Handbook of Graph Theory, Combinatorial Optimization, and Algorithms

CHAPMAN & HALL/CRC
COMPUTER and INFORMATION SCIENCE SERIES

Series Editor: Sartaj Sahni

PUBLISHED TITLES

ADVERSARIAL REASONING: COMPUTATIONAL APPROACHES TO READING THE OPPONENT'S MIND
Alexander Kott and William M. McEneaney

COMPUTER-AIDED GRAPHING AND SIMULATION TOOLS FOR AUTOCAD USERS
P. A. Simionescu

DELAUNAY MESH GENERATION
Siu-Wing Cheng, Tamal Krishna Dey, and Jonathan Richard Shewchuk

DISTRIBUTED SENSOR NETWORKS, SECOND EDITION
S. Sitharama Iyengar and Richard R. Brooks

DISTRIBUTED SYSTEMS: AN ALGORITHMIC APPROACH, SECOND EDITION
Sukumar Ghosh

ENERGY-AWARE MEMORY MANAGEMENT FOR EMBEDDED MULTIMEDIA SYSTEMS: A COMPUTER-AIDED DESIGN APPROACH
Florin Balasa and Dhiraj K. Pradhan

ENERGY EFFICIENT HARDWARE-SOFTWARE CO-SYNTHESIS USING RECONFIGURABLE HARDWARE
Jingzhao Ou and Viktor K. Prasanna

FUNDAMENTALS OF NATURAL COMPUTING: BASIC CONCEPTS, ALGORITHMS, AND APPLICATIONS
Leandro Nunes de Castro

HANDBOOK OF ALGORITHMS FOR WIRELESS NETWORKING AND MOBILE COMPUTING
Azzedine Boukerche

HANDBOOK OF APPROXIMATION ALGORITHMS AND METAHEURISTICS
Teofilo F. Gonzalez

HANDBOOK OF BIOINSPIRED ALGORITHMS AND APPLICATIONS
Stephan Olariu and Albert Y. Zomaya

HANDBOOK OF COMPUTATIONAL MOLECULAR BIOLOGY
Srinivas Aluru

HANDBOOK OF DATA STRUCTURES AND APPLICATIONS
Dinesh P. Mehta and Sartaj Sahni

HANDBOOK OF DYNAMIC SYSTEM MODELING
Paul A. Fishwick

HANDBOOK OF ENERGY-AWARE AND GREEN COMPUTING
Ishfaq Ahmad and Sanjay Ranka

HANDBOOK OF GRAPH THEORY, COMBINATORIAL OPTIMIZATION, AND ALGORITHMS
Krishnaiyan "KT" Thulasiraman, Subramanian Arumugam, Andreas Brandstädt, and Takao Nishizeki

PUBLISHED TITLES CONTINUED

HANDBOOK OF PARALLEL COMPUTING: MODELS, ALGORITHMS AND APPLICATIONS
Sanguthevar Rajasekaran and John Reif

HANDBOOK OF REAL-TIME AND EMBEDDED SYSTEMS
Insup Lee, Joseph Y-T. Leung, and Sang H. Son

HANDBOOK OF SCHEDULING: ALGORITHMS, MODELS, AND PERFORMANCE ANALYSIS
Joseph Y.-T. Leung

HIGH PERFORMANCE COMPUTING IN REMOTE SENSING
Antonio J. Plaza and Chein-I Chang

HUMAN ACTIVITY RECOGNITION: USING WEARABLE SENSORS AND SMARTPHONES
Miguel A. Labrador and Oscar D. Lara Yejas

IMPROVING THE PERFORMANCE OF WIRELESS LANs: A PRACTICAL GUIDE
Nurul Sarkar

INTEGRATION OF SERVICES INTO WORKFLOW APPLICATIONS
Paweł Czarnul

INTRODUCTION TO NETWORK SECURITY
Douglas Jacobson

LOCATION-BASED INFORMATION SYSTEMS: DEVELOPING REAL-TIME TRACKING APPLICATIONS
Miguel A. Labrador, Alfredo J. Pérez, and Pedro M. Wightman

METHODS IN ALGORITHMIC ANALYSIS
Vladimir A. Dobrushkin

MULTICORE COMPUTING: ALGORITHMS, ARCHITECTURES, AND APPLICATIONS
Sanguthevar Rajasekaran, Lance Fiondella, Mohamed Ahmed, and Reda A. Ammar

PERFORMANCE ANALYSIS OF QUEUING AND COMPUTER NETWORKS
G. R. Dattatreya

THE PRACTICAL HANDBOOK OF INTERNET COMPUTING
Munindar P. Singh

SCALABLE AND SECURE INTERNET SERVICES AND ARCHITECTURE
Cheng-Zhong Xu

SOFTWARE APPLICATION DEVELOPMENT: A VISUAL C++®, MFC, AND STL TUTORIAL
Bud Fox, Zhang Wenzu, and Tan May Ling

SPECULATIVE EXECUTION IN HIGH PERFORMANCE COMPUTER ARCHITECTURES
David Kaeli and Pen-Chung Yew

VEHICULAR NETWORKS: FROM THEORY TO PRACTICE
Stephan Olariu and Michele C. Weigle

Handbook of Graph Theory, Combinatorial Optimization, and Algorithms

Editor-in-Chief
Krishnaiyan "KT" Thulasiraman
University of Oklahoma
Norman, Oklahoma, USA

Edited by
Subramanian Arumugam
Kalasalingam University
Tamil Nadu, India

Andreas Brandstädt
University of Rostock
Rostock, Germany

Takao Nishizeki
Tohoku University
Sendai, Japan

CRC Press
Taylor & Francis Group
Boca Raton London New York

CRC Press is an imprint of the
Taylor & Francis Group, an **informa** business

A CHAPMAN & HALL BOOK

CRC Press
Taylor & Francis Group
6000 Broken Sound Parkway NW, Suite 300
Boca Raton, FL 33487-2742

© 2016 by Taylor & Francis Group, LLC
CRC Press is an imprint of Taylor & Francis Group, an Informa business

No claim to original U.S. Government works

Printed on acid-free paper
Version Date: 20151102

International Standard Book Number-13: 978-1-58488-595-5 (Hardback)

This book contains information obtained from authentic and highly regarded sources. Reasonable efforts have been made to publish reliable data and information, but the author and publisher cannot assume responsibility for the validity of all materials or the consequences of their use. The authors and publishers have attempted to trace the copyright holders of all material reproduced in this publication and apologize to copyright holders if permission to publish in this form has not been obtained. If any copyright material has not been acknowledged please write and let us know so we may rectify in any future reprint.

Except as permitted under U.S. Copyright Law, no part of this book may be reprinted, reproduced, transmitted, or utilized in any form by any electronic, mechanical, or other means, now known or hereafter invented, including photocopying, microfilming, and recording, or in any information storage or retrieval system, without written permission from the publishers.

For permission to photocopy or use material electronically from this work, please access www.copyright.com (http://www.copyright.com/) or contact the Copyright Clearance Center, Inc. (CCC), 222 Rosewood Drive, Danvers, MA 01923, 978-750-8400. CCC is a not-for-profit organization that provides licenses and registration for a variety of users. For organizations that have been granted a photocopy license by the CCC, a separate system of payment has been arranged.

Trademark Notice: Product or corporate names may be trademarks or registered trademarks, and are used only for identification and explanation without intent to infringe.

Visit the Taylor & Francis Web site at
http://www.taylorandfrancis.com

and the CRC Press Web site at
http://www.crcpress.com

Contents

Preface xi

Editors xiii

Contributors xv

Section I Basic Concepts and Algorithms

Chapter 1 ▪ Basic Concepts in Graph Theory and Algorithms 3
Subramanian Arumugam and Krishnaiyan "KT" Thulasiraman

Chapter 2 ▪ Basic Graph Algorithms 21
Krishnaiyan "KT" Thulasiraman

Chapter 3 ▪ Depth-First Search and Applications 59
Krishnaiyan "KT" Thulasiraman

Section II Flows in Networks

Chapter 4 ▪ Maximum Flow Problem 79
F. Zeynep Sargut, Ravindra K. Ahuja, James B. Orlin, and Thomas L. Magnanti

Chapter 5 ▪ Minimum Cost Flow Problem 113
Balachandran Vaidyanathan, Ravindra K. Ahuja, James B. Orlin, and Thomas L. Magnanti

Chapter 6 ▪ Multicommodity Flows 157
Balachandran Vaidyanathan, Ravindra K. Ahuja, James B. Orlin, and Thomas L. Magnanti

Section III Algebraic Graph Theory

Chapter 7 ▪ Graphs and Vector Spaces 177
Krishnaiyan "KT" Thulasiraman and M. N. S. Swamy

Chapter 8 ▪ Incidence, Cut, and Circuit Matrices of a Graph 191
Krishnaiyan "KT" Thulasiraman and M. N. S. Swamy

Chapter	9 ▪ Adjacency Matrix and Signal Flow Graphs	215
	Krishnaiyan "KT" Thulasiraman and M. N. S. Swamy	
Chapter	10 ▪ Adjacency Spectrum and the Laplacian Spectrum of a Graph	227
	R. Balakrishnan	
Chapter	11 ▪ Resistance Networks, Random Walks, and Network Theorems	247
	Krishnaiyan "KT" Thulasiraman and Mamta Yadav	

Section IV Structural Graph Theory

Chapter	12 ▪ Connectivity	273
	Subramanian Arumugam and Karam Ebadi	
Chapter	13 ▪ Connectivity Algorithms	291
	Krishnaiyan "KT" Thulasiraman	
Chapter	14 ▪ Graph Connectivity Augmentation	315
	András Frank and Tibor Jordán	
Chapter	15 ▪ Matchings	349
	Michael D. Plummer	
Chapter	16 ▪ Matching Algorithms	373
	Krishnaiyan "KT" Thulasiraman	
Chapter	17 ▪ Stable Marriage Problem	403
	Shuichi Miyazaki	
Chapter	18 ▪ Domination in Graphs	419
	Subramanian Arumugam and M. Sundarakannan	
Chapter	19 ▪ Graph Colorings	449
	Subramanian Arumugam and K. Raja Chandrasekar	

Section V Planar Graphs

Chapter	20 ▪ Planarity and Duality	475
	Krishnaiyan "KT" Thulasiraman and M. N. S. Swamy	
Chapter	21 ▪ Edge Addition Planarity Testing Algorithm	489
	John M. Boyer	
Chapter	22 ▪ Planarity Testing Based on PC-Trees	525
	Wen-Lian Hsu	

CHAPTER 23 ▪ Graph Drawing 537
MD. SAIDUR RAHMAN and TAKAO NISHIZEKI

SECTION VI Interconnection Networks

CHAPTER 24 ▪ Introduction to Interconnection Networks 587
S. A. CHOUDUM, LAVANYA SIVAKUMAR, and V. SUNITHA

CHAPTER 25 ▪ Cayley Graphs 627
S. LAKSHMIVARAHAN, LAVANYA SIVAKUMAR, and S. K. DHALL

CHAPTER 26 ▪ Graph Embedding and Interconnection Networks 653
S.A. CHOUDUM, LAVANYA SIVAKUMAR, and V. SUNITHA

SECTION VII Special Graphs

CHAPTER 27 ▪ Program Graphs 691
KRISHNAIYAN "KT" THULASIRAMAN

CHAPTER 28 ▪ Perfect Graphs 707
CHÍNH T. HOÀNG and R. SRITHARAN

CHAPTER 29 ▪ Tree-Structured Graphs 751
ANDREAS BRANDSTÄDT and FEODOR F. DRAGAN

SECTION VIII Partitioning

CHAPTER 30 ▪ Graph and Hypergraph Partitioning 829
SACHIN B. PATKAR and H. NARAYANAN

SECTION IX Matroids

CHAPTER 31 ▪ Matroids 879
H. NARAYANAN and SACHIN B. PATKAR

CHAPTER 32 ▪ Hybrid Analysis and Combinatorial Optimization 923
H. NARAYANAN

SECTION X Probabilistic Methods, Random Graph Models, and Randomized Algorithms

CHAPTER 33 ▪ Probabilistic Arguments in Combinatorics 945
C.R. SUBRAMANIAN

CHAPTER 34 ▪ Random Models and Analyses for Chemical Graphs 997
DANIEL PASCUA, TINA M. KOURI, and DINESH P. MEHTA

x ■ Contents

CHAPTER 35 ▪ Randomized Graph Algorithms: Techniques and Analysis	1011
SURENDER BASWANA and SANDEEP SEN	

SECTION XI Coping with NP-Completeness

CHAPTER 36 ▪ General Techniques for Combinatorial Approximation	1027
SARTAJ SAHNI	

CHAPTER 37 ▪ ε-Approximation Schemes for the Constrained Shortest Path Problem	1035
KRISHNAIYAN "KT" THULASIRAMAN	

CHAPTER 38 ▪ Constrained Shortest Path Problem: Lagrangian Relaxation-Based Algorithmic Approaches	1041
YING XIAO and KRISHNAIYAN "KT" THULASIRAMAN	

CHAPTER 39 ▪ Algorithms for Finding Disjoint Paths with QoS Constraints	1063
ALEX SPRINTSON and ARIEL ORDA	

CHAPTER 40 ▪ Set-Cover Approximation	1075
NEAL E. YOUNG	

CHAPTER 41 ▪ Approximation Schemes for Fractional Multicommodity Flow Problems	1079
GEORGE KARAKOSTAS	

CHAPTER 42 ▪ Approximation Algorithms for Connectivity Problems	1097
RAMAKRISHNA THURIMELLA	

CHAPTER 43 ▪ Rectilinear Steiner Minimum Trees	1115
TAO HUANG and EVANGELINE F. Y. YOUNG	

CHAPTER 44 ▪ Fixed-Parameter Algorithms and Complexity	1141
VENKATESH RAMAN and SAKET SAURABH	

Index	1197

Preface

Research in graph theory and combinatorial optimization has experienced explosive growth in the last three decades or so. Rapid technological advances such as those in telecommunication networks and large-scale integrated circuit design; emergence of new areas such as network science, which emphasizes applications in social networks and biological networks; and advances in theoretical computer science have all contributed to this explosion of interest and knowledge in graph theory and combinatorial optimization and related algorithmic issues. Therefore, it is no surprise that these disciplines have come to play a central role in engineering and computer science curricula. Several excellent textbooks dealing with graph theory or combinatorial optimization are now available. These books can be broadly classified into two categories. In the first category are the books that deal with all the essential topics in graph theory or combinatorial optimization. These books are intended to serve as textbooks for senior undergraduate students and beginning graduate students. In the second category are books that give an in-depth treatment of certain specific topics. They are appropriate for students who intend to pursue a research career in graph theory or combinatorial optimization. Since these disciplines have reached a certain level of maturity, we see a need for a book that gives a broader and an integrated treatment of both graph theory and combinatorial optimization. Such a book will help students and researchers equip themselves with techniques and tools that will strengthen their ability to see opportunities to apply graph theory and combinatorial optimization in solving problems they encounter in their applications. Our long years of experience in teaching and applying graph theory and combinatorial optimization have convinced us that while tools and techniques enhance one's ability to solve problems, a broader exposure to them will also help an individual see problems that will not be visible otherwise. This philosophy is the underlying motivating factor for undertaking this project.

A book that satisfies the above objective has to be necessarily a handbook with contributions from experts on the various topics to be covered. Size limitations also require that we make some sacrifices in the treatment of the topics. We decided to emphasize proofs of results and underlying proof techniques, since exposure to them would help enhance the analytical skills of students. So, the authors were requested to give proofs of all theorems unless they were too long, and limit the illustrations of theorems and algorithms to a minimum.

This book is organized into 11 sections, with each section consisting of chapters focusing on a specific theme. Overall there are 44 chapters. Roughly speaking, there are 21 chapters dealing exclusively with graph theory, 19 dealing exclusively with combinatorial optimization, and 24 dealing with algorithmic issues. We believe that this book will serve as a reference and also provide material to develop different courses according to the needs of the students.

The coverage of this book is by no means exhaustive. Advances in graph minors and extremal graph theory are obvious omissions. There is also room for including additional topics in combinatorial optimization, particularly, approximation algorithms and recent applications. We hope the survey section at the end of each chapter provides adequate pointers for exploring other related issues. We also hope a future edition will make the coverage more complete.

It has been a pleasure working with the editorial and production teams at Taylor & Francis Group. In particular we are thankful to Randi Cohen, senior acquisition editor; Joette Lynch, project editor; and Indumathi Sambantham, project management executive at Lumina Datamatics, for making the production process smooth, swift, and painless.

"KT" thanks Sartaj Sahni of the University of Florida, Gainesville, Florida, for the opportunity to undertake this project and graduate students Mamta Yadav (now at the University of Oklahoma, Norman, Oklahoma), Dr. Yuh-Rong Chen (now at Nanyang Technological University, Singapore), and Dr. Jincheng Zhuang (now at the Chinese Academy of Sciences, China) for their valuable help at different stages during the preparation of this book.

For "KT" it all started when he was introduced to graph theory during 1963–1964 by Professor Myril B. Reed of the University of Illinois who was then visiting the College of Engineering, Guindy (now Anna University, Chennai, India), under the USAID educational program. This marked the beginning of a career in exploring graph theoretic applications. "KT" gratefully dedicates this handbook to the memory of Professor Myril Reed whose inspirational teaching and works triggered all that happened to him in his academic life.

Krishnaiyan "KT" Thulasiraman
University of Oklahoma

Subramanian Arumugam
Kalasalingam University

Andreas Brandstädt
University of Rostock

Takao Nishizeki
Tohoku University

Editors

Krishnaiyan "KT" Thulasiraman, PhD, has been professor and Hitachi chair in computer science at the University of Oklahoma, Norman, Oklahoma, since 1994 and holds the professor emeritus position in electrical and computer engineering at Concordia University, Montreal, Québec, Canada. His prior appointments include professorships in electrical engineering and computer science at the Indian Institute of Technology Madras, Chennai, India (1965–1981) and in electrical and computer engineering at the Technical University of Nova Scotia, Halifax, Nova Scotia, Canada (1981–1982), and at Concordia University (1982–1994). He has held visiting positions at the University of Illinois, Champaign, Illinois; University of Waterloo, Waterloo, Ontario, Canada; University of Karlsruhe, Karlsruhe, Germany; Tokyo Institute of Technology, Meguro, Japan; and the National Chiao-Tung University, Hsinchu, Taiwan.

"KT" earned his bachelor's and master's degrees in electrical engineering from Anna University (formerly College of Engineering, Guindy), Chennai, India, in 1963 and 1965, respectively, and a PhD in electrical engineering from the Indian Institute of Technology Madras, Chennai, India, in 1968. His research interests have been in graph theory, combinatorial optimization, and related algorithmic issues with a specific focus on applications in electrical and computer engineering. He has published extensively in archival journals and has coauthored two textbooks entitled *Graphs, Networks and Algorithms* and *Graphs: Theory and Algorithms* published by Wiley-Interscience in 1981 and 1992, respectively. He has been professionally active within the IEEE, in particular, IEEE Circuits and Systems, Computer and Communications Societies, and the ACM.

"KT" has received several honors and awards, including the Distinguished Alumnus Award of the Indian Institute of Technology Madras; IEEE Circuits and Systems Society Charles Desoer Technical Achievement Award; IEEE Circuits and Systems Society Golden Jubilee Medal; senior fellowship of the Japan Society for Promotion of Science; and fellowship of the IEEE, AAAS, and the European Academy of Sciences.

Subramanian Arumugam, PhD, is currently senior professor (research) and director, National Centre for Advanced Research in Discrete Mathematics, Kalasalingam University, Krishnankoil, India. He was previously professor and head of the Department of Mathematics at Manonmaniam Sundaranar University, Tirunelveli, India. He is also a visiting professor at Liverpool Hope University, Liverpool, UK, and adjunct professor at Ball State University, Muncie, Indiana. He was conjoint professor at the University of Newcastle, Australia, from 2009 to 2012. His current area of research is graph theory and its applications. He has guided 35 PhD candidates and has published approximately 195 papers in national and international journals. He has 47 years of academic experience and has authored 32 books, including two textbooks in Tamil, which are widely used by undergraduate students of various universities in India. He has organized 27 conferences and workshops and has edited 12 proceedings. He is the founder editor-in-chief of *AKCE International Journal of Graphs and Combinatorics*, which has been indexed in SCOPUS and is to be published jointly with Elsevier from 2015 onward.

Andreas Brandstädt, PhD, has been a professor in computer science at the University of Rostock, Rostock, Germany, since 1994 (officially retired October 2014). His prior appointments include a professorship in computer science at the University of Duisburg, Germany (from 1991 to 1994), and assistant professorships in computer science at the University of Hagen, Hagen, Germany (from 1990 to 1991), and in mathematics at the University of Jena, Jena, (then) East Germany (from 1974 to 1990).

He has held various visiting professorships in France, for example, at the University of Amiens (thrice), University of Clermont-Ferrand (twice), University of Metz, as well as the University of Koper, Slovenia; and he has presented invited lectures at various international conferences.

Dr. Brandstädt earned his master's degree (diplom), his PhD (Dr. rer. nat.), and his habilitation (Dr. rer. nat. habil.) in mathematics from the University of Jena, East Germany, in 1974, 1976, and 1983, respectively.

His research interests have been in stochastics, complexity theory, formal languages, graph algorithms, graph theory, combinatorial optimization, and related algorithmic issues with a specific focus on efficient algorithms based on graph structure and graph classes with tree structure. He has published extensively in various international journals and conference proceedings, is the author of a textbook *Graphen und Algorithmen* (in German), and has coauthored a widely cited monograph, *Graph Classes: A Survey*.

He has been active within various program committees, such as the WG conferences and the ODSA conferences, as co-organizer of such conferences and coeditor of the corresponding conference proceedings.

Takao Nishizeki, PhD, was a student at Tohoku University, Japan, earning a bachelor's degree in 1969, a master's in 1971, and a PhD in 1974, all in electrical communication engineering. He continued at Tohoku as a faculty member, and became a full professor there in 1988. He retired in 2010, becoming a professor emeritus at Tohoku University, Japan, but continued teaching as a professor at Kwansei Gakuin University, Nishinomiya, Japan, over the period 2010–2015. He was also a visiting research mathematician at Carnegie-Mellon University, Pittsburgh, Pennsylvania, from 1977 to 1978.

Dr. Nishizeki has established himself, both nationally and internationally, as a world leader in computer science, in particular, algorithms for planar graphs, edge coloring, network flows, VLSI routing, graph drawing, and cryptology. His publication list includes 3 coauthored books, 5 edited books, and more than 300 technical papers in leading journals and prestigious conferences, such as *JACM*, *SIAM Journal on Computing*, *Algorithmica*, *Journal of Algorithms*, *Journal of Cryptology*, *STOC*, *ICALP*, and *SODA*.

Dr. Nishizeki is a fellow of many distinguished academic and scientific societies, including ACM, IEEE, IEICE of Japan, Information Processing Society of Japan, and Bangladesh Academy of Sciences. He served as advisory committee chair of ISAAC 1990–2009 and as steering committee member of Graph Drawing Conference 1993–2009.

For his great achievements in computer science, Professor Nishizeki has received many awards, including the Science and Technology Prize of the Japanese Ministry of Education, IEICE Achievement Award, ICF Best Research Award, Funai Information Science Promotion Award, TELECOM Technology Award, and Best Paper Awards of IEICE, JSIAM, IPSJ, ISAAC, FAW-AAIM, and WALCOM.

Contributors

Ravindra K. Ahuja
Department of Industrial and Systems Engineering
University of Florida
Gainesville, Florida

Subramanian Arumugam
National Centre for Advanced Research in Discrete Mathematics (n-CARDMATH)
Kalasalingam University
Krishnankoil, India
and
School of Electrical Engineering and Computer Science
The University of Newcastle
New South Wales, Australia
and
Department of Computer Science
Liverpool Hope University
Liverpool, United Kingdom
and
Department of Computer Science
Ball State University
Muncie, Indiana

R. Balakrishnan
Department of Mathematics
Bharathidasan University
Tiruchirappalli, India

Surender Baswana
Department of Computer Science and Engineering
Indian Institute of Technology Kanpur
Kanpur, India

John M. Boyer
IBM Canada
Victoria, British Columbia, Canada

Andreas Brandstädt
Department of Computer Science
University of Rostock
Rostock, Germany

K. Raja Chandrasekar
National Centre for Advanced Research in Discrete Mathematics (n-CARDMATH)
Kalasalingam University
Krishnankoil, India

S. A. Choudum
Department of Mathematics
Indian Institute of Technology Madras
Chennai, India

S. K. Dhall
School of Computer Science
University of Oklahoma
Norman, Oklahoma

Feodor F. Dragan
Department of Computer Science
Kent State University
Kent, Ohio

Karam Ebadi
National Centre for Advanced Research in Discrete Mathematics (n-CARDMATH)
Kalasalingam University
Krishnankoil, India

András Frank
Department of Operations Research
Eötvös University
and
MTA-ELTE Egerváry Research Group on Combinatorial Optimization
Budapest, Hungary

Chính T. Hoàng
Department of Physics and Computer Science
Wilfrid Laurier University
Waterloo, Ontario, Canada

Wen-Lian Hsu
Institute of Information Science
Academia Sinica
Taipei, Taiwan

Tao Huang
Synopsys, Silicon Valley
Sunnyvale, California

Tibor Jordán
Department of Operations Research
Eötvös University
and
MTA-ELTE Egerváry Research
  Group on Combinatorial
  Optimization
Budapest, Hungary

George Karakostas
Department of Computer Science
McMaster University
Hamilton, Ontario, Canada

Tina M. Kouri
Department of Computer Science
  and Engineering
University of South Florida
Tampa, Florida

S. Lakshmivarahan
School of Computer Science
University of Oklahoma
Norman, Oklahoma

Thomas L. Magnanti
Sloan School of Management
Massachusetts Institute of Technology
Cambridge, Massachusetts

Dinesh P. Mehta
Department of Electrical Engineering
  and Computer Science
Colorado School of Mines
Golden, Colorado

Shuichi Miyazaki
Academic Center for Computing and Media
  Studies
Kyoto University
Kyoto, Japan

H. Narayanan
Department of Electrical
  Engineering
Indian Institute of Technology
  Bombay
Mumbai, India

Takao Nishizeki
Graduate School of Information Sciences
Tohoku University
Sendai, Japan

Ariel Orda
Department of Electrical Engineering
Technion–Israel Institute
  of Technology
Haifa, Israel

James B. Orlin
Sloan School of Management
Massachusetts Institute of Technology
Cambridge, Massachusetts

Daniel Pascua
Department of Electrical Engineering
  and Computer Science
Colorado School of Mines
Golden, Colorado

Sachin B. Patkar
Department of Electrical Engineering
Indian Institute of Technology Bombay
Mumbai, India

Michael D. Plummer
Department of Mathematics
Vanderbilt University
Nashville, Tennesse

Md. Saidur Rahman
Department of Computer Science and Engineering
Bangladesh University of EngineeringInstitute of Mathematical and Technology
Dhaka, Bangladesh

Venkatesh Raman
The Institute of Mathematical Sciences
Chennai, India

Sartaj Sahni
Department of Computer and Information Sciences and Engineering
University of Florida
Gainesville, Florida

F. Zeynep Sargut
Department of Industrial Engineering
Izmir University of Economics
Izmir, Turkey

Saket Saurabh
The Institute of Mathematical Sciences
Chennai, India

Sandeep Sen
Department of Computer Science and Engineering
Indian Institute of Technology Delhi
New Delhi, India

Lavanya Sivakumar
SRM Research Institute (Mathematics)
SRM University
Chennai, India

Alex Sprintson
Department of Electrical and Computer Engineering
Texas A&M University
College Station, Texas

R. Sritharan
Department of Computer Science
The University of Dayton
Dayton, Ohio

C.R. Subramanian
The Institute of Mathematical Sciences
Chennai, India

M. Sundarakannan
Department of Mathematics
SSN College of Engineering
Chennai, India

V. Sunitha
Department of Information and Communication Technology
Dhirubhai Ambani Institute of Information and Communication Technology
Gandhinagar, India

M.N.S. Swamy
Department of Electrical and Computer Engineering
Concordia University
Montreal, Québec, Canada

Krishnaiyan "KT" Thulasiraman
School of Computer Science
University of Oklahoma
Norman, Oklahoma

Ramakrishna Thurimella
Department of Computer Science
University of Denver
Denver, Colorado

Balachandran Vaidyanathan
Operations Research
FedEx Corporation
Memphis, Tennessee

Ying Xiao
VT iDirect Inc.
Herndon, Virginia

Mamta Yadav
School of Computer Science
University of Oklahoma
Norman, Oklahoma

Evangeline F.Y. Young
Department of Computer Science and Engineering
The Chinese University of Hong Kong
Hong Kong, China

Neal E. Young
Department of Computer Science
University of California
Riverside, California

I
Basic Concepts and Algorithms

CHAPTER 1

Basic Concepts in Graph Theory and Algorithms

Subramanian Arumugam

Krishnaiyan "KT" Thulasiraman

CONTENTS

1.1	Introduction	3
1.2	Basic Concepts	4
1.3	Subgraphs and Complements	5
1.4	Connectedness	6
1.5	Operations on Graphs	7
1.6	Trees	8
1.7	Cutsets and Cuts	9
1.8	Eulerian Graphs	9
1.9	Hamiltonian Graphs	9
1.10	Graph Parameters	10
1.11	Directed Graphs	11
1.12	Paths and Connections in Digraphs	12
1.13	Directed Graphs and Relations	13
1.14	Directed Trees and Arborescences	13
1.15	Directed Eulerian Graphs	14
1.16	Directed Hamiltonian Graphs	14
1.17	Acyclic Directed Graphs	15
1.18	Tournaments	15
1.19	Computational Complexity and Completeness	16

1.1 INTRODUCTION

In this chapter we give a brief introduction to certain basic concepts and results in graph theory and algorithms, some of which are not explicitly defined in the remaining chapters of this book. For easy reference we also provide a list of commonly used symbols for graph parameters and concepts. For basic concepts in graphs, digraphs, and algorithms we refer to Bondy and Murty [1], Chartrand and Lesniak [2], Swamy and Thulasiraman [3], Thulasiraman and Swamy [4], and West [5].

1.2 BASIC CONCEPTS

A *graph* G consists of a pair (V, E) where V is a nonempty finite set whose elements are called *vertices* and E is a set of unordered pairs of distinct elements of V. The elements of E are called *edges* of the graph G. If $e = \{u, v\} \in E$, the edge e is said to join u and v. The vertices u and v are called the end vertices of the edge uv. We write $e = uv$ and we say that the vertices u and v are *adjacent*. We also say that the vertex u and the edge e are *incident* with each other. If two distinct edges e_1 and e_2 are incident with a common vertex, then they are called *adjacent edges*. A graph with n vertices and m edges is called a (n, m) *graph*. The number of vertices in G is called the *order* of G. The number of edges of G is called the *size* of G.

A graph is normally represented by a diagram in which each vertex is represented by a dot and each edge is represented by a line segment joining two vertices with which the edge is incident. For example, if $G = (V, E)$ is a graph where $V = \{a, b, c, d\}$ and $E = \{ab, ac, ad\}$, then G is a $(4, 3)$ graph and it is represented by the diagram given in Figure 1.1.

The definition of a graph does not allow more than one edge joining two vertices. It also does not allow any edge joining a vertex to itself. Such an edge joining a vertex to itself is called a *self-loop* or simply a *loop*. If more than one edge joining two vertices are allowed, the resulting object is called a *multigraph*. Edges joining the same pair of vertices are called *multiple edges*. If loops are also allowed, the resulting object is called a *pseudo graph*.

A graph in which any two distinct vertices are adjacent is called a *complete graph*. The complete graph on n vertices is denoted by K_n. A graph whose edge set is empty is called a *null graph* or a *totally disconnected graph*.

A graph G is called a *bipartite graph* if V can be partitioned into two disjoint subsets V_1 and V_2 such that every edge of G joins a vertex of V_1 to a vertex of V_2 and (V_1, V_2) is called a *bipartition* of G. If G contains every edge joining the vertices of V_1 to the vertices of V_2 then G is called a *complete bipartite graph*. If V_1 contains r vertices and V_2 contains s vertices, then the complete bipartite graph G is denoted by $K_{r,s}$. The graph given in Figure 1.1 is $K_{1,3}$.

A graph G is k-*partite*, $k \geq 1$, if it is possible to partition $V(G)$ into k subsets $V_1, V_2, \ldots, V_k$ (called *partite sets*) such that every element of $E(G)$ joins a vertex of V_i to a vertex of $V_j, i \neq j$.

A *complete k-partite graph* G is a k-partite graph with partite sets $V_1, V_2, \ldots, V_k$ having the additional property that if $u \in V_i$ and $v \in V_j, i \neq j$, then $uv \in E(G)$. If $|V_i| = n_i$, then this graph is denoted by $K(n_1, n_2, \ldots, n_k)$ or $K_{n_1, n_2, \ldots, n_k}$.

The *degree* of a vertex v_i in a graph G is the number of edges incident with v_i. The degree of v_i is denoted by $d_G(v_i)$ or $deg\ v_i$ or simply $d(v_i)$. A vertex v of degree 0 is called an *isolated vertex*. A vertex v of degree 1 is called a *pendant vertex*.

Theorem 1.1 *The sum of the degrees of the vertices of a graph G is twice the number of edges.* ∎

Corollary 1.1 *In any graph G the number of vertices of odd degree is even.* ∎

Figure 1.1 Example of a graph.

For any graph G, we define

$$\delta(G) = \min\{deg\ v : v \in V(G)\}$$

and

$$\Delta(G) = \max\{deg\ v : v \in V(G)\}.$$

If all the vertices of G have the same degree r, then $\delta(G) = \Delta(G) = r$ and in this case G is called a *regular graph* of degree r. A regular graph of degree 3 is called a *cubic graph*.

1.3 SUBGRAPHS AND COMPLEMENTS

A subgraph of a graph G is a graph H such that $V(H) \subseteq V(G)$ and $E(H) \subseteq E(G)$ and the assignment of end vertices to edges in H is the same as in G. We then write $H \subseteq G$ and say that G contains H. The graph H is called a proper subgraph of G if either $E(H)$ is a proper subset of $E(G)$ or $V(H)$ is a proper subset of $V(G)$. Also H is called a *spanning subgraph* of G if $V(H) = V(G)$. If $V_1 \subset V$, then the subgraph $G_1 = (V_1, E_1)$ is called an *induced subgraph* of G if G_1 is the maximal subgraph of G with vertex set V_1. Thus, if G_1 is an induced subgraph of G, then two vertices in V_1 are adjacent in G_1 if and only if they are adjacent in G. The subgraph induced by V_1 is denoted by $\langle V_1 \rangle$ or $G[V_1]$. It is also called a vertex-induced subgraph of G.

If $E_1 \subset E$, then the subgraph of G with edge set E_1 and having no isolated vertices is called the subgraph induced by E_1 and is denoted by $G[E_1]$. This is also called *edge-induced subgraph* of G.

Let $G = (V, E)$ be a graph. Let $v_i \in V$. The subgraph of G obtained by removing the vertex v_i and all the edges incident with v_i is called the *subgraph obtained by the removal of the vertex* v_i and is denoted by $G - v_i$. Clearly $G - v_i$ is an induced subgraph of G. If G is connected $S \subset V$ and $G - S$ is not connected, then S is called a vertex cut of G.

Let $e_j \in E$. Then $G - e_j = (V, E - \{e_j\})$ is called the subgraph of G obtained by the removal of the edge e_j. Clearly $G - e_j$ is a spanning subgraph of G which contains all the edges of G except e_j.

Let $G = (V, E)$ be a graph. Let v_i, v_j be two vertices which are not adjacent in G. Then $G + v_i v_j = (V, E \cup \{v_i, v_j\})$ is called the graph obtained by the *addition of the edge* $v_i v_j$ to G.

Two graphs $G_1 = (V_1, E_1)$ and $G_2 = (V_2, E_2)$ are said to be *isomorphic* if there exists a bijection $f : V_1 \to V_2$ such that u, v are adjacent in G_1 if and only if $f(u), f(v)$ are adjacent in G_2. If G_1 is isomorphic to G_2, we write $G_1 \cong G_2$. The map f is called an *isomorphism* from G_1 to G_2.

Let f be an isomorphism of the graph $G_1 = (V_1, E_1)$ to the graph $G_2 = (V_2, E_2)$. Let $v \in V_1$. Then $deg\ v = deg\ f(v)$.

An isomorphism of a graph G onto itself is called an *automorphism* of G. Let $\Gamma(G)$ denote the set of all automorphisms of G. Clearly the identity map $i : V \to V$ defined by $i(v) = v$ is an automorphism of G so that $i \in \Gamma(G)$. Further if α and β are automorphisms of G then $\alpha\beta$ and α^{-1} are also automorphisms of G. Hence $\Gamma(G)$ is a group and is called the *automorphism group* of G.

Let $G = (V, E)$ be a graph. The *complement* $\overline{G}$ of G is defined to be the graph which has V as its set of vertices and two vertices are adjacent in $\overline{G}$ if and only if they are not adjacent in G. The graph G is said to be a *self-complementary graph* if G is isomorphic to $\overline{G}$.

1.4 CONNECTEDNESS

A *walk* of a graph G is an alternating sequence of vertices and edges $v_0, x_1, v_1, x_2, v_2, \ldots, v_{n-1}, x_n, v_n$ beginning and ending with vertices such that each edge x_i is incident with v_{i-1} and v_i.

We say that the walk joins v_0 and v_n and it is called a $v_0 - v_n$ walk. Also v_0 is called the *initial vertex* and v_n is called the *terminal vertex* of the walk. The above walk is also denoted by $v_0, v_1, \ldots, v_n$, the edges of the walk being self evident. The number of edges in the walk is called the length of the walk.

A single vertex is considered as a walk of length 0. A walk is called a *trail* if all its edges are distinct and is called a *path* if all its vertices are distinct. A graph consisting of a path with n vertices is denoted by P_n. A $v_0 - v_n$ walk is called *closed* if $v_0 = v_n$. A closed walk $v_0, v_1, v_2, \ldots, v_n = v_0$ in which $n \geq 3$ and $v_0, v_1, \ldots, v_{n-1}$ are distinct is called a *cycle* (or *circuit*) of length n. The graph consisting of a cycle of length n is denoted by C_n.

Two vertices u and v of a graph G are said to be *connected* if there exists a $u - v$ path in G. A graph G is said to be *connected* if every pair of its vertices are connected. A graph which is not connected is said to be *disconnected*.

It is an easy exercise to verify that connectedness of vertices is an equivalence relation on the set of vertices V. Hence V can be partitioned into nonempty subsets $V_1, V_2, \ldots, V_n$ such that two vertices u and v are connected if and only if both u and v belong to the same set V_i.

Let G_i denote the induced subgraph of G with vertex set V_i. Clearly the subgraphs $G_1, G_2, \ldots, G_n$ are connected and are called the *components* of G. Clearly a graph G is connected if and only if it has exactly one component.

For any two vertices u, v of a graph we define the *distance* between u and v by

$$d(u,v) = \begin{cases} \text{the length of a shortest } u - v \text{ path if such a path exists} \\ \infty \text{ otherwise.} \end{cases}$$

If G is a connected graph, then $d(u,v)$ is always a nonnegative integer. In this case d is actually a *metric* on the set of vertices V.

Theorem 1.2 *A graph G with at least two vertices is bipartite if and only if all its cycles are of even length.* ∎

A *cutvertex* of a graph G is a vertex whose removal increases the number of components. A *bridge* of a graph G is an edge whose removal increases the number of components. Clearly if v is a cutvertex of a connected graph, then $G - v$ is disconnected.

Theorem 1.3 *Let v be a vertex of a connected graph G. The following statements are equivalent:*

1. *v is a cutvertex of G.*

2. *There exists a partition of $V - \{v\}$ into subsets U and W such that for each $u \in U$ and $w \in W$, the vertex v is on every $u - w$ path.*

3. *There exist two vertices u and w distinct from v such that v is on every $u - w$ path.* ∎

Theorem 1.4 *Let x be an edge of a connected graph G. The following statements are equivalent:*

1. x is bridge of G.

2. There exists a partition of V into two subsets U and W such that for every vertex $u \in U$ and $w \in W$, the edge x is on every $u - w$ path.

3. There exist two vertices u, w such that the edge x is on every $u - w$ path. ∎

Theorem 1.5 *An edge x of a connected graph G is a bridge if and only if x is not on any cycle of G.* ∎

A *nonseparable graph* is a connected graph with no cutvertices. All other graphs are *separable*. A *block* of a separable graph G is a maximal nonseparable subgraph of G.

Two $u - v$ paths are internally disjoint if they have no common vertices except u and v.

Theorem 1.6 *Every nontrivial connected graph contains at least two vertices that are not cutvertices.* ∎

Theorem 1.7 *A graph G with $n \geq 3$ vertices is a block if and only if any two vertices of G are connected by at least two internally disjoint paths.* ∎

A block G is also called *2-connected* or *biconnected* because at least two vertices have to be removed from G to disconnect it. The concepts of vertex and edge connectivities and generalized versions of the above theorem, called Menger's theorem, will be discussed further in Chapter 12 on connectivity.

1.5 OPERATIONS ON GRAPHS

Next we describe some binary operations defined on graphs. Let G_1 and G_2 be two graphs with disjoint vertex sets. The *union* $G = G_1 \cup G_2$ has $V(G) = V(G_1) \cup V(G_2)$ and $E(G) = E(G_1) \cup E(G_2)$. If a graph G consists of $k(\geq 2)$ disjoint copies of a graph H, then we write $G = kH$. The *join* $G = G_1 + G_2$ has $V(G) = V(G_1) \cup V(G_2)$ and $E(G) = E(G_1) \cup E(G_2) \cup \{uv | u \in V(G_1) \text{ and } v \in V(G_2)\}$.

A pair of vertices v_i and v_j in a graph G are said to be *identified* if the two vertices are replaced by a new vertex such that all the edges in G incident on v_i and v_j are now incident on the new vertex.

By *contraction* of an edge e we refer to the operation of removing e and identifying its end vertices. A graph G is *contractible* to a graph H if H can be obtained from G by a sequence of contractions.

The *Cartesian product* $G = G_1 \square G_2$ has $V(G) = V(G_1) \times V(G_2)$ and two vertices (u_1, u_2) and (v_1, v_2) of G are adjacent if and only if either $u_1 = v_1$ and $u_2 v_2 \in E(G_2)$ or $u_2 = v_2$ and $u_1 v_1 \in E(G_1)$.

The *n-cube* Q_n is the graph K_2 if $n = 1$, while for $n \geq 2$, Q_n is defined recursively as $Q_{n-1} \square K_2$. The n-cube Q_n can also be considered as that graph whose vertices are labeled by the binary n-tuples $(a_1, a_2, \ldots, a_n)$ (i.e., a_i is 0 or 1 for $1 \leq i \leq n$) and such that two vertices are adjacent if and only if their corresponding n-tuples differ at precisely one coordinate. The graph Q_n is an n-regular graph of order 2^n. The graph Q_n is called a *hypercube*.

The *strong product* $G_1 \boxtimes G_2$ of G_1 and G_2 is the graph with vertex set $V(G_1) \times V(G_2)$ and two vertices (u_1, u_2) and (v_1, v_2) are adjacent in $G_1 \boxtimes G_2$ if $u_1 = v_1$ and $u_2 v_2 \in E(G_2)$, or $u_1 v_1 \in E(G_1)$ and $u_2 = v_2$, or $u_1 v_1 \in E(G_1)$ and $u_2 v_2 \in E(G_2)$.

The *direct product* $G_1 \times G_2$ of two graphs G_1 and G_2 is the graph with vertex set $V(G_1) \times V(G_2)$ and two vertices (u_1, u_2) and (v_1, v_2) are adjacent in $G_1 \times G_2$ if $u_1 v_1 \in V(G_1)$ and $u_2 v_2 \in E(G_2)$.

The *lexicographic product* $G_1 \circ G_2$ of two graphs G_1 and G_2 is the graph with vertex set $V(G_1) \times V(G_2)$ and two vertices (u_1, u_2) and (v_1, v_2) are adjacent in $G_1 \circ G_2$ if $u_1 v_1 \in E(G_1)$ or $u_1 = v_1$ and $u_2 v_2 \in E(G_2)$.

1.6 TREES

The graphs that are encountered in most of the applications are connected. Among connected graphs trees have the simplest structure and are perhaps the most important ones. A tree is the simplest nontrivial type of a graph and in trying to prove a general result or to test a conjecture in graph theory, it is sometimes convenient to first study the situation for trees.

A graph that contains no cycles is called an *acyclic graph*. A connected acyclic graph is called a *tree*. Any graph without cycles is also called a *forest* so that the components of a forest are trees.

A tree of a graph G is a connected acyclic subgraph of G. A *spanning tree* of a graph G is a tree of G having all the vertices of G. A connected subgraph of a tree T is called a *subtree* of T.

The *cospanning tree* T^* of a spanning tree T of a graph G is the subgraph of G having all the vertices of G and exactly those edges of G that are not in T. Note that a cospanning tree may not be connected.

The edges of a spanning tree T are called the *branches* of T, and those of the corresponding cospanning tree T^* are called *links* or *chords*.

A spanning tree T uniquely determines its cospanning tree T^*. As such, we refer to the edges of T^* as the chords or links of T.

Theorem 1.8 *The following statements are equivalent for a graph G with n vertices and m edges:*

1. *G is a tree.*

2. *There exists exactly one path between any two vertices of G.*

3. *G is connected and $m = n - 1$.*

4. *G is acyclic and $m = n - 1$.*

5. *G is acyclic, and if any two nonadjacent vertices of G are connected by an edge, then the resulting graph has exactly one cycle.* ∎

Theorem 1.9 *A subgraph G' of an n-vertex graph G is a spanning tree of G if and only if G' is acyclic and has $n - 1$ edges.* ∎

Theorem 1.10 *A subgraph G' of a connected graph G is a subgraph of some spanning tree of G if and only if G' is acyclic.* ∎

A graph is trivial if it has only one vertex.

Theorem 1.11 *In a nontrivial tree there are at least two pendant vertices.* ∎

For a graph G with p components a spanning forest is a collection of p spanning trees, one for each component.

The *rank* $\rho(G)$ and *nullity* $\mu(G)$ of a graph G of order n and size m are defined by $\rho(G) = n - k$ and $\mu(G) = m - n + k$, where k is the number of components of G. Note that $\rho(G) + \mu(G) = m$.

The *arboricity* $a(G)$ of a graph G is the minimum number of edge disjoint spanning forests into which G can be decomposed.

1.7 CUTSETS AND CUTS

A *cutset* S of a connected graph G is a minimal set of edges of G such that $G - S$ is disconnected. Equivalently, a cutset S of a connected graph G is a minimal set of edges of G such that $G - S$ has exactly two components.

Let G be a connected graph with vertex set V. Let V_1 and V_2 be two disjoint subsets of V such that $V = V_1 \cup V_2$. Then the set S of all those edges of G having one end vertex in V_1 and the other in V_2 is called a *cut* of G. This is usually denoted by $[V_1, V_2]$.

Theorem 1.12 *A cut in a connected graph G is a cutset or union of edge-disjoint cutsets of G.* ∎

Several results connecting spanning trees, circuits, and cutsets will be discussed in Chapters 7 and 8.

1.8 EULERIAN GRAPHS

Let G be a connected graph. A closed trail containing all the edges of G is called an *Eulerian trail*. A graph G having an Eulerian trial is called an *Eulerian graph*.

The following theorem gives simple and useful characterizations of Eulerian graphs.

Theorem 1.13 *The following statements are equivalent for a connected graph G.*

1. *G is Eulerian.*

2. *The degree of every vertex in G is even.*

3. *G is the union of edge-disjoint circuits.* ∎

Corollary 1.2 *Let G be a connected graph with exactly $2k(k \geq 1)$ odd vertices. Then the edge set of G can be partitioned into k open trails.* ∎

Corollary 1.3 *Let G be a connected graph with exactly two odd vertices. Then G has an open trail containing all the edges of G.* ∎

1.9 HAMILTONIAN GRAPHS

A *Hamiltonian cycle* in a graph G is a cycle containing all the vertices of G. A *Hamiltonian path* in G is a path containing all the vertices of G. A graph G is defined to be *Hamiltonian* if it has a Hamiltonian cycle.

An Euler trail is a closed walk passing through each edge exactly once and a Hamiltonian cycle is a closed walk passing through each vertex exactly once. Thus there is a striking similarity between an Eulerian graph and a Hamiltonian graph. This may lead one to expect that there exists a simple, useful, and elegant characterization of a Hamiltonian graph, as in the case of an Eulerian graph. Such is not the case; in fact, development of such a characterization is a major unsolved problem in graph theory. However, several sufficient conditions have been established for a graph to be Hamiltonian.

We present several necessary conditions and sufficient conditions for a graph to be Hamiltonian. We observe that any Hamiltonian graph has no cutvertex.

Theorem 1.14 *If G is Hamiltonian, then for every nonempty proper subset S of $V(G)$, $\omega(G - S) \leq |S|$ where $\omega(H)$ denotes the number of components in any graph H.* ∎

A sequence $d_1 \leq d_2 \leq \cdots \leq d_n$ is said to be *graphic* if there is a graph G with n vertices $v_1, v_2, \ldots, v_n$ such that the degree $d(v_i)$ of v_i equals d_i for each i. Also $(d_1, d_2, \ldots, d_n)$ is then called the *degree sequence* of G.

If $S : d_1 \leq d_2 \leq \cdots \leq d_n$ and $S^* : d_1^* \leq d_2^* \leq \cdots \leq d_n^*$ are graphic sequences such that $d_i^* \geq d_i$ for $1 \leq i \leq n$, then S^* is said to *majorize* S.

The following result is due to Chavátal [6].

Theorem 1.15 *A simple graph $G = (V, E)$ of order n, with degree sequence $d_1 \leq d_2 \leq \cdots \leq d_n$ is Hamiltonian if $d_k \leq k < n/2 \Rightarrow d_{n-k} \geq n - k$.* ∎

Corollary 1.4 *A simple graph $G = (V, E)$ of order $n \geq 3$ with degree sequence $d_1 \leq d_2 \leq \cdots \leq d_n$ is Hamiltonian if one of the following conditions is satisfied:*

1. $1 \leq k \leq n \Rightarrow d_k \geq \frac{n}{2}$ [7].

2. $(u, v) \notin E \Rightarrow d(u) + d(v) \geq n$ [8].

3. $1 \leq k < \frac{n}{2} \Rightarrow d_k > k$ [6].

4. $j < k$, $d_j \leq j$ and $d_k \leq k - 1 \Rightarrow d_j + d_k \geq n$ [6]. ∎

The *closure* of a graph G with n vertices is the graph obtained from G by repeatedly joining pairs of nonadjacent vertices whose degree sum is at least n until no such pair remains. The closure of G is denoted by $c(G)$.

Theorem 1.16 *A graph is Hamiltonian if and only if its closure is Hamiltonian.* ∎

Corollary 1.5 *Let G be a graph with at least 3 vertices. If $c(G)$ is complete, then G is Hamiltonian.* ∎

1.10 GRAPH PARAMETERS

Several graph parameters relating to connectivity, matching, covering, coloring, and domination will be discussed in different chapters of this book. In this section we give a brief summary of some of the graph theoretic parameters.

Definition 1.1 *The distance $d(u, v)$ between two vertices u and v in a connected graph is defined to be the length of a shortest $u - v$ path in G. The eccentricity $e(v)$ of a vertex v is the number $\max_{u \in V(G)} d(u, v)$. Thus $e(v)$ is the distance between v and a vertex farthest from v. The radius $\operatorname{rad} G$ of G is the minimum eccentricity among the vertices of G, while the diameter $\operatorname{diam} G$ of G is the maximum eccentricity. A vertex v is a central vertex if $e(v) = rad(G)$ and the center $Cen(G)$ is the subgraph of G induced by its central vertices. A vertex v is a peripheral vertex if $e(v) = diam(G)$, while the periphery $Per(G)$ is the subgraph of G induced by its peripheral vertices.*

Definition 1.2 *If G is a noncomplete graph and t is a nonnegative real number such that $t \leq |S|/\omega(G-S)$ for every vertex-cut S of G, where $\omega(G-S)$ is the number of components of $G-S$, then G is defined to be t-tough. If G is a t-tough graph and s is a nonnegative real number such that $s < t$, then G is also s-tough. The maximum real number t for which a graph G is t-tough is called the toughness of G and is denoted by $t(G)$.*

Definition 1.3 *A subset S of vertices of a graph is called an independent set if no two vertices of S are adjacent in G. The number of vertices in a maximum independent is called the independence number of G.*

Definition 1.4 *The clique number $\omega(G)$ of a graph G is the maximum order among the complete subgraphs of G.*

We observe that the clique number of G is the independence number of its complement.

Definition 1.5 *The girth of a graph G having at least one cycle is the length of a shortest cycle in G. The circumference of G is the length of a longest cycle in G.*

Definition 1.6 *The vertex cover of a graph G is a set S of vertices such that every edge of G has at least one end vertex in S. An edge cover of G is a set L of edges such that every vertex of G is incident to some edge of L. The minimum size of a vertex cover is called the vertex covering number of G and is denoted by $\beta(G)$. The minimum size of an edge cover is called the edge covering number of G and is denoted by $\beta'(G)$.*

1.11 DIRECTED GRAPHS

A *directed graph* (or in short *digraph*) D is a pair (V, A) where V is a finite nonempty set and A is a subset of $V \times V - \{(x,x)/x \in V\}$. The elements of V and A are respectively called *vertices* and *arcs*. If $(u,v) \in A$ then the arc (u,v) is said to have u as its initial vertex (tail) and v as its terminal vertex (head). Also the arc (u,v) is said to join u to v.

As in the case of graphs a digraph can also be represented by a diagram. A digraph is represented by a diagram of its underlying graph together with arrows on its edges, each arrow pointing toward the head of the corresponding arc. For example $D = (\{1,2,3,4\}, \{(1,2),(2,3),(1,3),(3,1)\})$ is a digraph. The diagram representing D is given in Figure 1.2.

The *indegree* $d^-(v)$ of a vertex v in a digraph D is the number of arcs having v as its terminal vertex. The *outdegree* $d^+(v)$ of v is the number of arcs having v as its initial vertex. The ordered pair $(d^+(v), d^-(v))$ is called the *degree pair* of v.

The degree pairs of the vertices 1, 2, 3 and 4 of the digraph in Figure 1.2 are (2,1), (1,1), (1,2), and (0,0) respectively.

Theorem 1.17 *In a digraph D, the sum of the indegrees of all the vertices is equal to the sum of their outdegrees, each sum being equal to the number of arcs in D.* ∎

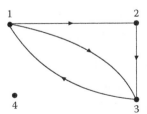

Figure 1.2 Example of a directed graph.

Subgraphs and *induced subgraphs* of a directed graph are defined as in the case of undirected graphs.

Let $D = (V, A)$ be a digraph. The *underlying graph* G of D is a graph having the same vertex set as D and two vertices u and w are adjacent in G whenever (u, w) or (w, u) is in A.

Similarly if we are given a graph G we can obtain a digraph from G by giving orientation to each edge of G. A digraph thus obtained from G is called an *orientation* of G.

The *converse* digraph D' of a digraph D is obtained from D by reversing the direction of each arc.

1.12 PATHS AND CONNECTIONS IN DIGRAPHS

A *walk* (*directed walk*) in a digraph is a finite alternating sequence $W = v_0 x_1 v_1 \ldots x_n v_n$ of vertices and arcs in which $x_i = (v_{i-1}, v_i)$ for every arc x_i. W is called a walk from v_0 to v_n or a $v_0 - v_n$ walk. The vertices v_0 and v_n are called *origin* and *terminus* of W, respectively, and $v_1, v_2, \ldots, v_{n-1}$ are called its *internal vertices*. The *length* of a walk is the number of occurrences of arcs in it. A walk in which the origin and terminus coincide is called a *closed walk*.

A *path* (*directed path*) is a walk in which all the vertices are distinct. A *cycle* (*directed cycle* or *circuit*) is a nontrivial closed path whose origin and internal vertices are distinct.

If there is a path from u to v then v is said to be *reachable* from u. A digraph is called *strongly connected* or *strong* if every pair of vertices are mutually reachable. A digraph is called *unilaterally connected* or *unilateral* if for every pair of vertices, at least one is reachable from the other. A digraph is called *weakly connected* or *weak* if the underlying graph is connected. A digraph is called *disconnected* if the underlying graph is disconnected.

Theorem 1.18 *The edges of a connected graph $G = (V, E)$ can be oriented so that the resulting digraph is strongly connected if and only if every edge of G is contained in at least one cycle.* ∎

Let $D = (V, A)$ be a digraph.

1. Let W_1 be a maximal subset of V such that for every pair of vertices $u, v \in W_1$, u is reachable from v and v is reachable from u. Then the subdigraph of D induced by W_1 is called a *strong component* of D.

2. Let W_2 be a maximal subset of V such that for every pair of vertices $u, v \in W_2$, either u is reachable from v or v is reachable from u. Then the subdigraph of D induced by W_2 is called a *unilateral component* of D.

3. Let W_3 be a maximal subset of V such that for every pair of vertices $u, v \in W_3$, u and v are joined by a path in the underlying graph of D. Then the subdigraph of D induced by W_3 is called a *weak component* of D.

We observe that each vertex of D is in exactly one strong component of D. An arc x lies in exactly one strong component if it lies on a cycle. There is no strong component containing an arc that does not lie on any cycle.

The *condensation* D^* of a digraph D has the strong components $S_1, S_2, \ldots, S_n$ of D as its vertices with an arc from S_i to S_j whenever there is at least one arc from S_i to S_j in D.

A directed graph is said to be *quasi-strongly connected* if for every pair of vertices v_1 and v_2 there is a vertex v_3 from which there is a directed path to v_1 and a directed path to v_2. Note that v_3 need not be distinct from v_1 or v_2.

1.13 DIRECTED GRAPHS AND RELATIONS

A *binary relation* R on a set $X = \{x_1, x_2, \ldots\}$ is a collection of ordered pairs of elements of X. If $(x_i, x_j) \in R$, then we write $x_i \, R \, x_j$. A most convenient way of representing a binary relation R on a set X is by a directed graph, the vertices of which stand for the elements of X and the edges stand for the ordered pairs of elements of X defining the relation R.

For example, the relation *is a factor of* on the set $X = \{2, 3, 4, 6, 9\}$ is shown in Figure 1.3.

Let R be a relation on a set $X = \{x_1, x_2, \ldots\}$. The relation R is called *reflexive* if $x_i R x_i$ for all $x_i \in X$. The relation R is said to be *symmetric* if $x_i R x_j$ implies $x_j R x_i$. The relation R is said to be *transitive* if $x_i R x_j$ and $x_j R x_k$ imply $x_i R x_k$. A relation R which is reflexive, symmetric, and transitive is called an *equivalence relation*. If R is an equivalence relation defined on a set S, then we can uniquely partition S into subsets $S_1, S_2, \ldots, S_k$ such that two elements x and y of S belong to S_i if and only if xRy. The subsets $S_1, S_2, \ldots, S_k$ are all called the *equivalence classes* induced by the relation R on the set S.

The directed graph representing a reflexive relation is called a *reflexive directed graph*. In a similar way *symmetric* and *transitive directed graphs* are defined. We now make the following observations about these graphs:

1. In a reflexive directed graph, there is a self-loop at each vertex.

2. In a symmetric directed graph, there are two oppositely oriented edges between any two adjacent vertices. Therefore, an undirected graph can be considered as representing a symmetric relation if we associate with each edge two oppositely oriented edges.

3. The edge (v_1, v_2) is present in a transitive graph G if there is a directed path in G from v_1 to v_2.

1.14 DIRECTED TREES AND ARBORESCENCES

A vertex v in a directed graph D is a *root* of D if there are directed paths from v to all the remaining vertices of D.

Theorem 1.19 *A directed graph D has a root if and only if it is quasi-strongly connected.* ∎

A directed graph D is a *tree* if the underlying undirected graph is a tree. A directed graph D is a *directed tree* or *arborescence* if D is a tree and has a root.

We present in the next theorem a number of equivalent characterizations of a directed tree.

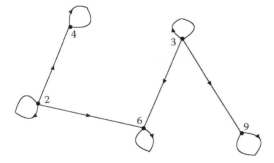

Figure 1.3 Directed graph representation of a relation on the set $X = \{2, 3, 4, 6, 9\}$.

Theorem 1.20 *Let D be a directed graph with $n > 1$ vertices. Then the following statements are equivalent:*

1. *D is a directed tree.*

2. *There exists a vertex r in D such that there is exactly one directed path from r to every other vertex of D.*

3. *D is quasi-strongly connected and loses this property if any edge is removed from it.*

4. *D is quasi-strongly connected and has a vertex r such that $d^-(r) = 0$.* ∎

Theorem 1.21 *A directed graph D has a directed spanning tree if and only if D is quasi-strongly connected.* ∎

1.15 DIRECTED EULERIAN GRAPHS

A *directed Euler trail* in a directed graph D is a closed directed trail that contains all the arcs of D. An *open directed Euler trail* is an open directed trail containing all the arcs of D. A directed graph possessing a directed Euler trail is called a *directed Eulerian graph*.

The following theorem gives simple and useful characterizations of directed Eulerian graphs.

Theorem 1.22 *The following statements are equivalent for a connected directed graph D.*

1. *D is a directed Eulerian graph.*

2. *For every vertex v of D, $d^-(v) = d^+(v)$.*

3. *D is the union of some edge-disjoint directed cycles.* ∎

Theorem 1.23 *A directed connected graph D possesses an open directed Euler trail if and only if the following conditions are satisfied:*

1. *In D there are two vertices v_1 and v_2, such that $d^+(v_1) = d^-(v_1) + 1$ and $d^-(v_2) = d^+(v_2) + 1$.*

2. *For every vertex v different from v_1 and v_2, we have $d^+(v) = d^-(v)$.* ∎

Theorem 1.24 *The number of directed Euler trails of a directed Eulerian graph D without self-loops is $\tau_d(D) \prod_{p=1}^{n} (d^-(v_p) - 1)!$, where n is the number of vertices in D and $\tau_d(D)$ is the number of directed spanning trees of G with v_1 as root.* ∎

Corollary 1.6 *The number of directed spanning trees of a directed Eulerian graph is the same for every choice of root.* ∎

1.16 DIRECTED HAMILTONIAN GRAPHS

A directed circuit in a directed graph D is a *directed Hamilton circuit* of D if it contains all the vertices of D. A directed path in D is a *directed Hamilton path* of G if it contains all the vertices of D. A digraph is a *directed Hamiltonian graph* if it has a directed Hamilton circuit. A directed graph D is *complete* if its underlying undirected graph is complete.

Theorem 1.25 *Let u be any vertex of a strongly connected complete directed graph with $n \geq 3$ vertices. For each $k, 3 \leq k \leq n$, there is a directed circuit of length k containing u.* ■

Corollary 1.7 *A strongly connected complete directed graph is Hamiltonian.* ■

Theorem 1.26 *Let D be a strongly connected n-vertex graph without parallel edges and self-loops. If for every vertex v in D, $d^-(v) + d^+(v) \geq n$, then D has a directed Hamilton circuit.* ■

Corollary 1.8 *Let D be a directed n-vertex graph without parallel edges or self-loops. If $\min(\delta^-, \delta^+) \geq n/2 > 1$, then D contains a directed Hamilton circuit.* ■

Theorem 1.27 *If a directed graph $D = (V, A)$ is complete, then it has a directed Hamilton path.* ■

1.17 ACYCLIC DIRECTED GRAPHS

A directed graph is *acyclic* if it has no directed circuits. Obviously the simplest example of an acyclic directed graph is a directed tree.

We can label the vertices of an n-vertex acyclic directed graph D with integers from the set $\{1, 2, \ldots, n\}$ such that the presence of the edge (i, j) in D implies that $i < j$. Note that the edge (i, j) is directed from vertex i to vertex j. Ordering the vertices in this manner is called *topological sorting*.

Theorem 1.28 *In an acyclic directed graph D there exists at least one vertex with zero indegree and at least one vertex with zero outdegree.* ■

Select any vertex with zero outdegree. Since D is acyclic, by Theorem 1.28, there is at least one such vertex in D. Label this vertex with the integer n. Now remove from D this vertex and the edges incident on it. Let D' be the resulting graph. Since D' is also acyclic, we can now select a vertex whose outdegree in D' is zero. Label this with the integer $n - 1$. Repeat this procedure until all the vertices are labeled. It is now easy to verify that this procedure results in a topological sorting of the vertices of D.

1.18 TOURNAMENTS

A *tournament* is an orientation of a complete graph. It derives its name from its application in the representation of structures of round-robin tournaments. In a round-robin tournament several teams play a game that cannot end in a tie, and each team plays every other team exactly once. In the directed graph representation of the round-robin tournament, vertices represent teams and an edge (v_1, v_2) is present in the graph if the team represented by the vertex v_1 defeats the team represented by the vertex v_2. Clearly, such a directed graph has no parallel edges and self-loops, and there is exactly one edge between any two vertices. Thus it is a tournament.

The teams participating in a tournament can be ranked according to their scores. The *score* of a team i is the number of teams it has defeated. This motivates the definition of the score sequence of a tournament.

The *score sequence* of an n-vertex tournament is the sequence $(s_1, s_2, \ldots, s_n)$ such that each s_i is the outdegree of a vertex of the tournament. An interesting characterization of a tournament in terms of the score sequence is given in the following theorem.

Theorem 1.29 *A sequence of nonnegative integers $s_1, s_2, \ldots, s_n$ is the score sequence of a tournament G if and only if*

1. $\sum_{i=1}^{n} s_i = \frac{n(n-1)}{2}$.

2. $\sum_{i=1}^{k} s_i = \frac{k(k-1)}{2}$ *for all $k < n$.* ∎

Suppose we can order the teams in a round-robin tournament such that each team precedes the one it has defeated. Then we can assign the integers $1, 2, \ldots, n$ to the teams to indicate their ranks in this order. Such a ranking is always possible since in a tournament there exists a directed Hamilton path and it is called *ranking* by a Hamilton path.

Note that ranking by a Hamilton path may not be the same as ranking by the score. Further, a tournament may have more than one directed Hamilton path. In such a case there will be more than one Hamilton path ranking. However, there exists exactly one directed Hamilton path in a transitive tournament. This is stated in the following theorem, which is easy to prove.

Theorem 1.30 *In a transitive tournament there exists exactly one directed Hamilton path.* ∎

1.19 COMPUTATIONAL COMPLEXITY AND COMPLETENESS

In assessing the efficiency of an algorithm two metrics are used; time complexity and space complexity. In this book we will be mainly concerned with time complexities of algorithms. The computational time complexity of an algorithm is a measure of the number of primitive operations (low level instructions) performed during the execution of the algorithm. We will use what is known as the random access machine (RAM) model in which it is assumed that the computer can perform any primitive operation in a constant number of steps which do not depend on the size of the input.

A function $t(n)$ is said to be $O(g(n))$ if there exist some constant c and some nonnegative integer n_0 such that

$$t(n) \leq c\, g(n) \text{ for all } n \geq n_0$$

This definition is referred to as the *big-Oh* notation. Usually it is pronounced as $t(n)$ *is big Oh of $g(n)$*. We can also say as $t(n)$ *is order $g(n)$*. For example $5n^2 + 100n$ is $O(n^2)$.

A function $t(n)$ is said to be $\Omega(g(n))$, if there exist some positive constant c and some nonnegative integer n_0 such that

$$t(n) \geq c\, g(n) \text{ for all } n \geq n_0$$

This definition is referred to as the *big-Omega* notation. Usually it is pronounced as $t(n)$ *is big-Omega of $g(n)$*. For example, n^3 is $\Omega(n^2)$.

A function $t(n)$ is said to be $\Theta(g(n))$, if $t(n)$ is $O(g(n))$ and $\Omega(g(n))$. It is pronounced as $t(n)$ *is big-Theta of $g(n)$*. For example, $(1/2)\, n(n-1)$ is $\Theta(n^2)$.

There are several texts that discuss the above and other asympototic notations as well as properties involving them. For example, see Levitin [9], Goodrich and Tamassia [10], and Cormen et al. [11].

An algorithm for a problem is efficient [12] if there exists a polynomial $p(k)$ such that an instance of the problem whose input length is k takes at most $p(k)$ elementary computational steps to solve. In other words any algorithm of polynomial time complexity is accepted to be efficient. There are several problems that defy polynomial time algorithms in spite of massive efforts to solve them efficiently. This family includes several important problems such as the traveling salesman problem, the graph coloring problem, the problem of simplifying Boolean functions, scheduling problems, and certain covering problems.

In this section we present a brief introduction to the theory of NP-completeness. A *decision* problem is one that asks only for a yes or no answer. For example the question *Can this graph be 5-colored?* is a decision problem. Many of the important optimization problems can be phrased as decision problems. Usually, if we find a fast algorithm for a decision problem, then we will be able to solve the corresponding original problem also efficiently. For instance, if we have a fast algorithm to solve the decision problem for graph coloring, by repeated applications (in fact, $n \log n$ applications) of this algorithm, we can find the chromatic number of an n-vertex graph.

A decision problem belongs to the class P if there is a polynomial time algorithm to solve the problem. A verification algorithm is an algorithm A which takes as input an instance of a problem and a candidate solution to the problem, called a *certificate* and verifies in polynomial time whether the certificate is a solution to the given problem instance. The class NP is the class of problems which can be verified in polynomial time.

The fundamental open question in computational complexity is whether the class P equals the class NP. By definition, the class NP contains all problems in class P. It is not known, however, whether all problems in NP can be solved in polynomial time.

In an effort to determine whether $P = NP$, Cook [13] defined the class of NP-complete problems. We say that a problem P_1 is polynomial-time reducible to a problem P_2, written $P_1 \leq_p P_2$, if

1. There exists a function f which maps any instance I_1 of P_1 to an instance of P_2 in such a way that I_1 is a *yes* instance of P_1 if and only if $f(I_1)$ is a *yes* instance of P_2.

2. For any instance I_1, the instance $f(I_1)$ can be constructed in polynomial time.

If P_1 is polynomial-time reducible to P_2, then any algorithm for solving P_2 can be used to solve P_1. We define a problem P to be NP-complete if (1) $P \in NP$, and (2) for every problem $P' \in NP, P' \leq_p P$. If a problem P can be shown to satisfy condition (2), but not necessarily condition (1), then we say that it is NP-hard. Let NPc denote the class of NP-complete problems.

The relation $\leq_p$ is transitive. Because of this, a method frequently used in demonstrating that a given problem is NP-complete is the following:

1. Show that $P \in NP$.

2. Show that there exists a problem $P' \in NPc$, such that $P' \leq_p P$.

It follows from the definition of NP-completeness that if any problem in NPc can be solved in polynomial time, then every problem in NPc can be solved in polynomial time and $P = NP$. On the other hand, if there is some problem in NPc that cannot be solved in polynomial time, then no problem in NPc can be solved in polynomial time.

Cook [13] proved that there is an NP-complete problem. The satisfiability problem is defined as follows:

Let $X = \{x_1, x_2, \ldots, x_n\}$ be a set of Boolean variables. A literal is either a variable x_i or its complement $\overline{x_i}$. Thus the set of literals is $L = \{x_1, x_2, \ldots, x_n, \overline{x_1}, \overline{x_2}, \ldots, \overline{x_n}\}$. A clause C is a subset of L. The *satisfiability problem* (SAT) is: Given a set of clauses, does there exist a set of truth values (T or F), one for each variable, such that every clause contains at least one literal whose value is T. Cook's proof that SAT is NP-complete opened the way to demonstrate the NP-completeness of a vast number of problems. The second problem to be

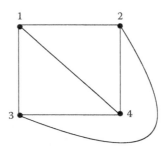

Figure 1.4 Graph for the illustration of the concept of reducibility.

proved NP-complete is 3-SAT the 3-*satisfiability problem*, which is the special case of SAT in that only three literals are permitted in each clause.

The list of NP-complete problems has grown very rapidly since Cook's work. Karp [14,15] demonstrated the NP-completeness of a number of combinatorial problems. Garey and Johnson [16] is the most complete reference on NP-completeness and is highly recommended. Other good textbooks recommended for this study include Horowitz and Sahni [17], Melhorn [18], and Aho et al. [19]. For updates on NP-completeness see the article titled "NP-completeness: An ongoing guide" in the *Journal of Algorithms*.

Since the concept of reducibility plays a dominant role in establishing the NP-completeness of a problem, we shall illustrate this with an example.

Consider the graph $G = (V, E)$ in Figure 1.4 and the decision problem *Can the vertices of G be 3-colored?* We now reduce this problem to an instance of SAT.

We define 12 Boolean variables $x_{i,j}$ ($i = 1,2,3,4; j = 1,2,3$) where the variable $x_{i,j}$ corresponds to the assertion that *vertex i has been assigned color j*. The clauses are defined as follows:

$$\begin{aligned}
C(i) &= \{x_{i,1}, x_{i,2}, x_{i,3}\}, 1 \leq i \leq 4; \\
T(i) &= \{\overline{x}_{i,1}, \overline{x}_{i,2}\}, 1 \leq i \leq 4; \\
U(i) &= \{\overline{x}_{i,1}, \overline{x}_{i,3}\}, 1 \leq i \leq 4; \\
V(i) &= \{\overline{x}_{i,2}, \overline{x}_{i,3}\}, 1 \leq i \leq 4; \\
D(e,j) &= \{\overline{x}_{u,j}, \overline{x}_{v,j}\}, \text{ for every } e = (u,v) \in E, 1 \leq j \leq 3.
\end{aligned}$$

Whereas, $C(i)$ asserts that each vertex i has been assigned at least one color, the clauses $T(i), U(i)$ and $V(i)$ together assert that no vertex has been assigned more than one color. The clauses $D(e,j)$'s guarantee that the coloring is proper (adjacent vertices have been assigned distinct colors).

Thus, the graph of Figure 1.4 is 3-colorable if and only if there exists an assignment of truth values T and F to the 12 Boolean variables $x_{1,1}, x_{1,2}, \ldots, x_{4,3}$ such that the each of the clauses contains at least one literal whose value is T.

References

[1] J. A. Bondy and U. S. R. Murty, *Graph Theory*, Springer, Berlin, Germany, 2008.

[2] G. Chartrand and L. Lesniak, *Graphs and Digraphs*, 4th Edition, CRC Press, Boca Raton, FL, 2005.

[3] M. N. S. Swamy and K. Thulasiraman, *Graphs, Networks and Algorithms*, Wiley-Interscience, New York, 1981.

[4] K. Thulasiraman and M. N. S. Swamy, *Graphs: Theory and Algorithms*, Wiley-Interscience, New York, 1992.

[5] D. B. West, *Introduction to Graph Theory*, 2nd Edition, Prentice Hall, Upper Saddle River, NJ, 2001.

[6] V. Chavátal, On Hamilton's ideals, *J. Comb. Th. B*, **12** (1972), 163–168.

[7] G. A. Dirac, Some theorems on abstract graphs, *Proc. London Math. Soc.*, **2** (1952), 69–81.

[8] O. Ore, Note on Hamilton circuits, *Amer. Math. Monthly*, **67** (1960), 55.

[9] A. Levitin, *Introduction to the Design and Analysis of Algorithms*, Pearson, Boston, MA, 2012.

[10] M. T. Goodrich and R. Tamassia, *Algorithm Design: Foundations, Analysis and Internet Examples*, John Wiley & Sons, New York, 2002.

[11] T. H. Cormen, C. E. Liercersona, and R. L. Rivest, *Introduction to Algorithms*, MIT Press, Cambridge, MA, 1990.

[12] J. Edmonds, Paths, trees and flowers, *Canad. J. Math.*, **17** (1965), 449–467.

[13] S. A. Cook, The complexity of theorem proving procedures, *Proc. 3rd ACM Symp. on Theory of Computing*, ACM, New York, 1971, 151–158.

[14] R. M. Karp, Reducibility among combinatorial problems, *Complexity of Computer Communications*, R. E. Miller and J. W. Thatcher, Eds., Plenum Press, New York, 1972, 85–104.

[15] R. M. Karp, On the computational complexity of combinatorial problems, *Networks*, **5** (1975), 45–68.

[16] M. R. Garey and D. S. Johnson, *Computers and Intractability: A Guide to the Theory of NP-Completeness*, Freeman, San Francisco, CA, 1979.

[17] E. Horowitz and S. Sahni, *Fundamentals of Computer Algorithms*, Computer Science Press, Potomac, MD, 1978.

[18] K. Melhorn, *Graph Algorithms and NP-Completeness*, Springer-Verlag, New York, 1984.

[19] A. V. Aho, J. E. Hopcroft, and J. D. Ullman *The Design and Analysis of Computer Algorithms*, Addison-Wesley, Reading, MA, 1974.

CHAPTER 2

Basic Graph Algorithms*

Krishnaiyan "KT" Thulasiraman

CONTENTS

2.1	Introduction	21
2.2	Minimum Weight Spanning Tree	22
2.3	Optimum Branchings	25
2.4	Transitive Closure	30
2.5	Shortest Paths	35
	2.5.1 Single Source Shortest Paths: Bellman–Ford–Moore Algorithm	36
	2.5.1.1 Negative Cycle Detection	38
	2.5.1.2 Shortest Path Tree	38
	2.5.2 Single Source Shortest Paths in Graphs with No Negative Length Edges: Dijkstra's Algorithm	39
	2.5.3 All Pairs Shortest Paths	42
2.6	Transitive Orientation	44

2.1 INTRODUCTION

Graphs arise in the study of several practical problems. The first step in such studies is to discover graph theoretic properties of the problem under consideration that would help us in the formulation of a method of solution to the problem. Usually solving a problem involves analysis of a graph or testing a graph for some specified property. Graphs that arise in the study of real-life problems are very large and complicated. Analysis of such graphs in an efficient manner, therefore, involves the design of efficient computer algorithms.

In this and the next chapter we discuss several basic graph algorithms. We consider these algorithms to be basic in the sense that they serve as building blocks in the design of more complex algorithms. While our main concern is to develop the theoretical foundation on which the design of the algorithms is based, we also develop results concerning the computational complexity of these algorithms.

The computational complexity of an algorithm is a measure of the running time of the algorithm. Thus it is a function of the size of the input. In the case of graph algorithms, complexity results will be in terms of the number of vertices and the number of edges in the graph. In the following function $g(n)$ is said to be $O(f(n))$ if and only if there exist constants c and n_0 such that $|g(n)| \leq c|f(n)|$ for all $n \geq n_0$. Furthermore, all our complexity results will be with respect to the worst-case analysis.

There are different methods of representing a graph on a computer. Two of the most common methods use the adjacency matrix and the adjacency list. Adjacency matrix representation is not a very efficient one in the case of sparse graphs. In the adjacency

---

*This chapter is an edited version of Sections 14.1, 14.2, 15.1, and 15.8 in Swamy and Thulasiraman [17].

list representation, we associate with each vertex a list that contains all the edges incident on it. A detailed discussion of data structures for representing a graph may be found in some of the references listed at the end of this chapter.

2.2 MINIMUM WEIGHT SPANNING TREE

Consider a weighted connected undirected graph G with a nonnegative real weight $w(e)$ associated with each edge e of G. The weight of a subgraph of G will refer to the sum of the weights of the edges of the subgraph. In this section we discuss the problem of constructing a minimum weight spanning tree of G. We present two algorithms for this problem, namely, Kruskal's algorithm [1] and the one due to Prim [2]. Theorems 2.1 through 2.3 provide the basis of these algorithms. In the following, $T + e$ denotes the subgraph $T \cup \{e\}$. Similarly, $T - e$ denotes the subgraph that results after deleting edge e from T. Also, contraction of an edge $e = (u, v)$ results in a new graph G' in which the vertices u and v are replaced by a single vertex w and all the edges in G that are incident on u or v are incident on vertex w in G'. The self loop on w that results because of the coalescing of u and v will not be present in G'.

Theorem 2.1 *Consider a vertex v in a weighted connected graph G. Among all the edges incident on v, let e be one of minimum weight. Then, G has a minimum weight spanning tree that contains e.*

Proof. Let $T_{\min}$ be a minimum weight spanning tree of G. If $T_{\min}$ does not contain e, then a circuit C is created when e is added to $T_{\min}$. Let e' be the edge of C that is adjacent to e. Clearly $e' \in T_{\min}$. Also $T' = T_{\min} - e' + e$ is a spanning tree of G. Since e and e' are both incident on v, we get $w(e) \leq w(e')$. But $w(e) \geq w(e')$ because $w(T') = w(T_{\min}) - w(e') + w(e) \geq w(T_{\min})$. So, $w(e) = w(e')$ and $w(T') = w(T_{\min})$. Thus we have found a minimum weight spanning tree, namely T', containing e. ∎

If the fundamental circuit with respect to a chord c of a spanning tree T contains branch b, then $T - b + c$ is also a spanning tree of G (see Chapter 7). Using this result we can prove the statements in Theorem 2.2 along the same lines as the proof of Theorem 2.1.

Theorem 2.2 *Let e be a minimum weight edge in a weighted connected graph G. Then*

1. *G has a minimum weight spanning tree that contains e.*

2. *If T_{min} is a minimum weight spanning tree of G, then for every chord c $w(b) \leq w(c)$, for every branch b in the fundamental circuit of T_{min} with respect to c.*

3. *If T_{min} is a minimum weight spanning tree of G, then for every branch $b \in T_{min}$ $w(b) \leq w(c)$, for every chord c in the fundamental cutset of T_{min} with respect to b.* ∎

Theorem 2.3 *Let T be an acyclic subgraph of a weighted connected graph G such that there exists a minimum weight spanning tree containing T. If G' denotes the graph obtained by contracting the edges of T, and T'_{min} is a minimum weight spanning tree of G', then $T'_{min} \cup T$ is a minimum weight spanning tree of G.*

Proof. Let $T_{\min}$ be a minimum weight spanning tree of G containing T. Let $T_{\min} = T \cup T'$. Then T' is a spanning tree of G'. Therefore

$$w(T') \geq w(T'_{\min}). \tag{2.1}$$

It is easy to see that $T'_{\min} \cup T$ is also a spanning tree of G. So

$$w(T'_{\min} \cup T) \geq w(T_{\min}) = w(T) + w(T'). \tag{2.2}$$

From this we get
$$w(T'_{\min}) \geq w(T'). \qquad (2.3)$$

Combining (2.1) and (2.3) we get $w(T'_{\min}) = w(T')$, and so $w(T'_{\min} \cup T) = w(T' \cup T) = w(T_{\min})$. Thus $T'_{\min} \cup T$ is a minimum weight spanning tree of G. ∎

We now present Kruskal's algorithm.

Algorithm 2.1 Minimum weight spanning tree (Kruskal)

Input: $G = (V, E)$ is the given nontrivial n-vertex weighted connected graph with m edges. The edges are ordered according to their weights, that is, $w(e_1) \leq w(e_2) \leq \cdots \leq w(e_m)$.
Output: A minimum weight spanning tree of G. The edges $e'_1, e'_2, \ldots, e'_{n-1}$ will be the required spanning tree.
begin
    $k \leftarrow 0$;
    $T_0 \leftarrow \phi$;
    for $i = 1$ to m **do**
        If $T_k + e_i$ is acyclic **then**
            begin
                $k \leftarrow k + 1$;
                $e'_k \leftarrow e_i$;
                $T_k \leftarrow T_{k-1} + e'_k$;
            end
end

Kruskal's algorithm essentially proceeds as follows. Edges are first sorted in the order of nondecreasing weights and then examined, one at a time, for inclusion in a minimum weight spanning tree. An edge is included if it does not form a circuit with the edges already selected.

Next we prove the correctness of Kruskal's algorithm.

Theorem 2.4 *Kruskal's algorithm constructs a minimum weight spanning tree of a weighted connected graph.*

Proof. Let G be the given nontrivial weighted connected graph. Clearly, when Kruskal's algorithm terminates, the edges selected will form a spanning tree of G. Thus the algorithm terminates with $k = n - 1$, and T_{n-1} as a spanning tree of G.

We next establish that T_{n-1} is indeed a minimum weight spanning tree of G by proving that every T_k, $k \geq 1$ constructed in the course of Kruskal's algorithm is contained in a minimum weight spanning tree of G. Our proof is by induction on k.

Clearly by Theorem 2.1, G has a minimum weight spanning tree that contains the edge $e'_1 = e_1$. In other words $T_1 = \{e_1\}$ is contained in a minimum weight spanning tree of G. As inductive hypothesis, assume that T_k, for some $k \geq 1$ is contained in a minimum weight spanning tree of G. Let G' be the graph obtained by contracting the edges of T_k. Then the edge e'_{k+1} selected by the algorithm will be a minimum weight edge in G'. So, by Theorem 2.2, the edge e'_{k+1} is contained in a minimum weight spanning tree $T'_{\min}$ of G'. By Theorem 2.3, $T_k \cup T'_{\min}$ is a minimum weight spanning tree of G. More specifically $T_{k+1} = T_k + e'_{k+1}$ is contained in a minimum weight spanning tree of G and the correctness of Kruskal's algorithm follows. ∎

We next present another algorithm due to Prim [2] to construct a minimum weight spanning tree of a weighted connected graph.

> **Algorithm 2.2 Minimum weight spanning tree (Prim)**
>
> **Input:** $G = (V, E)$ is the given nontrivial n-vertex weighted connected graph.
> **Output:** A minimum weight spanning tree of G. The edges $e_1, e_2, \ldots, e_{n-1}$ will form the required minimum spanning tree.
> **begin**
> $i \leftarrow 1$;
> $T_0 \leftarrow \phi$;
>   Select any vertex $v \in V$;
> $S \leftarrow \{v\}$;
>   **while** $i \leq n - 1$ **do**
>   **begin**
>   Select an edge $e_i = (p, q)$ of minimum weight such that e_i has exactly one end vertex, say p, in S;
> $T_i \leftarrow T_{i-1} + e_i$;
>   Add vertex q to the set S;
> $i \leftarrow i + 1$;
>   **end**
> **end**

As in Kruskal's algorithm, Prim's algorithm also constructs a sequence of acyclic subgraphs $T_1, T_2, \ldots$, and terminates with T_{n-1}, a minimum weight spanning tree of G. The subgraph T_{i+1} is constructed from T_i by adding an edge of minimum weight with exactly one end vertex in T_i. This construction ensures that all T_i's are connected. If G' denotes the graph obtained by contracting the edges of T_i, and W denotes the vertex of G', which corresponds to the vertex set of T_i, then e_{i+1} is in fact a minimum weight edge incident on W in G'. This observation and Theorems 2.1 and 2.3 can be used (as in the proof of Theorem 2.4) to prove that each T_i is contained in a minimum weight spanning tree of G. This would then establish that the spanning tree produced by Prim's algorithm is a minimum weight spanning tree of G.

Both Prim's and Kruskal's algorithms can be viewed as special cases of a more general version. Both these algorithms construct a sequence of acyclic subgraphs $T_1, T_2, \ldots, T_{n-1}$ with T_{n-1} as a minimum weight spanning tree. Each T_{i+1}, is constructed by adding an edge e_{i+1} to T_i. They differ in the way e_{i+1} is selected. As before, let G' denote the graph obtained by contracting the edges of T_i, and W denote the (super) vertex of G', which corresponds to the vertex set of T_i. Then the two algorithms select weight edge e_{i+1} as follows.

Prim: e_{i+1} is a minimum weight edge incident on W in G'.

>Note: This is a restricted application of Theorem 2.1 since this theorem is applicable to any vertex in G'.

Kruskal: e_{i+1} is a minimum weight edge in G'.

>Note: This is also a restricted application of Theorem 2.1 since this theorem only requires e_{i+1} to be the minimum weight edge incident on some vertex in G'.

Both these algorithms result in a minimum weight spanning tree since the edge e_{i+1} selected as above is in a minimum weight spanning tree of G' as proved in Theorems 2.1 and 2.3.

The following algorithm which unifies both Prim's and Kruskal's algorithms is slightly more general than both.

> **Algorithm 2.3 Minimum weight spanning tree (Unified version)**
>
> **Input:** $G = (V, E)$ is the given nontrivial n-vertex weighted connected graph.
> **Output:** A minimum weight spanning tree of G. The edges $e_1, e_2, \ldots, e_{n-1}$ will form the required minimum spanning tree.
> **begin**
> $i \leftarrow 1$;
> $T_0 \leftarrow \phi$;
> $G' \leftarrow G$;
>   **while** $i \leq n - 1$ **do**
>   **begin**
>   Select an edge $e_i = (p, q)$ such that e_{i+1} is a minimum weight edge incident on any vertex in G';
>   Contract the edge e_i in G' and let G' denote the contracted graph;
> $T_i \leftarrow T_{i-1} + e_i$;
> $i \leftarrow i + 1$;
>   **end**
> **end**

For complexity results relating to the minimum weight spanning tree enumeration problem see Kerschenbaum and Van Slyke [3], Yao [4], and Cheriton and Tarjan [5]. For sensitivity analysis of minimum weight spanning trees and shortest path trees see Tarjan [6]. See Papadimitriou and Yannakakis [7] for a discussion of the complexity of restricted minimum weight spanning tree problems. For a history of the minimum weight spanning tree problem see Graham and Hall [8]. The complexity of Kruskal's algorithm is $O(m \log n)$ (see Korte and Vygen [9]). Clearly the complexity of Prim's algorithm is $O(n^2)$. For more sophisticated implementations see [10–13]. See also [14] for an algorithm due to Jarnik.

2.3 OPTIMUM BRANCHINGS

Consider a weighted directed graph $G = (V, E)$. Let $w(e)$ be the weight of edge e. Recall that the weight of a subgraph of G is defined to be equal to the sum of the weights of all the edges in the subgraph.

A subgraph G_s of G is a branching in G if G_s has no directed circuits and the in-degree of each vertex of G_s is at most 1. Clearly each component of G_s is a directed tree*. A branching of maximum weight is called an *optimum branching*.

In this section we discuss an algorithm due to Edmonds [15] for computing an optimum branching of G. Our discussion here is based on Karp [16].

An edge $e = (i, j)$ directed from vertex i to vertex j is critical if

1. $w(e) > 0$.

2. $w(e) \geq w(e')$ for every edge $e' = (k, j)$ incident into j.

A spanning subgraph H of G is a critical subgraph of G if

1. Every edge of H is critical.

2. The in-degree of every vertex of H is at most 1.

---

*See Chapter 1 for the definition of a directed tree, also called an arborescence.

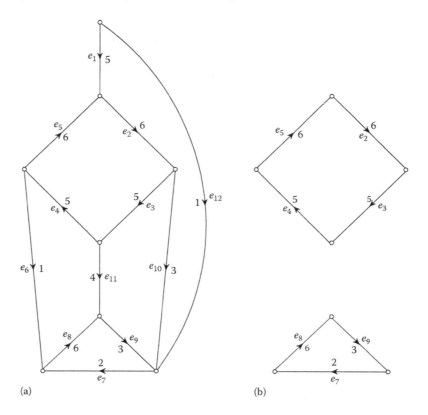

Figure 2.1 (a) A directed graph G. (b) A critical subgraph of G. (Data from M. N. Swamy and K. Thulasiraman, *Graphs, Networks and Algorithms*, Wiley-Interscience, 1981.)

A directed graph G and a critical subgraph H of G are shown in Figure 2.1. It is easy to see that

1. Each component of a critical subgraph contains at most one circuit, and such a circuit will be a directed circuit.

2. A critical subgraph with no circuits is an optimum branching of G.

Consider a branching B. Let $e = (i, j)$ be an edge not in B, and let e' be the edge of B incident into vertex j. Then e is eligible relative to B if

$$B' = (B \cup e) - e'$$

is a branching.

For example, the edges $\{e_1, e_2, e_3, e_4, e_7, e_8\}$ form a branching B of the graph of Figure 2.1. The edge e_6, not in B, is eligible relative to B since

$$(B \cup e_6) - e_7$$

is a branching of this graph.

Lemmas 2.1 and 2.2 are easy to prove, and they lead to Theorem 2.5, which forms the basis of Karp's proof of the correctness of Edmonds' algorithm. In the following the edge set of a subgraph H will also be denoted by H.

Lemma 2.1 *Let B be a branching, and let $e = (i, j)$ be an edge not in B. Then e is eligible relative to B if and only if in B there is no directed path from j to i.* ∎

Lemma 2.2 *Let B be a branching and let C be a directed circuit such that no edge of $C - B$ is eligible relative to B. Then $|C - B| = 1$.* ∎

Theorem 2.5 *Let H be a critical subgraph. Then there exists an optimum branching B such that, for every directed circuit C in H, $|C - B| = 1$.*

Proof. Let B be an optimum branching that, among all optimum branchings, contains a maximum number of edges of the critical subgraph.

Consider any edge $e \in H - B$ that is incident into vertex j, and let e' be the edge of B incident into j. If e were eligible, then

$$(B \cup e) - e'$$

would also be an optimum branching, containing a larger number of edges of H than B does; a contradiction. Thus no edge of $H - B$ is eligible relative to B. So, by Lemma 2.2, for each circuit C in H, $|C - B| = 1$. ∎

Let $C_1, C_2, \ldots, C_k$ be the directed circuits in H. Note that no two circuits of H can have a common edge. In other words, these circuits are edge-disjoint. Let e_i^0 be an edge of minimum weight in C_i, $i \geq 1$.

Corollary 2.1 *There exists an optimum branching B such that:*

1. *$|C_i - B| = 1$, $i = 1, 2, \ldots, k$.*

2. *If no edge of $B - C_i$ is incident into a vertex in C_i, $i = 1, 2, \ldots, k$, then*

$$C_i - B = e_i^0. \tag{2.4}$$

Proof. Among all optimum branchings that satisfy item 1, let B be a branching containing a minimum number of edges from the set $\{e_1^0, e_2^0, \ldots, e_k^0\}$. We now show that B satisfies item 2.

If not, suppose that, for some i, $e_i^0 \in B$, but that no edge of $B - C_i$ is incident into a vertex in C_i. Let $e = C_i - B$. Then $(B - e_i^0) \cup e$ is clearly an optimum branching that satisfies item 1 but has fewer edges than B from the set $\{e_1^0, e_2^0, \ldots, e_k^0\}$. This is a contradiction. ∎

This result is very crucial in the development of Edmonds' algorithm. It suggests that we can restrict our search for optimum branchings to those that satisfy (2.4).

Next we construct, from the given graph G, a simpler graph G' and show how to construct from an optimum branching of G' an optimum branching of G that satisfies (2.4).

As before, let H be the critical subgraph of G and let $C_1, C_2, \ldots, C_k$ be the directed circuits in H. The graph G' is constructed by contracting all the edges in each $C_i = 1, 2, \ldots, k$. In G', vertices of each circuit C_i are represented by a single vertex a_i, called a *pseudo-vertex*. The weights of the edges of G' are the same as those of G, except for the weights of the edges incident into the pseudo-vertices. These weights are modified as follows.

Let $e = (i, j)$ be an edge of G such that j is a vertex of some circuit C_r and i is not in C_r. Then in G', e is incident into the pseudo-vertex a_r. Define $\tilde{e}$ as the unique edge in C_r that is incident into vertex j. Then in G' the weight of e, denoted by $w'(e)$, is given by

$$w'(e) = w(e) - w(\tilde{e}) + w(e_r^0). \tag{2.5}$$

For example, consider the edge e_1 incident into the directed circuit $\{e_2, e_3, e_4, e_5\}$ of the critical subgraph of the graph G of Figure 2.1. Then $\tilde{e}_1 = e_5$, and the weight of e_1 in G' is given by

$$w'(e_1) = w(e_1) - w(e_5) + w(e_4)$$
$$= 5 - 6 + 5$$
$$= 4.$$

Note that e_4 is a minimum weight edge in the circuit $\{e_2, e_3, e_4, e_5\}$.

Let E and E', respectively, denote the edge sets of G and G'. We now show how to construct from a branching B' of G' a branching B of G that satisfies (2.4) and vice versa.

For any branching B of G that satisfies (2.4) it is easy to see that

$$B' = B \cap E' \tag{2.6}$$

is a branching of G'. Furthermore, B' as defined is unique for a given B.

Next consider a branching B' of G'. For each C_i, let us define C_i' as follows:

1. If the in-degree in B' of a pseudo-vertex a_i is zero, then

$$C_i' = C_i - e_i^0.$$

2. If the in-degree in B' of a_i is nonzero, and e is the edge of B' incident into a_i, then

$$C_i' = C_i - \tilde{e}.$$

Now it is easy to see that

$$B = B' \bigcup_{i=1}^{k} C_i'. \tag{2.7}$$

is a branching of G that satisfies (2.4). Furthermore, B as defined is unique for a given B'.

Thus we conclude that there is a one-to-one correspondence between the set of branchings of G that satisfy (2.4) and the set of branchings of G'. Also, the weights of the corresponding branchings B and B' satisfy

$$w(B) - w(B') = \sum_{i=1}^{k} w(C_i) - \sum_{i=1}^{k} w(e_i^0). \tag{2.8}$$

This property of B and B' implies that if B is an optimum branching of G that satisfies (2.4), then B' is an optimum branching of G' and vice versa. Thus we have proved the following theorem.

Theorem 2.6 *There exists a one-to-one correspondence between the set of all optimum branchings in G that satisfy (2.4) and the set of all optimum branchings in G'.* ∎

Edmonds' algorithm for constructing an optimum branching is based on Theorem 2.6 and is as follows:

Basic Graph Algorithms ■ 29

Algorithm 2.4 Optimum branching (Edmonds)

S1. From the given graph $G = G_0$ construct a sequence of graphs $G_0, G_1, G_2, \ldots, G_k$, where
1. G_k is the first graph in the sequence whose critical subgraph is acyclic and
2. G_i, $1 \leq i \leq k$, is obtained from G_{i-1} by contracting the circuits in the critical subgraph H_{i-1} of G_{i-1} and altering the weights as in (2.5).

S2. Since H_k is acyclic, it is an optimum branching in G_k. Let $B_k = H_k$.
Construct the sequence $B_{k-1}, B_{k-2}, \ldots, B_0$, where
1. B_i, $0 \leq i \leq k-1$, is an optimum branching of G_i.
2. B_i, for $i \geq 0$, is constructed by expanding, as in (2.7), pseudo-vertices in B_{i+1}.

As an example let G_0 be the graph in Figure 2.1a, and let H_0 be the graph in Figure 2.1b. H_0 is the critical subgraph of G_0. After contracting the edges of the circuits in H_0 and modifying the weights, we obtain the graph G_1 shown in Figure 2.2a. The critical subgraph H_1 of G_1 is shown in Figure 2.2b. H_1 is acyclic. So it is an optimum branching of G_1. An optimum branching of G_0 is obtained from H_1 by expanding the pseudo-vertices a_1 and a_2 (which correspond to the two directed circuits in H_0), and it is shown in Figure 2.2c.

The running time of Edmonds' optimum branching algorithm is $O(mn)$, where m is the number of edges and n is the number of vertices. Tarjan [18] gives an $O(mn \log n)$

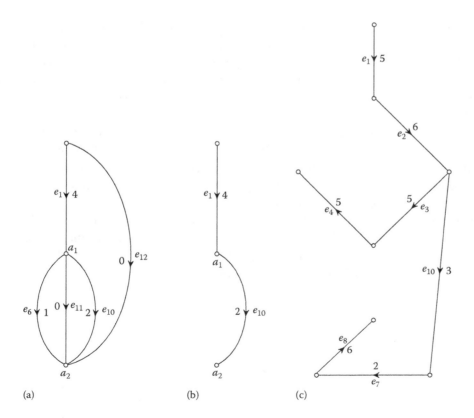

Figure 2.2 (a) Graph G_1. (b) H_1 a critical subgraph of G_1. (c) An optimum branching of graph G of Figure 2.1a. (Data from M. N. Swamy and K. Thulasiraman, *Graphs, Networks and Algorithms*, Wiley-Interscience, 1981.)

2.4 TRANSITIVE CLOSURE

A binary relation R on a set is a collection of ordered pairs of the elements of the set. If $(x, y) \in R$, then we say that x is related to y and denote this relationship by $x \, R \, y$. A relation R is transitive if $x \, R \, y$ whenever there exists a sequence

$$x_0 = x, x_1, x_2, \ldots, x_k = y$$

such that $k > 0$ and $x_0 \, R \, x_1$, $x_1 \, R \, x_2, \ldots$, and $x_{k-1} \, R \, x_k$.

The *transitive closure* of a binary relation R is a relation R^* defined as follows: $x \, R^* \, y$ if and only if there exists a sequence

$$x_0 = x, x_1, x_2, \ldots, x_k = y$$

such that $k > 0$ and $x_0 \, R \, x_1$, $x_1 \, R \, x_2, \ldots$, and $x_{k-1} \, R \, x_k$.

Clearly, if $x \, R \, y$, then $x \, R^* \, y$. Hence $R \subseteq R^*$. Further, it can be easily shown that R^* is transitive. In fact, it is the smallest transitive relation containing R. So if R is transitive, then $R^* = R$.

A binary relation R on a set S can be represented by a directed graph G in which each vertex corresponds to an element of S and (x, y) is a directed edge of G if an only if $x \, R \, y$. The directed graph G^* representing the transitive closure R^* of R is called the transitive closure of G. It follows from the definition of R^* that the edge (x, y), $x \neq y$, is in G^* if and only if there exists in G a directed path from the vertex x to the vertex y. Similarly the self-loop (x, x) at vertex x is in G^* if and only if there exists in G a directed circuit containing x. For example, the graph shown in Figure 2.3b is the transitive closure of the graph of Figure 2.3a.

Suppose that we define the *reachability matrix* of an n-vertex directed graph G as an $n \times n$ $(0, 1)$ matrix in which the (i, j) entry is equal to 1 if and only if there exists a directed path from vertex i to vertex j when $i \neq j$, or a directed circuit containing vertex i when $i = j$. In other words the (i, j) entry of the reachability matrix is equal to 1 if and only if vertex j is reachable from vertex i through a path of directed edges in G. The problem of constructing the transitive closure of a directed graph arises in several applications (e.g., see Gries [22]). In this section we discuss an elegant and computationally efficient algorithm due to Warshall [23] for computing the transitive closure of a directed graph. We also discuss a variation of Warshall's algorithm given by Warren [24].

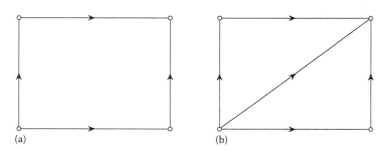

Figure 2.3 (a) Graph G. (b) G^*, transitive closure of G. (Data from M. N. Swamy and K. Thulasiraman, *Graphs, Networks and Algorithms*, Wiley-Interscience, 1981.)

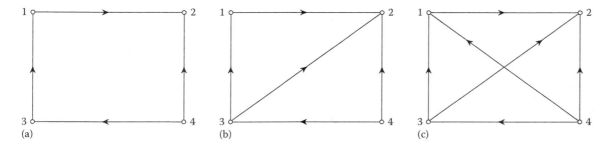

Figure 2.4 Illustration of Warshall's algorithm. (a) G^0. (b) $G^1 = G^2$. (c) $G^3 = G^4$. (Data from M. N. Swamy and K. Thulasiraman, *Graphs, Networks and Algorithms*, Wiley-Interscience, 1981.)

Let G be an n-vertex directed graph with its vertices denoted by the integers $1, 2, \ldots, n$. Let $G^0 = G$. Warshall's algorithm constructs a sequence of graphs so that $G^i \subseteq G^{i+1}$, $0 \leq i \leq n-1$, and G^n is the transitive closure of G. The graph G^i, $i \geq 1$, is obtained from G^{i-1} by processing vertex i in G^{i-1}. Processing vertex i in G^{i-1} involves addition of new edges to G^{i-1} as described next.

Let, in G^{i-1}, the edges (i, k), (i, l), $(i, m), \ldots$ be incident out of vertex i. Then for each edge (j, i) incident into vertex i, add to G^{i-1} the edges (j, k), (j, l), $(j, m), \ldots$ if these edges are not already present in G^{i-1}. The graph that results after vertex i is processed is denoted as G^i. Warshall's algorithm is illustrated in Figure 2.4.

It is clear that $G^i \subseteq G^{i+1}$, $i \geq 0$. To show that G^n is the transitive closure of G we need to prove the following result.

Theorem 2.7

1. *Suppose that, for any two vertices s and t, there exists in G a directed path P from vertex s to vertex t such that all its vertices other than s and t are from the set $\{1, 2, \ldots, i\}$. Then G^i contains the edge (s, t).*

2. *Suppose that, for any vertex s, there exists in G a directed circuit C containing s such that all its vertices other than s are from the set $\{1, 2, \ldots, i\}$. Then G^i contains the self-loop (s, s).*

Proof.

1. Proof is by induction on i.

    Clearly the result is true for G^1 since Warshall's construction, while processing vertex 1, introduces the edge (s, t) if $G^0(= G)$ contains the edges $(s, 1)$ and $(1, t)$.

    Let the result be true for all G^k, $k < i$.

    Suppose that i is not an internal vertex of P. Then it follows from the induction hypothesis that G^{i-1} contains the edge (s, t). Hence G^i also contains (s, t) because $G^{i-1} \subseteq G^i$.

    Suppose that i is an internal vertex of P. Then again from the induction hypothesis it follows that G^{i-1} contains the edges (s, i) and (i, t). Therefore, while processing vertex i in G^{i-1}, the edge (s, t) is added to G^i.

2. Proof follows along the same lines as in (1). ∎

As an immediate consequence of this theorem we get the following corollary.

Corollary 2.2 G^n is the transitive closure of G. ■

We next give a formal description of Warshall's algorithm. In this description the graph G is represented by its adjacency matrix M and the symbol $\vee$ stands for Boolean addition.

Algorithm 2.5 Transitive closure (Warshall)

Input: M is the adjacency matrix of G.
Output: Transitive closure of G.
begin
for $i = 1, 2, \ldots, n$ do
   begin
      for $j = 1, 2, \ldots, n$ do
         if $M(j, i) = 1$ then
            begin
               for $k = 1, 2, \ldots, n$ do
                  $M(j, k) \leftarrow M(j, k) \vee M(i, k)$;
            end
   end
end

A few observations are now in order:

1. Warshall's algorithm transforms the adjacency matrix M of a graph G to the adjacency matrix of the transitive closure of G by suitably overwriting on M. It is for this reason that the algorithm is said to work *in place*.

2. The algorithm processes all the edges incident into a vertex before it begins to process the next vertex. In other words it processes the matrix M column-wise. Hence, we describe Warshall's algorithm as *column-oriented*.

3. While processing a vertex no new edge (i.e., an edge that is not present when the processing of that vertex begins) incident into the vertex is added to the graph. This means that while processing a vertex we can choose the edges incident into the vertex in any arbitrary order.

4. Suppose that the edge (j, i) incident into the vertex i is not present while vertex i is processed, but that it is added subsequently while processing some vertex k, $k > i$. Clearly this edge was not processed while processing vertex i. Neither will it be processed later since no vertex is processed more than once. In fact, such an edge will not result in adding any new edges.

5. Warshall's algorithm is said to work in one pass since each vertex is processed exactly once.

Suppose that we wish to modify Warshall's algorithm so that it becomes *row-oriented*. In a row-oriented algorithm, while processing a vertex, all the edges incident out of the vertex are to be processed. The processing of the edge (i, j) introduces the edges (i, k) for every edge (j, k) incident out of vertex j. Therefore new edges incident out of a vertex may be added while processing a vertex *row-wise*. Some of these newly added edges may not be processed before the processing of the vertex under consideration is completed. If the processing of these edges is necessary for the computation of the transitive closure, then such a processing

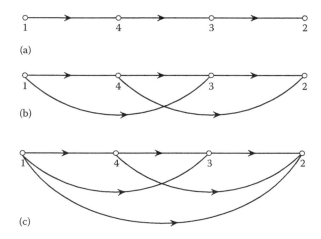

Figure 2.5 Example of row-oriented transitive closure algorithm: (a) G, (b) G', and (c) G^*. (Data from M. N. Swamy and K. Thulasiraman, *Graphs, Networks and Algorithms*, Wiley-Interscience, 1981.)

can be done only in a second pass. Thus, in general, a row-oriented algorithm may require more than one pass to compute the transitive closure.

For example, consider the graph G of Figure 2.5a. After processing row-wise the vertices of G we obtain the graph G' shown in Figure 2.5b. Clearly, G' is not the transitive closure of G since the edge (1,2) is yet to be added. It may be noted that the edge (1,3) is not processed in this pass because it is added only while processing the edge (1,4). The same is the case with the edge (4,2).

Suppose we next process the vertices of G'. In this second pass the edge (1,2) is added while processing vertex 1 and we get the transitive closure G^* shown in Figure 2.5c. Thus in the case of the graph of Figure 2.5a two passes of the row-oriented algorithm are required.

Now the question arises whether two passes always suffice. The answer is in the affirmative, and Warren [24] has demonstrated this by devising a clever two-pass row-oriented algorithm. In this algorithm, while processing a vertex, say vertex i, in the first pass only edges connected to vertices less than i are processed, and in the second pass only edges connected to vertices greater than i are processed. In other words the algorithm transforms the adjacency matrix M of the graph G to the adjacency matrix of G^* by processing in the first pass only entries below the main diagonal of M and in the second pass only entries above the main diagonal. Thus during each pass at most $n(n-1)/2$ edges are processed. A description of Warren's modification of Warshall's algorithm now follows.

Algorithm 2.6 Transitive closure (Warren)

Input: M is the adjacency matrix of G.
Output: Transitive closure of G.
begin
    for $i = 2, 3, \ldots, n$ do {**Pass 1 begins**}
        begin
            for $j = 1, 2, \ldots, i - 1$ do
                if $M(i,j) = 1$ **then**
                    begin
                        for $k = 1, 2, \ldots, n$ do
                            $M(i,k) \leftarrow M(i,k) \lor M(j,k)$;
        end

```
            end
        {Pass 1 ends}
    for i = 1, 2, ..., n − 1 do {Pass 2 begins}
        begin
            for j = i + 1, i + 2, ..., n do
                if M(i, j) = 1 then
                    begin
                        for k = 1, 2, ..., n do
                            M(i, k) ← M(i, k) ∨ M(j, k);
                    end
        end
        {Pass 2 ends}
end
```

As an example, consider again the graph shown in Figure 2.5a. At the end of the first pass of Warren's algorithm we obtain the graph shown in Figure 2.6a, and at the end of the second pass we get the transitive closure G^* shown in Figure 2.6b. The proof of correctness of Warren's algorithm is based on the following lemma.

Lemma 2.3 *Suppose that, for any two vertices s and t, there exists in G a directed path P from s to t. Then the graph that results after processing vertex s in the first pass of Warren's algorithm contains an edge (s, r), where r is a successor of s on P and either $r > s$ or $r = t$.*

Proof. Proof is by induction on s.

If $s = 1$, then the lemma is clearly true because all the successors of 1 on P are greater than 1. Assume that the lemma is true for all $s < k$ and let $s = k$. Suppose (s, i_1) is the first edge on P. If $i_1 > s$, then clearly the lemma is true.

If $i_1 < s$, then by the induction hypothesis the graph that results after processing vertex i_1 in the first pass contains an edge (i_1, i_2), where i_2 is a successor of i_1 on P and either $i_2 > i_1$ or $i_2 = t$.

If $i_2 \neq t$ and $i_2 < s$, then again by the induction hypothesis the graph that results after processing vertex i_2 in the first pass contains an edge (i_2, i_3) where i_3 is a successor of i_2 on P, and either $i_3 > i_2$ or $i_3 = t$.

If $i_3 \neq t$ and $i_3 < s$, we repeat the arguments on i_3 until we locate an i_m such that either $i_m > s$ or $i_m = t$. Thus the graph that we have before the processing of vertex s begins contains the edges $(s, i_1), (i_1, i_2), \ldots, (i_{m-1}, i_m)$ such that the following conditions are satisfied:

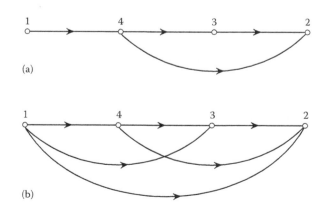

Figure 2.6 Illustration of Warren's algorithm. (Data from M. N. Swamy and K. Thulasiraman, *Graphs, Networks and Algorithms*, Wiley-Interscience, 1981.)

1. i_p is a successor of i_{p-1} on P, $p \geq 2$.
2. $i_{m-1} > i_{m-2} > i_{m-3} > \cdots > i_1$, and $i_k < s$ for $k \neq m$.
3. $i_m = t$ or $i_m > s$.

We now begin to process vertex s. Processing of (s, i_1) introduces the edge (s, i_2) because of the presence of (i_1, i_2). Since $i_2 > i_1$, the edge (s, i_2) is subsequently processed. Processing of this edge introduces (s, i_3) because of the presence of (i_2, i_3), and so on. Thus when the processing of s is completed, the required edge (s, i_m) is present in the resulting graph. ∎

Theorem 2.8 *Warren's algorithm computes the transitive closure of a directed graph G.*

Proof. We need to consider the following two cases.

Case 1 For any two distinct vertices s and t there exists in G a directed path P from s to t.

Let (i, j) be the first edge on P (as we proceed from s to t) such that $i > j$. Then it follows from Lemma 2.3 that the graph that we have, before the second pass of Warren's algorithm begins, contains an edge (i, k), where k is a successor of i on P and either $k = t$ or $k > i$. Thus after the first pass is completed there exists a path $P': s, i_1, i_2, \ldots, i_m, t$ such that $s < i_1 < i_2 < \cdots < i_m$ and each i_{j+1} is a successor of i_j on P.

When in the second pass we process vertex s, the edge (s, i_1) is first encountered. The processing of this edge introduces the edge (s, i_2) because of the presence of the edge (i_1, i_2). Since $i_2 > i_1$, the edge (s, i_2) is processed subsequently. This, in turn, introduces the edge (s, i_3), and so on. Thus when the processing of s is completed, we have the edge (s, t) in the resulting graph.

Case 2 There exists in G a directed circuit containing a vertex s.

In this case we can prove along the same lines as before that when the processing of vertex s is completed in the second pass, the resulting graph contains the self-loop (s, s). ∎

Clearly both Warshall's and Warren's algorithms have the worst-case complexity $O(n^3)$. Warren [24] refers to other row-oriented algorithms. For some of the other transitive closure algorithms (see [25–33]). Syslo and Dzikiewicz [34] discuss computational experiences with several of the transitive closure algorithms. Melhorn [35] discusses the transitive closure problem in the context of general path problems in graphs. Several additional references on this topic can also be found in [35].

2.5 SHORTEST PATHS

Let G be a connected directed graph in which each directed edge is associated with a finite real number called the length of the edge. The length of an edge directed from a vertex i to a vertex j is denoted by $w(i, j)$. If there is no edge directed from vertex i to vertex j, then $w(i, j) = \infty$. The length of a directed path in G is the sum of the lengths of the edges in the path. A minimum length directed $s - t$ path is called a shortest path from s to t. The length of a shortest directed $s - t$ path, if it exists, is called the distance from s to t, and it is denoted as $d(s, t)$. We assume that the vertices are denoted as $1, 2, \ldots, n$.

In this section we consider the following two problems:

1. *Single Source Shortest Paths Problem*: Find the shortest paths from a specified vertex s to all other vertices in G.

2. *All-Pairs Shortest Paths Problem*: Find the shortest paths between all the ordered pairs of vertices in G.

These two problems arise in several optimization problems. For example, finding a minimum cost flow in a transport network involves finding a shortest path from the source to the sink in the network [36] (see also Chapter 5).

2.5.1 Single Source Shortest Paths: Bellman–Ford–Moore Algorithm

We first make a few observations and assumptions that do not cause any loss of generality of the algorithms to be discussed in this section.

- Vertex 1 will serve as the source vertex. Algorithms for the shortest path problems start with an initial estimate of the distance of each vertex from the source vertex 1. We will denote this estimate for vertex i by $\lambda(i)$. Initially, $\lambda(1) = 0$ and $\lambda(i) = \infty$ for all $i \neq 1$. Algorithms repeatedly update these estimates until they all become the required distance values. That is, at termination, $d(1, i) = \lambda(i)$.

- We assume that all vertices are reachable from the source vertex. In other words, we assume that there is a directed path from the source vertex to every other vertex. Thus, at termination of the algorithms, $\lambda(i)$ values will be finite.

- Algorithms search for a shortest directed walk from the source vertex to every other vertex. We assume that the graph has no directed circuit of negative length. If such a circuit C were present, then a directed walk P of arbitrarily small length from the source vertex to every vertex in C can be found by repeatedly traversing the circuit C.

The following theorem is the basis of all shortest path algorithms that we will be discussing in this section. In this theorem and the subsequent discussions, an $i - j$ directed path refers to a directed path from vertex i to vertex j.

Theorem 2.9 (Optimality conditions) *Consider a connected directed graph $G = (V, E)$ with each edge $e \in E$ associated with a finite real number $w(e)$. Let $\lambda(i)$, $i \geq 1$ denote the length of a $1 - i$ directed path. Then for all $i \geq 1$, $\lambda(i)$ is the distance from vertex 1 to vertex i, that is $\lambda(i) = d(1, i)$, if and only if the following condition is satisfied.*

$$\lambda(i) + w(i, j) \geq \lambda(j), \textit{for every edge } e = (i, j). \tag{2.9}$$

Proof.
Necessity: Let $\lambda(i)$, $i \geq 1$ be the distance from vertex 1 to vertex i. Suppose condition (2.9) is violated for some edge $e = (i, j)$. Then, $\lambda(i) + w(e) < \lambda(j)$. Since there is no directed circuit of negative length, concatenating the edge e to a shortest path from vertex 1 to vertex i will give a path of length less than $\lambda(j)$, contradicting that $\lambda(j)$ is the length of a shortest path from vertex 1 to vertex i.

Sufficiency: Consider a directed path P from vertex 1 to vertex j and let

$$P : 1 = i_0, i_1, \ldots, i_{k-1}, i_k = j.$$

Assume that the $\lambda(i)$'s satisfy (2.9).
Then

$$\begin{aligned}
\lambda(j) = \lambda(i_k) &\leq \lambda(i_{k-1}) + w(i_{k-1}, i_k) \\
&\leq \lambda(i_{k-2}) + w(i_{k-2}, i_{k-1}) + w(i_{k-1}, i_k) \ldots \\
&\leq \lambda(i_0) + w(i_0, i_1) + w(i_1, i_2) + \cdots + w(i_{k-2}, i_{k-1}) + w(i_{k-1}, i_k) \\
&= w(i_0, i_1) + w(i_1, i_2) + \cdots + w(i_{k-2}, i_{k-1}) + w(i_{k-1}, i_k), \\
&\quad \text{since} \quad i_0 = 1 \text{ and } \lambda(1) = 0. \\
&= \text{length of path } P.
\end{aligned}$$

Since the length of every directed path P from vertex 1 to vertex j is at least $\lambda(j)$, and $\lambda(j)$ is the length of a directed path from vertex 1 to vertex j, it follows that $\lambda(j)$ is the length of a shortest path from vertex 1 to vertex j. ∎

We are now ready to present a shortest path algorithm due to Ford [37]. This algorithm is also attributed to Bellman [38] and Moore [39]. So we shall call this Bellman–Ford–Moore (in short, BFM) algorithm.

Algorithm 2.7 Shortest paths in graphs with negative length edges (Bellman–Ford–Moore)

Input: A connected graph $G = (V, E)$ with length $w(e) = w(i, j)$ for each edge $e = (i, j)$.
Output: Shortest paths and their lengths from vertex 1 to all other vertices.
begin
  $\lambda(1) \leftarrow 0$;
  $\lambda(i) \leftarrow \infty$, for each $i \neq 1$;
  PRED $(i) \leftarrow i$, for every vertex i;
  while there exists an edge $e = (i, j)$ satisfying $\lambda(i) + w(i, j) < \lambda(j)$ **do**
    begin
      $\lambda(j) \leftarrow \lambda(i) + w(i, j)$;
      PRED $(j) \leftarrow i$;
    end
end

Note that the algorithm starts by assigning $\lambda(1) = 0$ and $\lambda(j) = \infty$ for all $j \neq 1$. If there exists an edge (i, j) which violates the optimality condition (2.9) then the $\lambda(j)$ value is updated to $\lambda(j) = \lambda(i) + w(i, j)$ and the predecessor of j is set to i. If for every edge (i, j) the optimality condition is satisfied then the algorithm terminates.

A few observations are now in order.

- If at any iteration $\lambda(j)$ is finite and $\text{PRED}(j) = i$ then it means that vertex j received its current label $\lambda(j)$ while examining the edge (i, j) and $\lambda(j)$ is updated to $\lambda(i) + w(i, j)$. Since there are no directed circuits of negative length in the graph, a directed $1 - j$ path of length $\lambda(i) + w(i, j)$ can then be obtained by tracing the path backward from vertex j along the predecessors. For example, if $\text{PRED}(8) = 6$, $\text{PRED}(6) = 9$ and $\text{PRED}(9) = 1$, then the $1 - 8$ path along the edges $(1,9)$, $(9,6)$, and $(6,8)$ is of length $\lambda(8)$.

- Since there are no directed circuits of negative length, the number of times the label of a vertex j is updated is no more than the number of directed paths from vertex 1 to vertex j. Since the number of directed paths is finite, the BFM algorithm will terminate, and at termination the label values will satisfy (2.9).

This establishes the correctness of the BFM algorithm. We summarize this result in the following theorem.

Theorem 2.10 *For a directed graph G with no directed circuits of negative length, the Bellman–Ford–Moore Algorithm terminates in a finite number of steps, and upon termination, $\lambda(j) = d(1, j)$ for every vertex j.* ∎

Note that the number of directed paths from vertex 1 to another vertex could be exponential in the number of vertices of the graph. So, if in the implementation of Algorithm 2.7 edges are selected in an arbitrary manner, it is possible that the algorithm may perform an exponential number of operations before terminating. We next present an implementation which results in $O(mn)$ time complexity, where m denotes the number of edges and n denotes the number of vertices in the network.

Step 1: First arrange the edges in any specified order as follows: $e_1, e_2, e_3, \ldots, e_m$

Step 2: Scan the edges one by one and check if the condition $\lambda(j) > \lambda(i) + w(i,j)$ is satisfied for edge (i,j). If so, update $\lambda(j) = \lambda(i) + w(i,j)$.

Step 3: Stop if no $\lambda(j)$ is updated in step 2. Otherwise, repeat step 2.

In the above implementation Step 2 is called a *sweep* or a *phase*. During a phase, m edges are scanned and the $\lambda(j)$ values are updated if necessary. Thus a sweep takes $O(m)$ time. We now show that the algorithm terminates in atmost n sweeps. We first prove the following.

Theorem 2.11 *If there exists a shortest path from vertex 1 to vertex j having k edges, then $\lambda(j)$ will have reached its final value by the end of the kth sweep.*

Proof. Proof is by induction on k. Clearly the result is true for $k = 1$, because the vertex j will get its final $\lambda(j)$ value when the edge $(1,j)$ is scanned during the first sweep. As induction hypothesis, assume that the result is true for some $k \geq 1$. In other words, we assume that if there exists a shortest path from vertex 1 to vertex j having k edges, then at the end of the kth sweep, $\lambda(j)$ will have reached its final value.

Consider a vertex j which is connected to 1 by a shortest path having $(k+1)$ edges.

$$P: 1, i_1, i_2, i_3, \ldots, i_k, i_{k+1} = j$$

Since the subpath from 1 to i_k is a shortest path to i_k having k edges, then by the induction hypothesis $\lambda(i_k)$ will have reached its final value by the end of the kth sweep. During the $(k+1)$th sweep when edge (i_k, i_{k+1}) is scanned $\lambda(i_{k+1})$ will be updated to $\lambda(i_{k+1}) = \lambda(i_k) + w(i_k, i_{k+1})$ which is the length of P. Thus at the end of the $(k+1)$th sweep, $\lambda(i_{k+1}) = d(1,j)$ which is its final value. ∎

Now we can see that the BFM algorithm will terminate in at most n sweeps because every path has at most $(n-1)$ edges. So the complexity of the above implementation of the BFM algorithm is $O(mn)$. We call this implementation of the BFM as edge-based implementation.

In a similar manner, we can design a vertex-based implementation of the BFM algorithm. In this implementation, we first order the vertices as $1, 2, \ldots, n$. Recall that vertex 1 is the source vertex. During a sweep each vertex is examined in the order specified by the vertex ordering. Here, examining a vertex j involves examining all the edges incident on and directed away from j and updating the labels of the neighbors according to step 2. We can again show that no more than n sweeps will be required and that the algorithm terminates in $O(mn)$ time. We will call this vertex-based implementation.

2.5.1.1 Negative Cycle Detection

As we noted earlier, $\lambda(j)$'s satisfying the optimality conditions do not exist if there exists a directed circuit of negative length. So, if a negative directed were present, the BFM algorithm will not terminate. Suppose C denotes the maximum of the absolute values of all edge lengths, no path can have a cost smaller than $-nC$. Thus if the $\lambda(j)$ value falls below $-nC$ for some node j, then we can terminate the algorithm. The negative length circuit can be obtained by tracing the predecessor values starting at node j.

2.5.1.2 Shortest Path Tree

At the end of the shortest path algorithm each node j has a predecessor $\text{PRED}(j)$. The set of edges $(j, \text{PRED}(j))$ will form a tree called the shortest path tree. Each path from 1 to node j in the tree gives a shortest length path from 1 to j. Also, $\lambda(j) = \lambda(i) + w(i,j)$ for every edge (i,j) on this tree.

Sometimes we may be interested in getting the second, third, or higher shortest paths. These and related problems are discussed in Christofides [40], Dreyfus [41], Frank and Frisch [42], Gordon and Minoux [43], Hu [44], Lawler [45], Minieka [46], and Spira and Pan [47]. For algorithms designed for sparse networks see Johnson [48] and Wagner [49] (see also Edmonds and Karp [50], Fredman [51], and Johnson [52]). For a shortest path problem that arises in solving a special system of linear inequalities and its applications see Comeau and Thulasiraman [53], Lengauer [54], and Liao and Wong [55].

2.5.2 Single Source Shortest Paths in Graphs with No Negative Length Edges: Dijkstra's Algorithm

In this section we consider a special case of the shortest path problem where all the edge lengths are nonnegative. This restricted version of the shortest path problem admits a very efficient algorithm due to Dijkstra [56]. This algorithm may be viewed as a vertex-based implementation of the BFM algorithm. This implementation requires only one sweep of all the vertices, but in contrast to the BFM algorithm the order in which the vertices are selected for scanning cannot be arbitrary. This order depends on the edge lengths. Following is an informal description of Dijkstra's algorithm.

Like the BFM algorithm, Dijkstra's algorithm starts by assigning $\lambda(1) = 0$ and $\lambda(j) = \infty$ for all $j \neq 1$. Initially, all vertices are unlabeled.

A general iteration: In iteration i

- An unlabeled vertex i with minimum λ-value is labeled *Permanent*. Let this vertex be denoted as u_i, that is, among all unlabeled vertices vertex u_i has the smallest λ-value.

- Then every edge (u_i, j) where j is unlabeled is examined for violation of the optimality condition (2.9). If there exists such an edge (u_i, j), then the $\lambda(j)$ value is updated to $\lambda(j) = \lambda(u_i) + w(u_i, j)$ and the predecessor of j is set to u_i.

The algorithm terminates when all the vertices are permanently labeled. A formal description of Dijkstra's algorithm is given next. In this description, we use an array PERM to indicate which of the vertices are permanently labeled. If $\text{PERM}(v) = 1$, then v is a permanently labeled vertex. We start with $\text{PERM}(v) = 0$ for all v. PRED is an array that keeps a record of the vertices from which the vertices get permanently labeled. If a vertex v is permanently labeled, then v, $\text{PRED}(v)$, $\text{PRED}(\text{PRED}(v))$, ..., 1 are the vertices in a shortest directed 1-v path.

Algorithm 2.8 Shortest paths in graphs with non-negative edge lengths (Dijkstra)

Input: A connected graph $G = (V, E)$ with length $w(e) = w(i,j) \geq 0$ for each edge $e = (i,j)$.
Output: Shortest paths and their lengths from vertex 1 to all other vertices.
begin
  $\lambda(1) \leftarrow 0$;
  $\lambda(i) \leftarrow \infty$, for each $i \neq 1$;
  $\text{PRED}(i) \leftarrow i$, for every vertex i;
  $\text{PERM}(i) \leftarrow 0$, for every vertex i;
  $S_0 \leftarrow \phi$;
  for $k = 1, 2, \ldots, n$ **do**

```
        begin (iteration k begins)
            Let u_k be a vertex that is not yet labeled permanently and has minimum
            λ-value;
            PERM(u_k) ← 1;
            S_k ← S_{k-1} ∪ {u_k};
            for every edge e = (u_k, j) do
                If λ(j) > λ(u_k) + w(u_k, j), then λ(j) ← λ(u_k) + w(u_k, j) and
                PRED(j) ← u_k;
        end (iteration k ends)
end
```

Note that in a computer program ∞ is represented by as high a number as necessary. Further, if the final λ value of a vertex v is equal to ∞, then it means that there is no directed path from s to v.

To illustrate Dijkstra's algorithm, consider the graph G in Figure 2.7 in which the length of an edge is shown next to the edge. In Figure 2.8 we have shown the λ values of vertices and the entries of the PRED array.

For any i the circled entries correspond to the permanently labeled vertices. The entry with the mark * is the label of the latest permanently labeled vertex u_i. The shortest paths from s and the corresponding distances are obtained from the final λ values and the entries in the PRED array.

Theorem 2.12 *For $i \geq 1$, let u_i denote the vertex that is labeled permanently in iteration i of Dijkstra's algorithm, and $d(u_i)$ be the λ-value of u_i at the termination of the algorithm. Then*

1. $d(u_1) \leq d(u_2) \leq \cdots \leq d(u_n)$; and

2. $d(1, u_j) = d(u_i)$ for all $j \geq 1$.

Note: In Dijkstra's algorithm the λ-value of a vertex does not change once the vertex is labeled permanently.

Proof. Proof depends on the following observations based on the way labels are updated and how a vertex is selected for permanent labeling.

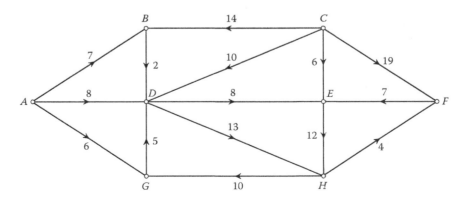

Figure 2.7 Graph for illustrating Dijkstra's algorithm. (Data from M. N. Swamy and K. Thulasiraman, *Graphs, Networks and Algorithms*, Wiley-Interscience, 1981.)

i	A	B	C	D	E	F	G	H
				Vertices				
1	⓪*	∞	∞	∞	∞	∞	∞	∞
2	⓪	7	∞	8	∞	∞	⑥*	∞
3	⓪	⑦*	∞	8	∞	∞	⑥	∞
4	⓪	⑦	∞	⑧*	∞	∞	⑥	∞
5	⓪	⑦	∞	⑧	⑯*	∞	⑥	21
6	⓪	⑦	∞	⑧	⑯	∞	⑥	㉑*
7	⓪	⑦	∞	⑧	⑯	㉕*	⑥	㉑
8	⓪	⑦	∞*	⑧	⑯	㉕	⑥	㉑

(a)

PRED(A) = A   PRED(E) = D
PRED(B) = A   PRED(F) = H
PRED(C) = C   PRED(G) = A
PRED(D) = A   PRED(H) = D

From	To	Shortest path
A	B	A, B
A	C	No Path
A	D	A, D
A	E	A, D, E
A	F	A, D, H, F
A	G	A, G
A	H	A, D, H

(b)

Figure 2.8 Illustration of Dijkstra's algorithm. The λ values are shown in (a). (Data from M. N. Swamy and K. Thulasiraman, *Graphs, Networks and Algorithms*, Wiley-Interscience, 1981.)

1. Let S_i denote the set of all the vertices permanently labeled at the end of iteration i. That is $S_i = \{u_1, u_2, \ldots, u_i\}$. Then, at the end of iteration i, $\lambda(j)$ of a vertex $j \neq S_i$, if it is finite, is the minimum of the labels considered for assignment to j while scanning the edges directed into j from the permanently labeled vertices $u_1, u_2, \ldots, u_i$. So,

$$\lambda(j) \leq d(u_k) + w(u_k, j), \text{ for all } j \neq S_i \text{ and } k \leq i \quad (2.10)$$

2. u_{i+1} is the vertex not in S_i with minimum λ-value. This means that

$$\lambda(u_{i+1}) = d(u_{i+1}) \leq d(u_k) + w(u_k, u_{i+1}), \text{ for all } k \leq i \quad (2.11)$$

PRED(u_{i+1}) is the vertex u_k for which the above minimum is achieved. So, if PRED(u_{i+1}) = u_a, then

$$d(u_{i+1}) = d(u_a) + w(u_a, u_{i+1}). \quad (2.12)$$

3. It follows from (2.10) that for $j > i$

$$d(u_j) \leq d(u_i) + w(u_i, u_j) \quad (2.13)$$

because λ-values of vertices do not increase from iteration to iteration.

Proof of (1): If $a = i$, then from (2.12) we get $d(u_{i+1}) = d(u_i) + w(u_i, u_{i+1})$ and so $d(u_i) \leq d(u_{i+1})$.

If $a < i$, then vertex u_{i+1} must have reached its final label value $d(u_{i+1}) = d(u_a) + w(u_a, u_{i+1})$ in iteration a.

So, in the iteration i when u_i with label $d(u_i)$ was selected for permanent labeling, the vertex u_{i+1} with label $d(u_{i+1}) = d(u_a) + w(u_a, u_{i+1})$ was also available for consideration. But vertex u_i was permanently labeled. So, $d(u_i) \leq d(u_{i+1})$.

Result (1) follows since the above argument is valid for all $i \geq 1$.

Proof of (2): We prove the result by showing that $d(u_1), d(u_2), \ldots, d(u_n)$ satisfy the optimality condition for shortest path lengths given in Theorem 2.9.

Consider any edge (u_i, u_j). If $i > j$, then $d(u_j) \leq d(u_i)$, by the result (1) of the theorem. So $d(u_i) + w(u_i, u_j) \geq d(u_j)$. Suppose that $i < j$. Then, we see from (2.13) that $d(u_i) + w(u_i, u_j) \geq d(u_j)$.

Thus, the labels $d(u_1), d(u_2), \ldots, d(u_n)$ at the termination of Dijkstra's algorithm satisfy the optimality condition for shortest path lengths and the result (2) follows. ∎

Dijkstra's algorithm requires examination of at most m edges and at most n minimum computations, leading to a complexity of $O(m + n \log n)$. In our discussions thus far we have assumed that all the lengths are nonnegative. Dijkstra's algorithm is not valid if some of the lengths are negative. (Why?)

2.5.3 All Pairs Shortest Paths

Suppose that we are interested in finding the shortest paths between all the $n(n-1)$ ordered pairs of vertices in an n-vertex directed graph. A straight-forward approach to get these paths would be to use the BFM algorithm n times. However, there are algorithms that are computationally more efficient than this. These algorithms are applicable when there are no negative-length directed circuits. Now we discuss one of these algorithms. This algorithm, due to Floyd [57], is based on Warshall's algorithm (Algorithm 2.5) for computing transitive closure.

Consider an n-vertex directed graph G with lengths associated with its edges. Let the vertices of G be denoted as $1, 2, \ldots, n$. Assume that there are no negative-length directed circuits in G. Let $W = [w_{ij}]$ be the $n \times n$ matrix of direct lengths in G, that is, w_{ij} is the length of the directed edge (i, j) in G. We set $w_{ij} = \infty$ if there is no edge (i, j) directed from i to j. We also set $w_{ii} = 0$ for all i.

Starting with the matrix $W^{(0)} = W$, Floyd's algorithm constructs a sequence $W^{(1)}, W^{(2)}, \ldots, W^{(n)}$ of $n \times n$ matrices so that the entry $w_{ij}^{(n)}$ in $W^{(n)}$ would give the distance from i to j in G. The matrix $W^{(k)} = [w_{ij}^{(k)}]$ is constructed from the matrix $W^{(k-1)} = [w_{ij}^{(k-1)}]$ according to the following rule:

$$w_{ij}^{(k)} = \min\left\{w_{ij}^{(k-1)}, w_{ik}^{(k-1)} + w_{kj}^{(k-1)}\right\} \tag{2.14}$$

Let $P_{ij}^{(k)}$ denote a path of minimum length among all the directed i–j paths, which use as internal vertices only those from the set $\{1, 2, \ldots, k\}$. The following theorem proves the correctness of Floyd's algorithm.

Theorem 2.13 *For $0 \leq k \leq n$, $w_{ij}^{(k)}$ is equal to the length of $P_{ij}^{(k)}$.*

Proof. Proof follows from the following observations.

1. If $P_{ij}^{(k)}$ does not contain vertex k then $w_{ij}^{(k)} = w_{ij}^{(k-1)}$.

2. If $P_{ij}^{(k)}$ contains vertex k then $w_{ij}^{(k)} = w_{ik}^{(k-1)} + w_{kj}^{(k-1)}$ because a subpath of a shortest path is a shortest path between the end vertices of the subpath.

Note: See the similarity between the proof of this theorem and the proof of correctness of Warshall's algorithm for transitive closure. ∎

Usually, in addition to the shortest lengths, we are also interested in obtaining the paths that have these lengths. Recall that in Dijkstra's algorithm we use the PRED array to keep a record of the vertices that occur in the shortest paths. This is achieved in Floyd's algorithm as described next.

As we construct the sequence $W^{(0)}, W^{(1)}, \ldots, W^{(n)}$, we also construct another sequence $Z^{(0)}, Z^{(1)}, \ldots, Z^{(n)}$ of matrices such that the entry $z_{ij}^{(k)}$ of $Z^{(k)}$ gives the vertex that immediately follows vertex i in $P_{ij}^{(k)}$. Clearly, initially we set

$$z_{ij}^{(0)} = \begin{cases} j, & \text{if } w_{ij} \neq \infty; \\ 0, & \text{if } w_{ij} = \infty. \end{cases} \qquad (2.15)$$

Given $Z^{(k-1)} = [z_{ij}^{(k-1)}]$, $Z^{(k)} = [z_{ij}^{(k)}]$ is obtained according to the following rule: Let

$$M = \min\left\{w_{ij}^{(k-1)}, w_{ik}^{(k-1)} + w_{kj}^{(k-1)}\right\}$$

Then

$$z_{ij}^{(k)} = \begin{cases} z_{ij}^{(k-1)}, & \text{if } M = w_{ij}^{(k-1)}; \\ z_{ik}^{(k-1)}, & \text{if } M < w_{ij}^{(k-1)}. \end{cases} \qquad (2.16)$$

It should be clear that the shortest i–j path is given by the sequence $i, i_1, i_2, \ldots, i_p, j$ of vertices, where

$$i_1 = z_{ij}^{(n)}, i_2 = z_{i_1 j}^{(n)}, i_3 = z_{i_2 j}^{(n)}, \ldots j = z_{i_p j}^{(n)}. \qquad (2.17)$$

Algorithm 2.9 Shortest paths between all pairs of vertices (Floyd)

Input: $W = [w_{ij}]$ is the $n \times n$ matrix of direct lengths in the given directed graph $G = (V, E)$. Here $w_{ii} = 0$ for all $i = 1, 2, \ldots, n$. The graph does not have negative-length directed circuits.
Output: Shortest directed paths and their lengths between every pair of vertices.
begin
    for every $(i, j) \in V \times V$ **do**
        if $(i, j) \in E$ **then** $z_{ij} \leftarrow j$ **else** $z_{ij} \leftarrow 0$;
    for $k = 1, 2, \ldots, n$ **do**
        for each $(i, j) \in V \times V$ **do**
        begin
        if $w_{ij} > w_{ik} + w_{kj}$ **then**
            begin
                $w_{ij} \leftarrow w_{ik} + w_{kj}$;
                $z_{ij} \leftarrow z_{ik}$;
            end
        end
end

Suppose the graph has some negative-length directed circuits. Then, during the execution of Floyd's algorithm, w_{ii} becomes negative for some i. This means that the vertex i is in some negative-length directed circuit and so the algorithm can be terminated at that stage.

It is easy to see that Floyd's algorithm is of complexity $O(n^3)$. This algorithm is also valid for finding shortest paths in a network with negative lengths provided the network does not have a directed circuit of negative length. Dantzig [58] proposed a variant of Floyd's algorithm that is also of complexity $O(n^3)$.

For some of the other shortest path algorithms see Tabourier [59], Williams and White [60], and Yen [61]. Deo and Pang [62] and Pierce [63] give exhaustive bibliographies of algorithms for the shortest path and related problems. A discussion of complexity results for shortest path problems can be found in Melhorn [35] and Tarjan [64]. Discussions of shortest path problems in a more general setting can be found in [43] and [35], Carré [65], and Tarjan [66]. For some other developments see Moffat and Takoka [67] and Frederickson [68].

2.6 TRANSITIVE ORIENTATION

An undirected graph G is *transitively orientable* if we can assign orientations to the edges of G so that the resulting directed graph is transitive. If G is transitively orientable, then $\vec{G}$ will denote a *transitive orientation* of G.

For example, the graph shown in Figure 2.9a is transitively orientable. A transitive orientation of this graph is shown in Figure 2.9b.

In this section we discuss an algorithm due to Pnueli et al. [69] to test whether a simple undirected graph G is transitively orientable and obtain a transitive orientation $\vec{G}$ if one exists. To aid the development and presentation of this algorithm, we introduce some notations:

1. $i \to j$ means that vertex i is connected to vertex j by an edge oriented from i to j.

2. $i \leftarrow j$ is similarly defined.

3. $i\text{---}j$ means that there is an edge connecting vertex i and vertex j.

4. $i \not\!\!\!-\!\! j$ means that there is no edge connecting vertex i and vertex j.

5. $i \twoheadrightarrow j$ means that either $i\not\!\!\!-\!\! j$ or $i \leftarrow j$ or the edge $i\text{---}j$ is not oriented.

$i \to j$, $i \leftarrow j$, and $i\text{---}j$ will also be used to denote the corresponding edges.

Consider now an undirected graph $G = (V, E)$ which is transitively orientable. Let $\vec{G} = (V, \vec{E})$ denote a transitive orientation of G.

Suppose that there exist three vertices $i, j, k \in V$ such that $i \to j$, $j\text{---}k$, and $i\not\!\!\!-\!\! k$, then transitivity of $\vec{G}$ requires that $j \leftarrow k$. Similarly, if $i, j, k \in V$, $i \to j$, $i\text{---}k$, and $j\not\!\!\!-\!\! k$, then transitivity of $\vec{G}$ requires that $i \to k$.

These two observations lead to the following simple rules which form the basis of Pnueli, Lempel and Even's algorithm.

Rule R_1 For $i, j, k \in V$, if $i \to j$, $j\text{---}k$, and $i\not\!\!\!-\!\! k$, then orient the edge $j\text{---}k$ as $j \leftarrow k$.

Rule R_2 For $i, j, k \in V$, if $i \to j$, $i\text{---}k$, and $j\not\!\!\!-\!\! k$, then orient the edge $i\text{---}k$ as $i \to k$.

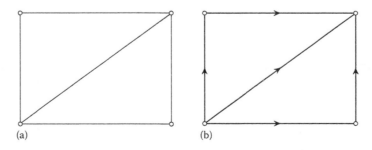

Figure 2.9 (a) Graph G. (b) A transitive orientation of G. (Data from M. N. Swamy and K. Thulasiraman, *Graphs, Networks and Algorithms*, Wiley-Interscience, 1981.)

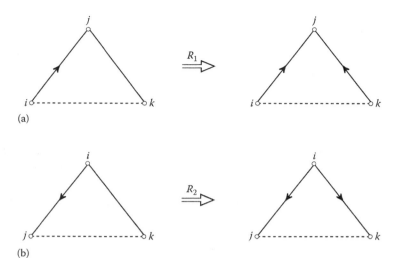

Figure 2.10 (a) Rule R_1. (b) Rule R_2. (Data from M. N. Swamy and K. Thulasiraman, *Graphs, Networks and Algorithms*, Wiley-Interscience, 1981.)

These two rules are illustrated in Figure 2.10, where a dashed line indicates the absence of the corresponding edge.

A description of the transitive orientation algorithm now follows.

Algorithm 2.10 Transitive orientation (Pnueli, Lempel, and Even)

S1. G is the given simple undirected graph $i \leftarrow 1$.

S2. (Phase i begins.) Select an edge e of the graph G and assign an arbitrary orientation to e. Assign, whenever possible, orientations to the edges in G adjacent to e, using Rule R_1 or Rule R_2. The directed edge e is now labeled *examined*.

S3. Test if there exists in G a directed edge which has not been labeled *examined*. If yes, go to step S4. Otherwise go to step S6.

S4. Let $i \to j$ be a directed edge in G which has not been labeled *examined*. Now do the following, whenever applicable for each edge in G incident on i or j, and then label $i \to j$ *examined*:

  Case 1 Let the edge under consideration be j—k.
   a. (Applicability of rule R_1.) If $i \not= k$ and the edge j—k is not oriented, then orient j—k as $j \leftarrow k$.
   b. (Contradiction of rule R_1.) If $i \not= k$ and the edge j—k is already oriented as $j \to k$, then a contradiction of rule R_1 has occurred. Go to step S9.

  Case 2 Let the edge under consideration be i—k.
   a. (Applicability of rule R_2.) If $j \not= k$ and the edge i—k is not oriented, then orient i—k as $i \to k$.
   b. (Contradiction of rule R_2.) If $j \not= k$ and edge i—k is already oriented as $i \leftarrow k$, then a contradiction of rule R_2 has occurred. Go to step S9.

S5. Go to step S3.

S6. (Phase i has ended successfully.) Test whether all the edges of G have been assigned orientations. If yes, go to step S8. Otherwise remove from G all its directed edges and let G' be the resulting graph.

> **S7.** $G \leftarrow G'$ and $i \leftarrow i+1$. Go to Step S2.
>
> **S8.** (All the edges of the given graph have been assigned orientations which are consistent with Rules R_1 and R_2. These orientations define a transitive orientation of the given graph.) HALT.
>
> **S9.** (The graph G is not transitively orientable.) HALT.

The main step in the above algorithm is S4. In this step we examine each edge adjacent to a directed edge, say, edge $i \to j$. If such an edge is already oriented, then we test whether its orientation and that of $i \to j$ are consistent with Rule R_1 or Rule R_2. If an edge under examination is not yet oriented, then we assign to it, if possible, an orientation using Rule R_1 or Rule R_2.

As we can see, the algorithm consists of different phases. Each phase involves execution of Step S2 and repeated executions of step S4 as more and more edges get oriented. If a phase ends without detecting any contradiction of Rule R_1 or Rule R_2, then it means that no more edges can be assigned orientations in this phase by application of the two rules, and that all the orientations assigned in this phase are consistent with these rules.

The algorithm terminates either (1) by detecting a contradiction of Rule R_1 or Rule R_2, or (2) by assigning orientations to all the edges of the given graph such that these orientations are consistent with Rules R_1 and R_2. In the former case the graph is not transitively orientable, and in the latter case the graph is transitively orientable with the resulting directed graph defining a transitive orientation.

The complexity of the algorithm depends on the complexity of executing step S4. This step is executed at most m times, where m is the number of edges in the given graph. Each execution of step S4 involves examining all the edges adjacent to an oriented edge. So the number of operations required to execute step S4 is proportional to 2Δ, where Δ is the maximum degree in the given graph. Thus the overall complexity of the algorithm is $O(2m\Delta)$.

Next we illustrate the algorithm with two examples. Consider first the graph G shown in Figure 2.11a.

Phase 1: We begin by orienting edge 7—2 as $7 \to 2$. By Rule R_2, $7 \to 2$ implies that $7 \to 4$ and $7 \to 5$, and $7 \to 4$ implies that $7 \to 6$. Rule R_1 is not applicable to any edge adjacent to $7 \to 2$.

By Rule R_2, $7 \to 5$ and $7 \to 6$ imply that $7 \to 1$ and $7 \to 3$, respectively. In this phase no more edges can be assigned orientations. We can also check that all the assigned orientations are consistent with Rules R_1 and R_2. Phase 1 now terminates, and the edges oriented in this phase are shown in Figure 2.11b.

Now remove from the graph G of Figure 2.11a all the edges which are oriented in Phase 1. The resulting graph G' is shown in Figure 2.11c. Phase 2 now begins, and G' is the graph under consideration.

Phase 2: We begin by orienting edge 1—2 as $1 \to 2$. This results in the following sequences of implications:

$$(1 \to 2) \underset{R_1}{\Rightarrow} (2 \leftarrow 3) \underset{R_2}{\Rightarrow} (3 \to 5) \underset{R_1}{\Rightarrow} (6 \to 5) \underset{R_1}{\Rightarrow} (4 \to 5)$$

$$(2 \leftarrow 3) \underset{R_2}{\Rightarrow} (3 \to 4)$$

$$(6 \to 5) \underset{R_2}{\Rightarrow} (6 \to 1)$$

$$(6 \to 5) \underset{R_2}{\Rightarrow} (6 \to 2)$$

Thus all the edges of G' have now been oriented as shown in Figure 2.11d.

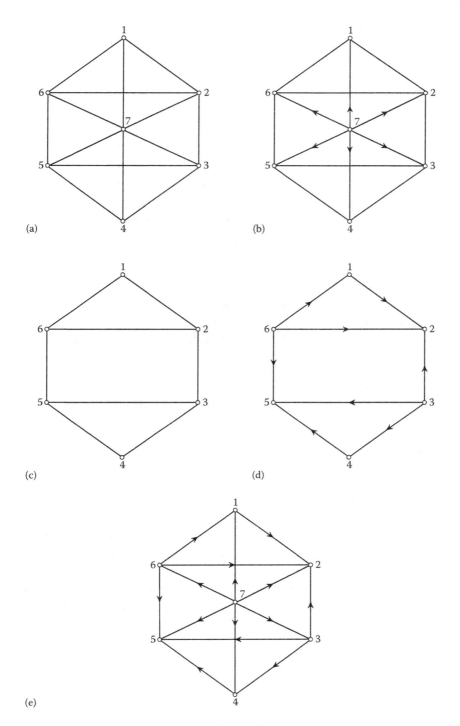

Figure 2.11 (a–e) Illustration of transitive orientation algorithm.

These orientations of G' are also consistent with the Rules R_1 and R_2.
Therefore phase 2 terminates successfully.
The resulting transitive orientation of G is shown in Figure 2.11e.
Consider next the graph shown in Figure 2.12. We begin by orienting edge 1—2 as $1 \to 2$. This leads to the following sequence of implications:

$$(1 \to 2) \underset{R_1}{\Rightarrow} (2 \leftarrow 3) \underset{R_2}{\Rightarrow} (3 \to 4) \underset{R_1}{\Rightarrow} (4 \leftarrow 5) \underset{R_2}{\Rightarrow} (5 \to 1) \underset{R_1}{\Rightarrow} (1 \leftarrow 2), \qquad (2.18)$$

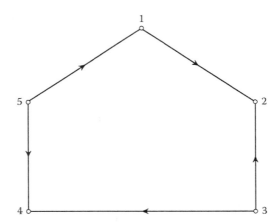

Figure 2.12 A non-transitively orientable graph.

which requires that 1—2 be directed as $1 \leftarrow 2$, contrary to the orientation we have already assigned to the edge 1—2. Thus a contradiction of Rule R_1 is observed. Hence the graph of Figure 2.12 is not transitively orientable.

We now proceed to prove the correctness of Algorithm 2.10. To do so we need to prove the following two assertions.

Assertion 2.1 *If Algorithm 2.10 terminates successfully (step S8), then the resulting directed graph is a transitive orientation of the given graph.*

Assertion 2.2 *If the given graph is transitively orientable, then Algorithm 2.10 terminates successfully.*

We first consider Assertion 1.

Given an undirected graph $G = (V, E)$. Suppose that the algorithm terminates successfully.

Consider now any two edges $i \to j$ and $k \to l$ which are assigned orientations in the same phase of the algorithm. Then we can construct a sequence of implications, which starts with the directed edge $i \to j$ and ends orienting the edge k—l as $k \to l$. Such a sequence will be called a *derivation chain* from $i \to j$ to $k \to l$. For example, in the directed graph of Figure 2.11d the following are two of the derivation chains from $2 \leftarrow 6$ to $3 \to 4$:

$$(2 \leftarrow 6) \Rightarrow (3 \to 2) \Rightarrow (3 \to 4)$$
$$(2 \leftarrow 6) \Rightarrow (5 \leftarrow 6) \Rightarrow (3 \to 5) \Rightarrow (3 \to 2) \Rightarrow (3 \to 4)$$

Thus it is clear that it is meaningful to talk about a shortest derivation chain between any pair of directed edges which are assigned orientations in the same phase of Algorithm 2.10. Proof of Assertion 1 is based on the following important lemma.

Lemma 2.4 *After a successful completion of phase 1 it is impossible to have three vertices i, j, k such that $i \to j$ and $j \to k$ with $i \not\to k$.*

Proof. Note that "$i \not\to k$" means that either $i \not\!-\! k$ or $i \leftarrow k$ or edge i—k is not oriented.

It is clear that there is an edge connecting i and k. For otherwise we get a contradiction because by Rule R_1, $i \to j$ implies that $j \leftarrow k$.

Now assume that forbidden situations of the type $i \to j$ and $j \to k$ with $i \not\to k$ exist after phase 1 of Algorithm 2.10. Then select, from among all the derivation chains which lead to a forbidden situation, a chain which is shortest with the minimum number of directed edges

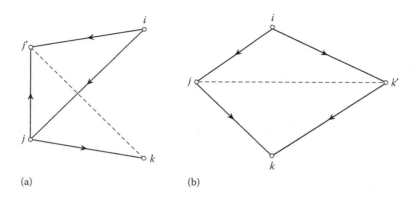

Figure 2.13 Illustration of the proof of Lemma 2.19.

incident into k. Clearly, any such chain must be of length at least 3. Let one such chain be as follows:

$$(i \to j) \Rightarrow (\alpha_1) \Rightarrow (\alpha_2) \Rightarrow \cdots \Rightarrow (\alpha_{p-1}) \Rightarrow (j \to k).$$

Now α_{p-1} must be either $j \to j'$ for some j' or $k' \to k$ for some k'. Thus we need to consider two cases.

Case 1 Let α_{p-1} be $j \to j'$.

Then the derivation $(\alpha_{p-1}) \Rightarrow (j \to k)$ requires that $j' \not\vdash k$. Further $i \to j'$, for otherwise the derivation chain

$$(i \to j) \Rightarrow (\alpha_1) \Rightarrow \cdots \Rightarrow (\alpha_{p-2}) \Rightarrow (j \to j')$$

would lead to the forbidden situation $i \to j$ and $j \to j'$ with $i \not\to j'$. But this chain is shorter than our chain, contradicting its minimality.

The situation arising out of the above arguments is depicted in Figure 2.13a, where a dashed line indicates the absence of the corresponding edge.

Now $i \to j'$ and $j' \not\vdash k$ imply that $i \to k$ by Rule R_2. But this contradicts our assumption that $i \not\to k$.

Case 2 Let α_{p-1} be $k' \to k$.

As in the previous case the derivation $(\alpha_{p-1}) \Rightarrow (j \to k)$ requires that $j \not\vdash k'$. Further $i\text{---}k'$, for otherwise $k' \to k$ would imply that $i \to k$, contrary to our assumption that $i \not\to k$. In addition, $i \to j$ and $j \not\vdash k'$ imply that $i \to k'$.

The situation resulting from the above arguments is depicted in Figure 2.13b.

Now the derivation chain

$$(i \to k') \Rightarrow (i \to j) \Rightarrow (\alpha_1) \Rightarrow \cdots \Rightarrow (\alpha_{p-2}) \Rightarrow (k' \to k),$$

leading to the forbidden situation $i \to k'$ and $k' \to k$ with $i \not\to k$ is of the same length as our chain, but with one less edge entering k. This is again a contradiction of the assumption we have made about the choice of our chain. ∎

Let E' be the set of edges which are assigned orientations in the first phase, and let $\vec{E'}$ be the corresponding set of directed edges. The following result is an immediate consequence of Lemma 2.4.

Theorem 2.14 *The subgraph $\vec{G'} = (V, \vec{E'})$ of the directed edges of the set $\vec{E'}$ is transitive.* ∎

Theorem 2.15 *If Algorithm 2.10 terminates successfully, then it gives a transitive orientation.*

Proof. Proof is by induction on the number of phases in the algorithm. If the algorithm orients all the edges of the given graph in phase 1, then, by Theorem 2.14, the resulting orientation is transitive.

Let the given graph $G = (V, E)$ be oriented in p phases. Let E' be the set of edges which are assigned orientations in the first phase. Then, by Theorem 2.14, the subgraph $\vec{G'} = (V, \vec{E'})$ is transitive. Further, since the subgraph $G'' = (V, E - E')$ is orientable in $p - 1$ phases, it follows from the induction hypothesis that $\vec{G''} = (V, \vec{E} - \vec{E'})$ is transitive. Now we prove that the directed graph $\vec{G} = (V, \vec{E})$ is transitive.

Suppose that $\vec{G}$ is not transitive, that is, there exist in $\vec{G}$ three vertices i, j, k such that $i \to j$ and $j \to k$, but $i \not\to k$. Then both $i \to j$ and $j \to k$ cannot belong to $\vec{E'}$ or to $\vec{E} - \vec{E'}$ because $\vec{G'}$ and $\vec{G''}$ are both transitive.

Without loss of generality, assume that $i \to j$ is in $\vec{E'}$ and $j \to k$ is in $\vec{E} - \vec{E'}$. Then there must be an edge i—k connecting i and k; for otherwise $i \to j$ would imply $j \leftarrow k$, by Rule R_1.

Suppose that the edge i—k is oriented as $i \leftarrow k$ in Phase 1. Then $\vec{G'}$ is not transitive, resulting in a contradiction. On the other hand, if it is oriented as $i \leftarrow k$ in a latter phase, then $\vec{G''}$ is not transitive, again resulting in a contradiction.

Thus it is impossible to have in G three vertices i, j, k such that $i \to j$ and $j \to k$, but $i \not\to k$. Hence $\vec{G}$ is transitive. ∎

Thus Assertion 1 is established. We next proceed to establish Assertion 2.

Consider a graph $G = (V, E)$ which is transitively orientable. It is clear that if we reverse the orientations of all the edges in any transitive orientation of G, then the resulting directed graph is also a transitive orientation of G.

Suppose that we pick an edge of G and assign an arbitrary orientation to it. Let this edge be $i \to j$. If we now proceed to assign orientations to additional edges using Rules R_1 and R_2, then the edges so oriented will have the same orientations in all possible transitive orientations in which the edge i—j is oriented as $i \to j$. This is because once the orientation of i—j is specified, the orientation derived by Rules R_1 and R_2 are necessary for transitive orientability. It therefore follows that if we apply Algorithm 2.10 to the transitively orientable graph G, then Phase 1 will terminate successfully without encountering any contradiction of Rule R_1 or Rule R_2. Further the edges oriented in the first phase will have the same orientations in some transitive orientation of G.

If we can prove that graph $G'' = (V, E - E')$, where E' is the set of edges oriented in the first phase, is also transitively orientable, then it would follow that the second phase and also all other phases will terminate successfully giving a transitive orientation of G. Thus proving Assertion 2 is the same as establishing the transitive orientability of $G'' = (V, E - E')$. Toward this end we proceed as follows.

Let the edges of the set E' be called *marked edges*, and the end vertices of these edges be called *marked vertices*. Let V' denote the set of marked vertices. Note that an unmarked edge may be incident on a marked vertex.

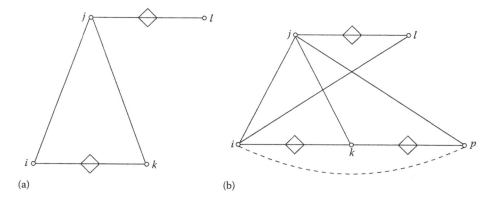

Figure 2.14 Illustration of the proof of Lemma 2.5.

Lemma 2.5 *It is impossible to have three marked vertices i, j, k such that edge i—j and j—k are unmarked, and edge i—k is marked.*

Proof. Assume that forbidden situations of the type mentioned in the lemma exist. In other words, assume that there exist triples of marked vertices i, j, k such that edges i—j and j—k are unmarked, and edge i—k is marked. For each such triple i, j, k there exists a marked edge j—l for some l, because j is a marked vertex. Therefore there exists a derivation chain from the marked edge i—k to the marked edge j—l.

Now select a forbidden situation with i, j, and k as the marked vertices such that there is a derivation chain P from i—k to j—l, which is a shortest one among all such chains that lead to a forbidden situation. This is shown in Figure 2.14a, where a diamond on an edge indicates that the edge is marked, and a dashed line indicates the absence of the corresponding edge.

The next marked edge after i—k in the shortest chain P is either i—p or k—p, for some p. We assume, without loss of generality, that it is k—p. Hence $i \not{-} p$, for otherwise edge k—p would not have been marked from edge i—k. Further there exists the edge i—l connecting i and l, for otherwise edge i—j would have been marked. Thus p and l are distinct. Also there exists the edge j—p, for otherwise edge j—k would have been marked. The relations established so far are shown in Figure 2.14b. Now j—p cannot be a marked edge, for marking it would result in marking i—j. Now we have a shorter derivation chain from edge k—p to edge j—l, leading to another forbidden situation where edges k—j and j—p are unmarked with edge k—p marked. A contradiction. ∎

Theorem 2.16 *If $G = (V, E)$ is transitively orientable, then $G''' = (V, E - E')$ is also transitively orientable.*

Proof. Since the Rules R_1 and R_2 mark only adjacent edges, it follows that the graph $G' = (V', E')$ is connected.

Consider now a vertex $v \in V - V'$. If v is connected to any vertex $v' \in V'$, then v should be connected to all the vertices of V' which are adjacent to v', for otherwise the edge $v - v'$ would have been marked. Since the graph G' is connected, it would then follow that v should be connected to all the vertices in V'.

Let $\vec{G}$ be a transitive orientation of G such that the orientations of the edges of $\vec{E'}$ agree in $\vec{G}$.

Next we partition the set $V - V'$ into four subsets as follows:

$$A = \{i | i \in V - V' \text{ and for all } j \in V', i \to j \text{ in } \vec{G}\},$$
$$B = \{i | i \in V - V' \text{ and for all } j \in V', j \to i \text{ in } \vec{G}\},$$
$$C = \{i | i \in V - V' \text{ and for all } j \in V', i \not{-} \text{ in } \vec{G}\},$$
$$D = V - (V' \cup A \cup B \cup C).$$

Note that D consists of all those vertices of $V - V'$ which are connected to all the vertices of V', but not all the edges connecting a vertex in D to the vertices in V' are oriented in the same direction.

Transitivity of $\vec{G}$ implies the following connections between the different subsets of V:

1. For every $i \in A$, $j \in D$, $k \in B$, $i \to j$, $j \to k$, and $i \to k$.

2. For all $i \in C$ and $j \in D$, $i \not{-} j$.

3. All edges connecting A and C are directed from A to C.

4. All edges connecting B and C are directed from C to B.

The situation so far is depicted in Figure 2.15a.

Now reverse the orientations of all the edges directed from V' to D so that all the edges connecting V' and D are directed from D to V'. The resulting orientation is as shown in Figure 2.15b. We now claim that this orientation is transitive.

To prove this claim we have to show that, in the graph of Figure 2.15b, if $i \to j$ and $j \to k$, then $i \to k$ for all i, j, and k. Clearly, this is true if none of these three edges is among those edges whose directions have just been reversed.

Thus we need to consider only the following four cases:

1. $i \in D$, $j \in V'$, and $k \in V'$.

2. $i \in D$, $j \in V'$, and $k \in B$.

3. $j \in D$, $k \in V'$, and $i \in A$.

4. $j \in D$, $k \in V'$, and $i \in D$.

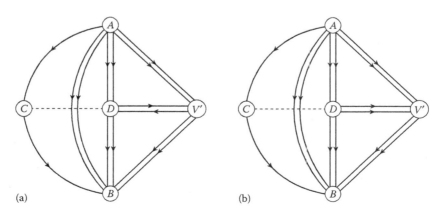

Figure 2.15 Illustration of the proof of Theorem 2.16.

In all these four cases $i \to k$ as shown in Figure 2.15b. Thus the orientation of Figure 2.15b is transitive.

Now remove from the graph of Figure 2.15b all the edges of E', namely, all the marked edges.

Suppose that in the resulting graph there exist vertices i, j, and k such that $i \to j$ and $j \to k$, but $i \not\to k$. Here $i \not\to k$ means only that $i \not\!\!-k$, for $i \leftarrow k$ would imply that the orientation in Figure 2.15b is not transitive. If edge $i\!-\!k$ is not in $E - E'$, it should be in E', for otherwise the orientation in Figure 2.15b is not transitive. Thus i and k are in V'.

Since there is no vertex outside V' which has both an edge into it from V' and an edge from it into V', it follows that j is also in V'.

Thus we have i, j, k marked, edges $i\!-\!j$ and $j\!-\!k$ unmarked, and edge $i\!-\!k$ marked. This is not possible by Lemma 2.5.

Therefore the directed graph which results after removing the edges of $\overrightarrow{E'}$ from the graph of Figure 2.15b is transitive. Hence $G'' = (V, E - E')$ is transitively orientable. ∎

Thus we have established Assertion 2 and hence the correctness of Algorithm 2.10.

It should now be clear that the transitive orientation algorithm discussed above is an example of an algorithm which is simple but whose proof of correctness is very much involved. For an earlier algorithm on this problem see Gilmore and Hoffman [70].

Pnueli et al. [69] and Even et al. [71] introduced permutation graphs and established a structural relationship between these graphs and transitively orientable graphs. They also discussed an algorithm to test whether a given graph is a permutation graph. For perhaps the most recent work on transitive orientation and related problems (see McConnell and Spinard [72]).

Certain graph problems which are in general very hard to solve become simple when the graph under consideration is transitively orientable. Problems of finding a maximum clique and a minimum coloration are examples of such problems. These problems arise in the study of memory relocation and circuit layout problems [71] (see also Liu [73] and Even [74]).

References

[1] J. B. Kruskal, Jr., On the shortest spanning subtree of a graph and the travelling salesman problem, *Proc. Am. Math. Soc.*, **7** (1956), 48–50.

[2] R. C. Prim, Shortest connection networks and some generalizations, *Bell Sys. Tech. J.*, **36** (1957), 1389–1401.

[3] A. Kerschenbaum and R. Van Slyke, Computing minimum spanning trees efficiently, *Proceedings of the 25th Annual Conference of the ACM*, 1972, 518–527.

[4] A. C. Yao, An $O(|E| \log \log |V|)$ algorithm for finding minimum spanning trees, *Inform. Process. Lett.*, **4** (1975), 21–23.

[5] D. Cheriton and R. E. Tarjan, Finding minimum spanning trees, *SIAM J. Comput.*, **5** (1976), 724–742.

[6] R. E. Tarjan, Sensitivity analysis of minimum spanning trees and shortest path trees, *Inform. Process. Lett.*, **14** (1982), 30–33.

[7] C. H. Papadimitriou and M. Yannakakis, The Complexity of restricted minimum spanning tree problems, *J. ACM*, **29** (1982), 285–309.

[8] R. L. Graham and P. Hall, On the history of the minimum spanning tree problem, *Mimeographed*, Bell Laboratories, Murray Hill, NJ, 1982.

[9] B. Korte and J. Vygen, *Combinatorial Optimization: Theory and Algorithms*, Springer, New York, 2000.

[10] B. Chazelle, A fast deterministic algorithm for minimum spanning trees, *Proceedings of the 38th Annual IEEE Symposium on Foundations of Computer Science*, 1997, 22–31.

[11] M. L. Fredman and R. E. Tarjan, Fibonacci heaps and their uses in improved network optimization problems, *J. ACM*, **34** (1987), 596–615.

[12] H. N. Gabow, Z. Galil, and T. Spencer, Efficient implementation of graph algorithms using contraction, *Proceedings of the 25th Annual IEEE Symposium on Foundations of Computer Science*, 1984, 338–346.

[13] H. N. Gabow, Z. Galil, T. Spencer, and R. E. Tarjan, Efficient algorithms for finding minimum spanning trees in undirected and directed graphs, *Combinatorica*, **6** (1986), 109–122.

[14] B. Korte and J. Nesetril, Vojtech Jarnik's work in combinatorial optimization, *Report No. 97855-0R*, Research Institute for Discrete Mathematics, University of Bonn, Germany, 1997.

[15] J. Edmonds, Optimum branchings, *J. Res. Nat. Bur. Std.*, **71B** (1967), 233–240.

[16] R. M. Karp, A simple derivation of Edmonds' algorithm for optimum branchings, *Networks*, **1** (1972), 265–272.

[17] M. N. Swamy and K. Thulasiraman, *Graphs, Networks and Algorithms*, Wiley-Interscience, New York, 1981.

[18] R. E. Tarjan, Finding optimum branchings, *Networks*, **7** (1977), 25–35.

[19] P. M. Camerini, L. Fratta, and F. Maffioli, A note on finding optimum branchings, *Networks*, **9** (1979), 309–312.

[20] F. C. Bock, An algorithm to construct a minimum directed spanning tree in a directed network, *Developments in Operations Research*, B. Avi-Itzak, Ed., Gordon & Breach, New York, 1971, 29–44.

[21] Y. Chu and T. Liu, On the shortest arborescence of a directed graph, *Scientia Sinica* [Peking], **4**, (1965), 1396–1400; *Math. Rev.*, **33**, 1245 (D. W. Walkup).

[22] D. Gries, *Compiler Construction for Digital Computers*, Wiley, New York, 1971.

[23] S. Warshall, A theorem on boolean matrices, *J. ACM*, **9** (1962), 11–12.

[24] H. S. Warren, A modification of Warshall's algorithm for the transitive closure of binary relations, *Comm. ACM*, **18** (1975), 218–220.

[25] V. L. Arlazarov, E. A. Dinic, M. A. Kronrod, and I. A. Faradzev, On economical construction of the transitive closure of a directed graph, *Soviet Math. Dokl.*, **11** (1970), 1209–1210.

[26] J. Eve and R. Kurki-Suonio, On computing the transitive closure of a relation, *Acta Inform.*, **8** (1977), 303–314.

[27] M. J. Fischer and A. R. Meyer, Boolean matrix multiplication and transitive closure, *Conference Record, IEEE 12th Annual Symposium on Switching and Automata Theory*, 1971, 129–131.

[28] M. E. Furman, Application of a method of fast multiplication of matrices in the problem of finding the transitive closure of a graph, *Soviet Math. Dokl.*, **11** (1970), 1252.

[29] I. Munro, Efficient determination of the transitive closure of a directed graph, *Inform. Process. Lett.*, **1** (1971), 56–58.

[30] P. E. O'Neil and E. J. O'Neil, A fast expected time algorithm for boolean matrix multiplication and transitive closure, *Inform. Control*, **22** (1973), 132–138.

[31] P. Purdom, A transitive closure algorithm, *BIT*, **10** (1970), 76–94.

[32] C. P. Schnorr, An algorithm for transitive closure with linear expected time, *SIAM J. Comput.*, **7** (1978), 127–133.

[33] V. Strassen, Gaussian elimination is not optimal, *Numerische Math.*, **13** (1969), 354–356.

[34] M. M. Syslo and J. Dzikiewicz, Computational experience with some transitive closure algorithms, *Computing*, **15** (1975), 33–39.

[35] K. Melhorn, *Graph Algorithm and NP-completeness*, Springer-Verlag, Berlin, Germany, 1984.

[36] L. R. Ford and D. R. Fulkerson, *Flows in Networks*, Princeton University Press, Princeton, NJ, 1962.

[37] L. R. Ford, Jr., Network flow theory, Paper P-923, RAND Corp., Santa Monica, CA, 1956.

[38] R. E. Bellman, On a routing problem, *Quart. Appl. Math.*, **16** (1958), 87–90.

[39] E. F. Moore, The shortest path through a maze, *Proceedings of the International Symposium on the Theory of Switching, Part II*, University Press, Cambridge, MA, 1957, 285–292.

[40] N. Christofides, *Graph Theory: An Algorithmic Approach*, Academic Press, New York, 1975.

[41] S. E. Dreyfus, An appraisal of some shortest-path algorithms, *Oper. Res.*, **17** (1969), 395–412.

[42] H. Frank and I. T. Frisch, *Communication, Transmission and Transportation Networks*, Addison-Wesley, Reading, MA, 1971.

[43] M. Gordon and M. Minoux, *Graphs and Algorithms* (trans. by S. Vajda), Wiley, New York, 1984.

[44] T. C. Hu, A decomposition algorithm for shortest paths in a network, *Oper. Res.*, **16** (1968), 91–102.

[45] E. L. Lawler, *Combinatorial Optimization: Networks and Matroids*, Holt, Rinehart & Winston, New York, 1976.

[46] E. Minieka, *Optimization Algorithms for Networks and Graphs*, Marcel Dekker, New York, 1978.

[47] P. M. Spira and A. Pan, On finding and updating spanning trees and shortest paths, *SIAM J. Comput.*, **4** (1975), 375–380.

[48] D. B. Johnson, Efficient algorithms for shortest paths in sparse networks, *J. ACM*, **24** (1977), 1–13.

[49] R. A. Wagner, A shortest path algorithm for edge-sparse graphs, *J. ACM*, **23** (1976), 50–57.

[50] J. Edmonds and R. M. Karp, Theoretical improvements in algorithmic efficiency for network flow problems, *J. ACM*, **19** (1972), 248–264.

[51] M. L. Fredman, New bounds on the complexity of the shortest path problem, *SIAM J. Comp.*, **5** (1976), 83–89.

[52] D. B. Johnson, A note on Dijkstra's shortest path algorithm, *J. ACM*, **20** (1973), 385–388.

[53] M. A. Comeau and K. Thulasiraman, Structure of the submarking-reachability problem and network programming, *IEEE Trans. Circuits Syst.*, **35** (1988), 89–100.

[54] T. Lengauer, On the solution of inequality systems relevant to IC layout, *J. Algorithms*, **5** (1984), 408–421.

[55] Y. Liao and C. K. Wong, An algorithm to compact VLSI symbolic layout with mixed constraints, *IEEE Trans. CAD. Circuits Syst.*, **2** (1983), 62–69.

[56] E. W. Dijkstra, A note on two problems in connection with graphs, *Numerische Math.*, **1** (1959), 269–271.

[57] R. W. Floyd, Algorithm 97: Shortest path, *Comm. ACM*, **5** (1962), 345.

[58] G. B. Dantzig, All shortest routes in a graph, *Theory of Graphs*, Gordon & Breach, New York, 1967, 91–92.

[59] Y. Tabourier, All shortest distances in a graph: An improvement to Dantzig's inductive algorithm, *Discrete Math.*, **4** (1973), 83–87.

[60] T. A. Williams and G. R. White, A note on Yen's algorithm for finding the length of all shortest paths in N-node nonnegative distance networks, *J. ACM*, **20** (1973), 389–390.

[61] J. Y. Yen, Finding the lengths of all shortest paths in N—node, nonnegative distance complete networks using $N^3/2$ additions and N^3 comparisons, *J. ACM*, **19** (1972), 423–424.

[62] N. Deo and C. Y. Pang, Shortest path algorithms-taxonomy and annotation, *Networks*, **14** (1984), 275–323.

[63] A. R. Pierce, Bibliography on algorithms for shortest path, shortest spanning tree and related circuit routing problems (1956–1974), *Networks*, **5** (1975), 129–149.

[64] R. E. Tarjan, Data structures and network algorithms, *CBMS-NSF Regional Conference Series in Applied Mathematics*, Vol. 44, Society for Industrial Applied Mathematics, Philadelphia, PA, 1983.

[65] B. Carré, *Graphs and Networks*, Clarendon Press, Oxford, 1979.

[66] R. E. Tarjan, A unified approach to path problems, *J. ACM.*, **28** (1981), 577–593.

[67] A. Moffat and T. Takoka, An all-pairs shortest path algorithm with expected time $O(n^2 \log n)$, *SIAM J. Comp.*, **16** (1987), 1023–1031.

[68] G. N. Frederickson, Fast algorithms for shortest paths in planar graphs with applications, *SIAM J. Comp.*, **16** (1987), 1004–1022.

[69] A. Pnueli, A. Lempel, and S. Even, Transitive orientation of graphs and identification of permutation graphs, *Canad. J. Math.*, **23** (1971), 160–175.

[70] P. C. Gilmore and A. J. Hoffman, A characterization of comparability graphs and of interval graphs, *Canad. J. Math.*, **16** (1964), 539–548.

[71] S. Even, A. Pnueli, and A. Lempel, Permutation graphs and transitive graphs, *J. ACM*, **19** (1972), 400–410.

[72] R. M. McConnell and J. P. Spinrad, Modular decomposition and transitive orientation, *Technical Report 475/1995*, Technische Universitat Berlin, Fachbereich Mathematik, 1995.

[73] C. L. Liu, *Introduction to Combinatorial Mathematics*, McGraw-Hill, New York, 1968.

[74] S. Even, *Algorithmic Combinatorics*, Macmillan, New York, 1973.

CHAPTER 3

Depth-First Search and Applications*

Krishnaiyan "KT" Thulasiraman

CONTENTS

3.1 Introduction 59
3.2 DFS of an Undirected Graph 59
3.3 DFS of a Directed Graph 63
3.4 Biconnectivity and Strong Connectivity Algorithms 66
    3.4.1 Biconnectivity Algorithm 66
    3.4.2 Strong Connectivity Algorithm 68
3.5 st-Numbering of a Graph 71

3.1 INTRODUCTION

In this chapter we describe a systematic method for exploring a graph. This method known as depth-first search, in short, DFS, has proved very useful in the design of several efficient algorithms [2,3]. We describe three of these application, namely, biconnectivity, strong connectivity and $s-t$ numbering algorithms.

3.2 DFS OF AN UNDIRECTED GRAPH

We first describe DFS of an undirected graph. To start with, we assume that the graph under consideration is connected. If the graph is not connected, then DFS would be performed separately on each component of the graph. We also assume that there are no self-loops in the graph. DFS of an undirected graph G proceeds as follows:

We choose any vertex, say v, in G and begin the search from v. The start vertex v, called the *root* of the DFS, is now said to be visited.

We then select an edge (v, w) incident on v and traverse this edge to visit w. We also orient this edge from v to w. The edge (v, w) is now said to be *examined* and is called a *tree edge*. The vertex v is called the *father* of w, denoted as FATHER(w).

In general, while we are at some vertex x, two possibilities arise:

1. If all the edges incident on x have already been examined, then we return to the father of x and continue the search from FATHER(x). The vertex x is now said to be *completely scanned*.

2. If there exists some unexamined edges incident on x, then we select one such edge (x, y) and orient it from x to y. The edge (x, y) is now said to be examined. Two cases need to be considered now:

---

*This chapter is an edited version of Sections 11.7, 11.8, and 11.10 in Thulasiraman and Swamy [1].

Case 1 If y has not been previously visited, then we traverse the edge (x, y), visit y, and continue the search from y. In this case (x, y) is a *tree edge* and $x = \text{FATHER}(y)$.

Case 2 If y has been previously visited, then we proceed to select another unexamined edge incident on x. In this case the edge (x, y) is called a *back edge*.

During the DFS, whenever a vertex x is visited for the first time, it is assigned a distinct integer $\text{DFN}(x)$ such that $\text{DFN}(x) = i$, if x is the ith vertex to be visited during the search. $\text{DFN}(x)$ is called the *depth-first number* (DFN) of x. Clearly, DFNs indicate the order in which the vertices are visited during DFS.

DFS terminates when the search returns to the root and all the vertices have been visited. We now present a formal description of the DFS algorithm. In this description the graph under consideration is assumed to be connected. The array MARK used in the algorithm has one entry for each vertex. To begin with we set $\text{MARK}(v) = 0$ for every vertex v in the graph, thereby indicating that no vertex has yet been visited. Whenever a vertex is visited for the first time, we set the corresponding entry in the MARK array equal to 1. We use an array SCAN that has one entry for each vertex in the graph. To begin with we set $\text{SCAN}(v) = 0$ for every vertex v, thereby indicating that none of the vertices is completely scanned. Whenever a vertex is completely scanned, the corresponding entry in the SCAN array is set to 1. The arrays DFN and FATHER are as defined before. TREE and BACK are two sets storing, respectively, the tree edges and the back edges as they are generated.

Algorithm 3.1 DFS of an undirected graph

Input: $G = (V, E)$ is a connected undirected graph. Vertex s is the start vertex of the depth-first search.

Output: Depth-first numbering of the vertices of G and the depth-first search tree with vertex s as the root vertex.

begin
    TREE $\leftarrow \phi$;
    BACK $\leftarrow \phi$;
    for every edge e in G, EXAMINED $(e) \leftarrow 0$;
    for vertex v in G
        do
            FATHER$(v) \leftarrow v$;
            MARK$(v) \leftarrow 0$;
            SCAN$(v) \leftarrow 0$;
        od
    MARK$(s) \leftarrow 1$;
    DFN$(s) \leftarrow 1$;
    $i \leftarrow 1$;
    $v \leftarrow s$;
    repeat
        while there exists an edge $e = (v, w)$ with EXAMINED$(e) = 0$
            do
                Orient the edge (v, w) from v to w;
                EXAMINED$(e) \leftarrow 1$;
                if MARK$(w) = 0$ **then**
                      begin
                          $i \leftarrow i + 1$;

```
                    DFN(w) ← i;
                    TREE ← TREE ∪ {(v, w)};
                    MARK(w) ← 1;
                    FATHER(w) ← v;
                    v ← w;
                end
            else BACK = BACK ∪ {(v, w)};
        od
    end while
    SCAN(v) ← 1;
    v ← FATHER(v);
    until v = s and SCAN(s) = 1;
end
```

As we can see from the preceding description, DFS partitions the edges of G into tree edges and back edges. It is easy to show that the tree edges form a spanning tree of G. DFS also imposes directions on the edges of G. The resulting directed graph will be denoted by $\hat{G}$. The tree edges with their directions imposed by the DFS will form a directed spanning tree of $\hat{G}$. This directed spanning tree will be called the *DFS tree*.

Note that DFS of a graph is not unique since the edges incident on a vertex may be chosen for examination in any arbitrary order.

As an example, we have shown in Figure 3.1 DFS of an undirected graph. In this figure tree edges are shown as continuous lines, and back edges are shown as dashed lines. Next to each vertex we have shown its DFN. We have also shown in the figure the list of edges incident on each vertex v. This list for a vertex v is called the *adjacency* list of v, and it gives the order in which the edges incident on v are chosen for examination.

Let T be a DFS tree of a connected undirected graph. As we mentioned before, T is a directed spanning tree of G. For further discussions, we need to introduce some terminology.

If there is a directed path in T from a vertex v to a vertex w, then v is called an *ancestor* of w, and w is called a *descendant* of v. Furthermore, if $v \neq w$, v is called a proper ancestor of w, and w is called a *proper descendant* of v. If (v, w) is a directed edge in T, then v is called the *father* of w, and w is called a *son* of v. Note that a vertex may have more than one son. A vertex v and all its descendants form a subtree of T with vertex v as the root of this subtree.

Two vertices v and w are *related* if one of them is a descendant of the other. Otherwise v and w are *unrelated*. If v and w are unrelated and $DFN(v) < DFN(w)$, then v is said to be to the *left* of w; otherwise, v is to the right of w. Edges of G connecting unrelated vertices are called *cross edges*. We now show that there are no cross edges in G.

Let v_1 and v_2 be any two unrelated vertices in T. Clearly then there are two distinct vertices s_1 and s_2 such that (1) $FATHER(s_1) = FATHER(s_2)$ and (2) v_1 and v_2 are descendants of s_1 and s_2, respectively (see Figure 3.2).

Let T_1 and T_2 denote the subtrees of T rooted at s_1 and s_2, respectively. Assume without loss of generality that $DFN(s_1) < DFN(s_2)$. It is then clear from the DFS algorithm that vertices in T_2 are visited only after the vertex s_1 is completely scanned.

Further, scanning of s_1 is completed only after all the vertices in T_1 are scanned completely. So there cannot exist an edge connecting v_1 and v_2. For if such an edge existed, it would have been visited before the scanning of s_1 is completed.

62 ■ Handbook of Graph Theory, Combinatorial Optimization, and Algorithms

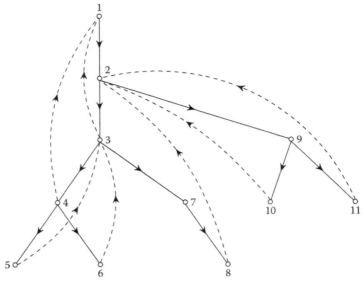

Vertex	Adjacency List
1	(1, 2), (1, 3), (1, 4)
2	(2, 1), (2, 3), (2, 8), (2, 9), (2, 10), (2, 11)
3	(3, 1), (3, 2), (3, 4), (3, 5), (3, 6), (3, 7)
4	(4, 1), (4, 3), (4, 5), (4, 6)
5	(5, 3), (5, 4)
6	(6, 3), (6, 4)
7	(7, 3), (7, 8)
8	(8, 2), (8, 7)
9	(9, 2), (9, 10), (9, 11)
10	(10, 2), (10, 9)
11	(11, 2), (11, 9)

Figure 3.1 DFS of an undirected graph.

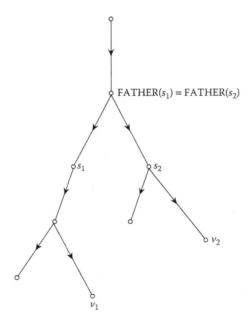

Figure 3.2 Illustration of the proof of Theorem 3.1.

Thus we have the following.

Theorem 3.1 *If (v, w) is an edge in a connected undirected graph G, then in any DFS tree of G either v is a descendant of w or vice versa. In other words, there are no cross edges.* ∎

The absence of cross edges in an undirected graph is an important property that forms the basis of an algorithm to be discussed in Section 3.4.1 for determining the biconnected components of a graph.

3.3 DFS OF A DIRECTED GRAPH

DFS of a directed graph is essentially similar to that of an undirected graph. The main difference is that in the case of a directed graph an edge is traversed only along its orientation. As a result of this constraint, edges in a directed graph G are partitioned into four categories (and not two as in the undirected case) by a DFS of G. An unexamined edge (v, w) encountered while at the vertex v would be classified as follows.

Case 1 w has not yet been visited.
In this case (v, w) is a tree edge.

Case 2 w has already been visited.

   a. If w is a descendant of v in the DFS forest (i.e., the subgraph of tree edges), then (v, w) is called a *forward edge*.

   b. If w is an ancestor of v in the DFS forest, then (v, w) is called a *back edge*.

   c. If v and w are not related in the DFS forest and $\mathrm{DFN}(w) < \mathrm{DFN}(v)$, then (v, w) is a *cross edge*. Note that there are no cross edges of the type (v, w) with $\mathrm{DFN}(w) > \mathrm{DFN}(v)$. The proof for this is along the same lines as that for Theorem 3.1.

A few useful observations are now in order:

  1. An edge (v, w), with $\mathrm{DFN}(w) > \mathrm{DFN}(v)$, is either a tree edge or a forward edge. During the DFS it is easy to distinguish between a tree edge and a forward edge because a tree edge always leads to a new vertex.

  2. An edge (v, w) with $\mathrm{DFN}(w) < \mathrm{DFN}(v)$ is either a back edge or a cross edge. Such an edge (v, w) is a back edge if and only if w is not completely scanned when the edge is encountered while examining the edges incident out of v.

  3. DFS forest, the subgraph of tree edges, may not be connected even if the directed graph under consideration is connected. The first vertex to be visited in each component of the DFS forest will be called the root of the corresponding component.

A description of the DFS algorithm for a directed graph is presented next. As we pointed out earlier, when we encounter an edge (v, w) with $\mathrm{DFN}(w) < \mathrm{DFN}(v)$, we shall classify it as a back edge if $\mathrm{SCAN}(w) = 0$; otherwise (v, w) is a cross edge. We also use two arrays, FORWARD and CROSS, that store respectively, forward and cross edges.

> **Algorithm 3.2 DFS of a directed graph**
>
> **Input:** $G = (V, E)$ is a connected directed graph.
> **Output:** Depth first numbering of the vertices of G.
> begin
>   TREE $\leftarrow \phi$; BACK $\leftarrow \phi$; FORWARD $\leftarrow \phi$; CROSS $\leftarrow \phi$;
>   **for** every edge e in G EXAMINED $(e) \leftarrow 0$;
>   **for** every vertex v in G
>   **do**
>   FATHER $(v) \leftarrow v$; MARK$(v) \leftarrow 0$; SCAN $(v) \leftarrow 0$;
>   ROOT$(v) \leftarrow 0$;
>   **od**
> $i \leftarrow 0$;
>   **repeat**
>   **while** there exists a vertex v with MARK$(v) = 0$;
>   MARK$(v) \leftarrow 1$; $i \leftarrow i + 1$; DFN$(v) \leftarrow i$; ROOT$(v) \leftarrow 1$;
>   **repeat**
>   **while** there exists an edge $e \leftarrow (v, w)$ with EXAMINED$(e) = 0$
>   **do**
>   Orient the edge (v, w) from v to w; EXAMINED$(e) \leftarrow 1$;
>   **if** MARK$(w) \leftarrow 0$ **then**
>   **begin**
> $i \leftarrow i + 1$;
>   DFN$(w) \leftarrow i$;
>   TREE $\leftarrow$ TREE $\cup \{(v, w)\}$;
>   MARK$(w) \leftarrow 1$;
>   FATHER$(w) \leftarrow v$;
> $v \leftarrow w$;
>   **end**
>   **else**
>   **if** DFN$(w) >$ DFN(v) **then**
>   FORWARD $\leftarrow$ FORWARD $\cup \{(v, w)\}$;
>   **else if** SCAN$(w) = 0$ **then**
>   BACK $\leftarrow$ BACK $\cup \{(v, w)\}$;
>   **else** CROSS $\leftarrow$ CROSS $\cup \{(v, w)\}$;
>   **od**
>   **end while**
>   SCAN$(v) \leftarrow 1$;
> $v \leftarrow$ FATHER(v);
>   **until** ROOT$(v) = 1$ and SCAN $(v) = 1$
>   **end while**
>   **until** $i = n$;
> end

As an example, DFS of a directed graph is shown in Figure 3.3a. Next to each vertex we have shown its DFN. The tree edges are shown as continuous lines, and the other edges are shown as dashed lines. The DFS forest is shown separately in Figure 3.3b.

We pointed out earlier that the DFS forest of a directed graph may not be connected, even if the graph is connected. This can also be seen from Figure 3.3b. This leads us to the problem of discovering sufficient conditions for a DFS forest to be connected. In the following we prove that the DFS forest of a strongly connected graph is connected. In fact, we shall be

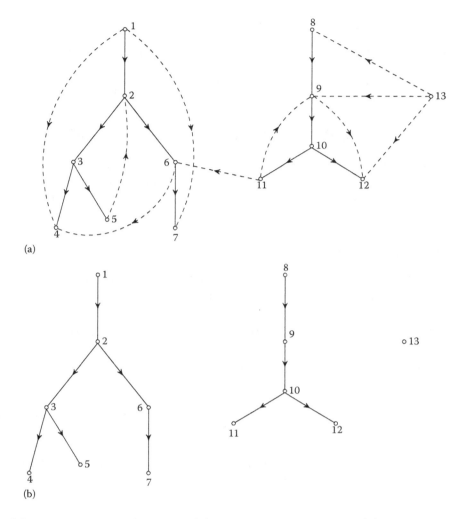

Figure 3.3 (a) DFS of a directed graph. (b) DFS forest of graph in (a).

establishing a more general result. Let T denote a DFS forest of a directed graph $G = (V, E)$. Let $G_i = (V_i, E_i)$, with $|V_i| \geq 2$, be a strongly connected component of G. Consider any two vertices v and w in G_i. Assume without loss of generality that $\text{DFN}(v) < \text{DFN}(w)$. Since G_i is strongly connected, there exists a directed path P in G_i from v to w. Let x be the vertex on P with the lowest DFN and let T_x be the subtree of T rooted at x. Note that cross edges and back edges are the only edges that lead out of the subtree T_x. Since these edges lead to vertices having lower DFNs than $\text{DFN}(x)$, it follows that once path P reaches a vertex in T_x, then all the subsequent vertices on P will also be in T_x. In particular, w also lies in T_x. So it is a descendant of x. Since $\text{DFN}(x) \leq \text{DFN}(v) < \text{DFN}(w)$, it follows from the DFS algorithm that v is also in T_x. Thus, any two vertices v and w in G_i have a common ancestor that is also in G_i.

We may conclude from this that all the vertices of G_i have a common ancestor r_i that is also in G_i. It may now be seen that among all the common ancestors in T of vertices in G_i vertex r_i has the highest DFN. Further, it is easy to show that if v is a vertex in G_i, then any vertex on the tree path from r_i to v will also be in G_i. So the subgraph of T induced by V_i is connected. Thus we have the following.

Theorem 3.2 *Let $G_i = (V_i, E_i)$ be a strongly connected component of a directed graph $G = (V, E)$. If T is a DFS forest of G, then the subgraph of T induced by V_i is connected.* ■

Following is an immediate corollary of Theorem 3.2.

Corollary 3.1 *The DFS forest of a strongly connected graph is connected.* ∎

It is easy to show that the DFS algorithms are both of complexity $O(n+m)$, where n is the number of vertices and m is the number of edges in a graph.

3.4 BICONNECTIVITY AND STRONG CONNECTIVITY ALGORITHMS

In this section we discuss algorithms due to Hopcroft and Tarjan [2] and Tarjan [3] for determining the biconnected components and the strongly connected components of a graph. These algorithms are based on DFS. We begin our discussion with the biconnectivity algorithm.

3.4.1 Biconnectivity Algorithm

First we recall that a biconnected graph is a connected graph with no cut-vertices. A maximal biconnected subgraph of a graph is called a biconnected component of the graph.*

A crucial step in the development of the biconnectivity algorithm is the determination of a simple criterion that can be used to identify cut-vertices as we perform a DFS. Such a criterion is given in the following two lemmas.

Let $G = (V, E)$ be a connected undirected graph. Let T be a DFS tree of G with vertex r as the root. Then we have the following.

Lemma 3.1 *Vertex $v \neq r$ is a cut-vertex of G if and only if for some son s of v there is no back edge between any descendant in T of s (including itself) and a proper ancestor of v.*

Proof. Let G' be the graph that results after removing vertex v from G. By definition, v is a cut-vertex of G if and only if G' is not connected.

Let $s_1, s_2, \ldots, s_k$ be the sons of v in T. For each i, $1 \leq i \leq k$, let V_i denote the set of descendants of s_i (including itself), and let G_i be the subgraph of G' induced on V_i. Further let $V'' = V' - \cup_{i=1}^{k} V_i$, where $V' = V - \{v\}$, and let G'' be the subgraph induced on V''. Note that all the proper ancestors of v are in V''.

Clearly, $G_1, G_2, \ldots, G_k$ and G'' are all subgraphs of G, which together contain all the vertices of G'. We can easily show that all these subgraphs are connected. Further, by Theorem 3.1 there are no edges connecting vertices belonging to different G_i's. So it follows that G' will be connected if and only if for every i, $1 \leq i \leq k$, there exists an edge (a, b) between a vertex $a \in V_i$ and a vertex $b \in V''$. Such an edge (a, b) will necessarily be a back edge, and b will be a proper ancestor of v. We may therefore conclude that G' will be connected if and only if for every son s_i of v there exists a back edge between some descendant of s_i (including itself) and a proper ancestor of v. The proof of the lemma is now immediate. ∎

Lemma 3.2 *The root vertex r is a cut-vertex of G if and only if it has more than one son.*

Proof. Proof in this case follows along the same line as that for Lemma 3.1. ∎

In the following we refer to the vertices of G by their DFNs. To embed into the DFS procedure the criterion given in Lemmas 3.1 and 3.2, we now define, for each vertex v of G,

$$\text{LOW}(v) = \min(\{v\} \cup \{w | \text{there exists a backedge } (x, w) \text{ such that } x \text{ is a descendant of } v, \text{ and } w \text{ is a proper ancestor of } v \text{ in } T\}). \quad (3.1)$$

---
*Note that a biconnected component is the same as a block defined in Chapter 1.

Using the LOW values, we can restate the criterion given in Lemma 3.1 as in the following theorem.

Theorem 3.3 *Vertex $v \neq r$ is a cut-vertex of G if and only if v has a son s such that $LOW(s) \geq v$.* ∎

Noting that $LOW(v)$ is equal to the lowest numbered vertex that can be reached from v by a directed path containing at most one back edge, we can rewrite (3.1) as

$$LOW(v) = \min(\{v\} \cup \{LOW(s) | s \text{ is a son of } v\} \cup \{w | (v,w) \text{ is a backedge}\})$$

This equivalent definition of $LOW(v)$ suggests the following steps for computing $LOW(v)$:

1. When v is visited for the first time during DFS, set $LOW(v)$ equal to the DFN of v.

2. When a back edge (v, w) incident on v is examined, set $LOW(v)$ to the minimum of its current value and the DFN of w.

3. When the DFS returns to v after completely scanning a son s of v, set $LOW(v)$ equal to the minimum of its current value and $LOW(s)$.

Note that for any vertex v, computation of $LOW(v)$ ends when the scanning of v is completed.

We next consider the question of identifying the edges belonging to a biconnected component. For this purpose we use an array STACK. To begin with STACK is empty. As edges are examined, they are added to the top of STACK.

Suppose DFS returns to a vertex v after completely scanning a son s of v. At this point computation of $LOW(s)$ will have been completed. Suppose it is now found that $LOW(s) \geq v$. Then, by Theorem 3.3, v is a cut-vertex. Further, if s is the first vertex with this property, then we can easily see that the edge (v, s) along with the edges incident on s, and its descendants will form a biconnected component. These edges are exactly those that lie on top of STACK up to and including (v, s). They are now removed from STACK. From this point on the algorithm behaves in exactly the same way as it would on the graph G', which is obtained by removing from G the edges of the biconnected component that has just been identified.

For example, a DFS tree of a connected graph may be as in Figure 3.4, where $G_1, G_2, \ldots, G_5$ are the biconnected components in the order in which they are identified.

A description of the biconnectivity algorithm now follows. This algorithm is essentially the same as Algorithm 3.1, with the inclusion of appropriate steps for computing $LOW(v)$ and identifying the cut-vertices and the edges belonging to the different biconnected components. Note that in this algorithm the root vertex r is treated as a cut-vertex, even if it is not one, for the purpose of identifying the biconnected component containing r.

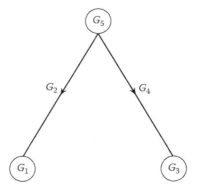

Figure 3.4 G_1, G_2, G_3, G_4, G_5—biconnected components of a graph.

Algorithm 3.3 Biconnectivity

Input: $G = (V, E)$ is a connected undirected graph.
Output: Biconnected components of G.
begin
    STACK $\leftarrow \phi$;
    for every edge e in G EXAMINED $(e) \leftarrow 0$;
    for vertex v in G
        do
            FATHER $(v) \leftarrow v$;
            MARK$(v) \leftarrow 0$;
            SCAN$(v) \leftarrow 0$;
    od
    Pick any vertex s with MARK$(s) \leftarrow 0$;
    MARK$(s) \leftarrow 1$;
    DFN$(s) \leftarrow 1$;
    LOW$(s) \leftarrow 1$;
    $i \leftarrow 1$;
    $v \leftarrow s$;
    repeat
        while there exists an edge $e = (v, w)$ with EXAMINED$(e) = 0$
            do
                EXAMINED$(e) \leftarrow 1$;
                STACK $\leftarrow$ STACK $\cup \{(v, w)\}$;
                if MARK$(w) \leftarrow 0$ **then**
                      begin
                          MARK$(w) \leftarrow 1$;
                          $i \leftarrow i + 1$;
                          DFN$(w) \leftarrow i$;
                          FATHER$(w) \leftarrow v$;
                          LOW$(w) \leftarrow i$;
                          $v \leftarrow w$;
                      end
                else LOW$(v) \leftarrow \min\{\text{LOW}(v), \text{DFN}(w)\}$;
            od
        end while
        SCAN$(v) \leftarrow 1$;
        if LOW$(v) \geq$ DFN(FATHER$(v))$ **then**
            remove all the edges from the top of the STACK up to and
            including the edge (FATHER$(v), v)$;
            LOW(FATHER$(v)) \leftarrow \min \{\text{LOW}(v), \text{LOW}(\text{FATHER}(v))\}$;
            $v \leftarrow$ FATHER(v);
    until $v = s$ and SCAN$(s) = 1$;
end

3.4.2 Strong Connectivity Algorithm

Recall from Chapter 1 that a graph is strongly connected if for every pair of vertices v and w there exists in G a directed path from v to w and a directed path from w to v; further a maximal strongly connected subgraph of a graph G is called a strongly connected component of the graph.

Consider a directed graph $G = (V, E)$. Let $G_1 = (V_1, E_1), G_2 = (V_2, E_2), \ldots, G_k = (V_k, E_k)$ be the strongly connected components of G. Let T be a DFS forest of G and $T_1, T_2, \ldots, T_k$ be the induced subgraphs of T on the vertex sets $V_1, V_2, \ldots, V_k$, respectively. We know from Theorem 3.2 that $T_1, T_2, \ldots, T_k$ are connected.

Let r_i, $1 \leq i \leq k$, be the root of T_i. If $i < j$, then DFS terminates at vertex r_i earlier than at r_j. Then we can see that for each $i < j$, either r_i is to the left of r_j or r_i is a descendant of r_j in T. Further G_i, $1 \leq i \leq k$, would consist of those vertices that are descendants of r_i, but are in none of $G_1, G_2, \ldots, G_{i-1}$.

The first step in the development of the strong connectivity algorithm is the determination of a simple criterion that can be used to identify the roots of strongly connected components as we perform a DFS. The following observations will be useful in deriving such a criterion. These observations are all direct consequences of the fact that there exist no directed circuits in the graph obtained by contracting all the edges in each one of the sets $E_1, E_2, \ldots, E_k$.

1. There is no back edge of the type (v, w) with $v \in V_i$ and $w \in V_j$, $i \neq j$. In other words all the back edges that leave vertices in V_i also end on vertices in V_i.

2. There is no cross edge of the type (v, w) with $v \in V_i$, and $w \in V_j$, $i \neq j$ and r_j is an ancestor of r_i. Thus for each cross edge (v, w) one of the following two is true:

   a. $v \in V_i$ and $w \in V_j$ for some i and j with $i \neq j$ and r_j to the left of r_i.

   b. For some i, $v \in V_i$ and $w \in V_i$.

Assuming that the vertices of G are named by their DFS numbers, we define for each v in G, LOWLINK$(v) = \min(\{v\} \cup \{w |$ there is a cross edge or a back edge from a descendant of v to w, and w is in the same strongly connected component as $v\})$.

Suppose $v \in V_i$. Then it follows from the above definition that LOWLINK(v) is the lowest numbered vertex in V_i that can be reached from v by a directed path that contains at most one back edge or one cross edge. From the observations that we have just made it follows that all the edges of such a directed path will necessarily be in G_i. As an immediate consequence we get

$$\text{LOWLINK}(r_i) = r_i \text{ for all } 1 \leq i \leq k. \tag{3.2}$$

Suppose $v \in V_i$ and $v \neq r_i$. Then there exists a directed path P in G_i from v to r_i. Such a directed path P should necessarily contain a back edge or a cross edge because $r_i < v$, and only cross edges and back edges lead to lower numbered vertices. In other words P contains a vertex $w < v$. So for $v \neq r_i$, we get

$$\text{LOWLINK}(v) < v. \tag{3.3}$$

Combining (3.2) and (3.3) we get the following theorem, which characterizes the roots of the strongly connected components of a directed graph.

Theorem 3.4 *A vertex v is the root of a strongly connected component of a directed graph G if and only if LOWLINK$(v) = v$.* ■

The following steps can be used to compute LOWLINK(v) as we perform a DFS.

1. On visiting v for the first time, set LOWLINK(v) equal to the DFS number of v.

2. If a back edge (v, w) is examined, then set LOWLINK(v) equal to the minimum of its current value and the DFS number of w.

3. If a cross edge (v, w) with w in the same strongly connected component as v is explored, set LOWLINK(v) equal to the minimum of its current value and the DFS number of w.

4. When the search returns to v after completely scanning a son s of v, set LOWLINK(v) to the minimum of its current value and LOWLINK(s).

To implement step 3 we need a test to check whether w is in the same strongly connected component as v. For this purpose we use an array STACK1 to which vertices of G are added in the order in which they are visited during the DFS. STACK1 is also used to determine the vertices belonging to a strongly connected component.

Let v be the first vertex during DFS for which it is found that LOWLINK$(v) = v$. Then by Theorem 3.4, v is a root and in fact it is r_1. At this point the vertices on top of STACK1 up to and including v are precisely those that belong to G_1. Thus, G_1 can easily be identified. These vertices are now removed from STACK1. From this point on the algorithm behaves in exactly the same way as it would on the graph G', which is obtained by removing from G the vertices of G_1.

As regards the implementation of step 3 in LOWLINK computation, let $v \in V_i$ and let (v, w) be a cross edge encountered while examining the edges incident on v. Suppose w is not in the same strongly connected component as v. Then it would belong to a strongly connected component G_j whose root r_j is to the left of r_i. The vertices of such a component would already have been identified, and so they would no longer be on STACK1. Thus, w will be in the same strongly connected component as v if and only if w is on STACK1.

A description of the strong connectivity algorithm now follows. This is the same as Algorithm 3.2 with the inclusion of appropriate steps for computing LOWLINK values and for identifying the vertices of the different strongly connected components. We use in this algorithm an array POINT. To begin with POINT$(v) = 0$ for every vertex v. This indicates that no vertex is on the array STACK1. POINT(v) is set to 1 when v is added to STACK1, and it is set to zero when v is removed from STACK1. We also use an array ROOT. ROOT$(v) = 1$ if it is the root vertex of a tree in the DFS forest. For example, in Figure 3.3b, 1, 8, and 13 are root vertices.

Algorithm 3.4 Strong connectivity

Input: $G = (V, E)$ is a connected directed graph.
Output: Strongly connected components of G.
begin
    STACK1 $\leftarrow \phi$.
    for every edge e in G, EXAMINED$(e) \leftarrow 0$;
    for every vertex v in G
        begin
            FATHER$(v) \leftarrow v$;
            MARK$(v) \leftarrow 0$;
            SCAN$(v) \leftarrow 0$;
            ROOT$(v) \leftarrow 0$;
            POINT$(v) \leftarrow 0$;
        end
    end for
    $i \leftarrow 0$;
    repeat
        while there exists a vertex v with MARK$(v) = 0$;
            MARK$(v) \leftarrow 1$; $i \leftarrow i + 1$; DFN$(v) \leftarrow i$; ROOT$(v) \leftarrow 1$;

```
                LOWLINK(v) ← i; STACK1 ← STACK1 ∪ {v}; POINT(v) ← 1;
            repeat
                while there exists an edge e ← (v, w) with EXAMINED(e) = 0
                    do
                        EXAMINED(v) ← 1;
                        if MARK(w) = 0 then
                            begin
                                i ← i + 1;
                                DFN(w) ← i;
                                MARK(w) ← 1;
                                LOWLINK(v) ← i;
                                FATHER(w) ← v;
                                v ← w;
                                STACK1 ← STACK1 ∪ {w};
                                POINT(w) ← 1;
                            end
                        else
                            if DFN(w) < DFN(v) and POINT(w) = 1 then
                                LOWLINK(v) ← min{LOWLINK(v), DFN(w)}.
                    od
                end while
                SCAN(v) ← 1;
                if LOWLINK(v) = DFN(v), then
                    begin
                        remove all the vertices from the top of STACK1 up to
                            and including v;
                        POINT(x) ← 0 for all such x removed from STACK1;
                    end
                if ROOT(FATHER(v)) = 0 then
                    LOWLINK(FATHER(v)) ← min{LOWLINK(FATHER(v)),
                        LOWLINK(v)};
                v ← FATHER(v);
            until ROOT(v) = 1 and SCAN(v) = 1;
        end while
    until i = n;
end
```

See Figure 3.5 for an illustration of this algorithm. In this figure, LOWLINK values are shown in parentheses. Strongly connected components are $\{3, 4, 5\}$, $\{6, 7, 8, 9, 10\}$, $\{2\}$, and $\{1, 11, 12, 13\}$.

3.5 *st*-NUMBERING OF A GRAPH

In this section we present yet another application of DFS-computing an *st*-numbering of a graph. For an application of $s - t$ numbering, see Reference [4].

Given an n-vertex biconnected graph $G = (V, E)$ and an edge (s, t) of G, a numbering of the vertices of G is called an *st*-numbering of G if the following conditions are satisfied, where $g(v)$ denotes the corresponding *st*-number of vertex v:

1. For all $v \in V$, $1 \leq g(v) \leq n$, and for $u \neq v$, $g(u) \neq g(v)$.

2. $g(s) = 1$.

72 ■ Handbook of Graph Theory, Combinatorial Optimization, and Algorithms

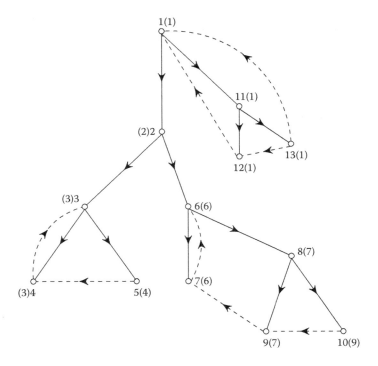

Figure 3.5 Illustration of Algorithm 3.4.

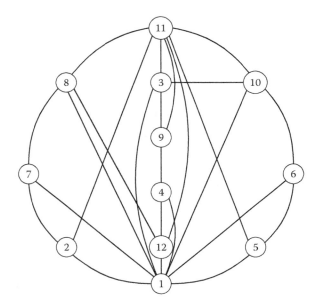

Figure 3.6 st-graph G [1].

3. $g(t) = n$.

4. For $v \in V - \{s, t\}$ there are adjacent vertices u and w such that $g(u) < g(v) < g(w)$.

A graph G and an st-numbering of G are shown in Figure 3.6. Lempel et al. [4] have shown that for every biconnected graph and every edge (s, t) there exists an st-numbering. The st-numbering algorithm to be discussed next is due to Even and Tarjan [5].

Given a biconnected graph G and an edge (s, t), the st-numbering algorithm of Even and Tarjan first performs a DFS of G with t as the start (root) vertex and (t, s) as the first edge. In other words $\mathrm{DFN}(t) = 1$, and $\mathrm{DFN}(s) = 2$. Recall that $\mathrm{DFN}(v)$ denotes the DFS number of vertex v. During the DFS, the algorithm also computes, for each vertex v, its DFN, FATHER(v) in the DFS tree, the low point LOW(v), and identifies the tree edges and back edges.

Next the vertices s and t and the edge (s, t) are marked *old*. All the other vertices and edges are marked *new*. An algorithm, called the *path finding algorithm*, is then invoked repeatedly (in an order to be described later) until all the vertices and edges are marked *old*.

The path finding algorithm when applied from an *old* vertex v finds a directed path into v or from v and proceeds as follows.

Algorithm 3.5 Path finding algorithm (applied from vertex v)

S1. Pick a *new* edge incident on v.

    i. If (v, w) is a back edge (DFN(w) < DFN(v)), then mark e *old* and HALT.
    Note: The path consists of the single edge e.

    ii. If (v, w) is a tree edge (DFN(w) > DFN(v)), then do the following:
    Starting from v traverse the directed path that defined LOW(w) and mark all edges and vertices on this path *old*. HALT.
    Note: The path here starts with the tree edge (v, t) and ends in the vertex u such that DFN(u) = LOW(w). This path has exactly one back edge.

    iii. If (w, v) is a back edge (DFN(w) > DFN(v)) do the following:
    Starting from v traverse the edge (w, v) backward and continue backward along tree edges until an *old* vertex is encountered. Mark all the edges and vertices on this path *old* and HALT.
    Note: The path in this case is directed into v.

S2. If all the edges incident on v are *old* HALT.
    Note: The path produced is empty.

The following facts hold true after each application of the path finding algorithm. Note that the algorithm is always applied from an *old* vertex.

1. All ancestors of an *old* vertex are old too. This is true before the first application of the algorithm since t is the only ancestor of s and it is *old*. This property remains true after any one of the applicable steps of the algorithm.

2. When the algorithm is applied from an *old* vertex v, it either produces an empty path or it produces a path that starts at v, passes through *new* vertices and edges and ends at another *old* vertex. This is obvious when (v, w) or (w, v) is a back edge (cases [i] and [iii] of S1). This is also true when (v, w) is a tree edge because in the biconnected graph G the vertex u defining LOW(w) is an ancestor of v and therefore u is *old*.

Even and Tarjan's *st*-numbering algorithm presented next uses a stack STACK that initially contains only t and s with s on top of t. In this description we do not explicitly include the details of DFS. We also assume that to start with t, s and the edge (t, s) are *old*.

Algorithm 3.6 *st*-Numbering (Even and Tarjan)

S1. $i \leftarrow 1$.

S2. Let v be the top vertex on STACK. Remove v from STACK. If $v = t$, then $g(v) \leftarrow i$ and HALT.

> **S3.** ($v \neq t$). Apply the path finding algorithm from v. If the path is empty, then $g(v) \leftarrow i$, $i \leftarrow i + 1$ and go to S2.
>
> **S4.** (The path is not empty.) Let the path be P: $v, u_1, u_2, \ldots, u_k, w$. Place $u_k, u_{k-1}, \ldots, u_1, v$ on STACK in this order (note that v comes on top of STACK) and go to S2.

Theorem 3.5 *Algorithm* 3.5 *computes an st-numbering for every biconnected graph* $G = (V, E)$.

Proof. The following facts about the algorithm are easy to verify:

1. No vertex appears in more than one place on STACK at any time.

2. Once a vertex v is placed on STACK, no vertex under v receives a number until v does.

3. A vertex is permanently removed from STACK only after all edges incident on v become *old*.

We now show that each vertex v is placed on STACK before t is removed. Clearly this is true for $v = s$ because initially t and s are placed on STACK with s on top of t.

Consider any vertex $v \neq s, t$. Since G is biconnected, there exists a directed path P of tree edges from s to v. Let P: $s, u_1, u_2, \ldots, u_k = v$. Let m be the first index such that u_m is not placed on STACK. Since u_{m-1} is placed on STACK, t can be removed only after u_{m-1} is removed (fact 2), and u_{m-1} is removed only after all edges incident on u_{m-1}, are *old* (fact 3). So u_m must be placed on STACK before t is removed.

We need to show that the numbers assigned to the vertices are indeed *st*-numbers. Since each vertex is placed on STACK and eventually removed, every vertex v gets a number $g(v)$.

Clearly all numbers assigned are distinct. Also $g(s) = 1$ and $g(t) = n$ because s is the first vertex and t is the last vertex to be removed. Every time a vertex $v \neq s, t$ is placed on STACK, there is an adjacent vertex placed above v and an adjacent vertex placed below v. By fact 2 the one above gets a lower number and the one below gets a higher number. ∎

Tarjan [6] gives a simplified version of the *st*-numbering algorithm. See Erbert [7] for another *st*-numbering algorithm.

Further Reading

A number of algorithms that use DFS as a building block have been reported in the literature. For example, see References [8–12]. See Chapter 27 for algorithms on program graphs that use DFS.

References

[1] K. Thulasiraman and M. N. S. Swamy, *Graphs: Theory and Algorithms*, Wiley-Interscience, 1992.

[2] J. Hopcroft and R. E. Tarjan, Efficient algorithms for graph manipulation, *Comm. ACM*, **16** (1973), 372–378.

[3] R. E. Tarjan, Depth-first search and linear graph algorithms, *SIAM J. Comput.*, **1** (1972), 146–160.

[4] A. Lempel, S. Even, and I. Cederbaum, An algorithm for planarity testing of graphs, *Theory of Graphs*, International Symposium, Rome, Italy, July 1966, P. Rosenstiehl, Ed., Gordon & Breach, New York, 1967, 215–232.

[5] S. Even and R. E. Tarjan, Computing an *st*-Numbering, *Th. Comp. Sci.*, New York, **2** (1976), 339–344.

[6] R. E. Tarjan, Two streamlined depth-first search algorithms, *Fund. Inform. IX* (1986), 85–94.

[7] J. Erbert, *st*-Ordering the vertices of biconnected graphs, *Computing*, **30** (1983), 19–33.

[8] H. de Fraysseix and P. Rosenstiehl, A depth-first search characterization of planarity, *Ann. Discrete Math.*, **13** (1982), 75–80.

[9] J. Hopcroft and R. E. Tarjan, Dividing a graph into triconnected components, *SIAM J. Comput.*, **2** (1973), 135–138.

[10] J. Hopcroft and R. E. Tarjan, Efficient planarity testing, *J. ACM*, **21** (1974), 549–568.

[11] S. Shinoda, W.-K. Chen, T. Yasuda, Y. Kajitani, and W. Mayeda, A necessary and sufficient condition for any tree of a connected graph to be a DFS-tree of one of its 2-isomorphic graphs, *IEEE ISCAS*, New Orleans, LA, 1990, 2841–2844.

[12] J. Valdes, R. E. Tarjan, and E. L. Lawler, The recognition of series parallel digraphs, *SIAM J. Comp.*, **11** (1982), 298–313.

II

Flows in Networks

CHAPTER 4

Maximum Flow Problem

F. Zeynep Sargut

Ravindra K. Ahuja

James B. Orlin

Thomas L. Magnanti

CONTENTS

4.1	Introduction	79
	4.1.1 Mathematical Formulation	81
	4.1.2 Assumptions	82
4.2	Preliminaries	84
	4.2.1 Residual Network	84
	4.2.2 Flow Across an s–t Cut	84
4.3	Augmenting Path Algorithms	85
	4.3.1 Generic Augmenting Path Algorithms	85
	4.3.2 Maximum Capacity and Capacity-Scaling Algorithms	89
	4.3.3 Shortest Augmenting Path Algorithm	91
	4.3.3.1 Worst-Case Improvements	96
	4.3.3.2 Improvement in Capacity-Scaling Algorithm	97
	4.3.3.3 Further Worst-Case Improvements	97
4.4	Preflow-Push Algorithms	98
	4.4.1 Generic Preflow-Push Algorithm	99
	4.4.2 FIFO Preflow-Push Algorithm	103
4.5	Blocking Flow Algorithms	104
	4.5.1 Blocking Flow Algorithm	106
	4.5.2 Malhotra, Kumar, and Maheshwari Algorithm	107
4.6	Maximum Flow on Unit Capacity Networks	107

4.1 INTRODUCTION

The maximum flow problem seeks the maximum possible flow in a capacitated network from a specified source node s to a specified sink node t without exceeding the capacity of any arc. A closely related problem is the minimum cut problem, which is to find a set of arcs with the smallest total capacity whose removal separates nodes s and t. The maximum flow and minimum cut problems arise in a variety of application settings as diverse as manufacturing, communication systems, distribution planning, and scheduling. These problems also arise as subproblems in the solution of more difficult network optimization problems. In this chapter, we study the maximum flow problem, introducing the underlying theory and algorithms. The book by Ahuja et al. [1] contains a wealth of additional material that amplifies this discussion.

We give the representative selection problem as an example of a maximum flow problem. Consider a town that has r residents $R_1, R_2, \ldots, R_r$; q clubs $C_1, C_2, \ldots, C_q$; and p political parties $P_1, P_2, \ldots, P_p$. Each resident is a member of at least one club and can belong to exactly one political party. Each club must nominate one of its members to represent it on the town's governing council so that the number of council members belonging to the political party P_k is at most u_k. Is it possible to find a council that satisfies this *balancing* property?

We consider a problem with $r = 7, q = 4, p = 3$ and formulate it as a maximum flow problem in Figure 4.1. The nodes $R_1, R_2, \ldots, R_7$ represent the residents; the nodes $C_1, C_2, \ldots, C_4$ represent the clubs; and the nodes P_1, P_2, P_3 represent the political parties. The network also contains a source node s and a sink node t. It contains an arc (s, C_i) for each node C_i denoting a club, an arc (C_i, R_j) whenever the resident R_j is a member of the club C_i, and an arc (R_j, P_k) if the resident R_j belongs to the political party P_k. Finally, we add an arc (P_k, t) for each $k = 1, 2, 3$ of capacity u_k; all other arcs have unit capacity.

We next find a maximum flow in this network. If the maximum flow value equals q, then the town has a balanced council; otherwise, it does not. The proof of this assertion is easy to establish by showing that (1) any flow of value q in the network corresponds to a balanced council, and that (2) any balanced council implies a flow of value q in the network. This type of model has applications in several resource-assignment settings. For example, suppose the residents are skilled craftsmen, the club C_i is the set of craftsmen with a particular skill, and

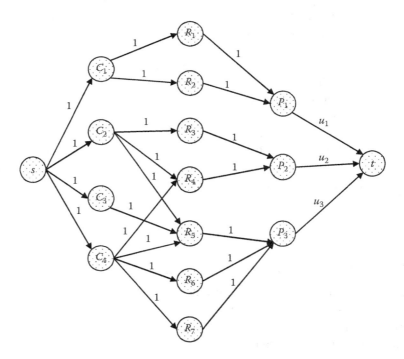

Figure 4.1 System of distinct representatives.

the political party P_k corresponds to a particular seniority class. In this instance, a balanced town council corresponds to an assignment of craftsmen to a union governing board so that every skill class has representation on the board and no seniority class has a dominant representation.

4.1.1 Mathematical Formulation

Let $G = (N, A)$ be a directed network defined by a set N of n nodes and a set A of m directed arcs. We refer to nodes i and j as endpoints of arc (i, j). A directed path $i_1-i_2-i_3-\cdots-i_k$ is a set of arcs $(i_1, i_2), (i_2, i_3), \ldots, (i_{k-1}, i_k)$. Each arc (i, j) has an associated capacity u_{ij} denoting the maximum possible amount of flow on this arc. We assume that each arc capacity u_{ij} is an integer, and let $U = \max\{u_{ij} : (i, j) \in A\}$. The network has two distinguished nodes, a source node s and a sink node t. To help in representing a network, we use the arc adjacency list $A(i)$ of node i, which is the set of arcs emanating from it; that is, $A(i) = \{(i, j) \in A : j \in N\}$. The maximum flow problem is to find the maximum flow from the source node s to the sink node t that satisfies the arc capacities and mass balance constraints at all nodes. We can state the problem formally as follows.

$$\text{Maximize } v$$

subject to

$$\sum_{\{j:(i,j)\in A\}}^{t} x_{ij} - \sum_{\{j:(j,i)\in A\}}^{t} x_{ji} = \begin{cases} v, & i = s \\ 0, & i \neq s \text{ or } t \\ -v, & i = t \end{cases} \quad i \in N \quad (4.1)$$

$$0 \leq x_{ij} \leq u_{ij} \text{ for all } (i, j) \in A \quad (4.2)$$

We refer to a vector $x = \{x_{ij}\}$ satisfying constraint sets 4.1 and 4.2 as a flow and the corresponding value of the scalar variable v as the value of the flow. The maximum flow problem has two constraints. We refer to the constraints 4.1 as the mass balance constraints, and we refer to the constraints 4.2 as the flow-bound constraints.

The maximum flow problem is a special type of minimum cost flow problem and can be formulated as a minimum cost flow problem by adding an arc from node t to s, if it does not exist. Moreover, the capacity of this arc is set to infinite. The minimum cost flow formulation of the maximum flow problem is given below and the network flow representation is given in Figure 4.2.

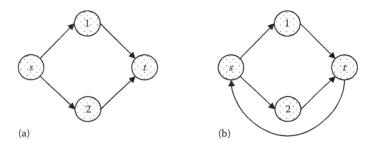

Figure 4.2 Representation of the maximum flow problem as a minimum cost flow problem.

$$\text{Maximize } x_{ts}$$
$$\text{subject to}$$
$$\sum_{\{j:(i,j)\in A\}}^{t} x_{ij} - \sum_{\{j:(j,i)\in A\}}^{t} x_{ji} = 0 \text{ for all } i \in N \tag{4.3}$$
$$0 \leq x_{ij} \leq u_{ij} \text{ for all } (i,j) \in A \tag{4.4}$$
$$x_{ts} \geq 0 \tag{4.5}$$

4.1.2 Assumptions

In examining the maximum flow problem, we impose the following assumptions.

Assumption 4.1. *Whenever the network contains arc (i,j), then it also contains arc (j,i).* This assumption is satisfied by adding the missing arcs with zero capacities.

Assumption 4.2. *The network is directed.* If the network contains an undirected arc (i,j) with capacity u_{ij}, it means that this arc permits flow from node i to node j and also from node j to node i, and the total flow (from node i to node j plus from node j to node i) has an upper bound u_{ij}.

The directed and undirected flow problems are closely related. A well-known reduction of Ford and Fulkerson [2] reduces the undirected problem to the directed problem with comparable size and capacity values.

Assumption 4.3. *The lower bounds on the arc flows are zero.* Sometimes the flow vector x might be required to satisfy lower bound constraints imposed upon the arc flows; that is, if l_{ij} specifies the lower bound on the flow on arc (i,j) A, we impose the condition $x_{ij} \geq l_{ij}$. We refer to this problem as the maximum flow problem with positive lower bounds.

Whereas the maximum flow problem with zero lower bounds always has a feasible solution (since the zero flow is feasible), the problem with nonnegative lower bounds could be infeasible. For example, consider the maximum flow problem given in Figure 4.3. This problem does not have a feasible solution because arc $(s,2)$ must carry at least 5 units of flow into node 2, and arc $(2,t)$ can take out at most 4 units of flow; therefore, we can never satisfy the mass balance constraint of node 2.

It is possible to transform a maximum flow problem with positive lower bounds into a maximum flow problem with zero lower bounds. As illustrated by this example, any maximum flow algorithm for problems with nonnegative lower bounds has two objectives: (1) to determine whether the problem is feasible or not, and (2) if so, to establish a maximum flow. It therefore comes as no surprise that most algorithms use a two-phase approach. The first phase determines a feasible flow, if one exists, and the second phase converts a feasible flow into a maximum flow. It can be shown that each phase essentially reduces to solving a maximum flow problem with zero lower bounds. Consequently, it is

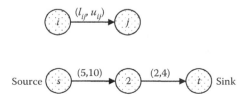

Figure 4.3 Maximum flow problem with no feasible solution.

possible to solve the maximum flow problem with positive lower bounds by solving two maximum flow problems, each with zero lower bounds.

Assumption 4.4. *There are single source and single sink.* If the problem is asking the maximum flow from the multiple sources to the multiple sinks, we can create a combined source that is connected with infinite capacity arcs to all source nodes and a combined sink that is connected with infinite capacity arcs to all sink nodes.

Arc capacities. There are two cases for the arc capacities: real valued and integral. Although some algorithms do not require the capacities to be integral, some algorithms are based on the assumption that the capacities are integral. Algorithms whose complexity bounds involve U assume integrality of the data. In reality, the integrality assumption is not a restrictive assumption because all modern computers store capacities as rational numbers and we can always transform rational numbers to integer numbers by multiplying them by a suitably large number.

We would like to design maximum flow algorithms that are guaranteed to be efficient in the sense that their worst-case running times, that is, the total number of multiplications, divisions, additions, subtractions, and comparisons in the worst case grow slowly in some measure of the problem's size. We say that a maximum flow algorithm is an $O(n^3)$ algorithm, or has a worst-case complexity of $O(n^3)$, if it is possible to solve any maximum flow problem using a number of computations that is asymptotically bounded by some constant times the term n^3. We say that an algorithm is a polynomial time algorithm if its worst-case running time is bounded by a polynomial function of the input size parameters. For a maximum flow problem, the input size parameters are n, m, and $\log U$ (the number of bits needed to specify the largest arc capacity). We refer to a maximum flow algorithm as a pseudopolynomial time algorithm if its worst-case running time is bounded by a polynomial function of n, m, and U. For example, an algorithm with a worst-case complexity of $O(nm \log U)$ is a polynomial time algorithm, but an algorithm with a worst-case complexity of $O(nmU)$ is a pseudopolynomial time algorithm.

For many years researchers attempted to find an algorithm faster than the decomposition barrier $\Omega(nm)$. This is a natural lower bound on algorithms that require flow decomposition and on algorithms that augment flow of one arc at a time. Cheriyan et al. [3] beat this algorithm for dense graphs, while Goldberg and Rao [4] further improve this lower bound.

The remainder of this chapter is organized as follows. In Section 4.2, we present some preliminary results concerning flows and cuts. We next discuss two important classes of algorithms for solving the maximum flow problem: (1) augmenting path algorithms and (2) preflow-push algorithms. Section 4.3 describes augmenting path algorithms, which augment flow along directed paths from the source node to the sink node. The proof of the validity of the augmenting path algorithm yields the well-known max-flow min-cut theorem, which states that the value of a maximum flow in a network equals the capacity of a minimum cut in the network. Moreover, we describe three polynomial-time implementations of the generic augmenting path algorithm. In Section 4.4, we study preflow-push algorithms that *flood* the network so that some nodes have excesses and then incrementally *relieve* the flow from nodes with excesses by sending flow from excess nodes forward toward the sink node or backward toward the source node. Additionally, we describe three polynomial implementations of the preflow-push type algorithms. In Section 4.5, we give the blocking flow type maximum flow algorithms. In Section 4.6, we describe the faster version of the preflow-push algorithms for the unit capacity networks. In Section 4.7, we give references for further reading.

4.2 PRELIMINARIES

In this section, we discuss some elementary properties of flows and cuts. We will use these properties to prove the celebrated max-flow min-cut theorem and to establish the correctness of the augmenting path algorithm described in Section 4.3.

4.2.1 Residual Network

The concept of residual network plays a central role in the development of maximum flow algorithms. Given a flow x, the residual capacity r_{ij} of any arc $(i,j) \in A$ is the maximum additional flow that can be sent from node i to node j using the arcs (i,j) and (j,i). (Recall our assumption from Section 4.1.2 that whenever the network contains arc (i,j), it also contains the arc (j,i).) The residual capacity r_{ij} has two components: (i) $u_{ij} - x_{ij}$, the unused capacity of arc (i,j), and (ii) the current flow x_{ji} on arc (j,i), which we can cancel to increase the flow from node i to node j. Consequently, $r_{ij} = u_{ij} - x_{ij} - x_{ji}$. We refer to the network $G(x)$ consisting of the arcs with positive residual capacities as the residual network (with respect to the flow x). Figure 4.4 gives an example of a residual network.

4.2.2 Flow Across an s–t Cut

Let x be a flow in the network. Adding the mass balance constraint 4.1 for the nodes in S, we obtain the following equation:

$$v = \sum_{i \in S} \left[\sum_{\{j:(i,j) \in A\}}^{t} x_{ij} - \sum_{\{j:(j,i) \in A\}}^{t} x_{ji} \right] = \sum_{\{(i,j) \in (S,\overline{S})\}}^{t} x_{ij} - \sum_{\{(j,i) \in (\overline{S},S)\}}^{t} x_{ij} \quad (4.6)$$

The second equality uses the fact that whenever both the nodes p and q belong to the node set S and (p,q) A, the variable x_{pq} in the first term within the bracket (for node $i = p$) cancels the variable $-x_{pq}$ in the second term within the bracket (for node $j = q$). The first expression in the right-hand side of Equation 4.6 denotes the amount of flow from the nodes in S to nodes in $\overline{S}$, and the second expression denotes the amount of flow returning from the nodes in $\overline{S}$ to the nodes in S. Therefore, the right-hand side denotes the total (net) flow across the cut, and 4.6 implies that the flow across any $s-t$ cut $[S,\overline{S}]$ equals v. Substituting $x_{ij} \leq u_{ij}$ in the first expression of 4.6 and $x_{ij} \geq 0$ in the second expression yields $v \leq \sum_{(i,j) \in [S,\overline{S}]} u_{ij} = u[S,\overline{S}]$, implying that the value of any flow can never exceed the capacity of any cut in the network. We record this result formally for future reference.

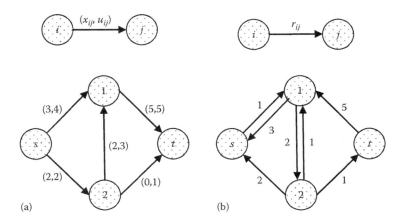

Figure 4.4 Residual network.

The minimum cut problem is a close relative of the maximum flow problem. A cut $[S, \overline{S}]$ partitions the node set N into two subsets S and $\overline{S} = N - S$. It consists of all arcs with one endpoint in S and the other in $\overline{S}$. We refer to the arcs directed from S to $\overline{S}$, denoted by $(S, \overline{S})$, as forward arcs in the cut and the arcs directed from $\overline{S}$ to S, denoted by $(\overline{S}, S)$, as backward arcs in the cut. The cut $[S, \overline{S}]$ is called an $s - t$ cut if $s \in S$ and $t \in \overline{S}$. We define the capacity of the cut $[S, \overline{S}]$, denoted as $u[S, \overline{S}]$, where

$$u[S, \overline{S}] = \sum_{(i,j) \in [S, \overline{S}]} u_{ji}$$

A *minimum cut* in G is an $s - t$ cut of minimum capacity. We will show that any algorithm that determines a maximum flow in the network also determines a minimum cut in the network.

Theorem 4.1 *The value of any flow can never exceed the capacity of any cut in the network. Consequently, if the value of some flow x equals the capacity of some cut $[S, \overline{S}]$, then x is a maximum flow and the cut $[S, \overline{S}]$ is a minimum cut.* ∎

The max-flow min-cut theorem, to be proved in the next section, states that the value of some flow always equals the capacity of some cut.

4.3 AUGMENTING PATH ALGORITHMS

In this section, we first describe one of the simplest and most intuitive algorithms for solving the maximum flow problem, an algorithm known as the augmenting path algorithm. Ford and Fulkerson [2] and Elias et al. [5] independently developed the basic augmenting path algorithm. We next describe four specific implementations of the augmenting path algorithm: maximum capacity, capacity scaling, and shortest path algorithms. Several studies show that prior to the push-relabel type algorithms, shortest path algorithm was superior to the other algorithms. We first explain the generic version of the augmenting path algorithm.

4.3.1 Generic Augmenting Path Algorithms

Let x be a feasible flow in the network G, and let $G(x)$ denote the residual network corresponding to the flow x. We refer to a directed path from the source to the sink in the residual network $G(x)$ as an augmenting path. We define the residual capacity $\delta(P)$ of an augmenting path P as the maximum amount of flow that can be sent along it; that is, $\delta(P) = \min\{r_{ij} : (i, j) \in P\}$. Since the residual capacity of each arc in the residual network is strictly positive, the residual capacity of an augmenting path is strictly positive. Therefore, we can always send a positive flow of δ units along it. Consequently, whenever the network contains an augmenting path, we can send additional flow from the source to the sink. (Sending an additional δ units of flow along an augmenting path decreases the residual capacity of each arc (i, j) in the path by δ units.) The generic augmenting path algorithm is essentially based upon this simple observation. The algorithm identifies augmenting paths in $G(x)$ and augments flow on these paths until the network contains no such path. We describe the generic augmenting path algorithm below.

Algorithm: Generic augmenting path

$x := 0$;

while $G(x)$ contains a directed path from node s to node t do

---

```
{
    identify an augmenting path P from node s to node t;
    δ(P) = min{r_{ij} : (i,j) ∈ P};
    augment δ(P) units of flow along path P and update G(x);
}
```

The performance of the algorithm depends on (1) the number of augmenting paths and (2) the time to identify an augmenting path. Both numbers depend on which augmenting path we choose at each step. We can identify an augmenting path P in $G(x)$ by using a graph search algorithm. A graph search algorithm starts at node s and progressively finds all nodes that are reachable from the source node using directed paths. Most search algorithms run in time proportional to the number of arcs in the network, that is, $O(m)$ time, and either identify an augmenting path or conclude that $G(x)$ contains no augmenting path; the latter happens when the sink node is not reachable from the source node.

For each arc $(i,j) \in P$, augmenting δ units of flow along P decreases r_{ij} by δ units and increases r_{ji} by δ units. The final residual capacities r_{ij} when the algorithm terminates specifies a maximum (arc) flow in the following manner. Since $r_{ij} = u_{ij} - x_{ij} + x_{ji}$, the arc flows satisfy the equality $x_{ij} - x_{ji} = u_{ij} - r_{ij}$. If $u_{ij} > r_{ij}$, we can set $x_{ij} = u_{ij} - r_{ij}$ and $x_{ji} = 0$; otherwise, we set $x_{ij} = 0$ and $x_{ji} = r_{ij} - u_{ij}$.

We use the maximum flow problem given in Figure 4.5 to illustrate the algorithm. Figure 4.5a shows the residual network corresponding to the starting flow $x = 0$, which is identical to the original network. The residual network contains three augmenting paths: 1-3-4, 1-2-4, and 1-2-3-4. Suppose the algorithm selects the path 1-3-4 for augmentation. The residual capacity of this path is $\delta = \min\{r_{13}, r_{34}\} = \min\{4, 5\} = 4$. This augmentation reduces the residual capacity of arc (1, 3) to zero (thus we delete it from the residual network) and

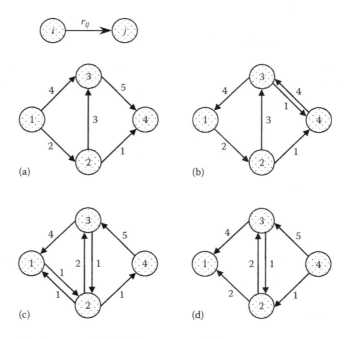

Figure 4.5 (a) The residual network $G(x)$ for $x = 0$. (b) The residual network after augmenting four units along the path 1-3-4. (c) The residual network after augmenting one unit along the path 1-2-3-4. (d) The residual network after augmenting one unit along the path 1-2-4.

increases the residual capacity of arc (3, 1) to 4 (so we add this arc to the residual network). The augmentation also decreases the residual capacity of arc (3, 4) from 5 to 1 and increases the residual capacity of arc (4, 3) from 0 to 4. Figure 4.5b shows the residual network at this stage. In the second iteration, the algorithm selects the path 1-2-3-4 and augments one unit of flow; Figure 4.5c shows the residual network after the augmentation. In the third iteration, the algorithm augments one unit of flow along the path 1-2-4. Figure 4.5d shows the corresponding residual network. Now the residual network contains no augmenting path and so the algorithm terminates.

Theorem 4.2 (Max-flow min-cut theorem) *The maximum value of the flow from a source node s to a sink node t in a capacitated network equals the minimum capacity among all $s - t$ cuts.*

Proof. The algorithm terminates when the search algorithm fails to identify a directed path in $G(x)$ from node s to node t, indicating that no such path exists (we prove later that the algorithm would terminate finitely). At this stage, let S denote the set of nodes in N that are reachable in $G(x)$ from the source node using directed paths, and $\overline{S} = N - S$. Clearly, $s \in S$ and $t \in \overline{S}$. Since the search algorithm cannot reach any node in $\overline{S}$ and it can reach each node in S, we know that $r_{ij} = 0$ for each $(i, j) \in (S, \overline{S})$. Recall that $r_{ij} = (u_{ij} - x_{ij}) + x_{ji}, x_{ij} \leq u_{ij}$, and $x_{ji} \geq 0$. If $r_{ij} = 0$, then $x_{ij} = u_{ij}$ and $x_{ji} = 0$. Since $r_{ij} = 0$ for each $(i, j) \in (S, \overline{S})$, by substituting these flow values in expression 4.6, we find that $v = [S, \overline{S}]$. Therefore, the value of the current flow x equals the capacity of the cut $[S, \overline{S}]$. Theorem 4.1 implies that x is a maximum flow and $[S, \overline{S}]$ is a minimum cut. This conclusion establishes the correctness of the generic augmenting path algorithm and, as a by-product, proves the max-flow min-cut theorem. ■

The proof of the max-flow min-cut theorem shows that when the augmenting path algorithm terminates, it also discovers a minimum cut $[S, \overline{S}]$, with S defined as the set of all nodes reachable from the source node in the residual network corresponding to the maximum flow. For our previous numerical example, the algorithm finds the minimum cut in the network, which is $[S, \overline{S}]$ with $S = \{1\}$. The augmenting path algorithm also establishes another important result given in Theorem 4.3.

Theorem 4.3 (Integrality theorem) *If all arc capacities are integer, then the maximum flow problem always has an integer maximum flow.*

Proof. This result follows from the facts that the initial (zero) flow is integer and all arc capacities are integer; consequently, all initial residual capacities will be integer. Since subsequently all arc flows change by integer amounts (because residual capacities are integer), the residual capacities remain integer throughout the algorithm. Further, the final integer residual capacities determine an integer maximum flow. The integrality theorem does not imply that every optimal solution of the maximum flow problem is integer. The maximum flow problem might have non-integer solutions and, most often, it has such solutions. The integrality theorem shows that the problem always has at least one integer optimal solution. ■

Theorem 4.4 *The generic augmenting path algorithm solves the maximum flow problem in $O(nmU)$ time.*

Proof. An augmenting path is a directed path in $G(x)$ from node s to node t. We have seen earlier that each iteration of the algorithm requires $O(m)$ time. In each iteration, the algorithm augments a positive integer amount of flow from the source node to the sink node. To bound the number of iterations, we will determine a bound on the maximum flow value. By definition, U denotes the largest arc capacity, and so the capacity of the cut $(\{s\}, S-\{s\})$ is at most nU. Since the value of any flow can never exceed the capacity of any cut in the network, we obtain a bound of nU on the maximum flow value and also on the number of iterations performed by the algorithm. Consequently, the running time of the algorithm is $O(nmU)$, which is a pseudopolynomial time bound. We summarize the preceding discussion with the following theorem. ∎

The generic augmenting path algorithm is possibly the simplest algorithm for solving the maximum flow problem. Empirically, the algorithm performs reasonably well. However, the worst-case bound on the number of iterations is not entirely satisfactory for large values of U. For example, if $U = 2^n$, the bound is exponential in the number of nodes. Moreover, the algorithm can indeed perform this many iterations, as the example given in Figure 4.6 demonstrates. For this example, the algorithm can select the augmenting paths $s-a-b-t$ and $s-b-a-t$ alternatively 10^6 times, each time augmenting unit flow along the path. This example illustrates one shortcoming of the algorithm.

The second drawback of the generic augmenting path algorithm is that if the capacities are irrational, the algorithm might not terminate. For some pathological instances of the maximum flow problem, the augmenting path algorithm does not terminate, and although the successive flow values converge, they converge to a value strictly less than the maximum flow value. (Note, however, that the max-flow min-cut theorem holds even if arc capacities are irrational.) Therefore, if the generic augmenting path algorithm is guaranteed to be effective, it must select augmenting paths carefully.

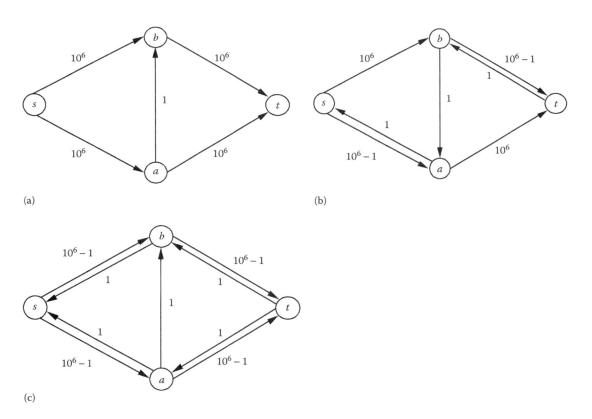

Figure 4.6 (a) Residual network for the zero flow. (b) Network after augmenting unit flow along the path $s-a-b-t$. (c) Network after augmenting unit flow along the path $s-b-a-t$.

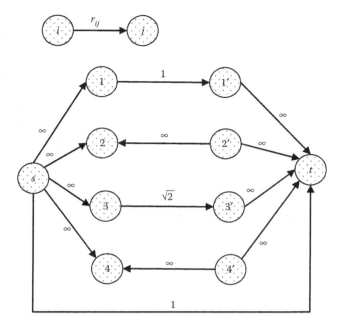

Figure 4.7 Another pathological instance for the generic augmenting path algorithm.

We give an example of this situation in Figure 4.7. Let us assume that we first choose the path $s-1-1'-t$ and then choose the path $s-t-1'-1-s-3-3'-t$. We can continue in this fashion and always identify a path from s to t. Therefore, it is possible that the generic augmenting path algorithm does not converge.

The third drawback of the generic augmenting path algorithm is its *forgetfulness*. In each iteration, the algorithm generates node labels that contain information about augmenting paths from the source node to other nodes. The implementation that we described erases the labels as it moves from one iteration to the next, even though much of this information might be valid in the next iteration. Erasing the labels therefore destroys potentially useful information. Ideally, we should retain a label when we can use it profitably in later computations.

Researchers have developed specific implementations of the generic augmenting path algorithms that overcome these drawbacks. Of these, the following four implementations are particularly noteworthy: (1) the maximum capacity augmenting path algorithm, which always augments flow along the augmenting path with the maximum residual capacity and can be implemented to run in $O(m^2 \log U)$ time; (2) the capacity-scaling algorithm, which uses a scaling technique on arc capacities and can be implemented to run in $O(nm \log U)$ time; and (3) the shortest augmenting path algorithm, which augments flow along a shortest path (as measured by the number of arcs) in the residual network and runs in $O(n^2 m)$ time. Next, we describe these three algorithms.

4.3.2 Maximum Capacity and Capacity-Scaling Algorithms

In this section, we describe two specific implementations of the generic augmenting path algorithm. The maximum capacity augmenting path algorithm always augments flow only along a path with the maximum residual capacity. Let x be any flow and let v be its flow value. As before, let v^* be the maximum flow value. It can be shown using the flow decomposition theory that in the residual network $G(x)$ we can find m or fewer directed paths from the source to the sink whose residual capacities sum to $(v^* - v)$. Therefore, the maximum capacity augmenting path has a residual capacity of at least $(v^* - v)/m$. Now consider a sequence of

$2m$ consecutive maximum capacity augmentations starting with the flow x. Suppose that v' is the new flow value. If in each of these augmentations we augment at least $(v^* - v)/2m$ units of flow, then we will establish a maximum flow within $2m$ or fewer iterations. However, if one of these $2m$ augmentations carries less than $(v^* - v)/2m$ units of flow, it implies that $(v^* - v')$ becomes less than $m(v^* - v)/2m = (v^* - v)/2$. This argument shows that within $2m$ consecutive iterations, the algorithm either establishes a maximum flow or reduces the residual capacity of the maximum capacity augmenting path by a factor of at least 2. Since the residual capacity of any augmenting path is at most $2U$ and at least 1, after $O(m \log U)$ iterations, the flow must be maximum.

As we have seen, the maximum capacity augmentation algorithm reduces the number of augmentations in the generic augmenting path algorithm from $O(nU)$ to $O(m \log U)$. However, the algorithm performs more computations per iteration, since it needs to identify an augmenting path with the maximum residual capacity and not just any augmenting path. Therefore, the running time is $O(Am \log U)$, where A is the time to obtain the maximum capacity augmenting path in the residual network.

We shall now suggest a variation of the maximum capacity augmenting path algorithm that does not perform more computations per iteration than the generic augmenting path algorithm and yet establishes a maximum flow within $O(m \log U)$ iterations. Since this algorithm scales the arc capacities implicitly, we refer to it as the capacity-scaling algorithm.

The essential idea underlying the capacity-scaling algorithm is conceptually quite simple: we augment flow along a path with a sufficiently large residual capacity instead of a path with the maximum augmenting capacity because we can obtain a path with a sufficiently large residual capacity fairly easily in $O(m)$ time. To define the capacity-scaling algorithm, let us introduce a parameter Δ and, with respect to a given flow x, define the Δ-residual network as a network containing arcs whose residual capacity is at least Δ. Let $G(x, \Delta)$ denote the Δ-residual network. Note that $G(x, 1) = G(x)$ and $G(x, \Delta)$ is a subgraph of $G(x)$. The capacity-scaling algorithm works as follows.

Algorithm: Capacity scaling

$x := 0$;

$\Delta := 2^{\log U}$;

while $\Delta \geq 1$ do

    while $G(x, \Delta)$ contains a path from node s to node t do

        {

        identify a path P in $G(x, \Delta)$;

        augment $\min\{r_{ij} : (i, j) \in P\}$ units of flow along P and update $G(x, \Delta)$;

        }

    $\Delta := \Delta/2$;

Theorem 4.5 *The capacity-scaling algorithm solves the maximum flow problem within $O(m \log U)$ augmentations and runs in $O(m^2 \log U)$ time.*

Proof. Let us refer to a phase of the algorithm during which Δ remains constant as a scaling phase and a scaling phase with a specific value of Δ as a Δ-scaling phase. Observe that in a Δ-scaling phase, each augmentation carries at least Δ units of flow. The algorithm starts with $\Delta = 2^{\log U}$ and halves its value in every scaling phase until $\Delta = 1$. Consequently, the algorithm performs $1 + \lfloor \log U \rfloor = O(\log U)$ scaling phases. In the last scaling phase, $\Delta = 1$

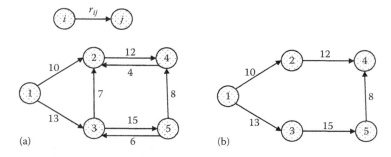

Figure 4.8 (a) Residual network $G(x)$. (b) Δ-residual network $G(x, \Delta)$ for $\Delta = 8$.

and so $G(x, \Delta) = G(x)$. This result shows that the algorithm terminates with a maximum flow. Figure 4.8 shows an example of the residual network when $\Delta = 1$ and $\Delta = 8$.

The efficiency of the algorithm depends upon the fact that it performs at most $2m$ augmentations per scaling phase. To establish this result, consider the flow at the end of the Δ-scaling phase. Let x' be this flow and let v' denote its flow value. Furthermore, let S be the set of nodes reachable from s in $G(x', \Delta)$. Since $G(x', \Delta)$ contains no augmenting path from the sources to the sink, $t \notin S$. Therefore, $[S, \overline{S}]$ forms an $s - t$ cut. The definition of S implies that the residual capacity of every arc in (S, Δ) is strictly less than Δ, and so the residual capacity of the cut $(S, \overline{S})$ is at most $m\Delta$. Consequently, $v^* - v' \leq m\Delta$. In the next scaling phase, each augmentation carries at least $\Delta/2$ units of flow, and so this scaling phase can perform at most $2m$ such augmentations. The algorithm described earlier requires $O(m)$ time to identify an augmenting path, and updating the Δ-residual network also requires $O(m)$ time. ∎

4.3.3 Shortest Augmenting Path Algorithm

The shortest augmenting path algorithm always augments flow along a shortest path (a path with the least number of arcs) in the residual network from the source to the sink. Since the minimum distance from any node i to the sink node t is monotonically nondecreasing over all augmentations, we obtain the average time per augmentation to $O(n)$.

The shortest augmenting path algorithm proceeds by augmenting flows along admissible paths. It constructs an admissible path incrementally by adding one arc at a time. The algorithm maintains a partial admissible path, that is, a path from s to some node i consisting solely of admissible arcs, and iteratively performs advance or retreat operations from the last node (i.e., the tip) of the partial admissible path, which we refer to as the current node. If the current node i has (i.e., is incident to) an admissible arc (i, j), then we perform an advance operation and add arc (i, j) to the partial admissible path; otherwise, we perform a retreat operation and backtrack one arc. We repeat these operations until the partial admissible path reaches the sink node, at which time we perform an augmentation. We repeat this process until the flow is maximum. Before presenting a formal description of the algorithm, we illustrate it on the numerical example given in Figure 4.9.

We first compute the initial distance labels by performing the backward breadth-first search of the residual network starting at the sink node. The numbers next to the nodes in Figure 4.9 specify these values of the distance labels. In this example, we adopt the convention of selecting the arc (i, j) with the smallest value of j whenever node i has several admissible arcs. We start at the source node with a null partial admissible path. The source node has several admissible arcs, so we perform an advance operation. This operation adds the arc $(1, 2)$ to the partial admissible path. We store this path using predecessor indices, so we set $\text{pred}(2) = 1$. Now node 2 is the current node and the algorithm performs an advance operation at node 2. In doing so, it adds arc $(2, 7)$ to the partial admissible path, which now

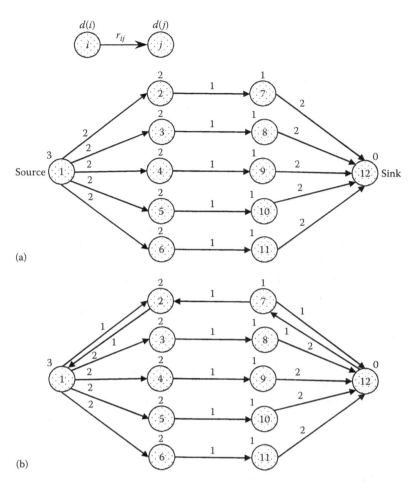

Figure 4.9 Illustrating the shortest augmenting path algorithm. (a) Original network. (b) Residual network after an augmentation.

becomes 1-2-7. We also set pred(7) = 2. In the next iteration, the algorithm adds arc (7, 12) to the partial admissible path obtaining 1-2-7-12, which is an admissible path to the sink node. We perform an augmentation of value $\min\{r_{12}, r_{27}, r_{7,12}\} = \min\{2, 1, 2\} = 1$, and thus saturate the arc (2, 7). Figure 4.9b specifies the residual network at this stage.

We again start at the source node with a null partial admissible path. The algorithm adds the arc (1, 2) and node 2 becomes the new current node. Now we find that node 2 has no admissible arcs. To create new admissible arcs, we must increase the distance label of node 2. We thus increase $d(2)$ to the value $\min\{d(j)+1 : (2,j) \in A(2), r_{2j} > 0\} = 4$. We refer to this operation as a relabel operation. We will later show that a relabel operation preserves the validity conditions. Observe that the increase in $d(2)$ causes arc (1, 2) to become inadmissible. Thus we delete arc (1, 2) from the partial admissible path, which again becomes a null path. In the subsequent operations, the algorithm identifies the admissible paths 1-3-8-12, 1-4-9-12, 1-5-10-12, and 1-6-11-12 and augments a unit flow on these paths. We encourage the reader to carry out the details of these operations. We now give the algorithmic description of the algorithm.

Algorithm: Shortest augmenting path

$x := 0$;

obtain the exact distance labels $d(i)$;

current node:= s;

```
while d(s) < n do
    {
    if current node has an admissible arc then advance(current node)
    else retreat(current node);
    if current node = t then augment;
    }
```

procedure advance(i)

```
{
let (i, j) be an admissible arc in A(i);
pred(j) := i and i := j;
}
```

procedure retreat(i)

```
{
d(i) := min{d(j) + 1 : (i, j) ∈ A(i) and r_{ij} > 0};
if i ≠ s then i := pred(i);
}
```

procedure augment

```
{
```

using the predecessor indices identify an augmenting path P from the source to the sink;

augment $\min\{r_{ij} : (i,j) \in P\}$ units of flow along path P;

```
}
```

We first show that the shortest augmenting path algorithm correctly solves the maximum flow problem.

Theorem 4.6 *The shortest augmenting path algorithm maintains valid distance labels at each step. Moreover, each relabel (or, retreat) operation strictly increases the distance label of a node.*

Proof. We show that the algorithm maintains valid distance labels at every step by performing induction on the number of augment and relabel operations. (The advance operation does not affect the admissibility of any arc because it does not change any residual capacity or distance label.) Initially, the algorithm constructs valid distance labels. Assume, inductively, that the distance labels are valid prior to an operation; that is, they satisfy the validity conditions. We need to check whether these conditions remain valid (1) after an augment operation and (2) after a relabel operation.

1. Although a flow augmentation on arc (i,j) might remove this arc from the residual network, this modification to the residual network does not affect the validity of the distance function for this arc. An augmentation on arc (i,j) might, however, create an additional arc (j,i) with $r_{ji} > 0$ and, therefore, also create an additional inequality $d(j) \leq d(i) + 1$ that the distance labels must satisfy. The distance labels satisfy this validity condition, though, since $d(i) = d(j) + 1$ by the admissibility property of the augmenting path.

2. The relabel operation modifies $d(i)$; therefore, we must show that the validity conditions are satisfied for each incoming and outgoing arc at node i with respect to the new distance labels, say $d'(i)$. The algorithm performs a relabel operation at node i when it has no admissible arc; that is, no arc $(i,j) \in A(i)$ satisfies the conditions $d(i) = d(j)+1$ and $r_{ij} > 0$. This observation, in light of the validity condition $d(i) \leq d(j) + 1$, implies that $d(i) < d(j) + 1$ for all arcs $(i,j) \in A$ with a positive residual capacity. Therefore, $d(i) < \min\{d(j) + 1 : (i,j) \in A(i), r_{ij} > 0\} = d'(i)$, which is the new distance label after the relabel operation. We have thus shown that relabeling preserves the validity condition for all arcs emanating from node i and that each relabel operation strictly increases the value of $d(i)$. Finally, note that every incoming arc (k,i) satisfies the inequality $d(k) \leq d(i) + 1$ (by the induction hypothesis). Since $d(i) < d'(i)$, the relabel operation again preserves the validity condition for the arc (k,i). ∎

Now, we will show that the shortest augmenting path algorithm runs in $O(n^2 m)$ time. We first describe a data structure used to select an admissible arc emanating from a given node. We call this data structure the current-arc data structure. We also use this data structure in almost all the maximum flow algorithms that we describe in subsequent sections. Therefore, we first review this data structure before proceeding further.

Recall that we maintain the arc list $A(i)$, which contains all the arcs emanating from node i. We can arrange the arcs in these lists arbitrarily, but the order, once decided, remains unchanged throughout the algorithm. Each node i has a current-arc, which is an arc in $A(i)$ and is the next candidate for admissibility testing. Initially, the current-arc of node i is the first arc in $A(i)$. Whenever the algorithm attempts to find an admissible arc emanating from node i, it tests whether the node's current-arc is admissible. If not, then it designates the next arc in the arc list as the current arc. The algorithm repeats this process until it either finds an admissible arc or reaches the end of the arc list.

Consider, for example, the arc list of node 1 in Figure 4.10. In this instance, $A(1) = \{(1,2),(1,3),(1,4),(1,5),(1,6)\}$. Initially, the current-arc of node 1 is arc $(1,2)$. Suppose the algorithm attempts to find an admissible arc emanating from node 1. It checks whether the node's current-arc, $(1,2)$, is admissible. Since it is not, the algorithm designates arc $(1,3)$ as the current-arc of node 1. The arc $(1,3)$ is also inadmissible, and so the current-arc becomes arc $(1,4)$, which is admissible. From this point on arc $(1,4)$ remains the current-arc of node 1 until it becomes inadmissible because the algorithm has increased the value of $d(4)$ or decreased the value of the residual capacity of arc $(1,4)$ to zero.

Let us consider the situation when the algorithm reaches the end of the arc list without finding any admissible arc. Can we say that $A(i)$ has no admissible arc? Yes, we can, because it is possible to show that if an arc (i,j) is inadmissible in previous iterations, then it remains inadmissible until $d(i)$ increases. So if we reach the end of the arc list, then we perform a relabel operation and again set the current-arc of node i to be the first arc in $A(i)$. The relabel operation also examines each arc in $A(i)$ once to compute the new distance label, which is same as the time it spends in identifying admissible arcs at node i in one scan of the arc list. We have thus established the following result.

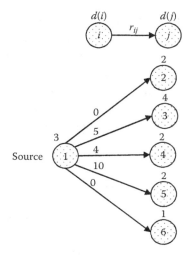

Figure 4.10 Selecting admissible arcs emanating from a node.

Theorem 4.7 *If the algorithm relabels any node at most k times, then the total time spent in finding admissible arcs and relabeling the nodes is $O(k \sum_{i \in N} |A(i)|) = O(km)$.* ∎

Theorem 4.8 *If the algorithm relabels any node at most k times, then the algorithm saturates arcs (i.e., reduces their residual capacity to zero) at most $km/2$ times.*

Proof. We show that between two consecutive saturations of an arc (i, j), both $d(i)$ and $d(j)$ must increase by at least 2 units. Since, by our hypothesis, the algorithm increases each distance label at most k times, this result would imply that the algorithm could saturate any arc at most $k/2$ times. Therefore, the total number of arc saturations would be $km/2$. Suppose that an augmentation saturates an arc (i, j). Since the arc (i, j) is admissible

$$d(i) = d(j) + 1.$$

Before the algorithm saturates the arc again, it must send back flow from node j to node i. At this time, the distance labels $d'(i)$ and $d'(j)$ satisfy the equality

$$d'(j) = d'(i) + 1.$$

In the next saturation of arc (i, j), we must have

$$d''(i) = d''(j) + 1.$$

Using the above equations, we obtain

$$d''(i) = d''(j) + 1 \geq d'(j) + 1 = d'(i) + 2 \geq d(i) + 2.$$

Similarly, it is possible to show that $d''(j) \geq d(j) + 2$. As a result, between two consecutive saturations of the arc (i, j), both $d(i)$ and $d(j)$ increase by at least 2 units. ∎

Theorem 4.9 *In the shortest augmenting path algorithm each distance label increases at most n times. Consequently, the total number of relabel operations is at most n^2. The number of augment operations is at most $nm/2$.*

Proof. Each relabel operation at node i increases the value of $d(i)$ by at least one unit. After the algorithm has relabeled node i at most n times, $d(i) \geq n$. From this point on, the algorithm never again selects node i during an advance operation since for every node k in the partial admissible path, $d(k) < d(s) < n$. Thus the algorithm relabels a node at most n times and the total number of relabel operations is bounded by n^2. The preceding result in view of Theorem 4.8 implies that the algorithm saturates at most $nm/2$ arcs. Since each augmentation saturates at least one arc, we immediately get a bound of $nm/2$ on the number of augmentations. ∎

Theorem 4.10 *The shortest augmenting path algorithm runs in $O(n^2m)$ time.*

Proof. Using Theorems 4.9 and 4.7, we find that the total effort spent in finding admissible arcs and in relabeling the nodes is $O(nm)$. Theorem 4.9 implies that the total number of augmentations is $O(nm)$. Each augmentation requires $O(n)$ time, this results in $O(n^2m)$ total effort for the augmentation operations. Each retreat operation relabels a node, so the total number of retreat operations is $O(n^2)$. Each advance operation adds one arc to the partial admissible path, and each retreat operation deletes one arc from it. Since each partial admissible path has length at most n, the algorithm requires at most $O(n^2 + n^2m)$ advance operations. The first term comes from the number of retreat (relabel) operations, and the second term from the number of augmentations. The combination of these bounds establishes the complexity of the algorithm. ∎

4.3.3.1 Worst-Case Improvements

The shortest augmenting path algorithm terminates when $d(s) \geq n$. This termination criterion is satisfactory for the worst-case analysis, but it might not be efficient in practice. Empirical investigations have revealed that the algorithm spends too much time in relabeling nodes and that a major portion of this effort is performed after the algorithm has established a maximum flow. This happens because the algorithm does not know that it has found a maximum flow. We next suggest a technique that is capable of detecting the presence of a minimum cut, and therefore the existence of a maximum flow, much earlier than $d(s) \geq n$ is satisfied. Incorporating this technique in the shortest augmenting path algorithm improves its performance substantially in practice.

We will illustrate this technique by applying it to the numerical example we used earlier to illustrate the shortest augmenting path algorithm. Figure 4.11 gives the residual network immediately after the last augmentation. Although the flow is now a maximum flow, since the source is not connected to the sink in the residual network, the termination criteria of $d(1) \geq 12$ is far from being satisfied. The reader can verify that after the last augmentation, the algorithm would increase the distance labels of nodes 6, 1, 2, 3, 4, 5, in the given order, each time by two units. Eventually, $d(1) \geq 12$ and the algorithm terminates. Observe that the node set S of the minimum cut $[S, \overline{S}]$ equals $\{6, 1, 2, 3, 4, 5\}$, and the algorithm increases the distance labels of all the nodes in S without performing any augmentation. The technique we describe essentially detects a situation like this one.

To implement this approach, we maintain an n-dimensional additional array, $numb$, whose indices vary from 0 to $(n-1)$. The value $numb(k)$ is the number of nodes whose distance label equals k. The algorithm initializes this array while computing the initial distance labels using the breadth-first search. At this point, the positive entries in the array numb are consecutive; that is, the entries $numb(0), numb(1), \ldots, numb(l)$ will be positive up to some index l and the remaining entries will all be zero. For example, the $numb$ array for the distance labels shown in Figure 4.11 is $numb(0) = 1, numb(1) = 5, numb(2) = 5, numb(3) = 1, numb(4) = 4$, and

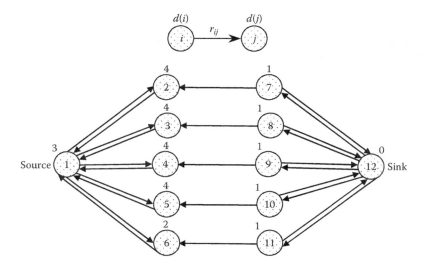

Figure 4.11 Bad example for the shortest augmenting path algorithm.

the remaining entries are zero. Subsequently, whenever the algorithm increases the distance label of a node from k_1 to k_2, it subtracts 1 from $numb(k_1)$, adds 1 to $numb(k_2)$, and checks whether $numb(k_1) = 0$. If $numb(k_1)$ does equal zero, then the algorithm terminates. As seen earlier, the shortest augmenting path algorithm augments unit flow along the paths 1-2-7-12, 1-3-8-12, 1-4-9-12, 1-5-10-12, and 1-6-11-12. At the end of these augmentations, we get the residual network shown in Figure 4.11. When we continue the shortest augmenting path algorithm from this point, then it constructs the partial admissible path 1-6. Next it relabels node 6, and its distance label increases from 2 to 4. The algorithm finds that $numb(2) = 0$ and it terminates.

4.3.3.2 Improvement in Capacity-Scaling Algorithm

In the previous section, we described the capacity-scaling algorithm for the maximum flow problem that runs in $O(m^2 \log U)$ time. We can improve the running time of this algorithm to $O(nm \log U)$ by using the shortest augmenting path as a subroutine in the capacity-scaling algorithm. Recall that the capacity-scaling algorithm performs a number of Δ-scaling phases and that in the Δ-scaling phase it sends the maximum possible flow in the Δ-residual network $G(x, \Delta)$ using the generic augmenting path algorithm as a subroutine. In the improved implementation, we use the shortest augmenting path algorithm to send the maximum possible flow from node s to node t. We accomplish this by defining the distance labels with respect to the network $G(x, \Delta)$ and augmenting flow along the shortest augmenting path in $G(x, \Delta)$. This algorithm yields a bound of $O(nm \log U)$ on the running time of the capacity-scaling algorithm.

4.3.3.3 Further Worst-Case Improvements

The idea of augmenting flows along the shortest paths is intuitively appealing and easy to implement in practice. The resulting algorithms identify at most $O(nm)$ augmenting paths, and this bound is tight; that is, on particular examples these algorithms perform $O(nm)$ augmentations. The only way to improve the running time of the shortest augmenting path algorithm is to perform fewer computations per augmentation.

The use of a sophisticated data structure, called dynamic trees, reduces the average time for each augmentation from $O(n)$ to $O(\log n)$. This implementation of the shortest

TABLE 4.1 Summary of the Augmenting Path Algorithms

	# Augment Iterations	Time to Find an Augmenting Path	Best Running Time
Generic [6,7]	$O(nU)$	$O(m)$	$O(nmU)$
Capacity scaling [8,9]	$O(m \log U)$	$O(n)$	$O(nm \log U)$
Shortest [8,10]	$O(nm)$	$O(\log n)$	$O(nm \log n)$

augmenting path algorithm runs in $O(nm \log n)$ time, and obtaining further improvements appears quite difficult except in very dense networks. See Table 4.1 for the summary of results in this section.

4.4 PREFLOW-PUSH ALGORITHMS

Another class of algorithms for solving the maximum flow problem, known as preflow-push algorithms, is more decentralized than augmenting path algorithms. Augmenting path algorithms send flow by augmenting along a path. This basic operation further decomposes into the more elementary operation of sending flow along individual arcs. Sending a flow of δ units along a path of k arcs decomposes into k basic operations of sending a flow of δ units along each of the arcs of the path. We shall refer to each of these basic operations as a push. The preflow-push algorithms push flows on individual arcs instead of on augmenting paths. Some theoretical time bounds for preflow-push type of algorithms are Goldberg and Tarjan [11] runs in $O(nm \log(n^2/m))$ time, an algorithm of King et al. [12] runs in $O(nm + n^{2+\varepsilon})$ time for any constant $\varepsilon > 0$, algorithms of Cheriyan et al. [3] runs in $O(n^3/\log n)$ time and $O(nm + (n \log n)^2)$ time with high probability, and an algorithm of Ahuja et al. [13] runs in $O(nm \log n/m\sqrt{\log U})$ time.

A path augmentation has one advantage over a single push: it maintains conservation of flow at all nodes. The preflow-push algorithms violate conservation of flow at all steps except at the very end and instead maintain a *preflow* at each iteration. A preflow is a vector x satisfying the flow-bound constraints and the following relaxation of the mass balance constraints:

$$\sum_{\{j:(i,j)\in A\}} x_{ij} - \sum_{\{j:(j,i)\in A\}} x_{ji} \geq 0 \qquad (4.7)$$

Each element of a preflow vector is either a real number or equals to $+\infty$. The preflow-push algorithms maintain a preflow at each intermediate stage. For a given preflow x, we define the excess for each node $i \in N - \{s,t\}$ as follows:

$$e(i) = \sum_{\{j:(i,j)\in A\}} x_{ij} - \sum_{\{j:(j,i)\in A\}} x_{ji} \qquad (4.8)$$

We refer to a node with positive excess as an active node. We adopt the convention that the source and sink nodes are never active. In a preflow-push algorithm, the presence of an active node indicates that the solution is infeasible. Consequently, the basic operation in this algorithm is to select an active node i and try to remove the excess by pushing flow out of it. When we push flow out of an active node, we need to do it carefully. If we just push flow to an adjacent node in an arbitrary manner and the other nodes do the same, then it is conceivable that some nodes would keep pushing flow among themselves, resulting in an infinite loop, which is not a desirable situation. Since ultimately we want to send the flow to the sink node, it seems reasonable for an active node to push flow to another node that is *closer* to the sink. If all nodes maintain this rule, then the algorithm could never encounter an infinite loop. The concept of distance labels defined next allows us to implement this algorithmic strategy.

The preflow-push algorithms maintain a distance label $d(i)$ with each node in the network. The distance labels are nonnegative (finite) integers defined with respect to the residual network $G(x)$. Let $d(i)$ be the valid distance label at node i; it is then easy to demonstrate that $d(i)$ is a lower bound on the length of any directed path (as measured by the number of arcs) from node i to node t in the residual network and thus is a lower bound on the length of the shortest path between nodes i and j. Let $i = i_1 - i_2 - i_3 - \cdots - i_k - t$ be any path of length k in the residual network from node i to node t. The validity conditions (5) imply that $d(i) = d(i_1) \leq d(i_2) + 1, d(i_2) \leq d(i_3) + 1, \ldots, d(i_k) \leq d(t) + 1 = 1$. Adding these inequalities shows that $d(i) \leq k$ for any path of length k in the residual network, and therefore any (shortest) path from node i to node t contains at least $d(i)$ arcs. Recall that we say that an arc (i,j) in the residual network is *admissible* if it satisfies the condition $d(i) = d(j) + 1$; we refer to all other arcs as inadmissible.

The basic operation in the preflow-push algorithm is to select an active node i and try to remove the excess by pushing flow to a node with a smaller distance label. (We will use the distance labels as estimates of the length of the shortest path to the sink node.) If node i has an admissible arc (i,j), then $d(j) = d(i) - 1$ and the algorithm sends flow on admissible arcs to relieve the node's excess. If node i has no admissible arc, then the algorithm increases the distance label of node i so that node i has an admissible arc. The algorithm terminates when the network contains no active nodes; that is, excess resides only at the source and sink nodes.

4.4.1 Generic Preflow-Push Algorithm

Now, we will explain the generic preflow-push algorithm.

Algorithm: Generic preflow-push

{

$x := 0$ and $d(j) := 0$ for all $j \in N$;

$x_{sj} = u_{sj}$ for each arc $(s,j) \in A(s)$;

$d(s) := n$;

while the residual network $G(x)$ contains an active node do

  {

  select an active node i;

  push/relabel(i);

  }

}

procedure push/relabel(i);

{

if the network contains an admissible arc (i,j) then

  push $\min\{e(i), r_{ij}\}$ units of flow from node i to node j

else replace $d(i)$ by $\min\{d(j) + 1 : (i,j) \in A(i) \text{ and } r_{ij} > 0\}$;

}

The algorithm first saturates all arcs emanating from the source node; then each node adjacent to node s has a positive excess so that the algorithm can begin pushing flow from active nodes. Since the preprocessing operation saturates all the arcs incident to node s, none of these arcs is admissible, and setting $d(s) = n$ will satisfy the validity condition. But then, since $d(s) = n$, and a distance label is a lower bound on the length of the shortest path from that node to node t, the residual network contains no directed path from s to t. The subsequent pushes maintain this property and drive the solution toward feasibility. Consequently, when there are no active nodes, the flow is a maximum flow.

A push of δ units from node i to node j decreases both the excess $e(i)$ of node i and the residual r_{ij} of arc (i, j) by δ units and increases both $e(j)$ and r_{ji} by δ units. We say that a push of δ units of flow on an arc (i, j) is saturating if $\delta = r_{ij}$ and is nonsaturating otherwise. A nonsaturating push at node i reduces $e(i)$ to zero. We refer to the process of increasing the distance label of a node as a relabel operation. The purpose of the relabel operation is to create at least one admissible arc on which the algorithm can perform further pushes.

It is instructive to visualize the generic preflow push algorithm in terms of a physical network: arcs represent flexible water pipes, nodes represent joints, and the distance function measures how far nodes are above the ground. In this network, we wish to send water from the source to the sink. We visualize flow in an admissible arc as water flowing downhill. Initially, we move the source node upward, and water flows to its neighbors. Although we would like water to flow downhill toward the sink, occasionally flow becomes trapped locally at a node that has no downhill neighbors. At this point, we move the node upward, and again water flows downhill toward the sink. Eventually, no more flow can reach the sink. As we continue to move nodes upward, the remaining excess flow eventually flows back toward the source. The algorithm terminates when all the water flows either into the sink or back to the source.

To illustrate the generic preflow-push algorithm, we use the example given in Figure 4.12. Figure 4.12a specifies the initial residual network. We first saturate the arcs emanating from

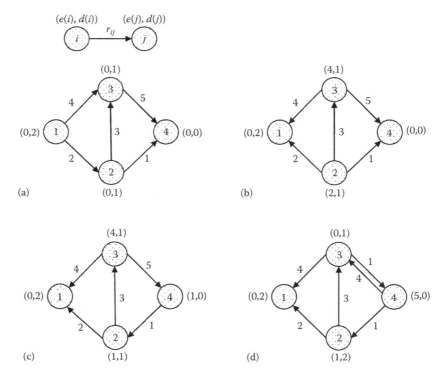

Figure 4.12 Example of generic preflow-push algorithm.

the source node, node 1, and set $d(1) = n = 4$. Figure 4.12b shows the residual graph at this stage. At this point, the network has two active nodes, nodes 2 and 3. Suppose that the algorithm selects node 2 for the push/relabel operation. Arc (2, 4) is the only admissible arc and the algorithm performs a saturating push of value $\delta = \min\{e(2), r_{24}\} = \min\{2, 1\} = 1$. Figure 4.12c gives the residual network at this stage. Suppose the algorithm again selects node 2. Since no admissible arc emanates from node 2, the algorithm performs a relabel operation and gives node 2 a new distance label, $d(2) = \min\{d(3) + 1, d(1) + 1\} = \min\{2, 5\} = 2$. The new residual network is the same as the one shown in Figure 4.12c except that $d(2) = 2$ instead of 1. Suppose this time the algorithm selects node 3. Arc (3, 4) is the only admissible arc emanating from node 3, and so the algorithm performs a nonsaturating push of value $\delta = \min\{e(3), r_{34}\} = \min\{4, 5\} = 4$. Figure 4.12d specifies the residual network at the end of this iteration. Using this process for a few more iterations, the algorithm will determine a maximum flow.

We now analyze the complexity of the algorithm. To begin with we establish one important result that distance labels are always valid and do not increase *too many* times. As in the shortest augmenting path algorithm, the preflow-push algorithm pushes flow only on admissible arcs and relabels a node only when no admissible arc emanates from it. The second conclusion follows from the following theorem.

Theorem 4.11 *At any stage of the preflow-push algorithm, each node i with positive excess is connected to node s by a directed path from i to s in the residual network.*

Proof. Notice that for a preflow $x, e(s) < 0$ and $e(i) \geq 0$ for all $i \in N - \{s\}$. Any preflow x can be decomposed with respect to the original network G into nonnegative flows along (i) paths from node s to node t, (ii) paths from node s to active nodes, and (iii) flows around directed cycles. Let i be an active node relative to the preflow x in G. Then the flow decomposition of x must contain a path P from node s to node i, since the paths from node s to node t and the flows around cycles do not contribute to the excess at node i. Then the residual network contains the reversal of P (P with the orientation of each arc reversed), and so a directed path from i to s. ∎

Theorem 4.12 *For each node $i \in N, d(i) < 2n$.*

Proof. The last time the algorithm relabeled node i, the node had a positive excess, and so the residual network contained a path P of length at most $n - 1$ from node i to node s. The fact that $d(s) = n$ and that $d(k) \leq d(l) + 1$ for every arc (k, l) in the path P implies that $d(i) \leq d(s) + |P| < 2n$. ∎

Since each time the algorithm relabels node i, $d(i)$ increases by at least one unit, we have established the following result.

Theorem 4.13 *Each distance label increases at most $2n$ times. Consequently, the total number of relabel operations is at most $2n^2$.* ∎

Theorem 4.14 *The algorithm performs at most nm saturating pushes.*

Proof. This result follows directly from Theorems 4.12 and 4.8. ∎

Theorems 4.7 and 4.13 imply that the total time needed to identify admissible arcs and to perform relabel operations is $O(nm)$. We next count the number of nonsaturating pushes performed by the algorithm.

Theorem 4.15 *The generic preflow-push algorithm performs $O(n^2m)$ nonsaturating pushes.*

Proof. We prove the theorem using an argument based on potential functions. Let I denote the set of active nodes. Consider the potential function $\Phi = \sum_{i \in I} d(i)$. Since $|I| \leq n$, and $d(i) \leq 2n$ for all $i \in I$, the initial value of Φ (after the preprocess operation) is at most $2n^2$. At the termination of the algorithm, Φ is zero. During the push/relabel(i) operation, one of the following two cases must apply. ∎

Case 1 The algorithm is unable to find an admissible arc along which it can push flow. In this case, the distance label of node i increases by $\varepsilon \geq 1$ units. This operation increases Φ by at most ε units. Since the total increase in $d(i)$ for each node i throughout the execution of the algorithm is bounded by $2n$, the total increase in Φ due to increases in distance labels is bounded by $2n^2$.

Case 2 The algorithm is able to identify an arc on which it can push flow, and so it performs a saturating push or a nonsaturating push. A saturating push on arc (i,j) might create a new excess at node j, thereby increasing the number of active nodes by 1, and increasing Φ by $d(j)$, which could be as much as $2n$ per saturating push, and so $2n^2m$ over all saturating pushes. Next note that a nonsaturating push on arc (i,j) does not increase $|I|$. The nonsaturating push will decrease Φ by $d(i)$ since i becomes inactive, but it simultaneously increases Φ by $d(j) = d(i) - 1$ if the push causes node j to become active; the total decrease in Φ being of value 1. If node j was active before the push, then Φ decreases by an amount $d(i)$. Consequently, net decrease in Φ is at least one unit per nonsaturating push.

We summarize these facts. The initial value of Φ is at most $2n^2$ and the maximum possible increase in Φ is $2n^2 + 2n^2m$. Each nonsaturating push decreases by at least one unit and always remains nonnegative. Consequently, the algorithm can perform at most $2n^2 + 2n^2 + 2n^2m = O(n^2m)$ nonsaturating pushes.

Finally, we indicate how the algorithm keeps track of active nodes for the push/relabel operations. The algorithm maintains a set LIST of active nodes. It adds to LIST those nodes that become active following a push and are not already in LIST, and deletes from LIST nodes that become inactive following a nonsaturating push. Several data structures (e.g., doubly linked lists) are available for storing LIST so that the algorithm can add, delete, or select elements from it in $O(1)$ time. Consequently, it is easy to implement the preflow-push algorithm in $O(n^2m)$ time. We have thus established the following theorem.

Theorem 4.16 *Generic preflow-push algorithm runs in $O(n^2m)$ time.* ∎

The preflow-push algorithm has several attractive features, particularly its flexibility and its potential for further improvements. Different rules for selecting active nodes for the push/relabel operations create many different versions of the generic algorithm, each with different worst-case complexity. As we have noted, the bottleneck operation in the generic preflow-push algorithm is the number of nonsaturating pushes, and many specific rules for examining active nodes can produce substantial reductions in the number of nonsaturating pushes. The following specific implementations of the generic preflow-push algorithms are noteworthy: (1) the first-in, first-out (FIFO) preflow-push algorithm examines the active nodes in the FIFO order and runs in $O(n^3)$ time; (2) the highest label preflow-push algorithm pushes flow from an active node with the highest value of a distance label and runs in $O(n^2\sqrt{m})$ time; and (3) the excess-scaling algorithm uses the scaling of arc capacities to attain a time bound of $O(nm + n^2 \log U)$. These algorithms are due to Goldberg and Tarjan [11], Cheriyan and Maheshwari [14], and Ahuja et al. [13], respectively. These preflow-push algorithms are more general, more powerful, and more flexible than augmenting path algorithms. The best preflow-push algorithms currently outperform the best augmenting path algorithms in theory as well as in practice (see, e.g., Ahuja et al. [15]).

4.4.2 FIFO Preflow-Push Algorithm

Before we describe the FIFO implementation of the preflow-push algorithm, we define the concept of a node examination. In an iteration, the generic preflow-push algorithm selects a node, say node i, and performs a saturating push or a nonsaturating push or relabels the node. If the algorithm performs a saturating push, then node i might still be active, but it is not mandatory for the algorithm to select this node again in the next iteration. The algorithm might select another node for the next push/relabel operation. However, it is easy to incorporate the rule that whenever the algorithm selects an active node, it keeps pushing flow from that node until either the node's excess becomes zero or the node is relabeled. Consequently, the algorithm might perform several saturating pushes followed either by a nonsaturating push or by a relabel operation. We refer to this sequence of operations as a node examination. We shall assume that every preflow-push algorithm adopts this rule for selecting nodes for the push/relabel operation.

The FIFO preflow-push algorithm examines active nodes in the FIFO order. The algorithm maintains the set LIST as a queue. It selects a node i from the front of LIST, performs pushes from this node, and adds newly active nodes to the rear of LIST. The algorithm examines node i until either it becomes inactive or it is relabeled. In the latter case, we add node i to the rear of the queue. The algorithm terminates when the queue of active nodes is empty.

We illustrate the FIFO preflow-push algorithm using the example shown in Figure 4.13a. The preprocess operation creates an excess of 10 units at each of the nodes 2 and 3. Suppose that the queue of active nodes at this stage is LIST = $\{2, 3\}$. The algorithm takes out node 2 from the queue and examines it. Suppose it performs a saturating push of 5 units on arc (2, 4) and a nonsaturating push of 5 units on arc (2, 5) (see Figure 4.13b). As a result of these pushes, nodes 4 and 5 become active and we add these nodes to the queue in this order, obtaining LIST = $\{3, 4, 5\}$. The algorithm next takes out node 3 from the queue. While examining node 3, the algorithm performs a saturating push of 5 units on arc (3, 5), followed by a relabel operation of node 3 (see Figure 4.13c). The algorithm adds node 3 to the queue, obtaining LIST = $\{4, 5, 3\}$.

We now analyze the worst-case complexity of the FIFO preflow-push algorithm. For the purpose of this analysis, we partition the total number of node examinations into different phases. The first phase consists of those node examinations for the nodes that become active during the preprocess operation. The second phase consists of the node examinations of all the nodes that are in the queue after the algorithm has examined the nodes in the first phase. Similarly, the third phase consists of the node examinations of all the nodes that are in the queue after the algorithm has examined the nodes in the second phase, and so on. For example, in the preceding illustration, the first phase consists of the node examinations of the set $\{2, 3\}$, and the second phase consists of the node examinations of the set $\{4, 5, 3\}$. Observe that the algorithm examines any node at most once during a phase.

We will now show that the algorithm performs at most $2n^2 + n$ phases. Each phase examines any node at most once, and each node examination performs at most one nonsaturating push. Therefore, a bound of $2n^2 + n$ on the total number of phases would imply a bound of $O(n^3)$ on the number of nonsaturating pushes. This result would also imply that the FIFO preflow-push algorithm runs in $O(n^3)$ time because the bottleneck operation in the generic preflow-push algorithm is the number of nonsaturating pushes.

To bound the number of phases in the algorithm, we consider the total change in the potential function $\Phi = \max\{d(i) : i \text{ is active}\}$ over an entire phase. By the total change, we mean the difference between the initial and final values of the potential function during a phase. We consider two cases.

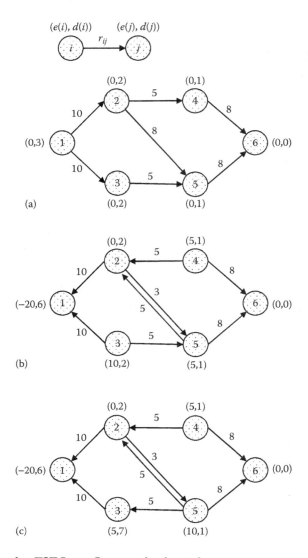

Figure 4.13 Illustrating the FIFO preflow-push algorithm.

Case 1 The algorithm performs at least one relabel operation during a phase. Then might increase by as much as the maximum increase in any distance label. Theorem 4.13 implies that the total increase in Φ over all the phases is $2n^2$.

Case 2 The algorithm performs no relabel operation during a phase. In this case, the excess of every node that was active at the beginning of the phase moves to nodes with smaller distance labels. Consequently, Φ decreases by at least one unit.

Combining Cases 1 and 2, we find that the total number of phases is at most $2n^2 + n$; the second term corresponds to the initial value of Φ which could be at most n. We have thus proved the following theorem.

Theorem 4.17 *The FIFO preflow-push algorithm runs in $O(n^3)$ time.* ∎

The summary of the algorithms in this section are given in Table 4.2.

4.5 BLOCKING FLOW ALGORITHMS

Instead of identifying augmenting paths one by one, Dinic [16] suggested identifying a set of augmenting paths at each iteration. Each iteration is called a blocking flow.

TABLE 4.2 Summary of the Preflow-Push Algorithms

	# Nonsaturating Pushes	Best Running Time
Generic [11]	$O(n^2 m)$	$O(n^2 m)$
FIFO [15]	$O(n^3)$	$O(n^3)$

Distance labels. A distance function $d : N \to Z^+ \cup \{0\}$ with respect to the residual capacities r_{ij} is a function from the set of nodes to the set of nonnegative integers. We say that a distance function is valid with respect to a flow x if it satisfies the following two conditions: $d(t) = 0$ and $d(i) \geq d(j) + 1$ for every arc (i,j) in the residual network $G(x)$.

The following observations show why the distance labels might be of use in designing network flow algorithms.

1. If the distance labels are valid, then the distance label $d(i)$ is a lower bound on the length of the shortest (directed) path from node i to node t in the residual network. To establish the validity of this observation, let $i = i_1 - i_2 - \cdots - i_k - i_{k+1} = t$ be any path of length k in the residual network from node i to t. The validity conditions imply the following:

$$d(i_k) \leq d(i_{k+1}) + 1 = d(t) + 1 = 1,$$
$$d(i_{k-1}) \leq d(i_k) + 1 \leq 2,$$
$$d(i_{k-2}) \leq d(i_{k-1}) + 1 \leq 3,$$
$$\ldots$$
$$d(i) = d(i_1) \leq d(i_2) + 1 \leq k$$

2. If $d(s) \geq n$, then the residual network contains no directed path from the source node to the sink node. The correctness of this observation follows from the facts that $d(s)$ is a lower bound on the length of the shortest path from s to t in the residual network and, therefore, no directed path can contain more than $(n-1)$ arcs. Therefore, if $d(s) \geq n$, then the residual network contains no directed path from node s to node t.

We say that the distance labels are exact if for each node i, $d(i)$ equals the length of the shortest path from node i to node t in the residual network. For example, in Figure 4.11, if node 1 is the source node and node 4 is the sink node, then $d = (0, 0, 0, 0)$ is a valid vector of distance labels, though $d = (3, 1, 2, 0)$ represents the vector of exact distance labels. We can determine exact distance labels for all nodes in $O(m)$ time by performing a backward breadth-first search of the network starting at the sink node.

We say that an arc (i,j) in the residual network is admissible if it satisfies the condition that $d(i) = d(j) + 1$; otherwise we refer that arc as inadmissible. We also refer to a path from node s to node t consisting entirely of admissible arcs as an admissible path. Later, we shall use the following property of admissible paths.

An admissible path is a shortest augmenting path from the source to the sink. Since every arc (i,j) in an admissible path P is admissible, the residual capacity of this arc and the distance labels of its end nodes satisfy the conditions (1) $r_{ij} > 0$; and (2) $d(i) = d(j) + 1$. The condition (1) implies that P is an augmenting path and condition (2) implies that if P contains k arcs, then $d(s) = k$. Since $d(s)$ is a lower bound on the length of any path from the source to the sink in the residual network, the path P must be a shortest augmenting path.

Dinic's algorithm constructs shortest path networks, called layered networks, and establishes blocking flows (to be defined later) in these networks. Now, we point out the relationship between layered networks and distance labels.

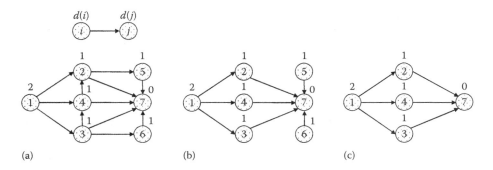

Figure 4.14 (a) Residual network. (b) Corresponding layered network. (c) Layered network after deleting redundant arcs.

Layered network. With respect to a given flow x, we define the layered network V as follows. We determine the exact distance labels d in $G(x)$. The layered network consists of those arcs (i, j) in $G(x)$ for which $d(i) = d(j) + 1$. For example, consider the residual network $G(x)$ given in Figure 4.14a. The number beside each node represents the exact distance label. Figure 4.14b shows the layered network of $G(x)$. Observe that, by definition, every path from the source to the sink in the layered network V is a shortest path in $G(x)$. Also observe that some arc in v might not be contained in any path from the source to the sink. For example, in Figure 4.14b, arcs (5, 7) and (6, 7) do not lie on any path in V from the source to the sink. Since these arcs do not participate in any flow augmentation, we typically delete them from the layered network and we obtain Figure 4.14c. In the resulting layered network, the nodes are partitioned into layers of nodes $V_0, V_1, V_2, \ldots, V_l$; and layer k contains the nodes whose distance labels are equal to k. Furthermore, for every arc (i, j) in the layered network, $i \in V_k$ and $j \in V_{k-1}$ for some k. Let the source node have the distance label l.

4.5.1 Blocking Flow Algorithm

A blocking flow is the list of augmentations each of which saturates (reduces the residual capacity to zero) at least one arc in the layered network. A blocking flow includes at most m augmenting paths. The flow will be maximum after $n-1$ blocking flow operations. A blocking flow can be identified in $O(nm)$ time by using depth first search or $O(m \log n)$ time by employing dynamic tree structure [16]. The total running time of these algorithms are $O(n^2 m)$ and $O(nm \log n)$, respectively. The algorithm is formally given below.

Blocking flow algorithm

$x := 0$;

while sink is reachable from the source do

{

    identify the current layered network;

    find the corresponding blocking flow f;

    augment x by f;

    update the residual network;

}

TABLE 4.3 Summary of the Blocking Flow Algorithms

Reference	# Blocking Flow	Time to Find a Blocking Flow	Best Running Time
[16]	$O(n)$	$O(nm)$	$O(n^2 m)$
[10]	$O(n)$	$O(m \log n)$	$O(nm \log n)$ Dynamic tree
[17]			$O(n^3)$

4.5.2 Malhotra, Kumar, and Maheshwari Algorithm

This algorithm relies on the idea of selecting the node with the lowest flow potential. The flow potential of node i, $p_x(i)$, is defined as the maximum amount of flow that can be forced through node i.

$$p_x(i) = \min\left\{ \sum_{(j,i)\in A} (u_{ji} - z_{ji}), \sum_{(i,k)\in A} (u_{ik} - x_{ik}) \right\}$$

Malhotra, Kumar, and Maheshwari (MPM) algorithm

$x := 0$;

while there are nodes in the network do

    { select node k with minimum $p_x(k)$;

    send $p_x(k)$ units of flow from node k toward the node s and t;

    augment x by $p_x(k)$, update $G(x)$;

    delete all saturated arcs, node k, and the nodes with only incoming or outgoing arcs; }

To get an efficient implementation of the MPM algorithm we adopt the following rules.

1. Push flow out of vertex v_k only after all the flows pushed into v_k from vertices in a lower layer have been received.

2. Push flow out of vertex v_k only after all the flows into v_k from vertices in an upper layer have been received.

3. While pushing a flow along an edge push the maximum possible flow allowed by the residual capacity at that step.

If we follow the above rules, the total number of saturating pushes during the entire algorithm is m. The number of non-saturating pushes during each execution of the while loop is at most n. The while loop will be executed at most n times since at least one node will be deleted during each execution. So the total number of non-saturating pushes during the entire algorithm is at most n^2 resulting in a overall time complexity of $O(m + n^2)$ which is $O(n^2)$ for the MPM algorithm.

In finding a maximum flow the blocking flow computations will be done at most n times. So the complexity of the maximum flow algorithm using the MPM algorithm for blocking computations is $O(n^3)$. The summary of the running times in the section are given in Table 4.3.

4.6 MAXIMUM FLOW ON UNIT CAPACITY NETWORKS

Certain combinatorial problems are naturally formulated as zero-one optimization models. When viewed as flow problems, these models yield networks whose arc capacities are all 1.

We will refer to these networks as unit capacity networks. Frequently, it is possible to solve flow problems on these networks more efficiently than those defined on general networks. In this section, we describe an efficient algorithm for solving the maximum flow problem on unit capacity networks. We subsequently refer to this algorithm as the unit capacity maximum flow algorithm.

In a unit capacity network, the maximum flow value is at most n since the capacity of the $s-t$ cut $[s, S-\{s\}]$ is at most n. The labeling algorithm, therefore, determines a maximum flow within n augmentations and requires $O(nm)$ effort. The shortest augmenting path algorithm also solves this problem in $O(nm)$ time since its bottleneck operation, which is the augmentation step, requires $O(nm)$ time instead of $O(n^2 m)$ time. The unit capacity maximum flow algorithm that we describe is a hybrid version of these two algorithms. This unit capacity maximum flow algorithm is noteworthy because by combining features of both algorithms, it requires only $O(\min\{n^{2/3}m, m^{3/2}\})$ time, which is consistently better than the $O(nm)$ bound of either algorithm by itself.

Theorem 4.18 *This unit capacity maximum flow algorithm runs in $O(\min\{n^{2/3}m, m^{3/2}\})$ time.*

Proof. The unit capacity maximum flow algorithm is a two-phase algorithm. In the first phase, it applies the shortest augmenting path algorithm, although not until completion; rather, this phase terminates whenever the distance label of the source node satisfies the condition $d(s) \geq d^* = \min\{2n^{2/3}, m^{1/2}\}$. Although the algorithm might terminate with a nonoptimal solution, the solution is provably near-optimal (its value is within d^* of the optimal flow value). In its second phase, the unit capacity maximum flow algorithm applies the labeling algorithm to convert this near-optimal flow into a maximum flow. As we will see, this two-phase approach works well for unit capacity networks because the shortest augmenting path algorithm obtains a near-optimal flow quickly (when augmenting paths are short) but then takes a long time to convert this solution into a maximum flow (when augmenting paths become long). It so happens that the labeling algorithm converts this near-optimal flow into a maximum flow far more quickly than does the shortest augmenting path algorithm.

Let us examine the behavior of the shortest augmenting path algorithm for $d^* = \min\{\lceil 2n^{2/3} \rceil, \lceil m^{1/2} \rceil\}$. Suppose the algorithm terminates with a flow vector x' with a flow value equal to v'. What can we say about $v^* - v'$? (Recall that v^* denotes the maximum flow value.) We shall answer this question in two parts: (1) when $d^* = \lceil 2n^{2/3} \rceil$ and (2) when $d^* = \lceil m^{1/2} \rceil$.

Suppose $d^* = \lceil 2n^{2/3} \rceil$. For each $k = 0, 1, 2, \ldots, d^*$, let V_k denote the set of nodes with a distance label equal to k; that is, $V_k = \{i \in N : d(i) = k\}$. We refer to V_k as the set of nodes in the kth layer of the network. Consider the situation when each of the sets $V_1, V_2, \ldots, V_{d^*}$ is nonempty. It can be shown that each arc (i,j) in the residual network $G(x')$ connects a node in the kth layer to a node in the $(k-1)$th layer for some k, for otherwise $d(i) > d(j)+1$, which contradicts the distance label validity conditions. Therefore, for each $k = 1, 2, \ldots, d^*$, the set of arcs joining the node sets V_k to V_{k-1} forms an $s-t$ cut in the residual network. In the case that one of the sets, say V_k, is empty, then the cut $[S, \overline{S}]$ defined by $S = V_{k+1} \bigcup V_k \bigcup \cdots \bigcup V_{d^*}$ is a minimum cut.

Note that $|V_1| + |V_2| + \cdots + |V_{d^*}| \leq n - 1$ because the sink node does not belong to any of these sets. We claim that the residual network contains at least two consecutive layers V_k and V_{k-1}, each with at most $n^{1/3}$ nodes. For, if not, then every alternate layer (say, $V_1, V_3, V_5, \ldots$) would have to contain more than $n^{1/3}$ nodes, and the total number of nodes in these layers would be strictly greater than $n^{1/3} d^*/2 \geq n$, leading to a contradiction. Consequently, $|V_k| \leq n^{1/3}$ and $|V_{k-1}| \leq n^{1/3}$ for some of the two layers V_k and V_{k-1}. The residual capacity of the $s-t$ cut defined by the arcs connecting V_k to V_{k-1} is at most

$|V_k||V_{k-1}| \leq n^{2/3}$ (since at most one arc of unit residual capacity joins any pair of nodes). Therefore, it can be shown that $v^* - v' \leq n^{2/3} \leq d^*$.

Next consider the situation when $d^* = \lceil m^{1/2} \rceil$. The layers of nodes $V_1, V_2, \ldots, V_{d^*}$ define d^* number of $s-t$ cuts, and these cuts are arc-disjoint. The sum of the residual capacities of these cuts is at most m since each arc contributes at most one to the residual capacity of any such cut. Thus some $s-t$ cut must have residual capacity at most $\lceil m^{1/2} \rceil$. This conclusion proves that $v^* - v' \leq \lceil m^{1/2} \rceil = d^*$.

In both the cases, we find that the first phase obtains a flow whose value differs from the maximum flow value by at most d^* units. The second phase converts this flow into a maximum flow in $O(d^*m)$ time since each augmentation requires $O(m)$ time and carries a unit flow. We now show that the first phase also requires $O(d^*m)$ time.

In the first phase, whenever the distance label of a node k exceeds d^*, this node never occurs as an intermediate node in any subsequent augmenting path since $d(k) < d(s) < d^*$. So the algorithm relabels any node at most d^* times. This observation gives a bound of $O(d^*n)$ on the number of retreat operations and a bound of $O(d^*m)$ on the time to perform the retreat operations. Consider next the augmentation time. Since each arc capacity is one, flow augmentation over an arc immediately saturates that arc. During two consecutive saturations of any arc (i, j), the distance labels of both the nodes i and j must increase by at least two units. Thus the algorithm can saturate any arc at most $\lfloor d^*/2 \rfloor$ times, giving an $O(d^*m)$ bound on the total time needed for flow augmentations. The total number of advance operations is bounded by the augmentation time plus the number of retreat operations and is again $O(d^*m)$. ∎

Another special case of unit capacity networks, called unit capacity simple networks, also arises in practice and is of interest to researchers. For this class of unit capacity networks, every node in the network, except the source and sink nodes, has at most one incoming arc or at most one outgoing arc. The unit capacity maximum flow algorithm runs even faster for this class of networks. We achieve this improvement by setting $d^* = n^{1/2}$ in the algorithm.

Theorem 4.19 *The unit capacity maximum flow algorithm establishes a maximum flow the in unit capacity simple networks in $O(\sqrt{n}m)$ time.*

Proof. Consider the layers of nodes $V_1, V_2, \ldots, V_{d^*}$ at the end of the first phase. Suppose layer V_h contains the smallest number of nodes. Then $|V_h| \leq \sqrt{n}$, since otherwise the number of nodes in all layers would be strictly greater than n. We define a cut Q of arcs as follows: we consider each node in V_h, and if the node has a unique incoming arc, then we add it to Q, and if the node has a unique outgoing arc, then we add it to Q. It can be easily shown that Q defines an $s-t$ cut. The residual capacity of the cut Q is at most $|V_h| \leq \sqrt{n}$. Consequently, at the termination of the first phase, the flow value differs from the maximum flow value by at most $\sqrt{n}$ units. Using arguments similar to those we have just used, we can now easily show that the algorithm would run in $O(\sqrt{n}m)$ time. ∎

The proof of Theorem 4.19 relies on the fact that only one unit of flow can pass through each node in the network (except the source and sink nodes). If we satisfy this condition but allow some arc capacities to be larger than one, then the unit capacity maximum flow algorithm would still require only $O(\sqrt{n}m)$ time. Theorem 4.19 has another by-product: it permits us to solve the maximum bipartite matching problem in $O(\sqrt{n}m)$ time since we can formulate this problem as a maximum flow problem on a unit capacity simple network.

Further Reading

In this section, we briefly mention other approaches to solve maximum flow problems. A new survey on maximum flows is given by Goldberg and Tarjan [18].

Maximum Adjacency Ordering Algorithm

The ordering of nodes $v_1, v_2, \ldots, v_n$ is called maximum adjacency (MA) ordering if for all i, v_i is the node in $N - \{v_1, v_2, \ldots, v_{i-1}\}$ that has the largest sum of residual capacities of the arcs to the nodes $v_1, v_2, \ldots, v_{i-1}$. MA ordering algorithm is proposed by Fujishige [19] and runs in $O(n(m + n \log n) \log nU)$ time and the scaling version runs is $O(nm \log U)$. Shioura [20] shows that the algorithm is not strongly polynomial by giving a real-valued instance for which algorithm does not terminate. Matsouka and Fujishige [21] propose an improved version using preflows. While the theoretical bound is the same as Fujishige [19], the algorithm is faster in practise.

Arc-Balancing Algorithm

Tarjan et al. [22] presents a round-robin arc-balancing algorithm that computes a maximum flow in $O(n^2 m \log(nU))$ time. Although this algorithm is slower than other known algorithms, it is very simple to implement this algorithm. The algorithm maintains a *pseudoflow* rather than a preflow. Let x be a pseudoflow then $x_{ij} = -x_{ji}$ and $x_{ij} \leq u_{ij}$ for all arcs $(i,j) \in A$. The balance at node i can be calculated as

$$b(i) = \sum_{(j,i) \in A} x_{ji}$$

where $b(i)$ can take both negative (deficit) and positive (excess) values.

Hochbaum's Pseudoflow Algorithm

Hochbaum proposes a pseudoflow algorithm that runs in $O(nm \log n)$ time. Chandran and Hochbaum shows that the highest label pseudoflow implementation is faster than the preflow-push type algorithms in most of the instances across many problem families. Hochbaum and Orlin show that the highest label version of the pseudoflow algorithm runs in $O(nm \log(n^2/m))$ and $O(n^3)$ time, with and without dynamic tree implementation.

Fastest Algorithm

The fastest strongly polynomial time algorithm is due to King et al. [12]. Its running time is $O(nm \log_{m/(n \log n)} n)$. When $m = \Omega(n^{1+\varepsilon})$ for any constant ε, the running time is $O(nm)$. Orlin [23] has shown that the maximum flow problem can be solved in $O(nm + m^{31/16} \log^2 n)$ time. When $m = O(n^{16/15-\varepsilon})$ this running time is $O(nm)$. Because the algorithm by [12] solves the max flow problem in $O(nm)$ time for $m > n^{1+\varepsilon}$, this work by Orlin establishes that the max flow problem can be solved in $O(nm)$ time for all n and m.

References

[1] R. K. Ahuja, T. L. Magnanti, and J. B. Orlin, *Network Flows: Theory, Algorithms, and Applications*, Prentice Hall, Englewood Cliffs, NJ, 1993.

[2] L. Ford and D. R. Fulkerson, *Flows in Networks*, Princeton University, Princeton, NJ, 1962.

[3] J. Cheriyan, T. Hagerup, and K. Mehlhorn, *Can a Maximum Flow Be Computed in o(nm) Time?* Berlin, Germany, Springer, 1990.

[4] A. V. Goldberg and S. Rao, Beyond the flow decomposition barrier, *J. ACM*, **45** (September 1998), 783–797.

[5] P. Elias, A. Feinstein, and C. E. Shannon, A note on the maximum flow through a network, *Inf. Theory, IRE Trans. on*, **2**(4) (1956), 117–119.

[6] L. R. Ford and D. R. Fulkerson, Maximal flow through a network, *Can. J Math.*, **8**(3) (1956), 399–404.

[7] P. Elias, A. Feinstein, and C. Shannon, A note on the maximum flow through a network, *Inf. Theory, IRE Trans. on*, **2**(4) (1956), 117–119.

[8] R. K. Ahuja and J. B. Orlin, Distance-directed augmenting path algorithms for maximum flow and parametric maximum flow problems, *Naval Res. Logist.*, **38**(3) (1991), 413–430.

[9] J. Edmonds and R. M. Karp, Theoretical improvements in algorithmic efficiency for network flow problems, *J. ACM*, **19**(2) (1972), 248–264.

[10] D. D. Sleator and R. E. Tarjan, A data structure for dynamic trees, *J. Comp. Syst. Sci.*, **26**(3) (1983), 362–391.

[11] A. V. Goldberg and R. E. Tarjan, A new approach to the maximum-flow problem, *J. ACM*, **35**(4) (1998), 921–940.

[12] V. King, S. Rao, and R. Tarjan, A faster deterministic maximum flow algorithm, *J. Algorithms*, **17**(3) (1994), 447–474.

[13] R. K. Ahuja, J. B. Orlin, and R. E. Tarjan, Improved time bounds for the maximum flow problem, *SIAM J. Comput.*, **18**(5) (1989), 939–954.

[14] J. Cheriyan and S. N. Maheshwari, Analysis of preflow push algorithms for maximum network flow, *SIAM J. Comput.*, **18**(6) (1989), 1057–1086.

[15] R. K. Ahuja, M. Kodialam, A. K. Mishra, and J. B. Orlin, Computational investigations of maximum flow algorithms, *Eur. J. Oper. Res.*, **97**(3) (1997), 509–542.

[16] E. A. Dinic, Algorithm for solution of a problem of maximum flow in networks with power estimation, *Soviet Math. Doklady*, **11**(5) (1970), 1277–1280.

[17] V. M. Malhotra, M. P. Kumar, and S. N. Maheshwari, An $O(v^3)$ algorithm for finding maximum flows in networks, *Inf. Process. Lett.*, **7**(6) (1978), 277–278.

[18] R. E. Tarjan and A. V. Goldberg, Efficient maximum flow algorithms, *Commun. ACM*, **57**(8) (August 2014), 82–89.

[19] S. Fujishige, A maximum flow algorithm using ma ordering, *Oper. Res. Lett.*, **31**(3) (2003), 176–178.

[20] A. Shioura, The ma-ordering max-flow algorithm is not strongly polynomial for directed networks, *Oper. Res. Lett.*, **32**(1) (2004), 31–35.

[21] Y. Matsuoka and S. Fujishige, Practical efficiency of maximum flow algorithms using ma orderings and preflows, *J. Oper. Res. Soc. Japan-Keiei Kagaku*, **48**(4) (2005), 297–307.

[22] R. Tarjan, J. Ward, B. Zhang, Y. Zhou, and J. Mao, Balancing applied to maximum network flow problems, In *Algorithms–ESA*, 612–623, Berlin, Germany, Springer, 2006.

[23] J. B. Orlin, Max flows in $O(nm)$ time, or better, *Proceedings of the 45th Annual ACM Symposium on Theory of Computing*, 2013, 765–774, New York. ACM.

CHAPTER 5

Minimum Cost Flow Problem

Balachandran Vaidyanathan

Ravindra K. Ahuja

James B. Orlin

Thomas L. Magnanti

CONTENTS

5.1	Introduction	114
	5.1.1 Notation and Assumptions	115
	5.1.2 Similarity Assumption	116
	5.1.3 Residual Network	116
	5.1.4 Tree, Spanning Tree, and Forest	117
5.2	Applications	117
	5.2.1 Distribution Problems	117
	5.2.2 Optimal Loading of a Hopping Airplane	118
	5.2.3 Scheduling with Deferral Costs	118
5.3	Optimality Conditions	120
	5.3.1 Negative Cycle Optimality Conditions	120
	5.3.2 Reduced Cost Optimality Conditions	121
	5.3.3 Complementary Slackness Optimality Conditions	122
5.4	Minimum Cost Flow Duality	123
5.5	Cycle-Canceling Algorithm	125
	5.5.1 Augmenting Flow in a Negative Cycle with Maximum Improvement	127
	5.5.2 Augmenting Flow along a Negative Cycle with Minimum Mean Cost	127
5.6	Successive Shortest Path Algorithm	127
5.7	Primal-Dual and Out-of-Kilter Algorithms	129
5.8	Network Simplex Algorithm	130
	5.8.1 Cycle Free and Spanning Tree Solutions	131
	5.8.2 The Network Simplex Algorithm	133
	5.8.2.1 Obtaining an Initial Spanning Tree Structure	133
	5.8.2.2 Maintaining a Spanning Tree Structure	134
	5.8.3 Computing Node Potentials and Flows	135
	5.8.4 Entering Arc	137
	5.8.5 Leaving Arc	138
	5.8.6 Updating the Tree	138

		5.8.7	Termination	140
		5.8.8	Strongly Feasible Spanning Trees	140
			5.8.8.1 Leaving Arc Rule	141
	5.9		Capacity-Scaling Algorithm	143

5.1 INTRODUCTION

Network flows is a problem domain that lies in the cusp between several fields of enquiry, including applied mathematics, computer science, engineering, management, and operations research. The minimum cost flow problem is a fundamental network flow problem. The objective of this problem is to send units of a good that reside at one or more points in a network (sources or supplies) with arc capacities to one or more other points in the network (sinks or demands), incurring minimum cost. In this chapter, we consider algorithmic approaches for solving the minimum cost flow problem and discuss a few applications. These algorithms are among the most efficient algorithms known in applied mathematics, computer science, and operations research for solving large-scale optimization problems.

All other network flow problems, such as the shortest path problem, maximum flow problem, assignment problem, and transportation problem, can be viewed as special cases of the minimum cost flow problem. The objective of the shortest path problem is to find a path of minimum cost on a network between a specified source node and a specified sink node; and the objective of the maximum flow problem is to find a solution that sends the maximum possible flow between a specified source node and a specified sink node on a network with arc capacities. The shortest path problem considers only the cost aspect of the minimum cost flow problem, and the maximum flow problem considers only arc capacities while neglecting costs. Consequently, many algorithms for solving the minimum cost flow problem combine ingredients of both the shortest path and the maximum flow algorithms. These algorithms solve the minimum cost flow problem by solving a sequence of shortest path problems. We consider four such algorithms in this chapter. The *cycle-canceling algorithm* uses shortest path computations to find augmenting cycles with negative flow costs; it then augments flows along these cycles and iteratively repeats these computations for detecting negative cost cycles and augmenting flows. The *successive shortest path algorithm* incrementally loads flow on the network from some source node to some sink node, each time selecting an appropriately defined shortest path. The *primal-dual* and *out-of-kilter algorithms* use a similar algorithmic strategy: At every iteration, they solve a shortest path problem and augment flow along one or more shortest paths. They differ, however, in their tactics. The primal-dual algorithm uses a maximum flow computation to simultaneously augment flow along several shortest paths. Unlike all the other algorithms, the out-of-kilter algorithm permits arc flows to violate their flow bounds. It uses shortest path computations to find flows that satisfy both the flow bounds and the cost- and capacity-based optimality conditions.

The simplex method for solving linear programming problems is perhaps the most powerful algorithm ever devised for solving constrained optimization problems. Since minimum cost flow problems define a special class of linear programs, we might expect the simplex method to be an attractive candidate solution procedure for solving these problems also. Then again, because network flow problems have considerable special structure, we might also ask whether the simplex method could possibly compete with other *combinatorial* methods, such as the many variants of the successive shortest path algorithm, that exploit the underling network structure. The general simplex method, when implemented in a way that does not exploit underlying network structure, is not a competitive solution procedure for solving minimum cost flow problems. Fortunately, however, if we interpret the core concepts of the simplex method appropriately as network operations, then when we apply it to the minimum cost flow problem, we can adapt and streamline the method to exploit the network structure, producing the *network simplex* algorithm, which is very efficient.

Scaling is a powerful idea that has produced algorithmic improvements to many problems in combinatorial optimization. We might view scaling algorithms as follows. We start with the conditions for optimality of the network flow problem we are examining, but instead of enforcing these conditions exactly, we generate an *approximate* solution that is permitted to violate one (or more) of the conditions by an amount Δ. Initially, by choosing Δ quite large, we will be able to easily find a starting solution that satisfies the relaxed optimality conditions. We then reset the parameter Δ to $\Delta/2$ and re-optimize so that the approximate solution now violates the optimality conditions by at most $\Delta/2$. We then repeat the procedure, re-optimizing again until the approximate solution violates the conditions by at most $\Delta/4$, and so forth. This solution strategy is quite flexible and leads to a different algorithm depending upon which of the optimality conditions we relax and how we perform the re-optimizations. In this chapter, we apply a scaling approach to the successive shortest path algorithm to develop a polynomial-time version. We relax two optimality conditions in the Δ-scaling phase: (1) we permit the solution to violate supply-demand constraints by an amount Δ and (2) we permit the residual network to contain negative cost cycles, but only if their capacities are less than Δ. The resulting *capacity-scaling algorithm* reduces the number of shortest path computations from pseudopolynomial to polynomial.

To begin our discussion of the minimum cost flow problem, we first lay out the notations, assumptions and underlying definitions that we use throughout the chapter.

5.1.1 Notation and Assumptions

Let $G = (N, A)$ be a directed network with a *cost* c_{ij} and a *capacity* u_{ij} associated with every arc $(i, j) \in A$. We associate with each node $i \in N$ a number $b(i)$ that indicates its supply or demand depending upon whether $b(i) > 0$ or $b(i) < 0$. The minimum cost flow problem can be defined as follows:

$$\text{Minimize} \sum_{(i,j) \in A} c_{ij} x_{ij} \quad (5.1)$$

subject to

$$\sum_{\{j:(i,j)\in A\}} x_{ij} - \sum_{\{j:(j,i)\in A\}} x_{ji} = b(i), \text{ for all } i \in N, \quad (5.2)$$

$$0 \leq x_{ij} \leq u_{ij}, \text{ for all } (i,j) \in A \quad (5.3)$$

Let C denote the largest magnitude of any arc cost. Further, let U denote the largest magnitude of any supply/demand or finite arc capacities. We assume that the lower bounds l_{ij} on arc flows are all zero. Let n denote the number of nodes and m the number of arcs. We further make the following assumptions:

Assumption 5.1 *All data (cost, supply/demand, and capacity) are integral.*

This assumption is not really restrictive in practice because computers work with rational numbers, which we can convert to integer numbers by multiplying by a suitably large number.

Assumption 5.2 *The network is directed.*

We can always fulfill this assumption by transforming any undirected network into a directed network by replacing each undirected arc by two arcs in opposite directions.

Assumption 5.3 *The supply/demands at the nodes satisfy the condition $\sum_{i \in N} b(i) = 0$, and the minimum cost flow problem has a feasible solution.*

We can determine whether the minimum cost flow problem has a feasible solution by introducing a preliminary step that involves solving a maximum flow problem (see Chapter 4). Introduce a source node s^* and a sink node t^*. For each node i with $b(i) > 0$, add a *source* arc (s^*, i) with capacity $b(i)$, and for each node i with $b(i) < 0$, add a *sink* arc (i, t^*) with capacity $-b(i)$. Now solve a maximum flow problem from s^* to t^*. If the maximum flow saturates all the source arcs, then the minimum cost flow problem is feasible; otherwise, it is infeasible.

Assumption 5.4 *We assume that the network G contains an uncapacitated directed path (i.e., each arc in the path has infinite capacity) between every pair of nodes.*

We impose this condition, if necessary, by adding artificial arcs $(1, j)$ and $(j, 1)$ for each $j \in N$ and assigning a large cost and infinite capacity to each of these arcs. No such arc would appear in an optimal minimum cost solution unless the problem has no feasible solution without artificial arcs.

Assumption 5.5 *All arc costs are nonnegative.*

This assumption imposes no loss of generality since a minimum cost flow problem with negative arc lengths can be transformed to one with nonnegative arc lengths. This transformation, however, works if all arcs have finite capacities. Suppose that some arcs are uncapacitated. We assume that the network contains no directed negative cost cycle of infinite capacity. The presence of such cycles indicates that the optimal value of the minimum cost flow problem is unbounded. In the absence of a negative cycle with infinite capacity, we can make each uncapacitated arc capacitated by setting its capacity equal to B, where B is the sum of all arc capacities and the supplies of all supply nodes.

5.1.2 Similarity Assumption

The assumption that each arithmetic operation takes one step might lead us to underestimate the asymptotic running time of arithmetic operations involving very large numbers since, in practice, a computer must store these numbers in several words of its memory. Therefore, to perform each operation on such numbers, a computer must access a number of words of data and thus take more than a constant number of steps. To avoid such systematic underestimation of running time, in comparing two running times, we sometimes assume that both C (i.e., the largest arc cost) and U (i.e., the largest arc capacity) are polynomially bounded in n (i.e., $C = O(n^k)$ and $U = O(n^k)$ for some constant k). We refer to this assumption as the *similarity assumption*.

5.1.3 Residual Network

Our algorithms rely on the concept of residual networks. The residual network $G(x)$ corresponding to a flow x is defined as follows. We replace each arc $(i, j) \in A$ by two arcs (i, j) and (j, i). The arc (i, j) has cost c_{ij} and residual capacity $r_{ij} = u_{ij} - x_{ij}$, and the arc (j, i) has cost $c_{ji} = -c_{ij}$ and residual capacity $r_{ji} = x_{ij}$. The residual network contains only arcs with positive residual capacity.

5.1.4 Tree, Spanning Tree, and Forest

A *tree* is a connected graph that contains no cycles. A graph that contains no cycles is called a *forest*. Alternatively, a forest is a collection of trees. A tree T is a spanning tree of G if T is a spanning sub-graph of G.

5.2 APPLICATIONS

Minimum cost flow problems arise in almost all industries, including agriculture, communications, defense, education, energy, health care, manufacturing, medicine, retailing, and transportation. Indeed, minimum cost flow problems are pervasive in practice. This section is intended to introduce some applications and to illustrate some of the possible uses of minimum cost flow problems in practice.

5.2.1 Distribution Problems

A large class of network flow problems centers on shipping and distribution applications. Suppose a firm has p plants with known supplies and q warehouses with known demands. It wishes to identify a flow that satisfies the demands at the warehouses with the available supplies at the plants while minimizing overall shipping costs. This problem is a well-known special case of the minimum cost flow problem, known as the *transportation problem*. We next describe in more detail a slight generalization of this model that also incorporates manufacturing costs at the plants.

A car manufacturer has several manufacturing plants and produces several car models at each plant that it then ships to geographically dispersed retail centers throughout the country. Each retail center requests a specific number of cars of each model. The firm must determine the production plan of each model at each plant and a shipping pattern that satisfies the demands of each retail center while minimizing the overall cost of production and transportation.

We describe this formulation through an example. Figure 5.1 illustrates a situation with two manufacturing plants, two retailers, and three car models. This model has four types of nodes: (1) *plant nodes*, representing various plants; (2) *plant/model nodes*, corresponding to each model made at a plant; (3) *retailer/model nodes*, corresponding to the models required by each retailer; and (4) *retailer nodes*, corresponding to each retailer. The network contains three types of arcs: (1) *Production arcs:* These arcs connect a plant node to a plant/model node; the cost of this arc is the cost of producing the model at that plant. We might place lower and upper bounds on these arcs to control for the minimum and maximum production of each particular car model at the plants. (2) *Transportation arcs:* These arcs connect plant/model nodes to retailer/model nodes; the cost of such an arc is the total cost of shipping one car from the manufacturing plant to the retail center. Any such arc might correspond to a complex distribution channel with, for example, three legs: (a) a delivery from a plant (by truck) to a rail system, (b) a delivery from the rail station to another rail station elsewhere in the system, and (c) a delivery from the rail station to a retailer (by a local delivery truck). The transportation arcs might have lower or upper bounds imposed upon their flows to model contractual agreements with shippers or capacities imposed upon any distribution channel. (3) *Demand arcs:* These arcs connect retailer/model nodes to the retailer nodes. These arcs have zero costs and positive lower bounds, which equal the demand of that model at that retail center. Clearly, the production and shipping schedules for the automobile company have a one-to-one correspondence with the feasible flows in this network model. Consequently, a minimum cost flow would yield an optimal production and shipping schedule.

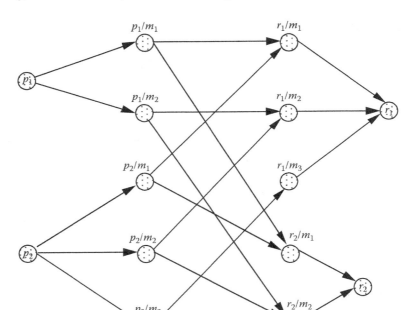

Figure 5.1 Production–distribution model.

5.2.2 Optimal Loading of a Hopping Airplane

A small commuter airline uses a plane with a capacity to carry at most p passengers on a *hopping flight*, as shown in Figure 5.2a. The hopping flight visits the cities 1, 2, 3, ..., n, in a fixed sequence. The plane can pick up passengers at any node and drop them off at any other node. Let b_{ij} denote the number of passengers available at node i who want to go to node j, and let f_{ij} denote the fare per passenger from node i to node j. The airline would like to determine the number of passengers that the plane should carry between the various origins and destinations in order to maximize the total fare per trip while never exceeding the plane capacity.

Figure 5.2b shows a minimum cost flow formulation of this hopping plane flight problem. The network contains data for only those arcs with nonzero costs and with finite capacities: any arc without an associated cost has a zero cost; any arc without an associated capacity has an infinite capacity. Consider, for example, node 1. Three types of passengers are available at node 1, those whose destination is node 2, node 3, or node 4. We represent these three types of passengers by the nodes 1–2, 1–3, and 1–4 with supplies b_{12}, b_{13}, and b_{14}. A passenger available at any such node, say, 1–3, either boards the plane at its origin node by flowing through the arc (1–3, 1) and thus incurring a cost of $-f_{13}$ units, or never boards the plane, which we represent by the flow through the arc (1–3, 3). There is a one-to-one correspondence between feasible flows on this network and feasible solutions to the hopping plane flight problem.

5.2.3 Scheduling with Deferral Costs

In some scheduling applications, jobs do not have any fixed completion times, but instead incur a deferral cost for delaying their completion. Some of these scheduling problems have

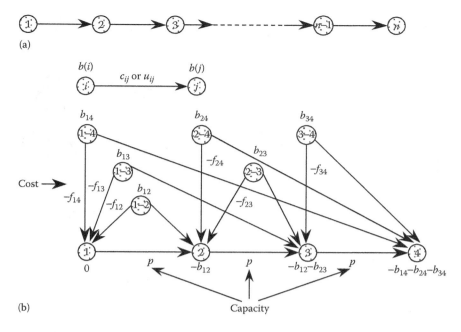

Figure 5.2 Formulating the hopping plane problem as a minimum cost flow problem. (a) Hopping plane problem. (b) Minimum cost formulation.

the following characteristics: one of q identical processors (machines) needs to process each of p jobs. Each job j has a fixed processing time α_j that does not depend upon which machine processes the job or which jobs precede or follow the job. Job j also has a *deferral cost* $c_j(\tau)$, which we assume is a monotonically nondecreasing function of τ, the completion time of the job. Figure 5.3a illustrates one such deferral cost function. We wish to find a schedule for the jobs, with completion times denoted by $\tau_1, \tau_2, \ldots, \tau_p$, that minimizes the total deferral cost $\Sigma_{j=1}^n c_j(\tau_j)$. This scheduling problem is difficult if the jobs have different processing times, but it can be modeled as a minimum cost flow problem for situations with uniform processing times, that is, $\alpha_j = \alpha$ for each $j = 1, \ldots, p$.

Since the deferral costs are monotonically nondecreasing with time, in some optimal schedule the machines will process the jobs one immediately after another, that is, the machines incur no *idle time*. As a consequence, in some optimal schedule the completion time of each job will be $k\alpha$ for some constant k. The first job assigned to every machine will have a completion time of α units, the second job assigned to every machine will have a completion time of 2α units, and so forth. This observation allows us to formulate the scheduling as a minimum cost flow problem in the network shown in Figure 5.3b.

Assume, for simplicity, that $r = p/q$ is an integer. This assumption implies that we will assign exactly r jobs to each machine. There is no loss of generality in imposing this assumption because we can add dummy jobs so that p/q becomes an integer. The network has p job nodes $1, 2, \ldots, p$, each with one unit of supply; it also has r position nodes, $\bar{1}, \bar{2}, \bar{3}, \ldots, \bar{r}$, each with a demand of q units, indicating that the position has the capability to process q jobs. The flow on each arc $(j, \bar{i})$ is 1 or 0, depending upon whether the schedule does or does not assign job j to the ith position on some machine. If we assign job j to the ith position on any machine, its completion time is $i\alpha$, and its deferral cost is $c_j(i\alpha)$. Therefore, arc $(j, \bar{i})$ has a cost of $c_j(i\alpha)$. Feasible schedules correspond, in a one-to-one fashion, with feasible flows in the network and both have the same cost. Consequently, a minimum cost flow will prescribe a schedule with the least possible deferral cost.

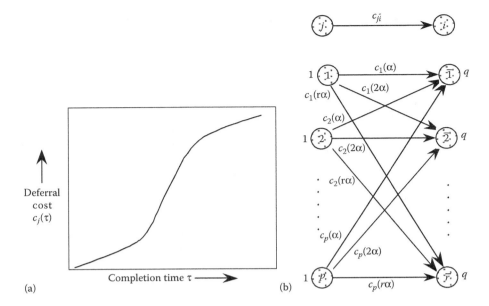

Figure 5.3 Formulating the scheduling problem with deferral costs.

5.3 OPTIMALITY CONDITIONS

Optimality conditions are useful in several respects. First, they give us a simple validation method to check whether a given feasible solution to the problem is indeed optimal. We simply check whether this solution satisfies the optimality conditions. Hence, the optimality conditions provide us with a *certificate* of optimality. A nice feature of the certificate is its ease of use. We need not invoke any complex algorithm to certify that a solution is optimal. The optimality conditions are also valuable for other reasons. They can help suggest or motivate algorithms for solving a problem and provide us with a mechanism for establishing the validity of algorithms for a given problem. To show that an algorithm correctly finds the optimal solution to a problem, we simply need to show that the solution it generates satisfies the optimality conditions.

Rather than launching immediately into a discussion of algorithms for the minimum cost flow problem, we first pause to describe a few different optimality conditions for this problem, which make our subsequent algorithmic descriptions in the following sections more understandable. All the optimality conditions that we state have an intuitive network interpretation. We consider three different (but equivalent) optimality conditions: (1) *negative cycle optimality condition*, (2) *reduced cost optimality condition*, and (3) *complementary slackness optimality condition*.

5.3.1 Negative Cycle Optimality Conditions

Theorem 5.1 (Negative cycle optimality conditions) *A feasible solution x^* is an optimal solution of the minimum cost flow problem if and only if it satisfies the following negative cycle optimality conditions: the residual network $G(x^*)$ contains no negative cost (directed) cycle.*

Proof. Suppose that x is a feasible flow and that residual network $G(x)$ contains a negative cycle. Then x cannot be an optimal flow, since by augmenting positive flow along the cycle we can improve the objective function value. Therefore, if x^* is an optimal flow, then $G(x^*)$ cannot contain a negative cycle. Now suppose that x^* is a feasible flow and that $G(x^*)$ contains no negative cycle. Let x^o be an optimal flow and $x^* \neq x^o$. The augmenting cycle property shows that we can decompose the difference vector $x^o - x^*$ into at most m augmenting cycles with respect to the flow x^* and the sum of the costs of flows on these cycles equals $cx^o - cx^*$. Since the lengths of all the cycles in $G(x^*)$ are nonnegative, $cx^o - cx^* \geq 0$, or $cx^o \geq cx^*$. Moreover, since x^o is an optimal flow, $cx^o \leq cx^*$. Thus, $cx^o = cx^*$, and x^* is also an optimal flow. This argument shows that if $G(x^*)$ contains no negative cycle then x^* must be optimal, and this conclusion completes the proof of the theorem. ∎

5.3.2 Reduced Cost Optimality Conditions

Suppose we associate a real number $\pi(i)$, unrestricted in sign, with each node $i \in N$. We refer to $\pi(i)$ as the *potential* of node i. We will show in Section 5.4 that $\pi(i)$ is the linear programming dual variable corresponding to the mass balance constraint of node i. For a given set of node potentials π, we define the *reduced* cost of an arc (i,j) as $c_{ij}^\pi = c_{ij} - \pi(i) + \pi(j)$. These reduced costs are applicable to both the original and the residual network. We now proceed to state and prove the following properties.

Property 5.1

a. *For any directed path P from node k to node l, $\sum_{(i,j) \in P} c_{ij}^\pi = \sum_{(i,j) \in P} c_{ij} - \pi(k) + \pi(l)$.*

b. *For any directed cycle W, $\sum_{(i,j) \in W} c_{ij}^\pi = \sum_{(i,j) \in W} c_{ij}$.*

Proof. Consider a directed path P from node k to node l.

$$\begin{aligned}
\sum_{(i,j) \in P} c_{ij}^\pi &= \sum_{(i,j) \in P} (c_{ij} - \pi(i) + \pi(j)) \\
&= \sum_{(i,j) \in P} c_{ij} - \sum_{(i,j) \in P} (\pi(i) - \pi(j)) \\
&= \sum_{(i,j) \in P} c_{ij} - \pi(k) + \pi(l)
\end{aligned}$$

All other terms in the summation $\sum_{(i,j) \in P} (\pi(i) - \pi(j))$ cancel out except the first and last terms. Property 5.1b can also be proved in a similar manner by considering a cycle W. ∎

Notice that this property implies that using the reduced costs c_{ij}^π instead of the actual costs c_{ij} does not change the shortest path between any pair of nodes k and l, since the potentials increase the length of every path by a constant amount of $\pi(l) - \pi(k)$. This property also implies that if W is a negative cycle with respect to c_{ij} as arc costs, then it is also a negative cycle with respect to c_{ij}^π. We can now provide an alternative form of the negative cycle optimality conditions, stated in terms of the reduced costs of the arcs.

Theorem 5.2 (Reduced cost optimality conditions) *A feasible solution x^* is an optimal solution of the minimum cost flow problem if and only if some set of node potentials π satisfy the following reduced cost optimality conditions:*

$$c_{ij}^\pi \geq 0 \text{ for every arc } (i,j) \in G(x^*). \tag{5.4}$$

Proof. We shall prove this result using Theorem 5.1. To show that the negative cycle optimality conditions are equivalent to the reduced cost optimality conditions, suppose solution x^* satisfies the latter conditions. Therefore, $\sum_{(i,j) \in W} c_{ij}^\pi \geq 0$ for every directed cycle W in $G(x^*)$. Consequently, by Property 5.1b, $\sum_{(i,j) \in W} c_{ij}^\pi = \sum_{(i,j) \in W} c_{ij} \geq 0$, so $G(x^*)$ contains no negative cycle.

To show the converse, assume that for the solution x^*, $G(x^*)$ contains no negative cycle. Let $d(\cdot)$ denote the shortest path distances from node 1 to all other nodes in $G(x^*)$ (refer to Chapter 4 for formal description of distance labels). If the network contains no negative cycle, then the distance labels $d(\cdot)$ are well defined and satisfy the conditions $d(j) \leq d(i) + c_{ij}$ for all (i,j) in $G(x^*)$. We can restate these inequalities as $c_{ij} + d(i) - d(j) \geq 0$, or $c_{ij}^\pi \geq 0$ if we define $\pi = -d$. Consequently, the solution x^* satisfies the reduced cost optimality conditions. ∎

In the preceding theorem, we characterized an optimal flow x as a flow that satisfied the conditions $c_{ij}^\pi \geq 0$ for all (i,j) in $G(x)$ for some set of node potentials π. In the same fashion, we could define *optimal node potentials* as a set of node potentials π that satisfy the conditions $c_{ij}^\pi \geq 0$ for all (i,j) in $G(x)$ for some feasible flow x.

The reduced cost optimality conditions have a convenient economic interpretation. Suppose we interpret c_{ij} as the cost of transporting one unit of a commodity from node i to node j through the arc (i,j), and we interpret $\mu(i) \equiv -\pi(i)$ as the cost of obtaining a unit of this commodity at node i. Then, $c_{ij} + \mu(i)$ is the cost of the commodity at node j if we obtain it at node i and transport it to node j. The reduced cost optimality condition, $c_{ij} - \pi(i) + \pi(j) \geq 0$, or equivalently $\mu(j) \leq c_{ij} + \mu(i)$, states that the cost of obtaining the commodity at node j is no more than the cost of the commodity if we obtain it at node i and incur the transportation cost in sending it from node i to j. The cost at node j might be smaller than $c_{ij} + \mu(i)$ because there might be a more cost-effective way to transport the commodity to node j via other nodes.

5.3.3 Complementary Slackness Optimality Conditions

Both Theorems 5.1 and 5.2 provide means for establishing optimality of solutions to the minimum cost flow problem by formulating conditions imposed upon the residual network; we shall now restate these conditions in terms of the original network.

Theorem 5.3 (Complementary slackness optimality conditions) *A feasible solution x^* is an optimal solution of the minimum cost flow problem if and only if for some set of node potentials π, the reduced costs and flow values satisfy the following complementary slackness optimality conditions for every arc $(i,j) \in A$:*

$$\text{if } c_{ij}^\pi > 0, \text{ then } x_{ij}^* = 0; \tag{5.5}$$
$$\text{if } 0 < x_{ij}^* < u_{ij}, \text{ then } c_{ij}^\pi = 0; \text{ and} \tag{5.6}$$
$$\text{if } c_{ij}^\pi < 0, \text{ then } x_{ij}^* = u_{ij}. \tag{5.7}$$

Proof. We show that the reduced cost optimality conditions are equivalent to (5.5 through 5.7). To establish this result, we first prove that if the node potentials π and the flow vector x satisfy the reduced cost optimality conditions, then they must satisfy (5.5 through 5.7). Consider the three possibilities for any arc $(i,j) \in A$.

Case 1. If $c_{ij}^\pi > 0$, then the residual network cannot contain the arc (j,i) because $c_{ji}^\pi = -c_{ij}^\pi < 0$ for that arc, contradicting (5.4). Therefore, $x_{ij}^* = 0$.

Case 2. If $0 < x_{ij}^* < u_{ij}$, then the residual network contains both the arcs (i,j) and (j,i). The reduced cost optimality conditions imply that $c_{ij}^\pi \geq 0$ and $c_{ji}^\pi \geq 0$. But since $c_{ji}^\pi = -c_{ij}^\pi$, these inequalities imply that $c_{ij}^\pi = c_{ji}^\pi = 0$.

Case 3. If $c_{ij}^\pi < 0$, then the residual network cannot contain the arc (i,j), because $c_{ij}^\pi < 0$ for that arc, contradicting (5.4). Therefore, $x_{ij}^* = u_{ij}$.

We have thus shown that if the node potentials π and the flow vector x satisfy the reduced cost optimality conditions, then they also satisfy the complementary slackness optimality conditions. Similarly we can also prove the converse result, that is, if the pair (x, π) satisfies the complementary slackness optimality conditions, then it also satisfies the reduced cost optimality conditions by doing a straightforward case-by-case analysis. ∎

5.4 MINIMUM COST FLOW DUALITY

For every linear programming problem, which we subsequently refer to as a *primal* problem, we can associate another intimately related linear programming problem, called its *dual*. For a minimization problem, the objective function value of *any* feasible solution of the dual is less than or equal to the objective function of any feasible solution of the primal. Furthermore, the maximum objective function value of the dual equals the minimum objective function of the primal. Duality theory is fundamental to the understanding of the theory of linear programming. In this section, we state and prove these duality theory results for the minimum cost flow problem, which is a special case of the linear programming problem.

While forming the dual of a (primal) linear programming problem, we associate a *dual variable* with every constraint of the primal except for the nonnegativity restriction on arc flows. For the minimum cost flow problem stated in (5.1), we associate the variable $\pi(i)$ with the mass balance constraint of node i and the variable α_{ij} with the capacity constraint of arc (i, j). In terms of these variables, the *dual minimum cost flow problem* can be stated as follows:

$$\text{Maximize } w(\pi, \alpha) = \sum_{i \in N} b(i)\pi(i) - \sum_{(i,j) \in A} u_{ij}\alpha_{ij} \quad (5.8)$$

subject to

$$\pi(i) - \pi(j) - \alpha_{ij} \leq c_{ij} \text{ for all } (i, j) \in A, \quad (5.9)$$

$$\alpha_{ij} \geq 0 \text{ for all } (i, j) \in A \text{ and } \pi(j) \text{ unrestricted for all } j \in N. \quad (5.10)$$

Our first duality result for the general minimum cost flow problem is known as the *weak duality theorem*.

Theorem 5.4 (Weak duality theorem) *Let $z(x)$ denote the objective function value of some feasible solution x of the minimum cost flow problem and let $w(\pi, \alpha)$ denote the objective function value of some feasible solution (π, α) of its dual. Then $w(\pi, \alpha) \leq z(x)$.*

Proof. We multiply both the sides of (5.9) by x_{ij} and sum these weighted inequalities for all $(i, j) \in A$, obtaining

$$\sum_{(i,j) \in A} (\pi(i) - \pi(j))x_{ij} - \sum_{(i,j) \in A} \alpha_{ij}x_{ij} \leq \sum_{(i,j) \in A} c_{ij}x_{ij}. \quad (5.11)$$

Notice that the first term on the left-hand side of (5.11) can be re-written as

$$\sum_{i \in N} \pi(i) \left(\sum_{\{j:(i,j) \in A\}} x_{ij} - \sum_{\{j:(j,i) \in A\}} x_{ji} \right) = \sum_{i \in N} \pi(i)b(i). \quad (5.12)$$

Next notice that replacing x_{ij} in the second term on the left-hand side of (5.11) by u_{ij} preserves the inequality because $x_{ij} \leq u_{ij}$ and $\alpha_{ij} \geq 0$. Consequently,

$$\sum_{i \in N} b(i)\pi(i) - \sum_{(i,j) \in A} \alpha_{ij}u_{ij} \leq \sum_{(i,j) \in A} c_{ij}x_{ij}. \quad (5.13)$$

Now notice that the left-hand side of (5.13) is the dual objective $w(\pi, \alpha)$ and the right-hand side is the primal objective, and so we have established the theorem. ∎

The weak duality theorem implies that the objective function value of *any* dual feasible solution is a lower bound on the objective function value of *any* primal feasible solution. One consequence of this result is immediate: if some dual solution (π, α) and a primal solution x have the same objective function value, then (π, α) must be an optimal solution of the dual problem and x must be an optimal solution of the primal problem. The *strong duality theorem*, proved next, shows that such a pair of solutions always exists.

We first eliminate the dual variables α_{ij}'s from the dual formation (5.8) using some properties of the optimal solution. We can rewrite the constraint (5.9) as

$$\alpha_{ij} \geq -c_{ij}^{\pi} \tag{5.14}$$

The coefficient associated with the variable α_{ij} in the dual objective (5.8) is $-u_{ij}$, and we wish to maximize the objective function value. Consequently, in any optimal solution we would assign the smallest possible value to α_{ij}. This observation, in view of (5.10) and (5.14), implies that

$$\alpha_{ij} = \max\{0, -c_{ij}^{\pi}\} \tag{5.15}$$

The condition (5.15) allows us to omit α_{ij} from the dual formulation. Substituting (5.15) in (5.8) yields

$$\text{Maximize } w(\pi) = \sum_{i \in N} b(i)\pi(i) - \sum_{(i,j) \in A} \max\{0, -c_{ij}^{\pi}\} u_{ij}. \tag{5.16}$$

We are now in the position to prove the strong duality theorem.

Theorem 5.5 (Strong duality theorem) *For any choice of problem data, the minimum cost flow problem always has a solution x^* and the dual minimum cost flow problem a solution π satisfying the property that $z(x^*) = w(\pi)$.*

Proof. We prove this theorem using the complementary slackness optimality conditions (5.5 through 5.7). Let x^* be an optimal solution of the minimum cost flow problem. Theorem 5.3 implies that x^* together with some vector π of node potentials satisfies the complementary slackness optimality conditions. We claim that this solution satisfies the condition

$$-c_{ij}^{\pi} x_{ij} = \max\{0, -c_{ij}^{\pi}\} u_{ij} \text{ for every arc } (i,j) \in A. \tag{5.17}$$

To establish this result, consider the following three cases: (1) $c_{ij}^{\pi} > 0$, (2) $c_{ij}^{\pi} = 0$, and (3) $c_{ij}^{\pi} < 0$. The complementary slackness conditions (5.5 through 5.7) imply that in the first two cases, both the left-hand side and right-hand side of (5.17) are zero, and in the third case both sides equal $c_{ij}^{\pi} u_{ij}$.

Substituting (5.17) in dual objective (5.16) yields

$$w(\pi) = \sum_{i \in N} b(i)\pi(i) + \sum_{(i,j) \in A} c_{ij}^{\pi} x_{ij}^* = \sum_{(i,j) \in A} c_{ij} x_{ij}^* = z(x^*). \tag{5.18}$$

The second equality follows from the fact that $\sum_{i \in N} b(i)\pi(i)$ can be re-written as $\sum_{(i,j) \in A} (\pi(i) - \pi(j)) x_{ij}$. Then, (5.18) proves the strong duality theorem. ∎

In Theorem 5.5, we showed that the complementary slackness optimality conditions imply strong duality. We next prove the converse result, namely that strong duality implies the complementary slackness optimality conditions.

Theorem 5.6 *If x is a feasible flow and π is an (arbitrary) vector satisfying the property that $z(x) = w(\pi)$, then the pair (x, π) satisfies the complementary slackness optimality conditions.*

Proof. Since $z(x) = w(\pi)$,

$$\sum_{(i,j) \in A} c_{ij} x_{ij} = \sum_{i \in N} b(i) \pi(i) - \sum_{(i,j) \in A} \max\{0, -c_{ij}^\pi\} u_{ij}. \tag{5.19}$$

Substituting $\sum_{i \in N} b(i) \pi(i) = \sum_{(i,j) \in A} (\pi(i) - \pi(j)) x_{ij}$ in (5.19), we get

$$\sum_{(i,j) \in A} \max\{0, -c_{ij}^\pi\} u_{ij} = \sum_{(i,j) \in A} -c_{ij}^\pi x_{ij} \tag{5.20}$$

Now observe that both the sides have m terms, and each term in the left-hand side is nonnegative and its value is an upper bound on the corresponding term in the right-hand side (because $\max\{0, -c_{ij}^\pi\} \geq -c_{ij}^\pi$ and $u_{ij} \geq x_{ij}$). Therefore, the two sides can be equal only when

$$\max\{0, -c_{ij}^\pi\} u_{ij} = -c_{ij}^\pi x_{ij} \text{ for every arc } (i,j) \in A. \tag{5.21}$$

We now consider three cases.

1. $c_{ij}^\pi > 0$. In this case, the left-hand side of (5.21) is zero, and the right-hand side can be zero only if $x_{ij} = 0$. This conclusion establishes (5.5).

2. $0 < x_{ij} < u_{ij}$. In this case, $c_{ij}^\pi = 0$; otherwise, the right-hand side of (5.21) is negative. This conclusion establishes (5.6).

3. $c_{ij}^\pi < 0$. In this case, the left-hand side of (5.21) is $-c_{ij}^\pi u_{ij}$ and, therefore, $x_{ij} = u_{ij}$. This conclusion establishes (5.7).

These results complete the proof of the theorem. ∎

The following result is an easy consequence of Theorems 5.5 and 5.6.

Property 5.2 *If x^* is an optimal solution of the minimum cost flow problem and π is an optimal solution of the dual minimum cost flow problem, then the pair (x^*, π) satisfies the complementary slackness optimality conditions.*

Proof. Theorem 5.5 implies that $z(x^*) = w(\pi)$ and Theorem 5.6 implies that the pair (x^*, π) satisfies complementary slackness optimality conditions. ∎

Minimum cost flow duality has several important implications. Since almost all algorithms for solving the primal problem also generate optimal node potentials $\pi(i)$ and the variables α_{ij}, solving the primal problem almost always solves both the primal and dual problems. Similarly, solving the dual problem typically solves the primal problem as well. Most algorithms for solving network flow problems explicitly or implicitly use properties of dual variables and of the dual linear program. In particular, the dual problem provides us with a certificate that if we can find a feasible dual solution that has the same objective function value as a given primal solution, then we know from the strong duality theorem that the primal solution must be optimal, *without* making additional calculations and without considering other potentially optimal primal solutions.

5.5 CYCLE-CANCELING ALGORITHM

Operations researchers, computer scientists, electrical engineers, and many others have extensively studied the minimum cost flow problem and proposed a number of different algorithms to solve this problem. Notable among these are the cycle-canceling, successive shortest path, primal-dual, out-of-kilter, and scaling-based algorithms. In this and the following sections,

```
algorithm cycle canceling;
begin
    establish a feasible flow x in the network;
    while G(x) contains a negative cycle do
    begin
        use some algorithm to identify a negative cycle W;
        δ := min{r_ij : (i,j) ∈ W};
        augment δ units of flow in the cycle W and update G(x);
    end
end
```

Figure 5.4 Cycle-canceling algorithm.

we discuss some of the important algorithms for the minimum cost flow problem and point out relationships among them. We first consider the cycle-canceling algorithm.

The negative cycle optimality conditions suggest one simple algorithmic approach for solving the minimum cost flow problem, which we call the *cycle-canceling algorithm*. This algorithm maintains a feasible solution and at every iteration attempts to improve its objective function value. The algorithm first establishes a feasible flow x in the network. Then it iteratively finds negative cost-directed cycles in the residual network and augments flows on these cycles. The algorithm terminates when the residual network contains no negative cost-directed cycle. Theorem 5.1 implies that when the algorithm terminates, it has found a minimum cost flow. Figure 5.4 specifies this generic version of the cycle-canceling algorithm.

A feasible flow in the network can be found by solving a maximum flow problem. One algorithm for identifying a negative cost cycle is the label correcting algorithm for the shortest path problem, which requires $O(nm)$ time to identify a negative cost cycle. A by-product of the cycle-canceling algorithm is the following important result.

Theorem 5.7 (Integrality property) *If all arc capacities and supply/demands of nodes are integer, then the minimum cost flow problem always has an integer minimum cost flow.*

Proof. We show this result by performing induction on the number of iterations. The algorithm first establishes a feasible flow in the network by solving a maximum flow problem (see Chapter 4). We assume that the maximum flow algorithm finds an integer solution since all arc capacities in the network are integer, and the initial residual capacities are also integer. The flow augmented by the cycle-canceling algorithm in any iteration equals the minimum residual capacity in the cycle canceled, which, by the inductive hypothesis, is integer. Hence, the modified residual capacities in the next iteration will again be integer. The result follows. ∎

Let us now calculate the number of iterations performed by the algorithm. For the minimum cost flow problem, mCU is an upper bound on the initial flow cost (since $c_{ij} \leq C$ and $x_{ij} \leq U$ for all $(i,j) \in A$), and $-mCU$ is a lower bound on the optimal flow cost (since $c_{ij} \geq -C$ and $x_{ij} \leq U$ for all $(i,j) \in A$). Each iteration of the cycle-canceling algorithm changes the objective function value by an amount $\left(\sum_{(i,j)\in W} c_{ij}\right)\delta$, which is strictly negative. Since we are assuming that all the data of the problem is integral, the algorithm terminates within $O(mCU)$ iterations and runs in $O(nm^2CU)$ time.

The generic version of the cycle-canceling algorithm does not specify the order for selecting negative cycles from the network. Different rules for selecting negative cycles produce different versions of the algorithm, each with different worst-case and theoretical behavior.

The network simplex algorithm, which is widely considered to be one of the fastest algorithms for the minimum cost flow problem in practice, can be considered a particular version of the cycle-canceling algorithm. The network simplex algorithm maintains information (a spanning tree solution and node potentials) that enables it to identify a negative cost cycle in $O(m)$ time. However, due to *degeneracy*, the algorithm cannot necessarily send a positive amount of flow along this cycle, and hence the most general implementation of the network simplex algorithm does not run in polynomial-time. The following two versions of the cycle-canceling algorithm are, however, polynomial-time implementations.

5.5.1 Augmenting Flow in a Negative Cycle with Maximum Improvement

Let x be any feasible flow and let x^* be an optimal flow. The improvement in the objective function value due to an augmentation along a cycle W is $-(\sum_{(i,j) \in W} c_{ij})(\min\{r_{ij} : (i,j) \in W\})$. It can be shown that x^* equals x plus the flow on at most m augmenting cycles with respect to x, and improvements in cost due to flow augmentations on these augmenting cycles sum to $cx - cx^*$. Consequently, at least one of these augmenting cycles with respect to x must decrease the objective function value by at least $(cx - cx^*)/m$. Therefore, if the algorithm always augments flow along a cycle giving the maximum possible improvement, then this implies that the method would obtain an optimal flow within $O(m \log(mCU))$ iterations. Finding a maximum improvement cycle is difficult (i.e., it is an *NP*-complete problem), but a modest variation of this approach yields a polynomial time algorithm for the minimum cost flow problem. We provide a reference of this algorithm in the bibliographic notes.

5.5.2 Augmenting Flow along a Negative Cycle with Minimum Mean Cost

We define the *mean cost* of a cycle as its cost divided by the number of arcs it contains. A *minimum mean cycle* is a cycle whose mean cost is as small as possible. It is possible to identify a minimum mean cycle in $O(nm)$ or $O(\sqrt{n}m \log(nC))$ time. Researchers have shown that if the cycle-canceling algorithm always augments flow along a minimum mean cycle, then it performs $O(\min\{nm \log(nC), nm^2 \log n\})$ iterations.

5.6 SUCCESSIVE SHORTEST PATH ALGORITHM

The cycle-canceling algorithm maintains feasibility of the solution at every step and attempts to achieve optimality. In contrast, the successive shortest path algorithm maintains optimality of the solution (as defined in Theorem 5.2) at every step and strives to attain feasibility. It maintains a solution x that satisfies the nonnegativity and capacity constraints, but it violates the mass balance constraints of the nodes. At each step, the algorithm selects a node s with excess supply (i.e., supply not yet sent to some demand node) and a node t with unfulfilled demand and sends flow from s to t along a shortest path in the residual network. The algorithm terminates when the current solution satisfies all the mass balance constraints.

In order to describe this algorithm, we first introduce the concept of *pseudoflows*. A *pseudoflow* is a function $x : A \to R^+$ satisfying only the capacity and nonnegativity constraints; it need not satisfy the mass balance constraints. For any pseudoflow x, we define the *imbalance* of node i as

$$e(i) = b(i) + \sum_{\{j:(j,i)\in A\}} x_{ji} - \sum_{\{j:(i,j)\in A\}} x_{ij} \text{ for all } i \in N.$$

If $e(i) > 0$ for some node i, then we refer to $e(i)$ as the *excess* of node i; if $e(i) < 0$, then we call $-e(i)$ the node's *deficit*. We refer to a node i with $e(i) = 0$ as *balanced*. Let E and D denote the sets of excess and deficit nodes in the network. Notice that $\sum_{i \in N} e(i) = \sum_{i \in N} b(i) = 0$, and hence $\sum_{i \in E} e(i) = -\sum_{i \in D} e(i)$. Consequently, if the network contains an excess node, then it must also contain a deficit node. The residual network corresponding to a pseudoflow is defined in the same way that we define the residual network for a flow.

Lemma 5.1 *Suppose a pseudoflow (or a flow) x satisfies the reduced cost optimality conditions with respect to some node potentials π. Let the vector d represent the shortest path distances from some node k to all other nodes in the residual network $G(x)$ with c_{ij}^π as the length of an arc (i, j). Then the following properties are valid:*

a. *The pseudoflow x also satisfies the reduced cost optimality conditions with respect to the node potentials $\pi' = \pi - d$.*

b. *The reduced costs $c_{ij}^{\pi'}$ are zero for all arcs (i, j) in a shortest path from node s to every other node.*

Proof. Since x satisfies the reduced cost optimality conditions with respect to π, we have $c_{ij}^\pi \geq 0$ for every arc (i, j) in $G(x)$. Furthermore, since the vector d represents shortest path distances with arc lengths c_{ij}^π, it satisfies the shortest path optimality conditions; that is

$$d(j) \leq d(i) + c_{ij}^\pi \text{ for all } (i,j) \in G(x) \tag{5.22}$$

Substituting $c_{ij}^\pi = c_{ij} - \pi(i) + \pi(j)$ in (5.22), we obtain $d(j) \leq d(i) + c_{ij} - \pi(i) + \pi(j)$. Alternatively, $c_{ij} - (\pi(i) - d(i)) + (\pi(j) - d(j)) \geq 0$, or $c_{ij}^{\pi'} \geq 0$. This conclusion establishes part (a) of the lemma.

Consider next a shortest path from node s to some node l. For each arc (i, j) in this path, $d(j) = d(i) + c_{ij}^\pi$. Substituting $c_{ij}^\pi = c_{ij} - \pi(i) + \pi(j)$ in this equation, we obtain $c_{ij}^{\pi'} = 0$. This conclusion establishes part (b) of the lemma. ∎

The following result is an immediate corollary of the preceding lemma.

Lemma 5.2 *Suppose a pseudoflow (or a flow) x satisfies the reduced cost optimality conditions and we obtain x' from x by sending flow along a shortest path from node s to some other node k; then x' also satisfies the reduced cost optimality conditions.*

Proof. Define the potentials π and π' as in Lemma 5.1. The proof of Lemma 5.1 implies that for every arc (i, j) in the shortest path P from node s to the node k, $c_{ij}^{\pi'} \geq 0$. Augmenting flow on any such arc might add its reversal (j, i) to the residual network. But since $c_{ij}^{\pi'} = 0$ for each arc $(i,j) \in P$, $c_{ij}^{\pi'} = 0$ and the arc (j, i) also satisfies the reduced cost optimality conditions. The lemma follows. ∎

We are now in a position to describe the successive shortest path algorithm. The node potentials play a very important role in this algorithm. In addition to using them to prove the correctness of the algorithm, we use them to maintain nonnegative arc lengths so that we can solve the shortest path problem more efficiently. Figure 5.5 gives a formal statement of the successive shortest path algorithm.

We now justify the successive shortest path algorithm. To initialize the algorithm, we set $x = 0$, which is a feasible pseudoflow. For the zero pseudoflow x, $G(x) = G$. Note that this solution together with $\pi = 0$ satisfies the reduced cost optimality conditions because $c_{ij}^\pi = c_{ij} \geq 0$ for every arc (i, j) in the residual network $G(x)$ (recall Assumption 5.5, which states that all arc costs are nonnegative). Observe that as long as any node has a nonzero imbalance, both E and D must be nonempty since the total sum of excesses equals the total sum of deficits. Thus, until all nodes are balanced, the algorithm always succeeds in identifying an excess node k and a deficit node l. Assumption 5.4 implies that the residual network contains

```
algorithm successive shortest path;
begin
    x := 0 and π := 0;
    e(i) := b(i) for all i ∈ N;
    initialize the sets E and D;
    while E ≠ ϕ do
    begin
        select a node k ∈ E and a node l ∈ D;
        determine shortest path distances d(j) from node k to all other nodes in G(x)
            with respect to the reduced costs $c_{ij}^\pi$;
        let P denote a shortest path from node s to node t;
        update π := π − d;
        δ := min[e(s), −e(t), min{$r_{ij}$ : (i, j) ∈ P}];
        augment δ units of flow along the path P;
        update x, G(x), E, D, and the reduced costs;
    end
end
```

Figure 5.5 Successive shortest path algorithm.

a directed path from node k to every other node, including node l. Therefore, the shortest path distances $d(\cdot)$ are well defined. Each iteration of the algorithm solves a shortest path problem with nonnegative arc lengths and strictly decreases the excess of some node (and, also, the deficit of some other node). Consequently, if U is an upper bound on the largest supply of any node, then the algorithm would terminate in at most nU iterations. If $S(n, m, C)$ denotes the time taken to solve a shortest path problem with nonnegative arc lengths, then the overall complexity of this algorithm is $O(nU\ S(n, m, C))$. Since the arc lengths are nonnegative, the shortest path problem at each iteration can be solved using Dijkstra's algorithm. Currently, the best known polynomial-time bound for the Dijkstra's algorithm is $O(m+n\ log\ n)$, and the best weakly polynomial bound is $O(\min\{m\ log\ log\ C, m+n\sqrt{\log C}\})$. The successive shortest path algorithm takes pseudopolynomial time since it is polynomial in n, m, and the largest supply U. This algorithm is, however, polynomial-time for the assignment problem, which is a special case of the minimum cost flow problem, for which $U = 1$. In Section 5.9, we describe a polynomial-time implementation of the successive shortest path algorithm in conjunction with scaling.

5.7 PRIMAL-DUAL AND OUT-OF-KILTER ALGORITHMS

The primal-dual algorithm for the minimum cost flow problem is similar to the successive shortest path algorithm in the sense that it also maintains a pseudoflow that satisfies the reduced cost optimality conditions and gradually converts it into a flow by augmenting flows along shortest paths. In contrast, however, instead of sending flow along one shortest path at a time, it solves a maximum flow problem that sends flow along all shortest paths.

The primal-dual algorithm generally transforms the minimum cost flow problem into a problem with a single excess node and a single deficit node. We transform the problem into this form by introducing a *source* node s and a *sink* node t. For each node i with $b(i) > 0$, we add a zero cost arc (s, i) with capacity $b(i)$, and for each node i with $b(i) < 0$, we add a zero cost arc (i, t) with capacity $-b(i)$. Finally, we set $b(s) = \sum_{i \in N: b(i) > 0} b(i)$, $b(t) = -b(s)$, and

$b(i) = 0$ for all $i \in N$. It is easy to see that a minimum cost flow in the transformed network gives a minimum cost flow in the original network.

At every iteration, the primal-dual algorithm solves a shortest path problem from the source to update the node potentials (i.e., as in the successive shortest path algorithm, $\pi(j)$ becomes $\pi(j) - d(j)$) and then solves a maximum flow problem to send the maximum possible flow from source to sink using only the arcs with zero reduced cost. The algorithm guarantees that the excess of some node strictly decreases at each iteration, and also assures that the node potential of the sink strictly decreases. The latter observation follows from the fact that after we have solved the maximum flow problem, the network contains no path from source to the sink in the residual network consisting entirely of arcs with zero reduced costs; consequently, in the next iteration $d(t) \geq 1$. These observations give a bound of $\min\{nU, nC\}$ on the number of iterations since the magnitude of each node potential is bounded by nC. This bound is better than that of the successive shortest path algorithm, but, of course, the algorithm incurs the additional expense of solving a maximum flow problem at each iteration. If $S(n, m, C)$ and $M(n, m, U)$ denote the solution times of shortest path and the maximum flow algorithms, then the primal-dual algorithm has an overall complexity of $O(\min\{nU\,S(n, m, C), nC\,M(n, m, U)\})$.

The successive shortest path and primal-dual algorithms maintain a solution that satisfies the reduced cost optimality conditions and the flow bound constraints but that violates the mass balance constraints. These algorithms iteratively modify arc flows and node potentials so that the flow at each step comes closer to satisfying the mass balance constraints. However, we could just as well have developed other solution strategies by violating other constraints at intermediate steps. The out-of-kilter algorithm satisfies only the mass balance constraints, and so intermediate solutions might violate both the optimality conditions and the flow bound restrictions. The basic idea is to drive the flow on an arc (i, j) to u_{ij} if $c_{ij}^\pi < 0$, drive the flow to zero if $c_{ij}^\pi > 0$, and to allow any flow between 0 and u_{ij} if $c_{ij}^\pi = 0$. The name *out-of-kilter algorithm* reflects the fact that arcs in the network either satisfy the reduced cost optimality conditions (are *in-kilter*) or do not (are *out-of-kilter*). We define the *kilter number* k_{ij} of each arc (i, j) in A as the magnitude of the change in x_{ij} required to make the arc an in-kilter arc while keeping c_{ij}^π fixed. Therefore, in accordance with the reduced cost optimality condition, if $c_{ij}^\pi > 0$, then $k_{ij} = |x_{ij}|$, and if $c_{ij}^\pi < 0$, then $k_{ij} = |u_{ij} - x_{ij}|$. If $c_{ij}^\pi = 0$ and $x_{ij} > u_{ij}$, then $k_{ij} = x_{ij} - u_{ij}$. If $c_{ij}^\pi = 0$ and $x_{ij} < 0$, then $k_{ij} = -x_{ij}$. The kilter number of any in-kilter arc is zero. The sum $\sum_{(i,j) \in A} k_{ij}$ of all kilter numbers provides us with a measure of how far the current solution is from optimality; the smaller the value of K, the closer the current solution is to being an optimal solution.

At each iteration, the out-of-kilter algorithm reduces the kilter number of at least one arc; it terminates when all arcs are in-kilter. Suppose the kilter number of arc (i, j) would decrease by increasing flow on the arc. The algorithm would obtain a shortest path P from node j to node i in the residual network and augment at least one unit of flow in the cycle $P \cup (i, j)$. The proof of correctness of this algorithm is similar to but more detailed than that of the successive shortest path algorithm.

5.8 NETWORK SIMPLEX ALGORITHM

The simplex method for solving linear programming problems is perhaps the most powerful algorithm ever devised for solving constrained optimization problems. Indeed, many members of the academic community view the simplex method as not only one of the principal computational engines of applied mathematics, computer science, and operations research, but also as one of the landmark contributions to computational mathematics of this century.

The algorithm has achieved this lofty status because of the pervasiveness of its applications throughout many problem domains, because of its extraordinary efficiency, and because it permits us to not only to solve problems numerically, but also to gain considerable practical and theoretical insight through the use of sensitivity analysis and duality theory.

In the introduction, we mentioned that minimum cost flow problems have considerable special structure, and applying the general simplex method, which does not exploit the problem structure on these problems gives algorithms that are not competitive when compared with the combinatorial methods described in the preceding sections. However, if we interpret the core concepts of the simplex method appropriately as network operations, then when we apply it to the minimum cost flow problem, we can adapt and streamline the method to exploit the network structure, producing an algorithm that is very efficient. Our purpose in this section is to develop this network-based implementation of the simplex method and show how to apply it to the minimum cost flow problem.

The central concept underlying the network simplex algorithm is the notion of spanning tree solutions, which are solutions that we obtain by fixing the flow on every arc not in a spanning tree either at value zero or the arc's flow capacity. As we show in this section, we can then solve uniquely for the flow on all the arcs in the spanning tree. We also show that the minimum cost flow problem always has at least one optimal spanning tree solution and that it is possible to find an optimal spanning tree solution by *moving* from one such solution to another, at each step introducing one new nontree arc into the spanning tree in place of one tree arc. This method is known as the *network simplex algorithm* because spanning trees correspond to the so-called basic feasible solutions of linear programming, and the movement from one spanning tree solution to another corresponds to a so-called pivot operation of the general simplex method.

In this section, we describe the network simplex algorithm in detail. We first define the concept of *cycle free* and *spanning tree* solutions and describe a data structure to store and manipulate the spanning tree. We then show how to compute arc flows and node potentials for any spanning tree solution. We next discuss how to perform various simplex operations such as obtaining a starting solution, and selection of entering arcs, leaving arcs, and pivots. Finally, we show how to guarantee the finiteness of the network simplex algorithm.

5.8.1 Cycle Free and Spanning Tree Solutions

Much of our development in the previous sections has relied upon a simple but powerful algorithmic idea: to generate an improving sequence of solutions to the minimum cost flow problem, we iteratively augment flows along a series of negative cycles and shortest paths. As one of these variants, the network simplex algorithm uses a particular strategy for generating negative cycles. In this section, as a prelude to our discussion of the method, we introduce some basic background material. We begin by examining two important concepts known as *cycle free solutions* and *spanning tree solutions*.

For any feasible solution x, we say that an arc (i, j) is a *free arc* if $0 < x_{ij} < u_{ij}$ and is a *restricted arc* if $x_{ij} = 0$ or $x_{ij} = u_{ij}$. Note that we can both increase and decrease flow on a free arc while honoring the bounds on arc flows. However, in a restricted arc (i, j) at its lower bound (i.e., $x_{ij} = 0$) we can only increase the flow. Similarly, in a restricted arc (i, j) at its upper bound (i.e., $x_{ij} = u_{ij}$) we can only decrease the flow. We refer to a solution x as a *cycle free solution* if the network contains no cycle composed only of free arcs. Note that in a cycle free solution, we can augment flow on any augmenting cycle in only a single direction since some arc in any cycle will restrict us from either increasing or decreasing that arc's flow. We also refer to a feasible solution x and an associated spanning tree of the network as a *spanning tree solution* if every nontree arc is a restricted arc. Notice that in a spanning tree solution, the tree arcs can be free or restricted.

In this section, we establish a fundamental result of network flows: minimum cost flow problems always have optimal cycle free and spanning tree solutions. The network simplex algorithm will exploit this result by restricting its search for an optimal solution to only spanning tree solutions.

Theorem 5.8 (Cycle free property) *If the objective function of a minimum cost flow problem is bounded from below over the feasible region, then the problem always has an optimal cycle free solution.*

Proof. We prove this result by showing that any feasible solution can be converted to a corresponding cycle free solution with lesser cost. Let the feasible solution contain at least one cycle having only free arcs. Consider one such cycle. We now have the following possibilities: (1) cost of the cycle is negative, (2) cost of the cycle is positive, and (3) cost of the cycle is zero. In case (1), we can obtain a better solution by sending maximum possible flow in the direction of the cycle; in case (2), we can obtain a better solution by sending maximum possible flow against the cycle; and in case (3), solution cost is unchanged if we send flow in either direction along the cycle. Since the cycle contains only free arcs, whenever we send maximum possible flow on the cycle, flow on at least one arc will hit its lower or upper bound and become restricted. This implies that the cycle will no longer contain only free arcs. The same procedure can be iteratively applied to obtain a cycle free solution. ∎

It is easy to convert a cycle free solution into a spanning tree solution. The free arcs in a cycle free solution define a forest (i.e., a collection of node-disjoint trees). If this forest is a spanning tree, then the cycle free solution is already a spanning tree solution. However, if this forest is not a spanning tree, since we assume that the underlying network is connected, we can add some restricted arcs and produce a spanning tree. So, we have established the following fundamental result:

Theorem 5.9 (Spanning tree property) *If the objective function of a minimum cost flow problem is bounded from below over the feasible region, then the problem always has an optimal spanning tree solution.* ∎

A spanning tree solution partitions the arc set A into three subsets: (1) T, the arcs in the spanning tree; (2) L, the nontree arcs whose flow is restricted to value zero; (3) U, the nontree arcs whose flow is restricted in value to the arcs' flow capacities. We refer to (T, L, U) as a *spanning tree structure*.

Just as we can associate a spanning tree structure with a spanning tree solution, we can also obtain a unique spanning tree solution corresponding to a given spanning tree structure (T, L, U). To do so, we set $x_{ij} = 0$ for all arcs $(i, j) \in L$, $x_{ij} = u_{ij}$ for all arcs $(i, j) \in U$ and then solve the mass balance equations to determine the flow values for arcs in T. We say a spanning tree structure is *feasible* if its associated spanning tree solution satisfies all of the arcs' flow bounds. In the special case in which every tree arc in a spanning tree solution is a free arc, we say that the spanning tree is *nondegenerate*; otherwise, we refer to it as a *degenerate* spanning tree. We refer to a spanning tree structure as *optimal* if its associated spanning tree solution is an optimal solution of the minimum cost flow problem. The following theorem states a sufficient condition for a spanning tree structure to be an optimal structure.

Theorem 5.10 (Minimum cost flow optimality conditions) *A spanning tree structure (T, L, U) is an optimal spanning tree structure of the minimum cost flow problem* *if it is feasible and for some choice of node potentials* π, *the arc reduced costs* c_{ij}^π *satisfy the following conditions:*

$$\text{a.} \quad c_{ij}^\pi = 0, \text{ for all } (i,j) \in T, \tag{5.23}$$

$$\text{b.} \quad c_{ij}^\pi \geq 0, \text{ for all } (i,j) \in L, \text{ and} \tag{5.24}$$

$$\text{c.} \quad c_{ij}^\pi \leq 0, \text{ for all } (i,j) \in U. \tag{5.25}$$

Proof. Let x^* be the solution associated with the spanning tree structure (T, L, U). We know that some set of node potentials π, together with the spanning tree structure (T, L, U), satisfies (5.23 through 5.25).

We need to show that x^* is an optimal solution of the minimum cost flow problem. Minimizing $\sum_{(i,j) \in A} c_{ij} x_{ij}$ is equivalent to minimizing $\sum_{(i,j) \in A} c_{ij}^\pi x_{ij}$. The conditions stated in (5.23 through 5.25) imply that for the given node potential π, minimizing $\sum_{(i,j) \in A} c_{ij}^\pi x_{ij}$ is equivalent to minimizing the following expression:

$$\text{Minimize} \sum_{(i,j) \in L} c_{ij}^\pi x_{ij} - \sum_{(i,j) \in U} |c_{ij}| x_{ij} \tag{5.26}$$

The definition of the solution x^* implies that for any arbitrary solution x, $x_{ij} \geq x_{ij}^*$ for all $(i,j) \in L$, and $x_{ij} \leq x_{ij}^*$ for all $(i,j) \in U$. The expression (5.26) implies that the objective function value of the solution x will be greater than or equal to that of x^*. ∎

The network simplex algorithm maintains a feasible spanning tree structure and moves from one spanning tree structure to another until it finds an optimal structure. At each iteration, the algorithm adds one arc to the spanning tree in place of one of its current arcs. The entering arc is a nontree arc violating its optimality condition. The algorithm (1) adds this arc to the spanning tree, creating a negative cycle (which might have zero residual capacity); (2) sends the maximum possible flow in this cycle until the flow on at least one arc in the cycle reaches its lower or upper bound; and (3) drops an arc whose flow has reached its lower or upper bound, giving us a new spanning tree structure. Because of its relationship to the primal simplex algorithm for the linear programming problem, this operation of moving from one spanning tree structure to another is known as a *pivot operation*, and the two spanning tree structures obtained in consecutive iterations are called *adjacent spanning tree structures*.

5.8.2 The Network Simplex Algorithm

The network simplex algorithm maintains a feasible spanning tree structure at each iteration and successively transforms it into an improved spanning tree structure until it becomes optimal. The following algorithmic description specifies the essential steps of the method (Figure 5.6).

In the following discussion, we describe in greater detail how the network simplex algorithm uses tree indices to efficiently perform the various steps in the algorithm. This discussion highlights the value of the tree indices in designing an efficient implementation of the algorithm.

5.8.2.1 *Obtaining an Initial Spanning Tree Structure*

Our connectedness assumption, Assumption 5.4, provides one way of obtaining an initial spanning tree structure. We have assumed that for every node $j \in N - \{1\}$, the network contains arcs $(1, j)$ and $(j, 1)$, possibly with sufficiently large costs and capacities. We construct the initial tree T as follows. We examine each node j, other than node 1, one by one. If $b(j) \geq 0$, we include arc $(j, 1)$ in T with a flow value of $b(j)$. If $b(j) < 0$, then we

algorithm *network simplex*;
begin
    determine an initial feasible tree structure (T, L, U);
    let x be the flow and π be the node potentials associated with this tree structure;
    while some non-tree arc violates the optimality conditions **do**
    begin
        select an entering arc (k, l) violating its optimality condition;
        add arc (k, l) to the tree and determine the leaving arc (p, q);
        perform a tree update and update the solutions x and π;
    end
end

Figure 5.6 Network simplex algorithm.

include arc $(1, j)$ in T with a flow value of $-b(j)$. The set L consists of the remaining arcs, and the set U is empty.

5.8.2.2 Maintaining a Spanning Tree Structure

Since the network simplex algorithm generates a sequence of spanning tree solutions, to implement the algorithm effectively, we need to be able to represent spanning trees conveniently in a computer so that the algorithm can perform its basic operations efficiently and can update the representation quickly when it changes the spanning tree. Over the years, researchers have suggested several ways for maintaining and manipulating a spanning tree structure. In this section, we describe one of the more popular representations.

We consider the tree as *hanging* from a specially designated node, called the *root*. Throughout this chapter, we assume that node 1 is the root node. Figure 5.7 gives an example of a tree. We associate three indices with each node i in the tree: a predecessor index, $pred(i)$; a depth index, $depth(i)$; and a thread index, $thread(i)$.

Predecessor index. Each node i has a unique path connecting it to the root. The index $pred(i)$ stores the first node in that path (other than node i). For example, the path 9–6–5–2–1 connects node 9 to the root; therefore, $pred(9) = 6$. By convention, we set the predecessor

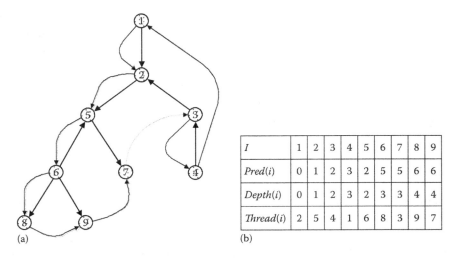

Figure 5.7 Example of tree indices: (a) rooted tree; (b) corresponding tree indices.

node of the root node, node 1, to zero. Figure 5.7 specifies these indices for the other nodes. Observe that by iteratively using the predecessor indices, we can enumerate the path from any node to the root.

A node j is called a successor of node i if $\text{pred}(j) = i$. For example, node 5 has two successors: nodes 6 and 7. A *leaf node* is a node with no successors. In Figure 5.7, nodes 4, 7, 8, and 9 are leaf nodes. The *descendants* of a node i are the node i itself, its successors, successors of its successors, and so forth. For example, in Figure 5.7, the node set {5, 6, 7, 8, 9} contains the descendants of node 5.

Depth index. We observed earlier that each node i has a unique path connecting it to the root. The index $\text{depth}(i)$ stores the number of arcs in that path. For example, since the path 9–6–5–2–1 connects node 9 to the root, depth $(9) = 4$. Figure 5.7 gives depth indices for the rest of the nodes in the network.

Thread index. The thread indices define a traversal of a tree; that is, a sequence of nodes that walks or threads its way through the nodes of a tree, starting at the root node, visiting nodes in a *top-to-bottom* order, and finally returning to the root. We can find thread indices by performing a depth-first search and setting the thread of a node to be the node in the depth-first search encountered just after the node itself. For our example, the depth-first traversal would read 1–2–5–6–8–9–7–3–4–1, and so $\text{thread}(1) = 2$, $\text{thread}(2) = 5$, $\text{thread}(5) = 6$, and so on.

The thread indices provide a particularly convenient means for visiting (or finding) all descendants of a node i. We simply follow the thread starting at that node and record the nodes visited until the depth of the visited node becomes at least as large as that of node i. For example, starting at node 5, we visit nodes 6, 8, 9, and 7 in order, which are the descendants of node 5, and then visit node 3. Since the depth of node 3 equals that of node 5, we know that we have left the *descendant tree* lying below node 5. We shall see later that finding the descendant tree of a node efficiently is an important step in developing an efficient implementation of the network simplex algorithm.

The network simplex method has two basic steps: (1) determining the node potentials of a given spanning tree structure and (2) computing the arc flows for the spanning tree structure. We now show how to perform these two steps efficiently using the tree indices.

5.8.3 Computing Node Potentials and Flows

We first consider the problem of computing the node potentials π for a given spanning tree structure (T, L, U). Note that we can set the value of one node potential arbitrarily because adding a constant k to each node potential does not alter the reduced cost of any arc; that is, for any constant k, $c_{ij}^\pi = c_{ij} - \pi(i) + \pi(j) = c_{ij} - [\pi(i) + k] + [\pi(j) + k]$. So for convenience, we henceforth assume that $\pi(1) = 0$. We compute the remaining node potentials using the fact that the reduced cost of every spanning tree arc is zero, that is,

$$c_{ij}^\pi = c_{ij} - \pi(i) + \pi(j) = 0 \text{ for every arc } (i,j) \in T. \tag{5.27}$$

The basic idea in the procedure is to start at node 1 and fan out along the tree arcs using the thread indices to compute other node potentials. The traversal of nodes using the thread indices ensures that whenever the procedure visits a node k, it has already evaluated the potential of its predecessor, and so it can compute $\pi(k)$ using (5.27). Figure 5.8 gives a formal statement of the procedure *compute-potentials*.

```
procedure compute-potentials;
begin
    π(1) := 0;
    j := thread(1);
    while j ≠ 1 do
    begin
        i := pred(j);
        if (i, j) ∈ A then π(j) := π(i) − c_{ij};
        if (j, i) ∈ A then π(j) := π(i) − c_{ji};
        j := thread(j);
    end
end
```

Figure 5.8 Procedure compute-potentials.

The procedure compute-potentials requires $O(1)$ time per iteration and performs $(n-1)$ iterations to evaluate the node potential of each node. Therefore, the procedure runs in $O(n)$ time.

One important consequence of the procedure compute-potentials is that the minimum cost flow problem always has integer optimal node potentials whenever all the arc costs are integer. To see this result, recall from Theorem 5.9 that the minimum cost flow problem always has an optimal spanning tree solution. The potentials associated with this tree constitute optimal node potentials, which we can determine using the procedure compute-potentials. The description of the procedure compute-potentials implies that if all arc costs are integer, then node potentials are integer as well (because the procedure performs only additions and subtractions). We refer to this integrality property of optimal node potentials as the *dual integrality property* since node potentials are the dual linear programming variables associated with the minimum cost flow problem.

Theorem 5.11 (Dual integrality property) *If all arc costs are integer, then the minimum cost flow problem always has optimal integer node potentials.* ∎

We next consider the problem of determining the flows on the tree arcs of a given spanning tree structure. A similar procedure will permit us to determine the flow on tree arcs. We proceed, however, in the reverse order: Start at a leaf node and move in toward the root using the predecessor indices while computing flows on arcs encountered along the way. The following procedure accomplishes this task (Figure 5.9).

The running time of the procedure *compute-flows* is easy to determine. Clearly, the initialization of flows and excesses requires $O(m)$ time. If we set aside the time to select leaf nodes of T, then each iteration requires $O(1)$ time, resulting in a total of $O(n)$ time. One way of identifying leaf nodes in T is to select nodes in the reverse order of the thread indices. We identify the reverse thread traversal of the nodes by examining the nodes in the order dictated by the thread indices, putting all the nodes into a stack in the order of their appearance, and then taking them out from the top of the stack one at a time. Therefore, the reverse thread traversal examines each node only after it has examined all of the node's descendants. We have thus the procedure *compute-flows* runs in $O(m)$ time.

The description of the procedure compute-flows implies that if the capacities of all the arcs and the supplies/demands of all the nodes are integer, then arc flows are integer as well (because the procedure performs only additions and subtractions). We state this result again because of its importance in network flow theory.

```
procedure compute-flows;
begin
    e(i) := b(i), for all i ∈ N;
    for each (i, j) ∈ L do set x_{ij} := 0;
    for each (i, j) ∈ U do
        set x_{ij} := u_{ij}, subtract u_{ij} from e(i) and add u_{ij} to e(j);
    T' := T;
    while T' ≠ {1} do
    begin
        select a leaf node j (other than node 1) in the sub-tree T';
        i := pred(j);
        if (i, j) ∈ T' then x_{ij} := -e(j);
        else x_{ji} := e(j);
        add e(j) to e(i);
        delete node j and the arc incident to it from T';
    end
end
```

Figure 5.9 Procedure compute-flows.

Theorem 5.12 (Primal integrality property) *If capacities of all the arcs and supplies/demands of all the nodes are integer, then the minimum cost flow problem always has an integer optimal flow.* ∎

5.8.4 Entering Arc

Two types of arcs are *eligible* to enter the tree: (1) any arc $(i, j) \in L$ with $c_{ij}^\pi < 0$ or (2) any arc $(i, j) \in U$ with $c_{ij}^\pi > 0$. For any eligible arc (i, j), we refer to $|c_{ij}^\pi|$ as its *violation*. An implementation that selects an arc that violates the optimality condition by the most might require the fewest number of iterations in practice, but it must examine each nontree arc in each iteration, which is very time consuming. On the other hand, examining the list cyclically and selecting the first eligible nontree arc may quickly find the entering arc, but it may require a relatively large number of iterations due to poor entering arc choice. One of the most successful implementation strategies uses a *candidate list* approach that strikes an effective compromise between these two strategies. This approach also allows sufficient opportunity for fine-tuning of special problem classes.

The algorithm maintains a candidate list of arcs that violate the optimality conditions, selecting arcs in a two-phase procedure consisting of *major* iterations and *minor* iterations. In a major iteration, we construct the candidate list. We examine arcs emanating from nodes, one node at a time, adding the arcs that violate the optimality conditions to the candidate list until we have either examined all nodes or the list has reached its maximum size. The next major iteration begins at the node where the previous major iteration ended. In other words, the algorithm examines nodes cyclically as it adds arcs emanating from them to the candidate list.

Once the algorithm has formed the candidate list in a major iteration, it performs minor iterations, scanning all the candidate arcs and selecting a nontree arc from this list that violates the optimality conditions by the most to enter the basis. As we scan the arcs, we update the candidate list by removing those arcs that no longer violate the optimality conditions. Once the list becomes empty or we have reached a specified limit on the number of minor iterations to be performed at each major iteration, we rebuild the list with another major iteration.

5.8.5 Leaving Arc

Suppose we select arc (k, l) as the entering arc. The addition of this arc to the tree T creates exactly one cycle W, which we refer to as the *pivot cycle*. The pivot cycle consists of the unique path in the tree T from node k to node l, together with arc (k, l). We define the orientation of the cycle W as the same as that of (k, l) if $(k, l) \in L$ and opposite of the orientation of (k, l) if $(k, l) \in U$. Let $\overline{W}$ and $\underline{W}$ denote the sets of *forward arcs* (i.e., those along the orientation of W) and *backward arcs* (those opposite to the orientation of W) in the pivot cycle. Sending additional flow around the pivot cycle W in the direction of its orientation strictly decreases the cost of the current solution at the per unit rate of $|c_{kl}^\pi|$. We augment the flow as much as possible until one of the arcs in the pivot cycle reaches its lower or upper bound. Notice that augmenting flow along W increases the flow on forward arcs and decreases flow on backward arcs. Consequently, the maximum flow change δ_{ij} on an arc $(i, j) \in W$ that satisfies the flow bound constraints is

$$\delta_{ij} = \left\{ \begin{array}{ll} u_{ij} - x_{ij} & \text{if } (i,j) \in \overline{W} \\ x_{ij} & \text{if } (i,j) \in \underline{W} \end{array} \right\}$$

To maintain feasibility, we can augment $\delta = \min\{\delta_{ij} : (i, j) \in W\}$ units of flow along W. We refer to any arc $(i, j) \in W$ that defines δ, that is, for which $\delta = \delta_{ij}$, as a *blocking arc*. We then augment δ units of flow and select an arc (p, q) with $\delta_{pq} = \delta$ as the leaving arc, breaking ties arbitrarily. We say that a pivot iteration is a *nondegenerate iteration* if $\delta > 0$ and is a *degenerate iteration* if $\delta = 0$. A degenerate iteration occurs only if T is a degenerate spanning tree. Observe that if two arcs tie in while determining the value of δ, then the next spanning tree will be degenerate.

The crucial step in identifying the leaving arc is to identify the pivot cycle. If $P(i)$ denotes the unique path in the tree from any node i to the root node, then this cycle consists of the arcs $\{(k,l)\} \cup P(k) \cup P(l) - (P(k) \cap P(l))$. In other words, W consists of the arc (k, l) and the disjoint portions of $P(k)$ and $P(l)$. Using the predecessor indices alone permits us to identify the cycle W as follows. First, we designate all the nodes in the network as unmarked. We then start at node k and, using the predecessor indices, trace the path from this node to the root and mark all the nodes in this path. Next we start at node l and trace the predecessor indices until we encounter a marked node, say w. The node w is the first common ancestor of nodes k and l; we refer to it as the *apex* of cycle W. The cycle W contains the portions of the paths $P(k)$ and $P(l)$ up to node w, together with the arc (k, l). This method identifies the cycle W in $O(n)$ time and therefore is efficient. However, it has the drawback of backtracking along those arcs of $P(k)$ that are not in W. If the pivot cycle lies *deep in the tree*, far from its root, then tracing the nodes back to the root will be inefficient. Ideally, we would like to identify the cycle W in time proportional to $|W|$. The simultaneous use of depth and predecessor indices, as indicated in Figure 5.10, permits us to achieve this goal.

This method scans the arcs in the pivot cycle W twice. During the first scan, we identify the apex of the cycle and also identify the maximum possible flow that can be augmented along W. In the second scan, we augment the flow. The entire flow change operation requires $O(n)$ time in the worst-case, but typically it examines only a small subset of nodes (and arcs).

5.8.6 Updating the Tree

When the network simplex algorithm has determined a leaving arc (p, q) for a given entering arc (k, l), it updates the tree structure. If the leaving arc is the same as the entering arc,

```
procedure identify-cycle;
begin
    i := k and j := l;
    while i ≠ j do
    begin
        if depth(i) > depth(j) then i := pred(i)
        else if depth(j) > depth(i) then j := pred(j)
        else i := pred(i) and j := pred(j);
    end
    w := i;
end
```

Figure 5.10 Procedure for identifying the pivot cycle.

which would happen when $\delta = \delta_{kl} = u_{kl}$, the tree does not change. In this instance, the arc (k, l) merely moves from the set $\boldsymbol{L}$ to the set $\boldsymbol{U}$, or vice versa. If the leaving arc differs from the entering arc, then the algorithm must perform more extensive changes. In this instance, the arc (p, q) becomes a nontree arc at its lower or upper bound depending upon whether (in the updated flow) $x_{pq} = 0$ or $x_{pq} = u_{pq}$. Adding arc (k, l) to the current spanning tree and deleting arc (p, q) creates a new spanning tree.

For the new spanning tree, the node potentials also change; we can update them as follows. The deletion of the arc (p, q) from the previous tree partitions the set of nodes into two sub-trees, one, T_1, containing the root node, and the other, T_2, not containing the root node. Note that the sub-tree T_2 hangs from node p or node q. The arc (k, l) has one endpoint in T_1 and the other in T_2. As is easy to verify, the conditions $\pi(1) = 0$ and $c_{ij} - \pi(i) + \pi(j) = 0$ for all arcs in the new tree imply that the potentials of nodes in the sub-tree T_1 remain unchanged, and the potentials of nodes in the sub-tree T_2 change by a constant amount. If $k \in T_1$ and $l \in T_2$, then all the node potentials in T_2 increase by $-c_{kl}^\pi$; if $l \in T_1$ and $k \in T_2$, they increase by the amount c_{kl}^π. Using the thread and depth indices, the method described in Figure 5.11 updates the node potentials quickly.

The final step in the updating of the tree is to re-compute the various tree indices. This step is rather involved and we refer the reader to the references given in the reference notes for the details. We do point out, however, that it is possible to update the tree indices in

```
procedure update-potentials;
begin
    if q ∈ T₂ then y := q else y := p
    if k ∈ T₁ then change := -c_{kl}^π else change := c_{kl}^π;
    π(y) := π(y) + change;
    z := thread(y);
    while depth(z) > depth(y) do
    begin
        π(z) := π(z) + change;
        z := thread(z);
    end
end
```

Figure 5.11 Updating node potentials in a pivot operation.

$O(n)$ time. In fact, the time required to update the tree indices is $O(|W| + \min\{|T_1|, |T_2|\})$, which is typically much less than n.

5.8.7 Termination

The network simplex algorithm, as just described, moves from one feasible spanning tree structure to another until it obtains a spanning tree structure that satisfies the optimality condition (5.23 through 5.27). If each pivot operation in the algorithm is nondegenerate, then it is easy to show that the algorithm terminates finitely. Recall that $|c_{kl}^\pi|$ is the net decrease in the cost per unit flow sent around the pivot cycle W. After a nondegenerate pivot (for which $\delta > 0$), the cost of the new spanning tree structure is $\delta|c_{kl}^\pi|$ units less than the cost of the previous spanning tree structure. Since any network has a finite number of spanning tree structures and every spanning tree structure has a unique associated cost, the network simplex algorithm will encounter any spanning tree structure at most once and, hence, will terminate finitely. Degenerate pivots, however, pose a theoretical difficulty: the algorithm might not terminate finitely unless we perform pivots carefully. We next discuss a special implementation, called the *strongly feasible spanning tree* implementation, which guarantees finite convergence of the network simplex algorithm even for problems that are degenerate.

5.8.8 Strongly Feasible Spanning Trees

The network simplex algorithm does not necessarily terminate in a finite number of iterations unless we impose some additional restriction on the choice of the entering and leaving arcs. Very small network examples show that a poor choice leads to *cycling*, that is, an infinite repetitive sequence of degenerate pivots. Degeneracy in network problems is not only a theoretical issue, but also a practical one. Computational studies have shown that as many as 90% of the pivot operations in commonplace networks can be degenerate. As we show next, by maintaining a special type of spanning tree called a *strongly feasible spanning tree*, the network simplex algorithm terminates finitely; moreover, it runs faster in practice as well.

Let $(\boldsymbol{T}, \boldsymbol{L}, \boldsymbol{U})$ be a spanning tree structure for a minimum cost flow problem with integral data. As before, we conceive of a spanning tree as a tree hanging from the root node. The tree arcs are either *upward pointing* (toward the root) or are *downward pointing* (away from the root). We now state two alternate definitions of a strongly feasible spanning tree.

1. A spanning tree $\boldsymbol{T}$ is *strongly feasible* if every tree arc with zero flow is upward pointing and every tree arc whose flow equals its capacity is downward pointing.

2. A spanning tree $\boldsymbol{T}$ is *strongly feasible* if we can send a positive amount of flow from any node to the root along the tree path without violating any flow bound.

If a spanning tree $\boldsymbol{T}$ is strongly feasible, then we also say that the spanning tree structure $(\boldsymbol{T}, \boldsymbol{L}, \boldsymbol{U})$ is strongly feasible.

To implement the network simplex algorithm so that it always maintains a strongly feasible spanning tree, we must first find an initial strongly feasible spanning tree. The method described earlier in this section for constructing the initial spanning tree structure always gives such a spanning tree. Note that a nondegenerate spanning tree is always strongly feasible; a degenerate spanning tree might or might not be strongly feasible. The network simplex algorithm creates a degenerate spanning tree from a nondegenerate spanning tree whenever two or more arcs are qualified as leaving arcs and we drop only one of these. Therefore, the algorithm needs to select the leaving arc carefully so that the next spanning tree is strongly feasible.

Suppose that we have a strongly feasible spanning tree and, during a pivot operation, arc (k, l) enters the spanning tree. We first consider the case when (k, l) is a nontree arc at its lower bound. Suppose W is the pivot cycle formed by adding arc (k, l) to the spanning tree and that node w is the apex of the cycle W; that is, w is the first common ancestor of nodes k and l. We define the orientation of the cycle W as compatible with that of arc (k, l). After augmenting δ units of flow along the pivot cycle, the algorithm identifies the *blocking arcs*, that is, those arcs (i, j) in the cycle that satisfy $\delta_{ij} = \delta$. If the blocking arc is unique, then we select it to leave the spanning tree. If the cycle contains more than one blocking arc, then the next spanning tree will be degenerate, that is, some tree arcs will be at their lower or upper bounds. In this case, the algorithm selects the leaving arc in accordance with the following rule.

5.8.8.1 Leaving Arc Rule

Select the leaving arc as the last blocking arc encountered in traversing the pivot cycle W along its orientation starting at the apex w.

To illustrate the leaving arc rule, we consider a numerical example. Figure 5.12 shows a strongly feasible spanning tree for this example. Let $(9, 10)$ be the entering arc. The pivot cycle is 10–8–6–4–2–3–5–7–9–10 and the apex is node 2. This pivot is degenerate because arcs $(2, 3)$ and $(7, 5)$ block any additional flow in the pivot cycle. Traversing the pivot cycle starting at node 2, we encounter arc $(7, 5)$ later than arc $(2, 3)$; so we select arc $(7, 5)$ as the leaving arc.

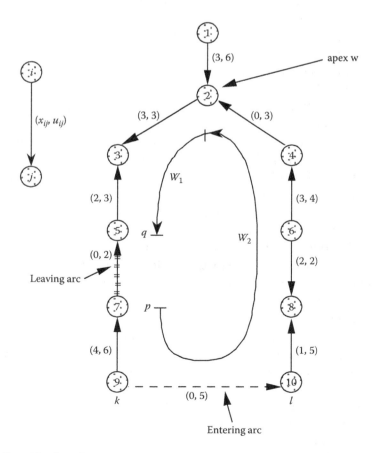

Figure 5.12 Selecting the leaving arc.

We show that the leaving arc rule guarantees that in the next spanning tree every node in the cycle W can send a positive amount of flow to the root node. Let (p, q) be the arc selected by the leaving arc rule. Let W_1 be the segment of the cycle W between the apex w and arc (p, q) when we traverse the cycle along its orientation. Let $W_2 = W - W_1 - \{(p, q)\}$. Define the orientation of segments W_1 and W_2 as compatible with the orientation of W. See Figure 5.12 for an illustration of the segments W_1 and W_2. We use the following property about the nodes in the segment W_2.

Property 5.3 *Each node in the segment W_2 can send a positive amount of flow to the root in the next spanning tree.* ∎

This observation follows from the fact that arc (p, q) is the last blocking arc in W; consequently, no arc in W_2 is blocking and every node in this segment can send a positive amount of flow to the root via node w along the orientation of W_2. Note that if the leaving arc does not satisfy the leaving arc rule, then not all nodes in the segment W_2 can send a positive amount of flow to the root; therefore, the next spanning tree will not be strongly feasible.

We next focus on the nodes contained in the segment W_1.

Property 5.4 *Each node in the segment W_1 can send a positive amount of flow to the root in the next spanning tree.* ∎

We prove this observation by considering two cases. If the previous pivot was a *nondegenerate* pivot, then the pivot augmented a positive amount of flow δ along the arcs in W_1; consequently, after the augmentation, every node in the segment W_1 can send a positive amount of flow back to the root opposite to the orientation of W_1 via the apex node w (each node can send at least δ units to the apex and then at least some of this flow to the root since the previous spanning tree was strongly feasible). If the previous pivot was a *degenerate* pivot, then W_1 must be contained in the segment of W between node w and node k because the property of strong feasibility implies that every node on the path from node l to node w can send a positive amount of flow to the root before the pivot and, thus, no arc on this path can be a blocking arc in a degenerate pivot. Now observe that before the pivot, every node in W_1 could send a positive amount of flow to the root, and, therefore, since the pivot does not change flow values, every node in W_1 must be able to send a positive amount of flow to the root after the pivot as well. This conclusion completes the proof that in the next spanning tree every node in the cycle W can send a positive amount of flow to the root node.

We next show that in the next spanning tree, nodes not belonging to the cycle W can also send a positive amount of flow to the root. In the previous spanning tree (before the augmentation), every node j could send a positive amount of flow to the root and if the tree path from node j does not pass through the cycle W, then the same path is available to carry a positive amount of flow in the next spanning tree. If the tree path from node j does pass through the cycle W, then the segment of this tree path to the cycle W is available to carry a positive amount of flow in the next spanning tree and once a positive amount of flow reaches the cycle W, then, as shown earlier, we can send it (or some of it) to the root node. This conclusion completes the proof that the next spanning tree is strongly feasible.

We now establish the finiteness of the network simplex algorithm. Since we have previously shown that each nondegenerate pivot strictly decreases the objective function value, the number of nondegenerate pivots is finite. The algorithm can, however, also perform degenerate pivots. We will show that the number of successive degenerate pivots between any two nondegenerate pivots is finitely bounded. Suppose arc (k, l) enters the spanning tree at its lower bound and in doing so it defines a degenerate pivot. In this case, the leaving arc belongs to the tree path from node k to the apex w. Now observe from the section on updating node potentials that node k lies in the sub-tree T_2 and the potentials of all nodes in T_2 change by an amount $-c_{kl}^p$. Since $c_{kl}^p < 0$, this degenerate pivot strictly decreases the sum of all node potentials (which, by our prior assumption, is integral). Since no node potential can fall below $-nC$, the number of successive degenerate pivots is finite.

So far we have assumed that the entering arcs are always at their lower bounds. If the entering arc (k, l) is at its upper bound, then we define the orientation of the cycle W as opposite to the orientation of arc (k, l). The criteria for selecting the leaving arc remains unchanged—the leaving arc is the last blocking arc encountered in traversing W along its orientation starting at the apex w. In this case, node l is contained in the sub-tree T_2 and, thus, after the pivot, the potentials of all the nodes in T_2 decrease by the amount $c_{kl}^\pi > 0$; consequently, the pivot again decreases the sum of the node potentials.

5.9 CAPACITY-SCALING ALGORITHM

Scaling techniques are among the most effective algorithmic strategies for designing polynomial time algorithms for the minimum cost flow problem. In this section, we describe an algorithm based on the capacity scaling technique applied to the pseudopolynomial successive shortest path algorithm described in Section 5.6. The resulting capacity-scaling algorithm is a polynomial time algorithm.

The successive shortest path algorithm is one of the fundamental algorithms for solving the minimum cost flow problem. An inherent drawback of this algorithm is that its augmentations might carry relatively small amounts of flow, resulting in a fairly large number of augmentations in the worst-case. By incorporating a scaling technique, the capacity-scaling algorithm described in this section guarantees that each augmentation carries *sufficiently large* flow and thereby reduces the number of augmentations substantially. This method permits us to improve the worst-case algorithmic performance from $O(nU \cdot S(n, m, C))$ to $O(m \log U \cdot S(n, m, C))$. (Recall that U is an upper bound on the largest supply/demand and the largest capacity in the network and $S(n, m, C)$ is the time required to solve a shortest path problem with n nodes, m arcs, and nonnegative arc costs whose values are no more than C). This algorithm not only improves upon the algorithmic performance of the successive shortest path algorithm, but also illustrates how small changes in an algorithm can produce significant algorithmic improvements (at least in the worst case).

The capacity-scaling algorithm applies to the general capacitated minimum cost flow problem. It uses a pseudoflow x and imbalances $e(i)$ as defined in Section 5.6. The algorithm maintains a pseudoflow satisfying the reduced cost optimality conditions and gradually converts this pseudoflow into a flow by identifying shortest paths from nodes with excesses to nodes with deficits and augmenting flows along these paths. It performs a number of scaling phases for different values of a parameter Δ. We refer to a scaling phase with a specific value of Δ as the Δ-*scaling phase*. Initially, $\Delta = 2^{\lfloor \log U \rfloor}$. The algorithm ensures that in the Δ-scaling phase each augmentation carries exactly Δ units of flow. When it is not possible to do so because no node has an excess of at least Δ or no node has a deficit of at least Δ, the algorithm reduces the value of Δ by a factor of 2 and repeats the process. Eventually, $\Delta = 1$ and at the end of this scaling phase, the solution becomes a flow. This flow must be an optimal flow because it satisfies the reduced cost optimality conditions.

For a given value of Δ, we define two sets $S(\Delta)$ and $T(\Delta)$ as follows: (1) $S(\Delta) = \{i : e(i) \geq \Delta\}$, and $T(\Delta) = \{i : e(i) \leq -\Delta\}$. In the Δ-scaling phase, each augmentation must start at a node in $S(\Delta)$ and end at a node in $T(\Delta)$. Moreover, the augmentation must take place on a path along which every arc has residual capacity of at least Δ. Therefore, we introduce another definition: the Δ-*residual network* $G(x, \Delta)$ is the sub-graph of $G(x)$ consisting of those arcs whose residual capacity is at least Δ. In the Δ-scaling phase, the algorithm augments flow from a node in $S(\Delta)$ to a node in $T(\Delta)$ along a shortest path in

```
algorithm capacity scaling;
begin
    x := 0, π := 0;
    Δ := 2^{⌊log U⌋};
    while Δ ≥ 1
    begin {Δ-scaling phase}
        for every arc (i, j) in the residual network G(x) do
        if r_{ij} ≥ Δ and c_{ij}^π < 0 then send r_{ij} units of flow along (i, j), update x and
            the imbalances e(.);
        S(Δ) := {i ∈ N : e(i) ≥ Δ};
        T(Δ) := {1 ∈ N : e(i) ≤ −Δ};
        while S(Δ) ≠ φ and T(Δ) ≠ φ do
        begin
            select a node k ∈ S(Δ) and a node l ∈ T(Δ);
            determine shortest path distances d(.) from node k to all other nodes in
                the Δ-residual network G(x, Δ) with respect to the reduced costs c_{ij}^π;
            let P denote a shortest path from node k to node l in G(x, Δ);
            update π := π − d;
            augment Δ units of flow along the path P;
            update x, S(Δ), T(Δ) and G(x, Δ);
        end
        Δ := Δ/2;
    end
end
```

Figure 5.13 Capacity-scaling algorithm.

$G(x, \Delta)$. The algorithm satisfies the property that every arc in $G(x, \Delta)$ satisfies the reduced cost optimality condition; those arcs in $G(x)$ but not in $G(x, \Delta)$ might violate this optimality condition. Figure 5.13 presents an algorithmic description of the capacity-scaling algorithm.

To establish the correctness of the capacity-scaling algorithm, observe that the 2Δ-scaling phase ends when $S(2\Delta) = \phi$ or $T(2\Delta) = \phi$. At that point, either $e(i) < 2\Delta$ for all $i \in N$ or $e(i) > -2\Delta$ for all $i \in N$. These conditions imply that the sum of the excesses (whose magnitude equals the sum of deficits) is bounded by $2n\Delta$. At the beginning of the Δ-scaling phase, the algorithm first checks whether all the arcs (i, j) in the Δ-residual network satisfy the reduced cost optimality condition $c_{ij}^\pi \geq 0$. The arcs introduced in the Δ-residual network at the beginning of the Δ-scaling phase (i.e., those arcs (i, j) for which $\Delta \leq r_{ij} < 2\Delta$) might not satisfy the optimality condition (since, conceivably $c_{ij}^\pi < 0$). Therefore, the algorithm immediately saturates those arcs (i, j) so that they drop out of the residual network; since the reversal of each of these arcs (j, i) satisfies the condition $c_{ji}^\pi = -c_{ij}^\pi > 0$, they satisfy the optimality condition. Notice that because $r_{ij} < 2\Delta$, saturating any such arc (i, j) changes the imbalance of its end points by at most 2Δ. As a result, after we have saturated all the arcs violating the reduced cost optimality condition, the sum of the excesses is at most $2n\Delta + 2m\Delta = 2(n + m)\Delta$.

In the Δ-scaling phase, each augmentation starts at a node $k \in S(\Delta)$, terminates at a node $l \in T(\Delta)$, and carries at least Δ units of flow. Note that Assumption 5.4 implies that the Δ-residual network contains a directed path from node k to node l, so we always succeed in identifying a shortest path from node k to node l. Augmenting flow along a shortest path in $G(x, \Delta)$ preserves the property that every arc satisfies the reduced cost optimality condition (see Section 5.6). When either $S(\Delta)$ or $T(\Delta)$ is empty, the Δ-scaling phase ends. At this point, we divide Δ by a factor of 2 and start a new scaling phase. Within $O(\log U)$ scaling phases, $\Delta = 1$, and by the integrality of data, every node imbalance will be zero at the end of this phase. In this phase, $G(x, \Delta) \equiv G(x)$ and every arc in the residual network satisfies the reduced cost optimality condition. Consequently, the algorithm will obtain a minimum cost flow at the end of this scaling phase.

As we have seen, the capacity-scaling algorithm is easy to state. Similarly, it is easy to analyze its running time. We have noted previously that in the Δ-scaling phase the sum of the excesses is bounded by $2(n+m)\Delta$. Since each augmentation in this phase carries at least Δ units of flow from a node in $S(\Delta)$ to a node in $T(\Delta)$, each augmentation reduces the sum of the excesses by at least Δ units. Therefore, a scaling phase can perform at most $2(n+m)$ augmentations. Since we need to solve a shortest path problem to identify each augmenting path, we have established the following result.

Theorem 5.13 *The capacity-scaling algorithm solves the minimum cost flow problem in $O(m \log U\, S(n, m, C))$ time.* ∎

Further Reading

In this section, we present reference notes on the topics covered in this chapter and also other progress made in minimum cost flows. This discussion has three objectives: (1) to review the important theoretical contributions in each topic, (2) to point out the interrelationships among different algorithms, and (3) to comment on the empirical aspects of the algorithms.

The minimum cost flow problem has a rich history. The classical transportation problem, a simple case of the minimum cost flow problem, was posed and solved (though incompletely) by Kantorovich [1], Hitchcock [2], and Koopmans [3]. Dantzig [4] developed the first complete solution procedure for the transportation problem by specializing his simplex algorithm for linear programming. He observed the spanning tree property of the basis and the integrality property of the optimal solution. Later, his development of the upper bounding techniques for linear programming led to an efficient specialization of the simplex algorithm for the minimum cost flow problem. Dantzig's book [5] discusses these topics.

Ford and Fulkerson [6,7] suggested the first combinatorial algorithms for the uncapacitated and capacitated transportation problem; these algorithms are known as primal-dual algorithms. Ford and Fulkerson [8] later generalized this approach for solving the minimum cost flow problem. Jewell [9], Iri [10], and Busaker and Gowen [11] independently developed the successive shortest path algorithm. These researchers showed how to solve the minimum cost flow problem as a sequence of shortest path problems with arbitrary arc lengths. Tomizava [12] and Edmonds and Karp [13] independently observed that if the computations use node potentials, then it is possible to implement these algorithms so that the shortest path problems have nonnegative arc lengths.

Minty [14] and Fulkerson [15] independently developed the out-of-kilter algorithm. Aashtiani and Magnanti [16] have described an efficient implementation of this algorithm. The cycle canceling algorithm is credited to Klein [17]. Three special implementations of the cycle canceling algorithms run in polynomial-time: the first, due to Barahona and Tardos [18] (which, in turn, modifies an algorithm by Weintraub [19]), augments flow along (negative) cycles with the maximum possible improvement; the second, due to Goldberg and Tarjan [20], augments flow along minimum mean cost (negative) cycles; and the third, due to Wallacher and Zimmermann [21], augments flow along minimum ratio cycles.

Zadeh [22,23] described families of minimum cost flow problems on which each of several algorithms—the cycle canceling algorithm, the successive shortest path algorithm, the primal-dual algorithm, and the out-of-kilter algorithm—perform an exponential number of iterations.

The fact that the same families of networks are bad for many network algorithms suggests an interrelationship among the algorithms. The insightful paper by Zadeh [24] points out that each of the algorithms we have just mentioned are indeed equivalent in the sense that they perform the same sequence of augmentations, which they obtained through shortest path computations, provided that we initialize them properly and break ties using the same rule.

The network simplex algorithm gained its current popularity in the early 1970s when the research community began to develop and test algorithms using efficient tree indices. Johnson [25] suggested the first tree-manipulating data structure for implementing the simplex algorithm. Srinivasan and Thompson [26] and Glover et al. [27] implemented these ideas; these investigations found the network simplex algorithm to be substantially faster than the existing codes that implemented the primal-dual and out-of-kilter algorithms. Subsequent research has been focused on designing improved tree indices and determining the best pivot rule. Glover et al. [28], Bradley et al. [29], and Barr et al. [30] subsequently discovered improved data structures. The book by Kennington and Helgason [31] is an excellent source for references and background concerning these developments. The book by Bazaraa et al. [32] also describes a method for updating tree indices.

Researchers have conducted extensive studies to determine the most effective pricing strategy; that is, selection of the entering variable. These studies show that the choice of pricing strategy has a significant effect on both the solution time and the number of pivots required to solve minimum cost flow problems. The candidate list strategy we described is due to Mulvey [33]. Goldfarb and Reid [34], Bradley et al. [29], Grigoriadis and Hsu [35], Gibby et al. [36], and Grigoriadis [37] described other pivot selection strategies that have been effective in practice. In a recent development, Sokkalingam et al. [38] considered a new pivot selection rule based on minimum cost-to-penalty ratio. This method gives rise to efficient primal simplex algorithms for shortest path and assignment problems.

Experience with solving large-scale minimum cost flow problems has established that for certain classes of problems, more than 90% of the pivots in the network simplex algorithm can be degenerate (see Bradley et al. [29], Gavish et al. [39], and Grigoriadis [37]). Thus, degeneracy is a computational as well as theoretical issue. The strongly feasible spanning tree technique, proposed by Cunningham [40], and independently by Barr et al. [41–43] has contributed on both fronts. Computational experiences have shown that maintaining a strongly feasible spanning tree basis substantially reduces the number of degenerate pivots. On the theoretical front, the use of this technique led to a finitely converging primal simplex algorithm. Orlin [44] showed, using a perturbation technique, that for integer data an implementation of the primal simplex algorithm that maintains a strongly feasible basis performs $O(nmCU)$ pivots when used with any arbitrary pricing strategy and $O(nmC \log(mCU))$ pivots when used with Dantzig's pricing strategy.

The strongly feasible spanning tree technique prevents cycling during a sequence of consecutive degenerate pivots, but the number of consecutive degenerate pivots can be exponential. This phenomenon is known as *stalling*. Cunningham [45] described an example of stalling and suggested several rules for selecting the entering variable to avoid stalling. One such rule is the LRC (least recently considered) rule, which orders the arcs in an arbitrary but fixed manner. The algorithm then examines arcs in the wrap-around fashion, each iteration starting at the place where it left off earlier, and introduces the first eligible arc into the basis. Cunningham showed that this rule admits at most nm consecutive degenerate pivots. Goldfarb et al. [46] have described more anti-stalling pivot rules for the minimum cost flow problem.

Orlin [47] developed a polynomial-time dual simplex algorithm; this algorithm performs $O(n^3 \log n)$ pivots for the uncapacitated minimum cost flow problem. Tarjan [48] and Goldfarb and Hao [49] have described polynomial-time implementations of a variant of the network simplex algorithm that permits pivots to increase value of the objective function.

A monotone polynomial-time implementation, in which the value of the objective function is nonincreasing, remained elusive to researchers until Orlin [50] settled this long-standing open problem by developing an $O(\min(n^2 m \log nC, n^2 m^2 \log n))$ time algorithm. They introduce a pseudopolynomial variant of the network simplex algorithm called the *pre-multiplier algorithm* and then develop a cost-scaling version of the pre-multiplier algorithm that solves the minimum cost flow problem in $O(\min(nm \log nC, nm^2 \log n))$ pivots. They also showed that the average time per pivot is $O(n)$, using simple data structures. Tarjan [51] further reduced the time per pivot to $O(\log n)$ using dynamic trees as search trees via Euler tours.

A number of empirical studies have extensively tested minimum cost flow algorithms for a wide variety of network structures, data distributions, and problem sizes. The most common problem generator is NETGEN, due to Klingman et al. [52], which is capable of generating assignment, capacitated or uncapacitated transportation, and minimum cost flow problems. Glover et al. [53] and Aashtiani and Magnanti [16] have tested the primal-dual and the out-of-kilter algorithms. Helgason and Kennington [54] and Armstrong et al. [55] have reported on extensive studies of the dual simplex algorithm. The primal simplex has been the subject of more rigorous investigation; studies conducted by Glover et al. [27,53], Bradley et al. [29], Mulvey [56], Grigoriadis and Hsu [35], and Grigoriadis [37] are noteworthy. Bertsekas and Tseng [57] presented computational results for the relaxation algorithm.

Bertsekas [58] suggested the relaxation algorithm for the minimum cost flow problems with integer data. Bertsekas and Tseng [59] developed the relaxation algorithm and conducted extensive computational investigations of it. The relaxation approach maintains a pseudoflow satisfying the optimality conditions. The algorithm operates so that each change in node potentials increases the dual objective function value and when it determines the optimal-dual objective function values, it has also obtained an optimal primal solution. A Fortran code of the relaxation algorithm appears in Bertsekas and Tseng [57]. Their study and ones conducted by Grigoriadis [37] and Kennington and Wang [60] indicate that the relaxation algorithm and the network simplex algorithm are the two fastest available algorithms for solving the minimum cost flow problem in practice. Previous computational studies conducted by Glover et al. [53] and Bradley et al. [29] have indicated that the network simplex algorithm is consistently superior to the primal-dual and out-of-kilter algorithms. Most of this computational testing has been done on random network flow problems generated by the well-known computer program NETGEN, suggested by Klingman et al. [52].

Polynomial Time Algorithms

Researchers have actively pursued the design of fast weakly polynomial and strongly polynomial time algorithms for the minimum cost flow problem. Recall that an algorithm is strongly polynomial time if its running time is polynomial in the number of nodes and arcs and does not involve terms containing logarithms of C and U. Table 5.1 summarizes these theoretical developments in solving the minimum cost flow problem. Ahuja et al. [61–63] and Goldberg et al. [64] provide more details concerning the development of this field.

The table reports running times for networks with n nodes and m arcs, m' of which are capacitated. It assumes that the integral cost coefficients are bounded in absolute value by C, and the integral capacities, supplies, and demands are bounded in absolute value by U. The term $S(.)$ is the running time for the shortest path problem and the term $M(.)$ represents the corresponding running time to solve a maximum flow problem.

Most of the available (combinatorial) polynomial time algorithms for the minimum cost flow problems use scaling techniques. Edmonds and Karp [13] introduced the scaling approach

TABLE 5.1 Polynomial Time Algorithms for the Minimum Cost Flow Problem

S.No	Developers	Running Time
Polynomial time combinatorial algorithms		
1	Edmonds and Karp [13]	$O((n+m')\log U\, S(n,m,C))$
2	Rock [65]	$O((n+m')\log U\, S(n,m,C))$
3	Rock [65]	$O(n\log C\, M(n,m,U))$
4	Bland and Jensen [66]	$O(n\log C\, M(n,m,U))$
5	Goldberg and Tarjan [67]	$O(nm\log(n^2/m)\log nC)$
6	Bertsekas and Eckstein [68]	$O(n^3\log nC)$
7	Goldberg and Tarjan [67]	$O(n^3\log nC)$
8	Gabow and Tarjan [69]	$O(nm\log n\log U\log nC)$
9	Goldberg and Tarjan [67,20]	$O(nm\log n\log nC)$
10	Ahuja et al. [70]	$O(nm(\log U/\log\log U)\log nC)$ and $O(nm\log\log U\log nC)$
11	Orlin [50]	$O(\min(n^2m\log nC, n^2m^2\log n))$
12	Goldfarb and Zhiying [71]	$O(m(m+n\log n)\log(B/(m+n)))$
Strongly polynomial-time combinatorial algorithms		
1	Tardos [72]	$O(m^4)$
2	Orlin [47]	$O((n+m')^2 S(n,m))$
3	Fujishige [73]	$O((n+m')^2 S(n,m))$
4	Galil and Tardos [74]	$O(n^3\log n\, S(n,m))$
5	Goldberg and Tarjan [67]	$O(nm^2\log n\log(n^2/m))$
6	Goldberg and Tarjan [20]	$O(nm^2\log^2 n)$
7	Orlin [75]	$O((n+m')\log n\, S(n,m))$

and obtained the first weakly polynomial time algorithm for the minimum cost flow problem. This algorithm used the capacity scaling technique. Rock [65] and, independently, Bland and Jensen [66] suggested a cost scaling technique for the minimum cost flow problem. This approach solves the minimum cost flow problem as a sequence of $O(n\log C)$ maximum flow problems. Orlin [75] developed the capacity-scaling algorithm presented in Section 5.9, which is a variant of Edmonds and Karp's capacity-scaling algorithm. Goldfarb and Zhiying [71] more recently developed a polynomial time algorithm based on Edmond and Karp's capacity scaling and Orlin's capacity-scaling algorithm. Scaling techniques yield many of the best (in the worst case) available minimum cost flow algorithms.

The pseudoflow push algorithms for the minimum cost flow problem use the concept of *approximate* optimality, introduced independently by Bertsekas [76] and Tardos [72]. Bertsekas [77] developed the first pseudoflow push algorithm. This algorithm was pseudopolynomial-time. Goldberg and Tarjan [67] used a scaling technique on a variant of this algorithm to obtain the generic pseudoflow push algorithm. Tarjan [78] proposed a wave algorithm for the maximum flow problem. The wave implementation of the minimum cost flow problem, which was developed independently by Goldberg and Tarjan [67] and Bertsekas and Eckstein [68], relies upon similar ideas. Using a dynamic tree data structure in the generic pseudoflow push algorithm, Goldberg and Tarjan [67] obtained a computational time bound of $O(nm\log n\log nC)$. They also showed that the minimum cost flow problem can be solved using $O(n\log nC)$ blocking flow computations. Using both *finger tree* (see Mehlhorn [79]) and *dynamic tree* data structures, Goldberg and Tarjan [67] obtained an $O(nm\log(n^2/m)\log nC)$ bound for the wave implementation. Goldberg [80] proposed several heuristics to further improve the real-life performance of this method.

These algorithms, except the wave algorithm, require sophisticated data structures that impose a very high computational overhead. Although the wave algorithm is very practical, its worst-case running time is not very attractive. This situation prompted researchers to investigate the possibility of improving the computational complexity of minimum cost flow algorithms without using any complex data structures. The first success in this direction was due to Gabow and Tarjan [69], who developed a triple scaling algorithm running in $O(nm \log n \log U \log nC)$ time. The second success was due to Ahuja et al. [70], who developed a double scaling algorithm. The double scaling algorithm runs in $O(nm \log U \log nC)$ time. Scaling costs by an appropriately large factor improves the algorithm to $O(nm(\log U/\log \log U) \log nC)$, and a dynamic tree implementation improves the bound further to $O(nm \log \log U \log nC)$. For problems satisfying the similarity assumptions, the double scaling algorithm is faster than all other algorithms for all but very dense networks; in these instances, algorithms by Goldberg and Tarjan appear more attractive.

Goldberg and Tarjan [20] and Barahona and Tardos [18] have developed other polynomial-time algorithms. Both algorithms are based on the cycle-canceling algorithm due to Klein [17]. Goldberg and Tarjan [20] showed that if the cycle-canceling algorithm always augments flow on a minimum-mean cycle (a cycle W for which $\sum_{(i,j)\in W} c_{ij}/|W|$ is minimum), then it is strongly polynomial-time. Goldberg and Tarjan described an implementation of this approach running in time $O(nm (\log n) \min\{\log nC, m \log n\})$. Barahona and Tardos [18], analyzing an algorithm suggested by Weintraub [19], showed that if the cycle-canceling algorithm augments flow along a cycle with maximum improvement in objective function, then it performs $O(m \log mCU)$ iterations. Since identifying a cycle with maximum improvement is difficult (i.e., NP-hard), they describe a method based on solving an auxiliary assignment problem to determine a set of disjoint augmenting cycles with the property that augmenting flows along these cycles improves the flow cost by at least as much as augmenting flow along any single cycle. Their algorithm runs in $O(m^2 \log(mCU)S(n, m, C))$ time. Sokkalingam et al. [81] developed a scaling version of the cycle-canceling algorithm that performs $O(m \log(nU))$ iterations and runs in $O(m(m + n \log n) \log (nU))$. They showed that this algorithm could be modified to obtain a strongly polynomial $O(m(m + n \log n)\min\{\log (nU), m \log n\})$ bound.

Edmonds and Karp [13] proposed the first polynomial time algorithm for the minimum cost flow problem and also highlighted the desire to develop a strongly polynomial time algorithm. This desire was motivated primarily by theoretical considerations. (Indeed, in practice, the terms $\log C$ and $\log U$ typically range from 1 to 20 and are sub-linear in n.) Strongly polynomial time algorithms are attractive for at least two reasons: (1) they might provide, in principle, network flow algorithms that can run on real-valued data as well as integer-valued data and (2) they might, at a more fundamental level, identify the source of underlying complexity in solving a problem; that is, are problems more difficult or equally difficult to solve as the values of the underlying data becomes increasingly larger?

The first strongly polynomial-time minimum cost flow algorithm is due to Tardos [72]. Subsequently, Orlin [47], Fujishige [73], Galil and Tardos [74], Goldberg and Tarjan [20,67], Orlin [75], Ervolina and McCormick [82], Sokkalingam et al. [81], and Fathabadi and Shirdel [83] provided other strongly polynomial time algorithms. Currently, the fastest strongly polynomial time algorithm is due to Orlin [75]. This enhanced capacity-scaling algorithm solves the minimum cost flow problem as a sequence of $O(\min(m \log U, m \log n))$ shortest path problems. Vygen [84] describe a new algorithm for the minimum cost flow problem that can be regarded as a variation of the enhanced capacity-scaling algorithm. Their algorithm can also be considered a variant of dual network simplex, and they showed that the best worst-case running time of dual simplex algorithms exceeds the running time of their algorithm by a factor of n.

Most of the strongly polynomial-time minimum cost flow algorithms use one of the following two ideas: *fixing arc flows* or *fixing node potentials*. Tardos [72] was the first investigator to propose the use of either of these ideas (her algorithm fixes arc flows). The minimum mean cycle-canceling algorithm due to Goldberg and Tarjan [20] also fixes arc flows. Goldberg and Tarjan [20] also presented several variants of the minimum mean cycle algorithm with improved worst-case complexity. Orlin [47] and Fujishige [73] independently developed the idea of fixing node potentials, which is the *dual* of fixing arc flows. Using this idea, Goldberg et al. [64] obtained the repeated capacity-scaling algorithm. The enhanced capacity-scaling algorithm achieves the best strongly polynomial-time to solve the minimum cost flow problem and is due to Orlin [75].

Interior point linear programming algorithms are another source of polynomial time algorithms for the minimum cost flow problem. Kapoor and Vaidya [85] have shown that Karmarkar's [86] algorithm, when applied to the minimum cost flow problem performs $O(n^{2.5}mK)$ operations, where $K = \log n + \log C + \log U$. Vaidya [87] suggested another algorithm for linear programming that solves the minimum cost flow problem in $O(n^{2.5}m^{0.5}K)$ time. Asymptotically, these bounds are worse than that of the double scaling algorithm.

Currently, the best available theoretical time bound for the minimum cost flow problem is $O(\min\{nm \log(n^2/m \log(nC)), nm (\log \log U) \log(nC), (m \log n)(m + n \log n)\})$; the three bounds in this expression are, respectively, due to Goldberg and Tarjan [67], Ahuja et al. [70], and Orlin [75].

References

[1] L. V. Kantorovich, Mathematical methods in the organization and planning of production, Publication House of the Leningrad University, Translated in *Manage. Sci.*, **6** (1939) (1960), 366–422.

[2] F. L. Hitchcock, The distribution of a product from several sources to numerous facilities, *J. Math. Phy.*, **20** (1941), 224–230.

[3] T. C. Koopmans, Optimum utilization of the transportation system, In *Proc. Int. Stat. Conf.*, Washington, DC. Also in *Econometrica*, **17** (1947) (1949).

[4] G. B. Dantzig, Application of the simplex method to a transportation problem, In *Activity analysis and production and allocation*, T. C. Koopmans, editor, John Wiley & Sons, New York, 1951, 359–373.

[5] G. B. Dantzig, *Linear programming and extensions*, Princeton University Press, Princeton, NJ, 1962.

[6] L. R. Ford and D. R. Fulkerson, Solving the transportation problem, *Manage. Sci.*, **3** (1956), 24–32.

[7] L. R. Ford and D. R. Fulkerson, A primal-dual algorithm for the capacitated Hitchcock problem, *Nav. Res. Log. Quart.*, **4** (1957), 47–54.

[8] L. R. Ford and D. R. Fulkerson, *Flows in networks*. Princeton University Press, Princeton, NJ, 1962.

[9] W. S. Jewell, *Optimal flow through networks. Interim Technical Report No. 8*, Operations Research Center, MIT, Cambridge, MA, 1958.

[10] M. Iri, A new method of solving transportation-network problems, *J. Oper. Res. Soc. Japan*, **3** (1960), 27–87.

[11] R. G. Busaker and P. J. Gowen, A procedure for determining minimal-cost network flow patterns, O. R. O. Technical Report No. 15, Operational Research Office, John Hopkins University, Baltimore, MD, 1961.

[12] N. Tomizava, On some techniques useful for solution of transportation network problems, *Netw.*, **1** (1972), 173–194.

[13] J. Edmonds and R. M. Karp, Theoretical improvements in algorithmic efficiency for network flow problems, *J. ACM*, **19** (1972), 248–264.

[14] G. J. Minty, Monotone networks, *Proc. R. Soc. London*, **257A** (1960), 194–212.

[15] D. R. Fulkerson, An out-of-kilter method for minimal cost flow problems, *SIAM J. Appl. Math.*, **9** (1961), 18–27.

[16] H. A. Aashtiani and T. L. Magnanti, *Implementing primal-dual network flow algorithms. Technical Report OR* 055–76, Operations Research Center, MIT, Cambridge, MA, 1976.

[17] M. Klein, A primal method for minimal cost flows, *Manage. Sci.*, **14** (1967), 205–220.

[18] F. Barahona and E. Tardos, Note on Weintraub's minimum cost circulation algorithm, *SIAM J. Comput.*, **18** (1989), 579–583.

[19] A. Weintraub, A primal algorithm to solve network flow problems with convex costs, *Manage. Sci.*, **21** (1974), 87–97.

[20] A. V. Goldberg and R. E. Tarjan, Finding minimum-cost circulations by canceling negative cycles, *Proc. 20th ACM Symp. Theory Comput.*, 388–397. Full paper in *J. ACM*, **36** (1988), 873–886.

[21] C. Wallacher and U. Zimmermann, A combinatorial interior point method for network flow problems, *Math. Prog.*, **56** (1992), 321–335.

[22] N. Zadeh, A bad network problem for the simplex method and other minimum cost flow algorithms, *Math. Prog.* **5** (1973a), 255–266.

[23] N. Zadeh, More pathological examples for network flow problems, *Math. Prog.*, **5** (1973b), 217–224.

[24] N. Zadeh, *Near equivalence of network flow algorithms. Technical Report No. 26*, Department of Operations Research, Stanford University, Stanford, CA, 1979.

[25] E. L. Johnson, Networks and basic solutions, *Oper. Res.*, **14** (1966), 619–624.

[26] V. Srinivasan and G. L. Thompson, Benefit-cost analysis of coding techniques for primal transportation algorithm, *J. ACM*, **20** (1973), 194–213.

[27] F. Glover, D. Karney, D. Klingman and A. Napier, A computational study on start procedures, basis change criteria, and solution algorithms for transportation problem, *Manage. Sci.*, **20** (1974), 793–813.

[28] F. Glover, D. Klingman and J. Stutz Augmented threaded index for network optimization, *INFOR*, **12** (1974), 293–298.

[29] G. Bradley, G. Brown and G. Graves, Design and implementation of large scale primal transshipment algorithms, *Manage. Sci.*, **21** (1977), 1–38.

[30] R. Barr, F. Glover and D. Klingman, Enhancement of spanning tree labeling procedures for network optimization, *INFOR*, **17** (1979), 16–34.

[31] J. L. Kennington and R. V. Helgason *Algorithms for network programming*, Wiley-Interscience, New York, 1980.

[32] M. S. Bazaraa, J. J. Jarvis and H. D. Sherali, *Linear programming and network flows*. Second Edition, John Wiley & Sons, New York, 1990.

[33] J. Mulvey, Pivot strategies for primal-simplex network codes, *J. ACM*, **25** (1978a), 266–270.

[34] D. Goldfarb and J. K. Reid, A practicable steepest edge simplex algorithm, *Math. Prog.*, **12** (1977), 361–371.

[35] M. D. Grigoriadis and Y. Hsu, The Rutgers minimum cost network flow subroutines, *SIGMAP Bull. ACM*, **26** (1979), 17–18.

[36] D. Gibby, F. Glover, D. Klingman and M. Mead, A comparison of pivot selection rules for primal simplex based network codes, *Oper. Res. Lett.* **2** (1983), 199–202.

[37] M. D. Grigoriadis, An efficient implementation of the network simplex method, *Math. Prog. Stud.*, **26** (1986), 83–111.

[38] P. T. Sokkalingam, R. K. Ahuja and P. Sharma, A new pivot selection rule for the network simplex algorithm, *Math. Prog.*, **78** (1997), 149–158.

[39] B. Gavish, P. Schweitzer and E. Shilfer, The zero pivot phenomenon in transportation problems and its computational implications, *Math. Prog.*, **12** (1977), 226–240.

[40] W. H. Cunningham, A network simplex method, *Math. Prog.*, **11** (1976), 105–116.

[41] R. Barr, F. Glover and D. Klingman, The alternative path basis algorithm for the assignment problem, *Math. Prog.*, **12** (1977a), 1–13.

[42] R. Barr, F. Glover and D. Klingman, A network augmenting path basis algorithm for transshipment problems, *Proc. Int. Symp. Extremal Methods Sys. Anal.* (1977b).

[43] R. Barr, F. Glover and D. Klingman, Generalized alternating path algorithms for transportation problems, *Eur. J. Oper. Res.*, **2** (1978), 137–144.

[44] J. B. Orlin, On the simplex algorithm for networks and generalized networks, *Math. Prog. Stud.*, **24** (1985), 166–178.

[45] W. H. Cunningham, Theoretical properties of the network simplex method, *Math. Oper. Res.*, **4** (1979), 196–208.

[46] D. Goldfarb, J. Hao and S. Kai, Anti-stalling pivot rules for the network simplex algorithm, *Networks*, **20** (1990), 79–91.

[47] J. B. Orlin, *Genuinely polynomial simplex and non-simplex algorithms for the minimum cost flow problem*, Technical Report No. 1615-84, Sloan School of Management, MIT, Cambridge, MA, 1984.

[48] R. E. Tarjan, Efficiency of the primal network simplex algorithm for the minimum-cost circulation problem, *Math. Oper. Res.*, **16** (1991), 272–291.

[49] D. Goldfarb and J. Hao, Polynomial-time primal simplex algorithms for the minimum cost network flow problem, *Algorithmica*, **8** (1992), 145–160.

[50] J. B. Orlin, A polynomial time primal simplex algorithm for minimum cost flows, *Math. Prog.*, **78** (1997), 109–129.

[51] R. E. Tarjan, Dynamic trees as search trees via Euler tours, applied to the network simplex algorithm, *Math. Prog.*, **78** (1997), 169–177.

[52] D. Klingman, A. Napier and J. Stutz, NETGEN: A program for generating large scale capacitated assignment, transportation, and minimum cost flow network problems, *Manage. Sci.*, **20** (1974), 814–821.

[53] F. Glover, D. Karney and D. Klingman, Implementation and computational comparisons of primal, dual and primal-dual computer codes for minimum cost network flow problem, *Netw.*, **4** (1974), 191–212.

[54] R. V. Helgason and J. L. Kennington, An efficient procedure for implementing a dual-simplex network flow algorithm, *AIIE Trans.*, **9** (1977), 63–68.

[55] R. D. Armstrong, D. Klingman and D. Whitman, Implementation and analysis of a variant of the dual method for the capacitated transshipment problem, *Eur. J. Oper. Res.*, **4** (1980), 403–420.

[56] J. Mulvey, Testing a large-scale network optimization problem, *Math. Prog.*, **15** (1978b), 291–314.

[57] D. P. Bertsekas and P. Tseng, Relaxation methods for minimum cost ordinary and generalized network flow problems, *Oper. Res.*, **36** (1988b), 93–114.

[58] D. P. Bertsekas, A unified framework for primal-dual methods in minimum cost network flow problems, *Math. Prog.*, **32** (1985), 125–145.

[59] D. P. Bertsekas and P. Tseng, The relax codes for linear minimum cost network flow problems *FORTRAN Codes for Network Optimization*, B. Simeone, P. Toth, G. Gallo, F. Maffioli and S. Pallottino, editors, *Ann. Oper. Res.*, **13** (1988a), 125–190.

[60] J. L. Kennington and Z. Wang, The shortest augmenting path algorithm for the transportation problem, Technical Report 90–CSE–10. Southern Methodist University, Dallas, TX, 1990.

[61] R. K. Ahuja, T. L. Magnanti and J. B. Orlin, Network flows. In *Handbooks in Operations Research and Management Science. Vol. 1: Optimization*, G. L. Nemhauser, A. H. G. Rinnooy Kan and M. J. Todd, editors, North-Holland, Amsterdam, the Netherlands, 211–369, 1989.

[62] R. K. Ahuja, T. L. Magnanti and J. B. Orlin, Some recent advances in network flows, *SIAM Rev.*, **33** (1991), 175–219.

[63] R. K. Ahuja, T. L. Magnanti and J. B. Orlin, *Network flows: Theory, algorithms, and applications*. Prentice Hall, Upper Saddle River, NJ, 1993.

[64] A. V. Goldberg, E. Tardos and R. E. Tarjan, *Network flow algorithms, Technical Report No. 860*, School of Operations Research and Industrial Engineering, Cornell University, Ithaca, NY, 1989.

[65] H. Rock, Scaling techniques for minimal cost network flows. In *Discrete Structures and Algorithms*, V. Page, editor, Carl Hansen, Munich, Germany, 181–191, 1980.

[66] R. G. Bland and D. L. Jensen, On the computational behavior of a polynomial-time network flow algorithm, *Math. Prog.*, **54** (1985), 1–39.

[67] A. V. Goldberg and R. E. Tarjan, Solving minimum cost flow problem by successive approximation, *Math. Oper. Res.*, **15** (1990), 430–466.

[68] D. P. Bertsekas and J. Eckstein, Dual coordinate step methods for linear network flow problems, *Math. Prog. B*, **42** (1988), 203–243.

[69] H. N. Gabow and R. E. Tarjan, Faster scaling algorithms for network problems, *SIAM J. Comput.*, **18** (1989), 1013–1036.

[70] R. K. Ahuja, A. V. Goldberg, J. B. Orlin and R. E. Tarjan, Finding minimum-cost flows by double scaling, *Math. Prog.*, **53** (1992), 243–266.

[71] D. Goldfarb and J. Zhiying A new scaling algorithm for the minimum cost network flow problem, *Oper. Res. Lett.*, **25** (1999), 205–211.

[72] E. Tardos, A strongly polynomial minimum cost circulation algorithm, *Combinatorica*, **5** (1985), 247–255.

[73] S. Fujishige, An $O(m^3 \log n)$ capacity-rounding algorithm for the minimum cost circulation problem: A dual framework of Tardos' algorithm, *Math. Prog.*, **35** (1986), 298–309.

[74] Z. Galil and E. Tardos, An $O(n^2(m+n\log n)\log n)$ min-cost flow algorithm, *J. ACM*, **35** (1987), 374–386.

[75] J. B. Orlin, A faster strongly polynomial minimum cost flow algorithm, *Oper. Res.*, **41** (1993), 377–387.

[76] D. P. Bertsekas, A distributed algorithm for the assignment problem, *Ann. Oper. Res.*, **14** (1979), 105–123.

[77] D. P. Bertsekas, Distributed relaxation methods for linear network flow problems, *Proc. 25th IEEE Conf. Decis. Control*, Athens, Greece, 1986.

[78] R. E. Tarjan, A simple version of Karzanov's blocking flow algorithm, *Oper. Res. Lett.*, **2** (1984), 265–268.

[79] K. Mehlhorn, *Data structures and algorithms, Vol. I: Searching and sorting.* Springer-Verlag, New York, 1984.

[80] A. V. Goldberg, An efficient implementation of a scaling minimum-cost flow algorithm. *Journal of Algorithms*, **22** (1997), 1–29.

[81] P. T. Sokkalingam, R. K. Ahuja and J. B. Orlin, New polynomial-time cycle-canceling algorithms for minimum-cost flows, *Networks*, **36** (2000), 53–63.

[82] T. R. Ervolina and S. T. McCormick, Two strongly polynomial cut canceling algorithms for minimum cost network flow, *Discrete Appl. Math.*, **46** (1993), 133–165.

[83] H. S. Fathabadi and G. H. Shirdel, An $O(nm^2)$ time algorithm for solving minimal cost network flow problems, *Asia-Pacific J. Oper. Res.*, **20** (2003), 161–175.

[84] J. Vygen, On dual minimum cost flow algorithms, *Math. Methods Oper. Res.*, **56** (2002), 101–126.

[85] S. Kapoor and P. Vaidya, Fast algorithms for convex quadratic programming and multicommodity flows, *Proc. 18th ACM Symp. Theor. Comput.*, 147–159, 1986.

[86] N. Karmarkar, A new polynomial-time algorithm for linear programming, *Combinatorica*, **4** (1984), 373–395.

[87] P. Vaidya, An algorithm for linear programming which requires $O(((m+n)n^2 + (m+n)^{1.5}n)L)$ arithmetic operations, *Proc. 19th ACM Symp. Theor. Comput.*, 29–38, 1987.

CHAPTER 6

Multicommodity Flows

Balachandran Vaidyanathan

Ravindra K. Ahuja

James B. Orlin

Thomas L. Magnanti

CONTENTS

6.1	Introduction	157
	6.1.1 Assumptions	158
6.2	Applications	159
	6.2.1 Application 1: Routing of Multiple Commodities	160
	6.2.1.1 Communication Networks	160
	6.2.1.2 Railroad Transportation Networks	160
	6.2.1.3 Distribution Networks	160
	6.2.2 Application 2: Multivehicle Tanker Scheduling	160
6.3	Optimality Conditions	161
	6.3.1 Multicommodity Flow Complementary Slackness Conditions	162
6.4	Lagrangian Relaxation	163
6.5	Dantzig-Wolfe Decomposition	164
	6.5.1 Reformulation with Path Flows	165
	6.5.2 Optimality Conditions	165
	6.5.3 Path Flow Complementary Slackness Conditions	166
6.6	Resource-Directive Decomposition	167

6.1 INTRODUCTION

The multicommodity flow problem is a generalization of the minimum cost flow problem, described in Chapter 5. Multicommodity flow problems are frequently encountered in several application domains. In many applications, several physical commodities, vehicles, or messages, each governed by their own network flow constraints, share the same network. For example, in telecommunications applications, telephone calls between specific node pairs in an underlying telephone network, each define a separate commodity, and all these commodities share common telephone line resources. In this chapter, we study the *multicommodity flow problem*, in which individual commodities share common arc capacities. That is, each arc has a capacity u_{ij} that restricts the total flow of all commodities on that arc. The objective of this problem is to send several commodities that reside at one or more points in a network (sources or supplies) with arc capacities to one or more points on the network (sinks or demands), incurring minimum cost. The arc capacities bind the flows of all commodities together. If the commodities do not interact with each other (common arc capacities constraints are relaxed), then the multicommodity flow problem can be solved as several independent single commodity problems using the techniques discussed in Chapter 5. We now proceed to give the mathematical formulation of this problem.

Let x_{ij}^k denote the flow of commodity k on arc (i, j), and let x^k and c^k denote the flow vector and per unit cost vector for commodity k. Using this notation, we can formulate the multicommodity flow problem as follows:

$$\text{Minimize} \sum_{1 \leq k \leq K} c^k x^k \tag{6.1}$$

subject to

$$\sum_{1 \leq k \leq K} x_{ij}^k \leq u_{ij}, \text{ for all } (i,j) \in A, \tag{6.2}$$

$$N x^k = b^k, \text{ for all } k = 1, 2, \ldots, K \tag{6.3}$$

$$0 \leq x_{ij}^k \leq u_{ij}^k, \text{ for all } (i,j) \in A, k = 1, 2, \ldots, K \tag{6.4}$$

This formulation has a collection of K ordinary mass balance constraints (6.3), one modeling the flow of each commodity $k = 1, 2, \ldots, K$. The *bundle* constraints (6.4) tie together the commodities by restricting the total flow of all the commodities on each arc (i, j) to at most u_{ij}. Note that we also impose individual flow bounds u_{ij}^k on the flow of commodity k on arc (i, j).

6.1.1 Assumptions

Note that the model (6.1 through 6.4) imposes capacities on the arcs but not on the nodes. This modeling assumption imposes no loss of generality, since by using the node splitting techniques we can use this formulation to model situations with node capacities as well. Three other features of the model are worth noting.

Homogeneous goods assumption. We are assuming that every unit flow of each commodity uses one unit of capacity of each arc. A more general model would permit the flow of each commodity k to consume a given amount ρ_{ij}^k of the capacity (or some other resource) associated with each arc and would replace the bundle constraint with a more general resource availability constraint $\sum_{1 \leq k \leq K} \rho_{ij}^k x_{ij}^k \leq u_{ij}$. With minor modifications, the solution techniques that we will be discussing in this chapter apply to this more general model as well.

No congestion assumption. We are assuming that we have a hard (i.e., fixed) capacity on each arc and that the cost on each arc is linear in the flow on that arc.

Indivisible goods assumption. The model (6.1 through 6.4) assumes that the flow variables can be fractional. In some applications encountered in practice, this assumption is appropriate; in other application contexts, however, the variables must be integer-valued. In these instances, the model that we are considering might still prove to be useful, since either the linear programming model might be a good approximation of the integer programming model, or we could use the linear programming model as a linear programming relaxation of the integer program and embed it within a branch and bound or some other type of enumeration approach.

We note that the integrality of solutions is one very important distinguishing feature between single and multicommodity flow problems. As we have seen several times in Chapter 5, one very nice feature of single commodity network flow problems is that they always have integer solutions whenever the supply/demand and capacity data are integer-valued. For multicommodity flow problems, however, this is not the case.

The three main approaches to solving multicommodity flow problems are (1) price-directive decomposition, (2) resource-directive decomposition, and (3) partitioning methods. Price-directive decomposition methods place Lagrangian multipliers (or prices) on the bundle constraints and bring them into the objective function so that the resulting problem decomposes into a separate minimum cost flow problem for each commodity k. These methods remove the capacity constraints and instead *charge* each commodity for the use of the capacity of each arc. These methods attempt to find appropriate prices so that some optimal solution to the resulting *pricing problem* or Lagrangian sub-problem also solves the overall multicommodity flow problem. Several methods are available for finding appropriate prices.

Resource-directive decomposition methods view the multicommodity flow problem as a capacity allocation problem. All the commodities are competing for the fixed capacity u_{ij} of every arc (i, j) of the network. Any optimal solution to the multicommodity flow problem will prescribe for each commodity a specific flow on each arc (i, j) that is the appropriate capacity to allocate to that commodity. If we started by allocating these capacities to the commodities and then solved the resulting (independent) single commodity flow problems, we would be able to solve the problem quite easily as a set of independent single commodity flow problems. Resource-directive methods provide a general solution approach for implementing this idea. They begin by allocating the capacities to the commodities and then use information gleaned from the solution to the resulting single commodity problems to reallocate the capacities in a way that improves the overall system cost.

Partitioning methods exploit the fact that the multicommodity flow problem is a specially structured linear program with embedded network flow problems. As we have seen in Chapter 5, to solve any minimum cost flow problem, we can use the network simplex method, which works by generating a sequence of improving spanning tree solutions. This observation raises the following questions: (1) can we adopt a similar approach for solving the multicommodity flow problem?; and (2) can we somehow use spanning tree solutions for the embedded network flow constraints $\mathcal{N}x^k = b^k$? The partitioning method is a linear programming approach that permits us to answer both of these questions affirmatively. It maintains a linear programming basis that is composed of spanning trees of the individual single commodity flow problems as well as additional arcs that are required to *tie* these solutions together to accommodate the bundle constraints.

We next describe some applications of the multicommodity flow problem before discussing solution techniques in more detail. We describe classical price-directive methods and resource-directive methods and direct the reader to suitable references for the basis partitioning methods.

6.2 APPLICATIONS

Multicommodity flow problems arise in a wide variety of application contexts. In this section, we consider some of these applications.

6.2.1 Application 1: Routing of Multiple Commodities

In many applications of the multicommodity flow problem, we distinguish commodities because they are different physical goods and/or because they have different points of origin and destination; that is, either (1) several physically distinct commodities (e.g., different manufactured goods) share a common network or (2) a single physical good (e.g., messages or products) flows on a network, but the good has multiple points of origin and destination defined by different pairs of nodes in the network that need to send the good to each other. This second type of application arises frequently in problem contexts such as communication systems or distribution/transportation systems. In this section, we briefly introduce several application domains of both types.

6.2.1.1 Communication Networks

In a communication network, nodes represent origin and destination stations for messages, and arcs represent transmission lines. Messages between different pairs of nodes define distinct commodities; the supply and demand for each commodity is the number of messages to be sent between the origin and destination nodes of that commodity. Each transmission line has a fixed capacity (in some applications the capacity of each arc is fixed; in others, we might be able to increase the capacity at a certain cost per unit). In this network, the problem of determining the minimum cost routing of messages is a multicommodity flow problem.

6.2.1.2 Railroad Transportation Networks

In a rail network, nodes represent yard and junction points and arcs represent track sections between the yards. The demand is measured by the number of cars (or, any other equivalent measure of tonnage) to be loaded on any train. Since the system incurs different costs for different goods, we divide traffic demand into different classes. Each commodity in this network corresponds to a particular class of demand between a particular origin-destination pair. The bundle capacity of each arc is the number of cars that we can load on the trains that are scheduled to be dispatched on that arc (over some period of time). The decision problem in this network is to meet the demands of cars at the minimum possible operating cost.

6.2.1.3 Distribution Networks

In distribution systems planning, we wish to distribute multiple (nonhomogeneous) products from plants to retailers using a fleet of trucks or railcars and using a variety of railheads and warehouses. The products define the commodities of the multicommodity flow problem, and the joint capacities of the plants, warehouses, rail yards, and the shipping lanes define the bundle constraints. Note that this application has important bundle constraints imposed upon the nodes (plants, warehouses) as well as the arcs.

6.2.2 Application 2: Multivehicle Tanker Scheduling

Suppose we wish to determine the optimal routing of fuel oil tankers required to achieve a prescribed schedule of deliveries: each delivery is a shipment of some commodity from a point of supply to a point of demand with a given delivery date. In the simplest form, this problem considers a single product (e.g., aviation gasoline or crude oil) to be delivered by a single type of tanker. The multivehicle tanker scheduling problem considers the scheduling and routing of a fixed fleet of nonhomogeneous tankers to meet a pre-specified set of shipments of multiple products. The tankers differ in their speeds, carrying capabilities, and operating costs.

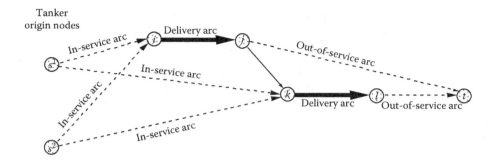

Figure 6.1 Multivehicle tanker scheduling problem.

To formulate the multivehicle tanker scheduling problem as a multicommodity flow problem, we let the different commodities correspond to different tanker types. Each distinct type of tanker originates at a unique source node s^k. This network has four types of arcs (see Figure 6.1 for a partial example with two tankers types): in-service, out-of-service, delivery, and return arcs. An in-service arc corresponds to the initial employment of a tanker type; the cost of this arc is the cost of deploying the tanker at the origin of the shipment. Similarly, an out-of-service arc corresponds to the removal of the tanker from service. A delivery arc (i,j) represents a shipment from origin i to destination j; the cost c_{ij}^k of this arc is the operating cost of carrying the shipment by a tanker of type k. A return arc (j,k) denotes the movement (*backhaul*) of an empty tanker, with an appropriate cost, between two consecutive shipments (i,j) and (k,l).

Each arc in the network has a capacity of one. The shipment arcs have a bundle capacity ensuring that at most one tanker type services that arc. Each shipment arc also has a lower flow bound of one unit, which ensures that the chosen schedule does indeed deliver the shipment. Some arcs might also have commodity-based capacities u_{ij}^k. For instance, if tanker type 2 is not capable of handling the shipment on arc (i,j), then we set $u_{ij}^2 = 0$. Moreover, if tanker type 2 can use the return arc (j,k), but the tanker type 1 cannot (because it is too slow to make the connection between shipments), then we set $u_{jk}^1 = 0$.

In the multivehicle tanker scheduling problem, we are interested in integer solutions of the multicommodity flow problem. The solutions obtained by the multicommodity flow algorithms to be described in this chapter need not be integral. Nevertheless, the fractional solution might be useful in several ways. For example, we might be able to convert the nonintegral solution into a (possibly, sub-optimal) integral solution by minor tinkering or, as we have noted earlier, we might use the nonintegral solution as a bound in solving the integer-valued problem by a branch and bound enumeration procedure.

6.3 OPTIMALITY CONDITIONS

For this discussion, we assume that the flow variables x_{ij}^k have no individual flow bounds; that is, each $u_{ij}^k = +\infty$ in the formulation (6.1 through 6.4). Since the multicommodity flow problem is a linear program, we can use linear programming optimality conditions to characterize optimal solutions to the problem. Since the linear programming formulation (6.1 through 6.4) of the problem has one bundle constraint for every arc (i,j) of the network and one mass balance constraint for each node-commodity combination, the dual linear program has two types of dual variables: a *price* w_{ij} on each arc (i,j) and a *node potential* $\pi^k(i)$ for each combination of commodity k and node i. Using these dual variables, we define the reduced cost $c_{ij}^{\pi,k}$ of arc (i,j) with respect to commodity k as follows:

$$c_{ij}^{\pi,k} = c_{ij}^k + w_{ij} - \pi^k(i) + \pi^k(j).$$

In the matrix notation, this definition is $c^{\pi,k} = c^k + w - \pi^k N$.

Note that if we consider a fixed commodity k, this reduced cost is similar to the reduced cost that we have used in Chapter 5 for the minimum cost flow problem; the difference is that we now add the arc price w_{ij} to the arc cost c_{ij}^k. Note that just as the bundle constraints provided a linkage between the otherwise independent commodity flow variables x_{ij}^k, the arc prices w_{ij} provide a linkage between the otherwise independent commodity reduced costs. Linear programming duality theory permits us to characterize optimal solutions to the multicommodity flow problem.

We first write the dual of the multicommodity flow problem (6.1 through 6.4) as follows:

$$\text{Maximize} - \sum_{(i,j) \in A} u_{ij} w_{ij} + \sum_{k=1}^{K} b^k \pi^k$$

subject to

$$c_{ij}^{\pi,k} = c_{ij}^k + w_{ij} - \pi^k(i) + \pi^k(j) \geq 0 \text{ for all } (i,j) \in A, k = 1, \ldots, K$$
$$w_{ij} \geq 0 \text{ for all } (i,j) \in A.$$

The optimality conditions for linear programming are called the *complementary slackness (optimality) conditions* and state that a primal feasible solution x and a dual feasible solution (w, π^k) are optimal to the respective problems if and only if the product of each primal (dual) variable and the slack in the corresponding dual (primal) constraint is zero. The complementary slackness conditions for the primal-dual pair of the multicommodity flow problem are as follows.

6.3.1 Multicommodity Flow Complementary Slackness Conditions

The commodity flows y_{ij}^k are optimal in the multicommodity flow problem (6.1 through 6.4) with each $u_{ij}^k = +\infty$ if and only if they are feasible and for some choice of arc prices w_{ij} and node potentials $\pi^k(i)$, the reduced costs and arc flows satisfy the following complementary slackness conditions:

a.
$$w_{ij} \left(\sum_{k=1}^{K} y_{ij}^k - u_{ij} \right) = 0 \text{ for all arcs } (i,j) \in A. \quad (6.5)$$

b.
$$c_{ij}^{\pi,k} \geq 0 \text{ for all } (i,j) \in A \text{ and all } k = 1, \ldots, K. \quad (6.6)$$

c.
$$c_{ij}^{\pi,k} y_{ij}^k = 0 \text{ for all } (i,j) \in A \text{ and } k = 1, \ldots, K. \quad (6.7)$$

We refer to any set of arc prices and node potentials that satisfy the complementary slackness conditions as *optimal arc prices* and *optimal node potentials*. The following theorem shows the connection between the multicommodity and single commodity flow problems.

Theorem 6.1 (Partial dualization result) *Let y_{ij}^k be optimal flows and let w_{ij} be optimal arc prices for the multicommodity flow problem (6.1 through 6.4). Then for each commodity k, the flow variables y_{ij}^k for $(i,j) \in A$ solve the following (uncapacitated) minimum cost flow problem:*

$$\text{Minimize} \left\{ \sum_{(i,j) \in A} (c_{ij}^k + w_{ij}) x_{ij}^k : Nx^k = b^k, x_{ij}^k \geq 0 \text{ for all } (i,j) \in A \right\} \quad (6.8)$$

Proof. Since y_{ij}^k are optimal flows and w_{ij} are optimal arc prices for the multicommodity flow problem (6.1), these variables together with some set of node potentials $\pi^k(i)$ satisfy the complementary slackness condition (6.5). Now notice that the conditions (6.6) and (6.7) are the optimality conditions for the uncapacitated minimum cost flow problem for commodity k with arc costs as $c_{ij}^k + w_{ij}$ (see Chapter 5, condition (5.3)). This observation implies that the flows y_{ij}^k solve the corresponding minimum cost flow problems.

This property shows that we can use a sequential approach for obtaining optimal arc prices and node potentials: we first find optimal arc prices and then attempt to find the optimal node potentials and flows by solving the single commodity minimum cost flow problems. In the next few sections, we use this observation to develop and assess algorithms for solving the multicommodity flow problem.

6.4 LAGRANGIAN RELAXATION

Lagrangian relaxation is a general solution strategy for solving mathematical programs that permits us to decompose problems to exploit their special structure. This solution approach is perfectly tailored for solving many problems with embedded network structure. The multicommodity flow problem is one such problem.

Lagrangian relaxation works by moving hard constraints into the objective function and penalizing their violation. In the multicommodity flow problem, the complicating constraints are the bundle arc capacity constraints (6.2). Therefore, to apply Lagrangian relaxation to the multicommodity flow problem, we associate nonnegative Lagrange multipliers w_{ij} with the bundle constraints (6.2), creating the following Lagrangian sub-problem:

$$L(w) = \min \sum_{1 \leq k \leq K} c^k x^k + \sum_{(i,j) \in A} w_{ij} \left(\sum_{1 \leq k \leq K} x_{ij}^k - u_{ij} \right) \tag{6.9}$$

Or, equivalently,

$$L(w) = \min \sum_{1 \leq k \leq K} \sum_{(i,j) \in A} \left(c_{ij}^k + w_{ij} \right) x_{ij}^k - \sum_{(i,j) \in A} w_{ij} u_{ij} \tag{6.10}$$

subject to

$$Nx^k = b^k \text{ for all } k = 1, \ldots, K, \tag{6.11}$$

$$x_{ij}^k \geq 0 \text{ for all } (i,j) \in A \text{ and all } k = 1, \ldots, K \tag{6.12}$$

Note that since the second term in the objective function of the Lagrangian sub-problem is a constant for any given choice of the Lagrange multipliers, we can ignore it. The resulting objective function for the Lagrangian sub-problem has a cost of $c_{ij}^k + w_{ij}$ associated with every flow variable x_{ij}^k. Since none of the constraints in this problem contains the flow variables for more than one of the commodities, the problem decomposes into separate minimum cost flow problems, one for each commodity.

For any value of Lagrange multipliers w, it can be shown that $L(w)$ is a lower bound on the optimal objective function of the original problem. Hence, to obtain the tightest possible lower bound, we need to solve the following optimization problem, which is referred to as the *Lagrangian multiplier problem*:

$$L^* = \max_w L(w)$$

The values w for which $L(w) = L^*$ are called the *optimal Lagrange multipliers*.

The following theorem, relating Lagrangian relaxation and linear programming, which we state without proof, is the basis of our solution methodology for the multicommodity flow problem.

Theorem 6.2 *Suppose we apply the Lagrangian relaxation technique to a linear programming problem. Then the optimal value L^* of the Lagrangian multiplier problem equals the optimal objective function value of the linear programming problem.* ∎

Lagrangian relaxation is a price-directive decomposition method to solve the multicommodity flow problem. From Theorem 6.2, it follows that by solving the Lagrangian multiplier problem, we obtain the optimal solution to the corresponding linear programming problem. Sub-gradient optimization is a commonly used method to solve Lagrangian multiplier problem. This works in the following method. We alternately do the following: (1) solve a set of minimum cost flow problems (for a fixed value of the Lagrange multipliers w) with the cost coefficients $c_{ij}^k + w_{ij}$ and (2) update the multipliers. In this case, if y_{ij}^k denotes the optimal solution to the kth minimum cost flow sub-problem when the Lagrange multipliers have the value w_{ij}^q at the qth iteration, the sub-gradient update formula can be written as

$$w_{ij}^{q+1} = \left[w_{ij}^q + \theta_q \left(\sum_{1 \leq k \leq K} y_{ij}^k - u_{ij} \right) \right]^+.$$

In this expression, the notation $[\alpha]^+$ denotes the positive part of α; that is, $\max(\alpha, 0)$. θ_q is a step size specifying how far we move from the current solution w_{ij}^q. Note that this update formula either increases the multiplier w_{ij}^q on arc (i,j) by the amount $\sum_{1 \leq k \leq K} y_{ij}^k - u_{ij}$ if the sub-problem solutions y_{ij}^k use more than the available capacity u_{ij} of that arc or it reduces the Lagrange multiplier of arc (i,j) by the amount $u_{ij} - \sum_{1 \leq k \leq K} y_{ij}^k$ if the sub-problem solutions y_{ij}^k use less than the available capacity of that arc. If, however, the decrease would cause the multiplier w_{ij}^{q+1} to become negative, then we reduce its value only to value zero. It can be proved that on a suitable choice of step size θ_q, the sub-gradient method converges in a finite number of steps to the optimal solution.

In the next section, we consider an alternate solution approach, known as *Dantzig-Wolfe decomposition*, for solving the Lagrangian multiplier problem. This approach requires considerably more work at each iteration for updating the Lagrange multipliers (the solution of a linear program) but has proved to converge faster than the sub-gradient optimization procedure for several classes of problems. Rather than describing the Dantzig-Wolfe decomposition procedure as a variant of Lagrangian relaxation, we will develop it from an alternate large-scale linear programming viewpoint that provides a somewhat different perspective on the approach.

6.5 DANTZIG-WOLFE DECOMPOSITION

To simplify our discussion in this section, we consider a special case of the multicommodity flow problem: we assume that each commodity k has a single source node s^k and a single sink node t^k, and a flow requirement of d^k units between these source and sink nodes. We also assume that we impose no flow bounds on the individual commodities other than the bundle constraints. Therefore, for each commodity k, the sub-problem constraints $\mathcal{N}x^k = b^k, x^k \geq 0$ define a shortest path problem: for this model, for any choice w_{ij} of the Lagrange multipliers for the bundle constraints, the Lagrangian relaxation requires the solution of a series of shortest path problems, one for each commodity.

6.5.1 Reformulation with Path Flows

To simplify our discussion even further, let us assume that the cost of every cycle W in the underlying network is nonnegative for every commodity. If we impose this nonnegative cycle cost condition, then in some optimal solution to the problem, the flow on every cycle is zero and so we can eliminate the cycle flow variables. Therefore, throughout this section, we assume that we can represent any potentially optimal solution as the sum of flows on directed paths.

For each commodity k, let P^k denote the collection of all directed paths from the source node s^k to the sink node t^k in the underlying network $G = (N, A)$. In the path flow formulation, each decision variable $f(P)$ is the flow on some path P, and for the kth commodity, we define this variable for every directed path P in P^k.

Let $\delta_{ij}(P)$ be an arc-path indicator variable; that is, $\delta_{ij}(P)$ equals 1 if arc (i,j) is contained in the path P and is 0 otherwise. Then, the flow decomposition theorem of network flows states that we can always decompose some optimal arc flow x_{ij}^k into path flows $f(P)$ as follows:

$$x_{ij}^k = \sum_{p \in P^k} \delta_{ij}(P) f(P)$$

By substituting the path variables in the multicommodity flow formulation, we obtain the following equivalent path flow formulation:

$$\text{Minimize} \sum_{1 \leq k \leq K} \sum_{P \in P^k} c^k(P) f(P) \tag{6.13}$$

subject to

$$\sum_{1 \leq k \leq K} \sum_{P \in P^k} \delta_{ij}(P) f(P) \leq u_{ij} \text{ for all } (i,j) \in A, \tag{6.14}$$

$$\sum_{P \in P^k} f(P) = d^k \text{ for all } k = 1, \ldots, K, \tag{6.15}$$

$$f(P) \geq 0 \text{ for all } k = 1, \ldots, K \text{ and all } P \in P^k. \tag{6.16}$$

Note that the path flow formulation of the multicommodity flow problem has a very simple constraint structure. The problem has a single constraint for each arc (i, j), which states the sum of the path flows passing through the arc is at most u_{ij}, the capacity of the arc. Further, the problem has a single constraint (6.15) for each commodity k, which states that the total flow on all the paths connecting the source node s^k and sink node t^k of commodity k must equal the demand d^k for this commodity. For a network with n nodes, m arcs, and K commodities, the path flow formulation contains $m + K$ constraints (in addition to the nonnegativity restrictions imposed on the path flow values).

6.5.2 Optimality Conditions

The revised simplex method of linear programming maintains a basis at every step, and using this basis determines a vector of simplex multipliers for the constraints. Since the path flow formulation (6.13) contains one bundle constraint for each arc and one demand constraint (6.15) for every commodity, the dual linear program has a dual variable w_{ij} for each arc and another dual variable σ^k for each commodity $k = 1, 2, \ldots, K$. With respect to these dual variables, the reduced cost $c_P^{\sigma, w}$ for each path flow variable $f(P)$ is

$$c_P^{\sigma, w} = c^k(P) + \sum_{(i,j) \in P} w_{ij} - \sigma^k$$

The complimentary slackness conditions (6.5 through 6.8) is discussed elaborately in the following section.

6.5.3 Path Flow Complementary Slackness Conditions

The commodity path flows $f(P)$ are optimal in the path flow formulation (6.13 through 6.16) of the multicommodity flow problem if and only if we can find (nonnegative) arc prices w_{ij} and commodity prices σ^k so that the reduced costs and arc flows satisfy the following complementary slackness conditions:

a.
$$w_{ij}\left[\sum_{1\leq k\leq K}\sum_{P\in P^k}\delta_{ij}(P)f(P)-u_{ij}\right]=0 \text{ for all } (i,j)\in A. \qquad (6.17)$$

b.
$$c_P^{\sigma,w}\geq 0 \text{ for all } k=1,\ldots,K \text{ and all } P\in P^k. \qquad (6.18)$$

c.
$$c_P^{\sigma,w}f(P)=0 \text{ for all } k=1,\ldots,K \text{ and all } P\in P^k. \qquad (6.19)$$

The Dantzig-Wolfe's decomposition method works in the following manner. Imagine that K different decision makers as well as one *coordinator* are solving the K-commodity flow problem. The coordinator's job is to solve the path formulation (6.13 through 6.16) of the problem, which we refer to as the *master or coordinating* problem. In general, the coordinator has on hand only a subset of the columns of the master problem. Since the coordinator can, at best, solve the linear program as restricted to this subset of columns, we refer to this smaller linear program as *the restricted master problem*. Each of these K decision makers plays a special role in solving the problem. The K decision makers, with guidance from the coordinator in the form of arc prices, generate entering variables or columns, with the kth decision maker generating the columns of the master problem corresponding to the kth commodity.

The path formulation of the multicommodity flow problem has $m+K$ constraints: (1) one for each commodity k, specifying that the flow of commodity k is d^k and (2) one for each arc (i,j), specifying that the total flow on that arc is at most u_{ij}. The coordinator solves the restricted master problem to optimality using any linear programming technique, such as the simplex algorithm, and then needs to determine whether the solution to the restricted master is optimal for the original problem, or if some another column has a negative reduced cost and can enter the basis. To this end, the coordinator broadcasts the optimal set of simplex multipliers (or prices) of the restricted master problem; that is, broadcasts an arc price w_{ij} associated with arc (i,j) and a path length σ^k associated with each commodity k.

After the coordinator has broadcast the prices, the decision maker for commodity k determines the least cost way of shipping d^k units from the source node s^k to the sink node t^k of commodity k, assuming that each arc (i,j) has an associated toll of w_{ij} in addition to its arc cost c_{ij}^k. If the cost of this shortest path is less than σ^k, this would imply that the corresponding reduced cost $c_P^{\sigma,w}$ is negative; then the kth decision maker will report this solution to the coordinator as an improving solution, and the decision variable corresponding to this path becomes a candidate to enter the basis. If the cost of this path equals σ^k, then the kth decision maker need not report anything to the coordinator. (The cost will never be less than σ^k because the current solution is optimal to the restricted master problem, and, hence, the coordinator is already using some path of cost σ^k for the kth commodity in the current solution.) The algorithm terminates when no decision maker finds a candidate path to enter the basis (complementary slackness conditions (6.17) are satisfied). Note that to price out the columns for commodity k, we need to solve the following shortest path sub-problem:

$$\text{Minimize } \sum_{(i,j)\in A}(c_{ij}^k+w_{ij})x_{ij}^k$$

subject to
$$Nx^k = b^k$$
$$x^k \geq 0$$

By solving this sub-problem, we generate the entering variable columns dynamically and employ the popular column generation technique that is used to solve large linear programming problems. Also, notice that the K sub-problems correspond to the relaxed Lagrangian problem with a multiplier of w_{ij} imposed upon each arc (i,j). Consequently, we could view the coordinator as setting the Lagrange multipliers and solving the Lagrangian multiplier problem. In fact, Dantzig-Wolfe decomposition is an efficient method for solving the Lagrangian multiplier problem if we measure efficiency by the number of iterations an algorithm performs. Unfortunately, in applying Dantzig-Wolfe decomposition, at each iteration the coordinator must solve a linear program with $m + K$ constraints, and this update step for the simplex multipliers is very expensive. It is far more time consuming to solve a linear program than to update the multipliers using sub-gradient optimization. Because each multiplier update for Dantzig-Wolfe decomposition is so expensive computationally, the Dantzig-Wolfe decomposition method has generally not proven to be an efficient method for solving the multicommodity flow problem; nevertheless, Dantzig-Wolfe decomposition has one important advantage that distinguishes it from other Lagrangian-based algorithms. It can be shown that the solution to the sub-problems provides us with a lower bound on the optimal value of the problem. Consequently, at each step we also have a bound on how far the current feasible solution is from optimal. Therefore, we can terminate the algorithm at any step not only with a feasible solution, but also with a guarantee of how far that solution is from optimality.

6.6 RESOURCE-DIRECTIVE DECOMPOSITION

Lagrangian relaxation and Dantzig-Wolfe decomposition are price-directive methods that decompose the multicommodity flow problem into single commodity network flow problems by placing tolls or prices on the complicating bundle constraints. The resource-directive method that we consider in this section takes a different approach. Instead of using prices to decompose the problem, it allocates the joint bundle capacity of each arc to the individual commodities. When applied to the problem formulation (6.1), the resource-directive approach allocates $r_{ij}^k \leq u_{ij}^k$ units of the bundle capacity u_{ij} of arc (i,j) to commodity k, producing the following *resource-directive problem*.

$$z = \min \sum_{1 \leq k \leq k} c^k x^k \tag{6.20}$$

subject to

$$\sum_{1 \leq k \leq K} r_{ij}^k \leq u_{ij} \text{ for all } (i,j) \in A, \tag{6.21}$$

$$Nx^k = b^k \text{ for } k = 1, \ldots, K, \tag{6.22}$$

$$0 \leq x_{ij}^k \leq r_{ij}^k \text{ for all } (i,j) \in A \text{ and all } k = 1, \ldots, K. \tag{6.23}$$

Note that the constraint (6.21) ensures that the total resource allocation for arc (i,j) does not exceed that arc's bundle capacity. Let $r = (r_{ij}^k)$ denote the vector of resource allocations. We now make the following elementary observations about the problem.

Property 6.1 *The resource-directive problem (6.20 through 6.23) is equivalent to the original multicommodity flow problem (6.1) in the sense that (i) if (x, r) is feasible in the resource-directive problem, then x is feasible for the original problem and has the same objective* *function value and (ii) if x is feasible in the original problem and we set $r = x$, then (x, r) is feasible and has the same objective function value in the resource-directive problem.* ∎

Now consider the following sequential approach for solving the resource-directive problem (6.20). Instead of solving the problem by choosing the vectors r and x simultaneously, let us choose them sequentially. We first fix the resource allocations r_{ij}^k and then choose the flows x_{ij}^k. Let $z(r)$ denote the optimal value of the resource-directive problem for a fixed value of the resource allocation r and consider the following derived *resource-allocation* problem:

$$\text{Minimize } z(r) \tag{6.24}$$

subject to

$$\sum_{1 \leq k \leq K} r_{ij}^k \leq u_{ij} \text{ for all } (i,j) \in A, \tag{6.25}$$

$$0 \leq r_{ij}^k \leq u_{ij}^k \text{ for all } (i,j) \in A \text{ and all } k = 1, \ldots, K. \tag{6.26}$$

The objective function $z(r)$ for this problem is complicated. We know its value only implicitly as the solution of an optimization problem in the flow variables x_{ij}^k. Moreover, note that for any fixed value of the resource variables r_{ij}^k, the resource-directive problem decomposes into a separate network flow sub-problem for each commodity. That is, $z(r) = \sum_{k \in K} z^k(r^k)$ with the value $z^k(r^k)$ of the kth sub-problem given by

$$z^k(x^k) = \min \; c^k x^k \tag{6.27}$$

subject to

$$Nx^k = b^k \text{ for all } k = 1, \ldots, K, \tag{6.28}$$

$$0 \leq x_{ij}^k \leq r_{ij}^k \text{ for all } (i,j) \in A \text{ and all } k = 1, \ldots, K. \tag{6.29}$$

We now make the following observation establishing the relationship between the resource-allocation and the resource-planning problem.

Property 6.2 *The resource-directive problem (6.20) is equivalent to the resource-allocation problem (6.24) in the sense that (1) if (x, r) is feasible in the resource-directive problem, then r is feasible in the resource-allocation problem and $z(r) \leq cx$, and (ii) if r is feasible in the resource-allocation problem, then for some vector x, (x, r) is feasible in the original problem and $cx = z(r)$.* ∎

Let us pause to consider the implications of Properties 6.1 and 6.2. They imply that rather than solving the multicommodity flow problem directly, we can decompose it into a resource-allocation problem with a very simple constraint structure with a single inequality constraint, but with a complex objective function $z(r)$. Although the overall structure of the objective function is complicated, it is easy to evaluate: to find its value for any choice of the resource-allocation vector r, we need merely to solve K single commodity flow problems.

Another way to view the objective function $z(r)$ is as the cost of the linear program (6.20) as a function of the right-hand side parameters r. That is, any value r for the allocation vector defines the values of right-hand side parameters for this linear program. A well-known result in linear programming shows us that the function has a special form. We state this result for a general linear programming problem that contains the multicommodity flow problem as a special case.

Property 6.3 *Let r denote the set of allocations for which the linear program minimize $\{cx : \mathcal{A}\,x = b, 0 \leq x \leq r\}$ is feasible. Let $z(r)$ denote the value of this linear program as a function of right-hand side parameter r. The objective function $z(r)$ is a piecewise linear convex function of r.*

Proof. To establish convexity of $z(r)$, we need to show that if $\bar{r}$ and $\hat{r}$ are any two values of the parameter r for which the given linear program is feasible and θ is any scalar, $0 \leq \theta \leq 1$, then $z(\theta \bar{r} + (1-\theta)\hat{r}) \leq \theta z(\bar{r}) + (1-\theta)\hat{r}$. Let $\bar{y}$ and $\hat{y}$ be optimal solutions to the linear program for the parameter choices $r = \bar{r}$ and $r = \hat{r}$. Note that $A\bar{y} = b$, $A\hat{y} = b$, $\bar{y} \leq \bar{r}$, and $\hat{y} \leq \hat{r}$. But then $A(\theta\bar{y} + (1-\theta)\hat{y}) = b$ and, hence, $\theta\bar{y} + (1-\theta)\hat{y} \leq \theta\bar{r} + (1-\theta)\hat{r}$. Therefore, the vector $\theta\bar{y} + (1-\theta)\hat{y}$ is feasible for the linear program with the parameter vector $r = \theta\bar{r} + (1-\theta)\hat{r}$, and so the optimal objective function value for this problem is at most $c(\theta\bar{y} + (1-\theta)\hat{y})$. Moreover, by our choice of $\bar{y}$ and $\hat{y}$, $z(\bar{r}) = c\bar{y}$ and $z(\hat{r}) = c\hat{y}$; therefore,

$$z(\theta\bar{r} + (1-\theta)\hat{r}) \leq c(\theta\bar{y} + (1-\theta)\hat{y}) = \theta z(\bar{r}) + (1-\theta)z(\hat{r})$$

and so $z(r)$ is a convex function.

The piecewise linearity of $z(r)$ follows from the optimal basis property of linear programs. That is, for any choice of the parameter r, the problem has a basic feasible optimal solution, and this basic feasible solution remains optimal for all values of r for which it remains feasible. Moreover, the objective function value of the linear program is linear in r for any given (optimal) basis. ■

A number of algorithmic approaches are available for solving the resource-directive models that we have introduced in this section. Since the function $z(r)$ is nondifferentiable (because it is piecewise linear), we cannot use gradient methods from nonlinear programming to solve the resource-allocation problem. We could, instead, use several other approaches. For example, we could search for local improvement in $z(r)$ using a heuristic method. As one such possibility, we could use an *arc-at-a-time approach* by adding 1 to $r_{pq}^{k'}$ and subtracting 1 from $r_{pq}^{k''}$ for some arc (p, q) for two commodities k' and k'', choosing the arc and commodities at each step using some criterion (e.g., the choices that give the greatest decrease in the objective function value at each step). This approach is easy to implement but does not ensure convergence to an optimal solution. Note that we can view this approach as changing the resource allocation at each step using the formula $r \leftarrow r + \theta\gamma$ with a step length of $\theta = 1$ and a movement direction of $\gamma = \gamma_{ij}^k$ given by $\gamma_{pq}^{k'} = 1$, $\gamma_{pq}^{k''} = -1$, and $\gamma_{ij}^k = 0$ for all other arc commodity combinations. Borrowing ideas from sub-gradient optimization, however, we could use an optimization approach by choosing the movement direction γ as a sub-gradient corresponding to the resource-allocation r. A natural approach would be to search for a sub-gradient or movement direction γ and step length θ that simultaneously maintains feasibility and ensures convergence to an optimal solution r of the resource-allocation problem (6.24 through 6.26). Since the scope of this discussion is purely mathematical and not related to network flows, we do not delve into details here but instead direct the reader to suitable references in the references section.

Further Reading

Researchers have proposed a number of basic approaches for solving the multicommodity flow problem. The following are the three basic approaches, all based upon exploiting network flow substructure, that we have mentioned in this chapter and selected references: (1) price-directive decomposition algorithm [1–7]; (2) resource-directive decomposition algorithm [4,8,9]; and (3) basis partitioning [10–12]. Ford and Fulkerson [13] and Tomlin [14] first suggested the column generation approach. The first of these papers was the forerunner to the general Dantzig and Wolfe [15] decomposition procedure of mathematical programming. The excellent survey papers by Assad [16] and by Kennington [17] describe all of these algorithms and several standard properties of multicommodity flow problems. The book by Kennington and Helgason [11] and the doctoral dissertation by Schneur [18] are other valuable references on this topic.

Most of the material discussed in this chapter is classical and dates from the 1960s and 1970s. Many of the standard properties of multicommodity flows (e.g., nonintegrality of optimal flows), are due to Fulkerson [19], Hu [20], and Sakarovitch [21]. The decomposition methods that we have considered in this chapter extend to other situations as long as we can represent any solution to a problem as a convex combination of other particularly *simple solutions*; in the text we have used shortest paths as the simple solutions. For some applications, we might use solutions to knapsack problems as the simple solutions, and in other cases, such as the general multicommodity flow problem with multiple sources and destinations for each commodity, the simple solutions might be spanning tree solutions.

Researchers have developed and tested several codes for multicommodity flow problems. Kennington [17] and Ali et al. [22] have described the results of some of these computational experiments. These results have suggested that price-directive and partitioning algorithms are the fastest algorithms for solving multicommodity flow problems. The best multicommodity flow codes are 2 to 5 times faster than a general-purpose linear programming code. Computational experience by Bixby [23] in solving large-scale network flow problems with side constraints has shown that the simplex method with an advanced starting basis technique can be very effective computationally.

Interior point algorithms provide another approach for solving multicommodity flow problems. Although these algorithms yield the only known polynomial-time bounds for these problems, an efficient and practical implementation of these algorithms has been the subject of research. The best time bound for the multicommodity flow problem is due to Vaidya [24]. Tardos' [25] algorithm solves the multicommodity flow problem in strongly polynomial time. Vaidya and Kapoor [26] show how to speed up Karmarkar's linear programming algorithm for the special case of multicommodity flows. Chardaire and Lisser [27] and Castro [28] developed other interior point methods.

Several researchers have suggested other algorithms for the multicommodity flow problem: Gersht and Shulman [29], Barnhart [30], Pinar and Zenios [31], Barnhart [32], Barnhart and Sheffi [33], Farvolden et al. [34], Frangioni and Gallo [35], and Detlefsen and Wallace [36]. Schneur [18] and Schneur and Orlin [37] studied scaling techniques for the multicommodity flow problem. Matsumoto et al. [38] gave a polynomial-time combinatorial algorithm for solving a multicommodity flow problem in $s-t$ planar networks. Radzig [39] developed approximation algorithms for the multicommodity flow problem. Barnhart et al. [40] developed a column generation model and an integer programming based branch-and-price-and-cut algorithm for integral multicommodity flow problems, and Brunetta et al. [41] investigated polyhedral approaches and branch-and-cut algorithms for the integral version see also Assad [58].

Multicommodity network flow models have wide applications in several domains. The routing of multiple commodities application has been adapted from Golden [42] and Crainic et al. [43], and the multivehicle tanker scheduling problem from Bellmore et al. [44]. Other applications are due to Kaplan [45], Evans [46], Geoffrion and Graves [8], Bodin et al. [47], Assad [4], Korte [48], Gautier and Granot [49], Lin and Yuan [50], Moz and Pato [51], Ouaja and Richards [52], Ahuja et al. [53], Vaidyanathan et al. [54], Vaidyanathan et al. [55], and Kumar et al. [56]. The book by Ahuja et al. [57] provides extensive coverage of applications of network and multicommodity flows.

References

[1] Cremeans, J.E., R.A. Smith, and G.R. Tyndall. Optimal multicommodity network flows with resource allocation. *Naval Research Logistics Quarterly* **17** (1970), 269–280.

[2] Swoveland, C. 1971. Decomposition algorithms for the multi-commodity distribution problem. Working Paper No. 184, Western Management Science Institute, University of California, Los Angeles, CA.

[3] Chen, H. and C.G. Dewald. A generalized chain labeling algorithm for solving multicommodity flow problems. *Computers & Operations Research* **1** (1974), 437–465.

[4] Assad, A.A. Models for rail transportation. *Transportation Research* **14A** (1980a), 205–220.

[5] Mamer, J.W. and R.D. McBride. A decomposition-based pricing procedure for large-scale linear programs: An application to the linear multicommodity flow problem. *Management Science* **46** (2000), 693–709.

[6] Holmberg, K. and D. Yuan. A multicommodity network-flow problem with side constraints on paths solved by column generation. *INFORMS Journal on Computing* **15** (2003), 42–57.

[7] Larsson, T. and D. Yuan. An augmented Lagragian algorithm for large scale multicommodity routing. *Computational Optimization and Applications* **27** (2004), 187–215.

[8] Goeffrion, A.M. and G.W. Graves. Multicommodity distribution system design by Benders decomposition. *Management Science* **20** (1974), 822–844.

[9] Kennington, J.L. and M. Shalaby. An effective subgradient procedure for minimal cost multicommodity flow problems. *Management Science* **23** (1977), 994–1004.

[10] Graves, G.W. and R.D. McBride. The factorization approach to large scale linear programming. *Mathematical Programming* **10** (1976), 91–110.

[11] Kennington, J.L. and R.V. Helgason. *Algorithms for Network Programming*. Wiley-Interscience, New York, 1980.

[12] Castro, J. and N. Nabona. An implementation of linear and nonlinear multicommodity network flows. *European Journal of Operational Research* **92** (1996), 37–53.

[13] Ford, L.R. and D.R. Fulkerson. A suggested computation for maximal multicommodity network flow. *Management Science* **5** (1958b), 97–101.

[14] Tomlin, J.A. A linear programming model for the assignment of traffic. *Proceedings of the 3rd Conference of the Australian Road Research Board* **3** (1966), 263–271.

[15] Dantzig, G.B. and P. Wolfe. Decomposition principle for linear programs. *Operations Research* **8** (1960), 101–111.

[16] Assad, A.A. Multicommodity network flows—A survey. *Networks* **8** (1978), 37–91.

[17] Kennington, J.L. Survey of linear cost multicommodity network flows. *Operations Research* **26** (1978), 209–236.

[18] Schneur, R. Scaling algorithms for multicommodity flow problems and network flow problems with side constraints. PhD Dissertation, Department of Civil Engineering, MIT, Cambridge, MA, 1991.

[19] Fulkerson, D.R. Flows in networks. In *Recent Advances in Mathematical Programming*, (eds.) R.L. Graves and P. Wolfe, McGraw-Hill, New York, 319–332, 1963.

[20] Hu, T.C. Multi-commodity network flows. *Operations Research* **11** (1963), 344–360.

[21] Sakarovitch, M. Two commodity network flows and linear programming. *Mathematical Programming* **4** (1973), 1–20.

[22] Ali, A.I., D. Barnett, K. Farhangian, J.L. Kennington, B. Patty, B. Shetty, B. McCarl, and P. Wong. Multicommodity network problems: Applications and computations. *A.I.I.E. Transactions* **16** (1984), 127–134.

[23] Bixby, R.E. The simplex method—It keeps getting better. Presented at the *14th International Symposium on Mathematical Programming*, Amsterdam, the Netherlands, 1991.

[24] Vaidya, P.M. Speeding up linear programming using fast matrix multiplication. In *Proceedings of the 30th Annual Symposium on the Foundations of Computer Science*, IEEE, 332–337, 1989.

[25] Tardos, E. A strongly polynomial algorithm to solve combinatorial linear programs. *Operations Research* **34** (1986), 250–256.

[26] Vaidya, P.M. and S. Kapoor. Speeding up Karmarkar's algorithm for multicommodity flows. *Mathematical Programming (Series A)* **73** (1996), 111–127.

[27] Chardaire, P. and A. Lisser. Simplex and interior point specialized algorithms for solving nonoriented multicommodity flow problems. *Operations Research* **50** (2002), 260–276.

[28] Castro, J. Solving difficult multicommodity problems with a specialized interior-point algorithm. *Annals of Operation Research* **124** (2003), 35–48.

[29] Gersht, A. and A. Shulman. A new algorithm for the solution of the minimum cost multicommodity flow problem. *Proceedings of the IEEE Conference on Decision and Control* **26** (1987), 748–758.

[30] Barnhart, C. A network-based primal-dual solution methodology for the multicommodity network flow problem. PhD Dissertation, Department of Civil Engineering, MIT, Cambridge, MA, 1988.

[31] Pinar, M.C. and S.A. Zenios. Parallel decomposition of multicommodity network flows using smooth penalty functions. Technical Report 90-12-06, University of Pennsylvania, Philadelphia, PA, 1990.

[32] Barnhart, C. Dual-ascent methods for large-scale multi-commodity flow problems. *Naval Research Logistics* **40** (1993), 305–324.

[33] Barnhart, C. and Y. Sheffi. A network-based primal-dual heuristic for the solution of multicommodity network flow problems. *Transportation Science* **27** (1993), 102–117.

[34] Farvolden, J.M., W.B. Powell, and I.J. Lustig. A primal partitioning solution for the arc-chain formulation of a multicommodity network flow problem. *Operations Research* **41** (1993), 669–693.

[35] Frangioni, A. and G. Gallo. A bundle type dual-ascent approach to linear multicommodity min-cost flow problems. *INFORMS Journal on Computing* **11** (1999), 370–393.

[36] Detlefsen, N.K. and S.W. Wallace. The simplex algorithm for multicommodity networks. *Networks* **39** (2002), 15–28.

[37] Schneur, R. and J.B. Orlin. A scaling algorithm for multicommodity flow problems. *Operations Research* **46** (1998), 231–246.

[38] Matsumoto, K., T. Nishizeki, and N. Saito. An efficient algorithm for finding multicommodity flows in planar networks. *SIAM Journal on Computing* **14** (1985), 289–302.

[39] Radzig, T. Fast deterministic approximation for the multicommodity flow problem. *Mathematical Programming* **78** (1997), 43–58.

[40] Barnhart, C., C.A. Hane, and P.H. Vance. Using branch-and-price-and-cut to solve origin-destination integer multicommodity flow problems. *Operations Research* **48** (2000), 318–326.

[41] Brunetta, L., M. Conforti, and M. Fischetti. A polyhedral approach to an integer multicommodity flow problem. *Discrete Applied Mathematics* **101** (2000), 13–36.

[42] Golden, B.L. A minimum cost multicommodity network flow problem concerning imports and exports. *Networks* **5** (1975), 331–356.

[43] Crainic, T., J.A. Ferland, and J.M. Rousseau. A tactical planning model for rail freight transportation. *Transportation Science* **18** (1984), 165–184.

[44] Bellmore, M., G. Bennington, and S. Lubore. A multivehicle tanker scheduling problem. *Transportation Science* **5** (1971), 36–47.

[45] Kaplan, S. Readiness and the optimal redeployment of resources. *Naval Research Logistics Quarterly* **20** (1973), 625–638.

[46] Evans, J.R. Some network flow models and heuristics for multiproduct production and inventory planning. *A.I.I.E. Transactions* **9** (1977), 75–81.

[47] Bodin, L.D., B.L. Golden, A.D. Schuster, and W. Rowing. A model for the blockings of trains. *Transportation Research* **14B** (1980), 115–120.

[48] Korte, B. Applications of combinatorial optimization. Technical Report No. 88541-OR. Institute für Okonometrie und Operations Research, Bonn, Germany, 1988.

[49] Gautier, A. and F. Granot. Forest management: A multicommodity flow formulation and sensitivity analysis. *Management Science* **41** (1995), 1654–1688.

[50] Lin, Y. and J. Yuan. On a multicommodity flow network reliability model and its application to a container-loading transportation problem. *Journal of Operations Research Society of Japan* **44** (2001), 366–377.

[51] Moz, M. and M.V. Pato. An integer multicommodity flow model applied to the rerostering of nurse schedules. *Annals of Operations Research* **119** (2003), 285–301.

[52] Ouaja, W. and B. Richards. A hybrid multicommodity routing algorithm for traffic engineering. *Networks* **43** (2004), 125–140.

[53] Ahuja, R.K., J. Liu, J.B. Orlin, D. Sharma, and L.A. Shughart. Solving real-life locomotive scheduling problems. *Transportation Science* **39** (2005), 503–517.

[54] Vaidyanathan, B., K.C. Jha, and R.K. Ahuja. Multi-commodity network flow approach to the railroad crew-scheduling problem. *IBM Journal of Research and Development* **51** (2007b), 325–344.

[55] Vaidyanathan, B., R.K. Ahuja, J. Liu, and L.A. Shughart. Real-life locomotive planning: New formulations and computational results. *Transportation Research B*, 2007a.

[56] Kumar, A., B. Vaidyanathan, and R.K. Ahuja. Railroad locomotive scheduling. In *Encyclopedia of Optimization*, 2nd Ed., (eds.) C.A. Floudas and P.M. Pardalos, Springer, New York, 2007.

[57] Ahuja, R.K., T.L. Magnanti, and J.B. Orlin. *Network Flows: Theory, Algorithms, and Applications*. Prentice Hall, NJ, 1993.

[58] Assad, A.A. Solving linear multicommodity flow problems. *Proceedings of the IEEE International Conference on Circuits and Computers*, 157–161, 1980b.

III

Algebraic Graph Theory

CHAPTER 7

Graphs and Vector Spaces*

Krishnaiyan "KT" Thulasiraman

M. N. S. Swamy

CONTENTS

7.1	Introduction	177
7.2	Fundamental Circuits and Cutsets	177
7.3	Spanning Trees, Circuits, and Cutsets	179
7.4	Circuit and Cutset Spaces of a Graph	182
7.5	Dimensions of Circuit and Cutset Subspaces	184
7.6	Relationship between Circuit and Cutset Subspaces	185
7.7	Orthogonality of Circuit and Cutset Subspaces	186

7.1 INTRODUCTION

Electrical circuit theory is one of the earliest applications of graph theory to a problem in physical science. The dynamic behavior of an electrical circuit is governed by three laws: Kirchhoff's voltage law, Kirchhoff's current law, and Ohm's law. Each element in a circuit is associated with two variables, namely, the current variable and the voltage variable. Kirchhoff's voltage law requires that the algebraic sum of the voltages around a circuit be zero, and Kirchhoff's current law requires that the algebraic sum of the currents across a cut be zero. Thus, circuits and cuts define a linear relationship among the voltage variables and a linear relationship among the current variables, respectively. It is for this reason that circuits, cuts, and the vector spaces associated with them have played a major role in the discovery of several fundamental properties of electrical circuits arising from the structure or the interconnection of the circuit elements. Several graph theorists and circuit theorists have immensely contributed to the development of what we may now call the structural theory of electrical circuits. The significance of the results to be presented in this section goes well beyond their application to circuit theory. They will bring out the fundamental duality that exists between circuits and cuts and the influence of this duality on the structural theory of graphs. Most of the results in this section are also relevant to the development of combinatorial optimization theory as well as matroid theory.

7.2 FUNDAMENTAL CIRCUITS AND CUTSETS

Consider a spanning tree T of a connected graph G. Let the branches of T be denoted by $b_1, b_2, \ldots, b_{n-1}$, and let the chords of T be denoted by $c_1, c_2, \ldots, c_{m-n+1}$, where n is the number of vertices in G and m is the number of edges in G.

---

*This chapter is an edited version of the Chapter 4 in Reference 1.

While T is acyclic, the graph $T \cup c_i$ contains exactly one circuit C_i. This circuit consists of the chord c_i and those branches of T which lie in the unique path in T between the end vertices of c_i. The circuit C_i is called the *fundamental circuit* of G with respect to the chord c_i of the spanning tree T.

The set of all the $m - n + 1$ fundamental circuits $C_1, C_2, \ldots, C_{m-n+1}$ of G with respect to the chords of the spanning tree T of G is known as the *fundamental set of circuits* of G with respect to T. The nullity $\mu(G)$ of a connected graph G is defined to be equal to $m - n + 1$. If G is not connected and has p components, then $\mu(G) = m - n + p$.

An important feature of the fundamental circuit C_i is that it contains exactly one chord, namely, chord c_i. Further, chord c_i is not present in any other fundamental circuit with respect to T.

A graph G and a set of fundamental circuits of G are shown in Figure 7.1. A cutset S of a connected graph G is a minimal set of edges of G such that its removal from G disconnects G, that is, the graph $G - S$ is disconnected.

We next define the concept of a cut which is closely related to that of a cutset. Consider a connected graph G with vertex set V. Let V_1 and V_2 be two mutually disjoint subsets of V such that $V = V_1 \cup V_2$; that is, V_1 and V_2 have no common vertices and together contain all the vertices of V. Then the set S of all those edges of G having one end vertex in V_1 and the other in V_2 be called a cut of G. This is usually denoted by $\langle V_1, V_2 \rangle$. Reed [2] refers to a cut as a seg (the set of edges segregating the vertex set V).

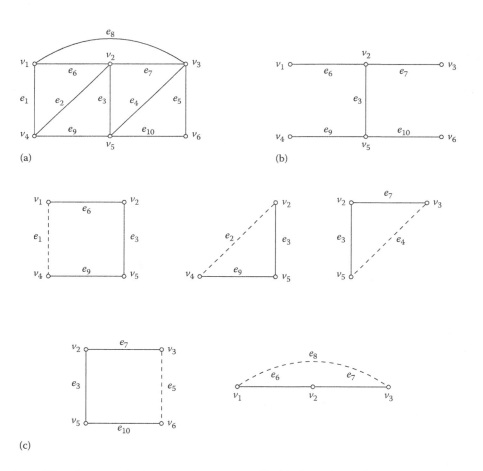

Figure 7.1 Set of fundamental circuits of a graph G. (a) Graph G. (b) Spanning tree T of G. (c) Set of five fundamental circuits of G with respect to T. (Chords are indicated by dashed lines.)

Note that the cut $\langle V_1, V_2 \rangle$ of G is a set of edges of G whose removal disconnects G into two graphs G_1 and G_2 which are induced subgraphs of G on the vertex sets V_1 and V_2. G_1 and G_2 may not be connected. If both these graphs are connected, then $\langle V_1, V_2 \rangle$ is also the minimal set of edges disconnecting G into exactly two components. Then, by definition, $\langle V_1, V_2 \rangle$ is a cutset of G.

Any cut $\langle V_1, V_2 \rangle$ in a connected graph G contains a cutset of G, since the removal of $\langle V_1, V_2 \rangle$ from G disconnects G. In fact, we can prove that a cut in a graph G is the union of some edge-disjoint cutsets of G. Formally, we state this in the following theorem.

Theorem 7.1 *A cut in a connected graph G is the union of some edge-disjoint cutsets of G.* ∎

Consider next a vertex v_1 in a connected graph. The set of edges incident on v_1 forms the cut $\langle v_1, V - v_1 \rangle$. The removal of these edges disconnects G into two subgraphs. One of these subgraphs containing only the vertex v_1 is, by definition, connected. The other subgraph is the induced subgraph G' of G on the vertex set $V - v_1$. Thus, the cut $\langle v, V - v_1 \rangle$ is a cutset if and only if G' is connected. However, G' is connected if and only if v_1 is not a cut-vertex. Thus we have the following theorem.

Theorem 7.2 *The set of edges incident on a vertex v in a connected graph G is a cutset of G if and only if v is not a cut-vertex of G.* ∎

We now show how a spanning tree can be used to define a set of fundamental cutsets.

Consider a spanning tree T of a connected graph G. Let b be a branch of T. Now, removal of the branch b disconnects T into exactly two components T_1 and T_2. Note that T_1 and T_2 are trees of G. Let V_1 and V_2, respectively, denote the vertex sets of T_1 and T_2. V_1 and V_2 together contain all vertices of G.

Let G_1 and G_2 be, respectively, the induced subgraphs of G on the vertex sets V_1 and V_2. It can be seen that T_1 and T_2 are, respectively, the spanning trees of G_1 and G_2. Hence, G_1 and G_2 are connected. This, in turn, proves that the cut $\langle V_1, V_2 \rangle$ is a cutset of G. This cutset is known as the fundamental cutset of G with respect to the branch b of the spanning tree T of G. The set of all the $n - 1$ fundamental cutsets with respect to the $n - 1$ branches of a spanning tree T of a connected graph G is known as the *fundamental set of cutsets* of G with respect to the spanning tree T. The rank $\rho(G)$ of a connected G is defined to be equal to $n - 1$. If G has p components, then $\rho(G) = n - p$.

Note that the cutset $\langle V_1, V_2 \rangle$ contains exactly one branch, namely, the branch b of T. All the other edges of $\langle V_1, V_2 \rangle$ are chords of T. This follows from the fact that $\langle V_1, V_2 \rangle$ does not contain any edge of T_1 or T_2. Further, branch b is not present in any other fundamental cutset with respect to T.

A graph G and a set of fundamental cutsets of G are shown in Figure 7.2.

7.3 SPANNING TREES, CIRCUITS, AND CUTSETS

In this section, we discuss some interesting results which relate cutsets and circuits to spanning trees and cospanning trees, respectively. These results will bring out the *dual* nature of circuits and cutsets. They will also lead to alternate characterizations of cutsets and circuits in terms of spanning trees and cospanning trees, respectively.*

It is obvious that removal of a cutset S from a connected graph G destroys all the spanning trees of G. A little thought will indicate that a cutset is a minimal set of edges whose removal from G destroys all spanning trees of G. However, the converse of this result

---
*See Section 1.6 for the definition of a cospanning tree.

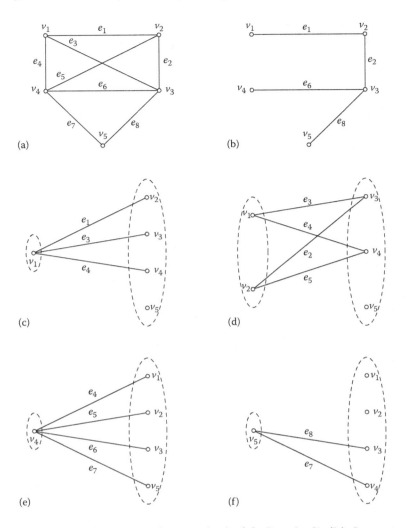

Figure 7.2 Set of fundamental cutsets of a graph G. (a) Graph G. (b) Spanning tree T of G. (c) Fundamental cutset with respect to branch e_1. (d) Fundamental cutset with respect to branch e_2. (e) Fundamental cutset with respect to branch e_6. (f) Fundamental cutset with respect to branch e_8.

is not so obvious. The first few theorems of this section discuss these questions and similar ones relating to circuits.

Theorem 7.3 *A cutset of a connected graph G contains at least one branch of every spanning tree of G.*

Proof. Suppose that a cutset S of G contains no branch of a spanning tree T of G. Then the graph $G - S$ will contain the spanning tree T and hence, $G - S$ is connected. This, however, contradicts that S is a cutset of G. ∎

Theorem 7.4 *A circuit of a connected graph G contains at least one edge of every cospanning tree of G.*

Proof. Suppose that a circuit C of G contains no edge of the cospanning tree T^* of a spanning tree T of G. Then the graph $G - T^*$ will contain the circuit C. Since $G - T^*$ is the same as spanning tree T, this means that the spanning tree T contains a circuit. However, this is contrary to the definition of a spanning tree. ∎

Theorem 7.5 *A set S of edges of a connected graph G is a cutset of G if and only if S is a minimal set of edges containing at least one branch of every spanning tree of G.*

Proof. Necessity: If the set S of edges of G is a cutset of G, then, by Theorem 7.3, it contains at least one branch of every spanning tree of G. If it is not a minimal such set, then a proper subset S' of S will contain at least one branch of every spanning tree of G. Then $G - S'$ will contain no spanning tree of G and it will be disconnected. Thus, removal of a proper subset S' of the cutset S of G will disconnect G. This, however, would contradict the definition of a cutset. Hence the necessity.

Sufficiency: If S is a minimal set of edges containing at least one branch of every spanning tree of G, then the graph $G - S$ will contain no spanning tree, and hence it will be disconnected. Suppose S is not a cutset, then a proper subset S' of S will be a cutset. Then, by the necessity part of the theorem, S' will be a minimal set of edges containing at least one branch of every spanning tree of G. This, however, will contradict that S is a minimal such set. Hence the sufficiency. ∎

The above theorem gives a characterization of a cutset in terms of spanning trees. We would like to establish next a similar characterization for a circuit in terms of cospanning trees.

Consider a set C of edges constituting a circuit in a graph G. By Theorem 7.4, C contains at least one edge of every cospanning tree of G. We now show that no proper subset C' of C has this property. It is obvious that C' does not contain a circuit. Hence, we can construct a spanning tree T that contains C'. The cospanning tree T^* corresponding to T has no common edge with C'. Hence for every proper subset C' of C, there exists at least one cospanning tree T^* which has no common edge with C'. In fact, this statement is true for every acyclic subgraph of a graph. Thus we have the following theorem.

Theorem 7.6 *A circuit of a connected graph G is a minimal set of edges of G containing at least one edge of every cospanning tree of G.* ∎

The converse of the above theorem follows next.

Theorem 7.7 *The set C of edges of a connected graph G is a circuit of G if it is a minimal set containing at least one edge of every cospanning tree of G.*

Proof. As shown earlier, the set C cannot be acyclic since there exists, for every acyclic subgraph G' of G, a cospanning tree not having in common any edge with G'. Thus C has at least one circuit C'. Suppose that C' is a proper subset of C. Then by Theorem 7.6, C' is a minimal set of edges containing at least one edge of every cospanning tree of G. This, however, contradicts the hypothesis that C is a minimal such set. Hence no proper subset of C is a circuit. Since C is not acyclic, C must be a circuit. ∎

Theorems 7.6 and 7.7 establish that a set C of edges of a connected graph G is a circuit if and only if it is a minimal set of edges containing at least one edge of every cospanning tree of G.

The new characterizations of a cutset and a circuit as given by Theorems 7.5 through 7.7 clearly bring out the dual nature of the concepts of circuits and cutsets. The next theorem relates circuits and cutsets without involving trees.

Theorem 7.8 *A circuit and a cutset of a connected graph have an even number of common edges.*

Proof. Let C be a circuit and S a cutset of a connected graph G. Let V_1 and V_2 be the vertex sets of the two connected subgraphs G_1 and G_2 of $G - S$.

If C is a subgraph of G_1 or of G_2, then obviously the number of edges common to C and S is equal to zero, an even number.

Suppose that C and S have some common edges. Let us traverse the circuit C starting from a vertex, say v_1, in the set V_1. Since the traversing should end at v_1, it is necessary that every time we meet with an edge of S leading us from a vertex in V_1 to a vertex in V_2, there must be an edge of S leading us from a vertex in V_2 back to a vertex in V_1. This is possible only if C and S have an even number of common edges. ∎

We would like to point out that the converse of Theorem 7.8 is not quite true. However, we show in Theorem 7.15 that a set S of edges of a graph G is a cutset (circuit) or the union of some edge-disjoint cutsets (circuits) if and only if S has an even number of edges in common with every circuit (cutset).

Fundamental circuits and fundamental cutsets of a connected graph have been defined with respect to a spanning tree of a graph. It is, therefore, not surprising that fundamental circuits and cutsets are themselves related as proved next.

Theorem 7.9

1. *The fundamental circuit with respect to a chord of a spanning tree T of a connected graph consists of exactly those branches of T whose fundamental cutsets contain the chord.*

2. *The fundamental cutset with respect to a branch of a spanning tree T of a connected graph consists of exactly those chords of T whose fundamental circuits contain the branch.*

Proof.

1. Let C be the fundamental circuit of a connected graph G with respect to a chord c_1 of a spanning tree T of G. Let C contain, in addition to the chord c_1, the branches $b_1, b_2, \ldots, b_k$ of T.

    Suppose $S_i, 1 \leq i \leq k$, is the fundamental cutset of G with respect to the branch $b_i, 1 \leq i \leq k$, of T. The branch b_i is the only branch common to both C and S_i. The chord c_1 is the only chord in C. Since C and S_i must have an even number of common edges, it is necessary that the fundamental cutset S_i contain c_1. Next we show that no other fundamental cutset of T contains c_1.

    Suppose the fundamental cutset S_{k+1} with respect to some branch b_{k+1} of T contains c_1. Then c_1 will be the only common edge between C and S_{k+1}. This will contradict Theorem 7.8. Thus the chord c_1 is present only in those cutsets defined by the branches $b_1, b_2, \ldots, b_k$.

2. Proof of this part is similar to that of part 1. ∎

7.4 CIRCUIT AND CUTSET SPACES OF A GRAPH

Let W_G be the collection of subsets of edges of a graph $G = (V, E)$ with n vertices and m edges. Denoting by $\oplus$ the ring sum (Exclusive-OR) operation, it is easy to verify that W_G is an m-dimensional vector space over $GF(2) = \{0, 1\}$.

Since, in this section, we are concerned only with edge-induced subgraphs, we refer to them simply as subgraphs without the adjective *edge-induced*. However, we may still use this adjective in some places to emphasize the edge-induced nature of the concerned subgraph.

We show that the following subsets of W_G are subspaces:

1. W_C, the set of all circuits (including the null graph ϕ) and unions of edge-disjoint circuits of G.

2. W_S, the set of all cutsets (including the null graph ϕ) and unions of edge-joint cutsets of G.

This result will follow once we show that W_C and W_S are closed under $\oplus$, the ring sum operation.

Theorem 7.10 *W_C, the set of all circuits and unions of edge-disjoint circuits of a graph G, is a subspace of the vector space W_G of G.*

Proof. A graph can be expressed as the union of edge-disjoint circuits if and only if every vertex in the graph is of even degree (i.e., G is Eulerian) (see Chapter 1). Hence we may regard W_C as the set of all edge-induced subgraphs of G in which all vertices are of even degree.

Consider any two distinct members C_1 and C_2 of W_C. C_1 and C_2 are edge-induced subgraphs with the degrees of all their vertices even. Let C_3 denote the ring sum of C_1 and C_2. To prove the theorem, we need only to show that C_3 belongs to W_C. In other words, we should show that in C_3 every vertex is of even degree.

Consider any vertex v in C_3. Obviously, this vertex should be present in at least one of the subgraphs C_1 and C_2. Let $X_i, i = 1, 2, 3$, denote the set of edges incident on v in C_i. Let $|X_i|$ denote the number of edges in X_i. Thus $|X_i|$ is the degree of the vertex v in C_i. Note that $|X_1|$ and $|X_2|$ are even and one of them may be zero. Further $|X_3|$ is nonzero.

Since $C_3 = C_1 \oplus C_2$, we get
$$X_3 = X_1 \oplus X_2.$$
Hence
$$|X_3| = |X_1| + |X_2| - 2|X_1 \cap X_2|.$$

It is now clear from the above equation that $|X_3|$ is even, because $|X_1|$ and $|X_2|$ are both even. In other words, the degree of vertex v in C_3 is even. Since this should be true for all vertices in C_3, it follows that C_3 belongs to W_C and the theorem is proved. ∎

Theorem 7.11 *The ring sum of any two cuts in a graph G is also a cut in G.*

Proof. Consider any two cuts $S_1 = \langle V_1, V_2 \rangle$ and $S_2 = \langle V_3, V_4 \rangle$ in a graph $G = (V, E)$. Note that
$$V_1 \cup V_2 = V_3 \cup V_4 = V$$
and
$$V_1 \cap V_2 = V_3 \cap V_4 = \varnothing$$
Let
$$A = V_1 \cap V_3,$$
$$B = V_1 \cap V_4,$$
$$C = V_2 \cap V_3,$$
$$D = V_2 \cap V_4.$$

It is easy to see that the sets $A, B, C,$ and D are mutually disjoint. Then

$$S_1 = \langle A \cup B, C \cup D \rangle$$
$$= \langle A, C \rangle \cup \langle A, D \rangle \cup \langle B, C \rangle \cup \langle B, D \rangle$$

and

$$S_2 = \langle A \cup C, B \cup D \rangle$$
$$= \langle A, B \rangle \cup \langle A, D \rangle \cup \langle C, B \rangle \cup \langle C, D \rangle.$$

Hence, we get

$$S_1 \oplus S_2 = \langle A, C \rangle \cup \langle B, D \rangle \cup \langle A, B \rangle \cup \langle C, D \rangle.$$

Since

$$\langle A \cup D, B \cup C \rangle = \langle A, C \rangle \cup \langle B, D \rangle \cup \langle A, B \rangle \cup \langle C, D \rangle,$$

we can write

$$S_1 \oplus S_2 = \langle A \cup D, B \cup C \rangle.$$

Because $A \cup D$ and $B \cup C$ are mutually disjoint and together include all the vertices in V, $S_1 \oplus S_2$ is a cut in G. Hence the theorem. ∎

Corollary 7.1 *The union of any two edge-disjoint cuts in a graph G is also a cut in G.* ∎

Since a cutset is also a cut, it is now clear from Corollary 7.1 that W_S is the set of all cuts in G.

Further, by Theorem 7.11, W_S is closed under the ring sum operation. Thus we get the following theorem.

Theorem 7.12 W_S, *the set of all cutsets and unions of edge-disjoint cutsets in a graph G, is a subspace of the vector space W_G of G.* ∎

W_S will be referred to as the *cutset subspace* of the graph G.

7.5 DIMENSIONS OF CIRCUIT AND CUTSET SUBSPACES

In this section, we show that the dimensions of the circuit and cutset subspaces of a graph are equal to the nullity and the rank of the graph, respectively. We do this by proving that the set of fundamental circuits and the set of fundamental cutsets with respect to some spanning tree of a connected graph are bases for the circuit and cutset subspaces of the graph, respectively.

Let T be a spanning tree of a connected graph G with n vertices and m edges. The branches of T will be denoted by $b_1, b_2, \ldots, b_{n-1}$ and the chords by $c_1, c_2, \ldots, c_{m-n+1}$. Let C_i and S_i refer to the fundamental circuit and the fundamental cutset with respect to c_i and b_i, respectively.

By definition, each fundamental circuit contains exactly one chord, and this chord is not present in any other fundamental circuit. Thus no fundamental circuit can be expressed as the ring sum of the other fundamental circuits. Hence the fundamental circuits $C_1, C_2, \ldots, C_{m-n+1}$ are independent. Similarly, the fundamental cutsets $S_1, S_2, \ldots, S_{n-1}$ are also independent, since each of these contains exactly one branch which is not present in the others.

To prove that $C_1, C_2, \ldots, C_{m-n+1}$ $(S_1, S_2, \ldots, S_{n-1})$ constitute a basis for the circuit (cutset) subspace of G, we need only to prove that every subgraph in the circuit (cutset) subspace of G can be expressed as a ring sum of C_i's (S_i's).

Consider any subgraph C in the circuit subspace of G. Let C contain the chords $c_{i_1}, c_{i_2}, \ldots, c_{i_r}$. Let C' denote the ring sum of the fundamental circuits $C_{i_1}, C_{i_2}, \ldots, C_{i_r}$. Obviously, the chords $c_{i_1}, c_{i_2}, \ldots, c_{i_r}$ are present in C', and C' contains no other chords of T. Since C also contains these chords and no others, $C' \oplus C$ contains no chords.

We now claim that $C' \oplus C$ is empty. If this is not true, then by the preceding arguments, $C' \oplus C$ contains only branches and hence has no circuits. On the other hand, being a ring sum of circuits, $C' \oplus C$ is, by Theorem 7.10, a circuit or the union of some edge-disjoint circuits. Thus the assumption that $C' \oplus C$ is not empty leads to a contradiction. Hence $C' \oplus C$ is empty. This implies that $C = C' = C_{i_1} \oplus C_{i_2} \oplus \ldots \oplus C_{i_r}$. In other words, every subgraph in the circuit subspace of G can be expressed as a ring sum of C_i's.

In an exactly similar manner we can prove that every subgraph in the cutset subspace of G can be expressed as a ring sum of S_i's. Thus we have the following theorem.

Theorem 7.13 *Let a connected graph G have m edges and n vertices. Then*

1. *The fundamental circuits with respect to a spanning tree of G constitute a basis for the circuit subspace of G, and hence the dimension of the circuit subspace of G is equal to $m - n + 1$, the nullity $\mu(G)$ of G.*

2. *The fundamental cutsets with respect to a spanning tree of G constitute a basis for the cutset subspace of G, and hence the dimension of the cutset subspace of G is equal to $n - 1$, the rank $\rho(G)$ of G.* ∎

It is now easy to see that in the case of a graph G which is not connected, the set of all the fundamental circuits with respect to the chords of a forest of G, and the set of all the fundamental cutsets with respect to the branches of a forest of G are, respectively, bases for the circuit and cutset subspaces of G. Thus we get the following corollary of the previous theorem.

Corollary 7.2 *If a graph G has m edges, n vertices, and p components, then*

1. *The dimension of the circuit subspace of G is equal to $m - n + p$, the nullity of G.*

2. *The dimension of the cutset subspace of G is equal to $n - p$, the rank of G.* ∎

7.6 RELATIONSHIP BETWEEN CIRCUIT AND CUTSET SUBSPACES

We establish in this section a characterization for the subgraphs in the circuit subspace of a graph G in terms of those in the cutset subspace of G.

Since every subgraph in the circuit subspace of a graph is a circuit or the union of edge-disjoint circuits, and every subgraph in the cutset subspace is a cutset or the union of edge-disjoint cutsets, we get the following as an immediate consequence of Theorem 7.8.

Theorem 7.14 *Every subgraph in the circuit subspace of a graph G has an even number of common edges with every subgraph in the cutset subspace of G.* ∎

In the next theorem, we prove the converse of the above.

Theorem 7.15

1. *A subgraph of a graph G belongs to the circuit subspace of G if it has an even number of common edges with every subgraph in the cutset subspace of G.*

2. *A subgraph of a graph G belongs to the cutset subspace of G if it has an even number of common edges with every subgraph in the circuit subspace of G.*

Proof.

1. We may assume, without any loss of generality, that G is connected. The proof when G is not connected will follow in an exactly similar manner.

    Let T be a spanning tree of G. Let $b_1, b_2, \ldots$ denote the branches of T and $c_1, c_2, \ldots$ denote its chords. Consider any subgraph C of G which has an even number of common edges with every subgraph in the cutset subspace of G. Without any loss of generality, assume that C contains the chords $c_1, c_2, \ldots, c_r$. Let C' denote the ring sum of the fundamental circuits $C_1, C_2, \ldots, C_r$ with respect to the chords $c_1, c_2, \ldots, c_r$.

    Obviously, C' consists of the chords $c_1, c_2, \ldots, c_r$ and no other chords. Hence $C' \oplus C$ consists of no chords.

    C', being the ring sum of some circuits of G, has an even number of common edges with every subgraph in the cutset subspace of G. Since C also has this property, so does $C' \oplus C$.

    We now claim that $C' \oplus C$ is empty. If not, $C' \oplus C$ contains only branches. Let b_i be any branch in $C' \oplus C$. Then b_i is the only edge common between $C' \oplus C$ and the fundamental cutset with respect to b_i. This is not possible since $C' \oplus C$ must have an even number of common edges with every cutset. Thus $C' \oplus C$ should be empty. In other words, $C = C' = C_1 \oplus C_2 \oplus \ldots \oplus C_r$, and hence C belongs to the circuit subspace of G.

2. The proof of this part follows in an exactly similar manner. ∎

7.7 ORTHOGONALITY OF CIRCUIT AND CUTSET SUBSPACES

Let $e_1, e_2, \ldots, e_m$ denote the m edges of a graph G. Suppose we associate each edge-induced subgraph G_i of G with an m-vector w_i such that the jth entry of w_i is equal to 1 if and only if the edge e_j is in G_i. Then the ring sum $G_i \oplus G_j$ of two subgraphs G_i and G_j will correspond to the m-vector $w_i + w_j$, the modulo 2 sum of w_i and w_j. It can now be seen that the association just described indeed defines an isomorphism between W_G and the vector space of all m-vectors over $GF(2)$. In fact, if we choose $\{e_1\}, \{e_2\}, \ldots, \{e_m\}$ as the basis vector for W_G, then the entries of w_i are the coordinates of G_i relative to this basis.

In view of this isomorphism, we again use the symbol W_G to denote the vector space of all the m-vectors associated with the subgraphs of the graph G. Also, W_C will denote the subspace of m-vectors representing the subgraphs in the circuit subspace of G and similarly W_S will denote the subspace of those representing the subgraphs in the cutset subspace of G.

Consider any two vectors w_i and w_j such that w_i is in W_C and w_j is in W_S. Because every subgraph in W_C has an even number of common edges with those in W_S, it follows that the dot product $\langle w_i, w_j \rangle$ of w_i and w_j is equal to the modulo 2 sum of an even number of 1's. This means $\langle w_i, w_j \rangle = 0$. In other words, the m-vectors in W_C are orthogonal to those in W_S. Thus we have the following theorem.

Theorem 7.16 *The cutset and circuit subspaces of a graph are orthogonal to each other.* ∎

Consider next the direct sum $W_C \boxplus W_S$. We know that

$$\dim(W_C \boxplus W_S) = \dim(W_C) + \dim(W_S) - \dim(W_C \cap W_S).$$

Since $\dim(W_C) + \dim(W_S) = m$, we get

$$\dim(W_C \boxplus W_S) = m - \dim(W_C \cap W_S).$$

Now the orthogonal subspaces W_C and W_S will also be orthogonal complements of W_G if and only if $\dim(W_C \boxplus W_S) = m$. In other words W_C and W_S will be orthogonal complements if and only if $\dim(W_C \cap W_S) = 0$, that is, $W_C \cap W_S$ is the zero vector whose elements are all equal to zero. Thus we get the following theorem.

Theorem 7.17 W_C *and* W_S, *the circuit and cutset subspaces of a graph are orthogonal complements if and only if* $W_C \cap W_S$ *is the zero vector.* ∎

Suppose W_C and W_S are orthogonal complements. Then it means that every vector in W_G can be expressed as $w_i + w_j$, where w_i is in W_C and w_j is in W_S. In other words, every subgraph of G can be expressed as the ring sum of two subgraphs, one belonging to the circuit subspace and the other belonging to the cutset subspace. In particular, the graph G itself can be expressed as above.

Suppose W_C and W_S are not orthogonal complements. Then, clearly, there exists a subgraph which cannot be expressed as the ring sum of subgraphs in W_C and W_S. The question then arises whether, in this case too, it is possible to express G as the ring sum of subgraphs from W_C and W_S. The answer is in the affirmative as stated in the next theorem.

Theorem 7.18 *Every graph G can be expressed as the ring sum of two subgraphs one of which is in the circuit subspace and the other is in the cutset subspace of G.* ∎

See Chen [3] and Williams and Maxwell [4] for a proof of this theorem.

A subgraph in the circuit subspace is called a *circ*. A subgraph that is in the intersection of the circuit and cutset subspaces of an undirected graph is called a *bicycle*. That is, a bicycle is a circ as well as a cut.

The subgraphs used in the decomposition presented in Theorem 7.18 may not be disjoint. That is, these subgraphs do not form a partition of the edge set of G. We next present two ways to partition a graph based on cuts and elements of W_c.

Rosenstiehl and Read [5] have proved the following. See also Parthasarathy [6].

Theorem 7.19 [5] *Any edge e of a graph G is of one of the following types:*

1. *e is in a circ that becomes a cut when e is removed from it.*

2. *e is in a cut that becomes a circ when e is removed from it.*

3. *e is in a bicycle.* ∎

The partition of the edges defined in Theorem 7.19 is called the *bicycle-based tripartition*.

The *tree distance*, $d(T_1, T_2)$, between any two spanning trees T_1 and T_2 is defined as $d(T_1, T_2) = |E(T_1) - E(T_2)| = |E(T_2) - E(T_1)|$. Two spanning trees T_1 and T_2 are *maximally distant* if $d(T_1, T_2) \geq d(T_i, T_j)$ for every pair of spanning trees T_i and T_j. The maximum distance between any two spanning trees of a connected graph is denoted by d_m.

Kishi and Kajitani [7] introduced the concept of principal partition of a graph using maximally distant spanning trees. Principal partition is also a tripartition of the edge set of G.

Summary and Related Works

An early paper on vector spaces associated with a graph is by Gould [8], where the question of constructing a graph having a specified set of circuits is also discussed. Chen [3] and Williams and Maxwell [4] are also recommended for further reading on this topic. Rosentiehl and Reed [5] have proved several interesting results relating to circuits and cuts and their relationship.

In electrical circuit analysis one is interested in solving for all the current and the voltage variables. The circuit method of analysis (also known as the loop analysis) requires solving for only $m - n + 1$ independent current variables. The remaining current variables and all the voltage variables can then be determined using these $m - n + 1$ independent current variables. The cutset method of analysis requires solving for only $n - 1$ independent voltage variables. A question that intrigued circuit theorists for a long time was whether one could use a hybrid method of analysis involving some current variables and some voltage variables and reduce the size of the system of equations to be solved to less than both $n - 1$ and $m - n + 1$, the rank and nullity of the graph of the circuit. Ohtsuki, Ishizaki, and Watanabe [9] studied this problem and showed that d_m, the maximum distance between any two spanning trees of the graph of the circuit is, in fact, the minimum number of variables required in the hybrid method of analysis. They also showed that the variables can be determined using the principal partition of the graph. The works by Kishi and Kajitani [7] on principal partition and by Ohtsuki, Ishizaki, and Watanabe [9] on the hybrid method of analysis are considered landmark results in electrical circuit theory. Swamy and Thulasiraman [1] give a detailed exposition of the principal partition concept and the hybrid and other methods of circuit analysis (see also Chapter 32).

Lin [10] presented an algorithm for computing the principal partition of a graph. Bruno and Weinberg [11] extended the concept of principal partition to matroids.

In the application of graph theory to the electrical circuit synthesis problem, one encounters a certain matrix of integers modulo 2 and seeks to determine if this matrix is the cutset or the circuit matrix of an undirected graph. The complete solution to this problem was given by Tutte [12]. Cederbaum [13] and Gould [8] considered this problem before Tutte provided the solution. Tutte [12] provided the necessary and sufficient conditions for the realizability of a matrix of integers modulo 2 as the circuit or the cutset matrix of an undirected graph. See Seshu and Reed [14] for a discussion of this topic.

Mayeda [15] gave an alternate proof of Tutte's realizability condition, shorter than Tutte's original proof, which is 27 pages long.

Early works on algorithms for constructing graphs having specified circuit or cutset matrices are in [16,17]. Bapeswara Rao [18] defined the tree-path matrix of an undirected graph which is essentially the nonunit submatrix of the fundamental circuit matrix and presented an algorithm for constructing a graph with a prescribed tree-path matrix. This is also an algorithmic solution to the cutset and the circuit matrix realization problems. A detailed presentation of Bapeswara Rao's algorithm is given in [1].

The circuit and cutset matrix realization problems arise in the design of multi-port resistance networks. It was in the context of this application that Cederbaum [13,19] encountered the realization problem. Interestingly, Bapeswara Rao [18] and Boesch and Youla [20] presented circuit-theoretic approaches to the realization of a matrix as the cutset or circuit matrix of a directed graph. Details of Bapeswara Rao's algorithm based on this approach may also be found in [1].

References

[1] M. N. S. Swamy and K. Thulasiraman, *Graphs, Networks and Algorithms*, Wiley-Interscience, New York, 1981.

[2] M. B. Reed, The seg: A new class of subgraphs, *IEEE Trans. Circuit Theory*, **CT-8** (1961), 17–22.

[3] W. K. Chen, On vector spaces associated with a graph, *SIAM J. Appl. Math.*, **20** (1971), 526–529.

[4] T. W. Williams and L. M. Maxwell, The decomposition of a graph and the introduction of a new class of subgraphs, *SIAM J. Appl. Math.*, **20** (1971), 385–389.

[5] P. Rosenstiehl and R. C. Reed, On the principal edge tripartion of a graph, *Ann. Discrete Math.*, **3** (1978), 195–226.

[6] K. R. Parthasarathy, *Basic Graph Theory*, Tata McGraw-Hill Publishing Company, New Delhi, India, 1994.

[7] G. Kishi and Y. Kajitani, Maximally distant trees and principal partition of a linear graph, *IEEE Trans. Circuit Theory*, **16** (1969), 323–330.

[8] R. Gould, Graphs and vector spaces, *J. Math. Phys.*, **37** (1958), 193–214.

[9] T. Ohtsuki, Y. Ishizaki, and H. Watanabe, Topological degrees of freedom and mixed analysis of electrical networks, *IEEE Trans. Circuit Theory*, **17** (1970), 491–499.

[10] P. M. Lin, An improved algorithm for principal partition of graphs, *Proc. IEEE Intl. Symp. Circuits and Systems*, Munich, Germany, 1976, 145–148.

[11] J. Bruno and L. Weinberg, The principal minors of a matroid, *Linar Algebra Its Appl.*, **4** (1971), 17–54.

[12] W. T. Tutte, Matroids and graphs, *Trans. Am. Math. Soc.*, **90** (1959), 527–552.

[13] I. Cederbaum, Matrices all of whose elements and subdeterminants are 1, −1 or 0, *J. Math. and Phys.*, **36** (1958), 351–361.

[14] S. Seshu and M. B. Reed, *Linear Graphs and Electrical Networks*, Addison-Wesley, Reading, MA, 1961.

[15] W. Mayeda, A proof of Tutte's realizability condition, *IEEE Trans. Circuit Theory*, **17** (1970), 506–511.

[16] W. T. Tutte, An algorithm for determining whether a given binary matroid is graphic, *Proc. Am. Math. Soc.*, **11** (1960), 905–917.

[17] W. T. Tutte, From matrices to graphs, *Can. J. Math.*, **56** (1964), 108–127.

[18] V. V. Bapeswara Rao, *The Tree-Path Matrix of a Network and Its Applications*, PhD thesis, Department of Electrical Engineering, Indian Institute of Technology, Madras, India, 1970.

[19] I. Cederbaum, Applications of matrix algebra to network theory, *IRE Trans. Circuit Theory*, **6** (1959), 127–137.

[20] F. T. Boesch and D. C. Youla, Synthesis of resistor n-port networks, *IEEE Trans. Circuit Theory*, **12** (1965), 515–520.

CHAPTER 8

Incidence, Cut, and Circuit Matrices of a Graph

Krishnaiyan "KT" Thulasiraman

M. N. S. Swamy

CONTENTS

8.1	Introduction	191
8.2	Incidence Matrix	191
8.3	Cut Matrix	193
8.4	Circuit Matrix	196
8.5	Orthogonality Relation	197
8.6	Submatrices of Cut, Incidence, and Circuit Matrices	199
8.7	Totally Unimodular Matrices	202
8.8	Number of Spanning Trees	203
8.9	Number of Spanning 2-Trees	205
8.10	Number of Directed Spanning Trees in a Directed Graph	207
8.11	Directed Spanning Trees and Directed Euler Trails	211

8.1 INTRODUCTION

In this chapter we introduce the incidence, circuit and cut matrices of a graph and establish several properties of these matrices which help to reveal the structure of a graph. The incidence, circuit, and cut matrices arise in the study of electrical networks because these matrices are the coefficient matrices of Kirchhoff's equations which describe a network. Thus the properties of these matrices and other related results to be established in this chapter have been used extensively in electrical circuit analysis.

Our discussions of incidence, circuit, and cut matrices are mainly with respect to directed graphs. However, these discussions become valid for undirected graphs too if addition and multiplication are in $GF(2)$, the field of integers modulo 2. For basic definitions and results on circuits, cutsets, and their relationship see Chapter 7. Also see Chapter 6.4 of Gross et al. [1].

8.2 INCIDENCE MATRIX

Consider a graph G with n vertices and m edges and having no self-loops. The *all-vertex incidence matrix* $A_c = [a_{ij}]$ of G has n rows, one for each vertex, and m columns, one for each edge. The element a_{ij} of A_c is defined as follows:

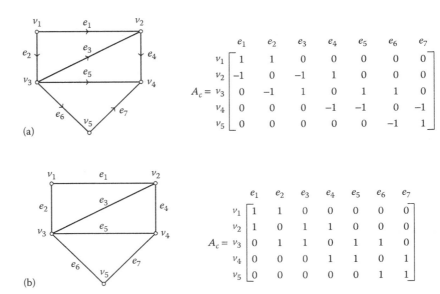

Figure 8.1 (a) Directed graph G and its all-vertex incidence matrix. (b) Undirected graph G and its all-vertex incidence matrix.

G is directed:

$$a_{ij} = \begin{cases} 1, & \text{if the } j\text{th edge is incident on the } i\text{th vertex and oriented away from it;} \\ -1, & \text{if the } j\text{th edge is incident on the } i\text{th vertex and oriented toward it;} \\ 0, & \text{if the } j\text{th edge is not incident on the } i\text{th vertex.} \end{cases}$$

G is undirected:

$$a_{ij} = \begin{cases} 1, & \text{if the } j\text{th edge is incident on the } i\text{th vertex;} \\ 0, & \text{otherwise.} \end{cases}$$

A row of A_c will be referred as an *incidence vector* of G. Two graphs and their all-vertex incidence matrices are shown in Figure 8.1a and 8.1b.

It should be clear from the preceding definition that each column of A_c contains exactly two nonzero entries, one $+1$ and one -1. Therefore we can obtain any row of A_c from the remaining $n - 1$ rows. Thus any $n - 1$ rows of A_c contain all the information about A_c. In other words, the rows of A_c are linearly dependent.

An $(n - 1)$-rowed submatrix A of A_c will be referred to as an *incidence matrix* of G. The vertex which corresponds to the row of A_c which is not in A will be called the *reference vertex* or *datum vertex* of A. Note that

$$\text{rank}(A) = \text{rank}(A_c) \leq n - 1. \tag{8.1}$$

Now we show that in the case of a connected graph, rank of A_c is in fact equal to $n-1$. This result is based on the following theorem.

Theorem 8.1 *The determinant of any incidence matrix of a tree is equal to ± 1.*

Proof. Proof is by induction on the number n of vertices in a tree.

Any incidence matrix of a tree on two vertices is just a 1×1 matrix with its only entry being equal to ± 1. Thus the theorem is true for $n = 2$. Note that the theorem does not arise for $n = 1$.

Let the theorem be true for $2 \leq n \leq k$. Consider a tree T with $k+1$ vertices. Let A denote an incidence matrix of T. T has at least two pendant vertices. Let the ith vertex of T be a pendant vertex, and let this not be the reference vertex of A. If the only edge incident on this vertex is the lth one, then in A

$$a_{il} = \pm 1 \quad \text{and} \quad a_{ij} = 0, \quad j \neq l.$$

If we now expand the determinant of A by the ith row, then

$$\det(A) = \pm(-1)^{i+l}\det(A'), \tag{8.2}$$

where A' is obtained by removing the ith row and the lth column from A.

Suppose T' is the graph that results after removing the ith vertex and the lth edge from T. Clearly T' is a tree because the ith vertex is a pendant vertex and the lth edge is a pendant edge in T. Further it is easy to verify that A' is an incidence matrix of T'. Since T' is a tree on $n-1$ vertices, we have by the induction hypothesis that

$$\det(A') = \pm 1. \tag{8.3}$$

This result in conjunction with (8.2) proves the theorem for $n = k+1$. ∎

Since a connected graph has at least one spanning tree, it follows from the above theorem that in any incidence matrix A of a connected graph with n vertices there exists a nonsingular submatrix of order $n-1$. Thus for a connected graph,

$$\text{rank}(A) = n - 1. \tag{8.4}$$

Since $\text{rank}(A_c) = \text{rank}(A)$, we get the following theorem.

Theorem 8.2 *The rank of the all-vertex incidence matrix of an n-vertex connected graph G is equal to $n-1$, the rank $\rho(G)$ of G.* ∎

An immediate consequence of the above theorem is the following.

Corollary 8.1 *If an n-vertex graph has p components, then the rank of its all-vertex incidence matrix is equal to $n-p$, the rank of G.* ∎

8.3 CUT MATRIX

To define the cut matrix of a directed graph we need to assign an orientation to each cut of the graph.

Consider a directed graph $G = (V, E)$. If V_a is a nonempty subset of V, then the set of edges connecting the vertices in V_a to those in $\bar{V}_a$ is a cut, and this cut is denoted as $\langle V_a, \bar{V}_a \rangle$. The orientation of $\langle V_a, \bar{V}_a \rangle$ may be assumed to be either from V_a to $\bar{V}_a$ or from $\bar{V}_a$ to V_a. Suppose we assume that the orientation is from V_a to $\bar{V}_a$. Then the orientation of an edge in $\langle V_a, \bar{V}_a \rangle$ is said to agree with the orientation of the cut $\langle V_a, \bar{V}_a \rangle$ if the edge is oriented from a vertex in V_a to a vertex in $\bar{V}_a$.

The *cut matrix* $Q_c = [q_{ij}]$ of a graph G with m edges has m columns and as many rows as the number of cuts in G. The entry q_{ij} is defined as follows:

G is *directed:*

$$q_{ij} = \begin{cases} 1, & \text{if the } j\text{th edge is in the } i\text{th cut and its orientation agrees with the cut orientation;} \\ -1, & \text{if the } j\text{th edge is in the } i\text{th cut and its orientation does not agree with the cut orientation;} \\ 0, & \text{if the } j\text{th edge is not in the } i\text{th cut.} \end{cases}$$

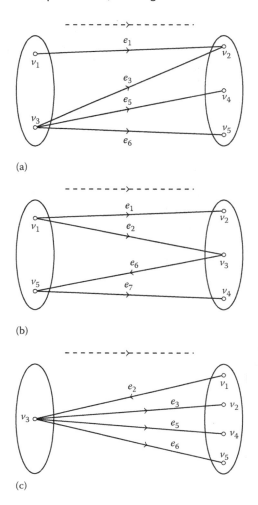

Figure 8.2 Some cuts of the graph of Figure 8.1a. (a) Cut 1, (b) Cut 2, (c) Cut 3.

G is *undirected*:

$$q_{ij} = \begin{cases} 1, & \text{if the } j\text{th edge is in the } i\text{th cut}; \\ 0, & \text{otherwise}. \end{cases}$$

A row of Q_c will be referred to as a *cut vector*.

Three cuts of the directed graph of Figure 8.1a are shown in Figure 8.2. In each case the cut orientation is shown in dashed lines. The submatrix of Q_c corresponding to these three cuts is as given below:

$$\begin{array}{c} \\ \text{cut 1} \\ \text{cut 2} \\ \text{cut 3} \end{array} \begin{array}{c} \begin{matrix} e_1 & e_2 & e_3 & e_4 & e_5 & e_6 & e_7 \end{matrix} \\ \left[\begin{matrix} 1 & 0 & 1 & 0 & 1 & 1 & 0 \\ 1 & 1 & 0 & 0 & 0 & -1 & 1 \\ 0 & -1 & 1 & 0 & 1 & 1 & 0 \end{matrix} \right] \end{array}$$

The corresponding submatrix in the undirected case can be obtained by replacing the -1's in the above matrix by $+1$'s.

Consider next any vertex v. The nonzero entries in the corresponding incidence vector represent the edges incident on v. These edges form the cut $\langle v, V - v \rangle$. If we assume that the orientation of this cut is from v to $V - v$, then we can see from the definitions of cut and incidence matrices that the row in Q_c corresponding to the cut $\langle v, V - v \rangle$ is the same as the row in A_c corresponding to the vertex v. Thus A_c is a submatrix of Q_c.

Next we show that the rank of Q_c is equal to that of A_c. To do so we need the following theorem.

Theorem 8.3 *Each row in the cut matrix Q_c can be expressed, in two ways, as a linear combination of the rows of the matrix A_c. In each case, the nonzero coefficients in the linear combination are all $+1$ or all -1.*

Proof. Let $\langle V_a, \bar{V}_a \rangle$ be the ith cut in a graph G with n vertices and m edges, and let q_i be the corresponding cut vector. Let $V_a = \{v_1, v_2, \ldots, v_r\}$ and $\bar{V}_a = \{v_{r+1}, v_{r+2}, \ldots, v_n\}$. For $1 \leq i \leq n$, let a_i denote the incidence vector corresponding to the vertex v_i.

We assume, without any loss of generality, that the orientation of $<V_a, \bar{V}_a>$ is from V_a to $\bar{V}_a$, and prove the theorem by establishing that

$$q_i = a_1 + a_2 + \cdots + a_r = -(a_{r+1} + a_{r+2} + \cdots + a_n). \tag{8.5}$$

Let v_p and v_q be the end vertices of the kth edge, $1 \leq k \leq m$. Let this edge be oriented from v_p to v_q so that

$$\begin{aligned} a_{pk} &= 1, \\ a_{qk} &= -1, \\ a_{jk} &= 0, \qquad j \neq p, q. \end{aligned} \tag{8.6}$$

Now four cases arise.

Case 1 $v_p \in V_a$ and $v_q \in \bar{V}_a$, that is, $p \leq r$ and $q \geq r+1$, so that $q_{ik} = 1$.

Case 2 $v_p \in \bar{V}_a$ and $v_q \in V_a$, that is, $p \geq r+1$ and $q \leq r$, so that $q_{ik} = -1$.

Case 3 $v_p, v_q \in V_a$, that is, $p, q \leq r$, so that $q_{ik} = 0$.

Case 4 $v_p, v_q \in \bar{V}_a$, that is, $p, q \geq r+1$, so that $q_{ik} = 0$.

It is easy to verify, using (8.6), that the following is true in each of these four cases.

$$\begin{aligned} q_{ik} &= (a_{1k} + a_{2k} + \cdots + a_{rk}) \\ &= -(a_{r+1,k} + a_{r+2,k} + \cdots + a_{nk}). \end{aligned} \tag{8.7}$$

Now (8.5) follows since the above equation is valid for all $1 \leq k \leq m$. Hence the theorem. ∎

An important consequence of Theorem 8.3 is that $\text{rank}(Q_c) \leq \text{rank}(A_c)$. However, $\text{rank}(Q_c) \geq \text{rank}(A_c)$, because A_c is a submatrix of Q_c. Therefore, we get

$$\text{rank}(Q_c) = \text{rank}(A_c).$$

Then Theorem 8.2 and Corollary 8.1, respectively, lead to the following theorem.

Theorem 8.4

1. *The rank of the cut matrix Q_c of an n-vertex connected graph G is equal to $n-1$, the rank of G.*

2. *The rank of the cut matrix Q_c of an n-vertex graph G with p components is equal to $n-p$, the rank of G.* ∎

As the above discussions show, the all-vertex incidence matrix A_c is an important submatrix of the cut matrix Q_c. Next we identify another important submatrix of Q_c.

We know from Chapter 7 that a spanning tree T of an n-vertex connected graph G defines a set of $n-1$ fundamental cutsets—one fundamental cutset for each branch of T. The submatrix of Q_c corresponding to these $n-1$ fundamental cutsets is known as the *fundamental cutset matrix* Q_f of G with respect to T.

Let $b_1, b_2, \ldots, b_{n-1}$ denote the branches of T. Suppose we arrange the columns and the rows of Q_f so that

1. For $1 \leq i \leq n-1$, the ith column corresponds to the branch b_i; and

2. The ith row corresponds to the fundamental cutset defined by b_i.

If, in addition, we assume that the orientation of a fundamental cutset is so chosen as to agree with that of the defining branch, then the matrix Q_f can be displayed in a convenient form as follows:

$$Q_f = [U | Q_{fc}] \tag{8.8}$$

where U is the unit matrix of order $n-1$ and its columns correspond to the branches of T.

For example, the fundamental cutset matrix Q_f of the connected graph of Figure 8.1a with respect to the spanning tree $T = \{e_1, e_2, e_6, e_7\}$ is as given below:

$$Q_f = \begin{array}{c} \\ e_1 \\ e_2 \\ e_6 \\ e_7 \end{array} \begin{array}{cccccccc} e_1 & e_2 & e_6 & e_7 & & e_3 & e_4 & e_5 \\ \left[\begin{array}{cccc|ccc} 1 & 0 & 0 & 0 & 1 & -1 & 0 \\ 0 & 1 & 0 & 0 & -1 & 1 & 0 \\ 0 & 0 & 1 & 0 & 0 & 1 & 1 \\ 0 & 0 & 0 & 1 & 0 & 1 & 1 \end{array}\right] \end{array} \tag{8.9}$$

It is clear from (8.8) that the rank of Q_f is equal to $n-1$, the rank of Q_c. Thus, every cut vector (which may be a cutset vector) can be expressed as a linear combination of the fundamental cutset vectors.

8.4 CIRCUIT MATRIX

A circuit can be traversed in one of two directions, clockwise or anticlockwise. The direction we choose for traversing a circuit defines its orientation.

Consider an edge e which has v_i and v_j as its end vertices. Suppose that this edge is oriented from v_i to v_j and that it is present in circuit C. Then we say that the orientation of e agrees with the orientation of the circuit if v_i appears before v_j when we traverse C in the direction specified by its orientation.

The *circuit matrix* $B_c = [b_{ij}]$ of a graph G with m edges has m columns and as may rows as the number of circuits in G. The entry b_{ij} is defined as follows:

G is *directed*:

$$b_{ij} = \begin{cases} 1, & \text{if the } j\text{th edge is in the } i\text{th circuit and its orientation agrees with the circuit orientation;} \\ -1, & \text{if the } j\text{th edge is in the } i\text{th circuit and its orientation does not agree with the circuit orientation;} \\ 0, & \text{if the } j\text{th edge is not in the } i\text{th circuit.} \end{cases}$$

G is *undirected*:

$$b_{ij} = \begin{cases} 1, & \text{if the } j\text{th edge is in the } i\text{th circuit;} \\ 0, & \text{otherwise.} \end{cases}$$

A row of B_c is called a *circuit vector* of G. Next we identify an important submatrix of B_c.

Consider any spanning tree T of a connected graph G having n vertices and m edges. Let $c_1, c_2, \ldots, c_{m-n+1}$ be the chords of T. We know that these $m - n + 1$ chords define a set of $m - n + 1$ fundamental circuits. The submatrix of B_c corresponding to these fundamental circuits is known as the *fundamental circuit matrix* B_f of G with respect to the spanning tree T (see Chapter 7).

Suppose we arrange the columns and rows of B_f so that

1. For $1 \leq i \leq m - n + 1$, the ith column corresponds to the chord c_i; and

2. The ith row corresponds to the fundamental circuit defined by c_i.

If, in addition, we choose the orientation of a fundamental circuit to agree with that of the defining chord, then the matrix B_f can be written as

$$B_f = [U | B_{ft}], \qquad (8.10)$$

where U is the unit matrix of order $m - n + 1$ and its columns correspond to the chords of T.

For example, the fundamental circuit matrix of the graph of Figure 8.1a with respect to the spanning tree $T = \{e_1, e_2, e_6, e_7\}$ is as given below:

$$B_f = \begin{array}{c} \\ e_3 \\ e_4 \\ e_5 \end{array} \begin{array}{cccccccc} e_3 & e_4 & e_5 & e_1 & e_2 & e_6 & e_7 \\ \left[\begin{array}{ccc|cccc} 1 & 0 & 0 & -1 & 1 & 0 & 0 \\ 0 & 1 & 0 & 1 & -1 & -1 & -1 \\ 0 & 0 & 1 & 0 & 0 & -1 & -1 \end{array}\right] \end{array} \qquad (8.11)$$

It is obvious from (8.10) that the rank of B_f is equal to $m - n + 1$, the nullity $\mu(G)$ of G. Since B_f is a submatrix of B_c, we get

$$\text{rank}(B_c) \geq m - n + 1. \qquad (8.12)$$

We show in the next section that the rank of B_c in the case of a connected graph is equal to $m - n + 1$.

8.5 ORTHOGONALITY RELATION

We showed in Chapter 7 that in the case of an undirected graph every circuit vector is orthogonal to every cut vector. Now we prove that this result is true in the case of directed graphs too. Our proof is based on the following theorem.

Theorem 8.5 *If a cut and a circuit in a directed graph have $2k$ edges in common, then k of these edges have the same relative orientations in the cut and in the circuit, and the remaining k edges have one orientation in the cut and the opposite orientation in the circuit.*

Proof. Consider a cut $\langle V_a, \bar{V}_a \rangle$ and a circuit C in a directed graph. Suppose we traverse C starting from a vertex in V_a. Then, for every edge e_1 which leads us from a vertex in V_a to a vertex in $\bar{V}_a$, there is an edge e_2 which leads us from a vertex in $\bar{V}_a$ to a vertex in V_a. The proof of the theorem will follow if we note that if $e_1(e_2)$ has the same relative orientation in the cut and in the circuit, then $e_2(e_1)$ has one orientation in the cut and the opposite orientation in the circuit. ∎

Now we prove the main result of this section.

Theorem 8.6 (Orthogonality Relation) *If the columns of the circuit matrix B_c and the cut matrix Q_c are arranged in the same edge order, then*

$$B_c Q_c^t = 0.$$

Proof. Consider a circuit and a cut which have $2k$ edges in common. The inner product of the corresponding circuit and cut vectors is equal to zero, since, by Theorem 8.5, it is the sum of k 1's and k -1's. The proof of the theorem now follows because each entry of the matrix $B_c Q_c^t$ is the inner product of a circuit vector and a cut vector. ∎

Now we use the orthogonality relation to establish the rank of the circuit matrix B_c.

Consider a connected graph G with n vertices and m edges. Let B_f and Q_f be the fundamental circuit and cutset matrices of G with respect to a spanning tree T. If the columns of B_f and Q_f are arranged in the same edge order, then we can write B_f and Q_f as

$$B_f = \begin{bmatrix} B_{ft} & U \end{bmatrix}$$

and

$$Q_f = \begin{bmatrix} U & Q_{fc} \end{bmatrix}.$$

By the orthogonality relation

$$B_f Q_f^t = 0,$$

that is,

$$\begin{bmatrix} B_{ft} & U \end{bmatrix} \begin{bmatrix} U \\ Q_{fc}^t \end{bmatrix} = 0,$$

that is,

$$B_{ft} = -Q_{fc}^t. \tag{8.13}$$

Let $\beta = [\beta_1, \beta_2, \ldots, \beta_\rho | \beta_{\rho+1}, \ldots, \beta_m]$, where ρ is the rank of G, be a circuit vector with its columns arranged in the same edge order as B_f and Q_f. Then, again by the orthogonality relation,

$$[\beta_1, \beta_2, \ldots, \beta_\rho | \beta_{\rho+1}, \ldots, \beta_m] \begin{bmatrix} U \\ Q_{fc}^t \end{bmatrix} = 0.$$

Therefore

$$[\beta_1, \beta_2, \ldots, \beta_\rho] = -[\beta_{\rho+1}, \beta_{\rho+2}, \ldots, \beta_m] Q_{fc}^t = [\beta_{\rho+1}, \beta_{\rho+2}, \ldots, \beta_m] B_{ft}.$$

Using the above equation we can write $[\beta_1, \beta_2, \ldots, \beta_m]$ as

$$[\beta_1, \beta_2, \ldots, \beta_m] = [\beta_{\rho+1}, \beta_{\rho+2}, \ldots, \beta_m][B_{ft} \ U] = [\beta_{\rho+1}, \beta_{\rho+2}, \ldots, \beta_m] B_f. \tag{8.14}$$

Thus any circuit vector can be expressed as a linear combination of the fundamental circuit vectors. So

$$\operatorname{rank}(B_c) \leq \operatorname{rank}(B_f) = m - n + 1.$$

Combining the above inequality with (8.12) establishes the following theorem and its corollary.

Theorem 8.7 *The rank of the circuit matrix B_c of a connected graph G with n vertices and m edges is equal to $m - n + 1$, the nullity $\mu(G)$ of G.* ∎

Corollary 8.2 *The rank of the circuit matrix B_c of a graph G with n vertices, m edges, and p components is equal to $m - n + p$, the nullity of G.* ∎

Suppose $\alpha = [\alpha_1, \alpha_2, \ldots, \alpha_\rho, \alpha_{\rho+1}, \ldots, \alpha_m]$ is a cut vector such that its columns are arranged in the same edge order as B_f and Q_f, then we can start from the relation

$$\alpha B_f^t = 0.$$

and prove that

$$\alpha = [\alpha_1, \alpha_2, \ldots, \alpha_\rho] Q_f, \tag{8.15}$$

by following a procedure similar to that used in establishing (8.14). Thus every cut vector can be expressed as a linear combination of the fundamental cutset vectors. Since $\text{rank}(Q_f) = n - 1$, we get

$$\text{rank}(Q_c) = \text{rank}(Q_f) = n - 1.$$

The above is thus an alternate proof of Theorem 8.4.

8.6 SUBMATRICES OF CUT, INCIDENCE, AND CIRCUIT MATRICES

In this section we characterize those submatrices of Q_c, A_c, and B_c which correspond to circuits, cutsets, spanning trees, and cospanning trees and discuss some properties of these submatrices.

Theorem 8.8

1. *There exists a linear relationship among the columns of the cut matrix Q_c which correspond to the edges of a circuit.*

2. *There exists a linear relationship among the columns of the circuit matrix B_c which correspond to the edges of a cutset.*

Proof.

1. Let us partition Q_c into columns so that

$$Q_c = [Q^{(1)}, Q^{(2)}, \ldots, Q^{(m)}].$$

Let $\beta = [\beta_1, \beta_2, \ldots, \beta_m]$ be a circuit vector. Then by the orthogonality relation, we have

$$Q_c \beta^t = 0$$

or

$$\beta_1 Q^{(1)} + \beta_2 Q^{(2)} + \cdots + \beta_m Q^{(m)} = 0. \tag{8.16}$$

If we assume, without loss of generality, that the first r elements of β are nonzero and the remaining ones are zero, then we have from (8.16)

$$\beta_1 Q^{(1)} + \beta_2 Q^{(2)} + \cdots + \beta_r Q^{(r)} = 0.$$

Thus there exists a linear relationship among the columns $Q^{(1)}, Q^{(2)}, \ldots, Q^{(r)}$ of Q_c which correspond to the edges of a circuit.

2. The proof in this case follows along the same lines as that for part 1. ∎

Corollary 8.3 *There exists a linear relationship among the columns of the incidence matrix which correspond to the edges of a circuit.*

Proof. The result follows from Theorem 8.8, part 1, because the incidence matrix is a submatrix of Q_c. ∎

Theorem 8.9 *A square submatrix of order $n-1$ of any incidence matrix A of an n-vertex connected graph G is nonsingular if and only if the edges which correspond to the columns of the submatrix form a spanning tree of G.*

Proof. Necessity: Consider the $n-1$ columns of a nonsingular submatrix of A. Since these columns are linearly independent, by Corollary 8.3, there is no circuit in the corresponding subgraph of G. Since this acyclic subgraph has $n-1$ edges, it follows that it is a spanning tree of G.

Sufficiency: This follows from Theorem 8.1. ∎

Thus the spanning trees of a connected graph are in one-to-one correspondence with the nonsingular submatrices of the matrix A.

Theorem 8.10 *Consider a connected graph G with n vertices and m edges. Let Q be a submatrix of Q_c with $n-1$ rows and of rank $n-1$. A square submatrix of Q of order $n-1$ is nonsingular if and only if the edges corresponding to the columns of this submatrix form a spanning tree of G.*

Proof. Necessity: Let the columns of the matrix Q be rearranged so that

$$Q = \begin{bmatrix} Q_{11} & Q_{12} \end{bmatrix},$$

with Q_{11} nonsingular. Since the columns of Q_{11} are linearly independent, by Theorem 8.8, part 1, there is no circuit in the corresponding subgraph of G. This acyclic subgraph has $n-1$ edges and is therefore a spanning tree of G.

Sufficiency: Suppose we rearrange the columns of Q so that

$$Q = \begin{bmatrix} Q_{11} & Q_{12} \end{bmatrix}$$

and the columns of Q_{11} correspond to the edges of a spanning tree T. Then the fundamental cutset matrix Q_f with respect to T is

$$Q_f = \begin{bmatrix} U & Q_{fc} \end{bmatrix}.$$

Since the rows of Q can be expressed as linear combinations of the rows of Q_f, we can write Q as

$$Q = \begin{bmatrix} Q_{11} & Q_{12} \end{bmatrix} = DQ_f$$
$$= D \begin{bmatrix} U & Q_{fc} \end{bmatrix}.$$

Thus

$$Q_{11} = DU = D.$$

Now D is nonsingular, because both Q and Q_f are of maximum rank $n-1$. So Q_{11} is nonsingular, and the sufficiency of the theorem follows. ∎

A dual theorem is presented next. Proof follows along the same lines as Theorem 8.10.

Theorem 8.11 *Consider a connected graph G with n vertices and m edges. Let B be a submatrix of the circuit matrix B_c of G with $m-n+1$ rows and of rank $m-n+1$. A square submatrix of B of order $m-n+1$ is nonsingular if and only if the columns of this submatrix correspond to the edges of a cospanning tree.* ∎

We conclude this section with the study of an interesting property of the inverse of a nonsingular submatrix of the incidence matrix.

Theorem 8.12 *Let A_{11} be a nonsingular submatrix of order $n-1$ of an incidence matrix A of an n-vertex connected graph G. Then the nonzero elements in each row of A_{11}^{-1} are either all 1 or all -1.*

Proof. Let A be the incidence matrix with v_r as the reference vertex. Assume that

$$A = \begin{bmatrix} A_{11} & A_{12} \end{bmatrix},$$

where A_{11} is nonsingular. We know from Theorem 8.9 that the edges corresponding to the columns of A_{11} constitute a spanning tree T of G. Then Q_f, the fundamental cutset matrix with respect to T will be

$$Q_f = \begin{bmatrix} U & Q_{fc} \end{bmatrix}.$$

By Theorem 8.3, each cut vector can be expressed as a linear combination of the rows of the incidence matrix. So we can write Q_f as

$$Q_f = \begin{bmatrix} U & Q_{fc} \end{bmatrix} = D \begin{bmatrix} A_{11} & A_{12} \end{bmatrix}.$$

Thus

$$D = A_{11}^{-1}.$$

Consider now the ith row q_i of Q_f. Let the corresponding cutset be $\langle V_a, \bar{V}_a \rangle$. Let

$$V_a = \{v_1, v_2, \ldots, v_k\}$$

and

$$\bar{V}_a = \{v_{k+1}, v_{k+2}, \ldots, v_n\}.$$

Suppose that the orientation of the cutset $\langle V_a, \bar{V}_a \rangle$ is from V_a to $\bar{V}_a$. Then we get from (8.5) that

$$q_i = a_1 + a_2 + \cdots + a_k \tag{8.17}$$
$$= -(a_{k+1} + a_{k+2} + \cdots + a_n), \tag{8.18}$$

where a_i is the ith row of A_c.

Note that row a_r corresponding to v_r will not be present in A. So if $v_r \in V_a$, then to represent q_i as a linear combination of the rows of A we have to write q_i as in (8.18). If $v_r \in \bar{V}_a$, then we have to write q_i as in (8.17). In both cases the nonzero coefficients in the linear combination are either all 1 or all -1.

Thus the nonzero elements in each row of $D = A_{11}^{-1}$ are either all 1 or all -1. ∎

The proof used in the above theorem suggests a simple procedure for evaluating A_{11}^{-1}. See Chapter 6 in [2].

8.7 TOTALLY UNIMODULAR MATRICES

A matrix is *totally unimodular* if the determinant of each of its square sub matrices is 1, −1, or 0.

We show in this section that the matrices A_c, Q_f, and B_f are all totally unimodular.

Theorem 8.13 *The incidence matrix A_c of a directed graph is totally unimodular.*

Proof. We prove the theorem by induction on the order of a square submatrix of A_c.

Obviously the determinant of every square submatrix of A_c of order 1 is 1, −1, or 0. Assume, as the induction hypothesis, that the determinant of every square submatrix of order less than k is equal to 1, −1, or 0.

Consider any nonsingular square submatrix of A_c of order k. Every column in this matrix contains at most two nonzero entries, one +1, and/or one −1. Since the submatrix is nonsingular, not every column can have both +1 and −1. For the same reason, in this submatrix there can be no column consisting of only zeros. Thus there is at least one column that contains exactly one nonzero entry. Expanding the determinant of the submatrix by this column and using the induction hypothesis, we find that the desired determinant is ±1. ∎

Let Q_f be the fundamental cutset matrix of an n-vertex connected graph G with respect to some spanning tree T. Let the branches of T be $b_1, b_2, \ldots, b_{n-1}$. Let G' be the graph which is obtained from G by identifying or short-circuiting the end vertices of one of the branches, say, the branch b_1. Then $T - \{b_1\}$ is a spanning tree of G'. Let us now delete from Q_f the row corresponding to branch b_1 and denote the resulting matrix as Q'_f. Then it is not difficult to show that Q'_f is the fundamental cutset matrix of G' with respect to the spanning tree $T - \{b_1\}$. Thus the matrix that results after deleting any row from Q_f is a fundamental cutset matrix of some connected graph. Generalizing this, we can state that each matrix formed by some rows of Q_f is a fundamental cutset matrix of some connected graph.

Theorem 8.14 *Any fundamental cutset matrix Q_f of a connected graph G is totally unimodular.*

Proof. Let Q_f be the fundamental cutset matrix of G with respect to a spanning tree T. Then

$$Q_f = \begin{bmatrix} U & Q_{fc} \end{bmatrix}.$$

Let an incidence matrix A of G be partitioned as $A = [A_{11} \; A_{12}]$ where the columns of A_{11} correspond to the branches of T. We know from Theorem 8.9 that A_{11} is nonsingular. Now we can write Q_f as

$$Q_f = \begin{bmatrix} U & Q_{fc} \end{bmatrix} = A_{11}^{-1} \begin{bmatrix} A_{11} & A_{12} \end{bmatrix}.$$

If C is any square submatrix of Q_f of order $n-1$, where n is the number of vertices of G, and D is the corresponding submatrix of A, then $C = A_{11}^{-1} D$. Since $\det(D) = \pm 1$ or 0 and $\det(A_{11}^{-1}) = \pm 1$, we get

$$\det(C) = \pm 1 \text{ or } 0. \tag{8.19}$$

Consider next any square submatrix H of Q_f of order less than $n-1$. From the arguments preceding this theorem we know that H is a submatrix of a fundamental cutset matrix of some connected graph. Therefore, $\det(H) = \pm 1$ or 0, the proof of which follows along the same lines as that used to prove (8.19).

Thus the determinant of every square submatrix of Q_f is ±1 or 0 and hence Q_f is totally unimodular. ∎

Next we show that B_f is totally unimodular.

Theorem 8.15 *Any fundamental circuit matrix B_f of a connected graph G is totally unimodular.*

Proof. Let B_f and Q_f be the fundamental circuit and cutset matrices of G with respect to a spanning tree T. If $Q_f = [U \; Q_{fc}]$, then we know from (8.13) that

$$B_f = \begin{bmatrix} -Q_{fc}^t & U \end{bmatrix}.$$

Since Q_{fc} is totally unimodular, Q_{fc}^t is also totally unimodular. It is now a simple exercise to show that $[-Q_{fc}^t \; U]$ is totally unimodular. ∎

8.8 NUMBER OF SPANNING TREES

We derive in this section a formula for counting the number of spanning trees in a connected graph. This formula is based on Theorems 8.9 and 8.13 and a result in matrix theory, known as the Binet–Cauchy theorem.

A major determinant or briefly a *major* of a matrix is a determinant of maximum order in the matrix. Let P be a matrix of order $p \times q$ and Q be a matrix of order $q \times p$ with $p \leq q$. The majors of P and Q are of order p. If a major of P consists of the columns $i_1, i_2, \ldots, i_p$ of P, then the corresponding major of Q is formed by the rows $i_1, i_2, \ldots, i_p$ of Q. For example, if

$$P = \begin{bmatrix} 1 & -1 & 3 & 3 \\ 2 & 2 & -1 & 2 \end{bmatrix}, \text{ and } Q = \begin{bmatrix} 1 & 2, \\ 2 & -1, \\ -3 & 1, \\ 1 & 2 \end{bmatrix},$$

then for the major

$$\begin{vmatrix} -1 & 3 \\ 2 & -1 \end{vmatrix}$$

of P,

$$\begin{vmatrix} 2 & -1 \\ -3 & 1 \end{vmatrix}$$

is the corresponding major of Q.

Theorem 8.16 (Binet–Cauchy) *If P is a $p \times q$ matrix and Q is a $q \times p$ matrix, with $p \leq q$, then*

$$\det(PQ) = \Sigma(\text{product of the corresponding majors of } P \text{ and } Q). \quad ∎$$

Proof of this theorem may be found in Hohn [4].

As an illustration, if the matrices P and Q are as given earlier, then applying the Binet–Cauchy theorem we get

$$\det(PQ) = \begin{vmatrix} 1 & -1 \\ 2 & 2 \end{vmatrix}\begin{vmatrix} 1 & 2 \\ 2 & -1 \end{vmatrix} + \begin{vmatrix} 1 & 3 \\ 2 & -1 \end{vmatrix}\begin{vmatrix} 1 & 2 \\ -3 & 1 \end{vmatrix} + \begin{vmatrix} 1 & 3 \\ 2 & 2 \end{vmatrix}\begin{vmatrix} 1 & 2 \\ 1 & 2 \end{vmatrix}$$

$$+ \begin{vmatrix} -1 & 3 \\ 2 & -1 \end{vmatrix}\begin{vmatrix} 2 & -1 \\ -3 & 1 \end{vmatrix} + \begin{vmatrix} -1 & 3 \\ 2 & 2 \end{vmatrix}\begin{vmatrix} 2 & -1 \\ 1 & 2 \end{vmatrix} + \begin{vmatrix} 3 & 3 \\ -1 & 2 \end{vmatrix}\begin{vmatrix} -3 & 1 \\ 1 & 2 \end{vmatrix}$$

$$= -167.$$

Theorem 8.17 *Let G be a connected undirected graph and A an incidence matrix of a directed graph which is obtained by assigning arbitrary orientations to the edges of G. Then*

$$\tau(G) = \det(AA^t),$$

where $\tau(G)$ is the number of spanning trees of G.

Proof. By the Binet–Cauchy theorem

$$\det(AA^t) = \Sigma(\text{product of the corresponding majors of } A \text{ and } A^t). \quad (8.20)$$

Note that the corresponding majors of A and A^t both have the same value equal to 1, -1, or 0 (Theorem 8.13). Therefore each nonzero term in the sum on the right-hand side of (8.20) has the value 1. Furthermore, a major of A is nonzero if and only if the edges corresponding to the columns of the major form a spanning tree.

Thus there is a one-to-one correspondence between the nonzero terms in the sum on the right-hand side of (8.20) and the spanning trees of G. Hence the theorem. ∎

Let $v_1, v_2, \ldots, v_n$ denote the vertices of an undirected graph G without self-loops. The *degree matrix** $K = [k_{ij}]$ is an $n \times n$ matrix defined as follows:

$$k_{ij} = \begin{cases} -p, & \text{if } i \neq j \text{ and there are } p \text{ parallel edges connecting } v_i \text{ and } v_j. \\ d(v_i), & \text{if } i = j. \end{cases}$$

We can easily see that $K = A_c A_c^t$ and is independent of our choice of the orientations for arriving at the all-vertex incidence matrix A_c. If v is the reference vertex of A, then AA^t is obtained by removing the rth row and the rth column of K. Thus the matrix AA^t used in Theorem 8.17 can be obtained by an inspection of the graph G.

It is clear from the definition of the degree matrix that the sum of all the elements in each row of K equals zero. Similarly the sum of all the elements in each column of K equals zero. A square matrix with these properties is called an *equi-cofactor matrix*. As its name implies, all the cofactors of an equi-cofactor matrix are equal [4]. Thus from Theorem 8.17 we get the following result, originally due to Kirchhoff [5].

Theorem 8.18 *All the cofactors of the degree matrix of a connected undirected graph have the same value equal to the number of spanning trees of G.* ∎

Next we derive a formula for counting the number of distinct spanning trees which can be constructed on n labeled vertices. Clearly this number is the same as the number of spanning trees of K_n, the complete graph on n labeled vertices.

Theorem 8.19 (Cayley) *There are n^{n-2} labeled trees on $n \geq 2$ vertices.*

Proof. In the case of K_n, the matrix AA^t is of the form

$$\begin{bmatrix} n-1 & -1 & \cdots & -1 \\ -1 & n-1 & \cdots & -1 \\ . & . & \cdots & . \\ . & . & \cdots & . \\ . & . & \cdots & . \\ -1 & -1 & \cdots & n-1 \end{bmatrix}.$$

---
*The *degree matrix* is also known as the graph Laplacian. See Chapters 10 and 11.

By Theorem 8.17, the determinant of this matrix gives the number of spanning trees of K_n, which is the same as the number of labeled trees on n vertices.

To compute $\det(AA^t)$, subtract the first column of AA^t from all the other columns of AA^t. Then we get

$$\begin{bmatrix} n-1 & -n & -n & \cdots & -n \\ -1 & n & 0 & \cdots & 0 \\ -1 & 0 & n & \cdots & 0 \\ . & . & . & \cdots & . \\ . & . & . & \cdots & . \\ . & . & . & \cdots & . \\ -1 & 0 & 0 & \cdots & n \end{bmatrix}.$$

Now adding to the first row of the above matrix every one of the other rows, we get

$$\begin{bmatrix} 1 & 0 & 0 & \cdots & 0 \\ -1 & n & 0 & \cdots & 0 \\ -1 & 0 & n & \cdots & 0 \\ . & . & . & \cdots & . \\ . & . & . & \cdots & . \\ -1 & 0 & 0 & \cdots & n \end{bmatrix}.$$

The determinant of this matrix is n^{n-2}. The theorem now follows since addition of any two rows or any two columns of a matrix does not change the value of the determinant of the matrix. ∎

Several proofs of Cayley's theorem [6] are available in the literature. See Moon [7] and Prüfer [8].

8.9 NUMBER OF SPANNING 2-TREES*

In this section we relate the cofactors of the matrix AA^t of a graph G to the number of spanning 2-trees of the appropriate type. For this purpose, we need symbols to denote 2-trees in which certain specified vertices are required to be in different components. We use the symbol $T_{ijk,\ldots,rst,\ldots}$ to denote spanning 2-trees in which the vertices $v_i, v_j, v_k, \ldots$ are required to be in one component and the vertices $v_r, v_s, v_t, \ldots$ are required to be in the other component of the 2-tree. The number of these spanning 2-trees in the graph G will be denoted by $\tau_{ijk,\ldots,rst,\ldots}$.

In the following, we denote by A an incidence matrix of the directed graph which is obtained by assigning arbitrary orientations to the edges of the graph G. However, we shall refer to A as an incidence matrix of G. We shall assume, without any loss of generality, that v_n is the reference vertex for A, and the ith row of A corresponds to vertex v_i. Δ_{ij} will denote the (i,j) cofactor of AA^t.

Let A_{-i} denote the matrix obtained by removing from A its ith row. If G' is the graph obtained by short-circuiting (contracting) the vertices v_i and v_n in G, then we can verify the following:

---

*An acyclic spanning subgraph of a connected graph G that contains exactly two trees is called a spanning 2-tree of G.

1. A_{-i} is the incidence matrix of G' with v_n as the reference vertex.

2. A set of edges form a spanning tree of G' if and only if these edges form a spanning 2-tree $T_{i,n}$ of G.

Thus, there exists a one-to-one correspondence between the nonzero majors of A_{-i} and the spanning 2-trees of the type $T_{i,n}$.

Theorem 8.20 *For a connected graph G,*

$$\Delta_{ii} = \tau_{i,n}.$$

Proof. Clearly $\Delta_{ii} = \det(A_{-i}A^t_{-i})$. Proof is immediate since the nonzero majors of A_{-i} correspond to the spanning 2-trees $T_{i,n}$ of G and vice versa. ■

Consider next the (i,j) cofactor Δ_{ij} of AA^t which is given by

$$\Delta_{ij} = (-1)^{i+j}\det(A_{-i}A^t_{-j}). \tag{8.21}$$

By the Binet–Cauchy theorem,

$$\det(A_{-i}A^t_{-j}) = \Sigma(\text{product of the corresponding majors of } A_{-i} \text{ and } A^t_{-j}). \tag{8.22}$$

Each nonzero major of A_{-i} corresponds to a spanning 2-tree of the type $T_{i,n}$, and each nonzero major of A_{-j} corresponds to a spanning 2-tree of the type $T_{j,n}$. Therefore the nonzero terms in the sum on the right-hand side of (8.22) correspond to the spanning 2-trees of the type $T_{ij,n}$. Each one of these nonzero terms is equal to a determinant of the type $\det(F_{-i}F^t_{-j})$, where F is the incidence matrix of a 2-tree of the type $T_{ij,n}$.

Theorem 8.21 *Let F denote the incidence matrix of a 2-tree $T_{ij,n}$ with v_n as the reference vertex. If the ith row of F corresponds to vertex v_i, then*

$$\det(F_{-i}F^t_{-j}) = (-1)^{i+j}.$$

Proof. Let T_1 and T_2 denote the two components of $T_{ij,n}$. Assume that v_n is in T_2. Then v_i and v_j will be in T_1. By interchanging some of its rows and the corresponding columns, we can write the matrix FF^t as

$$S = \begin{bmatrix} C & 0 \\ 0 & D \end{bmatrix},$$

where:
    C is the degree matrix of T_1
    D is obtained by removing from the degree matrix of T_2 the row and the column corresponding to v_n

Let row k' of S correspond to vertex v_k.

Interchanging some rows and the corresponding columns of a matrix does not alter the values of the cofactors of the matrix. So

$$\begin{aligned}(i',j') \text{ cofactor of } S &= (i,j) \text{ cofactor of } (FF^t) \\ &= (-1)^{i+j}\det(F_{-i}F^t_{-j}).\end{aligned} \tag{8.23}$$

By Theorem 8.18 all the cofactors of C have the same value equal to the number of spanning trees of T_1. So we have

$$(i',j') \text{ cofactor of } C = 1.$$

Furthermore,
$$\det(D) = 1.$$

So
$$(i', j') \text{ cofactor of } S = [(i', j') \text{ cofactor of } C] \det D = 1. \tag{8.24}$$

Now, from (8.23) and (8.24) we get
$$\det(F_{-i} F^t_{-j}) = (-1)^{i+j}. \qquad \blacksquare$$

Proof of the above theorem is due to Sankara Rao et al. [9]. The following result is due to Seshu and Reed [10].

Theorem 8.22 *For a connected graph G,*
$$\Delta_{ij} = \tau_{ij,n}.$$

Proof. Since each nonzero term in the sum on the right-hand side of (8.22) is equal to a determinant of the type given in Theorem 8.21, we get
$$\det(A_{-i}, A^t_{-j}) = (-1)^{i+j} \tau_{ij,n}.$$

So
$$\Delta_{ij} = (-1)^{i+j} \det(A_{-i}, A^t_{-j})$$
$$= \tau_{ij,n}. \qquad \blacksquare$$

8.10 NUMBER OF DIRECTED SPANNING TREES* IN A DIRECTED GRAPH

In this section we discuss a method due to Tutte [11] for computing the number of directed spanning trees in a given directed graph having a specified vertex as root. This method is in fact a generalization of the method given in Theorem 8.17 to compute the number of spanning trees of a graph, and it is given in terms of the in-degree matrix defined below.

The *in-degree matrix* $K = [k_{pq}]$ of a directed graph $G = (V, E)$ without self-loops and with $V = \{v_1, v_2, \ldots, v_n\}$ is an $n \times n$ matrix defined as follows:

$$k_{pq} = \begin{cases} -\omega, & \text{if } p \neq q \text{ and there are } \omega \text{ parallel edges directed from } v_p \text{ to } v_q. \\ d^-(v_p), & \text{if } p = q. \end{cases}$$

Let K_{ij} denote the matrix obtained by removing row i and column j from K. Tutte's method is based on the following theorem.

Theorem 8.23 *A direct graph $G = (V, E)$ with no self-loops and with $V = \{v_1, v_2, \ldots, v_n\}$ is a directed spanning tree with v_r as the root if and only if its in-degree matrix K has the following properties:*

---
*A spanning tree in a directed graph is a directed spanning tree with root V_r if in the tree there is a directed path from V_r to every vertex in the tree.

1. $k_{pp} = \begin{cases} 0, & \text{if } p = r, \\ 1, & \text{if } p \neq r. \end{cases}$

2. $\det(K_{rr}) = 1$.

Proof. Necessity: Suppose the given directed graph G is a directed spanning tree with v_r as the root. Clearly G is acyclic. So we can label the vertices of G with the numbers $1, 2, \ldots, n$ in such a way that (i, j) is a directed edge of G only if $i < j$. Such a labeling is called a topological sorting of the vertices of an acyclic graph. Then in such a numbering, the root vertex would receive the number 1. If the ith row and the ith column of the new in-degree matrix K' of G correspond to the vertex assigned the number i, then we can easily see that K' has the following properties:

$$k'_{11} = 0,$$
$$k'_{pp} = 1, \text{ for } p \neq 1,$$
$$k'_{pq} = 0, \text{ if } p > q.$$

Therefore

$$\det(K'_{11}) = 1.$$

The matrix K' can be obtained by interchanging some rows and the corresponding columns of K. Such an interchange does not change the value of the determinant of any submatrix of K. So

$$\det(K_{rr}) = \det(K'_{11}) = 1.$$

Sufficiency: Suppose the in-degree matrix K of the graph G satisfies the two properties given in the theorem. If G is not a directed tree, it contains a circuit C. The root vertex v_r cannot be in C, for this would imply that $d^-(v_r) > 0$ or that $d^-(v) > 1$ for some other vertex v in C, contradicting property 1. In a similar way, we can show that

1. C must be a directed circuit; and
2. No edge not in C is incident into any vertex in C.

Consider now the submatrix K_s of K consisting of the columns corresponding to the vertices in C. Because of the above properties, each row of K_s corresponding to a vertex in C has exactly one $+1$ and one -1. All the other rows in K_s contain only zero elements. Thus the sum of the columns of K_s is zero. In other words, the sum of the columns of K which correspond to the vertices in C is zero. Since v_r is not in C, this is true in the case of the matrix K_{rr} too, contradicting property 2. Hence the sufficiency. ∎

We now develop Tutte's method for computing the number τ_d of the directed spanning trees of a directed graph G having vertex v_r as the root. Assume that G has no self-loops.

For any graph g, let $K(g)$ denote its in-degree matrix, and let K' be the matrix obtained from K by replacing its rth column by a column of zeros. Denote by S the collection of all the subgraphs of G in each of which $d^-(v_r) = 0$ and $d^-(v_p) = 1$ for $p \neq r$. Clearly

$$|S| = \prod_{p=1}^{n} d^-(v_p).$$

Further, for any subgraph $g \in S$, the corresponding in-degree matrix satisfies property 1 given in Theorem 8.23.

It is well known in matrix theory that the determinant of a square matrix is a linear function of its columns. For example, if

$$P = [p_1, p_2, \ldots, p'_i + p''_i, \ldots, p_n]$$

is a square matrix with the columns $p_1, p_2, \ldots, p'_i + p''_i, \ldots, p_n$, then

$$\det(P) = \det[p_1, p_2, \ldots, p'_i, \ldots, p_n] + \det[p_1, p_2, \ldots, p''_i, \ldots, p_n]$$

Using the linearity of the determinant function and the fact that the sum of all the entries in each column of the matrix $K'(G)$ is equal to zero, we can write $\det K'(G)$ as the sum of $|S|$ determinants each of which satisfies the property 1 given in Theorem 8.23. It can be seen that there is a one-to-one correspondence between these determinants and the in-degree matrices of the subgraphs in S. Thus

$$\det(K'(G)) = \sum_{g \in S} \det(K'(g)).$$

So

$$\det(K'_{rr}(G)) = \sum_{g \in S} \det(K'_{rr}(g)).$$

Since

$$\det(K'_{rr}(G)) = \det(K_{rr}(G)).$$

and

$$\det(K'_{rr}(g)) = \det(K_{rr}(g)), \quad \text{for all } g \in S,$$

we get

$$\det(K_{rr}(G)) = \sum_{g \in S} \det(K'_{rr}(g)).$$

From Theorem 8.23 it follows that each determinant in the sum on the right-hand side of the above equation is nonzero and equal to 1 if and only if the corresponding subgraph in S is a directed spanning tree. Thus we have the following theorem.

Theorem 8.24 *Let K be the in-degree matrix of a directed graph G without self-loops. Let the ith row of K correspond to vertex v_i of G. Then the number τ_d of directed spanning trees of G having v_r as its root is given by*

$$\tau_d = \det(K_{rr}),$$

where K_{rr} is the matrix obtained by removing from K its rth row and its rth column. ∎

We now illustrate the above theorem and the arguments leading to its proof.

Consider the directed graph G shown in Figure 8.3. Let us compute the number of directed spanning trees with vertex v_1 as the root.

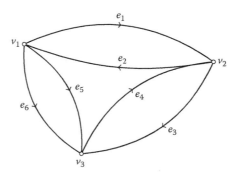

Figure 8.3 Illustration of the proof of Theorem 8.24.

The in-degree matrix K of G is

$$K = \begin{bmatrix} 1 & -1 & -2 \\ -1 & 2 & -1 \\ 0 & -1 & 3 \end{bmatrix}$$

and

$$K' = \begin{bmatrix} 0 & -1 & -2 \\ 0 & 2 & -1 \\ 0 & -1 & 3 \end{bmatrix}.$$

We can write $\det(K')$ as

$$\det(K') = \begin{vmatrix} 0 & -1 & -2 \\ 0 & 2 & -1 \\ 0 & -1 & 3 \end{vmatrix} = \begin{vmatrix} 0 & -1 & -2 \\ 0 & 1 & -1 \\ 0 & 0 & 3 \end{vmatrix} + \begin{vmatrix} 0 & 0 & -2 \\ 0 & 1 & -1 \\ 0 & -1 & 3 \end{vmatrix}$$

$$= \begin{vmatrix} 0 & -1 & -1 \\ 0 & 1 & 0 \\ 0 & 0 & 1 \end{vmatrix} + \begin{vmatrix} 0 & -1 & -1 \\ 0 & 1 & 0 \\ 0 & 0 & 1 \end{vmatrix} + \begin{vmatrix} 0 & -1 & 0 \\ 0 & 1 & -1 \\ 0 & 0 & 1 \end{vmatrix}$$

$$+ \begin{vmatrix} 0 & 0 & -1 \\ 0 & 1 & 0 \\ 0 & -1 & 1 \end{vmatrix} + \begin{vmatrix} 0 & 0 & -1 \\ 0 & 1 & 0 \\ 0 & -1 & 1 \end{vmatrix} + \begin{vmatrix} 0 & 0 & 0 \\ 0 & 1 & -1 \\ 0 & -1 & 1 \end{vmatrix}.$$

The six determinants on the right-hand side in the above equation correspond to the subgraphs on the following set of edges:

$$\{e_1, e_5\}, \{e_1, e_6\}, \{e_1, e_3\},$$
$$\{e_4, e_5\}, \{e_4, e_6\}, \{e_4, e_3\}.$$

Removing the first row and the first column from the above determinants, we get

$$\det(K'_{11}) = \begin{vmatrix} 2 & -1 \\ -1 & 3 \end{vmatrix} = \begin{vmatrix} 1 & 0 \\ 0 & 1 \end{vmatrix} + \begin{vmatrix} 1 & 0 \\ 0 & 1 \end{vmatrix} + \begin{vmatrix} 1 & -1 \\ 0 & 1 \end{vmatrix}$$

$$+ \begin{vmatrix} 1 & 0 \\ -1 & 1 \end{vmatrix} + \begin{vmatrix} 1 & 0 \\ -1 & 1 \end{vmatrix} + \begin{vmatrix} 1 & -1 \\ -1 & 1 \end{vmatrix}$$

$$= 5.$$

The five directed spanning trees with v_1 as root are

$$\{e_1, e_5\}, \{e_1, e_6\}, \{e_1, e_3\}, \{e_4, e_5\}, \{e_4, e_6\}.$$

8.11 DIRECTED SPANNING TREES AND DIRECTED EULER TRAILS

Let G be a directed Eulerian graph without self-loops. In this section we relate the number of directed Euler trails in G to the number of directed spanning trees of G.

Let $v_1, v_2, \ldots, v_n$ denote the vertices of G. Consider a directed Euler trail C in G. Let e_{j_1} be any edge of G incident into v_1. For every $p = 2, 3, \ldots, n$, let e_{j_p} denote the first edge on C to enter vertex v_p after traversing e_{j_1}.

Let H denote the subgraph of G on the edge set $\{e_{j_2}, e_{j_3}, \ldots, e_{j_n}\}$.

Lemma 8.1 *Let C be a directed Euler trail of a directed Eulerian graph G. The subgraph H defined as above is a directed spanning tree of G with root v_1.*

Proof. Clearly in H, $d^-(v_1) = 0$ and $d^-(v_p) = 1$ for all $p = 2, 3, \ldots, n$. Suppose H has a circuit C'. Then v_1 is not in C', for otherwise either $d^-(v_1) > 0$ or $d^-(v) > 1$ for some other vertex v on C'. For the same reason C' is a directed circuit. Since $d^-(v) = 1$ for every vertex v on C', no edge not in C' enters any vertex in C'. This means that the edge e which is the first edge of C to enter a vertex of C' after traversing e_{j_1} does not belong to H, contradicting the definition of H. Thus H has no circuits.

Now it follows that H is a directed spanning tree of G. ∎

Given a directed spanning tree H of an n-vertex directed Eulerian graph G without self-loops, let v_1 be the root of H and e_{j_1} an edge incident into v_1 in G. Let e_{j_p} for $p = 2, 3, \ldots, n$ be the edge entering v_p in H. We now describe a method for constructing a directed Euler trail in G.

1. Start from vertex v_1 and traverse backward on any edge entering v_1 other than e_{j_1} if such an edge exists, or on e_{j_1} if there is no other alternative.

2. In general on arrival at a vertex v_p, leave it by traversing backward on an edge entering v_p which has not yet been traversed and, if possible, other than e_{j_p}. Stop if no untraversed edges entering v_p exist.

In the above procedure every time we reach a vertex $v_p \neq v_1$, there will be an untraversed edge entering v_p because the in-degree of every vertex in G is equal to its out-degree. Thus this procedure terminates only at the vertex v_1 after traversing all the edges incident into v_1.

Suppose there exists in G an untraversed edge (u, v) when the above procedure terminates at v_1. Since the in-degree of u is equal to its out-degree, there exists at least one untraversed edge incident into u. If there is more than one such untraversed edge, then one of those will be the edge y entering u in H. This follows from step 2 of the procedure. This untraversed edge y will lead to another untraversed edge which is also in H. Finally we shall arrive at v_1 and shall find an untraversed edge incident into v_1. This is not possible since all the edges incident into v_1 will have been traversed when the procedure terminates at v_1.

Thus all the edges of G will be traversed during the procedure we described above, and indeed a directed Euler trail is constructed.

Since at each vertex v_p there are $(d^-(v_p) - 1)!$ different orders for picking the incoming edge (with e_{j_p} at the end), it follows that the number of distinct directed Euler trails that we can construct from a given directed spanning tree H and e_{j_1} is

$$\prod_{p=1}^{n}(d^-(v_p) - 1)!$$

Further, each different choice of H will yield a different e_{j_p} for some $p = 2, 3, \ldots, n$, which will in turn result in a different entry to v_p after traversing e_{j_1} in the resulting directed Euler trail.

Finally, since the procedure of constructing a directed Euler trail is the reversal of the procedure for constructing a directed spanning tree, it follows that every directed Euler trail can be constructed from some directed spanning tree.

Thus we have proved the following theorem due to Van Aardenne-Ehrenfest and de Bruijn [12].

Theorem 8.25 *The number of directed Euler trails of a directed Eulerian graph G without self-loops is*

$$\tau_d(G) \prod_{p=1}^{n} (d^-(v_p) - 1)!$$

where $\tau_d(G)$ is the number of directed spanning trees of G with v_1 as root. ∎

Since the number of directed Euler trails is independent of the choice of the root, we get the following.

Corollary 8.4 *The number of directed spanning trees of a directed Eulerian graph is the same for every choice of root.* ∎

Further Reading

Seshu and Reed [10], Chen [4], Mayeda [13], and Deo [14] are other references for the topics covered in this chapter. Chen [4] also gives historical details regarding the results presented here.

The textbook by Harry and Palmer [15] is devoted exclusively to the study of enumeration problems in graph theory, in particular, to those related to unlabeled graphs. See also Biggs [16].

The problem of counting spanning trees has received considerable attention in electrical network theory literature. Recurrence relations for counting spanning trees in special classes of graphs are available. For example, see Myers [17] and [18], Bedrosian [19], Bose et al. [20], and Swamy and Thulasiraman [21]. Berge [22] contains formulas for the number of spanning trees having certain specified properties. For algorithms to generate all the spanning trees of a connected graph see Gabow and Myers [23] and Jayakumar et al. [24] and [25].

Thulasiraman and Swamy [2] give a detailed discussion of applications of graphs in electrical network theory.

References

[1] J. L. Gross, J. Ellen, and P. F. Zhang, *Handbook of Graph Theory*, CRC Press, New York, 2013.

[2] M. N. S. Swamy and K. Thulasiraman, *Graphs, Networks and Algorithms*, Wiley-Interscience, New York, 1981.

[3] F. E. Hohn, *Elementary Matrix Algebra*, Macmillan, New York, 1958.

[4] W. K. Chen, *Applied Graph Theory*, North-Holland, Amsterdam, the Netherlands, 1971.

[5] G. Kirchhoff, Uber die Auflosung der Gleichungen, auf welche man bei der untersuchung der linearen verteilung galvanischer strome gefuhrt wird, *Ann. Phys. Chem.*, **72** (1847), 497–508.

[6] A. Cayley, A theorem on trees, *Quart. J. Math.*, **23** (1889), 376–378.

[7] J. W. Moon, Various proofs of Cayley's formula for counting trees, In F. Harary and L. W. Beinke, editors, *A Seminar on Graph Theory*, Holt, Rinehart & Winston, New York, 1967, 70–78.

[8] H. Prüfer, Neuer Beweis eines Satzes über Permutationen, *Arch. Math. Phys.*, **27** (1918), 742–744.

[9] K. Sankara Rao, V. V. Bapeswara Rao, and V. G. K. Murti, Two-tree admittance products, *Electron. Lett.*, **6** (1970), 834–835.

[10] S. Seshu and M. B. Reed, *Linear Graphs and Electrical Networks*, Addison-Wesley, Reading, MA, 1961.

[11] W. T. Tutte, The dissection of equilateral triangles into equilateral triangles, *Proc. Cambridge Phil. Soc.*, **44** (1948), 203–217.

[12] T. Van Aardenne-Ehrenfest and N. G. de Bruijn, Circuits and trees in oriented linear graphs, *Simon Stevin*, **28** (1951), 203–217.

[13] W. Mayeda, *Graph Theory*, Wiley-Interscience, New York, 1972.

[14] N. Deo, *Graph Theory with Applications to Engineering and Computer Science*, Prentice Hall, Englewood Cliffs, NJ, 1974.

[15] F. Harary and E. M. Palmer, *Graphical Enumeration*, Academic Press, New York, 1973.

[16] N. Biggs, *Algebraic Graph Theory*, Cambridge University Press, Cambridge, England, 1974.

[17] B. R. Myers, Number of trees in a cascade of 2-port networks, *IEEE Trans. Circuit Theory*, **CT-18** (1967), 284–290.

[18] B. R. Myers, Number of spanning trees in a wheel, *IEEE Trans. Circuit Theory*, **CT-18** (1971), 280–282.

[19] S. D. Bedrosian, Number of spanning trees in multigraph wheels, *IEEE Trans. Circuit Theory*, **CT-19** (1972), 77–78.

[20] N. K. Bose, R. Feick, and F. K. Sun, General solution to the spanning tree enumeration problem in multigraph wheels, *IEEE Trans. Circuit Theory*, **CT-20** (1973), 69–70.

[21] M. N. S. Swamy and K. Thulasiraman, A theorem in the theory of determinants and the number of spanning trees of a graph, *Can. Elec. Eng. J.*, **8** (1983), 147–152.

[22] C. Berge, *Graphs and Hypergraphs*, North-Holland, Amsterdam, the Netherlands, 1973.

[23] H. N. Gabow and E. W. Myers, Finding all spanning trees of directed and undirected graphs, *SIAM J. Comp.*, **7** (1978), 280–287.

[24] R. Jayakumar, K. Thulasiraman, and M. N. S. Swamy, Complexity of computation of a spanning tree enumeration algorithm, *IEEE Trans. Circuit and Systems*, **CAS-31** (1984), 853–860.

[25] R. Jayakumar, K. Thulasiraman, and M. N. S. Swamy, MOD-CHAR: An implementation of Char's spanning tree enumeration algorithm and its complexity analysis. *IEEE Trans. Circuit and Systems*, **CAS-36** (1989), 219–228.

CHAPTER 9

Adjacency Matrix and Signal Flow Graphs*

Krishnaiyan "KT" Thulasiraman

M. N. S. Swamy

CONTENTS

9.1 Introduction 215
9.2 Adjacency Matrix of a Directed Graph 215
9.3 Coates' Gain Formula 218
9.4 Mason's Gain Formula 222

9.1 INTRODUCTION

Signal flow graph theory is concerned with the development of a graph theoretic approach to solving a system of linear algebraic equations. Two closely related methods proposed by Coates [2] and Mason [3,4] have appeared in the literature and have served as elegant aids in gaining insight into the structure and nature of solutions of systems of equations. In this chapter we develop these two methods. Our development follows closely [5].

Coates' and Mason's methods may be viewed as generalizations of a basic theorem in graph theory due to Harary [6], which provides a formula for finding the determinant of the adjacency matrix of a directed graph. Thus, our discussion begins with the development of this theorem. An extensive discussion of this topic may be found in Chen [7].

9.2 ADJACENCY MATRIX OF A DIRECTED GRAPH

Consider a directed graph $G = (V, E)$ with no parallel edges. Let $V = \{v_1, \ldots, v_n\}$. The *adjacency matrix* $M = [m_{ij}]$ of G is an $n \times n$ matrix defined as follows:

$$m_{ij} = \begin{cases} 1, & \text{if } (v_i, v_j) \in E \\ 0, & \text{otherwise} \end{cases}$$

The graph shown in Figure 9.1 has the following adjacency matrix:

$$M = \begin{matrix} & \begin{matrix} v_1 & v_2 & v_3 & v_4 \end{matrix} \\ \begin{matrix} v_1 \\ v_2 \\ v_3 \\ v_4 \end{matrix} & \begin{pmatrix} 1 & 1 & 1 & 0 \\ 0 & 1 & 0 & 0 \\ 1 & 0 & 0 & 1 \\ 1 & 1 & 1 & 1 \end{pmatrix} \end{matrix}$$

---

*This chapter is an edited version of Chapter 8 in Reference 1.

In the following we shall develop a topological formula for $\det(M)$. Toward this end we introduce some basic terminology. A *1-factor* of a directed graph G is a spanning subgraph of G in which the in-degree and the out-degree of every vertex are both equal to 1. It is easy to see that a 1-factor is a collection of vertex-disjoint directed circuits. Because a self-loop at a vertex contributes 1 to the in-degree and 1 to the out-degree of the vertex, a 1-factor may have some self-loops. As an example, the three 1-factors of the graph of Figure 9.1 are shown in Figure 9.2.

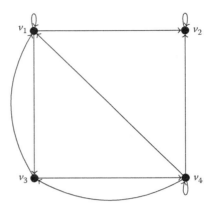

Figure 9.1 Graph G.

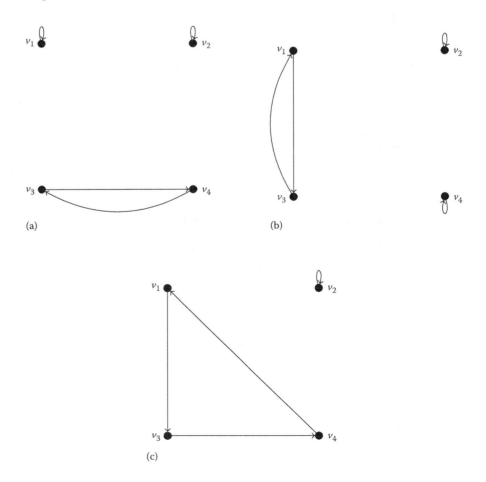

Figure 9.2 Three 1-factors of the graph of Figure 9.1.

A *permutation* $(j_1, j_2, \ldots, j_n)$ of integers $1, 2, \ldots, n$ is even (odd) if an even (odd) number of interchanges are required to rearrange it as $(1, 2, \ldots, n)$.

The notation
$$\begin{pmatrix} 1, 2, \ldots, n \\ j_1, j_2, \ldots, j_n \end{pmatrix}$$
is also used to represent the permutation $(j_1, j_2, \ldots, j_n)$. As an example, the permutation $(4, 3, 1, 2)$ is odd because it can be rearranged as $(1, 2, 3, 4)$ using the following sequence of interchanges:

1. Interchange 2 and 4.

2. Interchange 1 and 2.

3. Interchange 2 and 3.

For a permutation $(j) = (j_1, j_2, \ldots, j_n)$, $\varepsilon_{j_1, j_2, \ldots, j_n}$, is defined as equal to 1, if (j) is an even permutation; otherwise, $\varepsilon_{j_1, j_2, \ldots, j_n}$ is equal to -1.

Given an $n \times n$ square matrix $X = [x_{ij}]$, we note that $\det X$ is given by
$$\det(X) = \sum_{(j)} \varepsilon_{j_1, j_2, \ldots, j_n} x_{1j_1} \cdot x_{2j_2} \ldots x_{nj_n}$$
where the summation $\sum_{(j)}$ is over all permutations of $1, 2, \ldots, n$ [8].

The following theorem is due to Harary [6].

Theorem 9.1 *Let $H_i, i = 1, 2, \ldots, p$ be the 1-factors of an n-vertex directed graph G. Let L_i denote the number of directed circuits in H_i, and let M denote the adjacency matrix of G. Then*
$$\det(M) = (-1)^n \sum_{i=1}^{p} (-1)^{L_i}$$

Proof. From the definition of a determinant, we have
$$\det(M) = \sum_{(j)} \varepsilon_{j_1, j_2, \ldots, j_n} m_{1j_1} \cdot m_{2j_2} \ldots m_{nj_n} \tag{9.1}$$

Proof will follow if we establish the following:

1. Each nonzero term $m_{1j_1} \cdot m_{2j_2} \ldots m_{nj_n}$ corresponds to a 1-factor of G, and conversely, each 1-factor of G corresponds to a non-zero term $m_{1j_1} \cdot m_{2j_2} \ldots m_{nj_n}$.

2. $\varepsilon_{j_1, j_2, \ldots, j_n} = (-1)^{n+L}$ if the 1-factor corresponding to a nonzero $m_{1j_1} \cdot m_{2j_2} \ldots m_{nj_n}$ has L directed circuits.

A nonzero term $m_{1j_1} \cdot m_{2j_2} \ldots m_{nj_n}$ corresponds to the set of edges $(v_1, v_{j1}), (v_2, v_{j2}), \ldots, (v_n, v_{jn})$. Each vertex appears exactly twice in this set, once as an initial vertex and once as a terminal vertex of a pair of edges. Therefore, in the subgraph induced by these edges, for each vertex its in-degree and its out-degree are both equal to 1, and this subgraph is a 1-factor of G. In other words, each nonzero term in the sum in (9.1) corresponds to a 1-factor of G. The fact that each 1-factor of G corresponds to a nonzero term $m_{1j_1} \cdot m_{2j_2} \ldots m_{nj_n}$ is obvious.

As regards $\varepsilon_{j_1,j_2,\ldots,j_n}$, consider a directed circuit C in the 1-factor corresponding to $m_{1j_1} \cdot m_{2j_2} \ldots m_{nj_n}$. Without loss of generality, assume that C consists of the w edges

$$(v_1, v_2), (v_2, v_3), \ldots, (v_w, v_1)$$

It is easy to see that the corresponding permutation $(2, 3, \ldots, w, 1)$ can be rearranged as $(1, 2, \ldots, w)$ using $w - 1$ interchanges. If the 1-factor has L directed circuits with lengths $w_1, \ldots, w_L$, the permutation $(j_1, \ldots, j_n)$ can be rearranged as $(1, 2, \ldots, n)$ using

$$(w_1 - 1) + (w_2 - 1) + \cdots + (w_L - 1) = n - L$$

interchanges. So,

$$\varepsilon_{j_1,j_2,\ldots,j_n} = (-1)^{n+L}.$$
∎

As an example, for the 1-factors (shown in Figure 9.2) of the graph of Figure 9.1 the corresponding L_i are $L_1 = 3, L_2 = 3$, and $L_3 = 2$. So, the determinant of the adjacency matrix of the graph of Figure 9.1 is

$$(-1)^4[(-1)^3 + (-1)^3 + (-1)^2] = -1.$$

Consider next a weighted directed graph G in which each edge (v_i, v_j) is associated with a weight w_{ij}. Then we may define the *adjacency matrix* $M = [m_{ij}]$ of G as follows:

$$m_{ij} = \begin{cases} w_{ij}, & \text{if } (v_i, v_j) \in E \\ 0, & \text{otherwise} \end{cases}$$

Given a subgraph H of G, let us define the weight $w(H)$ of H as the product of the weights of all edges in H. If H has no edges, then we define $w(H) = 1$. The following result is an easy generalization of Theorem 9.1.

Theorem 9.2 *The determinant of the adjacency matrix of an n-vertex weighted directed graph G is given by*

$$\det(M) = (-1)^n \sum_H (-1)^{L_H} w(H),$$

where H is a 1-factor, $w(H)$ is the weight of H, and L_H is the number of directed circuits in H.
∎

9.3 COATES' GAIN FORMULA

Consider a linear system described by the equation

$$AX = Bx_{n+1} \tag{9.2}$$

where A is a nonsingular $n \times n$ matrix, X is a column vector of unknown variables $x_1, x_2, \ldots, x_n$; B is a column vector of elements $b_1, b_2, \ldots, b_n$; and x_{n+1} is the input variable. It is well known that

$$\frac{x_k}{x_{n+1}} = \frac{\sum_{i=1}^n b_i \Delta_{ik}}{\det A} \tag{9.3}$$

where Δ_{ik} is the (i, k) cofactor of A.

To develop Coates' topological formulas for the numerator and the denominator of (9.3), let us first augment the matrix A by adding $-B$ to the right of A and adding a row of zeroes at the bottom of the resulting matrix. Let this matrix be denoted by A'. The *Coates flow graph** $G_C(A')$, or simply the *Coates graph* associated with matrix A', is a weighted directed graph whose adjacency matrix is the transpose of the matrix A'. Thus, $G_C(A')$ has $n+1$ vertices $x_1, x_2, \ldots, x_{n+1}$, and if $a_{ji} \ldots \neq 0$, then $G_C(A')$ has an edge directed from x_i to x_j with weight a_{ji}. Clearly, the Coates graph $G_C(A)$ associated with matrix A can be obtained from $G_C(A')$ by removing the vertex x_{n+1}.

As an example, for the following system of equations

$$\begin{bmatrix} 3 & -2 & 1 \\ -1 & 2 & 0 \\ 3 & -2 & 2 \end{bmatrix} \begin{bmatrix} x_1 \\ x_2 \\ x_3 \end{bmatrix} = \begin{bmatrix} 3 \\ 1 \\ -2 \end{bmatrix} x_4 \qquad (9.4)$$

the matrix A' is

$$\begin{bmatrix} 3 & -2 & 1 & -3 \\ -1 & 2 & 0 & -1 \\ 3 & -2 & 2 & 2 \\ 0 & 0 & 0 & 0 \end{bmatrix}.$$

The Coates graphs $G_C(A')$ and $G_C(A)$ are shown in Figure 9.3.

Because a matrix and its transpose have the same determinant value and because A is the transpose of the adjacency matrix of $G_C(A)$, we obtain the following result from Theorem 9.2.

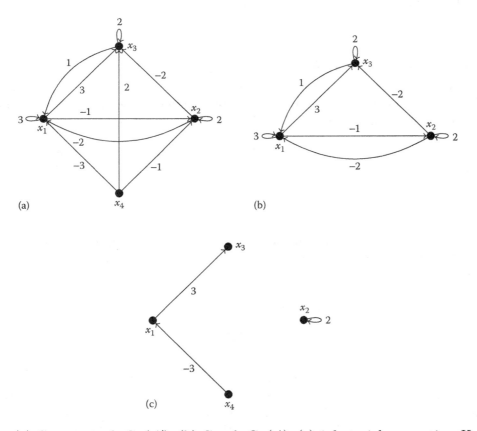

Figure 9.3 (a) Coates graph $G_C(A')$. (b) Graph $G_C(A)$. (c) 1-factorial connection $H_{4,3}$ of the graph $G_C(A)$.

---

*In networks and systems literature, the Coates graph is referred to as a *flow graph*.

Theorem 9.3 *If a matrix A is nonsingular, then*

$$\det(A) = (-1)^n \sum_H (-1)^{L_H} w(H) \qquad (9.5)$$

where H is a 1-factor of $G_C(A)$, $w(H)$ is the weight of H, and L_H is the number of directed circuits in H. ∎

To derive a similar expression for the sum in the numerator of (9.3), we first define the concept of *1-factorial connection*. A 1-factorial connection H_{ij} from x_i to x_j in $G_C(A)$ is a spanning subgraph of G which contains a directed path P from x_i to x_j and a set of vertex-disjoint directed circuits which include all the vertices of $G_C(A)$ other than those which lie on P. Similarly, a 1-factorial connection of $G_C(A')$ can be defined. As an example, a 1-factorial connection from x_4 to x_3 of the graph $G_C(A')$ of Figure 9.3a is shown in Figure 9.3c.

Theorem 9.4 *Let $G_C(A)$ be the Coates graph associated with an $n \times n$ matrix A. Then*

1. $\Delta_{ii} = (-1)^{n-1} \sum_H (-1)^{L_H} w(H)$
2. $\Delta_{ij} = (-1)^{n-1} \sum_{H_{ij}} (-1)^{L'_H} w(H_{ij}), \quad i \neq j$

where H is a 1-factor in the graph obtained by removing vertex x_i from $G_C(A)$, H_{ij} is a 1-factorial connection in $G_C(A)$ from vertex x_i to vertex x_j, and L_H and L'_H are the numbers of directed circuits in H and H_{ij}, respectively.

Proof.

1. Note that Δ_{ii} is the determinant of the matrix obtained from A by replacing its ith column by a column of zeros except for the element in row i, which is 1. Also, the Coates graph of the resulting matrix can be obtained from $G_C(A)$ by removing vertex x_i. Proof follows from these observations and Theorem 9.3.

2. Let A_α denote the matrix obtained from A by replacing its jth column by a column of zeros, except for the element in row i, which is 1. Then it is easy to see that

$$\Delta_{ij} = \det A_\alpha.$$

Now, the Coates graph $G_C(A_\alpha)$ can be obtained from $G_C(A)$ by removing all edges incident out of vertex x_j and adding an edge directed from x_j to x_i with weight 1. Then from Theorem 9.3, we get

$$\begin{aligned}\Delta_{ij} &= \det A_\alpha \\ &= (-1)^n \sum_{H_\alpha} (-1)^{L_\alpha} w(H_\alpha)\end{aligned} \qquad (9.6)$$

where H_α is a 1-factor of $G_C(A_\alpha)$ and L_α is the number of directed circuits in H_α.

Consider now a 1-factor H_α in $G_C(A_\alpha)$. Let C be the directed circuit of H_α containing x_i. Because in $G_C(A_\alpha)$, (x_j, x_i) is the only edge incident out of x_j, it follows that x_j also lies in C. If we remove the edge (x_j, x_i) from H_α we get a 1-factorial connection, $H_{ij} = H$. Furthermore, $L'_H = L_\alpha - 1$ and $w(H_{ij}) = w(H_\alpha)$ because (x_j, x_i) has weight equal to 1. Thus, each H_α corresponds to a 1-factorial connection H_{ij} of $G_C(A_\alpha)$ with $w(H_\alpha) = w(H_{ij})$

and $L'_H = L_\alpha - 1$. The converse of this is also easy to see. Thus in (9.6) we can replace H_α by H_{ij} and L_α by $(L'_H - 1)$. Then we obtain

$$\Delta_{ij} = (-1)^{n-1} \sum_{H_{ij}} (-1)^{L'_H} w(H_{ij})$$ ∎

Having shown that each Δ_{ij} can be expressed in terms of the weights of the 1-factorial connections H_{ij} in $G_C(A)$, we now show that $\sum b_i \Delta_{ik}$ can be expressed in terms of the weights of the 1-factorial connections $H_{n+1,k}$ in $G_C(A')$.

First, note that adding the edge (x_{n+1}, x_i) to H_{ik} results in a 1-factorial connection $H_{n+1,k}$, with $w(H_{n+1,k}) = -b_i w(H_{ik})$. Also, $H_{n+1,k}$ has the same number of directed circuits as H_{ik}. Conversely, from each $H_{n+1,k}$ that contains the edge (x_{n+1}, x_i) we can construct a 1-factorial connection H_{ik} satisfying $w(H_{n+1,k}) = -b_i w(H_{ik})$. Also, $H_{n+1,k}$ and the corresponding H_{ik} will have the same number of directed circuits. Thus, a one-to-one correspondence exists between the set of all 1-factorial connections $H_{n+1,k}$ in $G_C(A')$ and the set of all 1-factorial connections in $G_C(A)$ of the form H_{ik} such that each $H_{n+1,k}$ and the corresponding H_{ik} have the same number of directed circuits and satisfy the relation $w(H_{n+1,k}) = -b_i w(H_{ik})$. Combining this result with Theorem 9.4, we get

$$\sum_{i=1}^{n} b_i \Delta_{ik} = (-1)^n \sum_{H_{n+1,k}} (-1)^{L'_H} w(H_{n+1,k}). \tag{9.7}$$

where the summation is over all 1-factorial connections $H_{n+1,k}$ in $G_C(A')$, and L'_H is the number of directed circuits in $H_{n+1,k}$. From (9.5) and (9.7) we get the following theorem.

Theorem 9.5 *If the coefficient matrix A is nonsingular, then the solution of (9.2) is given by.*

$$\frac{x_k}{x_{n+1}} = \frac{\sum_{H_{n+1,k}} (-1)^{L'_H} w(H_{n+1,k})}{\sum_H (-1)^{L_H} w(H)} \tag{9.8}$$

for $k = 1, 2, \ldots, n$, where $H_{n+1,k}$ is a 1-factorial connection of $G_C(A')$ from vertex x_{n+1} to vertex x_k, H is a 1-factor of $G_C(A)$, and L'_H and L_H are the numbers of directed circuits in $H_{n+1,k}$ and H, respectively. ∎

Equation 9.8 is called the *Coates' gain formula*. We now illustrate Coates' method by solving the system (9.4) for x_2/x_4. First, we determine the 1-factors of the Coates graph $G_C(A)$ shown in Figure 9.3b. These 1-factors, along with their weights, are listed below. The vertices enclosed within parentheses represent a directed circuit.

1-Factor H	Weight $w(H)$	L_H
$(x_1)(x_2)(x_3)$	12	3
$(x_2)(x_1, x_3)$	6	2
$(x_3)(x_1, x_2)$	4	2
(x_1, x_2, x_3)	2	1

From the above we get the denominator in (9.8) as

$$\sum_H (-1)^{L_H} w(H) = (-1)^3 \cdot 12 + (-1)^2 \cdot 6 + (-1)^2 \cdot 4 + (-1)^1 \cdot 2 = -4.$$

To compute the numerator in (9.8) we need to determine the 1-factorial connections $H_{4,2}$ in the Coates graph $G_C(A')$ shown in Figure 9.3a. They are listed below along with their weights. The vertices in a directed path from x_4 to x_2 are given within parentheses.

1-Factor connection $H_{4,2}$	$w(H_{4,2})$	L'_H
$(x_4, x_1, x_2)(x_3)$	6	1
$(x_4, x_2)(x_1)(x_3)$	-6	2
$(x_4, x_2)(x_1, x_3)$	-3	1
(x_4, x_3, x_1, x_2)	-2	0

From the above we get the numerator in (9.8) as

$$\sum_{H_{4,2}} (-1)^{L'_H} w(H_{4,2}) = (-1)^1 \cdot 6 + (-1)^2 \cdot (-6) + (-1)^1 \cdot (-3) + (-1)^0 \cdot (-2) = -11.$$

Thus, we get

$$\frac{x_2}{x_4} = \frac{11}{4}.$$

9.4 MASON'S GAIN FORMULA

Consider again the system of equations

$$AX = Bx_{n+1}$$

We can rewrite the above as

$$x_j = (a_{jj} + 1)x_j + \sum_{\substack{k=1 \\ k \neq j}}^{n} a_{jk} x_k - b_j x_{n+1}, \quad j = 1, 2, \ldots, n, \tag{9.9}$$

$$x_{n+1} = x_{n+1}$$

Letting X' denote the column vector of the variables $x_1, x_2, \ldots, x_{n+1}$, and U_{n+1} denote the unit matrix of order $n+1$, we can write (9.9) in matrix form as follows:

$$(A' + U_{n+1})X' = X' \tag{9.10}$$

where A' is the matrix defined earlier in Section 9.3.

The Coates graph $G_C(A' + U_{n+1})$ is called the *Mason's signal flow graph* or simply the *Mason graph** associated with A', and it is denoted by $G_m(A')$. The Mason graph $G_m(A')$ and $G_m(A)$ associated with the system (9.4) are shown in Figure 9.4. Mason's graph elegantly represents the flow of variables in a system. If we associate each vertex with a variable and if an edge is directed from x_i to x_j, then we may consider the variable x_i as contributing $(a_{ji} x_i)$ to the variable x_j. Thus, x_j is equal to the sum of the products of the weights of the edges incident into vertex x_j and the variables corresponding to the vertices from which these edges emanate.

Note that to obtain the Coates graph $G_C(A)$ from the Mason graph $G_m(A)$ we simply subtract one from the weight of each self-loop. Equivalently, we may add at each vertex of the Mason graph a self-loop of weight -1. Let S denote the set of all such loops of weight -1 added to construct the Coates graph G_C from the Mason graph $G_m(A)$. Note that the Coates graph so constructed will have at most two and at least one self-loop at each vertex.

---

*In network and system theory literature Mason graphs are usually referred to as *signal flow graphs*.

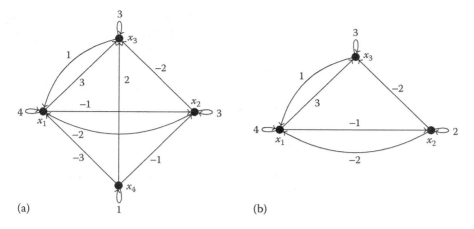

Figure 9.4 (a) Mason graph $G_m(A')$. (b) Mason graph $G_m(A)$.

Consider now the Coates graph G_C constructed as above and a 1-factor H in G_C having j self-loops from the set S. If H has a total of $L_Q + j$ directed circuits, then removing the j self-loops from H will result in a subgraph Q of $G_m(A)$ which is a collection of L_Q vertex disjoint directed circuits. Also,

$$w(H) = (-1)^j w(Q)$$

Then, from Theorem 9.3 we get

$$\det A = (-1)^n \sum_H (-1)^{L_Q+j} w(H)$$
$$= (-1)^n \left[1 + \sum_Q (-1)^{L_Q} w(Q)\right]. \tag{9.11}$$

So

$$\det A = (-1)^n \left[1 - \sum_j Q_{j1} + \sum_j Q_{j2} - \sum_j Q_{j3} \cdots\right]. \tag{9.12}$$

where each term in $\sum_j Q_{ji}$ is the weight of a collection of i vertex-disjoint directed circuits in $G_m(A)$.

Suppose we refer to $(-1)^n \det A$ as the determinant of the graph $G_m(A)$. Then, starting from $H_{n+1,k}$ and reasoning exactly as above we can express the numerator of (9.3) as

$$\sum_{i=1}^n b_i \Delta_{ik} = (-1)^n \sum_j w(P_{n+1,k}^j) \Delta_j \tag{9.13}$$

where $P_{n+1,k}^j$ is a directed path from x_{n+1} to x_k of $G_m(A')$ and Δ_j is the determinant of the subgraph of $G_m(A')$ which is vertex-disjoint from the path $P_{n+1,k}^j$. From (9.12) and (9.13) we get the following theorem.

Theorem 9.6 *If the coefficient matrix A in (9.2) is nonsingular, then*

$$\frac{x_k}{x_{n+1}} = \frac{\sum_j w(P_{n+1,k}^j) \Delta_j}{\Delta}, \quad k = 1, 2, \ldots, n \tag{9.14}$$

where $P_{n+1,k}^j$ is the jth directed path from x_{n+1} to x_k of $G_m(A')$, Δ_j is the determinant of the subgraph of $G_m(A')$ which is vertex-disjoint from the jth directed path $P_{n+1,k}^j$, and Δ is the determinant of the graph $G_m(A)$. ∎

Equation 9.14 is known as *Mason's gain formula*. In network and system theory $P_{n+1,k}^j$ is referred as a *forward path* from vertex x_{n+1} to x_k. The directed circuits of $G_m(A')$ are called the *feedback loops*.

We now illustrate Mason's method by solving the system (9.4) for x_2/x_4. To compute the denominator in (9.14) we determine the different collections of vertex-disjoint directed circuits of the Mason graph $G_m(A)$ shown in Figure 9.4. They are listed below along with their weights.

Collection of Vertex-Disjoint Directed Circuits of $G_m(A)$	Weight	No. of Directed Circuits
(x_1)	4	1
(x_2)	3	1
(x_3)	3	1
(x_1, x_2)	2	1
(x_1, x_3)	3	1
(x_1, x_2, x_3)	2	1
$(x_1)(x_2)$	12	2
$(x_1)(x_3)$	12	2
$(x_2)(x_3)$	9	2
$(x_2)(x_1, x_3)$	9	2
$(x_3)(x_1, x_2)$	6	2
$(x_1)(x_2)(x_3)$	36	3

From the above we obtain the denominator in (9.14) as

$$\Delta = 1 + (-1)^1[4 + 3 + 3 + 2 + 3 + 2] \\ + (-1)^2[12 + 12 + 9 + 9 + 6] + (-1)^3 36 = -4$$

To compute the numerator in (9.14) we need the forward paths in $G_m(A')$ from x_4 to x_2. They are listed below with their weights.

j	$P_{4,2}^j$	Weight
1	(x_4, x_2)	-1
2	(x_4, x_1, x_2)	3
3	(x_4, x_3, x_1, x_2)	-2

The directed circuits which are vertex-disjoint from $P_{4,2}^1$ are $(x_1), (x_3), (x_1, x_3)$.
Thus

$$\Delta_1 = 1 - (4 + 3 + 3) + 12 = 1 - 10 + 12 = 3$$

(x_3) is the only directed circuit which is vertex-disjoint from $P_{4,2}^2$. So,

$$\Delta_2 = 1 - 3 = -2.$$

No directed circuit is vertex-disjoint from $P_{4,2}^3$, so $\Delta_3 = 1$. Thus the numerator in (9.14) is

$$P_{4,2}^1 \Delta_1 + P_{4,2}^2 \Delta_2 + P_{4,2}^3 \Delta_3 = -3 - 6 - 2 = -11$$

and

$$\frac{x_2}{x_4} = \frac{11}{4}.$$

References

[1] W. K. Chen (EIC), *The Circuits and Filters Handbook*, CRC Press, Boca Raton, FL, 1995.

[2] C. L. Coates, Flow graph solutions of linear algebraic equations, *IRE Trans. Circuit Theory*, **CT-6** (1959), 170–187.

[3] S. J. Mason, Feedback theory: Some properties of signal flow graphs, *Proc. IRE*, **41** (1953), 1144–1156.

[4] S. J. Mason, Feedback theory: Further properties of signal flow graphs, *Proc. IRE*, **44** (1956), 920–926.

[5] K. Thulasiraman and M. N. S. Swamy, *Graphs: Theory and Algorithms*, Wiley-Interscience, New York, 1992.

[6] F. Harary, The determinant of the adjacency matrix of a graph, *SIAM Rev.*, **4** (1962), 202–210.

[7] W. K. Chen, *Applied Graph Theory*, North-Holland, Amsterdam, the Netherlands, 1971.

[8] F. E. Hohn, *Elementary Matrix Algebra*, Macmillan, New York, 1958.

CHAPTER 10

Adjacency Spectrum and the Laplacian Spectrum of a Graph

R. Balakrishnan

CONTENTS

10.1	Introduction	227
10.2	Spectrum of a Graph	228
10.3	Spectrum of the Complete Graph K_n	229
10.4	Spectrum of the Cycle C_n	229
	10.4.1 Coefficents of the Characteristic Polynomial of a Graph	230
10.5	Spectra of Regular Graphs	230
	10.5.1 Spectrum of the Complement of a Regular Graph	231
	10.5.2 Spectra of Line Graphs of Regular Graphs	232
10.6	Spectrum of the Complete Bipartite Graph $K_{p,q}$	234
10.7	Determinant of the Adjacency Matrix of a Graph	235
10.8	Spectra of Product Graphs	236
10.9	Laplacian Spectrum of a Graph	240
10.10	Algebraic Connectivity of a Graph	243

10.1 INTRODUCTION

In this chapter, we look at the properties of graphs from our knowledge of their eigenvalues. The set of eigenvalues (with their multiplicities) of a graph G is the spectrum of its adjacency matrix and it is the *spectrum* of G and denoted by $Sp(G)$. We compute the spectra of some well-known families of graphs—the family of complete graphs, the family of cycles, and so forth. We present Sachs' theorem on the spectrum of the line graph of a regular graph. We also obtain the spectra of product graphs—Cartesian product, direct product, and strong product of two graphs.

The properties of the spectra of graphs are based on [1]. For a comprehensive treatment of spectral graph theory see [2–5]. Some applications of the Laplacian Matrix in random walk routing and connections to electrical resistance networks are given in Chapter 11.

10.2 SPECTRUM OF A GRAPH

Let G be a simple graph of order n with vertex set $V = \{v_1, \ldots, v_n\}$. The *adjacency matrix* of G (with respect to this labeling of V) is the n by n matrix $A = (a_{ij})$, where

$$a_{ij} = \begin{cases} 1, & \text{if } v_i \text{ is adjacent to } v_j \text{ in } G \\ 0, & \text{otherwise} \end{cases}$$

Thus A is a real symmetric matrix of order n. Hence

i. The spectrum of A, that is, the multiset of its eigenvalues is real.

ii. $\mathbb{R}^n$ has an orthonormal basis of eigenvectors of A.

iii. The sum of the entries of the ith row (column) of A is $d(v_i) = $ degree of v_i in G.

The spectrum of A is called the *spectrum* of G and denoted by $Sp(G)$. We note that $Sp(G)$, as defined above, depends on the labeling of the vertex set V of G. We now show that it is independent of the labeling of G. Suppose we consider a new labeling of V. Let A' be the adjacency matrix of G with respect to this labeling. The new labeling can be obtained from the original labeling by means of a permutation π of $V(G)$. Any such permutation can be effected by means of a *permutation matrix* P of order n (got by permuting the rows of I_n, the identity matrix of order n. (For example if $n = 3$, the permutation matrix $P = \begin{pmatrix} 0 & 1 & 0 \\ 0 & 0 & 1 \\ 1 & 0 & 0 \end{pmatrix}$ takes v_1, v_2, v_3 to v_3, v_1, v_2, respectively, since $\begin{pmatrix} 1 & 2 & 3 \end{pmatrix} \begin{pmatrix} 0 & 1 & 0 \\ 0 & 0 & 1 \\ 1 & 0 & 0 \end{pmatrix} = \begin{pmatrix} 3 & 1 & 2 \end{pmatrix}$.)

Let $P = (p_{ij})$. Now given the new labeling of V, that is, given the permutation π on $\{1, 2, \ldots, n\}$, and the vertices v_i and v_j, there exist unique α_0 and β_0, such that $\pi(i) = \alpha_0$ and $\pi(j) = \beta_0$ or equivalently $p_{\alpha_0 i} = 1$ and $p_{\beta_0 j} = 1$ while for $\alpha \neq \alpha_0$ and $\beta \neq \beta_0$, $p_{\alpha i} = 0 = p_{\beta j}$. Thus the (α_0, β_0)th entry of the matrix $A' = PAP^{-1} = PAP^T$ (where P^T stands for the transpose of P) is

$$\sum_{k,l=1}^{n} p_{\alpha_0 k} a_{kl} p_{\beta_0 l} = a_{ij}.$$

Hence $v_{\alpha_0} v_{\beta_0} \in E(G)$ if and only if $v_i v_j \in E(G)$. This proves that the adjacency matrix of the same graph with respect to two different labelings are similar matrices. But then similar matrices have the same spectra.

We usually arrange the eigenvalues of G in their nondecreasing order: $\lambda_1 \geq \lambda_2 \geq \cdots \geq \lambda_n$. If $\lambda_1, \ldots, \lambda_s$ are the distinct eigenvalues of G, and if m_i is the multiplicity of λ_i as an eigenvalue of G, we write

$$Sp(G) = \begin{pmatrix} \lambda_1 & \lambda_2 & \ldots & \lambda_s \\ m_1 & m_2 & \ldots & m_s \end{pmatrix}.$$

Definition 10.1 *The characteristic polynomial of G is the characteristic polynomial of the adjacency matrix of G with respect to some labeling of G. It is denoted by $\chi(G; \lambda)$.*

Hence $\chi(G; \lambda) = det(xI - A) = det(P(xI - A)P^{-1}) = det(xI - PAP^{-1})$ *for any permutation matrix of P, and hence $\chi(G; \lambda)$ is also independent of the labeling of $V(G)$.*

Definition 10.2 *A circulant of order n is a square matrix of order n in which all the rows are obtainable by successive cyclic shifts of one of its rows (usually taken as the first row).*

For example, the circulant with first row $(a_1\ a_2\ a_3)$ is the matrix $\begin{pmatrix} a_1 & a_2 & a_3 \\ a_3 & a_1 & a_2 \\ a_2 & a_3 & a_1 \end{pmatrix}$.

Lemma 10.1 *Let C be a circulant matrix of order n with first row $(a_1\ a_2\ \ldots\ a_n)$. Then $Sp(C) = \{a_1 + a_2\omega + \cdots + a_n\omega^{n-1} : \omega =$ an nth root of unity$\} = \{a_1 + \zeta^r + \zeta^{2r} + \cdots + \zeta^{(n-1)r},\ 0 \leq r \leq n-1$ and $\zeta = $ a primitive nth root of unity$\}$.*

Proof. The characteristic polynomial of C is the determinant $D = \det(xI - C)$. Hence

$$D = \begin{vmatrix} x-a_1 & -a_2 & \cdots & -a_n \\ -a_n & x-a_1 & \cdots & -a_{n-1} \\ \vdots & \vdots & & \vdots \\ -a_2 & -a_3 & \cdots & x-a_1 \end{vmatrix}.$$

Let C_i denote the ith column of D, $1 \leq i \leq n$, and ω, an nth root of unity. Replace C_1 by $C_1 + C_2\omega + \cdots + C_n\omega^{n-1}$. This does not change D. Let $\lambda_\omega = a_1 + a_2\omega + \cdots + a_n\omega^{n-1}$. Then the new first column of D is $(x - \lambda_\omega, \omega(x - \lambda_\omega), \ldots, \omega^{n-1}(x - \lambda_\omega))^T$, and hence $x - \lambda_\omega$ is a factor of D. This gives $D = \prod_{\omega : \omega^n = 1}(x - \lambda_\omega)$, and $Sp(C) = \{\lambda_\omega : \omega^n = 1\}$. ∎

10.3 SPECTRUM OF THE COMPLETE GRAPH K_n

For K_n, the adjacency matrix A is given by $A = \begin{pmatrix} 0 & 1 & 1 & \cdots & 1 \\ 1 & 0 & 1 & \cdots & 1 \\ \vdots & \vdots & \vdots & & \vdots \\ 1 & 1 & 1 & \cdots & 0 \end{pmatrix}$, and so by Lemma 10.1,

$$\lambda_\omega = \omega + \omega^2 + \cdots + \omega^{n-1}$$
$$= \begin{cases} n-1, & \text{if } \omega = 1 \\ -1, & \text{if } \omega \neq 1. \end{cases}$$

Hence $Sp(K_n) = \begin{pmatrix} n-1 & -1 \\ 1 & n-1 \end{pmatrix}$.

10.4 SPECTRUM OF THE CYCLE C_n

Label the vertices of C_n as $0, 1, 2, \ldots, n-1$ in this order. Then i is adjacent to $i \pm 1 \pmod n$. Hence

$$A = \begin{bmatrix} 0 & 1 & 0 & 0 & \cdots & 0 & 1 \\ 1 & 0 & 1 & 0 & \cdots & 0 & 0 \\ \vdots & \vdots & \vdots & \vdots & \ddots & \vdots & \vdots \\ 1 & 0 & 0 & 0 & \cdots & 1 & 0 \end{bmatrix}$$

is the circulant with the first row $(0\ 1\ 0\ \ldots\ 0\ 1)$. Again by Lemma 10.1, $Sp(C_n) = \{\omega^r + \omega^{r(n-1)} : 0 \leq r \leq n-1$, where ω is a primitive nth root of unity$\}$. Taking $\omega = \cos(2\pi/n) + i\sin(2\pi/n)$, we get $\lambda_r = \omega^r + \omega^{r(n-1)} = (\cos(2\pi r/n) + i\sin(2\pi r/n)) + (\cos(2\pi r(n-1)/n) + i\sin(2\pi r(n-1)/n))$.

This simplifies to the following:

i. If n is odd, $Sp(C_n) = \begin{pmatrix} 2 & 2\cos\frac{2\pi}{n} & \cdots & 2\cos\frac{(n-1)\pi}{n} \\ 1 & 2 & \cdots & 2 \end{pmatrix}$.

ii. If n is even, $Sp(C_n) = \begin{pmatrix} 2 & 2\cos\frac{2\pi}{n} & \cdots & 2\cos\frac{(n-2)\pi}{n} & -2 \\ 1 & 2 & \cdots & 2 & 1 \end{pmatrix}$.

10.4.1 Coefficients of the Characteristic Polynomial of a Graph

Let G be a connected graph on n vertices, and let $\chi(G; x) = \det(xI_n - A) = x^n + a_1 x^{n-1} + a_2 x^{n-2} + \cdots + a_n$ be the characteristic polynomial of G. It is easy to check that $(-1)^r a_r =$ sum of the principal minors of A of order r. (Recall that a principal minor of order r of A is the determinant minor of A common to the same set of r rows and columns.)

Lemma 10.2 *Let G be a graph of order n and size m, and let $\chi(G; x) = x^n + a_1 x^{n-1} + a_2 x^{n-2} + \cdots + a_n$ be the characteristic polynomial of A. Then*

i. $a_1 = 0$

ii. $a_2 = -m$

iii. $a_3 = -(\text{twice the number of triangles in } G)$.

Proof.

i. Follows from the fact that all the entries of the principal diagonal of A are zero.

ii. A nonvanishing principal minor of order 2 of A is of the form $\begin{vmatrix} 0 & 1 \\ 1 & 0 \end{vmatrix}$ and its value is -1. Since any 1 in A corresponds to an edge of G, we get (ii).

iii. A nontrivial principal minor of order 3 of A can be one of the following three types:

$$\begin{vmatrix} 0 & 1 & 0 \\ 1 & 0 & 0 \\ 0 & 0 & 0 \end{vmatrix}, \begin{vmatrix} 0 & 1 & 1 \\ 1 & 0 & 0 \\ 1 & 0 & 0 \end{vmatrix}, \begin{vmatrix} 0 & 1 & 1 \\ 1 & 0 & 1 \\ 1 & 1 & 0 \end{vmatrix}$$

Of these, only the last determinant is nonvanishing. Its value is 2 and corresponds to a triangle in G. This proves (iii). ∎

10.5 SPECTRA OF REGULAR GRAPHS

In this section, we look at the spectra of some regular graphs.

Theorem 10.1 *Let G be a k-regular graph of order n. Then*

i. *k is an eigenvalue of G.*

ii. *If G is connected, every eigenvector corresponding to the eigenvalue k is a multiple of $\mathbf{1}$ (the all 1-column vector of length 1) and the multiplicity of k as an eigenvalue of G is one.*

iii. *For any eigenvalue λ of $G, |\lambda| \leq k$. (Hence $Sp(G) \subset [-k, k]$.)*

Proof.

i. We have $A\mathbf{1} = k\mathbf{1}$, and hence k is an eigenvalue of A.

ii. Let $\boldsymbol{x} = (x_1, \ldots, x_n)^T$ be any eigenvector of A corresponding to the eigenvalue k so that $A\boldsymbol{x} = k\boldsymbol{x}$. We may suppose that $\boldsymbol{x}$ has a positive entry (otherwise take $-\boldsymbol{x}$ in place of $\boldsymbol{x}$), and that x_j is the largest positive entry in $\boldsymbol{x}$. Let $v_{i_1}, v_{i_2}, \ldots, v_{i_k}$ be the k neighbors of v_j in G. Taking the innerproduct of the jth row of A with $\boldsymbol{x}$, we get $x_{i_1} + x_{i_2} + \cdots + x_{i_k} = kx_j$. This gives, by the choice of x_j, $x_{i_1} = x_{i_2} = \cdots = x_{i_k} = x_j$.

Now start at $v_{i_1}, v_{i_2}, \ldots, v_{i_k}$ in succession and look at their neighbors in G. As before, the entries x_p in $\boldsymbol{x}$ corresponding to these neighbors must all be equal to x_j. As G is connected, all the vertices of G are reachable in this way step by step. Hence $\boldsymbol{x} = x_j(1, 1, \ldots, 1)^T$, and every eigenvector $\boldsymbol{x}$ of A corresponding to the eigenvalue k is a multiple of $\mathbf{1}$. Thus, the space of eigenvectors of A corresponding to the eigenvalue k is one-dimensional, and therefore, the multiplicity of k as an eigenvalue of G is one.

iii. The proof is similar to (ii). Indeed if $A\boldsymbol{y} = \lambda\boldsymbol{y}, \boldsymbol{y} \neq 0$, and if y_j is the entry in $\boldsymbol{y}$ with the largest absolute value, we see that the equation $\sum_{p=1}^{k} y_{i_p} = \lambda y_i$ implies that $|\lambda||y_j| = |\lambda y_j| = |\sum_{p=1}^{k} y_{i_p}| \leq \sum_{p=1}^{k} |y_{i_p}| \leq k|y_j|$. Thus $|\lambda| \leq k$. ∎

Corollary 10.1 *If Δ denotes the maximum degree of G, then for any eigenvalue λ of G, $|\lambda| \leq \Delta$.*

Proof. Considering a vertex v_j of maximum degree Δ, and imitating the proof of (iii) above, we get $|\lambda||y_j| \leq \Delta|y_j|$. ∎

10.5.1 Spectrum of the Complement of a Regular Graph

Theorem 10.2 *Let G be a k-regular connected graph of order n with spectrum $\begin{pmatrix} k & \lambda_2 & \lambda_3 & \cdots & \lambda_s \\ 1 & m_2 & m_3 & \cdots & m_s \end{pmatrix}$. Then the spectrum of G^c, the complement of G, is given by*

$$Sp(G^c) = \begin{pmatrix} n-1-k & -\lambda_2 - 1 & -\lambda_3 - 1 & \cdots & -\lambda_s - 1 \\ 1 & m_2 & m_3 & \cdots & m_s \end{pmatrix}.$$

Proof. As G is k-regular, G^c is $n-1-k$ regular, and hence by Theorem 10.1 $n-1-k$ is an eigenvalue of G^c. Further the adjacency matrix of G^c is $A^c = J - I - A$, where J is the all 1 matrix of order n, I is the identity matrix of order n and A is the adjacency matrix of G. If $\chi(\lambda)$ is the characteristic polynomial of A, $\chi(\lambda) = (\lambda - k)\chi_1(\lambda)$. By Cayley–Hamilton theorem, $\chi(A) = 0$ and hence we have $A\chi_1(A) = k\chi_1(A)$. Hence every column vector of $\chi_1(A)$ is an eigenvector of A corresponding to the eigenvalue k. But by Theorem 10.1, the space of eigenvectors of A corresponding to the eigenvalue k is generated by $\mathbf{1}$, G being connected. Hence each column vector of $\chi_1(A)$ is a multiple of $\mathbf{1}$. But $\chi_1(A)$ is symmetric and hence $\chi_1(A)$ is a multiple of J, say, $\chi_1(A) = \alpha J$, $\alpha \neq 0$. Thus J and hence $J - I - A$ are polynomials in A (remember: $A^0 = I$). Let $\lambda \neq k$ be any eigenvalue of A (so that $\chi_1(\lambda) = 0$), and Y an eigenvector of A corresponding to λ. Then

$$\begin{aligned} A^c Y &= (J - I - A)Y \\ &= \left(\frac{\chi_1(A)}{\alpha} - I - A \right) Y \\ &= \left(\frac{\chi_1(\lambda)}{\alpha} - 1 - \lambda \right) Y \text{ (see note below)} \\ &= (-1 - \lambda)Y. \end{aligned}$$

Thus $A^c Y = (-1 - \lambda)Y$, and therefore $-1 - \lambda$ is an eigenvalue of A^c corresponding to the eigenvalue $\lambda (\neq k)$ of A.

Note: We recall that if $f(\lambda)$ is a polynomial in λ, and $Sp(A) = \begin{pmatrix} \lambda_1 & \lambda_2 & \ldots & \lambda_s \\ m_1 & m_2 & \ldots & m_s \end{pmatrix}$, then, $Sp(f(A)) = \begin{pmatrix} f(\lambda_1) & f(\lambda_2) & \ldots & f(\lambda_s) \\ m_1 & m_2 & \ldots & m_s \end{pmatrix}$.

10.5.2 Spectra of Line Graphs of Regular Graphs

We now establish Sachs' theorem which determines the spectrum of the line graph of a regular graph G in terms of $Sp(G)$. Recall that the line graph $L(G)$ of G is that graph whose vertex set is in $1-1$ correspondence with the edge set of G and two vertices of $L(G)$ are adjacent if and only if the corresponding edges of G are adjacent in G.

Let G be a labeled graph with vertex set $V(G) = \{v_1, \ldots, v_n\}$ and edge set $E(G) = \{e_1, \ldots, e_m\}$. With respect to these labelings, the incidence matrix $B = (b_{ij})$ of G, which describes the incidence structure of G as the m by n matrix $B = (b_{ij})$,

where

$$b_{ij} = \begin{cases} 1, & \text{if } e_i \text{ is incident to } v_j, \\ 0, & \text{otherwise.} \end{cases}$$

Lemma 10.3 *Let G be a graph of order n and size m with A and B as its adjacency and incidence matrices respectively. Let A_L denote the adjacency matrix of the line graph of G. Then*

i. $BB^T = A_L + 2I_m$, and

ii. *if G is k-regular, $B^T B = A + kI_n$.*

Proof. Let $A = (a_{ij})$ and $B = (b_{ij})$. We have

i. $(BB^T)_{ij} = \sum_{p=1}^{n} b_{ip} b_{jp}$

$= $ number of vertices v_p which are incident to both e_i and e_j
$= \begin{cases} 1, & \text{if } e_i \text{ and } e_j \text{ are adjacent} \\ 0, & \text{if } i \neq j \text{ and } e_i \text{ and } e_j \text{ are nonadjacent} \\ 2, & \text{if } i = j. \end{cases}$

ii. Proof of (ii) is similar. ■

Theorem 10.3 (Sachs' theorem) *Let G be a k-regular graph of order n. Then $\chi(L(G); \lambda) = (\lambda + 2)^{m-n} \chi(G; \lambda + 2 - k)$, where $L(G)$ is the line graph of G.*

Proof. Consider the two partitioned matrices U and V, each of order $n+m$ (where B stands for the incidence matrix of G):

$$U = \begin{bmatrix} \lambda I_n & -B^T \\ 0 & I_m \end{bmatrix}, \quad V = \begin{bmatrix} I_n & B^T \\ B & \lambda I_m \end{bmatrix}.$$

We have

$$UV = \begin{bmatrix} \lambda I_n - B^T B & 0 \\ B & \lambda I_m \end{bmatrix} \text{ and } VU = \begin{bmatrix} \lambda I_n & 0 \\ \lambda B & \lambda I_m - BB^T \end{bmatrix}.$$

Now $\det(UV) = \det(VU)$ gives:

$$\lambda^m \det(\lambda I_n - B^T B) = \lambda^n \det(\lambda I_m - BB^T). \tag{10.1}$$

Replacement of λ by $\lambda + 2$ in Equation 10.1 yields

$$(\lambda + 2)^{m-n} \det((\lambda + 2)I_n - B^T B) = \det((\lambda + 2)I_m - BB^T). \tag{10.2}$$

Hence, by Lemma 10.3,

$$\begin{aligned}
\chi(L(G); \lambda) &= \det(\lambda I_m - A_L) \\
&= \det((\lambda + 2)I_m - (A_L + 2I_m)) \\
&= \det((\lambda + 2)I_m - BB^T) \\
&= (\lambda + 2)^{m-n} \det((\lambda + 2)I_n - B^T B) \quad \text{(by Equation 10.2)} \\
&= (\lambda + 2)^{m-n} \det((\lambda + 2)I_n - (A + kI_n)) \quad \text{(by Lemma 10.3)} \\
&= (\lambda + 2)^{m-n} \det((\lambda + 2 - k)I_n - A) \\
&= (\lambda + 2)^{m-n} \chi(G; \lambda + 2 - k).
\end{aligned}$$
∎

Sachs' theorem implies the following: As $\chi(G; \lambda) = \prod_{1}^{n}(\lambda - \lambda_i)$, it follows that

$$\begin{aligned}
\chi(L(G); \lambda) &= (\lambda + 2)^{m-n} \prod_{i=1}^{n}(\lambda + 2 - k - \lambda_i) \\
&= (\lambda + 2)^{m-n} \prod_{i=1}^{n}(\lambda - (k - 2 + \lambda_i)).
\end{aligned}$$

Hence if

$$Sp(G) = \begin{pmatrix} k & \lambda_2 & \cdots & \lambda_s \\ 1 & m_2 & \cdots & m_s \end{pmatrix}$$

then

$$Sp(L(G)) = \begin{pmatrix} 2k-2 & k-2+\lambda_2 & \cdots & k-2+\lambda_s & -2 \\ 1 & m_2 & \cdots & m_s & m-n \end{pmatrix}.$$

It is easy to see that the Petersen graph P is isomorphic to $(L(K_5))^c$. Hence the spectrum of P can be obtained by using Theorems 10.2 and 10.3: $Sp(P) = \begin{pmatrix} 3 & 1 & -2 \\ 1 & 5 & 4 \end{pmatrix}$. We use this result to prove a well-known result on the Petersen graph.

Theorem 10.4 *The complete graph K_{10} cannot be decomposed into (i.e., expressed as an edge-disjoint union of) three copies of the Petersen graph.*

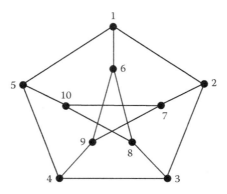

Figure 10.1 Petersen graph P.

Proof. [6] Assume the contrary. Suppose K_{10} is expressible as an edge-disjoint union of three copies, say, P_1, P_2, P_3 of the Petersen graph P (see Figure 10.1). Then

$$A(K_{10}) = J - I = A(P_1) + A(P_2) + A(P_3), \qquad (10.3)$$

where $A(H)$ stands for the adjacency matrix of the graph H, and J, the all 1-matrix of order 10. (Note that each P_i is a spanning subgraph of K_{10} and that the number of edges of P_i, namely 15, is a divisor of the number of edges of K_{10}, namely 45, and that the degree of any vertex v of P, namely 3, is a divisor of the degree of v in K_{10}, namely 9.) It is easy to check that 1 is an eigenvalue of P. Further, as P is 3-regular, the all 1-column vector $\mathbf{1}$ is an eigenvector of P. Now $\mathbb{R}^{10}$ has an orthonormal basis of eigenvectors of $A(P)$ containing $\mathbf{1}$. Again the null space of $(A(P) - 1.I) = A(P) - I$ is of dimension 5, and is orthogonal to $\mathbf{1}$. (For P with the labeling of Figure 10.1, one can check that for $A(P) - I$, the null space is spanned by the five vectors: $(1\ 0\ 0\ 0\ 0\ 1\ -1\ 0\ 0\ -1)^T, (0\ 1\ 0\ 0\ 0\ -1\ 1\ -1\ 0\ 0)^T, (0\ 0\ 1\ 0\ 0\ 0\ -1\ 1\ -1\ 0)^T, (0\ 0\ 0\ 1\ 0\ 0\ 0\ -1\ 1\ -1)^T, (0\ 0\ 0\ 0\ 1\ -1\ 0\ 0\ -1\ 1)^T$.) The orthogonal complement of $\mathbf{1}$ in $\mathbb{R}^{10}$ is of dimension $10 - 1 = 9$. Hence the null spaces of $A(P_1) - I$ and $A(P_2) - I$ must have a common eigenvector $\boldsymbol{x}$ orthogonal to $\mathbf{1}$. Applying the matrices on the two sides of Equation 10.3 to $\boldsymbol{x}$, we get $(J - I)\boldsymbol{x} = A(P_1)\boldsymbol{x} + A(P_2)\boldsymbol{x} + A(P_3)\boldsymbol{x}$, that is (as $J\boldsymbol{x} = 0$), $-\boldsymbol{x} = \boldsymbol{x} + \boldsymbol{x} + A(P_3)\boldsymbol{x}$. Thus $A(P_3)\boldsymbol{x} = -3\boldsymbol{x}$, and this means that -3 is an eigenvalue of the Petersen graph, a contradiction.

Various proofs of Theorem 10.4 are available in literature. For a second proof, see [7]. ∎

10.6 SPECTRUM OF THE COMPLETE BIPARTITE GRAPH $K_{p,q}$

We now determine the spectrum of the complete bipartite graph $K_{p,q}$.

Theorem 10.5 $Sp(K_{p,q}) = \begin{pmatrix} 0 & \sqrt{pq} & -\sqrt{pq} \\ p+q-2 & 1 & 1 \end{pmatrix}.$

Proof. Let $V(K_{p,q})$ have the bipartition (X, Y) with $|X| = p$ and $|Y| = q$. Then the adjacency matrix of $K_{p,q}$ is of the form:

$$A = \begin{pmatrix} 0 & J_{p,q} \\ J_{q,p} & 0 \end{pmatrix},$$

where $J_{r,s}$ stands for the all 1-matrix of size r by s. Clearly, $\text{rank}(A) = 2$, as the maximum number of independent rows of A is 2. Hence zero is an eigenvalue of A repeated $p + q - 2$

times (as the null space of A is of dimension $p + q - 2$). Thus the characteristic polynomial of A is of the form $\lambda^{p+q-2}(\lambda^2 + c_2)$.

(Recall that by Lemma 10.2, the coefficient of λ^{p+q-1} in $\chi(G; \lambda)$ is zero.) Further (again by the same Lemma), $-c_2 =$ *the number of edges of* $K_{p,q} = pq$. This proves the result. ∎

10.7 DETERMINANT OF THE ADJACENCY MATRIX OF A GRAPH

We now present the elegant formula given by Harary for the determinant of the adjacency matrix of a graph in terms of certain of its subgraphs.

Definition 10.3 *A linear subgraph of a graph G is a subgraph of G whose components are either single edges or cycles.*

Theorem 10.6 [8] *Let A be the adjacency matrix of a simple graph G. Then*
$$\det A = \sum_H (-1)^{e(H)} 2^{c(H)},$$
where the summation is over all the spanning linear subgraphs H of G, and $e(H)$ and $c(H)$ denote respectively the number of even components and the number of cycles in H.

Proof. Let G be of order n with $V = \{v_1, \ldots, v_n\}$, and $A = (a_{ij})$. A typical term in the expansion of $\det A$ is:
$$sgn(\pi) a_{1\pi(1)} a_{2\pi(2)} \cdots a_{n\pi(n)},$$
where π is a permutation on $\{1, 2, \ldots, n\}$ and $sgn(\pi) = 1$ or -1 according to whether π is an even or odd permutation. This term is zero if and only if for some $i, 1 \leq i \leq n, a_{i\pi(i)} = 0$, that is, if and only if $\pi(i) = i$ or $\pi(i) = j(\neq i)$ and $v_i v_j \notin E(G)$. Hence this term is nonzero if and only if the permutation π is a product of disjoint cycles of length at least 2, and in this case, the value of the term is $sgn(\pi).1.1\ldots 1 = sgn(\pi)$. Each cycle (ij) of length two in π corresponds to the single edge $v_i v_j$ of G while each cycle $(ij\ldots p)$ of length $r > 2$ in π corresponds to a cycle of length r of G. Thus each nonvanishing term in the expansion of $\det(A)$ gives rise to a linear subgraph H of G and conversely. Now for any cycle C of S_n, $sgn(C) = 1$ or -1 according to whether C is an odd or even cycle. Hence $sgn(\pi) = (-1)^{e(H)}$, where $e(H)$ is the number of even components of H (i.e., components which are either single edges or even cycles of the graph H). Moreover, any cycle of H has two different orientations. Hence each of the undirected cycles of H of length ≥ 3 yields two distinct even cycles in S_n. (For example, the 4-cycle $(v_{i_1} v_{i_2} v_{i_3} v_{i_4})$ gives rise to two cycles $(v_{i_1} v_{i_2} v_{i_3} v_{i_4})$ and $(v_{i_4} v_{i_3} v_{i_2} v_{i_1})$ in H.) This proves the result. ∎

Corollary 10.2 [9] *Let $\chi(G; x) = x^n + a_1 x^{n-1} + \cdots + a_n$ be the characteristic polynomial of G. Then*
$$a_i = \sum_H (-1)^{\omega(H)} 2^{c(H)},$$
where the summation is over all linear subgraphs H of order i of G, and $\omega(H)$ and $c(H)$ denote, respectively, the number of components and the number of cycle components of H.

Proof. Recall that $a_i = (-1)^i \sum_H \det A(H)$, where H runs through all the induced subgraphs of order i of G. But by Theorem 10.6,
$$\det A(H) = \sum_{H_i} (-1)^{e(H_i)} 2^{c(H_i)},$$
where H_i is a spanning linear subgraph of H and $e(H_i)$ stands for the number of even components of H_i while $c(H_i)$ stands for the number of cycles in H_i. The corollary follows from the fact that i and the number of odd components of H_i have the same parity. ∎

10.8 SPECTRA OF PRODUCT GRAPHS

In this section we determine spectra of the product graphs—Cartesian product, direct product, and strong product—in terms of the spectra of their factor graphs. Our approach is based on Cvetković [10] as described in [4]. We first recall the definitions of the Cartesian, direct, and strong products of two graphs.

Denote a general graph product of two *simple* graphs by $G * H$. We define the product in such a way that $G * H$ is also simple. Given graphs G_1 and G_2 with vertex sets V_1 and V_2, respectively, any product graph $G_1 * G_2$ has as its vertex set, the Cartesian product $V(G_1) \times V(G_2)$. For any two vertices $(u_1, u_2), (v_1, v_2)$ of $G_1 * G_2$, consider the following possibilities:

i. u_1 adjacent to v_1 in G_1 or u_1 nonadjacent to v_1 in G_1,

ii. u_2 adjacent to v_2 in G_2 or u_2 nonadjacent to v_2 in G_2,

iii. $u_1 = v_1$ and/or $u_2 = v_2$.

We use, with respect to any graph, the symbols E, N, and $=$ to denote adjacency (edge), nonadjacency (no edge), and equality of vertices, respectively. We then have the following structure table S for $G_1 * G_2$ where the rows of S correspond to G_1 and the columns to G_2 and

$$S : \begin{array}{c} \\ E \\ = \\ N \end{array} \begin{array}{ccc} E & = & N \\ \left[\begin{array}{ccc} \circ & \circ & \circ \\ \circ & = & \circ \\ \circ & \circ & \circ \end{array} \right] \end{array}$$

where each $\circ$ in the double array S is E or N according to whether a general vertex (u_1, u_2) of $G_1 * G_2$ is adjacent or nonadjacent to a general vertex (v_1, v_2) of $G_1 * G_2$. Since each $\circ$ can take two options, there are in all $2^8 = 256$ graph products $G_1 * G_2$ that can be defined using G_1 and G_2.

If $S = \begin{bmatrix} a_{11} & a_{12} & a_{13} \\ a_{21} & = & a_{23} \\ a_{31} & a_{32} & a_{33} \end{bmatrix}$, then the edge-nonedge entry of S will correspond to the nonedge-edge entry of the structure matrix of $G_2 * G_1$. Hence the product $*$ is commutative, that is, G_1 and G_2 commute under $*$ if and only if the double-array S is symmetric. Hence if the product is commutative, it is enough if we know the five circled positions in $\begin{bmatrix} \circ & \circ & \circ \\ & = & \circ \\ & & \circ \end{bmatrix}$ to determine S completely. Therefore there are in all $2^5 = 32$ commuting products.

We now give the matrix S for the Cartesian, direct, composition, and strong products.

Definition 10.4 *Cartesian product*, $G_1 \square G_2$.

$$S : \begin{array}{c} \\ E \\ = \\ N \end{array} \begin{array}{ccc} E & = & N \\ \left[\begin{array}{ccc} N & E & N \\ E & = & N \\ N & N & N \end{array} \right] \end{array}$$

Hence (u_1, u_2) and (v_1, v_2) are adjacent in $G_1 \square G_2$ if and only if either $u_1 = v_1$ and u_2 is adjacent to v_2 in G_2, or u_1 is adjacent to v_1 in G_1 and $u_2 = v_2$.

Definition 10.5 *Direct (or tensor or Kronecker) product, $G_1 \times G_2$.*

$$S := \begin{array}{c} E \\ E \\ N \end{array} \begin{bmatrix} E & = & N \\ E & N & N \\ N & = & N \\ N & N & N \end{bmatrix}$$

Hence (u_1, u_2) is adjacent to (v_1, v_2) in $G_1 \times G_2$ if and only if u_1 is adjacent to v_1 in G_1 and u_2 is adjacent to v_2 in G_2.

Definition 10.6 *Strong (or normal) product $G_1 \square G_2$.* By definition, $G_1 \square G_2 = (G_1 \square G_2) \cup (G_1 \times G_2)$. Hence its structure matrix is given by

$$S := \begin{array}{c} E \\ E \\ N \end{array} \begin{bmatrix} E & = & N \\ E & E & N \\ E & = & N \\ N & N & N \end{bmatrix}$$

These three products can be checked to be associative.

Example 10.1 Let G_1 and G_2 be the two graphs given in Figure 10.2. Then the Cartesian, direct, and strong products of G_1 and G_2 are given in Figure 10.3.

Let $\mathscr{B}$ be a set of binary n-tuples $(\beta_1, \beta_2, \ldots, \beta_n)$ not containing $(0, 0, \ldots, 0)$.

Definition 10.7 *Given a sequence of graphs $G_1, G_2, \ldots, G_n$, the NEPS (noncomplete extended P-sum) of $G_1, G_2, \ldots, G_n$, with respect to $\mathscr{B}$ is the graph G with $V(G) = V(G_1) \times V(G_2) \times \cdots \times V(G_n)$, and in which two vertices $(x_1, x_2, \ldots, x_n)$ and $(y_1, y_2, \ldots, y_n)$ are adjacent if and only if there exists an n-tuple $(\beta_1, \beta_2, \ldots, \beta_n) \in \mathscr{B}$ with the property that if $\beta_i = 1$, then $x_i y_i \in E(G_i)$ and if $\beta_i = 0$, then $x_i = y_i$.*

Example 10.2

i. $n = 2$ and $\mathscr{B} = \{(1, 1)\}$. Here the graphs are G_1 and G_2.

The vertices (x_1, x_2) and (y_1, y_2) are adjacent in the NEPS of G_1 and G_2 with respect to $\mathscr{B}$ if and only if $x_1 y_1 \in E(G_1)$ and $x_2 y_2 \in E(G_2)$. Hence $G = G_1 \times G_2$, the direct product of G_1 and G_2.

ii. $n = 2$ and $\mathscr{B} = \{(0, 1), (1, 0)\}$. Here G is the Cartesian product $G_1 \square G_2$.

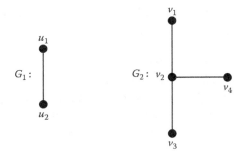

Figure 10.2 Graphs G_1 and G_2.

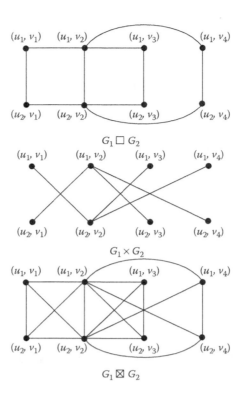

Figure 10.3 Product graphs of the graphs G_1 and G_2 given in Figure 10.2.

iii. $n = 2$ and $\mathscr{B} = \{(0,1), (1,0), (1,1)\}$. Here $G = (G_1 \square G_2) \cup (G_1 \times G_2) = G_1 \boxtimes G_2$, the strong product of G_1 and G_2.

Now given the adjacency matrices $G_1, \ldots G_n$, the adjacency matrix of the NEPS graph G with respect to the basis $\mathscr{B}$ is expressible in terms of the Kronecker product of matrices which we now define by the following definition.

Definition 10.8 *Let $A = (a_{ij})$ be an m by n matrix and $B = (b_{ij})$ be a p by q matrix. Then $A \otimes B$, the Kronecker product of A with B, is the mp by nq matrix got by replacing each entry a_{ij} of A by the double array $a_{ij}B$ (where $a_{ij}B$ is the matrix got by multiplying each entry of B by a_{ij}).*

It is well-known and easy to check that

$$(A \otimes B)(C \otimes D) = (AC \otimes BD), \tag{10.4}$$

whenever the matrix products AC and BD are defined. Clearly this can be extended to any finite product whenever the products are defined.

Remark 10.1 *Let us look more closely at the product $A_1 \otimes A_2$, where A_1 and A_2 are the adjacency matrices of the graphs G_1 and G_2 of orders n and t, respectively. To fix any particular entry of $A_1 \otimes A_2$, let us first label $V(G_1) = V_1$ and $V(G_2) = V_2$ as $V_1 = \{u_1, \ldots, u_n\}$, and $V_2 = \{v_1, \ldots, v_t\}$. Then to fix the entry in $A_1 \otimes A_2$ corresponding to $((u_i, u_j), (v_p, v_q))$, we look at the double array $(A_1)_{(u_i u_j)} A_2$ in $A_1 \otimes A_2$, where $(A_1)_{(u_i u_j)} := \alpha$ stands for the (i,j)th entry of A_1. Then the required entry is just $\alpha\{(p,q)$th entry of $A_2\}$. Hence it is 1 if and only if $(A_1)_{(u_i u_j)} = 1 = (A_2)_{(v_p v_q)}$, that is, if and only if, $u_i u_j \in E(G_1)$ and $v_p v_q \in E(G_2)$ and 0 otherwise. In other words, $A_1 \otimes A_2$ is the adjacency matrix of $G_1 \times G_2$. By associativity, $A(G_1) \otimes \cdots \otimes A(G_r)$ is the adjacency matrix of the graph product $G_1 \times \cdots \times G_r$.*

Our next theorem determines the adjacency matrix of the NEPS graph G in terms of the adjacency matrices of G_i, $1 \leq i \leq n$ for all the three products mentioned above.

Theorem 10.7 [10] *Let G be the NEPS of the graphs $G_1, \ldots, G_n$ with respect to the basis $\mathscr{B}$. Let A_i be the adjacency matrix of G_i, $1 \leq i \leq n$. Then the adjacency matrix A of G is given by*
$$A = \sum_{\beta=(\beta_1,\ldots,\beta_n)\in\mathscr{B}} A_1^{\beta_1} \otimes \cdots \otimes A_n^{\beta_n}.$$

Proof. Label the vertex set of each of the graphs G_i, $1 \leq i \leq n$, and order the vertices of G lexicographically. Form the adjacency matrix A of G with respect to this ordering. Then (by the description of Kronecker product of matrices given in Remark 10.1) we have $(A)_{(x_1,\ldots,x_n)(y_1,\ldots,y_n)} = \sum_{\beta\in\mathscr{B}}(A_1^{\beta_1})_{(x_1,y_1)}\cdots(A_n^{\beta_n})_{(x_n,y_n)}$, where $(M)_{(x,y)}$ stands for the entry in M corresponding to the vertices x and y. But by lexicographic ordering, $(M)_{(x_1,\ldots,x_n)(y_1,\ldots,y_n)} = 1$ if and only if there exists a $\beta = (\beta_1, \ldots, \beta_n) \in \mathscr{B}$ with $(A_i^{\beta_i})_{(x_i,y_i)} = 1$ for each $i = 1, \ldots, n$. This of course means that $x_iy_i \in E(G_i)$ if $\beta_i = 1$ and $x_i = y_i$ if $\beta_i = 0$ (the latter condition corresponds to $A_i^{\beta_i} = I$). ∎

We now determine the spectrum of the NEPS graph G with respect to the basis $\mathscr{B}$ in terms of the spectra of the factor graphs G_i.

Theorem 10.8 [10] *Let G be the NEPS of the graphs $G_1, \ldots, G_n$ with respect to the basis $\mathscr{B}$. Let k_i be the order of G_i and A_i, the adjacency matrix of G_i. Let $\{\lambda_{i1}, \ldots, \lambda_{ik_i}\}$ be the spectrum of G_i, $1 \leq i \leq n$. Then*
$$Sp(G) = \{\Lambda_{i_1 i_2 \ldots i_n} : 1 \leq i_j \leq k_j \text{ and } 1 \leq j \leq n\},$$
where $\Lambda_{i_1 i_2 \ldots i_n} = \sum_{\beta=(\beta_1,\ldots,\beta_n)\in\mathscr{B}} \lambda_{1i_1}^{\beta_1} \ldots, \lambda_{ni_n}^{\beta_n}, 1 \leq i_j \leq k_j$ and $1 \leq j \leq n$.

Proof. There exist vectors x_{ij} with $A_i x_{ij} = \lambda_{ij} x_{ij}$, $1 \leq i \leq n$; $1 \leq j \leq k_j$. Now consider the vector $x = x_{1i_1} \otimes \cdots \otimes x_{ni_n}$. Let A be the adjacency matrix of G. Then from Theorem 10.7 and Equation 10.4 (rather its extension),

$$\begin{aligned}
Ax &= \left(\sum_{\beta\in\mathscr{B}} A_1^{\beta_1} \otimes \cdots \otimes A_n^{\beta_n}\right)(x_{1i_1} \otimes \cdots \otimes x_{ni_n}) \\
&= \sum_{\beta\in\mathscr{B}} \left(A_1^{\beta_1} x_{1i_1} \otimes \cdots \otimes A_n^{\beta_n} x_{ni_n}\right) \\
&= \sum_{\beta\in\mathscr{B}} \left(\lambda_{1i_1}^{\beta_1} x_{1i_1} \otimes \cdots \otimes \lambda_{ni_n}^{\beta_n} x_{ni_n}\right) \\
&= \left(\sum_{\beta\in\mathscr{B}} \lambda_{1i_1}^{\beta_1} \ldots \lambda_{ni_n}^{\beta_n}\right) x \\
&= \Lambda_{i_1 i_2 \ldots i_n} x.
\end{aligned}$$

Thus $\Lambda_{i_1 i_2 \ldots i_n}$ is an eigenvalue of G. This yields $k_1 k_2 \ldots k_n$ eigenvalues of G and hence all the eigenvalues of G. ∎

Corollary 10.3 *Let A_1 and A_2 be the adjacency matrices of G_1 and G_2, respectively, and let $Sp(G_1) = \{\lambda_1, \ldots, \lambda_n\}$ and $Sp(G_2) = \{\mu_1, \ldots, \mu_t\}$.*

Then

    i. $A(G_1 \times G_2) = A_1 \otimes A_2$; *and*

$$Sp(G_1 \times G_2) = \{\lambda_i \mu_j : 1 \leq i \leq n, 1 \leq j \leq t\}.$$

    ii. $A(G_1 \square G_2) = (I_n \otimes A_2) + (A_1 \otimes I_t)$; *and*

$$Sp(G_1 \square G_2) = \{\lambda_i + \mu_j : 1 \leq i \leq n,\ 1 \leq j \leq t\}.$$

    iii. $A(G_1 \boxtimes G_2) = (A_1 \otimes A_2) + (I_n \otimes A_2) + (A_1 \otimes I_t)$; *and*

$$Sp(G_1 \boxtimes G_2) = \{\lambda_i \mu_j + \lambda_i + \mu_j : 1 \leq i \leq n,\ 1 \leq j \leq t\}. \blacksquare$$

10.9 LAPLACIAN SPECTRUM OF A GRAPH

In this section we discuss the Laplacian spectrum, that is, the spectrum of the Laplacian matrix of a graph. The Laplacian spectrum has several applications—in physics, electrical engineering, and computer science—to mention a few fields. For a comprehensive treatment of the spectrum of the Laplacian matrix, see [11]

Let G, as before, be a simple graph with $V = \{v_1, \ldots, v_n\}$ as its vertex set and $E = \{e_1, \ldots, e_m\}$ as its edge set. Let $D = D(G)$ be the n by n diagonal matrix $\text{diag}[d(v_1), \ldots, d(v_n)]$, where $d(v_i)$ stands for the degree of the vertex v_i in G.

Definition 10.9 *The Laplacian matrix of G is the matrix $Q(G) = D(G) - A(G)$, where $A(G)$ as usual stands for the adjacency matrix of G. The matrix $Q(G)$ is often called the Kirchoff matrix of G (see Theorem* 10.11 *below).*

Definition 10.10 *The Laplacian spectrum of G is the spectrum of the Laplacian matrix $Q(G)$ of G.*

We usually write it in the nonincreasing order: $\{\lambda_1 \geq \lambda_2 \geq \cdots \geq \lambda_{n-1} \geq \lambda_n\}$.

Definition 10.11 *The characteristic polynomial of $Q(G)$ is the polynomial $\mu(x) = \det(xI - Q)$.*

It is clear that as in the case of the adjacency matrix $A(G)$, the eigenvalues of $Q(G)$ are also real, $Q(G)$ being a real symmetric matrix. Hence $\mathbb{R}^n$ has an orthonormal basis consisting of eigenvectors of $Q(G)$.

Now orient G in some way and get the oriented graph $\vec{G}$. Let C be the $n \times m$ matrix (c_{ij}), where

$$c_{ij} = \begin{cases} 1, & \text{if } v_i \text{ is the head of the arc } e_j \text{ in } \vec{G}; \\ -1, & \text{if } v_i \text{ is the tail of the arc } e_j \text{ in } \vec{G}; \\ 0, & \text{if } v_i \text{ is nonincident to the arc } e_j \text{ in } \vec{G}. \end{cases} \quad (10.5)$$

If $R_1, \ldots, R_n$ are the row vectors of C, it is clear that

$$CC^T = \begin{bmatrix} R_1.R_1 & R_1.R_2 & \cdots & R_1.R_n \\ \vdots & \vdots & \vdots & \vdots \\ R_n.R_1 & R_n.R_2 & \cdots & R_n.R_n \end{bmatrix}.$$

where . stands for scalar product of vectors.

Now $R_i.R_i = d(v_i)$, while for $i \neq j$, $R_i.R_j = c_{i1}c_{j1} + c_{i2}c_{j2} + \cdots + c_{im}c_{jm}$

$$= \begin{cases} 0, & \text{if } v_i v_j \notin E(G); \\ -1, & \text{if } v_i v_j \in E(G). \end{cases} \quad (10.6)$$

(The last relation 10.6 is due to the fact that if $v_i v_j = e_k \in E(G)$, one of c_{ik}, c_{jk} is 1 and the other is -1 while if $p \neq k$, at least one of c_{ip} and c_{jp} is zero.) Thus we have proved the following theorem.

Theorem 10.9 $Q = CC^T$. ∎

A consequence of Theorem 10.9 is the following: Let $X \in \mathbb{R}^n$. Then the inner product $(QX, X) = (CC^T X, X) = (C^T X, C^T X) \geq 0$ and so (QX, X) is a positive semidefinite quadratic form. Consequently, all eigenvalues λ_i of Q are nonnegative. (Indeed, if λ is an eigenvalue of Q and x, a corresponding eigenvector, then $0 \leq (Qx, x) = (\lambda x, x) = \lambda(x, x)$ and so $\lambda \geq 0$.) Further, as each row of Q adds up to zero, $Q\mathbf{1} = 0$, where $\mathbf{1}$ is the all 1-column vector of length n and 0 is the zero vector of length n. Hence $\lambda_n = 0$.

It is also clear that

$$X^T Q X = X^T C C^T X$$
$$= \sum_{\substack{1 \leq i < j \leq n \\ (v_i, v_j) \in E(G)}} (x_i - x_j)^2, \quad (10.7)$$

where $X = (x_1, \ldots, x_n)^T$.

Example 10.3 Let G be the underlying undirected graph of the oriented graph $\vec{G}$ of Figure 10.4.

$$\text{Then } Q = D - A = \begin{bmatrix} 2 & -1 & -1 & 0 \\ -1 & 2 & -1 & 0 \\ -1 & -1 & 3 & -1 \\ 0 & 0 & -1 & 1 \end{bmatrix}, \text{ and}$$

$$C = \begin{array}{c} \\ v_1 \\ v_2 \\ v_3 \\ v_4 \end{array} \begin{array}{c} e_1 \quad e_2 \quad e_3 \quad e_4 \\ \begin{bmatrix} -1 & 0 & 1 & 0 \\ 1 & -1 & 0 & 0 \\ 0 & 1 & -1 & -1 \\ 0 & 0 & 0 & 1 \end{bmatrix} \end{array}.$$

It is easy to check that $Q = CC^T$ and also Equation 10.7.

We now proceed to establish Kirchoff's matrix-tree theorem. Before we do that, we need an auxiliary lemma which is known as Binet–Cauchy theorem.

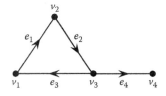

Figure 10.4 Oriented graph $\vec{G}$.

Theorem 10.10 (Binet–Cauchy theorem) *If P and Q are $l \times t$ and $t \times l$ matrices, respectively, with $l \leq t$, then $\det(PQ)$ is the sum of the products of the corresponding major determinants of P and Q.* ∎

For example, if $P = \begin{bmatrix} a_1 & a_2 & a_3 \\ b_1 & b_2 & b_3 \end{bmatrix}$ and $Q = \begin{bmatrix} c_1 & d_1 \\ c_2 & d_2 \\ c_3 & d_3 \end{bmatrix}$, then

$$\det PQ = \begin{vmatrix} a_1 & a_2 \\ b_1 & b_2 \end{vmatrix} \begin{vmatrix} c_1 & d_1 \\ c_2 & d_2 \end{vmatrix} + \begin{vmatrix} a_1 & a_3 \\ b_1 & b_3 \end{vmatrix} \begin{vmatrix} c_1 & d_1 \\ c_3 & d_3 \end{vmatrix} + \begin{vmatrix} a_2 & a_3 \\ b_2 & b_3 \end{vmatrix} \begin{vmatrix} c_2 & d_2 \\ c_3 & d_3 \end{vmatrix}.$$

Theorem 10.11 (Kirchoff's matrix-tree theorem) *Let G be a connected graph. Then the cofactors of all the entries of Q, the Kirchoff matrix of G, are equal and their common value is the number of spanning trees of G.*

Proof. Recall that the Kirchoff matrix of G is just the Laplacian matrix $Q = CC^T$, where C is the $n \times m$ matrix defined by (10.5).

Consider an $n \times (n-1)$ submatrix P of C defined by a set of $(n-1)$ columns of C. Then P is the incidence matrix of the oriented graph H defined by the $n-1$ edges corresponding to the $(n-1)$ columns of G.

Delete any row of P, say the row corresponding to the vertex v_k. This results in a square submatrix P_k of P of order $(n-1)$. We show that $|\det P_k|$ is 1 or 0 according as H is a tree or not.

If H is not a tree, as H has n vertices and $n-1$ edges, H must be disconnected. Further, one of the components of H does not contain v_k. The sum of the row vectors corresponding to the vertex subset of this component is zero. Hence $|\det P_k| = 0$.

We now assume that H is a tree. We now relabel the vertices and edges of H other than v_k as follows: Let $u_1 \neq v_k$ be an end vertex of H and let x_1 be the pendant edge incident at x_1 in H; let $u_2 \neq v_k$ be an end vertex of the subtree $H - u_1$ of H, and let x_2 be the pendant edge of H incident at u_2 and so on. This relabeling of the vertices and edges of H determines a new matrix P'_k which is lower triangular (i.e., if $P'_k = (p_{rs})$, then for $s > r$, $p_{rs} = 0$) and every diagonal entry is 1 or -1; hence $|\det P'_k| = 1$.

To complete the proof of the theorem, we apply the Binet–Cauchy theorem. We first note that if in a square matrix, each row and column adds up to zero, then all the cofactors of the entries of the matrix are equal. Hence this is true for Q since in Q each row and column adds up to zero. Therefore it suffices to show that the cofactor of the leading entry of Q is the number of spanning trees of G. Now the cofactor of the leading entry of Q is the determinant of the product $C_1 C_1^T$, where C_1 is the matrix got by deleting the first row of C. C_1 is then an $(n-1) \times m$ matrix. By Binet–Cauchy theorem, $\det(C_1 C_1^T)$ is equal to the sum of products of the corresponding major determinants of C_1 and C_1^T. A major determinant of C_1 has value 1 or 0 according as the columns of C_1 correspond to a spanning tree of G or not. Thus the sum of their products is exactly the number of spanning trees of G. ∎

Illustration Let G be the graph of Example 10.3. The cofactor of the leading entry of Q is $\begin{vmatrix} 2 & -1 & 0 \\ -1 & 3 & -1 \\ 0 & -1 & 1 \end{vmatrix} = 3$. Hence the number of spanning trees of this graph is 3.

Corollary 10.4 (Cayley) *The number of labeled trees with n vertices is n^{n-2}.*

Proof. The number of labeled trees with n vertices is just the number of spanning trees of a labeled complete graph K_n = the principal cofactor of $Q(K_n)$ =
$\begin{bmatrix} n-1 & -1 & \cdots & -1 \\ -1 & n-1 & \cdots & -1 \\ \vdots & \vdots & \ddots & \vdots \\ -1 & -1 & \cdots & n-1 \end{bmatrix}$. By Lemma 10.1, this number is easily seen to be n^{n-2}. ∎

10.10 ALGEBRAIC CONNECTIVITY OF A GRAPH

Since Q is singular, 0 is an eigenvalue of Q. What is the dimension of the eigenspace of 0? If G is connected, $\text{rank}(Q) = n-1$ by Theorem 10.11. It is clear that the eigenspace of Q is then generated by the all 1-vector **1**. In general, if G has ω components, the all 1-vector of relevant length is an eigenvector of each of the components and hence the ω binary vectors of length n, each having 1's in the positions corresponding to each of the components and zeros elsewhere form a basis for the eigenspace of 0. Thus the dimension of this eigenspace of G (namely, the eigenspace of the eigenvalue 0) is ω. Hence G is connected if and only if 0 is a simple eigenvalue of Q; that is, if and only if $\lambda_{n-1} \neq 0$. For this reason, λ_{n-1} was called by Fiedler [12] the *algebraic connectivity* of G. It is also called as the *Fiedler value* of G. Thus if the characteristic polynomial of Q, namely, $\mu(x) = x^n + a_1 x^{n-1} + \cdots + a_\omega x^\omega$, $a_\omega \neq 0$, then the number of components of G is ω.

Corollary 10.5 *Let $\lambda_1 \geq \lambda_2 \geq \cdots \geq \lambda_{n-1} > \lambda_n = 0$ be the eigenvalues of the Kirchoff matrix of a **connected** graph G of order n. Let δ and Δ be the minimum and maximum degrees of G respectively. Then*

i. $\sum_{i=1}^{n} \lambda_i = 2|E(G)|$.

ii. $\lambda_{n-1} \leq \frac{n}{n-1}\Delta$.

iii. $\lambda_1 \geq \frac{n}{n-1}\delta$.

iv. *The number of spanning trees $\tau(G)$ of G is $\frac{1}{n}\lambda_1 \lambda_2 \cdots \lambda_{n-1}$.*

Proof.

(i) follows from the fact that $\sum_{i=1}^{n} \lambda_i$ = trace of Q.

(ii) and (iii). From (i), $\lambda_1 + \cdots + \lambda_{n-1} + 0 = d_1 + \cdots + d_n$, where d_i is the degree of the vertex v_i. Since $\lambda_1 \geq \lambda_2 \geq \cdots \geq \lambda_{n-1}$,

$$(n-1)\lambda_{n-1} \leq \lambda_1 + \cdots + \lambda_{n-1} \leq (n-1)\lambda_1 \text{ and}$$

$$(n-1)\lambda_{n-1} \leq d_1 + \cdots + d_{n-1} + d_n \leq (n-1)\lambda_1$$

and so $(n-1)\lambda_{n-1} \leq n\Delta$ and $n\delta \leq (n-1)\lambda_1$.

(iv) As G is connected,

$$\mu(x) = x^n + a_1 x^{n-1} + \cdots + a_{n-1} x + 0$$

where $a_{n-1} \neq 0$. Hence $\lambda_1 \lambda_2 \cdots \lambda_{n-1} = (-1)^{n-1} a_{n-1}$. Now,

$$\begin{aligned} a_{n-1} &= \text{coefficient of } x \text{ in } \det(xI - Q) \\ &= (-1)^{n-1} \{\text{cofactor of } d_1 \text{ in } Q + \cdots + \text{ cofactor of } d_n \text{ in } Q\} \\ &= (-1)^{n-1}.n.\tau(G) \text{ (by Theorem 10.11)} \end{aligned}$$

Consequently, $\tau(G) = \frac{1}{n}(-1)^{n-1} a_{n-1} = \frac{1}{n} \lambda_1 \lambda_2 \cdots \lambda_{n-1}$. ∎

Lemma 10.4 *The Laplacian spectrum of the complete graph K_n is zero repeated once and n repeated $n-1$ times.*

Proof. Follows from the fact that

$$Q(K_n) = \begin{bmatrix} n-1 & -1 & -1 & \cdots & -1 \\ -1 & n-1 & -1 & \cdots & -1 \\ \vdots & \vdots & \vdots & \ddots & \vdots \\ -1 & -1 & -1 & \cdots & n-1 \end{bmatrix}$$

and by Lemma 10.1. ∎

Note:

1. Cayley's theorem (Corollary 10.4) is a consequence of Lemma 10.4 and (iv) of Corollary 10.5.

2. Lemma 10.4 implies that the Fiedler value of K_n is n.

Theorem 10.12 [13] *The Laplacian spectrum of any simple graph G of order n is contained in $[0, n]$. Further the multiplicity of n as a Laplacian eigenvalue of G is one less than the number of components of G^c, the complement of G.*

Proof. We have $Q(G) + Q(G^c) = Q(K_n)$ and $Q(G)\mathbf{1} = Q(G^c)\mathbf{1} = Q(K_n)\mathbf{1} = 0$. Let x be any nonzero vector orthogonal to $\mathbf{1}$ with $Q(G)x = \lambda x$ for some Laplacian eigenvalue λ of G. We have by Lemma 10.4, $Q(G^c)x = Q(K_n)x - Q(G)x = nx - \lambda x = (n - \lambda)x$ which means that $n - \lambda$ is a Laplacian eigenvalue of G^c. Hence $n - \lambda \geq 0$ and so $\lambda \leq n$. Moreover $\lambda = n$ if and only if $Q(G^c)x = 0$ and the dimension of the space spanned by such vectors is one less than the nullity of $Q(G^c)$. (Recall that $Q(G^c)\mathbf{1} = 0$, the vector x is orthogonal to $\mathbf{1}$ and the nullity of G^c is equal to the number of components of G^c.) ∎

Corollary 10.6 *If G is of order n and n is an eigenvalue of $Q(G)$, then G^c is disconnected.*

Proof. If G^c is connected, by Theorem 10.12, n can not be a Laplacian eigenvalue of G. ∎

We now improve upon the inequalities of Corollary 10.5 by invoking the Courant-Weyl inequalities [3,14].

Theorem 10.13 [3,14] *Let A and B be real symmetric (more generally Hermitian) matrices with spectra $\{\lambda_1(A) \geq \lambda_2(A) \geq \cdots \geq \lambda_n(A)\}$ and $\{\lambda_1(B) \geq \lambda_2(B) \geq \cdots \geq \lambda_n(B)\}$, respectively. Let the spectrum of $A + B$ be $\{\lambda_1(A+B) \geq \lambda_2(A+B) \geq \cdots \geq \lambda_n(A+B)\}$. Then, for $1 \leq i, j \leq n$, we have*

i. *If $i + j - 1 \leq n$, then $\lambda_{i+j-1}(A+B) \leq \lambda_i(A) + \lambda_j(B)$,*

ii. If $i + j - n \geq 1$, then $\lambda_i(A) + \lambda_j(B) \geq \lambda_{i+j-n}(A + B)$,

iii. If B is positive semidefinite, then $\lambda_i(A + B) \geq \lambda_i(A)$. ∎

Theorem 10.14 [14] *Let M be a positive semidefinite matrix with each row sum zero. Then the second smallest eigenvalue of M is given by $\min_{X \in W} X^T M X$, where W is the set of unit vectors X orthogonal to the all 1-vector 1.* ∎

An immediate consequence of Theorem 10.14 is the following:

Corollary 10.7 *Let G be a graph of order n with Laplacian spectrum $\{\lambda_1(G) \geq \lambda_2(G) \geq \cdots \geq \lambda_{n-1}(G) > 0\}$. Then*

$$\lambda_{n-1}(G) = \min_{X \in W} X^T Q(G) X, \ X \in \mathbb{R}^n,$$

where W is the set of unit vectors X orthogonal to the all 1-vector 1. ∎

A consequence of Corollary 10.7 is the following corollary.

Corollary 10.8 *If a graph G is the edge-disjoint union of two spanning subgraphs G_1 and G_2, then the Fiedler values of G_1, G_2, and G are related by*

$$\lambda_{n-1}(G_1) + \lambda_{n-1}(G_2) \leq \lambda_{n-1}(G).$$

Proof. Let W be the set of unit vectors $X \in \mathbb{R}^n$ orthogonal to the all 1-vector 1. Then,

$$\begin{aligned}
\lambda_{n-1}(G) &= \min_{X \in W} X^T Q(G) X \\
&= \min_{X \in W} X^T [Q(G_1) + Q(G_2)] X \\
&\geq \min_{X \in W} X^T Q(G_1) X + \min_{X \in W} X^T Q(G_2) X \\
&= \lambda_{n-1}(G_1) + \lambda_{n-1}(G_2).
\end{aligned}$$
∎

We now present an improvement of (ii) of Corollary 10.5 by showing that the maximum degree Δ of G can be replaced by the minimum degree δ of G on the right.

Theorem 10.15 [12] *Let G be a connected graph of order n with Laplacian spectrum $\{\lambda_1 \geq \lambda_2 \geq \cdots \geq \lambda_{n-1} > 0\}$. Then*

$$\lambda_{n-1} \leq \frac{n}{n-1} \delta.$$

Proof. By Corollary 10.7,

$$\lambda_{n-1} = \min_{X \in W} X^T Q X, \ Q = Q(G). \tag{10.8}$$

Now consider the matrix

$$\tilde{Q} = Q - \lambda_{n-1}(I - n^{-1} J).$$

Let $Y \in \mathbb{R}^n$ so that $Y = c_1 \mathbf{1} + c_2 X$, where $c_1, c_2 \in \mathbb{R}$ and $X \in W$. Then $Q\mathbf{1} = 0$ and $\tilde{Q}\mathbf{1} = 0$ as $(I - n^{-1} J)\mathbf{1} = 0$. Therefore

$$\begin{aligned}
Y^T \tilde{Q} Y &= c_2^2 X^T \tilde{Q} X = c_2^2 (X^T Q X - \lambda_{n-1}) \\
&\geq 0 \text{ (by Equation 10.8).}
\end{aligned}$$

Thus $\tilde{Q}$ is also a positive semidefinite matrix and hence all diagonal entries of $\tilde{Q}$ are nonnegative. This means that $\delta - \lambda_{n-1}(1 - n^{-1}) \geq 0$, that is,

$$\lambda_{n-1} \leq \frac{n}{n-1} \delta.$$
∎

In a similar manner, one can show that $\lambda_1 \geq (n/(n-1))\Delta$.

References

[1] R. Balakrishnan and K. Ranganathan. *A Textbook of Graph Theory*. Springer, New York, Second (Revised and Enlarged) edition, 2012.

[2] N.L. Biggs. *Algebraic Graph Theory*. Cambridge University Press, Second edition, 1993.

[3] A.E. Brouwer and W.H. Haemers. *Spectra of Graphs*. Springer, New York, 2011.

[4] D.M. Cvetković, M. Doob, and H. Sachs. *Spectra of Graphs: Theory and Application*. Wiley, New York, Third (revised and enlarged) edition, 1998.

[5] C.D. Godsil and G. Royle. *Algebraic Graph Theory*. Springer-Verlag, New York, 2001.

[6] A.J. Schwenk and O.P. Lossers. Solutions of advanced problems. *The American Mathematical Monthly*, **94** (1987), 885–887.

[7] D. Bryant. Another quick proof that $K_{10} \neq P + P + P$. *Bull. ICA*, **34** (2002), 86.

[8] F. Harary. The determinant of the adjacency matrix of a graph. *SIAM Rev.*, **4** (1962), 202–210.

[9] H. Sachs. Über teiler, Faktoren und Charakteristische Polynome von Graphen II. *Wiss. Z. Techn. Hochsch. Ilmenau*, **13** (1967), 405–412.

[10] D. Cvetković. *Graphs and Their Spectra*. PhD thesis, Univ. Beograd Publ. Elektrotehn. Fak., Ser. Mat. Fiz., 1971.

[11] J.J. Molitierno. *Applications of Combinatorial Matrix Theory to Laplacian Matrices of Graphs*. Chapman & Hall/CRC Press, 2012.

[12] M. Fiedler. Algebraic connectivity of graphs. *Czech. Math. J.*, **23** (1973), 298–305.

[13] W.N. Anderson and T.D. Morley. Eigenvalues of the Laplacian of a graph. *Linear Multilinear Algebra*, **18** (1985), 141–145.

[14] R.A. Horn and C.R. Johnson. *Matrix Analysis*. Cambridge University Press, Cambridge, 1985.

CHAPTER 11

Resistance Networks, Random Walks, and Network Theorems

Krishnaiyan "KT" Thulasiraman

Mamta Yadav

CONTENTS

11.1	Introduction	247
11.2	Resistance Networks	247
11.3	Topological Formulas for Resistance Network Functions	249
11.4	Random Walks	254
11.5	Kirchhoff Index of a Graph and the Graph Laplacian	258
	11.5.1 Formula for the Kirchhoff Index	259
	11.5.2 Kirchhoff Index Using Topological Formulas for Network Functions	262
11.6	Foster's Theorems	263
11.7	Arc-Coloring Theorem and the No-Gain Property	265

11.1 INTRODUCTION

In this chapter we first discuss some aspects of electrical network analysis which depend heavily on the theory of graphs. We then use these results in the discovery of certain fundamental properties of networks (not necessarily electrical networks). In doing so we view a weighted graph as an electrical resistance network with conductances (reciprocals of resistance values) as weights of the edges. The main topics considered are topological formulas for network functions, random walks, Kirchhoff Index of a graph, Foster's theorems, the arc-coloring lemma, and the no-gain property of resistance networks. Results to be presented make use of the theory developed in Chapter 8.

11.2 RESISTANCE NETWORKS

An electrical network is an interconnection of electrical network elements such as resistances, capacitances, inductances, and voltage and current sources. In this chapter, we will assume that all the network elements in the networks to be considered are resistances. However, unless explicitly stated all the results to be developed are applicable to any network of resistances, capacitances, and inductances. Each network element is associated with two variables—the voltage variable $v(t)$ and the current variable $i(t)$. We need to specify reference directions for these variables because they are functions of time and may take on positive and negative values in the course of time. This is done by assigning an arrow, called *orientation*, to each network element (Figure 11.1). This arrow means that $i(t)$ is positive whenever the current is in the direction of the arrow. Further we assume that the positive polarity of the voltage $v(t)$ is at the tail end of the arrow. Thus $v(t)$ is positive whenever the voltage drop in a network element is in the direction of the arrow.

Network elements are characterized by the physical relationships between the associated voltage and current variables. Ohm's law specifies the relationship between $v(t)$ and $i(t)$ as

$$v(t) = Ri(t)$$

where R is the resistance (in ohms) of the network element.

Note that for some of the network elements the voltage variables may be required to have specified values and for some others the current variables may be specified. Such elements are called, respectively, the *voltage* and *current sources*.

Two fundamental laws of network theory are *Kirchhoff's laws*, that are stated as follows.

Kirchhoff's Current Law (KCL) The algebraic sum of the currents flowing out of a vertex is equal to zero.

Kirchhoff's Voltage Law (KVL) The algebraic sum of the voltages around any circuit is equal to zero.

For instance, for the network shown in Figure 11.2a the KCL and KVL equations are as given below. In this figure element 5 is a voltage source and element 4 is a current source.

KCL equations

$$\begin{aligned} \text{vertex } a \quad & i_1 - i_5 + i_6 = 0, \\ \text{vertex } c \quad & -i_2 + i_4 - i_6 = 0, \\ \text{vertex } b \quad & -i_1 + i_2 + i_3 = 0. \end{aligned}$$

KVL equations

$$\begin{aligned} \text{circuit } \{1,3,5\} \quad & v_1 + v_3 + v_5 = 0, \\ \text{circuit } \{2,4,3\} \quad & v_2 + v_4 - v_3 = 0, \\ \text{circuit } \{1,6,2\} \quad & -v_1 + v_6 - v_2 = 0. \end{aligned}$$

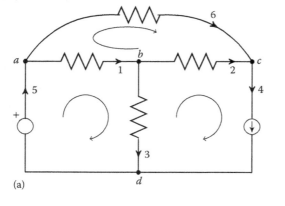
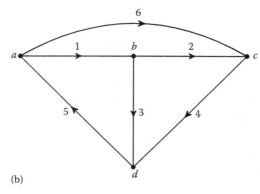

Figure 11.2 Directed graph representation of a network. (a) Network N. (b) Directed graph of N.

Given an electrical network N, the problem of network analysis is to determine the element voltages and currents that satisfy Kirchhoff's laws and the Ohm's law.

Notice that the equations which arise from an application of Kirchhoff's laws are algebraic in nature, and they depend only on the way the network elements are interconnected and not on the nature of the network elements. There are several properties of an electrical network which depend on the structure of the network. In studying such properties it will be convenient to treat each network element as a directed edge associated with the two variables $v(t)$ and $i(t)$. Thus, we may consider an electrical network as a directed graph in which each edge is associated with the two variables $v(t)$ and $i(t)$, which are required to satisfy Kirchhoff's laws and the Ohm's law.

For example, the directed graph corresponding to the network of Figure 11.2a is shown in Figure 11.2b.

It is now easy to see that KCL and KVL equations for a network N can be written, respectively, as

$$Q_c I_e = 0 \qquad (11.1)$$

and

$$B_c V_e = 0, \qquad (11.2)$$

where Q_c and B_c are the cut and circuit matrices of the directed graph associated with N, and I_e and V_e are, respectively, the column vectors of element currents and voltages of N.

Since the all-vertex incidence matrix A_c is a submatrix of Q_c and has the same rank as Q_c, we can use in Equation 11.1 the matrix A_c in place of Q_c. Thus KCL equations can be written as

$$A_c I_e = 0. \qquad (11.3)$$

Since the rank of A_c is $n-1$, we can remove any row from A_c and use the resulting matrix A called the incidence matrix (see Chapter 8). The vertex corresponding to the removed row is called the reference or datum vertex.

In all our discussions in this chapter we denote both an electrical network and the associated directed graph by the same symbol N. Most often a graph is also referred to as a network, and vice versa. We may also refer to a vertex as a *node*.

11.3 TOPOLOGICAL FORMULAS FOR RESISTANCE NETWORK FUNCTIONS

In this section we derive topological formulas for resistance network functions.

Consider first a 1-port resistance network N. Let the network N have $n+1$ nodes denoted by $0, 1, 2, \ldots, n$, and let the nodes 1 and 0 be, respectively, the positive and negative reference terminals of the port (Figure 11.3). Let us now excite the network by connecting a current source of value I_1 across the port. Let $V_1, V_2, \ldots, V_n$ denote the voltages of the nodes 1, 2, \ldots, n$ with respect to node 0. This means $V_0 = 0$ and V_i is the voltage between the nodes i and 0, (i.e., $V_i = V_i - V_0$) for $i \neq 0$. Also the A matrix does not contain the row corresponding to the vertex 0.

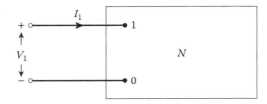

Figure 11.3 A 1-port network.

Then we have
$$AI_e = I, \qquad (11.4)$$
where
$$I = \begin{bmatrix} I_1 \\ 0 \\ 0 \\ \vdots \\ 0 \end{bmatrix}.$$

Let the network elements be labeled as $e_1, e_2, \ldots, e_m$ with r_i denoting the resistance value of element e_i. Then the conductance of e_i is given by $g_i = 1/r_i$.

Let G be the diagonal matrix with its (i,i) entry equal to g_i. Then we can write
$$I_e = GV_e. \qquad (11.5)$$

Suppose the end vertices of e_i are k and l. Then the voltage across this element (voltage drop from node k to node l) is given by $V_k - V_l$, assuming that the element is oriented from vertex k to vertex l. So we can write
$$V_e = A^t V \qquad (11.6)$$
where V is the vector of voltages $V_1, V_2, \ldots, V_n$.

Combining (11.4) through (11.6) we get the node equations
$$AGA^t V = I \qquad (11.7)$$
where
$$V = \begin{bmatrix} V_1 \\ V_2 \\ \vdots \\ V_n \end{bmatrix}.$$

Let
$$Y = AGA^t$$
so that
$$YV = I \qquad (11.8)$$

The matrix Y is called the node conductance matrix of the network with vertex 0 as the reference.

Solving (11.8) for V_1, we get
$$V_1 = \frac{\Delta_{11}}{\Delta} I_1,$$
where
$$\Delta = \det Y$$
and
$$\Delta_{11} = (1,1) \text{ cofactor of } Y.$$

So the driving-point resistance across vertices 1 and 0 is given by
$$z = \frac{V_1}{I_1} = \frac{\Delta_{11}}{\Delta}, \qquad (11.9)$$
and the driving-point conductance across 1 and 0 is given by
$$y = \frac{1}{z} = \frac{\Delta}{\Delta_{11}}. \qquad (11.10)$$

Figure 11.4 A 2-port network.

To illustrate certain principles of network analysis, consider next a 2-port network (see Figure 11.4). If the ports of N are excited by current sources of values I_1 and I_2, then the node equations of N can be written as

$$YV = I$$

where

$$I = \begin{bmatrix} I_1 \\ I_2 \\ -I_2 \\ \vdots \\ 0 \end{bmatrix},$$

Solving for the node voltages V_1, V_2, and V_3, we get

$$V_1 = \frac{1}{\Delta}(\Delta_{11}I_1 + \Delta_{21}I_2 - \Delta_{31}I_2),$$
$$V_2 = \frac{1}{\Delta}(\Delta_{12}I_1 + \Delta_{22}I_2 - \Delta_{32}I_2),$$
$$V_3 = \frac{1}{\Delta}(\Delta_{13}I_1 + \Delta_{23}I_2 - \Delta_{33}I_2).$$

From the above relations we get

$$\begin{bmatrix} V_1 \\ V_2 - V_3 \end{bmatrix} = \frac{1}{\Delta} \begin{bmatrix} \Delta_{11} & \Delta_{21} - \Delta_{31} \\ \Delta_{12} - \Delta_{13} & \Delta_{22} + \Delta_{33} - \Delta_{32} - \Delta_{23} \end{bmatrix} \begin{bmatrix} I_1 \\ I_2 \end{bmatrix} \quad (11.11)$$

$$= Z_{oc}I$$

Here Z_{oc} is called the open circuit resistance matrix of the 2-port network. This is because each element of Z_{oc} is obtained by setting one of the port currents equal to zero (i.e., open-circuiting the corresponding port).

Thus

$$z_{11} = \frac{V_1}{I_1}\bigg|_{I_2=0},$$
$$z_{12} = \frac{V_1}{I_2}\bigg|_{I_1=0},$$
$$z_{21} = \frac{V_2}{I_1}\bigg|_{I_2=0},$$
$$z_{22} = \frac{V_2}{I_2}\bigg|_{I_1=0}.$$

Here z_{11} and z_{22} are called driving point resistances across the respective ports and z_{12} and z_{21} are called transfer resistances between the ports. Note that since Y is symmetric, we have

$$\Delta_{ij} = \Delta_{ji}.$$

So

$$Z_{oc} = \frac{1}{\Delta}\begin{bmatrix} \Delta_{11} & \Delta_{12} - \Delta_{13} \\ \Delta_{12} - \Delta_{13} & \Delta_{22} + \Delta_{33} - 2\Delta_{23} \end{bmatrix}. \tag{11.12}$$

Now we recall from Chapter 8 that

$$\Delta = W$$
$$\Delta_{11} = W_{1,0}, \tag{11.13}$$

where W is the sum of the conductance products of all the spanning trees in N and $W_{1,0}$ is the sum of the conductance products of all the spanning 2-trees of the type $T_{1,0}$ (with 1 and 0 in separate trees of $T_{1,0}$). We also have from Chapter 8 that

$$\Delta_{ij} = W_{ij,0}$$

where $W_{ij,0}$ is the sum of the conductance products of all 2-trees $T_{ij,0}$ (i and j in one tree and 0 in the other tree). So

$$\Delta_{12} - \Delta_{13} = W_{12,0} - W_{13,0}. \tag{11.14}$$

Since each spanning 2-tree $T_{12,0}$ is either a spanning 2-tree $T_{12,30}$ or a spanning 2-tree $T_{123,0}$, we get

$$W_{12,0} = W_{12,30} + W_{123,0}. \tag{11.15}$$

Similarly,

$$W_{13,0} = W_{13,20} + W_{123,0}. \tag{11.16}$$

Then

$$\Delta_{12} - \Delta_{13} = W_{12,30} - W_{13,20}. \tag{11.17}$$

By a similar reasoning,

$$\begin{aligned}\Delta_{22} + \Delta_{33} - 2\Delta_{23} &= W_{2,0} + W_{3,0} - 2W_{23,0} \\ &= W_{23,0} + W_{2,30} + W_{23,0} + W_{3,20} - 2W_{23,0} \\ &= W_{2,30} + W_{3,20} \\ &= W_{2,3}.\end{aligned} \tag{11.18}$$

So we can write Z_{oc} as

$$Z_{oc} = \frac{1}{W}\begin{bmatrix} W_{1,0} & W_{12,30} - W_{13,20} \\ W_{12,30} - W_{13,20} & W_{2,3} \end{bmatrix}.$$

So the driving point resistance z_{11} across port 1 is given by

$$z_{1,0} = \frac{W_{1,0}}{W}$$

Similarly, the driving point resistance z_{22} across 2 and 3 is given by $W_{2,3}/W$. In general, the driving port resistance across any pair of nodes i and j is given by $W_{i,j}/W$. We shall denote by r_{ij} the driving port resistance across any pair of vertices i and j so that

$$r_{ij} = \frac{W_{i,j}}{W} \tag{11.19}$$

r_{ij} is also called the effective resistance across i and j.

We wish to emphasize that the formulas for z'_{ij}s in (11.11) are with respect to vertex 0 as reference. On the other hand the formula in (11.19) does not explicitly involve the reference vertex.

We conclude this section with the following facts that will be needed in the following sections. We assume that the vertices are labeled as $1, 2, \ldots, n$.

1. The degree matrix $K = [k_{ij}]$ of a simple undirected graph $G = (V, E)$ is defined as

$$k_{ii} = d(v_i), \quad \text{for all } i \in V$$
$$k_{ij} = -1, \quad \text{if } (i, j) \in E$$
$$= 0, \quad \text{otherwise}$$

   where $d(v_i)$ is the degree of vertex i.
   Then K can be written as
   $$K = A_c A_c^t$$
   where A_c is the all-vertex incidence matrix of G.

2. Let N be the resistance network N obtained by associating a 1 ohm resistance with each edge of G. Then in electrical engineering literature the matrix K is called the indefinite conductance matrix. In graph theory literature K is also known as the graph Laplacian. Also, if the conductances are defined by g_i, with G as the diagonal matrix of edge conductances, then the graph Laplacian of the corresponding weighted graph will be $A_c G A_c^t$. Here the degree of vertex i is the sum of the conductances incident on i.

3. Let K_{jj} be the matrix obtained by removing the jth row and the jth column from K. Then K_{jj} is the same as the matrix Y defined in (11.8) with vertex j as reference if all the resistances have 1 ohm value.

4. By Theorem 8.21 all cofactors of K are equal to the number of spanning trees of N. In particular
$$\det K_{jj} = W. \tag{11.20}$$

5. (i, i) cofactor of K_{jj} = Number of spanning 2-trees of the type $T_{i,j}$
$$= W_{i,j} \tag{11.21}$$

6. (i, k) cofactor of K_{jj} = Number of spanning 2-trees of the type $T_{ik,j}$
$$= W_{ik,j} \tag{11.22}$$

7. The effective resistance r_{ij} across i and j of N is given by

$$r_{ij} = \frac{(i,i) \text{ Cofactor of } Y}{\text{determinant of } Y}$$
$$= \frac{(i,i) \text{ Cofactor of } K_{jj}}{\text{determinant of } K_{jj}}$$
$$= \frac{(i,i) \text{ Cofactor of } K_{jj}}{W} \tag{11.23}$$
$$= \frac{W_{i,j}}{W}$$

11.4 RANDOM WALKS

A random v-walk on an undirected graph G is a walk that starts at vertex v, at each vertex selects uniformly (i.e., with equal probability) one of the edges incident on v, moves along that edge to the other end vertex of that edge, and repeats these actions at subsequent vertices. Note that the probability that the walk, while at v, selects an edge is equal to $1/(d(v))$ where the $d(v)$ is the degree of v. Random walks arise in several applications. For examples, see [1–3].

Two quantities of interest in the study of random walks are hitting time and commute time. A random v-walk is said to hit a vertex w when it reaches w. The hitting time H_{vw} is the expected number of steps taken by a random v-walk before hitting w. The commute time C_{vw} between v and w is defined by

$$C_{vw} = H_{vw} + H_{wv} \tag{11.24}$$

In this section, we establish certain connections between random walks and electrical resistance networks.

Theorem 11.1 *Let $G = (V, E)$ be a simple connected undirected graph with its vertices labeled as $1, 2, \ldots, n$. Consider any two distinct vertices say 1 and n. Let p_1 denote the probability that a random 1-walk on G hits n before returning to 1.*

Then

$$p_1 = \frac{1}{d(1)r_{1,n}}$$

Proof. Let p_x, for $x \neq 1$, be the probability that a random x-walk on G hits n before hitting 1. Clearly

$$p_n = 1 \tag{11.25}$$

and

$$p_i = \frac{1}{d(i)} \sum_{\substack{(i,j) \in E \\ j \neq 1}} p_j, \qquad i \neq n \tag{11.26}$$

(11.26) can be rewritten as

$$d(i)p(i) - \sum_{\substack{(i,j) \in E \\ j \neq 1}} p_j = 0, \qquad i \neq n \tag{11.27}$$

Then (11.25) and (11.27) can be written in matrix form as

$$K'P = \begin{bmatrix} 0 \\ 0 \\ \vdots \\ 0 \\ 1 \end{bmatrix}$$

where:

K' is the matrix obtained from the degree matrix K by setting to zero all nondiagonal elements of the first column and replacing the nth row by the row vector $[0 \quad 0 \quad \cdots \quad 0 \quad 1]$.

P is the column vector of probabilities $p_1, p_2, \ldots, p_n$.

The matrix K' will appear as

$$\begin{bmatrix} d(1) & \times & \times & \times & \times \\ 0 & & & & \times \\ 0 & & M & & \times \\ \vdots & & & & \vdots \\ 0 & 0 & 0 & 0 & 1 \end{bmatrix}$$

where M is the matrix obtained from the degree matrix K by removing its first and nth rows and columns. Also the conductance matrix (with n as reference) is given by $Y = K_{nn}$ and M is the same as the matrix obtained from Y by removing its first row and first column.

Then using Cramer's rule we can solve (11.28) for p_1 as

$$\begin{aligned} p_1 &= \frac{(n,1) \text{ cofactor of } K'}{\text{determinant of } K'}, \\ &= \frac{(n,1) \text{ cofactor of } K'}{d(1) \det M} \\ &= \frac{W}{d(1)(1,1) \text{ cofactor of } K_{nn}}, \qquad \text{from (11.20)} \\ &= \frac{W}{d(1)(1,1) \text{ cofactor of } Y} \\ &= \frac{W}{d(1)W_{1,n}} \\ &= \frac{1}{d(1)} \times \frac{1}{r_{1,n}}, \qquad \text{from (11.23)} \quad \blacksquare \end{aligned}$$

To illustrate the definitions of matrices used in the proof of Theorem 11.1, consider the undirected graph G shown in Figure 11.5. Here $n = 5$.

For G

$$K = \begin{bmatrix} 2 & -1 & -1 & 0 & 0 \\ -1 & 3 & -1 & -1 & 0 \\ -1 & -1 & 4 & -1 & -1 \\ 0 & -1 & -1 & 3 & -1 \\ 0 & 0 & -1 & -1 & 2 \end{bmatrix}$$

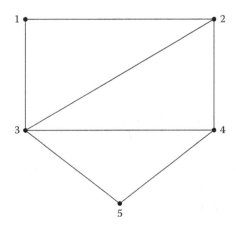

Figure 11.5 Graph G to illustrate the proof of Theorem 11.1.

$$K' = \begin{bmatrix} 2 & -1 & -1 & 0 & 0 \\ 0 & 3 & -1 & -1 & 0 \\ 0 & -1 & 4 & -1 & -1 \\ 0 & -1 & -1 & 3 & -1 \\ 0 & 0 & 0 & 0 & 1 \end{bmatrix}$$

$$M = \begin{bmatrix} 3 & -1 & -1 \\ -1 & 4 & -1 \\ -1 & -1 & 3 \end{bmatrix}, \quad Y = \begin{bmatrix} 2 & -1 & -1 & 0 \\ -1 & 3 & -1 & -1 \\ -1 & -1 & 4 & -1 \\ 0 & -1 & -1 & 3 \end{bmatrix}$$

$$W = \text{determinant of } Y$$
$$= 21$$
$$(1,1) \text{ cofactor of } Y = \det M$$
$$= 24$$

So
$$r_{1,n} = \frac{24}{21}$$

and
$$p_1 = \frac{1}{2} \times \frac{21}{24} = \frac{21}{48} = \frac{7}{16}.$$

Theorem 11.2 *In a simple connected undirected graph $G(V, E)$ the commute time between any two distinct vertices x and y is given by*

$$C_{xy} = 2mr_{x,y}$$

where m is the number of edges in G.

Proof. We assume that the vertices of G are labeled as $1, 2, \ldots, n$. For ease in presentation we take x and y as the first vertex and nth vertex, respectively. It is easy to see that the hitting times H_{ij} satisfy the following system of equations.

$$H_{in} = \sum_{\substack{(i,j) \in E \\ j \neq n}} \frac{1}{d(i)} (1 + H_{jn}) \qquad i \neq n \qquad (11.28)$$

This system of equations can be written as

$$d(i) H_{in} - \sum_{\substack{(i,j) \in E \\ j \neq n}} H_{jn} = d(i) \qquad i \neq n \qquad (11.29)$$

Equation 11.29 in matrix form becomes

$$K_{nn} H = D \qquad (11.30)$$

where K_{nn} is the matrix obtained from the degree matrix by removing its nth row and column, H is the column vector of $H_{1n}, H_{2n}, \ldots, H_{(n-1)n}$ and D is the column vector of degrees $d(1), d(2), \ldots, d(n-1)$.

Suppose we take the network N of 1 ohm resistances derived from G and inject a current of value $d(i)$ at each vertex $i \neq n$. If we let V_i denote the voltage from vertex i to vertex n, then the node equations of this network (see 11.7) will be

$$I = YV \qquad (11.31)$$

where
$$I = \begin{bmatrix} d(1) \\ d(2) \\ \vdots \\ d(n-1) \end{bmatrix} \quad \text{and} \quad V = \begin{bmatrix} V_1 \\ V_2 \\ \vdots \\ V_{(n-1)} \end{bmatrix}$$

and Y is the node conductance matrix with vertex n as the reference. Since Y is the same as K_{nn}, we can see that H and V both satisfy the same set of equations. Thus

$$V_i = H_{i,n} \quad \text{for all } i \neq n.$$

Solving (11.31) for V_1 we get

$$V_1 = \frac{1}{\Delta} \sum_{i=1}^{n-1} d(i) \Delta_{i,1}$$

So

$$\begin{aligned} H_{1n} &= \frac{1}{\Delta} \sum_{i=1}^{n-1} d(i) \Delta_{i,1} \\ &= \frac{1}{\Delta} \sum_{i=1}^{n-1} d(i) W_{i1,n} \end{aligned} \quad (11.32)$$

Similarly

$$H_{n1} = \frac{1}{\Delta} \sum_{i=2}^{n} d(i) W_{in,1}$$

So

$$\begin{aligned} C_{1n} &= H_{1n} + H_{n1}, \\ &= \frac{1}{\Delta}(d(1) W_{1,n} + d(n) W_{n,1}) + \frac{1}{\Delta} \sum_{i=2}^{n-1} (d(i) W_{i1,n} + d(i) W_{in,1}) \end{aligned}$$

Since $W_{1,n} = W_{i1,n} + W_{in,1}$
we get

$$\begin{aligned} C_{1n} &= \frac{W_{1,n}}{\Delta} \sum_{i=1}^{n} d(i) \\ &= 2m \frac{W_{1,n}}{\Delta} \\ &= 2m \frac{W_{1,n}}{W} \\ &= 2m r_{1,n} \end{aligned} \quad \blacksquare$$

If Y is the node conductance matrix with vertex n as reference then it follows from (11.32) that the hitting times $H_{k,n}$ for all $k \neq n$ can be obtained as

$$H_{k,n} = \frac{1}{\Delta} \sum_{i=1}^{n-1} d(i) \Delta_{i,k}$$

Note that $\Delta_{i,k}$ is the (i,k) cofactor of Y and is equal to $W_{ik,n}$.

In other words, one can get H'_{in}s for all $i \neq n$, from the elements of Y and the vertex degrees. H_{kn} is also called the first passage time taken by a random k-walk to hit n.

The cover time of G is defined as

$$C = \max\{C_v \in V\}$$

where C_v is the expected number of steps taken by a random v-walk to hit every vertex of G. Aleliunas et al. [4] have obtained a bound for C given by

$$C \leq 2m(n-1) \tag{11.33}$$

11.5 KIRCHHOFF INDEX OF A GRAPH AND THE GRAPH LAPLACIAN

Kirchhoff Index $Kf(G)$ of a connected undirected graph G is defined as

$$Kf(G) = \sum_{i<j} r_{ij} \tag{11.34}$$

Thus $Kf(G)$ is the sum of the effective resistances across all pairs of vertices of the 1-ohm resistance network obtained from G. This concept was introduced by Klein and Randić [5]. The problem of computing $Kf(G)$ has been studied extensively using what is known as the pseudoinverse of the degree matrix (Laplacian matrix) K defined in Section 11.3.

The Moore–Penrose pseudoinverse of Laplacian matrix $K(G)$ is denoted by $K^+(G)$ and has the following basic properties

1. $K(G)K^+(G)K(G) = K(G)$
2. $K^+(G)K(G)K^+(G) = K^+(G)$
3. $[K(G)K^+(G)]^t = K(G)K^+(G)$
4. $[K^+(G)K(G)]^t = K^+(G)K(G)$

The Moore–Penrose pseudoinverse $K^+(G)$ can be computed as follows [6]:

$$K^+(G) = \left(K(G) + \frac{J}{n}\right)^{-1} - \frac{J}{n} \tag{11.35}$$

where $J \in R^{n \times n}$ is a matrix of all 1's and n is the number of vertices of graph G.

The following properties of the Moore–Penrose pseudoinverse of the Laplacian matrix were established by several authors [6].

Lemma 11.1 *The Moore–Penrose pseudoinverse $K^+(G)$ of the Laplacian matrix $K(G)$ of a connected graph is a real and symmetric matrix.* ∎

Lemma 11.2 *The Laplacian matrix and its pseudoinverse satisfy the following relations*

$$K(G)J = JK(G) = 0$$
$$K^+(G)J = JK^+(G) = 0$$
∎

Lemma 11.3 *If $K(G)$ and $K^+(G)$ pertain to a connected graph G on n vertices, then*

$$K(G)K^+(G) = K^+(G)K(G) = I - \frac{J}{n}$$
∎

Theorem 11.3 *If G is a connected graph, then the inverse of the matrix $K(G) + J/n$ exists and is equal to $K^+(G) + J/n$.*

Proof. Using Lemmas 11.2 and 11.3, and the fact that $J^2 = nJ$, we have

$$\left(K(G) + \frac{J}{n}\right)\left(K^+(G) + \frac{J}{n}\right) = K(G)K^+(G) + \frac{J}{n}K^+(G) + \frac{1}{n}K(G)J + \frac{1}{n^2}J^2$$

$$= \left(I - \frac{J}{n}\right) + O + O + \frac{J}{n} = I. \blacksquare$$

It was proved by Klein and Randić [5] that the Kirchhoff Index can also be written as

$$Kf(G) = n\operatorname{tr}(K^+(G)) \tag{11.36}$$

where n is the number of vertices and $\operatorname{tr}(K^+(G))$ denotes the trace function which can be calculated by

$$\operatorname{tr}(K^+(G)) = \sum_{i=1}^{n} k_{ii}^+$$

Gutman and Mohar [7] demonstrated that it is possible to calculate the Kirchhoff Index without knowing the Moore–Penrose pseudoinverse of the Laplacian matrix. They obtained the Kirchhoff Index from the eigenvalues of the Laplacian matrix of a graph G as

$$Kf(G) = n\sum_{i=1}^{n-1}\frac{1}{\mu_i} \tag{11.37}$$

where μ_i's are the nonzero eigenvalues of the Laplacian matrix $K(G)$.

To avoid the computational efforts required to calculate the Moore–Penrose pseudoinverse of the Laplacian matrix, we next present a new formula for $Kf(G)$.

11.5.1 Formula for the Kirchhoff Index

Let K be the Laplacian matrix of a connected graph G and as before $K(\bar{i})$ be a submatrix obtained by deleting ith row and ith column of the Laplacian matrix K. Note that $K(\bar{i})$ is the same as the node conductance Y, if vertex i is chosen as reference.

Let Z be the inverse of $K(\bar{i})$, that is,

$$Z = K(\bar{i})^{-1}$$

See [8] for a proof of the following theorem.

Theorem 11.4 *Let K be the Laplacian matrix of a connected graph G with n vertices. Then*

$$K^+ = \frac{e^T Z e}{n^2} J + \left[\begin{array}{c|c} Z - (1/n)ZJ - (1/n)JZ & -(1/n)Ze \\ \hline -(1/n)e^T Z & O \end{array}\right] \tag{11.38}$$

where e is the left and right null vector to any Laplacian matrix and matrix Z is the inverse of a reduced Laplacian matrix obtained by deleting the last (nth) row and the last (nth) column, that is, $Z = K(\bar{n})^{-1} = Y^{-1}$. The new formula for computing Kirchhoff Index is given in the following theorem. $\blacksquare$

Theorem 11.5

$$Kf(G) = n\,Tr(Z) - \sum_{k,l} z_{kl} \qquad (11.39)$$

where Z is the inverse of the Laplacian matrix obtained by deleting any ith row and ith column, and $\sum_{k,l} z_{kl}$ is the sum of all the elements of matrix Z (note $Z = Y^{-1}$).

Proof. Using Equation 11.38 we can calculate the (i,j)th entry of pseudoinverse K^+ of the Laplacian matrix K in terms of the elements of the matrix Z.

$$k_{ij}^+ = \begin{cases} \frac{\sum_{k,l} z_{kl}}{n^2} + z_{ij} - \frac{1}{n}\sum_k z_{kj} - \frac{1}{n}\sum_l z_{il}, & i \neq n, j \neq n \\ \frac{\sum_{k,l} z_{kl}}{n^2} - \frac{1}{n}\sum_k z_{kj}, & i = n, j \neq n \\ \frac{\sum_{k,l} z_{kl}}{n^2} - \frac{1}{n}\sum_l z_{il}, & i \neq n, j = n \\ \frac{\sum_{k,l} z_{kl}}{n^2}, & i = n, j = n \end{cases} \qquad (11.40)$$

where:

$\sum_{k,l} z_{kl}$ is the sum of all the elements of the matrix Z

$\sum_k z_{kj}$ is the sum of the elements of the kth row of the matrix Z

$\sum_l z_{il}$ is the sum of the elements of the lth column of the matrix Z

Now using Equations 11.36 and 11.38, we get

$$Kf(G) = n\,\mathrm{Tr}(K^+) = n \sum_{i=1}^{n-1} (k_{ii}^+ + k_{nn}^+) \qquad (11.41)$$

The trace of the pseudoinverse K^+ of the Laplacian matrix satisfies

$$Tr(K^+) = \sum_{i=1}^{n-1} k_{ii}^+ + k_{nn}^+ \qquad (11.42)$$

From (11.40) we get

$$k_{ii}^+ = \frac{\sum_{k,l} z_{kl}}{n^2} + z_{ii} - \frac{2}{n}\sum_l z_{il} \qquad (11.43)$$

$$k_{nn}^+ = \frac{\sum_{k,l} z_{kl}}{n^2} \qquad (11.44)$$

Now using (11.42) through (11.44), we get

$$Tr(K^+) = \sum_{i=1}^{n-1} \left(\frac{\sum_{k,l} z_{kl}}{n^2} + z_{ii} - \frac{2}{n}\sum_l z_{il} \right) + \frac{\sum_{k,l} z_{kl}}{n^2} \qquad (11.45)$$

Note: $\sum_{i=1}^{n-1} \left(\sum_l z_{il} \right) = \sum_{k,l} z_{kl}$ (sum of all elements of matrix Z)
Thus,

$$Tr(K^+) = \frac{1}{n^2}(n-1)\sum_{k,l} z_{kl} + \sum_{i=1}^{n-1} z_{ii} - \frac{2}{n}\sum_{k,l} z_{kl} + \frac{\sum_{k,l} z_{kl}}{n^2}$$

After simplification we get

$$Tr(K^+) = \sum_{i=1}^{n-1} z_{ii} - \frac{\sum_{k,l} z_{kl}}{n} \qquad (11.46)$$

From (11.41) and (11.46), we get

$$Kf(G) = n \sum_{i=1}^{n-1} z_{ii} - \sum_{k,l} z_{kl} \qquad (11.47)$$

We know

$$\sum_{i=1}^{n-1} z_{ii} = Tr(Z) \qquad (11.48)$$

The required result follows from (11.41) and (11.42) as

$$Kf(G) = nTr(Z) - \sum_{k,l} z_{kl}$$

∎

The following example demonstrates the calculation of the Kirchhoff Index by first using the Moore–Penrose pseudoinverse and then by using our new formula.

Kirchhoff Index Using Moore–Penrose Pseudoinverse:

As an example, Figure 11.6 shows a graph G with six nodes and its Laplacian matrix.

First we find the Moore–Penrose pseudoinverse of Laplacian matrix K given in Figure 11.6b by using formula (11.35)

$$K^+ = \left(\begin{bmatrix} 2 & -1 & 0 & -1 & 0 & 0 \\ -1 & 2 & -1 & 0 & 0 & 0 \\ 0 & -1 & 3 & -1 & -1 & 0 \\ -1 & 0 & -1 & 3 & -1 & 0 \\ 0 & 0 & -1 & -1 & 3 & -1 \\ 0 & 0 & 0 & 0 & -1 & 1 \end{bmatrix} + \frac{1}{6} \begin{bmatrix} 1 & 1 & 1 & 1 & 1 & 1 \\ 1 & 1 & 1 & 1 & 1 & 1 \\ 1 & 1 & 1 & 1 & 1 & 1 \\ 1 & 1 & 1 & 1 & 1 & 1 \\ 1 & 1 & 1 & 1 & 1 & 1 \\ 1 & 1 & 1 & 1 & 1 & 1 \end{bmatrix} \right)^{-1}$$

$$-\frac{1}{6} \begin{bmatrix} 1 & 1 & 1 & 1 & 1 & 1 \\ 1 & 1 & 1 & 1 & 1 & 1 \\ 1 & 1 & 1 & 1 & 1 & 1 \\ 1 & 1 & 1 & 1 & 1 & 1 \\ 1 & 1 & 1 & 1 & 1 & 1 \\ 1 & 1 & 1 & 1 & 1 & 1 \end{bmatrix}$$

$$K^+ = \begin{bmatrix} 0.487 & 0.123 & -0.074 & 0.017 & -0.195 & -0.362 \\ 0.123 & 0.487 & 0.017 & -0.074 & -0.195 & -0.362 \\ -0.074 & 0.017 & 0.275 & 0.002 & -0.028 & -0.195 \\ 0.017 & -0.074 & 0.002 & 0.275 & -0.028 & -0.195 \\ -0.195 & -0.195 & -0.028 & -0.028 & 0.305 & 0.138 \\ -0.362 & -0.362 & -0.195 & -0.195 & 0.138 & 0.972 \end{bmatrix}$$

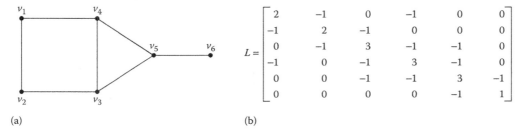

Figure 11.6 (a) Graph G with six nodes. (b) Laplacian matrix of graph G.

The trace of Moore–Penrose pseudoinverse is

$$Tr(K^+) = \sum_{i=1}^{n} k_{ii} = 2.801$$

Now using (11.36) we can calculate Kirchhoff Index $Kf(G)$ as

$$Kf(G) = 6*2.801 = 16.8$$

Let us next calculate Kirchhoff Index $Kf(G)$ by using matrix Z (i.e., $Z = K(\overline{n})^{-1}$). The matrix Z of graph G for Laplacian matrix K in Figure 11.6 is

$$Z = \begin{bmatrix} 2.182 & 1.818 & 1.455 & 1.545 & 1 \\ 1.818 & 2.182 & 1.545 & 1.455 & 1 \\ 1.455 & 1.545 & 1.636 & 1.364 & 1 \\ 1.545 & 1.455 & 1.364 & 1.636 & 1 \\ 1 & 1 & 1 & 1 & 1 \end{bmatrix}$$

In order to find the Kirchhoff Index $Kf(G)$, we calculate the trace of matrix Z and the sum of all the elements of matrix Z.

$$Tr(Z) = 8.63$$

$$\sum_{k,l} z_{kl} = 35$$

Using (11.39), the Kirchhoff Index $Kf(G)$ is

$$Kf(G) = 6*8.63 - 35 = 16.8$$

11.5.2 Kirchhoff Index Using Topological Formulas for Network Functions

As we have seen before

$$r_{ij} = \frac{W_{i,j}}{W}$$

But

$$W_{ij} = W_{i,nj} + W_{in,j}$$
$$= \{W_{i,n} - W_{ij,n}\} + \{W_{j,n} - W_{ij,n}\}$$
$$= W_{i,n} + W_{j,n} - 2W_{ij,n}$$

Dividing by W both sides of the above equation we get

$$\frac{W_{ij}}{W} = \frac{W_{i,n}}{W} + \frac{W_{j,n}}{W} - \frac{2W_{ij,n}}{W}$$

$$r_{i,j} = r_{i,n} + r_{j,n} - 2z_{ij}$$

Since each $r_{j,n}$ appears $(n-1)$ times on the right-hand side of the sum $\sum_{i,k>i} r_{i,k}$ we get

$$\sum_{i,k>i} r_{i,k} = (n-1) \sum_{j=1}^{n-1} r_{j,n} - 2 \sum_{i,k>i} z_{ik}$$

$$= (n-1) \sum_{j=1}^{n-1} r_{j,n} + \sum_{j=1}^{n-1} r_{j,n} - \left(\sum_{j=1}^{n-1} r_{j,n} + 2 \sum_{i,k>i} z_{ik} \right)$$

$$Kf(G) = n \sum_{j=1}^{n-1} r_{j,n} - \left(\sum_{j=1}^{n-1} r_{j,n} + 2 \sum_{i,k>i} z_{ik} \right)$$

The above is the same as

$$Kf(G) = n\sum_{i=1}^{n-1} z_{ii} - \left(\sum_{i,l} z_{il}\right)$$

11.6 FOSTER'S THEOREMS

Consider a resistance N. Let N have n nodes and m elements $e_1, e_2, \ldots, e_m$. The resistance and conductance of each e_i will be denoted by z_i and $y_i(=\frac{1}{z_i})$, respectively. Also, the two nodes of each e_i will be denoted by i_1 and i_2. If r_{i_1,i_2} denotes the effective resistance of N across the pair of nodes i_1 and i_2, then we have the following theorem due to Foster [9].

Theorem 11.6 (Foster's first theorem)

$$\sum_{i=1}^{m} y_i r_{i_1,i_2} = n - 1. \qquad (11.49)$$

Proof. Let T denote the set of all the spanning trees of N and, for each i, let T_i denote the set of all the spanning 2-trees of N separating the nodes i_1 and i_2. That is, T_i is the set of all the spanning trees of type T_{i_1,i_2}. Note that adding e_i to a spanning 2-tree separating i_1 and i_2 will generate a spanning tree. Further, let $w(t)$ denote the conductance product of spanning tree t and $w(t_i)$ denote the conductance product of a spanning 2-tree t_i separating i_1 and i_2. It is easy to see that if $t = t_i \cup e_i$ then

$$w(t) = y_i w(t_i).$$

If

$$W(T) = \sum_{t \in T} w(t)$$

and

$$W(T_i) = \sum_{t_i \in T_i} w(t_i)$$

then it is known (see 11.23) that

$$r_{i_1,i_2} = \frac{W(T_i)}{W(T)}.$$

Thus to prove the theorem, we need to show that

$$\sum_{i=1}^{m} y_i W(T_i) = (n-1)W(T) \qquad (11.50)$$

or

$$\sum_{i=1}^{m} y_i \sum_{t_i \in T_i} w(t_i) = (n-1)\sum_{t \in T} w(t).$$

Consider any tree conductance product $w(t)$. We may assume, without loss of generality, that the spanning tree t contains the elements $e_1, e_2, \ldots, e_{n-1}$. Then for every $i = 1, 2, \ldots, n-1$, $t - e_i$ is a spanning 2-tree t_i separating the nodes i_1 and i_2. So for every $i = 1, 2, \ldots, n-1$

$$w(t) = y_i w(t_i)$$

for some spanning 2-tree t_i. Thus, the conductance product $w(t)$ appears exactly once in each $y_i w(T_i), i = 1, 2, \ldots, n-1$. In other words, each $w(t)$ appears $(n-1)$ times in the sum

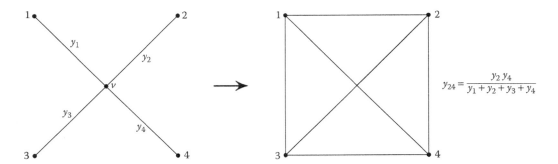

Figure 11.7 Star-delta transformation.

on the left-hand side of (11.50). The theorem follows since each $y_i w(t_i)$ corresponds to a unique $w(t)$. ∎

Next we state and prove Foster's second theorem. This theorem is based on the operation of star-delta transformation which we define as follows.

Consider a vertex v. Let $y_1, \ldots, y_k$ be the conductances of the edges incident on v, with $1, 2, \ldots, k$ denoting the other end vertices of these edges. Star-delta transformation at v is the operation of removing vertex v from N and adding a new element (i, j) with conductance $y_i y_j / d(v)$ for all $k \leq i, j \leq k$ (see Figure 11.7).

The following theorem is by Foster [10].

Theorem 11.7 (Foster's second theorem) *Consider a resistance network N. For any pair of conductances y_i and y_j incident on common vertex v, let r_{ij} denote the effective resistance across the two remaining vertices of y_i and y_j. Let $d(v)$ be the sum of the conductances of the elements incident on v. Then*

$$\sum r_{ij} y_{ij} = \sum r_{ij} \frac{y_i y_j}{d(v)} = n - 2 \tag{11.51}$$

where the sum is extended over all pairs of adjacent elements (i.e., elements incident on a common vertex v).

Proof. Consider any vertex v in N. Star-delta transformation at v results in a network N' with $n - 1$ vertices. Applying Foster's First theorem to N' we get

$$\sum_v r_{ij} y_{ij} + \sum z_k y_k = n - 2 \tag{11.52}$$

Here the first summation is over all pairs of elements of N' which reflect the original star of conductances incident on vertex v. The second summation is over all conductances of N that are not connected to v.

Note that y_k is a conductance and z_k is the effective resistance across the vertices of this conductance.

Summing (11.52) over all the n vertices in N, we get

$$\sum_v r_{ij} y_{ij} + \sum \sum z_k y_k = n(n-2)$$

The first sum is over all adjacent pairs of conductances on a common vertex v in N. The second sum is

$$\sum \sum z_k y_k = (n-2) \sum z_i y_i \tag{11.53}$$

because conductance y_k appears exactly $n-2$ times in the double summation. So

$$\sum r_{ij}y_{ij} = n(n-2) - (n-2)\sum z_i y_i$$
$$= n(n-2) - (n-2)(n-1), \quad \text{applying Foster's First theorem}$$
$$= (n-2)$$

This completes the proof. ∎

11.7 ARC-COLORING THEOREM AND THE NO-GAIN PROPERTY

We now derive a profound result in graph theory, the arc-coloring theorem for directed graphs, and discuss its application in establishing the no-gain property of resistance networks. In the special case of undirected graphs the arc-coloring theorem reduces to the *painting* theorem. Both of these theorems by Minty [11] are based on the notion of *painting a graph*. Other works of Minty relating to this are [12] and [13].

Given an undirected graph with edge set E, a painting of the graph is a partitioning of E into three subsets, R, G, and B, such that $|G| = 1$. We may consider the edges in the set R as being *painted red*, the edge in G as being *painted green* and the edges in B as being *painted blue*.

Theorem 11.8 (Painting theorem) *For any painting of a graph, there exists a circuit C consisting of the green edge and no blue edges, or a cutset C^* consisting of the green edge and no red edges.*

Proof. Consider a painting of the edge set E of a graph G. Assuming that there does not exist a required circuit, we shall establish the existence of a required cutset.

Let $E' = R \cup G$ and T' denote a spanning forest of the subgraph induced by E', containing the maximum number of red edges. (Note that the subgraph induced by E' may not be connected.) Then construct a spanning tree T of G such that $T' \subseteq T$.

Now consider any red edge y which is not in T'. y will not be in T for otherwise it would contradict the property that T' has the maximum number of red edges. Because the fundamental circuit of y with respect to T is the same as the fundamental circuit of y with respect to T', this circuit consists of no blue edges. Furthermore, this circuit will not contain the green edge, for otherwise a circuit consisting of the green edge and no blue edges would exist contrary to our assumption. Thus, the fundamental circuit of a red edge with respect to T does not contain the green edge. Then it follows from Theorem 7.9 that the fundamental cutset of the green edge with respect to T contains no red edges. Thus, this cutset satisfies the requirements of the theorem. ∎

A painting of a directed graph with edge set E is a partitioning of E into three sets R, G, and B, and the distinguishing of one element of the set G. Again, we may regard the edges of the graph as being colored red, green, or blue with exactly one edge of G being colored dark green. Note that the dark green edge is also to be treated as a green edge.

Next we state and prove Minty's arc-coloring theorem.

Theorem 11.9 (Arc-coloring theorem) *For any painting of a directed graph exactly one of the following is true:*

1. *A circuit exists containing the dark green edge, but no blue edges, in which all the green edges are similarly oriented.*

2. *A cutset exists containing the dark green edge, but no red edges, in which all the green edges are similarly oriented.*

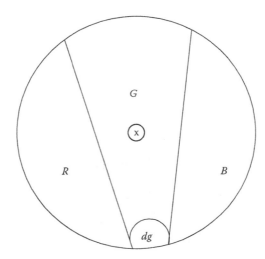

Figure 11.8 Painting of a directed graph.

Proof. Proof is by induction on the number of green edges. If only one green edge exists, then the result will follow from Theorem 11.8. Assume then that result is true when the number of green edges is $m \geq 1$. Consider a painting in which $m+1$ edges are colored green. Pick a green edge x other than the dark green edge (see Figure 11.8). Color the edge x red. In the resulting painting there are m green edges. If a cutset of type 2 is now found, then the theorem is proved. On the other hand if we color the edge x blue and in the resulting painting a circuit of type 1 exists, then the theorem is proved.

Suppose neither occurs. Then, using the induction hypothesis we have the following:

1. A cutset of type 2 exists when x is colored blue.

2. A circuit of type 1 exists when x is colored red.

Now let the corresponding rows of the circuit and cutset matrices be

	dg	R	B	G	x
Cutset	+1	0 0 . . . 0 0	1 −1 . . . 0 1	1 1 1 . . . 0	?
Circuit	+1	−1 1 . . . 0 −1	0 0 . . . 0 0	0 1 1 . . . 0	?

Here we have assumed, without loss of generality, that $+1$ appears in the dark green position of both rows.

By the orthogonality relation (Theorem 8.7) the inner product of these two row vectors is zero. No contribution is made to this inner product from the red edges or from the blue edges. The contribution from the green edges is a nonnegative integer p. The dark green edge contributes 1 and the edge x contributes an unknown integer q which is 0, 1, or -1. Thus we have $1 + p + q = 0$. This equation is satisfied only for $p = 0$ and $q = -1$. Therefore, in one of the rows, the question mark is $+1$ and in the other it is -1. The row in which the question mark is 1 corresponds to the required circuit or cutset. Thus, either statement 1 or 2 of the theorem occurs. Both cannot occur simultaneously because the inner product of the corresponding circuit and cutset vectors will then be nonzero. ∎

Theorem 11.10 *Each edge of a directed graph belongs to either a directed circuit or to a directed cutset but no edge belongs to both. (Note: A cutset is a directed cutset if all its edges are similarly oriented.)*

Proof. Proof will follow if we apply the arc-coloring theorem to a painting in which all the edges are colored green and the given edge is colored dark green. ∎

We next present an application of the arc-coloring theorem in the study of electrical networks. We prove what is known as the *no-gain property* of resistance networks. Our proof is the result of the work of Wolaver [14] and is purely graph theoretic in nature.

Theorem 11.11 *In a network of sources and (liner/nonliner) positive resistances the magnitude of the current through any resistance with nonzero voltage is not greater than the sum of the magnitudes of the currents through the sources.*

Proof. Let us eliminate all the elements with zero voltage by considering them to be short-circuits and then assign element reference directions so that all element voltages are positive.

Consider a resistance with nonzero voltage. Then, no directed circuit can contain this resistance, for if such a directed circuit were present, the sum of all the voltages in the circuit would be nonzero, contrary to Kirchhoff's voltage law. It then follows from Theorem 11.10 that a directed cutset contains the resistance under consideration.

Pick a directed cutset that contains the considered resistance. Let the current through this resistance be i_o. Let R be the set of all other resistances in this cutset and let S be the set of all sources. Then, applying Kirchhoff's current law to the cutset, we obtain

$$i_o + \sum_{k \in R} i_k + \sum_{s \in S} \pm i_s = 0 \tag{11.54}$$

Because all the resistances and voltages are positive, every resistance current is positive. Therefore, we can write the above equation as

$$|i_o| + \sum_{k \in R} |i_k| + \sum_{s \in S} \pm i_s = 0 \tag{11.55}$$

and so

$$|i_o| \leq \sum_{s \in S} \mp i_s \leq \sum_{s \in S} |i_s| \tag{11.56}$$

Thus follows the theorem. ∎

The following result is the dual of the above theorem. Proof of this theorem follows in an exactly dual manner, if we replace current with voltage, voltage with current, and circuit with cutset in the proof of the above theorem.

Theorem 11.12 *In a network of sources and (linear/nonlinear) positive resistances, the magnitude of the voltage across any resistance is not greater than the sum of the voltages across all the sources.* ∎

Schwartz [15] and Talbot [16] are early papers that have discussed the *no-gain* properties of resistance networks. Chua and Green [17] used the arc-coloring theorem to establish several properties of nonlinear networks and nonlinear multiport resistive networks.

Summary and Related Works

Swamy and Thulasiraman [18] give a detailed account of the graph theoretic foundation of electrical circuit analysis. Some of the early works on this are Seshu and Reed [19], Balabanian and Bickart [20], Chen [21], Kim and Chien [22], and Mayeda [23]. See also Chapters 7 through 9.

For a discussion of spectral graph theory see Biggs [24], Chung [25], Molitierno [8], and Chapter 10. Resistance distance and Kirchhoff Index of graph have been studied in several works. For example, see [5, 26–29]. For generalization of Foster's theorem and several related results see [30–33]. See also [34,35] for the relationship of Kirchhoff Index to network criticality. References [2,36,40] give a detailed discussion of the connection between electrical resistance networks and random graphs. Bollobas [37], Bondy and Murty [38], and West [39] give a good good introduction to random graphs and random walks. For an extensive treatment of random graphs see Bollobas [41]. Section 11.5 is based on Yadav and Thulasiraman [42].

References

[1] S. Bornholdt and H. G. Schuster, *Handbook of Graphs and Networks: From the Genome to the Internet*, Wiley-VCH, 2003.

[2] P. G. Doyle and J. L. Snell, *Random Walks and Electric Networks*, Mathematical Association of America, 1984.

[3] M. E. J. Newman, *Networks: An Introduction*. Oxford University Press, 2010.

[4] R. Aleliunas, R. M. Karp, R. J. Lipton, L. Lovász, and C. Rackoff, Random walks, universal traversal sequences, and the complexity of maze problems, in *IEEE 20th Annual Symposium on Foundations of Computer Science (San Juan, Puerto Rico)*, New York, 1979, 218–223.

[5] D. J. Klein and M. Randić, Resistance distance, *J. Math. Chem.*, **12** (1993), 81–95.

[6] I. Gutman and W. Xiao, Generalizeed inverse of the laplacian matrix and some applications, *Bull. Acad. Serbe Sci. Arts (Cl. Math. Natur.)*, **129** (2004), 15–23.

[7] I. Gutman and B. Mohar, The Quasi-Wiener and the Kirchhoff indices coincide, *J. Chem. Inf. Comput. Sci.*, **36** (1996), 982–985.

[8] J. J. Molitierno, *Applications of Combinatorial Matrix Theory to Laplacian Matrices of Graphs*, Chapman & Hall/CRC Press, 2012.

[9] K. Thulasiraman, R. Jayakumar, and M. N. S. Swamy, Graph theoretic proof of a network theorem and some consequences, *Proc. IEEE*, **71**, (1983), 771–772.

[10] R. M. Foster, An extension of a network theorem contribution to applied mechanics, *IRE Trans. Circ. Th.*, **8** (1961), 75–76.

[11] G. J. Minty, On the axiomatic foundations of the theories of directed linear graphs, electrical networks and network programming, *J. Math. and Mech.*, **15** (1966), 485–520.

[12] G. J. Minty, Monotone networks, *Proc. Roy. Soc., A*, **257** (1960), 194–212.

[13] G. J. Minty, Solving steady-state non-linear networks of 'Monotone' elements, *IRE Tras. Circuit Theory*, **CT-8** (1961), 99–104.

[14] D. H. Wolaver, Proof in graph theory of the 'No-Gain' property of resistor networks, *IEEE Trans. Circuit Theory*, **CT-17** (1970), 436–437.

[15] R. J. Schwartz, A note on the transfer ratio of resistive networks with positive elements, *Proc. IRE*, **43** (1955), 1670.

[16] A. Talbot, Some fundamental properties of networks without mutual inductance, *Proc. IEE (London).*, **102** (1955), 168–175.

[17] L. O. Chua and D. N. Green, Graph-theoretic properties of dynamic nonlinear networks, *IEEE Trans. Circuits Syst.*, **CAS-23** (1976), 292–312.

[18] M. N. Swamy and K. Thulasiraman, *Graphs, Networks and Algorithms*, Wiley-Interscience, 1981.

[19] S. Seshu and M. B. Reed, *Linear Graphs and Electrical Networks*, Addison-Wesley, Reading, MA, 1961.

[20] N. Balabanian and T. A. Bickart, *Electrical Network Theory*, Wiley, New York, 1969.

[21] W. K. Chen, *Applied Graph Theory*, North-Holland, Amsterdam, the Netherlands, 1971.

[22] W. H. Kim and R. T. Chien, *Topological Analysis and Synthesis of Communication Networks*, Columbia University Press, New York, 1962.

[23] W. Mayeda, *Graph Theory*, Wiley-Interscience, New York, 1970.

[24] N. Biggs, *Algebraic Graph Theory*, Cambridge University Press, Cambridge, 1993.

[25] F. R. K. Chung, *Spectral Graph Theory*, Wiley-Interscience, New York, 1970.

[26] D. Babić, D. J. Klein, I. Lukovits, S. Nikolić, and N. Trinajstić, Resistance-distance matrix: A computational algorithm and its application, *Int., J. Quantum Chem.*, **90**(1) (2002), 166–176.

[27] R. B. Bapat, I. Gutman, and W. Xiao, A simple method for computing resistance distance, *Z. Naturforsch*, **58a** (2003), 494–498.

[28] E. Bendito, A. Carmona, A. M. Encinas, and J. M. Gesto, A formula for the Kirchhoff index. *Int. J. Quantum Chem.*, **108** (2008), 1200–1206.

[29] J. L. Palacios, Closed-form formulas for Kirchhoff index. *Int. J. Quantum Chem.*, **81** (2001), 135–140.

[30] Z. Cinkir, *The Tau Constant of Metrized Graphs*, PhD Thesis, University of Georgia, Department of Mathematics, Athens, GA, 2007.

[31] J. L. Palacios, Foster's formulas via probability and the Kirchhoff index, *Method. Comp. Appl. Probab.*, **6** (2004), 381–387.

[32] P. Tetali, Random walks and effective resistance of networks, *J. Theor. Probab.*, **4** (1991), 101–109.

[33] P. Tetali, An extension of Foster's theorem, *Comb. Probab. Comput.*, **3** (1994), 421–427.

[34] A. Tizghadam and A. Leon-Garcia, A graph theoretical approach to traffic engineering and network control problem, *IEEE 21st Int. Teletraffic Cong.*, Paris, France, September 2009, 15–17.

[35] A. Tizghadam, *Autonomic Core Network Management System*, PhD Thesis, University of Toronto, School of Electrical and Computer Engineering, April 2009.

[36] J. L. Palacios and J. M. Renom, Sum rules for hitting times of markov chains, *Linear Algebra. Appl.*, **433** (2010),491–497.

[37] B. Bollobas, *Modern Graph Theory*, Springer, 1998.

[38] J. A. Bondy and U. S. R. Murty, *Graph Theory*, Springer, 2007.

[39] D. B. West, *Introduction to Graph Theory*, Prentice Hall, 1996.

[40] A. K. Chandra, P. Raghavan, W. L. Ruzzo, R. Smolensky, and P. Tiwari, The electrical resistance of a graph captures its commute and cover times, In *Proceedings of the 21st Annual ACM Symposium on Theory of Computing*, Seattle, WA, 574–586.

[41] B. Bollobas, *Random Graphs*, Cambridge University Press, 2001.

[42] M. Yadav, and K. Thulasiraman, Network science meets circuit theory: Kirchhoff index of a graph and the power of node-to-datum resistance matrix, In *Circuits and Systems, 2015 IEEE International Symposium on,* 854–857, 2015.

IV
Structural Graph Theory

CHAPTER 12

Connectivity

Subramanian Arumugam

Karam Ebadi

CONTENTS

12.1	Introduction	273
12.2	Cut-Vertices, Cut-Edges, and Blocks	273
12.3	Vertex Connectivity and Edge Connectivity	276
12.4	Structural Results	279
12.5	Menger's Theorem and Its Applications	280
12.6	Conditional Connectivity	283
12.7	Criticality and Minimality	287
12.8	Conclusion	288

12.1 INTRODUCTION

The interconnection network of a distributed computer system can be modeled as a graph in which the vertices are the processors and the edges are the communication links. It is generally expected that the system must be able to work even if some of its vertices or edges fail. This property of the system which is known as fault tolerance, requires that the graph must have high connectivity and edge connectivity. Further there are several nice min-max characterizations such as Menger's theorem and these results are closely related to several other key theorems in graph theory such as Ford and Fulkerson's max-flow-min-cut theorem and Hall's theorem on matching. Thus connectivity is one of the central concepts of graph theory both from theoretical and practical point of view. In this chapter we present basic results on these concepts. We limit ourselves to finite simple graphs.

12.2 CUT-VERTICES, CUT-EDGES, AND BLOCKS

We start with a brief review of the basic concepts such as cut-vertex, cut-edge and blocks.

Definition 12.1 *A cut-vertex of a graph G is a vertex whose removal increases the number of components. A cut-edge of a graph G is an edge whose removal increases the number of components.*

Clearly if v is a cut-vertex of a connected graph G, then $G - v$ is disconnected.

For the graph given in Figure 12.1, the vertices 1, 2, and 3 are cut-vertices. The edges $\{1,2\}$ and $\{3,4\}$ are cut-edges. The vertex 5 is a non-cut-vertex.

Theorem 12.1 *Let v be a vertex of a connected graph G. Then the following statements are equivalent.*

1. *The vertex v is a cut-vertex of G.*

Figure 12.1 Cut-vertices and cut-edges.

2. *There exists a partition of $V - \{v\}$ into subsets U and W such that for each $u \in U$ and $w \in W$, the vertex v is on every $u-w$ path.*

3. *There exist two vertices u and w distinct from v such that v is on every $u-w$ path.*

Proof. (1) $\Rightarrow$ (2): Since v is a cut-vertex of G, $G - v$ is disconnected. Hence $G - v$ has at least two components. Let U consist of the vertices of one of the components of $G - v$ and W consist of the vertices of the remaining components. Clearly $V - \{v\} = U \cup W$ is a partition of $V - \{v\}$.

Let $u \in U$ and $w \in W$. Then u and w lie in different components of $G - v$. Hence there is no $u-w$ path in $G - v$. Therefore every $u-w$ path in G contains v.

Trivially (2) $\Rightarrow$ (3).

(3) $\Rightarrow$ (1): Since v is on every $u-w$ path in G there is no $u-w$ path in $G - v$. Hence $G - v$ is not connected so that v is a cut-vertex of G. ∎

Theorem 12.2 *Let x be an edge of a connected graph G. Then the following statements are equivalent.*

1. *The edge x is a cut-edge of G.*

2. *There exists a partition of V into two subsets U and W such that for every vertex $u \in U$ and $w \in W$, the edge x is on every $u-w$ path.*

3. *There exist two vertices u, w such that the edge x is on every $u-w$ path.* ∎

The proof is analogous to that of Theorem 12.1.

Theorem 12.3 *An edge x of a connected graph G is a cut-edge if and only if x is not on any cycle of G.*

Proof. Let x be a cut-edge of G. Suppose x lies on a cycle C of G. Let w_1 and w_2 be any two vertices in G. Since G is connected, there exists a $w_1 - w_2$ path P in G. If x is not on P, then P is a path in $G - x$. If x is on P, replacing x by $C - x$, we obtain a $w_1 - w_2$ walk in $G - x$. This walk contains a $w_1 - w_2$ path in $G - x$. Hence $G - x$ is connected which is a contradiction. Hence x is not on any cycle on G.

Conversely, suppose $x = uv$ is not any cycle of G. Suppose $x = uv$ is not a cut-edge of G. Then $G - x$ is connected. Hence there is a $u - v$ path in $G - x$ and this path together with the edge x forms a cycle containing x, which is a contradiction. Hence x is a cut-edge of G. ∎

Theorem 12.4 *Every nontrivial connected graph G has at least two vertices which are not cut-vertices.*

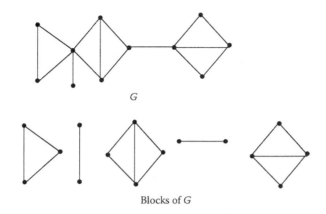

Figure 12.2 Graph G and its blocks.

Proof. Choose two vertices u and v such that $d(u,v)$ is maximum. We claim that u and v are not cut-vertices.

Suppose v is a cut-vertex. Hence $G - v$ has more than one component. Choose a vertex w in a component that does not contain u. Then v lies on every $u-w$ path and hence $d(u,w) > d(u,v)$ which is impossible. Hence v is not a cut-vertex of G. Similarly u is not a cut-vertex of G. ∎

Definition 12.2 *A connected nontrivial graph having no cut-vertex is a block. A block of a graph is a subgraph that is a block and is maximal with respect to this property.*

A graph and its blocks are given in Figure 12.2.

In the following theorem we give several equivalent conditions for a graph to be a block.

Theorem 12.5 *Let G be a connected graph with at least three vertices. Then the following statements are equivalent.*

1. *G is a block.*

2. *Any two vertices of G lie on a cycle.*

3. *Any vertex and any edge of G lie on a cycle.*

4. *Any two edges of G lie on a cycle.*

Proof. (1)⇒(2): Suppose G is a block.

We shall prove by induction on the distance $d(u,v)$ between u and v, that any two vertices u and v lie on a cycle.

Suppose $d(u,v) = 1$. Hence u and v are adjacent. By hypothesis, $G \neq K_2$ and G has no cut-vertices. Hence the edge $x = uv$ is not a cut-edge and hence by Theorem 12.3. x is on a cycle of G. Hence the vertices u and v lie on a cycle of G.

Now assume that the result is true for any two vertices at distance less than k and let $d(u,v) = k \geq 2$. Consider a $u - v$ path of length k. Let w be the vertex that precedes v on this path. Then $d(u,w) = k - 1$. Hence by induction hypothesis there exists a cycle C that contains u and w. Now since G is a block, w is not a cut-vertex of G and so $G-w$ is connected.

Hence there exists a $u-v$ path P not containing w.

Let v' be the last vertex common to P and C (see Figure 12.3). Since u is common to P and C, such a v' exists.

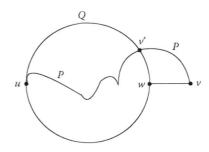

Figure 12.3 Figure for the proof of Theorem 12.5.

Now, let Q denote the $u - v'$ path along the cycle C not containing the vertex w. Then, Q followed by the $v' - v$ path along P the edge vw and the $w-u$ path along the cycle C edge disjoint from Q form a cycle that contains both u and v. This completes the induction.

Thus any two vertices of G lie on a cycle of G.

(2) $\Rightarrow$ (1): Suppose any two vertices of G lie on a cycle of G. Suppose v is a cut-vertex of G. Then there exist two vertices u and w distinct from v such that every u–w path contains v.

Now, by hypothesis u and w lie on a cycle and this cycle determines two u–w paths and at least one of these paths does not contain v which is a contradiction.

Hence G has no cut-vertices so that G is a block.

(2) $\Rightarrow$ (3): Let u be a vertex and vw an edge of G.

By hypothesis u and v lie on a cycle C. If w lies on C, then the edge vw together with the $v - w$ path of C containing u is the required cycle containing u and the edge vw.

If w is not on C, let C' be a cycle containing u and w. This cycle determines two $w - u$ paths and at least one of these paths does not contain v. Denote this path by P.

Let u' be the first vertex common to P and C (u' may be u itself). Then the edge vw followed by the $w - u'$ subpath of P and the $u' - v$ path in C containing u form a cycle containing u and the edge vw.

(3) $\Rightarrow$ (2) is trivial.

(3) $\Rightarrow$ (4): The proof is analogous to the proof of (2) $\Rightarrow$ (3).

(4) $\Rightarrow$ (3) is trivial. ∎

12.3 VERTEX CONNECTIVITY AND EDGE CONNECTIVITY

We define two parameters of a graph, its connectivity and edge connectivity which measure the extent to which it is connected.

Definition 12.3 *The connectivity $\kappa = \kappa(G)$ of a graph G is the minimum number of vertices whose removal results in a disconnected or trivial graph. The edge connectivity $\lambda = \lambda(G)$ of G is the minimum number of edges whose removal results in a disconnected graph.*

Example 12.1

1. The connectivity and edge connectivity of a disconnected graph is 0.

2. The connectivity of a connected graph with a cut-vertex is 1.

3. The edge connectivity of a connected graph with a cut-edge is 1.

4. The complete graph K_n cannot be disconnected by removing any number of vertices, but the removal of $n - 1$ vertices results in a trivial graph. Hence $\kappa(K_n) = n - 1$.

5. The n-dimensional hypercube Q_n is n-regular and it can be proved by induction on n that $\kappa(Q_n) = n$.

Theorem 12.6 *For any graph G, $\kappa \leq \lambda \leq \delta$.*

Proof. We first prove $\lambda \leq \delta$. If G has no edges, $\lambda = \delta = 0$. Otherwise removal of all the edges incident with a vertex of minimum degree results in a disconnected graph. Hence $\lambda \leq \delta$.

Now to prove $\kappa \leq \lambda$, we consider the following cases.

Case 1 G is disconnected or trivial.
In this case $\kappa = \lambda = 0$.

Case 2 G is a connected graph with a cut-edge x.
Then $\lambda = 1$. Further in this case $G = K_2$ or one of the vertices incident with x is a cut-vertex. Hence $\kappa = 1$ so that $\kappa = \lambda = 1$.

Case 3 $\lambda \geq 2$.
Then there exist λ edges, the removal of which disconnects the graph and the removal of $\lambda - 1$ of these edges results in a graph G with a cut-edge $x = uv$. For each of these $\lambda - 1$ edges select an incident vertex different from u or v. The removal of these $\lambda - 1$ vertices removes all the $\lambda - 1$ edges. If the resulting graph is disconnected, then $\kappa \leq \lambda - 1$. If not x is a cut-edge of this subgraph and hence the removal of u or v results in a disconnected or trivial graph. Hence $\kappa \leq \lambda$ and this completes the proof. ■

Remark 12.1 *The inequalities in Theorem 12.6 are often strict. For the graph given in Figure 12.4, $\kappa = 2, \lambda = 3$, and $\delta = 4$. In fact Chartrand and Harary [1] have proved that given three positive integers $a, b,$ and c with $0 < a \leq b \leq c$, there exists a graph G with $\kappa(G) = a, \lambda(G) = b,$ and $\delta(G) = c$.*

Remark 12.2 *There are several families of graphs for which $\lambda = \delta$. For graphs with diameter two and for graphs with $\delta \geq [(n-1)/2]$ we have $\lambda = \delta$. For any cubic graph, we have $\kappa = \lambda$.*

Remark 12.3 *Let G be a connected graph of order n and size m. Since $\delta \leq (2m/n)$, we have $\kappa \leq (2m/n)$.*

Definition 12.4 *A graph G is said to be k-connected if $\kappa(G) \geq k$ and k-edge connected if $\lambda(G) \geq k$.*

Thus a nontrivial graph is 1-connected if and only if it is connected. A nontrivial graph is 2-connected if and only if it is a block having more than one edge. Hence K_2 is the only block which is not 2-connected.

It follows from the Remark 12.3 that if G is a k-connected graph, then $m \geq \lceil kn/2 \rceil$. Harary [2] proved that this bound is best possible by constructing a family of k-connected

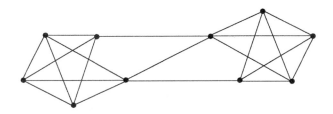

Figure 12.4 Graph with $\kappa = 2, \lambda = 3$, and $\delta = 4$.

graphs on n vertices with $m = \lceil kn/2 \rceil$ for all positive integers k and n with $k < n$. Given $k < n$, place n equally spaced vertices around a circle. If k is even, form $H_{k,n}$ by joining each vertex to the nearest $k/2$ vertices in each direction around the circle. If k is odd and n is even, form $H_{k,n}$ by joining each vertex to the nearest $(k-1)/2$ vertices in each direction and to the diametrically opposite vertex. If k and n are both odd, construct $H_{k,n}$ from $H_{k-1,n}$ by joining i and $i+(1/2)(n-1)$ for $0 \leq i \leq (1/2)(n-1)$, where $V(H_{k,n}) = \{0,1,2,\ldots,(n-1)\}$ and addition is modulo n. The graphs $H_{k,n}$ are called Harary graphs. The graph $H_{k,n}$ is a k-connected graph of order n with $m = \lceil kn/2 \rceil$.

Proposition 12.1 *Let $e = uv$ be any edge of a k-connected graph G and let $k \geq 3$. Then the subgraph $G - e$ is $(k-1)$-connected.*

Proof. Let $W = \{w_1, w_2, \ldots, w_{k-2}\}$ be any set of $k-2$ vertices in $G - e$. We claim that $(G - e) - W$ is connected. Let x and y be any two distinct vertices in $(G-e) - W$.

Suppose u or $v \in W$. Since $G - W$ is 2-connected there is an $x - y$ path P in $G - W$. Clearly P cannot contain e and hence P is an $x - y$ path in $(G - e) - W$. Now suppose that neither u or v is in W. Then there are two cases to consider.

Case 1 $\{x, y\} = \{u, v\}$.
Since G is k-connected, $|V(G)| \geq k+1$. Let $z \in G - \{w_1, w_2, \ldots, w_{k-2}, x, y\}$. Since G is k-connected there exists an $x - z$ path P_1 in $G - \{w_1, w_2, \ldots, w_{k-2}, y\}$ and a $z - y$ path P_2 in $G - \{w_1, w_2, \ldots, w_{k-2}, x\}$. Neither of these paths contain the edge e and therefore their concatenation is an $x - y$ walk in $G - \{w_1, w_2, \ldots, w_{k-2}\}$.

Case 2 $\{x, y\} \neq \{u, v\}$.
Suppose $x \neq u$. Since G is k-connected, $G - \{w_1, w_2, \ldots, w_{k-2}, u\}$ is connected. Hence there is an $x - y$ path P in $G - \{w_1, w_2, \ldots, w_{k-2}, u\}$. Clearly P is an $x - y$ path in $G - \{w_1, w_2, \ldots, w_{k-2}\}$ that does not contain u and hence does not contain the edge e. Thus P is an $x - y$ path in $(G - e) - \{w_1, w_2, \ldots, w_{k-2}\}$. ∎

Theorem 12.7 *Let G be a graph of order n with $V(G) = \{v_1, v_2, \ldots, v_n\}$ and let $d(v_1) \leq d(v_2) \leq \cdots \leq d(v_n)$. Then G is k-connected if $d(v_r) \geq r+k-1$ for $1 \leq r \leq n-1-d(v_{n-k+1})$.*

Proof. Suppose G satisfies the conditions of the theorem. If it is not k-connected, then there exists a disconnecting set S such that $|S| = s < k$.

Then the graph $G - S$ is not connected. Let H be a component of $G - S$ of minimum order h. Clearly the degree in H of each vertex of H is at most $h - 1$. Therefore, in G, the degree of each vertex of H is at most $h + s - 1$. Thus

$$d(v) \leq h + s - 1 < h + k - 1 \text{ for all } v \in V(H). \tag{12.1}$$

Therefore, by the conditions of the theorem,

$$h > n - 1 - d(v_{n-k+1}). \tag{12.2}$$

Since $G - S$ has $n - s$ vertices and H is a component of $G - S$ of minimum order, we have

$$h \leq n - s - h$$

or

$$h + s \leq n - h.$$

Therefore,

$$d(v) \leq h + s - 1 \leq n - h - 1 \text{ for all } v \in V(H). \tag{12.3}$$

Since every vertex $u \in V(G) - V(H) - S$ is adjacent to at most $n - h - 1$ vertices, we have

$$d(u) \leq n - h - 1 \text{ for all } u \in V(G) - V(H) - S. \tag{12.4}$$

From 12.3 and 12.4 we conclude that all the vertices of degrees exceeding $n - h - 1$ are in S. Thus there are at most s vertices of degrees exceeding $n - h - 1$. Therefore

$$d(v_{n-s}) \leq n - h - 1. \tag{12.5}$$

Using 12.2 in 12.5, we get $d(v_{n-s}) < d(v_{n-k+1})$. Therefore $n - s < n - k + 1$ or $s \geq k$, a contradiction. ∎

The Cartesian product $G \square H$ of two graphs G and H is connected if and only if both G and H are connected. Šacapan [3] proved that if G and H are two graphs with at least two vertices, then

$$\kappa(G \square H) = \min\{\kappa(G)|V(H)|, \kappa(H)|V(G)|, \delta(G) + \delta(H)\}.$$

Xu and Yang [4] proved a similar result for edge connectivity, namely, if G and H are two graphs with at least two vertices, then

$$\lambda(G \square H) = \min\{\lambda(G)|V(H)|, \lambda(H)|V(G)|, \delta(G) + \delta(H)\}.$$

12.4 STRUCTURAL RESULTS

We start with a structural characterization of 2-connected graphs.

Definition 12.5 *Let H be a graph and let $u, v \in V(H)$. Any nontrivial $u - v$ path P with $V(P) \cap V(H) = \{u, v\}$ is called a H-path.*

We observe that the edge of any H-path of length 1 is not an edge of H.

Proposition 12.2 *A graph G is 2-connected if and only if G can be constructed from a cycle by successively adding H-paths to graphs H already constructed.*

Proof. Clearly every graph constructed as described is 2-connected. Conversely, let G be a 2-connected graph. Then G contains a cycle and let H be a maximal subgraph of G constructible as above. Since any edge $uv \in E(G) - E(H)$ with $u, v \in V(H)$ defines a H-path in G, it follows that H is an induced subgraph of G. Now, suppose $H \neq G$. Since G is connected, there exist $v \in V(G) - V(H)$ and $w \in V(H)$ such that v and w are adjacent in G. Since G is 2-connected, $G - w$ is connected and hence there exists a $v - x$ H-path $P = (v, x_1, x_2, \ldots, x_k, x)$. Now $P_1 = (w, v, x_1, x_2, \ldots, x_k, x)$ is a H-path in G and hence $H \cup P_1$ is a constructible subgraph of G, contradicting the maximality of H. ∎

The above construction of 2-connected graphs can also be expressed in terms of ear decomposition.

An ear of a graph G is a maximal path whose internal vertices have degree 2 in G. An ear decomposition of G is a decomposition $P_0, \ldots, P_k$ such that P_0 is a cycle and P_i for $i \geq 1$ is an ear of $P_0 \cup \cdots \cup P_i$.

Whitney [5] has proved that a graph G is a 2-connected graph if and only if it has an ear decomposition. Furthermore, every cycle in a 2-connected graph is the initial cycle in some ear decomposition.

A closed ear in a graph G is a cycle C such that all vertices of C except one have degree 2 in G. A closed-ear decomposition of G is a decomposition $P_0, \ldots, P_k$ such that P_0 is a cycle and P_i for $i \geq 1$ is either an (open) ear or a closed ear in $P_0 \cup \cdots \cup P_i$.

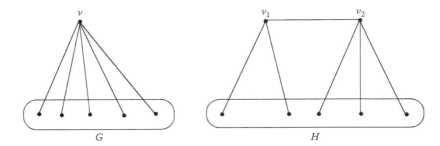

Figure 12.5 Expansion of a graph at a vertex.

A graph is 2-edge-connected if and only if it has a closed-ear decomposition, and every cycle in a 2-edge-connected graph is the initial cycle in some such decomposition.

Thomassen [6] established that any 3-connected graph G with $|V(G)| \geq 5$ contains an edge e such that G/e is 3-connected, where G/e is the graph obtained from G by contracting the edge e.

There is another graph operation which is in some sense an inverse to contraction. Let G be a 3-connected graph and let $v \in V(G)$ with $deg(v) \geq 4$. Replace v by two vertices v_1 and v_2, add the edge $e = v_1v_2$ and join v_1, v_2 to the neighbors of v in G in such a way that $deg(v_1) \geq 3$ and $deg(v_2) \geq 3$ in the resulting graph H. The graph H is called an expansion of G at v (see Figure 12.5).

Since there is some freedom in distributing the neighbors of v to v_1 and v_2, expansions are not uniquely determined. However, the contraction H/e is isomorphic to G.

Let G be a 3-connected graph, $v \in V(G)$ with $deg(v) \geq 4$ and H be an expansion of G at v. Then H is 3-connected. The above two results on contraction and expansion lead to the following structural characterization of 3-connected graphs. Given any 3-connected graph G, there exists a sequence $G_1, G_2, \ldots, G_k$ of graphs such that (i) $G_1 = K_4$, (ii) $G_k = G$, and (iii) G_{i+1} is obtained from G_i by adding an edge to G_i or by expanding G_i at a vertex of degree at least four.

In the above construction we cannot stay within simple graphs since the graph obtained from K_4 in the above construction is not a simple graph. Tutte [7] has shown that by starting with the class of all wheels, all simple 3-connected graphs can be constructed by means of the above two operations without creating parallel edges.

This recursive construction of 3-connected graphs have been used to prove many interesting theorems in graph theory.

12.5 MENGER'S THEOREM AND ITS APPLICATIONS

The following classical theorem of Menger is a cornerstone of the structural study of connectivity and plays a vital role in the theory of graphs and applications. It has been established that Menger's theorem implies Ford and Fulkerson's max-flow-min-cut theorem [8].

Theorem 12.8 (Menger) *The minimum number of vertices whose removal from a graph G disconnects two nonadjacent vertices s and t is equal to the maximum number of vertex-disjoint $s - t$ paths in G.*

Proof. We proceed by induction on $|E(G)| = m$. The theorem is trivially true when $m = 0$ or 1. Assume that the theorem holds for all graphs of size less than m, where $m \geq 2$, and let G be a graph of size m with $m \geq 3$. Let s and t be two nonadjacent vertices of G. If s and t belong to different components of G, then the result follows. So we may assume that s and t belong to the same component of G. Suppose that a minimum $s - t$ separating set consists of $k \geq 1$ vertices. We claim that G contains k internally disjoint $s-t$ paths. Since this is obviously true if $k = 1$, we may assume that $k \geq 2$. We now consider three cases.

Case 1 Some minimum $s-t$ separating set X in G contains a vertex x that is adjacent to both s and t.

Then $X - x$ is a minimum $s-t$ separating set in $G - x$ consisting of $k-1$ vertices. Since the size of $G - x$ is less than m, it follows by the induction hypothesis that $G - x$ contains $k-1$ internally disjoint $s-t$ paths. These paths together with the path $P = (s, x, t)$ produce k internally disjoint $s-t$ paths in G.

Case 2 For every minimum $s-t$ separating set S in G, either every vertex in S is adjacent to s and not to t or every vertex of S is adjacent to t and not to s.

In this case $d(s,t) \geq 3$. let $P = (s, x, y, \ldots, t)$ be a $s-t$ geodesic in G, where $e = xy$. Every minimum $s-t$ separating set in $G - e$ contains at least $k-1$ vertices. We show that every minimum $s-t$ separating set in $G - e$ contains k vertices. Suppose that there is some minimum $s-t$ separating set in $G - e$ with $k-1$ vertices, say $Z = \{z_1, z_2, \ldots, z_{k-1}\}$. Then $Z \cup \{x\}$ is a $s-t$ separating set in G and therefore is a minimum $s-t$ separating set in G. Since x is adjacent to s (and not to t), it follows that every vertex $z_i (1 \leq i \leq k-1)$ is also adjacent to s and not adjacent to t. Since $Z \cup \{y\}$ is also a minimum $s-t$ separating set in G and each vertex $z_i (1 \leq i \leq k-1)$ is adjacent to s but not to t, it follows that y is adjacent to s. This, however, contradicts the assumption that P is a $s-t$ geodesic. Thus k is the minimum number of vertices in a $s-t$ separating set in $G - e$. Since the size of $G - e$ is less than m, it follows by the induction hypothesis that there are k internally disjoint $s-t$ paths in $G - e$ and hence in G as well.

Case 3 There exists a minimum $s-t$ separating set W in G in which no vertex is adjacent to both s and t and containing at least one vertex not adjacent to s and at least one vertex not adjacent to t.

Let $W = \{w_1, w_2, \ldots, w_k\}$. Let G_s be the subgraph of G consisting of all $s - w_i$ paths in G in which $w_i \in W$ is the only vertex of the path belonging to W. Let G'_s be the graph constructed from G_s by adding a new vertex t' and joining t' to each vertex w_i for $1 \leq i \leq k$. The graphs G_t and G'_t are defined similarly. Since W contains a vertex that is not adjacent to s and a vertex that is not adjacent to t, the sizes of both G'_s and G'_t are less than m. So G'_s contains k internally disjoint $s - t'$ paths $A_i(1 \leq i \leq k)$, where A_i contains w_i. Also, G'_t contains k internally disjoint $t - t'$ paths $B_i(1 \leq i \leq k)$, where B_i contains w_i. Let A'_i be the $s - w_i$ subpath of A_i and let B'_i be the $w_i - t$ subpath of B_i for $1 \leq i \leq k$. The k paths constructed from A'_i and B'_i for each $i(1 \leq i \leq k)$ are internally disjoint $s - t$ paths in G. ∎

Whitney [5] proved that a graph G of order at least three is 2-connected if and only if there are two vertex disjoint $u - v$ paths for any two distinct vertices u and v of G. The following theorem gives a similar result for k-connected graphs.

Theorem 12.9 *Let G be a graph of order $n \geq k + 1$. Then G is k-connected if and only if there are k vertex-disjoint $s - t$ paths between any two vertices s and t of G.*

Proof. Obviously, the theorem is true for $k = 1$. So we need to prove the theorem for $k \geq 2$.

Necessity: If s and t are not adjacent, then the necessity follows from Theorem 12.8.

Suppose that s and t are adjacent and that there are at most $k - 1$ vertex-disjoint $s - t$ paths in G. Let $e = st$. Consider the graph $G' = G - e$. Since there are at most $k - 1$ vertex-disjoint $s - t$ paths in G, there cannot be more than $k - 2$ vertex-disjoint $s - t$ paths in G'. Thus there exists a set $A \subseteq V - \{s, t\}$ of vertices, with $|A| \leq k - 2$, whose removal disconnects s and t in G'. Then

$$|V - A| = |V| - |A| \geq k + 1 - (k - 2) = 3,$$

and therefore, there is a vertex u in $V - A$ different from s and t.

Now we show that there exists an $s - u$ path in G' that does not contain any vertex of A. Clearly, this is true if s and u are adjacent. If s and u are not adjacent, then there are k vertex-disjoint $s - u$ paths in G, and hence there are $k - 1$ vertex-disjoint $s - u$ paths in G'. Since $|A| \leq k - 2$ at least one of these $k - 1$ paths will not contain any vertex of A.

In a similar way we can show that in G' there exists a $u - t$ path that does not contain any vertex of A.

Thus there exists in G' an $s-t$ path that does not contain any vertex of A. This, however, contradicts that A is an $s - t$ disconnecting set in G'.

Hence the necessity.

Sufficiency: Since there are k vertex-disjoint paths between any two distinct vertices of G, it follows that G is connected. Further, at most one of these paths can be of length 1. The union of the remaining $k - 1$ paths must contain at least $k - 1$ distinct vertices other than s and t. Hence

$$|V| \geq (k - 1) + 2 > k.$$

Suppose in G there is a disconnecting set A with $|A| < k$. Then the subgraph G' of G on the vertex set $V - A$ contains at least two distinct components. If we select two vertices s and t from any two different components of G', then there are at most $|A| < k$ vertex-disjoint $s-t$ paths in G, which is a contradiction. ∎

Lemma 12.1 *Let G be a k-connected graph and let H be a graph obtained from G by adding a new vertex y and joining it to at least k vertices of G. Then H is also k-connected.*

Proof. The conclusion clearly holds if H is complete. Let S be a subset of $V(H)$ with $|S| = k - 1$. To complete the proof, it suffices to show that $H - S$ is connected.

Suppose first that $y \in S$. Then $H - S = G - (S \backslash \{y\})$. By hypothesis, G is k-connected and $|S \backslash \{y\}| = k - 2$. Hence $H - S$ is connected.

Now suppose that $y \notin S$. Since, by hypothesis, y has at least k neighbors in $V(G)$ and $|S| = k - 1$, there is a neighbor z of y which does not belong to S. Because G is k-connected, $G - S$ is connected. Furthermore, z is a vertex of $G - S$, and hence yz is an edge of $H - S$. It follows that $(G - S) + yz$ is a spanning connected subgraph of $H - S$. Hence $H - S$ is connected. ∎

The following useful property of k-connected graphs can be deduced from Lemma 12.1.

Proposition 12.3 *Let G be a k-connected graph and let X and Y be subsets of V of cardinality at least k. Then there exists in G a family of k pairwise disjoint (X, Y)-paths.*

Proof. Let H be the graph obtained from G by adding two vertices x and y and joining x to each vertex of X and y to each vertex of Y. By Lemma 12.1, H is k-connected. Therefore, by Menger's theorem, there exist k internally disjoint $x - y$-paths in H. Deleting x and y from each of these paths, we obtain k disjoint paths $Q_1, Q_2, \ldots, Q_k$ in G, each of which has its initial vertex in X and its terminal vertex in Y. Every path Q_i necessarily contains a segment, P_i with initial vertex in X, terminal vertex in Y and no internal vertex in $X \cup Y$. The paths $P_1, P_2, \ldots, P_k$ are pairwise disjoint (X, Y)-paths. ∎

A family of k internally disjoint (x, Y)-paths whose terminal vertices are distinct is referred to as a *k-fan* from x to Y. The following assertion is another consequence of Menger's theorem. Its proof is similar to the proof of Proposition 12.3.

Proposition 12.4 (Fan lemma) *Let G be a k-connected graph, let x be a vertex of G and let $Y \subseteq V \backslash \{x\}$ be a set of at least k vertices of G. Then there exists a k-fan in G from x to Y.* ∎

By Theorem 12.5 in a 2-connected graph any two vertices are connected by two internally disjoint paths; equivalently, any two vertices in a 2-connected graph lie on a cycle. Dirac [9] generalized this statement for k-connected graphs.

Theorem 12.10 *Let S be a set of k vertices in a k-connected graph G, where $k \geq 2$. Then there is a cycle in G which includes all the vertices of S.*

Proof. By induction on k. We have already observed that the assertion holds for $k = 2$. Now let $k \geq 3$. Let $x \in S$, and set $T = S \backslash x$. Since G is k-connected, it is $(k-1)$-connected. Therefore, by the induction hypothesis, there is a cycle C in G which includes T. Set $Y = V(C)$. If $x \in Y$, then C includes all the vertices of S. Thus we may assume that $x \notin Y$. If $|Y| \geq k$, the Fan lemma ensures the existence of a k-fan in G from x to Y. Because $|T| = k-1$, the set T divides C into $k-1$ edge-disjoint segments. By the Pigeonhole principle, two paths of the fan, P and Q, end in the same segment. The subgraph $C \cup P \cup Q$ contains three cycles, one of which includes $S = T \cup x$. If $|Y| = k - 1$, the Fan lemma yields a $(k-1)$-fan from x to Y in which each vertex of Y is the terminus of one path, and we conclude as before. ∎

We give another application of Menger's theorem. Let $A = \{A_1, \ldots, A_m\}$ be a collection of subsets of X with $\bigcup_{i=1}^{m} A_i = X$. A system of distinct representatives (SDR) is a set of distinct elements $x_1, \ldots, x_m$ such that $x_i \in A_i$. A necessary and sufficient condition for the existence of an SDR is that $|\bigcup_{i \in I} A_i| \geq |I|$ for all $I \subseteq \{1, \ldots, m\}$. This can be proved using Hall's theorem. In fact Hall's theorem was originally proved in the language of SDR and it is equivalent to Menger's theorem.

12.6 CONDITIONAL CONNECTIVITY

Harary [10] introduced the concept of conditional connectivity of a graph. Given a graph theoretic property $\mathcal{P}$, the conditional connectivity of G with respect to $\mathcal{P}$ is defined to be the minimum cardinality of a set S of vertices such that every component of $G - S$ has property $\mathcal{P}$ and it is denoted by $\kappa(G; \mathcal{P})$. Similarly we can define the conditional edge connectivity $\lambda(G; \mathcal{P})$ with respect to the property $\mathcal{P}$. The conditional connectivity and conditional edge connectivity with respect to several properties $\mathcal{P}$ have been investigated by several authors. We shall discuss some of these concepts in this section.

Definition 12.6 *Let $G = (V, E)$ be a nontrivial graph. A subset S of E is called a restricted edge-cut of G if $G - S$ is disconnected and contains no isolated vertices. The restricted edge-connectivity $\lambda'(G)$ is defined to be the minimum cardinality of a restricted edge-cut of G.*

The concept of restricted edge-connectivity was introduced by Esfahanian and Hakimi [11], which provides a more accurate measure of fault-tolerance of networks than the classical edge-connectivity (see [12]).

Definition 12.7 *The degree of an edge $e = uv$ in a graph G is defined by $\deg(e) = \deg u + \deg v - 2$. Also $\xi(G) = \min\{\deg(e) : e \in E\}$ is the minimum edge-degree of G.*

The following inequality has been proved in [11].

$$\lambda'(G) \leq \xi(G) \tag{12.6}$$

Any nontrivial graph G with $\lambda'(G) = \xi(G)$ is *optimal* and otherwise G is *nonoptional*. Xu and Xu [13] obtained several classes of optimal graphs.

For any two disjoint subsets X and Y of V, let $[X, Y]$ denote the set of all edges in E with one end in X and other end in Y. Let $\partial(X) = [X, \overline{X}]$ and $d(X) = |\partial(X)|$. A proper subset X of V is called a *fragment* of G if $\partial(X)$ is a λ'-cut of G. Clearly if X is a fragment of G, then $\overline{X}$ is also a fragment of G. Let

$$r(G) = \min\{|X| : X \text{ is a fragment of } G\}.$$

Then $2 \leq r(G) \leq (1/2)|V|$. A fragment X is called an *atom* if $|X| = r(G)$.

Theorem 12.11 [13] *A nontrivial graph G is optimal if and only if $r(G) = 2$.* ∎

Theorem 12.12 [13] *Let G be a nonoptimal graph. If G is k-regular, then $r(G) \geq k \geq 3$.* ∎

Theorem 12.13 [13] *Let G be a connected vertex-transitive graph with degree $k(\geq 3)$ and let X be an atom of G. If G is nonoptimal, then*

i. *$G[X]$ is a vertex-transitive subgraph of G with degree $k - 1$ containing a triangle.*

ii. *G has even order and there is a partition $\{X_1, X_2, \ldots, X_m\}$ of V such that $G[X_i] \cong G[X]$ for each $i = 1, 2, \ldots, m, m \geq 2$.* ∎

Theorem 12.14 [14] *Let G be a connected vertex-transitive graph. If it either contains no triangles or has odd order, then G is optimal.* ∎

Corollary 12.1 [12] *The k-cube Q_k is optimal.* ∎

The k-dimensional toroidal mesh $C(d_1, d_2, \ldots, d_k)$, is the Cartesian product $C_{d_1} \square C_{d_2} \square \cdots \square C_{d_k}$, where C_d is the cycle on d vertices.

Corollary 12.2 *The k-dimensional toroidal mesh $C(d_1, d_2, \ldots, d_k)$ is optimal if $d_i \geq 4$ for each $i = 1, 2, \ldots, k$.* ∎

A circulant graph, denoted by $G(n; a_1, a_2, \ldots, a_k)$, where $0 < a_1 < \cdots < a_k \leq (n/2)$, has vertices $0, 1, 2, \ldots, n - 1$ and edge ij if and only if $|j - i| \equiv a_t \pmod{n}$ for some $t, 1 \leq t \leq k$. If $a_k \neq (n/2)$, it is $2k$-regular. Otherwise, it is $(2k - 1)$-regular.

Corollary 12.3 [15] *Any connected circulant graph $G(n; a_1, a_2, \ldots, a_k), n \geq 4$, is optimal if either it contains no triangles or $a_k \neq (n/2)$.* ∎

The concept of k-restricted edge-connectivity was introduced by Fabrega and Fiol [16] and is a natural generalization of restricted edge-connectivity.

Definition 12.8 *An edge-cut S of a connected graph G is called a k-restricted edge-cut if every component of $G - S$ has at least k vertices. If G has at least one k-restricted edge-cut, then $\lambda_k(G) = \min\{|S| : S \subseteq E(G)$ is a k-restricted edge-cut of $G\}$ is called the k-restricted edge-connectivity of G.*

The concepts of minimum edge degree $\xi(G)$, fragment, atom, and optimal graphs can also be naturally extended in the context of k-restricted edge-cuts.

If $[X, \overline{X}]$ is a λ_k-cut, then X is called a k-fragment of G. Let $r_k(G) = \min\{|X| : X$ is a k-fragment of $G\}$. Obviously, $k \leq r_k(G) \leq (1/2)|V(G)|$.

The minimum k-edge degree $\xi_k(G)$ is defined by

$$\xi_k(G) = \min\{|[X,\overline{X}]| : |X| = k \text{ and } G[X] \text{ is connected}\}.$$

A λ_k-connected graph G with $\lambda_k(G) = \xi_k(G)$ is said to be λ_k-optimal.

We now present several recent results on k-restricted edge-connectivity and λ_k-optimal graphs.

Theorem 12.15 [17] *Let G be a connected triangle-free graph of order $n \geq 4$. If $d(u)+d(v) \geq 2\lfloor(n+2)/4\rfloor + 1$ for each pair u,v of vertices at distance 2, then G is λ_k-optimal.* ∎

Theorem 12.16 [17] *Let G be a λ_k-connected graph with $\lambda_k(G) \leq \xi_k(G)$ and let U be a k-fragment of G. If there is a connected subgraph H of order k in $G[U]$ such that $|[V(H), U\setminus V(H)]| \leq |[U\setminus V(H), \overline{U}]|$, then G is λ_k-optimal.* ∎

Theorem 12.17 [18] *Let G be a λ_2-connected and triangle-free graph with minimum degree $\delta \geq 1$. If G is not λ_2-optimal, then* ∎

$$r_2(G) \geq \max\left\{3, \frac{1}{\delta}((\delta-1)\xi_2(G) + 2\delta + 1)\right\}.$$

Corollary 12.4 [19] *Let G be a λ_2-connected and triangle-free graph with minimum degree $\delta \geq 2$. If G is not λ_2-optimal, then* ∎

$$r_2(G) \geq \begin{cases} 2\delta - 1 & \text{if } \delta \geq 3, \\ 4 & \text{if } \delta = 2. \end{cases}$$

Bonsma et al. [20] investigated the concept of 3-restricted edge-connectivity, λ_3-connected graphs and λ_3-optimal graphs.

Theorem 12.18 [20] *A graph G is λ_3-connected if and only if $n \geq 6$ and G is not isomorphic to the net N or to any graph of the family F in Figure 12.6.* ∎

Theorem 12.19 [20] *If G is a λ_3-connected graph, then $\lambda_3(G) \leq \xi_3(G)$.* ∎

Theorem 12.20 [18] *Let G be a connected triangle-free graph of order $n \geq 6$. If $d(u)+d(v) \geq 2\left\lceil\frac{n}{4}\right\rceil + 3$ for each pair u,v of non adjacent vertices, then G is λ_3-optimal.* ∎

A graph G with $|V(G)| \geq 2k$ is called a flower if it contains a cut vertex u such that every component of $G - u$ has order at most $k - 1$.

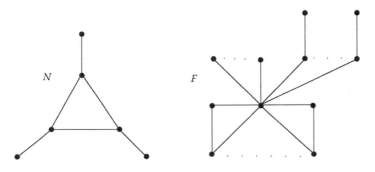

Figure 12.6 All graphs that are not λ_3-connected.

Theorem 12.21 [21] *Let G be a connected graph not isomorphic to a flower and k be a positive integer with $k \leq \delta(G) + 1$. Then G is λ_k-connected and $\lambda_k(G) \leq \xi_k(G)$.* ∎

Theorem 12.22 [22] *Let G be a λ_k-connected graph with minimum degree δ. If $\lambda_k(G) \leq \xi_k(G)$, then $r_k(G) \geq \max\{k+1, \delta - k + 1\}$.* ∎

Theorem 12.23 [18] *Let G be a λ_k-connected and triangle-free graph with minimum degree δ and $\lambda_k(G) \leq \xi_k(G)$. If G is not λ_k-optimal, then $r_k(G) \geq \max\{k+1, 2\delta - k + 1\}$.* ∎

Theorem 12.24 [23] *Let G be a bipartite graph of order $n \geq 2k$. If $\delta(G) \geq \lceil (n+2k)/4 \rceil$, then G is λ_k-optimal.* ∎

Corollary 12.5 [18] *Let G be a connected and triangle-free graph of order $n \geq 2k$. If $\delta(G) \geq (1/2)\left(\lceil n/2 \rceil + k\right)$, then G is λ_k-optimal.* ∎

Jianping [24] obtained a characterization of all connected graphs of order at least 10 which contain 4-restricted edge-cuts.

Definition 12.9 *Let $P_5 = uvxyz$ be a path of length 4. Let w be an arbitrary vertex of another connected graph H of order 3. Then Ω_4^8 is the collection of graphs obtained from one of the following two different ways.*

Method 1.

Step 1. *Join the graph H and path P_5 by adding two edges wv and wy to obtain a new graph N. This step results in three distinct graphs according to the choice of H and w.*

Step 2. *Add at most one of the two edges wx and vy to the graph N.*

Method 2.

Step 1. *When H is a path with w as one of its pendants, join x to the degree 2 vertex of H after performing step 1 of method 1.*

Step 2. *Add at most one edge between w and x.*

Definition 12.10 *Graphs in the collection Ω_3^8 can be obtained as follows. Let A and B be two connected graphs of order 3 and let C be an isolated edge. Take three arbitrary vertices, one each from these three graphs, and join them into a 3-cycle.*

Definition 12.11 *The collection Ω_3^9 consists of graphs obtained by joining three vertices, one each from three arbitrary connected graphs of order 3, into a 3-cycle.*

Theorem 12.25 *A connected graph G of order at least 8 contains no 4-restricted-edge-cut if and only if $G \in F_{n,4} \cup \Omega_4^8 \cup \Omega_3^8 \cup \Omega_3^9$, where $F_{n,4}$ is a flower.* ∎

Corollary 12.6 *Let G be a connected graph of order at least 10. Then G contains a 4-restricted-edge-cut if and only if G is not a flower.* ∎

12.7 CRITICALITY AND MINIMALITY

A graph G is said to be critically k-connected if G is k-connected and $G-v$ is not k-connected for all $v \in V$. Kaugars [25] has shown that $\delta(G) = 2$ for each critically 2-connected graph G. Chartrand et al. [26] obtained a generalization of the above result.

Theorem 12.26 [26] *If G is a critically k-connected graph with $k \geq 2$, then $\delta(G) < [(3k-1)/2]$.* ∎

Entringer and Slater [27] have proved that every critically 3-connected graph contains at least two vertices of degree 3 and that this result is best possible.

Maurer and Slater [28] generalized the concept of critically k-connected graphs. A graph G is called k^*-connected if $\kappa(G) = k$. Thus a k^*-connected graph is k-connected but not $(k+1)$-connected. A k^*-connected graph G is said to be s-critical if for any subset S of V with $|S| \leq s$, we have $\kappa(G - S) = k - |S|$.

If G is s-critical but not $(s+1)$-critical, then G is called s^*-critical. A graph which is k^*-connected and s^*-critical is called (k^*, s^*)-graph.

For example, the complete graph K_{n+1} is a (n^*, n^*)-graph and this is the only (n^*, n^*)-graph. The graph $C_t + K_{p-2}$ is a $(p^*, 1^*)$-graph if $p < t$ and is a $(p^*, 2^*)$-graph if $p = t$.

Maurer and Slater [28] obtained the following results on the existence and nonexistence of (k^*, s^*)-graphs.

Proposition 12.5 *The graph G_{2n} obtained from the complete graph K_{2n} by deleting the edges of a perfect matching is a $((2n)^*, n^*)$-graph.* ∎

Proposition 12.6 *If G is a (k^*, s^*)-graph and $1 \leq t \leq s$, then G has an induced subgraph which is a $((k-t)^*, (s-t)^*)$-graph.* ∎

Theorem 12.27 [28] *In any 3-connected graph G, with $G \neq K_n$, there are two adjacent vertices, u and v, for which $G - \{u, v\}$ is 2-connected.* ∎

Corollary 12.7 [28] *If $k \geq 3$, then there does not exist a $(k^*, (k-1)^*)$-graph.* ∎

Theorem 12.28 [28] *There does not exist a $(5^*, 3^*)$-graph.* ∎

Corollary 12.8 [28] *If $k \geq 5$, then there does not exist a $(k^*, (k-2)^*)$-graph.* ∎

Theorem 12.29 [28] *If $k \geq 7$, then there does not exist a $(k^*, (k-3)^*)$-graph.* ∎

Proposition 12.7 [28] *Suppose $H = G - v$ is a (k^*, s^*)-graph. The following are equivalent.*

  i. $\kappa(G) = k + 1$.

  ii. G is either a $((k+1)^*, s^*)$-graph or a $((k+1)^*, (s+1)^*)$-graph.

  iii. For each subset S of $V(H)$ with $|S| = s$, v is connected to each component of $H - S$. ∎

Proposition 12.8 [28] *Let G be a (k^*, s^*)-graph of order n, and let H be any disconnected graph of order $n - k$. Then $G + H$ is a $(n^*, (s+1)^*)$-graph.* ∎

Similar to the study of criticality and multiple criticality, Maurer and Slater [29] investigated the concept of minimality and multiple minimality with respect to connectivity and edge-connectivity. A graph G is s-minimal with respect to a graph parameter if the removal of any j edges, $1 \leq j \leq s$, reduces the value of that parameter by j. When $s = 1$, this corresponds to the well-known concept of minimality. Let G be a graph with edge-connectivity $\lambda(G) = k$. If G is s-minimal with respect to λ, then G is called a s-minimally k-edge connected graph, or simply a $(k, s)'$-graph. Let G be a $(k, s)'$-graph. If s is the largest integer for which G is s-minimal, then G is called a $(k^*, s^*)'$-graph.

For any graph G, let G^p be the graph obtained from G by replacing each edge of G by p parallel edges. Maurer and Slater [29] determined all $(k, s)'$-graphs.

Theorem 12.30 [29] *The only $(k, s)'$-graphs for $s \geq 2$ are C_m^p for $m \geq 3$ and K_2^p for $p \geq 2$. Further C_m^p is a $(2p^*, 2^*)'$-graph and K_2^p for $p \geq 2$ is a $(p^*, p^*)'$-graph.* ∎

Corollary 12.9 *The only simple $(k^*, s^*)'$-graphs with $s \geq 2$ are the cycles $C_m, m \geq 3$, and these are $(2^*, 2^*)'$-graphs.* ∎

In case of vertex connectivity it turns out that the only s-minimally k-connected graphs, $s \geq 2$, are the cycle $C_n, n \geq 3$, and these are 2-minimally 2-connected.

12.8 CONCLUSION

We have presented some of the key results dealing with connectivity of finite simple graphs. Several textbooks and surveys, (see, e.g., [30–32]) are available on the subject. The definitions of connectivity and edge-connectivity have straightforward extensions to directed graphs. It suffices to replace paths by directed paths throughout. The arc and vertex versions of Menger's theorem can be proved for directed graphs as well. Maurer and Slater [29] have extended the concept of multiply minimal connectedness to directed graphs and obtained a few basic results. Every 2-edge connected graph admits a strongly connected orientation. Nash-Williams [33] obtained a nice generalization of this result by proving that every $2k$-edge-connected graph has a k-arc-connected orientation.

References

[1] G. Chartrand and F. Harary, Graphs with prescribed connectivities, In P. Erdös and G. Katona, editors, *Theory of Graphs*, Akadémiai Kiadó, Budapest, Hungary, (1968), 61–63.

[2] F. Harary, The maximum connectivity of a graph, *Proc. Nat. Acad. Sci. U.S.A.*, **48** (1962), 1142–1146.

[3] S. Špacapan, Connectivity of Cartesian products of graphs, *Appl. Math. Lett.*, **21** (2008), 682–685.

[4] J. M. Xu and C. Yang, Connectivity of Cartesian product graphs, *Discrete Math.*, **306** (2006), 159–165.

[5] H. Whitney, Congruent graphs and the connectivity of graphs, *Am. J. Math.*, **54** (1932), 150–168.

[6] C. Thomassen, Kuratowski's theorem, *J. Graph Theory*, **5** (1981), 225–241.

[7] W. T. Tutte, A theory of 3-connected graphs, *Indag. Math.*, **23** (1961), 441–455.

[8] L. R. Ford and D. R. Fulkerson, Maximal flow through a network, *Can. J. Math.*, **8** (1956), 399–404.

[9] G. A. Dirac, In abstrakten graphen vorhandene vollständige 4-Graphen und ihre Unterteilungen, *Math. Nachr.*, **22** (1960), 61–85.

[10] F. Harary, Conditional connectivity, *Networks*, **13** (1983), 347–357.

[11] A. H. Esfahanian and S. L. Hakimi, On computing a conditional edge connectivity of a graph, *Inf. Process. Lett.*, **27** (1988), 195–199.

[12] A. H. Esfahanian, Generalized measure of fault tolerance with application to n-cube networks, *IEEE Trans. Comput.*, **38**(11) (1989), 1586–1591.

[13] J. M. Xu and K. L. Xu, On restricted edge-connectivity of graphs, *Discrete Math.*, **243** (2002), 291–298.

[14] J. M. Xu, Restricted edge-connectivity of vertex-transitive graphs, *Chin. J. Contemp. Math.*, **21**(4) (2000), 369–374.

[15] Q. L. Li and Q. Li, Reliability analysis of circulants, *Networks*, **31** (1998), 61–65.

[16] J. Fábrega and M. A. Fiol, On the extraconnectivity of graphs, *Discrete Math.*, **155** (1996), 49–57.

[17] J. Yuan and A. Liu, Sufficient conditions for λ_k-optimality in triangle-free graphs, *Discrete Math.*, **310** (2010), 981–987.

[18] A. Holtkamp, D. Meierling, and L. P. Montejano, k-restricted edge-connectivity in triangle-free graphs, *Discrete Appl. Math.*, **160** (2012), 1345–1355.

[19] N. Ueffing and L. Volkmann, Restricted edge-connectivity and minimum edge-degree, *Ars Combin.*, **66** (2003), 193–203.

[20] P. Bonsma, N. Ueffing, and L. Volkmann, Edge-cuts leaving components of order at least three, *Discrete Math.*, **256** (2002), 431–439.

[21] Z. Zhang and J. Yuan, A proof of an inequality concerning k-restricted edge-connectivity, *Discrete Math.*, **304** (2005), 128–134.

[22] Z. Zhang and J. Yuan, Degree conditions for restricted-edge-connectivity and isoperimetric-edge-connectivity to be optimal, *Discrete Math.*, **307** (2007), 293–298.

[23] J. Yuan, A. Liu, and S. Wang, Sufficient conditions for bipartite graphs to be super-k-restricted edge-connected, *Discrete Math.*, **309** (2009), 2886–2896.

[24] O. Jianping, 4-restricted edge cuts of graphs, *Australas. J. Combin.*, **30** (2004), 103–112.

[25] A. Kaugars, *A theorem on the removal of vertices from blocks*, Senior thesis, Kalamazoo College, 1968.

[26] G. Chartrand, A. Kaugars, and D. R. Lick, Critically n-connected graphs, *Proc. Amer. Math. Soc.*, **32** (1979), 63–68.

[27] R. C. Entringer and P. J. Slater, A theorem on critically 3-connected graphs, *Nanta Mathematica*, **11**(2) (1977), 141–145.

[28] S. Maurer and P. J. Slater, On k-critical, n-connected graphs, *Discrete Math.*, **20** (1977), 255–262.

[29] S. Maurer and P. J. Slater, On k-minimally n-edge-connected graphs, *Discrete Math.*, **24** (1978), 185–195.

[30] W. T. Tutte, *Connectivity in Graphs*, University of Toronto Press, London, 1966.

[31] A. Frank, Connectivity augmentation problems in network design, In J. R. Birge and K. G. Murty, editors, *Mathematical Programming: State of the Art 1994*, The University of Michigan, Ann Arbor, MI, (1994), 34–63.

[32] A. Frank, Connectivity and network flows, In R. Graham, M. Grötschel, and L. Lovász, editors, *Handbook of Combinatorics*, Elsevier Science B.V., North Holland, (1995), 111–177.

[33] C. S. J. A. Nash-Williams, On orientation, connectivity and odd-vertex pairings in finite graphs, *Can. J. Math.*, **12** (1960), 555–567.

CHAPTER 13

Connectivity Algorithms

Krishnaiyan "KT" Thulasiraman

CONTENTS

13.1	Introduction	291
13.2	Menger's Theorems and Maximum Flows in a Network	291
	13.2.1 Edge Version of Menger's Theorem	292
	13.2.2 Vertex Analog of Menger's Theorem	293
	13.2.3 Determining Connectivities in Undirected Graphs	294
13.3	Edge Connectivity of Graphs	295
	13.3.1 Edge Connectivity of Undirected Graphs: Matula's Algorithm	295
	13.3.2 Edge Connectivity of Directed Graphs	298
13.4	Global Minimum Cut in a Weighted Undirected Graph: Stoer and Wagner's Algorithm	298
13.5	Vertex Connectivity in Undirected Graphs	300
13.6	Gomory–Hu Tree	301
13.7	Graphs With Prescribed Degrees	308

13.1 INTRODUCTION

In this chapter we discuss algorithms to determine edge and vertex connectivities of a graph and related issues. We begin with Menger's theorems (see also Chapter 12) that relate maximum flow from a vertex s to a vertex t in a network to the maximum number of edge (vertex)-disjoint s–t paths as well as the minimum number of edges (vertices) to be deleted to disconnect s and t. This leads to natural maximum flow-based algorithms to determine vertex and edge connectivities. We then discuss efforts to develop connectivity algorithms that achieve better complexity results. In particular, we discuss Matula's algorithm for edge connectivity and Stoer and Wagner's algorithm to determine a minimum cut in a weighted graph. As regards vertex connectivity, we develop Even and Tarjan's algorithm. We conclude with a discussion of an algorithm for constructing the Gomory–Hu tree and results of interest in the design of graphs meeting certain connectivity requirements. In the course of our discussions we will develop certain results on connectivity that form the foundation of most of the connectivity algorithms.

13.2 MENGER'S THEOREMS AND MAXIMUM FLOWS IN A NETWORK

In this section we revisit Menger's theorems. In Chapter 12 graph theoretic proofs of these theorems were given. In this section we prove these results using maximum flow theory, thereby establishing the connection between these two areas, namely graph theory and network flow theory. We refer the reader to Chapter 4 for basic results and algorithms relating to the maximum flow problem.

13.2.1 Edge Version of Menger's Theorem

Given a network N with the underlying directed graph $G = (V, E)$. In the following, the terms network and graph will be used interchangeably. Each edge e in G is associated with a capacity $c(e) = 1$. A cut $K = \langle S, \overline{S} \rangle$ is the set of edges that have one vertex in S and the other in $\overline{S}$. K is called an s–t cut if $s \in S$ and $t \in \overline{S}$. The sum of the capacities of the edges in K directed from S to $\overline{S}$ is called the capacity of the cut denoted as $c(K)$. Since each capacity is equal to one, the capacity of K is equal to the number of edges directed from S to $\overline{S}$. An s–t cut with minimum capacity is called a minimum s–t cut. We would like to determine the minimum number of edges to be removed to disconnect all the directed s–t paths in G.

Let f^* be a maximum flow in G from vertex s to vertex t with $val(f^*)$ denoting the value of this flow. Let $K^* = \langle S^*, \overline{S}^* \rangle$ be a minimum s–t cut in N with $c(K^*)$ denoting the capacity of this cut. Let $\lambda(s,t)$ = minimum number of edges to be removed to destroy all directed s–t paths in G, and $l(s,t)$ = maximum number of edge-disjoint directed s–t path in G. We would like to prove that $\lambda(s,t) = l(s,t) = val(f^*) = c(K^*)$. By the max-flow min-cut theorem we know that

$$val(f^*) = c(K^*). \tag{13.1}$$

Consider any s–t cut $K = \langle X, \overline{X} \rangle$. Then the number of edges directed from X to $\overline{X}$ is the capacity $c(K)$. If we remove $c(K)$ edges from G, then all the directed s–t paths will be destroyed. So, we get

$$c(K^*) \geq \lambda(s,t). \tag{13.2}$$

Next, suppose that we pick a set of l edge-disjoint directed s–t paths. Then, we need to remove at least one edge from each one of these paths to destroy all directed s–t paths. So,

$$\lambda(s,t) \geq l(s,t). \tag{13.3}$$

Finally, any flow f in a unit capacity network can be decomposed into unit flows along $val(f)$ edge-disjoint directed s–t paths. So,

$$val(f^*) \leq l(s,t). \tag{13.4}$$

Combining (13.2)–(13.4) we get

$$val(f^*) \leq l(s,t) \leq \lambda(s,t) \leq c(K^*).$$

By (13.1), $val(f^*) = c(K^*)$. So,

$$val(f^*) = l(s,t) = \lambda(s,t) = c(K^*).$$

Thus, we have proved the following theorems.

Theorem 13.1 *Let N be a network with source s and sink t and in which each edge has unit capacity. Then*

1. *The value of a maximum s–t flow in N is equal to the maximum number of edge-disjoint directed s–t paths in N.*

2. *The minimum number of edges whose deletion destroys all directed s–t paths in G is equal to the capacity of a minimum s–t cut in N.* ∎

Theorem 13.2 [1] *Let s and t be two vertices in a directed graph G. Then, the maximum number of edge-disjoint directed s–t paths in G is equal to the minimum number of edges whose removal destroys all directed s–t paths in G.* ∎

For an undirected graph G, let $D(G)$ denote the directed graph obtained by replacing each edge e of G by a pair of oppositely oriented edges having the same end vertices as e. It can be shown that

1. There exists a one-to-one correspondence between the paths in G and the directed paths in $D(G)$.

2. For any two vertices s and t, the minimum number of edges whose deletion removes all s–t paths in G is equal to the minimum number of edges whose deletion destroys all directed s–t paths in $D(G)$.

The undirected version of Theorem 13.2 now follows from the above observations.

Theorem 13.3 [1] *Let s and t be two vertices in a graph G. Then the maximum number of edge-disjoint s–t paths in G is equal to the minimum number of edges whose removal destroys all s–t paths in G.* ∎

Suppose that we treat a given directed graph G as a network N with each edge having unit capacity. Given two vertices s and t, it follows from Theorems 13.1–13.3 that a maximum flow in N will also yield a maximum set of edge-disjoint s–t paths in N. Similarly a minimum s–t cut in N will determine a minimum set of edges whose deletion will destroy all s–t paths in N. Thus, Menger's theorem provides the main link between the maximum flow theory and graph theory. As a result, maximum flow algorithms provide the foundation for many algorithmic developments in graph theory.

13.2.2 Vertex Analog of Menger's Theorem

To develop the vertex analog of Menger's theorem we proceed as follows. Let s and t be any two nonadjacent vertices in a directed graph $G = (V, E)$. From G we construct a directed graph G' as follows:

1. Split each vertex $v \in V \setminus \{s, t\}$ into two vertices v' and v'' and connect them by a directed edge (v', v'').

2. Replace each edge of G having $v \in V \setminus \{s, t\}$ as terminal vertex by a new edge having v' as terminal vertex.

3. Replace each edge of G having $v \in V \setminus \{s, t\}$ as initial vertex by a new edge having v'' as initial vertex.

A graph G and the corresponding graph G' are shown in Figure 13.1. It is not difficult to see the following:

1. Each directed s–t path in G' corresponds to a directed s–t path in G that is obtained by contracting (i.e., identifying the vertices of) all the edges of the type (v', v''); conversely, each directed s–t path in G corresponds to a directed s–t path in G' obtained by splitting all the vertices other than s and t in the path.

2. Two directed s–t paths in G' are edge-disjoint if and only if the corresponding directed paths in G are vertex-disjoint.

3. The maximum number of edge-disjoint directed s–t paths in G' is equal to the maximum number of vertex-disjoint directed s–t paths in G.

4. The minimum number of edges whose deletion removes all directed s–t paths in G' is equal to the minimum number of vertices whose deletion destroys all directed s–t paths in G.

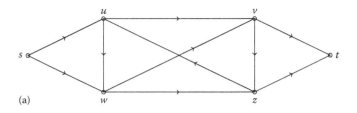

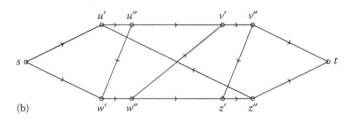

Figure 13.1 (a) Graph G. (b) Graph G' after vertex splitting.

From these observations we get the following vertex analogs of Theorems 13.2 and 13.3, respectively.

Theorem 13.4 [1] *Let s and t be two nonadjacent vertices in a directed graph G. Then the maximum number of vertex-disjoint directed $s-t$ paths in G is equal to the minimum number of vertices whose removal destroys all directed $s-t$ paths in G.* ∎

Theorem 13.5 [1] *Let s and t be two nonadjacent vertices in a graph G. Then the maximum number of vertex-disjoint $s-t$ paths in G is equal to the minimum number of vertices whose removal destroys all $s-t$ paths in G.* ∎

As we pointed out before, the maximum number of vertex-disjoint $s-t$ paths in a directed graph G can be computed by finding a maximum $s-t$ flow in a network N (with G' as the underlying graph) in which each edge has unit capacity. As regards the minimum number of vertices whose removal disconnects s and t in G cut in G, it can be computed by first finding a minimum $s-t$ cut in a network N (with G' as the underlying graph) in which each newly added edge is assigned unit capacity and all other edges are each assigned ∞ capacity. In such an $s-t$ cut $\langle S, \overline{S} \rangle$ all edges directed from S to $\overline{S}$ will have unit capacity. In other words, the edges in the cut will be of the form (v', v''), each one of which corresponding to vertex v in G. Removal of the edges in $\langle S, \overline{S} \rangle$ will disconnect s and t in G'. Equivalently, removal of the corresponding vertices in G will disconnect s and t in G.

13.2.3 Determining Connectivities in Undirected Graphs

A network is called a 0−1 network if each edge has unit capacity. Also, a 0−1 network is of type 1 if it has no parallel edges, and is of type 2 if the in-degree or the out-degree of each vertex is equal to one. In Chapter 4 the following complexity results for Dinic's maximum flow algorithm have been established.

i. $O(m^{3/2})$ if N is a 0−1 network.

ii. $O(n^{2/3}m)$ if N is of type 1.

iii. $O(n^{1/2}m)$ if N is of type 2.

As before let $\lambda(s,t)$ denote the minimum number of edges to be removed to destroy all s-t paths in an undirected graph. Let $\kappa(s,t)$ denote the minimum number of vertices whose removal destroys all s-t paths in G. As we discussed in Section 13.2.1, $\lambda(s,t)$ can be computed by applying Dinic's maximum flow algorithm on a 0-1 network, and $\kappa(s,t)$ can be computed by applying this algorithm on a 0-1 network of type 2. It follows from the complexity results presented above that $\lambda(s,t)$ and $\kappa(s,t)$ can be computed in $O(\min\{n^{2/3}m, m^{3/2}\})$ time and $O(n^{1/2}m)$ time, respectively.

The edge connectivity and vertex connectivity of a graph G denoted by $\lambda(G)$ and $\kappa(G)$ are defined, respectively, as follows:

$$\lambda(G) = \text{minimum } \{\lambda(s,t) | s, t \in V, s \neq t\} \text{ and}$$

$$\kappa(G) = \text{minimum } \{\kappa(s,t) | s, t \in V, s \neq t \text{ and nonadjacent}\}.$$

Since both these parameters require maximum flow computations over $O(n^2)$ pairs of vertices we get the following complexity result for computing $\lambda(G)$ and $\kappa(G)$.

Theorem 13.6

i. *The edge connectivity $\lambda(G)$ of an undirected graph can be computed in $O(n^2 \min\{n^{2/3}m, m^{3/2}\})$ time.*

ii. *The vertex connectivity $\kappa(G)$ of an undirected graph can be computed in $O(n^{5/2}m)$ time.* ∎

The question now arises if we can develop algorithms with better complexities for computing $\lambda(G)$ and $\kappa(G)$. We address this question in the following sections.

13.3 EDGE CONNECTIVITY OF GRAPHS

In this section we develop algorithms to compute the edge connectivity of directed and undirected graphs.

13.3.1 Edge Connectivity of Undirected Graphs: Matula's Algorithm

In this section we develop Matula's algorithm [2] for determining the edge connectivity of an undirected graph.

In the previous section we pointed out that the edge connectivity of an undirected graph can be computed using $O(n^2)$ maximum flow computations. We now show that with at most $n - 1$ maximum flow computations one can determine edge connectivity.

Given an undirected graph $G = (V, E)$, what we are looking for is a cut with the smallest number of edges. Such a cut will be called a minimum cut. Consider a minimum cut $\langle S, \overline{S} \rangle$ so that $\lambda(G) = |\langle S, \overline{S} \rangle|$. Then for any two vertices i and j such that $i \in S$ and $j \in \overline{S}$, we have $\lambda(i,j) = \lambda(G)$. This is because the minimum cut $\langle S, \overline{S} \rangle$ is also a smallest cut that separates i and j. So if we pick any vertex i in G then

$$\lambda(G) = \min\{\lambda(i,j) | j \in V \setminus \{i\}\}.$$

Thus, $\lambda(G)$ can be computed using at most $n - 1$ maximum flow computations, thereby reducing by a factor of n the complexity given in Theorem 13.6. That is, $\lambda(G)$ can be computed in $O(n \min\{n^{2/3}m, m^{3/2}\})$ time.

Next we develop Matula's algorithm [2] that reduces the complexity of computing $\lambda(G)$ to $O(mn)$.

A nonempty subset D of V is called a dominating set of G if every vertex $v \in V$ is either in D or adjacent to a vertex in D. Matula's algorithm is based on the following theorem where $\lambda(G)$ is simply represented as λ.

Theorem 13.7 *Let $G = (V, E)$ be a graph with a minimum cut $\langle A, \overline{A} \rangle$ such that $|\langle A, \overline{A} \rangle| = \lambda \leq \delta - 1$, where δ is the minimum degree in G. Then every dominating set of G contains vertices from both A and $\overline{A}$, that is, $|A \cap D| \neq \varnothing$ and $|\overline{A} \cap D| \neq \varnothing$.*

Proof. The sum of the degrees of the vertices in A satisfies the following:

$$|A|\delta \leq \sum_{v \in A} \text{degree}(v) \leq |A|(|A| - 1) + \lambda.$$

Rewriting the above we get

$$(|A| - \delta)(|A| - 1) \geq \delta - \lambda \geq 1.$$

This means that both the terms on the left-hand side are at least one. So $|A| \geq \delta + 1$ and A contains at least $\delta + 1$ vertices. But by assumption the cut $\langle A, \overline{A} \rangle$ has fewer than δ edges. So one of the vertices in A is not adjacent to any vertex in $\overline{A}$. So no dominating set can contain only members of A. Similarly no dominating set can contain only members of $\overline{A}$. ∎

We now present Matula's algorithm. We follow the presentation in [2]. The algorithm uses a partition S, T, and U of the vertex set V of a graph $G = (V, E)$. Sets T and U are defined as follows.

$T =$ Set of all vertices of $V \setminus S^*$ each of which is adjacent to a vertex of S.

$U =$ Set of all vertices of $V \setminus S$ that are not adjacent to any vertex in S.

The algorithm starts with $S = \{s_0\}$. At the beginning of iteration $i \geq 1$, let $S = \{s_0, s_1, \ldots, s_{i-1}\}$. The algorithm picks a vertex s_i that is in U. Note that s_i is not adjacent to any vertex in S. A min-cut $\langle A, \overline{A} \rangle$ is determined such that $S \subset A$ and $s_i \in \overline{A}$. That is, the cut $\langle A, \overline{A} \rangle$ separates s_i from all vertices of S. s_i is then moved to S and any adjacent vertices of s_i to T. When U becomes empty the algorithm returns the minimum cut obtained over all iterations as the required minimum cut of the graph G. A formal presentation of the algorithm is given in Figure 13.2.

Theorem 13.8 *For any connected graph G with n vertices and m edges, Algorithm 13.1 determines the edge connectivity $\lambda(G)$ in $O(mn)$ time.*

Proof. We need to prove that during some iteration of the algorithm a minimum cut of G is identified.

Suppose $\lambda = \delta$. Then the initial cut $\langle \{p\}, V \setminus \{p\} \rangle$ is a minimum cut and so the required cut is identified at the first iteration itself.

Consider then the case when $\lambda \leq \delta - 1$. Let $\langle A, \overline{A} \rangle$ be a minimum cut. Assume without loss of generality that $p \in A$. We first show that $S \subset A$ until a minimum cut is identified. This is true initially because $S = \{s_0\} = \{p\} \subset A$ by our assumption. Assume for $i \geq 1$ that the set $S = \{s_0, s_1, \ldots, s_{i-1}\} \subset A$. Then the vertex $k = s_i$ that is picked at this iteration is

---

*$V \setminus S$ is the same as $V - S$.

> **Algorithm 13.1 (Matula's Edge Connectivity Algorithm)**
> **Input**: Given a connected undirected graph $G = (V, E)$ with $n \geq 2$ vertices and m edges.
> **Output**: Edge connectivity λ of G and also a minimum cut $\langle A, \overline{A} \rangle$.
>   begin
>   Let p be a minimum degree vertex in G and δ its degree;
> $S \leftarrow \{p\}$ and $\lambda \leftarrow \delta$;
> $A \leftarrow \{p\}$ and $\overline{A} \leftarrow V \setminus \{p\}$;
> $T \leftarrow$ Set of all vertices of $V \setminus S$ each of which is adjacent to a vertex of S.
> $U \leftarrow V \setminus T \setminus S$;
>   while $U \neq \emptyset$ do
>   begin
>   Select a vertex $k \in U$;
>   Compute $\lambda(S, k)$ using the labeling algorithm for maximum flow
>   and the corresponding minimum cut $\langle B, \overline{B} \rangle$ with
> $S \subset B, k \in \overline{B}$;
> $S \leftarrow S \cup \{k\}$;
> $T \leftarrow T \cup \{w | w \text{ adjacent to } k\}$;
> $U \leftarrow U - T - k$.
>   If $\lambda > \lambda(S, k)$ then $\lambda \leftarrow \lambda(S, k)$ and $\langle A, \overline{A} \rangle \leftarrow \langle B, \overline{B} \rangle$;
>   end
>   end

Figure 13.2 Determining edge connectivity.

an element of A or an element of $\overline{A}$. If $s_i \in A$ then $S \cup \{s_i\} \subset A$. Otherwise $s_i \in \overline{A}$. Then the minimum cut $\langle B, \overline{B} \rangle$ separating S and vertex k is also a minimum cut of the graph. That is, $\lambda(S, k) = \lambda(G)$ and the required minimum cut is identified. This minimum cut identification must occur before termination, for otherwise, at termination the set $S \subset A$ is a dominating set containing no elements of $\overline{A}$ contradicting Theorem 13.7.

The complexity bound is dominated by the complexity of finding $\lambda(S, k)$. This requires finding a maximum $s - k$ flow in a network N constructed as follows: First construct the directed graph G' by replacing each edge of G by two oppositely oriented edges. Then construct a network N by adding a new vertex s, connecting s to all the vertices in S by edges oriented away from s and assigning unit capacity to all the edges in N. An $s - k$ maximum flow in N can be constructed by the iterated applications of the labeling algorithm using flow augmentation.

Let n_T be the number of vertices of T adjacent to the vertex k, and n_U be the number of vertices of U adjacent to k. Then n_T paths of length 2 from k to s can be identified in time proportional to n_T. The remaining $|\langle B, \overline{B} \rangle| - n_T$ flow augmenting paths can each be found in $O(m)$ time. Since each such path uses a distinct adjacent vertex of k in U, $|\langle B, \overline{B} \rangle| - n_T \leq n_U$. A cost of $O(m)$ is required to move vertex k to S. So an overall cost of $O((n_U + 1)m)$ is required to move k to S and n_U vertices from U to T. This implies a cost of $O(m)$ per vertex removed from U, yielding $O(mn)$ time for the algorithm. ∎

A graph G is k-edge connected if $\lambda(G) \geq k$. Matula has also given another elegant algorithm to determine if a graph G is k-edge connected. This algorithm is of time complexity $O(kn^2)$ and is presented in Figure 13.3. The proof of correctness and complexity analysis of the algorithm follow along the same lines as the proof of Theorem 13.8. The step which requires checking the existence of a cut separating S and vertex v of cardinality less than or equal to $k - 1$ can be accomplished by testing the existence $k - 1$ edge-disjoint paths from v to S.

> **Algorithm 13.2 (Testing k-edge connectivity)**
> **Input**: Given a connected undirected graph $G = (V, E)$ with $n \geq 2$ vertices and m edges.
> **Output**: Determine if G is k-edge connected.
>   **begin**
>   If $\delta \leq k - 1$ **then** Stop (The graph is not k-edge connected);
>   select a vertex v of maximum degree in G;
> $S \leftarrow \{v\}$;
> $T \leftarrow$ Set of all vertices of $V \setminus S$ each of which is adjacent to a vertex of S.
>   **while** $S \cup T \neq V$ **do**
>   **begin**
>   select a vertex $v \in \overline{S} = V \setminus S$ such that v has maximum degree in the
>   induced subgraph $\langle \overline{S} \rangle$;
>   **If** there is a cut separating S and v with $S \subset B, v \in \overline{B}$ and of
>   cardinality $\leq k - 1$ **then** Stop (The graph is not k- edge connected);
>   **else**
> $S \leftarrow S \cup \{v\}$;
> $T \leftarrow \{w | w \in V \setminus S$ adjacent to some vertex in $S\}$
>   **end**
>   **end** (The graph G is k-edge connected.)

Figure 13.3 Testing k-edge connectivity.

13.3.2 Edge Connectivity of Directed Graphs

Theorem 13.9 *Let $G = (V, E)$ be a directed graph with $V = \{v_1, v_2, \ldots, v_n\}$. Then*

$$\lambda(G) = \min\{\lambda(v_1, v_2), \lambda(v_2, v_3), \ldots, \lambda(v_n, v_1)\}.$$

Proof. Consider any two vertices u and v such that $\lambda(G) = \lambda(u, v)$. Then there is a set T of edges whose removal will destroy all directed paths from u to v. Note that $\lambda(u, v) = |T|$. Let X denote the set of all vertices w that are reachable from u by directed paths containing no edge of T. Let Y denote the set of all vertices w such that each directed path $u - w$ contains some edge of T. Then $\langle X, Y \rangle$ is a cut of G with $u \in X$ and $v \in Y$. Now T disconnects every $x \in X$ and $y \in Y$. So, $|T| = \lambda(G) \leq \lambda(x, y) \leq |T|$. Thus $\lambda(G) = \lambda(x, y)$. This means that there exists some index i such that $v_i \in X$ and $v_{i+1} \in Y$ and $\lambda(G) = \lambda(v_i, v_{i+1})$. ∎

The above theorem due to Schnorr [3] means that the edge connectivity of a directed graph can be computed using n maximum flow computations. The time complexity of such an algorithm is $O(n \min\{n^{2/3}m, m^{3/2}\})$. In fact Mansour an Schieber [4] have shown that this can be achieved in $O(mn)$ time.

13.4 GLOBAL MINIMUM CUT IN A WEIGHTED UNDIRECTED GRAPH: STOER AND WAGNER'S ALGORITHM

Consider a weighted undirected graph $G = (V, E)$ with each edge e associated with a capacity $c(e) > 0$. For ease of notation a cut $\langle S, \overline{S} \rangle$ will be denoted by $\delta(S)$, the set of edges that have exactly one vertex in the set S. Note that the capacity $c(\delta(S))$ of a cut $\delta(S)$ is the sum of the capacities of the edges in the cut. In this section we consider the problem of determining a cut with minimum capacity in an undirected graph. Such a cut is called a *global minimum cut* or simply, a minimum cut. Note that when the edge capacities are all equal to unity, then the edge connectivity $\lambda(G)$ is equal to the capacity of a minimum cut of G.

As pointed out in Section 13.2, a minimum cut can be obtained by picking a vertex s, computing $\lambda(s, v)$ for each $v \neq s$ and taking the minimum of the weights of these cuts. Such a maximum flow-based approach will have a complexity of $O(n^4)$. We now discuss a much simpler algorithm due to Stoer and Wagner [5]. This algorithm uses the operation of vertex identification defined below.

Given two distinct vertices v and w in a graph $G = (V, E)$, the graph G_{vw} resulting from the identification (contraction) of v and w has vertex set $V(G_{vw}) = V \setminus \{v, w\} \cup \{x\}$ where x is a new vertex and edge set $E(G_{vw}) = E \setminus \{\text{set of all parallel edges connecting } v \text{ and } w\}$. For each edge $e \in E(G_{vw})$ and end vertex $p \in V$, p is an end vertex of e in G_{vw}, if $p \neq v, w$; otherwise x is an end of e in G_{vw}. All edges in G_{vw} have the same capacities as in G. Note that G_{vw} will not have self-loops (x, x) even when there are parallel edges in G.

The following theorem is the basis of Store and Wagner's minimum cut algorithm.

Theorem 13.10 *Let s and t be two vertices of a graph G. Then $\lambda(G) = \min\{\lambda(G_{st}), \lambda(G; s, t)\}$, where $\lambda(G; s, t)$ is the capacity of a minimum s–t cut separating s and t in G, and $\lambda(G)$ is the capacity of a minimum cut of G.*

Note: For the sake of convenience we use $\lambda(G)$ to denote the capacity of a minimum cut of G.

Proof. The result follows because

1. If there is a minimum cut of G separating s and t then $\lambda(G) = \lambda(G; s, t)$;

2. Otherwise $\lambda(G) = \lambda(G_{st})$. ∎

This theorem suggests a recursive algorithm to compute $\lambda(G)$. To achieve an efficient algorithm, we need an algorithm to find two vertices s and t such that the computation of $\lambda(G; s, t)$ is easily achieved without the help of a maximum flow algorithm. This requires ordering the vertices by what is known as the MA (maximum adjacency) ordering. An MA ordering of G is an ordering of the vertices as $v_1, v_2, \ldots, v_n$ of G such that

$$c(\delta(V_{i-1}) \cap \delta(v_i)) \geq c(\delta(V_{i-1}) \cap \delta(v_j)) \text{ for } 2 \leq i < j \leq n,$$

where $V_i = \{v_1, v_2, \ldots, v_i\}$.

Basically, we choose any vertex as v_1 and at step i we select v_i to be the vertex that has the largest capacity of edges connecting it to the previously selected vertices. The time complexity to find an MA ordering is no more than the time complexity of Prim's minimum spanning tree algorithm. It is shown in [6] that an MA ordering can be found in $O(m + n \log n)$.

To establish the main result we need the following.

Lemma 13.1 *If $p, q, r \in V$, then $\lambda(G; p, q) \geq \min\{\lambda(G; p, r), \lambda(G; r, q)\}$.*

Proof. Consider a minimum $p - q$ cut $\delta(S)$ with $p \in S$. If $r \in S$, then $\delta(S)$ is an $r - q$ cut and so $c(\delta(S)) \geq \lambda(G; r, q)$. Otherwise, $\delta(S)$ is a $p - r$ cut and so $c(\delta(S)) \geq \lambda(G; p, r)$. Hence the result. ∎

In fact, using induction and Lemma 13.1 we can show the following.

Theorem 13.11 *For vertices $p_1, p_2, \ldots, p_k$,*

$$\lambda(G; p_1, p_k) \geq \min\{\lambda(G; p_1, p_2), \lambda(G; p_2, p_3), \ldots, \lambda(G; p_{k-1}, p_k)\}. \quad \blacksquare$$

We follow the treatment in [7] for the proof of the following main result of this section.

Theorem 13.12 *If $v_1, v_2, \ldots, v_n$ is an MA ordering of G, then $\delta(v_n)$ is a minimum $v_{n-1} - v_n$ cut of G.*

Proof. $c(\delta(v_n)) \geq \lambda(G; v_{n-1}, v_n)$ because $\delta(v_n)$ is a $v_{n-1} - v_n$ cut of G. So we just need to show that $c(\delta(v_n)) \leq \lambda(G; v_{n-1}, v_n)$. Proof is by induction on the number of edges and the number of vertices. The result is trivially true for $n = 2$ or $m = 0$.

We consider two cases. In the following δ' refers to the δ on graph G'.

Case 1 (v_{n-1}, v_n) is an edge of G. Let G' be the graph obtained after deleting edge e in G. It is easy to see that $v_1, v_2, \ldots, v_n$ is still an MA ordering of G'. So by the induction hypothesis $c(\delta'(v_n)) = \lambda(G'; v_{n-1}, v_n)$. Then $c(\delta(v_n)) = c(\delta'(v_n)) + c(e) = \lambda(G'; v_{n-1}, v_n) + c(e) = \lambda(G; v_{n-1}, v_n)$. Thus the result is true in this case.

Case 2 v_{n-1} and v_n are not adjacent in G. We will show that

$$c(\delta(v_n)) \leq \lambda(G; v_{n-2}, v_n) \tag{13.5}$$

and that

$$c(\delta(v_n)) \leq \lambda(G; v_{n-1}, v_{n-2}). \tag{13.6}$$

Combining these with the result of Lemma 13.1 will prove that $c(\delta(v_n)) \leq \lambda(G; v_{n-1}, v_n)$.

Consider first the inequality (13.5). Let $G' = G \setminus v_{n-1}$, be the graph obtained by deleting the vertex v_{n-1}. Again it is easy to see that $v_1, v_2, \ldots, v_{n-2}, v_n$ is an MA ordering of G'. Then by the induction hypothesis we have

$$c(\delta'(v_n)) = \lambda(G'; v_{n-2}, v_n).$$

So

$$c(\delta(v_n)) = c(\delta'(v_n)) = \lambda(G'; v_{n-2}, v_n) \leq \lambda(G; v_{n-2}, v_n).$$

To prove the inequality (13.6) we apply induction on $G' = G \setminus v_n$. Since $v_1, v_2, \ldots, v_{n-2}, v_{n-1}$ is an MA ordering of G' we get

$$c(\delta(v_n)) \leq c(\delta(v_{n-1})) = c(\delta'(v_{n-1})) \leq \lambda(G'; v_{n-2}, v_{n-1}) \leq \lambda(G; v_{n-1}, v_{n-2}). \blacksquare$$

Stoer and Wagner's algorithm for the minimum cut problem in undirected graphs is given in Figure 13.4. An example taken from [5] and shown in Figure 13.5 illustrates this algorithm. In Figure 13.5 s and t refer to the last two vertices in an MA ordering. For this graph, $\langle \{3, 4, 7, 8\}, \{1, 2, 5, 6\} \rangle$ is a min-cut with capacity $= 4$.

The correctness of this recursive algorithm follows from Theorems 13.10 and 13.12. The time complexity of the algorithm is dominated by the time required for at most n applications of an algorithm for determining an MA ordering. Since an MA ordering can be found in $O(m + n \log n)$ [6], the overall time complexity of the algorithm is $O(mn + n^2 \log n)$.

13.5 VERTEX CONNECTIVITY IN UNDIRECTED GRAPHS

It was pointed out in Theorem 13.6 that the vertex connectivity of a graph G can be computed in $O(n^{5/2} m)$. We can do much better than this. In this section we state and prove the correctness of an algorithm to compute $\kappa(G)$, the vertex connectivity of a simple n-vertex undirected graph G. This algorithm is due to Even and Tarjan [8]. Since the vertex connectivity of an n-vertex complete graph is $n - 1$, we assume that the graph under consideration is not complete.

Theorem 13.13 *Algorithm 13.4 terminates with vertex connectivity $\kappa(G) = \alpha$.*

> **Algorithm 13.3 (Stoer and Wagner's min-cut algorithm)**
> **Input**: A connected undirected graph $G = (V, E)$.
> **Output**: A minimum cut A of G.
>   begin
> $M \leftarrow \infty$ (a very large number);
>   while G has more than one vertex do
>   begin
>   find an MA ordering $v_1, v_2, \ldots, v_n$ of G;
>   if $c(\delta(v_n)) < M$ do $M \leftarrow c(\delta(v_n))$ and $A \leftarrow \delta(v_n)$;
> $G \leftarrow G_{v_{n-1}, v_n}$
>   end
>   end while
>   return A, a minimum cut of G.
>   end

Figure 13.4 Stoer–Wagner's min-cut algorithm.

Proof. Clearly after the first computation of $\kappa(v_1, v_j)$ for some v_j that is not adjacent to v_1, α satisfies

$$\kappa(G) \leq \alpha \leq n - 2. \tag{13.7}$$

From this point on, α can only decrease.

By definition $\kappa(G)$ is the minimum number of vertices whose removal disconnects two vertices in G. Let R denote such a set of vertices. Since $|R| = \kappa(G)$, and $p > \alpha \geq \kappa(G)$, at least one of the vertices $v_1, v_2, \ldots, v_p$ is not in R. Let v_i be one such vertex. Then removing from G the vertices in R will separate the remaining vertices into two sets such that each path from a vertex of one set to a vertex of another set passes through at least one vertex of R. So there exists a vertex v such that $\kappa(v_i, v) \leq |R| = \kappa(G)$. Therefore $\alpha \leq \kappa(G)$. Thus $\alpha = \kappa(G)$. ∎

Theorem 13.14 *The complexity of computing the vertex connectivity of an undirected G is $O(n^{1/2} m^2)$ where m and n are, respectively, the number of edges and the number of vertices of G.*

Proof. Each step of Algorithm 13.4 requires at most n maximum flow computations, each of complexity $O(n^{1/2} m)$. At most p steps will be performed. So the complexity of this algorithm is $O(\kappa(G) n^{3/2} m)$. But $\kappa(G) \leq 2m/n$. So the complexity of the algorithm is $O(n^{1/2} m^2)$. ∎

13.6 GOMORY–HU TREE

Consider a graph $G = (V, E)$ with a nonnegative capacity $c(e)$ associated with each edge e. Let T be a spanning tree on the vertex set V with a label f_e on each edge e in T. T is said to be flow equivalent to G, if for any pair of vertices u and v in G, the capacity of a minimum $u-v$ cut is equal to the minimum label f_e among the edges e in the path from u to v in T. Suppose edge e^* achieves this minimum for vertices u and v. Then the flow equivalent tree T is called a *Gomory–Hu* tree of G if the cut defined by the two trees obtained when e^* is deleted from T is a minimum $u-v$ cut in G. In this section we develop an algorithm to construct a *Gomory–Hu* tree [9].

First we establish certain conditions on the existence of a flow equivalent tree.

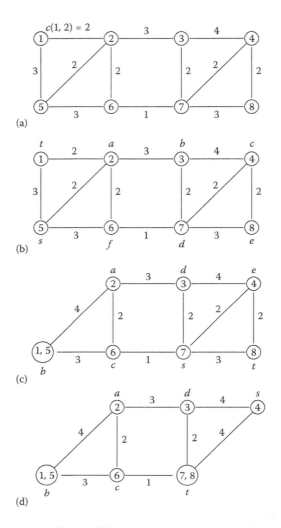

Figure 13.5 Example illustrating Stoer–Wagner's min-cut algorithm. (a) Graph G with edge capacities. (b) MA ordering $2,3,4,7,8,6,5,1$ of vertices of G. Here $s = 5$ and $t = 1$. Capacity of minimum s–t cut = 5. (c) Graph after contracting 1 and 5 in the graph (b). Now $s = 7$ and $t = 8$ in the MA ordering of the contracted graph. Capacity of minimum s–t cut = 5. (d) Graph after contracting s and t in the graph (c). Now $s = 4$ and $t = \{7,8\}$ in the MA ordering of the contracted graph. Capacity of minimum s–t cut = 7.

(*Continued*)

Theorem 13.15 *Given $f(x,y) = f(y,x)$ for any x and y, $1 \leq x \leq n$ and $1 \leq y \leq n$ and $x < y$. There exists an n-vertex connected graph $G = (V, E)$ such that $f(x,y)$ is the value of a maximum $x - y$ flow if and only if, for any three distinct vertices $p, q, r \in V$,*

$$f(p,q) \geq \min\{f(p,r), f(r,q)\}.$$

Proof. Necessary part of the theorem follows from Lemma 13.1.

Sufficiency: Treating $f(x,y)$'s as weights of the edges (x,y) of a complete graph, construct a maximum spanning tree T of G. For any two vertices x and y, let $q(x,y)$ denote the minimum of the weights of the edges that lie on the unique path from x to y in T. $q(x,y)$ will be called the capacity of this path. Then $q(x,y) \geq f(x,y)$ for otherwise one can get

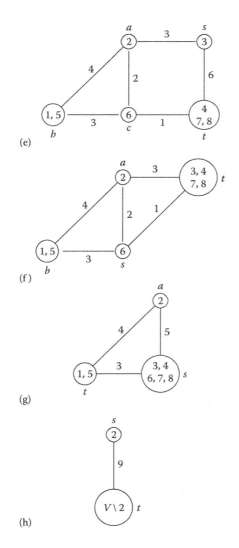

Figure 13.5 (Continued) Example illustrating Stoer–Wagner's min-cut algorithm. (e) Graph after contracting s and t in the graph of Figure 13.5d. Now $s = 3$ and $t = \{4, 7, 8\}$ in the MA ordering of the contracted graph. Capacity of minimum s–t cut = 7. (f) Graph after contracting s and t in the graph of Figure 13.5e. Now $s = 6$ and $t = \{3, 4, 7, 8\}$ in the MA ordering of the contracted graph. Capacity of minimum s–t cut = 4. (g) Graph after contracting s and t in the graph of Figure 13.5f. Now $s = \{3, 4, 6, 7, 8\}$ and $t = \{1, 5\}$ in the MA ordering of the contracted graph. Capacity of minimum s–t cut = 7. (h) Graph after contracting s and t in the graph of Figure 13.5g. Now $s = 2$ and $t = \{1, 3, 4, 5, 6, 7, 8\}$ in the MA ordering of the contracted graph. Capacity of minimum s–t cut = 9.

a spanning tree with larger weight by removing an edge with capacity $q(x, y)$ from T and adding edge (x, y). Also for any path $x = x_1, x_2, x_3, \ldots, x_k = y$,

$$f(x, y) = f(x_1, x_k) \geq \min\{f(x_1, x_2), f(x_2, x_3), \ldots, f(x_{k-1}, x_k)\}.$$

So, $f(x, y) \geq q(x, y)$ and $f(x, y) = q(x, y)$. Since the maximum $x - y$ flow in T is uniquely determined by the $x - y$ path in T, its value is equal to $q(x, y) = f(x, y)$, minimum of the weights of the edges in the $x - y$ path in T. Thus the tree T is the required graph realizing $f(x, y)$'s as maximum flows. ∎

> **Algorithm 13.4 (Even's vertex connectivity algorithm)**
> **Input**: A non-complete Undirected Graph $G = (V, E)$
> **Output**: Connectivity $\kappa(G)$ of G.
>   begin
>   Compute $\kappa(v_1, v_j)$ for every vertex v_j that is not adjacent to v_1. Let α be the minimum of these values;
>   Repeat this computation with $v_2, v_3, \ldots$, each time updating the value of α;
>   Terminate with v_p, once p exceeds the current value of α;
>   end /* At termination $\kappa(G) = \alpha$.*/

Figure 13.6 Even's vertex connectivity algorithm.

This theorem also demonstrates that a flow equivalent tree T can be constructed for a given G. Such a tree may not be a *Gomory–Hu* tree. An interesting corollary of this theorem is stated next.

Corollary 13.1 *The smallest two of the three maximum flow values $f(x, y)$, $f(y, z)$, and $f(z, x)$ in a graph are equal.* ∎

Next we proceed to develop an algorithm for constructing a Gomory–Hu tree. We follow the treatment in [7].

Lemma 13.2 *Let $\delta(A)$ and $\delta(B)$ be two cuts in a graph $G = (V, E)$ with a capacity $c(e) \geq 0$ associated with each edge e. Then*

$$c(\delta(A)) + c(\delta(B)) \geq c(\delta(A \cup B)) + c(\delta(A \cap B)).$$

Proof. Consider Figure 13.7 where the sets $(A \cap B), (A \cap \overline{B}), (\overline{A} \cap B)$, and $(\overline{A} \cap \overline{B})$ are shown as circles. The set of edges connecting two of these sets is shown by one single line with total capacity marked on this edge. For example, the sum of the capacities of the edges that connect $(A \cap B)$ and $(\overline{A} \cap B)$ is equal to a. So,

$$c(\delta(A)) = a + b + c + d, c(\delta(B)) = b + c + e + f, c(\delta(A \cap B)) = a + b + e$$

and $c(\delta(A \cup B)) = b + d + f$. So

$$c(\delta(A)) + c(\delta(B)) \geq c(\delta(A \cup B)) + c(\delta(A \cap B)).$$
∎

The inequality in Lemma 13.2 is called the *submodular inequality*.

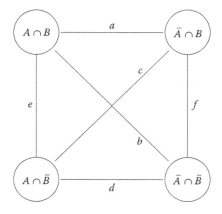

Figure 13.7 Illustration of the proof of Lemma 13.2.

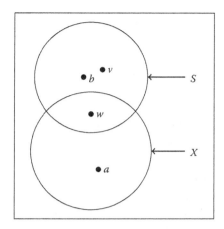

Figure 13.8 Illustration of Case 1 of Theorem 13.16.

Theorem 13.16 *Let $\delta(S)$ be a minimum $a - b$ cut for some vertices $a, b \in V$ in a graph $G = (V, E)$ with a capacity $c(e) \geq 0$ associated with each edge e. For any two vertices v and $w \in S$, there exists a minimum $v - w$ cut $\delta(T)$ such that $T \subset S$.*

Proof. Let $\delta(X)$ be a minimum $v - w$ cut. Suppose that both $X \cap S$ and $X \cap \overline{S}$ are nonempty. Without loss of generality assume that $b \in \overline{X}$.

Case 1 $a \in X$ (See Figure 13.8.) By the submodular inequality we have

$$c(\delta(S)) + c(\delta(\overline{X})) \geq c(\delta(S \cup \overline{X})) + c(\delta(S \cap \overline{X})). \tag{13.8}$$

Since $\delta(S \cup \overline{X})$ is an $a - b$ cut and $\delta(S)$ is a minimum $a - b$ cut

$$c(\delta(S \cup \overline{X})) \geq c(\delta(S)). \tag{13.9}$$

Combining (13.8) and (13.9) we get

$$c(\delta(S \cap \overline{X})) \leq c(\delta(\overline{X})) = c(\delta(X)).$$

So $\delta(S \cap \overline{X})$ is a minimum $v - w$ minimum cut. Here $T = S \cap \overline{X} \subset S$, as required.

Case 2 $a \in \overline{X}$. Following the same line of arguments as for case 1 we can show that $\delta(S \cap X)$ is a minimum $v - w$ minimum cut. ∎

This theorem has a very important implication. As in Theorem 13.16, let $\delta(S)$ be a minimum a–b cut in G. Let s and t be any two vertices in S. Let G' be the graph obtained after contracting the vertices in S to a single vertex $\{S\}$. Then by theorem 13.16 if $\{T \cup \{S\}\}$ is a minimum s–t cut in G' then $\{T \cup S\}$ is a minimum s–t cut in G. This result is the basis of the Gomory–Hu tree construction algorithm in Figure 13.9 (see steps 3–5 in this algorithm).

Next, we prove that at termination of Algorithm 13.5 T is a Gomory–Hu tree. Let T be the tree produced at some step in the algorithm, and let e be an edge in T joining sets R and S in T. Vertices $r \in R$ and $s \in S$ will be called representatives for e if $f_e = f(r, s)$, the capacity of a minimum $r - s$ cut in G.

Theorem 13.17 *At every stage in the Gomory–Hu Algorithm 13.5 there exist representatives for each edge in the tree T.*

> **Algorithm 13.5 (Gomory–Hu Tree Construction)**
>
> **Initialization**
> 1. Start with an empty tree T.
> 2. (a) Select any two vertices r and s in V, and determine a minimum $r - s$ cut $\delta(S) = \langle S, \overline{S} \rangle$ in G. Let $r \in R, s \in S$.
>   (b) Add to tree T vertices labeled as R and S and an edge e connecting R and S. Assign label $f_e = f(r,s)$, the capacity of $\delta(S)$. (Note: Each of R and S represents a group of vertices.)
>
> **General Step (Expansion of T)**
> 3. Pick any vertex A in the current tree T such that A represents at least two vertices of G. In G, identify (contract) the vertices in the subtree rooted on each neighbor of A in T. Let the resulting graph be denoted as G_A. Note that the subtrees will appear as contracted vertices in G_A (see Figure 13.10).
> 4. Select any two vertices x and y in A and determine a minimum $x - y$ cut $\delta(S)$ in G_A. Repeated application of Theorem 13.16 guarantees that the capacity of this minimum cut is also the capacity of a minimum $x - y$ cut in G. Also the set S may have some contracted vertices representing the subtrees.
> 5. Split A into two vertices X and Y with X representing all vertices of S except the contracted vertices contained in S. Add an edge e connecting X and Y with the capacity of the minimum $x - y$ cut as f_e. Use the following rule to reconnect the different subtrees to X and Y.
>   i. If a contracted vertex is in S, then connect the corresponding root to X in T;
>   ii. Otherwise, connect the root to Y.
>
> **Note**: While reconnecting the root the edge labels do not change.
>
> **End of General Step**
> 6. The general step is repeated until every vertex in the current tree T represents only one vertex.

Figure 13.9 Gomory–Hu tree construction.

Proof. Clearly this is true initially. The two vertices picked in the initialization step are the representatives of the first edge added to T.

In the general step vertex A is split into two sets X and Y given by an $x - y$ cut with $x \in X$ and $y \in Y$, and a tree edge connecting X and Y is added. The vertices x and y are clearly representatives for this newly added tree edge. So we need only show that the reconnected edges (whose vertices represent new sets) also have representatives. Consider the edge h connecting B (a root) to X. The case connecting B to Y can be handled in a similar manner. By assumption there exist representatives $a \in A$ and $b \in B$ such that $f_h = f(a,b)$. If $a \in X$ then a and b will continue to be representatives for h.

If $a \in Y$, we will show that x and b are representatives for h. That is, we will prove that $f(x,b) = f(a,b)$ (see Figure 13.11).

The minimum $a - b$ cut that was used to obtain the edge h also separates x and b. So, $f(x,b) \leq f(a,b)$. We will now show the reverse inequality.

Let G' be the graph obtained by identifying (contracting) all vertices in Y, and let V_Y denote the vertex in G' denoting the set Y. Then by Theorem 13.16 we have

$$f(x,b) = f'(x,b).$$

By Theorem 13.15,

$$f'(x,b) \geq \min\{f'(x,V_Y), f'(V_Y,b)\}.$$

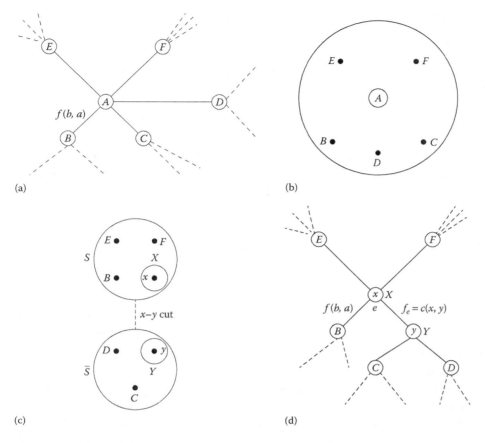

Figure 13.10 Example illustrating Gomory–Hu tree construction. (a) Tree T, $a \in A$, $b \in B$. (b) G_A. (c) $x-y$ cut $\delta(S)$, $x, y \in A$, $X \cup Y = A$. (d) Expansion of vertex A.

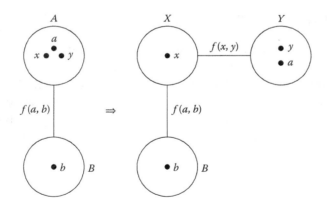

Figure 13.11 Illustration of proof of theorem 13.17.

Since $a \in Y$ we have
$$f'(V_Y, b) \geq f(a, b).$$

Also since the $x - y$ cut that splits A into X and Y also separates a and b, we have
$$f'(x, V_Y) \geq f(x, y) \geq f(a, b).$$

This leads to $f(x, b) \geq f(a, b)$. Thus $f(x, b) = f(a, b)$ and x and b are representatives for the reconnected edge h. ∎

Theorem 13.18 *The tree constructed by Algorithm 13.5 is a Gomory–Hu tree.*

Proof. Recall that a Gomory–Hu tree is defined as follows.

1. For any pair of vertices r and s in G, the capacity of a minimum $r-s$ cut is equal to the minimum label f_e among the edges e in the path from r to s in T.

2. Suppose edge e^* achieves this minimum for vertices r and s. Then the cut defined by the two trees obtained when e^* is deleted from T is a minimum $r-s$ cut in G.

To prove that the tree T produced by the algorithm is a Gomory–Hu tree, let $v_0, e_1, v_1, e_2, v_2, \ldots, e_k, v_k$ be the path in T from vertex r to vertex s. Then by Theorem 13.11 and the fact that $f_{e_i} = f(v_{i-1}, v_i)$ (by Theorem 13.17) we have

$$f(r,s) \geq \min\{f_{e_1}, f_{e_2}, \ldots, f_{e_k}\}.$$

Since each of the cuts corresponding to the edges $e_1, e_2, \ldots, e_k$ separate r and s, we have

$$f(r,s) \leq \min\{f_{e_1}, f_{e_2}, \ldots, f_{e_k}\}.$$

so $f(r,s) = \min\{f_{e_1}, f_{e_2}, \ldots, f_{e_k}\}$.

The second property follows because each edge e in T corresponds to a cut specified by the two trees obtained when e is deleted from T, and this cut has capacity equal to f_e. ∎

Gomory–Hu tree construction as in Algorithm 13.5 involves $n-1$ minimum cut computations and so is of complexity $O(n^4)$. The real difficulty in implementing the algorithm lies in performing the identification (contraction) operations. Gusfield [10] gives an approach that avoids this. This algorithm is simple, but its proof of correctness is very involved. If we are interested only in a flow equivalent tree (that may not be a Gomory–Hu tree), then see Gusfield [10] for a simpler algorithm. See also Jungnickel [11] for a discussion of this algorithm. Ahuja et al. [12] discuss another algorithm due to Cheng and Hu [13] for Gomory–Hu tree construction.

13.7 GRAPHS WITH PRESCRIBED DEGREES

A sequence $(d_1, d_2, \ldots, d_n)$ of nonnegative integers is graphic if there exists an n-vertex graph with vertices $v_1, v_2, \ldots, v_n$ such that vertex v_i has degree d_i.

In this section we first describe an algorithm to construct a simple graphs, if one exists, having a prescribed degree sequence. We then use this algorithm to establish Edmonds' theorem on the existence of k-edge-connected simple graphs having prescribed degree sequences.

Consider a graphic sequence $(d_1, d_2, \ldots, d_n)$ with $d_1 \geq d_2 \ldots \geq d_n$. Let d_i be the degree of vertex v_i. To *lay off* d_k means to connect the corresponding vertex v_k to the vertices

$$v_1, v_2, \ldots, v_{d_k}, \text{ if } d_k < k$$

or to the vertices

$$v_1, v_2, \ldots, v_{k-1}, v_{k+1}, \ldots, v_{d_k+1}, \text{ if } d_k \geq k.$$

The sequence

$$(d_1 - 1, \ldots, d_{d_k} - 1, d_{d_k+1}, \ldots, d_{k-1}, 0, d_{k+1}, \ldots, d_n), \text{ if } d_k < k$$

or

$$(d_1 - 1, \ldots, d_{k-1} - 1, 0, d_{k+1} - 1, \ldots, d_{d_k+1} - 1, d_{d_k+2}, \ldots, d_n), \text{ if } d_k \geq k$$

is called the *residual sequence* after laying off d_k or simply the residual sequence.

Hakimi [14] and Havel [15] have given an algorithm for constructing a simple graph, if one exists, having a prescribed degree sequence. This algorithm is based on a result that is a special case (where $k=1$) of the following theorem due to Wang and Kleitman [16].

Theorem 13.19 *If a sequence $(d_1, d_2, \ldots, d_n)$ with $d_1 \geq d_2 \geq \ldots \geq d_n$ is the degree sequence of a simple graph, then so is the residual sequence after laying off d_k.*

Proof. To prove the theorem we have to show that a graph having $(d_1, d_2, \ldots, d_n)$ as its degree sequence exists such that vertex v_k is adjacent to the first d_k vertices other than itself. If otherwise, select from among the graphs with the degree sequence $(d_1, d_2, \ldots, d_n)$ a simple graph G in which v_k is adjacent to the maximum number of vertices among the first d_k vertices other than itself. Let v_m be a vertex not adjacent to v_k in G such that

$$m \leq d_k, \text{ if } d_k < k$$

or

$$m \leq d_k + 1, \text{ if } k \leq d_k.$$

In other words v_m is among the first d_k vertices other than v_k. So v_k is adjacent in G to some vertex v_q that is not among these first d_k vertices. Then $d_m > d_q$ (if equality, the order of q and m can be interchanged), and hence v_m is adjacent to some vertex $v_t, t \neq q, t \neq m$, such that v_t and v_q are not adjacent in G. If we now remove the edges (v_m, v_t) and (v_k, v_q) and replace them by (v_m, v_k) and (v_t, v_q), we obtain a graph G' with one more vertex adjacent to v_k among the first d_k vertices other than itself violating the definition of G. ∎

From Theorem 13.19 we get the following algorithm which is a generalization of Hakimi's algorithm for realizing a sequence $D = (d_1, d_2, \ldots, d_n)$ with $d_1 \geq d_2 \geq \ldots \geq d_n$ by a simple graph. Choose any $d_k \neq 0$. Lay off d_k by connecting v_k to the first d_k vertices other than itself. Compute the residual degree sequence. Reorder the vertices so that the residual degrees in the resulting sequence are in nonincreasing order. Repeat this process until one of the following occurs:

1. All the residual degrees are zero. In this case the resulting graph has D as its graph sequence.

2. One of the residual degrees is negative. This means that the sequence D is not graphic.

To illustrate the preceding algorithm, consider the sequence

$$\begin{array}{cccccc} & v_1 & v_2 & v_3 & v_4 & v_5 \\ D = (& 4 & 3 & 3 & 2 & 2). \end{array}$$

After laying off d_3, we get the sequence

$$\begin{array}{cccccc} & v_1 & v_2 & v_3 & v_4 & v_5 \\ D' = (& 3 & 2 & 0 & 1 & 2), \end{array}$$

which, after reordering of the residual degrees, becomes

$$\begin{array}{cccccc} & v_1 & v_2 & v_5 & v_4 & v_3 \\ D_1 = (& 3 & 2 & 2 & 1 & 0). \end{array}$$

We next lay off the degree corresponding to v_5 and get

$$\begin{array}{cccccc} & v_1 & v_2 & v_5 & v_4 & v_3 \\ D_1' = (& 2 & 1 & 0 & 1 & 0). \end{array}$$

Reordering the residual degrees in D_1' we get

$$\begin{array}{cccccc} & v_1 & v_2 & v_4 & v_5 & v_3 \\ D_2 = (& 2 & 1 & 1 & 0 & 0). \end{array}$$

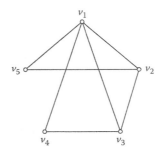

Figure 13.12 Graph with degree sequence $(4, 3, 3, 2, 2)$.

Now laying off the degree corresponding to v_1, we get

$$D'_2 = \begin{pmatrix} v_1 & v_2 & v_4 & v_5 & v_3 \\ 0 & 0 & 0 & 0 & 0 \end{pmatrix}.$$

The algorithm terminates here. Since all the residual degrees are equal to zero, the sequence (4, 3, 3, 2, 2) is graphic. The required graph (Figure 13.12) is obtained by the following sequence of steps, which corresponds to the order in which the degrees were laid off:

1. Connect v_3 to v_1, v_2, and v_4.

2. Connect v_5 to v_1 and v_2.

3. Connect v_1 to v_2 and v_4.

Erdös and Gallai [17] have given a necessary and sufficient condition (not of an algorithmic type) for a sequence to be graphic. See also Harary [18].

Suppose that in the algorithm we lay off at each step the smallest nonzero residual degree. Then using induction we can easily show that the resulting graph is connected if

$$d_i \geq 1 \quad \text{for all } i \tag{13.10}$$

and

$$\sum_{i=1}^{n} d_i \geq 2(n-1). \tag{13.11}$$

Note that (13.10) and (13.11) are necessary for the graph to be connected. In fact we prove in the following theorem a much stronger result. This result is due to Edmonds [19]. The proof given here is due to Wang and Kleitman [20].

Theorem 13.20 (Edmonds) *A necessary and sufficient condition for a graphic sequence $(d_1, d_2, \ldots, d_n)$ to be the degree sequence of a simple k-edge-connected graph, for $k \geq 2$, is that each degree $d_i \geq k$.*

Proof. Necessity is obvious.

We prove the sufficiency by showing that the algorithm we have just described results in a k-edge-connected graph when all the edges in the given graphic sequence are greater than or equal to k. Note that at each step of the algorithm, we should lay off the smallest nonzero residual degree.

Proof is by induction. Assume that the algorithm is valid for all sequences in which each degree $d_i \geq p$ with $p \leq k - 1$.

To prove the theorem we have to show that in the graph constructed by the algorithm every cutset $\langle A, \overline{A} \rangle$ has at least k edges. Proof is trivial if $|A| = 1$ or $|\overline{A}| = 1$. So assume that $|A| \geq 2$ and $|\overline{A}| \geq 2$.

We claim that in the connection procedure of the algorithm, one of the following three cases will eventually occur at some step, say, step r. (Step r means the step when the rth vertex is fully connected.)

Case 1 All the nonzero residual degrees are at least k.

Case 2 All the nonzero residual degrees are at least $k - 1$, and there is at least one edge constructed by steps $1, \ldots, r$ that lies in $\langle A, \overline{A} \rangle$.

Case 3 All the nonzero residual degrees are at least $k - 2$, and there are at least two edges constructed by steps $1, 2, \ldots, r$ that lie in $\langle A, \overline{A} \rangle$.

All these three cases imply that the cutset $\langle A, \overline{A} \rangle$ has at least k edges by induction.

We prove this claim as follows.

Let v_i be the vertex that is connected at step i. Without loss of generality we may assume that v_1 is in A. We now show that at some step in the connection procedure case 3 must occur if cases 1 and 2 never occur. At step 1, v_1 is fully connected. Then:

a. The smallest nonzero residual degrees are $k - 1$, because case 1 has not occurred.

b. The degrees of none of the vertices in $\overline{A}$ are decreased by 1 when v_1 is connected, because case 2 has not occurred.

Hence v_2, the vertex to be connected at step 2, must be in A. (All vertices in $\overline{A}$ must have degree at least k.) Now if v_2 is connected and case 1 does not occur and no edge connects A with $\overline{A}$, then we shall still have (a) and (b) as before. Hence the next vertex to be connected will still be in A.

Since the residual degrees are decreased at each step of the connection procedure, sooner or later there must exist an r with v_r in A, such that when v_r is connected, an edge will connect A with $\overline{A}$. If case 2 does not occur, then one of the nonzero residual degrees among the vertices of A not yet fully connected becomes $k - 2$. This means, by our connection procedure, that v_r must connect to every vertex in $\overline{A}$, because vertices in $\overline{A}$ all have residual degree equal to k, and since we connect v_r to the vertices of the largest residual degree. Since $|\overline{A}| \geq 2$, at this step case 3 occurs. ∎

Wang and Kleitman [16] have also established necessary and sufficient conditions for a graphic sequence to be the degree sequence of a simple k-vertex-connected graph.

Summary and Related Works

The book [6] by Nagomochi and Ibaraki gives a comprehensive treatment of various aspects of connectivity algorithms. Our treatment in Sections 13.4 and 13.6 follows closely the developments in the book [7] by Cook et al. Jungnickel [11], Korte and Vygen [21], and Ahuja et al. [12] are other excellent references for the material covered in this chapter. See Esfahanian [22] for a very readable exposition of several connectivity algorithms.

The idea of node contraction used by Stoer and Wagner [5] in their min-cut algorithm was earlier considered in Paderberg and Rinaldi [23]. In contrast to node contraction, Karger [24] used edge contraction to develop a randomized contraction algorithm for the min-cut problem. This is discussed in Chapter 35 of this book (see also Karger and Stein [25]). Nagamochi and Ibaraki [26] have discussed edge connectivity problem in multigraph and capacitated graphs.

As regards connectivity (i.e., vertex connectivity), algorithms for testing 2-connectivity and strong connectivity have been discussed in Chapter 3 of this book. Hopcroft and Tarjan [27] have developed an $O(n)$ time algorithm to test 3-connectivity of an undirected graph. For early algorithms to test if the connectivity of a graph is at least k, see Galil [28], Even [29], and Kleitman [30]. Cheriyan and Thurimella [31] presented distributed and parallel algorithms for the k-connectivity problem.

For a more detailed discussion of Gomory–Hu tree algorithm see Jungnickel [11], Ahuja et al. [12], and Korte and Vygen [21]. For a detailed discussion of the problem of synthesizing networks satisfying prescribed flow requests see Gomory and Hu [9]. See also Gondron and Minoux [32]. Korte and Vygen [21] discuss several related network design algorithms.

Boesch [33] has several papers on the design of graphs having specific connectivity and reliability properties. Some of the early papers that deal with this topic include Hakimi [34], Boesch and Thomas [35], and Amin and Hakimi [36]. Frank and Frisch [37] is one of the earliest works to discuss a variety of topics related to network design that are of great interest to researchers in the modern area of communication networks. For some more works see Krishnamoorthy et al. [38,39], Bermond et al. [40], Opatrny et al. [41], and Chung and Garey [42]. See Chapter 42 of this book for appropriate algorithm related to certain connectivity problem.

References

[1] K. Menger. Zur allgemeinen Kurventheorie. *Fund. Math.*, **10** (1927), 96–115.

[2] D. W. Matula. Determining edge connectivity in $O(mn)$. In *Proceedings of the 28th Annual Symposium on Foundations of Computer Science*, Los Angeles, CA, 249–251, 1987.

[3] C. P. Schnorr. Bottlenecks and edge connectivity in unsymmetrical networks. *SIAM J. Compu.*, **8** (1979), 265–274.

[4] Y. Mansour and B. Schieber. Finding the edge connectivity of directed graphs. *J. Algorithms*, **10** (1989), 76–85.

[5] M. Stoer and F. Wagner. A simple min-cut algorithm. *JACM*, **44**(4) (1997), 585–591.

[6] H. Nagamochi and T. Ibaraki. *Algorithmic Aspects of Graph Connectivity*. Cambridge University Press, Cambridge, 2008.

[7] W. J. Cook, W. H. Cunningham, W. R. Pulleyblank, and A. Schrijver. *Combinatorial Optimization*. Wiley-Interscience, New York, 1998.

[8] S. Even and R. E. Tarjan. Network flow and testing graph connectivity. *SIAM J. Comput.*, **4** (1975), 507–512.

[9] R. E. Gomory and T. C. Hu. Multi-terminal network flows. *J. SIAM*, **9** (1961), 551–570.

[10] D. Gusfield. Very simple methods for all pairs network flow analysis. *SIAM J. Comput.*, **19** (1990), 143–155.

[11] D. Jungnickel. *Graphs, Networks and Algorithms*. Springer-Verlag, Berlin, Germany, 2005.

[12] R. K. Ahuja, T. L. Magnanti, and J. B. Orlin. *Network Flows: Theory, Algorithms, and Applications.* Prentice Hall, Upper Saddle River, NJ, 1993.

[13] C. K. Cheng and T. C. Hu. Ancestor tree for arbitrary multi-terminal cut functions. In *Proceedings of the 1st Integer Programming and Combinatorial Optimization Conference*, Waterloo, Canada, 1990.

[14] S. L. Hakimi. On the realizability of a Set of integers as degrees of the vertices of a graph,. *SIAM J. Appl. Math.*, **10** (1962), 496–506.

[15] V. Havel. A remark on the existence of finite graphs. *Casopois Pest. Math.*, **80** (1955), 477–480.

[16] D. L. Wang and D. J. Kleitman. On the existence of n-connected graphs with prescribed degrees ($n \geq 2$). *Networks*, **3** (1973), 225–239.

[17] P. Erdös and T. Gallai. Graphs with prescribed degrees of vertices. *Mat. Lapok*, **11** (1960), 264–274.

[18] F. Harary. *Graph Theory*. Addison-Wesley, Reading, MA, 1969.

[19] J. Edmonds. Existence of k-edge-connected ordinary graphs with prescribed degrees. *J. Res. Nat. Bur. Stand. B.*, **68** (1964), 73–74.

[20] D. L. Wang and D. J. Kleitman. A note on n-edge connectivity. *SIAM J. Appl. Math.*, **26** (1974), 313–314.

[21] B. Korte and J. Vygen. *Combinatorial Optimization: Theory and Algorithms.* Springer, Berlin, Germany, 2010.

[22] A. Esfahanian, Connectivity algorithms. In Robin Wilson and Lowell Beineke, editors, *Structural Graph Theory*, pages 268–281. Cambridge University Press, Cambridge, 2013.

[23] M. Padberg and G. Rinaldi. An efficient algorithm for the minimum capacity cut problem. *Mathematical Programming*, **47** (1990), 19–36.

[24] D. R. Karger. Global min-cuts in RNC and other ramifications of a simple mincut algorithm. In *Proceedings of the 4th Annual ACM-SIAM Symposium on Discrete Algorithms*, Austin, TX, pages 84–93, 1993.

[25] D. R. Karger and C. Stein. An $\tilde{O}(n^2)$ algorithm for minimum cuts. In *Proc. 25th Symp. Theor. Comput.*, San Diego, CA, 757–765, 1993.

[26] H. Nagamochi and T. Ibaraki. Computing edge-connectivity in multigraphs and capacitated graphs. *SIAM J. on Discrete Maths*, **5** (1992), 54–66.

[27] J. E. Hopcroft and R. E. Tarjan. Dividing a graph into triconnected components. *SIAM J. Comput.*, **2** (1973), 135–158.

[28] Z. Galil. Finding the vertex connectivity of graphs. *SIAM J. on Comput.*, **9** (1980), 197–199.

[29] S. Even. An algorithm for determining whether the connectivity of a graph is at least k. *SIAM J. Comput.*, **4** (1975), 393–396.

[30] D. J. Kleitman. Methods for investigating the connectivity of large graphs. *IEEE Trans. Circuit Theory*, **16** (1969), 232–233.

[31] J. Cheriyan and R. Thurimella. Algorithms for parallel vertex connecitivity and sparse certificates. In *Proc. 24th ACM Symp. Theor. Comput.*, New Orleans, LA, 391–401, 1991.

[32] M. Gondron and N. Minoux. *Graphs and Algorithms*. Wiley, New York, 1984.

[33] F. T. Boesch, editor. *Large-Scale Networks: Theory and Design*. IEEE Press, New York, 1976.

[34] S. L. Hakimi. An algorihtm for construction of least vulnerable communication networks or the graph with the maximum connectivity. *IEEE Trans. Circuit Theory*, **CT-16** (1969), 229–230.

[35] F. T. Boesch and R. E. Thomas. On graphs of invulnerable communication nets. *IEEE Trans. Circuit Theory*, **CT-17** (1970), 183–192.

[36] A. T. Amin and S. L. Hakimi. Graphs with given connectivity and independence number or networks with given measures of vuleralility and survivability. *IEEE Trans. Circuit Theory*, **CT-20** (1973), 2–10.

[37] H. Frank and I. T. Frisch. *Communication, Transmission, and Transportation Networks*. Addison-Wesley, Reading, MA, 1971.

[38] V. Krishnamoorthy, K. Thulasiraman, and M. N. S. Swamy. Minimum-order graphs with prescribed diameter, connectivity and regularity. *Networks*, **19** (1989), 25–46.

[39] V. Krishnamoorthy, K. Thulasiraman, and M. N. S. Swamy. Incremental distance and diameter sequences of a graph: New measures of network performance. *IEEE Trans. Comput.*, **39** (1990), 230–237.

[40] J. C. Bermond, N. Homobono, and C. Peyrat. Large fault-tolerant inter-connection networks. *Graphs and Combinatorics*, **5** (1989), 107–123.

[41] J. Opatrny, N. Srinivasan, and V. S. Alagar. Highly fault-tolerant communication network models. *IEEE Trans. Circuits and Sys.*, **30** (1989), 23–30.

[42] F. R. K. Chung and M. R. Garey. Diameter bounds for altered graphs. *J. Graph Theory*, **8** (1984), 511–534.

CHAPTER 14

Graph Connectivity Augmentation

András Frank

Tibor Jordán

CONTENTS

14.1	Introduction	315
	14.1.1 Notation	316
14.2	Edge-Connectivity Augmentation of Graphs	316
	14.2.1 Degree-Specified Augmentations	318
	14.2.2 Variations and Extensions	320
14.3	Local Edge-Connectivity Augmentation of Graphs	320
	14.3.1 Degree-Specified Augmentations	321
	14.3.2 Node-to-Area Augmentation Problem	323
14.4	Edge-Connectivity Augmentation of Digraphs	323
	14.4.1 Degree-Specified Augmentations	325
14.5	Constrained Edge-Connectivity Augmentation Problems	325
14.6	Vertex-Connectivity of Graphs	329
14.7	Vertex-Connectivity Augmentation of Digraphs	333
	14.7.1 Augmenting ST-Edge-Connectivity	335
14.8	Hypergraph Augmentation and Coverings of Set Functions	337
	14.8.1 Detachments and Augmentations	340
	14.8.2 Directed Hypergraphs	341

14.1 INTRODUCTION

The problem of economically improving a network to meet given survivability requirements occurs in a number of areas. A straightforward problem of this type is concerned with creating more connections in a telephone or computer network so that it survives the failure of a given number of cables or terminals [1]. Similar problems arise in graph drawing [2], statics [3], and data security [4]. It is natural to model these networks by graphs or directed graphs and use graph connectivity parameters to handle the survivability requirements. This leads to the following quite general optimization problem.

Connectivity Augmentation Problem: Let $G = (V, E)$ be a (directed) graph and let $r : V \times V \to Z_+$ be a function on the pairs of vertices of G. Find a smallest (or cheapest) set F of new edges on vertex set V such that $\lambda(u, v; G') \geq r(u, v)$ (or $\kappa(u, v; G') \geq r(u, v)$) holds for all $u, v \in V$ in $G' = (V, E \cup F)$.

Here $\lambda(u,v;H)$ ($\kappa(u,v;H)$) denotes the local edge-connectivity (local vertex-connectivity, respectively) in graph H, that is, the maximum number of pairwise edge-disjoint (vertex-disjoint) paths from u to v in H. When the goal is to find a cheapest set of new edges, it is meant with respect to a given cost function on the set of possible new edges.

The connectivity augmentation problem includes a large number of different subproblems (e.g., G may be a graph or a digraph, or the requirements may involve edge- or vertex-connectivity), which are interesting on their own, and whose solutions require different methods of combinatorial optimization. Some special cases can be solved by using well-known techniques, even in the minimum cost version. For example, if there is only one vertex pair u,v for which $r(u,v)$ is positive then, after assigning zero costs to the edges in E, we can use shortest path or minimum cost flow algorithms to find an optimal solution. If $r(s,v) = k$ for some integer k for all $v \in V$, where s is a designated vertex and the graph is directed, then minimum cost s-arborescence or minimum cost matroid intersection algorithms can be used in a similar way. However, the minimum cost versions (even in the case when each edge cost is either one or ∞) are typically NP-hard. For example, a graph $H = (V, E)$ contains a Hamilton cycle if and only if the cheapest set of edges which makes the edgeless graph on V 2-edge-connected has cost $|V|$, where the cost function is defined so that the cost of each edge of H is equal to one, and all the other edge costs are ∞.

We shall focus on those efficiently solvable variants of the problem which are NP-hard in the minimum cost setting but for which polynomial time algorithms and minimax theorems are known when the addition of any new edge has the same unit cost. That is, we shall be interested in augmenting sets of minimum size. In most cases it will be possible to add (any number of parallel copies of) a new edge ab for all $a, b \in V$, but we shall also consider tractable versions of some constrained augmentation problems, where the set of possible new edges is restricted (e.g., where we wish to augment a bipartite graph preserving bipartiteness). A brief summary of generalizations to hypergraph augmentation problems and even more abstract versions will also be given. A detailed analysis of the corresponding efficient algorithms is not in the scope of this chapter. The reader is referred to the book of Nagamochi and Ibaraki [5] for more details on the algorithmic issues. Further related survey articles and book chapters can be found in Frank [6,7], Schrijver [8], and Szigeti [9]. NP-hard versions and approximation algorithms are discussed in Khuller [10] and Gupta and Könemann [11].

14.1.1 Notation

Let $G = (V, E)$ be a graph. We shall use $d_G(X)$ to denote the *degree* of a set X of vertices. The degree of single vertex is denoted by $d_G(v)$. For two disjoint subsets $X, Y \subseteq V$ the number of edges from X to Y in G is denoted by $d_G(X, Y)$. When we deal with a directed graph D we use $\rho_D(X)$ and $\delta_D(X)$ to denote the *in-degree* and the *out-degree* of X, respectively. We omit the subscript referrring to the (directed) graph when it is clear from the context. For simplicity the directed edges (arcs) in a directed graph will also be called edges. In this case the notation reflects the orientation of the edge, that is, a directed edge uv has a *tail* u and a *head* v.

A function f defined on the subsets of V is called *submodular* if it satisfies $f(X) + f(Y) \geq f(X \cap Y) + f(X \cup Y)$ for all pairs $X, Y \subset V$. We say that f is *supermodular* if $-f$ is submodular. The functions d, ρ, δ defined above are all submodular.

14.2 EDGE-CONNECTIVITY AUGMENTATION OF GRAPHS

The first papers on graph connectivity augmentation appeared in 1976, when Eswaran and Tarjan [12] and independently Plesnik [13] solved the 2-edge- (and 2-vertex-)connectivity augmentation problem. Although several graph synthesis problems (i.e., augmentation problems where the starting graph has no edges) had been solved earlier (see, e.g., Gomory and

Hu [14] and Frank and Chou [15]), these papers were the first to provide minimax theorems for arbitrary starting graphs. The size of a smallest augmenting set which makes a graph 2-edge-connected can be determined as follows.

Let $G = (V, E)$ be a graph. We say that a set $U \subseteq V$ is *extreme* if $d(X) > d(U)$ for all proper nonempty subsets X of U. For example, $\{v\}$ is extreme for all $v \in V$. It is not hard to see that the extreme sets of G form a laminar family. (A set family $\mathcal{F}$ is *laminar* if for each pair of members $X, Y \in \mathcal{F}$ we have that either $X \cap Y = \emptyset$ or one of them is a subset of the other.)

An extreme set with $d(U) \leq 1$ satisfying $d(X) \geq 2$ for all proper nonempty subsets X of U is called *2-extreme*. The subgraph G' induced by a 2-extreme set U is 2-edge-connected, for if there is a nonempty subset $X \subset U$ for which $d_{G'}(X) \leq 1$, then $2 + 2 \leq d_G(X) + d_G(U - X) = d_{G'}(X) + d_{G'}(U - X) + d_G(U) \leq 1 + 1 + 1$, which is not possible. Furthermore, the 2-extreme sets are pairwise disjoint, since $X \cap Y \neq \emptyset$ would imply $1 + 1 \geq d_G(X) + d_G(Y) \geq d_G(X - Y) + d_G(Y - X) \geq 2 + 2$. Let $t_0(G)$ and $t_1(G)$ denote the number of 2-extreme sets of degree 0 and 1 in G, respectively.

Theorem 14.1 [12] *The minimum number γ of new edges whose addition to a graph $G = (V, E)$ results in a 2-edge-connected graph is $t_0(G) + \lceil t_1(G)/2 \rceil$.*

Proof. Shrinking a 2-extreme subset into a single vertex does not affect the values t_0 and t_1, and the minimal γ remains unchanged, as well. Therefore, we can assume that every 2-extreme set is a singleton, and hence G is a forest. In this case, t_0 is the number of isolated vertices, while t_1 is the number of leaf vertices.

In a 2-edge-connected augmentation of G, there are at least 2 new edges incident to an isolated vertex of G, and at least 1 new edge incident to a leaf vertex of G. Therefore, the number of new edges is at least $\lceil 2t_0 + t_1/2 \rceil = t_0 + \lceil t_1/2 \rceil$.

To see that the graph can be made 2-edge-connected by adding $t_0 + \lceil t_1/2 \rceil$ new edges, it suffices to show by induction that there is a new edge e such that the addition of e to G decreases the value of $t_0 + \lceil t_1/2 \rceil$. Such an edge is said to be *reducing*.

Assume first that G is disconnected. Let u and v be two vertices of degree at most one belonging to distinct components. A simple case-checking—depending on the degrees of u and v—shows that the new edge $e = uv$ is reducing.

Therefore, we can assume that G is actually a tree (and hence $t_0 = 0$) which has at least 2 vertices (and hence $t_1 \geq 2$). When $t_1 = 2$, the tree is a path, and we obtain a 2-edge-connected graph (namely, a cycle) by adding one edge connecting the end-vertices of the path. In this case, $\gamma = 1 = t_0 + \lceil t_1/2 \rceil$.

If $t_1 = 3$, then the tree consists of three paths ending at a common vertex. Let a, b, and c denote the other end-vertices of these paths. By adding the two new edges ab and ac to the tree, we obtain a 2-edge-connected graph, and hence $\gamma = 2 = t_0 + \lceil t_1/2 \rceil$.

The remaining case is when $t_1 \geq 4$. There is a path P in the tree connecting two leaf vertices such that at least two edges leave $V(P)$. (For example, a longest path including a vertex of degree at least four or a longest path including two vertices of degree three will suffice.) By adding the new edge e between the two end-vertices of P, we obtain a graph G' in which the value of t_0 continues to be 0. Furthermore, the vertex-set of $P + e$ is not 2-extreme in G', since its degree is at least 2. Thus, the addition of e to G reduces the value of t_1 by exactly 2. Consequently, e is reducing. ∎

A different solution method is to (cyclically) order the 2-extreme sets as they are reached by a DFS and then connect opposite pairs by the new edges.

Next we consider the *k-edge-connectivity augmentation problem*, where the goal is to make the input graph k-edge-connected ($k \geq 2$) by adding a smallest set of new edges. Kajitani and Ueno [16] solved this problem for every $k \geq 1$ in the special case when the starting graph is a tree.

First we extend the lower bound used in Theorem 14.1 to general graphs and higher target connectivity as follows. A *subpartition* of V is a family of pairwise disjoint nonempty subsets of V. Let $G = (V, E)$ be a graph and $k \geq 2$. Let

$$\alpha(G, k) = \max\left\{\sum_1^t (k - d(X)) : \{X_1, X_2, \ldots, X_t\} \text{ is a subpartition of } V\right\} \quad (14.1)$$

and let $\gamma(G, k)$ be the size of a smallest augmenting set of G with respect to k-edge-connectivity. Every augmenting set must contain at least $k - d(X)$ edges entering X for every $X \subset V$ and every new edge can decrease this *deficiency* of at most two sets in any subpartition. Thus $\gamma(G, k) \geq \Phi(G, k)$, where $\Phi(G, k) = \lceil \alpha(G, k)/2 \rceil$. Theorem 14.1 implies that $\gamma(G, 2) = \Phi(G, 2)$.

Watanabe and Nakamura [17] were the first to prove that $\gamma = \Phi$ for all $k \geq 2$ by using extreme sets and constructing an increasing sequence of augmenting sets $F_1, F_2, \ldots, F_k$ such that for all $1 \leq i \leq k$ the set F_i is an optimal augmenting set of G with respect to i. Here we present a different method which was first employed by Cai and Sun [18] (and suggested already by Plesnik [13] for $k = 2$). This method is based on edge splitting. Let $H = (V+s, E)$ be a graph with a designated vertex s. By *splitting off* a pair of edges su, sv we mean the operation that replaces su, sv by a new edge uv. The resulting graph is denoted by H_{uv}. A *complete splitting* at s is a sequence of splittings which isolates s. A complete splitting exists only if $d_H(s)$ is even. The splitting off method adds a new vertex s and some new edges incident with s to the starting graph and constructs the augmenting set by splitting off all the edges from s. Frank [19] simplified and extended this method (and established the link between generalized polymatroids and augmentation problems).

Let $G = (V, E)$ be a graph. An *extension* $G' = (V + s, E')$ of G is obtained from G by adding a new vertex s and a set of new edges incident with s. An extension G' is said to be (k, s)-*edge-connected* if $\lambda(x, y; G') \geq k$ holds for every pair $x, y \in V$. G' is *minimally* (k, s)-*edge-connected* if $G' - e$ is no longer (k, s)-edge-connected for every edge e incident with s.

The following result of Lovász [20] is a key ingredient in this approach. Let $H = (V+s, E)$ be a (k, s)-edge-connected graph. We say that splitting off two edges us, sv is k-*admissible* if H_{uv} is also (k, s)-edge-connected. A *complete* k-*admissible splitting* at s is a sequence of k-admissible splittings which isolates s. Observe that the graph on vertex set V, obtained from H by a complete k-admissible splitting, is k-edge-connected.

Theorem 14.2 [20] *Let $H = (V + s, E)$ be a (k, s)-edge-connected graph for some $k \geq 2$ and suppose that $d_H(s)$ is even. Then (a) for every edge su there exists an edge sv such that the pair su, sv is k-admissible; (b) there exists a complete k-admissible splitting at s in H.* ∎

Note that (b) follows by $d_H(s)/2$ repeated applications of (a).

14.2.1 Degree-Specified Augmentations

Let $G = (V, E)$ be a graph and let $m : V \to Z_+$ be a function. We say that m is a k-*augmentation vector* if there exists a graph $H = (V, F)$ for which $G + H$ is k-edge-connected and $d_H(v) = m(v)$ for every vertex v. Here $G + H$ is the graph on vertex set V with edge set $E \cup F$. For a subset $X \subseteq V$ we put $m(X) = \sum_{v \in X} m(v)$.

Theorem 14.3 [19] *Let $G = (V, E)$ be a graph, $k \geq 2$ an integer, and let $m : V \to Z_+$. Then m is a k-augmentation vector if an only if $m(V)$ is even and*

$$m(X) \geq k - d_G(X) \text{ for every } \emptyset \subset X \subset V. \quad (14.2)$$

Furthermore, it suffices to require (14.2) for the extreme subsets of V.

Proof. If m is a k-augmentation vector then there is a graph $H = (V, F)$ for which $G + H$ is k-edge-connected and $d_H(v) = m(v)$ for every vertex v. We then have $k \leq d_{G+H}(X) = d_G(X) + d_H(X) \leq d_G(X) + \sum[d_H(v) : v \in X] = d_G(X) + m(X)$, from which (14.2) follows. Since m is a degree sequence, $m(V)$ must be even.

To see sufficiency, add a new vertex s to G and $m(v)$ parallel sv-edges for every vertex v of G. It follows from (14.2) that for all $X \subset V$ we have $d_{G'}(X) = d_G(X) + m(X) \geq k$ in the extended graph G'. Thus G' is (k, s)-edge-connected. Since $m(V)$ is even, $d_{G'}(s)$ is even. Hence we can apply Theorem 14.2 to G' and conclude that there is a complete k-admissible splitting at s resulting in a k-edge-connected graph on vertex-set V. This implies that $G + H$ is k-edge-connected, where $H = (V, F)$ is the graph whose edge set F consists of the edges arising from the edge splittings. Since $d_H(v) = m(v)$ for all $v \in V$, it follows that m is a k-augmentation vector.

The last part of the theorem follows by observing that every set $X \subseteq V$ has a nonempty extreme subset U with $d(U) \leq d(X)$. ∎

We define a set X ($\emptyset \subset X \subset V$) to be *tight* with respect to a function $m : V \to Z_+$ satisfying (14.2) if $m(X) = k - d_G(X)$.

Lemma 14.1 *Let $m : V \to Z_+$ be a function satisfying (14.2) and let T be the subset of vertices v for which $m(v) > 0$. Suppose that m is minimal in the sense that reducing $m(v)$ for any $v \in T$ destroys (14.2). Then there is a subpartition $\{X_1, \ldots, X_t\}$ of V consisting of tight extreme sets which cover T.*

Proof. The minimality of m implies that each vertex $v \in T$ belongs to a tight set. Let $T(v)$ be a minimal tight set containing v. We claim that $T(v)$ is extreme. For if not, then there is a proper subset $Z \subset T(v)$ with $d_G(Z) \leq d_G(T(v))$. Then $m(Z) \geq k - d_G(Z) \geq k - d_G(T(v)) = m(T(v)) = m(Z) + m(T(v) - Z)$, from which we obtain $m(Z) = k - d_G(Z)$ and $m(T(v) - Z) = 0$. Hence Z is tight and $v \in Z$, contradicting the minimal choice of $T(v)$.

Therefore each $v \in T$ belongs to tight extreme set. Let $\{X_1, \ldots, X_t\}$ denote the maximal tight extreme sets. These sets are disjoint and cover T since the extreme subsets of V form a laminar family. ∎

We are now ready to prove the following fundamental theorem, due to Watanabe and Nakamura [17]. The proof below appeared in [19].

Theorem 14.4 [17] *Let $G = (V, E)$ be a graph and $k \geq 2$ an integer. Then*

$$\gamma(G, k) = \Phi(G, k). \tag{14.3}$$

Proof. We have already observed that $\gamma(G, k) \geq \Phi(G, k)$. To see that equality holds choose a function $m : V \to Z_+$ for which (14.2) holds and for which $m(V)$ is as small as possible.

Claim 14.1 $m(V) \leq \alpha(G, k)$.

Proof. By the minimality of $m(V)$ we can apply Lemma 14.1 to obtain that there is a subpartition $\{X_1, \ldots, X_t\}$ of V consisting of tight extreme sets which cover every vertex v with $m(v) > 0$. Thus $m(V) = \sum_{i=1}^{t} m(X_i) = \sum_{i=1}^{t}[k - d_G(X_i)] \leq \alpha(G, k)$, as claimed. ∎

If $m(V)$ is odd, increase $m(v)$ by one for some $v \in V$ to make sure that $m(V)$ is even. Now m is a k-augmentation vector by Theorem 14.3 and hence there is a graph $H = (V, F)$ for which $G + H$ is k-edge-connected and $d_H(v) = m(v)$ for all $v \in V$. Since $|F| = m(V)/2 \leq \lceil \alpha(G, k)/2 \rceil = \Phi(G, k)$, the theorem follows. ∎

Note that the statement of the theorem of Watanabe and Nakamura fails to hold for $k = 1$. This can be seen by choosing G to be the edgeless graph on four vertices. It is also worth mentioning that the proof above gives rise to a polynomial algorithm since a minimally (k, s)-edge-connected extension and a complete k-admissible splitting can be computed in polynomial time by using maximum flow algorithms. For more efficient algorithms using maximum adjacency orderings, see Nagamochi and Ibaraki [5].

Naor et al. [21] came up with yet another proof (and algorithm) for Theorem 14.4. Their algorithm increases the edge-connectivity one by one and is based on extreme sets. They use the Gomory–Hu tree of G to find the extreme sets. They also show how to employ the cactus representation of minimum edge cuts [22] to find a smallest set F which increases the edge-connectivity by one. Benczúr and Karger [23] show how the so-called extreme set tree can be used to find a minimally (k, s)-edge-connected extension G' of G and a complete k-admissible splitting in G'.

14.2.2 Variations and Extensions

Frank [19] showed that, although finding an augmenting set with minimum total cost is NP-hard, the k-edge-connectivity augmentation problem with *vertex-induced* edge costs can be solved in polynomial time. In this version we are given a cost function $c' : V \to Z_+$ on the vertices of the input graph and the cost $c(uv)$ of a new edge uv is defined to be $c'(u) + c'(v)$. Several degree constrained versions of the problem are also dealt with in [19]. One of these results is the following characterization.

Theorem 14.5 [19] *Let $G = (V, E)$ be a graph, $k \geq 2$ an integer, and let $f \leq g$ be two nonnegative integer-valued functions on V. Then G can be made k-edge-connected by adding a set F of new edges so that $f(v) \leq d_F(v) \leq g(v)$ holds for every $v \in V$ if and only if $k - d(X) \leq g(X)$ for every $\emptyset \neq X \subset V$ and there is no partition $\mathcal{P} = \{X_0, X_1, \ldots, X_t\}$ of V, where only X_0 may be empty, for which $f(X_0) = g(X_0)$, $g(X_i) = k - d(X_i)$ for $1 \leq i \leq t$, and $g(V)$ is odd.* ∎

14.3 LOCAL EDGE-CONNECTIVITY AUGMENTATION OF GRAPHS

A function $r : V \times V \to Z_+$ is called a *local requirement function* on V. We shall only consider symmetric functions, that is we shall assume that $r(u, v) = r(v, u)$ for all $u, v \in V$ when we deal with undirected graphs. Given a local requirement function r, we say that a graph H on vertex set V is r-edge-connected if $\lambda(x, y; H) \geq r(x, y)$ for all $x, y \in V$. In the *local edge-connectivity augmentation problem* the goal is to find a smallest set F of new edges whose addition makes the input graph $G = (V, E)$ r-edge-connected.

The local edge-connectivity augmentation problem can also be solved by using the edge splitting method. In this version we need a stronger splitting result (Theorem 14.6 below, due to Mader), and a modified lower bound counting deficiencies of subpartitions.

Let $G = (V, E)$ be a graph and let r be a fixed local requirement function on V. We define a function R on the subsets of V as follows: we put $R(\emptyset) = R(V) = 0$ and let

$$R(X) = \max\{r(x, y) : x \in X, y \in V - X\} \text{ for all } \emptyset \neq X \subset V. \tag{14.4}$$

It is not difficult to check that R is *skew supermodular*, that is, for all $X, Y \subseteq V$ we have $R(X) + R(Y) \leq R(X \cap Y) + R(X \cup Y)$ or $R(X) + R(Y) \leq R(X - Y) + R(X - Y)$ (or both).

By Menger's theorem an augmented graph G' of G is r-edge-connected if and only if $d_{G'}(X) \geq R(X)$ for every $X \subseteq V$. Let $q(X) = R(X) - d_G(X)$ for $X \subseteq V$ and let

$$\alpha(G,r) = \max \left\{ \sum_{i=1}^{t} q(X_i) : \{X_1, X_2, \ldots, X_t\} \text{ is a subpartition of } V \right\}. \tag{14.5}$$

An argument analogous to that of the uniform case shows that $\gamma(G,r) \geq \Phi(G,r)$, where $\gamma(G,r)$ is the size of a smallest augmenting set and $\Phi(G,r) = \lceil \alpha(G,r)/2 \rceil$. Theorem 14.4 claims that this lower bound is achievable if $r \equiv k \geq 2$. In the local version this does not necessarily hold. For example, consider a graph with four vertices and no edges and let $r \equiv 1$. On the other hand, if

$$r(u,v) \geq 2 \tag{14.6}$$

for all $u,v \in V$ then, as we shall see, we do have the equality $\gamma = \Phi$. In the rest of this section we shall assume that (14.6) holds. In general G may contain some *marginal components* with respect to r which need to be taken care of before one may assume (14.6). This reduction is relatively easy but quite technical. Therefore we refer the reader to [19] for the details.

Let r be a local requirement function on V, let $G = (V, E)$ be a graph and let $G' = (V + s, E')$ be an extension of G. We say G' is (r,s)-*edge-connected* if $\lambda(x,y;G') \geq r(x,y)$ for every $x,y \in V$. Splitting off us, sv is r-admissible in G' if $\lambda(x,y;G'_{uv}) \geq r(x,y)$ for all $x,y \in V$.

Let $r_\lambda(x,y) = \lambda(x,y;G')$ be a special requirement function defined on pairs $x,y \in V$. Mader's [24] deep result, which extends Theorem 14.2 to local edge-connectivities, is as follows.

Theorem 14.6 [24] *Let $G' = (V + s, E')$ be a graph. Suppose that $d(s)$ is even and there is no cut-edge incident with s. Then there is a complete r_λ-admissible splitting at s.* ∎

14.3.1 Degree-Specified Augmentations

The next result is the local version of Theorem 14.3.

Theorem 14.7 [19] *Let $G = (V,E)$ be a graph, $m : V \to Z_+$ with $m(V)$ even, and let r be a local requirement function satisfying (14.6). There is a graph $H = (V, F)$ for which*

$$d_H(v) = m(v) \text{ for all } v \in V \tag{14.7}$$

and

$$\lambda_{G'}(x,y) \geq r(x,y) \text{ for all } x,y \in V, \tag{14.8}$$

where $G' = G + H$, if and only if

$$m(X) \geq R(X) - d_G(X) \text{ for every } X \subseteq V. \tag{14.9}$$

Proof. If there is a graph H for which G' satisfies (14.8), then $d_G(X) + d_H(X) = d_{G'}(X) \geq R(X)$, from which $m(X) \geq d_H(X) \geq R(X) - d_G(X)$, and hence (14.9) holds.

To prove sufficiency, add a new vertex s to G and $m(v)$ parallel sv-edges for every vertex $v \in V$. In the resulting graph G', $\lambda_{G'}(x,y) \geq r(x,y)$ holds for every pair $x,y \in V$ of vertices due to (14.9). Observe that there is no cut-edge of G' incident to s by (14.6). Therefore we can apply Theorem 14.6, which asserts that there is a complete splitting at s that preserves the local edge-connectivities in V. This means that $H = (V, F)$ satisfies the requirements of the theorem where F denotes the set of edges arising from the splittings. ∎

As in the global case, we can use the degree-specified version to deduce a min-max result on the size of a smallest augmenting set.

Theorem 14.8 [19] *Let $G = (V, E)$ be a graph and let r be a local requirement function on V satisfying (14.6). Then $\gamma(G, r) = \Phi(G, r)$.*

Proof. We have already observed that $\gamma(G, r) \geq \Phi(G, r)$. To prove that equality holds first recall that R is a skew supermodular function. Hence q is also skew supermodular, that is, for each pair of sets $X, Y \subset V$, at least one of the following two inequalities holds:

$$q(X) + q(Y) \leq q(X \cap Y) + q(X \cup Y), \tag{14.10}$$

$$q(X) + q(Y) \leq q(X - Y) + q(Y - X). \tag{14.11}$$

Let $m : V \to Z_+$ be chosen in such a way that (14.9) is satisfied and $m(V)$ is minimal in the sense that reducing any positive $m(v)$ by one destroys (14.9). ∎

Claim 14.2 $m(V) \leq \alpha(G, r)$.

Proof. By the minimality of $m(V)$ every vertex v with $m(v) > 0$ belongs to a tight set where a set X is *tight* if $m(X) = q(X)$. Let $\mathcal{F} = \{X_1, \ldots, X_t\}$ be a system of tight sets which covers each vertex v with $m(v) > 0$, in which $|\mathcal{F}|$ is minimal, and with respect to this, in which $\sum[|Z| : Z \in \mathcal{F}]$ is minimal.

Suppose that $\mathcal{F}$ contains two intersecting members X and Y. If (14.10) holds, then $X \cup Y$ is tight, in which case X and Y could be replaced by $X \cup Y$, contradicting the minimality of $|\mathcal{F}|$. Therefore (14.11) must hold, which implies

$$m(X) + m(Y) = q(X) + q(Y) \leq q(X - Y) + q(Y - X) \leq m(X - Y) + m(Y - X)$$
$$= m(X) + m(Y) - 2m(X \cap Y)$$

from which we can conclude that both $X - Y$ and $Y - X$ are tight and $m(X \cap Y) = 0$. That is, in $\mathcal{F}$ we could replace X and Y by $X - Y$ and $Y - X$, contradicting the minimality of $\sum[|Z| : Z \in \mathcal{F}]$.

Therefore $\mathcal{F}$ must be a subpartition. Then

$$m(V) = \sum_1^t m(X_i) = \sum_1^t q(X_i) \leq \alpha(G, r),$$

as claimed. ∎

If $m(V)$ is odd, increase $m(v)$ by one for some $v \in V$ to make sure that $m(V)$ is even. Now Theorem 14.7 applies and hence there is a graph $H = (V, F)$ for which $G + H$ is r-edge-connected and $d_H(v) = m(v)$ for all $v \in V$. Since $|F| = m(V)/2 \leq \lceil \alpha(G, r)/2 \rceil = \Phi(G, r)$, the theorem follows. ∎

One may also consider the fractional version of Theorem 14.8, in which edges with fractional capacities may be added and the goal is to find an augmenting set with minimum total capacity. It can be shown that the minimum total capacity is equal to $\alpha(G, r)/2$. Moreover, the fractional optimum can be chosen to be half-integral.

We say that an increasing sequence of local requirements $(r_1, r_2, \ldots, r_t)$ on V has the *successive augmentation property* if, for any starting graph $G = (V, E)$, there exists an increasing sequence $F_1 \subseteq F_2 \subseteq \ldots \subseteq F_t$ of sets of edges such that $G + F_i$ is an optimal augmentation of G with respect to r_i, for all $1 \leq i \leq t$. The proof of Theorem 14.4 by Watanabe and Nakamura [17] (and also Naor et al. [21]) implies that any increasing sequence of uniform requirements has the successive augmentation property in the edge-connectivity augmentation problem.

By using an entirely different approach, Cheng and Jordán [25] generalized this to sequences with the following property:

$$r_{i+1}(u,v) - 1 = r_i(u,v) \geq 2, \text{ for all } u,v \in V \text{ and}, 1 \leq i \leq t-1. \tag{14.12}$$

The proof is based on the fact that if $G' = (V+s, E')$ ($G'' = (V+s, E'')$) is a minimally r_{i-1}-edge-connected (minimally r_i-edge-connected, respectively) extension of G such that G' is a subgraph of G'', then any r_{i-1}-admissible splitting su, sv in G' is r_i-admissible in G''.

Theorem 14.9 [9] *Every increasing sequence $(r_1, r_2, \ldots, r_t)$ of local requirements satisfying (14.12) has the successive augmentation property in the edge-connectivity augmentation problem.* ∎

A *mixed graph* $D = (V, E \cup A)$ has edges as well as directed edges. Bang-Jensen et al. [26] extended Theorem 14.6 to mixed graphs and with the splitting off method, they generalized Theorem 14.8 to the case when the edge-connectivity of a mixed graph is to be increased by adding undirected edges only. See also [9] for a list of theorems of this type.

14.3.2 Node-to-Area Augmentation Problem

Let $G = (V, E)$ be a graph and let $\mathcal{W}$ be a family of subsets of V, called *areas*. Let $r : \mathcal{W} \to Z_+$ be a requirement function assigning a nonnegative integer to each area. The *node-to-area augmentation problem* is to find a smallest set F of new edges for which $G + F$ contains at least $r(W)$ edge-disjoint paths between v and W for all $v \in V$ and $W \in \mathcal{W}$. It generalizes the k-edge-connectivity augmentation problem (make each vertex v a one-element area W_v with requirement $r(W_v) = k$).

This problem was shown to be NP-hard when $r(W) = 1$ for all $W \in \mathcal{W}$ [27], but it turned out to be polynomially solvable in the uniform case when $r(W) = r \geq 2$ for all $W \in \mathcal{W}$ [28]. The most general result proved so far is due to Ishii and Hagiwara [29] who showed that the problem can be solved even if r is not uniform but satisfies $r(W) \geq 2$ for each $W \in \mathcal{W}$.

For a proper subset X of V let

$$p(X) = \max\{r(W) : W \in \mathcal{W}, X \cap W = \emptyset \text{ or } W \subseteq X\}.$$

It is easy to see, by using Menger's theorem, that an augmented graph $G' = G + F$ is a feasible solution if and only if $d_{G'}(X) \geq p(X)$ for all proper subsets X of V. Let $\alpha(G, \mathcal{W}, r) = \max\{\sum_1^t (p(X_i) - d(X_i)) : \{X_1, \ldots, X_t\}$ is a subpartition of $V\}$ and let $\Phi(G, \mathcal{W}, r) = \lceil \alpha(G, \mathcal{W}, r)/2 \rceil$. Then we have $\gamma(G, \mathcal{W}, r) \geq \Phi(G, \mathcal{W}, r)$. This inequality may be strict but the gap can be at most one, and the instances with strict inequality (call them *exceptional configurations*) can be completely characterized [29].

Theorem 14.10 *Let $G = (V, E)$ be a graph, let $\mathcal{W}$ be a family of subsets of V, and let $r : \mathcal{W} \to Z_+$ be a requirement function satisfying $r(W) \geq 2$ for all $W \in \mathcal{W}$. Then $\gamma(G, \mathcal{W}, r) = \Phi(G, \mathcal{W}, r)$, unless $G, \mathcal{W}$, and r form an exceptional configuration, in which case $\gamma(G, \mathcal{W}, r) = \Phi(G, \mathcal{W}, r) + 1$.* ∎

A shorter proof of Theorem 14.10 was given later by Grappe and Szigeti [30]. The case when the edge-connectivity requirement is given separately for each area-vertex pair remains open.

14.4 EDGE-CONNECTIVITY AUGMENTATION OF DIGRAPHS

Let $\gamma(D, k)$ denote the size of a smallest set F of new (directed) edges which makes a given directed graph D k-edge-connected. The first result on digraph augmentation is due to Eswaran and Tarjan [12], who solved the strong connectivity augmentation problem and gave the following minimax formula for $\gamma(D,1)$.

Let $D = (V, A)$ be a digraph and let D_c be obtained from D by contracting its strong components. D_c is acyclic and it is easy to see that $\gamma(D,1) = \gamma(D_c,1)$ holds. Thus we may focus on acyclic input graphs. A vertex v with $\rho(v) = 0$ ($\delta(v) = 0$) is a *source* (*sink*, respectively). For a set $X \subset V$ let X^+ and X^- denote the set of sources in X and the set of sinks in X, respectively. It is clear that to make D strongly connected we need at least $|V^+|$ ($|V^-|$) new arcs.

Eswaran and Tarjan [12] proved the following min-max theorem.

Theorem 14.11 [12] *Let $D_c = (V, E)$ be an acyclic digraph. Then $\gamma(D_c, 1) = \max\{|V^+|, |V^-|\}$.*

Proof. Let $k = |V^+|$, $l = |V^-|$, and let $m = \max\{k, l\}$. We may suppose that $m = l \geq k$. We need to show that D_c can be made strongly connected by adding l new edges. We can assume that there are no isolated vertices in D_c, since an isolated vertex v can be replaced by a single edge vv', where v' is a new vertex, without changing γ, the number of sources, or the number of sinks.

The proof is by induction on m. First suppose that there is a sink-vertex t which is not reachable from some source vertex s. Then by adding the new edge ts to D we obtain a digraph that is still acyclic. Moreover, m is decreased by one. Thus the theorem follows by induction.

Next suppose that each sink vertex is reachable from each source vertex. Let $s_1, \ldots, s_k$ denote the source vertices and $t_1, \ldots, t_\ell$ the sink vertices. By adding the edges $t_1 s_1, \ldots, t_k s_k$ along with the edges $t_{k+1} s_1, t_{k+2} s_1, \ldots, t_\ell s_1$ (altogether ℓ new edges), we obtain a strongly connected augmentation of D_c. This completes the proof. ∎

It follows that for arbitrary starting digraphs $\gamma(D, 1)$ equals the maximum number of pairwise disjoint sets $X_1, \ldots, X_t$ in D with $\rho(X_i) = 0$ for all $1 \leq i \leq t$ (or $\delta(X_i) = 0$ for all $1 \leq i \leq t$).

We note that there is another version of the strong connectivity augmentation problem which is nicely tractable. In this version only those edges uv are allowed to be added for which u is reachable from v in the initial digraph. The solution for this problem is based on a theorem of Lucchesi and Younger, see for example [7] and [31].

Kajitani and Ueno [16] solved the k-edge-connectivity augmentation problem for digraphs in the special case when D is a directed tree (but $k \geq 1$ may be arbitrary). The solution for arbitrary starting digraphs is due to Frank [19], who adapted the splitting off method to directed graphs and showed that, as in the undirected case, $\gamma(D, k)$ can be characterized by a subpartition-type lower bound. For a digraph $D = (V, A)$ let

$$\alpha_{in}(D, k) = \max\left\{\sum_1^t (k - \rho(X_i)) : \{X_1, \ldots, X_t\} \text{ is a subpartition of } V\right\},$$

$$\alpha_{out}(D, k) = \max\left\{\sum_1^t (k - \delta(X_i)) : \{X_1, \ldots, X_t\} \text{ is a subpartition of } V\right\}.$$

It is again easy to see that $\gamma(D, k) \geq \Phi(D, k)$, where $\Phi(D, k) = \max\{\alpha_{in}(D), \alpha_{out}(D)\}$. An *extension* $D' = (V + s, A')$ of a digraph $D = (V, A)$ is obtained from D by adding a new vertex s and a set of new edges, such that each new edge leaves or enters s. A digraph $H = (V + s, A)$ with a designated vertex s is (k, s)-*edge-connected* if $\lambda(x, y; H) \geq k$ for every $x, y \in V$. Splitting off two edges us, sv means replacing the edges us, sv by a new edge uv. Splitting off two edges us, sv is k-*admissible* in a (k, s)-edge-connected digraph H if H_{uv} is also (k, s)-edge-connected.

The next theorem, due to Mader [32], is the directed counterpart of Theorem 14.2.

Theorem 14.12 [32] *Let $D = (V+s, A)$ be a (k,s)-edge-connected digraph with $\varrho(s) = \delta(s)$. Then (a) for every edge us there exists an edge sv such that the pair us, sv is k-admissible, (b) there is a complete k-admissible splitting at s.* ∎

14.4.1 Degree-Specified Augmentations

With the help of Theorem 14.12 an optimal solution for the directed edge-connectivity augmentation problem can be obtained following the steps of the solution in the undirected case. Here we only state the key results and refer to [19] for the proofs.

Theorem 14.13 [19] *Let $D = (V, A)$ be a digraph and let $m_{in} : V \to Z_+$ and $m_{out} : V \to Z_+$ be in- and out-degree specifications. There is a digraph $H = (V, F)$ for which*

$$\varrho_H(v) = m_{in}(v) \text{ and } \delta_H(v) = m_{out}(v) \text{ for every } v \in V \qquad (14.13)$$

and $D + H$ is k-edge-connected if and only if $m_{in}(V) = m_{out}(V)$,

$$m_{in}(X) \geq k - \varrho_D(X) \quad \text{for } \emptyset \neq X \subset V \qquad (14.14)$$

and

$$m_{out}(X) \geq k - \delta_D(X) \quad \text{for } \emptyset \neq X \subset V. \qquad (14.15)$$

The directed counterpart of the theorem of Watanabe and Nakamura is as follows. ∎

Theorem 14.14 [19] *Let $D = (V, A)$ be a directed graph and let $k \geq 1$. Then $\gamma(D, k) = \Phi(D, k)$.* ∎

Frank [19] also showed that the minimum cost version with vertex-induced cost functions is also solvable in polynomial time. Cheng and Jordán [25] proved that the successive augmentation property holds for any increasing sequence of uniform requirements in the directed edge-connectivity augmentation problem as well.

The local edge-connectivity augmentation problem in directed graphs is NP-hard, even if $r(u, v) \in \{0, 1\}$ for all $u, v \in V$ [19]. Bang-Jensen et al. [26] generalized Theorem 14.14 to mixed graphs and special classes of local requirements. For instance, they showed that the local version is solvable for Eulerian digraphs (i.e., for digraphs where $\varrho(v) = \delta(v)$ for all $v \in V$). The proofs of these results rely on an edge splitting theorem, which is a common extension of Theorem 14.12 and a result of Frank [33] and Jackson [34] on splitting off edges in Eulerian digraphs preserving local edge-connectivities. A different version of the mixed graph augmentation problem was investigated by Gusfield [35].

14.5 CONSTRAINED EDGE-CONNECTIVITY AUGMENTATION PROBLEMS

In each of the augmentation problems considered so far it was allowed to add (an arbitrary number of parallel copies of) any edge connecting two vertices of the input graph. It is natural to consider (and in some cases the applications give rise to) variants where the set of new edges must meet certain additional constraints. In general such constraints may lead to hard problems. For example, Frederickson and Jaja [36] proved that, given a tree $T = (V, E)$ and a set J of edges on V, it is NP-hard to find a smallest set $F \subseteq J$ for which $T' = (V, E \cup F)$ is 2-edge-connected. This problem remains NP-hard even if J is the edge-set of a cycle on the leaves of T [37]. For some types of constraints, however, an optimal solution can be found in polynomial time. In this section we consider these tractable problems.

Motivated by a question in statics, Bang-Jensen et al. [3] solved the following partition-constrained problem. Let $G = (V, E)$ be a graph and let $\mathcal{P} = \{P_1, P_2, \ldots, P_r\}$, $r \geq 2$, be a partition of V. In the *partition-constrained k-edge-connectivity augmentation problem* the goal is to find a smallest set F of new edges, such that every edge in F joins two distinct members of $\mathcal{P}$ and $G' = (V, E \cup F)$ is k-edge-connected. If G is a bipartite graph with bipartition $V = A \cup B$ and $\mathcal{P} = \{A, B\}$ then the problem corresponds to optimally augmenting a bipartite graph while preserving bipartiteness. By a theorem of Bolker and Crapo [38] the solution of this bipartite version can be used to make a square grid framework highly redundantly rigid by adding a smallest set of new diagonal rods.

Let $\gamma(G, k, \mathcal{P})$ denote the size of a smallest augmenting set with respect to k and the given partition $\mathcal{P}$. Clearly, $\gamma(G, k, \mathcal{P}) \geq \gamma(G, k)$. The case $k = 1$ is easy, hence we assume $k \geq 2$. For $i = 1, 2, \ldots, r$ let

$$\beta_i = \max \left\{ \sum_{Y \in \mathcal{Y}} (k - d(Y)) : \mathcal{Y} \text{ is a subpartition of } P_i \right\}. \tag{14.16}$$

Since no new edge can join vertices in the same member P_i of $\mathcal{P}$, it follows that β_i is a lower bound for $\gamma(G, k, \mathcal{P})$ for all $1 \leq i \leq r$. By combining this bound and the lower bound of the unconstrained problem we obtain $\gamma(G, k, \mathcal{P}) \geq \Phi(G, k, \mathcal{P})$, where

$$\Phi(G, k, \mathcal{P}) = \max\{\lceil \alpha(G, k)/2 \rceil, \beta_1, \ldots, \beta_r\}. \tag{14.17}$$

Simple examples show that $\gamma \geq \Phi + 1$ may hold. Consider a four-cycle C_4 and let $\mathcal{P}$ be the natural bipartition of C_4. Here we have $\Phi(C_4, 3, \mathcal{P}) = 2$ and $\gamma(C_4, 3, \mathcal{P}) = 3$. Now consider a six-cycle C_6 and let $\mathcal{P} = \{P_1, P_2, P_3\}$, where the members of $\mathcal{P}$ contain pairs of opposite vertices. For this graph and partition we have $\Phi(C_6, 3, \mathcal{P}) = 3$ and $\gamma(C_6, 3, \mathcal{P}) = 4$.

On the other hand, Bang-Jensen et al. [3] proved that we always have $\gamma \leq \Phi + 1$ and characterized all graphs (and partitions) with $\gamma = \Phi + 1$. The proof employed the splitting off method. The first step was a complete solution of the corresponding constrained edge splitting problem. Let $H = (V + s, E)$ be a (k, s)-edge-connected graph and let $\mathcal{P} = \{P_1, P_2, \ldots, P_r\}$ be a partition of V. We say a splitting su, sv is *allowed* if it is k-admissible and respects the partition constraints, that is, u and v belong to distinct members of $\mathcal{P}$. If k is even, the following extension of Theorem 14.2(b) is not hard to prove.

Theorem 14.15 [3] *Let $H = (V + s, E)$ be a (k, s)-edge-connected graph, for some even integer k, let $\mathcal{P} = \{P_1, P_2, \ldots, P_r\}$ be a partition of V, and suppose that $d(s)$ is even. There exists a complete allowed splitting at s if and only if $d(s, P_i) \leq d(s)/2$ for all $1 \leq i \leq r$.*

For k odd, however, there exist more complicated *obstacles* that prevent a complete allowed splitting at s. Let S denote the set of neighbors of s.

A partition $A_1 \cup A_2 \cup B_1 \cup B_2$ of V is called a C_4-*obstacle* if it satisfies the following properties in H for some index i, $1 \leq i \leq r$:

i. $d(A_1) = d(A_2) = d(B_1) = d(B_2) = k$;

ii. $d(A_1, A_2) = d(B_1, B_2) = 0$;

iii. $S \cap (A_1 \cup A_2) = S \cap P_i$ or $S \cap (B_1 \cup B_2) = S \cap P_i$;

iv. $d(s, P_i) = d(s)/2$. ∎

C_4-obstacles exist only for k odd. It is not difficult to see that if H contains a C_4-obstacle, then there exists no complete allowed splitting at s. A more special family of obstacles, called C_6-*obstacles*, can be defined when $r \geq 3$, k is odd, and $d(s) = 6$, see [3]. These two families suffice to characterize when there is no complete allowed splitting. Note that in the bipartition constrained case only C_4-obstacles may exist.

Theorem 14.16 [3] *Let $H = (V + s, E)$ be a (k, s)-edge-connected graph with $d(s)$ even and let $\mathcal{P} = \{P_1, P_2, \ldots, P_r\}$ be a partition of V. There exists a complete allowed splitting at vertex s in G if and only if*

a. $d(s, P_i) \leq d(s)/2$ for $1 \leq i \leq r$,

b. H contains no C_4- or C_6-obstacle. ∎

Bang-Jensen et al. [3] show that there exists a (k, s)-edge-connected extension $G' = (V+s, E')$ of $G = (V, E)$ with $d(s) = 2\Phi(G, k, \mathcal{P})$ for which Theorem 14.16(a) holds. If G' satisfies Theorem 14.16(b), as well, a complete allowed splitting at s yields an optimal augmenting set (of size $\Phi(G, k, \mathcal{P})$). Since $\gamma \leq \Phi + 1$, it remains to characterize the exceptions, that is, those starting graphs G (and partitions) for which any extension G' with $d(s) = 2\Phi(G, k, \mathcal{P})$ contains an obstacle (and hence $\gamma(G, k, \mathcal{P}) = \Phi(G, k, \mathcal{P}) + 1$ holds).

Let $G = (V, E)$ be a graph. A partition X_1, X_2, Y_1, Y_2 of V is a C_4-*configuration* if it satisfies the following properties in G:

i. $d(A) < k$ for $A = X_1, X_2, Y_1, Y_2$;

ii. $d(X_1, X_2) = d(Y_1, Y_2) = 0$;

iii. There exist subpartitions $\mathcal{F}_1, \mathcal{F}_2, \mathcal{F}'_1, \mathcal{F}'_2$ of X_1, X_2, Y_1, Y_2 respectively, such that for A ranging over X_1, X_2, Y_1, Y_2 and $\mathcal{F}$ the corresponding subpartition of A, $k - d(A) = \sum_{U \in \mathcal{F}} (k - d(U))$. Furthermore for some $i \leq r$, P_i contains every set of either $\mathcal{F}_1 \cup \mathcal{F}_2$ or $\mathcal{F}'_1 \cup \mathcal{F}'_2$.

iv. $(k - d(X_1)) + (k - d(X_2)) = (k - d(Y_1)) + (k - d(Y_2)) = \Phi(G, k, \mathcal{P})$.

As with C_4-obstacles, k must be odd in a C_4-configuration. A C_6-*configuration* is more specialized, since it only exists in graphs with $r \geq 3$ and $\Phi = 3$, see [3].

Theorem 14.17 [3] *Let $k \geq 2$ and let $G = (V, E)$ be a graph with a partition $\mathcal{P} = \{P_1, \ldots, P_r\}$, $r \geq 2$ of V. Then $\gamma(G, k, \mathcal{P}) = \Phi(G, k, \mathcal{P})$ unless G contains a C_4- or C_6-configuration, in which case $\gamma(G, k, \mathcal{P}) = \Phi(G, k, \mathcal{P}) + 1$.* ∎

If each member of $\mathcal{P}$ is a single vertex then we are back at Theorem 14.4. The following special case solves the rigidity problem mentioned above. Let $G = (V, E)$ be a bipartite graph with bipartition $V = A \cup B$, let $\mathcal{P} = \{A, B\}$, and let

$$\beta'_1 = \sum_{v \in A} \max\{0, k - d(v)\},$$

$$\beta'_2 = \sum_{v \in B} \max\{0, k - d(v)\},$$

$$\Theta(G, k, \mathcal{P}) = \max\{\lceil \alpha(G, k, \mathcal{P})/2 \rceil, \beta'_1, \beta'_2\}.$$

Theorem 14.18 [3] *Let $G = (V, E)$ be a bipartite graph with bipartition $V = A \cup B$ and let $\mathcal{P} = \{A, B\}$. Then $\gamma(G, k, \mathcal{P}) = \Theta(G, k, \mathcal{P})$ unless k is odd and G contains a C_4-configuration, in which case $\gamma(G, k, \mathcal{P}) = \Theta(G, k, \mathcal{P}) + 1$.* ∎

The variant of the above problem, where the edges of the augmenting set F must lie within members of a given partition, is NP-hard [3]. The status of this variant is open if the number r of partition members is fixed, even if $r = 2$. The corresponding edge splitting problem, for $r = 2$, has been solved in [39].

A different application of Theorem 14.16 is concerned with permutation graphs. A *permutation graph* G^π of a graph G is obtained by taking two disjoint copies of G and adding a matching joining each vertex v in the first copy to $\pi(v)$ in the second copy, where π is a permutation of $V(G)$. Thus G has several permutation graphs. The edge-connectivity of any permutation graph of G is at most $\delta(G) + 1$, where $\delta(G)$ is the minimum degree of G. When does G have a k-edge-connected permutation graph for $k = \delta(G) + 1$?

Creating a permutation graph of G corresponds to performing a complete bipartition constrained splitting in G', where G' is obtained from $2G$ by adding a new vertex s and precisely one edge from s to each vertex of $2G$. (For some graph H we use $2H$ to denote the graph consisting of two disjoint copies of H.) If G is simple, it can be seen that G' is (k, s)-edge-connected. Thus Theorem 14.16 leads to the following characterization, due to Goddard et al. [40]. An extension to hypergraphs was given later by Jami and Szigeti [41].

Theorem 14.19 [40] *Let G be a simple graph without isolated vertices and let $k = \delta(G) + 1$. Then there is a k-edge-connected permutation graph of G unless $G = 2K_k$, and k is odd.* ∎

The partition constrained k-edge-connectivity augmentation problem has been investigated for digraphs as well. Although the general case of this problem is still open, Gabow and Jordán presented polynomial algorithms for several special cases, including the partition constrained strong connectivity augmentation problem [42–44].

It is also natrual to consider the *planarity-preserving k-edge-connectivity augmentation problem*. In this problem we are given a planar graph $G = (V, E)$ and the goal is to find a smallest set F of new edges for which $G' = (V, E \cup F)$ is k-edge-connected and planar. The complexity of this problem is still open, even for $k = 2$. (The corresponding problem for 2-vertex-connectivity is NP-hard [2].)

A typical positive result, due to Nagamochi and Eades [45], is as follows. For $k = 2$ it was proved earlier by Kant [2]. For a simpler proof see [46].

Theorem 14.20 [45] *Let $G = (V, E)$ be outer-planar and let k be even or $k = 3$. Then G can be made k-edge-connected and planar by adding $\Phi(G, k)$ edges.* ∎

Another natural constrained augmentation problem, which has been investigated by several authors, is the *simplicity preserving k-edge-connectivity augmentation problem*: given a simple graph $G = (V, E)$, find a smallest set F of new edges for which $G' = (V, E \cup F)$ is k-edge-connected and simple. Frank and Chou [15] solved this problem (even with local requirements) in the special case where the starting graph G has no edges. Some papers on arbitrary starting graphs G but with small target value k followed. Let us denote the size of a smallest simplicity-preserving augmenting set F by $\gamma(G, k, S)$. Clearly, we have $\gamma(G, k, S) \geq \gamma(G, k)$. Following the algorithmic proof of Theorem 14.1, it can be checked that if G is simple, so is the augmented graph G'. This proves $\gamma(G, 2, S) = \gamma(G, 2)$. Watanabe and Yamakado [47] proved that $\gamma(G, k, S) = \gamma(G, k)$ holds for $k = 3$ as well. Taoka et al. [48] pointed out that $\gamma(G, k, S) \geq \gamma(G, k) + 1$ may hold if $k \geq 4$, even if the starting graph G is $(k-1)$-edge-connected. On the other hand, they showed that for $(k-1)$-edge-connected starting graphs one has $\gamma(G, k, S) \leq \gamma(G, k) + 1$ for $k = 4, 5$. Moreover, in these special cases, we have $\gamma(G, k, S) = \gamma(G, k)$, provided $\gamma(G, k) \geq 4$. For general k, it was observed [21] that $\gamma(G, k, S) = \gamma(G, k)$ if G is $(k-1)$-edge-connected and the minimum degree of G is at least k.

Jordán [49] settled the complexity of the problem by proving that the simplicity-preserving k-edge-connectivity augmentation problem is NP-hard, even if the starting graph is $(k-1)$-edge-connected. For k fixed, however, the problem is solvable in polynomial time. This result of Bang-Jensen and Jordán [50] is based on the fact that if $\gamma(G, k)$ is large compared to k then $\gamma(G, k, S) = \gamma(G, k)$ holds.

Theorem 14.21 [50] *Let $G = (V, E)$ be a simple graph. If $\gamma(G, k) \geq 3k^4/2$ then $\gamma(G, k, S) = \gamma(G, k)$.* ∎

The algorithmic proof of Theorem 14.21 employed the splitting off method and showed that if $\gamma(G, k)$ is large then an optimal simplicity-preserving augmentation can be found in polynomial time, even if k is part of the input. It is also proved in [50] that for any graph G we have $\gamma(G, k, S) \leq \gamma(G, k) + 2k^2$. Using this fact and some additional structural properties lead to an $O(n^4)$ algorithm for k fixed. Most of these results have been extended to the local version of the simplicity-preserving edge-connectivity augmentation problem [50].

In the *reinforcement problem*, which is the opposite of the simplicity preserving problem in some sense, we are given a connected graph $G = (V, E)$ and an integer $k \geq 2$, and the goal is to find a smallest set F of new edges for which $G' = (V, E \cup F)$ is k-edge-connected and every edge of F is parallel to some edge in E. This problem is also NP-hard [49].

We close this section by mentioning a constrained problem of a different kind. In the *simultaneous edge-connectivity augmentation problem* we are given two graphs $G_1 = (V, E)$ and $G_2 = (V, I)$, and two integers $k, l \geq 2$ and the goal is to find a smallest set F of new edges for which $G'_1 = (V, E \cup F)$ is k-edge-connected and $G'_2 = (V, I \cup F)$ is l-edge-connected. For this problem Jordán [46] proved that the difference between a subpartition type lower bound and the optimum is at most one. Furthermore, if k and l are both even then we have equality. The status of the simultaneous augmentation problem is still open in the case when k or l is odd. Ishii and Nagamochi [51] solved a similar simultaneous augmentation problem where the goal is to make G_1 and G_2 k-edge-connected and 2-vertex-connected, respectively.

14.6 VERTEX-CONNECTIVITY OF GRAPHS

The vertex-connected versions of the augmentation problems are substantially more difficult than their edge-connected counterparts. This will be transparent by comparing the corresponding minimax theorems, the proof methods, as well as the hardness results and open questions. The following observation indicates that the k-vertex-connectivity augmentation problem, at least in the undirected case, has a different character. Suppose the goal is to make $G = (V, E)$ k-connected, optimally, where $k = |V| - 2$. Although this case may seem very special, it is in fact equivalent to the maximum matching problem. To see this observe that F is a feasible augmenting set if and only if the complement of $G + F$ consists of independent edges. Thus finding a smallest augmenting set for G corresponds to finding a maximum matching in its complement. The case $k = |V| - 3$, which is equivalent to finding a four-cycle free 2-matching of maximum size, is still open.

As in the edge-connected case, if k is small, the k-connectivity augmentation problem can be solved by considering the tree-like structure of the k-connected components of the graph. If $k = 2$, the familiar concept of 2-connected components, or blocks, and the block-cutvertex tree helps. For simplicity, suppose that $G = (V, E)$ is connected. Let $t(G)$ denote the number of end-blocks of G and let $b(G)$ denote the maximum number of components of $G - v$ over all vertices $v \in V$. Note that the end-blocks are pairwise disjoint. Since G' is 2-connected if and only if $t(G') = b(G') - 1 = 0$, and adding a new edge can decrease $t(G)$ by at most two and $b(G)$ by at most one, it follows that at least $\Psi(G) = \max\{\lceil t(G)/2 \rceil, b(G) - 1\}$ new edges are needed to make G 2-connected. Eswaran and Tarjan [12] and independently Plesnik [13] proved that this number can be achieved. See also Hsu and Ramachandran [52]. Finding two end-blocks X, Y for which adding a new edge xy with $x \in X$ and $y \in Y$ decreases $\Psi(G)$ by one can be done, roughly speaking, by choosing the end-blocks corresponding to the end-vertices of a longest path in the block-cutvertex tree of G.

Theorem 14.22 [12,13] *Let $G = (V, E)$ be a connected graph. Then G can be made 2-connected by adding $\max\{\lceil t(G)/2 \rceil, b(G) - 1\}$ edges.* ∎

The lower bounds used in Theorem 14.22 can be extended to $k \geq 3$ and arbitrary starting graphs G as follows. Let $N_G(X)$, or simply $N(X)$ if G is clear from the context, denote the set of neighbors of vertex set X in G. A nonempty subset $X \subset V$ is a *fragment* if $V - X - N(X) \neq \emptyset$. It is easy to see that every set of new edges F which makes G k-connected must contain at least $k - |N(X)|$ edges from X to $V - X - N(X)$ for every fragment X. By summing up these "deficiencies" over pairwise disjoint fragments, we obtain a subpartition-type lower bound, similar to the one used in the corresponding edge-connectivity augmentation problem. Let

$$t(G, k) = \max\left\{\sum_{i=1}^{r}(k - |N(X_i)|) : \{X_1, \ldots, X_r\} \text{ are pairwise disjoint fragments in } V\right\}. \tag{14.18}$$

Let $\gamma(G, k)$ denote the size of a smallest augmenting set of G with respect to k. Since an edge can decrease the deficiency $k - |N(X_i)|$ of at most two sets X_i, we have $\gamma(G, k) \geq \lceil t(G, k)/2 \rceil$. For $K \subset V$ let $b(K, G)$ denote the number of components in $G - K$. Let

$$b(G, k) = \max\{b(K, G) : K \subset V, |K| = k - 1\}. \tag{14.19}$$

Since the deletion of K from the augmented graph must leave a connected graph, we have that $\gamma(G, k) \geq b(G, k) - 1$. Thus we obtain $\gamma(G, k) \geq \Psi(G, k)$, where we put

$$\Psi(G, k) = \max\{\lceil t(G, k)/2 \rceil, b(G, k) - 1\}. \tag{14.20}$$

Theorem 14.22 implies that $\gamma(G, 2) = \Psi(G, 2)$. Watanabe and Nakamura [53] proved that this minimax equality is valid for $k = 3$, too. Hsu and Ramachandran [54] gave an alternative proof and a linear time algorithm, based on Tutte's decomposition theory of 2-connected graphs into 3-connected components. This method was further developed by Hsu [55], who solved the problem of making a 3-connected graph 4-connected by adding a smallest set of edges. His proof relies on the decomposition of 3-conneced graphs into 4-connected components.

This approach, however, which relies on the decomposition of a graph into its k-connected components, is rather hopeless for $k \geq 5$. While k-edge-connected components have a nice structure, k-connected components are difficult to handle. Furthermore, the successive augmentation property does not hold for vertex-connectivity augmentation [25].

Although $\Psi(G, k)$ suffices to characterize $\gamma(G, k)$ for $k \leq 3$, there are examples showing that $\gamma(G, k)$ can be strictly larger than $\Psi(G, k)$. Consider for example the complete bipartite graph $K_{k-1,k-1}$ with target k. For $k \geq 4$ this graph has $\Psi = k - 1$ and $\gamma = 2k - 4$, showing that the gap can be as large as $k - 3$. Jordán [56] showed that if the starting graph G is $(k-1)$-connected then this is the extremal case, that is, $\gamma(G, k) \leq \Psi(G, k) + k - 3$. A polynomial time algorithm to find an augmenting set with at most $k - 3$ surplus edges was also given in [56]. This gap was later reduced to $(k-1)/2$ with the help of two additional lower bounds [57]. Cheriyan and Thurimella [58] gave a more efficient algorithm with the same approximation gap and showed how to compute $b(G, k)$ in polynomial time if G is $(k-1)$-connected. A near optimal solution can be found efficiently even if G is not $(k-1)$-connected. This was proved by Ishii and Nagamochi [59] and, independently, by Jackson and Jordán [60]. The approximation gap in the latter paper is slightly smaller.

Theorem 14.23 [60] *Let $G = (V, E)$ be an l-connected graph. Then $\gamma(G, k) \leq \Psi(G, k) + (k-l)k/2 + 4$.* ∎

Jackson and Jordán [60] adapted the edge splitting method for vertex-connectivity. This method was subsequently employed to find an optimal augmentation in polynomial time, for k fixed.

Given an extension $G' = (V + s, E')$ of a graph $G = (V, E)$, define $\bar{d}(X) = |N_G(X)| + d'(s, X)$ for every $X \subseteq V$, where d' denotes the degree function in G'. We say that G' is (k, s)-*connected* if

$$\bar{d}(X) \geq k \text{ for every fragment } X \subset V, \tag{14.21}$$

and that it is *minimally (k, s)-connected* if the set of edges incident to s is inclusionwise minimal with respect to (14.21). The following result from [60] gives lower and upper bounds for $\gamma(G, k)$ in terms of $d'(s)$ in any minimally (k, s)-connected extension of G.

Theorem 14.24 [60] *Let $G' = (V + s, E')$ be a minimally (k, s)-connected extension of a graph G. Then $\lceil d'(s)/2 \rceil \leq \gamma(G, k) \leq d'(s) - 1$.* ∎

Let $G' = (V+s, E')$ be a minimally (k, s)-connected extension of G. Splitting off su and sv in G' is k-*admissible* if G'_{uv} also satisfies (14.21). Notice that if G' has no edges incident to s then (14.21) is equivalent to the k-connectivity of G. Hence, as in the case of edge-connectivity, it would be desirable to know, when $d(s)$ is even, that there is a sequence of k-admissible splittings which isolates s. In this case, using the fact that $\gamma(G, k) \geq d'(s)/2$ by Theorem 14.24, the resulting graph on V would be an optimal augmentation of G with respect to k. This approach works for the k-edge-connectivity augmentation problem but does not always work in the vertex connectivity case. The reason is that complete k-admissible splittings do not necessarily exist. On the other hand, the splitting off results in [60, 61] are 'close enough' to yield an optimal solution if k is fixed.

The obstacle for the existence of a k-admissible splitting can be described, provided $d'(s)$ is large enough compared to k. The proof of the following theorem is based on a new tripartite submodular inequality for $|N(X)|$, see [60].

Theorem 14.25 [60] *Let $G' = (V + s, E')$ be a minimally (k, s)-connected extension of $G = (V, E)$ and suppose that $d'(s) \geq k^2$. Then there is no k-admissible splitting at s in G' if and only if there is a set $K \subset V$ in G such that $|K| = k - 1$ and $G - K$ has $d'(s)$ components $C_1, C_2, \ldots, C_{d'(s)}$ (and we have $d'(s, C_i) = 1$ for $1 \leq i \leq d'(s)$).* ∎

Theorem 14.25 does not always hold if $d'(s)$ is small compared to k. To overcome this difficulty, Jackson and Jordán [61] introduced the following family of graphs. Let $G = (V, E)$ be a graph and k be an integer. Let X_1, X_2 be disjoint subsets of V. We say (X_1, X_2) is a k-*deficient pair* if $d(X_1, X_2) = 0$ and $|V - (X_1 \cup X_2)| \leq k - 1$. We say two deficient pairs (X_1, X_2) and (Y_1, Y_2) are *independent* if for some $i \in \{1, 2\}$ we have either $X_i \subseteq V - (Y_1 \cup Y_2)$ or $Y_i \subseteq V - (X_1 \cup X_2)$, since in this case no edge can simultaneously connect X_1 to X_2 and Y_1 to Y_2. We say G is k-*independence free* if G does not have two independent k-deficient pairs. Note that if G is $(k-1)$-connected and (X_1, X_2) is a k-deficient pair then $V - (X_1 \cup X_2) = N(X_1) = N(X_2)$. For example (a) $(k-1)$-connected chordal graphs and graphs with minimum degree $2k-2$ are k-independence free, (b) all graphs are 1-independence free and all connected graphs are 2-independence free, (c) a graph with no edges and at least $k + 1$ vertices is not k-independence free for any $k \geq 2$, (d) if G is k-independence free and H is obtained by adding edges to G then H is also k-independence free, and (e) a k-independence free graph is l-independence free for all $l \leq k$. In general, a main difficulty in vertex-connectivity problems is that vertex cuts can cross each other in many different ways. In the case of an independence free graph G these difficulties can be overcome.

Theorem 14.26 [61] *If G is k-independence free then $\gamma(G, k) = \Psi(G, k)$.* ∎

If G is not k-independence free but $t(G,k)$ is large, then the augmentation problem can be reduced to the independence free case by adding new edges. This crucial property is formulated by the next theorem.

Theorem 14.27 [61] *Let $G = (V, E)$ be $(k-1)$-connected and suppose that $t(G,k) \geq 8k^3 + 10k^2 - 43k + 22$. Then there exists a set of edges F for G such that $t(G+F,k) = t(G,k) - 2|F|$, $G + F$ is k-independence free, and $t(G + F, k) \geq 2k - 1$.* ∎

These results lead to the following theorem.

Theorem 14.28 [61] *Let G be $(k-1)$-connected. If $\gamma(G,k) \geq 8k^3 + 10k^2 - 43k + 21$ then*
$$\gamma(G,k) = \Psi(G,k).$$
∎

The min-max equality in Theorem 14.28 is not valid if we remove the hypothesis that G is $(k-1)$-connected. To see this consider the graph G obtained from the complete bipartite graph $K_{m,k-2}$ by adding a new vertex x and joining x to j vertices in the m set of the $K_{m,k-2}$, where $j < k < m$. Then $b(G,k) = m$, $t(G,k) = 2m + k - 2j$ and $\gamma(G,k) = m - 1 + k - j$. However, by modifying the definition of $b(G,k)$ slightly, an analogous minimax theorem can be obtained for augmenting graphs of arbitrary connectivity. For a set $K \subset V$ with $|K| = k-1$ let $\delta(K,k) = \max\{0, \max\{k - d(x) : x \in K\}\}$ and $b^*(K,G) = b(K,G) + \delta(K,k)$. Let $b^*(G,k) = \max\{b^*(K,G) : K \subset V, |K| = k-1\}$. It is easy to see that $\gamma(G,k) \geq b^*(G,k) - 1$. Let
$$\Psi^*(G,k) = \max\{\lceil t(G,k)/2 \rceil, b^*(G,k) - 1\}.$$

Theorem 14.29 [61] *Let $G = (V, E)$ be l-connected. If $\gamma(G,k) \geq 3(k-l+2)^3(k+1)^3$ then $\gamma(G,k) = \Psi^*(G,k)$.* ∎

The lower bounds, in terms of k, are certainly not best possible in Theorems 14.27 through 14.29. These bounds, however, depend only on k. This is the essential fact in the solution for k fixed. Note that by Theorem 14.24 one can efficiently decide whether $\gamma(G,k)$ (or $t(G,k)$) is large enough compared to k. The proofs of these results are algorithmic and give rise to an algorithm which solves the k-vertex-connectivity augmentation problem in polynomial time for k fixed. If $\gamma(G,k)$ is large, then the algorithm has polynomial running time even if k is not fixed. This phenomenon is similar to what we observed when we investigated the algorithm for the simplicity-preserving k-edge-connectivity augmentation problem.

Perhaps the most exciting open question of this area is the complexity of the k-vertex-connectivity augmentation problem, when k is part of the input. A recent result of Végh [62] settled the special case when the starting graph is $(k-1)$-connected. To state the result we need some new concepts.

Let $G = (V, E)$ be a $(k-1)$-connected graph. A *clump* of G is an ordered pair $C = (S, \mathcal{P})$ where $S \subset V$, $|S| = k-1$, and $\mathcal{P}$ is a partition of $V - S$ into nonempty subsets, called *pieces*, with the property that no edge of G joins two distinct pieces in C. (Note that a piece is not necessarily connected.) It can be seen that if $C = (S, \mathcal{P})$ is a clump of G then, in order to make G k-connected, we must add a set of at least $|\mathcal{P}| - 1$ edges between the pieces of C, where $|\mathcal{P}|$ is the number of pieces of $\mathcal{P}$. We shall say that a clump C *covers* a pair of vertices u, v of G if u and v belong to distinct pieces of C. A *bush* B of G is a set of clumps such that each pair of vertices of G is covered by at most two clumps in B. Thus, if B is a bush in G, then in order to make G k-connected, we must add a set of at least

$$def(B) = \left\lceil \frac{1}{2} \sum_{(S,P) \in B} (|\mathcal{P}| - 1) \right\rceil \quad (14.22)$$

edges between the pieces of the clumps in B. Two bushes B_1 and B_2 of G are *disjoint* if no pair of vertices of G is covered by clumps in both B_1 and B_2. Thus, if B_1 and B_2 are disjoint bushes, then the sets of edges which need to be added between the pieces of the clumps in B_1 and B_2 are disjoint. We can now state the theorem.

Theorem 14.30 [62] *Let G be a $(k-1)$-connected graph. Then the minimum number of edges which must be added to G to make it k-connected is equal to the maximum value of $\sum_{B \in D} def(B)$ taken over all sets of pairwise disjoint bushes D for G.* ∎

The local vertex-connectivity augmentation problem is NP-hard even in the special case when the goal is to find a smallest augmenting set which increases the local vertex-connectivity up to k within a given subset of vertices of a $(k-1)$-connected graph. However, there exist solvable subcases and some remaining open questions. For instance, Watanabe et al. [63] gave a linear-time algorithm for optimally increasing the connectivity to 2 within a specified subset. The special case when the starting graph has no edges is an interesting open problem.

We close this section by a different generalization of the connectivity augmentation problem. In some cases it is desirable to make the starting graph $G = (V, E)$ k-edge-connected as well as l-vertex-connected at the same time, by adding a new set of edges F. In this multiple target version l is typically small while k is arbitrary. We may always assume $l \leq k$. Hsu and Kao [64] solved a local version of this problem for $k = l = 2$. Ishii et al. [65–67] proved a number of results for $l \leq 3$ and presented near optimal polynomial time algorithms when l as well as k can be arbitrary.

A typical result is as follows. Let $k \geq 2$ and $l = 2$. By combining $\alpha(G, k)$ and $t(G, 2)$ define

$$\alpha'(G, k, 2) = \max \left\{ \sum_{i=1}^{p}(k - d(X_i)) + \sum_{i=p+1}^{t}(2 - |N(X_i)|) \right\},$$

where the maximum is taken over all subpartitions $\{X_1, \ldots, X_p, X_{p+1}, \ldots, X_t\}$ of V for which X_i is a fragment for $p+1 \leq i \leq t$. Clearly, $\lceil \alpha'(G, k, 2)/2 \rceil$ is a lower bound for this multiple target problem. By applying the splitting off method (and a new operation called edge switching), a common extension of Theorems 14.4 and 14.22 can be obtained.

Theorem 14.31 [66] *$G = (V, E)$ can be made k-edge-connected and 2-connected by adding γ new edges if and only if $\max\{\lceil \alpha'(G, k, 2)/2 \rceil, b(G, 2) - 1\} \leq \gamma$.* ∎

14.7 VERTEX-CONNECTIVITY AUGMENTATION OF DIGRAPHS

From several aspects, the directed k-edge-connectivity augmentation problem is less tractable than its undirected version. This may suggest that the *directed k-vertex-connectivity augmentation problem* is harder than the (still unsolved) undirected problem. Another sign of this is the fact that after the basic result of Eswaran and Tarjan [12] on the case $k = 1$ (Theorem 14.11) almost no results appeared for nearly twenty years. An exception was the following result of Masuzawa et al. [68] which solved the special case when the starting digraph $D = (V, A)$ is an arborescence (i.e., a directed tree with a root vertex r such that there is a directed path from r to every $v \in V$). Let $\gamma(D, k)$ denote the size of a smallest augmenting set with respect to the target vertex-connectivity k.

Theorem 14.32 [68] *Let $B = (V, A)$ be an arborescence. Then $\gamma(B, k) = \sum_{v \in V} \max\{0, k - \delta(v)\}$.* ∎

In spite of the above general feeling a complete solution for the directed version has been found. For arbitrary starting digraphs $D = (V, A)$ there is a natural subpartition-type lower bound for $\gamma(D, k)$, similar to $t(G, k)$. Let $N^-(X)$ and $N^+(X)$ denote the set of in-neighbors and out-neighbors of vertex set X in D, respectively. We say $X \subset V$ is an *in-fragment* if $V - X - N^-(X) \neq \emptyset$. If $V - X - N^+(X) \neq \emptyset$ then X is called an *out-fragment*. Let

$$t_{\text{in}}(D, k) = \max \left\{ \sum_{i=1}^{r} (k - |N^-(X)|) : X_1, \ldots, X_r \text{ are pairwise disjoint in-fragments in } V \right\},$$

$$t_{\text{out}}(D, k) = \max \left\{ \sum_{i=1}^{r} (k - |N^+(X)|) : X_1, \ldots, X_r \text{ are pairwise disjoint out-fragments in } V \right\},$$

and let

$$\Psi(D, k) = \max\{t_{\text{in}}(D, k), t_{\text{out}}(D, k)\}.$$

It is easy to see that $\gamma(D, k) \geq \Psi(D, k)$ holds. Theorem 14.11 shows that $\gamma(D, 1) = \Psi(D, 1)$ and Theorem 14.32 implies that $\gamma(B, k) = \Psi(B, k)$ for every arborescence B and every $k \geq 1$. For $k \geq 2$, however, Jordán [69] pointed out that $\gamma(D, k) \geq \Psi(D, k) + k - 1$ may hold, even if D is $(k-1)$-connected. On the other hand, for $(k-1)$-connected starting digraphs the gap cannot be larger than $k - 1$ [69].

A stronger lower bound can be obtained by considering deficient pairs of subsets of V rather than deficient in- or out-fragments. We say that an ordered pair (X, Y), $\emptyset \neq X, Y \subset V$, $X \cap Y = \emptyset$ is a *one-way pair* in a digraph $D = (V, A)$ if there is no edge in D with tail in X and head in Y. We call X and Y the *tail* and the *head* of the pair, respectively. The *deficiency* of a one-way pair, with respect to k-connectivity, is $def_k(X, Y) = \max\{0, k - |V - (X \cup Y)|\}$. Two pairs are *independent* if their tails or their heads are disjoint. For a family $\mathcal{F}$ of pairwise independent one-way pairs we define $def_k(\mathcal{F}) = \sum_{(X,Y) \in \mathcal{F}} def_k(X, Y)$. By Menger's theorem every augmenting set F with respect to k must contain at least $def_k(X, Y)$ edges from X to Y for every one-way pair (X, Y). Moreover, these arcs are distinct for independent one-way pairs. This proves $\gamma(D, k) \geq def_k(\mathcal{F})$ for all families $\mathcal{F}$ of pairwise independent one-way pairs. Frank and Jordán [70] solved the k-vertex-connectivity augmentation problem for digraphs by showing that this lower bound can be attained.

Theorem 14.33 [70] *A digraph $D = (V, A)$ can be made k-connected by adding at most γ new edges if and only if*

$$def_k(\mathcal{F}) \leq \gamma \qquad (14.23)$$

holds for all families $\mathcal{F}$ of pairwise independent one-way pairs. ∎

This result was obtained as a special case of a more general theorem on coverings of bi-supermodular functions, see Theorem 14.47. We present a more direct proof in the next subsection.

The minimax formula of Theorem 14.33 was later refined by Frank and Jordán [71]. Among others, it was shown that if $def_k(\mathcal{F}) \geq 2k^2 - 1$, then the tails or the heads of the pairs in $\mathcal{F}$ are pairwise disjoint. This implies that if $\gamma(D, k) \geq 2k^2 - 1$ then the simpler lower bound $\Psi(D, k)$ suffices. With the help of this refined version, one can deduce Theorem 14.32 from Theorem 14.33 as well. A related conjecture of Frank [6] claims that $\gamma(D, k) = \Psi(D, k)$ for every acyclic starting digraph D.

If D is strongly connected and $k = 2$ then a direct proof and a simplified minimax theorem was given in [72] by applying the splitting off method. In a strongly connected digraph $D = (V, A)$ there are two types of deficient sets with respect to $k = 2$: in-fragments X with $|N^-(X)| = 1$ (called *in-tight*) and out-fragments X with $|N^+(X)| = 1$ (called *out-tight*).

Theorem 14.34 [72] Let $D = (V, A)$ be strongly connected. Then $\gamma(D, 2) = \Psi(D, 2)$ holds, unless $\Psi(D, 2) = 2$ and there exist three in-tight (or three out-tight) sets B_1, B_2, B_3, such that

$$B_1 \cap B_2 \neq \emptyset, |B_3| = 1, \quad \text{and } V - (B_1 \cup B_2) = B_3.$$

In the latter case D can be made 2-connected by adding 3 arcs. ∎

14.7.1 Augmenting ST-Edge-Connectivity

Theorem 14.33 can also be deduced from the solution of a directed edge-connectivity augmentation problem involving local requirements of a special kind. We present this solution in detail.

Let $D = (V, A)$ be a digraph and let S and T be two nonempty (but not necessarily disjoint) subsets of V. One may be interested in an augmentation of D in which every vertex of T is reachable from every vertex of S. This generalizes the problem of making a digraph strongly connected ($S = T = V$) which was solved by Eswaran and Tarjan (Theorem 14.11). This generalization, however, leads to an NP-complete problem even in the special cases when $|S| = 1$ or $S = T$. See [7] for the proof.

On the other hand, we shall show that if only ST-edges are allowed to be added, then the augmentation problem is tractable even for higher edge-connectivity. An edge with tail s and head t is an ST-edge if $s \in S, t \in T$. Let A^* denote the set of all ST-edges, including loops, and let $m = |A^*|$. Clearly, $m = |S||T|$. We say that a subset X of vertices is ST-nontrivial, or nontrivial for short, if $X \cap T \neq \emptyset$ and $S - X \neq \emptyset$, which is equivalent to requiring that there is an ST-edge entering X. A digraph is k-ST-edge-connected if the number of edges entering $X \subseteq V$ is at least k for every nontrivial X. By Menger's theorem, this is equivalent to requiring the existence of k edge-disjoint st-paths for every possible choice of $s \in S$ and $t \in T$. Note that this property is much stronger than requiring only the existence of k edge-disjoint paths from S to T.

We say that two sets X and Y are ST-crossing if none of the sets $X \cap Y \cap T, S - (X \cup Y), X - Y$, and $Y - X$ is empty. In the special case when $S = T = V$ this coincides with the standard notion of crossing. A family $\mathcal{L}$ is ST-crossing if both the intersection and the union of any two ST-crossing members of $\mathcal{L}$ belong to $\mathcal{L}$. If $\mathcal{L}$ does not include two ST-crossing members, it is said to be ST-cross-free. A family $\mathcal{I}$ of sets is ST-independent or just independent if, for any two members X and Y of $\mathcal{I}$, at least one of the sets $X \cap Y \cap T$ and $S - (X \cup Y)$ is empty. Note that the relation between two sets can be of three types. Either they are ST-crossing, or one includes the other, or they are ST-independent. A set F of ST-edges (or the digraph (V, F)) covers $\mathcal{L}$ if each member of $\mathcal{L}$ is entered by a member of F.

For an initial digraph $D = (V, A)$ that we want to make k-ST-edge-connected, define the deficiency function h on sets as follows. For a real number x we put $x^+ = \max\{x, 0\}$. Let

$$h(X) = \begin{cases} (k - \varrho_D(X))^+ & \text{if } X \text{ is } ST\text{-nontrivial} \\ 0 & \text{otherwise} \end{cases} \quad (14.24)$$

Therefore, the addition of a digraph $H = (V, F)$ of ST-edges to D results in a k-ST-edge-connected digraph if and only if F covers h in the sense that $\varrho_H(X) \geq h(X)$ for every $X \subseteq V$. For a set-function h and a family $\mathcal{I}$ of sets we use $h(\mathcal{I})$ to denote $\sum[h(X) : X \in \mathcal{I}]$.

Theorem 14.35 [70] A digraph $D = (V, A)$ can be made k-ST-edge-connected by adding at most γ new ST-edges (or equivalently, h can be covered by γ ST-edges) if and only if

$$h(\mathcal{I}) \leq \gamma \text{ for every } ST\text{-independent family } \mathcal{I} \text{ of subsets of } V. \quad (14.25)$$

Equivalently, the minimum number $\tau_h = \tau_h(D)$ of ST-edges whose addition to D results in a k-ST-edge connected digraph is equal to the maximum $\nu_h = \nu_h(D)$ of the sum of h-values over all families of ST-independent sets.

Proof. We prove the second form. Since one ST-edge cannot cover two or more sets from an ST-independent family, we have $\nu_h \leq \tau_h$. For the reverse direction we proceed by induction on ν_h. If this value is 0, that is, there is if no deficient set, then D itself is k-ST-edge-connected and hence no new edge is needed. So we may assume that ν_h is positive.

First suppose that there is an edge $e \in A^*$ for which $\nu_h(D') \leq \nu_h(D) - 1$ for $D' = D + e$. Then it follows by induction that D' can be made k-ST-edge-connected by adding $\nu_h(D')$ ST-edges. But then the original D can be made k-ST-edge-connected by adding at most $\nu_h(D') + 1$ ST-edges and hence we have $\tau_h(D) \leq \tau_h(D') + 1 = \nu_h(D') + 1 \leq \nu_h(D) \leq \tau_h(D)$ from which equality follows throughout and, in particular, $\tau_h(D) = \nu_h(D)$. In this case we are done.

Thus, it remains to consider the case when for every ST-edge e, there is an ST-independent family $\mathcal{I}_e$ for which $h(\mathcal{I}_e) = \nu_h$ and e does not enter any member of $\mathcal{F}_e$. Let $\mathcal{J}'$ denote the union of all of these families $\mathcal{I}_e$ in the sense that as many copies of a set X are put into $\mathcal{J}'$ as the number of edges e for which X is in $\mathcal{I}_e$. Recall that m denotes the number of ST-edges. Then we have $h(\mathcal{J}') = m\nu_h$ and

$$\text{every } ST\text{-edge enters at most } m - 1 \text{ members.} \tag{14.26}$$

We may assume that the h-value of each member of $\mathcal{J}'$ is strictly positive since the members of zero h-values can be discarded. Now we apply the following uncrossing procedure as long as possible: if there are two ST-crossing members, replace them by their intersection and union. When a new member has zero h-value, remove it. The submodularity of the in-degree function implies that such an exchange operation preserves (14.26) and also that the h-value of the revised system is at least $h(\mathcal{J}')$.

The above uncrossing procedure terminates after a finite number of steps since the number of sets can never increase and hence it can decrease only a finite number of times. Moreover, when this number does not decrease, the sum of squares of the sizes in the family must strictly increase. Since the number of such steps is also finite, we can conclude that we arrive at an ST-cross-free family $\mathcal{J}$ after a finite number of uncrossing operations. Therefore we have $h(\mathcal{J}) \geq h(\mathcal{J}') = m\nu_h$.

We emphasize that a set $X \subseteq V$ can occur in several copies in $\mathcal{J}$. Let $s(X)$ denote the number of these copies. Evidently, the sum of the s-values over the subsets of V is exactly $|\mathcal{J}|$.

Claim 14.3 *The partial order on $\mathcal{J}$ defined by $X \subseteq Y$ admits no chain of s-weight larger than $m - 1$.*

Proof. For a contradiction suppose that there is a chain $\mathcal{C}$ of s-weight at least m. Then there are m (not necessarily distinct) members of $\mathcal{J}$ which are pairwise comparable. Since the members of a chain of ST-nontrivial sets can be covered by a single ST-edge, this contradicts property (14.26). ∎

We can apply the weighted polar-Dilworth theorem asserting that the maximum weight of a chain is equal to the minimum number of antichains covering each element as many times as its weight is. It follows that $\mathcal{J}$ contains $m - 1$ antichains such that $s(X)$ of them contain X for every $X \in \mathcal{J}$. Since $h(\mathcal{J}) \geq m\nu_h$, the h-sum of at least one of these antichains is larger than ν_h. However, $\mathcal{J}$ is ST-cross-free and hence this antichain is ST-independent, contradicting the definition of ν_h. This contradiction completes the proof of the theorem. ∎

Note that in the special case when $S = T = V$, the members of an ST-independent family $\mathcal{I}$ of nonempty proper subsets of V are either pairwise disjoint or pairwise co-disjoint. (To see this suppose, for a contradiction, that $\mathcal{I}$ has two members which are disjoint and has two members which are co-disjoint. This implies, since any two members of $\mathcal{I}$ are disjoint or co-disjoint, that there is an $X \in \mathcal{I}$ which is disjoint from some $Y \in \mathcal{I}$ and co-disjoint from some $Z \in \mathcal{I}$. But then we must have $Y \subseteq Z$, contradicting the independence of $\mathcal{I}$.) Thus in the special case $S = T = V$ Theorem 14.35 immediately implies Theorem 14.14. It is also possible to deduce Theorem 14.33, although the proof is a bit technical, see [71].

Although it was possible to obtain a polynomial algorithm from the original proof of Theorem 14.33, it was neither combinatorial nor very efficient. Using a different (but quite complicated) approach, Benczúr and Végh [73] developed a combinatorial polynomial time algorithm. In the special of the k-vertex-connectivity augmentation problem when the initial digraph is $(k-1)$-connected, [74] describes a much simpler algorithm. Enni [75] gave an algorithmic proof of Theorem 14.35 in the special case when $k = 1$. Note that no strongly polynomial algorithm is known for the capacitated version of Theorem 14.35.

14.8 HYPERGRAPH AUGMENTATION AND COVERINGS OF SET FUNCTIONS

Connectivity augmentation is about adding new edges to a graph or digraph so that it becomes sufficiently highly connected. Applying Menger's theorem, every augmentation problem has an equivalent formulation where the goal is to add new edges so that each cut receives at least as many new edges as its deficiency with respect to the given target. Cut typically means a subset of vertices, but it may also be a pair or collection of subsets of the vertex set. The deficiency function, say $k - \rho(X)$ or $R(X) - d(X)$, is determined by the input graph and the connectivity requirements. This leads to a more abstract point of view: given a function p on subsets of a ground-set V, find a smallest cover of p, that is, a smallest set of edges F such that at least $p(X)$ edges enter every subset $X \subset V$. Deficiency functions related to connectivity problems have certain supermodular properties. This motivates the study of minimum covers of functions of this type.

This is not just for the sake of proving more general minimax theorems. In some cases (e.g., in the directed k-connectivity augmentation problem) the only known way to the solution is via an abstract result. In other cases (e.g., in the k-edge-connectivity augmentation problem) generalizations lead to simpler proofs, algorithms, and extensions (to local requirements or vertex-induced cost functions) by showing the background of the problem.

An intermediate step toward an abstract formulation is to consider hypergraphs. A *hypergraph* is a pair $\mathcal{G} = (V, E)$ where V is a finite set (the set of *vertices* of $\mathcal{G}$) and E is a finite collection of *hyperedges*. Each hyperedge e is a set $Z \subseteq V$ with $|Z| \geq 2$. The *size* of e is $|Z|$. Thus (loopless) graphs correspond to hypergraphs with edges of size two only. A hyperedge of size two is called a *graph edge*. Let $d_\mathcal{G}(X)$ denote the number of hyperedges intersecting both X and $V - X$. A hypergraph is *k-edge-connected* if $d_\mathcal{G}(X) \geq k$ for all $\emptyset \neq X \subset V$. A *component* of $\mathcal{G}$ is a maximal connected subhypergraph of $\mathcal{G}$. Let $w(\mathcal{G})$ denote the number of components of $\mathcal{G}$.

One possible way to generalize the k-edge-connectivity augmentation problem is to search for a smallest set of graph edges whose addition makes a given hypergraph k-edge-connected. Cheng [76] was the first to prove a result in this direction. He determined the minimum number of new graph edges needed to make a $(k-1)$-edge-connected hypergraph $\mathcal{G}$ k-edge-connected, by invoking deep structural results of Cunningham [77] on decompositions of submodular functions. His result was soon extended to arbitrary hypergraphs by Bang-Jensen and Jackson [78]. They employed and extended the splitting off method to hypergraphs.

A hypergraph $\mathcal{H} = (V+s, E)$ is (k,s)-*edge-connected* if s is incident to graph edges only and $d_{\mathcal{H}}(X) \geq k$ for all $\emptyset \neq X \subset V$. Splitting off two edges su, sv is k-*admissible* if $\mathcal{H}_{uv}$ is also (k,s)-edge-connected. The extension of Theorem 14.2 to hypergraphs is as follows.

Theorem 14.36 [78] *Let $\mathcal{H} = (V+s, E)$ be a (k,s)-edge-connected hypergraph with $d_{\mathcal{H}}(s) = 2m$. Then exactly one of the following statements holds.*

i. *There is a complete k-admissible splitting at s, or*

ii. *There exists a set $A \subset E$ with $|A| = k-1$ and $w(\mathcal{H} - s - A) \geq m+2$.*

Let $\mathcal{G} = (V, E)$ be a hypergraph. Let

$$\alpha(\mathcal{G}, k) = \max\left\{\sum_{i=1}^{t}(k - d(X_i)) : \{X_1, \ldots, X_t\} \text{ is a subpartition of } V\right\} \quad (14.27)$$

$$c(\mathcal{G}, k) = \max\{w(\mathcal{G} - A) : A \subset E, |A| = k-1\}. \quad (14.28)$$

∎

As in the case of graphs, $\lceil \alpha(\mathcal{G}, k)/2 \rceil$ is a lower bound for the size of a smallest augmentation. Another lower bound is $c(\mathcal{G}, k) - 1$. Note that if $\mathcal{G}$ is a graph, $k \geq 2$, and $\mathcal{G} - A$ has $c(\mathcal{G}, k)$ components $C_1, C_2, \ldots, C_c$ for some $A \subset E$ with $|A| = k-1$ and $c \geq 2$ then $\sum_1^c (k - d(C_i)) = kc - \sum_1^c d(C_i) \geq kc - 2(k-1) = 2(c-1) + (k-2)(c-2) \geq 2(c-1)$, and hence $\alpha(\mathcal{G}, k)/2 \geq c(\mathcal{G}, k) - 1$.

Theorem 14.37 [78] *The minimum number of new graph edges which makes a hypergraph $\mathcal{G}$ k-edge-connected equals*

$$\max\{\lceil \alpha(\mathcal{G}, k)/2 \rceil, c(\mathcal{G}, k) - 1\}. \quad (14.29)$$

∎

Note that the successive augmentation property does not hold for hypergraphs [25]. Cosh [79] proved the bipartition constrained version of Theorem 14.37. The general partition constrained version was solved by Bernáth et al. [80]. Király et al. [81] proved that the local version of the hypergraph edge-connectivity augmentation problem (with graph edges) is NP-hard, even if the starting hypergraph is connected and the maximum requirement is two.

One may also want to augment a hypergraph by adding hyperedges. The minimum number of new hyperedges which make a given hypergraph $\mathcal{G} = (V, E)$ k-edge-connected is easy to determine: add l copies of the hyperedge containing all vertices of V, where $l = \max\{k - d_{\mathcal{G}}(X) : \emptyset \neq X \subset V\}$. So it is natural to either set an upper bound on the size of the new edges or to make the *cost* of a new hyperedge depend on its size. The following extension of Theorem 14.37 is due to T. Király. In a *t-uniform* hypergraph each hyperedge has size t.

Theorem 14.38 [82] *Let $\mathcal{H}_0 = (V, \mathcal{E}_0)$ be a hypergraph and $t \geq 2$, $\gamma \geq 0$ integers. There is a t-uniform hypergraph $\mathcal{H}$ on vertex-set V with at most γ hyperedges so that $\mathcal{H}_0 + \mathcal{H}$ is k-edge-connected ($k \geq 1$) if and only if*

$$\sum_{X \in \mathcal{P}}[k - d_{\mathcal{H}_0}(X)] \leq t\gamma \text{ for every subpartition } \mathcal{P} \text{ of } V,$$

$$k - d_{\mathcal{H}_0}(X) \geq \gamma \text{ for every } X \subset V,$$

$$w(\mathcal{H}_0 - \mathcal{E}_0') - 1 \leq (t-1)\gamma \text{ for every } \mathcal{E}_0' \subseteq \mathcal{E}_0, |\mathcal{E}_0'| = k-1.$$

∎

Szigeti [83] considered a different objective function and local requirements.

Theorem 14.39 [83] *Let $\mathcal{H}_0 = (V, \mathcal{E}_0)$ be a hypergraph and let $r(u,v)$ be a local requirement function. Then there is a hypergraph $\mathcal{H} = (V, \mathcal{E})$ so that $\lambda_{\mathcal{H}_0+\mathcal{H}}(u,v) \geq r(u,v)$ for every $u, v \in V$ and so that $\sum[|Z| : Z \in \mathcal{E}] \leq \gamma$ if and only if*

$$\sum_{X \in \mathcal{P}} [R(X) - d_{\mathcal{H}_0}(X)] \leq \gamma \text{ for every subpartition } \mathcal{P} \text{ of } V.$$ ∎

Benczúr and Frank [84] solved an abstract generalization of Theorem 14.37. Let V be a finite set and let $p : 2^V \to Z$ be a function with $p(\emptyset) = p(V) = 0$. p is *symmetric* if $p(X) = p(V - X)$ holds for every $X \subseteq V$. We say that p is *crossing supermodular* if it satisfies the following inequality for each pair of crossing sets $X, Y \subset V$:

$$p(X) + p(Y) \leq p(X \cup Y) + p(X \cap Y). \tag{14.30}$$

Recall that a set of edges F on V *covers* p if $d_F(X) \geq p(X)$ for all $X \subset V$.

Now suppose p is a symmetric crossing supermodular function on V and we wish to determine the minimum size $\gamma(p)$ of a cover of p consisting of graph edges. A subpartition-type lower bound is the following. Let

$$\alpha(p) = \max \left\{ \sum_{i=1}^{t} p(X_i) : \{X_1, \ldots, X_t\} \text{ is a subpartition of } V \right\}.$$

Since an edge can cover at most two sets of a subpartition, we have $\gamma(p) \geq \lceil \alpha(p)/2 \rceil$. This lower bound may be strictly less than $\gamma(p)$. To see this consider a ground set V with 4 elements and let $p \equiv 1$. Here $\alpha(p) = 4$ but, since every cover forms a connected graph on V, we have $\gamma(p) \geq 3$. This example leads to the following notions. We call a partition $\mathcal{Q} = \{Y_1, \ldots, Y_r\}$ of V with $r \geq 4$ *p-full* if

$$p(\cup_{Y \in \mathcal{Q}'} Y) \geq 1 \text{ for every non-empty subfamily } \mathcal{Q}' \subseteq \mathcal{Q}. \tag{14.31}$$

The maximum size of a p-full partition is called the *dimension* of p and is denoted by $\dim(p)$. If there is no p-full partition, then $\dim(p) = 0$. Since every cover induces a connected graph on the members of a p-full partition, we have $\gamma(p) \geq \dim(p) - 1$. Thus the minimum size of a cover is at least $\Phi(p) = \max\{\lceil \alpha(p)/2 \rceil, \dim(p) - 1\}$.

Theorem 14.40 [84] *Let $p : 2^V \to Z$ be a symmetric crossing supermodular function. Then $\gamma(p) = \Phi(p)$.* ∎

The proof of Theorem 14.40 yields a polynomial time algorithm to find a smallest cover, provided a polynomial time submodular function minimization oracle is available. The deficiency function of a (hyper)graph is symmetric and supermodular: Theorems 14.4 and 14.37 follow by taking $p(X) = k - d_\mathcal{G}(X)$ for all $X \subset V$ and $p(\emptyset) = p(V) = 0$, where $\mathcal{G}$ is the starting (hyper)graph, see [84]. The special case of Theorem 14.40 where $p(X) \in \{0, 1\}$ for all $X \subset V$ follows also from a result of Fleiner and Jordán [85]. Bernáth et al. [86] solved the partition-constrained version of this covering problem.

An abstract extension of Theorem 14.39 is as follows.

Theorem 14.41 [83] *Let $p : 2^V \to Z$ be a symmetric skew-supermodular function. Then*

$$\min \left\{ \sum_{e \in F} |e| : F \text{ is a cover of } p \right\} = \max \left\{ \sum_{1}^{t} p(X_i) : \{X_1, \ldots, X_r\} \text{ is a subpartition of } V \right\}.$$
$$\tag{14.32}$$
∎

Theorem 14.39 follows by taking $p(X) = q(X) = R(X) - d_{\mathcal{H}_0}(X)$. If p is an even valued skew-supermodular function, the minimum size of a cover consisting of graph edges can also be determined.

14.8.1 Detachments and Augmentations

We have seen that edge splitting results are important ingredients in solutions of connectivity augmentation problems and hence generalizations of some augmentation problems lead to extensions of edge splitting theorems. This works the other way round as well. Consider the following operation. Let $G = (V + s, E)$ be a graph with a designated vertex s. A *degree specification* for s is a sequence $\mathcal{S} = (d_1, \ldots, d_p)$ of positive integers with $\sum_{j=1}^{p} d_j = d(s)$. An $\mathcal{S}$-*detachment* of s in G is obtained by replacing s by p vertices $s_1, \ldots, s_p$ and replacing every edge su by an edge $s_i u$ for some $1 \leq i \leq p$ so that $d(s_i) = d_i$ holds in the new graph for $1 \leq i \leq p$. If $d_i = 2$ for all $1 \leq i \leq p$ then an $\mathcal{S}$-detachment corresponds to a complete splitting in a natural way. Given a local requirement function $r : V \times V \to Z_+$, an $\mathcal{S}$-detachment is called r-*admissible* if the detached graph G' satisfies $\lambda(x, y; G') \geq r(x, y)$ for every pair $x, y \in V$.

Extending an earlier theorem of Fleiner [87] on the case when $r \equiv k$ for some $k \geq 2$, Jordán and Szigeti [88] gave a necessary and sufficient condition for the existence of an r-admissible $\mathcal{S}$-detachment. We call r *proper* if $r(x, y) \leq \lambda(x, y; G)$ for every pair $x, y \in V$.

Theorem 14.42 [88] *Let r be a local requirement function for $G = (V + s, E)$ and suppose that G is 2-edge-connected and $r(u, v) \geq 2$ for each pair $u, v \in V$. Let $\mathcal{S} = (d_1, \ldots, d_p)$ be a degree specification for s with $d_i \geq 2$, $i = 1, \ldots, p$. Then there exists an r-admissible $\mathcal{S}$-detachment of s if and only if r is proper and*

$$\lambda(u, v; G - s) \geq r(u, v) - \sum_{i=1}^{p} \lfloor d_i/2 \rfloor \tag{14.33}$$

holds for every pair $u, v \in V$. ∎

Theorem 14.42 implies Theorem 14.6 by letting $r \equiv r_\lambda$ and $d_i \equiv 2$. It also gives the following extension of Theorem 14.8. By *attaching* a star of degree d to a graph $G = (V, E)$ we mean the addition of new vertex s and d edges from s to vertices in V. Let $G = (V, E)$ be a graph and suppose that we are given local requirements $r(u, v)$ for each pair $u, v \in V$ as well as a set of integers $d_1, \ldots, d_p$ ($d_j \geq 2$). Can we make G r-edge-connected by attaching p stars with degrees $d_1, d_2, \ldots, d_p$? Applying Theorem 14.42 to a minimally (r, s)-edge-connected extension of G gives the following necessary and sufficient condition. Recall the definition of $q(X)$ from Section 14.2. For simplicity, we again assume that r satisfies (14.6).

Theorem 14.43 [88] *Let $G = (V, E)$ be a graph and let $r(u, v)$, $u, v \in V$ be a local requirement function satisfying (6). Then G can be made r-edge-connected by attaching p stars with degrees $d_1, \ldots, d_p$ ($d_j \geq 2$, $1 \leq j \leq p$) if and only if*

$$\sum_{i=1}^{t} q(X_i) \leq \sum_{j=1}^{p} d_j \tag{14.34}$$

holds for every subpartition $\{X_1, \ldots, X_t\}$ of V and $\lambda(u, v; G) \geq r(u, v) - \sum_{j=1}^{p} \lfloor d_j/2 \rfloor$ for every pair $u, v \in V$. ∎

With the help of Theorem 14.43 it is easy to deduce a minimax formula for the following optimization problem: given G, r, and an integer $w \geq 2$, determine the minimum number γ for which G can be made r-edge-connected by attaching γ stars of degree w. If $w = 2$ then we are back at Theorem 14.8.

14.8.2 Directed Hypergraphs

In this subsection we consider extensions of some of the previous results on augmentations of directed graphs to directed hypergraphs and to directed covers of certain set functions.

A *directed hypergraph* (or *dypergraph*, for short) is a pair $\mathcal{D} = (V, A)$, where V is a finite set (the set of *vertices* of $\mathcal{D}$) and A is a finite collection of *hyperarcs*. Each hyperarc e is a set $Z \subseteq V$, $|Z| \geq 2$, with a specified *head vertex* $v \in Z$. We also use (Z, v) to denote a hyperarc on set Z and with head v. The *size* of e is $|Z|$. Thus a directed graph (without loops) is a dypergraph with hyperarcs of size two only. We say that a hyperarc (Z, v) *enters* a set $X \subset V$ if $v \in X$ and $Z - X \neq \emptyset$. Let $\rho(X)$ denote the number of hyperarcs entering X. A dypergraph $\mathcal{D} = (V, A)$ is k-*edge-connected* if $\rho(X) \geq k$ for every $\emptyset \neq X \subset V$. Berg et al. [89] extended Theorem 14.12 to dypergraphs and, among others, proved the following extension of Theorem 14.14.

Theorem 14.44 [89] *Let $\mathcal{D} = (V, A)$ be a dypergraph. Then $\mathcal{D}$ can be made k-edge-connected by adding γ new hyperarcs of size at most t if and only if*

$$\gamma \geq \sum_{1}^{r}(k - \rho(X_i)) \tag{14.35}$$

and

$$(t-1)\gamma \geq \sum_{1}^{r}(k - \delta(X_i)) \tag{14.36}$$

hold for every subpartition $\{X_1, X_2, \ldots, X_r\}$ of V. ∎

Directed covers of set functions have also been investigated. Frank [6] proved the directed version of Theorem 14.40. We say that a set F of directed edges on ground set V *covers* a function $p : 2^V \to \mathbb{Z}$ if $\rho_F(X) \geq p(X)$ for all $X \subset V$.

Theorem 14.45 [6] *Let $p : 2^V \to \mathbb{Z}$ be a crossing supermodular function. Then p can be covered by γ edges if and only if*

$$\gamma \geq \sum_{1}^{t} p(X_i) \tag{14.37}$$

and

$$\gamma \geq \sum_{1}^{t} p(V - X_i) \tag{14.38}$$

hold for every subpartition $\{X_1, X_2, \ldots, X_t\}$ of V. ∎

If $p(X) \in \{0, 1\}$ for all $X \subset V$, the problem corresponds to covering a *crossing family* of subsets of V by a smallest set of edges. Gabow and Jordán [43] solved the bipartition-constrained version of this special case.

Since $p(X) = k - \rho(X)$ is crossing supermodular, Theorem 14.45 implies Theorem 14.14. The following generalization of the directed k-edge-connectivity augmentation problem can also be solved by Theorem 14.45. Let $D = (V, A)$ be a directed graph with a specified root vertex $r \in V$ and let $k \geq l \geq 0$ be integers. D is called (k, l)-*edge-connected (from r)* if $\lambda(r, v; D) \geq k$ and $\lambda(v, r; D) \geq l$ for every vertex $v \in V - r$. Clearly, D is k-edge-connected if and only if D is (k, k)-edge-connected. The extension, due to Frank [90], is as follows. Let $p_{kl}(X) = \max\{k - \rho(X), 0\}$ for sets $\emptyset \neq X \subseteq V - r$ and let $p_{kl}(X) = \max\{l - \rho(X), 0\}$ for sets $X \subset V$ with $r \in X$.

Theorem 14.46 [90] *Let $D = (V, A)$ be a digraph and let $r \in V$. D can be made (k, l)-edge-connected from r by adding γ new arcs if and only if*

$$\gamma \geq \sum_1^t p_{kl}(X_i) \tag{14.39}$$

and

$$\gamma \geq \sum_1^t p_{kl}(V - X_i) \tag{14.40}$$

hold for every partition $\{X_1, X_2, \ldots, X_t\}$ of V. ∎

Finally we state a result on directed covers of pairs of sets, due to Frank and Jordán [70], which led to the solution of the k-vertex-connectivity augmentation problem for directed graphs.

Let V be a ground set and let $p(X, Y)$ be an integer-valued function defined on ordered pairs of disjoint subsets $X, Y \subset V$. We call p *crossing bi-supermodular* if

$$p(X, Y) + p(X', Y') \leq p(X \cap X', Y \cup Y') + p(X \cup X', Y \cap Y')$$

holds whenever $X \cap X', Y \cap Y' \neq \emptyset$. A set F of directed edges *covers* p if there are at least $p(X, Y)$ arcs in F with tail in X and head in Y for every pair $X, Y \subset V$, $X \cap Y = \emptyset$. Two pairs $(X, Y), (X', Y')$ are *independent* if $X \cap X'$ or $Y \cap Y'$ is empty.

Theorem 14.47 [70] *Let p be an integer-valued crossing bi-supermodular function on V. Then p can be covered by γ arcs if and only if $\sum_{(X,Y) \in \mathcal{F}} p(X, Y) \leq \gamma$ holds for every family $\mathcal{F}$ of pairwise independent pairs.* ∎

Let $D = (V, A)$ be a digraph. By taking $p(X, Y) = k - |(V - (X \cup Y)|$ for one-way pairs (X, Y) we can deduce Theorem 14.33. Furthermore, Theorem 14.47 implies Theorems 14.12, 14.14 and 14.35 as well as Edmonds' matroid partition theorem, a theorem of Győri on covering a rectilinear polygon with rectangles, and a theorem of Frank on $K_{t,t}$-free t-mathcings in bipartite graphs, see [70,91] for more details. A recent application to the jump number of two-directional orthogonal ray graphs can be found in [92].

The idea of abstract formulations may also lead to graph augmentation problems with somewhat different but still connectivity related objectives. A recent result of Frank and Király [93] solves the problem of optimally augmenting a graph G by adding a set F of edges so that $G + F$ is (k, l)-partition-connected. A graph $G = (V, E)$ is called (k, l)-partition-connected if the number of cross edges is at least $k(|\mathcal{P}| - 1) + l$ for all partitions $\mathcal{P}$ of V. With this definition (k, k)-partition-connectivity is equivalent to k-edge-connectivity while $(k, 0)$-partition-connectivity is equivalent, by a theorem of Tutte, to the existence of k-edge-disjoint spanning trees.

Acknowledgments

This work was supported by the MTA-ELTE Egerváry Research Group on Combinatorial Optimization and the Hungarian Scientific Research Fund grants K81472, K115483, CK80124, and K109240.

References

[1] M. Grötschel, C. L. Monma, and M. Stoer. Design of survivable networks. In *Network models*, volume 7 of *Handbooks in Operational Research and Management Science*, pages 617–672. M.O. Ball et al. (eds) North-Holland, Amsterdam, the Netherlands, 1995.

[2] G. Kant. Augmenting outerplanar graphs. *J. Algorithms*, **21**(1) (1996), 1–25.

[3] J. Bang-Jensen, H. N. Gabow, T. Jordán, and Z. Szigeti. Edge-connectivity augmentation with partition constraints. *SIAM J. Discrete Math.*, **12**(2) (1999), 160–207.

[4] M.-Y. Kao. Data security equals graph connectivity. *SIAM J. Discrete Math.*, **9**(1) (1996), 87–100.

[5] H. Nagamochi and T. Ibaraki. *Algorithmic aspects of graph connectivity*. Encyclopedia of Mathematics and Its Applications 123, Cambridge University Press, New York.

[6] A. Frank. Connectivity augmentation problems in network design. In *Mathematical Programming: State of the Art*, pages 34–63. J.R. Birge and K.G. Murty (eds.), University of Michigan, Ann Arbor, MI, 1994.

[7] A. Frank. *Connections in combinatorial optimization*, volume 38 of *Oxford Lecture Series in Mathematics and Its Applications*. Oxford University Press, Oxford, 2011.

[8] A. Schrijver. *Combinatorial Optimization-Polyhedra and Efficiency, Algorithms and Combinatorics*, 24, Springer, 2003.

[9] Z. Szigeti. On edge-connectivity augmentation of graphs and hypergraphs. In *Reearch Trends in Combinatorial Optimization*, pages 483–521. W. Cook, L. Lovász, J. Vygen (eds.), Springer, Berlin, Germany, 2009.

[10] S. Khuller. Approximation algorithms for finding highly connected subgraphs. In *Approximation Algorithms for NP-Hard Problems,* D. Hochbaum, (ed.), PWS Publishing Company, 1996.

[11] A. Gupta and J. Könemann. Approximation algorithms for network design: A survey. *Surv. Oper. Res. Manage. Sci.,* **16**(1) (January 2011), 3–20.

[12] K. P. Eswaran and R. E. Tarjan. Augmentation problems. *SIAM J. Comput.*, **5**(4) (1976), 653–665.

[13] J. Plesnik. Minimum block containing a given graph. *Arch. Math. (Basel)*, **27**(6) (1976), 668–672.

[14] R. E. Gomory and T. C. Hu. Multi-terminal network flows. *J. Soc. Ind. Appl. Math.*, **9** (1961), 551–570.

[15] H. Frank and W. Chou. Connectivity considerations in the design of survivable networks. *IEEE Trans. Circuit Theory*, **CT-17** (1970), 486–490.

[16] Y. Kajitani and S. Ueno. The minimum augmentation of a directed tree to a k-edge-connected directed graph. *Networks*, **16**(2) (1986), 181–197.

[17] T. Watanabe and A. Nakamura. Edge-connectivity augmentation problems. *J. Comput. Syst. Sci.*, **35**(1) (1987), 96–144.

[18] G. R. Cai and Y. G. Sun. The minimum augmentation of any graph to a K-edge-connected graph. *Networks*, **19**(1) (1989), 151–172.

[19] A. Frank. Augmenting graphs to meet edge-connectivity requirements. *SIAM J. Discrete Math.*, **5**(1) (1992), 25–53.

[20] L. Lovász. *Combinatorial Problems and Exercises*. North-Holland Publishing Co., Amsterdam, the Netherlands, 1979.

[21] D. Naor, D. Gusfield, and C. Martel. A fast algorithm for optimally increasing the edge connectivity. *SIAM J. Comput.*, **26**(4) (1997), 1139–1165.

[22] E. A. Dinic, A. V. Karzanov, and M. V. Lomonosov. The structure of a system of minimal edge cuts of a graph. In *Studies in Discrete Optimization (Russian)*, pages 290–306. A.A. Fridman (ed.), Izdat. Nauka, Moscow, Russia, 1976.

[23] A. A. Benczúr and D. R. Karger. Augmenting undirected edge connectivity in $\widetilde{O}(n^2)$ time. *J. Algorithms*, **37**(1) (2000), 2–36.

[24] W. Mader. A reduction method for edge-connectivity in graphs. *Ann. Discrete Math.*, **3** (1978), 145–164. *Advances in Graph Theory*, Cambridge Combinatorial Conf., Trinity College, Cambridge, 1977.

[25] E. Cheng and T. Jordán. Successive edge-connectivity augmentation problems. *Math. Program.*, **84**(3, Ser. B) (1999), 577–593. *Connectivity Augmentation of Networks: Structures and Algorithms*, Budapest, Hungary, 1994.

[26] J. Bang-Jensen, A. Frank, and B. Jackson. Preserving and increasing local edge-connectivity in mixed graphs. *SIAM J. Discrete Math.*, **8**(2) (1995), 155–178.

[27] H. Miwa and H. Ito. NA-edge-connectivity augmentation problems by adding edges. *J. Oper. Res. Soc. Jpn.*, **47**(4) (2004), 224–243.

[28] T. Ishii, Y. Akiyama, and H. Nagamochi. Minimum augmentation of edge-connectivity between vertices and sets of vertices in undirected graphs. *Algorithmica* **56**(4) (2010), 413–436.

[29] T. Ishii and M. Hagiwara. Minimum augmentation of local edge-connectivity between vertices and vertex subsets in undirected graphs. *Discrete Appl. Math.*, **154**(16) (November 1, 2006), 2307–2329.

[30] R. Grappe and Z. Szigeti. Covering symmetric semi-monotone functions. *Discrete Appl. Math.* **156** (2008), 138–144.

[31] A. Frank, How to make a digraph strongly connected, *Combinatorica* **1** (1981), 145–153.

[32] W. Mader. Konstruktion aller n-fach kantenzusammenhängenden Digraphen. *Eur. J. Combin.*, **3**(1) (1982), 63–67.

[33] A. Frank. On connectivity properties of Eulerian digraphs. In *Graph Theory in Memory of G. A. Dirac* (Sandbjerg, 1985), pages 179–194. L.D. Andersen et al. (eds.), North-Holland, Amsterdam, the Netherlands, 1989.

[34] B. Jackson. Some remarks on arc-connectivity, vertex splitting, and orientation in graphs and digraphs. *J. Graph Theory*, **12**(3) (1988), 429–436.

[35] D. Gusfield. Optimal mixed graph augmentation. *SIAM J. Comput.*, **16**(4) (1987), 599–612.

[36] G. N. Frederickson and J. Ja'Ja'. Approximation algorithms for several graph augmentation problems. *SIAM J. Comput.*, **10**(2) (1981), 270–283.

[37] J. Cheriyan, T. Jordán, and R. Ravi. On 2-coverings and 2-packings of laminar families. In *Algorithms—ESA '99* (Prague), pages 510–520. J. Nesetril (ed.), Springer, Berlin, Germany, 1999.

[38] E. D. Bolker and H. Crapo. Bracing rectangular frameworks. I. *SIAM J. Appl. Math.*, **36**(3) (1979), 473–490.

[39] J. Bang-Jensen and T. Jordán. Splitting off edges within a specified subset preserving the edge-connectivity of the graph. *J. Algorithms*, **37**(2) (2000), 326–343.

[40] W. Goddard, M. E. Raines, and P. J. Slater. Distance and connectivity measures in permutation graphs. *Discrete Math.*, **271**(1–3) (2003), 61–70.

[41] N. Jami and Z. Szigeti. Edge-connectivity of permutation hypergraphs. *Discrete Math.* **312**(17) (September 6, 2012), 2536–2539.

[42] H. N. Gabow and T. Jordán. How to make a square grid framework with cables rigid. *SIAM J. Comput.*, **30**(2) (2000), 649–680.

[43] H. N. Gabow and T. Jordán. Incrementing bipartite digraph edge-connectivity. *J. Comb. Optim.*, **4**(4) (2000), 449–486.

[44] H. N. Gabow and T. Jordán. Bipartition constrained edge-splitting in directed graphs. *Discrete Appl. Math.*, **115**(1–3) (2001), 49–62.

[45] H. Nagamochi and P. Eades. Edge-splitting and edge-connectivity augmentation in planar graphs. In *Integer Programming and Combinatorial Optimization* (Houston, TX, 1998), pages 96–111. R.E. Bixby, E.A. Boyd, and R.Z. Rios-Mercado (eds.), Springer, Berlin, Germany, 1998.

[46] T. Jordán. Constrained edge-splitting problems. *SIAM J. Discrete Math.*, **17**(1) (2003), 88–102.

[47] T. Watanabe and M. Yamakado. A linear time algorithm for smallest augmentation to 3-edge-connect a graph. *IEICE Trans. Fund. Jpn.*, **E76-A** (1993), 518–531.

[48] S. Taoka and T. Watanabe. The $(\sigma+1)$-edge-connectivity augmentation problem without creating multiple edges of a graphs. *Theoret. Comput. Sci.* **1872/2000** (2000), 169–185.

[49] T. Jordán. Two NP-complete augmentation problems. Tech. Rep. PP-1997-08, Odense University, Denmark, 1997.

[50] J. Bang-Jensen and T. Jordán. Edge-connectivity augmentation preserving simplicity. *SIAM J. Discrete Math.*, **11**(4) (1998), 603–623.

[51] T. Ishii and H. Nagamochi. Simultaneous augmentation of two graphs to an l-edge-connected graph and a biconnected graph. In *Algorithms and Computation* (Taipei, 2000), volume 1969 of *Lecture Notes in Computer Science*, pages 326–337. D.T. Lee and S.-H. Teng (eds.), Springer, Berlin, Germany, 2000.

[52] T.-S. Hsu and V. Ramachandran. Finding a smallest augmentation to biconnect a graph. *SIAM J. Comput.*, **22**(5) (1993), 889–912.

[53] T. Watanabe and A. Nakamura. A minimum 3-connectivity augmentation of a graph. *J. Comput. Syst. Sci.*, **46**(1) (1993), 91–128.

[54] T.-S. Hsu and V. Ramachandran. A linear time algorithm for triconnectivity augmentation. *Proc. 32nd Annu. Symp. Found. Comput. Sci.*, **22**(5) (1991), 548–559.

[55] T.-S. Hsu. On four-connecting a triconnected graph. *J. Algorithms*, **35**(2) (2000), 202–234.

[56] T. Jordán. On the optimal vertex-connectivity augmentation. *J. Combin. Theory Ser. B*, **63**(1) (1995), 8–20.

[57] T. Jordán. A note on the vertex-connectivity augmentation problem. *J. Combin. Theory Ser. B*, **71**(2) (1997), 294–301.

[58] J. Cheriyan and R. Thurimella. Fast algorithms for k-shredders and k-node connectivity augmentation. *J. Algorithms*, **33**(1) (1999), 15–50.

[59] T. Ishii and H. Nagamochi. On the minimum augmentation of an l-connected graph to a k-connected graph. In *Algorithm Theory—SWAT 2000* (Bergen), pages 286–299. Springer, Berlin, Germany, 2000.

[60] B. Jackson and T. Jordán. A near optimal algorithm for vertex connectivity augmentation. In *Algorithms and Computation* (Taipei, 2000), volume 1969 of *Lecture Notes in Comput. Sci.*, pages 313–325. D.T. Lee and S.-H. Teng (eds.), Springer, Berlin, Germany, 2000.

[61] B. Jackson and T. Jordán. Independence free graphs and vertex connectivity augmentation. *J. Combin. Theory Ser. B*, **94**(1) (2005), 31–77.

[62] L. A. Végh. Augmenting undirected node-connectivity by one. *SIAM J. Discrete Math.*, **25**(2) (2011), 695–718.

[63] T. Watanabe, Y. Higashi, and A. Nakamura. Graph augmentation problems for a specified set of vertices. In *Algorithms* (Tokyo, 1990), pages 378–387. Springer, Berlin, Germany, 1990.

[64] T.-S. Hsu and M.-Y. Kao. A unifying augmentation algorithm for two-edge connectivity and biconnectivity. *J. Comb. Optim.*, **2**(3) (1998), 237–256.

[65] T. Ishii, H. Nagamochi, and T. Ibaraki. Optimal augmentation of a 2-vertex-connected multigraph to an l-edge-connected and 3-vertex-connected multigraph. *J. Comb. Optim.*, **4**(1) (2000), 35–77.

[66] T. Ishii, H. Nagamochi, and T. Ibaraki. Multigraph augmentation under biconnectivity and general edge-connectivity requirements. *Networks*, **37**(3) (2001), 144–155.

[67] T. Ishii, H. Nagamochi, and T. Ibaraki. Augmenting a $(k-1)$-vertex-connected multigraph to an l-edge-connected and k-vertex-connected multigraph. *Algorithmica*, **44**(3) (2006), 257–280.

[68] T. Masuzawa, K. Hagihara, and N. Tokura. An optimal time algorithm for the k-vertex-connectivity unweighted augmentation problem for rooted directed trees. *Discrete Appl. Math.*, **17**(1–2) (1987), 67–105.

[69] T. Jordán. Increasing the vertex-connectivity in directed graphs. In *Algorithms—ESA '93* (Bad Honnef, 1993), pages 236–247. T. Lengauer (ed.), Springer, Berlin, Germany, 1993.

[70] A. Frank and T. Jordán. Minimal edge-coverings of pairs of sets. *J. Combin. Theory Ser. B*, **65**(1) (1995), 73–110.

[71] A. Frank and T. Jordán. Directed vertex-connectivity augmentation. *Math. Program.*, **84**(3, Ser. B) (1999), 537–553. *Connectivity augmentation of Networks: Structures and Algorithms* (Budapest, 1994).

[72] A. Frank and T. Jordán. How to make a strongly connected digraph two-connected. In *Integer Programming and Combinatorial Optimization* (Copenhagen, 1995), pages 414–425. E. Balas and J. Clausen (eds.), Springer, Berlin, Germany, 1995.

[73] A. A. Benczúr and L. Végh. Primal-dual approach for directed vertex connectivity augmentation and generalizations. *ACM Trans. Algorithms* **4**(2) (2008), 20.

[74] A. Frank and L. A. Végh. An algorithm to increase the node-connectivity of a digraph by one. *Discrete Optim.*, **5**(4) (2008), 677–684.

[75] S. Enni. A 1-(S,T)-edge-connectivity augmentation algorithm. *Math. Program.*, **84**(3, Ser. B) (1999), 529–535. *Connectivity Augmentation of Networks: Structures and Algorithms* (Budapest, 1994).

[76] E. Cheng. Edge-augmentation of hypergraphs. *Math. Program.*, **84**(3, Ser. B) (1999), 443–465. *Connectivity Augmentation of Networks: Structures and Algorithms* (Budapest, 1994).

[77] W.H. Cunningham. Decomposition of submodular functions. *Combinatorica*, (3) (1983), 53–68.

[78] J. Bang-Jensen and B. Jackson. Augmenting hypergraphs by edges of size two. *Math. Program.*, **84**(3, Ser. B) (1999), 467–481. *Connectivity Augmentation of Networks: Structures and Algorithms* (Budapest, 1994).

[79] B. Cosh. *Vertex Splitting and Connectivity Augmentation in Hypergraphs*. PhD Thesis, Goldsmiths College, London, 2000.

[80] A. Bernáth, R. Grappe, and Z. Szigeti. Augmenting the edge-connectivity of a hypergraph by adding a multipartite graph. *J. Graph Theory*, **72**(3) (March 2013), 291–312.

[81] Z. Király, B. Cosh, and B. Jackson. Local edge-connectivity augmentation in hypergraphs is NP-complete. *Discrete Appl. Math.*, **158**(6) (2010), 723–727.

[82] T. Király, Covering symmetric supermodular functions by uniform hypergraphs, *J. Comb. Theory, Ser. B*, **91**(2) (2004), 185–200.

[83] Z. Szigeti. Hypergraph connectivity augmentation. *Math. Program.*, **84**(3, Ser. B) (1999), 519–527. *Connectivity Augmentation of Networks: Structures and Algorithms* (Budapest, 1994).

[84] A. A. Benczúr and A. Frank. Covering symmetric supermodular functions by graphs. *Math. Program.*, **84**(3, Ser. B) (1999), 483–503. *Connectivity Augmentation of Networks: Structures and Algorithms* (Budapest, 1994).

[85] T. Fleiner and T. Jordán. Coverings and structure of crossing families. *Math. Program.*, **84**(3, Ser. B) (1999), 505–518. *Connectivity Augmentation of Networks: Structures and Algorithms* (Budapest, 1994).

[86] A. Bernáth, R. Grappe, and Z. Szigeti. Partition constrained covering of a symmetric crossing supermodular function by a graph. *Proc. of the 21st Annu. ACM-SIAM Symp. Discrete Algorithms*, (2010), 1512–1520.

[87] B. Fleiner. Detachment of vertices of graphs preserving edge-connectivity. *SIAM J. Discrete Math.*, **18**(3) (2004), 581–591.

[88] T. Jordán and Z. Szigeti. Detachments preserving local edge-connectivity of graphs. *SIAM J. Discrete Math.*, **17**(1) (2003), 72–87.

[89] A. R. Berg, B. Jackson, and T. Jordán. Edge splitting and connectivity augmentation in directed hypergraphs. *Discrete Math.*, **273**(1–3) (2003), 71–84. EuroComb'01 (Barcelona).

[90] A. Frank. Edge-connection of graphs, digraphs, and hypergraphs. In *More sets, graphs and numbers*, volume 15 of *Bolyai Society Mathematical Studies*, pages 93–141. E. Győri et al. (eds.), Springer, Berlin, Germany, 2006.

[91] A. Frank. Restricted t-matchings in bipartite graphs. *Discrete Appl. Math.*, **131**(2) (2003), 337–346. Submodularity.

[92] J. A. Soto and C. Telha. Jump number of two-directional orthogonal ray graphs. In *Proceedings of the IPCO*, LNCS 6655, pages 389–403, O. Günlük and G. Woeginger (eds.), Springer, New York, 2011.

[93] A. Frank and T. Király. Combined connectivity augmentation and orientation problems. *Discrete Appl. Math.*, **131**(2) (2003), 401–419. Submodularity.

CHAPTER 15

Matchings

Michael D. Plummer

CONTENTS

15.1	Introduction and Terminology	349
15.2	Bipartite Matching: Theorems of König, Hall, and Frobenius	351
15.3	Tutte's Theorem and Perfect Matching in General Graphs	354
15.4	Sufficient Conditions for Perfect Matchings	355
15.5	Maximum Matchings	356
15.6	Gallai–Edmonds Structure Theorem	357
15.7	Structure of Factor-Critical Graphs	359
15.8	Ear Decompositions of Matching-Covered Graphs	360
15.9	Brick Decomposition Procedure	362
15.10	Determinants, Permanents, and the Number of Perfect Matchings	364
15.11	Applications of Matching: Chemistry and Physics	366
15.12	Matching Extension	367
15.13	f- and k-Factors	368

15.1 INTRODUCTION AND TERMINOLOGY

A *matching* in a graph G is a set of independent edges; that is to say, a set of edges no two of which share a vertex. It is no surprise that the study of such a simple concept should have begun early in the history of graph theory. Nor is it surprising that the idea of a matching should arise in many different contexts as well. One can take the position that there are two main historical sources for the study of matchings. One can associate these sources with two individuals: the Dane Julius Petersen in the area of regular graphs and the Hungarian Dénes König in the area of bipartite graphs.

In 1891 Petersen [1] translated a problem in algebraic factorization due to David Hilbert into a factorization problem for regular graphs. He then proved that any graph regular of even degree can be *factored* (i.e., decomposed) into edge-disjoint spanning subgraphs regular of degree two. He proceded to point out that, in contrast, factorization into subgraphs regular of *odd* degree is a more difficult problem. He was able to show, however, that any connected 3-regular graph contains a perfect matching (i.e., a matching covering all of $V(G)$), provided it contains no more than two cut edges. He then pointed out that the assumption of no more than two cut edges was best possible by giving the example shown in Figure 15.1 below which he attributed to Silvester.

Somewhat later, Petersen published the famous graph which was to bear his name (the *Petersen* graph) as an example of a nonplanar cubic graph with no cut edges which could not be decomposed into three disjoint perfect matchings [2] (see Figure 15.2).

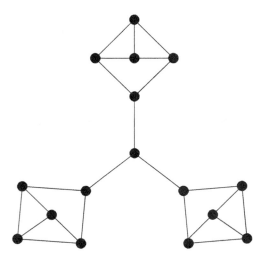

Figure 15.1 Sylvester graph.

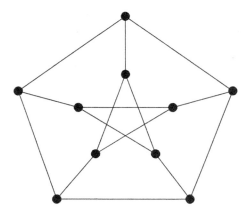

Figure 15.2 Petersen graph.

It is hard to overestimate the importance of this 10-vertex graph in the field of graph theory as it undoubtedly has provided counterexamples to more conjectures than any other graph. (See the book [3] and more recently the paper [4] in this context.)

König's study of matchings in bipartite graphs also arose from a problem in algebra. In particular, he translated a problem in the area of reducibility properties of determinants originally solved by the German Frobenius into a problem on bipartite graphs and in so doing provided a shorter proof [5]. He then extended this work in his twin papers [6,7]—one in Hungarian and the other in German—proving that any k-regular bipartite graph is the union of k disjoint perfect matchings.

Also in these two papers, König proved what has come to be known as *König's Edge Coloring Theorem* which states that in any bipartite graph the chromatic index is equal to the maximum degree of the graph.

In a second set of twin papers [8,9], König proved yet another theorem of at least as much importance in graph theory, in which he showed that in any bipartite graph the size of any maximum matching is equal to the size of any smallest set of vertices which together cover all the edges of the graph. This result, which is discussed below in Section 15.2, gave birth to the study of so–called *minimax* theorems in graph theory and allied areas.

Due to space limitations, we will be content with this necessarily brief historical introduction to matching theory. For a much more detailed treatment of the history in this area, see the Preface of [10,11].

Let us make a few comments on notation and terminology pertaining especially to this chapter on matching. A matching M in G is said to be *maximal* if it is not a proper subset of any other matching in G. A matching in G is said to be *maximum* if its cardinality is greatest among all matchings in G. The cardinality of any maximum matching in G is called the *matching number* of G and is denoted by $\nu(G)$. Of course, every maximum matching is maximal, but the converse is not true in general. A matching is said to be *perfect* if it covers all vertices of the graph and *near-perfect* if it covers all but exactly one vertex. A perfect matching is often called a *1-factor*. A set of edges in graph G is called an *edge cover* of G if it covers all the vertices of G. The cardinality of any smallest edge cover in G is called the *edge covering number* of G and is denoted by $\rho(G)$. A set of vertices in graph G is said to be *independent* if no two of its members are joined by an edge. The size of any largest independent set in G is called the *independence number* of G and is denoted by $\alpha(G)$. A set S of vertices in G is called a *vertex cover* of G if every edge of G has at least one end vertex in S. The size of any smallest vertex cover is called the *vertex covering number* of G and is denoted by $\tau(G)$.

These four parameters are not independent in general, but are related as shown in the following theorem.

Theorem 15.1 *If G is any graph, then*

i. $\alpha(G) + \tau(G) = |V(G)|$ *and*

ii. *If G has no isolated vertices, then $\nu(G) + \rho(G) = |V(G)|$.*

Proof. The proof of (i) is immediate upon realizing that the complement of any vertex cover is an independent set.

To prove (ii), let E be a smallest edge cover in G. Then by the minimality of E, it must consist of $|V(G)| - \rho(G)$ vertex-disjoint stars. But since there are no isolates, each of these stars contains at least one edge, so selecting one edge from each star, one obtains a matching M in G. Hence $\nu(G) \geq |M| = |V(G)| - \rho(G)$; that is, $\nu(G) + \rho(G) \geq |V(G)|$.

On the other hand, let us choose a maximum matching M in G and let U denote the set of vertices not covered by M. Then U is independent. Again since G has no isolated vertices, for each of the $|V(G)| - 2\nu(G)$ vertices of set U we may choose an edge covering it. If S denotes the collection of edges thus chosen, clearly $M \cup S$ is an edge cover for G. But then $\rho(G) \leq |M \cup S| = \nu(G) + |V(G)| - 2\nu(G) = |V(G)| - \nu(G)$ and hence $\nu G) + \rho(G) \leq |V(G)|$. The result then follows. ∎

The two equations in parts (i) and (ii) of the preceding theorem are often referred to as the *Gallai identities* (cf. [12]).

For any set $S \subseteq V(G)$, we shall denote by $N(S)$ the set of all vertices in G having a neighbor in S. For any graph G, we shall denote by $c(G)$ the number of components of G and by $c_o(G)$ the number of components of G of odd cardinality. The maximum degree in graph G will be denoted by $\Delta(G)$. Other terminology and symbols will be introduced as needed.

15.2 BIPARTITE MATCHING: THEOREMS OF KÖNIG, HALL, AND FROBENIUS

One of the earliest results on matching in bipartite graphs, and still one of the most important, is due to König [8,9].

Theorem 15.2 *If G is bipartite, then $\nu(G) = \tau(G)$.*

Proof. Since every cover must, in particular, cover every matching, we have $\tau(G) \geq \nu(G)$.

To obtain the reverse inequality, we delete edges from graph G as long as possible to obtain a subgraph G' with $\tau(G') = \tau(G)$. Hence $\tau(G' - e) < \tau(G)$ for every $e \in E(G')$.

We claim that G' is a matching. Suppose not. Then there must exist two edges e and f which share a common vertex x. Consider now the graph $G' - e$. By the minimality of graph G', there is a cover S_e in $G' - e$ covering the edges of $G' - e$ with $|S_e| < \tau(G') - 1$. Clearly neither end vertex of e lies in S_e. Similarly, there is a cover S_f in $G' - f$ containing neither end vertex of f with $|S_f| = \tau(G') - 1 = |S_e|$.

Now form the induced subgraph G'' of G' where $G'' = G'[\{x\} \cup (S_e \oplus S_f)]$, where "$\oplus$" denotes symmetric difference. Let $s = |S_e \cap S_f|$. Then $|V(G'')| = 2(\tau(G') - 1 - s) + 1$. But graph G'', being a subgraph of G, is bipartite and so if we let T be the smaller of the two partite sets for G'', T covers G'' and $|T| \leq \tau(G') - 1 - s$.

But $T' = T \cup (S_e \cap S_f)$ covers G'. To see this, suppose that g is any edge in G'. If $g \neq e$ and $g \neq f$, then g is covered by both S_e and S_f, for either it is covered by $S_e \cap S_f$ or else it joins $S_e - S_f$ to $S_f - S_e$. In the latter instance, g is an edge of G'' and hence covered by T.

So $\tau(G') \leq |T'| = |T \cup (S_e \cap S_f)| = |T| + |S_e \cap S_f| \leq \tau(G') - 1 - s + s = \tau(G') - 1$ and we have a contradiction. Thus G' is a matching as claimed.

But then $\tau(G) = \tau(G') = \nu(G') \leq \nu(G)$ and the proof is complete. ∎

Remark: König's theorem is an example of a so-called *minimax* theorem. Other examples include the max-flow-min-cut theorem and Menger's theorem (Chapters 4, 5, and 12).

König's minimax theorem, together with the following result of Philip Hall, can be said to be the twin pillars upon which bipartite matching is built.

Theorem 15.3 [13] $G = (A, B)$ *bipartite, then exists matching of A into B iff $|N(X)| \geq |X|$, for all $X \subseteq A$.*

Proof. Clearly, if G has a matching of A into B, that is, if $\nu(G) = |A|$, then for every $A' \subseteq A$, $|N(A')| \geq |A'|$.

To prove the converse, it is enough to prove that $\tau(G) = |A|$, by König's theorem. Suppose, to the contrary, that $\tau(G) < |A|$. Thus there is a minimum vertex cover of G consisting of a set of vertices $A' \cup B'$, where $A' \subseteq A$ and $B' \subseteq B$ and $|A' \cup B'| < |A|$. Hence $|B'| < |A| - |A'| = |A - A'|$. So $|N(A - A')| \leq |B'| < |A - A'|$, a contradiction. Thus $\tau(G) = |A|$. ∎

The statement that $|N(X)| \geq |X|$, for all $X \subseteq A$ in the statement of Hall's theorem above is often called *Hall's condition*.

Remark: There are many versions of Hall's theorem (cf. [10,11]). One immediate consequence of Hall's theorem is often called the *Marriage Theorem* or the *Frobenius Theorem*.

Corollary 15.1 [14] *Bipartite graph $G = (A, B)$ has a perfect matching iff $|A| = |B|$ and for each $X \subseteq A$, $|X| \leq |N(X)|$.*

Remark: It may be shown that the Marriage theorem is, in fact, equivalent to Hall's theorem and both are equivalent, in turn, to König's theorem.

Remark: For a bipartite graph $G = (A, B)$, the property of having a perfect matching is an *NP-property*. That is, given a set of edges in G, it may be verified in polynomial time that they are disjoint and cover all the vertices of G. Less obviously, the property of *not* having a perfect matching is also an NP-property. To see this, use Hall's theorem and simply exhibit a subset $A' \subseteq A$ such that $|N(A')| > |A'|$. For this reason, the Frobenius-Hall result is referred to as a *good characterization* of those bipartite graphs with perfect matchings. In contrast, consider the existence of a Hamilton cycle in G (i.e., a cycle through all vertices of G). Given a cycle it is polynomial to check that it contains all vertices of G. However, if a graph is *not* Hamiltonian, there is no known way to verify this in polynomial time. It is crucial to note that both in the matching and Hamiltonian cycle illustrations, how the candidate matching (respectively, the candidate cycle) is obtained does not concern us.

The area of study known as *computational complexity* deals with such concepts in much more depth than we can in our limited space here (cf. [10,11]).

Hall's theorem has immediate application to the problem of coloring the edges of a bipartite graph.

Theorem 15.4 *Let G be a regular bipartite graph. Then G contains a perfect matching.*

Proof. The regularity hypothesis immediately implies that Hall's Condition is satisfied. ∎

An *edge coloring* of a graph G is an assignment of colors to the edges such that edges sharing a common vertex must receive different colors. The *edge-chromatic number* $\chi_e(G)$ of the graph G is the smallest number of colors necessary to color the edges of G. The preceding theorem has the following immediate consequence.

Theorem 15.5 [6,7] *If G is a bipartite graph regular of degree Δ, then $\chi_e(G) = \Delta(G)$.*

Proof. The proof is by induction on $\Delta(G)$. If $\Delta(G) = 1$ the result is clear. So suppose the result holds for all bipartite graphs regular of degree less than $\Delta(G)$. By the preceding theorem, G contains a perfect matching M. Delete M from G to obtain a bipartite graph G' which is regular of degree $\Delta(G) - 1$. By the induction hypothesis, G' can be edge-colored in $\Delta(G) - 1$ colors. Then assigning a new color to the edges of the matching M we obtain the desired edge coloring for G in $\Delta(G)$ colors. ∎

This result was used by König to prove a more general edge coloring result for all bipartite graphs.

Theorem 15.6 [6,7] *If G is bipartite, then $\chi_e(G) = \Delta(G)$.*

Proof. The proof is by induction on $|E(G)|$. Trivially, the result holds if $|E(G)| = 0$, so suppose that $|E(G)| \geq 1$ and that the result holds for all graphs with fewer edges than G. Without loss of generality, let us suppose that if (A, B) is the bipartition of G, $|A| \leq |B|$. If $|A| = 1$, then clearly the result holds, so suppose $|A| \geq 2$. Then one can choose an edge $xy \in E(G)$ such that $\Delta(G - xy) = \Delta(G)$. So by the induction hypothesis, there is an edge coloring of $G - xy$ in $\Delta(G)$ colors. Let these colors be denoted $c_1, \ldots, c_\Delta$.

In the graph $G - xy$ each of x and y is incident with at most $\Delta - 1$ edges. Hence there must be two colors c_i and c_j such that x meets no c_i-edge and y meets no c_j-edge. If $c_i = c_j$, color the missing edge xy with this color and we have the desired coloring of $E(G)$ in $\Delta(G)$ colors. So assume that $c_i \neq c_j$. Without loss of generality, assume that vertex x is incident with a c_j-edge.

Starting with this edge, find a walk W consisting of edges colored alternately c_i and c_j of maximal length. Clearly such a walk exists and moreover, by the definition of edge coloring, it cannot use any vertex more than once; that is, W must be a path. We claim that W cannot use vertex y. Suppose, to the contrary, that it does use vertex y. Then, since G is bipartite, the edge of W incident with y must have color c_i. Thus W is even. But then if one adds the edge xy to W one obtains an odd cycle in the graph G, a contradiction.

Now interchange the colors c_i and c_j on the path W to obtain a new coloring of the edges of G. By the choice of color c_i and the fact that path W was chosen to be of maximal length, it remains true that adjacent edges of $G - xy$ are colored with different colors. But in this new coloring, neither vertex x nor y is incident with an edge colored c_j. So color edge xy with color c_j and we have obtained an edge coloring of $E(G)$ in $\Delta(G)$ colors, as desired. ∎

Other settings for Hall's theorem abound. In its original form, it was cast as a theorem from set theory dealing with *systems of distinct representatives*. In the language of partially ordered sets, it turns out to be equivalent to an important result known as Dilworth's theorem which states that in any finite poset, the size of any largest antichain equals the size of any smallest partition of the poset into chains. In yet another setting, the König-Hall result implies an important theorem in the theory of doubly stochastic matrices. A nonnegative square matrix M is said to be *doubly stochastic* if all row and column sums are equal to 1. A *permutation matrix* is a square matrix in which each row and each column contains exactly one entry 1, while the other entries are all 0. This *Birkhoff–von Neumann Theorem* says that any doubly stochastic matrix can be written as a convex combination of permutation matrices.

Two other extremely important graphical results are *Menger's Theorem* [15] which is central in the study of connectivity in graphs and the closely related *Max-Flow Min-Cut Theorem* [16,17] which is basic to the theory of network flows. Interestingly, Hall's theorem is also equivalent to each of these! We will not go into details here, but refer the reader to Chapters 4 and 12.

Remark: It is interesting to consider the generalization of Hall's condition to nonbipartite graphs. The condition then becomes

$$|S| \leq |N(S)| \text{ for all } S \subseteq V(G).$$

We give two applications of this condition. A graph is said to have the (*weak*) *odd cycle property* if every pair of odd cycles either share at least one vertex or are joined by an edge. Berge [18,19] proved the following result for graphs with this property. ∎

Theorem 15.7 *Let G be an even graph satisfying the odd cycle property. Then G contains a perfect matching if and only if $|S| \leq |N(S)|$, for all $S \subseteq V(G)$.* ∎

Tutte [20] obtained the following result involving a certain generalization of perfect matching called a perfect 2-matching. Given a graph G, a *perfect 2-matching* of G is a set of cycles and edges such that every two elements of this set are vertex-disjoint and together they cover $V(G)$.

Theorem 15.8 *A graph G has a perfect 2-matching if and only if $|S| \leq |N(S)|$, for every independent set of vertices $S \subseteq V(G)$.* ∎

15.3 TUTTE'S THEOREM AND PERFECT MATCHING IN GENERAL GRAPHS

For $S \subseteq V(G)$, let $c_o(G - S)$ denote the number of odd components of the graph $G - S$. The proof of the following classical matching theorem of Tutte [21] is due to Lovász [22].

Theorem 15.9 *A graph G contains a perfect matching if and only if for all $S \subseteq V(G)$, $c_o(G - S) \leq |S|$.*

Proof. ($\Longrightarrow$) Clear.

($\Longleftarrow$) Suppose that for all $S \subseteq V(G)$, $c_o(G - S) \leq |S|$, but G does not contain a perfect matching. If we set $S = \emptyset$, we have that $c_o(G - S) = 0$, so G is even. Moreover, $|S|$ and $c_o(G - S)$ must have the same parity.

Let us now successively add edges to G to obtain a graph G' which has no perfect matching, but for any edge $e \notin E(G')$, $G' + e$ does contain a perfect matching. (Clearly this is possible since G is a subgraph of the complete graph on $|V(G)|$ vertices.) Let V_1 consist of the set of vertices in $V(G)$ which are each adjacent to every other vertex in G and let $V_2 = V(G) - V_1$. Finally, let $G'' = G[V_2]$.

Claim: G'' consists of vertex-disjoint complete graphs.

Suppose not. Then there are three vertices x, y, and z in $V(G)$ such that x and y are adjacent, y and z are adjacent, but x and z are not adjacent. Furthermore, since $y \in V_2$, there must be a fourth vertex $w \notin \{x, y, z\}$ such that w is not adjacent to y.

Now by definition $G' + xz$ contains a perfect matching M_1 and $G' + yw$ contains a perfect matching M_2. Clearly $xz \in M_1$ and $yw \in M_2$, but $xz \notin M_2$ and $yw \notin M_1$. Moreover, $M_1 \cup M_2$ consists of disjoint even (in fact, alternating $M_1 - M_2$) cycles together with a set of independent edges which are just the edges of $M_1 \cap M_2$. Let C be a cycle in $M_1 \cup M_2$ which contains the edge xz.

We claim that C also contains the edge yw. For suppose not. Then construct a new perfect matching of G' from M_1 by exchanging the edges of M_1 in $E(C)$ for those of M_2 in $E(C)$. But then this new perfect matching contains neither xz nor yw and hence is a perfect matching in G', a contradiction.

Now since C is even, $C - xz$ is a path of odd length. Without loss of generality, suppose $C - xz = P_1 \cup P_2$, where P_2 contains the edge yw.

First suppose P_1 has odd length. Then form a new matching M_3 of $V(C)$ using edge xy and suitable edges from $E(G)$. Then $M_3 \cup (M_1 - E(C))$ is a perfect matching of G containing neither xz nor yw and hence is a perfect matching of G', again a contradiction.

If, on the other hand, P_1 has even length, form matching M_3 using edge yz and suitable edges of C. Then $M_3 \cup (M_1 - E(C))$ is a perfect matching of G', again a contradiction.

Thus the Claim is true; that is, G'' consists of a disjoint union of cliques.

Now form a perfect matching for G as follows. For each odd component of $G'[V_2]$, take a near-perfect matching together with a single edge joining the unmatched vertex of the component with a distinct vertex of V_1. To these matchings, add a perfect matching of each even component of $G''[V_2]$. Since $c_o(G - S) \leq |S|$ by our initial assumption, the matching so far constructed covers all components of $G''[V_2]$. Any remaining unmatched vertices of V_1 remaining can be matched pair-wise, since $|S|$ and $c_o(G - S)$ have the same parity. So G has a perfect matching, contrary to our initial assumption and the theorem is proved. ∎

15.4 SUFFICIENT CONDITIONS FOR PERFECT MATCHINGS

There are a number of known sufficient conditions for perfect matchings which involve other graphical properties. We give a brief sample.

An induced subgraph of any graph G which is isomorphic to the complete bipartite graph $K_{1,3}$ is called a *claw* in G. A graph containing no claws is said to be *claw-free*. Las Vergnas [23] and Sumner [24] independently proved the following theorem about matchings in claw-free graphs.

Theorem 15.10 *If G is a connected claw-free graph with an even number of vertices, then G contains a perfect matching.*

Proof. Suppose to the contrary that G has no perfect matching. Then by Tutte's theorem, there is a set $S \subseteq V(G)$ such that $c_o(G - S) > |S|$. Such a set S is often called a *Tutte set* in G. Among all Tutte sets in G let S_0 be a smallest one. Since G is connected, $S \neq \emptyset$, so choose a vertex $x \in S$. Then x must be adjacent to a vertex in an odd component of $G - S$, for otherwise $S - x$ would also be a Tutte set. Suppose x is adjacent to a vertex in exactly one odd component C of $G - S$. Then $S - x$ is a smaller Tutte set than S, a contradiction. Suppose x is adjacent to exactly two odd components of $G - S$; call these components C_1 and C_2. Then the subgraph induced by $V(C_1) \cup V(C_2) \cup \{x\}$ is odd and once again $S - x$ is a smaller Tutte set than S. So x is adjacent to vertices in at least three odd components of $G - S$ and the existence of a claw follows. ∎

Next we define the *toughness* $t(G)$ of a graph G as follows. If $G = K_n$, then $t(G) = +\infty$. If $G \neq K_n$ for all n, then $t(G) = \min\{|S|/c(G - S)\}$ where we take the minimum over all vertex cuts S in G. A graph G is said to be *t-tough* for all t such that $t \leq t(G)$. The proof of the next theorem follows immediately from Tutte's theorem.

Theorem 15.11 *If $|V(G)|$ is even and $t(G) \geq 1$, then G has a perfect matching.* ∎

Toughness deserves to be mentioned because of a well-known conjecture of Chvátal [25] which remains unsettled.

Conjecture: There exists a constant t_0 such that if G is any graph with $t(G) \geq t_0$, then G contains a Hamilton cycle.

It is known that $t_0 = 2$ is not sufficient here. In fact, we have the following result. A graph is said to be *traceable* if it contains a Hamilton path.

Theorem 15.12 [26] *For every $\epsilon > 0$, there exists a $(9/4) - \epsilon$-tough graph which is not traceable.* ∎

The *binding number* $\text{bind}(G)$ of graph G is the minimum over all S, $\emptyset \neq S \subseteq V(G)$ and $N(S) \neq V(G)$, of the quantity $|N(S)|/|S|$. Anderson [27] related the binding number to matchings via the following theorem.

Theorem 15.13 *If $\text{bind}(G) \geq 4/3$, then G has a perfect matching.* ∎

That toughness and binding number are really quite different is graphically pointed out by comparing Chvátal's unsolved conjecture above with the following result of Woodall [28].

Theorem 15.14 *If $\text{bind}(G) \geq 3/2$, then G contains a Hamilton cycle.* ∎

15.5 MAXIMUM MATCHINGS

Although a graph may or may not have a perfect matching, all graphs contain matchings of maximum size, so-called *maximum* matchings. So what can be said about the size of a maximum matching?

Note that König's minimax equation does not necessarily hold in the case of nonbipartite graphs. (Just consider $G = K_3$.) However, Berge [18] obtained a minimax result involving matchings for general graphs using the concept of *deficiency* of a graph. Given a graph G, the *deficiency* of G is defined as $\text{def}(G) = |V(G)| - 2\nu(G)$; that is, $\text{def}(G)$ is just the number of vertices in G left unsaturated by any maximum matching.

Theorem 15.15 (Berge's Deficiency Theorem): *For any graph G, $\text{def}(G) = \max\{c_o(G - S) - |S| | S \subseteq V(G)\}$.*

Proof. Define $d = \max\{c_o(G - S) - |S| | S \subseteq V(G)\}$. Setting $S = \emptyset$ shows that $d \geq 0$. Now consider the join of G and K_d; that is, join every vertex of a complete graph of size d to every vertex of G. Denote this graph by G'. We claim that G' satisfies Tutte's condition and therefore has a perfect matching.

Let us consider the subsets $S' \subseteq V(G')$. If $S' = \emptyset$, then $c_o(G' - S') \leq |S'| = 0$, for $|V(G')|$ is even and hence $c_o(G' - S') = 0$. Suppose next that $\emptyset \neq S'$, but S' does not contain all of K_d. Then $G' - S'$ is connected; that is, it consists of exactly one component and so $1 = c_o(G' - S') \leq |S'|$. Finally, suppose S' contains K_d. Let $S = S' - V(K_d)$. Then $G' - S' = G - S$ and so $c_o(G' - S') = c_o(G - S) \leq |S| + d = |S'|$. Hence G' contains a perfect matching as claimed.

Now let M' be such a perfect matching in G'. This matching can cover at most d vertices in G, so if we delete these edges from M', we are left with a matching M in G covering at least $|V(G)| - d$ vertices. That is, $2\nu(G) \geq |V(G)| - d$ or

$$\max\{c_o(G-S) - |S||S \subseteq V()\} \geq |V(G)| - 2\nu(G). \tag{15.1}$$

But for every $S \subseteq V(G)$, every matching fails in G to cover at least $c_o(G-S) - |S|$ vertices and hence every matching in G fails to cover at least $\max\{c_o(G-S) - |S||S \subseteq V(G)\}$ vertices. Thus, in particular, every *maximum* matching in G fails to cover at least $\max\{c_o(G-S) - |S||S \subseteq V(G)\}$ vertices. That is,

$$def(G) = |V(G)| - 2v(G) \geq \max\{c_o(G-S) - |S||S \subseteq V(G)\}. \tag{15.2}$$

But now combining (15.1) and (15.2), equality must hold. ∎

By virtue of Berge's deficiency theorem, sometimes the quantity def(G) (and therefore also $v(G)$) is said to be *well-characterized*. Suppose we are given a nonnegative integer $d \geq$ def(G). How can we demonstrate this to someone without knowing the value of def(G)? If our challenger knows Berge's theorem, we can proceed to convince him as follows. Let $r = (|V(G)| - d)/2$. Suppose we can provide him with a matching M with $|M| \geq r$. Then def(G) $= |V(G)| - 2v(G) \leq |V(G)| - 2r = d$ and he is convinced.

On the other hand, suppose $d \leq$ def(G). If we can produce a set $S \subseteq V(G)$ such that $c_o(G-S) - |S| \geq d$, then he knows, again, that def(G) $= \max\{c_o(G-S) - |S||S \subseteq V(G)\} \geq c_o(G-S) - |S| \geq d$ and again he is convinced.

15.6 GALLAI–EDMONDS STRUCTURE THEOREM

Suppose G does not have a perfect matching and hence by Tutte's theorem there is a set S such that $c_o(G-S) > |S|$. Recall such an S is called a *Tutte set* in G. There may be many different Tutte sets in G. Is there perhaps one Tutte set which is *best* in some sense? The aim of the next work is to show that indeed among all the Tutte sets, there is a unique such set which yields much information about the structure of G with respect to its maximum matchings. This beautiful canonical decomposition theorem for graphs in terms of their maximum matchings is independently due to Gallai [29,30] and Edmonds [31].

Let G be an arbitrary graph. We define three subsets of $V(G)$ as follows: $D(G)$ is the set of all vertices v such that G contains a maximum matching missing v. $A(G) = N(D(G)) - V(D(G))$ and $C(G) = V(G) - (A(G) \cup D(G))$. Clearly, $\{D(G), A(G), C(G)\}$ is a partition of $V(G)$. Now suppose that $G = (A, B)$ is bipartite. For every set $S \subseteq A$, define the *surplus* of S, $\sigma(G)$, as $|N(S)| - |S|$. We then define the *surplus* of graph G, $\sigma(G)$, by $\sigma(G) = \min\{\sigma(A)| A \neq \emptyset\}$.

In Figure 15.3 below we illustrate a Gallai–Edmonds partition $\{D(G), A(G), C(G)\}$.

Definition 15.1 *A graph G is said to be factor-critical (or hypomatchable) if $G - v$ contains a perfect matching for every $v \in V(G)$.*

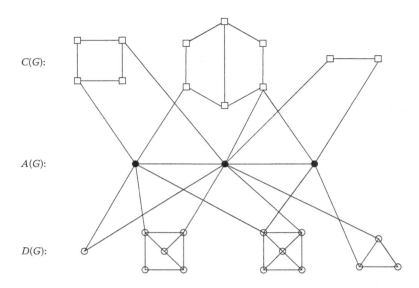

Figure 15.3 Example of the partition $\{D(G), A(G), C(G)\}$.

Theorem 15.16 (Gallai–Edmonds Structure Theorem): *Let G be an arbitrary graph with $D(G), A(G),$ and $C(G)$ defined as above. Then:*

a. *Each component of the subgraph induced by $D(G)$ is factor-critical;*

b. *The subgraph induced by $C(G)$ has a perfect matching;*

c. *If $B(G)$ is the bipartite graph obtained from G by deleting $C(G)$ and the edges spanned by $A(G)$ and then contracting each component of $D(G)$ to a separate single vertex, then $B(G)$ has positive surplus (when viewed from A); and*

d. *Every maximum matching of G is composed of a near-perfect matching of each component of the graph spanned by $D(G)$, a perfect matching of each component of the graph spanned by $C(G)$ and a complete matching of $A(G)$ into $D(G)$ each edge of which matches a vertex of $A(G)$ to a vertex in a different component of the graph spanned by $D(G)$, and*

e.
$$\nu(G) = \frac{1}{2}(|V(G)| - c(D(G)) + |A(G)|),$$

where $c(D(G))$ denotes the number of components of the graph spanned by $D(G)$. ∎

For two quite different proofs of this beautiful result, we refer the reader to [10,11] and to [32]. The latter of these derives the $\{D(G), A(G), C(G)\}$ partition using the blossom algorithm for maximum matching. (See Chapter 16 for details of the matching algorithm.) Since the blossom algorithm is known to be polynomial, it follows that the Gallai–Edmonds partition can be obtained in polynomial time.

Remark: The nontrivial parts of the proofs of Tutte's theorem and Berge's theorem are easy corollaries of the Gallai–Edmonds theorem.

Remark: It follows that $\mathrm{def}(G) = c(D(G)) - |A(G)|$, where $c(D(G))$ denotes the number of components of the graph spanned by $D(G)$.

Remark: It also follows that every edge incident with a vertex of $D(G)$ lies in some maximum matching of G, but no edge joining $A(G)$ to $C(G)$ belongs to any maximum matching of G. Moreover, no edge spanned by $A(G)$ belongs to any maximum matching either.

Although the Gallai–Edmonds result provides much useful information about the structure of any graph with respect to its maximum matchings, there are several degenerate cases which require further scrutiny. If G is factor-critical, then $V(G) = D(G)$ and hence $A(G) = C(G) = \emptyset$. On the other hand, if G contains a perfect matching, then $D(G) = A(G) = \emptyset$ and hence $V(G) = C(G)$. We will discuss these two cases later.

Question: Can one determine the number $\Phi(G)$ of maximum (or perfect) matchings in a graph G in an efficient way?

That this is highly unlikely has been demonstrated by Valiant [33] who showed that the problem of determining $\Phi(G)$ is #P-complete. Informally this means that if one could determine $\Phi(G)$ in polynomial time via some algorithm, then said algorithm could be converted into a polynomial algorithm for solving a large number of problems for which no *good* (i.e., polynomial) procedure is known.

If one recalls the Gallai–Edmonds decomposition $\{D(G), A(G), C(G)\}$, one might hope to obtain $\Phi(G)$ if—and this is a big *if*—one could count the number of near-perfect matchings in each component of $D(G)$ and the number of perfect matchings in each component of $C(G)$. But given the known #P-completeness of computing $\Phi(G)$ exactly, it is likely we are also doomed in this approach as well. So taking the *half a loaf is better than none* approach, one might ask if, for an arbitrary graph G, we can *bound* $\Phi(G)$ by deriving bounds for the number of near-perfect matchings in $D(G)$ and the number of perfect matchings in $C(G)$.

15.7 STRUCTURE OF FACTOR-CRITICAL GRAPHS

It is easy to see that every factor-critical graph is connected (in fact, 2-edge-connected), has minimum degree at least two and is odd. On the other hand, factor-critical graphs may contain cutvertices. (Consider two triangles sharing precisely one vertex.)

In fact, there is a nice iterative construction of any factor-critical graph. That such a so-called *ear structure* exists is guaranteed by the next theorem due to Lovász [34].

Theorem 15.17 *A graph is factor-critical if and only if it can be represented as $P_0 + P_1 + \cdots + P_r$, where P_0 is a single vertex and for each i, P_{i+1} is either a path of odd length having its two distinct end vertices in common with $P_0 + \cdots + P_i$ or else is an odd cycle having precisely one vertex in common with $P_0 + \cdots + P_i$.*

Proof. It is easy to see that a graph which possesses an ear decomposition as described must be factor-critical.

To prove the converse, let G be factor-critical. Select any vertex $v \in V(G)$ and set $P_0 = v$. Suppose u is a neighbor of v. Let M_u be a perfect matching of $G - u$ and M_v, a perfect matching of $G - v$. There must exist an $M_u - M_v$ alternating path P joining u and v and it must be of even length. Let $P_1 = P_0 + P$. (So P is our first *ear*.) If P_1 spans $V(G)$, then we add each edge of $E(G) - E(P)$ sequentially in any order as a separate single-edge ear. We note that each P_i in our ear sequence is factor-critical.

So suppose $P_1 = P + uv$ does not span $V(G)$. Since G is connected, there must exist an edge ab which has exactly one end vertex, say a, on P_1. Let M_b be a perfect matching of $G-b$. Then there must be an $M_v - M_b$ alternating path Q joining vertices b and v. Now traverse Q from vertex b until the first vertex of P_1 is reached. Call this vertex c. Let $P_2 = Q[b, c] + ab$ be our second ear.

We continue this process, maintaining matching M_v as a reference matching until each edge belongs to an ear. ∎

The following similar result holds for factor-critical graphs which are 2-connected. The proof is omitted. A path is said to be *open* if its end vertices are distinct.

Theorem 15.18 [10,11,35] *A 2-connected graph G is factor-critical of and only if it can be represented as $P_0 + P_1 + \cdots + P_r$, where P_0 is a single vertex, P_1 is an odd cycle and for each i, P_i is an open path of odd length. Moreover, the intermediate graphs $G_i = P_0 + \cdots + P_i$ are each 2-connected factor-critical for each $i = 1, \ldots, r$.* ∎

Note that ear decompositions of factor-critical graphs are no means necessarily unique. However, in any such ear decomposition, the *number* of ears is $|E(G)| - |V(G)|$ and hence is an invariant of the graph G.

If G is 2-connected and factor-critical, then the ear decomposition described in the preceding theorem can be used to derive a lower bound on the number of near-perfect matchings in G.

Theorem 15.19 [10,11,36] *Every 2-connected factor-critical graph contains at least $|E(G)|$ near-perfect matchings.*

Proof. Let $G = P_0 + \cdots + P_r$, $r \geq 0$, be the ear decomposition guaranteed by the preceding theorem. The proof is by induction on r. The result is obvious when $r = 0$. For each $i > 0$, let $G_i = P_o + \cdots + P_i$. Suppose the theorem holds for G_i and consider $G_{i+1} = G_i + P_{i+1}$. Let $P_{i+1} = x u_1 \cdots u_{l-1} y$ when l is odd and at least 3 or $P_{i+1} = xy$ when $l = 1$.

If M is any near-perfect matching of G_i, then $M' = M \cup \{u_1 u_2, \ldots, u_{l-2} u_{l-1}\}$ is a near-perfect matching of G_{i+1}, so by the induction hypothesis there are $|E(G_i)| = |E(G_{i+1}| - l$ near-perfect matchings in G_{i+1}. If M_j is a perfect matching of $G_{i+1} - u_j$, for each $j = 1, \ldots, l-1$, we have an additional $l-1$ near-perfect matchings of G_{i+1}.

It remains to construct one more near-perfect matching for G_{i+1} which is different from M' and from each of the M_js. This matching will cover both x and y and contain every second edge of ear P_{i+1}. Let M_x be a perfect matching of $G_i - x$ and suppose the edge of M_x covering y is denoted by yz which belongs to $E(G_i)$. Then $M_0 = (M_x - \{yz\}) \cup \{xu_1, u_2 u_3, \ldots, u_{l-1} y\}$ is a perfect matching of $G_{i+1} - z$ and therefore near-perfect in G_{i+1}. ∎

Remark: Let us express the matchings in G as binary vectors of length $|E(G)|$ in which there is a 1 in slot i if and only if the corresponding edge e_i belongs to the matching. Then Pulleyblank [36] proved that the $|E(G)|$ near-perfect matchings guaranteed by the preceding theorem are, in fact, linearly independent over $\Re$.

15.8 EAR DECOMPOSITIONS OF MATCHING-COVERED GRAPHS

One might hope, in view of the preceding theorem, that there might be an ear decomposition approach to counting the number of perfect matchings in a general graph. In fact, there is such a theory, but it turns out to be more complex than that for near-perfect matchings discussed here. We will only outline its main features.

Since we are motivated by the problem of counting the number of different perfect matchings, we do not really care about edges which lie in *no* perfect matching. Accordingly, let us define a graph G to be 1-*extendable* (or *matching-covered*) if every edge in G lies in some perfect matching of G.

Clearly, a graph is 1-extendable if and only if all of its components are 1-extendable. Henceforth, we will assume our 1-extendable graphs are connected. (In fact, it is easy to see that a connected 1-extendable graph must, in fact, be 2-connected.) There are several good references for ear decompositions of 1-extendable graphs (cf. [10,11,37,38]). First of all, one can build up any 1-extendable graph starting with an edge and adding a sequence of odd ears. However, it may not be the case that each intermediate graph in the sequence is also 1-extendable. On the other hand, if one is willing to allow the addition of two ears simultaneously, one can obtain a sequence of subgraphs each of which is itself 1-extendable.

Lemma 15.1 [32] *Let G be a connected 1-extendable graph and M a perfect matching in G. Then for any two distinct vertices x and y of G, there is an M-alternating path from x to y which starts with an edge of M.*

Proof. Choose an $x \in V(G)$ and define a set S_x to consist of x together with all vertices of G which can be reached via an M-alternating path from x. It will suffice to show that $S_x = V(G)$. Suppose to the contrary that there is an edge st with $s \in S_x$ and $t \notin S_x$.

If edge $st \in M$, then t belongs to S_x, a contradiction. Hence $st \notin M$. On the other hand, since G is 1-extendable, there is another perfect matching M' of G which contains edge st. Form the symmetric difference $M \oplus M'$. There must be an alternating cycle C_{st} in $M \oplus M'$ which contains edge st and contains vertices of S_x. Let P_x be an alternating path of minimum length from x to a vertex z of C_{st}, where, if $x \notin V(C)$, we begin this alternating path from x with an M-edge. Then one of the two paths in C_{st} joining z to t together with path P_x forms an M-alternating path from x to t. This $t \in S_x$, a contradiction. So $S_x = V(G)$ and the proof is complete. ∎

Definition 15.2 *Given a matching M in a graph G, an M-alternating path which begins and ends with edges of M is an M^+-path.*

Lemma 15.2 *Suppose x and y belong to $V(G)$. Then $G - x - y$ contains a perfect matching if and only if G has a perfect matching M such that x and y are the end vertices of some M^+-path.*

Proof. ($\Longleftarrow$) Let M be a perfect matching of G such that x and y are end vertices of some M^+-path P_{xy}. Let $M_0 = E(P_{xy}) - M$ and note that $M' = (M - E(P_{xy})) \cup M_0$ is the desired perfect matching of $G - x - y$.

($\Longrightarrow$) Conversely, suppose M_{xy} is a perfect matching of $G - x - y$. Then $M \oplus M_{xy}$ must consist of some $M - M_{xy}$-alternating (even) cycles and exactly one $M - M_{xy}$-alternating path joining x and y. ∎

Definition 15.3 *Let G be a 1-extendable graph. We define a relation on $V(G)$ by $u \sim v$ if $G - u - v$ does not contain a perfect matching.*

Lemma 15.3 *If G is 1-extendable, the relation $\sim$ is an equivalence relation on $E(G)$.*

Proof. Clearly we only need verify transitivity. Suppose M is a perfect matching in G and suppose that for some x, y, and $z \in V(G)$, $x \sim y$ and $y \sim z$, but $x \not\sim z$. Thus by Lemma 15.2, x and z are joined by an M^+-path P.

By Lemma 15.1 there exists an M-alternating path from y to x which starts with an edge of M. Choose a shortest such M-alternating P_0 from y to any vertex p of P. Then either $P_0 + P[p, x]$ or $P_0 + P[p, z]$ must be an M^+-path. But then either $y \sim x$ or $y \sim z$ and we have a contradiction. ∎

The partition of $V(G)$ induced by the equivalence relation $\sim$ will be denoted by $\mathcal{P}(G)$. (The above approach to the partition $\mathcal{P}(G)$ is that taken in [32]. This partition can be arrived at in a quite different manner, however. See Section 5.2 of [10,11].)

As in our discussion of 2-connected factor-critical graphs, an *ear* is a path (sometimes called an *open ear*). We now present the *Two Ear Theorem* for 1-extendable graphs.

Theorem 15.20 [10,11] *Let G be a 1-extendable graph with canonical partition $\mathcal{P}(G)$. Then there exists a sequence of 1-extendable subgraphs of G, $\{G_0, G_1, \ldots, G_m = G\}$ such that for each $i, 0 \leq i < m$, G_{i+1} arises from G_i by adding either one or two ears. Moreover, in the* case of a single-ear addition, the ear joins vertices in two different equivalence classes of $\mathcal{P}(G)$, while in a two-ear addition, one ear joins two vertices of one class of $\mathcal{P}(G)$, but the other joins two vertices in a different class of $\mathcal{P}(G)$.

Remark: That two-ear additions are sometimes necessary is shown by the complete graph K_4.

Remark: It should be noted that a 1-extendable graph can have more than one ear decomposition and, in contrast to the factor-critical case, it need not even be true that two different ear decompositions of the same graph contain the same number of ears.

Remark: In the construction of ear decompositions of a 1-extendable graph, each ear introduces a new perfect matching. Hence since the number of ears of any ear decomposition is a lower bound on the number of perfect matchings of the graph, it is of special interest to find *longest* ear decompositions.

Remark: On the other hand, the number d of double-ear additions in an ear decomposition also gives information on the number of perfect matchings.

Theorem 15.21 *If a 1-extendable graph G has an ear decomposition containing d double ears, then there exist $|E(G)| - |V(G)| + 2 - d$ perfect matchings in G the incidence vectors of which are linearly independent over $\Re$.* ∎

Hence we are also interested in ear decompositions which contain the *fewest* possible double ears.

15.9 BRICK DECOMPOSITION PROCEDURE

In this section we describe very briefly a second approach to decomposing 1-extendable graphs, the so called *Brick Decomposition Procedure*. The basic building blocks will be 1-extendable bipartite graphs and so-called *bicritical* graphs.

Definition 15.4 *A connected graph G is bicritical if $G - u - v$ contains a perfect matching for every choice of distinct vertices u and v in $V(G)$. It is easy to see that a bicritical graph must be 2-connected.*

Definition 15.5 *A 3-connected bicritical graph is called a brick.*

Examples of bricks include $K_4, \overline{C_6}$ (the triangular pyramid), the Petersen graph, and all even K_{2n}s, for all $n \geq 2$.

In this procedure too, the canonical partition $\mathcal{P}(G)$ plays an important role. More particularly, if $\mathcal{P}(G)$ contains a set S with $|S| \geq 2$, the set can be used to form a certain 1-extendable bipartite graph which serves as a frame for the decomposition.

Next, let the components of $G - S$ be $H_1, \ldots, H_k$. Form a new graph G'_S by contracting each H_i to a separate single vertex and deleting all edges joining vertices in $G[S]$. Now, for each $i = 1, \ldots, k$, form graph H'_i by contracting $V(G) - V(H_i)$ to a single vertex u_i. We then have the following result.

Theorem 15.22 *Let $G, S, G'_S, H_1, \ldots, H_k, H'_1, \ldots, H'_k$ be as described above. Then*

a. *G'_S is 1-extendable and bipartite,*

b. *H'_i is 1-extendable for each i, and*

c. *$\mathcal{P}(H'_i) = \{\{u_i\}\} \cup \{T \cap V(H_i) | T \in \mathcal{P}(G)\}.$* ∎

If some H_i' is not bicritical, this means that there must be some set $S_i \in \mathcal{P}(H_i')$ with $|S_i| \geq 2$ and we can repeat the process on H_i' starting with S_i. The goal is to repeat this process as long as possible until only bicritical graphs remain, keeping a census of the bicritical graphs formed in all steps. We illustrate one step in this procedure in Figure 15.4

If one of these bicritical graphs, call it K, is not 3-connected, but only 2-connected, we further decompose it into 3-connected bicritical graphs as follows. Let $\{x, y\}$ be a cutset in K and let the components of $K - x - y$ be denoted $K_1, \ldots, K_j$. Then if $e = xy$, each of the graphs $K_i \cup e$ is bicritical and we continue. If H' is not bicritical, the process is repeated on H' based upon its canonical decomposition $\mathcal{P}(H')$ and so on (see Figure 15.5). Finally, when we cannot proceed any further, we have assembled a list of bricks.

Although one may have a variety of paths to take to reach a final list of bricks for a given 1-extendable graph G, the following beautiful and somewhat surprising result holds.

Theorem 15.23 [39] *The final list of bricks in any brick decomposition of a given 1-extendable graph is unique, up to multiple edges.* ∎

Remark: There is a somewhat different decomposition procedure, called the *Tight Cut Decomposition Procedure*, in which, although the procedure is different and the intermediate graphs are different from the Brick Decomposition Procedure, the final catalog of bricks is the same (cf. [37,39]).

Finally, the above decomposition methods help to prove the following theorem.

Theorem 15.24

a. *If G is a 1-extendable graph with $b(G)$ bricks in its brick decomposition, then the number of linearly independent perfect matching vectors over $\Re = |E(G)| - |V(G)| + 2 - b(G)$ [39].*

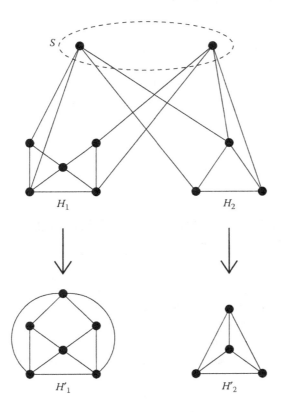

Figure 15.4 One step in the brick decomposition procedure.

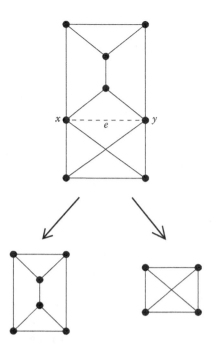

Figure 15.5 Decomposition of a bicritical graph into bricks.

b. *Every bicritical graph G has at least $|V(G)|/2 + 1$ perfect matchings* [40].

c. *Every cubic brick G has at least $|V(G)|/2 + 2$ perfect matchings* [37]. ∎

Bricks have proven to have a rich and deep structure that is not yet fully understood.

We close this section by observing that we have derived lower bounds for the number of perfect matchings in graphs which are bicritical. If one thinks a bit about the definition of a bicritical graph, one might conclude that such graphs are in general those with the largest number of perfect matchings. It therefore comes as something of a shock to find that, in a sense, the opposite is true! In particular, we present the following result without proof.

Theorem 15.25 [10,11] *Let G be a k-connected graph containing a perfect matching. Then, if G is not bicritical, G contains at least $k!$ perfect matchings.* ∎

15.10 DETERMINANTS, PERMANENTS, AND THE NUMBER OF PERFECT MATCHINGS

In this section we continue to pursue the problem of counting the number of perfect matchings in a graph. If the graphs in question belong to certain special classes, alternate tools for attacking this problem have been developed. In particular, we will address the problem for bipartite graphs and for regular graphs.

Let us begin with the bipartite case. If G is a simple balanced bipartite graph with bipartition (A, B), $A = \{u_1, \ldots, u_n\}$ and $B = \{w_1, \ldots, w_n\}$, let us define the so-called *biadjacency matrix* $A(G)$ of G by:

$$a_{ij} = \begin{cases} 1, & \text{if } u_i w_j \in E(G); \\ 0, & \text{otherwise.} \end{cases}$$

Note that every expansion term of det $A(G)$ which is different from 0 corresponds to a perfect matching in G. So if graph G has no perfect matching, it follows that det $A(G) = 0$. The problem is that the converse of this statement is false, for terms of a determinant may cancel each other as is well-known. However, if all the expansion terms had the same algebraic sign, these terms would indeed count the number of perfect matchings. This observation motivates the introduction of the permanent of a matrix. Let A be any $n \times n$ matrix. The *permanent* of A is defined by

$$\text{per } A = \sum a_{1\pi(1)} a_{2\pi(2)} \cdots a_{n\pi(n)},$$

where the summation is over all permutations π of the set $\{1, \ldots, n\}$. So then clearly if $A(G)$ is the biadjacency matrix of a balanced bipartite graph as defined above, it follows that $\text{per} A(G) = \Phi(G)$.

However, evaluating the permanent is more difficult in general than evaluating a determinant, as many of the standard tricks used to evaluate a determinant do not carry over to permanent evaluation. Nevertheless, there are some important results in permanent theory which help us to obtain new bounds for $\Phi(G)$. Perhaps the most famous of these is *van der Waerden's Conjecture* dating back to 1926, and not proved until some 50 years later by Falikman [41] and independently by Egoryčev [42].

Theorem 15.26 *Let A be a doubly stochastic $n \times n$ matrix. Then $\text{per} A \geq n!/n^n$, where equality holds if and only if every entry of A is $1/n$.* ∎

The next result follows immediately.

Corollary 15.2 *Suppose G is a k-regular bipartite graph on $2n$ vertices. Then*

$$\Phi(G) \geq n! \left(\frac{k}{n}\right)^n.$$ ∎

A second important lower bound was proved for $k = 3$ by Voorhoeve [43] and later for general k by Schrijver [44].

Theorem 15.27 *Suppose G is a k-regular bipartite graph on $2n$ vertices. Then*

$$\Phi(G) \geq \left(\frac{(k-1)^{(k-1)}}{k^{k-2}}\right)^n.$$ ∎

An upper bound for $\Phi(G)$ follows from another important result on permanents due to Brègman [45].

Theorem 15.28 *Suppose G is a k-regular simple bipartite graph on $2n$ vertices. Then*

$$\Phi(G) \leq (k!)^{n/k}.$$ ∎

Note that if one did not demand that the bipartite graph be simple, the best one could hope for as an upper bound would be k^n which is demonstrated by the bipartite graph with n components each consisting of two vertices joined by k parallel edges.

We would be remiss if we failed to mention one final approach to determining $\Phi(G)$ for arbitrary G, the so-called *Method of Pfaffians*.

Let G be any (not necessarily bipartite) graph with $n = |V(G)|$. Orient each edge of G in one of the two possible directions to obtain the oriented graph $\vec{G}$. Now define the *skew adjacency matrix* of G, $A_s(\vec{G})$ by $A_s(\vec{G}) = (a_{ij})_{x \times n}$, where

$$a_{ij} = \begin{cases} 1, & \text{if } (u_i, u_j) \in E(\vec{G}), \\ -1, & \text{if } (u_j, u_i) \in E(\vec{G}), \\ 0, & \text{otherwise.} \end{cases}$$

Note that the matrix $A_s(\vec{G})$ is skew-symmetric. We now define a certain function of any skew-symmetric matrix $B = (b_{ij})$ of size $2n \times 2n$ called the *Pfaffian* of B. Let $P = \{\{i_1, j_1\}, \ldots, \{i_n, j_n\}\}$ be a partition of the set of positive integers $\{1, \ldots, 2n\}$ into n (unordered) pairs. Now form the expression

$$b_P = \text{sgn} \begin{pmatrix} 1 & 2 & \cdots & 2n-1 & 2n \\ i_1 & j_1 & \cdots & i_n & j_n \end{pmatrix} b_{i_1 j_1} \cdots b_{i_n j_n}$$

where

$$\begin{pmatrix} 1 & 2 & \cdots & 2n-1 & 2n \\ i_1 & j_1 & \cdots & i_n & j_n \end{pmatrix}$$

denotes a permutation of the integers $1, \ldots, 2n$ and *sgn* is the sign of the permutation. (Note that b_P depends only upon the choice of the partition P.)

Now let us suppose that $|V(G)|$ is even; that is, that $|V(G)| = 2n$, for some n. Note that this assumption is a realistic one when we are trying to count perfect matchings, since a graph must have $|V(G)|$ even in order to have a perfect matching.

Finally, we define the *Pfaffian* of matrix B as pf $B = \sum_P b_P$. It is an old result from linear algebra (cf. [46,47]) that

Theorem 15.29 *If B is any skew-symmetric matrix, then* $\det B = (pf\ B)^2$. *But then since* $\det B$ *can be computed in polynomial time, so can* pf B. ∎

Moreover, it follows from the definition that for any orientation $\vec{G}$ of graph G, $|\text{pf}(A_s(\vec{G}))| \leq \Phi(G)$. Hence if one can find among all orientations of G one which realizes equality here, we can then use this orientation to compute *exactly* the number of perfect matchings in graph G and do so *in polynomial time*.

But how do we find such an orientation? Does such an orientation even exist? If the graph is planar, a Pfaffian orientation does always exist, and Kasteleyn [48,49] showed how to find it. For arbitrary graphs, however, no method of finding such an orientation efficiently (or even determining whether or not such an orientation exists) is known. In fact, the problem is known to be NP-complete. However, if the graph under consideration is bipartite, there is a polynomial algorithm to find such an orientation. This deep and important result is due to Robertson et al. [50] and, independently, to McCuaig [51] (see also [52]).

15.11 APPLICATIONS OF MATCHING: CHEMISTRY AND PHYSICS

Chemists have found that certain chemical compounds can be synthesized only when the graph of the compound has a perfect matching. (Chemists call perfect matchings *Kekulé structures*.) Moreover, it seems that in at least some cases the more perfect matchings the compound graph has, the more chemically stable it is. This has been shown to be true, for example, for the class of *benzenoid hydrocarbons* (cf. [53,54]). These can be modeled as follows. First, one suppresses the hydrogen atoms. The remaining graph consists of planar arrays of hexagons (rings of carbon atoms) such that any two of the hexagons either are vertex-disjoint or share exactly one edge.

Physicists too have brought to bear ideas from matching theory to study problems involving the Ising model of magnetic materials and the partition function from statistical physics. For further discussion of these applications, as well as others, the reader is referred to Chapter 8 of [10,11]. Kekulé structures are the subject of an entire book [55].

15.12 MATCHING EXTENSION

In this section of the chapter on matchings, we will be motivated by the following question:

Question: When can a *small* matching be extended to a perfect matching?

We have already met a version of this problem in Section 15.8 where we introduced the idea of a matching-covered graph; that is, a graph in which every edge belongs to a perfect matching. This idea admits a natural generalization.

Definition 15.6 *Let n be a nonnegative integer and G, a connected graph with at least $2n+2$ vertices. G is said to be n-extendable if* (a) *G contains a perfect matching when $n = 0$ and* (b) *for $n \geq 1$ every matching of size n extends to (i.e., is a subset of) a perfect matching.*

Thus 1-extendable graphs are exactly the matching-covered graphs introduced in Section 15.8.

We now present two basic properties of n-extendability.

Theorem 15.30 [56] *Suppose $n \geq 1$ and $|V(G)| \geq 2n+2$ is even. Then if G is n-extendable, it is also $(n-1)$-extendable.*

Proof. Suppose G is n-extendable, but not $(n-1)$-extendable. Hence there is a matching M_0 of size $n-1$ which lies in no perfect matching in G. Let M be any perfect matching of G. Then $M \oplus M_0$ consists of some even cycles together with at least two alternating paths each of which begins and ends with edges from M. Let P be the set of edges in one of these alternating paths. Then $P \oplus M_0$ is a matching of size n which can be extended to a perfect matching F. Moreover, F will contain at least one edge e which does not belong to $P \oplus M_0$, since $|P \oplus M_0| = n$ and $|V(G)| \geq 2n+2$. So $M_0 \cup e$ is a matching of size n which extends to a perfect matching containing M_0 and we have a contradiction. ∎

Remark: The reader may have wondered why we assumed that $|V(G)| \geq 2n+2$ in the definition for n-extendability of graph G. Is it not more natural to assume merely that $|V(G)| \geq 2n$? Let us consider the graph $K_4 - e$, the graph obtained by deleting any single edge from K_4. In this graph it is clearly true that any two independent edges belong to a perfect matching, but there is a single edge which does not belong to any perfect matching. That is, the statement in the above theorem fails to hold.

Theorem 15.31 [56] *Suppose n is a positive integer. Then if G is n-extendable, it is $(n+1)$-connected.*

Proof. The proof proceeds by induction on n. The proof for $n = 1$ is left to the reader.

So suppose $n \geq 2$ and that the result holds for all integers less than n. Let G be an n-extendable graph. By the preceding theorem, G is then $(n-1)$-extendable and hence by the induction hypothesis, G is n-connected.

Suppose, on the other hand, that G is not $(n+1)$-connected. Hence there must exist a vertex cutset S with $|S| = n$. Let the components of $G - S$ be denoted by $C_1, \ldots, C_k$, where $k \geq 2$. Note now that $|V(C_1)| + \cdots + |V(C_k)| = |V(G)| - |S| \geq 2n + 2 - n = n + 2 > n + 1$, and so we may apply a variation of Menger's theorem due to Dirac [57] which says that there must exist n vertex-disjoint paths each with one end vertex in S and the other in $\cup_{i=1}^{k} V(C_i)$. But then there must exist a set L of n independent edges each joining S to $\cup_{i=1}^{k} V(C_i)$.

Suppose first that there is a C_i with $|V(C_i)| \geq n$. Then Dirac's theorem says that there exist n vertex-disjoint paths in G joining S and $V(C_i)$ and hence there must be a set L of n independent edges joining C_i and S. But then these n edges cover S.

Now let e_1 be any edge of L, $e_1 = c_1 s_1$, where $c_1 \in V(C_1)$ and $s_1 \in S$. The set $S - \{s_1\}$ is not a cutset in G, so there must be a line e'_1 joining s_1 to a vertex in C_j for some $j \neq i$. Moreover, if $L' = (L - \{e_1\}) \cup \{e'_1\}$, then L' is also a set of n independent edges in G.

Let us first suppose that n is even. Then, since there is a perfect matching in G which contains L, $|V(C_i)|$ is also even. On the other hand, since L' also extends to a perfect matching in G, it follows that $|V(C_i)|$ is odd and we have a contradiction. A similar contradiction is reached if one assumes that n is odd.

So we may suppose that each of $|V(C_1)|, \ldots, |V(C_k)| \leq n - 1$. Next assume that for some i, $1 \leq i \leq k$, there is a C_i with $2 \leq |V(C_i)| \leq n - 1$. Let $V(C_i) = \{u_1, \ldots, u_m\}$. Let $R_1 = \{u_1, \ldots, u_{m-1}\}$. Then $|V(G) - S - V(C_i)| = |V(G)| - |S| - |V(C_i)| \geq 2n + 2 - n - m = n - m + 2$, so choose any set $R_2 \subseteq V(G) - S - V(C_i)$ such that $|R_2| = n - m + 1$. Then $|R_1 \cup R_2| = m - 1 + n - m + 1 = n$ and again using the result of Dirac, there are n disjoint paths in G joining S to $R_1 \cup R_2$. But then there must be a set L of n edges joining some $m - 1$ vertices of C_i and some $n - (m - 1)$ vertices of $V(G) - S - V(C_i)$ to S.

Let u denote the single vertex of C_i not covered by L. Then L covers S and extends to a perfect matching M of G. But then M cannot cover vertex u, a contradiction.

Thus for all $i, 1 \leq i \leq k$, $|V(C_i)| = 1$. But since G has a perfect matching, it follows that $k \leq n$. Hence $|V(G)| \leq 2n$, a contradiction of the assumption that $|V(G)| \geq 2n + 2$. ∎

In Section 15.9, we mentioned a method of decomposing a graph using its matchings which is called the tight cut procedure. The atoms of this decomposition turn out to be graphs of two types. Either they are bricks or 2-extendable bipartite graphs. (The latter are called *braces*.) It turns out that the family of all 2-extendable graphs partitions nicely into bricks and braces. We state this result without proof.

Theorem 15.32 [10,11] *Let G be a 2-extendable graph. Then G is either a brick or a brace.* ∎

The study of n-extendable graphs has blossomed quite dramatically over the past several decades. Space limitations dictate that we will not delve further into this area here, but instead refer the interested reader to several surveys of this topic, namely [58–60].

15.13 f- AND k-FACTORS

One can generalize the notion of a matching in many different ways. Two of the most widely studied are as follows. Given a graph G, let f be a function mapping $V(G)$ into the nonnegative integers. A spanning subgraph F of G is an $f-factor$ of G if at every vertex $v \in V(G)$, $\deg_F(v) = f(v)$. An important special case here is when $f(v) = k$, a constant. Necessary and sufficient conditions for a graph to admit an f-factor (k-factor) are known, but too complicated to go into here. Instead, we conclude this topic, and with it this chapter, by referring the reader to the survey articles [61,62], as well as the book [63].

References

[1] J. Petersen, Die Theorie der regulären Graphen, *Acta Math.*, **15** (1891), 193–220.

[2] J. Petersen, Sur le théorème de Tait, *L'Intermediaire des Mathematiciens*, **5** (1898), 225–227.

[3] D.H. Holton and J. Sheehan, *The Petersen graph, Australian Mathematical Society Lecture Series, 7*, Cambridge University Press, Cambridge, 1993.

[4] D. Nelson, M. Plummer, N. Robertson, and X. Zha, On a conjecture concerning the Petersen graph, *Electron. J. Comb.*, **18**(1) (2011), 20, 37.

[5] D. König, Line systems and determinants, *Math. Termész. Ért.*, **33** (1915), 221–229 (in Hungarian).

[6] D. König, Über Graphen und ihre Andwendung auf Determinantentheorie und Mengenlehre, *Math. Ann.*, **77** (1916), 453–465.

[7] D. König, Graphok és alkalmazásuk a determinánsok és a halmazok elméletére, *Math. Termész. Ért.*, **34** (1916), 104–119.

[8] D. König, Graphs and matrices, *Mat. Fiz. Lapok*, **38** (1931), 116–119 (in Hungarian).

[9] D. König, Über trennende Knotenpunkte in Graphen (nebst. Anwendungen auf Determinanten und Matrizen), *Acta Sci. Math. (Szeged)*, **6** (1933), 155–179.

[10] L. Lovász and M. Plummer, Matching theory. North-Holland Mathematics Studies, 121. Annals of Discrete Mathematics, 29. North-Holland Publishing Co., Amsterdam, the Netherlands; Akademiai Kiado (Publishing House of the Hungarian Academy of Sciences), Budapest, Hungary, 1986. xxvii+544 pp.

[11] L. Lovász and M. Plummer, Matching theory. Corrected reprint of the 1986 original, AMS Chelsea Publishing, Providence, RI, 2009. xxxiv+554 pp.

[12] T. Gallai, Über extreme Punkt- und Kantenmengen, *Ann. Univ. Sci. Budapest. Eötvös Sect. Math.*, **2** (1959), 133–138.

[13] P. Hall, On representatives of subsets, *J. London Math. Soc.*, **10** (1935), 26–30.

[14] G. Frobenius, Über zerlegbare Determinanten, *Sitzungsber. König. Preuss. Akad. Wiss.*, **XVIII** (1917), 274–277.

[15] K. Menger, Zur allgemeinen Kurventheorie, *Fund. Math.*, **10** (1927), 96–115.

[16] P. Elias, A. Feinstein, and C. Shannon, Note on maximum flow through a network, *IRE Trans. Inform. Theory*, **It-2** (1956), 117–119.

[17] L. Ford and D. Fulkerson, Maximal flow through a network, *Canad. J. Math.*, **8** (1956), 399–404.

[18] C. Berge, Sur le couplage maximum d'un graphe, *C.R. Acad. Sci. Paris Sér. I Math.*, **247** (1958), 258–259.

[19] C. Berge, *Graphs and hypergraphs*, North-Holland, Amsterdam, the Netherlands, 1973.

[20] W.T. Tutte, The 1-factors of oriented graphs, *Proc. Amer. Math. Soc.*, **4** (1953), 922–931.

[21] W.T. Tutte, The factorisation of linear graphs, *J. London Math. Soc.*, **22** (1947), 107–111.

[22] L. Lovász, Three short proofs in graph theory, *J. Combin. Theory Ser. B*, **19** (1975), 269–271.

[23] M. Las Vergnas, A note on matchings in graphs, *Colloque sur la Théorie des Graphes (Paris, 1974)*, Cahiers Centre Études Rech. Opér., **17** (1975), 257–260.

[24] D. Sumner, 1-factors and anti-factor sets, *J. London Math. Soc. Ser. 2*, **13** (1976), 351–359.

[25] V. Chvátal, Tough graphs and hamiltonian circuits, *Discrete Math.*, **5** (1973), 215–228.

[26] D. Bauer, H. Broersma, and E. Schmeichel, Toughness in graphs—A survey, *Graphs Combin.*, **22** (2006), 1–35.

[27] I. Anderson, Sufficient conditions for matching, *Proc. Edinburgh Math. Soc.* **18** (1973) 129–136.

[28] D. Woodall, The binding number of a graph and its Anderson number, *J. Combin. Theory Ser. B*, **15** (1973), 225–255.

[29] T. Gallai, Kritische Graphen II, *Magyar Tud. Akad. Mat. Kutató Int. Közl.*, **8** (1963), 373–395.

[30] T. Gallai, Maximale Systeme unabhängiger Kanten, *Magyar Tud. Akad. Mat. Kutató Int. Közl.*, **9** (1964), 401–413.

[31] J. Edmonds, Paths, trees and flowers, *Canad. J. Math.*, **17** (1965), 449–467.

[32] W.R. Pulleyblank, Matchings and extensions. *Handbook of Combinatorics*, Vol. 1, 2, Elsevier, Amsterdam, the Netherlands, 1995, 179–232.

[33] L. Valiant, The complexity of computing the permanent, *Theor. Comput. Sci.*, **8** (1979), 189–201.

[34] L. Lovász, A note on factor-critical graphs, *Studia Sci. Math. Hungar.*, **7** (1972), 279–280.

[35] G. Cornuéjols and W.R. Pulleyblank, Critical graphs, matchings and tours or a hierarchy of relations for the travelling salesman problem, *Combinatorica*, **3** (1983), 35–52.

[36] W.R. Pulleyblank, Faces of matching Polyhedra, University of Waterloo, Department of Combinatorics and Optimization, PhD Thesis, 1973.

[37] M.H. de Carvalho, C.L. Lucchesi, and U.S.R. Murty, *The Matching Lattice, Recent Advances in Algorithms and Combinatorics*, 125, CMS Books Math./Ouvrages Math. SMC, 11, Springer, New York, 2003.

[38] D. Naddef and W.R. Pulleyblank, Ear decompositions of elementary graphs and GF_2-rank of perfect matchings, *Bonn Workshop on Combinatorial Optimization (Bonn, 1980), Ann. Discrete Math.*, **16** (1982), 241–260.

[39] L. Lovász, The matching lattice, *J. Combin. Theory Ser. B*, **43** (1987), 187–222.

[40] J. Edmonds, L. Lovász, and W.R. Pulleyblank, Brick decompositions and the matching rank of graphs, *Combinatorica*, **2** (1982), 247–274.

[41] D. Falikman, A proof of the van der Waerden conjecture on the permanent of a doubly stochastic matrix, *Mat. Zametki*, **29** (1981), 931–938 (in Russian). (English translation: *Mathematical Notes of the Academy of Sciences of the USSR*, Consultants Bureau, New York, **29** (1981), 475–479.)

[42] G. Egoryčev, Solution of the van der Waerden problem for permanents, IFSO-13M, *Akad. Nauk. SSSR Sibirsk. Otdel. Inst. Fiz.*, Krasnoyarsk, preprint 1980 (in Russian). (English translation: *Soviet Math. Dokl.*, Amer. Math. Soc., Providence, RI, **23** (1982), 619–622.)

[43] M. Voorhoeve, A lower bound for the permanents of certain $(0,1)$ matrices, *Nederl. Akad. Wetensch. Indag. Math.*, **82** (1979), 83–86.

[44] A. Schrijver, Counting 1-factors in regular bipartite graphs, *J. Combin. Theory Ser. B*, **72** (1998), 122–135.

[45] L. Brègman, Certain properties of nonnegative matrices and their permanents, *Dokl. Akad. Nauk SSSR*, **211** (1973) 27–30 (in Russian). (English translation: *Soviet Math. Dokl.*, **14** (1973) 945–949.)

[46] T. Muir, *A Treatise on the Theory of Determinants*, Macmillan, London, 1882.

[47] T. Muir, *The Theory of Determinants*, Macmillan, London, 1906.

[48] P. Kasteleyn, Dimer statistics and phase transitions, *J. Math. Phys.*, **4** (1963), 287–293.

[49] P. Kasteleyn, Graph theory and crystal physics, *Graph Theory and Theoretical Physics*, Frank Harary (ed.), Academic Press, New York, 1967, 43–110.

[50] N. Robertson, P. Seymour, and R. Thomas, Permanents, Pfaffian orientations and even directed circuits, *Ann. of Math.*, **150** (1999), 929–975.

[51] W. McCuaig, Pólya's permanent problem, *Electron. J. Comb.*, **11** (2004) Research Paper, 79, 83.

[52] R. Thomas, A survey of Pfaffian orientations of graphs, *International Congress of Mathematicians*, Eur. Math. Soc., **III** Zürich (2006), 963–984.

[53] I. Gutman and B. Mohar, More difficulties with topological resonance energy, *Chem. Phys. Lett.*, **77** (1981), 567–570.

[54] I. Gutman, Topological properties of benzenoid molecules, *Bull. Societe Chemique Beograd*, **47** (1982), 453–471.

[55] S.J. Cyvin and I. Gutman, Kekulé structures in benzenoid hydrocarbons, *Lecture Notes in Chemistry*, **46** Springer-Verlag, Berlin, Germany, 1988. xv+348 pp.

[56] M. Plummer, On n-extendable graphs, *Discrete Math.*, **31** (1980), 201–210.

[57] G. Dirac, Généralisations du théoreme de Menger, *C.R. Acad. Sci. Paris*, **250** (1960), 4252–4253.

[58] M. Plummer, Extending matchings in graphs: a survey, *Discrete Math.*, **127** (1994), 277–292.

[59] M. Plummer, Extending matchings in graphs: An update, surveys in graph theory (San Francisco, CA, 1995), *Congr. Numer.*, **116** (1996), 3–32.

[60] M. Plummer, Recent progress in matching extension, *Building bridges, Bolyai Soc. Math. Stud.*, **19**, Springer, Berlin, Germany (2008), 427–454.

[61] J. Akiyama and M. Kano, Factors and factorizations of graphs—A survey, *J. Graph Theory*, **9** (1985), 1–42.

[62] M. Plummer, Graph factors and factorization: 1985–2003: A survey, *Discrete Math.*, **307** (2007), 791–821.

[63] J. Akiyama and M. Kano, Factors and factorizations of graphs, *Lecture Notes in Math.*, **2031**, Springer, Berlin, Germany, 2011. xii+353 pp.

CHAPTER 16

Matching Algorithms*

Krishnaiyan "KT" Thulasiraman

CONTENTS

16.1	Introduction	373
16.2	Berge's Alternating Chain Theorem	373
16.3	Maximum Matching in General Graphs	377
	16.3.1 Edmonds' Approach	377
	16.3.2 Gabow's Algorithm	379
16.4	Maximum Matchings in Bipartite Graphs	385
	16.4.1 Philosophy of Hopcroft and Karp's Approach	386
	16.4.2 Flow-Based Approach	388
16.5	Perfect Matching, Optimum Assignment, and Timetable Scheduling	388
	16.5.1 Perfect Matching	389
	16.5.2 Optimal Assignment	392
	16.5.3 Timetable Scheduling	394
16.6	Chinese Postman Problem	395

16.1 INTRODUCTION

In this chapter we discuss algorithms for constructing a maximum matching (i.e., a matching with the largest cardinality) in a graph and some related problems. We begin with Berge's alternating chain theorem which states that a matching M is maximum if and only if there is no alternating path between any two unsaturated vertices relative to M. This theorem is the basis of most maximum matching algorithms. We first discuss Edmonds' approach for constructing a maximum matching in general graphs followed by Gabow's $O(n^3)$ implementation of Edmonds' approach. We then discuss algorithms for constructing maximum matchings in bipartite graphs. Specifically, we discuss a result due to Hopcroft and Karp which provides a basis for evaluating the complexity of maximum matching algorithms. We conclude with a discussion of four related problems: perfect matchings in bipartite graphs, Kuhn–Munkres' algorithm for the optimum assignment problem, the timetable scheduling problem, and the Chinese Postman problem.

16.2 BERGE'S ALTERNATING CHAIN THEOREM

In this section we present a fundamental result in the theory of matchings, namely, Berge's alternating chain theorem which gives a necessary and sufficient condition for a matching to be maximum. We also establish some results relating to matchings in a general graph and bipartite graphs that will be of interest in the algorithms to be developed in this chapter.

---

*This chapter is an edited version of Sections 15.5 and 15.6 in Swamy and Thulasiraman [1].

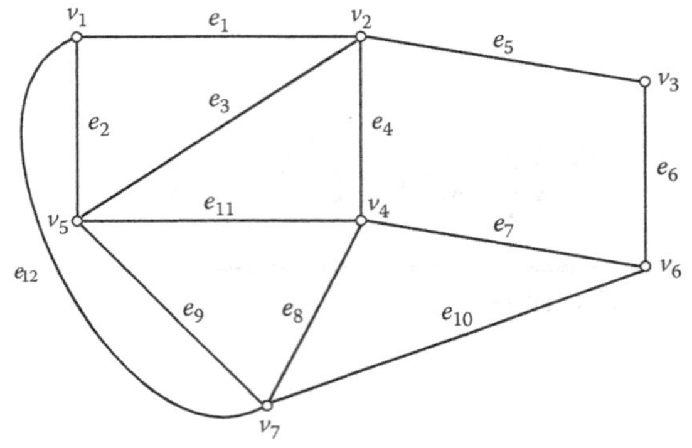

Figure 16.1 Graph for illustrating Theorem 16.1.

Consider a graph $G = (V, E)$ and a matching M in G. An *alternating chain* in G is a trail whose edges are alternately in M and $E - M$. For example, the sequence of edges e_1, e_2, e_3, e_4, e_7, e_6 is an alternating chain relative to the matching $M = \{e_2, e_4, e_6\}$ in the graph of Figure 16.1. The edges in the alternating chain that belong to M are called *dark edges* and those that belong to $E - M$ are called *light edges*. Thus e_1, e_3, e_7 are light edges, whereas e_2, e_4, e_6 are dark edges in the alternating chain considered above. Given a matching M, the vertices of the edges in M are said to be saturated in M. Other vertices are unsaturated vertices.

Theorem 16.1 *Let M_1 and M_2 be two matchings in a simple graph $G = (V, E)$. Let $G' = (V', E')$ be the induced subgraph of G on the edge set*

$$M_1 \oplus M_2 = (M_1 - M_2) \cup (M_2 - M_1).$$

Then each component of G' is of one of the following types:

1. *A circuit of even length whose edges are alternately in M_1 and M_2.*

2. *A path whose edges are alternately in M_1 and M_2 and whose end vertices are unsaturated in one of the two matchings.*

Proof. Consider any vertex $v \in V'$.

Case 1 $v \in V(M_1 - M_2)$ and $v \notin V(M_2 - M_1)$, where $V(M_i - M_j)$ denotes the set of vertices of the edges in $M_i - M_j$.

In this case v is the end vertex of an edge in $M_1 - M_2$. Since M_1 is a matching, no other edge of $M_1 - M_2$ is incident on v. Further, no edge of $M_2 - M_1$ is incident on v because $v \notin V(M_2 - M_1)$. Thus in this case the degree of v in G' is equal to 1.

Case 2 $v \in V(M_1 - M_2)$ and $v \in V(M_2 - M_1)$.

In this case a unique edge of $M_1 - M_2$ is incident on v and a unique edge of $M_2 - M_1$ is incident on v. Thus the degree of v is equal to 2.

Since the two cases considered are exhaustive, it follows that the maximum degree in G' is 2. Therefore the connected components will be of one of the two types described in the theorem. ∎

For example, consider the two matchings $M_1 = \{e_5, e_7, e_9\}$ and $M_2 = \{e_1, e_{10}, e_{11}\}$ of the graph G of Figure 16.1. Then

$$M_1 \oplus M_2 = \{e_1, e_5, e_7, e_9, e_{10}, e_{11}\},$$

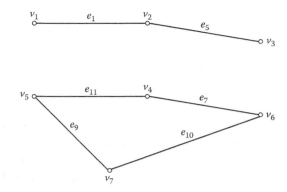

Figure 16.2 Illustration of Theorem 16.1.

and the graph G' will be as in Figure 16.2. It may be seen that the components of G' are of the two types described in Theorem 16.1.

In the following theorem we establish Berge's [2] characterization of a maximum matching in terms of an alternating chain.

Theorem 16.2 *A matching M is maximum if and only if there exists no alternating chain between any two unsaturated vertices.*

Proof. Necessity: Suppose there is an alternating chain P between two unsaturated vertices. Then replacing the dark edges in the chain by the light edges will give a matching M_1 with

$$|M_1| = |M| + 1,$$

contradicting that M is a maximum matching.

Note that $M_1 = (M - P) \cup (P - M)$.

For example, in the graph of Figure 16.1, consider the matching $M = \{e_2, e_4\}$. There is an alternating chain $e_5, e_4, e_{11}, e_2,$ and e_{12} between the unsaturated vertices v_3 and v_7. If we replace in M the dark edges e_4 and e_2 by the light edges $e_5, e_{11},$ and e_{12}, we get the matching $\{e_5, e_{11}, e_{12}\}$, which has one more edge than M.

Sufficiency: Suppose M satisfies the conditions of the theorem. Let M' be a maximum matching. Then it follows from the necessity part of the theorem that M' also satisfies the condition of the theorem, namely, there is no alternating chain between any vertices that are not saturated in M'. We now show that $|M| = |M'|$, thereby proving the sufficiency.

Since $M = (M \cap M') \cup (M - M')$ and $M' = (M \cap M') \cup (M' - M)$, it is clear that $|M| = |M'|$ if and only if $|M - M'| = |M' - M|$. Let G' be the graph on the edge set $M \oplus M' = (M - M') \cup (M' - M)$.

Consider first a circuit in G'. By Theorem 16.1 such a circuit is of even length, and the edges in this circuit are alternately in $M - M'$ and $M' - M$. Therefore each circuit in G' has the same number of edges from both $M - M'$ and $M' - M$.

Consider next a component of G' that is a path. Again, by Theorem 16.1, the edges in this path are alternately in $M - M'$ and $M' - M$. Further the end vertices of this path are unsaturated in M or M'. Suppose the path is of odd length, then the end vertices of the path will be both unsaturated in the same matching. This would mean that with respect to one of these two matchings there is an alternating chain between two unsaturated vertices. But this is a contradiction because both M and M' satisfy the condition of the theorem. So each component of G' that is a path has an even number of edges, and hence it has the same number of edges from $M - M'$ and $M' - M$.

Thus each component of G' has an equal number of edges from $M - M'$ and $M' - M$. Since the edges of G' constitute the set $(M - M') \cup (M' - M)$, we get

$$|M - M'| = |M' - M|,$$

and so, $|M| = |M'|$. ∎

Given a matching M in a graph G, let P be an alternating chain between any two vertices that are not saturated in M. Then as we have seen before, $M \oplus P$ is a matching with one more edge than M. For this reason the path P is called an *augmenting path* relative to M.

Next we prove two interesting results on bipartite graphs using the theory of alternating chains.

Consider a bipartite graph $G = (X, Y, E)$ with maximum degree Δ. Let X_1 denote the set of all the vertices in X of degree Δ. If G' is the bipartite graph $(X_1, \Gamma(X_1), E')$, where E' is the set of edges connecting X_1 and $\Gamma(X_1)$, then it can be seen from Hall's theorem (see Chapter 15) that there exists in G' a complete matching of X_1 into $\Gamma(X_1)$. Such a matching clearly saturates all the vertices in X_1. Thus there exists a matching in G that saturates all the vertices in X of degree Δ. Similarly, there exists a matching in G that saturates all the vertices in Y of degree Δ. The question that now arises is whether in a bipartite graph there exists a matching that saturates all the maximum degree vertices in both X and Y. To answer this question, we need the following result due to Mendelsohn and Dulmage [3].

Theorem 16.3 (Mendelsohn and Dulmage) *Let $G = (X, Y, E)$ be a bipartite graph, and let M_i be a matching that matches $X_i \subseteq X$ with $Y_i \subseteq Y$ ($i = 1, 2$). Then there exists a matching $M' \subseteq M_1 \cup M_2$ that saturates X_1 and Y_2.*

Proof. Consider the bipartite graph $G' = (X_1 \cup X_2, Y_1 \cup Y_2, M_1 \cup M_2)$. Each vertex of this graph has degree 1 or 2; hence each component of this graph is either a path or a circuit whose edges are alternatively in M_1 and M_2. (See proof of Theorem 16.1.)

Each vertex $y \in Y_2 - Y_1$ has a degree 1 in G'. So it is in a connected component that is a path P_y from y to a vertex $x \in X_2 - X_1$ or to a vertex $z \in Y_1 - Y_2$. In the former case the last edge of P_y is in M_2 and so $M_1 \oplus P_y$ matches $X_1 \cup \{x\}$ with $Y_1 \cup \{y\}$. In the latter case the last edge of P_y is in M_1 and so $M_1 \oplus P_y$ matches X_1 with $(Y_1 - z) \cup \{y\}$. In either case $M_1 \oplus P_y$ saturates $Y_1 \cap Y_2$. Thus $M_1 \oplus P_y$ saturates $y \in Y_2 - Y_1$, and all the vertices in X_1 and $Y_1 \cap Y_2$.

If we let

$$P = \bigcup_{y \in Y_2 - Y_1} P_y,$$

then we can see that $M_1 \oplus P$ is a matching that saturates X_1 and Y_2. This is a required matching $M' \subseteq M_1 \cup M_2$. ∎

Theorem 16.4 *In a bipartite graph there exists a matching that saturates all the maximum degree vertices.*

Proof. Consider a bipartite graph $G = (X, Y, E)$. Let $X' \subseteq X$ and $Y' \subseteq Y$ contain all the vertices of maximum degree in G. As we have seen before, there exists a matching M_1 that saturates all the vertices in X' and a matching M_2 that saturates all the vertices in Y'. So by Theorem 16.3, there exists a matching $M' \subseteq M_1 \cup M_2$ that saturates all the vertices in X' and Y'. This is a required matching saturating all the maximum degree vertices in G. ∎

Corollary 16.1 *The set of edges of a bipartite graph with maximum degree Δ can be partitioned into Δ matchings.*

Proof. Consider a bipartite graph $G = (X, Y, E)$ with maximum degree Δ. By Theorem 16.4 there exists a matching M_1 that saturates all the vertices of degree Δ. Then the bipartite graph $G' = (X, Y, E - M_1)$ has maximum degree $\Delta - 1$. This graph contains a matching M_2 that saturates every vertex of degree $\Delta - 1$. By repeating this process we can construct a sequence of disjoint matchings $M_1, M_2, \ldots, M_\Delta$ that form a partition of E. ∎

16.3 MAXIMUM MATCHING IN GENERAL GRAPHS

In this section we discuss the problem of constructing a maximum matching in a general graph. We first present a basic approach due to Edmonds [4] for constructing a maximum matching. We then describe Gabow's algorithm [5] which is an efficient implementation of Edmonds' algorithm.

16.3.1 Edmonds' Approach

Edmonds' algorithm is based on Berge's theorem (Theorem 16.2) which states that a matching is maximum if and only if there is no augmenting path relative to the matching. So, given a graph and an initial matching M, we may proceed as follows to get a maximum matching.

Find an augmenting path P with respect to M. Get the matching $M \oplus P$ which has one more edge than M. With respect to this new matching, find an augmenting path and proceed as before. Repeat this until we get a matching with respect to which there is no augmenting path. Then by Berge's theorem such a matching is maximum.

Thus the problem essentially reduces to finding an augmenting path relative to a given matching in an efficient way. The most important idea in this context is that of a *blossom* introduced by Edmonds, and this is described below.

To find an augmenting path relative to a matching M, we have to start our search necessarily at an unsaturated vertex, say u. If there exists an augmenting path P from u to u' (note that u' is also an unsaturated vertex), then, in P, u' is adjacent to either u or a saturated vertex v. Such a vertex v will be at an even distance from u in the path P; that is, there exists an alternating path of even length from u to v. This implies that the search for an augmenting path should be done only at a selected group of vertices, namely, those to which there are alternating paths of even length from u.

For example, let $v_1, v_2, \ldots, v_r$ be the vertices adjacent to u (Figure 16.3). If any one of them is unsaturated, then we have found an augmenting path. Otherwise let $u_1, u_2, \ldots, u_r$ be their respective mates in the matching M. At this stage the selected group consists of the vertices $u, u_1, u_2, \ldots, u_r$. We then pick a vertex, say u_1, from the selected group which is not yet examined. If u_1 has a neighbor which is unsaturated, then we have found an augmenting path.

Otherwise, suppose that u_1 is not adjacent to any vertex in the selected group. If $v'_1, v'_2, \ldots, v'_s$ are those vertices adjacent to u_1, such that $v'_i \neq v_j$ for all i and j, then their mates $u'_1, u'_2, \ldots, u'_s$ also join the selected group of vertices.

If we find, while searching a vertex in the selected group, that it is adjacent to some other vertex already in the selected group, then an odd circuit (i.e., a circuit of odd length) is created. This circuit, which is a closed alternating path of odd length, is called a *blossom*. For example, in Figure 16.4, the addition of the edge (u_9, u_7) creates a blossom $(u_2, v_6, u_6, v_9, u_9, u_7, v_7, u_2)$. Before the addition of this edge the selected group consisted of the vertices $u, u_1, u_2, \ldots, u_9$. But once the blossom is created, the vertices v_6, v_7, and v_9 also join the selected group, because we can now find alternating paths of even length from u to these vertices. For example, in Figure 16.4, $(u, v_2, u_2, v_7, u_7, u_9, v_9)$ is an alternating path of even

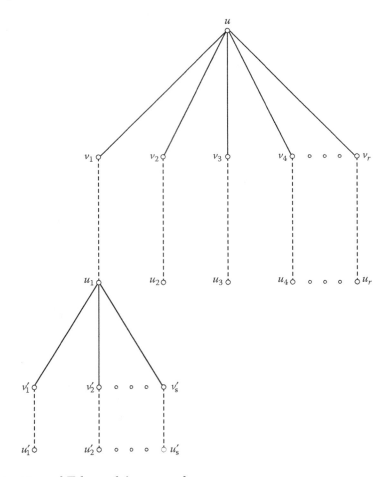

Figure 16.3 Illustration of Edmonds' approach.

length to v_9. So once a blossom is created, we find that all the vertices of that blossom join the selected group.

In Edmonds' algorithm, once a blossom is created all the vertices in the blossom are replaced by a single vertex called the pseudo vertex. All the edges that were adjacent to one or more of the vertices in the blossom will now be incident on the pseudo vertex. This is called the process of shrinking (also contracting) the blossom. Thus we get a reduced graph. We will continue the search in the reduced graph for an augmenting path from vertex u. If an augmenting path P is found then the current matching is updated to a new matching $M' = M \oplus P$ which has one more edge than M. Also, vertex u will be saturated in the new matching. Note that if an augmenting path is found in the reduced graph, then to find this path we have to trace back carefully by expanding the blossoms (pseudo vertices) found previously.

The above process is repeated until the searches have been completed starting from all unsaturated vertices.

The correctness of Edmonds' algorithm is based on the following facts:

1. Let M be a matching with vertex u unsaturated in M. Let P be an augmenting path relative to M and $M' = M \oplus P$. If there is no augmenting path relative to M starting at vertex u, then there is no augmenting path relative to M' starting at u.

2. Suppose, during a search at vertex u, there exists no augmenting path starting at u, then there exists no augmenting path starting at u in any of the subsequent iterations. Here an iteration refers to the step when an augmenting path relative to

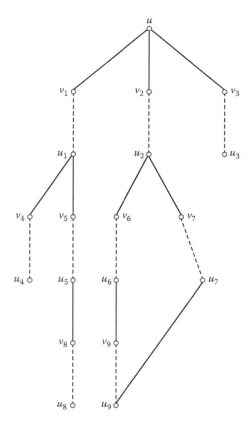

Figure 16.4 Blossom formation.

M is located and a new matching obtained using the augmenting path. This means that no unsaturated vertex needs to be examined more than once.

3. An augmenting path starting at a vertex u exists in the original graph G if and only if there is an augmenting path starting at u in the reduced graph after a blossom has been contracted.

The expansion and shrinking operations required in Edmonds' algorithm might lead to a complexity of $O(n^4)$ where n is the number of vertices in a graph.

Gabow [5] discussed an implementation that avoids the shrinking and expansion operations by recording the pertinent structure of blossoms using an efficient labeling technique and suitable arrays. This helps to achieve a complexity of $O(n^3)$. The labeling technique used by Gabow is similar to those in the matching algorithms of Balinski [6], Witzgall [7], and Kameda and Munro [8].

16.3.2 Gabow's Algorithm

First we discuss the basic strategy and define the different arrays used in Gabow's algorithm.

Let the given graph have n vertices and m edges. The algorithm begins by numbering the vertices and the edges of the graph. The vertices are numbered from 1 to n, and the edges are numbered as $n+2, n+4, \ldots, n+2m$. The number of edge (x, y) is denoted by $N(x, y)$. A dummy vertex numbered 0 is also used.

END is an array which has entries numbered from $n+1$ to $n+2m$. For each edge, there are two consecutive entries in END containing the numbers of the end vertices of the edge. Thus if edge (v, w) has number k (where $k = n + 2i$ for some $1 \leq i \leq m$), then $\text{END}(k-1) =$

v and END(k) = w. So given the number of an edge, its end vertices can easily be determined using this array.

Gabow's algorithm constructs a number of matchings starting with an initial matching which may be empty. It terminates with a maximum matching. A matching is stored in an array called MATE. This array has an entry for each vertex. Edge (v, w) is matched if MATE(v) = w and MATE(w) = v.

A vertex v is called *outer* with respect to a fixed unsaturated vertex u if and only if there exists an alternating path of even length from u to v. It is clear that this path $P(v)$ when traced from v to u starts with a matched edge. Thus $P(v) = (v, v_1, \ldots, u)$, where (v, v_1) is a matched edge.

If an edge joining an outer vertex v to an unsaturated vertex $u' \neq u$ is scanned, then the algorithm finds the augmenting path as

$$(u') * P(v) = (u', v, v_1, \ldots, u),$$

where * denotes concatenation. If no such edge is ever scanned, then the vertex u is not in an augmenting path.

LABEL is an array which has an entry for every vertex. The LABEL entry of an outer vertex v is used to find the alternating path $P(v)$.

The LABEL entry for an outer vertex is interpreted as a start label or a vertex label or an edge label.

Start Label. The start vertex u has a start label. LABEL(u) is set to 0 in this case. Now the alternating path $P(u) = (u)$.

Vertex Label. If LABEL(v) = i, where $1 \leq i \leq n$, then v is said to have a vertex label. In this case v is an outer vertex, and LABEL(v) is the number of another outer vertex. Path $P(v)$ is defined as $(v, \text{MATE}(v))*P(\text{LABEL}(v))$.

Edge Label. If LABEL(v) = $n + 2i$, $1 \leq i \leq m$, then v is said to have an edge label. Now v is an outer vertex, and LABEL(v) contains the number of an edge joining two outer vertices, say x and y. Thus LABEL(v) = $N(x, y)$. The edge label $N(x, y)$ of the vertex v indicates that there is an alternating path $P(v)$ of even length from v to the start vertex u, which passes through the edge (x, y). The path $P(v)$ can be defined in terms of paths $P(x)$ and $P(y)$. If v is in path $P(x)$, let $P(x, v)$ denote the portion of $P(x)$ from x to v along $P(x)$. Then $P(v) = \text{rev}P(x, v) * P(y)$, where the first term denotes the reverse of the path from x to v.

LABEL(v)< 0 when v is a nonouter vertex. To start with, all the vertices are nonouter and we assign -1 as LABEL value to all the vertices.

The algorithm also uses an array called FIRST. If v is an outer vertex, FIRST(v) is the first nonouter vertex in $P(v)$. If the path $P(v)$ does not contain a nonouter vertex, then FIRST(v) is set to 0. FIRST(v)= 0 if v is nonouter.

An array called OUTER is used to store the outer vertices encountered during the search for an augmenting path. The search graph is grown at the outer vertices in order of their appearance during the search. A breadth-first search is done at these outer vertices.

Gabow's algorithm (as presented below) consists of three procedures: PROC-EDMONDS, PROC-LABEL, and PROC-REMATCH.

PROC-EDMONDS is the main procedure. It starts a search for an augmenting path from each unsaturated vertex. It scans the edges of the graph, deciding to assign labels or to augment the matching.

When the presence of an augmenting path is detected (step E3 in Algorithm 16.1), PROC-REMATCH is invoked. This procedure computes a new matching which has one more edge than the current matching.

If a blossom is created (step E4) while scanning the edge (x, y), then PROC-LABEL(x, y) is invoked. Now x and y are outer vertices. PROC-LABEL performs the following:

1. The value of a variable JOIN is set to the first nonouter vertex which is in both $P(x)$ and $P(y)$.

2. All nonouter vertices preceding JOIN in $P(x)$ or $P(y)$ now become outer vertices. They are assigned the edge label $N(x, y)$. This edge label indicates that to each one of these vertices there is an alternating path of even length from the start vertex which passes through the edge (x, y).

3. Now JOIN is the first nonouter vertex in $P(x)$ as well as in $P(y)$. So the entries of the FIRST array corresponding to all the vertices which precede JOIN in $P(x)$ or $P(y)$ are set to JOIN.

A description of Gabow's algorithm now follows. In each step appropriate comments and explanations are given in parentheses.

Algorithm 16.1 Maximum Matching (Gabow)

PROC-EDOMONDS

- **E0.** (Initialize.) G is the given graph. Number the vertices of G from 1 to n and the edges as $n+2, n+4, \ldots, n+2m$. Create a dummy vertex 0. For $0 \leq i \leq n$, set LABEL$(i) = -1$, FIRST$(i) = 0$, and MATE$(i) = 0$. (To start with, all the vertices are nonouter and unsaturated.) Set $u = 0$.
- **E1.** (Find an unsaturated vertex.) Set $u = u+1$. If $u > n$, then HALT; now MATE contains a maximum matching. Otherwise if vertex u is saturated, repeat step E1. If u is unsaturated, add u to the OUTER array. Set LABEL$(u) = 0$. (Assign a start label to u and begin a new search.)
- **E2.** (Choose an edge.) Choose an edge (x, y) (where x is an outer vertex) which has not yet been examined at x. If no such edge exists, go to step E7. (**Note**: Edges (x, y) can be chosen in an arbitrary order. We adopt a breadth-first search: an outer vertex $x = x_1$ is chosen, and edges (x_1, y) are chosen in succeeding executions of step E2. When all such edges have been chosen, the vertex x_2 that was labeled immediately after x_1 is chosen, and the process is repeated for $x = x_2$. This breadth-first search requires maintaining a list of outer vertices $x_1, x_2, \ldots$ The OUTER array is used for this purpose.)
- **E3.** (Presence of an augmenting path is detected.) If y is unmatched and $y \neq u$, carry out PROC-REMATCH(x, y) and then go to setp E7.
- **E4.** (A blossom is created.) If y is outer, then carry out PROC-LABEL(x, y) and then go to step E2.
- **E5.** (Assign a vertex label.) Set $v = $ MATE(y). If v is outer, go to step E6. If v is nonouter, set LABEL$(v) = x$, FIRST$(v) = y$, and add v to the OUTER array. (Now y is encountered for the first time in this search; its mate v is a new outer vertex. This fact is noted in the OUTER array.) Then go to step E6.
- **E6.** (Get next edge.) Go to step E2. (A closed alternating path of even length is obtained; so edge (x, y) adds nothing.)
- **E7.** (Stop the search.) Set LABEL$(i) = -1$, for $0 \leq i \leq n$. Then go to step E1. (All the vertices are made nonouter for the next search.)

PROC-LABEL(x, y)

L0. (Initialize.) Set $r =$ FIRST(x) and $s =$ FIRST(y). If $r = s$, then go to step L6. (There is no nonouter vertex in the blossom.) Otherwise flag the vertices r and s. (Steps L1 and L2 find JOIN by advancing alternately along paths $P(x)$ and $P(y)$. Flags are assigned to nonouter vertices r in these paths. This is done by setting LABEL(r) to a negative edge number; that is, LABEL$(r) = -N(x, y)$. This way, each invocation of PROC-LABEL uses a distinct flag value.)

L1. (Switch paths.) If $s \neq 0$, interchange r and s. (r is a flagged nonouter vertex, alternately in $P(x)$ and $P(y)$.)

L2. (Get the nonouter vertex.) Set $r =$ FIRST(LABEL(MATE(r))). (r is set to the next nonouter vertex in $P(x)$ or $P(y)$.) If r is not flagged, flag r and go to step L1. Otherwise set JOIN$= r$. (We have found the JOIN.) Go to step L3.

L3. (Label vertices in $P(x)$, $P(y)$; that is, all nonouter vertices between x and JOIN or y and JOIN will be assigned edge labels, namely, $N(x, y)$.) Set $v =$ FIRST(x) and do step L4; then set $v =$ FIRST(y) and do step L4. Then go to step L5.

L4. (Label a nonouter vertex v.) If $v \neq$ JOIN, set LABEL$(v) = N(x, y)$ and FIRST$(v) =$ JOIN, and add v to the OUTER array. Then set $v =$ FIRST(LABEL(MATE(v))). (Get the next nonouter vertex.) Repeat step L4. Otherwise (i.e., $v =$ JOIN and hence we have assigned edge labels to all the nonouter vertices in the concerned path) continue as specified in step L3 (i.e., return to step L3).

L5. (Update FIRST.) For each outer vertex i, if FIRST(i) is outer, set FIRST$(i) =$ JOIN (i.e., JOIN is the new first nonouter vertex in $P(i)$).

L6. (Edge labeling is over.) End the procedure.

PROC-REMATCH(x, y)

R0. (Obtain the augmenting path.) Compute $P(x)$ as described below:

1. If x has an edge label $N(v, w)$, then compute $P(v)$ and $P(w)$. If x lies in $P(v)$, then
$$P(x) = (\text{rev} P(v, x)) * P(w).$$
Otherwise,
$$P(x) = (\text{rev} P(w, x)) * P(v).$$

2. If x has a vertex label, then
$$P(x) = (x, \text{MATE}(x)) * P(\text{LABEL}(x)).$$

The augmenting path P_a is then given by
$$P_a = (y) * P(x).$$

R1. (Augment the current matching.) Obtain a new matching by removing from the current matching all the matched edges in P_a and adding to it all the unmatched edges in P_a. (That is, if M is the current matching, then $M \oplus P_a$ is the new matching.) Modify suitably the entries in the MATE array and end the procedure.

It should be pointed out that in the above algorithm, a search for an augmenting path is made from a vertex only once. Suppose that the search from an unsaturated vertex u terminates without finding an augmenting path. Let S_u denote this search. Then *Hungarian subgraph* H for vertex u is the subgraph which consists of all the edges containing an outer

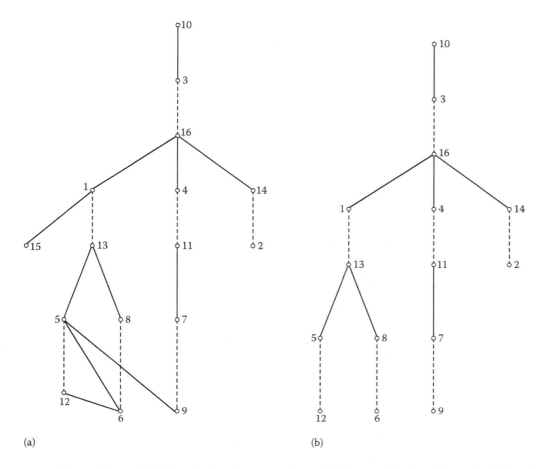

Figure 16.5 Illustration of Gabow's implementation of Edmonds' algorithm. (a) Graph to illustrate Gabow's algorithm. (b) Search graph at an intermediate step.

vertex of S_u and all the vertices in these edges. Edmonds [4] has shown that we can ignore the Hungarian subgraph H in searches after S_u. This suggests that we can modify Algorithm 16.1 by changing step E2 as follows:

E2′ (Choose an edge.) Choose an edge... If no such edge exists, go to step E1.

Step E2′ now causes step E7, which unlabels vertices to be skipped after S_u.

This modification speeds up the algorithm if the graph has no perfect matching. However, it does not change the worst-case complexity of $O(n^3)$.

For further discussions regarding the complexity and the proof of the correctness of Algorithm 16.1 (see [5]). Now we illustrate Gabow's algorithm.

For the graph in Figure 16.5a, with the initial matching shown is dashed lines, Gabow's algorithm proceeds as follows.

The start vertex is 10. After completing the search at the outer vertices 10, 16, 13, 11, and 2, the search graph will be as shown in Figure 16.5b. At this stage, the entries of the LABEL and FIRST arrays are shown in Table 16.1.

The OUTER array contains the vertices 10, 16, 13, 11, 2, 12, 6, and 9, in that order. When we search at vertex 12, we examine the edge (12, 6) and get the blossom (13, 5, 12, 6, 8, 13). Now the LABEL and FIRST entries get changed as follows:

$$\text{LABEL}(5) = \text{LABEL}(8) = N(12, 6);$$
$$\text{FIRST}(i) = 1, i = 5, 12, 6, 8.$$

Vertices 5 and 8 are now placed in the OUTER array.

TABLE 16.1 LABEL and FIRST Arrays at an Intermediate Step

Vertex Number	LABEL	FIRST
1	-1	0
2	16	14
3	-1	0
4	-1	0
5	-1	0
6	13	8
7	-1	0
8	-1	0
9	11	7
10	0	0
11	16	4
12	13	5
13	16	1
14	-1	0
15	-1	0
16	10	3

TABLE 16.2 LABEL and FIRST Arrays at an Intermediate Step

Vertex Number	LABEL	FIRST
1	$N(9,5)$	3
2	16	14
3	-1	0
4	$N(9,5)$	3
5	$N(12,6)$	3
6	13	3
7	$N(9,5)$	3
8	$N(12,6)$	3
9	11	3
10	0	0
11	16	3
12	13	3
13	16	3
14	-1	0
15	-1	0
16	10	3

Next, when we search at vertex 6, we examine the edge (6, 5), and this creates another blossom (6, 5, 12, 6). But all the vertices of this blossom are already outer, and so nothing gets changed.

Then we search at vertex 9. The edge (9, 5) is examined and the blossom (9, 7, 11, 4, 16, 1, 13, 8, 6, 12, 5, 9) is created. This again changes the LABEL and FIRST entries for some of the vertices. The resulting values are shown in Table 16.2.

Now the OUTER array contains the vertices 10, 16, 13, 11, 2, 12, 6, 9, 5, 8, 1, 7, and 4, in that order.

All the vertices in the OUTER array up to 9 have now been searched. Search continues with the remaining vertices. Searching at 5 and 8 does not add any new vertex to the

OUTER array. While searching at vertex 1, we examine the edge (1, 15) and find that 15 is unsaturated. So an augmenting path is noticed. The augmenting path is $(15)*P(1)$.

We now use the procedure given in step R0 to compute $P(1)$. Vertex 1 has the edge label $N(9,5)$. So to compute $P(1)$, we need $P(9)$ and $P(5)$. Further, since vertex 5 has the edge label $N(12,6)$ we need $P(12)$ and $P(6)$ to compute $P(5)$.

Vertex 12 has a vertex label. So

$$\begin{aligned} P(12) &= (12, \text{MATE}(12))^*P(\text{LABEL}(12)) \\ &= (12, 5)^*P(13) \\ &= (12, 5)^*(13, \text{MATE}(13))^*P(\text{LABEL}(13)) \\ &= (12, 5, 13, 1)^*P(16) \\ &= (12, 5, 13, 1)^*(16, \text{MATE}(16))^*P(\text{LABEL}(16)) \\ &= (12, 5, 13, 1, 16, 3, 10). \end{aligned}$$

Similarly
$$P(6) = (6, 8, 13, 1, 16, 3, 10)$$
and
$$P(9) = (9, 7, 11, 4, 16, 3, 10).$$

Since vertex 5 lies on $P(12)$,

$$\begin{aligned} P(5) &= (\text{rev}P(12,5))^*P(6) \\ &= (5, 12)^*(6, 8, 13, 1, 16, 3, 10) \\ &= (5, 12, 6, 8, 13, 1, 16, 3, 10). \end{aligned}$$

Now we find that vertex 1 lies on $P(5)$. Therefore

$$\begin{aligned} P(1) &= (\text{rev}P(5,1))^*P(9) \\ &= (1, 13, 8, 6, 12, 5)^*(9, 7, 11, 4, 16, 3, 10) \\ &= (1, 13, 8, 6, 12, 5, 9, 7, 11, 4, 16, 3, 10). \end{aligned}$$

Thus the augmenting path is

$$(15)^*P(1) = (15, 1, 13, 8, 6, 12, 5, 9, 7, 11, 4, 16, 3, 10).$$

After the augmentation, we get a new matching consisting of the edges (15, 1), (13, 8), (6, 12), (5, 9), (7, 11), (4, 16), (3, 10), and (14,2). Since all the vertices of the graph are saturated in this matching, it is a maximum matching (in fact, a perfect matching).

16.4 MAXIMUM MATCHINGS IN BIPARTITE GRAPHS

The problem of finding a maximum matching in a bipartite graph has a wide variety of applications. In view of these applications, the computational complexity of this problem is of great interest. Hopcroft and Karp [9] have shown how to construct a maximum matching in a bipartite graph in steps proportional to $n^{5/2}$. The philosophy of their approach is based on some interesting contributions they have made to the theory of matching. This is discussed in the following subsection.

16.4.1 Philosophy of Hopcroft and Karp's Approach

All maximum matching algorithms developed so far start with a matching (which may not be maximum) and obtain, if it exists, a matching of greater cardinality by locating an augmenting path. The choice of an augmenting path can be made in an arbitrary manner. The complexity of these algorithms is $O(n^3)$. Hopcroft and Karp have shown that if the augmentation is done along a shortest path, then a maximum matching can be obtained in $O(n^{5/2})$ phases, where each phase involves finding a maximal set of vertex-disjoint shortest augmenting paths relative to a matching. We now prove this result.

Let M be a matching. An augmenting path P is called a *shortest path relative to M* if P has the smallest length among all the augmenting paths relative to M.

Lemma 16.1 *Let M and N be two matchings in a graph G. If $|M| = s$ and $|N| = r$ with $r > s$, then $M \oplus N$ contains at least $r - s$ vertex-disjoint augmenting paths relative to M.*

Proof. Consider the induced subgraph G' of G on the edge set $M \oplus N$. By Theorem 16.1 each (connected) component of G' is either:

1. A circuit of even length, with edges alternately in $M - N$ and $N - M$; or

2. A path whose edges are alternately in $M - N$ and $N - M$.

Let the components of G' be $C_1, C_2, \ldots, C_k$, where each C_i has vertex set V_i and edge set E_i. Let

$$\delta(C_i) = |E_i \cap N| - |E_i \cap M|.$$

Then $\delta(C_i)$ is -1 or 0 or 1 for every i; and $\delta(C_i) = 1$ if and only if C_i is an augmenting path relative to M. Now

$$\sum_{i=1}^{k} \delta(C_i) = |N - M| - |M - N| = |N| - |M| = r - s.$$

Hence there are at least $r - s$ components of G', such that $\delta(C_i) = 1$. These components are vertex-disjoint, and each is an augmenting path relative to M. ∎

Lemma 16.2 *Let M be a matching. Let $|M| = r$ and suppose that the cardinality of a maximum matching is s. Then there exists an augmenting path relative to M of length at most*

$$2 \left\lfloor \frac{r}{s - r} \right\rfloor + 1$$

Proof. Let S be a maximum matching. Then by the previous lemma, $M \oplus S$ contains at least $s - r$ vertex-disjoint (and hence edge-disjoint) augmenting paths relative to M. Altogether, these paths contain at most r edges from M. So one of these paths will contain at most $\lfloor r/(s-r) \rfloor$ edges from M, and hence at most

$$2 \left\lfloor \frac{r}{s - r} \right\rfloor + 1$$

edges altogether. ∎

Lemma 16.3 *Let M be a matching, P a shortest augmenting path relative to M, and P' an augmenting path relative to $M \oplus P$. Then $|P'| \geq |P| + |P \cap P'|$.*

Proof. Let $N = M \oplus P \oplus P'$. Then N is a matching, and $|N| = |M| + 2$. So $M \oplus N$ contains two vertex-disjoint augmenting paths P_1 and P_2 relative to M.

Since $M \oplus N = P \oplus P'$, $|P \oplus P'| \geq |P_1| + |P_2|$. But $|P_1| \geq |P|$ and $|P_2| \geq |P|$, because P is a shortest augmenting path. So $|P \oplus P'| \geq |P_1| + |P_2| \geq 2|P|$. Then from the identity $|P \oplus P'| = |P| + |P'| - |P \cap P'|$, we get $|P'| \geq |P| + |P \cap P'|$. ∎

Suppose that we compute, starting with a matching $M_0 = \phi$, a sequence of matchings M_1, $M_2, \ldots, M_i, \ldots$, where $M_{i+1} = M_i \oplus P_i$ and P_i is a shortest augmenting path relative to M_i. Then from Lemma 16.3, $|P_{i+1}| \geq |P_i| + |P_i \cap P_{i+1}|$. Hence we have the following.

Lemma 16.4 $|P_i| \leq |P_{i+1}|$. ∎

Theorem 16.5 *For all i and j such that $|P_i| = |P_j|$, P_i and P_j are vertex-disjoint.*

Proof. Proof is by contradiction.

Assume that $|P_i| = |P_j|$, $i < j$, and P_i and P_j are not vertex-disjoint. Then there exist k and l such that $i \leq k < l \leq j$, P_k and P_l are not vertex-disjoint, and for each r, $k < r < l$, P_r is vertex-disjoint from P_k and P_l. Then P_l is an augmenting path relative to $M_k \oplus P_k$, so $|P_l| \geq |P_k| + |P_k \cap P_l|$. But $|P_l| = |P_k|$. So $|P_k \cap P_l| = 0$. Thus P_k and P_l have no edges in common. But if P_k and P_l had a vertex v in common, then they would have in common that edge incident on v which is in $M_k \oplus P_k$. Hence P_k and P_l are vertex-disjoint, and a contradiction is obtained. ∎

The main result of this section now follows.

Theorem 16.6 *Let s be the cardinality of a maximum matching. The number of distinct integers in the sequence $|P_0|, |P_1|, \ldots, |P_i|, \ldots$ is less than or equal to $2\lfloor\sqrt{s}\rfloor + 2$.*

Proof. Let $r = \lfloor s - \sqrt{s} \rfloor$. Then $|M_r| = r$, and by Lemma 16.2,

$$|P_r| \leq 2\lfloor s - \sqrt{s}\rfloor/(s - \lfloor s - \sqrt{s}\rfloor) + 1 \leq 2\lfloor\sqrt{s}\rfloor + 1.$$

Thus for each $i < r$, $|P_i|$ is one of the $\lfloor\sqrt{s}\rfloor + 1$ positive odd integers less than or equal to $2\lfloor\sqrt{s}\rfloor + 1$. Also $|P_{r+1}|, \ldots, |P_s|$ contribute at most $s - r = \lceil\sqrt{s}\rceil$ distinct integers, and so the total number of distinct integers in the sequence $|P_0|, |P_1|, \ldots$ is less than or equal to $\lfloor\sqrt{s}\rfloor + 1 + \lceil\sqrt{s}\rceil \leq 2\lfloor\sqrt{s}\rfloor + 2$. ∎

In view of Lemma 16.4 and Theorems 16.5 and 16.6, we may regard the computation of the sequence $M_0, M_1, M_2, \ldots$ as consisting of at most $2\lfloor\sqrt{s}\rfloor + 2$ phases, such that the augmenting paths found in each phase are vertex-disjoint and of the same length. Since all the augmenting paths in a phase are vertex-disjoint, they are also augmenting paths relative to the matching with which the phase is begun. This leads Hopcroft and Karp to suggest the following alternate way of describing the computation of a maximum matching.

Step 0 Start with a null matching M, that is, $M = \phi$.

Step 1 Let $l(M)$ be the length of a shortest augmenting path relative to M. Find a maximal set of paths $Q_1, Q_2, \ldots, Q_t$ with the following properties:

1. for each i, Q_i is an augmenting path relative to M, and $|Q_i| = l(M)$.
2. The Q_i are vertex-disjoint.

    HALT if no such path exists.

Step 2 Set $M = M \oplus Q_1 \oplus Q_2 \oplus \cdots \oplus Q_t$; go to step 1.

It is clear from our previous discussion that steps 1 and 2 of the above computation will be executed at most $2\lfloor\sqrt{s}\rfloor + 2$ times, that is, $O(n^{1/2})$ times. Further, the complexity of the computation depends crucially on the complexity of implementing step 1. In a general graph, implementing this step is quite involved, since it requires generation of all the augmenting paths relative to a given matching and then selecting from them a maximal set of shortest paths which are vertex-disjoint.

However, in the special case of bipartite graphs, an $O(n^2)$ implementation of step 1 is possible so that the complexity of the computation for such special graphs is $O(n^{5/2})$. See [9] for a discussion of such an algorithm.

This algorithm due to Hopcroft and Karp is similar to the flow based approach discussed in the following section.

16.4.2 Flow-Based Approach

In this subsection we show how to construct a maximum matching in a bipartite graph by finding a maximum flow in an appropriately constructed network.

Given a bipartite graph $G = (V, E)$ with bipartition (X, Y), let us construct a network $N = (V^*, E^*)$ as follows. N has vertex set $V^* = V \cup \{s, t\}$. And N has a directed edge (s, x) for each $x \in X$, a directed edge (y, t) for each $y \in Y$, and a directed edge (x, y) for each undirected edge $(x, y), x \in X, y \in Y$ of G. Also $c(x, y) = 1$ for every edge in N and s and t are, respectively, the source and the sink of N. Clearly N has the property that for every vertex except s and t, either the in-degree or the out-degree equals 1.

Consider now a matching M in G. Let us now define a flow f as follows. For each $(x, y) \in M$, let $f(x, y) = 1$, $f(s, x) = 1$ and $f(y, t) = 1$. Then we can see that $val(f) = |M|$. In other words each matching M in G defines an $s - t$ flow of value $|M|$.

Let f be an $s - t$ flow in N of value F. Then there are exactly F edges of the form (s, x) for which $f(s, x) = 1$. For each such edge (s, x) there is exactly one vertex $y \in Y$ such that $f(x, y) = 1$. Since $c(y, t) = 1$ and (y, t) is the only outgoing edge at y, it follows that $f(y, t) = 1$ and so there is no $x' \neq x$ such that $f(x', y) = 1$. In other words the edges (x, y) with $f(x, y) = 1$ define a matching M of cardinality F.

Summarizing, there is a one-to-one correspondence between the set of matchings in G and the set of $s - t$ flows in N. Thus a maximum matching in G corresponds to a maximum flow in N. Because of the special property of N mentioned above, we get the following. See Chapter 4 (Theorem 4.19).

Theorem 16.7 *A maximum matching in a bipartite graph $G = (V, E)$ with n vertices and m edges can be constructed in $O(mn^{1/2})$ times.* ∎

16.5 PERFECT MATCHING, OPTIMUM ASSIGNMENT, AND TIMETABLE SCHEDULING

The optimal assignment and the timetable scheduling problems, the study of which involves the theory of matching, are discussed in this section. Obtaining an optimal assignment requires as a first step the construction of a perfect matching in an appropriate bipartite graph. Recall that a matching which saturates all the vertices is called a perfect matching. With this in view, we first discuss an algorithm for constructing a perfect matching in a bipartite graph.

16.5.1 Perfect Matching

Consider the following personnel assignment problem in which n available workers are qualified for one or more of n available jobs, and we are interested to know whether we can assign jobs to all the workers, one job per worker, for which they are qualified. If we represent the workers by one set $X = \{x_1, x_2, \ldots, x_n\}$ of vertices and the jobs by the other set $Y = \{y_1, y_2, \ldots, y_n\}$ of vertices of a bipartite graph G, in which x_i is joined to y_j if and only if the worker x_i is qualified for the job y_j, then it is clear that the personnel assignment problem is to find whether the graph G has a perfect matching or not.

One method of finding a solution for this problem would be to apply a bipartite maximum matching algorithm and find a maximum matching. If this matching consists of n edges, then it shows that the graph has a perfect matching, and the maximum matching obtained is nothing but a perfect matching.

The main drawback of the above method is that if the graph does not have a perfect matching, then we will know this only at the end of the procedure. Now we discuss an algorithm which either finds a perfect matching of G or stops when it finds a subset S of X such that $|\Gamma(S)| < |S|$, where $\Gamma(S)$ is the set of vertices adjacent to those in S. Clearly, by Hall's theorem (see Chapter 15) there exists no perfect matching in the latter case.

The basic idea behind the algorithm is very simple. As usual, we start with an initial matching M. If M saturates all the vertices in X, then it is the one that we are looking for. Otherwise, as in the general case, we choose an unsaturated vertex u in X and systematically search for an augmenting path P starting from u. While looking for such a path P, we keep a count of the number of vertices selected from set X, the number of their neighbors, and the number of vertices selected from set Y.

The bipartite nature of the graph assures us that we can get no odd circuit during our search, and hence blossoms are not created. As we have seen in Section 16.3, a closed alternating path of even length is not of help in augmenting the given matching M. Hence the search graph which we develop is always a tree. This tree is called a *Hungarian tree*. At any stage, if we find an augmenting path, we perform the augmentation and get the new matching which saturates one more vertex in X and proceed as before. If such a path does not exist, then we would have obtained a set $S \subseteq X$, violating the necessary and sufficient condition for the existence of a perfect matching.

Let M be a matching in G, and let u be an unsaturated vertex in X. A tree H in G is called an M—*alternating tree* rooted at u if:

1. u belongs to the vertex set of H; and

2. For every vertex v of H, the unique path from u to v in H is an M—alternating path (i.e., an alternating path relative to M).

Let us denote by S the subset of vertices of X and by T the subset of vertices of Y which occur in H.

The alternating tree is grown as follows. Initially, H consists of only the vertex u. It is then grown in such a way that at any stage, there are two possibilities.

1. All the vertices of H except u are saturated (e.g., see Figure 16.6a).

2. H contains an unsaturated vertex different from u (e.g., see Figure 16.6b), in which case we have an augmenting path and hence we get a new matching.

In the first case, either $\Gamma(S) = T$ or $T \subset \Gamma(S)$.

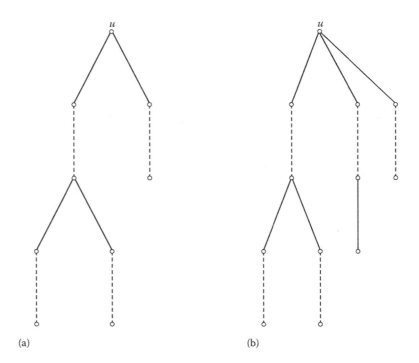

Figure 16.6 Examples of alternating trees.

1a. $\Gamma(S) = T$. Since $|S| = |T| + 1$ in the tree H, we get in this case $|\Gamma(S)| = |S| - 1$, and so the set S does not satisfy the necessary and sufficient condition required by Hall's theorem. Hence we conclude that there exists no perfect matching in G.

1b. $T \subset \Gamma(S)$. So there exists a vertex y in Y which does not occur in T, but which occurs in $\Gamma(S)$. Let this vertex y be adjacent to vertex x in S. If y is saturated with the vertex z as its mate, then we grow H by adding the vertices y and z and the edges (x, y) and (y, z). We are then back to the first case. If y is unsaturated, we grow H by adding the vertex y and the edge (x, y), resulting in the second case. The path from u to y in H is an augmenting path relative to M.

The method described above is presented in the following algorithm.

Algorithm 16.2 Perfect Matching

S1. Let G be a bipartite graph with bipartition (X, Y) and $|X| = |Y|$. Let M_0 be the null matching, that is, $M_0 = \phi$. Set $i = 0$.

S2. If all the vertices in X are saturated in the matching M_i, then HALT. (M_i is a perfect matching in G.) Otherwise pick an unsaturated vertex u in X and set $S = \{u\}$ and $T = \phi$.

S3. If $\Gamma(S) = T$, then HALT. (Now $|\Gamma(S)| < |S|$ and hence there is no perfect matching in G.) Otherwise select a vertex y from $\Gamma(S) - T$.

S4. If y is not saturated in M_i, go to step S5. Otherwise set $z =$ mate of y, $S = S \cup \{z\}$ and $T = T \cup \{y\}$, and then go to step S3.

S5. (An augmenting P path is found.) Set $M_{i+1} = M_i \oplus P$ and $i = i + 1$. Go to step S2.

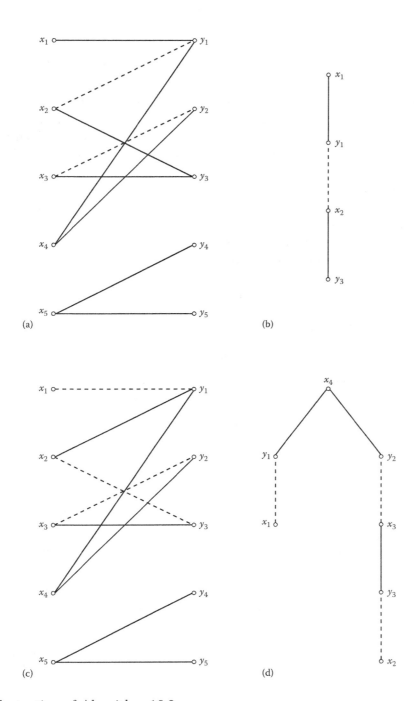

Figure 16.7 Illustration of Algorithm 16.2.

As an example, consider the bipartite graph G shown in Figure 16.7a. In this graph, the edges of an initial matching M are shown in dashed lines. The vertex x_1 is not saturated in M. The M-alternating tree rooted at x_1 is now developed. We terminate the growth of this tree as shown in Figure 16.7b when we locate the augmenting path x_1, y_1, x_2, y_3. We then augment M and obtain the new matching shown in Figure 16.7c.

Vertex x_4 is not saturated in this new matching. So we proceed to develop the alternating tree rooted at x_4, with respect to the new matching. This tree terminates as shown in Figure 16.7d. Further growth of this tree is not possible since at this stage $\Gamma(S) = T$, where $S = \{x_1, x_2, x_3, x_4\}$ and $T = \{y_1, y_2, y_3\}$. Hence the graph in Figure 16.7a has no perfect matching.

16.5.2 Optimal Assignment

Consider an assignment problem in which each worker is qualified for all the jobs. Here it is obvious that every worker can be assigned a job (of course, we assume, as before, that there are n workers and n jobs). In fact, any maximum matching performs this, and we have got $n!$ such matchings. A problem of interest in this case is to take into account the effectiveness of the workers in their various jobs, and then to make that assignment which maximizes the total effectiveness of the workers. The problem of finding such an assignment is known as the *optimal assignment problem*.

The bipartite graph for this problem is a complete one; that is, if $X = \{x_1, x_2, \ldots, x_n\}$ represents the workers and $Y = \{y_1, y_2, \ldots, y_n\}$ represents the jobs, then for all i and j, x_i is adjacent to y_j. Also, we assign a weight $w_{ij} = w(x_i, y_j)$ to every edge (x_i, y_j), which represents the effectiveness of worker x_i in job y_j (measured in some units). Then the optimal assignment problem corresponds to finding a maximum weight perfect matching in this weighted graph. Such a matching is referred to as an *optimal matching*.

We now discuss a $O(n^4)$ algorithm due to Kuhn [10] and Munkres [11] for the optimal assignment problem.

A *feasible vertex labeling* is a real-valued function f on the set $X \cup Y$ such that

$$f(x) + f(y) \geq w(x, y), \text{ for all } x \in X \text{ and } y \in Y.$$

$f(x)$ is then called the *label* of the vertex x.

For example, the following labeling is a feasible vertex labeling:

$$f(x) = \max\{w(x, y)\}, \text{ if } x \in X,$$
$$f(y) = 0, \text{ if } y \in Y.$$

From this it should be clear that there always exists a feasible vertex labeling irrespective of what the weights are.

For a given feasible vertex labeling f, let E_f denote the set of all those edges (x, y) of G such that $f(x) + f(y) = w(x, y)$. The spanning subgraph of G with the edge set E_f is called the *equality subgraph* corresponding to f. We denote this subgraph by G_f.

The following theorem relating equality subgraphs and optimal matchings forms the basis of the Kuhn–Munkres algorithm.

Theorem 16.8 *Let f be a feasible vertex labeling of a graph $G = (V, E)$. If G_f contains a perfect matching M^*, then M^* is an optimal matching in G.*

Proof. Suppose that G_f contains a perfect matching M^*. Since G_f is a spanning subgraph of G, M^* is also a perfect matching in G. Let $w(M^*)$ denote the weight of M^*, that is,

$$w(M^*) = \sum_{e \in M^*} w(e).$$

Since each edge $e \in M^*$ belongs to the equality subgraph and the vertices of the edges of M^* cover each vertex of G exactly once, we get

$$w(M^*) = \sum_{e \in M^*} w(e)$$
$$= \sum_{v \in V} f(v). \tag{16.1}$$

On the other hand, if M is any perfect matching in G, then

$$w(M) = \sum_{e \in M} w(e)$$
$$\leq \sum_{v \in V} f(v). \tag{16.2}$$

Now combining (16.1) and (16.2), we see that

$$w(M^*) \geq w(M).$$

Thus M^* is an optimal matching in G. ∎

In the Kuhn–Munkres algorithm, we first start with an arbitrary feasible vertex labeling f and find the corresponding G_f. We will choose an initial matching M in G_f and apply Algorithm 16.2. If a perfect matching is obtained in G_f, then, by Theorem 16.8, this matching is optimal. Otherwise Algorithm 16.2 terminates with a matching M' that is not perfect, giving an M'-alternating tree H that contains no M'-augmenting path and which cannot be grown further in G_f. We then modify f to a feasible vertex labeling f' with the property that both M' and H are contained in $G_{f'}$, and H can be extended in $G_{f'}$. We make such a modification in the feasible vertex labeling whenever necessary, until a perfect matching is found in some equality subgraph. Details of the Kuhn–Munkres algorithm are presented below:

Algorithm 16.3   Optimal Assignment (Kuhn and Munkres)

S1. G is the given complete bipartite graph with bipartition (X, Y) and $|X| = |Y|$. $W = [w_{ij}]$ is the given weight matrix. Set $i = 0$.

S2. Start with an arbitrary feasible vertex labeling f in G. Find the equality subgraph G_f and then select an initial matching M_i in G_f.

S3. If all the vertices in X are saturated in M_i, then M_i is a perfect matching, and hence by Theorem 16.8, it is an optimal matching. So HALT. Otherwise let u be an unsaturated vertex in X. Set $S = \{u\}$ and $T = \phi$.

S4. Let $\Gamma_f(S)$ be the set of vertices which are adjacent in G_f to the vertices in S. If $\Gamma_f(S) \supset T$, then go to step S5. Otherwise (i.e., if $\Gamma_f(S) = T$) compute

$$d_f = \min_{\substack{x \in S \\ y \notin T}} \{f(x) + f(y) - w(x,y)\} \qquad (16.3)$$

and get a new feasible vertex labeling f' given by

$$f'(v) = \begin{cases} f(v) - d_f, & \text{if } v \in S \\ f(v) + d_f, & \text{if } v \in T \\ f(v), & \text{otherwise} \end{cases} \qquad (16.4)$$

(Note that $d_f > 0$ and $\Gamma_f(S) = T$.)

Replace f by f' and G_f by $G_{f'}$.

S5. Select a vertex y from $\Gamma_f(S) - T$. If y is not saturated in M_i, go to step S6. Otherwise set z = mate of y in M_i, $S = S \cup \{z\}$ and $T = T \cup \{y\}$, and then go to step S4.

S6. (An augmenting path P is found.) Set $M_{i+1} = M_i \oplus P$ and $i = i + 1$. Go to step S3.

To illustrate the Kuhn–Munkres algorithm consider a complete bipartite graph G having the following weight matrix $W = [w_{ij}]$:

$$W = \begin{bmatrix} 4 & 4 & 1 & 3 \\ 3 & 2 & 2 & 1 \\ 5 & 4 & 4 & 3 \\ 1 & 1 & 2 & 2 \end{bmatrix}.$$

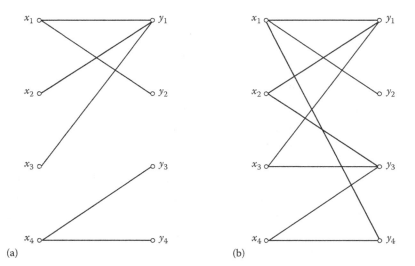

Figure 16.8 Illustration of Algorithm 16.3.

An initial feasible vertex labeling f and G may be chosen as follows:

$$f(x_1) = 4, f(x_2) = 3, f(x_3) = 5, f(x_4) = 2;$$
$$f(y_1) = f(y_2) = f(y_3) = f(y_4) = 0.$$

The equality subgraph G_f is shown in Figure 16.8a. Applying Algorithm 16.2, we find that G_f has no perfect matching because for the set $S = \{x_1, x_2, x_3\}$, $T = \Gamma(S) = \{y_1, y_2\}$. Using (16.3) we compute

$$d_f = 1.$$

The following new labeling f' is then obtained using (16.4):

$$f'(x_1) = 3, \ f'(x_2) = 2, \ f'(x_3) = 4, \ f'(x_4) = 2;$$
$$f'(y_1) = 1, \ f'(y_2) = 1, \ f'(y_3) = 0, \ f'(y_4) = 0.$$

The equality subgraph G_f is shown in Figure 16.8b. Using Algorithm 16.2 on G_f, we obtain the perfect matching M consisting of the edges (x_1, y_2), (x_2, y_1), (x_3, y_3), and (x_4, y_4). This matching is an optimal matching.

16.5.3 Timetable Scheduling

In a school there are p teachers $x_1, x_2, \ldots, x_p$ and q classes $y_1, y_2, \ldots, y_q$. Given that teacher x_i is required to teach class y_j for p_{ij} periods, we would like to schedule a timetable having the minimum possible number of periods. This is a special case of what is known as the *timetable scheduling* problem.

Suppose that we construct a bipartite graph $G = (X, Y)$ in which the vertices in X represent the teachers and those in Y represent classes, and vertex $x_i \in X$ is connected to vertex $y_j \in Y$ by p_{ij} parallel edges. Since in any one period each teacher can teach at most one class and each class can be taught by at most one teacher, it follows that a teaching schedule for one period corresponds to a matching in G, and conversely each matching corresponds to a possible assignment of teachers to classes for one period. Thus the timetable scheduling problem is to partition the edges of G into as few matchings as possible.

By Corollary 16.1, the edge set of a bipartite graph can be partitioned into Δ matchings. The proof of this theorem also suggests the following procedure for determining a partition having the smallest number of matchings:

Step 1 Let G be the given bipartite graph. Set $i = 0$ and $G_0 = G$.

Step 2 Construct a matching M_i of G_i that saturates all the maximum degree vertices in G_i.

Step 3 Remove M_i from G_i. Let G_{i+1} denote the resulting graph. If G_{i+1} has no edges, then $M_0, M_1, \ldots, M_i$ is a required partition of the edge set of G. Otherwise set $i = i + 1$ and go to step 2.

Clearly, the complexity of this procedure depends on the complexity of implementing step 2, which requires finding a matching that saturates all the maximum degree vertices in a bipartite graph $G = (X, Y)$. Such a matching may be found as follows. (See proof of Theorem 16.3.)

Let X_a denote the set of maximum degree vertices in X, and let Y_b denote the set of maximum degree vertices in Y. Let G_a denote the subgraph of G formed by the edges incident on the vertices in X_a. Similarly G_b is the subgraph on the edges incident on the vertices Y_b.

By Theorem 16.4 there is a matching M_a that saturates all the vertices in X_a. The matching M_a is a maximum matching in G_a. Similarly in G_b there is a maximum matching M_b that saturates all the vertices in Y_b. Following the procedure used in the proof of Theorem 16.3, we can find from M_a and M_b a matching M that saturates the vertices in X_a an Y_b. Then M is a required matching saturating all the maximum degree vertices in G.

The complexity of finding M_a and M_b is the same as the complexity of finding a maximum matching (see Theorem 16.7) which is $O(n^{2.5})$, where n is the number of vertices in the bipartite graph. It is easy to see that the complexity of constructing M from M_a and M_b is $O(n^2)$. Thus the overall complexity of implementing step 2 is $O(n^{2.5})$. This bound can be improved since all that we need in step 2 is a matching that saturates all the maximum degree vertices. Cole and Hopcroft [12] present such an algorithm of complexity $O(m \log n)$, where m is the number of edges in the bipartite graph. An earlier algorithm for this problem due to Gabow and Kariv [13] had complexity $O(\min \{n^2 \log n, m \log^2 n\})$.

Since step 2 will be repeated Δ times where Δ is the maximum degree in G and $\Delta \leq n$, it follows that the complexity of constructing the required timetable is $O(mn \log n)$.

A more general problem follows. Let us assume that only a limited number of classrooms are available. The question is: How many periods are now needed to schedule a complete timetable?

Suppose that l lessons are to be given and that they have to be scheduled in a p-period timetable. This timetable would require an average of $\lceil l/p \rceil$ lessons to be given per period. So it is clear that at least $\lceil l/p \rceil$ rooms will be needed in some one period. Interestingly one can always arrange l lessons in a p-period timetable so that at most $\lceil l/p \rceil$ rooms are occupied in any one period.

A discussion of the general form of the timetable scheduling problem and references related to this may be found in Even, Itai, and Shamir [14].

16.6 CHINESE POSTMAN PROBLEM

A postman picks up mail at the post office, delivers it along a set of streets, and returns to the post office. Of course, he must cover every street at least once, in either direction. The question is: What route would enable the postman to walk the shortest distance possible? This problem known as the *Chinese postman problem* was first proposed by the Chinese mathematician Kwan [15].

If G denotes the weighted connected graph representing the streets and their lengths, then the Chinese postman problem is simply that of finding a minimum weight closed walk that traverses every edge of G at least once. We shall refer to such a minimum weight closed walk as an *optimal Chinese postman tour*. Any closed walk traversing every edge at least once will be referred to as a *postman tour*.

If G is Eulerian, then clearly any Euler trail of G is an optimal postman tour. Suppose that G is not Eulerian. Then we can easily see that every postman tour of G corresponds to an Euler trail in the (Eulerian) graph G^* that has the same vertex set as G and that has as many copies of an edge (i, j) as the number of times it appears in the walk. Conversely, if G^* is an Eulerian graph constructed by adding to G appropriate numbers of additional copies of edges, then each Euler trail of G^* will correspond to a postman tour of G. We can consider G^* as consisting of two types of edges: the *original edges* (the edges of G) and the *pseudo-edges* (the parallel edges added to G to make it Eulerian). Our objective, therefore, is to obtain a G^* such that the sum of the weights of its pseudo-edges is as small as possible. Note here that the weight of a pseudo-edge is the same as that of the corresponding edge.

In order to gain an insight into the structure of G^*, consider a vertex v whose degree in G is odd. Clearly in G^* an odd number of pseudo-edges must be incident on v. Let (u, v) be one such pseudo-edge. If u is of even degree in G, then there must be a pseudo-edge (w, u) incident on w, for otherwise the degree of u in G^* will be odd. Continuing this argument, we can see that in G^* there is a path of pseudo-edges that starts at v and ends at a vertex whose degree in G is odd. Thus, if G has $2k$, $k \geq 1$ odd degree vertices, then we can group these vertices into k distinct pairs $(v_1, v_2), (v_3, v_4), \ldots, (v_{2k-1}, v_{2k})$ such that in G^* there is a path of pseudo-edges between the vertices of each pair. These observations suggest the following procedure to construct G^* from G.

Pick any two vertices, say u and v, whose degrees in G are odd. Find an $u - v$ path in G and add to G pseudo-edges along this path. Clearly, in the resulting graph G' both u and v will have even degrees. Repeat this procedure, picking any two odd degree vertices in G'. This procedure will terminate when we have obtained an Eulerian graph G^* in which all vertices have even degrees.

It should be easy to see that the sum W^* of the weights of pseudo-edges of the graph G^* (constructed as above) is equal to the sum of the weights of the paths chosen to add the pseudo-edges. So, in order to minimize W^* it is necessary that the paths selected must be shortest ones. Also, our choice of pairs of odd-degree vertices influences the value of W^*. In other words to minimize W^* we should group the odd-degree vertices into k pairs $(v_1, v_2), (v_3, v_4), \ldots, (v_{2k-1}, v_{2k})$ such that

$$W^* = \sum_{i=1}^{k} d(v_{2i-1}, v_{2i})$$

is as small as possible. Recall that $d(v_i, v_j)$ denotes the distance between v_i and v_j in G.

We can easily verify that the minimum value of W^* and the corresponding shortest paths can be obtained by solving a maximum weight perfect matching problem in a complete bipartite graph as described in S3 and S4 of the following algorithm due to Edmonds [16] and Edmonds and Johnson [17].

Algorithm 16.4 Optimal Chinese Postman Tour (Edmonds and Johnson)

S0. Given a weighted connected graph G, an optimal Chinese postman tour is required.

S1. Identify all the odd-degree vertices of G. Let these be $v_1, v_2, \ldots, v_{2k}, k \geq 1$. If there are no odd-degree vertices in G, set $G^* = G$ and go to S5.

S2. Compute the shortest paths between all pairs of odd-degree vertices. Let $d(v_i, v_j)$ denote the distance between v_i and v_j.

S3. Construct a complete bipartite graph G' with bipartition as follows:

$$X = \{x_1, x_2, \ldots, x_{2k}\},$$
$$Y = \{y_1, y_2, \ldots, y_{2k}\},$$
$$w(x_i, y_i) = 0,$$
$$w(x_i, y_j) = M - d(v_i, v_j) \text{ for } i \neq j,$$

where M is a large number.

S4. Find a maximum weight perfect matching in G'. For each (x_i, y_j) in this perfect matching, add pseudo-edges to G along a shortest $v_i - v_j$ path. Let G^* denote the resulting Eulerian graph.

S5. Construct an Euler trail of G^*. This trail defines an optimal Chinese postman tour. HALT.

It is easy to show that no edge can appear in more than one of the shortest paths identified in S4 in Algorithm 16.4. This means that no edge will be traversed more than twice in the optimal Chinese postman tour.

Note that S2 can be carried our by Floyd's algorithm for the all pairs shortest path problem (see Chapter 2). Step S4 can be carried out by Algorithm 16.3 for the optimal assignment problem. To carry out S5 we need an efficient algorithm to construct an Euler trail in an Eulerian graph. An algorithm due to Fleury described in Wilson [18], which achieves this, is described below.

Algorithm 16.5 Eulerian Trail (Fleury)

S0. Given an Eulerian graph $G = (V, E)$, an Euler trail of G is required.

S1. Let $i = 0$ and select an arbitrary vertex v_0 of G and define $T_0 : v_0$.

S2. Given that the trail $T_i : v_0, e_1, v_1, e_2, \ldots, e_i, v_i$ has been constructed, select an edge e_{i+1} from

$$E - \{e_1, e_2, \ldots, e_i\}$$

subject to the following conditions:

  a. e_{i+1} is incident on v_i.
  b. Unless there is no other choice, e_{i+1} is not a bridge of the graph
  $$G_i = G - \{e_1, e_2, \ldots, e_i\}.$$

If no such edge e_{i+1} exists, then HALT.

S3. Define $T_{i+1} : v_0, e_1, v_1, e_2, \ldots, e_{i+1}, v_{i+1}$, where $e_{i+1} = (v_i, v_{i+1})$.

S4. Set $i = i + 1$ and go to S2.

Theorem 16.9 *If G is Eulerian, then Fleury's algorithm constructs an Eulerian trail of G.*

Proof. Let $G = (V, E)$ be an Eulerian graph. Suppose that Fleury's algorithm starts with vertex v_0 of G and terminates with the trail

$$T_p : v_0, e_1, v_1, e_2, v_2, \ldots, e_p, v_p.$$

Note that whereas the vertices v_i's of T_p may not all be distinct, the edges e_i's are. We need to show that T_p is an Eulerian trail of G. For $i = 1, 2, \ldots, p$, let

$$G_i = G - \{e_1, e_2, \ldots, e_i\}.$$

First, we show that $v_p = v_0$. Since the algorithm terminates in v_p, the degree of v_p in G_p is zero. If $v_p \neq v_0$, then the degrees of v_p and v_0 in T_p will be odd because they are the terminal vertices of T_p. This would then imply that the degree of v_p in G is odd since the degree of a vertex in G is equal to the sum of its degrees in G_p and T_p. A contradiction because G is Eulerian and every vertex in G has even degree. Hence $v_p = v_0$ and T_p is a closed trail. Furthermore, this means that every vertex v_i has even degree in T_p and so every vertex of G_p must also have even degree.

We next show that T_p contains every edge of G. Suppose that this is not true. Let S denote the set of vertices of positive degree in G_p. Clearly S is nonempty and $v_p \in \bar{S}$, where $\bar{S} = V - S$. Let v_k be the last vertex of T_p that belongs to S. Since T_p terminates in $v_p \in \bar{S}$, it follows that the edge $e_{k+1} = (v_k, v_{k+1})$ is the only edge of the cut $\langle S, \bar{S} \rangle$ in G_k. In other words e_{k+1} is a bridge of G_k.

For $i = k, k+1, \ldots, p$, let G'_i denote the subgraph of G_i induced by the vertex set S. Since none of the edges $e_{k+1}, e_{k+2}, \ldots, e_p$ have both their end vertices in S, it follows that $G'_k = G'_{k+1} = \cdots = G'_p$. Also note that every vertex in G'_p has even degree because it is the subgraph of G_p induced by vertices of positive degree. So, G'_p is Eulerian. Now consider any edge $e \neq e_{k+1}$ of G_k incident on v_k. It follows from step S2 of the algorithm that e is also a bridge of G_k, for otherwise e_{k+1} would not have been picked at this step. Thus e is a bridge of G'_k and also of G'_p. But G'_p has no bridges because it is Eulerian and every edge of an Eulerian graph lies in a circuit of the graph.

Thus T_p is a closed trail and contains every edge of G. In other words Fleury's algorithm constructs an Eulerian trail of G. ∎

Summary and Related Works

In this chapter we have discussed algorithms for constructing a maximum matching (i.e., a matching with the largest cardinality) in a graph and some related problems. For a more detailed treatment of this topic, references Ahuja et al. [19], Applegate and Cook [20], Cook et al. [21], Jungnickel [22], Korte and Vygen [23], and the references therein are highly recommended. We have followed the treatment in Thulasiraman and Swamy [1,24].

Edmonds' algorithm [4] discussed in Section 16.3.1 was the first polynomial time algorithm for the maximum matching problem with a complextiy of $O(n^4)$. Several researchers subsequently improved this algorithm to achieve better complexity results. One such algorithm due to Gabow [5] and of complexity $O(n^3)$ is discussed in Section 16.3.2. The labeling technique used in this algorithm is similar to those used in the matching algorithms of Balinski [6], Witzgall and Zahn [7], and Kameda and Munro [8]. Witzgall and Zahn's algorithm is of complexity $O(n^2 m)$, and Kameda and Munro's algorithm is of complexity $O(mn)$. Another algorithm of complexity $O(n^3)$ using the primal approach is in Cunningham and Marsh [25]. Even and Kariv [26] gave an algorithm of complexity $O(n^{5/2})$. Micali and Vazirani [27] and Vazirani [28] gave an algorithm of complexity $O(n^{1/2} m)$. See also Ball and Derigs [29]. The theoretically fastest algorithm with complexity $O(n^{1/2} m \log (n^2/m) / \log n)$ is due to Fremuth-Paeger and Jungnickel [30].

Hopcroft and Karp [9] gave an $O(n^{5/2})$ algorithm for the maximum bipartite matching problem. This is based on their complexity result discussed in Section 16.4.2. The $O(n^{1/2} m)$ complexity of the maximum flow-based approach discussed in Section 16.4.3 follows from the complexity of the maximum flow algorithms for 0-1 networks given in [31] and also discussed in Chapter 4. The theoretically fastest algorithm for the maximum bipartite matching problem with complexity $O(n^{1/2} m \log (n^2/m) / \log n)$ is due to Feder and Motwani [32].

The Kuhn–Munkres algorithm [10,11,33] for the weighted bipartite matching problem is also known as the Hungarian algorithm. The best known complexity of the assignment problem is $O(mn + n^2 \log n)$. See Fredman and Tarjan [34]. Megiddo and Tamir [35] discuss an $O(n \log n)$ algorithm for a class of weighted matching problems that arise in certain applications relating to scheduling and optimal assignment. Some of the other works on perfect matchings and optimal assignments are [36–40]. See also References [41–43] for scaling and auction algorithms for the optimal assignment and other network problems.

Edmonds [44] gave the first polynomial time algorithm for the weighted matching problem in general graphs. Gabow [5] and Lawler [45] gave $O(n^3)$ implementations of this algorithm. An $O(n^2 m)$ implementation of the weighted matching algorithm is discussed in Applegate and Cook [20]. Currently the fastest algorithm for this problem is of complexity $O(mn + n^2 \log n)$ and due to Gabow [46]. See [47] for a scaling algorithm.

The complexity of the timetable scheduling algorithm discussed in Section 16.5.3 is $O(mn \log n)$. A discussion of the general form of the timetable scheduling problem and related references may be found in Even et al. [14].

The algorithm for the Chinese postman problem discussed in Section 16.6 is due to Edmonds and Johnson [16,17]. For a variant of the Chinese postman problem (called the rural Chinese postman) and its application in communication network protocol testing see Aho et al. [48]. For some generalization of this problem and related algorithms see Orloff [49], Papadimitriou [50], Lenstra and Rinooy-Kan [51], Frederickson [52], and Dror et al. [53].

A set J of vertices of a graph is a T−join if and only if the odd degree vertices of the subgraph (V, J) are exactly the elements of T. Several optimization problems can be viewed as special cases of the minimum weight T−join problem, including the Chinese postman problem. Hadlock [54] provides a polynomial time algorithm for the maximum planar cut problem by reducing the problem to an optimal T−join problem. An application of the maximum planar cut problem in 2-layer layout of VLSI circuits is discussed in Chen et al. [55] and Pinter [56]. A detailed exposition of several issues relating to T−join is given in Cook et al. [21] and Korte and Vygen [23].

References

[1] M.N.S. Swamy and K. Thulasiraman, *Graphs, Networks and Algorithms*, Wiley-Interscience, New York, 1981.

[2] C. Berge, Two theorems in graph theory, *Proc. Nat. Acad. Sci. USA*, **43** (1957), 842–844.

[3] N. S. Mendelsohn and A.L. Dulmage, Some generalizations of the problem of distinct representatives, *Can. J. Math.*, **10** (1958), 230–241.

[4] J. Edmonds, Paths, trees, and flowers, *Can. J. Math.*, **17** (1965), 449–467.

[5] H.N. Gabow, An efficient implementation of Edmonds' algorithm for maximum matchings on graphs, *J. ACM*, **23** (1976), 221–234.

[6] M.L. Balinski, Labelling to obtain a maximum matching, In *Combinatorial Mathematics and Its Applications*, R.C. Bose and T.A. Dowling, editors, University of North Carolina Press, Chappel Hill, NC, 585–602, 1967.

[7] D. Witzgall and C.T. Zahn, Modification of Edmond's algorithm for maximum matching of graphs, *J. Res. Nat. Bur. Std.*, **69B** (1965), 91–98.

[8] T. Kameda and I. Munro, A $O(|V|.|E|)$ algorithm for maximum matching of graphs, *Computing*, **12** (1974), 91–98.

[9] J.E. Hopcroft and R.M. Karp, An $n^{5/2}$ algorithm for maximum matching in bipartite graphs, *SIAM J. Comput.*, **2** (1971), 225–231.

[10] H.W. Kuhn, The Hungarian method for the assignment problem, *Nav. Res. Log. Quart.*, **2** (1955), 83–97.

[11] J. Munkres, Algorithms for the assignment and transportation problems, *SIAM J. Appl. Math.*, **5** (1957), 32–38.

[12] R. Cole and J. Hopcroft, On edge coloring bipartite graphs, *SIAM J. Comput.*, **11** (1982), 540–546.

[13] H.N. Gabow and O. Kariv, Algorithms for edge coloring bipartite graphs and multigraphs, *SIAM J. Comput.*, **11** (1982), 117–129.

[14] S. Even, A. Itai, and A. Shamir, On the complexity of time-table and multicommodity flow problems, *SIAM J. Comput.*, **15** (1976), 691–703.

[15] M.-K. Kwan, Graphic programming using even and odd points, *Chin. Math.*, **1** (1962), 273–277.

[16] J. Edmonds, The Chinese postman problem, *Oper. Res.*, **13**(Supplement 1) (1965), 373.

[17] J. Edmonds and E.L. Johnson, Matching, Euler tours, and the Chinese postman, *Math. Prog.*, **5** (1975), 88–124.

[18] R.J. Wilson, *Introduction to Graph Theory*, Oliver & Boyd, Edinburgh, 1972.

[19] R.K. Ahuja, T.L. Magnanti, and J.B. Orlin, *Network flows: Theory, Algorithms and Applications*, Prentice Hall, Upper Saddle River, NJ, 1993.

[20] D. Applegate and W. Cook, Solving large scale matching problems, In D.H. Johnson and C.C. McGeoch, editors, *Network Flows and Matching*, American Mathematical Society, Providence, pp. 557–576, *Networks*, **13** (1983), 475–493.

[21] W.J. Cook, W.H. Cunningham, W.R. Pulleyblank, and A. Schrijver, *Combinatorial Optimization*, Wiley-Interscience, New York, 1998.

[22] D. Jungnickel, *Graphs, Networks and Algorithms*, Springer-Verlag, Berlin, Germany, 2005.

[23] B. Korte and J. Vygen, *Combinatorial Optimization: Theory and Algorithms*, Springer-Verlag, New York, 1991.

[24] K. Thulasiraman and M.N.S. Swamy, *Graphs: Theory and Algorithms*, Wiley-Interscience, New York, 1992.

[25] W.H. Cunningham and A.B. Marsh, A primal algorithm for optimal matching, *Math. Prog. Stud.*, **8** (1978), 50–72.

[26] S. Even and O. Kariv, An $O(n^{2.5})$ algorithm for maximum matching in general graphs, *Proc. 16th Ann. Symp. Found. Comput. Sci.*, (1975), Berkeley, CA, 100–112.

[27] S. Micali and V.V. Vazirani, An $O(\sqrt{V} \cdot |E|)$ algorithm for finding maximum matchings in general graphs, *Proc. 21st Ann. Symp. Foundations Comput. Sci.*, IEEE, (1980), Syracuse, NY, 17–27.

[28] V.V. Vazirani, A theory of alternating paths and blossoms for proving correctness of the $O(n^{1/2}m)$ general graph matching algorithm, *Combinatorica*, **14** (1994), 71–109.

[29] M.O. Ball and U. Derigs, An analysis of alternative strategies for implementing matching algorithms, *Networks*, **13** (1983), 517–549.

[30] C. Fremuth-Paeger and D. Jungnickel, Balanced Network Flows VIII, A revised theory of phase ordered algorithms and the $O(n^{5/2}m \log (n^2/m)/\log n)$ bound for the non-bipartite cardinality matching problem, *Networks*, **41** (2003), 137–142.

[31] S. Even and R.E. Tarjan, Network flow and testing graph connectivity, *SIAM. J. Comput., Nav. Res. Log. Quart.*, **4** (1975), 507–518.

[32] T. Feder and R. Motwani, Clique partitions, graph compression, and speeding up algorithms, *J. Comput. Sys. Sci.*, **51** (1995), 261–272.

[33] H.W. Kuhn, Variants of the Hungarian method for the assignment problem, *Nav. Res. Log. Quart.*, **3** (1956), 253–258.

[34] M. L. Fredman and R.E. Tarjan, Fibonacci heaps and their uses on improved network optimization algorithms, *J. Ass. Comput. Mach.*, **34** (1987), 596–615.

[35] N. Megiddo and A. Tamir, An $O(n \log n)$ algorithm for a class of matching problems, *SIAM. J. Comput.*, **7** (1978), 154–157.

[36] R.M. Karp, An algorithm to solve the $m \times n$ assignment problem in expected time $O(mn \log n)$ *Netw.*, **10** (1980), 143–152.

[37] V. Derigs, A shortest augmenting path method for solving perfect matching problems, *Netw.*, **11** (1981), 379–390.

[38] M.D. Grigoriadis and B. Kalantari, A New class of heuristics for weighted perfect matching, *J. ACM*, **35** (1988), 769–776.

[39] D. Avis, A survey of heuristics for the weighted matching problem, *Networks*, **13** (1983), 475–493.

[40] D. Avis and C.W. Lai, The probabilistic analysis of heuristic for the assignment problem, *SIAM J. Comput.*, **17** (1988), 732–741.

[41] D.P. Bertsekas, The auction method: A distributed relaxation method for the assignment problem, *Ann. Oper. Res.*, **14** (1988), 105–123.

[42] H.N. Gabow and R.E. Tarjan, Faster scaling algorithms for general graph matching problems, *J. ACM*, **38** (1991), 815–853.

[43] J.B. Orlin and R.K. Ahuja, New scaling algorithms for the assignment and minimum cycle mean problems, *Math. Prog.*, **54** (1992), 41–56.

[44] J. Edmonds, Maximum matching and a polyhedron with 0, 1 vertices, *J. Res. Nat. Bur. Stand.*, **69 B** (1965), 125–130.

[45] E.L. Lawler, *Combinatorial Optimization: Networks and Matroids*, Holt, Rinehart & Winston, New York, 1976.

[46] H.N. Gabow, Data structures for weighted matching and nearest common ancestors with linking, *Proc. 1st Ann. ACM—SIAM Symp. Discrete Alg.*, SIAM, Philadelphia, PA (1990), 434–443.

[47] H.N. Gabow and R.E. Tarjan, Faster scaling algorithms for the network problems, *SIAM, J. Comput.*, **18** (1989), 1013–1036.

[48] A.V. Aho, A.T. Dahbura, D. Lee and M.U. Uyar, *Technique for protocol conformance test generation based on UIO sequences and rural Chinese postman tours*, in Protocol Specification, Testing and Verification, VIII , S. Aggarwal and K.K. Sabnani, Eds., Elsevier, North-Holland, New York, 1988, 75–86.

[49] C.S. Orloff, A fundamental problem in vehicle routing, *Networks*, **4** (1974), 35–64.

[50] C.H. Papadimitriou, On the complexity of edge traversing, *J. ACM*, **23** (1976), 544–554.

[51] J. Lenstra and A. Rinooy-Kan, On general routing problems, *Networks*, **6** (1976), 273–280.

[52] G.N. Frederickson, Approximation algorithms for some postman problems, *J. ACM*, **26** (1979), 538–554.

[53] M. Dror, H. Stern, and P. Trudeau, Postman tour on graphs with precedence relations on arcs, *Networks*, **17** (1987), 283–294.

[54] F.O. Hadlock, Finding a maximum cut in a planar graph in polynomial time, *SIAM J. Comput.*, **19** (1975), 221–225.

[55] R.W. Chen, Y. Kajitani, and S.P. Chan, A graph theoretic via minimization algorithm for two-layer printed circuit boards, *IEEE Transactions Circ. Sys.*, **30** (1983), 284–299.

[56] R.Y. Pinter, Optimal layer assignment for interconnect, *J. VLSI Comput. Sys.*, **1** (1984), 123–127.

CHAPTER 17

Stable Marriage Problem

Shuichi Miyazaki

CONTENTS

17.1	Introduction	403
17.2	Stable Marriage Problem	404
17.3	Gale–Shapley Algorithm	405
17.4	Extensions of Preference Lists	406
	17.4.1 Incomplete Preference Lists	406
	17.4.2 Preference Lists with Ties	407
	17.4.3 Incomplete Preference Lists with Ties	409
17.5	Optimal Stable Matchings	410
17.6	Stable Roommates Problem	411
17.7	Hospitals/Residents Problem	411
17.8	Other Variants	412
	17.8.1 HR Problem with Lower Quotas	412
	17.8.2 Student-Project Allocation Problem	413

17.1 INTRODUCTION

Consider a bipartite graph $G = (U, V, E)$, where U and V are sets of vertices and E is a set of edges. A matching M of G is a subset of E such that each vertex appears at most once in M. Bipartite matchings are sometimes characterized as the marriage between a man and a woman: U and V represent the sets of men and women, respectively, and the existence of an edge between $m \in U$ and $w \in V$ implies that m and w are acceptable to each other.

In the *stable marriage problem*, each person expresses not only the acceptability but also a preference order of the members of the opposite gender. Furthermore, an output matching must satisfy the *stability condition*, which intuitively means that there is no (man, woman)-pair in which both individuals have an incentive to elope (formal definitions are given in Section 17.2).

This problem was first introduced by Gale and Shapley in 1962 in their seminal paper [1], and it has received much attention in many areas, including economics, mathematics, and computer science, due to its inherent mathematical structure and rich applications to the real world. Among the various applications, the most famous is to assign residents (medical students) to hospitals according to the preference lists of both sides (via a suitable extension from one-to-one assignment to many-to-one) [2–6]. Other applications have included assigning students to schools [7–9], matching donors and recipients for kidney transplants [10–12], constructing a table of matches in a chess tournament [13], and designing routing algorithms in computer networks [14,15].

17.2 STABLE MARRIAGE PROBLEM

In this chapter, we give a brief survey of algorithmic results from the stable marriage problem and its variants. For a more comprehensive survey, refer to the major textbooks on this topic [16–19].

Although there are many variants, we will start with the simplest model. An instance of the stable marriage problem consists of sets of men and women of equal size n. Each person has a *preference list*, which orders all members of the opposite gender in order of the person's preference. Figure 17.1 is an example instance I_1 of the stable marriage problem, where $n = 5$, men are denoted by 1, 2, 3, 4, and 5, and women are denoted by a, b, c, d, and e. Each person's preference list is ordered from left to right. For example, man 3 most prefers woman a and least prefers woman b.

A *matching* is a set of n pairs of a man and a woman such that each person appears exactly once. For example, $M_1 = \{(1,a),(2,b),(3,c),(4,d),(5,e)\}$ illustrated in Figure 17.2 is a matching of I_1.

For a matching M, we denote $M(m)$ the partner of man m in M and, similarly, $M(w)$ the partner of woman w in M. In this case, we say that m is *matched with* w and w is *matched with* m in M. For example, $M_1(2) = b$ and $M_1(e) = 5$. Suppose that in a matching M, $M(m) \neq w$ (i.e., $(m,w) \notin M$), but m prefers w to $M(m)$ and w prefers m to $M(w)$. Then, (m,w) is called a *blocking pair* for M. For example, $(1,e)$ and $(3,d)$ are blocking pairs for M_1. Such pairs are undesirable because they create an incentive to deviate from the given matching. In this sense, a matching having a blocking pair is called *unstable*. A matching with no blocking pair is called a *stable matching*. The stable marriage problem is the problem of finding a stable matching for a given instance. For example, $M_2 = \{(1,e),(2,c),(3,a),(4,d),(5,b)\}$ illustrated in Figure 17.3 is a stable matching for I_1.

1:	c	e	a	d	b		a:	4	1	3	2	5
2:	a	b	c	d	e		b:	5	1	2	3	4
3:	a	d	c	e	b		c:	2	3	1	5	4
4:	b	e	d	a	c		d:	3	4	2	5	1
5:	d	a	e	b	c		e:	3	1	5	2	4

Figure 17.1 Instance I_1.

1:	c	e	ⓐ	d	b		a:	4	①	3	2	5
2:	a	ⓑ	c	d	e		b:	5	1	②	3	4
3:	a	d	ⓒ	e	b		c:	2	③	1	5	4
4:	b	e	ⓓ	a	c		d:	3	④	2	5	1
5:	d	a	ⓔ	b	c		e:	3	1	⑤	2	4

Figure 17.2 Matching M_1.

1:	c	ⓔ	a	d	b		a:	4	1	③	2	5
2:	a	b	ⓒ	d	e		b:	⑤	1	2	3	4
3:	ⓐ	d	c	e	b		c:	②	3	1	5	4
4:	b	e	ⓓ	a	c		d:	3	④	2	5	1
5:	d	a	e	ⓑ	c		e:	3	①	5	2	4

Figure 17.3 Matching M_2.

17.3 GALE–SHAPLEY ALGORITHM

In their paper [1], Gale and Shapley proved that for any instance there exists at least one stable matching, and they also proposed an efficient algorithm to find one in time $O(n^2)$ (recall that n is the number of men in an input). This algorithm is called the *Gale–Shapley algorithm* or *GS algorithm* (named for the authors) or the *Deferred Acceptance algorithm* (named for the algorithm's operation). Throughout this chapter, we simply call it the GS algorithm.

We first give a rough sketch of the GS algorithm. During the execution of this algorithm, each person takes one of two states: *engaged* (i.e., having a temporary partner) and *free* (i.e., being unattached). At the beginning, all persons are free. At any step of the algorithm, an arbitrary free man (say m) proposes to the woman (say w) at the top of his (current) preference list. The woman w accepts this proposal if she is currently free, in which case m and w become engaged. If w is currently engaged, she compares man m with her current partner (say m'). If w prefers m' to m, then w rejects m. In this case, m' and w remain engaged, and m remains free and deletes w from his list. Otherwise, if w prefers m to m', she accepts the proposal from m and rejects m'. In this case, m and w become newly engaged, m' becomes free, and m' deletes w from his list (note that w must be at the top of his current list). This procedure continues as long as there is still a man who can make a proposal (namely, being free and having a nonempty list); the algorithm terminates when there is no such man. Since men make and women receive proposals, this version of the GS algorithm is called the *man-oriented* GS algorithm. In the following, we give a pseudo-code of the GS algorithm.

Gale–Shapley algorithm

1: $M := \emptyset$ and let all people be free.
2: **while** there is a free man whose preference list is nonempty **do**
3:     Let m be an arbitrary free man with nonempty preference list.
4:     Let w be the woman at the top of m's current preference list.
5:     **if** w is free **then**
6:         $M := M \cup \{(m, w)\}$ and let m and w be engaged.
7:     **end if**
8:     **if** w is engaged **then**
9:         Let m' be the current partner of w, that is, $m' = M(w)$.
10:         **if** w prefers m' to m **then**
11:             Delete w from m's preference list.
12:         **else**
13:             $M := M \cup \{(m, w)\} \setminus \{(m', w)\}$. Let m' be free and m be engaged. Delete w from the preference list of m'.
14:         **end if**
15:     **end if**
16: **end while**
17: Output M.

It is not hard to see that the algorithm terminates in $O(n^2)$ steps, since no man proposes to the same woman twice. If we apply the GS algorithm to I_1 in Figure 17.1, we obtain the matching $M_3 = \{(1, c), (2, b), (3, a), (4, e), (5, d)\}$, which is a stable matching, unlike M_2. This implies that a stable matching is not necessarily unique. In fact, there are exponentially many stable matchings in general [16,20,21]. Now we show the correctness of the GS algorithm.

Theorem 17.1 *The matching M output by the Gale–Shapley algorithm is a stable matching.*

Proof. We first show that M is a matching. Clearly, no person is matched with two or more persons in M. Supposing that a man m has no partner in M, then m was rejected by all n women. Just after a woman w rejected m, w became engaged to a man better than m. After that, she may change partner but can never become free again. Hence, in M, all women are matched, while m is single; however, this is a contradiction, since there is an equal number of men and women. Therefore, every man must be matched in M. These observations prove that M is a one-to-one correspondence.

We then show the stability of M. Suppose, on the contrary, that M is unstable and there is a blocking pair (m, w) for M. Since m prefers w to $M(m)$, m must have been rejected by w at some point in the algorithm. Just after this rejection, w was matched with a man better than m. After this point, when w changes partner, she always changes to a better man. Therefore, it is impossible for w to prefer m to $M(w)$, contradicting the assumption that (m, w) is a blocking pair. ∎

Although there is arbitrarity in the order of proposals by men, any execution of the GS algorithm leads to the same stable matching, called the *man-optimal stable matching*, which has a special property [1]. In this man-optimal stable matching, each man is matched with the best woman he can obtain among all of the stable matchings, and thus the result is desirable for all of the men *simultaneously*.

Theorem 17.2 *The Gale–Shapley algorithm outputs the man-optimal stable matching.* ∎

Interestingly, the man-optimal stable matching is the *woman-pessimal stable matching*, that is, each woman is assigned the *worst* possible partner among all of the stable matchings [22]. Due to the symmetry of men and women, if we use the woman-oriented GS algorithm, then we obtain the woman-optimal (and man-pessimal) stable matching.

The $O(n^2)$ running time of the GS algorithm is optimal in the sense that any algorithm that always outputs a stable matching must know the contents of at least $\Omega(n^2)$ positions of the input preference lists [16,23].

17.4 EXTENSIONS OF PREFERENCE LISTS

Recall that in the stable marriage problem, each person's preference list must include *all* members of the opposite side. Also, a preference list must be *strictly* ordered. Apparently, this is inconvenient for large-scale matching systems such as nation-wide medical intern assignment. This motivates us to consider removing these restrictions and consider the use of *incomplete preference lists* and *ties*. In the following subsections, we see how the problem changes (or remains the same) when we allow each or both of these relaxations.

17.4.1 Incomplete Preference Lists

In this variant, each person's preference list may be incomplete, that is, one can exclude some members, with whom he/she does not want to be matched, from the preference list. We call this variant *SMI* (stable marriage with incomplete lists). If a person p's list includes a person q, we say that q is *acceptable* to p. Here, a *matching* is a set of pairs (m, w) such that m and w are acceptable to each other and each person appears at most once. Since here we are considering incomplete lists, a matching may no longer be a perfect matching. Therefore, we need to extend the definition of a blocking pair. If a person p is not included in a matching M, we say that p is *single* in M.

```
1: a  c  ⓑ           a: 2  ①  3  4  5
2: ⓒ  a              b: 2  ①
3: b  a              c: 1  ②
4: c  b  d  ⓔ        d: 3  1  4
5: c  d  b           e: ④  3
```

Figure 17.4 SMI instance I_2 and its (unstable) matching M_4.

```
1: ⓐ  c  b           a: 2  ①  3  4  5
2: ⓒ  a              b: 2  1
3: b  a              c: 1  ②
4: c  b  ⓓ  e        d: 3  1  ④
5: c  d  b           e: 4  3
```

Figure 17.5 Stable matching M_5.

For a matching M, (m, w) is a *blocking pair* if the following three conditions are met: (1) $M(m) \neq w$ but m and w are acceptable to each other, (2) m is single in M or prefers w to $M(m)$, and (3) w is single in M or prefers m to $M(w)$. The concept of a blocking pair is the same as before: Both m and w can improve the situation by forming a couple, assuming that a person prefers being matched with an acceptable partner to being single. As an example, let us consider an SMI instance I_2 and its (unstable) matching M_4 given in Figure 17.4. M_4 includes several types of blocking pairs, that is, $(1, c)$, $(4, d)$, and $(3, a)$.

As before, there is at least one stable matching in any instance, and one can be found by the GS algorithm. For example, the matching M_5 in Figure 17.5 is stable.

One important property of SMI is that we can partition the set of men (women) into two sets: the set of men (women) who have partners in all stable matchings and the set of men (women) who are single in all stable matchings [5,24,25]. (In Section 17.7, we revisit this property in the context of the more general Hospitals/Residents problem.) As an example, consider another stable matching M_6 of I_2 (Figure 17.6). Men 1, 2, and 4 and women a, c, and d are matched in both M_5 and M_6, while men 3 and 5 and women b and e are single in both matchings. This fact immediately implies that all stable matchings are of the same size.

17.4.2 Preference Lists with Ties

The other extension is to allow ties in preference lists, namely, two or more persons with the same preference may be tied in a preference list. We call this variant *SMT* (stable marriage with ties). We give an example instance I_3 of SMT in Figure 17.7. Two or more persons in the same tie are included in parentheses. For example, in man 5's preference list, women d and b are in the same tie, meaning that man 5 is indifferent between d and b.

```
1: a  ⓒ  b           a: ②  1  3  4  5
2: c  ⓐ              b: 2  1
3: b  a              c: ①  2
4: c  b  ⓓ  e        d: 3  1  ④
5: c  d  b           e: 4  3
```

Figure 17.6 Stable matching M_6.

1:	a	(c	b	d)	e	a:	2	1	3	4	5
2:	c	a	e	b	d	b:	(2	1)	4	5	3
3:	b	a	(e	d)	c	c:	1	2	3	5	4
4:	c	b	d	(e	a)	d:	(3	1	4)	(2	5)
5:	c	(d	b)	e	a	e:	4	3	1	2	5

Figure 17.7 SMT instance I_3.

1:	a	(c	b	ⓓ)	e	a:	②	1	3	4	5
2:	c	ⓐ	e	b	d	b:	(2	1)	4	⑤	3
3:	b	a	ⓔ	d)	c	c:	1	2	3	5	④
4:	ⓒ	b	d	(e	a)	d:	(3	①	4)	(2	5)
5:	c	(d	ⓑ)	e	a	e:	4	③	1	2	5

Figure 17.8 Matching M_7.

In SMT, there are three stability notions: *weak stability*, *strong stability*, and *super-stability*; these depend on *how stable* the matchings are. We give the definitions of these stabilities using a matching M_7 (Figure 17.8) for I_3. We say that w_1 *is at least as good as* w_2 *for* m, meaning that m prefers w_1 to w_2 or that m is indifferent between w_1 and w_2.

In the *weak stability*, a blocking pair of a matching M is defined as (m,w) such that (1) $M(m) \neq w$, (2) m prefers w to $M(m)$, and (3) w prefers m to $M(w)$. In our current example, $(5,c)$ is a blocking pair for M_7, but neither $(1,c)$ nor $(3,d)$ is a blocking pair. A matching with no such blocking pair is called a *weakly stable matching*.

In the *strong stability*, a blocking pair of a matching M is defined as a pair of p and q such that (1) $M(p) \neq q$, (2) p prefers q to $M(p)$, and (3) p is at least as good as $M(q)$ for q. Here, person p, who must be strictly improved, may be either a man or a woman. For example, $(5,c)$ and $(1,c)$ are blocking pairs for M_7, but $(3,d)$ is not. A matching with no such blocking pair is called a *strongly stable matching*.

Finally, in the *super-stability*, a blocking pair of a matching M is defined as (m,w) such that (1) $M(m) \neq w$, (2) w is at least as good as $M(m)$ for m, and (3) m is at least as good as $M(w)$ for w. In our example, $(5,c)$, $(1,c)$, and $(3,d)$ are all blocking pairs for M_7. A matching with no such blocking pair is called a *super-stable matching*.

Note that a blocking pair in the weak stability is also a blocking pair in the strong stability, and a blocking pair in the strong stability is a blocking pair in the super-stability. Consequently, a super-stable matching is also a strongly stable matching, and a strongly stable matching is also a weakly stable matching.

It is easy to see that a weakly stable matching always exists and can be found in polynomial time [26]: Given an instance I of SMI, we construct another instance I' by breaking all ties in I arbitrarily. As we have seen previously, I' (with no ties) has a stable matching, and one can be found in time $O(n^2)$ by the GS algorithm. It is not hard to see that this matching is a weakly stable matching for I.

In contrast, there is an instance that has no strongly stable matching (Figure 17.9 left) as well as an instance that has a strongly stable matching but no super-stable matching (Figure 17.9 right). However, there is an efficient way to confirm the existence of stable matchings: There is a polynomial time algorithm that tests whether a super-stable (or strongly stable) matching exists and finds one, if any, whose running time is $O(n^2)$ [26] (or $O(n^3)$ [27]).

1:	a	b		a:	(1	2)		1:	(a	b)
2:	a	b		b:	1	2		2:	(a	b)

a:	(1	2)
b:	(1	2)

Figure 17.9 SMT instance having no strongly stable matching (left) and SMT instance having no super-stable matching (right).

17.4.3 Incomplete Preference Lists with Ties

Another extension allows both incompleteness and ties in preference lists, and we call it *SMTI* (stable marriage with ties and incomplete lists). Definitions of blocking pairs (and hence stability) can be obtained by straightforwardly combining the definitions of Sections 17.4.1 and 17.4.2. Accordingly, we again have three stability notions: *super*, *strong*, and *weak* stabilities. For example, in the strong stability, a blocking pair of a matching M is defined as (p, q) such that (1) $M(p) \neq q$, (2) p is single in M or prefers q to $M(p)$, and (3) q is single in M or p is at least as good as $M(q)$ for q. (Here, we omit the formal definitions of the other two stabilities.)

For the super and the strong stabilities, similar results as SMT hold; namely, not all instances have a stable matching, but there is an algorithm that tests the existence of a stable matching and finds one if one exists. The running times of the algorithms are $O(a)$ for the super-stability [28] and $O(na)$ for the strong stability [27], where a is the total length of all preference lists (which is $2n^2$ if all preference lists are complete). Moreover, under both super and strong stabilities, SMTI has a similar property as SMI, that is, all stable matchings have the same size.

For the weak stability, there exists at least one stable matching for any instance, and one can be found in time $O(a)$. This time, however, one instance can have stable matchings of different sizes. For example, consider the example I_4 in Figure 17.10. All three of the matchings $\{(1, c), (4, d)\}$, $\{(1, a), (2, c), (4, d)\}$, and $\{(1, a), (2, c), (4, e), (5, d)\}$ are weakly stable for I_4, but their sizes are 2, 3, and 4, respectively.

When there are stable matchings of different sizes, it is natural to seek large ones. The cases of super and strong stabilities are easy to handle: If there exists at least one stable matching, one can be found in polynomial time, and trivially it is of maximum size because all stable matchings have the same size. In contrast, the problem of finding weakly stable matchings of the largest size, which we call *MAX SMTI*, is NP-hard [29,30].

Here, we briefly address the history of developing an approximation algorithm for MAX SMTI. First of all, it is easy to construct a 2-approximation algorithm: By the definition of stability, the sizes of minimum and maximum stable matchings differ by a factor of at most two, so returning any stable matching is a 2-approximation algorithm, and as mentioned previously, there is a polynomial time algorithm to find a weakly stable matching. There has been a chain of research improving the approximation ratio [31–35], and the current best approximation ratio is 1.5 [36–38]. On the negative side, there is no polynomial time

1:	(a	c)	(b	d)		a:	1			
2:	c	a	e			b:	(2	1)	4	
3:	b	a	(e | d)		c:	1	2	3	5	4
4:	c	(b	d	e	a)	d:	(5	3	1	4)
5:	c	(d	b)			e:	4			

Figure 17.10 SMTI instance I_4.

approximation algorithm with a ratio smaller than 33/29 unless P=NP, and no approximation algorithm with a ratio smaller than 4/3 if the unique games conjecture (UGC) is true [39].

In the special case where ties can appear on one side only, for example, women's preference lists may contain ties while men's preference lists must be strict, the situation is slightly better: We have a polynomial time 25/17-approximation algorithm [40]. On the negative side, the current best lower bound on the approximation ratio is 21/19 assuming P$\neq$NP, and is 5/4 assuming that UGC is true [41].

As a final remark in this subsection, Irving et al. [42] gave a border between the polynomially solvable case and the NP-hard case in terms of the lengths of preference lists: If all of the preference lists of one side are of length at most two, MAX SMTI is solvable in polynomial time, while it is NP-hard even if the preference lists of both sides are of lengths at most three.

17.5 OPTIMAL STABLE MATCHINGS

In this section, we assume that preference lists are strict and complete unless otherwise mentioned. As we mentioned in Section 17.3, there may be several stable matchings. However, the GS algorithm can find only two of them with extreme properties (Section 17.3), namely, one is desirable for men but undesirable for women while the other is desirable for women but undesirable for men. In this section, we consider how to find a *good* or *fair* stable matching.

There are many optimization criteria for the quality of stable matchings, but here we introduce just three of them. Let us define the *rank* of woman w in m's preference list, denoted $r_m(w)$, as the position of woman w in man m's preference list, or more precisely, one plus the number of women whom m prefers to w. Similarly, we define $r_w(m)$, the rank of m in w's list. Also, let X and Y be the sets of men and women, respectively. For a stable matching M, define the *regret cost* $r(M)$ as

$$r(M) = \max_{p \in X \cup Y} \{r_p(M(p))\},$$

which is the highest rank of the partner in M over all participants. Define the *egalitarian cost* $c(M)$ as

$$c(M) = \sum_{p \in X \cup Y} r_p(M(p)),$$

which is the sum of the ranks of the partner in M over all participants. Finally, define the *sex-equality cost* $d(M)$ as

$$d(M) = \sum_{m \in X} r_m(M(m)) - \sum_{w \in Y} r_w(M(w)).$$

The minimum regret stable marriage problem (or *the minimum egalitarian stable marriage problem* or *the sex-equal stable marriage problem*) is the problem of finding a stable matching M that minimizes $r(M)$ (or $c(M)$ or $|d(M)|$) [16].

Since the number of stable matchings for one instance grows exponentially in general (see Section 17.3), a naive algorithm that enumerates all of the stable matchings does not work efficiently. Nevertheless, for the first and second problems, respectively, Gusfield [43] and Irving et al. [44] proposed polynomial time algorithms by exploiting a partially ordered set of *rotations* that is of polynomial size but contains information of all stable matchings. In contrast, the sex-equal stable matching problem is NP-hard [45]. Iwama et al. gave approximation algorithms for the sex-equal stable matching problem and its variants [46]. Recently,

McDermid and Irving [47] considered the complexity of the sex-equal stable matching problem in terms of the lengths of preference lists, showing a border between P and NP-hard. They also presented an exact exponential time algorithm for it.

If we allow ties in preference lists, the problem of finding an optimal *weakly* stable matching in any of the above three problems becomes hard, even to approximate: For each problem, there exists a positive constant ϵ such that there is no polynomial time ϵn-approximation algorithm unless P = NP [48].

17.6 STABLE ROOMMATES PROBLEM

The *stable roommates problem* (*SR*) is a nonbipartite extension of the stable marriage problem, which is defined as follows: We are given an even number $2n$ of persons, each having a preference list over all of the other, $2n - 1$, persons. A *matching* is a set of n pairs in which each person appears exactly once. Persons p and q form a *blocking pair* for a matching M if they are not matched together in M but would both be improved if they were matched. A matching without a blocking pair is a *stable matching*. In addition to a direct application of assigning people to twin-rooms, SR has applications for pairing players in chess tournaments [13] and pairwise kidney exchange among patient–donor pairs who are incompatible [10–12].

In contrast to the case of stable marriage, there is an instance with no stable matching even for complete preference lists without ties. The instance I_5 in Figure 17.11, taken from earlier works [1,16], consists of four persons. There are three matchings, $\{(1,2),(3,4)\}$, $\{(1,3),(2,4)\}$, and $\{(1,4),(2,3)\}$, but in any of them, the person who is matched with 4 forms a blocking pair with one of the other two persons. For example, in the case of $\{(1,3),(2,4)\}$, $(1,2)$ is a blocking pair.

Irving [49] proposed a polynomial time algorithm to test whether there exists at least one stable matching and, if so, to find one. In the case where there is no stable matching, it is desirable to find a matching *as stable as possible*. In this context, the problem of finding a (not necessarily stable) matching with the minimum number of blocking pairs was introduced, and it was proved to be not only NP-hard but also hard to approximate [50].

If we allow ties in the lists, determining whether there is a weakly stable matching becomes NP-complete even for complete preference lists [51], while the same problems under the super-stability and the strong stability can still be solved in polynomial time [52,53].

17.7 HOSPITALS/RESIDENTS PROBLEM

The *Hospitals/Residents problem* (*HR*) is a many-to-one extension of the stable marriage problem, where the two sets are usually referred to as residents and hospitals. Each resident has a preference list over an acceptable set of hospitals, and each hospital has a preference over an acceptable set of residents (hence, preference lists are usually incomplete). Furthermore, each hospital h_j has a quota c_j, which specifies the maximum number of residents it can accept. Figure 17.12 shows an example instance I_6 of HR. There are six residents 1–6 and

1:	2	3	4
2:	3	1	4
3:	1	2	4
4:	1	2	3

Figure 17.11 SR instance I_5 having no stable matching.

1:	a			$a[2]$:	5	3	2	6	4
2:	a	c		$b[1]$:	2	6	3	5	
3:	b	a	c	$c[3]$:	2	1	3	6	4	5
4:	a	b	c						
5:	c	b	a						
6:	a	b	c						

Figure 17.12 HR instance I_6.

three hospitals a, b, and c that have quotas of 2, 1, and 3, respectively. (In this example, the sum of the quotas of all hospitals happens to be equal to the total number of residents, but in general these sums may be different.)

A *matching* is an assignment of residents to hospitals (possibly, leaving some residents unassigned). For a matching M, let $M(r)$ be the hospital to which the resident r is assigned (if any), and $M(h)$ be the *set* of residents assigned to the hospital h. A matching must satisfy $|M(h_j)| \leq c_j$ for each hospital h_j.

The definition of a blocking pair is an extension of that for SMI by regarding a hospital as *single* if the number of residents assigned to it is less than its quota; a resident r_i and a hospital h_j form a *blocking pair* for M if (1) r_i is not assigned to h_j in M but r_i and h_j are acceptable to each other; (2) r_i is unassigned in M or r_i prefers h_j to $M(r_i)$; and (3) $|M(h_j)| < c_j$ or h_j prefers r_i to one of the residents in $M(h_j)$.

As for the stable marriage case, there always exists at least one stable matching in any HR instance, and one can be found by the GS algorithm modified to fit the many-to-one case. Furthermore, similarly to the stable marriage problem, there are two versions of the GS algorithm, namely, the *resident-oriented* and the *hospital-oriented* versions, which find the *resident-optimal* and the *hospital-optimal* stable matchings, respectively.

Here, we give one remark concerning the property we mentioned in Section 17.4.1. This property can be generalized to HR, and it is known as the *Rural Hospitals Theorem* [5,24,25].

Theorem 17.3 *In HR, the following conditions hold:* (i) *Each hospital is assigned the same number of residents in all stable matchings,* (ii) *A hospital to which the number of assigned residents is strictly less than its quota is assigned the same set of residents in all stable matchings, and,* (iii) *The same set of residents are unassigned (equivalently, the same set of residents are assigned) in all stable matchings.* ∎

17.8 OTHER VARIANTS

Depending on the application, there are several different variants of the stable matching problem. In this last section, we briefly show two of them.

17.8.1 HR Problem with Lower Quotas

In HR, each hospital declares its quota to specify an upper bound on the number of residents assigned to it. In some cases, however, it may be convenient if we can declare not only an upper bound but also a *lower bound*. For example, the shortage of hospital doctors in rural areas is a critical issue, and it is sometimes necessary to guarantee a certain number of residents for such hospitals. Furthermore, when determining the supervisors of students in universities, it is quite common to expect the number of students assigned to each professor to be somehow balanced, which again can be achieved by specifying both upper and lower bounds on the number of students accepted by each professor. In this subsection, we show two different models that incorporate a quota for lower bounds in HR. In the following, the upper and the lower quotas of hospital h_j are denoted by u_j and l_j, respectively.

In the first model, proposed by Biró et al. [54], we allow some hospitals to be closed. In a matching M, each hospital h_j must satisfy either $|M(h_j)| = 0$ or $l_j \leq |M(h_j)| \leq u_j$. A hospital h_j is called *closed* if $|M(h_j)| = 0$, while it is *open* if $l_j \leq |M(h_j)| \leq u_j$.

A resident r_i and a hospital h_j form a blocking pair for a matching M if the following conditions are met: (1) r_i and h_j are acceptable to each other, (2) r_i is either unassigned or prefers h_j to $M(r_i)$, (3) h_j is open, and (4) either $|M(h_j)| < u_j$ or h_j prefers r_i to one of the residents in $M(h_j)$. A closed hospital h_j and l_j residents form a *blocking coalition* for M if each resident r_i is either unassigned or prefers h_j to $M(r_i)$. Note that all the residents in a blocking coalition have an incentive to collectively open the hospital h_j. A matching with neither a blocking pair nor a blocking coalition is *stable*.

Under this definition, there exist instances that have no stable matching, and deciding whether there is a stable matching for a given instance is NP-complete, even for restricted cases [54].

In the second model [55], every hospital must satisfy $l_j \leq |M(h_j)| \leq u_j$. In general, such a feasible matching need not exist; in their work [55], however, the authors imposed a restriction on inputs to guarantee the existence of a feasible matching. The definition of the blocking pair is the same as that for the original HR, and our aim in this problem is to seek a stable matching. As in the case of the first model, there are instances that admit no stable matching, but the existence of a stable matching can be determined in polynomial time.

Similarly to the problem we discussed in Section 17.6, we try to find a matching that is as stable as possible in case there is no stable matching. Hamada et al. [55] considered two problems: the problem of finding a matching with the minimum number of blocking pairs and the problem of finding a matching with the minimum number of *blocking residents* (i.e., the residents who are included in a blocking pair). They showed that both problems are NP-hard and that the first and second problems are approximable in polynomial time within a factor of $n_1 + n_2$ and $\sqrt{n_1}$, respectively, where n_1 and n_2 denote the number of residents and hospitals, respectively. They also showed that the first problem is hard to approximate within $(n_1 + n_2)^{1-\epsilon}$ for any constant ϵ unless P=NP, while the inapproximability of the second problem is established under a stronger assumption.

17.8.2 Student-Project Allocation Problem

In many universities, it is common to require students to take certain project courses provided by lecturers. Students have preferences over available projects and want to take a preferable one. Each project has its own *capacity* due to, for example, the limitation of available instruments, and also each lecturer has his/her own capacity. One lecturer may provide more than one projects, but the sum of the capacities of the projects she offers need not be equal to her capacity. In addition, each lecturer has a preference over students. Our goal is to assign students to projects while satisfying the capacity constraints of both lecturers and projects as well as achieving stability. This problem was modeled by Abraham et al. [56] as the *Student-Project Allocation problem* (*SPA*).

Here, we give only informal definitions. A *matching* is an assignment of students to projects satisfying all of the capacity constraints. The pair of a student s and a project p form a *blocking pair* if (1) s can improve the current situation if he is assigned to p, and (2) the lecturer who offers p has an incentive to accept s to p (possibly by rejecting a student currently assigned to one of her projects). A matching without a blocking pair is a *stable matching*. Abraham et al. [56] showed that there exists at least one stable matching for any instance, and that all of the stable matchings have the same size, by proving a variant of the Rural Hospitals Theorem. They also extended the GS algorithm to fit SPA, giving the *student-oriented* and the *lecturer-oriented* versions.

The *Student-Project Allocation problem with preferences over Projects* (SPA-P) [57] is a variant of SPA where lecturers have preference not over students but over the projects they offer. This problem models the situation where lecturers are indifferent to the set of students because they have no preliminary information on the students but they can rank projects they offer according to academic interests and their desire to have more students in higher-ranked projects. The rest of the input is the same as SPA: Students have preferences over projects, and each project and each lecturer has a capacity.

A *blocking pair* is a pair of a student s and a project p such that (1) s can improve the current situation if he is assigned to p, and (2) the lecturer who offers p has an incentive to accept s to p. Note that condition (2) in this case implies that the lecturer can increase the number of students assigned to her higher-ranked project. A set of two or more students form a *blocking coalition* if all of them can improve their situation by swapping their assigned projects. Note that since lecturers are indifferent to the set of students and are interested only in the number of students assigned to each project, they do not resist this swap. A matching having neither a blocking pair nor a blocking coalition is *stable*. Manlove and O'Malley [57] showed that as in the case of SPA, there always exists a stable matching, but here the sizes of stable matchings may differ. They showed that the problem of finding a maximum stable matching is APX-hard and gave a polynomial time 2-approximation algorithm. These upper and lower bounds for approximation were later improved to 1.5 and 21/19, respectively [58].

References

[1] D. Gale and L. S. Shapley, "College admissions and the stability of marriage," *The American Mathematical Monthly*, **69** (1962), 9–15.

[2] Canadian Resident Matching Service, http://www.carms.ca/.

[3] Japan Residency Matching Program, http://www.jrmp.jp/.

[4] National Resident Matching Program, http://www.nrmp.org/.

[5] A. E. Roth, "The evolution of the labor market for medical interns and residents: A case study in game theory," *Journal of the Political Economy*, **92**(6) (1984), 991–1016.

[6] Scottish Foundation Allocation Scheme, http://www.nes.scot.nhs.uk/sfas.

[7] A. Abdulkadiroğlu, P. A. Pathak, and A. E. Roth, "The New York City high school match," *American Economic Review*, **95**(2) (2005), 364–367.

[8] A. Abdulkadiroğlu, P. A. Pathak, A. E. Roth, and T. Sönmez, "The Boston public school match," *American Economic Review*, **95**(2) (2005), 368–371.

[9] C. P. Teo, J. V. Sethuraman, and W. P. Tan, "Gale-Shapley stable marriage problem revisited: Strategic issues and applications," *Management Science*, **47** (2001), 1252–1267.

[10] R. W. Irving, "The cycle roommates problem: A hard case of kidney exchange," *Information Processing Letters*, **103**(1) (2007), 1–4.

[11] A. E. Roth, T. Sönmez, and M. U. Unver, "Kidney exchange," *Quarterly Journal of Economics*, **119**(2) (2004), 457–488.

[12] A. E. Roth, T. Sönmez, and M. U. Unver, "Pairwise kidney exchange," *Journal of Economic Theory*, **125**(2) (2005), 151–188.

[13] E. Kujansuu, T. Lindberg, and E. Mäkinen, "The stable roommates problem and chess tournament pairings," *Divulgaciones Matemáticas*, **7**(1) (1999), 19–28.

[14] S. T. Chuang, A. Goel, N. McKeown, and B. Prabhakar, "Matching output queueing with a combined input output queued switch," *IEEE Journal on Selected Areas in Communications*, **17**(6) (1999), 1030–1039.

[15] G. Nong and M. Hamdi, "On the provision of quality-of-service guarantees for input queued switches," *IEEE Communications Magazine*, **38**(12) (2000), 62–69.

[16] D. Gusfield and R. W. Irving, *The Stable Marriage Problem: Structure and Algorithms*, MIT Press, Boston, MA, 1989.

[17] D. E. Knuth, *Mariages Stables*, Les Presses de l'Université Montréal, 1976. (Translated and corrected edition, *Stable Marriage and Its Relation to Other Combinatorial Problems*, CRM Proceedings and Lecture Notes, Vol. 10, American Mathematical Society, 1997.)

[18] D. F. Manlove, *Algorithmics of Matching Under Preferences*, World Scientific, 2013.

[19] A. E. Roth and M. Sotomayor, *Two-Sided Matching: A Study in Game-theoretic Modeling and Analysis*, Cambridge University Press, Cambridge, 1990.

[20] R. W. Irving and P. Leather, "The complexity of counting stable marriages," *SIAM Journal on Computing*, **15** (1986), 655–667.

[21] E. G. Thurber, "Concerning the maximum number of stable matchings in the stable marriage problem," *Discrete Mathematics*, **248** (2002), 195–219.

[22] D. McVitie and L. B. Wilson, "The stable marriage problem," *Communications of the ACM*, **14** (1971), 486–490.

[23] C. Ng and D. S. Hirschberg, "Lower bounds for the stable marriage problem and its variants," *SIAM Journal on Computing*, **19**(1) (1990), 71–77.

[24] D. Gale and M. Sotomayor, "Some remarks on the stable matching problem," *Discrete Applied Mathematics*, **11** (1985), 223–232.

[25] A. E. Roth, "On the allocation of residents to rural hospitals: A general property of two-sided matching markets," *Econometrica*, **54** (1986), 425–427.

[26] R. W. Irving, "Stable marriage and indifference," *Discrete Applied Mathematics*, **48** (1994), 261–272.

[27] T. Kavitha, K. Mehlhorn, D. Michail, and K. Paluch, "Strongly stable matchings in time $O(nm)$ and extension to the hospitals-residents problem," *ACM Transactions on Algorithms*, **3**(2), Article No. 15, (2007).

[28] D. F. Manlove, "Stable marriage with ties and unacceptable partners," Research Report, TR-1999-29, Computing Science Department, University of Glasgow, Glasgow, 1999.

[29] K. Iwama, D. F. Manlove, S. Miyazaki, and Y. Morita, "Stable marriage with incomplete lists and ties," *Proc. ICALP*, LNCS 1644, pp. 443–452, 1999.

[30] D. F. Manlove, R. W. Irving, K. Iwama, S. Miyazaki, and Y. Morita, "Hard variants of stable marriage," *Theoretical Computer Science*, **276**(1–2) (2002), 261–279.

[31] R. W. Irving and D. F. Manlove, "Approximation algorithms for hard variants of the stable marriage and hospitals/residents problems," *Journal of Combinatorial Optimization*, **16**(3) (2008), 279–292.

[32] K. Iwama, S. Miyazaki, and K. Okamoto, "A $(2 - c\log N/N)$-approximation algorithm for the stable marriage problem," *Proc. SWAT*, LNCS 3111, pp. 349–361, 2004.

[33] K. Iwama, S. Miyazaki, and N. Yamauchi, "A $(2 - c\frac{1}{\sqrt{N}})$-approximation algorithm for the stable marriage problem," *Algorithmica*, **51** (2008), 342–356.

[34] K. Iwama, S. Miyazaki, and N. Yamauchi, "A 1.875-approximation algorithm for the stable marriage problem," *Proc. SODA*, pp. 288–297, 2007.

[35] Z. Király, "Better and simpler approximation algorithms for the stable marriage problem," *Algorithmica*, **60**(1) (2011), 3–20.

[36] E. McDermid, "A 3/2-approximation algorithm for general stable marriage," *Proc. ICALP*, LNCS 5555, pp. 689–700, 2009.

[37] Z. Király, "Linear time local approximation algorithm for maximum stable marriage," *MDPI Algorithms*, **6**(3) (2013), 471–484.

[38] K. Paluch, "Faster and simpler approximation of stable matchings," *MDPI Algorithms*, **7**(2) (2014), 189–202.

[39] H. Yanagisawa, "Approximation algorithms for stable marriage problems," PhD thesis, Graduate School of Informatics, Kyoto University, Kyoto, Japan, 2007.

[40] K. Iwama, S. Miyazaki, and H. Yanagisawa, "A 25/17-approximation algorithm for the stable marriage problem with one-sided ties," *Algorithmica*, **68** (2014), 758–775.

[41] M. M. Halldórsson, K. Iwama, S. Miyazaki, and H. Yanagisawa, "Improved approximation results of the stable marriage problem," *ACM Transactions on Algorithms*, **3**(3), Article No. 30, (2007).

[42] R. W. Irving, D. F. Manlove, and G. O'Malley, "Stable marriage with ties and bounded length preference lists," *Proc. the 2nd Algorithms and Complexity in Durham Workshop*, Texts in Algorithmics, College Publications, 2006.

[43] D. Gusfield, "Three fast algorithms for four problems in stable marriage," *SIAM Journal on Computing*, **16**(1) (1987), 111–128.

[44] R. W. Irving, P. Leather, and D. Gusfield, "An efficient algorithm for the 'optimal' stable marriage," *Journal of the ACM*, **34** (1987), 532–543.

[45] A. Kato, "Complexity of the sex-equal stable marriage problem," *Japan Journal of Industrial and Applied Mathematics*, **10** (1993), 1–19.

[46] K. Iwama, S. Miyazaki, and H. Yanagisawa, "Approximation algorithms for the sex-equal stable marriage problem," *ACM Transactions on Algorithms*, **7**(1) Article No. 2, (2010).

[47] E. McDermid and R. W. Irving, "Sex-equal stable matchings: Complexity and exact algorithms," *Algorithmica*, **68** (2014), 545–570.

[48] M. M. Halldórsson, R. W. Irving, K. Iwama, D. F. Manlove, S. Miyazaki, Y. Morita, and S. Scott, "Approximability results for stable marriage problems with ties," *Theoretical Computer Science*, **306** (2003), 431–447.

[49] R. W. Irving, "An efficient algorithm for the 'stable roommates' problem," *Journal of Algorithms*, **6**(4) (1985), 577–595.

[50] D. J. Abraham, P. Biró, and D. F. Manlove, " 'Almost stable' matchings in the roommates problem," *Proc. WAOA*, LNCS 3879, pp. 1–14, 2005.

[51] E. Ronn, "NP-complete stable matching problems," *Journal of Algorithms*, **11** (1990), 285–304.

[52] R. W. Irving and D. F. Manlove, "The stable roommates problem with ties," *Journal of Algorithms*, **43**(1) (2002), 85–105.

[53] S. Scott, "A study of stable marriage problems with ties," PhD thesis, University of Glasgow, Glasgow, 2005.

[54] P. Biró, T. Fleiner, R. W. Irving, and D. F. Manlove, "The college admissions problem with lower and common quotas," *Theoretical Computer Science*, **411**(34–36) (2010), 3136–3153.

[55] K. Hamada, K. Iwama, and S. Miyazaki, "The hospitals/residents problem with lower quotas," *Algorithmica*, DOI 10.1007/s00453-014-9951-z.

[56] D. J. Abraham, R. W. Irving, and D. F. Manlove, "Two algorithms for the Student-Project Allocation problem," *Journal of Discrete Algorithms*, **5**(1) (2007), 73–90.

[57] D. F. Manlove and G. O'Malley, "Student-project allocation with preferences over projects," *Journal of Discrete Algorithms*, **6**(4) (2008), 553–560.

[58] K. Iwama, S. Miyazaki, and H. Yanagisawa, "Improved approximation bounds for the Student-Project Allocation problem with preferences over projects," *Journal of Discrete Algorithms*, **13** (2012), 59–66.

CHAPTER 18

Domination in Graphs

Subramanian Arumugam

M. Sundarakannan

CONTENTS

18.1	Introduction	419
18.2	Domination Number of a Graph—Basic Results and Bounds	420
18.3	Domination Chain	423
18.4	Various Types of Domination	426
18.5	Total Dominating Sets	427
18.6	Connected Dominating Sets	429
18.7	Paired-Dominating Sets	430
18.8	Equivalence Domination	432
18.9	Global Domination and Factor Domination	434
18.10	Other Types of Domination	435
18.11	Nordhaus–Gaddum Type Results	436
18.12	Domatic Number	437
18.13	Domination in Product Graphs	439
18.14	Algorithmic Aspects	441
18.15	Conclusion	442

18.1 INTRODUCTION

One of the fastest growing areas within graph theory is the study of domination and related subset problems such as independence, irredundance, and matching. The origin of the study of dominating sets in graphs is the following problem of dominating queens: What is the minimum number of queens to be placed on an $n \times n$ chessboard in such a way that every cell in the board is either occupied by a queen or dominated by a queen, where a queen dominates all those cells that can be reached in a single move? This number is the domination number of the graph G whose vertex set V is the set of n^2 cells of the $n \times n$ board and two vertices i and j are adjacent if the queen at i can reach j in a single move. The term domination was first introduced by Ore [1]. Domination and its several variants serve as natural models for many optimization problems. Domination theory has many and varied applications in diverse areas such as design and analysis of communication networks, social network theory, routing problems, kernels of games, operations research, bioinformatics, computational complexity, algorithm design, linear algebra, and optimization. The publication of a survey article by Cockayne and Hedetniemi [2] in 1977, led to the modern study of domination in graphs. Since then more than 1500 papers have been published on this topic. A comprehensive treatment of the fundamentals of domination is given in the book by Haynes et al. [3] and surveys of several advanced topics can be found in the book edited by Haynes et al. [4]. In this chapter we present some of the basic results in domination, recent developments, and directions for further research.

18.2 DOMINATION NUMBER OF A GRAPH—BASIC RESULTS AND BOUNDS

A vertex v in a graph G is said to dominate itself and each of its neighbors. Thus v dominates all the vertices in its closed neighborhood $N[v]$. A set S of vertices of G is a *dominating set* of G if every vertex of $V \setminus S$ is adjacent to at least one vertex of S. Obviously a set $S \subseteq V$ is dominating set if and only if S satisfies any one of the following equivalent conditions:

1. $N[S] = V$.

2. For every vertex $v \in V - S$, we have $d(v, S) \leq 1$.

3. For every vertex $v \in V - S$, there exists a vertex $u \in S$ such that v is adjacent to u.

4. For every vertex $v \in V$, we have $|N[v] \cap S| \geq 1$.

5. For every vertex $v \in V - S$, we have $|N(v) \cap S| \geq 1$.

We observe that any superset of a dominating set is also a dominating set and hence domination is a superhereditary property. Cockayne et al. [5] have proved that for any graph theoretic property P which is superhereditary (hereditary) 1-minimality (1-maximality) is equivalent to minimality (maximality). Thus a dominating set S of G is a minimal dominating set of G if and only if $S - \{v\}$ is not a dominating set of G for all $v \in S$.

Definition 18.1 *The domination number $\gamma(G)$ of a graph G is the minimum cardinality of a dominating set in G. The upper domination number $\Gamma(G)$ is the maximum cardinality of a minimal dominating set of G*

For the graph G given in Figure 18.1, the sets $S_1 = \{v_2, v_3, v_5\}, S_2 = \{v_1, v_4\}$, and $S_3 = \{v_1, v_3, v_5\}$ are minimal dominating sets, $\gamma(G) = 2$ and $\Gamma(G) = 3$. For the complete graph K_n, we have $\gamma = \Gamma = 1$. For the star $K_{1,n-1}$ we have $\gamma = 1$ and $\Gamma = n - 1$.

For any graph G of order n we have $1 \leq \gamma(G) \leq n$. Further $\gamma(G) = 1$ if and only if $\Delta(G) = n - 1$ and $\gamma(G) = n$ if and only if $G = \overline{K_n}$. The following result shows that the upper bound can be substantially improved for graphs having no isolated vertices.

Theorem 18.1 [1] *If a graph G has no isolated vertices, then $\gamma(G) \leq (n/2)$.*

Proof. We first prove that if S is a minimal dominating set of G, then $V - S$ is a dominating set of G. Suppose there exists a vertex $v \in S$ such that v is not dominated by any vertex in $V - S$. Since G has no isolated vertices, it follows that v is adjacent to a vertex w in S. Hence $S - \{v\}$ is a dominating set of G, contradicting the minimality of S. Thus $V - S$ is a dominating set of G. Hence $\gamma(G) \leq \min\{|S|, |V - S|\} \leq (n/2)$. ∎

The *corona* of two graphs G_1 and G_2 is the graph $G = G_1 \circ G_2$ formed from one copy of G_1 and $|V(G_1)|$ copies of G_2, where the ith vertex of G_1 is adjacent to every vertex in the ith copy of G_2. Walikar et al. [6] obtained a characterization of all graphs of even order n for which $\gamma = n/2$. Payen and Xuong [7] and Fink et al. [8] independently obtained the same result.

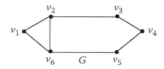

Figure 18.1 Graph with $\gamma = 2$ and $\Gamma = 3$.

Theorem 18.2 *For a graph G with even order n and no isolated vertices, $\gamma(G) = (n/2)$ if any only if the components of G are the cycle C_4 or the corona $H \circ K_1$ for any connected graph H.*

Proof. It is easy to verify that if the components of G are C_4 or $H \circ K_1$, then $\gamma(G) = n/2$. Conversely, suppose $\gamma(G) = (n/2)$. Without loss of generality, we assume that G is connected. Let $T = \{T_1, T_2, \ldots, T_p\}$ be a minimal set of stars which cover all the vertices of G. Since $\gamma(G) = n/2$, T must be a maximum matching and $|T| = p = n/2$. Let $T_i = u_i v_i, 1 \leq i \leq p$.

Assume $p \geq 3$. We claim that u_i or v_i is an end vertex. If not, there exists i such that both u_i and v_i have degree at least 2. Then G has a dominating set of cardinality $p - 1$. Thus $\gamma(G) = p - 1 < n/2$, a contradiction. This shows that G is of the form $H \circ K_1$ for some connected graph H. Now if $p \leq 2$, then G is K_2 or P_4 or C_4. Obviously $K_2 = K_1 \circ K_1$ and $P_4 = K_2 \circ K_1$. ∎

Cockayne et al. characterized graphs G of order n with $\gamma(G) = \lfloor n/2 \rfloor$, which includes Theorem 18.2 as a special case, when n is even. For details we refer to the book by Haynes et al. ([3], pages 42–48).

Remark 18.1 Ore's theorem states that if $\delta \geq 1$, then $\gamma \leq n/2$. Several authors have obtained improvements on the upper bound by restricting their attention to smaller families of graphs. McCuaig and Shepherd [9] have proved that if G is connected and $\delta \geq 2$, then $\gamma \leq 2n/5$, except for a family of 7 graphs. Reed [10] again improved the bound by proving that if G is connected and $\delta \geq 3$, then $\gamma \leq 3n/8$. In this connection, a natural conjecture is that for any connected graph G with $\delta \geq k, \gamma \leq kn/(3k-1)$. Caro and Roditty [11] have obtained a better bound, which settles the above conjecture for $\delta \geq 7$. The question still remains open for $\delta = 4, 5$, or 6. Cockayne et al. [12] improved the bounds for graphs with some forbidden graphs. They have proved that if G is a connected graph which is $K_{1,3}$-free and $K_3 \circ K_1$ free, then $\gamma \leq \lceil n/3 \rceil$.

The following theorem gives lower and upper bounds for γ in terms of the order and the maximum degree. The upper bound is attributed to Berge [13] and the lower bound to Walikar et al. [6].

Theorem 18.3 *For any graph G with order n and maximum degree Δ we have $\lceil n/(1+\Delta) \rceil \leq \gamma \leq n - \Delta$.*

Proof. Let S be a γ-set of G. Since every vertex v in S dominates at most $\Delta + 1$ vertices (including itself) it follows that $n \leq \gamma(\Delta+1)$. Hence $\gamma \geq \lceil n/(1+\Delta) \rceil$. Now, let v be a vertex of maximum degree Δ. Then $V - N(v)$ is a dominating set of cardinality $n - \Delta$ and hence $\gamma \leq n - \Delta$. ∎

It follows from the proof of Theorem 18.3 that $\gamma = n/(1+\Delta)$ if and only if G has a γ-set S such that $|N(v)| = \Delta$ for all $v \in S$ and $N[u] \cap N[v] = \emptyset$ for any two distinct vertices $u, v \in S$. For example, the collection of stars $tK_{1,\Delta}$ and cycles C_{3t} have $\gamma = n/(1+\Delta)$.

Vizing [14] obtained an upper bound for the size m of a graph in terms of its order n and domination number γ, which in turn gives a bound for γ in terms of m and n.

Theorem 18.4 [14] *If G is a graph for which $\gamma \geq 2$, then*

$$m \leq \frac{(n-\gamma)(n-\gamma+2)}{2}. \tag{18.1}$$

Proof. The proof is by induction on n. When $\gamma(G) = 2$, the right side of (1) is $(n-2)n/2$. Since $\gamma(G) = 2$, we have $\Delta(G) \leq n-2$. Hence $m = 1/2 \sum_{v \in V} deg(v) \leq \frac{1}{2} n \Delta(G) \leq \frac{1}{2} n(n-2)$ and the result follows.

Now, suppose $\gamma(G) \geq 3$. When $n = 3, G = \overline{K}_3$. Hence $m = 0$ and the inequality is trivially satisfied. We assume that the result is true for all graphs of order less than n. Let G be a graph of order n with $\gamma(G) \geq 3$. Let $v \in V(G)$ and $deg\, v = \Delta(G)$. Then by Theorem 18.3, $|N(v)| = \Delta(G) \leq n - \gamma(G)$. Let $|N(v)| = \Delta(G) = n - \gamma(G) - r$, where $0 \leq r \leq n - \gamma(G)$. Let $S = V - N[v]$. Then $|S| = n - \Delta(G) - 1 = n - (n - \gamma(G) + r) - 1 = \gamma(G) - r - 1$.

Let $m_1, m_2,$ and m_3 denote respectively the number of edges between $N(v)$ and S, $\langle S \rangle$, and $\langle N[v] \rangle$. Now, for any $u \in N(v)$, the set $S_1 = (S - N(u)) \cup \{u, v\}$ is a dominating set of G. Hence $\gamma(G) \leq |S - N(u)| + 2 \leq \gamma(G) + r - 1 - |S \cap N(u)| + 2$. Thus $|N(u) \cap S| \leq r + 1$ and hence $m_1 \leq \Delta(r+1)$.

Now if D is a minimum dominating set of $\langle S \rangle$, then $D \cup \{u\}$ is a dominating set of G. Hence $\gamma(G) \leq \gamma(\langle S \rangle) + 1$, so that $\gamma(\langle S \rangle) \geq \gamma(G) - 1 \geq 2$. By induction hypothesis, we have

$$m_2 \leq \left[\frac{1}{2}(|S| - \gamma(\langle S \rangle))(|S| - \gamma(\langle S \rangle) + 2\right]$$
$$\leq \left[\frac{1}{2}(\gamma(G) + r - 1 - (\gamma(G) - 1))(\gamma(G) + r - 1 - (\gamma(G) - 1) + 2\right]$$
$$= \frac{1}{2}r(r+2).$$

Now the vertex v is adjacent to $\Delta(G)$ vertices in $N[v]$ and for each vertex $u \in N(v)$ the number of edges between S and $N(v)$ incident with u is at most $r + 1$.

Hence $m_3 = |E(\langle N[v] \rangle)| \leq \Delta(G) + \frac{1}{2}\Delta(G)(\Delta(G) - r - 2)$.

Thus $m = m_1 + m_2 + m_3$
$$\leq \Delta(G)(r+1) + \frac{1}{2}r(r+2) + \Delta(G) + \frac{1}{2}\Delta(G)(\Delta(G) - r - 2)$$
$$= \Delta(G)(n - \gamma(G) - \Delta(G) + 1)$$
$$+ \frac{1}{2}(n - \gamma(G) - \Delta(G))(n - \gamma(G) - \Delta(G) + 2)$$
$$+ \Delta(G) + \frac{1}{2}\Delta(G)(2\Delta(G) - n + \gamma(G) - 2)$$
$$= \frac{1}{2}(n - \gamma(G))(n - \gamma(G) + 2) - \frac{1}{2}\Delta(G)(n - \gamma(G) - \Delta(G))$$
$$\leq \frac{1}{2}(n - \gamma(G))(n - \gamma(G) + 2).$$

Hence the proof is complete by induction. ∎

For graphs with $\gamma < n - \Delta$, Sanchis [15] improved the bound in Theorem 18.4.

Theorem 18.5 *Let G be a graph with $2 \leq \gamma \leq n - \Delta - 1$. Then*

$$m \leq \frac{1}{2}(n-\gamma)(n-\gamma+1).$$

∎

Theorem 18.6 *For any graph G,*

$$n - m \leq \gamma(G) \leq n + 1 - \sqrt{1 + 2m}.$$

Furthermore, $\gamma(G) = n - m$ if and only if each component of G is a star or an isolated vertex.

Proof. The inequality given in Theorem 18.4 can be rewritten as $(n-\gamma)(n-\gamma+2) - 2m \geq 0$. Solving for $n-\gamma$ and using the fact that $n - \gamma \geq 0$, we get

$$n - \gamma \geq -1 + \sqrt{1 + 2m}$$

which establishes the required upper bound.

Since $\gamma(G) \geq 1$, the lower bound is obvious for $m \geq n-1$. Assume that $m \leq n-1$. Then G is a graph with at least $n - m$ components. The domination number of each component of G is at least 1. Hence, $\gamma(G) \geq n - m$ with equality if and only of G has exactly $n - m$ components each with domination number equal to 1. Furthermore, G has $n-m$ components if and only if G is a forest. Hence $\gamma = n - m$ if and only if each component of G is a star or an isolated vertex. ∎

We now proceed to give a few more bounds for γ in terms of other graph theoretic parameters such as diameter, girth, minimum degree, and covering number.

Theorem 18.7 *For any connected graph G,*

$$\left\lceil \frac{diam(G) + 1}{3} \right\rceil \leq \gamma(G).$$

Proof. Let S be a γ-set of G. Let P be a path of length $diam(G)$. Clearly P includes at most two edges from the induced subgraph $\langle N[v] \rangle$ for each $v \in S$. Also since S is a γ-set, P includes at most $\gamma(G) - 1$ edges joining the neighborhood of the vertices of S. Hence $diam(G) \leq 2\gamma(G) + \gamma(G) - 1 = 3\gamma(G) - 1$ and the result follows. ∎

If G is a graph of diameter 2, then for any vertex $v \in V(G)$, the open neighborhood $N(v)$ is a dominating set of G and hence $\gamma(G) \leq \delta(G)$. Brigham et al. [16] have proved that if $\gamma(\overline{G}) \geq 3$, then $diam(G) \leq 2$. Further if G has no isolated vertices and $diam(G) \geq 3$, then $\gamma(\overline{G}) = 2$. The length of a shortest cycle in a graph G is called the girth of G and is denoted by $g(G)$. Brigham and Dutton [17] obtained the following bounds for $\gamma(G)$ in terms of $g(G)$.

If $\delta(G) \geq 2$ and $g(G) \geq 5$, then $\gamma(G) \leq \lceil n - \lfloor g(G)/3 \rfloor/2 \rceil$.

If $g(G) \geq 5$, then $\gamma(G) \geq \delta(G)$.

If $g(G) \geq 6$, then $\gamma(G) \geq 2(\delta(G) - 1)$.

If $\delta(G) \geq 2$ and $g(G) \geq 7$, then $\gamma(G) \geq \Delta(G)$.

18.3 DOMINATION CHAIN

In this section we discuss the relationship between domination, independence and irredundance, leading to an inequality chain of six graph parameters. This inequality chain, which was first observed by Cockayne et al. [18], has become one of the major focal points in the study of domination in graphs.

Let P be a graph theoretic property of a set of vertices S of a graph. Any subset S of V satisfying P is called a P-set and is otherwise called a $\overline{P}$-set. A property P is called *hereditary* (*superhereditary*) if whenever S is a P-set, then any subset (superset) of S is also a P-set. A P-set S in G is called a *minimal* (*maximal*) P-set if every proper subset (superset) of S is a $\overline{P}$-set. Also S is called 1-*minimal* (1-*maximal*) if for every vertex $v \in S, S - \{v\} (v \in V - S, S \cup \{v\})$ is a $\overline{P}$-set. For any hereditary (superhereditary) property, minimality (maximality) and 1-minimality (1-maximality) are equivalent.

Let $G = (V, E)$ be a graph. A subset S of V is called an *independent set* if no two vertices in S are adjacent. Since independence is a hereditary property, an independent set S is maximal if and only if S is 1-maximal. Thus, an independent set S is maximal if and only if $S \cup \{v\}$ is not an independent set for all $v \in V - S$, or equivalently v is adjacent to at least one vertex in S or equivalently S is a dominating set of G. Thus the maximality condition for independence is the definition of dominating set and we have the following basic result.

Proposition 18.1 *An independent set S is maximal independent if and only if it is independent and dominating.* ∎

Definition 18.2 *The minimum cardinality of a maximal independent set is called the independence domination number of G and is denoted by $i(G)$. The independence number $\beta_0(G)$ is the maximum cardinality of a maximal independence set in G.*

Proposition 18.2 *Every maximal independent set S in a graph G is a minimal dominating set of G.*

Proof. It follows from Proposition 18.1 that S is a dominating of G. Also if $v \in S$, then v is not adjacent to any vertex in $S - \{v\}$ and hence $S - \{v\}$ is not a dominating set of G. Hence S is 1-minimal and since domination is a superhereditary property, S is a minimal dominating set of G. ∎

Corollary 18.1 *For any graph G,*

$$\gamma(G) \leq i(G) \leq \beta_0(G) \leq \Gamma(G).$$

∎

We now proceed to investigate the condition for minimality of a dominating set. Let S be a minimal dominating set of G. Then $S - \{v\}$ is not a dominating set of G for all $v \in V - S$ and hence there exists a vertex $w \in V - (S - \{v\})$ such that w is not dominated by any vertex in $S - \{v\}$. Since w is dominated by S, it follows that $N[w] \cap S = \{v\}$. We observe that w may be equal to v in which case v is an isolated vertex in $\langle S \rangle$. This leads to the following definition.

Definition 18.3 *Let $G = (V, E)$ be a graph. Let $S \subseteq V$ and $u \in S$. A vertex v is called a private neighbor of u with respect to S if $N[v] \cap S = \{u\}$.*

We denote by $pn[u, S]$ the set of all private neighbors of u with respect to S. Any private neighbor of u with respect to S is called an *external private neighbor* if $u \in V \backslash S$. The set of all external private neighbors of u with respect to S is denoted by $epn(u, S)$. A dominating set S is a minimal dominating set if and only if $pn[v, S] \neq \emptyset$ for all $v \in S$ or equivalently every vertex in S has at least one private neighbor with respect to S.

Definition 18.4 *Let $G = (V, E)$ be a graph. A subset S of V is called an irredundant set if $pn[v, S] \neq \emptyset$ for all $v \in S$.*

Clearly the minimality condition for a dominating set is the definition of an irredundant set. Thus we have the following theorem.

Theorem 18.8 *A dominating set S is a minimal dominating set if and only if it is dominating and irredundant.* ∎

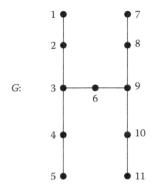

Figure 18.2 Graph with a maximal irredundant set which is not a dominating set.

Obviously, irredundance is a hereditary property and hence maximality of an irredundant set is equivalent to 1-maximality. Thus an irredundant set S is maximal if and only if for every vertex $w \in V - S$, there exists a vertex $v \in S \cup \{w\}$ such that $pn[v, S \cup \{w\}] = \emptyset$.

Theorem 18.9 *Every minimal dominating set in a graph G is a maximal irredundant set of G.*

Proof. It follows from Theorem 18.8 that every minimal dominating set S is irredundant. Now suppose S is not a maximal irredundant. Then there exists a vertex $u \in V - S$ for which $S_1 = S \cup \{u\}$ is irredundant. Hence $pn[u, S \cup \{u\}] \neq \emptyset$. Now, let w be a private neighbor of u with respect to $S \cup \{u\}$. Then w is not adjacent to any vertex in S and hence S is not a dominating set of G, a contradiction. Thus S is a maximal irredundant set of G. ∎

Example 18.1 The converse of Theorem 18.9 is not true. For the graph G given in Figure 18.2, $S = \{2, 3, 8, 9\}$ is a maximal irredundant set, but not a dominating set.

Since every minimal dominating set is a maximal irredundant set, we have the following inequality chain, which was first observed by Cockayne et al. [18].

Theorem 18.10 [18] *For any graph G,*
$$ir(G) \leq \gamma(G) \leq i(G) \leq \beta_0(G) \leq \Gamma(G) \leq IR(G).$$
∎

The above inequality chain is called the *domination chain* of G and has been the focus of more than 100 papers. The following theorem gives a relation between $\gamma(G)$ and $ir(G)$.

Theorem 18.11 *For any graph G,*
$$\frac{\gamma(G)}{2} < ir(G) \leq \gamma(G) \leq 2ir(G) - 1.$$

Proof. Let $ir(G) = k$ and let $S = \{v_1, v_2, \ldots, v_k\}$ be an ir-set of G. Let $u_i \in pn[v_i, S]$ and let $S' = \{u_1, u_2, \ldots, u_k\}$. We observe that u_i may be equal to v_i. Let $S'' = S \cup S'$. Clearly $|S''| \leq 2k$. We claim that S'' is a dominating set of G. If there exists $w \in V - S''$ such that w is not dominated by S'', then $w \in pn[w, S \cup \{w\}]$. Further $u_i \in pn[v_i, S \cup \{w\}]$ and hence $S \cup \{w\}$ is an irredundant set, which is a contradiction. Thus S'' is a dominating set of G. Now if $u_i = v_i$ for some i, then $\gamma(G) \leq |S''| \leq 2k - 1 = 2ir(G) - 1$. If $u_i \neq v_i, 1 \leq i \leq k$, then it follows from Theorem 18.9 that S'' is not a minimal dominating set of G. Hence $\gamma(G) < |S''| = 2k = 2ir(G)$. Thus $\gamma(G) \leq 2ir(G) - 1$ and $\gamma(G)/2 < ir(G)$. ∎

TABLE 18.1 Domination Sequence of Some Standard Graphs

Graph G	$ir(G)$	$\gamma(G)$	$i(G)$	$\beta_0(G)$	$\Gamma(G)$	$IR(G)$
K_n	1	1	1	1	1	1
Path P_n	$\lceil n/3 \rceil$	$\lceil n/3 \rceil$	$\lceil n/3 \rceil$	$\lceil n/2 \rceil$	$\lceil n/2 \rceil$	$\lceil n/2 \rceil$
Cycle C_n	$\lceil n/3 \rceil$	$\lceil n/3 \rceil$	$\lceil n/3 \rceil$	$\lfloor n/2 \rfloor$	$\lfloor n/2 \rfloor$	$\lfloor n/2 \rfloor$
Petersen graph $\mathbb{P}$	3	3	3	4	4	4
Complete bipartite $K_{s,t}, s,t \geq 2$	2	2	$\min\{s,t\}$	$\max\{s,t\}$	$\max\{s,t\}$	$\max\{s,t\}$

Definition 18.5 *An integer sequence $1 \leq a \leq b \leq c \leq d \leq e \leq f$ is called a domination sequence if there exists a graph G for which $ir(G) = a, \gamma(G) = b, i(G) = c, \beta_0(G) = d, \Gamma(G) = e$, and $IR(G) = f$.*

Cockayne and Mynhardt [19] obtained a complete characterization of domination sequences.

Theorem 18.12 [19] *A sequence a,b,c,d,e,f of positive integers is a domination sequence if and only if*

1. $a \leq b \leq c \leq d \leq e \leq f$,

2. $a = 1$ *implies that* $c = 1$,

3. $d = 1$ *implies that* $f = 1$, *and*

4. $b \leq 2a - 1$. ∎

Example 18.2 The domination chain of a few standard graphs is given in Table 18.1.

Observation 18.1 *The domination chain has been formed starting from the concept of independence as seed property. Haynes et al. ([3], page 286) have suggested that almost any property such as vertex cover, packing, $\langle S \rangle$ is acyclic, and so on can be used as a seed property to generate a similar inequality chain. In [20] the concept of vertex cover has been used as a seed property to form an inequality chain of six parameters which is called the covering chain of G.*

18.4 VARIOUS TYPES OF DOMINATION

Several domination parameters have been formed either by imposing a condition on the subgraph induced by a dominating set S or on the method by which the vertices in $V - S$ are dominated. In fact Haynes et al. [3] in the appendix have listed around 75 models of domination and many more models have been introduced since then. In the following sections we present basic results on some of the most fundamental models of domination. For any graph theoretic property P, Harary and Haynes [21] defined the conditional domination number $\gamma(G : P)$ to be the minimum cardinality of a dominating set $S \subseteq V$ such that the induced subgraph $\langle S \rangle$ has the property P.

If P is the property that $\langle S \rangle$ has no edges, then $\gamma(G; P)$ is the independent domination number $i(G)$, which we have covered in the previous section. For any graph G, we have $\gamma(G) \leq i(G)$. Allan and Laskar [22] obtained the following basic theorem giving a family of graphs for which $\gamma(G) = i(G)$. The graph $K_{1,3}$ is called a claw and a graph G is called claw-free if G does not contain $K_{1,3}$ as an induced subgraph.

Theorem 18.13 [22] *If a graph G is claw-free, then $\gamma(G) = i(G)$.*

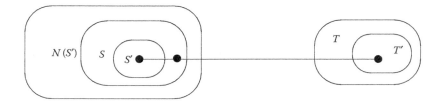

Figure 18.3 Sets S, S', T and T' in the proof of Theorem 18.13.

Proof. Let S be a γ-set in a claw-free graph G. Let S' be a maximal independent subset of S. Let $T = V - N(S')$ and let T' be a maximal independent subset of T. Since T' contains no neighbor of S', it follows that $S' \cup T'$ is independent. Since S' is maximal in S, we have $S \subseteq N(S')$. Since T' is maximal in T, T' dominates T. Hence $S' \cup T'$ is a dominating set of G. It remains to show that $|S' \cup T'| \leq \gamma(G)$ (see Figure 18.3).

Since S' is a maximal independent set in S, T' is independent, and G is claw-free, it follows that each vertex of $S - S'$ has at most one neighbor in T'. Since S is dominating, each vertex of T' has at least one neighbor in $S - S'$. Hence $|T'| \leq |S - S'|$, which gives $|S' \cup T'| \leq |S| = \gamma(G)$. ∎

Since $\gamma(G) \leq i(G)$ for any graph G, a proof of Allan and Laskar's result can be obtained by showing that in any claw-free graph, $i(G) \leq \gamma(G)$. Bollabas and Cockayne generalized this inequality in [23] as follows.

Theorem 18.14 [23] *If a graph G does not contain the star $K_{1,k+1}$, $k \geq 2$, as an induced subgraph, then*
$$i(G) \leq (k-1)\gamma(G) - (k-2).$$

Proof. Let S be a γ-set of G and let I be a maximal independent set in $\langle S \rangle$. Let X be the set of vertices in $V - S$ not dominated by I and let Y be a maximal independent set in $\langle X \rangle$. Since G is $K_{1,k+1}$-free, each $v \in S - I$ is adjacent to at most $k - 1$ vertices of Y. Therefore
$$|Y| \leq (k-1)|S - I|$$
$$\text{Hence } i(G) \leq |Y| + |I| \leq (k-1)(\gamma(G) - |I|) + |I|$$
$$= (k-1)\gamma(G) - (k-2)|I|$$
$$\leq (k-1)\gamma(G) - (k-2).$$
∎

18.5 TOTAL DOMINATING SETS

A solution to the famous five Queens problem inspired Cockayne et al. [24] to introduce the concept of total domination. They observed that in the solution shown in Figure 18.4, not only are the squares without queens dominated by queens, but each queen is dominated by another queen.

For total domination, a vertex v dominates just its open neighborhood $N(v)$ and not itself. Thus S is a total dominating set if $V = N(S)$ and the total domination number is $\gamma_t(G) = min\{|S| : S \text{ is a total dominating set of } G\}$. Thus a dominating set S is a total dominating set if and only if the subgraph induced by S has no isolated vertices.

For an application, we consider a computer network in which a core group of file servers has the ability to communicate directly with every computer outside the core group. In addition, each file server is directly linked to at least one other *back up* file server. A smallest core group with this property is a γ_t-set for the network.

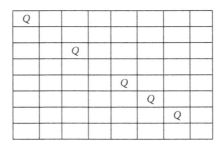

Figure 18.4 Total dominating set solution to the Five Queens Problem.

Clearly any graph G with no isolated vertices has a total dominating set. We note that $\gamma_t(G) = n$ if and only if $G \cong mK_2$. It is easy to see that $\gamma(G) \leq \gamma_t(G) \leq 2\gamma(G)$.

Example 18.3 For $n \geq 3$

$$\gamma_t(C_n) = \gamma_t(P_n) = \begin{cases} \dfrac{n}{2}, & \text{if } n \equiv 0 \pmod{4} \\ \left\lfloor \dfrac{n}{2} \right\rfloor + 1, & \text{otherwise} \end{cases}$$

Cockayne et al. [24] obtained an upper bound for $\gamma_t(G)$ in terms of its order for a connected graph G.

Theorem 18.15 [24] *If G is a connected graph with $n \geq 3$ vertices, then $\gamma_t(G) \leq 2n/3$. Further $\gamma_t(G) = 2n/3$ if and only if G is C_3, C_6, or the 2-corona of some connected graph.* ∎

Theorem 18.16

1. *If G has n vertices and no isolates, then $\gamma_t(G) \leq n - \Delta(G) + 1$.*

2. *If G is connected and $\Delta(G) < n - 1$, then $\gamma_t(G) \leq n - \Delta(G)$.* ∎

We remark that Archdeacon et al. [25] recently found an elegant one page graph theoretic proof of the upper bound of $n/2$ when $\delta \geq 3$. Two infinite families of connected cubic graphs with total domination number one half their orders are constructed in [26]. The result when $\delta \geq 2$ has recently been strengthened by Lam and Wei [27].

Theorem 18.17 *If G is a graph of order n with $\delta(G) \geq 2$ such that every component of the subgraph of G induced by its set of degree-2 vertices has size at most one, then $\gamma_t(G) \leq n/2$.* ∎

For an exhaustive treatment of recent results on total domination, we refer to the book by Henning and Yeo [30].

The known bounds for $\gamma_t(G)$ in terms of δ and n are given in Table 18.2.

TABLE 18.2 Bounds on the Total Domination Number of a Graph G

Graph	Constraint	Upper Bound	Reference
$n \geq 3$ and G is connected	$\delta(G) \geq 1$	$\gamma_t(G) \leq 2n/3$	[24]
$G \notin \{C_3, C_5, C_6, C_{10}\}$ and G is connected	$\delta(G) \leq 2$	$\gamma_t(G) \geq 4n/7$	[28]
G	$\delta(G) \geq 3$	$\gamma_t(G) \leq n/2$	[25]
G	$\delta(G) \geq 4$	$\gamma_t(G) \leq 3n/7$	[29]

18.6 CONNECTED DOMINATING SETS

Sampathkumar and Walikar [31] introduced the concept of *connected domination*.

Definition 18.6 *Let G be a connected graph. A dominating set S of G is called a connected dominating set of G if the induced subgraph $\langle S \rangle$ is connected. The minimum cardinality of a connected dominating set in G is called the connected domination number of G and is denoted by $\gamma_c(G)$.*

Obviously $\gamma(G) \leq \gamma_t(G) \leq \gamma_c(G)$ for any connected graph G with $\Delta(G) < n - 1$. For the complete bipartite graph $K_{r,s}, r, s \geq 2$, we have $\gamma = \gamma_t = \gamma_c = 2$. On the other hand for the cycle C_{12k} we have, $\gamma(C_{12k}) = 4k < \gamma_t(C_{12k}) = 6k < \gamma_c(C_{12k}) = 12k - 2$.

Sampathkumar and Walikar observed the following theorems.

Theorem 18.18 [31] *If H is a connected spanning subgraph of G, then $\gamma_c(G) \leq \gamma_c(H)$.* ∎

Theorem 18.19 [32] *If G is a connected graph and $n \geq 3$, then $\gamma_c(G) = n - \epsilon_T(G)$, where $\epsilon_T(G)$ is the maximum number of end vertices in any spanning tree T of G.*

Proof. Let T be a spanning tree of G with $\epsilon_T(G)$ end vertices and let L denote the set of end vertices of T. Then $T - L$ is a connected dominating set of G and hence $\gamma_c(G) \leq n - \epsilon_T(G)$. Conversely, let S be a γ_c-set of G. Since $\langle S \rangle$ is connected, $\langle S \rangle$ has a spanning tree T_S. Now since S is a dominating set of G, every vertex of $V - S$ is adjacent to at least one vertex in S. Let T be a spanning tree of G obtained by joining each vertex of $V - S$ to exactly one vertex in T_S. Clearly T has at least $n - \gamma_c(G)$ end vertices. Thus $\epsilon_T(G) \geq n - \gamma_c(G)$, so that $\gamma_c(G) \geq n - \epsilon_T(G)$. Hence, $\gamma_c(G) = n - \epsilon_T(G)$. ∎

Corollary 18.2 *Let G be a connected graph with $n \geq 3$. Then $\gamma_c(G) \leq n - 2$ and equality holds if and only if G is a path or a cycle.* ∎

Theorem 18.20 [31] *For any connected graph G,*

$$n/(\Delta(G) + 1) \leq \gamma_c(G) \leq 2m - n$$

with equality for the lower bound if and only if $\Delta(G) = n - 1$ and equality for the upper bound if and only if G is a path.

Proof. Since $n/(\Delta(G) + 1) \leq \gamma(G)$ and $\gamma(G) \leq \gamma_c(G)$, the lower bound follows. Further this lower bound is attained if any only if G has a vertex of degree $n - 1$. Now, $\gamma_c(G) \leq n - 2 = 2(n - 1) - n$ and since G is connected, $m \geq n - 1$. Hence $\gamma_c(G) \leq 2m - n$. If G is a path, then $\gamma_c(G) = n - 2 = 2(n - 1) - n = 2m - n$. Conversely, let $\gamma_c(G) = 2m - n$. Then it follows from Theorem 18.19 that $m \leq n - 1$. Since G is connected, $m = n - 1$ and G is a tree. Let $l(G)$ denote the number of leaves in G.

$$\text{Then } \gamma_c(G) = n - l(G)$$
$$= 2m - n$$
$$= 2(n - 1) - n$$
$$= n - 2.$$

Thus $l(G) = 2$ and G is a path. ∎

TABLE 18.3 Domination Parameters of Q_n

Parameter/Graph	Q_1	Q_2	Q_3	Q_4	Q_5	Q_6	Q_7	Q_n
γ	1	2	2	4	7	12	16	$\geq 2^{n-3} \ \forall \ n \geq 7$
i	1	2	2	4	8	12	?	$\leq 2^{n-2} \ \forall \ n \geq 3$
γ_t	2	2	4	4	8	14	?	$\leq 2^{n-2} - 2^{n-4} \ \forall \ n \geq 7$
γ_c	1	2	4	6	10	18	?	$\leq 2^{n-2} + 4 \ \forall \ n \geq 7$

Theorem 18.21 [32] *For any connected graph G,*
$$\gamma_c(G) \leq n - \Delta(G).$$

Proof. Let v be a vertex in G of maximum degree. Let T be a spanning tree of G in which v is adjacent to each of its neighbors in G. Then T has at least $\Delta(G)$ end vertices and by Theorem 18.19, $\gamma_c(G) \leq n - \Delta(G)$. ∎

Corollary 18.3 [32] *For any tree $T, \gamma_c(T) = n - \Delta(T)$ if and only if T has at most one vertex of degree three or more.* ∎

The concept of connected domination was extended to connected cut-free domination by Joseph and Arumugam [33] who required that $\langle S \rangle$ not only be connected, but be 2-connected. In [34] Joseph and Arumugam explored dominating sets with a required edge connectivity. A set S is a 2-edge connected dominating set if it dominates G and the subgraph $\langle S \rangle$ is 2-edge connected.

Remark 18.2 The n-cube Q_n is the graph whose vertex set is the set of all n-dimensional Boolean vectors, two vertices being adjacent if and only if they differ in exactly one coordinate. We observe that $Q_1 = K_2$ and $Q_n = Q_{n-1} \square K_2$ if $n \geq 2$. The n-cube has applications in coding theory. It admits a decomposition into Hamiltonian cycles if n is even and into a perfect matching and Hamiltonian cycles if n is odd. It has been successfully employed in the architecture of massively parallel computers. A dominating set in Q_n can be interpreted as a set of processors from which information can be passed on to all the other processors. The determination of the domination parameters of Q_n is a significant unsolved problem.

The known values of some of the domination parameters for Q_n are given in Table 18.3. Some of these results are given in [35].

18.7 PAIRED-DOMINATING SETS

A matching in a graph G is a set of independent edges in G. A perfect matching M in G is a matching in G such that every vertex of G is incident to an edge of M. A set $S \subseteq V$ is a paired-dominating set (PDS) if S is a dominating set and the induced subgraph $\langle S \rangle$ has a perfect matching. Paired-domination was introduced by Haynes and Slater [36,37] as a model for assigning backups to guards for security purposes. Every graph without isolated vertices has a PDS since the end vertices of any maximal matching form such a set. The *paired-domination* number of G denoted by $\gamma_{pr}(G)$ is the minimum cardinality of a PDS.

For example, for the graph Q_3 in Figure 18.5, $D_1 = \{v_1, v_2, v_3, v_4\}$ with $M_1 = \{v_1v_2, v_3v_4\}$ or D_1 with $M_2 = \{v_1v_4, v_2v_3\}$ are paired-dominating sets.

Both total domination and paired-domination require that there be no isolated vertices, and every paired dominating set is a total dominating set. Hence for any graph G without isolated vertices, we have $\gamma(G) \leq \gamma_t(G) \leq \gamma_{pr}(G)$. Haynes and Slater [36] obtained the following upper bound for the paired-domination number of a connected graph.

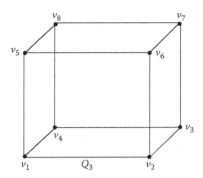

Figure 18.5 Hypercube $Q_3 : \gamma_{pr}(Q_3) = 4$.

Theorem 18.22 [36] *If G is a connected graph of order $n \geq 3$, then $\gamma_{pr}(G) \leq n-1$ with equality if and only if G is C_3, C_5, or a subdivided star.* ∎

Theorem 18.23 [36] *If a graph G has no isolated vertices, then*

$$\gamma_{pr}(G) \leq 2\gamma(G).$$

Proof. We first prove that there exists a γ-set S of G such that every vertex of S has a private neighbor in $V - S$. Let S be a γ-set of G such that the number of vertices in $\langle S \rangle$ having a private neighbor in $V - S$ is maximum. If there exists a vertex u in S having no private neighbor in $V - S$, then u is an isolated vertex in $\langle S \rangle$. Since G has no isolated vertices, u is adjacent to a vertex v in $V - S$. Now $S_1 = (S - \{u\}) \cup \{v\}$ is a γ-set of G in which v has u as its private neighbor in $V - S_1$, contradicting the maximality of the number of vertices in S having private neighbors in $V - S$. Thus every vertex of S has a private neighbor in $V - S$. For each $v \in S$, choose a private neighbor $v_1 \in V - S$. Then $D = S \cup \{v_1 : v \in S\}$ is a paired dominating set of G and hence $\gamma_{pr} \leq |D| = 2\gamma$. ∎

The above bound is sharp as can be seen with $K_n, K_{1,t}$, and graphs formed from a C_{3k} with vertices 0 to $3k-1$ by adding at least one leaf adjacent to each vertex whose label is congruent to 0 modulo 3. If every minimum dominating set is independent, we say that $\gamma(G)$ *strongly equals* $i(G)$ and we write $\gamma(G) \equiv i(G)$. For example, $\gamma(C_5) = i(C_5) = 2$ and each of the five γ-sets is also an i-set. Haynes and Slater defined the concept of *strong equality* of parameters and showed that graphs having $\gamma_{pr}(G) = 2\gamma(G)$ also have $\gamma(G) \equiv i(G)$.

Theorem 18.24 [36] *If a connected graph G has $n \geq 6$ and $\delta(G) \geq 2$, then $\gamma_{pr}(G) \leq 2n/3$.* ∎

The bound of Theorem 18.24 is sharp as can be seen with the cycle C_6. Although there is no known infinite family of graphs which achieves this upper bound, for the family of graphs shown in Figure 18.6, $\gamma_{pr}(G)$ approaches $2n/3$ for large n.

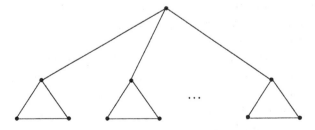

Figure 18.6 $\gamma_{pr}(G)$ approaches $2n/3$ for large n.

Observation 18.2 *The following inequality chains summarize the relation between the paired-domination number, domination number, and total domination number.*

i. $\gamma(G) \leq \gamma_t(G) \leq \gamma_{pr}(G) \leq 2\gamma(G) \leq 2i(G)$.

ii. $2 \leq \gamma_t(G) \leq \gamma_{pr}(G) \leq 2\gamma_t(G)$.

iii. *For any graph G without isolated vertices* $\gamma(G) \leq \gamma_t(G) \leq \gamma_{pr}(G) \leq 2\beta_1(G)$.

Theorem 18.25 [36] *If a graph G has no isolated vertices, then*

$$\gamma_{pr}(G) \leq 2\gamma_t(G) - 2.$$
∎

In [38], Chellali and Haynes showed that if T is a tree of order $n \geq 3$, then $\gamma_{pr}(T) \leq \gamma_t(T) + s - 1$, $\gamma_t(T) \leq n + s/2$, and $\gamma_{pr}(T) \leq n + 2s - 1/2$, where s is the number of support vertices of T.

In [39], Erfang Shan et al. provided a constructive characterization of those trees with equal total domination and paired-domination numbers and of those trees for which the paired-domination number is twice the matching number.

18.8 EQUIVALENCE DOMINATION

An *equivalence graph* is a vertex disjoint union of complete graphs. The equivalence covering number was first studied in [40]. Another concept which uses equivalence graph is subcoloring [41–43]. The concept of equivalence graph also arises naturally in the study of domination in claw-free graphs. It has been proved in [44] that if D is a minimal dominating set in a $K_{1,3}$-free graph, then D is a collection of disjoint complete subgraphs. Motivated by these observations, in [45], Arumugam and Sundarakannan introduced the concept of equivalence set and several new parameters using this concept. Further in [46] the concept of equivalence is used as a seed properly to form an inequality chain of six parameters which is called the *equivalence chain* of G.

A subset S of V is called an equivalence set if every component of the induced subgraph $\langle S \rangle$ is complete. The concept of equivalence set is a hereditary property and hence an equivalence set S is maximal if and only if it is 1-maximal. The maximum cardinality of an equivalence set of G is called the *equivalence number* of G and is denoted by $\beta_{eq}(G)$. The minimum cardinality of a maximal equivalence set of G is called the lower equivalence number of G and is denoted by $i_{eq}(G)$.

We observe that an equivalence set S is maximal if and only if for every $u \in V - S$, at least one component of the induced subgraph $\langle S \cup \{u\} \rangle$ is not complete. Hence for every $u \in V - S$, there exist two vertices $v, w \in S$ such that $\langle \{u, v, w\} \rangle$ is a path P_3. We use this maximality condition for the definition of equivalence dominating set (EDS) or eq-dominating set. A subset $S \subseteq V$ is said to be an *eq-dominating set* of G if for every $v \in V - S$, there exist two vertices $u, w \in S$ such that the induced subgraph $\langle \{u, v, w\} \rangle$ is isomorphic to P_3. The path P_3 may be formed in one of the two ways as shown in Figure 18.7.

Clearly any *eq*-dominating set of G is a dominating set of G. Further eq-domination is a superheriditary property and an *eq*-dominating set S is minimal if and only if S is 1-minimal. The maximum cardinality of a minimal *eq*-dominating set of G is called the *upper eq-domination number* of G and is denoted by $\Gamma_{eq}(G)$. The minimum cardinality of an *eq*-dominating set of G is called the *eq-domination number* of G and is denoted by $\gamma_{eq}(G)$.

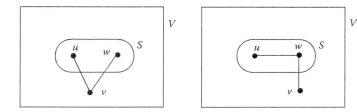

Figure 18.7 Structure of eq-dominating set.

An eq-dominating set S is minimal if and only if $S - \{u\}$ is not an eq-dominating set for every $u \in S$. Hence for any $u \in S$, there exists $v \in V - (S - \{u\})$ such that in the induced subgraph $\langle (S - \{u\}) \cup \{v\} \rangle$, the component containing v is complete. We use this minimality condition of an eq-dominating set for the definition of equivalence irredundant set or eq-irredundant set. A subset $S \subseteq V$ is said to be an eq-irredundant set of G if for each vertex $u \in S$, there exists $v \in V - (S - \{u\})$ such that in the induced subgraph $\langle (S - \{u\}) \cup \{v\} \rangle$, the component containing v is complete. Clearly eq-irredundance is a hereditary property and hence an eq-irredundant set S is maximal if and only if S is 1-maximal. The maximum cardinality of an eq-irredundant set of G is called the *upper eq-irredunance number* of G and is denoted by $IR_{eq}(G)$. The minimum cardinality of a maximal eq-irredundant set of G is called the *eq-irredundance number* of G and is denoted by $ir_{eq}(G)$.

Since the maximality condition for an equivalence set is the definition of an eq-dominating set and the minimality condition for an eq-dominating set is the definition of an eq-irredundant set, we have

$$ir_{eq}(G) \leq \gamma_{eq}(G) \leq i_{eq}(G) \leq \beta_{eq}(G) \leq \Gamma_{eq}(G) \leq IR_{eq}(G).$$

This inequality chain is called the *equivalence chain* of G.

For the complete graph $G = K_n$, we have $ir_{eq} = \gamma_{eq} = i_{eq} = \beta_{eq} = \Gamma_{eq} = IR_{eq} = n$. For the complete bipartite graph $K_{a,b}, 2 \leq a \leq b$, we have $ir_{eq} = \gamma_{eq} = i_{eq} = 2$ and $\beta_{eq} = \Gamma_{eq} = IR_{eq} = b$.

Haynes et al. [47] introduced the concept of H-forming set in graphs. In particular a subset S of V is a P_3-forming set of G if each vertex $u \in V - S$ is contained in a (not necessarily induced) path P_3 of $\langle S \cup \{u\} \rangle$. The P_3-forming number $\gamma_{\{P_3\}}(G)$ and the upper P_3-forming number $\Gamma_{\{P_3\}}(G)$ are defined to be the minimum and maximum cardinality of a minimal P_3-forming set respectively. For a graph G without isolated vertices, every total dominating set is a P_3-forming set, and so we have $\gamma_{\{P_3\}}(G) \leq \gamma_t(G)$.

Note that any eq-dominating set is a P_3-forming set with the added restriction that the 3-paths are induced. Hence, every eq-dominating set is a P_3-forming set and the following result is immediate.

Proposition 18.3 *For every graph* $G, \gamma_{\{P_3\}}(G) \leq \gamma_{eq}(G)$. ■

For any nontrivial tree $T, \gamma_t(T) \leq \gamma_{\{P_3\}}(T)$, which was proved by Haynes et al. in [47]. Chellali and Favaron [48] extended this result to chordal graphs.

Theorem 18.26 [48] *Every nontrivial connected chordal graph G satisfies* $\gamma_t(G) = \gamma_{\{P_3\}}(G)$. ■

Proposition 18.3 and Theorem 18.26 yield the following corollary.

Corollary 18.4 *For every nontrivial connected chordal graph G,* $\gamma_t(G) \leq \gamma_{eq}(G)$. ■

Theorem 18.27 [49] *Let G be a triangle-free graph without isolated vertices. Then every minimal total dominating set of G is a minimal eq-dominating set.*

Proof. Let D be a minimal total dominating set of G. Since G is triangle-free, for every $w \in V - D$, there exist two adjacent vertices w', w'' in D such that $\{w, w', w''\}$ induces a path P_3. Hence D is an eq-dominating set. Now, suppose D is not a minimal eq-dominating set. Then there is a vertex $u \in D$ such that $D' = D - \{u\}$ is an eq-dominating set. Since D' is an eq-dominating set, every vertex in $V - D'$ has a neighbor in D', implying that $epn(u, D) = \emptyset$. Since D is a minimal total dominating set, it follows that u is a private neighbor of a vertex, say x in D. Thus x is an isolate in D', and so u has at least two neighbors in D'. Then the minimality of D as a total dominating set implies that $epn(x, D) \neq \emptyset$. Let $z \in epn(x, D)$. Then z is not eq-dominated by D', a contradiction. Thus D is a minimal eq-dominating set of G. ∎

Corollary 18.5 *If G is a triangle-free graph without isolated vertices, then*

$$\gamma_{\{P_3\}}(G) = \gamma_{eq}(G) \leq \gamma_t(G) \leq \Gamma_t(G) \leq \Gamma_{eq}(G) = \Gamma_{\{P_3\}}(G).$$

∎

Remark 18.3 Since trees T are both triangle-free and chordal graphs, by Corollaries 18.4 and 18.5, we have that $\gamma_{eq}(T) = \gamma_t(T)$.

18.9 GLOBAL DOMINATION AND FACTOR DOMINATION

Sampathkumar [50] introduced the concept of global domination in graphs. Bringham and Dutton [51] introduced the concept of factor domination, which includes global domination as a particular case. A set $D \subseteq V$ is a *global dominating set* of G if D is a dominating set of both G and its complement $\overline{G}$. The *global domination number* $\gamma_g(G)$ of G is the minimum cardinality of a global dominating set of G. A *factor* of G is a spanning subgraph of G. A *k-factoring* of G is a set of k-factors $f = \{G_1, G_2, \ldots, G_k\}$ whose union is G. A *factor dominating set* with respect to f is a set of vertices D which is a dominating set in each factor G_i for $1 \leq i \leq k$. The minimum cardinality of a factor dominating set with respect to f is called the *factor domination number* of G and is denoted by $\gamma(G, f)$.

Factor domination has several interesting applications in many network communication problems. The communication network can be represented by a graph G where vertices of G correspond to nodes of the network and edges correspond to links joining nodes which can communicate directly and finally k-edge disjoint factors of G represent k private subnetworks. Therefore, the factor domination number represents the minimum number of nodes needed to send a message such that all other nodes receive the message in each such network independently in one hop.

The concept of global domination is a special case of factor domination of K_n with 2-factoring. The global domination number of a few standard graphs are given below:

i. $\gamma_g(K_n) = n$.

ii. $\gamma_g(C_n) = \begin{cases} 3, & \text{if } n = 3, 5 \\ \lceil n/3 \rceil, & \text{otherwise.} \end{cases}$

iii. $\gamma_g(W_n) = \begin{cases} 3, & \text{if } n = 4 \\ 4, & \text{otherwise.} \end{cases}$

iv. $\gamma_g(K_{n_1, n_2, \ldots, n_r}) = r$.

Notice that in all the above examples, $\gamma_g = \max\{\gamma, \overline{\gamma}\}$. This leads one to ask whether there might be other classes of graphs or other conditions under which $\gamma_g = \max\{\gamma, \overline{\gamma}\}$. In this regard we have the following theorem.

Theorem 18.28 [51] *If either G or $\overline{G}$ is a disconnected graph, then $\gamma_g = \max\{\gamma, \overline{\gamma}\}$.*

Proof. Assume that G is disconnected. Then G has at least two components and hence $\gamma(\overline{G}) = 2$. Also any dominating set of G is a dominating set of $\overline{G}$ since a dominating set of G must contain at least one vertex from each component. Therefore, $\gamma_g(G) = \gamma(G) = \max\{\gamma, \overline{\gamma}\}$. ∎

Because of the above theorem, we assume that both G and $\overline{G}$ are connected. We state without proof a few basic results on the global domination number.

Theorem 18.29 [51] *If both G and $\overline{G}$ are connected graphs, then*

a. $\gamma_g = \max\{\gamma, \overline{\gamma}\}$ *if* $diam(G) + diam(\overline{G}) \geq 7$.

b. $\gamma_g \leq \max\{3, \gamma, \overline{\gamma}\} + 1$ *if* $diam(G) + diam(\overline{G}) = 6$.

c. $\gamma_g \leq \max\{\gamma, \overline{\gamma}\} + 2$ *if* $diam(G) + diam(\overline{G}) = 5$.

d. $\gamma_g \leq \min\{\delta, \overline{\delta}\}$ *if* $diam(G) = diam(\overline{G}) = 2$. ∎

The global domination number is bounded from above by a variety of graphical invariants, including the minimum degree δ, maximum degree Δ, clique number ω, matching number β_1, and vertex covering number α_0.

Theorem 18.30 [51] *If $\gamma_g > \gamma$, then $\gamma_g \leq \Delta + 1$.* ∎

Theorem 18.31 [51] *Either $\gamma_g = \max\{\gamma, \overline{\gamma}\}$ or $\gamma_g \leq \min\{\Delta, \overline{\Delta}\} + 1$.* ∎

Theorem 18.32 [51] *For any graph G, $\gamma_g \leq \min\{\omega + \gamma, \overline{\omega} + \overline{\gamma}\} - 1$.* ∎

Corollary 18.6 [51] *If G is a triangle-free graph, then $\gamma \leq \gamma_g \leq \gamma + 1$.* ∎

18.10 OTHER TYPES OF DOMINATION

So far we have considered several types of domination obtained by imposing conditions on the dominating set D. Several other domination parameters have been investigated by imposing conditions on the dominated set $V - D$ or on the method in which vertices in $V - D$ are dominated.

Fink and Jacobson [52] introduced the concept of k-domination. Let $G = (V, E)$ be a graph and let k be a positive integer. A subset D of V is called a *k-dominating set* of G if every vertex $v \in V - D$ is dominated by at least k vertices of D. The minimum cardinality of a k-dominating set of G is called the *k-domination number* of G and is denoted by $\gamma_k(G)$. If $k = 1$, then $\gamma_k(G) = \gamma(G)$.

The distance version of dominating set is more applicable to modeling real-world problems. Let $G = (V, E)$ be a connected graph and let k be a positive integer. A subset D of V is called a *k-distance dominating set* of G if for each $u \in V - D$, there exists $v \in D$ such that $d(u, v) \leq k$. The minimum cardinality of a k-distance dominating set of G is called the *k-distance domination number* of G and is denoted by $\gamma_{\leq k}(G)$. We observe that $\gamma_{\leq k}(G)$ is $\gamma(G^k)$ where G^k is the kth power of G. For a survey of distance domination, the reader may refer to Chapter 12 of the book edited by Haynes et al. [3].

Sampathkumar and Pushpalatha [53] introduced the concept of strong and weak domination in graphs. Let u and v be two adjacent vertices of a graph G. We say that u *strongly dominates* v if $deg\ u \geq deg\ v$. Similarly v *weakly dominates* u if $deg\ v \leq deg\ u$. A subset D of V is said to be a strong (weak) dominating set if every vertex in $V - D$ is strongly (weakly) dominated by at least one vertex in D. The *strong (weak) domination number* of G is the minimum cardinality of a strong (weak) dominating set of G and is denoted by $\gamma_s(G)(\gamma_w(G))$. For further results on these concepts we refer to [53].

Another important area in domination is the study of fractional version of domination in graphs. For fractionalization of graph theoretic parameters one may refer to the book by Scheinerman and Ullman [54]. Hedetniemi et al. [55] introduced the fractional version of the concept of domination in graphs.

Let $G = (V, E)$ be a graph. Let $g : V \to \mathbb{R}$ be any function. For any subset S of V, let $g(S) = \sum_{v \in S} g(v)$. The *weight* of g is defined by $|g| = g(V) = \sum_{v \in V} g(v)$.

A function $g : V \to [0,1]$ is called a *dominating function* (DF) of the graph $G = (V,E)$ if $g(N[v]) = \sum_{u \in N[v]} g(u) \geq 1$ for all $v \in V$.

A DF g of a graph G is minimal (MDF) if for all functions $f : V \to [0,1]$ such that $f \leq g$ and $f(v) \neq g(v)$ for at least one $v \in V$, f is not a DF of G.

The fractional domination number $\gamma_f(G)$ and the upper fractional domination number $\Gamma_f(G)$ are defined as follows:

$$\gamma_f(G) = min\{|g| : g \text{ is a dominating function of } G\} \text{ and}$$

$$\Gamma_f(G) = max\{|g| : g \text{ is a minimal dominating function of } G\}.$$

For a detailed survey of Linear Programming formulation for fractional parameters we refer to Chapters 1 through 3 of the book edited by Haynes et al. [4].

18.11 NORDHAUS–GADDUM TYPE RESULTS

Nordhaus and Gaddum [56] obtained lower and upper bounds for the sum and product of the chromatic number $\chi(G)$ and the chromatic number $\chi(\overline{G})$, where $\overline{G}$ is the complement of G. In this section we present similar results for some of the domination parameters. Jager and Payan [57] obtained the following Nordhaus–Gaddum type result for domination number $\gamma(G)$. Cockayane and Hedetniemi [2] determined graphs for which the upper bound is attained.

Theorem 18.33 *Let G be a graph of order $n \geq 3$. Then*

  i. $3 \leq \gamma(G) + \gamma(\overline{G}) \leq n + 1$ *and*

  ii. $2 \leq \gamma(G) \cdot \gamma(\overline{G}) \leq n$.

Also $\gamma(G) + \gamma(\overline{G}) = n + 1$ if and only if $G = K_n$ or $\overline{K_n}$.

Proof. If $\gamma(G) = 1$, then $\gamma(\overline{G}) \geq 2$ and if $\gamma(\overline{G}) = 1$, then $\gamma(G) \geq 2$. Hence the lower bounds follow immediately. We now prove that $\gamma(G) + \gamma(\overline{G}) \leq n+1$ and $\gamma(G) \cdot \gamma(\overline{G}) \leq n$. If $\overline{G}$ has an isolated vertex, then $\gamma(G) = 1$ and $\gamma(\overline{G}) \leq n$. If both G and $\overline{G}$ have no isolated vertices, then by Theorem 18.1, $\gamma(G) \leq \lfloor n/2 \rfloor$ and $\gamma(\overline{G}) \leq \lfloor n/2 \rfloor$. Thus in all cases, $\gamma(G) + \gamma(\overline{G}) \leq n+1$. Also it follows from the above argument that $\gamma(G) + \gamma(\overline{G}) = n + 1$ if and only if either G or $\overline{G}$ has an isolated vertex. Hence either $\gamma(G) = 1$ and $\gamma(\overline{G}) = n$ or $\gamma(G) = n$ and $\gamma(\overline{G}) = 1$. Thus either $\overline{G} = K_n$ or $G = K_n$. The converse is obvious.

We now proceed to prove that $\gamma(G) \cdot \gamma(\overline{G}) \leq n$. For any subset X of V let
$D_e(X) = \{v \in V - X : v \text{ is adjacent to every vertex in } X\}$ and
$D_i(X) = \{u \in X : u \text{ is adjacent to all the other vertices in } X\}$.

Let $d_e(X) = |D_e(X)|$ and $d_i(X) = |D_i(X)|$. We observe that if $d_e(X) = 0$, then X is a dominating set of $\overline{G}$. Now, let $\gamma(G) = k$ and let $S = \{x_1, x_2, \ldots, x_k\}$ be a γ-set of G. Let $\{B_1, B_2, \ldots, B_k\}$ be a partition of V such that $x_j \in B_j$ and all the other vertices in B_j are adjacent to x_j where $1 \leq j \leq k$. We choose such a partition P for which $\sum_{j=1}^{k} d_i(B_j)$ is maximum. Suppose $d_e(B_j) \geq 1$. Then there exists a vertex $x \in B_r, r \neq j$, such that x is adjacent to every vertex in B_j. If $x \in D_i(B_r)$, then $S_1 = (S - \{x_j, x_r\}) \cup \{x\}$ is a dominating set of G and $|S_1| = |S| - 1 = \gamma - 1$, a contradiction. Hence $x \notin D_i(B_r)$.

Now $P' = \{B'_1, B'_2, \ldots, B'_k\}$, where $B'_\ell = B_l$ for all $l \neq j, r$, $B'_j = B_j \cup \{x\}$ and $B'_r = B_r - \{x\}$, is a partition of V and $\sum_{j=1}^{k} d_i(B'_j) > \sum_{j=1}^{k} d_i(B_j)$ contradicting the choice of P. Hence $d_e(B_j) = 0$ for all $j, 1 \leq j \leq k$, so that each B_j is a dominating set of $\overline{G}$. Therefore $\gamma(\overline{G}) \leq |B_j|$. Hence $n = \sum_{j=1}^{k} |B_j| \geq k\gamma(\overline{G}) = \gamma(G)\gamma(\overline{G})$. ■

Lasker and Peters [58] obtained an improvement in the upper bound for $\gamma(G) + \gamma(\overline{G})$ when both G and $\overline{G}$ are connected.

Theorem 18.34 [58] *If G and $\overline{G}$ are connected, then $\gamma(G) + \gamma(\overline{G}) \leq n$ and equality holds if and only if $G = P_4$.* ■

Joseph and Arumugam [59] obtained a substantial improvement when G and $\overline{G}$ have no isolated vertices.

Theorem 18.35 [59] *If G is a graph of order $n \geq 2$ such that neither G nor $\overline{G}$ has isolated vertices, then $\gamma(G) + \gamma(\overline{G}) \leq (n+4)/2$.*

Proof. It follows from Theorem 18.1 that $\gamma(G) \leq n/2$ and $\gamma(\overline{G}) \leq n/2$. Hence the result is obvious if either $\gamma(G) = 2$ or $\gamma(\overline{G}) = 2$. Now, suppose $\gamma(G) \geq 4$ and $\gamma(\overline{G}) \geq 4$. Since $\gamma(G) \cdot \gamma(\overline{G}) \leq n$, it follows that $\gamma(G) \leq n/4$ and $\gamma(\overline{G}) \leq n/4$. Hence $\gamma(G) + \gamma(\overline{G}) \leq n/2$. Hence we may assume that $\gamma(G) = 3$. Then $3 = \gamma(G) \leq n/2$ and so $n \geq 6$. Also $\gamma(\overline{G}) \leq n/\gamma(G) = n/3$. Hence $\gamma(G) + \gamma(\overline{G}) \leq 3 + n/3 \leq 2 + n/2$. ■

A property of most graphs attaining the bound given in the above theorem is given in [59]. Cockayne et al. [60] characterized all graphs which attain the bound.

The proof technique of Theorem 18.35 has been used by several authors to improve the upper bound for the sum of two parameters, in case the upper bound is n or $n+1$. For details we refer to Haynes et al. ([3], pages 251–253).

18.12 DOMATIC NUMBER

The word *domatic* was created from the words *dominating* and *chromatic*. The concept of domatic number $d(G)$ of a graph is analogus to the chromatic number and was introduced by Cockayne and Hedetniemi [2]. In fact $d(G)$ is related to the domination number $\gamma(G)$ in the same way as chromatic number to independence number.

Definition 18.7 *A domatic partition of a graph $G = (V, E)$ is a partition of V into dominating sets. The domatic number $d(G)$ of G is the maximum order of a domatic partition of G.*

Since every graph G admits a trivial domatic partition $\{V(G)\}$, it follows that $d(G)$ is always well defined. We can similarly define total domatic number $d_t(G)$ and connected domatic number $d_c(G)$. These concepts were introduced respectively by Cockayne et al. [4] and Hedetniemi and Laskar [32].

Observation 18.3 *For any graph G, $d \leq \delta + 1$. Let $v \in V(G)$ with $\deg v = \delta$ and let $P = \{S_1, S_2, \ldots, S_d\}$ be a domatic partition of maximum order. Since each S_i is a dominating set of G, it follows that S_i must contain at least one vertex of $N[v]$. Since $|N[v]| = \delta + 1$, it follows that $d = |P| \leq \delta + 1$.* ∎

By a similar argument we have $d_t(G) \leq \delta(G)$. Also if $\gamma(G) \geq 2$, then $d_c(G) \leq \delta$.

Definition 18.8 *A graph G is domatically full if $d(G) = \delta(G) + 1$.*

Similarly we can define total domatically full graph and connected domatically full graph as any graph which attains the upper bound given above.

Observation 18.4 *Let $\{D_1, D_2, \ldots, D_d\}$ be a domatic partition of G of maximum order. Since $|D_i| \geq \gamma(G)$ for each i, it follows that $n = |V| = \sum_{i=1}^{d} |D_i| \geq d(G)\gamma(G)$. Hence $d(G) \leq n/\gamma(G)$. Similarly $d_t(G) \leq n/\gamma_t(G)$ and $d_c(G) \leq n/\gamma_c(G)$.* ∎

Zelinka [61,62] obtained the following lower bounds for $d(G)$ and $d_t(G)$.

Theorem 18.36 *For any graph G, we have*

i. $d(G) \geq \lfloor n/(n - \delta(G)) \rfloor$ *and*

ii. $d_t(G) \geq \lfloor n/(n - \delta(G) + 1) \rfloor$.

Proof. Let D be any subset of V with $|D| \geq (n - \delta(G))$. Since $|N(v)| \geq \delta + 1$ for any $v \in V - D$, it follows that $N(v) \cap D \neq \emptyset$. Hence D is a dominating set of G. Clearly we can obtain $\lfloor n/(n - \delta(G)) \rfloor$ distinct subsets S of G with $|S| = n - \delta(G)$ and each such set S is a dominating set of G. Hence $d(G) \geq \lfloor n/(n - \delta(G)) \rfloor$.

By a similar argument we see that any subset D of G with $|D| \geq n - \delta(G) + 1$ is a total dominating set of G and hence (ii) follows. ∎

The following theorem gives a Nordhaus–Gaddum type result for the domatic number of a graph.

Theorem 18.37 [2] *Let G be any graph of order n. Then $d(G) + d(\overline{G}) \leq n + 1$ and equality holds if and only if $G = K_n$ or $\overline{K_n}$.*

Proof. Suppose $\gamma(G) = 1$. Then $d(G) \leq n$ and $\overline{G}$ has an isolated vertex v. Since every dominating set of $\overline{G}$ contains v, it follows that $d(\overline{G}) = 1$. Thus $d(G) + d(\overline{G}) \leq n + 1$. The proof is similar if $\gamma(\overline{G}) = 1$. If $\gamma(G) \geq 2$ and $\gamma(\overline{G}) \geq 2$, then by Observation 18.4, $d(G) \leq \lfloor n/2 \rfloor$ and $d(\overline{G}) \leq \lfloor n/2 \rfloor$. Hence $d(G) + d(\overline{G}) \leq n$. Further $d(G) + d(\overline{G}) = n + 1$ if and only if $\{d(G), d(\overline{G})\} = \{1, n\}$. Thus $d(G) + d(\overline{G}) = n + 1$ if and only if $G = K_n$ or $\overline{K_n}$. ∎

The cycle C_5 cannot be partitioned into independent dominating sets, whereas the cycle C_6 can be partitioned into three independent dominating sets. This naturally leads to the following definition.

Definition 18.9 [2] *A graph G is called idomatic if there exists at least one partition of V into independent dominating sets. The maximum order of a partition of V into independent dominating sets is called the idomatic number of G and is denoted by $id(G)$.*

Since any paired dominating set is of even order, it follows that the vertex set of any graph $G = (V, E)$ with $|V|$ odd cannot be partitioned into paired dominating sets.

For a survey of results on various types domatic partitions we refer to Chapter 13 of Haynes et al. [4].

18.13 DOMINATION IN PRODUCT GRAPHS

A natural problem for any graph invariant is to investigate how it behaves on graph products. Since any product of two graphs G and H is in some way related to the factors G and H, it is natural to expect that the value of the invariant on the product of G and H is related to its values on G and H. In this section we explore this relationship for domination parameters.

Vizing [63] posed the following conjecture giving a lower bound for the domination number of the Cartesian product $G \square H$ of two graphs G and H.

Vizing's Conjecture. For any two graph G and H,

$$\gamma(G \square H) \geq \gamma(G) \cdot \gamma(H).$$

Vizing's conjecture is one of the main open problems in the area of domination theory. In the absence of a proof of Vizing's conjecture the following question naturally arises.

Question 1. Given a graph G determine all graphs H for which the Vizing's conjecture is true?

The above question naturally leads to the following definition.

Definition 18.10 *A graph G is said to satisfy Vizing's conjecture if $\gamma(G \square H) \geq \gamma(G)\gamma(H)$ for every graph H.*

Most of the results supporting Vizing's conjecture are of the following two types.

i. If H is a graph related to G in some way, and if G satisfies Vizing's conjecture, then H also does.

ii. Let $\mathcal{P}$ be a graph property. If G satisfies $\mathcal{P}$, then G satisfies Vizing's conjecture.

We proceed to present a few results illustrating the above two types of theorems.

Theorem 18.38 [64] *Let K be an induced subgraph of G such that $\gamma(K) = \gamma(G)$. If G satisfies Vizing's conjecture, then K also satisfies Vizing's conjecture.*

Proof. The result is trivial if $K = G$. Hence we assume that K is a proper subgraph of G. Let $e \in E(G) - E(K)$. We claim that $G - e$ satisfies Vizing's conjecture. Since K is an induced subgraph of $G - e$ it follows that $\gamma(K) \geq \gamma(G - e) \geq \gamma(G)$. Further by hypothesis $\gamma(K) = \gamma(G)$. Hence $\gamma(G - e) = \gamma(G)$. Now for any graph H, $(G - e) \square H$ is an induced subgraph of $G \square H$ and hence

$$\gamma((G - e) \square H) \geq \gamma(G \square H)$$
$$\geq \gamma(G)\gamma(H) \text{(by hypothesis)}$$
$$= \gamma(G - e)\gamma(H).$$

Hence $G - e$ satisfies Vizing's conjecture. Since K can be obtained from $G - e$ by successively removing edges, it follows that K also satisfies Vizing's conjecture. ∎

Barcalkin and German [65] proved that Vizing's conjecture is true for a large class of graphs.

Definition 18.11 *Let G be a graph with $\gamma(G) = k$. If $V(G)$ can be partitioned into k subsets $C_1, C_2, \ldots, C_k$ such that each of the induced subgraphs $\langle C_i \rangle$ is complete, then the graph G is said to be decomposable.*

Theorem 18.39 [65] *Any spanning subgraph K of a decomposable graph G with $\gamma(K) = \gamma(G)$ satisfies Vizing's conjecture.* ∎

In particular it follows from the above theorem that all trees and all cycles satisfy Vizing's conjecture. Hartnell and Rall [66] posed the following natural question.

Problem 18.1 *Is there a constant $c > 0$ such that $\gamma(G \Box H) \geq c\gamma(G)\gamma(H)$?*

Theorem 18.40 [67] *For any two graphs G and H, $\gamma(G \Box H) \geq \frac{1}{2}\gamma(G)\gamma(H)$.* ■

Suen and Tarr [68] obtained an improvement of the above theorem by proving that $\gamma(G \Box H) \geq \frac{1}{2}\gamma(G)\gamma(H) + \frac{1}{2}\min\{\gamma(G), \gamma(H)\}$.

Brešar et al. [69] used Clark and Suen's approach to obtain the following theorem for claw-free graphs.

Theorem 18.41 [69] *Let G be a claw-free graph. Then for any graph H without isolated vertices, $\gamma(G \Box H) \geq \frac{1}{2}\beta_0(G)(\gamma(H) + 1)$ where $\beta_0(G)$ is the independence number of G.* ■

Since $\gamma(G) \leq \beta_0(G)$, it follows from Theorem 18.41 that if G is a claw-free graph, then for any graph H without isolated vertices,

$$\gamma(G \Box H) \geq \frac{1}{2}\gamma(G)(\gamma(H) + 1).$$

Another way to tackle Vizing's conjecture is to search for a counterexample. Suppose Vizing's conjecture is false. Then there exists a graph G such that $\gamma(G \Box H) < \gamma(G)\gamma(H)$ for some graph H. A graph G of smallest order with $\gamma(G \Box H) < \gamma(G)\gamma(H)$ for some graph H is called a minimal counterexample. Any minimal counterexample is connected and edge critical with respect to domination. Further it has been proved in [69] that if G is a minimal counterexample to Vizing's conjecture, then for any pair of distinct vertices u and v, $\gamma(G_{uv}) < \gamma(G)$, where G_{uv} is the graph obtained from G by identifying the vertices u and v and then removing any parallel edges.

There are several pairs of graphs G and H for which $\gamma(G \Box H) = \gamma(G)\gamma(H)$. For example $\gamma(G \Box C_4) = \gamma(G)\gamma(C_4)$. Additional results regarding pairs of graphs with equality in Vizing's conjecture are given in [70].

Versions of Vizing's conjecture for various domination-related parameters such as total, fractional, paired, and independent dominations have been studied by different authors. It was conjectured in [71] that $\gamma_t(G \Box H) \geq \frac{1}{2}\gamma_t(G)\gamma_t(H)$ for any two graphs G and H without isolated vertices. This conjecture was proved by Ho [72]. Henning and Rall [71] characterized graphs for which $\gamma_t(G \Box H) = \frac{1}{2}\gamma_t(G)\gamma_t(H)$ when at least one of G or H is a nontrivial tree.

Theorem 18.42 [71] *Let G be a nontrivial tree and let H be any graph without isolated vertices. Then $\gamma_t(G \Box H) = \frac{1}{2}\gamma_t(G)\gamma_t(H)$ if and only if $\gamma_t(G) = 2\gamma(G)$ and H consists of disjoint copies of K_2.* ■

The problem of characterizing all graphs G and H for which $\gamma_t(G \Box H) = \frac{1}{2}\gamma_t(G)\gamma_t(H)$ is open.

For fractional domination Fisher et al. [73] established the following theorem.

Theorem 18.43 [73] *For any two graphs G and H, $\gamma_f(G \Box H) \geq \gamma_f(G)\gamma_f(H)$.* ■

Brešar et al. [74] investigated Vizing-type result for paired-domination number $\gamma_{pr}(G)$. For any integer $k \geq 2$, a *k-packing* in a graph $G = (V, E)$ is a subset S of V such that $d(u, v) > k$ for all $u, v \in S$. The *k-packing number* $\rho_k(G)$ is the maximum cardinality of a k-packing in G. Brešar et al. [74] observed that the role played by 3-packing number for paired domination is similar to that of the packing number for domination and obtained the following theorem.

Theorem 18.44 [74] *Let G and H be graphs without isolated vertices. Then $\gamma_{pr}(G\square H) \geq \max\{\gamma_{pr}(G)\rho_3(H), \gamma_{pr}(H)\rho_3(G)\}$.* ∎

Further for any tree T, $\gamma_{pr}(T) = 2\rho_3(T)$ and hence we have the following corollary.

Corollary 18.7 [74] *Let T be any nontrivial tree and let H be any graph without isolated vertices. Then $\gamma_{pr}(T\square H) \geq \frac{1}{2}\gamma_{pr}(T)\gamma_{pr}(H)$ and this bound is sharp.* ∎

Vizing-like bounds for upper domination number and upper total domination number were established in [75] and [76], respectively.

Theorem 18.45 [75] *For any graphs G and H, $\Gamma(G\square H) \geq \Gamma(G)\Gamma(H)$.* ∎

Theorem 18.46 [76] *Let G and H be connected graphs of order at least 3 with $\gamma_t(G) \geq \Gamma_t(H)$. Then $\gamma_t(G\square H) \geq \frac{1}{2}\Gamma_t(G)(\Gamma_t(H) + 1)$ and this bound is sharp.* ∎

Domination in graph products, other than the Cartesian product, is an area that has not been fully explored.

18.14 ALGORITHMIC ASPECTS

Given a graph $G = (V, E)$ with $|V| = n$, the natural question is to design an algorithm for determining the domination number $\gamma(G)$. One possible method is to list all possible subsets of V in nondecreasing order of cardinality and check whether any of these subsets is a dominating set. The value of $\gamma(G)$ is simply the cardinality of the first dominating set that we get in this method. This algorithm requires $O(2^n)$ steps in the worst case and hence is of exponential time complexity. We shall prove that the decision problem corresponding to $\gamma(G)$ is NP-complete for arbitrary graphs and this indicates that it may not be possible to construct a polynomial time algorithm for this problem.

A well known NP-complete problem is 3-SAT .

3-SAT

> **INSTANCE.** A set $U = \{u_1, u_2, \ldots, u_n\}$ of Boolean variables and a set $\mathcal{C} = \{C_1, C_2, \ldots, C_m\}$ of 3-element sets where $|C_i| = 3$ and any element of C_i is either a variable u_i or its complement u'_i. Each C_i is called a clause.
>
> **QUESTION.** Does there exist an assignment of True or False to the variables in U such that at least one variable in each clause C_i is assigned the value True?

Such an assignment is called a satisfiable truth assignment.

The decision problem for the domination number of a graph takes the following form.

DOMINATING SET

> **INSTANCE.** A graph $G = (V, E)$ and a positive integer k.
>
> **QUESTION.** Does G have a dominating set of size $\leq k$?

Theorem 18.47 [77] *DOMINATING SET IS NP-complete.*

Proof. If $S \subseteq V$ and $|S| \leq k$, then it can be easily verified in polynomial time whether S is a dominating set of G. Hence DOMINATING SET $\in$ NP. We now construct a reduction from 3-SAT to DOMINATING SET. Give an instance $\mathcal{C}$ of 3-SAT, we construct an instance $G(\mathcal{C})$ of DOMINATING SET as follows:

For each variable u_i construct a triangle with vertices labeled u_i, u'_i, v_i. For each clause $C_j = \{u_i, u_k, u_e\}$ create a single vertex labeled C_j and add edges $u_i C_j, u_k C_j$, and $u_e C_j$. Also let $k = n$. Thus $G(\mathcal{C})$, n is an instance of DOMINATING SET. We claim that $\mathcal{C}$ has a satisfying truth assignment if and only if $G(\mathcal{C})$ has a dominating set S with $|S| \leq n$.

Suppose that $\mathcal{C}$ has a satisfying truth assignment. Let $S = \{u_i : u_i = True\} \cup \{u'_i : u_i = False\}$. Since exactly one of u_i, u'_i is in S for each $i, 1 \leq i \leq n$, we have $|S| = n$. Further S contains one vertex from each of the triangles $\{u_i, u'_i, v_i\}$ and have all the vertices of the triangle are dominated by S. Since each clause C_j contains at least one variable whose value is TRUE, it follows that each C_j is dominated by S. Hence S is a dominating set of $G(\mathcal{C})$ with $|S| = n$.

Conversely, suppose $G(\mathcal{C})$ has a dominating set S with $|S| \leq n$. Since each v_i is either in S or dominated by a vertex in S, each triangle $\{u_i, u'_i, v_i\}$ must contain a vertex of S. Thus $|S| \geq n$ and S contains no clause vertex C_j. Hence each C_j must be dominated by a vertex in S. Hence the truth assignment defined by

$$value\ of\ u_i = \begin{cases} True, & \text{if } u_i \in S \\ False, & \text{otherwise} \end{cases}$$

gives a satisfying truth assignment for $\mathcal{C}$.

Now the graph $G(\mathcal{C})$ has $3n+m$ vertices and $3n+3m$ edges. Hence $G(\mathcal{C})$ can be constructed from any instance of 3-SAT in polynomial time. ∎

Observation 18.5 *Since the dominating set S constructed in Theorem 18.47 is an independent dominating set, the INDEPENDENT DOMINATING SET problem is also NP-complete.*

Since the DOMINATING SET is NP-complete, the next natural question is to find whether the domination number of a graph can be computed in polynomial time when restricted to special classes of graphs. Several authors [78,79] have independently proved that DOMINATING SET remains NP-complete even when restricted to bipartite graphs.

Booth and Johnson [80] have shown that the DOMINATING SET remains NP-complete when restricted to chordal graphs. For details of NP-completeness of other domination related parameters such as irredundance number $ir(G)$, independence number $\beta_0(G)$, upper domination number $\Gamma(G)$, upper irredundance number $IR(G)$, connected domination number $\gamma_c(G)$, and total domination number $\gamma_t(G)$, for arbitrary graphs and the complexity status when restricted to special classes of graphs, we refer to Chapter 12 of Haynes et al. [3] and to Chapters 8 and 9 of the book edited by Haynes et al. [4].

18.15 CONCLUSION

In this chapter we have covered some of the basic concepts in domination. Since domination is a major area of current research in graph theory, we have not touched on several topics such as domination in directed graphs, domination in hypergraphs, chessboard domination problems, criticality, applications, and several other topics. The two books by Haynes et al. [3,4] give a comprehensive treatment of fundamental concepts and several advanced topics. A more recent book by Henning and Yeo [30] provides an up-to-date coverage of results on total domination in graphs.

References

[1] O. Ore, Theory of graphs, *Amer. Math. Soc. Colloq. Publ.*, **38** (Amer. Math. Soc, Providence, RI), 1962.

[2] E. J. Cockayne and S. T. Hedetniemi, Towards a theory of domination in graphs, *Networks*, **7** (1977), 247–261.

[3] T. W. Haynes, S. T. Hedetniemi and P. J. Slater, *Fundamentals of domination in graphs*, Marcel Dekker, New York, 1998.

[4] T. W. Haynes, S. T. Hedetniemi and P. J. Slater, *Domination in graphs—Advanced topics*, Marcel Dekker, 1998.

[5] E. J. Cockayne, J. H. Hatting, S. M. Hedetniemi, S. T. Hedetniemi and A. A. McRae, Using maximality and minimality conditions to construct inequality chain, *Discrete Math.*, **176**(1–3) (1997), 43–61.

[6] H. B.Walikar, B. D. Acharya and E. Sampathkumar, Recent developments in the theory of domination in graphs, *MRI Lecture Notes in Math.*, **1** (1976).

[7] C. Payan and N.H. Xuong, Domination-balanced graphs, *J. Graph Theory*, **6** (1982), 23–32.

[8] J. F. Fink, M. S. Jacobson, L. F. Kinch and J. Roberts, On graphs having domination number half their order, *Period. Math. Hungar*, **16** (1985), 287–293.

[9] W. McCuaig and B. Shepherd, Domination in graphs with minimum degree two, *J. Graph Theory*, **13** (1989), 749–762.

[10] B. Reed, Paths, stars and the number three, *Combin. Probab. Comput.*, **5** (1996), 277–295.

[11] Y. Caro and Y. Roditty, Improved bounds for the product of the domination and chromatic numbers of a graph, *Ars Combin.*, **56** (2000), 189–192.

[12] E. J. Cockayne, C. W. Ko and F. B. Shepherd, *Inequalities concerning dominating sets in graphs*, Technical Report DM-370-IR, Dept. Math., Univ. Victoria, Canada, 1985.

[13] C. Berge, *Theory of graphs and its applications*, Methuen, London, 1962.

[14] V. G. Vizing, A bound on the external stability number of a graph, *Dokl. Akad. Nauk SSSR*, **164** (1965), 729–731.

[15] L. A. Sanchis, Maximum number of edges in connected graphs with a given domination number, *Discrete Math.*, **87** (1991), 65–72.

[16] R. C. Brigham, P. Z. Chinn and R. D. Dutton, Vertex domination-critical graphs, *Networks*, **18** (1988), 173–179.

[17] R. C. Brigham and R. D. Dutton, Bounds on the domination number of a graph, *Quart. J. Math. Oxford Ser.* 2, **41** (1990), 269–275.

[18] E. J. Cockayne, S. T. Hedetniemi and D. J. Miller, Properties of hereditary graphs and middle graphs, *Canad. Math. Bull.*, **21** (1978), 461–468.

[19] E. J. Cockayne and C. M. Mynhardt, The sequence of upper and lower domination, independence and irredundance numbers of a graph, *Discrete Math.*, **122** (1993), 89–102.

[20] S. Arumugam, S. T. Hedetniemi, S. M. Hedetniemi and L. Sathikala, The covering chain of a graph, *Util. Math.*, (To appear).

[21] F. Harary and T. W. Haynes, Conditional graph theory IV: Dominating sets, *Util. Math.*, **40** (1995), 179–192.

[22] R. B. Allan and R. C. Laskar, On domination and independent domination numbers of a graph, *Discrete Math.*, **23** (1978), 73–76.

[23] B. Bollabas and E. J. Cockayne, Graph theoretic parameters concerning domination, independence and irredundance, *J. Graph Theory*, **3** (1979), 241–250.

[24] E. J. Cockayne, R. M. Dawes and S. T. Hedetniemi, Total domination in graphs, *Networks*, **10** (1980), 211–219.

[25] D. Archdeacon, J. Ellis-Managhan, D. Fischer, D. Froncek, P. C. B. Lam, S. Seager, B. Wei and R. Yuster, Some remarks on domination, *J.Graph Theory*, **46** (2004), 207–210.

[26] O. Favaron, M. A. Henning, C. M. Mynhardt and J. Puech, Total domination in graphs with minimum degree three, *J. Graph Theory*, **34** (2000), 9–19.

[27] P. C. B. Lam and B. Wei, On the total domination number of graphs, *Utilitas Math.*, **72** (2007), 223–240.

[28] M. A. Henning, Graphs with large total domination number, *J.Graph Theory*, **35** (2000), 21–45.

[29] A. Yeo, *Improved bound on the total domination in graphs with minimum degree four*, Manuscript, 2006.

[30] M. A. Henning and A. Yeo, *Total domination in graphs*, Springer, New York, 2013.

[31] E. Sampathkumar and H. B. Walikar, The connected domination numbers of a graph, *J. Math. Phys. Sci.*, **13** (1979), 607–613.

[32] S. T. Hedetniemi and R. C. Laskar, Connected domination graphs, In *Graph theory and combinatorics*, B. Bollabas, editors, Academic Press, London, 209–218, 1984.

[33] J. P. Joseph and S. Arumugam, On connected cut free domination in graphs, *Indian J. Pure Appl. Math.*, **23** (1992), 643–647.

[34] J. P. Joseph and S. Arumugam, On 2-edge connected domination in graphs, *Internat. J. Management Systems*, **12** (1996), 131–138.

[35] S. Arumugam and R. Kala, Domination parameters of hypercubes, *J. Indian Math. Soc.*, **65** (1998), 31–38.

[36] T. W. Haynes and P. J. Slater, Paired-domination in graphs, *Networks*, **32** (1998), 199–206.

[37] T. W. Haynes and P. J. Slater, Paired-domination and the paired-domatic number, *Congr. Numer.*, **109** (1995), 65–72.

[38] M. Chellali and T. W. Haynes, Total and paired-domination numbers of a tree, *AKCE Int. J. Graphs Combin.*, **2** (2004), 69–75.

[39] Erfang Shan, Liying Kang and M. A. Henning, A characterization of trees with equal total domination and paired-domination numbers, *Austr. Journal of Combin.*, **30** (2004), 31–39.

[40] P. Duchet, *Représentations, noyaus en théorie des graphes et hypergraphs*, Thése, Paris VI, France, 1979, 85–95.

[41] C. Mynhardt and I. Broere, Generalized colorings of graphs. In *Graph theory with applications to algorithms and computer science*, Y. Alavi, G. Chartrand, L. Lesniak, D. R. Lick and C. E. Wall, editors, Wiley, New York, 583–594, 1985.

[42] M. O. Albertson, R. E. Jamison, S. T. Hedetmemi and S. C. Locke, The subchromatic number of a graph, *Discrete Math.*, **74** (1989), 33–49.

[43] J. Gimbel and C. Hartman, Subcolorings and the subchromatic number of a graph, *Discrete Math.*, **272** (2003), 139–154.

[44] R. D. Dutton and R. C. Brigham, Domination in claw-free graphs, *Congr. Numer.*, **132** (1998), 69–75.

[45] S. Arumugam and M. Sundarakannan, Equivalence dominating sets in graphs, *Util. Math.*, **91** (2013), 231–242.

[46] S. Arumugam and M. Sundarakannan, The equivalence chain of a graph, *J. Combin. Math. Combin. Comput.*, **80** (2012), 277–288.

[47] T. W. Haynes, S. T. Hedetniemi, M. A. Henning and P. J. Slater, H-forming sets in graphs, *Discrete Math.*, **262** (2003), 159–169.

[48] M. Chellali and D. Favaron, On k-star forming sets in graphs, *J. Combin. Math. Combin. Comput.*, **68** (2009), 205–214.

[49] S. Arumugam, M. Chellai and T. W. Haynes, Equivalence domination in graphs, *Questiones Mathematicae*, **36** (2013), 331–340.

[50] E. Sampathkumar, The global domination number of a graph, *J. Math. Phys. Sci.*, **23** (1989), 377–385.

[51] R. C. Brigham and R. D. Dutton, Factor domination in graphs, *Discrete Math.*, **86** (1990), 127–136.

[52] J. F. Fink and M. S. Jacobson, n-domination in graphs, In *Graph theory with applications to algorithms and computer science*, Y. Alavi and A.J. Schwenk, editors, 283–300, 1985.

[53] E. Sampathkumar and L. Pushpalatha, Strong weak domination and domination balance in graph, *Discrete Math.*, **161** (1996), 235–242.

[54] E. R. Scheinerman and D. H. Ullman, *Fractional graph theory: A rational approch to the theory of graphs*, John Wiley & Sons, New York, 1997.

[55] S. M. Hedetniemi, S. T. Hedetniemi and T. V. Wimer, Linear time resource allocation algorithms for trees, Technical report URI-014, Department of Mathematics, Clemson University, SC, 1987.

[56] E. A. Nardhaus and J. W. Gaddum, On complementary graphs, *Amer. Math. Monthly*, **63** (1956), 175–177.

[57] F. Jaeger and C. Payen, Relation du type Nordhauss-Gaddam pour le nombre dábsorption dún graphe simple, *C.R. Acad. Sci. Ser. A*, **274** (1972), 728–730.

[58] R. Lasker and K. Peters, Vertex and edge domination parameters in graphs, *Congr. Numer.*, **48** (1985), 291–305.

[59] J. P. Joseph and S. Arumugam, A note on domination in graphs, *International Journal of Management and Systems*, **11** (1995), 177–182.

[60] Xu Baogen, E. J. Cockayne, T. W. Haynes, S. T. Hedetniemi and Zhou Shangchao, Extremal graphs for inequalities involving domination parameters, *Discrete Math.*, **216** (13) (2000), 1–10.

[61] B. Zelinka, On k-domatic numbers of graphs, *Czech. Math. J.*, **33** (1983), 309–313.

[62] B. Zelinka, Total domatic number and degrees of vertices of a graph, *Math. Slovaca*, **39** (1989), 7–11.

[63] V. G. Vizing, The Cartesian product of graphs, *Vychisl. Sistemy*, **9** (1963), 30–43.

[64] B. L. Hartnell and D. F. Rall, On Vizing's conjecture, *Congr. Numer.*, **82** (1991), 87–96.

[65] A. M. Barcalkin and L. F. German, The external stability number of the Cartesian product of graphs, *Bul. Akad, Stiince RSS Moldoven*, **1** (1979), 5–8.

[66] B. L. Hartnell and D. F. Rall, *Domination in Cartesian product: Vizing's conjecture*, In *domination in graph* s-Advanced topics, T.W. Haynes, S.T. Hedetniemi, and P.J. Slater, editors, New York: Dekker, 163, 189, 1998.

[67] W. E. Clark and S. Suen, An inequality related to Vizing's conjecture, *Electronic J. Combin.*, **7** (Note 4) (2000), 3p.

[68] S. Suen and J. Tarr, *An improved inequality related to Vizing's conjecture*, Manuscript (2010).

[69] B. Brešar, P. Dorbee, W. Goddard, B. L. Hartnell, M. A. Henning, S. Klavzar and D. F. Rall, Vizing's conjecture: A survey and recent results, *J. Graph Theory*, **69** (1) (2011), 46–76.

[70] S. M. Khamis and Kh. M. Nazzal, Equality in Vizing's conjecture fixing one factor of the Cartesian product, *Ars Combin.*, **96** (2010), 375–384.

[71] M. A. Henning and D. F. Rall, On the total domination number of Cartesian product of graphs, *Graphs Combin.*, **21**(1) (2005), 63–69.

[72] P. T. Ho, A note on total domination number, *Util. Math.*, **77** (2008), 97–100.

[73] D. C. Fisher, J. Ryan, G. Domke and A. Majumdar, Fractional domination of strong direct products, *Discrete Appl. Math.*, **50** (1) (1994), 89–91.

[74] B. Brešar, M. A. Henning and D. F. Rall, Paired domination of Cartesian products of graphs, *Util. Math.*, **73** (2007), 255–265.

[75] B. Brešar, On Vizing conjecture, *Discuss Math. Graph Theor.*, **21** (1) (2001), 5–11.

[76] P. Dorbec, M. A. Henning and D. F. Rall, On the upper total domination number of Cartesian products of graphs, *J. Comb. Optim.*, **16** (1) (2008), 68–80.

[77] M. R. Garey and D. S. Johnson, *Computers and intractability: A guide to the theory of NP-completeness*, Freeman, New York, 1979.

[78] A. A. Bertossi, Dominating set for split and bipartite graphs, *Inform. Process. Lett.*, **19** (1984), 37–40.

[79] G. J. Chang and G. L. Nemhauser, The k-domination and k-stability problems in sum-free chordal graphs, *SIAM J. Algebraic Discrete Methods*, **5** (1984), 332–345.

[80] K. S. Booth and J. H. Johnson, Dominating sets in chordal graphs, *SIAM J. Comput.*, **11** (1982), 191–199.

CHAPTER 19

Graph Colorings

Subramanian Arumugam

K. Raja Chandrasekar

CONTENTS

- 19.1 Introduction 449
- 19.2 Vertex Colorings 449
- 19.3 Edge Colorings 461
- 19.4 Other Variants of Graph Colorings 463
  - 19.4.1 Complete Colorings and Achromatic Number 463
  - 19.4.2 Grundy Colorings in Graphs 465
  - 19.4.3 Dominator Colorings 466
  - 19.4.4 List Colorings and Choosability 467
  - 19.4.5 Total Colorings 469
- 19.5 Conclusion 469

19.1 INTRODUCTION

Graph coloring theory has a central position in discrete mathematics. Graph coloring deals with the fundamental problem of partitioning a set into classes according to certain rules. Timetabling, sequencing, and scheduling problems are basically of this nature and hence graph coloring has interesting applications to several practical problems. Though many deep and interesting results have been obtained on graph coloring during the past 100 years, there are still many easily formulated, interesting, and challenging unsolved problems. In this chapter we present basic results on various graph coloring concepts such as vertex, edge, total, list, complete, and dominator colorings.

19.2 VERTEX COLORINGS

Definition 19.1 *A* proper coloring *of a graph G is an assignment of colors to the vertices of G in such a way that no two adjacent vertices receive the same color. The* chromatic number $\chi(G)$ *is the minimum number of colors required for a proper coloring of G. A graph G with chromatic number k is a k-chromatic graph.*

The chromatic numbers of some graphs can be easily determined. For example, $\chi(C_{2k}) = 2$, $\chi(C_{2k+1}) = 3, \chi(K_n) = n$, and $\chi(K_{n_1,n_2,\ldots,n_k}) = k$. Also a 3-chromatic graph is given in Figure 19.1.

In a given coloring of G, the set of all vertices which are assigned a particular color is called a *color class*. Thus, if $V_1, V_2, \ldots, V_k$ are the color classes of a k-coloring of G, then each V_i is independent and $\{V_1, V_2, \ldots, V_k\}$ forms a partition of V and the coloring is completely

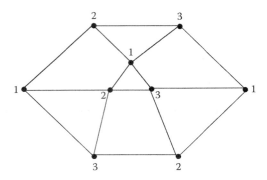

Figure 19.1 A 3-chromatic graph.

determined by this partition. Thus, the chromatic number of G is the minimum value k such that V can be partitioned into k independent sets.

Obviously if G is a k-partite graph with partite sets $V_1, V_2, \ldots, V_k$, then $\{V_1, V_2, \ldots, V_k\}$ is a k-coloring of G and hence $\chi(G) \leq k$. Further $\chi(G) = k$ if and only if G is k-partite and is not l-partite for any $l < k$. Consequently $\chi(G) = 1$ if and only if G is an empty graph and $\chi(G) = 2$ if and only if G is a nontrivial bipartite graph. However, for $k \geq 3$ the problem of deciding whether $\chi(G) \leq k$ is NP-complete ([1], page 191). We observe that if G is a disconnected graph with components $G_1, G_2, \ldots, G_r$, then $\chi(G) = max\ \{\chi(G_i) : 1 \leq i \leq r\}$ and if G is a separable graph with blocks $B_1, B_2, \ldots, B_s$, then $\chi(G) = max\ \{\chi(B_i) : 1 \leq i \leq s\}$. Hence, we can confine ourselves to 2-connected graphs while considering vertex colorings.

Coloring problems arise naturally in many practical applications. We give two examples of such problems.

Examination scheduling: An university has to conduct annual examination for all courses offered in the university. Obviously examinations in two different courses cannot be held simultaneously if the two courses have students in common. The problem of finding a schedule for the examination with minimum number of sessions can be reduced to a graph coloring problem. We define a graph G whose vertex set is the set of all courses and two courses are joined by an edge if they have students in common. The required minimum number of sessions for the schedule of the examination is the chromatic number of G.

Chemical storage: A set of n chemicals $c_1, c_2, \ldots, c_n$ are to be stored in a warehouse. Certain pairs of chemicals would cause explosion if they are in contact with each other and hence must be stored in different compartments of the warehouse. Then the least number of compartments into which the warehouse must be partitioned is the chromatic number of the graph G, where $V(G) = \{c_1, c_2, \ldots, c_n\}$, and $c_i c_j \in E(G)$ if and only if c_i and c_j cause explosion if they are in contact with each other.

We now proceed to give upper and lower bounds for the chromatic number of a graph. Two of the most elementary bounds for the chromatic number of a graph G involve the independence number $\alpha(G)$, which is the maximum cardinality of an independent set of vertices of G.

Theorem 19.1 *If G is a graph of order n, then $n/\alpha(G) \leq \chi(G) \leq n - \alpha(G) + 1$.*

Proof. Suppose $\chi(G) = k$. Let $\{V_1, V_2, \ldots, V_k\}$ be a χ-coloring of G. Then $n = |V(G)| = \sum_{i=1}^{k} |V_i| \leq k\ \alpha(G)$. Hence $n/\alpha(G) \leq \chi(G)$.

Now, let U be an independent set of G such that $|U| = \alpha(G)$. Then $\mathcal{C} = \{U\} \cup \{\{v\} : v \notin U\}$ is a proper coloring of G and hence $\chi(G) \leq 1 + |V(G) - U| = n - \alpha(G) + 1$. ■

Let $v_1, v_2, \ldots, v_n$ be an ordering of the vertices of G. We color the vertices in this order and assign to v_i the smallest positive integer not already used to the neighbors v_j of v_i with $j < i$. This method of coloring the vertices is called a *greedy coloring* of G.

Proposition 19.1 *For any graph G, $\chi(G) \leq \Delta(G) + 1$.*

Proof. In any vertex ordering each vertex has at most $\Delta(G)$ earlier neighbors and hence the greedy coloring uses at most $\Delta(G) + 1$ colors. Hence $\chi(G) \leq \Delta(G) + 1$. ■

For specific vertex ordering greedy coloring gives better bounds.

Proposition 19.2 [2] *Let G be a graph with degree sequence $d_1 \geq d_2 \geq \cdots \geq d_n$ and let $x_i = min(d_i, i-1)$. Then $\chi(G) \leq 1 + \max_i x_i$.*

Proof. We order the vertices of G in the nonincreasing order of the degrees. The greedy algorithm when applied to this ordering uses at most k colors, where $k = 1 + \max_i x_i$. ■

Proposition 19.3 *If G is a graph, then $\chi(G) \leq 1 + \max_{H \subseteq G} \delta(H)$.*

Proof. We order the vertices of G in such a way that $deg\ v_n = \delta$ and for $i < n$, v_i is a vertex of minimum degree in the induced subgraph $\langle G - \{v_{i+1}, v_{i+2}, \ldots, v_n\}\rangle$. Applying the greedy algorithm to this ordering, the result follows. ■

Definition 19.2 *A graph G is said to be k-critical if $\chi(G) = k$ and $\chi(G - v) = k - 1$ for all $v \in V$.*

Proposition 19.4 *If H is a k-critical graph, then $\delta(H) \geq k - 1$.*

Proof. Let u be a vertex of H. Since H is k-critical, $H - u$ is $(k-1)$-colorable. If $d_H(u) < k-1$, then there exists at least one color that does not appear on $N(u)$ and assigning this color to u, we obtain a proper $(k-1)$-coloring of H, which is a contradiction. Hence $d(u) \geq k - 1$ and $\delta(H) \geq k - 1$. ■

In fact Dirac [3] has proved that if G is k-critical, then $\kappa' \geq k - 1$, where κ' is the edge connectivity of G and since $\kappa' \leq \delta$, we get the above result as a corollary.

Definition 19.3 *Let S be a vertex cut of a connected graph G and let the components of $G - S$ have vertex sets $V_1, V_2, \ldots, V_r$. Then the induced subgraphs $G_i = G[V_i \cup S]$ are called the S-components of G.*

Theorem 19.2 *In a k-critical graph no vertex cut is a clique.*

Proof. Let G be a k-critical graph. Suppose G has a vertex cut S which is also a clique. Let $G_1, G_2, \ldots, G_r$ be the S-components of G. Since G is k-critical, it follows that each G_i is $(k-1)$-colorable and since S is a clique, the vertices of S receive distinct colors in any $(k-1)$-coloring of G. Hence, we can find $(k-1)$-colorings of $G_1, G_2, \ldots, G_r$ which agree on S. These coloring together give a $(k-1)$-coloring of G, which is a contradiction. ■

Corollary 19.1 *Every critical graph is a block.* ■

It follows from Theorem 19.2 that if G is k-critical and $S = \{u, v\}$ is a vertex cut of G, then u and v cannot be adjacent. An S-component G_1 of G is said to be of *type 1* if every $(k-1)$-coloring of G_1 assigns the same color to u and v and is of *type 2* if every $(k-1)$-coloring of G_1 assigns different colors to u and v.

Dirac proved the following theorem for k-critical graphs with a 2-vertex cut.

Theorem 19.3 [3] *Let G be a k-critical graph with a 2-vertex cut $\{u,v\}$. Then the following conditions are satisfied.*

  i. $G = G_1 \cup G_2$, where G_i is a $\{u,v\}$-component of type i ($i = 1, 2$).

  ii. *Both $G_1 + uv$ and $G_2 \cdot uv$ are k-critical, where $G_2 \cdot uv$ denotes the graph obtained from G_2 by identifying u and v.*

Proof.

  i. Since G is k-critical, each $\{u,v\}$-component of G is $(k-1)$-colorable. Further there cannot exist $(k-1)$-colorings of the $\{u,v\}$-components all of which agree on $\{u,v\}$, since such colorings together give a $(k-1)$-coloring of G. Hence, there exist two $\{u,v\}$-components G_1 and G_2 such that no $(k-1)$-coloring of G_1 agrees with any $(k-1)$-coloring of G_2.

  Hence one component say G_1 must be of type 1 and the other component say G_2 must be of type 2. Since G_1 and G_2 are of different types, the subgraph $G_1 \cup G_2$ of G is not $(k-1)$-colorable. Now, since G is critical, it follows that $G = G_1 \cup G_2$.

  ii. Now, let $H_1 = G_1 + uv$. Since G_1 is of type 1, $\chi(H_1) = k$. We shall prove that H_1 is k-critical, by showing that $\chi(H_1 - e) = k-1$ for every edge e of H_1. This is obviously true for $e = uv$. Now, let $e \in E(H_1) - uv$. Since G_2 is a subgraph of $G - e$, it follows that in any $(k-1)$-coloring of $G - e$, the vertices u and v must receive different colors. The restriction of such a coloring to the vertices of G_1 gives a $(k-1)$-coloring of $H_1 - e$. Thus H_1 is k-critical. We can prove by a similar argument that $G_2 \cdot uv$ is k-critical. ∎

The following theorem due to Brooks [4] shows that the bound given in Proposition 19.1 is attained only for complete graphs and odd cycles.

Theorem 19.4 [4] *If G is a connected graph other than a complete graph or an odd cycle, then $\chi(G) \leq \Delta(G)$.*

Proof. Let G be a connected graph that is neither an odd cycle nor a complete graph. Suppose that $\chi(G) = k$. We may assume that $k \geq 2$. Let H be a k-critical subgraph of G. Clearly H is nonseparable and $\Delta(H) \leq \Delta(G)$. Suppose that $H = K_k$ or that H is an odd cycle. Then $G \neq H$. Since G is connected, $\Delta(G) > \Delta(H)$. If $H = K_k$, then $\Delta(H) = k-1$ and $\Delta(G) \geq k$, so that $\chi(G) = k \leq \Delta(G)$. If H is an odd cycle, then $\Delta(G) > \Delta(H) = 2$, which implies that $\Delta(G) \geq 3 = k = \chi(G)$. Hence we may assume that H is a k-critical graph and is neither an odd cycle nor a complete graph. Then it follows that $k \geq 4$. Let H have order p. Since $\chi(H) = k \geq 4$ and H is not complete, it follows that $p \geq 5$. We now consider two cases depending on the connectivity of H.

Case i. H is 3-connected.

Let x and y be vertices of H such that $d_H(x,y) = 2$ and suppose that (x,w,y) is a path in H. Clearly the graph $H - \{x,y\}$ is connected. Let $x_1 = w, x_2, \ldots, x_{p-2}$ be the vertices of $H - \{x,y\}$, listed so that each vertex $x_i (2 \leq i \leq p-2)$ is adjacent to at least one vertex preceding it. Let $x_{p-1} = x$ and $x_p = y$. Now we assign the colors $1, 2, \ldots, \Delta(H)$ to the vertices of H as follows. Consider the sequence: $x_1 = w, x_2, \ldots, x_{p-2}, x_{p-1} = x, x_p = y$. Assign the color 1 to the vertices x_{p-1} and x_p. We successively color $x_{p-2}, x_{p-3}, \ldots, x_2$ with one of the colors $1, 2, \ldots, \Delta(H)$ that was not used in coloring adjacent vertices following it in the sequence. Such a color is available, since each x_i ($2 \leq i \leq p-2$) is adjacent to at most $\Delta(H)-1$ vertices following it in the sequence. Since $x_1 = w$ is adjacent to two vertices colored 1 (namely, x_{p-1} and x_p), a color is available for x_1. Hence $\chi(G) = \chi(H) \leq \Delta(H) \leq \Delta(G)$.

Case ii. $\kappa(H) = 2$.

Since H is k-critical, it follows by Proposition 19.4 that $\delta(H) \geq k-1$. Since $k \geq 4$, it follows that $\delta(H) \geq 3$. Since H is not complete, $\delta(H) \leq p-2$. Hence, there exists a vertex u in H such that $3 \leq deg_H(u) \leq p-2$. Suppose that $\kappa(H-u) = 2$. Let v be a vertex with $d_H(u,v) = 2$. We may let $x = u$ and $y = v$, and proceed as in Case (i).

On the other hand, if $\kappa(H-u) = 1$, then we consider two end-blocks B_1 and B_2 containing cut vertices w_1 and w_2, respectively, of $H - u$. Since H is 2-connected, there exist vertices u_1 in $B_1 - w_1$ and u_2 in $B_2 - w_2$ that are adjacent to u. Now let $x = u_1$ and $y = u_2$ and proceed as in Case (i). ∎

Our next result, due to Nordhaus and Gaddum [5] is the best known result on the sum and product of the chromatic numbers of a graph and its complement.

Theorem 19.5 [5] *If G is a graph of order n, then*

i. $2\sqrt{n} \leq \chi(G) + \chi(\overline{G}) \leq n+1$ *and*

ii. $n \leq \chi(G)\chi(\overline{G}) \leq \left(\frac{n+1}{2}\right)^2$.

Proof. Suppose that $\chi(G) = k$ and $\chi(\overline{G}) = l$. Let a k-coloring $\mathcal{C}$ of G and an l-coloring $\overline{\mathcal{C}}$ of $\overline{G}$ be given. Using these colorings, we obtain a coloring of K_n. With each vertex v of G, we associate the ordered pair $(\mathcal{C}(v), \overline{\mathcal{C}}(v))$. Since any two vertices of G are either adjacent in G or in $\overline{G}$, they are assigned different colors and thus this is a coloring of K_n using at most kl colors. Therefore,

$$n = \chi(K_n) \leq kl = \chi(G)\chi(\overline{G})$$

This establishes the lower bound in (ii). Since the geometric mean of two positive real numbers never exceeds their arithmetic mean, it follows that

$$\sqrt{n} \leq \sqrt{\chi(G)\chi(\overline{G})} \leq \frac{\chi(G) + \chi(\overline{G})}{2}$$

Consequently,

$$2\sqrt{n} \leq \chi(G) + \chi(\overline{G}),$$

which proves the lower bound in (i).

Now, to prove the upper bound in (i), let $p = \max_{H \subseteq G} \delta(H)$. Hence the minimum degree of every subgraph of G is at most p. By Proposition 19.3, $\chi(G) \leq 1 + p$.

We claim that the minimum degree of every subgraph of $\overline{G}$ is at most $n-p-1$. Assume, to the contrary, that there is a subgraph H of G such that $\delta(\overline{H}) \geq n-p$ for the subgraph $\overline{H}$ in $\overline{G}$. Thus every vertex of H has degree $p-1$ or less in G. Let F be a subgraph of G such that $\delta(F) = p$. So every vertex of F has degree p or more. This implies that no vertex of F belongs to H. Since the order of F is at least $p+1$, the order of H is at most $n - (p+1) = n-p-1$. This contradicts the fact that $\delta(\overline{H}) \geq n-p$. Thus, the minimum degree of every subgraph of $\overline{G}$ is at most $n-p-1$. By Proposition 19.3, $\chi(\overline{G}) \leq 1 + (n-p-1) = n-p$ and so

$$\chi(G) + \chi(\overline{G}) \leq (1+p) + (n-p) = n+1.$$

This gives the upper bound in (i). Now

$$\chi(G)\,\chi(\overline{G}) \leq \left(\frac{\chi(G) + \chi(\overline{G})}{2}\right)^2 \leq \left(\frac{n+1}{2}\right)^2.$$

∎

Definition 19.4 *If $\chi(G) = k$ and every k-coloring of G induces the same partition on $V(G)$, then G is called* uniquely k-colorable *or* uniquely colorable.

The complete graph K_n is uniquely n-colorable and $K_n - e$ where e is any edge of K_n, is uniquely $(n-1)$-colorable. Any connected bipartite graph is uniquely 2-colorable.

Theorem 19.6 *If G is uniquely k-colorable, then $\delta \geq k - 1$.*

Proof. Let v be any vertex of G. In any k-coloring of G, v must be adjacent with at least one vertex of every color different from that assigned to v. Otherwise, by recoloring v with a color which none of its neighbors is having, a different k-coloring can be achieved. Hence degree of v is at least $k - 1$ so that $\delta \geq k - 1$. ∎

Theorem 19.7 *Let G be a uniquely k-colorable graph. Then in any k-coloring of G, the subgraph induced by the union of any two color classes is connected.*

Proof. If possible, let C_1 and C_2 be two color classes in a k-coloring of G such that the subgraph induced by $C_1 \cup C_2$ is disconnected. We may assume that the vertices in C_1 are colored 1 and those in C_2 are colored 2. Let H be a component of the subgraph induced by $C_1 \cup C_2$. Obviously, no vertex of H is adjacent to a vertex in $V(G) - V(H)$ that is colored 1 or 2. Hence, interchanging the colors of the vertices in H and retaining the original colors for all other vertices, we get a different k-coloring for G. This gives a contradiction. ∎

Theorem 19.8 *Every uniquely k-colorable graph is $(k-1)$-connected.*

Proof. Let G be a uniquely k-colorable graph. Consider a k-coloring of G. If possible, let G be not $(k-1)$-connected. Hence there exists a set S of at most $k-2$ vertices such that $G - S$ is either trivial or disconnected. If $G - S$ is trivial, then G has at most $k - 1$ vertices, which is a contradiction. Hence $G - S$ has at least two components. In the considered k-coloring, there are at least two colors say c_1 and c_2 that are not assigned to any vertex of S.

If every vertex in a component of $G - S$ has color different from c_1 and c_2, then by assigning color c_1 to a vertex of this component, we get a different k-coloring of G. Otherwise, by interchanging the colors c_1 and c_2 in a component of $G - S$, a different k-coloring of G is obtained. In any case, G is not uniquely k-colorable, giving a contradiction.

Hence G is $(k-1)$-connected. ∎

Corollary 19.2 *In any k-coloring of a uniquely k-colorable graph G, the subgraph induced by the union of any r color classes, $2 \leq r \leq k$, is $(r-1)$-connected.*

Proof. If the subgraph H induced by the union of any r color classes, $2 \leq r \leq k$, had different r-colorings, then these r-colorings will induce different k-colorings for G giving a contradiction. Thus H is uniquely r-colorable. Hence H is $(r-1)$-connected. ∎

Theorem 19.9 *For any two graphs G and H, we have $\chi(G \square H) = \max(\chi(G), \chi(H))$, where $G \square H$ is the Cartesian product of G and H.*

Proof. Since the cartesian product $G\square H$ contains copies of G and H, it follows that $\chi(G\square H) \geq \max(\chi(G), \chi(H))$. Now, let $k = max(\chi(G), \chi(H))$. Let $g : V(G) \to \{1, 2, \ldots, \chi(G)\}$ and let $h : V(H) \to \{1, 2, \ldots, \chi(H)\}$ be proper colorings of G and H, respectively.

We now define $f : V(G) \times V(H) \to \{1, 2, \ldots, k\}$ by $f(u, v) = (g(u) + h(v))\ (mod\ k)$. We claim that f is a proper coloring of $G\square H$. Let (u, v) and (u', v') be two adjacent vertices. Without loss of generality we assume that $u = u'$ and $vv' \in E(H)$. Hence $g(u) = g(u')$ and $h(v) \neq h(v')$. Hence $(g(u) + h(v))\ (mod\ k) \neq (g(u') + h(v'))\ (mod\ k)$. Thus f is a proper coloring of $G\square H$ and hence $\chi(G\square H) = k$. ∎

The study of graph coloring problems has its root at the famous *four color conjecture,* which states that every map can be colored with four colors, subject to the usual condition that two regions which share a common boundary must be assigned different colors. This problem, from its first appearance in 1852 until its eventual solution in 1977, has been the motivation for the developments of several concepts of graph colorings. There are several nice historical accounts of the four color conjecture, see for example [6,7].

Given a map, we can define a graph G whose vertex set is the set of all regions of the map (including the exterior region) and two vertices are joined by an edge if the corresponding regions share a common boundary. Such a graph is obviously a planar graph and hence the four color conjecture is equivalent to the statement that every planar graph is 4-colorable. Several erroneous proofs of the four color problem have been reported in the literature. The error in the first proof of Kempe [8] was pointed out by Heawood [9].

The final successful proof of the four color theorem by Appel and Haken [10] is the first example of a mathematical proof relying heavily on the use of computers. Gardner [11] remarked that the proof of the four color theorem is an extraordinary achievement and however to most of the mathematicians the proof is deeply unsatisfactory. Upto this date we don't have a short proof of the four color theorem in which all the details can be checked by hand.

Heawood [9] showed that one can always color the vertices of a planar graph with at most five colors. This is known as the five color theorem.

Theorem 19.10 *Every planar graph is 5-colorable.*

Proof. We prove the theorem by induction on the number n of vertices. For any planar graph having $n \leq 5$ vertices, the result is obvious since the graph is n-colorable.

Now let us assume that all planar graphs with n vertices is 5-colorable for some $n \geq 5$. Let G be a planar graph with $n + 1$ vertices. Then G has a vertex v of degree 5 or less. By induction hypothesis, the planar graph $G - v$ is 5-colorable. Consider a 5-coloring of $G - v$ where c_i, $1 \leq i \leq 5$, are the colors used. If some color, say c_j is not used in coloring vertices adjacent to v, then by assigning the color c_j to v the 5-coloring of $G - v$ can be extended to a 5-coloring of G.

Hence we have to consider only the case in which $deg(v) = 5$ and all the five colors are used for coloring the vertices of G adjacent to v.

Let v_1, v_2, v_3, v_4 and v_5 be the vertices adjacent to v colored c_1, c_2, c_3, c_4 and c_5, respectively. Let G_{13} denote the subgraph of $G - v$ induced by those vertices colored c_1 or c_3. If v_1 and v_3 belong to different components of G_{13}, then a 5-coloring of $G - v$ can be obtained by interchanging the colors of vertices in the component of G_{13} containing v_1. (Since no vertex of this component is adjacent to a vertex with color c_1 or c_3 outside this component, this interchange of colors results in a coloring of $G - v$.) In this 5-coloring no vertex adjacent to v is colored c_1 and hence by coloring v with c_1, a 5-coloring of G is obtained.

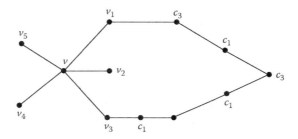

Figure 19.2 The graph in the proof of Theorem 19.10

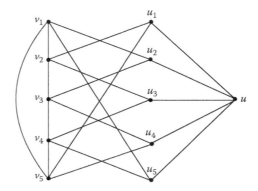

Figure 19.3 Mycielski's construction for the cycle C_5.

If v_1 and v_3 are in the same component of G_{13}, then in G there exists a $v_1 - v_3$ path all of whose vertices are colored c_1 or c_3. Hence there is no $v_2 - v_4$ path in $G - v$ all whose vertices are colored c_2, c_4 (refer Figure 19.2).

Hence if G_{24} denotes the subgraph of $G - v$ induced by the vertices colored c_2 or c_4, then v_2 and v_4 belong to different components of G_{24}. Hence if we interchange the colors of the vertices in the component of G_{24} containing v_2, a new 5-coloring $G - v$ results and in this, no vertex adjacent to v is colored c_2. Hence by assigning color c_2 to v, we get a 5-coloring of G. This completes the induction and the proof. ∎

Recall that the *clique number* $\omega(G)$ of a graph G is the order of a largest complete subgraph of G. Obviously for every graph G, $\chi(G) \geq \omega(G)$. Hence if G contains a triangle, then $\chi(G) \geq 3$. However, there exist triangle-free graphs with $\chi(G) \geq 3$. For example, the odd cycles C_{2k+1}, with $k \geq 2$, have chromatic number 3 and are, of course, triangle-free.

It may be surprising that there exist triangle-free graphs with arbitrarily large chromatic number. The following construction is due to Mycielski [12].

Theorem 19.11 *For every positive integer k, there exists a k-chromatic triangle-free graph.*

Proof. Since no graph with chromatic number 1 or 2 contains a triangle, the theorem is obviously true for $k = 1$ and $k = 2$. To verify the theorem for $k \geq 3$, we proceed by induction on k. Since $\chi(C_5) = 3$ and C_5 is triangle-free, the statement is true for $k = 3$.

Assume that there exists a triangle-free graph with chromatic number k, where $k \geq 3$. We show that there exists a triangle-free $(k+1)$-chromatic graph. Let H be a triangle-free graph with $\chi(H) = k$, where $V(H) = \{v_1, v_2, \ldots, v_n\}$. We construct a graph G from H by adding $n+1$ new vertices $u, u_1, u_2, \ldots, u_n$, joining u to each vertex $u_i (1 \leq i \leq n)$ and joining u_i to each neighbor of v_i in H (see Figure 19.3).

We claim that G is a triangle-free $(k+1)$-chromatic graph. First, we show that G is triangle-free. Since $S = \{u_1, u_2, \ldots, u_n\}$ is an independent set of vertices of G and u is adjacent to no vertex of H, it follows that u belongs to no triangle in G. Hence if there is a triangle T in G, then two of the three vertices of T must belong to H and the third vertex must belong to S, say $V(T) = \{u_i, v_j, v_k\}$. Since u_i is adjacent v_j and v_k, it follows that v_i is adjacent to v_j and v_k. Since v_j and v_k are adjacent, H contains a triangle, which is a contradiction. Hence G is triangle-free.

Next, we show that $\chi(G) = k + 1$. Since H is a subgraph of G and $\chi(H) = k$, it follows that $\chi(G) \geq k$. Let a k-coloring of H be given. Now assign the color to u_i the same color that is assigned to v_i for $1 \leq i \leq n$ and assign the color $k + 1$ to u. This gives a $(k + 1)$-coloring of G and so $\chi(G) \leq k + 1$. Hence either $\chi(G) = k$ or $\chi(G) = k + 1$. Suppose that $\chi(G) = k$. Then there is a k-coloring of G with colors $1, 2, \ldots, k$, where u is assigned the color k, say. Necessarily, none of the vertices $u_1, u_2, \ldots, u_n$ is assigned the color k; that is, each vertex of S is assigned one of the colors $1, 2, \ldots, k - 1$. Since $\chi(H) = k$, one or more vertices of H are assigned the color k. For each vertex v_i of H colored k, recolor it with the color assigned to u_i. This produces a $(k-1)$-coloring of H, which is a contradiction. Thus $\chi(G) = k + 1$. ∎

The above result has been extended significantly by Erdös [13] and Lovász [14].

Theorem 19.12 *For any two integers $k \geq 2$ and $l \geq 3$, there exists a k-chromatic graph G with $g(G) \geq l$, where $g(G)$ is the girth of G.* ∎

Birkhoff [15] introduced chromatic polynomials as a possible means of attacking the four color conjecture. This concept considers the number of ways of coloring a graph with a given number of colors.

Let G be a labeled graph. A λ-coloring of G is a coloring of G which uses λ or fewer colors. Two λ-colorings of G will be considered different if at least one of the labeled vertices is assigned different colors. Let $f(G, \lambda)$ denote the number of different λ-colorings of G.

For example, $f(K_1, \lambda) = \lambda$ and $f(\overline{K_2}, \lambda) = \lambda^2$. If one could prove that $f(G, 4) > 0$ for a planar graph, then this would give a positive answer to the four color problem.

Theorem 19.13 $f(K_n, \lambda) = \lambda(\lambda - 1) \cdots (\lambda - n + 1)$.

Proof. The first vertex in K_n can be colored in λ different ways (as there are λ colors). For each coloring of the first vertex, the second vertex can be colored in $\lambda - 1$ ways (as there are $\lambda - 1$ colors remaining). For each coloring of the first two vertices, the third vertex can be colored in $\lambda - 2$ ways and so on. Hence $f(K_n, \lambda) = \lambda(\lambda - 1) \cdots (\lambda - n + 1)$. ∎

Remark 19.1 $f(\overline{K_n}, \lambda) = \lambda^n$, *since each of the n vertices of $\overline{K_n}$ may be colored independently in λ ways.*

Theorem 19.14 *If G is a graph with k components $G_1, G_2, \ldots, G_k$, then $f(G, \lambda) = \prod_{i=1}^{k} f(G_i, \lambda)$.*

Proof. Number of ways of coloring G_i with λ colors is $f(G_i, \lambda)$. Since any choice of λ-colorings for $G_1, G_2, \ldots, G_k$ can be combined to give a λ-coloring for G, $f(G, \lambda) = \prod_{i=1}^{k} f(G_i, \lambda)$. ∎

Definition 19.5 *Let u and v be two vertices in a graph G. The graph obtained from G by the removal of u and v and the addition of a new vertex w adjacent to those vertices to which u or v was adjacent is called an* elementary homomorphism *of G. Thus an elementary homomorphism of G is obtained by identification of two vertices of G.*

Theorem 19.15 *If u and v are nonadjacent vertices in a graph G and hG denotes the elementary homomorphism of G which identifies u and v, then $f(G, \lambda) = f(G + uv, \lambda) + f(hG, \lambda)$ where $G + uv$ denotes the graph obtained from G by adding the edge uv.*

Proof.

$$\begin{aligned} f(G, \lambda) &= \text{number of } \lambda\text{-colorings of } G \\ &= (\text{number of } \lambda\text{-colorings of } G \text{ in which } u \text{ and } v \text{ get different colors}) \\ &\quad + (\text{number of } \lambda\text{-colorings of } G \text{ in which } u \text{ and } v \text{ get the same color}) \\ &= (\text{number of } \lambda\text{-colorings of } G + uv) + (\text{number of } \lambda\text{-colorings of } hG) \\ &= f(G + uv, \lambda) + f(hG, \lambda). \end{aligned}$$

∎

Corollary 19.3 *Let G be a graph and let $e = uv$ be an edge of G. Then $f(G, \lambda) = f(G - e, \lambda) - f(hG, \lambda)$.* ∎

Theorem 19.16 *Let G be a graph of order n and size m. Then $f(G, \lambda)$ is a polynomial in λ of degree n with leading coefficient 1. Further the coefficient of λ^{n-1} is $-m$, the constant term is zero and the coefficients of $f(G, \lambda)$ alternate in sign.*

Proof. The proof is by induction on m. If $m = 0$, then $G = \overline{K_n}$ and $f(G, \lambda) = \lambda^n$, which has the desired properties. We now assume that the result is true for all graphs whose size is less than m, where $m \geq 1$. Let G be a graph of size m and let $e = uv$ be an edge of G. By Corollary 19.3, we have

$$f(G, \lambda) = f(G - e, \lambda) - f(hG, \lambda). \tag{19.1}$$

Now by induction hypothesis we have

$$\begin{aligned} f(G - e, \lambda) &= \lambda^n - (m-1)\lambda^{n-1} + a_2\lambda^{n-2} - \cdots + (-1)^i a_i \lambda^{n-i} + \ldots \\ &\quad + (-1)^{n-1} a_{n-1} \lambda \end{aligned}$$

and

$$f(hG, \lambda) = \lambda^{n-1} - b_1 \lambda^{n-2} + \cdots + (-1)^{i-1} b_{i-1} \lambda^{n-i} + \cdots + (-1)^{n-2} b_{n-2} \lambda.$$

Hence from (19.1), we have

$$\begin{aligned} f(G, \lambda) &= \lambda^n - m\lambda^{n-1} + (a_2 + b_1)\lambda^{n-2} + \cdots + (-1)^i (a_i + b_{i-1})\lambda^{n-i} + \ldots \\ &\quad + (-1)^{n-1}(a_{n-1} + b_{n-2})\lambda. \end{aligned}$$

Thus $f(G, \lambda)$ has all the desired properties and the theorem follows by induction. ∎

The polynomial $f(G, \lambda)$ is called the *chromatic polynomial* of G. We observe that the chromatic polynomial of small graphs can be computed by repeated application of Theorem 19.15.

The following theorem shows that all trees of order n have the same chromatic polynomial.

Theorem 19.17 *Let G be a graph of order n, $n \geq 2$. Then G is a tree if and only if $f(G, \lambda) = \lambda(\lambda - 1)^{n-1}$.*

Proof. We prove the result by induction on n. For $n = 2$, $G = K_2$ and hence $f(G, \lambda) = f(K_2, \lambda) = \lambda(\lambda - 1)$ and the theorem holds. Assume that the chromatic polynomial of any tree with $n-1$ vertices is $\lambda(\lambda-1)^{n-2}$. Let G be a tree with n vertices. Let v be an end vertex of G and let u be the unique vertex of G adjacent to v. By hypothesis, the tree $G - v$ has $\lambda(\lambda - 1)^{n-2}$ for its chromatic polynomial. The vertex v can be assigned any color different from that assigned to u. Hence v may be colored in $\lambda - 1$ ways for each coloring of $G - v$. Thus, $f(G, \lambda) = (\lambda - 1)f(G - v, \lambda) = (\lambda - 1)\lambda(\lambda - 1)^{n-2} = \lambda(\lambda - 1)^{n-1}$.

Conversely, let G be a graph with $f(G, \lambda) = \lambda(\lambda - 1)^{n-1}$. It follows from Theorem 19.16 that the order of G is n and the size of G is $n - 1$. Further since λ^2 is not a factor of $f(G, \lambda)$, it follows that G is connected. Hence G is a tree. ∎

Two graphs are called *chromatically equivalent* if they have the same chromatic polynomial. It follows from Theorem 19.16 that two chromatically equivalent graphs must have the same order and same size. Also it follows from Theorem 19.17 that any two trees of the same order are chromatically equivalent. A graph G is *chromatically unique*, if $f(H, \lambda) = f(G, \lambda)$ implies that the graphs G and H are isomorphic. It is not known under what conditions two graphs are chromatically equivalent or a given graph is chromatically unique. Several authors have investigated the distribution of the roots of the chromatic polynomials both on the real line and in the complex plane. For a comprehensive treatment chromatically equivalent graphs, chromatically unique graphs and the distributions of the roots of chromatic polynomials we refer to the book by Dong et al. [16].

We have already seen that the clique number $\omega(G)$ is a lower bound for the chromatic number $\chi(G)$. There are many graphs such as K_n, $\overline{K_n}$, and bipartite graphs for which $\chi(G) = \omega(G)$. Also there are many graphs whose chromatic number is larger than the clique number. In fact Mycielski's theorem proves the existence of graphs with large chromatic number and having $\omega(G) = 2$. In this context Berge [17] introduced the concept of perfect graphs.

Definition 19.6 *A graph G is called* perfect *if $\chi(H) = \omega(H)$ for every induced subgraph H of G.*

Clearly the complete graph K_n and the graph $\overline{K_n}$ are perfect. Also if G is a bipartite graph and H is an induced subgraph of G then $\chi(H) = \omega(H)$. Hence G is perfect.

Berge conjectured that a graph is perfect if and only if its complement is perfect. Lovász [18] proved that this conjecture is true and the result is known as the perfect graph theorem.

Theorem 19.18 (Perfect graph theorem) *A graph is perfect if and only if its complement is perfect.* ∎

It follows from Theorem 19.18 that the complement of a bipartite graph is perfect. We now give another family of perfect graphs.

Definition 19.7 *A graph G with $V(G) = \{v_1, v_2, \ldots, v_n\}$ is an* interval graph *if there exists a collection of n intervals $[a_i, b_i]$, where $a_i < b_i$, such that v_i and v_j are adjacent if and only if $[a_i, b_i] \cap [a_j, b_j] \neq \emptyset$.*

Clearly every induced subgraph of an interval graph is an interval graph.

Theorem 19.19 *Every interval graph is perfect.*

Proof. Let G be an interval graph with $V(G) = \{v_1, v_2, \ldots, v_n\}$. Let $\{[a_i, b_i] : 1 \leq i \leq n\}$ be n closed intervals such that v_i is adjacent to v_j if and only if $I_i \cap I_j \neq \emptyset$, where $I_i = [a_i, b_i]$.

We order the intervals I_i in such a way that $a_1 \leq a_2 \leq \cdots \leq a_n$. Let k be the number of colors used for the vertex coloring of G by a greedy algorithm with respect to the above ordering of the vertices of G. If $k = 1$, then $G = \overline{K_n}$ and hence G is perfect.

Suppose $k \geq 2$. Let v_t be the vertex which has been assigned the color k. This means that the interval $I_t = [a_t, b_t]$ has nonempty intersection with $k-1$ intervals $I_{j_1}, I_{j_2}, \ldots, I_{j_{k-1}}$, where $1 \leq j_1 \leq j_2 \leq \cdots \leq j_{k-1} \leq t$. Hence $a_{j_1} \leq a_{j_2} \leq \cdots \leq a_{j_{k-1}} \leq a_t$. Since $I_{j_i} \cap I_t \neq \emptyset$, for $1 \leq i \leq k-1$, it follows that $a_t \in I_{j_1} \cap I_{j_2} \cap \cdots \cap I_{j_{k-1}} \cap I_t$. Hence the induced subgraph $G[U] = K_k$, where $U = \{v_{j_1}, v_{j_2}, \ldots, v_{j_{k-1}}, v_t\}$. Thus $\chi(G) \leq k \leq \omega(G)$. Since $\chi(G) \geq \omega(G)$, we have $\chi(G) = \omega(G)$. Further since any induced subgraph H of G is also an interval graph, we have $\chi(H) = \omega(H)$. Thus G is perfect. ∎

There is another important class of perfect graphs. A *chord* of a cycle C in a graph G is an edge that joins two nonconsecutive vertices of C. A graph is a *chordal graph* if every cycle of length at least 4 has a chord.

It has been proved independently by Hajnal and Surányi [19] and Dirac [20] that every chordal graph is perfect.

Lovász [21] proved that from a given perfect graph one can construct a class of perfect graphs.

Definition 19.8 *Let G be a graph and let $v \in V(G)$. The replication graph $R_v(G)$ of G with respect to v is the graph obtained from G by adding a new vertex v' and joining v' to every vertex in the closed neighborhood $N[v]$.*

Theorem 19.20 [21] *Let G be a graph and let $v \in V(G)$. If G is perfect, then $R_v(G)$ is perfect.* ∎

Berge proposed the following deeper conjecture.

Strong perfect graph conjecture: A graph G is perfect if and only if neither G nor $\overline{G}$ contains an induced odd cycle of length 5 or more.

Chudnovsky et al. [22] proved the above conjecture which is now known as the strong perfect graph theorem.

Though the complete graph K_k need not be present in a graph G with $\chi(G) = k$, it has been conjectured that K_k may be indirectly present in G.

Clearly K_k is present in a graph G with $\chi(G) = k$ for $k = 1, 2$. Also if $\chi(G) = 3$, then G contains an odd cycle which can be thought of as a subdivision of K_3. Dirac [23] showed that a similar result is true for graphs with $\chi(G) = 4$.

Theorem 19.21 *Any 4-chromatic graph contains a subdivision of K_4.*

Proof. Let G be a 4-chromatic graph. We may assume without loss of generality that G is critical. If $n = 4$, then $G = K_4$ and the theorem is trivially true. We now proceed by induction on n. We assume that the theorem is true for all 4-chromatic graphs of order less than n. Let G be a 4-chromatic graph of order n, with $n \geq 5$.

If G has a 2-vertex cut $\{u, v\}$, then by Theorem 19.3, G has two $\{u, v\}$-components G_1 and G_2 such that $G_1 + uv$ is 4-critical. By induction hypothesis $G_1 + uv$ contains a subdivision of K_4.

Now, let P be a $u - v$ path in G_2. Then $G_1 \cup P$ contains a subdivision of K_4. Since $G_1 \cup P$ is a subgraph of G, it follows that G contains a subdivision of K_4.

Now, suppose that G is 3-connected. Since $\delta \geq 3$, G has a cycle C of length at least 4. Let u and v be two nonconsecutive vertices on C. Since $G - \{u, v\}$ is connected, there exists a path P in $G - \{u,v\}$ connecting the two components of $C - \{u,v\}$. We assume that the origin x and terminus y are the only vertices of P on C. Similarly there exists a path Q in $G - \{x,y\}$. If P and Q have no common vertex, then $C \cup P \cup Q$ is a subdivision of K_4. Otherwise let w be the first vertex in P on Q and let P' be the $x - w$ section of P. Now $C \cup P' \cup Q$ is a subdivision of K_4. Hence in both cases G contains a subdivision of K_4. ∎

Thus it follows that if $2 \leq k \leq 4$, then every k-chromatic graph contains a subdivision of K_k. Hajós [24] proposed the following conjecture.

Hajós conjecture: If G is a k-chromatic graph where $k \geq 2$, then G contains a subdivision of K_k.

Catlin [25] showed that Hajós conjecture is false for $k \geq 7$. Hajós conjecture is still open for $k = 5$ or 6.

Thomassen [26] showed that there is a connection between perfect graphs and Hajós conjecture.

Theorem 19.22 *A graph G is perfect if and only if every replication of G satisfies Hajós conjecture.* ∎

Definition 19.9 *A graph H is called a* minor *of a graph G if H can be obtained from G by a sequence of contractions, edge deletions, and vertex deletions in any order.*

If a graph G contains a subdivision of a graph H, then H is a minor of G. In particular if G is a k-chromatic graph containing a subdivision of K_k, then K_k is a minor of G. In this context Hadwiger proposed the following conjecture much earlier than Hajós conjecture.

Hadwiger's conjecture: Every k-chromatic graph contains K_k as a minor.

This conjecture first appeared in [27]. Wagner [28] proved that Hadwiger's conjecture for $k = 5$ is equivalent to the four color conjecture and hence Hadwiger's conjecture can be considered as a generalization of the four color conjecture. Robertson et al. [29] have proved that Hadwiger's conjecture is true for $k = 6$ by using the four color theorem. The conjecture remains open for $k > 6$.

19.3 EDGE COLORINGS

In this section we consider the problem of coloring the edges of a graph.

Definition 19.10 *An* edge coloring *of a graph G is an assignment of colors to the edges of G such that adjacent edges are colored differently. A graph G is k-edge colorable if there exists a l-edge coloring of G for some $l \leq k$. The minimum k for which a graph G is k-edge colorable is called the* edge-chromatic number *(or* chromatic index*) of G and is denoted by $\chi'(G)$.*

It follows immediately from the definition that $\chi'(G) = \chi(L(G))$ where $L(G)$ is the line graph of G. Also if $\deg v = \Delta$, then the Δ edges incident at v must receive distinct colors in any edge coloring of G and hence $\chi'(G) \geq \Delta(G)$. The following fundamental theorem by Vizing [30] shows that $\chi'(G) = \Delta(G)$ or $\Delta(G) + 1$.

Let $\mathcal{C}$ be a k-edge coloring of G and let $v \in V$. We say that a color c is represented at v if an edge incident at v is assigned the color c; otherwise c is missing at v.

Theorem 19.23 [30] *For any simple graph G, $\chi'(G) \leq 1 + \Delta(G)$.*

Proof. The proof is by induction on m.

The result is trivial if $m = 1$. Assume that the result is true for all graphs G with $|E(G)| < m$ and let G be a graph with $|E(G)| = m \geq 2$ and let $\Delta(G) = \Delta$. Let uv_0 be an edge of G. By induction hypothesis $G - uv_0$ admits a $(\Delta + 1)$-edge coloring and let $\{1, 2, \ldots, \Delta + 1\}$ be the colors used. Since v_0 is incident to at most $\Delta - 1$ edges in $G - uv_0$, there exists a color c_1 missing at v_0. If c_1 is also missing at u, then by coloring uv_0 with c_1, we get a $(\Delta + 1)$-edge coloring of G. Suppose there exists an edge uv_1 with color c_1. Since v_1 is incident to at most Δ edges, a color c_2 is missing at v_1. If c_2 is missing at u, then recoloring uv_1 with c_2 and coloring uv_0 with c_1, we obtain a $(\Delta + 1)$-edge coloring of G. Hence we may assume that there exists an edge uv_2 with color c_2. Continuing this process, we obtain a sequence $v_1, v_2, \ldots$ of neighbors of u and a sequence of colors $c_1, c_2, \ldots$ such that uv_i is colored c_i and c_{i+1} is missing at v_i. Since the degree of u is finite, there exists a least positive integer l such that $c_{l+1} = c_k$ for some $k < l$. Now, for $0 \leq i \leq k - 1$, we recolor the edge uv_i with c_{i+1}. There exists a color c_0 missing at u and $c_0 \neq c_k$. Let P be a maximal path with origin v_{k-1} with edges alternately colored c_0 and c_k and we interchange colors c_0 and c_k on $P + uv_{k-1}$. If P does not contain v_k, we have a $(\Delta + 1)$-edge coloring of G. If P contains v_k, then recoloring the edge uv_i with c_{i+1} for $k \leq i \leq l$, we get a $(\Delta + 1)$-edge coloring of G. ∎

In view of Theorem 19.23, the set of all nonempty graphs can be divided naturally into two classes.

A nonempty graph G is said to be of *class 1* or *class 2* accordingly as $\chi'(G) = \Delta(G)$ or $\chi'(G) = \Delta(G) + 1$. The natural problem is to determine whether a given graph is of class 1 or class 2 and this problem is NP-complete [31]. However, there are classes of graphs for which we know if they are of class 1 or class 2. The cycle C_n is of class 1 if n is even and is of class 2 if n is odd. Also the complete graph K_n is of class 1 if n is even and is of class 2 if n is odd.

The following theorem due to König [32] shows that any nonempty bipartite graph is of class 1.

Theorem 19.24 (König's theorem) *If G is a nonempty bipartite graph, then $\chi'(G) = \Delta(G)$.*

Proof. Suppose that the theorem is false. Then among the counter examples, let G be one of minimum size. Thus G is a bipartite graph such that $\chi'(G) = \Delta(G) + 1$ and $\chi'(G - e) = \Delta(G - e)$ for all $e \in E(G)$. Now, let $e = uv$. Then $\Delta(G - e) = \Delta(G)$, for otherwise G is $\Delta(G)$-edge colorable. Let $\mathcal{C}$ be a $\Delta(G)$-edge coloring of $G - e$. If there exists a color c which is missing at both u and v, then c can be assigned to e, thus giving a $\Delta(G)$-edge coloring of G, a contradiction. Thus, every color is represented either at u or at v. Now, since $deg_{G-e}(u) < \Delta(G)$ and $deg_{G-e}(v) < \Delta(G)$, there is a color α and a color β such that α is missing at u and β is missing at v. Clearly, $\alpha \neq \beta$ and α is represented at v and β is represented at u.

Let P be a path of maximum length having initial vertex v whose edges are alternately colored α and β. The path P cannot contain u, for otherwise P has odd length, implying that the initial and terminal edges of P are both colored α, a contradiction. Now interchanging the colors α and β of the edges of P gives a new $\Delta(G)$-edge coloring $\mathcal{C}'$ of $G - e$ in which α is missing at both u and v. Now, assign the color α to the edge e in G and this gives a $\Delta(G)$-coloring of G, which is a contradiction. ∎

Beineke and Wilson [33] obtained a simple sufficient condition for a graph to be of class 2.

Theorem 19.25 *Let G be a graph of size m. If $m > \Delta(G)\beta_1(G)$, where $\beta_1(G)$ is the edge independence number of G, then G is of class two.*

Proof. Suppose G is of class 1, so that $\chi'(G) = \Delta(G)$. Let $\{E_1, E_2, \ldots, E_\Delta\}$ be an edge coloring of G. Since $|E_i| \leq \beta_1(G)$, it follows that $m \leq \Delta(G)\beta_1(G)$, which is a contradiction. Thus G is of class 2. ∎

A graph G of order n and size m is called *overfull* if $m > \Delta(G) \cdot \lfloor n/2 \rfloor$.

Since $\beta_1(G) \leq \lfloor n/2 \rfloor$, it follows from Theorem 19.25 that every overfull graph is of class two.

Hilton [34] and Chetwynd and Hilton [35] conjectured that a graph G of order n with $\Delta(G) > n/3$ is of class two if and only if G contains an overfull subgraph H with $\Delta(G) = \Delta(H)$.

As in the case of vertex colorings, the concept of minimal graphs with respect to edge colorings is quite useful. A graph G with at least two edges is χ'-*minimal* if $\chi'(G - e) = \chi'(G) - 1$ for every edge e of G. Vizing [36] obtained several fundamental results on χ'-minimal graphs.

Theorem 19.26 [36] *Let G be a connected graph of class two that is minimal with respect to edge-chromatic number. Then every vertex of G is adjacent to at least two vertices of degree $\Delta(G)$. In particular, G contains at least three vertices of degree $\Delta(G)$.* ∎

Theorem 19.27 [36] *Let G be a connected graph of class two that is minimal with respect to edge-chromatic number. If u and v are adjacent vertices with $\deg(u) = k$, then v is adjacent to at least $\Delta(G) - k + 1$ vertices of degree $\Delta(G)$.* ∎

Theorem 19.28 [36] *Let G be a connected graph with $\Delta(G) = d \geq 2$. Then G is minimal with respect to edge-chromatic number if and only if either*

i. *G is of class one and $G = K_{1,d}$ or*

ii. *G is of class two and $G - e$ is of class one for every edge e of G.* ∎

19.4 OTHER VARIANTS OF GRAPH COLORINGS

19.4.1 Complete Colorings and Achromatic Number

Let G be a graph with $\chi(G) = k$ and let $\{V_1, V_2, \ldots, V_k\}$ be a k-coloring of G. Then for any two distinct color classes V_i and V_j, there exist two adjacent vertices u and v such that $u \in V_i$ and $v \in V_j$; since otherwise the color classes V_i and V_j can be merged into a single color class, giving a $(k-1)$-coloring of G. This motivates the concept of complete coloring, which was introduced by Harary et al. [37].

Definition 19.11 *A* complete coloring *of a graph G is a proper coloring of G having the property that for any distinct colors i and j there exist adjacent vertices u, v of G such that u is colored i and v is colored j.*

A complete coloring using k-colors is called a complete k-coloring *of G. The largest positive integer k for which G has a complete k-coloring is called the* achromatic number *of G and its denoted by $\psi(G)$.*

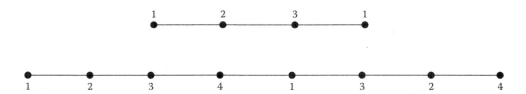

Figure 19.4 Complete 3-coloring of P_4 and a complete 4-coloring of P_8.

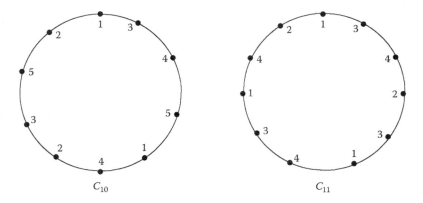

Figure 19.5 Complete colorings of the cycles C_{10} and C_{11}.

Trivially if $\chi(G) = k$, then any k-coloring of G is a complete coloring and hence $\psi(G) \geq \chi(G)$. For the complete graph K_n, we have $\psi(K_n) = \chi(K_n) = n$. However, a graph G may admit a complete k-coloring, where $k > \chi(G)$.

For example, a complete 3-coloring of P_4 and a complete 4-coloring of P_8 are given in Figure 19.4. Hell and Miller [38] determined the achromatic number of paths and cycles.

Theorem 19.29 [38] *For each $n \geq 2$, $\psi(P_n) = \max\{k : (\lfloor k/2 \rfloor)(k-2) + 2 \leq n\}$.* ∎

It follows from Theorem 19.29 that, $\psi(P_4) = 3$ and $\psi(P_8) = 4$.

Theorem 19.30 [38] *For each $n \geq 3$, $\psi(C_n) = \max\{k : k\lfloor k/2 \rfloor \leq n\} - s(n)$, where $s(n)$ is the number of positive integer solutions of $n = 2x^2 + x + 1$.* ∎

It follows from Theorem 19.30 that, $\psi(C_{10}) = 5$ and $\psi(C_{11}) = 4$ and the corresponding complete colorings are given in Figure 19.5.

Theorem 19.31 *For the complete bipartite graph $G = K_{r,s}$, we have $\psi(G) = 2$.*

Proof. Let $\psi(G) = k$. Clearly $k \geq 2$. Suppose $k \geq 3$. Then in any complete k-coloring of G, there exist two vertices in the same partite set receiving distinct colors i and j. Now no vertex in the other partite set can have color i or j and hence there does not exist adjacent vertices with colors i and j, a contradiction. ∎

There exists a close relation between the achromatic number and the concept of homomorphism. A *homomorphism* from a graph G to a graph G' is a function $\phi : V(G) \to V(G')$ such that $uv \in E(G)$ implies $\phi(u)\phi(v) \in E(G')$.

A graph H is called a *homomorphic image* of G if there exists a homomorphism ϕ of G onto H. There is an alternative way to obtain homomorphic images of a graph G. Obviously the only homomorphic image of the complete graph K_n is K_n. If G is not complete, then an elementary homomorphism of G is obtained by identifying two nonadjacent vertices u and

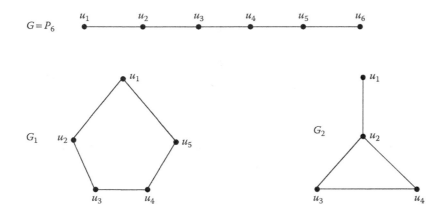

Figure 19.6 Homomorphic images of P_6.

v of G and the resulting graph G' is a homomorphic image of G. The vertex obtained by identifying u and v is denoted by either u or v.

Alternatively the mapping $\varepsilon : V(G) \to V(G')$ defined by

$$\varepsilon(w) = \begin{cases} w, & \text{if } w \in V(G) - \{u,v\} \\ v, & \text{if } w \in \{u,v\}. \end{cases}$$

is an elementary homomorphism from G to G'. Now any homomorphic image H of a graph G is obtained from G by a sequence of elementary homomorphisms. Some homomorphic images of the graph $G = P_6$ are given in Figure 19.6.

Since each homomorphic image of a graph G can be obtained from G by a sequence of elementary homomorphisms, it follows that each homomorphic image H of G can be obtained from a partition $\Pi = \{V_1, V_2, \ldots, V_k\}$ of $V(G)$ into independent sets such that $V(H) = \{v_1, v_2, \ldots, v_k\}$ and v_i is adjacent to v_j in H if and only if some vertex of V_i is adjacent to some vertex of V_j in G. In particular if Π is a complete k-coloring of G, then $H = K_k$. Thus, the chromatic number $\chi(G)$ and the achromatic number $\psi(G)$ are respectively the smallest and the largest positive integer k, such that K_k is a homomorphic image of G.

It is NP-hard to determine the achromatic number even for trees [39] but polynomial time solvable for trees of bounded degree [40].

19.4.2 Grundy Colorings in Graphs

The concepts introduced by Grundy [41] while dealing with combinatorial games led to the concept of Grundy coloring.

Definition 19.12 *A* Grundy coloring *of a graph G is a proper vertex coloring of G (the colors are taken as positive integers) having the property that for every two colors i and j with $i < j$, every vertex with color j has a neighbor with color i. The maximum positive integer k for which G has a Grundy k-coloring is called the* Grundy chromatic number *of G or simply the* Grundy number *of G and it is denoted by $\Gamma(G)$.*

Since any Grundy coloring is both a proper coloring and a complete coloring, it follows that $\chi(G) \leq \Gamma(G) \leq \psi(G)$. For the corona $G = C_5 \circ K_1$, a proper 3-coloring, a Grundy 4-coloring and a complete 5-coloring are given in Figure 19.7.

In fact $\chi(G) = 3$, $\Gamma(G) = 4$ and $\psi(G) = 5$.

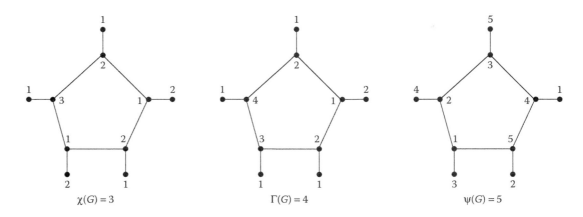

Figure 19.7 Complete and Grundy colorings.

For the complete bipartite graph $G = K_{r,s}$, we have $\Gamma(G) = 2$. In fact complete bipartite graphs are the only connected graphs with Grundy number 2.

The algorithmic complexity of finding the Grundy number has not been determined so far.

Christen and Selkow [42] have proved that given a graph G and an integer k with $\chi(G) \leq k \leq \Gamma(G)$, there is a Grundy k-coloring of G.

19.4.3 Dominator Colorings

Hedetniemi et al. [43] and Hedetniemi et al. [44] introduced the concepts of dominator partition and dominator coloring of a graph. A vertex v in a graph dominates itself and all vertices adjacent to v. We say that a vertex $v \in V$ is a *dominator* of a set $S \subseteq V$ if v dominates every vertex in S. A partition $\pi = \{V_1, V_2, \ldots, V_k\}$ is called a *dominator partition* if every vertex $v \in V$ is a dominator of at least one V_i. The *dominator partition number* $\pi_d(G)$ equals the minimum k such that G has a dominator partition of order k. If we further require that π be a proper coloring of G, then we have a *dominator coloring* of G. The *dominator chromatic number* $\chi_d(G)$ is the minimum number of colors required for a dominator coloring of G. Some basic results on dominator colorings are given in [45–49].

Since every vertex is a dominator of itself, the partition $\{\{v_1\}, \{v_2\}, \ldots, \{v_n\}\}$ into singleton sets is a dominator coloring. Thus, every graph of order n has a dominator coloring and therefore the dominator chromatic number $\chi_d(G)$ is well defined. A dominator coloring of G using $\chi_d(G)$ colors is called a χ_d-*coloring* of G.

In a graph G, any vertex v of degree 1 is called a *leaf* or a *pendant vertex* and the unique vertex u adjacent to v is called a *support vertex*. We observe that in any χ_d-coloring of G, either $\{u\}$ or $\{v\}$ must appear as a color class.

We start with an example to illustrate the concept of dominator coloring.

Example 19.1 For the graph G given in Figure 19.8, $\mathcal{C} = \{\{v_1\}, \{v_3\}, \{v_6\}, \{v_8\}, \{v_2, v_4, v_5, v_7\}\}$ is a dominator coloring of G and hence $\chi_d(G) \leq 5$. Since in any χ_d-coloring of G either a support vertex or a leaf adjacent to it appears as a singleton color class, it follows that $\chi_d(G) \geq 5$. Hence $\chi_d(G) = 5$.

Theorem 19.32 [47] *Let G be a connected graph of order $n \geq 2$. Then $\chi_d(G) = 2$ if and only if G is a complete bipartite graph of the form $K_{a,b}$, where $1 \leq a \leq b \leq n$ and $a + b = n$.* ∎

Theorem 19.33 [47] *Let G be a connected graph of order n. Then $\chi_d(G) = n$ if and only if G is the complete graph K_n.* ∎

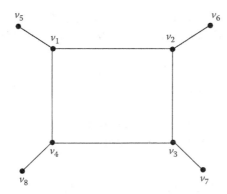

Figure 19.8 Graph with $\chi_d(G) = 5$.

Theorem 19.34 [47] *For the cycle C_n, we have*

$$\chi_d(C_n) = \begin{cases} \left\lceil \dfrac{n}{3} \right\rceil & \text{if } n = 4 \\ \left\lceil \dfrac{n}{3} \right\rceil + 1 & \text{if } n = 5 \\ \left\lceil \dfrac{n}{3} \right\rceil + 2 & \text{otherwise.} \end{cases}$$

∎

Theorem 19.35 [49] *For the path P_n, $n \geq 2$, we have*

$$\chi_d(P_n) = \begin{cases} \left\lceil \dfrac{n}{3} \right\rceil + 1 & \text{if } n = 2, 3, 4, 5, 7 \\ \left\lceil \dfrac{n}{3} \right\rceil + 2 & \text{otherwise.} \end{cases}$$

∎

If $\{V_1, V_2, \ldots, V_{\chi_d}\}$ is a χ_d-coloring of G and if $v_i \in V_i$, then $S = \{v_1, v_2, \ldots, v_{\chi_d}\}$ is a dominating set of G. Also if D is a γ-set of G, then $\mathcal{C} \cup \{\{v\} : v \in D\}$, where $\mathcal{C}$ is a proper coloring of $G - D$ gives a dominator coloring of G. These observations lead to the following bounds for $\chi_d(G)$.

Theorem 19.36 [47] *Let G be a connected graph. Then $\max\{\chi(G), \gamma(G)\} \leq \chi_d(G) \leq \chi(G) + \gamma(G)$.* ∎

Corollary 19.4 *For any connected bipartite graph G, we have $\gamma(G) \leq \chi_d(G) \leq \gamma(G) + 2$.* ∎

Chellali and Maffray [46] have obtained a characterization of split graphs G with $\chi_d(G) = \gamma(G) + 1$. Arumugam et al. [50] have proved that the dominator coloring problem is NP-complete even for split graphs.

19.4.4 List Colorings and Choosability

List coloring is a more general version of vertex coloring. This concept was introduced by Vizing [51] and independently by Erdös et al. [52].

Definition 19.13 *Let G be a graph and suppose for each $v \in V(G)$, there is associated a set $L(v)$ of permissible colors. The set $L(v)$ is called a* color list *for v. A* list coloring *of G is a proper coloring c of G such that $c(v) \in L(v)$ for each $v \in V$. A list coloring is also referred to as a* choice function. *Let $\mathcal{L} = \{L(v) : v \in V(G)\}$. If there exists a list coloring for $\mathcal{L}$, then G is said to be $\mathcal{L}$-choosable or $\mathcal{L}$-list colorable.*

A graph G is said to be *k-choosable* or *k-list colorable* if G is $\mathcal{L}$-choosable for every collection $\mathcal{L}$ with $|L(v)| \geq k$ for all $L \in \mathcal{L}$. The *list chromatic number* $\chi_l(G)$ is the least positive integer k such that G is k-choosable.

Clearly $\chi_l(G) \geq \chi(G)$. Further if $L(v) = \{1, 2, \ldots, \Delta + 1\}$, then a greedy algorithm produces a proper coloring of G and hence $\chi_l(G) \leq 1 + \Delta(G)$. Thus we have $\chi(G) \leq \chi_l(G) \leq 1 + \Delta(G)$ for every graph G.

Similarly we can define list edge coloring and the list chromatic index $\chi'_l(G)$. Clearly $\chi_{l'}(G) = \chi_l(L(G))$, where $L(G)$ is the line graph of G. We start with the following result for trees.

Theorem 19.37 *Every tree T is 2-choosable. Further for a vertex $u \in V(T)$ and for a collection $\mathcal{L} = \{L(v) : v \in V(T)\}$ of color lists of size 2 where $a \in L(u)$, there exists an $\mathcal{L}$-list coloring of T in which u is assigned the color a.*

Proof. We prove the result by induction on $n = |V(T)|$. The result is obvious for $n = 1$ or 2. We now assume that the result is true for all trees of order k, where $k \geq 2$. Let T be a tree of order $k+1$ and $\mathcal{L} = \{L(v) : v \in V(T)\}$ with $|L(v)| = 2$, for all $v \in V(T)$. Let $u \in V(T)$ and let $a \in L(u)$. Let x be an end vertex of T such that $x \neq u$ and let $\mathcal{L}' = \{L(v) : v \in V(T) - \{x\}\}$. Let y be the neighbor of x in T. By induction hypothesis there exists an $\mathcal{L}'$-list coloring c' of $T - \{x\}$ in which u is colored a.

Now, let $b \in L(x)$ and $b \neq c'(y)$. Then the coloring c defined by

$$c(v) = \begin{cases} b & \text{if } v = x \\ c'(v) & \text{if } v \neq x \end{cases}$$

is an $\mathcal{L}$-list coloring of T in which u is colored a. ∎

Graphs that are 2-choosable have been characterized. A *theta graph* consists of two vertices u and v connected by three internally disjoint u-v paths. We denote by $\Theta_{i,j,k}$ the theta graph whose three internally disjoint $u - v$ paths have lengths i, j, and k. The *core* of a graph G is obtained by successively removing end vertices until none remain.

Theorem 19.38 *A connected graph G is 2-choosable if and only if its core is K_1 or an even cycle or $\Theta_{2,2,2k}$, for some $k \geq 1$.* ∎

Vizing [51] and independently Erdös et al. [52] conjectured that any planar graph is 5-choosable. Thomassen [53] proved the above conjecture.

Theorem 19.39 [53] *Every planar graph is 5-choosable.* ∎

The above theorem gives a new proof of the five color theorem for planar graphs which avoids the recoloring technique of Kempe and the proof is conceptually simpler than all the previously known proofs.

For list edge coloring several researchers have independently posed the following conjecture.

List coloring conjecture: For any graph G,

$$\chi'_l(G) = \chi'(G).$$

This conjecture is given in Bollabás and Harris [54].

Galvin [55] proved the list coloring conjecture for bipartite graphs.

19.4.5 Total Colorings

We consider colorings that assign colors to both vertices and edges of the graph. Let $G = (V, E)$ be a graph. The *total graph* $T(G)$ of G is the graph with $V(T(G)) = V(G) \cup E(G)$ and two distinct vertices x and y of $T(G)$ are adjacent if x and y are adjacent vertices of G or adjacent edges of G or an incident vertex and edge.

The *total coloring* of a graph G is an assignment of colors to the vertices and the edges of G such that distinct colors are assigned to any two adjacent elements of $V(G) \cup E(G)$. The minimum number of colors required for a total coloring of G is called the *total chromatic number* of G and is denoted by $\chi''(G)$.

It follows from the definition that $\chi''(G) = \chi(T(G))$.

Obviously $\chi''(G) \geq 1 + \Delta(G)$, since the Δ-edges incident at a vertex v with $deg(v) = \Delta$ along with v must be assigned different colors. Behzad [56] and Vizing [30] independently conjectured an upper bound for the total chromatic number.

Total coloring conjecture: For every graph G,

$$\chi''(G) \leq 2 + \Delta(G).$$

This conjecture still remains open. However, we have the following theorem.

Theorem 19.40 *For any graph G, $\chi''(G) \leq 2 + \chi'_l(G)$, where $\chi'_l(G)$ is the list chromatic index of G.*

Proof. Let $\chi'_l(G) = k$. Then $\chi(G) \leq 1 + \Delta(G) \leq 1 + \chi'(G) \leq 1 + \chi'_l(G) < 2 + \chi'_l(G) = 2 + k$. Thus G is $(k+2)$-colorable. Let c be a $(k+2)$-coloring of G. For each edge $e = uv$ of G, let $L(e)$ be a list of $(k+2)$ colors and let $L'(e) = L(e) - \{c(u), c(v)\}$. Clearly $|L'(e)| \geq k$, for each edge e. Since $\chi'_l(G) = k$, it follows that there is a proper edge coloring c' of G such that $c'(e) \in L'(e)$. Hence $c'(e) \notin \{c(u), c(v)\}$. Now the total coloring c'' of G defined by

$$c''(x) = \begin{cases} c(x) & \text{if } x \in V(G) \\ c'(x) & \text{if } x \in E(G) \end{cases}$$

is a $(k+2)$-total coloring of G. Thus $\chi''(G) \leq 2 + k = 2 + \chi'_l(G)$. ∎

If the list coloring conjecture is true, then $\chi'_l(G) \leq 1 + \Delta(G)$ and so $\chi''(G) \leq 3 + \Delta(G)$.

19.5 CONCLUSION

In this chapter we have presented a few basic results on graph colorings. There are several other topics on graph colorings such as coloring graphs on surfaces, rainbow coloring, circular chromatic number, fractional coloring, harmonious coloring, $L(2,1)$-coloring, radio coloring, fall coloring, b-coloring, and so on. For more details on these topics we refer to the book by Chartrand and Zhang [57]. Jenson and Toft [58] have given an exhaustive list of unsolved problems on graph colorings. For survey papers on specific topics we refer to [59,60]. A recent monograph by Barenboim and Elkin [61] deals with distributed graph coloring. The main theme of this monograph are graph colorings and maximal independent sets. Several distributed graph coloring algorithms and a collection of open problems are presented.

References

[1] M.R. Garey and D.S. Johnson, *Computers and Intractability: A Guide to the Theory of NP-Completeness*, Freeman, San Francisco, CA, (1979).

[2] D.J.A. Welsh and M.B. Powell, An upper bound for the chromatic number of a graph an its application to timetabling problems, *The Computer Journal,* **10** (1967), 85–86.

[3] G.A. Dirac, The structure of k-chromatic graphs, *Fund. Math.*, **40** (1953), 42–55.

[4] R.L. Brooks, On coloring the nodes of a network, *Proc. Camb. Philos. Soc.*, **37** (1941), 194–197.

[5] E.A. Nordhaus and J.W. Gaddum, On complementary graphs, *Am. Math. Mon.*, **63** (1956), 175–177.

[6] R. Wilson, *Four Colors Suffice: How the Map Problem Was Solved*, Princeton University Press, Princeton, NJ, 2002.

[7] N.L. Biggs, E.K. Lloyd and R.J. Wilson, *Graph Theory 1736–1936*, Second edition, Clarendon Press, New York, 1986.

[8] A.B. Kempe, On the geographical problem of four colours, *Am. J. Math.*, **2** (1879), 193–200.

[9] P.J. Heawood, Map colour theorem, *Quart. J. Math.*, **24** (1890), 332–338.

[10] K. Appel and W. Haken, Every planar map is four colorable, Part I: Discharging, *Illinois J. Math.*, **21**, (1977), 429–490.

[11] M.Gardner, Mathematical games, *Sci. Am.*, **242**(2) (1980), 14–21.

[12] J. Mycielski, Sur le coloriage des graphes, *Colloq. Math.*, **3** (1955), 161–162.

[13] P. Erdös, Graph theory and probability, *Can. J. Math.*, **13** (1961), 346–352.

[14] L. Lovász, On chromatic number of finite set-systems, *Acta Mathematica Academiae Scientiarum Hungarica*, **79** (1967), 59–67.

[15] G.D. Birkhoff, A determinant formula for the number of ways of coloring a map, *Ann. Math.*, **14** (1912), 42–46.

[16] F.M. Dong, K.M. Koh and K.L. Teo, *Chromatic Polynomials and Chromaticity of Graphs*, World Scientific, Singapore, 2005.

[17] C. Berge, *Perfect Graphs, Six Papers on Graph Theory*, Indian Statistical Institute, Calcutta, India, 1963, 1–21.

[18] L. Lovász, A characterization of perfect graphs, *J. Comb. Theory Ser. B*, **13** (1972), 95–98.

[19] A. Hajnal and J. Surányi, Über die Auflösung von Graphen in vollständige Teilgraphen, *Ann. Univ. Sci. Budapest, Eötvös. Sect. Math.*, **1** (1958), 113–121.

[20] G.A. Dirac, On rigid circuit graphs, *Abh. Math. Sem. Univ. Hamburg*, **25** (1961), 71–76.

[21] L. Lovász, Normal hypergraphs and the perfect graph conjecture, *Discrete Math.*, **2** (1972), 253–267.

[22] M. Chudnovsky, N. Robertson, P. Seymour and R. Thomas, The strong perfect graph theorem, *Ann. Math.*, **164** (2006), 51–229.

[23] G.A. Dirac, A property of 4-chromatic graphs and some remarks on critical graphs, *J. Londan Math. Soc.*, **27** (1952), 85–92.

[24] G. Hajós, Über eine Konstruktion nicht n-färbbarer Graphen, *Wiss. Z. Martin – Luther – Univ. Halle – Wittenberg, Math. – Nat. Reihe.*, **10** (1961), 116–117.

[25] P.A. Catlin, Hajós' graph coloring conjecture: Variations and counter examples, *J. Comb. Theory Ser. B*, **26** (1979), 268–274.

[26] C. Thomassen, Some remarks on Hajós' conjecture, *J. Combin. Theor. Ser. B*, **93** (2005), 95–105.

[27] H. Hadwiger, Über eine Klassifikation der Streckenkomplexe, *Vierteljschr. Naturforsch. ges Zürich*, **88** (1943), 133–143.

[28] K. Wagner, Üer eine Eigenschaft der ebene Komplexe, *Math. Ann.*, **114** (1937), 570–590.

[29] N. Robertson, P. Seymour and R. Thomas, Hadwiger's conjecture for K_5-free graphs, *Combinatorica*, **14** (1993), 279–361.

[30] V.G. Vizing, On an estimate of the chromatic class of a p-graph, *Diskret. Analiz.*, **3** (1964), 25–30.

[31] I. Holyer, The NP-completeness of edge-coloring, *SIAM J. Computing*, **2** (1981), 225–231.

[32] D. König, Über Graphen ihre Anwendung auf Determinantentheorie und Mengenlehre, *Math. Ann.*, **77** (1916), 453–465.

[33] L.W. Beineke and R.J. Wilson, On the edge-chromatic number of a graph, *Discrete Math.*, **5** (1973), 15–20.

[34] A.J. Hilton, Recent progress in edge-coloring graphs, *Discrete Math.*, **64** (1987), 303–307.

[35] A.G. Chetwynd and A.J. Hilton, Star multigraphs with three vertices of maximum degree, *Math. Proc. Cambridge Philos. Soc.*, **100** (1986), 303–317.

[36] V.G. Vizing, Critical graphs with a given chromatic class, *Metody Diskret. Analiz.*, **5** (1965), 9–17.

[37] F. Harary, S.T. Hedetniemi and G. Prins, An interpolation theorem for graphical homomorphisms, *Portugal. Math.*, **26** (1967), 453–462.

[38] P. Hell and D.J. Miller, Graphs with given achromatic number, *Discrete Math.*, **16** (1976), 195–207.

[39] N. Cairnie and K. Edwards, Some results on the achromatic number, *J. Graph Theory*, **26** (1997), 129–136.

[40] N. Cairnie and K. Edwards, The achromatic number of bounded degree trees, *Discrete Math.*, **188** (1998), 87–97.

[41] P.M. Grundy, Mathematics and games, *Eureka*, **2** (1939), 6–8.

[42] C.A. Christen and S.M. Selkow, Some perfect coloring properties of graphs, *J. Combin. Theory Ser. B*, **27** (1979), 49–59.

[43] S.M. Hedetniemi, S.T. Hedetniemi, R. Laskar, A.A. McRae and C.K. Wallis, Dominator partitions of graphs, *J. Combin. Inform. System Sci.*, **34** (2009), 183–192.

[44] S.M. Hedetniemi, S.T. Hedetniemi, A.A. McRae and J.R.S. Blair, Dominator colorings of graphs, Preprint, 2006, (Unpublished).

[45] S. Arumugam, J. Bagga and K. Raja Chandrasekar, On dominator colorings in graphs, *Proc. Indian Acad. Sci. (Math. Sci.)*, Springer, **122**(4) (2012), 561–571.

[46] M. Chellali and F. Maffray, Dominator colorings in some classes of graphs, *Graphs Combin.*, **28** (2012), 97–107.

[47] R.M. Gera, On dominator coloring in graphs, *Graph Theory Notes NY.*, **LII** (2007), 25–30.

[48] R. Gera, On the dominator colorings in bipartite graphs, *Conference Proceedings of the 4th International Conference on Information Technology, IEEE Comp. Soc.*, (2007), 1–6.

[49] R. Gera, C. Rasmussen and S. Horton, Dominator colorings and safe clique partitions, *Congr. Numer.*, **181** (2006), 19–32.

[50] S. Arumugam, K. Raja Chandrasekar, N. Misra, G. Philip and S. Saurabh, Algorithmic aspects of dominator colorings in graphs, *Lecture Notes in Comput. Sci.*, Springer-Verlag, **7056** (2011), 19–30.

[51] V.G. Vizing, Coloring the vertices of a graph in prescribed colors (Russian), *Metody Diskret. Anal. v Teorii Kodov i Schem*, **29** (1976), 3–10.

[52] P. Erdös, A.L. Rudin and H. Taylor, Choosability in graphs, *Congr. Numer.*, **126** (1980), 125–157.

[53] C. Thomassen, 3-List coloring planar graphs of girth 5, *J. Combin. Theory Ser. B*, **64** (1995), 101–107.

[54] B. Bollobás and A.J. Harris, List colorings of graphs, *Graphs Combin.*, **1** (1985), 115–127.

[55] F. Galvin, The list chromatic index of a bipartite multigraph, *J. Combin. Theory Ser. B*, **63** (1995), 153–158.

[56] M. Behzad, Graphs and their chromatic numbers, PhD Thesis, Michigan State University, 1965.

[57] G. Chartrand and P. Zhang, *Chromatic Graph Theory*, Chapman & Hall, CRC Press, Boca Raton, FL, 2009.

[58] T.R. Jensen and B. Toft, *Graph Coloring Problems*, Wiley-Interscience, Hoboken, NJ, 1995.

[59] B. Randerath and I. Schiermeyer, Vertex colouring and forbidden subgraphs—A survey, *Graphs Combin.*, **20** (2004), 1–40.

[60] Z. Tuza, Graph colorings with local constraints—A survey, *Math. Graph Theory*, **17** (1997), 161–228.

[61] L. Barenboim and M. Elkin, *Distributed Graph Coloring—Fundamentals and Recent Developments*, Morgan & Claypool Publishers, San Rafael, CA, 2013.

V

Planar Graphs

CHAPTER 20

Planarity and Duality*

Krishnaiyan "KT" Thulasiraman

M. N. S. Swamy

CONTENTS

20.1	Introduction	475
20.2	Planar Graphs	475
20.3	Euler's Formula	477
20.4	Kuratowski's Theorem and Other Characterizations of Planarity	480
20.5	Dual Graphs	481
20.6	Planarity and Duality	485

20.1 INTRODUCTION

In this chapter we discuss two important concepts in graph theory, namely, planarity and duality. First we consider planar graphs and derive some properties of these graphs. Characterizations of planar graphs due to Kuratowski, Wagner, Harary and Tutte, and to MacLane are also discussed. We then discuss Whitney's definition of duality of graphs which is given in terms of circuits and cutsets and relate this concept to the seemingly unrelated concept of planarity.

Duality has been of considerable interest to electrical network theorists. This interest is due to the fact that the voltages and currents in an electrical network are dual variables. Duality of these variables arises as a result of Kirchhoff's laws. Kirchhoff's voltage law is in terms of circuits and Kirchhoff's current law is in terms of cutsets.

Many of the results developed in Chapter 7 on circuit and cutset spaces of a graph will be used in the developments of this chapter.

20.2 PLANAR GRAPHS

A graph G is said to be *embeddable* on a surface S if it can be drawn on S so that its edges intersect only at their end vertices. A graph is said to be *planar* if it can be embedded on a plane. Such a drawing of a planar graph G is called a *planar embedding* of G.

Two planar embeddings of a graph are shown in Figure 20.1. In one of these (Figure 20.1a) all the edges are drawn as straight line segments, while in the other (Figure 20.1b) one of the edges is drawn as a curved line. Note that the edge connecting vertices a and d in Figure 20.1b cannot be drawn as a straight line if all the remaining edges are drawn as shown.

Obviously, if a graph has self-loops or parallel edges, then in none of its planar embeddings all the edges can be drawn as straight line segments. This naturally raises the question

---

*This chapter is an edited version of Chapter 7 in K. Thulasiraman and M.N.S. Swamy [1].

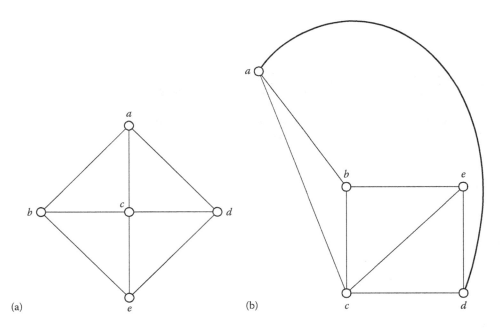

Figure 20.1 (a,b) Two planar embeddings of a graph.

whether for every simple planar graph G there exists a planar embedding in which all the edges of G can be drawn as straight line segments. The answer to this question is in the affirmative, as stated in the following theorem.

Theorem 20.1 *For every simple planar graph there exists a planar embedding in which all the edges of the graph can be drawn as straight line segments.* ∎

This result was proved independently by Wagner [2], Fary [3], and Stein [4].

If a graph is not embeddable on a plane, then it may be embeddable on some other surface. However, we now show that embeddability on a plane and embeddability on a sphere are equivalent; that is, if a graph is embeddable on a plane, then it is also embeddable on a sphere and vice versa. The proof of this result uses what is called the stereographic projection of a sphere onto a plane, which is described below.

Suppose that we place a sphere on a plane (Figure 20.2) and call the point of contact the south pole and the diametrically opposite point on the sphere the north pole N. Let P

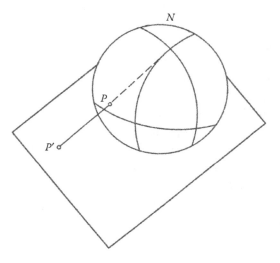

Figure 20.2 Stereographic projection.

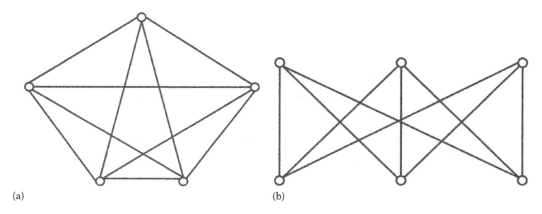

Figure 20.3 Basic nonplanar graphs (a) K_5 and (b) $K_{3,3}$.

be any point on the sphere. Then the point P' at which the straight line joining N and P, when extended, meets the plane, is called the *stereographic projection* of P onto the plane. It is clear that there is one-to-one correspondence between the points on a sphere and their stereographic projections on the plane.

Theorem 20.2 *A graph G is embeddable on a plane if and only if G is embeddable on a sphere.* ∎

Proof. Let G' be an embedding of G on a sphere. Place the sphere on the plane so that the north pole is neither a vertex of G' nor a point on an edge of G'.

Then the image of G' under the stereographic projection is an embedding of G on the plane because edges of G' intersect only at their end vertices and there is a one-to-one correspondence between points on the sphere and their images under stereographic projection. The converse is proved similarly. ∎

Two basic nonplanar graphs called *Kuratowski's graphs* are shown in Figure 20.3. One of these is K_5, the complete graph on five vertices, and the other is $K_{3,3}$. We call these graphs basic nonplanar graphs because they play a fundamental role in an important characterization of planarity due to Kuratowski (Section 20.4). The nonplanarity of these two graphs is established in the next section.

Before we conclude this section, we would like to point out that Whitney [5] has proved that a separable graph is planar if and only if its blocks are planar. So while considering questions relating to the embedding on a plane, it is enough if we concern ourselves with only nonseparable graphs.

20.3 EULER'S FORMULA

An embedding of a planar graph on a plane divides the plane into *regions*. A region is finite if the area it encloses is finite; otherwise it is infinite.

For example, in the planar graph shown in Figure 20.4, the hatched region f_5 is the infinite region; f_1, f_2, f_3, and f_4 are the finite regions.

Clearly, the edges on the boundary of a region contain exactly one circuit, and this circuit is said to enclose the region. For example, the edges e_1, e_8, e_9, e_{10} and e_{13} form the region f_1 in the graph of Figure 20.4, and they contain the circuit $\{e_1, e_8, e_9, e_{10}\}$.

Note that in any spherical embedding of a planar graph, every region is finite. Suppose we embed a planar graph on a sphere and place the sphere on a plane so that the north pole is inside any chosen region, say, region f. Then under the stereographic projection the image of f will be the infinite region. Thus a planar graph can always be embedded on a plane so that any chosen region becomes the infinite region.

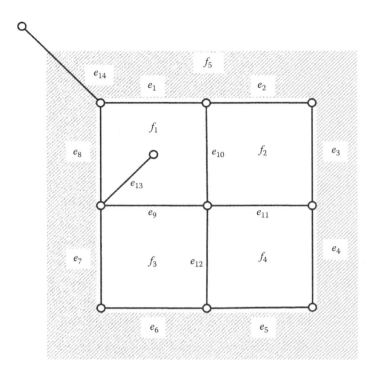

Figure 20.4 Regions of a planar embedding.

Let $f_1, f_2, \ldots, f_r$ be the regions of a planar graph with f_r as the infinite region. We denote by C_i, $1 \leq i \leq r$, the circuit on the boundary of region f_i. The circuits $C_1, C_2, \ldots, C_{r-1}$, corresponding to the finite regions, are called *meshes*. It is easy to verify that the ring sum (exclusive OR/symmetric difference denoted by $\oplus$) of any $k \geq 2$ meshes, say, $C_1, C_2, \ldots, C_k$, is a circuit or union of edge-disjoint circuits enclosing the regions $f_1, f_2, \ldots, f_k$. Since each mesh encloses only one region, it follows that no mesh can be obtained as the ring sum of some of the remaining meshes. Thus the meshes $C_1, C_2, \ldots, C_{r-1}$ are linearly independent.

Suppose that any element C of the circuit subspace of G encloses the finite regions $f_1, f_2, \ldots, f_k$. Then it can be verified that

$$C = C_1 \oplus C_2 \oplus \ldots \oplus C_k.$$

For example, in the graph of Figure 20.4, the set $C = \{e_1, e_2, e_3, e_4, e_5, e_6, e_7, e_8\}$ encloses the regions f_1, f_2, f_3, f_4. Therefore,

$$C = C_1 \oplus C_2 \oplus C_3 \oplus C_4.$$

Thus, every element of the circuit subspace of G can be expressed as the ring sum of some or all of the meshes of G. Since the meshes are themselves independent, we get the following theorem.

Theorem 20.3 *The meshes of a planar graph G form a basis of the circuit subspace of G.* ∎

Following is an immediate consequence of this theorem.

Corollary 20.1 (Euler's formula) *If a connected planar graph G has m edges, n vertices, and r regions, then*

$$n - m + r = 2.$$

Proof. The proof follows if we note that by Theorem 20.3, the nullity μ of G is equal to $r - 1$. ∎

In general it is not easy to test whether a graph is planar or not. We now use Euler's formula to derive some properties of planar graphs. These properties can be of help in detecting nonplanarity in certain cases, as we shall see soon.

Corollary 20.2 *If a connected simple planar graph G has m edges and $n \geq 3$ vertices, then*

$$m \leq 3n - 6.$$

Proof. Let $F = \{f_1, f_2, \ldots, f_r\}$ denote the set of regions of G.

Let the *degree $d(f_i)$ of region f_i* denote the number of edges on the boundary of f_i, bridges being counted twice. (For example, in the graph of Figure 20.4, the degree of region f_1 is 6.) Noting the similarity between the definitions of the degree of a vertex and the degree of a region, we get

$$\sum_{f_i \in F} d(f_i) = 2m.$$

Since G has neither parallel edges nor self-loops and $n \geq 3$, it follows that $d(f_i) \geq 3$, for all i, Hence

$$\sum_{f_i \in F} d(f_i) \geq 3r.$$

Thus $2m \geq 3r$, that is,

$$r \leq \frac{2}{3}m.$$

Using this inequality in Euler's formula, we get

$$n - m + \frac{2}{3}m \geq 2$$

or

$$m \leq 3n - 6.$$ ∎

Corollary 20.3 K_5 *is nonplanar.*

Proof. For K_5, $n = 5$ and $m = 10$. If it were planar, then by Corollary 20.2,

$$m = 10 \leq 3n - 6 = 9;$$

a contradiction. Thus K_5 must be nonplanar. ∎

Corollary 20.4 $K_{3,3}$ *is nonplanar.*

Proof. For $K_{3,3}$, $m = 9$ and $n = 6$. If it were planar, then by Euler's formula it has $r = 9 - 6 + 2 = 5$ regions.

In $K_{3,3}$ there is no circuit of length less than 4. Hence the degree of every region is at least 4. Thus,

$$2m = \sum_{i=1}^{r} d(f_i) \geq 4r$$

or

$$r \leq \frac{2m}{4},$$

that is, $r \leq 4$; a contradiction. Hence $K_{3,3}$ is nonplanar. ∎

Corollary 20.5 *In a simple planar graph G there is at least one vertex of degree less than or equal to 5.*

Proof. Let G have m edges and n vertices. If every vertex of G has degree greater than 5, then
$$2m \geq 6n$$
or
$$m \geq 3n.$$
But by Corollary 20.2,
$$m \leq 3n - 6.$$
These two inequalities contradict one another. Hence the result. ∎

20.4 KURATOWSKI'S THEOREM AND OTHER CHARACTERIZATIONS OF PLANARITY

Characterizations of planarity given by Kuratowski, Warner, Harary and Tutte, and MacLane are presented in this section.

To explain Kuratowski's characterization, we need the definition of the concept of homeomorphism between graphs.

The two edges incident on a vertex of degree 2 are called *series edges*. Let $e_1 = (u, v)$ and $e_2 = (v, w)$ be the series edges incident on a vertex v. Removal of vertex v and replacing e_1 and e_2 by a simple edge (u, w) is called *series merger* (Figure 20.5a).

Adding a new vertex v on an edge (u, w) thereby creating the edges (u, v) and (v, w), is called *series insertion* (Figure 20.5b).

Two graphs are said to be *homeomorphic* if they are isomorphic or can be made isomorphic by repeated series insertions and/or mergers.

It is clear that if a graph G is planar, then any graph homeomorphic to G is also planar, that is, planarity of a graph is not affected by series insertions or mergers.

We proved in the previous section that K_5 and $K_{3,3}$ are nonplanar. Therefore, a planar graph does not contain a subgraph homeomorphic to K_5 or $K_{3,3}$. It is remarkable that Kuratowski [6] could prove that the converse of this result is also true. In the following

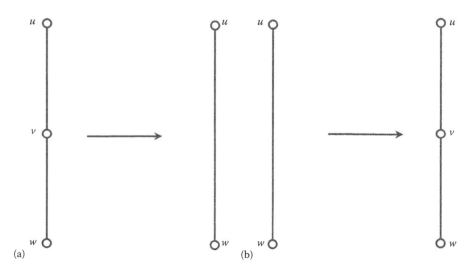

Figure 20.5 (a) Series merger. (b) Series insertion.

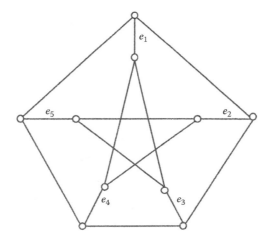

Figure 20.6 Petersen graph.

theorem, we state this celebrated characterization of planarity. Proof of this may be found in Harary [7].

Theorem 20.4 *[6] A graph is planar if and only if it does not contain a subgraph homeomorphic to K_5 or $K_{3,3}$.* ∎

See also Tutte [8] for a constructive proof of this theorem.

We now present another characterization of planarity independently proved by Wagner [9] and by Harary and Tutte [10].

Theorem 20.5 *A graph is planar if and only if it does not contain a subgraph contractible to K_5 or $K_{3,3}$.* ∎

Consider now the graph (known as the *Petersen graph*) shown in Figure 20.6. This graph does not contain any subgraph isomorphic to K_5 or $K_{3,3}$, but is known to be nonplanar. So if we wish to use Kuratowski's criterion to establish the nonplanar character of the Petersen graph, then we need to locate a subgraph homeomorphic to K_5 or $K_{3,3}$. However, the nonplanarity of the graph follows easily from the above characterization, because the graph reduces to K_5 after contracting the edges e_1, e_2, e_3, e_4, and e_5. MacLane's characterization of planar graphs is stated next.

Theorem 20.6 *A graph G is planar if and only if there exists in G a set of basis circuits such that no edge appears in more than two of these circuits.* ∎

We know that the meshes of a planar graph form a basis of the circuit subspace of the graph, and that no edge of the graph appears in more than two of the meshes. This proves the necessity or Theorem 20.6. The proof of the sufficiency may be found in MacLane [11].

Another important characterization of planar graphs in terms of the existence of dual graphs is discussed in Section 20.6.

20.5 DUAL GRAPHS

A graph G_1 is a *dual* of a graph G_2 if there is a one-to-one correspondence between the edges of G_1 and those of G_2 such that a set of edges in G_1 is a circuit vector of G_1 if and only if the corresponding set of edges in G_2 is a cutset vector of G_2. Duals were first defined by Whitney [12], though his original definition was given in a different form.

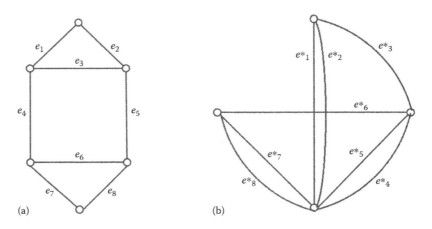

Figure 20.7 Dual graphs: (a) graph G_1 and (b) graph G_2.

Clearly, to prove that G_1 is a dual of G_2 it is enough if we show that the vectors forming a basis of the circuit subspace of G_1 correspond to the vectors forming a basis of the cutset subspace of G_2.

For example, consider the graphs G_1 and G_2 shown in Figure 20.7. The edge e_i of G_1 corresponds to the edge e_i^* of G_2. It may be verified that the circuits $\{e_1, e_2, e_3\}$, $\{e_3, e_4, e_5, e_6\}$, and $\{e_6, e_7, e_8\}$ form a basis of the circuit subspace of G_1, and the corresponding sets of edges $\{e_1^*, e_2^*, e_3^*\}$, $\{e_3^*, e_4^*, e_5^*, e_6^*\}$, and $\{e_6^*, e_7^*, e_8^*\}$ form a basis of the cutset subspace of G_2. Thus G_1 is a dual of G_2.

We next study some properties of the duals of a graph.

Theorem 20.7 *Let G_1 be a dual of a graph G_2. Then a circuit in G_1 corresponds to a cutset in G_2 and vice versa.*

Proof. Let C^* be a circuit in G_1 and C be the corresponding set of edges in G_2.

Suppose that C is not a cutset in G_2. Then it follows from the definition of a dual that C must be the union of disjoint cutsets $C_1, C_2, \ldots, C_k$.

Let $C_1^*, C_2^*, \ldots, C_k^*$ be the sets of edges in G_1 which correspond to the cutsets $C_1, C_2, \ldots, C_k$. Again from the definition of a dual it follows that $C_1^*, C_2^*, \ldots, C_k^*$ are circuits or unions of disjoint circuits.

Since C^* is the union of $C_1^*, C_2^*, \ldots, C_k^*$, it is clear that C^* must contain more than one circuit. However, this is not possible, since C^* is a circuit, and no proper subset of a circuit is a circuit. Thus $k = 1$, or in other words, C is a cutset of G_2.

In a similar way we can show that each cutset in G_2 corresponds to a circuit in G_1. ∎

Theorem 20.8 *If G_1 is a dual of G_2, then G_2 is a dual of G_1.*

Proof. To prove the theorem we need to show that each circuit vector of G_2 corresponds to a cutset vector of G_1 and vice versa.

Let C be a circuit vector in G_2, with C^* denoting the corresponding set of edges in G_1. Since every circuit and every cutset have an even number of common edges, C has an even number of common edges with every cutset vector of G_2. Since G_1 is a dual of G_2, C^* has an even number of common edges with every circuit vector of G_1. Therefore, C^* is a cutset vector of G_1 (see Theorem 7.17). In a similar way, we can show that each cutset vector of G_1 corresponds to a circuit vector of G_2. Hence the theorem. ∎

In view of the above theorem we refer to graphs G_1 and G_2 as simply duals if any one of them is dual of the other.

The following result is a consequence of Theorem 20.8 and the definition of a dual.

Theorem 20.9 *If G_1 and G_2 are dual graphs, then the rank of one is equal to the nullity of the other; that is*

$$\rho(G_1) = \mu(G_2)$$

and

$$\rho(G_2) = \mu(G_1).$$ ∎

Suppose a graph G has a dual. Then the question arises whether every subgraph of G has a dual. To answer this question we need the following result.

Theorem 20.10 *Consider two dual graphs G_1 and G_2. Let $e = (v_1, v_2)$ be an edge in G_1, and $e^* = (v_1^*, v_2^*)$ be the corresponding edge in G_2. Let G_1' be the graph obtained by removing the edge e from G_1; let G_2' be the graph obtained by contracting e^* in G_2. Then G_1' and G_2' are duals, the one-to-one correspondence between their edges being the same as in G_1 and G_2.*

Proof. Let C and C^* denote corresponding sets of edges in G_1 and G_2, respectively.

Suppose C is a circuit in G_1'. Since it does not contain e, it is also a circuit in G_1. Hence C^* is a cutset, say, $\langle V_a^*, V_b^* \rangle$, in G_2. Since C^* does not contain e^*, the vertices v_1^* and v_2^* are both in V_a^* or in V_b^*. Therefore C^* is also a cutset in G_2'. Thus every circuit in G_1' corresponds to a cutset in G_2'.

Suppose C^* is a cutset in G_2'. Since C^* does not contain e^*, it is also a cutset in G_2. Hence C is a circuit in G_1. Since it does not contain e, it is also a circuit in G_1'. Thus every cutset in G_2' corresponds to a circuit in G_1'. ∎

In the view of this theorem, we may say, using the language of electrical network theory, that *open-circuiting* an edge in a graph G corresponds to *short-circuiting* the corresponding edge in a dual of G.

A useful corollary of Theorem 20.10 now follows.

Corollary 20.6 *If a graph G has a dual, then every edge-induced subgraph of G also has a dual.*

Proof. The result follows from Theorem 20.10, if we note that every edge-induced subgraph H of G can be obtained by removing from G the edges not in H. ∎

To illustrate the above corollary, consider the two dual graphs G_1 and G_2 of Figure 20.7. The graph G_1' shown in Figure 20.8a is obtained by removing from G_1 the edges e_3 and e_6. The graph G_2' of Figure 20.8b is obtained by contracting the edges e_3^* and e_6^* of G_2. It may be verified that G_1' and G_2' are duals.

Observing that series edges in a graph G correspond to parallel edges in a dual of G, we get the following corollary of Theorem 20.10.

Corollary 20.7 *If a graph G has a dual, then every graph homeomorphic to G also has a dual.* ∎

We now proceed to develop an equivalent characterization of a dual.

Let G be an n-vertex graph. We may assume without loss of generality that it is connected. Let K^* be the subgraph of G, and let G' be the graph obtained by contracting the edges of K^*. Note that G' is also connected.

If K^* has n^* vertices and p connected components then G' will have $n - (n^* - p)$ vertices. Therefore, the rank of G' is given by

$$\begin{aligned}\rho(G') &= n - (n^* - p) - 1 \\ &= \rho(G) - \rho(K^*).\end{aligned} \quad (20.1)$$

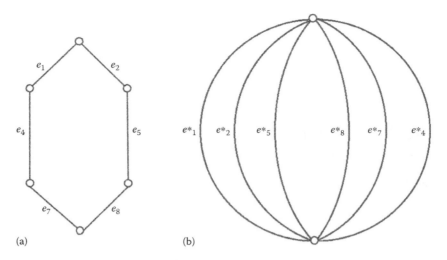

Figure 20.8 Dual graphs: (a) graph G_1' and (b) graph G_2'.

Theorem 20.11 *Let G_1 and G_2 be two graphs with a one-to-one correspondence between their edges. Let H be any subgraph of G_1 and H^* the corresponding subgraph in G_2. Let K^* be the complement of H^* in G_2. Then G_1 and G_2 are dual graphs if and only if*

$$\mu(H) = \rho(G_2) - \rho(K^*). \tag{20.2}$$

Proof. Necessity: Let G_1 and G_2 be dual graphs. Let G_2' be the graph obtained from G_2 by contracting the edges of K^*. Then by Theorem 20.10, H and G_2' are dual graphs. Therefore, by Theorem 20.9,

$$\mu(H) = \rho(G_2').$$

But by (20.1),

$$\rho(G_2') = \rho(G_2) - \rho(K^*).$$

Hence

$$\mu(H) = \rho(G_2) - \rho(K^*).$$

Sufficiency: Assume that (20.2) is satisfied for every subgraph H of G_1. We now show that each circuit in G_1 corresponds to a cutset in G_2 and vice versa.

Let H be a circuit in G_1. Then $\mu(H) = 1$. Therefore by (20.2),

$$\rho(K^*) = \rho(G_2) - 1.$$

Since H is a minimal subgraph of G_1 with nullity equal to 1, and K^* is the complement of H^* in G_2, it is clear that K^* is a maximal subgraph of G_2 with rank equal to $\rho(G_2) - 1$. It now follows from the definition of a cutset that H^* is a cutset in G_2.

In a similar way, we can show that a cutset in G_2 corresponds to a circuit in G_1. Thus G_1 and G_2 are dual graphs. ∎

Whitney's [12] original definition of duality was stated as in Theorem 20.11.

To illustrate this definition, consider the dual graphs G_1 and G_2 of Figure 20.7. A subgraph H of G_1 and the complement K^* of the corresponding subgraph in G_2 are shown in Figure 20.9. We may now verify that

$$\mu(H) = \rho(G_2) - \rho(K^*).$$

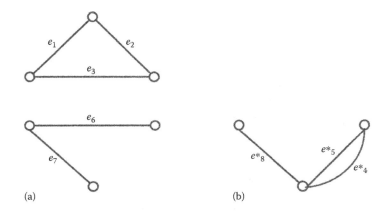

Figure 20.9 Illustration of Whitney's definition of duality. (a) Graph $H, \mu(H) = 1$. (b) Graph $K^*, \rho(k^*) = 2$.

20.6 PLANARITY AND DUALITY

In this section we characterize the class of graphs which have duals. While doing so, we relate the two seemingly unrelated concepts, planarity and duality.

First we prove that every planar graph has a dual. The proof is based on a procedure for constructing a dual of a given planar graph.

Consider a planar graph and let G be a planar embedding of this graph. Let $f_1, f_2, \ldots, f_r$ be the regions of G. Construct a graph G^* defined as follows:

1. G^* has r vertices $v_1^*, v_2^*, \ldots, v_r^*$, vertex v_i, $1 \leq i \leq r$, corresponding to region f_i.

2. G^* has as many edges as G has.

3. If an edge e of G is common to the regions f_i and f_j (not necessarily distinct), then the corresponding edge e^* in G^* connects vertices v_i^* and v_j^*. (Note that each edge e of G is common to at most two regions, and it is possible that an edge may be in exactly one region.)

A simple way to construct G^* is to first place the vertices $v_1^*, v_2^*, \ldots, v_r^*$, one in each region of G. Then, for each edge e common to regions f_i and f_j, draw a line connecting v_i^* and v_j^* so that it crosses the edge e. This line represents the edge e^*.

The procedure for constructing G^* is illustrated in Figure 20.10. The continuous lines represent the edges of the given planar graph G and the dashed lines represent those of G^*.

We now prove that G^* is a dual of G.

Let $C_1, C_2, \ldots, C_{r-1}$ denote the meshes of G, and $C_1^*, C_2^*, \ldots, C_{r-1}^*$, denote the corresponding sets of edges in G^*. It is clear from the procedure used to construct G^* that the edges in C_i^* are incident on the vertex v_i^* and form a cut whose removal will separate v_i^* from the remaining vertices of G^*.

By Theorem 20.3, $C_1, C_2, \ldots, C_{r-1}$ form a basis of the circuit subspace of G, and we know that the incidence vectors $C_1^*, C_2^*, \ldots, C_{r-1}^*$ form a basis of the cutset subspace of G^*. Since there is a one-to-one correspondence between C_i's and $C_i^{*'}$s, G and G^* are dual graphs. Thus we have the following theorem.

Theorem 20.12 *Every planar graph has a dual.* ∎

The question that immediately arises now is whether a nonplanar graph has a dual. The answer is *no* and it is based on the next two lemmas.

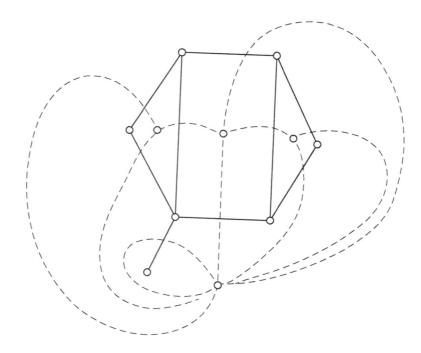

Figure 20.10 Construction of a dual.

Lemma 20.1 $K_{3,3}$ *has no dual.*

Proof. First observe that

1. $K_{3,3}$ has no cutsets of two edges.

2. $K_{3,3}$ has circuits of length four or six only.

3. $K_{3,3}$ has nine edges.

Suppose $K_{3,3}$ has a dual G. Then these observations would, respectively, imply the following for G:

1. G has no circuits of two edges; that is, G has no parallel edges.

2. G has no cutsets with less than four edges. Thus every vertex in G is of degree at least 4.

3. G has nine edges.

The first two of the above imply that G has at least five vertices, each of degree at least 4. Thus G must have at least $(1/2) \times 5 \times 4 = 10$ edges. However, this contradicts observation 3. Hence $K_{3,3}$ has no dual. ∎

Lemma 20.2 K_5 *has no dual.*

Proof. First observe that

1. K_5 has no circuits of length one or two.

2. K_5 has cutsets with four or six edges only.

3. K_5 has 10 edges.

Suppose K_5 has a dual G. Then by observation 2, G has circuits of lengths four and six only. In other words, all circuits of G are of even length. So G is bipartite.

Since a bipartite graph with six or fewer vertices cannot have more than nine edges, it is necessary that G has at least seven vertices. But by observation 1 the degree of every vertex of G is at least 3. Hence G must have at least $(1/2) \times 7 \times 3 > 10$ edges. This, however, contradicts observation 3. Hence K_5 has no dual. ∎

The main result of this section now follows.

Theorem 20.13 *A graph has a dual if and only if it is planar.*

Proof. The sufficiency part of the theorem is the same as Theorem 20.12.

We can prove the necessity by showing that a nonplanar graph G has no dual. By Kuratowski's theorem, G has a subgraph H homeomorphic to $K_{3,3}$ or K_5. If G has a dual, then by Corollary 20.6, H has a dual. But then, by Corollary 20.7, $K_{3,3}$ or K_5 should have a dual. This, however, will contradict the fact that neither of these graphs has a dual. Hence G has no dual. ∎

The above theorem gives a characterization of planar graphs in terms of the existence of dual graphs and was originally proved by Whitney. The proof given here is due to Parsons [13]. Whitney's original proof, which does not make use of Kuratowski's theorem, may be found in [14].

From the procedure given earlier in this section it is clear that different (though isomorphic) planar embeddings of a planar graph may lead to nonisomorphic duals. The following theorem presents a property of the duals of a graph.

Theorem 20.14 *All duals of a graph G are 2-isomorphic; every graph 2-isomorphic to a dual of G is also a dual of G.* ∎

The proof of this theorem follows from the definition of a dual and a result by Whitney on 2-isomorphic graphs [15].

Summary and Related Works

Whitney [5,12,15,16], and books by Seshu and Reed [14], Ore [17], Harary [7], Bondy and Murty [18], and West [19] are recommended for further reading on planar graphs.

Two properties of a nonplanar graph G which are of interest are:

1. The minimum number of planar subgraphs whose union is G; this is called the *thickness* of G.

2. The minimum number of crossings (or intersections) in order to draw a graph on a plane; this is called the *crossing number* of G.

For several results on the thickness and the crossing numbers of a nonplanar graph, see Harary [7], Bondy and Murty [18], and West [19]. See also Bose and Prabhu [20].

References

[1] K. Thulasiraman and M.N.S. Swamy, *Graphs: Theory and Algorithms*, Wiley-Interscience, New York, 1992.

[2] K. Wagner, Bemerkungen zum Vierfarbenproblem, *Üeber. Deutsch. Math.-Verein*, **46** (1936), 26–32.

[3] I. Fary, On straight line representation of planar graphs, *Acta Sci. Math. Szeged*, **11** (1948), 229–233.

[4] S.K. Stein, Convex maps, *Proc. Am. Math. Soc.*, **2** (1951), 464–466.

[5] H. Whitney, Non-separable and planar graphs, *Trans. Am. Math. Soc.*, **34** (1932), 339–362.

[6] C. Kuratowski, Sur le Problème des courbes gauches en topologie, *Fund. Math.*, **15**, (1930), 271–283.

[7] F. Harary, *Graph Theory*, Addison-Wesley, Reading, MA, 1969.

[8] W.T. Tutte, How to draw a graph, *Proc. Lond. Math.Soc.*, **13** (1963), 743–767.

[9] K. Wagner, Über eine Eigneschaft der ebenen Komplexe, *Math. Ann.*, **114** (1937), 570–590.

[10] F. Harary and W.T. Tutte, A dual form of Kuratowski's theorem, *Can. Math. Bull.*, **8** (1965), 17–20.

[11] S. MacLane, A structural characterization of planar combinatorial graphs, *Duke Math. J.*, **3** (1937), 340–372.

[12] H. Whitney, Planar graphs, *Fund. Math.*, **21** (1933), 73–84.

[13] T.D. Parsons, On planar graphs, *Am. Math. Monthly*, **78** (1971), 176–178.

[14] S. Seshu and M. B. Reed, *Linear Graphs and Electrical Networks*, Addison-Wesley, Reading, MA, 1961.

[15] H. Whitney, 2-Isomorphic graphs, *Am. J. Math.*, **55** (1933), 245–254.

[16] H. Whitney, A set of topological invariants for graphs, *Am.J. Math.* **55** (1933), 231–235.

[17] O. Ore, *The Four Colour Problem*, Academic Press, New York, 1967.

[18] J.A. Bondy and U.S.R. Murty, *Graph Theory*, Springer, Berlin, Germany, 2008.

[19] D.B.West, *Introduction to Graph Theory*, Prentice Hall, Upper Saddle River, NJ, 2001.

[20] N.K. Bose and K. A. Prabhu, Thickness of graphs with degree constrained vertices, *IEEE Trans. Circ. Sys.*, **CAS-24** (1975), 184–190.

CHAPTER 21

Edge Addition Planarity Testing Algorithm

John M. Boyer

CONTENTS

21.1	Introduction	489
21.2	Overview of Edge Addition Planarity	493
21.3	Walkdown	495
	21.3.1 Example of Walkdown Processing	497
21.4	Proof of Correctness	499
21.5	Efficient Implementation	502
	21.5.1 Graph Storage and Manipulation	502
	21.5.2 Sorted Child Lists and Forward Edge Lists	503
	21.5.3 Systematic Walkdown Invocation and Nonplanarity Detection	503
	21.5.4 Pertinence Management with Walkup	504
	21.5.5 Future Pertinence Management	505
	21.5.6 Merging, Flipping, and Embedding Recovery	505
	21.5.7 External Face Management	509
	21.5.8 Coda	510
21.6	Isolating an Obstruction to Planarity	511
21.7	Drawing a Visibility Representation of a Planar Graph	514
	21.7.1 Computing Vertical Positions of Vertices	514
	21.7.1.1 Localized Sense of Up and Down	515
	21.7.1.2 Making the Localized Settings	515
	21.7.1.3 Postprocessing to Generate the Vertex Order	516
	21.7.2 Computing Horizontal Positions of Edges	518
	21.7.3 Correctness and Performance	519
21.8	Conclusion	522

21.1 INTRODUCTION

A *graph* is a data structure comprising a set V of vertices and a set E of edges, each edge corresponding to a pair of vertices called its *endpoints*. The number of vertices is denoted n, and the number of edges is denoted m. An edge is *incident* to its endpoints, and the *degree* of a vertex is the number of edges incident to the vertex. A *walk* is a sequence of vertices $(v_0, v_1, \ldots, v_k)$ and the edges (v_{i-1}, v_i) for $0 < i \leq k$. A *path* is a walk with no repeated vertex. A *cycle* is a walk with no repeated vertex except $v_0 = v_k$. A vertex v is *adjacent* to a vertex w, and has w as a *neighbor*, if there exists an edge (v, w) in E that associates v with w. A graph is *connected* if, for every pair of vertices u and v, there exists a path from u to

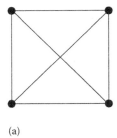 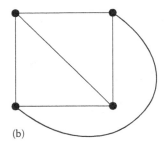

(a)   (b)

Figure 21.1   (a) Nonplanar embedding of K_4 and (b) planar embedding of K_4.

v in the graph, and a *connected component* of a graph is a maximal connected subgraph. A *cut vertex* of a graph is a vertex whose removal, along with its incident edges, increases the number of connected components in the graph. A graph with no cut vertices is *biconnected*, and a *biconnected component* of a graph is a maximal biconnected subgraph. A *directed* edge (v, w) is an edge in which v is adjacent to w but w is not adjacent to v, and an edge is *undirected* if both its endpoints are neighbors. A graph is *undirected* if, for each neighbor w of each vertex v, the vertex v is also a neighbor of w. A graph is *simple* if it contains no duplicate edges and no loops (edges with both endpoints at the same vertex). In this chapter, graphs are assumed to be simple and undirected since edge direction, duplicate edges, and loops are not significant concerns with respect to planarity and since input graphs can easily be converted to simple undirected graphs.

A graph can be embedded in a plane by placing the vertices at distinct locations in the plane, such as points, and placing a continuous curve in the plane, such as a line, to connect the vertex endpoints of each edge. The embedding is *planar* if none of the edge embeddings intersect, except at their common endpoints. A *planar graph* is a graph for which a planar embedding exists. A planar embedding of a graph divides the plane into connected regions, called *faces*, each bounded by edges of the graph. A region of finite area is called a *proper face* and is bound by a cycle. The *external face* is bound by a walk and comprises the plane less than the union of the proper faces, plus its boundary. The *interior region* of a face is its region less its boundary. Figure 21.1a depicts an embedding of K_4, a graph with four vertices and six edges, and Figure 21.1b shows a planar embedding of K_4.

The planar graph embedding problem is typically divided into a geometric component and a combinatorial component. A *planar drawing* is a geometric planar embedding that associates locations (point sets) of the plane with vertices and edges. There are many kinds of planar drawings with different geometric characteristics, and the underlying combinatorial component of planar graph embedding is independent of these differences. A *combinatorial planar embedding* is a consistent edge ordering for each vertex that determines the bounding edges of the planar embedding faces. The sequential edge order of a vertex is called its *orientation* because it corresponds to a clockwise (or counterclockwise) order for the edges in a geometric planar embedding of the graph. Hence, the combinatorial planar embedding is an equivalence class of planar drawings described by the orientations of each vertex ([1], p. 7). For any consecutive pair of edges e_i and e_j incident to a vertex v, the entire boundary of the face containing the corner (e_i, v, e_j) can be traversed by starting at $e = e_j$ and $w = v$ then iteratively obtaining the next w as the neighbor indicated by e and then the next e as the successor of e in the edge order of w. The iteration ends successfully when it returns to v via the e becoming equal to e_i. The boundaries of all faces can be obtained in this way, and the edges can be marked as they are visited to ensure each edge is used in only two face boundaries and only once per face boundary. Lastly, the number of faces f produced by all m edges and n vertices in a connected component is governed by the

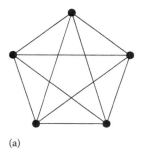 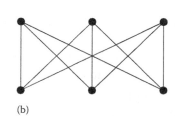

Figure 21.2 Planar obstructions (a) K_5 and (b) $K_{3,3}$.

well-known Euler formula $f = m - n + 2$. Thus, a combinatorial planar embedding provides a testable certification of the planarity of a graph.

It is also desirable to have a simple certificate of nonplanarity for a nonplanar graph, in part so that the output of a planarity algorithm implementation can be checked for correctness regardless of the input. Due to Euler's formula above, it is known that any graph with more than $3n - 6$ edges is not planar since the boundary of each face must contain at least three edges. For this reason, planarity processing is typically restricted on input to a subgraph of at most $3n - 5$ edges, but an excess of edges does not certify nonplanarity as there are many nonplanar graphs with $3n - 6$ or fewer edges. The first characterization of planarity by Kuratowski [2] showed that it is always possible to create a simple certificate of nonplanarity for nonplanar graphs. A *series reduction* replaces a degree two vertex v with an edge connecting the two neighbors of v. An *edge subdivision* is the inverse of a series reduction. A graph G has a *subgraph homeomorphic* to H if vertex and edge deletions can be performed on G so that the remaining subgraph can be converted to H by performing any sequence of series reductions or edge subdivisions. Kuratowski proved that a graph is planar if and only if it contains no subgraph homeomorphic to either of two graphs, which are denoted K_5 or $K_{3,3}$ and depicted in Figure 21.2. Essentially, any nonplanar graph must contain a subgraph that structurally matches K_5 or $K_{3,3}$, except that the edges of the K_5 or $K_{3,3}$ could be paths in the original graph.

In the subsequent planarity characterization by Wagner [3], the series reduction and edge subdivision operations are replaced by the *edge contraction* operation, which removes an edge $e = (u, v)$ and replaces u and v and their incident edges with a single vertex w whose neighbors are the vertices that were neighbors of u and v. Duplicate edges between w and the common neighbors of u and v can be easily detected and reduced to single edges so that the graph remains simple. A graph G contains a *minor* of a graph H, that is an H minor, if a subgraph of G can be converted to H by applying zero or more edge contractions. Wagner proved that a graph is planar if and only if it contains neither a K_5 minor nor a $K_{3,3}$ minor.

The edge addition planarity algorithm [4] described in this chapter constructs a combinatorial planar embedding by adding each edge from the input graph while preserving planarity. If the algorithm becomes unable to embed an edge while preserving planarity, then the proof of correctness shows that the input graph contains a K_5 minor or a $K_{3,3}$ minor. Graph minors are used in the proof because they allow the structural components leading to nonplanarity to be more succinctly depicted. The proof provides an alternative planarity characterization those of Kuratowski and Wagner because it not only presents a set of five nonplanar graph minors, but it also proves by contradiction that the set is complete. For a nonplanar graph, a post-processing step can be used to isolate a subgraph homeomorphic to K_5 or $K_{3,3}$. This process is typically called Kuratowski subgraph isolation. Although graph minors are more powerful for the mathematical characterization of the correctness of the algorithm, a subgraph homeomorphic to K_5 or $K_{3,3}$ is a more easily validated certificate of nonplanarity for a graph.

The first linear-time graph planarity test is due to Hopcroft and Tarjan [5] and was based on a *path addition* technique. A particular cycle C is embedded first, and then each path P is attached to C, either on the inside or the outside of C in order to avoid crossing previously embedded paths whose attachment points to C may overlap those of P.

A conceptually simpler *vertex addition* planarity test had its start with Lempel et al. [6]. A *flip* operation on a planar embedding of a biconnected component reverses the orientations of the vertices in the biconnected component. The Lempel–Even–Cederbaum algorithm begins by creating an *st*-numbering for a biconnected graph, which is a vertex ordering wherein every vertex has a path of lower numbered vertices that connect to the least numbered vertex s and a path of higher numbered vertices leading to the last vertex t. Thus, for each value of k from s to t, there must exist an embedding $\tilde{G}_k$ of the first k vertices such that the remaining vertices can be embedded in a single face of $\tilde{G}_k$. The adjustment of a partial embedding $\tilde{G}_k$ to obtain $\tilde{G}_{k+1}$ consists of permutations at cut vertices and flips of biconnected components of $\tilde{G}_k$. The Lempel–Even–Cederbaum planarity test was optimized to linear time by a pair of contributions. Even and Tarjan [7] optimized *st*-numbering. Booth and Lueker [8] developed a comprehensive data structure called a PQ-tree that could efficiently track which parts of $\tilde{G}_k$ could be permuted or flipped in order to add a new vertex while preserving planarity.

Achieving linear-time performance for either the Hopcroft–Tarjan method or the Lempel–Even–Cederbaum method is considered to be quite complex (e.g., see [9,10]). There are other planarity characterizations that also lead to complex linear-time planarity algorithms, such as [11–13]. However, recent advancements in combinatorial planar embedding techniques have yielded substantially simpler new vertex addition methods in which the *st*-numbering and PQ-tree are replaced by depth-first search (DFS) and simpler data structures (see [14,15] and [16,17]). These new vertex addition methods were rationalized with PQ-tree operations in [18].

The planarity algorithm [4] described in this chapter is not a vertex addition method, though its topmost level is guided by some of the same principles as the vertex addition method in [14]. However, the processing model and proof of correctness are defined in terms of *edge addition* as the atomic operation of combinatorial planar embedding. By comparison, the path and vertex addition methods may be regarded as batch methods that have greater complexity because they must determine whether a whole path or a vertex and all incident edges can be added while preserving planarity. In this regard, the edge addition method is more closely aligned to the task of planar embedding, which is fundamentally about ensuring that an edge does not cross any other edge when it is embedded.

The remaining sections of this chapter provide a detailed exposition of the Boyer–Myrvold edge addition planarity algorithm as follows. Section 21.2 provides the key definitions and describes the top-level processing, while Section 21.3 presents the core method of the algorithm, called the **Walkdown**. Section 21.4 provides the proof of correctness, Section 21.5 describes techniques for efficient implementation of the algorithm, and Section 21.6 provides additional techniques for postprocessing a nonplanar graph to obtain a minimal subgraph obstructing planarity, that is, a subgraph homeomorphic to $K_{3,3}$ or K_5. Section 21.7 describes an augmentation to the edge addition planarity algorithm that produces a particular kind of planar drawing for planar graphs. Lastly, Section 21.8 provides concluding remarks, including information about a reference implementation and several related problems that can be solved by the edge addition planarity method.

21.2 OVERVIEW OF EDGE ADDITION PLANARITY

The edges of a simple undirected input graph G are added one at a time to a combinatorial planar embedding $\tilde{G}$ in such a way that planarity is preserved with each edge addition. Throughout the process, $\tilde{G}$ is managed as a collection of planar embeddings of the biconnected components that develop as each edge is embedded. Initially, a DFS is performed [19] to number each vertex according to its visitation order and to distinguish a spanning tree called a *DFS tree* in each connected component. Each undirected edge in a DFS tree is called a *tree edge*, and each undirected edge not in a DFS tree is called a *back edge*. The DFS tree of each connected component establishes parent, child, ancestor, and descendant relationships among the vertices in the component. Each vertex has a lower DFS number than its children and descendants and a higher number than its parent and ancestors, except the DFS *tree root* is a vertex with no parent or ancestors. The vertex endpoints of a DFS back edge share the ancestor–descendant relationship. A cut vertex r in $\tilde{G}$ separates at least one DFS child c of r from the DFS ancestors (and any other DFS children) of r. A *virtual vertex* is an extra vertex in $\tilde{G}$ (but not in G) that is used to represent r in the separate biconnected component containing c. The virtual vertex is denoted r^c, or simply r' when the child identity is unimportant. The virtual vertex r^c is the *root* of the biconnected component B_{r^c} that contains c.

For each DFS tree edge (v, c) of G, a singleton biconnected component (v^c, c) is added to $\tilde{G}$. Then, each back edge of G is added to $\tilde{G}$ in an order that is partially organized into steps based on the depth-first index (DFI) order of the vertices. For each vertex v, each back edge between v and a DFS descendant of v that can be added while preserving planarity is embedded in $\tilde{G}$. Each DFS tree of vertices is processed in a bottom-up fashion by simply using reverse DFI order. Thus, the back edges between a vertex v and its DFS ancestors are embedded in the future steps in which those ancestors are processed.

A single new back edge (v, w) has the potential to eliminate cut vertices in $\tilde{G}$. To add a back edge (v, w) to $\tilde{G}$, any previously separable biconnected components are first merged, as is shown in the example in Figure 21.3. In the diagram, r is no longer a cut vertex once an edge representing (v, w) is added, so the child biconnected component rooted by the virtual vertex r' is merged with the parent biconnected component containing the nonvirtual vertex r. Then, the back edge (v^c, w) is embedded to complete the biconnection. A back edge (v, w) is always added incident to a virtual vertex v^c, where c is a DFS ancestor of w, because the back edge does not biconnect w and the parent of v. In some future step when a back edge is added that biconnects w and the parent of v, then v^c will be merged with v, and the edge (v^c, w) will become (v, w) due to that merge operation, just as the edges incident to r' were transferred to r in the merge operation depicted in Figure 21.3.

The adjacency list nodes for a back edge (v^c, w) are added to either the start or end of the adjacency lists of v^c and w so that the new edge appears along the external face of the biconnected component. The new edge also forms a new proper face that includes any cut vertex merge points and selected external face paths between v^c and w. In order to select the paths that form the new proper face, one or more biconnected component embeddings may need to be flipped during the merge operation. The paths selected are those that do not contain certain vertices that must remain on the external face boundary when the new back edge is embedded. For example, in Figure 21.3, the child biconnected component is flipped so that vertex y stays on the external face boundary. For efficiency, this operation is performed in amortized constant time using a method described in Section 21.5.

The following are some operational definitions that are used to determine when a vertex w must be kept on the external face boundary of a biconnected component in $\tilde{G}$. A vertex

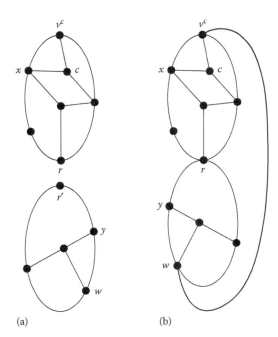

Figure 21.3 (a) Parent biconnected component with cut vertex r and child biconnected component rooted by a virtual vertex r'. (b) Adding the back edge (v, w) and merging biconnected components.

w is *active* if there is an unembedded back edge e for which w is either the descendant endpoint or a cut vertex in $\tilde{G}$ on the DFS tree path between the endpoints of e. A vertex w is *inactive* if it is not active, and it can become inactive during the embedding of any single back edge. Since each back edge is embedded incident to a virtual vertex, extending into the interior region of the external face region, and then incident to the descendant endpoint, the algorithm keeps all active vertices on the external face boundaries of biconnected components in $\tilde{G}$ so that a newly embedded edge does not cross an edge of an external face boundary.

A vertex w is *pertinent* in step v of edge addition processing if it is active due to an unembedded back edge e incident to v. In this definition, the descendant endpoint of edge e may be w or a descendant d of w, in which case w has a child biconnected component rooted by a virtual vertex w^c, where c is the root of a DFS subtree containing d. Similarly, a vertex w is *future pertinent* in step v if it is active due to an unembedded back edge e incident to a DFS ancestor u of v. Again, the descendant endpoint of edge e may be w or a descendant d of w, in which case w has a child biconnected component rooted by a virtual vertex w^c, where c is the root of a DFS subtree containing d. A vertex is *only pertinent* if it is pertinent but not future pertinent, and it is *only future pertinent* if it is future pertinent but not pertinent.

In [4], a future pertinent vertex was called *externally active* because it had to be kept on the external face throughout all edge additions in step v. This equivalence makes sense for planarity, but extension algorithms can benefit from having a definition of externally active that differs from future pertinence. For example, the edge addition planarity method becomes an outerplanarity method by simply making all vertices externally active, and some extensions that are based on the outerplanarity method still make use of future pertinence as a separate phenomenon. Under both definitions of the term externally active, a *stopping vertex* is externally active but not pertinent; this name refers to the fact that a stopping vertex terminates an external face traversal of the main edge adding method called the **Walkdown**, which is discussed in Section 21.3.

> **Algorithm:** Edge Addition Planarity
> 1: Initialize embedding $\tilde{G}$ based on input graph G
>
> 2: For each vertex v from $n-1$ down to 0
> 3:   Establish pertinence for step v within $\tilde{G}$
> 4:   For each successive DFS child c of v
> 5:   Embed the tree edge (v,c) as a singleton biconnected component (v^c, c)
> 6:   Perform **Walkdown** to embed back edges from v^c to descendants of c
> 7:   if any back edge from v to a descendant of c was not embedded
> 8:   Isolate planarity obstruction and return NONPLANAR
> 9: Postprocess planar embedding and return PLANAR

Figure 21.4 Top-level processing model of the edge addition planarity algorithm.

The definitions related to activity and pertinence are applicable to vertices, but not to virtual vertices, which are automatically kept on the external face boundaries until they are merged with the cut vertices they represent. There are, however, analogous definitions for activity and pertinence for each whole biconnected component. A biconnected component is *active, pertinent,* or *future pertinent* if it contains an active, pertinent, or future pertinent vertex, respectively. Efficient methods of implementing these definitions are discussed in Section 21.5. These definitions support the correct operation of the top-level edge addition processing model presented in Figure 21.4.

Initialization includes performing the DFS, arranging the vertices into DFI order in linear time, and calculating the least ancestor and lowpoint [19] of each vertex. The edge embedding process is performed for each vertex v in reverse DFI order, beginning with establishing pertinence for step v. The DFS tree edges can be embedded either all at once during initialization (as in [4]) or a DFS tree edge (v, c) can be embedded as a singleton biconnected component (v^c, c) immediately before embedding the back edges between v and descendants of c.

The back edges between v and its descendants are added systematically for each child c by traversing the external faces of (v^c, c) and its descendant biconnected components. This traversal is performed by a method called **Walkdown**. The planarity of $\tilde{G}$ is preserved for each edge addition that the **Walkdown** performs, and if the **Walkdown** is unable to traverse to the descendant endpoint of any back edge, then the input graph is not planar [4] according to the proof in Section 21.4.

If the **Walkdown** successfully embeds all back edges for each child c of each vertex v, then a planar embedding is produced. For efficiency, especially of biconnected component flip operations, the consistency of vertex orientation in $\tilde{G}$ is relaxed so that vertices can have either a clockwise or counterclockwise orientation. Section 21.5 describes efficient techniques for relaxed orientation and external face management, biconnected component flipping and merging, and the postprocessing to recover a planar embedding. On the other hand, if the **Walkdown** is unable to traverse to the descendant endpoint of any back edge during a step v, then a subgraph homeomorphic to $K_{3,3}$ or K_5 is obtained using the techniques described in Section 21.6.

21.3 WALKDOWN

To embed the back edges from a vertex v to its descendants in a DFS subtree rooted by child c, the **Walkdown** is invoked on a biconnected component B_{v^c} rooted by the virtual vertex v^c. The **Walkdown** performs two traversals of the external face of B_{v^c}, corresponding to the two opposing external face paths emanating from v^c. The traversals perform the same operations and are terminated by the same types of conditions, so the method of traversal will only be described once.

A traversal begins at v^c and proceeds in a given direction from vertex to vertex along the external face boundary in search of the descendant endpoints of back edges. Whenever a vertex is found to have a pertinent child biconnected component, the **Walkdown** descends to its root and proceeds with the search. Once the descendant endpoint w of a back edge is found, the biconnected component roots visited along the way must be merged (and the biconnected components flipped as necessary) before the back edge (v^c, w) is embedded. An initially empty *merge stack* is used to help keep track of the biconnected component roots to which the **Walkdown** has descended as well as information that helps determine whether each biconnected component must be flipped when it is merged.

A **Walkdown** traversal terminates either when it returns to v^c or when it encounters a stopping vertex. If the **Walkdown** were to proceed to embed an edge after traversing past such a stopping vertex, then the vertex would not remain on the external face. In planarity testing, this would be problematic because the stopping vertex is future pertinent, and embedding an edge corresponding to that future pertinence would cause a path from the stopping vertex to cross the external face boundary. By comparison, a future pertinent vertex that is also pertinent does not stop processing because the **Walkdown** can embed an edge directly to the vertex or descend to one of its pertinent child biconnected components and merge it without removing the vertex from the external face.

Observe that if a child biconnected component $B_{w'}$ is only pertinent, then after its root is merged with w, the **Walkdown** traversal eventually visits the entire external face boundary of $B_{w'}$ and returns to w. By comparison, once the **Walkdown** descends to a pertinent child biconnected component that is also future pertinent, the traversal encounters a stopping vertex before returning to w. To avoid prematurely encountering a stopping vertex, the **Walkdown** enforces Rule 21.1.

Rule 21.1 *When vertex w is encountered, first embed a back edge to w (if needed) and then descend to all of its child biconnected components that are only pertinent (if any) before descending to a pertinent child biconnected component that is also future pertinent.*

A similar argument governs how the **Walkdown** chooses a direction from which to exit a biconnected component root r^s to which it has descended. Both external face paths emanating from r^s are searched to find the first active vertices x and y in each direction. The path along which traversal continues is then determined by Rule 21.2.

Rule 21.2 *When selecting an external face path from the root r^s of a biconnected component to the next vertex, preferentially select the path to a vertex that is only pertinent, if one exists, and select an external face path to a pertinent vertex otherwise.*

If both external face paths from r^s lead to vertices that are only future pertinent, then both are stopping vertices and the entire **Walkdown** (not just the current traversal) can be immediately terminated due to a nonplanarity condition.

In general, the **Walkdown** fails to embed a back edge if it is blocked from traversing to a pertinent vertex w by stopping vertices x and y appearing along each of the two external face paths emanating from the root of a biconnected component. In the nonplanarity condition above, the biconnected component is rooted by r', where r is a descendant of v. In this case, the merge stack will be nonempty, and r' will be on the top of the stack. Otherwise, the two **Walkdown** traversals were terminated at distinct stopping vertices x and y, and the biconnected component B_{v^c} still contains a pertinent vertex w. This second nonplanarity

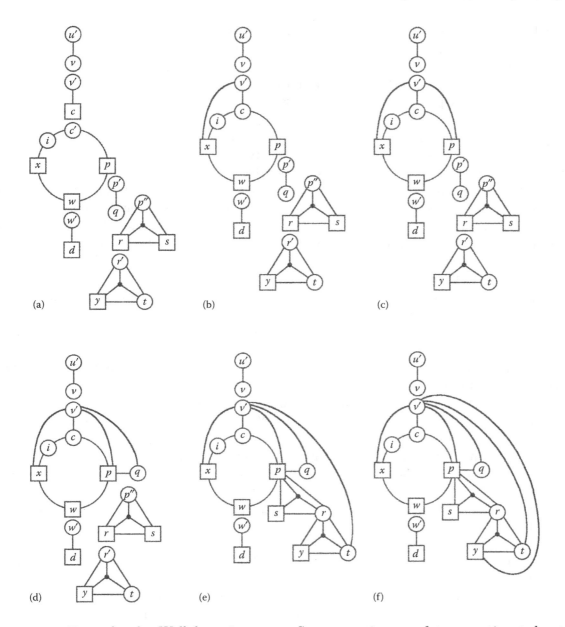

Figure 21.5 Example of a Walkdown in step v. Square vertices are future pertinent due to unembedded back edges (u,d), (u,s), (u,x), and (u,y). Back edges (v,p), (v,q), (v,t), (v,x), and (v,y) are to be added in step v. (a) Embedding at the start of step v. (b) Merge at c to add (v,x), then stop counterclockwise traversal. (c) Clockwise traversal visits p and embeds (v,p). (d) Merge p and p' and embed (v,q). (e) Flip biconnected component rooted by p'', merge p and p'', merge r and r', then embed (v,t). (f) Embed (v,y) and stop clockwise traversal.

condition is detected when the merge stack is empty and the list of unembedded back edges still contains an edge to a descendant of c. In Section 21.4, the proof of correctness associates these two nonplanarity conditions with five minors of $K_{3,3}$ and K_5.

21.3.1 Example of Walkdown Processing

This section presents an example that demonstrates the key processing rules of the **Walkdown** method. Figure 21.5a presents a partial embedding of a graph at the beginning of step

v and with the following edges still to embed: (u,d), (u,s), (u,x), (u,y), (v,p), (v,q), (v,t), (v,x), and (v,y). Note that the vertex i is inactive and the biconnected component rooted by w' is not pertinent. The *square vertices* are future pertinent; some are also pertinent, such as p, and others are only future pertinent, such as x and y. Now we will discuss the actions performed by the **Walkdown** to embed the back edges from v to its descendants.

The first traversal in the counterclockwise direction begins at v', travels to c, then descends to c'. The first active vertices along the two external face paths are x and p, both of which are future pertinent and pertinent. The decision to proceed in the direction of x is therefore made arbitrarily. At x, there is a back edge to embed, so the **Walkdown** first merges c and c' with no flip operation since the traversal direction was consistently counterclockwise when entering c and exiting c'. After this merge, c becomes nonpertinent because it has no more pertinent child biconnected components, and it is no longer future pertinent because it has no separated DFS children with back edge connections to ancestors of v. Figure 21.5b shows the result of the merge and the embedding of (v,x).

Once the back edge to x has been embedded, the **Walkdown** determines that x is a stopping vertex, so the second **Walkdown** traversal commences in a clockwise direction from v' to c. In this example, c became inactive in the first traversal, so the second traversal proceeds beyond c in the clockwise direction to p.

At p, the back edge (v,p) is embedded first, as shown in Figure 21.5c. Then, the **Walkdown** descends to p', rather than p'', because $B_{p'}$ is only pertinent. Both paths lead to q, which is only pertinent, so p and p' are merged and the back edge (v,q) is embedded as shown in Figure 21.5d. Since q becomes inactive, the **Walkdown** proceeds to its successor on the external face, which is p.

In this second visitation of p, the **Walkdown** again tests whether a back edge to p must be embedded, but since the back edge has already been embedded, the result is negative. The **Walkdown** again tests for pertinent child biconnected components, but this time there are none which are only pertinent, so the **Walkdown** descends to p''. The two external face paths from p'' lead to future pertinent vertices r and s, but r is pertinent and s is not, so the **Walkdown** selects the counterclockwise direction from p'' to r. This is contrary to the clockwise direction by which the **Walkdown** entered p, so the indication of a flip operation is pushed onto the merge stack, along with p''.

The **Walkdown** proceeds to r, where it finds r has no back edge to embed, but r does have a pertinent child biconnected component, so the **Walkdown** descends to r'. The two external face paths from r' lead to y and t. While y is pertinent, it is also future pertinent, whereas t is only pertinent. The clockwise path to t is selected, in opposition to the counterclockwise direction used to enter r. Thus, r' and a flip indicator are pushed onto the merge stack.

At t, the **Walkdown** determines that a back edge must be embedded. First, the merge stack is processed. $B_{r'}$ is flipped, and r' is merged with r. Then, p'' is popped and the component comprised of $B_{p''}$ merged with $B_{r'}$ is flipped. Finally, the back edge (v,t) is embedded. Notice that $B_{r'}$ is logically flipped a second time, restoring its original orientation. All such double flips are effectively eliminated using an efficient implementation technique described in Section 21.5. The logical result of these operations is shown in Figure 21.5e.

The clockwise traversal then continues from vertex t, which is now inactive, to vertex y. The back edge (v,y) is embedded as shown in Figure 21.5f. Once the back edge to y is embedded, y is no longer pertinent since it has no pertinent child biconnected components. Thus, y is a stopping vertex that terminates the second **Walkdown** traversal.

21.4 PROOF OF CORRECTNESS

In this section, we prove that the edge addition planarity algorithm described in Sections 21.2 and 21.3 correctly distinguishes between planar and nonplanar graphs. It is clear that the algorithm maintains planarity of the biconnected components in $\tilde{G}$ during the addition of each edge. Thus, a graph G is planar if all of its edges are added to $\tilde{G}$, and we focus on showing that if the **Walkdown** is unable to embed a back edge between a virtual vertex v^c and a descendant of c, then the graph G is nonplanar.

The **Walkdown** halts when a traversal descends to the root of a pertinent biconnected component if both external face paths are blocked by stopping vertices, as depicted in Figure 21.6a. Otherwise, if any back edge from v^c to a descendant of c is not embedded, then both **Walkdown** traversals were blocked by stopping vertices from reaching a pertinent vertex in the biconnected component B_{v^c}, as depicted in Figure 21.6b.

These configurations omit unimportant details such as virtual vertices, and unimportant structure is eliminated using edge contraction and deletion. For example, in Figure 21.6, all ancestors of v are edge contracted into u. Similarly, the future pertinence of x and y and the pertinence of w may involve descendant biconnected components in addition to unembedded back edges, and these descendants are edge contracted into x, y, and w. Hence, the dashed edge (v,w) represents a back edge connection that the **Walkdown** was blocked from completing. The dashed edges (u,x) and (u,y) represent the unembedded back edge connections from x and y to ancestors of v. The edge (u,v) represents the DFS tree path from v to the ancestors needed to complete the connections with x and y. Similarly, the edge (v,r) in Figure 21.6a represents the DFS tree path between v and the root of the pertinent biconnected component containing the stopping vertices x and y.

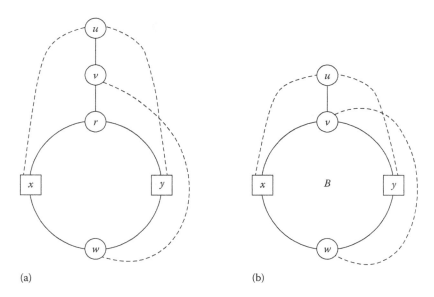

Figure 21.6 Unembedded back edge configurations for the **Walkdown**. (a) A $K_{3,3}$ minor occurs if the **Walkdown** descends from v to the root r of a biconnected component, but stopping vertices x and y block the external face paths to a pertinent vertex w. (b) Both **Walkdown** traversals of the external face of a biconnected component B rooted by v encountered stopping vertices, x and y, and could not reach the pertinent vertex w. Theorem 21.1 proves that this configuration is accompanied by one of four additional structures that help to form a $K_{3,3}$ minor or K_5 minor.

Since the graph minor Figure 21.6a is a $K_{3,3}$, the input graph is clearly nonplanar. In the case of the graph minor in Figure 21.6b, it is natural to ask why did the **Walkdown** halt rather than embedding (v, w) before embedding (v, x) or (v, y) so that (v, w) is inside the bounding cycle of B. In short, Rules 21.1 and 21.2 ensure that additional structure exists either in the subgraph represented by (v, w) or within B that prevents the **Walkdown** from embedding the connection from w to v inside B. An examination of the possibilities yields additional nonplanarity minors depicted in Figure 21.7(b–e). Theorem 21.1 argues that one of these five nonplanarity minors must exist if the **Walkdown** fails to embed a back edge, and the absence of the conditions that give rise to these nonplanarity minors contradicts the assumption that the **Walkdown** failed to embed a back edge.

Theorem 21.1 *Given a biconnected component B with root v^c, if the **Walkdown** fails to embed a back edge from v to a descendant of c, then the input graph G is not planar.*

Proof. By contradiction, suppose the input graph G is planar but the **Walkdown** halts without embedding a back edge from v to a descendant of c. To do so, the **Walkdown** must encounter two distinct stopping vertices, denoted x and y, that prevent traversal to a pertinent vertex, denoted w. The stopping vertices and pertinent vertex cannot be in a biconnected component rooted by a descendant r of v, since in that case the graph is nonplanar due to the $K_{3,3}$ minor depicted in Figure 21.7a. Hence, assume the **Walkdown** halted on stopping vertices in the biconnected component containing v^c. ∎

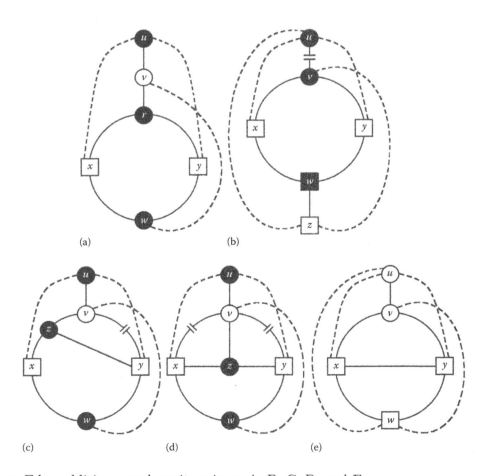

Figure 21.7 Edge addition nonplanarity minors A, B, C, D, and E.

Figure 21.7b depicts the relevant additional structure that exists if w has a pertinent child biconnected component that is also future pertinent. Embedding the pertinent connection from w to v would place a future pertinent descendant of w, or one of x or y, inside the bounding cycle of B. However, this condition cannot arise in a planar graph since the minor in Figure 21.7b contains a $K_{3,3}$.

Therefore, we consider conditions related to having an obstructing path inside B that contains only internal vertices of B except for two points of attachment along the external face: one along the path $v, \ldots, x, \ldots, w$, and the other along the path $v, \ldots, y, \ldots, w$. The obstructing path, which is called an x–y path, contains neither v nor w. If such an x–y path exists, then the connection from w to v would cross it if the connection were embedded inside B. We use p_x and p_y to denote the points of attachment of the obstructing x–y path.

Figure 21.7c represents the condition in which the x–y path has a *high* point of attachment, that is closer to v than the stopping vertex. The diagram depicts p_x attached closer to v than x, but symmetrically, p_y could be attached closer to v than y. Given one high point of attachment, the opposing point of attachment can be edge contracted into the stopping vertex. None of these cases can occur in a planar graph since the graph minor in Figure 21.7c, which characterizes these cases, contains a $K_{3,3}$.

Thus, we consider the cases in which both x–y path attachment points are at or below the respective stopping vertices. Without loss of generality, the stopping vertices x and y are edge contracted into the respective points of attachment, p_x and p_y, of the x–y path.

Figure 21.7d depicts the condition of having a second path of vertices attached to v that (other than v) contains vertices internal to B that lead to an attachment point z along the x–y path. This second path cannot exist in a planar graph since the minor in Figure 21.7d contains a $K_{3,3}$.

In Figure 21.7e, a future pertinent vertex exists along the lower external face path strictly between p_x and p_y. Without loss of generality, this vertex can be edge contracted into w. This condition cannot occur in a planar graph due to the K_5 minor in Figure 21.7e.

Since the input graph is assumed to be planar, the above nonplanarity conditions must be absent. Due to the absence of the condition of Figure 21.7a, the two **Walkdown** traversals must have ended on stopping vertices along external face paths in the biconnected component B rooted by v^c. Conjunctive to the contradictive assumption, B has a pertinent vertex w along the lower external face path strictly between stopping vertices x and y. We address two cases based on whether or not there is an obstructing x–y path.

If no obstructing x–y path exists, then at the start of step v all paths between x and y in $\tilde{G}$ contain w. Thus, w is a DFS ancestor of x or y (or both), and it becomes a merge point when its descendants (x or y or both) are incorporated into B. Due to processing Rule 21.1, when the **Walkdown** first visits w, it first embeds a direct back edge from w to v if one is required and processes the active child biconnected components that are only pertinent. Thus, w must be pertinent due to a child biconnected component that is also future pertinent. Yet, this contradicts the pertinence of w since otherwise the input graph is not planar due to Figure 21.7b.

On the other hand, suppose there is an obstructing x–y path, but the conditions of nonplanarity minors C, D, and E do not apply. The *highest x–y path* is the obstructing x–y path that would be contained by a proper face cycle starting at v^c if the internal edges to v^c were removed, along with any resulting separable components. The highest x–y path and the lower external face path from its attachment points, p_x to p_y, formed the external face of a biconnected component at the beginning of step v. Let r_1 denote whichever of p_x or p_y was the root of that biconnected component, and let r_2 denote the other of p_x or p_y such that $r_1 \neq r_2$. Figure 21.8 illustrates an example in which p_x is r_1.

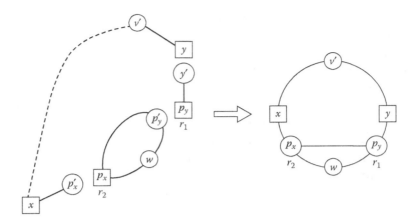

Figure 21.8 Example of contradiction to assumption of pertinence of w. If there had been a vertex w that was only pertinent, then **Walkdown** processing Rule 21.2 would not have selected the path from p'_y to p_x since p_x was not only pertinent but also future pertinent.

Since the condition of Figure 21.7c does not exist, r_2 is equal to or an ancestor of one of x or y and was therefore future pertinent when the **Walkdown** descended to r'_1. Moreover, the first active vertex along the path that is now the highest x–y path was r_2 because the condition of Figure 21.7d does not exist. Descending from r'_1 along the path that is now the lower external face path between p_x and p_y, the existence of a pertinent vertex w implies that there are no future pertinent vertices along the path due to the absence of the condition of Figure 21.7e. Thus, we reach a contradiction to the pertinence of w since processing Rule 21.2 stipulates that the **Walkdown** would have preferentially selected the path to w, which is only pertinent, rather than r_2. ■

Theorem 21.2 *Given a simple undirected graph G, the edge addition planarity algorithm correctly determines that G is nonplanar or generates a combinatorial planar embedding if G is planar.*

Proof. If G is not planar, then eventually a **Walkdown** invocation must fail to embed an edge because no mutation of $\tilde{G}$ perturbs planarity, nor the property that $\tilde{G}$ is a combinatorial planar embedding, as follows. Adding a tree edge produces a single edge biconnected component. Back edges are added into the interior region of the external face and to endpoints on the external face boundaries of biconnected components. The flip of a planar embedding of a biconnected component is also a planar embedding. A merge operation adds the adjacency list of a virtual vertex to its nonvirtual counterpart; both are on external face boundaries, and the resulting adjacency list order preserves planarity. Due to Theorem 21.1, the **Walkdown** fails to embed an edge only if G is not planar. Thus, if G is planar, then all of its edges are added to $\tilde{G}$, and $\tilde{G}$ is a combinatorial planar embedding of G. ■

21.5 EFFICIENT IMPLEMENTATION

21.5.1 Graph Storage and Manipulation

The graph is represented by storing the list of neighbors, or the adjacency list, of each vertex as a doubly linked list, with the vertex containing links to both the head and the tail of the adjacency list. The adjacency list can be efficiently traversed and mutated regardless of the vertex orientation. Each adjacency list node, or *arc*, indicates a neighbor vertex. The pair of arcs representing an edge (v, w) are called *twin arcs*, and the one in v's list indicates w while the other in w's list indicates v. The twin arcs are stored in consecutive array locations so it takes constant time to traverse an edge since the location of an arc can be calculated given the location of its twin arc.

During embedding, each cut vertex is represented by a virtual vertex in each biconnected component containing one of its DFS children. An efficient implementation relies on constant time traversal from a virtual vertex r^c to its nonvirtual counterpart r. Since each virtual vertex r^c is uniquely associated with a specific DFS child c of the cut vertex r, the virtual vertex can be stored at location c in an array of virtual vertices. In fact, since a virtual vertex has an adjacency list whose arcs are the twins of arcs in nonvirtual vertices, it is expedient to store the virtual vertices in the same array as the nonvirtual vertices by using positions 0 to $n-1$ for vertices and positions n to $2n-1$ for virtual vertices. Then, given r^c, the vertex r is obtained by subtracting n to get c, then obtaining the DFS parent value associated with c.

When a vertex is on the external face boundary of a biconnected component, it is convenient to have immediate access to the two arcs that attach the vertex to the external face boundary. If the adjacency list is noncircular, then a vertex can be equipped with indicators of both the first and last arcs of its adjacency list. Then, given a vertex w entered via arc w_{in}, traversing to the next vertex x on the external face consists of exiting from the opposing arc w_{out} from among w's first and last arcs, then obtaining x_{in} as the twin arc of w_{out} and obtaining x from the neighbor field of w_{out}. In Section 21.5.7, an optimization of this technique is presented that helps the **Walkdown** to avoid repeatedly traversing through paths of inactive vertices.

21.5.2 Sorted Child Lists and Forward Edge Lists

Several operations are facilitated by equipping each vertex with a *sorted DFS child list* and a *forward arc list* sorted by DFIs of the descendant endpoints. In the initial DFS, it is easy to construct a sorted list of DFS children since the DFS children of each vertex are visited in ascending DFI order. Similarly, when a back arc to an ancestor is found by DFS, the companion forward arc can be added to the forward arc list of the ancestor endpoint. Again, since vertices are visited in DFI order, the forward arc list of each vertex will be formed in ascending order of the descendant DFIs.

The child list is a list of the DFS children, not a list of arcs indicating the children, so the child list is not changed during embedding. The forward arc list contains the actual forward arcs of the vertex, so when the **Walkdown** embeds a back edge from v^c to a descendant d, the forward arc indicating d is removed from the forward arc list of v in order to place it in the adjacency list of v^c. Note that the remaining elements of the forward arc list are still sorted by the DFIs of the descendant endpoints.

21.5.3 Systematic Walkdown Invocation and Nonplanarity Detection

The sorted DFS child list provides the direct ability to perform the systematic iteration of the DFS children indicated on Line 4 of the pseudo-code in Figure 21.4. The following Lines 5, 6, and 7 perform a constant time tree edge embedding, the invocation of the **Walkdown**, and a test for nonplanarity.

Nonplanarity can be detected in constant time because the forward arc list is sorted and only contains the forward arcs of unembedded edges. After the **Walkdown** has finished processing for a particular child c_i, there are two cases to consider. If there is no next DFS child, then the graph is nonplanar if there are any forward arcs left in the forward arc list. Otherwise, if there is a next DFS child c_j, then it is compared to the descendant endpoint

d of the first element of the sorted forward arc list. If $d < c_j$, then the graph is nonplanar because d is in the DFS subtree of c_i.

21.5.4 Pertinence Management with Walkup

To help manage pertinence, each vertex is equipped with a few extra data members. For each back edge (v, d) to be embedded in step v, the *pertinentEdge* member of the descendant d is used to store the forward arc from v to d. This member has the value NIL when not in use, so when the **Walkdown** traverses to each vertex, a constant time query of the **pertinentEdge** member detects whether to embed a back edge and provides the forward arc of the edge to embed. The forward arc of (v, d) is removed from the forward arc list of v, embedded as (v^c, d), and then the **pertinentEdge** member of d is set to NIL.

Each vertex is also equipped with a member called *pertinentRoots*, which is used in each cut vertex to store a list of the roots of the pertinent child biconnected components. The roots of biconnected components that are only pertinent are stored at the front of the list, and those that are both pertinent and future pertinent are stored at the back of the list. Thus, after a **Walkdown** traversal processes the **pertinentEdge** member of a vertex w, it need only obtain the first element, if any, of its **pertinentRoots** list. If the list is empty, then w has become inactive and the **Walkdown** proceeds to the next vertex on the external face. If the list is nonempty, then the first element will be the root of an only pertinent biconnected component unless there are none.

The settings for **pertinentEdge** and **pertinentRoots** for each vertex are made in Line 3 of the pseudo-code in Figure 21.4. For each arc e in the forward arc list, a process called **Walkup** is performed. The **Walkup** first sets the **pertinentEdge** member to e in the descendant endpoint d indicated by e. Then, for each cut vertex r in $\tilde{G}$ between d and the current vertex v, there is a child c of r that is ancestor to d. The **Walkup** ensures that r^c is stored in the **pertinentRoots** list of r.

The **Walkup** arrives at a first vertex w within a biconnected component in one of two ways. Either w is the descendant endpoint d of a forward arc from v, or w is a cut vertex r in $\tilde{G}$ along the DFS tree path between v and d. Either way, starting at a vertex w, the **Walkup** traverses the external face of the biconnected component containing w in search of the root r'. When it is found, it is placed in the **pertinentRoots** list of r, at the end if the biconnected component is future pertinent and at the beginning otherwise. Then, the **Walkup** reiterates this process by setting w equal to r, except it terminates when the biconnected component root found by this process is v'. Two optimization techniques are used to ensure that **Walkup** processing takes linear time in total.

The first optimization technique is *visitation detection*. As the **Walkup** visits vertices along the external face paths of biconnected components, each is marked as visited in step v. Then, if a **Walkup** invocation encounters a vertex w that is marked visited in step v, then that invocation is terminated because a prior **Walkup** invocation for another forward arc of v has already recorded all of the pertinent roots for all ancestors of w.

The second optimization technique is *parallel face walking*. When the **Walkup** starts at a vertex w in a biconnected component, it simultaneously traverses both external face paths emanating from w in search of the root of the biconnected component containing w. A single loop advances two vertex variables so that the root r' is found by the shorter external face path. Later, when the **Walkdown** embeds the forward arc (v', d), either the shorter or the longer external face path will become part of a proper face in the embedding. Therefore, when combined with visitation detection, this technique ensures that the total cost of all **Walkup** processing is within a constant factor of the sum of the bounding cycle sizes of proper faces in the embedding.

21.5.5 Future Pertinence Management

During embedding initialization, two additional values are calculated and stored for each vertex. The *least ancestor* of a vertex v is the ancestor u with the lowest DFI that is directly adjacent to v by a back edge. This value can be calculated during the initial DFS as each back edge is identified. The *lowpoint* is the well-known value originally invented by Tarjan [19] for helping to identify cut vertices in a graph. The lowpoint of a vertex is the minimum of its least ancestor value and the lowpoint values of its DFS children. The lowpoint values of vertices can be computed with a bottom-up traversal of the DFS tree (i.e., post-order visitation). Furthermore, since outer loop of the edge addition planarity algorithm performs a bottom-up traversal of the DFS tree, the initial calculation of the lowpoint value for each vertex v could optionally be deferred until the edge addition planarity algorithm processing begins for vertex v. This is typically more efficient for nonplanar graphs since planarity is a rarity, and nonplanarity is often discovered without processing the whole input graph.

Given the root r^c of a biconnected component B_{r^c}, where r is a descendant of the current vertex v, the biconnected component is determined to be future pertinent in constant time if the lowpoint of c is less than v. Since some vertex in the DFS subtree rooted by c has a back edge to an ancestor of v, and all back edges for ancestors of v are unembedded, B_{r^c} must contain a future pertinent vertex and is therefore future pertinent. With this operation, it is a constant time decision whether to place r^c at the front or back of the **pertinentRoots** list of r when the **Walkup** visits r^c.

A vertex w is future pertinent in step v if its least ancestor value is less than v or if w has a DFS child c for which there is a future pertinent biconnected component rooted by w^c in $\tilde{G}$. This operation is reduced to a constant time cost via the following optimization.

Each vertex is equipped with a *futurePertinentChild* member that is initially set equal to the first child in the DFS child list of the vertex. This member is updated for a vertex at two times during the embedding process. First, when a root w^c is merged with its nonvirtual counterpart w during a back edge embedding, if the **futurePertinentChild** value of w is c, then the **futurePertinentChild** is advanced to the next child in the DFS child list of w (or becomes NIL if there is no successor). Second, immediately before testing the future pertinence of w, if the **futurePertinentChild** member of w indicates a child whose lowpoint is not less than v, then the **futurePertinentChild** is advanced along the DFS child list of w until a child with a lowpoint less than v is found or until the end of the child list is found, in which case the **futurePertinentChild** member becomes NIL. The total cost of initializing and updating the **futurePertinentChild** members of all vertices is linear. Yet, if a vertex w has any DFS child c not in the same biconnected component (due to the first type of update) with a lowpoint less than v (due to the second type of update), then it is indicated by the **futurePertinentChild** member. Thus, a vertex is deemed future pertinent if its least ancestor value is less than v or if it has a non NIL **futurePertinentChild** value c for which the lowpoint of c is less than v.

21.5.6 Merging, Flipping, and Embedding Recovery

When adding a back edge (v', w) to the embedding $\tilde{G}$, one or more descendant biconnected components may be merged into the biconnected component $B_{v'}$, and some may be flipped in order to keep their future pertinent vertices on the external face boundary of $B_{v'}$. Merging a biconnected component root r^c with its nonvirtual counterpart r has a cost commensurate with the degree of r^c, since each must be redirected to indicate r. If the edge used to enter r does not form a proper face corner with the edge used to exit r^c, then a flip operation is performed to invert the orientation of r^c and form the proper face corner. Then, the adjacency list of r^c is added to r in a constant time list union operation.

As described in Section 21.1, a combinatorial planar embedding provides a consistent edge ordering for each vertex. To preserve a consistent orientation of vertices in $\tilde{G}$, the adjacency lists of all vertices in the biconnected component could literally be inverted when the root is inverted. However, it is easy to create graphs that would require $O(n)$ vertices to be inverted $O(n)$ times with such a direct biconnected component flip operation. Instead, the edge ordering in $\tilde{G}$ is relaxed such that the adjacency list order of each vertex is either the desired forward edge ordering or a reversed edge ordering. Furthermore, when a biconnected component rooted by r^c is flipped, only the virtual vertex r^c is inverted and otherwise the DFS tree edge (r, c) is marked with an inversion sign of -1. The initial inversion sign for each tree edge is $+1$. Thus, at any time in the embedding process, and particularly at Line 9 of the pseudo-code in Figure 21.4, a consistent orientation for the adjacency lists of each vertex in each connected component can be obtained in a single DFS tree traversal that takes the product of the inversion signs along the ancestor tree path for the vertex and inverts the edge order of the vertex if the product is -1.

Via this scheme, the embedding $\tilde{G}$ can be regarded as providing a consistent edge order for each vertex indirectly through the combination of its adjacency list order and the markings on its ancestor tree edges. For efficiency, the edge marking could be implemented with a single bit flag, in which case the multiply operation is replaced with exclusive-or. It is also worth noting that marking the tree edge (r, c) conceptually flips not only the biconnected component that was rooted by r^c but also the entire DFS subtree rooted by c. If done literally, there would be no harm since the descendant biconnected components would still be planar embeddings, so there is also no harm in flipping the whole subtree indirectly with a constant time cost for edge marking. Further, since each biconnected component root is merged only once, it can be flipped at most once, after which its edges are added to a nonvirtual vertex and do not participate in any further flip and merge operations. Hence, there is an amortized constant cost per edge for all flip and merge operations performed during embedding, plus a single linear-time cost for the post-processing step to eliminate the edge markings and impart a consistent edge ordering on each vertex. Illustrations of the details of these processes are provided in Figures 21.9 through 21.11.

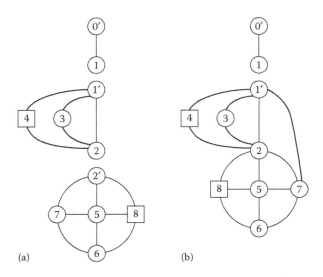

Figure 21.9 (a,b) Overview of data structures for flip operation.

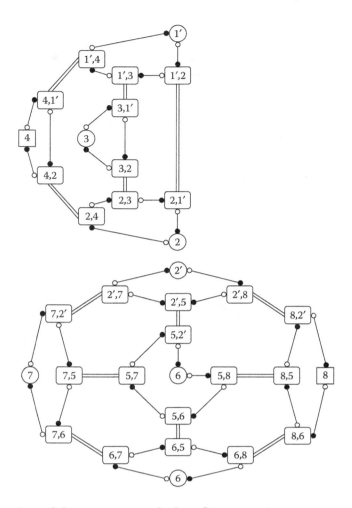

Figure 21.10 Elaboration of data structures before flip operation.

Figure 21.9 presents an overview of the embedding of back edge (1, 7). Figure 21.9a shows the state of the data structures during step 1 after embedding back edges (1, 3) and (1, 4). Because vertex 4 is future pertinent, the first **Walkdown** traversal returns and the second **Walkdown** traversal begins at 1′ such that back edge (1, 7) will be embedded around the right-hand side of the diagram. However, since vertex 8 is also future pertinent, the biconnected component rooted at 2′ must be flipped so that vertex 8 remains on the external face boundary when edge (1, 7) is embedded. The result is shown in Figure 21.9b.

An elaboration of Figure 21.9a appears in Figure 21.10. The rounded rectangles represent adjacency list arcs, or edge records, and the double lines represent the constant-time connection between twin arcs. The circles represent vertex structures, except that vertices 4 and 8 are represented by squares to indicate their future pertinence. The single lines with black and white dots for endpoints represent the links that implement the doubly linked adjacency list format. At this point of the embedding, all vertices still have the same orientation, that is the adjacency list of each vertex can be traversed in counterclockwise order by traversing the black dot links to exit the vertex structure and each edge record in the adjacency list. On the last arc, the black dot link is NIL, but this indirectly indicates the containing vertex, which can be obtained in constant time using the neighbor member of the twin arc.

As stated previously, the first **Walkdown** traversal embeds edges (1, 3) and (1, 4), then stops at vertex 4. The second **Walkdown** traversal starts along the right side of edge (1′, 2),

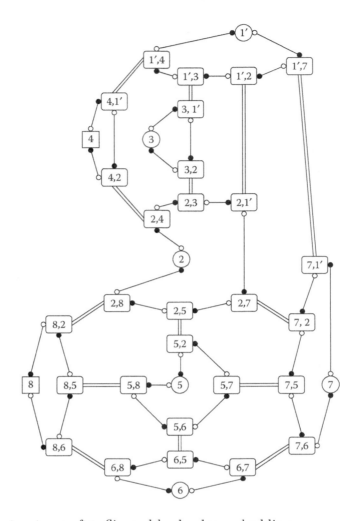

Figure 21.11 Data structures after flip and back edge embedding.

and it descends to the pertinent child biconnected component rooted at $2'$. Since vertex 7 is only pertinent, the biconnected component must be flipped before merging so that vertex 8 remains on the external face boundary.

The results of the flip and merge operations can be seen in Figure 21.11. The merge operation begins by changing all neighbor values for the twin arcs of the arcs in the adjacency list of $2'$ so that they contain 2. Next, the black dot and white dots links in the arcs of $2'$ are inverted. Then, the adjacency list of $2'$ is attached into the adjacency list of vertex 2. The arc $(2, 7)$ is joined with $(2, 1)$, and arc $(2, 8)$ is receives a NIL since it now attaches vertex 2 to the external face boundary.

The diagram of Figure 21.11 shows that the biconnected component has been flipped relative to Figure 21.10, but looking carefully at the sequence of arcs produced by any set of black dot or white dot links shows that no work has been done on the adjacency lists of vertices 5, 6, 7, and 8. For example, in vertex 5, following the black dot links produces the same sequency of adjacency nodes to the neighbors 2, 7, 6, and then 8. Instead of inverting the adjacency lists of vertices 5, 6, 7, and 8, the inversion sign mark is affixed to the tree edge $(2, 5)$.

The final change made in Figure 21.11 is the addition of the back edge $(1, 7)$. The **Walkdown** traversal exited vertex $1'$ using arc $(1', 2)$, the new arc $(1', 7)$ is added between vertex structure $1'$ and arc $(1', 2)$. Since the **Walkdown** entered vertex 7 using the arc $(7, 2')$, the new arc $(7, 1')$ is added between the vertex structure for 7 and the arc $(7, 2)$. Thus, the new edge was added along the the external face boundary and has formed a new proper face in the embedding that includes the corner formed by arc $(2, 1')$, vertex 2, and arc $(2, 7)$.

21.5.7 External Face Management

Based on the detail in Figure 21.11, it is evident that the optimized biconnected component flipping technique has affected the structure of the external face boundary. A simple counterclockwise walk of the external face boundary rooted at $1'$ begins with vertex $1'$ then vertices 4, 2, 8, 6, 7 and back to $1'$. In detail, the counterclockwise traversal can exit from black dot links to proceed from vertex $1'$ to vertex 4 and then vertex 2. This is because the adjacency lists of vertices $1'$, 4, and 2 have the same orientation. However, to continue in the counterclockwise direction, the traversal must exit vertex 8 using the white dot link because the adjacency list of vertex 8 has not been inverted.

To solve this problem, we switch focus from the link used to exit a vertex w, denoted w_out and instead maintain the link used to enter its successor s, denoted s_in. Then, when it is necessary to proceed past s on the external face boundary, the traversal simply selects s_out as the opposing link from s_in. For example, if s_in was stored as the first arc in s, then the last arc in the adjacency list of s is chosen to be s_out.

Generally, during embedding, only the two edges that affix a vertex to the external face boundary are important, and the remainder of the adjacency list is preserved only for post-processing to recover a combinatorial planar embedding or a minimal subgraph obstructing planarity (see Section 21.6). When a new edge is added incident to a vertex, it is placed between the two external face edges and becomes one of the two external face edges. It is possible to represent the external face boundary using a pair of *external face links* for each vertex and virtual vertex to indicate the next and previous vertices on the external face boundary. These links provide an alternative to storing the two arcs of the edges that affix a vertex to the external face boundary. In [4], special *short-circuit* edges were described rather than external face linkages. Either way, a method of optimizing the traversal of the external face is needed in order to avoid traversing paths of inactive vertices multiple times.

When the **Walkdown** descends to the root of a pertinent biconnected component, it must obtain the first active vertices along both external face paths emanating from the root in order to decide in which direction to continue the traversal. The path along the selected direction will become part of a proper face, so the cost of traversing that path can be associated as an additional amortized constant cost per edge of the proper faces of the embedding. The opposing path remains on the external face boundary. It is possible to construct graphs in which such a path contains $O(n)$ inactive vertices and is traversed $O(n)$ times, only to have the **Walkdown** select the opposing path each time. Therefore, the **Walkdown** path selection operation must be optimized to constant time, which can be done by maintaining external face links throughout the edge addition process.

For a biconnected component with root r^c, at the end of the **Walkdown** of r^c in step r, the stopping vertices x and y are known, along with the entry direction x_in and y_in for each. The external face links for r^c are set directly to x and y, which enables a future **Walkdown** that descends to r^c to immediately access x and y without traversing any intervening inactive vertices. Likewise, the vertex r^c is stored in the x_in external face link of x and the y_in external face link of y, which eliminates the inactive vertices from the external face boundary in both traversal directions (from r^c to x or y and from x or y to r^c).

Finally, in the special case where x and y are equal, the above technique is amended to avoid creating an external face boundary with only two vertices. This is necessary because it is challenging to determine whether a vertex is inversely oriented relative to the root r^c if both of its external face links indicate r^c. However, one of the two paths from r^c contains an inactive vertex adjacent to the stopping vertex. If it is the first **Walkdown** traversal path from r^c, then we let x be the neighboring inactive vertex rather than the stopping vertex y. Otherwise, we let y be the inactive vertex neighbor of the stopping vertex x. Then, the external face boundary will contain three vertices. Despite the inactive vertex, a future **Walkdown** that descends to r^c will still make the correct decision to proceed in the direction of the active vertex, and it will still do so in constant time.

21.5.8 Coda

The optimization techniques described in this section support the following theorem.

Theorem 21.3 *The edge addition planarity algorithm produces a combinatorial planar embedding of a simple undirected planar graph G in worst case $O(n)$ time.*

Proof. The initializations in Line 1 of Figure 21.4 are performed with well-known linear-time methods, such as the DFS and lowpoint calculations. The iteration of vertices in Line 2 is a total linear cost, as is the total cost of embedding the DFS tree edges. There is a total linear cost for pre-computing the sorted DFS child lists and forward edge lists according to the methods in Section 21.5.2. Various low-level operations are clearly constant time per operation, such as traversing an edge (due to the twin arc storage mechanism) and determining the exit edge of a vertex on the external face given the entry edge.

The pertinence in step v is efficiently established via the pre-computed forward edge list and the **Walkup** method of Section 21.5.4. At constant cost per back edge, that is linear in total, each forward edge to a descendant d of v is obtained and used as the starting point for an invocation of the **Walkup**. The parallel face-walking ensures that the root of each biconnected component is found at a cost no greater than twice the length of the path that will become part of a proper face of $\tilde{G}$ when the edge (v, d) is embedded. Vertex visitation by the **Walkup** is tracked in each step v to ensure that a biconnected component root r' is not added to the **pertinentRoots** list of r more than once but also to ensure that paths in the biconnected components containing r and its ancestors are not traversed more than a constant number of times in step v. Each setting of **pertinentEdge** is a constant-time cost associated with a unique back edge, and each addition of a biconnected component root r' to a **pertinentRoots** list can only occur once and is associated with a unique DFS child of r.

Using the pre-computed sorted DFS child lists, each DFS child c of v is obtained and used to initiate the invocation of each **Walkdown** on v^c in constant time per DFS tree edge, for a total linear cost. During a vertex visitation by the **Walkdown**, testing the **pertinentEdge** setting and clearing the setting if an edge is embedded are constant-time operations. Determining a child biconnect component to which a traversal must descend is a constant-time decision because the **Walkup** stores the roots of only pertinent biconnected components at the front of the list. Deciding the direction of traversal from the root of a biconnected component is constant-time because the external face management technique ensures the active vertices are immediate neighbors of the root on the external face boundary. All stack manipulations related to descending to a child biconnected component are constant time. Once the descendant endpoint of a back edge is found, the stack pop operations are constant time, and the merge and flip operations are constant time due to the methods in Section 21.5.6. Future pertinence testing and updating is held constant per test or update by the methods in Section 21.5.5. Overall, once a back edge is embedded, constant work is associated with each of the vertices and edges that become part of the newly formed proper face.

The **Walkdown** traversals have additional cost to proceed beyond the descendant endpoint of the last back edge embedded to find a stopping vertex. However, the cost of traversing these paths can be associated with the formation of virtual proper faces formed by updating the external face links, which removes the paths from the external face.

Finally, the sort orders of the DFS child list and the forward edge list ensure there is a constant time test for whether the **Walkdown** embedded all of the back edges between v and descendants of c. After all **Walkdown** operations have been performed, there is a final linear cost to resolve the relaxed vertex orientations and produce a combinatorial planar embedding. Thus, the overall performance for all operations is worse case $O(m)$, where $m \leq 3n - 6$ since G is planar. ∎

21.6 ISOLATING AN OBSTRUCTION TO PLANARITY

The nonplanarity minors of Figure 21.7 can be used to find a Kuratowski subgraph in a nonplanar graph (or a subgraph with at most $3n - 5$ edges). The first step is to determine which nonplanarity minor to use. Minors A–D can be used directly to find a subgraph homeomorphic to $K_{3,3}$. Minor E is a K_5 minor, so a few further tests are performed afterward to determine whether a subgraph homeomorphic to $K_{3,3}$ or K_5 can be obtained.

Once the input graph is found to be nonplanar, the vertices of $\tilde{G}$ can be consistently oriented as described in Section 21.5, and the short-circuiting external face links described in Section 21.5.7 can be discarded. When the nonplanarity condition is detected, the **Walkdown** will have just returned from processing the biconnected component B_{v^c}, and the merge stack will be empty except in the case of nonplanarity minor A, wherein the root r' of a blocked descendant biconnected component will be at the top of the merge stack.

Based on the desired biconnected component root, r' or v^c, the two external face paths from the selected root are searched for the stopping vertices x and y, then the lower external face path $(x, \ldots, y)$ is searched for a pertinent vertex w that the **Walkdown** could not reach.

If the merge stack was nonempty, it is now possible to invoke the minor A isolator (the isolators are described below). If the merge stack is empty, then we must choose one of minors B, C, D, and E. If the last element of the **pertinentRoots** list of w exists and is the root of a future pertinent biconnected component, then the minor B isolator is invoked. Otherwise, the highest x–y path is obtained by temporarily deleting the internal edges incident to v^c, then traversing the proper face bordered by v^c and its two remaining edges. Due to the removal of edges, the boundary of the face will contain cut vertices, which can be easily recognized and eliminated as their cut vertices are visited more than once during the face walk. Once the x–y path is obtained, the internal edges incident to v^c are restored.

If either p_x or p_y is attached high, then the minor C isolator is invoked. Otherwise, nonplanarity minor D is detected by scanning the internal vertices of the x–y path for a vertex z whose x–y path edges are not consecutive above the x–y path. If it exists, such a vertex z may be directly incident to v^c or it may have become a cut vertex during the x–y path test. Either way, the nonplanarity minor D isolator if z is found, and otherwise the nonplanarity minor E isolator is invoked.

Each isolator marks the vertices and edges to be retained, then deletes unmarked edges and merges the biconnected components. The edges are added and marked to complete the pertinent path from w to v and the future pertinence paths from x and y to ancestors of v. Minors B and E also require an additional edge to complete the future pertinence path for z. Finally, the tree path is added from v to the ancestor of least DFI associated with the future pertinence of x, y, and (for minors B and E) z. Otherwise, we mark previously embedded edges along DFS tree paths, the x–y path and v–z path, and the external face of the biconnected component containing the stopping vertices.

To exemplify marking a future pertinence path, we consider the one attached to x (in any of the nonplanarity minors). If the least ancestor directly attached to x by a back edge (a value obtained during the lowpoint calculation) is less than v, then let u_x be that least ancestor, and let d_x be x. Otherwise, let χ indicate the child of x with the least lowpoint for which $\tilde{G}$ still contains the root x^χ. Then, let u_x be the lowpoint of χ, and let d_x be the least descendant endpoint greater than χ from among the endpoints of the forward arcs of u_x. Mark for inclusion the DFS tree path from d_x to x, then add and mark the edge (u_x, d_x). Future pertinence paths for y and (if needed) z are obtained in the same way.

Marking the pertinent path is similar, except that minor B requires the path to come from the pertinent child biconnected component containing z. In the other cases, if the **pertinentEdge** of w is not NIL, then let d_w be w. If the **pertinentEdge** of w is clear or in the case of minor B, then we obtain the last element w^χ in the **pertinentRoots** list of w, then scan the unembedded forward arc list of v for the neighbor d_w with least DFI greater than χ. Finally, mark the DFS tree path d_w to w and add and mark the edge (v, d_w).

To conclude the $K_{3,3}$ isolation for minor A, we mark the DFS tree path from v to the least of u_x and u_y, and we mark the external face boundary of the biconnected component rooted by r. For minor B, we mark the external face boundary of the biconnected component rooted by v^c and the DFS tree path from $\max(u_x, u_y, u_z)$ to $\min(u_x, u_y, u_z)$. The path from v to $\max(u_x, u_y, u_z)$, excluding endpoints, is not marked because the edge (u, v) in minor B is not needed to form a $K_{3,3}$. For the same reason, minors C and D omit parts of the external face boundary of the biconnected component rooted by v^c, but both require the tree path v to $\min(u_x, u_y)$. Minor C omits the short path from p_x to v if p_x is attached high, and otherwise it omits the short path from p_y to v. Minor D omits the upper paths $(x, \ldots, v)$ and $(y, \ldots, v)$. In all cases, the endpoints of the omitted paths are not omitted.

For, the minor E isolator, we must determine whether to isolate a $K_{3,3}$ homeomorph or a K_5 homeomorph. Four simple tests are applied, the failure of which implies that minor E can be used to isolate a K_5 homeomorph based on the techniques described above. The first test to succeed implies the ability to apply the corresponding minor from Figure 21.12. Minor E_1 occurs if the pertinent vertex w is not future pertinent (i.e., a second vertex z is future pertinent along the lower external face path strictly between p_x and p_y). If this condition fails, then $w = z$. Minor E_2 occurs if the future pertinence connection from w to an ancestor u_w of v is a descendant of u_x and u_y. Minor E_3 occurs if u_x and u_y are distinct and at least one is a descendant of u_w. Minor E_4 occurs if either $p_x \neq x$ or $p_y \neq y$.

As with minors A–D, there are symmetries to handle and edges that are not required to form a $K_{3,3}$. For minors E_1 and E_2 it is easier to handle the edge omissions (and symmetries) because they reduce to minors C and A, respectively. Minor E_3 does not require (x, w) and (y, v) to form a $K_{3,3}$, and minor E_4 does not require (u, v) and (w, y) to form a $K_{3,3}$. Moreover, note that the omission of these edges must account for the fact that p_x or p_y may have been edge contracted into x or y in the depiction of the minor (e.g., eliminating (w, y) in minor E_4 corresponds to eliminating the path $(w, \ldots, p_y)$ but not $(p_y, \ldots, y)$). As for symmetries, minor E_1 in Figure 21.12a depicts z between x and w along the path $(x, \ldots, z, \ldots, w, \ldots, y)$, but z may instead appear between w and y. Also, Figure 21.12c depicts minor E_3 with u_x an ancestor of u_y, but u_y could instead be an ancestor of u_x. For minor E_4, Figure 21.12d depicts p_x distinct from x (and p_y can be equal to or distinct from y), but if $p_x = x$, then p_y must be distinct from y. Finally, in the symmetric cases, the edges to delete to isolate the $K_{3,3}$ homeomorph are different but analogous to those indicated above.

The techniques described in this section support the following theorem.

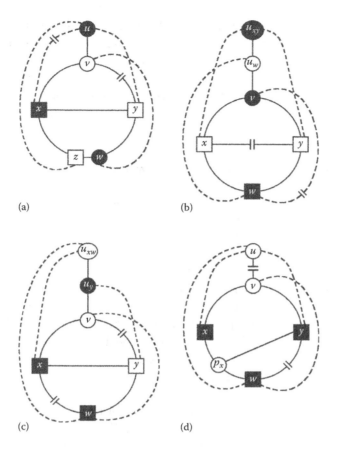

Figure 21.12 Additional $K_{3,3}$ minors from the K_5 minor E in Figure 21.7: (a) minor E_1, (b) minor E_2, (c) minor E_3, and (d) minor E_4.

Theorem 21.4 *Given a simple undirected nonplanar graph G with n vertices, the edge addition planarity algorithm produces a subgraph homeomorphic to one of $K_{3,3}$, or K_5 in worst case $O(n)$ time.*

Proof. Due to Theorem 21.2, G is correctly determined to be nonplanar in some step v due to the **Walkdown** failing to embed at least one back edge while processing some biconnected component B within the partial embedding $\tilde{G}$. As a corollary of Theorem 21.3, the performance is linear in the number of edges embedded up to and including those embedded in step v since planarity of $\tilde{G}$ is maintained until nonplanarity is discovered. A one-time additional linear cost can be assumed for establishing pertinence for back edges not embedded in the step v in which nonplanarity is discovered. The straightforward techniques above then isolate a subgraph homeomorphic to $K_{3,3}$ or K_5. Thereafter, an additional one-time linear cost is associated with isolating the Kuratowski subgraph as follows.

The stopping vertices x and y, the pertinent vertex w, the points of attachment p_x and p_y of the x–y path (for minors C, D, and E), and the v–z path (for minor D) are all found with a cost linear in the size of B. Determining the nonplanarity minor is then a set of constant time decisions. In the case of minor E, a further set of constant-time decisions determines which of minors E_1 to E_4 will be used or whether a subgraph homeomorphic to K_5 will be isolated. For an additional linear cost, we find and embed an edge incident to v and at most three new edges incident to ancestors of v in order to complete the pertinent and future pertinence path connections, and the DFS tree path is also embedded from v to the least of the ancestors in the future pertinence paths. There is an additional linear cost to mark all the DFS tree paths and paths within B that are associated with the particular Kuratowski subgraph being isolated and to delete the unmarked edges previously embedded but not needed to form the Kuratowski subgraph. Thus, the overall performance for all operations is worse case $O(m)$. Since any graph with more than $3n - 6$ edges is nonplanar, by restricting the input of G to a subgraph of $O(n)$ edges with m at least $min(m, 3n - 5)$, the worse case $O(n)$ performance is obtained. ∎

21.7 DRAWING A VISIBILITY REPRESENTATION OF A PLANAR GRAPH

A *visibility representation* of a planar graph is a general-purpose planar drawing in which each vertex is mapped to a horizontal segment at a given vertical position, and each edge is mapped to a vertical segment that is at a given horizontal position and that terminates at the vertical positions associated with the endpoint vertices of the edge ([1], p. 25). This type of planar drawing has sometimes been called a *horvert diagram* [20].

The *dual* of a combinatorial planar embedding $\tilde{G}$ is a combinatorial planar embedding $\tilde{H}$ that contains a vertex corresponding to each face of $\tilde{G}$ and an edge for each edge e of $\tilde{G}$ to connect the pair of vertices in $\tilde{H}$ that represent the pair of faces in $\tilde{G}$ that include e in their boundaries. The first linear-time algorithms for producing visibility representations were presented by Tamassia and Tollis [21] and by Rosenstiehl and Tarjan [22]. These algorithms take as input an st-numbering, a combinatorial planar embedding, and its dual. Jayakumar et al. [20] augmented the PQ-tree [8] planarity algorithm to also produce a visibility representation, the stated benefits of which were removing the necessity of pre-computing the dual, reusing the st-numbering, and using the PQ-tree information available while computing the combinatorial planar embedding rather than having a separate algorithm. Although this showed that it was beneficial to extend an existing planarity algorithm and reuse the information it produces during planar embedding, the edge addition planarity algorithm derives simplicity in part from eliminating st-numbering and the PQ-tree, so a different augmentation is required.

This section presents an extension to the core edge addition planarity method that enables simple, efficient generation of visibility representations [23]. The method does not pre-compute an st-numbering nor a planar graph dual, and the augmentations are simple and geometrically intuitive because they are not encumbered by a batch *vertex addition* processing model (e.g., [8,17]). Instead, information is cached at key steps related to adding an edge or traversing an external face edge. The cached information indicates whether a vertex shall appear *locally* above or below its DFS parent relative to a particular DFS ancestor. In postprocessing, these local markings are converted into a global sense of up and down, resulting in a vertex order. Then, the vertex positions are used to guide a computational geometry sweep over the combinatorial planar embedding to assign horizontal positions to edges. In a final trivial step, the vertex positions are used to assign spanning ranges to the edges, and the edge positions are used to assign spanning ranges to the vertices.

21.7.1 Computing Vertical Positions of Vertices

Vertices are represented by horizontal segments at vertical positions in the visibility representation. The horizontal range for each vertex is computed later once the horizontal positions of the edges are known (see Section 21.7.2). The vertical positions of the vertices are computed using geometric positioning information collected during planar embedding. To maintain a linear-time bound, the information collected is *relaxed* to be only relative positioning of each vertex compared with its DFS parent and some DFS ancestor. A post-processing step resolves the relative positioning information into a vertex order.

21.7.1.1 Localized Sense of Up and Down

To see the geometric intuition for vertex placement, consider the localized, relational positioning characteristics, for example in Figure 21.5. In a localized sense, v and its ancestors are *above* the descendants of v. Then, the descendants that are future pertinent remain *below* (in a localized sense) the vertices that are only pertinent or inactive. This is because the back edges being embedded for v tend to surround vertices that are only pertinent or inactive as the **Walkdown** traverses toward stopping vertices, whereas the back edges being embedded do not surround future pertinent vertices. As vertices become surrounded by edges, they are placed closer to v, whereas future pertinent vertices are placed farther away from v because they will be extended horizontally to attach to back edges from the ancestors of v that will be embedded in future steps of the planar embedder.

Of course, in a future step that processes an ancestor u of v, the biconnected component containing v and a child c may be regarded as only pertinent (if it contains only back edge connections to u), so it may be turned upside down, reversing the localized sense of *above* and *below* for relative positions assigned in step v. In fact, such reversals of vertical geometric orientation can occur any number of additional times before the planar embedder finishes processing. If all reversals of the vertical direction were immediately and directly processed, a total linear-time bound could not be achieved. Instead, the terms *above* and *below* are relaxed during the planar embedding. A vertex whose relative positioning is determined in step v is a descendant of v, and it is marked as being either *between* its DFS parent and v or *beyond* its DFS parent relative to v.

21.7.1.2 Making the Localized Settings

Initially, each vertex is marked *beyond* its DFS parent relative to the DFS tree root. During the planar embedding, each vertex is eventually positioned either *between* or *beyond* its DFS parent relative to some ancestor, specifically the vertex v whose back edges are being embedded at the time the decision is made. The decision is based on whether or not (*beyond* or *between*) the parent is removed from the external face before all vertices in the DFS subtree rooted by the given child vertex. To make this decision efficiently, we cache some information that allows us to exploit some structural characteristics of biconnected components and the result of merging them together.

Figure 21.13 depicts two DFS children f and g of a vertex r while they are still in separated child biconnected components rooted by virtual copies of r. Note that the virtual vertex root of a biconnected component has two edges emanating from it to other vertices on the external face of the biconnected component. When a biconnected component is merged, one of those edges becomes surrounded, but the other remains on the external face. For example, when (v', w) is added, the virtual vertex r' is merged with r, transforming the edges incident to r' into edges incident to r. The edge (r, w) no longer appears on the external face, but (r, x) does. This edge is marked with the identity of the DFS child of the biconnected component root vertex (f in this case). This marking helps to make a final determination later in processing about the vertex placement of the child (f in this case) relative to its parent (r in this case) and some ancestor of the parent.

In the example of Figure 21.13, the **Walkdown** embeds edge (v', w) and then proceeds along the external face until the stopping vertex x is encountered. The **Walkdown** then proceeds down the right-hand side of the biconnected component from v' to r and merges the biconnected component containing r'' and g so that the back edge (v', z) can be embedded. At this time, the identity of the DFS child g of r is cached on the external face edge (r, y) and then **Walkdown** traversal continues until the stopping vertex y is encountered.

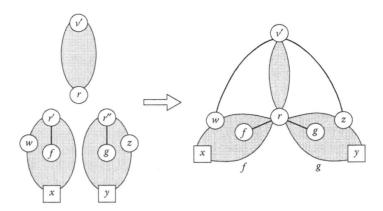

Figure 21.13 Child biconnected components rooted by r' and r'' are merged when back edges (v', w) and (v', z) are added. During the merge, the edge incident to the virtual vertex root that will remain on the external face is marked with the identity of the DFS child. Above, (r, x) is marked with f, and (r, y) is marked with g. Initially, f and g are placed *beyond* r, but one or both may be changed to *between* r and some ancestor of r when these marks are resolved. Note that (r, x) and (r, y) may be edges or paths on the external face, so these segments are marked using parallel data fields to those used in external face management.

Observe that in a planar graph r cannot have another pertinent child biconnected component while both x and y are still future pertinent since it is easy to isolate a subgraph homeomorphic to $K_{3,3}$ in that case [14]. Another way to view this fact is that it is not possible in a future **Walkdown** for an ancestor of v to reach r again without traversing either the edge (r, x) or (r, y). If, during step u, the **Walkdown** traverses a marked edge from descendant (x or y) to ancestor (r), then the DFS child (f or g, obtained from the marking on the edge) is determined to be *between* its parent (r) and the current vertex (u). Similarly, if the **Walkdown** traversal goes from the parent to the descendant, then the DFS child is placed *beyond* the parent relative to the current vertex. Figure 21.14 depicts the main vertex placement cases corresponding to the example of Figure 21.13.

Without loss of generality, Figure 21.14a depicts the resolution of the future pertinence of x (and r, if any) occurring before that of y. The opposing case, in which x is future pertinent after both r and y, is symmetric. The embedder traverses the edge (r, x) in the direction from x to r, so the marking f on that edge is used to place f between r and u_x. Assuming r is not still future pertinent in step u_y, then the edge (r, y) is traversed from r to y, so the marking g on that edge is used to place g beyond r relative to u_x. Figure 21.14b illustrates the case in which r is future pertinent after both x and y. In this case, **Walkdown** traversals will reach r as a stopping vertex in both directions, so that both children f and g are placed *between* their parent r and the ancestor u.

As a final component of making the localized setting, when a vertex such as f is marked as being *between* or *beyond* its parent r relative to the current vertex v, the DFS child c of v whose subtree contains f is also stored in f. This is needed to help achieve linear-time operation of the post-processing step that generates the vertex ordering (see Section 21.7.1.3).

21.7.1.3 Postprocessing to Generate the Vertex Order

The vertical *vertex order* is a list of size n that provides an absolute positioning for the vertices, that is vertices appearing earlier in the list appear above those later in the list. The vertex order is constructed as a doubly linked list to allow $O(1)$ insertion before or after any

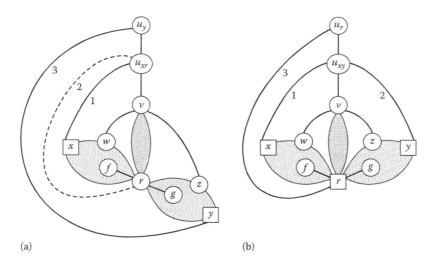

Figure 21.14 (a) In embedding step u_x, traversing from the descendant x of f to the parent r of f results in placing f between its parent r and the ancestor u_x. Traversing from the parent r of g to the descendant y of g results in placing g beyond parent r relative to ancestor u_x. (b) Future pertinence at r can result in both children f and g being placed between r and some ancestor.

node. Each vertex of the input graph receives a pointer to its node in the vertex order when it is added to the vertex order.

The localized vertex placements collected for each biconnected component during planar embedding are converted to the vertical vertex order using a pre-order traversal of each DFS tree. When a vertex is visited to add it to the vertex order, all of its ancestors have already been added to the vertex order.

The localized information includes a marking of *between* the DFS parent and a given ancestor or *beyond* the DFS parent relative to the given ancestor. We convert the marking into a decision to insert the vertex immediately *above* or *below* its DFS parent in the vertex order. This conversion has one challenging aspect. Although the parent and the ancestor are both in the vertex order, their relative positions in the vertex order cannot be determined in constant time.

Three observations help to solve this problem. First, all ancestors of a vertex have been marked *above* or *below* their parents. Second, a child vertex is added immediately above or below its parent. Inductively, this means that the entire DFS subtree rooted by a child vertex appears above or below the parent in the vertex order. Third, as the **Walkdown** operates over a biconnected component B rooted by a virtual copy of v, the DFS child c of v in B is known. When the **Walkdown** makes a localized vertex placement, this child vertex is also stored (see Section 21.7.1.2).

Hence, the localized vertex placement information known for a vertex are (1) the DFS parent, (2) a drawing flag set to *between* or *beyond*, (3) a DFS ancestor, and (4) the DFS child of the ancestor whose DFS subtree contains the vertex. The key to positioning a vertex is that the drawing flag of the ancestor's child has already been set to *above* or *below* the ancestor due to pre-order DFS traversal, and this setting controls whether the DFS parent of a given vertex is above or below the ancestor. Thus, a vertex position can be assigned as follows. If a vertex is *between* its DFS parent and the given ancestor, then the vertex is inserted before (above) the parent if the parent is below the ancestor, and inserted after (below) the parent otherwise. Symmetrically, if a vertex is *beyond* its DFS parent relative to the given ancestor, then the vertex is inserted after (below) the parent if the parent is below the ancestor, and inserted before (above) the parent otherwise.

21.7.2 Computing Horizontal Positions of Edges

The *edge order* is a list of size m that provides absolute horizontal edge positions. Edges earlier in the list are to the left of those later in the list. The edge order is determined by a geometric sweep over the combinatorial planar embedding. The edge order list develops in the horizontal sweep line as it advances through the vertex order.

The method keeps track of a *generator edge* for each vertex, which is the first edge incident to the vertex that is added to the edge order list. The generator edge provides the insertion point along the horizontal sweep line for the edges emanating from the vertex to the vertices that are below it (which have a greater vertex position number). The generator edge for each vertex is initially *nil*.

To begin, the edges incident to each DFS tree root r are added to the edge order list in the order of r's adjacency list. For each edge $e = (r, w)$, e is the generator edge of w. Then, for each vertex v below a DFS tree root in vertex position order, we obtain the generator edge e as the starting point of a cyclic traversal of v's adjacency list, and the set of edges from v to vertices with greater positions (i.e., below v) are added to the edge order list immediately after e. For each edge (v, w), if w has no generator edge, then (v, w) is the generator edge of w. Figure 21.15 illustrates the method on a sample graph whose combinatorial planar embedding and vertex positions have already been determined.

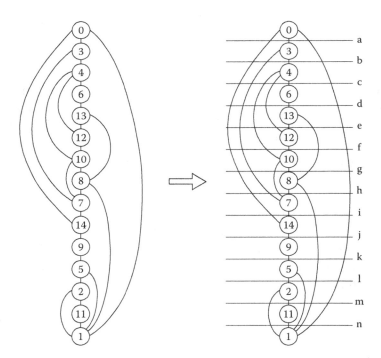

Figure 21.15 Combinatorial planar embedding with vertices numbered by DFI and vertically placed according to positions computed by the method of Section 21.7.1.3. The horizontal edge positions are computed by a geometric sweep over the planar embedding guided by vertex positions. At each step, edges from a vertex v to succeeding vertices in the vertex order are added after the generator edge for v. If an edge added to the sweep line is the first one incident to a vertex w, it becomes the generator edge for w.

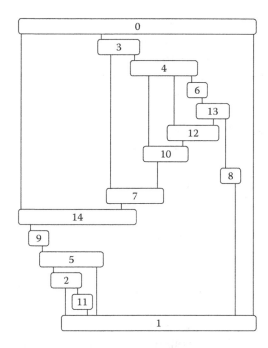

Figure 21.16 Planar drawing of the resulting visibility representation.

Some of the steps of the edge positioning method on the example graph in Figure 21.15 are as follows. First, in step (a), the edge order list is initialized to the following edges $[(0, 14)(0, 3)(0, 1)]$. These edges are the generator edges of the vertices labeled 14, 3, and 1. In step (b), vertex 3 is visited, and the edges incident to later vertices, that is those with higher vertex positions, are added after the generator edge for vertex 3. After step (b), the edge order list is $[(0, 14)(0, 3)(3, 7)(3, 4)(0, 1)]$. Note that the numeric vertex labels are based on the initial labeling and are unrelated to the vertex position order. Thus, because vertex 10 has a later vertex position number than vertex 12, the processing of vertex 12 in step (f) consists of obtaining the edge list location of the previously saved generator edge $(4, 12)$ and then adding the edge $(12, 10)$ after it. The edge list at the end of step (f) is $[(0, 14)(0, 3)(3, 7)(3, 4)(4, 10)(4, 12)(10, 12)(4, 6)(6, 13)(12, 13)(8, 13)(0, 1)]$.

Given the edge positions, and the previously computed vertex positions, the visibility representation is easily completed by assigning a vertical range to each edge based on the span of its endpoints and a horizontal range to each vertex based on the span of the positions of its incident edges. Figure 21.16 depicts a planar drawing of a visibility representation.

21.7.3 Correctness and Performance

The Edge Addition Planar Graph Drawing Algorithm generates visibility representations by augmenting the processing of the Edge Addition Planarity Algorithm. Relative vertex placement information is collected within each biconnected component during embedding, and a postprocessing step determines exact vertical positions for all vertices. The key to correctness is the vertex positioning method and its association to the combinatorial planar embedding of the biconnected components, since obvious methods are then employed to rationalize the vertex positions in the biconnected components, to derive the horizontal edge positions based on the vertex positions and the combinatorial planar embedding, and then to determine the horizontal ranges of vertices and the vertical ranges of edges.

For the purpose of assessing correctness, there is no need to defer absolute vertex positioning to postprocessing. As such, correct vertex positioning can be established as a property of all biconnected components that is held invariant by the operations of the Edge Addition Planar Graph Drawing Algorithm. Vertex positioning within a biconnected component is deemed correct if the vertex positions can be used with the combinatorial planar embedding to determine edge positions such that vertical edge lines intersect only the horizontal lines corresponding to their vertex endpoints. Thus, we have the following theorem.

Theorem 21.5 *Given a simple undirected planar graph G, the Edge Addition Planar Graph Drawing Algorithm correctly generates a visibility representation for G.*

Proof. The inductive anchor for the correct vertex positioning property is established in the initial embedding of each DFS tree edge as a singleton biconnected component with vertex position 0 for the root and 1 for the DFS child. Hence, we focus on the only other algorithm actions defined to modify the state of the planar embedding or the vertex positioning: adding a back edge and traversing an edge of the external face.

There are four atomic operations to consider. Adding a back edge (v', w) decomposes into a composite operation of merging a biconnected component and the final atomic operation of embedding the back edge along the external face. Merging a biconnected component $B_{r'}$ decomposes into two atomic operations: optionally flipping $B_{r'}$ and attaching the biconnected component root r' to its nonvirtual counterpart r in the parent biconnected component of $B_{r'}$. Finally, a last atomic operation is traversal of an external face edge marked f, which may cause the vertex f to be changed so that it is placed *between* its DFS parent and the current vertex v whose back edges are being embedded.

Flipping a biconnected component $B_{r'}$ does not change vertex positions. It does change the horizontal edge positions within $B_{r'}$, the changes directly correspond to changes made to the combinatorial planar embedding of $B_{r'}$ to logically invert its vertex orientations (which, for assessing correctness, is assumed to occur but which, for efficiency, is mostly deferred to postprocessing by the core edge addition planarity algorithm).

When merging the biconnected component root r' with r, the relative vertex locations within $B_{r'}$ are not changed, and the DFS child endpoint of the biconnected component root edge is not changed from the initial setting of *beyond* its parent. In the overall view of absolute vertex positions, though, $B_{r'}$ may be inverted if there is an odd number of *between* markings on vertices on the path from r to the root of the parent biconnected component containing r. However, this inversion corresponds to a half rotation within the plane that remains in accord with the rotation scheme (adjacency list orders of vertices) of the combinatorial planar embedding for $B_{r'}$. Thus, the horizontal positions of edges $B_{r'}$ determined by the sweep method will simply be reversed. As for attaching the adjacency list of vertex r' to r, this places the vertices of $B_{r'}$ immediately above or below the position of r. The only effect on vertex positions in $B_{r'}$ is a simple additive adjustment. The horizontal positions of edges within $B_{r'}$ are unaltered, and their placement within the parent biconnected component is in accord with the combinatorial planar embedding. Lastly, for vertices below r, and also r if $B_{r'}$ is placed above it, the absolute positions are shifted downward to accommodate the positions of vertices in $B_{r'}$. However, this only lengthens edges and does not perturb the positions that would be determined for them by the sweep of the combinatorial planar embedding.

The direction of traversing an external face edge marked f enables the drawing algorithm to determine whether a vertex f should be placed *between* its DFS parent r and v or *beyond* its DFS parent r relative to v. Only for efficiency is this determination made when the external face edge is traversed. To preserve the correct vertex positioning property, the traversal direction can be regarded as equivalent in effect to being known and accommodated in the earlier merge operation of B_{r_f} onto r, which is covered above.

Hence, all operations prior to visiting the descendant endpoint of the back edge (v', w) produce only vertex and edge position shifts and transpositions and widening of lines for merge point vertices, but they do not perturb the property that every edge position is such that the edge intersects only the lines for its vertex endpoints. The final operation of embedding the back edge (v', w) also makes no vertex position adjustments and embeds the new edge along the external face in accord with its position in the combinatorial planar embedding. The lines for the two vertex endpoints of the edge are horizontally extended to meet the new vertical edge line. The vertex v' is topmost in the biconnected component, so its horizontal extension does not cross any vertical lines for edges. The vertex positioning operations performed during the **Walkdown** traversal to the descendant endpoint w ensure that preceding vertices along the external face are above w such that horizontal extension of w also does not cross any edge lines.

Thus, for each successive embedding of G, beginning with the embedding of the DFS tree edges and held invariant for each back edge addition, the correct vertex positioning property is maintained for each biconnected component such that the vertex positions and the embedding determine the edge positions needed for a correct visibility representation of each biconnected component of G. ∎

In terms of run-time performance, the core edge addition planarity algorithm is $O(n)$ and the drawing method adds constant time operations during the planar embedding process while deferring expensive work to linear-time post-processing steps. Thus we have the following theorem.

Theorem 21.6 *Given a simple undirected planar graph G with n vertices, the Edge Addition Planar Graph Drawing Algorithm operates in worst case $O(n)$ time.*

Proof. The relative vertex placements are determined in total $O(n)$ time as follows. When a biconnected component root r' is attached to its nonvirtual copy r, a constant time operation is performed to mark the edge e incident to r' that will remain on the external face. Later, when the **Walkdown** traverses e, a constant time operation is performed to mark a DFS child of r as either *between* r and v or *beyond* r relative to v. Once an edge e is marked by the first operation, it is never incident to a biconnected component root again, and once the **Walkdown** traverses that edge, it is removed from the external face.

The postprocessing to generate the absolute vertex positioning is $O(n)$ as follows. First, when a vertex f is placed relative to its parent r and the ancestor v, the child c of v whose DFS subtree contains r is also stored, which is a constant-time cost since it is known in the **Walkdown**. Then, the vertex order is computed in a linear-time postprocessing via a pre-order traversal of the DFS tree, which uses constant time per vertex visitation to assess the four pieces of information stored during relative vertex placement and then insert a node for the vertex above or below the node for its DFS parent in a doubly linked list. As each vertex is placed in the vertex order list, its representative node in the list is stored in the vertex so that the node for a parent can be obtained in constant time when each of its DFS children is processed.

The horizontal edge positions are computed in linear time via a vertex ordered sweep over the combinatorial planar embedding. For each vertex, all of its edges are traversed and each is processed in constant time. Those edges leading to lower numbered vertices (processed earlier in the sweep) are ignored. Those leading to higher numbered vertices are added in constant time to the sweep line after a generator edge for the vertex, which is obtained in constant time. As a last constant time operation per edge, for each edge leading to a vertex with no generator edge, the edge is stored as the generator edge of the vertex.

Finally, determining the horizontal and vertical ranges of the vertices and edges is a trivial $O(n)$ post-processing step. Thus, the Edge Addition Planar Graph Drawing Algorithm adds worst case $O(n)$ processing time to the core edge addition planarity algorithm. ∎

21.8 CONCLUSION

This chapter has presented the main ideas and optimization techniques of the Boyer–Myrvold edge addition planarity algorithm [4]. A reference implementation is available from an online open source code project [24]. It has been tested on all graphs on 12 or fewer vertices as well as billions of randomly generated graphs. For each result, an integrity check is performed. For a planar embedding, a face walk is performed to ensure that the number of faces is correct and that each edge is used in only two face boundaries. For a Kuratowski subgraph, the result is checked to ensure that there are only five vertices of degree four or six of degree 3, that the remaining vertices are degree two or zero, that the degree two vertices appear along paths between the higher degree vertices, that the paths form a K_5 or $K_{3,3}$ homeomorph, and that the result is a subgraph of the input graph. An earlier version of the reference implementation was found to be typically the fastest among several planarity algorithm implementations [15]. The current reference implementation of the edge addition planarity algorithm [24] is nearly twice as fast now that it has more of the low-level implementation optimizations that were included in some of the other algorithm implementations analyzed in [15].

The definitions and techniques of the edge addition planariy algorithm provide an extensible framework for solving several planarity-related problems. The open source project provides implementations of several such extensions, including the method for generating visibility representation drawings of planar graphs [23] described in this chapter as well as methods for outerplanar graph embedding and obstruction isolation and for a number of subgraph homeomorphism problems [25].

References

[1] G. Di Battista, P. Eades, R. Tamassia, and I. G. Tollis. *Graph Drawing: Algorithms for the Visualization of Graphs*. Prentice Hall, Upper Saddle River, NJ, 1999.

[2] K. Kuratowski. Sur le problème des courbes gauches en topologie. *Fundamenta Mathematicae*, **15** (1930), 271–283.

[3] K. Wagner. Über einer eigenschaft der ebener complexe. *Mathematische Annalen*, **14** (1937), 570–590.

[4] J. M. Boyer and W. J. Myrvold. On the cutting edge: Simplified $O(n)$ planarity by edge addition. *Journal of Graph Algorithms and Applications*, **8**(3) (2004), 241–273. DOI: 10.7155/jgaa.00091; http://jgaa.info/08/91.html.

[5] J. Hopcroft and R. Tarjan. Efficient planarity testing. *Journal of the Association for Computing Machinery*, **21**(4) (1974), 549–568.

[6] A. Lempel, S. Even, and I. Cederbaum. An algorithm for planarity testing of graphs. In P. Rosenstiehl, editor, *Theory of Graphs*, pages 215–232, New York, 1967. Gordon & Breach.

[7] S. Even and R. E. Tarjan. Computing an *st*-numbering. *Theoretical Computer Science*, **2** (1976), 339–344.

[8] K. S. Booth and G. S. Lueker. Testing for the consecutive ones property, interval graphs, and graph planarity using PQ-tree algorithms. *Journal of Computer and Systems Sciences*, **13** (1976), 335–379.

[9] N. Chiba, T. Nishizeki, A. Abe, and T. Ozawa. A linear algorithm for embedding planar graphs using PQ-trees. *Journal of Computer and Systems Sciences*, **30** (1985), 54–76.

[10] M. Jünger, S. Leipert, and P. Mutzel. Pitfalls of using PQ-trees in automatic graph drawing. In G. Di Battista, editor, *Proceedings of the 5th International Symposium on Graph Drawing*, volume 1353 of Lecture Notes in Computer Science, pages 193–204. Springer-Verlag, September 1997.

[11] H. de Fraysseix. Trémaux trees and planarity. *Electronic Notes in Discrete Mathematics*, **31** (2008), 169–180.

[12] H. de Fraysseix, P. Ossona de Mendez, and P. Rosenstiehl. Trémaux trees and planarity. *International Journal of Foundations of Computer Science*, **17**(5) (2006), 1017–1029.

[13] H. de Fraysseix and P. Rosenstiehl. A characterization of planar graphs by trémaux orders. *Combinatorica*, **5**(2) (1985), 127–135.

[14] J. Boyer and W. Myrvold. Stop minding your P's and Q's: A simplified $O(n)$ planar embedding algorithm. *Proceedings of the 10th Annual ACM-SIAM Symposium on Discrete Algorithms*, pages 140–146, 1999.

[15] J. M. Boyer, P. F. Cortese, M. Patrignani, and G. Di Battista. Stop minding your P's and Q's: Implementing a fast and simple DFS-based planarity testing and embedding algorithm. In G. Liotta, editor, *Proceedings of the 11th International Symposium on Graph Drawing 2003*, volume 2912 of Lecture Notes in Computer Science, pages 25–36. Springer-Verlag, Perugia, Italy, 2004.

[16] J. M. Boyer. Additional PC-tree planarity conditions. In J. Pach, editor, *Proceedings of the 12th International Symposium on Graph Drawing 2004*, volume 3383 of Lecture Notes in Computer Science, pages 82–88. Springer-Verlag, New York, 2005.

[17] W.-K. Shih and W.-L. Hsu. A new planarity test. *Theoretical Computer Science*, **223** (1999), 179–191.

[18] Bernhard Haeupler and Robert E. Tarjan. Planarity algorithms via PQ-trees. In Patrice Ossona de Mendez, Michel Pocchiola, Dominique Poulalhon, Jorge Luis Ramrez Alfonsín, and Gilles Schaeffer, editors, *International Conference on Topological and Geometric Graph Theory*, volume 31 of Electronic Notes in Discrete Mathematics, pages 143–149. ScienceDirect, Strasbourg, France, 2008.

[19] R. E. Tarjan. Depth-first search and linear graph algorithms. *SIAM Journal of Computing*, **1**(2) (1972), 146–160.

[20] R. Jayakumar, K. Thulasiraman, and M. N. S. Swamy. Planar embedding: Linear-time algorithms for vertex placement and edge ordering. *IEEE Transactions on Circuits and Systems*, **35**(3) (1988), 334–344.

[21] R. Tamassia and I. G. Tollis. A unified approach to visibility representations of planar graphs. *Discrete and Computational Geometry*, **1**(4) (1986), 321–341.

[22] P. Rosenstiehl and R. Tarjan. Rectilinear planar layouts and bipolar orientations of planar graphs. *Discrete and Computational Geometry*, **1**(4) (1986), 343–353.

[23] J. M. Boyer. A new method for efficiently generating planar graph visibility representations. In P. Eades and P. Healy, editors, *Proceedings of the 13th International Symposium on Graph Drawing 2005*, volume 3843 of Lecture Notes in Computer Science, pages 508–511. Springer-Verlag, Limerick, Ireland, 2006.

[24] J. M. Boyer. Edge Addition Planarity Suite, version 3.0.0.3. July 2015, https://github.com/graph-algorithms/edge-addition-planarity-suite/.

[25] J. M. Boyer. Subgraph homeomorphism via the edge addition planarity algorithm. *Journal of Graph Algorithms and Applications*, **16**(2) (2012), 381–410. DOI: 10.7155/jgaa.00268; http://jgaa.info/16/268.html.

CHAPTER 22

Planarity Testing Based on PC-Trees

Wen-Lian Hsu

CONTENTS

22.1 Introduction 525
22.2 Overview of S&H Planarity Test 526
22.3 Creating the First C-Node 527
22.4 Creating C-Nodes in General 528
22.5 Embedding Algorithm 529
22.6 Linear-Time Implementation 529

22.1 INTRODUCTION

A planar graph is one which can be drawn on the plane without any crossing edge. Given an undirected graph G, the planarity test is to determine whether there exists a clockwise edge ordering around each vertex, such that the graph G can be drawn on the plane without any crossing edge. Linear-time planarity test was first established by Hopcroft and Tarjan [1] based on a *path addition approach*. The *vertex addition approach*, originally developed by Lempel et al. [2], was later improved by Booth and Lueker [3] (hereafter, referred to as B&L) to run in linear time using a data structure called *PQ-tree*. Several other approaches have also been developed for simplifying the planarity test (see e.g., [4–8]) and the embedding algorithm [9,10]. Shih and Hsu [11] (hereafter referred to as S&H) developed a linear-time test based on PC-trees (a generalization of P-trees), which did not use any template. In fact, based on this idea, Hsu [12] and Hsu and McConnell [13] further eliminated the template operations of the original PQ-tree for the consecutive ones test, and used PC-tree for the circular ones test directly. An earlier version [14] of Shih and Hsu [8] has been referred to as the simplest linear-time planarity test by Thomas in his lecture notes [15]. In this chapter we shall describe a PC-tree–based planarity test, which is much simpler than any previous version (c.f. [16]) of S&H. A software for our planarity test is available at https://github.com/x1213/planarity-algorithms/ in *GitHub*.

In S&H algorithm, a data structure called PC-tree was introduced, in which a P-node denotes a node whose neighbors can be permuted arbitrarily, whereas the neighbors of a C-node must observe a cyclic order or its reversal. One can associate a PC-tree to a planar graph, in which a P-node is an original node of the graph; a C-node represents a biconnected component C_w whose neighbors are the P-nodes on the boundary of C_w. Intuitively, a PC-tree represents the relationships between biconnected components and regular nodes in planar graphs as shown in Figure 22.1.

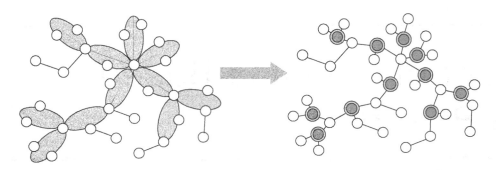

Figure 22.1 Biconnected components and their corresponding C-nodes in the PC-tree.

We regard the left diagram of Figure 22.1 as the *cycle view* of the planar graph and the right diagram as its *tree view*. PC-tree enables us to provide a more streamlined algorithm in dealing with the biconnected components that are created, traversed, and merged during the planarity test. As demonstrated in Theorem 22.1 in Section 22.4, C-nodes in many ways can be treated as P-nodes conceptually, which allows us to quickly grasp the essence of planar graphs and to easily identify Kuratowski subgraphs for non-planar graphs. Not only can PC-tree be used in planarity test, it can also be extended to finding maximal planar subgraphs in linear time [17] (a more updated version is coming up). PC-tree provides the needed data structure in complexity reduction for both computer implementation and conceptual understanding.

This chapter is arranged as follows. In Section 22.2 we give an overview of the S&H algorithm. A more concrete description of our PC-tree algorithm is carried out in Section 22.3 when the first C-node is created. Section 22.4 discusses two important properties of planar graphs as revealed by S&H algorithm and gives the recognition algorithm for the general case. Section 22.5 is devoted to the embedding algorithm. A linear-time implementation of the algorithm is provided in Section 22.6. Proofs of the theorems are included in Appendix 22A.

22.2 OVERVIEW OF S&H PLANARITY TEST

In this section we give an overview of the planarity algorithm based on PC-trees. Since the inception of S&H algorithm [11], there have been a few simplifications [13,16]. The version presented here is by far the simplest. The basic notations, such as terminal path, essential node, representative boundary cycle, stay the same. But the algorithms to identify them are completely renovated. Some notations, such as partial, full nodes, are no longer used.

Let n be the number of vertices and m, the number of edges of the given graph G. Construct a depth-first search (DFS) tree T for G. Note that every nontree edge of G must be a back edge from a vertex to one of its ancestors. To simplify our discussion and proofs, assume the given graph G is biconnected. This is certainly not a restriction since we can split the graph into biconnected components along articulation vertices, which can be identified in the DFS tree. Let $1, \ldots, n$ be the vertex order resulting from a postorder traversal of T. So the order of a child is always less than that of its parent. Denote the subtree of T with root k by T_k. Initially, we include all edges of T, namely the DFS tree, in the embedding. All nodes are P-nodes. At each iteration k, we add all back edges from descendants in T_k to k. Such a back edge addition will produce biconnected components (abbreviated as *components* later), which are then replaced by C-nodes. Components are created, traversed, and merged, which are all managed by PC-tree operations. The creation of C-nodes for G_r is discussed in two stages. In the next section, we consider the creation of the first C-node.

22.3 CREATING THE FIRST C-NODE

Before any back edge is added, every node in the current tree is a P-node. To introduce the first C-node in the tree, let us consider the first iteration, say i, that there exists a back edge to i. At this iteration, we shall add all back edges from the descendants to node i, and update the embedding. A neighbor v of a vertex i is called its *back neighbor* if the edge (v, i) is a back edge. A descendant k of i is called an *i-descendant* if T_k has a back neighbor of i. Note that there may be several i-children and we could test the embedding of each i-child subtree and related back edges independently for planarity. Let r denote an i-child. Below, we concentrate on embedding all back edges to i from nodes in T_r.

In general, we regard the graph as *undirected*. Only when we refer to the parent–child relationship the edge direction becomes important. Let u be a back neighbor of i in T_r. A back edge traversal (BET) initiated from (u,i) is a traversal starting from i, through (u,i), and following the parent pointers from u to r. A path from i formed through a BET is called an *i-path*. Thus, there is at least one i-path from i to each i-descendant. In contrast, we shall refer to any path from i to r, then to a node in T_r through tree edges an *i_r-path*.

Every time a vertex or an edge is traversed, it is assigned the label i. Perform a BET for every back edge of T_r from i. Terminate a traversal whenever it encounters a node previously labeled i. After all BETs are performed, let the traversed node set be V_r^*, which is exactly the set of i-descendants in T_r, and the traversed edge set be E_r^*. The traversed subgraph $T_r^* = (V_r^*, E_r^*)$ forms a subtree of T_r. The set of back edges from the nodes of T_r^* to node i is denoted by $\beta_{r,i}$. Since each node of V_r^* is in a traversed cycle containing the edge (r, i), the subgraph $G_r = (V_r^* \cup i, E_r^* \cup (r, i) \cup \beta_{r,i})$ must be biconnected, which is called an *i-component*. If G is planar, then G_r has a planar embedding with a boundary cycle.

The most important step in our recognition algorithm is to identify some nodes that should be on the boundary cycle of G_r. Define a traversed node v in T_r^* to be *essential* if it is incident (in the undirected graph) to an un-traversed edge (v, v'). An essential node is connected to an ancestor t of i through a t-path. Since an essential node cannot be embedded inside G_r, it must lie on the boundary cycle. The proof of the following Lemma 22.1 will be provided in Theorem 22.1.

Lemma 22.1 *If G is planar and every node is a P-node in T_r, then all essential nodes must lie on a path of the PC-tree.* ∎

The shortest path including all essential nodes is called the *terminal path*, denoted by F, which can be identified efficiently by finding the least common ancestor of all essential nodes. The two end nodes of F must be essential, referred to as the *terminal nodes*. When we embed G_r, there could be many paths that can be used as part of the boundary path to connect the terminal nodes to i. Exactly which path is used is immaterial for our recognition algorithm. To avoid such trivial variations, we connect terminal nodes to i with two artificial edges. These two artificial edges together with the terminal path F and node i form a cycle BC, referred to as the *boundary cycle* for G_r. An important property is that BC is uniquely determined. In the degenerate case, there could be only one essential node and the cycle BC is reduced to an edge connecting this unique terminal node to i.

BC divides the plane into inside and outside. Since each node in T_r is a P-node, we can flip all traversed subtrees to the inside of BC and flip the un-traversed ones to the outside. Since the entire tree T_r^* can be flipped to the inside, one can easily form a planar embedding for G_r by connecting i to r and all its back neighbors in T_r^*. We call this the *internal embedding* of the component G_r with the boundary cycle BC.

Since only essential nodes are related to future embedding, we could delete the internal nodes; extract a *representative boundary cycle* (RBC) from BC by contracting all nonessential nodes. The edge connections in the RBC are called *links*, to be distinguished from the tree edges or back edges. Define the corresponding C-nodes w_r for G_r as follows: The parent of w_r is i, referred to as the *head* of this C-node, and the children of w_r are the remaining nodes in its RBC following their cyclic order. This completes the C-node construction for the subtree T_r. Note that a tree path from u to v containing some C-nodes always contains a path of P-nodes in the original graph by traversing one side of the boundary cycle of each C-node on the tree path.

22.4 CREATING C-NODES IN GENERAL

Assume now the graph contains some C-nodes. We shall abuse the notation a little by still assuming the current iteration is i, and the current subtree is T_r, which has a back edge to i, except that now there could be some C-nodes in T_r. As far as node traversal is concerned, C-nodes are no different from P-nodes. Hence, we can still perform all BETs as before. Denote the new i-component by G_r. Define the essential nodes, the terminal path just as before. In order to successfully embed G_r, we need to prove the following two theorems, which allow us to find the boundary cycle efficiently through the PC-tree. Since the proofs essentially amount to identifying Kuratowski subgraphs for nonplanar graphs (in linear time), we describe them in Appendix 22A for interested readers.

Theorem 22.1 *If G is planar, essential nodes of an i-component must lie on a path of the PC-tree.* ∎

Definition 22.1 *The shortest path including all essential nodes in the PC-tree is called the terminal path, denoted by F. The separating nodes of an intermediate C-node w on the terminal path F are the two neighbors of w on F.* ∎

Theorem 22.2 *Let w be a C-node on the terminal path F. If G is planar, the traversed neighbors of w form a consecutive path in the RBC of w. Furthermore, if w is an intermediate node in F, such a path must end in the two separating nodes of w.* ∎

By connecting i to the two ends of F with two artificial links, we obtain the *boundary cycle* BC for G_r. Note that BC could consist of both P-nodes and C-nodes. By Theorem 22.2, the RBC of each C-node in F can be divided into the traversed side and the un-traversed side. Obtain the *external boundary path* F_r of G_r by following path F through all its P-nodes and the un-traversed side of its C-nodes. In Figure 22.2, we illustrate the formation of F_r by showing how to extract it from F. Now, P-nodes of G_r not in BC will be embedded inside; un-traversed nodes not in F_r will be embedded outside. Delete all inside nodes and C-nodes in BC. Extract an RBC from F_r by contracting all nodes of degree 2 so that each remaining node in F_r must be incident to an un-traversed node not in BC. Only those nodes on the RBC are relevant for future embedding. Edge connections in the RBC are referred to as *links*. Note that RBC, useful for the cycle view, is not part of the PC-tree. We associate an RBC to each C-node for implementation purpose.

Create the new C-node w_r for G_r as follows: The parent of w_r is i, called its *head*, and the children of w_r are the remaining nodes in the RBC following their cyclic order. This completes the operations related to T_r in the ith iteration. If there is no violation to the properties stated in both Theorems 22.1 and 22.2 at each iteration, the algorithm can continue successfully until the end of the nth iteration, at which point a single biconnected component is formed and the graph is declared planar.

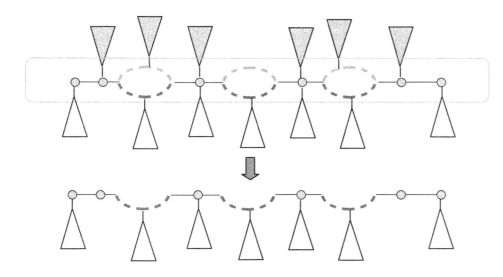

Figure 22.2 Boundary path F_r and its un-traversed children subtrees.

Figure 22.3 Internal boundary path F'_r and its traversed children subtrees.

22.5 EMBEDDING ALGORITHM

The internal embedding of the C-node w_r can be carried out alongside the recognition algorithm. Similar to the extraction of the external boundary path in the section above, one can obtain the internal boundary path F'_r of G_r by following path F through all its P-nodes and the traversed side of its C-nodes as illustrated in Figure 22.3. Path F'_r together with all their traversed children form a PC-tree T_{G_r}. Embed T_{G_r} inside BC by embedding the tree nodes observing the C-node constraint (note that each C-node can be embedded in two possible ways). Finally, connect i to r and its back descendants in T_{G_r}. This completes an internal embedding of G_r within BC. Store such an embedding for G_r. At the end of the nth iteration, we get a single biconnected component and its internal embedding. From this embedding, we can trace back and paste all stored sub-embeddings back (by identifying the boundary cycle) recursively to form the final planar embedding of G.

22.6 LINEAR-TIME IMPLEMENTATION

The recognition algorithm described above could take $O(n^2)$ time in a brute-force implementation (as was probably carried out in [4]). This is because C-nodes can be merged and the parent pointers could be changed quite often ($O(n^2)$ time in the worst case). A trick similar to the one used in Booth and Lueker [3] that reduces the original $O(n^2)$ PQ-tree operations to linear time can be applied here, which requires no parent pointers: drop the parent–child relations for all C-nodes and consider the cycle view of the PC-tree. For each C-node w with parent i, associate w to the two artificial links incident to i in its RBC. Whenever the algorithm traverses to a node u that is on an RBC, we find its owner w' using a *parallel search* along the links of its RBC cycle: start traversing links of the cycle from the two neighbors of u in both directions in parallel; whichever search first reaches an artificial link associated with w' will terminate both searches. In other words, it is useful to think that each node of the PC-tree has a parent pointer in the algorithm and the proofs; but for efficient implementation, we need to go around the RBC of each C-node to find the head of w' instead. Note that every P-node can be the heads of several (child) C-nodes; but it can belong to at most one RBC (of its parent C-node if any), a property of the PC-tree.

That such an implementation takes linear time can be argued below.

It is clear that the running time can be divided into two parts:

a. The time spent in the BETs, which is bounded by (1) the number of edges traversed, and (2) the number of links traversed and

b. The time it takes to create the new C-node.

In part (a)(1), every tree edge traversed will be included in a biconnected component. If it is embedded inside, then we never have to look at it again; otherwise, this edge is on the terminal path, which is also deleted and replaced by links in the RBC. So every tree edge can be traversed at most once, and the total number of tree edges traversed is bounded by $O(n)$.

For part (a)(2), since the number of links in an RBC is no more than the number of edges in a terminal path, the total number of links created is bounded by the total number of tree edge traversed, which is $O(n)$. Now, these links are traversed to identify the parent C-node in a parallel search. Suppose it takes $2d$ link traversals to achieve that. Since one of the two traversed link paths will be embedded inside, at least $d-1$ nodes and d links will be embedded inside and deleted. Hence, the total number of links traversed is at most $2n$. Therefore, the total time spent in (a) is $O(n)$.

In part (b), it suffices to show that finding the boundary cycle BC and contracting all nonessential nodes take time proportional to the number of tree nodes in the terminal path since the latter is bounded above by the time spent in the BET of (a). Now, for each C-node s in BC, we only need to connect from its separating nodes to their neighboring un-traversed P-nodes in the RBC of s to form that part of BC (there is no need to traverse the un-traversed P-nodes in between). So connecting through each C-node takes constant time. Hence, part (b) takes $O(n)$ time in total.

Therefore, the entire recognition algorithm takes $O(n)$ time. It is easy to see from Section 22.5 that, through backtracking, the embedding algorithm also takes $O(n)$ time. Furthermore, the reader can check that, in case the graph is nonplanar, finding Kuratowski subgraphs also takes $O(n)$ time from the proofs of Theorems 22.1 and 22.2.

APPENDIX 22A

Proof of Theorem 22.1. Suppose not. Assume there are three essential nodes v_1, v_2, and v_3 such that v_3 is not on the tree path between v_1 and v_2. Let q be the node on this path closest to v_3 in the PC-tree. Then there are three node-disjoint tree paths from q to v_1, v_2, and v_3, respectively. Now, there are also three node-disjoint paths from i to v_1, v_2, and v_3, respectively, through i-paths or an i_r-path (in case a node $v_k, k \in 1, 2, 3$, is located on the i_r-path from i to q; note there could be at most one such v_k) as shown in Figure 22A.1a. Consider the following two cases:

Case 1 G_r contains only P-nodes: Rearrange the structure in Figure 22A.1a as the basic structure in Figure 22A.1b. Let T_{v_1}, T_{v_2}, and T_{v_3} have back edges to t_1, t_2, and t_3, respectively, with $t_1 \leq t_2 \leq t_3$. There exist three node-disjoint paths through un-traversed nodes and

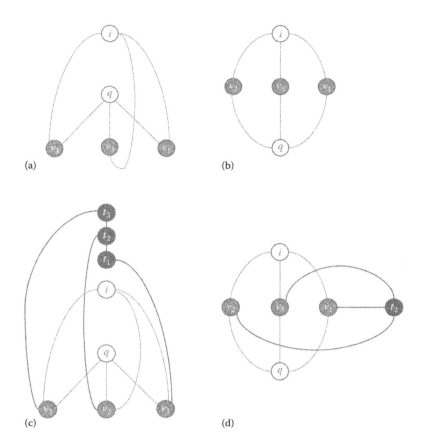

Figure 22A.1 (a–d) $K_{3,3}$ for Case 1.

these back edges from t_2 to v_1, v_2, and v_3 as shown in Figure 22A.1c. Together, they form a graph homeomorphic to $K_{3,3}$ with v_1, v_2, and v_3 on one side, and i, q, t_2 on the other as in Figure 22A.1d.

Case 2 G_r contains both C-nodes and P-nodes: We can further divide this into the following subcases depending on whether the v's and q are C-nodes. Note, however, these subcases are not mutually exclusive.

Case 2.1 Suppose an essential node, say v_1, is a C-node that has an un-traversed child n_1 and two traversed children n_2 and n_3 on the path from q to i as shown in Figure 22A.2a. Since these children are on the RBC for v_1, there exists a path from n_2, through n_1, to n_3 as shown in Figure 22A.2b. In this way, an essential C-node for v_1, v_2, or v_3 can be regarded as an essential P-node. We can then follow the arguments in Case 1 for essential P-nodes to find a subgraph homeomorphic to $K_{3,3}$.

Case 2.2 q is a C-node as illustrated in Figure 22A.3a. Let t_1, t_2, and t_3 be three back nodes from nodes in T_{v_1}, T_{v_2}, and T_{v_3}, respectively with $t_1 \leq t_2 \leq t_3$, similar to Case 1. Consider the following subcases:

Case 2.2.1 At least one of v_1, v_2, and v_3 is not a neighbor of q. Without loss of generality, assume v_1 is not. Let u be the neighbor of q on the tree path from v_1 to i. There is a basic structure as shown in Figure 22A.3b, where u plays the role

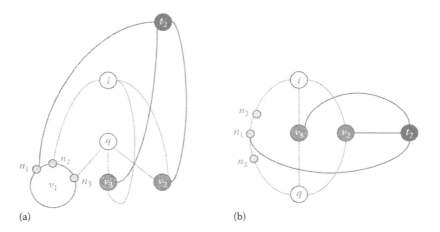

Figure 22A.2 (a,b) $K_{3,3}$ for Case 2.1.

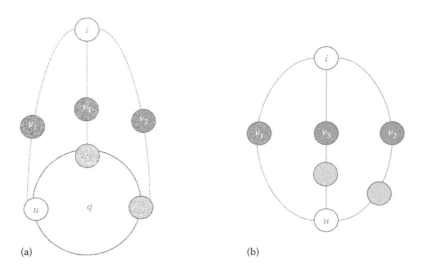

Figure 22A.3 (a,b) Basic structure in which q is a C-node in Case 2.2.

of q as in Case 1 Figure 22A.4d. We can similarly find a subgraph homeomorphic to $K_{3,3}$ with v_1, v_2, v_3 on one side and i, u, t_2 on the other.

Case 2.2.2 v_1, v_2, and v_3 are neighbors of q. If $t_1 = t_2$, there are three node-disjoint paths from t_2 to v_1, v_2, and v_3 through un-traversed nodes as in Figure 22A.4a. Together with the tree path from t_2 to i, they form a subgraph homeomorphic to K_5 in Figure 22A.4b. Otherwise, $t_1 < t_2$ and there are three node-disjoint paths emanating from t_2: one from t_2 to t_1 and two from t_2 to v_2 and v_3, respectively, through un-traversed nodes as in Figure 22A.5a. They form a subgraph homeomorphic to $K_{3,3}$ with t_1, v_2, and v_3 on one side, i, v_1, and t_2 on the other as in Figure 22A.5b. ∎

Proof of Theorem 22.2. Suppose not. Consider the following two cases.

Case 1 The traversed neighbors are not consecutive. Then there are at least two traversed neighbors, say u_1 and u_2, of w separated by two un-traversed neighbors v_1 and v_2 as in Figure 22A.6a. Let T_{v_1} and T_{v_2} have back edges to t_1 and t_2, respectively. Without loss of

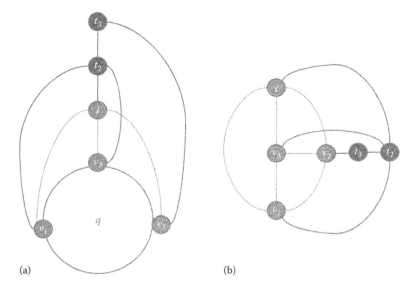

Figure 22A.4 (a,b) K_5 in Case 2.2.2.

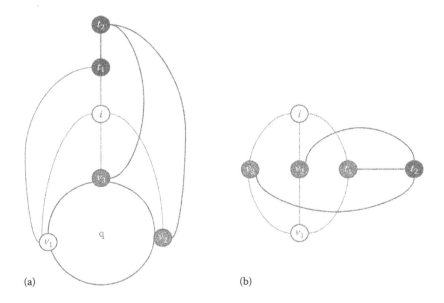

Figure 22A.5 (a,b) $K_{3,3}$ in Case 2.2.2.

generality, assume $t_1 \leq t_2$. Then we get a subgraph homeomorphic to $K_{3,3}$ with i, v_1, v_2 on one side and t_1, u_1, u_2 on the other as in Figure 22A.6b.

Note that, the same holds for the case where w is the only terminal node. From now on, we assume there are two terminal nodes.

Case 2 The traversed neighbors of w form a consecutive path, say H, in its RBC, but the separating nodes are not the two end nodes of H. Consider the following two sub-cases:

Case 2.1 w has only one separating node d (w is a terminal node of the terminal path in the i-component). Node d is traversed but not at the end of H. Let u_1 and u_2 be the two traversed neighbors of d. Pick any un-traversed node v_1 in the RBC. Let v_2 be the

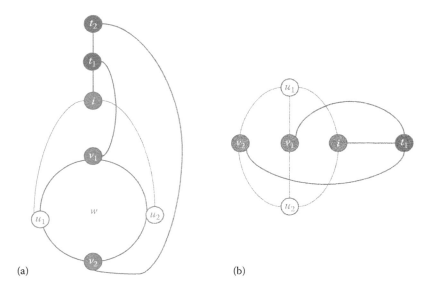

Figure 22A.6 (a,b) $K_{3,3}$ in Case 1.

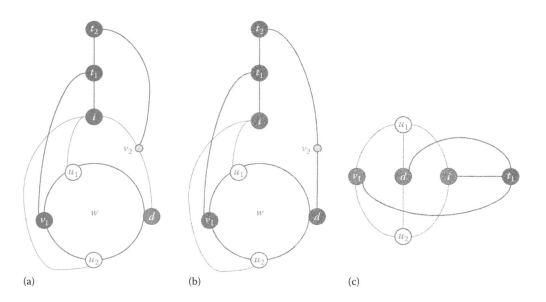

Figure 22A.7 (a,b) A $K_{3,3}$ in Case 2.1.

other terminal node. Without loss of generality, assume T_{v_1} and T_{v_2} have back-edges to t_1 and t_2, respectively, with $t_1 \leq t_2$ as in Figure 22A.7a. From t_1, there are two node-disjoint paths through these back edges, one to v_1 and the other to d through t_2, v_2, and tree edges as in Figure 22A.7b. Then we get a subgraph homeomorphic to $K_{3,3}$ with i, d, v_1 on one side and t_1, u_1, u_2 on the other as in Figure 22A.7c.

Case 2.2 w has two separating nodes d_1 and d_2, which are not both at the ends of H. We could assume d_1 is not an end of H. Let u_1 and u_2 be the two traversed neighbors of d_1 in the RBC. Let v_1 and v_2 be the two terminal nodes of this i-component as shown in Figure 22A.8a. (In case v_k is a C-node, we could use one of its un-traversed neighbors instead.) Without loss of generality, assume T_{v_1} and T_{v_2} have back-edges to t_1 and t_2, respectively, and $t_1 \leq t_2$. Similar to case 2.1, there are two node-disjoint

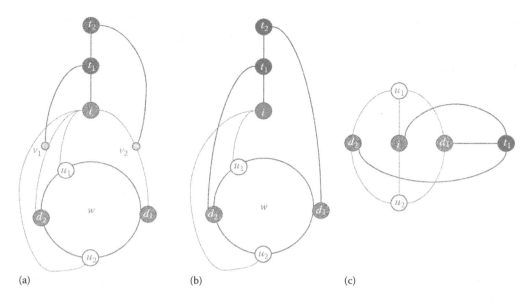

Figure 22A.8 (a–c) A $K_{3,3}$ in Case 2.2.

paths from t_1 to d_1 and d_2, through v_1 and v_2, respectively, as in Figure 22A.8b. Then we get a subgraph homeomorphic to $K_{3,3}$ with i, d_1, d_2 on one side and t_1, u_1, u_2 on the other as in Figure 22A.8c. ∎

References

[1] J. Hopcroft and R. Tarjan, Efficient planarity testing, *J. ACM*, **21**(4) (1974), 549–568.

[2] A. Lempel, S. Even, and I. Cederbaum, An algorithm for planarity testing of graphs, In P. Rosenstiel (ed.) *Theory of Graphs: International Symposium*, Gordon & Breach, New York, 1967, volume 67, 215–232.

[3] K. S. Booth and G. S. Lueker, Testing for the consecutive ones property, interval graphs, and graph planarity using pq-tree algorithms, *J. Comp. Syst. Sci.*, **13**(3) (1976), 335–379, December 1976.

[4] J. Boyer and W. Myrvold, Stop minding your p's and q's: A simplified $O(n)$ planar embedding algorithm, In *Proc. 10th Ann. ACM-SIAM Symp. Discrete Algorithms*, pages 140–146, Philadelphia, PA, 1999. Society for Industrial and Applied Mathematics.

[5] J. M. Boyer and W. J. Myrvold, On the cutting edge: Simplified $O(n)$ planarity by edge addition, *J. Graph Algorithms and Appl.*, **8** (2004), 241–273.

[6] J. Small, A unified approach of testing, embedding and drawing planar graphs, In *Proc. ALCOM Int. Workshop Graph Draw.*, Sevre, France, 1993.

[7] H. Stamm-Wilbrandt, A simple linear-time algorithm for embedding maximal planar graphs, In *ALCOM Inte. Workshop Graph Draw.*, 92, 1993.

[8] S. G. Williamson, Depth-first search and kuratowski subgraphs, *J. ACM*, **31**(4) (1984), 681–693.

[9] N. Chiba, T. Nishizeki, S. Abe, and T. Ozawa, A linear algorithm for embedding planar graphs using *PQ*-trees, *J. of Comput. Sys. Sci.*, **30**(1) (1985), 54–76.

[10] K. Mehlhorn and P. Mutzel, On the embedding phase of the hopcroft and Tarjan planarity testing algorithm, *Algorithmica*, **16**(2) (1996), 233–242.

[11] W.-K. Shih and W.-L. Hsu, A new planarity test, *Theoretical Comput. Sci.*, **223**(1–2) (1999), 179–191.

[12] W.-L. Hsu, PC-trees vs. PQ-trees, *Lecture Notes Comput. Sci.*, **2108** (2001), 207–217.

[13] W.-L. Hsu and R. McConnell, PQ trees, PC trees and planar graphs, In Dinesh P. Mehta and Sartaj Sahni (eds.) *Handbook of Data Structures and Applications*, 2004.

[14] W.-K. Shih and W.-L. Hsu, A simple test for planar graphs, In *Proc. Int. Workshop on Discrete Math. Algorithms*, 110–122. University of Hong Kong, 1993.

[15] R. Thomas, Planarity in linear time, *Lecture Notes*, Georgia Institute of Technology, 1997. http://www.math.gatech.edu/~thomas/planarity.ps.

[16] W.-L. Hsu, An efficient implementation of the PC-tree algorithm of Shih & Hsu's planarity test, Technical report, Institute of Information Science, Academia Sinica, 2003.

[17] W.-L. Hsu, A linear time algorithm for finding a maximal planar subgraph based on PC-trees, *Lecture Notes Comput. Sci.*, **3595** (2005), 787–797.

CHAPTER 23

Graph Drawing

Md. Saidur Rahman

Takao Nishizeki

CONTENTS

23.1	Introduction	537
	23.1.1 Drawing Styles	538
	23.1.2 Applications of Graph Drawing	542
23.2	Straight Line Drawing	543
	23.2.1 Canonical Ordering	544
	23.2.2 Shift Algorithm	546
23.3	Convex Drawing	552
	23.3.1 Canonical Decomposition	553
	23.3.2 Algorithm for Convex Grid Drawing	556
23.4	Rectangular Drawing	560
	23.4.1 Rectangular Drawing and Matching	562
	23.4.2 Linear Algorithm	564
23.5	Orthogonal Drawing	567
	23.5.1 Orthogonal Drawing and Network Flow	570
	23.5.2 Linear Algorithm for Bend-Optimal Drawing	575

23.1 INTRODUCTION

A drawing of a graph can be thought of as a diagram consisting of a collection of objects corresponding to the vertices of the graph together with some line segments corresponding to the edges connecting the objects. People are using diagrams from ancient time to represent abstract things like ideas, concepts, and so forth as well as concrete things like maps, structures of machines, and so on. A graph may be used to represent any information which can be modeled as objects and relationship between those objects. A drawing of a graph is a sort of visualization of information represented by the graph. The graph in Figure 23.1a represents eight components and their interconnections in an electronic circuit, and Figure 23.1b depicts a drawing of the graph. Although the graph in Figure 23.1a correctly represents the circuit, the representation is messy and hard to trace the circuit for understanding and troubleshooting. Furthermore, in this representation one cannot lay the circuit on a single layered PCB (printed circuit board) because of edge crossings. On the other hand, the drawing of the graph in Figure 23.1b looks better and it is easily traceable. Furthermore one can use the drawing to lay the circuit on a single layered PCB, since it has no edge crossing. Thus, the objective of graph drawing is to obtain a nice representation of a graph such that the structure of the graph is easily understandable, and moreover the drawing should satisfy some criteria that arises from the application point of view.

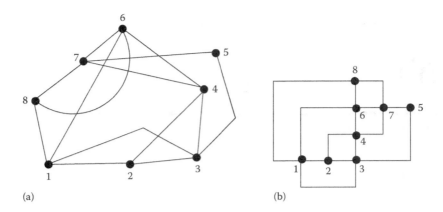

Figure 23.1 (a,b) Example of graph drawing in circuit schematics.

The industrial need for graph drawing algorithms arose in the late 1960s when a large number of elements in complex circuit designs made hand drawing too complicated [1–4]. Algorithms were developed to aid circuit design; an overview can be found in the book of Lengauer [5]. The field of graph drawing with the objective of producing aesthetically pleasing pictures became of interest in the late 1980s for presenting information of engineering and production process [6,7].

The field of graph drawing has flourished very much in the last two decades. Recent progress in computational geometry, topological graph theory, and order theory has considerably affected the evolution of this field, and has widened the range of issues being investigated. A comprehensive bibliography on graph drawing algorithms [8] cites more than 300 papers written before 1993. From 1993, an international symposium on graph drawing is being held annually in different countries and the proceedings of the symposium are published by Springer-Verlag in the LNCS series [9–20]. Several special issues of journals dedicated to graph drawing have been assembled [21–26]. A few books on graph drawing have also been published [4,27–30].

23.1.1 Drawing Styles

In this section we introduce some important drawing styles and related terminologies [29].

Various graphic standards are used for drawing graphs. Usually, vertices are represented by symbols such as points or boxes, and edges are represented by simple open Jordan curves connecting the symbols that represent the associated vertices. From now on, we assume that vertices are represented by points if not specified. We now introduce the following drawing styles.

Planar drawing: A drawing of a graph is *planar* if no two edges intersect in the drawing. Figure 23.2 depicts a planar drawing and a nonplanar drawing of the same graph. It is preferable to find a planar drawing of a graph if the graph has such a drawing. Unfortunately not all graphs admit planar drawings. A graph which admits a planar drawing is called a *planar graph*.

If one wants to find a planar drawing of a given graph, first he/she needs to test whether the given graph is planar or not. If the graph is planar, then he/she needs to find a planar embedding of the graph, which is a data structure representing adjacency lists: in each list the edges incident to a vertex are ordered, all clockwise or all counterclockwise, according to the planar embedding. Kuratowski [31] gave the first complete characterization of planar graphs. Unfortunately, the characterization does not lead to an efficient algorithm for planarity testing. Linear-time algorithms for this

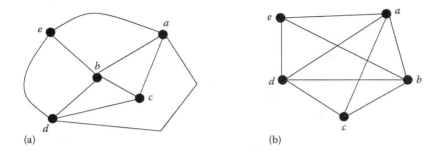

Figure 23.2 (a) Planar drawing and (b) nonplanar drawing of the same graph.

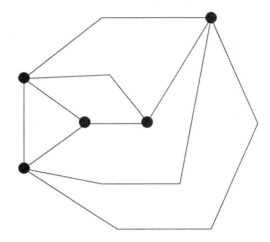

Figure 23.3 Polyline drawing of a graph.

problem have been developed by Hopcroft and Tarjan [32] and Booth and Lueker [33]. Chiba et al. [34] and Mehlhorn and Mutzel [35] gave linear-time algorithms for finding a planar embedding of a planar graph. Shih and Hsu [36] gave a simple linear-time algorithm which performs planarity testing and finds a planar embedding of a planar graph simultaneously. A planar graph with a fixed planar embedding is called a *plane graph*.

Polyline drawing: A *polyline drawing* is a drawing of a graph in which each edge of the graph is represented by a polygonal chain. A polyline drawing of a graph is shown in Figure 23.3. A point at which an edge changes its direction in a polyline drawing is called a *bend*. Polyline drawings provide great flexibility since they can approximate drawings with curved edges. However, it may be difficult to follow edges with more than two or three bends by the eye. Several interesting results on polyline drawings can be found in [37–39].

Straight line drawing: A *straight line drawing* is a drawing of a graph in which each edge of the graph is drawn as a straight line segment, as illustrated in Figure 23.4. A straight line drawing is a special case of a polyline drawing, where edges are drawn without bend. Wagner [40], Fáry [41], and Stein [42] independently proved that every planar graph has a straight line drawing. A straight line drawing of a plane graph G is called a *convex drawing* if the boundaries of all faces of G are drawn as convex polygons, as illustrated in Figure 23.4b [6,43–47].

Orthogonal drawing: An *orthogonal drawing* is a drawing of a plane graph in which each edge is drawn as a chain of horizontal and vertical line segments, as illustrated

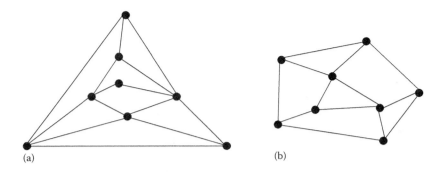

Figure 23.4 (a) Straight line drawing and (b) convex drawing.

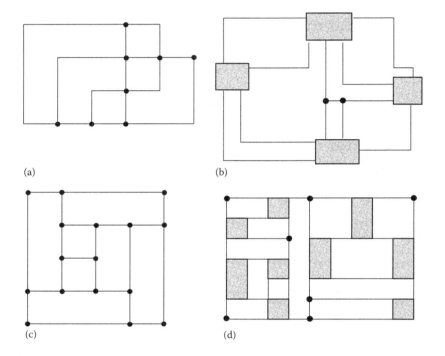

Figure 23.5 (a) Orthogonal drawing, (b) box-orthogonal drawing, (c) rectangular drawing, and (d) box-rectangular drawing.

in Figure 23.5a. Many results have been published in recent years on both planar orthogonal drawings [45,48–56] and nonplanar orthogonal drawings [57–59]. An orthogonal drawing is called an *octagonal drawing* if the outer cycle is drawn as a rectangle and each inner face is drawn as a rectilinear polygon of at most eight corners [60]. Conventionally, each vertex in an orthogonal drawing is drawn as a point, as illustrated in Figure 23.5a. Clearly a graph having a vertex of degree five or more has no orthogonal drawing, because at most four edges can be incident to a vertex in an orthogonal drawing. A *box-orthogonal drawing* of a graph is a drawing such that each vertex is drawn as a (possibly degenerate) rectangle, called a *box*, and each edge is drawn as a sequence of alternate horizontal and vertical line segments, as illustrated in Figure 23.5b. Every plane graph has a box-orthogonal drawing. Several results are known for box-orthogonal drawings [61–63].

Rectangular drawing: A *rectangular drawing* of a plane graph G is a drawing of G in which each vertex is drawn as a point, each edge is drawn as a horizontal or vertical line segment without edge-crossings, and each face is drawn as a rectangle, as illustrated in Figure 23.5c [64–67]. A *box-rectangular drawing* of a plane graph G is a drawing of G on the plane such that each vertex is drawn as a (possibly degenerate) rectangle, called a *box*, and the contour of each face is drawn as a rectangle, as illustrated in Figure 23.5d [68–70].

Grid drawing: A drawing of a graph in which vertices and bends are located at grid points of an integer grid as illustrated in Figure 23.6 is called a *grid drawing*. The *size* of an integer grid required for a grid drawing is measured by the size of the smallest rectangle on the grid which encloses the drawing. The *width* W of the grid is the width of the rectangle and the *height* H of the grid is the height of the rectangle. The grid size is usually described as $W \times H$. The grid size is sometimes described by the *half perimeter* $W + H$ or the *area* $W \cdot H$ of the grid.

Visibility drawing: A *visibility drawing* of a plane graph G is a drawing of G where each vertex is drawn as a horizontal line segment and each edge is drawn as a vertical line segment. The vertical line segment representing an edge must connect points on the horizontal line segments representing the end vertices [71–73]. Figure 23.7b depicts a visibility drawing of the plane graph G in Figure 23.7a. A *2-visibility drawing* is a generalization of a visibility drawing where vertices are drawn as boxes and edges are drawn as either a horizontal line segment or a vertical line segment [62]. Figure 23.7c depicts a 2-visibility drawing of the plane graph G in Figure 23.7a.

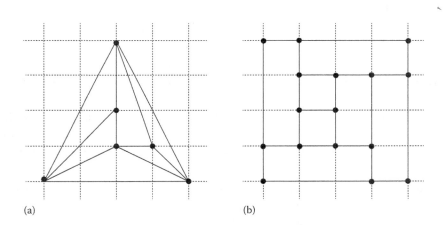

Figure 23.6 (a) Straight line grid drawing and (b) rectangular grid drawing.

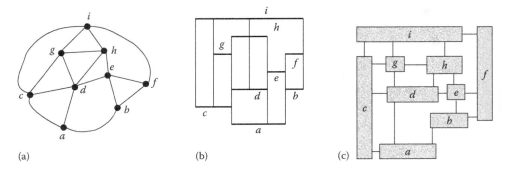

Figure 23.7 (a) Plane graph G, (b) visibility drawing of G, and (c) 2-visibility drawing of G.

23.1.2 Applications of Graph Drawing

Graph drawings have applications in almost every branch of science and technology [4,28,29,74]. In [29] applications of graph drawings in computer networks, circuit schematics, floorplanning, VLSI layout sofware engineering, bioinformatics, and so forth are illustrated. In this section we will illustrate an application of graph drawing in VLSI floorplanning as well as architectural floorplanning [29]. In a VLSI floorplanning problem, an input is a plane graph F as illustrated in Figure 23.8a; F represents the functional entities of a chip, called *modules*, and interconnections among the modules; each vertex of F represents a module, and an edge between two vertices of F represents the interconnections between the two corresponding modules. An output of the problem for the input graph F is a partition of a rectangular chip area into smaller rectangles as illustrated in Figure 23.8d; each module is assigned to a smaller rectangle, and furthermore, if two modules have interconnections, then their corresponding rectangles must be adjacent, that is, must have a common boundary. A similar problem may arise in architectural floorplanning also. When building a house, the owner may have some preference; for example, a bed room should be adjacent to a reading room. The owner's choice of room adjacencies can be easily modeled by a plane graph F, as illustrated in Figure 23.8a; each vertex represents a room and an edge between two vertices represents the desired adjacency between the corresponding rooms.

A rectangular drawing of a plane graph may provide a suitable solution of the floorplanning problem described above. First, obtain a plane graph F' by triangulating all inner faces of F as illustrated in Figure 23.8b, where dotted lines indicate new edges added to F. Then obtain a dual-like graph G of F' as illustrated in Figure 23.8c, where the four vertices of degree 2 drawn by white circles correspond to the four corners of the rectangular area. Finally, by finding a rectangular drawing of the plane graph G, obtain a possible floorplan for F as illustrated in Figure 23.8d.

In the floorplan above, two rectangles are always adjacent if the modules corresponding to them have interconnections in F. However, two rectangles may be adjacent even if the modules corresponding to them have no interconnections. For example, modules e and f have no interconnection in F, but their corresponding rectangles are adjacent in the floorplan in Figure 23.8d. Such unwanted adjacencies are not desirable in some other floorplanning problems. In floorplanning of a multichip module (MCM), two chips generating excessive heat should not be adjacent, or two chips operating on high frequency should not be adjacent to avoid malfunctioning due to their interference [75,76]. Unwanted adjacencies may cause a dangerous situation in some architectural floorplanning too [77]. For example, in a chemical industry, a processing unit that deals with poisonous chemicals should not be adjacent to a cafeteria.

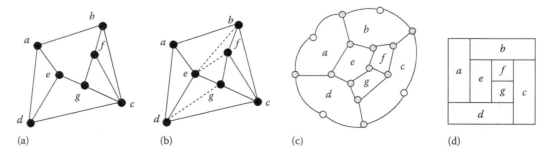

Figure 23.8 (a) Graph F, (b) triangulated graph F', (c) dual-like graph G, and (d) rectangular drawing of G. (Figure taken from Nishizeki, T. and Rahman, M. S., *Planar Graph Drawing*, World Scientific, Singapore, 2004.)

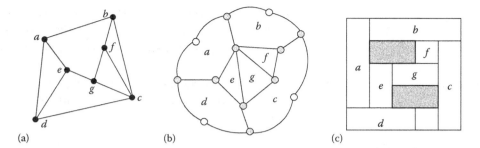

Figure 23.9 (a) F, (b) G, and (c) box-rectangular drawing of G. (Figure taken from Nishizeki, T. and Rahman, M S., *Planar Graph Drawing*, World Scientific, Singapore, 2004.)

We can avoid the unwanted adjacencies if we obtain a floorplan for F by using a box-rectangular drawing instead of a rectangular drawing, as follows. First, without triangulating the inner faces of F, find a dual-like graph G of F as illustrated in Figure 23.9b. Then, by finding a box-rectangular drawing of G, obtain a possible floorplan for F as illustrated in Figure 23.9c. In Figure 23.9c rectangles e and f are not adjacent although there is a dead space corresponding to a vertex of G drawn by a rectangular box. Such a dead space to separate two rectangles in floorplanning is desirable for dissipating excessive heat in an MCM or for ensuring safety in a chemical industry.

23.2 STRAIGHT LINE DRAWING

A *straight line drawing* of a plane graph is a drawing in which each edge is drawn as a straight line segment without edge-crossings, as illustrated in Figure 23.4. Wagner [40], Fáry [41], and Stein [42] independently proved that every planar graph G has a straight line drawing. Their proofs immediately yield polynomial time algorithms to find a straight line drawing of a given plane graph. However, the area of a rectangle enclosing a drawing on an integer grid obtained by these algorithms is not bounded by any polynomial in the number n of vertices in G. In fact, it remained as an open problem for long time to obtain a drawing of area bounded by a polynomial. In 1990, de Fraysseix et al. [78] and Schnyder [79] showed by two different methods that every planar graph of $n \geq 3$ vertices has a straight line drawing on an integer grid of size $(2n - 4) \times (n - 2)$ and $(n - 2) \times (n - 2)$, respectively. The two methods can be implemented as linear-time algorithms, and are well known as the *shift method* and the *realizer method*, respectively [29].

A natural question arises: what is the minimum size of a grid required for a straight line drawing? de Fraysseix et al. showed that, for each $n \geq 3$, there exists a plane graph of n vertices, for example nested triangles, which needs a grid of size at least $\lceil 2(n-1)/3 \rceil \times \lceil 2(n-1)/3 \rceil$ for any grid drawing [78,80]. It has been conjectured that every plane graph of n vertices has a grid drawing on a $\lceil 2n/3 \rceil \times \lceil 2n/3 \rceil$ grid, but it is still an open problem. On the other hand, a restricted class of graphs has a more compact grid drawing. For example, if G is a 4-connected plane graph, then G has a more compact grid drawing [81,82].

In this section, we describe a constructive proof for the theorem by de Fraysseix et al. [78] that every plane graph G of $n \geq 3$ vertices has a straight line grid drawing of size $(2n - 4) \times (n - 2)$, and present a linear-time implementation of an algorithm for finding such a drawing [83]. If G is not triangulated, then we obtain a triangulated plane graph G' from G by adding dummy edges to G. From a straight line grid drawing of G' we can immediately obtain a straight line grid drawing of G by deleting the dummy edges. Therefore it is sufficient to prove that a triangulated plane graph G of n vertices has a straight line grid drawing of size $(2n-4) \times (n-2)$. To construct such a drawing, de Fraysseix et al. introduced an ordering of vertices called a *canonical ordering* and installed vertices one by one in the drawing according to the ordering.

In Section 23.2.1 we present a canonical ordering, and in Section 23.2.2 we present the algorithm of de Fraysseix et al. and a linear time implementation of the algorithm.

23.2.1 Canonical Ordering

For a cycle C in a graph, an edge joining two nonconsecutive vertices in C is called a *chord* of C. For a 2-connected plane graph G, we denote by $C_o(G)$ the *outer cycle* of G, that is, the boundary of the outer face of G. A vertex on $C_o(G)$ is called an *outer vertex* and an edge on $C_o(G)$ is called an *outer edge*. A plane graph is *internally triangulated* if every inner face is a triangle.

Let $G = (V, E)$ be a triangulated plane graph of $n \geq 3$ vertices, as illustrated in Figure 23.10. Since G is triangulated, there are exactly three vertices on $C_o(G)$. One may assume that these three vertices, denoted by v_1, v_2 and v_n, appear on $C_o(G)$ counterclockwise in this order. Let $\pi = (v_1, v_2, \ldots, v_n)$ be an ordering of all vertices in G. For each integer k, $3 \leq k \leq n$, we denote by G_k the plane subgraph of G induced by the k vertices $v_1, v_2, \ldots, v_k$. Then $G_n = G$. We call π a *canonical ordering* of G if the following conditions (co1)–(co3) hold for each index $k, 3 \leq k \leq n$:

(co1) G_k is 2-connected and internally triangulated;

(co2) (v_1, v_2) is an outer edge of G_k; and

(co3) if $k + 1 \leq n$, then vertex v_{k+1} is located in the outer face of G_k, and all neighbors of v_{k+1} in G_k appear on $C_o(G_k)$ consecutively.

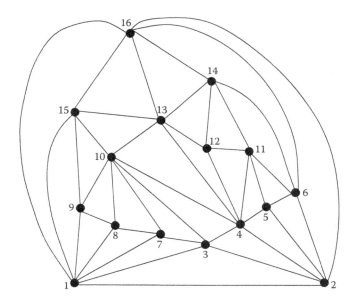

Figure 23.10 Canonical ordering of a triangulated plane graph of $n = 16$ vertices. (Figure taken from Nishizeki, T. and Rahman, M. S. *Planar Graph Drawing*, World Scientific, Singapore, 2004.)

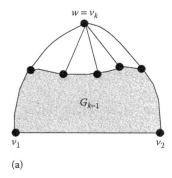
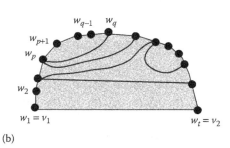

Figure 23.11 (a,b) Graph G_k and chords.

An example of a canonical ordering is illustrated for a triangulated plane graph of $n = 16$ vertices in Figure 23.10.

We now have the following lemma.

Lemma 23.1 *Every triangulated plane graph G has a canonical ordering.*

Proof. Obviously G has a canonical ordering if $n = 3$. One may thus assume that $n \geq 4$. Since $G = G_n$, clearly (co1)–(co3) hold for $k = n$. We then choose the $n - 3$ inner vertices $v_{n-1}, v_{n-2}, \ldots, v_3$ in this order, and show that (co1)–(co3) hold for $k = n-1, n-2, \ldots, 3$.

Assume for inductive hypothesis that the vertices $v_n, v_{n-1}, \ldots, v_{k+1}$, $k + 1 \geq 4$, have been appropriately chosen, and that (co1)–(co3) hold for k. If one can choose as v_k a vertex $w \neq v_1, v_2$ on the cycle $C_o(G_k)$ which is not an end of a chord of $C_o(G_k)$, as illustrated in Figure 23.11a, then clearly (co1)–(co3) hold for $k - 1$ since $G_{k-1} = G_k - v_k$. Thus it suffices to show that there is such a vertex w.

Let $C_o(G_k) = w_1, w_2, \ldots, w_t$, where $w_1 = v_1$ and $w_t = v_2$. If $C_o(G_k)$ has no chord, then any of the vertices $w_2, w_3, \ldots, w_{t-1}$ is such a vertex w. One may thus assume that $C_o(G_k)$ has a chord. Then G_k has a *minimal* chord (w_p, w_q), $p + 2 \leq q$, such that none of the vertices $w_{p+1}, w_{p+2}, \ldots, w_{q-1}$ is an end of a chord, as illustrated in Figure 23.11b where chords are drawn by thick lines. Then any of the vertices $w_{p+1}, w_{p+2}, \ldots, w_{q-1}$ is such a vertex w. ∎

The following algorithm computes a canonical ordering of a triangulated plane graph $G = (V, E)$. For each vertex v, we keep the following variables:

- $mark(v) = true$ if v has been added to the ordering, and *false* otherwise;

- $out(v) = true$ if v is an outer vertex of a current plane graph, and *false* otherwise; and

- $chords(v) =$ the number of chords of the outer cycle whose end vertex is v.

The algorithm is as follows.

Algorithm Canonical-Ordering(G)
begin
1   Let v_1, v_2, and v_n be the vertices appearing on the outer cycle counterclockwise in this order;
2   Set $chords(x) = 0$, $out(x) = false$, and $mark(x) = false$ for all vertices $x \in V$;
3   Set $out(v_1) = true$, $out(v_2) = true$, and $out(v_n) = true$;

```
4       for k = n down to 3 do
        begin
5           Choose any vertex x such that mark(x) = false,
            out(x) = true, chords(x) = 0, and x ≠ v_1, v_2;
6           Set v_k = x and mark(x) = true;
7           Let C_o(G_{k-1}) = w_1, w_2, ..., w_t, where w_1 = v_1 and w_t = v_2;
8           Let w_p, w_{p+1}, ..., w_q be the neighbors of v_k which
            have mark(w_i) = false;
            {They are consecutive on C_o(G_{k-1}), as illustrated in Figure 23.12.}
9           For each vertex w_i, p < i < q, set out(w_i) = true, and
            update the variable chords for w_i and its neighbors.
        end
end
```

The following lemma holds for Algorithm *Canonical-Ordering* [29].

Lemma 23.2 *Algorithm* Canonical-Ordering(G) *computes a canonical ordering of a triangulated plane graph G in time* $O(n)$. ∎

23.2.2 Shift Algorithm

In this section we describe the shift algorithm given by de Fraysseix et al. [78]. The algorithm embeds G, one vertex at a time in a canonical order $\pi = (v_1, v_2, \ldots, v_n)$ at each stage, adjusting the current partial embedding. With each vertex v_i, a set of vertices need to be moved whenever the position of v_i is adjusted. We denote by $L(v_i)$ the set of such vertices. Note that $v_i \in L(v_i)$.

We denote the current position of a vertex v by $P(v)$; $P(v)$ is expressed by its x- and y-coordinates as $(x(v), y(v))$. If $P_1 = (x_1, y_1)$ and $P_2 = (x_2, y_2)$ are two grid points whose Manhattan distance is even, then the straight line with slope $+1$ through P_1 and the straight line with slope -1 through P_2 intersects at a grid point, which is denoted by $\mu(P_1, P_2)$. Clearly

$$\mu(P_1, P_2) = \left(\frac{1}{2}(x_1 - y_1 + x_2 + y_2), \frac{1}{2}(-x_1 + y_1 + x_2 + y_2)\right) \quad (23.1)$$

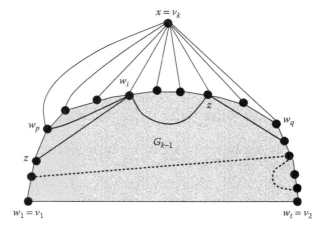

Figure 23.12 G_k.

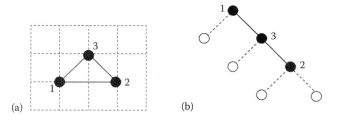

Figure 23.13 (a) Graph G_3 and (b) binary tree T for G_3.

We are now ready to describe the drawing algorithm.

First we draw G_3 by a triangle as follows. Set $P(v_1) = (0,0), P(v_2) = (2,0), P(v_3) = (1,1)$, and $L(v_i) = \{v_i\}$ for $i = 1, 2, 3$ (see Figure 23.13a).

Assume that $k - 1 \geq 3$ and we have embedded G_{k-1} in such a way that the following conditions hold:

(e1) $P(v_1) = (0,0)$ and $P(v_2) = (2k - 6, 0)$;

(e2) $x(w_1) < x(w_2) < \cdots < x(w_t)$, where $C_o(G_{k-1}) = w_1, w_2, \ldots, w_t$, $w_1 = v_1$ and $w_t = v_2$; and

(e3) each edge (w_i, w_{i+1}) on $C_o(G_{k-1})$ is drawn by a straight line having slope either $+1$ or -1, as illustrated in Figure 23.14a.

We now explain how to install v_k to a drawing of G_{k-1}. Let $w_p, w_{p+1}, \ldots, w_q$ be the neighbors of v_k on $C_o(G_{k-1})$, as illustrated in Figure 23.14a. We say that the vertex v_k covers the vertices $w_{p+1}, w_{p+2}, \ldots, w_{q-1}$. By (e3) the Manhattan distance between w_p and w_q is

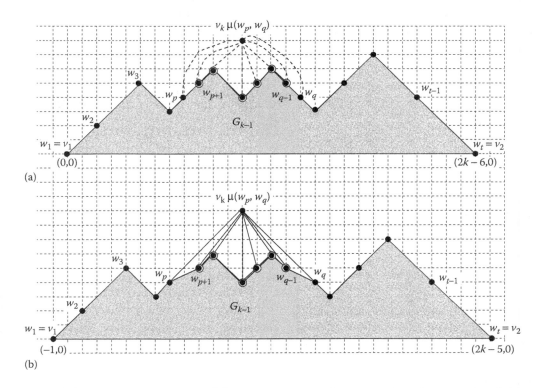

Figure 23.14 (a) G_k before shift and (b) G_k after shift. (Figure taken from Nishizeki, T. and Rahman, M. S., *Planar Graph Drawing*, World Scientific, Singapore, 2004.)

even, and hence $\mu(w_p, w_q)$ is a grid point. However, if we installed v_k at $\mu(w_p, w_q)$, then the straight line segment $w_p v_k$ would overlap with $w_p w_{p+1}$, because $w_p w_{p+1}$ may have slope $+1$ as illustrated in Figure 23.14a. We thus shift vertices $w_1(=v_1), w_2, \ldots, w_p$ together with some inner vertices to the left by one, as illustrated in Figure 23.14b. Similarly we shift vertices $w_q, w_{q+1}, \ldots, w_t(=v_2)$ together with some inner vertices to the right by one. We then install v_k at the grid point $\mu(w_p, w_q)$ for the new positions of w_p and w_q. More precisely, we execute the following Steps 1–4.

Step 1: for each $v \in \bigcup_{i=1}^{p} L(w_i)$ do $x(v) = x(v) - 1$;

Step 2: for each $v \in \bigcup_{i=q}^{t} L(w_i)$ do $x(v) = x(v) + 1$;

Step 3: $P(v_k) = \mu(w_p, w_q)$

Step 4: $L(v_k) = \{v_k\} \cup (\bigcup_{i=p+1}^{q-1} L(w_i))$

Figure 23.14a depicts a drawing of G_{k-1}, and Figure 23.14b depicts a drawing of G_k obtained by Steps 1–4. The Manhattan distance between w_p and w_q was even in the drawing of G_{k-1}. Vertex w_p is moved to the left by one by Step 1, and w_q is moved to the right by one by Step 2. Therefore the Manhattan distance between w_p and w_q is even in the drawing of G_k, and hence $\mu(w_p, w_q)$ is a grid point as in Figure 23.14b. Vertices $w_1, w_2, \ldots, w_p$ are moved to the left by one, and $w_q, w_{q+1}, \ldots, w_t$ are moved to the right by one. However, the positions of all vertices $w_{p+1}, w_{p+2}, \ldots, w_{q-1}$ are unchanged by Steps 1 and 2; they are indicated by double circles in Figure 23.14. Therefore the slopes of all edges $(w_{p+1}, w_{p+2}), (w_{p+2}, w_{p+3}), \ldots, (w_{q-2}, w_{q-1})$ have absolute value 1; these edges are drawn by thick solid lines in Figure 23.14. The slopes of edges (w_p, w_{p+1}) and (w_{q-1}, w_q) have absolute values smaller than 1 in the drawing of G_k as illustrated in Figure 23.14b. Thus all the vertices $w_p, w_{p+1}, \ldots, w_q$ are visible from the point $\mu(w_p, w_q)$, and hence one can draw all edges $(v_k, w_p), (v_k, w_{p+1}), \ldots, (v_k, w_q)$ by straight line segments without edge crossings as illustrated in Figure 23.14b.

Clearly $P(v_1) = (-1, 0)$ in Figure 23.14b. Replace Steps 1 and 2 above by the following Steps 1' and 2' to make $P(v_1) = (0, 0)$ by translating the drawing in Figure 23.14b to the right by one.

Step 1': for each $v \in \bigcup_{i=p+1}^{q-1} L(w_i)$ do $x(v) = x(v) + 1$;

Step 2': for each $v \in \bigcup_{i=q}^{t} L(w_i)$ do $x(v) = x(v) + 2$.

Then (e1), (e2), and (e3) hold for G_k.

Figure 23.15a illustrates $L(w_i)$ for all outer vertices w_i of G_{15} for the graph G in Figure 23.10.

The following lemma ensures that G_{k-1}, $3 \le k \le n$, remains to be a straight line grid drawing after Steps 1' and 2' are executed and hence G_k is a straight line grid drawing [29].

Lemma 23.3 *Let G_k, $3 \le k \le n$, be straight line grid embedded as described above. Let $C_o(G_k)$ contain t' vertices, and let $\delta_1 \le \delta_2 \le \cdots \le \delta_{t'}$ be any nondecreasing sequence of t' nonnegative integers. If, for each i, we shift the vertices in $L(w_i)$ by δ_i to the right, then we again obtain a straight line grid embedding of G_k.* ∎

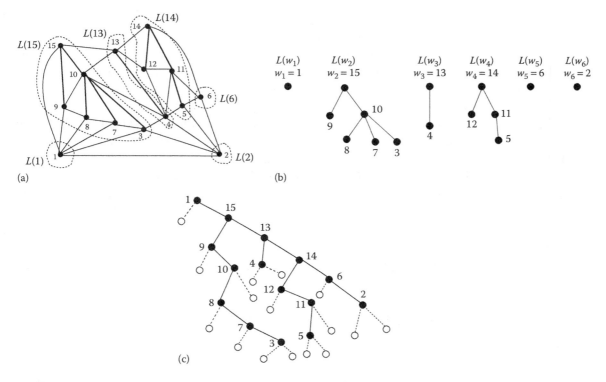

Figure 23.15 (a) Graph G_{15} for G in Figure 23.10, (b) forest F, and (c) binary tree T. (Figure taken from Nishizeki, T. and Rahman, M. S., *Planar Graph Drawing*, World Scientific, Singapore, 2004.)

So, in the end we have a straight line embedding of $G = G_n$ such that $P(v_1) = (0,0)$ and $P(v_2) = (2n-4, 0)$. By (e3), $P(v_n) = (n-2, n-2)$. Therefore, the whole graph G is drawn in a $(2n-4) \times (n-2)$ grid, as illustrated in Figure 23.16.

It is easy to implement the drawing algorithm described above in time $O(n^2)$. In the remainder of this section we describe a linear-time implementation of the straight line drawing algorithm in Section 23.2.2 [83].

We assume that G is already triangulated and embedded in the plane, and that a canonical ordering $\pi = (v_1, v_2, \ldots, v_n)$ of G is given. We view the family of sets $L(w_1), L(w_2), \ldots, L(w_t)$

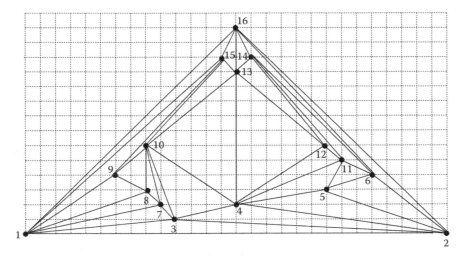

Figure 23.16 Grid drawing of the plane graph in Figure 23.10. (Figure taken from Nishizeki, T. and Rahman, M. S., *Planar Graph Drawing*, World Scientific, Singapore, 2004.)

for the outer vertices $w_1, w_2, \ldots, w_t$ of graph G_k as a forest F in G_k consisting of trees $L(w_1), L(w_2), \ldots, L(w_t)$ rooted at the vertices $w_1, w_2, \ldots, w_t$. For G_{15} in Figure 23.15a the forest F is drawn by thick solid lines in Figure 23.15a and is depicted also in Figure 23.15b. The children of root w_i of a tree $L(w_i)$ are the vertices that w_i covers, that is, its neighbors that leave the outer cycle when w_i is installed. The forest F is represented by a binary tree T as illustrated in Figure 23.15c. The root of T is $w_1(=v_1)$. The w_1's right child is w_2, he w_2's right child is w_3, and so on. The set $L(w_i)$ consists of w_i and all nodes in the w_i's left subtree in T. Thus, the subtree of T rooted at w_i consists of the vertices in $\cup_{j \geq i} L(w_j)$. In the left subtree of T rooted at w_i, the left child of w_i is the w_i's leftmost child in tree $L(w_i)$ (if any), the left child's right child in T is its next sibling to the right in tree $L(w_i)$ (if any), the left child's right child's right child in T is its next next sibling to the right in tree $L(w_i)$ (if any), and so on. For G_3 in Figure 23.13a, T is illustrated in Figure 23.13b.

Since v_k is embedded at a point $\mu(w_p, w_q)$, by Equation 23.1 we have

$$x(v_k) = \frac{1}{2}\{x(w_q) + x(w_p) + y(w_q) - y(w_p)\}, \tag{23.2}$$

$$y(v_k) = \frac{1}{2}\{[x(w_q) - x(w_p)] + y(w_q) + y(w_p)\} \tag{23.3}$$

and hence

$$x(v_k) - x(w_p) = \frac{1}{2}\{[x(w_q) - x(w_p)] + y(w_q) - y(w_p)\}. \tag{23.4}$$

The crucial observation is that, when we embed v_k, it is not necessary to know the exact position of w_p and w_q. If we know only their y-coordinates and their relative x-coordinates, that is, $x(w_q) - x(w_p)$, then by Equation 23.3 we can compute $y(v_k)$ and by Equation 23.4 we can compute the x-coordinate of v_k relative to w_p, that is, $x(v_k) - x(w_p)$.

For each vertex $v \neq v_1$, the *x-offset of v* is defined as $\Delta x(v) = x(v) - x(w)$, where w is the parent of v in T. More generally, if w is an ancestor of v, then the *x-offset between w and v* is $\Delta x(w, v) = x(v) - x(w)$.

With each vertex v we store the following information:

- $left(v) = $ the left child of v in T;
- $right(v) = $ the right child of v in T;
- $\Delta x(v) = $ the x-offset of v from its parent in T; and
- $y(v) = $ the y-coordinate of v.

The algorithm consists of two phases. In the first phase, we add new vertices one by one, and each time we add a vertex we compute its x-offset and y-coordinate, and update the x-offsets of one or two other vertices. In the second phase, we traverse the tree T and compute the final x-coordinates by accumulating offsets.

The first phase is implemented as follows. First we initialize the values stored at v_1, v_2, and v_3 as follows (see Figure 23.13):

- $\Delta x(v_1) = 0$; $y(v_1) = 0$; $right(v_1) = v_3$; $left(v_1) = nil$;
- $\Delta x(v_3) = 1$; $y(v_3) = 1$; $right(v_3) = v_2$; $left(v_3) = nil$; and
- $\Delta x(v_2) = 1$; $y(v_2) = 0$; $right(v_2) = nil$; $left(v_2) = nil$.

We then embed the other vertices, one by one, as follows.

```
1    for k = 4 to n do
         begin
2            Let w_1, w_2, ..., w_t be the outer cycle C_o(G_{k-1}) of G_{k-1};
             {See Figure 23.17a.}
3            Let w_p, w_{p+1}, ..., w_q be the neighbors of v_k on C_o(G_{k-1});
4            Increase offset of w_{p+1} and w_q by one; {cf. Steps 1' and 2'.}
5            Calculate Δx(w_p, w_q) as
                 Δx(w_p, w_q) = Δx(w_{p+1}) + Δx(w_{p+2}) + ··· + Δx(w_q);
6            Calculate Δx(v_k) as
                 Δx(v_k) = ½{Δx(w_p, w_q) + y(w_q) − y(w_p)};
             {cf. Equation 23.4.}
7            Calculate y(v_k) as y(v_k) = ½{Δx(w_p, w_q) + y(w_q) + y(w_p)};
             {cf. Equation 23.3.}
8            Calculate Δx(w_q) as Δx(w_q) = Δx(w_p, w_q) − Δx(v_k);
9            if p + 1 ≠ q then
10               Calculate Δx(w_{p+1}) as Δx(w_{p+1}) = Δx(w_{p+1}) − Δx(v_k);
11           Set right(w_p) = v_k and right(v_k) = w_q;
12           if p + 1 ≠ q then
13               Set left(v_k) = w_{p+1} and right(w_{q-1}) = nil
14           else
15               Set left(v_k) = nil;
16       end
```

Figure 23.17 illustrates the construction of T for G_k from T for G_{k-1} by the algorithm above.

In the second phase, we compute the x-coordinate $x(v_i)$ for each vertex v_i in G. Let Q be the path from the root v_1 to v_i in tree T. Then $x(v_i) = \sum\{\Delta(x)|$ vertex x is on $Q\}$. One can compute $x(v_i)$ for all vertices v_i by invoking *Accumulate-Offset*(v_1,0); procedure *Accumulate-Offset* is as follows.

```
procedure Accumulate-Offset(v:vertex; δ:integer);
    begin
        if v ≠ nil then begin
            Set Δx(v) = Δx(v) + δ;
            Accumulate-Offset(left(v); Δx(v));
            Accumulate-Offset(right(v); Δx(v));
        end
    end
```

Clearly $x(v_i) = \Delta x(v_i)$ for each vertex v_i in G.

The first phase takes linear time, since adding a vertex v_k takes at most time $O(d(v_k))$. The second phase, that is, *Accumulate-Offset*, takes time proportional to the number nodes in T. Thus, the algorithm takes linear time in total.

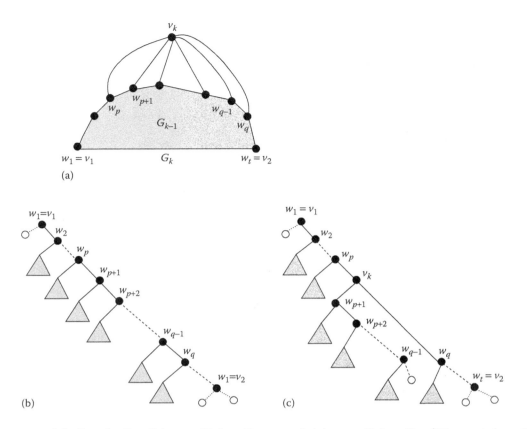

Figure 23.17 (a) Graph G_k, (b) tree T for G_{k-1}, and (c) tree T for G_k. (Figure taken from Nishizeki, T. and Rahman, M. S., *Planar Graph Drawing*, World Scientific, Singapore, 2004.)

23.3 CONVEX DRAWING

Some planar graphs can be drawn in such a way that each edge is drawn as a straight line segment and each face is drawn as a convex polygon, as illustrated in Figures 23.4b and 23.18b. Such a drawing is called a *convex drawing*. The drawings in Figures 23.18d and f are not convex drawings. Although not every planar graph has a convex drawing, Tutte showed that every 3-connected planar graph has a convex drawing, and obtained a necessary and sufficient condition for a plane graph to have a convex drawing [84]. Furthermore, he gave a *barycentric mapping* method for finding a convex drawing of a plane graph, which requires solving a system of $O(n)$ linear equations [85]. The system of equations can be solved either in $O(n^3)$ time and $O(n^2)$ space using the ordinary Gaussian elimination method, or in $O(n^{1.5})$ time and $O(n \log n)$ space using the sparse Gaussian elimination method [86]. Thus the barycentric mapping method leads to an $O(n^{1.5})$ time convex drawing algorithm for plane graphs.

Chiba et al. gave two linear algorithms for the convex drawing problem of planar graphs: drawing and testing algorithms [44]. One of them finds a convex drawing of a given *plane* graph G if there is; it extends a given convex polygonal drawing of the outer cycle of G into a convex drawing of G. The other algorithm tests the possibility for a given *planar* graph. That is, it examines whether a given planar graph has a *plane embedding* which has a convex drawing.

A convex drawing is called a *convex grid drawing* if it is a grid drawing. Every 3-connected plane graph has a convex grid drawing on an $(n-2) \times (n-2)$ grid, and there is an algorithm to find such a grid drawing in linear time [43].

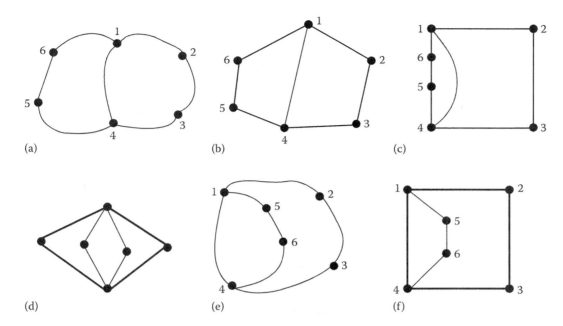

Figure 23.18 (a–f) Plane graphs and drawings.

One may expect that the size of an integer grid required by a convex grid drawing will be smaller than $(n-2) \times (n-2)$ for 4-connected plane graphs. Miura et al. presented an algorithm which finds in linear time a convex grid drawing of any given 4-connected plane graph G on an integer grid such that $W + H \leq n - 1$ if G has four or more outer vertices [87]. Since $W + H \leq n - 1$, the area of the grid satisfies $W \cdot H \leq n^2/4$.

In this section, we describe a linear algorithm for finding a convex grid drawing of a 3-connected plane graph on an $(n-2) \times (n-2)$ grid [43]. The algorithm is based on a *canonical decomposition* of a 3-connected plane graph, which is a generalization of a canonical ordering described in Section 23.2.1. In Section 23.3.1 we present a canonical decomposition, and in Section 23.3.2 we present the algorithm.

23.3.1 Canonical Decomposition

We say that a plane graph G is *internally 3-connected* if G is 2-connected and, for any separation pair $\{u, v\}$ of G, u and v are outer vertices and each connected component of $G - \{u, v\}$ contains an outer vertex. In other words, G is internally 3-connected if and only if it can be extended to a 3-connected graph by adding a vertex in an outer face and connecting it to all outer vertices. If a 2-connected plane graph G is not internally 3-connected, then G has a separation pair $\{u, v\}$ of one of the three types illustrated in Figure 23.19, where the

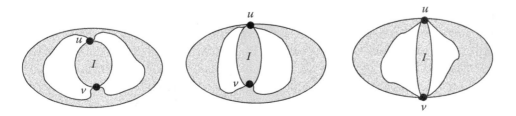

Figure 23.19 Biconnected plane graphs which are not internally 3-connected.

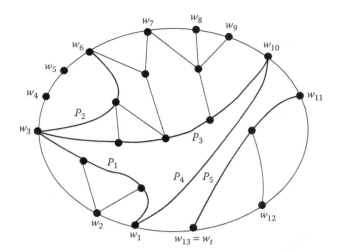

Figure 23.20 Plane graph with chord-paths $P_1, P_2, \ldots, P_5$.

split component I contains a vertex other than u and v. If an internally 3-connected plane graph G is not 3-connected, then G has a separation pair of outer vertices and hence G has a "chord-path" when G is not a single cycle.

We now define a "chord-path". Let G be a 2-connected plane graph, and let $w_1, w_2, \ldots, w_t$ be the vertices appearing clockwise on the outer cycle $C_o(G)$ in this order, as illustrated in Figure 23.20. We call a path P in G a *chord-path* of the cycle $C_o(G)$ if P satisfies the following (i)–(iv):

i. P connects two outer vertices w_p and w_q, $p < q$;

ii. $\{w_p, w_q\}$ is a separation pair of G;

iii. P lies on an inner face; and

iv. P does not pass through any outer edge and any outer vertex other than the ends w_p and w_q.

The plane graph G in Figure 23.20 has six chord-paths $P_1, P_2, \ldots, P_5$ drawn by thick lines. A chord-path P is *minimal* if none of $w_{p+1}, w_{p+2}, \ldots, w_{q-1}$ is an end of a chord-path. Thus, the definition of a minimal chord-path depends on which vertex is considered as the starting vertex w_1 of $C_o(G)$. P_1, P_2, and P_5 in Figure 23.20 are minimal, while P_3 and P_4 are not minimal.

Let $\{v_1, v_2, \ldots, v_p\}$, $p \geq 3$, be a set of three or more outer vertices consecutive on $C_o(G)$ such that $d(v_1) \geq 3$, $d(v_2) = d(v_3) = \cdots = d(v_{p-1}) = 2$, and $d(v_p) \geq 3$. Then we call the set $\{v_2, v_3, \ldots, v_{p-1}\}$ an *outer chain* of G. The graph in Figure 23.19 has two outer chains $\{w_4, w_5\}$ and $\{w_8\}$.

We are now ready to define a canonical decomposition. Let $G = (V, E)$ be a 3-connected plane graph of $n \geq 4$ vertices like one in Figure 23.22. For an ordered partition $\Pi = (U_1, U_2, \ldots, U_l)$ of set V, we denote by G_k, $1 \leq k \leq l$, the subgraph of G induced by $U_1 \cup U_2 \cup \cdots \cup U_k$, while we denote by $\overline{G_k}$, $0 \leq k \leq l-1$, the subgraph of G induced by $U_{k+1} \cup U_{k+2} \cup \cdots \cup U_l$. Clearly $G_k = G - U_{k+1} \cup U_{k+2} \cdots \cup U_l$, and $G = G_l = \overline{G_0}$. Let (v_1, v_2) be an outer edge of G. We then say that Π is a *canonical decomposition* of G (for an outer edge (v_1, v_2)) if Π satisfies the following conditions (cd1)–(cd3).

(cd1) U_1 is the set of all vertices on the inner face containing edge (v_1, v_2), and U_l is a singleton set containing an outer vertex $v_n \notin \{v_1, v_2\}$.

(cd2) For each index k, $1 \le k \le l$, G_k is internally 3-connected.

(cd3) For each index k, $2 \le k \le l$, all vertices in U_k are outer vertices of G_k and the following conditions hold:

   a. if $|U_k| = 1$, then the vertex in U_k has two or more neighbors in G_{k-1} and has at least one neighbor in $\overline{G_k}$ when $k < l$, as illustrated in Figure 23.21a; and

   b. If $|U_k| \ge 2$, then U_k is an outer chain of G_k, and each vertex in U_k has at least one neighbor in $\overline{G_k}$, as illustrated in Figure 23.21b.

Figure 23.22 illustrates a canonical decomposition $\Pi = (U_1, U_2, \ldots, U_8)$ of a 3-connected plane graph of $n = 15$ vertices. We now have the following lemma on a canonical decomposition [29].

Lemma 23.4 *Every 3-connected plane graph G of $n \ge 4$ vertices has a canonical decomposition Π, and Π can be found in linear time.* ∎

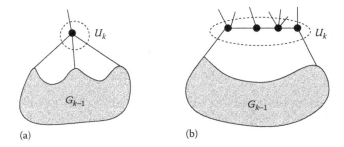

Figure 23.21 (a,b) G_k with some edges joining U_k and $\overline{G_k}$.

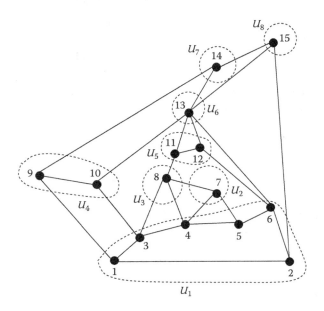

Figure 23.22 A canonical decomposition of a 3-connected plane graph. (Figure taken from Nishizeki, T. and Rahman, M. S., *Planar Graph Drawing*, World Scientific, Singapore, 2004.)

23.3.2 Algorithm for Convex Grid Drawing

In this section, we describe a linear algorithm for finding a convex grid drawing of a 3-connected plane graph [43].

Let G be a 3-connected plane graph, and let $\Pi = (U_1, U_2, \ldots, U_l)$ be a canonical decomposition of G. The algorithm will add to a drawing the vertices in set U_k, one by one, in the order $U_1, U_2, \ldots, U_l$, adjusting the drawing at every step. Before giving the detail of the algorithm we need some preparation.

We say that a vertex $v \in U_k$, $1 \le k \le l$, has *rank* k. Let $2 \le k \le l$, and let $C_o(G_{k-1}) = w_1, w_2, \ldots, w_t$, where $w_1 = v_1$ and $w_t = v_2$. The definition of a canonical decomposition implies that there is a pair of indices a and b, $1 \le a < b \le t$, such that each of w_a and w_b has a neighbor in $\overline{G_{k-1}}$ but any vertex w_i, $a < i < b$, has no neighbor in $\overline{G_{k-1}}$ and is an inner vertex of G, as illustrated in Figure 23.23. (see also Figure 23.21). Then the path $w_a, w_{a+1}, \ldots, w_b$ is a part of an inner facial cycle F of G; F also contains two edges connecting w_a and w_b with $\overline{G_{k-1}}$, plus possibly some edges in $\overline{G_{k-1}}$. Let c, $a \le c < b$, be an index such that w_c has the smallest rank among the vertices $w_a, w_{a+1}, \ldots, w_{b-1}$. If there are two or more vertices with the smallest rank, then let w_c be the leftmost one, that is, let c be the smallest index of these vertices. Intuitively, the algorithm will work in such a way that, for any such pair of indices a and b, either the vertex w_c or w_{c+1} will have the smallest y-coordinate among the vertices on the face F (see Figure 23.23). We denote the index c for a and b by $\mu_k^+(a)$ and $\mu_k^-(b)$. Thus $c = \mu_k^+(a) = \mu_k^-(b)$. The superscript $+$ indicates $a \le c$, while the superscript $-$ indicates $c < b$. We often omit the subscript k for simplicity. Note that if $b = a + 1$ then $a = \mu_k^+(a) = \mu_k^-(b)$.

We denote the current position of a vertex v by $P(v)$; $P(v)$ is expressed by its x- and y-coordinates as $(x(v), y(v))$. With each vertex v, a set of vertices need to be moved whenever the position of v is adjusted. We denote by $L(v)$ the set of such vertices.

We are now ready to describe the drawing algorithm.

First we draw $C_o(G_1) = w_1, w_2, \ldots, w_t$ as follows. Set $P(w_1) = (0,0)$, $P(w_t) = (t-1, 0)$ and $P(w_i) = (i-1, 0)$ for all indices $i = 2, 3, \ldots, t-1$, as illustrated in Figure 23.24a for the graph in Figure 23.22. Also set $L(w_i) = \{w_i\}$ for each index $i = 1, 2, \ldots, t$.

Then, for each index $k = 2, 3, \ldots, l$, we do the following. Let $C_o(G_{k-1}) = w_1, w_2, \ldots, w_t$ be the outer cycle of G_{k-1} where $w_1 = v_1$ and $w_t = v_2$. Let $U_k = \{u_1, u_2, \ldots, u_r\}$. U_k is either a singleton set or an outer chain of G_k, but in the algorithm we will treat both cases uniformly.

Let w_p and w_q be the leftmost and rightmost neighbors of U_k in G_{k-1} as illustrated in Figures 23.25 and 23.26. Let $\alpha = \mu^+(p)$, and let $\beta = \mu^-(q)$. If U_k is an outer chain, then

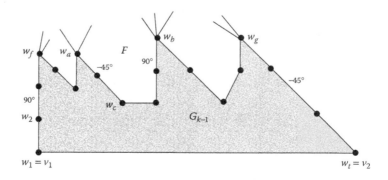

Figure 23.23 G_{k-1} with some edges connecting G_{k-1} and $\overline{G_{k-1}}$. (Figure taken from Nishizeki, T. and Rahman, M. S., *Planar Graph Drawing*, World Scientific, Singapore, 2004.)

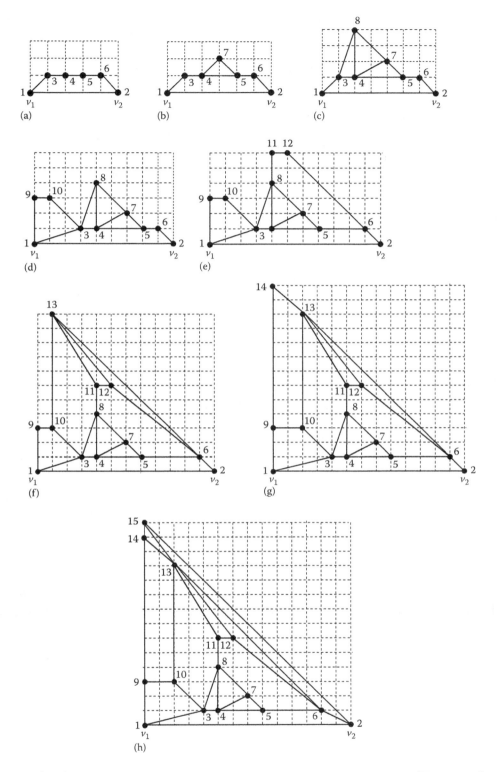

Figure 23.24 (a–h) Drawing process of the plane graph in Figure 23.22. (Figure taken from Nishizeki, T. and Rahman, M. S., *Planar Graph Drawing*, World Scientific, Singapore, 2004.)

all vertices $w_{p+1}, w_{p+2}, \ldots, w_{q-1}$ belong to the same inner face of G_k and none of them has a neighbor in $\overline{G_k}$ and hence $\alpha = \beta$, as illustrated in Figure 23.26a. If U_k is a singleton set of a vertex u_1 having three or more neighbors in G_{k-1}, then at least one of the vertices $w_{p+1}, w_{p+2}, \ldots, w_{q-1}$ has a neighbor in $\overline{G_k}$ and hence $\alpha < \beta$; in fact, w_α and w_β will

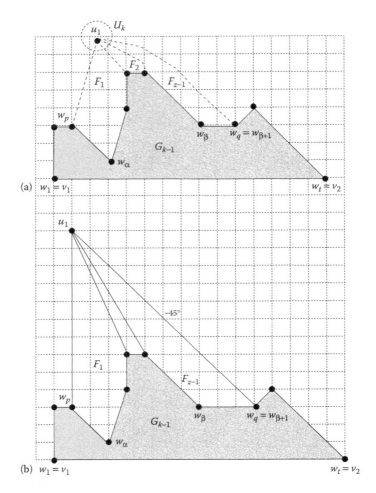

Figure 23.25 (a) G_k before shift and (b) G_k after shift and install with $\epsilon = 0$. (Figure taken from Nishizeki, T. and Rahman, M. S., *Planar Graph Drawing*, World Scientific, Singapore, 2004.)

belong to two different inner faces of G_k, to the first and last faces among those that are created when adding u_1 to G_{k-1}, as illustrated in Figure 23.25. We thus execute the following steps.

Update
Set $L(w_p) = \bigcup_{i=p}^{\alpha} L(w_i)$;
Set $L(w_q) = \bigcup_{i=\beta+1}^{q} L(w_i)$;
Set $L(u_1) = \{u_1\} \cup (\bigcup_{i=\alpha+1}^{\beta} L(w_i))$;
Set $L(u_i) = \{u_i\}$ for each index i, $2 \leq i \leq r$;

Shift: For each vertex $v \in \bigcup_{i=q}^{t} L(w_i)$, set $x(v) = x(v) + r$;

Install U_k: Let ϵ be 0 if w_p has no neighbor in $\overline{G_k}$ and 1 otherwise. For each $i = 1, 2, \ldots, r$, we set $x(u_i) = x(w_p) + i - 1 + \epsilon$, and set $y(u_i) = y(w_q) + x(w_q) - x(w_p) - r + 1 - \epsilon$. In other words, we draw U_k horizontally in such a way that the slope of the segment $u_r w_q$ is $-45°$.

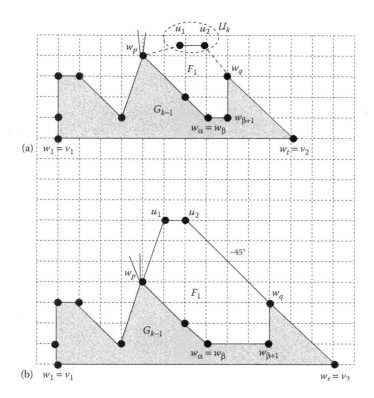

Figure 23.26 (a) G_k before shift, and (b) G_k after shift and install with $\epsilon = 1$. (Figure taken from Nishizeki, T. and Rahman, M. S., *Planar Graph Drawing*, World Scientific, Singapore, 2004.)

Vertex u_1 is placed above w_p if w_p has no neighbor in $\overline{G_k}$, and at the next x-coordinate otherwise. Note that in the last equation we use the new updated value of $x(w_q)$.

We call the algorithm above Algorithm *Convex-Grid-Drawing*. Figure 23.24 illustrates the execution of Algorithm *Convex-Grid-Drawing* for the plane graph in Figure 23.22. The linear-time implementation of *Convex-Grid-Drawing* can be achieved by using a data structure presented in Section 23.2.2.

We now verify the correctness of Algorithm *Convex-Grid-Drawing*. Let $2 \leq k \leq l$, and let $C_o(G_{k-1}) = w_1, w_2, \ldots, w_t$, where $w_1 = v_1$ and $w_t = v_2$. Then by induction on k it can be proved that in the drawing of G_{k-1}, $P(v_1) = (0,0)$, $P(v_2) = (|\cup_{i=1}^{k-1} U_i| - 1, 0)$, and any line segment $w_i w_{i+1}$, $1 \leq i \leq t-1$, has slope in $\{-45°, 0°\} \cup [45°, 90°]$ as illustrated in Figure 23.23, where $[45°, 90°]$ denotes the set of all angles θ, $45° \leq \theta \leq 90°$. More specifically, the following properties (a)–(c) hold on the slopes of line segments on $C_o(G_{k-1})$ in the drawing of G_{k-1} [43].

a. Let w_f, $1 \leq f \leq t$, be the first vertex on $C_o(G_{k-1})$ which has a neighbor in $\overline{G_{k-1}}$, then the slope of each line segment on the path $w_1, w_2, \ldots, w_f$ is $90°$.

b. Let w_g, $1 \leq g \leq t$, be the last vertex on $C_o(G_{k-1})$ which has a neighbor in $\overline{G_{k-1}}$, then the slope of each line segment on the path $w_g, w_{g+1}, \ldots, w_t$ is $-45°$.

c. For any triple of indices a, b and c as defined earlier and illustrated in Figure 23.23, each of the first $c - a$ line segments on the path $w_a, w_{a+1}, \ldots, w_b$ has slope $-45°$, while each of the last $b - c - 1$ segments has slope $90°$. The remaining line segment $w_c w_{c+1}$ has slope in $\{-45°, 0°\} \cup [45°, 90°]$, and the slope is not $90°$ if $c = a$.

One can observe that, after the shift operation while adding U_k, all neighbors of U_k on $C_o(G_{k-1})$ are visible from the vertices in U_k. Hence the edges joining vertices in U_k and vertices in $C_o(G_{k-1})$ do not intersect themselves or edges on $C_o(G_{k-1})$. Thus, adding U_k does not distroy the planarity. One can observe also that the newly created inner faces are convex polygons.

What remains to show is that we do not destroy the planarity and convexity when we apply the shift operation. This is shown in the following lemma, which is similar to Lemma 23.3 for straight line drawings [43]. We call a drawing of a plane graph *internally convex* if all inner faces are drawn as convex polygons.

Lemma 23.5 *(a) Each graph G_k, $1 \leq k \leq l$, is straight-line embedded and internally convex. (b) Suppose that $C_o(G_k) = w'_1, w'_2, \ldots, w'_{t'}$, $w'_1 = v_1$ and $w'_{t'} = v_2$, and that s is any index, $1 \leq s \leq t'$, and δ is any nonnegative integer. If we shift all vertices in $\cup_{i=s}^{t'} L(w'_i)$ by δ to the right, then G_k remains straight-line embedded and internally convex.* ∎

The conditions (a) and (b) on the slopes of line segments on $C_o(G_k)$ imply that the outer cycle $C_o(G)$ is drawn as an isosceles right triangle in the final drawing of G, as illustrated in Figure 23.24h. Clearly $x(v_1) = 0, x(v_2) = n-1$, and $y(v_n) = n-1$, and hence Algorithm Convex-Grid-Drawing produces a convex drawing of a 3-connected plane graph G on an $(n-1) \times (n-1)$ grid.

Using a canonical decomposition such that v_n is adjacent to v_2 and fixing the position of v_n carefully, one can obtain a convex grid drawing on an $(n-2) \times (n-2)$ grid.

23.4 RECTANGULAR DRAWING

A *rectangular drawing* of a plane graph G is a drawing of G in which each vertex is drawn as a point, each edge is drawn as a horizontal or vertical line segment without edge-crossings, and each face is drawn as a rectangle. Thus, a rectangular drawing is a special case of a convex drawing. Figure 23.27b illustrates a rectangular drawing of the plane graph in Figure 23.27a. In Section 23.1.2 we have seen applications of a rectangular drawing to VLSI floorplanning and architectural floorplanning. In a rectangular drawing of G, the outer cycle $C_o(G)$ is drawn as a rectangle and hence has four convex corners such as a, b, c and d drawn by white circles in Figure 23.27. Such a convex corner is an outer vertex of degree two and is called a *corner of the rectangular drawing*. Not every plane graph G has a rectangular drawing. Of course, G must be 2-connected and the maximum degree Δ of G is at most four if G has

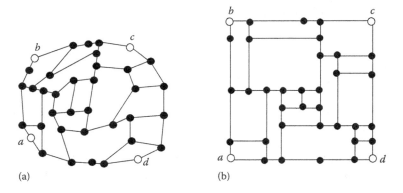

Figure 23.27 (a) Plane graph and (b) its rectangular drawing for the designated corners a, b, c, and d.

a rectangular drawing. Miura et al. recently showed that a plane graph G with $\Delta \leq 4$ has rectangular drawing D if and only if a new bipartite graph constructed from G has a perfect matching, and D can be found in time $O(n^{1.5}/\log n)$ whenever G has D [88].

Since a planar graph with $\Delta \leq 3$ often appears in many practical applications, much works have been devoted on rectangular drawings of planar graphs with $\Delta \leq 3$. Thomassen gave a necessary and sufficient condition for a plane graph G with $\Delta \leq 3$ to have a rectangular drawing when four outer vertices of degree two are designated as the corners [64]. Based on this characterization Rahman et al. gave a linear-time algorithm to obtain a rectangular drawing with the designated corners [66]. The problem of examining whether a plane graph has a rectangular drawing becomes difficult when four outer vertices are not designated as the corners. Rahman et al. gave a necessary and sufficient condition for a plane graph with $\Delta \leq 3$ to have a rectangular drawing for some quadruplet of outer vertices appropriately chosen as the corners, and present a linear time algorithm to find such a quadruplet [67].

Kozminski and Kinnen [89] established a necessary and sufficient condition for the existence of a *rectangular dual* of an inner triangulated plane graph, that is, a rectangular drawing of the dual graph of an inner triangulated plane graph, and gave an $O(n^2)$ algorithm to obtain it. Based on the characterization of [89], Bhasker and Sahni [65], and Xin He [90] developed linear-time algorithms to find a rectangular dual. Kant and Xin He [91] presented two more linear-time algorithms. Xin He [92] presented a parallel algorithm for finding a rectangular dual. Lai and Leinwand [93] reduced the problem of finding a rectangular dual of an inner triangulated plane graph G to a problem of finding a perfect matching of a new bipartite graph constructed from G.

The outer face boundary must be rectangular in a rectangular drawing, as illustrated in Figure 23.27b. However, the outer boundary of a VLSI chip or an architectural floor plan is not always rectangular, but is often a rectilinear polygon of L-shape, T-shape, Z-shape, and so forth as illustrated in Figures 23.28a–c. We call such a drawing of a plane graph G an *inner rectangular drawing* if every inner face of G is a rectangle although the outer face boundary is not always a rectangle. Miura et al. [88] reduced the problem of finding an inner rectangular drawing of a plane graph G with $\Delta \leq 4$ to a problem of finding a perfect matching of a new bipartite graph constructed from G.

A planar graph may have many embeddings. We say that *a planar graph G has a rectangular drawing* if at least one of the plane embeddings of G has a rectangular drawing. Since a planar graph may have an exponential number of embeddings, it is not a trivial problem to examine whether a planar graph has a rectangular drawing or not. Rahman et al. gave a linear-time algorithm to examine whether a planar graph G with $\Delta \leq 3$ has a rectangular drawing or not, and find a rectangular drawing of G if it exists [94].

In Section 23.4.1 we present the result of Miura et al. on rectangular drawings of plane graphs with $\Delta \leq 4$ [29,88]. In Section 23.4.2 we present the characterization of Thomassen [64] an outline of the linear-time algorithm of Rahman et al. [66].

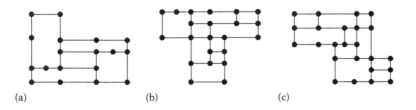

Figure 23.28 Inner rectangular drawings of (a) L-shape, (b) T-shape, and (c) Z-shape.

23.4.1 Rectangular Drawing and Matching

In this section, we consider rectangular drawings of plane graphs with $\Delta \leq 4$. We show that a plane graph G with $\Delta \leq 4$ has rectangular drawing D if and only if a new bipartite graph G_d constructed from G has a perfect matching, and D can be found in time $O(n^{1.5}/\log n)$ if D exists [29,88]. G_d is called a decision graph.

We may assume without loss of generality that G is 2-connected and $\Delta \leq 4$, and hence every vertex of G has degree two, three or four.

An angle formed by two edges e and e' incident to a vertex v in G is called an *angle of* v if e and e' appear consecutively around v. An angle of a vertex in G is called an *angle of* G. An angle formed by two consecutive edges on a boundary of a face F in G is called an *angle of* F. An angle of the outer face is called an *outer angle* of G, while an angle of an inner face is called an *inner angle*.

In any rectangular drawing, every inner angle is 90° or 180°, and every outer angle is 180° or 270°. Consider a labeling Θ which assigns a label 1, 2, or 3 to every angle of G, as illustrated in Figure 23.29b. Labels 1, 2, and 3 correspond to angles 90°, 180°, and 270°, respectively. Therefore each inner angle has label either 1 or 2, exactly four outer angles have label 3, and all other outer angles have label 2.

We call Θ a *regular labeling* of G if Θ satisfies the following three conditions (a)–(c):

a. For each vertex v of G, the sum of the labels of all the angles of v is equal to 4;

b. The label of any inner angle is 1 or 2, and every inner face has exactly four angles of label 1; and

c. The label of any outer angle is 2 or 3, and the outer face has exactly four angles of label 3.

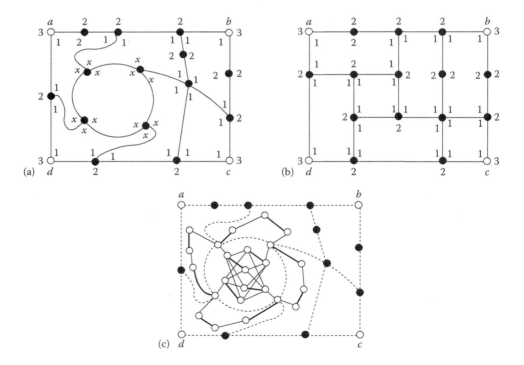

Figure 23.29 (a) Plane graph G, (b) rectangular drawing D and regular labeling Θ of G, and (c) decision graph G_d. (Figure taken from Nishizeki, T. and Rahman, M. S., *Planar Graph Drawing*, World Scientific, Singapore, 2004.)

A regular labeling Θ of the plane graph in Figure 23.29a and a rectangular drawing D corresponding to Θ are depicted in Figure 23.29b. A regular labeling is a special case of an orthogonal representation of an orthogonal drawing presented in [54].

Conditions (a) and (b) implies the following (i)–(iii):

i. If a noncorner vertex v has degree two, that is, $d(v) = 2$, then the two labels of v are 2 and 2.

ii. If $d(v) = 3$, then exactly one of the three angles of v has label 2 and the other two have label 1.

iii. If $d(v) = 4$, then all the four angles of v have label 1.

If G has a rectangular drawing, then clearly G has a regular labeling. Conversely, if G has a regular labeling, then G has a rectangular drawing, as can be proved by means of elementary geometric considerations. We thus have the following fact.

Fact 23.1 *A plane graph G has a rectangular drawing if and only if G has a regular labeling.*

We now assume that four outer vertices $a, b, c,$ and d of degree two are designated as corners. Then the outer angles of $a, b, c,$ and d must be labeled with 3, and all the other outer angles of G must be labeled with 2, as illustrated in Figure 23.29a. Some of the inner angles of G can be immediately determined, as illustrated in Figure 23.29a. If v is a noncorner outer vertex of degree two, then the inner angle of v must be labeled with 2. The two angles of any inner vertex of degree two must be labeled with 2. If v is an outer vertex of degree three, then the outer angle of v must be labeled with 2 and both of the inner angles of v must be labeled with 1. We label all the three angles of an inner vertex of degree three with x, because one cannot determine their labels although exactly one of them must be labeled with 2 and the others with 1. We label all the four angles of each vertex of degree four with 1. Label x means that x is either 1 or 2, and exactly one of the three labels x's attached to the same vertex must be 2 and the other two must be 1 (see Figures 23.29a and b).

We now present how to construct a decision graph G_d of G. Let all vertices of G attached label x be vertices of G_d. Thus, all the inner vertices of degree three are vertices of G_d, and none of the other vertices of G is a vertex of G_d. We then add to G_d a complete bipartite graph inside each inner face F of G, as illustrated in Figure 23.30 where G_d is drawn by

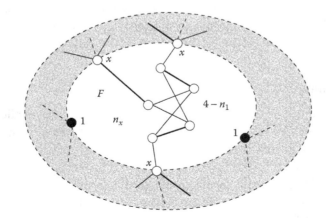

Figure 23.30 Construction of G_d for an inner face F of G. (Figure taken from Nishizeki, T. and Rahman, M. S., *Planar Graph Drawing*, World Scientific, Singapore, 2004.)

solid lines and G by dotted lines. Let n_x be the number of angles of F labeled with x. For example, $n_x = 3$ in Figure 23.30. Let n_1 be the number of angles of F which have been labeled with 1. Then n_1 is the number of vertices v on F such that either v is a corner vertex or $d(v) = 4$. Thus, $n_1 = 2$ for the example in Figure 23.30. One may assume as a trivial necessary condition that $n_1 \leq 4$; otherwise, G has no rectangular drawing. Exactly $4 - n_1$ of the n_x angles of F labeled with x must be labeled with 1 by a regular labeling. We add a complete bipartite graph $K_{(4-n_1), n_x}$ in F, and join each of the n_x vertices in the second partite set with one of the n_x vertices on F whose angles are labeled with x. Repeat the operation above for each inner face F of G. The resulting graph is a *decision graph* G_d of G. The decision graph G_d of the plane graph G in Figure 23.29a is drawn by solid lines in Figure 23.29c, where G is drawn by dotted lines. The idea of adding a complete bipartite graph originates from Tutte's transformation for finding an f-factor of a graph [95].

A *matching* of G_d is a set of pairwise nonadjacent edges in G_d. A *maximum matching* of G_d is a matching of the maximum cardinality. A matching M of G_d is called a *perfect matching* if an edge in M is incident to each vertex of G_d. A perfect matching is drawn by thick solid lines in Figures 23.29c and 23.30.

Each edge e of G_d incident to a vertex v attached a label x corresponds to an angle α of v labeled with x. A fact that e is contained in a perfect matching M of G_d means that the label x of α is 2. Conversely, a fact that e is not contained in M means that the label x of α is 1.

We now have the following theorem.

Theorem 23.1 *Let G be a plane graph with $\Delta \leq 4$ and four outer vertices a, b, c and d be designated as corners. Then G has a rectangular drawing D with the designated corners if and only if the decision graph G_d of G has a perfect matching. D can be found in time $O(n^{1.5}/\log n)$ whenever G has D.*

Proof. We only show a proof for the time complexity; the proof for necessity and sufficiency can be found in [29].

Clearly, G_d is a bipartite graph, and $4 - n_1 \leq 4$. Obviously, n_x is no more than the number of edges on face F. Let m be the number of edges in G, then we have $2m \leq 4n$ since $\Delta \leq 4$. Therefore the sum $2m$ of the numbers of edges on all faces is at most $4n$. One can thus know that both the number n_d of vertices in G_d and the number m_d of edges in G_d are $O(n)$. Since G_d is a bipartite graph, a maximum matching of G_d can be found either in time $O(\sqrt{n_d} m_d) = O(n^{1.5})$ by an ordinary bipartite matching algorithm [96–98] or in time $O(n^{1.5}/\log n)$ by a recent pseudoflow-based bipartite matching algorithm using boolean word operations on $\log n$-bit words [99,100]. One can find a regular labeling Θ of G from a perfect matching of G_d in linear time. It is easy to find a rectangular drawing of G from Θ in linear time. ∎

23.4.2 Linear Algorithm

In this section, we present Thomassen's theorem on a necessary and sufficient condition for a plane graph G with $\Delta \leq 3$ to have a rectangular drawing when four outer vertices of degree two are designated as the corners [64], and give a linear-time algorithm to find a rectangular drawing of G if it exists [66].

Before presenting Thomassen's theorem we recall some definitions. An edge of a plane graph G is called a *leg* of a cycle C if it is incident to exactly one vertex of C and located outside C. The vertex of C to which a leg is incident is called a *leg-vertex* of C. A cycle in G is called a *k-legged cycle* of G if C has exactly k legs in G and there is no edge which joins two vertices on C and is located outside C. Figure 23.31a illustrates 2-legged cycles C_1, C_2, C_3,

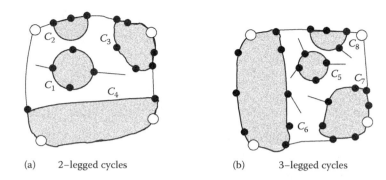

(a) 2-legged cycles (b) 3-legged cycles

Figure 23.31 Good cycles C_4, C_6, and C_7 and bad cycles C_1, C_2, C_3, C_5, and C_8. (Figure taken from Nishizeki, T. and Rahman, M. S., *Planar Graph Drawing*, World Scientific, Singapore, 2004.)

and C_4, while Figure 23.31b illustrates 3-legged cycles C_5, C_6, C_7, and C_8, where corners are drawn by white circles.

If a 2-legged cycle contains at most one corner like C_1, C_2, and C_3 in Figure 23.31a, then some inner face cannot be drawn as a rectangle and hence G has no rectangular drawing. Similarly, if a 3-legged cycle contains no corner like C_5 and C_8 in Figure 23.31b, then G has no rectangular drawing. One can thus observe the following fact.

Fact 23.2 *In any rectangular drawing D of G, every 2-legged cycle of G contains two or more corners, every 3-legged cycle of G contains one or more corner, and every cycle with four or more legs may contain no corner, as illustrated in Figure 23.32.*

The necessity of the following Thomassen's theorem [64] is immediate from Fact 23.2.

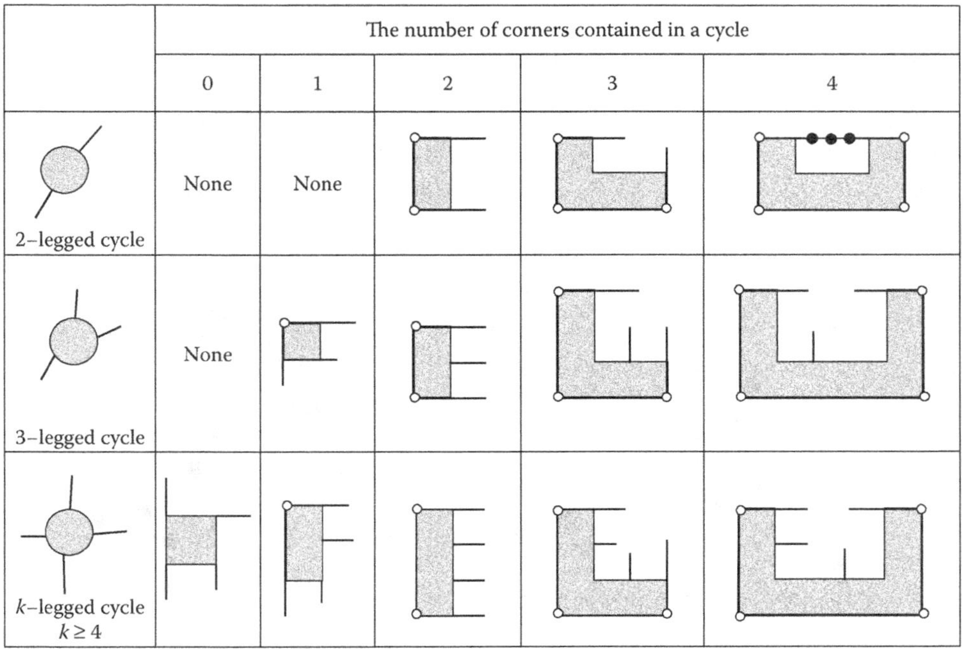

Figure 23.32 Numbers of corners in drawings of cycles. (Figure taken from Nishizeki, T. and Rahman, M. S., *Planar Graph Drawing*, World Scientific, Singapore, 2004.)

Theorem 23.2 *Assume that G is a 2-connected plane graph with $\Delta \leq 3$ and four outer vertices of degree two are designated as the corners a, b, c, and d. Then G has a rectangular drawing if and only if*

(r1) *Any 2-legged cycle contains two or more corners, and*

(r2) *Any 3-legged cycle contains one or more corners.*

A cycle of type (r1) or (r2) is called *good*. Cycles C_4, C_6, and C_7 in Figure 23.31 are good cycles; the 2-legged cycle C_4 contains two corners, and the 3-legged cycles C_6 and C_7 contain one or two corners. On the other hand, a 2-legged or 3-legged cycle is called *bad* if it is not good. Thus, 2-legged cycles C_1, C_2, and C_3 and 3-legged cycles C_5 and C_8 are bad cycles. Thus, Theorem 23.2 can be rephrased as follows: G has a rectangular drawing if and only if G has no bad cycle. In particular, a 2-legged bad cycle is called a *bad corner* if it contains exactly one corner like C_3.

In the rest of this section we present an outline of a constructive proof of the sufficiency of Theorem 23.2 [66]. The proof leads to a linear-time algorithm which we call Algorithm *Rectangular-Draw*.

The *union* $G = G' \cup G''$ of two graphs G' and G'' is a graph $G = (V(G') \cup V(G''), E(G') \cup E(G''))$.

In a given 2-connected plane graph G, four outer vertices of degree two are designated as the corners a, b, c, and d. These four corners divide the outer cycle $C_o(G)$ of G into four paths, the north path P_N, the east path P_E, the south path P_S, and the west path P_W, as illustrated in Figure 23.33a. We will draw the north and south paths on two horizontal straight line segments and the east and west paths on two vertical line segments. We thus fix the embedding of $C_o(G)$ as a rectangle. We call a rectangular embedding of $C_o(G)$ an *outer rectangle*.

A graph of a single edge, not in the outer cycle $C_o(G)$, joining two vertices in $C_o(G)$ is called a $C_o(G)$-*component* of G. A graph which consists of a connected component of $G - V(C_o(G))$ and all edges joining vertices in that component and vertices in $C_o(G)$ is also called a $C_o(G)$-*component*. The outer cycle $C_o(G)$ of the plane graph G in Figure 23.33a is drawn by thick lines, and the $C_o(G)$-components J_1, J_2, and J_3 of G are depicted in Figure 23.33b. Clearly the following lemma holds. ∎

Lemma 23.6 *Let $J_1, J_2, \ldots, J_p$ be the $C_o(G)$-components of a plane graph G, and let $G_i = C_o(G) \cup J_i$, $1 \leq i \leq p$, as illustrated in Figure 23.34. Then G has a rectangular drawing with corners a, b, c, and d if and only if, for each index i, $1 \leq i \leq p$, G_i has a rectangular drawing with corners a, b, c, and d.* ∎

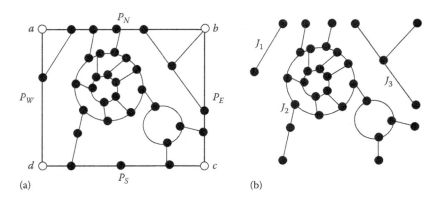

Figure 23.33 (a) Plane graph G and (b) $C_o(G)$-components. (Figure taken from Nishizeki, T. and Rahman, M. S., *Planar Graph Drawing*, World Scientific, Singapore, 2004.)

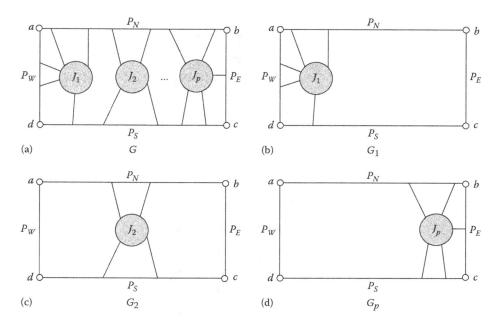

Figure 23.34 (a) G, (b) G_1, (c) G_2, and (d) G_p. (Figure taken from Nishizeki, T. and Rahman, M. S., *Planar Graph Drawing*, World Scientific, Singapore, 2004.)

In the remainder of this section, because of Lemma 23.6, we may assume that G has exactly one $C_o(G)$-component J.

We now outline the proof of the sufficiency of Theorem 23.2. Assume that G has no bad cycle. We divide G into two subgraphs having no bad cycle by slicing G along one or two paths. For example, the graph G in Figure 23.35a is divided into two subgraphs G_1 and G_2, each having no bad cycle, by slicing G along a path drawn by thick lines, as illustrated in Figure 23.35b. We then recursively find rectangular drawings of the two subgraphs as illustrated in Figure 23.35c, and obtain a rectangular drawing of G by patching them, as illustrated in Figure 23.35d. However, the problem is not so simple, because, for some graphs having no bad cycles like one in Figure 23.36a, there is no such path that the resulting two subgraphs have no bad cycle. For any path, one of the resulting two subgraphs has a bad 3-legged cycle C, although C is not a bad cycle in G, as illustrated in Figure 23.36b where a bad cycle C in a subgraph is indicated by dotted lines. For such a case, we split G into two or more subgraphs by slicing G along two paths P_c and P_{cc} having the same ends on P_N and P_S. For example, as illustrated in Figure 23.36c, the graph G in Figure 23.36a is divided into three subgraphs G_1, G_2, and G_3, each having no bad cycle, by slicing G along path P_c indicated by dotted lines and path P_{cc} drawn by thick lines in Figure 23.36a. We then recursively find rectangular drawings of G_1, G_2, and G_3 as illustrated in Figure 23.36d, and slightly deform the drawings of G_1 and G_2, as illustrated in Figure 23.36e. We finally obtain a rectangular drawing of G by patching the drawings of the three subgraphs as illustrated in Figure 23.36f.

The algorithm *Rectangular-Draw* (G) finds only the directions of all edges in G. From the directions the integer coordinates of vertices in G can be determined in linear time.

23.5 ORTHOGONAL DRAWING

An *orthogonal drawing* of a plane graph G is a drawing of G, with the given embedding, in which each vertex is mapped to a point, each edge is drawn as a sequence of alternate horizontal and vertical line segments, and any two edges do not cross except at their common end, as illustrated in Figure 23.37. Orthogonal drawings have numerous practical applications in

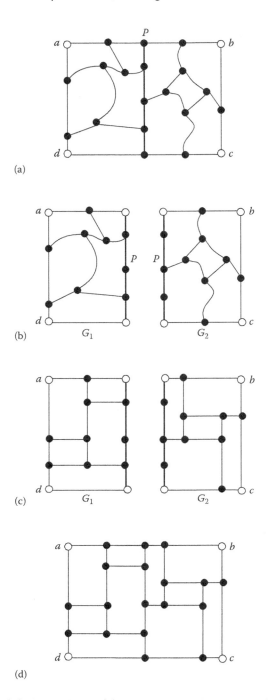

Figure 23.35 (a) G and P, (b) G_1 and G_2, (c) rectangular drawings of G_1 and G_2, and (d) rectangular drawing of G. (Figure taken from Nishizeki, T. and Rahman, M. S., *Planar Graph Drawing*, World Scientific, Singapore, 2004.)

circuit schematics, data flow diagrams, entity relationship diagrams, and so forth [29]. Clearly the maximum degree Δ of G is at most four if G has an orthogonal drawing. Conversely, every plane graph with $\Delta \leq 4$ has an orthogonal drawing, but may need *bends*, that is, points where an edge changes its direction in a drawing. For the cubic plane graph in Figure 23.37a, two orthogonal drawings are depicted in Figures 23.37b and c, which have six and five bends, respectively. If a graph corresponds to a VLSI circuit, then one may be interested in an orthogonal drawing such that the number of bends is as small as possible, because bends

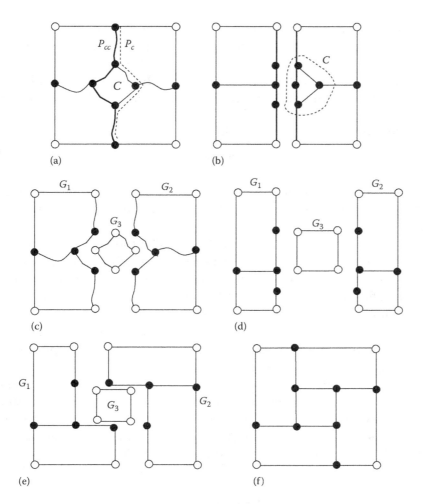

Figure 23.36 (a) G, (b) splitting G along a single path P_{cc}, (c) splitting G along two paths P_{cc} and P_c, (d) rectangular drawings of three subgraphs, (e) deformation, and (f) rectangular drawing of G. (Figure taken from Nishizeki, T. and Rahman, M. S., *Planar Graph Drawing*, World Scientific, Singapore, 2004.)

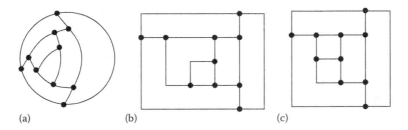

Figure 23.37 (a) Plane graph G, (b) orthogonal drawing of G with 6 bends, and (c) orthogonal drawing of G with 5 bends.

increase the manufacturing cost of a VLSI chip. However, for a given planar graph G, if one is allowed to choose its planar embedding, then finding an orthogonal drawing of G with the minimum number of bends is NP-complete [101]. On the other hand, Tamassia [54] and Garg and Tamassia [102] presented algorithms which find an orthogonal drawing of a given plane graph G with the minimum number $b(G)$ of bends in time $O(n^2 \log n)$ and $O(n^{7/4}\sqrt{\log n})$ respectively, where G has a fixed planar embedding and one is not allowed to alter the planar embedding. Such a drawing is called a *bend-optimal* orthogonal drawing of a plane graph G. They reduce the problem of finding a bend-optimal orthogonal drawing of G to a minimum cost flow problem. Rahman et al. [51] gave a linear algorithm to find a bend-optimal orthogonal drawing for 3-connected cubic plane graphs and Rahman and Nishizeki [103] gave a linear algorithm to find a bend-optimal orthogonal drawing for plane graphs with $\Delta \leq 3$.

23.5.1 Orthogonal Drawing and Network Flow

In this section we describe a network flow model for finding a bend-optimal orthogonal drawing of a plane graph with $\Delta \leq 4$ [54,102].

We first introduce some definitions and terminologies related to an orthogonal representation of a plane graph. In this section, we describe a network flow model for an orthogonal drawing of a plane graph.

Let G be a plane connected graph with $\Delta \leq 4$. The topological structure of G can be described by listing edges that appear on the contour of each face, and by specifying the outer face. A *planar representation* P of a plane graph G is a set of circularly ordered edge lists $P(F)$, one for each face F. Edges in a list $P(F)$ appear as they are encountered when going around the contour of F in the *positive* direction, that is, having the face at one's right. Note that every edge of G appears exactly twice in lists. If this happens in the same list $P(F)$, the edge is called a *bridge*. For the plane graph G in Figure 23.38a, a planar representation P is depicted in Figure 23.38c, and edge e_7 is a bridge in G.

Let D be an orthogonal drawing of G like one in Figure 23.38b. Then each face of G is drawn in D as a rectilinear polygon. Note that a facial polygon is not always a simple polygon. There are two types of angles in D. We call an angle formed by two edges incident to a vertex a *vertex-angle*, and call an angle formed by two line segments at a bend a *bend-angle*. Clearly both a vertex-angle and a bend-angle are $k \cdot 90°$ for some integer k, $1 \leq k \leq 4$. We now have the following two facts.

Fact 23.3 *The sum of the vertex-angles around any vertex is $360°$.*

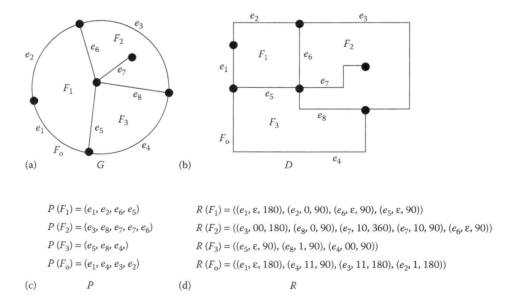

Figure 23.38 (a) Plane graph G, (b) orthogonal drawing D of G, (c) planar representation P of G, and (d) orthogonal representation R of D. (Figure taken from Nishizeki, T. and Rahman, M. S., *Planar Graph Drawing*, World Scientific, Singapore, 2004.)

Fact 23.4 *The sum of the angles inside any facial polygon is $(2p - 4)90°$, and the sum of the angles of the outer facial polygon is $(2p + 4)90°$, where p is the number of angles of the polygon.*

We now introduce a concept of an orthogonal representation R of an orthogonal drawing D in terms of bends occurring along edges and of angles formed by edges. This orthogonal representation is obtained by enriching the lists of the planar representation with information about bends and angles formed by edges. An *orthogonal representation* R of D is a set of circularly ordered lists $R(F)$, one for each face F of G. Each element r of a list is a triple (e_r, s_r, a_r); e_r is an edge, s_r is a bit string, and a_r is an integer in the set $\{90, 180, 270, 360\}$. The bit string s_r provides information about the bends along edge e_r; the kth bit of s_r describes the kth bend on the right side of e_r; bit 0 indicates a 90° bend, and bit 1 indicates a 270° bend. An empty string ϵ is used to characterize a straight line edge. The number a_r specifies the angle formed in face F by edges e_r and $e_{r'}$, where r' is the element following r in the circular list $R(F)$. Figure 23.38d depicts an orthogonal representation R of the orthogonal drawing D in Figure 23.38b. Clearly R preserves only the shape of D without considering lengths of line segments, and hence describes actually an equivalence class of orthogonal drawings of G with *similar shape*, that is, with the same lists of triples r for the edges of G.

For a set R of circular lists to be an orthogonal representation of an orthogonal drawing D of a plane graph G, the following properties are necessary and sufficient, as can be proved by means of elementary geometric considerations.

(p1) There is some planar graph whose planar representation is given by the e-fields of the lists in R.

(p2) For each pair of elements r and r' in R with $e_r = e_{r'}$, string $s_{r'}$ can be obtained by applying bitwise negation to the reversion of s_r.

(p3) For each element r in R, define the rotation $\rho(r)$ as follows:

$$\rho(r) = |s_r|_0 - |s_r|_1 + \left(2 - \frac{a_r}{90}\right) \tag{23.5}$$

where $|s_r|_0$ is the number of zeros in string s_r and $|s_r|_1$ the number of ones. Then

$$\sum_{r \in R(F)} \rho(r) = \begin{cases} +4 & \text{if } F \text{ is an inner face;} \\ -4 & \text{if } F \text{ is the outer face } F_o. \end{cases} \tag{23.6}$$

(p4) For each vertex $v \in V$, the sum of the vertex-angles around v given by the a-fields in R is equal to 360°.

Property (p2) means that each edge must have consistent descriptions in the faces in which it appears. Fact 23.4 implies Property (p3), which means that every face described by R is a rectilinear polygon. Fact 23.3 implies Property (p4).

We say that an orthogonal drawing D of a plane graph G *realizes* an orthogonal representation R if R is a valid description for the shape of D. Figure 23.38d depicts an orthogonal representation R of the plane graph G in Figure 23.38a, and Figure 23.38b depicts an orthogonal drawing D realizing R. Note that the number $b(R)$ of bends in any orthogonal drawing D that realizes R is

$$b(R) = \frac{1}{2} \sum_{F \in \mathcal{F}} \sum_{r \in R(F)} |s_r|, \tag{23.7}$$

where $|s_r|$ is the number of bits in string s_r and $\mathcal{F}$ is the set of all faces in G.

We now introduce notations and terminologies related to flow networks. A *flow network* $\mathcal{N}$ is a directed graph such that $\mathcal{N}$ has two disjoint nonempty sets of distinguished nodes called its *sources* and *sinks*, and each arc e of $\mathcal{N}$ is labeled with three nonnegative integers

- A *lower bound* $\lambda(e)$,

- A *capacity* $\mu(e)$, and

- A *cost* $c(e)$.

A *flow* ϕ in $\mathcal{N}$ associates a nonnegative integer $\phi(e)$ with each arc e; $\phi(e)$ is called a *flow of arc* e. The flow $\phi(e)$ of each arc e must satisfy $\lambda(e) \leq \phi(e) \leq \mu(e)$. Furthermore, ϕ must satisfy the so-called conservation law as follows. For each node u of $\mathcal{N}$ that is neither a source nor a sink, the sum of the flows of the outgoing arcs from u must be equal to the sum of the flows of the incoming arcs to u. Each source u has a *production* $\sigma(u) \geq 0$ of flow, and each sink u has a *consumption* $-\sigma(u) \geq 0$ of flow. That is, for each u of the sources and sinks, the sum of the flows of the outgoing arcs from u minus the sum of the flows of the incoming arcs to u must be equal to $\sigma(u)$.

The total amount of production of the sources is equal to the total amount of consumption of the sinks.

The *cost* $COST(\phi)$ *of a flow* ϕ in $\mathcal{N}$ is the sum of $c(e)\phi(e)$ over all the arcs e of $\mathcal{N}$. The *minimum cost flow problem* is stated as follows. Given a network $\mathcal{N}$, find a flow ϕ in $\mathcal{N}$ such that the cost of ϕ is minimum.

In the remainder of this section we present a flow network $\mathcal{N}$ for an orthogonal drawing of a plane graph G [29,54,102]. All angles in a drawing of G are viewed as commodities that are produced by the vertices, are transported between faces by the edges through their bends, and are eventually consumed by the faces in $\mathcal{N}$. The nodes of $\mathcal{N}$ are vertices and faces of G. Since all angles we deal with have measure $k \cdot 90°$ with $1 \leq k \leq 4$, we establish the convention that a unit of flow represents an angle of $90°$. We shall see Facts 23.3 and 23.4 express the conservation of flow at vertices and faces, respectively. The formal description of $\mathcal{N}$ is given below.

Let $G = (V, E)$ be a plane graph of maximum degree $\Delta \leq 4$ with face set $\mathcal{F}$. We construct a flow network $\mathcal{N}$ from G as follows. The nodes of $\mathcal{N}$ are the vertices and faces of G. That is, the node set U of $\mathcal{N}$ is $U = U_\mathcal{F} \cup U_V$. Each node $u_F \in U_\mathcal{F}$ corresponds to a face F of G, while each node $u_v \in U_V$ corresponds to a vertex v of G. Each node $u_v \in U_V$ is a source and has a production

$$\sigma(u_v) = 4. \tag{23.8}$$

Each node $u_F \in U_\mathcal{F}$ is a sink and has a consumption

$$-\sigma(u_F) = \begin{cases} 2p(F) - 4 & \text{if } F \text{ is an inner face;} \\ 2p(F) + 4 & \text{if } F \text{ is the outer face } F_o \end{cases} \tag{23.9}$$

where $p(F)$ is the number of vertex-angles inside face F (see Figures 23.39a and b). Thus, every node in $\mathcal{N}$ is either a source or a sink. Clearly the total production is $4n$, and the total consumption is

$$\sum_{F \neq F_o} (2p(F) - 4) + 2p(F_o) + 4 = 4m - 4f + 8$$

where n, m, and f are the numbers of vertices, edges, and faces of G, respectively. The total consumption is equal to the total production $4n$, according to Euler's formula.

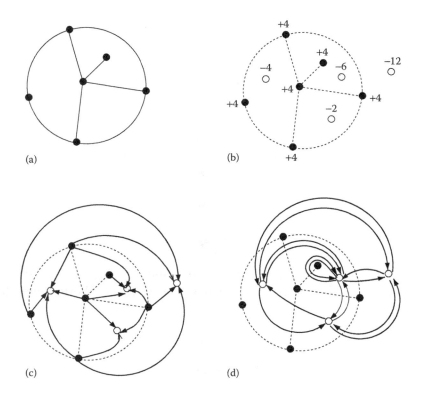

Figure 23.39 (a) Plane graph G, (b) nodes of $\mathcal{N}$ with their productions and consumptions, (c) arcs in A_V, and (d) arcs in $A_\mathcal{F}$. (Figure taken from Nishizeki, T. and Rahman, M. S., *Planar Graph Drawing*, World Scientific, Singapore, 2004.)

The arc set of network $\mathcal{N}$ is $A = A_V \cup A_\mathcal{F}$.

i. A_V consists of all arcs of type (u_v, u_F) such that vertex v is on face F (see Figure 23.39c); the flow $\phi(u_v, u_F)$ in arc (u_v, u_F) represents the sum of vertex-angles at vertex v inside face F, the lower bound $\lambda(u_v, u_F)$ is equal to the number of vertex-angles at v inside face F, the capacity is $\mu(u_v, u_F) = 4$, and the cost is $c(u_v, u_F) = 0$ (see Figure 23.40); and

ii. $A_\mathcal{F}$ consists of all arcs of type $(u_F, u_{F'})$ such that face F shares an edge with face F' (see Figure 23.39d); the flow $\phi(u_F, u_{F'})$ in arc $(u_F, u_{F'})$ represents the number of bends with an angle of 90° inside face F along the edges which are common to F and F', and the lower bound is $\lambda(u_F, u_{F'}) = 0$, the capacity is $\mu(u_F, u_{F'}) = +\infty$, and the cost is $c(u_F, u_{F'}) = 1$ (see Figure 23.41).

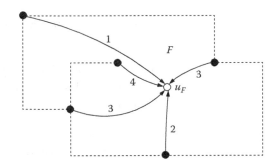

Figure 23.40 Face F and flows in arcs in A_V. (Figure taken from Nishizeki, T. and Rahman, M. S., *Planar Graph Drawing*, World Scientific, Singapore, 2004.)

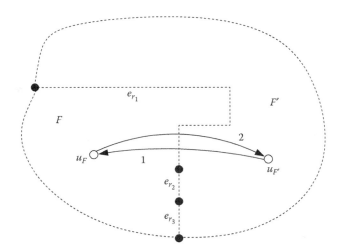

Figure 23.41 Faces F and F', and arcs $e = (u_F, u_{F'})$ and $e' = (u_{F'}, u_F)$ in $A_{\mathcal{F}}$. (Figure taken from Nishizeki, T. and Rahman, M. S., *Planar Graph Drawing*, World Scientific, Singapore, 2004.)

The conservation rule implies that for each source $u_v \in U_V$

$$\sum_{F \in \mathcal{F}} \phi(u_v, u_F) = 4 \tag{23.10}$$

and for each sink $u_F \in U_{\mathcal{F}}$

$$\sum_{F' \in \mathcal{F}} \phi(u_F, u_{F'}) - \left(\sum_{F' \in \mathcal{F}} \phi(u_{F'}, u_F) + \sum_{v \in V} \phi(u_v, u_F) \right) = \sigma(u_F). \tag{23.11}$$

A plane graph G together with a transformation into a network $\mathcal{N}$ is illustrated in Figure 23.39. The intuitive interpretation of the assignment above to the arcs is as follows.

1. Each unit of flow in network $\mathcal{N}$ represents an angle of 90°; for each arc $(u_v, u_F) \in A_V$, flow $\phi(u_v, u_F)$ represents the sum of the vertex-angles formed inside face F by the edges incident to v, which is given by $\phi(u_v, u_F) \cdot 90°$ (see Figure 23.40); for each arc $(u_F, u_{F'}) \in A_{\mathcal{F}}$, flow $\phi(u_F, u_{F'})$ represents the number of bends with an angle of 90° inside face F that appear along the edges separating face F from face F' (see Figure 23.41).

2. The conservation rule at a vertex-node, Equation 23.10 means that the sum of vertex-angles around each vertex must be equal to 360°. The conservation rule at a face-node, Equation 23.11, means that each face must be a rectilinear polygon.

3. The cost $COST(\phi)$ of the flow ϕ is equal to the number of bends of an orthogonal representation corresponding to ϕ.

It is easy to see that every orthogonal representation R of G yields a feasible flow ϕ in network $\mathcal{N}$. Conversely, every feasible flow ϕ can be used to construct an orthogonal representation R of G as in the following theorem.

Theorem 23.3 *Let G be a plane graph, and let $\mathcal{N}$ be the network constructed from G. For each integer flow ϕ in network $\mathcal{N}$, there is an orthogonal representation R that represents an orthogonal drawing D of G and whose number of bends is equal to the cost of the flow ϕ. In particular, the minimum cost flow can be used to construct a bend-optimal orthogonal drawing of G.*

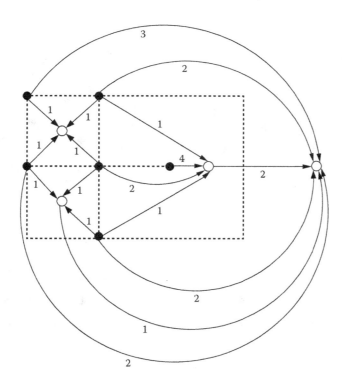

Figure 23.42 Minimum cost flow in network N associated with G in Figure 23.39a and the corresponding orthogonal grid drawing of G. Arcs with zero flow are omitted. (Figure taken from Nishizeki, T. and Rahman, M. S., *Planar Graph Drawing*, World Scientific, Singapore, 2004.)

Garg and Tamassia [102] have shown that the minimum cost flow problem in this specific network can be solved in time $O(n^{7/4}\sqrt{\log n})$. Figure 23.42 depicts a minimum cost flow in the network constructed in Figure 23.39 and a realizing grid embedding for the derived bend-optimal orthogonal representation.

23.5.2 Linear Algorithm for Bend-Optimal Drawing

One can find a bend-optimal orthogonal drawing of a plane graph with $\Delta \leq 4$ in time $O(n^{7/4}\sqrt{\log n})$ as explained in Section 23.5.1. In a VLSI floorplanning problem, an input is often a plane graph with $\Delta \leq 3$ [5]. In this section, we present a linear algorithm to find a bend-optimal orthogonal drawing of a 3-connected cubic plane graph G [51], which depicts the key idea behind a linear algorithm for plane graphs with $\Delta \leq 3$ [103].

Let G be a plane 3-connected cubic graph. We assume for simplicity' sake that G has four or more outer edges. Since G is 3-connected, G has no 1- or 2-legged cycle. In any orthogonal drawing of G, every cycle C of G is drawn as a rectilinear polygon, and hence has at least four convex corners, i.e., polygonal vertices of inner angle 90°. Since G is cubic, such a corner must be a bend if it is not a leg-vertex of C. Thus we have the following facts for any orthogonal drawing of G.

Fact 23.5 *At least four bends must appear on the outer cycle $C_o(G)$ of G.*

Fact 23.6 *At least one bend must appear on each 3-legged cycle in G.*

The algorithm is outlined as follows.

For a cycle C in a plane graph G, we denote by $G(C)$ the plane subgraph of G inside C (including C).

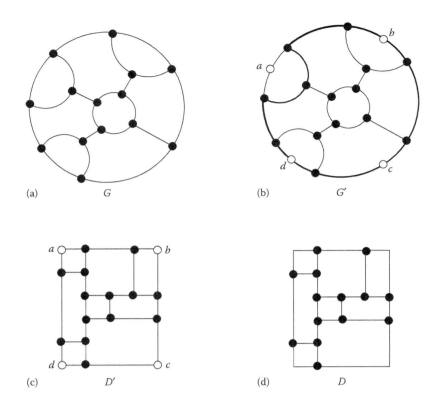

Figure 23.43 (a) G, (b) G', (c) D', and (d) D. (Figure taken from Nishizeki, T. and Rahman, M. S., *Planar Graph Drawing*, World Scientific, Singapore, 2004.)

Let G' be a graph obtained from G by adding four dummy vertices a, b, c, and d of degree two, as corners, on any four distinct outer edges, one for each. Then the resulting graph G' has exactly four outer vertices of degree two designated as corners, and all other vertices of G' have degree three. Figure 23.43b illustrates G' for the graph G in Figure 23.43a. If G' has a rectangular drawing D' with the designated corners a, b, c, and d as illustrated in Figure 23.43c, that is, G' satisfies the condition in Theorem 23.2, then from D' one can immediately obtain an orthogonal drawing D of G with exactly four bends by replacing the four dummy vertices with bends at the corners a, b, c, and d as illustrated in Figure 23.43d. By Fact 23.5 D is a bend-optimal orthogonal drawing of G.

One may thus assume that G' does not satisfy the condition in Theorem 23.2. Then G' has a bad cycle, that is, a 2-legged cycle containing at most one corner or a 3-legged cycle containing no corner. Since G is 3-connected, G has no 2-legged cycle. However, G' has four 2-legged cycles, each passing through all outer vertices except one of the four corners. (One of them is drawn by thick lines in Figure 23.43b.) Clearly all these four 2-legged cycles in G' are not bad, because each of them contains three corners. Thus every bad cycle in G' is a 3-legged cycle containing no corner. A bad cycle C in G' is defined to be *maximal* if C is not contained in the subgraph $G'(C')$ of G' inside C' for any other bad cycle C' in G'. In Figure 23.44a $C_1, C_2, \ldots, C_6$ are the bad cycles, $C_1, C_2, \ldots, C_4$ are the maximal bad cycles in G', and C_5 and C_6 are not maximal bad cycles since they are contained in $G'(C_4)$. The 3-legged cycle C_7 indicated by a dotted line in Figure 23.44a is not a bad cycle in G' since it contains a corner a. We say that cycles C and C' in G' are *independent* if $G'(C)$ and $G'(C')$ have no common vertex. Since G is a plane 3-connected cubic graph, all maximal bad cycles in G' are independent of each other. Let $C_1, C_2, \ldots, C_l$ be the maximal bad cycles in G'. Let G'' be the graph obtained from G' by contracting $G'(C_i)$ into a single vertex v_i for each maximal bad cycle C_i,

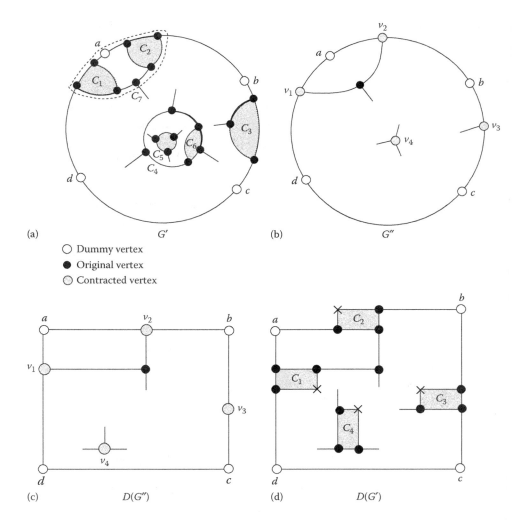

Figure 23.44 (a) G', (b) G'', (c) rectangular drawing $D(G'')$ of G'', and (d) orthogonal drawing $D(G')$ of G'. (Figure taken from Nishizeki, T. and Rahman, M. S., *Planar Graph Drawing*, World Scientific, Singapore, 2004.)

$1 \leq i \leq l$, as illustrated in Figure 23.44b. Clearly G'' has no bad cycle, and hence by Theorem 23.2 G'' has a rectangular drawing. We first find a rectangular drawing of G'', and then recursively find a suitable orthogonal drawing of $G'(C_i)$, $1 \leq i \leq l$, with the minimum number of bends, called a *feasible drawing*, and finally patch them to get an orthogonal drawing of G (see Figure 23.44).

Further Reading

Due to numerous practical applications of graph drawing, researchers have worked on graph drawing with various view-points, and as a result, the field of graph drawing has become very wide. Although we have focused mainly on some important results on planar graph drawing [29], the field has been enriched with many recent results on 3D drawing, dynamic drawing, proximity drawing, symmetric drawing, simultaneous embedding, map labeling, and so on [4,27,28,30]. Different classes of graphs such as planar graphs, hierarchical graphs, clustered graphs, interval graphs, and so forth are investigated to produce graph drawings having some desired properties. Aesthetic drawing styles for nonplanar graphs, such as confluent drawing

[104,105], right angle crossings drawing [106,107], bar 1-visibility drawing [108], and so forth have also been studied in recent years.

References

[1] T. C. Biedl, *Orthogonal graph visualization: The three-phase method with applications*, PhD Thesis, RUTCOR, Rutgers University, New Brunswick, NJ, 1997.

[2] G. Kant, *algorithms for drawing planar graphs*, PhD Thesis, Faculty of Information Science, Utrecht University, the Netherlands, 1993.

[3] M. S. Rahman, *Efficient algorithms for drawing planar graphs*, PhD Thesis, Graduate School of Information Sciences, Tohoku University, Sendai, Japan, 1999.

[4] K. Sugiyama, *Graph Drawing and Applications for Software and Knowledge Engineers*, World Scientific, Singapore, 2002.

[5] T. Lengauer, *Combinatorial Algorithms for Integrated Circuit Layout*, John Wiley & Sons, Chichester, 1990.

[6] N. Chiba, K. Onoguchi, and T. Nishizeki, Drawing planar graphs nicely, *Acta Informatica*, **22** (1985), 187–201.

[7] R. Tamassia, G. Di Battista, and C. Batini, Automatic graph drawing and readability of diagrams, *IEEE Trans. Syst. Man Cybern.* **SMC-18** (1988), 61–79.

[8] G. Di Battista, P. Eades, R. Tamassia, and I. G. Tollis, Algorithms for drawing graphs: An annotated bibliography, *Comp. Geom. Theory Appl.*, **4** (1994), 235–282.

[9] R. Tamassia and I. G. Tollies (editors), *Graph Drawing (Proc. of GD '94)*, Lect. Notes in Computer Science, Springer, 894, 1995.

[10] F. J. Brandenburg (editor), *Graph Drawing (Proc. of GD '95)*, Lect. Notes in Computer Science, Springer, 1027, 1996.

[11] S. North (editor), *Graph Drawing (Proc. of GD '96)*, Lect. Notes in Computer Science, Springer, 1190, 1997.

[12] G. Di Battista (editor), *Graph Drawing (Proc. of GD '97)*, Lect. Notes in Computer Science, Springer, 1353, 1997.

[13] S. H. Whiteside (editor), *Graph Drawing (Proc. of GD '98)*, Lect. Notes in Computer Science, Springer, 1547, 1998.

[14] J. Kratochvíl (editor), *Graph Drawing (Proc. of GD '99)*, Lect. Notes in Computer Science, Springer, 1731, 1999.

[15] J. Marks (editor), *Graph Drawing (Proc. of GD'00)*, Lect. Notes in Computer Science, Springer, 1984, 2001.

[16] M. T. Goodrich and S. G. Kobourov (editors), *Graph Drawing (Proc. GD '02)*, Lect. Notes in Computer Science, Springer, 2528, 2002.

[17] G. Liotta (editor), *Graph Drawing (Proc. of GD '03)*, Lect. Notes in Computer Science, Springer, 2912, 2004.

[18] S. Hong, T. Nishizeki, and W. Quan (editors), *Graph Drawing (Proc. of GD '07)*, Lect. Notes in Computer Science, Springer, 4875, 2008.

[19] U. Brandes and S. Cornelsen (editors), *Graph Drawing (Proc. of GD '10)*, Lect. Notes in Computer Science, Springer, 6502, 2011.

[20] S. Wismath and A. Wolf (editors), *Graph Drawing (Proc. of GD '13)*, Lect. Notes in Computer Science, Springer, 8242, 2013.

[21] I. F. Cruz and P. Eades (editors), Special issue on graph visualization, *J. Vis. Lang. Comput.*, **6**(3) (1995).

[22] G. Di Battista and R. Tamassia (editors), Special issue on graph drawing, *Algorithmica*, **16**(1) (1996).

[23] G. Di Battista and R. Tamassia (editors), Special issue on geometric representations of graphs, *Comput. Geom. Theory Appl.*, **9**(1–2) (1998).

[24] G. Di Battista and P. Mutzel (editors), New trends in graph drawing: special issue on selected papers from the 1997 symposium on graph drawing, *J. Graph Algorithms Appl.*, **3**(4) (1999).

[25] G. Liotta and S. H. Whitesides (editors), Special issue on selected papers from the 1998 symposium on graph drawing, *J. Graph Alg. Appl.*, **4**(3) (2000).

[26] M. Kaufmann (editor), Special issue on selected papers from the 2000 symposium on graph drawing, *J. Graph Alg. Appl.*, **6**(3) (2002).

[27] G. Di Battista, P. Eades, R. Tamassia, and I. G. Tollis, *Graph Drawing: Algorithms for the Visualization of Graphs*, Prentice Hall, Upper Saddle River, NJ, 1999.

[28] M. Kaufmann and D. Wagner (editors), *Drawing Graphs: Methods and Models*, Lect. Notes in Compt. Sci., Springer, 2025, Berlin, Germany, 2001.

[29] T. Nishizeki and M. S. Rahman, *Planar Graph Drawing*, World Scientific, Singapore, 2004.

[30] R. Tamassia (editor), *Handbook of Graph Drawing and Visualization*, CRC Press, 2014.

[31] C. Kuratowski, Sur le probléme des courbes gauches en topologie, *Fund. Math.*, **15** (1930), 271–283.

[32] J. E. Hopcroft and R. E. Tarjan, Efficient planarity testing, *J. Assoc. Comput. Mach.*, **21** (1974), 549–568.

[33] K. S. Booth and G. S. Lueker, Testing the consecutive ones property, interval graphs, and graph planarity using PQ-tree algorithms, *J. Comput. Syst. Sci.*, **13** (1976), 335–379.

[34] N. Chiba, T. Nishizeki, S. Abe, and T. Ozawa, A linear algorithm for embedding planar graphs using PQ-trees, *J. Comput. Syst. Sci.*, **30** (1985), 54–76.

[35] K. Mehlhorn and P. Mutzel, On the embedding phase of the Hopcroft and Tarjan planarity testing algorithm, *Algorithmica*, **16** (1996), 233–242.

[36] W. K. Shih and W.-L. Hsu, A new planarity test, *Theoret. Comput. Sci.*, **223** (1999), 179–191.

[37] N. Bonichon, B. L. Saëc, and M. Mosbah, *Optimal Area Algorithm for Planar Polyline Drawings (Proc. of WG '02)*, Lect. Notes in Computer Science, Springer, 2573, pp. 35–46, 2002.

[38] E. Di Giacomo, W. Didimo, G. Liotta, and S. K. Wismath, *Drawing Planar Graphs on a Curve (Proc. WG '03)*, Lect. Notes in Computer Science, Springer, 2880, pp. 192–204, 2003.

[39] C. Gutwenger and P. Mutzel, *Planar Polyline Drawings with Good Angular Resolution (Proc. GD '98)*, Lect. Notes in Compt. Sci., Springer, 1547, pp. 167–182, 1998.

[40] K. Wagner, Bemerkungen zum vierfarbenproblem, *Jahresber. Deutsch. Math-Verien.*, **46** (1936), 26–32.

[41] I. Fáry, *On Straight Line Representations of Planar Graphs*, Acta Sci. Math. Szeged, **11** (1948), 229–233.

[42] K. S. Stein, Convex maps, *Proc. Am. Math. Soc.*, **2** (1951), 464–466.

[43] M. Chrobak and G. Kant, Convex grid drawings of 3-connected planar graphs, *Inter. J. Comput. Geom. Appl.*, **7**(3) (1997), 211–223.

[44] N. Chiba, T. Yamanouchi, and T. Nishizeki, Linear algorithms for convex drawings of planar graphs, J. A. Bondy and U. S. R. Murty (editors), *Progress in Graph Theory*, Academic Press Canada, pp. 153–173, 1984.

[45] G. Kant, Drawing planar graphs using the canonical ordering, *Algorithmica*, **16** (1996), 4–32.

[46] X. Zhou and T. Nishizeki, Convex drawings of internally triconnected plane graphs on $0(n^2)$ grids, *Discrete Math. Algorithms Appl.*, **2**(3) (2010), 347–362.

[47] D. Mondal, R. I. Nishat, S. Biswas, and M. S. Rahman, Minimum-segment convex drawings of 3-connected cubic plane graphs, *J. Comb. Optim.*, **25**(3) (2013), 460–480.

[48] T. C. Biedl, *New Lower Bounds for Orthogonal Graph Drawings (Proc. GD '95)*, Lect. Notes in Compt. Sci., Springer, 1027, pp. 28–39, 1996a.

[49] T. C. Biedl, Optimal orthogonal drawings of triconnected plane graphs, *Proc. Scandinavian Workshop on Algorithm Theory, SWAT '96*, Lect. Notes in Compt. Sci., Springer, 1097, pp. 333–344, 1996b.

[50] G. Di Battista, G. Liotta, and F. Vargiu, Spirality and optimal orthogonal drawings, *SIAM J. Comput.*, **27**(6) (1998), 1764–1811.

[51] M. S. Rahman, S. Nakano, and T. Nishizeki, A linear algorithm for bend-optimal orthogonal drawings of triconnected cubic plane graphs, *J. Graph Algorithms Appl.*, **3**(4) (1999), 31–62, http://jgaa.info.

[52] M. S. Rahman, T. Nishizeki, and M. Naznin, Orthogonal drawings of plane graphs without bends, *J. Graph Algorithms Appl.*, **7**(4) (2003), 335–362, http://jgaa.info.

[53] J. A. Storer, On minimal node-cost planar embeddings, *Networks*, **14** (1984), 181–212.

[54] R. Tamassia, On embedding a graph in the grid with the minimum number of bends, *SIAM J. Comput.*, **16**(3) (1987), 421–444.

[55] R. Tamassia, I. G. Tollis, and J. S. Vitter, Lower bounds for planar orthogonal drawings of graphs, *Inf. Proc. Lett.*, **39** (1991), 35–40.

[56] X. Zhou and T. Nishizeki, Orthogonal drawings of series–parallel graphs with minimum bends, *SIAM J. Discrete Math.*, **22**(4) (2008), 1570–1604.

[57] T. C. Biedl and G. Kant, A better heuristic for orthogonal graph drawings, *Comput. Geom. Theory Appl.*, **9** (1998), 159–180.

[58] A. Papakostas and I. G. Tollis, *Improved Algorithms and Bounds for Orthogonal Drawings (Proc. GD '94)*, Lect. Notes in Computer Science, Springer, 894, pp. 40–51, 1995.

[59] A. Papakostas and I. G. Tollis, *A Pairing Technique for Area Efficient Orthogonal Drawings (Proc. GD '96)*, Lect. Notes in Computer Science, Springer, 1190, pp. 355–370, 1997.

[60] M. S. Rahman, K. Miura, and T. Nishizeki, Octagonal drawings of plane graphs with prescribed face areas, *Comput. Geom. Theory Appl.*, **42**(3) (2009), 214–230.

[61] T. C. Biedl and M. Kaufmann, Area-efficient static and incremental graph drawings, *Proc. 5th European Symposium on Algorithms*, Lect. Notes in Compt. Sci., Springer, 1284, pp. 37–52, 1997.

[62] U. Fößmeier, G. Kant, and M. Kaufmann, *2-visibility drawings of plane graphs (Proc. GD '96)*, Lect. Notes in Computer Science, Springer, 1190, pp. 155–168, 1997.

[63] A. Papakostas and I. G. Tollis, Efficient orthogonal drawings of high degree graphs, *Algorithmica*, **26** (2000), 100–125.

[64] C. Thomassen, Plane representations of graphs, J. A. Bondy and U. S. R. Murty (editors), *Progress in Graph Theory*, Academic Press Canada, pp. 43–69, 1984.

[65] J. Bhasker and S. Sahni, A linear algorithm to find a rectangular dual of a planar triangulated graph, *Algorithmica*, **3** (1988), 247–278.

[66] M. S. Rahman, S. Nakano, and T. Nishizeki, Rectangular grid drawings of plane graphs, *Comput. Geom. Theory Appl.*, **10**(3) (1998), 203–220.

[67] M. S. Rahman, S. Nakano, and T. Nishizeki, Rectangular drawings of plane graphs without designated corners, *Comput. Geom. Theory Appl.*, **21**(3) (2002), 121–138.

[68] X. He, A simple linear time algorithm for proper box rectangular drawings of plane graphs, *J. Algorithms*, **40**(1) (2001), 82–101.

[69] M. S. Rahman, S. Nakano, and T. Nishizeki, Box-rectangular drawings of plane graphs, *J. Algorithms*, **37** (2000), 363–398.

[70] M. M. Hasan, M. S. Rahman, and M. R. Karim, Box-rectangular drawings of planar graphs, *J. Graph Algorithms Appl.*, **17**(6) (2013), 629–646.

[71] G. Kant, A more compact visibility representation, *Int. J. Comput. Geom. Appl.*, **7**(3) (1997), 197–210.

[72] X. He, J. Wang, and H. Zhang, Compact visibility representation of 4-connected plane graphs, *Theor. Comput. Sci.*, **447** (2012), 62–73.

[73] J. Wang and X. He, Visibility representation of plane graphs with simultaneous bound for both width and height, *J. Graph Algorithms Appl.*, **16**(2) (2012), 317–334.

[74] M. Jünger and P. Mutzel (editors), *Graph Drawing Software*, Springer, Berlin, Germany, 2004.

[75] N. Sherwani, *Algorithms for VLSI Physical Design Automation*, 2nd edition, Kluwer Academic Publishers, Boston, MA, 1995.

[76] S. M. Sait and H. Youssef, *VLSI Physical Design Automation: Theory and Practice*, World Scientific, Singapore, 1999.

[77] R. L. Francis and J. A. White, *Facility Layout and Location*, Prentice Hall, New Jersey, 1974.

[78] H. de Fraysseix, J. Pach, and R. Pollack, How to draw a planar graph on a grid, *Combinatorica*, **10** (1990), 41–51.

[79] W. Schnyder, Embedding planar graphs on the grid, *Proc. 1st ACM-SIAM Symp. on Discrete Algorithms*, San Francisco, CA, pp. 138–148, 1990.

[80] M. Chrobak and S. Nakano, Minimum-width grid drawings of plane graphs, *Comput. Geom. Theory Appl.*, **11** (1998), 29–54.

[81] X. He, Grid embedding of 4-connected plane graphs, *Discrete Comput. Geom.*, **17** (1997), 339–358.

[82] K. Miura, S. Nakano, and T. Nishizeki, Grid drawings of 4-connected plane graphs, *Discrete Comput. Geom.*, **26**(1) (2001), 73–87.

[83] M. Chrobak and T. H. Payne, A linear-time algorithm for drawing a planar graph on a grid, *Inf. Process. Lett.*, **54** (1995), 241–246.

[84] W. T. Tutte, Convex representations of graphs, *Proc. Lond. Math. Soc.*, **10** (1960), 304–320.

[85] W. T. Tutte, How to draw a graph, *Proc. Lond. Math. Soc.*, **13** (1963), 743–768.

[86] R. J. Lipton, D. J. Rose, and R. E. Tarjan, Generalized nested dissections, *SIAM J. Numer. Anal.*, **16**(2) (1979), 346–358.

[87] K. Miura, S. Nakano, and T. Nishizeki, Convex grid drawings of four-connected plane graphs, *Proc. 11th Annual International Symposium on Algorithms and Computation*, Lect. Notes in Compt. Sci., Springer, 1969, pp. 254–265, 2000.

[88] K. Miura, H. Haga, and T. Nishizeki, Inner rectangular drawings of plane graphs, *Proc. of ISAAC*, Lect. Notes in Compt. Sci., Springer, 3341, pp. 693–704, 2004.

[89] K. Kozminski and E. Kinnen, An algorithm for finding a rectangular dual of a planar graph for use in area planning for VLSI integrated circuits, *Proc. 21st DAC*, Albuquerque, NM, pp. 655–656, 1984.

[90] X. He, On finding the rectangular duals of planar triangular graphs, *SIAM J. Comput.*, **22**(6) (1993), 1218–1226.

[91] G. Kant and X. He, Regular edge labeling of 4-connected plane graphs and its applications in graph drawing problems, *Theoret. Comput. Sci.*, **172** (1997), 175–193.

[92] X. He, An efficient parallel algorithm for finding rectangular duals of plane triangulated graphs, *Algorithmica*, **13** (1995), 553–572.

[93] Y.-T. Lai and S. M. Leinwand, A theory of rectangular dual graphs, *Algorithmica*, **5** (1990), 467–483.

[94] M. S. Rahman, T. Nishizeki, and S. Ghosh, Rectangular drawings of planar graphs, *J. Algorithms*, **50** (2004), 62–78.

[95] W. T. Tutte, A short proof of the factor theorem for finite graphs, *Can. J. Math.*, **6** (1954), 347–352.

[96] J. E. Hopcroft and R. M. Karp, An $n^{5/2}$ algorithm for maximum matching in bipartite graphs, *SIAM J. Comput.*, **2** (1973), 225–231.

[97] S. Micali and V. V. Vazirani, An $O(\sqrt{|V|} \cdot |E|)$ algorithm for finding maximum matching in general graphs, *Proc. 21st Annual Symposium on Foundations of Computer Science*, pp. 17–27, 1980.

[98] C. H. Papadimitriou and K. Steiglitz, *Combinatorial Optimization*, Prentice Hall, Englewood Cliffs, NJ, 1982.

[99] D. S. Hochbaum and B. G. Chandran, *Further below the flow decomposition barrier of maximum flow for bipartite matching and maximum closure*, Working paper, 2004.

[100] D. S. Hochbaum, *Faster pseudoflow-based algorithms for the bipartite matching and the closure problems*, Abstract, CORS/SCRO-INFORMS Joint Int. Meeting, Banff, Canada, p. 46, 2004.

[101] A. Garg and R. Tamassia, On the computational complexity of upward and rectilinear planarity testing, *SIAM J. Comput.*, **31**(2) (2001), 601–625.

[102] A. Garg and R. Tamassia, A new minimum cost flow algorithm with applications to graph drawing, *Proc. of Graph Drawing '96*, Lect. Notes in Compt. Sci., Springer, 1190, pp. 201–216, 1997.

[103] M. S. Rahman and T. Nishizeki, Bend-minimum orthogonal drawings of plane 3-graphs, *Proc. International Workshop on Graph Theoretic Concepts in Computer Science*, Lect. Notes in Computer Science, Springer, 2573, pp. 367–378, 2002.

[104] M. Dickerson, D. Eppstein, M. T. Goodrich, and J. Y. Meng, Confluent drawings: visualizing non-planar diagrams in a planar way, *J. Graph Algorithms Appl.*, **9**(1) (2005), 31–52.

[105] G. Quercini and M. Ancona, Confluent drawing algorithms using rectangular dualization, *Proc. of GD 2010*, Lect. Notes in Computer Science, Springer, 6502, pp. 341–352, 2011.

[106] W. Didimo, P. Eades, and G. Liotta, Drawing graphs with right angle crossings, *Theoret. Comput. Sci.*, **412**(39) (2011), 5156–5166.

[107] E. N. Argyriou, M. A. Bekos, and A. Symvonis, The straight-line RAC drawing problem is NP-hard, *J. Graph Algorithms Appl.*, **16**(2) (2012), 569–597.

[108] S. Sultana, M. S. Rahman, A. Roy, and S. Tairin, Bar 1-visibility drawings of 1-planar praphs, *Proc. of ICAA*, Lect. Notes in Computer Science, Springer, 8321, pp. 62–76, 2014.

[109] T. H. Cormen, C. E. Leiserson, and R. L. Rivest, *Introduction to Algorithms*, Cambridge, MIT Press, MA, 1990.

[110] A. Lubiw, Some NP-complete problems similar to graph isomorphism, *SIAM J. Comput.*, **10**(1) (1981), 11–21.

[111] T. Nishizeki and N. Chiba, *Planar Graphs: Theory and Algorithms*, North-Holland, Amsterdam, the Netherlands, 1988.

VI

Interconnection Networks

CHAPTER 24

Introduction to Interconnection Networks

S. A. Choudum

Lavanya Sivakumar

V. Sunitha

CONTENTS

24.1	Interconnection Networks		588
	24.1.1	Computing the Sum of n Numbers	588
	24.1.2	Computing the Product of Two Matrices	589
24.2	Introduction to Hypercubes		591
	24.2.1	Alternative Definitions of Hypercubes	592
		24.2.1.1 Cartesian Product of Graphs	592
		24.2.1.2 Basic Algebraic Notions	592
		24.2.1.3 Cayley Graph	593
	24.2.2	Basic Properties of Hypercubes	594
	24.2.3	Characterizations of Hypercubes	599
		24.2.3.1 Characterizations through Splitting	599
		24.2.3.2 Characterizations through (0,2)-Graphs	600
		24.2.3.3 Characterizations through Intervals	602
		24.2.3.4 Characterizations through Medians	605
		24.2.3.5 Characterizations through Projections	607
		24.2.3.6 Characterizations through Convex Sets	608
		24.2.3.7 Characterizations through Some Monotone Properties	609
		24.2.3.8 Characterizations through Edge Colorings	612
24.3	Hypercube-Like Interconnection Networks		612
	24.3.1	Twisted Cube TQ_n	612
	24.3.2	k-Skip Enhanced Cube $Q_{n,k}$	614
	24.3.3	Möbius Cube $0MQ_n, 1MQ_n$	615
	24.3.4	Shuffle Cube ShQ_n	615
	24.3.5	Fibonacci Cube Γ_n	615
	24.3.6	k-Ary n-Cube Q_n^k	616
	24.3.7	Augmented Cube AQ_n	617
	24.3.8	Hamming Graph/Generalized Base-b Cube $H(b,n)$	618
	24.3.9	de Bruijn Graph $DG(d,k)$	618
	24.3.10	Cube-Connected Cycles Graph CCC_n	619
	24.3.11	Butterfly and Wrapped Butterfly Graph UBF_n, WBF_n	619
	24.3.12	Degree Four Cayley Graph G_n	620

24.3.13 k-Valent Cayley Graph $G_{k,n}$ 621
24.3.14 Star Graph $S(n)$ 622

24.1 INTERCONNECTION NETWORKS

Parallel computing, distributed computing, and cloud computing are evolutions of serial computing. They use multiple resources to solve a single problem. These new paradigms of computing are faster and cost effective. Their applications extend to various fields such as global weather forecasting, genetic engineering, quantum chemistry, relativistic physics, computational fluid dynamics, and turbulence.

At the basic level, a collection of processors and memory units interconnected by a network of links is called a *multiprocessor computing system*. The network is called the *interconnection network*. These systems are further classified into parallel computing systems, distributed computing systems, and cloud computing systems depending on the paradigms employed in computing. This chapter is devoted to parallel computing systems and algorithms to have a definite focus.

Parallel algorithms are so designed that they concurrently utilize the various units of a parallel computer. During the computations, processors communicate among themselves by passing messages through the links of the interconnection network. So, the interconnection network plays a central role in boosting the performance rate of a parallel computer. There are two types of interconnection networks, namely direct/static and indirect/dynamic, depending on whether the communicating devices are connected directly node to node or via switches. The engineering aspects of the interconnection network and parallel computing are extensively covered in the books [1–6]. In this chapter we give a brief survey of various interconnection networks from a graph theoretical perspective.

Before we formally define these network models, let us look at the working of two examples of parallel algorithms to get motivated.

24.1.1 Computing the Sum of n Numbers Using a Complete Binary Tree of Height $\log_2 n$ as a Data Structure

The usual sequential algorithm requires $n-1$ steps to compute the sum of n numbers. However, using a parallel algorithm as illustrated in Figure 24.1, the sum can be computed in $\log_2 n$ steps. Initially, the data is stored in the processors located at the leaves of the complete binary tree. Two numbers are stored in each of the leaf processors. Each of these processors computes the sum of the two numbers and routes the sum to its parent. The parents act identically as their children, under the same instruction. The computation and routing are repeated until the total sum reaches the output device. The processors located at each level compute the

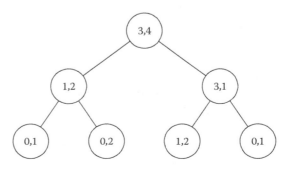

Figure 24.1 Addition using complete binary tree.

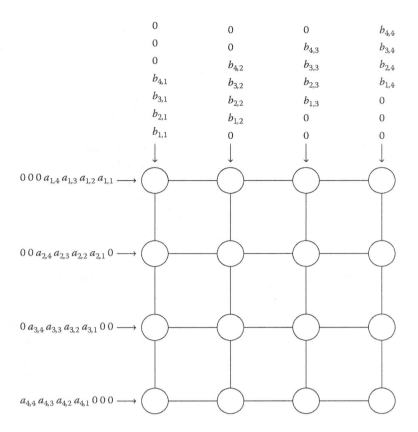

Figure 24.2 Matrix product using two-dimensional mesh.

sums parallely/concurrently. The synchronization of computation and routing are vital for the successful implementation of the algorithm. The reader may also note that by changing the strategy in data storage and routing, one can reduce the number of processors utilized.

24.1.2 Computing the Matrix Product $C = [c_{i,j}]$ of Two $n \times n$ Matrices $A=[a_{i,j}]$, $B=[b_{i,j}]$ Using an $n \times n$ Mesh as a Data Structure

A sequential algorithm can be easily designed to compute AB in $O(n^3)$ steps, by directly using the definition of the matrix product. However, using an $n \times n$ mesh one can calculate AB in $3n - 2$ steps; see Figure 24.2. The ith row $a_{i,1}, a_{i,2}, \ldots, a_{i,n}$ is processed through the ith row of the mesh, and the jth column $b_{1,j}, b_{2,j}, \ldots, b_{n,j}$ is processed through the jth column of the mesh as shown in the figure. At each step k, the (i,j)th processor receives (i) $a_{i,k}$ and $b_{k,j}$ concurrently; (ii) computes the product $a_{i,k}b_{k,j}$; (iii) adds the product to the sum $a_{i,1}b_{1,j} + a_{i,2}b_{2,j} + \cdots + a_{i,k-1}b_{k-1,j}$, retains the consequent sum; and (iv) sends $a_{i,k}$ to the right processor and $b_{k,j}$ to the processor one below it. At the end of the computation, $c_{i,j}$ is available at the (i,j)th processor. The computation of $c_{i,j}$ is completed at step $i+j+n-2$ and so, it is easy to see that C is computed after $3n - 2$ steps.

Graphs serve as natural mathematical models to represent the interconnection networks, where the vertices denote the processors and the edges denote the physical links between the processors. So, the terminology of graphs is used for the interconnection networks too. Many networks have been proposed because of their favorable graph theoretic properties. Among these, mesh networks, torus networks, and hypercube networks have been widely studied. Some of the well-known commercial parallel computers and their interconnection networks are given in Table 24.1.

TABLE 24.1 Commercial Parallel Computers and Their Interconnection Networks

Parallel Computers	Interconnection Network
Intel Delta	Linear Array
Intel Paragon, Transputers	2-D Mesh
MIT J-Machine	3-D Mesh
ILLIAC IV	Torus
Intel/CMU iWarp	2-D Torus
Cray/SGI T3D,T3E	3-D Torus
CM5 (Thinking Machines)	4-ary Hypertree
nCUBE 10, nCUBE-2 (NCUBE), iPSC/862(Intel), Cosmic cube(Caltech), CM200(Thinking Machines)	Binary hypercube

The problems arising in the study of interconnection network can be broadly classified as follows:

1. *Design and analysis of networks.* In this problem, a network is designed and its suitability for various tasks are then studied. To evaluate the performance of a network, many graph theoretic concepts like diameter, connectivity, parallel paths, symmetry, and recursive construction are used. A designer of interconnection networks has to take into account the following criteria.

   a. *Communication speed*—There should be fast communication among processors. In graph theoretical terminology, this goal is achieved by keeping the diameter small.

   b. *High robustness*—The network should be least vulnerable to any disruption. This goal is achieved by providing large connectivity.

   c. *Rich structure*—The network must contain many types of subnetworks. This demand is to be satisfied for the implementation of various kinds of algorithms.

   d. *Fixed degree*—The network should have small average degree or fixed degree for easy scalability.

2. *Routing messages in a network.* The process of sending data/messages in a network from a source to a destination is called *routing*. One of the problem is finding optimal routing paths in the designed network.

3. *Design of algorithms.* Under this topic, one designs the parallel algorithms and analyzes their properties stated above.

4. *Implementation of algorithms in a network.* The computations involved in a parallel algorithm A are represented by a graph $G(A)$ called the *computation graph* (or *algorithmic graph*), where a vertex represents the data set allotted to the processor and an edge represents the computation involving the two data sets. So, in graph theoretic terms, the task of executing a parallel algorithm A in a parallel computer P is equivalent to finding an embedding of $G(A)$ into the interconnection network $N(P)$ of the parallel computer. If $G(A)$ is a subgraph of $N(P)$ then the algorithm performs very efficiently without any communication delay. So, the performance of the algorithm depends on the qualities of the embedding. We cover this significant topic in Chapter 26.

5. *Fault tolerance.* The ability of a system to continue operations correctly in the presence of failures in one or many of its components is known as *fault tolerance*.

The reader may easily notice that some of the above demands are conflicting. So some trade-off is necessary in designing the network. The graphs of some of the networks are described and analyzed with much detail by Hayes [3], Hsu and Lin [4], Hwang and Briggs [5], Leighton [6], Quinn [2], and Xu [7,8]. Thus a large literature on interconnection networks is already available. However, there seems to be no single source which contains information on the various variations of the hypercube. In the next section, we will give various definitions and properties of hypercubes. This is followed by a section on variants and generalizations of hypercubes.

24.2 INTRODUCTION TO HYPERCUBES

Hypercubes and their various generalizations which include Hamming graphs are fundamental objects in communication and coding theory; see [1,9]. In this section, we define hypercubes in various ways and state their properties which are used in the following sections.

> **Notation:** Q_n, hypercube of dimension n; n is always assumed to be at least one. Hypercubes are also called binary cube, n-cube, cube-connected network, cosmic cube, binary n-cube, and Boolean n-cube.

Definition 24.1 *For any integer $n \geq 1$, the* hypercube Q_n *of dimension n is the graph with vertex set*

$$V(Q_n) = \{X = x_1 x_2 \ldots x_n : x_i \in \{0,1\}, 1 \leq i \leq n\}$$

and edge set

$$E(Q_n) = \{(X,Y) : X = x_1 x_2 \ldots x_n, Y = y_1 y_2 \ldots y_n \text{ and } x_i \neq y_i \text{ for exactly one } i, 1 \leq i \leq n\}.$$

See Figure 24.3 for hypercubes of small dimensions.

So each vertex of Q_n is a binary string $X = x_1 x_2 \ldots x_n$ with n bits $x_i = 0$ or 1. Throughout this article, we will follow this convention of denoting the vertices of Q_n by capital letters and the binary strings by the respective small letters. The complement of x_i is denoted by

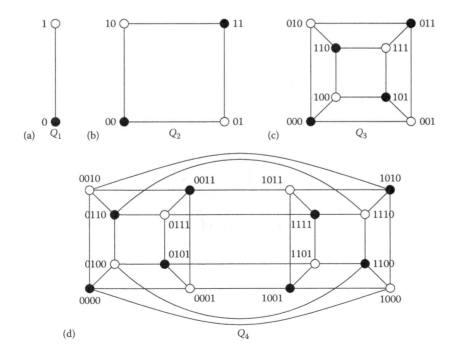

Figure 24.3 Hypercubes of dimensions (a) one, (b) two, (c) three, and (d) four.

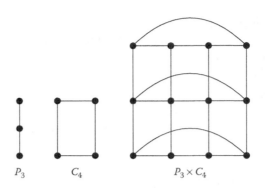

Figure 24.4 Cartesian product of P_3 and C_4.

$\overline{x_i}$, so $\overline{0} = 1$ and $\overline{1} = 0$. An edge (X, Y), where Y differs at position i from X, that is $Y = x_1 x_2 \ldots x_{i-1} \overline{x_i} x_{i+1} \ldots x_n$, is called an edge of dimension i. In this case, we use X_i to denote Y. Let $N(X; Q_n) = \{X_i : 1 \leq i \leq n\}$ denote the set of all neighbours of X. Two edges of the same dimension are called *parallel edges*. Similarly, $X_{i_1, i_2, \ldots, i_k}$ denotes the vertex which differs from X at positions $i_1, i_2, \ldots, i_k$. Furthermore, $\overline{X}$ denotes $X_{1,2,\ldots,n} = \overline{x}_1 \overline{x}_2 \ldots \overline{x}_n$ and $X_\emptyset$ denotes X. Clearly, if π is any permutation of $\{i_1, i_2, \ldots, i_k\}$ then $X_{i_1, i_2, \ldots, i_k} = X_{\pi(i_1), \pi(i_2), \ldots, \pi(i_k)}$.

24.2.1 Alternative Definitions of Hypercubes

Before we proceed to give alternative definitions of Q_n, we require some terminology from graph theory and set theory.

24.2.1.1 Cartesian Product of Graphs

Definition 24.2 *Given p arbitrary graphs $G_1, G_2, \ldots, G_p$, the* Cartesian product *of $G_1, G_2, \ldots, G_p$, denoted by $G_1 \times G_2 \times \cdots \times G_p$, is the graph with vertex set $V(G_1) \times V(G_2) \times \ldots \times V(G_p)$. Two vertices $(u_1, u_2, \ldots, u_p)$ and $(v_1, v_2, \ldots, v_p)$ are adjacent in the product graph if and only if for exactly one i, $1 \leq i \leq p$, $u_i \neq v_i$, and $(u_i, v_i) \in E(G_i)$.*

Figure 24.4 shows the Cartesian product of two graphs.

24.2.1.2 Basic Algebraic Notions

Let S be a nonempty set with n elements and let 2^S denote the set of all subsets of S. For $X, Y \subseteq S$, let $X \Delta Y$ denote the *symmetric difference* of X and Y, that is, $X \Delta Y := (X \cup Y) \setminus (X \cap Y) (= (X \setminus Y) \cup (Y \setminus X))$. It is easily verified that $(2^S, \Delta)$ is a commutative group with $\emptyset$ as the identity, and the inverse of any X is X itself.

Next, let $\mathbb{Z}_2 = \{0, 1\}$. Let $(\mathbb{Z}_2, \oplus)$ be the group of integers modulo 2, that is, for $X, Y \in \mathbb{Z}_2$,

$$X \oplus Y = \begin{cases} 0, & \text{if } X = Y \\ 1, & \text{if } X \neq Y. \end{cases}$$

Let $\mathbb{Z}_2^n$ denote the n-fold Cartesian product of $\mathbb{Z}_2$. If $X = (x_1, x_2, \ldots, x_n)$ and $Y = (y_1, y_2, \ldots, y_n)$ are two elements of $\mathbb{Z}_2^n$, let $X \oplus Y$ denote the binary string $(x_1 \oplus y_1, x_2 \oplus y_2, \ldots, x_n \oplus y_n)$. It is again easily verified that $(\mathbb{Z}_2^n, \oplus)$ is a commutative group with $(0, 0, \ldots, 0)$ as the identity, and the inverse of any X is X itself.

If $S = \{s_1, s_2, \ldots, s_n\}$ then the function $f : 2^S \to \mathbb{Z}_2^n$ defined by $f(X) = (x_1, x_2, \ldots, x_n)$ where

$$x_i = \begin{cases} 0, & \text{if } s_i \in X \\ 1, & \text{if } s_i \notin X \end{cases}$$

for all $X \in 2^S$ is a group homomorphism, that is, $f(X \Delta Y) = f(X) \oplus f(Y)$, for all $X, Y \in 2^S$.

24.2.1.3 Cayley Graph

Definition 24.3 *Let Γ be a finite group and S be a set of generators of Γ satisfying the following properties:*

1. *S does not contain the identity.*

2. *If $g \in S$, then $g^{-1} \notin S$.*

The Cayley graph $G(\Gamma, S)$ is the undirected graph with vertex set consisting of the elements of Γ. Two vertices g, h are joined by an edge iff $g^{-1}h \in S$, and it is labeled $g^{-1}h$.

Every vertex in Cayley graph $G(\Gamma, S)$ has degree $|S|$. Many well-known graphs can be shown to be isomorphic with Cayley graphs by choosing appropriate group Γ and the set of generators S. For more such properties and importance of Cayley graphs we refer to Chapter 25. We are now ready to give alternative definitions of hypercubes.

Proposition 24.1

1. Let S be a nonempty set with n elements and let G be the graph with vertex set 2^S, where two vertices X and Y are adjacent if and only if $|X \Delta Y| = 1$. Then $G \cong Q_n$.

2. $Q_n \cong \underbrace{K_2 \times K_2 \times \cdots \times K_2}_{n \text{ times}}$.

3. For $1 \leq k \leq n-1$, $Q_n \cong Q_k \times Q_{n-k}$; in particular, $Q_n \cong K_2 \times Q_{n-1}$.

4. Q_n is the Cayley graph $G(\mathbb{Z}_2^n, S)$ of the group $(\mathbb{Z}_2^n, \oplus)$ with $S = \{10\ldots0, 010\ldots0, \ldots, 00\ldots01\}$ as the generating set.

Proof.

1. The map f, defined above in the introductory remarks, is a graph isomorphism too, since $|X \Delta Y| = 1$ if and only if $f(X)$ and $f(Y)$ differ in exactly one position.

2. Label the vertices of K_2 with $0, 1 \in \mathbb{Z}_2$. Then $V(K_2 \times K_2 \times \cdots \times K_2) = \mathbb{Z}_2^n$, and by the definition of the Cartesian product of graphs, two vertices $(x_1, x_2, \ldots, x_n)$ and $(y_1, y_2, \ldots, y_n)$ are adjacent in $K_2 \times K_2 \times \cdots \times K_2$ if and only if for exactly one i $(1 \leq i \leq n)$ (i) $x_i \neq y_i$ and (ii) (x_i, y_i) is an edge in the ith copy K_2 (a redundant demand). Hence (2) holds.

3. A consequence of (2).

4. By definition, $V(Q_n) = \mathbb{Z}_2^n$. Let $X, Y \in V(Q_n)$. Suppose that X and Y are adjacent in Q_n where $Y = X_i$, for some i, $1 \leq i \leq n$. So, $X \oplus Y^{-1} = X \oplus Y = (0\ldots010\ldots0)$. So, $X \oplus Y^{-1} \in S$. Conversely, suppose $X \oplus Y^{-1} \in S$. Then $X \oplus Y^{-1} = X \oplus Y = (0\ldots010\ldots0)$. Hence $Y = X_i$. So, X and Y are adjacent in Q_n. ∎

In view of Proposition 24.1(1), the vertices of Q_n may be labeled by the subsets of an n-set and work with subsets rather than binary strings, for example, see Figure 24.5.

Proposition 24.1(3), is a useful recursive construction of Q_n. For $n \geq 2$, Q_n is constructed from Q_{n-1} in three steps as follows:

a. Take two copies of Q_{n-1}, say Q_{n-1}^0 and Q_{n-1}^1.

b. Prefix each vertex of Q_{n-1}^0 with 0 and each vertex of Q_{n-1}^1 by 1.

c. Join $0X$ of Q_{n-1}^0 and $1X$ of Q_{n-1}^1 for every $X \in Q_{n-1}$.

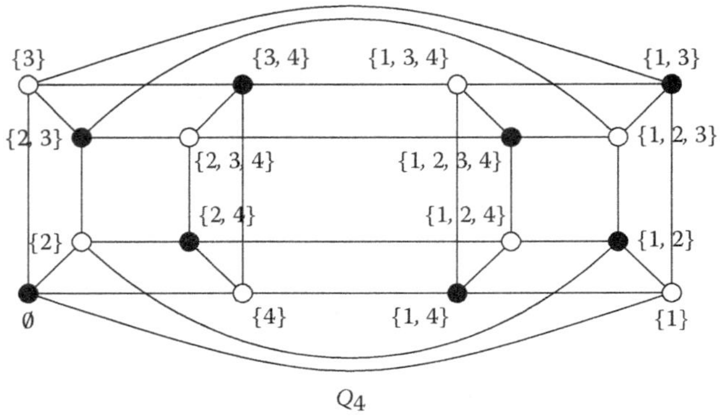

594 ■ Handbook of Graph Theory, Combinatorial Optimization, and Algorithms

Figure 24.5 Hypercube of dimension four with vertices labeled by the subsets of $\{1, 2, 3, 4\}$.

Remark 24.1 The hypercubes Q_2, Q_3, and Q_4 shown in Figure 24.3 are constructed following this procedure. Moreover, this construction is reversible, in the sense that there is a perfect matching of 2^{n-1} edges in Q_n whose deletion results in a graph with two components, each isomorphic with Q_{n-1}. This operation of splitting is called a *canonical decomposition of Q_n* and is denoted by $Q_n = Q_{n-1}^0 \ominus Q_{n-1}^1$. Splitting and the recursive construction is often used in proving the structural properties of Q_n by induction on n.

Further, in view of Proposition 24.1(3), Q_n can be recursively constructed by adding a 0 and a 1 at any position k in the corresponding copies of Q_{n-1}. We denote this construction by $Q_n = Q_{n-1}^{0,k} \ominus Q_{n-1}^{1,k}$.

In a graph G, let $d(X, Y; G)$ denotes the distance between two vertices X and Y in G and let $\text{diam}(G)$ denote the diameter of G.

24.2.2 Basic Properties of Hypercubes

Proposition 24.2 *The hypercube Q_n has the following properties:*

1. $|V(Q_n)| = 2^n$.

2. Q_n is n-regular.

3. $|E(Q_n)| = n2^{n-1}$.

4. Q_n is connected.

5. Q_n is bipartite where each part contains exactly 2^{n-1} vertices.

6. For any i, $1 \leq i \leq n$, no two edges of dimension i are adjacent; hence they are called *parallel edges*.

7. For any i, $1 \leq i \leq n$, there are 2^{n-1} edges of dimension i.

Proof. The statement (1) is obvious, and (3) is a consequence of (1) and (2). So, we prove the remaining assertions.
2: For any vertex X, $N(X; Q_n) = \{X_i : 1 \leq i \leq n\}$. Therefore, $deg(X) = |N(X; Q_n)| = n$.
4: Let X and Y be any two vertices in Q_n, where $Y = X_{i_1, i_2, \ldots, i_k}$. Then

$$P = \langle X, X_{i_1}, X_{i_1, i_2}, \ldots, X_{i_1, i_2, \ldots, i_k} = Y \rangle$$

is an (X, Y)-path in Q_n.

5: Define $S = \{X \in V(Q_n) :$ the number of 1's in X is even$\}$ and $T = \{X \in V(Q_n) :$ the number of 1's in X is odd$\}$. Clearly, no two vertices in S (and T) are adjacent and hence (S,T) is a bipartition of Q_n. Further, it can be easily seen that $|S| = |T| = 2^{n-1}$.

6: If (X, X_i) and (Y, Y_i) are two distinct edges of dimension i, then their end-vertices either differ in at least two positions or the given edges are connected by an edge of dimension not equal to i.

7: For any fixed i, $1 \leq i \leq n$, $\{(x_1 \ldots x_{i-1}\ 0\ x_{i+1} \ldots x_n, x_1 \ldots x_{i-1}\ 1\ x_{i+1} \ldots x_n) : \{x_j \in \{0,1\}, 1 \leq j \neq i \leq n\}$ is the set of dimension i edges. This set contains exactly 2^{n-1} edges. ∎

Remark 24.2 The path P described in the proof of (4) is a shortest (X,Y)-path, since any two successive vertices in any (X,Y)-path differ in exactly one position and so contains at least $k-1$ internal vertices.

Definition 24.4 *Let U, V be binary strings of equal length. The* Hamming distance *$H(U,V)$ between U and V is the number of positions at which U and V differ.*

Proposition 24.3 *The following distance properties hold.*

1. *For any $X, Y \in V(Q_n)$, $d(X, Y; Q_n) = H(X, Y)$,*

2. *$diam(Q_n) = n$.*

Proof.

1. Let k be the number of positions at which X and Y differ. Since the path P described in the proof of Proposition 24.2(4), is a shortest (X,Y)-path, we have
$$d(X, Y; Q_n) = k = H(X, Y)$$

2. By (1), for any two vertices $X, Y \in V(Q_n)$, $d(X, Y; Q_n) = H(X, Y) \leq n$. Therefore, $diam(Q_n) \leq n$. Since $d(X, \overline{X}; Q_n) = H(X, \overline{X}) = n$, we deduce that $diam(Q_n) = n$. ∎

Proposition 24.4

1. *In Q_n, any two adjacent edges belong to exactly one cycle of length four.*

2. *$K_{2,3}$ is not a subgraph of Q_n.*

Proof.

1. Let (X, Y) and (Y, Z) be two adjacent edges in Q_n. Then $Y = X_i$ for some i, $1 \leq i \leq n$ and $Z = Y_j = X_{i,j}$, $1 \leq j \leq n$, $i \neq j$. If A denotes X_j, then $(X, A), (Z, A) \in E(Q_n)$ and hence $C : \langle X, Y, Z, A, X \rangle$ is a C_4 containing (X,Y) and (Y,Z) in Q_n (Figure 24.6).

   Next, if possible, let $C^* : \langle X, Y, Z, B, X \rangle$ be any other C_4 containing (X, Y) and (Y, Z) in Q_n. Now, since $(X, B) \in E(Q_n)$ and $B \neq Y, A$ we have,
$$B = X_p, \text{ for some } p, 1 \leq p \leq n, \text{ where } p \neq i, j. \tag{24.1}$$

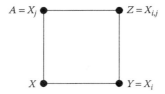

Figure 24.6 Illustration of existence of C_4 containing two adjacent edges in Q_n.

Since $(Z, B) \in E(Q_n)$ and $B \neq Y, A$, we deduce that, $B = Z_k = X_{i,j,k}$ where i, j, k are distinct integers. This is a contradiction to Equation (24.1).

2. A consequence of (1). ∎

Proposition 24.5 [10] *Let X, Y be any two adjacent vertices in Q_n. Then, (i) $|N(X; Q_n)| = |N(Y; Q_n)|$, (ii) every vertex of $N(X; Q_n)$ is adjacent to exactly one vertex of $N(Y; Q_n)$, and (iii) every vertex of $N(Y; Q_n)$ is adjacent to exactly one vertex of $N(X; Q_n)$.*

Proof. Let $Y = x_1 x_2 \ldots x_{k-1} \overline{x}_k x_{k+1} \ldots x_n$; w.l.g. let $x_k = 0$. Then, $X \in Q_{n-1}^{0,k}$ and $Y \in Q_{n-1}^{1,k}$. Further, $N(X; Q_n) \setminus \{Y\} \subseteq Q_{n-1}^{0,k}$ and $N(Y; Q_n) \setminus \{X\} \subseteq Q_{n-1}^{1,k}$. The vertices of $N(X; Q_n)$ and $N(Y; Q_n)$ can be put in a one-to-one correspondence as required, by mapping X_i to Y_i, for $1 \leq i \neq k \leq n$. ∎

Proposition 24.6 *Let X, Y be any two vertices of Q_n. Then there exist $d(X, Y)!$ paths of length $d(X, Y)$ (shortest paths/geodesics) between X and Y in Q_n.*

Proof. Let $Y = X_{i_1, i_2, \ldots, i_k}$, where $k = d(X, Y)$. Let Π_j be a permutation of $\{i_1, i_2, \ldots, i_k\}$, $1 \leq j \leq k!$. The path $P_j = \langle X, X_{\Pi_j(i_1)}, X_{\Pi_j(i_1), \Pi_j(i_2)}, \ldots, X_{\Pi_j(i_1), \Pi_j(i_2), \ldots, \Pi_j(i_k)} = Y \rangle$ is an (X, Y)-path of length k. ∎

Definition 24.5 *In a graph G, two (x, y)-paths are said to be parallel if they have no common internal vertices.*

In a multiprocessor computing system, computing involves exchange of data among several of its processors. The data is transmitted from one processor to another through a sequence of interlinked processors. Obviously, the transmission is faster if there are very few intermediate processors and a large number of alternative parallel paths are available. Equivalently, the demand is that in the graph of the interconnection network, the distance between two given vertices be small and there be a large number of parallel paths connecting any two vertices. Hypercube has these desirable properties.

Proposition 24.7 [10] *Let X, Y be any two vertices of Q_n. Then there exist $d(X, Y)$ parallel paths of length $d(X, Y)$ between X and Y in Q_n.*

Proof. Let $d(X, Y) = k$ and $Y = X_{i_1, i_2, \ldots, i_k}$. The following are k parallel (X, Y)-paths each of length k.

$$P_1 : \langle X, X_{i_1}, X_{i_1, i_2}, \ldots, X_{i_1, i_2, \ldots, i_k} = Y \rangle$$
$$P_2 : \langle X, X_{i_2}, X_{i_2, i_3}, \ldots, X_{i_2, i_3, \ldots, i_k}, X_{i_2, i_3, \ldots, i_k, i_1} = Y \rangle$$
$$P_3 : \langle X, X_{i_3}, X_{i_3, i_4}, \ldots, X_{i_3, i_4, \ldots, i_k}, X_{i_3, i_4, \ldots, i_k, i_1}, X_{i_3, i_4, \ldots, i_k, i_1, i_2} = Y \rangle$$
$$\vdots$$
$$P_k : \langle X, X_{i_k}, X_{i_k, i_1}, X_{i_k, i_1, i_2}, \ldots, X_{i_k, i_1, i_2, \ldots, i_{k-1}} = Y \rangle$$
∎

Proposition 24.8 [10] *Let X, Y be any two vertices of Q_n. Then there exist n parallel paths between X and Y, each of length at most $d(X, Y) + 2$.*

Proof. If $d(X, Y) = n$, the result follows by Proposition 24.7. Next let $d(X, Y) = k < n$, and let $Y = X_{i_1, i_2, \ldots, i_k}$. Let $\{1, 2, \ldots, n\} \setminus \{i_1, i_2, \ldots, i_k\} = \{j_1, j_2, \ldots, j_{n-k}\}$. That is, $j_1, j_2, \ldots, j_{n-k}$

are the positions at which X and Y coincide. In addition to the k vertex disjoint paths described in Proposition 24.7, we have the following $n-k$ parallel paths between X and Y.

$$P_{k+1}: \langle X, X_{j_1}, X_{j_1,i_1}, X_{j_1,i_1,i_2}, \ldots, X_{j_1,i_1,i_2,\ldots,i_k}, X_{i_1,i_2,\ldots,i_k} = Y \rangle$$
$$P_{k+2}: \langle X, X_{j_2}, X_{j_2,i_2}, X_{j_2,i_2,i_3}, \ldots, X_{j_2,i_2,i_3,\ldots,i_k}, X_{j_2,i_2,i_3,\ldots,i_k,i_1}, X_{i_2,i_3,\ldots,i_k,i_1} = Y \rangle$$
$$P_{k+3}: \langle X, X_{j_3}, X_{j_3,i_3}, X_{j_3,i_3,i_4}, \ldots, X_{j_3,i_3,i_4,\ldots,i_k}, X_{j_3,i_3,i_4,\ldots,i_k,i_1}, X_{j_3,i_3,i_4,\ldots,i_k,i_1,i_2},$$
$$X_{i_3,i_4,\ldots,i_k,i_1,i_2} = Y \rangle$$
$$\vdots$$
$$P_{k+n-k}: \langle X, X_{j_{n-k}}, X_{j_{n-k},i_k}, X_{j_{n-k},i_k,i_1}, X_{j_{n-k},i_k,i_1,i_2}, \ldots, X_{j_{n-k},i_k,i_1,i_2,\ldots,i_{k-1}},$$
$$X_{i_k,i_1,i_2,\ldots,i_{k-1}} = Y \rangle$$

■

Proposition 24.9 *The vertex connectivity of Q_n is n.*

Proof. Follows by Menger's theorem. ■

Definition 24.6 *A graph G is said to be*

1. Vertex-symmetric *if given any two vertices u, v there exists an automorphism f of G such that $f(u) = v$,*

2. Edge-symmetric *if given any two edges (u, v), (x, y) there exists an automorphism f of G such that $f(u) = x$ and $f(v) = y$,*

3. P_n-symmetric *if give any two paths $\langle u_1, u_2, \ldots, u_n \rangle$ and $\langle v_1, v_2, \ldots, v_n \rangle$, there exists an automorphism f of G such that $f(u_i) = v_i$, $1 \le i \le n$,*

4. C_n-symmetric *if given any two cycles $\langle u_1, u_2, \ldots, u_n, u_1 \rangle$ and $\langle v_1, v_2, \ldots, v_n, v_1 \rangle$, there exists an automorphism f of G such that $f(u_i) = v_i$, $1 \le i \le n$,*

5. Distance-symmetric *if given any two pairs of vertices $\{u, v\}$ and $\{x, y\}$ such that $d(u, v) = d(x, y)$, there exists an automorphism f of G such that $f(u) = x$ and $f(v) = y$.*

Proposition 24.10 *The following symmetric properties hold for Q_n.*

1. Let X, Y be any two vertices of Q_n and let Π be any permutation of $\{1, 2, \ldots, n\}$. Then there exists an automorphism f of Q_n such that

    a. $f(X) = Y$,

    b. $f(X_i) = Y_{\Pi(i)}$, $1 \le i \le n$.

That is, given any two vertices X, Y (not necessarily distinct), there exists an automorphism which maps X onto Y, and maps the neighbors of X onto the neighbors of Y, with an a priori given ordering of the neighbors.

2. Q_n is vertex-symmetric.

3. Q_n is edge-symmetric.

4. Q_n is P_3-symmetric.

5. Q_n is C_4-symmetric.

6. Q_n is distance-symmetric.

Proof.
(1): Define $f_{X,Y,\Pi} = f : V(Q_n) \to V(Q_n)$ by $f(A) = Y_{\Pi(i_1),\Pi(i_2),...,\Pi(i_k)}$ if $A = X_{i_1,i_2,...,i_k}$. It is easily verified that f is an automorphism satisfying the required properties.
(2), (3), (4): Follow by (1).
(5): Follows by (4) and Proposition 24.4.
(6): Let X, Y, U, V be vertices with $d(X, Y) = k = d(U, V)$. Let $Y = X_{i_1,i_2,...,i_k}$ and $V = U_{j_1,j_2,...,j_k}$. Define the permutation $\Pi : \{1, 2, ..., n\} \to \{1, 2, ..., n\}$ by $\Pi(i_l) = j_l, 1 \le l \le k$ and $\Pi(m) = m, m \in \{1, 2, ..., n\} \setminus \{i_1, i_2, ..., i_k\}$. With this Π, the map $f = f_{X,U,\Pi}$ defined in (1) is an automorphism of Q_n such that $f(X) = U$ and $f(Y) = V$. ∎

Despite the properties (1) to (6), Q_n is not P_m-symmetric for $m > 3$. This follows as there can be no automorphism which maps a path $\langle A, B, C, ..., D \rangle$ with $d(A, D) = m(> 3)$ onto a path $\langle U, V, X, ..., Y \rangle$ with $d(U, Y) = 1$.

The level decomposition of hypercubes has been fruitfully used by many researchers to characterize hypercubes and derive their properties.

Definition 24.7 *Let G be a simple connected graph with $diam(G)=k$ and let $x \in V(G)$. The level decomposition of G with respect to x is the partition $(N_0(x), N_1(x), ..., N_k(x))$ of $V(G)$ where $N_i(x) = \{y \in V(G) : d(x,y) = i\}, 0 \le i \le k$. The vertex sets $N_i(x)$ are called the levels.*

Readers familiar with search algorithms will immediately recognize that $N_0(x), N_1(x), ..., N_k(x)$ are the sets generated by the breadth search algorithm.

We now list some of the properties of such a level decomposition. Often, we will be using these properties without actually referring to the list.

Proposition 24.11 *For any vertex x, the level decomposition of G with respect to x has the following properties.*

1. *$N_0(x) = \{x\}$ and $N_1(x) = N(x)$.*

2. *Every vertex in $N_{i+1}(x)$ is adjacent with some vertex in $N_i(x)$, $0 \le i \le k-1$.*

3. *No vertex in $N_{i+2}(x)$ is adjacent with a vertex in $N_i(x)$, $0 \le i \le k-2$.*

4. *Two vertices in $N_i(x)$ may or may not be adjacent, $0 \le i \le k$.*

5. *G is bipartite if and only if no two vertices in $N_i(x)$ are adjacent, for every i, $0 \le i \le k$.*

6. *G is distance-symmetric if and only if it is vertex symmetric and every automorphism of G which fixes x permutes the elements of $N_i(x)$, for every i, $1 \le i \le k$.* ∎

The level decomposition of a hypercube has a few more additional properties. Here, for a subset of vertices U of the graph G, we denote the subgraph of G induced by the vertices in U by $[U]$.

Proposition 24.12 *For any $X \in V(Q_n)$ and $0 \le i \le n$, the following hold:*

1. *No two vertices in $[N_i(X)]$ are adjacent.*

2. *Every vertex $Y \in N_i(X)$ is adjacent with exactly i vertices in $N_{i-1}(X)$.*

3. *Every vertex $Y \in N_i(X)$ is adjacent with exactly $n-i$ vertices in $N_{i+1}(X)$.*

4. *Any two vertices $Y, Z \in N_i(X)$ are mutually adjacent with at most one vertex $N_{i-1}(X)$.*

5. Any two vertices $Y, Z \in N_i(X)$ are mutually adjacent with at most one vertex in $N_{i+1}(X)$.

6. $|N_i(X)| = \binom{n}{i}$.

7. Every 4-cycle in Q_n intersects exactly three levels.

Proof.
(1): Follows since Q_n is bipartite.
(2): Let $Y = X_{j_1, j_2, \ldots, j_i} \in N_i(X)$. Then

$$N(Y) \cap N_{i-1}(X) = \{X_{j_1, j_2, \ldots, j_{i-1}}, X_{j_1, j_2, \ldots, j_{i-2}, j_i}, X_{j_1, j_2, \ldots, j_{i-3}, j_{i-1}, j_i}, \ldots, X_{j_2, j_3, \ldots, j_i}\}.$$

(3): A consequence of (1),(2) and that Q_n is n-regular.
(4), (5): Let $Y, Z \in N_i(X)$ be adjacent to a vertex in $N_{i-1}(X)$. Then $d(Y, Z) = 2$. W.l.o.g, let $Y = X_{j_1, j_2, \ldots, j_i}$ and $Z = X_{j_1, j_2, \ldots, j_{i-1}, j_{i+1}}$. By Proposition 24.4, Y, Z are mutually adjacent with exactly two vertices, namely $X_{j_1, j_2, \ldots, j_{i-1}}$ and $X_{j_1, j_2, \ldots, j_i, j_{i+1}}$. Among these two vertices, $X_{j_1, j_2, \ldots, j_{i-1}}$ lies in $N_{i-1}(X)$ and $X_{j_1, j_2, \ldots, j_i, j_{i+1}}$ lies in $N_{i+1}(X)$. So if Y, Z have a common neighbor then they are mutually adjacent with exactly one vertex in $N_{i-1}(X)$ and with exactly one vertex in $N_{i+1}(X)$.
(6): $N_i(X) = \{Y \in V(Q_n) : Y = X_{j_1, j_2, \ldots, j_i}, \text{ where} \{j_1, j_2, \ldots, j_i\} \subseteq \{1, 2, \ldots, n\}\}$. Hence $|N_i(X)| = \binom{n}{i}$.
(7): A restatement of (4) and (5). ■

24.2.3 Characterizations of Hypercubes

In the previous section, we observed many properties of hypercubes on bipartition, existence of short parallel paths, level decomposition, absence of $K_{2,3}$ and symmetry. So, all these properties are necessary conditions for a graph to be a hypercube. Over the years, many characterizations have been proved which show that a combination of these necessary conditions and a few more additional conditions are sufficient for an arbitrary graph to be a hypercube. In this section, we present some of these characterizations, some with proofs and some without proofs. We have classified these characterizations into four broad types.

24.2.3.1 Characterizations through Splitting

Theorem 24.1 [10] *A graph G is a hypercube if and only if*

1. $|V(G)| = 2^n$, for some n,

2. G is n-regular,

3. G is connected, and

4. Any two adjacent vertices x and y of G are such that (i) $|N(x; G)| = |N(y; G)|$, (ii) every vertex of $N(x; G) \setminus \{y\}$ is adjacent to exactly one vertex of $N(y; G) \setminus \{x\}$, and (iii) every vertex of $N(y; G) \setminus \{x\}$ is adjacent to exactly one vertex of $N(x; G) \setminus \{y\}$.

Proof. $\Longrightarrow$: Follows by Propositions 24.2 and 24.5.
$\Longleftarrow$: The converse is proved by induction on n. The only graph for $n = 1$ satisfying (1) to (4) is $K_2 \cong Q_1$. Assume that the implication is true for $n - 1$, and let G be a graph satisfying (1) to (4). Consider any two adjacent vertices in G. Color them with red and blue and name them as r and b. By (4), $N(r; G)$ and $N(b; G)$ are linked in a one-to-one fashion (and hence $N(r; G) \cap N(b; G) = \emptyset$). Color the vertices in $N(r; G)$ in red (except b) and the vertices in $N(b; G)$ in blue (except r). Select a new edge whose ends are colored red and blue. As above,

color the neighbors of its ends. Continue this process of selecting a new edge and coloring the neighbors of its ends till all the vertices have been colored. In the colored graph, the vertices and edges have the following properties.

1. All the vertices of G have been colored either blue or red, since G is connected.

2. Since $N(r; G)$ and $N(b; G)$ are linked in a one-to-one fashion, exactly half the vertices have received the color red and the other half have received the color blue.

3. By construction, each vertex of red color is connected to r. So, the red colored vertices induce a connected subgraph say G_r of G. Similarly, let G_b be the connected subgraph of G induced by the blue colored vertices.

4. On the removal of the edges of G which have one end vertex colored red and the other blue, each vertex of G looses a degree and results in a spanning disconnected subgraph of G with components G_r and G_b. The graph G_r satisfies the following properties:

    a. $|V(G_r)| = 2^{n-1}$,
    b. G_r is regular of degree $n-1$,
    c. G_r is connected, and
    d. G_r satisfies (4); for let $(x, y) \in E(G_r)$ and let x_b (or y_b) be the unique vertex of blue color adjacent to x (or y) in G. Then $(x, y) \in E(G) \Rightarrow N(x; G)$ and $N(y; G)$ are linked in a one-to-one fashion $\Rightarrow N(x; G) \setminus \{x_b\} = N(x; G_r)$ and $N(y; G) \setminus \{y_b\} = N(y; G_r)$ are linked in a one-to-one fashion.

All these properties hold for G_b as well. So, by induction hypothesis, $G_r \cong Q_{n-1} \cong G_b$. Hence $G = G_r \ominus G_b \cong Q_n$. ∎

24.2.3.2 Characterizations through (0,2)-Graphs

The class of $(0,2)$-graphs is a subclass of strongly regular graphs widely studied in the theory of combinatorial designs. Characterizations of hypercubes contained in this section state that a (0,2)-graph satisfying a few more properties is a hypercube.

Definition 24.8 *A (0,2)-graph is a connected graph in which any two vertices have zero or exactly two common neighbors.*

For example, while C_4 is a $(0,2)$-graph, $K_4 - e$ is not a $(0,2)$-graph.

Proposition 24.13 *A graph G is a $(0,2)$-graph if and only if any two adjacent edges in G belong to exactly one cycle of length four.*

Proof. $\Rightarrow$: Assume that G is a $(0,2)$-graph and let (x, y), (y, z) be two adjacent edges. Since G is a $(0,2)$-graph and vertices x, z have y as a common neighbor, there exists exactly one more vertex, say a which is adjacent to both x and z. Thus $\langle x, y, z, a, x \rangle$ is a unique 4-cycle containing the edges (x, y) and (y, z).

$\Leftarrow$: Let vertices x, z of G have a common neighbor, say y. Then $(x, y), (y, z)$ being adjacent edges of G lie in a unique 4-cycle. Let a be the vertex which completes the unique 4-cycle, containing (x, y) and (y, z). Then $N(x) \cap N(z) = \{y, a\}$ and hence G is a $(0, 2)$-graph. ∎

Proposition 24.14 *Every $(0, 2)$-graph is regular.*

Proof. Let G be a $(0, 2)$-graph. We first show that if x and y are two adjacent vertices in G then $deg(x) = deg(y)$.

Let a be a vertex in $N(x) \setminus \{y\}$. The vertices a and y have a common neighbor x. Since G is a $(0,2)$-graph, there is a unique vertex $b \in N(y) \setminus \{x\}$ which is adjacent with a. Hence $|N(y) \setminus \{x\}| \geq |N(x) \setminus \{y\}|$. Similarly, the reverse inequality holds. So, $deg(x) = deg(y)$.

Next, if x and y are nonadjacent vertices, let $\langle x = x_1, x_2, \ldots, x_p = y \rangle$ be an (x,y)-path. Then $deg(x_1) = deg(x_2) = \cdots = deg(x_p)$. ∎

Proposition 24.15 *For $n \geq 2$, Q_n is a $(0,2)$-graph.*

Proof. A consequence of Propositions 24.4 and 24.13. ∎

Theorem 24.2 *A graph G is a hypercube if and only if*

1. *G is connected,*
2. *G is a $(0,2)$-graph (with degree of regularity, say, n),*
3. *For some $x \in V(G)$ and for every $i \geq 0$,*
    a. *No two vertices in $N_i(x)$ are adjacent.*
    b. *every vertex in $N_i(x)$ is adjacent with exactly i vertices in $N_{i-1}(x)$.*

Proof. $\Longrightarrow$: Follows by Propositions 24.2, 24.12, and 24.15.
$\Longleftarrow$: In view of Theorem 24.1, it is enough if we show that $|V(G)| = 2^n$ and that Theorem 24.1(4) holds.

At the outset, observe that G is bipartite, since no two vertices in $N_i(x)$ are adjacent. Moreover, every vertex in $N_i(x)$ is adjacent with exactly $n - i$ vertices in $N_{i+1}(x)$, since G is n-regular, and it satisfies (b).

Claim 1 $|N_i(x)| = \binom{n}{i}$

We prove the claim by induction on i. Since $|N_0(x)| = |\{x\}| = 1$, the claim holds for $i = 0$. Assuming that the claim is true for i, we prove it for $i + 1$. Let F be the set of edges in $[N_i(x), N_{i+1}(x)]$. We count the number of edges emanating from $N_i(x)$ to $N_{i+1}(x)$ to obtain

$$|F| = |N_i(x)|(n - i).$$

Similarly, we count the number of edges emanating from $N_{i+1}(x)$ to $N_i(x)$, to obtain

$$|F| = |N_{i+1}(x)|(i + 1).$$

Combining the above two equations, we get

$$|N_{i+1}(x)| = |N_i(x)| \frac{n - i}{i + 1}$$
$$= \binom{n}{i} \frac{n - i}{i + 1} \quad \text{(using induction hypothesis)}$$
$$= \binom{n}{i + 1}$$

We have completed the proof of Claim 1. It implies that

a. $|V(G)| = \sum_{i \geq 0} |N_i(x; G)| = \sum_{i=0}^{n} \binom{n}{i} = 2^n$.
b. The diameter of G is n, since $N_{n+1} = \emptyset$ and $N_n \neq \emptyset$.

Claim 2 *G satisfies Theorem 24.1(4).*

Let y, z be two adjacent vertices of G. Since G is bipartite, y, z lie in two consecutive levels, say $N_i(x)$ and $N_{i-1}(x)$, respectively. Let $\{z_1, z_2, \ldots, z_{i-1}\}$ be the neighbors of z in N_{i-2}. Since

G is a $(0,2)$-graph and z is a common neighbor of y and z_j, $1 \leq j \leq i-1$, there exists y_j (say) in N_{i-1} adjacent with y and z_j. By Theorem 24.2(3a) it follows that $z, y_1, y_2, \ldots, y_{i-1}$ are the only neighbors of y in $N_{i-1}(x)$. Let $y_{i+1}, y_{i+2}, \ldots, y_n$ be the neighbors of y in $N_{i+1}(x)$, and $y, z_{i+1}, z_{i+2}, \ldots, z_n$ be the neighbors of z in $N_i(x)$. Since y is a common neighbor of z and y_k, $i+1 \leq k \leq n$, we conclude as before that z_k and y_k are adjacent. Hence the claim. So, G is a hypercube by Theorem 24.1. ∎

In the following we list other characterizations which involve $(0,2)$-graphs. We omit the proofs.

Theorem 24.3 [11] *A graph G is a hypercube if and only if*

1. *G is connected (with diameter n, say),*

2. *G is bipartite, and*

3. *for any two vertices x and y in G, the number of shortest (x,y)-paths is $d(x,y)$!* ∎

Theorem 24.4 [12] *A graph G is a hypercube if and only if*

1. *G is connected (with $\delta(G) = n$ say),*

2. *every pair of adjacent edges lies in exactly one 4-cycle, and*

3. *$|V(G)| = 2^n$.* ∎

Proposition 24.16 [13] *A graph G is a hypercube if and only if*

1. *G is connected,*

2. *G is a $(0,2)$-graph (with degree of regularity n, say), and*

3. *$|V(G)| = 2^n$.* ∎

Proposition 24.17 [13] *A graph G is a hypercube if and only if*

1. *G is connected (with diameter n, say),*

2. *G is a $(0,2)$ graph, and*

3. *G has a level decomposition such that every 4-cycle intersects exactly three levels.* ∎

24.2.3.3 Characterizations through Intervals

In this section, characterizations do not forbid $K_{2,3}$ but add a few distance properties.

Definition 24.9 *For any two vertices x, y of a graph G, the* interval *between x and y is the set*
$$I_G(x,y) := \{z \in V(G) : \text{there exists a shortest } (x,y)\text{-path containing } z\}.$$
Clearly,

a. If x, y are connected, then $I_G(x,y) = \{z \in V(G) : d(x,z) + d(z,y) = d(x,y)\}$, and $x, y \in I_G(x,y)$. In particular, if $x = y$ then $|I_G(x,y)| = 1$.

b. If x, y are not connected then $I_G(x,y) = \emptyset$.

Proposition 24.18 *For $X, Y \in V(Q_n)$,*
$$|N(X) \cap I(X,Y)| = d(X,Y) = |N(Y) \cap I(X,Y)|.$$

Proof. See Proposition 24.7 and its proof. ∎

Theorem 24.5 [14] *A graph G is a hypercube if and only if*

1. G is connected (with diameter n, say),

2. G is bipartite,

3. G is a $(0,2)$-graph, and

4. If θ and $\hat{\theta}$ are two diametrical vertices of G, then for all $u \in V(G)$
$$|N(u) \cap I(\theta, u)| = d(\theta, u). \tag{24.2}$$

and
$$|N(u) \cap I(u, \hat{\theta})| = d(u, \hat{\theta}). \tag{24.3}$$

Proof. $\Longrightarrow$: Follows by Propositions 24.2, 24.15, and 24.18.
$\Longleftarrow$: We verify that G satisfies the hypothesis of Theorem 24.2.

Claim 1 $N(\theta) \subseteq I(\theta, \hat{\theta})$.

Let $x \in N(\theta)$. Since $d(\theta, \hat{\theta}) = diam(G) = n$, we have, $d(x, \hat{\theta}) = n - 1$ or n.

Case 1 $d(x, \hat{\theta}) = n - 1$.

Let P be a shortest $(x, \hat{\theta})$-path in G. Then the path $\langle \theta, x, P(x, \hat{\theta}) \rangle$ is a shortest $(\theta, \hat{\theta})$-path containing x. Thus, $x \in I(\theta, \hat{\theta})$.

Case 2 $d(x, \hat{\theta}) = n$.

Then $\langle P(\theta, \hat{\theta}), Q(\hat{\theta}, x), \theta \rangle$ is a closed walk of length $2n+1$ in G, where P and Q are shortest paths. But every closed walk of odd length contains an odd cycle, a contradiction to (2). So, this case does not arise.

Claim 2 $deg(u) = n$, for all $u \in V(G)$.

We have
$$deg(\theta) = |N(\theta)| = |N(\theta) \cap I(\theta, \hat{\theta})|, \ (by \ Claim \ 1)$$
$$= d(\theta, \hat{\theta}) \ (by \ substituting \ u = \theta \ in \ (24.3))$$
$$= n.$$

Since every $(0,2)$-graph is a regular graph, we conclude that G is n-regular.

Next consider a level decomposition of G w.r.t. θ. By (24.2), it follows that every vertex v in $N_i(\theta)$ is adjacent with exactly i vertices in $N_{i-1}(\theta)$. So, by (24.3) and Claim 2, every vertex v in $N_i(\theta)$ is adjacent with exactly $n - i$ vertices in $N_{i+1}(\theta)$. We now appeal to Theorem 24.2 and conclude that G is a hypercube. ∎

Theorem 24.6 [11] *A graph G is a hypercube if and only if*

1. G is connected (with diameter n, say),

2. G is bipartite,

3. For all $x, y \in V(G)$, $|N(x) \cap I(x,y)| = d(x,y)$.

Proof. $\Longrightarrow$: Follows by Propositions 24.2 and 24.18.
$\Longleftarrow$: Follows by Theorem 24.5. ■

Next characterizations involve concepts associated with an interval in a graph. As before, we first observe a necessary condition for a graph G to be a hypercube.

Proposition 24.19 *Every interval in a hypercube induces a sub hypercube.*

Proof. Let $X, Y \in V(Q_n)$ and let $Y = X_{i_1, i_2, \ldots, i_k}$. Then $I(X, Y) = \{X_T : T \subseteq \{i_1, i_2, \ldots, i_k\}\}$ and so $[I(X, Y)] \cong Q_k$. ■

Definition 24.10 *Let G be a connected graph and let u, v be vertices. An interval $I(u, v)$ is said to be* end-regular, *if both u and v have $d(u, v)$ neighbors in the subgraph induced by $I(u, v)$. The graph G is said to be* interval-regular *if every interval is end-regular.*

It follows by Proposition 24.19 that hypercubes are interval regular.

Proposition 24.20 *A connected graph G is interval-regular if and only if for any two vertices u and v of G at least one of u and v has $d(u, v)$ neighbors in $I(u, v)$.*

Definition 24.11 *A graph G is called* antipodal *if for every vertex u there exist a vertex u' such that $I(u, u') = V(G)$. The vertex u' is called the* antipode *of u.*

Clearly:

- For every vertex $x \in I(u, v)$, $d(u, x; G) = d(u, x; [I(u, v)])$.
- An antipodal graph is necessarily connected.
- In an antipodal graph, given u, its antipode u' is unique.
- Hypercubes are antipodal.

Proposition 24.21 *If $I(u, v)$ is an interval in an antipodal graph G and $x, x' \in I(u, v)$ are antipodal then $d(u, v) = d(x, x')$.*

Proof. Since $x, x' \in [I(u, v)]$, we have $d(u, x) + d(x, v) = d(u, v)$, and $d(u, x') + d(x', v) = d(u, v)$. Since x, x' are antipodal, we have $[I(x, x')] = [I(u, v)]$, $d(x, u) + d(u, x') = d(x, x')$, and $d(x, v) + d(v, x') = d(x, x')$. These equations imply that $d(u, v) = d(x, x')$. ■

In the following characterizations, the statements 24.7(2) to 24.7(6) are due to Bandelt and Mulder [15], and the statement 24.7(7) is due to Wenzel [16].

Theorem 24.7 *For a connected bipartite graph G, the following statements are equivalent.*

1. *G is a hypercube.*
2. *Every interval in G induces a hypercube.*
3. *G is interval regular.*
4. *Every interval in G induces a $(0, 2)$-graph.*
5. *Every interval $I(u, v)$ in G contains exactly $2^{d(u,v)}$ vertices.*
6. *Every interval $I(u, v)$ in G induces a graph with exactly $d(u, v) 2^{d(u,v)-1}$ edges.*
7. *Every interval in G is antipodal.*

Proof. We have already shown that (1) implies all other statements. So, we prove the reverse implications.

(2) $\Rightarrow$ (1): In view of Theorem 24.3, it is enough if we show that G satisfies 24.3(3). So, let $u, v \in V(G)$. Let H be the hypercube induced by $I(u, v)$. Then,

the number of shortest (u, v)-paths in G = the number of shortest (u, v)-paths in H
$$= d(u, v; H)!, \text{ by Theorem 24.3}$$
$$= d(u, v; G)!$$

(3) $\Leftrightarrow$ (1): This equivalence is Theorem 24.6.

(4) $\Rightarrow$ (3): We would like to use Proposition 24.20. So, we show that for any two vertices u, v of G at least one of u and v has degree $d(u, v)$ in $[I(u, v)]$, by induction on $d(u, v)$. If $d(u, v)=1$, $[I(u, v)] \cong K_2$ and so the claim holds. Next we proceed to the induction step assuming that $d(u, v) = d$. Let w be a neighbor of u in $[I(u, v)]$. Then $d(w, v) = d - 1$. Therefore by induction hypothesis, $deg(w; [I(w, v)]) = d - 1$ or $deg(v; [I(w, v)]) = d - 1$. However, $[I(w, v)]$ is regular being a $(0, 2)$-graph. So, $deg(w; [I(w, v)]) = d - 1 = deg(v; [I(w, v)])$ and hence $deg(w; [I(u, v)]) = d$. Therefore $deg(u; [I(u, v)]) = d$, since $[I(u, v)]$ is regular.

(5) $\Rightarrow$ (3): We show that every interval $[I(u, v)]$ is end-regular by induction on $d(u, v)$. For $d(u, v) \leq 2$, the assertion is obvious. So, let $d(u, v) = d \geq 3$. Consider a level decomposition of $[I(u, v)]$ w.r.t. u and let $x \in N_i(u)$, where $1 \leq i \leq d - 1$; $d(u, x) = i$ and $d(x, v) = d - i$. We apply our induction hypothesis to the intervals $[I(u, x)]$ and $[I(x, v)]$ and infer that x has exactly i neighbors in $N_{i-1}(u)$ and exactly $d - i$ neighbors in $N_{i+1}(u)$. Hence counting the edges between $N_{i-1}(u)$ and $N_i(u)$ in two ways, we get

$$|N_{i-1}(u)|(d - i + 1) = |N_i(u)|i, \quad 1 \leq i \leq d - 1.$$

Putting $i = 1$ in the above equation, we get $|N_0|d = |N_1|$. But $N_0 = \{u\}$ and $|N_1| = deg(u; [I(u, v)])$. Therefore the claim is proved and we conclude that G is interval-regular.

(6) $\Rightarrow$ (3): Proof is exactly as above.

(7) $\Rightarrow$ (3): We prove the implication by induction on $d(u, v) = d$. For $d = 1$, the claim is obvious. We next prove the claim for $d = 2$. Let x' be the antipode of x in $[I(u, v)]$. By Proposition 24.21, $d(x, x') = d(u, v) = 2$. So, $[\{u, v, x, x'\}]$ is a 4-cycle. We next assert that $[I(u, v)]$ is the 4-cycle $\{u, v, x, x'\}$. On the contrary, if $z \in [I(u, v)] - \{u, v, x, x'\}$, we arrive at a contradiction as follows: while $(z, u), (z, v) \in E([I(u, v)])$, $(z, x), (z, x') \notin E([I(u, v)])$ since G is bipartite. But then there is no shortest (x, x')-path containing z in $[I(x, x')]$; so $z \notin I(x, x') = I(u, v)$, a contradiction.

We now proceed to the anchor step. Let x be a neighbor of u in $[I(u, v)]$ and let x' be its antipode. By induction hypothesis $deg(u; [I(u, x')]) = d - 1$ and so $deg(u; [I(x, x')]) = d = deg(u; [I(u, v)])$. ∎

24.2.3.4 Characterizations through Medians

Definition 24.12 *A simple graph G is called a* median graph *if G is connected, and if for any three vertices u, v, and w of G there exists a unique vertex x, called the* median *of u, v, and w, such that*

$$d(u, x; G) + d(x, v; G) = d(u, v; G)$$
$$d(v, x; G) + d(x, w; G) = d(v, w; G)$$
$$d(w, x; G) + d(x, u; G) = d(w, u; G)$$

In other words, a simple connected graph is a median graph if for any three vertices u, v, and w in G, there exists a unique vertex that lies simultaneously on a shortest (u, v)-path, a shortest (v, w)-path, and a shortest (w, u)-path.

It is easily seen that (i) a median graph is bipartite, (ii) every tree is a median graph, (iii) cycles of length greater than 4 are not median, and (iv) $K_{2,3}$ is not median.

Proposition 24.22 *For $n \geq 2$, Q_n is a median graph.*

Proof. Let $U, V, W \in Q_n$, and let $V = U_S$ and $W = U_T$ where $S, T \subseteq \{1, 2, \ldots, n\}$. Then, $d(V, W) = |S \Delta T|$.

Claim $X = U_{S \cap T}$ is the unique median of U, V, W.

Let

$$S \cap T = \{i_1, i_2, \ldots, i_\alpha\},$$
$$S \setminus (S \cap T) = \{j_1, j_2, \ldots, j_\beta\},$$
$$T \setminus (S \cap T) = \{k_1, k_2, \ldots, k_\gamma\}.$$

So, we can write:

$$S = \{i_1, i_2, \ldots, i_\alpha, j_1, j_2, \ldots, j_\beta\}$$
$$T = \{i_1, i_2, \ldots, i_\alpha, k_1, k_2, \ldots, k_\gamma\}.$$

Then, $d(U, X) = \alpha, d(U, V) = \alpha + \beta, d(U, W) = \alpha + \gamma, d(V, W) = \beta + \gamma$. Moreover,

$$\langle U, U_{i_1}, U_{i_1,i_2}, \ldots, U_{i_1,i_2,\ldots,i_\alpha} = X, U_{i_1,i_2,\ldots,i_\alpha,j_1}, U_{i_1,i_2,\ldots,i_\alpha,j_1,j_2}, \ldots, U_{i_1,i_2,\ldots,i_\alpha,j_1,j_2,\ldots,j_\beta} = V \rangle$$

is a shortest (U, V)-path,

$$\langle U, U_{i_1}, U_{i_1,i_2}, \ldots, U_{i_1,i_2,\ldots,i_\alpha} = X, U_{i_1,i_2,\ldots,i_\alpha,k_1}, U_{i_1,i_2,\ldots,i_\alpha,k_1,k_2}, \ldots, U_{i_1,i_2,\ldots,i_\alpha,k_1,k_2,\ldots,k_\gamma} = W \rangle$$

is a shortest (U, W)-path, and $\langle V = U_{i_1,i_2,\ldots,i_\alpha,j_1,j_2,\ldots,j_\beta}, U_{i_1,i_2,\ldots,i_\alpha,j_1,j_2,\ldots,j_{\beta-1}}, \ldots, U_{i_1,i_2,\ldots,i_\alpha,j_1}, U_{i_1,i_2,\ldots,i_\alpha} = X, U_{i_1,i_2,\ldots,i_\alpha,k_1}, U_{i_1,i_2,\ldots,i_\alpha,k_1,k_2}, \ldots, U_{i_1,i_2,\ldots,i_\alpha,k_1,k_2,\ldots,k_\gamma} = W \rangle$ is a shortest (V, W)-path. Clearly, X is unique. ∎

The following theorem characterizes median graphs; please refer [14,17,18] for its proof.

Theorem 24.8 [14] *A graph G is a median graph if and only if G is a connected induced subgraph of a hypercube Q such that for any three vertices of G their median in Q is also a vertex of G.* ∎

Theorem 24.9 *The following statements are equivalent for a graph G.*

1. *G is a hypercube.*

2. *G is median and contains two diametrical vertices at least one of which has maximum degree (say n) [17].*

3. G is median and regular (with regularity say n) [17].

4. G is median and diametrical (with diameter say n) [17]. *(A connected graph G is diametrical if each vertex of G has a unique diametrical vertex.)*

5. G is median and a $(0,2)$-graph [15].

Proof. We only prove (1) $\Leftrightarrow$ (2). The other characterizations easily follow.
$\Longrightarrow$: Follows by Proposition 24.22 and the fact that G is regular.
$\Longleftarrow$: Let G satisfy (2). Let $\theta, \hat{\theta}$ be two diametrical vertices of G with $deg(\theta) = n$. Embed G in an m-cube Q ($m \geq n$) as in Theorem 24.8. Assume that the vertices of Q are labeled by using the subsets of $\{1, 2, \ldots, m\}$ as in Proposition 24.1. Consider a level decomposition $(N_0, N_1, \ldots, N_n)$ of Q with respect to the vertex labeled $\emptyset$. W.l.g, assume that θ of G is the vertex $\emptyset$ of Q and that $\{1\}, \{2\}, \ldots, \{n\}$ are the neighbors of θ. We prove that every subset of $\{1, 2, \ldots, n\}$ is a vertex of G, by induction on the cardinality of the subset. It then follows that $G \cong Q_n \subseteq Q$.

For the inductive anchor, assume that all subsets $A \subseteq \{1, 2, \ldots, n\}$ with $|A| \leq i$ are vertices of G, and that $d(A, \hat{\theta}; G) = diam(G) - |A|$, for all such sets. Note that when $i = 1$, these assumptions are true.

Let A be a $(i+1)$-element subset of $\{1, 2, \ldots, n\}$. Let B, C be two distinct i-element subsets of A. Then

$$B \cup C = A, \quad |B \cap C| = i - 1, \quad B, C \in N_i, \quad B \cap C \in N_{i-1} \quad \text{and } A \in N_{i+1}.$$

By induction hypothesis, $B, C, B \cap C \in V(G)$. Clearly, A is the median of B, C, and $\hat{\theta}$. Therefore, by Theorem 24.8, $A \in G$ and moreover

$$d(A, \hat{\theta}; G) = d(B, \hat{\theta}; G) - 1 = (diam(G) - |B|) - 1 = diam(G) - |A|. \quad \blacksquare$$

24.2.3.5 Characterizations through Projections

As in the last few sections, in this section too, the characterizations depend on the distance properties.

Definition 24.13 *Let G be a simple connected graph and S be a subset of vertices. The* projection *of a vertex x over S is the set*

$$P(x; S) := \{s \in S : d(x, s) \leq d(x, s'), \text{ for every } s' \in S\}.$$

Definition 24.14 *Let G be a simple connected graph and S be a subset of vertices. The* antiprojection *of a vertex x over S is the set*

$$AP(x; S) := \{s \in S : d(x, s) \geq d(x, s'), \text{ for every } s' \in S\}.$$

Proposition 24.23 *For all X, Y, Z in $V(Q_n)$, $|AP(Z; I(X,Y))| = 1$.*

Proof. By Proposition 24.19, $[I(X,Y)] \cong Q_d$, where $d = d(X, Y)$. W.l.g, let $I(X,Y) = \{X \in V(Q_n) : x_j = 0, d+1 \leq j \leq n\}$. Then $Z' = (\overline{z_1}\overline{z_2}\ldots\overline{z_d}00\ldots0)$ is the unique vertex in $AP(Z; I(X,Y))$. $\blacksquare$

Proposition 24.24 *If G is a connected graph such that*

$$|AP(z; I_G(x,y))| = 1, \text{ for all vertices } x, y, \text{ and } z. \tag{24.4}$$

Then

1. *G is bipartite.*

2. *$K_{2,3}$ is not a subgraph of G.*

3. *G is a $(0,2)$-graph.*

4. *For any $x \in V(G)$, every 4-cycle intersects exactly three levels of the level decomposition w. r. t. x.*

Proof.
(1): If $\langle x_1, x_2, \ldots, x_{2p+1}, x_1 \rangle$ is a shortest odd cycle in G, then $d(x_1, x_{p+1}) = p = d(x_1, x_{p+2})$, and $I(x_{p+1}, x_{p+2}) = \{x_{p+1}, x_{p+2}\}$. So, $AP(x_1; I(x_{p+1}, x_{p+2})) = \{x_{p+1}, x_{p+2}\}$, a contradiction to (24.4).
(2): On the contrary suppose $[\{x, y\}, \{a, b, c\}] = K_{2,3} \subseteq G$. Since $d(x, y) = 2$ and G is bipartite, $\{b, c\} \subseteq AP(a; I(x, y))$, a contradiction to (24.4).
(3): Let z be a common neighbor of x, y in G. Then there exists another common neighbor, say r, of x, y; else $I(x, y) = \{x, y, z\}$ and so $AP(z; I(x, y)) = \{x, y\}$, a contradiction to (24.4). By (2), r is unique.
(4): Since G is bipartite by (1), every 4-cycle intersects two or three levels. If possible, let $\langle a, b, c, d, a \rangle$ be a 4-cycle intersecting only two levels, say $N_i(x)$ and $N_{i+1}(x)$. W.l.g let $a \in N_i(x)$. Since G is bipartite, $b, d \in N_{i+1}(x)$ and $c \in N_i(x)$. But then since G is a $(0,2)$-graph by (3), $I(a, c) = \{a, b, c, d\}$ and $AP(x; I(a, c)) = \{b, d\}$, a contradiction to (24.4). ∎

Theorem 24.10 [13] *A connected graph G is a hypercube if and only if $|AP(z; I_G(x,y))| = 1$, for all vertices x, y and z in G.*

Proof. Follows by Propositions 24.23, 24.17, and 24.24. ∎

24.2.3.6 Characterizations through Convex Sets

Definition 24.15 *Let G be a simple connected graph. A set of vertices C is said to be* convex *if for any two vertices x, y of C the interval $I(x, y)$ is contained in C.*

The next proposition reveals a connection between convex sets and intervals in hypercubes. It says that the intervals are the only convex sets in hypercubes. This property and a few more obvious necessary conditions characterize hypercubes.

Proposition 24.25 *A set of vertices C in a hypercube is convex if and only if C is an interval.*

Proof. $\Rightarrow$: Let $X, Y \in C$ be two vertices such that $d(X, Y) = diam([C]) = d$ (say). We claim that $C = I(X, Y)$. Since C is convex, $I(X, Y) \subseteq C$. If there is a $Z \in C \setminus I(X, Y)$, we arrive at a contradiction as follows: By Proposition 24.19, $H := [I(X,Y)] \cong Q_d$. W.l.g, let $V(H) = \{X \in V(Q_n) : x_j = 0, d+1 \leq j \leq n\}$. Then $z_j = 1$ for some j, $d+1 \leq j \leq n$. Now $Z' = (\overline{z_1 z_2} \ldots \overline{z_d} 00 \ldots 0) \in V(H)$ and $d(Z, Z') > d$, a contradiction to the choice of X, Y. Therefore $C = I(X, Y)$.
$\Leftarrow$: Let $I(X, Y)$ be an interval in Q_n. Let $U, V \in I(X, Y)$. By Proposition 24.19, $I(X, Y)$ induces a sub-hypercube Q_d of Q_n of dimension $d = d(X, Y) \leq n$. Again by Proposition 24.19, $I(U, V)$ induces a sub-hypercube in Q_d. ∎

It can be easily verified that if $U, V \in H := [I(X,Y)] \subseteq Q_n$, then $I_H(U, V) = I_{Q_n}(U, V)$.

Proposition 24.26 *For every convex set C and every vertex X in Q_n, $|AP(X;C)| = 1$.*

Proof. Follows by Propositions 24.23 and 24.25. ∎

Proposition 24.27 *If G is a connected graph such that*

$$|AP(x;C)| = 1, for\ every\ convex\ set\ C\ and\ every\ vertex\ x\ in\ G, \qquad (24.5)$$

then G is bipartite.

Proof. If possible, let $\langle x_1, x_2, \ldots, x_{2p+1}, x_1 \rangle$ be a shortest odd cycle in G. Since $(x_{p+1}, x_{p+2}) \in E(G)$, $\{x_{p+1}, x_{p+2}\}$ is a convex set. Moreover, $AP(x_1; x_{p+1}, x_{p+2}) = \{x_{p+1}, x_{p+2}\}$, a contradiction to (24.5). ∎

Theorem 24.11 [13] *A graph G is a hypercube if and only if*

1. *G is connected,*
2. *G does not contain $K_{2,3}$ as an induced subgraph, and*
3. *For every convex set C and every vertex x in G, $|AP(x; C)| = 1$.*

Proof. $\Longrightarrow$: A consequence of Propositions 24.2, 24.4, and 24.26.
$\Longleftarrow$: We verify that G satisfies the conditions of Proposition 24.17, and conclude that G is a hypercube.

Claim 1 G is a $(0,2)$-graph.

Let z be a common neighbor of x, y in G. Then there exists another common neighbor, say r, of x, y; else $\{x, y, z\}$ is a convex set and so $AP(z; \{x, y, z\}) = \{x, y\}$—a contradiction to (24.5). Since $K_{2,3}$ is forbidden in G, r and z are the only common neighbors of x and y. This proves the claim. Consider a level decomposition of G with respect to some $x \in V(G)$.

Claim 2 Every 4-cycle intersects exactly three levels.

Since G satisfies (24.5), it is bipartite and so every 4-cycle intersects two or three levels. If possible, let $\langle a, b, c, d, a \rangle$ be a 4-cycle intersecting only two levels, say $N_i(x)$ and $N_{i+1}(x)$. W.l.o.g let $a \in N_i(x)$. Since G is bipartite, $b, d \in N_{i+1}(x)$ and $c \in N_i(x)$. Also, since G is a $(0,2)$-graph, $I(a,c) = \{a,b,c,d\}$ is a convex set and so $AP(x; \{a,b,c,d\}) = \{b,d\}$, a contradiction to (24.5). ∎

24.2.3.7 Characterizations through Some Monotone Properties

Definition 24.16 *Let G be a simple connected graph. An interval $I(x,y)$ is said to be* closed *if for all $z \in V(G) \setminus I(x,y)$ there exists $z' \in I(x,y)$ such that $d(z,z') > d(x,y)$.*

Remark 24.3 *Let $I(x,y)$ be a closed interval and z be a vertex in G. If for every $u \in I(x,y)$, $d(z,u) \leq d(x,y)$, then $z \in I(x,y)$.*

Definition 24.17 *A simple graph G is said to be* distance monotone *if all its intervals are closed.*

Remark 24.4 *Any distance monotone graph G is connected; for if G is not connected, then it contains an empty interval which is not closed.*

Proposition 24.28 *Every distance monotone graph G is bipartite.*

Proof. If possible, let $\langle x_1, x_2, \ldots, x_{2p+1}, x_1 \rangle$ be a shortest odd cycle in G. Then $d(x_1, x_{p+1}) = p = d(x_1, x_{p+2})$. So, $x_{p+2} \notin I(x_1, x_{p+1})$. Further, for all $u \in I(x_1, x_{p+1})$, we have $d(x_{p+2}, u) \leq p$. Therefore, $I(x_1, x_{p+1})$ is not closed, a contradiction. ∎

Proposition 24.29 *The hypercube Q_n is distance monotone.*

Proof. Let $I(X, Y)$ be an interval in Q_n. If $d(X, Y) = n$, then clearly $I(X, Y)(= V(Q_n))$ is closed. Next assume that $d(X, Y) < n$. By Theorem 24.19, $H := [I(X, Y)] \cong Q_d$ where $d = d(X, Y)$. W.l.o.g, let $V(H) = \{X \in V(Q_n) : x_j = 0, d+1 \leq j \leq n\}$. Let $Z \in V(Q_n) \setminus V(H)$. Then $z_j = 1$ for some j, $d+1 \leq j \leq n$. Now $Z' = (\overline{z_1}\overline{z_2}\ldots\overline{z_d}00\ldots0) \in V(H)$ and further, $d(Z, Z') > d$. ∎

Definition 24.18 *A connected graph G is said to be* interval monotone *if each interval in G is convex.*

Proposition 24.30 *Q_n is interval monotone.*

Proof. Follows by Proposition 24.25. ∎

Proposition 24.31 *An interval monotone graph G does not contain $K_{2,3}$ as an induced subgraph.*

Proof. If possible, let $K_{2,3} \subseteq G$ and let $[\{a, b\}, \{x, y, z\}]$ be the bipartition of $K_{2,3}$. Then $I(x, y)$ is not convex: for $a, b \in I(x, y)$, $z \notin I(x, y)$ whereas $z \in I(a, b)$ and so $I(a, b) \not\subseteq I(x, y)$. ∎

Theorem 24.12 [19] *A connected graph G with minimum degree $\delta(G) = n \geq 3$ is a hypercube if and only if*

1. *G is distance monotone, and*
2. *G is interval monotone.*

Proof. $\Longrightarrow$: Follows by Propositions 24.29 and 24.30.

$\Longleftarrow$: Claim 1: For any four vertices x, a, b, c such that a, b, c are neighbors of x, there exists $y \in V(G)$ such that y is adjacent to both a and b but not to c.

Since G is bipartite (being a distance monotone graph), $d(a, b) = d(a, c) = d(b, c) = 2$. Therefore, $c \notin I_G(a, b)$. Since G is distance monotone, $I(a, b)$ is closed. So, there exists $y \in I(a, b)$ such that $d(c, y) > d(a, b) = 2$. Now, since $y \in I(a, b)$ and $d(a, b) = 2$, we deduce that $(a, y), (b, y) \in E(G)$. Further, $d(c, y) > 2$ and so $(c, y) \notin E(G)$. The claim is proved.

Using this claim, we next show that for any two vertices u, v, the interval $I(u, v)$ induces a $(0, 2)$-graph, so that G is a hypercube by Theorem 24.7. Let G' denote the subgraph induced by $I(u, v)$.

If $d(u, v; G') = 1$, then $G' \cong K_2$ with $V(G') = \{u, v\}$ and so it is a $(0, 2)$-graph. Next, let $d(u, v) \geq 2$. Let $a, b \in V(G')$ have a common neighbor x in G'. Since $deg_G(x) \geq 3$, there exists $c \in V(G) \setminus \{a,b\}$ adjacent to x. Then, by Claim 1, there exists $y \in V(G)$ such that $(a,y), (b,y) \in E(G)$ and $(c,y) \notin E(G)$. Therefore $y \in I_G(a,b)$. Since G' is convex (G being interval monotone), $I_G(a,b) \subseteq V(G')$. So, $y \in V(G')$. Therefore, x, y are common neighbors of a, b in G'. These are the only common neighbors since $K_{2,3}$ is forbidden in G and hence in G'. That is, G' is a $(0,2)$-graph. ∎

Definition 24.19 *A simple connected graph G is said to be* interval distance monotone *if for any two vertices x, y in G, the interval $I(x,y)$ induces a distance monotone graph.*

Proposition 24.32 *Q_n is an interval distance monotone graph.*

Proof. Follows by Propositions 24.19 and 24.29. ∎

Theorem 24.13 [20] *A graph G with minimum degree $\delta(G) \geq 3$ is a hypercube if and only if*

1. *G is distance monotone, and*
2. *G is interval distance monotone.*

Proof. $\Longrightarrow$: Follows by Propositions 24.29 and 24.32.
$\Longleftarrow$: Our aim is to show that every interval in G is antipodal and then appeal to Theorem 24.7. Let $u, v \in V(G)$.

Claim 1 *For any $w \in I(u,v)$, there exists a $\tilde{w} \in I(u,v)$ such that $d(w, \tilde{w}) = d(u,v)$.*

Clearly, $\tilde{u} = v$ and $\tilde{v} = u$. So the claim holds if $w \in \{u,v\}$. Next, let $w \notin \{u,v\}$; so $d(u,v) \geq 2$.

Case 1 $d(u,w) = 1$. (If $d(v,w) = 1$, a similar proof holds.)

Let $\langle u, w, w_1, \ldots, v \rangle$ (it is possible that $w_1 = v$) be a shortest (u,v)-path. Since $\delta(G) \geq 3$, there exists $a \in V(G) \setminus \{u, w_1\}$ adjacent to w. Then as in Claim 1 of Theorem 24.12, we can prove that there exists $b \in V(G)$ adjacent to both u and w_1 but not to a. Then $\langle u, b, w_1, \ldots, v \rangle$ is also a shortest (u,v)-path, hence $b \in I(u,v)$. Clearly, $w \notin I(b,v)$; since $d(b,v) = d(u,v) - 1 = d(w,v)$. Since $I(b,v)$ is closed ($[I(u,v)]$ being distance monotone), there exists $\tilde{w} \in I(b,v)$ such that $d(w, \tilde{w}) > d(b,v) = d(u,v) - 1$. Thus $d(w, \tilde{w}) \geq d(u,v)$. Also, $d(w, \tilde{w}) \leq d(w,u) + d(u,b) + d(b, \tilde{w}) \leq 1 + 1 + (d(u,v) - 2) = d(u,v)$.

Also, note that in this case, $v, \tilde{w}$ are adjacent.

Case 2 $d(u,w) \geq 2$ and $d(v,w) \geq 2$.

Let $\langle u, u_1, \ldots, w, \ldots, v_1, v \rangle$ be a shortest (u,v)-path. By Case 1, there exists $\tilde{v}_1 \in I(u,v)$ adjacent to u such that $d(v_1, \tilde{v}_1) = d(u,v)$. Clearly, $\tilde{v}_1 \notin I(u,w)$ and $w \notin I(v, \tilde{v}_1)$. Since $I(v, \tilde{v}_1)$ is closed, there exists a $\tilde{w} \in I(v, \tilde{v}_1)$ such that $d(w, \tilde{w}) > d(v, \tilde{v}_1) = d(u,v) - 1$.

The proof of Claim 1 is now complete. Next, let $H = [I_G(u,v)]$.

Claim 2 *$I_H(w, \tilde{w}) = V(H)$, for any $w, \tilde{w} \in V(H)$ such that $d(w, \tilde{w}; H) = d(u,v; H) = diam(H)$.*

Clearly, $I_H(w, \tilde{w}) \subseteq V(H)$. If possible, let $y \in V(H) \setminus I_H(w, \tilde{w})$. Then there exists $y' \in I_H(w, \tilde{w})$ such that $d(y, y'; H) > d(w, \tilde{w}; H) = diam(H)$, a contradiction.

By Claims 1 and 2, it follows that H is antipodal. We have thus proved that every interval in G is antipodal. Therefore by Theorem 24.7, G is a hypercube. ∎

Theorem 24.14 [20] *A graph G with minimum degree $\delta(G) \geq 3$ is a hypercube if and only if*

1. *G is bipartite, and*

2. *G is interval distance monotone.*

Proof. $\implies$: Follows by Propositions 24.2 and 24.32.

$\impliedby$: Just notice that in the proof of the above theorem we have used the hypothesis that G is distance monotone in only one instance that too to deduce that G is bipartite. ∎

24.2.3.8 Characterizations through Edge Colorings

Unlike the earlier characterizations, this does not use the distance concept and absence of $K_{2,3}$. It depends on edge colorings. Recall that, in Q_n if we color an ith dimensional edge with color i, we obtain a proper n-edge-coloring. Buratti [21] identified few more properties of this edge-coloring to obtain the following interesting characterization. We omit its proof.

Theorem 24.15 [21] *A graph G is a hypercube if and only if*

1. *G is connected,*

2. *G is regular (with degree of regularity, say n),*

3. *G admits a proper n-edge-coloring satisfying the following conditions:*

   C1: *any two-colored path $\langle u, v, x, y, z \rangle$ of length four is closed; that is, it induces a cycle $\langle u, v, x, y, z = u \rangle$ of length four, and*

   C2: *any path whose edges have pairwise distinct colors is open.* ∎

24.3 HYPERCUBE-LIKE INTERCONNECTION NETWORKS

In this section, we first present the definitions and properties of the variants and generalizations of the hypercube. Following this we summarize the topological properties of these networks in Table 24.2. A large number of interconnection networks can be defined using Cartesian product of elementary graphs, like paths, cycles, complete graphs and complete k-ary trees The Cartesian product of paths is called a *mesh*, the Cartesian product of cycles is called a *torus*. See Figure 24.7 for a mesh and a torus.

Yet another general method of constructing interconnection networks is through Cayley graphs. In fact hypercubes and several variations of hypercubes are Cayley graphs widely studied in algebra and appearing in the theory of manifolds. They have several graph theoretical properties which are ideal for the design of interconnection networks. These include large number of vertices, small vertex-regularity, vertex-symmetry and small diameter. They also admit simple routing techniques. We end this section by defining some interconnection networks which have fixed vertex degree and/or Cayley graphs.

24.3.1 Twisted Cube TQ_n

The twisted cubes are constructed by applying an operation called twist to some of the edges of the binary hypercubes [22]. The twist is an operation defined on two edges (u, v) and (x, y) which have no nodes in common. The operation consists of adding two new edges (u, x) and (v, y) and deleting the edges (u, v) and (x, y).

TABLE 24.2 Topological Properties of the Various Interconnection Networks

Network	$|V|$	$|E|$	deg and κ	Diameter	Symmetry	Cayley
TQ_n	2^n	$n2^{n-1}$	n	$\left\lceil \dfrac{n+1}{2} \right\rceil$	No	No
Q_n^T, Q_n^C	2^n	$n2^{n-1}$	n	$\left\lceil \dfrac{n+1}{2} \right\rceil$	No	No
T_nQ	2^n	$n2^{n-1}$	n	$n-1$	No	No
CQ_n	2^n	$n2^{n-1}$	n	$\left\lceil \dfrac{n+1}{2} \right\rceil$	No	No
GQ_n	2^n	$n2^{n-1}$	n	$\left\lceil \dfrac{2n}{3} \right\rceil$	No	No
$Q_{n,k}$	2^n	$(n+1)2^{n-1}$	$n+1$	$\left\lceil \dfrac{n}{2} \right\rceil$	Yes	Yes
$0MQ_n$	2^n	$n2^{n-1}$	n	$\left\lceil \dfrac{n+2}{2} \right\rceil$	No	No
$1MQ_n$	2^n	$n2^{n-1}$	n	$\left\lceil \dfrac{n+1}{2} \right\rceil$	No	No
ShQ_n	2^n	$n2^{n-1}$	n	$\simeq \dfrac{n}{4}$	Yes	No
Γ_n	f_n	$\dfrac{2(n-1)f_n - nf_{n-1}}{5}$	$\Delta = n-2$, $\lfloor \dfrac{n}{8} \rfloor \leq \kappa \leq \lfloor \dfrac{n-2}{3} \rfloor$	$n-2$	No	No
Q_n^k ($k \geq 3$)	k^n	nk^n	$2n$	$n\left\lfloor \dfrac{k}{2} \right\rfloor$	Yes	Yes
AQ_n	2^n	$(2n-1)2^{n-1}$	$2n-1$	$\left\lceil \dfrac{n}{2} \right\rceil$	Yes	Yes
$H(b,n)$	b^n	$\tfrac{1}{2}(b-1)nb^n$	$(b-1)n$	n	Yes	Yes
$DG(d,k)$	d^k	d^{k+1}	$2d$	k	Yes	No
CCC_n	$n2^n$	$3n2^{n-1}$	3	$2n + \lfloor \tfrac{n}{2} \rfloor - 2$ ($n \geq 4$)	Yes	Yes
UBF_n	$(n+1)2^n$	$n2^{n+1}$	$\Delta = 4$, $\kappa = 2$	$2n$	No	No
$G_n \cong WBF_n$	$n2^n$	$n2^{n+1}$	4	$\left\lfloor \dfrac{3n}{2} \right\rfloor$	Yes	Yes
$G_{k,n}$	$n(k-1)^n$	$\dfrac{nk(k-1)^n}{2}$	k	$\lfloor \tfrac{5n}{2} \rfloor - 2$ ($k \geq 6, n \geq 4$)	Yes	Yes
$S(n)$	$n!$	$\dfrac{n!(n-1)}{2}$	$n-1$	$\left\lfloor \dfrac{3(n-1)}{2} \right\rfloor$	Yes	Yes

Note: $|V|$, number of vertices; $|E|$, number of edges; *deg*, degree of regularity (if regular); Δ, maximum degree (if not regular); κ, vertex connectivity; symmetry, whether vertex symmetric or not; Cayley, whether a Cayley graph or not; f_n, the nth Fibonacci number.

Definition 24.20 (Recursive) *Define $TQ_1 = Q_1$ and $TQ_2 = Q_2$. For $n \geq 3$, TQ_n is obtained by taking four copies of the twisted cube TQ_{n-1}, denoted by TQ_{n-2}^{00}, TQ_{n-2}^{01}, TQ_{n-2}^{10}, and TQ_{n-2}^{11} and adding edges between these copies as follows: For $\alpha, \beta \in \{0,1\}$, let*

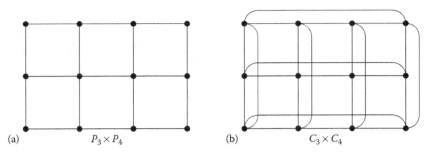

Figure 24.7 (a) Mesh and (b) torus.

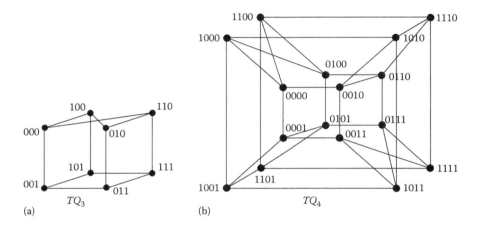

Figure 24.8 Twisted cubes of dimension (a) three and (b) four.

$V(TQ_{n-2}^{\alpha\beta}) = \{\alpha\beta x_3 x_4 \ldots x_n : x_i \in \{0,1\}, 3 \leq i \leq n\}$. *A vertex* $\alpha\beta x_3 x_4 \ldots x_n$ *is connected to* $\overline{\alpha\beta} x_3 x_4 \ldots x_n$, *and* $\overline{\alpha}\beta x_3 x_4 \ldots x_n$ *if* $\sum_{i=3}^{n} x_i = 0 \pmod{2}$ *and is connected to* $\alpha\overline{\beta} x_3 x_4 \ldots x_n$, *and* $\overline{\alpha}\beta x_3 x_4 \ldots x_n$ *if* $\sum_{i=3}^{n} x_i = 1 \pmod{2}$.

Twisted cubes of dimension 3 and 4 are given in Figure 24.8.

The following variants of the hypercube are some variations of the twisted cubes:

- X hypercube (twisted hypercube) Q_n^T, Q_n^C defined by Sung [23].

- Twisted n-cube T_nQ defined by Esfahanian et al. [24].

- Multiply-twisted cube (crossed cube) CQ_n defined by Efe [25,26].

- Generalized twisted cube GQ_n defined by Chedid and Chedid [27].

24.3.2 k-Skip Enhanced Cube $Q_{n,k}$

An efficient and widely used variant of the hypercube is the k-skip-enhanced cube, wherein an edge is added between any two vertices $X = x_1 x_2 \ldots x_k\, x_{k+1} \ldots x_n$ and $Y = y_1\, y_2 \ldots y_k\, y_{k+1} \ldots y_n$ provided $x_i = y_i$ for $1 \leq i \leq k$ and $x_i = \overline{y}_i$ for $k+1 \leq i \leq n$ [28]. This process reduces the diameter of the network by a factor of 2. When $k = 1$, the 1-skip enhanced cube was independently defined by El-Amawy and Latifi [29] and they called as *folded hypercube* FQ_n.

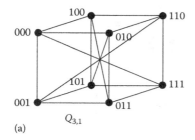 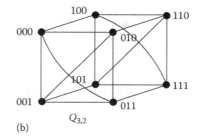

Figure 24.9 (a) 1- and (b) 2-skip-enhanced cubes of dimension three.

Definition 24.21 *For $1 \leq k \leq n-1$, let $E_k = \{(x_1 x_2 \ldots x_n, x_1 x_2 \ldots x_{k-1}\overline{x_k}\overline{x_{k+1}} \ldots \overline{x_n}) : x_i \in \{0,1\}, 1 \leq i \leq n\}$. Then the n-dimensional k-skip enhanced cube $Q_{n,k}$, has the edge set $E(Q_{n,k}) = E(Q_n) \cup E_k$.*

Enhanced cubes of dimension 3 for $k = 1$ and $k = 2$ are given in Figure 24.9.

24.3.3 Möbius Cube $0MQ_n$, $1MQ_n$

There are two types of Möbius cubes called the 0-Möbius cube and 1-Möbius cube [30].

Definition 24.22 *Two vertices $X = x_1 x_2 \ldots x_n$ and $Y = y_1 y_2 \ldots y_n$ are joined in $0MQ_n$ iff $y_1 y_2 \ldots y_n = \overline{x_1} x_2 \ldots x_n$ or there exists an integer i, $2 \leq i \leq n$ such that $y_1 \ldots y_{i-1} y_i y_{i+1} \ldots y_n = x_1 \ldots x_{i-1} \overline{x_i} x_{i+1} \ldots x_n$. Similarly, two vertices $X = x_1 x_2 \ldots x_n$ and $Y = y_1 y_2 \ldots y_n$ are joined in $1MQ_n$ iff $y_1 y_2 \ldots y_n = \overline{x_1 x_2 \ldots x_n}$ or there exists an integer i, $2 \leq i \leq n$ such that $y_1 \ldots y_{i-1} y_i y_{i+1} \ldots y_n = x_1 \ldots x_{i-1} \overline{x_i} \overline{x_{i+1} \ldots x_n}$.*

Mobius Cubes of dimension 4 are shown in Figure 24.10.

24.3.4 Shuffle Cube ShQ_n

Let $X = x_1 x_2 \ldots x_n$ where $x_i \in \{0,1\}$, $1 \leq i \leq n$. Let $p_j(X)$ denote the j-prefix of X, that is, $p_j(X) = x_1 x_2 \ldots x_j$, and $s_i(X)$ denote the i-suffix of X, that is, $s_i(X) = x_{n-i+1} x_{n-i+2} \ldots x_n$. Let $V_{00} = \{0000, 0001, 0010, 0011\}$, $V_{01} = \{0100, 0101, 0110, 0111\}$, $V_{10} = \{1000, 1001, 1010, 1011\}$ and $V_{11} = \{1100, 1101, 1110, 1111\}$ [31].

Definition 24.23 (Recursive) *Define $ShQ_1 = Q_1$, $ShQ_2 = Q_2$, $ShQ_3 = Q_3$, and $ShQ_4 = Q_4$. For $n \geq 5$, ShQ_n is obtained by taking sixteen copies of ShQ_{n-4}, say $ShQ_{n-4}^{i_1 i_2 i_3 i_4}$ with $i_j \in \{0,1\}$ for $1 \leq j \leq 4$, and adding edges between these copies as follows: Let $V(ShQ_{n-4}^{i_1 i_2 i_3 i_4}) = \{i_1 i_2 i_3 i_4 x_5 x_6 \ldots x_n : x_i \in \{0,1\}, 5 \leq i \leq n\}$, that is, for $i_j \in \{0,1\}$, $1 \leq j \leq 4$, $p_4(X) = i_1 i_2 i_3 i_4$ for all vertices X in $ShQ_{n-4}^{i_1 i_2 i_3 i_4}$. Vertices $X = x_1 x_2 \ldots x_n$ and $Y = y_1 y_2 \ldots y_n$ in different subcubes of dimension $n-4$ are adjacent in ShQ_n iff (1) $s_{n-4}(X) = s_{n-4}(Y)$, and (2) $p_4(X) \oplus_2 p_4(Y) \in V_{s_2(X)}$, that is, (1)' $x_5 x_6 \ldots x_n = y_5 y_6 \ldots y_n$, and (2)' $(x_1 x_2 x_3 x_4) \oplus_2 (y_1 y_2 y_3 y_4) \in V_{x_{n-1} x_n}$.*

An illustration of Shuffle cube of dimension 6 is shown in Figure 24.11.

24.3.5 Fibonacci Cube Γ_n

The Fibonacci cubes can have arbitrary number of vertices as against the other variations of the hypercube which have 2^n vertices. Further, these cubes are not regular [32]. Fibonacci numbers are well studied in number theory. The Fibonacci numbers are recursively defined by $f_0 = 0$, $f_1 = 1$, and, for $n \geq 2$, $f_n = f_{n-1} + f_{n-2}$. It is known that any natural number can be uniquely represented as a sum of Fibonacci numbers as follows: Let $n \geq 3$. Let i be

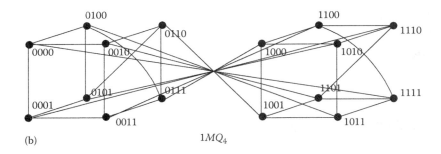

Figure 24.10 (a,b) Möbius cubes of dimension four.

an integer and $0 \leq i \leq f_n - 1$. Then $i = \sum_{j=2}^{n-2} b_j f_j$, where $b_j \in \{0,1\}$ for $2 \leq j \leq n-1$. The sequence $(b_{n-1},\ldots,b_3,b_2)_F$ is called the order-n Fibonacci code (or simply Fibonacci code) of i. For example, the numbers from 0 to $4 = f_5 - 1$ are expressed in the Fibonacci code of order-5 as $0 = (000)_F, 1 = (001)_F, 2 = (010)_F, 3 = (100)_F, 4 = (101)_F$.

Definition 24.24 *Let N denote an integer, where $1 \leq N \leq f_n$ for some n. Let I_F and J_F denote the Fibonacci codes of i and j, respectively, where $0 \leq i, j \leq N - 1$. The Fibonacci cube of size N is a graph with vertex set $\{0, 1, \ldots N - 1\}$ with vertices i and j adjacent iff I_F and J_F differ in exactly one position.*

Fibonacci cubes of dimensions one to five are shown in Figure 24.12.

Definition 24.25 *The Fibonacci cube Γ_n of order n is a Fibonacci cube with f_n nodes.*

We now give a method for constructing Fibonacci cube Γ_n using binary strings.

Definition 24.26 *For $n \geq 3$, the Fibonacci cube Γ_n of order n is a simple graph with f_n nodes. The nodes are labeled with binary strings of length $n - 2$ with no consecutive 1's. Two nodes of Γ_n are adjacent if and only if their labels differ in exactly one position.*

24.3.6 *k*-Ary *n*-Cube Q_n^k

A popular generalization of the hypercube is the k-ary n-cube Q_n^k because many other networks-like rings, meshes, tori, hypercubes, and Omega networks are all isomorphic to a k-ary n-cube [33]. The vertices of Q_n^k are labeled using n-tuples with components from the set $\{0, 1, \ldots, k-1\}$. There exists an edge between two vertices of Q_n^k if and only if their labels differ by $1 \mod k$ in exactly one position. Consequently, a 2-ary n-cube/binary n-cube is Q_n, k-ary 1-cube is a ring on k vertices and a k-ary 2-cube is a $k \times k$ torus. A formal definition of Q_n^k is as follows.

Definition 24.27 *The vertex set of Q_n^k is $\{x_1 x_2 \ldots x_n : x_i \in \{0, 1, \ldots, k-1\}, 1 \leq i \leq n\}$ and the edge set is $\{(x_1 x_2 \ldots x_n, y_1 y_2 \ldots y_n) : x_i \equiv (y_i - 1) \mod k$ or $x_i \equiv (y_i + 1) \mod k$, for some i, and $x_j = y_j$, for all $j \neq i\}$.*

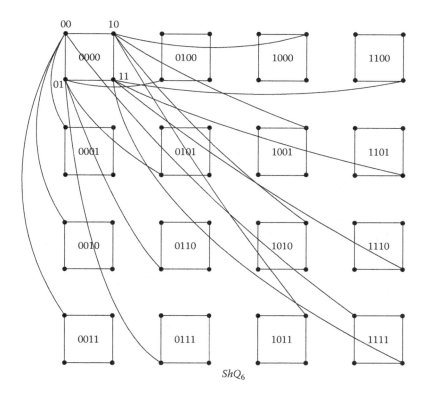

Figure 24.11 Shuffle cube of dimension six. For clarity, the edges incident with vertices in ShQ_2^{0000} alone are drawn.

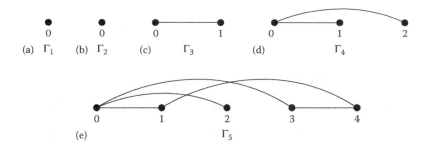

Figure 24.12 Fibonacci cubes of order (a) one, (b) two, (c) three, (d) four, and (e) five.

The k-ary n-cube Q_n^k has k^n vertices each with degree n for $k = 2$ and $2n$ for $k \geq 3$. So, it has nk^{n-1} edges of $k = 2$ and nk^n edges for $k \geq 3$. Its connectivity is n or $2n$ according as $k = 2$ or $k \geq 3$. It is a Cayley graph and so is vertex-symmetric. It is also edge-symmetric.

24.3.7 Augmented Cube AQ_n

One of the recently proposed variant of the hypercube is the Augmented cube [34].

Definition 24.28 *Augmented cube (AQ_n) of dimension n is defined recursively as follows: $AQ_1 = K_2$ and AQ_n is obtained from two copies of AQ_{n-1}, denoted by AQ_{n-1}^0 and AQ_{n-1}^1 and adding $2 \times 2^{n-1}$ edges as follows. Let $V(AQ_{n-1}^0) = \{0u_{n-1} \ldots u_2 u_1 : u_i = 0 \text{ or } 1\}$ and $V(AQ_{n-1}^1) = \{1u_{n-1} \ldots u_2 u_1 : u_i = 0 \text{ or } 1\}$. A vertex $U = 0u_{n-1} \ldots u_2 u_1$ of AQ_{n-1}^0 is joined to a vertex $V = 1v_{n-1} \ldots v_2 v_1$ in AQ_{n-1}^1 if and only if either (i) $u_i = v_i$, for all i, $1 \leq i \leq n-1$ or (ii) $u_i = \overline{v_i}$, for all i, $1 \leq i \leq n-1$.*

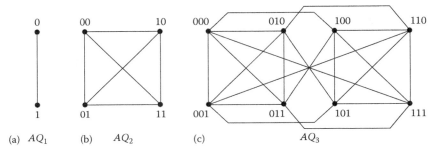

Figure 24.13 Augmented cubes of dimension (a) one, (b) two, and (c) three.

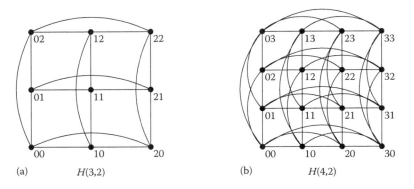

Figure 24.14 Hamming graphs of dimensions (a) three and (b) four with base two.

Augmented cubes of dimensions 1, 2, and 3 are shown in Figure 24.13.
In a manner analogous to the extension of Q_n to AQ_n, there exists an extension of Q_n^k to the augmented k-ary n-cube [35].

24.3.8 Hamming Graph/Generalized Base-b Cube $H(b,n)$

The Hamming graphs or generalized base-b cubes were introduced in [36].

Definition 24.29 *For any two positive integers b and n, the Hamming graph $H(b,n)$ is the Cartesian product of n complete graphs each with b vertices. So, $H(b,n)$ has vertex set $\{X : X = x_1 x_2 \cdots x_n,$ where $x_i \in \{0, 1, \ldots, b-1\},$ for $1 \leq i \leq n\}$, and edge set $\{(X,Y) : X$ and Y differ in exactly one position$\}$.*

Hamming graphs are also Cayley graphs and hence they possess various symmetric properties such as vertex symmetry, edge symmetry, distance symmetry, and distance regularity. Figure 24.14 shows Hamming graphs $H(3,2)$ and $H(4,2)$. In a recent work, Huang and Fang [37] have shown that Hamming graphs are Hamiltonian connected and pancyclic, for $b \geq 3$.

We now proceed to defining some more interconnection networks other than the generalization and variants of hypercubes which have fixed vertex degree and/or Cayley graphs.

24.3.9 de Bruijn Graph $DG(d,k)$

Definition 24.30 *The de Bruijn network $DG(d,k)$ is an undirected graph with $V(DG(d,k)) = \{X = a_{k-1}a_{k-2}\cdots a_1 a_0 : a_i \in \{0, 1, \ldots, d-1\}, 0 \leq i \leq k-1\}$ and $E(DG(d,k)) = \{(X,Y) : X = 2Y + j \pmod{N}$ or $Y = 2X + j \pmod{N}$, where $N = d^k$ and for $0 \leq j \leq d-1\}$.*

$DG(2,k)$ is called the binary de Bruijn multiprocessor network [38]. See Figure 24.15 for the graph $DG(2,3)$.

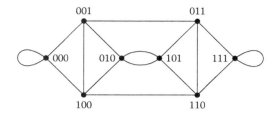

Figure 24.15 Binary de Bruijn network of dimension three.

The de Bruijn network $DG(d,k)$ has d^k vertices and diameter k. It is a regular graph with degree $2d$ and connectivity $2d$.

24.3.10 Cube-Connected Cycles Graph CCC_n

An interconnection network which was proposed combining the principles of both parallelism and pipelining are the cube-connected cycles [39]. Here the underlying topology is the hypercube of dimension n and every vertex of the Q_n is replaced by a cycle on n vertices.

Definition 24.31 *Let $n \geq 3$. The cube-connected cycles CCC_n of dimension n is obtained from Q_n by replacing its vertices by cycles of length n. So it has $n2^n$ vertices. It is constructed as follows:*

1. *The vertex $x_1 x_2 \ldots x_n$ of Q_n is replaced by the n vertices $(1, x_1 x_2 \ldots x_n), (2, x_1 x_2 \ldots x_n),$ $\ldots (n, x_1 x_2 \ldots x_n)$.*

2. *For $i = 1, 2, \ldots, n$, the vertex $(i, x_1 x_2 \ldots x_n)$ is joined with the three vertices $((i-1) \bmod n, x_1 x_2 \ldots x_n), ((i+1) \bmod n, x_1 x_2 \ldots x_n)$ and $(i, x_1 x_2 \ldots x_{i-1} \overline{x}_i x_{i+1} \ldots x_n)$.*

The cube-connected cycles graph CCC_n of dimension n is a vertex-symmetric regular graph of degree 3. It contains $n2^n$ vertices and $3n2^{n-1}$ edges. Its diameter is 6 if $n = 3$, and it is $2n + \lfloor \frac{n}{2} \rfloor - 2$, if $n \geq 4$. The cube-connected cycle, CCC_3, is shown in Figure 24.16.

In a manner analogous to the construction of CCC_n from Q_n, one can construct a three regular graph from AQ_n.

24.3.11 Butterfly and Wrapped Butterfly Graph UBF_n, WBF_n

Definition 24.32 *Let $n \geq 3$. The butterfly graph UBF_n of dimension n is the graph with vertex set $\{(h, x_1 x_2 \ldots x_n) : 0 \leq h \leq n,$ and $x_i \in \{0, 1\}\}$. Vertices $(i, x_1 x_2 \ldots x_n)$ and $(j, y_1 y_2 \ldots y_n)$ are adjacent if and only if $j = i + 1$ and either $y_1 y_2 \ldots y_n = x_1 x_2 \ldots x_n$ or $y_1 y_2 \ldots y_n = x_1 x_2 \ldots \overline{x}_j x_{j+1} \ldots x_n$.*

It has $(n+1)2^n$ vertices and $n2^{n+1}$ edges. Vertices labeled $(0, x_1 x_2 \ldots x_n)$ or $(n, x_1 x_2 \ldots x_n)$ have degree 2 and the remaining vertices have degree 4. Its diameter is $2n$. Its vertex connectivity is 2 and edge connectivity is 2.

A wrapped butterfly graph WBF_n of dimension n is obtained from UBF_n by merging the vertex labeled $(0, x_1 x_2 \ldots x_n)$ with the vertex labeled $(n, x_1 x_2 \ldots x_n)$. Consequently, WBF_n is a graph on $n2^n$ vertices, $n2^{n+1}$ edges and each vertex has degree 4. Its diameter is $n + \lfloor n/2 \rfloor$. Its vertex connectivity as well as edge connectivity is 4. It is a Cayley graph [40] and hence vertex symmetric. See Figure 24.17 for UBF_3 and WBF_3.

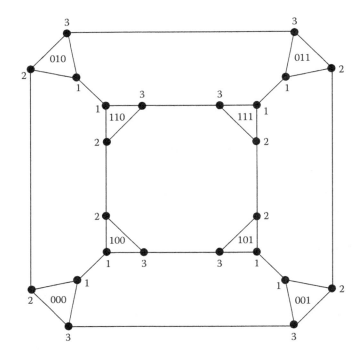

Figure 24.16 Cube-connected cycle of dimension three.

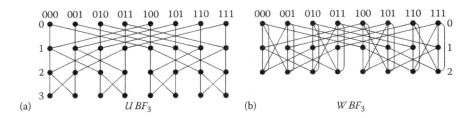

Figure 24.17 (a) Butterfly and (b) wrapped butterfly graphs of dimension three.

24.3.12 Degree Four Cayley Graph G_n

A new class of interconnection networks was proposed as an alternative to the de Bruijn graphs for VLSI implementation in terms of regularity [41].

Definition 24.33 *The degree four Cayley graph G_n is defined as follows: Each vertex is represented by a circular permutation of n symbols in lexicographic order where each symbol may be present in either uncomplemented or complemented form. Let t_k, $1 \leq k \leq n$ denote the kth symbol in the set of n symbols. Let t_k^* denote that is, $V(G_n) = \{t_1^* t_2^* \ldots t_n^* : t_k^* \in \{t_k, \bar{t}_k\}, 1 \leq k \leq n\}$. The edges of G, are defined by the following four generators in the graph:*

- $g(a_1 a_2 \cdots a_n) = a_2 a_3 \cdots a_n a_1$
- $f(a_1 a_2 \cdots a_n) = a_2 a_3 \cdots a_n \bar{a}_1$
- $g^{-1}(a_1 a_2 \cdots a_n) = a_n a_1 \cdots a_{n-1}$
- $f^{-1}(a_1 a_2 \cdots a_n) = \bar{a}_n a_1 \cdots a_{n-1}$

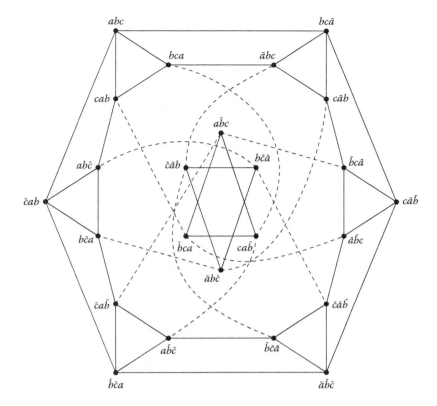

Figure 24.18 Example of G_3 on 24 nodes.

The degree four Cayley graph G_3, is shown in Figure 24.18. G_n is a symmetric regular graph with degree 4 on $n2^n$ vertices, $n2^{n+1}$ edges. Its connectivity is 4 and diameter is $\lfloor 3n/2 \rfloor$. It has a higher connectivity than cube-connected cycles. Further the constant degree of these networks were considered practically important when compared to the logarithmic degree of hypercubes. Later in 1999, it was showed that this class of graph is isomorphic to the wrapped butterfly by Wei et al. [40].

24.3.13 k-Valent Cayley Graph $G_{k,n}$

Definition 24.34 [42] *A k-valent Cayley graph $G_{k,n}$ is an undirected graph with $N = n(k-1)^n$ vertices for any integers $n \geq 2$ and $k \geq 3$. Each node v of $G_{k,n}$ has the form $s_0 s_1 \ldots s_{m-1} \tilde{s}_m s_{m+1} \ldots s_{n-1}$ corresponding to a string of n symbols selected from $\{0, 1, \ldots, k-2\}$ such that exactly one symbol $\tilde{s}_m$ is in marked form and the others are in unmarked form. We sometimes use v^m for representing a node v with the marked symbol on position m. Let $s_i^* = s_i$ or $\tilde{s}_i$. Each edge is of the type $(v, \delta(v))$, where $\delta \in \{f, f^{-1}, g_1, g_2, \ldots, g_{k-2}\}$ is a generator defined as follows:*

- $f(u^m) = v^{(m-1) \pmod n}$, where $u^m = s_0^* s_1^* \ldots s_{n-1}^*$, $v^{(m-1) \pmod n} = s_1^* s_2^* \ldots s_{n-1}^* \alpha^*$, and $\alpha = (s_0 + 1) \pmod{k-1}$.

- $f^{-1}(u^m) = v^{(m+1) \pmod n}$, where $u^m = s_0^* s_1^* \ldots s_{n-1}^*$, $v^{(m+1) \pmod n} = \beta^* s_1^* s_2^* \ldots s_{n-2}^*$, and $\beta = (s_{n-1} - 1) \pmod{k-1}$.

- $g_i(u^m) = v^m$, where $u^m = s_0^* s_1^* \ldots s_{n-1}^*$, $v^m = s_0^* s_1^* \ldots s_{n-2}^* \gamma^*$, and $\gamma = (s_{n-1} + i) \pmod{k-1}$ for $1 \leq i \leq k-2$.

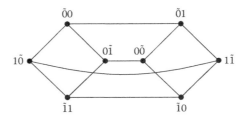

Figure 24.19 Three valent Cayley graph on two symbols.

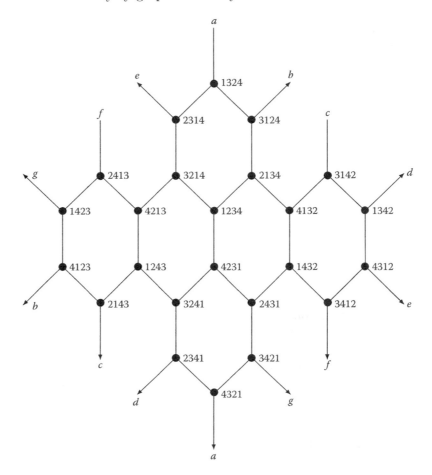

Figure 24.20 Star graph of dimension four.

The graph $G_{k,n}$ is a symmetric regular graph with degree k on $n(k-1)^n$ vertices and $1/2nk(k-1)^n$ edges. Its vertex connectivity is k. The diameter is at most $\max\{2n, \lfloor 5n/2 \rfloor - 2\}$. Moreover, when $k \geq 6$ and $n \geq 4$ the diameter of $G_{k,n}$ is $\lfloor 5n/2 - 2 \rfloor$ and for $k \geq 6$ and $n = 2, 3$ the diameter of $G_{k,n}$ is $2n$. For $k = 3$, it can be easily verified that $G_{k,n}$ is isomorphic to the trivalent Cayley graph of Vadapalli and Srimani [43]. See Figure 24.19 for the graph $G_{3,2}$.

24.3.14 Star Graph $S(n)$

Let $n \geq 2$. A permutation $\pi = a_1 a_2 \ldots a_n$ of an n-set $1, 2, \ldots n$ is called a transposition if for some pair (i,j), $i \neq j$, $\pi(i) = j$, $\pi(j) = i$, and $\pi(k) = k$, for every other k. It is denoted by (i,j).

Definition 24.35 [44] *The star graph of dimension n, $S(n)$ has $n!$ vertices, each labeled by a permutation $a_1 a_2 \ldots a_n$. The vertex $a_1 a_2 \ldots a_n$ is joined to $n-1$ transpositions $(a_1, a_2), (a_1, a_3), \ldots, (a_1, a_n)$.*

The star graph $S(n)$ of dimension n has $n!(n-1)/2$ edges. Each vertex has degree $n-1$. Its diameter is $\lfloor 3(n-1)/2 \rfloor$. Its vertex connectivity is $n-1$ and edge connectivity is $n-1$. It is a Cayley graph. The star graph S(4) is shown in Figure 24.20.

Akers and Krishnamurthy [44] also constructed bubble sort graphs and pancake graphs as Cayley graphs on the set of permutations of an n-set. A remarkable feature of these networks is that they have a large number ($n!$) of vertices but small degree ($n-1$) of regularity. A distinguishing feature of these networks is their diameter. The bubble sort graph of dimension n has diameter $n(n-1)/2$. The diameter of the pancake graph of dimension n is $O(n)$ and the exact value for large values of n is still unknown.

Some topologies like the alternating group graphs [45], split-stars [46] share some features with the star graph. For instance, each of these graphs have a vertex set of size $O(n!)$, vertex degree $O(n)$, and diameter $O(n)$. A drawback of these topologies is the big difference between the number of vertices of consecutive dimensions. The arrangement graphs [47], the (n, k)-star [48] were some of the topologies that were introduced to circumvent the above mentioned short comings. Any interested reader may refer to [49] to know about developments with respect to these networks.

References

[1] J. Duato, S. Yalamanchili, and L.M. Ni. *Interconnection Networks: An Engineering Approach*. Morgan Kaufmann, San Francisco, CA, 1st edition, 2002.

[2] M.J. Quinn. *Parallel Computing—Theory and Practice*. Tata McGraw-Hill, New Delhi, 2nd edition, 1994.

[3] J.P. Hayes. *Computer Architecture and Organisation*. McGraw-Hill, Singapore, 1988.

[4] L.-H. Hsu and C.K. Lin. *Graph Theory & Interconnection Networks*. Taylor & Francis Group, New York, 2008.

[5] K. Hwang and F.A. Briggs. *Computer Architecture and Parallel Processing*. McGraw-Hill, New York, 1984.

[6] F.T. Leighton. *Introduction to Parallel Algorithms and Architectures: Arrays, Trees, Hypercubes*. Morgan Kauffmann, San Mateo, CA, 1992.

[7] J.-M. Xu. *Topological Structure and Analysis of Interconnection Networks*. Kluwer Academic Publishers, Dordrecht/Boston/London, 2001.

[8] J.-M. Xu. *Theory and Application of Graphs*, Network Theory and Applications Series, Volume 10, Springer US, Springer Science + Business Media, New York, 2003.

[9] W. Imrich and S. Klavžar. *Product Graphs: Structure and Recognition*. Wiley-Interscience Series in Discrete Mathematics and Optimization, USA, 2000.

[10] Y. Saad and M.H. Schultz. Topological properties of hypercubes. *IEEE Transactions on Computers*, **37**(7) (1988), 867–872.

[11] S. Foldes. A characterization of hypercubes. *Discrete Mathematics*, **17**(2) (1977), 155–159.

[12] J.-M. Laborde and S.P. Rao Hebbare. Another characterization of hypercubes. *Discrete Mathematics*, **39** (1982), 161–166.

[13] A. Berrachedi and M. Mollard. Median graphs and hypercubes: Some new characterizations. *Discrete Mathematics*, **208–209** (1999), 71–75.

[14] H.M. Mulder. n-cubes and median graphs. *Journal of Graph Theory*, **4** (1980), 107–110.

[15] H.-J. Bandelt and H.M. Mulder. Infinite median graphs, $(0,2)$-graphs and hypercubes. *Journal of Graph Theory*, **7** (1983), 487–497.

[16] W. Wenzel. A sufficient condition for a bipartite graph to be a cube. *Discrete Mathematics*, **259** (2002), 383–386.

[17] H.M. Mulder. The structure of median graphs. *Discrete Mathematics*, **24**(2) (1978), 197–204.

[18] H.M. Mulder. $(0,\lambda)$-graphs and n-cubes. *Discrete Mathematics*, **28**(2) (1979), 179–188.

[19] G. Burosch, I. Havel, and J.-M. Laborde. Distance monotone graphs and a new characterization of hypercubes. *Discrete Mathematics*, **110**(1–3) (1992), 9–16.

[20] M. Aïder and M. Aouchiche. Distance monotonicity and a new characterization of hypercubes. *Discrete Mathematics*, **245** (2002), 55–62.

[21] M. Buratti. Edge-colourings characterizing a class of Cayley graphs and a new characterization of hypercubes. *Discrete Mathematics*, **161**(1–3) (1996), 291–295.

[22] P.A.J. Hilbers, M.R.J. Koopman, and J.L.A. van de Snepscheut. The twisted cube. In *Proceedings of the Parallel Architectures and Languages Europe, Volume I: Parallel Architectures*, pages 152–159. Springer-Verlag, 1987.

[23] Y.Y. Sung. X-hypercube: A better interconnection network. *Proceedings of the 26th Annual Southeast Regional ACM Conference*, 1988.

[24] A.-H. Esfahanian, L.M. Ni, and B.E. Sagan. The twisted n-cube with application to multiprocessing. *IEEE Transactions on Computers*, **40**(1) (1991), 88–93.

[25] K. Efe. A variation of the hypercube with lower diameter. *IEEE Transactions on Computers*, **40**(1) (1991), 1312–1316.

[26] K. Efe. The crossed cube architecture for parallel computation. *IEEE Transactions on Parallel and Distributed Systems*, **3**(5) (1992), 513–524.

[27] F.B. Chedid and R.B. Chedid. A new variation on hypercubes with smaller diameter. *Information Processing Letters*, **46**(6) (1993), 275–280.

[28] N.-F. Tzeng and S. Wei. Enhanced hypercubes. *IEEE Transactions on Computers*, **40**(3) (1991), 284–294.

[29] A. El-Amawy and S. Latifi. Properties and performance of folded hypercubes. *IEEE Transactions on Parallel and Distributed Systems*, **2**(1) (1991), 31–42.

[30] P. Cull and S.M. Larson. The mobius cubes. *IEEE Transactions on Computers*, **44**(5) (1995), 647–659.

[31] T.-K. Li, J.J.M. Tan, L.-H. Hsu, and T.-Y. Sung. The shuffle-cubes and their generalization. *Information Processing Letters*, **77**(1) (2001), 35–41.

[32] W.J. Hsu. Fibonacci cubes-a new interconnection technology. *IEEE Transactions on Parallel and Distributed Systems*, **4**(1) (1993), 3–12.

[33] W.J. Dally. Performance analysis of k-ary n-cube interconnection networks. *IEEE Transactions on Computers*, **39**(6) (1990), 775–785.

[34] S.A. Choudum and V. Sunitha. Augmented cubes. *Networks*, **40** (2002), 71–84.

[35] Y. Xiang and I.A. Stewart. Augmented k-ary n-cubes. *Information Sciences*, **181**(1) (2011), 239–256.

[36] S. Lakshmivarahan and S.K. Dhall. A new hierarchy of hypercube interconnection schemes for parallel computers. *Journal of Supercomputing*, **2** (1988), 81–108.

[37] C.-H. Huang and J.-F. Fang. The pancyclicity and the Hamiltonian-connectivity of the generalized base-b hypercube. *Computers and Electrical Engineering*, **34**(4) (2008), 263–269.

[38] M.R. Samatham and D.K. Pradhan. A multiprocessor network suitable for single-chip vlsi implementation. *ACM SIGARCH Computer Architecture News*, **12**(3) (1984), 328–339.

[39] F.P. Preparata and J. Vuillemin. The cube-connected cycles: A versatile network for parallel computation. *Communications of the ACM*, **24**(5) (1981), 300–309.

[40] D.S.L. Wei, F.P. Muga, and K. Naik. Isomorphism of degree four Cayley graph and wrapped butterfly and their optimal permutation routing algorithm. *IEEE Transactions on Parallel and Distributed Systems*, **10**(11) (1999), 1290–1298.

[41] P. Vadapalli and P.K. Srimani. A new family of Cayley graph interconnection networks of constant degree four. *IEEE Transactions on Parallel and Distributed Systems*, **7** (1996), 26–32.

[42] S.-Y. Hsieh and T.-T. Hsiao. The k-valent graph: A new family of Cayley graphs for interconnection networks. *International Conference on Parallel Processing*, **1** (2004), 206–213.

[43] P. Vadapalli and P.K. Srimani. Trivalent Cayley graphs for interconnection networks. *Information Processing Letters*, **54** (1995), 329–335.

[44] S. Akers and B. Krishnamurthy. A group-theoretic model for symmetric interconnection networks. *IEEE Transactions on Computers*, **38** (1989), 555–566.

[45] J.S. Jwo, S. Lakshmivarahan, and S.K. Dhall. A new class of interconnection networks based on the alternating group. *Networks*, **23**(4) (1993), 315–326.

[46] E. Cheng, M.J. Lipman, and H.A. Park. An attractive variation of the star graphs: split-stars. Technical report, Oakland University, Rochester, MI, 1998. Technical Report no. 98-3.

[47] K. Day and A. Tripathi. Arrangement graphs: A class of generalized star graphs. *Information Processing Letters*, **42** (1992), 235–241.

[48] W.K. Chiang and R.J. Chen. The star graph: A generalized star graph. *Information Processing Letters*, **56** (1995), 259–264.

[49] E. Cheng and M.J. Lipman. Basic structures of some interconnection networks. *Electronic Notes in Discrete Mathematics*, **11** (2002), 140–156.

CHAPTER 25

Cayley Graphs

S. Lakshmivarahan

Lavanya Sivakumar

S. K. Dhall

CONTENTS

25.1	Groups	627
	25.1.1 Definitions and Examples	628
	25.1.2 Subgroup	633
	25.1.3 Homomorphism, Isomorphism, and Automorphism	634
	25.1.4 Operations on Groups	634
	25.1.5 Generators of a Group	635
25.2	Cayley Graphs	636
25.3	Symmetry in Cayley Graphs	642
	25.3.1 Graph Symmetry	643
	25.3.2 Symmetry and Cayley Graphs	647
25.4	Consequences of Vertex Transitivity	648
25.5	Conclusion	649

THEORY OF CAYLEY graphs provides a mathematical basis for the design of simple, undirected, uniform scalable families of interconnection networks that constitute the backbone of distributed memory parallel architectures. In this chapter, we provide a comprehensive introduction to the properties of Cayley graphs that directly affect the design of basic algorithms for performing various communication tasks such as point-to-point, broadcast, personalized communication, and gossip, to name a few.

Since the theory of Cayley graphs is deeply rooted in the theory of finite groups, in Section 25.1 we provide a short introduction to this latter theory. Cayley graphs are defined in Section 25.2 and their properties are developed in Section 25.3.

25.1 GROUPS

In this section, we provide an elementary discussion of the concept of the important algebraic system called groups along with several properties of interest that are critical to the analysis of Cayley graphs. For more details on the group theoretic concept, we refer the reader to [1–3].

25.1.1 Definitions and Examples

Let Γ be a *set* and $* : \Gamma \times \Gamma \to \Gamma$ be a *binary operation* called the *product*, defined on Γ. That is, for $a, b \in \Gamma$, $a * b \in \Gamma$. The pair $(\Gamma, *)$ is called a group if the following three properties hold:

$G1$: For $a, b, c \in \Gamma$, $a * b * c = (a * b) * c = a * (b * c)$, that is, $*$ is *associative*.

$G2$: There exists a distinguished element $e \in \Gamma$ such that $a * e = e * a = a$, for all $a \in \Gamma$. This element e is called the *unit* element.

$G3$: For every $a \in \Gamma$, there exists a $b \in \Gamma$ such that $a * b = b * a = e$. Such an element b is called *inverse* of a and is often denoted by a^{-1}, that is, $a * a^{-1} = a^{-1} * a = e$.

In addition, if $a * b = b * a$, for all $a, b \in \Gamma$, then it is called a *commutative (or abelian) group*, otherwise it is a noncommutative (or nonabelian) group. The *number* $|\Gamma|$ of elements in Γ is called the *order* of the group. Thus, a group $(\Gamma, *)$ is called a *finite* or *infinite* group when $|\Gamma| < \infty$ or $|\Gamma| = \infty$.

The concept of group is all pervasive and it arises in various shapes and forms. Here are a few examples of interest.

Example 25.1 (Integer additive groups) Let $Z = \{\ldots, -2, -1, 0, 1, 2, \ldots\}$ be the set of all integers and consider the usual integer addition operation $+$. Clearly, $(Z, +)$ is an infinite, commutative, additive group. In this group, 0 (the number zero) is the unit element and for any integer a, the *negative of a,* denoted by $-a$, is the inverse of a.

The pair $(Z_m, +_m)$ where $Z_m = \{0, 1, 2, \ldots, m-1\}$ and $+_m$ denote the (mod m) addition of integers in Z_m is a finite, commutative, additive group. Here, 0 is the unit element.

Finite groups can be easily represented as a table. Examples of group tables for $(Z_2, +_2)$ and $(Z_3, +_3)$ are given in Tables 25.1 and 25.2.

Example 25.2 (Integer multiplicative groups) Let $R^+ = \{a > 0 | a \in R\}$ be the set of all positive real numbers with the usual operation of the multiplication of real numbers denoted by $\times$. Clearly, $(R^+, \times)$ is an infinite, commutative, multiplicative group. For this group, the number one, denoted by 1, is the unit element and the *reciprocal* of a, denoted by $(1/a) = a^{-1}$ is the inverse of a.

Let p be a *prime* integer and $(Z_p, \times_p)$ be such that $Z_p = \{1, 2, \ldots, p-1\}$ and $\times_p$ denotes the product (mod p) of integers in Z_p. This is an example of a finite, multiplicative, commutative group. The group table for $(Z_5, \times_5)$ is given in Table 25.3.

Example 25.3 (Matrix groups) Let $GL(2)$ denote the set of 2×2 real *nonsingular* matrices and $\times$ denote the usual matrix–matrix product operation. Then $(GL(2), \times)$ is an example of an infinite, noncommutative multiplicative group with the identity matrix, denoted by I, as its unit element and A^{-1}, the inverse of the matrix A in $GL(2)$ is the inverse of A.

As another example, consider $O(2)$ the set of all 2×2 orthogonal matrices given by

$$A = \begin{bmatrix} \cos\theta & \sin\theta \\ -\sin\theta & \cos\theta \end{bmatrix} \text{ for } 0 \leq \theta < 2\pi.$$

It can be verified that $O(2)$ is an example of an infinite, commutative, multiplicative matrix group. It is well known that the action of the matrix A on a vector X is to rotate it by an angle θ in the anticlockwise direction while keeping its length constant.

TABLE 25.1 Group table for $(Z_2, +_2)$. 0 is the unit element and each element is its own inverse, (i.e.) $1 +_2 1 \equiv 0 (\mod 2)$. This $+_2$ operation is also called *exclusive-or* operation

$+_2$	0	1
0	0	1
1	1	0

TABLE 25.2 Group table for $(Z_3, +_3)$. 0 is the unit element and $1^{-1} = 2$ and $2^{-1} = 1$

$+_3$	0	1	2
0	0	1	2
1	1	2	0
2	2	0	1

TABLE 25.3 Group table for $(Z_5, \times_5)$. 1 is the unit element and $2^{-1} = 3$, $3^{-1} = 2$, $4^{-1} = 4$, and $1^{-1} = 1$

$\times_5$	1	2	3	4
1	1	2	3	4
2	2	4	1	3
3	3	1	4	2
4	4	3	2	1

TABLE 25.4 Group table for $(B_2, +)$. 00 is the unit element and each element is its own inverse

+	00	01	10	11
00	00	01	10	11
01	01	00	11	10
10	10	11	00	01
11	11	10	01	00

Example 25.4 (Binary groups) Let B_n denote the set of all n bit binary strings. Clearly $|B_n| = 2^n$. Let $X = X_n X_{n-1} \ldots X_2 X_1$ and $Y = Y_n Y_{n-1} \ldots Y_2 Y_1$ be two elements of B_n. Let $Z = Z_n Z_{n-1} \ldots Z_2 Z_1$ and define $Z = X + Y$ where $Z_i = X_i +_2 Y_i$, for $1 \leq i \leq n$ with $+_2$ defined as in Example 25.1 is called the bit-wise (mod 2) addition or exclusive-or addition. It can be verified that the string $0^n = \underbrace{00\ldots0}_{n \text{ times}}$ is the unit element and each element is its own inverse. An example of $(B_2, +)$ is given in Table 25.4.

An important and a useful concept is the notion of the Hamming weight of a binary string which is defined as the number of 1 bits in the string. Similarly, the Hamming distance between two strings is the number of bit positions in which they differ. Thus, the Hamming weight of 0110 is 2 and Hamming distance between 0110 and 1011 is 3. Accordingly, we define the *parity* of a binary string to be odd or even depending on if its Hamming weight is odd or even. Thus, 0110 is of even parity.

Example 25.5 (Symmetric permutation groups) Let $A = \{1, 2, \ldots, n\}$ and $p: A \to A$ be a bijective (i.e., one-to-one and onto) function of A on itself. Such a map p is called a *permutation* and has the property that if $i \neq j$, then $p(i) \neq p(j)$. Let S_n denote the set of all permutations of A. Clearly $|S_n| = n!$. A permutation p is usually denoted by a two-dimensional array

$$p = \begin{pmatrix} 1 & 2 & 3 & \ldots & i & \ldots & n \\ p_1 & p_2 & p_3 & \ldots & p_i & \ldots & p_n \end{pmatrix}$$

where $p_i = p(i)$ for $1 \leq i \leq n$. Let $p, q \in S_n$, then define the *product* operation, denoted by $\cdot$, as $p \cdot q(i) = p(q(i))$, $1 \leq i \leq n$, which is simply the *composition* of the two bijective functions in the right to left order. It can be verified that $(S_n, \cdot)$ is a finite noncommutative group with the identity permutation I as its unit element and

$$p^{-1} = \begin{pmatrix} p_1 & p_2 & p_3 & \ldots & p_i & \ldots & p_n \\ 1 & 2 & 3 & \ldots & i & \ldots & n \end{pmatrix}$$ called symmetric group.

Remark 25.1 One could have defined the product as $p \cdot q(i) = q(p(i))$, $1 \leq i \leq n$ in the left to right order and could develop a parallel theory of permutation group. Since the resulting permutations are *isomorphic*, without loss of generality we adopt the natural right to left composition.

Permutations can be represented in different ways. For example, p can be represented as a single array given by

$$p = \boxed{\begin{array}{|c|c|c|c|c|c|c|} \hline 1 & 2 & 3 & \ldots & i & \ldots & n \\ \hline p_1 & p_2 & p_3 & \ldots & p_i & \ldots & p_n \\ \hline \end{array}} = p_1 p_2 p_3 \ldots p_n.$$

Similarly, permutations can be represented using the so called cycle notation which is illustrated below. Let

$$p = \begin{pmatrix} 1 & 2 & 3 & 4 & 5 & 6 & 7 & 8 & 9 & 10 \\ 3 & 4 & 2 & 1 & 7 & 5 & 6 & 8 & 10 & 9 \end{pmatrix} \qquad (25.1)$$

Using this p, we can define a graph with a directed edge from i to p_i for all $1 \leq i \leq n$. The graph of this p is given in Figure 25.1 which is clearly the union of 4 subgraphs each of which is a cycle. Accordingly, p is expressed as a product of 4 disjoint cycles $p = (1\ 3\ 2\ 4)(5\ 7\ 6)(8)(9\ 10)$ where $C_1 = (1\ 3\ 2\ 4)$, $C_2 = (5\ 7\ 6)$, $C_3 = e_1 = (8)$ and $C_4 = (9\ 10)$. The number of elements in a cycle C_i is denoted by $|C_i|$.

If $|C_i| = k$, then C_i is called a k-cycle. In the special case, when a cycle is of length 2 then it is called a *transposition*. If a cycle is of length one, then it denotes a cycle from i to i and is called an invariant under p and denoted by e (not to be confused with the unit element of the group). Thus, the above permutation p is the product of two cycles—a 4-cycle C_1 and

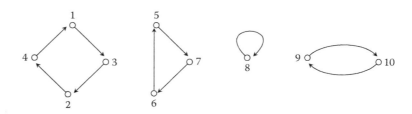

Figure 25.1 Graphical representation of a permutation $p = (1\ 3\ 2\ 4)(5\ 7\ 6)(8)(9\ 10)$.

TABLE 25.5 Group table for $(S_3, \cdot)$ where the entries are computed as $p \cdot q(i) = p(q(i)), 1 \leq i \leq 3$

p	q					
	I	(1 2)	(1 3)	(2 3)	(1 2 3)	(1 3 2)
I	I	(1 2)	(1 3)	(2 3)	(1 2 3)	(1 3 2)
(1 2)	(1 2)	I	(1 3 2)	(1 2 3)	(2 3)	(1 3)
(1 3)	(1 3)	(1 2 3)	I	(1 3 2)	(1 2)	(2 3)
(2 3)	(2 3)	(1 3 2)	(1 2 3)	I	(1 3)	(1 2)
(1 2 3)	(1 2 3)	(1 3)	(2 3)	(1 2)	(1 3 2)	I
(1 3 2)	(1 3 2)	(2 3)	(1 2)	(1 3)	I	(1 2 3)

TABLE 25.6 Group table for the parity group $S = \{1, -1\}$ under multiplication. 1 is the unit element and each element is its own inverse

×	1	−1
1	1	−1
−1	−1	1

a 3-cycle C_2, one transposition C_4 and one invariant C_3. In general, a permutation is given by $p = C_1 C_2 \ldots C_k e_1 e_2 \ldots e_l$ as a product of k disjoint cycles and l invariants. From this, we obtain a basic relation

$$n = \sum_{i=1}^{k} |C_i| + l. \tag{25.2}$$

When representing permutations using the cycle notation, it is customary to drop the invariants. Thus the above permutation in Figure 25.1 is often denoted as $p = (1\ 3\ 2\ 4)(5\ 7\ 6)(9\ 10)$. The group table for $(S_3, \cdot)$ using the cycle notation is given in Table 25.5.

Let p be a permutation and consider two pairs (i, p_i) and (j, p_j). If $p_i > p_j$, for $i < j$, then these two pairs are said to constitute an inversion in p. Thus, for the p in (25.1), the pairs $(3, 2)$ and $(4, 1)$ are inversions with respect to the first pair $(1, 3)$. Likewise, with respect to the pair $(4, 1)$ there is no inversion in p. If the total number of inversion in a permutation p is an *odd* or *even* integer, then p is called an *odd* or *even* permutation. It can be verified that the permutation p in (25.1) has 6 inversions and hence is an even permutation. Define a function called the *parity*, $PAR : S_n \to \{1, -1\}$ where

$$PAR(p) = \begin{cases} 1, & \text{if } p \text{ is an even permutation} \\ -1, & \text{if } p \text{ is an odd permutation} \end{cases} \tag{25.3}$$

Clearly, PAR partitions S_n into two subsets consisting of all even and odd permutations, equivalently of all even and odd parity. It can be verified that the set $S = \{1, -1\}$ under the standard integer multiplication is a group described by the Table 25.6. It can be verified that if p and q are permutations, then $PAR(pq) = PAR(p)PAR(q)$.

For later reference, let $A_n = \{p \in S_n | PAR(p) = 1\}$ denote the subset of all permutations of even parity. It can be verified that $(A_n, \cdot)$ also forms a group under the usual product of permutations. This group is known as the *alternating group*. It can be verified that S_3 represented in Table 25.5 is the union of A_3 and $\{(1\ 2), (1\ 3), (2\ 3)\}$, where $A_3 = \{I, (1\ 2\ 3), (1\ 3\ 2)\}$.

Permutation groups arise naturally in the context of rigid body rotation. Consider an equilateral triangle with vertices labeled inside the triangle as

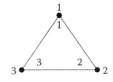

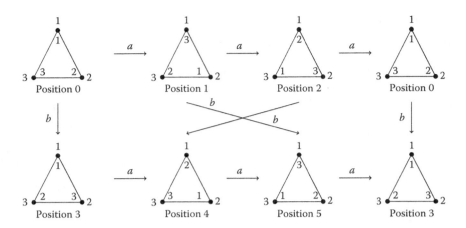

Figure 25.2 Permutation group as rigid body rotation and reflection of an equilateral triangle. The rotation a and reflection b generate a group called dihedral group, $D_3 = \langle a, b | a^3 = I, b^2 = I, \text{ and } ba = a^2 b \rangle$.

The three fixed positions labeled 1, 2, and 3 occur outside the triangle and initially the triangle is located in such a way that inside node labels and the labels of the fixed outside positions are the same. This arrangement corresponds to the initial position 0 in the Figure 25.2, and is denoted by

$$e = \begin{pmatrix} 1 & 2 & 3 \\ 1 & 2 & 3 \end{pmatrix} = I$$

which is the identity position.

Let the operation a denote the clockwise rotation of the triangle by 60° with respect to the vertical axis passing through the centroid of the triangle. Clearly, this operation a corresponds to a permutation

$$a = \begin{pmatrix} 1 & 2 & 3 \\ 3 & 1 & 2 \end{pmatrix}.$$

Similarly, let the operation b denote the reflection of the triangle with respect to the vertical axis perpendicular to the base passing through the vertex 1. Then b can be denoted by a permutation

$$b = \begin{pmatrix} 1 & 2 & 3 \\ 1 & 3 & 2 \end{pmatrix}.$$

The rotation a and reflection b generate a group called dihedral group

$$D_3 = \langle a, b | a^3 = I, b^2 = I \text{ and } ba = a^2 b \rangle.$$

Clearly $D_3 = \{e, a, a^2, b, ab, a^2 b\}$ and $D_3 = S_3$, the symmetric permutation group over 3 objects. The sequence of configuration of the equilateral triangle by repeated application of a and b are given in Figure 25.2. By referring to the Figure 25.2, it is obvious that $a^3 = I$ and $b^2 = I$. We can also derive other relations. For example, starting from position 0 at the left end of the first row we can reach, say position 5, in two ways. First applying the sequence $a^2 b$ (right to left) or by ba. Hence we get the relation $ba = a^2 b$. Similarly, we can verify that $ba^2 = ab$ and $ba^{-1} = ab$. We invite the reader to discuss other relations and verify the group table for D_3 given in Table 25.7.

TABLE 25.7 Group table for the dihedral group D_3

·	e	a	a^2	b	ab	a^2b
e	e	a	a^2	b	ab	a^2b
a	a	a^2	e	ab	a^2b	b
a^2	a^2	e	a	a^2b	b	ab
b	b	a^2b	ab	e	a^2	a
ab	ab	b	a^2b	a	e	a^2
a^2b	a^2b	ab	b	a^2	a	e

25.1.2 Subgroup

Let $(\Gamma, *)$ be a group and let $\mathcal{G} \subseteq \Gamma$ be a subset. If $(\mathcal{G}, *)$ satisfy the conditions $G1 - G3$, then $(\mathcal{G}, *)$ is a group in its own right and is called the subgroup of $(\Gamma, *)$.

Example 25.6 (Subgroups) Referring to Example 25.3 given above, it can be verified that the set $O(2)$ of all 2×2 orthogonal matrices form an infinite subgroup of the general linear group $GL(2)$, under the usual matrix–matrix multiplication.

Again referring to the Example 25.5, it can be checked that the set A_n consisting of all even permutations form a finite subgroup of the symmetric group S_n.

Let $(\Gamma, *)$ be a finite group and let $a \in \Gamma$ such that $a \neq e$. Define $a^2 = a * a$, $a^3 = a^2 * a$, and inductively $a^j = a^{j-1} * a$. Notice that $a^j \in \Gamma$ for each j. Since Γ is finite, a little reflection reveals that there exists an integer k such that $a^k = e$. Thus clearly, the subset $\mathcal{G} = \{a, a^2, \ldots, a^k = e\}$ forms a subgroup where the index of the powers are computed (mod k). This subgroup generated by an element $a \neq e$, is called the *cyclic* subgroup of Γ. The smallest integer k for which $a^k = e$ is called the order of a and is denoted by $ord_\Gamma(a) = k$. Clearly, each element of the group has a unique order.

Example 25.7 (Cyclic subgroup) Consider the finite multiplicative group $(Z_5, \times_5)$ in Example 25.2. The order of each element is given below.

a	1	2	3	4
$ord(a)$	1	4	4	2

We now state several facts without proof. Let Γ be a group and $\mathcal{G}$ be a subgroup of Γ.

SG1: The order of a subgroup divides the order of the group, that is $|\mathcal{G}| \mid |\Gamma|$, where $m|n$ denotes the fact that n is a multiple of m.

SG2: $ord_\Gamma(a) \mid |\Gamma|$.

SG3: Define a *left coset* of $\mathcal{G}$ as

$$x\mathcal{G} = \{y = xa \mid a \in \mathcal{G} \text{ and } x \in \Gamma\}.$$

Similarly, $\mathcal{G}x$ is called the *right coset*. It can be verified that $x_1\mathcal{G} \neq x_2\mathcal{G}$ for $x_1 \neq x_2$ and

$$\Gamma = \bigcup_{x \in \Gamma} x\mathcal{G}$$

that is, Γ is disjoint union of the left cosets generated by the subgroup $\mathcal{G}$ of Γ. Similar statement holds for right cosets as well.

SG4: If $x\mathcal{G} = \mathcal{G}x$ for all x, then $\mathcal{G}$ is called the *normal* subgroup of Γ.

25.1.3 Homomorphism, Isomorphism, and Automorphism

Let $(\Gamma_1, *_1)$ and $(\Gamma_2, *_2)$ be two (finite) groups, and let $h : \Gamma_1 \to \Gamma_2$ be a function from Γ_1 to Γ_2. If h satisfies the condition

$$h(a *_1 b) = h(a) *_2 h(b)$$

then h is called a *homomorphism* from Γ_1 to Γ_2. If the function h is *one-to-one*, then it is called a *monomorphism*, and if it is *onto*, then it is called *epimorphism*. If h is one-to-one and onto, then it is called an *isomorphism*. The function h is called an *automorphism* if it defines an isomorphism of a group onto itself.

Example 25.8 (Homomorphism) Referring to the Example 25.5, the parity function $PAR : S_n \to S = \{-1, 1\}$ is a *homomorphism*.

Let $g \in \Gamma$ then the function $h_g : \Gamma \to \Gamma$ is defined by

$$h_g(x) = gxg^{-1}$$

is an automorphism of Γ. For, if $x, y \in \Gamma$, then

$$\begin{aligned} h_g(xy) &= g(xy)g^{-1} \\ &= (gxg^{-1})(gyg^{-1}) \\ &= h_g(x)h_g(y). \end{aligned}$$

Facts about homomorphisms $h : \Gamma_1 \to \Gamma_2$.

H1: If e_1 and e_2 are unit elements of Γ_1 and Γ_2, respectively, then $h(e_1) = e_2$.
H2: $h(x^{-1}) = [h(x)]^{-1}$, for $x \in \Gamma_1$.
H3: $h(\Gamma_1) = \{y | y = h(x), \text{ for } x \in \Gamma_1\}$ called the *image* of Γ_1 is a subgroup of Γ_2.
H4: $Ker(h) = \{x \in \Gamma_1 | h(x) = e_2\}$ is the set of preimage of e_2 in Γ_1, called the *kernel* of h, is a subgroup of Γ_1.

Set $A(\Gamma)$ of all automorphisms of a group Γ, itself forms a group called the automorphism group of Γ.

25.1.4 Operations on Groups

Given two groups, say $(\Gamma_1, *_1)$ and $(\Gamma_2, *_2)$, we can obtain a new group $(\Gamma, *)$ by defining binary operations on the two given groups. In this section we introduce the reader to two types of group operations called the *direct product* and *semidirect product* of groups.

Direct product Let $\Gamma = \Gamma_1 \times \Gamma_2 = \{(a, b) | a \in \Gamma_1 \text{ and } b \in \Gamma_2\}$, the Cartesian product of the sets Γ_1 and Γ_2. Clearly, $|\Gamma| = |\Gamma_1| \times |\Gamma_2|$. We now define the product group, $(\Gamma, *)$ as follows: Let (a_1, b_1) and (a_2, b_2) be in Γ. Then

$$(a_1, b_1) * (a_2, b_2) = (a_1 *_1 a_2, b_1 *_2 b_2)$$

defines the group operation on Γ. It can be verified that (e_1, e_2) is the identity in Γ, where e_1 and e_2 are the unit elements of Γ_1 and Γ_2 respectively, and $(a, b)^{-1} = (a^{-1}, b^{-1})$ where a^{-1} and b^{-1} are the inverses of a and b in $(\Gamma_1, *_1)$ and $(\Gamma_2, *_2)$, respectively.

Example 25.9 (Product group) Let $(Z_5, +_5)$ and $(Z_7, +_7)$ be two groups. Then

$$Z = \{(a,b) | 0 \leq a < 5 \text{ and } 0 \leq b < 7\}$$

and $|Z| = 35$. $(0,0)$ is the identity in Z and $(2,3) * (4,5) = (2 +_5 4, 3 +_7 5) = (1,1)$ and $(2,3)^{-1} = (3,4)$.

As another example consider $(Z_5, +_5)$ and $(Z_7, \times_7)$ with $Z = \{(a,b) | 0 \leq a < 5 \text{ and } 1 \leq b < 7\}$ and $|Z| = 30$. Clearly $(0,1)$ is the identity, and $(3,2) * (3,5) = (3 +_5 3, 2 \times_7 5) = (1,3)$ and $(3,2)^{-1} = (2,4)$.

Semidirect product We now define the *semidirect product* of two groups $(\Gamma_1, *_1)$ and $(\Gamma_2, *_2)$. To this end, let $A(\Gamma_2)$ be an automorphism group of the (second) group $(\Gamma_2, *_2)$. Since identity is always an automorphism, clearly $A(\Gamma_2)$ is always nonempty. Let $f: \Gamma_1 \to A(\Gamma_2)$ be a (group) homomorphism from the group $(\Gamma_1, *_1)$ into the automorphism group $A(\Gamma_2)$. That is, for any $a \in \Gamma_1$, $f_a \in A(\Gamma_2)$ is such that $f_a: \Gamma_2 \to \Gamma_2$ is an automorphism of Γ_2 onto itself, that is, $f_a(b) \in \Gamma_2$ for $b \in \Gamma_2$.

Let $\Gamma = \Gamma_1 \times \Gamma_2$ be the Cartesian product defined above. Define a new operation on Γ as follows. For (a_1, b_1) and (a_2, b_2) in Γ, let

$$(a_1, b_1) * (a_2, b_2) = (a_1 *_1 a_2, f_{a_2^{-1}}(b_1) *_2 b_2).$$

It can be verified that Γ with this operation forms a group called the semidirect product of Γ_1 and Γ_2 and denoted by $\Gamma_1 \ltimes \Gamma_2$. Clearly (e_1, e_2) is the identity and the inverse of (a,b) is given by $(a^{-1}, f_a(b^{-1}))$. For since f_a is an automorphism on Γ, $(a,b) * (a^{-1}, f_a(b^{-1})) = (a *_1 a^{-1}, f_a(b) *_2 f_a(b^{-1})) = (e_1, e_2)$, see [4].

Example 25.10 (Semidirect product) Consider three integers (c, k, l) and let $\Gamma_1 = Z_{ck}$ and $\Gamma_2 = Z_{c^2 l}$. For $(a, b), (u, v)$ in $\Gamma = \Gamma_1 \times \Gamma_2$ define

$$(a, b) * (u, v) = (a + u, (1 - ucl)b + v)$$

where the first component is computed (mod ck) and the second (mod $c^2 l$). Thus, $(a,b)^{-1} = (-a, -b(1+ucl))$ and $(0,0)$ is the identity.

25.1.5 Generators of a Group

Let $\Omega \in \Gamma$ be such that every element of Γ can be expressed as product of elements of Ω, then Ω is called a *generator* of Γ and is denoted by $\Gamma = \langle \Omega \rangle$. Clearly $\Omega = \Gamma$ is a generator and if Ω is a generator, and $\Omega \subset \Omega_1$, then Ω_1 is also a generator. This fact gives rise to the notion of minimal generator of a group.

Example 25.11 (Generators)

a. $\Omega = \{1\}$ is a minimal generator of the additive group $(Z_6, +_6)$ since $2 = 1+1$, $3 = 1+1+1$, $4 = 1+1+1+1$, ... and $0 = 1+1+1+1+1+1$. It can be verified that $-1 \equiv 5 \pmod{6}$ is also a generator for $(Z_6, +_6)$ and hence $\Omega = \{1, -1\}$ is a generator for this group.

b. $\Omega = \{2\}$ is a generator for $(Z_5, \times_5)$, the multiplicative group (mod 5). Since $3 \equiv 2^{-1} \pmod 5$ is a generator for this group, it follows $\Omega = \{2, 3\}$ is also a generator for this group.

c. Define $g_i = 0^i 10^{n-i+1}$, the binary string that has one 1-bit at position i and zero bits at other positions and let $\Omega = \{g_i | 1 \leq i \leq n\}$. It can be verified that Ω generates B_n, the binary group.

d. Consider the permutation group S_n. Recall any $p \in S_n$ can be expressed as $p = C_1 C_2 \ldots C_k e_1 e_2 \ldots e_l$. Let $C = (a_1\ a_2\ a_3\ \ldots\ a_r)$ be a cycle of length r. Then it can be verified that this cycle C can be expressed as the product of $(r-1)$ transpositions as follows:

$$C = (a_1\ a_2\ a_3\ \ldots\ a_r) = (a_1\ a_r)(a_1\ a_{r-1}) \ \ldots\ (a_1\ a_3)(a_1\ a_2)$$

Consequently, the permutation p in Example 25.5 can be factored using transpositions as follows: Since 8 is an invariant,

$$p = (1\ 3\ 2\ 4)(5\ 7\ 6)(9\ 10)$$
$$= (1\ 4)\ (1\ 2)\ (1\ 3)\ (5\ 6)\ (5\ 7)\ (9\ 10).$$

Let

$$\Omega_{CT} = \{(i\ j) | 1 \leq i < j \leq n\}$$

be the set of all transpositions and $|\Omega_{CT}| = n(n-1)/2$. From the above discussion it readily follows that Ω_{CT} is a generator set for S_n, that is, $S_n = \langle\ \Omega_{CT}\ \rangle$ where the subscript CT denotes the complete transposition set.

We now define two other smaller generator sets for S_n. Define [5,6]

$$\Omega_{ST} = \{(1\ j) | 1 < j \leq n\}$$

and

$$\Omega_{BS} = \{(i\ i+1) | 1 \leq i < n\}.$$

Clearly, $|\Omega_{ST}| = |\Omega_{BS}| = n - 1$. It can be verified that any arbitrary transposition $(i\ j)$ can be expressed as a product of the members of Ω_{ST} as follows:

$$(i\ j) = (1\ i)\ (1\ j)\ (1\ i) = (1\ j)\ (1\ i)\ (1\ j)$$

Thus, every element in Ω_{CT} can be expressed as a product of elements in Ω_{ST}. Hence it follows

$$S_n = \langle\ \Omega_{CT}\ \rangle = \langle\ \Omega_{ST}\ \rangle.$$

Similarly it can be verified that any $(1\ i)$ can be expressed as product of the elements of Ω_{BS} and hence $S_n = \langle\ \Omega_{BS}\ \rangle$. For later reference Ω_{ST} is called the *star generator* and Ω_{BS} is called the *bubble sort generator*.

25.2 CAYLEY GRAPHS

In this section we provide the definition, examples and extensions of the concept of Cayley graphs. For basic terminology of graphs, we refer to [7–9].

Let $(\Gamma, *)$ be a finite group with e as its unit element, and Ω be its generator set, that is, $\Gamma = \langle\ \Omega\ \rangle$. In the following, for similarity in notations, we refer to the group $(\Gamma, *)$ simply as Γ. For reasons stemming from network design considerations, we first restrict our attention to an important subclass of generators satisfying the following two conditions:

(c1) $e \notin \Omega$ and

(c2) If $g \in \Omega$, then $g^{-1} \in \Omega$, that is, Ω is closed under the inverse.

Given the pair (Γ, Ω), define the Cayley graph $G(\Gamma, \Omega) = (V, E)$ where $V = \Gamma$ and $E = \{(x,y)_g | x, y \in V \text{ and } g \in \Omega\}$. That is, the Cayley graph has $|V| = |\Gamma|$ number of nodes labeled by the elements of the group Γ. Further, any two nodes x and y are neighbors if and only if $y = x * g$ for some $g \in \Omega$. Thus, if $\Omega = \{g_1, g_2, \ldots g_k\}$, then each node x has exactly k neighbors $y_i = x * g_i$, $1 \leq i \leq k$. Further, from $y = x * g$ we readily obtain $x = y * g^{-1}$, that is, if g leads to y from x, then g^{-1} leads to x from y. Thus, each generator g defines a two way edge, which is represented by an undirected edge labeled by g.

From the above definition it follows that a Cayley graph is a uniform graph of degree $|\Omega|$. Since $e \notin \Omega$, there are no self loops at any node, and since $g_i \in \Omega$ are all distinct, between any two nodes there is at most one edge labeled by g_i. Hence a Cayley graph defined above is a simple, undirected, vertex and edge labeled uniform graph.

Referring to Section 25.1, indeed there are unending choices for groups. Further, given a group, there are many generators satisfying the conditions $(c1) - (c2)$. By picking a family of groups indexed by n such as $(Z_n, +_n)$, $(Z_p, \times_p)$, $(B_n, *)$, $(S_n, *)$ to name a few, we can ensure scalability as well.

Remark 25.2 While every Cayley graph is a uniform graph, the converse is not always true. To this end, consider the set $\mathcal{T}$ of ten 2-element subsets of a given set S with five elements. Thus, if $S = \{1, 2, 3, 4, 5\}$, then $\mathcal{T} = \{12, 13, 14, 15, 23, 24, 25, 34, 35, 45\}$ where $ij = \{i, j\}$ for simplicity in notation. Now construct a graph $G = (V, E)$ with $|V| = |\mathcal{T}|$ where the ten nodes are labeled by the subsets in $\mathcal{T}$. Thus any two are neighbors if and only if the subsets corresponding to their label do not have any common element. The resulting graph is called the Petersen* graph with 10 nodes and 15 edges given in Figure 25.3.

If the Petersen graph were to be a Cayley graph, we would have to start with a group Γ of order 10 with a generator set Ω such that $|\Omega| = 3$. It can be shown [2] that the only abelian group of order 10 is the product group $(Z_2, +_2) \times (Z_5, +_5)$ which is isomorphic to the cyclic group (denoted by C_{10}) of order 10. Further, the only nonabelian group of order 10 is the Dihedral group D_5. This latter group is generated by the following relation:

$$D_5 = \langle\, a, b | a^5 = e, b^2 = e, ba = a^4 b \,\rangle$$

Next we show that this C_{10} admits a generator set Ω with $|\Omega| = 3$.

Let the elements of C_{10} be labeled by the pair (i, j) denoted by ij, for $0 \leq i < 2$ and $0 \leq j < 5$. That is, $C_{10} = \{00, 01, 02, 03, 04, 10, 11, 12, 13, 14\}$. The generators are then given

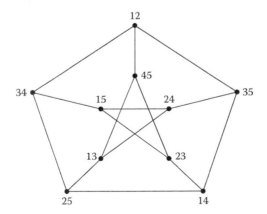

Figure 25.3 Petersen graph.

---

*Julius Petersen (1839–1910) was a Danish mathematician. Petersen graph was constructed by him in 1898. This graph provides a counter example to many claims in graph theory.

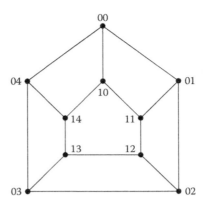

Figure 25.4 Cayley Graph of the abelian group $C_{10} \equiv (Z_2, +_2) \times (Z_5, +_5)$.

by $\Omega = \{01, 04, 10\}$. It can be easily verified that Ω is indeed a generator of size 3. Thus, the Cayley graph obtained by the 3 element generator of C_{10} is given by Figure 25.4. As will be seen in the next section (Example 25.17), the Petersen graph is not isomorphic to this Cayley graph of C_{10}. Similarly, one can easily see that the Cayley graph of D_5 is again not isomorphic to the Petersen graph. Thus, we conclude that the Petersen graph which is a uniform graph is not a Cayley graph.

We now provide a number of examples of Cayley graphs.

Example 25.12 (n-node ring) Consider the finite additive group $(Z_n, +_n)$ with $\Omega = \{1, -1\}$ as its generator. Setting $V = Z_n$, the n-nodes are labeled with integers 0 through $n - 1$. Since $i \pm 1 = (i \pm 1) \pmod n$, a node i is connected to node $i + 1$ and $i - 1$. Thus, every n-node ring or cycle is a Cayley graph. Refer to Figure 25.5 for an example. Similarly, one can define a p-node ring for p a prime integer using $(Z_p, \times_p)$ by using a generator Ω with two elements satisfying the condition $(c1 - c2)$, see [10].

Example 25.13 (nm toroid) Consider a product group $(Z_n, +_n) \times (Z_m, +_m)$ with $\Omega = \{(1, 0), (-1, 0), (0, 1), (0, -1)\}$, see [10]. The nm nodes are labeled by the pair $(i, j) = ij$ for $0 \leq i < n$ and $0 \leq j < m$. Since $\Omega = 4$, there are exactly four neighbors (when $n, m > 2$) for each node; see Figure 25.6.

The east/west neighbors E and W are obtained by using the generators $(1, 0)$ and $(-1, 0)$ and the north/south neighbors N and S are obtained by using the generators $(0, 1)$ and $(0, -1)$ respectively. The resulting Cayley graph of degree 4 on nm nodes is called a (two-dimensional) toroid. Refer to Figure 25.7 for examples.

Remark 25.3 To visualize a toroid, hold a rectangular sheet of paper and roll it longitudinally and form a tube. Then roll the tube to join them at their open ends. The resulting

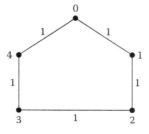

Figure 25.5 5-Node ring as a Cayley graph of $(Z_5, +_5)$.

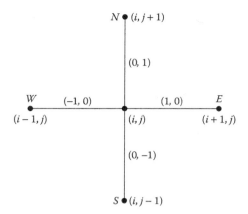

Figure 25.6 Neighbors of two dimensional toroid.

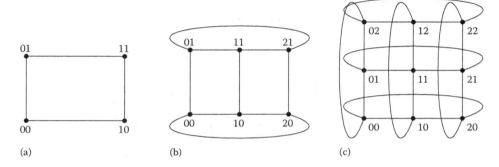

Figure 25.7 Examples of two-dimensional toroids.

object has a donut shape. The two-dimensional nm toroid is a graph that is drawn on the surface of this donut. Also notice that the nm grid, a close relative of the nm toroid, is not a Cayley graph.

Remark 25.4 (d-dimensional toroid) The concept of a toroid can be readily extended to higher dimensions. For $d = 3$, let $(Z_{n_1}, +_{n_1})$, $(Z_{n_2}, +_{n_2})$, and $(Z_{n_3}, +_{n_3})$ be three additive groups where $n_i \geq 2$, for $1 \leq i \leq 3$. Let $Z = Z_{n_1} \times Z_{n_2} \times Z_{n_3} = \{(i,j,k) | 0 \leq i < n_1, 0 \leq j < n_2, 0 \leq k < n_3\}$ with $|Z| = n_1 n_2 n_3 = n$. Let $\Omega = \{(\pm 1, 0, 0), (0, \pm 1, 0), (0, 0, \pm 1)\}$, where $|\Omega| = 6$. Thus, each node labeled by a triple (i, j, k) has six neighbors. The east/west neighbors $(i \pm 1, j, k)$ are obtained by using the generators $(\pm 1, 0, 0)$, the north/south neighbors $(i, j \pm 1, k)$ obtained by using the generators $(0, \pm 1, 0)$ and the back/front neighbors $(i, j, k \pm 1)$ obtained by using the generators $(0, 0, \pm 1)$, respectively.

A very popular CRAY-research parallel process T-D3 is based on the 3-dimensional toroidal network.

Example 25.14 (Supertoroid) Referring to the Example 25.10, consider the triple $(c, k, l) = (4, 1, 1)$. Let $\Gamma_1 = Z_{ck} = Z_4$ and $\Gamma_2 = Z_{c^2 l} = Z_{16}$. consider the generator $\Omega = \{(1, 0), (-1, 0), (0, 1), (0, -1)\}$. Let (a, b) be a node where $0 \leq a < 4$ and $0 \leq b < 16$, the four neighbors of (a, b) are then given by, see [4].

$$(a, b) * (\pm 1, 0) = (a \pm 1, (1 \mp cl)b) = (a + 1, -3b) \text{ and } (a - 1, 5b)$$

and

$$(a, b) * (0, \pm 1) = (a, b \pm 1) = (a, b + 1) \text{ and } (a, b - 1).$$

As an illustration consider the node $(1, 1)$. Its four neighbors are shown in Figure 25.8.

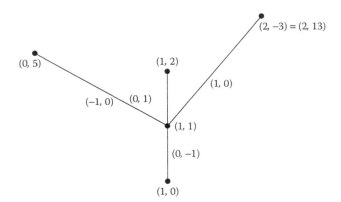

Figure 25.8 Four neighbors of supertoroid.

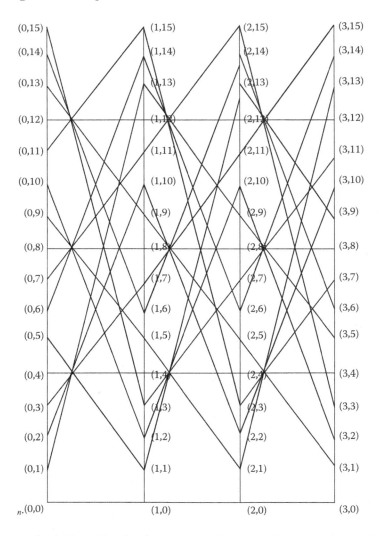

Figure 25.9 Supertoroid which is the Cayley graph of the semidirect product. The wrap around edges are not shown for simplicity.

The resulting Cayley graph on $4 \times 16 = 64$ nodes of degree 4 is shown in Figure 25.9.

Example 25.15 (Binary hypercubes) Consider the graph $(B_n, *)$ in Example 25.4 with $\Omega = \{g \in B_n |$ the Hamming weight of $g = 1\}$. Consider a graph with $|V| = 2^n$ nodes each labeled by an n-bit binary string. Let $X = x_n x_{n-1} \ldots x_2 x_1$ and $Y = y_n y_{n-1} \ldots y_2 y_1$ be two node labels. Then

$$E = \{(X,Y)_g | H_2(X,Y) = 1\}$$

where $H_2(X,Y)$ denotes the Hamming distance between X and Y defined by

$$H_2(X,Y) = \sum_{i=1}^{n} |x_i - y_i|$$

where $|a|$ is the absolute value of a. That is, two nodes are neighbors if and only if their node labels differ in only one bit position. For a given X, since Y can differ in each of the n-bits, the degree of this graph is n. The resulting Cayley graph, called the binary hypercube of dimension n, has $N = 2^n$ nodes with degree $n = \log_2 N$. Thus unlike the ring and toroid which are constant degree Cayley graphs, in the binary hypercube the degree is logarithmic in the size of the graph see [10–12]. Examples of binary hypercubes are given in Figure 25.10.

Remark 25.5 (Recursive structure of the binary hypercube) The binary hypercube admits a nice recursive definition as well. Let us denote the n-dimensional binary hypercube as $(N, 2, n)$-cube where $N = 2^n$. This $(N, 2, n)$-cube can be obtained by taking two copies of $(N/2, 2, n-1)$-cubes. The node labels in each of the two $n-1$ dimensional subcubes are $(n-1)$-bit strings. Let $X = x_{n-1} x_{n-2} \ldots x_2 x_1$. Then there is one node in each of these subcubes labeled X. To obtain $(N, 2, n)$-cube from these two $(N/2, 2, n-1)$-cubes, first add a prefix 0 to X to make it an n-bit string $0X = 0 x_{n-1} x_{n-2} \ldots x_2 x_1$ in one cube and add a prefix 1 to X to make it an n-bit string in the other cube. Since $H_2(0X, 1X) = 1$, we join $0X$ and $1X$ by a new edge. By repeating this process to each of the 2^{n-1} nodes, we obtain the $(N, 2, n)$-cube. This process is illustrated in Figure 25.11.

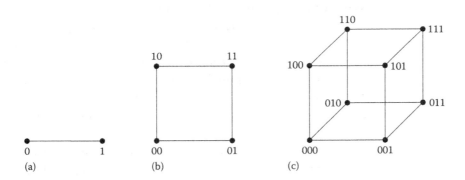

Figure 25.10 (a-c) Examples of d-dimensional binary hypercube for $1 \leq d \leq 3$.

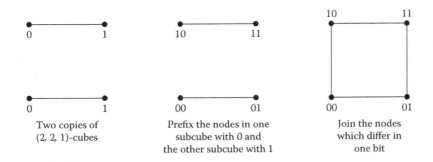

Figure 25.11 Demonstration of the recursive construction of $(4, 2, 2)$-cube.

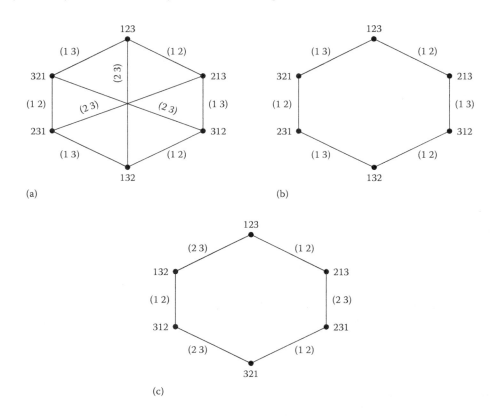

Figure 25.12 Examples of Cayley graphs of permutation graphs. (a) Complete transposition graph on S_3 using $\Omega_{CT} = \{(1,2),(1,3),(2,3)\}$. (b) Star graph on S_3 using $\Omega_{ST} = \{(1,2),(1,3)\}$. (c) Bubble sort graph on S_3 using $\Omega_{BS} = \{(1,2),(2,3)\}$

Example 25.16 (Cayley graphs of permutation groups)
  a. Complete transposition graph: Consider the symmetric group of $n!$ permutations. From Example 25.5 and Example 25.11(d), we know $\Omega_{CT} = \{(i,j) | 1 \leq i < j \leq n\}$ is a generator set for S_n. Since transpositions are their own inverse, Ω_{CT} satisfy the conditions $(c1)-(c2)$. The resulting Cayley graph of uniform degree $= n(n-1)/2$ is called the *complete transposition graph*, see [6,13]. For $n = 3$, $S_3 = \{I, (1\,2), (1\,3), (2\,3), (1\,2\,3), (1\,3\,2)\}$ and $\Omega_{CT} = \{(1\,2),(1\,3),(2\,3)\}$. The complete transposition graph is given in Figure 25.12a.
  b. Consider S_n with $\Omega_{ST} = \{(1\,j) | 2 \leq j \leq n\}$. The resulting Cayley graph is called the *star graph*. An example of the star graph for $n = 3$ is given in Figure 25.12b.
  c. Consider S_n with $\Omega_{BS} = \{(i\,i+1) | 1 \leq i < n\}$. The resulting Cayley graph is called the *bubble sort graph*. An example of the bubble sort graph on S_3 is given in Figure 25.12c.

Remark 25.6: These Cayley graphs on permutation group readily admit a recursive structure. Thus, a complete transposition graph on S_4 is built of 4-copies of S_3. While each of four subgraphs are generated using the generators $\{(1\,2),(1\,3),(2\,3)\}$, the four subgraphs are interconnected using the generators $\{(1\,4),(2\,4),(3\,4)\}$ resulting in a 24 node degree 6 graph called the complete transposition graph on S_4. Similarly one can recursively construct star graphs and bubble sort graphs on S_4.

25.3 SYMMETRY IN CAYLEY GRAPHS

In Section 25.1, we described the basic concepts of group isomorphism and group automorphism. In this section, we develop analogous concepts of isomorphisms and automorphisms for graphs in general and Cayley graphs in particular.

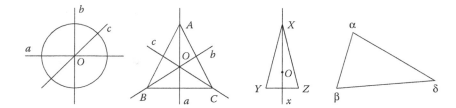

Figure 25.13 Examples of objects with varying symmetry.

Intuitively speaking, isomorphic objects—be it graphs or groups—have identical structures. However, an automorphism of an object onto itself relates to underlying structural symmetry of the object in question. Symmetry of an object relates to the ability to map the object onto itself using simple operations including rotation, reflection, to name a few. For example, consider a set of simple geometric figures—a circle, an equilateral triangle, an isosceles triangle, and an arbitrary triangle as in Figure 25.13.

Anticlockwise rotation with respect to an axis through its center O perpendicular to the plane of the circle by an angle θ, $0 \leq \theta < 2\pi$ maps the circle with center O onto itself. Similarly, reflection of the circle with respect to any line such as a, b, c, and so on that passes through the center O maps the circle onto itself. Stated in other words, the circle admits an infinitely many automorphisms $A_1 = \{$rotation by $\theta, 0 \leq \theta < 2\pi$, reflection with respect to any axis$\}$. An anticlockwise rotation P of the equilateral triangle ABC by an angle $2\pi/3 = 120°$ with respect to an axis passing through the center O and perpendicular to the plane of the triangle maps the equilateral triangle onto itself. The operation $P^2 = P \cdot P$ denotes the rotation by $240°$ and $P^3 = P^2 \cdot P$ denotes the rotation by $360°$. Thus, $P^3 = I$, the identity mapping of the triangle onto itself. Reflection R of the triangle ABC with respect to any one of the axes a, b, or c that passes through the center O and a vertex also maps the triangle onto itself. It can be verified that $R^2 = I$. Thus, the set $A_2 = \{P, P^2, P^3 = I, PR, P^2R, P^3R = R\}$ denotes a set of six automorphisms of the equilateral triangle ABC. Considering the isosceles triangle XYZ, it can be verified the reflection R with respect to the vertical axis x passing through the center O and the vertex X maps the triangle XYZ onto itself. In this case, $A_3 = \{R, R^2 = I\}$ are the two automorphisms. Similarly, it can be verified that $A_4 = \{I\}$, the identity mapping is the only set of automorphism of the general triangle $\alpha\beta\delta$.

The above exercise naturally leads to an inescapable conclusion namely the more the number of ways in which an object can be mapped onto itself, the more symmetric it is. Thus, a circle is more symmetric compared to an equilateral triangle which is more symmetric compared to an isosceles triangle which in turn is more symmetric compared to an arbitrary triangle. That is, larger the set of automorphisms of an object more symmetric the underlying object is.

Analysis of symmetry has a long and cherished history in science and engineering. Examples include analysis and classification of symmetry in crystals in solid-state physics and insistence of symmetry in engineering design and construction of Eiffel tower in Paris, St. Louis arch, Golden Gate bridge in San Francisco, commercial aircrafts such as Boeing 747, ocean liners, great cathedrals, and temples all over the world, to name a few.

In this section we develop important concepts and tools needed to describe the inherent symmetry in graphs, in particular Cayley graphs.

25.3.1 Graph Symmetry

We begin by introducing the notion of graph isomorphism. Let $G_1 = (V_1, E_1)$ and $G_2 = (V_2, E_2)$ be two simple, regular connected graphs on same number of vertices, that is,

$|V_1| = |V_2|$. We say G_1 is isomorphic to G_2 if there exists a function f satisfying the following two conditions.

I_1: There exists a bijective function $\alpha : V_1 \to V_2$, that is, α is a permutation of the node labels of the two graphs and

I_2: $(x, y) \in E_1$ if and only if $(\alpha(x), \alpha(y)) \in E_2$, that is, α in addition to being a permutation, also preserves the neighborhood.

Example 25.17 Referring to Figure 25.14, the graphs in Figure 25.14a are isomorphic to each other by the bijection α as defined in the table, while the graphs in Figure 25.14b are not isomorphic.

An isomorphism of a graph $G = (V, E)$ onto itself is called *(graph) automorphism*. Stated in other words, a graph automorphism is a permutation of its nodes that preserves the neighborhood. Let $A(G)$ denote the set of all automorphisms of a given graph G. Under the usual operation of composition of bijective functions (which is the same as the composition of permutations described in Section 25.2), this set $A(G)$ forms a group called the automorphism group of the graph G. We say an automorphism group is *regular* if $|A(G)| \geq |V|$.

Example 25.18 Referring to equilateral triangle ABC in Figure 25.13, it can be verified that
$$A(G) = \{(ABC), (ACB), (BC), (AB), (CA), I\}$$
is a regular automorphism group.

Now considering the isoceles triangle XYZ its automorphism group $A(G) = \{I, (YZ)\}$ and for the arbitrary triangle $A(G) = \{I\}$ are not regular.

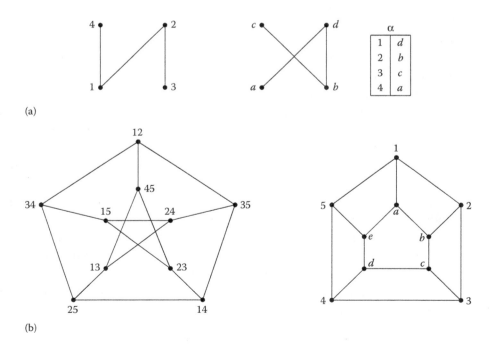

Figure 25.14 Examples of isomorphic and nonisomorphic graphs. (a) An example of isomorphic graphs. (b) An example of nonisomorphic graphs.

Remark 25.7: An automorphism group of the circle with center at O is given by

$$A(G) = \left\{ T = \begin{bmatrix} \cos\theta & \sin\theta \\ -\sin\theta & \cos\theta \end{bmatrix}, 0 \leq \theta < 2\pi \right\}$$

which is the orthogonal group $O(2)$. This infinite group is a continuous group. This is an example of the Lie group which in addition to the group structure also admits a rich topological structure. Thus one can talk about the derivative of T with respect to θ. However, for the other three objects in Figure 25.13, $A(G)$ is finite, which represents the discrete symmetry of these objects.

Given $G = (V, E)$ and its $A(G)$ we now characterize several important notions of graph symmetry.

Vertex symmetry A graph G is said to be *vertex symmetric* or *vertex transitive* if for every pair u, v of vertices, there exists an automorphism α (depending on u and v) such that $\alpha(u) = v$. Refer to Figure 25.15 for an illustration. In mapping u to v, the automorphism α also maps every neighbor x_i of u onto some neighbor y_i of v.

Example 25.19 Consider a four node graph of degree 2 given by Figure 25.16.

Let $P = (1234)$ denote the rotation by $2\pi/4 = 90°$ degrees and $R = (24)$ denote the reflection with respect to the diagonal connecting the vertices 1 and 3. Thus, $A(G) = \{P, P^2, P^3, P^4 = I, PR, P^2R, P^3R, P^4R = R\}$ is a regular automorphism group of this graph. Indeed, this graph is vertex transitive. For example, if $u = 1$ and $v = 3$, then $\alpha = P^2$ maps $u = 1$ to $v = 3$. In this process, the neighbors 2 and 4 of u are mapped to 4 and 2, respectively, which are the neighbors of $v = 3$.

Remark 25.8: It is immediate that while every vertex-transitive graph is an uniform (regular) graph, the converse is not always true as the following example shows.

Example 25.20 Consider the regular graph on 12 vertices with degree 3 as shown in Figure 25.17. It can be easily deduced that, this is an uniform graph but not vertex transitive.

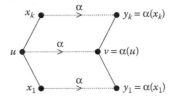

Figure 25.15 An illustration of vertex transitivity. Dotted lines refer to the map and solid lines are edges in the graphs.

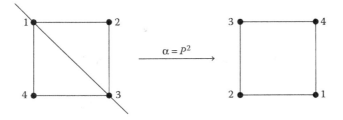

Figure 25.16 Example of vertex symmetry.

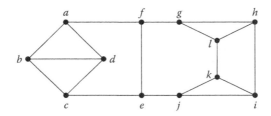

Figure 25.17 Uniform (regular) graph on 12 vertices but not vertex transitive.

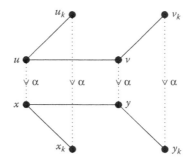

Figure 25.18 Illustration of edge transitivity.

Edge symmetry A graph G is *edge transitive* or *edge symmetric* if given any pair of edges $e_1 = (u, v)$ and $e_2 = (x, y)$, there exists a (vertex) automorphism $\alpha \in A(G)$ such that $\alpha(e_1) = e_2$, that is, $\alpha(u) = x$ and $\alpha(v) = y$. That is, there exists an automorphism α that simultaneously maps u to x and v to y. In so doing, α also maps all the neighbors of u onto the neighbors of x. Refer to Figure 25.18 for an illustration.

Example 25.21 The four node graph of degree 2 in Figure 25.16 is also edge transitive. Thus, if $e_1 = (1, 2)$ and $e_2 = (1, 4)$, thus $\alpha = R = (2\ 4)$ is such that $\alpha(1) = 1$ and $\alpha(2) = 4$, that is, $\alpha(e_1) = \alpha(e_2)$.

Remark 25.9: It is to be recognized that edge transitivity neither implies nor implied by vertex transitivity, that is these are two independent attributes of a graph. We illustrate this with the following example.

Example 25.22 Referring to Figure 25.19, the graph in Figure 25.19a, with three nodes has $A(G) = \{I, (1\ 2)\}$ and is edge transitive, since $\alpha = (1\ 2)$ maps the edge $(1, a)$ onto $(2, a)$. Since it is not uniform, it is not vertex transitive. The graph in Figure 25.19b on ten nodes has $A(G) = \{P, P^2, P^3, P^4, P^5 = I, PR, P^2R, P^3R, P^4R, P^5R = R\}$ where $P = (1\ 2\ 3\ 4\ 5)(a\ b\ c\ d\ e)$ and $R = (1\ a)(2\ b)(3\ c)(4\ d)(5\ e)$ and clearly is vertex transitive. However, the edge $e_1 = (1, a)$ cannot be mapped onto $e_2 = (c, d)$. For the edge (c, d) is a part of the 5-cycle a, b, c, d, e but the edge $(1, a)$ is *not* a part of any 5-cycle. Hence it is not edge transitive.

Remark 25.10: The following properties of the automorphisms are easily verified. Let $\alpha \in A(G)$. Then

A_1: $1 \leq |A(G)| \leq n!$ where $n = |V|$, that is the order of the automorphism group can vary widely.

A_2: $\text{degree}(u) = \text{degree}(\alpha(u))$, where $u \in V$.

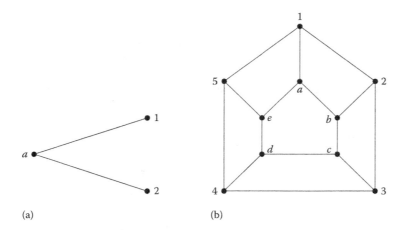

Figure 25.19 Edge transitivity. (a) Edge transitive. (b) Non-edge transitive.

A_3: Let $(u_0, u_1, u_2, \ldots, u_k)$ be a simple path/cycle in G. Then $(\alpha(u_0), \alpha(u_1), \alpha(u_2), \ldots, \alpha(u_k))$ is also a simple path/cycle.

A_4: $d(u, v) = d(\alpha(u), \alpha(v))$ where $d(x, y)$ denotes the length of the shortest path (measured by the number of edges) between x and y.

Distance symmetry Let D be the diameter of the graph which is defined as $D = \max\{d(x, y) |\ x, y \in V \text{ and } x \neq y\}$, that is, D is the maximum of the shortest distance between any pair of nodes in G. Let $1 \leq k \leq D$. The graph G is said to be k-*distance transitive* if given a set of four vertices u, v, x, and y such that $d(u, v) = d(x, y) = k$, then there exists $\alpha \in A(G)$ such that $x = \alpha(u)$ and $y = \alpha(v)$. If a graph is k-distance transitive for every k, $1 \leq k \leq D$, then it is called *distance transitive*.

Remark 25.11: It can be verified that vertex transitivity is simply zero-distance transitivity and 1-distance transitivity implies edge transitivity and not vice versa.

25.3.2 Symmetry and Cayley Graphs

We now move onto analyzing the symmetry in Cayley graphs.

Let $G = (V, E)$ be a Cayley graph of (Γ, Ω). We can exploit the properties of the group to define vertex automorphisms of the associated Cayley graphs. Let $a \in \Gamma$ be arbitrary but fixed. Define a function $L_a: \Gamma \to \Gamma$ where $f_a(x) = ax$, for all $x \in \Gamma$, called the left multiplication of the elements of the group by a. Similarly, one can define the right multiplication $R_a: \Gamma \to \Gamma$ where $f_a(x) = xa$, for all $x \in \Gamma$. It can be verified both L_a and R_a define a permutation of the elements of the group. From the definition of the Cayley graph, for each $g \in \Omega$, we have $(x, y)_g \in E$ only if $y = x.g$, that is $g = x^{-1}y$. From this using a series of algebraic manipulation we get

$$g = x^{-1}y = x^{-1}(a^{-1}a)y = (ax)^{-1}(ay) \text{ or } (ay) = (ax) \cdot g.$$

That is, if $(x, y)_g \in E$, then $(ax, ay)_g \in E$. Stated in other words left multiplication by an element of the group is a graph automorphism. The above analysis naturally leads to the following.

Fact: Every Cayley graph is vertex transitive. For, let u and v be two arbitrary nodes in the Cayley graph G. Define $a = vu^{-1}$. Then, $L_a(u) = v$. Let, $(u, x)_g \in E$. Then $g = u^{-1}x = u^{-1}(a^{-1}a)x = (au)^{-1}(ax)$ and hence $(au, ax)_g \in E$. From the arbitrariness of u, v and x, the claim follows.

Example 25.23

1. The binary hypercube is vertex transitive, edge transitive, and is also distance transitive.

2. Petersen graph is vertex transitive, edge transitive and distance transitive but it is not a Cayley graph.

3. The graph shown below

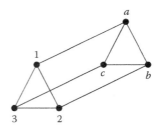

is vertex transitive but not edge transitive.

Another result of interest is stated below:

Let h be a homomorphism, of the group Γ generated by S into itself such that $h(S) = S$. Then h is a graph automorphism of the Cayley graph $G(\Gamma, S)$.

25.4 CONSEQUENCES OF VERTEX TRANSITIVITY

Intuitively a vertex-transitive graph looks the same when viewed through any vertex. This property has important consequences for developing many communication algorithms such as shortest path, broadcast, and so forth.

Shortest path problem Let $u, v \in V$ and $d(u, v) = k$, where $1 \leq k \leq D$, D the diameter of the Cayley graph $G = (V, E)$. This implies there exists a sequence of generators $g_{i_1}, g_{i_2}, \ldots, g_{i_k} \in \Omega$ such that the shortest path is given by

$$u = u_0 \xrightarrow{g_{i_1}} u_1 \xrightarrow{g_{i_2}} u_2 \to \cdots \xrightarrow{g_{i_{k-1}}} u_{k-1} \xrightarrow{g_{i_k}} u_k = v$$

where

$$u_j = u_0 \cdot g_{i_1} \cdot g_{i_2} \cdots g_{i_j}$$

for $1 \leq j \leq k$. That is

$$v = u \cdot g_{i_1} \cdot g_{i_2} \cdots g_{i_k}$$

from which we readily obtain

$$u^{-1}v = e \cdot g_{i_1} \cdot g_{i_2} \cdots g_{i_k}.$$

That is, for every shortest path specified by the sequence $g_{i_1}, g_{i_2}, \ldots, g_{i_k}$ between u and v, there exists a corresponding shortest path from e to $u^{-1}v$ also specified by the same sequence $g_{i_1}, g_{i_2}, \ldots, g_{i_k}$ and vice versa. Since $u^{-1}v$ span Γ as u and v are varied through Γ, it is immediate that instead of developing the $n(n-1)/2$ shortest paths between all possible pairs of nodes, we only need to find $(n-1)$ shortest paths from e to every other node in G. Hence the complexity of the shortest path is $O(n)$ in Cayley graph instead of $O(n^2)$ in general graphs. Further, we could use the same algorithm at each node.

Example 25.24 Referring to Example 25.16, consider the complete transposition graph on S_3 with Ω_{CT} as the generator set. Let $u = 123$ and $v = 231$. A shortest path P_1 between u and v is then given by

$$P_1 < u = 123 \xrightarrow{(1\ 3)} 321 \xrightarrow{(1\ 2)} 231 = v >.$$

We also observe that there are two more vertex-disjoint shortest paths, when we start with another generator.

(Two paths are said to be *vertex-disjoint*, if the paths have no common internal vertices.)

That is, $P_2 < 123 \xrightarrow{(1\ 2)} 213 \xrightarrow{(2\ 3)} 231 >$ and $P_3 < 123 \xrightarrow{(2\ 3)} 132 \xrightarrow{(1\ 3)} 231 >$ are also shortest paths between u and v.

While in the star graph, P_1 is the only shortest path between u and v. Similarly, P_2 is the only shortest path between u and v in the bubble sort graph.

Remark 25.12: In addition to determining a shortest path between pairs of vertices, finding vertex-disjoint paths in networks helps in reducing the congestion in network. Establishing vertex-disjoint paths by itself is an interesting and daunting problem.

Example 25.25 As another example, consider a three-dimensional torus given by $Z_4 \times Z_3 \times Z_4$ on 48 nodes. Let $u = (1, 2, 1)$ and $v = (3, 1, 0)$. A shortest path of length 4 between u and v is given by

$$P :< (1,2,1) \xrightarrow{(+1,0,0)} (2,2,1) \xrightarrow{(+1,0,0)} (3,2,1) \xrightarrow{(0,-1,0)} (3,1,1) \xrightarrow{(0,0,-1)} (3,1,0) >.$$

Alternatively, we also arrive at a shortest path by interchanging the generator $(+1, 0, 0)$ and the generator $(0, 0, -1)$. This technique of finding shortest paths in a d-dimensional toroid is well known as *one-step greedy algorithm*, where we move one step at a time along a particular dimension and obtain the desired shortest path.

Broadcasting The broadcast problem is one wherein a single node, say e, in a Cayley graph $G = (V, E)$ wants to send the same piece of message to every other node in an optimal time. A little reflection reveals that this can be accomplished by defining a minimum depth spanning tree of the Cayley graph G. In principle, such a tree can be obtained by a breadth-first search, where the depth of the resulting tree is D, the diameter of the graph.

Another interesting result is stated next. This result follows as a corollary to Propositions 24.7 and 24.8 given in Chapter 24.

> Every pair of vertices in a n-dimensional hypercube is connected by $n - 1$ paths of length at most n and a path of length at most $n + 1$, such that all these paths are vertex-disjoint.

25.5 CONCLUSION

In this chapter after a brief review of some key concepts of finite groups, we have described a large class of uniform, scalable graphs called Cayley graphs and their natural extension called Cayley coset graphs [4,11,14]. These Cayley graphs enjoy several symmetry properties which form the basis for the development of various communications algorithms; see [5,13]. Design of efficient Cayley graphs to serve as an interconnection network of parallel computing systems also serves as an interesting application; see [10,12,14–17]. For more details about interconnection networks, refer [8,9,18–21].

A recent development in the study of Cayley graphs is the generation of Cayley graph expanders. Expanders are graphs which are sparse but nevertheless highly connected.

Expanders have been useful in solving many fundamental problems in computer science (including network design, complexity theory, coding theory, and cryptography) and in various pure mathematics (such as game theory, measure theory, group theory). Construction of expanders has been a very daunting problem. Recently, it was proved that Cayley graphs serve as an efficient class of graphs in efficiently (polynomial time) constructing expanders; see [22]. In light of this work, new Cayley expanders (both deterministic and random) have come into existence. For more details, we refer to [23,24].

References

[1] M.A. Armstrong, *Groups and Symmetry*, Springer-Verlag, New York, 1988.

[2] C.F. Gardiner, *A First Course In Group Theory*, Springer-Verlag, New York, 1980.

[3] S. Lang, *Algebra*, Third edition, Springer Science, New York, 1993.

[4] F.L. Wu, S. Lakshmivarahan and S.K. Dhall, Routing in a class of Cayley graphs of semidirect products of finite groups, *J. Parallel Dist. Comput.*, **60**(5) (2000), 539–565.

[5] S. Lakshmivarahan, J.S. Jwo and S.K. Dhall, Symmetry in interconnection networks based on Cayley graphs of permutation groups: A survey, *Parallel Computing*, **19**(4) (1993), 361–407.

[6] S. Akers and B. Krishnamurthy, A group-theoretic model for symmetric interconnection networks, *IEEE Trans. Comput.*, **38** (1989), 555–566.

[7] J.A. Bondy and U.S.R. Murty, *Graph theory with applications*, Elsevier, North-Holland, 1976.

[8] W. Imrich and S. Klavžar, *Product Graphs: Structure and Recognition*, Wiley-Interscience Series in Discrete Mathematics and Optimization, New York, 2000.

[9] J.-M. Xu, *Topological Structure and Analysis of Interconnection Networks*, Kluwer Academic Publishers, Dordrecht/Boston/London, 2001.

[10] S. Lakshmivarahan and S.K. Dhall, Ring, torus and hypercube architectures/algorithms for parallel computing, *Parallel Comput.*, **25**(13–14) (1999), 1877–1906.

[11] S. Lakshmivarahan and S.K. Dhall, *Analysis and Design of Parallel Algorithms: Arithmetic and Matrix Problems*, McGraw-Hill, New York, 1990.

[12] S. Lakshmivarahan and S.K. Dhall, A new hierarchy of hypercube interconnection schemes for parallel computers, *J. Supercomput.*, **2** (1988), 81–108.

[13] M.-C.H. Bertrand and D. Bertrand, *Cayley graphs and interconnection networks*, Kluwer Academic Publishers, Boston, MA, 1997.

[14] J.-P. Huang, S. Lakshmivarahan and S.K. Dhall, Analysis of interconnection networks based on simple Cayley coset graphs. In *Parallel and Distributed Processing. Proc. of the 5th IEEE Symp.*, 150–157, December 1993.

[15] P. Vadapalli and P.K. Srimani, Trivalent Cayley graphs for interconnection networks, *Inform. Process. Lett.*, **54** (1995), 329–335.

[16] P. Vadapalli and P.K. Srimani, A new family of Cayley graph interconnection networks of constant degree four, *IEEE Trans. Parallel Dist. Sys.*, **7** (1996), 26–32.

[17] J.S. Jwo, S. Lakshmivarahan and S.K. Dhall. A new class of interconnection networks based on the alternating group. *Networks*, **23**(4) (1993), 315–326.

[18] J. Duato, S. Yalamanchili and L.M. Ni, *Interconnection networks: An engineering approach*, First edition, Morgan Kaufmann, San Francisco, CA, 2002.

[19] S.-Y. Hsieh and T.-T. Hsiao, The k-valent graph: A new family of Cayley graphs for interconnection networks, *Int. Conf. Parallel Process.*, **1** (2004), 206–213.

[20] L.-H. Hsu and C.K. Lin, *Graph Theory and Interconnection Networks*, Taylor & Francis Group, New York, 2008.

[21] D.S.L. Wei, F.P. Muga and K. Naik, Isomorphism of degree four Cayley graph and wrapped butterfly and their optimal permutation routing algorithm, *IEEE Trans. Parallel Dist. Sys.*, **10**(11) (1999), 1290–1298.

[22] N. Alon and Y. Roichman, Random Cayley graphs and expanders, *Random Struct. Algorithms*, **5** (1997), 271–284.

[23] E. Rozenman, A. Shalev and A. Wigderson, A new family of Cayley expanders, In *Proc. 36th Ann. ACM Symp. Theor. Comput.*, 445–454, New York, 2004. ACM.

[24] E. Rozenman, A. Shalev and A. Wigderson, Iterative construction of Cayley expander graphs, *Theor. Comput.*, **2**(1) (2006), 91–120.

CHAPTER 26

Graph Embedding and Interconnection Networks

S.A. Choudum

Lavanya Sivakumar

V. Sunitha

CONTENTS

26.1	Embeddings and Quality Measures	653
26.2	Embedding into Hypercubes	656
	26.2.1 Cubical Graphs	657
	26.2.2 Trees in Hypercubes	662
	26.2.3 Collection of Cycles, Paths, and Trees in Hypercubes	670
26.3	Embedding into Variants of Hypercubes	674
	26.3.1 Embedding into Augmented Cubes AQ_n	674
	26.3.2 Embedding into Crossed Cubes CQ_n	680
	26.3.3 Embedding into Twisted Cubes TQ_n	681
	26.3.4 Embedding into Enhanced Cubes $Q_{n,k}$	681
	26.3.5 Embedding into Hamming Graphs $H_{b,n}$	681
26.4	Summary	682

GRAPH EMBEDDINGS are a subject of study in mathematics and computer science. In one of the applications, they are mathematical models capturing the issues involved in the implementation of parallel algorithms on a parallel computer. Theoretically too they are significant: they generalize the fundamental concepts of graph isomorphism and subgraph relation. This chapter is a brief report on fundamental problems and their progress with emphasis on embeddings of various computational graphs into hypercubes and variants of hypercubes.

26.1 EMBEDDINGS AND QUALITY MEASURES

We assume that the reader is familiar with the contents of Chapter 24.

Definition 26.1 *Let $\wp(G)$ denote the set of all paths in a graph G. An* embedding *of a guest graph $G(V, E)$ into a* host *graph $H(W, F)$ is a pair of functions (f, ρ) where $f : V \to W$, $\rho : E \to \wp(H)$, and ρ maps an edge (u, v) of G onto a path connecting $f(u)$ and $f(v)$ in H.*

It is not necessary that the functions f or ρ be injective or surjective. However, we will be often seeking the functions with these properties. They are practically more useful. An extensively studied special case of (f, ρ) is one in which ρ maps (u, v) to a shortest $(f(u), f(v))$-path. We write $\rho(u, v)$, instead of $\rho((u, v))$ for the image of (u, v) under ρ. For brevity, an embedding (f, ρ) may be denoted by f alone, when ρ is clear from the context, this is especially so when ρ maps (u, v) onto a shortest $(f(u), f(v))$-path.

Several parameters are associated with an embedding (f, ρ) to measure its qualities. In the following definitions, (f, ρ) is an embedding of a graph $G(V, E)$ into a graph $H(W, F)$.

Definition 26.2 *The* load *of a vertex w in H is the number of vertices of G mapped onto w; that is, $load(w) = |f^{-1}(w)|$. The* load *of an embedding f is then defined as*

$$load(f) := \max\{|f^{-1}(w)| : w \in V(H)\}.$$

In the context of parallel processing, an embedding which maps a set P of distinct nodes of the guest graph onto a single node p of the host graph will reduce the parallel processing intended to be performed by the distinct processors of P in the guest graph to a sequential operation to be performed by the single processor p in the host graph in actual processing. So, one aims to minimize the load factor.

Definition 26.3 *The* dilation *of an edge $e(u, v)$ in G is the length of the path $\rho(u, v)$. It is denoted by $dil(e)$. The* dilation *of the embedding (f, ρ) is then defined as*

$$dil(f, \rho) := \max\{dil(e) : e \in E(G)\}.$$

When ρ is a map which associates an edge (u, v) with a shortest $(f(u), f(v))$-path, the above definition of dilation can be rewritten as

$$dil(f) := \max\{d(f(u), f(v); H) : (u, v) \in E(G)\}.$$

Clearly, there is a close relationship between dilation and computational latency in the host graph. If t is the time required to pass a message from u to v through a link $e(u, v)$ in the guest graph, then $t \cdot dil(e)$ is the time required to pass the message from $f(u)$ to $f(v)$ in the host graph.

A load-1 embedding is an injective map and a dilation-1 embedding preserves adjacency. *So, there is a load-1, dilation-1 embedding of G into H if and only if G is isomorphic to a subgraph of H.* Hence it is convenient to say that G is a subgraph of H whenever there is a load-1, dilation-1 embedding of G into H.

Definition 26.4 *The* edge congestion *of an edge e' of H is defined as*

$$ec(e') := |\{e \in E(G) : e' \in \rho(e)\}|.$$

The edge congestion *of (f, ρ) is then defined as*

$$ec(f, \rho) := \max\{ec(e') : e' \in E(H)\}.$$

Edge-congestion again leads to computational latency. For example, if a link $(s,t) \in E(H)$ is on two paths $P_1 = P(u,v)$ and $P_2 = P(x,y)$, and if the messages to be passed through P_1 and P_2 reach the node s at the same time, then one of the messages has to wait until the other passes through the link (s,t). (In such cases, the priority protocol is assigned a priori.) The edge congestion may also lead to deadlock as the same link in the host graph may have to serve the demands of a number of message routings. Clearly, a load-1, dilation-1 embedding has edge congestion-1.

When one aims to minimize the congestion $ec(f,\rho)$, the map ρ need not necessarily associate an edge (u,v) with a shortest $(f(u),f(v))$-path.

Definition 26.5 *The node congestion of a node $x \in V(H)$ is defined as*

$$nc(x) := |\{e \in E(G) : x \in \rho(e)\}|.$$

The node congestion of (f,ρ) is then defined as

$$nc(f,\rho) := \max\{nc(x) : x \in V(H)\}.$$

Again it is easy to see that higher the node congestion, higher the computational latency. An illustration of embedding is given in Figure 26.1. Here $v_i \in V(G)$ is mapped into $V_i \in V(H)$.

Definition 26.6 *The expansion of an embedding f is defined as*

$$exp(f) := \frac{|V(H)|}{|V(G)|}.$$

Unlike the previous four measures, this is not related to the computational latency. Rather it gives a measure of the number of unutilized processors by f in H. So, given f, it can be used to select an appropriate H to bring down the overhead cost.

Clearly, a load-1 embedding has expansion at least 1. From a practical point of view, one prefers to have an embedding with load-1, dilation-1, and expansion ε, where $1 \leq \varepsilon < 2$. However, in many cases such an embedding may not exist. Note that there exists a load-1, dilation-1, and expansion-1 embedding of a graph G into a graph H if and only if G is a spanning subgraph of H.

These measures are worst-case measures. On the same lines average measures are also defined. A central problem on graph embeddings can now be formulated.

> **Meta problem:** Given two graphs G and H, find an embedding that maps G into H with minimum load, minimum dilation, minimum node-congestion, minimum edge-congestion, and minimum expansion.

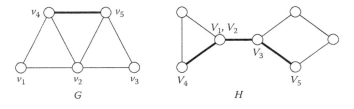

Figure 26.1 Embedding (f,ρ) of G into H with load 2, dilation 3, edge congestion 3, and node congestion 2.

This is too general a problem to hope for a neat solution. For arbitrary graphs G and H, there need not exist an embedding with all the parameters having minimum value. The following is a well-known NP-complete decision problem which is related to the above meta problem.

Problem 26.1 (Subgraph problem)

Input: *Two graphs G and H.*
Question: *Is G a subgraph of H?*

Linear arrays (i.e., paths) and cycles being most trivial topologies, their embeddings into networks have gained a lot of importance. Through Sekanina's [1,2] and Karaganis's [3] result on Hamiltonian-connectedness of cubes of connected graphs, it becomes evident that cycles and hence linear arrays can be embedded with constant measures into any network. Below, we present this result using the graph embedding terminologies.

Theorem 26.1 [1,2] *A cycle C on n vertices can be embedded into an n-node connected graph G with load-1, dilation ≤ 3, and edge-congestion ≤ 2.*

Proof. Consider any spanning tree of G and traverse the tree in a depth-first order (prepostorder). This traversal produces a listing of the vertices of G in a cyclic order $(v_0, v_1, \ldots, v_{n-1})$ such that the distance between v_k and $v_{(k+1) \bmod n}$ is at most 3, for $0 \leq k < n$. So, the load and the dilation of the resulting embedding are 1 and ≤ 3, respectively. Since any edge of G is traversed at most twice (once in each direction), the edge-congestion of the embedding is at most two. ∎

26.2 EMBEDDING INTO HYPERCUBES

In this section, we discuss embeddings in which the hypercubes are host graphs. We begin with an embedding of an arbitrary graph into a hypercube.

Theorem 26.2 [4,5] *If G is a graph on $n + 1$ vertices, then there exists a subdivision G' of G such that G' is isomorphic to a subgraph of Q_n. In particular, K_{n+1} can be embedded into Q_n with dilation 2.*

Proof. Since every graph G on $n+1$ vertices is a spanning subgraph of K_{n+1}, in order to prove the theorem, it is enough to prove that there exists a subdivision of K_{n+1} which is isomorphic to a subgraph of Q_n. Let $v_1, v_2, \ldots, v_{n+1}$ be the vertices of K_{n+1}. For $1 \leq i < j \leq n$, subdivide the edge (v_i, v_j) of K_{n+1} exactly once by introducing a new vertex $v_{i,j}$ on it. Let K' denote this subdivision of K_{n+1}. Note that $V(K') = \{v_i : 1 \leq i \leq n+1\} \cup \{v_{i,j} : 1 \leq i < j \leq n\}$. We use the following notation to denote some of the vertices of Q_n:

$$0^n := \underbrace{0 \ldots 0}_{n \text{ times}}, \quad 0^n_x := \underbrace{0 \ldots 0}_{x-1 \text{ times}} 1 \underbrace{0 \ldots 0}_{n-x \text{ times}}, \quad 0^n_{x,y} := \underbrace{0 \ldots 0}_{x-1 \text{ times}} 1 \underbrace{0 \ldots 0}_{y-x-1 \text{ times}} 1 \underbrace{0 \ldots 0}_{n-y \text{ times}}$$

We now define a bijection f between $V(K')$ and the set $\{0^n\} \cup \{0^n_i : 1 \leq i \leq n\} \cup \{0^n_{i,j} : 1 \leq i \neq j \leq n\}$ as

$$f(v_{n+1}) = 0^n, f(v_i) = 0^n_i, 1 \leq i \leq n, f(v_{i,j}) = 0^n_{i,j}, 1 \leq i < j \leq n.$$

Clearly, this bijection yields an isomorphism between K' and a subgraph of Q_n and in turn a dilation-2 embedding from K_{n+1} to Q_n. ∎

Although, from the above theorem, we have an embedding of any arbitrary graph into a hypercube, the expansion $(2^n/(n+1))$ is too large to be of any practical use. Therefore, one aims at finding better embeddings of graphs into hypercubes.

From the recursive construction of hypercubes, it follows that, there exists an embedding from Q_m to Q_n with load-1 and dilation-1 whenever $m \leq n$. The following result is about embedding from Q_m to Q_n when $m > n$.

Theorem 26.3 [6] *Hypercube Q_m can be embedded into hypercube Q_n with dilation-1 and an optimal congestion of 2^{m-n}, when $m > n$.* ∎

Definition 26.7 *Given a graph G, let n be the smallest integer such that $2^n \geq |V(G)|$. Then*

- Q_n *is called the* optimal hypercube *of G.*
- Q_{n+1} *is called the* next-to-optimal hypercube *of G.*

Remark 26.1 The expansion of any embedding of a graph G into its optimal hypercube will lie in the interval [1,2).

26.2.1 Cubical Graphs

Definition 26.8 *A graph G is said to be* cubical *if it is a subgraph of Q_n, for some n.*

It is easy to see that every even cycle is cubical and that every tree is cubical. A standard technique to show that such an elementary graph G, with a certain property P, is cubical is by induction on the number of vertices in G. It involves the following four steps:

i. Delete certain edges of a matching M in G to obtain two smaller subgraphs G_1 and G_2, both having property P. It is not always easy to find such an M.

ii. Choose a hypercube Q_n of large dimension n and canonically decompose it into two smaller hypercubes say Q_{n-1}^0 and Q_{n-1}^1.

iii. Embed G_1 into Q_{n-1}^0 and G_2 into Q_{n-1}^1. Here one uses symmetric properties of the hypercubes, for convenience these properties are stated under Proposition 24.10 as given in Chapter 24.

iv. Finally combine G_1 and G_2 by utilizing the edges of the perfect matching which exists between Q_{n-1}^0 and Q_{n-1}^1.

As an illustration of this technique we prove the following result that every tree is cubical [7].

Theorem 26.4 *Every tree T_n on n vertices is a subgraph of Q_{n-1}, where $n \geq 2$.*

Proof. We prove the result by induction on n. Clearly, T_2 is contained in Q_1. So we proceed to the induction step. Let (v, w) be an edge in T_n where $n \geq 3$. By deleting (v, w), we obtain two smaller trees, say T^1 and T^2, where $v \in T^1$ and $w \in T^2$. Decompose Q_{n-1} into smaller hypercubes Q_{n-2}^0 and Q_{n-2}^1. By induction hypothesis, T^1 is a subgraph of Q_{n-2}^0 and T^2 is a subgraph of Q_{n-2}^1. Since hypercubes are vertex-symmetric (see Proposition 24.10 as given in Chapter 24), we can assume that v is mapped onto a vertex $0X$ of Q_{n-2}^0 and that w is mapped onto the vertex $1X$ of Q_{n-2}^1. Since $0X$ and $1X$ are adjacent in Q_{n-1}, we conclude that T_n is isomorphic to a subgraph of Q_{n-1}. ∎

Observe that, embedding a graph G into Q_n with load-1 and dilation-1 is equivalent to labeling the vertices of G using n-bit binary strings such that the labels of the adjacent vertices differ in exactly one position. For example, any path or an even cycle admits such a labeling.

The next theorem states a few necessary conditions for a graph to be cubical; its proof follows since hypercubes have these properties.

Theorem 26.5 *If G is cubical then the following hold:*

  i. *G is bipartite.*

 ii. *$K_{2,3}$ is not a subgraph of G.*

iii. *If G has bipartition $[X, Y]$ and it is subgraph of Q_n, for some n, then $|X| \leq 2^{n-1}$, $|Y| \leq 2^{n-1}$.*

 iv. *$\Delta(G) \leq n$.*

  v. *Given any vertex v of G, the number of vertices at distance at most d from v is $\leq \binom{n}{0} + \binom{n}{1} + \cdots + \binom{n}{d}$.* ∎

However, the problem of deciding whether a given graph G is cubical is NP-complete.

Theorem 26.6 [8] *Given an arbitrary undirected graph G with maximum degree at most four, it is NP-complete to decide whether G is cubical.* ∎

If the maximum degree of G is at most two, then every component of G is a path or a cycle and so, one can decide in linear time whether G is cubical. However, the time complexity of the following decision problem is unknown.

Problem 26.2

Input: *A graph G with maximum degree three.*
Question: *Is G cubical?*

Theorem 26.7 [9] *Given an undirected graph G and positive integers k and n, it is NP-complete to decide whether G is embeddable into Q_n with load-1 and dilation-k. The problem is NP-complete even when $k = 1$. That is, given n, the problem of deciding whether G is a subgraph of Q_n is NP-complete.* ∎

For similar results on NP-completeness of congestion-1 embeddings we refer to [10]. Following theorem characterizes cubical graphs.

Theorem 26.8 [11] *A graph G is a subgraph of Q_n if and only if there exists a proper n-edge coloring of G such that:*

1. *In every open path of G, at least one color appears an odd number of times; (open path := origin and terminus are nonadjacent in G).*

2. *In every cycle of G, no color appears an odd number of times.*

Proof. Assume, without loss of generality, that G is connected.

First let $G \subseteq Q_n$; for $x \in V(G)$, let X denote the corresponding vertex in Q_n. Define $C : E(G) \to \{1, 2, 3, \ldots, n\}$ by $C(e(x,y)) = i$, if X and Y differ in position i. Clearly, C is a proper n-edge coloring of G. Moreover, if $P := \langle v_1, v_2, \ldots, v_t \rangle$ is a path in G and $\theta(P) = \{i :$ there are odd number of edges on P colored $i\}$, then the vertices $(x_1 x_2 \ldots x_n)$ and $(z_1 z_2 \ldots z_n)$ are related as follows:

$$z_i = \begin{cases} \overline{x_i}, & \text{if } i \in \theta(P), \\ x_i, & \text{if } i \in \{1, 2, \ldots, n\} \setminus \theta(P). \end{cases}$$

It now easily follows that C satisfies (1) and (2).

Next to prove the converse, let $C : E(G) \to \{1, 2, 3, \ldots, n\}$ be a proper n-edge coloring of G satisfying (1) and (2). Define $f : E(G) \to V(Q_n)$ by $f(e) = (0 \ldots 010 \ldots 0)$, where 1 appears in the jth position if $C(e) = j$. Choose a vertex x_0 in G and define $\phi : V(G) \to V(Q_n)$ by

$$\phi(x_0) = (00 \ldots 0),$$

and

$$\phi(x) = f(e_1) \oplus f(e_2) \oplus \cdots \oplus f(e_r),$$
$$\text{if } \langle x_0 e_1 x_1 e_2 \ldots x_{r-1} e_r x_r (= x) \rangle \text{ is a } (x_0, x) - \text{path } P \text{ in } G.$$

Observe that if $\theta(P) = \{j : j$ appears an odd number of times on the edges of $P\}$, then $\phi(x) = (z_1 z_2 \ldots z_n)$ where

$$z_i = \begin{cases} 1, & \text{if } i \in \theta(P), \\ 0, & \text{if } i \in \{1, 2, \ldots, n\} \setminus \theta(P). \end{cases}$$

Using this observation it is straight forward to verify that $\phi(x)$ is independent of the choice of (x_0, x)-path (i.e., ϕ is well defined) and that ϕ is a embedding of G into Q_n with load-1 and dilation-1. ∎

See Livingston and Stout [9] for more details on cubical graphs.

Theorem 26.9 [12] *If a graph G on n vertices is cubical, then $G \subseteq Q_{n-1}$.*

Proof. Let p be the smallest integer such that $G \subseteq Q_p$. For any i, there is at least one edge of G in $[Q_{p-1}^{0,i}, Q_{p-1}^{1,i}]$; else $G \subseteq Q_{p-1}^{0,i}$ (or $Q_{p-1}^{1,i}$), a contradiction to the minimality of p. So, by deleting the edges of G which are along the dimension i, we increase the number of components. By successively deleting the edges of dimension $i = 1, 2, \ldots, n-1$, every time we increase the number of components in the residual of G. So, after at most $n-1$ steps, the number of components in the residual graph H of G is n, that is no edge of G remains. Hence, every edge of G is an edge of dimension i in Q_p, for some $i \in \{1, 2, \ldots, n-1\}$. We conclude that $G \subseteq Q_{n-1}$. ∎

The above theorem motivates the following concept.

Definition 26.9 *If k is the smallest integer such that G is a subgraph of Q_k, then k is called the* cubical dimension *of G and it is denoted by $cd(G)$.*

The problem of finding $cd(G)$ does not get easier with the knowledge that G is cubical.

Theorem 26.10 [9] *The problem of finding the cubical dimension of an arbitrary cubical graph is NP-complete.* ∎

Clearly, a graph is cubical if and only if its components are cubical. However, if a cubical graph G has components $G_1, G_2, \ldots, G_k$ then the problem of expressing $cd(G)$ as a function of $cd(G_i)$, $1 \leq i \leq k$, seems to be difficult. However, one can use the recursive definitions of a hypercube to derive rough bounds.

In the next theorem, we list a few known results on cubical dimension.

Theorem 26.11 *For any cubical graph G on p vertices, the following inequalities hold.*

1. $\lceil \log p \rceil \leq cd(G) \leq p - 1$.

2. *If G is a connected cubical graph (with $\delta(G) \geq 2$), then $cd(G) \leq 2(p-1)/3$ [9].*

3. *If G is a 2-connected cubical graph then $cd(G) \leq p/2$ [8].* ∎

Note that, if a graph G on p vertices is a subgraph of Q_n, then $2^n \geq p$, that is $n \geq \lceil \log p \rceil$, and hence the lower bound of Theorem 26.11(1) follows immediately. Further, the upper bound of Theorem 26.11(1) follows by Theorem 26.9.

Remark 26.2 Each of the bounds given by the above theorem is tight in the following sense.

1. There are graphs whose cubical dimensions attain the bounds of Theorem 26.11(1). For example, $cd(P_p) = \lceil \log p \rceil$ and $cd(K_{1,p-1}) = p - 1$.

   A tighter lower bound can be obtained by looking at the bipartite sets X, Y of G: If $G \subseteq Q_n$, then $|X| \leq 2^{n-1}$ and $|Y| \leq 2^{n-1}$. For example, for any connected bipartite graph $G[X, Y]$ where $|X| = 9$, $|Y| = 6$, we deduce that $cd(G) \geq 5 = \lceil \log p \rceil + 1$.

2. There are cubical graphs on p nodes with cubic dimension $2(p-1)/3$. However, a tighter bound can be given for graphs with no cut-vertices.

3. There are 2-connected graphs on p nodes with cubic dimension $p/2$.

Both Havel and Morávek [11] and Garey and Graham [12] independently introduced *critical graphs* to obtain structural description of cubical graphs. A few results and open problems are found in both.

Definition 26.10 *A graph G is said to be Q-critical if G is not cubical but every proper subgraph of G is cubical.*

Clearly, every odd cycle is Q-critical. Some examples of bipartite Q-critical graphs are shown in Figure 26.2.

Theorem 26.12 [11,12] *For any integer g, there exists a bipartite Q-critical graph with smallest cycle having length greater than g.* ∎

Problem 26.3 [12] *Does every Q-critical graph have a vertex of degree 2?*

Problem 26.4 [12] *How to combine two Q-critical graphs to obtain a new Q-critical graph?*

In a load-1 and dilation-1 embedding, we demand that the map preserve the adjacency. We can further demand that the map preserve the distance between any two vertices, as in *isometric embeddings* in classic analysis.

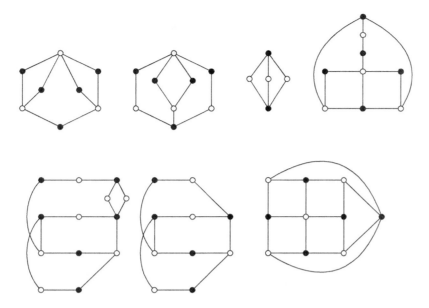

Figure 26.2 Examples of bipartite Q-critical graphs.

Definition 26.11 A graph G is said to be isometrically embeddable into Q_n if there exists a one-one function $f : V(G) \to V(Q_n)$ such that $dist(u, v; G) = dist(f(u), f(v); Q_n)$, for all $u, v \in V(G)$. A graph isometrically embeddable into a hypercube of some dimension is called a partial hypercube.

Several characterizations of isometrically embeddable graphs are known. We list some in Theorem 26.13. These in turn have led to algorithms for their recognition (see [13]).

Theorem 26.13 Let G be a connected bipartite graph. Then the following statements are equivalent.

a. G is isometrically embeddable in a hypercube.

b. For a pair of vertices v_1, v_2 in G, let $c(v_1, v_2) = \{x \in V(G) : dist(x, v_1) < dist(x, v_2)\}$. Then for every edge (v_1, v_2) of G and for all $x, z \in c(v_1, v_2)$, $d(x, y) + d(y, z) = d(x, z) \Rightarrow y \in c(v_1, v_2)$; that is, for every edge e with end vertices v_1, v_2, both $c(v_1, v_2)$ and $c(v_2, v_1)$ are convex [14].

c. The relation $\theta(G)$, defined on $E(G)$ as follows, is transitive. If $e = (x, y)$ and $f = (u, v)$ are two edges, then $e \, \theta \, f$ iff $d(x, u) + d(y, v) \neq d(x, v) + d(y, u)$. (In general, θ is reflexive and symmetric but need not be transitive) [15].

d. The distance matrix of G has exactly one positive eigenvalue [16]. ∎

Clearly, every isometrically embeddable graph is cubical. However, the converse is not true; that is, there are cubical graphs which are not partial hypercubes. For instance, the graph shown in Figure 26.3 is cubical as is evident through the labeling but it is not isometrically embeddable into a hypercube because its distance matrix has more than one positive eigenvalue.

The strict hierarchy of hypercubes to bipartite graphs is shown in Figure 26.4 and the time-complexity for recognizing them is shown in Table 26.1.

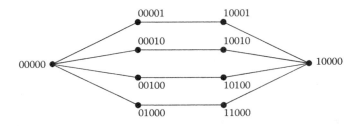

Figure 26.3 Cubical graph which is not a partial hypercube.

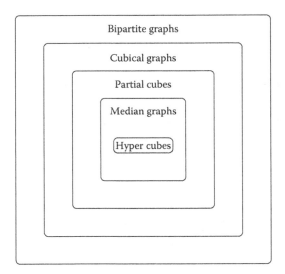

Figure 26.4 Hierarchy of graphs.

TABLE 26.1 Time Complexity for Recognition for the Hierarchy of Graphs

Class of Graphs with n Vertices and m Edges	Time Complexity for Recognition
Hypercubes	$O(m+n)$ [13]
Median graphs	$O((m \log n)^{1.41})$ [17]
Partial cubes	$O(mn)$
Cubical graphs	NP-complete
Bipartite graphs	$O(m+n)$

26.2.2 Trees in Hypercubes

Tree embeddings have received maximum attention. From a programming perspective, trees form an important class of computational structures. They naturally arise in the design of sequential and parallel algorithms which require basic operations like merging, sorting, and searching. Despite the easy proof that every tree is cubical (see Theorem 26.4), the following problems are open and consequent decision problems are known to be NP-complete.

Problem 26.5 *Which trees are subgraphs of Q_n?*

Theorem 26.14 [18] *Given a tree T and an integer n, it is NP-complete to decide if T is a subgraph of Q_n.*

Problem 26.6 [18] *Given k (≥ 3) and n, find the complexity of the problem of embedding a tree T with $\Delta(T) = k$ into Q_n. The problem is open even for $k = 3$.*

To illustrate the difficulties involved in embedding trees in hypercubes we state and prove a few basic theorems. Before that we introduce some necessary definitions and notations.

Definition 26.12 *A bipartite graph is called* equibipartite *or* equipartite *if its two parts are of equal order.*

Hypercubes are equibipartite graphs; see Proposition 24.2 as given in Chapter 24.

Definition 26.13

1. *A* rooted tree *is a tree in which a vertex is arbitrarily chosen and designated as its* root.

2. *A* k-ary tree *is a rooted tree with root having degree at most k, and every other vertex having degree at most $k + 1$.*

3. *A vertex of degree one is called a* leaf. *All other vertices are called* internal *vertices.*

4. *The maximum distance between the root and a leaf is called the* height *of the tree.*

5. *The* complete k-ary tree *is a k-ary tree in which the root has degree k, every internal vertex has degree $k + 1$, and all the leaves are at the same distance from the root.*

6. *If $\langle v_0, v_1, \ldots \rangle$ is a path in a k-ary tree where v_0 is the root, then v_{i+1} is called a* child *of v_i, v_{i+t} is called a* descendent *of v_i, and v_i is called an* ancestor *of v_{i+t}.*

Any tree T with $\Delta(T) \leq k + 1$ is a k-ary tree; designate a vertex of degree at most k as its root. In a k-ary tree every vertex has at most k children. A complete k-ary tree of height h has $1 + k + k^2 + \cdots + k^h = (k^{h+1} - 1)/(k - 1)$ vertices ($k \neq 1$).

A 2-ary tree is often called a binary tree. In the following, CBT_n denotes a complete binary tree; it has $2^n - 1$ vertices and height $n - 1$. Although CBT_n has $2^n - 1$ vertices and Q_n has 2^n vertices, it is not a subgraph of Q_n. The proof uses the fact that Q_n is equibipartite where as CBT_n is not equibipartite.

Theorem 26.15 [19] CBT_n *is not a subgraph of Q_n, for every $n \geq 3$.*

Proof. Consider the level decomposition $[N_0(r), N_1(r), \ldots, N_{n-1}(r)]$ of CBT_n where $N_i(r)$ denotes the set of vertices at distance i from the root r. Then, $S = N_0(r) \cup N_2(r) \cup \cdots$ and $T = N_1(r) \cup N_3(r) \cup \cdots$ are the two bipartite sets of CBT_n with one part containing more than 2^{n-1} vertices. Hence $CBT_n \not\subseteq Q_n$. See Figure 26.5 for an illustration showing $CBT_3 \not\subseteq Q_3$. ∎

Theorem 26.16 [19] CBT_n *is a subgraph of Q_{n+1}, for $n \geq 1$.*

Proof. Let $G(n; r, u, v)$ be the tree on $2^n + 1$ vertices obtained from CBT_n with root r, by adding two new vertices u, v and two new edges $(r, u), (u, v)$. We prove that this supergraph $G(n; r, u, v)$ of CBT_n is a subgraph of Q_{n+1}, by induction on n.
The case $n = 1$ is obvious. Assume that the theorem holds for $n - 1$ and consider a canonical decomposition $Q'_n \ominus Q''_n$ of Q_{n+1}. By induction hypothesis, $G(n-1; r', u', v')$ is a subgraph of Q'_n. Let the path $<r', u', v'>$ of the tree be mapped onto the path $<R', U', V'>$ in Q'_n. Using induction hypothesis and P_3-symmetry of Q''_n, we can embed another copy of $G(n-1; r', u', v')$

in Q_n'' where the path $<r', u', v'>$ is mapped onto the path $<U'', V'', W>$ in Q_n'', such that U'' and V'' are the unique vertices of Q_n'' that are adjacent to U' and V' in Q_{n+1}. Now, we combine these two embeddings by adding the edges (U', U''), (V', V'') and deleting the edges (U'', V'') and (V'', W) to obtain $G(n; r, u, v)$ where $<r, u, v>$ is mapped onto the path $<U', V', V''>$; refer to Figure 26.6 for an illustration of the proof. ∎

Corollary 26.1 *For $n \geq 3$, $cd(CBT_n) = n + 1$.*

Proof. Follows from Theorems 26.15 and 26.16 ∎

Definition 26.14 *In CBT_n let r be the root and x, y be its children. The tree obtained by subdividing the edge (r, y) with a new vertex s is called a* double rooted complete binary tree with roots r and s. *It is denoted by $D(n; x, r, s, y)$ or $D(n)$; see Figure 26.7.*

Clearly, every $D(n)$ is equibipartite and contains 2^n vertices. It has been discovered several times that $D(n)$ spans Q_n.

Theorem 26.17 *For every $n \geq 1$, the double rooted complete binary tree on 2^n vertices spans Q_n.*

Proof. Theorem is proved by induction on n. The base case of $n = 1$ can be easily verified. Let $D(n-1; e, f, g, h)$ be a vertex disjoint copy of $D(n-1; a, b, c, d)$. Let $Q_n = Q'_{n-1} \ominus Q''_{n-1}$ be a canonical decomposition of Q_n. If A' is a vertex of Q'_{n-1}, then we shall denote by A'', the unique vertex of Q''_{n-1} adjacent to A' in Q_n. By induction hypothesis $D(n-1; e, f, g, h)$ and

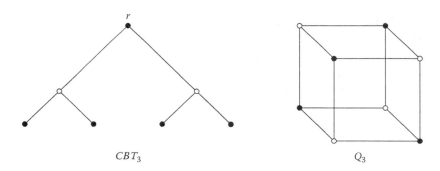

Figure 26.5 CBT_3 is not a subgraph of Q_3, since one part of CBT_3 has five vertices.

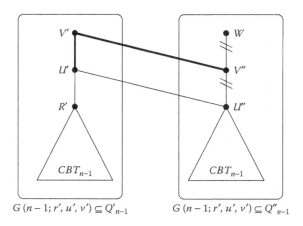

Figure 26.6 Illustration to explain the induction step of the proof of Theorem 26.16.

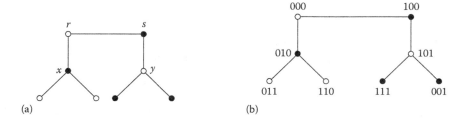

Figure 26.7 (a) Double rooted complete binary tree $D(3)$ and (b) an embedding showing $D(3) \subseteq Q_3$.

$D(n-1; a, b, c, d)$ are embeddable into Q'_{n-1} and Q''_{n-1}, respectively. Let the path $\langle e, f, g, h \rangle$ be mapped onto a path $\langle E', F', G', H' \rangle$ of Q'_{n-1}. By the P_3-symmetry of Q''_{n-1}, one can choose an embedding of $D(n-1; a, b, c, d)$ into Q''_{n-1} such that the path $\langle a, b, c, d \rangle$ is mapped onto the path $\langle F'', G'', H'', Y \rangle$ of Q''_{n-1}.

Deleting the edges (G', H'), (F'', G'') and adding the edges (F', F''), (G', G''), (H', H'') we get a double rooted complete binary tree on 2^n vertices with roots G', G''; see Figure 26.8 for an illustration of the induction step. ∎

Corollary 26.2 CBT_n has a load-1 and dilation-2 embedding into Q_n, where exactly one edge has dilation-2.

Proof. The construction of $D(n)$ from CBT_n shows that CBT_n has a load-1 and dilation-2 embedding into $D(n)$, with only one edge with dilation 2. So, appealing to Theorem 26.17 we have the corollary. ∎

Theorem 26.18 [20] CBT_n has a load-2 and dilation-1 embedding into Q_{n-1} where $n \geq 2$.

Proof. Let $f : D(n; a, b, c, d) \to Q_n$ be the load-1 and dilation-1 embedding described in the proof of Theorem 26.17. Without loss of generality, let $f(b) = 0^n$ and $f(c) = 0^{j-1} 1 0^{n-j}$; there is no loss of generality here, since Q_n is edge-symmetric. Let $p : Q_n \to Q_{n-1}$ be the embedding defined by $p(X) = X^{(-j)}$, for every X in Q_n, where $X^{(-j)}$ denotes the binary string obtained by deleting the jth co-ordinate of X. It is now easy to verify that $p \circ f : D(n) \setminus \{c\} \to Q_{n-1}$ is a load-2 and dilation-1 embedding of CBT_n into Q_{n-1}. ∎

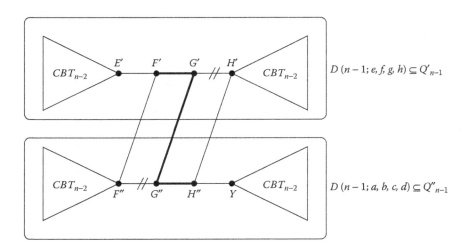

Figure 26.8 Illustration to explain the induction step in the proof of Theorem 26.17.

These and similar results lead to a long standing conjecture of Havel.

Conjecture 26.1 [21] *Every equipartite binary tree on 2^n vertices spans Q_n. Equivalently, every binary tree with bipartition $[X, Y]$ such that $|X| \leq 2^{n-1}$ and $|Y| \leq 2^{n-1}$ is a subgraph of Q_n.*

In support of the conjecture, several kinds of binary trees with additional constraints have been shown to be embeddable into hypercubes. Two kinds of techniques have been employed to obtain these embeddings.

Technique 1: Given a binary tree T on 2^n vertices (satisfying a certain property P), first show that there exists a small set S of independent edges whose deletion yields a tree T' on 2^{n-1} vertices satisfying P, and a forest T'' on 2^{n-1} vertices. Embed T' into Q'_{n-1} by induction and embed T'' into Q''_{n-1}. Some effort is required to find a proper S and to embed (inductively) the forest T'' into Q''_{n-1}.

Finally, combine T' and T'' by using the edges of the perfect matching that joins Q'_{n-1} and Q''_{n-1} to obtain a copy of T in Q_n. Here one uses the symmetric properties of hypercubes. See the proof of Theorem 26.4.

Technique 2: Instead of showing that T is a spanning subgraph of Q_n, identify a spanning subgraph G' of Q_n where G' is a supergraph of T. Some effort is required to guess this intermediate graph G' and then further efforts are required to show that $G' \subseteq Q_n$. See the proof of Theorem 26.16.

To illustrate these techniques we outline proofs of two more theorems. First theorem is proved using Technique 1, and the second theorem is proved using Technique 2. Both the theorems prove Havel's conjecture for special classes of trees called *caterpillars*. There are various definitions of caterpillars in the literature. In the following, we unify these.

Definition 26.15 *A k-caterpillar C is a tree having maximum degree k and a path P such that all the vertices of degree k lie on P, and by deleting the vertices of P we obtain a collection of paths. If such a path Q_i is incident with the vertex v_i of P, then $L_i = <v_i, Q_i>$ is called a leg of C. The path P is called a* spine *of C. A 3-caterpillar is called a* binary caterpillar*; see Figure 26.9 for a binary caterpillar.*

Theorem 26.19 [22] *Let $n \geq 2$. Every equipartite binary caterpillar on 2^n vertices in which every leg has length at most one is a spanning subgraph of Q_n.*

Proof. The proof uses induction on n and the canonical decomposition $Q_n = Q'_{n-1} \ominus Q''_{n-1}$. Crucial assertions are the following:

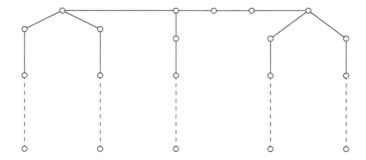

Figure 26.9 Binary caterpillar.

1. On the spine of C there exist at most two edges whose deletion results in a balanced caterpillar T' on 2^{n-1} vertices, and a balanced forest T'' on 2^{n-1} vertices with at most two components.

2. T' can be embedded in Q'_{n-1} such that the ends of its spine are at distance at most two in Q'_{n-1}.

3. T'' can be embedded in Q''_{n-1} such that the ends attached to T' in C are at distance at most two in Q''_{n-1}.

4. Using the symmetric properties of Q'_{n-1} and Q''_{n-1}, the embeddings of T' and T'' can be combined to get an embedding of C in Q_n. ∎

Definition 26.16 *Take two paths $\langle a_1, a_2, \ldots, a_k \rangle$ and $\langle b_1, b_2, \ldots, b_k \rangle$ and join each pair of vertices a_i, b_i, $1 \leq i \leq k$, with a new path P_i. The resulting graph is called a* ladder, *and the paths P_i are called its* rungs; *see Figure 26.10.*

Theorem 26.20 [23] *Every binary caterpillar C on 2^n vertices in which every leg has even length is a spanning subgraph of Q_n.*

Outline of the Proof:

1. At the outset observe that C is equipartite which is a necessary condition for embeddability of C in a hypercube.

2. It is first shown that every ladder L on 2^n vertices in which every rung has even length spans Q_n.

   (This is done by induction on n. L is carefully divided into two ladders L_1 and L_2 each on 2^{n-1} vertices, by deleting certain edges and adding certain edges. Consider the canonical decomposition of Q_n into two subcubes Q'_{n-1} and Q''_{n-1}. By induction, embed L_1 into Q'_{n-1} and L_2 into Q''_{n-1}. Retrieve the deleted edges of L from the perfect matching that combines Q'_{n-1} and Q''_{n-1}.)

3. Join the end vertices of the legs of C by a path to obtain a ladder L.

4. It follows by (2) that C is a spanning subgraph of Q_n. ∎

We next address the following problem.

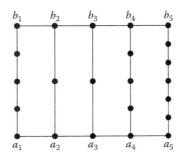

Figure 26.10 Ladder with rungs consisting of 5, 3, 3, 5, 7 vertices.

Problem 26.7 *If a binary tree tree T is not a subgraph of Q_n, then what is the minimum load, minimum dilation, minimum congestion, and minimum expansion required to embed T into Q_n.*

A central theorem on this topic is the following.

Theorem 26.21 [24] *Every binary tree on N nodes, $2^n < N \leq 2^{n+1}$, can be embedded into an $(n+1)$-dimensional hypercube with load-1 and every other parameter, namely dilation, node congestion, edge congestion, bounded by a constant.* ∎

Theorem 26.22

  a. *Every binary tree can be embedded with load-1, dilation-8, and constant node congestion into its optimal hypercube* [25].

  b. *Every binary tree can be embedded with load-1 and dilation-6 into its optimal hypercube* [26].

  c. *Every binary tree can be embedded in a hypercube with load-1, dilation-3, and expansion $O(1)$* [26].

  d. *Every binary tree can be dynamically embedded (i.e., recursively level by level) in a hypercube with load-1, dilation-9, constant node congestion, constant edge congestion, and nearly optimal expansion* [27]. ∎

Theorem 26.23 [28] *Every N-node binary tree is a subgraph of the hypercube with $O(N \log N)$ nodes.* ∎

This is an improvement over an earlier result of Afrati et al. [8], who showed that every N-node binary tree is a subgraph of the hypercube with $O(N^2)$ nodes.

We next list more open problems in this area.

Conjecture 26.2 [29] *Any binary tree is embeddable into its optimal hypercube with dilation at most 2.*

Remark 26.3 *Any binary tree can be embedded into (optimal) equibipartite binary tree with load-1 and dilation-2. So, if Conjecture 26.1 is true, then the above Conjecture 26.2 is true.*

Conjecture 26.3 [30] *Any binary tree is a subgraph of its next-to-optimal hypercube.*

Conjecture 26.4 [24] *Every N-node binary tree is a subgraph of an $O(N)$-node hypercube.*

Conjecture 26.5 [31] *Any tree T with $\Delta(T) \leq 5$ can be embedded into its optimal hypercube with load-1 and dilation-2.*

In the above conjecture, the condition $\Delta(T) \leq 5$ is essential. For any $n \geq 5$, $n \neq 6$, Dvořák et al. [24] have constructed a tree T_n with $\Delta(T_n) = 6$, $|V(T_n)| = 2^n$ which has no embedding in Q_n with load-1 and dilation-2.

Definition 26.17 *A rooted binary tree is said to be a* height-balanced tree *or* AVL-tree *if for every nonleaf vertex v, the heights of the subtrees rooted at the left and right child of v differ by 0 or 1. This difference is called the* balance-factor *of v.*

In the design of efficient algorithms, an appropriate choice of a data structure often reduces the time complexity of algorithms. When a data structure with a hierarchical relationship among its various elements is represented as an AVL-tree, several operations like search, insertion and deletion can be performed with least time complexity. Clearly, an AVL-tree in which each nonleaf vertex has balance-factor 0 is a complete binary tree. An AVL-tree in which every nonleaf vertex has balance-factor 1 is called a Fibonacci tree.

The problem of embedding AVL-trees in hypercubes was initiated by Choudum and Indhumathi in [32] and several subclasses of AVL-trees were optimally embedded in [32] and [33]. However, the problem of optimally embedding an arbitrary AVL-tree into a hypercube remains open.

We conclude this section with a catalogue of results proved in support of above conjectures.

1. For any k-ary tree T of height h, $kh/e \leq cd(T) \leq k(h+1)/2 + h - 1$ [34]. (If $h = 2$, then $cd(T) = (3k+1)/2$.)

2. Every equipartite tree on 2^n vertices with at most one vertex of degree greater than two spans Q_n [35]. These trees are called *starlike trees*.

3. Every equipartite binary tree on 2^n vertices with exactly two vertices of degree three spans Q_n [36].

4. Every equipartite caterpillar on 2^n vertices with exactly two vertices on the spine spans Q_n [37]. These trees are called *double starlike trees*.

5. Every equipartite binary caterpillar on 2^n vertices with every leg of length at most two spans Q_n [38].

6. Every equipartite binary caterpillar on 2^n vertices with at most $n-3$ vertices on the spine spans Q_n [39].

7. Any caterpillar is embeddable into its optimal hypercube with dilation-2 [31,40].

8. There exists a load-1, dilation-2 embedding of a level-1 hierarchical caterpillar into its optimal hypercube [40].

9. An m-sequential k-ary tree can be embedded into its optimal hypercube with dilation at most 2 [41].

10. Complete ternary (3-ary) tree of height h can be embedded with load-1, dilation-3, and congestion-3 into $Q_{d(h)}$, where $d(h) = \lceil 1.6h \rceil + 1$ [42].

11. Complete ternary tree of height h can be embedded with load-1, dilation-2, congestion-2 into $Q_{d(h)}$, where [43]

$$d(h) = \begin{cases} \lceil (1.6)h \rceil, & \text{if } h = 2 \pmod 5 \text{ or } h = 4 \pmod 5 \\ \lceil (1.6)h \rceil + 1, & \text{if } h = 0 \pmod 5 \text{ or } h = 1 \pmod 5 \text{ or } h = 3 \pmod 5 \end{cases}$$

For various particular values of h, this result proves one of the conjectures of Havel [36] specific to the class of complete ternary trees.

26.2.3 Collection of Cycles, Paths, and Trees in Hypercubes

In a multiprocessor computing system, computing involves exchange of data among several of its processors. The data is transmitted from one processor to another through a sequence/collection of interlinked processors. Obviously, the transmission is faster if a large number of alternative cycles/paths/trees are available. Equivalently, the demand is that in the graph of the interconnection network, there be a large number of distinct cycles/paths/trees. The presence of multiple copies can be viewed as embedding a collection. The availability of such alternatives may also help in circumventing faults of vertices and/or edges. In this section, we give some results in support of the claim that hypercube has these desirable properties.

It is well known that Q_n is Hamiltonian. However, it may come as a surprise to know that Q_4 has 1344 distinct Hamilton cycles. In fact, there exists at least $n!$ distinct Hamilton cycles in Q_n [44]. The exact number of distinct Hamilton cycles in Q_n, for an arbitrary n, is still open; see [45] and the references there in to know the bounds. However, the counts for the number of distinct Hamilton cycles are $1, 1, 6, 1344, 906545760, 14754666598334433250560$ for hypercubes of dimension $1, 2, 3, 4, 5, 6$; see Sequence A0066037 in *The On-Line Encyclopedia of Integer Sequences* [46]. Knowing that there exist a large number of distinct Hamilton cycles in hypercubes, a natural question that arises is "How many (edge) disjoint Hamilton cycles can one find in a hypercube?" The answer to this question is as follows.

Theorem 26.24 [47] *In Q_n, there exist $\lfloor n/2 \rfloor$ disjoint Hamiltonian cycles.*

Proof. The proof is dependent on the following results on Hamiltonian decomposition of cartesian product of cycles.

- There exist two disjoint Hamilton cycles in $C_p \times C_q$ [48].
- There exist three disjoint Hamilton cycles in $C_p \times C_q \times C_r$ [49].

One can now use induction on m and prove that $C_{i_1} \times C_{i_2} \times \ldots \times C_{i_m}$ contains m disjoint Hamilton cycles.

We now prove the result on existence of $\lfloor n/2 \rfloor$ disjoint Hamilton cycles in Q_n. One can easily verify the result for small values of n. We therefore assume that $n \geq 4$.

Case 1 n is even.

In this case, $Q_n = \underbrace{C_4 \times C_4 \times \ldots \times C_4}_{n/2 \text{ times}}$, so Q_n has $n/2$ disjoint Hamilton cycles.

Case 2 n is odd.

In this case, consider $Q_n = Q'_{n-1} \ominus Q''_{n-1}$. Then by Case 1, each of the Q_{n-1} contains $(n-1)/2$ disjoint Hamilton cycles. Remove an edge from each Hamilton cycle of Q'_{n-1} in such a way that the removed edges do not have any vertex in common. Remove the corresponding edges from each Hamilton cycle of Q''_{n-1}. Now, use the perfect matching that exists between Q'_{n-1} and Q''_{n-1} to obtain $(n-1)/2 = \lfloor n/2 \rfloor$ Hamilton cycles in Q_n. ∎

Note that, from the above proof we only learn that there exist disjoint Hamilton cycles in hypercubes. But for practical purposes, one would be interested in simple algorithms to construct these cycles; see [50–52] for such algorithms.

When one considers non-Hamilton cycles/paths, in addition to distinctness and/or (edge) disjointness, one can also ask about existence of vertex-disjoint copies. We next discuss the presence of vertex-disjoint copies of paths in hypercubes.

Definition 26.18 *In a graph G, two (x, y)-paths are said to be* parallel *if they have no common internal vertices. In general, any two paths are said to be* parallel *or* vertex-disjoint *if they have no common vertices.*

Proposition 26.1 [53] *Let X, Y be any two vertices of Q_n. Then there exist $d(X, Y)$ parallel paths of length $d(X, Y)$ between X and Y in Q_n.* ∎

Proposition 26.2 [53] *Let X, Y be any two vertices of Q_n. Then there exist n parallel paths between X and Y, each of length at most $d(X, Y) + 2$.* ∎

In the context of routing in parallel computers, finding vertex disjoint paths between a given set of source–destination pairs of vertices is another interesting study that has attracted various researchers.

Problem 26.8 *Given a set $U = \{u_1, u_2, \ldots, u_k\}$ of nodes and a set $A = \{a_1, a_2, \ldots, a_k\}$ of non-negative integers, find sufficient conditions the sets U and A should satisfy for the existence of a family $F = \{P_{(1)}, P_{(2)}, \ldots, P_{(k)}\}$ of vertex disjoint paths, where each $P_{(i)}$ is a u_i-path with a_i vertices.*

The initial results were proved in [21] for $k = 2$ and [35] when A is a set of even integers. Here, an u-path of order p refers to a path with origin u and containing p number of vertices. The terminus of a path P is denoted by $t(P)$.

Theorem 26.25 [21] *Given any two distinct vertices u_1, u_2 in Q_n, and any two positive integers a_1, a_2 such that $a_1 + a_2 = 2^n$ and let u_1, u_2 be such that $dist(u_1, u_2)$ is odd, whenever a_1, a_2 are odd, there exist two vertex disjoint paths $P_{(1)}$ and $P_{(2)}$, where $P_{(i)}$ is an u_i-path of order a_i $(i = 1, 2)$.* ∎

Nebeský [35] proved a similar result on the existence of paths of even order as follows.

Theorem 26.26 [35] *Let k and n be integers such that (1) $1 \leq k \leq n$, if $1 \leq n \leq 3$, and (2) $1 \leq k < n$, if $n \geq 4$. Let:*

i. *$u_1, u_2, \ldots, u_k$ be distinct vertices of Q_n,*

ii. *$a_1, a_2, \ldots, a_k$ be positive even integers with $a_1 + a_2 + \cdots + a_k = 2^n$, and*

iii. *$W_1, W_2, \ldots, W_k$ be subsets of $V(Q_n)$ such that $|W_i| \leq n - k$, $1 \leq i \leq k$.*

Then there exist vertex-disjoint paths $P_{(1)}, P_{(2)}, \ldots, P_{(k)}$ in Q_n, where $P_{(i)}$ is an u_i-path of order a_i and $t(P_{(i)}) \notin W_i$, $i = 1, 2, \ldots, k$. ∎

Recently, Choudum et al. [54] generalized both these results to the existence of paths both of even and odd order.

Theorem 26.27 [54] *Let n and k be integers as in Theorem 26.26 and let m be an integer such that $m \leq \lfloor k/2 \rfloor$. Let*

i. *$u_1, u_2, \ldots, u_k$ be distinct vertices of Q_n such that $(u_1, u_2), (u_3, u_4), \ldots, (u_{2m-1}, u_{2m})$ are edges of the same dimension,*

ii. *$a_1, a_2, \ldots, a_{2m}$ be positive odd integers and $a_{2m+1}, \ldots, a_k$ be positive even integers such that $a_1 + a_2 + \cdots + a_k = 2^n$,*

iii. $W_1, W_2, \ldots, W_k$ be subsets of $V(Q_n)$ such that $|W_i| \leq n-k$, and if $a_i = 1$, then $u_i \notin W_i$, for $1 \leq i \leq k$.

Then there exist vertex-disjoint paths $P_{(1)}, P_{(2)}, \ldots, P_{(k)}$ in Q_n, where $P_{(i)}$ is an u_i-path of order a_i and $t(P_{(i)}) \notin W_i$, $1 \leq i \leq k$. ∎

Remark 26.4 *In Theorem 26.27, it is assumed that the edges $(u_1, u_2), \ldots, (u_{2m-1}, u_{2m})$ have the same dimension. A generalization of the theorem under a weaker assumption that $dist(u_{2i-1}, u_{2i})$ $(1 \leq i \leq m)$, is odd would be more useful for applications.*

The above remark with the weaker assumption has been proved for $m = 1$ in [54]. In this direction, a more general problem can also be formulated as follows.

Problem 26.9 *Given two sets of vertices $S = \{u_1, u_2, \ldots, u_k\}$ and $T = \{v_1, v_2, \ldots, v_k\}$ in Q_n and a set $A = \{a_1, a_2, \ldots, a_k\}$ of integers, find sufficient conditions that S, T, and A should satisfy for the existence of a family $F = \{P_{(1)}, P_{(2)}, \ldots, P_{(k)}\}$ of vertex disjoint paths, where each $P_{(i)}$ is a (u_i, v_i)-path with a_i vertices.*

In this spirit, there already exist various results. We state a few. Here the conditions on the sets S, T, and A are stated.

Theorem 26.28

1. *For $S = \{u_1, u_2, \ldots, u_k\}$ and $T = \{v_1, v_2, \ldots, v_k\}$ with $k = \lceil n/2 \rceil$ and each $a_i \in A$ is such that $a_i \leq 2n$ [55].*

2. *For $S = \{u_1, u_2, \ldots, u_k\}$ and $T = \{v_1, v_2, \ldots, v_k\}$ with $k = \lceil n/2 \rceil$ and each $a_i \in A$ is such that $a_i \leq n + \log_2 n + 1$ [56].*

3. *For $S = \{u_1, u_2, \ldots, u_k\}$, $T = \{v_1, v_2, \ldots, v_k\}$ and $W = \{w_1, w_2 \ldots w_q\}$ such that $W \cap S = \emptyset$, $W \cap T = \emptyset$, $k \leq n/2$ and $2k + q \leq n + 1$ [57].*

4. *For $S = \{u_1, u_2, \ldots, u_k\}$ and $T = \{v_1, v_2, \ldots, v_k\}$ such that $col(u_i) \neq col(v_i)$ and $k \leq n - 1$. Here the paths $P_{(i)}$ can be of arbitrary length [58].* ∎

Recently Caha and Koubek [59] characterized the existence of vertex disjoint paths as follows.

Theorem 26.29 [59] *Given two sets of nodes $S = \{u_1, u_2, \ldots, u_k\}$ and $T = \{v_1, v_2, \ldots, v_k\}$ in Q_n, there exists a family $F = \{P_{(1)}, P_{(2)}, \ldots, P_{(k)}\}$ of vertex disjoint paths, where each $P_{(i)}$ is a (u_i, v_i)-path if and only if $col(u_i) \neq col(v_i)$ for every $i, 1 \leq i \leq k$, $k \leq n - 1$.* ∎

The complexity issue related to the k-pairwise edge disjoint shortest paths problem was studied by Gu and Peng [56] where they give an $O(n^2 \log^* n)$ time algorithm. Further Gonzalez and Serena in [60,61] presented an efficient algorithm for the case when every source point is at a distance at most two from its destination, and for pairs at a distance at most three they showed that the problem is NP-complete.

We next consider the availability of a collection of spanning trees in hypercubes. We know that the number of edges in Q_n is $n2^{n-1}$ and the number of edges in a spanning tree of Q_n is $2^n - 1$. Consequently, there can exist at most $\lfloor n2^{n-1}/(2^n - 1) \rfloor = \lfloor n/2 \rfloor$ edge-disjoint spanning trees. A construction for identifying n edge-disjoint spanning trees in Q_{2n} is available through the following result while no such construction is as yet available for Q_{2n-1}.

Theorem 26.30 [62] *There exists an embedding of n edge-disjoint spanning trees in Q_{2n}, with the remaining n edges forming a path.* ∎

One can also consider imposing some kind of vertex-disjointness on the collection of spanning trees. Independent spanning trees are one such collection.

Definition 26.19 *Two spanning trees T_1 and T_2 of a connected graph G are said to be* independent *if they are rooted at the same vertex, say r, and for each vertex $v \neq r$, the two paths $P_{(1)} \subseteq T_1$ and $P_{(2)} \subseteq T_2$ from r to v are parallel. A set of spanning trees of G is said to be independent if they are pairwise independent.*

In [63], it is shown that Q_n has n independent spanning trees rooted at any vertex. Note that this result is optimal since Q_n is n-regular.

It can now be asked whether these disjoint trees are isomorphic. That is, can Q_n be decomposed into T where T is a tree? The following results are available in this direction.

Theorem 26.31 [64,65] *If T is any tree having n edges, then Q_n can be decomposed into 2^{n-1} edge-disjoint induced subgraphs, each of which is isomorphic to T.* ∎

The above result has been generalized as follows.

Theorem 26.32 [66] *Q_n can be decomposed into 2^{n-1} arbitrary $(p+q)$-trees with $p+q = n+1$, where a $(p+q)$-tree is a tree which has a bipartition with p vertices in one part and q vertices in the other.* ∎

If we consider $K_{1,k}$, that is the star with k edges, then the following holds.

Theorem 26.33 [67] *Q_n can be decomposed into $K_{1,k}$ if and only if $k \leq n$ and k divides the number of edges of Q_n.* ∎

When we progress from trees with n edges to more than n edges, we have the following result.

Theorem 26.34 [68] *Q_n decomposes into copies of any tree with $n+6$ edges.* ∎

A related result by Horak et al. [69] is that Q_n can be decomposed into any graph with n edges each of whose blocks is either an even cycle or an edge. We next look at decomposing Q_n into a spanning tree.

Definition 26.20 *A binomial tree of height 0, B_0, is a single vertex. For all $h > 0$, a binomial tree of height h, B_h, is a tree formed by joining the roots of two binomial trees of height $h-1$ with a new edge and designating one of these roots to be the root of the new tree.*

It is clear that B_n has 2^n vertices. Often, broadcasting of messages in hypercubes is done through binomial trees. This is possible because of the following theorem.

Theorem 26.35 *B_n is a spanning subgraph of Q_n.*

Proof. The result can be easily proved using induction along with the vertex-symmetry property of hypercube. ∎

Johnsson and Ho investigated the presence of multiple copies of binomial trees in hypercubes and have concluded the following theorem.

Theorem 26.36 [70] *There exists an edge-disjoint embedding of n spanning binomial trees in Q_n.* ∎

Note that, the above result is based on the understanding that two spanning tree are edge-disjoint if they share no common directed edges (directed from parent to child). However, if the direction is dropped then we have the following theorem.

Theorem 26.37 [71] *Q_n can be decomposed into n edge-disjoint trees on 2^{n-1} edges, all of which are isomorphic to BT_n, where BT_n is the tree obtained by adding a new vertex to B_{n-1} and providing an edge between this new vertex and the root of B_{n-1}.* ∎

26.3 EMBEDDING INTO VARIANTS OF HYPERCUBES

In this section, we provide some results on embedding into some variants of hypercubes. In particular, the role of computation graphs (such as paths, cycles, meshes, and trees) as guest graphs has received maximum attention from the researchers.

26.3.1 Embedding into Augmented Cubes AQ_n

Out of our own interest in augmented cubes and the fact that many researchers have studied augmented cubes, we present an exhaustive coverage of the embedding results on augmented cubes.

From the definition of the augmented cubes, it easily follows the following theorem.

Theorem 26.38 *For $n \geq 1$,*

1. *Q_n is a spanning subgraph of AQ_n.*

2. *$Q_{n,k}$ is a spanning subgraph of AQ_n for every k.* ■

The following result can lead to identifying some good embeddings into hypercubes via the augmented cubes.

Theorem 26.39 *There exists a dilation 2 embedding of AQ_n into Q_n.* ■

Definition 26.21 *The graph obtained from a complete binary tree by adding edges to connect consecutive vertices on the same level of the tree is called an X-tree.*

Definition 26.22 *The one-dimensional multigrid also called the* linear multigrid, *consists of $2^n - 1$ nodes arranged in n levels, with the ith level containing a 2^i-node linear array (the top level is level 0 and the last level is $n-1$). In addition, the jth node on level i is connected to the 2jth node on level $i+1$ for every j, $1 \leq j \leq 2^i$, $0 \leq i \leq n-2$.*

Theorem 26.40 [38,72]

1. *For $n \geq 3$, AQ_n contains two edge disjoint complete binary trees on $2^n - 1$ vertices both rooted at a given vertex.*

2. *For $n \geq 3$, two copies of the binomial tree of height n can be embedded into AQ_n such that one of the following properties is satisfied.*

    a. *The roots of the copies are mapped onto distinct vertices and the embeddings are edge disjoint.*

    b. *The roots of the copies are mapped onto a single vertex of AQ_n and the embeddings have exactly one edge in common.*

3. *For $n \geq 2$, the X-tree on $2^n - 1$ vertices and hence the linear multigrid on $2^n - 1$ vertices is a subgraph of AQ_n.*

4. *For $n \geq 3$, AQ_n contains $n-1$ edge disjoint spanning trees.* ■

Conjecture 26.6 *Any binary tree is a subgraph of its optimal augmented cube.*

If the above conjecture is true, then in view of Theorem 26.39, it follows that Conjecture 26.2 is also true.

We now look at embedding linear arrays and cycles in AQ_n. It is well known that hypercubes are Hamiltonian. Combining this with the fact that Q_n is a spanning subgraph of AQ_n, it follows that AQ_n is also Hamiltonian. The Hamiltonicity property of AQ_n has been strengthened by Hsu et al. [73] by assuming that some vertices and/or edges can become faulty.

Theorem 26.41 [73]

1. AQ_2 is 1-fault-tolerant Hamiltonian and not k fault-tolerant Hamiltonian for $k \geq 2$.

2. AQ_3 is 2 fault-tolerant Hamiltonian and not k fault-tolerant Hamiltonian for $k \geq 3$.

3. For $n \geq 4$, the fault-tolerant Hamiltonicity of AQ_n is exactly $2n - 3$. ∎

If it is assumed that only edges become faulty, then AQ_3 can tolerate up to 3 faulty edges and still contain a Hamilton cycle [74]. This is an improvement over the above result for AQ_3. In addition to edge-faults, if it is assumed that each vertex is incident to at least two fault-free edges then Hsieh and Cian [75] have shown that AQ_n can tolerate up to $4n - 8$ faulty edges to retain Hamiltonicity; they also give a worst-case scenario in which there is no Hamilton cycle in AQ_n when there are $4n - 7$ faulty edges.

Theorem 26.42 [75] *Under the conditional fault-assumption, AQ_n ($n \geq 3$) with faulty edges up to $4n - 8$ contains a fault-free Hamilton cycle. Moreover, this is optimal with respect to the number of faulty edges tolerated.* ∎

Though AQ_n as well as Q_n are Hamiltonian and hence contain a Hamilton path, one distinguishing feature is their Hamiltonian connectedness—existence of a Hamilton path between every pair of vertices. Hypercube being bipartite, there cannot exist a Hamilton path between two vertices belonging to the same part. In this respect, AQ_n is superior due to the following result.

Theorem 26.43 [76] *For any three distinct vertices x, y and z of AQ_n ($n \geq 2$) and for any l, where $d_{AQ_n}(x, y) \leq l \leq 2^n - 1 - d_{AQ_n}(y, z)$, there exists a Hamiltonian path $R(x, y, z; l)$ from x to z such that $d_{R(x,y,z;l)}(x, y) = l$.* ∎

Note that the above result not only shows that there exists in AQ_n a Hamiltonian path between any pair of distinct vertices x, z (i.e., AQ_n is Hamiltonian connected) but it also shows that a third vertex y can be introduced into the Hamiltonian path at a required position. The Hamiltonian connected property of AQ_n has been strengthened as shown by the following theorem.

Theorem 26.44

1. AQ_3 is 1-fault-tolerant Hamiltonian connected and is not 2-fault Hamiltonian connected [73].

2. For $n \geq 4$, AQ_n is $2n - 4$-fault-tolerant Hamiltonian connected and is not $2n - 3$-fault Hamiltonian connected [73].

3. AQ_3 is 2-edge-fault-tolerant Hamiltonian connected [74]. ∎

Owing to bipartiteness, there exist no odd cycles in Q_n and $Q_{n,k}$ when $n \equiv k \pmod{2}$.

Theorem 26.45 [38,72] *For $n \geq 2$, AQ_n is pancyclic.*

Proof. Since Q_n contains a $2k$ cycle, say C_{2k}, for every k, $2 \leq k \leq 2^{n-1}$ and since Q_n is a spanning subgraph of AQ_n, it follows that AQ_n also contains C_{2k}. Further, these cycles are induced by n-bit $2k$-element Gray code sequences. The first three elements in these sequences are 0^n, $0^{n-1}1$ and $0^{n-2}11$. In AQ_n, 0^n and $0^{n-2}11$ are adjacent. So, $C_{2k} - \{0^{n-1}1\}$ is a $2k-1$ cycle in AQ_n. ∎

Some results on pancyclicity of augmented cubes under faults are the following.

Theorem 26.46

1. AQ_3 is not 3-fault-tolerant pancyclic but is 2-fault-tolerant pancyclic [77].

2. AQ_n is $(2n - 3)$-faults tolerant pancyclic for $n \geq 4$. Further, the number of faults tolerated is optimal [77].

3. For $n \geq 2$, AQ_n contains cycles of all length from 3 to 2^n even when any $(2n-3)$ edges are deleted from AQ_n [74]. ∎

In Theorem 26.45, we have shown that AQ_n is pancyclic. Since AQ_n is vertex symmetric it follows that AQ_n is vertex-pancyclic. An alternate proof for this result is provided in [78] using the following property.

Theorem 26.47 [78] *For any two distinct vertices u, v, there exists a uv-path of length $2^n - 2$ in AQ_n, where $n \geq 3$.* ∎

When it comes to vertex-pancyclicity in the presence of faults, we have the following theorem.

Theorem 26.48 [79] *For $n \geq 2$, every vertex in AQ_n lies on a fault-free cycle of every length from 3 to 2^n, even if there are up to $n - 1$ edge faults.* ∎

It has also been shown that this result is optimal. That is, AQ_n is not n-edge fault-tolerant vertex-pancyclic. Consequently, AQ_n is not n fault-tolerant vertex-pancyclic.

Lai et al. [80] have shown that AQ_n is panconnected (Theorem 26.49).

Theorem 26.49 [80] *For any two distinct vertices $x, y \in AQ_n$, there exists a path $P_l(x, y)$ of length l joining x and y for every l satisfying $d(x, y) \leq l \leq 2^n - 1$.* ∎

As an extension to the above theorem, one would also like to see if the path $P_l(x, y)$ can be further extended by including the vertices not in $P_l(x, y)$ into a Hamiltonian path from x to a fixed vertex z or a Hamiltonian cycle. This possibility has been explored in [76] with the outcome being available in the form of Theorem 26.43 and Theorem 26.50.

Theorem 26.50 [76] *There exists a Hamiltonian cycle $S(x, y; l)$ such that $d_{S(x,y;l)}(x, y) = l$ for any two distinct vertices x and y and for any $d_{AQ_n}(x, y) \leq l \leq 2^{n-1}$.* ∎

Definition 26.23 *A graph G is* panpositionably Hamiltonian *if for any two distinct vertices x and y of G, it contains a Hamiltonian cycle C such that $d_C(x, y) = l$ for any interger l satisfying $d_G(x, y) \leq l \leq |V(G)|/2$.*

Corollary 26.3 AQ_n *is panpositionably Hamiltonian.*

Ma et al. [74] rediscovered the panconnectedness property of AQ_n while studying its edge-fault-tolerant pancyclicity. On panconnectedness of AQ_n in the presence of faults we have the following theorem.

Theorem 26.51 [81] *For $n \geq 3$, $AQ_n - f$, is panconnected for any vertex $f \in AQ_n$.* ∎

The above result is a weaker form of the following result.

Theorem 26.52 [82] *For $n \geq 3$, $AQ_n - f$ is panconnected if f is a vertex or an edge of AQ_n.* ∎

In addition to the above result, Chan [82] has also shown that $AQ_n - F$ is not panconnected if $|F| \geq 2$ where $F \subseteq V(AQ_n) \cup E(AQ_n)$. A related result is the following theorem.

Theorem 26.53 [83] *If AQ_n ($n \geq 3$) has at most $2n - 5$ faulty vertices and/or edges, then for any two distinct fault-free vertices u and v with distance d in AQ_n, there exist fault-free uv-paths of every length from $d+2$ to $2^n - f - 1$, where f is the number of faulty vertices in AQ_n.* ∎

The above result is the best possible in the sense that

a. In AQ_n ($n \geq 4$), there exist a pair u, v of vertices with distance d (in particular, $d = 1$) and a set F of faulty vertices with $|F| \leq 2n - 5$ such that there exists no uv-path of length $d + 1$, and

b. In AQ_n, if F is a set of faulty vertices with $|F| = 2n - 4$, then there exist two distinct fault-free vertices u and v with distance $d = 1$, such that there exists no fault-free uv-path of length l for some $l \in \{d+2, d+3, \ldots, 2^n - f - 1\}$.

Note that the optimality of the above result is due to $d = 1$. Excluding this case, for $d \geq 2$ or $n \geq 4$, it would be worthwhile to investigate whether AQ_n is $(2n - 4)$-fault-tolerant panconnected for some large $d \geq 2$ or $n \geq 4$. Also note that this will not violate the optimality of Theorem 26.52 because there again the optimality is based on adjacent vertices.

The panpositionable panconnected property of a graph—a refinement of the panconnected property—is defined as follows.

Definition 26.24 *Let $x, y,$ and z be any three distinct vertices in a graph G. Then G is said to be* panpositionably panconnected *if for any l_1, where $d_G(x, z) \leq l_1 \leq |V(G)| - d_G(y, z) - 1$, it contains a path P such that x is the beginning vertex of P, z is the (l_1+1)th vertex of P, and y is the (l_1+l_2+1)th vertex of P for any integer l_2 satisfying $d_G(x, y) \leq l_2 \leq |V(G)| - l_1 - 1$.*

Theorem 26.54 [84]

1. AQ_2 is panpositionably panconnected.

2. AQ_n, for $n \geq 3$, is almost panpositionably panconnected. That is, for any three distinct vertices x, y, z, for any $d(x, z) \leq l_1 \leq 2^n - d(y, z) - 1$, and for any $d(y, z) \leq l_2 \leq 2^n - l_1 - 1$, there exists an xy-path P in AQ_n, such that x is the beginning vertex of P, z is the $(l_1 + 1)$th vertex of P, and y is the $(l_1 + l_2 + 1)$th vertex of P, except for the case that $l_1 = l_2 = 2$ when there just exists a vertex w distinct from x, y, z with $\{y, w\} = N(x) \cap N(z)$ and $\{x, w\} = N(y) \cap N(z)$. ∎

Corollary 26.4 [84] *Let $n \geq 2$.*

1. Let u, v, w be any three distinct vertices in AQ_n. Then $d_{AQ_n - \{w\}}(u, v) = d_{AQ_n}(u, v)$.

2. AQ_n is panconnected.

3. Let w be any vertex of AQ_n. Then $AQ_n - \{w\}$ is panconnected.

4. Let (u_1, v_1) and (u_2, v_2) be any two distinct edges in AQ_n. Then there exists a Hamiltonian cycle C in AQ_n such that $(u_1, v_1) \in E(C)$ and $(u_2, v_2) \in E(C)$.

5. Let (u_1, v_1) and (u_2, v_2) be any two vertex-disjoint edges of AQ_n. Let l_1, l_2 be any two integers such that $1 \leq l_1, l_2 \leq 2^n - 3$ with $l_1 + l_2 \leq 2^n - 2$. Then AQ_n has two vertex-disjoint paths, P_1 and P_2, such that P_1 is a u_1v_1-path of length l_1 and P_2 is a u_2v_2-path of length l_2 except for the following two cases: (a) $l_1 = 2$ or $l_2 = 2$ with $\{u_2, v_2\} = N(u_1) \cap N(v_1)$; (b) $4 \leq l_1 \leq 5$ or $4 \leq l_2 \leq 5$ with $\{u_2, v_2\} = V(AQ_n) - (N(u_1) \cup N(u_2))$.

It is well known that a panconnected graph is also edge-pancyclic. Thus, as a consequence of Theorem 26.49, AQ_n is edge pancyclic. Hsu et al. [85] give a constructive proof for the edge-pancyclicity of AQ_n by giving an algorithm to generate cycles of length from 3 to 2^n containing a pre-chosen edge $(u, v) \in AQ_n$. Using Theorem 26.51 we can conclude that AQ_n is 1-vertex fault-tolerant edge-pancyclic for $n \geq 3$. As a consequence of Theorem 26.48, we find that AQ_n is not n-edge fault-tolerant edge-pancyclic.

The proofs for most of the results stated above rely on the following $2H$ property of AQ_n discovered by Hsu et al. [73].

Definition 26.25 [73] *A graph G has* property 2H, *if for any two pairs of $\{w, x\}$ and $\{y, z\}$ of four distinct vertices of G, there exist two disjoint paths P_1 and P_2 of G such that (1) P_1 joins w to x, (2) P_2 joins y to z, and (3) $P_1 \cup P_2$ span G.*

Theorem 26.55 [73] *Let n be a positive integer with $n \geq 4$. Then AQ_n has property 2H.* ∎

Balanced-pancyclicity, geodesic-pancyclicity are some graph properties arising out of embeddings of paths and cycles on interconnection networks. Not many interconnection networks possess these strong properties. The following discussions are about these properties holding for augmented cubes.

Whenever we have a cycle passing through two vertices u and v of a graph, it would be of interest to see if the lengths of two disjoint paths between u and v in this cycle are as equal as possible. The balanced-pancyclic property of a graph captures the existence of such cycles. It would also be of interest to see if this cycle contains a shortest uv-path. The weakly geodesic pancyclic and geodesic pancyclic properties of a graph capture the existence of such cycles. That is, geodesic pancyclic properties are an enhancement of cycle embedding using shortest path(s) as a part of the cycle.

Definition 26.26 *A graph G is called* balanced pancyclic *if for every pair u, v of distinct vertices and for every integer l satisfying $\max\{2d_G(u, v), 3\} \leq l \leq |V(G)|$, there exists a cycle C of length l with $d_C(u, v) = \lfloor l/2 \rfloor$.*

Theorem 26.56 [86] *AQ_n is balanced pancyclic for $n \geq 2$.* ∎

Definition 26.27

1. *A graph G is said to be* weakly geodesic pancyclic *if for each pair u, v of distinct vertices and for each integer l with $\max\{2d_G(u, v), 3\} \leq l \leq |V(G)|$, there exists a cycle of length l that contains a shortest uv-path.*

2. *A graph G is said to be* geodesic pancyclic *if for each pair u, v of distinct vertices, every shortest uv-path lies on every cycle of length l where $\max\{2d_G(u, v), 3\} \leq l \leq |V(G)|$.*

Note that a geodesic pancyclic graph is also weakly geodesic pancyclic but the converse is not true. Further, a weakly geodesic pancyclic graph is edge pancyclic.

Theorem 26.57 [86] AQ_n is weakly geodesic pancyclic for $n \geq 2$. ∎

Note that, in [86], the term geodesic pancyclic has been used to refer to the concept of weakly geodesic pancyclicity.

Theorem 26.58 [81] The augmented cube AQ_n, $n \geq 2$, is geodesic pancyclic. ∎

We now consider the embedding of copies of linear arrays in AQ_n.

Since AQ_n, for $n \geq 4$, is $(2n-1)$ connected, it follows that there exist $2n-1$ parallel paths between any pair of vertices. Chen et al. [87] studied the existence of parallel paths in the presence of faults.

Theorem 26.59 [87] Let $n \geq 4$ and $F \subset V(AQ_n)$.

1. If $|F| \leq 4n-9$, then $AQ_n - F$ has a connected component containing at least $2^n - |F| - 1$ vertices.

2. If $|F| \leq 2n - 7$, then each pair of vertices u and v in $AQ_n - F$ is connected by $\min\{\deg_{AQ_n-F}(u), \deg_{AQ_n-F}(v)\}$ vertex-disjoint paths in $AQ_n - F$. ∎

Through the connectivity of a graph, we get to learn about the existence of parallel paths in the graph. It would be interesting to see if these parallel paths span the graph. The spanning connectivity of a graph measures the existence of such parallel paths.

Definition 26.28

1. A w-container $C(u,v)$ between vertices u and v of a κ-connected graph G is a set of w internally disjoint paths between u and v.

2. A w-container $C(u,v)$ of G is a w^*-container if it contains all the vertices of G.

3. A graph G is w^*-connected if there exists a w^*-container between any two distinct vertices.

4. The spanning connectivity, $\kappa^*(G)$, of a w^*-connected graph G ($w \leq \kappa$) is the largest integer k such that G is i^*-connected, for all i, $1 \leq i \leq k$.

5. A graph G is super spanning connected if $\kappa^*(G) = \kappa(G)$, where $\kappa(G)$ is the vertex connectivity of G.

Remark 26.5

1. A graph G is 1^*-connected if and only if it is Hamiltonian connected,

2. A graph G is 2^*-connected if it is Hamiltonian,

3. An 1^*-connected graph, except K_1 and K_2, is 2^*-connected.

Theorem 26.60 [88] AQ_n is super spanning connected if and only if $n \neq 3$. ∎

The above theorem says that, for $n \neq 3$, AQ_n contains $2n - 1$ internally disjoint paths $P_{(1)}(u,v), P_{(2)}(u,v), \ldots, P_{(2n-1)}(u,v)$ between any pair of vertices u, v such that $P_{(1)}(u,v) \cup P_{(2)}(u,v) \cup \ldots \cup P_{(2n-1)}(u,v) = V(AQ_n)$.

Though the connectivity of a connected graph allows us to determine the minimum number of vertices/edges to be removed to disconnect the graph, we cannot determine whether the resultant components have an isolated vertex or not. The super connectivity measures the extent to which vertices/edges have to be removed to ensure that there are no isolated vertices.

Definition 26.29 *Let G be a graph.*

1. *A subset $S \subset V(G)$ (respectively, $F \subset E(G)$) is called a* super vertex-cut *(respectively,* super edge-cut*) if $G - S$ (respectively, $G - F$) is not connected and every component contains at least two vertices.*

2. *The* super connectivity *(respectively,* super edge-connectivity*) of G is the minimum cardinality over all super vertex-cuts (respectively, super edge-cuts) in G if any, and, $+\infty$ otherwise.*

It has been shown that for $n \geq 3$, the super connectivity as well as the super edge-connectivity of Q_n is $2n - 2$.

Theorem 26.61 [89,90]

1. *Super connectivity of AQ_n is $4n - 8$, for $n \geq 6$.*

2. *Super edge-connectivity of AQ_n is $4n - 4$, for $n \geq 5$.* ∎

From the above theorem it follows, for $n \geq 6$ (respectively, $n \geq 5$) at least $4n - 8$ vertices (respectively, $4n - 4$ edges) of AQ_n are to be removed to get a disconnected graph that contains no isolated vertices.

The discussions above are only about the existence of parallel paths in AQ_n and do not provide any insight into the length of these paths. To permit some embeddings, it may be important to know the lengths of these paths. We have the $2H$ property of AQ_n (see Theorem 26.55) and the following result in this direction.

Theorem 26.62 [72] *For $n \neq 3$, between any two vertices $x, y \in AQ_n$, there exist $2n - 1$ vertex disjoint (x, y)-paths of length at most $\lceil n/2 \rceil + 1$.* ∎

26.3.2 Embedding into Crossed Cubes CQ_n

In this section, we present some of the recent results on embeddings into crossed cubes. Yang and Megson [91], studied the existence of cycles between two chosen vertices in the crossed cubes.

Theorem 26.63 [91] *For any two distinct vertices X and Y on CQ_n, such that $dist(X, Y) = d$ and for each integer l satisfying $2d + 6 \leq l \leq 2^n$, CQ_n contains a cycle of length l that passes through the two vertices.* ∎

Recently, Wang [92] studied the embedding of Hamiltonian cycle and presented an algorithm to generate Hamiltonian cycles for a given link permutation.

The embeddability of meshes into crossed cubes as studied by Fan and Jia [93] lead to the following result.

Theorem 26.64 [93] *A mesh of size $2 \times 2^{n-1}$ is embeddable into an n-dimensional crossed cube with unit dilation and unit expansion. Further, two disjoint meshes each of size $4 \times 2^{n-3}$ are embeddable into an n-dimensional crossed cube with unit dilation and unit expansion.* ■

Dong et al. [94] studied the embeddability of 3-dimensional meshes into crossed cubes.

Theorem 26.65 [94] *For $n \geq 4$, a family of two disjoint 3-dimensional meshes of size $2 \times 2 \times 2^{n-3}$ can be embedded in an n-dimensional crossed cube with unit dilation and unit expansion. Further, for $n \geq 6$ a family of four disjoint 3-dimensional meshes of size $4 \times 2 \times 2^{n-5}$ can be embedded in an n-dimensional crossed cube with unit dilation and unit expansion.* ■

26.3.3 Embedding into Twisted Cubes TQ_n

In a recent paper, Fan et al. [95] studied the optimal embedding of paths of all possible lengths between two arbitrary distinct nodes in twisted cubes.

Theorem 26.66 [95] *Let $n \geq 3$. Given any two distinct vertices U and V in TQ_n, there exists a path of length l between U and V, where l is any integer satisfying $dist(U,V) + 2 \leq l \leq 2^n - 1$. Further, for $n \geq 3$, there exist two nodes U and V such that no path of length $dist(U,V)$ can be embedded between U and V with dilation 1.* ■

Yang [96] presents an algorithm to construct n edge-disjoint spanning trees in TQ_n. Furthermore $\lceil n/2 \rceil$ of these trees are independent. Note that, here the edge-disjointness is with respect to edges of the trees being directed from parent to child when rooted at the same vertex. Since TQ_n is n-regular, this result is optimal with respect to the number of edge-disjoint spanning trees.

26.3.4 Embedding into Enhanced Cubes $Q_{n,k}$

Embeddings into a particular class of enhanced cubes, namely the folded hypercubes ($Q_{n,1}$), has gained maximum attention. Sunitha [38] studied the embeddability of the complete binary trees into $Q_{n,n-1}$—another particular class.

Theorem 26.67 [38] *A complete binary tree on $2^n - 1$ vertices can be embedded into the $n-1$-skip enhanced cube $Q_{n,n-1}$.* ■

Choudum and Usha Nandhini [97], studied the embeddability of the complete binary trees into the general class of enhanced cubes $Q_{n,k}$.

Theorem 26.68 [97] *Complete binary tree on $2^n - 1$ vertices is a subgraph of enhanced cube $Q_{n,k}$, when $n \not\equiv k \pmod{2}$.* ■

In particular, this result settles a conjecture of Wang [98] that the complete binary tree on $2^n - 1$ vertices is a subgraph of the folded cube $Q_{n,1}$, when n is even.

26.3.5 Embedding into Hamming Graphs $H_{b,n}$

In this section, we will present some of the results relating embedding into the generalized hypercubes (or the base-b cubes or Hamming graphs). The authors in [99] studied the embeddability of linear arrays, cycles, and meshes into the base-b cube, $H(b,n)$.

Theorem 26.69 [99] *The following classes of graphs are known to be embeddable in generalized hypercubes with dilation-1:*

1. *Linear arrays of length l, $1 \leq l \leq b^n$ can be embedded into $H(b,n)$.*

2. *A 2-dimensional mesh on $b^t \times b^m$ vertices can be embedded onto $H(b,n)$ where $n = t+m$.*

3. *Complete binary tree on $2^{2k} - 1$ vertices can be embedded onto a base-4 k-cube.* ∎

Recently, the authors in [100] have studied the embeddablity of complete k-ary trees into these cubes.

Theorem 26.70 [100] *Complete k-ary tree of height h is a subgraph of base-k cube of dimension $h + 1$, for $k \geq 3$.* ∎

For the embeddability of complete binary trees the following results are known.

Theorem 26.71 [100]

1. *Complete binary tree of height h is a subgraph of the base-3 cube of dimension n, where $\lceil (\log_3 2)(h+1) \rceil \leq n \leq \lceil (2/3)(h+1) \rceil$.*

2. *In general, complete binary tree of height h can be embedded as a subgraph into the base-b cube, whose dimension $n(b,k,h)$ is given by: $\lceil (h+1)/log_2 b \rceil \leq n(b,2,h) \leq \lceil (h+1)/\lfloor \log_2 b \rfloor \rceil$ for every $b \neq 2^l$, and $\lceil (h+1)/\log_2 b \rceil \leq n(b,2,h) \leq \lceil (h+2)/\log_2 b \rceil$ for every $b = 2^l$.*

3. *Further, a complete binary tree of height h can be embedded with load-1, dilation-2 into its optimal base-b cube, $H(b,n)$, where $n = \begin{cases} \lceil (h+1)/\lfloor \log_2 b \rfloor \rceil & \text{if } b \text{ is not a power of 2,} \\ \lceil (h+2)/log_2 b \rceil, & \text{if } b \text{ is a power of 2.} \end{cases}$* ∎

Apart from these classes of computation graphs, the embeddability of other classes of graphs into generalized hypercubes remains open.

26.4 SUMMARY

In this chapter, we have surveyed some of the results on embedding various classes of computational structures, such as meshes and trees, into interconnection networks mainly hypercubes and its variants. There are also studies related to embedding into many other classes of host graphs like star graphs, alternating group graphs, matching composition networks, de Bruijn graphs, and Mobiüs cubes.

The problem of embedding has also been studied in the context of fault tolerance of networks. Here, given a set of faulty vertices/edges in the host graph, the problem is to embed the guest graph into the host graph avoiding the fault set. The fault-tolerant embedding of various guest graphs such as paths, cycles, meshes, torus, and trees are also studied in detail by many authors. A recent survey article by Xu and Ma [101], also covers many results in the fault-tolerant embedding of paths and cycles in hypercubes.

References

[1] M. Sekanina. On an ordering of the set of vertices of a connected graph. *Publications of the Faculty of Science,* University of Brno, **412** (1960), 137–142.

[2] M. Sekanina. On an algorithm for ordering of graphs. *Canadian Mathematical Bulletin,* **14** (1971), 221–224.

[3] J.J. Karaganis. On the cube of a graph. *Canadian Mathematical Bulletin,* **11** (1968), 295–296.

[4] J. Hartman. The homeomorphic embedding of K_n in the m-cube. *Discrete Mathematics*, **16** (1976), 157–160.

[5] M. Winkler. Proof of the squashed cube conjecture. *Combinatorica*, **3** (1983), 135–139.

[6] A.K. Gupta, A.J. Boals, N.A. Sherwani, and S.E. Hambrusch. A lower bound on embedding large hypercubes into small hypercubes. *Congressus Numerantium*, **78** (1990), 141–151.

[7] V.V. Firsov. On isometric embedding of a graph into a boolean cube. *Cybernetics*, **1** (1965), 112–113.

[8] F.N. Afrati, C.H. Papadimitriou, and G. Papageorgiou. The complexity of cubical graphs. *Information and Control*, **66** (1985), 53–60.

[9] M. Livingston and Q.F. Stout. Embedding in hypercubes. *Mathematical and Computer Modelling*, **11** (1988), 222–227.

[10] Y.M. Kim and T.H. Lai. The complexity of congestion-1 embedding in a hypercube. *Journal of Algorithms*, **12**(2) (1991), 246–280.

[11] I. Havel and J. Morávek. B-valuations of graphs. *Czechoslovak Mathematical Journal*, **22**(97) (1972), 338–351.

[12] M.R. Garey and R.L. Graham. On cubical graphs. *Journal of Combinatorial Theory, Series B*, **18**(1) (1975), 84–95.

[13] W. Imrich and S. Klavžar. On the complexity of recognizing Hamming graphs and related classes of graphs. *European Journal of Combinatorics*, **17** (1996), 209–221.

[14] D.Z. Djokovic. Distance-preserving subgraphs of hypercubes. *Journal of Combinatorial Theory, Series B*, **14**(3) (1973), 263–267.

[15] P.M. Winkler. Isometric embedding in products of complete graphs. *Discrete Applied Mathematics*, **7**(2) (1984), 221–225.

[16] R.L. Roth and P.M. Winkler. Collapse of the metric hierachy for bipartite graphs. *European Journal of Combinatorics*, **7**(4) (1986), 371–375.

[17] W. Imrich and S. Klavžar. *Product Graphs: Structure and Recognition*. Wiley-Interscience Series in Discrete Mathematics and Optimization, 2000.

[18] A. Wagner and D.G. Corneil. Embedding trees in a hypercube is NP-complete. *SIAM Journal on Computing*, **19**(3) (1990), 570–590.

[19] I. Havel and P. Liebl. On imbedding the dichotomic tree into the cube. *Časopis pro Pěstování Matematiky*, **97** (1972), 201–205.

[20] B. Monien and H. Sudbourough. Embedding one interconnection network into another. *Computing Supplementary*, **7** (1990), 257–282.

[21] I. Havel. On Hamilton circuits and spanning trees of hypercubes. *Časopis pro Pěstování Matematiky*, **109** (1984), 135–152.

[22] I. Havel and P. Liebl. One-legged caterpillars span hypercubes. *Journal of Graph Theory*, **10** (1986), 69–77.

[23] S. Bezrukov, B. Monien, W. Unger, and G. Wechsung. Embedding ladders and caterpillars into the hypercube. *Discrete Applied Mathematics*, **83** (1998), 21–29.

[24] S.N. Bhatt, F.R.K. Chung, F.T. Leighton, and A.L. Rosenberg. Efficient embeddings of trees in hypercubes. *SIAM Journal of Computing*, **21** (1992), 151–162.

[25] V. Heun and E.W. Mayr. A new efficient algorithm for embedding an arbitrary binary tree into its optimal hypercube. *Journal of Algorithms*, **20**(2) (1996), 375–399.

[26] B. Monien and H. Sudbourough. Simulating binary trees on hypercubes. In *Proceedings of the 3rd Aegean Workshop on Computing*, volume 319 of *Lecture Notes in Computer Science*, pages 170–180, 1988. Technical Report, tr-rsfb-99-064 ed.: University of Paderborn, Germany, 1999.

[27] V. Heun and E.W. Mayr. Efficient dynamic embeddings of binary trees into hypercubes. *Journal of Algorithms*, **43** (2002), 51–84.

[28] A.S. Wagner. Embedding arbitrary binary trees in a hypercube. *Journal of Parallel and Distributed Computing*, **7**(3) (1989), 503–520.

[29] S.N. Bhatt and I.C.F. Ipsen. How to embed trees in hypercubes. Technical report, Yale University, 1985.

[30] S.N. Bhatt, F.R.K. Chung, F.T. Leighton, and A.L. Rosenberg. Optimal simulations of tree machines. *Proceedings of the 27th Annual IEEE Symposium of Foundations of Computer Science*, pages 274–282, 1986.

[31] T. Dvořák, I. Havel, J.M. Laborde, and P. Liebl. Generalized hypercubes and graph embeddings with dilation. *Rostocker Mathematisches Kolloqium*, **39** (1990), 13–20.

[32] S.A. Choudum and R. Indhumathi. On embedding subclasses of height-balanced trees in hypercubes. *Information Sciences*, **179**(9) (2009), 1333–1347.

[33] S.A. Choudum and R. Indhumathi. Embedding height balanced trees and Fibonacci trees in hypercubes. *Journal of Applied Mathematics and Computing*, **30** (2009), 39–52.

[34] I. Havel and P. Liebl. Embedding the polytomic tree into the n-cube. *Časopis pro Pěstování Matematiky*, **98** (1973), 307–314.

[35] L. Nebeský. Embedding m-quasistars into n-cubes. *Czechoslovak Mathematical Journal*, **38** (1988), 705–712.

[36] I. Havel. On certain trees in hypercube. In R. Bodendick and R. Henn, editors, *Topics in Combinatorics and Graph Theory*, pages 353–358. Physica-Verlag, Heidelberg, Germany, 1990.

[37] M. Kobeissi and M. Mollard. Disjoint cycles and spanning graphs of hypercubes. *Discrete Mathematics*, **288** (2004), 73–87.

[38] V. Sunitha. *Augmented cube: A new interconnection network*. PhD thesis, Department of Mathematics, IIT (Madras), India, 2002.

[39] S.A. Choudum and S. Lavanya. Embedding a subclass of trees into hypercubes. *Discrete Mathematics*, **311** (2011), 866–871.

[40] V. Sunitha. Embedding some hierarchical caterpillars into hypercube. *Electronic Notes in Discrete Mathematics*, **22** (2005), 385–389.

[41] I. Rajasingh, B. Rajan, and R.S. Rajan. On embedding of m-sequential k-ary trees into hypercubes. *Applied Mathematics*, **3** (2010), 499–503.

[42] A.K. Gupta, D. Nelson, and H. Wang. Efficient embeddings of ternary trees into hypercubes. *Journal of Parallel and Distributed Computing*, **63**(6) (2003), 619–629.

[43] S.A. Choudum and S. Lavanya. Embedding complete ternary trees into hypercubes. *Discussiones Mathematicae Graph Theory*, **28** (2008), 463–476.

[44] M.S. Chen and K.G. Shin. Processor allocation in an n-cube multiprocessor using gray codes. *IEEE Transactions of Computers*, **36** (1987), 1396–1407.

[45] T. Feder and C. Subi. Nearly tight bounds on the number of Hamiltonian cycles of the hypercube and generalizations. *Information Processing Letters*, **109** (2009), 267–272.

[46] John Tromp. Number of Hamiltonian cycles in the binary n-cube, or the number of cyclic n-bit gray codes. *http://oeis.org/A0066037*, 2010.

[47] B. Alspach, J.C. Bermond, and D. Sotteau. Decomposition into cycles I: Hamilton decompositions. In Gena Hahn, ed., *Cycles and Rays*, pages 9–18. Kluwer Academic, Dordrecht, Holland, 1990.

[48] A. Kotzig. Every cartesian product of two circuits is decomposable into two Hamiltonian circuits. Technical report, Centre de Recherches Mathematiques, Montreal, Canada, 1973.

[49] M.F. Foregger. Hamiltonian decompositions of product of cycles. *Discrete Mathematics*, **24** (1978), 251–260.

[50] D.W. Bass and I.H. Sudborough. Hamilton decompositions and (n/2)-factorizations of hypercubes. *Journal of Graph Algorithms and Applications*, **6** (2002), 174–194.

[51] K. Okuda and S.W. Song. Revisiting Hamiltonian decomposition of the hypercube. In *Proceedings of the 13th Symposium on Integrated Circuits and Systems Design,* Manaus, Brazil, pages 18–24, 2000.

[52] S.W. Song. Towards a simple construction method for Hamiltonian decomposition of the hypercube. *DIMACS series in Discrete Mathematics and Theoretical Computer Science*, **21** (1995), 297–306.

[53] Y. Saad and M.H. Schultz. Topological properties of hypercubes. *IEEE Transactions on Computers*, **37**(7) (1988), 867–872.

[54] S.A. Choudum, S. Lavanya, and V. Sunitha. Disjoint paths in hypercubes with prescribed origins and lengths. *International Journal of Computer Mathematics*, **87**(8) (2010), 1692–1708.

[55] S. Madhavapeddy and I.H. Sudbourough. A topological property of hypercubes: Node disjoint paths. In *Proceedings of the 2nd IEEE Symposium on Parallel and Distributed Processing*, Dallas, TX, pages 532–539, 1990.

[56] Q.-P. Gu and S. Peng. An efficient algorithm for the k-pairwise disjoint paths problem in hypercubes. *Journal of Parallel and Distributed Computing*, **60** (2000), 764–774.

[57] T.F. Gonzalez and D. Serena. n-cube network: Node disjoint shortest paths for maximal distance pairs of vertices. *Parallel Computing*, **30**(8) (2004), 973–998.

[58] T. Dvořák, P. Gregor, and V. Koubek. Spanning paths in hypercubes. In *DMTCS Proceedings*, pages 363–368, 2005.

[59] R. Caha and V. Koubek. Spanning multi-paths in hypercubes. *Discrete Mathematics*, **301** (2007), 2053–2066.

[60] T.F. Gonzalez and D. Serena. Complexity of k-pairwise disjoint shortest paths in the undirected hypercubic network and related problems. In *Proceedings of the International Conference on Parallel and Distributed Computing and Systems*, pages 61–66, 2002.

[61] T.F. Gonzalez and D. Serena. Pairwise edge disjoint shortest paths in the n-cube. *Theoretical Computer Science*, **369**(1–3) (2006), 427–435.

[62] B. Barden, R.L. Hadas, J. Davis, and W. Williams. On edge-disjoint spanning trees in hypercubes. *Information Processing Letters*, **70** (1999), 13–16.

[63] K. Obokata, Y. Iwasaki, F. Bao, and Y. Igarashi. Independent spanning trees of product graphs and their construction. *IEICE Transactions on Fundamentals of Electronics, Communications and Computer Science*, **E79**-A (1996), 184–193.

[64] J.F. Fink. On the decomposition of n-cubes into isomorphic trees. *Journal of Graph Theory*, **14** (1990), 405–411.

[65] M. Ramras. Symmetric edge-decompositions of hypercubes. *Graphs and Combinatorics*, **7** (1991), 65–87.

[66] M.S. Jacobson, M. Truszczynski, and Z. Tuza. Decompositions of regular bipartite graphs. *Discrete Mathematics*, **89** (1991), 17–27.

[67] D.E. Bryant, S. El-Zanati, C.V. Eynden, and D.G. Hoffman. Star decompositions of cubes. *Graphs and Combinatorics*, **17** (2001), 55–59.

[68] S. El-Zanati, C.V. Eynden, and S. Stubbs. On the decomposition of Cayley graphs into isomorphic trees. *Australasian Journal of Combinatorics*, **22** (2000), 13–18.

[69] P. Horak, J. Sira, and W. Wallis. Decomposing cubes. *Journal of the Australian Mathematical Society*, Series A **61** (1996), 119–128.

[70] S.L. Johnsson and C.T. Ho. Optimum broadcasting and personalized communication in hypercubes. *IEEE Transactions on Computers*, **38** (1989), 1249–1268.

[71] S. Wagner and M. Wild. Partitioning the hypercube Q_n into n isomorphic edge-disjoint trees. Preprint, http://hdl.handle.net/10019.1/16108, 2011.

[72] S.A. Choudum and V. Sunitha. Augmented cubes. *Networks*, **40** (2002), 71–84.

[73] H.C. Hsu, L.C. Chiang, J.J.M. Tan, and L.H. Hsu. Fault Hamiltonicity of augmented cubes. *Parallel Computing*, **31** (2005), 131–145.

[74] M. Ma, G. Liu, and J.M. Xu. Panconnectivity and edge-fault-tolerant pancyclicity of augmented cubes. *Parallel Computing*, **33** (2007), 36–42.

[75] S.Y. Hsieh and Y.R. Cian. Conditional edge-fault Hamiltonicity of augmented cubes. *Information Sciences*, **180** (2010), 2596–2617.

[76] C.M Lee, Y.H. Teng, J.J.M. Tan, and L.H. Hsu. Embedding Hamiltonian paths in augmented cubes with a required vertex in a fixed position. *Computers and Mathematics with Applications*, **58** (2009), 1762–1768.

[77] W.W. Wang, M.J. Ma, and J.M. Xu. Fault-tolerant pancyclicity of augmented cubes. *Information Processing Letters*, **103** (2007), 52–56.

[78] S.Y. Hsieh and J.Y. Shiu. Cycle embedding of augmented cubes. *Applied Mathematics and Computation*, **191** (2007), 314–319.

[79] J.S. Fu. Edge-fault-tolerant vertex-pancyclicity of augmented cubes. *Information Processing Letters*, **110** (2010), 439–443.

[80] P.L. Lai, J.W. Hsue, J.J.M. Tan, and L.H. Hsu. On the panconnected properties of the augmented cubes. In *Proceedings of the International Computer Symposium*, Taipei, Taiwan, pages 1249–1251, 2004.

[81] H.C. Chan, J.M. Chang, Y.L. Wang, and S.J. Horng. Geodesic-pancyclicity and fault-tolerant panconnectivity of augmented cubes. *Applied Mathematics and Computation*, **207** (2009), 333–339.

[82] H.C. Chan. *Geodesic-pancyclic graphs*. PhD thesis, Department of Computer Science and Information Technology, National Taiwan University of Science and Technology, Taiwan, 2006.

[83] H. Wang, J. Wang, and J.M. Xu. Fault-tolerant panconnectivity of augmented cubes. *Frontiers of Mathematics in China*, **4** (2009), 697–719.

[84] T.L. Kung, Y.H. Teng, and L.H. Hsu. The panpositionable panconnectedness of augmented cubes. *Information Sciences*, **180** (2010), 3781–3793.

[85] H.C. Hsu, P.L. Lai, C.H. Tsai, and T.K. Li. Efficient algorithms for embedding cycles in augmented cubes. In *Proceedings of the 10th International Symposium on Pervasive Systems, Algorithms and Networks*, Kaohsiung, Taiwan, pages 596–600, 2009.

[86] H.C. Hsu, P.L. Lai, and C.H. Tsai. Geodesic pancyclicity and balanced pancyclicity of augmented cubes. *Information Processing Letters*, **101** (2007), 227–232.

[87] Y.C. Chen, M.H. Chen, and J.J.M. Tan. Maximally local connectivity on augmented cubes. In *Proceedings of the 9th International Conference on Algorithms and Architectures for Parallel Processing*, pages 121–128, Springer-Verlag, Berlin, Germany, 2009.

[88] T.Y. Ho, C.K. Lin, and L.H. Hsu. The super spanning connectivity of the augmented cubes. *Proceedings of the 23rd Workshop on Combinatorial Mathematics and Computation Theory*, pages 53–61, 2006.

[89] M. Ma, G. Liu, and J.M. Xu. The super connectivity of augmented cubes. *Information Processing Letters*, **106** (2008), 59–63.

[90] M. Ma, X. Tan, J.M. Xu, and G. Liu. A note on "the super connectivity of augmented cubes." *Information Processing Letters*, **109** (2009), 592–593.

[91] X. Yang and G.M. Megson. On the double-vertex-cycle-connectivity of crossed cubes. *International Journal of Parallel, Emergent and Distributed Systems*, **19**(1) (2004), 11–17.

[92] D. Wang. On embedding Hamiltonian cycles in crossed cubes. *IEEE Transactions on Parallel and Distributed Systems*, **19**(3) (2008), 334–346.

[93] J. Fan and X. Jia. Embedding meshes into crossed cubes. *Information Sciences*, **177**(15) (2007), 3151–3160.

[94] Q. Dong, X. Yang, J. Zhao, and Y.Y. Tang. Embedding a family of disjoint 3D meshes into a crossed cube. *Information Sciences*, **178**(11) (2008), 2396–2405.

[95] J. Fan, X. Jia, and X. Lin. Optimal embeddings of paths with various lengths in twisted cubes. *IEEE Transactions on Parallel and Distributed Systems*, **18**(4) (2007), 511–521.

[96] M.C. Yang. Constructing edge-disjoint spanning trees in twisted cubes. *Information Sciences*, **180** (2010), 4075–4083.

[97] S.A. Choudum and R. Usha Nandini. Complete binary trees in folded and enhanced cubes. *Networks*, **43**(4) (2004), 266–272.

[98] D. Wang. On embedding binary trees into folded cubes. *Congressus Numerantium*, **134** (1998), 89–97.

[99] S. Lakshmivarahan and S.K. Dhall. A new hierarchy of hypercube interconnection schemes for parallel computers. *Journal of Supercomputing*, **2** (1988), 81–108.

[100] S.A. Choudum and S. Lavanya. Complete k-ary trees and Hamming graphs. *Australasian Journal of Combinatorics*, **45** (2009), 15–24.

[101] J.-M. Xu and M. Ma. Survey on path and cycle embedding in some networks. *Frontiers of Mathematics in China* (2009).

VII

Special Graphs

CHAPTER 27

Program Graphs*

Krishnaiyan "KT" Thulasiraman

CONTENTS

27.1 Introduction 691
27.2 Program Graph Reducibility 691
27.3 Dominators in a Program Graph 698

27.1 INTRODUCTION

A program graph is a directed graph G with a distinguished vertex s such that there is a directed path from s to every other vertex of G. In other words, every vertex in G is reachable from s. The vertex s is called the start vertex of G. We assume that there are no parallel edges in a program graph. The flow of control in a computer program can be modeled by a program graph in which each vertex represents a block of instructions which can be executed sequentially. Such a representation of computer programs has proved very useful in the study of several questions relating to what is known as the *code-optimization* problem. For many of the code-optimization methods to work, the program graph must have a special property called *reducibility* (see [1–9]). Two other concepts of interest in the study of program graphs are: dominators and dominator tree. The dominator tree of a program graph provides information about what kinds of code motion are safe. In this chapter we discuss algorithms to test the reducibility of a program graph and to generate a dominator tree. These algorithms make extensive use of depth-first search (DFS). For background information on DFS see Chapter 3.

27.2 PROGRAM GRAPH REDUCIBILITY

Reducibility of a program graph G is defined in terms of the following two transformations on G:

S_1 Delete self-loop (v,v) in G.

S_2 If (v,w) is the only edge incident into w, and $w \neq s$, delete vertex w. For every edge (w,x) in G add a new edge (v,x) if (v,x) is not already in G. (This transformation is called *collapsing* vertex w into vertex v.)

For example collapsing vertex 5 into vertex 4 in the program graph of Figure 27.1a results in the graph shown in Figure 27.1b.

A program graph is *reducible* if can be transformed into a graph consisting of only the vertex s by repeated applications of the transformations S_1 and S_2. For example, the graph in Figure 27.1a is reducible. It can be verified that this graph can be reduced by collapsing the vertices in the order 5, 8, 4, 3, 10, 9, 7, 6, 2.

---
*This chapter is an edited version of Sections 14.5 and 14.6 in M.N.S. Swamy and K. Thulasiraman, *Graphs, Networks and Algorithms*, Wiley-Interscience, 1981.

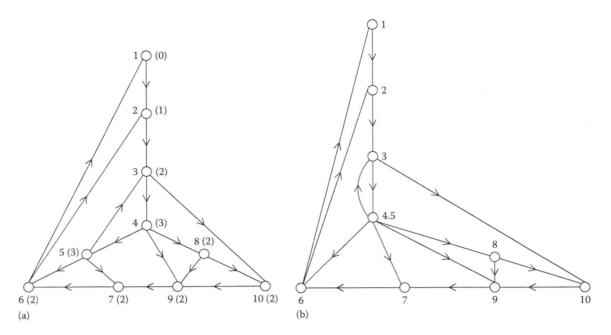

Figure 27.1 (a) Rreducible program graph G. HIGHPT1 values are given in parentheses. (b) Graph obtained after collapsing in G vertex 5 into vertex 4.

Cocke [5] and Allen [4] were the original formulators of the notion of reducibility, and their definition is in terms of a technique called *interval analysis*. The definition given above is due to Hecht and Ullman [10], and it is equivalent to that of Cocke and Allen.

If a graph G is reducible, then it can be shown [10] that any graph G' obtained from G by one or more applications of the transformations S_1 and S_2 is also reducible. Thus the order of applying transformations does not matter in a test for reducibility. Further, some interesting classes of programs such as *go-to-less* programs give rise to graphs which are necessarily reducible [10], and most programs may be modeled by a reducible graph using a process of *node splitting* [11].

Suppose we wish to test the reducibility of a graph G. This may be done by first deleting self-loops using transformation S_1 and then counting the number of edges incident into each vertex. Next we may find a vertex w with only one edge (v, w) incident into it and apply transformation S_2, collapsing w into v. We may then repeat this process until we reduce the graph entirely or discover that it is not reducible. Clearly each application of S_2 requires $O(n)$ time, where n is the number of vertices in G and reduces the number of vertices by 1. Thus the complexity of this algorithm is $O(n^2)$. Hopcroft and Ullman [12] have improved this to $O(m \log m)$, where m is the number of edges in G. Tarjan [13,14] has subsequently given an algorithm which compares favorably with that of Hopcroft and Ullman.

Hecht and Ullman [10,15] have given several useful structural characterizations of program graphs. One of these is given in the following theorem.

Theorem 27.1 *Let G be a program graph with start vertex s. G is reducible if and only if there do not exist distinct vertices $v \neq s$ and $w \neq s$, directed paths P_1 from s to v and P_2 from s to w, and a directed circuit C containing v and w, such that C has no edges and only one vertex in common with each of P_1 and P_2 (see Figure 27.2).*

Proof. Proof of the above theorem may be found in [16] and [10]. ∎

We discuss in this section Tarjan's algorithm for testing the reducibility of a program graph. This algorithm uses DFS and is based on a characterization of reducible program graphs which we shall prove using Theorem 27.1. Our discussion here is based on [13].

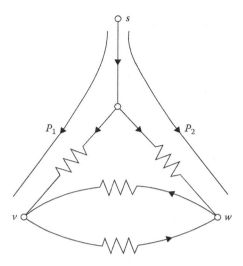

Figure 27.2 Basic non-reducible graph.

Let G be a program graph with start vertex s. Let T be a DFS tree of G with s as the root. Henceforth we refer to vertices by their depth-first numbers.

Theorem 27.2 *G is reducible if and only if G contains no directed path P from s to some vertex v such that v is a proper ancestor in T of some other vertex on P.*

Proof. Suppose G is not reducible. Then there exist vertices v and w, directed paths P_1 and P_2, and circuit C which satisfy the condition in Theorem 27.1. Assume that $v < w$. Let C_1 be the part of C from v to w. Then C_1 contains some common ancestor u of v and w. Then the directed path consisting of P_2 followed by the part of C from w to u satisfies the condition in the theorem.

Conversely, suppose there exists a directed path P which satisfies the condition in the theorem. Let then v be the first vertex on P which is a proper ancestor of some earlier vertex on P. Suppose w is the first vertex on P which is a descendant of v. Let P_1 be the part of T from s to v, and let P_2 be the part of P from s to w. Also let C be the directed circuit consisting of the part of P from w to v followed by the path of tree edges from v to w. Then we can see that v, w, P_1, P_2 and C satisfy the condition in Theorem 27.1. So G is not reducible. ∎

For any vertex v, let HIGHPT1(v) be the highest numbered proper ancestor of v such that there is a directed path P from v to HIGHPT1(v) and P includes no proper ancestors of v except HIGHPT1(v). We define HIGHPT1$(v) = 0$ if there is no directed path from v to a proper ancestor of v. As an example, in Figure 27.1a we have indicated in parentheses the HIGHPT1 values of the corresponding vertices.

Note that in HIGHPT1(v) calculation we may ignore forward edges since if P is a directed path from v to w and P contains no ancestors of v except v and w, we may substitute for each forward edge in P a path of tree edges or a part of it and still have a directed path from v to w which contains no ancestors of v except v and w. Tarjan's algorithm is based on the following DFS characterization of program graphs.

Theorem 27.3 *G is reducible if and only if there is no vertex v with an edge (u, v) incident into v such that $w < $ HIGHPT1(v), where w is the highest numbered common ancestor of u and v.*

Proof. Suppose G is not reducible. Then by Theorem 27.2 there is a directed path P from s to v with v a proper ancestor of some other vertex on P. Choose P as short as possible. Let w be the first vertex on P which is a descendant of v. Then all the vertices except v which follow w on P are descendants of v in T. In other words, the part of P from w to v includes no proper ancestors of w except v. So $\text{HIGHPT1}(w) \geq v$. Thus w, $\text{HIGHPT1}(w)$, and the edge of P incident into w satisfy the condition in the theorem.

Conversely, suppose the condition in the theorem holds. Then the edge (u, v) is not a back edge because in such a case the highest numbered common ancestor w of u and v is equal to v, and so $w = v \geq \text{HIGHPT1}(v)$. Thus (u, v) is either a forward edge or a cross edge. Let P_1 be a directed path from v to $\text{HIGHPT1}(v)$ which passes through no proper ancestors of v except $\text{HIGHPT1}(v)$. Then the directed path consisting of the tree edges from s to u followed by the edge (u, v) followed by P_1 satisfies the condition in Theorem 27.2. So G is not reducible. ∎

Now it should be clear that testing reducibility of a graph G using the above theorem involves the following main steps:

1. Perform a DFS of G with s as the root.

2. Calculate $\text{HIGHPT1}(v)$ for each vertex v in G.

3. For cross edges, check the condition in Theorem 27.3 during the HIGHPT1 calculation.

4. For forward edges, check the condition in Theorem 27.3 after the HIGHPT1 calculation.

Note that, as we observed earlier, forward edges may be ignored during the HIGHPT1 calculation. Further as we observed in the proof of Theorem 27.3, back edges need not be tested for the condition in Theorem 27.3.

To calculate HIGHPT1 values, we first order the back edges (u, v) by the number of v. Then we process the back edges in order, from highest to lowest v. Initially all vertices are unlabeled. To process the back edge (u, v), we proceed up the tree path from u to v, labeling each currently unlabeled vertex with v. (We do not label v itself.) If a vertex w gets labeled, we examine all cross edges incident into w. If (z, w) is such a cross edge (see Figure 27.3), we proceed up the tree path from z to v, labeling each unlabeled vertex with v. If z is

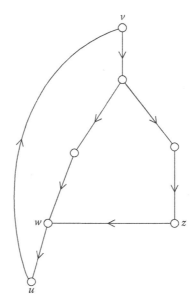

Figure 27.3 Illustration of HIGHPT1 calculation.

not a descendant of v, then G is not reducible by Theorem 27.3, and the calculation stops. We continue labeling until we run out of cross edges incident into just labeled vertices; then we process the next back edge. When all the back edges are processed, the labels give the HIGHPT1 values of the vertices. Each unlabeled vertex has HIGHPT1 equal to zero.

We now describe Tarjan's algorithm for testing reducibility. In this algorithm we use n queues, called buckets, one for each vertex. The bucket BUCKET(w) corresponding to vertex w contains the list of back edges (u, w) incident into vertex w. While processing back edge (u, w) we need to keep track of certain vertices from which u can be reached by directed paths. The set CHECK is used for this purpose.

Algorithm 27.1 Program graph reducibility (Tarjan)

S1. Perform a DFS of the given n-vertex program graph G. Denote vertices by their DFS numbers. Order the back edges (u, v) of G by the number of v.

S2. For $i = 1, 2, \ldots, n$
HIGHPT1(i) $\leftarrow$ 0,
BUCKET (i) $\leftarrow$ the empty list.

S3. Add each back edge (u, w) to BUCKET (w).

S4. $w \leftarrow n - 1$.

S5. Test if BUCKET(w) is empty. If yes, go to step S6; otherwise go to step S7.

S6. $w \leftarrow w - 1$. If $w < 1$, go so step S16; otherwise go to step S5.

S7. (Processing of a new back edge begins.) Delete a back edge (x, w) from BUCKET(w) and CHECK $\leftarrow \{x\}$.

S8. Test if CHECK is empty. If yes (processing of a back edge is over.) go to step S5; otherwise go to step S9.

S9. Delete u from CHECK.

S10. Test if u is a descendant of w. If yes, go to step S11; otherwise go to step S17.

S11. Test if $u = w$. If yes, go to step S8; otherwise go to step S12.

S12. Test if HIGHPT1(u) = 0. If yes, go to step S13; otherwise go to step S15.

S13. HIGHPT1(u) $\leftarrow w$.

S14. For each cross edge (u, v) add v to CHECK.

S15. $u \leftarrow$ FATHER(u). Go to step S11.

S16. If $u \geq$ HIGHPT1(v) for each forward edge (u, v) (the graph is reducible), then HALT. Otherwise go to step S17.

S17. HALT. The graph is not reducible.

Note that step S10 in Algorithm 27.2 requires that we be able to determine whether a vertex w is a descendant of another vertex u. Let ND(u) be the number of descendants of vertex u in T. Then we can show that w is a descendant of u if and only if $u \leq w < u + $ND($u$). We can calculate ND($u$) during the DFS in a straightforward fashion.

The efficiency of Algorithm 27.2 depends crucially on the efficiency of the HIGHPT1 calculation. To make the HIGHPT1 calculation efficient, in Algorithm 27.2 we need to avoid examining vertices which have already been labeled. Tarjan suggests a procedure to achieve this. The following observation forms the basis of this procedure:

Suppose at step S12 we are examining a vertex u for labeling. Let at this stage, u' be the highest unlabeled proper ancestor of u. This means that all the proper ancestors of u except u', which lie between u' and u in T, have already been labeled. Thus u' is the next vertex to be examined. Furthermore when u is labeled, then all the vertices for which u is the highest unlabeled proper ancestor will have u' as their highest unlabeled proper ancestor.

To implement the method which follows from the above observation, we shall use sets numbered 1 to n. A vertex $w \neq 1$ will be in the set numbered v, that is, SET(v), if v is the highest numbered unlabeled proper ancestor of w. Since vertex 1 never gets labeled, each vertex is always in a set. Initially a vertex is in the set whose number is its father in T. Thus initially add i to SET(FATHER(i)) for $i = 2, 3, \ldots, n$.

To carry out step S15, we find the number u' of the set containing u and let that be the new u. Further when u becomes labeled, we shall combine the sets numbered u and u' to form a new set numbered u'. Thus u' becomes the highest numbered unlabeled proper ancestor of all the vertices in the old SET(u).

It can now be seen that replacing steps S12 through S15 in Algorithm 27.2 by the following sequence of steps will implement the method described above. Of course, SETs are initialized as explained earlier.

S12′. $u' \leftarrow$ the number of the set containing u.

S13′. Test if HIGHPT1(u) = 0. If yes, go to step S14′; otherwise go to step S16′.

S14′. (a) HIGHPT1(u) $\leftarrow w$; and

(b) SET(u') $\leftarrow$ SET(u) $\cup$ SET(u')

S15′. For each cross edge (v, u) add v to CHECK.

S16′. $u \leftarrow u'$. Go to step S11.

It can be verified that the HIGHPT1 calculations modified as above require $O(n)$ set unions, $O(m+n)$ executions of step S12′, and $O(m+n)$ time exclusive of set operations. If we use the algorithm described in Fischer [17] and Hopcroft and Ullman [18] for performing disjoint set unions and for performing step S12′, then it follows from the analysis given in Tarjan [19] that the reducibility algorithm has complexity $O(m\alpha(m,n))$, where $\alpha(m,n)$ is a very slowly growing function which is related to a functional inverse of Ackermann's functions $A(p,q)$, and it is defined as follows:

$$\alpha(m,n) = \min\left\{z \geq 1 \,\middle|\, A\left(z, 4\left\lceil \frac{m}{n} \right\rceil\right) > \log 2n\right\}.$$

The definition of Ackermann's function is

$$A(p,q) = \begin{cases} 2q, & p = 0 \\ 0, & q = 0 \text{ and } p \geq 1 \\ 2, & p \geq 1 \text{ and } q = 1 \\ A(p-1, A(p, q-1)), & p \geq 1 \text{ and } q \geq 2 \end{cases} \quad (27.1)$$

Note that Ackermann's function is a very rapidly growing function. It it easy to see that $A(3,4)$ is a very large number, and it can be shown that $\alpha(m,n) \leq 3$ if $m \neq 0$ and $\log_2 n < A(3,4)$. The algorithm is also described in Aho et al. [20] and Horowitz and Sahni [21].

Algorithm 27.2 is nonconstructive, that is, it does not give us the order in which vertices have to be collapsed to reduce a reducible graph. However, as we shall see now, this information can be easily obtained as this algorithm progresses.

During DFS let us assign to the vertices, numbers called SNUMBERs, from n to 1 in the order in which scanning at a vertex is completed. It can be easily verified that

1. If (v,w) is a tree edge, then $\text{SNUMBER}(v) < \text{SNUMBER}(w)$.

2. If (v,w) is a cross edge, then $\text{SNUMBER}(v) < \text{SNUMBER}(w)$.

3. If (v,w) is a back edge, then $\text{SNUMBER}(v) > \text{SNUMBER}(w)$.

4. If (v,w) is a forward edge, then $\text{SNUMBER}(v) > \text{SNUMBER}(w)$.

As an example, in Figure 27.4 we have shown in parentheses the SNUMBERs of the corresponding vertices of the graph in Figure 27.1a.

Suppose we apply the reducibility algorithm, and each time we label a vertex v let us associate with it a pair $(\text{HIGHPT1}(v), \text{SNUMBER}(v))$. When the algorithm is finished, let us order the vertices so that a vertex labeled (x_1, y_1) appears before a vertex labeled (x_2, y_2) if and only if $x_1 > x_2$ or $x_1 = x_2$ and $y_1 < y_2$. This order of vertices is called *reduction order*. Note that an unlabeled vertex v is associated with the pair $(0, \text{SNUMBER}(v))$.

Suppose $v_a, v_b, v_c, \ldots$ is a reduction order for a reducible program graph G. Let T be a DFS tree of G with vertex s as the root. Using Theorem 27.3 and the properties of SNUMBERs outlined earlier, it is easy to show that the tree edge (u, v_a) is the only edge incident into v_a. Suppose we collapse v_a into u and let G' be the resulting reducible graph. Also let T' be the tree obtained from T by contracting the edge (u, v_a). Then clearly T' is a DFS tree of G'. Further

1. A cross edge of G corresponds to nothing or to a cross edge or to a forward edge of G'.

2. A forward edge of G corresponds either to nothing or to a forward edge of G'.

3. A back edge of G corresponds either to nothing or to a back edge of G'.

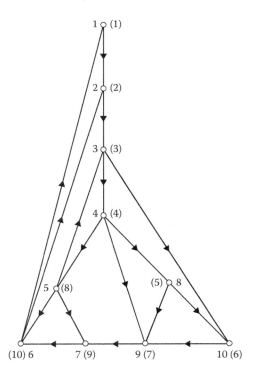

Figure 27.4 Graph of Figure 27.1a with SNUMBER values in parentheses.

It can now be verified that the relative HIGHPT1 values of the vertices in G' will be the same as those in G. This is also true for the SNUMBERs. Thus $v_b, v_c, \ldots$ will be a reduction order for G'. Repeating the above arguments, we get the following theorem.

Theorem 27.4 *If a program graph G is reducible, then we may collapse the vertices of G in the reduction order using the transformation S_2 (interspersed with the applications of S_1).* ∎

For example, it can be verified from the HIGHPT1 values given in Figure 27.1a and the SNUMBER values given in Figure 27.4 that the sequence 4, 5, 3, 8, 10, 9, 7, 6, 2 is a reduction order for the graph in Figure 27.1a.

27.3 DOMINATORS IN A PROGRAM GRAPH

Let G be a program graph with start vertex s. If, in G, vertex v lies on every directed path from s to w, then v is called a *dominator* of w and is denoted by $\text{DOM}(w)$. If v is a dominator of w and every other dominator of w also dominates v, then v is called the *immediate dominator* of w, and it is denoted by $\text{IDOM}(w)$. For example, in the program graph G shown in Figure 27.5a vertex 1 is the immediate dominator of vertex 9.

It can be shown that every vertex of a program graph $G = (V, E)$, except for the start vertex s, has a unique immediate dominator. The edges $\{(\text{IDOM}(w), w) | w \in V - \{s\}\}$ form

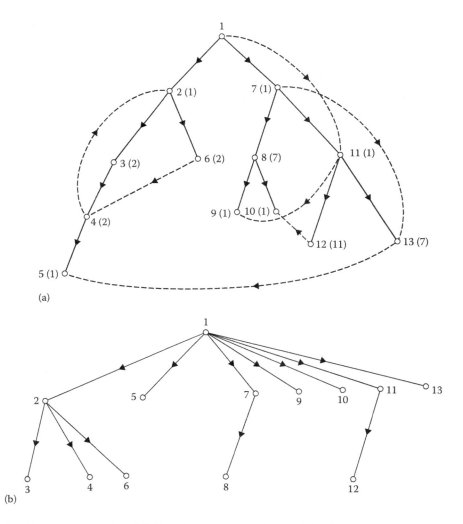

Figure 27.5 (a) Program graph. (b) Dominator tree of graph in (a).

a directed tree rooted at s, called the *dominator tree* of G, such that v dominates w if and only if v is a proper ancestor of w in the dominator tree. If G represents the flow of control in a computer program, then the dominator tree provides information about what kinds of code motion are safe. The dominator tree of the program graph in Figure 27.5a is shown in Figure 27.5b.

We now develop an algorithm due to Lengauer and Tarjan [22] for finding the dominator tree of a program graph. This algorithm is a simpler and faster version of an algorithm presented earlier by Tarjan [23].

Let G be a program graph with start vertex s. Let T be a DFS tree of G. In the following we shall identify the vertices of G by their DFS numbers. Furthermore, the notation $x \xrightarrow{*} y$ means that x is an ancestor of y in T, $x \xrightarrow{+} y$ means that $x \xrightarrow{*} y$ in T and $x \neq y$, and $x \rightarrow y$ means that x is the father of y in T.

The following two lemmas are crucial in the development of the algorithm.

Lemma 27.1 *If v and w are vertices of G such that $v < w$, than any directed path from v to w must contain a common ancestor of v and w in T.* ∎

Lemma 27.2 *Let $w \neq s$, $v \xrightarrow{*} w$, and P be a directed path from s to w. Let x be the last vertex on P such that $x < v$, and let y be the first vertex following x on P and satisfying $v \xrightarrow{*} y \xrightarrow{*} w$. If $Q : x = v_0, v_1, v_2, \ldots, v_k = y$ is the part of P from x to y, then $v_i > y$ for $1 \leq i \leq k - 1$.* ∎

Let us now define for each vertex $w \neq s$,

$\text{SDOM}(w) = \min \{v |$ there is directed path $v = v_0, v_1, v_2, \ldots, v_k = w$ such that $v_i > w$ for $1 \leq i \leq k - 1\}$.

$\text{SDOM}(w)$ will be called the *semi-dominator* of w. It easily follows from the above definition that

$$\text{SDOM}(w) < w. \tag{27.2}$$

In Figure 27.5a the continuous edges are tree edges and the dashed edges are nontree edges with respect to a DFS. The semi-dominator of each vertex is shown in parentheses next to the vertex.

As a first step, Lengauer and Tarjan's algorithm computes semi-dominators of all the vertices. The semi-dominators are then used to compute the immediate dominators of the vertices.

The following theorem provides a way to compute semi-dominators.

Theorem 27.5 *For any vertex $w \neq s$,*

$$\text{SDOM}(w) = \min(\{v|(v,w) \in E \text{ and } v < w\} \cup \{SDOM(u) | u > w \text{ and there is an edge}$$
$$(v,w) \text{ such that} (u \xrightarrow{*} v)\}) \tag{27.3}$$

Proof. Let x equal the right-hand side of (27.3). It can be shown using the definition of semi-dominators that $\text{SDOM}(w) \leq x$.

To prove that $\text{SDOM}(w) \geq x$, let $y = \text{SDOM}(w)$, and let $y = v_0, v_1, \ldots, v_k = w$ be a directed path such that $v_i > w$ for $1 \leq i \leq k - 1$. If $k = 1$, then $(y, w) \in E$, and $y < w$ by (27.2). Thus $\text{SDOM}(w) = y \geq x$. Suppose $k > 1$. Let j be minimum such that $j \geq 1$ and $v_j \xrightarrow{*} v_{k-1}$. Such a j exists since $k - 1$ is a candidate for j. We now claim $v_i > v_j$ for $1 \leq i \leq j - 1$.

Suppose to the contrary that $v_i < v_j$ for some $1 \leq i \leq j-1$. Then choose the i such that $1 \leq i \leq j-1$ and v_i is minimum. Then by Lemma 27.1 $v_i \xrightarrow{*} v_j$, contradicting the choice of j. This proves the claim.

The claim implies that $\mathrm{SDOM}(w) = y \geq \mathrm{SDOM}(v_j)$. Since $v_j > w$, $v_j \xrightarrow{*} v_{k-1}$ and $(v_{k-1}, w) \in E$ it follows from (27.3) that $\mathrm{SDOM}(v_j) \geq x$. So $\mathrm{SDOM}(w) \geq x$. Thus whether $k = 1$ or $k > 1$, we have $\mathrm{SDOM}(w) \geq x$, and the theorem is proved. ∎

We now need a way to compute immediate dominators from semi-dominators. Toward this end we proceed as follows.

The next three lemmas are easy to prove. Proof of Lemma 27.4 uses Lemma 27.1.

Lemma 27.3 *For any vertex $w \neq s$, $\mathrm{IDOM}(w) \xrightarrow{+} w$.* ∎

Lemma 27.4 *For any vertex $w \neq s$, let $v = \mathrm{SDOM}(w)$. Then $v \xrightarrow{+} w$.* ∎

Lemma 27.5 *For any vertex $w \neq s$, let $v = \mathrm{SDOM}(w)$. Then $\mathrm{IDOM}(w) \xrightarrow{*} v$.* ∎

Lemma 27.6 *Let vertices v, w satisfy $v \xrightarrow{*} w$. Then either $v \xrightarrow{*} \mathrm{IDOM}(w)$ or $\mathrm{IDOM}(w) \xrightarrow{*} \mathrm{IDOM}(v)$.*

Proof. Let x be any proper descendant of $\mathrm{IDOM}(v)$ which is also a proper ancestor of v. Then there is a directed path from s to v which avoids x. By concatenating this path with the path in T from v to w, we obtain a directed path from s to w which avoids x. Thus $\mathrm{IDOM}(w)$ must be either a descendant of v or an ancestor of $\mathrm{IDOM}(v)$. ∎

Using the foregoing lemmas we next prove two results which provide a way to compute immediate dominators and semi-dominators.

Theorem 27.6 *Let $w \neq s$ and let $v = \mathrm{SDOM}(w)$. Suppose every u for which $v \xrightarrow{+} u \xrightarrow{*} w$ satisfies $\mathrm{SDOM}(u) \geq \mathrm{SDOM}(w)$. Then $\mathrm{IDOM}(w) = v$.*

Proof. By Lemma 27.5, $\mathrm{IDOM}(w) \xrightarrow{*} v$. So to prove that $\mathrm{IDOM}(w) = v$, it suffices to show that v dominates w.

Consider any directed path P from s to w. Let x be the last vertex on this path such that $x < v$. If there is no such x, then $v = s$ dominates w. Otherwise let y be the first vertex following x on the path and satisfying $v \xrightarrow{*} y \xrightarrow{*} w$. Let $Q: x = v_0, v_1, v_2, \ldots, v_k = y$ be the part of P from x to y. Then by Lemma 27.2 $v_i > y$ for $1 \leq i \leq k-1$. This together with the definition of semi-dominators implies that $\mathrm{SDOM}(y) \leq x < v = \mathrm{SDOM}(w)$. So $\mathrm{SDOM}(y) < \mathrm{SDOM}(w)$.

By the hypothesis of the theorem, $\mathrm{SDOM}(u) \geq \mathrm{SDOM}(w)$ for every u satisfying $v \xrightarrow{+} u \xrightarrow{*} w$. So y cannot be a proper descendant of v. Since y satisfies $v \xrightarrow{*} y \xrightarrow{*} w$, it follows that $y = v$, and v lies on P. Since the choice of P was arbitrary, v dominates w. ∎

Theorem 27.7 *Let $w \neq s$, and let $v = \mathrm{SDOM}(w)$. Let u be a vertex for which $\mathrm{SDOM}(u)$ is minimum among vertices u satisfying $v \xrightarrow{+} u \xrightarrow{*} w$. Then $\mathrm{SDOM}(u) \leq \mathrm{SDOM}(w)$ and $\mathrm{IDOM}(u) = \mathrm{IDOM}(w)$.*

Proof. Let z be the vertex such that $v \to z \xrightarrow{*} w$. Then $\mathrm{SDOM}(u) \leq \mathrm{SDOM}(z) \leq v = \mathrm{SDOM}(w)$.

By Lemma 27.5, $\mathrm{IDOM}(w)$ is an ancestor of v and hence a proper ancestor of u. Thus by Lemma 27.6, $\mathrm{IDOM}(w) \xrightarrow{*} \mathrm{IDOM}(u)$. To prove that $\mathrm{IDOM}(u) = \mathrm{IDOM}(w)$, it suffices to show that $\mathrm{IDOM}(u)$ dominates w.

Consider any directed path P from s to w. Let x be the last vertex on P satisfying $x < \text{IDOM}(u)$. If there is no such x, then $\text{IDOM}(u) = s$ dominates w. Otherwise let y be the first vertex following x on P and satisfying $\text{IDOM}(u) \xrightarrow{*} y \xrightarrow{*} w$. As in the proof of Theorem 27.6, we can show using Lemma 27.2 that $\text{SDOM}(y) \leq x$. Since by Lemma 27.5 $\text{IDOM}(u) \leq \text{SDOM}(u)$, we have $\text{SDOM}(y) \leq x < \text{IDOM}(u) \leq \text{SDOM}(u)$. So $\text{SDOM}(y) < \text{SDOM}(u)$.

Since u has a minimum semi-dominator among the vertices on the tree path from z to w, y cannot by a proper descendant of v. Furthermore, y cannot be both a proper descendant of $\text{IDOM}(u)$ and an ancestor of u, for if this were the case, the directed path consisting of the tree path from s to $\text{SDOM}(y)$ followed by a path $\text{SDOM}(y) = v_0, v_1, v_2, v_3, \ldots, v_k = y$ such that $v_i > y$, for $1 \leq i \leq k-1$, followed by the tree path from y to u would avoid $\text{IDOM}(u)$, but no path from s to u avoids $\text{IDOM}(u)$.

The only remaining possibility is that $\text{IDOM}(u) = y$. Thus $\text{IDOM}(u)$ lies on the directed path P from s to w. Since the choice of P was arbitrary, $\text{IDOM}(u)$ dominates w. ∎

The following main result is an immediate consequence of Theorems 27.6 and 27.7.

Theorem 27.8 *Let $w \neq s$ and let $v = SDOM(w)$. Let u be the vertex for which $SDOM(u)$ is minimum among vertices u satisfying $v \xrightarrow{+} u \xrightarrow{*} w$. Then*

$$\text{IDOM}(w) = \begin{cases} v, & \text{if } \text{SDOM}(u) = \text{SDOM}(w) \\ \text{IDOM}(u), & \text{otherwise.} \end{cases}$$ ∎

We are now ready to describe the dominator algorithm of Lengauer and Tarjan.

Following are the main steps in this algorithm.

Algorithm 27.2 Dominators (Lengauer and Tarjan)

S1. Carry out a DFS of the given program graph $G = (V, E)$ with the start vertex as the root.

S2. Compute the semi-dominators of all vertices by applying Theorem 27.5. Carry out the computation vertex by vertex in the decreasing order of their DFS numbers.

S3. Implicitly define the immediate dominator of each vertex by applying Theorem 27.8.

S4. Explicitly define the immediate dominator of each vertex, carryisg out the computation vertex by vertex in the increasing order of their DFS numbers.

Implementation of Step S1 is straightforward. In the following we denote vertices by their DFS numbers assigned in Step S1.

In our description of steps S2, S3, and S4 we use the arrays FATHER, SEMI, BUCKET, and DOM, defined below.

SEMI(w):

1. Before the semi-dominator of w is computed,
$$\text{SEMI}(w) = w.$$

2. After the semi-dominator of w is computed,
$$\text{SEMI}(w) = \text{SDOM}(w).$$

BUCKET(w): It is a set of vertices whose semi-dominator is w.

DOM(w):

1. Afters step S3, if the semi-dominator of w is its immediate dominator, then DOM(w) is the immediate dominator of w. Otherwise DOM(w) is a vertex v such that $v < w$ and the immediate dominator of v is also the immediate dominator of w.

2. After step S4, DOM(w) is the immediate dominator of w.

After carrying out step S1, the algorithm carries out steps S2 and S3 simultaneously, processing the vertices $w \neq 1$ in the decreasing order of their DFS numbers. During this computation, the algorithm maintains a forest contained in the DFS tree of G. The forest consists of vertex set V and edge set $\{(\text{FATHER}(w), w) \mid \text{vertex } w \text{ has been processed}\}$. The algorithm uses one procedure to construct the forest and another to extract information from it. These procedures are:

LINK(v, w): Add edge (v, w) to the forest.

EVAL(v):

1. If v is the root of a tree in the forest, then EVAL(v) = v.

2. Otherwise, let r be the root of the tree in the forest which contains v, and let u be a vertex for which SEMI(u) is minimum among vertices satisfying $r \xrightarrow{+} u \xrightarrow{*} v$. Then EVAL($v$) = u.

To process a vertex w, the algorithm computes the semi-dominator of w by applying Theorem 27.5. Thus the algorithm assigns

$$\text{SEMI}(w) = \min\{\text{SEMI}(\text{EVAL}(v)) | (v, w) \in E\}.$$

After this computation, SEMI(w) is a semi-dominator of w. This follows from Theorem 27.5 and the definition of EVAL(v).

After computing SEMI(w), the algorithm adds w to BUCKET(SEMI(w)) and adds a new edge to the forest using LINK(FATHER(w),w). This completes step S2 for w.

The algorithm then carries out step S3 by considering each vertex in BUCKET (FATHER(w)).

Let v be such a vertex. The algorithm implicitly computes the immediate dominator of v by applying Theorem 27.8. Let $u = \text{EVAL}(v)$. Then u is the vertex satisfying FATHER(w) $\xrightarrow{+} u \xrightarrow{*} v$ whose semi-dominator is minimum. If SEMI(u) = SEMI(v), then FATHER(w) is the immediate dominator of v and the algorithm assigns DOM(v) = FATHER(w). Otherwise u and v have the same immediate dominator, and the algorithm assigns DOM(v) = u. This completes step S3 for v.

In step S4 the algorithm examines vertices in the increasing order of their DFS numbers, filling in the immediate dominators not explicitly computed in step S3.

Thus step S4 is as follows:

For each $i = 2, 3, \ldots, n$, if DOM(i) $\neq$ SEMI(i), then let
$$\text{DOM}(i) = \text{DOM}(\text{DOM}(i)).$$

For an illustration of the dominator algorithm consider the program graph shown in Figure 27.5a. Just before vertex 11 is processed, the forest will be as in Figure 27.6a. The entries of the SEMI array at this stage are shown in parentheses next to the corresponding vertices. Let us now process vertex 11.

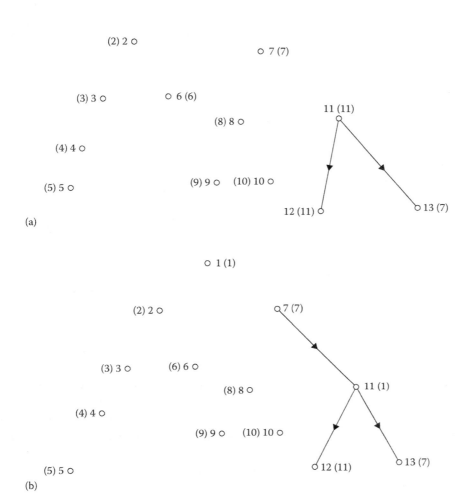

Figure 27.6 Illustration of Algorithm 27.2.

The edges (1,11) and (7,11) are incident into 11. So

$$\text{SEMI}(11) = \min\{\text{SEMI}(\text{EVAL}(1)), \text{SEMI}(\text{EVAL}(7))\}$$

Now EVAL(1) = 1 because vertex 1 is a tree root in the forest. For the same reason, EVAL(7) = 7. Thus

$$\begin{aligned}\text{SEMI}(11) &= \min\{\text{SEMI}(1), \text{SEMI}(7)\} \\ &= \min\{1, 7\} \\ &= 1.\end{aligned}$$

The algorithm now adds the edge (7, 11) to the forest and the vertex 11 to BUCKET(SEMI(11)) = BUCKET(1). The new forest with the SEMI array entries is shown in Figure 27.6b. This completes step S2 for vertex 11.

The algorithm now considers BUCKET(FATHER(11)) = BUCKET(7). Vertex 13 is the only one whose semi-dominator is equal to 7. So BUCKET(7) = {13}. Now EVAL(13) = 11, since SEMI(11) is minimum along vertices u satisfying $7 \xrightarrow{+} u \xrightarrow{*} 13$ (see Figure 27.6b). Since SEMI(13) $\neq$ SEMI(11), the algorithm sets DOM(13) = 11. This completes step S3 for vertex 11.

After steps S2 and S3 have been carried out for every vertex $w \neq 1$, the semi-dominaters of all the vertices will be available. At this stage, the entries of the DOM array and the semi-dominators will be as given below:

Vertex	DOM	Semi-Dominator(SEMI)
2	1	1
3	2	2
4	2	2
5	1	1
6	2	2
7	1	1
8	7	7
9	1	1
10	1	1
11	1	1
12	11	11
13	11	7

For every vertex $w \neq 13$, $\text{DOM}(w) = \text{SEMI}(w)$. So for all vertices w except 13, $\text{IDOM}(w) = \text{DOM}(w)$. For vertex 13, we compute

$$\text{DOM}(13) = \text{DOM}(\text{DOM}(13))$$
$$= \text{DOM}(11)$$
$$= 1.$$

So $\text{IDOM}(13) = 1$. This completes step S4 of the algorithm and we get the dominator tree shown in Figure 27.5b.

Clearly, the complexity of the above algorithm depends crucially on the implementation of LINK and EVAL instructions. Tarjan [24] discusses two methods which use path compression. One of these is described below.

To represent the forest built by the LINK instructions, the algorithm uses two arrays, ANCESTOR and LABEL. Initially $\text{ANCESTOR}(v) = 0$ and $\text{LABEL}(v) = v$ for each v. In general, $\text{ANCESTOR}(v) = 0$ only if v is a tree root in the forest; otherwise, $\text{ANCESTOR}(v)$ is ancestor of v in the forest.

The algorithm maintains the labels so that they satisfy the following property. Let v be any vertex, let r be the root of the tree in the forest containing v, and let $v = v_k, v_{k-1}, \ldots, v_0 = r$ be such that $\text{ANCESTOR}(v_i) = v_{i-1}$ for $1 \leq i \leq k$. Let x be a vertex such that $\text{SEMI}(x)$ is minimum among vertices $x \in \{\text{LABEL}(v_i) | 1 \leq i \leq k\}$. Then we have the following property.

x is a vertex such that $\text{SEMI}(x)$ is minimum among vertices x satisfying

$$r \xrightarrow{+} x \xrightarrow{*} v.$$

To carry out $\text{LINK}(v, w)$, the algorithm assigns $\text{ANCESTOR}(w) = v$. To carry out $\text{EVAL}(v)$, the algorithm follows ancestor pointers and determines the sequence $v = v_k, v_{k-1}, \ldots, v_0 = r$ such that $\text{ANCESTOR}(v_i) = v_{i-1}$ for $1 \leq i \leq k$. If $v = r$, then the algorithm sets $\text{EVAL}(v) = v$. Otherwise the algorithm sets $\text{ANCESTOR}(v_i) = r$ for $2 \leq i \leq k$, simultaneously updating labels as follows (to maintain the property mentioned before):

If
$\text{SEMI}(\text{LBBEL}(v_{i-1})) < \text{SEMI}(\text{LABEL}(v_i))$,
then
$\text{LABEL}(v_i) \leftarrow \text{LABEL}(v_{i-1})$.

Then the algorithm sets EVAL(v) = LABEL(v).

Tarjan [24] has shown that the complexity of implementing $(n-1)$ LINKs and $(m+n-1)$ EVALs using the method described above is $O(m \log n)$. If we use the more sophisticated implementation of LINK and EVAL instructions, also described in [24], then the algorithm would require $O(m\alpha(m,n))$ time, where $\alpha(m,n)$ is the functional inverse of Ackermann's function defined in the previous section.

For other dominator algorithms see Aho and Ullman [25] and Purdom and Moore [26].

References

[1] A.V. Aho and J.D. Ullman, *The Theory Of Parsing, Translation And Compiling, Volume II: Compiling*, Prentice Hall, Englewood Cliffs, NJ, 1973.

[2] A.V. Aho, J.E. Hopcroft, and J.D. Ullmann, On finding the least common ancestors in trees, *SIAM J. Comput.*, **5** (1976), 115–132.

[3] F.E. Allen, *Program Optimization. Annual Review in Automatic Programming*, Vol. 5, Pergamon, New York, 1969.

[4] F.E. Allen, Control flow analysis, *SIGPLAN Notices*, **5** (1970), 1–19.

[5] J. Cocke, Global common subexpression elimination, *SIGPLAN Notices*, **5** (1970), 20–24.

[6] M.S. Hecht, *Flow Analysis of Computer Programs*, Elsevier, New York, 1977.

[7] K. Kennedy, A global flow analysis algorithm, *Int. J. Comput. Math.*, **3** (1971), 5–16.

[8] M. Schaefer, *A Mathematical Theory of Global Program Optimization*, Prentice Hall, Englewood Cliffs, NJ, 1973.

[9] J.D. Ullman, Fast algorithms for the elimination of common subexpressions, *Acta Inf.*, **2** (1973), 191–213.

[10] M.S. Hecht and J.D. Ullman, Flow graph reducibility, *SIAM J. Comput.*, **1** (1972), 188–202.

[11] J. Cocke and R.E. Miller, Some analysis techniques for optimizing computer programs, In *Proc. 2nd Int. Conf. Sys. Sci.*, Honolulu, HI, 1969.

[12] J.E. Hopcroft and J.D. Ullman, An $n \log n$ algorithm for detecting reducible graphs, In *Proc. 6th Ann. Princeton Conf. Inform. Sci. Sys.*, Princeton, NJ, 119–122, 1972.

[13] R.E. Tarjan, Testing flow graph reducibility, *Proc. 5th Ann. ACM Symp. Theor. Comput.*, 96–107, 1973.

[14] R.E. Tarjan, Testing flow graph reducibility, *J. Comput. Sys.Sci.*, **9** (1974), 355–365.

[15] M.S. Hecht and J.D. Ullman, Characterizations of reducible flow graphs, *J. ACM*, **21** (1974), 367–375.

[16] J.M. Adams, J.M. Phelan, and R.H. Stark, A note on the Hecht-Ullman characterization of non-reducible flow graphs. *SIAM J. Comput.*, **3** (1974), 222–223.

[17] M. Fischer, Efficiency of equivalence algorithm, In *Complexity of Computer Computations*, R.E. Miller and J.W. Thatcher, editors, Plenum Press, New York, 153–168, 1972.

[18] J.E. Hopcroft and J.D. Ullman, Set merging algorithms, *SIAM J. Comput.*, **2** (1973), 294–303.

[19] R.E. Tarjan, On the efficiency of a good but not linear set union algorithm, *J. ACM*, **22** (1975), 215–225.

[20] A.V. Aho, J.E. Hopcroft, and J.D. Ullman, *The Design and Analysis of Computer Algorithms*, Addison-Wesley, Reading, MA, 1974.

[21] E. Horowitz and S. Sahni, *Fundamentals of Data Structures*, Computer Science Press, Potomac, MD, 1976.

[22] T. Lengauer and R.E. Tarjan, A fast algorithm for finding dominators in a flow graph, *Trans. on Prog. Lang. and Sys.*, **1** (1979), 121–141.

[23] R.E. Tarjan, Finding dominators in directed graphs, *SIAM J. Comput.*, **3** (1974), 62–89.

[24] R.E. Tarjan, Applications of path compression on balanced trees, *J. ACM*, **26** (1979), 690–715.

[25] A.V. Aho and J.D. Ullman, *Principles of Compiler Design*, Addison-Wesley, Reading, MA, 1977.

[26] P.W. Purdom and E.F. Moore, Algorithm 430: Immediate predominators in a directed graph, *Comm. ACM.*, **15** (1972), 777–778.

CHAPTER 28

Perfect Graphs

Chính T. Hoàng*

R. Sritharan[†]

CONTENTS

28.1	Introduction	708
28.2	Notation	710
28.3	Chordal Graphs	710
	28.3.1 Characterization	710
	28.3.2 Recognition	712
	28.3.3 Optimization	715
28.4	Comparability Graphs	715
	28.4.1 Characterization	715
	28.4.2 Recognition	718
	28.4.2.1 Transitive Orientation Using Modular Decomposition	720
	28.4.2.2 Modular Decomposition	720
	28.4.2.3 From the Modular Decomposition Tree to Transitive Orientation	721
	28.4.2.4 How Quickly Can Comparability Graphs Be Recognized?	722
	28.4.3 Optimization	725
28.5	Interval Graphs	726
	28.5.1 Characterization	727
	28.5.2 Recognition	728
	28.5.3 Optimization	728
28.6	Weakly Chordal Graphs	729
	28.6.1 Characterization	729
	28.6.2 Recognition	731
	28.6.3 Optimization	732
	28.6.4 Remarks	733
28.7	Perfectly Orderable Graphs	733
	28.7.1 Characterization	735
	28.7.2 Recognition	735
	28.7.3 Optimization	738
28.8	Perfectly Contractile Graphs	741
28.9	Recognition of Perfect Graphs	742
28.10	χ-Bounded Graphs	744

---

*Acknowledges support from NSERC of Canada.
[†]Acknowledges support from the National Security Agency, Fort Meade, Maryland.

28.1 INTRODUCTION

This chapter is a survey on perfect graphs with an algorithmic flavor. Our emphasis is on important classes of perfect graphs for which there are fast and efficient recognition and optimization algorithms. The classes of graphs we discuss in this chapter are chordal, comparability, interval, perfectly orderable, weakly chordal, perfectly contractile, and χ-bound graphs. For each of these classes, when appropriate we discuss the complexity of the recognition algorithm and algorithms for finding a minimum coloring, and a largest clique in the graph and in its complement.

In the late 1950s, Berge [1] started his investigation of graphs G with the following properties: (i) $\alpha(G) = \theta(G)$, that is the number of vertices in a largest stable set is equal to the smallest number of cliques that cover $V(G)$ and (ii) $\omega(G) = \chi(G)$, that is the number of vertices in a largest clique is equal to the smallest number of colors needed to color G. At about the same time, Shannon [2] in his study of the zero-error capacity of communication channels asked: (iii) what are the minimal graphs that do not satisfy (i)?, and (iv) what is the zero-error capacity of the chordless cycle on five vertices? In today's language, the graphs G all of whose induced subgraphs satisfy (ii) are called *perfect*.

In 1959, it was proved [3] that chordal graphs (graphs such that every cycle of length at least four has a chord) satisfy (i), that is complements of chordal graphs are perfect. In 1960, it was proved [1] that chordal graphs are perfect. These two results led Berge to propose two conjectures which after many years of work by the graph theory community were proved to hold.

Theorem 28.1 (Perfect graph theorem) *If a graph is perfect, then so is its complement.* ∎

Theorem 28.2 (Strong perfect graph theorem) *A graph is perfect if and only if it does not contain an odd chordless cycle with at least five vertices, or the complement of such a cycle.* ∎

Perfect graphs are prototypes of min-max characterizations in combinatorics and graph theory. The theory of perfect graphs can be used to prove well known theorems such as the Dilworth's theorem on partially ordered sets [4], or the König's theorem on edge coloring of bipartite graphs [5]. On the other hand, algorithmic considerations of perfect graphs have given rise to techniques such as clique cutset decomposition, and modular decomposition. Question (iv) was answered completely in [6]; in the process of doing so, the so-called Lovász's theta function Θ were introduced. Theta function satisfies $\omega(G) \leq \Theta(\overline{G}) \leq \chi(G)$ for any graph G. Thus, a perfect graph G has $\omega(G) = \Theta(\overline{G}) = \chi(G)$. Subsequently, [7] gave a polynomial time algorithm based on the ellipsoid method to compute $\Theta(G)$ for any graph G. As a consequence, a largest clique and an optimal coloring of a perfect graph can be found in polynomial time. Furthermore, the algorithm of [7] is *robust* in the sense of [8]: given the input graph G, it finds a largest clique and an optimal coloring, or says correctly that G is not perfect; [7] is also the first important paper in the now popular field of semidefinite programming (see [9]).

This paper is a survey on perfect graphs with an algorithmic flavor. Even though there are now polynomial time algorithms for recognizing a perfect graph and for finding an optimal coloring—and a largest clique—of such a graph, they are not considered fast or efficient. Our emphasis is on important classes of perfect graphs for which there are fast and efficient recognition and optimization algorithms. The purpose of this survey is to discuss these classes of graphs, named below, together with the complexity of the recognition problem and the optimization problems. The reader is referred to [10–12] for background on perfect graphs.

Chordal graphs form a class of graphs among the most studied in graph theory. Besides being the impetus for the birth of perfect graphs, chordal graphs have been studied in contexts such as matrix computation and database design. Chordal graphs have given rise to well known search methods such as lexicographic breadth-first search and maximum cardinality search. We discuss chordal graphs in Section 28.3.

Comparability graphs (the graphs of partially order sets) are also among the earliest known classes of perfect graphs. The well-known Dilworth's theorem—stating that in a partially ordered set, the number of elements in a largest anti-chain is equal to the smallest number of chains that cover the set—is equivalent to the statement that complements of comparability graphs are perfect. Early results of [13] and [14] imply polynomial time algorithms for comparability graph recognition. But despite much research, there is still no linear-time algorithm for the recognition problem. It turns out that recognizing comparability graphs is equivalent to testing for a triangle in a graph, via an $O(n^2)$ time reduction. We discuss comparability graphs in Section 28.4.

Interval graphs are the intersection graphs of intervals on a line. Besides having obvious application in scheduling, interval graphs have interesting structural properties. For example, interval graphs are precisely the chordal graphs whose complements are comparability graphs. We discuss interval graphs is Section 28.5.

Weakly chordal graphs are graphs without chordless cycles with at least five vertices and their complements. This class of graphs generalizes chordal graphs in a natural way. For weakly chordal graphs, there are efficient, but not linear time, algorithms for the recognition and optimization problems. We discuss weakly chordal graphs in Section 28.6.

An order on the vertices of a graph is perfect if the greedy (sequential) coloring algorithm delivers an optimal coloring on the graph and on its induced subgraphs. A graph is perfectly orderable if it admits a perfect order. Chordal graphs and comparability graphs admit perfect orders. Complements of chordal graphs are also perfectly orderable. Recognizing perfectly orderable graphs is NP-complete; however, there are many interesting classes of perfectly orderable graphs with polynomial time recognition algorithms. We discuss perfectly orderable graphs in Section 28.7.

An even-pair is a set of two nonadjacent vertices such that all chordless paths between them have an even number of edges. If a graph G has an even-pair, then by contracting this even-pair we obtain a graph G' satisfying $\omega(G) = \omega(G')$ and $\chi(G) = \chi(G')$. Furthermore, if G is perfect, then so is G'. Perfectly contractile graphs are those graphs G such that, starting with any induced subgraph of G by repeatedly contracting even-pairs we obtain a clique. Weakly chordal graphs and perfectly orderable graphs are perfectly contractile. We discuss perfectly contractile graphs in Section 28.8.

Recently, a polynomial time algorithm for recognizing perfect graphs was given in [15]. We give a sketch of this algorithm in Section 28.9.

A graph G is χ-bound if there is a function f such that $\chi(G) \leq f(\omega(G))$. Perfect graphs are χ-bound. Identifying sufficient conditions for a graph to be χ-bound is an interesting problem. It is proved in [16] that a graph is χ-bound if it does not contain an even chordless cycle. One many ask a similar question for odd cycles [17]: Is it true that a graph is χ-bound if it does not contain an odd chordless cycle with at least five vertices? In Section 28.10, we discuss this question and related conjectures.

We give the definitions used in this chapter in Section 28.2.

28.2 NOTATION

For graph $G = (V, E)$ and $x \in V$, $N_G(x)$ is the neighborhood of x in G; we omit the subscript G when the context is clear. Let $d(x)$ denote $|N(x)|$. For $S \subseteq V$, $G[S]$ denotes the subgraph of G induced by S, and $G - S$ denotes $G[V - S]$; for $x \in V$, we use $G - x$ for $G - \{x\}$. $\omega(G)$ is the number of vertices in a largest clique in G. $\alpha(G)$ is the number of vertices in a largest stable set in G. $\chi(G)$ is the chromatic number of G. $\theta(G)$ is the smallest number of cliques that cover the vertices of G. A clique is *maximal* if it is not a proper subset of another clique. For $A, B \subseteq V$ such that $G[A]$ and $G[B]$ are connected, $S \subseteq V$ is a *separator* for A and B provided A and B belong to different components of $G - S$. Further, S is a *minimal separator* for A and B if no proper subset of S is also a separator for A and B. We will also call a set C of vertices a *cutset* if C is a separator for some sets A, B of V; C is a *minimal cutset* if no proper subset of C is a cutset.

We use n to refer to $|V|$ and m to refer to $|E|$.

In a bipartite graph $G = (X, Y, E)$, X and Y are the parts of the partition of the vertex-set and E is the set of edges. A *matching* is a set of pairwise non-incident edges.

A set C of V is *anti-connected* if C spans a connected subgraph in the complement $\overline{G}$ of G. For a set $X \subset V$, a vertex v is X-*complete* if v is adjacent to every vertex of X. An edge is X-complete if both its endpoints are X-complete. A vertex v is X-*null* if v has no neighbor in X.

C_k denotes the chordless cycle with k vertices. A *hole* is the C_k with $k \geq 4$. An *anti-hole* is the complement of a hole. P_k denotes the chordless path with k vertices. K_t denotes the clique on t vertices. The K_3 is sometimes called a *triangle*. The complement of a C_4 is denoted by $2K_2$. The *claw* is the tree on four vertices with a vertex of degree 3.

For problems A and B, $A \preceq B$ *via an* $f(m, n)$ *time reduction* means that an instance of problem A can be reduced to an instance of problem B using an algorithm with the worst case complexity of $f(m, n)$; $A \equiv B$ *via* $f(m, n)$ *time reductions* means that we have $A \preceq B$ as well as $B \preceq A$ via $f(m, n)$ time reductions.

Let $O(n^\alpha)$ be the complexity of the current best algorithm to multiply two $n \times n$ matrices. It is currently known that $\alpha < 2.376$ [18].

28.3 CHORDAL GRAPHS

Definition 28.1 *A graph is chordal (or, triangulated) if it does not contain a chordless cycle with at least four vertices.*

Chordal graphs can be used to model various combinatorial structures. For example, they are the intersection graphs of subtrees of a tree as we will see later. See [19] for applications of chordal graphs to sparse matrix computations. Chordal graphs are among the earliest known classes of perfect graphs [3,20,21]. We will now discuss the combinatorial structures of chordal graphs.

28.3.1 Characterization

Definition 28.2 *A vertex is simplicial if its neighborhood is a clique.*

Theorem 28.3 [21] *A graph G is chordal if and only if each of its induced subgraphs is a clique or contains two nonadjacent simplicial vertices.* ∎

To prove Theorem 28.3, we need the following two lemmas.

Lemma 28.1 *Any minimal cutset of a chordal graph G is a clique.*

Proof. Suppose C is a minimal cutset of G and A_1, A_2 are two distinct components of $G - C$. Further, suppose for $x \in C$ and $y \in C$, $xy \notin E(G)$. As C is a minimal cutset of G, each of x, y has a neighbor in A_i, $i = 1, 2$. Let P_i, $i = 1, 2$, be a shortest path connecting x and y in $G[A_i \cup C]$ such that all the internal vertices of P_i lie in A_i. Then, $G[V(P_1) \cup V(P_2)]$ is a hole, a contradiction. ∎

Lemma 28.2 *Let G be a graph with a clique cutset C. Consider the induced subgraphs G_1, G_2 with $G = G_1 \cup G_2$ and $G_1 \cap G_2 = C$. Then, G is chordal if and only if G_1, G_2 are both chordal.*

Proof. If G is chordal, then as G_1 and G_2 are induced subgraphs of a chordal graph, they themselves are chordal; this proves the *only if* part. For the *if* part, suppose each of G_1, G_2 is chordal, but G has a hole L. Then, L must involve a vertex from each of $G_1 - C$, $G_2 - C$. Therefore, C contains a pair of nonadjacent vertices from L, contradicting C being a clique. ∎

Proof of Theorem 28.3. The *if* part is easy: If G is a graph and x is a simplicial vertex of G, then G is chordal if and only if $G - x$ is. Now, we prove the *only if* part by induction on the number of vertices. Let G be a chordal graph. We may assume G is connected, for otherwise by the induction hypothesis, each component of G is a clique or contains two nonadjacent simplicial vertices, and so G contains two nonadjacent simplicial vertices. Let C be a minimal cutset of G. By Lemma 28.1, C is a clique. Thus, G has two induced subgraphs G_1, G_2 with $G = G_1 \cup G_2$ and $G_1 \cap G_2 = C$. By the induction hypothesis, each G_i has a simplicial vertex $v_i \in G_i - C$ (since C is a clique, it cannot contain two nonadjacent simplicial vertices). The vertices v_1, v_2 remain simplicial vertices of G, and they are nonadjacent. ∎

Definition 28.3 *For a graph G and an ordering $v_1 v_2 \cdots v_n$ of its vertices, let G_i denote $G[\{v_i, \cdots, v_n\}]$. An ordering $\sigma = v_1 v_2 \cdots v_n$ of vertices of G is a perfect elimination scheme (p.e.s.) for G if each v_i is simplicial in G_i.*

Theorem 28.4 [21,22] *G is chordal if and only if G admits a perfect elimination scheme.*

Proof. For any vertex v in a chordal graph G, $G - v$ is also chordal; this together with Theorem 28.3 prove the *only if* part. Since no hole has a simplicial vertex, the *if* part follows. ∎

Corollary 28.1 *A chordal graph G has at most n maximal cliques whose sizes sum up to at most m.*

Proof. By induction on the number of vertices of G. Let x be a simplicial vertex of G. Then, $\{x\} \cup N(x)$ is the only maximal clique of G containing x. By the induction hypothesis, $G - x$ has at most $n - 1$ maximal cliques whose sizes sum up to at most $m - d(x)$. Then, the result follows. ∎

Definition 28.4 *Let $\mathcal{F}$ be a family of nonempty sets. The intersection graph of $\mathcal{F}$ is the graph obtained by identifying each set of $\mathcal{F}$ with a vertex, and joining two vertices by an edge if and only if the two corresponding sets have a nonempty intersection.*

Theorem 28.5 [23,24] *A graph is chordal if and only if it is the intersection graph of subtrees of a tree.*

Proof. By induction on the number of vertices. We prove the *if* part first. Let $G = (V, E)$ be a graph that is the intersection graph of a set $\mathcal{S}$ of subtrees of a tree T, that is, every vertex v of V is a subtree T_v of T, and two vertices $v, u \in V$ are adjacent if and only if T_v and T_u intersect. We may assume G is connected, for otherwise, we are done by the induction hypothesis. By Lemma 28.2, and the induction hypothesis, we only need prove G is a clique, or contains a clique cutset. We may assume G is not a clique, and let u, v be two nonadjacent vertices of G. Then, $T_u \cap T_v = \emptyset$. Let $P = x_1, \ldots, x_p$ be the path in T with $x_1 \in T_u, x_p \in T_v$ such that all interior vertices of P are not in $T_u \cup T_v$. Since T is a tree, P is unique; furthermore, all paths with one endpoint in T_u and the other endpoint in T_v must contain all vertices of P. Thus, $x_1 x_2$ is a cut-edge of T. Let $\mathcal{S}'$ be the set of all subtrees of $\mathcal{S}$ that contains the edge $x_1 x_2$. Then in G, the set C of vertices that corresponds to the subtrees of $\mathcal{S}'$ forms a clique. We claim C is a cutset of G. In G, consider a path from u to v; let the vertices of this path be $u = t_1, t_2, \ldots, v = t_k$. Some subtree T_{t_i} must contain the edge $x_1 x_2$ (because it is the cut-edge of T). Thus, the vertex that corresponds to T_{t_i} is in C. We have established the *if* part.

Now, we prove the *only if* part. Let $G = (V, E)$ be a chordal graph. We will prove that there is a tree T and a family $\mathcal{S}$ of subtrees of T such that (i) the vertices of T are the maximal cliques of G, and (ii) for each $v \in V$, the set of maximal cliques of G containing v induces a subtree of T. The proof is by induction on the number of vertices. Suppose that G is disconnected. Then, the induction hypothesis implies for each component C_i of G, there is a tree T_i satisfying (i) and (ii). Construct the tree T from the trees T_i by adding a new *root* vertex r and joining r to the root of each T_i. It is easy to see that T satisfies (i) and (ii). So, G is connected. We may assume G is not a clique, for otherwise we are easily done. Consider a simplicial vertex v of G. As v is simplicial in G, it is not a cut vertex of G and therefore, $G - v$ is connected. By the induction hypothesis, the graph $G - v$ is the intersection graph of a set $\mathcal{B}$ of subtrees of a tree $T_\mathcal{B}$ satisfying (i) and (ii). Let K be a maximal clique of $G - v$ containing $N_G(v)$ and let t_k be the vertex of $T_\mathcal{B}$ that corresponds to K. If $K = N_G(v)$, then we simply add v to t_K to get the tree T from $T_\mathcal{B}$. Otherwise, let $K' = N_G(v) \cup \{v\}$. Let T be the tree obtained from $T_\mathcal{B}$ by adding a new vertex $t_{K'}$ and the edge $t_K t_{K'}$. Let T_K be the subtree formed by the single vertex $t_{K'}$. We construct $\mathcal{S}$ as follows. Add T_K to $\mathcal{S}$; for each tree $T_u \in \mathcal{B}$, if T_u corresponds to a vertex in $N_G(v)$, then add the tree $T_u \cup \{t_K t_{K'}\}$; otherwise, add T_u to $\mathcal{S}$. It is seen that (i) and (ii) hold for T and $\mathcal{S}$. ∎

28.3.2 Recognition

Given G, an approach to testing whether G is chordal is: first generate an ordering σ of vertices of G that is guaranteed to be a perfect elimination scheme for G when G is chordal; then, verify whether σ is indeed a perfect elimination scheme for G. The first linear-time algorithm to generate a perfect elimination scheme of a chordal graph is given in [25]; it uses the lexicographic breadth-first search (LexBFS). We present the maximum cardinality search algorithm for the same purpose.

The maximum cardinality search algorithm (MCS), introduced in [26], is used to construct an ordering of vertices of a given graph; the ordering is constructed incrementally right to left (if a comes before b in the order, then we consider a to be to the left of b). An arbitrary vertex is chosen to be the last in the ordering. In each remaining step, from the vertices still not chosen (unlabeled vertices), one with the most neighbors among the already chosen vertices (labeled vertices) is picked with the ties broken arbitrarily.

> **Algorithm 28.1** MCS
>
> **input:** graph G
> **output:** ordering $\sigma = v_1 v_2 \cdots v_n$ of vertices of G
>
> $v_n \leftarrow$ an arbitrary vertex of G;
> **for** $i \leftarrow n-1$ **downto** 1 **do**
> $v_i \leftarrow$ unlabeled vertex adjacent to the most in $\{v_{i+1}, \cdots, v_n\}$;
> **end for**

Theorem 28.6 [26] *Algorithm MCS can be implemented to run in $O(m+n)$ time.*

Proof. We keep the array $set[0] \cdots set[n-1]$ where $set[j]$ is a doubly linked list of all the unlabeled vertices that are adjacent to exactly j labeled vertices. Thus, initially every vertex belongs to $set[0]$. For each vertex, we maintain the array index of the set it belongs to as well as a pointer to the node containing it in the $set[i]$ lists. Finally, we maintain *last*, the largest index such that $set[last]$ is nonempty. In the ith iteration of the algorithm, a vertex in $set[last]$ is taken to be v_i and v_i is deleted from $set[last]$. For every unlabeled neighbor w of v_i, if w belongs to $set[i]$, then we move w from $set[i]$ to $set[i+1]$. As each set is implemented as a doubly linked list, a single addition or deletion can be done in constant time, and hence all of the above operations can be done in $O(d(v_i))$ time. Finally, in order to update the value of *last*, we increment *last* once and then we repeatedly decrement the value of *last* until $set[last]$ is nonempty. As *last* is incremented at most n times and its value is never less than -1, the overall time spent manipulating *last* is $O(n)$ and we have the claimed complexity. ∎

Definition 28.5 *For vertices x, y of graph G and an ordering σ of vertices of G, $x <_\sigma y$ denotes that x precedes y in σ.*

Lemma 28.3 [26] *Let σ be the output of algorithm MCS on chordal graph G. Then, G does not have a chordless path $P = (x = u_0) u_1 \cdots u_{k-1} (u_k = y)$ with $k \geq 2$ such that $u_i <_\sigma x$, $1 \leq i \leq k-1$, and $x <_\sigma y$.*

Proof. Suppose such a path existed; from all such chordless paths, pick P so that the position of x in σ is as much to the right as possible. Given the logic of the algorithm MCS, as $u_{k-1} <_\sigma x <_\sigma y$, $u_{k-1} y \in E(G)$, and $xy \notin E(G)$, there must exist a vertex z such that $x <_\sigma z$, $xz \in E(G)$, and $u_{k-1} z \notin E(G)$). Let j be the largest index less than $k-1$ such that $u_j z \in E(G)$; such a j exists as $xz \in E(G)$. Let P' be the path $z u_j \cdots u_{k-1} y$. As G is chordal and P' has at least four vertices, $zy \notin E(G)$. Now, whether $x <_\sigma z <_\sigma y$ holds or $x <_\sigma y <_\sigma z$ holds, existence of the chordless path P' violates the choice of P, a contradiction. ∎

Theorem 28.7 [26] *If G is chordal, then the output $\sigma = v_1 v_2 \cdots v_n$ produced by the algorithm MCS is a perfect elimination scheme for G.*

Proof. Suppose not, and let i be the smallest such that v_i is not simplicial in G_i. Then, there exist v_j and v_k such that $v_i <_\sigma v_j <_\sigma v_k$, $v_i v_j \in E(G)$, $v_i v_k \in E(G)$, and $v_j v_k \notin E(G)$. Then, the chordless path $P = v_j v_i v_k$ contradicts Lemma 28.3. ∎

> **Algorithm 28.2** chordal-recognition
>
> **input:** graph G
> **output:** *yes* when G is chordal and *no* otherwise
>
> Run algorithm MCS on G to get $\sigma = v_1 v_2 \cdots v_n$;
> **if** σ is a perfect elimination scheme for G **then**
>   output *yes*
> **else**
>   output *no*
> **end if**

Next, we discuss how to verify in linear time [25] whether $\sigma = v_1 v_2 \cdots v_n$ is a perfect elimination scheme for G. The key idea in [25] is that part of the work involved in checking whether v_i is simplicial in G_i can be handed over to an appropriate vertex v_j such that $v_i <_\sigma v_j$. In particular, let v_j be the smallest neighbor of v_i such that $v_i <_\sigma v_j$. Let $L(v_i) = \{v_k \mid v_j <_\sigma v_k$ and $v_i v_k \in E(G)\}$. In other words, $L(v_i)$ is the set of those neighbors of v_i that follow v_j in σ.

If v_j is simplicial in G_j and v_j is adjacent to every vertex in $L(v_i)$, then v_i is simplicial in G_i. On the other hand, if either v_j is not simplicial in G_j or v_j is not adjacent to some vertex in $L(v_i)$ (making v_i not simplicial in G_i), then σ is not a perfect elimination scheme for G. Further, part of the work involved in checking whether v_j is simplicial in G_j can likewise be deferred to a later vertex.

In the following, the list $bba(v_k)$ is the list of vertices that v_k better be adjacent to; it is the concatenation of the $L(v_i)$ lists handed over to v_k by the v_i's preceding it in σ.

> **Algorithm 28.3** pes-verification
>
> **input:** graph G and ordering $\sigma = v_1 v_2 \cdots v_n$ of vertices of G
> **output:** *yes* when σ is a p.e.s. for G and *no* otherwise
>
> **for** $i \leftarrow 1$ **to** n **do**
>   Initialize $bba(v_i)$ to an empty list;
> **end for**
> **for** $i \leftarrow 1$ **to** $n-1$ **do**
>   **if** v_i is not adjacent to some vertex in $bba(v_i)$ **then**
>   output *no*;
>   stop
>   **end if**
>   Let v_j be the smallest neighbor of v_i such that $v_i <_\sigma v_j$;
> $L(v_i) \leftarrow \{v_k \mid v_j <_\sigma v_k$ and $v_i v_k \in E(G)\}$;
>   Append $L(v_i)$ to $bba(v_j)$
> **end for**
> output *yes*

Theorem 28.8 [25] *Algorithm pes-verification can be implemented to run in $O(m+n)$ time.*

Proof. Assume that the array $v[1] \cdots v[n]$ stores σ. In order to check whether v_i is adjacent to every vertex in $bba(v_i)$: use a boolean array $flag[1] \cdots flag[n]$ that is initialized in the first step of the entire algorithm. Now, mark the neighbors of v_i in the array $flag$. Then, traverse the list $bba(v_i)$ and check for each member of $bba(v_i)$ whether the corresponding entry in $flag$ is marked. Finally, unmark the neighbors of v_i in $flag$. Thus, this operation takes $O(|bba(v_i)| + d(v_i))$ time. As a vertex v_k hands over an $L(v_k)$ list at most once, the total size of all bba lists is $O(m+n)$ and the overall time spent on this operation is $O(m+n)$. The rest of the operations can easily be implemented in $O(m+n)$ time. ∎

28.3.3 Optimization

For a chordal graph, a largest clique and an optimal coloring can be found in linear time using the combined results in [25,27]. Even the weighted versions of these problems can be solved efficiently. This will be discussed in the context of the more general class of perfectly orderable graphs in Section 28.7.

The known optimization algorithms for chordal graphs use the clique cutset property. For a general graph, there are polynomial time algorithms [28,29] to find a clique cutset if one exists in the graph. [28,30] discuss optimization algorithm for classes of graphs, more general than chordal, using the clique cutset decomposition.

28.4 COMPARABILITY GRAPHS

Definition 28.6 *A graph $G = (V, E)$ is a* comparability *graph if there is a partially ordered set $(P, \prec)$ such that $V = P$ and two vertices of G are adjacent if and only if the corresponding elements of P are comparable in the relation $\prec$.*

Definition 28.7 *An orientation of a graph is* transitive *if whenever $a \to b, b \to c$ are arcs, $a \to c$ is an arc.*

An ordered graph $(G, \prec)$ corresponds to an orientation in a natural way: for vertices a, b, we orient $a \to b$ if $a \prec b$. Now, we can redefine the notion of a comparability graph as follows.

Definition 28.8 *A graph is a comparability graph if it admits an orientation that is both acyclic and transitive.*

28.4.1 Characterization

Several theorems on comparability graphs have become folklore. We start with a classical theorem of [13] that as we will see later implies a polynomial time algorithm to recognize a comparability graph.

Theorem 28.9 [13] *If a graph admits a transitive orientation, then it admits an acyclic and transitive orientation.* ∎

Definition 28.9 *A subset M of vertices of a graph $G = (V, E)$ is a* module *if any vertex outside of M is either adjacent to every vertex in M or adjacent to no vertex in M. Trivially, $\{x\}$ for any $x \in V$, and V are modules. Module M is nontrivial if $|M| \geq 2$ and $M \subset V$.*

To prove Theorem 28.9, we need the following.

Theorem 28.10 [13] *If a graph admits a cyclic transitive orientation, then it contains a nontrivial module.*

Proof. Let G be a graph and let $\overrightarrow{G}$ be transitive orientation of G containing a directed cycle C. We may assume C is a shortest cycle and thus chordless. Since $\overrightarrow{G}$ is transitive, C has length three. We may assume G has at least four vertices, for otherwise the theorem is trivially true. Let the vertices of C be a, b, c in the cyclic order, with $a \to b, b \to c, c \to a$. A vertex x

outside C cannot have exactly one neighbor in C, for otherwise x and some two vertices in C violate the transitivity of $\vec{G}$. There must be a vertex v adjacent to exactly two vertices of C, for otherwise C is a nontrivial module of G. We may assume v is adjacent to b, c. Let X be the set of vertices that are adjacent to b, c such that X is anti-connected, $a, v \in X$, and X is maximal with respect to this property. Since X is anti-connected, and $a \to b, c \to a$, it follows that every $x \in X$ has $x \to b, c \to x$. We may assume X is not a module of G, for otherwise we are done. Thus, there is a vertex $u \notin X$ such that $A = N(u) \cap X$ and $B = X - A$ are not empty. As X is anti-connected, there are vertices $x \in A, x' \in B$ with $xx' \notin E(G)$. Vertex u must be adjacent to b, or c, for otherwise $\{u, x, b, c\}$ violate the transitivity of $\vec{G}$. The maximality of X means u cannot be adjacent to both b and c. We may assume $ub \in E(G)$, $uc \notin E(G)$. Now, $\{u, b, x'\}$ or $\{u, b, c\}$ violates the transitivity of $\vec{G}$. ∎

Lemma 28.4 *Let G be a graph with a nontrivial module X and x be a vertex in X. Let G_1 be the subgraph of G induced by $(V(G) - X) \cup \{x\}$, let G_2 be the subgraph of G induced by X. Then G is a comparability graph if and only if both G_1 and G_2 are.*

Proof. We obviously need only to prove the *if* part. Assume both G_1 and G_2 admit acyclic transitive orientations $\vec{G_1}$ and $\vec{G_2}$. An acyclic transitive orientation $\vec{G}$ of G can be constructed as follows. Consider adjacent vertices a, b of G. If $a \to b$ is an arc in G_1 or G_2, then let $a \to b$ be an arc of $\vec{G}$. Otherwise, we may assume $a \in G_1 - x$, $b \in X - x$. If $a \to x$ is an arc of G_1, then let $a \to b$ be an arc of $\vec{G}$, else let $b \to a$ be an arc of $\vec{G}$. It is easy to verify that $\vec{G}$ is an acyclic transitive orientation. ∎

Lemma 28.4 implies the following.

Corollary 28.2 *A minimally noncomparability graph cannot contain a nontrivial module.* ∎

Proof of Theorem 28.9. We prove by contradiction. Let G be a graph such that every transitive orientation of G is cyclic. Therefore, G is not a comparability graph, and so G contains an induced subgraph H that is minimally noncomparability. Therefore, every transitive orientation of H is cyclic. By Theorem 28.10, H contains a proper module, contradicting Corollary 28.2. ∎

Definition 28.10 *Let $G = (V, E)$ be a graph. The corresponding* knotting *graph is given by $K[G] = (V_K, E_K)$ where V_K and E_K are defined as follows. For each vertex v of G there are copies $v_1, v_2, \ldots, v_{i_v}$ in V_K, where i_v is the number of components of $\overline{G}[N(v)]$. For each edge vw of E, there is an edge $v_i w_j$ in E_K, where v is contained in the jth component of $\overline{G}[N(w)]$) and w is contained in the ith component of $\overline{G}[N(v)]$.*

An illustration of the knotting relation is shown in Figure 28.1. It is easy to see that if G is a comparability graph, then its knotting graph $K(G)$ is bipartite. The converse is also true.

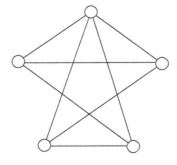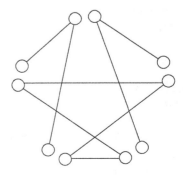

Figure 28.1 Graph and its knotting graph.

Theorem 28.11 [14] *A graph is a comparability graph if and only if its knotting graph is bipartite.* ∎

A characterization of comparability graphs by forbidden induced subgraphs is given in [14] (see [31] for an English translation of [14]).

Definition 28.11 *A sequence $\sigma = \{y_1 W_1 y_2 \ldots y_{2n+1} W_{2n+1} y_1\}$ is an* asteroid, *more exactly a $(2n+1)$-asteroid, if the y_i are pairwise distinct vertices, each W_i is a path with endpoints y_i, y_{i+1}, and y_i has no neighbor in W_{i+n} (subscripts are taken modulo $2n+1$).*

Theorem 28.12 [14] *A graph G is a comparability graph if and only if its complement $\overline{G}$ contains no asteroid.* ∎

By characterizing all minimal asteroids, a list of all minimal non-comparability graphs can be found.

Theorem 28.13 [14] *A graph G is a comparability graph if and only if G does not contain as induced subgraphs any of the four graphs shown in Figure 28.2 or the complements of the 14 graphs shown in Figure 28.3.* ∎

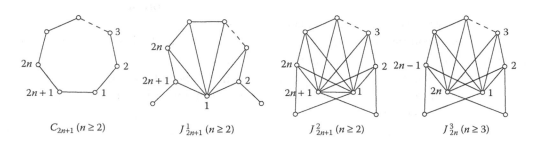

Figure 28.2 Four graphs with non-bipartite knotting graphs.

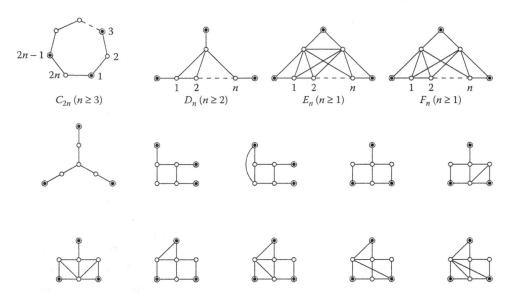

Figure 28.3 Fourteen graphs containing a 3-asteroid.

The reader may verify that the graphs in Figure 28.2 have nonbipartite knotting graphs, and the graphs in Figure 28.3 contain a 3-asteroid.

Definition 28.12 *Given a partial order $(P, \prec)$, a* chain *is a set of pairwise comparable elements, an* anti-chain *is a set of pairwise incomparable elements.*

A proof of the following well-known theorem is presented later.

Theorem 28.14 [4] *In a partially ordered set $(P, \prec)$, the size of a largest anti-chain is equal to the smallest number of chains needed to cover all elements of P.* ∎

Let $(P, \prec)$ be a partial order, and let $\vec{G}$ be the transitive orientation of the comparability graph G of $(P, \prec)$. Because of transitivity, a directed path of $\vec{G}$ induces a clique. Thus, a chain of P corresponds to a clique of G. And, an anti-chain of P corresponds to a stable set of G. Thus, Theorem 28.14 is equivalent to the statement that the complements of comparability graphs are perfect.

28.4.2 Recognition

Consider the problem of determining whether a given graph G is a comparability graph. Equivalently, the problem asks if G can be oriented so that the resulting directed graph is acyclic and transitive. First, we consider an algorithm for the problem with the complexity of $O(mn)$. Then, we discuss a more efficient algorithm.

Suppose G is a comparability graph, xy is an edge of G, and some transitive orientation $\vec{G}$ of G contains $x \to y$. Then, reversing the direction of every arc in $\vec{G}$ also yields a transitive orientation of G. Therefore, if we were to test whether G admits a transitive orientation, it is enough to pick an arbitrary edge xy of G and determine whether there exists a transitive orientation of G that contains $x \to y$.

Suppose xyz is a P_3 of G. If a transitive orientation of G contains $x \to y$, then it must contain $z \to y$ also; in this situation, we say that $x \to y$ forces $z \to y$. Now, the forced choice of $z \to y$ might in turn force the orientation of some other edges. The *implication class* of $x \to y$ consists of all the arcs that are forced, in one or more steps, by the initial choice of $x \to y$. Clearly, for some edge uv, if the implication class of $x \to y$ contains $u \to v$ as well as $v \to u$, then G cannot be a comparability graph. Conversely, it can be shown [11] that if the implication class of $x \to y$ does not contain $u \to v$ as well as $v \to u$, for any edge uv, then all the edges oriented thus far can be deleted from G, and the process can be repeated on the remaining graph until it has no edges left.

Theorem 28.15 [11] *Algorithm comparability-recognition-1 is correct and it can be implemented to run in $O(mn)$ time.* ∎

Algorithm comparability-recognition-1 produces an acyclic transitive orientation when the input graph is a comparability graph. Since the proof of its correctness is involved, we will not give it here. In this context, we note Theorem 28.9 already implies a simple polynomial time algorithm for recognizing comparability graphs: a graph G is a comparability graph if and only if for each edge xy, the implication class of $x \to y$ does not contain both $u \to v$ and $v \to u$ for some vertices u, v. Since the number of P_3 of a graph is $O(nm)$ (each edge can be extended to at most n P_3), it is not difficult to see that all implication classes of G can be enumerated in $O(nm)$ time, and so this simple algorithm runs in $O(nm)$ time.

Algorithm 28.4 comparability-recognition-1

input: graph G
output: *yes* when G is a comparability graph and *no* otherwise

$i = 1$;
while G has edges left **do**
  Pick edge xy and orient it $x \to y$;
  Enumerate the implication class D_i of $x \to y$;
  if some $u \to v$ and $v \to u$ are in D_i **then**
    output *no*;
    stop
  end if
  Let E_i be the set of underlying edges of members of D_i;
  $G = G - E_i$;
  $i = i + 1$
end while
output *yes*

Suppose we had an algorithm that can transitively orient a given comparability graph. Then, we can combine that with an algorithm to verify whether a given orientation of a graph is acyclic and transitive to obtain an algorithm to recognize comparability graphs. This is the basis for the algorithm comparability-recognition-2.

Algorithm 28.5 comparability-recognition-2

input: graph G
output: *yes* when G is a comparability graph and *no* otherwise

Run on G an algorithm for transitively orienting a comparability graph to obtain the directed graph H;
if H is acyclic and transitive **then**
  output *yes*
else
  output *no*
end if

First, we consider the second step of the algorithm comparability-recognition-2, where it is verified whether a given directed graph H is acyclic and transitive. The acyclicity of H can be verified in linear time using standard search algorithms. Having done that, by considering each P_3 of H, one can easily verify in $O(nm)$ time whether H is transitive. A faster algorithm can be derived using multiplication of Boolean matrices. The following is folklore.

Theorem 28.16 *It can be verified in $O(n^\alpha)$ time whether a given directed acyclic graph G is transitive.*

Proof. Let A be the adjacency matrix of G. Set each entry on the main diagonal of A to 1. Then, G is transitive if and only if $A = A^2$, where A^2 is computed via multiplication of Boolean matrices. ∎

In contrast to the verification step, a given comparability graph can be transitively oriented in linear time [32]. Next, we discuss the ideas behind the algorithm.

28.4.2.1 Transitive Orientation Using Modular Decomposition

The overall idea of the algorithm is to first decompose the given comparability graph using a technique called modular decomposition, store the result of the decomposition using a unique tree structure, and then orient the edges of the graph via a post order traversal of the decomposition tree. We note that modular decomposition of graphs in general has many other applications.

Suppose M is a nontrivial module in graph $G = (V, E)$. Then, G can be decomposed into $G_1 = G[V - M \cup \{x\}]$ and $G_2 = G[M]$, where x is any vertex in M. By Lemma 28.4, G is a comparability graph if and only if G_1 and G_2 are. Therefore, the notion of modules is directly relevant to the problems of recognizing comparability graphs and finding a transitive orientation of a comparability graph. Lemma 28.4 shows when G is a comparability graph, it is easy to construct a transitive orientation of G from transitive orientations of G_1 and G_2. Therefore, when G is a comparability graph that has a nontrivial module, one can find a transitive orientation of G by recursively solving the problem on G_1 and G_2; thus, the problem essentially reduces to finding a transitive orientation of a comparability graph that has no nontrivial modules. In this case, the problem is solved using the fact [14] that such a graph admits a unique transitive orientation (i.e., the transitive orientation and its reversal are the only possible ones). The notion of modular decomposition of a graph, described next, is a systematic procedure to decompose a graph into modules and record the result as a unique tree structure.

28.4.2.2 Modular Decomposition

The graph is decomposed recursively into subsets of vertices each of which is a module of the graph. The procedure stops when every subset has a single vertex. The result is represented as a tree.

Definition 28.13 *A module which induces a disconnected subgraph in the graph is a parallel module. A module which induces a disconnected subgraph in the complement of the graph is a series module. A module which induces a connected subgraph in the graph as well as in the complement of the graph is a neighborhood module.*

If the current set Q of vertices induces a disconnected subgraph, Q is decomposed into its components. A node labeled P (for parallel) is introduced, each component of Q is decomposed recursively, and the roots of the resulting subtrees are made children of the P node. If the complement of the subgraph induced by current set Q is disconnected, Q is decomposed into the components of the complement. A node labeled S (for series) is introduced, each component of the complement of Q is decomposed recursively, and the roots of the resulting subtrees are made children of the S node. Finally, if the subgraph induced by the current set Q of vertices and its complement are connected, then Q is decomposed into its maximal proper submodules (a proper submodule M of Q is maximal if there does not exist module M' of Q such that $M \subset M' \subset Q$); it is known [14] that in this case, each vertex of Q belongs to a unique maximal proper submodule of Q. A node labeled N (for neighborhood) is introduced, each maximal proper submodule of Q is decomposed recursively, and the roots of the resulting subtrees are made children of the N node. A graph and its modular decomposition tree are shown in Figure 28.4.

Theorem 28.17 [32] *The modular decomposition tree of a graph is unique and it can be constructed in $O(m + n)$ time.* ∎

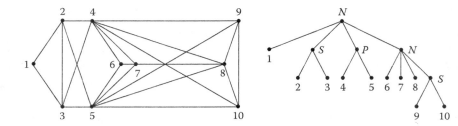

Figure 28.4 Graph and its modular decomposition tree.

28.4.2.3 From the Modular Decomposition Tree to Transitive Orientation

Definition 28.14 *Let M be the module corresponding to a node of the modular decomposition tree. The quotient graph of M is the graph obtained as follows: take a representative vertex of the graph from the subtree rooted at each child of M in the decomposition tree, and then construct the subgraph induced by the set of chosen vertices.*

We note that the choice of the representative vertex is irrelevant. The reader is referred to Figure 28.5 where the quotient graph of the root node of the decomposition tree in Figure 28.4 is shown. Vertex v_i corresponds to the subtree containing the representative vertex i of the graph.

Let us now consider the problem of transitively orienting a comparability graph, given its modular decomposition tree T. We do a post order traversal of T. Suppose we are at node D of T and all the subtrees of D have already been processed (and hence any edge of the graph with both endpoints in the same subtree of D is already oriented), our goal is to orient any edge of the graph whose endpoints are in different subtrees of D. In order to accomplish this, we construct the quotient graph H of D. We then transitively orient H. Suppose x, y are vertices of the graph that are in different subtrees of D such that v_i corresponds to the subtree of D containing x while v_j corresponds to the subtree containing y. We add $x \to y$ to the transitive orientation of the graph if and only if $v_i \to v_j$ is in the transitive orientation of H.

For example, consider the transitive orientation of the quotient graph shown in Figure 28.5. As it contains $v_4 \to v_3$, each of $4 \to 2$, $5 \to 2$, $4 \to 3$, and $5 \to 3$ will be added to the transitive orientation of the graph.

The remaining issues to be addressed are construction of the quotient graphs and finding a transitive orientation of each of the quotient graphs. It is easily seen that the sum of the sizes of all the quotient graphs is $O(m+n)$. However, this does not automatically imply that they can all be constructed efficiently. It is shown in [32] that all the required quotient graphs can be constructed in $O(m+n)$ time. Now, let us consider the problem of transitively orienting a quotient graph. The quotient graph of an S node is a complete graph; in this case, we can take any permutation R of the vertices and orient the edges so that R is a topological sort of the resulting orientation. The quotient graph of a P node has no edges.

Now, let H be the quotient graph of an N node. Clearly, H itself does not have any nontrivial modules. Therefore, as noted earlier, H admits a unique transitive orientation.

Figure 28.5 Quotient graph of the module corresponding to the root of the tree in Figure 28.4 and its transitive orientation. v_1 represents $\{1\}$, v_3 represents $\{2, \mathbf{3}\}$, v_4 represents $\{\mathbf{4}, 5\}$, and v_{10} represents $\{6, 7, 8, 9, \mathbf{10}\}$.

The idea of *vertex partitioning* is employed in [32] to transitively orient H in linear time and we explain this next. Suppose we are given a partition of $V(H)$ such that for blocks X and Y of the partition every edge of H with an endpoint in X and another in Y is already oriented in a way consistent with some transitive orientation of H (however, an edge with both endpoints inside a block may not yet be oriented). Now, suppose $u \in X$ is adjacent to some vertices in Y and also is nonadjacent to some vertices in Y. Then, we can split Y into Y_1 (neighbors of u) and Y_2 (nonneighbors of u) and replace the block Y of the partition with Y_1 and Y_2. Further, for $v \in Y_1$ and $w \in Y_2$ such that v and w are adjacent, as uvw is a P_3 and the edge uv is already oriented, orientation of the edge vw is forced. In other words, we can now orient every edge of H with an endpoint each in Y_1 and Y_2. As a result, we would have more blocks in the partition satisfying the property that any edge with endpoints in two different blocks of the partition is already oriented (and any edge with both endpoints in the same block may not yet be oriented). Observe that if a block Y had more than one vertex, then there must be a vertex in a block different from Y that splits Y; for otherwise, Y will be a nontrivial module in H. Therefore, as H contains no nontrivial modules, the process will terminate with each block containing exactly one vertex and all the edges in H will be oriented. The only remaining issue is finding the initial partition. It is shown in [32] that a source vertex s of a transitive orientation of H can be found in linear time, again, using a version of vertex partitioning. Once s is found, we can start with $X = \{s\}$ and $Y = V(H) - X$ as the blocks of the initial partition, with any edge incident on s oriented away from s.

Theorem 28.18 [32] *A transitive orientation of a comparability graph can be found in $O(m + n)$ time.* ∎

Corollary 28.3 *Comparability graphs can be recognized in $O(n^\alpha)$ time.*

28.4.2.4 How Quickly Can Comparability Graphs Be Recognized?

Next, we consider the feasibility of recognizing comparability graphs in better time than $O(n^\alpha)$.

Definition 28.15 *A dag is a directed acyclic graph.*

An h2dag $G = (X, Y, Z, E)$ is a dag (of height two) in which $\{X, Y, Z\}$ is a partition of the set of vertices of G, E is the set of arcs of G, each of X, Y, Z is a stable set, arcs between X and Y are oriented X to Y, arcs between Y and Z are oriented Y to Z, and arcs between X and Z are oriented X to Z. Further, $X = \{x_i \mid 1 \le i \le |X|\}$, $Y = \{y_i \mid 1 \le i \le |Y|\}$, and $Z = \{z_i \mid 1 \le i \le |Z|\}$.

In a tripartite graph $G = (X, Y, Z, E)$, $\{X, Y, Z\}$ is a partition of the set of vertices of G, E is set of edges of G, and each of X, Y, Z is a stable set.

Consider the following problems:

Problem-Comparability
   Instance: Graph G.
   Question: Is G a comparability graph?

Problem-Transitivity
   Instance: dag G.
   Question: Is G transitively oriented?

Problem-h2Transitivity

Instance: h2dag G.

Question: Is G transitively oriented?

Problem-Triangle

Instance: Graph G.

Question: Does G contain a triangle?

Problem-tripartiteTriangle

Instance: Tripartite graph G.

Question: Does G contain a triangle?

Lemma 28.5 [32] *Problem-Comparability $\preceq$ Problem-Transitivity via an $O(m+n)$ time reduction.*

Proof. Follows from Theorem 28.18. ∎

Lemma 28.6 [33] *Problem-Transitivity $\preceq$ Problem-Comparability via an $O(m+n)$ time reduction.*

Proof. Let $G = (V, E)$ be the given dag with $|E| \geq 1$. Construct graph H as follows: let $X = \{x_i \mid i \in V\}, Y = \{y_i \mid i \in V\}$, and $Z = \{z_i \mid i \in V\}$. Then, $V(H) = \{t\} \cup X \cup Y \cup Z \cup \{s\}$ and $E(H) = \{tx_i \mid x_i \in X\} \cup \{z_i s \mid z_i \in Z\} \cup \{x_i y_j \mid i \to j \in E\} \cup \{y_i z_j \mid i \to j \in E\} \cup \{x_i z_j \mid i \to j \in E\}$. ∎

In other words, H has two special vertices t and s and a copy in each of X, Y, and Z for every vertex $i \in V$. Corresponding to every arc $i \to j$ in G, H has three edges. Finally, t is adjacent to every vertex in X and s is adjacent to every vertex in Z. Next, we verify that G is transitive if and only if H is a comparability graph.

Suppose G is transitively oriented. Construct an orientation of H as follows: for every x_i, add the arc $x_i \to t$. For every z_i, add the arc $s \to z_i$. If $i \to j$ is an arc in G, then add the arcs $x_i \to y_j$, $y_i \to z_j$, and $x_i \to z_j$. If the resulting orientation had a violation of transitivity, then we must have $x_i \to y_j \to z_k$ (as only a vertex in Y can have an incoming as well as an outgoing arc), but no $x_i \to z_k$. This would then imply that G has $i \to j \to k$ but no $i \to k$, making it not transitive. Thus, the resulting orientation of H is transitive and therefore, H is a comparability graph.

Now, suppose H is a comparability graph and consider a transitive orientation of H. As the reversal of a transitive orientation is also a transitive orientation, we can assume that for some x_i, we have the arc $x_i \to t$. This forces the arc $x_j \to t$, for every $x_j \in X$. This in turn forces every edge between X and Y to be oriented from X to Y and also forces every edge between X and Z to be oriented from X to Z. As $|E| \geq 1$, there must be some edge $x_i z_j$ in H and hence the arc $x_i \to z_j$ must be in the transitive orientation of H. This forces the arc $s \to z_j$, which in turn forces the arc $s \to z_i$, for every $z_i \in Z$. Finally, as there cannot be a directed path with two arcs from s to a vertex in Y, every edge between Y and Z is oriented from Y to Z. In order to verify that G must be transitive, suppose G had $i \to j \to k$. Then, H has the P_3 $x_i y_j z_k$ and given the discussion above, the transitive orientation of H has $x_i \to y_j \to z_k$, and hence has the arc $x_i \to z_k$ also. Therefore, H has the edge $x_i z_k$, and given the construction of H, G has the arc $i \to k$.

Corollary 28.4 *Problem-Comparability $\equiv$ Problem-Transitivity via $O(m + n)$ time reductions.* ∎

Lemma 28.7 [33] *Problem-Transitivity $\preceq$ Problem-h2Transitivity via an $O(m + n)$ time reduction.*

Proof. Let $G = (V, E)$ be the given dag. Construct h2dag $H = (X, Y, Z, F)$ as follows: $X = \{x_i \mid i \in V\}$, $Y = \{y_i \mid i \in V\}$, $Z = \{z_i \mid i \in V\}$, and $F = \{x_i \to y_j \mid i \to j \in E\} \cup \{y_i \to z_j \mid i \to j \in E\} \cup \{x_i \to z_j \mid i \to j \in E\}$. It is seen that G has violation $i \to j \to k$ of transitivity if and only if H has violation $x_i \to y_j \to z_k$ of transitivity. ∎

Note that we trivially have Problem-h2Transitivity $\preceq$ Problem-Transitivity.

Lemma 28.8 [34] *Problem-Triangle $\preceq$ Problem-tripartiteTriangle via an $O(m + n)$ time reduction.*

Proof. Given $G = (V, E)$ construct the tripartite graph $H = (X, Y, Z, F)$ as follows: $X = \{x_i \mid i \in V\}$, $Y = \{y_i \mid i \in V\}$, $Z = \{z_i \mid i \in V\}$, and $F = \{x_i y_j, x_j y_i \mid ij \in E\} \cup \{y_i z_j, y_j z_i \mid ij \in E\} \cup \{x_i z_j, x_j z_i \mid ij \in E\}$. As H is a tripartite graph, any triangle of H must involve a vertex from each of X, Y, and Z. It is then seen that $\{i, j, k\}$ form a triangle in G if and only if $\{x_i, y_j, z_k\}$ form a triangle in H. ∎

Note that we trivially have Problem-tripartiteTriangle $\preceq$ Problem-Triangle.

Lemma 28.9 [34] *Problem-h2Transitivity $\preceq$ Problem-tripartiteTriangle via an $O(n^2)$ time reduction.*

Proof. Let $G = (X, Y, Z, E)$ be the given h2dag. Construct tripartite graph $H = (X, Y, Z, F)$ where $F = \{x_i y_j \mid x_i \to y_j \in E\} \cup \{y_i z_j \mid y_i \to z_j \in E\} \cup \{x_i z_j \mid x_i \to z_j \notin E\}$. It is seen that $x_i \to y_j \to z_k$ is a violation of transitivity in G if and only if $\{x_i, y_j, z_k\}$ form a triangle in H. ∎

Lemma 28.10 [34] *Problem-tripartiteTriangle $\preceq$ Problem-h2Transitivity via an $O(n^2)$ time reduction.*

Proof. Let $G = (X, Y, Z, E)$ be the given tripartite graph. Construct the h2dag $H = (X, Y, Z, F)$ where $F = \{x_i \to y_j \mid x_i y_j \in E\} \cup \{y_i \to z_j \mid y_i z_j \in E\} \cup \{x_i \to z_j \mid x_i z_j \notin E\}$. It is seen that $\{x_i, y_j, z_k\}$ form a triangle in G if and only if $x_i \to y_j \to z_k$ is a violation of transitivity in H. ∎

Corollary 28.5 *Problem-tripartiteTriangle $\equiv$ Problem-h2Transitivity via $O(n^2)$ time reductions.* ∎

Thus, we have the following theorem.

Theorem 28.19 *Problem-Comparability $\equiv$ Problem-Transitivity $\equiv$ Problem-Triangle via $O(n^2)$ time reductions.* ∎

We note that the current best algorithm to test for a triangle in a graph with $\Omega(n^2)$ edges runs in $O(n^\alpha)$ time.

28.4.3 Optimization

In this section, we consider the problems of finding a largest clique, a minimum coloring, a largest stable set, and a minimum clique cover of a comparability graph.

Theorem 28.20 *A largest clique and a minimum coloring of a comparability graph G can be computed in $O(m+n)$ time.*

Proof. Let $\vec{G}$ be a transitive orientation of G; from Theorem 28.18, $\vec{G}$ can be computed in $O(m+n)$ time. Observe that a directed path of $\vec{G}$ corresponds to a clique of G and vice versa.

For a vertex v of $\vec{G}$, let $height(v) = 0$ if there is no arc in $\vec{G}$ leaving v; otherwise, $height(v) = 1 + max\{height(w) \mid v \to w \text{ is an arc in } \vec{G}\}$. Now, $height(v)$ can be computed in $O(m+n)$ time for all the vertices in $\vec{G}$ as follows: compute a topological sort R of $\vec{G}$, then process the vertices of $\vec{G}$ by scanning R once from right to left (from largest to smallest), and compute $height(v)$ when vertex v is processed. During that computation, for every vertex v that has an arc leaving it in $\vec{G}$, we also record $next(v) = $ vertex w such that $height(v) = 1 + height(w)$.

Then, a longest directed path in $\vec{G}$, which corresponds to a largest clique of G, can be found starting from a vertex v of largest height, following to vertex $next(v)$, and repeating the process. Further, by assigning color h to all the vertices with height h, a minimum coloring of G can also be found. That the coloring found is optimal follows from the fact that the number of colors used equals the size of a largest clique of G. ■

Consider the following problems:

Problem-bipartiteStable

    Instance: Bipartite graph G and positive integer k.

    Question: Is there a stable set of size at least k in G?

Problem-bipartiteMatching

    Instance: Bipartite graph G and positive integer k.

    Question: Is there a matching of size at least k in G?

Problem-comparabilityStable

    Instance: Comparability graph G and positive integer k.

    Question: Is there a stable set of size at least k in G?

Theorem 28.21 [5,35] *In a bipartite graph, the size of a largest matching equals the size of a smallest vertex cover.* ■

The proof of the following theorem is adopted from [36].

Theorem 28.22 [37,38] *Let $\vec{G}$ be a transitive orientation of the comparability graph $G = (V, E)$. Construct bipartite graph $B = (X, Y, F)$ where $X = \{x' \mid x \in V\}$, $Y = \{x'' \mid x \in V\}$, and $F = \{x'y'' \mid x \to y \text{ is an arc in } \vec{G}\}$. Suppose M is a largest matching in B. Then, $\alpha(G) = \theta(G) = n - |M|$ where $n = |V|$.*

Proof [36,38]. For $x, y \in V$ with $x'y'' \in M$, refer to y as *successor* of x, and to x as *predecessor* of y. As M is a matching, every $x \in V$ has at most one predecessor and at most one successor.

Every $u \in V$ defines a unique sequence $K_u = u_{-p}, \ldots, u_{-2}, u_{-1}, u = u_0, u_1, u_2, \ldots, u_s$ where u_{i+1} is successor of u_i, u_{i-1} is predecessor of u_i, u_{-p} has no predecessor, and u_s has no successor. It then follows from transitivity of $\vec{G}$ that whenever $i < j$, we have the arc $u_i \to u_j$ in $\vec{G}$. This in turn implies that no two elements of K_u are the same.

Clearly, every $x'y'' \in M$ appears as $u_i'u_{i+1}''$ for some u_i, u_{i+1} in a specific sequence K_u. Let the total number of such sequences be k and the length of the ith sequence be r_i. Then, $\sum_{i=1}^{k} r_i = n$ and $\sum_{i=1}^{k}(r_i - 1) = |M|$. It then follows that $k = n - |M|$. As each K_u is a chain in $\vec{G}$, and hence corresponds to a clique of G, we have that $\theta(G) \leq k$.

In order to show that $\theta(G) \geq k$ also holds, we construct a stable set in G of size k based on M. From Theorem 28.21, B has a vertex cover R of size $|M|$. Let $S = \{x \in V \mid x' \notin R$ and $x'' \notin R\}$. Note that for $x, y \in S$, the arc $x \to y$ cannot be in $\vec{G}$; otherwise, as $x' \notin R$ and $y'' \notin R$, $x'y'' \in F$ is not covered by R. Therefore, S is a stable set of G and hence $\theta(G) \geq |S|$. However, as each $x \in R$ prevents only one vertex of G from being a member of S, $|S| \geq n - |R| = n - |M|$, and therefore $|S| \geq k$. Thus, we have $\theta(G) \geq |S| \geq k$ also. Finally, as $\theta(G) \geq \alpha(G)$ and $\alpha(G) \geq |S|$ also hold, we have $k \geq \theta(G) \geq \alpha(G) \geq |S| \geq k$, and we conclude that $\theta(G) = \alpha(G) = k = n - |M|$. ∎

Theorem 28.23 *Problem-bipartiteStable $\equiv$ Problem-bipartiteMatching $\equiv$ Problem-comparabilityStable via $O(m + n)$ time reductions.*

Proof. That Problem-bipartiteStable $\equiv$ Problem-bipartiteMatching follows from Theorem 28.21. As every bipartite graph is a comparability graph, we have Problem-bipartiteStable $\preceq$ Problem-comparabilityStable. That Problem-comparabilityStable $\preceq$ Problem-bipartiteMatching follows from Theorem 28.22. ∎

Given the current best time bounds of $O(n^{1.5}\sqrt{m/\log n})$ [39] and $O(n^{2.5}/\log n)$ [40] for computing a largest matching in a bipartite graph, we have the following:

Corollary 28.6 *A largest stable set and a smallest clique cover of a comparability graph can be computed in $O(\min(n^{1.5}\sqrt{m/\log n}, n^{2.5}/\log n))$ time.* ∎

Now, we present a proof of Theorem 28.14.

Proof of Theorem 28.3 [38]. Construct transitive orientation $\vec{G}$ of the comparability graph G of $(P, \prec)$ by adding arc $x \to y$ to $\vec{G}$ if and only if $x \prec y$. As a chain of $(P, \prec)$ corresponds to a clique of G and an anti-chain of $(P, \prec)$ corresponds to a stable set of G, the proof follows from Theorem 28.22. ∎

28.5 INTERVAL GRAPHS

Definition 28.16 *Graph $G = (V, E)$ is an* interval *graph if every $v \in V$ can be mapped to an interval I_v on the real line such that $xy \in E$ if and only if $I_x \cap I_y \neq \emptyset$. When G is an interval graph, the collection $\{I_v \mid v \in V\}$ is an* interval model *for G. For $v \in V$, v_L and v_R denote the left and right endpoints, respectively, of I_v.*

It is known that in an interval model for an interval graph, the endpoints can be assumed to be distinct. Thus, the $2n$ endpoints can be represented by the integers 1 through $2n$. Further, for a cost of $O(n)$ using bin-sort, one can assume the endpoints are given in increasing order.

28.5.1 Characterization

Theorem 28.24 [41] *For a graph $G = (V, E)$ the following statements are equivalent:*

i. G *is an interval graph.*

ii. G *is chordal and $\overline{G}$ is a comparability graph.*

iii. *There is an ordering $\mathcal{R}$ of the maximal cliques of G such that for every $v \in V$, the maximal cliques containing v are consecutive in $\mathcal{R}$.*

Proof.

(i) $\Rightarrow$ (ii) Let $\{I_v \mid v \in V\}$ be an interval model for G. Suppose $v_1 v_2 v_3 \cdots v_k$, $k \geq 4$ is a chordless cycle in G. For $1 \leq i \leq k-1$, let p_i be a point in $I_{v_i} \cap I_{v_{i+1}}$. Given that $v_1 v_2 \cdots v_{k-1}$ is a chordless path, we can assume $p_1 < p_2 < \cdots < p_{k-1}$. Then, it is impossible for I_{v_1} to intersect I_{v_k}. Therefore, G is chordal.

For $x, y \in V$, $xy \notin E$ if and only if either $x_R < y_L$ holds or $y_R < x_L$ holds. For $xy \notin E$, orient $x \to y$ in $\overline{G}$ if $x_R < y_L$. It is easily verified that the resulting orientation is acyclic and transitive. Therefore, $\overline{G}$ is a comparability graph.

(ii) $\Rightarrow$ (iii) Suppose A and B are distinct maximal cliques of G. Then, there must exist $x \in A$ and $y \in B$ such that $xy \notin E$; otherwise, $A \cup B$ is also a clique of G. Now, consider a transitive orientation of $\overline{G}$. For $w, x \in A$ and $y, z \in B$ such that $xy \notin E$ and $wz \notin E$, if we have $x \to y$ in $\overline{G}$, then we must have $w \to z$ in $\overline{G}$. Suppose not, and we have $x \to y$ and $z \to w$ in $\overline{G}$. Clearly, $w \neq x$ and $y \neq z$ or else, there is a violation of transitivity in $\overline{G}$. Further, as G is chordal, either $xz \notin E$ or $wy \notin E$; say, $xz \notin E$. Then, there is no way to orient the edge xz in $\overline{G}$ to avoid a violation of transitivity. Thus, the edges of $\overline{G}$ that go across A, B are all oriented either from A to B, or from B to A.

Now, for distinct maximal cliques A, B, and C of G and $w \in A$, $x, y \in B$, and $z \in C$, suppose we have $w \to x$ and $y \to z$ in $\overline{G}$. Then, we claim $wz \notin E$ and $w \to z$ in $\overline{G}$. Suppose not. As $\overline{G}$ is transitively oriented, we can assume $x \neq y$. Further, $xz \in E$ and $wy \in E$; otherwise, we have $x \to z$ or $w \to y$, and the transitivity of $\overline{G}$ is violated. Now, $wyxz$ is a chordless cycle in G. So, we have $wz \notin E$. Now, we must have $w \to z$ in $\overline{G}$ or else, $z \to w \to x$ is a violation of transitivity in $\overline{G}$.

Now, consider the ordering $\mathcal{R}$ of the maximal cliques of G where $A < B$ in $\mathcal{R}$ if there exist $x \in A$ and $y \in B$ such that we have $x \to y$ in $\overline{G}$; from the claim above, such a total ordering exists. In order to verify that $\mathcal{R}$ is the required ordering: for maximal cliques A, B, and C with $A < B < C$ in $\mathcal{R}$, suppose $x \in A$, $x \in C$, but $x \notin B$. As B is a maximal clique and $x \notin B$, there must exist $y \in B$ such that $xy \notin E$. As $A < B$ in $\mathcal{R}$, we must have $x \to y$ in $\overline{G}$. However, this contradicts $B < C$ which dictates that we have $y \to x$ in $\overline{G}$.

(iii) $\Rightarrow$ (i) Consider an ordering $\mathcal{R} = K_1 K_2 \cdots K_p$ of the maximal cliques of G as stated in the theorem. For $v \in V$, let K_{v_L} be the left most maximal clique in $\mathcal{R}$ that contains v. Similarly, let K_{v_R} be the right most maximal clique in $\mathcal{R}$ that contains v. Set $I_v = [v_L, v_R]$. It is easily verified that $\{I_v \mid v \in V\}$ is an interval model for G. ∎

Definition 28.17 *A set $\{x, y, z\}$ of pair-wise nonadjacent vertices of G is an* asteroidal triple *if there exists a path between any two of them that does not involve a neighbor of the third.*

Theorem 28.25 [42] *G is an interval graph if and only if G is chordal and G does not contain an asteroidal triple.* ∎

28.5.2 Recognition

As chordal graphs and complements of comparability graphs can be recognized in polynomial time, a direct consequence of Theorem 28.24 is that interval graphs can be recognized in polynomial time; further, an interval model for an interval graph can also be constructed in polynomial time. The first $O(m+n)$ time algorithm to recognize interval graphs was given in [43] and we describe the ideas employed there next. Given input graph $G = (V, E)$, we first test whether G is chordal (recall that every interval graph is chordal). If G is chordal, then we use the algorithms in [25,27] to generate all the maximal cliques of G; by Corollary 28.1 G has at most n maximal cliques whose sizes sum up to at most m. The remaining task is to determine whether an ordering of all the maximal cliques of G, as stipulated in Theorem 28.24, exists. In [43] the data structure *PQ-tree* was used to solve the following problem in $O(m+n)$ time: given a finite set X with $|X| = n$ and a collection $S_1, \cdots, S_k$ of subsets of X with $|S_1| + \cdots + |S_k| = m$, determine if there is an ordering of members of X such that for each S_i the members of S_i occur consecutively in the ordering. In order to use this algorithm for the recognition of interval graphs, we just have to let $X = V$ and let the set of maximal cliques of the chordal graph G to be the collection S_i of subsets.

Subsequently, several linear-time algorithms have been designed to recognize interval graphs; some of these algorithms employ some variation of *PQ-trees* where as the rest avoid the use of such data structures. In [44], the algorithm from [43] is simplified with the use of modified PQ-trees. An algorithm that relies on modular decomposition of chordal graphs is given in [45]. We remarked in the section on chordal graphs that the algorithm LexBFS [25] can be used to generate a perfect elimination scheme of a chordal graph. An algorithm to recognize interval graphs using LexBFS is given in [46]. The final algorithm that we comment on relies on the following characterization of interval graphs which has been observed by multiple researchers.

Theorem 28.26 [47–49] *$G = (V, E)$ is an interval graph if and only if vertices of G can be ordered $v_1 v_2 \cdots v_n$ such that for v_i, v_j, v_k with $i < j < k$, if $v_i v_k \in E$ then $v_j v_k \in E$.*

Proof. For an interval graph G with an interval model where the endpoints are distinct, an ordering of vertices of G according to the right endpoints of their intervals gives the desired ordering. Conversely, given such an ordering, one can derive an interval model for G by taking the interval for v_i to be $[v_{i_f}, v_i]$ where v_{i_f} is the left most neighbor of v_i in the ordering. ∎

In [50], a (very complicated) linear-time algorithm is given which employs six passes of LexBFS with various rules for breaking ties when choices have to be made. When the input is an interval graph, the algorithm is guaranteed to produce an ordering satisfying the conditions of Theorem 28.26. In order to test whether a given graph is an interval graph, we run the algorithm in [50] to get an ordering of vertices, and then verify if the ordering satisfies the conditions of Theorem 28.26.

28.5.3 Optimization

As interval graphs are chordal, given the adjacency lists for an interval graph, each of a largest clique, a largest stable set, an optimal vertex coloring, and a smallest vertex cover, as will be discussed in Section 28.7, can be computed in $O(m+n)$ time. However, when the interval model for an interval graph is given as input, it is possible to solve the problems more efficiently. Next, we illustrate this with algorithms for computing a largest clique and an optimal vertex coloring.

We will assume that the $2n$ endpoints in the interval model of the given interval graph $G = (V, E)$ are distinct and they are given in sorted order; recall that the endpoints can be sorted in $O(n)$ time. The algorithms scan the endpoints of the intervals from left to right (i.e., from the smallest to the largest). We *open* an interval when its left endpoint is scanned and we *close* it when its right endpoint is scanned. Further, an interval itself is *open* if its left endpoint has been scanned and its right endpoint is yet to be scanned.

First, we consider the problem of computing a largest clique. As a set of pair-wise intersecting intervals must share a common point, the problem reduces to considering each endpoint and computing how many intervals contain that endpoint. In order to do this efficiently, we scan the endpoints from left to right keeping track of the set K of intervals open at any point. The set K can be recorded in a boolean vector of size n. For a vertex v, when we scan v_L, I_v is added to K and it is deleted from K when v_R is scanned. This provides the set up to compute $\omega(G)$ in $O(n)$ time. One can then scan the endpoints again from left to right stopping when $|K| = \omega(G)$. The set K at this point corresponds to a maximum clique of G. Thus, a maximum clique of G can be found in $O(n)$ time.

Next, we consider the problem of optimal vertex coloring. We scan the endpoints from left to right and color a vertex v when v_L is scanned. Let k, initially set to zero, record the number of colors used at any point. The list *freed-colors* contains colors assigned to intervals that have already closed, that is, those whose right endpoints have already been scanned; initially, *freed-colors* is empty. For a vertex v, when v_L is scanned, if *freed-colors* is nonempty, then we remove any color c from *freed-colors* and assign it to v. If *freed-colors* is empty, then we increase k by 1 and assign the color k to v (i.e., v is given a new color). When v_R is scanned, the color assigned to v is added to *freed-colors*.

It is easily seen that the coloring is proper and that the algorithm can be implemented to run in $O(n)$ time. In order to verify that the coloring is optimal, observe that every time a new color k is assigned to a vertex v, as *freed-colors* is empty, each of the colors 1 through $k-1$ has been assigned to an interval that is currently open. Hence each of those $k-1$ open intervals contains v_L and v belongs to a clique of size k in the graph.

The reader is referred to [51] for a detailed exposition on interplay between representation of graphs and complexity of algorithms.

28.6 WEAKLY CHORDAL GRAPHS

A *long hole* is a chordless cycle with at least five vertices and a *long anti-hole* is the complement of a long hole.

Definition 28.18 *A graph is* weakly chordal (*also called* weakly triangulated) *if it does not contain any long holes or long anti-holes.*

It is seen from the definition that the complement of a weakly chordal graph is also weakly chordal. Further, the class of weakly chordal graphs is a proper generalization of the class of chordal graphs.

28.6.1 Characterization

Definition 28.19 *Let G be a graph and x, y be nonadjacent vertices of G. $\{x, y\}$ is a* two-pair *of G if either every induced path between x and y has exactly two edges or x and y belong to different components of G. A* co-pair *of a graph is a two-pair of the complement of the graph.*

Weakly chordal graphs were characterized [52] via the presence of two-pairs. As weakly chordal graphs are closed under complementation, the presence of a co-pair also characterizes weakly chordal graphs.

Theorem 28.27 [52] *G is a weakly chordal graph if and only if for every induced subgraph H of G, either H induces a stable set or H has a co-pair.* ∎

To prove Theorem 28.27, we will need to establish a preliminary result. We first start with a definition.

Definition 28.20 *A handle in a graph G is a proper vertex-subset H with size at least two such that $G[H]$ is connected, some component $J \neq H$ of $G - N(H)$ satisfies $N(J) = N(H)$, and each vertex of $N(H)$ is adjacent to at least an endpoint of each edge of $G[H]$. J is called a co-handle of H.* ∎

Note that $N(H)$ is a minimal separator of H and J.

Theorem 28.28 [53,54] *A graph has a handle if and only if the graph has a $\overline{P_3}$, and a handle and its co-handle can be found in polynomial time.* ∎

When vertex-subset H of G with $|H| \geq 2$ induces a component of G, as $N(H) = \emptyset$, H is trivially a handle of G; any other component of G can be considered a co-handle of H. In this case, it is easily seen that when G is a weakly chordal graph, any co-pair of $G[H]$ is a co-pair of G also. Next, we prove that this holds for any handle H of G when G is a weakly chordal graph.

Lemma 28.11 [55] *Suppose H is a handle of a weakly chordal graph G and $\{x, y\}$ is a co-pair of $G[H]$. Then, $\{x, y\}$ is a co-pair of G.*

Proof. Let J be a co-handle of H in G, $I = N(H) = N(J)$, and $R = V(G) - H - I$. Suppose $\{x, y\}$ is a co-pair of $G[H]$ but not a co-pair of G.

Then, there exists an induced path $P = x \ldots y$ with at least four vertices in $\overline{G}$. As each vertex in R is adjacent to both x and y in $\overline{G}$, P does not involve any vertex in R; therefore, P has at least a vertex from I. Now, P cannot have a segment uvw such that u and w are in H but v is in I, for otherwise, vertex v of I is not adjacent in G to any endpoint of the edge uw of $G[H]$, contradicting H being a handle of G. Thus, at least two consecutive vertices of P are in I and P involves at least an edge of $\overline{G}$ with both endpoints in I.

In $\overline{G}$, consider a segment $P' = x_2 x_3 x_4 \ldots x_r$ of P with $r \geq 4$ such that x_2 and x_r are in H but x_3 through x_{r-1} are in I. Observe that x_3 is not adjacent to x_4 in G. Since I is a minimal separator for H and J in G, and G has no long holes, in G every two nonadjacent vertices of I must have a common neighbor in J. In particular, x_3 and x_4 are adjacent in G to some vertex x_1 of J. Thus in $\overline{G}$ x_1 is adjacent to x_2, x_1 is not adjacent to x_3, x_1 is not adjacent to x_4, and $x_1 x_2 x_3 x_4$ is a P_4. Let x_k be the first vertex in P' after x_4 such that x_1 is adjacent to x_k in $\overline{G}$; such an x_k exists as x_1 is adjacent to x_r in $\overline{G}$. Then, $\{x_1, x_2, \ldots, x_k\}$ induces a long hole in $\overline{G}$, contradicting G being weakly chordal. ∎

Proof of Theorem 28.27. For one direction, if G is not weakly chordal, then it contains induced subgraph H such that H induces either a long hole or a long anti-hole. It is seen that neither does H induce a stable set nor it contains a co-pair of H.

For the other direction, as an induced subgraph of a weakly chordal graph is also weakly chordal, it suffices to prove the theorem for the given weakly chordal graph G. Let G be a weakly chordal graph with at least one edge. Let $G = H_0, H_1, \cdots, H_p, p \geq 0$, be a sequence of subsets of $V(G)$ such that H_i is a handle of $G[H_{i-1}]$, for $1 \leq i \leq p$, and $G[H_p]$ has no handle. Then, by Theorem 28.28, $G[H_p]$ has no $\overline{P_3}$, and is a complete multipartite graph. Therefore, every edge of $G[H_p]$ induces a co-pair of $G[H_p]$. Then, by Lemma 28.11, every edge of $G[H_p]$ induces a co-pair of $G[H_{p-1}]$, since H_p is a handle of $G[H_{p-1}]$. Continuing this argument, every edge of $G[H_p]$ induces a co-pair of G.

The current best recognition and optimization algorithms for weakly chordal graphs exploit the presence of two-pairs and co-pairs.

28.6.2 Recognition

An algorithm to test for the presence of a long hole in a graph is to check whether a P_3 of a graph extends into a long hole. As all the P_3's of a graph can be generated in $O(nm)$ time, this can be implemented to run in $O(nm^2)$ time. By running this algorithm on the graph and then on the complement, weakly chordal graphs can be recognized in $O(n^5)$ time. Later, we discuss more efficient algorithms for the same problem.

More generally, whether a P_k, $k \geq 2$, of a graph extends into a hole of size at least $k+3$ can be tested in $O(n^\alpha)$ time [56], where $O(n^\alpha)$ refers to the current best complexity of multiplying two $n \times n$ Boolean matrices, by testing whether an auxiliary directed graph in transitive. The algorithm is as follows: given the P_k $T = v_1 \cdots v_k$ of G, first we discard from G all the neighbors of v_2 through v_{k-1} that are not on T. Now, let $A = N(v_1) - N(v_k) - V(T)$, $B = N(v_k) - N(v_1) - V(T)$, and $D_1, \cdots, D_r$ be the components of $G - (A \cup B \cup V(T))$. Let M be the set formed by adding a vertex m_i corresponding to each D_i. Now, construct the directed graph H on the vertex-set $A \cup M \cup B$. For $x \in A$, add the directed edge $x \rightarrow m_i$ provided x is adjacent in G to some vertex in D_i. Similarly, for $x \in B$, add the directed edge $m_i \rightarrow x$ provided x is adjacent in G to some vertex in D_i. Finally, for $x \in A$ and $y \in B$, add the directed edge $x \rightarrow y$ provided x and y are adjacent in G. It can be seen that G has a hole of size at least $k+3$ through T if and only if H is not transitive. As whether a directed acyclic graph is transitive can be tested in $O(n^\alpha)$ time (cf. Theorem 28.16) we get the desired result. Thus, as the number of P_k's in a graph is $O(n^k)$, we can check whether a graph has a hole of size at least t, $t \geq 5$, in time $O(n^{t-3+\alpha})$.

Using the above mentioned algorithm on the graph and then on the complement of the graph, weakly chordal graphs can be recognized in $O(n^{2+\alpha})$ time which is currently $O(n^{4.376})$ [18]. For the specific case of finding long holes in a graph, an $O(m^2)$ time algorithm is known [57]. By using this on the graph and then on the complement, weakly chordal graphs can be recognized in $O(n^4)$ time. The current best algorithms to recognize weakly chordal graphs run in $O(m^2)$ time [55,58]. However, one of them requires $O(m^2)$ space [58] while the other [55] uses linear amount of space.

Lemma 28.12 [59] *Suppose $\{x, y\}$ is a co-pair of graph G. Let H be the graph obtained from G by deleting the edge xy but not its endpoints. Then, G is weakly chordal if and only if H is weakly chordal.* ∎

Algorithm 28.6 wc-recognition

input: graph G
output: *yes* when G is weakly chordal and *no* otherwise

found ← true;
while *found* and G has at least one edge **do**
  if G has co-pair $\{x, y\}$ **then**
    Delete edge xy from G
  else
    found ← false
  end if
end while
if G has no edges **then**
  output *yes*
else
  output *no*
end if

Theorem 28.29 [55] *Algorithm wc-recognition can be implemented to run in $O(m^2)$ time using $O(m+n)$ space.* ∎

28.6.3 Optimization

Definition 28.21 *For a graph G and a pair $\{x,y\}$ of nonadjacent vertices in G, the graph G/xy is obtained from G by* contracting *the pair $\{x,y\}$ as follows: delete vertices x and y and introduce vertex (xy) and edges $(xy)u$ for all u in $N_G(x) \cup N_G(y)$.*

Definition 28.22 *Two nonadjacent vertices x,y in a graph G form an* even-pair *if every induced path between them has an even number of edges.*

Our interest in even-pairs is motivated by the following two observations.

Lemma 28.13 [60] *Let G be any graph with an even-pair $\{x,y\}$. Then*

  i. $\omega(G/xy) = \omega(G)$;

  ii. $\chi(G/xy) = \chi(G)$.

Proof. We will establish (i) first. Let K be clique in G/xy. For simplicity, write $z = (xy)$. If $z \notin K$, then K is also a clique in G. Suppose $z \in K$. Then, either x or y must be adjacent in G to every vertex in $K - \{z\}$. Otherwise, there exist $u,v \in K$ such that $xu \in E(G)$, $xv \notin E(G)$, $yu \notin E(G)$, and $yv \in E(G)$ so that $xuvy$ is a P_4 in G; this contradicts $\{x,y\}$ being an even-pair of G. Thus, G also has a clique of size $|K|$ and $\omega(G/xy) \leq \omega(G)$. Now suppose K is a clique in G. Clearly, at most one of $x \in K$, $y \in K$ holds. Further, if $x \in K$ ($y \in K$), then $K - \{x\} \cup \{z\}$ ($K - \{y\} \cup \{z\}$) is a clique in G/xy. Therefore, $\omega(G) \leq \omega(G/xy)$ also holds and $\omega(G) = \omega(G/xy)$.

To prove (ii), consider a coloring of G/xy. It gives a coloring of G by assigning to x,y the color of (xy). So, we have $\chi(G/xy) \geq \chi(G)$. Now, we will prove $\chi(G/xy) \leq \chi(G)$. Consider a coloring of G. If x,y have the same color, then this color can be assigned to (xy), and we are done. So, assume x has color 1 and y has color 2. Let B be the bipartite graph induced by vertices of colors 1 and 2. x and y must belong to different components of B, for otherwise there is an induced odd path in B between the two vertices, a contradiction to the assumption that $\{x,y\}$ is an even-pair. Interchange colors 1 and 2 in the component of B containing x. In the new coloring, x and y have the same color, implying as above, that $\chi(G/xy) \leq \chi(G)$. ∎

The proof of Lemma 28.13 gives a simple algorithm that given a largest clique of G/xy produces a largest clique of G, and given a coloring of G/xy with k colors, produces a coloring of G with k colors. If, on subsequent graphs, we can always find an even-pair to contract until we obtain a clique, we could produce a largest clique and an optimal coloring of the original graph. The following lemma shows this is indeed the case for weakly chordal graphs.

Lemma 28.14 [52] *Suppose G is a weakly chordal graph and $\{x,y\}$ is a two-pair of G. Then, G/xy is weakly chordal. Further, $\omega(G) = \omega(G/xy)$ and $\chi(G) = \chi(G/xy)$.*

Proof. We show that if G/xy is not weakly chordal, then G is not weakly chordal. Clearly, G/xy cannot have a long hole or long anti-hole that does not involve $z = (xy)$. Suppose $zv_2 \cdots v_k$, for $k \geq 5$, is a long hole in G/xy. Then, as G is weakly chordal and given the construction of G/xy, neither x nor y is adjacent in G to each of v_2, v_k. Also, each of v_2, v_k

is adjacent in G to at least one of x, y. Without loss of generality, assume that $xv_2 \in E(G)$, $xv_k \notin E(G)$, $yv_k \in E(G)$, and $yv_2 \notin E(G)$. Then, $xv_2 \cdots v_k y$ is chordless path in G with at least five edges, contradicting $\{x, y\}$ being a two-pair of G.

Suppose $zv_2 \cdots v_k$, is a long anti-hole in G/xy where the ordering of the vertices corresponds to the cyclic ordering of the vertices along the long hole in the complement. As C_5 is isomorphic to $\overline{C_5}$, we can assume $k \geq 6$. One of x, y must be adjacent in G to each of v_3, v_4. Otherwise, given the construction of G/xy, we can assume $xv_3 \in E(G)$, $xv_4 \notin E(G)$, $yv_4 \in E(G)$, and $yv_3 \notin E(G)$. Then, $xv_3v_kv_4y$ a chordless path in G with four edges, contradicting $\{x, y\}$ being a two-pair of G. Assume x is adjacent in G to each of v_3, v_4 and let r be the smallest index such that $r \geq 5$ and $xv_r \notin E(G)$; such an r exists as $xv_k \notin E(G)$. Then, $xv_2v_3v_4 \cdots v_r$ is a long anti-hole in G, a contradiction. Since two-pairs are even-pairs, the rest of the lemma follows from Lemma 28.13. ∎

Algorithm 28.7 wc-optimization

  input: weakly chordal graph G
  output: $\chi(G)$ and $\omega(G)$

  while G is not a complete graph **do**
    find two-pair $\{x, y\}$ of G;
    replace G by G/xy
  end while
  $\chi(G) = |V(G)|$;
  $\omega(G) = |V(G)|$;
  output $\chi(G)$ and $\omega(G)$

Theorem 28.30 [55] *Algorithm wc-optimization can be implemented to run in $O(mn)$ time using $O(m+n)$ space.* ∎

For a weakly chordal graph G, $\alpha(G)$ and $\theta(G)$ can be computed by running the algorithm *wc-optimization* on $\overline{G}$.

28.6.4 Remarks

An $O(m^2)$ time algorithm to find a long hole in a given graph is given in [57]. An $O(m^2)$ time algorithm to recognize weakly chordal graphs using $O(m^2)$ space is given in [58]; unlike the algorithm described here, the one in [58] does not use the idea of a two-pair at all. The weighted versions of the clique, coloring, stable set, and clique cover problems can be solved on weakly chordal graphs in $O(n^4)$ time [52,59]. A consequence of algorithm wc-recognition is that graph G is a weakly chordal if and only if an empty graph can be derived from G by repeatedly removing a co-pair. As an interesting contrast, it is proved in [61] that graph G is chordal if and only if G can be derived from an empty graph by repeatedly adding an edge between vertices that form a two-pair. Efficient algorithms for finding a two-pair in a graph are given in [62] and [63]. The fact that weakly chordal graphs are perfect was first established in [64].

28.7 PERFECTLY ORDERABLE GRAPHS

A natural way to color a graph is to impose an order $<$ on its vertices and then scan the vertices in this order, assigning to each vertex v_i the smallest positive integer not assigned to a neighbor v_j of v_i with $v_j < v_i$. This method, referred to as the *greedy algorithm*, does not necessarily produce an optimal coloring of the graph (i.e., one using the smallest possible number of colors). However, on a *perfectly ordered* graph, the algorithm does produce an optimal coloring.

Definition 28.23 *Given an ordered graph* $(G, <)$, *the ordering* $<$ *is called* perfect *if for each induced ordered subgraph* $(H, <)$ *the greedy algorithm produces an optimal coloring of H. The graphs admitting a perfect ordering are called* perfectly orderable. *An* obstruction *in an ordered graph is a chordless path with vertices a, b, c, d, edges ab, bc, cd with $a < b$ and $d < c$.*

Several well known classes of graphs (in particular, chordal and comparability graphs) are perfectly orderable. It is easy to see that a perfectly ordered graph cannot contain an obstruction. It was shown [65] that this condition is also sufficient.

Theorem 28.31 [65] *A graph is perfectly orderable if and only if it admits an obstruction-free ordering.* ∎

We will need the following lemma.

Lemma 28.15 *Let G be a graph and let C be a clique of G such that each $w \in C$ has a neighbor $p(w) \notin C$ such that the set S consisting of the vertices $p(w)$ form a stable set of G. If there is an obstruction-free order $<$ such that $p(w) < w$ for all $w \in C$, then some $p(w)$ is C-complete.*

Proof. By induction on the number of vertices in C. The induction hypothesis implies that, for each $w \in C$, there is a vertex $f(w) \in C$ such that the vertex $p(f(w))$ is adjacent to all of C, except possibly w. In fact, we may assume $p(f(w))$ is not adjacent to w, for otherwise we are done. Thus, the mapping f is one-to-one and therefore onto, that is f is a bijection. Let v be the smallest vertex in C in the order $<$. There are vertices a, b such that $v = f(b)$ and $b = f(a)$. Now, $p(v), a, b, p(b)$ form an obstruction, a contradiction. ∎

Proof of Theorem 28.31. The 'only if' part is trivial. We will prove the 'if' part by induction on the number of vertices. Let G be a graph with an obstruction-free order $<$. By the induction hypothesis, we only need to prove the greedy algorithm delivers an optimal coloring on G. Let k be the number of colors used on G. We will prove G contains a clique on k vertices. This obviously shows the coloring produced by the greedy algorithm is optimal. Let i be the smallest integer such that there is a clique C on vertices $v_{i+1}, \ldots, v_k$ such that each v_j has color j, for $j = i+1, \ldots, k$. We may assume $i > 0$, for otherwise we are done. Properties of the greedy algorithm imply that each v_j has a neighbor $p(v_j)$ with color i with $p(v_j) < v_j$, for each $v_j \in C$. But Lemma 28.15 implies some $p(v_j)$ is C-complete, a contradiction to our choice of i. ∎

The proof of Theorem 28.31 shows that perfectly orderable graphs are perfect. In studying perfectly orderable graphs, the following two problems arise naturally: to decide on the complexity of recognizing perfectly orderable graphs and to find a subgraph characterization of perfectly orderable graphs (by *subgraph characterization*, we mean *characterization by minimal forbidden induced subgraphs*). The subgraph characterization problem is open but appears to be very difficult. It was proved in [66] that the problem of recognizing perfectly orderable graph is NP-complete. However, many classes of perfectly orderable graphs, together with their polynomial recognition algorithms, have been found. We will discuss some of these classes in this chapter. For a survey on perfectly orderable graphs, see [67].

28.7.1 Characterization

As mentioned before, there is no known characterization by forbidden induced subgraphs of perfectly orderable graphs. We will discuss several subclasses of perfectly orderable graphs that have been much studied.

Definition 28.24 *For a P_4 with vertices a, b, c, d, edges ab, bc, cd, the vertices a, d are endpoints, c, d are midpoints of the P_4. A vertex is soft if it is not a midpoint or an endpoint of a P_4. A graph G is brittle if each of its induced subgraphs contains a soft vertex.*

Observation 28.1 *Brittle graphs are perfectly orderable.*

Proof. By induction on the number of vertices. Let G be a brittle graph with a soft vertex v. Let $v_1 < v_2 < \ldots < v_{n-1}$ be a perfect order of $G - v$. If v is not the endpoint of a P_4, then $v < v_1 < v_2 < \ldots < v_{n-1}$ is a perfect order of G. If v is not a midpoint of a P_4, then $v_1 < v_2 < \ldots < v_{n-1} < v$ is a perfect order of G. ∎

Corollary 28.7 *Chordal graphs, their complements, and comparability graphs are perfectly orderable.*

Proof. Observe that a simplicial vertex is soft and that a soft vertex of a graph remains soft in the complement. Thus, chordal graphs are brittle; by Observation 28.1, they and their complements are perfectly orderable. Since a transitive orientation of a graph contains no obstruction, comparability graphs are perfectly orderable. ∎

28.7.2 Recognition

It is proved in [66] that the problem of recognizing perfectly orderable graphs is NP-complete. We have seen that chordal graphs and their complements are perfectly orderable. Since weakly chordal graphs are a generalization of these two classes, it is of interest to investigate the complexities of recognizing weakly chordal perfectly orderable graphs. In [68], it is shown that this problem is NP-complete by modifying the argument of [66]. Since [68] is an unpublished technical report, we will reproduce the proof here.

Theorem 28.32 *It is NP-complete to determine if a weakly chordal graph is perfectly orderable.*

Proof. We will reduce the 3SAT problem to our problem. Given a 3SAT formula E with clauses $C_0, C_1, \ldots, C_{m-1}$ and variables $v_0, v_1, \ldots, v_{n-1}$ where each clause C_i contains literals c_{i0}, c_{i1}, c_{i2}, we construct a weakly chordal graph $G(E)$ such that E is satisfiable if and only if $G(E)$ is perfectly orderable.

For each clause $C_j = (c_{j0}, c_{j1}, c_{j2})$, we define the *clause graph* $G(C_j)$ as in shown in Figure 28.6. For each variable v_i, we define the *variable graph* $G(v_i)$ as shown in Figure 28.7. In the graph $G(v_i)$, the chordless path between A_i and B_i has $2m$ vertices $v(i, j, 1)$ for $j = 0, 1, 2, \ldots, 2m - 1$.

Next, we obtain the graph $G'(v_i)$ (see Figure 28.8) from $G(v_i)$ by

- If C_j contains v_i, adding vertices $v(i, 2j, 2), v(i, 2j, 3)$ and edges $v(i, 2j, 1)v(i, 2j, 2)$, $v(i, 2j, 2)v(i, 2j, 3)$.

- If C_j contains $\overline{v_i}$, adding vertices $v(i, 2j+1, 2), v(i, 2j+1, 3)$ and edges $v(i, 2j+1, 1)v(i, 2j+1, 2)$, $v(i, 2j+1, 2)v(i, 2j+1, 3)$.

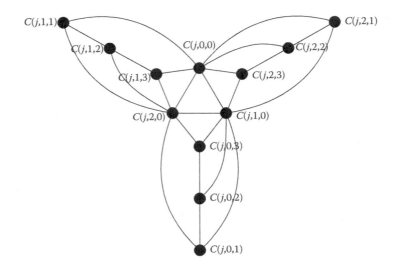

Figure 28.6 Clause graph $G(C_j)$.

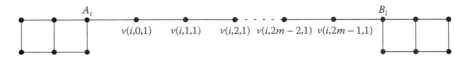

Figure 28.7 Graph $G(v_i)$.

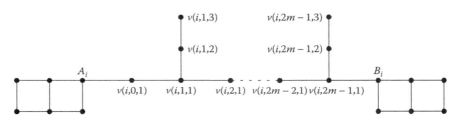

Figure 28.8 Graph $G'(v_i)$.

The graph $G(E)$ is obtained by

i. Taking m disjoint $G(C_j), 0 \leq j \leq m-1$;

ii. Taking n disjoint $G'(v_i), 0 \leq i \leq n-1$;

iii. For $k = 1, 2, 3$,

   identifying $v(i, 2j, k)$ with $c(j, l, k)$ if $c_{jl} = v_i$;

   identifying $v(i, 2j+1, k)$ with $c(j, l, k)$ if $c_{jl} = \overline{v_i}$;

   for each $c(j, l, 0), 0 \leq j \leq m-1, l = 0, 1, 2$, adding the edge $xc(j, l, 0)$ for all vertices x not in $G(C_j)$.

A vertex is of type k if it is of the form $c(j, l, k)$ for some j and some l. We denote by V_k the set of vertices of type k, $0 \leq k \leq 3$. Our construction is similar to [66], except that $G(v_i)$ is a chordless cycle in [66]. Figure 28.9 shows the interaction between a clause graph and a variable graph; for clarity we do not show all edges coming out of the vertices of type 0.

Remark 28.1 *A vertex $c(j, l, 0)$ (of type 0) is nonadjacent to exactly four vertices of $G(E)$: they are $c(j, l, k), 1 \leq k \leq 3$ and $c(j, l+1 \bmod 3, 2)$).*

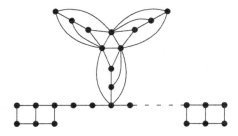

Figure 28.9 A portion of the graph $G(E)$.

Figure 28.10 Obstruction.

It is a routine but tedious matter to prove that $G(E)$ is weakly chordal. For detail, see [68].

For the rest of the proof, we will show that $G(E)$ is perfectly orderable if and only if E is satisfiable. It will be more convenient to work with orientations instead of orders. For an ordered graph, we may construct an oriented graph on the same vertex set as follows: If ab is an edge and $a < b$, then we add the arc $a \to b$. Thus, an obstruction is a $\vec{P_4}$ with vertices a, b, c, d and arcs $a \to b, b \to c, d \to c$ (see Figure 28.10). An orientation $\vec{G}$ of a graph G is *perfect* if it is acyclic and does not contain an induced obstruction. It is a routine matter to verify the following observation.

Observation 28.2 *The graph $G(v_j)$ admits a perfect orientation, but any perfect orientation of $G(v_j)$ is alternating on the path from A_j to B_j.* ∎

From now on, the argument of [66] carries through, for the sake of completeness we will complete the proof.

Claim 28.1 *If $G(E)$ admits a perfect orientation, then E is satisfiable.*

Proof. For each $i, 0 \leq i \leq n-1$, if the vertex $v(i, 0, 1)$ is a source in $G(v_i)$, then the variable v_i is assigned value true; otherwise, it is assigned value false. Note that, by Observation 28.2, $v(i, 0, 1)$ being a source (resp., sink) in $G(v_i)$ implies all $v(i, 2j, 1)$ are sources (resp., sink) in $G(v_i)$.

Consider the graph $G(C_j)$ with $C_j = (c_{j0}, c_{j1}, c_{j2})$. If all three vertices $c(j, l, 1), 0 \leq l \leq 2$, are sinks in the three corresponding graphs $G(v_i)$ where $c_{jl} = v_i$, or $c_{jl} = \overline{v}_i$, then we have $c(j, l, 2) \to c(j, l, 3)$, and thus $c(j, l, 0) \to c(j, l+1 \bmod 3, 0)$ for $0 \leq l \leq 2$; but then $\vec{G}$ is not acyclic, a contradiction. Thus, some $c(j, l, 1)$ is a source in $G(v_i)$ with $c_{jl} = v_i$ or $c_{jl} = \overline{v}_i$. If $c_{jl} = v_i$, then $c(j, l, 1) = v(i, 2j, 1)$ implying $v(i, 0, 1)$ is a source in $G(v_i)$, and thus v_i is true. Similarly, if $c_{jl} = \overline{v}_i$, then $v(i, 0, 1)$ is a sink, and thus v_i is false. In both cases, C_j is satisfied. ∎

Claim 28.2 *If E is satisfiable, then $G(E)$ admits a perfect orientation.*

Proof. Suppose there is a truth assignment of the variables $v_0, v_1, \ldots, v_{n-1}$ that satisfies E. For each variable graph $G(v_i)$, we assign a perfect orientation such that $v(i, 0, 1)$ is a source if and only if v_i is true. Such orientation exists by Observation 28.2.

Consider a clause graph $G(C_j)$ with $C_j = (c_{j0}, c_{j1}, c_{j2})$. Suppose c_{jl} is the ith variable, that is $c_{jl} = v_i$ or $\overline{v}_i$ ($0 \leq j \leq 2$). Then $c(j, l, 1) = v(i, 2j, 1)$ or $v(i, 2j+1, 1)$. If $c(j, l, 1)$ is a source in $G(v_i)$, then direct $c(j, l, 3) \to c(j, l, 2)$; otherwise, direct $c(j, l, 2) \to c(j, l, 3)$, and $c(j, l-1 \bmod 3, 0) \to c(j, l, 0)$. Since C_j contains a true literal, some $c(j, l, 0)$ is a source, and it follows that V_0 contains no directed cycle. Extend the partial orientation of V_0 into an acyclic orientation.

Now, for each edge ab, we direct $a \to b$ if $a \in V_0, b \notin V_0$; or if $a \in V_1, b \in V_2$. Every edge of G has been directed. Call the resulting directed graph $\vec{G}$. It is easy to see that $\vec{G}$ is acyclic.

Suppose $\vec{G}$ contains an obstruction P with vertices a, b, c, d and arcs $a \to b, b \to c, d \to c$. Because V_0 is a clique, P contains at most two vertices of type 0.

If P contains no vertex of type 0, then P must lie entirely in some $G'(v_i)$ because V_0 is a cutset of $G(E)$. But, clearly the orientation of every $G'(v_i)$ is perfect, a contradiction. Suppose P contains one vertex of type 0. The arcs $a \to b, d \to c$ imply $b, c \notin V_0$ by our construction. So, we may assume that $a \in V_0$ (for the rest of the proof, we will not argue on the direction of the arc $b \to c$). This means $a = c(j, l, 0)$ for some j and l. Since cd is an edge, we have $\{c, d\} \subset \{c(j, l, k) \mid 1 \le k \le 3\}$. Therefore, $b \in G(v_i)$ for some i such that v_i or $\bar{v}_i$ is a literal of the clause C_j. Thus, b is the vertex next to $c(j, l, 1) = v(i, r, 1)$ ($r = 2j$, or $r = 2j + 1$) on the path from A_i to B_i of $G(v_i)$. It follows that $c = c(j, l, 1), d = c(j, l, 2)$. But our construction implies $c(j, l, 1) \to c(j, l, 2)$, a contradiction.

Now, we may assume that P contains two vertices of type 0. Since V_0 is a clique, one of the two middle vertices of P must be of type 0. We may assume $b \in V_0$. Since $a \to b$, a must be in V_0. From Remark 28.1, P is the P_4 (i) $c(j, l, 0)c(j, l-1 \bmod 3, 0)c(j, l, 1)c(j, l, 2)$, or (ii) $c(j, l, 0)c(j, l-1 \bmod 3, 0)c(j, l, 3)c(j, l, 2)$. In case (i), our construction implies $c(j, l, 1) \to c(j, l, 2)$, a contradiction. In case (ii), the arc $c(j, l, 2) \to c(j, l, 3)$ implies $c(j, l, 1)$ is a sink in $G'(v_i)$ (for some appropriate i), and our construction implies $c(j, l-1 \bmod 3, 0) \to c(j, l, 0)$, a contradiction. ∎

28.7.3 Optimization

In this section, we consider the problems of finding a largest clique, a minimum coloring, a largest stable set, and a minimum clique-cover of perfectly ordered graphs. We note that these four problems (even in their weighted versions) for perfect graphs have been solved in [7]. This algorithm does not exploit the combinatorial structure of a perfect graph, instead it uses deep properties of the ellipsoid method. Thus, it is of interest to optimize the graphs discussed in this chapter by using combinatorial structures.

Theorem 28.33 [69] *Given a graph G and a perfect order on G, one can find in $O(n+m)$ time a minimum coloring and a largest clique of G.*

Proof. Let the vertices of G be $v_1, \ldots, v_n$ and the perfect order be $v_1 < \ldots < v_n$. We will show that the greedy coloring algorithm can be implemented in linear time on G. Vertices are colored in the order given by $<$. Suppose we are about to process vertex v_j. We find the smallest integer t such that no neighbor x of v_j has color t, and assign color t to v_j. The index t can be computed by traversing the adjacency list of v_j and computing the number a_i of neighbors of v_j with color i; t is the smallest index such that $a_t = 0$ (we may assume all the a_i are initially set to 0). At most $d(v_j)$ number a_i are modified in computing t. After v_j is colored, we reset these a_i to 0. So, the cost of coloring v_j is $O(d(v_j))$. Thus, we can color G in time $O(n + m)$.

From the proof of Theorem 28.31, we can extract a largest clique of G in linear time. Let k be the number of colors used by the greedy algorithm. We will show how to find a clique C with k vertices. Start with a vertex x of color k, put x in C. We go backward in $<$ to enlarge C. Suppose C contains vertices $w_i, w_{i+1}, \ldots, w_k$ with $i > 1$ and w_j having color j, $j = i, \ldots, k$. Let S_{i-1} be the set of vertices of color $i-1$. The proof of Theorem 28.31 implies there is a vertex $s \in S_{i-1}$ that is C-complete and so can be added to C. Such vertex can be found by scanning the adjacency list of every vertex x in S_{i-1} and computing the number of neighbors of x in C. The adjacency list of each vertex of G is scanned at most once, so the algorithm runs in linear time. ∎

Theorem 28.34 [69] *Given a graph G and a perfect order on its complement $\overline{G}$, one can find in $O(n+m)$ time a largest stable set and a minimum clique cover of G.*

Proof. Let the vertices of G be $v_1, \ldots, v_n$ and the perfect order on the complement of G be $v_1 < \ldots < v_n$. To stay within the linear-time bound, we will obviously not construct $\overline{G}$. We process the vertices in this order and produce a coloring of $\overline{G}$. Let the variable b_i count the number of vertices of color i. Suppose we are processing vertex v_j. Then v_j can be colored i if in $\overline{G}$, v_j is not adjacent to any vertex of color i, that is, in G, v_j is adjacent to b_i vertices of color i. This condition can be tested by scanning the adjacency list of v_j. If such color i exist, then we would choose the smallest such i for v_j; otherwise, we color v_j with a new color. The cost of coloring v_j is $O(d(v_j))$, so we can color $\overline{G}$ in $O(n+m)$ time. This coloring is a partition of G with a minimum number of cliques.

Now we show how to find a largest stable set of G. Let k be the number of colors used on $\overline{G}$ by the greedy algorithm. We will show how to find a stable set S of G with k vertices. Start with a vertex x of color k, put x in S. We go backward in $<$ to enlarge S. Suppose S contains vertices $w_i, w_{i+1}, \ldots, w_k$ with $i > 1$ and w_j having color j, $j = i, \ldots, k$. Let S_{i-1} be the set of vertices of color $i-1$. The proof of Theorem 28.31 implies there is a vertex $s \in S_{i-1}$ that is S-null and so can be added to S. Such vertex s can be found by scanning the adjacency list of every vertex s in S_{i-1}. The adjacency list of each vertex of G is scanned at most once, so the algorithm runs in linear time. ∎

Several classes $\mathcal{C}$ of perfectly orderable graphs have the property that if G is in $\mathcal{C}$ then not only that G is perfectly orderable, but its complement $\overline{G}$ also is (for example, brittle graphs, and therefore chordal graphs). Theorem 28.34 is useful for optimizing these graphs.

Corollary 28.8 [27] *There is a linear-time algorithm for finding a largest clique, a minimum coloring, a largest stable set, and a minimum clique cover for a chordal graph.*

Proof. Let G be a chordal graph with a perfect elimination scheme $<$. Then $<$ is a perfect order on $\overline{G}$, and the reverse of $<$ is a perfect order on G. The result follows from Theorems 28.33 and 28.34. ∎

A linear-time algorithm to recognize a co-chordal graph (complement of a chordal graph) and to construct a perfect order of such a graph is given in [70]. Thus, we have the following corollary.

Corollary 28.9 [70] *There is a linear-time algorithm for finding a largest clique, a minimum coloring, a largest stable set, and a minimum clique cover for a co-chordal graph.* ∎

Actually, for a perfectly ordered graph, there are algorithms to solve more general optimization problems. Consider the following.

Minimum weighted coloring. Given a weighted graph G such that each vertex x has a weight $w(x)$ which is a positive integer. Find stable sets $S_1, S_2, \ldots, S_k$ and integers $I(S_1), \ldots, I(S_k)$ such that for each vertex x we have $w(x) \leq \Sigma_{x \in S_i} I(S_i)$ and that the sum of the numbers $I(S_i)$ is minimized. This sum is called the weighted chromatic number and denoted by $\chi_w(G)$.

Maximum weighted clique. Given a weighted graph G such that each vertex x has a weight $w(x)$ which is a positive integer. Find a clique C such that $\Sigma_{x \in C} w(x)$ is maximized. This sum is called the weighted clique number and denoted by $\omega_w(G)$.

Definition 28.25 *A stable set of a graph G is* strong *if it meets all maximal cliques of G. (Here, as usual, Maximal is meant with respect to set-inclusion, and not size. In particular, a maximal clique may not be a largest clique.) A graph is* strongly perfect *if each of its induced subgraphs contains a strong stable set.*

Theorem 28.35 [65] *Perfectly orderable graphs are strongly perfect. And if a perfect order on G is given, then a strong stable set of G can be found in linear time.*

Proof. By induction on the number of vertices. We only need to prove that a graph G with a perfect order $<$ contains a strong stable set. Let S be the set of vertices colored with color 1 by the greedy algorithm. Assume that S is not a strong stable set, for otherwise we are done. So, consider a maximal clique C such that no vertex in C has color 1. Properties of the greedy algorithm implies each vertex $w \in C$ has a neighbor $p(w)$ of color 1 with $p(w) < w$. But then Lemma 28.15 implies some $p(w)$ is C-complete, a contradiction. The fact that S can be found in linear time follows from Theorem 28.33. ∎

Theorem 28.36 [71] *If there is a polynomial time algorithm A to find a strong stable set of a strongly perfect graph then there is a polynomial time algorithm B to find a minimum weighted coloring and maximum weighted clique of a strongly perfect graph. If algorithm A runs in time $O(f(n))$ then algorithm B runs in time $O(nf(n))$. Moreover if algorithm A is strongly polynomial then so is algorithm B.*

Proof. For a perfect graph G, it is known that $\chi_w(G) = \omega_w(G)$. Let G be a strongly perfect (and therefore, perfect) graph with a weight function w on its vertices. We will show the problem on G can be transformed to the problem on a smaller graph G' with an $O(f(n))$-time reduction. Suppose we can find a strong stable set S of G in $O(f(n))$ time. Let x be a vertex in S with the smallest weight among all vertices of S. Define a new weigh function $w'(v) = w(v) - w(x)$ for each $v \in S$, and $w'(v) = w(v)$ for each $v \in G - S$. Let $X = \{v | w'(v) = 0\}$. Since $x \in X$, X is not empty. Consider the graph $G' = G - X$. Since every maximal clique of G meets S, we have $\omega_w(G) = \omega_{w'}(G') + w(x)$, and thus, $\chi_w(G) = \chi_{w'}(G') + w(x)$. Suppose $S_1, \ldots, S_k$ is a minimum weighted coloring of G' with weights $I(S_i)$. Then $S_1, \ldots, S_k, S$ is a minimum weighted coloring of G with weights $I(S_i)$ for $i = 1, \ldots, k$, and $I(S) = w(x)$. Similarly, if C' is a maximum weighted clique of G', then a maximum weighted clique of G can be found as follows. If $C' \cap (S - X) \neq \emptyset$, then $C = C'$; otherwise, $C = C' \cup \{y\}$ where y is a vertex in X that is (C')-complete, y exists because S is a strong stable set (note that for C, we use the original weight function w).

We may recursively apply the above reduction until we get a trivial graph in at most n steps. Since the complexity of our procedure does not depend on the size of the number $w(v)$, the reduction is strongly polynomial. ∎

Theorems 28.35 and 28.36 implies the following.

Corollary 28.10 *Given a graph G and a perfect order on G, maximum weighted clique and minimum weighted coloring can be solved in $O(nm)$ time.* ∎

For comparability and chordal graphs, these two problems can be solved even faster.

Theorem 28.37 [71] *If G is a comparability graph or a chordal graph, then maximum weighted clique and minimum weighted coloring can be solved in $O(n^2)$ time.* ∎

Space-efficient algorithms for maximum weighted clique and minimum weighted coloring of co-chordal graphs are given in [70]. Theorem 28.36 shows that the problem of finding a strong stable set of a strongly perfect graph is of some consequence. However, no polynomial algorithm for solving this problem is known. Finding a strong stable set of an arbitrary graph is NP-hard [71].

28.8 PERFECTLY CONTRACTILE GRAPHS

Recall the definition of an even-pair in Section 28.6. Even-pairs play a central role in the study of perfect graphs, as illustrated by the following two results.

Lemma 28.16 [60] *Let G be a perfect graph with an even-pair $\{x, y\}$. Then G/xy is perfect.* ∎

Lemma 28.17 [72] *No minimal imperfect graph contains an even-pair.* ∎

From the above, it is of interest to know which perfect graphs contain even-pairs.

Definition 28.26 *A graph G is* even-contractile *if there is a sequence $G_0 = G, G_1, \ldots, G_k$ such that G_k is a clique, and for $i \leq k-1$, G_{i+1} is obtained from G_i by a contraction of some even-pair of G_i.*

An even-contractile graph G has $\chi(G) = \omega(G)$ by Lemma 28.13. But this class seems to be difficult to characterize; perhaps because the class is not hereditary. Now, consider the following definition from [73].

Definition 28.27 *A graph is* perfectly contractile *if each of its induced subgraphs is even-contractile.*

By Lemma 28.13, perfectly contractile graphs are perfect. Most classes of graphs discussed in this chapter are perfectly contractile. Lemma 28.14 implies the following.

Theorem 28.38 [52] *Weakly chordal graphs are perfectly contractile.* ∎

A graph is called a *Meyniel* graph if each of its odd cycle with at least five vertices has two chords. Perfection of Meyniel graphs was established in [74]. Note that chordal graphs are Meyniel graphs.

Theorem 28.39 [75] *Meyniel graphs are perfectly contractile.* ∎

Theorem 28.40 [76] *Perfectly orderable graphs perfectly contractile.* ∎

Definition 28.28 *A* prism *is a graph that consists of two vertex-disjoint triangles (cliques of size three) and three vertex-disjoint paths, each of length at least one and having an endpoint in each triangle, with no other edge than those in the two triangles and in the three paths. A prism is odd if all three paths are odd.*

The following beautiful and challenging conjecture was proposed in [77].

Conjecture 28.1 [77] *A graph is perfectly contractile if and only if it contains no odd hole, no anti-hole, and no odd prism.*

Definition 28.29 *A graph is an* Artemis *graph if it contains no odd hole, no anti-hole, and no prism.*

Validity of Conjecture 28.1 was partially established by the following remarkable result.

Theorem 28.41 [78] *Artemis graphs are perfectly contractile.* ∎

An $O(n^2 m)$ time algorithm to color an Artemis graph is given in [79]. Note that weakly chordal graphs and perfectly orderable graphs are Artemis graphs. An $O(n^9)$ time algorithm for recognizing an Artemis graph is given in [80].

28.9 RECOGNITION OF PERFECT GRAPHS

In this section, we give a sketch of a polynomial time algorithm to recognize a perfect graph. By the strong perfect graph theorem, the problem is equivalent to determining if a graph is Berge (graphs with no odd holes and no odd anti-holes). A polynomial time algorithm to solve this problem is given in [15]. The algorithm can be divided into three phases. In the first phase, given a graph G, the algorithm looks for one of five configurations. Each of these five configurations can be detected in time $O(n^9)$ or faster. If G contains one of these, then G is not Berge; otherwise, every shortest odd hole of G has a special property called *amenable*. Given an odd hole C of length at least seven, a set X of vertices is a near-cleaner if it contains all vertices that have two neighbors of distance at least three in C and $X \cap C$ is a subset of the vertex set of some path of length three of C. Amenable odd holes are those odd holes such that all near-cleaners have some special adjacency property (definitions not given here will be given later). If the first phase does not produce an odd hole or odd antihole, the second phase will generate $O(n^5)$ sets that are guaranteed to contain all near-cleaners of some amenable odd hole if one exists. Finally, the third phase provides an $O(n^4)$ algorithm that given a graph and a near-cleaner for a shortest odd hole finds an odd hole. Now, we describe the algorithm in more detail.

Definition 28.30 *A* pyramid *is an induced subgraph formed the union of a triangle* $\{b_1, b_2, b_3\}$, *a fourth vertex* a, *and three induced paths* P_1, P_2, P_3, *satisfying:*

- *For $i = 1, 2, 3$, the endpoints of P_i are a, b_i.*

- *For $i \leq i < j \leq 3$, a is the only vertex in both P_i, P_j, and $b_i b_j$ is the only edge between $P_i - a$ and $P_j - a$.*

- *a is adjacent to at most one of b_1, b_2, b_3.*

Definition 28.31 *A* jewel *is the graph formed by a cycle with vertices $v_1, v_2, \ldots, v_5$ and edges $v_i v_{i+1}$ (with the subscript taken modulo 5) and an induced path P such that $v_1 v_3, v_2 v_4, v_1 v_4$ are nonedges, v_1, v_4 are the endpoints of P, and there is no edges between $\{v_2, v_3, v_5\}$ and the interior vertices of P.*

Definition 28.32 *A configuration of type $\mathcal{T}_1$ is the hole on five vertices.*

Definition 28.33 *A configuration of type $\mathcal{T}_2$ is a sequence v_1, v_2, v_3, v_4, P, X such that*

- v_1, v_2, v_3, v_4 *induce a P_4 with endpoints v_1, v_4,*

- X *is an anticomponent of the set of all $\{v_1, v_2, v_4\}$-complete vertices,*

- P *is an induced path in $G \setminus (X \cup \{v_2, v_3\})$ between v_1, v_4, and no interior vertex of P is X-complete or adjacent to v_2 or adjacent to v_3.*

Definition 28.34 *A configuration of type $\mathcal{T}_3$ is a sequence $v_1, \ldots, v_6, P, X$ such that*

- $v_1, \ldots, v_6$ *are distinct vertices*

- $v_1 v_2, v_3 v_4, v_1 v_4, v_2 v_3, v_3 v_5, v_4 v_6$ *are edges, and $v_1 v_3, v_2 v_4, v_1 v_5, v_2 v_5, v_1 v_6, v_2 v_6, v_4 v_5$ are nonedges*

- X *is an anticomponent of the set of all $\{v_1, v_2, v_5\}$-complete vertices, and v_3, v_4 are not X-complete*

- P is an induced path of $G \setminus (X \cup \{v_1, v_2, v_3, v_4\})$ between v_5, v_6, and no interior vertex of P is X-complete or adjacent to v_1 or adjacent to v_2
- If $v_5 v_6$ is an edge, then v_6 is not X-complete

In [15], it is shown that a pyramid can be detected in $O(n^9)$ time, a jewel in $O(n^6)$ time, a configuration of type $\mathcal{T}_1$ in $O(n^5)$ time (obviously), a configuration of type $\mathcal{T}_2$ or $\mathcal{T}_3$ in $O(n^6)$ time.

Theorem 28.42 [15] *If G or $\overline{G}$ contains a pyramid, a jewel, or a configuration of type $\mathcal{T}_1$, $\mathcal{T}_2$, or $\mathcal{T}_3$, then G is not Berge.* ∎

Given a hole C of length at least seven, a vertex x is C-*major* if x has two neighbors in C whose distance in C is at least three. A hole C of G is *amenable* if (i) C is a shortest odd hole of length at least seven of G, and (ii) for every anticonnected set X of C-major vertices, there is an X-complete edge in C.

Theorem 28.43 [15] *If G contains no pyramid, and no configuration of type $\mathcal{T}_1$, $\mathcal{T}_2$, or $\mathcal{T}_3$, and both $G, \overline{G}$ contains no jewel, then every shortest odd hole of G is amenable.* ∎

Recall that a set X of vertices is a *near-cleaner* for an odd hole C of length at least seven if it contains all C-major vertices, and $X \cap C$ is a subset of the vertex set of some path of length three of C.

Theorem 28.44 [15] *There is an $O(n^5)$ algorithm which given a graph G outputs $O(n^5)$ subsets of $V(G)$ such that if C is an amenable odd hole of G, then one of the subsets is a near-cleaner for C.* ∎

Theorem 28.45 [15] *There is an $O(n^4)$ algorithm which given a graph G containing no pyramid or jewel, and a subset X of $V(G)$ outputs an odd hole, or determines that there is no shortest odd hole C of G such that X is a near-cleaner for C.* ∎

The steps needed to recognize a perfect graph are described in Algorithm 28.8. There are two bottlenecks to making the algorithm run faster than $O(n^9)$ time.

Algorithm 28.8 perfect graph recognition

input: graph G
output: a determination that G is Berge or not

(1) Determine if G or $\overline{G}$ contains a pyramid, or a jewel, or a configuration of type $\mathcal{T}_1$, $\mathcal{T}_2$, or $\mathcal{T}_3$. If it does, output G *is not Berge*, and stop
(2) Produce $O(n^5)$ subsets X of $V(G)$ using Theorem 28.44. These subsets contain all near-cleaners of some odd hole of G, if such an odd hole exists
(3) For each subset X of (2), run the algorithm of Theorem 28.45. If an odd hole is produced, output G *is not Berge*, and stop
(4) Run (2) and (3) with G replaced by $\overline{G}$
(5) Output G *is Berge*

The first one is that as of present, there is no algorithm to detect a pyramid in time faster than $O(n^9)$. The second involves the near-cleaners. It is not known if given a near-cleaner, one can find an odd hole in time faster than $O(n^4)$. It is also not known if a graph can have fewer than $O(n^5)$ near-cleaners.

28.10 χ-BOUNDED GRAPHS

Definition 28.35 *A graph G is χ-bounded if there is a function f such that $\chi(G) \leq f(\omega(G))$.*

We have seen that perfect graphs are χ-bounded. One may wonder about sufficient conditions on the holes of a graph for it to be χ-bounded. Some interesting conditions have been found.

Theorem 28.46 [16] *If a graph G is even hole-free, then G contains a vertex whose neighborhood can be partitioned into two cliques. In particular, G satisfies $\chi(G) \leq 2\omega(G) - 1$.* ∎

References [81–83] give two different polynomial time algorithms (of high complexity) for finding an even hole in a graph.

It is reasonable to expect that graphs without odd holes have bounded chromatic number. Before discussing this matter, we will need a definition.

Definition 28.36 *A k-division of a graph G with at least one edge is a partition of $V(G)$ into k sets $V_1, \ldots, V_k$ such that no V_i contains a clique with $\omega(G)$ vertices. A graph is k-divisible if each induced subgraph of G with at least one edge admits a k-division.*

It is easy to see the following.

Lemma 28.18 *A k-divisible graph G has $\chi(G) \leq k^{\omega(G)-1}$.* ∎

Consider the following conjectures.

Conjecture 28.2 [84] *A graph is 2-divisible if and only if it is odd hole-free.*

The above conjecture implies that an odd hole-free graph G has $\chi(G) \leq 2^{\omega(G)-1}$, and thus is χ-bounded. The conjecture is known to hold for claw-free graphs [84], $2K_2$-free graphs [17], and K_4-free graphs [85]. The problem of recognizing odd hole-free graphs is open.

We now mention a number of conjectures related to χ-bounded graphs and forbidden subgraphs.

Conjecture 28.3 [84] *Let F be any forest on k vertices. Then any graph G that does not contain F as induced subgraph is k-divisible.*

It is not known if Conjecture 28.3 holds for claw-free graphs.

Definition 28.37 *Let G be a graph with at least one hole. The hole number $h(G)$ of G is the length of the longest hole in G.*

Conjecture 28.4 [84] *Let G be a graph with at least one hole. Then G is $(h(G)-2)$-divisible.*

The following special case of Conjecture 28.4 is still open.

Conjecture 28.5 [84] *If G is a triangle-free graph with at least one hole, then $\chi(G) \leq h(G) - 2$.*

References

[1] C. Berge. Les problèmes de colorations en théorie des graphes. *Publications de l'Institut de Statistique de l'Université de Paris*, **IX** (1960), 123–160.

[2] C. E. Shannon. The zero-error capacity of a noisy channel. *IRE Trans. Inform. Th*, **2** (1956), 8–19.

[3] A. Hajnal and J. Surányi. Uber die auflösung von graphen in vollständige teilgraphen. *Ann. Univ. Sci. Budapest Eötvös. Sect. Math*, **1** (1958), 113–121.

[4] R. P. Dilworth. A decomposition theorem for partially ordered sets. *Annals of Mathematics*, **51** (1950), 161–166.

[5] D. König. Graphs and matrices. *Mat. Lapok*, **38** (1931), 116–119.

[6] L. Lovász. On the shannon capacity of a graph. *IEEE Trans. Inform. Th. IT*, **25** (1979), 1–7.

[7] M. Grötschel, L. Lovász, and A. Schrijver. Polynomial algorithms for perfect graphs. In Berge and Chvátal, editors, *Topics on Perfect Graphs*, pages 325–356. North-Holland Mathematics Studies, 1984.

[8] V. Raghavan and J. Spinrad. Robust algorithms for restricted domains. *Journal of Algorithms*, **48** (2003), 160–172.

[9] B. A. Reed. A gentle introduction to semi-definite programming. In *Perfect Graphs*, pages 67–92.

[10] A. Brandstädt, V. B. Le, and J. P. Spinrad. *Graph Classes: A Survey*. SIAM Monographs on Discrete Mathematics and Applications, Society for Industrial and Applied Mathematics, 1999.

[11] M. C. Golumbic. *Algorithmic Graph Theory and Perfect Graphs*. Academic Press, New York, 1980.

[12] J. L. Ramírez-Alfonsín and B. A. Reed, editors. *Perfect Graphs*. Wiley, 2001.

[13] A. Ghouila-Houri. Caractérisation des graphes non orientés dont on peut orienter les arêtes de manière à obtenir le graphe d'une relation d'ordre. *C. R. Acad. Sci. Paris*, **254** (1962), 1370–1371.

[14] T. Gallai. Transitiv orientierbare graphen. *Acta Math. Acad. Sci. Hungar*, **18** (1967), 25–66.

[15] M. Chudnovsky, G. Cornuéjols, X. Liu, P. Seymour, and K. Vuskovic. Recognizing berge graphs. *Combinatorica*, **25** (2005), 143–186.

[16] L. Addario-Berry, M. Chudnovsky, F. Havet, B. Reed, and P. Seymour. Bisimplicial vertices in even-hole-free graphs. *Journal of Combininatorial Theory Series B*, **98** (2008), 1119–1164.

[17] C. T. Hoàng and C. McDiarmid. A note on the divisibility of graphs. In *Congressus Numerantium 136*, pages 215–219. Proceedings of the 30th Southeastern International Conference on Combinatorics, Graph Theory, and Computing, 1999.

[18] D. Coppersmith and S. Winograd. Matrix multiplication via arithmetic progressions. *Journal of Symbolic Computation*, **9** (1990), 251–280.

[19] J. R. S. Blair and B. W. Peyton. An introduction to chordal graphs and clique trees. In *Graph Theory and Sparse Matrix Computation*, pages 1–29. Springer, New York, 1993.

[20] P. Buneman. A charactarization of rigid circuit graphs. *Discrete Mathematics*, **9** (1990), 205–212.

[21] G. A. Dirac. On rigid circuit graphs. *Abh. Math. Sem. Univ. Hamburg*, **25** (1961), 71–76.

[22] D. R. Fulkerson and O. A. Gross. Incidence matrices and interval graphs. *Pacific Journal of Mathematics*, **15** (1965), 835–855.

[23] F. Gavril. The intersection graphs of subtrees in trees are exactly the chordal graphs. *Journal of Combinatorial Theory Series B*, **16** (1974), 47–56.

[24] J. R. Walter. Representations of chordal graphs as subtrees of a tree. *Journal of Graph Theory*, **2** (1978), 265–267.

[25] D. J. Rose, R. E. Tarjan, and G. S. Leuker. Algorithmic aspects of vertex elimination on graphs. *SIAM Journal on Computing*, **5** (1976), 266–283.

[26] R. E. Tarjan and M. Yannakakis. Simple linear-time algorithms to test chordality of graphs, test acyclicity of hypergraphs, and selectively reduce acyclic hypergraphs. *SIAM Journal on Computing*, **13** (1984), 566–579.

[27] F. Gavril. Algorithms for minimum coloring, maximum clique, minimum cover by cliques and maximum independent set of a chordal graphs. *SIAM Journal on Computing*, **1** (1972), 180–187.

[28] R. E. Tarjan. Decomposition by clique separators. *Discrete Mathematics*, **55**(2) (1985), 221–232.

[29] S. H. Whiteside. An algorithm for finding clique cut-sets. *Information Processing Letters*, **12**(1) (1981), 31–32.

[30] S. H. Whitesides. A method for solving certain graph recognition and optimization problems, with applications to perfect graphs. In *Topics on Perfect Graphs. Annals of Discrete Mathematics*, pages 281–297.

[31] F. Maffray and M. Preissmann. A translation of gallai's paper:'transitiv orientierbare graphen'. In *Perfect Graphs*, pages 25–66.

[32] R. M. McConnell and J. P. Spinrad. Modular decomposition and transitive orientation. *Discrete Mathematics*, **201** (1999), 189–241.

[33] J. P. Spinrad. On comparability and permutation graphs. *SIAM Journal on Computing*, **14** (1985), 658–670.

[34] J. P. Spinrad. Problems (14a) and (14b). In *Efficient Graph Representations*. 2003.

[35] E. Egerváry. On combinatorial properties of matrices. *Mat. Lapok*, **38** (1931), 16–28.

[36] V. Chvátal. *Linear programming*. W. H. Freeman and Company, New York, 1983.

[37] G. B. Dantzig and D. R. Fulkerson. Minimizing the number of tankers to meet a fixed schedule. *Naval Research Logistics Quarterly*, **1** (1954), 217–222.

[38] D. R. Fulkerson. Note on dilworth's decomposition theorem for partially ordered sets. In *Proceedings of the American Mathematical Society*, pages 701–702, 1956.

[39] H. Alt, N. Blum, K. Mehlhorn, and M. Paul. Computing a maximum cardinality matching in a bipartite graph in time $o(n^{1.5}\sqrt{\frac{m}{\log n}})$. *Information Processing Letters*, **37** (1991), 237–240.

[40] T. Feder and R. Motwani. Clique partitions, graph compression, and speeding up algorithms. In *Proceedings of the 23rd Annual ACM Symposium on Theory of Computing*, pages 123–133, 1991.

[41] P. C. Gilmore and A. J. Hoffman. A characterization of comparability graphs and interval graphs. *Canadian Journal of Mathematics*, **16** (1964), 539–548.

[42] C. G. Lekkerkerker and J. Boland. Representation of a finite graph by a set of intervals on the real line. *Fund. Math*, **51** (1962), 45–64.

[43] K. S. Booth and G. Lueker. Testing for the consecutive ones property, interval graphs, and graph planarity using pq-tree algorithms. *Journal of Computer and System Science*, **13** (1976), 335–379.

[44] N. Korte and R. H. Möhring. An incremental linear-time algorithm for recognizing interval graphs. *SIAM Journal on Computing*, **18** (1989), 68–81.

[45] W. L. Hsu and T. H. Ma. Fast and simple algorithms for recognizing chordal comparability graphs and interval graphs. *SIAM Journal on Computing*, **28** (1999), 1004–1020.

[46] M. Habib, R. McConnell, C. Paul, and L. Viennot. LEX_BFS and partition refinement, with applications to transitive orientation, interval graph recognition, and consecutive ones testing. *Theoretical Computer Science*, **234** (2000), 59–84.

[47] S. Olariu. An optimal greedy heuristic to color interval graphs. *Information Processing Letters*, **37** (1991), 65–80.

[48] G. Ramalingam and C. Pandurangan. A uniform approach to domination problems on interval graphs. *Information Processing Letters*, **27** (1988), 271–274.

[49] A. Raychaudhuri. On powers of interval and unit interval graphs. *Congressus Numerantium*, **59** (1987), 235–242.

[50] D. G. Corneil, S. Olariu, and L. Stewart. The LBFS structure and recognition of interval graphs. *SIAM Journal on Discrete Mathematics*, **23** (2009/10), 1905–1953.

[51] J. P. Spinrad. *Efficient Graph Representations*. Fields Institute Monographs, American Mathematical Society, 2003.

[52] R. B. Hayward, C. T. Hoàng, and F. Maffray. Optimizing weakly triangulated graphs. *Graphs and Combinatorics*, **5** (1989), 339–349.

[53] R. B. Hayward. Meyniel weakly triangulated graphs i. Co-perfect orderability. *Discrete Applied Mathematics*, **73** (1997), 199–210.

[54] R. B. Hayward. Meyniel weakly triangulated graphs ii: A theorem of dirac. *Discrete Applied Mathematics*, **78** (1997), 283–289.

[55] R. B. Hayward, J. P. Spinrad, and R. Sritharan. Improved algorithms for weakly chordal graphs. *ACM Transactions on Algorithms*, **3**(2) (2007).

[56] J. P. Spinrad. Finding large holes. *Information Processing Letters*, **39** (1991), 227–229.

[57] S. D. Nikolopoulos and L. Palios. Hole and antihole detection in graphs. In *Proceedings of the 15th Annual ACM-SIAM Symposium on Discrete Algorithms*, pages 843–852, 2004.

[58] A. Berry, J. P. Bordat, and P. Heggernes. Recognizing weakly triangulated graphs by edge separability. *Nordic Journal on Computing*, **7** (2000), 164–177.

[59] J. P. Spinrad and R. Sritharan. Algorithms for weakly triangulated graphs. *Discrete Applied Mathematics*, **19** (1995), 181–191.

[60] J. Fonlupt and J. P. Uhry. Transformations which preserve perfectness and h-perfectness of graphs. *Annals of Discrete Mathematics*, **16** (1982), 83–85.

[61] A. Berry, A. Sigayret, and C. Sinoquet. Maximal sub-triangulation as improving phylogenetic data. Technical report, RR-02-02, LIMOS, Clermont-Ferrand, France, 2002.

[62] S. Arikati and C. Rangan. An efficient algorithm for finding a two-pair, and its applications. *Discrete Applied Mathematics*, **31** (1991), 71–74.

[63] D. Kratsch and J. P. Spinrad. Between o(mn) and o(n^α). In *Proceedings of the 14th Annual ACM-SIAM Symposium on Discrete Algorithms*, pages 158–167, 2003.

[64] R. B. Hayward. Weakly triangulated graphs. *Journal of Combinatorial Theory Series B*, **39** (1985), 200–209.

[65] V. Chvátal. Perfectly ordered graphs. In C. Berge and V. Chvátal, editors, *Topics on Perfect Graphs*, pages 63–65, Annals of Discrete Mathematics, Vol. 21, 1984.

[66] M. Middendorf and F. Pfeiffer. On the complexity of recognizing perfectly orderable graphs. *Discrete Mathematics*, **80** (1990), 327–333.

[67] C. T. Hoàng. Perfectly orderable graphs: a survey. In J. L. Ramirez Alfonsin and B. A. Reed, editors, *Perfect Graphs*, pages 139–166. John Wiley & Sons, 2001.

[68] C. T. Hoàng. The complexity of recognizing weakly triangulated graphs that are perfectly orderable. Technical Report Report No. 90638, Institute for Discrete Mathematics, University of Bonn, Germany, 1990.

[69] V. Chvátal, C. T. Hoàng, N. V. R. Mahadev, and D. deWerra. Four classes of perfectly orderable graphs. *Journal of Graph Theory*, **11** (1987), 481–495.

[70] C. T. Hoàng. Recognition and optimization algorithms for co-triangulated graphs. Technical report, Institute for Discrete Mathematics, University of Bonn, Germany, Report No. 90637, 1990.

[71] C. T. Hoàng. Efficient algorithms for minimum weighted colouring of some classes of perfect graphs. *Discrete Applied Mathematics*, **55** (1994), 133–143.

[72] H. Meyniel. A new property of critical imperfect graphs and some consequences. *European Journal of Combinatorics*, **8** (1987), 313–316.

[73] M. E. Bertschi. Perfectly contractile graphs. *Journal of Combinatorial Theory Series B*, **50** (1990), 222–230.

[74] H. Meyniel. On the perfect graph conjecture. *Discrete Mathematics*, **16**(4) (1976), 339–342.

[75] A. Hertz. A fast algorithm for coloring meyniel graphs. *Journal of Combinatorial Theory Series B*, **50** (1990), 231–240.

[76] A. Hertz and D. de Werra. Perfectly orderable graphs are quasi-parity graphs: A short proof. *Discrete Mathematics*, **68** (1988), 111–113.

[77] H. Everett, C. M. H. de Figueiredo, C. Linhares-Sales, F. Maffray, O. Porto, and B. Reed. Even pairs. In *Perfect Graphs*, pages 67–92.

[78] F. Maffray and N. Trotignon. A class of perfectly contractile graphs. *Journal of Combinatorial Theory Series B*, **96** (2006), 1–19.

[79] B. Lévêque, F. Maffray, B. Reed, and N. Trotignon. Coloring artemis graphs. *Theoretical Computer Science*, **410** (2009), 2234–2240.

[80] F. Maffray and N. Trotignon. Algorithm for perfectly contractile graphs. *SIAM Journal on Discrete Mathematics*, **19** (2005), 553–574.

[81] M. Chudnovsky, K. Kawarabayashi, and P. Seymour. Detecting even holes. *Journal of Graph Theory*, **48** (2005), 85–111.

[82] M. Conforti, G. Cornuéjols, A. Kapoor, and K. Vušković. Even-hole-free graphs, part i: Decomposition theorem. *Journal of Graph Theory*, **39** (2002), 6–49.

[83] M. Conforti, G. Cornuéjols, A. Kapoor, and K. Vušković. Even-hole-free graphs, part ii: Recognition algorithm. *Journal of Graph Theory*, **40** (2002), 238–266.

[84] C. T. Hoàng and C. McDiarmid. On the divisibility of graphs. *Discrete Mathematics*, **242** (2002), 145–156.

[85] M. Chudnovsky, N. Robertson, P. Seymour, and R. Thomas. K_4-free graphs with no odd holes. *Journal of Combinatorial Theory Series B*, **100** (2010), 313–331.

[86] C. Berge and V. Chvátal, editors. *Topics on Perfect Graphs*. Annals of Discrete Mathematics, Vol. 21. North Holland, Amsterdam, the Netherlands, 1984.

[87] M. Chudnovsky, N. Robertson, P. Seymour, and R. Thomas. The strong perfect graph theorem. *Annals of Mathematics*, **64** (2006), 51–229.

[88] L. Lovász. Normal hypergraphs and the perfect graph conjecture. *Discrete Math*, **2** (1972), 253–267.

CHAPTER 29

Tree-Structured Graphs

Andreas Brandstädt

Feodor F. Dragan

CONTENTS

29.1	Graphs with Tree Structure, Related Graph Classes, and Algorithmic Implications	752
29.2	Chordal Graphs and Variants	753
	29.2.1 Chordal Graphs	753
	29.2.2 Some Subclasses of Chordal Graphs	756
29.3	α-Acyclic Hypergraphs and Their Duals	757
	29.3.1 Motivation from Relational Database Theory	757
	29.3.2 Some Basic Hypergraph Notions	759
	29.3.3 Hypergraph 2-Coloring	763
	29.3.4 Kőnig Property	764
	29.3.5 α-Acyclic Hypergraphs and Tree Structure	764
	29.3.6 Graham's Algorithm, Running Intersection Property, and Other Desirable Properties Equivalent to α-Acyclicity	767
	29.3.7 Dually Chordal Graphs, Maximum Neighborhood Orderings, and Hypertrees	770
	29.3.8 Bipartite Graphs, Hypertrees, and Maximum Neighborhood Orderings	773
	29.3.9 Further Matrix Notions	775
29.4	Totally Balanced Hypergraphs and Matrices	776
	29.4.1 Totally Balanced Hypergraphs versus β-Acyclic Hypergraphs	776
	29.4.2 Totally Balanced Matrices	778
29.5	Strongly Chordal and Chordal Bipartite Graphs	779
	29.5.1 Strongly Chordal Graphs	779
	29.5.1.1 Elimination Orderings of Strongly Chordal Graphs	779
	29.5.1.2 Γ-Free Matrices and Strongly Chordal Graphs	782
	29.5.1.3 Strongly Chordal Graphs as Sun-Free Chordal Graphs	783
	29.5.2 Chordal Bipartite Graphs	786
29.6	Tree Structure Decomposition of Graphs	788
	29.6.1 Cographs	788
	29.6.2 Optimization on Cographs	790
	29.6.3 Basic Module Properties	791
	29.6.4 Modular Decomposition of Graphs	793
	29.6.5 Clique Separator Decomposition of Graphs	794
29.7	Distance-Hereditary Graphs, Subclasses, and γ-Acyclicity	794
	29.7.1 Distance-Hereditary Graphs	794
	29.7.2 Minimum Cardinality Steiner Tree Problem in Distance-Hereditary Graphs	799

	29.7.3	Important Subclasses of Distance-Hereditary Graphs	801
	29.7.3.1	Ptolemaic Graphs and Bipartite Distance-Hereditary Graphs	801
	29.7.3.2	Block Graphs	802
	29.7.3.3	γ-Acyclic Hypergraphs	802
29.8	Treewidth and Clique-Width of Graphs		803
	29.8.1	Treewidth of Graphs	803
	29.8.2	Clique-Width of Graphs	805
29.9	Complexity of Some Problems on Tree-Structured Graph Classes		807
29.10	Metric Tree-Like Structures in Graphs		808
	29.10.1	Tree-Breadth, Tree-Length, and Tree-Stretch of Graphs	808
	29.10.2	Hyperbolicity of Graphs and Embedding Into Trees	810

29.1 GRAPHS WITH TREE STRUCTURE, RELATED GRAPH CLASSES, AND ALGORITHMIC IMPLICATIONS

The aim of this chapter is to present various aspects of tree structure in graphs and hypergraphs and its algorithmic implications together with some important graph classes having nice and useful tree structure. In particular, we describe the hypergraph background and the tree structure of chordal graphs (introduced in Chapter 28) and some graph classes which are closely related to chordal graphs such as chordal bipartite graphs, dually chordal graphs, and strongly chordal graphs as well as important subclasses.

As already defined in Chapter 28, a graph is *chordal* if each of its induced cycles has only three vertices (i.e., each cycle with at least four vertices has a so-called *chord*). The study of chordal graphs goes back to [1], and the many aspects of chordal graphs are described in surveys and monographs such as [2–5] and others. The interest in chordal graphs and related classes comes from applications in computer science, in particular, relational database schemes [6,7], matrix analysis, models in biology, statistics, and others. Chordal graphs are closely related to the famous concept of *treewidth* introduced by Robertson and Seymour [8] but appears also under the name of *partial k-trees* in [9, 10] (see, e.g., [11]). The notion of treewidth plays a central role in algorithmic and complexity aspects on graphs.

Chordal graphs appear in the literature under different names such as *triangulated graphs* (Chapter 4 of [4]), *rigid-circuit graphs*, *perfect elimination graphs* and others. Most of the applications are due to the tree structure of chordal graphs which can be described in terms of so-called *clique trees* (arranging the maximal cliques of the graph in a tree).

The hypergraph-theoretical background of chordal graphs is given by α-acyclic hypergraphs which play an important role in the theory of relational database schemes. Various desirable properties of such schemes can be expressed in terms of various levels of acyclicity of hypergraphs [6,7]: Chordal graphs correspond to α-acyclic hypergraphs, dually chordal graphs correspond to the dual hypergraphs of α-acyclic hypergraphs, strongly chordal graphs correspond to β-acyclic hypergraphs (which are equivalent to totally balanced hypergraphs), ptolemaic graphs correspond to γ-acyclic hypergraphs, and block graphs correspond to Berge-acyclic hypergraphs. Actually, tree structure of hypergraphs was captured as *arboreal hypergraphs* by Berge [12,13]; a hypergraph is α-acyclic if and only its dual is arboreal.

We discuss also another width parameter of graphs, namely clique-width, and its relationship to treewidth as well as its algorithmic applications. Very similar to treewidth, it is known that whenever a problem is expressible in a certain kind of Monadic Second-Order Logic, and one deals with a class of graph whose clique-width is bounded by a constant then the problem is efficiently solvable on this class. This is one of the main reasons for the great interest in treewidth and clique-width of (special) graphs. In general, it is NP-hard to determine the clique-width of a graph, and for many important graph classes, the clique-width is unbounded. For some interesting classes, however, clique-width is bounded.

Finally, we discuss some other graph parameters, namely, the tree-length and the tree-breadth of a graph, the tree-distortion and the tree-stretch of a graph, the Gromov's hyperbolicity of a graph. All these parameters try to capture and measure tree likeness of a graph from a metric point of view. The smaller such a parameter is for a graph, the closer graph is to a tree metrically. Graphs for which such parameters are bounded by small constants have many algorithmic advantages; they allow efficient approximate solutions for a number of optimization problems. Note also that recent empirical and theoretical work has suggested that many real-life complex networks and graphs arising in Internet applications, in biological and social sciences, in chemistry and physics have tree-like structures from a metric point of view.

29.2 CHORDAL GRAPHS AND VARIANTS

In this section, we collect some notions and well-known facts on chordal graphs which are described in Chapter 28 (see also the monograph [4] and the survey [3] as well as [5] for details). In order to make this section self-contained, we briefly repeat some of the basic definitions and properties. Throughout this section, let $G = (V, E)$ be a finite undirected graph which is simple (i.e., loop-free and without multiple edges).

29.2.1 Chordal Graphs

Definition 29.1 *A graph is* chordal *if it does not contain any chordless cycle with at least four vertices.*

Obviously, trees and forests are chordal since they are cycle-free for any cycle length. Chordal graphs have a nice separator property which was found by Dirac [14].

Definition 29.2

  i. *The vertex set $S \subseteq V$ is a* separator *(or* cutset*) for nonadjacent vertices $a, b \in V$ (a–b-separator) if a and b are in different connected components in $G[V \setminus S]$.*

  ii. *S is a* minimal *a–b-separator if S is an a–b-separator and no proper subset of S is an a–b-separator.*

  iii. *S is a (*minimal*) separator if there are vertices a, b such that S is a (*minimal*) a–b-separator.*

Theorem 29.1 [14] *A graph G is chordal if and only if every minimal separator in G induces a clique.* ∎

Definition 29.3

  i. *A vertex $v \in V$ is* simplicial *in G if $N(v)$ induces a clique in G.*

  ii. *An ordering $(v_1, \ldots, v_n)$ of the vertices of V is a* perfect elimination ordering *(p.e.o.) of G if for all $i \in \{1, \ldots, n\}$, the vertex v_i is simplicial in the remaining subgraph $G_i := G[\{v_i, \ldots, v_n\}]$.*

Obviously, the notion of a simplicial vertex generalizes leaves in trees.

Lemma 29.1 [14] *Every chordal graph with at least one vertex contains a simplicial vertex. If G is not a clique then G contains at least two nonadjacent simplicial vertices.* ∎

Corollary 29.1 [14,15] *G is chordal if and only if G has a perfect elimination ordering. Moreover, every simplicial vertex of a chordal graph G can be the first vertex of a perfect elimination ordering of G.*

For a collection $\mathcal{T}$ of subtrees of a tree T, let the *vertex intersection graph* $G_\mathcal{T}$ of $\mathcal{T}$ be the graph having the elements of $\mathcal{T}$ as its vertices, and two subtrees t and t' from $\mathcal{T}$ are adjacent in $G_\mathcal{T}$ if they share a vertex in T.

Proposition 29.1 *The vertex intersection graph of a collection of subtrees in a tree is chordal.*

Proof. Let $G = (V, E)$ be the vertex intersection graph of a collection of subtrees in a tree T. Suppose G contains a chordless cycle $(v_0, v_1, \ldots, v_{k-1}, v_0)$ with $k > 3$ corresponding to the sequence of subtrees $T_0, T_1, \ldots, T_{k-1}, T_0$ of the tree T; that is, $T_i \cap T_j \neq \emptyset$ if and only if i and j differ by at most one modulo k. All arithmetic will be done modulo k.

Choose a point a_i from $T_i \cap T_{i+1}$ ($i = 0, \ldots, k-1$). Let b_i be the last common point on the (unique) simple paths from a_i to a_{i-1} and a_i to a_{i+1}. These paths lie in T_i and T_{i+1}, respectively, so that b_i also lies in T_i and T_{i+1}. Let P_{i+1} be the simple path connecting b_i to b_{i+1} in T. Clearly $P_i \subseteq T_i$, so $P_i \cap P_j = \emptyset$ for i and j differing by more than 1 mod k. Moreover, $P_i \cap P_{i+1} = \{b_i\}$ for $i = 0, \ldots, k-1$. Thus, $\bigcup_i P_i$ is a simple cycle in T, contradicting the definition of a tree. ∎

The tree structure of chordal graphs is described in terms of so-called *clique trees* of the maximal cliques of the graph; see Theorem 29.2. Let $\mathcal{C}(G)$ denote the family of $\subseteq$-maximal cliques of G. A *clique tree* T of G has the maximal cliques of G as its nodes, and for every vertex v of G, the maximal cliques containing v form a subtree of T. This property will be generalized in the hypergraph chapter; it can be taken for defining α-acyclicity of a hypergraph (see Definition 29.17). The existence of a clique tree characterizes chordal graphs:

Theorem 29.2 [16–18] *A graph is chordal if and only if it has a clique tree.*

Proof. "$\Longleftarrow$": Assume that G has a clique tree T. If T has only one node then G is a clique and thus chordal. Now let T have $k > 1$ nodes and assume as induction hypothesis that the assertion is true for clique trees with less than k nodes. Let C be a leaf node in T, let C' be its neighbor in T, let V_C be the subset of G vertices occuring only in C, and let T' be the clique tree restricted to $V \setminus V_C$. ∎

V_C must be nonempty since otherwise, $C \subset C'$ which is impossible by maximality of the cliques. Now start a p.e.o. of G with the vertices of V_C and then continue with a p.e.o. for $G - V_C$ which must exist since T' has less nodes than T.

"$\Longrightarrow$": For this direction, we use a version described by Spinrad in [19]: Assume that G is chordal and let $\sigma = (v_1, \ldots, v_n)$ be a p.e.o. of G. We construct a clique tree for the subgraph $G_i = G[v_i, \ldots, v_n]$ for all vertices, starting with $i = n$ and ending with $i = 1$. Let C_i be the clique consisting of v_i and all neighbors v_j of v_i, $j > i$. After each vertex v_i is processed, v_i is given a pointer to the clique C_i in the tree. We note that vertices may be added to this clique later in the algorithm, but v_i will always point to a clique which contains C_i.

Let v_i be the next vertex considered, and assume we know the clique tree on the graph induced by vertices $v_{i+1}, \ldots, v_n$. We need to add C_i to the clique tree. Let v_j be the first (i.e., leftmost) vertex of C_i on the right of v_i in σ. If $|C_i| = |C_j| + 1$, and the clique pointed to by v_j is equal to C_j then we add v_i to this clique; in other words, C_i replaces C_j in the tree. Otherwise, add C_i as a new node of the tree. Connect C_i to the tree by adding an edge from C_i to the clique pointed to by v_j.

To see that the algorithm is correct, it is sufficient to look at two cases. Either C_j is a maximal clique in $G_{i+1} = G[\{v_{i+1}, \ldots, v_n\}]$ or it is not. If C_j is a maximal clique, it clearly must be replaced by C_i if C_j is contained in C_i, which occurs if $C_i = C_j \cup \{v_i\}$, and the algorithm does this correctly. If C_j is not a maximal clique in G_{i+1} or C_i does not contain C_j, then C_i cannot contain any maximal clique of G_{i+1}, and must be added as a new node. All elements of $C_i - v_i$ are in the clique pointed to by v_j, so the subtrees generated by the occurrences of all vertices remain connected. ∎

A consequence of Theorem 29.2 and Proposition 29.1 is as follows.

Corollary 29.2 [16–18] *A graph is chordal if and only if it is the intersection graph of certain subtrees of a tree.*

Since a p.e.o. of a chordal graph can be determined in linear time (see, e.g., [4,20]), the proof of Theorem 29.2 implies the following.

Theorem 29.3 *Given a chordal graph $G = (V, E)$, a clique tree of G can be constructed in linear time $\mathcal{O}(|V| + |E|)$.* ∎

Interestingly, a clique tree of a chordal graph G gives also the minimal separators of G.

Lemma 29.2 [21,22] *Let $G = (V_G, E_G)$ be a chordal graph with clique tree $T = (\mathcal{C}(\mathcal{G}), E_T)$. Then $S \subseteq V_G$ is a minimal separator in G if and only if there are maximal cliques Q_i, Q_j of G with $Q_i Q_j \in E_T$ such that $S = Q_i \cap Q_j$.* ∎

The specific structure of chordal graphs allows to solve various problems efficiently which is well described in [4]; as another example we give here a linear-time algorithm by András Frank [23] for maximum weight independent set (MWIS) on chordal graphs.

Let $G = (V, E)$ be a chordal graph with perfect elimination ordering $(v_1, \ldots, v_n)$ of G and $\omega : V \longrightarrow R^+$ a nonnegative weight function on V. The algorithm of Frank efficiently constructs a maximum weight stable set $\mathcal{I}$ of G in the following way:

---

(0) $\mathcal{I} := \emptyset$; all vertices in V are *unmarked*

(1) **for** $i := 1$ **to** n **do**

    if $\omega(v_i) > 0$ **then** mark v_i and let $\omega(u) := \max(\omega(u) - \omega(v_i), 0)$ for all vertices $u \in N_i(v_i)$.

(2) **for** $i := n$ **downto** 1 **do**

    if v_i is marked **then** let $\mathcal{I} := \mathcal{I} \cup \{v_i\}$ and unmark all vertices $u \in N(v_i)$.

Theorem 29.4 [23] *The algorithm described above is correct and runs in linear time.* ∎

It is clear that the algorithm runs in linear time. For the correctness, we need the following (inductive) argument: As in the algorithm, let $(v_1, \ldots, v_n)$ be a p.e.o. of G and ω a weight function on V. Now let ω' be the weight function resulting from step (1) of the algorithm for the simplicial vertex v_1. We claim the following proposition.

Proposition 29.2 $\alpha_\omega(G) = \alpha_{\omega'}(G - v_1) + \omega(v_1)$.

This is clear by the following argument: If v_1 is in a maximum weight stable set S in G then none of its neighbors are in S, and the claim holds. Otherwise, if $v_1 \notin S$ then exactly one of its neighbors, say v_i, $i > 1$, is in S (otherwise S would not be a maximal stable set), and now $\omega'(v_i) = \omega(v_i) - \omega(v_1)$ holds.

29.2.2 Some Subclasses of Chordal Graphs

As mentioned in Chapter 28, interval graphs are a very important subclass of chordal graphs. Here is another subclass of chordal graphs which plays an important role in various contexts:

Definition 29.4 *A graph is a* split graph *if its vertex set can be partitioned into a clique and a stable set. Such a partition is called a* split partition.

It is easy to see that the complement of a split graph is a split graph as well, and split graphs are chordal. In what follows, we say a vertex x *sees* a vertex y if x is adjacent to y; otherwise we say x *misses* y.

Theorem 29.5 [24] *The following conditions are equivalent:*

i. G *is a split graph.*

ii. G *and* $\overline{G}$ *are chordal.*

iii. G *contains no induced* $2K_2 = \overline{C_4}$, C_4, C_5 *(i.e., G is $(2K_2, C_4, C_5)$-free).*

Proof. "(ii) $\iff$ (iii)": If G and $\overline{G}$ are chordal then obviously G contains no induced $2K_2$, C_4 and C_5. In the other direction, note that for every $k \geq 6$, C_k contains a $2K_2$, and $\overline{C_5} = C_5$. Thus, if G contains no induced $2K_2$, C_4 and C_5 then G and $\overline{G}$ are chordal.

"(i) $\implies$ (ii)": If the vertex set V of G has a partition into a clique Q and a stable set S then obviously, every vertex in S is simplicial in G. Thus, a p.e.o. of G can start with all vertices of S and finish with all vertices of Q. Similar arguments hold for $\overline{G}$, and thus G and $\overline{G}$ are chordal.

"(i) $\impliedby$ (ii)": Suppose that G and $\overline{G}$ are chordal (or, equivalently, G contains no induced $2K_2$, C_4, and C_5).

If there is a vertex $v \in V$ which is simplicial in G and $\overline{G}$ then $N[v]$ is a clique and $\overline{N[v]}$ is a stable set giving the desired split partition.

If there is a vertex $v \in V$ which is neither simplicial in G nor simplicial in $\overline{G}$ then let $a, b \in N(v)$ be vertices with $ab \notin E$ and let $c, d \in \overline{N[v]}$ with $cd \in E$. Since G is $2K_2$-free, a sees c or d, and similarly, b sees c or d but since G is C_4-free, a and b do not have a common neighbor in c, d. Thus, say, a sees c but not d and vice versa for b but now v, a, b, c, d induce a C_5 in G which is a contradiction.

Thus, every vertex $v \in V$ is either simplicial in G or simplicial in $\overline{G}$. Let $V_1 := \{v \in V \mid v$ is simplicial in $G\}$ and $V_2 := \{v \in V \mid v$ is simplicial in $\overline{G}\}$. Note that $V = V_1 \cup V_2$ is a partition of V. Now, if V_1 is a stable set and V_2 is a clique then this gives the desired split partition. Suppose to the contrary that V_1 contains an edge $xy \in E$. Then since G is $2K_2$-free, the set of nonneighbors of x and y form a stable set, and since x and y are simplicial, the set of neighbors of x and y form a clique which gives the desired split partition. ∎

Theorem 29.5 does not immediately give a linear-time recognition of split graphs. The following nice characterization of split graphs in terms of their degree sequence leads to linear-time recognition of split graphs:

Theorem 29.6 [25,26] *Let G have the degree sequence $d_1 \geq d_2 \geq \ldots \geq d_n$ and $\omega := \max\{i \mid d_i \geq i - 1\}$. Then G is a split graph if and only if $\Sigma_{i=1}^{\omega} d_i = \omega(\omega - 1) + \Sigma_{i=\omega+1}^{n} d_i$.* ∎

See [25,26] for more details.

Finally, another interesting subclass of chordal graphs should be mentioned which will be discussed in more detail in the section on strongly chordal graphs and on β-acyclicity. Assume that G is a chordal graph. A chord $x_i x_j$ in a cycle $C = (x_1, x_2, \ldots, x_{2k}, x_1)$ of even length $2k$ is an *odd chord* if the distance in C between x_i and x_j is odd.

Farber [27] defined strongly chordal graphs in terms of strong elimination orderings rather than odd chords in even cycles (see Definition 29.33), but he showed that chordal graphs having odd chords in even cycles are exactly the strongly chordal graphs (see Theorem 29.34).

Chordal graphs can be generalized in a natural way by placing a variety of restrictions on the number and type of chords with respect to a cycle. A fairly general scheme is given in the following definition (which was motivated by relational database schemes).

Definition 29.5 [28] *For $k \geq 4$ and $\ell \geq 1$, a graph G is (k, ℓ)-chordal if each cycle in G of length at least k contains at least ℓ chords.*

Thus chordal graphs are the (4,1)-chordal graphs. Further conditions can be placed on the parity of the cycles (chords in odd cycles), the parity of the cycle distance of the end vertices of chords (odd chords), requiring crossing and/or parallel chords, requiring all these conditions for G and $\overline{G}$, and requiring these conditions in bipartite graphs (where all cycles are of even length). Thus, for example, the (5,2)-odd-crossing-chordal graphs are the graphs such that every odd cycle of length at least five has at least two crossing chords.

See [3] for more details and Theorem 29.45 for a characterization of (5,2)-chordal graphs.

29.3 α-ACYCLIC HYPERGRAPHS AND THEIR DUALS

29.3.1 Motivation from Relational Database Theory

Fagin [7] gives a very nice introduction into acyclic database schemes (of various degrees, namely α-, β-, and γ-acyclicity) and their equivalence to desirable properties of relational databases. Since Fagin's introduction is mostly informal and we need some definitions, we follow the presentation in papers such as [29] for this subsection.

A (relational) *database scheme* as introduced by Codd [30] can be thought of as a collection of table skeletons, or, alternatively, as a set of subsets of *attributes*, or column names in the tables. These attribute subsets form the hyperedges of a finite hypergraph. A *relational database* corresponds to a family of relations over the attributes.

Let $V = \{v_1, \ldots, v_n\}$ be a finite set of distinct symbols called *attributes* or *column names* (*name, first name, age, birthday, citizenship, married, home address, telephone number*, etc).

Let $Y \subseteq V$. A Y-*tuple* is a mapping that associates a value (from a certain universe U) with each attribute in Y. For instance, if $Y = \{name, age, citizenship, married\}$ then a Y-tuple is a 4-tuple such as $(Higgins, 48, Canada, no)$.

If $X \subseteq Y$ and t is a Y-tuple, then *the projection* $t[X]$ denotes the X-tuple obtained by restricting t to X. For instance, if $X = \{name, citizenship\}$ and $t = (Higgins, 48, Canada, no)$ then $t[X] = (Higgins, Canada)$.

A Y-*relation* is a finite set of Y-tuples. If r is a Y-relation and $X \subseteq Y$ then by the *projection* $r[X]$ of r onto X, we mean the set of all tuples $t[X]$, where $t \in r$.

If V is a set of attributes, then we define a *relational database scheme* (*database scheme* for short) $\mathcal{E} = \{E_1, \ldots, E_m\}$ to be a set of subsets of V, that is, $(V, \mathcal{E})$ is a hypergraph over vertex set V.

Intuitively, for each i, the set E_i of attributes is considered to be the set of column names for a relation; the E_i's are called *relation schemes*. If $r_1, \ldots, r_m$ are relations, where r_i is a relation over E_i, $i \in \{1, \ldots, m\}$, then we call $\{r_1, \ldots, r_m\}$ a *database over $\mathcal{E}$*.

The *join* $r_1 \bowtie r_2$ of two relations r_1 and r_2 with attribute sets E_1 and E_2, respectively, is the set of all tuples t with attribute set $E_1 \cup E_2$ for which the projection $t[E_i]$ is in r_i, $i = 1, 2$.

Example 29.1

$r_1:$	A	B		$r_2:$	B	C		$r_3:$	A	C
	0	0			0	0			0	1
	1	1			1	1			1	0

$r_1 \bowtie r_2:$	A	B	C		$r_1 \bowtie r_3:$	A	B	C		$r_2 \bowtie r_3:$	A	B	C
	0	0	0			0	0	1			1	0	0
	1	1	1			1	1	0			0	1	1

More generally, the *join* $r_1 \bowtie \ldots \bowtie r_m$ of the relations $r_1, \ldots, r_m$, $m \geq 2$, with attribute sets $E_1, \ldots, E_m$, respectively, is the set of all tuples t with attribute set $E_1 \cup \ldots \cup E_m$, such that for each $i \in \{1, \ldots, m\}$, the projection $t[E_i]$ of tuple t onto attributes E_i fulfills $t[E_i] \in r_i$.

The join of all three relations in Example 29.1 is empty: $r_1 \bowtie r_2 \bowtie r_3 = \emptyset$.

We say that a relation r with attributes $E_1 \cup \ldots \cup E_m$ obeys the *join dependency* $\bowtie \{E_1, \ldots, E_m\}$ if $r = r_1 \bowtie \ldots \bowtie r_m$, where $r_i = r[E_i]$ for each $i \in \{1, \ldots, m\}$.

A highly desirable property of a relational database $r_1, \ldots, r_m$, $m \geq 2$, is that the entries in it are conflict-free. In general, the attribute sets are not pairwise disjoint, and it easily might happen that an entry in one of the relations is updated while the same entry in another relation is not. Pairwise consistency captures conflict-freeness for every two of the relations, and global consistency, roughly saying, means that all of them together are conflict-free. If the relations are globally consistent then they are pairwise consistent but not vice versa as Example 29.1 shows; surprisingly, it turns out that the equivalence of pairwise and of global consistency corresponds to a hypergraph acyclicity property of the underlying attribute sets.

More formally, let r and s be relations with attributes R and S, respectively, and let $Q = R \cap S$, that is, Q is precisely the set of attributes that r and s have in common. We say that *r and s are consistent* if $r[Q] = s[Q]$, that is, the projections of r and s onto their common attributes are the same.

Example 29.2

$r_1:$	A	B	C		$r_2:$	A	D	E
	0	1	2			0	3	4
	1	2	3			0	5	6
	2	3	4			3	4	5

$r_1 \bowtie r_2:$	A	B	C	D	E
	0	1	2	3	4
	0	1	2	5	6

In Example 29.2, r_1 and r_2 have only A as common attribute, and the projection $r_1[A]$ is $\{0, 1, 2\}$ while the projection $r_2[A]$ is $\{0, 3\}$; thus, r_1 and r_2 are not consistent.

In Example 29.1, each pair r_i, r_j of relations, $i, j \in \{1, 2, 3\}$, is consistent.

Definition 29.6 *Let $\{r_1, \ldots, r_m\}$ be an arbitrary database over $\mathcal{E} = \{E_1, \ldots, E_m\}$.*

  i. *$\{r_1, \ldots, r_m\}$ is* pairwise consistent *if for all $i, j \in \{1, \ldots, m\}$, r_i and r_j are consistent.*

  ii. *$\{r_1, \ldots, r_m\}$ is* globally consistent *if there is a relation r over the attribute set $E_1 \cup \ldots \cup E_m$ such that for each $i \in \{1, \ldots, m\}$, $r_i = r[E_i]$. Then r is called* universal *for $\{r_1, \ldots, r_m\}$.*

Thus, $\{r_1, \ldots, r_m\}$ is globally consistent if and only if there is a (universal) relation r such that each r_i is the projection of r onto the corresponding attribute set of r_i. Such a universal relation need not be unique, but it is known that if there is such a universal relation r, then also $r_1 \bowtie \ldots \bowtie r_m$ is such a universal relation.

Lemma 29.3 *If r is a universal relation for $r_1, \ldots, r_m$ with attribute sets $E_1, \ldots, E_m$ then $r \subseteq r[E_1] \bowtie \ldots \bowtie r[E_m] = r_1 \bowtie \ldots \bowtie r_m$.*

It is clear that if $\{r_1, \ldots, r_m\}$ is globally consistent then it is pairwise consistent but in general, the converse is false as the relations r_1, r_2, r_3 in Example 29.1 show which are pairwise consistent but not globally consistent.

Honeyman et al. [31] have shown the following theorem.

Theorem 29.7 [31] *The global consistency of a relational database is an NP-complete problem.* ∎

In [29], it is shown that for a relational database scheme, pairwise consistency implies global consistency if and only if it is α-acyclic (see Theorem 29.17).

29.3.2 Some Basic Hypergraph Notions

A pair $H = (V, \mathcal{E})$ is a (finite) *hypergraph* if V is a finite vertex set and $\mathcal{E}$ is a collection of subsets of V (the *edges* or *hyperedges* of H). Hypergraphs are a natural generalization of undirected graphs; unlike edges, hyperedges are not necessarily two-elementary. In many cases, hyperedges containing exactly one vertex (so-called *loops*) are excluded. Equivalently, a hypergraph $H = (V, \mathcal{E})$ with $V = \{v_1, \ldots, v_n\}$ and $\mathcal{E} = \{e_1, \ldots, e_m\}$ can be described by its $n \times m$ vertex-hyperedge incidence matrix $M(H)$ with entries $m_{ij} \in \{0, 1\}$ and $m_{ij} = 1 \iff v_i \in e_j$ for $i \in \{1, \ldots, n\}$ and $j \in \{1, \ldots, m\}$.

Subsequently, we collect some basic notions and properties—see, for example, [13].

Definition 29.7 *A hypergraph $H = (V, \mathcal{E})$ is* simple *if it has no repeated edges. Moreover, if no hyperedge $e \in \mathcal{E}$ is properly contained in another hyperedge $e' \in \mathcal{E}$ then H is called a* Sperner family *or* clutter.

In the database community (see, e.g., [29]), clutters are called *reduced hypergraphs*.

Definition 29.8 *Let $H = (V, \mathcal{E})$ be a finite hypergraph.*

i. *The* subhypergraph *induced by the subset $A \subseteq V$ is the hypergraph $H[A] = (A, \mathcal{E}_A)$ with edge set $\mathcal{E}_A = \{e \cap A \mid e \in \mathcal{E}\}$.*

ii. *The* partial hypergraph *given by the edge subset $\mathcal{E}' \subseteq \mathcal{E}$ is the hypergraph with the vertex set $\bigcup \mathcal{E}'$ and the edge set $\mathcal{E}'$.*

Note that both restrictions $A \subset V$ and $\mathcal{E}' \subset \mathcal{E}$ can be combined in a subhypergraph $H'[A] = (A, \mathcal{E}'_A)$ with edge set $\mathcal{E}'_A = \{e \cap A \mid e \in \mathcal{E}' \subset \mathcal{E}\}$ called *partial subhypergraph* in [13].

The partial hypergraphs [13] are called *subhypergraphs* in [6]. Since this may cause confusion, we also use the name *edge-subhypergraphs* for partial hypergraphs and *vertex-subhypergraphs* in case (i).

Dualization is a classical concept which is well-known from geometry; there, points and hyperplanes exchange their role. Here, dualization means that vertices and hyperedges exchange their role.

Definition 29.9 *Let $H = (V, \mathcal{E})$ be a finite hypergraph. For $v \in V$, let $\mathcal{E}_v = \{e \in \mathcal{E} \mid v \in e\}$. The* dual hypergraph *$H^* = (\mathcal{E}, \mathcal{E}^*)$ of H has vertex set $\mathcal{E}$ and hyperedge set $\{\mathcal{E}_v \mid v \in V\}$.*

If the hypergraph H is given in terms of its incidence matrix $M(H)$ then the incidence matrix of the dual of H is the transposal of $M(H)$: $M(H^*) = (M(H))^T$.

Evidently, the dual of the dual of H is isomorphic to H itself since the twofold transposal of a matrix is the matrix itself.

Proposition 29.3 $(H^*)^* \sim H$.

Graphs and hypergraphs are closely related to each other. The next definition represents two examples.

Definition 29.10 *Let $H = (V, \mathcal{E})$ be a finite hypergraph.*

  i. *The* 2-section graph *$2SEC(H)$ of H has the vertex set V, and two vertices u, v are adjacent if u and v are contained in a common hyperedge: $\exists e \in \mathcal{E}$ such that $u, v \in e$.*

  ii. *The* line graph *$L(H) = (\mathcal{E}, F)$ is the intersection graph of $\mathcal{E}$, that is, for any $e, e' \in \mathcal{E}$ with $e \neq e'$, $ee' \in F \iff e \cap e' \neq \emptyset$.*

The 2-section graph of H is denoted by $[H]_2$ in [13]; the line graph is also called *representative graph* in [13]. Again, these notions have different names in different communities; the 2-section graph is also called *adjacency graph* in [32], *primal graph* [33], or *Gaifman graph* [34] and has no name but is denoted by $G(H)$ in [29]. The line graph is also called *dual graph* in [35].

The following isomorphism is easy to see.

Proposition 29.4 $2SEC(H) \sim L(H^*)$.

A subfamily $\mathcal{E}' \subseteq \mathcal{E}$ is called *pairwise intersecting* if for all $e, e' \in \mathcal{E}'$, $e \cap e' \neq \emptyset$.

Definition 29.11 *Let $H = (V, \mathcal{E})$ be a hypergraph.*

  i. *H is* conformal *if every clique C in $2SEC(H)$ is contained in a hyperedge $e \in \mathcal{E}$.*

  ii. *H has the* Helly *property if every pairwise intersecting subfamily $\mathcal{E}' \subseteq \mathcal{E}$ has nonempty total intersection: $\bigcap \mathcal{E}' \neq \emptyset$.*

The following is easy to see.

Proposition 29.5 *H has the Helly property if and only if H^* is conformal.*

The next theorem gives a polynomial time criterion for testing the Helly property of a hypergraph. It is closely related to an earlier criterion for conformality given by Gilmore which will be mentioned in Theorem 29.9.

For a hypergraph $H = (V, \mathcal{E})$ and for any 3-elementary set $A = \{a_1, a_2, a_3\} \subseteq V$, let $\mathcal{E}_A$ denote the set of all hyperedges $e \in \mathcal{E}$ such that $|e \cap A| \geq 2$.

Theorem 29.8 [13,36] *A hypergraph $H = (V, \mathcal{E})$ has the Helly property if and only if for all 3-elementary sets $A = \{a_1, a_2, a_3\} \subseteq V$, the total intersection of all hyperedges containing at least two vertices of A is nonempty: $\bigcap \mathcal{E}_A \neq \emptyset$.*

Proof. "$\Longrightarrow$": Let H be a hypergraph with the Helly property, and let $\{e_1, \ldots, e_k\} \subseteq \mathcal{E}$ be the hyperedges for which $|e_i \cap A| \geq 2$, $i \in \{1, \ldots, k\}$. Then for all $i \neq j$, $i, j \in \{1, \ldots, k\}$, $e_i \cap e_j$ is nonempty and thus, their total intersection is nonempty since H has the Helly property.

"$\Longleftarrow$": Now assume that $\{e_1, \ldots, e_\ell\} \subseteq \mathcal{E}$ is a collection of pairwise intersecting hyperedges. If $\ell = 2$ then obviously their total intersection is nonempty; thus let $\ell > 2$. We assume inductively that the assertion of nonempty total intersection is true for less than ℓ hyperedges with pairwise nonempty intersection.

Then by the induction hypothesis, $e_1 \cap \ldots \cap e_{\ell-1} \neq \emptyset$; let $a_1 \in e_1 \cap \ldots \cap e_{\ell-1}$. Moreover, $e_2 \cap \ldots \cap e_\ell \neq \emptyset$; let $a_2 \in e_2 \cap \ldots \cap e_\ell$. Finally $e_1 \cap e_\ell \neq \emptyset$; let $a_3 \in e_1 \cap e_\ell$.

Let $A := \{a_1, a_2, a_3\}$. It is easy to see that in the case $|A| < 3$ we are done. Now let $|A| = 3$. Thus every e_i, $i = 1, \ldots, \ell$, contains at least two elements from the 3-elementary set A, and by the assumption, their total intersection is nonempty. ∎

An obvious consequence of Theorem 29.8 is as follows:

Corollary 29.3 *Testing the Helly property for a given hypergraph can be done in polynomial time.*

Corollary 29.4 *Every collection of subtrees of a tree has the Helly property.*

Proof. Let T be a tree with at least three vertices (otherwise the assertion is obviously fulfilled), and let a, b, c be any three vertices in T. We consider the set of all subtrees of T containing at least two of the vertices a, b, c. Let $P(x, y)$ denote the uniquely determined path in the tree T between x and y. Let x_0 denote the last vertex in $P(a, b) \cap P(b, c)$ (this intersection contains at least vertex b). Then $P(a, c)$ consists of $P(a, x_0)$ followed by $P(x_0, c)$. Thus the three paths $P(a, b)$, $P(b, c)$ and $P(a, c)$ have vertex x_0 in common, that is, x_0 is contained in every subtree of T which contains at least two of the vertices a, b, c. Thus, by Theorem 29.8, every system of subtrees has the Helly property. ∎

A nice inductive proof of Corollary 29.4 is given in a script by Alexander Schrijver: The induction is on $|V(T)|$. If $|V(T)| = 1$ then the assertion is trivial. Now assume $|V(T)| \geq 2$, and let $\mathcal{S}$ be a collection of pairwise intersecting subtrees of T. Let t be a leaf of T. If there exists a subtree of T consisting only of t, the assertion is trivial. Hence we may assume that each subtree in $\mathcal{S}$ containing t also contains the neighbor of t in T. So, after deleting t from T and from all subtrees in $\mathcal{S}$, this collection is still pairwise intersecting, and the assertion follows by induction.

Actually, Theorem 29.8 is formulated in a more general way in [13]; there are various interesting generalizations of the Helly property.

According to Proposition 29.5, Theorem 29.8 can be dualized as follows.

Theorem 29.9 (Gilmore, see [13]) *Let $H = (V, \mathcal{E})$ be a hypergraph. H is conformal if and only if for all 3-elementary edge sets $A = \{e_1, e_2, e_3\} \subseteq \mathcal{E}$ of hyperedges, there is a hyperedge $e \in \mathcal{E}$ with $(e_1 \cap e_2) \cup (e_1 \cap e_3) \cup (e_2 \cap e_3) \subseteq e$.*

Proof. "$\Longrightarrow$": Obviously, $(e_1 \cap e_2) \cup (e_1 \cap e_3) \cup (e_2 \cap e_3)$ is a clique in the 2-section graph $2SEC(H)$ of H. By conformality, there is a hyperedge e with $(e_1 \cap e_2) \cup (e_1 \cap e_3) \cup (e_2 \cap e_3) \subseteq e$.

"$\Longleftarrow$": Let $A = \{e_1, e_2, e_3\} \subseteq \mathcal{E}$ and let $\mathcal{E}_u$ be a hyperedge in H^* containing at least two of e_1, e_2, e_3. Then $u \in (e_1 \cap e_2) \cup (e_1 \cap e_3) \cup (e_2 \cap e_3)$ and thus also $u \in e$. Thus, e is in the total intersection of all hyperedges $\mathcal{E}_u$ which contain at least two of e_1, e_2, e_3. Then by Theorem 29.8, H^* has the Helly property and thus, by Proposition 29.5, H is conformal. ∎

There is a third type of graphs derived from a hypergraph $H = (V, \mathcal{E})$, namely the bipartite vertex-edge incidence graph $\mathcal{I}(H)$ (which is a reformulation of the incidence matrix of H in terms of a bipartite graph). The two color classes of $\mathcal{I}(H)$ are the sets V and $\mathcal{E}$, respectively, and a vertex v and an edge e are adjacent if and only if $v \in e$. More formally:

Definition 29.12 *Let $H = (V, \mathcal{E})$ be a finite hypergraph. In the bipartite incidence graph $\mathcal{I}(H) = (V, \mathcal{E}, I)$ of H, $v \in V$ and $e \in \mathcal{E}$ are adjacent if and only if $v \in e$.*

In the other direction, namely from graphs to hypergraphs, the most basic constructions are the following:

Definition 29.13 *Let $G = (V, E)$ be a graph.*

i. *The clique hypergraph $\mathcal{C}(G)$ consists of the $\subseteq$-maximal cliques of G.*

ii. *The neighborhood hypergraph $\mathcal{N}(G)$ consists of the closed neighborhoods $N[v]$ of all vertices v in G.*

iii. *The disk hypergraph $\mathcal{D}(G)$ consists of the iterated closed neighborhoods $N^i[v]$, $i \geq 1$, of all vertices v in G, where $N^1[v] := N[v]$ and $N^{i+1}[v] := N[N^i[v]]$.*

Note that in general, the neighborhood hypergraph $\mathcal{N}(G)$ is not simple since different vertices can have the same closed neighborhood in G. The following is easy to see.

Proposition 29.6 $\mathcal{N}(G)$ *is self-dual, that is,* $(\mathcal{N}(G))^* \sim \mathcal{N}(G)$.

Moreover, the 2-section graph of $\mathcal{C}(G)$ is isomorphic to G and thus, $\mathcal{C}(G)$ is conformal. Note that a hypergraph uniquely determines its 2-section graph but not vice versa.

Lemma 29.4 *Every conformal Sperner hypergraph $H = (V, \mathcal{E})$ is the clique hypergraph of its 2-section graph $2SEC(H)$: $H = \mathcal{C}(2SEC(H))$.*

Proof. Let H be conformal and Sperner. We show:

1. For every $e \in \mathcal{E}$, e is a maximal clique in $2SEC(H)$:

   Obviously, e is a clique in $2SEC(H)$ and thus, there is a maximal clique C' in $2SEC(H)$ with $e \subseteq C'$. Since H is conformal, there is an $e' \in \mathcal{E}$ with $C' \subseteq e'$, that is, $e \subseteq C' \subseteq e'$ and since H is Sperner, $e = C' = e'$ follows.

2. For every maximal clique C in $2SEC(H)$, $C \in \mathcal{E}$ holds:

   By conformality of H, there is $e \in \mathcal{E}$ with $C \subseteq e$, and since e is a clique in $2SEC(H)$, there is a maximal clique C' in $2SEC(H)$ with $e \subseteq C'$, that is, $C \subseteq e \subseteq C'$. By maximality of C, $C = e = C'$ follows. ∎

For a graph $G = (V, E)$, let $G^2 = (V, E^2)$ with $xy \in E^2$ for $x \neq y$ if and only if $d_G(x, y) \leq 2$, that is, either $xy \in E$ or there is a common neighbor z of x and y.

The following is easy to see.

Proposition 29.7 $G^2 \sim L(\mathcal{N}(G))$.

For graph $G = (V, E)$, let $B(G) = (V', V'', F)$ denote the bipartite graph with two disjoint copies V' and V'' of V, and for $v' \in V'$ and $w'' \in V''$, $v'w'' \in F$ if and only if either $v = w$ or $vw \in E$.

The following is easy to see.

Proposition 29.8 $B(G) \sim \mathcal{I}(\mathcal{N}(G))$.

The line graph of $\mathcal{C}(G)$ is the classical *clique graph operator* in graph theory.

Definition 29.14 *Let G be a graph.*

  i. *The* clique graph $K(G)$ *of G is defined as* $K(G) = L(\mathcal{C}(G))$.

  ii. *G is a* clique graph *if there is a graph G' such that G is the clique graph of G', that is, $G = K(G')$.*

Theorem 29.10 [37] *A graph G is a clique graph if and only if some class of complete subgraphs of G covers all edges of G and has the Helly property.* ∎

See [3,5] and in particular the survey [38] by Szwarcfiter for more details on clique graphs. Recognizing whether a graph is a clique graph is NP-complete [39].

29.3.3 Hypergraph 2-Coloring

A hypergraph $H = (V, \mathcal{E})$ is *2-colorable* if its vertex set V has a partition $V = V_1 \cup V_2$ such that every hyperedge $e \in \mathcal{E}$ has at least one vertex from each of the sets V_1 and V_2. See [13] for the more general notion of hypergraph coloring. The *Hypergraph 2-Coloring Problem* (also called *Bicoloring Problem, Set Splitting Problem* [SP4] in [40]) is the question whether a given hypergraph is 2-colorable.

Lovász [41] has shown that the Hypergraph 2-Coloring Problem is NP-complete even for hypergraphs whose hyperedges have size at most 3 (see [40]); the original reduction in [41] is from the graph coloring problem (which has been shown to be NP-complete in [42]) to hypergraph 2-coloring.

The following nice reduction from the satisfiability problem SAT to the hypergraph 2-coloring problem was given in [43].

Let $F = C_1 \wedge \ldots \wedge C_m$ be a Boolean expression in conjunctive normal form (CNF for short) with clauses $C_1, \ldots, C_m$ and variables $x_1, \ldots, x_n$. Each clause consists of a disjunction of literals, that is, unnegated or negated variables.

Let $H_F = (V_F, \mathcal{E}_F)$ be the following hypergraph for F:

> The vertex set $V_F = \{x_1, \ldots, x_n\} \cup \{\neg x_1, \ldots, \neg x_n\} \cup \{f\}$ where f is a new symbol different from the variable symbols.
>
> The edge set $\mathcal{E}_F$ of H_F consists of the following edges:
>
>   i. For all $i \in \{1, \ldots, n\}$, let $X_i = \{x_i, \neg x_i\}$,
>
>   ii. For all $j \in \{1, \ldots, m\}$, let Y_j be the set of all literals in C_j plus, additionally, the element f.

We show that F is satisfiable if and only if H_F is 2-colorable:

Given a truth assignment which satisfies F, we associate with it the following 2-coloring $V_1 \cup V_2$. If x_i has truth value 1 then $x_i \in V_1$ and $\neg x_i \in V_2$ and vice versa if x_i has truth value 0. The element f belongs to V_2. Now, for each $i \in \{1, \ldots, n\}$, the edge $\{x_i, \neg x_i\}$ intersects both V_1 and V_2. An edge Y_i intersects V_2 on f and intersects V_1 since it has a true literal.

On the other hand, given a 2-coloring $V_1 \cup V_2$ of H_F, with, say $f \in V_2$ we assign true to each x_i in V_1 and false to those in V_2. This gives a truth assignment since the edges $\{x_i, \neg x_i\}$ meet both V_1 and V_2. The edge Y_j of every clause C_j meets V_1 on an element other than f which ensures that every clause is satisfied. This shows the following theorems.

Theorem 29.11 [41] *The 2-coloring problem for hypergraphs is NP-complete.* ∎

Based on Theorem 29.11, in [41], Lovász has shown the following theorem.

Theorem 29.12 [41] *The 3-coloring problem for graphs is NP-complete.* ∎

See [44] for another proof of the NP-completeness of the 3-coloring problem.

29.3.4 Kőnig Property

The following definition generalizes the fundamental notions of matching and vertex cover in graphs to the corresponding notions in hypergraphs.

Definition 29.15 *Let $H = (V, \mathcal{E})$ be a hypergraph.*

   i. *An edge set $\mathcal{E}' \subseteq \mathcal{E}$ is called* matching *if the edges of $\mathcal{E}'$ are pairwise disjoint. The* matching number $\nu(H)$ *is the maximum number of pairwise disjoint hyperedges of H. This parameter $\nu(H)$ is also frequently called* packing number *of H.*

   ii. *A* transversal *of $\mathcal{E}$ is a subset $U \subseteq V$ such that U contains at least one vertex of every $e \in \mathcal{E}$. The* transversal number $\tau(H)$ *is the minimum number of vertices in a transversal of H.*

   iii. *H has the* Kőnig Property *if $\nu(H) = \tau(H)$.*

Note that for every hypergraph, $\nu(H) \leq \tau(H)$ holds. A well-known theorem of Kőnig states that for bipartite graphs G, $\nu(G) = \tau(G)$ holds. This justifies the name *Kőnig property* and is closely related to the celebrated max-flow min-cut theorem by Ford and Fulkerson.

29.3.5 α-Acyclic Hypergraphs and Tree Structure

Unlike the case of graphs, there is a bewildering diversity of cycle notions in hypergraphs, and some of them play an important role in connection with desirable properties of relational database schemes [6,7,29,32,45]. Thus, for example, the desirable property of a relational database scheme that pairwise consistency should imply global consistency turns out to be equivalent to α-acyclicity of the scheme [6,29]; as shown in Theorems 29.16 and 29.17, a relational database scheme has this property if and only if it is α-acyclic. Moreover, α-acyclicity is equivalent to many other desirable properties of such schemes. The most important property of an α-acyclic hypergraph for applications in databases and other fields seems to be the existence of a join tree for α-acyclic hypergraphs:

Definition 29.16 *Let $H = (V, \mathcal{E})$ be a hypergraph.*

   i. *Tree T is a* join tree *of H if the node set of T is the set of hyperedges $\mathcal{E}$ and for every vertex $v \in V$, the set $\mathcal{E}_v$ of hyperedges containing v forms a subtree in T.*

   ii. *H is α-acyclic if H has a join tree.*

Note that in this way, α-acyclicity of a hypergraph is defined without referring to any cycle notion in hypergraphs.

Tarjan and Yannakakis [20] gave a linear-time algorithm for testing α-acyclicity of a given hypergraph.

Tree structure in hypergraphs has been captured in the hypergraph community as *arboreal hypergraphs* [13] (as well as its dual version, the *co-arboreal hypergraphs*) and *tree-hypergraphs* in [46]. We call arboreal hypergraphs *hypertrees*.

Definition 29.17 *A hypergraph $H = (V, \mathcal{E})$ is a* hypertree *if there is a tree T whose set of nodes is V and such that every hyperedge $e \in \mathcal{E}$ induces a subtree in T.*

Note that in [33], Gottlob et al. define the notion of hypertrees in a completely different way.

The following properties are easy to see:

Proposition 29.9 *Let $H = (V, \mathcal{E})$ be a hypergraph.*

  i. *H is a hypertree if and only if its dual H^* is α-acyclic.*

  ii. *If H is a hypertree then every edge-subhypergraph of H is a hypertree as well but not necessarily every vertex-subhypergraph of H.*

  iii. *If H is α-acyclic then every vertex-subhypergraph of H is α-acyclic as well but not necessarily every edge-subhypergraph of H.*

The fact that α-acyclic hypergraphs may contain hyperedge cycles of a certain kind (there are various cycle definitions in hypergraphs), and the fact that edge-subhypergraphs of α-acyclic hypergraphs are not necessarily α-acyclic are somewhat counterintuitive in comparison with cycles in graphs and led Goodman and Shmueli [32] to the name *tree schema* for α-acyclic hypergraphs (see also [47] for a discussion).

The following theorem gives an important characterization of hypertrees (α-acyclic hypergraphs, respectively).

Theorem 29.13 [48–50] *A hypergraph H is a hypertree if and only if H has the Helly property and its line graph $L(H)$ is chordal.*

Proof. "$\Longrightarrow$": Let $H = (V, \mathcal{E})$ be a hypertree and let T be a tree with vertex set V such that for all $e \in \mathcal{E}$, $T[e]$ induces a subtree in T. By Corollary 29.4, every hypertree H has the Helly property. By Proposition 29.1, $L(H)$ is chordal.

"$\Longleftarrow$": A dual variant of the assertion is the following: If H is conformal and $2SEC(H)$ is chordal then H is α-acyclic. Without loss of generality we may assume that no hyperedge of H is contained in another one. By Lemma 29.4, H is the clique hypergraph of its 2-section graph, and by Theorem 29.2, the chordal graph $2SEC(H)$ has a clique tree. Thus, H is α-acyclic. ∎

By Propositions 29.4 and 29.5, Theorem 29.13 can also be formulated in the following equivalent way.

Corollary 29.5 *H is α-acyclic if and only if H is conformal and $2SEC(H)$ is chordal.*

See Definition 29.15 for the Kőnig property. As a consequence of Theorem 29.13, we obtain.

Corollary 29.6 *Hypertrees have the Kőnig property.*

Proof. Let $H = (V, \mathcal{E})$ be a hypertree. Then by Theorem 29.13, H has the Helly property and there is a p.e.o. $(e_1, \ldots, e_m)$ of the edge set $\mathcal{E}$ of $L(H)$. Since e_1 is simplicial in $L(H)$, the set $\mathcal{E}_1$ of hyperedges intersecting e_1 is pairwise intersecting. By the Helly property, there is a vertex v in the intersection of $\mathcal{E}_1$. Now assume inductively that the hypergraph $H' = (V, \mathcal{E} \setminus \mathcal{E}_1)$ fulfills already the condition $\tau(H') = \nu(H')$. A maximum packing of H consists of a packing of H' and one additional hyperedge from $\mathcal{E}_1$, and a minimum transversal of H consists of a minimum transversal of H' and additionally the vertex v. Thus, also $\tau(H) = \nu(H)$ holds. ∎

The α-acyclicity of a hypergraph H can also be characterized in terms of an inequality concerning the weighted line graph of H. This was shown by Acharya and Las Vergnas in the hypergraph community (see Theorem 29.14) but was also discovered by Bernstein and Goodman [51] in the database community.

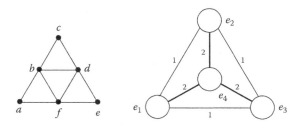

Figure 29.1 A 3-sun with its four maximal cliques $e_1 = \{a,b,f\}$, $e_2 = \{b,c,d\}$, $e_3 = \{d,e,f\}$ and $e_4 = \{b,d,f\}$ and the weighted line graph of them.

Definition 29.18 *Given a hypergraph $H = (V, \mathcal{E})$ with $\mathcal{E} = \{e_1, \ldots, e_m\}$, let $L_w(H)$ denote the weighted line graph of H whose nodes are the hyperedges of H which are pairwise connected and the edges are weighted by $w(e_i e_j) = |e_i \cap e_j|$. For any edge set F of $L(H)$, let $w(F)$ denote the sum of all edge weights in F. Let w_H denote the maximum weight of a spanning tree in $L_w(H)$. For a spanning tree T of $L_w(H)$, let T_v denote the subgraph of T induced by the hyperedges containing v, let $N(T_v)$ denote its node set and $E(T_v)$ its edge set.*

If T is a spanning tree of $L_w(H)$ then obviously for every vertex $v \in V$, the following inequality holds (Figure 29.1):

$$1 \leq |N(T_v)| - |E(T_v)|. \tag{29.1}$$

Since any tree with $k \geq 2$ nodes has $k-1$ edges, equality holds in (29.1) exactly when T_v is a subtree of T. The following lemma summarizes what is implicitly contained in Theorem 29.14.

Lemma 29.5 [52] *Let $H = (V, \mathcal{E})$ be a hypergraph with $\mathcal{E} = \{e_1, \ldots, e_m\}$ and let $L_w(H)$ be as in Definition 29.18. Then a spanning tree T of $L_w(H)$ is a join tree of H if and only if*

$$|V| = \sum_{j=1}^{m} |e_j| - \sum_{ij \in E(T)} |e_i \cap e_j|. \tag{29.2}$$

Proof. Suppose T is a spanning tree of $L_w(H)$. For each $v \in V$, the subgraph T_v consisting of all hyperedges containing v satisfies $1 \leq |N(T_v)| - |E(T_v)|$ as described in (29.1), with equality if and only if, for all $v \in V$, T_v is connected. Summing over all $v \in V$ in (29.1) proves that inequality

$$|V| \leq \sum_{j=1}^{m} |e_j| - \sum_{ij \in E(T)} |e_i \cap e_j| \tag{29.3}$$

holds, and equality holds in (29.3) if and only if the spanning tree T is a join tree. ∎

Note that the result of summing the right hand side of (29.1) is $\sum_{j=1}^{m} |e_j| - w(T)$ for the spanning tree T of $L_w(H)$. Thus also

$$|V| \leq \sum_{j=1}^{m} |e_j| - \max\{w(T) \mid T \text{ spanning tree of } L_w(H)\} = \sum_{j=1}^{m} |e_j| - w_H \tag{29.4}$$

with equality in (29.4) if and only if H has a join tree.

Inequality (29.4) led to the following parameter (see [53–55]):

Definition 29.19 *Let $H = (V, \mathcal{E})$ be a hypergraph and w_H as in Definition 29.18. The cyclomatic number $\mu(H)$ of H is defined as*

$$\mu(H) = \sum_{j=1}^{m} |e_j| - |V| - w_H.$$

Note that the cyclomatic number of a hypergraph can be efficiently determined by any maximum spanning tree algorithm. Now, the following theorem is a simple corollary of Lemma 29.5.

Theorem 29.14 [53] *A hypergraph H satisfies $\mu(H) = 0$ if and only if H is α-acyclic.* ∎

Note that Lemma 29.5 respectively Theorem 29.14 suggests a way how to find a join tree of an α-acyclic hypergraph, namely, taking any maximum spanning tree (determined, e.g., by Kruskal's greedy algorithm) of the weighted line graph $L_w(H)$. Independently, this has been discovered in the database community by Bernstein and Goodman [51] and rediscovered several times; see Chapter 2 of the monograph by McKee and McMorris [5].

However, this is not the most efficient way to construct a clique tree of a given chordal graph; Theorem 29.3 gives a linear-time algorithm for constructing a clique tree.

29.3.6 Graham's Algorithm, Running Intersection Property, and Other Desirable Properties Equivalent to α-Acyclicity

In this subsection, we collect some properties which are equivalent to α-acyclicity of a hypergraph. Some of these conditions are desirable properties of relational database schemes as mentioned in the introduction. Beeri et al. [29] give a long list of such equivalences; we mention here only some of them and give a few proofs which might be suitable for a first glance at this field of research.

In Corollary 29.1 we have seen: A graph G is chordal if and only if G has a p.e.o.

A generalization of this for α-acyclic hypergraphs is known under the name *Graham's Algorithm* (or *Graham Reduction*):

Definition 29.20 [56,57] *Let $H = (V, \mathcal{E})$ be a hypergraph.*

  i. Graham's Algorithm *on H applies the following two operations to H repeatedly as long as possible:*

      1. *If a vertex $v \in V$ is contained in exactly one hyperedge $e \in \mathcal{E}$ then delete v from e.*

      2. *If a hyperedge e is contained in another hyperedge e' then delete e.*

  ii. Graham's Algorithm *succeeds on H if repeatedly applying the two operations leads to empty hypergraph, that is, to $\mathcal{E} = \{\emptyset\}$.*

Graham's algorithm is also called *GYO algorithm* since Yu and Ozsoyoglu [57] came to exactly the same algorithm. Vertices which occur in only one edge are frequently called *ear vertices* (*isolated vertices* in [29]) and edges containing such a vertex are frequently called *ears* (*knobs* in [29]). Note that any ear node in H is simplicial in the 2-section graph of H.

Theorem 29.15 [29,32] *H is α-acyclic if and only if Graham's algorithm succeeds on H.*

Proof. "$\Longrightarrow$": Let H be α-acyclic, that is, by Corollary 29.5, H is conformal and $2SEC(H)$ is chordal. If H is not Sperner then the (possibly repeated) application of rule (2) leads to a Sperner hypergraph H' which is conformal and for which $2SEC(H')$ is chordal. By Lemma 29.4, H' is isomorphic to the maximal clique hypergraph $\mathcal{C}(2SEC(H'))$.

Let $(v_1, \ldots, v_n)$ be a p.e.o. of $2SEC(H')$. Then v_1 is simplicial and thus contained in only one hyperedge of H', that is, v_1 can be deleted by rule (1). Now the same argument can be repeated and shows the assertion.

"$\Longleftarrow$": Assume that Graham's Algorithm succeeds on H. Then, the repeated application of rules (1) and (2) defines a vertex ordering $\sigma = (v_1, \ldots, v_n)$ of V (i.e., the ordering in which by rule (1), the vertices get deleted). We claim that σ is a p.e.o. Indeed, for each $i \in \{1, \ldots, n\}$, when (1) is applicable to v_i, this vertex is contained in only one hyperedge and thus is simplicial in the remaining 2-section graph.

We finally show that H is conformal: Let C be a clique in $2SEC(H)$ and let v_i be its leftmost element in σ. Then, when eliminating this vertex by rule (1), v_i is contained in only one hyperedge, say e, and all its neighbors in the 2-section graph are in e including C, that is, $C \subseteq e$ which means that H is conformal. ∎

Note that Graham's algorithm produces a perfect elimination ordering of the 2-section graph of H if H is α-acyclic.

The *Running Intersection Property* is another notion from the database community which turns out to be equivalent to α-acyclicity of a hypergraph:

Definition 29.21 [29] *Let $H = (V, \mathcal{E})$ be a hypergraph. H has the* running intersection property *if there is an ordering $(e_1, e_2, \ldots, e_m)$ of $\mathcal{E}$ such that for all $i \in \{2, \ldots, m\}$, there is a $j < i$ such that $e_i \cap (e_1 \cup \ldots \cup e_{i-1}) \subseteq e_j$.*

Theorem 29.16 [29,32] *A hypergraph is α-acyclic if and only if it has the running intersection property.*

Proof. "$\Longrightarrow$": If the hypergraph $H = (V, \mathcal{E})$ is α-acyclic then it has a join tree T. Select a root for T. Let $(e_1, \ldots, e_m)$ be an ordering of $\mathcal{E}$ by increasing depth. Thus, if e_j is the parent of e_i, then $j < i$. Clearly, each path from e_i to any of $e_1, \ldots, e_{i-1}$ must pass through e_i's parent e_j. Now if $v \in V$ is a vertex in $e_i \cap e_k$ for some $k < i$, then all hyperedges along the T-path between e_i and e_k contain v. Since this path passes through e_j, it follows that $v \in e_j$ which implies $e_i \cap (e_1 \cup \ldots \cup e_{i-1}) \subseteq e_j$. Thus, H has the running intersection property.

"$\Longleftarrow$": Let $H = (V, \mathcal{E})$ be a hypergraph and let $(e_1, \ldots, e_m)$ be an ordering of $\mathcal{E}$ fulfilling the running intersection property. The proof is by induction on the number m of hyperedges. The basis $m = 2$ is trivial. $(e_1, \ldots, e_{m-1})$ also has the running intersection property, and by induction hypothesis, there is a join tree T' for $e_1, \ldots, e_{m-1}$. Let T be obtained from T' by adding node e_m and edge $e_m e_j$ for a j such that $e_m \cap (e_1 \cup \ldots \cup e_{m-1}) \subseteq e_j$. Obviously, T is a join tree for $e_1, \ldots, e_m$. ∎

For the next theorem, we need a few more definitions.

A *path* between two vertices $u, v \in V$ in hypergraph $H = (V, \mathcal{E})$ is a sequence of $k \geq 1$ edges $e_1, \ldots, e_k \in \mathcal{E}$ such that $u \in e_1$, $v \in e_k$ and for all $i = 1, \ldots, k-1$, $e_i \cap e_{i+1} \neq \emptyset$.

H is *connected* if for all pairs $u, v \in V$, there is a path between u and v in H.

The *connected components* of H are the maximal connected vertex-subhypergraphs of H.

For a reduced hypergraph $H = (V, \mathcal{E})$ and two edges $e, e' \in \mathcal{E}$, $e \cap e'$ is an *edge-intersection-separator*, *e.i.-separator* for short (called an *articulation set* in [29]) if the reduced vertex-subhypergraph $H[V \setminus (e \cap e')]$ has more connected components than H.

A hypergraph $H = (V, \mathcal{E})$ is *edge-intersection-separable*, *e.i.-separable* for short (called *acyclic* in [29]) if for each $U \subseteq V$, if the reduction of $H[U]$ is connected and has more than one edge (i.e., is nontrivial) then it has an edge-intersection-separator.

A hyperedge subset $\mathcal{F} \subseteq \mathcal{E}$ is *closed* if for each $e \in \mathcal{E}$, there is an edge $f \in \mathcal{F}$ such that $e \cap \bigcup \mathcal{F} \subseteq f$.

A reduced hypergraph $H = (V, \mathcal{E})$ is *closed-e.i.-separable* (called *closed-acyclic* in [29]) if for each $U \subseteq V$, if $H[U]$ is connected and has more than one edge and its set of edges is closed then it has an e.-i.-separator. A hypergraph is *closed-e.i.-separable* if its reduction is.

Note that in this definition, separators are always intersections of edges.

In [58], it is shown that a hypergraph is acyclic if and only if it is closed-acyclic, that is, e.i.-separable and closed-e.i.-separable are equivalent notions. This has the advantage that it is not necessary to deal with partial edges that are not edges.

Recall that in Section 29.3.1, pairwise and global consistency, semijoins and full reducers, monotone join expressions, and monotone sequential join expressions are defined.

Apparently, there is a close connection between the Helly property of a hypergraph and the equivalence between pairwise and global consistency of a relational database scheme (see Definition 29.6): A relational database $r_1, \ldots, r_m$ over scheme $\mathcal{E} = \{e_1, \ldots, e_m\}$ is pairwise consistent if for every pair $i, j \in \{1, \ldots, m\}$, r_i, r_j is consistent. Let R_i, $i \in \{1, \ldots, m\}$, denote the set of relations over at least the attributes e_i such that the projection to e_i is r_i. In other words, pairwise consistency means that for all $i, j \in \{1, \ldots, m\}$, the intersection $R_i \cap R_j$ is nonempty. Global consistency means that the intersection $\bigcap_{i=1}^m R_i$ is nonempty.

The next theorem is part of the main theorem in [29] which contains various other conditions. See the same paper for a detailed discussion of other papers where parts of these equivalences were shown.

Theorem 29.17 [29] *Let $\mathcal{E}$ be a hypergraph. The following conditions are equivalent:*

i. $\mathcal{E}$ *has the running intersection property.*

ii. $\mathcal{E}$ *has a monotone sequential join expression.*

iii. $\mathcal{E}$ *has a monotone join expression.*

iv. *every pairwise consistent database over $\mathcal{E}$ is globally consistent.*

v. $\mathcal{E}$ *is closed-e.i.-separable.*

vi. *every database over $\mathcal{E}$ has a full reducer.*

vii. *the GYO reduction algorithm succeeds on $\mathcal{E}$.*

viii. $\mathcal{E}$ *has a join tree (i.e., $\mathcal{E}$ is α-acyclic).*

Proof. In Theorems 29.16 and 29.17, it is already shown that conditions (i), (vii), and (viii) are equivalent.

(i) $\Longrightarrow$ (ii): Assume that $\mathcal{E}$ has the running intersection property. Let $(e_1, e_2, \ldots, e_m)$ be an ordering of $\mathcal{E}$ such that for all $i \in \{2, \ldots, m\}$, there is a $j_i < i$ such that $e_i \cap (e_1 \cup \ldots \cup e_{i-1}) \subseteq e_{j_i}$. Now we show that $(\ldots((e_1 \bowtie e_2) \bowtie e_3) \ldots \bowtie e_m)$ is a monotone, sequential join expression: If $r = \{r_1, \ldots, r_m\}$ is a pairwise consistent database over $\mathcal{E} = \{e_1, e_2, \ldots, e_m\}$, then the join $r_1 \bowtie \ldots \bowtie r_i$ (which we abbreviate as q_i) is consistent with r_{i+1} ($1 \leq i < n$).

An easy inductive argument shows that $r_k = q_i[e_k]$ whenever $k \leq i$. In particular, let $k = j_{i+1}$, and let $V := e_{i+1} \cap (e_1 \cup \ldots \cup e_i)$. Since $V \subseteq e_m$, it follows that $r_k[V] = q_i[V]$. But also $r_{i+1}[V] = r_k[V]$ since r_{i+1} and r_k are consistent. Hence $r_{i+1}[V] = q_i[V]$. So r_{i+1} is consistent with q_i which was to be shown.

(ii) $\Longrightarrow$ (iii): This is immediate since every monotone sequential join expression is a monotone join expression.

(iii) $\implies$ (iv): Assume that $\mathcal{E}$ has a monotone join expression. We must show that every pairwise consistent database over $\mathcal{E}$ is globally consistent. Let r be a pairwise consistent database over $\mathcal{E}$. It is not hard to see that since no tuples are lost in joining together the relations in r as dictated by the monotone join expression, it follows that every member of r is a projection of the final result $\bowtie r$. Hence r is globally consistent, which was to be shown.

(iv) $\implies$ (v): Details are described in [29].

(v) $\implies$ (i): Details are described in [29]—GYO reduction succeeds.

(iv) $\implies$ (vi): Assume that every pairwise consistent database over $\mathcal{E}$ is globally consistent. Let $r_1, \ldots, r_m$ be a database over $\mathcal{E}$. We have to show that $r_1, \ldots, r_m$ has a full reducer, that is, after finitely many semijoins $r_i \ltimes r_j$, we obtain a globally consistent database. Note that when further semijoin operations do not change anything, the resulting relations are pairwise consistent. By assumption, these are also globally consistent which means that we have a full reducer.

(vi) $\implies$ (iv): Assume that every database over $\mathcal{E}$ has a full reducer. Let $r_1, \ldots, r_m$ be a pairwise consistent database over $\mathcal{E}$. By assumption, it has a full reducer but in the case of pairwise consistent relations, the input and output of the full reducer is the same, that is, the result of the full reducer is the database $r_1, \ldots, r_m$ itself, and the result of a full reducer is guaranteed to be globally consistent. Thus, $r_1, \ldots, r_m$ is globally consistent. ∎

29.3.7 Dually Chordal Graphs, Maximum Neighborhood Orderings, and Hypertrees

Theorem 29.2 says that a graph is chordal if and only if it has a clique tree, that is, a graph G is chordal if and only if its hypergraph $\mathcal{C}(G)$ of maximal cliques is α-acyclic (or co-arboreal). The dual variant of this means that $\mathcal{C}(G)$ is a hypertree; the corresponding graph class called *dually chordal graphs* was studied in [59–61] and has remarkable properties. In particular, the notion of *maximum neighbor* and *maximum neighborhood ordering* (used in [60,62,63]) has many consequences for algorithmic applications and is somehow dual to the notion of a simplicial vertex. For the next definition, we need the notation of neighborhood in the remaining subgraph.

Let $G = (V, E)$ be a graph and $(v_1, \ldots, v_n)$ be a vertex ordering of G. For all $i \in \{1, \ldots, n\}$, let $G_i := G[\{v_i, \ldots, v_n\}]$ and $N_i[v]$ be the neighborhood of v in G_i: $N_i[v] := N[v] \cap \{v_i, \ldots, v_n\}$.

Definition 29.22 *Let $G = (V, E)$ be a graph.*

i. *A vertex $u \in N[v]$ is a maximum neighbor of v if for all $w \in N[v]$, $N[w] \subseteq N[u]$, that is, $N^2[v] = N[u]$. (Note that possibly $u = v$ in which case v sees all vertices of G.)*

ii. *A vertex ordering $(v_1, v_2, \ldots, v_n)$ of V is a maximum neighborhood ordering of G if for all $i \in \{1, \ldots, n\}$, v_i has a maximum neighbor in G_i, that is, there is a vertex $u_i \in N_i[v_i]$ such that for all $w \in N_i[v_i]$, $N_i[w] \subseteq N_i[u_i]$ holds.*

iii. *A graph is dually chordal if it has a maximum neighborhood ordering.*

Note that dually chordal graphs are not a hereditary class; adding a universal vertex makes every graph dually chordal. The following characterization of dually chordal graphs shows that these graphs are indeed dual (in the hypergraph sense) with respect to chordal graphs:

Theorem 29.18 [59,61] *For a graph G, the following conditions are equivalent:*

   i. *G has a maximum neighborhood ordering.*

   ii. *There is a spanning tree T of G such that every maximal clique of G induces a subtree of T.*

   iii. *There is a spanning tree T of G such that every disk of G induces a subtree of T.*

   iv. *$\mathcal{N}(G)$ is a hypertree.*

   v. *$\mathcal{N}(G)$ is α-acyclic.*

Proof. Let $G = (V, E)$ be a graph.

(i) $\Longrightarrow$ (ii): By induction on $|V|$. Let x be the leftmost vertex in a maximum neighborhood ordering of G and let y be a maximum neighbor of x, that is, $N^2[x] = N[y]$. If $x = y$, that is, $N^2[x] = N[x]$, then x sees all other vertices of G; let T be a star with central vertex x which fulfills (ii). Now assume that $x \neq y$; by induction hypothesis, there is a spanning tree of the graph $G - x$ fulfilling (ii) for $G - x$. Among all such spanning trees, choose a tree T in which y is adjacent to a maximum number of vertices from $N(x)$.

Claim 29.1 *In T, y sees all vertices of $N(x) \setminus \{y\}$.*

Proof of Claim 29.1. Assume to the contrary that there is a vertex $z \in N(x) \setminus \{y\}$ which is nonadjacent to y in T. Consider the T-path $y - \ldots - v - z$ connecting y and z. Let T_v (T_z, respectively) be the connected component of T obtained by deleting the T-edge vz such that T_v contains v (T_z contains z, respectively). Adding to these subtrees T_v, T_z a new edge yz, we obtain the tree T'. Since y and z are adjacent in $G - x$, T' is a spanning tree of $G - x$. Now we show that T' fulfills condition (ii) as well.

Let Q be a maximal clique of $G - x$. If $z \notin Q$ then Q is completely contained in one of the subtrees T_v, T_z, that is, Q induces one and the same subtree in both T and T'. Now suppose that $z \in Q$. Since $N[z] \subseteq N[y] = N^2[x]$, we have $y \in Q$ by maximality of Q. Let u_1, u_2 be any two vertices of Q. If both belong to the same subtree T_v or T_z then u_1 and u_2 are connected by the same path in T and T', and we are done. Now let u_1 be in T_v and u_2 be in T_z. In T_v, the vertices y and u_1 are connected by a T-path P_1 consisting of vertices from Q. In a similar way, the vertices z and u_2 are connected by a T-path P_2 in T_z. Gluing together these paths P_1 and P_2 with the edge yz, we obtain a T'-path connecting u_1 and u_2 in T'. Hence any maximal clique Q of $G - x$ induces a subtree in T', that is, T' satisfies condition (ii) as well. This, however, contradicts to the choice of T; thus, in T, y sees all vertices of $N(x) \setminus \{y\}$ which shows Claim 29.1.

Now let T be a spanning tree fulfilling the claim for $G - x$. Let T^* be the tree obtained from T by adding a leaf x adjacent to y. Obviously, T^* fulfills condition (ii).

(ii) $\Longrightarrow$ (iii): Let T be a spanning tree of G such that every clique of G induces a subtree in T. We claim that every disk $N^r[z]$ induces a subtree in T as well. In order to prove this, it is sufficient to show that the vertex z and every vertex $v \in N^r[z]$ are connected by a T-path consisting of vertices from $N^r[z]$. Let $v = v_1 - v_2 - \ldots - v_k - v_{k+1} = z$ be a shortest G-path between v and z. By Q_i we denote a maximal clique of G containing the edge $v_i v_{i+1}$, $i \in \{1, \ldots, k\}$. From the choice of T, it follows that v_i and v_{i+1} are connected by a T-path $P_i \subseteq Q_i$. The vertices of $P = \bigcup_{i=1}^{k} P_i$ induce a subtree $T[P]$ of T. Thus, v and z are connected by a T-path p. Since for all vertices $w \in Q_i$, for the G-distances $d(z, w) \leq d(z, v_i) \leq r$ holds,

every clique Q_i is contained in the disk $N^r[z]$. Thus, the claim follows from the obvious inclusion
$$p \subseteq P \subseteq \bigcup_{i=1}^{k} Q_i \subseteq N^r[z]$$

(iii) $\implies$ (iv) is obvious.

(iv) $\iff$ (v) is obvious by the self-duality of the neighborhood hypergraph $\mathcal{N}(G)$ and the duality between hypertree and α-acyclicity.

(iv) $\implies$ (i): Let $\mathcal{N}(G)$ be a hypertree. Then by Theorem 29.13, $\mathcal{N}(G)$ has the Helly property and $L(\mathcal{N}(G))$ is chordal. Let $\sigma = (e_1, \ldots, e_m)$ be a perfect elimination ordering of $L(\mathcal{N}(G))$. Since the hyperedges e_i of $\mathcal{N}(G)$ are the closed neighborhoods, $\sigma = (N[v_1], \ldots, N[v_n])$. Suppose inductively that there is a maximum neighborhood ordering for $G - v_1$. It suffices to show that v_1 has a maximum neighbor u_1. Since $N[v_1]$ is simplicial in $L(\mathcal{N}(G))$, the closed neighborhoods intersecting $N[v_1]$ are pairwise intersecting. By the Helly property of $\mathcal{N}(G)$, there is a vertex u_1 in the intersection of all such closed neighborhoods including $N[v_1]$ itself, that is, there is a vertex u_1 with $N^2[v_1] = N[u_1]$. Thus, u_1 is a maximum neighbor of v_1.

The equivalence of (i) and (iv) can be shown in an easy direct way as follows: We know already $(iv) \implies (i)$. By Theorem 29.13, we can also show the other direction

$(i) \implies (iv)$: Let G have the maximum neighborhood ordering $\sigma = (v_1, \ldots, v_n)$. We have to show that $\mathcal{N}(G)$ has the Helly property and $L(\mathcal{N}(G))$ is chordal.

Let $N[x_1], \ldots, N[x_k]$ be a collection of pairwise intersecting closed neighborhoods in G. Without loss of generality, let x_1 be the leftmost vertex of $x_1, \ldots, x_k$ in σ. Then x_1 has a maximum neighbor u_1, that is, there is a vertex u_1 for which $N^2[x_1] = N[u_1]$. Then $u_1 \in \bigcap_{i=1}^{k} N[x_i]$, and thus, $\mathcal{N}(G)$ has the Helly property.

Now we show that $N[v_1]$ is simplicial in $L(\mathcal{N}(G))$: Let $N[x]$ and $N[y]$ be closed neighborhoods intersecting $N[v_1]$. Let v_1 have a maximum neighbor u_1, that is, $N^2[v_1] = N[u_1]$. Since $x, y \in N^2[v_1]$, it follows that $u_1 \in N[x] \cap N[y]$ and thus, $N[v_1]$ is simplicial in $L(\mathcal{N}(G))$. Inductively, it follows that $\sigma = (N[v_1], \ldots, N[v_n])$ is a perfect elimination ordering of $L(\mathcal{N}(G))$. ∎

Since $L(\mathcal{N}(G))$ is isomorphic to G^2 (recall Proposition 29.7), Theorem 29.18 implies the following corollary.

Corollary 29.7 *Graph G is dually chordal if and only if G^2 is chordal and $\mathcal{N}(G)$ has the Helly property.*

Another characterization which follows from the basic properties is the following.

Corollary 29.8 *Graph G is dually chordal if and only if $G = L(H)$ for some α-acyclic hypergraph H.*

As a corollary of Theorem 29.18, dually chordal graphs can be recognized in linear time since α-acyclicity of $\mathcal{N}(G)$ can be tested in linear time [20]. Parts of Theorem 29.18 were found also by Szwarcfiter and Bornstein [64] and later again by Gutierrez and Oubiña [65]; in particular, it was shown in [64] that dually chordal graphs are the clique graphs of intersection graphs of paths in a tree. This implies that dually chordal graphs are the clique graphs of chordal graphs (in the sense of Definition 29.14). See [63,66] for algorithmic applications of maximum neighborhood orderings and [3] for more structural details. In [19], a linear-time algorithm for constructing a special (canonical) maximum neighborhood ordering for a dually chordal graph is described.

New characterizations of dually chordal graphs in terms of separator properties are given by De Caria and Gutierrez in [67–70]. Another new characterization was found by Leitert in [71].

In [72], Moscarini introduced the concept of *doubly chordal graphs*, that is, the graphs which are chordal and dually chordal. This class was introduced for efficiently solving the Steiner problem (motivated by database theory); this can be done, however, also for the larger class of dually chordal graphs (see [63]) and also for the class of homogeneously orderable graphs which contain the dually chordal graphs [73]:

$$\text{doubly chordal} \subset \text{dually chordal} \subset \text{homogeneously orderable}$$

29.3.8 Bipartite Graphs, Hypertrees, and Maximum Neighborhood Orderings

For bipartite graphs $B = (X, Y, E)$, the one-sided neighborhood hypergraphs are of fundamental importance. Let

$$\mathcal{N}^X(B) = \{N(y) \mid y \in Y\} \text{ as well as}$$

$$\mathcal{N}^X(B) = \{N(x) \mid x \in X\}.$$

Note that $(\mathcal{N}^X(B))^* = \mathcal{N}^Y(B)$ and vice versa.

Motivated by database schemes, the following concepts were introduced.

Definition 29.23 [28] *Let $B = (X, Y, E)$ be a bipartite graph.*

i. *B is X-conformal if for all $S \subseteq Y$ with the property that all vertices of S have pairwise distance 2, there is an $x \in X$ with $S \subseteq N(x)$.*

ii. *B is X-chordal if for every cycle C in B of length at least 8, there is a vertex $x \in X$ which is adjacent to at least two vertices of C whose distance in C is at least 4.*

Analogously, define Y-conformal and Y-chordal for bipartite graphs.

These notions are justified by the following simple facts.

Proposition 29.10 [28] *Let $B = (X, Y, E)$ be a bipartite graph.*

i. *B is X-conformal $\iff \mathcal{N}^Y(B)$ is conformal $\iff \mathcal{N}^X(B)$ has the Helly property.*

ii. *B is X-chordal $\iff 2SEC(\mathcal{N}^Y(B))$ is chordal $\iff L(\mathcal{N}^X(B))$ is chordal.*

Corollary 29.9 *The following conditions are equivalent:*

i. *B is X-chordal and X-conformal;*

ii. *$\mathcal{N}^Y(B)$ is α-acyclic;*

iii. *$\mathcal{N}^X(B)$ is a hypertree.*

Maximum neighborhood orderings can be defined for bipartite graphs as well. For this we need the following notations: Let $B = (X, Y, E)$ be a bipartite graph, and let $(y_1, \ldots, y_n)$ be a vertex ordering of Y. Then let $B_i^Y = B[X \cup \{y_i, y_{i+1}, \ldots, y_n\}]$ and let $N_i(x)$ denote the neighborhood of x in the remaining subgraph B_i^Y.

Definition 29.24 *Let $B = (X, Y, E)$ be a bipartite graph.*

i. *For $y \in Y$, a vertex $x \in N(y)$ is a maximum neighbor of y if for all $x' \in N(y)$, $N(x') \subseteq N(x)$ holds.*

ii. A *linear ordering* $(y_1, \ldots, y_n)$ *of* Y *is a* maximum X-neighborhood ordering *of* B *if for all* $i \in \{1, \ldots, n\}$, *there is a maximum neighbor* $x_i \in N_i(y_i)$ *of* y_i: *for all* $x \in N(y_i)$, $N_i(x) \subseteq N_i(x_i)$ *holds.*

Analogously, maximum Y-neighborhood orderings *are defined.*

Theorem 29.19 [59] *Let* $B = (X, Y, E)$ *be a bipartite graph. The following conditions are equivalent:*

i. *B has a maximum X-neighborhood ordering;*

ii. *B is X-conformal and X-chordal.*

Moreover, $(y_1, \ldots, y_n)$ *is a maximum X-neighborhood ordering of B if and only if* $(y_1, \ldots, y_n)$ *is a p.e.o. of* $2SEC(\mathcal{N}^Y(B))$.

Proof. (i) $\Longrightarrow$ (ii): Let $\sigma = (y_1, \ldots, y_n)$ be a maximum X-neighborhood ordering of Y.

(a) B is X-conformal: Assume that the vertices in $S \subseteq Y$ have pairwise distance 2. Let $y_j \in S$ be the leftmost vertex of S in σ and let x be a maximum neighbor of y_j in B_j^Y. Since every $y' \in S$ has a common neighbor $x' \in X$ with y_j, also x is adjacent to y' which implies $S \subseteq N(x)$. Thus, B is X-conformal.

(b) B is X-chordal: Let $C = (x_{i_1}, y_{i_1}, \ldots, x_{i_k}, y_{i_k})$, $k \geq 4$, be a cycle in B. If C has a chord then it has an X-vertex which fulfills the condition. Now assume that C is a chordless cycle. Let $y_{i_1} = y_j$ be the leftmost Y-vertex of C in $(y_1, \ldots, y_n)$. Since $y_{i_k} \in N_j(x_{i_1}) \setminus N_j(x_{i_2})$ and $y_{i_2} \in N_j(x_{i_2}) \setminus N_j(x_{i_1})$, the sets $N_j(x_{i_1})$ and $N_j(x_{i_2})$ are incomparable with respect to set inclusion. Thus, neither x_{i_1} nor x_{i_2} are maximum neighbors of y_{i_1}. Let x be a maximum neighbor of $y_{i_1} = y_j$. Then $y_{i_1}, y_{i_2}, y_{i_k} \in N_j(x)$. Thus, B is X-chordal.

(ii) $\Longrightarrow$ (i): Let B be X-conformal and X-chordal. Then by Proposition 29.10, the line graph $G' = L(\mathcal{N}^X(B))$ is chordal and $\mathcal{N}^X(B)$ has the Helly property. Let $(y_1, \ldots, y_n)$ be a p.e.o. of G'. Thus $N_{G'}[y_1]$ is a a clique, that is, for all $y, y' \in N_{G'}[y_1]$, $N(y) \cap N(y') \neq \emptyset$. By the Helly property of $\mathcal{N}^X(B)$, the total intersection of all $N(y)$ such that $N(y) \cap N(y_1) \neq \emptyset$ is nonempty: there is a vertex $x \in X$ in all these neighborhoods. Now, x is a maximum neighbor of y_1, and the same argument can be repeated with the smaller graph $B - y_1$. ∎

Corollary 29.10 *Let* $B = (X, Y, E)$ *be a bipartite graph. The following conditions are equivalent:*

i. *B has a maximum X-neighborhood ordering.*

ii. $\mathcal{N}^X(B)$ *is a hypertree.*

iii. $\mathcal{N}^Y(B)$ *is α-acyclic.*

Theorems 29.18 and 29.19 imply the following connection between maximum neighborhood orderings in graphs and in bipartite graphs.

Corollary 29.11 [59] *A graph G has a maximum neighborhood ordering if and only if $B(G)$ has a maximum X-neighborhood ordering (maximum Y-neighborhood ordering, respectively).*

Proof. Recall that by Proposition 29.8, $B(G)$ is isomorphic to the bipartite incidence graph of $\mathcal{N}(G)$. By Theorem 29.18, G has a maximum neighborhood ordering if and only if $\mathcal{N}(G)$ is a hypertree. Now, it is easy to see that the underlying tree of $\mathcal{N}(G)$ immediately leads to the fact that $\mathcal{N}^X(B(G))$ is a hypertree as well, and for symmetry reasons the same happens for $\mathcal{N}^Y(B(G))$. Conversely, if $\mathcal{N}^X(B(G))$ is a hypertree then the underlying tree immediately leads to an underlying tree for $\mathcal{N}(G)$. ∎

29.3.9 Further Matrix Notions

As already mentioned, a hypergraph $H = (V, \mathcal{E})$ can be described by its incidence matrix. The notion of a hypertree (see Definition 29.17) is also close to what is called *subtree matrix* in [74].

Definition 29.25

i. *A Γ matrix has the form*

$$\begin{array}{|cc|} \hline 1 & 1 \\ 1 & 0 \\ \hline \end{array}$$

ii. *A subtree matrix is the incidence matrix of a collection of subtrees of a tree, that is, it is a $(0,1)$-matrix with rows indexed by vertices of a tree T and columns indexed by some subtrees of T and with an entry of 1 if and only if the corresponding vertex is in the corresponding subtree.*

iii. *An ordered $(0,1)$-matrix M is supported Γ if for every pair $r_1 < r_2$ of row indices and $c_1 < c_2$ of column indices whose entries form a Γ, there is a row index $r_3 > r_2$ with $M(r_3, c_1) = M(r_3, c_2) = 1$. One says that row r_3 supports the Γ.*

Theorem 29.20 [74] *A $(0,1)$-matrix is a subtree matrix if and only if it is a matrix with supported Γ ordering.*

Proof. "$\Longrightarrow$": Let M be a subtree matrix for a collection $\mathcal{S}$ of subtrees of a tree T. Pick a vertex r of T and order the vertices of T by decreasing distance from r (breaking ties arbitrarily). The distance between r and a subtree S is the minimum distance between r and any vertex from S. Also order the subtrees from $\mathcal{S}$ by decreasing distance from r.

We claim that this is a supported Γ ordering of M, for suppose vertices $v_1 < v_2$ and subtrees $t_1 < t_2$ form a Γ in M: For $i \in \{1, 2\}$, let r_i be the vertex of t_i closest to r. Then $r_1 \geq v_2$ since v_2 is in t_1. We claim that r_1 supports the Γ: Since r_1 is in t_1, $M(r_1, t_1) = 1$. We have to show that r_1 is also in t_2, that is, $M(r_1, t_2) = 1$. If $r_1 = v_1$ or $r_1 = r_2$, we are done. Now suppose that $r_1 \neq v_1$ and $r_1 \neq r_2$.

Since $t_1 < t_2$, r_2 is closer to r than r_1 but t_1 and t_2 contain a common vertex v_1, and thus also r_1 is on the T path between v_1 and r_2, that is, r_1 is in subtree t_2 and supports the Γ.

"$\Longleftarrow$": If the ordered $n \times m$ matrix M is supported Γ, create a tree T on vertex set $\{1, 2, \ldots, n\}$ by setting for $i \in \{1, 2, \ldots, n-1\}$

$$f(i) = \begin{cases} \min\{k \mid M(i, k) = 1\} & \text{if there exists } j > i, M(j, k) = 1 \\ \text{not defined} & \text{otherwise} \end{cases}$$

and $b(i) = \max\{j \mid M(j, f(i)) = 1\}$ and creating the edges $(i, b(i))$ if $f(i)$ exists and (i, n) otherwise. We claim that T defined in this way is a tree: Since $b(i) > i$ when $f(i)$ (and thus also $b(i)$) exists, and the edges (i, n) otherwise, T is obviously cycle-free, and for the same reason, T is connected.

Finally, we show that each column of M is the incidence vector of a subtree of T. It suffices to show that for $i < j$, $M(i, k) = M(j, k) = 1$ implies $M(b(i), k) = 1$: If this were not true then $f(i) < k$ and rows $i, b(i)$ and columns $f(i), k$ would form an unsupported Γ in M. ∎

Note that Theorem 29.20 gives a characterization of hypertrees and of α-acyclic hypergraphs in terms of a matrix property. This also gives corresponding characterizations of chordal as well as of dually chordal graphs.

Definition 29.26 *A $(0,1)$-matrix M is doubly lexically ordered if the rows (columns, respectively) form a lexicographically increasing sequence from top to bottom (from left to right, respectively) where for rows (columns, respectively), the rightmost position (lowest position, respectively) has highest priority.*

Theorem 29.21 [74] *Every $(0,1)$-matrix M can be doubly lexically ordered by some suitable permutations of rows and columns.*

Proof. Let $M = (M_{ij})$ be an $m \times n$ matrix. We form a $m \cdot n$ vector $d(M)$ as follows: The entries of M will be ordered with respect to $i+j$, and for the same $i+j$ with respect to j:

$$d(M) = M_{11}, M_{21}, M_{12}, M_{31}, M_{22}, M_{13}, M_{41}, \ldots, M_{mn}$$

$$\begin{vmatrix} d_1 & d_3 & d_6 & \cdot & \ldots \\ d_2 & d_5 & \cdot & & \\ d_4 & \cdot & & & \\ \cdot & & & & \\ \ldots & & & & d_{m \cdot n} \end{vmatrix}$$

Claim 29.2 *If two rows (columns, respectively) of M are permuted which do not appear in lexical order then the result $d(M)$ will be lexically larger (with highest priority at $d_{m \cdot n}$).*

Proof. of Claim 29.2. Let k, l be row indices of M with $k < l$ and the property that the kth row is lexically larger than the lth row.

Let $j \in \{1, \ldots, n\}$ be the largest index for which $M_{kj} \neq M_{lj}$; then $M_{kj} > M_{lj}$. After permuting the kth and lth row, the part of $d(M)$ which was M_{lj} becomes M_{kj} and the parts on its right hand side do not change their value, and analogously for columns. This shows Claim 29.2.

By Claim 29.2, an ordering of M which maximizes $d(M)$, is a doubly lexical ordering of M. ∎

Theorem 29.21 holds also for other ordered matrix entries instead of $\{0,1\}$.

There is an efficient way for finding a doubly lexical ordering: Let $L := n + m +$ number of 1's in a $(0,1)$-matrix M.

Theorem 29.22 [75] *A doubly lexical ordering of an $m \times n$ matrix M over entries $\{0,1\}$ can be determined in $O(L \log L)$ steps.* ∎

29.4 TOTALLY BALANCED HYPERGRAPHS AND MATRICES

29.4.1 Totally Balanced Hypergraphs versus β-Acyclic Hypergraphs

Fagin [6,7] defined β-acyclic hypergraphs in connection with desirable properties of relational database schemes. Recall that for α-acyclic hypergraphs, edge-subhypergraphs are not necessarily α-acyclic.

Definition 29.27 [6,7] *A hypergraph $H = (V, \mathcal{E})$ is β-acyclic if each of its edge-subhypergraphs is α-acyclic, that is, for all $\mathcal{E}' \subseteq \mathcal{E}$, $\mathcal{E}'$ is α-acyclic.*

Fagin [6] gives a variety of equivalent notions of β-acyclicity in terms of certain forbidden cycles in hypergraphs (one of them goes back to Graham [56]) which Fagin in [6] shows to be equivalent.

Actually, β-acyclic hypergraphs appear under the name *totally balanced hypergraphs* much earlier in hypergraph theory (as it will turn out in Theorems 29.25 and 29.27).

Definition 29.28 [12,76] *Let $H = (V, \mathcal{E})$ be a hypergraph.*

i. *A sequence $C = (v_1, e_1, v_2, e_2, \ldots, v_k, e_k)$ of distinct vertices $v_1, v_2, \ldots, v_k$ and distinct hyperedges $e_1, e_2, \ldots, e_k$ is a* special cycle *(or* chordless cycle *or* induced cycle *or, in* [77], unbalanced circuit*) if $k \geq 3$ and for every i, $1 \leq i \leq k$, $v_i, v_{i+1} \in e_i$ (index arithmetic is done modulo k) and $e_i \cap \{v_1, \ldots, v_k\} = \{v_i, v_{i+1}\}$. The* length *of cycle C is k.*

ii. *H is* balanced *if it has no special cycles of odd length $k \geq 3$.*

iii. *H is* totally balanced *if it has no special cycles of any length $k \geq 3$.*

Special cycles are called *weak β-cycles* by Fagin in [6], and a hypergraph is called *β-acyclic* if it has no weak β-cycles. Actually, [6] mentions four other conditions and shows that all five are equivalent.

We will see in Theorem 29.27 that a hypergraph is β-acyclic if and only if it is totally balanced. Totally balanced hypergraphs are a natural generalization of trees.

Balanced hypergraphs are a natural generalization of bipartite graphs. See the monograph of Berge [13] for many properties and characterizations of balanced hypergraphs, and in particular, the following theorems.

Theorem 29.23 [13] *A hypergraph is balanced if and only if its vertex-subhypergraphs are 2-colorable.* ■

Theorem 29.24 [78] *A hypergraph is balanced if and only if its vertex- and edge-subhypergraphs have the Kőnig property.* ■

Corollary 29.12 [13] *Balanced hypergraphs have the Helly property and are conformal.*

In this subsection, we focus on totally balanced hypergraphs.

Proposition 29.11 *Let $H = (V, \mathcal{E})$ be a totally balanced hypergraph. Then the following holds:*

i. *The dual H^* of H and any vertex- or edge-subhypergraph of H are totally balanced.*

ii. *H has the Helly property.*

iii. *$L(H)$ is a chordal graph.*

Proof. Let $H = (V, \mathcal{E})$ be a totally balanced hypergraph.

i. By definition, it immediately follows that the dual H^* and any vertex- or edge-subhypergraph of a totally balanced hypergraph H is totally balanced.

ii. Since the Helly property is satisfied by any hypergraph without special cycle of length three (see Theorem 29.8), H must have the Helly property.

iii. $L(H)$ is a chordal graph since H contains no special cycle. ■

Recall that vertex-subhypergraphs of hypertrees are not necessarily hypertrees. The next theorem gives a characterization of totally balanced hypergraphs in terms of hypertrees.

Theorem 29.25 [36,46] *A hypergraph H is totally balanced if and only if every vertex-subhypergraph of H is a hypertree.*

Proof. Let $H = (V, \mathcal{E})$ be a hypergraph. By Proposition 29.11, we have:

i. The dual of H and any vertex- or edge-subhypergraph are totally balanced.

ii. H has the Helly property.

iii. $L(H)$ is a chordal graph.

Now by Theorem 29.13, H (and every vertex-subhypergraph of H) must be a hypertree and vice versa. ∎

Actually, Lehel [46] gives a complete structural characterization of totally balanced hypergraphs in terms of certain tree sequences. Lehel's result implies the following characterization of totally balanced hypergraphs which was originally found by Brouwer and Kolen [79] and nicely corresponds to the existence of simple vertices in strongly chordal graphs.

Theorem 29.26 [46,79] *A hypergraph H is totally balanced if and only if every vertex-subhypergraph H' has a vertex v (a so-called* nested point*) such that the hyperedges of H' containing v are linearly ordered by inclusion.* ∎

By simple duality arguments, the next theorem follows immediately from Theorem 29.25.

Theorem 29.27 [80] *A hypergraph H is totally balanced if and only if H is β-acyclic.* ∎

29.4.2 Totally Balanced Matrices

Definition 29.29 *Let B_k denote the $k \times k$ square $(0,1)$-matrix consisting of rows with exactly two consecutive 1's beginning with $10\ldots 01$, then $110\ldots$ and so on; $00\ldots 11$ is the last row.*

For example, B_4 is the following matrix:

$$\begin{array}{|cccc|} \hline 1 & 0 & 0 & 1 \\ 1 & 1 & 0 & 0 \\ 0 & 1 & 1 & 0 \\ 0 & 0 & 1 & 1 \\ \hline \end{array}$$

Thus, a hypergraph is totally balanced if and only if its incidence matrix contains no square submatrix B_k for $k \geq 3$ (in any row and column order), and correspondingly for balanced hypergraphs and odd $k \geq 3$. Obviously, the dual of a balanced (totally balanced, resp.) hypergraph is balanced (totally balanced, respectively).

Lubiw in [74] defines totally balanced matrices in the following way.

Definition 29.30

i. *For $n \geq 3$, a* cycle matrix *is an $n \times n$ $(0,1)$-matrix with no identical rows and columns which has exactly two 1's in each row and in each column such that no proper submatrix has this property.*

ii. *A* totally balanced matrix *is a $(0,1)$-matrix with no cycle submatrices.*

Recall Definition 29.25 for the notion of a Γ submatrix.

Definition 29.31 *An ordered $(0,1)$-matrix M is Γ-free if M has no Γ submatrix.*

Theorem 29.28 [81–83] *A $(0,1)$-matrix is totally balanced if and only if it has a Γ-free ordering.* ∎

This is shown in [74] as a consequence of the existence of doubly lexical orderings and the following.

Observation 29.1 *If a $(0,1)$-matrix has a cycle submatrix then for any ordering of the matrix there is a Γ submatrix (formed by a topmost, leftmost 1 of the cycle submatrix; the other 1 in its row in the cycle; and the other 1 in its column in the cycle).*

Theorem 29.29 [74] *Any doubly lexical ordering of a totally balanced matrix is Γ-free.* ∎

For example, the matrix B_4 from Definition 29.29 has the following doubly lexical ordering which is not Γ-free:

$$\begin{bmatrix} 1 & 1 & 0 & 0 \\ 1 & 0 & 1 & 0 \\ 0 & 1 & 0 & 1 \\ 0 & 0 & 1 & 1 \end{bmatrix}$$

(resulting from the matrix B_4 by first permuting rows 1 and 2 and then permuting columns 3 and 4).

29.5 STRONGLY CHORDAL AND CHORDAL BIPARTITE GRAPHS

29.5.1 Strongly Chordal Graphs

The subsequently defined *strongly chordal graphs* are an important subclass of chordal graphs for many reasons. Originally, they were introduced by Farber [27] as a subclass of chordal graphs for which the domination problem ([GT2] in [40]), which remains NP-complete for chordal graphs and even for split graphs [84], can be solved efficiently. Chang and Nemhauser [85,86] independently studied the same class and also showed that some problems such as domination can be solved efficiently. Later on, this has been extended to larger classes and other problems (see, e.g., [63,66,73,87]).

The motivation from the database community is the fact that strongly chordal graphs are the 2-section graphs of β-acyclic hypergraphs (as it will turn out in Theorem 29.32 as a consequence of Theorem 29.27).

29.5.1.1 Elimination Orderings of Strongly Chordal Graphs

Farber [27] defined strongly chordal graphs in terms of so-called *strong elimination orderings* which are closely related to neighborhood matrices of these graphs:

Definition 29.32 *Let $\sigma = (v_1, \ldots, v_n)$ be an ordering of the vertex set V of G. The neighborhood matrix $N_\sigma(G)$ ($N(G)$ if σ is understood) of G is the $n \times n$ matrix with entries*

$$n_{ij} = \begin{cases} 1 & \text{if } v_i \in N[v_j] \\ 0 & \text{otherwise} \end{cases}$$

Note that this matrix is symmetric and the main diagonal has values 1:

$$v_i \in N[v_j] \iff v_j \in N[v_i]$$

($N(G)$ is the incidence matrix of the closed-neighborhood hypergraph $\mathcal{N}(G)$).

The subsequent Definition 29.33 must be read as follows: If in the $(0,1)$ neighborhood matrix of graph G, for $i < j$ and $k < \ell$, the entries in row i and column k, in row i and column ℓ as well as in row j and column k are 1, then the entry in row j and column ℓ must be 1 as well (i.e., rows $i < j$ and columns $k < \ell$ do not form a Γ—see Definition 29.25).

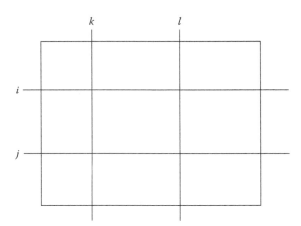

Figure 29.2 Selected rows and columns.

Definition 29.33 [27] *Let $G = (V, E)$ be a graph.*

i. *A vertex ordering $(v_1, \ldots, v_n)$ of G is a* **strong elimination ordering** *(st.e.o.) if for all i, j, k and ℓ with $i < j, k < \ell$ for $v_k, v_\ell \in N[v_i]$ holds: if $v_j \in N[v_k]$ then also $v_j \in N[v_\ell]$.*

ii. *G is* **strongly chordal** *if G has a st.e.o.*

Obviously, every st.e.o. is also a p.e.o. (let $i = k$ in condition (i)); thus, strongly chordal graphs are chordal.

Observation 29.2 *Let $\sigma = (v_1, \ldots, v_n)$ be an ordering of the vertex set V of graph G. Then σ is a st.e.o. of G if and only if the corresponding neighborhood matrix $N_\sigma(G)$ is Γ-free.*

Proof. Let $(v_1, \ldots, v_n)$ be a st.e.o. We consider the ith and the jth row as well as the kth and the lth column of the matrix, $i < j$, $k < l$.

Figure 29.2 schematically indicates the selected rows and columns.

If $n_{ik} = 1, n_{il} = 1$ and $n_{jk} = 1$ then $v_k \in N[v_i]$, $v_l \in N[v_i]$, $v_j \in N[v_k]$ and thus also $v_j \in N[v_l]$, that is, $n_{jl} = 1$.

Conversely, if the submatrix $\begin{smallmatrix} 11 \\ 10 \end{smallmatrix}$ is forbidden then obviously $(v_1, \ldots, v_n)$ is a st.e.o.

Observation 29.2 describes an important matrix aspect of strongly chordal graphs. We will also show that strongly chordal graphs are the hereditarily dually chordal graphs. For this, we need the following notion:

Definition 29.34 [27] *Let $G = (V, E)$ be a graph.*

i. *A vertex $v \in V$ is called* **simple** *if the set of closed neighborhoods $\{N[u] \mid u \in N[v]\}$ is linearly ordered with respect to set inclusion.*

ii. *A vertex ordering $(v_1, \ldots, v_n)$ of V is a* **simple elimination ordering** *(si.e.o.) if for all $i \in \{1, \ldots, n\}$, v_i is simple in $G_i = G[\{v_i, \ldots, v_n\}]$.*

It is easy to see that every simple vertex is also simplicial, that is, whenever a graph has a simple elimination ordering, it is chordal.

For proving Theorem 29.30, we need the following property.

Lemma 29.6 *Let v be simple in $G = (V, E)$ and $u_0 \in N[v]$ be a vertex with smallest neighborhood $N[u_0]$. Then also u_0 is simple in G.*

Proof. Assume that u_0 is not simple. Then let $x, y \in N[u_0]$ be two vertices with incomparable neighborhoods $N[x], N[y]$. Since $\{N[u] \mid u \in N[v]\}$ is linearly ordered with respect to $\subseteq$, for all $u \in N[v]$ $N[u_0] \subseteq N[u]$ holds, in particular for $u = v$, $N[u_0] \subseteq N[v]$. Thus, v has two neighbors with incomparable neighborhood—contradiction. ∎

Theorem 29.30 [27] *A graph G has a st.e.o. if and only if every induced subgraph of G contains a simple vertex.*

Proof. "$\Longrightarrow$": If G has a st.e.o. $(v_1, \ldots, v_n)$ then also every induced subgraph of G has such an ordering by Definition 29.33. We show that v_1 is simple.

Let $v_k, v_l \in N[v_1]$ with $k < l$ and $v_j \in N[v_k]$ with $1 < j$. By Definition 29.33, it follows immediately that $v_j \in N[v_l]$. Thus, $N[v_k] \subseteq N[v_l]$, and v_1 is simple (which means that the st.e.o. is also a si.e.o.).

"$\Longleftarrow$": Assume that every induced subgraph of G contains a simple vertex. We recursively construct a st.e.o. $(v_1, \ldots, v_n)$ of G as follows: For every $1 \leq i \leq n$, choose in $G_i = G(\{v_i, \ldots, v_n\})$ a simple vertex v_i with smallest $|N_i[v_i]|$.

We claim that this ordering is a st.e.o. Since the vertex v_i is simple in G_i, that is, for their neighbors from $N_i[v_i]$, the neighborhoods are linearly ordered with respect to $\subseteq$, we have the following.

The vertices from $N_i[v_i]$ appear in $(v_1, \ldots, v_n)$ in the same order (this follows by Lemma 29.6 for G_i). Now, for $i < j$ and $k < l$ let $v_k, v_l \in N[v_i]$ and $v_j \in N[v_k]$.

Case 1 $i < k$. Then v_i is simple in G_i and $v_k, v_l \in N_i[v_i]$ with $k < l$. Thus, $N_i[v_k] \subseteq N_i[v_l]$ and therefore also $v_j \in N[v_l]$.

Case 2 $i = k$. In this case, the assertion is fulfilled since any simple vertex is simplicial.

Case 3 $i > k$. Then v_k is simple in G_k and $v_i, v_j \in N_k[v_k], i < j$. Thus, $N_k[v_i] \subseteq N_k[v_j]$. From $v_l \in N[v_i]$, $l > k$, it follows that $v_l \in N_k[v_i]$, thus also $v_l \in N_k[v_j]$ and finally $v_j \in N[v_l]$. ∎

Corollary 29.13 *The following conditions are equivalent:*

  i. *G is strongly chordal.*

  ii. *G has a si.e.o.*

  iii. *G is hereditarily dually chordal, that is, every induced subgraph of G is dually chordal.*

  iv. *$\mathcal{N}(G)$ is β-acyclic (i.e., by Theorem 29.27, totally balanced).*

Proof. Theorem 29.30 shows the equivalence of (i) and (ii).

For the equivalence of (ii) and (iii), assume first that G has a si.e.o. Then every induced subgraph of G has a si.e.o. as well, and note that a si.e.o. is also a maximum neighborhood ordering which means that every induced subgraph of G is dually chordal. Conversely, let G be a hereditarily dually chordal graph. Let v_1 have a maximum neighbor u_1, that is, the neighborhood of u_1 is largest among all $N[u]$, $u \in N[v_1]$. Then a straightforward discussion shows that also the subgraph of G induced by $N[v_1] - u_1$ has a maximum neighborhood ordering and so on which leads to a linear ordering of neighborhoods w.r.t. v_1, that is, v_1 is simple. Now the same can be repeated for $G[\{v_2, \ldots, v_n\}]$ which shows the equivalence.

The equivalence of (iii) and (iv) is a simple consequence of Theorem 29.18. ∎

The equivalence of (i) and (iv) has been obtained independently by Iijima and Shibata [88]; they showed that a graph is sun-free chordal (see Theorem 29.33) if and only if its neighborhood matrix is totally balanced.

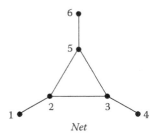

Figure 29.3 The *net*.

29.5.1.2 Γ-Free Matrices and Strongly Chordal Graphs

Definition 29.33 implies a useful characterization of strongly chordal graphs by matrices.

Observation 29.2 leads to the fastest known recognition algorithms for strongly chordal graphs by using doubly lexical orderings of matrices as given in Definition 29.26 (which permute rows and columns in a suitable way)—see the subsequent Theorem 29.31 and Corollary 29.14.

Example 29.3 Take the graph G from Figure 5.1 with vertices $1, \ldots, 6$, edges $12, 23, 25, 34, 35, 56$ and vertex ordering $\sigma_1 = (1, 2, 3, 4, 5, 6)$ (Figure 29.3).

The adjacency matrix M of this graph corresponding to σ_1:

	1	2	3	4	5	6
1	1	1	0	0	0	0
2	1	1	1	0	1	0
3	0	1	1	1	1	0
4	0	0	1	1	0	0
5	0	1	1	0	1	1
6	0	0	0	0	1	1

M is not Γ-free and not doubly lexically ordered (e.g., the third and fifth row together with the second and fourth column form a Γ, and likewise the fourth and fifth row together with the third and fourth column) and not doubly lexically ordered (e.g., the third column from the left is larger than the fourth column).

A strong elimination ordering for G is $\sigma_2 = (1, 4, 6, 2, 3, 5)$. The adjacency matrix M' of G resulting from this ordering σ_2 is as follows:

	1	4	6	2	3	5
1	1	0	0	1	0	0
4	0	1	0	0	1	0
6	0	0	1	0	0	1
2	1	0	0	1	1	1
3	0	1	0	1	1	1
5	0	0	1	1	1	1

M' is doubly lexically ordered.

Theorem 29.21 holds also for other ordered matrix entries instead of $\{0, 1\}$.

Recall Theorem 29.22 for an efficient way for finding a doubly lexical ordering. An efficient (but not linear-time) recognition of strongly chordal graphs results from the following property.

Theorem 29.31 [74] *A graph G is strongly chordal if and only if any doubly lexical ordering of its neighborhood matrix $N(G)$ is Γ-free.* ■

The connection to Γ-free matrices has been used by Paige and Tarjan in [75] as well as by Spinrad in [89] (see also [19]) to design fast (but not linear-time) recognition algorithms for strongly chordal graphs.

Corollary 29.14 *Recognition of strongly chordal graphs can be done in time $O(m \cdot \log n)$.*

It is an open problem whether strongly chordal graphs can be recognized in linear time.

Recall that a hypergraph is defined to be totally balanced if it contains no special cycle (Definition 29.28), and recall Corollary 29.13; this has been expressed in terms of totally balanced matrices.

Recall also (see Definition 29.30) that a $(0,1)$-matrix M is *totally balanced* if M contains no submatrix which is the vertex-edge incidence matrix of a cycle of length ≥ 3 of an undirected graph.

Example 29.4 The vertex-edge incidence matrix of C_3 with vertices v_1, v_2, v_3 and edges $e_1 = \{v_1, v_2\}, e_2 = \{v_2, v_3\}, e_3 = \{v_1, v_3\}$ is

	v_1	v_2	v_3
e_1	1	1	0
e_2	0	1	1
e_3	1	0	1

By Corollary 29.13, G is strongly chordal if and only if its closed neighborhood hypergraph $\mathcal{N}(G)$ is totally balanced. Thus, the next theorem is no surprise:

Theorem 29.32 [27] *A graph G is strongly chordal if and only if its neighborhood matrix $N(G)$ is totally balanced.* ■

29.5.1.3 Strongly Chordal Graphs as Sun-Free Chordal Graphs

Strongly chordal graphs have a variety of different characterizations, among them one in terms of forbidden induced subgraphs. Recall that we say a vertex x *sees* a vertex y if x is adjacent to y; otherwise we say x *misses* y.

Definition 29.35

i. A *k-sun* is a chordal graph G with $2k$ vertices, $k \geq 3$, whose vertex set is partitioned into two sets $W = \{w_0, \ldots, w_{k-1}\}$ and $U = \{u_0, \ldots, u_{k-1}\}$, such that $U = \{u_0, \ldots, u_{k-1}\}$ induces a cycle (the *central clique of the sun*), W is a stable set and for all $i \in \{0, \ldots, k-1\}$, w_i sees exactly u_i and u_{i+1} (index arithmetic modulo k).

ii. A *complete k-sun* is a k-sun where $G[U]$ is a clique.

See, for example, Figure 29.4 for 3-sun and complete 4-sun.

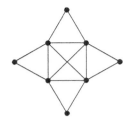

Figure 29.4 3-sun and complete 4-sun.

As shown in [27,85], the following holds.

Lemma 29.7 *In a chordal graph, every k-sun contains a complete k'-sun for some $k' \leq k$.*

Proof. [85] Let $U = \{u_1, \ldots, u_n\}$ and $W = \{w_1, \ldots, w_n\}$ describe the partition of the vertex set of an n-sun G. Since G is chordal and the degree of every vertex w_i in G is 2, its two neighbors u_i and u_{i+1} are adjacent. The proof is by induction on n. If $n = 3$ and $n = 4$ then the claim is obviously fulfilled. Suppose that $n > 4$ and that Lemma 29.7 holds for all suns on fewer than $2n$ vertices and suppose that G is an n-sun which is not complete. Let u_1 miss u_j for some j with $1 < j < n$. Since u_1 sees u_2 and u_n, there exist k and l such that u_1 sees u_k and u_l but misses u_p for any p with $k < p < l$. In that case,

$$G' := G[\{u_1, u_k, u_{k+1}, \ldots, u_l, w_k, \ldots, w_{l-1}\}]$$

is a smaller sun for which $U' := \{u_k, u_{k+1}, \ldots, u_l\}$ and $W' := \{u_1, w_k, w_{k+1}, \ldots, w_{l-1}\}$ gives the required partition. By induction, G' (and hence G) contains a complete sun. ∎

Lemma 29.8 [90] *Let $p \geq 3$ be an integer and suppose G is a graph in which every cycle of length k, for $4 \leq k \leq 2p$, has a chord. Then, $\mathcal{N}(G)$ has an induced special cycle C_p if and only if G has an induced p-sun.*

Proof. Clearly, if K is some induced subgraph of G, $\mathcal{N}(K)$ is isomorphic to an induced partial subhypergraph of $\mathcal{N}(G)$. Thus, the *if* part of Lemma 29.8 is easy and left to the reader.

The converse is proved by contradiction: Suppose that every cycle of G with length k, $4 \leq k \leq 2p$, has a chord and suppose G has no induced sun while $\mathcal{N}(G)$ has an induced special cycle C_p with p vertices and p hyperedges. Thus, by definition, there exists a set $A = \{a_1, \ldots, a_p\}$ and a set $B = \{b_1, \ldots, b_p\}$ with the following properties:

1. $(a_1, N[b_1], \ldots, a_p, N[b_p])$ is a special cycle in $\mathcal{N}(G)$.

2. $N[b_j] \cap A = \{a_j, a_{j+1}\}$ for every $j, 1 \leq j \leq p$ (index arithmetic modulo p). (2) is clearly equivalent to

3. For $j \neq i$ or $i + 1$, $a_i \neq b_j$ and $a_i b_j$ is not an edge of G.

(Note that so far, we do not know whether $A \cap B = \emptyset$.)

Claim 29.3 *If $(v_1, v_2, \ldots, v_q)$ is a cycle C of G $(4 \leq q \leq 2p)$, then either $v_2 v_q$ is a chord of C or C has a chord of the form $v_1 v_k$ for some $k, 3 \leq k \leq q - 1$*

The proof easily follows from the assumption that every cycle of length k, for $4 \leq k \leq 2p$, has a chord.

Claim 29.4 *G contains an edge of the form $a_i a_j$ $(i \neq j)$.*

Otherwise, by (1), $A \cap B = \emptyset$. Thus $(a_1, b_1, a_2, b_2, \ldots, a_p, b_p)$ is a cycle of length $2p$ in G which must have a chord. Claim 29.3 together with (3) implies that such a chord is an edge between two vertices from B, and since every cycle of length k, for $4 \leq k \leq 2p$, has a chord, it turns out that $b_k b_{k+1}$ is a chord of this cycle for each k $(1 \leq k \leq p)$. Hence $A \cup B$ induces a (chordal) subgraph of G which is a sun of order p: B is the central clique, and A is the stable set. The contradiction proves Claim 29.4.

Claim 29.5 *If $a_i a_j$ is an edge of G, then $a_i a_{i+1}$ is also an edge of G.*

By symmetry, we may suppose $i = 1$. Let j be the smallest integer for which a_1 sees a_j. If $j > 2$, the vertices a_1, b_1, a_2, a_j are different. $(a_2, b_2, a_3, b_3, \ldots, a_{j-1}, b_{j-1}, a_j)$ is a walk not passing through a_1 or b_1, by (3). This walk induces a minimal path, say P, from a_2 to a_j. By (3) and the definition of j, the cycle (a_1, b_1, P) with length ≥ 4 has no chord containing a_1. Hence b_1 must see a_j (Claim 29.3), in contradiction with (3). So, $j = 2$.

Claim 29.6 *$(a_1, \ldots, a_p)$ is a cycle of G.*

Claim 29.6 is an easy consequence of the previous claim.

Claim 29.7 *$a_i \neq b_j$ for all i, j, that is, $A \cap B = \emptyset$.*

Otherwise $N[b_j]$ would contain a_{i-1}, a_i and a_{i+1} (Claims 29.5 and 29.6), in contradiction with (2) (i.e., the definition of a special cycle).

Claim 29.8 *G contains some edge $b_i b_j$.*

Otherwise, $A \cup B$ would induce a p-sun with central clique A and stable set B.

For obtaining the final contradiction, we observe that in the last claim i and j play a symmetrical role. So, we may assume without loss of generality that G contains an edge of the form $b_1 b_j$ with $j \neq 2$ (the arguments for $j = 2$ are similar). Then G has the following cycle:

$$(b_1, a_2, a_3, a_4, \ldots, a_j, b_j)$$

and, by Claim 29.3, some edge $b_1 a_i (3 \leq i \leq j)$ or the edge $b_j a_2$ must exist, contradicting (3).

Theorem 29.33 [27] *A graph G is strongly chordal if and only if G is sun-free chordal.*

Proof. Theorem 29.33 follows by Lemma 29.8 and Corollary 29.13. ∎

A similar characterization of dually chordal graphs was obtained in [60]: A graph G is dually chordal if and only if G is a Helly graph containing no isometric complete k-suns for $k \geq 4$. Recall that G is a *Helly graph* if its disk hypergraph $\mathcal{D}(G)$ is Helly. A subgraph S of G is *isometric* if $d_S(x, y) = d_G(x, y)$ for all vertices x and y of S.

An *odd chord* $v_i v_j$ in an even cycle $(v_1, \ldots, v_{2k})$ is a chord with odd $|i - j|$.

Theorem 29.34 [27] *A graph G is strongly chordal if and only if it is chordal and every even cycle of length at least 6 in G has an odd chord.*

Proof. Let G be a chordal graph. If every even cycle of length at least 6 has an odd chord, then G contains no induced k-sun, $k \geq 3$. Thus, by Theorem 29.33, G is strongly chordal.

Conversely, we use Theorem 29.32: If G is strongly chordal then its neighborhood matrix $N(G)$ is totally balanced. If there is a cycle $(v_1, v_2, \ldots, v_{2k})$ in G without odd chord then the submatrix of $N(G)$ consisting of the rows corresponding to $v_1, v_3, \ldots, v_{2k-1}$ and the columns corresponding to $v_2, v_4, \ldots, v_{2k}$ is precisely the incidence matrix of a cycle of length k. Consequently, G is not strongly chordal. ∎

Finally, we give yet another characterization.

Corollary 29.15 [27] *Graph G is strongly chordal if and only if $\mathcal{C}(G)$ is totally balanced.*

Proof. By Theorems 29.18 and 29.25 and Corollary 29.13, G is strongly chordal if and only if $\mathcal{C}(G)$ is totally balanced. ∎

It follows from the basic properties that G is strongly chordal if and only if $G = L(H)$ for some totally balanced hypergraph H.

Recall that for a clique tree T of G, the intersections $Q \cap Q'$ of maximal cliques for which $QQ' \in E(T)$ form the minimal vertex separators in G. Let $\mathcal{S}(G)$ denote the separator hypergraph of G. It can be considered as the first derivative of a chordal graph. McKee [91] discusses this concept in detail.

Theorem 29.35 [92] *Graph G is strongly chordal if and only if G is chordal and its separator hypergraph $\mathcal{S}(G)$ is totally balanced.* ∎

29.5.2 Chordal Bipartite Graphs

The most natural variant of chordality for bipartite graphs is the following.

Definition 29.36 [93] *A bipartite graph is* chordal bipartite *if each of its cycles of length at least six has a chord.*

In the terminology of Definition 29.5, this means that a bipartite graph is chordal bipartite if and only if it is $(6, 1)$-chordal. Note that chordal bipartite does not mean *chordal and bipartite* (as the name might suggest); if graph G is chordal and bipartite, G is a forest, whereas C_4 is chordal bipartite.

Thus, a better name for chordal bipartite graphs would have been *weakly chordal bipartite* since graph G is chordal bipartite if and only if it is bipartite and *weakly chordal*, that is, every cycle in G and in $\overline{G}$ of length at least five has a chord. See Chapter 28 and [94] for the important class of weakly chordal graphs and their perfection.

Chordal bipartite graphs have various characterizations in terms of elimination orderings and tree structure properties of related hypergraphs; see for example Theorem 29.36 and [3,4] for more details. They are closely related to strongly chordal graphs.

Theorem 29.36 *A bipartite graph $B = (X, Y, E)$ is $(6, 1)$-chordal (i.e., chordal bipartite) if and only if every induced subgraph of B is X-conformal, Y-conformal and X-chordal, Y-chordal.*

Proof. "$\Longrightarrow$": Let $B = (X, Y, E)$ be bipartite $(6, 1)$-chordal. Then every induced subgraph B' of B is also bipartite $(6, 1)$-chordal.

We first show that B is X- and Y-chordal. If C is a cycle of length at least 8 in B' then C has a chord $\{x,y\}, x \in X, y \in Y$. Let $x_1, x_2 \in X$ be the neighbors of y in C and let $y_1, y_2 \in Y$ be the neighbors of x in C. Let C_1, C_2 denote the subcycles defined by the chord $\{x,y\}$ subdividing C. Without loss of generality, assume $|C_1| \leq |C_2|$. Moreover, assume without loss of generality that x_2, y_2 are in C_2. Then y und y_2 have distance at least 4 in C and x is a neighbor of both vertices. Likewise, x and x_2 have distance at least 4 in C and y is a neighbor of both vertices. Thus, B' is X-chordal and Y-chordal.

Now we show that B is X- and Y-conformal. Let $S \subseteq Y$ be a vertex set with pairwise distance 2 in B'. We show inductively the existence of a vertex $x \in X$ with $S \subseteq N(x)$: For $|S| = 2$ and $|S| = 3$, the assertion is obviously fulfilled (for $|S| = 3$, the existence of a chord in any cycle of length 6 is used).

Now, by induction hypothesis, let the assertion be fulfilled for all S', $|S'| \leq k$, with pairwise distance 2 and let $S \subseteq Y$, $|S| = k+1$ be a vertex set with pairwise distance 2. Then for every k-elementary subset $S_i \subseteq S$, $i \in \{1, \ldots, \binom{k+1}{k}\}$ (note $\binom{k+1}{k} = k+1$) there is a vertex x_i for which $S_i \subseteq N(x_i)$. If there is an i with $S \subseteq N(x_i)$ then the assertion is fulfilled. Otherwise, we can assume that the vertices $x_1, \ldots, x_{k+1}$ have exactly one nonneighbor in S: Without loss of generality, let

$$x_i \notin N\left(y_{i+2 (\mathrm{mod}\ k+1)}\right)$$

Now there is a C_6 $(x_1, y_1, x_2, y_3, x_k, y_4)$—contradiction. Thus, there is an index i such that $S \subseteq N(X_i)$. Analogously, one shows Y-conformality of B'.

"$\Longleftarrow$": If every induced subgraph of B is X-conformal, Y-conformal, X-chordal, and Y-chordal then B cannot contain chordless cycles of length at least 6 since chordless cycles of length 6 are neither X- nor Y-conformal and chordless cycles of length at least 8 are neither X- nor Y-chordal. ∎

By Corollary 29.9, Theorem 29.36 implies the following.

Corollary 29.16 *A bipartite graph $B = (X, Y, E)$ is $(6,1)$-chordal (i.e., chordal bipartite) if and only if $\mathcal{N}^X(B)$ and $\mathcal{N}^Y(B)$ are β-acyclic.*

Strongly chordal graphs are closely related to chordal bipartite graphs.

Definition 29.37 *Let $G = (V, E)$ be a graph.*

  i. *The bipartite copy $B(G) = (V', V'', F)$ of G is defined as follows: For every vertex $v \in V$, there are two copies $v' \in V'$ and $v'' \in V''$, and $x'y'' \in F$ if either $x = y$ or $xy \in E$.*

  ii. *$B_C(G)$ denotes the bipartite incidence graph $\mathcal{I}(\mathcal{C}(G))$.*

Note that $B(G)$ is isomorphic to the bipartite incidence graph $\mathcal{I}(\mathcal{N}(G))$. It follows from the basic properties that a graph is chordal bipartite if and only if it is the bipartite incidence graph of a totally balanced hypergraph.

Lemma 29.9 [27] *A graph G is strongly chordal if and only if $B_C(G)$ is chordal bipartite.*

Proof. Lemma 29.9 is an obvious consequence of Theorem 29.36 and Corollary 29.15. ∎

A similar connection is given in the following lemma.

Lemma 29.10 [95] *A graph G is strongly chordal if and only if $B(G)$ is chordal bipartite.*

Proof. Lemma 29.10 is an obvious consequence of Corollary 29.13, Corollary 29.11, Corollary 29.10, and Theorem 29.36. ∎

For a bipartite graph $B = (X, Y, E)$, let $split_X(B)$ denote the one-sided completion when X becomes a clique. Another relation between chordal bipartite and strongly chordal graphs is the following (see also [59]).

Lemma 29.11 [96] *A bipartite graph $B = (X, Y, E)$ is chordal bipartite if and only if $split_X(B)$ is strongly chordal.* ∎

Lemma 29.11 is a simple consequence of the following more general property.

Proposition 29.12 [59] *Let $B = (X, Y, E)$ be a bipartite graph. Then*

i. $\mathcal{N}^X(B)$ *has the Helly property if and only if* $\mathcal{C}(split_X(B))$ *has the Helly property;*

ii. $L(\mathcal{N}^X(B))$ *is chordal if and only if* $L(\mathcal{C}(split_X(B)))$ *is chordal.*

Spinrad [19] gives simple direct proofs of Lemmas 29.10 and 29.11 in terms of Γ-free matrices and discusses the relationship between fast recognition of strongly chordal graphs and fast recognition of chordal bipartite graphs; linear time for recognizing chordal bipartite graphs would imply linear time for recognizing strongly chordal graphs but not vice versa (a linear-time algorithm for recognizing chordal bipartite graphs as claimed in [97] turned out to contain a flaw). Linear-time recognition of strongly chordal graphs (chordal bipartite graphs, respectively) is still an open problem. See [98] for other characterizations of chordal bipartite graphs in terms of intersection graphs of compatible subtrees, and [99] for a relationship between dismantlable lattices and chordal bipartite graphs.

29.6 TREE STRUCTURE DECOMPOSITION OF GRAPHS

Various kinds of decomposition of graphs such as modular decomposition and clique separator decomposition lead to decomposition trees and algorithmic applications. In this section, we first describe cographs and modular decomposition (cographs are the completely decomposable graphs with respect to modular decomposition) and then mention some other decompositions, and in particular clique separator decomposition.

29.6.1 Cographs

In this subsection, we describe an auxiliary class, the *cographs*, which occur in many places and which are fundamental for distance-hereditary graphs and for clique-width. See [3] for additional information.

For disjoint vertex sets $A, B \subseteq V$, the *join operation* (denoted by ①) adds edges between all pairs x, y with $x \in A$, $y \in B$, and the *co-join operation* (denoted by ⓪) adds nonedges between all pairs x, y with $x \in A$ and $y \in B$. These notions are closely related to connectedness of a graph and its complement: G is disconnected if and only if G is decomposable into the co-join of two subgraphs, and $\overline{G}$ is disconnected if and only if G is decomposable into the join of two subgraphs. Subsequently we also use ① and ⓪ in order to denote the relationship between disjoint vertex sets.

Definition 29.38 *Graph G is a* cograph *(complement-reducible graph) if G can be constructed from single vertices by a finite number of join and co-join operations.*

See [3,100–102] for properties of this graph class.

Theorem 29.37 *G is a cograph if and only if G is P_4-free.*

Proof. "$\Longrightarrow$": By induction on the number of vertices in G. For single vertices, the assertion is obviously true. Now, let $G = G_1 \oplus G_2$ and G_1, G_2 being P_4-free. If G would contain a P_4 $P = abcd$ then P has vertices from G_1 and G_2. Assume first that P has exactly one vertex from G_1. If $a \in V(G_1), b, c, d \in V(G_2)$ then $ac \notin E$ contradicts to the join between G_1 and G_2, if $b \in V(G_1), a, c, d \in V(G_2)$ then $bd \notin E$ contradicts to the join between G_1 and G_2. Now assume that P has exactly two vertices from each of G_1, G_2. If $a, d \in V(G_1), b, c \in V(G_2)$ then $ac \notin E$ contradicts to the join between G_1 and G_2, if $b, d \in V(G_1), a, c \in V(G_2)$ then $ad \notin E$ contradicts to the join between G_1 and G_2, and if $a, b \in V(G_1), c, d \in V(G_2)$ then $ac \notin E$ contradicts to the join between G_1 and G_2. In every case, there is a nonedge of the P_4 between G_1 and G_2, and thus $G = G_1 \oplus G_2$ is again P_4-free.

In the same way one can show that $G = G_1 \otimes G_2$ is again P_4-free if G_1 and G_2 are P_4-free.

"$\Longleftarrow$": Let G be a P_4-free graph. We will show that then G is decomposable with respect to the operations $\oplus, \otimes$ into subgraphs G_1, G_2, that is, either G or $\overline{G}$ is disconnected. Assume that not every P_4-free graph would have this property. Then let $G = (V, E)$ be a smallest P_4-free graph not having this property, that is, G is P_4-free connected and co-connected but for every $v \in V$, either $G - v$ is disconnected or $\overline{G - v}$ is disconnected. Note that in this case, G has at least four vertices.

Case 1 $G - v$ is disconnected. Let $H_1, \ldots, H_k$, $k \geq 2$, be the connected components of $G - v$, that is, there are no edges between H_i and H_j for $i \neq j$, $i, j \in \{1, \ldots, k\}$ but since G is connected, v has edges to each of $H_1, \ldots, H_k$. Let x_i be a neighbor of v in H_i. Since G is also co-connected, v has at least one nonneighbor in $V \setminus \{v\}$. Without loss of generality, let $y \in H_1$ be a vertex with $vy \notin E$. Since H_1 is connected, there is a path $P_{x_1 y}$ between x_1 and y in H_1. Let $x'y'$ be the first edge on this path for which $vx' \in E$ but $vy' \notin E$ holds. Since $vx_1 \in E$, $vy \notin E$, the existence of such an edge is guaranteed. But now the vertices x_2, v, x', y' induce a P_4 in G—contradiction.

Case 2 The case that $\overline{G - v}$ is disconnected can be handled in the same way as the previous case. ∎

Theorem 29.37 implies that the property of being a cograph is a hereditary property, that is, if G is a cograph then every induced subgraph G' of G is a cograph as well.

The recursive generation of cographs by the two operations join and co-join is described in a tree structure—the *cotree*. This tree has the vertices of the graph as its leaves, and the internal nodes are labeled with $\oplus$ and $\otimes$ according to the operations. If $G = G_1 \oplus G_2$ ($G = G_1 \otimes G_2$, respectively) then the root vertex of the cotree of G carries the label $\oplus$ ($\otimes$, respectively), and its two children are the root nodes of G_1, G_2, respectively.

A cotree is not necessarily a binary tree; for example, a clique with k vertices is represented by one $\oplus$ node with the k vertices as its children.

In [102], it is described how to recognize in linear time $\mathcal{O}(n + m)$ whether a given input graph G is a cograph; starting with a single vertex, the algorithm tries to incrementally construct a cotree T of G, that is, in every step, a new vertex is added and the new cotree is constructed if the graph is still a cograph; otherwise, an induced P_4 in G is given as output. The algorithm is performed by a complicated marking procedure which cannot be described here. However, it has a remarkable property: It does not only give the correct *Yes/No* answer to the recognition problem; if the answer is *Yes* then the algorithm gives a certificate namely a cotree, and it is easily checkable whether the cotree indeed represents the graph, and if the answer is *No*, it gives a certificate for this answer, that is, in the case of cograph recognition a P_4 in the input graph. Such recognition algorithms are called *certified algorithms* and are known for various graph classes [103].

A simpler recognition of cographs is described in [104] which time bound, however is $\mathcal{O}(n+ m\log n)$ (and not linear). In [105], a simple multisweep LexBFS algorithm for recognizing cographs in linear time is given.

29.6.2 Optimization on Cographs

Various algorithmic graph problems being NP-complete in general, can be solved efficiently in a bottom-up procedure along the cotree of a cograph. As examples, we describe this for the problems MAXIMUM STABLE SET and MAXIMUM CLIQUE.

Let $G = (V, E)$ be a graph. A vertex set $U \subseteq V$ is *stable* (*independent*) if for all $x, y \in U$, $xy \notin E$. $U \subseteq V$ is *a clique* if U is stable in $\overline{G}$. If $G = (V, E)$ is a graph with vertex weight function w then for $U \subseteq V$, $w(U) := \Sigma_{x \in U} w(x)$.

Let $\alpha_w G$ be the maximum weight of a stable set in G, and let $\omega_w G$ be the maximum weight of a clique in G. Now, obviously the values of $\alpha_w(G)$ and $\omega_w(G)$ can be computed recursively for $G = G_1 \oplus G_2$ and $G = G_1 \otimes G_2$:

- If $G = G_1 \oplus G_2$ then
$$\omega_w(G) = \omega_w(G_1) + \omega_w(G_2)$$
and
$$\alpha_w(G) = \max(\alpha_w(G_1), \alpha_w(G_2)).$$

- If $G = G_1 \otimes G_2$ then
$$\omega_w(G) = \max(\omega_w(G_1), \omega_w(G_2))$$
and
$$\alpha_w(G) = \alpha_w(G_1) + \alpha_w(G_2).$$

This implies linear-time algorithms for these two problems on cographs.

As a further example, we show how to color cographs in an optimal way. The coloring problem of a graph is how to assign a minimum number of colors to the vertices such that adjacent vertices get different colors. The *chromatic number* $\chi(G)$ of the graph G is the minimum number of colors needed to color G. Obviously, for every graph G, $\omega(G) \leq \chi(G)$ holds. A graph is called χ-*perfect* if for every induced subgraph G' of G (including G itself), $\omega(G') = \chi(G')$ holds. Let $\kappa(G) = \chi(\overline{G})$. Obviously, $\alpha(G) \leq \kappa(G)$ holds. A graph is called κ-*perfect* if for every induced subgraph G' of G (including G itself), $\alpha(G') = \kappa(G')$ holds. The following theorem is a celebrated result by Laszló Lovász (see, e.g., Chapter 28):

Theorem 29.38 (Perfect graph theorem) *A graph is χ-perfect if and only if it is κ-perfect.*

Now, G is called *perfect* if G is χ-perfect (κ-perfect).

Corollary 29.17 *A graph G is perfect if and only if its complement graph $\overline{G}$ is perfect.*

Corollary 29.18 *Cographs are perfect.*

Proof. We show inductively on the number of vertices that cographs are perfect. For one-vertex graphs, the claim is obviously fulfilled. Now assume first that $G = G_1 \oplus G_2$ and $\omega(G_i) = \chi(G_i)$ holds for $i \in \{1, 2\}$. Since there is a join between G_1 and G_2, $\chi(G) = \chi(G_1) + \chi(G_2) = \omega(G_1) + \omega(G_2) = \omega(G)$ which shows the claim.

Now assume that $G = G_1 ⓪ G_2$ and $\omega(G_i) = \chi(G_i)$ holds for $i \in \{1,2\}$. Since there is a co-join between G_1 and G_2, $\chi(G) = \max(\chi(G_1), \chi(G_2)) = \max(\omega(G_1), \omega(G_2)) = \omega(G)$ which again shows the claim. Thus, cographs are perfect. ∎

See Chapter 28 for many other important subclasses of perfect graphs.

Another remarkable property of cographs is the fact that they are transitively orientable. Hereby a graph $G = (V, E)$ is called *transitively orientable* if its edge set E can be oriented as E' in such a way that for all oriented edges $(x, y), (y, z) \in E'$, $(x, z) \in E'$ holds. One can easily show by induction that cographs have this property. Hereby, for $G = G_1 ① G_2$, the edges of the join are oriented from G_1 to G_2 – this obviously gives again a transitive orientation if it is assumed that G_1 and G_2 are already transitively oriented—and for a co-join, there is nothing to show.

Subsequently, the modular decomposition of arbitrary graphs is described which generalizes cographs and cotrees and gives a strong algorithmic tool for many problems. See [106] for the connection between transitive orientation, cographs and modular decomposition.

29.6.3 Basic Module Properties

Let $G = (V, E)$ be a graph. A vertex set $M \subseteq V$ is a *module* in G if its vertices are indistinguishable from outside M. More formally: For all $u \in V \setminus M$, either $\{u\} ⓪ M$ or $\{u\} ① M$. Sets A and B *overlap* if $A \setminus B \neq \emptyset$, $B \setminus A \neq \emptyset$, and $A \cap B \neq \emptyset$.

Theorem 29.39 (Basic module properties) *Let G be a graph and let $\mathcal{M}(G)$ denote the set of modules in G. Then the following properties hold:*

i. $\emptyset$, V and $\{v\}$ for all $v \in V$ are modules (the *trivial* modules);

ii. If $M_1, M_2 \in \mathcal{M}(G)$ then $M_1 \cap M_2 \in \mathcal{M}(G)$;

iii. If $M_1, M_2 \in \mathcal{M}(G)$ and $M_1 \cap M_2 \neq \emptyset$ then $M_1 \cup M_2 \in \mathcal{M}(G)$;

iv. If M_1 and M_2 are overlapping modules then $M_1 \setminus M_2 \in \mathcal{M}(G)$, $M_2 \setminus M_1 \in \mathcal{M}(G)$, $(M_1 \setminus M_2) \cup (M_2 \setminus M_1) \in \mathcal{M}(G)$;

v. If M is a module in G and $U \subseteq V$ then $M \cap U$ is a module in $G[U]$.

Proof.
i. This property is obviously fulfilled.

ii. Let M_1 and M_2 be modules in G. If their intersection is empty then due to (i), the assertion is fulfilled. Now assume that $M_1 \cap M_2 \neq \emptyset$. If $M_1 \subseteq M_2$ or $M_2 \subseteq M_1$ then again the assertion holds true. Now assume that M_1 and M_2 are overlapping modules. Vertices outside $M_1 \cap M_2$ cannot distinguish two vertices from $M_1 \cap M_2$: if a vertex $x \notin M_1 \cup M_2$ would distinguish vertices $a, a' \in M_1 \cap M_2$, that is, $xa \in E$, $xa' \notin E$ then this would contradict to the module property of M_1 (M_2, respectively); if a vertex $x \in M_1 \setminus M_2$ would distinguish vertices $a, a' \in M_1 \cap M_2$, that is, $xa \in E$, $xa' \notin E$ then this would contradict the module property of M_2, and the same holds for $x \in M_2$.

iii. Let $M_1 \cap M_2 \neq \emptyset$ with $a \in M_1 \cap M_2$. If $M_1 \subseteq M_2$ or vice versa then the assertion is trivial. Now assume that M_1 and M_2 are overlapping modules. Due to condition (ii), vertices in $M_1 \cap M_2$ cannot be distinguished from outside. The same holds for two vertices in M_1 (M_2, respectively). Now assume that vertices $a' \in M_1 \setminus M_2$ and $a'' \in M_2 \setminus M_1$ could be distinguished by $x \notin M_1 \cup M_2$: $xa' \in E$ and $xa'' \notin E$. Since $xa' \in E$ and $a, a' \in M_1$, also $xa \in E$ holds but since $a, a'' \in M_2$, it follows that $xa'' \in E$—contradiction. Thus $M_1 \cup M_2$ is a module.

iv. We first show that $M_1 \setminus M_2$ is a module. Assume to the contrary that there are vertices $a, a' \in M_1 \setminus M_2$ and $x \notin M_1 \setminus M_2$ such that $ax \in E$, $a'x \notin E$. Then $x \in M_1$ since M_1 is a module, that is, $x \in M_1 \cap M_2$. Let $b \in M_2$. Since $x, b \in M_2$ and M_2 is a module, also $ab \in E$ and $a'b \notin E$ holds but now $b \notin M_1$ is a vertex outside M_1 distinguishing vertices $a, a' \in M_1$—contradiction. Analogously, M_2 is a module.

Now we show that $\Delta := (M_1 \setminus M_2) \cup (M_2 \setminus M_1)$ is a module: Let $a, a' \in \Delta$. Since $M_1 \setminus M_2$ ($M_2 \setminus M_1$) is a module, we can assume that $a \in M_1 \setminus M_2$ and $a' \in M_2 \setminus M_1$. Due to (iii), $M_1 \cup M_2$ is a module. Thus, a and a' cannot be distinguished from outside $M_1 \cup M_2$. Assume that there is a vertex $x \notin \Delta$, $x \in M_1 \cap M_2$ such that $xa \in E$, $xa' \notin E$. Since $x, a \in M_1$, $a' \notin M_1$ and M_1 a module, $aa' \notin E$ holds. Since $x, a' \in M_2$, $a \notin M_2$ and M_2 a module, $aa' \in E$ holds—contradiction.

v. If $M \subseteq U$ then the assertion is obviously fulfilled. Assume now that $M \setminus U \neq \emptyset$. If $M \cap U = \emptyset$ then again the assertion is obviously fulfilled. Now assume that $M \cap U \neq \emptyset$ and $M \cap U$ is no module in $G[U]$, that is, there are vertices $a, a' \in M \cap U$ and a vertex $v \in U \setminus M$ distinguishing a and a' from outside M but then M is no module—contradiction. ∎

Theorem 29.40 *In a connected and co-connected graph G, the nontrivial $\subseteq$-maximal modules are pairwise disjoint.*

Proof. Let M_1 and M_2 be nontrivial modules in G being maximal with respect to set inclusion and assume that $M_1 \cap M_2 \neq \emptyset$. This implies that they are overlapping modules. Then according to Theorem 29.39 (iii), $M_1 \cup M_2$ is a module. If $M_1 \cup M_2 \neq V$ then M_1 and M_2 are not maximal—thus $M_1 \cup M_2 = V$. Note that vertices from $M_1 \setminus M_2$ are either completely adjacent to M_2 or completely nonadjacent to M_2, and the same holds for vertices from $M_2 \setminus M_1$. Let $M_1^+ := \{x : x \in M_1 \setminus M_2 \text{ and } x \text{ has a join to } M_2\}$, $M_1^- := \{x : x \in M_1 \setminus M_2$ and x has a cojoin to M_2, and define the sets M_2^+ and M_2^- in a completely analogous way. Obviously, $M_1 \setminus M_2 = M_1^+ \cup M_1^-$ and $M_2 \setminus M_1 = M_2^+ \cup M_2^-$. If one of the sets M_1^+, M_1^-, M_2^+, and M_2^- is empty then G is not connected or not co-connected. Thus, all of these sets are nonempty. Now let $x \in M_1^+$, $x' \in M_1^-$ and $y \in M_2^+$. The fact that $xy \in E$ and M_1 is a module implies that $x'y \in E$ but now x' is adjacent to a vertex from M_2—contradiction. ∎

A graph is *prime* if it contains no nontrivial module. The *characteristic graph G^** of G is the graph obtained by contracting the maximal modules of G to one vertex.

Theorem 29.41 *The characteristic graph G^* of a connected and co-connected graph G is prime.*

Proof. By Theorem 29.40, the maximal nontrivial modules in $G = (V, E)$ are pairwise disjoint and thus define a partition of V into equivalence classes. Let v^* denote the equivalence class of a vertex v. Let $G^* = (V^*, E^*)$ and $U \subseteq V^*$ and denote by K_x the equivalence class in V belonging to $x \in V^*$. Then the *expansion $E(U)$* of U is the union of the equivalence classes belonging to U, that is, the vertex set $E(U) = \bigcup_{x \in U} K_x$. We first claim that for a module M in G^*, its expansion $E(M)$ is a module in G. Assume to the contrary that there are $a, b \in E(M)$ and $x \notin E(M)$ such that $ax \in E$ and $bx \notin E$. Then obviously, a and b are not in the same class in $E(M)$ since the classes are modules. This means that $a^* \neq b^*$, $a^*, b^* \in M$ and $x^* \notin M$ but now M is no module—contradiction. This shows the claim.

Now assume that M is a nontrivial module in G^*. If M consists only of vertices whose classes are one-elementary then $E(M) = M$ and M is a module in G; thus, after shrinking the modules in G, M cannot have more than one element. If M contains at least one vertex u whose class U is a nontrivial module in G then $U \subset E(M)$ but U is a maximal module in G and $E(M)$ is a module in G—contradiction. Thus, G^* is a prime graph. ∎

29.6.4 Modular Decomposition of Graphs

Theorems 29.39 and 29.40 lead to the following tree structure of a given graph G: Every vertex in G is contained in a unique (possibly one-elementary) maximal module different from V, and these modules define a partition of V. The *modular decomposition tree* has V as its root and the maximal modules smaller than V are the children of V in the tree. Then the children of an inner vertex M are the maximal modules in $G[M]$ smaller than M. Thus, if the inner vertex M of the modular decomposition tree has the partition $M_1, M_2, \ldots, M_k$ into its maximal modules then $M_1, M_2, \ldots, M_k$ are the children of M. Note that the leaf descendants of M are the vertices of M, and the edges in M between M_i and M_j are given by a sequence of join and co-join operations between the modules M_i and the vertices outside M_i. The graphs being completely decomposable by join and co-join are the cographs.

The following decomposition theorem is implicitly contained in the seminal paper by Tibor Gallai [107].

Theorem 29.42 (Modular decomposition theorem) *Let $G = (V, E)$ be an arbitrary graph. Then precisely one of the following conditions is satisfied:*

1. *G is disconnected (i.e., decomposable by the co-join operation);*
2. *$\overline{G}$ is disconnected (i.e., decomposable by the join operation);*
3. *G and $\overline{G}$ are connected: There is some $U \subseteq V$ and a unique partition $\mathcal{P}$ of V such that*
   a. *$|U| \geq 4$,*
   b. *$G[U]$ is a maximal prime subgraph of G, and*
   c. *for every class S of the partition $\mathcal{P}$, S is a module and $|S \cap U| = 1$.*

Each vertex of G forms a leaf of the decomposition tree. Each module M of G occuring as a node in the tree contains exactly the vertices that are leaves of the subtree rooted at M. According to the Decomposition Theorem, the tree has three kinds of nodes:

- Parallel nodes (co-join operation);
- Series nodes (join operation);
- Prime nodes.

Linear-time algorithms for finding the modular decomposition tree are given in [108,109] and in [106]. See [110,111] for simpler linear-time algorithms.

The modular decomposition is of crucial importance in many algorithmic applications; see [112] for many aspects of modular decomposition. Since for many algorithmic problems the operations join and co-join are easy to handle (cf. the case of cographs), it is important to look at prime graphs. There are some cases where prime graphs have *simple structure*.

A nice example for a graph class having simple prime graphs with respect to modular decomposition are P_4-sparse graphs.

A graph $G = (V, E)$ is P_4-*sparse* [113] if every five vertices induce at most one P_4 in G. Thus, cographs are P_4-sparse, and the only one-vertex extensions of a P_4 in a P_4-sparse graph G are the bull, gem and co-gem, that is, G is P_4-sparse if and only if all the other seven one-vertex extension (such as P_5, C_5, etc.) are forbidden induced subgraphs in G. Obviously, the complement of a P_4-sparse graph is P_4-sparse.

A graph is a *thin spider* if its vertex set can be partitioned into a clique Q and a stable set S such that the edges between Q and S form a matching, every vertex in S has exactly one neighbor in Q, and at most one vertex in Q has no neighbor in S (the *head of the spider*). Obviously, thin spiders are prime graphs and P_4-sparse. A graph is a *thick spider* if it is the complement of a thin spider; it is a *spider* if it is a thin or thick spider (these graphs were called *turtles* in [113]).

Theorem 29.43 [113] *A graph is P_4-sparse if and only if its prime graphs are spiders.*

Various structural and algorithmic consequences are given in [113–117].

A lot of research has been done in generalizing, refining and modifying modular decomposition. *Split* (or *join*) *decomposition* was introduced and studied by Cunningham [118]. A graph is *split decomposable* if its vertex set has a partition into A_1, A_2 and B_1, B_2 such that $A = A_1 \cup A_2, B = B_1 \cup B_2$, and the set of all edges between A and B forms a join $A_1 \oplus B_1$. The decomposition is discussed in detail in the monograph [19] by Spinrad, mentioning the linear-time algorithm for split decomposition by Dahlhaus [119]. A simplified linear-time algorithm for split decomposition is given in [120].

The class of graphs such that every induced subgraph on at least four vertices is decomposable by the join decomposition is of particular interest. It turns out that these are exactly the distance-hereditary graphs which are the central topic of the next section.

Another interesting concept is the homogeneous decomposition where a third operation is added which is a combination of join and co-join. This approach is based on a different kind of connectedness—the *p-connectedness*—and is described in [121].

29.6.5 Clique Separator Decomposition of Graphs

A *clique separator* of a graph G is a separator of G which is a clique in G. For a chordal graph G which is not a clique and a simplicial vertex v in G, obviously $N(v)$ is a clique separator of G. Clique separator decomposition of a graph is generalizing chordal graphs by repeatedly choosing a clique separator in G until there is no longer a clique separator in the resulting subgraphs; such subgraphs are called *atoms* of G. Note that such decomposition trees are not uniquely determined. Obviously, chordal graphs are those graphs whose atoms are cliques. This kind of decomposition was introduced in [122,123] and has a number of algorithmic applications described in [122] among them efficiently solving the MWIS problem on a graph class whenever it is efficiently solvable on the atoms of the class. This refers to the weight modification approach described in the algorithm of Frank for the same problem on chordal graphs—see Theorem 29.4.

Various examples of such classes were studied: In [124], a subclass of hole-free graphs, namely hole- and paraglider-free graphs, is characterized by the structure of their atoms. Among others, this is motivated by a result of Alekseev [125] showing that atoms of (P_5,paraglider)-free graphs are $3K_2$-free which implies polynomial time for MWIS on this class. For P_5-free graphs, the complexity of the MWIS problem was open for a long time; meanwhile, it has been shown by Lokshtanov et al. [126] that it is polynomially solvable for P_5-free graphs. For hole-free graphs, the complexity of the MWIS problem is open.

29.7 DISTANCE-HEREDITARY GRAPHS, SUBCLASSES, AND γ-ACYCLICITY

29.7.1 Distance-Hereditary Graphs

Distance-hereditary graphs are another fundamental generalization of trees. They are closely related to γ-acyclic hypergraphs (see Definition 29.44) and have bounded clique-width. Originally, they were defined via a distance property.

Definition 29.39 [127] *A graph G is* distance hereditary *if for each connected induced subgraph F of G, the distance functions d_G in G and d_F in F coincide.*

Definition 29.40 *A u-v-geodesic is a u-v-path α such that $l(\alpha) = d_G(u,v)$. Let Φ be a cycle of G. A path α is an* essential part *of Φ if $\alpha \subset \Phi$ and $1/2 l(\Phi) < l(\alpha)$.*

Theorem 29.44 [127] *The following conditions are equivalent:*

i. *G is distance hereditary.*

ii. *Every induced path of G is geodesic.*

iii. *No essential part of a cycle of G is induced.*

iv. *Each cycle of G of length ≥ 5 has at least two chords, and each 5-cycle of G has a pair of crossing chords.*

v. *Each cycle of G of length ≥ 5 has a pair of crossing chords.*

Proof. Howorka [127] has shown that (i) $\iff$ (ii) $\iff$ (iii) $\implies$ (iv) $\implies$ (v) $\implies$ (iii); here, we give his proof.

(i) $\implies$ (ii): Let α be an induced path of a distance-hereditary graph G and let u and v be the endpoints of α. Then $d_G(u,v) = d_\alpha(u,v) = l(\alpha)$. Hence α is a geodesic.

(ii) $\implies$ (i): Suppose that F is a connected induced subgraph of G. Let u,v be arbitrary vertices of F, and let α be a u-v-geodesic of F. Thus naturally, α is an induced path of F and, consequently, also an induced path of G. Hence, by assumption, α is a u-v-geodesic of G. Thus $d_F(u,v) = l(\alpha) = d_G(u,v)$. This proves that G is distance hereditary.

(ii) $\implies$ (iii): Since an essential part of a cycle cannot be a geodesic, (ii) clearly implies (iii).

(iii) $\implies$ (ii): Let G be a graph satisfying (iii). Let $u \neq v$ be vertices of G and assume that $\alpha = (u = a_0, a_1, \ldots, a_m = v)$ is a u-v-path of G which is not a geodesic. Consider any u-v-geodesic $\beta = (u = b_0, b_1, \ldots, b_n = v)$, $n < m$. Let i be the largest index for which $b_i = a_i$, $0 \leq i < n$. Let t be the least index $> i$ for which $b_t \in \alpha$. Thus $b_t = a_j$ for some $j > t$. Consequently, the path $\delta = (a_i, a_{i+1}, \ldots, a_j)$ is an essential part of the cycle $(a_i, a_{i+1}, \ldots, a_j = b_t, b_{t-1}, \ldots, b_i = a_i)$. By assumption, δ is not induced. Hence α is not induced. This completes the proof.

(iii) $\implies$ (iv): Let $\Phi = (a_0, a_1, \ldots, a_n = a_0)$, $n \geq 5$, be a cycle of a graph G satisfying (iii). By considering any essential part of Φ of length $\leq n-2$, we see from (iii) that Φ must have at least one chord, say $a_i a_j$. Since, in turn, $(a_{i+1}, a_{i+2}, \ldots, a_{i-1})$ is an essential part of Φ, then Φ must have a chord distinct from $a_i a_j$. This proves that each cycle of G of length ≥ 5 has at least two chords. An easy verification shows that if (iii) holds then a 5-cycle of G must have a pair of crossing chords.

(iv) $\implies$ (v): Assume that G satisfies (iv). We will prove by induction that each n-cycle of G, $n \geq 5$, has a pair of crossing chords. By assumption, the assertion is true for $n = 5$. Let $n > 5$ and suppose that each cycle of length m, $5 \leq m < n$, has a pair of crossing chords. Consider an n-cycle $\Phi = (a_0, a_1, \ldots, a_n = a_0)$ and let $a_i a_j$ and $a_r a_s$ be two distinct chords of Φ. If they do not cross one another, we may assume without loss of generality that $0 \leq i \leq j \leq r \leq s \leq n$. Consider the cycles $(a_i, a_j, a_{j+1}, \ldots, a_i)$ and $(a_r, a_s, a_{s+1}, \ldots, a_r)$. Since $n \geq 6$, at least one of them has length ≥ 5 and hence, by induction hypothesis, it must have a pair of crossing chords. This same pair is, of course, a pair of crossing chords of Φ. This completes the proof.

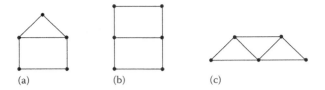

Figure 29.5 House (a), domino (b), and gem (c) are not distance-hereditary.

(v) $\Longrightarrow$ (iii): Let G be a graph satisfying (v). We will prove by induction on n that no essential part of an n-cycle of G is induced. This is trivially true if $n = 3$ or $n = 4$. Assume that $n > 4$ and that the assertion is true for all cycles of length $< n$. Let $\Phi = (a_0, a_1, \ldots, a_n = a_0)$ be a cycle of G and let α be an essential part of Φ, say $\alpha = (a_0, a_1, \ldots, a_k)$, where $n/2 < k < n$. Let $a_i a_j$ and $a_r a_s$ be a pair of crossing chords of Φ. Without loss of generality we may assume that $0 \leq i < j < n$, $0 \leq r < s < n$ and $i < r$. If $j \leq k$ then $a_i a_j$ joins two vertices of α; hence α is not induced. If $i \geq k$ then α is an essential part of the cycle $(a_0, a_1, \ldots, a_k, \ldots, a_i, a_j, \ldots, a_n = a_0)$ of length $< n$. Hence, by induction hypothesis, α is not induced. We may assume therefore that $0 \leq i < k < j < n$. Applying the same argument to $a_r a_s$, we obtain $0 \leq r < k < s < n$. Since the chords $a_i a_j$ and $a_r a_s$ cross one another, it follows that $0 \leq i < r < k < j < s < n$. Denote $\alpha' = (a_0, a_1, \ldots, a_r)$ and $\alpha'' = (a_i, a_{i+1}, \ldots, a_k)$. We claim that either α' is an essential part of the cycle $(a_s, a_{s+1}, \ldots, a_r, a_s)$ or α'' is an essential part of the cycle $(a_i, a_{i+1}, \ldots, a_j, a_i)$. We have indeed: $l(\alpha') + l(\alpha'') \geq k + 1 \geq n - k + 2 > n - s + j - k + 2 = (n - s + 1) + (j - k + 1)$ and so, either $l(\alpha') > n - s + 1$, or $l(\alpha'') > j - k + 1$, which proves our claim. It follows now from an induction hypothesis that either α' or α'' is not an induced path. Hence α cannot be induced. This completes the inductive step and proves the theorem. ∎

Chordless cycles with at least five vertices are called *holes*. Obviously, holes are not distance hereditary. Recall that, in connection with relational database schemes, (k, l)-chordal graphs were defined (see Definition 29.5) (see Figure 29.5 for house, domino, and gem).

Theorem 29.45 *Let G be a graph.*

i. *G is $(5, 2)$-chordal if and only if G is (house, hole, domino)-free.*

ii. *G is distance-hereditary if and only if G is (house, hole, domino, gem)-free* [128].

Proof. (i): Obviously, every $(5, 2)$-chordal graph is (house, hole, domino)-free. For the other direction, let G be (house, hole, domino)-free, and let $C = (x_1, \ldots, x_k)$, $k \geq 5$, be a cycle in G. If $k = 5$ then C is no C_5 and no house, that is, C must have at least two chords. If $k = 6$ then C is no C_6 and no domino, and since G is C_5-free, C must have at least two chords. If $k \geq 7$ then C has a chord $x_i x_j$ since G is hole-free. A cycle C' consisting of an essential part of C together with the chord $x_i x_j$ has length at least 5 and thus has another chord (since G is hole-free) which shows the assertion.

(ii): Obviously, every distance-hereditary graph is (house, hole, domino, gem)-free. For the other direction, let G be (house, hole, domino, gem)-free, and let $C = (x_1, \ldots, x_k)$, $k \geq 5$, be a cycle in G. By Theorem 29.44, (v), it is sufficient to show that C has two crossing chords. If $k = 5$ then C is no C_5, no house and no gem, that is, C must have two crossing chords. If $k = 6$ then C is no C_6 and no domino, and since G is C_5- and gem-free, C must have two crossing chords. If $k \geq 7$ then C has a chord $x_i x_j$ since G is hole-free. A cycle C' consisting of an essential part of C together with the chord $x_i x_j$ has length at least 5 and thus, by an induction hypothesis, has two crossing chords which shows the assertion. ∎

For most of the algorithmic applications, a characterization of distance-hereditary graphs in terms of three simple operations is crucial which is described in the next theorem:

Theorem 29.46 [128] *A connected graph G is distance-hereditary if and only if G can be generated from a single vertex by repeatedly adding a pendant vertex, a false twin or a true twin.*

Proof. Assume first that graph G can be generated from a single vertex by repeatedly adding a pendant vertex, a false twin or a true twin. Then it can easily be seen that G must be (house, hole, domino, gem)-free.

For the other direction, we give the short proof of Theorem 29.46 contained in [129]. Actually, [128] is claiming more namely that every distance-hereditary graph with at least two vertices contains either a pair of twins or two pendant vertices. In [130], an even slightly stronger version is given (and an incorrectness of the proof in [128] is corrected).

Let G be a distance-hereditary graph, thus having crossing chords in each cycle of length at least 5. It suffices to show that G has a pendant vertex or a pair of twins since every induced subgraph of G is again distance hereditary. This is trivially fulfilled if G is a disjoint union of cliques. We may assume that some component H of G is not a clique. Let Q be a minimal cutset of H and $R_1, \ldots, R_m$ be the components of $H - Q$. Suppose that $|Q| \geq 2$; we show that Q is a homogeneous set. If not, there are two vertices $p, q \in Q$ and a vertex $r \in V(H) - Q$ with $rp \in E$ and $rq \notin E$. Let $r \in R_1$. Since Q is a minimal cutset of H, vertex q has a neighbor $s \in R_1$. Note that there is an r-s-path P_1 in R_1. We choose s so that P_1 is as short as possible. Similarly p has a neighbor $t \in R_2$ and q has a neighbor $u \in R_2$. We choose t and u so that a shortest t-u-path P_2 in R_2 has smallest length (possibly $t = u$). The vertices s, q, u, t, p, r and the paths P_1 and P_2 form a cycle C of length at least 5. The only possible chords of C join p to q or to some vertices of P_1. Thus, C has no crossing chords, a contradiction.

Now if x is any vertex in R_1 which is adjacent to Q, it must be adjacent to all vertices of Q and thus Q is P_4-free (otherwise, G has a gem). We know that a nontrivial P_4-free graph has a pair of twins. They will also be twins in G because Q is homogeneous.

Now suppose that every minimal cutset contains only one vertex. Let R be a terminal block of H, that is, a maximal 2-connected subgraph of H that contains just one cut-vertex, say x, of H. If $|R| = 2$, the vertex in $R - x$ is a pendant vertex of G. If $|R| \geq 3$ and $R - x \subseteq N(x)$, the set $R - x$ must induce a P_4-free subgraph. So R contains a pair of twins, and clearly they are also twins in G.

If $R \setminus N(x) \neq \emptyset$, $N(x) \cap R$ is a cutset of H and so it contains a minimal cutset of size one but then R is not 2-connected, a contradiction which proves the theorem. ∎

For a distance-hereditary graph G, a *pruning sequence* of G describes how G can be generated (dismantled, respectively) by repeatedly adding (deleting, respectively) a pendant vertex, a false twin or a true twin. Pruning sequences and pruning trees are a fundamental tool for most of the efficient algorithms on distance-hereditary graphs. There is a more general way, however, to efficiently solve problems on graph classes captured in the notion of clique-width described in the section on clique-width.

Definition 29.41 *Let G be a graph with vertices $v_1, \ldots, v_n$, and let $S = (s_2, \ldots, s_n)$ be a sequence of tuples of the form $((v_i, v_j), type)$, where $j < i$ and $type \in \{leaf, true, false\}$. S is a pruning sequence for G, if for all i, $2 \leq i \leq n$, the subgraph of G induced by $\{v_1, \ldots, v_i\}$ is obtained from the subgraph of G induced by $\{v_1, \ldots, v_{i-1}\}$ by adding vertex v_i and making it adjacent only to v_j if $type = leaf$, making it a true twin of v_j if $type = true$, and making it a false twin of v_j if $type = false$.*

By Theorem 29.46, a graph is distance hereditary if and only if it has a pruning sequence.

Definition 29.42 *Let G be a graph with vertices $v_1, \ldots, v_n$, and let $S = (s_2, \ldots, s_n)$ be a pruning sequence for G. The* pruning tree *corresponding to S is the labeled ordered tree T constructed as follows:*

1. *Set T_1 as the tree consisting of a single root vertex v_1, and set $i := 1$.*

2. *Set $i := i + 1$. If $i > n$ then set $T := T_n$ and stop.*

3. *Let $s_i = ((v_i, v_j), leaf)$ (respectively, $s_i = ((v_i, v_j), true)$, or $s_i = ((v_i, v_j), false)$), then set T_i as the tree obtained from T_{i-1} by adding the new vertex v_i and making it a rightmost son of the vertex v_j, and labeling the edge connecting v_i to v_j by* leaf *(respectively by* true *or* false*).*

4. *Go back to step (2) above.*

A linear-time recognition algorithm for distance-hereditary graphs using pruning sequences was claimed already in [129]; however, their algorithm contained a flaw. Damiand et al. [104] used the following characterization given by Bandelt and Mulder for linear-time recognition of distance-hereditary graphs.

Theorem 29.47 [128] *Let G be a connected graph and $L_1, \ldots, L_k$ be the distance levels of a hanging from an arbitrary vertex v of G. Then G is a distance-hereditary graph if and only if the following conditions hold for any $i \in \{1, \ldots, k\}$:*

   i. *If x and y belong to the same connected component of $G[L_i]$ then $L_{i-1} \cap N(x) = L_{i-1} \cap N(y)$.*

   ii. *$G[L_i]$ is a cograph.*

   iii. *If $u \in L_i$ and vertices x and y from $L_{i-1} \cap N(u)$ are in different connected components X and Y of $G[L_{i-1}]$ then $X \cup Y \subseteq N(u)$ and $L_{i-2} \cap N(x) = L_{i-2} \cap N(y)$.*

   iv. *If x and y are in different connected components of $G[L_i]$ then sets $L_{i-1} \cap N(x)$ and $L_{i-1} \cap N(y)$ are either disjoint or comparable with respect to set inclusion.*

   v. *If $u \in L_i$ and vertices x and y from $L_{i-1} \cap N(u)$ are in the same connected component C of $G[L_{i-1}]$ then the vertices of C which are nonadjacent to u are either adjacent to both x and y or to none of them.*

The next theorem gives yet another characterization of distance-hereditary graphs. It will be used in the following subsection.

Theorem 29.48 [128,131] *For a graph G, the following conditions are equivalent:*

1. *G is distance-hereditary,*

2. *For each vertex v of G and every pair of vertices $x, y \in L_i(v)$, that are in the same connected component of the graph $G[V \setminus L_{i-1}(v)]$, we have*

$$N(x) \cap L_{i-1}(v) = N(y) \cap L_{i-1}(v).$$

Here, $L_1(v), \ldots, L_k(v)$ are the distance levels of a hanging from vertex v of G.

For many other graph classes defined in terms of metric properties in graphs, related convexity properties and connections to geometry, see the recent survey by Bandelt and Chepoi [132].

29.7.2 Minimum Cardinality Steiner Tree Problem in Distance-Hereditary Graphs

For a given graph $G = (V, E)$ and a set $S \subseteq V$ (of target vertices), a *Steiner tree* $T(S, G)$ is a tree with the vertex set $S \cup S'$ (i.e., $T(S, G)$ spans all vertices of S) and the edge set E' such that $S' \subseteq V$ and $E' \subseteq E$. The *minimum cardinality Steiner tree problem* asks for a Steiner tree with minimum $|S \cup S'|$.

An $O(|V||E|)$ time algorithm for the minimum cardinality Steiner tree problem on distance-hereditary graphs was presented in [131]. Later, in [133], a linear-time algorithm was obtained as a consequence of a linear-time algorithm for the connected r-domination problem on distance-hereditary graphs. Here, we present a direct linear-time algorithm for the minimum cardinality Steiner tree problem.

Algorithm ST-DHG (Find a minimum cardinality Steiner tree in a distance-hereditary graph)

Input: A distance-hereditary graph $G = (V, E)$ and a set $S \subseteq V$ of target vertices.
Output: A minimum cardinality Steiner tree $T(S, G)$.
    begin
        pick an arbitrary vertex $s \in S$ and build in G the distance levels $L_1(s), \ldots, L_k(s)$
        of a hanging from vertex s;
        for $i = k, k-1, \ldots, 2$ **do**
           if $S \cap L_i(s) \neq \emptyset$ **then**
              find the connected components $A_1, A_2, \ldots, A_p$ of $G[L_i(s)]$;
              in each component A_j pick an arbitrary vertex x_j;
              order these components in nondecreasing order with respect to $d'(A_j) = |N(x_j) \cap L_{i-1}(s)|$;
              for all components A_j taken in nondecreasing order with respect to $d'(A_j)$ **do**
                  set $B := N(x_j) \cap L_{i-1}(s)$;
                  if $(S \cap A_j \neq \emptyset$ and $S \cap B = \emptyset)$ **then**
                      add an arbitrary vertex y from B to set S;
        $T(S, G) :=$ a spanning tree of a subgraph $G[S]$ of G induced by vertices S;
    end

Clearly, this is a linear-time algorithm. The correctness proof is based on Theorem 29.47, Theorem 29.48 and the following claims.

Let $G = (V, E)$ be a distance-hereditary graph, $S \subseteq V$ be a set of target vertices, and $s \in S$ be an arbitrary vertex from S.

Claim 29.9 *There exists a minimum cardinality Steiner tree $T(S, G)$ such that $d_{T(S,G)}(x, s) = d_G(x, s)$ for any vertex x of $T(S, G)$.*

Proof. Let $L_1(s), \ldots, L_k(s)$ be the distance levels of a hanging of G from vertex $s \in S$. It is enough to show that there exists a minimum cardinality Steiner tree $T(S, G)$ such that if $T(S, G)$ is rooted at s then for any vertex x of $T(S, G)$ the following property holds:

(P^*) if x belongs to $L_i(s)$ $(i \in \{1, \ldots, k\})$ then its parent x^* in $T(S, G)$ belongs to $L_{i-1}(s)$.

Let $T(S, G)$ be a minimum cardinality Steiner tree with maximum number of vertices satisfying property (P^*) and let x be a vertex of $T(S, G)$ not satisfying (P^*) and with maximum $d_G(x, s)$. Assume x belongs to $L_i(s)$. Consider the (x, s)-path $P(x, s)$ in $T(S, G)$ and let

$y \in P(x, s)$ be the vertex closest to x in $P(x, s)$ with $y \in L_i(s)$ and $y^* \in L_{i-1}(s)$, where y^* is the parent of y in $T(S, G)$. From choices of vertices x and y, we conclude that the subpath of $P(x, s)$ between vertices x and y lays entirely in $L_i(s) \cup L_{i+1}(s)$. By Theorem 29.48, vertices x and y^* must be adjacent in G. Hence, we can modify tree $T(S, G)$ by removing edge xx^* and adding edge xy^*. The new tree obtained spans all vertices of S and has the same vertex-set. Since $T(S, G)$ was chosen to have maximum number of vertices satisfying property (P^*), such a vertex $x \in L_i(s)$ with $x^* \notin L_{i-1}(s)$ cannot exist, proving the claim. ∎

Let $A_1, A_2, \ldots, A_p$ be the connected components of $G[L_i(s)]$. By Theorem 29.48, $N(x) \cap L_{i-1}(s) = N(y) \cap L_{i-1}(s)$ for every pair of vertices $x, y \in A_j$, $j \in \{1, \ldots, p\}$. Hence, $N(A_j) \cap L_{i-1}(s) = N(x_j) \cap L_{i-1}(s)$ for any vertex $x_j \in A_j$. Denote $d'(A_j) := |N(A_j) \cap L_{i-1}(s)|$. Assume, without loss of generality, that $d'(A_1) \leq d'(A_2) \leq \cdots \leq d'(A_p)$. Let $B_j := N(u) \cap L_{i-1}(s) = N(A_j) \cap L_{i-1}(s)$, where u is an arbitrary vertex of A_j.

Claim 29.10 *For any vertices $x, y \in B_j$, $N(x) \setminus (B_j \cup A_1 \cup \cdots \cup A_{j-1}) = N(y) \setminus (B_j \cup A_1 \cup \cdots \cup A_{j-1})$.*

Proof. We have $u \in L_i(s)$, $x, y \in L_{i-1}(s) \cap N(u)$ and every vertex of A_j is adjacent to both x and y. By Theorem 29.48, any vertex $z \in L_{i-2}(s)$ either adjacent to both x and y or to none of them. Since $d'(A_j) \leq d'(A_{j'})$ for $j' > j$, by Theorem 29.47(iv), any vertex from $A_{j+1} \cup \cdots \cup A_p = L_i(s) \setminus (A_1 \cup \cdots \cup A_j)$ is adjacent to both or neither one of x and y. Assume now that there is a vertex $z \in L_{i-1}(s) \setminus B_j$ which is adjacent to x but not to y. Since path (z, x, u, y) lays in $L_i(s) \cup L_{i-1}(s)$, by Theorem 29.48, there must exist a vertex w in $L_{i-2}(s)$ adjacent to all y, x, z. But then, it is easy to see that the vertices u, x, y, z, w induce either a house or a gem in G, which is impossible. ∎

Let now i be the largest number such that $L_i(s) \cap S \neq \emptyset$ and, as before, $A_1, A_2, \ldots, A_p$ be the connected components of $G[L_i(s)]$ with $d'(A_1) \leq d'(A_2) \leq \cdots \leq d'(A_p)$. Let also j be the smallest number such that $A_j \cap S \neq \emptyset$. Set $B := N(A_j) \cap L_{i-1}(s)$. We know that any vertex of $A_j \cap S$ is adjacent to all vertices of B.

Claim 29.11 *Let $S \cap B \neq \emptyset$, $x \in S \cap A_j$ and $y \in S \cap B$. T' is a minimum cardinality Steiner tree of G for target set $S \setminus \{x\}$ if and only if T, obtained from T' by adding vertex x and edge xy, is a minimum cardinality Steiner tree of G for target set S.*

Proof. By Claim 29.9, for G and target set S, there exists a minimum cardinality Steiner tree T where vertex x is a leaf and its neighbor x^* in T belongs to $L_{i-1}(s)$, that is, to B. If $x^* \neq y$, we can get a new minimum cardinality Steiner tree for G and target set S by replacing edge xx^* in T with edge xy. We can do that since vertex y is in T and vertices x and y are adjacent in G. ∎

Claim 29.12 *Let $S \cap B = \emptyset$, $x \in S \cap A_j$, and y is an arbitrary vertex from B. T' is a minimum cardinality Steiner tree of G for target set $S \cup \{y\} \setminus \{x\}$ if and only if T, obtained from T' by adding vertex x and edge xy, is a minimum cardinality Steiner tree of G for target set S.*

Proof. By Claim 29.9, for G and target set S, there exists a minimum cardinality Steiner tree T such that $d_T(v, s) = d_G(v, s)$ for any vertex v of T. In particular, vertex x is a leaf and its neighbor x^* in T belongs to $L_{i-1}(s)$, that is, to B. Furthermore, any neighbor of x^* in T must belong to $A_j \cup A_{j+1} \cup \cdots \cup A_p$ or to $L_{i-2}(s)$. If $x^* \neq y$, we can get a new minimum cardinality Steiner tree for G and target set S by replacing in T vertex x^* with y and any edge ux^* of T with edge uy. We can do that since, by Claim 29.10, vertex y is adjacent in G to every vertex u to which vertex x^* was adjacent in T (recall, $u \in A_j \cup A_{j+1} \cup \cdots \cup A_p \cup L_{i-2}(s)$). ∎

Thus, we have the following theorem.

Theorem 29.49 [133] *The minimum cardinality Steiner tree problem in distance-hereditary graphs can be solved in linear $O(|V|+|E|)$ time.* ∎

29.7.3 Important Subclasses of Distance-Hereditary Graphs

29.7.3.1 Ptolemaic Graphs and Bipartite Distance-Hereditary Graphs

In this subsection, we describe the chordal and distance-hereditary graphs.

The *ptolemaic inequality* $(*)$ in metric spaces is defined as follows.

Definition 29.43 [134] *A connected graph G is ptolemaic if, for any four vertices u, v, w, x of G,*

$$(*) \quad d(u,v)d(w,x) \leq d(u,w)d(v,x) + d(u,x)d(v,w).$$

Theorem 29.50 [135] *Let G be a graph. The following conditions are equivalent:*

  i. *G is ptolemaic.*
  ii. *G is distance hereditary and chordal.*
  iii. *G is chordal and does not contain an induced gem.*
  iv. *For all distinct nondisjoint cliques P and Q of G, $P \cap Q$ separates $P \setminus Q$ and $Q \setminus P$.*

The equivalence of (ii) and (iii) follows from Theorem 29.45: If G is distance-hereditary then obviously G is gem-free. Conversely, if G is gem-free chordal then G is (house, hole, domino, gem)-free and by Theorem 29.45, it is distance-hereditary.

Ptolemaic graphs are characterized in various other ways; see, for example, [136] where the laminar structure of maximal cliques of ptolemaic graphs is described. This is closely related to Bachman Diagrams as described in [6].

Recall that G is chordal if and only if $\mathcal{C}(G)$ is α-acyclic and G is strongly chordal if and only if $\mathcal{C}(G)$ is β-acyclic. A similar fact holds for ptolemaic graphs (see Definition 29.44 for γ-acyclicity).

Theorem 29.51 [80] *Graph G is ptolemaic if and only if the hypergraph $\mathcal{C}(G)$ of its maximal cliques is γ-acyclic.* ∎

Theorems 29.45 and 29.44 imply the following corollary.

Corollary 29.19 *A graph is bipartite distance-hereditary if and only if it is bipartite $(6,2)$-chordal.*

Proof. Obviously, bipartite $(6,2)$-chordal graphs are (house, hole, domino, gem)-free and thus, by Theorem 29.45, are distance-hereditary. Conversely, let G be a bipartite distance-hereditary graph. Then, by Theorem 29.45, every cycle of length at least 5 has two (crossing) chords which shows the assertion. ∎

29.7.3.2 Block Graphs

There is an even more restrictive subclass of chordal distance-hereditary graphs, namely the *block graphs* which can be defined as the connected graphs whose blocks (i.e., 2-connected components) are cliques. Let $K_4 - e$ denote the clique of four vertices minus an edge (also called *diamond*).

Buneman's four-point condition $(**)$ for distances in connected graphs requires that for every four vertices u, v, x and y the following inequality holds:

$$(**) \quad d(u,v) + d(x,y) \leq \max\{d(u,x) + d(v,y),\ d(u,y) + d(v,x)\}.$$

It characterizes the metric properties of trees as Buneman [137] has shown. A connected graph is a tree if and only if it is triangle-free and fulfills Buneman's four-point condition $(**)$.

Theorem 29.52 [138] *Let G be a connected graph. The following conditions are equivalent:*

i. *G is a block graph.*

ii. *G is $(K_4 - e)$-free chordal.*

iii. *G fulfills Buneman's four-point condition $(**)$.* ∎

Theorem 29.53 [13] *G is a block graph if and only if $\mathcal{C}(G)$ is Berge-acyclic.* ∎

There are various other characterizations of block graphs—see for example [3] for a survey.

29.7.3.3 γ-Acyclic Hypergraphs

The basic subject of this subsection are γ-acyclic hypergraphs. Fagin [6,7] gives various equivalent definitions of γ-acyclicity.

Definition 29.44 [6,7] *Let $H = (V, \mathcal{E})$ be a hypergraph.*

i. *A γ-cycle in a hypergraph $H = (V, \mathcal{E})$ is a sequence $C = (v_1, E_1, v_2, E_2, \ldots, v_k, E_k)$, $k \geq 3$, of distinct vertices $v_1, v_2, \ldots, v_k$ and distinct hyperedges $E_1, E_2, \ldots, E_k$ such that for all i, $1 \leq i \leq k$, $v_i \in E_i \cap E_{i+1}$ holds and for all i, $1 \leq i < k$, $v_i \notin E_j$ for $j \neq i, i+1$ holds (index arithmetic modulo k).*

ii. *A hypergraph is γ-acyclic if it has no γ-cycle.*

Note that the only difference to special cycles is the condition $1 \leq i < k$ instead of $1 \leq i \leq k$. Fagin [6] gives some other variants of γ-acyclicity and shows that all these conditions are equivalent. A crucial property among them is the following separation property:

Theorem 29.54 *A hypergraph $H = (V, \mathcal{E})$ is γ-cyclic if and only if there is a nondisjoint pair E, F of hyperedges such that in the hypergraph that results by removing $E \cap F$ from every edge, what is left of E is connected to what is left of F.* ∎

This leads to the following tree structure of separators in γ-acyclic hypergraphs (it has been rediscovered under various names in subsequent papers on ptolemaic graphs, e.g., in [136]).

Definition 29.45 [6,139–141] *For a hypergraph $H = (V, \mathcal{E})$, we define:*

  i. *Bachman (H) is the hypergraph obtained by closing $\mathcal{E}$ under intersection, that is, S is in Bachman(H) if it is the intersection of some hyperedges from H (including the hyperedges from $\mathcal{E}$ themselves).*

  ii. *The Bachman diagram of H is the following undirected graph with Bachman (H) as its node set, and with an edge between two nodes S, T if S is a proper subset of T, that is, $S \subset T$ and there is no other W in Bachman (H) with $S \subset W \subset T$.*

  iii. *A Bachman diagram is* loop-free *if it is a tree.*

The tree property of the Bachman diagram is closely related to uniqueness properties in data connections; see [6] for a detailed discussion of various properties which are equivalent to γ-acyclicity and related work on desirable properties of relational database schemes.

The main theorem on γ-acyclicity is the following:

Theorem 29.55 [6] *Let $H = (V, \mathcal{E})$ be a connected hypergraph. The following are equivalent:*

1. *H is γ-acyclic.*

2. *Every connected join expression over H is monotone.*

3. *Every connected, sequential join expression over H is monotone.*

4. *The join dependency $\bowtie H$ implies that every connected subset of H has a lossless join.*

5. *There is a unique relationship among each set of attributes for each consistent database over H.*

6. *The Bachman diagram of H is loop-free.*

7. *H has a unique minimal connection among each set of its nodes.* ∎

29.8 TREEWIDTH AND CLIQUE-WIDTH OF GRAPHS

29.8.1 Treewidth of Graphs

Treewidth of a graph measures the tree-likeness of a graph. Treewidth of trees has value one, and if the treewidth of a graph class is bounded by a constant, this has important consequences for the efficient solution of many problems on the class. Treewidth was introduced by Robertson and Seymour in the famous graph minor project by Robertson and Seymour (see, e.g., [142–145] and is one of the most important concepts of algorithmic graph theory. It also came up as partial k-trees which have many applications (see e.g., [9]). A good survey is given by Bodlaender [11] and Kloks [146].

We first define k-trees recursively.

Definition 29.46 *Let $k \geq 1$ be an integer. The following graphs are k-trees:*

  i. *Any clique K_k with k vertices is a k-tree.*

  ii. *Let $G = (V, E)$ be a k-tree, let $x \notin V$ be a new vertex and let $C \subseteq V$ be a clique with k vertices. Then also $G' = (V \cup \{x\}, E \cup \{ux \mid u \in C\})$ is a k-tree.*

  iii. *There are no other k-trees.*

It is easy to see that for $k = 1$, the k-trees are exactly the trees, and for any k, k-trees are chordal with maximum clique size $k + 1$ if the graph is no clique. More exactly, all maximal cliques have size $k + 1$ in this case. See [147] for simple characterizations of k-trees.

Definition 29.47 *Graph $G' = (V, E')$ is a* partial k-tree *if there is a k-tree $G = (V, E)$ with $E' \subseteq E$.*

Obviously, every graph with n vertices is a partial n-tree, and every k-tree is a partial k-tree. The following parameter is of tremendous importance for the efficient solution of algorithmic problems on graphs.

Definition 29.48 *The* treewidth $tw(G)$ *of a given graph G is the minimum value k for which G is a partial k-tree.*

Determining the treewidth of a graph is NP-hard [9].

Treewidth was defined in a different way by Robertson and Seymour (see, e.g., [142–145]) via tree decompositions of graphs:

Definition 29.49 *A tree decomposition of a graph $G = (V, E)$ is a pair $D = (S, T)$ with the following properties:*

   i. *$S = \{V_i \mid i \in I\}$ is a finite collection of subsets of vertices (sometimes called bags).*

   ii. *$T = (I, F)$ is a tree with one node for each subset from S.*

   iii. *$\bigcup_{i \in I} V_i = V$.*

   iv. *For all edges $(v, w) \in E$, there is a subset (i.e., a bag) $V_i \in S$ such that both v and w are contained in V_i.*

   v. *For each vertex $x \in V$, the set of tree nodes $\{i \mid x \in V_i\}$ forms a subtree of T.*

Condition (v) corresponds to the join tree condition of α-acyclic hypergraphs and to the clique tree condition of chordal graphs. Thus, a graph is chordal if and only if it has a tree decomposition into cliques.

The *width* of a tree decomposition is the maximum bag size minus one. It is not hard to see that the following holds (see, e.g., [146]):

Lemma 29.12 *The treewidth of a graph equals the minimum width over all of its tree decompositions.* ∎

The fundamental importance of treewidth for algorithmic applications is twofold: First of all, many problems can be solved by dynamic programming in a bottom-up way along a tree decomposition (or equivalently, an embedding into a k-tree) of the graph, and the running time is *quite good* for *small k*. The literature [10,148] give many examples for this approach. Second, there is a deep relationship to Monadic Second-Order Logic described in various papers by Courcelle [149] (and in many other papers of this author; see also Bodlaender's tourist guide [11]). Roughly speaking, the following holds.

Whenever a problem Π is expressible in Monadic second-order logic and $\mathcal{C}$ is a graph class of bounded treewidth (with given tree decomposition for each input graph) then problem Π can be efficiently solved on every input graph from $\mathcal{C}$.

As an example, consider 3-colorability of a graph (which is well known to be NP-complete):

$$\exists W_1 \subseteq V \exists W_2 \subseteq V \exists W_3 \subseteq V \forall v \in V (v \in W_1 \vee v \in W_2 \vee v \in W_3) \wedge \forall v \in V \forall w \in V (vw \in E \Rightarrow (\neg(v \in W_1 \wedge w \in W_1) \wedge \neg(v \in W_2 \wedge w \in W_2) \wedge \neg(v \in W_3 \wedge w \in W_3))).$$

The detour via logic, however, leads to astronomically large constant factors in the running time of such algorithms. Therefore it is of crucial importance to have a tree decomposition of the input graph with very small width. We know already that the problem of determining treewidth is NP-complete.

Theorem 29.56 [150] *For each integer $k \geq 1$ there is a linear-time algorithm which for given graph G either determines that $tw(G) > k$ holds or otherwise finds a tree decomposition with width k.* ■

Some classes of graphs (cactus graphs, series-parallel graphs, Halin graphs, outerplanar graphs, etc.) have bounded treewidth. See [11] for more information.

Thorup [151] gives important examples of small treewidth in computer science applications.

Another closely related graph parameter called *tree-length* is proposed by Dourisboure and Gavoille [152]. It measures how close a graph is to being chordal. The tree-length of G is defined using tree decompositions of G (see Definition 29.49). Graphs of tree-length k are the graphs that have a tree decomposition where the distance in G between any pair of vertices that appear in the same bag of the tree decomposition is at most k. We discuss this and related parameters in Section 29.10.

29.8.2 Clique-Width of Graphs

The notion of *clique-width* of a graph, defined by Courcelle et al. (in the context of graph grammars) in [153], is another fundamental example of a width parameter on graphs which leads to efficient algorithms for problems expressible in some kind of Monadic second-order logic.

More formally, the clique-width $cw(G)$ of a graph G is defined as the minimum number of different integer labels which allow to generate graph G by using the following four kinds of operations on vertex-labeled graphs:

  i. Creation of a new vertex labeled by integer l.

  ii. Disjoint union of two (vertex-labeled and vertex-disjoint) graphs (i.e., co-join).

  iii. Join between the set of all vertices with label i and the set of all vertices with label j for $i \neq j$ (i.e., all edges between the two sets are added).

  iv. Relabeling of all vertices of label i by label j.

A *k-expression* for a graph G of clique-width k describes the recursive generation of G by repeatedly applying these operations using at most k different labels.

Obviously, any graph with n vertices can be generated using n labels (for each vertex a specific one). Thus $cw(G) \leq n$ if G has n vertices.

Clique-width is more powerful than treewidth in the sense that if a class of graphs has bounded treewidth then it also has bounded clique-width but not vice versa [154]—the clique-width of cliques of arbitrary size is two whereas their treewidth is unbounded. In particular, an upper bound for the clique-width of a graph is obtained from its treewidth as follows.

Theorem 29.57 [155] *For any graph G, $cw(G) \leq 3 \cdot 2^{tw(G)-1}$.* ∎

Similarly as for treewidth, the concept of clique-width of a graph has attracted much attention due to the fact that there is a similarly close connection to Monadic second-order logic. In [156], Courcelle et al. have shown that every graph problem definable in LinMSOL(τ_1) (a variant of Monadic second-order logic using quantifiers on vertex sets but not on edge sets) is solvable in linear time on graphs with bounded clique-width if a k-expression describing the input graph is given.

The problems maximum weight stable set, maximum weight clique, k-coloring for fixed k, Steiner tree, and domination are examples of LinMSOL(τ_1) definable problems whereas coloring and Hamiltonian circuit are not.

Theorem 29.58 [156] *Let $\mathcal{C}$ be a class of graphs of clique-width at most k such that there is an $\mathcal{O}(f(|E|,|V|))$ algorithm, which for each graph G in $\mathcal{C}$, constructs a k-expression defining it. Then for every LinMSOL(τ_1) problem on $\mathcal{C}$, there is an algorithm solving this problem in time $\mathcal{O}(f(|E|,|V|))$.* ∎

Moreover, for some other problems which are not expressible in this way, there are polynomial time algorithms for classes of bounded clique-width [157–159].

It is not hard to see that the class of cographs is exactly the class of graphs having clique-width at most 2, and a 2-expression can be found in linear time along the cotree of a cograph:

Proposition 29.13 *The clique-width of graph G is at most 2 if and only if G is a cograph.*

Clique-width is closely related to modular decomposition as the following proposition shows:

Proposition 29.14 [154,156] *The clique-width of a graph G is the maximum of the clique-width of its prime subgraphs, and the clique-width of the complement graph $\overline{G}$ is at most twice the clique-width of G.*

It is easy to see that the clique-width of thin spiders is at most 4. Thus, a simple consequence of Proposition 29.14 is that the clique-width of P_4-sparse graphs is bounded.

The fact that the clique-width of distance-hereditary graphs is at most three (which, at first glance, does not seem to be surprising but the proof is quite technical) is based on pruning sequences (see Theorem 29.46).

Theorem 29.59 [160] *The clique-width of distance-hereditary graphs is at most 3, and corresponding 3-expressions can be constructed in linear time.*

In the same paper [160] it is shown that unit interval graphs have unbounded clique-width. For very similar reasons, bipartite permutation graphs have unbounded clique-width [161]. Various other classes of bounded and unbounded clique-width are described in [162–167] and many other papers. See [168] for recent results on graph classes of bounded clique-width.

In [169], Fellows et al. show that determining clique-width is NP-complete. The recognition problem for graphs of clique-width at most three is solvable in polynomial time [170]. For any fixed $k \geq 4$, the problem of recognizing all graphs with clique-width at most k in polynomial time is open.

The notion of NLC-width introduced by Wanke [171] is closely related to clique-width. The NLC-width of a graph is not greater than its clique-width, and the clique-width of a graph is twice its NLC-width [172]. Computing the NLC-width of a graph is NP-complete [173]. The graphs of NLC-width 1 are the cographs, and the class of graphs of NLC-width at most 2 can be recognized in polynomial time [174]. Similarly as for clique-width (with $k \geq 4$), recognition of NLC-width at most k is open for $k \geq 3$.

Oum and Seymour [175,176] investigated the important concept of rank-width and its relationship to clique-width, treewidth and branchwidth. Oum showed that a graph has rank-width 1 if and only if it is distance hereditary.

29.9 COMPLEXITY OF SOME PROBLEMS ON TREE-STRUCTURED GRAPH CLASSES

The most prominent classes with tree structure in this chapter are chordal and dually chordal graphs, strongly chordal graphs and chordal bipartite graphs as well as distance-hereditary graphs. In the following, we describe a variety of complexity results for some problems on these classes. See also [19] for a final chapter on such results.

Recall that the recognition problem for chordal and dually chordal graphs is solvable in linear time, while the recognition of strongly chordal and of chordal bipartite graphs can be done in time $\mathcal{O}(\min(n^2, m \log n))$ (see [19]). Recall also that distance-hereditary graphs can be recognized in linear time [104,129].

The graph isomorphism problem was shown to be isomorphism-complete, that is as hard as in the general case, for strongly chordal graphs and chordal bipartite graphs [177]. The graph isomorphism problem for distance-hereditary graphs is solvable in linear time [136] (a first step for this was done in [178]); see also [179].

The four basic problems independent set [GT20], clique [GT19], chromatic number [GT4], and partition into cliques [GT15] (see [40]), are known to be polynomial-time solvable for perfect graphs [180,181] and thus for chordal graphs as well as strongly chordal graphs and chordal bipartite graphs. In some cases, there are better time bounds using prefect elimination orderings and similar tools. For dually chordal graphs, however, these four problems are NP-complete [63].

Hamiltonian circuit ([GT37] of [40]) is NP-complete for strongly chordal graphs and for chordal bipartite graphs [182] (and thus it is NP-complete for chordal as well as for dually chordal graphs).

Dominating set [GT2] and Steiner tree [ND12] [40] are solvable in linear time for dually chordal graphs [63] and thus for strongly chordal graphs while they are NP-complete for chordal graphs (even for split graphs [84]) and for chordal bipartite graphs [183].

For a given graph $G = (V, E)$, the *maximum induced matching problem* asks for a maximum set of edges having pairwise distance at least 2. While it is well known that the maximum matching problem is solvable in polynomial time, the maximum induced matching problem was shown to be NP-complete even for bipartite graphs [184,185]. For chordal graphs and for chordal bipartite graphs, however, it is solvable in polynomial time [184,186] and for chordal graphs, it is solvable in linear time [187]. It is NP-complete for dually chordal graphs [188]. Maximum induced matching can be generalized to hypergraphs and is solvable in polynomial time for α-acyclic hypergraphs but NP-complete for hypertrees [188].

For a given hypergraph $H = (V, \mathcal{E})$, the *exact cover problem* ([SP2] of [40]) asks for the existence of a subset $\mathcal{E}' \subseteq \mathcal{E}$ such that every vertex of V is in exactly one of the sets in $\mathcal{E}'$. The exact cover problem is NP-complete even for 3-regular hypergraphs [42]. In [188], it is shown that the exact cover problem is NP-complete for α-acyclic hypergraphs but solvable in linear time for hypertrees.

For a given graph $G = (V, E)$, the *efficient domination problem* asks for the existence of a set of closed neighborhoods of G forming an exact cover of V; thus, the efficient domination problem for G corresponds to the Exact Cover problem for the closed neighborhood hypergraph of G. It was introduced by Biggs [189] under the name *perfect code*.

The efficient domination problem is NP-complete for chordal graphs [190] and for chordal bipartite graphs [191]. In [188], it is shown that the efficient domination problem is solvable in linear time for dually chordal graphs.

For a given graph $G = (V, E)$, the *efficient edge domination problem* is the efficient domination problem for the line graph $L(G)$. It appears under the name *dominating-induced matching problem* in various papers; see for example [192]. The efficient edge domination problem is solvable in linear time for chordal graphs [188] and for dually chordal graphs [188] as well as for chordal bipartite graphs (and even solvable in polynomial time for hole-free graphs) [194].

For distance-hereditary graphs, there is a long list of papers showing that certain problems are efficiently solvable on this class. Most of these papers were published before the clique-width aspect was found. Theorem 29.58 covers many of these problems; on the other hand, it might be preferable to have direct dynamic programming algorithms using the tree structure of distance-hereditary graphs since the constant factors in algorithms using Theorem 29.58 are astronomically large (and similarly for graphs of bounded treewidth). However, various problems such as Hamilton cycle (HC) and variants cannot be expressed in MSOL; see also the algorithm for Steiner tree on distance-hereditary graphs.

The four basic problems can be solved in time $\mathcal{O}(n)$ if a pruning sequence of the input graph is given [129].

HC was shown to be solvable in time $\mathcal{O}(n^3)$ [195,196], in time $\mathcal{O}(n^2)$ [197] and finally in time $\mathcal{O}(n + m)$ for the HC problem [198,199] for HC and variants giving a unified approach. In [199], a detailed history of the complexity results for HC on distance-hereditary graphs is given. For the subclass of bipartite distance-hereditary graphs, a linear-time algorithm for HC was given already in [200].

The dominating set problem was solved in linear time in [201,202] for distance-hereditary graphs. The efficient domination and efficient edge domination problems are expressible in MSOL and thus efficiently solvable for distance-hereditary graphs.

29.10 METRIC TREE-LIKE STRUCTURES IN GRAPHS

There are few other graph parameters measuring tree likeness of a (unweighted) graph from a metric point of view. Two of them are also based on the notion of tree-decomposition of Robertson and Seymour [145] (see Definition 29.49).

29.10.1 Tree-Breadth, Tree-Length, and Tree-Stretch of Graphs

The *length* of a tree-decomposition T of a graph G is $\lambda := \max_{i \in I} \max_{u,v \in V_i} d_G(u, v)$ (i.e., each bag V_i has diameter at most λ in G). The *tree-length* of G, denoted by $tl(G)$, is the minimum of the length over all tree-decompositions of G [152]. As chordal graphs are exactly those graphs that have a tree decomposition where every bag is a clique [16–18], we can see that tree-length generalizes this characterization and thus the chordal graphs are exactly the graphs with tree-length 1. Note that tree-length and treewidth are not related to each other graph parameters. For instance, a clique on n vertices has tree-length 1 and treewidth $n - 1$, whereas a cycle on $3n$ vertices has treewidth 2 and tree-length n. One should also note that many graph classes with unbounded treewidth have bounded tree-length, such as chordal, interval, split, AT-free, and permutation graphs [152]. Analysis of a number of real-life networks, taken from different domains like Internet measurements, biological datasets, web graphs, social and collaboration networks, performed in [203,204] shows that those networks have sufficiently large treewidth but their tree-length is relatively small.

The *breadth* of a tree-decomposition T of a graph G is the minimum integer r such that for every $i \in I$ there is a vertex $v_i \in V$ with $V_i \subseteq N^r[v_i]$ (i.e., each bag V_i can be covered by a disk $N^r[v_i] := \{u \in V(G) : d_G(u, v_i) \leq r\}$ of radius at most r in G). Note that vertex v_i does not need to belong to V_i. The *tree-breadth* of G, denoted by $tb(G)$, is the minimum of the breadth over all tree-decompositions of G [205]. Evidently, for any graph G, $1 \leq tb(G) \leq tl(G) \leq 2tb(G)$ holds. Hence, if one parameter is bounded by a constant for a graph G then the other parameter is bounded for G as well.

Note that the notion of *acyclic (R,D)-clustering of a graph* introduced in [206] combines tree-breadth and tree-length into one notion. Graphs admitting acyclic (D,D)-clustering are exactly graphs with tree-length at most D, and graphs admitting acyclic $(R, 2R)$-clustering are exactly graphs with tree-breadth at most R. Hence, all chordal, chordal bipartite, and dually chordal graphs have tree-breadth 1 [206].

In view of tree-decomposition T of G, the smaller parameters $tl(G)$ and $tb(G)$ of G are, the closer graph G is to a tree metrically. Unfortunately, while graphs with tree-length 1 (as they are exactly the chordal graphs) can be recognized in linear time, the problem of determining whether a given graph has tree-length at most λ is NP-complete for every fixed $\lambda > 1$ (see [207]). Judging from this result, it is conceivable that the problem of determining whether a given graph has tree-breadth at most ρ is NP-complete, too. 3-Approximation algorithms for computing the tree-length and the tree-breadth of a graph are proposed in [152,204,205].

Proposition 29.15 [152] *There is a linear-time algorithm that produces for any graph G a tree-decomposition of length at most $3tl(G) + 1$.*

Proposition 29.16 [204,205] *There is a linear-time algorithm that produces for any graph G a tree-decomposition of breadth at most $3tb(G)$.*

It follows from results of [208] and [152] also that any graph G with small tree-length or small tree-breadth can be embedded to a tree with a small additive distortion.

Proposition 29.17 *For any (unweighted) connected graph $G = (V, E)$ there is an unweighted tree $H = (V, F)$ (on the same vertex set but not necessarily a spanning tree of G) for which the following is true:*

$$\forall u, v \in V, \quad d_H(u,v) - 2 \leq d_G(u,v) \leq d_H(u,v) + 3\, tl(G) \leq d_H(u,v) + 6\, tb(G).$$

Such a tree H can be constructed in $O(|E|)$ time.

Previously, these type of results were known for chordal graphs and dually chordal graphs [209], k-chordal graphs [210], and δ-hyperbolic graphs [211].

Graphs with small tree-length or small tree-breadth have many other nice properties. Every n-vertex graph with tree-length $tl(G) = \lambda$ has an additive 2λ-spanner with $O(\lambda n + n \log n)$ edges and an additive 4λ-spanner with $O(\lambda n)$ edges, both constructible in polynomial time [212]. Every n-vertex graph G with $tb(G) = \rho$ has a system of at most $\log_2 n$ collective additive tree $(2\rho \log_2 n)$-spanners constructible in polynomial time [213]. Those graphs also enjoy a 6λ-additive routing labeling scheme with $O(\lambda \log^2 n)$ bit labels and $O(\log \lambda)$ time routing protocol [214], and a $(2\rho \log_2 n)$-additive routing labeling scheme with $O(\log^3 n)$ bit labels and $O(1)$ time routing protocol with $O(\log n)$ message initiation time (by combining results of [213] and [215]). See appropriate papers for more details.

Here we elaborate a little bit more on a connection established in [205] between the tree-breadth and the tree-stretch of a graph (and the corresponding tree t-spanner problem).

The *tree-stretch* $ts(G)$ of a graph $G = (V, E)$ is the smallest number t such that G admits a *spanning* tree $T = (V, E')$ with $d_T(u, v) \leq t d_G(u, v)$ for every $u, v \in V$. T is called a *tree t-spanner* of G and the problem of finding such tree T for G is known as the *tree t-spanner problem*. Note that as T is a spanning tree of G, necessarily $d_G(u, v) \leq d_T(u, v)$ and $E' \subseteq E$. It is known that the tree t-spanner problem is NP-hard [216]. The best known approximation algorithms have approximation ratio of $O(\log n)$ [205,217].

The following two results were obtained in [205].

Proposition 29.18 [205] *For every graph G, $tb(G) \leq \lceil ts(G)/2 \rceil$ and $tl(G) \leq ts(G)$.*

Proposition 29.19 [205] *For every n-vertex graph G, $ts(G) \leq 2tb(G) \log_2 n$. Furthermore, a spanning tree T of G with $d_T(u, v) \leq (2tb(G) \log_2 n) \, d_G(u, v)$, for every $u, v \in V$, can be constructed in polynomial time.*

Proposition 29.19 is obtained by showing that every n-vertex graph G with $tb(G) = \rho$ admits a tree $(2\rho \log_2 n)$-spanner constructible in polynomial time. Together with Proposition 29.18, this provides a $\log_2 n$-approximate solution for the tree t-spanner problem in general unweighted graphs.

29.10.2 Hyperbolicity of Graphs and Embedding Into Trees

δ-Hyperbolic metric spaces have been defined by Gromov [218] in 1987 via a simple 4-point condition: for any four points u, v, w, x, the two larger of the distance sums $d(u, v) + d(w, x), d(u, w) + d(v, x), d(u, x) + d(v, w)$ differ by at most 2δ. They play an important role in geometric group theory, geometry of negatively curved spaces, and have recently become of interest in several domains of computer science, including algorithms and networking. For example, (a) it has been shown empirically in [219] (see also [220]) that the Internet topology embeds with better accuracy into a hyperbolic space than into an Euclidean space of comparable dimension, (b) every connected finite graph has an embedding in the hyperbolic plane so that the greedy routing based on the virtual coordinates obtained from this embedding is guaranteed to work (see [221]).

A connected graph $G = (V, E)$ equipped with standard graph metric d_G is δ-*hyperbolic* if the metric space (V, d_G) is δ-hyperbolic. More formally, let G be a graph and u, v, w and x be its four vertices. Denote by S_1, S_2, S_3 the three distance sums, $d_G(u, v) + d_G(w, x)$, $d_G(u, w) + d_G(v, x)$ and $d_G(u, x) + d_G(v, w)$ sorted in nondecreasing order $S_1 \leq S_2 \leq S_3$. Define the *hyperbolicity of a quadruplet* u, v, w, x as $\delta(u, v, w, x) = \frac{S_3 - S_2}{2}$. Then the *hyperbolicity* $\delta(G)$ *of a graph* G is the maximum hyperbolicity over all possible quadruplets of G, that is,

$$\delta(G) = \max_{u,v,w,x \in V} \delta(u, v, w, x).$$

δ-Hyperbolicity measures the local deviation of a metric from a tree metric; a metric is a tree metric if and only if it has hyperbolicity 0. Note that chordal graphs have hyperbolicity at most 1 [222], while k-chordal graphs have hyperbolicity at most $k/4$ [223].

The best known algorithm to calculate hyperbolicity has time complexity of $O(n^{3.69})$, where n is the number of vertices in the graph; it was proposed in [224] and involves matrix multiplications. Authors of [224] also propose a 2-approximation algorithm for calculating hyperbolicity that runs in $O(n^{2.69})$ time and a $2 \log_2 n$-approximation algorithm that runs in $O(n^2)$ time.

According to [211], if a graph G has small hyperbolicity then it can be embedded to a tree with a small additive distortion.

Proposition 29.20 [211] *For any (unweighted) connected graph $G = (V, E)$ with n vertices there is an unweighted tree $H = (V, F)$ (on the same vertex set but not necessarily a spanning tree of G) for which the following is true:*

$$\forall u, v \in V, \quad d_H(u, v) - 2 \leq d_G(u, v) \leq d_H(u, v) + O(\delta(G) \log n).$$

Such a tree H can be constructed in $O(|E|)$ time.

Thus, the distances in n-vertex δ-hyperbolic graphs can efficiently be approximated within an additive error of $O(\delta \log n)$ by a tree metric and this approximation is sharp (see [211,218,225]). An earlier result of Gromov [218] established similar distance approximations, however Gromov's tree is weighted, may have Steiner points and needs $O(n^2)$ time for construction.

It is easy to show that every graph G admitting a tree T with $d_G(x, y) \leq d_T(x, y) \leq d_G(x, y) + r$ for any $x, y \in V$ is r-hyperbolic. So, the hyperbolicity of a graph G is an indicator of an embedability of G in a tree with an additive distortion.

Graphs and general geodesic spaces with small hyperbolicities have many other algorithmic advantages. They allow efficient approximate solutions for a number of optimization problems. For example, Krauthgamer and Lee [226] presented a PTAS for the traveling salesman problem when the set of cities lie in a hyperbolic metric space. Chepoi and Estellon [227] established a relationship between the minimum number of balls of radius $r + 2\delta$ covering a finite subset S of a δ-hyperbolic geodesic space and the size of the maximum r-packing of S and showed how to compute such coverings and packings in polynomial time. Chepoi et al. gave in [211] efficient algorithms for fast and accurate estimations of diameters and radii of δ-hyperbolic geodesic spaces and graphs. Additionally, Chepoi et al. showed in [228] that every n-vertex δ-hyperbolic graph has an additive $O(\delta \log n)$-spanner with at most $O(\delta n)$ edges and enjoys an $O(\delta \log n)$-additive routing labeling scheme with $O(\delta \log^2 n)$ bit labels and $O(\log \delta)$ time routing protocol.

The following relations between the tree-length and the hyperbolicity of a graph were established in [211].

Proposition 29.21 [211] *For every n-vertex graph G, $\delta(G) \leq tl(G) \leq O(\delta(G) \log n)$.*

Combining this with results from [205] (see Propositions 29.18 and 29.19), one gets the following inequalities.

Proposition 29.22 [229] *For any n-vertex graph G, $\delta(G) \leq ts(G) \leq O(\delta(G) \log^2 n)$.*

This proposition says, in particular, that every δ-hyperbolic graph G admits a tree $O(\delta \log^2 n)$-spanner. Furthermore, such a spanning tree for a δ-hyperbolic graph can be constructed in polynomial time (see [205]).

The problem of approximating a given graph metric by a *simpler* metric is well motivated from several different perspectives. A particularly simple metric of choice, also favored from the algorithmic point of view, is a tree metric, that is, a metric arising from shortest path distance on a tree containing the given points. In recent years, a number of authors considered problems of minimum distortion embeddings of graphs into trees (see [208,230–232]), most popular among them being a noncontractive embedding with minimum multiplicative distortion.

Let $G = (V, E)$ be a graph. The (multiplicative) *tree-distortion* $td(G)$ of G is the smallest number α such that G admits a tree (not necessarily a spanning tree, possibly weighted and with Steiner points) with

$$\forall u, v \in V, \quad d_G(u, v) \leq d_T(u, v) \leq \alpha \, d_G(u, v).$$

The problem of finding, for a given graph G, a tree $T = (V \cup S, F)$ satisfying $d_G(u,v) \leq d_T(u,v) \leq td(G)d_G(u,v)$, for all $u, v \in V$, is known as the *problem of minimum distortion noncontractive embedding of graphs into trees*. In a noncontractive embedding, the distance in the tree must always be larger that or equal to the distance in the graph, that is, the tree distances *dominate* the graph distances.

It is known that this problem is NP-hard, and even more, the hardness result of [230] implies that it is NP-hard to approximate $td(G)$ better than γ, for some small constant γ. The best known 6-approximation algorithm using layering partition technique was recently given in [208]. It improves the previously known 100-approximation algorithm from [232] and 27-approximation algorithm from [231].

The following interesting result was presented in [208].

Proposition 29.23 [208] *For any (unweighted) connected graph $G = (V, E)$ with n vertices there is an unweighted tree $H = (V, F)$ (on the same vertex set but not necessarily a spanning tree of G) for which the following is true:*

$$\forall u, v \in V, \quad d_H(u,v) - 2 \leq d_G(u,v) \leq d_H(u,v) + 3\, td(G).$$

Such a tree H can be constructed in $O(|E|)$ time.

Surprisingly, a multiplicative distortion is turned into an additive one. Moreover, while a tree $T = (V \cup S, F)$ satisfying $d_G(u,v) \leq d_T(u,v) \leq td(G)d_G(u,v)$, for all $u, v \in V$, is NP-hard to find, tree H of Proposition 29.23 is constructible in $O(|E|)$ time. Furthermore, H is unweighted and has no Steiner points.

By adding at most $n = |V|$ new Steiner points to tree H and assigning proper weights to edges of H, the authors of [208] achieve a good noncontractive embedding of a graph G into a tree.

Proposition 29.24 [208] *For any (unweighted) connected graph $G = (V, E)$ there is a weighted tree $H'_\ell = (V \cup S, F)$ for which the following is true:*

$$\forall u, v \in V, \quad d_G(x,y) \leq d_{H'_\ell}(x,y) \leq 3td(G)(d_G(x,y) + 1).$$

Such a tree H'_ℓ can be constructed in $O(|V||E|)$ time.

As pointed out in [208], tree H'_ℓ provides a 6-approximate solution to the problem of minimum distortion noncontractive embedding of an unweighted graph into a tree.

We conclude this section with one more chain of inequalities establishing relations between the tree-stretch, the tree-length, and the tree-distortion of a graph.

Proposition 29.25 [229] *For every n-vertex graph G, $tl(G) \leq td(G) \leq ts(G) \leq 2td(G)\log_2 n$.*

Proposition 29.25 says that if a graph G is noncontractively embeddable into a tree with distortion $td(G)$ then it is embeddable into a spanning tree with stretch at most $2td(G)\log_2 n$. Furthermore, a spanning tree with stretch at most $2td(G)\log_2 n$ can be constructed for G in polynomial time.

References

[1] A. Hajnal and J. Surányi, Über die Auflösung von Graphen in vollständige Teilgraphen, *Ann. Univ. Sci. Budapest, Eötvös Sect. Math.* **1** (1958), 113–121.

[2] J.R.S. Blair and B. Peyton, An introduction to chordal graphs and clique trees, In *Graph Theory and Sparse Matrix Computation*, A. George, J.R. Gilbert, and J.W.H. Liu (Eds.), Springer, New York, 1993, 1–29.

[3] A. Brandstädt, V.B. Le, and J.P. Spinrad, Graph classes: A survey, *SIAM Monographs on Discrete Math. Appl.*, Vol. 3, SIAM, Philadelphia, PA, 1999.

[4] M.C. Golumbic, *Algorithmic Graph Theory and Perfect Graphs*, Academic Press, New York 1980; 2nd edition: *Ann. Discrete Math.* 57, Elsevier Science B.V., Amsterdam, the Netherlands, 2004.

[5] T.A. McKee and F.R. McMorris, Topics in intersection graph theory, *SIAM Monographs on Discrete Math. and Appl.* Vol. 2, Society for Industrial and Applied Mathematics, Philadelphia, PA, 1999.

[6] R. Fagin, Degrees of Acyclicity for hypergraphs and relational database schemes, *Journal ACM* **30** (1983), 514–550.

[7] R. Fagin, Acyclic database schemes (of various degrees): A painless introduction, *Proc. CAAP83 8th Colloquium on Trees in Algebra and Programming*, G. Ausiello and M. Protasi (Eds.), Springer LNCS 159 (1983), pp. 65–89.

[8] N. Robertson and P.D. Seymour, Graph minors. I. Excluding a forest, *J. Comb. Theory (B)* **35** (1983), 39–61.

[9] S. Arnborg, D.G. Corneil, and A. Proskurowski, Complexity of finding embeddings in a k-tree, *SIAM J. Alg. Discr. Meth.* **8** (1987), 277–284.

[10] S. Arnborg and A. Proskurowski, Linear time algorithms for NP-hard problems restricted to partial k-trees, *Discrete Applied Math.* **23** (1989), 11–24.

[11] H.L. Bodlaender, A tourist guide through treewidth, *Acta Cybernetica* **11** (1993), 1–23.

[12] C. Berge, *Graphs and Hypergraphs*, American Elsevier Publishing Co., North-Holland, 1973.

[13] C. Berge, *Hypergraphs*, Elsevier Publishing Co., North-Holland, 1989.

[14] G. Dirac, On rigid circuit graphs, *Abhandl. Math. Seminar Univ. Hamburg* **25** (1961), 71–76.

[15] D.R. Fulkerson and O.A. Gross, Incidence matrices and interval graphs, *Pacific J. Math.* **15** (1965), 835–855.

[16] A. Buneman, A characterization of rigid circuit graphs, *Discrete Math.* **9** (1974), 205–212.

[17] F. Gavril, The intersection graphs of subtrees in trees are exactly the chordal graphs, *J. Comb. Theory (B)* **16** (1974), 47–56.

[18] J.R. Walter, *Representations of Rigid Cycle Graphs*, PhD dissertation, Wayne State University, Detroit, MI, 1972.

[19] J.P. Spinrad, *Efficient Graph Representations, Fields Institute Monographs*, American Mathematical Society, Providence, RI, 2003.

[20] R.E. Tarjan and M. Yannakakis, Simple linear-time algorithms to test chordality of graphs, test acyclicity of hypergraphs, and selectively reduce acyclic hypergraphs, *SIAM J. Computing* **13** (1984), 566–579; Addendum *SIAM J. Computing* **14** (1985), 254–255.

[21] W.W. Barrett, C.R. Johnson, and M. Lundquist, Determinantal formulae for matrix completions associated with chordal graphs, *Linear Algebra Appl.* **121** (1989), 265–289.

[22] C.-W. Ho and R.C.T. Lee, Counting clique trees and computing perfect elimination schemes in parallel, *Inf. Proc. Letters* **31** (1989), 61–68.

[23] A. Frank, Some polynomial algorithms for certain graphs and hypergraphs, In *Proceedings of the 5th British Combinatorial Conference* (1975), *Congressus Numerantium* **XV** (1976), 211–226.

[24] S. Földes and P.L. Hammer, Split graphs, In *8th South–Eastern Conf. on Combinatorics, Graph Theory and Computing*, F. Hoffman, L. Lesniak-Foster, D. McCarthy, R.C. Mullin, K.B. Reid, and R.G. Stanton (Eds.), Louisiana State University, Baton Rouge, LA (1977), *Congressus Numerantium* **19** (1977), 311–315.

[25] P.L. Hammer and B. Simeone, The splittance of a graph, *Combinatorica* **1** (1981), 275–284.

[26] R.I. Tyshkevich, O.I. Melnikow, and V.M. Kotov, On graphs and degree sequences: The canonical decomposition (in Russian), *Kibernetika* **6** (1981), 5–8.

[27] M. Farber, Characterizations of strongly chordal graphs, *Discrete Math.* **43** (1983), 173–189.

[28] G. Ausiello, A. D'Atri, and M. Moscarini, Chordality properties on graphs and minimal conceptual connections in semantic data models, *J. Comput. Syst. Sci.* **33** (1986), 179–202.

[29] C. Beeri, R. Fagin, D. Maier, and M. Yannakakis, On the desirability of acyclic database schemes, *J. ACM* **30** (1983), 479–513.

[30] E.F. Codd, A relational model of data for large shared data banks, *Communications of the ACM* **13** (1970), 377–387.

[31] P. Honeyman, R. E. Ladner, and M. Yannakakis, Testing the universal instance assumption, *Inf. Proc. Letters* **10** (1980), 14–19.

[32] N. Goodman and O. Shmueli, Syntactic characterization of tree database schemas, *J. ACM* **30** (1983), 767–786.

[33] G. Gottlob, N. Leone, and F. Scarcello, Hypertree decompositions: A survey, In *Proc. MFCS 2001*, J. Sgall, A. Pultr, and P. Kolman, (Eds.), LNCS 2136, Springer, Mariánské Lázně, Czech Republic, 2001, 37–57.

[34] H. Gaifman, On local and nonlocal properties, In *Logic Colloquium'81* (J. Stern ed.,) Elsevier, North-Holland, Amsterdam, the Netherlands, 1982, 105–135.

[35] D. Maier, *The Theory of Relational Databases*, Computer Science Press, Rockville, MD, 1983.

[36] H.J. Ryser, Combinatorial configurations, *SIAM J. Appl. Math.* **17** (1969), 593–602.

[37] F.S. Roberts and J. H. Spencer, A characterization of clique graphs, *J. Comb. Theory (B)* **10** (1971), 102–108.

[38] J. L. Szwarcfiter, A survey on clique graphs, In *Recent Advances in Algorithmic Combinatorics*, C. Linhares-Sales and B. Reed (Eds.), CMS Books in Mathematics, Springer, 2003, 109–136.

[39] L. Alcón, L. Faria, C.M.H. de Figueiredo, and M. Gutierrez, Clique graph recognition is NP-complete, F. Fomin (ed.), WG 2006, *Lecture Notes in Comp. Sci.* **4271** (2006), 269–277; full version in: The complexity of clique graph recognition. *Theor. Comp. Sci.* **410**(21–23) (2009), 2072–2083.

[40] M.R. Garey and D.S. Johnson, *Computers and Intractability: A Guide to the Theory of NP-Completeness*, Freeman & Co., San Francisco, CA, 1979.

[41] L. Lovász, Coverings and colorings of hypergraphs, In *Proc. 4th Southeastern Conf. on Combinatorics, Graph Theor. Comput., Util. Math. Publ.*, Congr. Numerantium **VIII** (1973), 3–12.

[42] R.M. Karp, Reducibility among combinatorial problems, In *Complexity of Computer Computations*, R.E. Miller and J.W. Thatcher (Eds.), Plenum Press, New York, 1972, 85–103.

[43] N. Linial and M. Tarsi, Deciding hypergraph 2-colorability by H-resolution, *Theor. Comp. Sci.* **38** (1985), 343–347.

[44] M.R. Garey, D.S. Johnson, and L. Stockmeyer, Some simplified NP-complete graph problems, *Theor. Comp. Sci.* **1** (1976), 237–267.

[45] R. Fagin and M.Y. Vardi, The theory of data dependencies—A survey, *Mathematics of Information Processing*, In *Proc. Symp. Appl. Math.*, M. Anshel, W. Gewirtz, (Eds.) Vol. **34** (1986), 19–71, American Mathemaical Society, Providence, RI.

[46] J. Lehel, A characterization of totally balanced hypergraphs, *Discrete Math.* **57** (1985), 59–65.

[47] M.C. Golumbic, Algorithmic aspects of intersection graphs and representation hypergraphs, *Graphs and Combinatorics* **4** (1988), 307–321.

[48] P. Duchet, Propriété de Helly et problèmes de représentation, *Colloqu. Internat. CNRS 260, Problemes Combinatoires et Theorie du Graphs, Orsay, France* (1976), 117–118.

[49] C. Flament, Hypergraphes arborés, *Discrete Math.* **21** (1978), 223–226.

[50] P.J. Slater, A characterization of SOFT hypergraphs, *Canad. Math. Bull.* **21** (1978), 335–337.

[51] P.A. Bernstein and N. Goodman, Power of natural semijoins, *SIAM J. Comput.* **10** (1981), 751–771.

[52] T.A. McKee, How chordal graphs work, *Bull. ICA* **9** (1993), 27–39.

[53] B.D. Acharya and M. las Vergnas, Hypergraphs with cyclomatic number zero, triangulated graphs, and an inequality, *J. Comb. Theor. (B)* **33** (1982), 52–56.

[54] P. Hansen and M. Las Vergnas, On a property of hypergraphs with no cycles of length greater than two, In *Hypergraph Seminar, Lecture Notes in Math.* **411** (1974), 99–101.

[55] M. Lewin, On hypergraphs without significant cycles, *J. Comb. Theor. (B)* **20** (1976), 80–83.

[56] M.H. Graham, On the universal relation, *Tech. Report*, University of Toronto, Ontario, Canada, 1979.

[57] C.T. Yu and M.Z. Ozsoyoglu, An algorithm for tree-query membership of a distributed query, *Proc. 1979 IEEE COMPSAC*, IEEE, New York, 1979, 306–312.

[58] R. Fagin, A.O. Mendelzon, and J.D. Ullman, A simplified universal relation assumption and its properties, *ACM Trans. Database Syst.* **7** (1982), 343–360.

[59] A. Brandstädt, F.F. Dragan, V.D. Chepoi, and V.I. Voloshin, Dually chordal graphs, Technical report SM-DU-225, University of Duisburg 1993; extended abstract in: Proceedings of WG 1993, LNCS 790, 237–251, 1993; full version in *SIAM J. Discr. Math.* **11** (1998), 437–455.

[60] F.F. Dragan, HT-graphs: Centers, connected r-domination and steiner trees, *Comp. Sci. J. Moldova* **1** (1993), 64–83.

[61] F.F. Dragan, C.F. Prisacaru, and V.D. Chepoi, Location problems in graphs and the Helly property (in Russian) (1987) (appeared partially in *Diskretnaja Matematika* **4** (1992), 67–73).

[62] H. Behrendt and A. Brandstädt, Domination and the use of maximum neighborhoods, Technical report SM-DU-204, University of Duisburg, Germany, 2002.

[63] A. Brandstädt, V.D. Chepoi, and F.F. Dragan, The algorithmic use of hypertree structure and maximum neighbourhood orderings, Technical report SM-DU-244, University of Duisburg 1994; extended abstract in: Proceedings of WG 1994, LNCS 903, 65–80, 1994; full version in *Discrete Applied Math.* **82** (1998), 43–77.

[64] J.L. Szwarcfiter and C.F. Bornstein, Clique graphs of chordal and path graphs, *SIAM J. Discrete Math.* **7** (1994) 331–336.

[65] M. Gutierrez and L. Oubiña, Metric characterizations of proper interval graphs and tree-clique graphs, *J. Graph Theor.* **21** (1996), 199–205.

[66] A. Brandstädt, V.D. Chepoi, and F.F. Dragan, Clique r-domination and clique r-packing problems on dually chordal graphs, *SIAM J. Discrete Math.* **10** (1997), 109–127.

[67] P. De Caria, *A Joint Study of Chordal and Dually Chordal Graphs*, PhD thesis, Universidad Nacional de la Plata, Argentina, 2012.

[68] P. De Caria and M. Gutierrez, On minimal vertex separators of dually chordal graphs: Properties and characterizations, *Discrete Appl. Math.* **160** (2012), 2627–2635.

[69] P. De Caria and M. Gutierrez, Comparing trees characteristic to chordal and dually chordal graphs, *Electronic Notes in Discrete Math.* **37** (2011), 33–38.

[70] P. De Caria and M. Gutierrez, On the correspondence between tree representations of chordal and dually chordal graphs, *Discrete Appl. Math.* **164** (2014), 500–511.

[71] A. Leitert, Das Dominating Induced Matching Problem für azyklische Hypergraphen, Diploma thesis, University of Rostock, Germany, 2012.

[72] M. Moscarini, Doubly chordal graphs: Steiner trees and connected domination, *Netw.* **23** (1993), 59–69.

[73] A. Brandstädt, F.F. Dragan, and F. Nicolai, Homogeneously orderable graphs, *Theor. Comput. Sci.* **172** (1997), 209–232.

[74] A. Lubiw, Doubly lexical orderings of matrices, *SIAM J. Comput.* **16** (1987), 854–879.

[75] R. Paige and R.E. Tarjan, Three partition refinement algorithms, *SIAM J. Comput.* **16** (1987), 973–989.

[76] L. Lovász, *Combinatorial Problems and Exercises*, North-Holland, Amsterdam, the Netherlands, 1979.

[77] L. Lovász and M.D. Plummer, *Matching Theory*, North-Holland, Amsterdam, the Netherlands, Math. Studies Vol. 29, 1986.

[78] C. Berge and M. Las Vergnas, Sur un théorème du type Kőnig pour hypergraphes, *Annals NY Acad. Sci.* **175** (1970), 32–40.

[79] A.E. Brouwer and A. Kolen, A super-balanced hypergraph has a nest point, Report ZW 146/80, Mathematisch Centrum, Amsterdam, the Netherlands, 1980.

[80] A. D'Atri and M. Moscarini, On hypergraph acyclicity and graph chordality, *Inf. Proc. Letters* **29** (1988), 271–274.

[81] R.P. Anstee and M. Farber, Characterizations of totally balanced matrices, *J. Algorithms* **5** (1984), 215–230.

[82] A.J. Hoffman, A.W.J. Kolen, and M. Sakarovitch, Totally balanced and greedy matrices, *SIAM J. Alg. Discrete Meth.* **6** (1985), 721–730.

[83] A. Lubiw, Γ-*free matrices*, Master's thesis, Department of Combinatorics and Optimization, University of Waterloo, Canada, 1982.

[84] A.A. Bertossi, Dominating sets for split graphs and bipartite graphs, *Inf. Proc. Letters* **19** (1984), 37–40.

[85] G.J. Chang, Labeling algorithms for domination problems in sun-free chordal graphs, *Discrete Appl. Math.* **22** (1988), 21–34.

[86] G.J. Chang and G.L. Nemhauser, The k-domination and k-stability problem on sun-free chordal graphs, *SIAM J. Alg. Discrete Meth.* **5** (1984), 332–345.

[87] G.J. Chang, M. Farber, and Z. Tuza, Algorithmic aspects of neighbourhood numbers, *SIAM J. Discrete Math.* **6** (1993), 24–29.

[88] K. Iijima and Y. Shibata, A bipartite representation of a triangulated graph and its chordality, Deptartment of Computer Science, Gunma University, Maebashi, Japan, CS 79-1, 1979.

[89] J.P. Spinrad, Doubly lexical ordering of dense 0–1 matrices, *Inf. Proc. Letters* **45** (1993), 229–235.

[90] A.E. Brouwer, P. Duchet, and A. Schrijver, Graphs whose neighborhoods have no special cycles, *Discrete Math.* **47** (1983), 177–182.

[91] T.A. McKee, Strong clique trees, neighborhood trees, and strongly chordal graphs, *J. Graph Theor* **33** (2000), 125–183.

[92] S. Ma and J. Wu, Characterizing strongly chordal graphs by using minimal relative separators, *Combinatorial Designs and Applications*, W.D. Wallis, H. Shen, W. Wei, and L. Shu (Eds.): Lecture Notes in Pure and Applied Mathematics 126, Marcel Dekker, New York, 1990, 87–95.

[93] M.C. Golumbic and C.F. Goss, Perfect elimination and chordal bipartite graphs, *J. Graph Theor.* **2** (1978), 155–163.

[94] R.B. Hayward, Weakly triangulated graphs, *J. Comb. Theory (B)* **39** (1985), 200–208.

[95] A. Brandstädt, Classes of bipartite graphs related to chordal graphs, *Discrete Appl. Math.* **32** (1991), 51–60.

[96] E. Dahlhaus, Chordale Graphen im besonderen Hinblick auf parallele Algorithmen, Habilitation Thesis, Universität Bonn, Germany, 1991.

[97] R. Uehara, Recognition of chordal bipartite graphs, Proceedings of ICALP, *Lecture Notes in Comp. Sci.* **2380** (2002), 993–1004.

[98] J. Huang, Representation characterizations of chordal bipartite graphs, *J. Combinatorial Theor. B* **96** (2006) 673–683.

[99] A. Berry and A. Sigayret, Dismantlable lattices in the mirror, *Proc. ICFCA* 2013, 44–59.

[100] D.G. Corneil, H. Lerchs, and L. Stewart-Burlingham, Complement reducible graphs, *Discrete Appl. Math.* **3** (1981), 163–174.

[101] D.G. Corneil, Y. Perl, and L.K. Stewart, Cographs: Recognition, applications, and algorithms, *Congressus Numer.* **43** (1984), 249–258.

[102] D.G. Corneil, Y. Perl, and L.K. Stewart, A linear recognition algorithm for cographs, *SIAM J. Comput.* **14** (1985), 926–934.

[103] D. Kratsch, R.M. McConnell, K. Mehlhorn, and J. Spinrad, Certifying algorithms for recognizing interval graphs and permutation graphs, *SIAM J. Comput.* **36**(2) (2006), 326–353.

[104] G. Damiand, M. Habib, and Ch. Paul, A simple paradigm for graph recognition: Application to cographs and distance-hereditary graphs, *TCS* **263** (2001), 99–111.

[105] A. Bretscher, D.G. Corneil, M. Habib, and C. Paul, A simple linear time LexBFS cograph recognition algorithm, *Conference Proceedings of International Workshop on Graph-Theoretic Concepts in Computer Science*, In *Lecture Notes in Comp. Sci.* 2880, Hans L. Bodlaender (Ed.), Elspeet, the Netherlands, 2003, 119–130.

[106] R.M. McConnell and J.P. Spinrad, Modular decomposition and transitive orientation, *Discrete Math.* **201** (1999), 189–241.

[107] T. Gallai, Transitiv orientierbare Graphen, *Acta Math. Acad. Sci. Hung.* **18** (1967), 25–66.

[108] A. Cournier and M. Habib, A new linear algorithm for modular decomposition, *LIRMM, University Montpellier* (1995), Preliminary version in: *Trees in Algebra and Programming—CAAP*, LNCS **787** (1994), 68–84.

[109] E. Dahlhaus, J. Gustedt, and R.M. McConnell, Efficient and practical modular decomposition, *J. Algorithms* **41**(2) (2001), 360–387.

[110] M. Habib, F. de Montgolfier, and C. Paul, A simple linear time modular decomposition algorithm for graphs, using order extension, *Proc. 9th Scandinav. Workshop on Algorithm Theory, Lecture Notes in Comp. Sci.* **3111** (2004), 187–198.

[111] M. Tedder, D.G. Corneil, M. Habib, and C. Paul, Simpler linear-time modular decomposition via recursive factorizing permutations, *35th International Colloquium on Automata, Languages and Programming, Lecture Notes in Comput. Sci.* **5125** (2008), 634–645.

[112] R.H. Möhring and F.J. Radermacher, Substitution decomposition for discrete structures and connections with combinatorial optimization, *Annals of Discrete Math.* **19** (1984), 257–356.

[113] C.T. Hoàng, *A class of perfect graphs*, MSc Thesis, School of Computer Science, McGill University, Montreal, Canada, 1983.

[114] B. Jamison and S. Olariu, A tree representation for P_4-sparse graphs, *Discrete Appl. Math.* **35**(2) (1992), 115–129.

[115] B. Jamison and S. Olariu, Recognizing P_4-sparse graphs in linear time, *SIAM J. Comput.* **21**(2) (1992), 381–406.

[116] B. Jamison and S. Olariu, Linear time optimization algorithms for P_4-sparse graphs, *Discrete Appl. Math.* **61**(2) (1995), 155–175.

[117] C.T. Hoàng, *Perfect graphs*, PhD thesis, School of Computer Science, McGill University, Montreal, Canada, 1985.

[118] W.H. Cunningham, Decomposition of directed graphs, *SIAM J. Algebraic and Discrete Meth.* **3** (1982), 214–228.

[119] E. Dahlhaus, Parallel algorithms for hierarchical clustering and applications to split decomposition and parity graph recognition, *J. Algorithms* **36** (2000), 205–240.

[120] P. Charbit, F. de Montgolfier, and M. Raffinot, A simple linear time split decomposition algorithm of undirected graphs, CoRR abs/0902.1700, 2009.

[121] L. Babel and S. Olariu, On the p-connectedness of graphs—A survey, *Discrete Appl. Math.* **95** (1999), 11–33.

[122] R.E. Tarjan, Decomposition by clique separators, *Discrete Math.* **55** (1985), 221–232.

[123] S.H. Whitesides, A method for solving certain graph recognition and optimization problems, with applications to perfect graphs, In *Topics on Perfect Graphs*, Berge, C. and V. Chvátal (Eds.), North-Holland, Amsterdam, the Netherlands, 1984.

[124] A. Brandstädt, V. Giakoumakis, and F. Maffray, Clique-separator decomposition of hole-free and diamond-free graphs, *Discrete Appl. Math.* **160** (2012), 471–478.

[125] V.E. Alekseev, On easy and hard hereditary classes of graphs with respect to the independent set problem, *Discrete Appl. Math.* **132** (2004), 17–26.

[126] D. Lokshtanov, M. Vatshelle, and Y. Villanger, Independent set in P_5-free graphs in polynomial time, Tech. Report 2013, *Proceedings of the ACM-SIAM Symposium on Discrete Algorithms.*

[127] E. Howorka, A characterization of distance-hereditary graphs, *Quart. J. Math. Oxford Ser.* **2**(28) (1977), 417–420.

[128] H.-J. Bandelt and H.M. Mulder, Distance-hereditary graphs, *J. Combin. Theor. (B)* **41** (1986), 182–208.

[129] P.L. Hammer and F. Maffray, Completely separable graphs, *Discrete Appl. Math.* **27** (1990), 85–99.

[130] H.-J. Bandelt, A. Henkmann, and F. Nicolai, Powers of distance-hereditary graphs, *Discrete Math.* **145** (1995), 37–60.

[131] A. D'Atri and M. Moscarini, Distance-Hereditary Graphs, Steiner Trees, and Connected Domination, *SIAM J. Comput.* **17** (1988), 521–538.

[132] H.-J. Bandelt and V.D. Chepoi, Metric graph theory and geometry: A survey, *Contemporary Mathematics* 453, 2006, *Surveys on Discrete and Computational Geometry, Twenty Years Later, AMS-IMS-SIAM Joint Summer Conference*, Snowbird, UT, J.E. Goodman, J. Pach, and R. Pollack (Eds.), American Mathematical Society, June 18–22, 2006, 49–86.

[133] A. Brandstädt and F.F. Dragan, A linear-time algorithm for connected r-domination and Steiner Tree on distance-hereditary graphs, *Netw.* **31** (1998), 177–182.

[134] D.C. Kay and G. Chartrand, A characterization of certain ptolemaic graphs, *Canad. J. Math.* **17** (1965), 342–346.

[135] E. Howorka, A characterization of ptolemaic graphs, *J. Graph Theor.* **5** (1981), 323–331.

[136] R. Uehara and T. Uno, Laminar structure of ptolemaic graphs and its applications, Proc. ISAAC 2005, X. Deng, D. Du (Eds.), *Lecture Notes in Comp. Sci.* **3827** (2005), 186–195; *Discrete Appl. Math.* **157** (2009), 1533–1543.

[137] A. Buneman, A note on the metric properties of trees, *J. Comb. Theory (B)* **1** (1974), 48–50.

[138] E. Howorka, On metric properties of certain clique graphs, *J. Comb. Theory (B)* **27** (1979), 67–74.

[139] C.W. Bachman, Data structure diagrams, *Data Base* **1**(2) (1969), 4–10.

[140] Y.E. Lien, On the equivalence of database models, *J. ACM* **29**(2) (1982), 333–363.

[141] M. Yannakakis, Algorithms for Acyclic Database Schemes, In *Proc. of Int. Conf. on Very Large Data Bases*, C. Zaniolo, C. Delobel (Eds.), Cannes, France, 1981, 82–94.

[142] N. Robertson and P.D. Seymour, Graph minors. III. Planar tree-width, *J. Comb. Theor.(B)* **36** (1984), 49–64.

[143] N. Robertson and P.D. Seymour, Graph width and well-quasi ordering: A survey, *Progress in Graph Theory*, J. Bondy and U. Murty (Eds.), Academic Press, New York, 1984, 399–406.

[144] N. Robertson and P.D. Seymour, Graph minors—A survey, *Surveys in Combinatorics*, I. Anderson (Ed.), London Mathematical Society, Lecture Note Series 103, Invited papers for the 10th British Combinatorial Conference, Cambridge University Press, 1985, 153–171.

[145] N. Robertson and P.D. Seymour, Graph minors. II. Algorithmic aspects of tree width, *J. Algorithms* **7** (1986), 309–322.

[146] T. Kloks, Treewidth—Computations and approximations, *Lecture Notes in Comput. Sci.* **842** (1994), 1–209.

[147] D.J. Rose, On simple characterizations of k-trees, *Discrete Math.* **7** (1974), 317–322.

[148] S. Arnborg, J. Lagergren, and D. Seese, Easy problems for tree-decomposable graphs, *J. Algorithms* **12** (1991), 308–340.

[149] B. Courcelle, The monadic second-order logic of graphs III: Tree-decompositions, minor and complexity issues, *Informatique Theorique et Applications* **26** (1992), 257–286.

[150] H.L. Bodlaender, A linear time algorithm for finding tree-decompositions of small treewidth, *SIAM J. Comput.* **25** (1996), 1305–1317.

[151] M. Thorup, All structured programs have small tree width and good register allocation, *Information and Computation* **142**(2) (1988), 159–181.

[152] Y. Dourisboure and C. Gavoille, Tree-decompositions with bags of small diameter, *Discrete Math.* **307** (2007), 2008–2029.

[153] B. Courcelle, J. Engelfriet, and G. Rozenberg, Handle-rewriting hypergraph grammars, *J. Comput. Syst. Sci.* **46** (1993), 218–270.

[154] B. Courcelle and S. Olariu, Upper bounds to the clique width of graphs, *Discrete Appl. Math.* **101** (2000), 77–114.

[155] D.G. Corneil and U. Rotics, On the relationship between clique-width and treewidth, *Internat. Workshop on Graph-Theoretic Concepts in Computer Science, Lecture Notes in Comput. Sci.* **2204** (2001), 78–90; *SIAM J. Computing* **34** (2005), 825–847.

[156] B. Courcelle, J.A. Makowsky, and U. Rotics, Linear time solvable optimization problems on graphs of bounded clique width, *Theor. Comput. Syst.* **33** (2000), 125–150.

[157] W. Espelage, F. Gurski, and E. Wanke, How to solve NP-hard graph problems on clique-width bounded graphs in polynomial time, *Internat. Workshop on Graph-Theoretic Concepts in Computer Science, Lecture Notes in Comput. Sci.* **2204** (2001), 117–128.

[158] D. Kobler and U. Rotics, Edge dominating set and colorings on graphs with fixed clique-width, *Discrete Appl. Math.* **126** (2002), 197–221.

[159] M.U. Gerber and D. Kobler, Algorithms for vertex partitioning problems on graphs with fixed clique-width, *Theor. Comput. Sci.* **1–3** (2003), 719–734.

[160] M.C. Golumbic and U. Rotics, On the clique-width of perfect graph classes, *Int. J. Foundations Comput. Sci.* **11** (2000), 423–443.

[161] A. Brandstädt and V.V. Lozin, On the linear structure and clique width of bipartite permutation graphs, RUTCOR Research Report, Rutgers University, New Brunswick, NJ, 29–2001 (2001); *Ars Combinatoria* (2003) 273–281.

[162] A. Brandstädt, F.F. Dragan, H.-O. Le, and R. Mosca, New graph classes of bounded clique width, *Theor. Comput. Syst.* **38** (2005), 623–645.

[163] A. Brandstädt, J. Engelfriet, H.-O. Le, and V.V. Lozin, Clique-width for four-vertex forbidden subgraphs, *Theor. Comput. Syst.* **39** (2006), 561–590.

[164] A. Brandstädt, Hoàng-Oanh Le, and R. Mosca, Gem- and co-gem-free graphs have bounded clique width, *Internat. J. Foundat. Computer Science* **15** (2004), 163–185.

[165] A. Brandstädt, Hoàng-Oanh Le, and R. Mosca, Chordal co-gem-free graphs and (P_5,gem)-free graphs have bounded clique width, *Discrete Appl. Math.* **145** (2005), 232–241.

[166] H.-O. Le, Contributions to clique-width of graphs, Dissertation, University of Rostock, Germany, 2003.

[167] J.A. Makowsky and U. Rotics, On the clique-width of graphs with few P_4's, *Int. J. Foundat. Comput. Sci.* **10** (1999), 329–348.

[168] M. Kamiński, V.V. Lozin, and M. Milanič, Recent developments on graphs of bounded clique-width, *Discrete Appl. Math.* **157** (2009), 2747–2761.

[169] M.R. Fellows, F.A. Rosamond, U. Rotics, and S. Szeider, Clique-width is NP-complete, *SIAM J. Alg. Discr. Math.* **23**(2) (2009), 909–939.

[170] D.G. Corneil, M. Habib, J.M. Lanlignel, B. Reed, and U. Rotics, Polynomial time recognition of clique-width ≤ 3 graphs, *Proceedings of LATIN, Lecture Notes in Comput. Sci.* **1776** (2000), 126–134.

[171] E. Wanke, k-NLC graphs and polynomial algorithms, *Discrete Appl. Math.* **54** (1994), 251–266.

[172] Ö. Johansson, Clique decomposition, NLC decomposition, and modular decomposition—Relationships and results for random graphs, *Congressus Numerantium* **132** (1998), 39–60.

[173] F. Gurski and E. Wanke, Line graphs of bounded clique-width, *Discrete Math.* **307** (2007), 2734–2754.

[174] V. Limouzy, F. de Montgolfier, and M. Rao, NLC_2 recognition and isomorphism, Proceedings of WG 2007, *Lecture Notes in Comput. Sci.* **4769** (2007), 86–98.

[175] S.-I. Oum, Approximating rank-width and clique-width quickly, *ACM Transactions on Algorithms* **5** (2008), 1–20.

[176] S.-I. Oum and P.D. Seymour, Approximating clique-width and branch-width, *J. Comb. Theory (B)* **96** (2006), 514–528.

[177] R. Uehara, S. Toda, and T. Nagoya, Graph isomorphism completeness for chordal bipartite graphs and strongly chordal graphs, *Discr. Appl. Math.* **145** (2005), 479–482.

[178] A. D'Atri, M. Moscarini, and H.M. Mulder, On the isomorphism problem for distance-hereditary graphs, Econometric Institute Tech. Report EI9241, A Rotterdam School of Economics, 1992.

[179] S.-I. Nakano, R. Uehara, and T. Uno, A new approach to graph recognition and applications to distance-hereditary graphs, *J. Comput. Sci. Technol.* **24**(3) (2009), 517–533.

[180] M. Grötschel, L. Lovász, and A. Schrijver, Polynomial algorithms for perfect graphs, *Ann. Discr. Math.* **21** (1984), 325–356.

[181] M. Grötschel, L. Lovász, and A. Schrijver, *Geometric Algorithms and Combinatorial Optimization*, Springer, 1988.

[182] H. Müller, Hamilton circuits in chordal bipartite graphs, *Discr. Math.* **156** (1996), 291–298.

[183] P. Damaschke, H. Müller, and D. Kratsch, Domination in convex and chordal bipartite graphs, *Inf. Proc. Letters* **36** (1990), 231–236.

[184] K. Cameron, Induced matchings, *Discr. Appl. Math.* **24** (1989), 97–102.

[185] L.J. Stockmeyer and V.V. Vazirani, NP-completeness of some generalizations of the maximum matching problem, *Inform. Process. Lett.* **15** (1982), 14–19.

[186] K. Cameron, R. Sritharan, and Y. Tang, Finding a maximum induced matching in weakly chordal graphs, *Discrete Math.* **266** (2003), 133–142.

[187] A. Brandstädt and C.T. Hòang, Maximum induced matching for chordal graphs in linear time, *Algorithmica* **52**(4) (2008), 440–447.

[188] A. Brandstädt, A. Leitert, and D. Rautenbach, Efficient dominating and edge dominating sets for graphs and hypergraphs, extended abstract in: *Proceedings of ISAAC*, Taiwan, 2012; LNCS 7676, 267–277.

[189] N. Biggs, Perfect codes in graphs, *J. Combinatorial Theor. (B)* **15** (1973), 289–296.

[190] C.-C. Yen and R.C.T. Lee, The weighted perfect domination problem and its variants, *Discrete Appl. Math.* **66** (1996), 147–160.

[191] C.L. Lu and C.Y. Tang, Weighted efficient domination problem on some perfect graphs, *Discr. Appl. Math.* **117** (2002), 163–182.

[192] D.M. Cardoso, N. Korpelainen, and V.V. Lozin, On the complexity of the dominating induced matching problem in hereditary classes of graphs, *Discr. Appl. Math.* **159** (2011), 521–531.

[193] C.L. Lu, M.-T. Ko, and C.Y. Tang, Perfect edge domination and efficient edge domination in graphs, *Discr. Appl. Math.* **119** (2002), 227–250.

[194] A. Brandstädt, C. Hundt, and R. Nevries, Efficient edge domination on hole-free graphs in polynomial time, Extended abstract in: *Conference Proceedings LATIN*, LNCS **6034** (2010), 650–661.

[195] F. Nicolai, Strukturelle und algorithmische Aspekte distanz-erblicher Graphen und verwandter Klassen, Dissertation, Gerhard-Mercator-Universität Duisburg, Germany, 1994.

[196] F. Nicolai, Hamiltonian problems on distance-hereditary graphs, Schriftenreihe des Fachbereichs Mathematik der Universität Duisburg SM-DU-255 (1994); corrected version 1996.

[197] R.-W. Hung, S.-C. Wu, and M.-S. Chang, Hamiltonian cycle problem on distance-hereditary graphs, *J. Inform. Sci. Engg.* **19** (2003), 827–838.

[198] S.-Y. Hsieh, C.-W. Ho, T.-S. Hsu, and M.-T. Ko, The Hamiltonian problem on distance-hereditary graphs, *Discr. Appl. Math.* **154** (2006), 508–524.

[199] R.-W. Hung and M.-S. Chang, Linear-time algorithms for the Hamiltonian problems on distance-hereditary graphs, *Theor. Comput. Sci.* **341** (2005) 411–440.

[200] H. Müller and F. Nicolai, Polynomial time algorithms for the Hamiltonian problems on bipartite distance-hereditary graphs, *Inf. Proc. Lett.* **46** (1993), 225–230.

[201] M.S. Chang, S.C. Wu, G.J. Chang, and H.G. Yeh, Domination in distance-hereditary graphs, *Discrete Appl. Math.* **116** (2002), 103–113.

[202] F. Nicolai and T. Szymczak, Homogeneous sets and domination: A linear time algorithm for distance-hereditary graphs, Schriftenreihe des Fachbereichs Mathematik der Universität Duisburg SM-DU-336 (1996); *Netw.* **37** (2001), 117–128.

[203] F. de Montgolfier, M. Soto, and L. Viennot, Treewidth and hyperbolicity of the Internet, *Proceedings of the 10th IEEE International Symposium on Networking Computing and Applications*, NCA 2011, August 25–27, 2011, Cambridge, MA. IEEE Computer Society, 2011, 25–32.

[204] M. Abu-Ata and F.F. Dragan, Metric tree-like structures in real-life networks: An empirical study, Manuscript 2013.

[205] F.F. Dragan and E. Köhler, An approximation algorithm for the tree t-spanner problem on unweighted graphs via generalized chordal graphs, approximation, randomization, and combinatorial optimization. algorithms and techniques. *Proceedings of the 14th International Workshop, APPROX* 2011, and *15th International Workshop, RANDOM*, Princeton, NJ, August 17–19, 2011, *Lecture Notes in Computer Science* 6845, Springer, 171–183; *Algorithmica* (in print 2014).

[206] F.F. Dragan and I. Lomonosov, On compact and efficient routing in certain graph classes, *Discrete Appl. Math.* **155** (2007), 1458–1470.

[207] D. Lokshtanov, On the complexity of computing tree-length, *Discrete Appl. Math.* **158** (2010), 820–827.

[208] V.D. Chepoi, F.F. Dragan, I. Newman, Y. Rabinovich, and Y. Vaxes, Constant approximation algorithms for embedding graph metrics into trees and outerplanar graphs, *Discrete & Computational Geometry* **47** (2012), 187–214.

[209] A. Brandstädt, V.D. Chepoi, and F.F. Dragan, Distance approximating trees for chordal and dually chordal graphs, *J. Algorithms* **30** (1999), 166–184.

[210] V.D. Chepoi and F.F. Dragan, A note on distance approximating trees in graphs, *European J. Combin.* **21** (2000), 761–766.

[211] V.D. Chepoi, F.F. Dragan, B. Estellon, M. Habib, and Y. Vaxes, Diameters, centers, and approximating trees of δ-hyperbolic geodesic spaces and graphs, *Proceedings of the 24th Annual ACM Symposium on Computational Geometry*, June 9–11, 2008, College Park, MD, pp. 59–68.

[212] Y. Dourisboure, F.F. Dragan, C. Gavoille, and C. Yan, Spanners for bounded tree-length graphs, *Theor. Comput. Sci.* **383** (2007) 34–44.

[213] F.F. Dragan and M. Abu-Ata, Collective additive tree spanners of bounded tree-breadth graphs with generalizations and consequences, *SOFSEM: Theory and Practice of Computer Science, Lecture Notes in Comput. Sci.* **7741** (2013), 194–206.

[214] Y. Dourisboure, Compact routing schemes for generalised chordal graphs, *J. Graph Algorithms Appl.* **9** (2005), 277–297.

[215] F.F. Dragan, C. Yan, and I. Lomonosov, Collective tree spanners of graphs, *SIAM J. Discrete Math.* **20** (2006), 241–260.

[216] L. Cai and D.G. Corneil, Tree spanners, *SIAM J. Discrete Math.* **8** (1995), 359–387.

[217] Y. Emek and D. Peleg, Approximating minimum max-stretch spanning trees on unweighted graphs, *SIAM J. Comput.* **38** (2008), 1761–1781.

[218] M. Gromov, Hyperbolic groups, In *Essays in Group Theory*, S.M. Gersten (Ed.), MSRI Series **8** (1987), 75–263.

[219] Y. Shavitt and T. Tankel, On internet embedding in hyperbolic spaces for overlay construction and distance estimation, In *INFOCOM*, 2004.

[220] I. Abraham, M. Balakrishnan, F. Kuhn, D. Malkhi, V. Ramasubramanian, and K. Talwar, Reconstructing approximate tree metrics, *Proceedings of the 26th Annual ACM Symposium on Principles of Distributed Computing*, Portland, OR, August 12–15, 2007, ACM, pp. 43–52.

[221] R. Kleinberg, Geographic routing using hyperbolic space, In *INFOCOM*, 2007, pp. 1902–1909.

[222] G. Brinkmann, J. Koolen, and V. Moulton, On the hyperbolicity of chordal graphs, *Ann. Comb.* **5** (2001), 61–69.

[223] Y. Wu and Ch. Zhang, Hyperbolicity and chordality of a graph, *Electr. J. Comb.* **18** (2011), P43.

[224] H. Fournier, A. Ismail, and A. Vigneron, Computing the Gromov hyperbolicity of a discrete metric space, *CoRR abs/1210.3323* (2012), http://arxiv.org/abs/1210.3323.

[225] C. Gavoille and O. Ly, Distance labeling in hyperbolic graphs, In *ISAAC*, 2005, pp. 171–179.

[226] R. Krauthgamer and J.R. Lee, Algorithms on negatively curved spaces, In *Proceedings of the 47th FOCS*, Berkeley, CA, 2006, pp. 119–132.

[227] V. Chepoi and B. Estellon, Packing and covering δ-hyperbolic spaces by balls, In *APPROX-RANDOM*, 2007, pp. 59–73.

[228] V.D. Chepoi, F.F. Dragan, B. Estellon, M. Habib, Y. Vaxès, and Y. Xiang, Additive spanners and distance and routing labeling schemes for δ-hyperbolic graphs, *Algorithmica* **62** (2012), 713–732.

[229] F.F. Dragan, Tree-like structures in graphs: A metric point of view, In *Graph-Theoretic Concepts in Computer Science—39th International Workshop*, Lübeck, Germany, June 19–21, 2013, Springer, *Lecture Notes in Comp. Sci.* **8165**, 1–4.

[230] R. Agarwala, V. Bafna, M. Farach, B. Narayanan, M. Paterson, and M. Thorup, On the approximability of numerical taxonomy (fitting distances by tree metrics), *SIAM J. Comput.* **28** (1999), 1073–1085.

[231] M. Bǎdoiu, E.D. Demaine, M.T. Hajiaghayi, A. Sidiropoulos, and M. Zadimoghaddam, Ordinal embedding: Approximation algorithms and dimensionality reduction, In *Proceedings of the 11th International Workshop on Approximation Algorithms for Combinatorial Optimization Problems*, Boston, MA, August 25–27, 2008, Springer, *Lecture Notes in Computer Science* 5171, 21–34.

[232] M. Bǎdoiu, P. Indyk, and A. Sidiropoulos, Approximation algorithms for embedding general metrics into trees, In *Proceedings of the 18th Annual ACM-SIAM Symposium on Discrete Algorithms*, New Orleans, LA, January 7–9, 2007, ACM/SIAM, 512–521.

VIII
Partitioning

CHAPTER 30

Graph and Hypergraph Partitioning

Sachin B. Patkar

H. Narayanan

CONTENTS

30.1	Overview of the Partitioning Problem	830
	30.1.1 Some Definitions	831
30.2	Class of Graph Partition Problems and Combinatorial Approaches	832
	30.2.1 Bipartitioning Algorithm of Stoer, Wagner, and Frank	833
	30.2.1.1 Justification	833
	30.2.2 Multiway Cut	834
	30.2.3 Minimum Cost Multicut Problem	835
30.3	Submodular Optimization-Based Partitioning Algorithms	837
	30.3.1 Finding Densest Cluster	838
	30.3.2 Principal Lattice of Partitions	839
	30.3.3 Improving Ratio-Cut Using Principal Partition	842
	30.3.3.1 Algorithm to Improve Ratio-Cut	843
	30.3.3.2 Justification of Algorithm	844
30.4	Iterative and Multilevel Partitioning Algorithms	845
	30.4.1 K–L Algorithm	845
	30.4.1.1 Some Definitions	845
	30.4.1.2 Important Features of the K–L Algorithm	845
	30.4.2 F–M Algorithm	847
	30.4.2.1 Drawback of Iterative Move-Based Algorithms	847
	30.4.3 Multilevel Partitioning	848
	30.4.3.1 Multilevel Steps	848
	30.4.3.2 Coarsening	849
	30.4.3.3 V-Cycle Refinement	849
	30.4.3.4 Connectivity-Based Clustering Algorithm	849
30.5	Spectral Approaches for Partitioning	849
	30.5.1 Hall's Approach and Its Variations	849
	30.5.1.1 Spectral Bipartitioning Algorithm of Hall	854
	30.5.2 Another View of Spectral Approach to Graph Bisection	854
	30.5.2.1 Dealing with Fixed Vertices	854
	30.5.3 Partitioning into k Blocks: Barnes' Approach	854
	30.5.4 Ratio-Cut Bipartitioning Using Eigenspectrum	857
	30.5.5 Spectral Bound Involving Multiple Eigenvectors	857
	30.5.6 Simple Eigenvector-Based Clustering	859
	30.5.6.1 Algorithm	859

		30.5.6.2	Theoretical Motivation	860
	30.5.7		Graph Partitioning Using Multiple Eigenvectors	860
	30.5.8		Multilevel Spectral Partitioning	861
		30.5.8.1	Multilevel Spectral Partitioning Algorithm	861
	30.5.9		Eigenvalue-Based Hypergraph Reordering and Partitioning	862
30.6	Simulated Annealing and Graph Bisection			864
30.7	Specific Variations of Graph/Hypergraph Partitioning			866
	30.7.1		Efficient Network Flow-Based Minimum Cost Balanced Partitioning	867
		30.7.1.1	Modeling a Net in a Flow-Network	867
		30.7.1.2	Optimal Minimum Netcut Algorithm	868
		30.7.1.3	Most r-Balanced Minimum Netcut Bipartition: An NP-Complete Problem	868
		30.7.1.4	Minimum Cost Balanced Bipartition	868
		30.7.1.5	Efficient Implementation	869
	30.7.2		Optimal Replication for Minimum k-Cut Partitioning	869
		30.7.2.1	Vertex Replication	870
		30.7.2.2	Minimum k-Cut Replication Problem	870

30.1 OVERVIEW OF THE PARTITIONING PROBLEM

It is often desirable to break up a large problem into parts in such a way as to distribute the workload in a balanced manner and also to optimize the *volume* of communication between the processes carrying out the computation. Often this gets modeled as a graph/hypergraph partitioning problem.

Two examples from critical diverse domains are the following. The graph partitioning problem is of interest in areas such as CAD for VLSI circuits (computer-aided design for very large-scale integrated circuits) [1–4], and efficient parallel implementations of finite element methods [5–7]. Both these kind of problems present massive-sized instances and therefore a divide and conquer style of approach needs to be taken for scalability. Specifically in case of automated placement of VLSI netlist, the netlist is partitioned in such a way that it is effective to perform the placement on the parts which are of manageable size. Further the cost of patching up the placement of these sublists into that of the original is attempted to be minimized. The cost of the partitioning in such a problem is related to wirelengths.

On the other hand, in the case of finite element method the core workhorse performs a large number of invocations of the multiplication of a large sparse matrix with a vector. And this multiplication (SpMxV) is typically solved in a distributed fashion after partitioning the domain variables in such a manner the volume of communication between the processes collaborating on the SpMxV is minimized. Furthermore a realistic constraint is that the computational load should be balanced across multiple processes.

In this chapter, we aim to describe a wide variety of ideas and algorithms for graph and hypergraph partitioning problems. Given the vast literature available on this subject, it is impossible to cover a full spectrum of the ideas in this chapter, and therefore the reader is also referred to several survey articles which together cover a wide range of themes and paradigms in this subject [1,3,4,8–11]. Forthcoming sections are organized according to different major paradigms to which the graph/hypergraph partitioning approaches belong. Section 30.2 describes a couple of approaches based on combinatorial techniques or on mathematical programming techniques. Essential ideas underlying techniques based on submodular function optimization [12–15] are brought out in Section 30.3. In Section 30.4, the classical iterative algorithms of Kernighan–Lin (K–L) and Fiduccia–Mattheyses (F–M) for graph and hypergraph partitioning are outlined. Further the same section also briefly outlines multilevel approaches, which contribute most efficient and practical algorithms for very large-scale partitioning. Spectral graph theory contributes a large variety of techniques for graph partitioning. Furthermore, there are interesting applications of ideas based on eigenspectra of matrices that are relevant to hypergraph partitioning too. These spectral ideas are described in Section 30.5. Section 30.6 outlines simulated annealing approach to graph partitioning and Section 30.7 describes a couple of approaches which deal with certain practical variants of graph/hypergraph partitioning problems which occur in the domain of VLSI CAD.

We assume reader's familiarity with standard concepts, definitions and notation in graph theory, algorithms, complexity analysis, and combinatorial optimization [16,17]. Similarly for linear/matrix algebra and computations relevant to this chapter, the reader may refer to any standard text of choice (e.g., [18]) and for more specific topics, texts such as [19] are recommended. Also the concept of a *hypergraph* is of primary relevance to this chapter (quite simply put, a hypergraph is a generalization of the concept of graph, wherein rather than edges which connect at most two vertices, a hyperedge may connect an arbitrary number of vertices).

Given a graph or a hypergraph H, k-partition of H is an assignment of all the nodes (vertices) of H to k disjoint nonempty blocks. A 2-partition is also called a *bipartition*. The k-partitioning problem seeks to minimize a specified cost function of such a partitioning. Often this cost function simply measures the weight of edges or hyperedges that have been cut across the blocks of the partition. One also typically imposes constraints on the partitions for example, fixing vertices in specific blocks, ensuring weight of each block to be within certain bounds, and so forth. This increases the complexity of the problem further. As an illustration of the point, note that for the case of $k = 2$, the graph 2-partitioning problem is the well known min-cut (minimum 2-cut) problem that has elegant solutions (network flow based and otherwise). But when one demands that the 2 parts be equal in size, the resulting problem turns out to be NP-complete [20,21]. However due to its criticality in several practical applications, heuristic algorithms for such graph/hypergraph partitioning problems have been developed possessing near-linear runtime performance. The standard graph partitioning problem (partitioning vertex set of an edge weighted graph into k equal sized parts, such that the weight of the cut edges is minimum) was shown to be NP-complete [20,21] for fixed $k \geq 3$. Due to this intractability researchers have concentrated on heuristics and approximation algorithms for this problems, that is, algorithms that often find good quality but not necessarily optimal solutions.

30.1.1 Some Definitions

A *hypergraph* $H(V, E)$ is described by a collection V of *vertices* (or *nodes*) and, the collection E of *hyperedges*. Each hyperedge $e \in E$ is itself a subset of vertices from V. A graph is a special case of the concept of hypergraph where an hyperedge is merely an edge described by its (at most!) 2 endpoints. Hypergraphs are very useful for modeling VLSI circuit. A VLSI circuit is usually described by its netlist. A *netlist* is a representation of the vlsi *cells* and the signal *nets* connecting them. Each signal net listed in the netlist can naturally be represented as a hyperedge incident on the vertices representing the *cells* connected by the signal net. A netlist can thus be represented as a hypergraph, a cell becomes a node and the nets are hyperedges of the corresponding hypergraph. *Partitioning procedure* divides a given circuit into a number of divisions to meet some specified objective. A net is said to be *cut* if not all the vertices connected by this net are in the same part of the partition.

Network flow theory is of paramount importance in combinatorial optimization [16,17, 22,23]. A multitude of graph (and hypergraph) partitioning approaches rely significantly on network flow ideas, and many such ideas are described in this chapter. We assume familiarity of the reader with the concepts of flow (and its maximization) in a capacitated flow network, the notion of (source–sink) cuts in the flow network and their (forward) capacity. The celebrated maximum flow, minimum cut duality in network flow theory is often employed in solving several problems efficiently. One word of caution: when we refer to a *cut* in the context of a flow network, we mean the set of flow network arcs which when removed separate the source s and the sink t (usual notation!). Often we would associate *cuts* in a graph/hypergraph with $(s-t)$ *cuts* in a flow network.

Submodular function optimization [13,14] plays an important role in graph partitioning. We need the following definitions for later use. Let S be a finite set and let $f : 2^S \longrightarrow \Re$. The function f is said to be a *submodular* function iff $f(X) + f(Y) \geq f(X \cup Y) + f(X \cap Y) \forall X, Y \subseteq S$. If the inequality is reversed, one gets the definition of *supermodular* functions. In other words, f is supermodular *iff* $-f$ is submodular. And the function f is *modular* if the inequality is satisfied as equality.

30.2 CLASS OF GRAPH PARTITION PROBLEMS AND COMBINATORIAL APPROACHES

In this section, we aim to illustrate through a few important general variations of graph partitioning problems, important combinatorial solution approaches. Some of them are based on clever ordering schemes, some on simple approximation algorithms for such NP-complete problems. On the other hand, several approaches are based on mathematical programming and duality theory. Our choice of the algorithms to highlight this, is not canonical. There are several other equally interesting and powerful ideas elsewhere in the vast literature on this subject. A few of the other notable combinatorial approaches are due to Nagamochi and Ibaraki [24], Karger [25,26], Leighton and Rao [27], Cheng and Hu [28], Gomory and Hu [29], and due to Goldschmidt and Hochbaum [30].

Definition 30.1 (Minimum k-cut) *A k-cut is a set of edges whose removal disconnects graph into at least k connected components. Minimum k-cut is a k-cut of least weight.*

Definition 30.2 (Minimum multiway cut) *A multiway cut is a set of edges whose removal separates the given set of terminals $\{s_1, s_2, \ldots, s_i, \ldots, s_k\}$ from each other.*

It is NP-complete to find minimum weight multiway cut for $k \geq 3$ (note that for $k = 2$ it reduces to max-flow-min-cut [16,31] problem that is in P). However the problem of finding minimum k-cut, though NP-complete when k is part of the input, is solvable in time polynomial in n for fixed k (specifically the result of [30] gives an $O(n^{k^2/2})$ algorithm). In this sense minimum k-cut is an easier problem than minimum weight multiway cut problem.

Definition 30.3 (Minimum multicut) *A multicut is a set of edges whose removal separates the terminals s_i from t_i for all $i = 1, 2, \ldots, k$ (note that s_i's need not be distinct and also t_i's need not be distinct). Minimum multicut is a such a multicut of smallest weight.*

It is easy to observe that minimum multiway cut problem may be formulated as a minimum multicut separating each pair of vertices among the terminals listed for the multiway cut problem. The number of pairs whose vertices are to be separated in the multicut instance is quadratically many compared to that in the corresponding original multiway cut problem. However as the multiway problem is NP-complete for even fixed $k \geq 3$, so is the multicut problem.

30.2.1 Bipartitioning Algorithm of Stoer, Wagner, and Frank

The problem of finding a *bipartition* (i.e., a 2-partition) of vertex set of a graph with minimum cost of the cut edges has been well researched. In the absence of constraints on the sizes of the block sizes (excepting that the blocks are non-empty), it is well known that the problem is solvable using $|V|$ invocations of max-flow-min-cut algorithm which builds a Gomory–Hu cut-tree [29]. However currently there are more efficient algorithms available. One of the radically different one to appear was due to Nagamochi & Ibaraki [24]. Their formidable algorithm was simplified independently by Stoer and Wagner [32] and Frank.

Algorithm 30.1 The Stoer–Wagner and Frank algorithm

Input An undirected edge-weighted graph.
Output A minimum 2-cut

- Select a vertex v_0 at random.
- In the kth iteration, do the following. Let $v_1, \ldots, v_{k-1}$ be the sequence of vertices selected so far.
    - Find v_k, a vertex that maximizes weighted connectivity with the set $\{v_1, \ldots, v_{k-1}\}$.
- When this *phase* terminates the vertices of the current graph would be enumerated as $v_1, \ldots, v_{n-1}, v_n$. The set of edges incident on v_n (*star of v_n*) is a cut that we call *the cut of the phase*.
- Shrink v_n, v_{n-1} into a single node.
- Repeat the above steps for the resulting graph. Let the new sequence of vertices be $\hat{v}_1, \ldots, \hat{v}_{n-2}, \hat{v}_{n-1}$.
- The cut of this phase is the one that separates $\hat{v}_{n-1}$ from the earlier set of vertices. Proceed in this manner for $(n-1)$ phases.
- Minimum among the cuts of phases is the desired minimum 2-cut.

30.2.1.1 Justification

We claim that the cut of a phase has the minimum weight among all those separating the last vertex of the phase from the last but one vertex. For instance the first phase gives a cut of minimum weight which separates v_n from v_{n-1}. We defer the proof of this claim. If there exists a minimum 2-cut of the graph that also separates v_n from v_{n-1} then cut of this phase is one such. Otherwise any min 2-cut of the original graph would persist as a minimum 2-cut even after fusing v_n and v_{n-1} into a single vertex. The algorithm now has to work on a graph with $(n-1)$ vertices. The algorithm is clearly valid when there are only two vertices. Therefore by induction, the algorithm finds a minimum 2-cut, provided the above claim about the cut of the phase is true.

Claim 30.1 *Among all cuts which separate v_{n-1} and v_n, the cut of the phase has minimum weight.*

Proof. Let C be any cut separating v_{n-1} and v_n. We will use the following notation. $w(C)$ denotes the cost of the edges in C, and $w(X,p)$ denote the weight of the edges connecting vertex p to a set of vertices X, furthermore let A_u denote the set of vertices appearing before u in the sequence generated by the current phase. Note that when we select v_n in the phase, we just crossed the cut C (i.e., v_{n-1} and v_n are on opposite sides of C). To use induction, we will consider, in the sequence of vertices generated by the phase, those vertices whenever the sequence *just crosses the cut* C. We call such vertices *active* (more formally, a vertex v_i is said to be active w.r.t. C, if the previous vertex v_{i-1} in the sequence generated by the phase is on the other side of C). Using induction on the sequence of active vertices, we will show that $w(A_u, u) \leq w(C_u)$, where C_u denotes the cut induced by C on the subgraph on $A_u \cup \{u\}$. Clearly, for the first active vertex, say u_1, $w(A_{u_1}, u_1) = w(C_{u_1})$. To prove it for arbitrary active vertex, say v, let u be the preceding active vertex. $w(A_v, v) = w(A_v \backslash A_u, v) + w(A_u, v)$. But $w(A_u, v) \leq w(A_u, u)$ due to the scheme of ordering the vertices. Furthermore the set of edges connecting $A_v \backslash A_u$ and v is disjoint from C_u and are both contained in C_v. Therefore $w(C_v) \geq w(A_v \backslash A_u, v) + w(C_u)$. Combining these observations, we obtain $w(A_v, v) \leq w(C_v)$. Hence the proof of the claim. ∎

Queyranne [33] generalized this to arbitrary symmetric submodular functions. An equivalent description may also be found in [14]. This allows us to handle hypergraph extension of min-cut problem. Similar algorithm to find hypergraph min-cut has also been proposed by [14,34].

A brief idea of the approach [33] based on symmetric submodular function minimization is described next.

Let $B \equiv (V_L, V_R, E)$ be a bipartite graph. Let X, Y be disjoint subsets of V_L. We define $C(X, Y)$ as follows

$$C(X,Y) \equiv \frac{1}{2}[|\Gamma(X)| + |\Gamma(Y)| - |\Gamma(X \cup Y)|] \tag{30.1}$$

(here $\Gamma(X)$, for $X \subseteq V_L$, denotes the set of nodes of V_R that are neighbors of X). The min-cut problem in this situation is to find $X \subset V_L$ s.t. $C(X, V_L - X)$ is minimum. A line-by-line translation of Stoer–Wagner algorithm works in this case (weight of edges between disjoint vertex sets X, Y is replaced in this case by $C(X, Y)$).

30.2.2 Multiway Cut

Here we describe the approach of Saran and Vazirani [35] for the variation of graph partitioning problem called *Multiway Cut*. This approach illustrates how simple ideas can be used to provide approximation algorithm for NP-complete problems. The presentation follows that of [35].

Let $S = \{s_1, s_2, \ldots s_k\} \subseteq V$ be a set of given terminals. A multiway cut is defined as the set of edges whose removal disconnects the terminals from each other. The problem is to find such a set of minimum weight. The isolating cut for a terminal s_i is one whose removal disconnects s_i from the other terminals. The problem of finding a minimum weight multiway cut is NP-complete for $k \geq 3$ [20,21] (clearly for $k = 2$, the problem is the polynomial time solvable max-flow-min-cut problem [16,31]).

Algorithm to Find 2-Approximate Multiway Cut

For $i = 1, 2, \ldots k$. find a minimum weight isolating cut C_i for s_i.
Discard the heaviest of these cuts. The union of the remaining, say C, is output
  as the desired multiway cut guaranteed to be within twice the optimal.

The first step of the algorithm can be done by fusing terminals except s_i into a single node, and then, using maximum flow, find a min-cut separating s_i from the (super-)node created by this fusion.

Theorem 30.1 (Saran–Vazirani multiway cut) *The above algorithm achieves an approximation guarantee of $(2 - 2/k)$ compared to the optimal multiway cut.*

Proof. Suppose A is an optimal multiway cut in G. Consider A as the union of k cuts corresponding to the k connected components containing one terminal each. Let A_i be the cut of the component containing s_i. Therefore $A = \bigcup A_i$ Noting that each edge of A_i belongs to two of the cuts A_i, we get

$$\sum_{i=1}^{k} w(A_i) = 2w(A). \tag{30.2}$$

Since A_i is an isolating cut for s_i and C_i is the minimum weight isolating cut for s_i, we therefore have $w(C_i) \leq w(A_i)$. The algorithm discards the heaviest of the cuts C_i. Hence

$$w(C) \leq \frac{k-1}{k} \sum_{i=1}^{k} w(C_i) \leq \frac{k-1}{k} \sum_{i=1}^{k} w(C_i) \tag{30.3}$$

that is,

$$w(C) \leq 2(1 - \frac{1}{k})w(A). \tag{30.4}$$

∎

Example 30.1 Let there be $2k$ vertices, k of which form a cycle, say C, and that there is a distinct vertex attached to each vertex of the cycle C. For $e \in C$, $w(e) = 1$, else $w(e) = 2-\varepsilon$, where $0 < \varepsilon < 1$. Then the optimal multiway cut is given by the cycle edges, having weight k, whereas the cut C by the above algorithm has weight $(k-1)(2-\varepsilon)$. Please refer Figure 30.1. Here we have considered $k = 4$.

30.2.3 Minimum Cost Multicut Problem

Here we aim to find a smallest cost subset C of edges of an edge-weighted graph so that, for a given collection of k pairs of terminals (s_i, t_i) $1 \leq i \leq k$, s_i and t_i are separated into different components after removal of C.

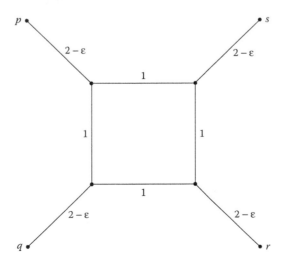

Figure 30.1 Example.

Following discussion is mainly based on the result of Garg et al. [36]. Other notable work on similar problems include that of [27,28,37] (also see [38]).

Let $\mathcal{P}_i$ denote the set of paths between s_i and t_i. The following can be easily seen to be an integer programming formulation of this problem [36].

$$\min \sum_{e \in E} c_e d_e$$

subject to

$$\sum_{e \in P} d_e \geq 1 \qquad P \in \mathcal{P}_i \text{ and } 1 \leq i \leq k$$
$$d_e \in \{0,1\} \qquad \forall e \in E \qquad (30.5)$$

The edges with $d_e = 1$ are taken as cut edges. The constraints ensure that every path between each terminal pair has at least one edge with $d_e = 1$. Therefore such edges, when removed, disconnect each pair of terminals, and hence form a multicut.

We shall consider the Linear Programming relaxation of the problem obtained by relaxing the constraint $d_e \in \{0,1\}$ to $0 \leq d_e \leq 1$.

The peculiar thing about the above formulation is the potentially exponentially many constraints. The reason one considers this formulation is for developing insights into an approximation approach due to its naturalness in describing the problem and its dual in terms of paths and edges. Furthermore using the famous result of Groetschel, Lovasz, and Schrijver, polynomial solvability is still guaranteed due to the existence of an efficient separation oracle for this problem [15]. Note that the constraints simply say that the shortest path between any pair (s_i, t_i) of vertices to be disconnected, should be at least 1. Checking feasibility of these exponentially many constraints is therefore easily done using a shortest path algorithm. Furthermore, in case of infeasibility, one can obtain as a proof of it, obtain a pair for which the shortest path is of length less than one. This describes a separation oracle for the above formulation.

A feasible solution, d, of the linear programming relaxation can be interpreted as length of the edges (not the costs c that are specified in the input). The constraints ensure that the distance between any two vertices, s_i and t_i, to be separated is at least one.

One can think in terms of dealing with a pipe system with edges as pipes and the vertices as the junctions. The cost c_e can be interpreted as cross-sectional area and their length equal to the value of d_e. Then the optimum cost of the relaxed solution equals the minimum total volume of the pipes under the constraint that each pair of terminals are separated by distance at least 1.

Let *ropt* denote the optimum value of the relaxed multicut problem, and let *opt* denote the minimum cost of a multicut. Clearly

$$ropt \leq opt. \qquad (30.6)$$

The approximation algorithm of Garg et al. [36] finds a multicut with cost $\leq \beta \, ropt \leq \beta \, opt$, for some $\beta = O(\log k)$.

This algorithm uses a variation of the idea of pipe volumes. It proceeds to iteratively find disjoint spherical regions centered on terminals in a manner such that the volume of the pipes contained fully or partially in the spherical region and the volume associated with the central terminal is large compared to a multiple of the cost of the cut pipes (which is just cross-sectional areas of the pipes cut). Let $\text{Vol}(B(s_i, r))$ denote the volume of the ball of radius r centered at a chosen terminal, w.l.o.g, say s_i. Precise definition follows.

$$\text{Vol}(B(s_i, r)) = \sum_{e=\{u,v\}, u,v \in B(s_i,r)} c(u,v)d(u,v)$$
$$+ \sum_{e=(u,v), u \in B(s_i,r), v \notin B(s_i,r)} \frac{c(u,v)d(u,v)(r - d_i(u))}{d_i(v) - d_i(u)}$$
$$+ \frac{ropt}{k}$$

here $d_i(u)$ denotes the shortest distance from terminal s_i and u, according to the solution d of the relaxed linear program above. Note that the center could also be t_i. This definition of volume is intuitively justified (volume of the pipes completely or partly contained in the ball of radius r), except the artifact $ropt/k$. We will indicate the reason later.

Interestingly, the volume of the spherical region as a function of the radius is piecewise differentiable and has its derivative lower bounded by the cost of the edges cut by the region.

$$\frac{d\text{Vol}(B(s_i, r))}{dr} \geq \sum_{e \in C(r)} c_e * d_e \qquad (30.7)$$

(here $C(r)$ denotes the set of edges/pipes cut by the ball of radius r).

In [36], the authors use this relationship to show existence of radius less than $1/2$ at which derivative of the volume is bounded by a chosen multiple of the volume itself. As the cuts of such spherical regions (of the above radii, less than $1/2$) together would form the required multicut, its cost is therefore bounded by the same multiple of the total volume of the pipe system (including the artificial volumes of the centers of the regions). The terminal node volumes are set to $ropt/k$, and therefore the cost of the multicut produced with the help of the above disjoint spherical regions is bounded above by a multiple $ropt$, say $\beta * ropt$, which is what needs to be shown. The choice of β should be such as to guarantee the existence of $\hat{r} < 1/2$ for which

$$\frac{d\text{Vol}(B(s_i, r))}{dr}\Big|_{r=\hat{r}} \leq \beta \text{Vol}(B(s_i, r)). \qquad (30.8)$$

The need of artificial volume $ropt/k$ associated with the centers of the regions is also for the purpose of establishing existence of such $\hat{r}$.

Note that a ball of radius less than half, centered at an s_i, can not contain its mate t_i, for the reason that the distance separating each pair of terminals is at least 1. Therefore such spherical regions grown by the algorithm of [36] form a multicut within $O(\log k)$ approximation factor of the optimal one. Details may be found in [36,38].

30.3 SUBMODULAR OPTIMIZATION-BASED PARTITIONING ALGORITHMS

In this section, a few applications of submodular function optimization to partitioning domain are presented. Submodular, supermodular and modular functions are defined in section 30.1.

Some well known examples of submodular functions are given below:

1. The function, $|\mathcal{V}|(.) : 2^E \longrightarrow \Re$, defined as $|\mathcal{V}|(X) \equiv$ the cardinality of the set of endpoints of the given subset of edges X, is a submodular function.

2. Similarly $|\delta|(.) : 2^V \longrightarrow \Re$, where $|\delta|(U)$ is the size of the edges crossing U, is also submodular.

We begin with the NP-complete problem of finding dense enough clusters in a graph. This problem is a well-studied one [28,39,40]. The NP-completeness of this problem is clearly associated with the NP-completeness of the problem of finding maximum complete subgraph (cliques) in a given graph. The following idea is similar to the one in [40]. This illustrates typical use of network flow techniques to optimize certain natural submodular functions. The paper of Picard and Queyeranne [41] is a good source of useful ideas involving network flows and submodular function optimization.

30.3.1 Finding Densest Cluster

Let $\mathcal{E}(.) : 2^V \longrightarrow \mathbf{R}$ is defined as $\mathcal{E}(U)$ = the set of edges having both the endpoints in the vertex set U. Let $wt(e)$ be the (non-negative) weight of edge e and let $wt(X)$ denote the sum of the weights of edges in $X \subseteq E$. The function $|\mathcal{E}(.)| : 2^V \longrightarrow \mathbf{R}$ is a supermodular function. The function $wt(\mathcal{E}(.)) : 2^V \longrightarrow \mathbf{R}$ (which represents the weight of all the edges having both the endpoints in the given vertex subset) is also a supermodular function as weights are nonnegative.

Given an (nonnegative) edge-weighted graph $G(U, F, wt(.))$ and $\mu \geq 0$, we define a flow network $FL(U, F, wt(.), \mu)$ as follows: it consists of source s, sink t and a directed bipartite graph (U, F, A), (where U and F are the left and right node sets of the bipartite graph and A is the following set of arcs (ie. directed edges) $A = \{(u, e) | u \in U, e \in F, e \text{ incident on } u \text{ in } G\})$, along with the following set of additional arcs, $\{(s, u) \mid u \in U\} \cup \{(e, t) \mid e \in F\}$. The capacities on arcs are as follows, $cap(s, u) = 1$ ...if $u \in U$, $cap(e, t) = \mu * wt(e)$ $\forall e \in F$ and $cap(u, e) = \infty$ $\forall (u, e) \in A$.

Any $(s-t)$-cut partitions the node set of the flow network in two parts one of which has the source as one of its members and the other, the sink as one of its members. The set of nodes containing the source (sink) is called s-part (t-part) of the $(s-t)$-cut.

Canonical $(s-t)$-cut: Given $U' \subseteq U$ in $FL(U, F, wt(.), \mu)$, we define *canonical_cut(U')* as the one whose t-part is $\{t\} \cup U' \cup \mathcal{E}(U')$. We call such a cut a *canonical $(s-t)$-cut*.

Note that the capacity of a canonical cut, say *canonical_cut(U')* $U' \subseteq U$, is $\mu * |F| - \mu * wt(\mathcal{E}(U')) + |U'|$. Using this and typical arguments about the properties of a minimum cut, it is not difficult to see that there exists a *canonical_cut($\hat{U}$)* that is a minimum $(s-t)$-cut, and indeed $\hat{U}$ solves $\max_{U' \subseteq U}(\mu * wt(\mathcal{E}(U')) - |U'|)$.

As $f(U') = (\mu * wt(\mathcal{E}(U')) - |U'|)$, $\mu \geq 0$, is a supermodular function, it follows from the properties of submodular and supermodular functions that the collection of subsets which maximize the function $f(.)$ form a lattice under the usual operations of union and intersection [13,14]. In particular there exist unique minimum (smallest) and maximum (largest) sets in such a collection.

Algorithm to Find a Cluster of Largest Density

We wish to compute a subset $\hat{U}$ of the set $\hat{V}$, that solves

$$\max_{W \subseteq \hat{V}, W \neq \emptyset} \frac{wt(\mathcal{E}(W))}{|W|}$$

(i.e., $\hat{U}$ has maximum *density*).

Optimization of $wt(\mathcal{E}(U)) - k * |U|$ over the subsets of the given set of nodes can be performed using the network flow model described earlier.

The above scheme generalizes in a straightforward manner to hypergraphs too.

> **Algorithm FindDensestCluster**
> Initialize $U_0 = \hat{V}$ and $i = 0$
> **repeat**
> $d_i = \frac{wt(\mathcal{E}(U_i))}{|U_i|}$
>   Find largest $U_{i+1} \subseteq U_i$ s.t.
> $wt(\mathcal{E}(U_{i+1})) - d_i|U_{i+1}| = \max_{W \subseteq U_i, W \neq \emptyset}(wt(\mathcal{E}(W)) - d_i|W|)$
> **until** $U_{i+1} = U_i$
> Output U_i as the required $\hat{U}$

Theorem 30.2 $\hat{U}$ output by the above algorithm is a nonempty subset of $\hat{V}$ that maximizes

$$\frac{wt(\mathcal{E}(\hat{W}))}{|W|}$$

over the nonempty subsets of $\hat{V}$.

Proof. Let i be such that $U_i \neq \emptyset$ and $U_{i+1} = U_i$. According to the algorithm, $\hat{U} = U_i$. We have, $wt(\mathcal{E}(\hat{U}))/|\hat{U}| = d_i$. We need to prove that $d_i = \max_{W \subseteq \hat{V}, W \neq \emptyset} wt(\mathcal{E}(W))/|W|$. Suppose the contrary. Let $d = \max_{W \subseteq \hat{V}, W \neq \emptyset} wt(\mathcal{E}(W))/|W|$. Clearly, $d > d_{i+1}$. Let $\hat{W}$ be a subset of $\hat{V}$ such that

$$\frac{wt(\mathcal{E}(\hat{W}))}{|\hat{W}|} = d. \tag{30.9}$$

Note that $\hat{V} = U_0 \supseteq U_1 \supseteq U_2 \supseteq \cdots \supseteq U_i = \hat{U}$.

Let j be the smallest index such that $\hat{W} \not\subseteq U_j$. Note that $j \leq i$ and therefore $d > d_{j-1}$. By the supermodularity of $wt(\mathcal{E}(.)) - d_{j-1}|.|$,

$$wt(\mathcal{E}(\hat{W} \cup U_j)) - d_{j-1}|\hat{W} \cup U_j| + (wt(\mathcal{E}(\hat{W} \cap U_j)) - d_{j-1}|\hat{W} \cap U_j|)$$
$$\geq wt(\mathcal{E}(\hat{W})) - d_{j-1}|\hat{W}| + wt(\mathcal{E}(U_j)) - d_{j-1}|U_j| \tag{30.10}$$

But $wt(\mathcal{E}(\hat{W})) - d|\hat{W}| \geq wt(\mathcal{E}(\hat{W} \cap U_j)) - d|\hat{W} \cap U_j|$ due to optimality of $\hat{W}$. Therefore

$$wt(\mathcal{E}(\hat{W})) - wt(\mathcal{E}(\hat{W} \cap U_j)) \geq d(|\hat{W}| - |\hat{W} \cap U_j|) > d_{j-1}(|\hat{W}| - |\hat{W} \cap U_j|).$$

Hence $wt(\mathcal{E}(\hat{W} \cup U_j)) - d_{j-1}|\hat{W} \cup U_j| > wt(\mathcal{E}(U_j)) - d_{j-1}|U_j|$ which is a contradiction, since $\hat{W} \cup U_j \subseteq U_{j-1}$. ∎

30.3.2 Principal Lattice of Partitions

In [42], the authors use strategies related to the principal lattice of partitions of a submodular function, for solving a generic partitioning problem (see also [14]).

A natural way of associating a set function with a function defined over partitions of the underlying set of the former is as follows: Let $f(\cdot)$ be a real-valued set function on the subsets of S. The *partition associate* of $f(\cdot)$ denoted by $\overline{f}(\cdot)$, is defined as $\overline{f}(\Pi) \equiv \sum_{N_i \in \Pi} f(N_i)$. We often need to consider functions such as $f - \lambda$, which are defined as $(f - \lambda)(X) = f(X) - \lambda$. Note that the partition associate of such a biased function satisfies, by definition, $\overline{f - \lambda}(\Pi) = \overline{f}(\Pi) - \lambda * |\Pi|$.

Partitioning problems generally involve the maximization or minimization of the partition associate of a set function. Quite often they are NP-complete. However, submodularity or supermodularity of the underlying function often makes these problems tractable.

The collection of partitions that minimize $\overline{(f-\lambda)}(\cdot)$ for some $\lambda \in \Re$ is called the *principal lattice of partitions of* $f(\cdot)$.

For describing the *principal lattice of partitions of* $f(\cdot)$ we need the following additional defintions. We say $\Pi_1 \geq \Pi_2$ (equivalently, Π_1 is *coarser* than Π_2 or Π_2 is *finer* than Π_1) iff every block of Π_2 is contained in some block of Π_1. The partition $\Pi_1 \vee \Pi_2$ ($\Pi_1 \wedge \Pi_2$) is the least upper bound (greatest lower bound) of the partitions Π_1 and Π_2 in the partial order ($\geq$). In other words, $\Pi_1 \vee \Pi_2$ is the unique finest partition, coarser than both Π_1 and Π_2. Similarly we can describe $\Pi_1 \wedge \Pi_2$ to be the unique coarsest partition that is finer than both Π_1 and Π_2. We use the notation Π_0 to denote the partition into singleton sets.

i. **Property PLP1**

   The collection of partitions that are optimum for a particular λ is closed under join ($\vee$) and meet ($\wedge$) operations naturally defined on the partitions. Thus there a unique maximal (coarsest) and a unique minimal (finest) partition minimizing $\overline{(f-\lambda)}(\cdot)$ denoted by Π^λ, Π_λ respectively.

ii. **Property PLP2**

   If $\lambda_1 > \lambda_2$, then $\Pi^{\lambda_1} \leq \Pi_{\lambda_2}$.

iii. **Property PLP3**

   We need the following definition.

Definition 30.4 *A number λ for which more than one partition minimizes $\overline{(f-\lambda)}(\cdot)$ is called a critical PLP value of* $f(\cdot)$.

This property states that the number of critical PLP values of $f(\cdot)$ is bounded by $|S|$.

iv. **Property PLP4**

   If $\lambda_1 > \cdots > \lambda_t$ is the complete decreasing sequence of critical PLP values of $f(\cdot)$, then, $\Pi^{\lambda_i} = \Pi_{\lambda_{i+1}}$ for $i = 1, \ldots, t-1$.

v. **Property PLP5**

   Let $\lambda_1 > \cdots > \lambda_t$ be the complete decreasing sequence of critical values of the PLP. If $\lambda_i > \sigma > \lambda_{i+1}$. Then $\Pi^{\lambda_i} = \Pi^\sigma = \Pi_\sigma = \Pi_{\lambda_{i+1}}$.

The sequence $\Pi_0 = \Pi_{\lambda_1}, \Pi_{\lambda_2}, \ldots, \Pi_{\lambda_t}, \Pi^{\lambda_t} = \{S\}$ is called the *principal sequence of partitions of* $f(\cdot)$.

The importance of the PLP for the 'partition associate minimization with number of blocks specified' problem lies in the following easy fact.

Let Π be a partition of S in the PLP of a submodular function $f(\cdot)$. If Π' is any other partition of S with the same number of blocks as Π then $\bar{f}(\Pi) \geq \bar{f}(\Pi')$, the equality holding *if and only if* Π' is also a partition in the PLP of $f(\cdot)$.

We now give a simple strategy for producing an approximation algorithm for finding a k-block partition that minimizes $\overline{|\delta|}(.)$ function, that is, to find a k-block partition of the vertex set that approximately minimizes the cost of the cut induced.

First build the principal sequence of partitions of the cut-size function $|\delta|(.)$, which is a well known submodular function (see [14,42] for the algorithms to find this sequence). The main subroutine is to minimize $|\delta|(X) - \lambda|X|$ among all subsets of a given set which have a specified element as a member. This problem can be converted easily to a network flow problem.

Let $\Pi_1 = \{N_1, N_2, \ldots N_t, S_2, S_3, \ldots S_m\}$, $\Pi_2 = \{S_1, S_2, \ldots S_m\}$ be partitions minimizing $\overline{|\delta|}(.) - \lambda|.|$ for some λ, such that $|\Pi_2| \leq k \leq |\Pi_1|$ (Π_1 is a refinement of Π_2).

Let for simplicity of presentation, $|\delta|_\lambda$ denote the function $|\delta(.)| - \lambda|.|$.

Consider the subgraph $\mathcal{G}_1$ of $\mathcal{G}_{fus.\Pi_1}$ (i.e., the graph obtained from $\mathcal{G}$ by fusing the vertices within each of the blocks of Π_1), induced on its supernodes corresponding to $\{N_1, N_2, \ldots N_t\}$. Select $k - m$ vertices of the least degree in $\mathcal{G}_1$ (without loss of generality, assume that these are $\{N_1, N_2, \ldots N_{k-m}\}$). Fuse the remaining vertices of $\mathcal{G}_1$ into a single vertex M. Select the partition

$$\Pi \equiv \{N_1, N_2, \ldots N_{k-m}, M, S_2, S_3 \ldots S_m\} \quad (30.11)$$

as the required *approximate* optimal partition. We now prove a desirable property of this partition.

Theorem 30.3 (2-approximate partition) *Let Π_{opt} be the k-block partition that minimizes $\overline{|\delta|}(.)$. Then*

$$\overline{|\delta|}(\Pi_{opt}) - \overline{|\delta|}(\Pi_2) \geq \frac{k-m}{t-1}(\overline{|\delta|}(\Pi_1) - \overline{|\delta|}(\Pi_2)) \quad (30.12)$$

$$\frac{\overline{|\delta|}(\Pi) - \overline{|\delta|}(\Pi_2)}{\overline{|\delta|}(\Pi_{opt}) - \overline{|\delta|}(\Pi_2)} \leq \frac{2(t-1)}{t} \quad (30.13)$$

Proof. Recall Π_1, Π_2 both minimize $\overline{|\delta|_\lambda}(.)$. So $\overline{|\delta|}(\Pi_{opt}) - \lambda k \geq \overline{|\delta|}(\Pi_2) - \lambda m$. That is

$$\overline{|\delta|}(\Pi_{opt}) - \overline{|\delta|}(\Pi_2) \geq \lambda(k-m). \quad (30.14)$$

But $\overline{|\delta|}(\Pi_1) - \lambda|\Pi_1| = \overline{|\delta|}(\Pi_2) - \lambda|\Pi_2|$. Thus, $\lambda = (\overline{|\delta|}(\Pi_1) - \overline{|\delta|}(\Pi_2))/(t-1)$. Substituting in Equation 30.14 yields Equation 30.12.

To prove Equation 30.13, let $\mathcal{G}_2$ be the graph with vertex set $\{N_1, N_2, \ldots N_{k-m}, M\}$ and note that $|E(\mathcal{G}_2)| \leq 2|E(\mathcal{G}_1)|(k-m)/t$ (from the choice of $N_1, N_2, \ldots N_{k-m}$). Now we note that $\overline{|\delta|}(\Pi_1) - \overline{|\delta|}(\Pi_2) = |E(\mathcal{G}_1)|$ and $\overline{|\delta|}(\Pi) - \overline{|\delta|}(\Pi_2) = |E(\mathcal{G}_2)|$. Using this in Equation 30.12, we obtain the required result. ∎

Corollary 30.1

$$\frac{\overline{|\delta|}(\Pi)}{\overline{|\delta|}(\Pi_{opt})} \leq \frac{2(n-1)}{n} \quad (30.15)$$

Proof. We have

$$\frac{\overline{|\delta|}(\Pi)}{\overline{|\delta|}(\Pi_{opt})} = \frac{\overline{|\delta|}(\Pi) - \overline{|\delta|}(\Pi_2) + \overline{|\delta|}(\Pi_2)}{\overline{|\delta|}(\Pi_{opt}) - \overline{|\delta|}(\Pi_2) + \overline{|\delta|}(\Pi_2)} \quad (30.16)$$

We obtain the result of Equation 30.15 using Equation 30.13 and noting that $t \leq n$. ∎

30.3.3 Improving Ratio-Cut Using Principal Partition

In [43], optimization of submodular functions is employed to design new heuristics and approximate algorithms for ratio-cut problem. Later such ideas were explored further in [44,45].

Although typically the bipartitions are required to be balanced (in the sense of both the parts having similar weights of their vertex sets), very often it may be desired to have a possibly unbalanced partition that is more *natural* (in the sense of the small value of ratio of the cut value of a block to its weight). The above mentioned *naturalness* is captured by the well-known concept of *ratio-cut* (see [26,46,47]).

The ratio-cut for $\emptyset \neq W \subset V$ is defined as follows:

$$ratio - cut(W) = \frac{w_e(\delta(W))}{w_v(W) * w_v(V - W)}$$

Note that the notion of the *ratio-cut* as defined above is really associated with the corresponding bipartition. We also call *ratio-cut(W)* as the *ratio-cut* of the corresponding bipartitions, that is $(W, V - W)$ or $(V - W, W)$. In [46], the authors use the concept of *scaled cost* which specializes to the concept of *ratio-cut* for 2-partition.

The approach of [43] formulates a ratio-optimization problem and solves it using theory of principal partition. Ratio-optimization problems have also been widely researched in several different contexts (see [13,48–51]).

In [43] the authors use an intuitive generalization of the concept of $Gain(.)$ to arbitrary subsets of V_1 or V_2. Given a graph $G(V, E, w_v, w_e)$, and a bipartition (V_1, V_2) and a subset U of either V_1 or V_2, $GGain(U) = $ the amount of reduction in the cut value if U is moved across to the other block. The concept of $GGain$ (read, *generalized-gain*) leads to a formulation involving the tractable problem of minimization of submodular functions. Main ideas used in this work are from the theory of Principal Partition [13,14] and from the literature on *fractional programming* [52].

Let $g : 2^S \longrightarrow \mathbf{R}$ be a submodular function and let $w(\cdot)$ be a weight function on S. Then the principal partition of (g, w) (or of g, if w is clear from the context), is the collection of all sets which minimize $g(X) - \lambda * w(X), X \subseteq S$, λ real. (See [12–14] for more details.)

Note that, for $U \subseteq V_1$, $GGain(U) = w_e(\delta(V_1)) - w_e(\delta(V_1 - U))$. Similarly for $U \subseteq V_2$, $GGain(U) = w_e(\delta(V_2)) - w_e(\delta(V_2 - U))$. We define *averageGain* in a natural way as follows:

$$averageGain(U) = GGain(U)/w_v(U).$$

Consider the natural problem of finding nonempty $\hat{U} \subseteq V_1$ that maximizes the *averageGain* over all nonempty subsets of V_1. and similarly for V_2.

Let μ_1 denote the maximum averageGain attained over the nonempty subsets of V_1, similarly μ_2.

Since the above problem is a ratio-optimization problem, ideas from [52] and [13] tell us to consider the following problem. For each real λ, find subsets that solve

$$\min_{U \subseteq V_1} (w_e(\delta(U)) - \lambda * w_v(U)).$$

This is the problem of computing the *principal partition of* $(w_e(\delta(.)), w_v(.))$ The case of V_2 is identical.

Proposition 30.1 $\hat{\lambda} = \mu_1$ iff $\hat{\lambda}$ is such that $\exists$ a proper subset $Z \subset V_1$ satisfying

$$\min_{U \subseteq V_1} (w_e(\delta(U)) - \hat{\lambda} * w_v(U))$$
$$= w_e(\delta(V_1)) - \hat{\lambda} * w_v(V_1)$$
$$= w_e(\delta(Z)) - \hat{\lambda} * w_v(Z).$$

Next one notes the significance of the concept of maximum averageGain in the context of the ratio-cut.

Proposition 30.2 *Let $G(V, E, w_v, w_e)$ be a graph with a given bipartition $(V_1, V - V_1)$. Let $\hat{U}$ be a nonempty proper subset of V_1 such that $\mu_1 = \frac{w_e(\delta(V_1)) - w_e(\delta(V_1 - \hat{U}))}{w_v(\hat{U})}$. Then ratio-cut$(V_1 - \hat{U}) <$ ratio-cut(V_1).*

Proof.
$$\frac{w_e(\delta(V_1)) - w_e(\delta(V_1 - \hat{U}))}{w_v(\hat{U})} \geq \frac{w_e(\delta(V_1)) - w_e(\delta(V_1 - V_1))}{w_v(V_1)}.$$

This may be rewritten as
$$\frac{w_e(\delta(V_1)) - w_e(\delta(V_1 - \hat{U}))}{w_v(V_1) - w_v(V_1 - \hat{U})} \geq \frac{w_e(\delta(V_1))}{w_v(V_1)}.$$

The above implies $\frac{w_e(\delta(V_1 - \hat{U}))}{w_v(V_1 - \hat{U})} \leq \frac{w_e(\delta(V_1))}{w_v(V_1)}$. This clearly implies that ratio_cut$(V_1 - \hat{U}) <$ ratio_cut(V_1), as $w_v(V - V_1) < w_v(V - (V_1 - \hat{U}))$. ∎

30.3.3.1 Algorithm to Improve Ratio-Cut

As the algorithm is based on principal partition, we need some relevant properties of it.
Some relevant properties of principal partition. Consider, in particular, the principal partition of $(w_e(\delta(.)), w_v(.))$ (with the function $w_e(\delta(.))$ defined on the subsets of V_1). Let f_λ denotes the function $w_e(\delta(.)) - \lambda * w_v(.)$, defined on subsets of V_1.

i. There is a unique maximal set X^λ and a unique minimal set X_λ at which f_λ reaches the minimum over the subsets of V_1. We call these sets *critical sets* in the principal partition of $(w_e(\delta(.)), w_v)$ for $G(V, E, w_v, w_e)$ and $V_1 \subseteq V$. Thus every subset that minimizes f_λ is contained in X^λ and contains X_λ.

ii. If $\lambda_1 < \lambda_2$, it can be shown that $X^{\lambda_1} \subseteq X_{\lambda_2}$. Thus all the critical sets form a *nested* sequence w.r.t. inclusion.

iii. For not more than $|V_1|$ values of λ, $X_\lambda \neq X^\lambda$. Such values are called *critical values* in the above principal partition of $(w_e(\delta(.)), w_v)$.

iv. Let $\lambda_1 < \lambda_2 < \lambda_3 \cdots < \lambda_t$ be the sequence of all critical values.
Then $X^{\lambda_i} = X_{\lambda_{i+1}}$ for $i = 1, 2, \ldots, t-1$, and $X_{\lambda_1} = \emptyset$, $X^{\lambda_t} = V_1$. The sequence $X_{\lambda_1} \subset X_{\lambda_2} \subset \ldots \subset X_{\lambda_t} \subset X^{\lambda_t}$ is called the *principal sequence* of function $w_e(\delta(.))$ for $G(V, E, w_v, w_e)$ and $V_1 \subseteq V$.

The definition of the largest critical value in the above mentioned principal sequence yields the following corollary to Proposition 30.1.

Corollary 30.2 *μ_1 is equal to the largest critical value in the Principal Partition of $(w_e(\delta(.)), w_v(.))$ for $G(V, E, w_v, w_e)$ and $V_1 \subseteq V$.*

Also note the the following useful and simple fact: All the critical values associated with the principal partition of $(w_e(\delta(.)), w_v(.))$ for $G(V, E, w_v, w_e)$ and $V_1 \subseteq V$ are nonnegative.

A high level description of the algorithm that attempts to improve ratio-cut is as follows:

i. Let V_1, V_2 denote the two blocks of the given bipartition.

   a. Compute the Principal Sequence of $w_v(\delta(.))$ for the given $G(V, E, w_v, w_e)$ and $V_1 \subseteq V$.

   b. Compute the Principal Sequence of $w_v(\delta(.))$ for the given $G(V, E, w_v, w_e)$ and $V_2 \subseteq V$.

ii. For each of the nonempty subsets, say X, in either of the above two principal sequences, if $(X, V - X)$ is not the same as (V_1, V_2) and (V_2, V_1) then *output* $(X, V - X)$ as a bipartition whose ratio-cut is strictly lower (i.e., better) than that of (V_1, V_2).

iii. For each pair of nonempty subsets (one from each of the above two principal sequences), compute a minimum cut separating these two subsets. Let $(Y, V - Y)$ be the minimum cut obtained. *Output* $(Y, V - Y)$ if it is not already output and is different from (V_1, V_2) and (V_2, V_1). The ratio-cut of the above bipartition $(Y, V - Y)$ is strictly lower than that of (V_1, V_2).

Note that it could happen that above algorithm may fail to find an improved bipartition, as there exists a possibility of nonexistence of any nonempty proper subsets (of V_1 and V_2, respectively) in either of the two principal sequences used in the above algorithm.

30.3.3.2 Justification of Algorithm

One can show that any critical set that is nonempty and a proper subset of V_1 will *help us* in improving the ratio-cut.

Lemma 30.1 *Let $G(V, E, w_v, w_e)$ be a given graph and let $V_1 \subseteq V$. Let U' be a subset that minimizes $w_e(\delta(.)) - \lambda * w_v(.)$ over the subsets of V_1. If $\emptyset \neq U' \subset V_1$ then, $ratio\text{-}cut(U') < ratio\text{-}cut(V_1)$.*

Proof. We have

$$0 = w_e(\delta(\emptyset)) - \lambda * w_v(\emptyset) \geq w_e(\delta(U')) - \lambda * w_v(U') \leq w_e(\delta(V_1)) - \lambda * w_v(V_1) \quad (30.17)$$

From the above we get,

$$\left(\lambda - \frac{w_e(\delta(U'))}{w_v(U')}\right) \geq 0. \quad (30.18)$$

Rewriting part of equation 30.17 we get,

$$\left(\lambda - \frac{w_e(\delta(U'))}{w_v(U')}\right) * w_v(U') \geq \left(\lambda - \frac{w_e(\delta(V_1))}{w_v(V_1)}\right) * w_v(V_1) \quad (30.19)$$

As $(\lambda - \frac{w_e(\delta(U'))}{w_v(U')}) \geq 0$ and $w_v(V_1) > w_v(U') \geq 0$ (due to positivity of weights), we get

$$\left(\lambda - \frac{w_e(\delta(U'))}{w_v(U')}\right) \geq \left(\lambda - \frac{w_e(\delta(V_1))}{w_v(V_1)}\right) \quad (30.20)$$

thus implying,

$$\frac{w_e(\delta(U'))}{w_v(U')} \leq \frac{w_e(\delta(V_1))}{w_v(V_1)}. \quad (30.21)$$

Now, as U' is a proper subset of V_1, and the weights are positive, $ratio\text{-}cut(U') < ratio\text{-}cut(V_1)$. ∎

30.4 ITERATIVE AND MULTILEVEL PARTITIONING ALGORITHMS

Among the earliest algorithms for graph and hypergraph partitioning problems were those of iterative and move-based flavor. We briefly ilustrate, in this section, the most well-known among these. Further, this section outlines the multilevel paradigm that has yielded practical, efficient algorithms for large-scale instances of partitioning problems.

30.4.1 K–L Algorithm

Arguably the most well-known and the earliest algorithm for graph partitioning is the algorithm of Kernighan and Lin (1970) [53] (usually called K–L algorithm). The algorithm is a clever improvement on the basic local search procedure that is typically successful in finding significantly better quality bipartitions.

The K–L algorithm also formed the basis for the practical improvised algorithm of Fiduccia and Mattheyses [54], whose data structures supported a modification of K–L algorithm to facilitate linear-time execution (per pass) and applicability to a more general setting of hypergraphs.

The K–L algorithm starts with an arbitrary bipartition (usually bisection, i.e., a balanced bipartition) of the vertices of the given undirected graph. The algorithm proceeds in a series of *passes*. At every basic step of each pass, it determines a pair of vertices, one from each block, whose *swap* (exchange) across the bipartition results in the largest decrease or the smallest increase (in case there is no cost decrease) in the cost of the edges cut. The vertices in this pair are then *locked* for this *pass* and listed in a data structure. The locking is necessary to prevent this move-based process from looping indefinitely. The *pass* continues till all the vertices are locked. The best cut encountered during the sequence of these swap steps in the pass is then *committed*.

30.4.1.1 Some Definitions

Let (A, B) denote the partition of the vertices of the graph. We wish to improve upon this given partition.

- *External edge cost.* $Ext(a)$ measures the cost of the connection from a vertex a to the block not containing the vertex a. That is, for $a \in A$, $Ext(a) = \sum_{y \in B} c(a, y)$. Here $c(a, y)$ denotes the cost of edge (a, y).

- *Internal edge cost.* $Int(a)$ measures the cost of the connection from a vertex a within the block containing the vertex a. Hence we define for $a \in A$, $\text{Int}(a) = \sum_{z \in A} c(a, z)$.

- *Gain* in the cost when a vertex x moves to the other side is denoted by $\text{gain}(x)x$, so $\text{gain}(x) = \text{Ext}(x) - \text{Int}(x)$. The gain when a pair of vertices $a \in A$ and $b \in B$ are swapped, equals $\text{gain}(a, b) = \text{gain}(a) + \text{gain}(b) - 2^*c(a, b)$.

30.4.1.2 Important Features of the K–L Algorithm

Some important charecteristics of this algorithm are:

- During each pass of the algorithm, every vertex moves exactly once, either from A to B or from B to A.

- At the beginning of a pass, each vertex is *unlocked*, meaning that it is free to be swapped; after a vertex is swapped it becomes *locked*.

- K–L algorithm iteratively swaps the pair of unlocked vertices with the highest gain.

- *Pretend* to swap the highest gain pair of unlocked nodes a_i, b_i even if the gain is not positive.

- The swapping process is iterated until all nodes become locked, and the lowest-cost bipartition observed during the pass is returned. More precisely, we find j such that $\sum_{1 \leq i \leq j} \text{gain}(a_i, b_i)$ is maximum. The swaps until the jth step are *committed* that is, the bipartition (A, B) is updated. The sum $\sum_{1 \leq i \leq j} \text{gain}(a_i, b_i)$ is the total gain during this *pass* of the algorithm.

- Another pass is then executed using this bipartition as its starting bipartition; the algorithm terminates when a pass fails to find a solution with lower cost than its starting solution.

- Due to the basic step of swapping of vertices, typically K–L algorithm is used for balanced bipartitions (or bisections) of graphs.

The Algorithm
K–L Algorithm
  Input An initial balanced bipartition of $G(V, E)$
  Output Balanced bipartition A and B with *small* cut cost
  Bipartition G into A and B such that $|A| = |B|$, $A \cap B = \emptyset$, and $A \cup B = V$.
  repeat
    Compute $gain(v), \forall v \in V$
    for all $i = 1$ to n **do**
      Find a pair of unlocked vertices $v_{a_i} \in A$ and $v_{b_i} \in B$
        whose exchange makes it the largest decrease or smallest increase in the cut cost.
      Mark v_{a_i} and v_{b_i} as locked
        store the gain $gain(a_i, b_i)$ and compute updated values of $gain(v)$ for all unlocked $v \in V$
    end for
    Find k, such that
      $G_k = \sum_{i=1}^{k} gain(a_i, b_i)$ is maximized
    if $G_k > 0$ **then**
      Swap the pairs $(v_{a_1}, v_{b_1}), (v_{a_2}, v_{b_2}), \ldots (v_{a_k}, v_{b_k})$
    end if
  until $G_k \leq 0$

A simple illustration of K–L algorithm is shown in Figure 30.2. Notice that each edge has unit weight in the example but the weights can be different in general. Table 30.1 displays the moves executed according to the algorithm. From the table, one notes that we get a minimum cost cut for $j = 2$. That is, $\sum_{1 \leq i \leq j} \text{gain}(a_i, b_i)$ is maximum.

It is important to note the ability of the K–L heuristic to climb out of local minima. This is due to the fact that it swaps the pair of nodes with highest gain even if this gain is negative. However, if we consider all solutions reachable within a single pass of the algorithm to be *neighbors* of the current solution, the K–L algorithm is still seen to be greedy. A simple (Figure 30.2) implementation of K–L requires $O(n^3)$ time per pass since finding the highest-gain swap involves evaluating $O(n^2)$ swaps.

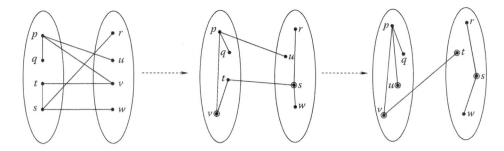

Figure 30.2 Simple example for K–L algorithm.

TABLE 30.1 Simple Example: Moves according to K–L Algorithm

Step	Vertex pair	Cost reduction	Cut cost
0	–	0	5
1	$\{s,v\}$	3	2
2	$\{t,u\}$	1	1
3	$\{q,w\}$	−2	3
4	$\{p,r\}$	−2	5

30.4.2 F–M Algorithm

The F–M algorithm is a near linear-time heuristic [54] with respect to the size of the hypergraph (equivalently, a netlist in the context of VLSI) which improve upon a given initial k-partition [8,54].

The F–M method applies a sequence of linear-time *passes* to iteratively improve a given initial partition [10]. The initial solution may even be produced by a simple randomized algorithm. The algorithm is not typically sensitive to choice of initial partition. All the vertices are *moved* exactly once in the whole process (one at a time, and not an exchange of a pair), that is, they are *locked* once they are moved from one part to the other. An important distinction between K–L and F–M algorithms is that the former is based on *swaps* and the latter on *moves* of the vertices. At the initial stage of the algorithm all vertices are free to move. For every move the change in cost which is called as *gain*, is calculated efficiently with the help of a clever bucket-based data structures employed for book-keeping of gain values [54]. The move with the highest gain is chosen. During the moves it is ensured that the partition does not get too unbalanced. Otherwise, all the vertices would move to one of the parts, since that is the trivial best solution with minimum cut if we ignore balancing constraint to such a ridiculous extent. Also note that movement of vertices across the blocks of partition might lead to changes in the gain of the adjacent vertices, therefore gains are to be updated after every iteration for all the affected vertices.

For details regarding the implementation and the data structures used for the F–M algorithm, the reader is referred to [8,54].

30.4.2.1 Drawback of Iterative Move-Based Algorithms

A major drawback of the traditional iterative procedures like K–L and F–M heuristics is that the performance of these procedures degrades as the size of the circuit increases [8]. Therefore, for large circuit sizes, multilevel partitioning approaches have become popular.

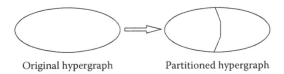

Figure 30.3 Schematic: Flat partitioning of a hypergraph.

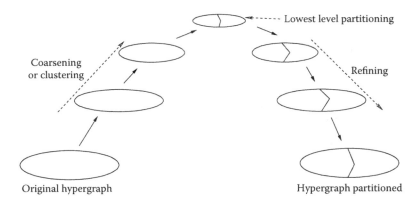

Figure 30.4 Schematic: multilevel approach.

30.4.3 Multilevel Partitioning

The evidence suggests that multilevel hypergraph partitioning schemes often gives high quality solutions to the partitioning problem for large graph/hypergraph partitioning problems [10] as compared to other heuristic approaches. The state-of-the-art fast hypergraph partitioner packages like hmetis and UCLAPack are based on the multilevel approach [7].

The multilevel partitioning algorithms have become popular because of the good quality of partitions they give in shorter execution times [7,55].

The idea of multilevel approach is to represent a very large hypergraph (for which the conventional iterative algorithms like F–M, K–L may execute for significantly long time) through a series of coarser hypergraphs. The hypergraph are coarsened in such a way that the information of the parent graph is inherited/encapsulated/embedded in the series of children hypergraphs. At the end a good partition of the coarsest hypergraph is computed and used as an initial partition for the parent hypergraph to refine it further, with the help of, say, a variant of F–M algorithms. This process is applied in an iterative way till one reaches back the original hypergraph through the refinement process.

The schematic in Figure 30.4 shows the multilevel approach and in contrast to the flat partitioning approach shown in Figure 30.3.

30.4.3.1 Multilevel Steps

The multilevel approach has, in general, three steps/phases described below.

i. *Coarsening phase.* In this phase the original hypergraph is reduced to a coarser graph iteratively until a small hypergraph of size less than a few hundred vertices is obtained.

ii. *Partitioning.* In this step the coarsest hypergraph which inherits the traits of the original hypergraph, is then partitioned using the algorithms which are effective on small sized graphs/hypergraphs [8].

iii. *Refinement.* In this step the partitioning of the coarsest hypergraph is projected to the hypergraph one level up (the finer hypergraph), and then the projected partition is refined using variations of move-based iterative partitioning algorithms (in a few cases, certain spectral techniques are used too [56]). Thus a better partition is obtained for the hypergraph, one level above, is obtained. This procedure is applied iteratively till one obtains a refined partitioning solution for original hypergraph.

30.4.3.2 Coarsening

Coarsening or clustering procedure is one of the critical steps in the multilevel approach and different clustering approach may lead to different quality of partitions. Also the performance of the partitioning algorithm as a whole depends upon the clustering approach used.

The coarsening is done by merging of the vertices, chosen intelligently with concerns for time efficiency, resulting in a smaller, coarser hypergraph. A typical procedure for coarsening is based on *maximal matching* of the vertex set [4,7,16].

30.4.3.3 V-Cycle Refinement

The V-cycle refinement idea aims to further increase the quality of the partition obtained from the multilevel approach. The procedure runs primarily in two phases, namely, the coarsening and the uncoarsening phase.

The complete multilevel partition is run for the hypergraph to get an initial solution for the hypergraph this solution works as an initial partition for the second run of the multilevel. In the coarsening phase the vertices coarsened belong to only one of the partitions. This can be termed as *restricted* coarsening [7] and this basically preserves the initial partition. The uncoarsening phase is the same as of the multilevel approach in which local improvement methods are used to improve upon the given partition.

30.4.3.4 Connectivity-Based Clustering Algorithm

Typically in multilevel partitioning approaches the clustering approach uses the *edge-coarsening* based on collapsing pairs of vertices (e.g., the endpoints of a maximal matching). However, the drawback is that it is likely to ignore the natural clusters in a graph. Figure 30.5 explains the idea of naturally occurring clusters in a graph and how the edge coarsening which gives importance to an unclustered vertex with the highest connectivity gives a poorer result. In their approach [57], the authors propose to use clustering approach (first step of the multilevel approach) that is based on finding *natural clusters* in the graph, that is parts of the graph which are heavily connected.

30.5 SPECTRAL APPROACHES FOR PARTITIONING

In this section we describe various elegant approaches to graph/hypergraph partitioning problems based on eigenvalues and eigenvectors of matrices associated with graphs and hypergraphs. Due to limitation of space many other interesting ideas of similar flavor could not be presented here.

30.5.1 Hall's Approach and Its Variations

Among the very early use of spectral ideas in graph partitioning and related areas were those of Hall and its variations. They made interesting use of the eigenspectrum of the adjacency matrix A and the Laplacian matrix $L = D - A$ (where D is the diagonal matrix

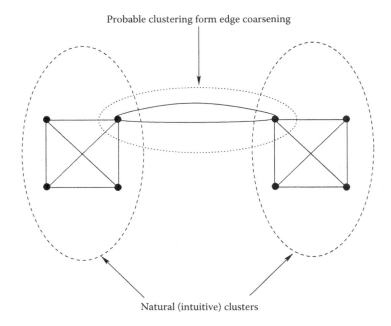

Figure 30.5 Edge coarsening versus coarsening of natural clusters.

containing degrees of the corresponding nodes). Hall [58] showed that the eigenspectrum of the Laplacian of graph G can be used to solve 1-dimensional quadratic cost placement problem. This problem requires us to find a nontrivial placement of nodes of G along x-axis such that, sum of the squares of the distance between the locations of nodes that are neighbors in G is minimized. That is,

$$\min \tfrac{1}{2} \sum_{i=1}^{n} \sum_{j=1}^{n} a_{ij}(x_i - x_j)^2$$
$$\text{subject to } \sum_{i=1}^{n} x_i^2 = 1 \qquad (30.22)$$

Here, x_i denotes the location of ith node. And the constraint $\sum_{i=1}^{n} x_i^2 = 1$ is to disallow trivial solution to the optimization problem (in fact, we will require additional constraint for obtaining nontrivial solution, as will be seen later!). In the matrix notation, this problem can be rephrased as follows.

$$\min x^T L x$$
$$\text{such that, } x^T x = 1 \qquad (30.23)$$

The following theorem describes certain properties of the Laplacian matrix $L = D - A$, where A is the adjacency matrix and D is the diagonal matrix containing the degrees of the nodes.

Theorem 30.4 (Properties of Laplacian) *Given an undirected graph G, its node-arc incidence matrix N and Laplacian matrix L have following properties.*

1. *L is symmeteric matrix and hence, posseses real eigenvalues and real, orthogonal eigenvectors.*

2. *$L\underline{e} = 0$, where $\underline{e}$ denotes $[1, \ldots, 1]^T$,*

3. *$L = NN^T$ independent of signs chosen in columns of N.*

4. If $\underline{v}$ is an eigenvector of L for eigenvalue λ, then

$$\lambda = \frac{\|N^T \underline{v}\|_2^2}{\|\underline{v}\|_2^2}$$

$$= \frac{\sum_{e=(i,j)} (\underline{v}_i - \underline{v}_j)^2}{\sum_i \underline{v}_i^2}$$

5. Eigenvalues of L are nonnegative. $0 \leq \lambda_1 \ldots \leq \lambda_n$.

6. The number of connected components of G equals the multiplicity of 0 eigenvalue. In particular, $\lambda_2 \neq 0$ if and only if G is connected.

Proof.

i. Symmetry follows from the definition of L: since G is an undirected graph, (i,j) is an edge iff (j,i) is an edge. From standard matrix algebra, we know that the eigenspectrum of real symmetric matrix is real valued and the eigenvectors form an orthonormal basis.

ii. The ith entry of $L\underline{e}$ is just the sum of the entries of the ith row of L. This is due to cancellation of the contribution of the degree of node i, that is L_{ii}, by terms $L_{ij} = -1$ together for each edge incident on i.

iii.

$$(NN^T)_{ii} = \sum_{e:\text{ edge } e \text{ incident on } i} (\pm 1)^2$$

$$= \text{degree of node } i$$

$$\text{and } (NN^T)_{ij} = \sum_{\text{all edges } e = (i,j)} (-1) \times (+1)$$

$$= -1 \quad \text{if an edge } e = (i,j) \text{ exists}$$

iv.

$$L\underline{v} = \lambda \underline{v}$$
$$\underline{v}^T L \underline{v} = \lambda \underline{v}^T \underline{v}, (\underline{v} \neq 0)$$
$$\underline{v}^T N N^T \underline{v} = \lambda \underline{v}^T \underline{v}$$
$$\lambda = \frac{\|N^T \underline{v}\|_2^2}{\|\underline{v}\|_2^2}$$
$$\lambda = \frac{\sum (\underline{v}_i - \underline{v}_j)^2}{\sum_i \underline{v}_i^2}$$

v. Nonnegativity of eigenvalues of L clearly follows from above.

vi. $\lambda = 0$ implies $\underline{v}_i = \underline{v}_j$ for each edge $e = (i,j)$. Hence, for any node k reachable from i, $\underline{v}_k = \underline{v}_i = c$. Therefore, within each connected component of graph, the components of the eigenvector $\underline{v}$ are constant. Therefore it is easy to see that there are exactly d independent eigenvectors, corresponding to these d connected components. Next, due to L being symmetric, it possesses full eigenspectrum, and therefore the number of independent eigenvectors corresponding to $\lambda = 0$ equals the multiplicity of 0 as an eigenvalue. Hence, if G is connected then then number of connected component is 1 and consequently only one eigenvalue should be zero. Therefore, $\lambda_2 \neq 0$. ■

Let U be the $n \times n$ matrix whose columns are the orthonormal eigenvectors of L. Therefore our problem can be restated as

$$\min x^T U \Lambda U^T x$$
$$\text{such that, } x^T U U^T x = 1 \qquad (30.24)$$

(here Λ denotes the diagonal matrix with eigenvalues of L). So we need to solve

$$\min y^T \Lambda y$$
$$\text{such that, } y^T y = 1 \qquad (30.25)$$

That is,

$$\min \sum_{i=1}^{n} \lambda_i y_i^2$$
$$\text{such that, } \sum_{i=1}^{n} y_i^2 = 1 \qquad (30.26)$$

Clearly the solution to above is $[1, 0, 0, \ldots, 0]^T$, and the value at this optimum is λ_1 which is zero. This is clearly not of any interest.

To make things slightly more interesting, we impose one more constraint, namely, $x^T[1, 1, \ldots, 1]^T = 0$ in other words, $\sum x_i = 0$. This transforms the problem as follows.

$$\min x^T L x$$
$$\text{such that, } x^T x = 1 \text{ and } \sum x_i = 0 \qquad (30.27)$$

Note that, $\sum x_i = 0$ can be written as $x^T \underline{u_1} = 0$ where, $\underline{u_1} = 1/\sqrt{n}[1, 1, \ldots, 1]^T$ is the normalized eigenvector of L corresponding to $\lambda_1 = 0$, Therefore the problem can be rephrased as

$$\min \left(\frac{v^T B v}{v^T v} \right)$$
$$\text{such that, } v \neq 0 \text{ and } v^T \underline{u_1} = 0 \qquad (30.28)$$

Following theorem solves this rephrased version of the problem.

Theorem 30.5 (Second smallest eigenvalue of Laplacian) *Let $\lambda_1 \leq \lambda_2 \ldots \leq \lambda_n$ be the eigenvalues of a real symmetric matrix B with $v_1, v_2, \ldots, v_n$ as corresponding orthonormal eigenvectors. Then,*

$$\lambda_2 = \min \left(\frac{v^T B v}{v^T v} \right) \qquad (30.29)$$
such that, $v \neq 0$ and $v^T \underline{u_1} = 0$

Thus we have solved the following minor variation

$$\min x^T L x \text{ such that, } \sum_{i=1}^{n} x_i^2 = n, \sum_{i=1}^{n} x_i = 0 \qquad (30.30)$$

Interestingly, if we were to restrict x_i values to only $+1$ and -1, then we have,

$$\min x^T L x \qquad (30.31)$$
$$\text{such that, } \sum_{i=1}^{n} x_i^2 = n, \sum_{i=1}^{n} x_i = 0 \qquad (30.32)$$
$$\text{and } x_i \in \{-1, +1\} \ \forall i = 1, \ldots, n \qquad (30.33)$$

Equivalently, we have

$$\min x^T L x \qquad (30.34)$$
$$\text{such that, } \sum_{i=1}^{n} x_i = 0 \qquad (30.35)$$
$$\text{and } x_i \in \{-1, +1\} \ \forall i = 1, \ldots, n \qquad (30.36)$$

which is a discrete optimization problem, a continuous relaxation of which was solved above using the eigenspectrum idea. The reason for our interest in this discrete optimization problem is that it is indeed graph bisection problem.

Theorem 30.6 *Let x be a vector representing a bipartition $\{V_+, V_-\}$ of the node set of graph G $x_i = 1$ iff $i \in V_+$ and -1 iff $i \in V_-$, then the number of edges cut across $\{V_+, V_-\}$, denoted by, say E_c, equals $\frac{1}{4} x^T L x = \frac{1}{4} \sum_{(i,j) \in E} (x_i - x_j)^2$.*

Proof. Consider:

$$\begin{aligned}
x^T L x &= \sum_i \sum_j L_{ij} x_i x_j \\
&= \sum_{i=j} L_{ii} x_i^2 + \sum_{i \neq j} L_{ij} x_i x_j \\
&= \sum_{i=j} L_{ii} + \sum_{\substack{i \neq j \\ i,j \in V_+}} L_{ij} x_i x_j + \sum_{\substack{i \neq j \\ i,j \in V_-}} L_{ij} x_i x_j + \sum_{\substack{i \neq j \\ i \in V_+ \\ j \in V_-}} L_{ij} x_i x_j \\
&= \sum_i degree(i) + \sum_{\substack{i \neq j \\ i,j \in V_+}} -1 + \sum_{\substack{i \neq j \\ i,j \in V_-}} -1 + \sum_{\substack{i \neq j \\ i \in V_+ \\ j \in V_-}} 1 \\
&= \begin{cases} 2(\text{ number of edges in G }) \\ -2(\text{number of edges connecting nodes in } V_+ \text{ to nodes in } V_+) \\ -2(\text{ number of edges connecting nodes in } V_- \text{ to nodes in } V_-) \\ +2(\text{number of edges connecting nodes in } V_- \text{ to nodes in } V_+) \end{cases} \\
&= 4(\text{ number of edges connecting nodes in } V_- \text{ to nodes in } V_+)
\end{aligned}$$

Hence,

$$E_c = \frac{1}{4} x^T L x$$

Further,

$$\begin{aligned}
x^T L x &= x^T N N^T x \\
&= \sum_{e=(i,j)} ((N^T x)_e)^2 \\
&= \sum_{e=(i,j)} (x_i - x_j)^2
\end{aligned}$$

where e denotes an edge of the graph and N is the node-arc incidence matrix of the graph. ∎

The above discrete optimization problem is a well-known NP-complete problem [20,21] and therefore the insights and the lower bound provided by the above continuous relaxation (which is variation of Hall's quadratic 1-dim placement problem) are of importance.

30.5.1.1 Spectral Bipartitioning Algorithm of Hall

The *spectral bipartitioning algorithm* based on this analysis is as follows: Compute eigenvector $\underline{\mu}_2$ corresponding to the second smallest eigenvalue of L and sets $\underline{x}_i$ to one if $\underline{\mu}_2(i) \geq 0$ and $\underline{x}_i$ to be zero, otherwise. This vector $\underline{x}$ may be interpreted as the one which maximizes $\underline{x} \cdot \underline{\mu}_2$. In other words, $\underline{x}$ is the indicator vector *projecting maximally* onto $\underline{\mu}_2$. The nodes of graph are separated into 2 blocks depending upon the $0-1$ values of the corresponding components of the indicator vector $\underline{x}$. The quality of the bipartition can be compared with the lower bound described in the theorem below, which summarizes the prior discussion based around the second smallest eigenvalue of the Laplacian.

Theorem 30.7 (Second smallest eigenvalue) *Minimum number of edges cut by a balanced partition of the node set of G is bounded below by $\frac{n}{4}\lambda_2(L)$.* ∎

30.5.2 Another View of Spectral Approach to Graph Bisection

The graph bisection problem may be stated as

$$\min x^T L x \tag{30.37}$$

such that,

$$x_i = \pm 1 \text{ and } \sum_i^{|V|} x_i = 0 \tag{30.38}$$

This discrete problem is NP-complete, hence it makes sense to study a tractable continuous relaxation of it, for instance the following one: min $x^T L x$ such that $x^T x = 1$, and $x^T \underline{e} = 0$ ($\underline{e}$ denotes the vector with all components 1). This continuous constrained optimization problem can be solved using Lagrangean technique. The following Lagrangean is to be considered for the stationarity analysis of its saddle point

$$F(x; \lambda) = x^T L x - \lambda(x^T x - 1).$$

The usual analysis involving taking partial derivatives of the above w.r.t. x and equating them to 0 yields necessary conditions for optimal solution as $\lambda x = Lx$ that is λ is an eigenvalue of L and x is corresponding eigevector. The cost of the original objective function turns out to be $x^T L x = \lambda$ under the constraint that x is of unit length. Therefore we seek smallest eligible eigenvalue and corresponding eigenvector of L. Since $\lambda_1 = 0$ gives $x = \underline{e}$ which violates the constraint that $x^T \underline{e} = 0$, the second smallest eigenvalue λ_2 gives valid optimal solution for the above problem. Note the equivalence of this approach with the earlier analysis that also results in the technique and lower bound for the graph bisection that is based on the second smallest eigenvalue and the corresponding eigenvector of L.

30.5.2.1 Dealing with Fixed Vertices

Hendrickson et al. [59] extended this standard bisection method to include information about fixed vertices. Fixed vertices are the vertices which may not be assigned to any other than the specified block. Their Lagarangean formulation results in a generalized eigenvalue problem whose instances tend to get solved effectively by Lanczos-like approaches [59].

30.5.3 Partitioning into k Blocks: Barnes' Approach

In general one considers the problem of finding a partition of the vertex set of G into k blocks. Sometimes block sizes are also specified, say $m_1 \geq m_2 \ldots \geq m_k$. Donath and Hoffman [60] derived a lower bound on the size of the set of edges cut by a k-partition with prescribed block sizes.

Theorem 30.8 (Donath–Hoffman bound): *Let $\lambda_1 \leq \lambda_2 \ldots \leq \lambda_n$ be the eigenvalues of the Laplacian. Let $m_1 \geq m_2 \ldots \geq m_k$ be prescribed block sizes. Then the number of edges cut by a k-partition that satisfies the prescribed block sizes is bounded below by $\frac{1}{2} \sum_{i=1}^{k} \lambda_i m_i$.* ∎

Later Bopanna (see [4]) and Rendl and Wolkowitz [61] were among those who improved Donath–Hoffman's bound. Barnes [62] worked with adjacency matrix directly rather than Laplacian. He relates minimum k-cut (with specified block sizes) problem to that of approximating the adjacency matrix by an $n \times n$ partition indicator matrix in the Frobenius norm. Barnes' technique involves use of k eigenvectors corresponding to first k eigenvectors of the adjacency matrix.

We outline here Barnes' approach to graph partitioning. Let Π be 0-1 matrix with k columns and n rows, each column indicating the nodes belonging to the block of a k-partition it represents. Thus Π represents the k-partition of the node set of the graph. $P = \Pi \Pi^T$ is $n \times n$ matrix representing the same partition. P has interesting and simple structure due to which its eigenspectrum is easy to find. Indeed the n eigenvalues are $m_i \geq m_2 \geq \ldots \geq m_k \geq 0 \ldots \geq 0$ where, $m_1, \ldots, m_k$ are block sizes. It is interesting to consider the Frobenius distance between the adjacency matrix and the $n \times n$ partition matrix P.

$$\begin{aligned}
&\|A - P\|_F^2 \\
&= trace((A-P)^T(A-P)) \\
&= trace(A^T A) + trace(P^T P) - 2\, trace(P^T A) \\
&= \sum_{i=1}^{n} \lambda_i^2 + \sum_{h=1}^{k} m_h^2 - 2\, trace(P^T A) \\
&= \sum_{i=1}^{n} \lambda_i^2 + \sum_{h=1}^{k} m_h^2 - 2 \sum_{i=1}^{n} (P^i . A^i)
\end{aligned}$$

Here P^i and A^i denote ith column of P and A, respectively,

$$= \sum_{i=1}^{n} \lambda_i^2 + \sum_{h=1}^{k} m_h^2 - 4(\text{number of uncut edges})$$

Thus we see that finding a partition matrix that is closest, in Frobenius norm, to the adjacency matrix is equivalent to solving the problem of finding k-partition with the prescribed block sizes, that minimizes the number of cut edges. Additional interesting information about the Frobenius distance between A and P is obtained using the well-known Hoffman–Wielandt's inequality, which is stated as follows.

Theorem 30.9 (Hoffman–Wielandt inequality): *Let A and B be real, symmetric $n \times n$ matrices with eigenvalues $\alpha_1 \geq \alpha_2 \ldots \geq \alpha_n$ and $\beta_1 \geq \beta_2 \ldots \geq \beta_n$. Then $\|A - P\|_F^2 \geq \sum_{i=1}^{n} (\alpha_i - \beta_i)^2$*

We use Hoffman–Wielandt inequality for the adjacency matrix A and a partition matrix P, to get

$$\|A - P\|_F^2 \geq \sum_{i=1}^{k} (\lambda_i - m_i)^2 + \sum_{i=k+1}^{n} \lambda_i^2$$

where, $\lambda_1 \geq \lambda_2 \ldots \geq \lambda_n$ are eigenvalues of A. Using this and relationship between the Frobenius distance and the number of uncut edges we get the following Donath–Hoffman's upper bound on the number of uncut edges.

$$4(|E| - |E_c|) \leq 2\sum_{i=1}^{k} \lambda_i m_i$$

where E_c denotes the set of edges cut across the partition.

Therefore this gives a lower bound on the number of cut edges in terms of eigenvalues of the adjacency matrix.

Next we analyze $||A - P||_F^2$ in terms of eigenspectra of A and P.

$$A = U\Lambda U^T \text{ and } P = VMV^T$$
$$\Lambda = \text{diag}(\lambda_1, \lambda_2, \ldots, \lambda_n)$$
$$M = \text{diag}(m_1, m_2, \ldots, m_k)$$
$$U = \text{matrix with orthonormal eigenvectors of } A \text{ as columns}$$
$$V = \text{matrix with orthonormal eigenvectors of } P \text{ as columns}$$

P, being of simple structure allows eigenvectors to be found by inspection. $P = \Pi\Pi^T = \Pi M^{-1/2} M M^{-1/2}\Pi^T$ and $m_1, m_2, \ldots, m_k, 0, \ldots, 0$ are the eigenvalues of P. Therefore, $\Pi M^{-1/2}$ is the matrix whose columns are the orthonormal eigenvectors of P. Thus, $V = \Pi M^{-1/2}$. Further noting that Frobenius norm is invariant under orthogonal linear transformation and that U and its transpose are orthogonal matrices, we get the following.

$$\begin{aligned} ||A - P||_F^2 &= ||U\Lambda U^T - VMV^T||_F^2 \\ &= ||\Lambda - U^T V M V^T U||_F^2 \end{aligned}$$

If V were such that $U^T V = [I_k\ 0]^T$, then it can be seen that Donath–Hoffman lower bound is attained. Therefore, we would like to find V such that $||U^T V - [I_k\ 0]^T||_F^2$ is as small as possible.

$$\begin{aligned} ||U^T V - [I_k\ 0]^T||_F^2 &= ||V - U[I_k\ 0]^T||_F^2 \\ &= \sum_{j=1}^{k} ||\underline{v_j} - \underline{u_j}||^2 \\ &= \sum_{j=1}^{k}(2 - 2\sum_{i=1}^{n} v_{ij} u_{ij}) \\ &= 2k - 2\sum\sum v_{ij} u_{ij} \end{aligned}$$

where, u_{ij} and v_{ij} are the ith components of the jth column vectors, $\underline{u_j}$ and $\underline{v_j}$, of U and V respectively. Clearly, jth column of V is

$$\underline{v_j} = \pm \frac{1}{\sqrt{m_j}}[x_{1j}, x_{2j}, \ldots, x_{nj}]^T$$

where $[x_{1j}, x_{2j}, \ldots, x_{nj}]^T$ denotes jth column of Π. Therefore V that minimizes $||U^T V - [I_k\ 0]^T||_F^2$ is obtained by

$$\min -\sum_{j=1}^{k} sgn(\underline{v_j}) \sum_{i=1}^{n} \left(\frac{u_{ij}}{\sqrt{m_j}}\right) x_{ij}$$

subject to,

$$\sum_{i=1}^{n} x_{ij} = m_j \quad j = 1, \ldots, k$$

$$\sum_{j=1}^{k} x_{ij} = 1 \quad i = 1, \ldots, n$$

$$x_{ij} \geq 0 \quad i = 1, \ldots, n; j = 1, \ldots, k$$

This is the transportation problem [16]. Note that the term $sgn(v_j)$ in the objective function of the above transportation problem is the sign chosen for v_j.

Since, $m_1, m_2, \ldots, m_k$ are integers, an extreme integer valued solution to the above transportation problem will be integer valued [16]. Also $\sum_{j=1}^{k} x_{ij} = 1$ implies $x_{ij} \in \{0, 1\}$. Thus we will get a partition matrix P from VMV^T. But how are the signs in the $v_j = \pm 1/\sqrt{m_j}[x_{1j}, x_{2j}, \ldots, x_{nj}]^T$ to be chosen? Clearly the most desirable choice is one for which the corresponding transportation problem has smallest optimum. We need to therefore solve 2^k such instances of transportation problems.

30.5.4 Ratio-Cut Bipartitioning Using Eigenspectrum

The following interesting result of Hagen and Kahng [63] relates second smallest eigenvalue of the Laplacian to a cost metric called ratio-cut. This measure, 2-ratio-cut is defined as follows. Ratio-cut cost of a bipartition H, K is the ratio $|E_c|/|H||K|$. Here E_c denotes the set of edges cut across the bipartition.

Theorem 30.10 [63] *Given a graph $G(V, E)$ with A as its adjacency matrix and $L = D - A$ as its Laplacian the 2nd smallest eigenvalue of the Laplacian gives a lower bound on the ratio-cut cost of any bipartition.*

$$\frac{|E_c|}{|H||K|} \geq \frac{\lambda_2(L)}{|V|}$$

where, E_c is the set of edges cut by the bipartition H, K of V and $\lambda_2(L)$ is the second smallest eigenvalue of the Laplacian.

Proof. Let H, K be an optimum bipartition of V in the sense of the above ratio-cut metric. Let $p = |H|/|V|$ and $q = |K|/|V|$. Thus, $p, q \geq 0$ and $p + q = 1$. Define a vector x with $x_i = q$ if $i \in H$ and $x_i = -p$ if $i \in K$. Then $x \perp [1 \ldots 1]^T$, that is, x is perpendicular to the first eigenvector (i.e., corresponding to the smallest eigenvalue of L). $x^T L x$ can be seen to equal $|E_c|$, the number of edges cut in H, K bipartition. Furthermore, $|H||K| = pq|V|^2 = |V|||x||_2^2$. Therefore, $\lambda_2(L) = \min_{y \perp [1\ldots1]^T, y \neq 0} y^T L y / y^T y \leq |E_c||V|/|H||K|$. Therefore, $|E_c|/|H||K| \geq \lambda_2(L)/|V|$. ∎

Using this result one develops intuition of use of the eigenvector corresponding to the second smallest eigenvector of the Laplacian to design heuristic for finding good bipartition in the sense of the above defined ratio-cut.

30.5.5 Spectral Bound Involving Multiple Eigenvectors

Chan et al. [64] extended the ideas of Hall [58], Pothen et al. [65], and Hagen et al. [63] using the following main theoretical result.

Theorem 30.11 (k-smallest eigenvalues of Laplacian) *Subject to the condition that k columns of $n \times k$ matrix Y are orthonormal, trace of $Y^T L Y$ is minimized, when the k columns* *of Y are the eigenvectors of L corresponding to the k-smallest eigenvalues of L. And the minimum value is $\sum_{i=1}^{k} \lambda_i$ where $\lambda_1 \leq \lambda_2 \ldots \leq \lambda_n$ are the eigenvalues of L.*

Proof. Let U denote $n \times n$ matrix with the normalized eigenvectors of L as columns in order of $\lambda_1 \leq \lambda_2 \ldots \leq \lambda_n$. Let U_k denote the $n \times k$ submatrix of U consisting of first k columns.

$$trace(U_k^T L U_k) = trace(U_k^T U \Lambda U^T U_K)$$
$$= trace(\begin{bmatrix} I_k & 0 \end{bmatrix} \Lambda \begin{bmatrix} I_k \\ 0 \end{bmatrix})$$
$$= \sum_{i=1}^{k} \lambda_i$$

We prove by contradiction. Let Z be $n \times k$ matrix that violates the theorem, that is $trace(Z^T L Z) < \sum_{i=1}^{k} \lambda_i$ and let k be smallest such integer. Clearly k cannot be 1 as $x^T L x \geq 0 = \lambda_1$ due to positive semidefiniteness of L. So, $k \geq 2$ and will be able to use inductive assertion for $k-1$ and below. Let $y \in sp(Z)$ be a unit vector (in the linear span of columns of Z) that maximizes $y^T L y$. Extend y to an orthonormal basis, say Y that is $Y^T Y = I_k = Z^T Z$ and $sp(Y) = sp(Z)$. We now show that there exists orthonormal W such that $Y = ZW$. This is seen as follows. We begin with assumption that W exists and show that it is orthogonal matrix. Let $\hat{Z}$ be orthogonal basis of the complement of $sp(Y)$ in $\Re^n$. Therefore, $[\ Y\ \hat{Z}\ ]$ is orthonormal basis columns of $\Re^n$ and so is $[\ Z\ \hat{Z}\ ]$. Clearly,

$$[\ Y\ \hat{Z}\ ] = [\ Z\ \hat{Z}\ ] \begin{bmatrix} W & 0 \\ 0 & I \end{bmatrix}$$

Therefore, $\begin{bmatrix} W & 0 \\ 0 & I \end{bmatrix}$ is orthogonal and hence W is also orthogonal $k \times k$ matrix. Due to similarity transformation $trace(Z^T L Z) = trace(Y^T L Y)$, therefore $trace(Y^T L Y) < \sum_{i=1}^{k} \lambda_i$. Let $\overline{Y_y}$ be Y with column y removed.

$$trace(Y^T L Y) = trace(\overline{Y_y}^T L \overline{Y_y}) + y^T L y$$

So, $\sum_{i=1}^{k} \lambda_i > \sum_{i=1}^{k-1} \lambda_i + y^T L y$.

Therefore, $\lambda_k > y^T L y = \max_{x \in sp(Z) = sp(Y), ||x||=1} x^T L x$.

Therefore, $\lambda_k > y^T L y = \max_{x \in sp(Z)} \frac{x^T L x}{x^T x} \geq \min_{s: k\text{-dim}} \max_{x \in S} \frac{x^T L x}{x^T x} = \lambda_k$.

The last equality is due to Courant–Fischer min-max theorem. Hence, we get a contradiction. Having shown that if W exists then it is orthogonal, to complete the proof we need to show that such W indeed exists. Due to independence of columns of Y there exists $\tilde{W}_1$ matrix composed of elementary column operations, such that $Y\tilde{W}_1 = \begin{bmatrix} I_k \\ Y_1 \end{bmatrix}$. Similarly for $\tilde{W}_2$ i.e $Z\tilde{W}_2 = \begin{bmatrix} I_k \\ Z_1 \end{bmatrix}$. Furthermore $Y_1 = Z_1$ since column spaces of $\begin{bmatrix} I_k \\ Y_1 \end{bmatrix}$ and $\begin{bmatrix} I_k \\ Z_1 \end{bmatrix}$ are identical to column spaces of Y and Z, respectively. So $\begin{bmatrix} I_k \\ Y_1 \end{bmatrix} \hat{W} = \begin{bmatrix} I_k \\ Z_1 \end{bmatrix}$ for some invertible matrix $\hat{W}$, which must be I_k therefore $Y_1 = Z_1$ And hence $Y\tilde{W}_1 = Z\tilde{W}_2$ and therefore $Y = ZW$, where $W = \tilde{W}_2 \tilde{W}_1^{-1}$. This completes the proof. ∎

For completeness, we include a version of Courant–Fischer min-max theorem with a proof.

Theorem 30.12 (Courant–Fischer min-max theorem) *If $A \in \Re^{n \times n}$ is symmetric, then*

$$\lambda_k(A) = \max_{dim(S)=k} \min_{0 \neq y \in S} \frac{y^T A y}{y^T y}$$

for $k = 1, \ldots, n$.

Proof. Let $Q^T A Q = diag(\lambda_i)$ be the Schur decompsition of A. Hence, $\lambda_k = \lambda_k(A)$ and $Q = [\underline{q_1}, \underline{q_2}, \ldots, \underline{q_n}]$. We define vector space $S_k = span(\{\underline{q_1}, \underline{q_2}, \ldots, \underline{q_k}\})$.

$$\max_{dim(S)=k} \min_{0 \neq y \in S} \frac{y^T A y}{y^T y} \geq \min_{0 \neq y \in S_k} \frac{y^T A y}{y^T y} = \underline{q_k}^T A \underline{q_k} = \lambda_k(A)$$

We establish reverse inequality to show that they are actually equal. Let S be any k-dimensional subspace and note that it must intersect the $n - k + 1$-dim subspace $span(\{\underline{q_k}, \ldots, \underline{q_n}\})$ nontrivially. If $\overline{y} = \alpha_k \underline{q_k} + \cdots + \alpha_n \underline{q_n}$ is in this intersection, then

$$\min_{0 \neq y \in S} \frac{y^T A y}{y^T y} \leq \frac{\overline{y}^T A \overline{y}}{\overline{y}^T \overline{y}} \leq \lambda_k(A)$$

Since this holds for all k-dimensional subspaces,

$$\max_{dim(S)=k} \min_{0 \neq y \in S} \frac{y^T A y}{y^T y} \leq \lambda_k(A)$$

Hence the proof. ∎

Motivated by the above characterization of eigenvectors of k-smallest eigenvalues of the Laplacian yielding columns of a suitable constrained minimizer Y of $trace(Y^T L Y)$, Chan et al. [64] proposed a new metric for k-ratio-cuts as follows: $k\_ratio\_cut(\Pi) = \sum_{h=1}^{k} w(E_h)/|V_h|$, where $w(E_h)$ denotes the weight of edges crossing the kth cluster/block of vertex partition $\Pi = \{V_1, V_2, \ldots, V_k\}$. They showed that the sum of the k smallest eigenvalues of the Laplacian is a lower bound on the optimal such ratio cut cost over k-block partitions of the vertex set. The same authors extended these ideas in the setting of partitioning for multiple FPGAs [66].

30.5.6 Simple Eigenvector-Based Clustering

30.5.6.1 Algorithm

Alpert and Kahng [67] proposed a simple but effective algorithm based on spectral methods for clustering hypergraphs. For a given hypergraph $H(V, E)$ over the n nodes $V = \{v_1, v_2, \ldots, v_n\}$, each hyperedge $e \in E$ can be regarded as a subset of two or more nodes from V, with $|e|$ representing the number of nodes incident on e. Then $n \times n$ adjacency matrix $A = (a_{ij})$ of the H can be defined using *clique* net model. If v_i and v_j both are in hyperedge (net) e then for entry a_{ij}, $f(|e|)$ is added. A typical choice for a weight function is : $f(|e|) = 6/|e|(|e| + 1)$. The weighted degree matrix D is defined as $d_{ii} = \sum_{j=1}^{n} a_{ij}$, that is degree of the ith node. First d eigenvectors (corresponding to d smallest eigenvalues) $\mu_1, \mu_2, \ldots, \mu_d$ of *Laplacian matrix* $L = D - A$, are computed, where 2^d is number of clusters in which H is to be partitioned. Based on these d eigenvectors d-digit binary code is calculated for each node v_i. Let μ_{ij} denote the ith entry of the jth eigenvector of L. The bit j of $code[i]$

is set to 1 if $\mu_{ij} \geq 0$ and 0 otherwise. Finally, nodes having same binary code are assigned the same cluster.

30.5.6.2 Theoretical Motivation

The idea of this algorithm is sort of generalization of well-known *spectral bipartitioning algorithm*. The *spectral bipartitioning algorithm* computes eigenvector $\underline{\mu_2}$ corresponding to the second smallest eigenvalue of L and uses it to define a $0-1$ vector $\underline{x}$ to describe a bipartition. To be precise, the spectral bipartitioning algorithm of Hall sets ith component of $\underline{x}$ to one if the corresponding component of $\underline{\mu_2}$ is ≥ 0 and to zero, otherwise. Recall that $\underline{x}$ is the indicator vector *projecting maximally* onto $\underline{\mu_2}$. The optimality of $\underline{\mu_2}$ It is well-known that $\underline{\mu_2}$ is an optimal vector y for the following quadratic optimization problem.

$$f(x) = \sum_{i=1}^{n} \sum_{j=1}^{n} a_{ij}(y_i - y_j)^2 \qquad (30.39)$$

under the constraints that $||y||^2 = 1$ and $\sum_{i=1}^{n} y_i = 0$. And the $\underline{x}$ computed above by Hall's algorithm is the nearest legal discrete solution to this problem. Similarly Chan et al. [64] showed the generalized result that the d best nondiscrete solutions to the above equation are given by $\underline{\mu_1}, \underline{\mu_2}, \ldots, \underline{\mu_d}$ under the constraint that all solutions are mutually orthogonal. Indeed the indicator vector for these vectors are the binary code in this algorithm. Let $\underline{x_h}$ be the indicator vector for block V_h of the clustering (partition). It is easy to see that the clustering obtained by the above simple algorithm maximizes $\sum_{h=1}^{k} \sum_{j=1}^{d} |\underline{x_h} \cdot \underline{\mu_j}|$, where k is the number of blocks (clusters). That is the clustering found will maximize the *cumulative* projection of the indicator vectors of the clusters onto the eigenvectors corresponding to the d smallest nonzero eigenvalues of the Laplacian L.

Hence, the above algorithm may be regarded as a natural extension of the simple *spectral bipartitioning algorithm* due to Hall [58].

30.5.7 Graph Partitioning Using Multiple Eigenvectors

Various researchers including Alpert et al. [67] as well as Frankle and Karp [68] investigated the approach of using multiple eigenvectors of the Laplacian to map the graph partitioning problem to a clustering or a partitioning problem on the collection of vectors obtained using eigenvectors (not eigenvectors themselves necessarily, though!).

Since, the collection of eigenvectors $\{\underline{\mu_j} : j = 1 \ldots n\}$, of the Laplacian L of a graph form an orthonormal n-dimensional basis, any $0-1$ vector indicating a cluster (i.e. a subset of nodes) can be expressed as a sum of projections on these eigenvectors.

$$\underline{x} = \sum_{j=1,\ldots,n} (\underline{\mu_j}^T \underline{x}) \underline{\mu_j}$$

Given any k-partition $\Pi = \{V_1, V_2, \ldots, V_k\}$ with $m_1(=|V_1|), m_2(=|V_2|)$ and so forth, the cost of the partition satisfies, for suitably large number H (to be clarified later!),

$$H * n - (2(\text{cost of partition } \Pi)) \\ = H \sum_{h=1,\ldots,k} m_h - \sum_{h=1,\ldots,k} \underline{x_h}^T L \underline{x_h}$$

where, $\underline{x_h}$ is $0-1$ indicator vector for V_h that is $\underline{x_h}(i) = 1$ if $i \in V_h$ and 0 otherwise.

$$H * n - 2(\text{cost}(\Pi)) = H \sum_{h=1,\ldots,k} (\sum_{j=1,\ldots,n} (\underline{\mu_j} \cdot \underline{x_h})^2) - \sum_{h=1,\ldots,k} \underline{x_h}^T U \Lambda U^T \underline{x_h}$$

$$= H \sum_{h=1,\ldots,k} (\sum_{j=1,\ldots,n} (\underline{\mu_j} \cdot \underline{x_h})^2) - \sum_{h=1,\ldots,k} (U^T \underline{x_h})^T \Lambda (U^T \underline{x_h})$$

$$= \sum_{h=1,\ldots,k} (\sum_{j=1,\ldots,n} H(\underline{\mu_j} \cdot \underline{x_h})^2) - \sum_{h=1,\ldots,k} \sum_{j=1,\ldots,n} (\sqrt{\lambda_j} \underline{\mu_j} \cdot \underline{x_h})^2$$

$$= \sum_{h=1,\ldots,k} (\sum_{j=1,\ldots,n} (H - \lambda_j)(\underline{\mu_j} \cdot \underline{x_h})^2)$$

$$= \sum_{h=1,\ldots,k} (\sum_{j=1,\ldots,n} (\sqrt{H - \lambda_j} \underline{\mu_j} \cdot \underline{x_h})^2$$

$$= \sum_{h=1,\ldots,k} \|M^T \underline{x_h}\|_2^2$$

where, the jth column of M is the scaled version (scaled by $\sqrt{H-\lambda_j}$), of the jth eigenvector of L (i.e., $\underline{\mu_j}$). Denoting jth column of M by M_j, we know $M_j = \sqrt{H-\lambda_j}\underline{\mu_j}$. Therefore the problem of finding a k-partition is equivalent to the following.

Find a k-partition Π for which the $n \times k$ matrix X, consisting of columns which are the indicators of the blocks of Π, maximizes $\|M^T X\|_F$, that is the Frobenius norm of $M^T X$. The equivalence is evident from the following.

$$\|M^T X\|_F^2 = \sum_{h=1,\ldots,k} \|M^T \underline{x_h}\|_2^2$$

$$= \sum_{h=1,\ldots,k} \|\sum_{q \in V_h} (\text{row } q \text{ of } M)\|_2^2$$

Thus, we have an equivalent vector partitioning problem on the collection of vectors which are the rows of M. Recall that, $H = U(\sqrt{diag(H) - \Lambda})$.

30.5.8 Multilevel Spectral Partitioning

Plenty of evidence points to efficiency and quality of multilevel partitioning schemes for large graphs. The methodology of multilevel partitioning has been described earlier Section 30.4.3.

Now we present brief outline of an adaptation of multilevel paradigm in the context of spectral technique from the work of Barnard and Simon [56]. They use a coarsening scheme based around maximal independent subsets.

30.5.8.1 Multilevel Spectral Partitioning Algorithm

- Coarsen the graph using *Maximal independent sets*, and partition the coarsened graph.
- Project the partition of the coarsened graph back to that of the original and *refine* using *Rayleigh quotient iterations*.

An independent set of a graph $G(V, E)$ is defined as a subset V_I of V such that, no two nodes in V_I are connected by an edge. A maximal independent set of a graph is a maximal such set. A simple greedy algorithm can be used to compute maximal independent set.

Coarsening can be done using the generated maximal independent set V_I. This is done by building *domains* D_i around each node in $i \in V_I$. The edges E_c are then the edges which connects such domains. The algorithm for selecting E_c is as follows.

```
$E_c = \{\emptyset\}$. Unmark all edges in $E$.
for nodes $i = 1$ to $|V_I|$ do
    $D_i = (\{i\}, \emptyset)$
end for
repeat
    Choose an unmarked edge $e = (i, j)$ from $E$
    if exactly one of $i$ and $j$ (say $i$) is in some $D_k$ then
        Mark $e$. Add $j$ and $e$ to $D_k$
    else if $i$ and $j$ are in different $D_k$'s (say $D_{k_i}$ and $D_{k_j}$ then
        Mark $e$. Add an edge $(k_i, k_j)$ to $E_c$
    else if both $i$ and $j$ are in the same $D_k$ then
        Mark $e$. Add it to $D_k$
    else
        Leave $e$ unmarked
    end if
until No unmarked edges
```

This coarsened graph G_c is then partitioned using spectral methods. The eigenvector $\underline{\mu}_c$, associated with the second smallest eigenvalue of Laplacian of G_c, is then used to approximate eigenvector $\underline{\mu}$ of Laplacian of G. For nodes which are in V_I, corresponding components of $\underline{\mu}_c$ are used for $\underline{\mu}$. For the nodes which are not in V_I, average of corresponding components of neighbors that are in V_I is taken.

This averaging method is not guaranteed to yield good results. Hence, further refinement is necessary. Natural idea is to use *Rayleigh quotient iterations*, which will refine the eigenvector so that it approximates actual eigenvector more closely. The algorithm is as follows: ρ is called as *Rayleigh quotient*, which is a good approximation of eigenvalue of the Laplacian L of G. ρ would converge to an eigenvalue of L and $\underline{\mu}$ would become the corresponding eigenvector.

30.5.9 Eigenvalue-Based Hypergraph Reordering and Partitioning

Otten [69] proposed an elegant scheme for reordering nodes and hyperedges of a hypergraph in such a way that layout of the resulting hypergraph would be more efficient. This ordering would naturally provide a heuristic for partitioning the nodes of the hypergraph.

```
Rayleigh quotient iterations
    $i = 0$
    Choose starting vetcor $\underline{\mu}_0 = \underline{\mu}_c$ (after expansion and normalization)
    repeat
        $i = i + 1$; $\rho_i = \underline{\mu}_{i-1}^T L \underline{\mu}_{i-1}$;
        $\underline{\mu}_i = (L - \rho_i I)^{-1} \underline{\mu}_{i-1}$; $\underline{\mu}_i = \frac{\underline{\mu}_i}{\|\underline{\mu}_i\|_2}$;
    until convergence
```

This scheme does not require us to model the hypergraph (representing typically a VLSI netlist) as a graph. Instead it works on the genuine representation of the hypergraph itself. Indeed, the scheme uses a *net connectivity matrix* Q, an $m \times n$ matrix where m is the number of nodes and n is the number of nets or hyperedges of the given hypergraph. If a node i is connected to net (hyperedge) j, then, the entry q_{ij} is marked 1, otherwise 0. In [70], this scheme of Otten is used at the core of a multilevel partitioning approach.

It may be intuitively noted that if the rows and columns of the net connectivity matrix Q are independently permuted such that the nonzeros of Q tend to be clustered near its diagonal region, then the placement of the nodes according to the permutation of the rows would yield an effective placement of nodes in terms of the total span of the nets connecting these nodes. This ordering of nodes also yields heuristically good quality partitioning consistent with the ordering itself.

Otten [69] related this problem of matrix (or hypergraph) reordering, to that of maximizing $\mathbf{r}^T Q \mathbf{k}$, over a collection of permutation vectors $\mathbf{r}$ and $\mathbf{k}$ of size m and n respectively (a *permutation vector* is a vector of length n whose entries form a permutation of $\{1,\ldots,n\}$).

Example 30.2 To illustrate the above idea, an example taken from [70] is taken Consider a matrix Q given as

$$Q = \begin{bmatrix} 1 & 0 & 0 & 1 \\ 1 & 1 & 1 & 0 \\ 1 & 0 & 1 & 1 \end{bmatrix}$$

then $\mathbf{r}$ and $\mathbf{k}$ are later shown to be derived from

$$r = \begin{bmatrix} 0.59 & -0.29 & -0.16 \end{bmatrix}^T$$

$$k = \begin{bmatrix} 0.05 & -0.29 & -0.22 & 0.59 \end{bmatrix}^T$$

In order to sort $\mathbf{r}$ and $\mathbf{k}$ from smallest to largest, $\mathbf{r}$ is reordered according to coordinates $[2\ 3\ 1]^T$ and $\mathbf{k}$ according to coordinates $[2\ 3\ 1\ 4]^T$. If the rows can columns of $\mathbf{Q}$ are reordered based on the above reordering index, the reordered matrix will look like

$$Q = \begin{bmatrix} 1 & 1 & 1 & 0 \\ 0 & 1 & 1 & 1 \\ 0 & 0 & 1 & 1 \end{bmatrix}$$

and the reordered matrix product $(\mathbf{r}^{\mathbf{opt}})^T \mathbf{Q}^{\mathbf{opt}} \mathbf{k}^{\mathbf{opt}}$ is 0.54 which is observed to be the maximum value.

Problem formulation. The matrix reordering technique can be considered as the following optimization problem

$$\max_{r,k} \mathbf{r}^T \mathbf{Q} \mathbf{k}$$
$$\text{such that} (\mathbf{C}_\mathbf{r}^{-1} \mathbf{r})^T (\mathbf{C}_\mathbf{r}^{-1} \mathbf{r}) = 1$$

To solve this problem a method based on eigenvalues is proposed in [69] First define diagonal matrices $\mathbf{C_k}$, $\mathbf{C_r}$

$$C_{k_{jj}} = \frac{1}{\sqrt{\sum_{i=1}^{m} q_{ij}}} \qquad j = 1, 2 \ldots, n$$

$$C_{r_{ii}} = \frac{1}{\sqrt{\sum_{j=1}^{m} q_{ij}}} \qquad i = 1, 2 \ldots, n$$

Note that the square of the reciprocals the diagonal entries of C_k are the number of nodes of the nets and for C_r, the reciprocal of the square of ith diagonal entry represents the number of nets incident on ith node (or a cell, as in VLSI netlist).

Also, **E** and **F** are defined as
$$E = C_r Q C_k$$
$$F = C_r^2 Q C_k^2 Q^T = C_r E E^T C_r^{-1}$$

Otten argued that the eigenvector solutions of $\mathbf{Fv} = \lambda \mathbf{v}$ satisfy certain *iterative averaging process* which aims to identify good **r** and **k**. It was also noted in [2,69] that, in order to maximize $\mathbf{r}^T \mathbf{Q} \mathbf{k}$, we should work with largest or near-largest eigenvalue λ of F.

As F and EE^T are similar, that is $F = C_r E E^T C_r^{-1}$, they have the same eigenvalues and related eigenvectors (v is an eigenvector of F if and only if $C_r^{-1} v$ is an eigenvector of EE^T). Due to symmetric, positive-definite nature of EE^T, the eigenvalues of F are real and nonnegative too. Furthermore, it can be easily seen that row sums of F equal to 1, and therefore, the nonnegative matrix F has 1 as its dominant eigenvalue. However, this dominant eigenvalue has trivial eigenvector (all 1's) and hence of no use for a permutation to be inferred from it.

Premultiply the above eigenvalue equation for F by $\mathbf{C_r^{-1}}$, we get

$$\mathbf{C_r^{-1} F r} = \lambda \mathbf{C_r^{-1} r}$$

Substituting $\mathbf{F} = \mathbf{C_r^2 Q C_k^2 Q^T} = \mathbf{C_r E E^T C_r^{-1}}$, the equation becomes

$$\mathbf{E E^T C_r^{-1} r} = \lambda \mathbf{C_r^{-1} r}$$

If we restrict r such that $C_r^{-1} r$ is of unit magnitude, multiplying the previous equation by $(\mathbf{C_r^{-1} r})^\mathbf{T}$ leads to

$$(\mathbf{C_r^{-1} r})^\mathbf{T} \mathbf{E E^T C_r^{-1} r} = (\mathbf{C_r^{-1} r})^\mathbf{T} \lambda (\mathbf{C_r^{-1} r}) = \lambda$$
$$= (\mathbf{r^T}) \mathbf{Q}(\mathbf{C_k^2 Q^T r}) = \lambda.$$

This has form
$$\mathbf{r^T Q k}$$
where
$$\mathbf{k} = \mathbf{C_k^2 Q^T r}$$

Note that **k** here is derived from **r** which in turn is obtained from solving the eigenvalue problem and obtaining the eigenvector corresponding to the appropriate eigenvalue of **F**. So for reordering **r** and **k** the second largest eigenvector of F is chosen and the rows and columns of the matrix are reordered according to the **r** and **k** found with respect to the eigenvector corresponding to the second largest eigenvalue.

30.6 SIMULATED ANNEALING AND GRAPH BISECTION

Simulated annealing is a Monte Carlo-based metaheuristic for optimization. It is based on the manner in which metals recrystallize (settles to an *optimum* configuration) in the process of annealing. Annealing is analogous to combinatorial problems; the state of the metal is like the current configuration and the energy function to be minimized is the objective function.

Often in optimization problems, local search based technique tends to get trapped at a local minima. Simulated annealing technique, inspired by the metropolis method [71], provides a way to escape from them. The main idea of simulated annealing is roughly as follows: Generate (sample) one random neighboring solution. If the new solution is better, go to this solution (as one would in the ordinary local search). However even if the new solution is worse, give it a chance with a certain probability, accept the worse solution for the sake of hoping to avoid getting trapped at a local optimum. The probability of accepting the neighboring solution that is worse than the current should indeed be governed by the amount of degradation certainly as the intuition would guide us. This apparently counterintuitive idea allows the algorithm to from a local optimum.

Over the time, the role of the above counterintuitive step should be curbed as we hope to converge to a global optimum. This policy is implemented using a parameter called *temperature*, which is progressively decreased according to a *cooling schedule*. The probability to accept a worse solution is gradually reduced as the temperature is reduced. Analogous to physical phenomenon, at a high temperature more random movements are likely, however with lowering of temperature things tend to settle down in a *downhill groove*, as would in a typical local search. It is hoped that algorithm finds some region around a global optimum while the temperature is hot, and fine-tune the rough choice of solution during the later cooler phase. For a good discussion of Simulation Annealing, reader is referred to Aarts and Korst [72].

Let us now describe in some detail the adaptation of metropolis algorithm at the core of simulated annealing. In general metropolis algorithm generates a sample according to an arbitrary probability distribution, however for simulated annealing we use it for sampling from Boltzmann probability distribution (probability of x, the cost or energy, is proportional to $exp(-x/(kT))$. Therefore the generated sample is very highly likely to be the minimum cost (or energy) solution.

Each solution (configuration) in the metropolis algorithm can be regarded as a state in a Markov chain. The transition probabilities are determined by the probability of generating the neighboring solution (for example, uniform distribution) and the probability of acceptance $exp(-\Delta/kT)$. This Markov chain turns out to be *ergodic*. The metropolis algorithm simulates the above ergodic Markov chain. Being ergodic, its simulation would yield its stationary distribution. In fact Boltzmann distribution happens to be the stationary distribution of this ergodic Markov chain. This can be proved using the observation that the so called *detailed balance* conditions are satisfied by it (Boltzmann's distribution) for this chain.

Simulated annealing works by running the metropolis algorithm while gradually decreasing the value of T over the course of the execution. The exact way in which T is updated is called a cooling schedule. The key idea of simulated annealing is to start at a high temperature, and slowly lower the temperature. At a given temperature, we run Metropolis for a while.

Metropolis algorithm
  Choose arbitrarily an initial solution S
  repeat
    Generate a new random neighboring solution S'
    Let $\Delta = cost(S') - cost(S)$.
    if $\Delta < 0$ /* downhill move */ **then**
      set $S = S'$.
    else
      /* uphill move */, with probability proportional to $exp(-\frac{\Delta}{kT})$, set $S = S'$.
    end if
  until convergence

There are some interesting facts about annealing. For example, you can prove that if you lower the temperature slowly enough, you will find the global minimum with arbitrarily high probability. However these optimal annealing schedules may note be very useful in practice.

```
Simulated annealing
  Get an initial solution S
  Get an initial temperature T > 0
  while not yet frozen do
    repeat
      Pick a random neighbor S' of S
      Let Δ = cost(S') − cost(S)
      if Δ ≤ 0 /* (downhill move) */ then
        Set S = S'
      else
        /* (uphill move) */ Set S = S' with probability $exp(-\frac{\Delta}{kT})$
      end if
    until L times
    Set T = rT (reduce temperature)
  end while
  Return S
```

Next we describe the straightforward scheme of adapting simulated annealing metaheuristic to graph bisection problem.

A solution/configuration would be any bipartition $\{V_1, V_2\}$ of the vertex set (not necessarily a vertex bisection). Neighborhood is defined as follows: two bipartitions are neighbors of each other if one can be obtained from the other through transfer of a single vertex from one of the parts of to the other (rather than by exchanging two vertices). Finally the cost of a bipartition $\{V_1, V_2\}$ is defined to be the size of cut edges cut across the bipartition and α times the square of the difference in the sizes of V_1 and V_2. The parameter α is called the imbalance factor. We may note that this scheme does allow for unbalanced partitions as solutions, however they are penalized according to the square of the imbalance. Such a penalty function approach is common and effective in adaptations of simulated annealing. This could be attributed to the intuition that the extra solutions that get facilitated provide new escapes out of local optima. Furthermore for our graph partitioning problem the penalty based relaxation approach working with possibly unbalanced bipartitions has another benefit. Neighborhoods are smaller (n neighbors vs. $n^2/4$ as would be the case if we were to explore only balanced bipartitions through exchange of a pair of vertices). The experiments reported by Johnson et al. [73] indicate that, under usual cooling rates such as $r = 0.95$, temperature lengths that are significantly smaller than the neighborhood size tend to give poor results. Therefore smaller neighborhood size may imply shorter execution time.

Several results relating to convergence of simulated annealing under suitable cooling schedules are available, for example in Gelfand and Mitter [74,75] or Hajek [76]. Essentially, convergence is guaranteed by a cooling schedule where the temperature T_k is decreased as $\Gamma/\log k$, for sufficiently large Γ. However due to practical computational time constraints, the cooling is done much faster than this suggestion.

30.7 SPECIFIC VARIATIONS OF GRAPH/HYPERGRAPH PARTITIONING

In this section we illustrate how specific variants of the generic graph/hypergraph partitioning algorithms are treated. Such variants result from technological and other practical constraints. A large volume of research has been done for such problems [1,3]. We discuss two such problems and proposed approaches for them.

30.7.1 Efficient Network Flow-Based Minimum Cost Balanced Partitioning

30.7.1.1 Modeling a Net in a Flow-Network

A *logic netlist* $N(V, S)$ has nodes V which represent *cells* (e.g., gates and registers), and the *nets* S which represent *signal* connections. More formally, a (directed) net $n = (v : v_1, v_2 \ldots v_l)$ consists of a driving cell v and the driven cells $\{v_1, v_2, \ldots, v_l\}$. In Figure 30.6, net a is $\{p_1 : q_1, q_2\}$, driven by cell p_1 and driving q_1 and q_2. In our subsequent description of essence of the flow-based bisection approach of Yang and Wong [77], we shall always assume that the cuts separate specified source and sink cells.

The $netcut(X, \overline{X})$ of the bipartition $\{X, \overline{X}\}$ (here $\overline{X}$ denotes complement of the subset X of cells) is the set of those nets incident on nodes (cells) both in X and in $\overline{X}$. A $netcut(X, \overline{X})$ is a minimum netcut if $|netcut(X, \overline{X})|$ is minimum. In Figure 30.6, (the nets driven by $p_1, p_2, q_1, q_2,$ and q_3 are denoted by $a, b, c, d,$ and e, respectively), we see that $netcut(X, \overline{X}) = \{b, e\}$, $netcut(Y, \overline{Y}) = \{c, a, b, e\}$. $(X, \overline{X})$ in this case happens to be a minimum netcut.

Yang and Wong [77] described how the problem of finding a minimum netcut in a given netlist N can be transformed to the problem of finding a cut of minimum capacity in a suitably defined capacitated flow-network.

One constructs the flow network $FN = (V', A)$ from a given netlist $N(V, S)$ as follows: V' contains all nodes in V, along with two special nodes called *split* nodes n_1 and n_2 for each net $n \in S$. The set of arcs of FN is formed by the *infinite capacity* arcs (u, n_1) and (n_2, u) for each cell u incident on the net n. In addition, for each net $n \in S$, *unit capacity bridging* arc (n_1, n_2) is added to FN. Two distinguished nodes s, t respectively denote the specified source and sink of FN. The node weights in FN are same as the weights of the corresponding nodes in N. However for the *split* nodes of FN, which represent the nets of N, zero weights are assigned.

It is easy to see that $|V'| < 3|V|$ and $|A| < 2|S| + 3|V|$. Also note that the directions of the nets (drivers and driven cells) are not considered in the construction of FN. That is, one may as well think of signal nets as ordinary undirected *hyperedges*.

We will now argue that the problem of finding a minimum netcut in N can be reduced to the problem of finding a cut with minimum capacity in FN.

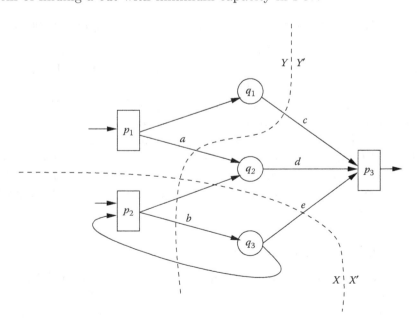

Figure 30.6 Logic circuit and its netcuts.

Lemma 30.2

i. *For every subset X of cells of N (ie. $X \subseteq V$), FN has a $cut(X', \overline{X'})$ of capacity $|netcut(X, \overline{X})|$. In fact $X' = X \bigcup delXN_1$, where $delXN_1$ is the collection of first of split nodes, n_1, for the nets $n \in netcut(X, \overline{X})$.*

ii. *For any cut $(X', \overline{X'})$ in FN of finite capacity C, then there exists $X \subseteq S$ for which $|netcut(X, \overline{X})| \leq C$.* ∎

Corollary 30.3 *Let $(X', \overline{X'})$ be a cut of minimum capacity C in FN, and let $X = X' \bigcap V$. Then $netcut(X, \overline{X})$ is a minimum netcut in N and $|netcut(X, \overline{X})| = C$.*

30.7.1.2 Optimal Minimum Netcut Algorithm

The above lemma and the corollary, cleary justify the following algorithm for finding a bipartition of a netlist $N = (V, S)$ w.r.t. two specified nodes (cells) s and t, that minimizes the number of crossing nets.

Algorithm MinNetCut
1. Construct the flow network $FN = (V', A)$ as illustrated earlier.
2. Find a maximum flow in FN from s to t.
3. Find a cut $(X', \overline{X'})$ of minimum capacity in FN
   as described in the maxflow-mincut theorem.
4. Set $X = X' \bigcap V$, Return $netcut(X, \overline{X})$ as a minimum netcut in N.

Using simple incremental Ford-Fulkerson style max-flow algorithm [31] we can guarantee the following.

Theorem 30.13 *Algorithm MinNetCut finds the minimum netcut in $O(|V||S|)$ time.*

30.7.1.3 Most r-Balanced Minimum Netcut Bipartition: An NP-Complete Problem

Minimum netcut bipartition may often yield unbalanced components. So one is naturally interested in finding a minimum netcut that is the most r-balanced among all minimum netcuts defined by a maximum flow, that is, among all possible minimum netcuts $(X, \overline{X})$ find one such that $|w(X) - r * W|$ is as close to 0 as possible, where W is the total weight. It is not difficult to show that this problem is also NP-complete.

Further note that this problem is different from the more familiar NP-complete minimum cost balanced bipartition. While the latter fixes the weights of the two partitions and tries to minimize the number of crossing nets, the first one requires that the number of crossing nets is minimum and tries to find such a (minimum netcut) partition $\{X, \overline{X}\}$ for which $w(X)$ is as close to rW as possible.

30.7.1.4 Minimum Cost Balanced Bipartition

Yang and Wong [77] proposed the heuristic flow-balanced-bipartition (FBB), for finding an r-balanced bipartition minimizing the number of crossing nets. Given a netlist $N = (V, S)$, FBB randomly picks a pair of nodes s and t in N, and then tries to find an r-balanced bipartition that separates s and t, minimizing the number of crossing nets. Let W be the total weight of the nodes in N. For the two blocks of the required bipartition, a deviation in weight from $(1 - \varepsilon)rW$ to $(1 + \varepsilon)rW$ is allowed.

> **Algorithm FBB**
> 1. Randomly pick a pair of nodes s and t in N.
>   Create the flow network FN for netlist N.
> **repeat**
> 2. Maximize flow in the flow network. Find mincut $\{\tilde{X}', \overline{\tilde{X}'}\}$ as follows.
>   Let $\tilde{X}'$ be the nodes reachable from s through the search for augmenting paths in the flow network, and $\overline{\tilde{X}'}$ be the rest of nodes.
>   Let X' denote the subset of nodes V' of original flow network FN, corresponding to $\tilde{X}'$.
>   Let $X = X' \cap V$, that is the cells of N inside X'.
> **if** $(1-\varepsilon)rW \leq w(X) \leq (1+\varepsilon)rW$ **then**
> 3.a. Stop and return X as the answer. **break**
> **end if**
> **if** $w(X) < (1-\varepsilon)rW$ **then**
> 4.a. Collapse all nodes in $\tilde{X}'$ to s.
> 4.b. Collapse to s a node $v \in \overline{\tilde{X}'}$ incident on a forward arc across $\{\tilde{X}', \overline{\tilde{X}'}\}$.
> **continue**
> **end if**
> **if** $w(X) > (1+\varepsilon)rW$ **then**
> 5.a. Collapse all nodes in $\overline{\tilde{X}'}$ to t.
> 5.b. Collapse to t a node $v \in \tilde{X}'$ incident on a forward arc across $\{\tilde{X}', \overline{\tilde{X}'}\}$.
> **continue**
> **end if**
> **until true**

30.7.1.5 Efficient Implementation

Instead of repeatedly applying Algorithm MinNetCut in step 2 of Algorithm FBB to compute a max-flow and a minimum netcut from scratch everytime, Yang and Wong [77] employ incremental flow computations to keep the total complexity almost of the same order as that of a single max-flow-min-cut computation. They utilize the maximum flow found in step 2 of Algorithm FBB as an initial flow in the modified (collapsed) flow network in the next iteration. The required bounds on the sizes of the max-flows and the number of collapsing steps are consequences of $|V'| < 3|V|$ and $|A| < 2|S| + 3|V|$.

30.7.2 Optimal Replication for Minimum k-Cut Partitioning

The idea of vertex replication can be used to reduce the size of a cut in an already partitioned graph. That is, we can improve upon the given partition in the sense of cost of the cut, however at the expense of more area (block sizes). This work is motivated by the increase in mapping large logic networks into multiple FPGAs. Often the number of pins on an FPGA is not large enough to permit high utilization of its gate capacity after partitioning. Vertex replication technique has the following potential benefits.

1. Replication can reduce the number of FPGAs required to implement a design.
2. It can reduce the number of wires interconnecting the FPGAs.
3. It can reduce the number of inter-chip wires along a path in a design, resulting in increased performance.

We describe below main ideas of [78,79] for tackling optimum vertex replication problem.

30.7.2.1 Vertex Replication

Let $G = (V, E)$ be a directed graph and $\Pi = \{V_1, V_2, \ldots, V_k\}$ be a partition of V into k disjoint subsets.

Vertex replication is a transformation on directed graphs. For any vertex $u \in V_i$, the replication of u into V_j yields the transformed graph $G' = (V', E')$, with $V' = V \cup u_j$. The transformed graph is obtained by adding an vertex u_j to V_j and E' is obtained starting with E performing the following modifications.

1. Every cut edge (u, v), where $v \in V_j$ is replaced by an edge (u_j, v).

2. $\forall$ edge (v, u), E' contains a new edge (v, u_j).

Subsequent replication of u_j into V_i and further replications of u into V_j are defined to be null operations.

We define
$$in(V_i) = |(u, v) \in E : u \notin V_i, v \in V_i| \tag{30.40}$$

that is, the number of directed edges entering V_i. As each cut edge is incident into exactly one component, we have

$$cutsize(\Pi) = \sum_i in(V_i) \tag{30.41}$$

30.7.2.2 Minimum k-Cut Replication Problem

Given a directed graph $G = (V, E)$, and a k-partition $\Pi = \{V_1, V_2 \ldots V_k\}$ of V, the problem is to determine a collection $\{V_{ij}^* : 1 \leq i, j \leq k\}$, that minimizes the cutsize(Π^*), where Π^* is the partition that results when V_{ij}^* is replicated from V_i to V_j for all i and j (Figure 30.7).

We present the essential ideas proposed for this problem by Hwang and El Gamal [78]. The core idea can be effectively illustrated using the case of $k = 2$, that is, replication for improving a bipartitioning.

Our objective is to improve a given bipartition (V_1, V_2) using replication technique. The following approach will also generalize to a more general optional replication problem in the settings of the k-partitions.

Suppose we need to choose $V_{12} \subset V_1$ (note the notation for proper subset rather than just a subset) and $V_{21} \subset V_2$ such that in the graph, that results after the replication of V_{12} in V_2 and V_{21} in V_1, the altered cut is minimum.

Note that the resulting graph differs from the original exactly in the removal of arcs belonging to the sets $out(V_{12}, V_2)$ and $out(V_{21}, V_1)$ from the original graphs (recall that

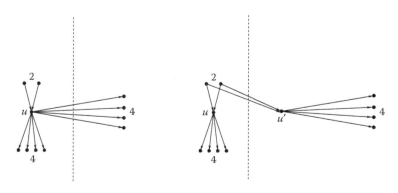

Figure 30.7 Replication reduces the cut size.

$out(X, Y)$ denotes the set of arcs starting at a node in X and terminating at a node Y). The resulting cut consists of the disjoint subsets of arcs, namely, those entering the block $V_2 \bigcup V_{12}$ (i.e. V_2 after replication of V_{12}) and those entering $V_1 \bigcup V_{21}$. The former can be seen to be identical to the subset $in(V_2 \bigcup V_{21})$ of original graph G and the arcs in the latter are in $1-1$ correspondence with those belonging to $in(V_1 \bigcup V_{21})$ of G. Then our task is to find $V_{12} \subset V_1$ and $V_{21} \in V_2$ which minimize $|in(V_2 \bigcup V_{12})| + |in(V_1 \bigcup V_{21})|$. Noting the separability, the above is equivalent to

$$\min_{V_{12} \subset V_1} |in(V_2 \bigcup V_{12})| + \min_{V_{21} \subset V_2} |in(V_1 \bigcup V_{21})|$$

(here the $in(.)$ function is defined for G). Therefore the problem decouples into independent problems. Let us analyze without loss of generality, just one of them, say $\min_{V_{12} \in V_1} |in(V_2 \bigcup V_{12})|$.

It is quite apparent that this problem can be solved by employing max-flow-min-cut technique [16,31]. Indeed the problem

$$\min_{V_{12} \subset V_1} |in(V_2 \bigcup V_{12})|$$

can be easily seen to be equivalent to

$$\min_{u \in V_1}[\text{ minimum capacity cut in flow network } FG_u]$$

where FG_u is as follows. The nodes of FG_u are those of G along with two new special nodes s and t (source and sink of the flow network). The arcs of FG_u consist of those of G each given unit capacity and the arc (s, u) as well as (v, t) for each $v \in V_2$, all with infinite capacity each.

The justification of the above equivalence rests on the simple observation that minimum cut in FG_u must be of finite capacity and therefore arcs crossing from the s-side to t-side. As a result, any mincut of FG_u must contain u on the source-side and whole of V_2 on the sink side. Hence the capacity of such a mincut must be $|in(V' \bigcup V_2)|$ for some $V' \subset V_1$, $u \notin V'$. Therefore repeated applications of max-flow-min-cut algorithms would solve the optimal replication problem for the case of 2-partitions. In fact, this approach readily generalizes to the setting of k-partitions. Details may be found in [78].

References

[1] C.J. Alpert and A.B. Kahng, Recent Directions in Netlist Partitioning: A Survey, *Integration VLSI J.*, **19**(1–2) (1995), 1–81.

[2] L. Behjat, D. Kucar and A. Vannelli, A Novel Eigenvector Technique for Large Scale Combinatorial Problems in VLSI Layout, *J. Comb. Optim.* **6**(3) (2002), 271–286.

[3] S-J. Chen and C-K. Cheng, Tutorial on VLSI Partitioning, *VLSI Design*, **11**(3) (2000), 175–218.

[4] T. Lengauer, *Combinatorial Algorithms for Integrated Circuit Layout*, John Wiley & Sons, New York, 1990.

[5] B. Hendrickson and T.G. Kolda, Graph Partitioning Models for Parallel Computing, *Parallel Comput.*, **26** (2000), 1519–1534.

[6] B. Hendrickson and R. Leland, An Improved Graph Partitioning Algorithm for Mapping Parallel Computations, *SIAM J. Sci. Comput.*, **16**(2) (1995), 452–469.

[7] G. Karypis and V. Kumar, Multilevel k-way Hypergraph Partitioning, *Proc. Design Automat. Conf.*, (1999), 343–348.

[8] P.-O. Fjallstrom, Algorithms for Graph Partitioning: A Survey, *Linkoping Elect. Art. Comput. Inform. Sci.*, **3**(10) (1998).

[9] D. Kucar, S. Areibi and A. Vannelli, Hypergraph Partitioning Techniques, *Special Issue of Dynam. Contin., Discrete Impulsive Syst. J.*, **11**(2–3) (2004), 341–369.

[10] D.A. Papa and I.L. Markov, Hypergraph Partitioning and Clustering, in *Approximation Algorithms and Metaheuristics*, T. Gonzalez, ed., CRC Press, 2007, pp. 61-1–61-19.

[11] A. Pothen, Graph Partitioning Algorithms with Applications to Scientific Computing, *Parallel Numerical Algorithms*, Kluwer Academic Press, 1997, pp. 323–368.

[12] J. Edmonds, Submodular Functions, Matroids and Certain Polyhedra, *Proc. Calgary Intl. Conf. Combinatorial Structures*, (1970), 69–87.

[13] S. Fujishige, *Submodular Functions and Optimization, Annals of Discrete Mathematics*, North Holland, Amsterdam, the Netherlands, 1991.

[14] H. Narayanan, *Submodular Functions and Electrical Networks, Annals of Discrete Mathematics-54*, North Holland, 1997.

[15] A. Schrijver, *Combinatorial Optimization: Polyhedra and Efficiency*, Springer, 2003.

[16] E.L. Lawler, *Combinatorial Optimization: Networks and Matroids*, Holt, Rinehart & Winston, New York, 1976.

[17] C.H. Papadimitriou and K. Steiglitz, *Combinatorial Optimization: Algorithms and Complexity*, Prentice Hall, Upper Saddle River, NJ, 1982.

[18] G.H. Golub and C.F. van Loan, *Matrix Computations*, Johns Hopkins Studies in Mathematical Sciences, 1996.

[19] F. Chung, *Spectral Graph Theory*, vol. 92, in CBMS Regional Conference Series in Mathematics, American Mathematical Society, 1997.

[20] M.R. Garey and D.S. Johnson, *Computers and Intractability: A Guide to the Theory of NP-Completeness*, W. H. Freeman & Co., New York, 1979.

[21] M.R. Garey, D.S. Johnson and L. Stockmeyer, Some Simplified NP-complete Problems, *Proc. ACM Symp. Theor. Comput.*, 1974, pp. 47–63.

[22] R.K. Ahuja, T.L. Magnanti and J.B. Orlin, *Network Flows: Theory, Algorithms, and Applications*, Prentice Hall, Englewood Cliffs, NJ, 1993.

[23] A.V. Goldberg and R.E. Tarjan, A New Approach to the Maximum Flow Problem, *J. Assoc. Comput. Mach.*, **35**(4) (1988), 921–940.

[24] H. Nagamochi and T. Ibaraki, Computing Edge-connectivity in Multigraphs and Capacitated Graphs, *SIAM J. Discrete Math.*, **5**(1) (1992), 54–66.

[25] D.R. Karger, Global Min-cuts in RNC, and Other Ramifications of a Simple Min-out Algorithm, *Proc. 4th Ann. ACM-SIAM Symp. Discrete Algorithms*, 1993, pp. 21–30.

[26] D.R. Karger and C. Stein, A New Approach to the Minimum Cut Problem, *J. ACM*, **43**(4) (1996), 601–640.

[27] F.T. Leighton and S. Rao, Multicommodity Max-Flow Min-Cut Theorems and Their Use in Designing Approximation Algorithms, *J. Assoc. Comput. Mach.*, **46**(6) (1999), 787–832.

[28] C.-K. Cheng and T.C. Hu, Maximum Concurrent Flows and Minimum Cuts, *Algorithmica*, **8**(3) (1992), 233–249.

[29] R. Gomory and T.C. Hu, Multi-Terminal Network Flows, *J. SIAM*, **9** (1961), 551–570.

[30] O. Goldschmidt and D. Hochbaum, Polynomial Algorithm for the k-Cut Problem, *29th Ann. Symp. Found. Comput. Sci.*, 1988, pp. 444–451.

[31] L.R. Ford, Jr. and D.R. Fulkerson, *Flows in Networks*, Princeton University Press, Princeton, NJ, 1962.

[32] M. Stoer and F. Wagner, A Simple Min-Cut Algorithm, *J. ACM*, **44**(4) (1997), 585–591.

[33] M. Queyranne, A Combinatorial Algorithm for Minimizing Symmetric Submodular Functions, *Proc. 6th Ann. ACM-SIAM Symp. Discrete Algorithms*, 1995, pp. 98–101.

[34] W.K. Mak and D.F. Wong, A Fast Hypergraph Min-Cut Algorithm for Circuit Partitioning. *Integration, VLSI J.*, **30**(1) (2000), 1–11.

[35] H. Saran and V.V. Vazirani, Finding k-Cuts within Twice the Optimal, *SIAM J. Comput.*, **24**(1) (1995), 101–108.

[36] N. Garg, V.V. Vazirani and M. Yannakakis, Approximate Max-Flow Min-(Multi)Cut Theorems and Their Applications, *SIAM J. Comput.*, **25**(2) (1996), 235–251.

[37] Y. Aumann and Y. Rabani, An $O(\log k)$ Approximate Min-Cut Max-Flow Theorem and Approximation Algorithm, *SIAM J. Comput.*, **27** (1998), 291–301.

[38] V.V. Vazirani, *Approximation Algorithms*, Springer, 2004.

[39] Y. Asahiro, K. Iwama and H. Tamaki, Greedily Finding a Dense Subgraph, *J. Alg.*, **34** (2000), 203–221.

[40] D.J.-H. Huang and A.B. Kahng, When Clusters Meet Partitions: New Density-Based Methods for Circuit Decomposition, *Proc. European Design Test Conf.*, 1995, pp. 60–64.

[41] J.C. Picard and M. Queyeranne, Selected Applications of Minimum Cuts in Networks, *INFOR-Canada J. Oper. Res. Inform. Process.*, **20**(4) (1982), 394–422.

[42] H. Narayanan, S. Roy and S.B. Patkar, Approximation Algorithms for Min-k-Overlap Problems, Using the Principal Lattice of Partitions Approach, *J. Algorithms*, **21**(2) (1996), 306–330.

[43] S.B. Patkar and H. Narayanan, Improving Graph Partitions Using Submodular Functions, *Discrete Appl. Math.*, **131**(2) (2003), 535–553.

[44] R. Andersen and K. Lang, An Algorithm for Improving Graph Partitions, *Proc. 19th ACM-SIAM Symp. SODA*, 2008, pp. 651–660.

[45] M. Narsimhan and J. Bilmes, Local Search for Balanced Submodular Clustering, *IJCAI*, (2007), 981–986.

[46] C.J. Alpert, S-Z. Yao, and A.B.Kahng, Spectral Partitioning with Multiple Eigenvectors, *Discrete Appl. Math.*, **90** (1999), 3–26.

[47] E. Cheng and W. Cunningham, A Faster Algorithm for Computing the Strength of a Network, *Inform. Process. Lett.*, **49**(4) (1994), 209–212.

[48] H.N. Gabow, Algorithms for Graphic Polymatroids and Parametric S-Sets, *J. Algorithms*, **26** (1998), 48–86.

[49] Y.C. Wei and C.K. Cheng, Towards Efficient Hierarchical Designs by Ratio Cut Partitioning, *Proc. Int. Conf. Comput.-Aided Design*, 1989, pp. 298–301.

[50] D. Gusfield, Computing the Strength of a Graph, *SIAM J. Comput.*, **20** (1991), 639–654.

[51] S.B. Patkar and H. Narayanan, Fast On-Line/Off-Line Algorithms for Optimal Reinforcement of a Network and Its Connections with Principal Partition, *Proc. 20th Ann. Found. Softw. Tech. Theoretical Comput. Sci., LNCS-1974*, Springer, 2000, pp. 94–105.

[52] W. Dinkelbach, On Nonlinear Fractional Programming, *Manage. Sci.*, **13** (1967), 492–498.

[53] B. Kernighan and S. Lin, An Effective Heuristic Procedure for Partitioning Graphs, *Bell Syst. Tech. J.*, (1970), 291–308.

[54] C. M. Fiduccia and R. M. Mattheyses, A Linear-Time Heuristic for Improving Network Partitions, *Proc. 19th Conf. Design Automation*, 1982, pp. 175–181.

[55] Y. Saab, An Effective Multilevel Algorithm for Bisecting Graphs and Hypergraphs, *IEEE Trans. Comput.*, **53**(6) (2004), 641–652.

[56] S.T. Barnard and H.D. Simon, Fast Multilevel Implementation of Recursive Spectral Bisection for Partitioning Unstructured Problems, *Concurrency—Practice and Experience*, **6**(2) (1994), 101–117.

[57] J. Li and L. Behjat, A Connectivity Based Clustering Algorithm with Application to VLSI Circuit Partitioning, *IEEE Trans. Circ. Syst.-II: Express Briefs* **53**(5) (2006), 384–388.

[58] M. Hall, An r-Dimensional Quadratic Placement Algorithm, *Manage. Sci.*, **17** (1970), 219–229.

[59] B. Hendrickson, R. Leland and R. van Driessche, Enhancing Data Locality Using Terminal Propagation, *Proc. 29th Ann. Hawaii Int. Conf. Syst. Sci.*, 1996, pp. 565–574.

[60] W.E. Donath and A.J. Hoffman, Lower Bounds for the Partitioning of Graphs, *IBM J. Res. Dev.* **17** (1973), 420–425.

[61] F. Rendl and H. Wolkowicz, A Projection Technique for Partitioning the Nodes of a Graph, *Ann. Oper. Res.* **58** (1995), 155–179.

[62] E. R. Barnes, An Algorithm for Partitioning the Nodes of a Graph, *Siam J. Alg. Discrete Methods*, **3**(4) (1982), 541–550.

[63] L. Hagen and A.B. Kahng New Spectral Methods for Ratio Cut Partitioning and Clustering, *IEEE Trans. CAD*, **11**(9) (1992), 1074–1085.

[64] P.K. Chan, M.D.F. Schlag and J.Y. Zien, Spectral K-Way Ratio-Cut Partitioning and Clustering, *IEEE Trans. CAD*, **13**(9) (1994), 1088–1096.

[65] A. Pothen, H.D. Simon and K.-P. Liou, Partitioning Sparse Matrices with Eigenvectors of Graphs, *SIAM J. Matrix Anal. Appl.*, **11**(3) (July 1990), 430–452.

[66] P.K. Chan, M.D.F. Schlag and J.Y. Zien, Spectral-Based Multiway FPGA Partitioning, *IEEE Trans. Comput.-Aided Design Integrat. Circ. Syst.*, **15**(5) (1996), 554–560.

[67] C.J. Alpert and A.B. Kahng, Simple Eigenvector-Based Circuit Clustering Can Be Effective, *IEEE Int. Symp. on Circ. Syst.*, **IV** (1996), 683–686

[68] J. Frankle and R.M. Karp, Circuit Placements and Cost Bounds by Eigenvector Decomposition, *IEEE Conf. Comput. Aided Design*, 1986, pp. 414–417.

[69] R.H.J.M. Otten, Eigensolutions in Top-Down Layout Design, *IEEE Symp. Circ. Syst.*, (1982), 1017–1020.

[70] B. Schiffner, J. Li and L. Behjat, A Multilevel Eigenvalue Based Circuit Partitioning Technique, *IWSOC*, 2005, pp. 312–316.

[71] N. Metropolis, A. Rosenbluth, M. Rosenbluth, A. Teller and E. Teller, Equation of State Calculations by Fast Computing Machines, *J. Chem. Phys.*, **21**(6) (1953), 1087–1092.

[72] E. Aarts and J. Korst, *Simulated Annealing and the Boltzmann Machine*, John Wiley & Sons, New York, 1990.

[73] D.S. Johnson, C.R. Aragon, L.A. McGeoch and C. Schevon, Optimization by Simulated Annearling: An Experimental Evaluation, Part I, Graph Partitioning, *Oper. Res.*, **37** (1989), 865–892.

[74] S.B. Gelfand and S.K. Mitter, Analysis of Simulated Annealing for Optimization, *Proc. 24th IEEE Conf. Decision Contr.*, **2** (1985), 779–786.

[75] S.B. Gelfand and S.K. Mitter, Simulated Annealing with Noisy or Imprecise Measurements, *J. Optim. Theor. Appl.*, **69** (1989), 49–62.

[76] B. Hajek, Cooling Schedules for Optimal Annealing. *Math. Oper. Res.*, **13** (1988), 311–329.

[77] H. Yang and D.F. Wong, Efficient Network Flow Based Min-Cut Balanced Partitioning, *IEEE Trans. CAD Integr. Circ. Syst.*, **15**(12) (1996), 1533–1540.

[78] J. Hwang and A. El Gamal, Optimal Replication for Min-Cut Partitioning, *Proc. IEEE Int. Conf. Comput. Aided Design*, 1992, pp. 432–435.

[79] J. Hwang and A. El Gamal, Min-Cut Replication in Partitioned Networks, *IEEE Trans. Comput.-Aided Design*, **14** (January 1995), 96–106.

IX
Matroids

CHAPTER 31

Matroids

H. Narayanan

Sachin B. Patkar

CONTENTS

31.1	Introduction	880
31.2	Axiom systems for Matroids	880
	31.2.1 Independence and Base Axioms	880
	31.2.2 Examples of Matroids	881
	31.2.3 Base Axioms	882
	31.2.4 Rank Axioms	884
	31.2.5 Circuit Axioms	886
31.3	Dual Matroid	887
31.4	Minors of Graphs, Vector Spaces, and Matroids	889
	31.4.1 Restriction and Contraction of Graphs	889
	31.4.2 Restriction and Contraction of Vector Spaces	890
	31.4.3 Minors of Dual Vector Spaces	891
	31.4.4 Representative Matrices of Minors of Vector Spaces	892
	31.4.5 Minors of Matroids	893
	31.4.6 Notes	896
31.5	Convolution	896
	31.5.1 Introduction	896
	31.5.2 Polymatroid Rank Functions	897
	31.5.3 Formal Properties of the Convolution Operation	897
	31.5.4 Connectedness for $f*g$	899
31.6	Principal Partition	901
	31.6.1 Introduction	901
	31.6.2 Basic Properties of Principal Partition	901
	31.6.3 Principal Partition of Contraction and Restriction	904
	31.6.4 Principal Partition of the Dual	905
	31.6.5 Principal Partition and the Density of Sets	905
	31.6.6 Outline of Algorithm for Principal Partition	906
	31.6.7 Notes	906
31.7	Matroid Union	907
	31.7.1 Matroid Union Algorithm	907
	31.7.2 Complexity of the Matroid Union Algorithm	910
	31.7.3 Matroid Union Theorem	911
	31.7.4 Fundamental Circuits and Coloops of $\mathcal{M}_1 \vee \mathcal{M}_2$	912
	31.7.5 Union of Matroids and the Union of Dual Matroids	912
	31.7.6 Matroid Union and Matroid Intersection	914

	31.7.7	Applications of Matroid Union and Matroid Intersection	915
		31.7.7.1 Representability of Matroids	915
		31.7.7.2 Decomposition of a Graph into Minimum Number of Subforests	915
		31.7.7.3 Rado's Theorem	916
	31.7.8	Algorithm for Construction of the Principal Sequence of a Matroid Rank Function	916
	31.7.9	Example	918
		31.7.9.1 Principal Sequence of $(r(\cdot), |\cdot|)$ Where $r(\cdot)$ Is the Rank Function of a Graph	918

31.1 INTRODUCTION

Matroids are important combinatorial structures both from the point of view of theory and from that of applications. Whitney [1] introduced matroids as a generalization of the concept of linear independence in the context of matrices. The idea was arrived at independently also by Van der Waerden in [2]. Matroid theory is one of the areas that straddles across several branches of discrete mathematics such as combinatorics, graph theory, finite fields, algebra, and coding theory. One of the subjects to which applications were found early was electrical network theory [3]. In this chapter we give a brief sketch of the theory with electrical networks in mind.

31.2 AXIOM SYSTEMS FOR MATROIDS

A *matroid* can be defined in several equivalent ways. Each of these is based on an axiom system. The primitive objects of each axiom system can be identified with either the primitive or some derived objects of every other axiom system. We restrict ourselves to finite underlying sets even though it is possible to define infinite matroids. The concept of maximality and minimality (with respect to set inclusion) is often used in matroid theory.

We may note that, in general, maximal and minimal members of a collection of sets may not be largest or smallest in terms of size.

Example: Consider the collection of sets $\{\{a,b,c\},\{g\},\{e,f\},\{a,b,c,d,e,f\}\}$. The minimal members of this collection are $\{a,b,c\},\{g\},\{e,f\}$, that is, these do not contain proper subsets which are members of this collection. The maximal members of this collection are $\{g\},\{a,b,c,d,e,f\}$, that is, these are not proper subsets of other sets which are members of this collection.

However, in the key concept of collection of independent sets in matroids, maximality implies the maximum size property.

31.2.1 Independence and Base Axioms

Independent sets of a matroid correspond to subforests (or, dually, subcoforests) of graphs and to independent sets of columns of matrices.

I. Independence Axioms

Definition 31.1 (Matroid: Independence Axioms) *A matroid $\mathcal{M}$ on S is a pair $(S, \mathcal{I})$, where S is a finite set and $\mathcal{I}$ is a family of subsets of S, always containing the empty subset, called* independent sets, *that satisfies the following:*

I1 *if $I_1 \in \mathcal{I}$ and $I_2 \subseteq I_1$, then $I_2 \in \mathcal{I}$.*

I2 *maximally independent sets contained in a subset of S have the same cardinality. (equivalently axiom I2 can be replaced by the following)*

I2' *if X, X' are independent and $|X'| > |X|$, then there exists $e' \in (X' - X)$ such that $X \cup e'$ is independent.*

A *base* of the matroid $\mathcal{M} \equiv (S, \mathcal{I})$, is a maximally independent subset of $\mathcal{M}$ contained in S. The complement, relative to S, of a base is called a *cobase* of $\mathcal{M}$.

It follows immediately from the above that all bases of a matroid have the same cardinality. Indeed, as we shall soon see, matroids can also be defined using an axiom system that describes the collection of bases.

31.2.2 Examples of Matroids

Theorem 31.1 (Polygon and bond matroids) *Let $\mathcal{G}(V, E)$ be an undirected graph. Let $\mathcal{I}_f$ be the collection of subforests and let $\mathcal{I}_c$ be the collection of subcoforests of $\mathcal{G}$ (denoting by forest, a spanning forest of the graph). Then $(E, \mathcal{I}_f), (E, \mathcal{I}_c)$ are matroids. Further, the bases of either matroid are cobases of the other.*

Proof. We will only prove that the collection $\mathcal{I}_f$ satisfies independence axioms. The proofs of the statements for $\mathcal{I}_c$ will follow directly from the definition of dual matroid that we will encounter later. That the independence axioms are satisfied by $\mathcal{I}_f$, follows from the facts given below:

- Maximal intersection of a forest of $\mathcal{G}$ with $T, T \subseteq E$, is a forest of the subgraph of $\mathcal{G}$ on T.
- All forests of the subgraph of $\mathcal{G}$ on T have the same cardinality. ∎

The matroid $(E, \mathcal{I}_f)$ is called the *polygon matroid* of $\mathcal{G}$, and is denoted by $\mathcal{M}(\mathcal{G})$. The matroid $(E, \mathcal{I}_c)$ is called the *bond matroid* of $\mathcal{G}$, and is denoted by $\mathcal{M}^*(\mathcal{G})$.

Let $\mathcal{V}$ be a vector space over the field of real numbers with components indexed by the elements of S. Let $\mathbf{R}, \mathbf{R}^*$ be representative matrices (i.e., their rows constitute bases) of $\mathcal{V}, \mathcal{V}^\perp$, respectively. Since the column dependence structure of all representative matrices of a vector space is the same (due to non-singularity of elementary row operations), it makes sense to consider an arbitrary representative matrix to study linear independence of columns. Let $\mathcal{I}$ be the collection of independent column sets of $\mathbf{R}$ (identified with corresponding subsets of S) and let $\mathcal{I}^*$ be the collection of independent column sets of $\mathbf{R}^*$.

Theorem 31.2 (Vectorial matroid) *Let $(S, \mathcal{I}), (S, \mathcal{I}^*)$ be matroids. Further, bases of each matroid are cobases of the other.*

Proof. $(S, \mathcal{I})$ satisfies independent axioms, which from the fact that maximally independent subsets of columns of any submatrix of $\mathbf{R}$ ($\mathbf{R}^*$) have the same cardinality. Further if $(\mathbf{I}:\mathbf{K})$ is a standard representative matrix of $\mathcal{V}$ then $(-\mathbf{K}^T : \mathbf{I})$ is a standard representative matrix of $\mathcal{V}^\perp$. Since there is a standard representative matrix corresponding to each maximally independent subset of columns, we conclude that bases of either of $(S, \mathcal{I}), (S, \mathcal{I}^*)$ are cobases of the other. ∎

We say that $(S, \mathcal{I})$ $((S, \mathcal{I}^*))$ is the *matroid (dual matroid) associated with* $\mathcal{V}$ and denote it by $\mathcal{M}(\mathcal{V})$ $(\mathcal{M}^*(\mathcal{V}))$.

Theorem 31.3 (Matroid union) [4,5] *Let $\mathcal{G}(V, E)$ be an undirected graph and let k be a positive integer. Let $\mathcal{I}_{k\cup}$ be the collection of unions of k forests (not necessarily edge-disjoint) of $\mathcal{G}$. Then $(E, \mathcal{I}_{k\cup})$ is a matroid.*

Proof. Let $\mathcal{I}_{12}$ be the collection of all sets of the form $X \cup Y$, where X_1, X_2 are independent respectively in the matroids $\mathcal{M}_1, \mathcal{M}_2$. We prove that $\mathcal{M}_{12} \equiv (S, \mathcal{I}_{12})$ is a matroid. The hereditary axiom (Axiom I1) is clearly satisfied. Therefore it suffices to prove Axiom $I2'$ to conclude that $\mathcal{M}_{12} \equiv (S, \mathcal{I}_{12})$ is a matroid. (This result [as Theorem 31.38] is discussed in detail later. The present proof follows the one in [6].) The theorem would then follow immediately since by Theorem 31.1 the collection of forests of a graph constitute the independent sets of a matroid. Let $Z_a = X_1 \cup X_2$ and $Z_b = Y_1 \cup Y_2$ with X_1, Y_1 independent in $\mathcal{M}_1$ and X_2 and Y_2 independent in $\mathcal{M}_2$. Further let the division of Z_a into $X_1 \cup X_2$ be such that the *cross sum* $|X_1 \cap Y_2| + |X_2 \cap Y_1|$ is a minimum among all such divisions. Now if $|Z_b| > |Z_a|$, either $|Y_1| > |X_1|$ or $|Y_2| > |X_2|$, let $|Y_2| > |X_2|$. Then there exists $e \in (Y_2 - X_2)$ such that $e \cup X_2$ is independent in $\mathcal{M}_2$. If $e \in X_1$ then the division of Z_a as $(X_1 - e) \cup (X_2 \cup e)$ would yield a lower cross sum than the division $X_1 \cup X_2$, a contradiction. Hence $e \notin X_1$ and therefore $e \in (Z_b - Z_a)$. Thus we have $e \cup Z_a = X_1 \cup (X_2 \cup e)$ is in the collection $\mathcal{I}_{12}$ confirming Axiom $I2'$. ∎

Let $\mathcal{G}(V, E)$ be an undirected graph. A matching of $\mathcal{G}$ is a subset of edges no two of which have a common endpoint. We say that a set of vertices U and a matching X *meet iff* the set of the endpoints of the edges in X include all the vertices of U. Let $\mathcal{I}_m$ be the collection of subsets of those vertices which meet a matching.

Theorem 31.4 [7] (**Matching matroid**) $(V, \mathcal{I}_m)$ *is a matroid.*

Proof. Let $T \subseteq V(\mathcal{G})$ and let I_1, I_2 be two maximal subsets of T in the collection $\mathcal{I}_m$. We will show that $|I_2| = |I_1|$. There exist matchings M_1, M_2 which meet I_1, I_2. Consider the subgraph of $\mathcal{G}$ on $M_1 \cup M_2$ (the vertex set of this subgraph may contain vertices outside T). Each component of this subgraph is either a circuit or a path. If a node of T has degree two in the above subgraph, then it must belong to both I_1 and I_2 (using the fact that both these are maximal subsets of T in the collection $\mathcal{I}_m$). So vertices in T, which are in components as circuits, are in $I_1 \cap I_2$. Now let $|I_2| > |I_1|$. Then in one of the components, that is a path, the subset of nodes of I_2 is of larger size than the subset of nodes of I_1 in the same component. Once again, the *middle* T nodes, in such components, must belong to $I_1 \cap I_2$. This means one of the end nodes, say v, of the path is in $I_2 - I_1$ and the other end node is not in $I_1 - I_2$. If we modify M_1 by dropping the edges of M_1 in this path and adding the edges of M_2, the new matching M_1' will meet T in $I_1 \cup v$. Hence, $I_1 \cup v \in \mathcal{I}_m$, a contradiction. Therefore $|I_1| = |I_2|$. ∎

31.2.3 Base Axioms

In this subsection, we characterize matroids through an axiom system for *bases* (i.e., *maximally independent sets*). We also consider an abstraction of the concept of a circuit. Note that a circuit of a graph is not contained in any forest and is the minimal such subset of edges. This motivates us to define a *circuit* of a matroid $(S, \mathcal{I})$ to be a minimal subset of S not contained in any independent set, equivalently, we could say that a circuit is a *minimal dependent (or non-independent)* subset of the matroid.

We will now arrive at a characterizing property of bases of a matroid. Let B_1, B_2 be two bases of the matroid $\mathcal{M} \equiv (S, \mathcal{I})$. Let $e \in B_2 - B_1$. Therefore $B_1 \cup e$, not being independent, contains a circuit (a minimal dependent set). We will prove uniqueness of this circuit. By contrast, let C_1, C_2 be two circuits contained in $B_1 \cup e$. Clearly $e \in C_1 \cap C_2$ and $\{e\}$ is not a circuit as e is an element inside a base. A maximally independent subset of $C_1 \cup C_2$ containing e, cannot have the cardinality exceed $|C_1 \cup C_2| - 2$. On the other hand $C_1 \cup C_2 - e$ is stated to be independent. This violates the Axiom I2 for $C_1 \cup C_2$. Therefore

$C_1 = C_2$. (The unique circuit contained in $e \cup B_1$ is referred to as the *fundamental circuit (f-circuit) of e with respect to B_1* and denoted by $L(e, B_1)$.) Now $L(e, B_1)$ has a nonempty intersection with $B_1 - B_2$ since $\{e\}$ itself is not a circuit and since B_2 is independent, $L(e, B_1)$ cannot be contained in it. Let $e' \in L(e, B_1) \cap (B_1 - B_2)$. Let $B_1' \equiv e \cup (B_1 - e')$. Clearly B_1' is also a base as it contains no circuit and has the same cardinality as B_1. On the other hand for any $e'' \in B_1$, $(B_1 - e'') \cup e$ is independent provided $e'' \in L(e, B_1)$. We therefore have the following theorem.

Theorem 31.5 *Let B_1, B_2 be bases of a matroid $\mathcal{M}$ on S. Let $e \in B_2 - B_1$. Then,*

1. *$e \cup B_1$ contains a unique circuit $L(e, B_1)$. This circuit has a nonempty intersection with $B_1 - B_2$.*
2. *If $e' \in B_1$, then $(B_1 - e') \cup e$ is a base of $\mathcal{M}$ iff $e' \in L(e, B_1)$.* ∎

This consequence of independent axioms for collection of bases would, as we shall soon see, indeed characterize the collection of bases of a matroid. In fact, the following dual observation also leads to an alternative axiomatic characterization for the collection of bases.

Let B_1, B_2 be bases of a matroid and let $e_1 \in B_1 - B_2$. B_2 is a maximally independent subset of $B_1 \cup B_2 - e_1$. Further as $|B_1| = |B_2|$, $B_1 - e_1$ is not maximally independent in this set. Hence, by Axiom $I2'$, there exists $e_2 \in B_2 - B_1$ so that $(B_1 - e_1) \cup e_2$ is independent and, therefore, a base of $\mathcal{M}$. We therefore have the following.

Theorem 31.6 *Let B_1, B_2 be bases of $\mathcal{M}$. and let $e_1 \in B_1 - B_2$. Then there exists $e_2 \in B_2 - B_1$, such that $(B_1 - e_1) \cup e_2$ is a base of $\mathcal{M}$.*

As remarked earlier, Theorems 31.5 and 31.6 can be used to generate axiom systems for matroids.

Theorem 31.7 (Base axioms) *Let a collection $\mathcal{B}$ of subsets of S satisfy the following equivalent axioms*

> **Axiom B** *If B_1, B_2 are members of $\mathcal{B}$ and if $e_2 \in B_2 - B_1$, then there exists $e_1 \in B_1 - B_2$ so that $(B_1 - e_1) \cup e_2$ is a member of $\mathcal{B}$.*
> **Axiom B'** *If B_1, B_2 are in $\mathcal{B}$ and if $e_1 \in B_1 - B_2$ then there exists $e_2 \in B_2 - B_1$ so that $(B_1 - e_1) \cup e_2$ is in $\mathcal{B}$.*

Then the collection $\mathcal{I}$ of subsets of the sets in $\mathcal{B}$ is a collection of independent subsets of a matroid $(S, \mathcal{I})$. (In other words $\mathcal{B}$ is the collection of bases of a matroid.)

Proof. When Axiom B is satisfied, to prove that $(S, \mathcal{I})$ is a matroid, we only need to prove that maximal subsets in $\mathcal{I}$ contained in a given subset T (denoted say $\mathcal{B}_T$) of S have the same cardinality. If $\mathcal{B}_{T1}, \mathcal{B}_{T2}$ are two such subsets we take an element $e_2 \in \mathcal{B}_{T2} - \mathcal{B}_{T1}$ and add it to $\mathcal{B}_{T1}$ and drop a suitable element in $\mathcal{B}_{T1} - \mathcal{B}_{T2}$ as in the argument to prove Theorem 31.5. The resulting set is also in $\mathcal{B}_T$ and has the same cardinality as $\mathcal{B}_{T1}$. Repeatedly using the argument will finally give us a set in $\mathcal{B}_T$ containing $\mathcal{B}_{T2}$ with the same cardinality as $\mathcal{B}_{T1}$. Clearly this must be $\mathcal{B}_{T2}$ itself.

Let us now consider the (Axiom B') case. Let $\mathcal{B}$ satisfy Base Axioms B'. We need to only show that maximal subsets, from the collection $\mathcal{I}$, contained in $T \subseteq S$ have the same cardinality.

Case 1 $T = S$. If B_1, B_2 are bases (i.e., members of $\mathcal{B}$) and $e \in B_1 - B_2$, we can find an $e' \in B_2 - B_1$ so that $(B_1 - e) \cup e'$ is a base. If we repeat this procedure we would finally get a base $B_k \subseteq B_2$ so that $|B_k| = |B_1|$. But one base cannot properly contain another. So $B_k = B_2$ and $|B_2| = |B_1|$.

Case 2 $T \subset S$. Suppose $X \equiv \{x_1, \ldots, x_k\}$ and $Y \equiv \{y_1, \ldots, y_m\}$ are maximal subsets of T, from the collection $\mathcal{I}$, with $k < m$. Let X be contained in a base B_x and Y in a base B_y. Let

$$B_x \equiv \{x_1, \ldots, x_k, p_{k+1}, \ldots, p_r\}$$
$$B_y \equiv \{y_1, \ldots, y_m, q_{m+1}, \ldots, q_r\}$$

(Note that p_i's and q_j's are outside T.) Since $k < m$, there exists $p_t \in B_x - B_y$. Therefore, there exists z in $B_y - B_x$ so that $(B_x - p_t) \cup z$ is a base. Clearly z cannot be one of the y_i for otherwise it would violate maximality of X. Say, $z = q_s$. We thus have a new base $B'_x \equiv (B_x - p_t) \cup q_s$. Note the progress, $(B_y - B'_x) \cap (S - T) \subset (B_y - B_x) \cap (S - T)$. Repeating this procedure we would finally arrive at a base B_x^f so that $B_x^f \cap (S - T) \supset B_y \cap (S - T)$. Any further attempted exchange using one of the remaining p elements from B_x^f would result in violation of maximality of X as a subset of T from the collection $\mathcal{I}$. The only way to avoid this contradiction is to have $k = m$. Therefore we conclude that the maximal subsets of T, belonging to the collection $\mathcal{I}$, have the same cardinality.

31.2.4 Rank Axioms

In this subsection we discuss characterization of matroids through an axiom system for the rank function.

Let $\mathcal{M}$ be a matroid on S. The *rank* of a subset $T \subseteq S$ is defined as the cardinality of the maximally independent set contained in T. This number is well defined since all maximally independent subsets of T have the same cardinality. The rank of T is denoted by $r(T)$. $r(S)$ is also called the rank of $\mathcal{M}$ and is denoted also by $r(\mathcal{M})$.

Clearly $r(\cdot)$ takes value 0 on $\emptyset$. Moreover the rank function is clearly an integral, increasing function on subsets of S. Also $r(X \cup e) - r(X) \leq 1 \; \forall \; X \subseteq S, e \in S$. We have the following properties of rank function which will motivate axiom systems for matroids in terms of rank function.

Theorem 31.8 *Let $r(\cdot)$ be the rank function of a matroid on S, $X \subseteq S$ and $e_1, e_2 \in S$.*

1. $r(X \cup e_1) = r(X \cup e_2) = r(X)$, *implies* $r(X \cup e_1 \cup e_2) = r(X)$.

2. $r(X \cup e) - r(X) \geq r(Y \cup e) - r(Y)$ *whenever* $X \subseteq Y \subseteq S - e$.

3. $r(\cdot)$ *is submodular, that is,*

$$r(X) + r(Y) \geq r(X \cup Y) + r(X \cap Y) \; \forall \; X, Y \subseteq S.$$

Proof.

i. Let B_X be a maximal independent subset of X. Clearly B_X is also a maximally independent subset of $X \cup e_1$ as well as of $X \cup e_2$ as their ranks are same. Suppose B_X is not a maximally independent subset of $X \cup e_1 \cup e_2$. This means either $B_X \cup e_1$ or $B_x \cup e_2$ must be independent. But this is a contradiction.

ii. Consider any maximally independent subset B_X of X. One can grow it into a maximally independent subset B_Y of Y. $r(X \cup e) - r(X) \geq 0$ and $r(Y \cup e) - r(Y)$ can be seen to be 0 or 1. It suffices to consider the case where $r(Y \cup e) - r(Y) = 1$. Then B_Y is not maximally independent in $Y \cup e$. Therefore $B_Y \cup e$ must be independent, which implies $B_x \cup e$ must also be independent (by Axiom I1). Therefore $r(X \cup e) - r(X) = 1$. Thus rank function always satisfies $r(X \cup e) - r(X) \geq r(Y \cup e) - r(Y)$.

iii. Let $Y - X = \{e_1, \ldots, e_k\}$. We then have

$$\begin{aligned}
r(Y) - r(X \cap Y) &= r((X \cap Y) \cup e_1) - r(X \cap Y) \\
&\quad + r((X \cap Y) \cup e_1 \cup e_2) - r((X \cap Y) \cup e_1) + \cdots \\
&\quad + r((X \cap Y) \cup e_1 \cdots \cup e_k) - r((X \cap Y) \cup e_1 \cup \cdots \cup e_{k-1}) \\
&\geq r(X \cup e_1) - r(X) + \cdots \\
&\quad + r(X \cup e_1 \cup \cdots \cup e_k) - r(X \cup e_1 \cup \cdots \cup e_{k-1}) \\
&\geq r(X \cup Y) - r(X).
\end{aligned}$$
∎

Example 31.1 For the polygon matroid $\mathcal{M}(\mathcal{G})$ (independent set ≡ subforest), the rank function of the matroid is the same as the rank function of the graph. For the bond matroid $\mathcal{M}^*(\mathcal{G})$, the rank function is the nullity function $\nu(\cdot)$ of the graph, where $\nu(A), A \subseteq E(\mathcal{G})$, is the number of edges in a coforest of $\mathcal{G} \times A$ (the graph obtained by fusing the endpoints of edges outside A and removing them).

An important property of matroid rank functions is that they are submodular. Next we describe and justify an axiom system for matroids based on the rank function. The axioms will all be properties of rank function that follow from being a matroid rank function. We show that certain combinations of these properties as axioms ensure matroid-ness.

Theorem 31.9 (Rank axioms) *Let S be a finite set and let $r(\cdot)$ be an integer valued submodular function on subsets of S satisfying in addition*

$$r(\emptyset) = 0$$
$$0 \leq r(X \cup e) - r(X) \leq 1 \ \forall \ X \subseteq S, e \in S.$$

Define $\mathcal{I}$ to be the collection of all subsets X of S satisfying $r(X) = |X|$. Let members of $\mathcal{I}$ be called independent. *Then $(S, \mathcal{I})$ is a matroid (satisfying the Independent Axioms).*

Proof.

i. Let $Y \in \mathcal{I}$ and $X \subseteq Y$. We need to show that $X \in \mathcal{I}$.
   We are given that $r(\emptyset) = 0$ and $r(A \cup e) \leq r(A) + 1 \ \forall \ A \in S$. Therefore, $r(X) \leq |X|$ and $r(Y) - r(X) \leq |Y| - |X|$. So $r(X) < |X|$, implies $r(Y) < |Y|$, which is a contradiction. Thus $r(X) = |X|$, that is, $X \in I$.

ii. Let B_1, B_2 be two maximal members of $\mathcal{I}$ contained in a subset T of S. We need to establish that $|B_1| = |B_2|$

For each $e_i \in T - B_1$, $r(B_1) \leq r(B_1 \cup e_i) < |B_1| + 1$, since B_1 is a maximal subset of T such that $r(B_1) = |B_1|$. Therefore, $r(B_1 \cup e_i) = r(B_1) \ \forall_i \in T - B_1$. For any U, V satisfying $r(B_1 \cup U) = r(B_1 \cup V) = r(B_1)$, we have, using submodularity of $r(\cdot)$ that

$$r(B_1 \cup U) + r(B_1 \cup V) \geq r(B_1 \cup U \cup V) + r(B_1 \cup (U \cap V)).$$

Note that LHS above is $2r(B_1)$ and RHS is greater or equal to $2r(B_1)$ ($r(\cdot)$ is an increasing function), therefore

$$r(B_1 \cup U \cup V) = r(B_1 \cup (U \cap V)) = r(B_1).$$

Using this inductively we can prove that

$$r(B_1 \cup e_1 \cup \cdots \cup e_k) = r(B_1),$$

where $\{e_1, \ldots, e_k\} = T - B_1$. Hence, $r(B_1) = r(T)$.
  Similarly $r(B_2) = r(T)$.
  Therefore $|B_1| = r(B_1) = r(B_2) = |B_2|$.
∎

We call a set function $r(\cdot)$, satisfying the properties stated in Theorem 31.9, a *matroid rank function*.

31.2.5 Circuit Axioms

Circuits of matroids satisfy the conditions given in the following theorem. We will use these conditions to define an axiom system for matroids using circuits as primitive objects.

Theorem 31.10 *Let $\mathcal{M} \equiv (S, \mathcal{I})$ be a matroid and let C_1, C_2 be non-disjoint circuits of $\mathcal{M}$ with e_c as one of the common elements. There exists an $e_1 \in C_1 - C_2$, (as circuits cannot be contained in one another, due to their minimality). Then there exists a circuit $C_3 \subseteq C_1 \cup C_2 - e_c$ so that $e_1 \in C_3$.*

The following lemma will be used for the proof of the theorem.

Lemma 31.1 *Let $\mathcal{M}, C_1, C_2$ be as in Theorem 31.10. Then there exists a circuit $C_3' \subseteq C_1 \cup C_2 - e_c$.*

Proof. We use properties of rank function of the matroid. Recall that a set X is independent iff $r(X) = |X|$, and by the definition of a circuit, $r(C_i) = |C_i| - 1, i = 1, 2$.
$r(C_1 \cup C_2) + r(C_1 \cap C_2) \leq r(C_1) + r(C_2) = |C_1| - 1 + |C_2| - 1$.
Therefore
$$r(C_1 \cup C_2) + |C_1 \cap C_2| \leq |C_1 \cup C_2| + |C_1 \cap C_2| - 2,$$
since $r(C_1 \cap C_2) = |C_1 \cap C_2|$. This implies $r(C_1 \cup C_2) \leq |C_1 \cup C_2| - 2$, therefore $r(C_1 \cup C_2 - e_c) \leq r(C_1 \cup C_2) \leq |C_1 \cup C_2 - e_c| - 1$, which proves that there exists a circuit inside $C_1 \cup C_2 - e_c$. ∎

Proof of Theorem 31.10 The result is clearly true when the union of the two circuits has size 3 and, trivially, when it is 2. We now use induction on the size of the union of the two circuits. Suppose that the result is true when the size of the union of the two circuits is less than n.

Let $|C_1 \cup C_2| = n$, $e_c \in C_1 \cap C_2$ and $e_1 \in C_1 - C_2$. By the above lemma (Lemma 31.1), one is guaranteed existence of a circuit $C_3' \subseteq C_1 \cup C_2 - e_c$. If $e_1 \in C_3'$ we are done.

So assume that $e_1 \notin C_3'$. We have, $C_3' \not\subseteq C_1$ and $C_3' \subseteq C_1 \cup C_2$. So $C_3' \cap C_2 \not\subseteq C_1 \cap C_2$. Let $e_2 \in C_3' \cap C_2 - C_1 \cap C_2$.

Consider $C_2 \cup C_3'$. We aim to use induction hypothesis on it, for which we need to show that its size is less than n. But this follows from the observation that $e_1 \notin C_2 \cup C_3'$. Further $e_2 \in C_3' \cap C_2$ and $e_c \in C_2 - C_3'$. By the induction hypothesis there is a circuit $C_2'' \subseteq C_2 \cup C_3' - e_2$ so that $e_c \in C_2''$.

Now consider $C_1 \cup C_2''$. We have $e_c \in C_2'' \cap C_1$ and $e_1 \in C_1 - C_2''$. Further $e_2 \notin C_1 \cup C_2''$ so that $|C_1 \cup C_2''| < n$ and we can apply induction hypothesis. Therefore, there exists a circuit $C_3 \subseteq C_1 \cup C_2'' - e_c$ so that $e_1 \in C_3$. ∎

Example 31.2 For the polygon matroid $\mathcal{M}(\mathcal{G})$ of $\mathcal{G}$ (independent set ≡ subforest), a circuit of the matroid is the same as a circuit of the graph. For the bond matroid $\mathcal{M}^*(\mathcal{G})$ (independent set ≡ subcoforest), a circuit of the matroid is the same as a cutset of the graph. For the vectorial matroid (associated with the columns of a representative matrix), a circuit is a minimal dependent set of columns.

Theorem 31.11 (Circuit axioms)

Let S be a finite set. Let $\mathcal{C}$ denote a family of subsets (called circuits) of S satisfying the following axioms.

Axiom C_1 *No member of $\mathcal{C}$ is a proper subset of another.*

Axiom C₂ *Let $C_1, C_2 \in \mathcal{C}$ and let $e_c \in C_1 \cap C_2$ and $e_1 \in C_1 - C_2$. Then there exists $C_3 \in \mathcal{C}$ so that $C_3 \subseteq C_1 \cup C_2 - e_c$ and $e_1 \in C_3$. Let $\mathcal{I}$ be the class of subsets of S that do not contain a member of $\mathcal{C}$. Then $(S, \mathcal{I})$ satisfies the axioms I1 and I2 of a matroid. (This also justifies denoting a matroid $\mathcal{M}$ as a pair $(S, \mathcal{C})$ describing collection of circuits as primitive objects.)*

Proof. We will show that maximal subsets of $T \subseteq S$ that are in $\mathcal{I}$ (i.e., that do not contain a circuit) have the same cardinality. For readability, during the course of the proof we will call members of $\mathcal{I}$ independent and maximal members, bases although this terminology is justified only after the proof is complete.

Let B_1, B_2 be two maximal independent sets contained in T. If $B_1 \neq B_2$, clearly $B_1 \not\supseteq B_2$. Let $e_2 \in B_2 - B_1$. Then $e_2 \cup B_1$ contains a circuit. Claim is that this circuit is unique. For otherwise if C_1, C_2 are two such circuits since both have e_2 as a member, then by the circuit axioms there exists a circuit $C_3 \subseteq C_1 \cup C_2 - e_2$. This is impossible as it would imply that a circuit C_3 is a subset of the base B_1.

Define $L(e_2, B_1)$ to be the unique circuit contained in $e_2 \cup B_1$. Since $\{e_2\}$ is not a circuit (element e_2 is inside the base B_2) we must have $L(e_2, B_1) \cap B_1$ is nonempty. Also $L(e_2, B_1)$ is not a subset of B_2. Therefore, there exists an $e_1 \in L(e_2, B_1) \cap (B_1 - B_2)$. Then $B_1' \equiv e_2 \cup B_1 - e_1$ is independent.

We claim that B_1' is also maximally independent. For, let $e' \in T - B_1'$. For the case $e' = e_1$, we note that $e' \cup B_1'$ contains $L(e_2, B_1)$, thereby not violating maximality of B_1'. So consider $e' \neq e_1$. Now $e' \cup B_1$ contains a circuit $L(e', B_1)$. If this circuit does not contain e_1, we have $L(e', B_1) \subseteq e' \cup B_1'$, again not violating maximality of B_1'. Suppose it contains e_1. Then $L(e_2, B_1)$ and $L(e', B_1)$ have the element e_1 in common. Hence, $L(e_2, B_1) \cup L(e', B_1) - e_1$ contains a circuit. This circuit is contained in $e' \cup B_1'$.

Therefore, we conclude that $e' \cup B_1'$ is not independent for every $e' \in T - B_1'$. We now have a maximally independent subset B_1' of T which has the same cardinality as B_1. However B_1' also satisfies $\mid B_2 - B_1' \mid < \mid B_2 - B_1 \mid$. Repeating this procedure we would arrive at a maximally independent subset B_k of T that has the same cardinality as B_1 and also contains B_2. But this implies $B_k = B_2$ and hence, $\mid B_1 \mid = \mid B_2 \mid$. ∎

31.3 DUAL MATROID

Matroids occur naturally in pairs. This pairing is analogous to that of complementary orthogonal vector spaces.

Consider a vector space $\mathcal{V}$ on S. We remind the reader that *dot product* of two vectors $\mathbf{f}, \mathbf{g}$ on S denoted by $<\mathbf{f}, \mathbf{g}>$ over a field $\mathcal{F}$ is defined by $<\mathbf{f}, \mathbf{g}> \equiv \sum_{e \in S} \mathbf{f}(e).\mathbf{g}(e)$. We say $\mathbf{f}, \mathbf{g}$ are *orthogonal* if their dot product is zero. $\mathcal{V}^\perp$, the space *complementary orthogonal* to $\mathcal{V}$, is the collection of all vectors on S that are orthogonal to every vector in $\mathcal{V}$. Let $\mathcal{M}(\mathcal{V})$ denote the matroid on S whose independent sets are the linearly independent column sets of a representative matrix (i.e., rows constitute a basis) of $\mathcal{V}$. Note that column dependence structure of all representative matrices of $\mathcal{V}$ is identical. Whenever we have a set of columns as a base of this matroid we know that we can build a *standard* representative matrix with identity matrix corresponding to this set as below.

$$\mathbf{R} \equiv \begin{matrix} B & S-B \\ \begin{bmatrix} \mathbf{I} & \vdots & \mathbf{K} \end{bmatrix} \end{matrix}. \qquad (31.1)$$

Then we know that

$$\mathbf{R}^* \equiv \begin{bmatrix} B & S-B \\ -\mathbf{K}^T & \vdots & \mathbf{I} \end{bmatrix}, \quad (31.2)$$

is a representative matrix of $\mathcal{V}^\perp$. It is thus clear that the bases of $\mathcal{M}(\mathcal{V}^\perp)$ are cobases of $\mathcal{M}(\mathcal{V})$ and vice versa. This situation can be shown to hold also for arbitrary matroids and the pairs of matroids are said to be dual to each other.

Theorem 31.12 *Let $\mathcal{M} \equiv (S, \mathcal{I})$ be a matroid. Then, $\mathcal{M}^* \equiv (S, \mathcal{I}^*)$, where $X \in \mathcal{I}^*$ iff $S - X$ contains a base of $\mathcal{M}$, is a matroid. $\mathcal{M}^*$ is called the dual matroid. Further, $(\mathcal{M}^*)^* = \mathcal{M}$.*

Proof. It suffices to prove that the collection of complements of the bases of $\mathcal{M}$ satisfy either of the equivalent base Axioms B or B'. This is indeed easy to see due to the duality evident between the Axioms B and B'. ∎

Example 31.3 If $\mathcal{V}$ is a vector space on S then $(\mathcal{M}(\mathcal{V}))^* = \mathcal{M}(\mathcal{V}^\perp)$. If $\mathcal{M}(\mathcal{G})$ is the polygon matroid associated with the graph $\mathcal{G}$, then we know that $\mathcal{M}(\mathcal{G}) = \mathcal{M}(\mathcal{V}_v(\mathcal{G}))$, where $\mathcal{V}_v(\mathcal{G})$ is the space of all vectors spanned by the rows of the incidence matrix of the directed graph $\mathcal{G}$ and $(\mathcal{M}(\mathcal{G}))^* = \mathcal{M}((\mathcal{V}_v(\mathcal{G}))^\perp)$. Thus matroid $(\mathcal{M}(\mathcal{G}))^*$ is the bond matroid associated with $\mathcal{G}$, for which the independent subsets are subcoforests. When the graph $\mathcal{G}$ is planar there exists a graph $\mathcal{G}^*$ so that $\mathcal{V}_v(\mathcal{G}^*) = (\mathcal{V}_v(\mathcal{G}))^\perp$. It would follow that $\mathcal{M}(\mathcal{G}^*) = (\mathcal{M}(\mathcal{G}))^*$.

The circuit of the matroid $\mathcal{M}^*$ is called a *bond* of $\mathcal{M}$. The following theorem gives some characterizations of a bond.

Theorem 31.13 *Let $\mathcal{M} \equiv (S, \mathcal{I})$ be a matroid. A subset $K \subseteq S$ is a bond of $\mathcal{M}$ iff any of the following equivalent conditions hold:*

1. *K is a circuit of $\mathcal{M}^*$.*

2. *K is a minimal set intersecting every base of $\mathcal{M}$.*

3. *K is a minimal set which meets no circuit of $\mathcal{M}$ in just a single element.*

We need the following lemma to prove the theorem.

Lemma 31.2 *Let L be independent in $\mathcal{M} \equiv (S, \mathcal{I})$ and let K be independent in $\mathcal{M}^*$. Further let $L \cap K = \emptyset$. Then there exists a base of $\mathcal{M}$ that contains L and does not intersect K.*

Proof. Independence of K in $\mathcal{M}^*$ implies $S - K$ contains a base B of $\mathcal{M}$. Now, L is given to be independent in $\mathcal{M}$ and is also a subset of $S - K$. So there exists a subset B', that is maximally independent in $\mathcal{M}$ containing L and also a subset of $S - K$. Axiom I2 tells us that $|B'| = |B|$. Hence B' is the desired base of $\mathcal{M}$. ∎

Proof of Theorem 31.13. Condition (i) is the definition of a bond. We will show that each of the conditions (ii) and (iii) is equivalent to (i).

(i) ⇔ (ii): K is a minimal set that is not contained in any base of $\mathcal{M}^*$, equivalently, that intersects every base of $\mathcal{M}$ (since bases of $\mathcal{M}$ are complements of bases of $\mathcal{M}^*$).

(i) ⇔ (iii): First we will show (i) implies (iii). We begin by showing the minimality (with respect to the intersection property of [iii]) of a circuit of the dual. Suppose K is a circuit of $\mathcal{M}^*$. Let $X \subset K$. Then X is independent in $\mathcal{M}^*$. Let B_X^* be a base of $\mathcal{M}^*$ containing X. Let B_X be the complement of B_X^*. B_X is a base of $\mathcal{M}$ and $B_X \cap X = \emptyset$. Let $e \in X$ and let

$L(e, B_X)$ be the unique circuit of $\mathcal{M}$ contained in $e \cup B_X$. This circuit intersects X in $\{e\}$. Thus every proper subset of K meets some circuit of $\mathcal{M}$ in exactly a single element.

Now we show that when a circuit of the dual meets a circuit of the original matroid, they intersect in at least two elements. Suppose K meets a circuit C of $\mathcal{M}$. Let $e \in C \cap K$. $C - e$ and $K - e$ are independent in the matroid and its dual, respectively. Suppose C and K intersect in just the element e. This implies $(C-e) \cap (K-e) = \emptyset$. Now, by Lemma 31.2, there exists a base B of $\mathcal{M}$ that contains $C - e$ but does not intersect $K - e$. We must have either $e \in B$ or $e \in (S - B)$. This would imply either $C \subseteq B$ or in the latter case, $K \subseteq (S - B)$ contradicting the independence of B in $\mathcal{M}$ and $S - B$ in $\mathcal{M}^*$ respectively. Therefore $C - e$ and $K - e$ must intersect, implying $|C \cap K| > 1$.

It remains to show that (iii) implies (i). Let $K \subseteq S$ be such that it does not meet any circuit of $\mathcal{M}$ in just a single element. Such K cannot be contained in a cobase of $\mathcal{M}$, for, if $e \in S - B$, for a base B of $\mathcal{M}$, then the fundamental circuit of matroid $\mathcal{M}$, formed by e with B intersects $(S - B)$ only in e. Hence, K is dependent in $\mathcal{M}^*$. We only need to show that K is a minimally dependent subset of the dual. Suppose the contrary, that is, there exists a proper subset K' of K that is a circuit of the dual matroid. However, as already seen, any circuit K' of $\mathcal{M}^*$ does not meet any circuit of $\mathcal{M}$ in just a single element. This would contradict minimality (with respect to the intersection property) of K. Therefore K is a circuit of $\mathcal{M}^*$. ∎

31.4 MINORS OF GRAPHS, VECTOR SPACES, AND MATROIDS

Given a matroid, there are some natural ways of deriving matroids on subsets of the underlying sets which we will call minors of the original matroid. In this section we motivate and define this notion by first studying the most important instances of graphs and vector spaces.

31.4.1 Restriction and Contraction of Graphs

Let $\mathcal{G}(V, E)$ be a graph and let $T \subseteq E$.

Definition 31.2 *The graph $\mathcal{G}$ open $(E - T)$ is the subgraph of $\mathcal{G}$ with T as the set of edges and the whole $V(\mathcal{G})$ as the vertex set. That is, to obtain $\mathcal{G}$ open $(E - T)$ we remove (delete) edges in $E - T$, however, leaving their endpoints in place.*

The restriction of $\mathcal{G}$ to T, denoted by $\mathcal{G} \cdot \mathbf{T}$, is the subgraph of $\mathcal{G}$ obtained by deleting isolated vertices from $\mathcal{G}$ open $(E - T)$. Thus, $\mathcal{G} \cdot T$ is the subgraph of $\mathcal{G}$ on T. In case of a directed graph $\mathcal{G}$, we retain original directions.

Definition 31.3 *The graph $\mathcal{G}$ short $(E - T)$, is built by first building $\mathcal{G}$ open T. We then get connected components. Let $V_1, \ldots, V_k$ be the vertex sets of the connected components of $\mathcal{G}$ open T. The set $\{V_1, \ldots, V_k\}$ is the vertex set and T is the edge set of $\mathcal{G}$ short $(E-T)$. (The reader may imagine $\{V_1, \ldots, V_k\}$ as a set of supernodes enclosed by surfaces.) An edge $e \in T$ would have V_i, V_j as its endpoints in $\mathcal{G}$ short $(E - T)$ iff the endpoints of e in $\mathcal{G}$ lie in V_i, V_j. If $\mathcal{G}$ is directed, V_i, V_j would be the positive and negative endpoints of e in $\mathcal{G}$ short $(E - T)$ provided the positive and negative endpoints of e in $\mathcal{G}$ lie in V_i, V_j, respectively.*

(In other words, $\mathcal{G}$ short $(E - T)$ is obtained from $\mathcal{G}$ by short circuiting the edges in $(E - T)$ (fusing their end points) and removing them.)

The contraction of $\mathcal{G}$ to T, denoted by $\mathcal{G} \times T$, is obtained from $\mathcal{G}$ short $(E - T)$ by deleting the isolated vertices of the latter.

An immediate consequence of the above definitions is that the union of a forest of $\mathcal{G}$ short $(E - T)$ (i.e., of $\mathcal{G} \times T$) and a forest of $\mathcal{G}$ open T (i.e., of $\mathcal{G} \cdot (E - T)$) is a forest of $\mathcal{G}$. We therefore have the following theorem.

Theorem 31.14 $r(\mathcal{G}) = r(\mathcal{G} \times T) + r(\mathcal{G} \cdot (E - T))$.

We denote $(\mathcal{G} \times T_1) \cdot T_2$, $T_2 \subseteq T_1 \subseteq E(\mathcal{G})$ by $\mathcal{G} \times T_1 \cdot T_2$ and $(\mathcal{G} \cdot T_1) \times T_2$. $T_2 \subseteq T_1 \subseteq E(\mathcal{G})$ by $\mathcal{G} \cdot T_1 \times T_2$. Graphs denoted by such expressions are called *minors* of $\mathcal{G}$. It is easy to see that when we short a set $A \subseteq E(\mathcal{G})$ and open a disjoint set $B \subseteq E(\mathcal{G})$, then the final graph does not depend on the order in which these operations are carried out. Also note that $\mathcal{G} \times T(\mathcal{G} \cdot T)$ differs from $\mathcal{G}$ short $(E-T)$ ($\mathcal{G}$ open $(E-T)$) only in that the isolated vertices are omitted. We therefore have the following theorem. (In the statement below equality refers to isomorphism.)

Theorem 31.15 *Let $\mathcal{G}$ be a graph and $X_2 \subseteq X_1 \subseteq E(G)$. Then*

1. $\mathcal{G} \times X_1 \times X_2 = \mathcal{G} \times X_2$,

2. $\mathcal{G} \cdot X_1 \cdot X_2 = \mathcal{G} \cdot X_2$,

3. $\mathcal{G} \times X_1 \cdot X_2 = \mathcal{G} \cdot (E - (X_1 - X_2)) \times X_2$.

Proof. To prove each statement of the above theorem we only need to note that the graph minors on both LHS and RHS are obtained by shorting and opening the same sets. In statement (i), on both sides $E - X_2$ is shorted. In the statement (ii) $E - X_2$ is opened. In the more interesting statement of (iii), on both sides, the shorted set is $E - X_1$ and the subset of edges opened is $X_1 - X_2$. ∎

The following result shows that construction of minors is essentially a two step process. The proof is by a routine application of Theorem 31.15.

Theorem 31.16 *Any minor of the form $\mathcal{G} \times X_1 \cdot X_2 \times X_3 \ldots X_n, X_1 \supseteq \ldots \supseteq X_n$ (the graph being obtained by starting from $\mathcal{G}$ and performing the operations from left to right in succession), can be simplified to a minor of the form $\mathcal{G} \cdot X' \times X_n$ or $\mathcal{G} \times X' \cdot X_n$.* ∎

The next two results are about circuits and cutsets of minors in terms of the corresponding subsets in the original graph. The routine proofs are omitted.

Theorem 31.17

1. *$C \subseteq T$ is a circuit of $\mathcal{G} \cdot T$ iff C is a circuit of $\mathcal{G}$.*

2. *$C \subseteq T$ is a circuit of $\mathcal{G} \times T$ iff C is a minimal intersection of circuits of $\mathcal{G}$ with T (equivalently, iff C is an intersection of a circuit of $\mathcal{G}$ with T but no proper subset of C is such an intersection).*

Theorem 31.18

1. *$B \subseteq T$ is a cutset of $\mathcal{G} \cdot T$ iff it is a minimal intersection of cutsets of $\mathcal{G}$ with T.*

2. *A subset B of T is a cutset of $\mathcal{G} \times T$ iff it is a cutset of $\mathcal{G}$.*

31.4.2 Restriction and Contraction of Vector Spaces

There are natural operations on vector spaces that are analogous to the operations of opening and shorting edges in a graph. We describe them now. Let $\mathcal{V}$ be a vector space on S and let $T \subseteq S$.

Definition 31.4 *The* restriction *of $\mathcal{V}$ to T, denoted by $\mathcal{V}.T$, is defined as follows:*

$$\mathcal{V}.T \equiv \{\mathbf{f}_T : \mathbf{f}_T = \mathbf{f}/T, \mathbf{f} \in \mathcal{V}\}.$$

The contraction *of $\mathcal{V}$ to T, denoted by $\mathcal{V} \times T$, is defined as follows:*

$$\mathcal{V} \times T \equiv \{\mathbf{f}'_T : \mathbf{f}'_T = \mathbf{f}/T, \mathbf{f} \in \mathcal{V} \text{ and } \mathbf{f}/(S-T) = \mathbf{0}\}.$$

It is easily seen that $\mathcal{V} \cdot T, \mathcal{V} \times T$ are vector spaces.

We denote $(\mathcal{V} \times T_1) \cdot T_2$ by $\mathcal{V} \times T_1 \cdot T_2$, as in the case of graphs. Such vector spaces are called *minors* of $\mathcal{V}$. We say we open T when we restrict $\mathcal{V}$ to $(S-T)$ and say we short T when we contract $\mathcal{V}$ to $(S-T)$.

The order in which we open and short disjoint sets of elements is unimportant. This is stated formally below.

Theorem 31.19 *Let $T_2 \subseteq T_1 \subseteq S$. Then*

1. $\mathcal{V} \cdot T_1 \cdot T_2 = \mathcal{V} \cdot T_2$,

2. $\mathcal{V} \times T_1 \times T_2 = \mathcal{V} \times T_2$,

3. $\mathcal{V} \times T_1 \cdot T_2 = \mathcal{V} \cdot (S - (T_1 - T_2)) \times T_2$.

Proof of 3. $LHS \subseteq RHS$

Let $\mathbf{f}_{T_2} \in \mathcal{V} \times T_1 \cdot T_2$. Then there exists a vector $\mathbf{f}_{T_1} \in \mathcal{V} \times T_1$ such that $\mathbf{f}_{T_1}/T_2 = \mathbf{f}_{T_2}$ and a vector $\mathbf{f} \in \mathcal{V}$ with $\mathbf{f}/(S - T_1) = \mathbf{0}$ such that $\mathbf{f}/T_1 = \mathbf{f}_{T_1}$. Now let $\mathbf{f}'$ denote $\mathbf{f}/(S - (T_1 - T_2))$. Clearly $\mathbf{f}' \in \mathcal{V} \cdot (S - (T_1 - T_2))$. Now $\mathbf{f}'/(S - T_1) = \mathbf{0}$. Hence, $\mathbf{f}'/T_2 \in \mathcal{V} \cdot (S - (T_1 - T_2)) \times T_2$. Thus, $\mathcal{V} \times T_1 \cdot T_2 \subseteq \mathcal{V} \cdot (S - (T_1 - T_2)) \times T_2$. The reverse containment is similarly proved. ∎

Remark 31.1 Observe that a typical vector of both LHS and RHS is obtained by restricting a vector of $\mathcal{V}$, that takes zero value on $S - T_1$, to T_2. We now have as in the case of graphs.

Theorem 31.20 *Any minor of the form $\mathcal{V} \times T_1 \cdot T_2 \times T_3 \ldots T_n, T_1 \supseteq T_2 \supseteq \ldots \supseteq T_n$, can be simplified to a minor of the form*

$$\mathcal{V} \cdot T' \times T_n \text{ or } \mathcal{V} \times T' \cdot T_n.$$

31.4.3 Minors of Dual Vector Spaces

We now relate the minors of $\mathcal{V}$ to the minors of the complementary orthogonal space $\mathcal{V}^\perp$. We remind the reader that $\mathcal{V}^\perp \equiv \{\mathbf{g} :< \mathbf{g}, \mathbf{f} >= 0, \mathbf{f} \in \mathcal{V}\}$, and that for any finite dimensional vector space $\mathcal{V}'$ $(\mathcal{V}'^\perp)^\perp = \mathcal{V}'$. In the following results we see that the contraction (restriction) of a vector space corresponds to the restriction (contraction) of the orthogonal complement. We say that contraction and restriction are *(orthogonal) duals* of each other.

Theorem 31.21 *Let $\mathcal{V}$ be a vector space on S and let $T \subseteq S$. Then,*

1. $(\mathcal{V} \cdot T)^\perp = \mathcal{V}^\perp \times T$.

2. $(\mathcal{V} \times T)^\perp = \mathcal{V}^\perp \cdot T$.

Proof. i. Let $\mathbf{g}_T \in (\mathcal{V} \cdot T)^\perp$. For any $\mathbf{f}$ on S let $\mathbf{f}_T$ denote $\mathbf{f}/T$. Now if $\mathbf{f} \in \mathcal{V}$, then $\mathbf{f}_T \in \mathcal{V} \cdot T$ and $< \mathbf{g}_T, \mathbf{f}_T > = 0$.

Let $\mathbf{g}$ on S be defined by $\mathbf{g}/T \equiv \mathbf{g}_T$, $\mathbf{g}/S - T \equiv \mathbf{0}$. If $\mathbf{f} \in \mathcal{V}$ we have

$$< \mathbf{f}, \mathbf{g} > = < \mathbf{f}_T, \mathbf{g}_T > + < \mathbf{f}_{S-T}, \mathbf{g}_{S-T} > = 0 + < \mathbf{f}_{S-T}, \mathbf{0}_{S-T} > = 0.$$

Thus $\mathbf{g} \in \mathcal{V}^\perp$ and therefore, $\mathbf{g}_T \in \mathcal{V}^\perp \times T$. Hence, $(\mathcal{V} \cdot T)^\perp \subseteq \mathcal{V}^\perp \times T$.

Next let $\mathbf{g}_T \in \mathcal{V}^\perp \times T$. Then there exists $\mathbf{g} \in \mathcal{V}^\perp$ so that $\mathbf{g}/S - T = \mathbf{0}$ and $\mathbf{g}/T = \mathbf{g}_T$. Let $\mathbf{f}_T \in \mathcal{V} \cdot T$. There exists $\mathbf{f} \in \mathcal{V}$ so that $\mathbf{f}/T = \mathbf{f}_T$. Now $0 = <\mathbf{f}, \mathbf{g}> = <\mathbf{f}_T, \mathbf{g}_T> + <\mathbf{f}_{S-T}, \mathbf{0}_{S-T}> = <\mathbf{f}_T, \mathbf{g}_T>$. Hence, $\mathbf{g}_T \in (\mathcal{V} \cdot T)^\perp$. We conclude that $\mathcal{V}^\perp \times T \subseteq (\mathcal{V} \cdot T)^\perp$. This proves that $(\mathcal{V} \cdot T)^\perp = \mathcal{V}^\perp \times T$.

ii. We have $(\mathcal{V}^\perp \cdot T)^\perp = (\mathcal{V}^\perp)^\perp \times T$.

For any finite dimensional vector space $\mathcal{V}'$ we know that $(\mathcal{V}'^\perp)^\perp = \mathcal{V}'$. Hence, $((\mathcal{V}^\perp \cdot T)^\perp)^\perp = \mathcal{V}^\perp \cdot T$ and $(\mathcal{V}^\perp)^\perp = \mathcal{V}$. Hence, $\mathcal{V}^\perp \cdot T = (\mathcal{V} \times T)^\perp$. ∎

The following corollary is immediate.

Corollary 31.1 $(\mathcal{V} \times P \cdot T)^\perp = \mathcal{V}^\perp \cdot P \times T, T \subseteq P \subseteq S.$

31.4.4 Representative Matrices of Minors of Vector Spaces

As defined earlier, the *representative matrix* $\mathbf{R}$ of a vector space $\mathcal{V}$ on S has the vectors of a basis of $\mathcal{V}$ as its rows. We now describe how to construct a representative matrix containing representative matrices of $\mathcal{V} \cdot P$ and $\mathcal{V} \times (S - P)$ as its submatrices. In such a case, $\mathcal{V} \cdot P$ and $\mathcal{V} \times (S - P)$ are said to become *visible* in $\mathbf{R}$.

Theorem 31.22 *Let $\mathcal{V}$ be a vector space on S. Let $P \subseteq S$. Let $\mathbf{R}$ be a representative matrix as shown below*

$$\mathbf{R} = \begin{bmatrix} \mathbf{R}_{PP} & \mathbf{R}_{P2} \\ \mathbf{0} & \mathbf{R}_{22} \end{bmatrix} \quad \begin{matrix} P & S-P \end{matrix} \tag{31.3}$$

where the rows of $\mathbf{R}_{PP}$ are linearly independent. Then $\mathbf{R}_{PP}$ is a representative matrix for $\mathcal{V} \cdot P$ and $\mathbf{R}_{22}$, a representative matrix for $\mathcal{V} \times (S - P)$.

Proof. The rows of $\mathbf{R}_{PP}$ are restrictions of vectors on S to P. If $\mathbf{f}_P$ is any vector in $\mathcal{V} \cdot P$ there exists a vector $\mathbf{f}$ in $\mathcal{V}$ so that $\mathbf{f}/P = \mathbf{f}_P$. Now $\mathbf{f}$ is a linear combination of the rows of $\mathbf{R}$. Hence, $\mathbf{f}/P (= \mathbf{f}_P)$ is a linear combination of the rows of $\mathbf{R}_{PP}$. Further it is given that the rows of $\mathbf{R}_{PP}$ are linearly independent. It follows that $\mathbf{R}_{PP}$ is a representative matrix of $\mathcal{V} \cdot P$.

It is clear from the structure of $\mathbf{R}$ (the zero in the second set of rows) that any linear combination of the rows of $\mathbf{R}_{22}$ belongs to $\mathcal{V} \times (S - P)$. Further if $\mathbf{f}$ is any vector in $\mathcal{V}$ so that $\mathbf{f}/P = \mathbf{0}$ then $\mathbf{f}$ must be a linear combination only of the second set of rows of $\mathbf{R}$. For, if the first set of rows are involved in the linear combination, since rows of $\mathbf{R}_{PP}$ are linearly independent, $\mathbf{f}/P$ cannot be zero. We conclude that if $\mathbf{f}/(S - P)$ is a vector in $\mathcal{V} \times (S - P)$, it is linearly dependent on the rows of $\mathbf{R}_{22}$. Now rows of $\mathbf{R}$ are linearly independent. We conclude that $\mathbf{R}_{22}$ is a representative matrix of $\mathcal{V} \times P$. ∎

The following corollary is immediate.

Corollary 31.2 $r(\mathcal{V}) = r(\mathcal{V} \cdot P) + r(\mathcal{V} \times (S - P)), P \subseteq S.$

31.4.5 Minors of Matroids

In this subsection we generalize the notion of minors of graphs and vector spaces to matroids.

Let $\mathcal{M} \equiv (S, \mathcal{I})$ be a matroid and $X \subseteq S$. The *restriction* (or *reduction*) of $\mathcal{M}$ to X, denoted by $\mathcal{M} \cdot X$, is the matroid on the ground set X whose independence family is the collection of all subsets of X which are members of $\mathcal{I}$. We define the *contraction* of $\mathcal{M}$ to X (which we shall denote by $\mathcal{M} \times X$), as the matroid on X whose independent sets are precisely those $Y \subseteq X$ which satisfy the property that $Y \cup B_{S-X} \in \mathcal{I}$ whenever B_{S-X} is a base of $\mathcal{M} \cdot (S - X)$. Indeed it is clear from the definition that $\mathcal{M} \cdot X$ is a matroid. That $\mathcal{M} \times X$ is also a matroid needs to be proved and this we do below.

Before we proceed to its proof, we define the notion of *minor* of a matroid. A *minor* of $\mathcal{M}$ is a matroid of the form $(\mathcal{M} \times X_1) \cdot X_2$ or $(\mathcal{M} \cdot X_1) \times X_2, X_2 \subseteq X_1 \subseteq S$. Since there is no room for confusion we omit the bracket while denoting minors. We need the following useful lemma.

Lemma 31.3 *Let $\mathcal{M}$ be a matroid on S and let $Y \subseteq X \subseteq S$. Suppose B'_1, B'_2 are two bases of $\mathcal{M} \cdot (S - X)$ and $Y \cup B'_1$ is independent in $\mathcal{M}$. Then so is $Y \cup B'_2$.*

Proof. Suppose the contrary. Then there exist bases B'_1, B'_2 of $\mathcal{M} \cdot (S - X)$ so that $Y \cup B'_1$ is independent, but $Y \cup B'_2$ is dependent and $|B'_1 - B'_2|$ is a minimum for these conditions. For $e \in B'_2 - B'_1$, $e \cup B'_1$ contains the unique circuit $L(e, B'_1)$ of $\mathcal{M} \cdot (S - X)$. There must exist e' of $(B'_2 - B'_2)$ inside $L(e, B'_1)$. So $B'_3 = (B'_1 - e') \cup e$ is a base of $\mathcal{M} \cdot (S - X)$. There exists a base B_1 of $\mathcal{M}$ containing $Y \cup B'_1$. Therefore $e \cup B_1$ contains the unique circuit $L(e, B_1)$. Observe that $L(e, B_1)$ is the same as $L(e, B'_1)$ (this follows as circuits of $\mathcal{M} \cdot (S - X)$ are the same as circuits of $\mathcal{M}$ contained in $(S - X)$). As $L(e, B_1) = L(e, B'_1)$, it follows that $B_3 \equiv (B_1 - e') \cup e$ is a base of $\mathcal{M}$. Now, $Y \cup B'_3$ is independent in $\mathcal{M}$, being a subset of B_3. But note that B'_3 and B'_2 violate the minimum size assumption about B'_1 and B'_2 (since $|B'_3 - B'_2| < |B'_1 - B'_2|$). We conclude therefore that $Y \cup B'_2$ is independent in $\mathcal{M}$. ∎

To prove that $\mathcal{M} \times X$, given by $(S, \mathcal{I}'_X)$, is a matroid, we will verify that it satisfies the independent axioms. If $Y \in \mathcal{I}'_X$ and $Z \subseteq Y$ it is clear from the definition of $\mathcal{I}'_X$ that $Z \in \mathcal{I}'_X$. Therefore Axiom I1 holds. Now we prove that Axiom I2 holds. Let $X_1 \subseteq X$ and let Z_1, Z_2 be maximal members of $\mathcal{I}'_X$ which are subsets of X_1. Then Z_1, Z_2 are maximal with respect to the property that $Z_1, Z_2 \subseteq X_1$ and $Z_1 \uplus B', Z_2 \uplus B'$ are independent in $\mathcal{M}$, for each base B' of $\mathcal{M} \cdot (S - X)$. We would be done if we show that $|Z_1 \uplus B'| = |Z_2 \uplus B'|$ (from which $|Z_1| = |Z_2|$ would follow). It suffices to show that $Z_1 \uplus B', Z_2 \uplus B'$ are both maximally independent subsets of $X_1 \uplus (S - X)$ in $\mathcal{M}$. Suppose without loss of generality, let $Z_1 \uplus B'$ not be a maximally independent subset of $X_1 \uplus (S - X)$ in $\mathcal{M}$. Let W be a proper superset of it that is independent in $\mathcal{M}$ and is as well a subset of $X_1 \uplus (S - X)$. Due to maximality of B' and independence of W, $W \cap (S - X) = B'$, and therefore $W \cap X_1$ would be a proper superset of Z_1. But $(W \cap X_1) \cup B'$ is independent which implies that $W \cap X_1$ is also a member of $\mathcal{I}'_X$ (note that we use Lemma 4.1 which assures us testing with any one base B' of $\mathcal{M} \cdot (S - X)$ is adequate). This would however violate maximality of Z_1 as a member of $\mathcal{I}'_X$. Therefore $Z_1 \uplus B'$ and similarly $Z_2 \uplus B'$ are maximally independent subsets of $X_1 \uplus (S - X)$ in $\mathcal{M}$. Thus, $|Z_1 \uplus B'| = |Z_2 \uplus B'|$ and therefore, $|Z_1| = |Z_2|$ as required.

Theorem 31.23 *Let $\mathcal{M}$ be a matroid on S and let $X \subseteq S$. Then*

1. *The union of a base of $\mathcal{M} \times X$ and a base of $\mathcal{M} \cdot (S - X)$ is a base of $\mathcal{M}$.*

2. $r(\mathcal{M} \times X) + r(\mathcal{M} \cdot (S - X)) = r(\mathcal{M})$.

Proof. i and ii. Let B_1 be a base of $\mathcal{M} \times X$ and let B_2 be a base of $\mathcal{M} \cdot (S - X)$. By the definition of $\mathcal{M} \times X$, $B_2 \cup B_1$ is independent in $\mathcal{M}$. Hence, $r(\mathcal{M} \times X) \leq r(\mathcal{M}) - r(\mathcal{M} \cdot (S-X))$.

Next, $B_2 \cup B_1$ can be extended to a base B of $\mathcal{M}$. By the definition of $\mathcal{M} \cdot (S - X)$, we must have $B_2 = B \cap (S - X)$. As $(B \cap X) \cup B_2$ is independent in $\mathcal{M}$, by the definition of $\mathcal{M} \times X$ and Lemma 31.3, $B \cap X$ is independent in $\mathcal{M} \times X$. Hence, $r(\mathcal{M} \times X) \geq r(\mathcal{M}) - r(\mathcal{M} \cdot (S - X))$. Therefore that $r(\mathcal{M} \times X) = r(\mathcal{M}) - r(\mathcal{M} \cdot (S - X))$ and $B_1 \cup B_2$ is a base of $\mathcal{M}$. ∎

We next study the relation between primitive notions (such as bases, circuits, bonds) associated with a matroid and those associated with restrictions and contractions of a matroid. We begin with bases.

Theorem 31.24 *Let $\mathcal{M}$ be a matroid on S and $X \subseteq S$. Then*

1. *B_X is a base of $\mathcal{M} \cdot X$ iff it is a maximal intersection of a base of $\mathcal{M}$ with X.*

2. *B'_X is a base of $\mathcal{M} \times X$ iff it is a minimal intersection of a base of $\mathcal{M}$ with X.*

Proof.

i. B_X is a maximal intersection of a base of $\mathcal{M}$ with X iff it is a maximal subset of X that is independent in $\mathcal{M}$, that is, iff B_X is a base of $\mathcal{M} \cdot X$.

ii. Let B_{S-X} be a base of $\mathcal{M} \cdot (S - X)$. By the definition of $\mathcal{M} \times X$ and Lemma 31.3 B_X is a base of $\mathcal{M} \times X$ iff $B_X \cup B_{S-X}$ is a base of $\mathcal{M}$, that is, iff a base B of $\mathcal{M}$ intersects X in B_X and intersects $(S - X)$ maximally among all bases of $\mathcal{M}$, that is, iff a base B of $\mathcal{M}$ intersects X in B_X and this intersection is minimal among all bases of $\mathcal{M}$. ∎

We next characterize circuits of minors.

Theorem 31.25 *Let $\mathcal{M}$ be a matroid on S and let $X \subseteq S$. Then*

1. *C_X is a circuit of $\mathcal{M} \cdot X$ iff it is a circuit of $\mathcal{M}$ contained in X.*

2. *C_X is circuit of $\mathcal{M} \times X$ iff it is a minimal nonvoid intersection of a circuit of $\mathcal{M}$ with X.*

Proof.

i. Independent sets of $\mathcal{M} \cdot X$ are just the independent sets of $\mathcal{M}$ contained in X. Hence, C_X is a minimal dependent set of $\mathcal{M} \cdot X$ iff it is a minimal dependent set of $\mathcal{M}$ contained in X.

ii. We use the following observation. If C is a circuit of $\mathcal{M}$ intersecting X, then $C \cap X$ is dependent in $\mathcal{M} \times X$. (This follows as otherwise, the disjoint union $(C \cap X) \cup (C \cap (S - X))$ which is indeed C, would be independent in $\mathcal{M}$.)

Let C_X be a circuit of $\mathcal{M} \times X$. Let B' be a base of $\mathcal{M} \cdot (S - X)$. Therefore $C_X \cup B'$ is dependent in $\mathcal{M}$, but $(C_X - e) \cup B'$ is independent in $\mathcal{M}$ for any $e \in C_X$. Let C be the unique circuit of $\mathcal{M}$ containing e, contained in $e \cup (C_X - e) \cup B'$. Note that $C \cap X \subseteq C_X$.

But $C \cap X$ is dependent in $\mathcal{M} \times X$ by the above observation. Therefore due to C_X being a circuit of $\mathcal{M} \times X$, we conclude that $C_X = C \cap X$.

We still need to prove that C_X is a *minimal* nonvoid intersection of a circuit of $\mathcal{M}$ with X. If $C_X = C \cap X$ were not minimal such, then there would exist nonvoid $C'_X = C' \cap X$, C' a circuit of $\mathcal{M}$ such that $C'_X \subset C_X$ (proper!). But then C'_X would be dependent in $\mathcal{M} \times X$ by the above observation, contradicting that C_X is a circuit of $\mathcal{M} \times X$.

Next we prove the other implication, that is, a minimal nonvoid intersection $C \cap X$ of a circuit C of $\mathcal{M}$ with X is a circuit of $\mathcal{M} \times X$. Suppose the contrary, then there exists a circuit C'_X of $\mathcal{M} \times X$, that is properly contained in $C \cap X$. But then by the above there exists a circuit C' of $\mathcal{M}$ such that $C' \cap X = C'_X$, contradicting that $C \cap X$ is a minimal nonvoid intersection of a circuit of $\mathcal{M}$ with X. ■

The next result speaks of the rank function of minors. The routine proof is omitted (the expression for rank function of contraction follows from Lemma 31.3).

Theorem 31.26 *Let $\mathcal{M}$ be a matroid on S and let $X \subseteq S$. Let $r(\cdot), r_r(\cdot), r_c(\cdot)$ be the rank functions of $\mathcal{M}, \mathcal{M} \cdot X, \mathcal{M} \times X$ respectively. Then*

1. $r_r(Y) = r(Y), Y \subseteq X,$
2. $r_c(Y) = r(Y \cup (S - X)) - r(S - X), Y \subseteq X.$ ■

A minor of a general form could be obtained from the original matroid by a sequence of restrictions and contractions. As in the case of graphs we can simplify these operations to a single contraction followed by a single restriction or vice versa. The following result is needed for such simplification.

Theorem 31.27 *Let $\mathcal{M}$ be a matroid on S and let $Y \subseteq X \subseteq S$. Then,*

1. $\mathcal{M} \cdot X \cdot Y = \mathcal{M} \cdot Y.$
2. $\mathcal{M} \times X \times Y = \mathcal{M} \times Y.$
3. $\mathcal{M} \times X \cdot Y = \mathcal{M} \cdot (S - (X - Y)) \times Y.$

Proof.

i. Immediate from the definition of restriction.
ii. A base of $\mathcal{M} \times Y$ is a minimal intersection of a base of $\mathcal{M}$ with Y, while a base of $\mathcal{M} \times X \times Y$ is a minimal intersection of a base of $\mathcal{M} \times X$ with Y. To construct the former base, one could begin with a maximally independent set of $\mathcal{M}$ within $S - Y$ and extend it to a base of $\mathcal{M}$ using a set B_Y of elements from Y. B_Y would then be a base of $\mathcal{M} \times Y$. But this could have been done by first choosing a maximally independent set B_{S-X} within $S - X$ of $\mathcal{M}$, extending it to a maximally independent set B_{S-Y} of $\mathcal{M}$ within $S - Y$, and then growing it further using B_Y. Thus B_Y is a base of $\mathcal{M} \times X \times Y$. On the other hand, starting with a base B_Y of $\mathcal{M} \times X \times Y$, one can see that its union with a maximally independent set B_{S-Y} of $\mathcal{M}$ within $S - Y$ gives a base of $\mathcal{M}$. Therefore B_Y is a minimal intersection of a base of $\mathcal{M}$ with Y and therefore a base of $\mathcal{M} \times Y$.

A more routine proof uses Theorem 31.28 proved below.

$$(\mathcal{M}^* \cdot Y)^* = (\mathcal{M}^* \cdot X \cdot Y)^* = (\mathcal{M}^* \cdot X)^* \times Y = (\mathcal{M}^*)^* \times X \times Y = \mathcal{M} \times X \times Y.$$

But $(\mathcal{M}^* \cdot Y)^* = (\mathcal{M}^*)^* \times Y = \mathcal{M} \times Y.$ We conclude that $\mathcal{M} \times X \times Y = \mathcal{M} \times Y.$

iii. Let Z be an independent set of $\mathcal{M} \times X \cdot Y$. Then, by the definition of restriction, $Z \subseteq Y$ and Z is independent in $\mathcal{M} \times X$. Let B_{S-X} be a base of $\mathcal{M} \cdot (S - X)$. Then by the definition of contraction, $Z \cup B_{S-X}$ is independent in $\mathcal{M}$. By the definition of restriction $Z \cup B_{S-X}$ must be independent in $\mathcal{M} \cdot (S - (X - Y))$. Now B_{S-X} is a base of $\mathcal{M} \cdot (S - X) = \mathcal{M} \cdot (S - (X - Y)) \cdot (S - X)$. Hence, Z is independent in $\mathcal{M} \cdot (S - (X - Y)) \times Y$ (note that $(S - X) \cup Y = S - (X - Y)$ since $Y \subseteq X$). It is easy to see that the above sequence of implications can be reversed. Hence, if Z is independent in $\mathcal{M} \cdot (S - (X - Y)) \times Y$ then Z is also independent in $\mathcal{M} \times X \cdot Y$. Thus,
$$\mathcal{M} \times X \cdot Y = \mathcal{M} \cdot (S - (X - Y)) \times Y.$$
■

Let $\mathcal{M}$ be a matroid on S and let $S \supseteq X_1 \supseteq X_2 \supseteq \cdots \supseteq X_n$. From Theorem 31.27, it is clear that $\mathcal{M} \times X_1 \cdot X_2 \times X_3 \ldots X_n$ can be written in the form $\mathcal{M} \times P \cdot X_n$ for a suitable $P \supseteq X_n$.

Now we relate the minors of the dual matroid to the duals of the minors of the original matroid.

Theorem 31.28 *Let $\mathcal{M}$ be a matroid on S. Let $X \subseteq S$. Then*

1. $(\mathcal{M} \times X)^* = \mathcal{M}^* \cdot X$,

2. $(\mathcal{M} \cdot X)^* = \mathcal{M}^* \times X$.

3. *The rank of X in $\mathcal{M}^*$ equals $|X| - r(\mathcal{M} \times X) = |X| - (r(\mathcal{M}) - r(\mathcal{M} \cdot (S - X)))$.*

Proof.

i. B_X is a base of $\mathcal{M} \times X$ iff it is a minimal intersection of a base of $\mathcal{M}$ with X, that is, iff it is the complement of a maximal intersection of a cobase of $\mathcal{M}$ with X, that is, iff it is the complement of a maximal intersection of a base of $\mathcal{M}^*$ with X, that is, iff it is the complement of a base of $\mathcal{M}^* \cdot X$.

ii. By the definition of dual, dual of the dual of a matroid is the original matroid itself. Hence, using (i) above, $(\mathcal{M}^* \times X)^* = (\mathcal{M}^*)^* \cdot X = \mathcal{M} \cdot X$

i.e., $\mathcal{M}^* \times X = (\mathcal{M}^* \times X)^{**} = (\mathcal{M} \cdot X)^*$.

iii. Immediate from (i) above.
■

31.4.6 Notes

The reader interested in making a serious study of matroid theory would do well to begin by referring to [8] where the key foundation papers in matroid theory are reproduced. A terse but readable account of matroid theory, as seen in the 1960s by a master, is available in [9]. Matroids seen from the unifying perspective of combinatorial geometries is available in [10]. The first comprehensive textbook on matroids is [11]. More recent books are [12–14].

31.5 CONVOLUTION

31.5.1 Introduction

The convolution operation popularized by Edmonds [15] is fundamental to the study of submodular functions and is extremely useful for both studying the structure of matroids and generating new matroids. Indeed, the well-known concept of principal partition associated with submodular functions in terms of a given weight function and the union matroid of two matroids, to name just two important instances, are best understood using convolution. In this section we begin with a description of polymatroid rank functions and use this as a framework for presenting results on convolution, and principal partition.

31.5.2 Polymatroid Rank Functions

It is convenient to state the basic results for convolution in terms of polymatroid rank functions which are a simple generalization of matroid rank functions.

We remind the reader that a function $f(\cdot) : 2^S \longrightarrow \Re$ is said to be *submodular* if for every pair of subsets X, Y of S we have $f(X) + f(Y) \geq f(X \cup Y) + f(X \cap Y)$. It is *supermodular* if the inequality is reversed and *modular* if the inequality is replaced by an equality. A *weight* function $g(\cdot)$ on subsets of S satisfies $g(X) \equiv \sum_{e_i \in X} g(e_i), X \subseteq S$, with $g(\emptyset) = 0$. A modular function can be seen to differ from a weight function only in that, for the latter, the value on null set is zero. Indeed if $g(\cdot)$ is modular we have $g(X) = \sum_{e \in X}(g(e) - g(\emptyset)) + g(\emptyset)$.

We saw in Section 31.2.4 that a matroid rank function $r(\cdot)$ on subsets of S is increasing, submodular with $r(\emptyset) = 0$, integral and satisfies $r(e) = 0$ or 1 for all $e \in S$. A polymatroid rank function $f(\cdot)$ on subsets of S is increasing, submodular with $r(\emptyset) = 0$, the remaining conditions being omitted.

We define restriction, contraction, and dual of a polymatroid rank function essentially as in the case of a matroid rank function. The restriction $f(\cdot)/T$ of $f(\cdot)$ to $T, T \subseteq S$, is defined by $f/T(X) \equiv f(X), X \subseteq T$. The contraction $f \diamond T(\cdot)$ of $f(\cdot)$ to $T, T \subseteq S$, is defined by $f \diamond T(X) \equiv f(X \cup (S - T)) - f(S - T), X \subseteq T$.

The dual $f^*(\cdot)$ of $f(\cdot)$ on subsets of S is defined relative to a positive weight function $g(\cdot)$ on subsets of S just as $r^*(\cdot)$ is defined relative to the $|\cdot|$ function (see Theorem 31.28):

$$f^*(X) \equiv g(X) - [f(S) - f(S - X)], X \subseteq S.$$

It is easily verified that $f^{**}(\cdot) = f(\cdot)$. Further, if $f(e) \leq g(e), \forall e \in S$, it will follow that $f^*(\cdot)$ is a polymatroid rank function whenever $f(\cdot)$ is.

Just as in the case of matroid rank functions, contraction and restriction operations turn out to be duals of each other for polymatroid rank functions.

We have, for $X \subseteq T$,

$$\begin{aligned}(f \diamond T)^*(X) &= g(X) - [f \diamond T(T) - f \diamond T(T - X)] \\ &= g(X) - [f(S) - f(S - T) - f((S - T) \cup (T - X)) + f(S - T)] \\ &= g(X) - [f(S) - f((S - T) \cup (T - X))] \\ &= g(X) - [f(S) - f(S - X)] = f^*/T(X).\end{aligned}$$

Since $f^{**}(\cdot) = f(\cdot)$, it will follow that $(f/T)^*(X) = (f^* \diamond T)(X)$.

31.5.3 Formal Properties of the Convolution Operation

Definition 31.5 *Let $f(\cdot), g(\cdot) : 2^S \longrightarrow \Re$. The lower convolution of $f(\cdot)$ and $g(\cdot)$, denoted by $f^*g(\cdot)$, is defined by*

$$f^*g(X) \equiv min_{Y \subseteq X}[f(Y) + g(X - Y)].$$

The collection of subsets Y, at which $f(Y) + g(X - Y) = f^*g(X)$, is denoted by $\mathcal{B}_{f,g}(X)$. But if $X = S$, we will simply write $\mathcal{B}_{f,g}$.

It is clear that $f^*g(\cdot) = g^*f(\cdot)$. We now have the following elementary but important result.

Theorem 31.29 *If $f(\cdot)$ is submodular on subsets of S and $g(\cdot)$ is modular, then $f^*g(\cdot)$ is submodular.*

We need the following lemma for the proof of the theorem.

Lemma 31.4 *Let $g(\cdot)$ be a modular function on the subsets of S. Let $A, B, C, D \subseteq S$ such that $A \cup B = C \cup D$, $A \cap B = C \cap D$. Then*

$$g(A) + g(B) = g(C) + g(D).$$

Proof. Both LHS and RHS in the statement of the lemma are equal to

$$\sum_{e \in A \cup B} (g(e) - g(\emptyset)) + \sum_{e \in A \cap B} (g(e) - g(\emptyset)) + 2(g(\emptyset)). \qquad \blacksquare$$

Proof of Theorem 31.29. Let $X, Y \subseteq S$. Further let

$$f^*g(X) = f(Z_X) + g(X - Z_X), f^*g(Y) = f(Z_Y) + g(Y - Z_Y).$$

Then,

$$f^*g(X) + f^*g(Y) = f(Z_X) + g(X - Z_X) + f(Z_Y) + g(Y - Z_Y).$$

We observe that, since $Z_X \subseteq X, Z_Y \subseteq Y$,

$$(X - Z_X) \cup (Y - Z_Y) = (X \cup Y - (Z_X \cup Z_Y)) \cup (X \cap Y - (Z_X \cap Z_Y))$$

and

$$(X - Z_X) \cap (Y - Z_Y) = ((X \cup Y) - (Z_X \cup Z_Y)) \cap (X \cap Y - (Z_X \cap Z_Y)),$$

we must have, by Lemma 31.4,

$$g(X - Z_X) + g(Y - Z_Y) = g(X \cup Y - (Z_X \cup Z_Y)) + g(X \cap Y - (Z_X \cap Z_Y)).$$

Hence, $f^*g(X) + f^*g(Y)$
$\geq f(Z_X \cup Z_Y) + f(Z_X \cap Z_Y) + g(X \cup Y - (Z_X \cup Z_Y)) + g(X \cap Y - (Z_X \cap Z_Y))$. Thus,

$$f^*g(X) + f^*g(Y) \geq f^*g(X \cup Y) + f^*g(X \cap Y),$$

which is the desired result. $\blacksquare$

Remark 31.2 It is clear that if $g(\cdot)$ is not modular, but only submodular, then $g(X - Z_X) + g(Y - Z_Y)$ need not be greater or equal to $g(X \cup Y - (Z_X \cup Z_Y)) + g(X \cap Y - (Z_X \cap Z_Y))$. Thus the above proof would not hold if $g(\cdot)$ is only submodular. Indeed the following counterexample shows the convolution of two submodular functions need not be submodular. Let B_1, B_2 be bipartite graphs on $V_L \equiv \{a, b, c\}, V_R \equiv \{a', b', c', d'\}$ with adjacency functions Γ_1, Γ_2 defined as follows:

$$\Gamma_1(a) = \{a', b', d'\}, \Gamma_1(b) = \{a', b', d'\}, \Gamma_1(c) = \{b', c', d'\},$$
$$\Gamma_2(a) = \{a', b', c'\}, \Gamma_2(b) = \{a', b', d'\}, \Gamma_2(c) = \{b', c'\}.$$

It may be verified that

$$|\Gamma_1|^*|\Gamma_2|(a) = 3, |\Gamma_1|^*|\Gamma_2|(a, b) = 3, |\Gamma_1|^*|\Gamma_2|(a, c) = 3, |\Gamma_1|^*|\Gamma_2|(a, b, c) = 4.$$

Hence

$$|\Gamma_1|^*|\Gamma_2|(a, b, c) - |\Gamma_1|^*|\Gamma_2|(a, c) > |\Gamma_1|^*|\Gamma_2|(a, b) - |\Gamma_1|^*|\Gamma_2|(a).$$

This shows that $|\Gamma_1|^*|\Gamma_2|(\cdot)$ is not submodular. But it is easily seen that $|\Gamma_1(\cdot)|, |\Gamma_2(\cdot)|$ are submodular.

Theorem 31.30 *Let $f(\cdot), g(\cdot)$ be arbitrary set functions on subsets of S.*

1. *Then $f^*g(X \cup e) - f^*g(X)$*
   $\leq \min[\max_{Y \subseteq X}(f(Y \cup e) - f(Y)), \max_{Y \subseteq X}(g(Y \cup e) - g(Y))], \quad X \subseteq S, e \in S$.

2. *Let $f(\cdot), g(\cdot)$ be increasing. Then $f^*g(\cdot)$ is increasing.*

3. *Let $f(\cdot), g(\cdot)$ be integral. Then so is $f^*g(\cdot)$.*

4. *Let $f(\cdot)$ be an integral polymatroid rank function and let $g(\cdot) = |\cdot|$. Then $f^*g(\cdot)$ is a matroid rank function* [15].

Proof.

i. Let $f^*g(X) = f(Z) + g(X - Z)$, where $Z \subseteq X$. Then
$$f^*g(X \cup e) \leq \min[f(Z \cup e) + g(X - Z), f(Z) + g((X - Z) \cup e)].$$
The proof is now immediate.

ii. Let, without loss of generality
$$f^*g(X \cup e) = f(Z \cup e) + g(X - Z), Z \subseteq X, e \in (S - X).$$
But then
$$f^*g(X) \leq f(Z) + g(X - Z) \leq f(Z \cup e) + g(X - Z).$$

iii. The proof is immediate from the definition of convolution.

iv. We need to show that $f^*g(\cdot)$ is an integral polymatroid rank function that takes value atmost one on singletons. We have, $f(\cdot), g(\cdot)$ are increasing, integral, submodular, taking value zero on the null set and further $g(\cdot)$ is a weight function with $g(e) = 1 \ \forall e \in S$. From Theorem 31.29 it follows that $f^*g(\cdot)$ is submodular. It is clear that $f^*g(\emptyset) = 0$. The remaining properties for being a matroid rank function follow from the preceding sections of the present theorem. ∎

Theorem 31.31 *Let $\rho(\cdot)$ be an integral polymatroid rank function on subsets of S. A set $X \subseteq S$ is independent in the matroid whose rank function is $\rho^* | \cdot |$ iff $\rho(Y) \geq |Y| \ \forall Y \subseteq X$.*

Proof. Let $r(\cdot) \equiv \rho^* | \cdot |$. A set $X \subseteq S$ is independent iff $r(X) = |X|$, that is, iff $(\rho^* | \cdot |)(X) = |X|$, that is, iff
$$\min_{Y \subseteq X}(\rho(Y) + |X - Y|) = |X|.$$
Clearly this would happen iff $\rho(Y) \geq |Y| \ \forall Y \subseteq X$. ∎

31.5.4 Connectedness for f^*g

Let $f(\cdot)$ be submodular on subsets of S with $f(\phi) = 0$. We say that T is a separator for S iff
$$f(T) + f(S - T) = f(S).$$

We then have the following result.

Theorem 31.32 *Let $f(\cdot)$ be submodular on subsets of S with $f(\phi) = 0$ and let T be a separator of $f(\cdot)$. Suppose $X \subseteq T, Y \subseteq S - T$. Then, $f(X) + f(Y) = f(X \cup Y)$.*

By submodularity,
$$f(Y) + f(T) \geq f(Y \cup T)$$
$$\text{i.e., } f(Y) + f(S) - f(S - T) \geq f(Y \cup T)$$
$$\text{i.e., } f(Y \cup T) + f(S - T) \leq f(Y) + f(S).$$

But by submodularity again,
$$f(Y \cup T) + f(S - T) \geq f(S) + f(Y).$$

We conclude that $f(Y \cup T) + f(S - T) = f(S) + f(Y)$, that is, $f(Y \cup T) = f(S) - f(S - T) + f(Y) = f(T) + f(Y)$. We could now repeat the argument working with X in place of Y, Y in place of T and obtain
$$f(X \cup Y) = f(X) + f(Y). \qquad \blacksquare$$

Thus when $f(\cdot)$ is submodular with $f(\phi) = 0$ on subsets of S and T is a separator, $f(\cdot)$ behaves as though it is the direct sum of its restrictions on subsets of T and subsets of $S - T$ since
$$f(X) = f(X \cup T) + f(X \cap (S - T)).$$

The above discussion is particularly relevant when we consider $f*g$, where $f(\cdot)$ is submodular with $f(\phi) = 0$ and $g(\cdot)$ is a positive weight function on subsets of S.

Suppose
$$f*g(S) = f(T) + g(S - T).$$

We claim $f*g(S) = f*g(T) + f*g(S - T)$. To see this suppose $f*g(T) = f(T_1) + g(T - T_1)$, $T_1 \subseteq T$ and $f*g(S - T) = f(T_2) + g(S - T - T_2)$, $T_2 \subseteq S - T$. We then have
$$f(T) + g(S - T) \geq f(T_1) + g(T - T_1) + f(T_2) + g(S - T - T_2), \ldots (*).$$

By submodularity of $f(\cdot)$,
$$f(T) + g(S - T) \geq f(T_1 \cup T_2) + g((T - T_1) \cup (S - T - T_2)) = f(T_1 \cup T_2) + g(S - (T_1 \cup T_2)).$$

However,
$$f(T) + g(S - T) = \min_{X \subseteq S} f(X) + g(S - X).$$

Hence, $f(T) + g(S - T) = f(T_1 \cup T_2) + g(S - (T_1 \cup T_2))$, and the inequality $(*)$ above is an equality. The only way this can happen is if $f(T) = f(T_1) + g(T - T_1)$ and $g(S - T) = f(T_2) + g(S - T - T_2)$. It follows that $f*g(T) = f(T)$ and $f*g(S - T) = g(S - T)$ and therefore
$$f*g(S) = f*g(T) + f*g(S - T),$$

proving the claim. Thus, $T, S - T$ are separators of $f*g$.

When $f(\cdot)$ is an integral polymatroid rank function and $g(\cdot) = |\cdot|$, if $f*g(S) = f(T) + g(S - T)$, the matroid $\mathcal{M}_{f*g}$ whose rank function is $f*g$ has $T, S - T$ as separators. Further $f*g(S - T) = |S - T|$ so that $(S - T)$ is independent. Now consider any base b of $\mathcal{M}_{f*g}$. We have $b = b \cap T \cup (b \cap (S - T))$. However
$$f*g(b \cap T \cup (S - T)) = f*g(b \cap T) + f*g(S - T)) = |b \cap T| + |S - T|.$$

It is thus clear that $b \cap T \cup (S - T)$ is independent. Since b is maximally independent in $\mathcal{M}_{f*g}$, we conclude that $b \cap (S - T) = S - T$. Thus $S - T$ is a subset of every base of $\mathcal{M}_{f*g}$, that is, $S - T$ is a set of coloops (elements which do not belong to any circuit) of $\mathcal{M}_{f*g}$.

On the other hand, suppose $(S-K)$ is the set of all coloops of $\mathcal{M}_{f*g}$. It is clear that

$$f*g(S) = f*g(K) + |S-K| = f*g(K) + f*g(S-K).$$

Thus $K, (S-K)$ are separators of $f*g$. We claim $f*g(K) = f(K)$. For, if $f*g(K) = f(K_1) + g(K-K_1)$ and $K_1 \subset K$, we have $f*g(S) = f(K_1) + g((S-K) \cup (K-K_1))$. But this means $(S-K) \cup (K-K_1)$ is a set of coloops which contradicts the fact $(S-K)$ is the set of all coloops of $\mathcal{M}_{f*g}$.

We summarize the above discussion in the following theorem.

Theorem 31.33 *Let $f*g(S) = f(T) + g(S-T), T \subset S$.*

1. *If $f(\cdot)$ is submodular on subsets of S with $f(\phi) = 0$, then $T, S-T$ are separators of $f*g(\cdot)$.*

2. *If $f(\cdot)$ is an integral polymatroid rank function and $g(\cdot) = |\cdot|$, the matroid $\mathcal{M}_{f*g}$ whose rank function is $f*g$ has $T, S-T$ as separators and $S-T$ as a set of coloops. Also, if $S-T$ is the set of all coloops of $\mathcal{M}_{f*g}$, then $f*g(S) = f(T) + g(S-T)$.*

31.6 PRINCIPAL PARTITION

31.6.1 Introduction

The *principal partition of a graph* was defined by Kishi and Kajitani in their seminal paper [16] and was originally an offshoot of their work on maximally distant trees. A graph was decomposed into three minors according to how strongly subsets of edges can be covered by unions of two trees or two cotrees. The extensions of this concept can be in two directions: toward making the partition finer or toward making the functions involved more general. Our present description favors the former approach and is mainly aimed at describing the principal partition of a matroid [17,18]. However, the results are best stated in terms of convolution of polymatroid rank functions with positive weight functions. Essentially, we study the collection of all minimizing T such that $\lambda f*g(S) = \lambda f(T) + g(S-T), \lambda \geq 0$ where $f(\cdot)$ is a polymatroid rank function and $g(\cdot)$, a positive weight function on subsets of S. The algorithms for building this structure for a matroid rank function are based on the matroid union algorithm which finds the maximal union of bases from two different matroids.

31.6.2 Basic Properties of Principal Partition

Definition 31.6 *Let $f(\cdot), g(\cdot)$ be a polymatroid rank function and a positive weight function respectively on the subsets of a set S. The collection of all sets in $\mathcal{B}_{\lambda f,g}$ (i.e., the collection of sets $X \subseteq S$ which minimize $\lambda f(X) + g(S-X)$ over subsets of S) $\forall \lambda, \lambda \geq 0$, is called the principal partition (PP) of $(f(\cdot), g(\cdot))$.*

We denote $\mathcal{B}_{\lambda f,g}$ by $\mathcal{B}_\lambda$ when $f(\cdot), g(\cdot)$ are clear from the context. We denote the maximal and minimal members of $\mathcal{B}_\lambda$ by X^λ, X_λ, respectively.

We now list the important properties of the principal partition of $(f(\cdot), g(\cdot))$.

Property PP1
The collection $\mathcal{B}_{\lambda f,g}, \lambda \geq 0$, is closed under union and intersection and thus has a unique maximal and a unique minimal element.

Property PP2
If $\lambda_1 > \lambda_2 \geq 0$, then $X^{\lambda_1} \subseteq X_{\lambda_2}$.

Definition 31.7 *A non-negative value λ for which $\mathcal{B}_\lambda$ has more than one subset as a member is called a critical value of $(f(\cdot), g(\cdot))$.*

Property PP3
The number of critical values of $(f(\cdot), g(\cdot))$ is bounded by $|S|$.

Property PP4
Let $(\lambda_i), i = 1, \ldots, t$ be the decreasing sequence of critical values of $(f(\cdot), g(\cdot))$. Then, $X^{\lambda_i} = X_{\lambda_{i+1}}$ for $i = 1, \ldots, t-1$.

Property PP5
Let (λ_i) be the decreasing sequence of critical values. Let $\lambda_i > \sigma > \lambda_{i+1}$. Then $X^{\lambda_i} = X^\sigma = X_\sigma = X_{\lambda_{i+1}}$.

Definition 31.8 *Let $f(\cdot)$ be a polymatroid rank function and let $g(\cdot)$ be a positive weight function on subsets of S. Let $(\lambda_i), i = 1, \ldots, t$ be the decreasing sequence of critical values of $(f(\cdot), g(\cdot))$. Then the sequence $X_{\lambda_1}, X_{\lambda_2}, \ldots, X_{\lambda_t}, X^{\lambda_t} = S$ is called the principal sequence of $(f(\cdot), g(\cdot))$. A member of $\mathcal{B}_\lambda$ would be alternatively referred to as a minimizing set corresponding to λ in the principal partition of $(f(\cdot), g(\cdot))$.*

Remark 31.3 A word about the terminology is in order. For convenience, we have defined the principal partition to be the collection of all minimizing sets of the expression $\lambda f(X) + g(S - X)$. Literally speaking, this is not a partition. However, there is a natural associated partition which is simply the collection of all minimal sets of the form $X_1 - X_2$, where X_1, X_2 minimize $\lambda f(X) + g(S - X)$, for a critical value λ. The coarser partition associated with the principal sequence is $X_{\lambda_1}, X_{\lambda_2} - X_{\lambda_1}, \ldots, X^{\lambda_t} - X_{\lambda_t}$.

Proof of the properties of the principal partition.

i. *PP1:* Define $h(X) \equiv \lambda f(X) + g(S - X) \forall X \subseteq S, \lambda \geq 0$. Observe that the function $g'(\cdot)$, defined through $g'(X) \equiv g(S - X) \forall X \subseteq S$, is submodular. Thus $h(\cdot)$ is the sum of two submodular functions and is therefore submodular. The collection of sets on which this function reaches a minimum is called the principal structure of $h(\cdot)$ [19]. If T_1, T_2 minimize $h(\cdot)$, since $h(T_1) + h(T_2) \geq h(T_1 \cup T_2) + h(T_1 \cap T_2)$, it follows that $T_1 \cup T_2, T_1 \cap T_2$ also minimize $h(\cdot)$. Thus the principal structure of $h(\cdot)$ is closed under union and intersection and therefore has a unique minimal and a unique maximal set. The principal structure of $h(\cdot)$ is precisely the same as $\mathcal{B}_\lambda$.

ii. *PP2:* Observe that minimizing $\lambda_i f(X) + g(S - X) \forall X \subseteq S, \lambda_i \geq 0, i = 1, 2$, is equivalent to minimizing $f(X) + (\lambda_i)^{-1} g(S - X) \quad \forall X \subseteq S, \lambda_i \geq 0, i = 1, 2$. (Here $+\infty \times 0$, corresponding to $\lambda_i = 0$ and $g(\emptyset) = 0$, is treated as zero.) So we may take the sets which minimize the latter expression to be the sets in $\mathcal{B}_{\lambda_i}, i = 1, 2$. Define $p_i(X) \equiv f(X) + (\lambda_i)^{-1} g(S - X) \quad \forall X \subseteq S, \lambda_i \geq 0, i = 1, 2$. As in the case of $h_i(\cdot), p_i(\cdot), i = 1, 2$ is also submodular. Let Z_1 minimize $p_1(\cdot)$. We will now show that $p_2(Z_1) < p_2(Y) \quad \forall Y \subset Z_1$. Let $Y \subset Z_1$. We have,

$$p_2(Z_1) = p_1(Z_1) + ((\lambda_2)^{-1} - (\lambda_1)^{-1}) g(S - Z_1)$$

and

$$p_2(Y) = p_1(Y) + ((\lambda_2)^{-1} - (\lambda_1)^{-1}) g(S - Y).$$

Since $g(\cdot)$ is a positive weight function, $S - Z_1 \subset S - Y$ and $((\lambda_2)^{-1} - (\lambda_1)^{-1}) > 0$, we must have $((\lambda_2)^{-1} - (\lambda_1)^{-1}) g(S - Z_1) < ((\lambda_2)^{-1} - (\lambda_1)^{-1})$

$g(S - Y)$. Since $p_1(Y) \geq p_1(Z_1)$, it follows that $p_2(Y) > p_2(Z_1)$. Now let $Z \subseteq S$ minimize $p_2(\cdot)$. Applying the submodular inequality for p_2 on Z, Z_1, it would follow that if $Z \cup Z_1$ is not the same as Z, then $p_2(Z \cap Z_1) \leq p_2(Z_1)$, with $(Z \cap Z_1) \neq Z_1$ which would be a contradiction. It follows that $Z \supseteq Z_1$.

iii. *PP3:* If $\mathcal{B}_\lambda$ has more than one set as a member then $|X^\lambda| > |X_\lambda|$. So if λ_1, λ_2 are critical values and $\lambda_1 > \lambda_2$, by Property PP2, we must have $|X_{\lambda_1}| < |X_{\lambda_2}|$. Thus the sequence X_{λ_i}, where (λ_i) is the decreasing sequence of critical values cannot have more than $|S|$ elements.

iv. *PP4:* We need the following lemma.

Lemma 31.5 *Let $\lambda > 0$. Then, for sufficiently small $\epsilon > 0$, the only set that minimizes $\lambda - \epsilon$ is X^λ.*

Proof. Since there are only a finite number of $(f(X), g(S - X))$ pairs, for sufficiently small $\epsilon > 0$ we must have the value of $(\lambda - \epsilon)f(X) + g(S - X)$ lower on the members of $\mathcal{B}_\lambda$ than on any other subset of S. We will now show that, among the members of $\mathcal{B}_\lambda, X^\lambda$ takes the least value of $(\lambda - \epsilon)f(X) + g(S - X), \epsilon > 0$. This would prove the required result. If λ is not a critical value this is trivial. Let λ be a critical value and let X_1, X^λ be two distinct sets in $\mathcal{B}_\lambda$. Since $X_1 \subset X^\lambda$, we have, $g(S - X_1) > g(S - X^\lambda)$. But, $\lambda f(X_1) + g(S - X_1) = \lambda f(X^\lambda) + g(S - X^\lambda)$. So, $\lambda f(X_1) < \lambda f(X^\lambda)$. Since $\lambda > 0$, we must have, $-\epsilon f(X_1) > -\epsilon f(X^\lambda), \epsilon > 0$. It follows that, $(\lambda - \epsilon)f(X_1) + g(S - X_1) > (\lambda - \epsilon)f(X^\lambda) + g(S - X^\lambda)$. ∎

Proof of PP4: By Lemma 31.5, for sufficiently small values of $\epsilon > 0$, X^{λ_i} would continue to minimize $(\lambda_i - \epsilon)f(X) + g(S - X)$. As ϵ increases, because there are only a finite number of $(f(X), g(S - X))$ pairs, there would be a least value, say σ, at which X^{λ_i} and atleast one other set minimize $(\lambda_i - \sigma)f(X) + g(S - X)$. Clearly, the next critical value $\lambda_{i+1} = \lambda_i - \sigma$. Since $\lambda_i > \lambda_i - \sigma$, by Property PP2, we must have $X^{\lambda_i} \subseteq X_{\lambda_i - \sigma}$. Hence we must have, $X^{\lambda_i} = X_{\lambda_i - \sigma} = X_{\lambda_{i+1}}$, as desired.

Proof of PP5: This is clear from the above arguments.

Informally, the situation is as follows. Suppose we start with $\lambda = +\infty$. Here X_∞ would be the null set and X^∞, the set of elements e for which $f(e)$ is zero. As we reduce λ, a point would be reached, say when $\lambda = \lambda_1$, where X^{λ_1} becomes a proper superset of X^∞. This would be the next critical value. Between ∞ and λ_1, $X^\lambda = X_\lambda = X^\infty$. As we lower λ further, $X^\lambda = X_\lambda = X^{\lambda_1}$, till the next critical value λ_2 is reached. The last critical value λ_k will be such that $X^{\lambda_k} = S$. When $\lambda = 0$, it is clear that the minimum of $\lambda f(X) + g(S - X)$ is reached only at S. It follows that all critical values have to be positive.

A characterization of principal partition would be useful for justifying algorithms for its construction. We will describe one such characterization in Theorem 31.34 below. This is a routine restatement of the properties of principal partition (PP).

Theorem 31.34 *Let $f(\cdot)$ be a polymatroid rank function on subsets of S and let $g(\cdot)$ be a positive weight function on subsets of S. Let $\mathcal{B}_\lambda$ denote $\mathcal{B}_{\lambda f, g}$. Let $\lambda_1, \ldots, \lambda_t$ be a strictly decreasing sequence of numbers such that*

1. *Each $\mathcal{B}_{\lambda_i}, i = 1, \ldots, t$ has atleast two members,*

2. *$\mathcal{B}_{\lambda_i}, \mathcal{B}_{\lambda_{i+1}}, i = 1, \ldots, t - 1$ have atleast one common member set,*

3. *$\emptyset$ belongs to $\mathcal{B}_{\lambda_1}$, while S belongs to $\mathcal{B}_{\lambda_t}$.*

Then $\lambda_1, \ldots, \lambda_t$ is the decreasing sequence of critical values of $(f(\cdot), g(\cdot))$ and therefore the collection of all the sets which are member sets in all the $\mathcal{B}_{\lambda_i}, i = 1, \ldots, t$ is the principal partition of $(f(\cdot), g(\cdot))$.

Proof. We note that, by definition, $\lambda_1, \ldots, \lambda_t$ are some of the critical values and, in the present case, $\emptyset = X_{\lambda_1}, X_{\lambda_2}, \ldots, X_{\lambda_t}, X^{\lambda_t} = S$ is a subsequence of the principal sequence. Let $\lambda'_1, \ldots, \lambda'_k$ be the critical values and let $Y_0, \ldots, Y_k = S$ be the principal sequence of $(f(\cdot), g(\cdot))$. Since the principal sequence is increasing, it follows that $Y_0 = \emptyset$. By Property PP2 of $(f(\cdot), g(\cdot))$, the only member set in $\mathcal{B}_\lambda$, when $\lambda > \lambda'_1$, is Y_0. Further when $\lambda < \lambda'_1$, Y_0 is not in $\mathcal{B}_\lambda$. Hence $\lambda_1 = \lambda'_1$. Next by Property PP5, when $\lambda'_1 > \lambda > \lambda'_2$, the only member in $\mathcal{B}_\lambda$ is Y_1 which is the maximal set in $\mathcal{B}_{\lambda'_1}$. Since $\mathcal{B}_{\lambda_2}$ has atleast two sets we conclude that $\lambda_2 \leq \lambda'_2$. We know that $\mathcal{B}_{\lambda_1}$ and $\mathcal{B}_{\lambda_2}$ have a common member which by Property PP2 can only be Y_1. But for $\lambda < \lambda'_2$, by Property PP5, Y_1 cannot be a member of $\mathcal{B}_\lambda$. Hence $\lambda_2 = \lambda'_2$. By repeating this argument, we see that t must be equal to k and $\lambda_i = \lambda'_i, i = 1, \ldots, t$. ∎

31.6.3 Principal Partition of Contraction and Restriction

There is a simple relationship between the principal partition of a polymatroid rank function and its restrictions and contractions relative to sets in the principal partition.

If the function is $f \diamond T(\cdot), S - T \subseteq X^{\lambda_1}$ then the principal partition for $\lambda < \lambda_1$ remains essentially that of $f(\cdot)$ except that we have $X - (S - T)$ as minimizing set in place of X, a superset of $S - T$. On the other hand if the function is $f(\cdot)/T$, where $T \supseteq X^{\lambda_1}$, then the principal partition for $\lambda \geq \lambda_1$, is identical to that of $f(\cdot)$. We formalize these ideas below.

Theorem 31.35 *Let $f(\cdot)$ be a polymatroid rank function and $g(\cdot)$ a positive weight function on subsets of S.*

1. *Let $(S - T) \subseteq X^{\lambda_1}$. Let $\lambda < \lambda_1$. Then for $\hat{X} \supseteq X^{\lambda_1}$, $\lambda f(\hat{X}) + g(S - \hat{X}) = \min_{X \subseteq S} \lambda f(X) + g(S - X)$ iff $\lambda f \diamond T(\hat{X} \cap T) + g(T - \hat{X} \cap T) = \min_{Y \subseteq T} \lambda f \diamond T(Y) + g(T - Y)$.*

2. *Let $T \supseteq X^{\lambda_1}$. Let $\lambda \geq \lambda_1$. Then, $\lambda f(\hat{X}) + g(S - \hat{X}) = \min_{X \subseteq S} \lambda f(X) + g(S - X)$ iff $\lambda f/T(\hat{X}) + g(T - \hat{X}) = \min_{Y \subseteq T} \lambda f/T(Y) + g(T - Y)$.*

Proof.

i. For $\lambda < \lambda_1$, we have $X_\lambda \supseteq X^{\lambda_1}$. We now have, for $\hat{X} \supseteq X^{\lambda_1} \supseteq (S - T)$,
$\lambda f \diamond T(\hat{X} \cap T) + g(T - \hat{X} \cap T) = \lambda[f(\hat{X} \cap T \cup (S - T)) - f(S - T)] + g(S - \hat{X})$
$= \lambda f(\hat{X}) + g(S - \hat{X}) - \lambda f(S - T)$.

Next we have, $\min_{Y \subseteq T} \lambda f \diamond T(Y) + g(T - Y) = \min_{Y \subseteq T} \lambda[f(Y \cup (S - T)) - f(S - T)] + g(S - (Y \cup (S - T)))] = \min_{Y \subseteq T} \lambda f(Y \cup (S - T)) + g(S - (Y \cup (S - T))) - \lambda f(S - T)] = \min_{X \subseteq S} \lambda f(X) + g(S - X) - \lambda f(S - T)$ (noting that $\lambda < \lambda_1$ implies $X_\lambda \supseteq X^{\lambda_1} \supseteq (S - T)$).

Thus the LHS of the two equations as well as the RHS of the equations differ by $\lambda f(S - T)$, which proves the result.

ii. We note that if $\lambda \geq \lambda_1$, $X^\lambda \subseteq X^{\lambda_1}$. So the first equation holds for $\hat{X}$, only if $\hat{X} \subseteq X^{\lambda_1} \subseteq T$. The result follows by noting that the LHS of the two equations differ by $g(S - T)$ and so do the RHS of the two equations.

31.6.4 Principal Partition of the Dual

We next study the principal partition of the dual. We have the following result which summarizes the relation between the PP of $(f(\cdot), g(\cdot))$ and that of $(fj(\cdot), g(\cdot))$. Essentially, critical values of the dual are of the form $\lambda^* \equiv (1 - (\lambda)^{-1})^{-1}$, where λ are the critical values of the original function and the minimizing sets in the dual corresponding to λ^* are complements of those corresponding to λ in the original.

Theorem 31.36 *Let $f(\cdot)$ be a submodular function on the subsets of S and let $g(\cdot)$ be a positive weight function on subsets of S. Let $\mathcal{B}_\lambda, \mathcal{B}^*_\lambda$ denote respectively the collection of minimizing sets corresponding to λ in the principal partitions of $(f(\cdot), g(\cdot)), (f^*(\cdot), g(\cdot))$, where $f^*(\cdot)$ denotes the dual of $f(\cdot)$ with respect to $g(\cdot)$. Let λ^* denote $(1 - (\lambda)^{-1})^{-1}$ $\forall \lambda \in \Re$.
Then*

1. *A subset X of S is in $\mathcal{B}_\lambda$ iff $S - X$ is in $\mathcal{B}^*_{\lambda^*}$,*

2. *If $\lambda_1, \ldots, \lambda_t$ is the decreasing sequence of critical values of $(f(\cdot), g(\cdot))$, then $\lambda^*_t, \ldots, \lambda^*_1$ is the decreasing sequence of critical values of $(f^*(\cdot), g(\cdot))$,*

3. *If the principal sequence of $(f(\cdot), g(\cdot))$ is $\emptyset = X_0, \ldots, X_t = S$, then the principal sequence of $(f^*(\cdot), g(\cdot))$ is $\emptyset = S - X_t, \ldots, S - X_0 = S$.*

Proof.

i. We will show that Y minimizes $\lambda f(X) + g(S-X)$ iff $S-Y$ minimizes $\lambda^* f^*(X) + g(S-X)$. We have

$$\lambda^* f^*(X) + g(S - X) = \lambda^*[g(X) - (f(S) - f(S - X))] + g(S - X)$$
$$= \lambda^* f(S - X) + (\lambda^* - 1)g(X) - \lambda^* f(S) + g(S).$$

Minimizing this expression is equivalent to minimizing the expression $\lambda^*(\lambda^*-1)^{-1} f(S-X) + g(X)$. Noting that $\lambda^*(\lambda^* - 1)^{-1} = \lambda$ we get the desired result. (We note that when one of λ, λ^* is 1, the other is to be taken as $+\infty$.) ∎

The remaining sections of the theorem are now straightforward.

31.6.5 Principal Partition and the Density of Sets

The principal partition gives information about which subsets are densely packed relative to $(f(\cdot), g(\cdot))$. Let us define the *density* of $X \subseteq S$ relative to $(f(\cdot), g(\cdot))$ to be $g(X)/f(X)$, taking the value to be $+\infty$ when $f(X)$ is zero. For instance if $f(\cdot)$ is the rank function of a graph and $g(X) \equiv |X|$, the sets of the highest density correspond to subgraphs where we can pack the largest (fractional) number of disjoint forests. As we see below, the sets of the highest density will be the sets in $\mathcal{B}_{\lambda_1}$, where λ_1 is the highest critical value.

The problem of finding a subset T of S of highest density for a given $g(T)$ value would be NP hard even for very simple submodular functions.

Example: Let $f(\cdot) \equiv$ rank function of a graph, $g(X) \equiv |X|$. In this case $g(T) = |T|$ and if we could find a set of branches of given size and highest density we can solve the problem of finding the maximal clique subgraph of a given graph. However, as we show below in Theorem 31.37, every set in the principal partition has the highest density for its $g(T)$ value and further is easy to construct. This apparent contradiction is resolved when we note that there may be no set of the given value of $g(T)$ in the principal partition.

Theorem 31.37 *Let $f(\cdot), g(\cdot)$ be polymatroid rank functions on subsets of S with $g(\cdot)$, a positive weight function. Let T be a set in the principal partition of $(f(\cdot), g(\cdot))$. If $T' \subseteq S$ so that $g(T) = g(T')$ and T' not in the principal partition, then the density of $T \subseteq S$ is greater than that of T'.*

Proof. Suppose otherwise. Let λ be the density of T. We must have $g(T') - \lambda f(T') \geq 0 = g(T) - \lambda f(T)$. Hence, $g(S-T) + \lambda f(T) \geq g(S-T') + \lambda f(T')$. But $g(S-T) = g(S-T')$. Hence, $f(T) \geq f(T')$, since $\lambda > 0$. Let $T \in \mathcal{B}_\sigma$. Then T minimizes the expression $g(S-X) + \sigma f(X) \ \forall X \subseteq S$. But since $\sigma > 0$ ($\sigma = 0$ minimizes the expression $g(S-X) + \sigma f(X)$ only at $X = S$), $g(S-T) + \sigma f(T) \geq g(S-T') + \sigma f(T')$, a contradiction, since T' is given to be not a set in the principal partition (and therefore, not in $\mathcal{B}_\sigma$). ∎

31.6.6 Outline of Algorithm for Principal Partition

We now present an informal algorithm for building the principal partition of a polymatroid rank function $f(\cdot)$ on subsets of S with respect to a positive weight function $g(\cdot)$.

We assume that we have a subroutine $\int_\lambda (f, g, P)$ for finding all sets $\hat{X}$ which minimize $\lambda f(X) + g(P - X), X \subseteq P$, where $f(\cdot), g(\cdot)$ are as above, $\lambda \geq 0$ and P is the underlying set. By property PP1 (Subsection 31.6.2), such sets are closed under union and intersection. This enables them to be represented through a partial order on a suitable partition of P. (Details may be found in [20].)

Step 1. Take $\lambda =$ the *density* $g(S)/f(S)$ and apply $\int_\lambda (f, g, S)$. We obtain X_λ. Output the family $\mathcal{F}_\lambda$ of sets $\hat{X} - X_\lambda$, where $\hat{X}$ is in the family of sets output by $\int_\lambda (f, g, S)$. By Theorem 31.35, the sets in this family minimize $\lambda \hat{f}(X) + g(S - X_\lambda - X), X \subseteq S - X_\lambda$, where $\hat{f}(\cdot) \equiv f \diamond (S - X_\lambda)$. If $X_\lambda = \emptyset$ and $X^\lambda = S$, (i.e., if $\lambda f(S) = g(S)$) we stop.

Step 2. Now work with $f_1 \equiv f/X_\lambda, f_2 \equiv f \diamond (S - X_\lambda), g_1 \equiv g/X_\lambda, g_2 \equiv g/(S - X_\lambda)$. Repeat with $(f_1(\cdot), g_1(\cdot)), (f_2(\cdot), g_2(\cdot))$ using $\int_{\lambda_1}(f_1, g_1, X_\lambda), \int_{\lambda_2}(f_2, g_2, S - X_\lambda)$, respectively, where λ_1, λ_2 are the corresponding densities.

The only λ's in the above sequence of steps which are critical values are those for which the sets X^λ, X_λ $(f_j(\cdot), g_j(\cdot))$ are the full set (at that stage of the algorithm) and the null set, respectively.

At the end of the algorithm, we will have a number of families $\mathcal{F}_\lambda$. In the process, we will have a number of disjoint sets $K_1, K_2, \ldots K_j, \ldots$, and a corresponding sequence of critical values such that $\int_{\lambda_j}(f_j, g_j, K_j)$ yields $K_j, \emptyset$, as the maximal and minimal minimizing sets for λ_j. Let the critical values λ_j be reordered as a decreasing sequence λ^j, and let the corresponding sets be $K^1, K^2, \ldots K^j, \ldots$.

The principal sequence then is $K^1, K^1 \cup K^2, K^1 \cup K^2 \cup K^3, \ldots, S$ and the critical values are $\lambda^1, \lambda^2, \ldots$.

We have X^j as a minimizing set corresponding to λ^j, for $f^j(\cdot), g^j(\cdot), K^j$, where $f^j(\cdot) \equiv f \diamond (S - [\bigcup_{i \leq (j-1)} K^i])/K^j, g^j(\cdot) \equiv g/K^j(\cdot)$, iff $X^j \cup [\bigcup_{i \leq (j-1)} K^i]$ is a minimizing set for $f(\cdot), g(\cdot), S$ corresponding to λ^j.

If we could take $\int_\lambda(f, g, P)$ to output only the minimal set minimizing $\lambda f(X) + g(P - X), X \subseteq P$, the above algorithm would construct only the principal sequence instead of the complete principal partition. The algorithm is justified through the use of Theorem 31.35.

31.6.7 Notes

An excellent overview of submodular functions (and therefore polymatroid rank functions) is available in [21]. The principal partition of the polymatroid rank function can be generalized to that of a pair of them. Details might be found in [22]. If the principal partitions of two matroids on the same underlying set have common sets then this goes over also to the union of the matroids under simple conditions. Details may be found in [20]. For the matroid case, applications of the *structural solvability* kind may be found in the following representative references [23–30], An up to date survey of principal partition and related ideas may be found in [31].

31.7 MATROID UNION

In this section we give a self contained description of the matroid union concept and link it to the principal partition of a matroid. We first give an informal algorithm for building a maximal union of bases, one from a matroid $\mathcal{M}_1$ and the other from the matroid $\mathcal{M}_2$. We show that in the process, we really are constructing the base of another matroid, which could aptly be called the *union* of the two matroids. The algorithm is due to Edmonds [32]. It can be easily modified to give the maximum size common independent set of two matroids. It also allows us to discuss the structure of various standard *objects* associated with the matroid union namely, f-circuit, the set of coloops, and so on.

31.7.1 Matroid Union Algorithm

In this subsection we give an informal description of the matroid union algorithm and justify it.

Let b_1, b_2 be bases of matroids $\mathcal{M}_1, \mathcal{M}_2$, respectively, on S. We aim to make $b_1 \cup b_2$ a maximal such union (equivalently make b_1, b_2 *maximally distant*). If $b_1 \cup b_2 = S$ there is nothing to be done. Otherwise let $e \in S - (b_1 \cup b_2)$. Now let $L_1(e, b_1), L_2(e, b_2)$ be the unique fundamental circuits that e forms with b_1, b_2 in the matroid $\mathcal{M}_1, \mathcal{M}_2$, respectively. The elements in $L_1(e, b_1)$ which do not intersect b_2 in $\mathcal{M}_2$ form fundamental circuits with b_2 in $\mathcal{M}_2$ and similarly elements in $L_2(e, b_2)$ with b_1 in $\mathcal{M}_1$. Let the set of all elements in the fundamental circuits obtained by repeatedly performing these operations be $R(e, b_1, b_2)$ which let us call $\widehat{R}$ temporarily. It is clear that $b_i \cap \widehat{R}$ spans the set $\widehat{R}$ in $\mathcal{M}_i.\widehat{R}$ since all elements in $\widehat{R} - b_i$ form fundamental circuits with it. If $b_1 \cap b_2 \cap \widehat{R}$ is not null we will show that $b_1 \cup b_2$ can be enlarged.

Let $e_c \in b_1 \cap b_2 \cap \widehat{R}$. We then must have a sequence $e, e_1, e_2, \ldots, e_k = e_c$ with property that $e_1 \in L_1(e, b_1), e_2 \in L_2(e_1, b_2), \ldots, e_k \in L_i(e_{k-1}, b_i)$ where $i = 1$ or 2 depending on whether k is odd or even. We may, without loss of generality, assume that if e_r is in the sequence it does not occur in a fundamental circuit of $e_j, j < r - 1$. Now we update the bases b_1, b_2 as follows. (Let $e_k \in L_1(e_{k-1}, b_1)$ for notational convenience.)

$$b_1^1 = b_1 - e_k + e_{k-1}$$

$$b_2^1 = b_2 - e_{k-1} + e_{k-2}$$

$$b_1^2 = b_1^1 - e_{k-2} + e_{k-3}$$

$$\vdots$$

The claim now is that each of the sets b_1^j, b_2^j is actually a base of $\mathcal{M}_1, \mathcal{M}_2$, respectively, for every j.

It is clear that b_1^1, b_2^1 are indeed such bases. Consider b_1^2. This would be the base of $\mathcal{M}_1$ provided $e_{k-2} \in L_1(e_{k-3}, b_1^1)$. But this is so because in $L_1(e_{k-3}, b_1^1)$ we know that e_k does not lie so that $L_1(e_{k-3}, b_1^1) = L_1(e_{k-3}, b^1)$.

Repeating this argument it is clear that b_1^j, b_2^j is actually a base of $\mathcal{M}_1, \mathcal{M}_2$, respectively, for j.

Suppose finally, $b_1^t = b_1^{t-1} - e_1 + e$ and $b_2^{t-1} = b_2^{t-2} - e_2 + e_1$,

$$b_1^t \cup b_2^{t-1} = b_1 \cup b_2 \cup e.$$

Thus $b_1 \cup b_2$ has been enlarged, by including e and making e_c belong only to one of the bases b_1^t, b_2^{t-1}.

We repeat this procedure, which we shall call updating using reachability, with every element e outside $b_1 \cup b_2$ and stop when we can proceed no further, that is, till a stage is reached where no element in $b_1 \cap b_2$ can be reached from an element outside $b_1 \cup b_2$. (We will call the resulting bases b_1, b_2 *maximally distant*.) This is the matroid union algorithm.

Let b_1, b_2 be maximally distant and let R be the set of all such elements reachable from elements of $S - b_1 \cup b_2$ by using fundamental circuits in the matroids $\mathcal{M}_i, i = 1, 2$ with b_1, b_2 repeatedly as above. We then have the following lemma which also contains a justification for the matroid union algorithm.

Lemma 31.6

1. *For the matroids $\mathcal{M}_i.R, i = 1, 2$, $b_i \cap R, i = 1, 2$, respectively, are disjoint bases.*

2. *If b_1, b_2 are the output of the matroid union algorithm (i.e., are maximally distant), then $|b_1 \cup b_2|$ has the maximum size among all unions of bases from $\mathcal{M}_1, \mathcal{M}_2$, respectively, and this number is $r_1(R) + r_2(R) + |S - R|$ which is $\min_{X \subseteq S} r_1(X) + r_2(X) + |S - X|$.*

3. *Given an element e in R, it is possible to find maximally distant bases b_1, b_2 of $\mathcal{M}_i$, $i = 1, 2$, respectively, such that $e \notin b_1 \cup b_2$.*

Proof. All elements in $R - (b_i \cap R)$ form fundamental circuits with $b_i \cap R$ in the matroid $\mathcal{M}_i.R, i = 1, 2$ (equivalently in $\mathcal{M}_i$, $i = 1, 2$). When the algorithm terminates, the set of all elements reachable from outside $b_1 \cup b_2$, by taking repeated fundamental circuit operations with respect to the two matroids, does not contain any element of $b_1 \cap b_2$. This is so since, otherwise, by using the algorithm, we can enlarge $b_1 \cup b_2$ by adding the external element, from which the common element can be reached, to the union of the bases (making the common element not a part of one of the bases). This proves that $b_i \cap R, i = 1, 2$ are disjoint bases of $\mathcal{M}_i.R, i = 1, 2$, respectively.

We have $|b_1 \cup b_2| = r_1(R) + r_2(R) + |S - R|$. If b_1', b_2' are bases of $\mathcal{M}_i$, $i = 1, 2$, then $b_i' \cap X, i = 1, 2, X \subseteq S$, are contained in bases of $\mathcal{M}_i.X, i = 1, 2$ and $(b_1' \cup b_2') \cap (S - X) \subseteq (S - X)$. So $|b_1' \cup b_2'| \leq r_1(X) + r_2(X) + |S - X|$, for every subset X of S. This proves that $|b_1 \cup b_2|$ is the maximum possible and $r_1(R) + r_2(R) + |S - R| = \min_{X \subseteq S} r_1(X) + r_2(X) + |S - X|$.

Next, every element e_{in} in $b_i \cap R, i = 1, 2$ is reachable from some element e_{out} outside $b_1 \cup b_2$. If we use the updating using reachability procedure, e_{out} would move into the union of the updated bases and e_{in} would move out. This proves that any element in R lies outside some maximally distant pair of bases. ∎

Example 31.4 In Figure 31.1, consider the trees t_1, t_2 of the graph $\mathcal{G}$. We illustrate the algorithm using two trees of $\mathcal{G}$ and obtaining a maximally distant pair by using the matroid union algorithm.

In the present case both the matroids are the same being the polygon matroids of $\mathcal{G}$ (i.e., independent set ≡ circuit free set). We have,

$$L(e_0, t_1) = \{e_0, e_1, e_2\}$$

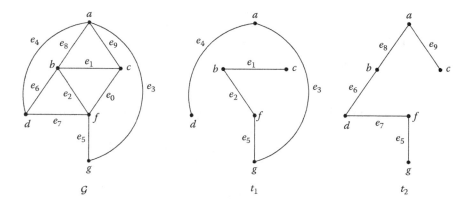

Figure 31.1 Example for matroid union algorithm.

$$L(e_1, t_2) = \{e_1, e_8, e_9\}$$
$$L(e_8, t_1) = \{e_8, e_3, e_5, e_2\}.$$

Note that e_5 belongs to both the trees.

So, we update the trees by,

$$t_1^1 = t_1 - e_5 + e_8$$
$$t_2^1 = t_2 - e_8 + e_1$$
$$t_1^2 = t_1^1 - e_1 + e_0.$$

Observe that $t_1^2 \cup t_2^1 = t_1 \cup t_2 \cup e_0$. Further $e_5 \in t_2^1$ and $e_5 \notin t_1^2$.

For completeness, we give a formal description of the matroid union algorithm below. We make use of a directed graph, $G(b_1, b_2)$, associated with bases b_1, b_2 of matroids $\mathcal{M}_1, \mathcal{M}_2$ respectively defined on S. The *graph* $G(b_1, b_2)$ *is built as follows:* S is the vertex set of the directed graph. Let v_1, v_2 be vertices. Then there is an edge (v_1, v_2, i) directed from v_1 to v_2 iff $v_2 \in L_i(v_1, b_i)$, that is, iff v_2 lies in the fundamental circuit of v_1 with respect to b_i in the matroid $\mathcal{M}_i$. If $v_1 \in b_i$ there is no edge of the kind (v_1, v_2, i). The notational difference between the informal algorithm and the present description is that elements of S are here denoted by v and the edges of the graph $G(b_1, b_2)$ by e.

Algorithm 31.1 Matroid union algorithm

INPUT   Matroids $\mathcal{M}_1, \mathcal{M}_2$ on S. Bases b_1, b_2 of $\mathcal{M}_1, \mathcal{M}_2$, respectively.

OUTPUT
1. Bases b_1^f, b_2^f of $\mathcal{M}_1, \mathcal{M}_2$ respectively such that $b_1^f \cup b_2^f$ has maximum size (i.e., is a base of $\mathcal{M}_1 \vee \mathcal{M}_2$).

2. The set R of all element reachable from $S - b_1^f \cup b_2^f$ in $G(b_1^f, b_2^f)$.

Initialize $j \leftarrow 0$
(COMMENT: j describes the current index of the base set.)
$b_1^j \leftarrow b_1, b_2^j \leftarrow b_2$.

STEP 1   Construct $G(b_1^j, b_2^j)$. If $S = b_1^j \cup b_2^j$, GOTO STEP 7.

> **STEP 2** Mark all vertices which belong to both of the b_i^j.
> For each $v \in S - b_1^j \cup b_2^j$ in $G(b_1^j, b_2^j)$, do
> Starting from v do a bfs (breadth first search) and find the set of all vertices reachable through directed paths from v.
> (COMMENT: The directed edges (v_a, v_b, p), (v_c, v_d, q), $p \neq q$ may be in the same directed path.)
> If no marked vertex is reachable from v in $G(b_1^j, b_2^j)$ call v good. Otherwise v is bad.
>
> **STEP 3** If all $v \in S - b_1^j \cup b_2^j$ are good, GOTO STEP 7.
>
> **STEP 4** Let v be a bad vertex of $G(b_1^j, b_2^j)$ and let v_m be a marked vertex reachable from v. Let $v = v_o, e_1, v_1, \ldots, e_m, v_m$ be the shortest directed path from v to v_m (where e_i is the directed edge from v_{i-1} to v_i).
> For $i = 0$ to $m - 1$, do
> If $e_{m-i} \equiv (v_{m-i-1}, v_{m-i}, q)$
> $b_q^j \leftarrow (b_q^j \cup v_{m-i-1}) - v_{m-i}$
> (COMMENT: The union of the updated bases has size one more than the union of the original bases since v_o has moved into the union by pushing v_m out of one of the bases to which it belonged.)
>
> **STEP 5** For $i = 1, 2$, do
> $b_i^{j+1} \leftarrow b_i^j$
>
> **STEP 6** $j \leftarrow j + 1$. GOTO STEP 1.
>
> **STEP 7** Declare: $b_1^f = b_1^j, b_2^f = b_2^j$
> and R to be the set of all vertices reachable in $G(b_1^f, b_2^f)$ from $S - b_1^f \cup b_2^f$.
>
> **STOP**

31.7.2 Complexity of the Matroid Union Algorithm

It is convenient to discuss the complexity of the algorithm in terms of the directed graph, $G(b_1, b_2)$, associated with bases b_1, b_2 of matroids $\mathcal{M}_1, \mathcal{M}_2$, respectively, defined on S.

Let us suppose that the matroids are available through the independence oracle which would declare, once per call, whether a particular subset of S is independent in the specified matroid $\mathcal{M}_i, i = 1, 2$.

How many calls do we require to build $G(b_1, b_2)$? This requires the knowledge of the f-circuits of an element outside b_i with respect to it in the matroid $\mathcal{M}_i$. To build $L_i(v, b_i)$ we check for each $v' \in b_i$ whether $v \cup b_i - v'$ is independent in the matroid $\mathcal{M}_i$. This requires atmost $r(\mathcal{M}_i)$ calls to the independence oracle. Thus the total number of calls to the independence oracle to build $G(b_1, b_2)$ is atmost $\mid S - b_1 \mid\mid b_1 \mid + \mid S - b_2 \mid\mid b_2 \mid$.

Finding the reachable set from $S - (b_1 \cup b_2)$ requires $O \mid E(G(b_1, b_2)) \mid$ elementary steps, where $\mid E(G(b_1, b_2)) \mid$ is the number of edges in $G(b_1, b_2)$ treating parallel edges as a single edge.

We may have started with (in the worst case) $b_1 = b_2$ and end with a base of the union of size atmost $|b_1| + |b_2|$. The graph $G(b_1, b_2)$ has to be rebuilt after each update. Let us call such graphs $\mathcal{G}_j$. So the overall complexity is $O|b_1|(|b_1||S - b_1| + |b_2||S - b_2|)$ calls to the independence oracle. If r, r' are the maximum and minimum of the ranks of the matroids this simplifies to $O(r^2(|S| - r'))$ calls to the independence oracle.

There are $O(r|E(\mathcal{G}_j)|)$ elementary steps involved in building the reachable set for all the $\mathcal{G}_j$. Let us simplify $|E(\mathcal{G}_j)|$ to $O(|S|^2)$.

So the *time complexity of the matroid union algorithm* is $O(r^2(|S| - r'))$ calls to the independence oracle $+O(r|S|^2)$ elementary operations.

The space requirement is that of storing the updated version of the graph $G(b_1, b_2)$. This has at most $|S|^2$ edges. So the *space complexity of the matroid union algorithm* is $O(|S|^2)$.

31.7.3 Matroid Union Theorem

We now state and prove the matroid union theorem. Note that a less illuminating but brief proof has been given in Theorem 31.3.

Theorem 31.38 *Let $\mathcal{M}_1 \equiv (S, \mathcal{I}_1)$, $\mathcal{M}_2 \equiv (S, \mathcal{I}_2)$ be matroids with rank functions $r_1(.)$, $r_2(.)$, respectively, and let $\mathcal{I}_1 \vee \mathcal{I}_2$ be the collection of all sets X such that $X = X_1 \cup X_2$, where X_1, X_2 are independent sets respectively in $\mathcal{M}_1, \mathcal{M}_2$. Then $\mathcal{M}_1 \vee \mathcal{M}_2 \equiv (S, \mathcal{I}_1 \vee \mathcal{I}_2)$ is a matroid with rank function*

$$r_v(K) = min_{X \subseteq K}(r_1 + r_2)(X) + |K - X|, i.e., r_v(\cdot) = (r_1 + r_2)^*|\cdot|.$$

Proof. It is clear that subsets of a set $X \in \mathcal{I}_1 \vee \mathcal{I}_2$ also belong to $\mathcal{I}_1 \vee \mathcal{I}_2$. We will verify that maximal subsets belonging to $\mathcal{I}_1 \vee \mathcal{I}_2$ which are contained in a given subset $K \subseteq S$ have the same size.

We first observe that if b_1, b_2 are bases of $\mathcal{M}_1.K, \mathcal{M}_2.K$ respectively then for any set $X \subseteq K$,

$$|(b_1 \cup b_2) \cap X| \leq r_1(X) + r_2(X)$$

$$|(b_1 \cup b_2) \cap (K - X)| \leq |K - X|.$$

Hence,

$$|b_1 \cup b_2| \leq min_{X \subseteq K}[(r_1 + r_2)(X) + |K - X|].$$

We will now construct a subset R of K where we have equality.

If we use the matroid union algorithm on bases of $\mathcal{M}_1.K$ and $\mathcal{M}_2.K$ we will finally reach bases b_1, b_2 respectively of these matroids which are maximally distant. At this stage one of the following two situations will occur.

1. $b_1 \cup b_2 = K$
   In this case $(r_1 + r_2)(\phi) + |K| = |b_1 \cup b_2|$.

2. $K - (b_1 \cup b_2) = T \neq \phi$.

By using the repeated fundamental circuit operation in $\mathcal{M}_1, \mathcal{M}_2$ starting from each element $e \in T$, it should be impossible to reach any element $e' \in b_1 \cap b_2$ (i.e., there is no directed path in $G(b_1, b_2)$ from the vertex e to the vertex e'), since otherwise we can enlarge $b_1 \cup b_2$.

Let R be the set of all such elements reachable from elements of T by using b_1, b_2. By Lemma 7.1, $b_i \cap R$ are disjoint bases in the matroids $\mathcal{M}_i.R$, $i = 1, 2$, respectively. (Note that $\mathcal{M}_i.K.R = \mathcal{M}_i.R$ and the rank function of the matroids $\mathcal{M}_i.R$ and $\mathcal{M}_i$ coincide on subsets of R.) Thus

$$|(b_1 \cup b_2) \cap R| = r_1(R) + r_2(R).$$

The elements in $K - R$ are covered by $(b_1 \cup b_2)$. Hence

$$|(b_1 \cup b_2) \cap (S - R)| = |K - R|.$$

Thus the size of the maximal union of independent sets is

$$= (r_1 + r_2)(X) + |K - X| \quad \text{for } X = R.$$

Since we have already seen that it is less than or equal to $\min_{X \subseteq K}(r_1 + r_2)(X) + |K - X|$, it follows that the size of any maximal union of independent sets of $\mathcal{M}_1, \mathcal{M}_2$ contained in K equals $\min_{X \subseteq K}(r_1 + r_2)(X) + |K - X|$ and is therefore always the same as required. Further it is clear that $r_v(K) = \min_{X \subseteq K}(r_1 + r_2)(X) + |K - X|$.

31.7.4 Fundamental Circuits and Coloops of $\mathcal{M}_1 \vee \mathcal{M}_2$

Let us now understand details of the matroid union algorithm in the context of the fact that $\mathcal{M}_1 \vee \mathcal{M}_2$ is a matroid. In particular we obtain a picture of fundamental circuits and coloops (i.e., elements not in any circuit) of the matroid $\mathcal{M}_1 \vee \mathcal{M}_2$.

First if b_1, b_2 is a pair of maximally distant bases of $\mathcal{M}_1, \mathcal{M}_2$ (say as output by the matroid union algorithm), then $b_1 \cup b_2$ is a base of $\mathcal{M}_1 \vee \mathcal{M}_2$. Consider the set R of all elements reachable from elements of $S - b_1 \cup b_2$ in the graph $G(b_1, b_2)$ (equivalently by the process of taking repeated fundamental circuits relative to b_1, b_2 in the matroids $\mathcal{M}_1, \mathcal{M}_2$, respectively). By Lemma 31.6 we know that for each $e \in R$, there exist some pair of maximally distant bases b_1', b_2' such that $e \notin b_1' \cup b_2'$, that is, there exists a base of $\mathcal{M}_1 \vee \mathcal{M}_2$ which does not contain e. So R contains no coloops of $\mathcal{M}_1 \vee \mathcal{M}_2$. Lemma 31.6 assures us that $r_1(R) + r_2(R) + |S - R| = \min_{X \subseteq S} r_1(X) + r_2(X) + |S - X|$. By Theorem 31.33, this means that $|S - R|$ is a set of coloops of the matroid whose rank function is $(r_1 + r_2)^* | \cdot |$, that is, of the matroid $\mathcal{M}_1 \vee \mathcal{M}_2$,

Thus the set R that we encounter in the matroid union algorithm is the set of all noncoloops of the matroid $\mathcal{M}_1 \vee \mathcal{M}_2$ and is independent of the pair of maximally distant bases b_1, b_2. Further, again by Theorem 31.33, R is the unique minimal set that minimizes $r_1(X) + r_2(X) + |S - X|, X \subseteq S$.

Next let $b_1 \cup b_2$ be a base of $\mathcal{M}_1 \vee \mathcal{M}_2$ and let $e \notin b_1 \cup b_2$. Consider the set R_e of all elements reachable from e in $G(b_1, b_2)$. It is clear that all the elements of R_e are spanned by $b_i \cap R_e, i = 1, 2$ in the matroid $\mathcal{M}_i.R_e$ and further that $b_i \cap R_e, i = 1, 2$ are disjoint. So the union of no pair of maximally distant bases can contain R_e. On the other hand, given any e' in $b_i \cap R_e, i = 1, 2$, by using the updating through reachability process in the algorithm, we can build a pair of maximally distant bases b_1', b_2' such that $b_1' \cup b_2' = b_1 \cup b_2 \cup e - e'$. We conclude, using Theorem 31.5, that R_e is the fundamental circuit of e with respect to the base $b_1 \cup b_2$ of $\mathcal{M}_1 \vee \mathcal{M}_2$. Note that if $b_1 \cup b_2 = b_1'' \cup b_2''$, R_e would be the same using $G(b_1, b_2)$ or $G(b_1'', b_2'')$.

31.7.5 Union of Matroids and the Union of Dual Matroids

It is natural to examine the relation between $\mathcal{M}_1 \vee \mathcal{M}_2$ and $\mathcal{M}_1^* \vee \mathcal{M}_2^*$. We show in this section that the complements of coloops of these matroids do not intersect and that this gives a natural partition of S relative to $\mathcal{M}_1, \mathcal{M}_2$

Theorem 31.39
Let $\mathcal{M}_1, \mathcal{M}_2$ be matroids on S and let $\mathcal{M}_1^*, \mathcal{M}_2^*$ be their duals. Let $r_1(\cdot), r_2(\cdot), r_1{}^*(\cdot), r_2{}^*(\cdot)$ be the rank of functions of $\mathcal{M}_1, \mathcal{M}_2, \mathcal{M}_1^*, \mathcal{M}_2^*$, respectively. Let R, R^* be the minimal sets that minimize $(r_1 + r_2)(X) + |S - X|, X \subseteq S$ and $(r_1{}^* + r_2{}^*)(X) + |S - X|, X \subseteq S$, respectively. Then,

1. b_1, b_2 are maximally distant bases of $\mathcal{M}_1, \mathcal{M}_2$, respectively, iff $S - b_1, S - b_2$, are maximally distant cobases of the same matroids (equivalently maximally distant cobases of $\mathcal{M}_1^*, \mathcal{M}_2^*$).

2. A set $K \subseteq S$ minimizes $(r_1 + r_2)(X) + \mid S - X \mid, X \subseteq S$ iff $S - K$ minimizes $(r_1^* + r_2^*)(X) + \mid S - X \mid, X \subseteq S$.

3. $S - R^*, S - R$ are the maximal sets that minimize

$$(r_1 + r_2)(X) + \mid S - X \mid, X \subseteq S,$$

$$(r_1^* + r_2^*)(X) + \mid S - X \mid, X \subseteq S,$$

respectively.

4. R is the collection of non-coloops of $\mathcal{M}_1 \vee \mathcal{M}_2$ and is disjoint from R^* which is the collection of non-coloops of $\mathcal{M}_1^* \vee \mathcal{M}_2^*$.

5. The set $S - (R \cup R^*)$ can be covered by disjoint bases of $\mathcal{M}_i.(S - R^*) \times (S - (R \cup R^*))$, $i = 1, 2$ (equivalently by those of $\mathcal{M}_i^*.(S - R) \times (S - (R \cup R^*))$, $i = 1, 2$).

Proof.

i. This follows essentially by noting that $b_1 \cup b_2$ is of maximum size iff $b_1 \cap b_2$ is of minimum size.

ii. We have
$$(r_1^* + r_2^*)(X) + \mid S - X \mid$$
$$= (2 \mid X \mid - (r_1(S) + r_2(S) - r_1(S - X) - r_2(S - X)) + \mid S - X \mid$$
$$= ((r_1 + r_2)(S - X) + \mid S - (S - X) \mid) + (\mid S \mid - (r_1 + r_2)(S)).$$

It is thus clear that K minimizes $(r_1 + r_2)(Y) + \mid S - Y \mid, Y \subseteq S$ iff $(S - K)$ minimizes $(r_1^* + r_2)^*(Y) + \mid S - Y \mid, Y \subseteq S$.

iii. This is an immediate consequence of the above result when we note that R, R^* are the minimal sets which minimize respectively the expressions $(r_1 + r_2)(X) + \mid S - X \mid$, $(r_1^* + r_2^*)(X) + \mid S - X \mid$.

iv. We saw in subsection 31.7.4 that the collection of non-coloops of $\mathcal{M}_1 \vee \mathcal{M}_2$ ($\mathcal{M}_1^* \vee \mathcal{M}_2^*$) is the minimal set R (R^*) that minimizes $(r_1 + r_2)(X) + \mid S - X \mid, X \subseteq S((r_1^* + r_2^*)(X) + \mid S - X \mid, X \subseteq S)$.

However, the second part (above) shows that $S - R \supseteq R^*$.

v. From Lemma 31.6 we know that maximally distant bases of $\mathcal{M}_1, \mathcal{M}_2$, respectively, must intersect any set T which minimizes the expression $(r_1 + r_2)(X) + \mid S - X \mid, X \subseteq S$ in disjoint bases of $\mathcal{M}_1.T, \mathcal{M}_2.T$, respectively, and the corresponding (maximally distant) cobases must cover T. Similarly maximally distant bases of $\mathcal{M}_1^*, \mathcal{M}_2^*$, respectively, must intersect any set P which minimizes the expression $(r_1^* + r_2^*)(X) + \mid S - X \mid$, $X \subseteq S$ in disjoint bases of $\mathcal{M}_1^*.P, \mathcal{M}_2^*.P$, respectively, and the corresponding (maximally distant) cobases must cover P. It follows that set $P \cap T$ is covered by any pair of maximally distant bases of $\mathcal{M}_1, \mathcal{M}_2$ as well as $\mathcal{M}_1^*, \mathcal{M}_2^*$. Since by (i) above $S - P, S - T$, respectively, minimize the expressions $(r_1 + r_2)(X) + \mid S - X \mid, X \subseteq S, (r_1^* + r_2^*)(X) + \mid S - X \mid, X \subseteq S$, (since minimizing sets are closed under intersection) so do $(S - P) \cap T, (S - T) \cap P$, respectively. But this means, whenever b_1, b_2 are maximally distant bases of $\mathcal{M}_1, \mathcal{M}_2$ respectively they intersect $(S-P) \cap T$ in disjoint bases of $\mathcal{M}_1.((S-P) \cap T), \mathcal{M}_2.((S-P) \cap T)$, respectively, intersect T in disjoint bases of $\mathcal{M}_1.T, \mathcal{M}_2.T$, respectively, and cover $P \cap T$. It follows that $b_1 \cap P \cap T, b_2 \cap P \cap T$ are disjoint bases of $\mathcal{M}_1.T \times (P \cap T), \mathcal{M}_2.T \times (P \cap T)$ which cover $P \cap T$. The result now follows substituting $S - R$ for P and $S - R^*$ for T. The dual result follows by working with dual matroids. Kishi and Kajitani's principal partition for graphs [16] is essentially the partition $R, S - R \cup R^*, R^*$ where $\mathcal{M}_1 = \mathcal{M}_2 = \mathcal{M}(\mathcal{G})$, $\mathcal{G}$ being the given graph.

31.7.6 Matroid Union and Matroid Intersection

The problem of matroid intersection (find the maximum size common independent set of two given matroids) and its solution has received more attention in the literature than matroid union. This is probably because Lawler based his well-known book [33] on matroid intersection. In this subsection we consider the relation between the two problems. These results are due essentially to Edmonds [15].

Theorem 31.40 *Let $\mathcal{M}_1, \mathcal{M}_2$ be matroids on S. Let b_{12} be the largest set independent in $\mathcal{M}_1$ as well as in $\mathcal{M}_2$. Then*

1. *b_{12} can be represented as $b_{12*} - b_2^*$ where b_{12*} is a base of $\mathcal{M}_1 \vee \mathcal{M}_2^*$ which is the union of a base b_1 of $\mathcal{M}_1$ and a base b_2^* of $\mathcal{M}_2^*$.*

2. *Every set of the form $b_{12*} - b_2^*$ is a common independent set of $\mathcal{M}_1, \mathcal{M}_2$ of maximum size.*

3. *$|b_{12}| = r(\mathcal{M}_1 \vee \mathcal{M}_2^*) - r(\mathcal{M}_2^*) = \min_{X \subseteq S} r(\mathcal{M}_1.X) + r(\mathcal{M}_2.(S-X))$.*

Proof.

i. Let b_1, b_2 be bases of $\mathcal{M}_1, \mathcal{M}_2$ so that $b_1 \cap b_2 = b_{12}$. Let $b_2^* \equiv S - b_2$. Then $b_{12} = b_1 \cup b_2^* - b_2^*$. Next $b_1 \cup b_2^*$ is independent in $\mathcal{M}_1 \vee \mathcal{M}_2^*$, since b_2^* is a base of $\mathcal{M}_2^*$. Let this be contained in the base $b_{1n} \cup b_{2n}^*$ of $\mathcal{M}_1 \vee \mathcal{M}_2^*$. Now $b_{1n} \cup b_{2n}^* - b_{2n}^*$ is independent in $\mathcal{M}_1$ and $\mathcal{M}_2$. Further,

$$|b_{1n} \cup b_{2n}^* - b_{2n}^*| \geq |b_1 \cup b_2^* - b_2^*| = |b_{12}|.$$

But b_{12} is the largest common independent set of $\mathcal{M}_1$ and $\mathcal{M}_2$. We conclude that $|b_{12}| = |b_{1n} \cup b_{2n}^* - b_{2n}^*|$, and $|b_1 \cup b_2^*| = |b_{1n} \cup b_{2n}^*|$. Therefore $b_1 \cup b_2^*$ is a base of $\mathcal{M}_1 \vee \mathcal{M}_2^*$ and the result follows.

ii. If b_{12*} is a base of $\mathcal{M}_1 \vee \mathcal{M}_2^*$ with $b_{12*} = b_1 \cup b_2^*$, where b_1, b_2^* are bases of $\mathcal{M}_1, \mathcal{M}_2^*$ respectively, then $b_{12*} - b_2^*$ is independent in $\mathcal{M}_1$ as well as in $\mathcal{M}_2$ and further its size equals $r(\mathcal{M}_1 \vee \mathcal{M}_2^*) - r(\mathcal{M}_2^*)$.

iii. It is clear from the above that $|b_{12}| = r(\mathcal{M}_1 \vee \mathcal{M}_2^*) - r(\mathcal{M}_2^*) = \min_{X \subseteq S}(r(\mathcal{M}_1.X) + r(\mathcal{M}_2^*.X) + |S - X|) - r(\mathcal{M}_2^*)$. Now $r(\mathcal{M}_2^*.X) - r(\mathcal{M}_2^*) = r(\mathcal{M}_2.(S - X)) - |S - X|$ and the result follows. Note that the collection of maximal common independent sets of two matroids do not form the bases of a matroid since they do not always have the maximum size. ∎

We next show how to convert matroid union algorithm to an algorithm for finding the maximum size common independent set of two matroids $\mathcal{M}_1, \mathcal{M}_2$.

We begin with two bases b_1, b_2 of matroids $\mathcal{M}_1, \mathcal{M}_2$ respectively on S. Let $b_2 = S - b_2^*$, where b_2^* is a base of $\mathcal{M}_2^*$. We now try to push updated versions of b_1, b_2^* apart. However, we would like to work with f-circuits of $\mathcal{M}_2$ rather than with f-circuits of $\mathcal{M}_2^*$. For this it suffices to observe that $v_p \in L_2^*(v_q, b_2^*)$ iff $v_q \in L_2(v_p, b_2)$, where $L_2^*(\cdot, \cdot), L_2(\cdot, \cdot)$ denote f-circuits of $\mathcal{M}_2^*, \mathcal{M}_2$, respectively. So while constructing $G(b_1, b_2^*)$ it is convenient to build edges of the type $(v_p, v_q, 2)$ at the node v_q directed into v_q (rather than at v_p directed away from v_p). If b_1, b_2^* are maximally distant bases of $\mathcal{M}_1, \mathcal{M}_2^*$, then $b_1 \cap b_2$ is a common independent set of $\mathcal{M}_1, \mathcal{M}_2$ of maximum size as we saw in Theorem 31.40 above.

31.7.7 Applications of Matroid Union and Matroid Intersection

31.7.7.1 Representability of Matroids

Horn [34] showed that k independent sets of columns can cover the set of all columns of a matrix iff there exists no subset A of columns such that $|A| > kr(A)$. He conjectured that this might be correct only for representable matroids (i.e., for matroids which are associated with column sets of matrices over fields). If the conjecture had been true then there would have been a nice characterization of representability. It is clear that the problem for matroids is to check if S is independent in the matroid $\mathcal{M}_1 \vee \cdots \vee \mathcal{M}_k$, where all the $\mathcal{M}_i$ are the same matroid $\mathcal{M}$. From Theorem 31.38, this happens iff $\min_{A \subseteq S} kr(A) + |S - A| = |S|$, that is, iff $|A| \leq kr(A), \forall A \subseteq S$. So the result is true for arbitrary matroids.

31.7.7.2 Decomposition of a Graph into Minimum Number of Subforests

Tutte and Nash-Williams [35,36] characterized graphs which can be decomposed into k disjoint subforests as those which satisfy $kr(X) \geq |X|, \forall X \subseteq E(\mathcal{G})$. This condition again fits into the matroid union framework as described above.

We need some preliminary definitions to describe the following results. Let $B \equiv (V_L, V_R, E)$ be a bipartite graph, which has all edges (members of E) with one endpoint in (left vertex set) V_L and another in (right vertex set) V_R. If $X \subseteq V_L, (X \subseteq V_R)$, then $\Gamma_L(X)(\Gamma_R(X))$ denotes the set of vertices adjacent to X. A matching is a subset of edges with no two incident on the same vertex. A cover is a subset of vertices containing atleast one end point of every edge in E. We give below some fundamental results about bipartite graph matching and derive them using matroid union or matroid intersection. Given a family $S_1, \ldots, S_k$ of S, a transversal of the family is a set $\{v_1, \ldots, v_k\}$ of k elements such that $v_i \in S_i$. (Note that the definition of a family permits $S_i = S_j$ even if $i \notin j$.) In a bipartite graph $V_L(V_R)$ can be regarded as a family of subsets of $V_R(V_L)$, by identifying a vertex in $V_L(V_R)$ with the subset of vertices of $V_R(V_L)$ it is adjacent to. Thus we could say V_R has a transversal whenever there is a subset T of V_L such that a matching has T as its left end points and V_R as its right end points.

Transversal matroids [7]. For each vertex $v \in V_R$, we define a matroid $\mathcal{M}_v$ on the set V_L. In this matroid the set of all vertices, say $v_{l1}, \ldots, v_{lk}$, which are adjacent to v, has rank one and contains no selfloops (rank zero elements). The complementary subset of vertices of V_L are all selfloops. Let its rank function be denoted by r_v. The union of all the matroids $\mathcal{M}_v, v \in V_R$ is a matroid which has, as independent sets, the subsets of V_L which are endpoints of matchings. This matroid is called the transversal matroid $\mathcal{M}_{tr}$. Its rank function, by using Theorem 31.38, can be seen to be $r_{tr}(X) \equiv \min_{Y \subseteq X}(\sum_{v \in V_R} r_v(Y) + |X - Y|) = \min_{Y \subseteq X}(|\Gamma_L(Y)| + |X - Y|), X \subseteq V_L$.

König's theorem [37]. Let $B \equiv (V_L, V_R, E)$ be a bipartite graph. A cover meets the edges of every matching. No two edges of any matching can meet the same vertex of any cover and therefore the size of a matching can never exceed the size of any cover. The following result is therefore remarkable.

Theorem 31.41 *In a bipartite graph the sizes of a maximum matching and a minimum cover are equal* [37].

This follows naturally from the matroid intersection result in Theorem 31.40 part (ii). We have the following matroids defined on E: $\mathcal{M}_L$ where the independent sets are subsets of E which do not meet any vertex of V_L in more than one edge and $\mathcal{M}_R$ where the independent sets are subsets of E which do not meet any vertex of V_R in more than one edge. ($\mathcal{M}_L, \mathcal{M}_R$ are easily seen to be matroids.) A subset of E is a matching iff it is independent in both matroids. The size of the maximum matching is therefore $min_{X \subseteq E}(r_L(X) + r_R(E - X))$, where $r_L(\cdot)$, $r_R(\cdot)$ are the rank functions respectively of $\mathcal{M}_L, \mathcal{M}_R$. Now $r_L(X)$ ($r_R(E-X)$) is the size of the left vertex subset $v_L(X)$ ($v_R(E-X)$) meeting X ($E-X$). It is clear that $v_L(X) \cup v_R(E-X)$ is a cover. The result now follows.

31.7.7.3 Rado's Theorem

Theorem 31.42 [38]
Let $B \equiv (V_L, V_R, E)$ be a bipartite graph. Let $\mathcal{M}$ be a matroid on V_L with rank function $f(\cdot)$. Then V_R has a transversal that is independent in $\mathcal{M}$ iff

$$f(\Gamma_R(Z)) \geq |Z| \quad \forall Z \subseteq V_R.$$

Here we consider the intersection of two matroids on V_L, namely, $\mathcal{M}$ and the above mentioned transversal matroid $\mathcal{M}_t$. By Theorem 31.40 part (ii), the maximum size of a common independent set in the two matroids is

$$\begin{aligned} & min_{X \subseteq V_L}(r_t(X) + f(V_L - X)) \\ = & min_{X \subseteq V_L}(min_{Y \subseteq X}(|\Gamma_L(Y)| + |X - Y|) + f(V_L - X))) \\ = & min_{X \subseteq V_L}(|\Gamma_L(X)| + f(V_L - X)) \end{aligned}$$

where we have used the fact that $f(K) \leq |K|, K \subseteq V_L$. Clearly the maximum size of common independent set in the two matroids must become $|V_R|$ for V_R to have an independent transversal. This will happen iff $min_{X \subseteq V_L}(|\Gamma_L(X)| + f(V_L - X)) = |V_R|$, that is, iff $min_{X \subseteq V_L}(|\Gamma_L(X)| + f(V_L - X)) - |V_R| = 0$ that is, iff $f(V_L - X)) \geq |V_R| - |\Gamma_L(X)|, \forall X \subseteq V_L$. (*) We claim that the condition (*) is equivalent to $f(\Gamma_R(Z)) \geq |Z|, \forall Z \subseteq V_R$. (**)

To see this, first observe since $\Gamma_R(V_R - \Gamma_L(X)) \subseteq V_L - X$, and $f(\cdot)$ is an increasing function, it follows that (**) implies (*). Next, define $\overline{Z} \equiv V_R - \Gamma_L(V_L - \Gamma_R(Z))$. If (*) is true, taking $X \equiv V_L - \Gamma_R(Z)$, we have $f(\Gamma_R(Z)) \geq |\overline{Z}|, \forall Z \subseteq V_R$. But $\overline{Z} \supseteq Z$ and $\Gamma_R(\overline{Z}) = \Gamma_R(Z)$. So (**) is true.

31.7.8 Algorithm for Construction of the Principal Sequence of a Matroid Rank Function

In this subsection we outline an algorithm for building the principal sequence of a matroid rank function with respect to a positive rational weight function. The main subroutine is the matroid union algorithm. The algorithm for the complete principal partition is along the same lines and may be found, for instance, in [20]. This algorithm is elementary and handles the weight function in a naive manner. The case of real weight function may be tackled by using the methods in [39,40].

We need some preliminary ideas about parallel elements in a matroid for describing our algorithm.

For a matroid $\mathcal{M}$ on S, two elements e_1, e_2 are in parallel iff $\{e_1, e_2\}$ is a circuit or e_1, e_2 are both selfloops. It is immediate that if $e_1 \in I$, where I is independent in $\mathcal{M}$ and e_1, e_2 are in parallel, then $(I - e_1) \cup e_2$ is also independent. Given a matroid $\mathcal{M}$ on S, and $e \in S$, we can create a new matroid $\mathcal{M}'$ on $S \cup e', e' \notin S$, by making e, e' parallel. The independent sets of this new matroid are simply all the independent sets of $\mathcal{M}$ and in addition sets of the form $(I - e) \cup e'$, where $e \in I$, I independent in $\mathcal{M}$. This process can be repeated by adding more than one element in parallel with a given element. In particular, we could replace each element e of $\mathcal{M}$ by k parallel elements $\{e^1, \ldots, e^k\}$. The resulting matroid $\mathcal{M}_k$ is on $S^k \equiv \{\bigcup_{e_j \in S} P^k(e_j)\}$, where $P^k(e_j) \equiv \{e_j^1, \ldots, e_j^k\}$. The sets $P^k(e_j)$ constitute a partition of S^k. We denote by $P^k(T), T \subseteq S$, the set $\bigcup_{e_j \in T} P^k(e_j)$. If $r(\cdot), r_k(\cdot)$, are the rank functions of $\mathcal{M}, \mathcal{M}_k$, then $r(T) = r_k(P^k(T)), T \subseteq S$. More generally, given a positive integral weight function $g(\cdot)$ on subsets of S, we can build the g-copy $\mathcal{M}_g$ on S^g, of $\mathcal{M}$ on S, with each element $e \in S$ replaced by the set $P^g(e)$ of $g(e)$ parallel elements in S^g. Here again $r(T) = r_g(P^g(T)), T \subseteq S$, where $P^g(T)$ is defined as $\bigcup_{e_j \in T} P^g(e_j)$ and $r_g(\cdot)$ is the rank function of the matroid $\mathcal{M}^g$.

The principal partition of $r(\cdot), g(\cdot)$ is essentially the same as that of $r_g(\cdot), |\cdot|$. This situation basically does not change even if $g(\cdot)$ were divided by a positive integer. We formalize these ideas in the following theorem.

Theorem 31.43 *Let $\mathcal{M}$ be a matroid with rank function $r(\cdot)$. Let q be a positive and λ, a nonnegative number.*

1. *Let $g(\cdot)$ be a positive integral weight function. Then, $\min_{X \subseteq S} \lambda r(X) + g(S - X)$ occurs at K iff $\min_{Y \subseteq S^g} \lambda r_g(Y) + |S^g - Y|$ occurs at $P^g(K)$.*

2. *Let $qg(\cdot)$ be a positive integral weight function. Then, $\min_{X \subseteq S}(\lambda/q) r(X) + g(S - X)$ occurs at K iff $\min_{X \subseteq S} \lambda r(X) + qg(S - X)$ occurs at K, equivalently, $\min_{Y \subseteq S^g} \lambda r_{qg}(Y) + |S^{qg} - Y|$ occurs at $P^{qg}(K)$.*

Proof. Let e_i, e_j be in parallel and let $e_i \in X, e_j \notin X$. Then $\lambda r(X) + g(S - X) > \lambda r(X \cup e_j) + g(S - (X \cup e_j))$. Thus a minimizing set of $\lambda r(X) + g(S - X)$ must contain all elements parallel to e if it contains e.

Now consider the principal partition of $(r(\cdot), g(\cdot))$. If we replace every element e by $g(e)$ parallel elements $\lambda r(X) + g(S - X) = \lambda r_g(P^g(X)) + |S^g - P^g(X)|$. Next if $\hat{X}$ minimizes $\min_{Y \subseteq S^g} \lambda r_g(Y) + |S^g - Y|$, it must be of the form $P^g(X)$ for some $X \subseteq S$.
This proves (i).
(ii) is a routine consequence. ∎

To build the principal sequence of $(r(\cdot), g(\cdot))$, where $r(\cdot)$ is a matroid rank function and $g(\cdot)$, a positive integral weight function, the key step is the construction of $f_\lambda(r, g, Q)$, which we will take as outputting the minimal minimizing set for $\lambda r(X) + g(Q - X)$. Since $g(\cdot), r(\cdot)$ are integral, the λs for which we need $f_\lambda(r, g, Q)$, are of the form p/q, where p, q are positive integers. By Theorem 31.43, we need to build $f_p(r_{qg}(\cdot), |\cdot|, P^{qg}(Q))$. The output of this subroutine is simply the set of noncoloops of the matroid $\mathcal{M}_{qg}^p$, which is the union of $\mathcal{M}_{qg}$ with itself p times. As we have seen in Theorem 31.43 this set, since it minimizes $pr_{qg}(Y) + |Q^{qg} - Y|$ must have the form $P^{qg}(K)$, for some set $K \subseteq P$. The set K is the minimal minimizing set for $\lambda r(X) + g(Q - X)$ and therefore the desired output of $f_\lambda(r, g, Q)$.

Algorithmically speaking, parallel elements can be handled by using just one of the elements and simply remembering how many elements are in parallel to it. So the underlying size of set does not go up in an essential way.

918 ■ Handbook of Graph Theory, Combinatorial Optimization, and Algorithms

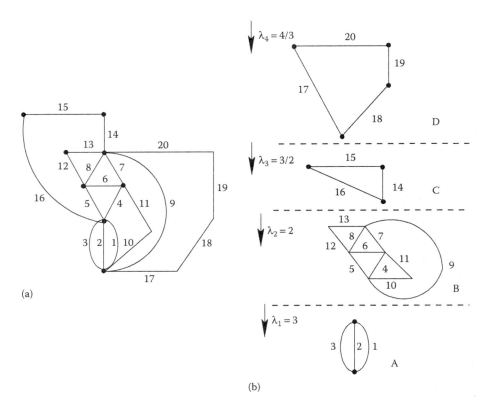

Figure 31.2 Example of principal sequence for a graph.

31.7.9 Example

31.7.9.1 Principal Sequence of $(r(\cdot), |\cdot|)$ Where $r(\cdot)$ Is the Rank Function of a Graph

Consider the graph $\mathcal{G}$ in Figure 31.2. We have, $E(\mathcal{G}) \equiv \{1, \ldots, 20\}$. We need to compute the principal sequence of $(r(\cdot), |\cdot|)$. We trace the steps of the algorithm of Subsection 31.6.6 as specialized in Subsection 31.7.8 for this purpose.

First we compute the density $\lambda \equiv E(\mathcal{G})/r(\mathcal{G})$. This is 20/11. So we use the subroutine $f_\lambda(r, |\cdot|, E(\mathcal{G}))$ which is essentially $f_{20}(r_{11}, |\cdot|, P^{11}(E(\mathcal{G}))) \equiv f_{20}(r_{11}, |\cdot|, E(\mathcal{G}_{11}))$, where $\mathcal{G}_{11}$ is obtained by putting in place of each edge of $\mathcal{G}$, 11 parallel edges. The subroutine does its job by computing the set of noncoloop elements of the matroid $\mathcal{M}_{11}^{20}$ (where $\mathcal{M}_{11}$ is the polygon matroid of $\mathcal{G}_{11}$ and $\mathcal{M}_{11}^{20}$ is the union of $\mathcal{M}_{11}$ with itself 20 times).

This set is the parallel copy of the subset $\{1, \ldots, 13\}$. So we build the graphs $\mathcal{G}_\alpha \equiv \mathcal{G} \cdot \{1, \ldots, 13\}$ and $\mathcal{G}_\beta \equiv \mathcal{G} \times \{14, \ldots, 20\}$ and repeat the algorithm on $\mathcal{G}_\alpha, \mathcal{G}_\beta$. For $\mathcal{G}_\alpha$ the density is 13/6 and for $\mathcal{G}_\beta$, it is 7/5. So we build the parallel 6– copy of $\mathcal{G}_\alpha$ and parallel 5– copy of $\mathcal{G}_\beta$ and find the set of noncoloops of $(\mathcal{M}_\alpha)_6^{13}, (\mathcal{M}_\beta)_5^7$. This yields the appropriate parallel copies of sets $\{1, 2, 3\}, \{14, 15, 16\}$, respectively.
We are now left with the graphs

$$\mathcal{G}_{\alpha 1} \equiv \mathcal{G} \cdot \{1, 2, 3\}, \mathcal{G}_{\alpha 2} \equiv \mathcal{G} \cdot \{1, \ldots, 13\} \times \{4, \ldots, 13\},$$
$$\mathcal{G}_{\beta 1} \equiv \mathcal{G} \times \{14, \ldots, 20\} \cdot \{14, 15, 16\}, \mathcal{G}_{\beta 2} \equiv \mathcal{G} \times \{17, \ldots, 20\}.$$

On these graphs when we apply our subroutine $f_\lambda(r, |\cdot|, E(\mathcal{G}'))$, ($\mathcal{G}'$ being the appropriate graph), we find the set of noncoloops is the null set which means that the null set minimizes $\lambda r(X) + |E(\mathcal{G}') - X|$. Further, we find that the full set also minimizes $\lambda r(X) + |E(\mathcal{G}') - X|$ since $\lambda r(\emptyset) + |E(\mathcal{G}')| = \lambda r(E(\mathcal{G}')) + |\emptyset|$. So at this stage we get the sets $A \equiv \{1, 2, 3\}, B \equiv \{4, \ldots, 13\}, C \equiv \{14, 15, 16\}, D \equiv \{17, \ldots, 20\}$ with the corresponding critical values $3, 2, 3/2, 4/3$ (being the densities of the graphs $\mathcal{G} \cdot \{1,2,3\}, \mathcal{G} \cdot \{1,\ldots,13\} \times \{4,\ldots,13\}, \mathcal{G} \times \{14,\ldots,20\} \cdot \{14,15,16\}, \mathcal{G} \times \{17,\ldots,20\}$). Thus the principal sequence is $E_0 \equiv \emptyset = X_{\lambda_1}, E_1 = X_{\lambda_2} = A, E_2 = X_{\lambda_3} = A \cup B, E_3 = X_{\lambda_4} = A \cup B \cup C, E_4 = X^{\lambda_4} = A \cup B \cup C \cup D \equiv E(\mathcal{G})$. The critical values are $\lambda_1 = 3, \lambda_2 = 2, \lambda_3 = 3/2, \lambda_4 = 4/3$.

Further Reading

A good way of approaching matroid union is through submodular functions induced through a bipartite graph [11]. Related material may be found in the survey paper by Brualdi [41]. A book that emphasizes the algorithmic uses of matroid intersection is [33]. An important class of applications of the matroid union, intersection, and its generalizations is in the structural solvability of systems [30,42].

References

[1] H. Whitney: On the abstract properties of linear dependence. *American Journal of Mathematics* **57** (1935), 509–533.

[2] B.L. van der Waerden: *Moderne Algebra* (2nd ed.), Springer, Berlin, Germany, 1937.

[3] S. Seshu and M.B. Reed: *Linear Graphs and Electrical Networks*, Addison-Wesley, Reading, MA, 1961.

[4] J. Edmonds: Matroid partition. *Mathematics of the Decision Sciences, Part I.* Lectures in Applied Mathematics. **11** (1968), 335–345.

[5] C. St. J.A. Nash-Williams: An application of matroids to graph theory. In: *Theory of Graphs Proceedings of the International Symposium*, Rome, Italy, 1966, P. Rosenstiehl, ed., Gordon & Breach, New York, 1967, 263–265.

[6] L. Mirsky: *Transversal Theory*, Academic Press, London, 1971.

[7] J. Edmonds and D.R. Fulkerson: Transversals and matroid partition. *Journal of Research of the National Bureau of Standards* **69B** (1965), 147–157.

[8] J.P.S. Kung: *A Source Book in Matroid Theory*, Birkhäuser, Boston, MA, 1986.

[9] W.T. Tutte: Lectures on matroids. *Journal of Research of the National Bureau of Standards* **69B** (1965), 1–48.

[10] H.H. Crapo and G.C. Rota: *On the Foundations of Combinatorial Theory—Combinatorial Geometry*, MIT Press, Cambridge, MA, 1970.

[11] D.J.A. Welsh: *Matroid Theory*, Academic Press, Cambridge, 1976.

[12] N.L. White: *Theory of Matroids*, Cambridge University Press, Cambridge, 1986.

[13] N. White: *Combinatorial Geometries*, N. White, ed., Encyclopedia of Mathematics and Its Applications **29**, Cambridge University Press, 1987.

[14] J.G. Oxley: *Matroid Theory*, Oxford University Press, New York, 1992.

[15] J. Edmonds: Submodular functions, matroids, and certain polyhedra. In: *Proceedings of the Calgary International Conference on Combinatorial Structures and Their Applications*, R. Guy, H. Hanani, N. Sauer, and J. Schönheim, eds., Gordon & Breach, New York, 1970, 69–87.

[16] G. Kishi and Y. Kajitani: Maximally distant trees and principal partition of a linear graph. *IEEE Transactions on Circuit Theory* **CT-16** (1969), 323–329.

[17] H. Narayanan: Theory of Matroids and Network Analysis, PhD Thesis, Department of Electrical Engineering, Indian Institute of Technology, Bombay, India, February 1974.

[18] N. Tomizawa: Strongly irreducible matroids and principal partition of a matroid into irreducible minors (in Japanese). *Transactions of the Institute of Electronics and Communication Engineers of Japan* **59A** (1976), 83–91.

[19] S. Fujishige: Principal structures of submodular systems. *Discrete Applied Mathematics* **2** (1980), 77–79.

[20] H. Narayanan: Submodular Functions and Electrical Networks. *Annals of Discrete Mathematics* **54** North Holland, London, New York, Amsterdam, the Netherlands, 1997. Revised version at http://www.ee.iitb.ac.in/hn/book/.

[21] L. Lovász: Submodular functions and convexity. In: *Mathematical Programmming—The State of the Art*, A.Bachem. M. Grötschel, and B.Korte, eds., Springer, Berlin, Germany, 1983, 235–257.

[22] S. Fujishige: *Submodular Functions and Optimization*, Annals of Discrete Maths **47**, North Holland, Amsterdam, New York, Oxford,Tokyo, 1991.

[23] T. Ozawa: Topological conditions for the solvability of active linear networks. *International Journal of Circuit Theory and Its Applications* **4** (1976), 125–136.

[24] K. Sugihara and M. Iri: A mathematical approach to the determination of the structure of concepts. *Matrix and Tensor Quarterly* **30** (1980), 62–75.

[25] M. Iri, J. Tsunekawa, and K. Murota: Graph theoretical approach to large-scale systems—Structural solvability and block-triangularization. *Transactions of Information Processing Society of Japan* **23** (1982), 88–95.

[26] M. Iri: Applications of matroid theory. In: *Mathematical Programming—The State of the Art*, A. Bachem, M. Grötschel, and B. Korte, eds., Springer, Berlin, Germany, 1983, 158–201.

[27] K. Sugihara: A unifying approach to descriptive geometry and mechanisms. *Discrete Applied Mathematics* **5** (1983), 313–328.

[28] K. Sugihara: *Machine Interpretion of Line Drawings*, MIT Press, Cambridge, MA, 1986.

[29] K. Murota and M. Iri: Structural solvability of systems of equations—A mathematical formulation for distinguishing accurate and inaccurate numbers in structural analysis of systems. *Japan Journal of Applied Mathematics* **2** (1985), 247–271.

[30] K. Murota: *Systems Analysis by Graphs and Matroids—Structural Solvability and Controllability*, Algorithms and Combinatorics **3**, Springer, 1987.

[31] S. Fujishige: Theory of principal partitions revisited. In: *Research Trends in Combinatorial Optimization*, W. J. Cook, L. Lovász, and J. Vygen, eds., Springer, 2009, pp. 127–162.

[32] J. Edmonds: Minimum partition of a matroid into independent subsets. *Journal of Research of the National Bureau of Standards* **69B** (1965), 67–72.

[33] E.L. Lawler: *Combinatorial Optimization—Networks and Matroids*, Holt, Rinehart & Winston, New York, 1976.

[34] A. Horn: A characterization of unions of linearly independent sets. *Journal of the London Mathematical Society* **30** (1955), 494–496.

[35] W.T. Tutte: On the problem of decomposing a graph into n-connected factors. *Journal of the London Mathematical Society* **36** (1961), 221–230.

[36] C. St. J.A. Nash-Williams: Edge-disjoint spanning trees of finite graphs. *Journal of the London Mathematical Society* **36** (1961), 445–450.

[37] D. König: *Theorie der Endlichen und Unendlichen Graphen*, Leipzig, Germany, 1936, Reprinted New York, Chelsea, 1950.

[38] R. Rado: A theorem on independence relations. *Quarterly Journal of Mathematics*, Oxford **13** (1942), 83–89.

[39] W.H. Cunningham: Testing membership in matroid polyhedra. *Journal of Combinatorial Theory* **B36** (1984), 161–188.

[40] H. Narayanan: A rounding technique for the polymatroid membership problem. *Linear Algebra and Its Applications* **221** (1995), 41–57.

[41] R.A. Brualdi: Matroids induced by directed graphs–a survey. In: *Recent Advances in Graph Theory*, Proceedings of the Symposium, Prague, Czech Republic, Academia Praha, June 1974, 115–134.

[42] A. Recski: *Matroid Theory and Its Applications in Electric Network Theory and in Statics*, Springer-Verlag, Berlin, Heidelberg, New York, London, Paris, Tokyo, 1989.

CHAPTER 32

Hybrid Analysis and Combinatorial Optimization

H. Narayanan

CONTENTS

32.1 Introduction 923
32.2 Preliminaries 924
32.3 Topological Hybrid Analysis Procedure 928
32.4 Proofs for Topological Hybrid Analysis 931
32.5 Principal Partition Problem 932
    32.5.1 Topological Degree of Freedom of an Electrical Network 932
    32.5.2 Shannon Switching Game 933
    32.5.3 Maximum Distance between Two Forests 933
    32.5.4 Forest of Minimum Size Hybrid Representation 933
    32.5.5 Maximum Rank of a Cobase Submatrix 933
32.6 Building Maximally Distant Forests 934
    32.6.1 Algorithm Maximally Distant Forests 934
32.7 Network Analysis Through Topological Transformation 937
    32.7.1 Fusion–Fission Method 937
    32.7.2 Solution of the Node Fusion–Fission Problem 939

32.1 INTRODUCTION

In this chapter we discuss the hybrid analysis problem and sketch one of its natural generalizations. Focusing attention on these naturally leads to the study of fundamental combinatorial optimization problems, which can be solved using the matroid union operation (see Chapter 31) and the Dilworth truncation operation (see Section 32.7).

Electrical network analysis is the process of finding pairs of vectors (v, i), such that v satisfies Kirchhoff's voltage law (KVL) for the graph $\mathcal{G}$ of the network, i satisfies Kirchhoff's current law (KCL) for $\mathcal{G}$, and the pair (v, i) satisfies the device characteristic of the network. For ease of discussion, we will assume that the network is static (no derivative terms in the constraints) and has a unique solution, that is, a unique (v, i) pair satisfies the constraints of the network. The basic methods of analysis reduce the constraints (KCL, KVL, device characteristic) of the network to a more compact form involving node voltages or loop currents. In the former case, once the node voltages are obtained, by the use of KVL the branch voltage vector can be obtained uniquely and thence using device characteristic, the branch current vector. In the latter case, from the loop currents, the branch current vector can be obtained through use of KCL and thence using device characteristic, the branch voltage vector. A natural generalization of these procedures is to pick as unknowns some voltages and some currents from which by the use of KCL and KVL either the voltage or current of every branch can be obtained, after which the use of device characteristic will enable us to obtain the remaining variable of the branch. This class of methods where the unknowns are a mixture of current and voltage variables is called *hybrid analysis*.

Hybrid analysis is originally due to Kron [1,2] and was simplified by Branin [3]. A description very accessible to the general reader is available in Brameller et al. [4]. The development presented here is however based on a topological version reported in Narayanan [5]. It has the advantage of greater flexibility in the choice of unknowns and also advantages in storage. The manner in which voltage and current variables are chosen in hybrid analysis can be viewed in a general way as a process of transforming the given network through the operations of node fusions and fissions into another simpler network. Such a general transformation can be called *topological transformation of electrical networks* [6] and has applications particularly in the parallel processing of network analysis.

Natural questions that arise are on how to choose the unknowns minimally. In the case of hybrid analysis this leads to the question of *principal partition* and in the more general case of topological transformations this leads to the *principal lattice of partitions* problem. The former is discussed in Section 32.5 and the latter is sketched in Section 32.7.

32.2 PRELIMINARIES

We need a few preliminary definitions and results before we move on to a discussion of the methods.

We assume familiarity with the notions of graph, subgraph, directed graph, path, circuit, cutset, connectedness, connected components, (spanning) tree, cotree, and so on. A forest of the graph is obtained by taking a tree for each component and a coforest is its complement. For a graph $\mathcal{G}$, which for us will invariably be directed, $V(\mathcal{G})$ and $E(\mathcal{G})$ denote the vertex set, and edge set, respectively. The number of edges in a forest of $\mathcal{G}$ is its rank, denoted by $r(\mathcal{G})$, and that in a coforest is its nullity, denoted by $\nu(\mathcal{G})$. A separator of a graph is a subset of edges with the property that there is no circuit of the graph containing an edge inside and an edge outside the subset. Minimal separators are called elementary separators. Any connected graph can be uniquely decomposed into subgraphs (called two-connected components) on elementary separators, which will be linked to each other at *hinges* or *cut vertices*. A hinge is the only vertex where two-connected subgraphs, whose edge sets are complements of each other, meet. Electrically speaking, that is, in terms of KCL and KVL, it is as though these subgraphs are disconnected.

Vectors are treated as functions from a set to a field, invariably that of real numbers. Examples are voltage and current vectors defined on the set of edges of a graph and potential vector defined on the set of nodes of a graph. If f is a vector on S and $T \subseteq S$, the *restriction* of f to T denoted f/T, is the vector on T, whose values agree with the values of f on T. Given a collection $\mathcal{K}$ of vectors on S which includes the zero vector, and a subset $T \subseteq S$, the collection $\mathcal{K} \cdot T$ is made up of all restrictions of vectors in $\mathcal{K}$ and the collection $\mathcal{K} \times T$ is the subset of $\mathcal{K} \cdot T$ where each vector is the restriction of a vector that is zero on $S - T$. If f, g are both vectors on a set S, the *dot product* $<f,g> \equiv \sum_{e \in S} f(e).g(e)$. The vectors f, g are said to be *orthogonal* if their dot product is zero. The collection of all vectors on S orthogonal to vectors in $\mathcal{K}$, is denoted as $\mathcal{K}^\perp$. When $\mathcal{K}$ is a vector space and $T \subseteq S$, it can be shown directly that $(\mathcal{K} \cdot T)^\perp = \mathcal{K}^\perp \times T$. Using the fact that when S is finite and $\mathcal{K}$ is a vector space, we have $\mathcal{K}^{\perp\perp} = \mathcal{K}$, it would then follow that $(\mathcal{K} \times T)^\perp = \mathcal{K}^\perp \cdot T$ (see Chapter 31).

The incidence matrix (usually denoted by A), of a directed graph, has one row per node and one column per edge, with the (i,j) entry being $+1(-1)$ if edge j is directed away (toward) node i and zero otherwise. The matrix obtained from the incidence matrix, by omitting one row per component of the graph, is called the reduced incidence matrix and has the same row space as the incidence matrix. For a graph $\mathcal{G}$, a *current vector* i is a vector on $E(\mathcal{G})$ that is orthogonal to the rows of the incidence matrix of $\mathcal{G}$, equivalently, that satisfies Kirchhoff's current equations (KCEs): $Ax = 0$.

A *voltage vector* v of $\mathcal{G}$ is a vector on $E(\mathcal{G})$ that is linearly dependent on the rows of the incidence matrix of $\mathcal{G}$, that is, $v^T = \lambda^T A$ for some vector λ.

The vector λ assigns a value to each node of $\mathcal{G}$ and s called a *potential vector*. We say v is *derived* from the node potential vector λ.

Voltage vectors and current vectors form vector spaces denoted by $\mathcal{V}_v(\mathcal{G}), \mathcal{V}_i(\mathcal{G})$, and called voltage space of $\mathcal{G}$ and current space of $\mathcal{G}$, respectively. An immediate consequence of the definition of the voltage and current spaces is the celebrated *Tellegen's theorem*, which states that $(\mathcal{V}_v(\mathcal{G}))^\perp = \mathcal{V}_i(\mathcal{G})$.

It is clear from the definition of voltage vector that we can assign, for the edges of any tree of a connected graph, arbitrary voltage values and this would uniquely determine cotree voltages. For, if tree voltages are given, we can assign a reference potential to some node and by traversing the tree, assign to all other nodes an appropriate unique potential. Thus, the tree voltages uniquely fix difference of potential between any pair of nodes and thence fix all cotree voltages. In particular, we can assign to one branch e of the tree t, value 1 and to all others in the tree, value 0. Let $v^e \equiv (v_t^e | v_{\bar{t}}^e)$ be the corresponding voltage vector. Note that when the branch e is removed from the tree, the latter splits into two connected pieces with vertex sets V_{e+}, V_{e-}, say, with V_{e+} being the vertex set where the tail of e is incident. The set of edges in the original graph between these two vertex sets is called the fundamental cutset of e with respect to cotree $\bar{t}$ and denoted as $L^*(e, \bar{t})$. The vector v^e has nonzero values only on the edges in $L^*(e, \bar{t})$ with the value being $+1(-1)$ if the tail is in $V_{e+}(V_{e-})$. The matrix, which has as rows the voltage vectors constructed in the above manner for each edge in a tree t, is called the fundamental cutset matrix $Q_{\bar{t}}$, with respect to the cotree $\bar{t}$. Given any voltage vector $\hat{v} \equiv (\hat{v}_t | \hat{v}_{\bar{t}})$, we observe that the vector $\hat{v} - \sum_{e \in t}(\hat{v}(e)v^e)$ is a voltage vector but has zero value on all tree branches. The above traversal through tree branches shows that it must be a zero vector. It follows that the rows of the fundamental cutset matrix form a basis for $\mathcal{V}_v(\mathcal{G})$.

Now if $i \equiv (i_t | i_{\bar{t}})$ is any current vector, it is clear, since v^e, i are orthogonal, that $<v_{\bar{t}}^e, i_{\bar{t}}> = -<v_t^e, i_t> = -i(e)$. Thus cotree current values uniquely determine tree current values and, if all cotree current values are zero, so will all tree currents be. In particular, we can assign to one branch c of the cotree $\bar{t}$, value 1 and to all others in the cotree, value 0. Let $i^c \equiv (i_t^c | i_{\bar{t}}^c)$ be the corresponding current vector. Note that when the branch c is added to the tree t, exactly one circuit is formed, called the fundamental circuit of c with respect to the tree t denoted by $L(c, t)$. The vector i^c has non zero values only on the edges in $L(c, t)$, with the value being $+1(-1)$ if the orientation of the edge in the circuit agrees with (opposes) that of c. The matrix, which has as rows the current vectors constructed in the above manner for each edge in a cotree, is called the fundamental circuit matrix B_t with respect to the tree t. By using the argument that we used for voltage vectors, it follows that the rows of the fundamental circuit matrix form a basis for $\mathcal{V}_i(\mathcal{G})$.

Let $\mathcal{G}$ be a graph, with $E(\mathcal{G}) \equiv E$ and let $T \subseteq E$. We now define some useful derived graphs natural to circuit theory.

The graph $\mathcal{G}_{\text{open}}(E - T)$ has the same vertex set as $\mathcal{G}$ but with the edges of $(E - T)$ removed. The graph $\mathcal{G} \cdot T$ is obtained from $\mathcal{G}_{\text{open}}(E - T)$ by removing isolated (with no edges incident) vertices.

The vertex set of $\mathcal{G}_{\text{short}}(E-T)$ is the set $\{V_1, V_2, \ldots V_n\}$, where V_i is the vertex set of the ith component of $\mathcal{G}_{\text{open}}\, T$, an edge $e \in T$ being directed from V_i to V_j in $\mathcal{G}_{\text{short}}(E-T)$, if it is directed from $a \in V_i$ to $b \in V_j$ in $\mathcal{G}$. The graph $\mathcal{G} \times T$ is obtained from $\mathcal{G}_{\text{short}}(E-T)$ by removing isolated vertices. (The two are the same if $\mathcal{G}$ is connected.)

From the above construction of $\mathcal{G} \cdot (E-T)$, $\mathcal{G} \times T$, we have the following theorem.

Theorem 32.1

1. *Maximal intersection of forest (coforest) of $\mathcal{G}$ with T is a forest of $\mathcal{G} \cdot T$ (coforest of $\mathcal{G} \times T$).*

2. *Union of forests of $\mathcal{G} \cdot (E-T)$ and $\mathcal{G} \times T$ is a forest of $\mathcal{G}$.*

3. $r(\mathcal{G}) = r(\mathcal{G} \cdot (E-T)) + r(\mathcal{G} \times T)$.

4. $\nu(\mathcal{G}) = \nu(\mathcal{G} \cdot (E-T)) + \nu(\mathcal{G} \times T)$.   ∎

Graphs obtained from $\mathcal{G}$, by opening some edges and shorting others, are called *minors of $\mathcal{G}$*. We note that, when some edges are shorted and others open-circuited, the order in which these operations are performed does not affect the resulting graph. So we have the following theorems.

Theorem 32.2 *Let $T_1 \subseteq T_2 \subseteq E$. Then*

1. $\mathcal{G} \cdot T_2 \cdot T_1 = \mathcal{G} \cdot T_1$.

2. $\mathcal{G} \times T_2 \times T_1 = \mathcal{G} \times T_1$.

3. $\mathcal{G} \times T_2 \cdot T_1 = \mathcal{G} \cdot (E - (T_2 - T_1)) \times T_1$.   ∎

We have the following important results on the voltage and current spaces associated with minors of graphs.

Theorem 32.3

1. $\mathcal{V}_v(\mathcal{G} \cdot T) = (\mathcal{V}_v(\mathcal{G})) \cdot T$.

2. $\mathcal{V}_v(\mathcal{G} \times T) = (\mathcal{V}_v(\mathcal{G})) \times T$.

Proof.

i. Let $v_T \in \mathcal{V}_v(\mathcal{G} \cdot T)$. Now $\mathcal{V}_v(\mathcal{G} \cdot T) = \mathcal{V}_v(\mathcal{G}_{\text{open}}(E-T))$.
Thus, $v_T \in \mathcal{V}_v(\mathcal{G}_{\text{open}}(E-T))$. Let v_T be derived from the potential vector λ of $\mathcal{G}_{\text{open}}(E-T)$. Now for any edge $e \in T$, $v_T(e) = \lambda(a) - \lambda(b)$, where a, b are the positive and negative end points of e. However, λ is also a potential vector of $\mathcal{G}$. Let the voltage vector v of $\mathcal{G}$ be derived from λ. For the edge $e \in T$, we have, as before, $v(e) = \lambda(a) - \lambda(b)$. Thus, $v_T = v/T$ and therefore, $v_T \in (\mathcal{V}_v(\mathcal{G})) \cdot T$. Hence $\mathcal{V}_v(\mathcal{G} \cdot T) \subseteq (\mathcal{V}_v(\mathcal{G})) \cdot T$. The reverse containment is proved similarly.

ii. Let $v_T \in \mathcal{V}_v(\mathcal{G} \times T)$. We have $\mathcal{V}_v(\mathcal{G} \times T) = \mathcal{V}_v(\mathcal{G}_{\text{short}}(E-T))$.
The vertex set of $\mathcal{G}_{\text{short}}(E-T)$ is, say, the set $\{V_1, V_2, \ldots, V_n\}$, where V_i is the vertex set of the ith component of $\mathcal{G}_{\text{open}}\, T$, an edge $e \in T$ being directed from V_i to V_j in $\mathcal{G}_{\text{short}}(E-T)$, if it is directed from $a \in V_i$ to $b \in V_j$ in $\mathcal{G}$.
Now, $v_T \in \mathcal{V}_v(\mathcal{G}_{\text{short}}(E-T))$. Let v_T be derived from the potential vector $\hat{\lambda}$ in $\mathcal{G}_{\text{short}}(E-T)$. The vector $\hat{\lambda}$ assigns to each of the V_i, the value $\hat{\lambda}(V_i)$. Define a potential vector λ on the nodes of $\mathcal{G}$ as follows: $\lambda(n) \equiv \hat{\lambda}(V_i), n \in V_i$. Since $\{V_1, \ldots, V_k\}$ is a

partition of $V(\mathcal{G})$, it is clear that λ is well defined. Let v be the voltage vector derived from λ in $\mathcal{G}$. Whenever $e \in E - T$ we must have $\mathbf{v}(e) = 0$ since both end points must belong to the same V_i.

Next, whenever $e \in T$ we have $\mathbf{v}(e) = |(a) - |(b)$ where a is the positive end point of e and b, the negative endpoint. Let $a \in V_a$, $b \in V_b$, where $V_a, V_b \in V(\mathcal{G}_{short}(E-T))$. Then the positive endpoint of e in $\mathcal{G}_{short}(E-T)$ is V_a and the negative end point, V_b. By definition $\lambda(a) - \lambda(b) = \hat{\lambda}(V_a) - \hat{\lambda}(V_b)$. Thus $\mathbf{v}/T = \mathbf{v}_T$. Hence, $\mathbf{v}_T \in (\mathcal{V}_v(\mathcal{G})) \times T$. Thus, $\mathcal{V}_v(\mathcal{G} \times T) \subseteq (\mathcal{V}_v(\mathcal{G})) \times T$.

The reverse containment is proved similarly, but using the idea, that if a voltage vector is zero on all elements of $E - T$, then a potential vector from which it is derived, must have the same value on all vertices of each V_i, since these are vertex sets of components of $\mathcal{G}_{open} T$. ∎

Using duality we can now prove the following theorem.

Theorem 32.4 *Let $\mathcal{G}$ be a directed graph on edge set E. Let $T \subseteq E$. Then,*

1. $\mathcal{V}_i(\mathcal{G} \cdot T) = (\mathcal{V}_i(\mathcal{G})) \times T$.

2. $\mathcal{V}_i(\mathcal{G} \times T) = (\mathcal{V}_i(\mathcal{G})) \cdot T$.

Proof.

i. $\mathcal{V}_i(\mathcal{G} \cdot T) = (\mathcal{V}_v(\mathcal{G} \cdot T))^\perp$ by Tellegen's theorem. By Theorem 32.3, $\mathcal{V}_v(\mathcal{G} \cdot T) = (\mathcal{V}_v(\mathcal{G})) \cdot T$. Hence, $\mathcal{V}_i(\mathcal{G} \cdot T) = ((\mathcal{V}_v(\mathcal{G})) \cdot T)^\perp = (\mathcal{V}_v(\mathcal{G}))^\perp \times T = \mathcal{V}_i(\mathcal{G}) \times T$.

ii. The proof is similar. ∎

It is useful to note some elementary facts about coloops and self-loops. A coloop, by definition, does not belong to any circuit and therefore must belong to every forest. Dually, a self-loop, by definition, does not belong to any cutset and therefore must belong to every coforest. The fundamental circuit and cutset matrices that result when coloop (self-loop) edges are shorted or open circuited are the same. So we have the following theorems.

Theorem 32.5 *Let $T \subseteq E$ be a set of edges composed entirely of self-loops and coloops. Then $\mathcal{V}_v(\mathcal{G} \cdot (E - T)) = (\mathcal{V}_v(\mathcal{G} \times (E - T))$ and $\mathcal{V}_i(\mathcal{G} \cdot (E - T)) = (\mathcal{V}_i(\mathcal{G} \times (E - T))$.* ∎

We are now in a position to state and prove a result, which will enable us to give a topological version of hybrid analysis.

Theorem 32.6 *Let (A, B) be a partition of $E(\mathcal{G})$. Let K be a forest and L_A, a coforest of $\mathcal{G} \cdot A$ and t_B be a forest and L, a coforest of $\mathcal{G} \times B$. Let $\mathcal{G}_{AL}$ be the graph $\mathcal{G} \times (A \cup L)$ and let $\mathcal{G}_{BK}$ be the graph $\mathcal{G} \cdot (B \cup K)$. Then*

1. *$i_K | i_{L_A} | i_{t_B} | i_L$ is a current vector of $\mathcal{G}$, iff there exist current vectors $i_K | i_{L_A} | i_L$ of $\mathcal{G}_{AL}$ and $i'_K | i_{t_B} | i_L$ of $\mathcal{G}_{BK}$.*

2. *$v_K | v_{L_A} | v_{t_B} | v_L$ is a voltage vector of $\mathcal{G}$, iff there exist voltage vectors $v_K | v_{L_A} | v'_L$ of $\mathcal{G}_{AL}$ and $v_K | v_{t_B} | v_L$ of $\mathcal{G}_{BK}$.* ∎

For ease of readability we relegate the proof of this result to Section 32.4.

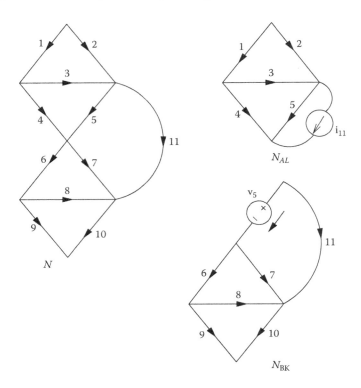

Figure 32.1 To illustrate the $\mathcal{N}_{AL} - \mathcal{N}_{BK}$ method.

32.3　TOPOLOGICAL HYBRID ANALYSIS PROCEDURE

In this section we present a topological version of the hybrid analysis procedure. Essentially, the network is decomposed into two subnetworks, whose simultaneous analysis, matching certain *boundary conditions*, is equivalent to the analysis of the original network. The validity of this procedure rests on Theorem 32.6 and requires only that the device characteristic of the subsets of edges of the derived networks appear decoupled in the original network. In the special case of resistive networks we could write nodal equations for one of the networks and loop equations for the other and match boundary conditions. This yields hybrid analysis equations with greater freedom in the choice of unknowns, which could translate to good properties such as sparsity for the coefficient matrix.

Let $\mathcal{N}$ be a network on the graph $\mathcal{G}$. (The reader may use Figure 32.1 to illustrate the procedure.) Let (A, B) be a partition of $E(\mathcal{G})$ such that in the device characteristic of $\mathcal{N}$ the devices in A, B are independent of, that is, decoupled from, each other. (In Figure 32.1, $A \equiv \{1, 2, 3, 4, 5\}$ and B, the complement.) Let K be a forest and L_A, a coforest of $\mathcal{G} \cdot A$ and t_B be a forest and L, a coforest of $\mathcal{G} \times B$. ($K \equiv \{1, 2, 5\}, L_A \equiv \{3, 4\}, t_B \equiv \{6, 8, 9\}, L \equiv \{7, 10, 11\}$.) Let $\mathcal{G}_{AL}$ be the graph $\mathcal{G} \times (A \cup L)$ and let $\mathcal{G}_{BK}$ be the graph $\mathcal{G} \cdot (B \cup K)$. (In the present example, $\mathcal{G} \times (A \cup L)$ would have $7, 10$ as self-loops. In the figure, the graph shown with caption $\mathcal{N}_{AL}$ is this graph with the self-loops omitted. The graph $\mathcal{G} \cdot (B \cup K)$ would have $1, 2$ as coloops. In the figure, the graph shown with caption $\mathcal{N}_{BK}$ is this graph with the coloops omitted.)

We now build two networks $\mathcal{N}_{AL}$ and $\mathcal{N}_{BK}$ as follows: $\mathcal{N}_{AL}$ has graph $\mathcal{G}_{AL}$ with edge set $A \cup L$ built from $\mathcal{G}$ by short-circuiting (fusing the end points of) edges in t_B and removing them. The devices in A have the same characteristics as in $\mathcal{N}$ and L has no device characteristic constraints. $\mathcal{N}_{BK}$ has graph $\mathcal{G}_{BK}$ with edge set $B \cup K$ built from $\mathcal{G}$ by open circuiting edges (removing the edges but leaving the end points in place) in L_A. The devices in B have the same characteristics as in $\mathcal{N}$ and K has no device characteristic constraints. (Note that the L, K edges are present in both networks.) Theorem 32.6 implies that solving $\mathcal{N}$ is equivalent to solving $\mathcal{N}_{AL}$ and $\mathcal{N}_{BK}$ simultaneously keeping i_L, v_K the same in both networks [5]. It may be noted that, if branches in L contain selfloops in the graph $\mathcal{G}_{AL}$, they may be deleted or contracted (endpoints fused and edge removed) from that graph and the current matching between $\mathcal{N}_{AL}$ and $\mathcal{N}_{BK}$ be confined to the remaining L branches. Similarly, if branches in K contain coloops in the graph $\mathcal{G}_{BK}$, they may be deleted or contracted from that graph and the voltage matching between $\mathcal{N}_{AL}$ and $\mathcal{N}_{BK}$ be confined to the remaining K branches.

Consider now the case where the device characteristic for the edges has the form

$$(i - \mathcal{J}) = G(v - \mathcal{E}), \tag{32.1}$$

where G is a block diagonal matrix with principal diagonal submatrices G_A, G_B, where we take G_B to be invertible. In this case, hybrid analysis equations can be written as follows:

1. Write nodal analysis equations for $\mathcal{N}_{AL}$ treating branches in L as current sources of value i_L.

2. Write loop analysis equations for $\mathcal{N}_{BK}$ treating branches in K as voltage sources of value v_K.

3. Force the constraints that i_L is the same in both networks and v_K is the same in both networks.

We now proceed formally. The reader is invited to refer to Figure 32.1.

Let $[A_{rA} A_{rL}] = [A_{rK} A_{r(A-K)} A_{rL}]$ be a reduced incidence matrix of $\mathcal{G}_{AL}$. Let the device characteristic of the edges in A be expressible as

$$(i_A - \mathcal{J}_A) = G_A(v_A - \mathcal{E}_A). \tag{32.2}$$

We then have, (since $i \in \mathcal{V}_i(\mathcal{G})$),

$$A_{rA} i_A + A_{rL} i_L = 0 \tag{32.3}$$

that is,

$$A_{rA}(i_A - \mathcal{J}_A) + A_{rL} i_L = -A_{rA} \mathcal{J}_A \tag{32.4}$$

that is,

$$A_{rA} G_A (v_A - \mathcal{E}_A) + A_{rL} i_L = -A_{rA} \mathcal{J}_A \tag{32.5}$$

that is,

$$A_{rA} G_A v_A + A_{rL} i_L = -A_{rA} \mathcal{J}_A + A_{rA} G_A \mathcal{E}_A. \tag{32.6}$$

Now,

$$\begin{bmatrix} v_A \\ v_L \end{bmatrix} = \begin{bmatrix} A_{rA}^T \\ A_{rL}^T \end{bmatrix} v_{nA}, \tag{32.7}$$

for some v_{nA} (since $\begin{bmatrix} v_A \\ v_L \end{bmatrix} \in \mathcal{V}_v(\mathcal{G}_{AL})$). We thus have,

$$(A_{rA} G_A A_{rA}^T) v_{nA} + A_{rL} i_L = -A_{rA} \mathcal{J}_A + A_{rA} G_A \mathcal{E}_A. \tag{32.8}$$

These are the nodal analysis equations of $\mathcal{N}_{AL}$. Note that we could have used any matrix which has as its rows a basis of $\mathcal{V}_v(\mathcal{G}_{AL})$, in place of $(A_{rA}|A_{rL})$, resulting in a valid set of equations with voltage type of unknowns.

Next for $\mathcal{N}_{BK}$, we choose the forest t for building the fundamental circuit matrix of $\mathcal{G}_{BK}$. Let $[B_K B_B] \equiv [B_K B_{t \cap B} B_L] \equiv [B_K B_{t \cap B} I_L]$ be the fundamental circuit matrix of $\mathcal{G}_{BK}$ with respect to forest t (where I_L denotes the identity matrix with its columns corresponding to L).

Let the device characteristic in B be expressible as $v_B - \mathcal{E}_B = R_B(i_B - \mathcal{J}_B)$. We then have,

$$B_K v_K + B_B v_B = 0 \tag{32.9}$$

that is,

$$B_K v_K + B_B(v_B - \mathcal{E}_B) = -B_B \mathcal{E}_B \tag{32.10}$$

that is,

$$B_K v_K + B_B R_B(i_B - \mathcal{J}_B) = -B_B \mathcal{E}_B \tag{32.11}$$

that is,

$$B_K v_K + B_B R_B i_B = -B_B \mathcal{E}_B + B_B R_B \mathcal{J}_B. \tag{32.12}$$

We have,

$$\begin{bmatrix} i_B \\ i_K \end{bmatrix} = \begin{bmatrix} B_B^T \\ B_K^T \end{bmatrix} y \tag{32.13}$$

for some y, since $\begin{bmatrix} i_B \\ i_K \end{bmatrix} \in \mathcal{V}_i(\mathcal{G}_{BK})$. Hence,

$$B_K v_K + B_B R_B B_B^T y = -B_B \mathcal{E}_B + B_B R_B \mathcal{J}_B. \tag{32.14}$$

Here again we could have used any matrix which has as its rows a basis of $\mathcal{V}_i(\mathcal{G}_{BK})$, in place of $(B_B | B_K)$, resulting in a valid set of equations with current type of unknowns. Now we impose the condition that v_K is the same in both networks and so is i_L. But this means,

$$A_{rK}^T v_{nA} = v_K \tag{32.15}$$

and,

$$I_L^T y = i_L. \tag{32.16}$$

So we get the hybrid equations,

$$A_{rA} G_A A_{rA}^T v_{nA} + A_{rL} i_L = -A_{rA} \mathcal{J}_A + A_{rA} G_A \mathcal{E}_A \tag{32.17}$$

$$B_K A_{rK}^T v_{nA} + B_B R_B B_B^T i_L = -B_B \mathcal{E}_B + B_B R_B \mathcal{J}_B. \tag{32.18}$$

The matrix $\begin{bmatrix} A_{rA} G_A A_{rA}^T & A_{rL} \\ B_K A_{rK}^T & B_B R_B B_B^T \end{bmatrix}$ is positive definite if G_A, R_B are positive definite. This matrix will usually not be very sparse unless $\mathcal{G}_{BK}$ has a suitable basis for its current space which makes $B_B R_B B_B^T$ sparse. In practice this may often be possible. The real power of these methods, however, is revealed when we try to use iterative methods. We show in Section 32.4 that $B_K A_{rK}^T = -A_{rL}^T$. This fact is computationally useful. Indeed, this means that we can use a variation of the conjugate gradient method to solve Equations 32.17 and 32.18 [7,8]. The advantage of such methods is that the matrix need not be stored explicitly storing $\mathcal{G}_{AL}$, $\mathcal{G}_{BK}$ and the device characteristic is adequate. The basic subroutine for the conjugate gradient method only requires multiplication of the coefficient matrix by a given vector, which arises at each iteration. This process can be carried out entirely by breaking it down into graph theoretic operations and multiplication by the device characteristic matrix (which would often be nearly diagonal).

We note that if in $\mathcal{G}_{AL}$, there are self-loops in L and in $\mathcal{G}_{BK}$ there are coloops in K, then in Equations 32.17 and 32.18, the matrix entries corresponding respectively to the current and voltage variables associated with such branches would be zero. We would therefore be justified in open circuiting or short-circuiting such edges before we write equations.

These methods were originally derived for parallelization of network analysis by G.Kron and were called *Diakoptics* [1]. They exploit the fact that when $\mathcal{G}_{AL}$, $\mathcal{G}_{BK}$ have several two-connected components, the matrix

$$\begin{bmatrix} A_{rA} G_A A_{rA}^T & A_{rL} \\ B_K A_{rK}^T & B_B R_B B_B^T \end{bmatrix}$$

will have block diagonal structure within $A_{rA} G_A A_{rA}^T$ and $B_B R_B B_B^T$.

32.4 PROOFS FOR TOPOLOGICAL HYBRID ANALYSIS

We need a preliminary lemma for the proof of Theorem 32.6.

Lemma 32.1 *Let A, B, L, K be defined as in Theorem 32.6. Then,*

1. $\mathcal{V}_i(\mathcal{G}_{AL} \cdot A) = \mathcal{V}_i(\mathcal{G} \cdot A)$ *and dually,* $\mathcal{V}_v(\mathcal{G}_{AL} \cdot A) = \mathcal{V}_v(\mathcal{G} \cdot A)$.
   $\mathcal{V}_i(\mathcal{G}_{BK} \times B) = \mathcal{V}_i(\mathcal{G} \times B)$ *and dually* $\mathcal{V}_v(\mathcal{G}_{BK} \times B) = \mathcal{V}_v(\mathcal{G} \times B)$.

2. $r(\mathcal{G}_{AL}) = r(\mathcal{G} \cdot A); \nu(\mathcal{G}_{BK}) = \nu(\mathcal{G} \times B)$.

Proof. It can be seen that $K \cup t_B$ is a forest and $L_A \cup L$, a coforest of $\mathcal{G}$. It follows that $K \cup t_B$ is a forest of $\mathcal{G} \cdot (A \cup t_B)$ and $L_A \cup L$ is a coforest of $\mathcal{G} \times (B \cup L_A)$. Since $K \cup t_B$ is a forest of $\mathcal{G} \cdot (A \cup t_B)$ and K is a forest of $\mathcal{G} \cdot (A \cup t_B) \cdot A(= \mathcal{G} \cdot A)$, it follows that (using Theorem 32.1) $r(\mathcal{G} \cdot (A \cup t_B) \times t_B) = r(\mathcal{G} \cdot (A \cup t_B)) - r(\mathcal{G} \cdot A) = |t_B|$. So $\nu(\mathcal{G} \cdot (A \cup t_B) \times t_B) = 0$. Thus the edges of t_B are not part of any circuit in $\mathcal{G} \cdot (A \cup t_B)$. Similarly, since $L_A \cup L$ is a coforest of $\mathcal{G} \times (B \cup L_A)$ and L is a coforest of $\mathcal{G} \times (B \cup L_A) \times B(= \mathcal{G} \times B)$, it follows that (using Theorem 32.1) $\nu(\mathcal{G} \times (B \cup L_A) \cdot L_A) = \nu(\mathcal{G} \times (B \cup L_A)) - \nu(\mathcal{G} \times B) = |L_A|$. So $r(\mathcal{G} \times (B \cup L_A) \cdot L_A) = 0$. Thus the edges of L_A are not part of any cutset in $\mathcal{G} \times (B \cup L_A)$.

We observe that $\mathcal{G}_{AL} \cdot A \equiv \mathcal{G} \times (A \cup L) \cdot A = \mathcal{G} \cdot (A \cup t_B) \times A$. But in the graph $\mathcal{G} \cdot (A \cup t_B)$, t_B is a set of coloops. Hence shorting or opening these edges will not affect the KCL or KVL constraints of the resulting graph. Thus, $\mathcal{V}_i(\mathcal{G}_{AL} \cdot A) = \mathcal{V}_i(\mathcal{G} \times (A \cup L) \cdot A) = \mathcal{V}_i(\mathcal{G} \cdot (A \cup t_B) \times A) = \mathcal{V}_i(\mathcal{G} \cdot (A \cup t_B) \cdot A) = \mathcal{V}_i(\mathcal{G} \cdot A)$ and dually, $\mathcal{V}_v(\mathcal{G}_{AL} \cdot A) = \mathcal{V}_v(\mathcal{G} \cdot A)$.

Further, we note that $\mathcal{G}_{AL}$ is obtained by shorting the branches t_B in the forest $K \cup t_B$ of $\mathcal{G}$. Hence K, which is a forest of $\mathcal{G} \cdot A$ is also a forest of $\mathcal{G}_{AL}$. This proves that $r(\mathcal{G}_{AL}) = r(\mathcal{G} \cdot A)$.

Next, $\mathcal{G}_{BK} \times B \equiv \mathcal{G} \cdot (B \cup K) \times B = \mathcal{G} \times (B \cup L_A) \cdot B$. But in the graph $\mathcal{G} \times (B \cup L_A)$, L_A is a set of self-loops. Hence shorting or opening these edges will not affect the KCL or KVL constraints of the resulting graph. Thus, $\mathcal{V}_i(\mathcal{G}_{BK} \times B) = \mathcal{V}_i(\mathcal{G} \cdot (B \cup K) \times B) = \mathcal{V}_i(\mathcal{G} \times (B \cup L_A) \cdot B) = \mathcal{V}_i(\mathcal{G} \times (B \cup L_A) \times B) = \mathcal{V}_i(\mathcal{G} \times B)$ and dually, $\mathcal{V}_v(\mathcal{G}_{BK} \times B) = \mathcal{V}_v(\mathcal{G} \times B)$.

Further, we note that $\mathcal{G}_{BK}$ is obtained by opening the branches L_A in the coforest $L \cup L_A$ of $\mathcal{G}$. Hence L, which is a coforest of $\mathcal{G} \times B$ is also a coforest of $\mathcal{G}_{BK}$. This proves that $\nu(\mathcal{G}_{BK}) = \nu(\mathcal{G} \times B)$. ∎

Proof of Theorem 32.6.

1. $i_K | i_{L_A} | i_{t_B} | i_L$ is a current vector of $\mathcal{G}$, iff there exist current vectors $i_K | i_{L_A} | i_L$ of $\mathcal{G}_{AL}$ and $i'_K | i_{t_B} | i_L$ of $\mathcal{G}_{BK}$.

2. $v_K|v_{L_A}|v_{t_B}|v_L$ is a voltage vector of $\mathcal{G}$, iff there exist voltage vectors $v_K|v_{L_A}|v'_L$ of $\mathcal{G}_{AL}$ and $v_K|v_{t_B}|v_L$ of $\mathcal{G}_{BK}$.

Let $i_K|i_{L_A}|i_{t_B}|i_L$ be a current vector of $\mathcal{G}$. From Theorem 32.4, it follows that $i_K|i_{L_A}|i_L$ is a current vector of $\mathcal{G}_{AL}$ and $i_{t_B}|i_L$ is a current vector of $\mathcal{G} \times B$. But $\mathcal{V}_i(\mathcal{G}_{BK} \times B) = (\mathcal{V}_i(\mathcal{G}_{BK})) \cdot B = \mathcal{V}_i(\mathcal{G} \times B)$. So there exists a current vector $i'_K|i_{t_B}|i_L$ of $\mathcal{G}_{BK}$. Next, suppose there exist current vectors $i_K|i_{L_A}|i_L$ of $\mathcal{G}_{AL}$ and $i'_K|i_{t_B}|i_L$ of $\mathcal{G}_{BK}$. Since $i'_K|i_{t_B}|i_L$ is a current vector of $\mathcal{G}_{BK}$, it follows that $i_{t_B}|i_L$ is a current vector of $\mathcal{G}_{BK} \times B$ and therefore of $\mathcal{G} \times B$. Now $L \cup L_A$ is a coforest of $\mathcal{G}$. Hence for any arbitrary vector $i_{L_A}|i_L$, there exists a unique current vector $i''_K|i_{L_A}|i''_{t_B}|i_L$ of $\mathcal{G}$. But then $i''_K|i_{L_A}|i_L$ is a current vector of $\mathcal{G}_{AL}$ and $i''_{t_B}|i_L$ is a current vector of $\mathcal{G} \times B$, that is, of $\mathcal{G}_{BK} \times B$. Now L is a coforest of $\mathcal{G} \times B$ and $L \cup L_A$ is a coforest of $\mathcal{G}_{AL}$. So for any arbitrary vector $i_{L_A}|i_L$, there is a unique current vector $i^3_K|i_{L_A}|i_L$ of $\mathcal{G}_{AL}$ and $i^3_{t_B}|i_L$ of $\mathcal{G} \times B$. But we already have seen that there exist current vectors $i_K|i_{L_A}|i_L$ of $\mathcal{G}_{AL}$ and $i_{t_B}|i_L$ of $\mathcal{G} \times B$. It follows that $i_K = i''_K$ and $i_{t_B} = i''_{t_B}$ and therefore $i_K|i_{L_A}|i_{t_B}|i_L$ is a current vector of $\mathcal{G}$. The voltage part of the theorem is dual to the above, that is, in the proof above we interchange voltage and current, $\mathcal{G}_{AL}$ and $\mathcal{G}_{BK}$ '$\cdot$' and '$\times$', K and L, t_B and L_A. ∎

Proof of claim $B_K A^T_{rK} = -A^T_{rL}$.

Observe that the rows of $(B_K|B_{t_B}|I_L)$ span the space $\mathcal{V}_i(\mathcal{G}_{BK})$. Hence the rows of $(B_K|I_L)$ span the space $(\mathcal{V}_i(\mathcal{G}_{BK})) \cdot (K \cup L) = \mathcal{V}_i(\mathcal{G}_{BK} \times (K \cup L))$, by Theorem 32.3. But $\mathcal{G}_{BK} = \mathcal{G} \cdot (B \cup K)$. Hence the rows of $(B_K|I_L)$ span the space $\mathcal{V}_i(\mathcal{G} \cdot (B \cup K) \times (K \cup L)) = \mathcal{V}_i(\mathcal{G} \times (A \cup L) \cdot (K \cup L))$. Next, we note that the rows of $(A_{rK}|A_{rL})$ span the space $(\mathcal{V}_v(\mathcal{G}_{AL})) \cdot (K \cup L) = \mathcal{V}_v(\mathcal{G}_{AL} \cdot (K \cup L)) = \mathcal{V}_v(\mathcal{G} \times (A \cup L) \cdot (K \cup L))$. The claim follows since rows of $(B_K|I_L)$ and $(A_{rK}|A_{rL})$ are orthogonal.

32.5 PRINCIPAL PARTITION PROBLEM

In this section we relate the hybrid analysis problem to the principal partition of graphs. The latter has played a fundamental role in the development of combinatorial optimization in the context of matroids (Chapter 31) and submodular functions [9]. All the results in the present section are best understood in a unified way as applications of the matroid union theorem and the principal partition for polymatroid rank functions discussed in Chapter 31. However, for readability we give a self-contained, graph based, treatment here while pointing out connections to results of that chapter.

When the network $\mathcal{N}$ is linear, if we write nodal equations for $\mathcal{N}_{AL}$ and loop equations for $\mathcal{N}_{BK}$ (defined as in Section 32.3), the total number of equations would be $r(\mathcal{G} \cdot A) + \nu(\mathcal{G} \times B)$. So one could ask for the partition A, B for which the above expression reaches a minimum value. More generally, given a partition (A, B) of $E(\mathcal{G})$, by Theorems 32.3 and 32.4 we know that there exists a current vector $i_A|i_B$ of $\mathcal{G}$ iff i_B is a current vector of $\mathcal{G} \times B$ and that there exists a voltage vector $v_A|v_B$ of $\mathcal{G}$ iff v_A is a voltage vector of $\mathcal{G} \cdot A$. Thus v_A, i_B of $\mathcal{G}$ can be uniquely determined using only KVL and KCL from the forest voltage vector v_{t_A} of $\mathcal{G} \cdot A$ and the coforest current vector i_{L_B} of $\mathcal{G} \times B$. Thus we have the first formulation of the hybrid rank problem.

Given a graph $\mathcal{G}$, partition $E(\mathcal{G})$ into A and B such that $r(\mathcal{G} \cdot A) + \nu(\mathcal{G} \times B)$ is minimized. Historically, the following problems, related to the first formulation, were solved at about the same time. After stating them we give their solution in brief. Detailed solution may be found in the references cited therein as well as in the works of Narayanan [9].

32.5.1 Topological Degree of Freedom of an Electrical Network

Please refer to [10]. This problem was posed by G. Kron. Select a minimum-sized set of branch voltages and branch currents from which, by using Kirchoff's voltage equations and KCEs, we can find either the voltage or the current associated with each branch. This minimum size is called the *topological degree of freedom of the network*, equivalently, the *hybrid rank* of the graph.

32.5.2 Shannon Switching Game

Please refer to [11]. $\mathcal{G}$ is a graph with one of its edges say e_M marked. There are two players a cut player and a short player. The cut player, during his turn, deletes (opens) an edge leaving the end points in place. The short player, during his turn, contracts an edge, that is, fuses its end points and removes it. Neither player is allowed to touch e_M. The cut player wins if all the paths between the end points of e_M are destroyed (equivalently, all circuits containing e_M are destroyed). The short player wins if the end points of e_M get fused (equivalently, all cutsets containing e_M are destroyed by shorting of edges). The problem is to analyze this game and characterize situations where the cut or short player, playing second, can always win and to determine the winning strategy.

32.5.3 Maximum Distance between Two Forests

Please refer to [12]. Define distance between two forests t_1 and t_2 as $|t_1 - t_2|$. Find two forests in a given graph which have the maximum distance between them, that is, the size of their union is the largest possible.

32.5.4 Forest of Minimum Size Hybrid Representation

Please refer to [12]. Let a forest t be represented by a pair of sets (A_t, B_t) where $A_t \subseteq t, t \cap B_t = \emptyset$ such that $(A_{t_1}, B_{t_1}) = (A_{t_2}, B_{t_2})$ iff $t_1 = t_2$. Note that we can represent the same forest by several pairs, for instance $(t, \emptyset), (\emptyset, E(\mathcal{G}) - t)$ both represent t. We call $|A_t \cup B_t|$ the *size of the representation* (A_t, B_t). Find a forest, which has the representation of minimum size.

32.5.5 Maximum Rank of a Cobase Submatrix

Please refer to [13]. For a rectangular $(m \times n)$ matrix with linearly independent rows, let us call an $m \times (n - m)$ submatrix a *cobase* submatrix iff the remaining set of columns are from an identity matrix. The *term rank* of a matrix is the maximum number of nonzero entries in the matrix, which belong to distinct rows and distinct columns. Find a cobase matrix of maximum rank, and a cobase matrix of minimum term rank among all matrices row equivalent to the given matrix.

For the above five problems the solution involves essentially the same strategy: Find a set A (or a minimal set $A_{\min}$ or a maximal set $A_{\max}$) which minimizes $2r(\mathcal{G} \cdot A) + |E(\mathcal{G}) - A|$. That these sets are unique is proved in Theorems 32.7 and 32.8. The partition of the graph into $A_{\min}, A_{\max} - A_{\min}, E(\mathcal{G}) - A_{\max}$ was called the *Principal Partition of $\mathcal{G}$* by Kishi and Kajitani [12]. This has been discussed in detail for the more general case of polymatroid rank functions in Chapter 31. The essential ideas will be repeated for graphs in Section 32.6 of this chapter. Below we have given a sketch of the solutions to the five problems. More details may be found in the works of Narayanan [9].

Let t_A be a forest of the subgraph on A. Let $L_{\overline{A}}$ be a coforest of the graph on $\mathcal{G} \times (E(\mathcal{G}) - A)$. Select the branch voltages of t_A and the branch currents of $L_{\overline{A}}$ as the desired set of variables.

If $e_M \in A_{\max}$, the short player can always win. If $e_M \in (E(\mathcal{G}) - A_{\max})$ the cut player can always win. If $e_M \in A_{\max} - A_{\min}$, whoever plays first can always win. The winning strategies involve the construction of appropriate maximally distant forests during every turn.

Kishi and Kajitani gave an algorithm for building a pair of maximally distant forests which is essentially the well-known algorithm for building a base of the union of two matroids (see [14]) for the case where the matroids are identical–essentially the same algorithm works for the general case.

Select a forest t which has maximal intersection with A. The representation $(t \cap A, (E(\mathcal{G})-t) \cap (E(\mathcal{G}) - A))$ has the least size among all representations of all forests. As is easily seen, the minimum size among all representations of forests of $\mathcal{G}$ is also the same as the topological degree of freedom of $\mathcal{G}$ and the above maximum distance.

The solution is similar for the last problem. Let S be the set of all columns and let $r(\cdot)$ be the rank function on the collection of subsets of S. Then the maximum rank of a cobase matrix = the minimum term rank of a cobase matrix. Select two maximally distant bases (bases $\equiv$ maximally independent columns). In this case the matroid union algorithm described in Chapter 31, essentially a generalization of the algorithm for building maximally distant forests, has to be used. Perform row operations so that an identity matrix appears coresponding to one of these. The submatrix corresponding to the complement of this base is the desired cobase matrix, which has both maximum rank as well as minimum term rank.

32.6 BUILDING MAXIMALLY DISTANT FORESTS

The five problems stated in the previous section were solved originally without reference to the matroid union theorem. Indeed, the problems of finding maximally distant forests, of minimal representation of forests and the topological degree of freedom were solved graph theoretically. The algorithm for building maximally distant forests of a graph is essentially the same as the above-mentioned matroid union algorithm except that both the matroids whose union is sought are the polygon matroids of the same graph. We sketch this algorithm informally in this section and also relate it to the principal partition of a graph as described by Kishi and Kajitani [12]. We hope that this treatment helps in better visualization of the more general matroid ideas.

We begin with a simple but useful observation in the following lemma.

Lemma 32.2 *Let $\mathcal{G}$ be a graph. Let t_1, t_2 be two forests and let $\overline{t_1}, \overline{t_2}$ be the corresponding coforests of $\mathcal{G}$. Then the following are equivalent.*

1. t_1, t_2 are maximally distant;
2. $|t_1 \cup t_2|$ is the maximum possible;
3. $|t_1 \cap t_2|$ is the minimum possible;
4. $\overline{t_1}, \overline{t_2}$ are maximally distant.

Proof. Since the sizes of all forests are the same, maximizing $|t_1 - t_2|$ is the same as maximizing $|t_1 \cup t_2|$ and minimizing $|t_1 \cap t_2|$. Sizes of all coforests are the same. So these hold also for coforests. But maximizing $|t_1 \cup t_2|$ is the same as minimizing $|\overline{t_1} \cap \overline{t_2}|$. So t_1, t_2 being maximally distant is the same as $\overline{t_1}, \overline{t_2}$ being maximally distant. ∎

32.6.1 Algorithm Maximally Distant Forests

Let t_1, t_2 be two forests of $\mathcal{G} = (V, E)$. Define a directed graph $G(t_1, t_2)$ with E as the set of vertices and with directed edges as described below.

Whenever in $\mathcal{G}, e \notin t_j, j = 1, 2$, draw, in $G(t_1, t_2)$, directed edges from the vertex e to all vertices which are edges of $\mathcal{G}$ in the fundamental circuit $L(e, t_j)$ (formed when e of $\mathcal{G}$ is added to tree t_j) and mark each directed edge as a t_j edge. From a given vertex e_s in $G(t_1, t_2)$, it is easy to determine the set of all vertices that can be reached through directed paths by using breadth-first search. In the process, the shortest path (in terms of number of edges in the path), from e_s to every vertex in $G(t_1, t_2)$, can be determined.

The present algorithm starts from some pair of forests t_1, t_2 (which could even be the same forest) and tries to build another pair t'_1, t'_2 for which the distance $|t'_1 - t'_2|$ is greater than the distance $|t_1 - t_2|$. Clearly, if $t_1 \cup t_2$ covers all edges in E, the forests are maximally distant. Suppose $e_{\text{out}} \notin t_1 \cup t_2$. In $G(t_1, t_2)$, we find the set of all vertices (which are edges of $\mathcal{G}$) reachable from vertex e_{out} through directed paths. If no vertex in this set corresponds to an edge common to both t_1 and t_2, we repeat the process with another such e_{out} from which an edge e_{com} common to both t_1 and t_2 may be reached. If no such vertex e_{out} exists we stop and output the current t_1, t_2 as maximally distant.

Let $e_{\text{out}}, (t_1), e_1, (t_2), \ldots, e_i, (t_j), e_{i+1}, \ldots, e_k, (t_j), e_{k+1}, \ldots, e_{n-1}, (t_m), e_{\text{com}}$ be a shortest path from e_{out} to e_{com}, where $e_r, (t_p), e_{r+1}$ indicates that from the vertex e_r there is a directed edge to e_{r+1}, which is marked t_p, where t_p could be either t_1 or t_2 (the indices used for trees and edges being unrelated). We will use this path to alter t_1, t_2. The key idea we use is the following: $L(e_i, t_j)$ does not have e_{k+1} as a member as otherwise we could have shortened the path to $e_{\text{out}}, (t_1), e_1, (t_2), \ldots, e_i, (t_j), e_{k+1}, \ldots, e_{n-1}, (t_m), e_{\text{com}}$. Therefore in the forest $\hat{t}_j = t_j \cup e_k - e_{k+1}$, the fundamental circuit $L(e_i, \hat{t}_j)$ will be the same as $L(e_i, t_j)$.

We can therefore alter t_1 and t_2 as

$$t_m \leftarrow t_m \cup e_{n-1} - e_{\text{com}}$$
$$\ldots$$
$$t_j \leftarrow t_j \cup e_i - e_{i+1}$$
$$\ldots$$
$$t_1 \leftarrow t_1 \cup e_{\text{out}} - e_1$$

The result of the above alteration would be that e_{out} would now move into $t_1 \cup t_2$ while $e_{\text{com}} \notin t_1 \cap t_2$. Therefore the size $|t_1 \cup t_2|$ and the distance $|t_1 - t_2|$ would have increased.

We repeat the above step until from none of the e_{out} we can reach any e_{com} and output the current t_1, t_2 as maximally distant.

By Lemma 32.2, equivalently, $\overline{t_1}, \overline{t_2}$ may be output as maximally distant. More directly, the above algorithm can be converted into one which finds maximally distant coforests by working with coforest $\overline{t_i}$ in place of forest t_i and replacing $L(e_i, t_j)$ wherever it occurs, by the fundamental cutset $L^*(e_i, \overline{t_j})$.

We know that $e_k \in L^*(e_i, \overline{t_j})$ iff $e_i \in L(e_k, t_j)$. This yields the relationship between $G(t_1, t_2)$ and $G(\overline{t_1}, \overline{t_2})$ stated in the following lemma.

Lemma 32.3 *Reversing the direction of arrows in $G(t_1, t_2)$ and marking t_j edges by $\overline{t_j}$ yields $G(\overline{t_1}, \overline{t_2})$.*

We justify the above algorithm through the following theorem. Here $r(A), \nu(A)$ denote, respectively, $r(\mathcal{G} \cdot A), \nu(\mathcal{G} \times A)$. ∎

Theorem 32.7 *Let t_1, t_2 denote forests of $\mathcal{G}$ on edge set E and let $A \subseteq E$. Then,*

1. *$|t_1 \cup t_2| \leq 2r(A) + |E - A|$ and the inequality becomes an equality only if t_1, t_2 are maximally distant and $t_i \cap A, i = 1, 2$, are disjoint forests of $\mathcal{G} \cdot A$ and $(t_1 \cup t_2) \cap (E - A) = (E - A)$.*

2. *The pair of forests t_1, t_2 output by maximally distant forests algorithm and the set of all edges $A_{\min}$ in E corresponding to vertices in $G(t_1, t_2)$ which can be reached from $E - (t_1 \cup t_2)$ satisfy $|t_1 \cup t_2| = 2r(A_{\min}) + |E - A_{\min}|$ and hence t_1, t_2 are maximally distant.*

3. $\max |t_1 \cup t_2| = \min 2r(A) + |E - A|$, *where t_1, t_2 are forests of $\mathcal{G}(V, E)$ and $A \subseteq E$.*

4. *If $\bar{t}_1, \bar{t}_2$ denote coforests of $\mathcal{G}$, then $|\bar{t}_1 \cup \bar{t}_2| \leq 2\nu(A) + |E - A|$ and the inequality becomes an equality iff $\bar{t}_1, \bar{t}_2$ are maximally distant, $\bar{t}_i \cap A, i = 1, 2$, are disjoint coforests of $\mathcal{G} \times A$ and $(\bar{t}_1 \cup \bar{t}_2) \cap (E - A) = (E - A)$.*

Proof.

i. We have $t_i \cap A, i = 1, 2$, as a subforest of $\mathcal{G} \cdot A$ and $(t_1 \cup t_2) \cap (E - A) \subseteq (E - A)$. So $|t_1 \cup t_2| \leq 2r(A) + |E - A|$. It is clear that if the inequality becomes an equality, the corresponding t_1, t_2 must be such that $|t_1 \cup t_2|$ is a maximum and therefore be maximally distant. The inequality becomes an equality iff the set A on the right-hand side is such that $|(t_1 \cup t_2) \cap A| = 2r(A)$, that is, $t_i \cap A, i = 1, 2$, are disjoint forests of $\mathcal{G} \cdot A$ and $(t_1 \cup t_2) \cap (E - A) = (E - A)$.

ii. If the t_1, t_2 output by the algorithm are such that $t_1 \cup t_2 = E$, the inequality will be satisfied as an equality by taking $A_{\min}$ to be the null set. If on the other hand, the algorithm outputs t_1, t_2 such that $t_1 \cup t_2 \neq E$, then $A_{\min}$ has the following properties. First, it contains $E - (t_1 \cup t_2)$, or equivalently, $(t_1 \cup t_2) \supseteq (E - A_{\min})$ and $A_{\min}$ does not contain any edge in $t_1 \cap t_2$. Next in $\mathcal{G} \cdot A_{\min}$, every edge in $t_i \cap A_{\min}, i = 1, 2$, can be reached from an edge outside $t_1 \cup t_2$ by repeatedly taking fundamental circuits with respect to the two forests. Thus $t_i \cap A_{\min}, i = 1, 2$, span both each other as well as edges in $E - (t_1 \cup t_2)$. So $t_i \cap A_{\min}, i = 1, 2$, are both forests of $\mathcal{G} \cdot A_{\min}$ and, further, have no intersection since $A_{\min}$ does not contain any edge in $t_1 \cap t_2$. Thus we have, $|t_1 \cup t_2| = 2r(A_{\min}) + |E - A_{\min}|$.

iii. Is now immediate from (i) and (ii).

iv. The proof for the coforest case is dual, that is, by replacing in the above argument, forests by coforests, $G(t_1, t_2)$ by $G(\bar{t}_1, \bar{t}_2)$, $A_{\min}$ by $A^*_{\min}$, $\mathcal{G} \cdot A_{\min}$ by $\mathcal{G} \times A^*_{\min}$ and $r(\mathcal{G} \cdot A_{\min})$ by $\nu(\mathcal{G} \times A^*_{\min})$.

Theorem 32.8 *Let t_1, t_2 ($\bar{t}_1, \bar{t}_2$) be maximally distant forests (coforests) of graph $\mathcal{G}$ on edge set E. Let $A_{\min}$ ($B_{\min}$) denote the set of all edges in E corresponding to vertices in $G(t_1, t_2)$ ($G(\bar{t}_1, \bar{t}_2)$) which can be reached from $E - (t_1 \cup t_2)$ ($E - (\bar{t}_1 \cup \bar{t}_2)$).*

1. *$A_{\min}(B_{\min})$ is the unique minimal set that minimizes $2r(A) + |E - A|$ ($2\nu(A^*) + |E - A^*|$).*

2. *$\hat{A} \subseteq E$ minimizes $2r(A) + |E - A|$ iff $E - \hat{A}$ minimizes $2\nu(A^*) + |E - A^*|$.*

3. *$E - B_{\min}$ ($E - A_{\min}$) is the unique maximal set that minimizes $2r(A) + |E - A|$ ($2\nu(A^*) + |E - A^*|$).*

4. *An edge e belongs to A_{min} (e belongs to B_{min}) iff there exist maximally distant forests t_1, t_2 (coforests $\bar{t}_1, \bar{t}_2$) s.t. $e \in (E - (t_1 \cup t_2))$, ($e \in (E - (\bar{t}_1 \cup \bar{t}_2)))$.*

Proof.

i. We know by Theorem 32.7, that a subset $\hat{A}$ minimizes $2r(A)+ \mid E - A \mid, A \subseteq E$ iff for every pair of maximally distant forests t_1, t_2 $|t_1 \cup t_2| = 2r(\hat{A}) + |E - \hat{A}|$ and that this happens iff $t_i \cap \hat{A}, i = 1, 2$, are disjoint forests of $\mathcal{G} \cdot \hat{A}$ and $(t_1 \cup t_2) \cap (E - \hat{A}) = (E - \hat{A})$, that is, $\hat{A} \supseteq E - (t_1 \cup t_2)$. Thus in $G(t_1, t_2)$, we see that $\hat{A}$ contains all vertices corresponding to $E - (t_1 \cup t_2)$ and further it is not possible to reach outside $\hat{A}$ from within since each $t_i \cap \hat{A}, i = 1, 2$, is a forest of $\mathcal{G} \cdot \hat{A}$. Thus $A_{\min} \subseteq \hat{A}$. However $A_{\min}$ itself minimizes $2r(A)+ \mid E - A \mid, A \subseteq E$ and so is the unique minimal minimizing set.

The proof for the dual statement follows by arguing with coforests $\bar{t}_1, \bar{t}_2$ and $G(\bar{t}_1, \bar{t}_2)$.

ii. We have, $t \cap A$ is a forest of $\mathcal{G} \cdot A$ iff $\bar{t} \cap (E - A)$ is a coforest of $\mathcal{G} \times (E - A)$ (Theorem 32.1 of Preliminaries). Next, $t_i \cap A, i = 1, 2$, are disjoint iff $(E - A) \supseteq t_1 \cap t_2$, that is, iff $(E - A) \supseteq (E - (\bar{t}_1 \cup \bar{t}_2))$ and $\bar{t}_i \cap (E - A), i = 1, 2$, are disjoint iff $A \supseteq \bar{t}_1 \cap \bar{t}_2$, that is, iff $A \supseteq E - (t_1 \cup t_2)$. From Theorem 32.8 we know that the expression $2r(A)+ \mid E - A \mid$ reaches a minimum iff $t_i \cap A, i = 1, 2$, are disjoint forests of $\mathcal{G} \cdot A$ and $A \supseteq E - (t_1 \cup t_2)$ and the expression $2\nu(A^*)+ \mid E - A^* \mid$ reaches a minimum iff $\bar{t}_i \cap (E - A), i = 1, 2$, are disjoint coforests of $\mathcal{G} \times (E - A)$ and $(E - A) \supseteq (E - (\bar{t}_1 \cup \bar{t}_2))$. The result follows.

iii. This follows immediately from (i) and (ii) above.

iv. Observe that every $e \in (t_1 \cup t_2) \cap A_{\min}$ can be reached from some $e_{\text{out}} \in A_{\min} - (t_1 \cup t_2)$. In $G(t_1, t_2)$ we therefore have a shortest path from e_{out} to e, say $e_{\text{out}}, (t_1), e_1, (t_2), \ldots, e_i, (t_j), e_{i+1}, \ldots, e_{n-1}, (t_m), e$. Modifying t_1, t_2 as in Algorithm *Maximally Distant Forests* would give us a new pair of maximally distant forests t'_1, t'_2, which would not contain e as a member. The result for $B_{\min}$ follows by using $G(\bar{t}_1, \bar{t}_2)$. ∎

32.7 NETWORK ANALYSIS THROUGH TOPOLOGICAL TRANSFORMATION

We say that we use topological transformation while analyzing networks, if at intermediate stages of the analysis, we modify the topology of the network. Instances of such transformations are the construction of two derived networks during topological hybrid analysis, multiport decomposition, and so on. In this section we consider a fairly general class of transformations which we may call the fusion fission method, and also sketch certain optimization problems which arise naturally during this study and which generalize the principal partition problem. Detailed description of these ideas may be found in the works of Narayanan [9,15,16].

32.7.1 Fusion–Fission Method

Consider the network in Figure 32.2. Four subnetworks have been connected together to make up the network. Assume that the devices in the subnetworks are decoupled. Clearly the networks in Figures 32.2 and 32.3 are equivalent, provided the current through the additional unknown voltage source and the voltage across the additional unknown current source are set equal to zero. But the network in Figure 32.3 is equivalent to that in Figure 32.4 under the additional conditions

$$i_{v1} + i_{v2} + i = 0$$

$$v_{i3} + v_{i4} - v = 0.$$

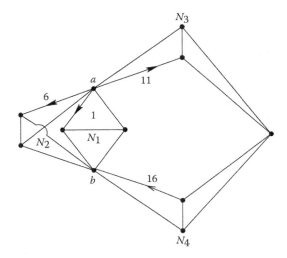

Figure 32.2 Network $\mathcal{N}$ to illustrate the fusion–fission method.

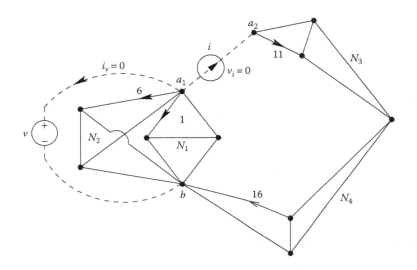

Figure 32.3 Network equivalent to $\mathcal{N}$ with virtual sources.

(The variables $i_{v1}, i_{v2}, v_{i3}, v_{i4}$ can be expressed in terms of currents and voltages of the original graph so that the additional constraints will involve only old current and voltage variables and the new variables i, v.)

As can be seen, the subnetworks of Figure 32.4 are decoupled except for the common variables v and i and the additional conditions.

A natural optimization problem here is the following:

Given a partition of the edges of a graph into $E_1, \ldots, E_k$, what is the minimum size set of node pair fusions and node fissions by which all circuits (equivalently cutsets) passing through more than one E_i are destroyed?

In the present example the optimal set of operations is to fuse nodes a and b and cut node a into a_1, a_2 as in Figure 32.3. Artificial voltage sources are introduced across the node pairs to be fused and artificial current sources are introduced between two halves of a split node.

The above formulation can be stated in a more convenient form as *node fusion–fission problem*. Let $\mathcal{G}$ be a graph and let Π_s be a specified partition of $E(\mathcal{G})$ so that $\mathcal{G} \cdot N_i$ is connected for each $N_i \in \Pi_s$. Find a minimum length sequence of node pair fusions and node

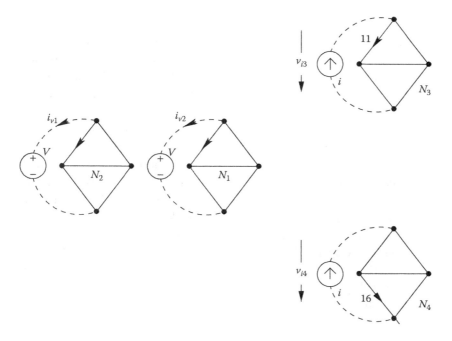

Figure 32.4 Network $\mathcal{N}$ decomposed by the fusion–fission method

fissions which, when performed on $\mathcal{G}$, result in a graph $\mathcal{G}_{new}$ in which each circuit intersects only one of the blocks of Π_s (equivalently each cutset intersects only one of the blocks of Π_s).

Assuming each $\mathcal{G} \cdot N_i$ to be connected is reasonable electrically speaking since, if they are not, we can usually be working with their connected components rather than with themselves while analyzing.

We will show that this problem generalizes the hybrid rank problem (see Section 32.5). We first note that the result of a node fission followed by a node fusion can always be achieved by a node fusion followed by a node fission. Thus whenever we have a sequence of node fissions and fusions needed for converting a graph into another, the result can always be achieved by a sequence of node fusions followed by a sequence of node fissions.

We think of the original network as being made up of a number of single edge networks and the problem is to decouple them in the above manner. At the end of the fusions and fissions no edges belonging to different networks should belong to the same circuit. Therefore every edge should have become a self-loop or a coloop. Let A, B be the subsets of edges, which are self-loops and coloops, respectively. Every edge in A has its endpoints fused. This is exactly equivalent to fusing the endpoints of the edges of a forest of $\mathcal{G} \cdot A$. Thus with $r(\mathcal{G} \cdot A)$ node fusions all edges in A would have become self-loops. After these node fusions the resulting graph would be $\mathcal{G} \times B$. In any graph, making all edges into coloops can be achieved with minimum number of operations by cutting each coforest edge at one of its end points. Therefore making all edges in B into coloops requires $\nu(\mathcal{G} \times B)$ node fissions. Thus the minimizing of $r(\mathcal{G} \cdot A) + \nu(\mathcal{G} \times B)$ over all partitions A, B is equivalent to finding a minimum length sequence of node fusions and fissions which will make every edge in the graph into a self-loop or coloop.

32.7.2 Solution of the Node Fusion–Fission Problem

In this subsection we give a sketch of the ideas involved in the solution of the node fusion-fission problem. Let $\mathcal{G}$ be a graph and Π_s, a partition of $E(\mathcal{G})$. The *fusion rank of $\mathcal{G}$ relative* to Π_s is the minimum length of a sequence of node pair fusions needed to destroy every circuit that intersects more than one block of Π_s. The *fission rank of $\mathcal{G}$ relative to* Π_s is the minimum length of a sequence of node fissions needed to destroy every circuit that intersects more than one block of Π_s. The *hybrid rank of $\mathcal{G}$ relative to* Π_s is the minimum length of a sequence of node pair fusions and node fissions needed to destroy every circuit that intersects more than one block of Π_s.

Now consider the situation where we use both fusions and fissions, with all the fusions occurring first. Any sequence of node pair fusions would ultimately fuse certain groups of nodes into single nodes. Hence, as far as the effect of these node pair fusions on the graph is concerned, we may identify them with a partition of $V(\mathcal{G})$ (with singleton blocks being permitted) each block of which would be reduced to a single node by the fusions. The number of node pair fusions required to convert a set of nodes V to a single node is $(|V|-1)$. Hence, if Π is a partition of $V(\mathcal{G})$, the number of node pair fusions required to go from $\mathcal{G}$ to the graph obtained from $\mathcal{G}$ by fusing blocks of Π into single vertices, which we shall denote by $\mathcal{G}_{\text{fus} \cdot \Pi}$, is $|V(\mathcal{G})|-|\Pi|$. This number we would henceforth call, the *fusion number of* Π. The fission rank of $\mathcal{G}_{\text{fus} \cdot \Pi}$ relative to a partition Π_s of $E(\mathcal{G})$ would be called the *fission number of Π relative to* Π_s. The sum of the fusion number and the fission number of Π relative Π_s would be called the *fusion–fission number of Π relative to* Π_s. Our task is to find a partition of $V(\mathcal{G})$ which minimizes this number.

We now define a bipartite graph which relates Π_s to $V(\mathcal{G})$. Let $B_\mathcal{G}$ be the bipartite graph associated with $\mathcal{G}$, with left vertices $V_L \equiv V(\mathcal{G})$ and right vertices $V_R \equiv E(\mathcal{G})$, with $e \in V_R$ adjacent to $v \in V$ iff edge e is incident on v in $\mathcal{G}$. Let $B(\Pi_s)$ be the bipartite graph obtained from $B_\mathcal{G}$ by fusing the right vertices in the blocks of Π_s and replacing parallel edges by single edges.

Let $X \subseteq V(\mathcal{G})$. Let $|\Gamma_L|(X)$ denote the size of the set of right vertices adjacent to vertices in X, $(|\Gamma_L|-\lambda)(X)$ denote $|\Gamma_L|(X)-\lambda$. Let Π be a partition of $V(\mathcal{G})$. We define $\overline{(|\Gamma_L|-\lambda)}(\Pi)$ to be the sum of the values of $(|\Gamma_L|-\lambda)$ on the blocks of Π. We then have the following result whose proof we omit in the interest of brevity.

Theorem 32.9 *Let $\mathcal{G}$ be a connected graph. Let Π_s be a partition of $E(\mathcal{G})$ so that $\mathcal{G} \cdot N_i$ is connected for each $N_i \in \Pi_s$. Let Π be a partition of $V(\mathcal{G})$. Then*

1. *The fusion–fission number of Π relative to Π_s equals*

$$\overline{(|\Gamma_L|-2)}(\Pi)+|V(\mathcal{G})|-|\Pi_s|+1.$$

2. *The hybrid rank of $\mathcal{G}$ relative to Π_s equals*

$$min(\overline{(|\Gamma_L|-2)}(\Pi)+|V(\mathcal{G})|-|\Pi_s|+1),$$

   Π a partition of $V(\mathcal{G})$. ∎

The problem of minimizing $\overline{(|\Gamma_L|-2)}(\Pi)$ over partitions of $V(\mathcal{G})$ falls under computing *Dilworth truncation* of a submodular function. The principal lattice of partitions problem is that of computing all partitions which minimize $\overline{(|\Gamma_L|-\lambda)}(\cdot)$ for some λ. Strongly polynomial algorithms are available for this purpose. Details are available in the works of Narayanan [9,15,16].

There are strong analogies between the principal partition (minimize $r(X)-\lambda|X|$, $r(\cdot)$ submodular, over subsets of a given set) and the principal lattice of partitions (minimize $\overline{(r-\lambda)}(\cdot)$, $r(\cdot)$ submodular, over partitions of a given set) problems. In the case of principal partition, if $\lambda_1 > \lambda_2$, a minimizing set corresponding to the former is always a subset of any minimizing set corresponding to the latter. In the case of principal lattice of partitions, if $\lambda_1 > \lambda_2$, a minimizing partition corresponding to the former is always finer than any minimizing partition corresponding to the latter [9]. In addition, in the case of graphs, the principal partition problem of Kishi–Kajitani (on the edge set of a graph) can be posed, as described earlier, as that of finding an optimal node fusion–fission problem and therefore can actually be solved as a principal lattice of partitions problem on the vertex set of the graph [9]. Indeed the current fastest principal partition algorithm for graphs is actually of this type [17].

Further Reading

The hybrid analysis notion is peculiar to network theory giving rise naturally to the hybrid rank problem. This problem and its generalizations can be regarded as unifiers for large parts of combinatorial optimization including the theory of submodular functions. This theme has been enlarged in the works of Narayanan [9]. The principal partition and principal lattice of partitions have many practical applications particularly in building partitioners for large-scale systems (see Chapter 30) and in structural solvability of systems [18,19] (see Chapter 31 for more references).

References

[1] G. Kron. *Diakoptics—Piecewise Solution of Large Scale Systems* (McDonald, London, 1963).

[2] G. Kron. *Tensor Analysis of Networks* (John Wiley & Sons, New York, 1939).

[3] F.H. Branin Jr. The relationship between Kron's method and the classical methods of network analysis. *Matrix and Tensor Quarterly* **12** (1962), 69–105.

[4] A. Brameller, M. John, and M. Scott. *Practical Diakoptics for Electrical Networks* (Chapman & Hall, London, 1969).

[5] H. Narayanan. A theorem on graphs and its application to network analysis. *Proceedings of the IEEE International Symposium on Circuits and Systems* (1979), 1008–1011.

[6] H. Narayanan. Topological transformations of electrical networks. *International Journal of Circuit Theory and Its Applications* **15** (1987), 211–233.

[7] H. Narayanan. Mathematical programming and electrical network analysis II: Computational linear algebra through network analysis, *International Symposium on Mathematical Programming for Decision Making: Theory and Applications*, ISI Delhi, India, January 10–11, 2007.

[8] V. Siva Sankar, H. Narayanan, and S.B. Patkar: Exploiting Hybrid Analysis in Solving Electrical Networks, *22nd International Conference on VLSI Design*, New Delhi, India, January 5–9, 2009, 206–210.

[9] H. Narayanan. Submodular functions and electrical networks. *Annals of Discrete Mathematics* **54** North Holland (London, New York, Amsterdam) (1997). (Revised version at http://www.ee.iitb.ac.in/ hn/book/.)

[10] T. Ohtsuki, Y. Ishizaki, and H. Watanabe. Topological degrees of freedom and mixed analysis of electrical networks. *IEEE Transactions on Circuit Theory* **CT–17** (1970), 491–499.

[11] J. Edmonds. Lehman's switching game and a theorem of Tutte and Nash-Williams. *Journal of Research of the National Bureau of Standards* **69B** (1965), 73–77.

[12] G. Kishi and Y. Kajitani. Maximally distant trees and principal partition of a linear graph. *IEEE Transactions on Circuit Theory* **CT–16** (1969), 323–329.

[13] M. Iri. The maximum rank minimum term rank theorem for the pivotal transformations of a matrix. *Linear Algebra and Its Applications* **2** (1969), 427–446.

[14] J. Edmonds. Minimum partition of a matroid into independent subsets. *Journal of Research of the National Bureau of Standards* **69B** (1965), 67–72.

[15] H. Narayanan. On the minimum hybrid rank of a graph relative to a partition of its edges and its application to electrical network analysis. *International Journal of Circuit Theory and Its Applications* **18** (1990), 269–288.

[16] H. Narayanan. The principal lattice of partitions of a submodular function. *Linear Algebra and Its Applications* **144** (1991), 179–216.

[17] S. Patkar and H. Narayanan. Fast algorithm for the principal partition of a graph. *Proceedings of the 11th Annual Symposium on Foundations of Software Technology and Theoretical Computer Science* **LNCS–560** (1991), 288–306.

[18] T. Ozawa. Topological conditions for the solvability of active linear networks. *International Journal of Circuit Theory and Its Applications* **4** (1976), 125–136.

[19] T. Ozawa and Y. Kajitani. Diagnosability of linear active networks. *IEEE Transactions on Circuits and Systems* **CAS–26** (1979), 485–489.

X

Probabilistic Methods, Random Graph Models, and Randomized Algorithms

CHAPTER 33

Probabilistic Arguments in Combinatorics*

C.R. Subramanian

CONTENTS

33.1	Introduction	946
33.2	Tools from Discrete Probability	947
33.3	Basic Philosophy	950
33.3.1	R1. Upper Bounding $\mathbf{Pr}(E^c)$: $\mathbf{Pr}(E^c) < 1 \Rightarrow \mathbf{Pr}(E) > 0$	950
33.3.2	R2. Lower Bounding $\mathbf{Pr}(E)$: Show Directly That $\mathbf{Pr}(E) > 0$	951
33.3.3	R3. Determining the Expectation: $E[X] \geq b \Rightarrow \mathbf{Pr}(X \geq b) > 0$	951
33.3.4	R4. Expectation and Tail Probabilities: Apply Bounds on Tail Probabilities to Show $\mathbf{Pr}(X \geq b) > 0$	951
33.4	Applications to Graph Theory and Number Theory	951
33.4.1	Lower Bound on Maximum Cut	951
33.4.2	Lower Bounds on Independence Number	953
33.4.3	Small Dominating Sets in Graphs	954
33.4.4	Number of Minimum Cuts in Multigraphs	955
33.4.4.1	Near-Minimum Cuts	956
33.4.5	List Coloring of Graphs	957
33.4.6	High Girth and High Chromatic Number	960
33.4.7	Global Coloring and Local Coloring	961
33.4.8	Tournaments of Specified Type	962
33.4.9	Bounds on Oriented Chromatic Numbers	964
33.4.10	Bounds on Constrained Colorings	965
33.4.10.1	Frugal Colorings	965
33.4.10.2	Acyclic Colorings	966
33.4.10.3	Other Constrained Colorings	967
33.4.11	Waring Bases	968
33.4.12	Sum-Free Subsets	970
33.5	Random Graphs	971
33.5.1	Existence of Triangles	972
33.5.2	Being Connected	973
33.5.3	Emergence of a Giant Component in the Vicinity of $1/n$	974
33.5.4	Diameter of Random Graphs	977
33.5.5	Concentration of Invariants	979
33.5.5.1	Concentration of $diam(G)$	979
33.5.5.2	Concentration of $\omega(G)$ and $\alpha(G)$	979

*Dedicated to my elder brother Dr. C.R. Seshan.

	33.5.5.3	Concentration of $\chi(G)$	981
	33.5.5.4	Concentration of Induced Paths and Induced Trees	982
33.6	Random Digraphs		984
	33.6.1	Induced Acyclic Tournaments	984
	33.6.2	Induced Acyclic Subgraphs	985
	33.6.3	Induced Tournaments	986
	33.6.4	Being Strongly Connected	987
	33.6.5	Emergence of a Giant Strongly Connected Component Around $1/n$	988
33.7	Conclusions		989

USING PROBABILISTIC arguments to prove statements in combinatorics is gaining popularity. A number of examples illustrate how this approach can be a powerful tool in proving mathematical statements in combinatorics in general, and graph theory in particular. This approach is based on applying (often simple) ideas, tools, and techniques from probability theory to obtain and prove a mathematical statement. On the other hand, several specific applications of this approach have also motivated and led to the development of powerful results in probability theory, in particular, with respect to discrete probability spaces.

In this expository chapter, we illustrate this approach with some specific applications from graph theory. A powerful tool employed is the notion of random graphs. We will also elaborate on this topic with specific examples and their probabilistic analyses. We do not necessarily present the best results (obtained using probabilistic arguments) since the main purpose is to give an introduction to the beauty and elegance of the approach. Many of the examples and the results are already known and published in the literature and have also been improved further. Some of the results we present are new ones not published before in the literature.

33.1 INTRODUCTION

Randomness is an important phenomenon that occurs in nature and in fact, scientists believe that randomness is an inherent part of the nature of physical reality. Probability theory is an axiomatic approach to model and infer conclusions about random phenomena. While the theory may have its inspiraton from nature, it has been found to be a powerful tool in obtaining and proving mathematical statements in combinatorics.

Combinatorics is the study of finite or countable structures in which we are interested in questions on the existence, number, and interplay between structures satisfying some property. The common idea underlying the application of probabilistic arguments is as follows:

Suppose $\mathcal{U}$ is a finite or countable set and let $\mathcal{P} \subseteq \mathcal{U}$ be a property possibly satisfied by some of the elements. Often, $\mathcal{P}$ is specified implicitly and we want to determine if $\mathcal{P} \neq \emptyset$. Then, we introduce a suitable probability measure P so that $(\mathcal{U}, P)$ becomes a probability space. Let $\omega \in \mathcal{U}$ be a random element drawn according to P. It follows that $\mathcal{P} \neq \emptyset$ if one can show that $\mathbf{Pr}(u \in \mathcal{P}) > 0$. If we also assume that P is uniform over $\mathcal{U}$, then $|\mathcal{P}|$ can be determined exactly or approximately from an exact or approximate knowledge of $\mathbf{Pr}(u \in \mathcal{P})$ and $|\mathcal{U}|$.

While this might look like nothing more than rephrasing the questions in the language of probability theory, it becomes easier now because probability theory has a rich wealth of tools, results and paradigms which can be employed on the rephrased questions to obtain a resolution. Often the proof turns out to be very simple and leads to the resolution quickly. This paradigm is known as the probabilistic method. There have been instances where this method has been successfully employed before (see, for instance, the probabilistic proof obtained by Bernstein [1,2] for Weierstrass approximation theorem). But Paul Erdös was the first person who fully understood the power of this method and pioneered its usage by applying it to a number of problems in number theory and graph theory with surprising conclusions. In that respect, one can as well call this as Erdös method. Also, he (in collaboration with Alfred Renyi) founded the theory of random graphs.

In what follows, we illustrate this approach with examples from graph theory and number theory. Some of the illustrations are based on random graphs the study of which is now a theory by itself since random graphs exhibit several interesting and curious properties which warrant a deep and detailed study by itself. With this view, we present an exposition on random graphs also.

The chapter is organized as follows. In Section 33.2, we provide a brief exposition of basic and necessary tools from discrete probability. In Section 33.3, we present and explain the basic philosophy underlying the probabilistic method. In Section 33.4, we present several applications of probabilistic arguments to graph theory and number theory. In Section 33.5, we present an exposition of random graphs. In Section 33.6, we present an exposition of random digraphs. Finally, in Section 33.7, we conclude with remarks. For a good introduction to probabilistic method, see the book authored by Alon and Spencer [2]. For a comprehensive introduction to random graphs, see the book authored by Bollobas [3] and also the one authored by Janson et al. [4].

33.2 TOOLS FROM DISCRETE PROBABILITY

We first recall some basic facts from probability theory. We focus only on discrete probability spaces. A discrete probability space (Ω, P) consists of a finite or countable set Ω of sample points and a probability or distribution function $P : \Omega \to [0, 1]$ which satisfies $\sum_{x \in \Omega} P(x) = 1$. An event E is any subset of Ω and its probability is defined as $\sum_{w \in E} P(w)$. Also, for any finite or countable collection $\mathcal{E} = \{E_1, E_2, \ldots\}$ of events, it follows from the Union Law of Probabilities that $\mathbf{Pr}(\cup_{E \in \mathcal{E}} E) \leq \sum_{E \in \mathcal{E}} \mathbf{Pr}(E)$ and the inequality becomes equality if the events in $\mathcal{E}$ are pairwise disjoint.

A real valued random variable X is a function $X : \Omega \to \mathcal{R}$ defined over Ω. Any distribution over Ω naturally induces a distribution on $Range(X)$. We only consider real valued random variables. From now on, for the sake of simplicity, we use $\mathcal{R}$ to denote $Range(X)$ and not the set of real numbers. Two random variables X and Y are said to be *identically distributed* if $\mathbf{Pr}(X = a) = \mathbf{Pr}(Y = a)$ for each $a \in \mathcal{R}$.

For a random variable X, the *expectation or mean* of X, denoted by $E[X]$ or μ_X, is defined to be $E[X] = \sum_{x \in \mathcal{R}} x \cdot \mathbf{Pr}(X = x)$ whenever the sum is well-defined. An important and very useful property of expectations is the following well-known fact.

Fact 33.1 *(Linearity of expectation) For any two real valued random variables X and Y defined over (Ω, P), we have $E(X + Y) = E(X) + E(Y)$.*

By induction, this holds for the sum of any finite number of variables. Note that this fact does not depend on any assumption about the variables.

Two random variables X and Y defined over Ω are *independent* if

$$\mathbf{Pr}(X = x, Y = y) = \mathbf{Pr}(X = x) \cdot \mathbf{Pr}(Y = y), \quad \forall x, y \in \mathcal{R}$$

or equivalently, for any two events $\mathcal{E}_X$ (which depends only on X) and $\mathcal{E}_Y$ (which depends only on Y), we have

$$\mathbf{Pr}(\mathcal{E}_X \cap \mathcal{E}_Y) = \mathbf{Pr}(\mathcal{E}_X) \cdot \mathbf{Pr}(\mathcal{E}_Y).$$

A collection $\mathcal{X} = \{X_1, \ldots, X_n\}$ of random variables are *mutually or totally independent* if

$$\mathbf{Pr}(X_1 = x_1, \ldots, X_n = x_n) = \prod_{1 \leq j \leq n} \mathbf{Pr}(X_j = x_j), \ \forall x_1, \ldots, x_n \in \mathcal{R}.$$

$\mathcal{X}$ is *pairwise independent* if $\mathbf{Pr}(X_i = x_i, X_j = x_j) = \mathbf{Pr}(X_i = x_i) \cdot \mathbf{Pr}(X_j = x_j)$ for every $i \neq j$. More generally, for $2 \leq k \leq n$, $\mathcal{X}$ is *k-wise independent* if, for every $1 \leq i_1 < \ldots < i_k \leq n$ and $x_{i_1}, \ldots, x_{i_k} \in \mathcal{R}$, we have

$$\mathbf{Pr}(X_{i_1} = x_{i_1}, \ldots, X_{i_k} = x_{i_k}) = \prod_{1 \leq j \leq k} \mathbf{Pr}(X_{i_j} = x_{i_j}).$$

Note that total independence implies k-wise independence for every k.

The *variance* of X, denoted by $\mathrm{Var}(X)$, is defined to be $\mathrm{Var}(X) = E((X - \mu_X)^2) = E(X^2) - E(X)^2$. The positive square root of $\mathrm{Var}(X)$ is known as the *standard deviation* of X and is denoted by σ_X. The following fact about independent random variables will be useful later.

Fact 33.2 *For any collection $\mathcal{X} = \{X_1, \ldots, X_n\}$ of n random variables, we have*

- $E(\prod_j X_j) = \prod_j E(X_j)$ *if $\mathcal{X}$ is totally independent.*
- $\mathrm{Var}(\sum_j X_j) = \sum_j \mathrm{Var}(X_j)$ *if $\mathcal{X}$ is pairwise independent.*

For a random variable X, its lower tail probabilities are estimates of the form $\mathbf{Pr}(X \leq a)$ for some $a \in \mathcal{R}$ and its upper tail probabilities estimates are of the form $\mathbf{Pr}(X \geq a)$ for some $a \in \mathcal{R}$. We recall some well-known upper bounds on such tail probabilities.

Fact 33.3 *(Markov inequality) For any non-negative real valued random variable X and any $t > 0$, we have $\mathbf{Pr}(X \geq t) \leq E(X)/t$.*

Fact 33.4 *(Chebyschev inequality) For any random variable X with mean μ_X and standard deviation σ_X and any $t > 0$, we have $\mathbf{Pr}(|X - \mu_X| \geq t) \leq Var(X)/t^2$. In particular, for any $t > 0$, we have $\mathbf{Pr}(|X - \mu_X| \geq t\sigma_X) \leq 1/t^2$.*

An inequality of this type which bounds the probability of deviating from the mean on either side is called a concentration inequality.

A random variable X which only takes values from $\{0, 1\}$ is called an indicator random variable. For an indicator variable X which takes the value 1 with probability p, it follows that $E(X) = p$ and $Var(X) = p(1 - p)$.

A sequence $(X_1, \ldots, X_n)$ of n mutually independent and identically distributed (iid) indicator random variables is said to form a Bernoulli trial. A Poisson trial is a sequence of independent (need not be identically distributed) indicator random variables.

If $(X_1, \ldots, X_n)$ forms a Poisson trial with respective means $E(X_i) = p_i$, then for their sum $X = X_1 + \cdots + X_n$, we have

$$\mu = E(X) = \sum_{1 \leq j \leq n} p_j; \ \ Var(X) = \sum_{1 \leq j \leq n} p_j(1 - p_j) \leq \mu.$$

As a result, for any $\epsilon > 0$, it follows by applying Chebyschev's inequality that

$$\mathbf{Pr}(|X - \mu| \geq \epsilon\mu) \leq \frac{1}{\epsilon^2 \mu}.$$

Thus the probability of deviating from the mean decreases in an inversely linear fashion. A relatively recent bound due to Chernoff and Hoeffding is tighter and the bound decreases inversely exponentially in μ. This exponential dependence will play an important role in our analyses later.

Fact 33.5 (*Chernoff–Hoeffding [CH] bounds*) *If X is the sum of the variables of a Poisson trial $(X_1, \ldots, X_n)$ with respective means $E(X_i) = p_i$ and $\mu = E(X)$, then for any $0 \leq a \leq \mu$, we have*

$$\Pr(X \leq \mu - a) \leq e^{-\frac{a^2}{2\mu}}$$
$$\Pr(X \geq \mu + a) \leq e^{-\frac{a^2}{3\mu}}$$
$$\Pr(|X - \mu| \geq a) \leq 2e^{-\frac{a^2}{3\mu}}$$

Note that the dependence on μ is inversely exponential. Applying this to the sum of n variables of a Bernoulli trial having a common mean of p, for $0 \leq \epsilon \leq 1$, CH bounds show that

$$\Pr(|X - np| \geq \epsilon np) \leq 2e^{-\epsilon^2 \frac{(np)}{3}}$$

whereas Chebyschev inequality only shows that

$$\Pr(|X - np| \geq \epsilon np) \leq \frac{1}{\epsilon^2 (np)}.$$

The CH-bounds provide tight concentration results but require that X is the sum of a finite collection of independent 0/1 variables. There are situations where X cannot be expressed in this way. For more general situations, another inequality based on the notion of martingales is known.

A sequence $(X_0, X_1, \ldots)$ of variables forms a *martingale* sequence if

$$E[X_i | X_{i-1}, \ldots, X_0] = X_{i-1}, \quad \forall i > 0.$$

It follows that $E[X_i] = E[X_0]$ for each i by an inductive argument.

Example 33.1 Suppose $X = X_1 + \cdots + X_n$ where $\Pr(X_i = 1) = \Pr(X_i = -1) = 1/2$ for each i. Define $Y_i = \sum_{j \leq i} X_j$. Then, $Y_i = Y_{i-1} + X_i$ and hence $E[Y_i | y_{i-1}, \ldots, y_0] = y_{i-1} + E[X_i] = y_{i-1}$. Thus $(Y_0, \ldots, Y_n)$ forms a martingale sequence.

Example 33.2 Suppose $(X_1, \ldots, X_n)$ forms a Poisson trial with $\Pr(X_i = 1) = p_i$ for each i. Let $X = X_1 + \cdots + X_n$. By an exposure of X_i, we refer to setting $X_i = b_i$ for some specific $b_i \in \{0, 1\}$. Define $Y_0 = E[X]$ and for $i \geq 1$, define Y_i as the expectation of $E[X]$ conditioned on the exposure of the first i variables $X_1, \ldots, X_i$. It is easy to see that Y_js are random variables and also that each Y_j is determined uniquely by the exposures of $(X_1, \ldots, X_j)$. Also, Y_n is exactly X. It can also be verified that

$$\begin{aligned} Y_{j-1} &= E[X | X_{j-1}, \ldots, X_1] \\ &= p_j E[X | X_j = 1, X_{j-1}, \ldots, X_1] + (1 - p_j) E[X | X_j = 0, X_{j-1}, \ldots, X_1] \\ &= E_{X_j}[E[X | X_j, X_{j-1}, \ldots, X_1]] = E[Y_j | Y_{j-1}, \ldots, Y_1] \end{aligned}$$

Thus $(Y_0, \ldots, Y_n)$ forms a martingale sequence.

The following result bounds tail probabilities of variables from a martingale sequence with bounded difference between consecutive variables. It was obtained by Azuma [2].

Fact 33.6 (*Azuma's bounds*) *If $c = Y_0, \ldots, Y_m$ form a martingale with $|Y_{i+1} - Y_i| \leq 1$ for each $i < m$. Then, for any $\lambda > 0$, we have*

$$\Pr(Y_m - c \geq \lambda \sqrt{m}) \leq e^{-\frac{\lambda^2}{2}}$$
$$\Pr(Y_m - c \leq -\lambda \sqrt{m}) \leq e^{-\frac{\lambda^2}{2}}$$
$$\Pr(|Y_m - c| \geq \lambda \sqrt{m}) \leq 2e^{-\frac{\lambda^2}{2}}$$

In the context of Example 33.2, $c = Y_0 = E(X)$, $m = n$, $Y_m = X$ and hence the three inequalities provide bounds on the lower and upper tail probabilities. Even though Y (in this example) is the sum of variables of a Poisson trial, we will later see applications of Azuma's result where this assumption is not true, yet Azuma's bound provides tight concentration results.

Unlike the previous tools, where we typiclly focus on proving that a certain event occurs almost surely, the following lemma is a very powerful tool in proving the occurrence of an event which occurs very rarely, provided it is implied by the simultaneous non-occurrence of so-called *bad* events which are almost independent. The following lemma provides sufficient conditions which guarantee the simultaneous non-occurrence of bad events. This lemma is known as Lovász local lemma (LLL). We use the following non-symmetric form of LLL (see [2,5]).

Lemma 33.1 [5] *Let $\mathcal{A} = \{A_1, A_2, ..., A_n\}$ be a family of events in an arbitrary probability space such that each event A_i is mutually independent of $\mathcal{A} \setminus (\{A_i\} \cup \mathcal{D}_i)$ for some $\mathcal{D}_i \subset \mathcal{A}$. Then if there are reals $0 < y_i < 1$ such that for all i,*

$$\mathbf{Pr}(A_i) \leq y_i \prod_{A_j \in \mathcal{D}_i} (1 - y_j)$$

then

$$\mathbf{Pr}(\cap (\overline{A_i})) \geq \prod_{i=1}^{n} (1 - y_i) > 0$$

so that with positive probability no event A_i occurs. ∎

For a more elaborate discussion on the probability tools presented in this section, the reader is referred to any of the following references [2,6].

33.3 BASIC PHILOSOPHY

The main argument underlying the application of probabilistic arguments in combinatorics is very simple. Suppose we want to show that a set Ω has a member w satisfying some property $E \subseteq \Omega$. In other words, given some description of E, we want to show that $E \neq \emptyset$. Introduce a probability distribution $P(x)$ over Ω. Then, using some of the tools of probability theory, show that $\mathbf{Pr}(E) > 0$. This implies that there is a point $x \in \Omega$ satisfying E. If not, we will have $\mathbf{Pr}(E) = 0$. Note that the choice of the distribution $P(.)$ is ours and also the choice of the probabilistic tools we employ are also ours. We describe below some rules which help us in this process. The rules themselves are straightforward but when applied they become a very powerful tool.

33.3.1 R1. Upper Bounding $\mathbf{Pr}(E^c)$: $\mathbf{Pr}(E^c) < 1 \Rightarrow \mathbf{Pr}(E) > 0$

E^c is the complement event $\Omega \setminus E$. It suffices to show that $\mathbf{Pr}(E^c) < 1$. One can do this by choosing events $F_1, F_2, \ldots$ such that whenever E^c holds, one of the F_i's holds. Hence $\mathbf{Pr}(E^c) \leq \sum_i \mathbf{Pr}(F_i)$. If this sum can be shown to be strictly less than 1, we are done.

Example: If n and k are such that $f(n,k) = \binom{n}{k} 2^{1-\binom{k}{2}} < 1$, then the edges of the complete graph K_n can be 2-colored in such a way that there is there is no subgraph K_k (of K_n) all of whose edges are colored with the same color. Here, Ω is the set of all 2-colorings of $E(K_n)$.

$P(.)$ is the uniform distribution over Ω. Equivalently, color each edge of K_n independently with (equal probability) one of the two colors $\{R, B\}$. E^c is the event that some $E(K_k)$ is monochromatically colored. We define an event E_S for every $S \subseteq V(K_n)$ with $|S| = k$. E_S is holds if the random coloring makes $E(S)$ monochromatic and to 0 otherwise. We have $\binom{n}{k}$ such events. Note that E^c holds if and only if E_S holds for some S. But $\mathbf{Pr}(E_S) = 2/2^{\binom{k}{2}}$ for every S. Hence $\mathbf{Pr}(E^c) \leq \sum_{S,|S|=k} \mathbf{Pr}(E_S) = \binom{n}{k} 2^{1-\binom{k}{2}} < 1$. Hence $\mathbf{Pr}(E) > 0$ and hence there is a 2-coloring with no monochromatic K_k.

33.3.2 R2. Lower Bounding $\mathbf{Pr}(E)$: Show Directly That $\mathbf{Pr}(E) > 0$

Choose a finite collection $\{F_1, \ldots, F_m\}$ events such that if none of the F_i's hold, then E holds. Hence it suffices to show that $\mathbf{Pr}(\wedge_i F_i^c) > 0$ since $\mathbf{Pr}(E) \geq \mathbf{Pr}(\wedge_i F_i^c)$. We can use powerful tools like LLL or Janson inequalities to establish this. We will see examples of this approach later.

33.3.3 R3. Determining the Expectation: $E[X] \geq b \Rightarrow \mathbf{Pr}(X \geq b) > 0$

If the property to be established can be stated as (or be implied by) a statement of the form "there exists a point $w \in \Omega$ with $X(w)$ at least (or at most) b" for some random variable X, then we can do this by determining $E[X]$. If $E[X] \geq b$, then there is indeed a point $w \in \Omega$ with $X(w) \geq b$ and $P(w) > 0$. Otherwise, $E[X] = \sum_{w \in \Omega} P(w) X(w) < b$. Similarly, if $E[X] \leq b$, then $\mathbf{Pr}(X \leq b) > 0$.

Example: We again look at the example of 2-colorings $E(K_n)$. By associating with each event E_S an indicator variable X_S which is set to 1 if E_S holds and to 0 otherwise, we see that E^c holds if and only if $X = \sum_S X_S \geq 1$. In other words, E holds if and only if $X < 1$. To show that $\mathbf{Pr}(X < 1) > 0$, we calculate $E[X]$ and notice that $E[X] = f(n, k) < 1$ and deduce our statement.

33.3.4 R4. Expectation and Tail Probabilities: Apply Bounds on Tail Probabilities to Show $\mathbf{Pr}(X \geq b) > 0$

Suppose the property to be established is equivalent to establishing $X \geq b$ (or $X \leq b$) for some random variable X. This is same as showing that $\mathbf{Pr}(X < b) = \mathbf{Pr}(X < \mu + (b - \mu)) < 1$ where $\mu = E[X]$. One can achieve this by calculating $E[X]$ and using some of the tail probability results we have mentioned before. We will see examples of this later.

33.4 APPLICATIONS TO GRAPH THEORY AND NUMBER THEORY

We present some interesting observations mostly from graph theory and also a few ones from number theory which are established by simple applications of probabilistic arguments. The purpose is to illustrate the power and simplicity of probabilistic arguments in arriving at interesting conclusions. Since this is an expository write-up, we do not bother to present the best known results but focus more on the method.

33.4.1 Lower Bound on Maximum Cut

A simple and direct application is the following example. We study the maximum size (= number of edges) of any spanning bipartite subgraph of a given undirected graph. More precisely, we establish that: Every graph $G = (V, E)$ with $|V| = n$ and $|E| = m$ has a vertex bipartition $V = A \cup B$ such that $e(A, B) \geq m/2$ where $e(A, B) = |E(A, B)| = |\{(u, v) \in E : u \in A, v \in B\}|$. One can easily prove this as follows. Choose uniformly at random a bipartition $V = A \cup B$. Equivalently, include each $u \in V$ independently and with equal probability into one of the two parts. For each $e = \{u, v\} \in E$, let X_e denote the indicator variable which takes 1 if $e \in E(A, B)$ and 0 otherwise. For each $e \in E$, $E(X_e) = 1/2$ since 2 of the four possible ways of including u and v into A or B will put the two vertices in different parts. Let $X = \sum_{e \in E} X_e$ denote the number of edges in $E(A, B)$. Then $E[X] = m/2$ and hence by R3, we have $\mathbf{Pr}(\exists A, B : V = A \cup B, e(A, B) \geq m/2) > 0$.

Consider the following deterministic argument: Start with an arbitrary bipartition $V = A \cup B$ and repeat the following rule as long as possible. If there exists $u \in A$ such that u has more neighbors in A than in B, then shift u from A to B. Do similarly for B. Each application of this rule takes us to another bipartition with at least one more edge in $E(A, B)$ than the previous bipartition. Hence this rule fails to be applicable after some iterations. At that moment, if $V = A \cup B$ denotes the current bipartition, then each vertex has at least as many neighbors in the other part than it has in its part. Hence, if we ignore edges falling within a part, we deduce that the final bipartition has at least $m/2$ edges. Even though this argument establishes a strengthening, it is not as simple and direct as the probabilistic arguments given before. It actually proves something stronger.

In fact, one can show a stronger lower bound than $m/2$. Let $\mu(G)$ denote the maximum size of any matching in G. Let $M = \{e_1, \ldots, e_\mu\}$ where $e_i := (u_i, v_i)$ for each $i = 1, \ldots, \mu$, be any such matching. Consider the following random experiment. For each $e_i \in M$, with equal probability, we choose one of the two possibilities: either $u_i \in A, v_i \in B$ or $u_i \in B, v_i \in A$. For the remaining vertices of V, we assign them randomly to either A or B with probability $1/2$. The choices are totally independent of each other. It follows that for each i, $\mathbf{Pr}(e_i \in E(A, B)) = 1$. For every other edge $e \in E(G) \setminus M$, we have $\mathbf{Pr}(e \in E(A, B)) = 0.5$. Hence, $E[X] = \mu + m - (\mu/2) = m + (\mu/2)$. Hence by R3, we have an edge-cut of size at least $m + \mu/2$.

Further improvements on the smaller additive term have been explored and tight estimates have been obtained. For a graph G, let $f(G)$ denote the size of the largest spanning bipartite subgraph of G. Let $f(m)$ denote the least value of $f(G)$ as G ranges over all graphs having m edges. Edwards [7,8] established that $f(m) \geq (m/2) + (\sqrt{8m+1} - 1)/8$ for every m. Erdös [9] conjectured that the gap between the $f(m)$ and the above lower bound can become arbitrarily large for large values of m. This was affirmatively settled by Noga Alon [10] (using probabilistic arguments) by establishing that $f(m) \geq m/2 + \sqrt{m/8} + cm^{1/4}$ (for some positive constant c), for infinitely many values of m. It was also established in [10] that $f(m) \leq m/2 + \sqrt{m/8} + c'm^{1/4}$ (for some positive constant c'), for every m. Thus, $f(m)$ has been essentially determined (upto constant multiplicative factors for the third order small term) for infinitely many values of m. The cute proof of the upper bound was based on exhibiting for every m a graph G_m defined as follows: G_m is the disjoint union of k complete subgraphs K_{n_i} ($i = 1, \ldots, k$) where n_i's are defined as follows: Let $n_0 = 1$. n_i is the largest positive integer such that $\binom{n_i}{2} \leq m - \sum_{j < i} \binom{n_j}{2}$. It is easy to verify that $f(G_m) \leq (m/2) + \sqrt{(m/8)} + O\left(m^{1/4}\right)$.

What happens if we focus only on triangle-free graphs? Let $g(m)$ be the same as $f(m)$ except that G ranges over all triangle-free graphs having m edges. Improving the earlier lower bounds of Erdös and Lovasz [11] and Poljak and Tuza [12], shearer [13] established that $g(m) \geq m/2 + cm^{3/4}$ (for some positive constant c) for every m. Alon [10] improves this further and shows that $m/2 + c_1 m^{4/5} \leq g(m) \leq m/2 + c_2 m^{4/5}$ for every m, where c_1, c_2 are some suitable positive constants. Shortly, we now know that $g(m) = m/2 + \Theta\left(m^{4/5}\right)$.

33.4.2 Lower Bounds on Independence Number

For a graph $G = (V, E)$ on n vertices, let $\alpha(G)$ denote the maximum size of an independent set in G. If Δ denotes the maximum degree of any vertex in G, then any maximal solution (and hence a maximum solution) should be of size at least $n/(\Delta+1)$. Hence $\alpha(G) \geq n/(\Delta+1)$.

A better bound can be obtained by a simple and elegant probabilistic argument. Let $d = (\sum_u d_u)/n$ denote the average degree of a vertex in G with $d > 0$. Choose a random subset S of V by including each u independently with probability $p = 1/d$. Then, $|S| = \sum_u X_u$ where X_u indicates $u \in S$. Hence, we have $E[|S|] = np$. Let t denote the number of edges in $G[S]$. Then, $t = \sum_{e \in E(G)} Y_e$ where Y_e (for $e = \{u,v\}$) indicates $u, v \in S$. Also, $E[Y_e] = p^2$ since the vertices defining e are included into S independently. Hence, we have $E[t] = |E(G)| \cdot p^2 = ndp^2/2$. Now, for each edge $e \in E(G[S])$, pick an arbitrary endpoint of e and remove it. Let I denote the resulting set. I is clearly an independent set. Also, $|I| \geq |S| - t$ always. Hence $E[|I|] \geq E[|S|] - E[t] = np - ndp^2/2 = n/2d$. Hence, by R3, it follows that $\Pr(|I| \geq n/2d) > 0$ and hence $\alpha(G) \geq n/2d$. Equivalently, since $nd = 2m$ where m denotes the number of edges in G, it follows that $\alpha(G) \geq n^2/4m$.

This later lower bound is at least as good as the first one whenever $\Delta + 1 \geq 2d$. Also, there are many examples where this gives a much better bound. For example, $\Delta(K_{1,n}) = n$ and hence we get $\alpha(K_{1,n}) \geq (n+1)/(n+1) = 1$ from the first bound. But, $d(K_{1,n}) = 2n/(n+1)$ and the second bound gives $\alpha(G) \geq (n+1)^2/4n \geq n/4$, which is much better than the first bound. While we know that $\alpha(K_{1,n}) = n$, this comes from our knowledge of the structure of $K_{1,n}$. We are aiming to get as strong a lower bound on $\alpha(G)$ as possible in the context of possessing very little information about G.

Later, Caro [14] and Wei [15] independently established that $\alpha(G) \geq g(G) := \sum_{u \in V} 1/(d_u + 1)$ where d_u denotes the degree of a vertex u. This bound is always at least as good as the previous two bounds (for connected graphs) and often yields much better results also. Applying this to $K_{1,n}$, we get $\alpha(K_{1,n}) \geq (n/2) + (1/(n+1)) \geq n/2$.

We present below the beautiful probabilistic proof of this bound of Caro and Wei [14,15] due to Boppana [2,11]. Without loss of generality, assume that $V = \{1, \ldots, n\}$. Choose uniformly at random a total order '<' of V. There are $n!$ such orders. Based on '<,' define a random set I as follows. $I = \{u \in V : u < v, \forall (u,v) \in E\}$. I is the set of those vertices which come before their neighbors. Clearly, for every $e = (u,v) \in E$, either $u \notin I$ or $v \notin I$. Hence I is always an independent set in G. For every $u \in V$, define I_u to be the indicator random variable for $u \in I$. Then $|I| = \sum_{u \in v} I_u$ and hence $E[|I|] = \sum_{u \in V} E[I_u]$. But, for any u, $E[I_u] = \Pr(I_u = 1) = (d_u)!/(d_u + 1)! = 1/(d_u + 1)$. The second equality follows by noting that the required probability depends only on the relative ordering of u and its neighbors and of these there are exactly $(d_u)!$ orderings which enable u to become a member of I. Hence, by R3, it follows that $\Pr(|I| \geq g(G)) > 0$ and hence there exists an ordering for which $|I| \geq g(G)$. This implies that $\alpha(G) \geq g(G)$.

Recently, this bound has been suitably generalized for k-uniform hypergraphs by Caro and Tuza [17]. A k-uniform hypergraph on $H = (V, E)$ is a collection $E \subseteq \binom{V}{k}$ of distinct k-element subsets of V. A subset $S \subseteq V$ is independent in H if $e \not\subseteq S$ for every $e \in E$. Caro and Tuza [17] established that

$$\alpha(H) \geq \sum_{u \in V} \binom{d_u + \frac{1}{t}}{d_u}^{-1} \quad \text{where } t := k - 1 \ldots \ldots (A).$$

For graphs (corresponding to $k = 2$), this bound specializes to the Caro–Wei–Boppana bound mentioned before. The proof of this bound is based on carefully chosen counting and structural arguments. An easy consequence of (A) is the following result.

Theorem 33.1 [17,18] *For every $k \geq 3$, there exists $d_k > 0$ such that every k-uniform hypergraph H has*

$$\alpha(H) \geq d_k \cdot \sum_{u \in V} \frac{1}{(d_u + 1)^{1/t}} \ldots \ldots (B).$$ ∎

The simple probabilistic proof of Boppana (outlined before), together with some additional ideas, can also be shown to yield a new short proof of (B) as has been done by Dutta et al. [18].

33.4.3 Small Dominating Sets in Graphs

Let $G = (V, E)$ be a graph whose minimum degree is δ. A subset $D \subseteq V$ is a dominating set in G if, for every $u \in V$, either $u \in D$ or $v \in D$ for some neighbor v of u. Let $\gamma(G)$ denote the minimum size of any dominating set in G. For graphs with minimum degree δ, using probabilistic arguments, one can obtain (as shown in [2]) an upper bound on $\gamma(G)$ as $\gamma(G) \leq h(G) = n[(1 + \ln(\delta + 1))/(\delta + 1)]$.

The proof is as follows. For $p = \ln(\delta + 1)/(\delta + 1)$, choose a random subset S of V by including independently each u with probability p. Now, let S' denote the set of vertices $u \notin S$ such that u has no neighbor in S. Let $D = S \cup S'$. Clearly, D is always a dominating set in G. For every $u \in V$, let X_u denote the indicator variable for $u \in S$ and let Y_u denote the variable for $u \in S'$. Note that $X_u = 1 \Rightarrow Y_u = 0$ for each u. Also, $|S| = \sum_u X_u$ and $|S'| = \sum_u Y_u$. Hence $|D| = \sum_u X_u + \sum_u Y_u$. Also, $E[|S|] = np$ and $E[|S'|] = \sum_u E[Y_u]$. For any u, $E[Y_u] = \Pr(Y_u = 1) = (1-p)^{d_u+1} \leq (1-p)^{\delta+1} \leq e^{-p(\delta+1)} = 1/(\delta+1)$. Hence $E[|D|] \leq np + n/(\delta+1) = n[(1 + \ln(\delta+1))/(\delta+1)]$. Hence, by R3, $\Pr(|D| \leq h(G)) > 0$. Hence $\gamma(G) \leq h(G)$. It follows that larger the value of δ is, the smaller is the upper bound on the size of a minimum dominating set. The proof yields a linear-time randomized algorithm for computing such a D. It can also be derandomized to get an efficient, deterministic algorithm.

Small size dominating sets play a useful role in designing faster algorithms for some NP-hard problems. For example, linear minimum degree guarantees the existence of a dominating set of size $O(\log n)$ and its efficient computation will help us decide 3-colorability (a NP-complete problem) in polynomial time. Recently, Narayanaswamy and Subramanian [19] exploited the presence of dominating sets of size guaranteed above to design faster exact and exponential time algorithms for the 3-colorability problem. The results are also shown to be *tight* in view of a widely held belief (known as exponential time hypothesis) that rules out the existence of sub-exponential time algorithms for 3-colorability.

The above bound can also be interpreted as an upper bound on $\gamma(G, d)$ (the minimum size of a set which dominates all vertices of degree at least d in an arbitrary graph). It follows that $\gamma(G, \sqrt{n}) = O(\sqrt{n}(\ln n))$. A dominating set D of such a size can be used to compute a shortest path (within an additive error of 2) between each pair of vertices in an unweighted, undirected graph. The running time is $O(n^{2.5}(\ln n))$. This led to the first combinatorial algorithm for computing 2-additive approximations to shortest paths. For details, see the work of Aingwork et al. [20].

A natural and related problem is to find upper bounds on minimum size $\gamma_c(G)$ of a connected dominating set in a connected graph G. It was shown by Duchet and Meyniel [21]) that $\gamma_c(G) \leq 3\gamma(G) - 2$. This was further sharpened (by employing a more careful probabilistic analysis of the random experiment described before and also some simple structural arguments) by Caro et al. [22] to $\gamma_c(G) \leq n[(1.45 + 0.5\sqrt{\ln \delta_1} + \ln \delta_1)/\delta_1]$ where δ_1 stands for $\delta + 1$. Combining this result with the observations in [19], one can now deduce that 3-colorability can be tested in $O((1.245)^n)$ time if $\delta \geq 15$ and in $O((1.0905)^n)$ time if $\delta \geq 50$.

33.4.4 Number of Minimum Cuts in Multigraphs

Let $G = (V, E)$ be a multigraph (without self-loops). G is allowed to have parallel edges (i.e., multiple copies of an edge $\{u, v\}$, where $u \neq v$) but has no self-loop. An edge-cut is any subset $F \subseteq E$ such that $G \setminus F$ is disconnected. Every edge cut F is precisely the set of edges joining vertices in different pieces of some bipartition $V = A \cup B$ into two non-empty sets. It is possible that there is more than one such bipartition. We will often specify an edge cut by one of the corresponding bipartitions. Given an edge cut (A, B), let $e(A, B)$ denote the number of edges in this cut. A minimum cut (or shortly a min-cut) of G is a cut (A, B) for which $e(A, B)$ is minimum. We describe below a proof (based on probabilistic arguments) of the following claim. The proof is probabilistic and algorithmic in nature and follows from the works of Karger and Stein [23–25].

Claim 33.1 *The number of minimum cuts in a connected multi-graph G on n vertices is at most $\binom{n}{2}$.*

Let m denote the number of all edges (counting the parallel ones also). Let c denote the size of any minimum cut in G. It follows that $\delta(G) \geq c$ where $\delta(G)$ denotes the minimum degree of any vertex in G with the convention that the degree counts the number of edges (including all parallel ones) incident at any vertex. This is because if $d_u < c$ for some vertex u, then $(\{u\}, V \setminus \{u\})$ is a cut of size less than c contradicting the optimality of c. As a result, it follows that $m = (\sum_u d_u)/2 \geq (nc/2)$. Before we proceed further, we describe some terminology which play an important role in our proof arguments.

By contracting an edge e (joining u and v) in a graph, we mean (1) removing u and v, (2) add a new vertex labeled as, say, uv, and (3) for every edge $f \neq e$ which joins either u or v with some $w \neq u, v$, we add the edge $\{w, uv\}$. The addition of edges preserves multiplicities also. For example, if there are 3 copies of the edge $\{u, w\}$ and 2 copies of the edge $\{v, w\}$, then we add 5 copies of the edge $\{uv, w\}$. Also, all multiple edges between u and v get destroyed. The result is a graph with one vertex less. The newly added vertex represents the subset $\{u, v\}$. It follows by an inductive argument that the result of repeated applications of the contraction operation on a graph G is a multigraph G' (without self-loops) such that

1. Each $u \in V(G')$ represents a subset S_u of $V(G)$ inducing a connected subgraph of G,
2. Each edge in G' represents an edge in G joining the corresponding induced subgraphs,
3. The subsets $\{S_u : u \in V(G')\}$ represented by $V(G')$ forms a partition of $V(G)$,
4. G' is connected if G is connected,
5. The minimum size of a cut in G' is at least the size of a minimum cut in G. Hence, if no edge of a minimum cut (A, B) of G is chosen for contraction, then (A, B) is also a cut in G' in the sense that $S_u \subseteq A$ or $S_u \subseteq B$ for each $u \in V(G')$ and hence a minimum cut in G' also. In that case, we say that (A, B) has survived the contraction process.

Hence, the result of $n - 2$ applications of the contraction operation starting from a connected multigraph on n vertices is a connected multigraph on 2 vertices which admits a unique minimum cut.

The proof of the claim follows by analyzing the the following random process : Initialize $H = G$. Repeatedly (for $n - 2$ times) pick uniformly (using independent choices) at random an edge e of the current graph H and contract e to get a new graph H' with one vertex less. Reset H to be H'.

Let G_0 denote the initial graph G stated in the Claim. For $i \leq n - 2$, let G_i denote the graph resulting after the first i random contractions. Fix an arbitrary minimum cut (C, D) of G. Consider any $1 \leq i \leq n - 2$. Suppose it is given that (C, D) has survived upto the first $i-1$ contractions and is a minimum cut in G_{i-1}. Then, $\delta(G_{i-1}) \geq c$ and hence $|E(G_{i-1})| \geq (n-i+1)c/2$. Hence,

$$\mathbf{Pr}(\text{ some } e \in E(C,D) \text{ is used in the } i^{th} \text{ contraction step }) = \frac{c}{|E(G_{i-1})|} \leq \frac{2}{n-i+1} \text{ and}$$

$$\mathbf{Pr}((C,D) \text{ survives the } i^{th} \text{contraction}) \geq 1 - \frac{2}{n-i+1} = \frac{n-i-1}{n-i+1}$$

For each $i = 0, \ldots, n-2$, let $\mathcal{E}_i$ denote the event that (C,D) has survived the first i random contractions and hence is a minimum cut in G_i. From the previous inequality it follows that $\mathbf{Pr}(\mathcal{E}_i|\mathcal{E}_{i-1}) \geq (n-i-1)/(n-i+1)$. Also, $\mathbf{Pr}(\mathcal{E}_0) = 1$ trivially. Hence,

$$\begin{aligned}\mathbf{Pr}((C,D) \text{ survives each of the } n{-}2 \text{ contractions}) &= \mathbf{Pr}(\mathcal{E}_{n-2}) \\ &= \mathbf{Pr}(\mathcal{E}_{n-2}|\mathcal{E}_{n-3})\ldots \mathbf{Pr}(\mathcal{E}_1|\mathcal{E}_0) \cdot \mathbf{Pr}(\mathcal{E}_0) \\ &\geq \frac{1}{3} \cdot \frac{2}{4} \cdot \frac{3}{5} \cdots \frac{n-2}{n} \\ &= \frac{2}{n(n-1)} = \binom{n}{2}^{-1}\end{aligned}$$

Clearly, if (C,D) survives the random process in some execution, then it is the only minimum cut of G that survives in this execution. In other words, the events $\mathcal{E}_{n-2}$ corresponding to different minimum cuts of G are mutually exclusive and hence all these probabilities should sum to at most 1. But each of these probabilities is at least $\binom{n}{2}^{-1}$. This implies that the number of minimum cuts is at most $\binom{n}{2}$.

Summary: What we have established is a statement in structural graph theory. But the proof of this statement was accomplished by employing probabilistic arguments. We first obtained a lower bound on the probability that any fixed (but arbitrary) minimum cut surviving a sequence of random contractions. We also noticed that these events (for different minimum cuts) are pairwise disjoint. Then we applied the union law on probabilities of events to obtain the desired conclusion.

33.4.4.1 Near-Minimum Cuts

Suppose a real $\alpha \geq 1$ is a half-integer, that is, 2α is an integer. An edge cut (A,B) is an α-min cut if $e(A,B) \leq \alpha c$ where c denotes the size of the minimum cut. The proof (based on probabilistic arguments) presented above can in fact be adapted to establish that the number of α-min cuts in G is at most $\binom{n}{2\alpha} 2^{2\alpha} \leq n^{2\alpha}$. If $c = 1$, then number of α-min cuts is at most $\binom{n^2}{\lfloor \alpha \rfloor} \leq n^{2\alpha}$. Hence, for the rest of the section, we assume that $c \geq 2$. The idea is to continue the edge contractions only as long as the current multigraph has more than $k := 2\alpha$ vertices. There are exactly $n-k$ contractions before we reach a graph H on k vertices after which we choose uniformly at random one of the $2^{k-1}-1$ non-trivial bipartitions of $V(H)$. Fix an α-min cut (C,D). Using arguments given before, it can be verified that the conditional probability that (C,D) survives the ith contraction given that it has survived the first $(i-1)$ contractions is at least $(n-i+1-k)/(n-i+1)$.

$$\begin{aligned}\mathbf{Pr}((C,D) \text{ survives the } n-k \text{ contractions and chosen}) &= \mathbf{Pr}(\mathcal{E}_{n-k}) \cdot \frac{1}{2^{k-1}-1} \\ &\geq \mathbf{Pr}(\mathcal{E}_{n-k}|\mathcal{E}_{n-k-1})\ldots \mathbf{Pr}(\mathcal{E}_0) \cdot \frac{1}{2^k} \\ &\geq \frac{1}{k+1} \cdots \frac{n-k}{n} \cdot \frac{1}{2^k} \\ &= \binom{n}{k}^{-1} \cdot 2^{-k}\end{aligned}$$

Since the events corresponding to different α-min cuts surviving and being chosen are mutually exclusive, we deduce that the number of α-min cuts is at most $\binom{n}{2\alpha}2^{2\alpha} \leq n^{2\alpha}$. What we have established is a pure structural result on the number of α-min cuts using probabilistic arguments. The union law on probabilities applied to mutually exclusive events helped us obtain this upper bound.

This upper bound can be established even if $\alpha \geq 1$ is any real using generalized binomial coefficients. Such a bound and enumeration of all such α-min cuts play an important role in approximately determining the reliability of a dynamic network. For more details, see [26–28].

33.4.5 List Coloring of Graphs

Here, we obtain upper bounds on the choice number (also known as list chromatic number) of undirected graphs. Given an undirected graph $G = (V, E)$, a list-assignment $\mathcal{L} = \{L_u : u \in V\}$ of colors is a map assigning (for every u) a list L_u of colors. We say that G is $\mathcal{L}$-colorable if there exists a proper coloring which assigns (for every u) a color $f(u) \in L_u$ such that $f(u) \neq f(v)$ for every $(u, v) \in E$. Otherwise, it is not $\mathcal{L}$-colorable. The choice number $ch(G)$ (also denoted by $\chi_l(G)$) is the minimum value of k such that G is $\mathcal{L}$-colorable for every list assignment $\mathcal{L}$ satisfying $|L_u| \geq k$ for every $u \in V$. When we restrict $\mathcal{L}$ to those list-assignments in which $L_u = L_v$ for every $u, v \in V$, the choice number specializes to the standard chromatic number $\chi(G)$. Hence it follows that $ch(G) \geq \chi(G)$.

For example, $ch(K_{3,3}) = 3$, whereas $\chi(K_{3,3}) = 2$. Suppose $G = K_{3,3}$ joins completely the sets $U = \{u_1, u_2, u_3\}$ and $V = \{v_1, v_2, v_3\}$ and label each u_i and v_i with the set $S_i = \{1, 2, 3\} \setminus \{i\}$ and consider the list assignment $\mathcal{L}$ where $L_{u_i} = L_{v_i} = S_i$ for every i, then it can be verified easily that $K_{3,3}$ is not $\mathcal{L}$-colorable. and hence $ch(G) \geq 3$. To prove $ch(G) \leq 3$, consider any list-assignment $\mathcal{L}$ where we assume, without loss of generality, that $|L_{u_i}| = |L_{v_i}| = 3$ for every i. Suppose there exist c and $i \neq j$ such that $c \in L_{v_i} \cap L_{v_j}$, then we assign $f(v_i) = f(v_j) = c$ and $f(v_k)$ ($k \neq i, j$) to be any $d \in L_{v_k}$, then $L_{u_l} \setminus \{c, d\} \neq \emptyset$ for every l, there by leading to a proper coloring. One can argue analogously by exchanging u_l's by v_l's. Hence, we assume that $L_{u_i} \cap L_{u_j} = L_{v_i} \cap L_{v_j} = \emptyset$ for every $i \neq j$. Of the 27 elements in $\prod_{i=1}^{3} L_{v_i}$, at most three of them can equal some L_{u_j} for some j. Any other 3-tuple can be extended to a proper $\mathcal{L}$-coloring of $K_{3,3}$. This establishes that $ch(K_{3,3}) = 3$.

Some natural questions that come to one's mind are: (i) does there exist a function $f : \mathcal{N} \to \mathcal{N}$ such that $ch(G) \leq f(\chi(G))$ for every G? (ii) how big can $ch(G)$ be when compared to $\chi(G)$? First, we prove that the Question (i) has a negative answer. The following example illustrates this. It is a generalization of the arguments for $ch(K_{3,3}) \geq 3$ and is provided in detail below.

For any $k \geq 1$, denote by $[2k + 1] = \{1, 2, \ldots, 2k + 1\}$. Let $\mathcal{S} = \binom{[2k+1]}{k+1}$ denote the set of all $(k+1)$-sized subsets of $[2k+1]$. Let n denote $\binom{2k+1}{k+1}$. Consider the $K_{n,n}$ which joins completely the sets $U = \{u_s | s \in \mathcal{S}\}$ and $V = \{v_s | s \in \mathcal{S}\}$ and the list-assignment $\mathcal{L}$ where $L_{u_s} = L_{v_s} = s$ for every $s \in \mathcal{S}$. We claim that $K_{n,n}$ is not $\mathcal{L}$-colorable. Suppose, on the contrary, there exists a proper coloring admitted by $\mathcal{L}$ with C_U and C_V denoting respectively the colors used by this coloring on the vertices of U and V. It follows that C_U and C_V are disjoint . However, $|C_U| \geq k + 1$ since $C_U \cap s \neq \emptyset$ for each $s \in \mathcal{S}$. Analogously, we also have $|C_V| \geq k + 1$. Since C_U and C_V are disjoint, this is not possible. This contradiction establishes that $K_{n,n}$ is not $\mathcal{L}$-colorable. This establishes that $ch(K_{n,n}) \geq k + 2 = \Omega(\log n)$.

In view of this negative answer to Question (i), we wonder if there exists a function $f(n, \chi)$ such that $ch(G) \leq f(n, \chi(G))$ for every G with $n = |V(G)|$?. Below, we present (by simple probabilistic arguments) the following bound arrived at independently by several researchers.

Theorem 33.2 *For every G on n vertices, we have $ch(G) \leq \lceil f(n, \chi(G)) \rceil$ where $f(n, \chi) = 2\chi(\ln n)$.*

Here $\ln(.)$ refers to the natural logarithmic function. By focussing on bipartite graphs (corresponding to $\chi(G) = 2$), the previous arguments on $ch(K_{n,n})$ imply that this upper bound is tight upto a constant multiplicative factor for bipartite graphs.

Proof. Let $G = (V, E)$ be a graph on n vertices with $k = \chi(G)$. Fix an arbitrary k-coloring $(C_1, C_2, \ldots, C_k)$ of G. Let $\mathcal{L} = \{L_u : u \in V\}$ be any list-assignment of colors to vertices with the assumption that $|L_u| \geq f(n, k)$ for each u. Without loss of generality, we assume that $f(n, k)$ is an integer and also that $|L_u| = f(n, k)$ for every u. Define $S = \cup_{u \in V} L_u$ be the union of all color lists in $\mathcal{L}$. Choose a mapping $f : S \to [k]$ uniformly at random. This random choice can be thought of as $|S|$ independent random choices (one for each $c \in S$) where each random choice is uniform over $[k]$. Now we use this random choice to define a random truncation of each L_u as follows.

Define (for each $u \in V$) $K_u = L_u \cap f^{-1}(j)$ where j is the index of the unique color class C_j containing u. Now, for each u, choose an arbitrary member (if any) c_u of K_u to define a coloring $\{c_u\}_u$. Obviously, this is a proper coloring (if well-defined) admitted by $\mathcal{L}$ for all choices of f. It is admitted by $\mathcal{L}$ because, for each u, $c_u \in K_u \subseteq L_u$. It is proper because, for every $(u, v) \in E$ with $u \in V_i$ and $v \in V_j$ (for some $i \neq j$), we have $K_u \cap K_v \subseteq f^{-1}(i) \cap f^{-1}(j) = \emptyset$ and hence $c_u \neq c_v$. This shows that there exists a proper coloring admitted by $\mathcal{L}$ for any $\mathcal{L}$ with each list having $f(n, k)$ colors and hence $ch(G) \leq f(n, k)$.

By **R2**, it only remains to establish that, with positive probability, this is a well-defined coloring. For any fixed (but arbitrary) u, we have

$$\mathbf{Pr}(K_u = \emptyset) = \left(1 - \frac{1}{k}\right)^{f(n,k)} \leq e^{-2(\ln n)} \leq \frac{1}{n^2}$$

and hence $\quad \mathbf{Pr}(\exists u \in V : K_u = \emptyset) \leq \dfrac{1}{n}$

Thus, with positive probability, $K_u \neq \emptyset$ for every $u \in V$ and hence $\{c_u\}_u$ will be a well-defined coloring. ∎

This bound is tight within a constant (i.e., independent of n) multiplicative factor for bipartite graphs as can be seen from the example of $K_{n,n}$. The tightness of this bound (for the class of graphs with $\chi(G) \geq n^\epsilon$, ϵ fixed but arbitrary) follows from the bounds (particularly, the lower bound) obtained by Alon [29]. Below, we adopt the following notation: For $r, m \geq 1$, let K_{m*r} denote the complete r-partite graph with m vertices in each part.

Theorem 33.3 [29] *There exist two positive constants c_1 and c_2 such that for every $m > 2$ and for every $r \geq 2$,*

$$c_1 r(\ln m) \leq ch(K_{m*r}) \leq c_2 r(\ln m).$$

Proof. (sketch) We provide a sketch of the proof of the upper bound. The lower bound is established by proving the existence of an explicit labeled graph K_{m*r} and also by proving that its choice number is at least the stated lower bound. For the upper bound, the proof of Theorem 33.3 is based on probabilistic arguments and proceeds as follows: Assume, without loss of generality, that r is a power of 2. Let $(C_1, \ldots, C_r)$ be the unique r-coloring of K_{m*r}. Let $\mathcal{L} = \{L_u\}_u$ (with $|L_u| \geq c_2 r(\ln m)$ for each u) be any list-assignment. If $r \leq m$, then $m \geq \sqrt{n}$ and hence we can apply Theorem 33.2 to deduce the upper bound.

Hence, we assume that $r \geq m$. Define S as in the proof of Theorem 33.2 and choose uniformly randomly a bipartition of $S = S_1 \cup S_2$. For each $u \in \cup_{j \leq r/2} C_j$, define $K_u = L_u \cap S_1$ and for each $u \in \cup_{j > r/2} C_j$, define $K_u = L_u \cap S_2$. It can be shown that, with high probability, $|K_u| \geq |L_u|[1 - o(1)]/2$ for each $u \in V$. Since the subgraphs induced by the sets $T_1 = \cup_{j \leq r/2} C_j$ and $T_2 = \cup_{j > r/2} C_j$ are isomorphic, it reduces to proving the result for $K_{m*r/2}$ with the list-assignment $\{K_u\}_u$. Repeating this argument inductively, it reduces to proving the theorem for $K_{m*r'}$ for some $r' \leq m$ which (as explained before) is taken care of by applying Theorem 33.2. The cumulative effect of the $[1 - o(1)]$ factors introduced (in the lower bounds on list sizes) in successive inductive steps can be taken care of by initially choosing c_2 sufficiently large. ∎

In the language of n and χ, the above theorem can be stated as $c_1\chi(\ln(n/\chi)) \leq ch(K_{(n/\chi)*\chi}) \leq c_2\chi(\ln(n/\chi))$. The lower bound of Theorem 33.3 establishes that the upper bound of Theorem 33.2 is tight within a constant multiplicative factor for any class of k-colorable graphs as long as $k = k(n)$ satisfies $k = O(n^\delta)$ for some fixed $\delta < 1$. However, when χ is nearly linear (i.e., $\chi(G) \geq n^{1-o(1)}$), Theorem 33.3 provides an asymptotically superior upper bound for the special graph $K_{n/r*r}$, as against the bound of Theorem 33.2 which works for any graph.

This naturally leads us to the question of whether the upper bound of Theorem 33.3 can be extended to any arbitrary graph. Subramanian [30] established that this is indeed possible even for a more general notion of list multicoloring (also known as list set coloring) and is stated below. Below, $ch(G,b)$ denotes the b-analogue of $ch(G)$ in which each vertex has to be given b distinct colors from its list so that adjacent vertices do not share any color in common.

Theorem 33.4 [30] *There exists a constant $c > 0$ such that for every graph G and for each $b \geq 1$,*

$$ch(G, b) \leq c(b\chi)\left[\ln\left(\frac{n}{\chi}\right) + 1\right]$$

$$\text{In particular, } ch(G) \leq c\chi\left[\ln\left(\frac{n}{\chi}\right) + 1\right]$$

∎

For graphs G whose chromatic number is lower bounded by a linear function of n, the upper bound of the previous theorem shows that $ch(G) = \Theta(\chi)$. The proof of this theorem is also based on probabilistic arguments and its broad outline is the same as that of the proof of Theorem 33.3. However, certain new complications arise when we consider an arbitrary G. For example, an arbitrary G can have more than one $\chi(G)$-coloring (unlike the case of K_{m*r}) and the size of the two subgraphs induced by T_1 and T_2 need not be of equal sizes and can differ very widely. Hence the random bipartition cannot be simply a uniform choice but a careful random choice which ensures that sizes of the randomly truncated lists K_u have sizes which are at least the respective guarantees required for further inductive reasoning. There are few other issues which are skipped here and [30] addresses all these issues and establishes Theorem 33.4. It also shows how these proof arguments can be made constructive leading to an efficient algorithm which, given an arbitrary graph G together with an optimal or nearly optimal coloring and a suitable list assignment $\mathcal{L}$, produces a proper coloring of G admitted by $\mathcal{L}$.

In a related work, Subramanian [31] continues further and generalizes the results of Theorem 33.4 in another direction to hereditary choice numbers. A hereditary property $\mathcal{P}$ is a class of labeled graphs which is closed under isormorphism and which also satisfies the property that for every $G \in \mathcal{P}$ and for every induced subgraph H of G, H is also in $\mathcal{P}$. A $(k, \mathcal{P})$-coloring of G is a vertex coloring of G using k colors such that each color class induces a subgraph of G which is a member of $\mathcal{P}$. The $\mathcal{P}$-chromatic number (denoted by $\chi_\mathcal{P}(G)$) is the minimum k such that G admits a $(k, \mathcal{P})$-coloring. The $\mathcal{P}$-choice number (denoted by $ch_\mathcal{P}(G)$) is the list analogue of $\chi_\mathcal{P}(G)$. Generalizing the proof arguments of [30], the following bound on $\mathcal{P}$-choice numbers was obtained in [31].

Theorem 33.5 [31] *There exists a constant $c > 0$ such that for every hereditary property $\mathcal{P}$ and for every graph G with $n = |V(G)|$,*

$$ch_{\mathcal{P}}(G) \leq c\chi_{\mathcal{P}} \left(\ln\left(\frac{n}{\chi_{\mathcal{P}}}\right) + 1 \right)$$

where $\chi_{\mathcal{P}} = \chi_{\mathcal{P}}(G)$ denotes the $\mathcal{P}$-chromatic number of G. Also, given lists $\mathcal{L} = \{L_u\}_u$ of this size and also an optimal $\mathcal{P}$-coloring of G, one can efficiently find a $\mathcal{P}$-coloring permitted by $\mathcal{L}$ in polynomial time. ∎

Other related extensions and generalizations of Theorems 33.4 and 33.5 to hypergraphs and hereditary properties are being presented in [32].

33.4.6 High Girth and High Chromatic Number

Some of the impressive applications of probabilistic arguments have been toward proving the existence of counterexamples to conjectures which are proposed on the basis of known facts and intuitive deductions. It follows from the definition of $\chi(G)$ that $\chi(G) \geq \omega(G)$ where $\omega(G)$ denotes the maximum size of a clique (i.e., a complete subgraph) of G. Odd cycles C_{2k+1} are examples showing that $\chi(G)$ can be larger than $\omega(G)$. If we define G to be the graph obtained by taking n vertex disjoint odd cycles and joining any two vertices from different cycles, then it is easy to verify that $\omega(G) = 2n$ and $\chi(G) = 3n$. This shows that $\chi(G) - \omega(G)$ cannot be bounded by any fixed constant. However, cliques being among the most dense subgraphs, it was believed for a long time and was also conjectured that there exists a function $f : \mathcal{N} \to \mathcal{N}$ such that $\chi(G) \leq f(\omega(G))$ for every G.

Erdös [33] refuted this conjecture using probabilistic arguments and showed the existence, for every $k \geq 3$, of triangle-free graphs (i.e., graphs for which $\omega(G) \leq 2$) having chromatic number at least k. The simple proof of this statement is described below. For a graph G, we use $\alpha(G)$ to denote the maximum size of an independent set in G.

For $n \in \mathcal{N}$ and $p \in [0, 1]$, let $\mathcal{G}(n, p)$ denote the model of a random labeled graph G on n vertices $V = [n]$ where each undirected edge $\{i, j\} \in \binom{V}{2}$ is chosen to be in $E(G)$ with probability p and the random choices for the $\binom{n}{2}$ 2-sets are totally independent. Hence, for any fixed graph H over V, $\mathbf{Pr}(G = H) = p^m \cdot (1-p)^{\binom{n}{2}-m}$ where $m = |E(H)|$. For the proof below, we fix $p = n^{-\frac{5}{6}}$.

There are $\binom{n}{3}$ 3-sets and for each 3-set $S = \{u, v, w\} \in \binom{V}{3}$, define an indicator random variable X_S as follows: $X_S = 1$ if $G[S]$ is a trianlge in G and $X_S = 0$ otherwise. $E[X_S] = \mathbf{Pr}(X_S = 1) = p^3$. Let $X = \sum_S X_S$ denote the number of triangles in G. Clearly, $E[X] = \sum_S E[X_S] = \binom{n}{3}p^3 \approx n^{1/2}/6$. By applying Markov Inequality (Fact 2.3), we have $\mathbf{Pr}(X > n^{3/4}) \leq n^{-0.25}/6 = o(1)$. Hence, $X \leq n^{0.75}$ with probability at least $1 - o(1)$. Let T be a canonically defined minimum sized set which intersects with each of the X triangles in G. Clearly, $|T| \leq X$. Hence, **(C1):** with probability $1 - o(1)$, $V \setminus T$ induces a triangle-free subgraph of G and has at least $n - n^{0.75} \geq n/2$ vertices.

Define $a := \lceil 4(\ln n)/p \rceil$. For every sufficiently large n, we have $a \geq 3(\ln n)/p + 1$. For any fixed set S of a vertices, the probability that S induces an independent set in G is exactly $(1-p)^{\binom{a}{2}} \leq e^{-pa(a-1)/2} \leq e^{-3a(\ln n)/2} = n^{-3a/2}$. Hence,

$$\mathbf{Pr}(\alpha(G) \geq a) = \mathbf{Pr}(\exists S \subseteq V : |S| = a, G[S] \text{ is an independent set in } G)$$
$$\leq \binom{n}{a} n^{-\frac{3a}{2}} \leq \left(n \cdot n^{-\frac{3}{2}}\right)^a \leq \left(n^{-0.5}\right)^a = o(1)$$

Hence, **(C2)** with probability at least $1 - o(1)$, G (and hence every induced subgraph of G) has no independent set of size at least a. Combining **(C1)** and **(C2)**, we have with probability $1 - o(1)$,

$$\chi(G[V \setminus T]) \geq \frac{\frac{n}{2}}{a-1} \geq \frac{np}{8(\ln n)} = \frac{n^{\frac{1}{6}}}{8(\ln n)}.$$

Thus, with positive probability, for every $k \geq 3$, there exists an infinite family of graphs, each of which is triangle-free and has chromatic number at least k.

In fact, Erdös [33] proved something stronger: for every $l, k \geq 3$, there are graphs G having no cycle of length less than l, and having chromatic number at least k. For this claim, X would be defined as the number of all cycles of length less than l and p would also be defined suitably.

33.4.7 Global Coloring and Local Coloring

Another natural and seemingly correct hypothesis to conjecture is as follows: There exists a positive constant k such that any graph G is k-colorable provided *every* induced subgraph of G on $n/(\ln n)$ vertices is 3-colorable. The guarantee of 3-colorability of *every* $n/(\ln n)$-sized subgraph might lead one to hope that this is possible perhaps through a sequence of local color changes. The following theorem of Erdös [34] rules out such possibilities even if we employ stronger assumptions about 3-colorability of induced subgraphs. This shows that chromatic number cannot be determined from local properties. More precisely, the following result is established.

Theorem 33.6 [34] *For every k, there exists an $\epsilon > 0$ such that for every sufficiently large n, there exists a graph G on n vertices such that $\chi(G) > k$ but $\chi(G[S]) \leq 3$ for every S with $|S| \leq \epsilon n$.*

Proof. Let k be any fixed (but arbitrary) positive integer. The proof is based on random graphs. Consider a random graph G drawn from the model $\mathcal{G}(n, p)$ with $p = \frac{c}{n}$ where c is a sufficiently large positive constant. We assume that $\epsilon > 0$ is a sufficiently small constant.

Claim 33.2 $\chi(G) > k$ with probability $1 - o(1)$.

Proof. Suppose, on the contrary, that $\chi(G) \leq k$. Then G has an independent set of size at least n/k. For the sake of simplicity, we assume that n/k is an integer. Then, for each sufficiently large n,

$$\begin{aligned}
\mathbf{Pr}(\chi(G) \leq k) &\leq \mathbf{Pr}(\alpha(G) \geq n/k) \\
&\leq \mathbf{Pr}(\exists S : |S| = n/k, S \text{ is independent in } G) \\
&\leq \binom{n}{n/k} \cdot (1-p)^{\binom{n/k}{2}} \leq (ek)^{n/k} \cdot e^{-\frac{pn(n-k)}{2k^2}} \\
&\leq e^{\frac{n(\ln k+1)}{k}} \cdot e^{-\frac{cn}{3k^2}} \leq e^{-n\left(\frac{c-3k(\ln k+1)}{3k^2}\right)} = o(1)
\end{aligned}$$

In the last line, we have used our assumptions: (1) k is fixed; (2) c and n are sufficiently large. ∎

Claim 33.3 *Every graph F with $\chi(F) > l$ has an induced subgraph H with $\delta(H) \geq l$.*

Proof. Suppose not. We construct a linear ordering of the vertices of F by repeatedly removing an arbitrary vertex of minimum degree in the current remaining graph. Every vertex will have at most $l-1$ neighbors among vertices that come after it. Hence, a simple inductive argument starting from the last vertex establishes that F is l colorable contradicting our assumption.

Hence existence of an induced subgraph F on at most ϵn vertices with $\chi(F) > 3$ would imply the existence of an induced subgraph H on at most $t \leq \epsilon n$ vertices with $\delta(H) \geq 3$ and consequently, having at least $3t/2$ edges. As a result,

$$\begin{aligned}
\mathbf{Pr}(\exists S : |S| \leq \epsilon n, \chi(G[S]) > 3) &\leq \mathbf{Pr}(\exists T : |T| \leq \epsilon n, |E(G[T])| \geq 3|T|/2) \\
&\leq \sum_{t \leq \epsilon n} \binom{n}{t} \left[\binom{\binom{t}{2}}{\frac{3t}{2}}\right] \left(\frac{c}{n}\right)^{\frac{3t}{2}} \\
&\leq \sum_{t \leq \epsilon n} \left[\frac{en}{t} \cdot \left(\frac{et}{3} \cdot \frac{c}{n}\right)^{\frac{3}{2}}\right]^t \\
&= \sum_{t \leq \epsilon n} \left[\frac{e^{\frac{5}{2}} c^{\frac{3}{2}}}{3^{\frac{3}{2}}} \cdot \left(\frac{t}{n}\right)^{\frac{1}{2}}\right]^t = \sum_{t \leq \epsilon n} X^t
\end{aligned}$$

where X denotes the expression within the parentheses. For $t \leq n^{1/2}$, we have $X = O(1/n^{1/4})$ and hence $\sum_{t \leq n^{1/2}} X^t = o(1)$. For $n^{1/2} \leq t \leq \epsilon n$, by choosing ϵ sufficiently small, we have $X \leq 1/2$ and hence $\sum_{n^{1/2} \leq t \leq \epsilon n} X^t = O(2^{-n^{1/2}}) = o(1)$. Hence, for every sufficiently large n, with probability $1 - o(1)$, the random graph G is such that $\chi(G) > k$ and $\chi(G[S]) \leq 3$ for every S with $|S| \leq \epsilon n$. ∎

33.4.8 Tournaments of Specified Type

An oriented graph is obtained by orienting each edge of an undirected graph. A tournament $T = (V, A)$ is an oriented graph obtained by orienting the edges of complete undirected graph on V. They can be used to model outcomes of games between players in a tournament. Each vertex represents a player and each edge represents the outcome of the play between the players. $u \to v$ means that u has won v in the unique game played between u and v. A natural question that arises and which was raised by Schuttle, is : given a k, is it true that there exists a tournament in which for every set of k players, there is a player who beats each of them? Erdös [35] has shown that this question can be answered quite easily using probabilistic arguments. For a given k, we say that a tournament T *satisfies property* S_k if, for every set of k players, there is some player who defeats each of the k players.

Theorem 33.7 [35] *For every k, there exists a N such that for every $n \geq N$, there is a tournament on n vertices satisfying S_k.*

Proof. Consider a random tournament T on $V = [n]$ obtained by orienting independently each undirected edge $\{i, j\}$ in one of the two directions with equal probability. Fix an arbitrary $S \subset V$ of size k. An arbitrary $v \notin S$ beats each member of S with probability 2^{-k}. Equivalently, v is won over by some $u \in S$ with probability $1 - 2^{-k}$. The events (of losing to some member of S) associated with different $v \in V \setminus S$ are independent since the edge sets determining any two such events are pairwise disjoint. Thus, for any fixed S,

$$\mathbf{Pr}(\text{no player can defeat each } u \in S) = \mathbf{Pr}(\forall v \in V \setminus S, \exists u \in S : u \to v) = \left(1 - 2^{-k}\right)^{n-k}$$

Hence,

$$\begin{aligned}
\mathbf{Pr}(T \text{ does not satisfy } S_k) &= \mathbf{Pr}(\exists S, |S| = k, \text{no player can defeat each } u \in S) \\
&\leq \binom{n}{k} \cdot \left(1 - 2^{-k}\right)^{n-k} := t(n, k)
\end{aligned}$$

For a given k, the ratio $t(n+1,k)/t(n,k) = (1 + k/(n-k+1)) \cdot (1 - 2^{-k}) \leq 1 - 2^{-(k+1)}$ for every sufficiently large n. This implies that for some suitably large N, $t(n,k) < 1$ for every $n \geq N$. By (R1), the theorem is proved. ∎

We can significantly strengthen the definition of the above property and also prove the existence of tournaments satisfying the strengthening. A special case of such a strengthening and the associated existence was in fact employed by Kostochka et al. [36] in proving their upper bounds on oriented chromatic numbers of undirected graphs. A more careful analysis of these arguments by Aravind and Subramanian [37] led to further improved bounds. See the subsection on oriented chromatic numbers for further details. We recall the following notation from [36]. For an oriented graph $G = (V, A)$, and a subset $I = \{x_1, \ldots, x_i\}$ of V and a vertex $v \notin I$ such that v is adjacent to each x_j, we use $F(I, v, G)$ to denote the vector $a = (a_1, \ldots, a_i)$ where, for each $j \leq i$, $a_j = 1$ if $(v, x_j) \in A$ and $a_j = -1$ if $(x_j, v) \in A$. We have assumed and used the ordering of I induced by the natural ordering on $[n]$.

Given $d, s \geq 1$, we say that a tournament $T = (V, A)$ satisfies property $T_{d,s}$ if, for every $I \subseteq V, |I| = i \leq d$ and for every $a \in \{-1, 1\}^i$, there are more than s vertices $v \in V \setminus I$ with $F(I, v, T) = a$. We prove (using probabilistic arguments) the following theorem not known before.

Theorem 33.8 *There exists a $N = N(d, s)$ such that for every $d, s \geq 1$, there is a tournament $T = (V, A)$ on n vertices satisfying the property $T_{d,s}$, for every $n \geq N(d, s)$.*

Proof. Consider a random tournament $T = (V, A)$ on n vertices obtained by randomly and independently orienting each edge of K_n in one of the two directions with equal probability. We assume that $n \geq 2d$ so that we can write n as $n = d2^d X + d$ where $X \geq 0$ is a real. We will specify later how big X should be. It follows that $(n - d)/2^d = dX$ and $n/d = 2^d X + 1 \leq 2^{d+1} X$.

Fix an $i \leq d$ and fix any $I \subseteq V$ of size i. Also, fix a vector $a \in \{1, -1\}^i$. Define the random variable
$$X_{I,a} = |\{u \in V \setminus I : F(I, u, T) = a\}|.$$
It is easy to verify that $X_{I,a}$ is the sum of $n - i$ independent and identically distributed indicator random variables each having the common expectation 2^{-i}. Hence it follows that
$$\mu_{I,a} = E(X_{I,a}) = (n - i)2^{-i} \geq (n - d)2^{-d}.$$
Also, by the well-known Chernoff–Hoeffding bounds (see Section 2), it follows from choosing X so large that $(n - d)2^{-d} = dX \geq 10s \geq (2s)/(2 - \sqrt{3})$,
$$\begin{aligned}
\mathbf{Pr}(X_{I,a} \leq s) &= \mathbf{Pr}(X_{I,a} - \mu_{I,a} \leq s - \mu_{I,a}) \\
&\leq e^{-\mu_{I,a}(1 - s/\mu_{I,a})^2/3} \leq e^{-\mu_{I,a}/4} \leq e^{-(n-d)/(4 \cdot 2^d)}
\end{aligned}$$
Hence, for the event $\mathcal{E}$ defined by $\mathcal{E} = \exists I, a : |I| \leq d, X_{I,a} \leq s$, we have
$$\begin{aligned}
\mathbf{Pr}(\mathcal{E}) &\leq d \cdot \binom{n}{d} \cdot 2^d \cdot e^{-(n-d)/(4 \cdot 2^d)} \\
&\leq e^{-\frac{n-d}{4 \cdot 2^d} + d(\ln 2e) + \ln d + d\left(\ln \frac{n}{d}\right)} \\
&\leq e^{-\frac{dX}{4} + d(\ln 2e) + \ln d + d(d+1)(\ln 2) + d(\ln X)} \\
&\leq e^{-d\left(\frac{X}{4} - (\ln 2e) - \frac{\ln d}{d} - (d+1)(\ln 2) - (\ln X)\right)} < 1
\end{aligned}$$
where the last strict inequality follows by assuming that X is suitably large, say, $X \geq 64d$. Thus, for $X_0 = \max\{64d, 10s/d\}$, we infer that with positive probability, there exists a tournament on n vertices (for every $n \geq d2^d X_0 + d$) satisfying $T_{d,s}$. We have not made any attempt to optimize the value X_0 and it can be brought down further. ∎

33.4.9 Bounds on Oriented Chromatic Numbers

An oriented graph is a graph obtained from a simple undirected graph by orienting each edge in one of the two possible ways. Thus, from an undirected graph on m edges, we get 2^m oriented graphs. Sopena, in [38], introduced the notion of oriented chromatic number for oriented graphs. Let $\vec{G}$ and $\vec{H}$ be two oriented graphs. A *homomorphism* of $\vec{G}$ to $\vec{H}$ is a mapping ϕ from $V(\vec{G})$ to $V(\vec{H})$ such that for every arc (x,y) in $E(\vec{G})$, $(\phi(x), \phi(y))$ is an arc in $E(\vec{H})$. This mapping is also called an oriented coloring of $\vec{G}$ using $\vec{H}$.

The oriented chromatic number of an oriented graph $\vec{G}$ is the smallest order $|V(\vec{H})|$ of a $\vec{H}$ for which a homomorphism $\phi : \vec{G} \to \vec{H}$ exists. Note that we can assume, without loss of generality, that $\vec{H}$ is a tournament (an oriented graph in which there is an arc between any two distinct vertices). Equivalently, $\chi_o(\vec{G})$ is the smallest $k \geq 1$ such that there is a proper k-coloring $(V_1, \ldots, V_k)$ of $V(\vec{G})$ such that for every $i \neq j$, all edges joining V_i and V_j are oriented in the same way.

Sopena [38] also extended this definition to undirected graphs. The oriented chromatic number $\chi_o(G)$ of an undirected graph G is the maximum value of $\chi_o(\vec{G})$ where the maximum is over all orientations $\vec{G}$ of G. Upper and lower bounds for the oriented chromatic number have been obtained in terms of the maximum degree and upper bounds have been obtained for various special families of graphs such as trees, planar graphs, partial k-trees, grid graphs [39], and so on.

We present below an upper bound on $\chi_o(G)$ in terms of its maximum degree $\Delta(G)$. Even though better bounds are known, we present this one because it can be established by simple structural arguments not seen before. It is not straightforward to see that such a bound exists. For a graph G, we use $\chi_2(G)$ to denote its distance-2 chromatic number, that is, the minimum k such that G admits a proper k-coloring f such that $f(u) \neq f(v)$ whenever u and v share a common neighbor. Note that $\chi_2(G) \leq \min\{\Delta^2, n\}$ always.

Theorem 33.9 *For any undirected G with $\Delta(G) = d$ and $\chi_2(G) = k$, $\chi_o(G) \leq k2^{k+d}$ if $k \leq 2d$ and $\chi_o(G) \leq kd\binom{k}{d}2^d$ if $k > 2d$. Both of these can be upper bounded as $\chi_o(G) \leq d^3(6d)^d$.*

Proof. Fix any orientation $\vec{G}$ of G. Let $(V_1, \ldots, V_k)$ be a distance-2 coloring of G. For every $i \neq j$, each $u \in V_i$ has at most one neighbor in V_j. We capture this neighbor information alongwith the direction of edges by associating a k-vector α_u with each u as follows. Suppose $u \in V_i$ has $l \leq d$ neighbors one each in $V_{j_1}, \ldots, V_{j_l}$. Then, α_u is defined to be $\alpha_u(j) = 0$ if u has no neighbor in V_j, $\alpha_u(j) = 1$ if u has a neighbor $v \in V_j$ and $\{u,v\}$ is directed as $u \to v$ in $\vec{G}$ and $\alpha_u(j) = -1$ if u has a neighbor $v \in V_j$ where $\{u,v\}$ is directed as $v \to u$. The total number L of possible values for α_u are at most $\sum_{0 \leq l \leq d} \binom{k}{l} 2^l$ which itself is upper bounded by 2^{k+d} if $k \leq 2d$ and $d\binom{k}{d}2^d$ if $k > 2d$. For every possible k-vector α such that $\alpha = \alpha_u$ for some $u \in V$, define $C_\alpha = \{u \in V : \alpha_u = \alpha\}$. Now, for each V_i, color each $u \in V_i$ with $f(u) = (i, \alpha_u)$. For each i and for each of the L possible vectors a, define $C_{i,a} = \{u : f(u) = (i,a)\}$. Since each $C_{i,a}$ is a subset of V_i, it is independent.

Claim 33.4 $\{C_{i,a}\}_{(i,a)}$ *forms an oriented coloring of $\vec{G}$.*

Proof. Suppose there are $(i,a) \neq (j,b)$, $x, w \in C_{i,a}$, $y, z \in C_{j,b}$ such that $x \to y, z \to w \in E(\vec{G})$. Clearly, $i \neq j$. We have $\alpha_x = \alpha_w = a$ and $\alpha_y = \alpha_z = b$. We also have $\alpha_x(j) = \alpha_z(i) = 1$ and $\alpha_y(i) = \alpha_w(j) = -1$. Since $\alpha_x = \alpha_w$, this is a contradiction. This establishes the claim.

The total number of different colors used is kL. Using the upper bounds on L, we get the desired bounds stated before. ∎

The above bound is based on counting and structural aspects of graphs. However, the worst-case value of this bound has a d^d factor and this can be brought down to a 2^d factor by employing probabilistic arguments. By setting $d = \Delta$ and $s = \Delta^2$, Theorem 33.8 guarantees the existence of a tournament on $O(\Delta^2 2^\Delta)$ vertices satisfying T_{Δ,Δ^2}. As mentioned before, we did not optimize on the constant hidden in $O()$ notation and in fact, one can show the existence of such a tournament T (as was shown in [36]) on at most $2\Delta^2 2^\Delta$ vertices. Kostochka et al. [36] demonstrated how to obtain a coloring of any orientation of G using vertices of T thereby establishing that $\chi_o(G) \leq 2\Delta^2 2^\Delta$ for any graph G.

We present below a brief sketch of the proof of this bound. Fix any orientation $\vec{G}$ of G. Choose an arbitrary linear ordering $v_n, \ldots, v_2, v_1$ of the vertices. For each i, let G_i denote the induced subgraph $G[\{v_1, \ldots, v_i\}]$. Color the vertices inductively in the order $v_1, v_2, \ldots, v_n$ always maintaining the the following property after every coloring of a vertex.

Suppose vertices $v_1, \ldots, v_m$ have been colored with $f(v_1), \ldots f(v_m) \in V(T)$, respectively. Then,

1. f restricted to $\vec{G}_m$ is an oriented coloring of $\vec{G}_m$.

2. For any $j > m$, all neighbors of v_j in $\vec{G}_m$ are colored with distinct colors.

Now consider $v = v_{m+1}$ and let $I = \{v_{j_1}, \ldots, v_{j_i}\}$ ($i \leq \Delta$) be the neighbors of v in $\vec{G}_{m+1}$ and let $\vec{a}$ denote $F(I, v, \vec{G}_{m+1})$. Let $I_T = \{f(v_{j_1}), \ldots, f(v_{j_i})\}$. Since T satisfies T_{Δ,Δ^2}, there are at least $\Delta^2 + 1$ vertices $w \in V(T)$ such that $F(I_T, w, T) = \vec{a}$. Among these, at most Δ^2 vertices u are such that if v_{m+1} is colored with u, (ii) might be violated for some $j > m+1$. Hence, there exists a color $u \in V(T) \setminus I_T$ such that $f(v_{m+1}) = u$ extends the partial oriented coloring of $\vec{G}_m$ to a partial oriented coloring of $\vec{G}_{m+1}$. Continuing this, we get an oriented coloring of $\vec{G}$ using $V(T)$. Since $\vec{G}$ was arbitrary, this establishes that $\chi_o(G) \leq 2\Delta^2 2^\Delta$.

The proof of this bound was further refined by Aravind and Subramanian [37] to yield a better bound. Here, you set $d = d(G)$ and $s = d\Delta$ where $d(G)$ denotes the degeneracy of G, that is, the maximum value of the minimum degree of any induced subgraph of G. This led to an upper bound of $16\Delta d 2^d$ on $\chi_o(G)$. Note that this replaces a factor $\Delta 2^\Delta$ in the previous bound by $d 2^d$ and will result in a significant improvement for those G having $d \ll \Delta$. As a consequence, we get an $O(\Delta)$ bound on graphs of bounded degeneracy, like planar graphs.

33.4.10 Bounds on Constrained Colorings

A proper k-coloring of an undirected $G = (V, E)$ is a labeled partition $(V_1, \ldots, V_k)$ of V such that each V_i is an independent set. V_i's are known as color classes. The minimum value of k for which G admits such a k-coloring is the chromatic number of G and is denoted by $\chi(G)$. We look at the application of probabilistic arguments to obtain bounds on colorings with some constraints added between color classes. We study a few examples of such colorings.

33.4.10.1 Frugal Colorings

Suppose, for a proper k-coloring, we also require that no vertex has more than b neighbors in any color class. Such a coloring is called a b-frugal k-coloring. The minimum k for which such a coloring exists is known as the b-frugal chromatic number of G and is denoted by $\chi_b^{fr}(G)$. A 1-frugal coloring is also known as a distance-2 coloring since any two vertices which are either adjacent or share a common neighbor are colored distinctly. The following bound follows from a simple application of LLL.

Theorem 33.10 *There exists an absolute constant $c > 0$ such that $\chi_b^{fr}(G) \leq cd^{(b+1)/b}$ for any integer $b \geq 1$ and for any graph G with maximum degree d.*

Proof. We assume that c is sufficiently large. Let $C = cd^{(b+1)/b}$. We assume that C is an integer for simplifying the arguments. If not, we redefine C to the ceiling of the expression on right-hand side. Ignoring the ceiling can be easily justified. Choose uniformly at random a coloring $f : V \to [C]$. We will show that f is a proper, b-frugal coloring with positive probability, thereby showing the existence of such a coloring. This, in turn, establishes the claim of the theorem. We define the following collection of bad events and provide their associated probabilities.

Type 1: For every $uv \in E$, let E_{uv} denote the event that $f(u) = f(v)$; $\mathbf{Pr}(E_{uv}) = 1/C$.

Type 2: For every u and for every $S \subseteq N(u)$ with $|S| = b+1$, let $E_{u,S}$ denote the event that $f(v) = f(w)$ for all $v, w \in S$. $\mathbf{Pr}(E_{u,S}) = 1/C^b$.

f is a b-frugal coloring if and only if no event of each of Types 1 and 2 occurs. Also, each event E of any of the two types is mutually independent of all other events which do not share any vertex with event E. Thus, E is mutually independent of all other events but at most kd events of Type 1 and $kd\binom{d}{b}$ events of Type 2 where k denotes the number of vertices defining E. We now define the reals y_E required by the LLL as follows: $y_E = (2/C)^{k-1}$ where k was defined before. It now only suffices to verify the associated set of inequalities and this is achieved by verifying that

$$1 \leq 2 \cdot \left(1 - \frac{2}{C}\right)^{2d} \cdot \left(1 - \frac{2^b}{C^b}\right)^{2d\binom{d}{b}}.$$

Using $(1 - \frac{1}{x})^x \geq 1/4$ for all $x \geq 2$, this is true if

$$1 \leq 2 \cdot 4^{-\frac{4}{cd^{1/b}}} \cdot 4^{-\frac{2^{b+1}}{c^b}}$$

which is clearly true if we choose $c = 16$. ∎

We have not tried to optimize the constant c. When $b = 1$, $\chi_1^{fr}(G) \leq d^2 + 1$ by a simple inductive way of coloring. The notion of frugal colorings was introduced by Hind et al. [40] as a tool to bound the total chromatic number of a graph. They also obtained the result of Theorem 33.10 but with a weaker constant $c = e^3$. It was also established in [40] that the bound is nearly tight in the following sense: For any $\beta \geq 1$ and for infinitely many values of d, there are explicit bipartite graphs (based on a $(\beta+2)$-dimensional projective geometry) with maximum degree d having β-frugal chromatic number at least $d^{(\beta+1)/\beta}/(2\beta)$. It was also established that any graph G with a sufficiently large maximum degree d admits a $(\ln d)^8$-frugal coloring using $d + 1$ colors. In their subsequent work on total colorings, Hind et al. [41] improved the frugality to $(\ln d)^5$ again using $d + 1$ colors.

33.4.10.2 Acyclic Colorings

Suppose, in a proper k-coloring $(V_1, \ldots, V_k)$ of G, we require that the induced subgraph $G[V_i \cup V_j]$ be acyclic for every $i \neq j$. Equivalently, we require that there is no 2-colored cycle. The minimum k such that G admits such a k-coloring is known as its acyclic chromatic number and is denoted by $a(G)$. It might be tempting to conclude that an acyclic coloring can be obtained from a proper coloring by performing suitable local operations to take care of 2-colored cycles and thereby establish that $a(G)$ is close to $\chi(G)$. The futility of this approach is illustrated by $K_{n,n} := (A \cup B, A \times B)$. In any acyclic coloring of $K_{n,n}$, either each vertex of A is a color class by itself or each vertex of B is a color class by itself. Hence $a(K_{n,n}) \geq n+1$. In fact $a(K_{n,n}) = n+1$ by keeping each vertex of A as a color class and B as another color class. This shows that $a(G)$ cannot bounded by a function of $\chi(G)$. However, a 1-frugal coloring of G is also an acyclic coloring and hence $a(G) \leq d^2 + 1$ for any graph with maximum degree d. This raises the question of how tightly one can bound $a(G)$ as a function of d. In fact, Alon et al. [42] have shown that there is a positive constant $c > 0$ such that $a(G) \leq cd^{4/3}$ for any G. The proof of this bound is based on an application of LLL. They also show that this upper bound is nearly tight in the sense that for each of infinitely many values of d, there is a graph G with maximum degree d and $a(G) \geq cd^{4/3}/(\ln d)^{1/3}$ where c is a positive constant independent of d. To keep the arguments simple, we present the proof of a weaker upper bound in what follows.

Theorem 33.11 *There exists a positive constant $c > 0$ such that : $a(G) \leq cd^{3/2}$ for any graph G with maximum degree d.*

Proof. Let $C = cd^{3/2}$ and as before assume, without loss of generality, that C is an integer. Let $f : V \to [C]$ be uniformly chosen. Consider the following bad events with their probabilities.

Type 1: For every $uv \in E$, let E_{uv} denote the event that $f(u) = f(v)$; $\mathbf{Pr}(E_{uv}) = 1/C$.

Type (2,k): For every $k \geq 2$ and for every cycle C of length $2k$ in G, let $E_{C,2k}$ denote the event that C is properly 2-colored. $\mathbf{Pr}(E_{C,2k}) \leq 1/C^{2k-2}$.

f is an acyclic coloring if no event of each of Types 1 and 2 occurs. Each event E is mutually independent of all other events but at most ld events of Type 1 and ld^{2k-1} events of Type $(2,k)$ where l denotes the number of vertices defining E. We now define the reals y_E required by the LLL as follows : $y_E = 2/C$ if E is of Type 1 and $y_E = (2/C)^{2k-2}$ if E is of Type $(2,k)$. To prove the theorem, it only suffices to verify the associated set of inequalities and this is achieved by verifying that

$$1 \leq 2 \cdot \left(1 - \frac{2}{C}\right)^{2d} \cdot \prod_{j \geq 2}\left(1 - \left(\frac{2}{C}\right)^{2j-2}\right)^{2d^{2j-1}} \text{ and}$$

$$1 \leq 2 \cdot \left(1 - \frac{2}{C}\right)^{(kd)/(k-1)} \cdot \prod_{j \geq 2}\left(1 - \left(\frac{2}{C}\right)^{2j-2}\right)^{(kd^{2j-1})/(k-1)}$$

Using $(1 - \frac{1}{x})^x \geq 1/4$ for all $x \geq 2$, this is true if

$$1 \leq 2 \cdot 4^{-\frac{4}{c\sqrt{d}}} \cdot 4^{-\sum_{j \geq 2} \frac{2^{2j-1}d}{c^{2j-2}d^{j-1}}}$$

which is clearly true if we choose $c = 16$. ∎

33.4.10.3 Other Constrained Colorings

Likewise, one can also require that $G[V_i \cup V_j]$ is a star forest for every $i \neq j$. This is a more restricted version of acyclic coloring and is known as a *star coloring* of G. The associated chromatic number is known as the *star chromatic number* of G (denoted by $\chi_s(G)$). Obviously, $\chi_s(G) \geq a(G)$ but it is also upper bounded by $16d^{3/2}$ (as can be seen from the proof of Theorem 33.11). This bound is also tight within a multiplicative factor of $\sqrt{\ln d}$ as was shown by Fertin et al. [43]. The notions of acyclic coloring and star coloring were introduced by Grünbaum [44].

The above two bounds raise the question of whether one can obtain similar upper bounds for other types of constrained colorings. This was raised and studied in a generalized framework by Aravind and Subramanian in [45]. For a family $\mathcal{F}$ of graphs, we say that G is $\mathcal{F}$-free if G does not have any subgraph which is isomorphic to some $H \in \mathcal{F}$. Given $j \geq 2$ and a family $\mathcal{F}$ of connected j-colorable graphs H on more than j vertices each, a $(j, \mathcal{F})$-coloring of G is a proper vertex coloring such that the union of any j color classes induces a $\mathcal{F}$-free subgraph. The minimum number of colors required in any such coloring is known as the $(j, \mathcal{F})$-chromatic number of G and is denoted by $\chi_{j,\mathcal{F}}(G)$. Reference [45] establishes that this number is upper bounded by $cd^{(k-1)/(k-j)}$ where $c > 0$ is a constant depending only on j and $\mathcal{F}$. A star coloring is a $(2, \mathcal{F})$-coloring with $\mathcal{F} = \{P_4\}$ (P_4 is a path on four vertices) and a b-frugal coloring is a $(2, \mathcal{F})$-coloring where $\mathcal{F} = \{K_{1,b+1}\}$. As a result, the respective upper bounds (provided before) on these numbers are derived as special cases of this general result. Bounds on other interesting variants of constrained colorings can also be derived and the details can be found in [45]. For example, one can obtain bounds on the chromatic numbers with two or more restrictions required to be satisfied simultaneously.

Aravind and Subramanian also studied the edge analogues of constrained colorings in [46] and obtained bounds on the corresponding chromatic indices. Again, we study proper edge colorings satisfying that union of any j color classes (matchings here) is $\mathcal{F}$-free. For example, acyclic edge coloring is a proper edge coloring with no 2-colored cycle. The acyclic chromatic index $a'(G)$ is upper bounded by $16d$ for any graph. This is in contrast with $a(G)$ which can become nearly $d^{4/3}$. Reference [46] presents upper bounds on such chromatic indices. As a consequence of these bounds, some interesting conclusions like the following are also presented: With $O(d)$ colors, any graph G (of maximum degree d) can be properly edge colored so that the coloring satisfies simultaneously (1) union of any three color classes is an outerplanar graph, (2) union of any four color classes is a partial 2-tree, (3) union of any five color classes is planar, and so on. For more details, see [46].

All of these bounds (for both vertex and edge colorings) were obtained using probabilistic arguments. Analogous derivations which are as simple as the probabilistic ones and which are based on structural arguments are currently not known. These bounds and their derivations illustrate the power and applicability of the probabilistic method.

33.4.11 Waring Bases

This is an example which involves an uncountable probability space and a probability measure. Still, we present it because of the simplicity of the solution obtained by Erdös. Let $\mathcal{N}$ denote the set of all positive integers. A subset $X \subseteq \mathcal{N}$ is an *asymptotic basis of order k* if every sufficiently large $n \in \mathcal{N}$ can be represented as an ordered sum of k distinct elements from X. We use $R_X^k(n)$ to denote the number of such representations of n using elements of X. A trivial upper bound is n^{k-1}. There are bases which achieve this bound within a constant multiplicative factor. For example, set $X = \mathcal{N}$. For the ease of presentation, assume that k divides n and let $m := n/k$ (otherwise, define $m := \lfloor n/k \rfloor$). Now consider the set $\mathcal{R} := \{(n_i)_{i=1}^k : 1 \leq n_1 < n_2 < \cdots < n_{k-1} \leq m, \ n_k = n - (n_1 + \cdots + n_{k-1})\}$. Since $|\mathcal{R}| = \binom{m}{k-1} = \Theta(n^{k-1})$, we have that $R_X^k(n) = \Theta(n^{k-1})$, for every fixed $k \geq 2$.

An interesting question to ask is: are there bases having fewer representations? In particular, one wishes to know if there is a basis X with $R_X^k(n) \leq n^{o(k)}$. In 1932, Sidon posed the question of whether there exists a basis X of order 2 for which $R_X^2(n) \leq n^{o(1)}$, to Erdös. Erdös hoped to solve it within a few days but it took more than two decades to answer it positively. The wait was worth since the solution obtained by Erdös [47] was much stronger and was a very simple probabilistic one. We present below this beautiful application of probabilistic method.

Theorem 33.12 [47] *There exists a subset $X \subseteq \mathcal{N}$ such that $R_X^2(n) = \Theta(\ln n)$.*

Proof. For some sufficiently large constant $c > 0$ (whose value can be chosen easily later) and for any $m \in \mathcal{N}$, define $p_m := c\sqrt{\ln m/m}$ if the RHS expression is less than 1 and

$p_m := 1$ otherwise. For all sufficiently large m, $p_m < 1$. Now choose a random subset $X \subseteq \mathcal{N}$ by including each m into X with probability p_m. The choices are independent for different m's. Let t_m denote the indicator variable for $m \in X$. It follows that $X_n := R_X^2(n) = \sum_{i+j=n, i<j} t_i t_j$. Each $1 \leq i \leq n-1$ is part of exactly one pair and hence X_n is a sum of $\lfloor (n-1)/2 \rfloor$ independent indicator variables each with expectation $E[t_i t_j] = p_i p_j$. Hence, $\mu_n := E[X_n] = \sum_{i+j=n, i<j} c^2 \sqrt{(\ln i)(\ln j)/ij} + o(1)$. Using standard approximations, it can be shown easily that $\mu_n = \Theta(\ln n)$. By choosing c sufficiently large, we can ensure that $\mu_n/12 \geq 2(\ln n)$.

Applying Chernoff bounds (Fact 2.5 of Section 33.2) to X_n, we deduce that $\mathbf{Pr}(|X_n - \mu_n| \geq \mu_n/2) \leq 2e^{-\mu_n/12} \leq 2/n^2$. Thus, for every sufficiently large n, with probability at least $1 - 2n^{-2}$, we have $\mu_n/2 \leq X_n \leq 3\mu_n/2$. Since $\sum_n 2n^{-2}$ converges, an application of Borel-Cantelli Lemma (see [3]) shows that $X_n \in [\mu_n/2, 3\mu_n/2]$ for every sufficiently large n, with probability 1 (with respect to the probability measure governing X). This proves the existence of a Waring basis X such that $R_X^2(n) = \Theta(\ln n)$. ∎

For $k = 3$, by redefining $p_m := c(\ln m/m^2)^{1/3}$, we have $X_n := R_X^3(n) = \sum_{i<j<k: i+j+k=n} t_i t_j t_k$ is the number of representations of n as a sum of terms from X. We can again show that $\mu_n := E[R_X^3(n)] = \Theta(\ln n)$ but X_n is no longer a sum of independent indicator variables, it is a sum of $O(n^2)$ variables which are dependent. Hence, Chernoff bounds cannot be applied as before. However, Erdös and Tetali found a way of analyzing Y_n and established that Y_n is concentrated it around its mean with a very small failure probability. As a result, arguing as before, one can show that $R_X^3(n) = \Theta(\ln n)$. In fact, the same was shown for any fixed $k \geq 3$. See [48] for more details.

By interpreting $R_X^k(n)$ as a homogeneous degree-k polynomial in variables $t_1, \ldots, t_n$ and applying the concentration results for such variables, Vu [49] provided an alternative proof of the $\Theta(\ln n)$ bound of [48]. In fact, Vu obtained something stronger in the sense that the k integers from X (whose sum equals n) need not be distinct and the same number can be used more than once. The number $T_X^k(n)$ of such representations is certainly as large as $R_X^k(n)$ but for some X, we have $T_X^k(n) = \Theta(\ln n)$. A still further generalization was also established in [49]. Suppose $a_1, \ldots, a_k$ are k fixed positive integers such that $gcd(a_1, \ldots, a_k) = 1$. Let $Q_X^k(n)$ be the number of representations such that $n = a_1 x_1 + \cdots + a_k x_k$ where $x_i \in X$ for each i. Again, Vu [49] established by probabilistic arguments the existence of a $X \subseteq \mathcal{N}$ such that $Q_X^k(n) = \Theta(\ln n)$.

For a subset $X \subseteq \mathcal{W}$ and a positive integer s, we say that X is a *basis of order s* if $T_X^s(n) \geq 1$ for every $n \geq 1$. $\mathcal{W}$ denotes the set of all nonnegative integers. Another generalization considered by Vu is related to the classical Hilbert–Waring theorem (see the survey by Vaughan and Wooley [50]) which asserts that for every fixed k, there exists a $g = g(k)$ such that $\mathcal{W}^k$ is a basis of order g (and hence of order s for every $s \geq g$). Here, $\mathcal{W}^k$ denotes the set $\{n^k : n \in \mathcal{W}\}$. The theorem was first conjectured by Waring in 1770 and was proved by Hilbert [51] in 1909. Exact determination of $g(k)$ has been done for all values of k (see [50] for details). For example, it is now known that $g(2) = 4$, $g(3) = 9$, $g(4) = 19$ [52], $g(5) = 37$, $g(6) = 73$ [53], and so on. Vinogradov and also others (see [54]) established the asymptotic order to be $T_{\mathcal{W}^k}^s(n) = \Theta\left(n^{\frac{s}{k}-1}\right)$ for every $s \geq g(k)$. Thus, the number of representations is quite large for the standard basis $\mathcal{W}^k$.

Any subset $X \subseteq \mathcal{W}^k$ satisfying $T_X^s(n) \geq 1$ (for every n) is referred to as a *subbasis of* $\mathcal{W}^k$ *of order s*. A natural question that arises is whether, for every sufficiently large s, there exists a subbasis X of $\mathcal{W}^k$ of order s admitting *fewer* representations. Vu [55] establishes the existence of such a subbasis : $\forall k \geq 2 \; \exists g(k) \; : \; \forall s \geq g(k)$, there exists a subbasis (of order s) $X \subseteq \mathcal{W}^k$ for which $T_X^s(n) = \Theta(\ln n)$. The proof is based on studying a random sequence $X = (a_i)_i$ of kth powers obtained by choosing each $n^k \in \mathcal{W}^k$ with probability

$p_n = cn^{-1+\frac{k}{s}}(\ln n)^{1/s}$ for some suitable constant $c > 0$ and $p_0 = 0$. Analyzing this random sequence for the number of representations, using recent concentration results (obtained by Vu himself) on random variables which are degree-s polynomials over indicator variables for various m^k, the above-stated claim is established in [55].

In another direction, one can also focus on the density $Y(n)$ of a basis Y, instead of the number of representations. The density $Y(n)$ (for every n) is defined as $|\{m \leq n : m \in Y\}|$. As noticed in [55], for the random sequence X mentioned before, by applying a simple double counting argument (based on the $\Theta(\ln n)$ bound), it follows that $X(n) = O\left(n^{1/s}(\ln n)^{1/s}\right)$ answering affirmatively a question of Nathanson [56] who asked if there exists a subbasis Y of $\mathcal{W}^k$ of order s for which $Y(m) = O(m^{\frac{1}{s}+o(1)})$. It is also easy to see that the density of any such subbasis $Y \subseteq \mathcal{W}^k$ of order s should satisfy $Y(n) = \Omega(n^{1/s})$ thereby establishing that density of X is best possible upto the $n^{o(1)}$ factor.

33.4.12 Sum-Free Subsets

This is again an application involving a probability space which is not finite or countable. We present it for the sake of its beauty. A set A of integers is *sum-free* if $(A + A) \cap A = \emptyset$ where $A + A := \{a + b : a, b \in A\}$. That is, no two (not necessarily distinct) members of A add up to another member of A. For a set A of non-zero integers, let $s(A)$ denote the maximum size of a sum-free subset of A. Using simple probabilistic arguments, Erdös [57] obtained a lower bound on $s(A)$ in terms of $|A|$. We present below this bound and a very simple proof argument of this (due to [58]).

Theorem 33.13 *For any A of n non-zero integers, $s(A) \geq n/3$.*

Proof. For a real x, let $f(x)$ denote the fractional part of x, that is, $f(x) = x - \lfloor x \rfloor = x \pmod 1$. Choose $\mu \in [0, 1)$ uniformly at random. Define $\mu A := \{a \in A : 1/3 \leq f(\mu a) < 2/3\}$. We claim that μA is sum-free. Suppose there exist $a, b, c \in \mu A$ with $a + b = c$. Then,

$$f(\mu c) = f(\mu(a+b)) = f(\mu a + \mu b) = f(\mu a) + f(\mu b)$$

where the last addition is a (mod 1) addition. Since this is impossible by the choice of a, b, c, it follows that μA is sum-free. Since $\mathbf{Pr}(a \in \mu A) = 1/3$ for any $a \in A$, the expected size of μA is $n/3$. Hence, for some $\mu \in [0, 1)$, we have $|\mu A| \geq n/3$. This establishes the lower bound. ∎

For every fixed A (of size n), $f(\epsilon a) \notin [1/3, 2/3]$ for every $a \in A$ if $\epsilon \leq 1/5m_A$ or if $\epsilon \geq 1 - 1/5m_A$ where $m_A := \max\{|a| : a \in A\}$. This implies that $\mathbf{Pr}(\mu A = \emptyset) \geq 2/5m_A$. Since m_A is a positive constant for a fixed A and since $E[|\mu A|] = \frac{n}{3}$, this implies that there exists an $\epsilon > 0$ such that $|\epsilon A| > n/3$. Since $|\epsilon A|$ is integer valued, this implies the existence of a sum-free subset of size at least $(n+1)/3$. Thus, we actually have $s(A) \geq (n+1)/3$ for any A as pointed out by Alon and Kleitman [59]. The currently best known lower bound is due to Bourgain [60] who proved that $s(A) \geq (n+2)/3$ for any $A \subseteq \mathcal{N}$.

Define $f(n) := \min\{s(A) : |A| = n\}$. As pointed out in [58], $f(n)$ is subadditive (meaning $f(m+n) \leq f(m) + f(n)$) which is known to imply that $\lim_{n \to \infty} f(n)/n$ exists and equals $\inf_n f(n)/n$. We denote this limit (known as the sum-free subset constant) by δ. Erdös [57] obtained that $\delta \leq 3/7$. That $\delta \leq 1/2$ can be seen by considering the sets $[n] = \{1, \ldots, n\}$. Further improved estimates of δ were obtained later, like $12/29$ by Alon and Kleitman [59] and $11/28$ by Lewko [61]. Recently, Eberhard et al. [58] established that $\delta = 1/3$ thereby determining the constant exactly. Even so, it is still not known if $f(n) \geq (n/3) + \omega$ for some $\omega \to \infty$ and remains a challenging problem.

Alon and Kleitman [59] also generalized the problem to sets of non-zero elements of finite abelian groups. Precisely, it was shown that any subset A of non-zero elements of a finite abelian group admits a subset $A' \subseteq A$ such that (i) A' is sum-free and (ii) $|A'| > 2|A|/7$. They also observed (based on a result of Rhemtulla and Street [62]) that the constant $2/7$ is optimal. In fact, [59] even extends the bound $1/3$ to measurable subsets of one-dimensional torus T (the group of reals $[0, 1)$ under (mod 1) addition). It is shown that for every $\epsilon > 0$ and for every measurable $A \subseteq T$, there is a measurable and sum-free $A' \subseteq A$ such that $\mu(A') > (1/3 - \epsilon)\mu(A)$.

Bourgain [60] also looks at the k-sum-free extension, for arbitrary but fixed k but for sets of positive integers. A set of positive integers A is k-sum-free if for any multi-set of k elements from A, their sum is not in A. $s_k(A)$ is the maximum size of a k-sum-free subset of A. One can generalize the probabilistic arguments of Theorem 33.13 to obtain that $s_k(A) > |A|/(k+1)$. Using tools from harmonic analysis, [60] establishes that $s_3(A) > |A|/4 + c \log |A|/\log \log |A|$ for some positive constant c. It is also established that for any choice of δ_k satisfying $s_k(A) > \delta_k |A|$ (for every finite A), it should be that $Lim_{k \to \infty} \delta_k = 0$.

33.5 RANDOM GRAPHS

Random graphs have become a powerful tool in the hands of researchers to study the properties of a *typical* graph. They can also be used as test inputs to compare different algorithms for their performances. They are also used to model the random and dynamic nature of real-life networks like communication networks. The study of random graphs was inspired by the work of Erdös and Renyi [63] which is generally considered to be the most influential work in the area of random graphs. This work brought out several surprising and interesting properties of random graphs and helped in inspiring further work in this area. In this exposition, we will focus mainly on a model introduced by Gilbert [64] and which is often referred to as Erdös–Renyi model of random graphs and is denoted by $\mathcal{G}(n, p)$.

$\mathcal{G}(n, p)$ *model*. Assume, without loss of gnerality, that $V = \{1, 2, \ldots, n\}$. In this model, a random graph $G = (V, E)$ is drawn as follows. For each $e \in \binom{V}{2}$, include $e \in E$ with probability p, that is $\mathbf{Pr}(e \in E) = p$. The random choices are independent for different $e \in \binom{V}{2}$. We use $G \in \mathcal{G}(n, p)$ to denote a random graph drawn in this way. For any $F \subseteq \binom{V}{2}$ with $|F| = m$, the probability that G equals (V, F) is exactly $p^m \cdot (1-p)^{\binom{n}{2} - m}$.

While studying random graphs, we are interested in the typical behavior of a n-vertex random graph in the asymptotic (as $n \to \infty$) scenario. We consider a fixed but arbitrary infinite sequence of random graphs on n vertices for each $n \geq 1$. Precisely, for any $p = p(n) : \mathcal{N} \to [0, 1]$, consider the infinite sequence of random graphs $\{G_n \in \mathcal{G}(n, p) | n \geq 1\}$. One is usually interested in knowing the typical properties, a phrase which captures properties which hold with a probability *very close* to 1. Also, one is usually interested in the asymptotic scenario, that is, we wish to study the typical properties possessed by random graphs on all sufficiently large number of vertices. An example of such a question would be : Given $\{G_n \in \mathcal{G}(n, p)\}_n$, is it true that $\mathbf{Pr}(\chi(G) \leq 3) \to 1$ as $n \to \infty$. Sometimes, it is possible that the limit may not exist or it may approach a value other than 1, say 0 or 0.75. This is still an asymptotic study but is not about a typical property. For every n, an event E_n is any set of graphs over $V = [n]$. For a sequence $\{E = E_n\}_n$ of events, we say that $G \in \mathcal{G}(n, p)$ satisfies E asymptotically almost surely (a.a.s.) if $\mathbf{Pr}(G_n \text{ satisfies } E_n) \to 1$ as $n \to \infty$.

The classic work [63] was mostly focussed on another related and equivalent (in a quantifiable sense) model, denoted by $\mathcal{G}(n, m)$, defined below.

$\mathcal{G}(n, m)$ *model*. $V = \{1, \ldots, n\}$. Choose uniformly at random $E \subseteq \binom{V}{2}$ with $|E| = m$. For any $F \subseteq \binom{V}{2}$ with $|F| = m$, the probability that $G = (V, E)$ equals (V, F) is exactly $\binom{\binom{n}{2}}{m}^{-1}$.

One can also view this model as $\mathcal{G}(n,p)$ model conditioned on $|E(G)| = m$. Again, one can fix an infinite sequence of models $\{G \in \mathcal{G}(n,m)\}_n$ for some $m = m(n) : \mathcal{N} \to \mathcal{N}$ with $m(n) \leq \binom{n}{2}$ always and analyze this sequence for the asymptotic behavior of its members. The two models are related to each other and often proving statements for one model is reduced to proving similar statements for the other model. Below, we focus mostly on the $\mathcal{G}(n,p)$ model for analyzing some graph properties.

The following technical observation will be often applied in several of derivations below.

Lemma 33.2 *Suppose $X_1, \ldots, X_m$ be a sequence of indicator variables with respective means $\mu_1, \ldots, \mu_m$ and let $X = X_1 + \cdots + X_m$. Let $\mu := E[X] = \sum_i \mu_i$. Then,*

$$Var(X) = E[(X-\mu)^2] = \sum_{i,j} E[(X_i - \mu_i)(X_j - \mu_j)] = \sum_i Var(X_i) + \sum_{i \neq j} E[X_i X_j] - \mu_i \mu_j.$$

Here, the second summation is over ordered pairs (i,j) with $i \neq j$. ∎

In what follows, we present (with formal justifications) some interesting properties possessed by random graphs. For more details and for a comprehensive treatment of the mathematical theory of random graphs, the reader is referred to the following excellent books [2–4]. For an exposition of different models of random graphs, the reader is referred to the book by Rick Durrett [65].

33.5.1 Existence of Triangles

Assume that $p = p(n)$ is such that $p = x/n$ where $x = x(n)$ is an arbitrary function. We want to determine the asymptotic limit (if it exists) of the probability of G_n containing a triangle K_3. For every $S \in \binom{V}{3}$, let X_S denote the indicator variable for S inducing a K_3 in G. We have $\mu_S := E[X_S] = p^3$ for every S. Let $X = \sum_S X_S$ denote the number of K_3's in G. Then, G has a triangle if and only if $X > 0$. We have $\mu := E[X] = \sum_S E[X_S] = \sum_S p^3 = \binom{n}{3} p^3 \approx (np)^3/6$. If $x(n) \to 0$ then $\mu \to 0$ and hence $\mathbf{Pr}(G \text{ has a } K_3) \leq \mu \to 0$ by Markov's inequality. Hence, with probability approaching 1 (as $n \to \infty$), G has no triangle.

Suppose $x(n) \to \infty$. By Chebyschev Inequality, it follows that $\mathbf{Pr}(X = 0) \leq Var(X)/\mu^2$. We also have, by applying Lemma 33.2,

$$Var(X) = \sum_S Var(X_S) + \sum_{S \neq T} E(X_S X_T) - \mu_S \mu_T$$

$$\leq \sum_S \mu_S + \sum_S \mu_S \left(\sum_{T : |S \cap T| = 2} E(X_T | X_S = 1) \right)$$

$$= \mu + \mu \cdot \left(\sum_{T : |S \cap T| = 2} E(X_T | X_S = 1) \right)$$

$$= \mu + 3\mu \cdot (n-3) p^2$$

The first inequality uses the fact that when S and T share at most one vertex, X_S and X_T are independent and hence $E[X_S X_T] = \mu_S \mu_T$. The second last equality relies on the homogeneous nature of the model, that is, the parenthesized quantity in the second summation is uniformly the same for all S. This quantity is the sum of $3(n-3)$ (one for each pair of an edge in S and a vertex outside S) identically distributed indicator variables with common expectation p^2. Thus, $Var(X)/\mu^2 \leq \mu^{-1} + 3\mu^{-1} np^2 = o(1)$ since $\mu \to \infty$ and $\mu^{-1} np^2 \approx 6/(n^2 p) = o(1)$ using our assumption $p = x(n)/n$. This shows that with probability approaching 1 (as $n \to \infty$), we have $X > 0$ and hence G has a triangle.

Note 1: (i) For a graph property $\mathcal{P}$, a function $q(n) \in [0,1]$ is a threshold function for satisfying $\mathcal{P}$ if it satisfies: (i) $\mathbf{Pr}(G \in \mathcal{G}(n,p))$ satisfies $\mathcal{P}) \to 0$ for any $p(n) = o(q(n))$ and (ii) $\mathbf{Pr}(G \in \mathcal{G}(n,p))$ satisfies $\mathcal{P}) \to 1$ for any $p(n) = \omega(q(n))$, or vice versa. The previous analysis shows that $q = 1/n$ is a threshold for G having a K_3.

Note 2: A slight modification of the previous arguments actually prove the following: If $p \geq \omega(n)/n$, then for every fixed $\epsilon > 0$, with probability $1 - o(1)$, the number X of triangles in G satisfies $X \in [(1-\epsilon)\mu, (1+\epsilon)\mu]$. This is an example of a typical property satisfied by G_n in the asymptotic setting.

33.5.2 Being Connected

Below, we show that there are probability functions $q = q(n)$ and $r = r(n)$ such that a.a.s.: either G is not connected if $p \leq q - r$ or G is connected if $p \geq q + r$. $q(n)$ is known as a threshold probability for G being connected. q is considered as a sharp threshold if $r(n) = o(q(n))$.

Theorem 33.14 *Let $\omega = \omega(n) \to \infty$ be any sufficiently slowly growing function. Define $q(n) = (\ln n)/n$ and $r(n) = \omega(n)/n$. Then, with probability $1 - o(1)$, the following holds:*

1. *If $p \geq q + r$, then G is connected.*
2. *If $p \leq q - r$, then G is not connected.*

Moreover, $q(n)$ is a sharp threshold since $\omega(n)$ can be arbitrarily small compared to $\ln n$.

Proof. Without loss of generality, we assume that $p = q + r$ (or $p = q - r$) depending on whether we want to prove (1) (or (2)). We first prove (1). Equivalently, we show that $\mathbf{Pr}(G \text{ is not connected}) \to 0$. G is not connected if and only if there exists a $S \subseteq V$, $|S| \leq n/2$, such that there is no edge in G between S and $V \setminus S$. For $1 \leq k \leq n/2$, let $\mathcal{E}_k$ denote the event that there exists such a S with $|S| = k$. We have

$$\mathbf{Pr}(\mathcal{E}_k) \leq \binom{n}{k}(1-p)^{k(n-k)} \leq \left(\frac{en}{k}(1-p)^{n-k}\right)^k \leq \left(\frac{en}{k}e^{-np(1-\frac{k}{n})}\right)^k$$

$$\leq \left(\frac{en}{k}e^{-(\ln n+\omega)(1-\frac{k}{n})}\right)^k \leq \left(\frac{en^{\frac{k}{n}}}{k}e^{-\omega(1-\frac{k}{n})}\right)^k \leq \left(e\left(n^{\frac{k}{n}-\frac{\ln k}{\ln n}}\right)e^{-\frac{\omega}{2}}\right)^k \leq \left(\frac{1}{\omega}\right)^k$$

In the above derivation, we have used (i) $\binom{n}{k} \leq (en/k)^k$, (ii) $k \leq n/2$ and (iii) $\ln m/m$ is a decreasing function of m. As a result, we have

$$\mathbf{Pr}(G \text{ is not connected}) = \mathbf{Pr}(\exists k \leq n/2 : \mathcal{E}_k) \leq \sum_{1 \leq k \leq \frac{n}{2}} \omega^{-k} = o(1)$$

We now prove (2). For each $u \in V$, let X_u denote the indicator variable for u being isolated in G. Then, for any u, $\mu_u := E[X_u] = (1-p)^{n-1}$. If $X := \sum_u X_u$ denotes the number of isolated vertices, then $\mu := E[X] = n(1-p)^{n-1} \geq n(1-p)^n$. We have

$$\mu \geq n(1-p)^n = ne^{-np[1+O(p)]} \geq ne^{-(\ln n - \omega)[1+O(p)]} \geq n^{-O(p)}e^\omega \geq [1-o(1)] \cdot e^\omega \to \infty.$$

Also,

$$\mathrm{Var}(X) = \sum_u \mathrm{Var}(X_u) + \sum_{u \neq v} E(X_u X_v) - \mu_u \mu_v$$

$$\leq \sum_u \mu_u + n^2\left((1-p)^{2n-3} - (1-p)^{2n-2}\right)$$

$$= \mu + \mu^2 \cdot \left((1-p)^{-1} - 1\right) = \mu + \mu^2 \cdot \left(\frac{p}{1-p}\right)$$

Hence,

$$\mathbf{Pr}(G \text{ has no isolated vertex}) = \mathbf{Pr}(X = 0) \leq \frac{\text{Var}(X)}{\mu^2} \leq \mu^{-1} + \frac{p}{1-p} = o(1)$$

using $\mu \to \infty$ and $p/1 - p = o(1)$ due to our assumption about p. This shows that

$$\mathbf{Pr}(G \text{ is not connected}) \geq \mathbf{Pr}(G \text{ has an isolated vertex}) = 1 - o(1).$$

Thus, $\ln n / n$ is a sharp threshold for being connected. ∎

33.5.3 Emergence of a Giant Component in the Vicinity of $1/n$

In their seminal paper, Erdös and Renyi [63] studied the asymptotical evolution of $G \in \mathcal{G}(n, m)$ as $m = m(n)$ goes from 0 to $N := \binom{n}{2}$. They discovered that there are several critical regions during which a sudden and surprising change occurs. Using standard reductions, it follows that similar phenomena occur as $G \in \mathcal{G}(n, p)$ evolves from $p(n) = 0$ to $p(n) = 1$. We focus on the $\mathcal{G}(n, p)$ model. As an example, for every fixed $\epsilon > 0$, with probability $1 - o(1)$, there is a sudden jump in the structure of G during the interval $[(1 - \epsilon)/n, (1 + \epsilon)/n]$ from having no connected component of size more than $O(\log n)$ to having a unique connected component of size $\Theta(n)$ with every other component of size at most $O(\log n)$. The sudden emergence of a *giant component* (in the language of Erdös–Renyi) around $p = 1/n$ is similar to the sudden change in the physical form of water around freezing point or around its boiling point and hence is often referred to as a phase transition. This sudden emergence or vanishing of a property around specific choices of $p(n)$ is one of the most interesting and curious phenomena about random graphs.

In what follows, we provide a formal derivation of this phase transition around $p = 1/n$. On account of the importance of this phenomenon, several proofs (some of these based on Galton–Watson branching processes) have been obtained before but we present below a simple and elegant proof due to Krivelevich and Sudakov [66] which exploits the basic properties of a well-known graph exploration heuristic known as depth-first search (DFS) whose usage was pioneered by Hopcroft and Tarjan to design efficient graph algorithms (see, e.g., [67]).

DFS. W.l.o.g. assume that $V = \{1, 2, \ldots, n\}$. The DFS is an algorithm which, starting from vertex 1, traverses the edges of the graph and visits all vertices reachable from 1 and thereby discovers the component containing 1. If there are vertices still unvisited, it picks the smallest of these and repeats the procedure again and continues until all vertices have been visited. At any point of time, DFS maintains three sets of vertices S (vertices whose exploration is over), T (vertices which have not been visited yet) and $U = V \setminus (S \cup T)$ (vertices which have been visited but the exploration is not complete yet). U is maintained as a stack (last-in and first-out data structure which maintains elements in a top-down fashion in the reverse order of their arrivals and access is only for the top element always). T is always maintained in the sorted order. DFS starts with $S = U = \emptyset$ and $T = V$ and goes on until $T \cup U = \emptyset$. The algorithm works in rounds. During each round, if u is the last vertex added to U, it finds the smallest neighbor v (in T) of u and shifts it from T to top of the stack. If u has no neighbor in T, then u is shifted from U to S. This completes a round and the algorithm proceeds to the next round. When U becomes empty, it starts again from the smallest vertex in T and shifts it to U. The following observations can be inferred about DFS.

1. Consider the time period between two instants when a vertex is about to be added to an empty U and the next time it has become empty. We call this an *epoch* of the execution. The sets of vertices which have been shifted from T to U and then from U to S during an epoch constitute a connected component of G.

2. The set of vertices in U at any instant are part of some connected component of G. Also, they constitute a path (in the order in which they were added).

3. At any instant, the sets S and T are such that there is no edge joining a vertex in S with a vertex in T.

To find the smallest neighbor (in T) of u, the algorithm needs to make queries about edges (in the form of neighbor queries) joining u and vertices in T until it obtains a YES answer. The random outcomes of these queries are independent with each outcome being a YES with probability p. We capture these random answers by a $N = \binom{n}{2}$-length vector $(X_i)_1^N$ of identical and independent random bits. The ith bit corresponds to the ith query posed by the algorithm. By studying the properties of such a vector of random bits, we can infer properties about the random graph. Since every YES answer to a query results in a vertex being moved from T to U, we notice that **(A)** after the first t queries, we have $|S \cup U| \geq \sum_{i=1}^t X_i$. The inequality is in fact strict since the first vertex of each component discovered so far has been added to $S \cup U$ without posing any query. It also follows that **(B)** $|U| \leq 1 + \sum_{i=1}^t X_i$ after t queries. The analysis exploits the following simple observation about $(X_i)_i$ whose proof (based on an application of Chernoff bounds and Chebyschev inequality) can be found in [66].

Lemma 33.3 [66] *Let $\epsilon > 0$ be a sufficiently small constant. The sequence $(X_i)_{i=1}^N$ of iid indicator variables each with probability p satisfies the following:*

1. *Let $p = (1 - \epsilon)/n$. Then, for $k = k(n) := 7(\ln n)\epsilon^{-2}$, with probability $1 - o(1)$, for every subsequence of kn consecutive variables in $(X_i)_i$, less than k of these are set to 1.*

2. *Let $p = (1 + \epsilon)/n$ and $N_0 = \epsilon n^2/2$. Then, with probability $1 - o(1)$,*

   i. $\sum_{i=1}^{n^{7/4}} X_i \leq n^{5/6}$, *and*

   ii. *For every $t : n^{7/4} \leq t \leq N_0$, $|\sum_{i=1}^t X_i - (1+\epsilon)t/n| \leq n^{2/3}$.*

The above lemma leads to the following threshold for the existence of a giant (whose size is linear in n) component.

Theorem 33.15 [66] *Let $\epsilon > 0$ be fixed and sufficiently small. Let $G \in \mathcal{G}(n,p)$. Define $q_l(n) = (1-\epsilon)/n$ and $q_h(n) = (1+\epsilon)/n$. $k = k(n)$ is as defined in Lemma 33.3. Then, with probability $1 - o(1)$, the following holds:*

1. *If $p \leq q_l$, then every connected component of G is of size at most $k(n)$.*

2. *If $p \geq q_h$, then G contains a connected component on at least $\epsilon n/2$ vertices.*

Proof. For each of the cases (1) and (2), one can assume, without loss of generality, that $p = q_l$ and $p = q_h$ respectively. Also, we assume that the sequence $(X_i)_i$ which defines $G \in \mathcal{G}(n,p)$ and guides the DFS satisfies respectively the conclusions (i) and (ii) of Lemma 33.3.

We first prove (1). Suppose that G contains a component C of size more than k. Consider the epoch which discovers C and consider the time instant just before the $(k+1)$-st addition of a vertex of C. Each (except the first) of the k vertices of C which have already made to $S \cup U$ have been added on account of a YES answer to a query to $(X_i)_i$. Each such query (successful or not) corresponds to an edge incident at one of these k vertices and the number of such edges is at most $\binom{k}{2} + k(n-k) < kn$. This implies that $(X_i)_i$ has a subsequence of at most kn consecutive bits in which there are at least k 1's present, a contradiction to (i) of Lemma 33.3.

We now prove (2). Applying Lemma 33.3, we obtain the following observations:

O1: Applying **(B)** and (ii) of Lemma 33.3, we deduce that $|U| \leq 3\epsilon n/4$ throughout the execution till we reach the first N_0 queries.

O2: $|S|$ cannot decrease as the execution of DFS evolves. Also, $|S| < n/3$ just after the first N_0 queries. Otherwise, at the time instant immediately after $|S|$ becomes $n/3$, we have $|T| = n - |S| - |U| \geq n/3$ (since ϵ is sufficiently small) and hence DFS would have queried all of the $|S||T| \geq n^2/9 > N_0$ potential edges within the first N_0 queries, a contradiction.

O3: Applying **(A)** and (ii) of Lemma 33.3, we have $|S \cup U| \geq (1+\epsilon)t/n - n^{2/3}$ just after the first t queries, for every $n^{7/4} \leq t \leq N_0$.

O4: For every $n^{7/4} \leq t \leq N_0$, U cannot be empty just after the first t queries because then that would imply that $|S| = |S \cup U|$ and also (using $|S| < n/3$) that DFS will have made at least

$$|S||T| = |S|(n-|S|) \geq \left(\frac{(1+\epsilon)t}{n} - n^{\frac{2}{3}}\right)\left(n - \frac{(1+\epsilon)t}{n} + n^{\frac{2}{3}}\right)$$

$$\geq (1+\epsilon)t - (1+\epsilon)^2 \frac{t^2}{n^2} - 2n^{\frac{5}{3}} \geq \left(1 + \frac{\epsilon}{2} - \epsilon^2 - \frac{\epsilon^3}{2}\right)t - 2n^{\frac{5}{3}} > t$$

queries within the first t queries, a contradiction.

Hence, U is not empty after every $t \in [n^{7/4}, N_0]$ queries. This means that vertices added to U during this interval are part of the same connected component and the number of such vertices is at least

$$\sum_{i=n^{\frac{7}{4}}}^{N_0} X_i \geq (1+\epsilon)\frac{N_0}{n} - n^{\frac{2}{3}} - n^{\frac{5}{6}} \geq \frac{\epsilon n}{2} \quad \text{for all large } n.$$

This establishes that, with probability $1 - o(1)$, there is some component on at least $\epsilon n/2$ vertices. ∎

The dependence on ϵ in both regimes (i) $\Theta(\epsilon^{-2})(\ln n)$ being the order of the maximum size of any connected compoennt in the regime $p \leq (1-\epsilon)/n$ and (ii) $\Theta(\epsilon)n$ being the order of the size of the largest connected component in the regime $p \geq (1+\epsilon)/n$, are both of correct order of magnitude (see [3] and [4]). Also, for $p \geq (1+\epsilon)/n$, there is a path on at least $\epsilon^2 n/5$ vertices in G and this also follows from analyzing the DFS algorithm as shown in [66].

What happens when $p = (1 \pm o(1))/n$? Alon and Spencer [2] (based on the extensive published work on this topic like those of [68–72]) identify within the range $\Theta(1/n)$ five subregions of $p = c/n$ where the connected components of G exhibit some interesting phenomena. In particular, they focus on L_1 where L_k (for $k \geq 1$) stands for the size of the kth largest connected component of G. They also study the complexities of the components where the complexity of a component is the excess number of edges it has in addition to the edges of any spanning tree. A component is simple if it is either a tree or is unicyclic. Below we present a very brief sketch of these subregions and state their properties (which hold with probability $1 - o(1)$). For more details, the reader is referred to [2] and the related references therein.

1. *Very subcritical*: $c = 1 - \epsilon$, ϵ is any sufficiently small positive constant. In this regime, $L_1 = \Theta\left(\ln n/\epsilon^2\right)$ (upper bound shown in Theorem 33.15) and all components are simple.

2. *Barely subcritical*: $c = 1 - \epsilon$ where $\epsilon = \lambda n^{-1/3}$ for some $0 \leq \lambda = o(n^{1/3})$ and $\lambda \to \infty$. In this regime, all components are simple, $L_1 = \Theta(n^{2/3}\lambda^{-2} \ln \lambda)$ and also $L_k \approx L_1$ for every fixed k.

3. *Critical*: $c = 1 \pm \epsilon$ where $\epsilon = \lambda n^{-1/3}$ for some constant real λ. In this regime, for every fixed $k \geq 1$, $L_k = \Theta(n^{2/3})$.

4. *Barely supercritical*: $c = 1 + \epsilon$ where $\epsilon = \lambda n^{-1/3}$ for some $0 \leq \lambda = o(n^{1/3})$ and $\lambda \to \infty$. In this regime, every component other than the largest is simple and the largest component has a complexity approaching infinity. Also, $L_1 \approx 2\lambda n^{2/3}$ and $L_2 = \Theta(n^{2/3}\lambda^{-2}(\ln \lambda))$.

5. *Very supercritical*: $c = 1+\epsilon$, ϵ is any sufficiently small positive constant. In this regime, $L_1 = \Theta(\epsilon n)$ and $L_2 = \Theta(\ln n)$. Further, every component (except the largest) is simple and the largest component has a complexity approaching infinity.

It is interesting to note that the largest component has size $\Theta(n^{2/3})$ throughout the critical regime on both sides of the threshold $1/n$. See also [73] for a discussion on the barely supercritical regime.

33.5.4 Diameter of Random Graphs

The random graph $G \in \mathcal{G}(n,p)$ has a small diameter. If $p(n) \geq 4(\ln n)/n$, then $diam(G) \leq c(\ln n)/\ln \ln n$ with probability $1 - o(1)$ for some positive constant c, as shown by Subramanian in [74]. For *denser* random graphs, the diameter is at most 2 a.a.s. Below, we state and prove a sharp threshold for the property of $diam(G) \leq 2$.

Theorem 33.16 *Let $\omega = \omega(n) \to \infty$ be any suitably growing function. Define $q_h(n) = \sqrt{2(\ln n) + \omega}/\sqrt{n}$ and $q_l(n) = \sqrt{2(\ln n) - \omega}/\sqrt{n}$. Then, with probability $1 - o(1)$, the following holds:*

1. *If $p \geq q_h$, then $diam(G) \leq 2$.*

2. *If $p \leq q_l$, then $diam(G) \geq 3$.*

Moreover, $q(n) = \sqrt{2(\ln n)}/\sqrt{n}$ is a sharp threshold since $q_h - q_l$ can be made arbitrarily small (by appropriately choosing $\omega(n)$) compared to $q(n)$.

Proof. We first prove (1). Without loss of generality, assume that $p = q_h$. For every $u \neq v$, let $\mathcal{E}_{u,v}$ denote the event that u and v do not share a common neighbor. We have

$$\mathbf{Pr}(\mathcal{E}_{u,v}) = (1 - p^2)^{n-2} \approx (1 - p^2)^n \leq e^{-np^2} = e^{-2(\ln n) - \omega} \leq \frac{1}{n^2 e^\omega}$$

for every $u \neq v$. Hence,

$$\mathbf{Pr}(diam(G) \geq 3) \leq \mathbf{Pr}(\exists u \neq v : \mathcal{E}_{u,v}) \leq \binom{n}{2} \frac{1}{n^2 e^\omega} \to 0$$

Thus, when $p \geq q_h$, $diam(G) \leq 2$ with probability $1 - o(1)$.

We now prove (2). Without loss of generality, assume that $p = q_l$. For every $u \neq v$, let X_{uv} denote the indicator variable for the event that u and v are not joined by a path on at most 2 edges. Then, $\mu_{uv} := E[X_{uv}] = (1-p)(1-p^2)^{n-2} \approx e^\omega/n^2$ for every $u \neq v$. Define $X = \sum_{u \neq v} X_{uv}$ denote the number of such pairs. It follows that $diam(G) \geq 3$ if and only if $X > 0$. Also, $E[X] = \sum_{u \neq v} \mu_{uv} = \binom{n}{2}\mu_{12} \approx \frac{e^\omega}{2} \to \infty$ as $n \to \infty$.

To calculate the variance, fix any $u \neq v$. Consider any $w \neq x$ such that $\{u,v\} \neq \{w,x\}$. There are two cases, namely, $|\{u,v\} \cap \{w,x\}| = 0$ or 1. We first consider the case $\{u,v\} \cap \{w,x\} = \emptyset$. The number of such $\{w,x\}$'s is $\binom{n-2}{2}$. We have

$$E_2 := E[X_{uv}X_{wx}] = (1-p)^2 \cdot (1-p^2)^{2n-8} \cdot \left((1-p)^4 + 4p(1-p)^3 + 2p^2(1-p)^2\right)$$

$$E_1^2 := E[X_{uv}]E[X_{wx}] = (1-p)^2 \cdot (1-p^2)^{2n-4}$$

$$E_2 - E_1^2 = (1-p)^2 \cdot (1-p^2)^{2n-8} \cdot \left(4p^3 - 7p^4 + 4p^6 - p^8\right)$$

$$\approx (1-p)^2 \cdot (1-p^2)^{2n-8} \cdot 4p^3 = O\left(\frac{p^3 e^{2\omega}}{n^4}\right)$$

We now consider the case $\{u,v\} \cap \{w,x\} \neq \emptyset$. The number of such $\{w,x\}$'s is $2(n-2)$. Assume, without loss of generality, that $w = u$. We have

$$E_2 := E[X_{uv}X_{ux}] = (1-p)^2 \cdot \left((1-p) + p(1-p)^2\right)^{n-3}$$

$$= (1-p)^2 \cdot \left(1 - 2p^2 + p^3\right)^{n-3}$$

$$\approx (1-2p^2)^n \cdot \left(1 + O(np^3)\right) \cdot (1 - O(p))$$

$$= \frac{e^{2\omega}}{n^4} \cdot \left(1 + O(np^3)\right)$$

$$E_1^2 := E[X_{uv}]E[X_{ux}] = (1-p)^2 \cdot (1-p^2)^{2n-4} \approx (1-p^2)^{2n} \cdot (1 - O(p))$$

$$= \frac{e^{2\omega}}{n^4} \cdot (1 - O(p))$$

$$E_2 - E_1^2 = \frac{e^{2\omega}}{n^4} \cdot O(np^3) = O\left(\frac{p^3 e^{2\omega}}{n^3}\right)$$

Summing $E_2 - E_1^2$ over all possible choices of $\{u,v\} \neq \{w,x\}$, we obtain that

$$\sum_{\{u,v\} \neq \{w,x\}} E_2 - E_1^2 \leq n^4 \cdot O\left(\frac{p^3 e^{2\omega}}{n^4}\right) + n^3 \cdot O\left(\frac{p^3 e^{2\omega}}{n^3}\right) = O\left(p^3 e^{2\omega}\right)$$

Hence, $\mathbf{Pr}(X = 0) \leq \frac{\text{Var}(X)}{\mu^2} \leq \mu^{-1} + O\left(p^3\right) \to 0$ since $\mu \to \infty$.

It follows that $diam(G) \geq 3$ with probability $1 - o(1)$. ∎

Similar thresholds can be shown to exist for (every fixed $l \geq 2$) for $G \in \mathcal{G}(n,p)$ to have diameter at most l. This follows from the works of [3,75]. We state the result without proof. See [76] for an earlier work on weaker results on diameter of random graphs.

Theorem 33.17 *Let $\omega = \omega(n) \to \infty$ be any growing function. For any fixed $l \geq 2$, define $q_H^l(n) = ((2(\ln n) + \omega)/n^{l-1})^{1/l}$ and $q_L^l(n) = ((2(\ln n) - \omega)/n^{l-1})^{1/l}$. Then, with probability $1 - o(1)$, the following holds:*

1. *If $p \geq q_H^l$, then $diam(G) \leq l$.*

2. *If $p \leq q_L^l$, then $diam(G) \geq l + 1$.*

Moreover, $q^l(n) = (2(\ln n)/n^{l-1})^{1/l}$ is a sharp threshold since $q_H^l - q_L^l$ can be made arbitrarily small (by appropriately choosing $\omega(n)$) compared to $q^l(n)$.

33.5.5 Concentration of Invariants

Another type of phenomenon that repeatedly occurs is the concentration of the *likely* values of a graph invariant in a narrow band. Let $f(G) \in \mathcal{N}$ be a nonnegative integral valued graph invariant. Examples of such invariants are: (1) the maximum size $\omega(G)$ of a clique, (2) the maximum size $\alpha(G)$ of an independent set in G, (3) the chromatic number $\chi(G)$, (4) the maximum size $p(G)$ of an induced path, (5) the maximum size $T(G)$ of an induced tree, (6) the diameter $diam(G)$, and so on. Even though the set of values that an invariant (say $\omega(G)$) potentially takes can be large (like $\omega(G) \in \{1, \ldots, n\}$), the actual values are highly concentrated in a very small band of consecutive integers (like $\omega(G) \in \{k, k+1\}$ for some $k = k(n, p)$), with probability $1 - o(1)$. This is known as the concentration phenomena. This is another interesting and surprising fact about random graphs whose study has been influenced by powerful tools from probability theory and which has also motivated the development of new tools in probability theory. In what follows, we illustrate this phenomenon with few examples.

33.5.5.1 Concentration of diam(G)

It follows from Theorem 33.17 that for every fixed $l \geq 2$, $diam(G)$ is exactly l (with probability $1 - o(1)$) provided p satisfies $q_H^l(n) \leq p \leq q_L^{l-1}(n)$. For p satisfying $q_L^l(n) < p < q_H^l(n)$, with probability $1 - o(1)$, $diam(G) \in \{l, l+1\}$. As a consequence, it follows that for every fixed $\epsilon > 0$, there exists a definition of $f(n) = 1 - o(1)$ such that for every sufficiently large n, $diam(G \in \mathcal{G}(n,p))$ is a single value with probability at least $f(n)$ as long as $p \in I_n \subseteq [n^{-1+\epsilon}, 1]$ where I_n is a set whose Lebseque measure approaches 1 asymptotically.

As for sparser random graphs are concerned, we set $p = n^{-1+o(1)}$. Bollobas [77] showed that $diam(G)$ is concentrated on at most four values provided $np = (\ln n) + \omega$ and $\omega \to \infty$. He further strengthened (see [3], Chapter 10) and showed that the value is in fact concentrated on at most two values provided $np = \omega \cdot (\ln n)$ for some $\omega \to \infty$. In a later work, Chung and Lu [78] established that diameter is concentrated on at most three values provided $np = c(\ln n)$ for some constant $c > 2$. It was also shown that the concentration is in a band of two values if $c > 8$. If $np = c(\ln n)$ for $\delta \leq c \leq 1$ (where $\delta > 0$ is any positive constant), then G is not necessarily connected and we redefine $diam(G)$ to be the diameter of its largest connected component. In this case, it is shown in [78] that $diam(G)$ is concentrated in a band of at most $2\lfloor \delta^{-1} \rfloor + 4$ values. Note that this regime corresponds to the scenario where G is not connected but has a unique giant component of linear size. For some related work on diameter when $np < 1$, see the work of Luczak [79].

33.5.5.2 Concentration of ω(G) and α(G)

We first focus on $\omega(G)$. Consider $G \in \mathcal{G}(n, 1/2)$. Consider any $1 \leq k \leq n$. For each k-subset S, let X_S be the indicator variable for S inducing a complete subgraph of G. Let $X = \sum_S X_S$. For each S, we have $\nu_S := E[X_S] = 2^{-\binom{k}{2}}$ and $\mu_k := E[X] = \sum_S \nu_S = \binom{n}{k} 2^{-\binom{k}{2}}$.

Let $\epsilon > 0$ be any small constant. Define $k_h := \lceil 2(\log_2 n) - 2(\log_2 \log_2 n) + 2\epsilon + 2(\log_2 e) - 1 \rceil$. Also, define $k_l := \lfloor 2(\log_2 n) - 2(\log_2 \log_2 n) - 2\epsilon + 2(\log_2 e) - 1 \rfloor$. We have, for $k = k_h$,

$$\mu_k \leq \left(\frac{en}{(2 - o(1))(\log_2 n)} \cdot 2^{-(\log_2 n) + (\log_2 \log_2 n) - \epsilon - \log_2 e + 1} \right)^k$$
$$\leq \left(2^{-\epsilon/2} \right)^k \to 0.$$

Hence, $\mathbf{Pr}(\omega(G) \geq k_h) = \mathbf{Pr}(X > 0) \leq \mu_k \to 0$.

But for $k = k_l$,

$$\mu_k \geq \frac{1-o(1)}{\sqrt{2\pi k}} \cdot \left(\frac{en}{(2-o(1))(\log_2 n)} \cdot 2^{-(\log_2 n)+(\log_2 \log_2 n)+\epsilon - \log_2 e + 1}\right)^k$$

$$\geq \frac{1-o(1)}{\sqrt{2\pi k}} \cdot 2^{k\epsilon} \to \infty$$

We now establish that $\mathbf{Pr}(\omega(G) < k_l) \to 0$. As in the case of triangles (discussed before), it reduces to establishing that the common value (i.e., independent of $S \in \binom{V}{k}$) of $\Delta_S := \sum_{T: 2 \leq |S \cap T| \leq k-1} E(X_T | X_S = 1)$ satisfies $\Delta_S = o(\mu_k)$. Note that

$$\Delta_S = \sum_{2 \leq l \leq k-1} \binom{k}{l}\binom{n-k}{k-l} 2^{-\binom{k}{2}+\binom{l}{2}}$$

$$\leq \mu_k \cdot \left(\sum_{2 \leq l \leq k-1} \frac{(k)_l^2}{(n)_l \cdot l!} \cdot 2^{\binom{l}{2}}\right)$$

$$= \mu_k \cdot [1 + o(1)] \cdot \left(\sum_{2 \leq l \leq k-1} A_l\right) \quad \text{where } A_l := \frac{(k)_l^2}{n^l \cdot l!} \cdot 2^{\binom{l}{2}}$$

Defining (as is done in [80])

$$t_l := \frac{A_{l+1}}{A_l} = \frac{(k-l)^2 \cdot 2^l}{n(l+1)} \quad \text{for } 2 \leq l \leq k-2$$

$$s_l := \frac{t_{l+1}}{t_l} = \left(\frac{k-l-1}{k-l}\right)^2 \cdot \frac{l+1}{l+2} \cdot 2 \quad \text{for } 2 \leq l \leq k-3$$

it can be verified that $s_l \geq 1$ for $2 \leq l \leq k-4$ and hence $\max\{A_l : 2 \leq l \leq k-2\} = \max\{A_2, A_{k-2}\}$ and also that $t_{k-2} > 1$. This implies that $\max_l A_l = \max\{A_2, A_{k-1}\} = o(n^{-\epsilon})$. As a result, we have $\Delta_S = o(\mu_k)$. Hence, with probability $1 - o(1)$, $k_l \leq \omega(G) < k_h$. The number of integers in the range $[k_l, k_h)$ is exactly $k_h - k_l < 4\epsilon + 2$. By choosing ϵ sufficiently small, we note that $k_h - k_l$ is at most 2. This shows that $\omega(G)$ is concentrated in a band of at most 2 integers, even though it can potentially take one of n possible values. It can also be shown that the set of values of n for which $[k_l, k_h)$ consists of just one integer (and hence $\omega(G)$ takes a unique value) is a subset of density 1 in the set of natural numbers. In an ongoing work [81], it has been shown that a similar two-point concentration result can be established for any $p = p(n)$ as long as $p \leq 1 - n^{-\epsilon}$ where $\epsilon < 1/3$ is any constant.

Since the complement of $G \in \mathcal{G}(n, p)$ is distributed as $\mathcal{G}(n, 1-p)$, we deduce (from our discussion on $\omega(G)$) that $\alpha(G)$ (for $G \in \mathcal{G}(n, p)$) is concentrated in at most two values provided $p \geq n^{-\epsilon}$ where $\epsilon < 1/3$ is any constant. However, for smaller values of p (like $p = n^{-0.75}$), $\alpha(G)$ has not been shown to be sharply concentrated but has only been shown to be concentrated in a band of $\Theta(p^{-1})$. It is shown in the work of Alan Frieze [82] that $\alpha(G)$ is concentrated as follows.

Theorem 33.18 *For every $\epsilon > 0$, there exists a w_ϵ such that for any p satisfying $w_\epsilon \leq w := np = o(n)$, we have with probability $1 - o(1)$,*

$$\left|\alpha(G) - \frac{2}{p}(\ln w - \ln \ln w - \ln 2 + 1)\right| \leq \frac{\epsilon}{p}.$$

∎

This result establishes $\alpha(G)$ is concentrated in a band of $\Theta(1/p)$ for a wide range of p.

It can also be established that with high probability $\alpha(G)$ is concentrated on a single value k_0 provided $k = k_0$ satisfies

$$k = O(n^{1/3}), \quad \binom{n}{k}(1-p)^{\binom{k}{2}} \to \infty \quad \text{and} \quad \binom{n}{k+1}(1-p)^{\binom{k+1}{2}} \to 0.$$

Thus, we see that $\alpha(G)$ is concentrated in a very narrow band for *large* enough values of p but is not sharply concentrated for *small* values of p.

In the case of sparse random graphs (those with $d = np$ being a constant), following can be said: For every $d \in (0, \infty)$, there exists $\alpha_d > 0$ satisfying $\mathbf{Pr}(\alpha(G) \geq \beta n) \to 1$ for every $\beta < \alpha_d$ and $\mathbf{Pr}(\alpha(G) \geq \beta n) \to 0$ for every $\beta > \alpha_d$. The work of Frieze [82] determines α_d to within a $o(1/d)$ additive factor, provided d is sufficiently large. For further work in this direction, the reader is referred to the work of Dani and Moore [83].

33.5.5.3 Concentration of $\chi(G)$

Let $G \in \mathcal{G}(n, 1/2)$. Since $\chi(H) \geq n/\alpha(H)$ for any graph H, it follows from upper bounds on $\alpha(G)$ that

$$\chi(G) \geq \frac{n}{2(\log_2 n) - \log_2 \log_2 n} = \frac{n}{2(\log_2 n)[1 - o(1)]}.$$

Grimmett and McDiarmid [84] proved that the analysis of a simple greedy algorithm yields an upper bound $\chi(G) \leq n/(\log_2 n)[1 - o(1)]$. while this established that $\chi(G)$ is concentrated in a band of sub-linear (i.e., $o(n)$) size, the upper is still nearly twice the lower bound.

It was suspected by many researchers that the upper bound should be actually closer to the lower bound. The first improvement in this direction was obtained by Matula [85] who proved that $\chi(G) \leq (2n/3(\log_2 n))[1 + o(1)]$. This was finally settled by Bollobas [85] who proved that $\chi(G) \leq (n/2(\log_2 n))[1 + o(1)]$. The proof of this result required obtaining exponentially low upper bounds on the probability of not having an independent set of size which is only away from $\alpha(G)$ by a constant. The powerful Azuma's inequality based on martingales was employed for this purpose. The proof of Bollobas was in fact motivated by the proof of a result of Shamir and Spencer [87] which (by employing Azuma's inequality) shows that $\chi(G)$ (for arbitrary p) is concentrated around its mean (even though no knowledge of the mean was employed) within a band of size $\omega(n)\sqrt{n}$ (where $\omega \to \infty$ is arbitrary). In addition, for $p = n^{-5/6-\epsilon}$, $\epsilon > 0$ fixed, Shamir and Spencer showed that there exists $u = u(n, p)$ such that $u \leq \chi(G) \leq u + 4$ with probability $1 - o(1)$. Thus, $\chi(G)$ is concentrated in five consecutive integers. It is interesting that one can obtain a very sharp concentration without knowing the location of the most likely values of $\chi(G)$. This is consistent with a similar nature of Azuma's inequality employed in the proof. Below, we provide a sketch of the weaker upper bound (due to [84]) for the sake of exposition.

Fix the ordering $\sigma = (1, 2, \ldots, n)$ on V. Consider the greedy algorithm which starts with coloring vertex 1 with color 1 and colors vertices as per σ. Suppose colors from $[c] = \{1, \ldots, c\}$ have been used when it comes to coloring vertex i. It tries to use an already used color (if possible), otherwise it borrows a new color $c + 1$ and colors i with $c + 1$. Let $\chi_g(G)$ be the number of colors used by the greedy algorithm on G. We establish the following bound (with the justifiable assumption that the expression used is an integer for the ease of exposition). The exposition is essentially the one presented by Krivelevich [88].

Theorem 33.19 [84] $\chi_g(G) \leq u := n/(\log_2 n - 2(\log_2 \log_2 n)) = n/((\log_2 n)[1 - o(1)])$ *with probability* $1 - o(1)$.

Proof. Let E denote the event that greedy uses more than u colors. If E occurs, then there is a special vertex on which color $u + 1$ is used for the first time. For every $i > u$, let E_i

denote the event that i is the first vertex on which $u+1$ is used. Then, E occurs if and only if E_i occurs for some $i > u$. Also, events E_i's are pairwise mutually exclusive. Hence, $\mathbf{Pr}(E) = \sum_{i>u} \mathbf{Pr}(E_i)$. It suffices to establish that $\mathbf{Pr}(E_i) = o(1/n)$ for any $i > u$.

Fix any $i > u$. Suppose that each of the colors in $[u]$ is used on at least one vertex from $\{1,\ldots,i-1\}$. Fix any such coloring $(C_1,\ldots,C_u)$ obtained by exposing only edges between vertices in $[i-1]$. For E_i to occur, it is necessary that i has a neighbor in each C_j. Thus, conditioned on the choice of $(C_j)_j$,

$$\mathbf{Pr}(E_i|(C_j)_j) = \prod_{j=1}^{u}\left(1-2^{-|C_j|}\right) \leq \left(1-2^{-\frac{1}{u}\sum_{j=1}^{u}|C_j|}\right)^u$$
$$\leq \left(1-2^{-\frac{i-1}{u}}\right)^u \leq exp\left(-u 2^{-n/u}\right)$$
$$= exp\left(-\left(\frac{n}{\log_2 n - 2\log_2\log_2 n}\right) 2^{-\log_2 n + 2\log_2\log_2 n}\right)$$
$$\leq exp(-(1+o(1))(\log_2 n)) = o\left(\frac{1}{n}\right)$$

Hence, $\mathbf{Pr}(E_i) = o\left(\frac{1}{n}\right)$

Thus, $\mathbf{Pr}(\chi_g(G) > u) = \mathbf{Pr}(E) = o(1)$ as required to be proved. ∎

The sharp 5-point concentration presented in [87] was strengthened to a 2-point concentration (for $p = n^{-5/6-\epsilon}$, $\epsilon > 0$ fixed) by Luczak [89]. The 2-point concentration result was further extended to $p = n^{-1/2-\epsilon}$ by Alon and Krivelevich [90]. It is still not known if similar sharp concentration can be established for denser random graphs (like for $p = n^{-1/2+\epsilon}$), perhaps it is not possible. However, for general p, the value of $\chi(G)$ (upto asymptotically smaller additive terms) was determined by Luczak [91] who established that

$$\frac{d}{2(\ln d)}\left(1 + \frac{\ln\ln d - 1}{\ln d}\right) \leq \chi(G) \leq \frac{d}{2(\ln d)}\left(1 + \frac{30(\ln\ln d)}{\ln d}\right)$$

with probability $1 - o(1)$, provided $p \geq C/n$ for a suitable constant C. Here d is a short notation for np.

As for the concentration width of sparse random graphs (those characterized by $p = d/n$ for some arbitrary but fixed $d \in (0,\infty)$) are concerned, there was a large gap between what is guaranteed (namely, a 2-point concentration due to [89,90]) and what is explicitly achieved (namely, a width (due to [91]) of $29d(\ln\ln d)/2(\ln d)^2$ provided $d \geq C$). This gap was finally resolved by Achlioptas and Naor [92] who established that: For every $d \in (0,\infty)$, with probability $1 - o(1)$, $\chi(G(n,d/n)) \in \{k_d, k_d+1\}$ where k_d is the smallest k satisfying $d < 2k(\ln k)$. In addition, it is also shown that: for every $k \geq 2$, if $d \in [(2k-1)(\ln k), 2k(\ln k))$, with probability $1 - o(1)$, $\chi(G(n,d/n)) = k+1$. This, in turn, is shown to imply that $\chi(G)$ is exactly determined for every $d \in S$ where $S \subseteq (0,\infty)$ is a subset of asymptotic density $1/2$. In a recent work, Coja-Oghlan and Vilenchik [93] strengthen further by showing that $\chi(G)$ is determined exactly for every $d \in S'$ where S' is a set of asymptotic density 1, by sharpening the thresholds for k-colorability.

33.5.5.4 Concentration of Induced Paths and Induced Trees

Another parameter which has been well-studied and closely related to independence number is the maximum size of an induced path in a graph G. We denote this parameter by $mip(G)$. Similarly, we use $h(G)$ to denote the maximum size of a hole (an induced cycle) in G. We also use $T(G)$ to denote the maximum size of an induced tree in G. From definition, it follows that $T(G) \geq mip(G)$ always. While each of these parameters has been studied in great detail for very sparse random graphs (those having $p = c/n$ for some fixed but arbitrary $c > 1$), there has been little (comparatively) attention paid to the case of dense random graphs (those having *large p*).

In a recent work [94], Dutta and Subramanian studied $mip(G)$ for dense random graphs and obtained a 2-point concentration for $mip(G)$. Precisely, it was established that: For $G \in \mathcal{G}(n,p)$ with $p \geq n^{-1/2}(\ln n)^2$, with probability $1 - o(1)$, $mip(G) \in \{b^*, b^* + 1\}$ where b^* is the largest b such that $\mu_b \geq np/(\ln \ln n)$. Here μ_b denotes the expected number of induced paths on b vertices. It is also shown that $\lfloor 2(\log_q np) + 2\rfloor \leq b^* \leq \lceil 2(\log_q np) + 3\rceil$ where $q = (1-p)^{-1}$.

As a corollary, it follows that $T(G) \geq b^*$ a.a.s. When this is combined with an improved upper bound obtained from a more careful analysis of the proof of an earlier bound of Erdös and Palka [95], it is established that $T(G) = 2(\log_q np) + O(1/\ln q)$ a.a.s. This significantly improves the gap $O(\ln n/\ln q)$ between earlier (and 25-year old) lower and upper bounds obtained in [95]. The precise statement of the result of [95] is: For every $\epsilon > 0$ and for every fixed p, with probability $1 - o(1)$, $G \in \mathcal{G}(n,p)$ satisfies $(2-\epsilon)(\log_q np) \leq T(G) \leq (2+\epsilon)(\log_q np)$.

Similarly, it is also established in [94] that: $G \in \mathcal{G}(n,p)$ satisfies (with probability $1 - o(1)$) $h(G) \in \{h^*, h^* + 1\}$ provided $p \geq n^{-0.5}(\ln n)^2$. The proofs for the bounds on $mip(G)$ and $h(G)$ are based on bounding the variance. While it might look like it will be essentially the standard proof for $\alpha(G)$, several complications arise when handling-induced paths and holes while computing the second moment. For example, the intersection of two independent sets is also independent. But the intersection of two induced paths need not be an induced path but will only be an induced linear forest.

For very sparse random graphs, Erdös and Palka [95] conjectured that for every $c > 1$, $G \in \mathcal{G}(n, \frac{c}{n})$ contains an induced tree of size $\alpha(c)n$ where $\alpha(c)$ depends only on c. This was affirmatively established de la Vega [96] and several others including Frieze and Jackson [97], Kucera and Rödl [98], and Luczak and Palka [99]. In particular, de la Vega [96] established that G almost surely contains an induced tree of size $\alpha_c n[1 - o(1)]$ where α_c is the positive root of the equation $\alpha c = \ln(1 + \alpha c^2)$. It can be verified that $\alpha_c = [1 - o(1)](\ln c)/c$. Here, $o(1)$ is with respect to increasing c. Later on, de la Vega [100] established that $T(G) \geq \beta_c n$ where $\beta_c := (2/c)(\ln c - \ln \ln c - 1)$. This bound is nearly tight (for large c) in view of an assertion established by Luczak and Palka [99] that: for every fixed $\epsilon > 0$, for every sufficiently large c, with probability $1 - o(1)$, $T(G) \leq 2(1+\epsilon)(\ln c)n/c$. In this context, we note that Palka and Rucinski [101] have obtained the following: If $p = c(\ln n)/n$ where $c \geq e$ is any constant and $G \in \mathcal{G}(n,p)$, then for any fixed $\epsilon > 0$, $(1/c - \epsilon)n(\ln \ln n)/\ln n \leq T(G) \leq (2/c + \epsilon)n(\ln \ln n)/\ln n$ with probability $1 - o(1)$.

As for induced paths and holes, Frieze and Jackson [102] studied the maximum size of a hole and obtained that: For every $\epsilon > 0$ and for every sufficiently large $c > 0$, $G \in \mathcal{G}(n,p)$ with $p = c/n$ contains a hole of size at least $(n/4c)\left(1 - c^6 e^{-c}\right)\left(1 - q\ln(1+q^{-1}) - \epsilon\right)$ with probability $1 - o(1)$. Here, $q := (8c-3)(8c-2)$. Note that for large c, we can approximate the above expression as $n/O(c^3)$. Since a hole of size m contains an induced path on $m-1$ vertices, this also yields a lower bound on $mip(G)$. In a related work, Suen [103] also studied induced paths and obtained the following significant improvement : For $G \in \mathcal{G}(n,p)$ with $p = c/n$ (where $c > 1$ is any fixed constant) and for any $\epsilon > 0$, a.a.s. G has an induced path of size at least $(1-\epsilon)h(c)n$, where

$$h(c) = c^{-1} \int_1^c \frac{1 - y(\zeta)}{\zeta} d\zeta$$

and $y(\zeta)$ is the smallest positive root of $y = e^{\zeta(y-1)}$. As $c \to \infty$, $h(c) \to (\ln c)/c$ and hence a.a.s., $mip(G) \geq (1-\epsilon)(n \ln c)/c$. Suen [103] also establishes that G contains a hole of size at least $(1-\epsilon)(n \ln c)/c$.

33.6 RANDOM DIGRAPHS

A digraph $D = (V, A)$ is a graph in which we orient each edge and also allow both edges of the form $i \to j$ and $j \to i$. Also, there is at most one copy of each directed edge. It is simple if at most one directed edge is allowed between any unordered pair of vertices. Simple digraphs can also be thought of as the result of orienting the edges of an undirected graph. Random models of digraphs have also been studied (for both simple and non-simple digraphs). It is possible to have directed 2-cycles in a non-simple digraph while this is impossible in the case of a simple digraph. We use $\mathcal{D}(n,p)$ and $\mathcal{D}_2(n,p)$ to denote respectively the random models for simple and non-simple digraphs over $V = [n]$. Let $N := \binom{n}{2}$.

$\mathcal{D}(n,p)$ *model.* The model is defined for every $p \leq 0.5$. Choose a random $G \in \mathcal{G}(n, 2p)$ and then orient each $e \in E(G)$ independently in one of the two directions equiprobably. This results in a $D \in \mathcal{D}(n,p)$. For any $F = (V, A)$ having m edges, we have $\mathbf{Pr}(D = F) = \binom{N}{m} p^m (1-2p)^{N-m}$.

$\mathcal{D}_2(n,p)$ *model.* The model is defined for every $p \leq 1$. For each of the $n(n-1)$ ordered pairs (u, v) (with $u \neq v$), include $u \to v$ in $E(D)$ with probability p. The choices are independent for different pairs. For any $F = (V, A)$ having m edges, we have $\mathbf{Pr}(D = F) = \binom{2N}{m} p^m (1-p)^{2N-m}$.

The $\mathcal{D}(n,p)$ model was introduced by Subramanian in [104] to study the size of maximum induced acyclic subgraphs and the $\mathcal{D}_2(n,p)$ model was introduced by Karp in [105] to study the maximum size of a strongly connected component. There are other models for random digraphs like $\mathcal{D}_{k\text{-out}}(n)$ (introduced by Fenner and Frieze in [106]) in which each $u \in V$ uniformly randomly and independently chooses k out-neighbors from other vertices. We focus only on $\mathcal{D}(n,p)$ and $\mathcal{D}_2(n,p)$ models.

33.6.1 Induced Acyclic Tournaments

Given a digraph $D = (V, A)$, let $mat(D)$ denote the maximum size (meaning $|U|$) of $U \subseteq V$ such that U induces an acyclic tournament in D. The problem of determining this invariant was posed and studied by Dutta and Subramanian [80] for a random digraph D (drawn from either of the $\mathcal{D}(n,p)$ and $\mathcal{D}_2(n,p)$ models) and they obtained a 2-point concentration on its size (for every p). In particular, the following is established: If $D \in \mathcal{D}(n,p)$ with $p \geq 1/n$, then (with probability $1 - o(1)$) $mat(D) \in \{b^*, b^* + 1\}$ where $b^* = \lfloor 2(\log_r n) + 0.5 \rfloor$ and $r = p^{-1}$. Also, sufficient conditions were obtained which guarantee that $mat(D)$ actually takes just one value. As a consequence, it is also shown that for every fixed p and every definition of $f(n) = 1 - o(1)$, $mat(D)$ takes just one value with probability at least f, for every $n \in \mathcal{N}_{f,p}$ where $\mathcal{N}_{f,p}$ is a subset of natural numbers having asymptotic density 1. Analogous results are also established if $D \in \mathcal{D}_2(n,p)$.

For the sake of exposition, we present below the simpler derivation of 2-point concentration of $mat(D)$ (where $D \in \mathcal{D}(n,p)$) for the case $p = 1/2$. When $p = 1/2$, D becomes a random tournament on V chosen uniformly.

Theorem 33.20 *If $D = (V, A)$ is a uniformly chosen tournament, then, with probability $1 - o(1)$, $mat(D) \in \{b^*, b^* + 1\}$ where $b^* = \lfloor 2(\log_2 n) + 0.5 \rfloor$.*

Proof. Let $b = b^* + 2$. For each $S \subseteq V$ of size b, let X_S indicate if $D[S]$ is acyclic or not. We have $E[X_S] = b! \cdot 2^{-\binom{b}{2}}$. Hence if $X_b = \sum_S X_S$ denotes the number of acyclic subtournaments of size b, then

$$\mathbf{Pr}(X_b > 0) \leq \mu_b := E[X_b] = \binom{n}{b} \cdot b! \cdot 2^{-\binom{b}{2}} \leq \left(n \cdot 2^{-\frac{b-1}{2}}\right)^b \leq \left(2^{-1/4}\right)^b \to 0 \text{ as } n \to \infty$$

Thus, $mat(D) \leq b^* + 1$ a.a.s. Now we set $b = b^*$ and note that

$$E[X_b] = [1 - o(1)] \cdot \left(n \cdot 2^{-\frac{b-1}{2}}\right)^b \geq \left(2^{1/4}\right)^b \to \infty \text{ as } n \to \infty$$

We now establish that $\mathbf{Pr}(mat(D) < b) \to 0$. As in the case of cliques (discussed before), it reduces to establishing that the common value (i.e., independent of $S \in \binom{V}{b}$) of $\Delta_S := \sum_{T : 2 \leq |S \cap T| \leq b-1} E(X_T | X_S = 1)$ satisfies $\Delta_S = o(\mu_b)$. Note that

$$\Delta_S = \sum_{2 \leq l \leq b-1} \binom{b}{l} \binom{n-b}{b-l} \frac{b!}{l!} \cdot 2^{-\binom{b}{2} + \binom{l}{2}}$$

$$= \mu_b \cdot \left(\sum_{2 \leq l \leq b-1} \frac{b!}{(n)_b \cdot l!} \cdot \frac{(b)_l}{l!} \cdot \frac{(n-b)_{b-l}}{(b-l)!} \cdot 2^{\binom{l}{2}} \right)$$

$$\leq \mu_b \cdot \left(\sum_{2 \leq l \leq b-1} \binom{b}{l}^2 \cdot \frac{2^{\binom{l}{2}}}{(n)_l} \right)$$

$$= \mu_b \cdot [1 + o(1)] \cdot \left(\sum_{2 \leq l \leq b-1} A_l \right) \text{ where } A_l := \binom{b}{l}^2 \cdot \frac{2^{\binom{l}{2}}}{n^l}$$

Defining

$$t_l := \frac{A_{l+1}}{A_l} = \frac{(b-l)^2 \cdot 2^l}{n(l+1)^2} \text{ for } 2 \leq l \leq b-2$$

$$s_l := \frac{t_{l+1}}{t_l} = \left(\frac{b-l-1}{b-l} \cdot \frac{l+1}{l+2}\right)^2 \cdot 2 \text{ for } 2 \leq l \leq b-3$$

it can be verified that $s_l \geq 1$ for $2 \leq l \leq b-4$ and hence $\max\{A_l : 2 \leq l \leq b-2\} = \max\{A_2, A_{b-2}\}$ and also that $t_{b-2} > 1$. This implies that $\max_l A_l = \max\{A_2, A_{b-1}\} = o(n^{-1})$. As a result, we have $\Delta_S = o(\mu_b)$. Hence, with probability $1 - o(1)$, $b^* \leq mat(D) \leq b^* + 1$. ∎

33.6.2 Induced Acyclic Subgraphs

For a given digraph $H = (V, A)$, let $mas(H)$ denote the maximum size ($|U|$) of a subset U of V such that U induces an acyclic subgraph of H. Subramanian [104] analyzed this invariant $mas(D)$ for a random graph $D \in \mathcal{D}(n, p)$ and obtained that $mas(D) \leq \lceil 2(\ln_q n) + 1 \rceil$ for any $p \leq 0.5$ where $q := (1-p)^{-1}$. It was also observed that $\mathbf{Pr}(mas(D) \geq b) \geq \mathbf{Pr}(\alpha(G) \geq b)$ for any $b \geq 1$ and for $G \in \mathcal{G}(n, p)$. Combining this fact with the lower bounds on $\alpha(G)$ presented in [82,107,108], it was established that, for any $\epsilon > 0$, $mas(D) \geq (2/\ln q)(\ln w - \ln \ln w - O(1))$ provided $w = np \geq w_\epsilon$ for some w_ϵ depending only on ϵ. While for larger values of p, the upper and lower bounds are the same (upto constant multiplicative factors), for $p = n^{-1+o(1)}$, the lower bound is asymptotically smaller than the above upper bound. It was conjectured in [104] that the upper bound is also essentially the lower bound (upto asymptotically negligible additive terms) and this was affirmatively established by Spencer and Subramanian in [109] by improving the upper bound to $(2/\ln q)(\ln w + 3e)$. As a result, we now know that: for every $p \leq 0.5$ satisfying $w = np = \omega(1)$, with probability $1 - o(1)$, we have $mas(D) = (2(\ln w)/\ln q)[1 \pm o(1)]$.

Still, the gap between lower and upper bounds is at least $(2(\ln \ln w) + 6e)/\ln q$. A simple application of Azuma's inequality (based on the fact that $mas(D)$ is a 1-Lipschitz function) establishes that $|mas(D) - \mu| \leq \omega\sqrt{n}$ (for any $\omega \to \infty$) for every choice of $p = p(n)$. Here, μ denotes the expectation of $mas(D)$. But this does not tell us the location of the concentration and is also weak (when compared to the gap obtained from known bounds) for $p \geq n^{-1/2+\epsilon}$. Similarly, an application of Talagrand's inequality establishes that $|mas(D) - m| \leq \omega\sqrt{\log_q np}$ (for any $\omega \to \infty$) for every choice of $p = p(n)$. Here, m denotes any median value of $mas(D)$. This gives a better concentration gap (when compared to that guaranteed by Azuma's inequality) and even beats the gap between known bounds for all p such that $p = o((\ln \ln n)^2/\ln n)$. Again, the location of the concentration is not known.

Recently, Dutta and Subramanian [110–112] improved the lower bound as follows: $mas(D) \geq \lfloor 2(\log_q np) - X \rfloor$ where $X = 1$ if $p \geq n^{-1/3+\epsilon}$ ($\epsilon > 0$ is any constant) and $X = W/\ln q$ if $p \geq C/n$ where $W > 4$ is any constant (and $C = C(W)$ is any suitably large constant). While this is a significant improvement (of the second-order additive term) over the previous lower bound, the asymptotics of the gap between known and best lower and upper bounds remain as they are. The first improvement to $X = 1$ was based on analyzing the total number of acyclic orderings of induced sub-digraphs of the stated size and is based on bounding the variance.

When p becomes smaller, the variance becomes larger and the variance-based approach does not work. Hence, the authors of [111,112] employ an approach which exploits the following fact: For every $b_1 < b_2$, Talagrand's inequality leads to an exponentially low upper bound on the product $\mathbf{Pr}(mas(D) \leq b_1) \cdot \mathbf{Pr}(mas(D) \geq b_2)$ where the exponent directly depends on $(b_2 - b_1)^2$ and inversely depends on the sparsity of the certifiability of $mas(D)$. If $mas(D) \geq b$ is certified by the set of edges incident on at most $f(b)$ vertices, then we say that $mas(D)$ is f-certifiable. It can be seen that $mas(D)$ is linearly certifiable. We combine this fact with a weaker lower bound on $\mathbf{Pr}(mas(D) \geq b_2)$ to get an $o(1)$ upper bound on $\mathbf{Pr}(mas(D) < b_1)$. By choosing b_1 and b_2 appropriately, [112] establishes the second improvement mentioned above. The weak lower bound on the second term of the product was established using the well-known Paley–Zigmund inequality which says that $\mathbf{Pr}(X > 0) \geq E[X]^2/E[X^2]$ for any nonnegative random variable X.

33.6.3 Induced Tournaments

For a digraph D, let $\omega(D)$ denote the maximum size of an induced tournament in D. Clearly, $\omega(D) \geq mat(D)$ for any D. It follows from the definitions of the respective random models that $\omega(D_1)$ (for $D_1 \in \mathcal{D}(n,p)$) and $\omega(G_1)$ (for $G_1 \in \mathcal{G}(n, 2p)$) are identically distributed. Similarly, $\omega(D_2)$ (for $D_2 \in \mathcal{D}_2(n,p)$) and $\omega(G_2)$ (for $G_2 \in \mathcal{G}(n, 2p(1-p))$) are identically distributed. Since $2p(1-p) \leq 0.5$ for any p, it follows that $\omega(G_2)$ and hence $\omega(D_2)$ are concentrated in two consecutive values for any p. This follows from the two-point concentration of $\omega(G)$ discussed in Section 33.5.5.2.

However, $\omega(D_1)$ is known to be two-point concentrated only for $p \leq 0.5 - n^{-\delta}$ where δ is a suitably small positive constant. For larger values of p which are closer to 0.5 like $p = 0.5 - 100n^{-1}$, $\omega(G_1)$ and hence $\omega(D_1)$ are only known to be concentrated in a band of size $\Theta((1-2p)^{-1})$. Thus, the concentration behaviors of $\omega(D_1)$ and $\omega(D_2)$ differ. Recall that both $mat(D_1)$ and $mat(D_2)$ are two-point concentrated for any $p = p(n)$. For a more detailed discussion on this topic, we refer the reader to [80].

33.6.4 Being Strongly Connected

A digraph $D = (V, A)$ is strongly connected if, for every ordered pair (u, v) of vertices, there is a directed path from u to v. Strongly connected components of a digraph are analogues of connected components of an undirected graph. Given a digraph $D = (V, A)$, a strongly connected component (shortly, a strong component) of D is a maximal (with respect to vertex inclusion) induced subdigraph which is also strongly connected. They are precisely the equivalence classes of an equivalence relation R over V where R is the relation: for every $u, v \in V$, $(u, v) \in R$ if and only if u and v are reachable from each other by a directed path in D.

Below, we present and prove a sharp threshold function for $D \in \mathcal{D}_2(n, p)$ being strongly connected.

Theorem 33.21 *Let $\omega = \omega(n) \to \infty$ be any sufficiently slowly growing function. Define $q(n) = (\ln n)/n$ and $r(n) = \omega(n)/n$. For $D \in \mathcal{D}_2(n, p)$, with probability $1 - o(1)$, the following holds:*

1. *If $p \geq q + r$, then D is strongly connected.*

2. *If $p \leq q - r$, then D is not strongly connected.*

Moreover, $q(n)$ is a sharp threshold since $r(n)$ can be arbitrarily small compared to $q(n)$.

Proof. The arguments are similar to those employed in the proof of Theorem 33.14. Without loss of generality, we assume that $p = q + r$ (or $p = q - r$) depending on whether we want to prove (1) (or (2)). We first prove (1). We show that $\mathbf{Pr}(D$ is not strongly connected$) \to 0$. D is not strongly connected if and only if there exists a $S \subseteq V$, $|S| \leq n/2$, such that there is no edge in one of the two directions $(S \to V \setminus S, V \setminus S \to S)$. For $1 \leq k \leq n/2$, let $\mathcal{E}_k$ denote the event that there exists such a S with $|S| = k$. We have

$$\mathbf{Pr}(\mathcal{E}_k) \leq 2\binom{n}{k}(1-p)^{k(n-k)} \leq 2\left(\frac{en}{k}(1-p)^{n-k}\right)^k \leq 2\left(\frac{en}{k}e^{-np(1-\frac{k}{n})}\right)^k$$

$$\leq 2\left(\frac{en}{k}e^{-(\ln n + \omega)(1-\frac{k}{n})}\right)^k \leq 2\left(\frac{en^{\frac{k}{n}}}{k}e^{-\omega(1-\frac{k}{n})}\right)^k \leq 2\left(e\left(n^{\frac{k}{n}-\frac{\ln k}{\ln n}}\right)e^{-\omega/2}\right)^k \leq 2\left(\frac{1}{\omega}\right)^k$$

In the above derivation, we have used (i) $\binom{n}{k} \leq (en/k)^k$, (ii) $k \leq n/2$, and (iii) $\ln m / m$ is a decreasing function of m. As a result, we have

$$\mathbf{Pr}(D \text{ is not strongly connected}) = \mathbf{Pr}(\exists k \leq \frac{n}{2} : \mathcal{E}_k) \leq 2\left(\sum_{1 \leq k \leq \frac{n}{2}} \omega^{-k}\right) = o(1)$$

We now prove (2). For each $u \in V$, let X_u denote the indicator variable for u being isolated in D. We say that u is isolated if either there is no edge of the form $u \to v$ or there is no edge of the form $v \to u$ in D. Then, for any u, $\mu_u := E[X_u] = 2(1-p)^{n-1} - (1-p)^{2n-2} \approx 2(1-p)^{n-1}$. If $X := \sum_u X_u$ denotes the number of isolated vertices, then $\mu := E[X] \approx 2n(1-p)^{n-1} \geq 2n(1-p)^n$. We have

$$\mu \geq 2n(1-p)^n = 2ne^{-np[1+O(p)]} \geq 2ne^{-(\ln n - \omega)[1+O(p)]} \geq 2n^{-O(p)}e^{\omega} \geq 2[1-o(1)] \cdot e^{\omega} \to \infty.$$

Also,

$$Var(X) = \sum_u Var(X_u) + \sum_{u \neq v} E(X_u X_v) - \mu_u \mu_v$$

$$\leq \mu + n^2 \left(4(1-p)^{2n-2} + 2p(1-p)^{2n-3} - 4(1-p)^{2n-2} - (1-p)^{4n-4} + 4(1-p)^{3n-3}\right)$$

$$\approx \mu + \frac{\mu^2}{2} \cdot \left(\frac{p}{1-p} - \frac{(1-p)^{2n-2}}{2} + 2(1-p)^{n-1}\right)$$

$$= \mu + \frac{\mu^2}{2} \cdot p[1 + o(1)]$$

Hence,

$$\mathbf{Pr}(D \text{ has no isolated vertex}) = \mathbf{Pr}(X=0) \leq \frac{Var(X)}{\mu^2} \leq \mu^{-1} + p[1+o(1)] = o(1)$$

using $\mu \to \infty$ and $p = o(1)$. This shows that

$$\mathbf{Pr}(D \text{ is not strongly connected}) \geq \mathbf{Pr}(D \text{ has an isolated vertex}) = 1 - o(1).$$

Thus, $\ln n/n$ is a sharp threshold for being strongly connected. ∎

Remark 33.1 Theorem 33.21 holds true even if we replace our assumption $D \in \mathcal{D}_2(n,p)$ to $D \in \mathcal{D}(n,p)$. The same proof arguments work with this assumption also except the following two changes:

$$\mu_u = 2(1-p)^{n-1} - (1-2p)^{n-1}, \ \forall u; \quad E[X_u X_v] \leq (4-6p)(1-p)^{2n-4}, \ \forall u \neq v.$$

33.6.5 Emergence of a Giant Strongly Connected Component Around $1/n$

What happens when $p = c/n$ with $c = 1 \pm \epsilon$ where ϵ is a small positive constant or $\epsilon = o(1)$? Phase transition phenomena, similar to those which occur in the case of random graphs, are witnessed in the case of random digraphs also. In particular, the size of the largest strongly connected component makes a huge jump from being *very small* to a giant size as p makes a transtion from being less than $1/n$ to being more than $1/n$. Palasti [113] was the first to study the strongly connectedness property of random directed graphs for the $\mathcal{D}(n,m)$ model (the directed analogue of the $\mathcal{G}(n,m)$ model). Here, a directed graph on $[n]$ with m edges is chosen uniformly randomly with probability $\binom{n(n-1)}{m}^{-1}$. By establishing an equivalence (in a quantifiable sense) between the $\mathcal{D}(n,m)$ and $\mathcal{D}_2(n,p)$ models, Graham and Pike [114] adapted the work of Palasti to the $\mathcal{D}_2(n,p)$ model and established $p = 1/n$ as a threshold for strong connectivity. A detailed study of the phase transition phenomenon was carried out (for giant-size strong components of $D \in \mathcal{D}_2(n,p)$) by Richard Karp (in [105]), by Luczak (in [115]), and by Luczak and Seierstad (in [116]). We present a brief and consolidated overview of the interesting observations presented in the above mentioned references.

As before, the focus is on L_1 where L_1 stands for the size of the largest strong component of $D \in \mathcal{D}_2(n,p)$ with $p = c/n$. Below we identify some subregions of p and state the behavior of L_1 (which holds with probability $1 - o(1)$) without any formal justification. For a constant $c > 1$, let $\Theta = \Theta(c)$ be the unique root in $(0,1)$ of the equation $1 - x - e^{-cx} = 0$.

1. *Very subcritical.* $c = 1 - \epsilon$ where ϵ is a positive constant. In this regime, each of the strong components is of size at most $C(\epsilon)(\ln n)$ where $C(\epsilon)$ is a constant depending only on ϵ. In addition, if $D \in \mathcal{D}(n,m)$ where $m = (1-\epsilon)n$, then for every $\omega(n) \to \infty$, every strong component of D is a cycle of length less than ω.

2. *Barely subcritical.* $c = 1 - \epsilon$ for some ϵ such that $\epsilon \to 0$ and $\epsilon n^{1/3} \to \infty$. Then, for every $\omega \to \infty$, each strong component of D is either an isolated vertex or or a directed cycle of length at most $\omega \cdot \epsilon^{-1}$. We have $L_1 = o(n^{1/3})$ throughout this regime.

3. *Critical.* $c = 1 \pm \epsilon$ where $\epsilon = \lambda n^{-1/3}$ for some constant real λ. No detailed study seems to have been done for this regime.

4. *Barely supercritical.* $c = 1 + \epsilon$ where $\epsilon \to 0$ and $\epsilon n^{1/3} \to \infty$. Then, for every $\omega \to \infty$, D has a unique strong component of size $(4 + o(1))\epsilon^2 n$ and every other component is of size at most $\omega \cdot \epsilon^{-1}$. We have $L_1 = \omega(n^{2/3})$ throughout this regime.

5. *Very supercritical.* $c = 1 + \epsilon$ where ϵ is a positive constant. For some constant $A = A(c)$, for every $\omega = \omega(n) \to \infty$, we have $|L_1 - \Theta^2 n| \leq \omega\sqrt{n(\ln n)}$ and every other strong component is of size at most $A(\ln n)$. Thus, there exists a unique giant strong component of linear size. In the recent work [66], by analyzing the DFS algorithm (as was done for the $\mathcal{G}(n,p)$ model), it was established that $D \in \mathcal{D}_2(n,p)$ has in fact a directed cycle of length $\Theta(\epsilon^2)n$. Also, for every $\epsilon > 0$, there exists a $\alpha = \alpha(\epsilon)$ such that for every $\omega \to \infty$, $D \in \mathcal{D}(n,m)$ (where $m = (1 + \epsilon)n$) contains a unique strong component of size larger than αn and every other component is a cycle length less than ω. If $m/n \to \infty$, then D contains a unique strong component of size $(1 - o(1))n$.

33.7 CONCLUSIONS

Several interesting applications of probabilistic techniques to obtain and prove results on combinatorial structures have been presented or introduced. For several of these applications, there is no known proof other than the probabilistic one which first established it. For example, there is no known constructive proof of Theorem 33.12 except the probabilistic one obtained by Erdös nearly 60 years ago. This illustrates the power of this approach to resolving questions in combinatorics. There are a number of other interesting applications which have not been introduced in this exposition and the interested reader can find them in books (and their references) and other references listed below.

While the applications of probabilistic arguments in combinatorics have mostly been in the context of obtaining existence proofs, many of these applications can actually be translated to constructive proofs which actually produce an object of the type whose existence is being sought to be established. Also, under some assumptions about the random experiment, the constructive proof can be made efficient. The method of conditional expectations and pessimistic estimators are some of the techniques that can be applied to successfully produce a desired structure. See the books [2] and [6] for a comprehensive introduction to these approaches. Recently, Moser and Tardos [117] have presented a randomized polynomial time algorithm which (under some assumptions about the underlying probability space) finds a desired object guaranteed by LLL.

References

[1] S.N. Bernstein. Demonstration du theoreme de weierstrass fondee sur le calcul des probabilites. *Comm. Soc. Math. Kharkov*, **13** (1912), 1–2.

[2] Joel H. Spencer and Noga Alon. *The Probabilistic Method, Third Edition*. John Wiley & Sons, New York, 2008.

[3] Bela Bollobas. *Random Graphs, Second Edition*. Cambridge University Press, United Kingdom, 2001.

[4] Tomasz Luczak, Svante Janson, and Andrzej Rucinski. *Random Graphs*. John Wiley & Sons, New York, 2000.

[5] Paul Erdös and Laszlo Lovasz. Problems and results on 3-chromatic hypergraphs and some related questions, in: *Infinite and Finite Series* (A. Hajnal, R. Rado, and V.T. Sos, eds.), pages 609–628, 1975.

[6] Prabhakar Raghavan and Rajeev Motwani. *Randomized Algorithms*. Cambridge University Press, United Kingdom, 1995.

[7] C.S. Edwards. Some extremal properties of bipartite subgraphs. *Canadian Journal of Mathematics*, **3** (1973), 475–485.

[8] C.S. Edwards. An improved lower bound for the number of edges in a largest bipartite subgraph. *Proceedings of the 2nd Czechoslovak Symposium on Graph Theory,* Prague, Czech Republic, 167–181, 1975.

[9] Paul Erdös. Some recent problems in combinatorics and graph theory. *Proceedings of the 26th Southeastern International Conference on Graph Theory, Combinatorics and Computing,* Boca Raton, FL, 1995.

[10] Noga Alon. Bipartite subgraphs. *Combinatorica*, **16** (1996), 301–311.

[11] P. Erdös. Problems and results in graph theory combinatorial analysis. *Proceedings of the Conference on Graph Theory and Related Topics,* Waterloo, Canada, 153–163, 1979.

[12] S. Poljak Zs. Tuza. Bipartite subgraphs of triangle-free graphs. *SIAM Journal of Discrete Mathematics*, **37** (1988), 130–143.

[13] J.B. Shearer. A note on bipartite subgraphs of triangle-free graphs. *Random Structures and Algorithms*, **3** (1992), 223–226.

[14] Y. Caro. New Results on the Independence Number. *Technical Report*, 1979.

[15] V.K. Wei. A lower bound on the stability number of a simple graph. *Technical Report*, 1981.

[16] R. Bopanna. comment #1. *Comment on Lance Fortnow's blog*, 2010.

[17] Z. Tuza and Y. Caro. Improved lower bounds on k-independence. *Journal of Graph Theory*, **15** (1991), 99–107.

[18] Kunal Dutta, Dhruv Mubayi, and C.R. Subramanian. New lower bounds for the independence number of sparse graphs and hypergraphs. *SIAM Journal on Discrete Mathematics*, **26**(3) (2012), 1134–1147.

[19] N.S. Narayanaswamy and C.R. Subramanian. Dominating set based exact algorithms for 3-coloring. *Information Processing Letters*, **111**(6) (2011), 251–255.

[20] D. Aingworth, C. Chekuri, P. Indyk, and R. Motwani. Fast estimation of diameter and shortest paths (without matrix multiplication). *SIAM Journal on Computing*, **82** (1999), 1167–1181.

[21] P. Duchet and H. Meyniel. On hadwiger's number and stability numbers. *Annals of Discrete Mathematics*, **13** (1982), 71–74.

[22] Y. Caro, D.B. West, and R. Yuster. Connected domination and spanning trees with many leaves. *SIAM Journal on Discrete Mathematics*, **13**(2) (2000), 202–211.

[23] D. Karger. Global min-cuts in *rnc* and other ramifications of a simple min-cut algorithm. *Proceedings of the 4th ACM-SIAM Symposium on Discrete Algorithms*, Austin, TX, 21–30, 1993.

[24] D. Karger and C. Stein. An $O(n^2)$ algorithm for minimum cuts. *Proceedings of the 25th ACM Symposium on Theory of Computing*, San Diego, CA, 757–765, 1993.

[25] D. Karger and C. Stein. A new approach to the minimum cut problem. *Journal of the ACM*, **43**(4) (1996), 601–640.

[26] Vijay V. Vazirani. *Approximation Algorithms*. Springer-Verlag, Berlin, Germany, 2001.

[27] D. Karger. A randomized fully polynomial approximation scheme for the all terminal network reliability problem. *Proceedings of the 27th ACM Symposium on Theory of Computing*, Las Vegas, NV, 11–17. ACM, 1995.

[28] D. Karger. A randomized fully polynomial approximation scheme for the all terminal network reliability problem. *SIAM Journal on Computing*, **29**(2) (1999), 492–514.

[29] Noga Alon. Choice numbers of graphs: A probabilistic approach. *Combinatorics, Probability and Computing*, **1**(2) (1992), 107–114.

[30] C.R. Subramanian. List set coloring: Bounds and algorithms. *Combinatorics, Probability and Computing*, **16**(1) (2007), 145–158.

[31] C.R. Subramanian. List hereditary colorings. *Proceedings of the 2nd International Conference on Discrete Mathematics,* India, June 6–10, 2008, RMS Lecture Note Series No. 13, Ramanujan Mathematical Society, 191–205, 2010.

[32] C.R. Subramanian. List hereditary colorings of graphs and hypergraphs. *Manuscript*, 2014.

[33] Paul Erdös. Graph theory and probability. *Canadian Journal of Mathematics*, **11** (1959), 34–38.

[34] Paul Erdös. On circuits and subgraphs of chromatic graphs. *Mathematika*, **9** (1962), 170–175.

[35] Paul Erdös. On a problem in graph theory. *Mathematical Gazette*, **47** (1963), 220–223.

[36] A.V. Kostochka, E. Sopena, and X. Zhu. Acyclic and oriented chromatic numbers of graphs. *Journal of Graph Theory*, **24**(4) (1997), 331–340.

[37] N.R. Aravind and C.R. Subramanian. Forbidden subgraph colorings and the oriented chromatic number. *European Journal of Combinatorics*, **34** (2013), 620–631.

[38] E. Sopena. The chromatic number of oriented graphs. *Journal of Graph Theory*, **25**(2) (1997), 191–205.

[39] G. Fertin, A. Raspaud, and A. Roychowdhury. On the oriented chromatic numbers of grids. *Information Processing Letters*, **85**(5) (2003), 261–266.

[40] H. Hind, M. Molloy, and B. Reed. Colouring a graph frugally. *Combinatorica*, **17**(4) (1997), 469–482.

[41] H. Hind, M. Molloy, and B. Reed. Total coloring with $\delta + poly(\log \delta)$ colors. *SIAM Journal on Computing*, **28**(3) (1998), 816–821.

[42] Noga Alon, Colin McDiarmid, and Bruce Reed. Acyclic coloring of graphs. *Random Structures and Algorithms*, **2**(3) (1991), 277–288.

[43] G. Fertin, A. Raspaud, and B. Reed. Star coloring of graphs. *Journal of Graph Theory*, **47**(3) (2004), 163–182.

[44] B. Grünbaum. Acyclic colorings of planar graphs. *Israel Journal of Mathematics*, **14**(3) (1973), 390–408.

[45] N.R. Aravind and C.R. Subramanian. Bounds on vertex colorings with restrictions on the union of color classes. *Journal of Graph Theory*, **66**(3) (2011), 213–234.

[46] N.R. Aravind and C.R. Subramanian. Bounds on edge colorings with restrictions on the union of color classes. *SIAM Journal of Discrete Mathematics*, **24**(3) (2010), 841–852.

[47] Paul Erdös. Problems and results in additive number theory. *Colloque sur le Theorie des Nombres* (CBRM, Bruselles, Belgium), 127–137, 1956.

[48] Paul Erdös and Prasad Tetali. Representations of integers as the sum of k terms. *Random Structures and Algorithms*, **1**(3) (1990), 245–261.

[49] V.H. Vu. On the concentration of multivariate polynomials with small expectation. *Random Structures and Algorithms*, **16** (2000), 344–363.

[50] R.C. Vaughan and T.D. Wooley. Waring's problem: A survey. *www.personal.psu.edu/rcv4/Waring.pdf*, pages 1–40.

[51] D. Hilbert. Beweis für di darstellbarkeit der ganzen zahlen durch eine feste anzahl n-ter potenzen (waringsches probelm). *Mathematische Annalen*, **67** (1909), 281–300.

[52] R. Balasubramanian, J.-M. Deshouillers, and F. Dress. Problme de waring pour les bicarrs. i. schma de la solution (French. English summary) [Waring's problem for biquadrates. i. sketch of the solution]. *C.R. Academy of Science Paris Series of Mathematics I*, **303**(4) (1986), 85–88.

[53] S.S. Pillai. On Waring's problem. $g(6) = 73$. *Proceedings of the Indian Academy of Sciences*, 12A:30–40, 1940.

[54] M. Nathanson. *Additive Number Theory: Classical Bases*. Graduate Texts in Mathematics 164, Springer, New York, 1996.

[55] V.H. Vu. On a refinement of waring's problem. *Duke Mathematical Journal*, **105**(1) (2000), 107–134.

[56] M.B. Nathanson. Waring's problem for sets of density zero, in: *Analytic Number Theory, Lecture Notes in Mathematics 899*, Temple University, PA, pages 301–310, 1981.

[57] Paul Erdös. Extremal problems in number theory. *Proceedings of the Symosium in Pure Mathematics*, American Mathematical Society, **VIII** (1965), 181–189.

[58] S. Eberhard, B. Green, and F. Manners. Sets of integers with no large sum-free subsets. *arXiv:1301.4579v2 [math.CO]*, **31**, 2013.

[59] Noga Alon and D.J. Kleitman. Sum-free subsets. *Proceedings of a Tribute to Paul Erdös*, 13–26. Cambridge University Press, 1990.

[60] Jean Bourgain. Estimates related to sum-free subsets of sets of integers. *Israel Journal of Mathematics*, **97**(1) (1997), 71–92.

[61] Mark Lewko. An improved upper bound for the sum-free subset constant. *Journal of Integer Sequences*, **13**(8) (2010), Article 10:8:3.

[62] A.H. Rhemtulla and Ann Penfold Street. Maximum sum-free sets in elementary abelian p-groups. *Canadian Mathematical Bulletin*, **14** (1971), 73–80.

[63] Paul Erdös and Alfred Renyi. On the evolution of random graphs. *Publications of the Mathematical Institute of the Hungarian Academy of Sciences*, **5** (1960), 17–61.

[64] Edgar Gilbert. Random graphs. *Annals of Mathematical Statistics*, **30**(4) (1959), 1141–1144.

[65] Rick Durrett. *Random Graph Dynamics*. Cambridge University Press, United Kingdom, 2007.

[66] M. Krivelevich and B. Sudakov. The phase transition in random graphs: A simple proof. *Random Structures and Algorithms*, **43** (2013), 131–138.

[67] John Hopcroft and Robert E. Tarjan. Efficient planarity testing. *Journal of the ACM*, **21**(4) (1974), 549–568.

[68] B. Bollobas. The evolution of random graphs. *Transactions of the American Mathematical Society*, **286**(1) (1984), 257–274.

[69] R. van der Hofstad and Joel H. Spencer. Counting connected graphs asymptotically. *European Journal of Combinatorics*, **27**(8) (2006), 1294–1320.

[70] T. Luczak. Component behavior near the critical point of the random graph process. *Random Structures and Algorithms*, **1**(3) (1990), 287–310.

[71] S. Janson, D.E. Knuth, T. Luczak, and B. Pittel. The birth of the giant component. *Random Structures and Algorithms*, **4**(3) (1993), 233–358.

[72] E.M. Wright. The number of connected sparsely edged graphs. *Journal of Graph Theory*, **1**(4) (1977), 317–330.

[73] B. Bollobas and O. Riordan. A simple branching process approach to the phase transition in $G(n,p)$. *The Electronic Journal of Combinatorics*, **19**(4) (2012), #P21.

[74] C.P. Schnorr and C.R. Subramanian. Almost optimal (on the average) algorithms for boolean matrix product witnesses, computing diameter. *Proceedings of the 2nd International Workshop on Randomization and Approximation Techniques in Computer Scienc*, Barcelona, Spain, October 1998, LNCS 1518, 218–231. Springer-Verlag, Germany, 1998.

[75] B. Bollobas. The diameter of random graphs. *Transactions of the American Mathematical Society*, **267**(1) (1981), 41–52.

[76] V. Klee and D. Larman. Diameters of random graphs. *Canadian Journal of Mathematics*, **XXXIII**(3) (1981), 618–640.

[77] B. Bollobas. The evolution of sparse graphs. *Graph Theory and Combinatorics*, 35–57, Univeristy of Cambridge, 1984.

[78] F. Chung and L. Lu. The diameter of sparse random graphs. *Advances in Applied Mathematics*, **26** (2001), 257–279.

[79] T. Luczak. Random trees and random graphs. *Random Structures and Algorithms*, **13**(13) (1998), 485–500.

[80] Kunal Dutta and C.R. Subramanian. Induced acyclic tournaments in random digraphs: Sharp concentration, thresholds and algorithms. *Discussiones Mathematicae Graph Theory*, **34**(3) (2014), 467–495.

[81] C.R. Subramanian. Sharp concentration of independence number in random graphs. *Manuscript*, 2014.

[82] A.M. Frieze. On the independence number of random graphs. *Discrete Mathematics*, **81**(2) (1990), 171–175.

[83] Varsha Dani and Cristopher Moore. Independent sets in random graphs from the weighted second moment method. In *Proceedings of the 15th International Workshop RANDOM,* Princeton, NJ, LNCS Series No. 6845, 472–482. Springer-Verlag, Germany, 2011.

[84] G.R. Grimmett and C.J.H. McDiarmid. On colouring random graphs. *Mathematical Proceedings of the Cambridge Philosophical Society*, **77**(2) (1975), 313–324.

[85] D. Matula. Expose-and-merge exploration and the chromatic number of a random graph. *Combinatorica*, **7**(3) (1987), 275–284.

[86] B. Bollobas. The chromatic number of random graphs. *Combinatorica*, **8**(1) (1988), 49–55.

[87] E. Shamir and Joel H. Spencer. Sharp concentration of the chromatic number on random graphs $g_{n,p}$. *Combinatorica*, **7**(1) (1987), 121–129.

[88] M. Krivelevich. Topics in random graphs. *Lecture Notes, ETH Zurich*, 2010.

[89] T. Luczak. A note on the sharp concentration of the chromatic number of random graphs. *Combinatorica*, **11**(3) (1991), 295–297.

[90] N. Alon and M. Krivelevich. The concentration of the chromatic number of random graphs. *Combinatorica*, **17**(3) (1997), 303–313.

[91] T. Luczak. The chromatic number of random graphs. *Combinatorica*, **11**(1) (1991), 45–54.

[92] D. Achlioptas and A. Naor. The two possible values of the chromatic number of a random graph. *Annals of Mathematics*, **162**(3) (2005), 1335–1351.

[93] A. Coja-Oghlan and D. Vilenchik. Chasing the k-colorability threshold. *Proceedings of the 54th IEEE Symposium on Foundations of Computer Science*, 380–389, Berkeley, CA, 2013.

[94] Kunal Dutta and C.R. Subramanian. On induced paths, holes and trees in random graphs. *Manuscript*, 23, 2014.

[95] Paul Erdös and Z. Palka. Trees in random graphs. *Discrete Mathematics*, **46** (1983), 145–150.

[96] W. Fernandez and de la Vega. Induced trees in random graphs. *Graphs and Combinatorics*, **2**(1) (1986), 227–231.

[97] A.M. Frieze and B. Jackson. Large induced trees in sparse random graphs. *Journal of Combinatorial Theory, Series B*, **42** (1987), 181–195.

[98] L. Kucera and V. Rödl. Large trees in random graphs. *Commentationes Mathematicae Universitatis Carolinae*, **28** (1987), 7–14.

[99] T. Luczak and Z. Palka. Maximal induced trees in sparse random graphs. *Discrete Mathematics*, **72** (1988), 257–265.

[100] W. Fernandez and de la Vega. The largest induced tree in a sparse random graph. *Random Structures and Algorithms*, **9**(1–2) (1996), 93–97.

[101] Z. Palka and A. Rucinski. On the order of the largest induced tree in a random graph. *Discrete Applied Mathematics*, **15** (1986), 75–83.

[102] A.M. Frieze and B. Jackson. Large holes in sparse random graphs. *Combinatorica*, **7** (1987), 265–284.

[103] W.C. Suen. On large induced trees and long induced paths in sparse random graphs. *Journal of Combinatorial Theory, Series B*, **56**(2) (1992), 250–262.

[104] C.R. Subramanian. Finding induced acyclic subgraphs in random digraphs. *The Electronic Journal of Combinatorics*, **10** (2003), #R46.

[105] R.M. Karp. The transitive closure of a random digraph. *Random Structures and Algorithms*, **1**(1) (1990), 73–93.

[106] T.I. Fenner and A.M. Frieze. On the connectivity of random m-orientable graphs and digraphs. *Combinatorica*, **2** (1982), 347–359.

[107] B. Bollobas and P. Erdös. Cliques in random graphs. *Mathematical Proceedings of the Cambridge Philosophical Society*, **80**(3) (1976), 419–427.

[108] D. Matula. The largest clique size in a random graph. *Technical Report*, 1976.

[109] J.H. Spencer and C.R. Subramanian. On the size of induced acyclic subgraphs in random digraphs. *Discrete Mathematics and Theoretical Computer Science*, **10**(2) (2008), 47–54.

[110] Kunal Dutta and C.R. Subramanian. Induced acyclic subgraphs in random digraphs: Improved bounds. *DMTCS Proceedings of the 21st International Meeting on Probabilistic, Combinatorial and Asymptotic Methods for the Analysis of Algorithms*, 159–174, 2010.

[111] Kunal Dutta and C.R. Subramanian. On induced acyclic subgraphs in sparse random digraphs. *Electronic Notes in Discrete Mathematics, Proceedings of the 6th European Conference on Combinatorics, Graph Theory and Applications*, **38** (2011), 319–324.

[112] Kunal Dutta and C.R. Subramanian. Improved bounds on induced acyclic subgraphs in random digraphs. Accepted by *SIAM Journal of Discrete Mathematics*, **17**, 2014.

[113] I. Palasti. On the strong connectedness of directed random graphs. *Studia Scientiarum Mathematicarum Hungarica*, **1** (1966), 205–214.

[114] A.J. Graham and D.A. Pike. The critical behavior of random digraphs. *Atlantic Electronic Journal of Mathematics*, **3**(1) (2008), 1–5.

[115] T. Luczak. The phase transition in the evolution of random digraphs. *Journal of Graph Theory*, **14**(2) (1990), 217–223.

[116] T. Luczak and T.G. Seierstad. The critical behavior of random digraphs. *Random Structures and Algorithms*, **35**(3) (2009), 271–293.

[117] R.A. Moser and G. Tardos. A constructive proof of the general Lovász local lemma. *Journal of the ACM*, **57**(2) (2010), Article no. 11, 15 pp.

CHAPTER 34

Random Models and Analyses for Chemical Graphs

Daniel Pascua

Tina M. Kouri

Dinesh P. Mehta

CONTENTS

34.1	Introduction	997
34.2	Background	998
	34.2.1 Graph Representation of Molecules	998
	34.2.2 Graph Isomorphism and Canonical Labeling	999
	34.2.3 Random Graph Models	999
	34.2.4 Random Graph Isomorphism	1000
	34.2.5 Practical Canonical Naming Algorithms for Chemical Graphs	1000
34.3	Random Graph Models for Chemistry	1002
	34.3.1 Definitions	1002
	34.3.2 Random Pair Model	1003
	34.3.3 Molecule Build Model	1004
34.4	Bound on Failure Rate of Canonical Labeling Algorithms	1005
	34.4.1 Class of Hydrocarbons $\mathcal{C}$	1005
	34.4.2 Reasonable Canonical Labeling Algorithms	1005
34.5	Discussion	1008

THIS CHAPTER[*] describes a random model for chemical graphs that captures the notion of valence along with algorithms to generate chemical graphs using this model. The model is also used to provide theoretical bounds on the accuracy of a class of canonical labeling algorithms for a class of hydrocarbons.

34.1 INTRODUCTION

Biological processes involve reactions between molecules. While many computational biology problems are able to abstract out knowledge of chemicals (e.g., a DNA sequence may be viewed as a string with an alphabet of size four even though the underlying A, C, T, and

---

[*]This chapter is an edited version of [1].

G are chemicals), others such as pharmaceutical applications depend on the knowledge of chemical structure of the drug and its interaction with biomolecules. This chapter focuses on the fundamental entity of these chemical processes, the chemical graph, and proposes a randomized framework for working with chemical graphs.

Randomized techniques have been traditionally used to enhance *simplicity* and *speed*. Motwani and Raghavan [2] state that "For many applications, a randomized algorithm is the simplest algorithm available, or the fastest, or both." In a well-designed randomized algorithm, the price for the gains in simplicity and speed is that there is a (very small) probability that the algorithm is incorrect or inefficient. This is considered to be a reasonable trade-off for many applications. Further, analyses of randomized algorithms are closely related to probabilistic analyses of related deterministic algorithms. Probabilistic analyses have been used to explain the phenomenon that provably difficult problems (e.g., NP-hard problems), while intractable for pathological inputs, usually admit fast solutions on most real-world inputs.

Although there is a significant body of work in random graphs, existing models in the random graph literature such as $G_{n,p}$, $G_{[n,N]}$ [3], and scale-free networks [4] do not satisfactorily model chemical graphs. Our first contribution (Section 34.3) is to define a random model to model chemical graphs. More precisely, we describe two intuitive graph models (random pair and molecule build) that respect atom valences and show that these two models are mathematically equivalent.

Our focus in this chapter is on a fundamental problem in cheminformatics; that of computing a *canonical label* or name for a molecule [5]. The objective is to uniquely and efficiently represent a chemical graph using a linear text string, which can then be conveniently used to determine whether two molecules are the same or to find a molecule in a database of molecules. If a unique canonical label can be computed efficiently in general, this would represent a significant theoretical breakthrough because it would imply that an efficient (polynomial time) solution exists for graph isomorphism (an open problem for which no polynomial time algorithm has yet been found). The literature indicates that graph isomorphism/canonical labeling can be solved in polynomial time for the special case of *chemical graphs* [6], but no practical software that takes polynomial time has been implemented [7]. In practice, however, the cheminformatics community uses several naming algorithms in software systems and databases [7–13]. These algorithms usually fall in one of two categories: Type 1—the naming algorithm is efficient, but names are not always unique (i.e., two different molecules can have the same name) or Type 2—the naming algorithm is inefficient for pathological inputs but results in unique names. In both scenarios, the likelihood of failure (non-unique name in the former and an inefficient algorithm in the latter) has been empirically found to be low. We are able to show through a probabilistic analysis based on our random model that reasonable Type 1 canonical naming algorithms generate non-unique names with exponentially small probabilities on a class of hydrocarbons. We begin with a review of the technical background for this chapter in the next section.

34.2 BACKGROUND

34.2.1 Graph Representation of Molecules

A chemical graph may be used to represent a molecule where the vertices and edges represent atoms and bonds, respectively. Definition 34.1 provides a formal definition of a chemical graph.

In Figure 34.1, the H_2O molecule has two bonds. Each bond is formed by joining the oxygen atom with a hydrogen atom. A chemical graph has bounded degree or valence since each atom has a limited number of valence electrons, which is a constant value for each atom.

Figure 34.1 H_2O molecule.

Definition 34.1 (chemical graph) *A chemical graph is a graph $G(V,E)$, where each vertex v has an associated label $l(v) \in [a_1, \ldots, a_t]$. Each label denotes a chemical atom (e.g., C, O, H). Each edge in E corresponds to a chemical bond* [14].

A *chemical multigraph* may have more than one edge between vertices to represent multiple bonds between atoms (e.g., atoms may be joined by a double or triple bond). In this chapter, we focus on chemical multigraphs (and use the term *chemical graphs* to mean chemical multigraphs).

34.2.2 Graph Isomorphism and Canonical Labeling

No efficient algorithm has been found to determine if two general graphs are isomorphic [15]. There are special cases for which graph isomorphism can be solved in polynomial time including triconnected planar graphs [16], planar graphs [17], interval graphs [18], and trees [19]. Chemical graphs may not meet these special cases and therefore we cannot use these algorithms.

The problem of finding a canonical name for a graph is closely related to the graph isomorphism problem.

Definition 34.2 (canonical name) *A canonical naming is a function over all graphs mapping a graph G to a canonical name $CN(G)$ such that for any two graphs, G_1 and G_2, G_1 is isomorphic to G_2 if and only if $CN(G_1) = CN(G_2)$.*

If canonical names can be found for two graphs, the graphs can easily be checked for isomorphism by comparing their canonical names [5]. The problem of determining whether two graphs are isomorphic can be performed at least as fast as the problem of finding a canonical name for a graph. Algebraic methods for testing for graph isomorphism involve determining the automorphic groups of vertices. Babai and Luks [5] bridged the gap between the canonical naming problem and the graph isomorphism problem by showing that knowledge of automorphic groups of vertices can lead to a canonical name of a graph.

Luks provide a theoretical polynomial time solution for graphs of bounded valence [6], but does not describe an algorithm for determining graph isomorphism. Fürer et al. provide a polynomial time solution for graphs of bounded valence with a worst-case time complexity of $O\left(n^{\tau(d)}\right)$, for a suitable integer $\tau(d)$, where d is the valence of the graph [20]. Although these polynomial time solutions exist for graphs of bounded valence, no practical, polynomial time algorithm has yet been implemented [7].

34.2.3 Random Graph Models

A random graph model is used to generate a graph using a random process. Two of the most well-known random graph models are the $G_{n,p}$ and $G_{[n,N]}$ models. Random graph models provide a probabilistic setting for studying a variety of graph problems [3].

Definition 34.3 *In the $G_{n,p}$ random graph model, start with an undirected graph with n vertices and no edges. Consider each of the $\binom{n}{2}$ possible edges and add it to the undirected graph with probability p* [3].

Definition 34.4 *In the $G_{[n,N]}$ random graph model, we consider all undirected graphs on n vertices with exactly N edges. Each graph with exactly N edges is equally likely to be chosen* [3].

Scale-free networks are characterized by their degree distributions $P(k)$, which describes the proportion of nodes in a graph with degree k.

Definition 34.5 *In a scale-free network, $P(k) = ck^{-\lambda}$, where λ is a constant generally between 2 and 3 and c is the normalization constant for the distribution.*

Barabasi and Albert have shown that many real-life networks can be modeled by scale-free networks [4]. Some examples include internet graphs, scientific collaboration graphs, social networking graphs, and paper citation graphs. A key property of scale-free networks is that the majority of nodes have a low degree; however, unlike $G_{n,p}$, the probability of a node with a large degree does not decrease exponentially, and as such, there are likely to be a few nodes (called hubs) that have a large degree.

Chemical graphs have no nodes with degree greater than a constant (typically 4), making all of the models above (all of which permit nodes with high degrees) unsuitable for modeling chemical graphs. Note that these models also preclude the generation of double and triple bonds.

34.2.4 Random Graph Isomorphism

Simple polynomial time graph canonical labeling algorithms have been developed for determining isomorphism on random graphs. These algorithms have been proven to produce a unique canonical label for most graphs generated under the $G_{n,p}$ random graph model. The algorithms are designed with failure criteria to ensure that only unique canonical labels are generated.

Babai et al. present a simple isomorphism testing algorithm based on vertex degree distributions [21]. They use degree distributions to determine a canonical name for the input graph. The worst-case complexity of their algorithm is quadratic in the number of vertices. The algorithm fails if the largest vertex degrees are not unique. This approach cannot be used for chemical graphs because many vertices will have the same largest degree.

Babai and Kucera present an algorithm (which the authors state is not meant to be practicable) for random graphs that are selected from the uniform distribution over all $2^{\binom{n}{2}}$ graphs on a set of n labeled vertices [22]. (We note that this is a special case of the $G_{n,p}$ model with $p = 0.5$.) The algorithm, which is also based on classifying vertices based on their degree distributions, improves upon the result of Babai et al. [21] by reducing the probability of failure. Similarly, Czajka and Pandurangan present a linear-time algorithm for the canonical labeling of a random graph, which is invariant under isomorphism [23]. The basic idea of their algorithm is to distinguish the vertices of a graph using the degree of the neighbors. The algorithm fails if the degree neighborhoods (DNs) are not distinct. Many vertices in chemical graphs can have identical DNs making this approach inapplicable as well.

34.2.5 Practical Canonical Naming Algorithms for Chemical Graphs

Several methods for labeling molecules have been developed in the literature for use in cheminformatics systems. These methods may also be used to solve the chemical graph isomorphism problem. Some graph isomorphism algorithms are designed for simple chemical graphs, but can be modified to support chemical multigraphs, either directly or by using Faulon's algorithm to convert a chemical multigraph into a simple chemical graph in polynomial time [24].

Morgan's algorithm: One of the first canonical labeling algorithms for chemical graphs was proposed by H.L. Morgan [8]. The algorithm is based on node connectivity and the creation of unambiguous strings, which describe a molecule. Morgan's algorithm may fail on highly regular graphs since it results in oscillatory behavior [25].

Nauty: One of the most well-known and fastest algorithms for determining chemical graph isomorphism is Nauty [26], which is based on finding the automorphism groups of a graph [9]. The worst-case complexity of Nauty was analyzed and found to be exponential [27], but in practice Nauty is much faster.

Bliss: The authors of Bliss improve the Nauty algorithm using an advanced data structure and incremental computations. The worst-case complexity of Bliss remains exponential, but the authors have shown that Bliss performs better than Nauty on benchmark tests [10].

Signature: Another well-known canonical naming algorithm for chemical graph isomorphism is Signature [7], which finds a canonical name using extended valence sequences. The authors state that the algorithm has exponential worst-case complexity, but in practice it appears to run much faster.

SMILES: SMILES is a chemical notation system commonly used to describe the structure of a molecule [11–13]. However, the SMILES string used to describe a molecule is not necessarily unique [28].

DN: DN is a simple labeling algorithm that does not guarantee uniqueness, that is, two non-isomorphic molecules may have the same DN name (i.e., there may be a Type 1 error) [29,30]. DN assigns each atom a name based on its label and degree and the label and degree of each of its neighbors. The names of each atom are then used to assign a name to the molecule. For example, consider the CH_3O molecule in Figure 34.2a. We first label each atom, using its symbol and degree (Figure 34.2b). We then add to each atom's name the symbol and degree of its neighbors lexicographically (Figure 34.2c). Now that each atom is named, we lexicographically sort the atom names to create the name for the molecule. The resulting name is therefore:

$$[[C4][H1H1H1O1]][[H1][C4]][[H1][C4]][[H1][C4]][[O1][C4]]$$

Algorithm 34.1: Degree neighborhood canonical labeling

Input: Molecule (M) as an adjacency matrix and list of n atoms
Output: Canonical Name (C)

1 **for** $i = 1$ to n **do**
2 $\quad$ $labels[i] = $ LabelVertexWithSymbolAndDegree(i);
3 **end**
4 **for** $i = 1$ to n **do**
5 $\quad$ $dLabels[i] = $ CreateArrayOfConnectedLabels(i);
6 $\quad$ Sort($degreeLabels[i]$)
7 $\quad$ $canonicalVertexLabels[i] = $ "[["$+labels[i]+$"][" $+$ ArrayToString($dLabels[i]$) $+$ "]]"
8 **end**
9 Sort($canonicalVertexLabels$)
10 $C = $ ArrayToString($canonicalVertexLabels$)
11 Return C

Although the DN algorithm is not 100% accurate, it is a very practical algorithm that may be used as a preliminary check when determining if two molecules are isomorphic. DN was used to speed up automated reaction mapping (an important tool in bioinformatics and

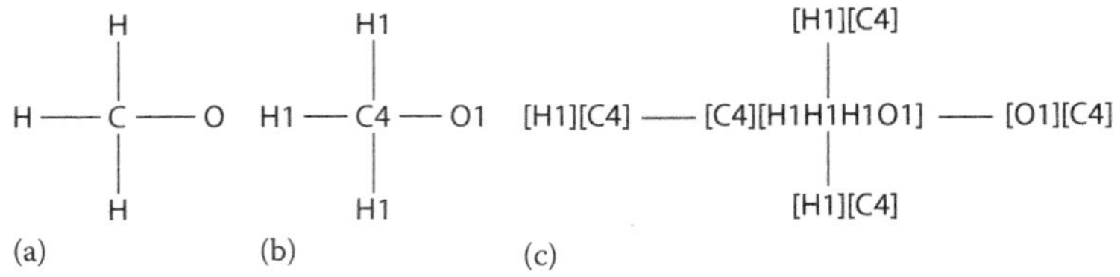

Figure 34.2 (a–c) Degree neighborhood canonical labeling.

TABLE 34.1 Automated Reaction Mapping Type 1 Error Probability

Mechanism	Type 1 Error Probability
CSM	0.14%
LLNL	0.02%
KEGG/LIGAND	0.02%

cheminformatics) computations [29–31]. The automated reaction mapping algorithms, which utilize the DN were tested on a variety of mechanisms, including a Colorado School of Mines (CSM) mechanism, a Lawrence Livermore National Laboratory (LLNL) [32] mechanism, and the KEGG/LIGAND v57 database [33]. In practice we found that the DN algorithm was able to correctly distinguish isomorphic and non-isomorphic molecules over 98% of the time. Table 34.1 summarizes the results and shows the probability of Type 1 error for each mechanism tested.

34.3 RANDOM GRAPH MODELS FOR CHEMISTRY

In this section, we introduce two random graph models as a means to model chemical graphs. We then show that these two random graph models are equivalent and are guaranteed to produce chemically valid graphs (defined later). Note that these models do not necessarily capture the chemistry perfectly, but do so better than the previous random graph models described earlier.

34.3.1 Definitions

Definition 34.6 (atom type) *An atom type refers to each element type (e.g., hydrogen, carbon, and oxygen are different atom types).*

Definition 34.7 (half-edge) *A half-edge is a potential bond for an atom. The number of half-edges an atom has is determined by its number of valence electrons (e.g., hydrogen has one half-edge, oxygen has two half-edges, and carbon has four half-edges). A bond joins two half-edges to create an edge in the resulting graph.*

Definition 34.8 (chemically valid graph) *In a chemically valid graph no atom has more edges than its number of available half-edges.*

Definition 34.9 (equally-likely bond assumption) *Given a collection of atoms and their half-edges, the equally-likely bond assumption is that the probability that any pair of half-edges forms a bond is equal. If there are $2e$ half-edges, this means that the probablility that a given half-edge forms a bond with one of the remaining $2e - 1$ half-edges is $1/(2e - 1)$.*

Definition 34.10 (molecule M's probability) *A molecule M's probability, $Pr(M)$, is the probability that the half-edges in a collection of atoms combine to form molecule M under the equally-likely bond assumption.*

34.3.2 Random Pair Model

The random pair model starts with the set of all available half-edges and repeatedly chooses (and removes from the set) two half-edges at random until there are no half-edges left in the set. The pair of half-edges removed at each stage are joined to form an edge. Clearly the random pair model satisfies the equally-likely bond assumption and may be implemented with a worst-case running time of $O(e)$, where e is the number of edges in the graph. Note that the resulting graph may not be connected. It is easy to see that the random pair model creates a chemically valid graph. Note that it may be possible for an atom to have fewer than its maximum number of bonds since its half-edges may bond with each other.

An alternative (equivalent) implementation of the random pair model is obtained by generating a uniform random permutation of the list of half-edges and pairing half-edges from left to right to form edges.

We present pseudocode for two algorithms, which may be used to implement the random pair model: `PairHalfEdges` (Algorithm 34.2) and `PermuteHalfEdges` (Algorithm 34.3).

Algorithm 34.2: PairHalfEdges

Input: the set of atoms, A
Output: chemical graph of a given atoms, G

1. $availableHalfEdges =$ GetHalfEdges(A)
2. $e =$ NumEdges(A) /* The number of edges in resulting graph */
3. **for** $i = 1$ to e **do**
4. $index1 =$ Random($availableHalfEdges$)
5.   Remove($availableHalfEdges, index1$)
6. $index2 =$ Random($availableHalfEdges$)
7.   Remove($availableHalfEdges, index2$)
8.   AddEdge($G, index1, index2$)
9. **end**
10. Return G

Algorithm 34.3: PermuteHalfEdges

Input: the set of atoms, A
Output: chemical graph of a given atoms, G

1. $availableHalfEdges =$ GetHalfEdges(A)
2. RandomPermute($availableHalfEdges$)
3. $e =$ NumEdges(A) /* The number of edges in resulting graph */
4. **for** $i = 1$ to e **do**
5. $index1 =$ NextAvailable($availableHalfEdges$)
6. $index2 =$ NextAvailable($availableHalfEdges$)
7.   AddEdge($G, index1, index2$)
8. **end**
9. Return G

34.3.3 Molecule Build Model

The molecule build model starts with an arbitrarily chosen atom.* Each half-edge of this atom is then connected to a randomly chosen half-edge from the list of available half-edges. When all half-edges associated with the current atom are used up, another atom with available half-edges from the connected molecule built so far is arbitrarily chosen and the process repeated. If the molecule has no remaining half-edges and there are atoms remaining with unpaired half-edges, a new molecule is started.

It is easy to see that the molecule build model creates a chemically valid graph. (The maximum number of half-edges available for each atom is its maximum number of connections. It is not possible for an atom to exceed its maximum number of connections.) Note again that it may be possible for an atom to have fewer than its maximum number of bonds since it may bond with itself.

We present pseudocode for the molecule build model in Algorithm 34.4.

Algorithm 34.4: MoleculeBuild

Input: the set of atoms, A
Output: chemical graph of a given atoms, G

1 $availableHalfEdges=$ GetHalfEdges(A)
2 $index1 =$ GetArbHalfEdge($availableHalfEdges$)
3 $e =$ NumEdges(A) /* The number of edges in resulting graph */
4 **for** $i = 1$ to e **do**
5     $index2 =$ Random($availableHalfEdges$)
6   AddEdge($G, index1, index2$)
7   **if** AllEdgesUsedInConnectedMolecule($index1$)
8         $index1 =$ GetArbHalfEdge($availableHalfEdges$)
9   **else**
10        $index1 =$ GetArbHalfEdgeFromConnectedMolecule($availableHalfEdges$)
11   **end**
12 **end**
13 **Return** G

Theorem 34.1 *Algorithm MoleculeBuild satisfies the equally-likely bond assumption.*

Proof. The behavior of the molecule build algorithm can be modeled as a sequence of e rounds that act on a set of $2e$ objects. In each round, one object is chosen arbitrarily while the other is chosen uniformly at random from the remaining objects. We now show by induction that P_e, the probability that a pair of objects A and B are chosen in the same round in one of the e rounds, is $1/(2e-1)$.

Basis: When $e = 1$, the two objects A and B are guaranteed to be chosen; so $P_1 = 1 = 1/[2(1) - 1]$.

We next consider two cases when $e > 1$.

Case 1. A is arbitrarily chosen in Round 1. The probability that B is chosen in the random phase of Round 1 is the probability of choosing one object uniformly at random from among $2e - 1$ objects and is given by $1/(2e - 1)$. (The case where B was arbitrarily chosen in Round 1 is identical.)

---

*Note that an arbitrary choice is different from a uniform random choice in that the former can be deterministic or biased.

Case 2. Neither A nor B was chosen in the arbitrary phase of Round 1. The probability that A and B are chosen in the same round is the product of (1) the probability that neither A nor B are chosen in the random phase of Round 1 multiplied by (2) the probability that A and B are in the same round in one of the remaining $e-1$ rounds. The first probability is given by $(2e-3)/(2e-1)$. The second probability is P_{e-1} which, by induction, is $1/(2e-3)$. Multiplying gives $1/(2e-1)$, proving the result. ∎

Corollary 34.1 *The molecule build model generates a molecule with the same probability as the random pair model.*

Proof. We need to only show that any two arbitrary half-edges have the same probability of being joined as any other half-edges in both models. This follows from Theorem 34.1. ∎

Note that Theorem 34.1 implies that the random pair model and the molecule build model are equivalent. Although all three algorithms described above ultimately implement equivalent models, it is more practical to use MoleculeBuild than the two RandomPair algorithms outlined above because it is easier to manage data structures when joining half-edges.

34.4 BOUND ON FAILURE RATE OF CANONICAL LABELING ALGORITHMS

In this section, we provide a theoretical bound on the failure rate of a class of canonical labeling algorithms (such as the DN naming algorithm we have described previously) on a class of hydrocarbon molecules under the RandomPair model. In our analysis, we consider two molecules. Molecule M_1 belongs to the class of hydrocarbons $\mathcal{C}$ defined below. Molecule M_2 is a chemical graph generated by our random graph model *with the same set of atoms as M_1*. We will show that the probability that the canonical names are the same ($CN(M_1) = CN(M_2)$) decreases exponentially with size.

34.4.1 Class of Hydrocarbons $\mathcal{C}$

Connected hydrocarbons have two properties:

1. Molecules are of the form $C_nH_{\Theta(n)}$, $n > 0$.
2. Molecules are of the form C_nH_{2n-2i}, where $-1 \leq i < n$.

The first property simply requires that the number of H atoms grows linearly with the number of C atoms. The second property includes a number of familiar hydrocarbons. For example, when $i = -1$, we obtain molecules of the form C_nH_{2n+2}, which is the class of acyclic alkanes (or saturated hydrocarbons). This includes linear chains (e.g., methane, ethane, propane, etc.) and branched chains. When $i = 0$, we have the general formula C_nH_{2n} for alkenes (unsaturated hydrocarbons containing one double-bond), which includes ethylene and propylene *or* cycloalkanes (saturated hydrocarbons with a cycle). When $i > 0$, we further include hydrocarbons with one or more triple bonds or two or more double bonds. *Although our proof below is restricted to the class of hydrocarbons described here to simplify the presentation, the underlying mathematics can be extended to other classes of organic compounds.*

34.4.2 Reasonable Canonical Labeling Algorithms

We define a *reasonable* canonical labeling algorithm to be one whose (1) runtime is at least linear in the size of the chemical graph (i.e., $\Omega(|G|)$) and (2) one that generates a canonical name such that it is possible to determine whether the chemical graph is connected in linear time with respect to the length of the name. Our justification for condition (1) is that any reasonable naming algorithm must visit each vertex and edge in the chemical graph at least once. Whether condition (2) is satisfied depends on details of the canonical naming algorithm. However, we argue that condition (2) can be trivially satisfied by modifying the naming algorithm so that it adds a prefix of 1 to the name if the chemical graph is connected and 0 otherwise. Such a connectivity check can be accomplished in $\Theta(|G|)$ time and clearly does not violate condition (1). Based on these conditions, we make the following observations.

Observation 34.1 *$CN(M_1)$ always begins with a 1, while $CN(M_2)$ begins with a 1 if it is connected and with a 0 if it is not.*

Observation 34.2 *The probability of failure ($M_1 \neq M_2$, but $CN(M_1) = CN(M_2)$) is bounded by the probability that M_2 is connected.*

The failure rate of a randomized algorithm is typically bounded by an expression of probability of the form $1/f(n)$, where n is the size of the input and $f(n)$ is a function that can take many forms; that is, $f(n)$ could be a constant greater than 1 (e.g., 1000), a polynomial in n (e.g., n^2), or exponential in n (e.g., e^n). Clearly, the larger (asymptotically faster growing) $f(n)$ is, the better randomized algorithm. In the following we show that the failure probability does indeed decrease exponentially with respect to n; that is, there is a constant $c > 1$ such that the probability of failure is bounded by c^{-n}. We make the following additional observations.

Observation 34.3 *If there are one or more H–H bonds in M_2, then M_2 is disconnected. However, the converse is not true (i.e., a disconnected graph does not imply that there are one or more H–H bonds in the chemical graph).*

Observation 34.4 *$Pr\ (M_2\ has > 0\ H\text{–}H\ bonds) \leq Pr\ (M_2\ is\ disconnected)$ or equivalently $Pr\ (M_2\ is\ connected) \leq Pr\ (M_2\ has\ 0\ H\text{–}H\ bonds)$.*

Theorem 34.2 *Under the random-pair model, the failure rate of a reasonable canonical labeling algorithm on M_2 whose composition is of the form in $\mathcal{C}$ decreases exponentially in n.*

Proof. We have already shown that the failure rate is bounded by the probability of no H–H bonds.

Recall that we are considering hydrocarbons of the form C_nH_{2n-2i}. The probability of 0 H–H bonds is the probability that each hydrogen half-edge bonds with a carbon half-edge. The first hydrogen atom has $4n$ carbon half-edges to choose from out of $6n - 2i - 1$ half-edges, the second has $4n - 1$ out of $6n - 2i - 3$, and so on, until the last hydrogen atom has $4n - (2n - 2i) + 1$ carbon half-edges to choose from out of $(6n - 2i) - 2 * (2n - 2i) + 1$ total half-edges left, which is $2n + 2i + 1$ out of $2n + 2i + 1$. The probability of all of these occurring can be written as a product of the individual probabilities, so the probability of no H–H bonds is

$$\frac{4n}{6n - 2i - 1} \frac{4n - 1}{6n - 2i - 3} \cdots \frac{2n + 2i + 1}{2n + 2i + 1}$$
$$= \frac{(4n)!/(2n + 2i)!}{(6n - 2i - 1)!!/(2n + 2i - 1)!!}$$
$$= \frac{(4n)!(2n + 2i - 1)!!}{(2n + 2i!)(6n - 2i - 1)!!}$$

where the double factorial of x for odd x, written $x!!$, is the product over all positive odd integers less than or equal to x. The double factorials can be changed to single factorials, as $x!! = x!/\left\{((x-1)/2)!2^{(x-1)-2}\right\}$. Using this fact gives the probability of no H–H bonds as

$$\frac{(4n)!}{(2n+2i)!}\frac{(2n+2i-1)!}{2^{n+i-1}(n+i-1)!}\frac{2^{3n-i-1}(3n-i-1)!}{(6n-2i-1)!}$$

Some terms can be canceled:

$$= \frac{2^{2n-2i}}{2n+2i}\frac{(4n)!(3n-i-1)!}{(n+i-1)!(6n-2i-1)!}$$

and, multiplying by $(2n+2i)/2^*(n+i)$:

$$= 2^{2n-2i-1}\frac{(4n)!(3n-i-1)!}{(n+i)!(6n-2i-1)!}$$

Now, using Stirling's inequality $\sqrt{2\pi}n^{n+(1/2)}e^{-n} \leq n! \leq e*n^{n+(1/2)}e^{-n}$ to provide an upper bound for the terms in the numerator and a lower bound for the terms in the denominator, an upper bound for the desired probability is:

$$\leq \frac{2^{2n-2i-1}e^2}{2\pi}\frac{e^{-7n+i+1}}{e^{-7n+i+1}}\frac{(4n)^{4n+(1/2)}(3n-i-1)^{3n-i-(1/2)}}{(n+i)^{n+i+(1/2)}(6n-2i-1)^{6n-2i-(1/2)}}$$

$$= \frac{2^{2n-2i-2}e^2}{\pi}4^{4n+(1/2)}n^{4n+(1/2)}\frac{(3n-i-1)^{3n-i-(1/2)}}{(n+i)^{n+i+(1/2)}(6n-2i-1)^{6n-2i-(1/2)}}$$

Rearranging terms and combining the powers of 2:

$$= \frac{2^{10n-2i-1}e^2n^{4n+(1/2)}}{\pi}\left(\frac{(3n-i-1)}{(n+i)(6n-2i-1)}\right)^{n+i+(1/2)}\left(\frac{(3n-i-1)}{(6n-2i-1)}\right)^{2n-2i-1}\left(\frac{1}{(6n-2i-1)}\right)^{3n-i}$$

Upper bounding by decreasing the denominator:

$$\leq \frac{2^{10n-2i-1}e^2n^{4n+(1/2)}}{\pi}\left(\frac{(3n-i-1)}{(n+i)(6n-2i-2)}\right)^{n+i+(1/2)}\left(\frac{(3n-i-1)}{(6n-2i-2)}\right)^{2n-2i-1}\left(\frac{1}{(6n-2i-1)}\right)^{3n-i}$$

$$= \frac{2^{10n-2i-1}e^2n^{4n+(1/2)}}{\pi}\left(\frac{1}{2*(n+i)}\right)^{n+i+(1/2)}\left(\frac{1}{2}\right)^{2n-2i-1}\left(\frac{1}{(6n-2i)*\{1-(1/(6n-2i))\}}\right)^{3n-i}$$

$$= \frac{2^{7n-i-(1/2)}e^2n^{4n+(1/2)}}{\pi}\left(\frac{1}{n+i}\right)^{n+i+(1/2)}\left(\frac{1}{(6n-2i)*\{1-(1/(6n-2i))\}}\right)^{3n-i}$$

The term

$$\frac{1}{1-(1/(6n-2i))} = \frac{6n-2i}{6n-2i-1} = 1 + \frac{1}{6n-2i-1}$$

Using $1+x \leq e^x$ for all x, $1+(1/(6n-2i-1)) \leq (6n-2i-1)$:

$$\leq \frac{2^{7n-i-(1/2)}e^2n^{4n+(1/2)}}{\pi}\left(\frac{1}{n+i}\right)^{n+i+(1/2)}\left(\frac{1}{2*(3n-i)}\right)^{3n-i}e^{(3n-i)/(6n-2i-1)}$$

$e^{(3n-i)/(6n-2i-1)}$ is roughly $e^{1/2}$, but can be bounded by the constant e.

$$\leq \frac{2^{4n-(1/2)}e^2n^{4n+(1/2)}}{\pi}(n+i)^{-(n+i+(1/2))}(3n-i)^{-(3n-i)}e$$

$$= \frac{2^{4n}e^3n^{4n+(1/2)}}{\pi\sqrt{2}}(n+i)^{-(n+i+(1/2))}(3n-i)^{-(3n-i)}$$

Let us assume that the number of hydrogen atoms grows linearly as $k*n$ ($0 \leq k \leq 2$), where n is the number of carbon atoms. This means that $2n - 2i = kn$ and $i = (1 - (k/2))*n$. Then, the bound on the probability of no H–H bonds becomes

$$\frac{2^{4n}e^3 n^{4n+(1/2)}}{\sqrt{2\pi}} ((2-k/2)n)^{-((2-k/2)n+(1/2))} ((2+k/2)n)^{-((2+k/2)n)}$$

$$= \frac{e^3}{\sqrt{2*(2-k/2)\pi}} \left\{ \frac{2^4}{(2-(k/2))^{2-(k/2)}(2+(k/2))^{2+(k/2)}} \right\}^n$$

For $k > 0$, define

$$\frac{1}{c} = \frac{2^4}{(2-(k/2))^{2-(k/2)}(2+(k/2))^{2+(k/2)}} < 1$$

which means that as long as the number of hydrogen atoms grows as the molecule increases in size, then the probability of having 0 H–H bonds will decrease exponentially. ∎

34.5 DISCUSSION

We have described a model for the random generation of graphs that are significantly closer to chemical graphs than previous random graph models. Three practical algorithms for generating molecules based on these models are provided, giving the programmer ample flexibility based on the nature of the underlying data structures used in a system. We use this model to prove a theoretical result related to graph isomorphism similar in spirit to results by Babai et al. [21] and Czajka and Pandurangan [23].

We also contrast this work with that of Goldberg and Jerrum [34], which presents a theoretical (i.e., no implementation is provided and the algorithm appears to be too complicated to implement) polynomial time algorithm that generates and selects a connected isomer from a set of isomers with equal probability. In contrast, our approach does not aim to generate an isomer with equal probability. Instead, we present a simple model that is easy to implement and to use or to perform probabilistic analyses. Our approach may be seen as similar (but chemically more relevant than) the $G_{n,p}$ model, which is used for probabilistic analysis but does not generate all graphs with equal probability.

References

[1] T.M. Kouri, D. Pascua, and D.P. Mehta. Random models and analyses for chemical graphs. *Int. J. Found. Comput. Sci.*, **26** (2015), 269–291.

[2] R. Motwani and P. Raghavan. *Randomized Algorithms*. Cambridge University Press, New York, 1995.

[3] M. Mitzenmacher and E. Upfal. *Probability and Computing: Randomized Algorithms and Probabilistic Analysis*. Cambridge University Press, New York, 2005.

[4] A.-L. Barabasi and R. Albert. Emergence of scaling in random networks. *Science*, **286**(5439) (1999), 509–512.

[5] L. Babai and E. Luks. Canonical labeling of graphs. In *Proceedings of the 15th Annual ACM Symposium on Theory of Computing*, New York, 1983.

[6] E.M. Luks. Isomorphism of graphs of bounded valence can be tested in polynomial time. *J. Comput. Syst. Sci.*, **25**(1) (1982), 42–65.

[7] J.-L. Faulon, M. Collins, and R. Carr. The signature molecular descriptor. 4. Canonizing molecules using extended valence sequences. *J. Chem. Inf. Model.*, **44**(2) (2004), 427–436.

[8] H.L. Morgan. The generation of a unique machine description for chemical structures—A technique developed at chemical abstracts service. *J. Chem. Doc.*, **5**(2) (1965), 107–113.

[9] B. McKay. Practical graph isomorphism. *Congr. Numer.*, **30** (1981), 45–87.

[10] T. Junttila and P. Kaski. Engineering an efficient canonical labeling tool for large and sparse graphs. In *Proceedings of the 9th Workshop on Algorithm Engineering and Experiments and the 4th Workshop on Analytic Algorithms and Combinatorics*, D. Applegate, G.S. Brodal, D. Panario, and R. Sedgewick, editors, SIAM, New Orleans, LA, pp. 135–149, 2007.

[11] D. Weininger. Smiles—A chemical language and information system. 1. Introduction to methodology and encoding rules. *J. Chem. Inf. Comp. Sci.*, **28**(1) (1988), 31–36.

[12] D. Weininger, A. Weininger, and J.L. Weininger. Smiles. 2. Algorithm for generation of unique smiles notation. *J. Chem. Inf. Comp. Sci.*, **29**(2) (1989), 97–101.

[13] D. Weininger. Smiles. 3. DEPICT—Graphical depiction of chemical structures. *J. Chem. Inf. Comp. Sci.*, **30**(3) (1990), 237–243.

[14] J. Crabtree and D. Mehta. Automated reaction mapping. *J. Exp. Algorithmics*, **13** (2009), 15:1.15–15:1.29.

[15] S. Pemmaraju and S. Skiena. *Computational Discrete Mathematics: Combinatorics and Graph Theory with Mathematica*. Cambridge University Press, New York, 2003.

[16] J.E. Hopcroft and R.E. Tarjan. A $V \log V$ algorithm for isomorphism of triconnected planar graphs. *J. Comput. Syst. Sci.*, **7** (1973), 323–331.

[17] J.E. Hopcroft and J.K. Wang. Linear time algorithm for isomorphism of planar graphs (preliminary report). In *Proceedings of the 6th Annual ACM Symposium on Theory of Computing*, New York, pp. 172–184, 1974.

[18] M.R. Garey and D.S. Johnson. *Computers and Intractability: A Guide to the Theory of NP-Completeness*. W.H. Freeman & Co., New York, 1990.

[19] A.V. Aho, J.E. Hopcroft, and J.D. Ullman. *The Design and Analysis of Computer Algorithms*. Addison-Wesley, Reading, MA, 1974.

[20] M. Fürer, W. Schnyder, and E. Specker. Normal forms for trivalent graphs and graphs of bounded valence. In *Proceedings of the 15th Annual ACM Symposium on Theory of Computing*, New York, pp. 161–170, 1983.

[21] L. Babai, P. Erdös, and S. Selkow. Random graph isomorphism. *SIAM J. Comput.*, **9**(3) (1980), 628–635.

[22] L. Babai and L. Kucera. Canonical labelling of graphs in linear average time. In *Proceedings of the 20th Annual Symposium on Foundations of Computer Science*, Washington, DC, pp. 39–46, 1979.

[23] T. Czajka and G. Pandurangan. Improved random graph isomorphism. *J. Discrete Algorithms*, **6**(1) (2008), 85–92.

[24] J.-L. Faulon. Isomorphism, automorphism partitioning, and canonical labeling can be solved in polynomial-time for molecular graphs. *J. Chem. Inf. Comput. Sci.*, **38**(3) (1998), 432–444.

[25] J. Gasteiger and T. Engel, editors. *Chemoinformatics: A Textbook*. Wiley, New York, 2003.

[26] B. McKay. No automorphisms, yes? http://cs.anu.edu.au/ bdm/nauty/.

[27] T. Miyazaki. The complexity of Mckay's canonical labeling algorithm. *Groups and Computation II, DIMACS Ser. Discret. M.*, **28** (1997), 239–256.

[28] G. Neglur, R.L. Grossman, and B. Liu. Assigning unique keys to chemical compounds for data integration: Some interesting counter examples. In *Data Integration in the Life Sciences, Lecture Notes in Computer Science 3615*, B. Ludascher and L. Raschid, editors, Springer, Berlin, Germany, pp. 145–157, 2005.

[29] T. Kouri and D. Mehta. Improved automated reaction mapping. In *Experimental Algorithms, Lecture Notes in Computer Science 6630*, P.M. Pardalos and S. Rebennack, editors, Springer, Berlin, Germany, pp. 157–168, 2011.

[30] T.M. Kouri and D.P. Mehta. Faster reaction mapping through improved naming techniques. *J. Exp. Algorithmics*, **18** (2013), 2.5:2.1–2.5:2.32.

[31] T.M. Kouri, M. Awale, J.K. Slyby, J.-L. Reymond, and D.P. Mehta. Social network of isomers based on bond count distance: Algorithms. *Journal of Chemical Information and Modeling*, **54**(1) (2014), 57–68.

[32] H.J. Curran, P. Gaffuri, W.J. Pitz, and C.K. Westbrook. A comprehensive modeling study of n-heptane oxidation. *Combust. Flame*, **114**(1/2) (1998), 149–177.

[33] S. Goto, T. Nishioka, and M. Kanehisa. LIGAND: Chemical database of enzyme reactions. *Bioinformatics*, **14**(7) (1998), 591–599.

[34] L.A. Goldberg and M. Jerrum. Randomly sampling molecules. *SIAM J. Computing*, **29**(3) (2000), 834–853.

CHAPTER 35

Randomized Graph Algorithms: Techniques and Analysis

Surender Baswana

Sandeep Sen

CONTENTS

- 35.1 Introduction 1011
- 35.2 Linear Time MST 1012
  - 35.2.1 Algorithm 1013
- 35.3 Global Min-Cut 1014
  - 35.3.1 Contraction Algorithm 1014
- 35.4 Estimating the Size of Transitive Closure 1016
  - 35.4.1 Idea of Randomization 1016
  - 35.4.2 Monte Carlo Algorithm 1017
  - 35.4.3 How Accurate the Estimate $\hat{\tau}(v)$ Is to $\tau(v)$? 1018
- 35.5 Decremental Algorithm for Maintaining SCCs 1019
  - 35.5.1 Handling Deletion of Edges 1019
- 35.6 Approximate Distance Oracles 1021
  - 35.6.1 3-Approximate Distance Oracle 1021
    - 35.6.1.1 Expected Size of the Data Structure 1022
    - 35.6.1.2 Answering a Distance Query 1022
  - 35.6.2 Recent and Related Work on Approximate Distance Oracles 1023

35.1 INTRODUCTION

Although the first significant application of randomized techniques in algorithm design happened in mid-1970s in the celebrated primality testing algorithms of Miller–Rabin and Solovay–Strassen, its impact in graph algorithms was felt more than a decade later. Arguably, the first non-trivial applications were in the area of parallel graph algorithms like connectivity [1], depth-first search (DFS) [2], and matching [3]. Luby's work [4] on parallel maximal matching inspired a new line of work in derandomization. Seidel's work [5] in shortest paths was soon followed by some very innovative approaches to global min-cuts and minimum spanning trees (MSTs) by Karger et al. [6,7] that consolidated the use of randomization in the area of mainstream graph algorithms. Subsequently, randomization has been used very effectively in dynamic graph algorithms such as connectivity [8], transitive closure, and shortest path maintenance problems, some of which have no matching deterministic counterparts.

The above line of work must be clearly distinguished from a related but a fundamentally different model, namely, the random graph model of Erdős–Rényi that has a very long and rich history. However, these results are obtained by analyzing the expected performance of an algorithm on a *random* instance of a graph in the $G_{n,p}$, a class of graphs with n vertices where each of the $\binom{n}{2}$ edges occur with probability p. Our focus is on algorithms that use *random choices* and work for any input instance. Traditionally randomized algorithms come in two flavors, namely *Monte Carlo* and *Las Vegas*. The former denotes a class of algorithms whose output may contain errors (with probability bounded below 1/2) whereas the Las Vegas algorithms always output the correct result. While Monte Carlo algorithms terminate in some predictable number of steps, the running time of a Las Vegas is a random variable whose expected value must be analyzed. For efficiency consideration, we are interested in polynomial time randomized algorithms (in case of Las Vegas algorithms the expected running time is bounded by a polynomial in size of the input). The class of polynomial time Monte Carlo algorithms are called RP (one-sided error* randomized polynomial), whereas the (expected) polynomial time Las Vegas algorithms are known as ZPP (zero error probabilistic polynomial). As the reader might have suspected, the Las Vegas algorithms is the more desirable kind and it is known that the complexity classes are related as $ZPP \subseteq RP$, but not known in the reverse direction.

In this chapter we will focus on Las Vegas randomized algorithms that always produce the correct answer, but manifest some variation in the running times. The most common measure is the expected running time of the algorithm, where the expectation is over the choice of random bits in the algorithm and not the input distribution. The natural concern is what is an *acceptable* probability of deviating from the expected running time by a certain amount—say a small constant factor. Roughly speaking, the expected running time implies that the probability of exceeding twice the expected time is 1/2. If the hardware itself is subjected to a similar scrutiny, there is empirical evidence that it has a failure probability as high as $1/2^k$ where k is a constant. Therefore it is not unreasonable to aim for a success probability for randomized algorithms that is about an order of magnitude smaller than this. Over the past decade, the notion of *high probability* bound has gained wide acceptance. An algorithm is said to have a running time $f(n)$ with high probability (whp), if the following holds

$$\Pr\left(T(n) > c\alpha f(n)\right) \leq \frac{1}{n^\alpha}$$

for some constant c and any α, where $T(n)$ is the running time for an input of size n. This is also referred to as *inverse polynomial* probability of success. It must be noted that there is an implicit trade-off between the success probability and the running time. An inverse-*exponential* probability is even better, namely,

$$\Pr\left(T(n) > cf(n)\right) \leq \frac{1}{2^n}$$

However, randomized algorithms that succeed with high probability are considered as reliable as any deterministic algorithm.

In this chapter, we shall discuss five problems on graph algorithms and present randomized algorithms for them. The particular choice of these problems has been due to the following reasons. These problems are simple as well as some of the well-researched problems. Moreover, they employ simple randomization ideas in a powerful way to achieve efficiency in the time complexity.

35.2 LINEAR TIME MST

The basic idea for obtaining a faster algorithm for MST is very intuitive. We first randomly sample a set of edges $S \subset E$ and construct a minimum spanning forest F (may not be

---

*There is a more general class called BPP that allows two-sided errors.

connected). Using F, we can filter some of the edges using the *red rule*, that is, the heaviest edge in any cycle cannot belong to the final MST. However, the technical lemma that achieves a provable bound on the number of edges that can be eliminated based on the sampling is quite subtle.

Lemma 35.1 *Given an undirected graph $G = (V, E)$, with a weight function $w : E \to \mathcal{R}$, let $G(p) = (V, E(p))$ denote the subgraph where each edge of G is included in $G(p)$ independently with probability p. If $F(p)$ is a minimum spanning forest of $G(p)$, we define an edge $(u, v) \in E - E(p)$ to be heavy if each edge on the path from u to v in $F(p)$ has weights smaller than $w(u, v)$. Since the heavy edges cannot be part of the MST of G, we retain only the remaining edges, that are called light and this set is denoted by L. Then $E[L] = O(|V|/p)$.* ∎

The proof is based on an useful observation called principle of deferred decision. We pretend that the sampling is happening online on a predetermined sequence using independent Bernoulli trials (with success probability p). This principle states that the sequence in which we do the Bernoulli trials is indistinguishable from independent offline sampling and therefore the bounds that we prove using some convenient online ordering will hold for the offline sampling.

We first sort the edges in increasing order of their weights and divide the sampling into $|V| - 1$ phases—in each phase we discover an additional edge of F. Note that F may not be connected and may have less than $|V| - 1$ edges. We will prove that in each phase, the expected number of *light* edges is $O(1/p)$. We argue using induction. Suppose we are currently in the ith phase, that is, the corrent size of F is $i - 1$ and E_i is the remaining sequence for which we have not yet sampled. The (partial) F categorizes E_i into two distinct classes—those edges that are light (on the basis of the current F) and those that are not, call them *heavy*. The outcome of the sampling on the *heavy* edges of E_i is not relevant, since they will be discarded. So we only focus on the subsequence E_i^ℓ of the *light* edges. The first edge that is selected from E_i^ℓ terminates the ith phase. What is the expected number of edges in E_i^ℓ that are not sampled before the first one? This is upperbounded by a geometric random variable that has expectation $1/p$. In fact, we observe that the number of light edges in each phase is independent and so we can obtain concentration bounds (that holds with probability approaching 1) on the total number of light edges over all the phases.

35.2.1 Algorithm

The algorithm first applies two rounds of Boruvka's algorithm, that is, each vertex v chooses its nearest neighbor $u = N(v)$ and includes (u, v) in the final MST*. Now contract these edges, and the resultant graph has no more than $|V|/2$ vertices (after contraction, every merged group has at least two vertices). Therefore after two rounds, the contracted graph, which is denoted by G_2 has at most $|V|/4$ vertices and each round takes $O(|E|)$ time. In G_2 we sample every edge with probability $1/2$ and in the resultant graph $G_2(1/2)$, we construct the MST of $G_2(1/2)$ *recursively*, denote this by $F(G_2(1/2))$. Using $F(G_2(1/2))$ to filter out those edges that cannot contribute to $F(G_2)$, we are left with a set of edges E' where the expected size of E' is $\leq 2 \cdot |V|/4 = |V|/2$. We again run the algorithm recursively on this graph $G' = (V', E)$ and output the edges in $F(G')$.

The running time of the algorithm depends on the procedure by which we detect the *light* edges. Let the time complexity of this procedure be $L(m, n)$. The entire algorithm can be viewed in terms of the (binary) recursion tree where the left nodes correspond to the sampling and filtering step and the right node corresponds to the second recursive. Let us first bound the cost of applying Boruvka's algorithm to all nodes of the recursion tree. For this, it suffices to count the total number of vertices and edges in all the subproblems. Since the number of vertices decrease by a factor of 4 at each level, the total number of nodes can be bounded by

$\sum_{d=0}^{\infty} (n/4^d) \cdot 2^d \leq \sum_d (n/2^d) \leq 2n$. The sum total of edges can be bound by looking at the (disjoint) union of the edges belonging to all the maximal length left paths from each node. Since the expected number of edges in successive left nodes decrease by a factor of 2, the expected total number of edges is no more than a factor of 2 of the number of edges in the starting node. At the root level, there is only one starting node that contributes $2m$ edges. In level d, there are 2^{d-1} starting nodes (that are right children), each with $2(n3h^d)$ edges, that yields a total of $\sum_d (n/2^d)$ edges in these nodes and therefore the maximal left paths starting from these have a total of twice this quantity, that is, $2n$. So the overall (expected) number of edges in all subproblems combined is $O(m+n)$ which is also the asymptotic cost of Boruvka's iterations. Moreover if $L(m,n)$ is $O(m+n)^*$, then, the expected running time of the algorithm is $O(m+n)$.

35.3 GLOBAL MIN-CUT

A *cut* of a given (connected) graph $G = (V, E)$ is a set of edges that when removed disconnects the graph. An $s-t$ cut must have the property that the designated vertices s and t should be in separate components. A *min-cut* is the minimum number of edges that disconnects a graph and is sometimes referred to as *global* min-cut to distinguish it from $s-t$ min-cut. The weighted version of the min-cut problem is the natural analogue when the edges have non-negative associated weights. A cut can also be represented by a set of vertices S where the cut edges are the edges connecting S and $V - S$.

It was believed for a long time that the min-cut is a harder problem to solve than the $s-t$ min-cut—in fact the earlier algorithms for min-cuts determined the $s-t$ min-cuts for all pairs $s, t \in V$. The $s-t$ min-cut can be determined from the $s-t$ max-flow algorithms and over the years, there have been improved reductions of the global min-cut problem to the $s-t$ flow problem, such that it can now be solved in one computation of $s-t$ flow.

In a remarkable departure from this line of work, first Karger and Stein [7], followed by Karger [9] developed faster algorithms (than max-flow) to compute the min-cut with *high probability*. The algorithms produce a cut that is very likely the min-cut, that is, these are Monte Carlo algorithms. Unfortunately, there is yet no known matching verification algorithms. We will describe an algorithm that runs in time $O(n^2 \text{polylog}(n))$, $(n = |V|)$ which is nearly best possible for dense graphs.* This algorithm exploits some properties of branching processes.

35.3.1 Contraction Algorithm

The basis of the algorithm is the procedure contraction described below. The fundamental operation $contract(v_1, v_2)$ replaces vertices v_1 and v_2 by a new vertex v and assigns the set of edges incident on v by the union of the edges incident on v_1 and v_2. We do not merge edges from v_1 and v_2 with the same end point but retain them as multiple edges. Notice that by definition, the edges between v_1 and v_2 disappear.

Procedure Contraction(t)
Input: A multigraph $G = (V, E)$
Output: A t partition of V

Repeat until t vertices remain

    choose an edge (v_1, v_2) at random
    contract (v_1, v_2)

---

*A more recent algorithm of Karger improves this to $O(|E|polylog(n))$ using a more sophisticated Monte Carlo algorithm.

Procedure *Contraction*(2) produces a cut. Using the observation that, in an n-vertex graph with a min-cut value k, the minimum degree of a vertex is k, the following can be shown quite easily.

Lemma 35.2 *The probability that a specific min-cut $\mathcal{C}$ survives at the end of Contraction(t) is at least $(t(t-1)/n(n-1))$.* ∎

Therefore *Contraction*(2) produces a min-cut with probability $\Omega\left(1/n^2\right)$.

Lemma 35.3 *A single iteration of the procedure contraction can be carried out in $O(n)$ steps.* ∎

This is done by using an adjacency graph representation (see Karger and Stein [7] for details). Therefore using the procedure contract to produce min-cut is somewhat expensive since we need to repeat it about n^2 times. Instead, we run procedure Contraction ($\sqrt{n}/2$) twice independently and repeat it recursively on the contracted graphs. The algorithm is described below.

Algorithm Fastmincut
Input: A multigraph $G = (V, E)$
Output: A cut $\mathcal{C}$

1. Let $n := |V|$.
2. If $n \leq 6$ then compute mincut of G directly else

    2.1 $t := \lceil 1 + n/\sqrt{2} \rceil$.

    2.2 Call *Contraction*(t) twice (independently) to produce to graphs H_1 and H_2.

    2.3 Let $C_1 =$ **Fastmincut** (H_1) and $C_2 =$ **Fastmincut** (H_2).

    2.4 $\mathcal{C} = \min\{C_1, C_2\}$

The running time of algorithm *Fastmincut* satisfies the following recurrence

$$T(n) = 2T\left(\lceil 1 + n/\sqrt{2} \rceil\right) + O(n^2)$$

which yields $T(n) = O(n^2 \log n)$. Perhaps a more interesting question is to ascertain the probability with which *Fastmincut* returns a min-cut. The probability that a min-cut survives in H_1 after Step 2.2 is

$$\frac{(\lceil 1 + n/\sqrt{2} \rceil)(\lceil 1 + n/\sqrt{2} \rceil - 1)}{n(n-1)} \geq \frac{1}{2}$$

from Lemma 35.2. The same argument applies to H_2 independently. Therefore, we can view the recursive algorithm as a branching process where any node can have zero, one, or two children depending on the fact if the min-cut survived in zero, one, or two children. The distribution function at each node can be approximated by a binomial distribution with two trials, each with success probability greater than $1/2$ (i.e., mean $\mu \geq 1$). Since the algorithm has roughly $2 \log n$ levels of recursion, we can restate the survival probability of the min-cut as the complement of the extinction probability at the $2 \log n$ generation.

The extinction probability of a branching process* with mean $\mu > 1$ converges to the solution of $x = P(x)$ where P is the generating function of the probability distribution. Here, we can approximate $P(s)$ by $(1/4) + (1/2)\,s^2 + (1/4)\,s^2$. Solving for x yields $x = 1$, which is an asymptotic solution but does not give us much information about the rate of convergence. For this, we need to solve the recurrence

$$x_n = P(x_{n-1})$$

where x_i is the extinction probability of the ith generation. Substituting our generating function and simplifying yields

$$x_n = \frac{1}{4}(1 + x_{n-1})^2$$

The solution to this recurrence is $x_n = \Theta(1/n)$. So the survival probability of the min-cut (after $2\log n$ levels of recursion) is $\Omega(1/\log n)$. Repeating the procedure (with independently chosen random bits) $m \log n$ times increases the probability of finding the min-cut to $1 - \exp^{-m}$.

35.4 ESTIMATING THE SIZE OF TRANSITIVE CLOSURE

Let $G = (V, E)$ be a directed graph on $n = |V|$ vertices and $m = |E|$ edges. For each vertex v, let $\tau(v)$ be the number of vertices reachable from v in the given graph. Consider the problem of computing a very accurate estimate of $\tau(v)$ for each $v \in V$. We can compute τ trivially by executing a breadth-first or depth-first traversal from each vertex and this will take $O(mn)$ time. We can also use repeated squaring of the adjacency matrix and this approach will achieve $O(n^\omega \log n)$ time, where ω is the exponent of the best-known algorithm for multiplying two $n \times n$ integer matrices. Currently best bound on ω is 2.317 due to Coppersmith and Winograd [10]. These are the only two deterministic algorithms known for this problem. We shall now discuss an $O(m \log n)$ time randomized Monte Carlo algorithm by Cohen [11] which, for any given constant $c_1, c_2 > 1$, computes $\hat{\tau}(v)$ for each vertex $v \in V$ satisfying

$$\frac{\tau(v)}{c_1} < \hat{\tau}(v) < c_2 \tau(v)$$

with high probability. Here c_1, c_2 can be chosen arbitrarily close to 1. Notice that the constant in $O(m \log n)$ running time will depend upon the values c_1, c_2 and the desired probability of success.

In addition to being a problem of independent theoretical interest, this problem has applications in databases. Before answering a database query, one would like to estimate the size of the *query–answer set* to be reported. This prior knowledge may sometimes help in optimizing query processing time.

35.4.1 Idea of Randomization

Suppose we select k numbers uniformly independently from the interval $[0, 1]$. Let X be the random variable defined as the smallest of these k numbers.

Lemma 35.4 *Expected value of X is $1/(k+1)$.*

Proof. Notice that selecting k numbers splits the interval $[0, 1]$ into $k + 1$ intervals. Taking this viewpoint, X is equal to the length of the leftmost of the $k + 1$ intervals formed.

---

*The reader is referred to standard textbooks, for example, Feller [12] for theoretical analysis of branching process.

To calculate $\mathbf{E}[X]$, we shall pursue this viewpoint. Consider a circle of circumference 1. Suppose we select $k+1$ points randomly uniformly and independently from the circumference of this circle. This will split the circumference into $k+1$ intervals. Exploiting the uniformity in sampling the points and the symmetry of the circle conclude that the lengths of these $k+1$ intervals have identical probability distribution (though they are not independent). This implies that the expected length of any interval would be $1/(k+1)$. Let us straighten the circle to form a line interval $[0,1]$ by cutting the circle at any of the $k+1$ selected point. This creates an instance of our original experiment of selecting k points from interval $[0,1]$. Hence $\mathbf{E}[X] = 1/(k+1)$. ∎

The fact that the expected value of the smallest number among the k numbers selected uniformly independently from $[0,1]$ is $1/(k+1)$ conveys the following important observation: the number of random variables k is related to $\mathbf{E}[X]$ very closely. We can use X to infer the number of random variables, which define it. In particular, if X takes value a, we may return $1/a - 1$ as the estimate of the number of random variables. This randomization idea is the underlying idea of the algorithm for estimating the value $\tau(v)$ for each $v \in V$. However, to improve the accuracy of our estimate and the associated probability, we shall use the idea of multiple sampling.

35.4.2 Monte Carlo Algorithm

Though there is no deterministic algorithm to compute $\tau(v)$ for all $v \in V$ in $O(m)$ time, there are many problems on directed graphs which can be solved in $O(m)$ time (e.g., computing the strongly connected components [SCC] of the graph). One such problem is the following. Let each vertex stores some key which is a real number and the aim is to compute, for each $v \in V$: the key of the smallest key vertex reachable from v. There is a deterministic $O(m)$ time algorithm for this problem based on DFS of the graph. We shall use this algorithm and the randomization idea described above to solve the problem of estimating $\tau(v)$ for each $v \in V$.

The algorithm will perform ℓ iterations. In ith iteration, we assign a key to each vertex in the graph by selecting a number uniformly randomly from the interval $[0,1]$. After this we execute $O(m)$ time algorithm which computes, for each $v \in V$, $k_i[v]$: the key of the smallest key vertex reachable from v. At the end of l iterations, we shall have a set $\{k_1[v], \ldots, k_\ell[v]\}$ of ℓ such labels for each $v \in V$. It follows from Lemma 35.4 that the expected value of any element of the set is $1/\tau(v)$. (Actually, it is $1/(\tau(v)+1)$, but for simplicity and clarity of exposition we ignore the additive term of 1 from the denominator.) In order to accurately estimate $\tau(v)$, we should select that element from $\{k_1[v], \ldots, k_\ell[v]\}$ which is going to be closest to $1/\tau(v)$ most likely. For this purpose, we state Lemma 35.5 whose proof is elementary.

Lemma 35.5 *If we select j numbers uniformly independently from interval $[0,1]$, the probability that the smallest number is greater than c is $(1-c)^j$ for any $0 < c < 1$.* ∎

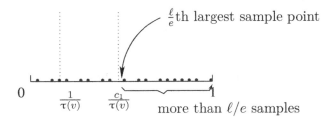

Figure 35.1 The event when (ℓ/e)th largest element happens to be greater than $c_1/\tau(v)$

It follows from Lemma 35.5 that the probability $k_i[v]$, for any $i \leq \ell$, takes value greater than $1/\tau(v)$ and is given by

$$\mathbf{P}\left(k_i(v) > \frac{1}{\tau(v)}\right) = \left(1 - \frac{1}{\tau(v)}\right)^{\tau(v)} \approx 1/e \qquad (35.1)$$

The above equation implies that out of ℓ iterations of the algorithm, roughly ℓ/e times the sample will be greater than $1/\tau(v)$. Therefore, from the set $\{k_1(v), \ldots, k_\ell(v)\}$ the (ℓ/e)th largest element is most likely to be closest to $1/\tau(v)$. Let this element be denoted as k_v^*. So the algorithm finally reports $1/k_v^*$ as the estimate $\hat{\tau}(v)$ of $\tau(v)$.

35.4.3 How Accurate the Estimate $\hat{\tau}(v)$ Is to $\tau(v)$?

We shall try to get a bound on the probability for the event $\tau(v)/c_1 < \hat{\tau}(v) < c_2\tau(v)$ for any given constants $c_1, c_2 > 1$. For this purpose we need to calculate a bound on the probability for the event $\hat{\tau}(v) < \tau(v)/c_1$ and a bound on the probability for the event $\hat{\tau}(v) > c_2\tau(v)$ separately. Let us focus on bounding the probability of $\hat{\tau}(v) < \tau(v)/c_1$.

$\hat{\tau}(v) < \tau(v)/c_1$ means that the (ℓ/e)th largest element from $\{k_1(v), k_2(v), \ldots, k_\ell(v)\}$ happened to be greater than $c_1/\tau(v)$. This means that at least ℓ/e of the elements from $\{k_1(v), k_2(v), \ldots, k_t(v)\}$ turned out to be greater than even $c_1/\tau(v)$ (see Figure 35.1).

Let us introduce random variables $X_i, i \leq t$ at this moment. X_i takes value 1 if $k_i[v]$ is greater than $c_1/\tau(v)$ and zero otherwise. Note that each of X_i, $i \leq \ell$ chooses its value independent of other X_j's $j \neq i$ since every iteration of the algorithm assigns labels to vertices independent of other iterations. Let $X = \sum_i X_i$. So we can see that the event $X > \ell/e$ is necessary for happening of the event $\hat{\tau}(v) < \tau(v)/c_1$. So we shall now calculate the probability of event $X > \ell/e$.

Using Lemma 35.5 and linearity of expectation, it is easy to observe that $\mathbf{E}[X]$ is $(1 - c_1/\tau(v))^{\tau(v)}\ell$, which is at most $e^{-c_1}l$ since $1 + \alpha < e^\alpha$, $\forall \alpha > 0$. So the situation is the following. Compared to its expected value of ℓ/e^{c_1}, the random variable X took value greater than ℓ/e which is quite large (depending upon c_1).

Let us compute the probability for this event using Chernoff bound. Convince yourself that X fulfills all requirements for applying Chernoff bound. Let us, for the sake of clarity of exposition, assume the value of c_1 such that $e^{-c_1} = (1/2e)$ (the value of c_1 is close to 1.7). So the expected value of X is $\ell/(2e)$. Since the value taken by X is at least ℓ/e, it implies that $\delta > 1$. Hence applying the Chernoff bound

$$\mathbf{P}\left(X > \frac{\ell}{e}\right) < e^{-[(\ell/2e)/4]} = e^{-(\ell/8e)}$$

If we choose $\ell = 24e \ln n$, that is, if we repeat the main iteration of the algorithm more than $24e \ln n$ times, the probability that $\tau(v)$ takes value less than $\tau(v)/1.7$ is less than $1/n^3$. Using Boole's inequality (union theorem) we can thus state Lemma 35.6.

Lemma 35.6 *If we repeat the main iteration of the algorithm at least $24e \ln n$ times, the probability that $\tau(v)$ is less than $\tau(v)/1.7$ for any v is less than $1/n^2$.* ∎

In a similar fashion, we can calculate the probability of the event $\hat{\tau}(v) > c_2\tau(v)$. We can thus conclude with the following theorem.

Theorem 35.1 *Given a directed graph G on n vertices and m edges, and any two constants $c_1, c_2 > 1$, there is a Monte Carlo randomized algorithm which takes $O(m \log n)$ time to compute $\hat{\tau}(v)$ for each v such that with probability exceeding $1 - 1/n^2$,*

$$\frac{\tau(v)}{c_1} \leq \hat{\tau}(v) \leq c_2\tau(v)$$

∎

35.5 DECREMENTAL ALGORITHM FOR MAINTAINING SCCs

Given a directed graph $G = (V, E)$, two vertices $u, v \in V$ are said to be strongly connected if there is a path from u to v as well as a path from v to u. A SCC in a graph is a maximal subset of strongly connected vertices. The problem of maintaining SCCs under deletion of edges can be stated formally as follows:

There is an online sequence of edge deletions interspersed with the queries *are u and v strongly connected* for any $u, v \in V$. The aim is to maintain a data structure which can answer each query in $O(1)$ time and can be updated efficiently upon any edge deletion.

In addition to being a problem of independent interest, an efficient algorithm for this problem provides efficient decremental algorithm for maintaining all-pairs reachability.

Let us start with a simple-minded solution for this problem. This solution will use the well-known static $O(m)$ time algorithm to compute SCCs of a given graph. In order to answer any query in $O(1)$ time, we may keep an array A such that $A[u] = A[v]$ if u and v belong to the same SCC. For this purpose, we may select a unique vertex $\text{REP}(c)$ called *representative* vertex of c. For each $v \in c$, $A[v]$ store $\text{REP}(c)$. This ensures $O(1)$ query time. In order to handle any edge deletion, we may execute the static $O(m)$ time algorithm to recompute SCC, and hence update A, after each edge deletion. This is a trivial decremental algorithm for maintaining SCCs while ensuring $O(1)$ query time. However, this algorithm will take a total of $O(m^2)$ update time to process any sequence of edge deletions.

We shall now discuss a very simple randomized Las Vegas algorithm by Roditty and Zwick [13] for maintaining SCCs under deletion of edges. The expected time taken by this algorithm to process any arbitrary sequence of edge deletions will be $O(mn \log n)$ only. Let us introduce a couple of notations here. For a SCC c, we shall use $G(c)$ to denote the subgraph of G induced by c and $E(c)$ to denote the edges of $G(c)$. Let G^r denote the graph obtained by reversing all the edge directions in G.

We basically need an efficient mechanism to maintain the array A under deletion of edges which bypasses the need of recomputing SCCs after each edge deletion. In particular, whenever an edge is deleted, we need to determine if it has indeed split an SCC into multiple SCCs. For this objective, for each SCC c, the following data structure is maintained:

- $T_{\text{out}}(c)$: the breadth-first search (BFS) tree rooted at $\text{REP}(c)$ in graph $G(c)$.

- $T_{\text{in}}(c)$: the BFS tree rooted at $\text{REP}(c)$ in graph $G^r(c)$.

We now provide an overview of a simple algorithm for maintaining a BFS tree rooted at a vertex under deletion of edges. This algorithm maintains the level of each vertex. As the edges are being deleted, vertices may fall from their level to lower levels. Upon deletion of an edge, this algorithm takes $O(1)$ time to determine if it has caused fall of one or more vertices. The algorithm computes new levels of each vertex which falls. In doing so, for each vertex x which has fallen from its level i to level j, the algorithm incurs $O((j-i)\deg(x))$ computation cost in processing x. For the sake of clarity of our analysis, we *charge* the total computation performed during any sequence of edge deletions to the respective vertices, which the algorithm processes. Since a vertex can only fall during edge deletions, and the lowest level is n, the total computation cost charged to x during any sequence of edge deletion will be $O(n \deg(x))$. This implies a total of $O(mn)$ update time for maintaining a rooted BFS tree in a graph. Next, we describe the procedure for handling deletion of an edge.

35.5.1 Handling Deletion of Edges

Consider deletion of an edge (u, v). If u and v belong to different SCCs, nothing needs to be done except deletion of (u, v) from the graph. But, if u and v belong to the same SCC, say c, we update the data structures associated with c. If the deletion leads to any vertex leaving $T_{\text{in}}(c)$ or $T_{\text{out}}(c)$, it implies that SCC c has split. We execute $O(|E(c)|)$ time algorithm to determine the new SCCs. For all vertices which leave the old SCC c, we delete them and all their edges from $T_{\text{in}}(c)$ and $T_{\text{out}}(c)$. For each new SCC c', we select the representative $REP(c')$, and build BFS trees $T_{\text{in}}(c')$ and $T_{\text{out}}(c')$. We now state an important observation.

Observation 35.1 *Suppose deletion of an edge splits a SCC c into SCCs $c_1, \ldots, c_k$ with $REP(c)$ belonging to some SCC, say c_1. Each vertex of c_1 will not be charged any additional computation cost. But any vertex $x \in c_i, i > 1$ will be charged a computation cost of $O(|c_i| \deg(x))$ in building and maintaining the data structure associated with c_i.* ∎

The algorithm described above is deterministic and its worst-case running time can still be $O(m^2)$. To achieve efficiency, we now add the following simple randomization ingredient to this algorithm. Each SCC c, at the moment of its creation, selects its $REP(c)$ randomly uniformly from c. It will turn out that this simple randomization leads to an efficient decremental algorithm for maintaining SCC. Let us analyze its running time.

There are two major computational tasks, which are performed by the algorithm for any sequence of edge deletions. The first task is the execution of the static algorithm for determining new SCC. This task is executed whenever some SCC gets split and hence will be executed at most $n - 1$ times during any sequence of edge deletions. A single execution of this algorithm takes $O(m)$ time, so overall $O(mn)$ computation time is spent in this task. Another computation task is associated with maintaining the data structures for various SCCs that get created during a sequence of edge deletions. To analyze this major task, we shall take vertex-centric approach.

Consider any arbitrary sequence of edge deletions and focus on any vertex v. Let vertex v changed its SCC a total of t times and let $c_1, c_2, \ldots, c_t$ be the respective connected components. Notice that $c_i \subset c_{i-1}$ for all $1 < i \leq t$. Let $n_i = |c_i|$ and $n_1 = n$. We shall show that the expected computation cost charged to v while maintaining the data structures associated with $c_1, \ldots, c_t$ will be $O(n \deg(v))$ only. Recall Observation 35.1. Let X_i be the random variable which takes value 1 if during transition from c_i to c_{i+1}, vertex v is charged $O(|n_i| \deg(v))$ computation cost, and zero otherwise.

Note that $X_i = 1$ if $REP(c_i)$ belonged to $c_i \setminus c_{i+1}$. Since $REP(c_i)$ was selected randomly uniformly from c_i, hence

$$\mathbf{P}(X_i = 1) = \frac{n_i - n_{i+1}}{n_i}$$

So the expected computation cost charged to v during any sequence of edge deletion is of the order of

$$\sum_{i=1}^{t-1} \mathbf{P}(X_i = 1) n_i \deg(v) = \sum_{i=1}^{t-1} \frac{n_i - n_{i+1}}{n_i} n_i \deg(v)$$

$$= \deg(v) \sum_{i=1}^{t-1} (n_i - n_{i+1}) = \deg(v)(n-1) = O(n \deg(v)).$$

So the expected total computation performed by the decremental algorithm of SCC for processing any sequence of edge deletions is $O(mn)$.

Theorem 35.2 *For any directed graph $G = (V, E)$, there is a randomized decremental algorithm for maintaining SCCs with $O(1)$ query time and expected $O(mn)$ total update time.* ∎

35.6 APPROXIMATE DISTANCE ORACLES

The all-pairs shortest paths problem is one of the most fundamental algorithmic graph problem. This problem is commonly phrased as follows: *Given a graph on n vertices and m edges, compute shortest-paths/distances between each pair of vertices.*

In many applications the aim is not to compute *all* distances, but to have a mechanism (data structure) through which we can extract distance/shortest-path for any pair of vertices efficiently. Therefore, the following is a useful alternate formulation of the all pairs shortest paths (APSP) problem.

Preprocess a given graph efficiently to build a data structure that can answer a shortest-path query or a distance query for any pair of vertices.

The objective is to construct a data structure for this problem such that it is efficient both in terms of the space and the preprocessing time. There is a lower bound of $\Omega(n^2)$ on the space requirement of any data structure for APSP problem, and space requirement of all the existing algorithms for APSP match this bound. However, this quadratic bound on the space is too large for many graphs, which appear in various large-scale applications. In most of these graphs, it is usual to have $m = n^2$, hence a table of $\Theta(n^2)$ size is too large to be kept. This has motivated researchers to design a subquadratic space data structures, which may report approximate instead of exact distance between any two vertices. A path from u to v in a graph is said to be t-approximate if its length is at most t times the length of the shortest path from u to v. Thorup and Zwick [14] presented a novel data structure for all-pairs approximate shortest paths, called approximate distance oracles. They showed that any given weighted undirected graph can be preprocessed in subcubic time to build a data structure of subquadratic size for answering a distance query with stretch 3 or more. Note that 3 is also the least stretch for which we can achieve subquadratic space for APSP (see [15]). There are two very impressive features of their data structure. First, the trade-off between stretch and the size of data structure is essentially optimal assuming a 1963 girth lower bound conjecture of Erdös [16] and second, in spite of its subquadratic size their data structure can answer any distance query in *constant* time, hence the name *oracle*. In precise words, Thorup and Zwick [14] achieved the following result.

Theorem 35.3 [14] *For any integer $k \geq 1$, an undirected weighted graph on n vertices and m edges can be preprocessed in expected $O(kmn^{1/k})$ time to build a data structure of size $O(kn^{1+1/k})$ that can answer any $(2k-1)$-approximate distance query in $O(k)$ time.* ∎

We provide below description of 3-approximate distance oracle.

35.6.1 3-Approximate Distance Oracle

In order to achieve subquadratic space, the 3-approximate distance oracle is based on the following idea.

From each vertex v, we store distance and shortest paths to a small set of vertices lying within small vicinity of v. This will take care of distance queries from v to its nearby vertices. For querying distance to a vertex, say w, lying outside the vicinity of v, we may do the following. We may have a small set S of special vertices and we store distance between each special vertex and every vertex of the graph. In order to report (approximate) distance between v and w, we may report the sum of the distances from some (carefully defined) special vertex to v and w.

The above idea, though looks intuitively appealing, appears difficult to materialize. In particular, how do we define *vicinity* around a vertex and ensure a finite stretch, while having only a small set of special vertices? Randomization in selecting S plays a crucial role to achieve all these goals simultaneously. The following terminology captures the notion of vicinity of a vertex in terms of the special vertices.

Definition 35.1 *Given a graph $G = (V, E)$, a vertex $v \in V$, and any subset $S \subset V$ of vertices, we define $Ball(v, V, S)$ as a set in the following way.*

$$Ball(v, V, S) = \{x \in V | \delta(v, x) < \delta(v, Y)\}$$

Let $p(v, S)$ denote the vertex from set S which is nearest to v. In simple words, $Ball(v, V, S)$ consists of all those vertices of the graph whose distance from v is less than the distance of $p(v, S)$ from v. The 3-approximate distance oracle is built as follows:

1. Let $S \subseteq V$ be formed by selecting each vertex uniformly and independently with probability $1/\sqrt{n}$.

2. From each vertex S, store distance to all the vertices.

3. From each vertex $v \in V \backslash S$, compute $p(v, S)$ and build a hash table which stores all the vertices of $Ball(v, V, S)$ and their distance from v.

See Figure 35.2 to get a better picture of the 3-approximate distance oracle.

35.6.1.1 Expected Size of the Data Structure

The expected size of the data structure computed as described above depends upon the expected size of $Ball(v, V, S)$ which turns out to be $O(\sqrt{n})$ as follows. Consider the sequence $\langle (v=)x_0, x_1, x_2, \cdots \rangle$ of vertices V arranged in non-decreasing order of their distances from v. The vertex x_i will belong to $Ball(v, V, S)$ only if none of $x_1, \cdots, x_i$ are selected in the set S. Therefore, $x_i \in Ball(v, V, S)$ with probability at most $(1 - 1/\sqrt{n})^i$. Hence using linearity of expectation, the expected number of vertices in $Ball(v, V, S)$ is at most $\sum_i (1 - 1/\sqrt{n})^i < \sqrt{n}$. Hence the expected size of the data structure will be $O(n\sqrt{n})$.

35.6.1.2 Answering a Distance Query

Let $u, v \in V$ be any two vertices whose approximate distance is to be computed. First it is determined whether or not $u \in Ball(v, V, S)$, and if so, the exact distance $\delta(u, v)$ is reported. Note that $u \notin Ball(v, V, S)$ would imply $\delta(v, p(v, S)) \leq \delta(v, u)$. In this case, report the

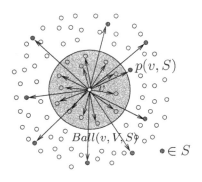

Figure 35.2 The oracle keeps distance information between v and all the vertices pointed by arrows.

distance $\delta(v, p(v, S)) + \delta(u, p(v, S))$, which is at least $\delta(u, v)$ (using the triangle inequality) and upperbounded by $3\delta(u, v)$ as shown below.

$$\begin{aligned}\delta(v, p(v, S)) + \delta(u, p(v, S)) &\leq \delta(v, p(v, S)) + \{\delta(u, v) + \delta(v, p(v, S))\} \\ &= 2\delta(v, p(v, S)) + \delta(u, v) \\ &\leq 2\delta(u, v) + \delta(u, v) = 3\delta(u, v)\end{aligned} \quad (35.2)$$

35.6.2 Recent and Related Work on Approximate Distance Oracles

Thorup and Zwick [14] presented an expected $O(kmn^{1/k})$ time algorithm for computing a $(2k - 1)$ spanner. There has been a lot of work [17,18] in designing faster algorithms for computing a $(2k - 1)$-approximate distance oracle. Currently, the best bounds for constructing these oracles is $O(n^2)$. Interestingly, these fast algorithms also employ simple randomization ideas.

An equally interesting question is to explore the possibility of approximate distance oracles for stretch better than 3. Recently Patrascu and Roditty [19] presented a positive answer to this question. They designed a 2-approximate distance oracle of size $O(n^{5/3})$ for unweighted graphs. They also generalize it for weighted graphs, though the size bounds depend on the number of edges of the graph: For graphs with edges n^2/α, they design a 2-approximate distance oracle with $O(n^2/\sqrt[3]{\alpha})$.

A related structure in graph theory, similar to the approximate distance oracles, is a graph spanner. A spanner is a subgraph which is sparse and yet preserves all-pairs distances approximately: Given an undirected graph $G = (V, E)$ and positive integer k, a $(2k - 1)$-spanner is a subgraph (V, E_S), such that the distance between any pair of vertices u, v in the subgraph is at most $(2k - 1)\delta(u, v)$. An interesting by-product of the $(2k - 1)$-approximate distance oracle of Thorup and Zwick [14] is a $(2k - 1)$-spanner with $O(kn^{1+1/k})$ edges. In fact 3-approximate distance oracle described above leads to 3-spanner as follows. From each special vertex $v \in S$, include all the edges of the shortest path tree rooted at v in the spanner. For every nonspecial vertex $v \in V \backslash S$, just store the edges of the shortest path tree from v to only those vertices that belong to $Ball(v, V, S)$. It follows that the subgraph defined by these edges is indeed a 3-spanner. Interestingly, there has been a Las Vegas randomized algorithm [20] to compute a $(2k - 1)$-spanner with $O(kn^{1+1/k})$ edges in expected $O(km)$ time.

References

[1] H. Gazit. An optimal randomized parallel algorithm for finding connected components in a graph. *SIAM J. Comput.* **20**(6) (1991), 1046–1067.

[2] A. Aggarwal, R. J. Anderson, and M.-Y. Kao. Parallel depth-first search in general directed graphs. *SIAM J. Comput.*, **19**(2) (1990), 397–409.

[3] K. Mulmuley, U. V. Vazirani, and V. V. Vazirani. Matching is as easy as matrix inversion. *Combinatorica*, **7**(1) (1987), 105–113.

[4] M. Luby. A simple parallel algorithm for the maximal independent set problem. *SIAM J. Comput.*, **15**(4) (1986), 1036–1053.

[5] R. Seidel. On the all-pairs-shortest-path problem in unweighted undirected graphs. *J. Comput. Syst. Sci.*, **51**(3) (1995), 400–403.

[6] D. R. Karger, P. N. Klein, and R. E. Tarjan. A randomized linear-time algorithm to find minimum spanning trees. *J. ACM*, **42**(2) (1995), 321–328.

[7] D. R. Karger and C. Stein. A new approach to the minimum cut problem. *J. ACM*, **43**(4) (1996), 601–640.

[8] M. R. Henzinger and V. King. Randomized fully dynamic graph algorithms with polylogarithmic time per operation. *J. ACM*, **46**(4) (1999), 502–516.

[9] D. R. Karger. Minimum cuts in near-linear time. *J. ACM*, **47**(1) (2000), 46–76.

[10] D. Coppersmith and S. Winograd. Matrix multiplication via arithmetic progressions. *J. Symb. Comput.*, **9**(3) (1990), 251–280.

[11] E. Cohen. Size-estimation framework with applications to transitive closure and reachability. *J. Comput. Syst. Sci.*, **55**(3) (1997), 441–453.

[12] W. Feller. *An Introduction to Prabability Theory and Its Applications*, John Wiley & Sons, 1957.

[13] L. Roditty and U. Zwick. Improved dynamic reachability algorithms for directed graphs. *SIAM J. Comput.*, **37**(5) (2008), 1455–1471.

[14] M. Thorup and U. Zwick. Approximate distance oracles. *J. ACM*, **52**(1) (2005), 1–24.

[15] E. Cohen and U. Zwick. All-pairs small-stretch paths. *J. Algorithms*, **38**(2) (2001), 335–353.

[16] P. Erdös. Extremal problems in graph theory. In *Theory of Graphs and Its Applications*, Publishing House of the Czechoslovak Academy of Sciences, Prague, Czech Republic, pp. 29–36, 1964.

[17] S. Baswana and S. Sen. Approximate distance oracles for unweighted graphs in expected $O(n^2)$ time. *ACM Trans. Algorithms*, **2**(4) (2006), 557–577.

[18] S. Baswana and T. Kavitha. Faster algorithms for all-pairs approximate shortest paths in undirected graphs. *SIAM J. Comput.*, **39**(7) (2010), 2865–2896.

[19] M. Patrascu and L. Roditty. Distance oracles beyond the Thorup-Zwick bound. In *FOCS*, Las Vegas, NV, IEEE, pp. 815–823, 2010.

[20] S. Baswana and S. Sen. A simple and linear time randomized algorithm for computing sparse spanners in weighted graphs. *Random Struct. Algorithms*, **30**(4) (2007), 532–563.

XI
Coping with NP-Completeness

CHAPTER 36

General Techniques for Combinatorial Approximation*

Sartaj Sahni

CONTENTS

36.1 Introduction 1027
36.2 Rounding 1029
36.3 Interval Partitioning 1031
36.4 Separation 1032

36.1 INTRODUCTION

Many combinatorial optimization problems are known to be NP-complete[†] [1–4]. An NP-complete problem can be solved in polynomial time iff all other NP-complete problems can also be solved in polynomial time. The class of NP-complete problems includes such difficult problems as the traveling salesman, multicommodity network flows, integer programming with bounded variables, set cover, and node cover problems. There is no known polynomial time algorithm for any of these problems. Moreover, mounting empirical evidence (i.e., the identification of more and more NP-complete problems) suggests that it is very likely that no polynomial time algorithms exist for any of these problems. This realization has led many researchers to develop polynomial time approximation algorithms for some NP-complete optimization problems. An approximation algorithm for an optimization problem is generally a heuristic that attempts to obtain a solution whose value is *close* to the optimal value. For many problems the data themselves are only an estimate. The exact values may be slightly different from these estimates. In such case it is probably just as meaningful to find a solution whose value is close to the optimal value as it is to find an optimal value (as the optimal solution may not remain so once the exact data values are known). For the case of NP-complete problems, the study of approximation algorithms derives an even stronger motivation from the fact that all known optimization algorithms for these problems require an exponential amount of time (measured as a function of problem size) and the expectation that these problems will never be solvable by polynomial time algorithms. Even on the fastest computers, exponential time algorithms are feasible only for relatively small problem sizes. It is better to be able to obtain an approximately optimal solution than no solution at all.

---

*This chapter is an edited version of the paper, Sartaj Sahni, "General Techniques for Combinatorial Approximation," *Operations Research*, **25**(6) (1977), 920–936.

[†]We use the term NP-complete somewhat loosely here, as strictly speaking only decision problems may be NP-complete and their optimization counterparts are NP-hard.

Many researchers have presented heuristics that obtain reasonably good solutions. The performance of these heuristics is usually measured only through computational tests. A major thrust of research on approximation algorithms has been the insistence on performance guarantees for heuristics. This involves the establishment of bounds on maximum difference between the value of an optimal solution and the value of the solution generated by the heuristic. While performance bounds were obtained as early as 1966 by Graham [5] for certain scheduling heuristics, the development of the class of NP-complete problems has led to an increased interest in approximation methods that are guaranteed to get solutions with value within some specified fraction of the optimal solution value.

Definition 36.1 *An algorithm will be said to be an ε-approximate algorithm for a problem P iff it is the case that for every instance I of P, $|F^*(I) - \hat{F}(I)|/F^*(I) \leq \varepsilon$. $F^*(I) > 0$ is the value of an optimal solution to I, $\hat{F}(I)$ is the value of the solution generated by the algorithm, and ε is some constant. For a maximization problem we require $0 \leq \varepsilon < 1$ and for a minimization problem $\varepsilon \geq 0$.*

Many known polynomial time approximation algorithms are also ε-approximation algorithms. Thus while it seems necessary to spend, in the worst case, a nonpolynomial amount of time to obtain optimal solutions to NP-complete problems, we can get to within an ε factor of the optimal for some such problems in polynomial time. For yet others [4] it is known that the problem of obtaining ε-approximate solutions is also NP-complete.

In some cases it is possible to obtain an approximation algorithm that for every $\varepsilon > 0$ generates solutions that are ε-approximate [6–10]. In the terminology of Garey and Johnson [11,12] such an algorithm is called an approximation scheme. Definitions 36.2 through 36.4 are due to Garey and Johnson.

Definition 36.2 *An approximation scheme for a problem P is an algorithm that, given an instance I and a desired degree of accuracy $\varepsilon > 0$, constructs a problem solution with value $\hat{F}(I)$, such that, if $F^*(I) > 0$ is the value of an optimal solution to I, then $|F^*(I) - \hat{F}(I)|/F^*(I) \leq \varepsilon$.*

Definition 36.3 *An approximation scheme is a polynomial time approximation scheme if for every fixed $\varepsilon > 0$ it has a polynomial computing time. (References [6–10] present such schemes for some NP-complete problems.)*

While the existence of polynomial time approximation schemes for NP-complete problems may appear surprising, for some problems one can in fact obtain approximation schemes with a computing time polynomial in both the input size and $1/\varepsilon$. Such schemes are called *fully polynomial time approximation schemes*.

Definition 36.4 *A fully polynomial time approximation scheme is a polynomial time approximation scheme whose computing time is a polynomial in both the input size and $1/\varepsilon$.*

Garey and Johnson [11] present an annotated bibliography of research on combinatorial approximation. Ibarra and Kim [7] present an $O(n/\varepsilon^2 + n \log n)$ fully polynomial time approximation scheme for the 0-1 knapsack problem. Sahni [10] and Horowitz and Sahni [6] present $O(n^2/\varepsilon)$ fully polynomial time approximation schemes for several machine-scheduling problems. In this chapter, we present a tutorial on two of the techniques used in [6,7,10] to obtain these schemes. These techniques are called *rounding* [7] and *interval partitioning* [10]. Both techniques are very general and applicable to a wide variety of optimization problems. Next we present a third technique, called *separation*, which is a modification of interval partitioning. This technique is as general as the others. While it results in approximation schemes with the same worst-case complexity as those obtained when interval partitioning is used, intuition backed by experimental results indicates that it performs better than interval partitioning. These three general methods for combinatorial optimization have the added advantage that heuristics, which could be applied to the exact algorithm, can also be used with the approximation, scheme.

36.2 ROUNDING

We state the three approximation techniques in terms of maximization problems. The extension to minimization problems is immediate. We shall assume the maximization problem to be of the form

$$\begin{aligned} \max\ & \sum_{i=1}^{n} c_i x_i \\ & \sum_{i=1}^{n} a_{ij} x_i \leq b_j,\ 1 \leq j \leq m \\ & x_i = 0\ \text{or}\ 1,\ 1 \leq i \leq n, \end{aligned} \qquad (36.1)$$

where $c_i, a_{ij} \geq 0$ for all i and j. Without loss of generality, we will assume that $a_{ij} \leq b_j, 1 \leq i \leq n$, and $1 \leq j \leq m$.

If $1 \leq k \leq n$, then the assignment $x_i = y_i, 1 \leq i \leq k$ will be said to be a *feasible assignment* iff there exists at least one feasible solution to (36.1) with $x_i = y_i, 1 \leq i \leq k$. A *completion* of a feasible assignment $x_i = y_i, 1 \leq i \leq k$, is any feasible solution to (36.1) with $x_i = y_i, 1 \leq i \leq k$. Let $x_i = y_i, 1 \leq i \leq k$, and $x_i = z_i, 1 \leq i \leq k$, be two feasible assignments such that for at least one $j, 1 \leq j \leq k, y_j \neq z_j$. Let $\sum c_i y_i = \sum c_i z_i$. We shall say that $y_1, \ldots, y_k$ *dominates* $z_1, \ldots, z_k$ iff there exists a completion $y_1, \ldots, y_k, y_{k+1}, \ldots, y_n$ such that $\sum_{i=1}^{n} c_i y_i$ is greater than or equal to $\sum_{i=1}^{n} c_i z_i$ for all completions $z_1, \ldots, z_n$ of $z_1, \ldots, z_k$. The approximation techniques to be discussed will apply to those problems that can be formulated as (36.1) and for which simple rules can be found to determine when one feasible assignment dominates another. Such rules exist, for example, for many problems solvable by dynamic programming [12–14].

One way to solve problems stated as above is to systematically generate all feasible assignments starting from the null assignment. Let $S^{(i)}$ represent the set of all feasible assignments for $x_1, \ldots, x_i$. Then $S^{(0)}$ represents the null assignment and $S^{(n)}$ the set of all completions. The answer to our problem is an assignment in $S^{(n)}$ that maximizes the objective function. The solution approach is then to generate $S^{(i+1)}$ from $S^{(i)}, 1 \leq i < n$. If an $S^{(i)}$ contains two feasible assignments $y_1, \ldots, y_i$ and $z_1, \ldots, z_i$ such that $\sum c_j y_j = \sum c_k z_k$, then use of the dominance rules enables us to discard that assignment which is dominated. (In some cases the dominance rules may permit the discarding of a feasible assignment even when $\sum c_j y_j \neq \sum c_k z_k$. This happens, for instance, in the knapsack problem [2,15].) Following the use of the dominance rules, it is the case that for each feasible assignment in $S^{(i)}$ $\sum_{j=1}^{i} c_j x_j$ is distinct. However, despite this, it is possible for each $S^{(i)}$ to contain twice as many feasible assignments as $S^{(i-1)}$. This results in a worst-case computing time that is exponential in n. The approximation methods we are about to discuss will restrict the number of distinct $\sum_{i=1}^{n} c_j x_j$ to be only a polynomial function of n. The error introduced will be within some prespecified bound. The methods are computationally efficient only when there exist efficient dominance rules to eliminate all but one of a set of assignments that yield the same profit (i.e., objective function value $\sum c_j x_j$).

The aim of rounding is to start from a problem instance I, formulated as in (36.1), and to transform it to another problem instance I' that is easier to solve. This transformation is carried out in such a way that the optimal solution value of I' is *close* to the optimal solution value of I. In particular, if we are provided with a bound ε on the fractional difference between the exact and approximate solution values, then we require that $|(F^*(I) - F^*(I'))/F^*(I)| \leq \varepsilon$, where $F^*(I)$ and $F^*(I')$ represent the optimal solution values of I and I', respectively. The transformation from I to I' is carried out in the following way: First we observe that the feasible solutions to I are independent of the c_i, $1 \leq i \leq n$. Thus if I' has the same constraints as I and the c_i' differ from the c_i by *small amounts* the optimal solution to I will be a feasible solution to I'. In addition, the solution value in I' will be close to that in I. For example, if the c_i in I have the values $(c_1, c_2, c_3, c_4) = (1.1, 2.1, 1001.6, 1002.3)$, then if we construct I' with $(c_1', c_2', c_3', c_4') = (0, 0, 1000, 1000)$ it is easy to see that the value of any solution in I is at most 7.1 more than the value of the same solution in I'. This worst-case difference is achieved only when $x_i = 1, 1 \leq i \leq 4$, is a feasible solution for I (and hence also for I'). Since $a_{ij} \leq b_j, 1 \leq i \leq n$ and $1 \leq j \leq m$, it follows that $F^*(I) \geq 1002.3$ (as one feasible solution is $x_1 = x_2 = x_3 = 0$ and $x_4 = 1$). But $F^*(I) - F^*(I') \leq 7.1$ and so $(F^*(I) - F^*(I'))/F^*(I) \approx 0.007$.

Solving I by using the procedure outlined above, the feasible assignments in $S^{(i)}$ could have the following distinct profit values: $S^{(0)} = \{0\}$, $S^{(1)} = \{0, 1.1\}$, $S^{(2)} = \{0, 1.1, 2.1, 3.2\}$, $S^{(3)} = \{0, 1.1, 2.1, 3.2, 1001.6, 1002.7, 1003.7, 1004.8\}$, and $S^{(4)} = \{0, 1.1, 2.1, 3.2, 1001.6, 1002.3, 1002.7, 1003.4, 1003.7, 1004.4, 1004.8, 1005.5, 2003.9, 2005, 2006, 2007.1\}$. Thus, barring any elimination of feasible assignments resulting from the dominance rules or from any heuristic, the solution I using the procedure outlined above would require the computation of $\sum_{i=0}^{n} |S^{(i)}| = 31$ feasible assignments. The feasible assignments for I' have the following values: $S^{(0)} = \{0\}$, $S^{(1)} = \{0\}$, $S^{(2)} = \{0\}$, $S^{(3)} = \{0, 1000\}$, $S^{(4)} = \{0, 1000, 2000\}$. Note that $\sum_{i=0}^{n} |S^{(i)}|$ is only 8. Hence I' can be solved in about one-fourth the time needed for I. An inaccuracy of at most 0.7% is introduced.

Given the c_i's and an ε, what should the c_i''s be so that $(F^*(I) - F^*(I'))/F^*(I) \leq \varepsilon$ and $\sum_{i=0}^{n} |S^{(i)}| \leq p(n)$ for some polynomial in n and $1/\varepsilon$? Once we can figure this out we will have a fully polynomial time approximation scheme for our problem since it is possible to go from $S^{(i-1)}$ to $S^{(i)}$ in time proportional to $O(|S^{(i-1)}|)$. (We shall see this in greater detail in the examples.)

Let LB be an estimate for $F^*(I)$ such that $F^*(I) \geq$ LB. Clearly, we may assume LB $\geq \max_i\{c_i\}$. If $\sum_{i=1}^{n} |c_i - c_i'| \leq \varepsilon F^*(I)$, then it is clear that $(F^*(I) - F^*(I'))/F^*(I) \leq \varepsilon$. Define $c_i' = c_i - \text{rem}(c_i, (\text{LB}\varepsilon)/n)$, where $\text{rem}(a, b)$ is the remainder of a/b, that is, $a - \lfloor a/b \rfloor b$ (e.g., rem(7,6)=1/6 and rem(2.2,1.3)=0.9).* Since $\text{rem}(c_i, \text{LB}\varepsilon/n) < \text{LB}\varepsilon/n$, it follows that $\sum |c_i - c_i'| < \text{LB}\varepsilon \leq F^*\varepsilon$. Hence, if the optimal solution to I' is used as an optimal solution for I, the fractional error is less than ε. In order to determine the time required to solve I' exactly, it is useful to introduce another problem I'' with c_i'', $1 \leq i \leq n$, as its objective function coefficients. Define $c_i'' = \lfloor (c_i n)/(\text{LB}\varepsilon) \rfloor$, $1 \leq i \leq n$. It is easy to see that $c_i'' = (c_i' n)/(\text{LB}\varepsilon)$. Clearly, the $S^{(i)}$'s corresponding to the solutions of I' and I'' will have the same number of tuples. (p_1, t_1) is a tuple in an $S^{(i)}$ for I' iff $[(p_1 n)/(\text{LB}\varepsilon), t_1]$ is a tuple in the $S^{(i)}$ for I''. Hence the time needed to solve I' is the same as that needed to solve I''. Since $c_i \leq$ LB, it follows that $c_i'' \leq \lfloor n/\varepsilon \rfloor$. Hence $|S^{(i)}| \leq 1 + \sum_{j=1}^{i} c_j'' \leq 1 + i\lfloor n/\varepsilon \rfloor$ and so $\sum_{i=0}^{n-1} |S^{(i)}| \leq n + \sum_{i=0}^{n-1} i\lfloor n/\varepsilon \rfloor = O(n^3/\varepsilon)$. Thus, if we can go from $S^{(i-1)}$ to $S^{(i)}$ in $O(|S^{(i-1)}|)$ time, then I'' and hence I' can be solved in $O(n^3/\varepsilon)$ time. Moreover, the solution for I' would be an ε-approximate solution for I and we would thus have a fully polynomial

---

*$\lfloor x \rfloor$ is the largest integer not greater than x.

time approximation scheme. When using rounding, we will actually solve I'' and use the resulting optimal solution as the solution to I.

Example 36.1 The 0-1 knapsack problem is formulated as

$$\max \sum_{i=1}^{n} c_i x_i$$
$$\sum_{i=1}^{n} w_i x_i \leq M$$
$$x_i = 0 \text{ or } 1, \ 1 \leq i \leq n.$$

While solving this problem by successively generating $S^{(0)}, S^{(1)}, \ldots, S^{(n)}$, the feasible assignments for $S^{(i)}$ may be represented by tuples of the form (p, t), where $p = \sum_{j=1}^{i} c_j x_j$ and $t = \sum_{j=1}^{i} w_j x_j$. One may easily verify the validity of the following dominance rule [2,16]: The assignment corresponding to (p_1, t_1) dominates that corresponding to (p_2, t_2) iff $t_1 \leq t_2$ and $p_1 \geq p_2$.

Let us solve the following instance of the 0-1 knapsack problem: $n = 5, M = 1112$ and $(c_1, c_2, c_3, c_4, c_5) = (w_1, w_2, w_3, w_4, w_5) = \{1, 2, 10, 100, 1000\}$. Since $c_i = w_i, 1 \leq i \leq 5$, the tuples (p, t) in $S^{(i)}, 0 \leq i \leq 5$ will have $p = t$. Consequently, it is necessary to retain only one of the two coordinates p, t. The $S^{(i)}$ obtained for this instance are $S^{(0)} = \{0\}$; $S^{(1)} = \{0, 1\}$; $S^{(2)} = \{0, 1, 2, 3\}$; $S^{(3)} = \{0, 1, 2, 3, 10, 11, 12, 13\}$; $S^{(4)} = \{0, 1, 2, 3, 10, 11, 12, 13, 100, 101, 102, 103, 110, 111, 112, 113\}$; and $S^{(5)} = \{0, 1, 2, 3, 10, 11, 12, 13, 100, 101, 102, 103, 110, 111, 112, 113, 1000, 1001, 1002, 1003, 1010, 1011, 1012, 1013, 1100, 1101, 1102, 1103, 1110, 1111, 1112\}$. The optimal solution has value $\sum c_i x_i = 1112$.

Now let us use rounding on the above problem instance to find an approximate solution with value at most 10% less than the optimal value. We thus have $\varepsilon = 1/10$. Also, we know that $F^*(I) \geq \text{LB} \geq \max\{c_i\} = 1000$. The problem I' to be solved is $n = 5, M = 1112, (c_1'', c_2'', c_3'', c_4'', c_5'') = (0, 0, 0, 5, 50)$ and $(w_1, w_2, w_3, w_4, w_5) = (1, 2, 10, 100, 1000)$. Hence $S^{(0)} = S^{(1)} = S^{(2)} = S^{(3)} = \{(0,0)\}$; $S^{(4)} = \{(0,0), (5, 100)\}$; $S^{(5)} = \{(0,0), (5, 100), (50, 1000), (55, 1100)\}$.

The optimal solution is $(x_1, x_2, x_3, x_4, x_5) = (0, 0, 0, 1, 1)$. Its value in I'' is 55 and in the original problem 1100. The error $(F^*(I) - \hat{F}(I))/F^*(I)$ is therefore $12/1112 < 0.011 < \varepsilon$. At this time we see that the solution may be improved by setting either $x_1 = 1$ or $x_2 = 1$ or $x_3 = 1$.

Rounding as described in its full generality results in $O(n^3/\varepsilon)$ time approximation schemes. It is possible to specialize this technique to the specific problem being solved. Thus Ibarra and Kim [7] obtain an $O(n/\varepsilon^2 + n \log n)$ ε-approximate algorithm for the 0-1 knapsack problem and an $O(n/\varepsilon^2)$ ε-approximate algorithm for the unrestricted nonnegative integer knapsack problem. Both their algorithms use rounding. See also Hassin [17] and Lorenz and Raz [18] where ε-approximation algorithms have been developed for the constrained shortest path problem. For further details see Chapter 37.

36.3 INTERVAL PARTITIONING

Unlike rounding, interval partitioning does not transform the original problem instance into one that is easier to solve. Instead, an attempt is made to solve the problem instance I by generating a restricted class of the feasible assignments for $S^{(0)}, S^{(1)}, \cdots, S^{(n)}$. Let P_i be the maximum $\sum_{j=1}^{i} c_j x_j$ among all feasible assignments generated for $S^{(i)}$. Then the profit interval $[0, P_i]$ is divided into subintervals each of size $P_i \varepsilon / n$ (except possibly the last interval, which may be a little smaller). All feasible assignments in $S^{(i)}$ with $\sum_{j=1}^{i} c_j x_j$ in the same subinterval are regarded as having the same $\sum_{j=1}^{i} c_j x_j$ and the dominance rules are used to discard all but one of them. The $S^{(i)}$ resulting from this elimination is used in the generation of $S^{(i+1)}$. Since the number of subintervals for each $S^{(i)}$ is at most $\lceil n/\varepsilon \rceil + 1$, $|S^{(i)}| \leq \lceil n/\varepsilon \rceil + 1$. Hence $\sum_{i=1}^{n} |S^{(i)}| = O(n^2/\varepsilon)$. The error introduced in each feasible assignment due to this elimination in $S^{(i)}$ is less than the subinterval length. This error may, however, propagate from $S^{(1)}$ up through $S^{(n)}$. However, the error is additive. If $\hat{F}(I)$ is the value of the optimal generated by interval partitioning and $F^*(I)$ is the value of a true optimal, it follows that $F^*(I) - \hat{F}(I) \leq (\varepsilon \sum_{i=1}^{n} P_i)/n$. Since $P_i \leq F^*(I)$, it follows that $(F^*(I) - \hat{F}(I))/F^*(I) \leq \varepsilon$, as desired.

In many cases the algorithm may be speeded up by starting with a good estimate, LB for $F^*(I)$ such that $F^*(I) \geq \text{LB}$. The subinterval size is then $\text{LB}\varepsilon/n$ rather than $P_i\varepsilon/n$. When a feasible assignment with value greater than LB is discovered, the subinterval size can be chosen as described above.

Example 36.2

Consider the same instance of the 0-1 knapsack problem as in Example 36.1. $\varepsilon = 1/10$ and $F^* \geq \text{LB} \geq 1000$. We can start with a subinterval size of $\text{LB}\varepsilon/n = 1000/50 = 20$. Since all tuples (p,t) in $S^{(i)}$ have $p = t$, only p will be explicitly retained. The intervals are $[0, 20), [20, 40), [40, 60), \cdots$ and so on. Using interval partitioning we obtain $S^{(0)} = S^{(1)} = S^{(2)} = S^{(3)} = \{0\}$; $S^{(4)} = \{0, 100\}$; $S^{(5)} = \{0, 100, 1000, 1100\}$.

The optimal generated by interval partitioning is $(x_1, x_2, x_3, x_4, x_5) = (0, 0, 0, 1, 1)$ and the value $\hat{F}(I) = 1100$. $(F^*(I) - \hat{F}(I))/F^*(I) = 12/1112 < 0.011 < \varepsilon$. Again, the solution value may be improved by using a heuristic to change some of the x_i's from 0 to 1.

See Hassin [17] for an application of interval partitioning in developing an ε-approximation algorithm for the constrained shortest path problem.

36.4 SEPARATION

Assume that in solving a problem instance I, we have obtained an $S^{(i)}$ with feasible solutions having the following values of $\sum_{j=1}^{i} c_j x_j$: $0, 3.9, 4.1, 7.8, 8.2, 11.9, 12.1$. Further assume that the interval size $P_i\varepsilon/n$ is 2. Then the subintervals are $[0, 2), [2, 4), [4, 6), [6, 8), [8, 10), [10, 12)$, and $[12, 14)$. Each value falls in a different subinterval and so no feasible assignments are eliminated. However, there are three pairs of assignments with values within $P_i\varepsilon/n$. If the dominance rules are used for each pair, only four assignments will remain. The error introduced is at most $P_i\varepsilon/n$. More formally, let $a_0, a_1, a_2, \ldots, a_r$ be the distinct values of $\sum_{j=1}^{i} c_j x_j$ in $S^{(i)}$. Let us assume $a_0 < a_1 < a_2 \cdots < a_r$. We will construct a new set J from $S^{(i)}$ by making a left-to-right scan and retaining a tuple only if its value exceeds the value of the last tuple in J by more than $P_i\varepsilon/n$. This is described by the following algorithm:

```
J ← assignment corresponding to a_0; XP ← a_0
  for j ← 1 to r do
    if a_j > XP + P_i ε/n then
      [put assignment corresponding to a_j into J, XP ← a_j]
    end if
  end for
```

The analysis for this strategy is the same as that for interval partitioning. The same comments regarding the use of a good estimate for $F^*(I)$ hold here too.

Intuitively, one may expect separation always to work better than interval partitioning. The following example illustrates that this need not be the case. However, empirical studies with one problem (see [19]) indicate that interval partitioning is inferior in practice.

Example 36.3

Using separation on the data of Example 36.1 yields the same $S^{(i)}$ as obtained using interval partitioning. We have already seen an instance where separation performs better than interval partitioning. Now we shall see an example where interval partitioning does better than separation. Assume that the subinterval size $\text{LB}\varepsilon/n$ is 2. Then the intervals are $[0, 2), [2, 4), [4, 6) \ldots$. Assume further that $(c_1, c_2, c_3, c_4, c_5) = (3, 1, 5.1, 5.1, 5.1)$. Then, following the use of interval partitioning, we have $S^{(0)} = \{0\}$; $S^{(1)} = \{0, 3\}$; $S^{(2)} = \{0, 3, 4\}$; $S^{(3)} = \{0, 3, 4, 8.1\}$; $S^4 = \{0, 3, 4, 8.1, 13.2\}$; and $S^{(5)} = \{0, 3, 4, 8.1, 13.2, 18.3\}$.

Using separation with $\text{LB}\varepsilon/n = 2$ we have $S^{(0)} = \{0\}$; $S^{(1)} = \{0, 3\}$; $S^{(2)} = \{0, 3\}$; $S^{(3)} = \{0, 3, 5.1, 8.1\}$; $S^{(4)} = \{0, 3, 5.1, 8.1, 10.2, 13.2\}$; and $S^{(5)} = \{0, 3, 5.1, 8.1, 10.2, 13.2, 15.3, 18.3\}$. The three methods for obtaining fully polynomial time approximation schemes can be applied to a wide variety of problems. Some of these problems are 0-1 knapsack problem [7]; integer knapsack problem [7]; job sequencing with deadlines [10]; minimizing weighted mean finish time [6,10]; finding an optimal SPT schedule [10]; and finding minimum finish time schedules on identical, uniform, and nonidentical machines [6,10].

Summary

Over the past two decades, there have been a number of significant advances in the area of approximation algorithms for combinatorial optimization. See Vazirani [20] and Hochbaum [21].

References

[1] M. Karp. Reducibility among combinatorial problems. In R. E. Miller and J. W. Thatcher, editors, *Complexity of Computer Computations*, pp. 85–103. Plenum Press, New York, 1972.

[2] G. Nemhauser and Z. Ullman. Discrete dynamic programming and capital allocation. *Management Sci.*, **15** (1969), 494–505.

[3] S. Sahni. Computationally related problems *SIAM J. Computing*, **3** (1974), 277–292.

[4] S. Sahni and T. Gonazalez. P-complete approximation problems. *J. Assoc. Comput. Machinery*, **23** (1976), 555–556.

[5] R. Graham. Bounds for certain multiprocessing anomalies. *Bell Systems Tech.*, **4** (1966), 1563–1581.

[6] E. Horowitz and S. Sahni. Exact and approximate algorithms for scheduling nonidentical processors. *J. Assoc. Comput. Machinery*, **23** (1976), 317–327.

[7] O. Ibarra and C. Kim. Fast approximation algorithms for the knapsack and sum of subsets problems. *J. Assoc. Comput. Machinery*, **22** (1975), 463–468.

[8] D. Johnson. Approximation algorithms for combinatorial problems. *J. Comput. Sys. Sci.*, **9** (1974), 256–278.

[9] S. Sahni. Approximate algorithms for the 0/1 knapsack problem. *J. Assoc. Comput. Machinery*, **22** (1975), 115–124.

[10] S. Sahni. Algorithms for scheduling independent tasks. *J. Assoc. Comput. Machinery*, **23** (1976), 114–127.

[11] M. Garey and D. Johnson. Approximation algorithm for combinatorial problems: An annotated bibliography. In J. Traub, editor *Algorithms and Complexity*, pp. 41–52. Academic Press, New York, 1976.

[12] M. Garey and D. Johnson. Approximation algorithms for combinatorial problems: Prospects and limitations. Lecture by D. Johnson presented at the *Symposium on Algorithms and Complexity*, Carnegie Mellon Institute, Pittsburgh, PA, 1976.

[13] R. Bellman and S. Dreyfus. *Applied Dynamic Programming*. Princeton University Press, Princeton, NJ, 1962.

[14] G. Nemhauser. *Introduction to Dynamic Programming*. John Wiley & Sons, New York, 1966.

[15] E. Horowitz and S. Sahni. Computing partitions with applications to the knapsack problem. *J. Assoc. Comput. Machinery*, **21** (1974), 277–292.

[16] E. Horowitz and S. Sahni. *Fundamentals of Data Structures*. Computer Science Press, Woodland Hills, CA, 1976.

[17] R. Hassin. Approximation schemes for the restricted shortest path problem. *Math. Oper. Res.*, **17** (1992), 36–42.

[18] D. H. Lorenz and Danny Raz. A simple efficient approximation scheme for the restricted shortest path problem. *Operations Research Letters*, **28** (2001), 213–219.

[19] S. Sahni. General techniques for combinatorial approximation. *Operations Research*, **25(6)** (1977), 920–936.

[20] V. V. Vazirani. *Approximation Algorithms*. Springer, Berlin, Germany, 2003.

[21] D. S. Hochbaum. *Approximation Algorithms for NP-Hard Problems*. PWS Publishing, Boston, MA, 1998.

CHAPTER 37

ε-Approximation Schemes for the Constrained Shortest Path Problem

Krishnaiyan "KT" Thulasiraman

CONTENTS

37.1 Introduction 1035
37.2 Lorenz–Raz Approach 1035

37.1 INTRODUCTION

In this chapter we present an ε-approximation algorithm for the constrained shortest path (CSP) problem. Recall from Chapter 36 that the CSP problem is to determine, from among all $s-t$ paths, a minimum cost path that has a delay less than or equal to a specified value. Also, an ε-approximation algorithm for the CSP problem constructs a path with cost $\tilde{c}$ such that $|(c^*-\tilde{c})|/c^* \leq \varepsilon$ for any $\varepsilon > 0$, where c^* is the cost of the optimum solution. Warburton [1] was the first to develop an ε-approximation algorithm for the CSP problem when the given graph is acyclic. This result was improved later by Hassin [2] that resulted in a fully polynomial time approximation algorithm with complexity of $O(mn^2/\varepsilon \log(n/\varepsilon))$ using scaling and rounding of edge costs, a dynamic programming-based test procedure to determine if the optimum value OPT $\geq v$ or not for a given value of v and binary search in a logarithmic scale. Hassin first developed an approximation algorithm of complexity $O((mn/\varepsilon)\log\log(\text{UB}/\text{LB}))$ where UB and LB are valid upper and lower bounds, respectively for the optimum solution. He also showed how to refine this algorithm resulting in a fully polynomial time approximation algorithm. To achieve this Hassin used the interval partitioning technique [3] (see also Chapter 35). Subsequently Lorenz and Raz [4] achieved an algorithm which improves Hassin's ε-approximation algorithm by a factor of n. This is done by showing how to find UB and LB such that $UB/LB \leq n$. We call this improved algorithm as Hassin–Lorenz–Raz (HLR) algorithm. Through Section 37.2 the HLR algorithm is developed. We follow the treatment in Lorenz and Raz [4].

37.2 LORENZ–RAZ APPROACH

Consider a directed graph $G = (V, E)$ with each edge $e \in E$ associated with a cost $c_e \geq 0$ and a delay $d_e \geq 0$. Let T be a positive integer and $s, t \in V$ be the source and destination nodes. The CSP problem is to find a path p from s to t such that delay(p) on this path is less than or equal to T and the cost $c(p)$ of p is minimum. We denote the optimum cost

Scaled Pseudopolynomial Plus [SPPP] $(G(V,E), \{d_l, c_l\}_{l \in E}, T, L, U, \varepsilon)$

1: $S \leftarrow \frac{L\varepsilon}{n+1}$
2: for each $l \in E$
3:     define $\tilde{c}_l \equiv \lfloor c_l/S \rfloor + 1$
4: $\tilde{U} \leftarrow \lfloor U/S \rfloor + n + 1$
5: for all $v \neq s$
6:     $D(v, 0) \leftarrow \infty$
7: $D(s, 0) \leftarrow 0$
8: for $i = 1, 2, \ldots, \tilde{U}$
9:     for $v \in V$
10:       $D(v, i) \leftarrow D(v, i-1)$
11:       for $l \in \{(u,v) \mid \tilde{c}_{(u,v)} \leq i\}$
12:         $D(v, i) \leftarrow \min\{D(v, i), d_l + D(u, i - \tilde{c}_l)\}$
13:     if $D(t, i) \leq T$
14:       return the corresponding path and cost
15: return FAIL

Figure 37.1 Scaled pseudopolynomial plus algorithm. (Data from D. H. Lorenz and Danny Raz. A simple efficient approximation scheme for the restricted shortest path problem. *Operations Research Letters*, **28** [June 2001], 213–219.)

by c^* and an optimum path by p^*. The main idea in Lorenz and Raz's approach is to scale the costs and then apply the dynamic programming-based algorithm in Figure 37.1 to find a path with the smallest delay for each cost. We denote by $D(v, i)$ the minimum delay on a path from s to node v with cost bounded by i. Note that the scaled cost $\tilde{c}_l \geq 1$ for every edge $l \in E$. Also, $\tilde{c}(p)$ will denote the cost of p with scaled values.

Lemma 37.1 *For any path p and the scaled costs $\tilde{c}_s$ defined in Figure 37.1, $c(p) \leq \tilde{c}(p)S \leq c(p) + nS$.*

Proof. It follows from the definition of scaled costs that for each $l \in E$

$$c_l/S \leq \tilde{c}_l \leq c_l/S + 1. \tag{37.1}$$

So

$$c_l \leq \tilde{c}_l S, \tag{37.2}$$

$$\tilde{c}_l S \leq c_l + S, \tag{37.3}$$

and

$$c(p) = \sum_{l \in p} c_l$$

$$\leq S \sum_{l \in p} \tilde{c}_l, \text{ from } (37.2)$$

$$= \tilde{c}(p)S$$

$$\leq c(p) + nS, \text{ from } (37.3). \blacksquare$$

Lemma 37.2 *Any path p returned by the Scaled Pseudopolynomial Plus (SPPP) algorithm satisfies*

$$c^* \leq c(p) \leq U + (n+1)S = U + L\varepsilon.$$

Proof. Since c^* is the optimum cost, $c^* \leq c(p)$. Since p is returned by SPPP algorithm, $\tilde{c}(p) \leq \tilde{U}$. So

$$\tilde{c}(p)S \leq \tilde{U}S$$
$$\leq U + (n+1)S, \text{ from the definition of } \tilde{U} \qquad (37.4)$$
$$= U + L\varepsilon, \text{ from the definition of } S.$$

Since $c(p) \leq \tilde{c}(p)S$, by Lemma 37.1, we have $c(p) \leq U + L\varepsilon$ by (37.4). ∎

Theorem 37.1 *If $\tilde{U} \geq c^*$ SPPP algorithm returns a feasible path p with cost $c(p) \leq c^* + L\varepsilon$. The complexity of the algorithm is $O(m(n/\varepsilon)(U/L))$, if $\varepsilon \leq 1$ and $U > L$.*

Proof. For each $l \in E$, $\tilde{c}_l \leq c_l/S + 1$. So

$$\tilde{c}(p^*) \equiv \sum_{l \in p^*} \tilde{c}_l \leq \frac{c^*}{S} + |p^*|$$
$$\leq \frac{U}{S+n} \qquad (37.5)$$
$$\leq \tilde{U},$$

where $|p^*|$ is the number of edges in p^*. If $\tilde{c}(p^*) < \tilde{c}(p)$, it follows from (37.5) that SPPP algorithm will encounter p^* earlier than p and would have returned p^*, instead of p, contradicting our assumption. So $\tilde{c}(p^*) \geq \tilde{c}(p)$. Then

$$c(p) \leq \tilde{c}(p)S, \text{ by Lemma 37.1}$$
$$\leq \tilde{c}(p^*)S$$
$$\leq c(p^*) + nS, \text{ by Lemma 37.1}$$
$$\leq c^* + L\varepsilon, \text{ from the definition of } S.$$

As regards the complexity of SPPP algorithm, each edge is examined at most once during each iteration i, $1 \leq i \leq \tilde{U}$. So the overall complexity of the algorithm is $O(m\tilde{U}) = O(m(n/\varepsilon)(U/L) + n)$. If $\varepsilon \leq 1$ and $U > L$, the overall complexity of SPPP algorithm is $O(m(n/\varepsilon)(U/L))$. ∎

We are now ready to define $T(\varepsilon, B)$, a test procedure to determine if $c^* \geq B$ for a given value of $\varepsilon > 0$. For an instance of the CSP problem, an ε-test procedure $T(\varepsilon, B)$ for a given value B and $\varepsilon > 0$ is defined as follows:

1. If $T(\varepsilon, B)$ answers Yes, then $c^* \geq B$.
2. If $T(\varepsilon, B)$ answers No, then $c^* \leq B(1+\varepsilon)$.

Theorem 37.2 *The test*

$$T(1,B) = \begin{cases} YES, & \text{if } SPPP(T,B,B,1) \text{ returns FAIL} \\ NO, & \text{otherwise} \end{cases}$$

is a 1-test and requires $O(mn)$ steps.

Proof. If $c^* \leq B$, then by Theorem 37.1, $SPPP(T,B,B,1)$ returns a feasible path. So if $SPPP(T,B,B,1)$ returns FAIL, $c^* > B$. In other words, if $T(1,B)$ answers YES, then $c^* > B$. If $c^* \leq B$, then $SPPP(T,B,B,1)$ returns a feasible path p with $c(p) \leq U + L\varepsilon \leq 2B$. Thus $T(1,B)$ as defined satisfies the requirement of a 1-test and has complexity $O(mn)$ because $L = B$ and $\varepsilon = 1$. ∎

```
Hassin–Lorenz–Raz $(G(V, E), \{d_l, c_l\}_{l \in E}, T, LB, UB, \varepsilon)$
 1:   $B_L \leftarrow LB$
 2:   $B_U \leftarrow \lceil UB/2 \rceil$
 3:   while $B_U/B_L > 2$
 4:       $B \leftarrow (B_L \cdot B_U)^{1/2}$;
 5:       if $T(1, B) =$ YES then $B_L \leftarrow B$
 6:          else $(T(1, B) =$ NO) $B_U \leftarrow B$;
 7:   return SPPP$(G(V, E), \{d_l, c_l\}_{l \in E}, T, B_L, 2B_U, \varepsilon)$
```

Figure 37.2 Hassin–Lorenz–Raz algorithm. (Data from D. H. Lorenz and Danny Raz. A simple efficient approximation scheme for the restricted shortest path problem. *Operations Research Letters*, **28** [June 2001], 213–219.)

Lorenz and Raz (LR algorithm) ε-approximation algorithm for the CSP problem combines the $T(1, B)$ test with SPPP algorithm, and is given in Figure 37.2. We will call this HLR algorithm since many of its ideas have their origins in Hassin's work [2].

Theorem 37.3 *The HLR algorithm is an ε-approximation algorithm for the CSP problem with time complexity $O((mn/\varepsilon) + mn \log \log(\text{UB}/\text{LB}))$, where LB and UB are, respectively, valid lower and upper bounds on the cost of the optimum solution.*

Proof. The HLR algorithm in Figure 37.2 starts with valid lower and upper bounds LB and UB, respectively. The algorithm looks for a lower bound B_L and a value for B_U such that $B_U/B_L \leq 2$ and $2B_U$ is a valid upper bound. This is achieved by a binary search on $\log B_L$, $\log B_U$ in the range $\log \text{LB} \ldots \log(\text{UB}/2)$. In each iteration a value $B = (B_U B_L)^{1/2}$ is selected and the test $T(1, B)$ is applied. Depending on the outcome of the test, the search interval is updated. It is easy to show that the value of $\log(B_U/B_L)$ after each iteration is half its value in the previous iteration. So the binary search will terminate after $\log \log(\text{UB}/\text{LB})$ tests and the time complexity of the binary search is $O(mn \log \log(\text{UB}/\text{LB}))$.

The $T(1, B)$ test guarantees that at each iteration B_L is a valid lower bound and $2B_U$ is a valid upper bound on c^* (see lines 5 and 6 in the algorithm). After the binary search is completed, the algorithm applies SPPP$(T, B_L, 2B_U, \varepsilon)$ and determines a feasible path p with cost $c(p) \leq c^* + B_L \varepsilon \leq c^*(1 + \varepsilon)$ (Theorem 37.1). So HLR algorithm is an ε-approximation algorithm. Since $U/L = 2B_U/B_L = O(1)$ in this case, SPPP$(T, B_L, 2B_U, 1)$ (line 7) requires $O(mn/\varepsilon)$ steps (see Theorem 37.1). So the overall complexity of this algorithm is $O((mn/\varepsilon) + mn \log \log(\text{UB}/\text{LB}))$. ∎

The complexity of the HLR algorithm involves UB and LB and so is not a fully polynomial time approximation scheme (FPTAS). To achieve a FPTAS, Lorenz and Raz propose a method to find UB and LB such that UB/LB $= n$. Let $c_1 < c_2 < \ldots < c_l$ be all the distinct costs of edges in the given graph $G(V, E)$. Let G_j be the induced subgraph of all the edges with cost no greater than c_j and G_0 defined as $\emptyset$. We have $G_i \subset G_{i+1}$ for $0 \leq i \leq l - 1$. Note that $G_l = G$. Then there is a unique j such that $G_i, i < j$ has no feasible $s - t$ path and G_j has a feasible $s - t$ path. Since G_j has a feasible $s - t$ path, the cost of such a path will be at most nc_j. So $c^* \leq nc_j$. Since G_{j-1} has no feasible $s - t$ path, every $s - t$ feasible path must have at least one edge with cost greater than or equal to c_j. So $c^* \geq c_j$. So, once c_j is determined, we can apply HLR algorithm with LB $= c_j$ and UB $= nc_j$. Then we would get a fully polynomial time approximation time algorithm of complexity $O((mn/\varepsilon) + mn \log \log n)$. Note that c_j can be determined using binary search and applying a shortest path algorithm

> **Simple Efficient Approximation [SEA]** (G, T, ε)
> 1: $low \leftarrow 0; high \leftarrow l;$
> 2:   while $low < high - 1$
> 3: $j \leftarrow \lfloor (high + low)/2 \rfloor$
> 4:   if $shortest_{s,t}path(G_j) < T$ then $high \leftarrow j;$
> 5:   else $low \leftarrow j$
> 6: $UB = nc_{high}; LB \leftarrow c_{high};$
> 7:   return Hassin–Lorenz–Raz $(G, \{d_l, c_l\}_{l \in E}, T, LB, UB, \varepsilon)$

Figure 37.3 Simple efficient approximation (SEA) algorithm. (Data from D. H. Lorenz and Danny Raz. A simple efficient approximation scheme for the restricted shortest path problem. *Operations Research Letters*, **28** [June 2001], 213–219.)

at each iteration (see Figure 37.3). The complexity of such a search is $O(\log m)$ times the complexity of a shortest path algorithm, which is $O(n \log n + m)$. So the HLR algorithm remains the dominant part of the simple efficient algorithm in Figure 37.3.

Theorem 37.4 *Algorithm simple efficient approximation (Figure 37.3) is a FPTAS for the CSP problem with time complexity* $O(mn(\log \log n + 1/\varepsilon))$.

We conclude by drawing attention to the work by Cynthia Phillips [5]. The work uses Dijkstra's shortest path algorithm as a building block to find an ε-approximation solution. The complexity of this algorithm is $O((mn/\varepsilon) + (n^2/\varepsilon) \log(n^2/\varepsilon) \log \log(UB/LB))$. Goel et al. [6] deal with another variant of the CSP problem. Xue et al. [7] discuss approximation schemes for the CSP problem under multiple additive constraints.

References

[1] A. Warburton. Approximation of Pareto Optima in Multiple-Objective Shortest Path Problems. *Oper. Res.*, **35** (1987), 70–79.

[2] R. Hassin. Approximation Schemes for the Restricted Shortest Path Problem. *Math. Oper. Res.*, **17** (1992), 36–42.

[3] S. Sahni. General Techniques for Combinatorial Approximations. *Oper. Res.*, **25** (1977), 920–936.

[4] D. H. Lorenz and D. Raz. A Simple Efficient Approximation Scheme for the Restricted Shortest Path Problem. *Operations Research Letters*, **28** (June 2001), 213–219.

[5] C. Phillips. The Network Inhibition Problem. In *Proceedings of the 25th Annual Symposium on Theory of Computing*, San Diego, CA, pp. 776–785, May 1993.

[6] A. Goel, K. G. Ramakrishnan, D. Kataria, and D. Logothetis. Efficient Computation of Delay-sensitive Routes from One Source to All Destinations. In *Proceedings of IEEE International Conference on Computer Communications*, Anchorage, AK, pp. 854–858, 2001.

[7] G. Xue, A. Sen, W. Zhang, J. Tang, and K. Thulasiraman. Finding a Path Subject to Many Additive QoS Constraints. *IEEE/ACM Trans. Netw.*, **15**(1) (2007), 201–211.

CHAPTER 38

Constrained Shortest Path Problem: Lagrangian Relaxation-Based Algorithmic Approaches

Ying Xiao

Krishnaiyan "KT" Thulasiraman

CONTENTS

38.1	Introduction	1041
38.2	CSP Problem and Generality of the LARAC Algorithm	1042
	38.2.1 CSP and DUAL-RELAX CSP Problems	1042
	38.2.2 LARAC Algorithm	1043
	38.2.3 MCRT Problem	1044
	38.2.4 Equivalence of LARAC and MCRT Algorithms	1045
38.3	An Algebraic Study of the RELAX-CSP Problem and Its Generalization	1046
38.4	LARAC-BIN: A Binary Search-Based Approach to the DUAL-RELAX CSP Problem	1049
38.5	GEN-LARAC: A Generalized Approach to the CSP Problem Under Multiple Additive Constraints	1052
	38.5.1 Formulation of the CSP(k) Problem and Its Relaxation	1052
	38.5.2 A Strongly Polynomial Time Approximation Algorithm for CSP(1) Problem	1054
	38.5.3 GEN-LARAC for the CSP(k) Problem	1056
	38.5.3.1 Optimality Conditions	1056
	38.5.3.2 GEN-LARAC: Coordinate Ascent Method	1056
	38.5.3.3 Verification of Optimality of Λ	1057

38.1 INTRODUCTION

Shortest path, minimum cost flow, and maximum flow computation are fundamental problems in operations research. Though interesting in their own right, algorithms for these problems also serve as building blocks in the design of algorithms for complex problems encountered in large-scale industrial applications. So, over the years an extensive literature on various aspects of these problems emerged, which helped in problem solving in polynomial time. However, adding one or more additional additive constraints makes them intractable.

In this chapter, we focus on the constrained shortest path (CSP) problem. This problem requires determination of a minimum cost path from a source node to a destination node of a network subject to the condition that the total delay of the path be less than or equal to a specified value. The CSP problem has attracted considerable attention from different research communities: operations research, computer science, and telecommunication networks. The interest from the telecommunications community arises from a great deal of emphasis on the need to design communication protocols that deliver certain performance guarantees. This need, in turn, is the result of an explosive growth in high bandwidth real-time applications that require stringent quality of service guarantees.

This chapter is organized as follows. In Section 38.2, we present the CSP problem and the general class of optimization problems, namely, the minimum cost-restricted time combinatorial optimization (MCRT) problem [1]. We also present the Lagrangian Relaxation-based Aggregated Cost (LARAC) algorithm of [2] for the CSP and MCRT problems. We point out the equivalence of the LARAC and the MCRT algorithms. In Section 38.3 we present an algebraic study of the integer relaxation of the CSP problem. In view of the equivalence of the LARAC and the MCRT algorithms, the results to be presented, although originally intended for the CSP problem, hold true for the MCRT problem too. We establish these results and certain new results for the general case without involving the properties of shortest paths. In Section 38.4, we present a binary search-based approach for the CSP problem and also show that both the LARAC algorithm and this algorithm can be embedded with a tuning parameter whose value can be specified in advance depending on the allowable deviation of the cost of the path produced from the optimal cost. In Section 38.5, we develop a strongly polynomial time algorithm for the integer relaxation of the CSP problem. This is based on the parametric search approach developed by Megiddo [3] for fractional combinatorial optimization problems. In Sections 38.6 through 38.8 we develop GEN-LARAC, a generalization of the LARAC algorithm for the CSP problem with multiple additive constraints.

38.2 CSP PROBLEM AND GENERALITY OF THE LARAC ALGORITHM

First we give a formal definition of the CSP problem. We use the terms links and nodes for edges and vertices, respectively, following the convention in the networking literature.

38.2.1 CSP and DUAL-RELAX CSP Problems

Consider a directed network $G(V, E)$. Each link $(u, v) \in E$ is associated with two weights $c_{uv} > 0$ (say, cost) and $d_{uv} > 0$ (say, delay). Also are given two distinguished nodes s and t and a real number $\Delta > 0$. Let P_{st} denote the set of all directed $s - t$ paths and for any $s - t$ path p, define

$$c(p) = \sum_{(u,v) \in p} c_{uv} \text{ and } d(p) = \sum_{(u,v) \in p} d_{uv}.$$

All paths considered in this chapter are directed paths. Let $P_{st}(\Delta)$ be the set of all the $s - t$ paths p such that $d(p) \leq \Delta$. A path in the set $P_{st}(\Delta)$ is called a feasible path. The CSP problem is to find a path $p^* = arg \min\{c(p) |\ p \in P_{st}(\Delta)\}$. In other words, the CSP problem is to find a minimum cost feasible $s - t$ path. It can be formulated as the following integer linear program.

CSP:

$$\text{Minimize} \sum_{(u,v) \in E} c_{uv} x_{uv}$$

subject to $\forall u \in V$,

$$\sum_{\{v|(u,v) \in E\}} x_{uv} - \sum_{\{v|(v,u) \in E\}} x_{vu} = \begin{cases} 1, & \text{for } u = s \\ -1, & \text{for } u = t \\ 0, & \text{otherwise} \end{cases}$$

$$\sum_{(u,v) \in E} -d_{uv} \cdot x_{uv} - w = -\Delta, w \geq 0$$

$$x_{uv} = 0 \text{ or } 1, \forall (u,v) \in E.$$

The CSP problem is known to be NP-hard [4,5]. The main difficulty lies with the integrality condition that requires that the variables x_{uv} be 0 or 1. Removing or relaxing this requirement from the above integer linear program and letting $x_{uv} \geq 0$ leads to RELAX-CSP, the relaxed CSP problem. It is often convenient to solve the dual of the relaxed form of the CSP problem which we present below.

The dual involves $s - t$ paths and a variable $\lambda \geq 0$. For each link (u,v), let the aggregated cost c_λ be defined as $c_{uv} + \lambda\, d_{uv}$. For a given λ, let $c_\lambda(p)$ denote the aggregated cost of the path p. Finally define $L(\lambda)$ as:

$$L(\lambda) = \min\{c_\lambda(p) | p \in P_{st}\} - \lambda \Delta. \tag{38.1}$$

Note that in the above, $\min\{c_\lambda(p)| p \in P_{st}\}$ is the same as the minimum aggregated cost of an $s - t$ path with respect to a given value of λ. This can be easily obtained by applying Dijkstra's algorithm using aggregated link costs. Let the $s - t$ path which has minimum aggregated cost with respect to a given λ be denoted as p_λ. Then $L(\lambda) = c_\lambda(p_\lambda) - \lambda \Delta$ and the dual of the RELAX CSP can be presented in the following form.

DUAL-RELAX CSP:

Find $L^* = \max\{L(\lambda) \mid \lambda \geq 0\}$.

We note that the problem of maximizing $L(\lambda)$ as above is also called the Lagrangian dual problem. The value of λ that achieves the maximum $L(\lambda)$ in DUAL-RELAX CSP will be denoted by λ^*. Note that L^*, the optimum value of DUAL-RELAX CSP is a lower bound on the optimum cost of the path solving the corresponding CSP problem. The key issue in solving DUAL-RELAX CSP is how to search for the optimal λ and determining the termination condition for the search. The LARAC algorithm of [2] presented in Figure 38.1 is one such efficient search procedure.

38.2.2 LARAC Algorithm

In the LARAC algorithm $Dijkstra\,(s, t, c)$, $Dijkstra\,(s, t, d)$, and $Dijkstra\,(s, t, c_\lambda)$ denote, respectively, Dijkstra's shortest path algorithm using link costs, link delays, and aggregated link costs with respect to the multiplier λ.

1. In the first step, the algorithm calculates the shortest path on link costs. If the path found meets the delay constraint, this is surely the optimal path. Otherwise, the algorithm stores the path as the latest infeasible path, simply called the p_c path. Then it determines the shortest path on link delays denoted as p_d. If p_d is infeasible, there is no solution to this instance.

> **Procedure** LARAC (s, t, d, Δ)
>
> $p_c := Dijkstra\ (s, t, c)$
>
> **If** $d(p_c) \leq \Delta$ **then** return p_c
>
> $p_d := Dijkstra\ (s, t, d)$
>
> **if** $d(p_d) > \Delta$ **then** return *there is no solution*
>
> **repeat**
>
> $\lambda := \dfrac{c(p_c) - c(p_d)}{d(p_d) - d(p_c)}$
>
> $r := Dijkstra\ (s, t, c_\lambda)$
>
> **if** $c_\lambda(r) = c_\lambda(p_c)$ **then** return p_d
>
> **else if** $d(r) \leq \Delta$ **then** $p_d := r$ **else** $p_c := r$
>
> **end repeat**
>
> **end procedure**

Figure 38.1 LARAC algorithm.

2. Set $\lambda = (c(p_c) - c(p_d))/(d(p_d) - d(p_c))$. With this value of λ, we can find a new c_λ-minimal path r. If $c_\lambda(r) = c_\lambda(p_c)(= c_\lambda(p_d))$, we have obtained the optimal λ according to Theorem 38.2 to be proved in Section 38.3. Otherwise, set r as the new p_c or p_d according to whether r is infeasible or feasible.

We next define the minimum cost restricted time (MCRT) problem studied in [1].

38.2.3 MCRT Problem

Given a finite set P, a finite collection S of subsets of P, a non-negative threshold h, and two non-negative real-valued functions $y : P \to R^+$ (say, cost) and $x : P \to R^+$ (say, delay). The MCRT problem is to seek a solution $F^* = \arg\min\{y(F)|\ F \in S, x(F) \leq h\}$, where $z(G) = \sum_{g \in G} z(g)$ for $z \in \{x, y\}$ and $G \in S$.

Evidently, the CSP problem is a special case of the MCRT problem and so the MCRT problem is also NP-hard. Therefore, we consider solving the integer relaxation of the MCRT problem. This is achieved by the MCRT algorithm given in [1] and presented in Figure 38.2. In this algorithm, it is assumed that there is an effective algorithm $A(a, b)$ for the corresponding minimum cost problem with respect to $ax(p) + by(p)$, $p \in S$, where a, b are the multipliers. For instance, in the case of the CSP problem, Dijkstra's algorithm for the minimum cost path problem can play the role of algorithm A. In Figure 38.2, algorithm $A(a, b)$ returns $p = \arg\min\{ax(r) + by(r)|r \in S\}$.

```
Procedure MCRT (h)

    F := A(0,1)

    if x(F) ≤ h then return F.

    H := A(1,0)

    if x(H) > h then return no solution

    repeat

        a := y(H) − y(F)

        b := x(F) − x(H)

        c := x(F)y(H) − x(H)y(F)    (a)

        G := A(a,b)

        if c = ax(G) + by(G) then    (b)

            if x(G) ≤ h then return G else return H

        if c > ax(G) + by(G) then    (c)

            if x(G) ≤ h then H := G else F := G

    end repeat

end procedure
```

Figure 38.2 MCRT algorithm.

38.2.4 Equivalence of LARAC and MCRT Algorithms

Following the definition of the variables in Figures 38.1 and 38.2, it can be seen that H corresponds to p_d while F corresponds to p_c and λ corresponds to a/b because

$$\frac{a}{b} = \frac{y(H) - y(F)}{x(F) - x(H)}.$$

Furthermore,

$$\frac{c}{b} = \frac{x(F)y(H) - x(H)y(F)}{x(F) - X(H)} = \frac{y(H) - y(F)}{x(F) - x(H)}x(F) + y(F) = y(F) + \frac{a}{b}x(F).$$

If the expressions (a)–(c) in procedure MCRT are scaled by b, the MCRT algorithm reduces to the LARAC algorithm. In view of the equivalence of the LARAC algorithm and the MCRT algorithm, in the rest of the chapter we shall refer to both these algorithms as simply LARAC.

38.3 AN ALGEBRAIC STUDY OF THE RELAX-CSP PROBLEM AND ITS GENERALIZATION

The LARAC algorithm was originally intended for the CSP problem. In view of its generality as discussed in the previous section, one would expect that the claims in [2] (stated without proof) on which the LARAC algorithm is based do not depend on the properties of shortest paths. In other words, we would like to establish these claims without invoking properties of shortest paths. Furthermore, in the following section we also establish certain other new results that throw much insight into the structure of the solutions of the DUAL-RELAX CSP problem. Though our proofs below do not involve shortest paths or their properties, we have decided to retain the terms such as *minimal path* whose interpretation in the general context should be obvious.

Claim 38.1 [2] *Let $L(\lambda) = \min\{c_\lambda(p) | \ p \in P_{st}\} - \lambda \Delta$. Then $L(\lambda)$ is a lower bound to the optimum objective of the CSP problem for any $\lambda \geq 0$.*

Claim 38.2 [2] *L is a concave piecewise linear function, namely, the minimum of the linear functions $c(p) + \lambda(d(p) - \Delta)$ for all $p \in P_{st}$.*

Claim 38.3 [2] *For any $\lambda \geq 0$ and c_λ-minimal path $p_\lambda, d(p_\lambda)$ is a subgradient of L in the point λ.*

Theorem 38.1 [2] *If $\lambda < \lambda^*$, then $d(p_\lambda) \geq \Delta$, and if $\lambda > \lambda^*$, then $d(p_\lambda) \leq \Delta$ for each c_λ-minimal path p_λ.*

Proof. Let p and p^* denote a c_λ-minimal path and a c_{λ^*}-minimal path, respectively.

$$L(\lambda^*) = c(p^*) + \lambda^* d(p^*) - \lambda^* \Delta \leq c(p) + \lambda^* d(p) - \lambda^* \Delta = L(\lambda) + (\lambda^* - \lambda)(d(p) - \Delta).$$

Since $L(\lambda^*) \geq L(\lambda), (\lambda^* - \lambda)(d(p) - \Delta) \geq 0$.

Therefore, if $\lambda < \lambda^*$ then $d(p_\lambda) \geq \Delta$, and if $\lambda > \lambda^*$ then $d(p_\lambda) \leq \Delta$ for each c_λ-minimal path p_λ. ∎

Theorem 38.2 [2] *A value $\lambda > 0$ maximizes the function $L(\lambda)$ if and only if there are paths p_c and p_d which are both c_λ-minimal and for which $d(p_c) \geq \Delta$ and $d(p_d) \leq \Delta$. (p_c and p_d can be the same, in this case $d(p_d) = d(p_c) = \Delta$.)*

Proof.

a. Proof of *only if* part: Suppose λ is the optimal value that maximizes $L(\lambda)$. Let p be the corresponding c_λ-minimal path and thus $L(\lambda) = c(p) + \lambda(d(p) - \Delta)$. Without loss of generality, we only consider the case $d(p) > \Delta$. If the λ is slightly increased to $\lambda'(> \lambda), c(p) + \lambda(d(p) - \Delta)$ is also increased. Since $L(\lambda)$ is optimal, p cannot be the $c_{\lambda'}$-minimal path any more; otherwise $L(\lambda') > L(\lambda)$. Let p' be the new $c_{\lambda'}$-minimal path. If $|\lambda - \lambda'|$ is small enough, p' is also the c_λ−minimal path because there are only a finite number of paths. It follows that $c(p') + \lambda'(d(p') - \Delta) = L(\lambda') \leq L(\lambda) = c(p') + \lambda(d(p') - \Delta)$.

Hence $\lambda'(d(p') - \Delta) \leq \lambda(d(p') - \Delta) \Rightarrow d(p') \leq \Delta$ since $\lambda' > \lambda$. Let $p_c = p$ and $p_d = p'$ completing the proof of the *only if* part.

b. Proof of *if* part: Let p_c and p_d be two c_λ-minimal paths and $d(p_c) \geq \Delta$ and $d(p_d) \leq \Delta$. Without loss of generality, assume λ^* maximizes the function $L(\lambda^*)$ and $\lambda^* > \lambda$.

Since $\lambda < \lambda^*, d(p_c) \geq \Delta$ and $d(p_d) \leq \Delta$, it follows that $d(p_d) = \Delta$.

Let p^* denote the c_{λ^*}-minimal path. Then,

$$L(\lambda^*) = c(p^*) + \lambda^* d(p^*) - \lambda^* \Delta \leq c(p_d) + \lambda^* d(p_d) - \lambda^* \Delta$$
$$= L(\lambda) + (\lambda^* - \lambda)(d(p_d) - \Delta) \leq L(\lambda)$$

Therefore, $L(\lambda) = L(\lambda^*)$, which proves that λ maximizes $L(\lambda)$. ∎

Theorem 38.3 [2] *Let $0 \leq \lambda_1 < \lambda_2$, and $p_{\lambda_1}, p_{\lambda_2} \in P_{st}$ be c_{λ_1}-minimal and c_{λ_2}-minimal paths. Then $c(p_{\lambda_1}) \leq c(p_{\lambda_2})$ and $d(p_{\lambda_1}) \geq d(p_{\lambda_2})$.*

Proof. Note that $c_\lambda(p) = c(p) + \lambda d(p)$.
Because $p_{\lambda_1}, p_{\lambda_2} \in P_{st}$ are c_{λ_1}-minimal and c_{λ_2}-minimal paths

$$c_{\lambda_1}(p_{\lambda_1}) \leq c_{\lambda_1}(p_{\lambda_2}) \Leftrightarrow c(p_{\lambda_1}) + \lambda_1 d(p_{\lambda_1}) \leq c(p_{\lambda_2}) + \lambda_1 d(p_{\lambda_2}),$$

and

$$c_{\lambda_2}(p_{\lambda_1}) \geq c_{\lambda_2}(p_{\lambda_2}) \Leftrightarrow c(p_{\lambda_1}) + \lambda_2 d(p_{\lambda_1}) \geq c(p_{\lambda_2}) + \lambda_2 d(p_{\lambda_2}).$$

Then

$$(\lambda_1 - \lambda_2) d(p_{\lambda_1}) \leq (\lambda_1 - \lambda_2) d(p_{\lambda_2}) \Rightarrow d(p_{\lambda_1}) \geq d(p_{\lambda_2}),$$

and

$$c(p_{\lambda_1}) \leq c(p_{\lambda_2}) + \lambda_1 [d(p_{\lambda_2}) - d(p_{\lambda_1})] \leq c(p_{\lambda_2}).$$

Hence the theorem. ∎

The convergence of the LARAC algorithm is guaranteed by the following result.

Theorem 38.4 [2]. *Let $p_c^1, p_c^2, p_c^3, \ldots$ and $p_d^1, p_d^2, p_d^3, \ldots$ denote the sequences of paths generated by the LARAC algorithm. Then*

$$d(p_c^1) > d(p_c^2) > d(p_c^3) > \cdots > \Delta \text{ and } d(p_d^1) < d(p_d^2) < d(p_d^3) < \cdots \leq \Delta.$$

Proof. Suppose p_c and p_d are the current paths in the LARAC algorithm with λ_c and λ_d as the corresponding λ values. Suppose that neither of these two λ values is the maximizing value.

Let $\lambda = (c(p_c) - c(p_d))(d(p_d) - d(p_c))$ and p_λ be the corresponding c_λ-minimal path. Evidently, $c_\lambda(p_c) = c_\lambda(p_d)$ (recalling that $c_\lambda(p) = c(p) + \lambda d(p)$).

Suppose λ is not the maximizing value either; otherwise, the algorithm stops immediately. We also have

$$c(p_c) + \lambda_c d(p_c) \leq c(p_d) + \lambda_c d(p_d),$$
$$c(p_c) + \lambda_d d(p_c) \geq c(p_d) + \lambda_d d(p_d).$$

In fact, the equality cannot hold because neither λ_c nor λ_d is the maximizing multiplier.
So

$$\lambda_c < \frac{c(p_c) - c(p_d)}{d(p_d) - d(p_c)} = \lambda < \lambda_d.$$

Consider 2 cases:

1. $d(p_\lambda) \leq \Delta$: In this case, because $d(p_\lambda) \geq d(p_d)$ by Theorem 38.3, it suffices to show that $d(p_\lambda) \neq d(p_d)$.

   Assume $d(p_\lambda) = d(p_d)$. Consider the following inequalities.

   $$c(p_\lambda) + \lambda d(p_\lambda) \leq c(p_d) + \lambda d(p_d) \text{ and } c(p_\lambda) + \lambda_d d(p_\lambda) \geq c(p_d) + \lambda_d d(p_d).$$

Because $d(p_\lambda) = d(p_d)$, it follows that $c(p_\lambda) = c(p_d)$. Hence $c_\lambda(p_c) = c_\lambda(p_d) = c_\lambda(p)$, which implies that λ is the maximizing value. This contradiction establishes the theorem.

2. $d(p_\lambda) > \Delta$: Proof in this case follows along the same lines as above. ∎

Theorem 38.5 *Consider the problem:*

$$\text{Minimize} \quad yc(p_d) + (1-y)c(p_c) \tag{38.2}$$

$$\text{subject to} \quad yd(p_d) + (1-y)d(p_c) = \Delta \text{ and } 0 \leq y \leq 1, \tag{38.3}$$

where p_c and p_d are two $s-t$ paths such that $d(p_d) > \Delta$ and $d(p_c) < \Delta$.
Let $\lambda = (c(p_d) - c(p_c))/(d(p_c) - d(p_d))$ and suppose that for all $s-t$ path p, $d(p) \neq \Delta$. Then p_d and p_c minimize (38.2) if and only if they are both c_λ-minimal.

Proof. First, we prove that

$$yc(p_d) + (1-y)c(p_c) \geq L(\xi), \quad \xi \in R^+. \tag{38.4}$$

In fact,

$$L(\xi) = \min\{c_\xi(p) | p \in P_{st}\} - \xi \Delta$$
$$\leq y c_\xi(p_d) + (1-y) c_\xi(p_c) - \xi(y\, d(p_d) + (1-y) d(p_c))$$
$$= y(c_\xi(p_d) - \xi d(p_d)) + (1-y)(c_\xi(p_c) - \xi d(p_c))$$
$$= y c(p_d) + (1-y) c(p_c).$$

Using (38.3), (38.2) can be rewritten as:

$$yc(p_d) + (1-y)c(p_c) = c(p_c) + \lambda(d(p_c) - \Delta) = c(p_d) + \lambda(d(p_d) - \Delta). \tag{38.5}$$

Evidently, $d(p_c) \neq \Delta$ and $d(p_d) \neq \Delta$.

a. Proof of the *if* part: Suppose p_d and p_c are c_λ-minimal paths. Then

$$L(\lambda) = c(p_c) + \lambda(d(p_c) - \Delta) = yc(p_d) + (1-y)c(p_c),$$

where $yd(p_d) + (1-y)d(p_c) = \Delta, 0 \leq y \leq 1$. So (38.2) is minimized.

b. Proof of the *only if* part: Suppose p_d and p_c minimize (38.2) or rather (38.5). Assume p is a c_λ-minimal path and p_d and p_c are not c_λ-minimal. Consider the case when p is infeasible (if p is feasible, the theorem can be proven similarly). We have

$$c(p) + \lambda d(p) < c(p_d) + \lambda d(p_d). \tag{38.6}$$

Then

$$\lambda' = \frac{c(p_d) - c(p)}{d(p) - d(p_d)} > \lambda.$$

Thus

$$y'c(p_d) + (1-y')c(p) = c(p_d) + \lambda'(d(p_d) - \Delta)$$
$$< c(p_d) + \lambda(d(p_d) - \Delta) = y\,c(p_d) + (1-y)c(p_c),$$

where $y'd(p_d) + (1-y')d(p) = y\,d(p_d) + (1-y)d(p_c) = \Delta$.

The contradiction above proves that p_c and p_d are c_λ-minimal paths. ∎

From the above proof, it can be shown that the value of λ defined by the optimal solution p_c and p_d of (38.2) is equal to the maximizing λ searched by LARAC algorithm. Also the optimum value of RELAX-CSP is equal to the optimum value $L(\lambda^*)$ of DUAL-RELAX CSP.

There may be more than one maximizing λ. Assume that there is some multiplier λ such that the delay of the corresponding path p_λ is equal to the delay bound. In this case, an interval will serve as the maximizing multiplier and we can find the actual optimal path for the original CSP problem with that λ, recalling that $c(p_\lambda) = L(\lambda)$ which is the lower bound on the cost of the actual optimal path.

Theorem 38.6 *If $\exists \lambda$ and the corresponding path p_λ such that $d(p_\lambda) = \Delta$, the maximizing λ is one unique interval (maybe just one point); Otherwise, the maximizing λ^* is unique.*

Proof. This is a direct consequence of the concavity of the function $L(\lambda)$ as stated in Claim 38.2. ∎

Summarizing the discussions thus far we have the following theorem.

Theorem 38.7 *Given λ_1, λ_2 such that $d(p_{\lambda 1}) > \Delta \geq d(p_{\lambda 2})$. If we start the LARAC algorithm by initializing p_c and p_d as $p_{\lambda 1}$ and $p_{\lambda 2}$, respectively, then the LARAC algorithm finds a maximizing multiplier λ^* satisfying $\lambda_1 < \lambda^* \leq \lambda_2$.* ∎

38.4 LARAC-BIN: A BINARY SEARCH-BASED APPROACH TO THE DUAL-RELAX CSP PROBLEM

In this section we present a new algorithm called LARAC-BIN that uses the binary search technique to find the maximizing multiplier. LARAC-BIN as presented in Figure 38.3 stops when $L(\lambda^*) - L(\lambda) < \tau$, for a given value of τ. The parameter τ serves as a tuning parameter and can be specified in advance depending on the allowable deviation of the cost of the produced solution from the optimum value. We also establish an optimality condition. This criterion can be used to terminate the algorithm and at termination the optimum value of $L(\lambda)$ will be obtained.

In effect, the goal of LARAC-BIN is to find the minimum λ with which we can obtain a feasible path because the smaller the λ, the smaller the cost of the path obtained. This goal is compatible with that of the LARAC algorithm searching for the maximizing λ^* and $L(\lambda^*)$. To put it formally, we have the following theorem.

Theorem 38.8 *Let λ^* denote the smallest maximizing value for $L(\lambda)$ and p_λ denote a path corresponding to λ. Then $c(p_{\lambda^*}) \leq c(p_\lambda)$ for all λ such that $d(p_\lambda) \leq \Delta$.*

Proof. According to Theorem 38.3, if $\lambda^* \leq \lambda, c(p_{\lambda^*}) \leq c(p_\lambda)$. So assume $\lambda^* > \lambda$. In this case, $d(p_\lambda) \leq \Delta$ implies $d(p_\lambda) = \Delta$ by Theorem 38.1. Hence $L(\lambda) = L(\lambda^*)$ according to Theorem 38.2, which is impossible because λ^* is the smallest maximizing value for $L(\lambda)$.

The above contradiction proves the theorem. ∎

The initial values of λ_{begin} and λ_{end} in Figure 38.3 are to be selected such that p_{begin} is infeasible and p_{end} is feasible. We can initialize λ_{end} as in the following theorem.

Theorem 38.9 *If $\lambda = \frac{c(p_d) - c(p_c)}{\Delta - d(p_d)}$, $d(p_d) < \Delta$ and $c(p_d) > c(p_c)$, then the c_λ-minimal path is feasible, where p_c and p_d are the minimal cost and minimal delay path, respectively.*

> **Procedure LARAC-BIN** (s, t, Δ, τ)
>
> $\quad p_c : Dijkstra\,(s, t, c)$
>
> $\quad$ **if** $d(p_c) \leq \Delta$ **then return** p_c
>
> $\quad p_d := Dijkstra\,(s, t, d)$
>
> $\quad$ **if** $d(p_d) > \Delta$ **then return** *there is no solution*
>
> $\quad$ **if** $d(p_d) = \Delta$ **or** $c(p_d) = c(p_c)$ **then return** p_d
>
> $\quad \lambda_{\text{begin}} := 0,\ \lambda_{\text{end}} := (c(p_d) - c(p_c))/(\Delta - \boldsymbol{d(p_d)})$
>
> $\quad$ **while** $(\lambda_{\text{end}} - \lambda_{\text{begin}})(\Delta - \boldsymbol{d(p_d)}) > \tau$
>
> $\quad\quad \lambda := (\lambda_{\text{begin}} + \lambda_{\text{end}})/2$
>
> $\quad\quad r := Dijkstra\,(s, t, c_\lambda)$
>
> $\quad\quad$ **if** $d(r) = \Delta$ **then return** r
>
> $\quad\quad$ **else if** $d(r) < \Delta$ **then** $\lambda_{\text{end}} := \lambda$ **else** $\lambda_{\text{begin}} := \lambda$
>
> $\quad$ **end while**
>
> $\quad$ **return** $r := Dijkstra\,(s, t, c_{\lambda_{\text{end}}})$
>
> **end procedure**

Figure 38.3 LARAC-BIN algorithm.

Proof. Assume that p is a c_λ-minimal path and $d(p) > \Delta$. It follows that

$$c(p_d) + \lambda d(p_d) \geq c(p) + \lambda d(p).$$

Then

$$0 \leq c(p_d) - c(p) - \frac{c(p_d) - c(p_c)}{\Delta - d(p_d)}(d(p) - d(p_d))$$
$$< c(p_d) - c(p) - (c(p_d) - c(p_c)) = c(p_c) - c(p) \leq 0$$

The above contradiction proves the theorem. ∎

Theorem 38.10 *Let λ^* denote the smallest maximizing Lagrangian multiplier of $L(\lambda)$ and p^* be the resulting path. Let p_{begin} and p_{end} be the minimal aggregated cost paths with respect to λ_{begin} and λ_{end}, where λ_{begin} and λ_{end} are as defined in the LARAC-BIN algorithm in Figure 38.3. Here p_{begin} is infeasible and p_{end} is feasible. Then*

$$0 \leq L(\lambda^*) - L(\lambda_{\text{end}}) \leq (\lambda_{\text{end}} - \lambda_{\text{begin}})(\Delta - d(p_{\text{end}})).$$

Proof. The left inequality holds because $L(\lambda^*)$ is the maximum value.
Evidently, $d(p_{\text{end}}) \leq \Delta, \lambda_{\text{begin}} \leq \lambda^* \leq \lambda_{\text{end}}$, and

$$c(p^*) + \lambda^* d(p^*) \leq c(p_{\text{end}}) + \lambda^* d(p_{\text{end}}).$$

It follows that

$$\begin{aligned} L(\lambda^*) - L(\lambda_{\text{end}}) &= c(p^*) + \lambda^* d(p^*) - \lambda^* \Delta - [c(p_{\text{end}}) + \lambda_{\text{end}} d(p_{\text{end}}) - \lambda_{\text{end}} \Delta] \\ &= \{c(p^*) + \lambda^* d(p^*) - [c(p_{\text{end}}) + \lambda^* d(p_{\text{end}})]\} - (\lambda_{\text{end}} - \lambda^*) d(p_{\text{end}}) \\ &\quad + (\lambda_{\text{end}} - \lambda^*) \Delta \\ &\leq (\lambda_{\text{end}} - \lambda^*)[\Delta - d(p_{\text{end}})] \leq (\lambda_{\text{end}} - \lambda_{\text{begin}})[\Delta - d(p_{\text{end}})]. \quad \blacksquare \end{aligned}$$

Note that we have used the result of the above theorem in the termination of the LARAC-BIN algorithm (Figure 38.3).

Since a number of optimization problems only involve integer values (integer problems) or can be converted to integer problems, we now derive a termination condition for the LARAC-BIN algorithm when all the link costs and delays are integers. If terminated according to this condition, the algorithm computes the maximizing λ^* with polynomial time complexity.

Consider the set of rational numbers $Q(D) = \{p/q \mid GCD(p,q) = 1, q \leq D$, and $p, q, D \in N^+\}$. Define the density of $Q(D)$ as $DENS(Q(D)) = \min\{|x_1 - x_2| : x_1, x_2 \in Q(D)$ and $x_1 \neq x_2\}$. It is easy to show that $DENS(Q(D)) = 1/D^2$ and that for $x, y \in Q(D), x = y$ if $|x - y| < DENS(Q(D))$.

Suppose that we modify LARAC-BIN so that it terminates when $|\lambda_{\text{begin}} - \lambda_{\text{end}}| < 1/D^2$ and that the paths at termination are p_{end} and p_{begin}, where $D = |d(p_{\text{begin}}) - d(p_{\text{end}})|$. Let

$$\lambda = \lambda' = \frac{c(p_{\text{end}}) - c(p_{\text{begin}})}{d(p_{\text{begin}}) - d(p_{\text{end}})}.$$

Theorem 38.11 *λ' defined as above is a maximizing multiplier.*

Proof. Consider $Q(D)$, where $D = |d(p_{\text{begin}}) - d(p_{\text{end}})|$. Because

$$c(p_{\text{begin}}) + \lambda_{\text{begin}} d(p_{\text{begin}}) \leq c(p_{\text{end}}) + \lambda_{\text{begin}} d(p_{\text{end}})$$

and

$$c(p_{\text{begin}}) + \lambda_{\text{end}} d(p_{\text{begin}}) \geq c(p_{\text{end}}) + \lambda_{\text{end}} d(p_{\text{end}}),$$

$$\lambda_{\text{begin}} \leq \lambda' = \frac{c(p_{\text{end}}) - c(p_{\text{begin}})}{d(p_{\text{begin}}) - d(p_{\text{end}})} \leq \lambda_{\text{end}}.$$

Suppose that $\lambda_{\text{begin}} \leq \lambda^* \leq \lambda_{\text{end}}$, where λ^* is the maximizing multiplier obtained by LARAC-BIN algorithm initialized with $p_c = p_{\text{begin}}$ and $p_d = p_{\text{end}}$.

Clearly $\lambda^* = (c(p_{\lambda 1}) - c(p_{\lambda 2}))/(d(p_{\lambda 2}) - d(p_{\lambda 1}))$ for some paths $p_{\lambda 1}$ and $p_{\lambda 2}$ with respect to the multipliers λ_1 and λ_2. Following the similar argument as above it can be seen that λ_1 and $\lambda_2 \in [\lambda_{\text{begin}}, \lambda_{\text{end}}]$. Hence $|d(p_{\lambda 2}) - d(p_{\lambda 1})| \leq D$ according to Theorem 38.3, that is, $\lambda^* \in Q(D)$.

Evidently $|d(p_{\text{begin}}) - d(p_{\text{end}})| = D \leq D$ and thus $\lambda \in Q(D)$.

Because $|\lambda' - \lambda^*| < |\lambda_{\text{begin}} - \lambda_{\text{end}}| < 1/D^2 = \text{DENS}(Q(D))$, the only possibility is that $\lambda' = \lambda^*$. $\blacksquare$

For the CSP problem, the size of D is bounded as $D \leq n \max\{d_{ij} \mid (i,j) \in E\}$, where n is the number of nodes in the network. If the LARAC-BIN algorithm is terminated using the condition given above, then we have the following complexity result.

Theorem 38.12 *LARAC-BIN terminates in $O((m+n\log n)(\log(COST \times DELAY^2)))$ time where COST is the cost of the minimum delay path and DELAY is the delay of the minimum cost path in the network.* ∎

38.5 GEN-LARAC: A GENERALIZED APPROACH TO THE CSP PROBLEM UNDER MULTIPLE ADDITIVE CONSTRAINTS

In this section we study the CSP(k) problem that requires determination of $s-t$ paths that satisfy $k > 1$ additive constraints. We develop a new approach using Lagrangian relaxation. We use the LARAC algorithm as a building block in the design of this approach.

38.5.1 Formulation of the CSP(k) Problem and Its Relaxation

Consider a directed graph $G(V, E)$ where V is the set of nodes and E is the set of links in G. Each link (u, v) is associated with a set of $k + 1$ additive non-negative integer weights $C_{uv} = (c_{uv}, w_{uv}^1, w_{uv}^2, \ldots, w_{uv}^k)$. Here c_{uv} is called the cost of link (u, v) and w_{uv}^i is called the ith delay of (u, v). Given two nodes s and t, an $s-t$ path in G is a directed simple path from s to t. Let P_{st} denote the set of all $s-t$ paths in G. For an $s-t$ path p define

$$c(p) \equiv \sum_{(u,v) \in p} c_{uv} \text{ and } d_i(p) \equiv \sum_{(u,v) \in p} w_{uv}^i, \; i = 1,\ldots k.$$

The value $c(p)$ is called the cost of path p, and $d_i(p)$ is called the ith delay of path p. Given k positive integers $r_1, r_2 \ldots, r_k$, an $s-t$ path is called feasible (*respectively*, strictly feasible) if $d_i(p) \leq r_i$ (*respectively*, $d_i(p) < r_i$), for all $i = 1, 2 \ldots k$ (r_i is called the bound on the ith delay of a path).

The CSP(k) problem is to find a minimum cost feasible $s-t$ path. An instance of the CSP(k) problem is strictly feasible if all the feasible paths are strictly feasible. Without loss of generality, we assume that the problem under consideration is always feasible. In order to guarantee strict feasibility, we do the following transformation.

For $i = 1, 2 \ldots, k$, transform the ith delay of each link (u, v) such that the new weight vector C'_{uv} is given by

$$C'_{uv} = \left(c_{uv}, 2w_{uv}^1, 2w_{uv}^2, \ldots, 2w_{uv}^k\right).$$

Also transform the bounds r_i's so that the new vector of bounds R' is given by

$$R' = (2r_1 + 1, 2r_2 + 1, \ldots, 2r_k + 1).$$

In the rest of the section, we only consider the transformed problem. Thus all link delays are even integers, and delay bounds are odd integers. We will use symbols with capital or bold letters to represent vectors. Also, for a matrix A, A^T denotes its transpose. For simplicity of presentation, we will use C_{uv} and R instead of C'_{uv} and R' to denote the transformed weight vector and the vector of bounds.

Two immediate consequences of this transformation are stated below.

Lemma 38.1 $\forall\, p \in P_{st}, \forall i \in \{1, 2 \ldots, k\}, d_i(p) \neq r_i$ *in the transformed problem.* ∎

Lemma 38.2 *An $s-t$ path in the original problem is feasible (resp. optimal) iff it is strictly feasible (resp. optimal) in the transformed problem.* ∎

Starting with an integer linear programming (LP) formulation of the CSP(k) problem and relaxing the integrality constraints we get the RELAX-CSP(k) problem below. In this formulation, for each $s-t$ path p, we introduce a variable x_p.

RELAX-CSP(k)

$$\text{Minimize} \sum_p c(p) x_p \tag{38.7}$$

$$\text{subject to} \sum_p x_p = 1 \tag{38.8}$$

$$\sum_p d_i(p) x_p \leq r_i, i = 1, \ldots, k \tag{38.9}$$

$$x_p \geq 0, \forall p \in P_{st} \tag{38.10}$$

The Lagrangian dual of RELAX-CSP(k) is given below.

DUAL-RELAX CSP(k)

$$\text{Maximize } w - \lambda_1 r_1 \ldots - \lambda_k r_k \tag{38.11}$$

$$\text{subject to } w - d_1(p)\lambda_1 \ldots - d_k(p)\lambda_k \leq c(p), \forall p \in P_{st} \tag{38.12}$$

$$\lambda_i \geq 0, i = 1, \ldots, k \tag{38.13}$$

In the above dual problem $\lambda_1, \lambda_2 \ldots, \lambda_k$ and w are the dual variables, with w corresponding to (38.8) and each λ_i corresponding to the ith constraint in (38.9).

It follows from (38.12) that $w \leq c(p) + d_1(p)\lambda_1 \ldots + d_k(p)\lambda_k \, \forall \, p \in P_{st}$. Since we want to maximize (38.11), the value of w should be as large as possible, that is,

$$w = \min_{p \in P_{st}} \{c(p) + d_1(p)\lambda_1 + \cdots + d_k(p)\lambda_k\}.$$

With the vector Λ defined as $\Lambda = (\lambda_1, \lambda_2, \ldots, \lambda_k)$, define

$$L(\Lambda) = \min_{p \in P_{st}} \{c(p) + \lambda_1(d_1(p) - r_1) \cdots + \lambda_k(d_k(p) - r_k)\} \tag{38.14}$$

Notice that $L(\Lambda)$ is called the Lagrangian function in the literature and is a concave continuous function of Λ [6].

Then DUAL-RELAX CSP(k) can be written as follows.

$$\begin{aligned} \text{Maximize} \quad & L(\Lambda) \\ \text{subject to} \quad & \Lambda \geq 0 \end{aligned} \tag{38.15}$$

The Λ^* that maximizes (38.15) is called the maximizing multiplier and is defined as

$$\Lambda^* = \arg\max_{\Lambda \geq 0} L(\Lambda) \tag{38.16}$$

Claim 38.4 *If an instance of the CSP(k) problem is feasible and a path p_{opt} is an optimal path, then $\forall \Lambda \geq 0, L(\Lambda) \leq c(p_{opt})$.* ∎

We shall use $L(\Lambda)$ as a lower bound of $c(p_{opt})$ to evaluate the quality of the approximate solution obtained by our algorithm. Given $p \in P_{st}$ and Λ, define

$$C(p) \equiv (c(p), d_1(p), d_2(p), \ldots, d_k(p)),$$
$$D(p) \equiv (d_1(p), d_2(p), \ldots, d_k(p)),$$
$$R \equiv (r_1, r_2, \ldots, r_k),$$
$$c_\Lambda(p) \equiv c(p) + d_1(p)\lambda_1 + \cdots + d_k(p)\lambda_k,$$

and
$$d_\Lambda(p) \equiv d_1(p)\lambda_1 + \cdots + d_k(p)\lambda_k.$$

Here $c_\Lambda(p)$ and $d_\Lambda(p)$ are called the aggregated cost and the aggregated delay of path p, respectively. We shall use P_Λ to denote the set of $s-t$ paths attaining the minimum aggregated cost with respect to Λ. A path $p_\Lambda \in P_\Lambda$ is called a Λ-minimal path.

38.5.2 A Strongly Polynomial Time Approximation Algorithm for CSP(1) Problem

The key issue now is to search for the maximizing multiplier and determine termination conditions. If there is only one delay constraint, that is, $k=1$, we have the following claim from Theorem 38.2 repeated below for ease of reference.

A value $\lambda > 0$ maximizes the function $L(\lambda)$ if and only if there are paths p_c and p_d which are both c_λ-minimal and for which $d(p_c) \geq r$ and $d(p_d) \leq r$. (p_c and p_d can be the same. In this case $d(p_d) = d(p_c) = r$.)

Theorem 38.13 *DUAL-RELAX CSP(1) is solvable in $O((m+n\log n)^2)$ time.*

Proof. We prove this theorem by presenting an algorithm with $O((m+n\log n)^2)$ time complexity. The algorithm called PSCSP (parametric search-based CSP) algorithm is based on a methodology first proposed by Megiddo [3] to solve fractional combinatorial optimization problems.

Assume node 1 is the source and node n is the target. In Figure 38.4, we present algorithm PSCSP for computing a CSP using lexicographic order on a pair of link weights $(l_{uv}, c_{uv})\ \forall (u,v) \in E$ and based on parametric search, where $l_{uv} = c_{uv} + \lambda^* d_{uv}$ and λ^* is unknown. The algorithm is the same as the BFM (Bellman–Ford–Moore) algorithm except for Step 4 which needs special care. We use BFM algorithm here because it is easy to explain. Actually we use Dijkstra's algorithm for better time complexity results.

In Figure 38.4, we need extra steps (Oracle test) to evaluate the Boolean expression in the if statement in Step 4 since $\lambda^* \geq 0$ is unknown. If $x_v = \infty, y_v = \infty$, then the inequality holds. Assume x_v and y_v are finite (non-negative) values. Then it suffices to evaluate the following Boolean expression.

$$p + q\lambda^* \leq 0?, \text{ where } p = x_u + c_{uv} - x_v \text{ and } q = (y_u + d_{uv} - y_v).$$

If $p \cdot q \geq 0$, then it is trivial to evaluate the Boolean expression. Without loss of generality, assume $p \cdot q < 0$, that is, $-p/q > 0$. The Oracle test algorithm is presented in Figure 38.5.

The time complexity of the Oracle test is $O(m + n\log n)$. On the other hand, we can revise the algorithm in Figure 38.4 using Dijkstra's algorithm and the resulting algorithm will have time complexity $O((m+n\log n)^2)$.

Next, we show how to compute the value of λ^* and $L(\lambda^*)$. The algorithm in Figure 38.4 computes a λ^*-minimal path p with minimal cost. Similarly, we can compute a λ^*-minimal path q with minimal delay. Then the value of λ^* is given by the following equation: $c(p) + \lambda^* d(p) = c(q) + \lambda^* d(q)$ and $L(\lambda^*) = c(p) + \lambda^*(d(p) - \Delta)$, where Δ is the path delay constraint (here $k=1$). Notice that $d(q) \neq d(p)$ is guaranteed by the transformation in Section 38.5.1. ∎

Because PSCSP and LARAC algorithms are based on the same methodology and obtain the same solution, we shall also call PSCSP as LARAC. In the rest of the chapter, we shall

Step 1. $M_v = (X_v, Y_v) = (+\infty, +\infty)$ for $v = 2, 3, \ldots, n$ and $M_1 = (0, 0)$

Step 2. $i \leftarrow 1$

Step 3. $u \leftarrow 1$

Step 4. $\forall v, (u, v) \in E$, if $(X_v + \lambda^* Y_v > X_u + \lambda^* Y_v + C_{uv} + \lambda^* d_{uv})$ or

$(X_v + \lambda^* Y_v = X_u + \lambda^* Y_u + C_{uv} + \lambda^* d_{uv})$ and $(X_v > X_u + C_{uv}))$

$M_v \leftarrow (X_v + C_{uv}, Y_u + d_{uv})$

Step 5. $u \leftarrow u + 1$ and if $u \leq n$, go to Step 4.

Step 6. $i \leftarrow i + 1$ and if $i \leq n$, go to Step 3.

Figure 38.4 PSCSP algorithm for CSP(1) problem.

Δ : Path delay constraint

Step 1. Let $\lambda = -p/q > 0$ for each link $(u, v) \in E$,

define its length $l_{uv} = C_{uv} + \lambda d_{uv}$.

Step 2. Compute two shortest paths p_c and p_d using the lexicographic order on

(l_{uv}, C_{uv}) and (l_{uv}, d_{uv}), respectively.

Step 3. Obviously, $d(p_c) \geq d(p_d)$. Only four cases are possible:

a. $d(p_c) > \Delta$ and $d(p_d) > \Delta$: By Theorem 38.1, $\lambda < \lambda^*$

and thus $p + q\lambda^* < 0$ if $q < 0$ and $p + q\lambda^* > 0$ otherwise.

b. $d(p_c) < \Delta$ and $d(p_d) < \Delta$: By Theorem 38.1, $\lambda < \lambda^*$

and thus $p + q\lambda^* > 0$ if $q < 0$ and $p + q\lambda^* > 0$ otherwise.

c. $d(p_c) > \Delta$ and $d(p_d) < \Delta$: By Theorem 38.1, $\lambda < \lambda^*$ and $p + q\lambda^* = 0$.

d. $d(p_c) > \Delta$ or $d(p_d) < \Delta$: By Lemma 38.1, this is impossible.

Figure 38.5 Oracle test algorithm.

discuss how to extend it for $k > 1$. In particular we develop an approach that combines the LARAC algorithm as a building block with certain techniques in mathematical programming. We shall call this new approach as GEN-LARAC.

38.5.3 GEN-LARAC for the CSP(k) Problem

38.5.3.1 Optimality Conditions

Theorem 38.14 *Given an instance of a feasible CSP(k) problem, a vector $\Lambda \geq 0$ maximizes $L(\Lambda)$ if and only if the following problem in the variables u_j is feasible.*

$$\sum_{p_j \in P_\Lambda} u_j \cdot d_i(p_j) = r_i, \ \forall i, \lambda_i > 0 \tag{38.17}$$

$$\sum_{p_j \in P_\Lambda} u_j \cdot d_i(p_j) \leq r_i, \ \forall i, \lambda_i = 0 \tag{38.18}$$

$$\sum_{p_j \in P_\Lambda} u_j = 1 \tag{38.19}$$

$$u_j \geq 0, \ \forall p_j \in P_\Lambda \tag{38.20}$$

Proof. Sufficiency: Let $x = (u_1 \ldots, u_r, 0, 0 \ldots)$ be a vector of size $|P_{st}|$, where $r = |P_\Lambda|$. Obviously, x is a feasible solution to RELAX-CSP(k). It suffices to show that x and Λ satisfy the complementary slackness conditions.

According to (38.12), $\forall p \in P_{st}, w \leq c(p) + d_1(p) \lambda_1 \ldots + d_k(p) \lambda_k$. Since we need to maximize (38.11), the optimal $w = c(p_\Lambda) + d_1(p_\Lambda) \lambda_1 \ldots + d_k(p_\Lambda) \lambda_k$ $\forall p_\Lambda \in P_\Lambda$. For all other paths p, $w - c(p) + d_1(p) \lambda_1 \ldots + d_k(p) \lambda_k. < 0$. So x satisfies the complementary slackness conditions. By (38.17) and (38.18), Λ also satisfies complementary slackness conditions.

Necessary: Let x^* and (w, Λ) be the optimal solution to RELAX-CSP(k) and DUAL-RELAX CSP(k), respectively. It suffices to show that we can obtain a feasible solution to (38.17) through (38.20) from x^*.

We know that all the constraints in (38.12) corresponding to paths in $P_{st} - P_\Lambda$ are strict inequalities, and $w = c(p_\Lambda)_+ d_1(p_\Lambda) \lambda_1 \ldots + d_k(p_\Lambda)\lambda_k$ $\forall p_\Lambda \in P_\Lambda$. So, from complementary slackness conditions we get $x_p = 0, \forall p \in P_{st} - P_\Lambda$.

Now let us set u_j corresponding to path p in P_Λ equal to x_p, and set all other u_j's corresponding to paths not in P_Λ equal to zero. The $u_i's$ so elected will satisfy (38.17) and (38.18) since these are complementary conditions satisfied by (w, Λ). Since x_i's satisfy (38.8), u_j's satisfy (38.19). Thus we have identified a solution satisfying (38.17) through (38.20). ∎

38.5.3.2 GEN-LARAC: Coordinate Ascent Method

Our approach for the CSP(k) problem is based on the coordinate ascent method called GEN-LARAC (Figure 38.6) and proceeds as follows. Given a multiplier Λ, in each iteration we try to improve the value of $L(\Lambda)$ by updating one component of the multiplier vector. If the objective function is not differentiable, the coordinate ascent method may get stuck at a corner Λ_s not being able to make progress by only changing one component. We call Λ_s a pseudo optimal point which requires updates of at least two components to achieve improvement in the solution. Our simulations show that the objective value attained at pseudo optimal points is usually very close to the maximum value of $L(\Lambda)$.

Step 1: $\Lambda^0 \leftarrow (0, 0..., 0); t \leftarrow 0; flag \leftarrow \text{true}; B \leftarrow 0$

Step 2: (Coordinate ascent steps)

  while $(flag)$

    $flag \leftarrow \text{false}$

    for $i = 1$ to k

      $\gamma \leftarrow \arg\max_{\xi \geq 0} L(\lambda_1^t..., \lambda_{i-1}^t, \xi, \lambda_{i-1}^t..., \lambda_k^t).$

  If $(\gamma \neq \lambda_i^t)$ **then**

    $flag \leftarrow \text{true}$

    $\lambda_j^{t+1} = \begin{cases} \gamma & j = i, \\ \lambda_j^t & j \neq i. \end{cases}, j = 1, 2..., k$

    $t \leftarrow t + 1$

  end if

  end for

  end while

Step 3: If Λ^t is optimal then return Λ^t.

Step 4: $B \leftarrow B + 1$ and go to Step 5 if $B < B_{\max}$ ($B_{\max}$ is the maximum number of iteration allowed); Otherwise, stop.

Step 5: Compute a new vector Λ^* such that $L(\Lambda^*) > L(\Lambda^t)$.

Step 6: $t \leftarrow t + 1$, $\Lambda^t \leftarrow \Lambda^*$, and go to Step 2.

Figure 38.6 GEN-LARAC: Coordinate ascent algorithm for the CSP(k) problem.

38.5.3.3 Verification of Optimality of Λ

In Step 3 we need to verify if a given Λ is optimal. We show that this can be accomplished by solving the following LP problem, where $P_\Lambda = \{p_1, p_2, \ldots, p_r\}$ is the set of Λ-minimal paths.

$$\text{Maximize } 0 \tag{38.21}$$

$$\text{subject to } \sum_{p_j \in P_\Lambda} u_j \cdot d_i(p_j) = r_i, \forall i, \lambda_i > 0 \tag{38.22}$$

$$\sum_{p_j \in P_\Lambda} u_j \cdot d_i(p_j) \leq r_i, \forall i, \lambda_i = 0 \tag{38.23}$$

$$\sum_{p_j \in P_\Lambda} u_j = 1 \tag{38.24}$$

$$u_j \geq 0, \forall p_j \in P_\Lambda \tag{38.25}$$

By Theorem 38.14, if the above linear program is feasible then the multiplier Λ is a maximizing multiplier.

Let $(y_1 \ldots, y_k, \delta)$ be the dual variables corresponding to the above problem. Let $Y = (y_1, y_2, \ldots, y_k)$. The dual of (38.21) through (38.25) is as follows

$$\text{Minimize} \quad RY^T + \delta \quad (38.26)$$
$$\text{subject to} \quad D(p_i)Y^T + \delta \geq 0, i = 1, 2, \ldots, r \quad (38.27)$$
$$y_i \geq 0, \forall i, \lambda_i > 0 \quad (38.28)$$

Evidently the LP problem (38.26) through (38.28) is feasible. From the relationship between primal and dual problems, it follows that if the linear program (38.21) through (38.25) is infeasible, then the objective of (38.26) is unbounded $(-\infty)$. Thus, if the optimum objective of (38.26) through (38.28) is 0, then the linear program (38.21) through (38.25) is feasible and by Theorem 38.14 the corresponding multiplier Λ is optimal. In summary, we have the following lemma.

Lemma 38.3 *If (38.21) through (38.25) is infeasible, then $\exists Y = (y_1, y_2, \ldots, y_k)$ and δ satisfying (38.27) through (38.28) and $RY^T + \delta < 0$.* ∎

The Y in Lemma 38.3 can be identified by applying any LP solver on (38.26) through (38.28) and terminating it once the current objective value becomes negative.

Let Λ be a non-optimal Lagrangian multiplier and $\Lambda(s, Y) = \Lambda + Y/s$ for $s > 0$.

Theorem 38.15 *If a multiplier $\Lambda \geq 0$ is not optimal, then*

$$\exists M > 0, \forall s > M, \ L(\Lambda(s, Y)) > L(\Lambda).$$

Proof. If M is big enough, $P_\Lambda \cap P_\Lambda(s, Y) \neq \emptyset$. Let $p_j \in P_\Lambda \cap P_\Lambda(s, V)$.

$$L[\Lambda(s, Y)] = c(p_j) + (D(p_j) - R)(\Lambda + Y/s)^T$$
$$= c(p_j) + (D(p_j) - R)\Lambda^T + (D(p_j) - R)(Y/s)^T$$
$$= L(\Lambda) + (D(p_j)Y^T - RY^T)/s.$$

Since $D(p_j)Y^T + \delta \geq 0$ and $RY^T + \delta < 0$, $D(p_j)Y^T - RY^T > 0$.

Hence $L(\Lambda(s, Y)) > L(\Lambda)$. ∎

We can find the proper value of M by binary search after computing Y. The last issue is to compute P_Λ. It can be expected that the size of P_Λ is usually very small. In our experiments, $|P_\Lambda|$ never exceeded 4 even for large and dense networks. The k-shortest path algorithm can be adapted easily to computing P_Λ.

A detailed convergence analysis of the GEN-LARAC algorithm may be found in [20].

Summary and Related Works

The literature on the CSP problem is vast. So in this section we survey only a subset of published works in this area. It has been shown in [4,5] that the CSP problem is NP-hard even for acyclic networks. So, in the literature, heuristic approaches and approximation algorithms have been proposed. Heuristics, in general, do not provide performance guarantees on the quality of the solution produced, though they are usually fast in practice. On the other hand, ε-approximation algorithms (subject matter of Chapter 37) deliver solutions with cost within $(1 + \varepsilon)$ time the optimal cost for all $\varepsilon > 0$, but are usually very slow in practice because they guarantee the quality of the solutions.

As regards heuristics, a number of them have appeared in the literature providing different levels of performance with regard to the quality of the solution as well as the computation time required. For instance, the LHWHM algorithm [7] is a very simple heuristic which is very fast requiring only two invocations of Dijkstra's shortest path algorithm for a feasible problem. Reference [8] also discusses further enhancements of the LHWHM algorithm as well as a heuristic based on the BFM algorithm for the shortest path problem. It should be emphasized that in all these cases, only simulations are used to evaluate the performance of the algorithms. A comprehensive overview of a number of quality of service (QoS) routing algorithms may be found in [9].

There are heuristics that are based on sound theoretical foundations. These algorithms are based on solutions to the integer relaxation or the dual of the integer relaxation of the CSP problem. To the best of our knowledge, the first such algorithm was reported in [10] by Handler and Zhang. This is based on the geometric approach (also called the hull approach [11,12]). More recently, in an independent work, Jüttner et al. [2] developed the LARAC algorithm which solves the Lagrangian relaxation of the CSP problem. In another independent work, Blokh and Gutin [1] defined a general class of combinatorial optimization problems (that are called the MCRT problems, namely, MCRT problems) of which the CSP problem is a special case, and proposed an approximation algorithm to this problem. Xiao et al. [13] drew attention to the fact that the algorithms in [1] and [2] are equivalent. Mehlhorn and Ziegelmann [11] and Ziegelmann [12] have also observed this equivalence and have developed several insightful results. In view of this equivalence, we refer to these algorithms as the LARAC algorithm. The work in [13] also establishes certain results using the algebraic approach. These results also hold true in the case of the general optimization problem considered in [1]. In another independent work, Xue [14] also arrived at the LARAC algorithm using the primal-dual method of LP. Usually, approximation algorithms are developed using the dual of the relaxed version of the CSP problem. Xiao et al. [15] described an efficient approximation algorithm for the CSP problem using the primal simplex method of LP. In [16], Jüttner proved the strong polynomiality of the LARAC algorithm, both for the general case and for the CSP problem. He has used certain results from the general area of fractional combinatorial optimization. Jüttner [17] gave a general method to solve budgeted optimization problems in strongly polynomial time. An application of the parametric search method to the general class of combinatorial optimization problems involving two additive parameters may be found in [18]. Radzik [19] gives an excellent exposition of approaches to fractional combinatorial optimization problems. The LARAC-BIN algorithm discussed in this chapter was developed in [13]. The GEN-LARAC algorithm also discussed in this chapter was developed in [20]. Multiconstrained routing problem and the constrained disjoint path problem have been considered in [14,21–26]. Several interesting algorithms related to the CSP problem and motivated by applications have appeared in the literature. For examples, see [27–36].

References

[1] D. Blokh and G. Gutin, "An approximation algorithm for combinatorial optimization problems with two parameters," *Australasian Journal of Combinatorics*, **14** (1996), 157–164.

[2] A. Jüttner, B. Szviatovszki, I. Mécs, and Z. Rajkó, "Lagrange relaxation based method for the QoS routing problem," in *Proceedings of the IEEE International Conference on Computer Communications*, Anchorage, AK, (2001), 859–868.

[3] N. Megiddo, "Combinatorial optimization with rational objective functions," *Mathematics of Operations Research*, **4**(4) (1979), 1–12.

[4] M.R. Garey and D.S. Johnson, *Computers and Intractability: A Guide to the Theory of NP-Completeness*. Freeman Press, San Francisco, CA, 1979.

[5] Z. Wang and J. Crowcroft, "Quality-of-Service routing for supporting multimedia applications," *IEEE JSAC*, **14**(7) (1996), 1228–1234.

[6] S. Boyd and L. Vandenberghe, *Convex Optimization*, Cambridge University Press, Cambridge, 2003.

[7] G. Luo, K. Huang, C. Hobbs, and E. Munter, "Multi-QoS constraints based routing for IP and ATM Networks," in *Proceedings of the IEEE Workshop on QoS Support for Real Time Internet Applications*, Vancouver, Canada, June 1999.

[8] R. Ravindran, K. Thulasiraman, A. Das, K. Huang, G. Luo, and G. Xue, "Quality of service routing: heuristics and approximation schemes with a comparative evaluation," in *Proceedings of the IEEE International Symposium on Circuits and Systems*, 2002, pp. 775–778.

[9] F.A. Kuipers, T. Korkmaz, M. Krunz, and P. Van Mieghem, "An overview of constraint-based path selection algorithms for QoS routing," *IEEE Communications Magazine*, **40** (2002), 50–55.

[10] G. Handler and I. Zhang, "A dual algorithm for the constrained shortest path problem," *Networks*, Saarbrücken, Germany, **10** (1980), 293–310.

[11] K. Mehlhorn and M. Ziegelmann, "Resource constrained shortest path," in *Proceedings of the 8th European Symposium on Algorithms*, Saarbrücken, Germany, 2000, pp. 326–337.

[12] M. Ziegelmann, "Constrained shortest paths and related problems," PhD thesis, Max-Planck-Institut für Informatik, Saarbrücken, Germany, 2001.

[13] Y. Xiao, K. Thulasiraman, and G. Xue, "The constrained shortest path problem: algorithmic approaches and an algebraic study with generalization," *AKCE International Journal of Graphs and Combinatorics*, **2**(2) (2005), 63–86.

[14] G. Xue, "Minimum-cost QoS multicast and unicast routing in communication networks," *IEEE Transactions on Communications*, **51** (2003), 817–827.

[15] Y. Xiao, K. Thulasiraman, and G. Xue, "QoS routing in communication networks: Approximation algorithms based on the primal simplex method of linear programming," *IEEE Transactions on Computers*, **55** (2006), 815–829.

[16] A. Jüttner, "On resource constrained optimization problems," in *4th Japanese-Hungarian Symposium on Discrete Mathematics and Its Applications*, Budapest, Hungary, 2005.

[17] A. Jüttner, "On budgeted optimization problems," *SIAM Journal on Discrete Mathematics*, **20**(4) (2006), 880–892.

[18] A. Jüttner. "Optimization with additional variables and constraints," *Operations Research Letters*, **33**(3) (2005), 305–311.

[19] T. Radzik, Fractional combinatorial optimization, in *Handbook of Combinatorial Optimization*, Editors, DingZhu Du and Panos Pardalos, vol. 1, Kluwer Academic Publishers, Dordrecht, the Netherlands, 1998.

[20] Y. Xiao, K. Thulasiraman, G. Xue, and M. Yadav, "QoS routing under additive constraints: A generalization of the LARAC algorithm," in *IEEE Transactions on Emerging Topics in Computing*, May 2015.

[21] J. M. Jaffe, "Algorithms for finding paths with multiple constraints," *Networks*, **14** (1984), 95–116.

[22] T. Korkmaz and M. Krunz, "Multi-constrained optimal path selection," *Proceedings of the IEEE INFOCOM*, **2** (2001), 834–843.

[23] T. Korkmaz, M. Krunz, and S. Tragoudas, "An efficient algorithm for finding a path subject to two additive constraints," *Computer Communications Journal*, **25**(3) (2002), 225–238.

[24] G. Xue, A. Sen, W. Zhang, J. Tang, and K. Thulasiraman, "Finding a path subject to many additive QoS constraints," *IEEE/ACM Transactions on Networking*, **15** (2007), 201–211.

[25] G. Xue, W. Zhang, J. Zhang, and K. Thulasiraman, "Polynomial time algorithms for multiconstrained QoS routing," *IEEE/ACM Transactions on Networking*, **16** (2008), 656–669.

[26] X. Yuan, "Heuristic algorithms for multiconstrained quality-of-service routing," *IEEE/ACM Transactions on Networking*, **10** (2003), 244–256.

[27] Y. Bejerano, Y. Breitbart, A. Orda, R. Rastogi, and A. Sprintson, "Algorithms for computing QoS paths with restoration," *IEEE Transactions on Networking*, **13**(3) (2005), 648–661.

[28] A. Chakrabarti and G. Manimaran, "Reliability constrained routing in QoS networks," *IEEE Transactions on Networking*, **13**(3) (2005), 662–675.

[29] T. Korkmaz and M. Krunz, "Bandwidth-delay constrained path selection under inaccurate state information," *IEEE/ACM Transactions on Networking*, **11** (2003), 384–398.

[30] P. Van Mieghem, H. De Neve, and F. Kuipers, "Hop-by-hop quality of service routing," *Computer Networks*, **37**(3/4) (2001), 407–423.

[31] P. Van Mieghem and F. Kuipers, "Concepts of exact QoS routing algorithms," *IEEE/ACM Transactions on Networking*, **12** (2004), 851–864.

[32] H. De Neve and P. Van Mieghem, "TAMCRA: A tunable accuracy multiple constraints routing algorithm," *Computer Communications*, **23** (2000), 667–679.

[33] J. L. Sobrinho, "Algebra and algorithms for QoS path computation and hop-by-hop routing in the internet," *IEEE/ACM Transactions on Networking*, **10**(4) (2002), 541–550.

[34] A. Warburton, "Approximation of pareto optima in multiple-objective shortest path problems," *Operations Research*, **35** (1987), 70–79.

[35] Y. Xiao, K. Thulasiraman, X. Fang, D. Yang, and G. Xue, "Computing a most probable delay constrained path: NP-hardness and approximation schemes," *IEEE Transactions on Computers*, **61** (2012), 738–744.

[36] Y. Xiao, K. Thulasiraman, and G. Xue, "Constrained shortest link-disjoint paths selection a network programming based approach," *IEEE Transactions on Circuits & Systems* I, **53** (2006), 1174–1187.

CHAPTER 39

Algorithms for Finding Disjoint Paths with QoS Constraints*

Alex Sprintson

Ariel Orda

CONTENTS

39.1	Introduction	1063
39.2	Model and Problem Definition	1064
	39.2.1 Minimum Cost Disjoint Paths Problem	1064
	39.2.2 Restricted Shortest Paths Problem	1064
	39.2.3 Restricted Disjoint Paths Problem	1065
39.3	Network Flows	1065
39.4	Algorithm for the Minimum Cost Disjoint Paths Problem	1066
39.5	Disjoint Paths with a Delay Constraint	1067
39.6	Minimizing Path Delays	1068
39.7	Algorithm Analysis	1069
39.8	Minimizing the Computational Complexity	1071
39.9	Computing Upper and Lower Bounds	1072
39.10	Conclusion	1073

39.1 INTRODUCTION

Finding a set of disjoint paths is one of the basic problems in graph theory. This problem appears in many settings in different areas such as communication networks, circuit design, transportation, and many others. Disjoint paths can be used to increase throughput, mitigate failures of network elements, and minimize congestion. Network operators can use disjoint paths to improve resilience to failures, spread the traffic evenly across the network, and maximize the rate of data transfer. For example, to minimize packet loss during an edge failure event, a copy of each packet can be sent along two disjoint paths, which allows the receiver to recover at least one copy of the packet if the failure occurs [2].

The need to find disjoint paths that satisfy quality of service (QoS) constraints leads to a broad range of interesting problems. The QoS constrains can be of *additive* type, such as delay or jitter, or of *bottleneck type* such as bandwidth. For example, a network operator might be interested in finding two disjoint paths such that both of the paths satisfy a given delay requirement. Another interesting problem in this domain is to find a set of disjoint paths that satisfy a given QoS constraint at minimum cost. In general, additive QoS constraints are harder to handle than bottleneck constraints. The problems that involve multiple criteria such as cost and delay pose additional algorithmic challenges.

---

*This chapter closely follows the treatment of the topic in [1].

While the basic problem of finding k disjoint paths can be solved in polynomial time, that is, through Suurballe–Tarjan algorithm [3], many of the problems related to finding disjoint paths that satisfy QoS constraints are intractable. In particular, the problem of finding two disjoint paths that satisfy a given delay constraint is NP-hard [4]. The problems that include multiple criteria, such as cost and delay, are typically NP-hard as well. Indeed, even the basic problem of finding a single path of minimum cost that satisfies an additive QoS constraint is NP-hard [5]. Accordingly, for many problems approximation algorithms are of interest. An approximation algorithm may relax the cost optimality constraint, relax the QoS constraints, or both. With the former approach, an approximation algorithm returns a set of disjoint paths that satisfy the QoS constraints at a cost that is slightly higher than the optimum. Alternatively, an algorithm can return a set of paths that slightly violate the QoS constraint, but whose cost is less than or equal to the optimum. This approach is acceptable for applications that tolerate slight delay violations. In addition, there is a trade-off between the quality of the paths and the computational complexity of the algorithm, since typically the quality of the paths can be improved through successful iterations.

The goal of this chapter is to present several algorithmic techniques for finding disjoint paths. We begin by introducing the classical Suurballe–Tarjan algorithm that finds a required number of disjoint paths of minimum total cost. Then, we present solutions for bi-criteria settings with an additive QoS constraint and cost minimization requirements.

39.2 MODEL AND PROBLEM DEFINITION

We focus on the problem of finding a set of disjoint paths between a source node s and a destination node t in a graph $G(V, E)$. For clarity, we assume that the graph $G(V, E)$ is a directed graph, however, all algorithms discussed in this chapter can be used for undirected graphs as well. We say that edges e' and e'' are *interlacing* if they connect the same endpoints in opposite directions, that is, $e' = (u, v)$ and $e'' = (v, u)$. We assume that each edge $e \in E$ is associated with a positive cost c_e and a QoS parameter d_e. The cost c_e of an edge e typically captures the amount of network resources consumed if e is selected by the algorithm. The QoS parameter captures the guarantee provided by the edge in terms of the QoS metric. For clarity, we refer to d_e as an edge delay, but it can be any other additive QoS metric. An (s,t)-*path* is a sequence of distinct nodes $P = (s = v_0, v_1, \ldots, t = v_n)$, such that, for $0 \leq i \leq n-1$, $(v_i, v_{i+1}) \in E$. A path W with identical source and destination nodes is referred to as a *cycle*. We denote by $E(P)$ the set of edges that belong to path P. The total cost $C(P)$ of a path P is equal to the sum of the costs of the individual edges along the path, that is, $C(P) = \sum_{e \in P} c_e$. Similarly, the total delay $D(P)$ of P is defined as $D(P) = \sum_{e \in P} d_e$. We also define the delay $D(\hat{E})$ of a set $\hat{E} \subseteq E$ to be the total delay of all edges in $\hat{E}$. In this chapter we focus on the following problems.

39.2.1 Minimum Cost Disjoint Paths Problem

An instance of the minimum cost disjoint paths (MCDP) problem includes a graph $G(V, E)$, a source node s, a destination node t, and the number k of disjoint paths. The goal is to find a set of k edge disjoint (s,t)-paths $\mathbf{P} = \{P_1, P_2, \ldots, P_k\}$ of minimum total cost $\sum_{P_i \in \mathbf{P}} C(P_i)$.

39.2.2 Restricted Shortest Paths Problem

An instance of the restricted shortest path (RSP) problem includes a graph $G(V, E)$, a source node s, a destination node t, and a delay constraint D. The goal is to find an (s,t)-path P with minimum cost $C(P)$ that satisfies the delay constraint $D(P) \leq D$.

The RSP problem is in general NP-hard, but it admits a *fully polynomial approximation scheme* (FPAS), that is, it is possible to find a polynomial time algorithm whose cost is at most $(1+\varepsilon)$ times more than the optimum. There are several exact and approximation algorithms proposed for the RSP problem [6–10]. The fastest approximation algorithm, due to Raz and Lorentz [8], has a computational complexity of $O(|E||V|((1/\varepsilon) + \log\log|V|))$.

39.2.3 Restricted Disjoint Paths Problem

An instance of the RSP problem includes a graph $G(V,E)$, a source node s, a destination node t, and a delay constraint D. The goal is to find two edge disjoint paths P_1 and P_2 of minimum total cost, $C(P_1) + C(P_2)$, that satisfy the delay constraint, that is, $D(P_1) \leq D$ and $D(P_2) \leq D$.

Restricted disjoint paths (RDP) problem is hard since it generalizes the RSP problem as well as the problem of finding two disjoint paths with minimum maximal weight, which was shown to be NP-hard by Li et al. [11] (it is straightforward to construct a reduction from either of these problems). The reduction from the latter problem also implies that it is NP-hard to find a solution to the RDP problem that satisfies the delay constraint. Accordingly, our goal is to find an (α, β)-approximate solution that both relaxes the delay constraint as well as the cost optimality condition. With an (α, β)-approximate solution, our goal is to identify two paths, P_1 and P_2, that satisfy the following conditions:

- The delay of each of the paths P_1 and P_2 is at most α times the delay constraint, that is, $D(P_1) \leq \alpha D$ and $D(P_2) \leq \alpha D$.

- The total cost of paths P_1 and P_2 is at most β times the optimum, that is, $C(P_1) + C(P_2) \leq \beta\mathrm{OPT}$.

39.3 NETWORK FLOWS

Disjoint paths in networks are closely related to *network flows*. An (s,t)-flow f is a mapping that assigns each edge $e \in E$ a value f_e. The flow value f_e cannot exceed the capacity of the edge e. In this chapter we assume that each edge has unit capacity and that flow values are either zero or one, that is, $f_e \in \{0,1\}$. In addition, the flow values satisfy *flow conservation constraints*, that is, for each $v \in V \setminus \{s,t\}$, it holds that

$$\sum_{(w,v) \in E} f_{(w,v)} = \sum_{(v,w) \in E} f_{(v,w)}.$$

The *value* of a flow f is defined as the sum of values of the outgoing edges of s:

$$|f| = \sum_{(s,v) \in E} f_{(s,v)}. \tag{39.1}$$

Note that $|f|$ is also equal to the sum of flow values of the incoming edges of t.

The cost $C(f)$ and delay $D(f)$ of a flow f are defined as

$$C(f) = \sum_{e \in E} c_e \cdot f_e \tag{39.2}$$

and

$$D(f) = \sum_{e \in E} d_e \cdot f_e, \tag{39.3}$$

respectively.

An (s,t)-flow can be decomposed into a set of disjoint paths between s and t and a set of cycles by using the *flow decomposition algorithm* [12]. Note that a minimum cost flow only contains a set of disjoint paths. Indeed, removal of a cycle from the flow minimizes its cost, but does not affect the value of the flow.

Note that the problem of finding a minimum cost set of k disjoint paths is related to the problem finding a minimum cost (s,t)-flow of value k.

39.4 ALGORITHM FOR THE MINIMUM COST DISJOINT PATHS PROBLEM

In this section we focus on the MCDP problem. This problem can be solved by using the algorithm due to Suurballe and Tarjan [3]. The detailed description of the algorithm is presented in Figure 39.1.

In Step 1, we find a shortest path P' between the source s and destination t and form a set $\hat{E} \subseteq E$ that includes the edges of P'. The goal of Step 2 is to compute, for each node $v \in V$, the minimum cost $\pi(v)$ of the shortest path between s and v in G. The values $\{\pi(v)\}$ are then used in Step 3 to modify or *reduce* the cost $c(e)$ for each edge $e \in E$. The reduced edge costs have several important properties: the reduced cost of each edge $e \in E$ is non-negative and the cost of each edge that belongs to the shortest path P' is zero. Note that the cost of any (s,t)-path with respect to the reduced costs is $\pi(t)$ units lower than its cost with respect to the original edge costs. Thus, the ordering of the paths with respect to reduced costs is the same as with respect to the original edge costs. The first two steps can be accomplished by a single invocation of a shortest path algorithm, such as Dijkstra's.

The goal of Step 4 is to construct an auxiliary graph G' that will be used for finding the second (s,t)-path, P''. Graph G' is formed from G by reversing the edges in $\hat{E}$. This way, a path in G' can use edges of P'' in the reverse direction. In Step 5, we identify a shortest path P'' between s in t in the auxiliary graph G'. Finally, in Step 6 we combine $\hat{E}$ and $E(P)$, excluding the interlacing edges of both sets. We refer to this step as *augmenting* set $\hat{E}$ along P''.

The algorithm can be extended to a k-disjoint paths algorithm by performing additional iterations. In each iteration, we find a shortest path P'' in the auxiliary graph formed from G by reversing edges in $\hat{E}$ and then augmenting set $\hat{E}$ along P''.

Algorithm MCDP $(G(V,E), s, t)$

1: Find a minimum cost path P' between source s and destination t;
   form a subset $\hat{E}$ of E that includes all edges in P',
   i.e., $\hat{E} \leftarrow E(P')$
2: **For** each node $v \in V$ **do**
   determine the minimum cost $\pi(v)$ of a shortest path in G that connects s and v;
3: **For** each edge $e(u,v) \in E$ **do**
   modify the cost $c(e)$ as follows:
   $$c(e) \leftarrow c(e) + \pi(u) - \pi(v)$$
4: Construct an auxiliary graph G' formed from G by reversing all edges that belong to $\hat{E}$
5: Find a minimum cost (s,t)-path P'' in G'
6: **For** each edge (u,v) of P'': **do**
   if there exists an interlacing edge $(v,u) \in \hat{E}$ **then** remove (v,u) from $\hat{E}$,
   otherwise add (u,v) to $\hat{E}$

Figure 39.1 Algorithm for minimum cost disjoint paths (MCDP) problem.

This algorithm belongs to the general class of *successive shortest paths algorithms* for the minimum cost flow problems [12]. Note that the set $\hat{E}$ corresponds to an (s,t)-flow f that assigns the value $f_e = 1$ for each edge $e \in \hat{E}$ and $f_e = 0$ otherwise. Each iteration of the algorithm increases the value of the flow by one unit. Therefore, at the end of the algorithm, set $\hat{E}$ can be decomposed into k edge-disjoint (s,t)-paths, for example, using the flow decomposition algorithm [12]. The algorithm requires k invocations of Dijkstra's shortest path algorithm, hence its complexity is $O(k(|E| + |V|\log(|V|)))$.

The following lemma establishes the correctness of the MCDP algorithm. We present a sketch of the proof that is slightly different from the standard approach. We will use a similar proof technique for establishing the correctness of approximation algorithms for the RDP problem below.

Lemma 39.1 *The algorithm finds an optimal solution to the minimum cost disjoint paths problem.*

Proof. (sketch) Let P_1^{opt} and P_2^{opt} be an optimal pair of disjoint (s,t)-paths and let OPT be their total cost. Since the cost of each of those paths with respect to the reduced costs is $C(P')$ units lower than that of the original cost, the total cost of these two paths with respect to the reduced cost is $\text{OPT} - 2C(P')$. Now, let us consider the subgraph G'' of G' induced by edges in $E(P_1^{\text{opt}})$, $E(P_2^{\text{opt}})$, as well as edges of P' in the reverse direction. It is easy to verify that G'' contains an (s,t)-path whose reduced cost is bounded by $\text{OPT} - 2C(P')$. Since G'' is a subgraph of G', it holds that G' has an (s,t)-path whose reduced cost is at most $\text{OPT} - 2C(P')$. Since P'' is a shortest path in G''', its cost with respect to the reduced edge costs is also at most $\text{OPT} - 2C(P')$, hence its cost with respect to the original costs is at most $\text{OPT} - C(P')$. This implies that the total cost of all edges in $\hat{E}$ (with respect to the original costs) is bounded by OPT. Note that the cost of some of the edges in P'' can be negative with respect to the original costs, but these edges do not appear in $\hat{E}$ because they are canceled by interlacing edges in P'. As mentioned above, $\hat{E}$ can be decomposed into two disjoint (s,t)-paths. We conclude that the total cost of these paths is bounded by OPT. ∎

Node-disjoint paths algorithms. With a small extension, the MCDP algorithm can be used for finding two node-disjoint paths. The basic idea is to construct an auxiliary graph G' formed from G by splitting each node $v \in V$ into two nodes, v' and v'' connected by an edge (v', v''). Each incoming edge (w, v) of v in G will be substituted by an edge (w, v') in G'; every outgoing edge (v, w) of v in G will be substituted by edge (v'', w) in G'. It is easy to verify that for every pair of node disjoint (s,t)-paths in G there exists a pair of edge disjoint (s,t)-paths in G' and vice versa.

39.5 DISJOINT PATHS WITH A DELAY CONSTRAINT

In this section we focus on the approximation solutions for the RDP problem, for different values of (α, β). We begin with a simple algorithm, referred to as Algorithm RDP1, depicted in Figure 39.2.

Similar to the MCDP algorithm we use the path augmentation approach. The first step is to identify an (s,t)-path P' that satisfies a given delay constraint D, whose cost is at most $(1+\varepsilon)$ more than the optimum. In Step 2, we construct an auxiliary graph G' formed from G by reversing all edges in $E(P')$ and setting their delay and cost to zero. The auxiliary graph is then used in Step 3 for finding an (s,t)-path P''. To that end we invoke Algorithm RSP to find a path P'' that satisfies the delay constraint $2D$. Next we augment set $\hat{E}$ along P'' by constructing a set $\hat{E}$ that includes all edges in $E(P')$ and $E(P'')$, except for the interlacing edges.

> **Algorithm RDP1** $(G(V,E), s, t, D)$
> 1: Invoke Algorithm RSP to find an (s,t)-path P' that satisfies the delay constraint D
> 2: Construct an auxiliary graph G' by reversing all edges in G that belong to P' and assigning them zero delay and zero cost
> 3: Invoke Algorithm RSP to find an (s,t)-path P'' that satisfies the delay constraint $2D$
> 4: Construct a set of edges $\hat{E}$ by combining $E(P')$ and $E(P'')$ and removing all interlacing edges from the resulting set
> 5: Decompose the set $\hat{E}$ into two disjoint paths P_1 and P_2

Figure 39.2 Algorithm RDP1 for restricted disjoint paths problem.

Lemma 39.2 *Algorithm RDP1 identifies a $(3, 1.5(1+\varepsilon))$ solution to the RDP problem.*

Proof. First, we note that $D(P') \leq D$ and that the cost $C(P')$ of P' is less than or equal to $(1+\varepsilon)\min\{C(P_1^{\text{opt}}), C(P_2^{\text{opt}})\} \leq (1+\varepsilon)(\text{OPT}/2)$.

Next, let P_1^{opt} and P_2^{opt} be an optimal solution to the RDP problem, that is, $D(P_1^{\text{opt}}) \leq D$, $D(P_2^{\text{opt}}) \leq D$, and $C(P_1^{\text{opt}}) + C(P_2^{\text{opt}}) = \text{OPT}$. Let G'' be a subgraph of G that includes edges in $E(P_1^{\text{opt}})$, $E(P_2^{\text{opt}})$, as well as edges of P' in reverse direction. It is easy to verify that graph G'' contains an (s,t)-path whose delay is at most $D(P_1^{\text{opt}}) + D(P_2^{\text{opt}}) \leq 2D$ and whose cost is at most OPT. This implies that the cost of path P'' is at most $(1+\varepsilon)\text{OPT}$.

We conclude that the total delay of the edges in $\hat{E}$ is at most $3D$ and total cost of at most $1.5(1+\varepsilon)\text{OPT}$. Since both P_1 and P_2 are constructed using the edges in $\hat{E}$, it holds that $D(P_1) \leq 3D$, $D(P_2) \leq 3D$, and

$$C(P_1) + C(P_2) \leq 1.5(1+\varepsilon)\text{OPT}.$$

We conclude that paths P_1 and P_2 constitute a $(3, 1.5(1+\varepsilon))$ solution to the RDP problem. ∎

We observe that Lemma 39.2 implies a stronger result, namely, that the total delay $D(P_1) + D(P_2)$ of two paths, P_1 and P_2 is bounded by $3D$. Thus, the minimum delay path among P_1 and P_2 violates the delay constraint by at most 50%. The algorithm includes two invocations of the RSP algorithm, hence its computational complexity is bounded by the standard notation is $O(|E||V|((1/\varepsilon) + \log\log|V|))$.

39.6 MINIMIZING PATH DELAYS

Algorithm RDP1 presented in the previous section identifies two paths P_1 and P_2 whose total delay is at most $3D$ and whose cost is at most $1.5(1+\varepsilon)$ times more than the optimum. In this section we show how to reduce the delay of the disjoint paths through an iterative procedure that includes augmentation along negative delay cycles in the auxiliary graph. In particular, we present Algorithm RDP2 that identifies two disjoint paths whose total delay is at most $2D(1+\varepsilon)$, and with cost at most $2.5 + 1.5\varepsilon + 2\log(1/2\varepsilon)\text{OPT}$. The parameter ε captures the trade-off between the accuracy of the algorithm (in terms of delay violation), the cost of the paths, and its running time. We refer to this algorithm as Algorithm RDP2 and present its detailed description in Figure 39.3.

In Step 1 we invoke Algorithm RDP1 to find two disjoint (s,t)-paths P_1 and P_2 that satisfy $D(P_1) + D(P_2) \leq 3D$. We also denote by $\hat{E}$ the set of edges that belong to P_1 and P_2. In Step 2, we construct an auxiliary graph G' by reversing edges in $\hat{E}$, and setting the cost of each of the reversed edges to zero. The delay of a reverse edge $\hat{e}$ is set to $d(\hat{e}) = -d(e)$, where $d(e)$ is the delay of the original edge. Thus, all of the reversed edges have negative delays.

Algorithm RDP2 $(G(V, E), s, t, D, \varepsilon)$

1: Invoke Algorithm RDP1 with parameter ε to find two disjoint (s,t)-paths P_1 and P_2, define $\hat{E} = E(P_1) \cup E(P_2)$;
2: Construct an auxiliary graph G' from G by reversing the edges of $\hat{E}$, i.e., for each edge $e(u,v) \in \hat{E}$, add an edge $\hat{e}(v,u)$ to G' with cost $c(\hat{e}) = 0$ and delay $d(\hat{e}) = -d(e)$
3: Find a negative delay cycle W in G' that minimizes $\frac{D(W)}{C(W)}$
4: Augment $\hat{E}$ along W, i.e., add edges in $E(W)$ to $\hat{E}$ and then remove all interlacing edges
5: **If** $\sum_{e \in \hat{E}} d(e) \leq 2D(1+\varepsilon)$ **then** proceed to the next step,
   else repeat steps 2–5
6: Decompose $\hat{E}$ into two edge disjoint paths P_1' and P_2'

Figure 39.3 Algorithm RDP2 for restricted disjoint paths problem.

The intuition behind this assignment is that if a reversed edge is included in the augmenting cycle, then the original edge will be excluded from $\hat{E}$, which will result in reducing the value of $D(\hat{E})$.

The key idea of the algorithm is to find cycles in G' that have negative delays and then use these cycles to minimize the total delay of edges in $\hat{E}$. Since we maintain the requirement that the edges in $\hat{E}$ can always be decomposed into two disjoint (s,t)-paths, we are able to iteratively minimize the delay violation until the total delay of the paths is less than or equal to $2D(1+\varepsilon)$.

More specifically, in Step 3, we find a negative delay cycle W that minimizes the ratio $D(W)/C(W)$. This cycle can be determined by using the *minimum cost-to-time ratio cycle algorithm* [12]. In Step 4 we augment set $\hat{E}$ along cycle W. This is accomplished by adding the set of edges in $E(W)$ to $\hat{E}$ and then removing all pairs of interlacing edges. Since for each edge $e \in W$ with a negative delay $d(e)$, we remove an edge from $\hat{E}$ whose delay is equal to $|d(e)|$, after Step 4 the total delay of all edges in $\hat{E}$ decreases by $|D(W)|$. It is easy to verify that after the set of disjoint paths is augmented by a cycle or a set of cycles, the resulted set of edges can still be decomposed into two disjoint (s,t)-paths.

We perform several iterations until the total delay of edges in $\hat{E}$ is less than or equal to $2D(1+\varepsilon)$. Finally, set $\hat{E}$ is decomposed into two edge disjoint paths P_1' and P_2' whose total delay $D(P_1') + D(P_2')$ is less than or equal to $2D(1+\varepsilon)$.

39.7 ALGORITHM ANALYSIS

In this section we present a detailed analysis of Algorithm RDP2. Our analysis is based on the *augmenting cycle theorem* [12]. Let $\hat{E}$ be the set of edges maintained by algorithm and let E^{opt} be a set of edges that corresponds to the optimal solution to Problem RDP. Note that $\hat{E}$ and E^{opt} correspond to two (s,t)-flows of value two. Also, let G' be an auxiliary graph formed from G by reversing edges in $\hat{E}$. The theorem implies that there exists a set of cycles $\mathbf{W} = \{W_1, W_2, \ldots\}$ in G' such that E^{opt} corresponds to an (s,t)-flow obtained by augmenting $\hat{E}$ along cycles in $\mathbf{W}$. The set $\mathbf{W}$ has the following properties:

$$\sum_{W_i \in \mathbf{W}} D(W_i) = D(E^{\text{opt}}) - D(\hat{E}) \leq -(D(\hat{E}) - 2D);$$

$$\sum_{W_i \in \mathbf{W}} C(W_i) = C(E^{\text{opt}}) - C(\hat{E}) \leq \text{OPT}.$$

Note that the total delay of cycles in **W** is negative (since $D(\hat{E}) \geq 2D$) and its total cost is positive. Thus, the ratio of the total delay to the total cost for all cycles in **W** is negative and bounded by

$$\frac{\sum_{W_i \in \mathbf{W}} D(W_i)}{\sum_{W_i \in \mathbf{W}} C(W_i)} \leq -\frac{D(\hat{E}) - 2D}{\text{OPT}}.$$

By the averaging argument, there must be a cycle $\bar{W} \in \mathbf{W}$ for which it holds that

$$\frac{D(\bar{W})}{C(\bar{W})} \leq -\frac{D(\hat{E}) - 2D}{\text{OPT}}. \tag{39.4}$$

Since the cycle W identified in Step 3 of the algorithm minimizes $D(W)/C(W)$, it holds that

$$\frac{D(W)}{C(W)} \leq -\frac{D(\hat{E}) - 2D}{\text{OPT}}. \tag{39.5}$$

We proceed to bound the number of iterations performed by the algorithm. Note that the cost $C(W)$ of every augmenting cycle W is at least one, hence by Equation 39.5, each iteration decreases $D(\hat{E})$ by at least $\left(D(\hat{E}) - 2D\right)/\text{OPT}$ units. The algorithm makes an improvement in the objective value, which is proportional to the difference between the objective value at the current operation and the optimal solution. We use the following theorem, which is an extension of the *geometric improvement theorem* [12].

Theorem 39.1 *Suppose that z^0 is the objective value in the beginning of a minimization algorithm, z^i is the objective value at the ith iteration of the algorithm, and z^* is the minimum objective value. Suppose that the cost of each iteration is c^i. Furthermore, suppose that the algorithm guarantees that for every iteration i,*

$$z^i - z^{i+1} \geq \zeta c^i (z^i - z^*) \tag{39.6}$$

for some constant $\zeta, 0 < \zeta < 1$. For a given integer $k \geq 2$, let i^* be the first iteration for which it holds that

$$\sum_{i=0}^{i^*} c^i \geq \frac{2k}{\zeta}.$$

Then, the objective value z^{i^*} at iteration i^* satisfies

$$z^{i^*+1} - z^* \leq \frac{z^0 - z^*}{2^k}.$$

Proof. For the analysis purposes, we divide each iteration into c^i steps, each step decreases the objective value of the function by $(z^i - z^{i+1})/c^i$. We denote by y^j the value of the objective function at the jth step. Note that the possible improvement $y^j - y^{j+1}$ at each step, is lower bounded by

$$y^j - y^{j+1} = \frac{z^{i_j} - z^{i_j+1}}{c^i} \geq \zeta(z^{i_j} - z^*) \geq \zeta(y^j - z^*).$$

where i_j is the iteration that includes step j.

Then, we apply the *geometric improvement theorem* [12] to show that after at most $2/\zeta$ steps, the algorithm reduces the value of the total possible improvement by a factor of a least 2. This implies, in turn, that after $2k/\zeta$ steps, that is, $j \geq 2k/\zeta$, it holds that $y^j - z^* \leq (z^0 - z^*)/2^k$. Since the first i^* iterations include at least $2k/\zeta$ steps it holds that $z^{i^*} - z^* \leq (z^0 - z^*)/2^k$. ∎

Let z^0 be the value of $D(\hat{E})$ in the beginning of the algorithm. Note that $z^0 \leq 3D$ since the total delay of the paths returned by Algorithm RDP1 is bounded by $3D$. We also let $\zeta = (1/\text{OPT})$, and $z^* = 2D$. Equation 39.5 implies that at each iteration the algorithm improves the objective function by at least $\left(C(W)(D(\hat{E}) - 2D)\right)/\text{OPT}$. Thus, the conditions of Theorem 39.13 are satisfied when $\zeta = (1/\text{OPT})$, $z^* = 2D$, and when z^i is equal to the value of $D(\hat{E})$ at iteration i.

Let i^* be the first iteration for which it holds that the cost of all augmenting cycles identified in iterations $1, \ldots, i^*$ is greater than

$$\frac{2}{\zeta} \log\left(\frac{1}{2\varepsilon}\right) = 2\text{OPT} \log\left(\frac{1}{2\varepsilon}\right).$$

For $k = \log(1/2\varepsilon)$, the total cost of all augmenting cycles identified at or before iteration i^* is at least $2k/\zeta$, hence the theorem implies that

$$D(\hat{E}) - 2D \leq \frac{D}{2^k} = 2D\varepsilon,$$

or, equivalently,

$$D(\hat{E}) \leq 2D + 2D\varepsilon = 2D(1+\varepsilon).$$

Note that the total additional cost incurred at all iterations of Algorithm RDP2 is at most $\text{OPT}(2\log(1/2\varepsilon) + 1)$. Since the initial cost of $C(\hat{E})$ is bounded by $1.5(1+\varepsilon)\text{OPT}$, the total cost of the solution is bounded by $(2.5 + 1.5\varepsilon + 2\log(1/2\varepsilon))\text{OPT}$.

Since the cost of the cycle identified at each iteration is at least one, the algorithm performs $O(\text{OPT}\log(1/\varepsilon))$ iterations. The running time for each iteration is dominated by the time required to find a cycle W that minimizes the delay to cost ratio $D(W)/C(W)$. The complexity of this procedure is equal to $O(|E||V|\log(CD))$, where C is the maximum cost of an edge in E. We conclude that the computational complexity of Algorithm RDP2 is $O(|E||V|\log(CD)\text{OPT}\log(1/\varepsilon))$.

We conclude that Algorithm RDP2 provides a $(2(1+\varepsilon), 2.5 + 1.5\varepsilon + 2\log(1/2\varepsilon))$ solution to Problem RDP. We note that the delay of the minimum path is at most $D(1+\varepsilon)$ so it only slightly violates the delay constraint.

39.8 MINIMIZING THE COMPUTATIONAL COMPLEXITY

While Algorithm RDP2 provides good quality paths, its computational complexity is proportional to the cost OPT of the optimal solution to Problem RDP. Thus, the algorithm can only be applicable for settings in which the costs the edges in the networks edge are limited to small values. However, the complexity of the algorithm can be high for settings with large cost and delay values of network edges. More precisely, the complexity of Algorithm RDP2 is not polynomial in the size of the input. Algorithm RDP2 belongs to the class of *pseudo-polynomial* algorithms, since its complexity depends on the values of the edge parameters (delays and costs).

To reduce the computational complexity of the algorithm, we apply the *cost scaling* approach [7]. With this approach, the cost of each edge is scaled by *scaling factor* Δ. Specifically, we reduce the cost $c(e)$ of each edge $e \in E$, replacing it by

$$c'(e) = \left\lfloor \frac{c(e)}{\Delta} \right\rfloor + 1. \tag{39.7}$$

The scaling approach requires to obtain sufficiently tight bounds L and U on the value OPT of the optimal solution. In Section 39.9 below we show how to identify bounds L and U that

> **Algorithm RDP3** $(G(V,E), s, t, D, \varepsilon)$
> 1: Obtain lower and upper bounds, L and U, on the cost of the optimal solution to Problem RDP
> 2: $\Delta \leftarrow \frac{L\varepsilon}{2|V|}$
> 3: **For** each edge $e \in E$ **do**
> $c'(e) \leftarrow \left\lfloor \frac{c(e)}{\Delta} \right\rfloor + 1.$
> 4: Invoke Algorithm RDP2 with respect to scaled costs to find two disjoint (s,t)-paths P_1 and P_2

Figure 39.4 Algorithm RDP3 for the restricted disjoint paths problem.

satisfy $(U/L) \leq 2|V|$ through a simple procedure. The lower bound L is then used in order to determine the scaling factor Δ,

$$\Delta = \frac{L\varepsilon}{2|V|}.$$

With this assignment, the scaling factor is large enough to reduce the edge costs, but it also ensures that the scaling procedure does not result in a significant increase in the overall cost of the solution. The detailed description of Algorithm RDP3 appears in Figure 39.4.

We proceed to show that the scaling procedure increases the cost of each pair P_1 and P_2 by a factor of at most $(1+\varepsilon)$. Equation 39.7 implies that, the cost of each edge increases by at most Δ. Since two disjoint paths include at most $2|V|$ edges, the total cost of P_1 and P_2 with respect to scaled edge cost is bounded by

$$C(P_1) + C(P_2) + 2|V|\Delta = C(P_1) + C(P_2) + L\varepsilon \leq (C(P_1) + C(P_2))(1+\varepsilon).$$

We conclude that the approximation solution provided by Algorithm RDP3 is similar to that provided by Algorithm RDP2, with a slightly increased cost (by a factor of $(1+\varepsilon)$).

We proceed to analyze the computational complexity of the algorithm. Let P_1^{opt} and P_2^{opt} be two disjoint (s,t)-paths that constitute an optimal solution to Problem RDP. Since P_1^{opt} and P_2^{opt} contain at most $2|V|$ edges, their cost OPT' with respect to scaled edge cost is bounded by

$$\text{OPT}' \leq \frac{2|V| \cdot \text{OPT}}{\varepsilon \cdot L} + 2|V|.$$

Since $(U/L) \leq 2|V|$, it holds that

$$L \geq \frac{U}{2|V|} \geq \frac{\text{OPT}}{2|V|},$$

hence

$$\text{OPT}' \leq \frac{2|V|^2}{\varepsilon} + 2|V| = O\left(\frac{|V|^2}{\varepsilon}\right).$$

By applying Algorithm RDP2 with respect to scaled costs we can reduce its complexity to $O((1/\varepsilon)\log(1/\varepsilon)|E||V|^3\log(CD))$. A further improvement in the computational complexity is discussed in [1].

39.9 COMPUTING UPPER AND LOWER BOUNDS

We denote by $c^1 < c^2 < \cdots < c^r$ the distinct costs values of the edges. Our goal is to find the maximum cost value $c^* \in \{c^i\}$ such that the graph G' derived from G by omitting all edges whose cost is greater than c^*, does not contain two disjoint paths P_1 and P_2 such that

$D(P_1) + D(P_2) \leq 2D$. Clearly, a feasible solution contains at least one edge whose cost is c^* or more, hence c^* is a lower bound on OPT. In addition, there exists a feasible solution P_1 and P_2 with $D(P_1) + D(P_2) \leq 2D$ that comprises edges whose cost is c^* or less. Since P_1 and P_2 include at most $2|V|$ edges, it holds that $2|V| \cdot c^*$ is an upper bound on OPT

We perform a binary search on the values $c^1, c^2, \cdots, c^r$. At each iteration, we need to check whether $c \leq c^*$, where c is the current estimate of c^*. For this purpose, we remove from G all edges whose cost is more than c, and assign the unit cost to the remaining edges. Then, we find a pair of disjoint paths of minimum delay (e.g., using the MCDP algorithm) in the resulting graph. If this algorithm returns a feasible flow, then we set $c \geq c^*$; otherwise, $c < c^*$. This procedure requires $O(\log|V|)$ iterations.

39.10 CONCLUSION

This chapter focused on two basic disjoint paths problems, namely, finding a set of disjoint paths of minimum total cost and finding a pair of edge disjoint paths that satisfy a given delay constraint at minimum cost. For the minimum cost disjoint paths problem we described the classical Suurballe–Tarjan algorithm. For the problem of finding QoS paths with a delay constraint, we presented three approximation algorithms. The first algorithm is simple and efficient, but it provides only a loose bound on the delay requirement. The second algorithm computes paths with tight delay guarantees at the price of higher cost. The third algorithm allows to compute the paths in a computationally efficient manner.

Finding a set of disjoint paths that satisfy QoS constraints have received a significant attention from the research community. The related studies focused on finding disjoint paths subject to multiple additive QoS constraints [13,14], improving network reliability [15,16], the hardness of the disjoint paths problem [17], extending the problem to the realm of disjoint spanning trees [18], and the trade-off between reliability and QoS constraints [19,20].

References

[1] A. Orda and A. Sprintson. Efficient Algorithms for Computing Disjoint QoS Paths. In *23rd Annual Joint Conference of the IEEE Computer and Communications Societies*, Hong Kong, China, March 2004.

[2] E. Mannie and D. Papadimitriou (editors). Recovery (Protection and Restoration) Terminology for Generalized Multi-Protocol Label Switching (GMPLS). Internet draft, Internet Engineering Task Force, May 2003.

[3] J. Suurballe and R. Tarjan. A Quick Method for Finding Shortest Pairs of Disjoint Paths. *Networks*, **14** (1984), 325–336.

[4] A. Itai, Y. Perl, and Y. Shiloach. The Complexity of Finding Maximum Disjoint Paths with Length Constraints. *Networks*, **12** (1982), 277–286.

[5] M.R. Garey and D.S. Johnson. *Computers and Intractability*. Freeman, San Francisco, CA, 1979.

[6] F. Ergun, R. Sinha, and L. Zhang. An Improved FPTAS for Restricted Shortest Path. *Information Processing Letters*, **83** (2002), 237–293.

[7] R. Hassin. Approximation Schemes for the Restricted Shortest Path Problem. *Mathematics of Operations Research*, **17** (1992), 36–42.

[8] D.H. Lorenz and D. Raz. A Simple Efficient Approximation Scheme for the Restricted Shortest Path Problem. *Operations Research Letters*, **28** (2001), 213–219.

[9] Y. Xiao, K. Thulasiraman, and G. Xue. QoS Routing in Communication Networks: Approximation Algorithms Based on the Primal Simplex Method of Linear Programming. *IEEE Transactions on Computers*, **55** (2006), 815–829.

[10] Y. Xiao, K. Thulasiraman, and G. Xue. Constrained Shortest Link-Disjoint Paths Selection: A Network Programming Based Approach. *IEEE Transactions on Circuits and Systems*, **53** (2006), 1174–1187.

[11] C.L. Li, T. McCormick, and D. Simchi-Levi. The Complexity of Finding Two Disjoint Paths with Min-Max Objective Function. *Discrete Applied Mathematics*, **26** (1990), 105–115.

[12] R.K. Ahuja, T.L. Magnanti, and J.B. Orlin. *Networks Flows*. Prentice Hall, Upper Saddle River, NJ, 1993.

[13] G. Xue, A. Sen, W. Zhang, J. Tang, and K. Thulasiraman. Finding a Path Subject to Many Additive QoS Constraints. *IEEE/ACM Transactions on Networking*, **15** (2007), 201–211.

[14] G. Xue, W. Zhang, J. Tang, and K. Thulasiraman. Polynomial Time Approximation Algorithms for Multi-Constrained QoS Routing. *IEEE/ACM Transactions on Networking*, **16** (2008), 656–669.

[15] G. Kuperman, E. Modiano, and A. Narula-Tam. Network Protection with Multiple Availability Guarantees. In *IEEE International Conference on Communications*, pp. 6241–6246. IEEE, Ottawa, ON, 2012.

[16] R. Chze Loh, S. Soh, and M. Lazarescu. Addressing the Most Reliable Edge-Disjoint Paths with a Delay Constraint. *IEEE Transactions on Reliability*, **60** (2011), 88–93.

[17] M. Andrews and L. Zhang. Hardness of the Undirected Edge-Disjoint Paths Problem. In *Proceedings of the 7th annual ACM Symposium on Theory of Computing*, pp. 276–283. ACM, Baltimore, MD, 2005.

[18] J. Yallouz, O. Rottenstreich, and A. Orda. Tunable Survivable Spanning Trees. *SIGMETRICS Performance Evaluation Review*, **42** (2014), 315–327.

[19] R. Banner and A. Orda. The Power of Tuning: A Novel Approach for the Efficient Design of Survivable Networks. *IEEE/ACM Transactions on Networking*, **15** (2007), 737–749.

[20] J. Yallouz and A. Orda. Tunable QoS-Aware Network Survivability. In *Proceedings of the INFOCOM*, pp. 944–952. IEEE, Turin, Italy, April 2013.

CHAPTER 40

Set-Cover Approximation

Neal E. Young

CONTENTS

40.1	Greedy Set-Cover Algorithm	1075
40.2	Analysis	1075
40.3	Vertex Cover	1076
40.4	Generalizations and Variants	1077

THE *Weighted Set-Cover Problem*: given a collection S of sets over a universe U, and a weight $w_s \geq 0$ for each set $s \in S$, find a collection $C \subseteq S$ of the sets whose union is U—a *set cover*—of minimum weight $\sum_{s \in C} w_s$.

40.1 GREEDY SET-COVER ALGORITHM

This algorithm, due to Johnson [1], Lovász [2], and Chvátal [3], computes an approximately optimal cover as follows: It chooses a set s to minimize the *price per element*—the weight w_s divided by the number of elements in s not yet covered by chosen sets. It repeats this step until the chosen sets cover all elements, then stops and returns the chosen sets. The algorithm is in Figure 40.1.

40.2 ANALYSIS

The standard linear-program relaxation for weighted set cover, and its dual, are shown in Figure 40.2. Let $H_k = 1 + (1/2) + (1/3) + \cdots + (1/k) \sim \ln k$ denote the kth harmonic number.

Theorem 40.1 *Fix any instance (S, w) of weighted set cover and any solution x to the linear-program relaxation. Let C be the set cover returned by the* greedy-set-cover (S, w). *Then C has weight at most $\sum_s H_{|s|} w_s x_s$.* ∎

Before we prove the theorem, observe that the standard performance guarantee for the algorithm follows as a corollary:

Corollary 40.1 [1–3] *Greedy set-cover (S, w) returns a set cover C of weight at most H_k times the minimum weight of any set cover, where $k = \max_{s \in S} |s|$ is the maximum set size.*

Proof of Theorem 40.1. For each element $e \in U$, define y_e to be the price per element (as defined in Figure 40.1) during the iteration of the algorithm that covers e. Then sum $\sum_{e \in U} y_e$ is the cost of the cover C returned by the algorithm. To complete the proof we show that $\sum_{e \in U} y_e$ is at most $\sum_s H_{|s|} w_s x_s$.

```
Greedy-set-cover(S, w)
1. Initialize C ← ∅. Define f(C) ≐ |∪_{s∈C} s|.
2. Repeat until f(C) = f(S):
3.     Choose s ∈ S minimizing the price per element

            w_s / [f(C ∪ {s}) − f(C)].

4.     Let C ← C ∪ {s}.
5. Return C.
```

Figure 40.1 Greedy set-cover algorithm.

$$\text{Minimize } \sum_{s\in S} w_s x_s \text{ subject to} \quad \Big| \quad \text{Maximize } \sum_{e\in U} y_e \text{ subject to}$$
$$\sum_{s\ni e} x_s \geq 1 \quad \forall e \in U, \quad \Big| \quad \sum_{e\ni s} y_e \leq w_e \quad \forall s \in S,$$
$$x_s \geq 0 \quad \forall s \in S. \quad \Big| \quad y_e \geq 0 \quad \forall e \in U.$$

Figure 40.2 Linear-program relaxation for set cover, and its dual.

We claim that the vector y satisfies $\sum_{e\ni s} y_e \leq H_{|s|} w_s$ for all $s \in S$. Indeed, consider any such set $s = \{e_k, e_{k-1}, \ldots, e_1\} \in S$, where e_k was the first element covered by the algorithm, e_{k-1} was the second, and so on (breaking ties arbitrarily). Since greedy-set-cover minimizes the price per element, when an element e_i was covered, the price per element y_{e_i} was at least w_s/i. Summing over i proves the claim.

Since $\sum_{e\ni s} y_e \leq H_{|s|} w_s$ for all $s \in S$, the vector y is feasible for the dual of the linear program with modified cost vector w' where $w'_s = H_{|s|} w_s$. Hence, by weak duality, $\sum_{e\in U} y_e$ is at most $\sum_{s\in S} w'_s x_s = \sum_{s\in S} H_{|s|} w_s x_s$, as desired. ∎

40.3 VERTEX COVER

The *weighted vertex-cover problem* is: given an undirected graph $G = (V, E)$ and a weight $w_v \geq 0$ for each vertex $v \in V$, find a vertex cover (a collection $C \subseteq V$ of the vertices whose union touches all edges of the graph) of minimum weight $\sum_{v\in C} w_v$.

Weighted vertex cover reduces to set cover by creating a set s_v for each vertex v, containing the edges incident to v, and having weight w_v. Applied to this instance, greedy-set-cover yields an H_k-approximate solution, where k is the maximum degree of any vertex in G.

For applications such as this, where the set sizes can be large but no element occurs in many sets (at most two for vertex cover), the algorithm in Figure 40.3 gives a better approximation. The algorithm chooses any maximal solution y to the dual of the linear-program relaxation, then returns the sets whose constraints are tight in the dual.

```
Set-cover-2(S, w)
1. Let y be any maximal solution to the dual of the linear-program relaxation. (For
   example, initialize each y_e equal to zero, then consider the elements e ∈ U in any
   order, and raise y_e as much as possible without violating some dual constraint.)
2. Return the cover C = {s ∈ S : ∑_{e∈s} y_e = w_s}.
```

Figure 40.3 Other set-cover algorithm.

Theorem 40.2 [4,5] *Set-cover-2(S,w) returns a cover of cost at most Δ times the minimum weight of any set cover, where Δ is the maximum number of sets that any element is contained in, $\Delta = \max\{|\{s \in S : e \in s\}| : e \in U\}$.*

For vertex cover, $\Delta = 2$.

Proof of Theorem 40.2. First observe that the set C returned by the algorithm is a cover, because otherwise the vector y would not be maximal (if $e \in U$ is not covered by S, then y_e is not in any tight constraint, so y_e could be raised).

Next observe that the cost of the cover C is at most $\Delta \sum_{e \ni U} y_e$:

$$\sum_{s \in C} w_s = \sum_{s \in C} \sum_{e \in s} y_e \leq \Delta \sum_{e \ni U} y_e \qquad (40.1)$$

(The last inequality holds because each element e occurs in at most Δ sets.)

Since y is a feasible dual solution of cost $\sum_{e \in U} y_e$, weak duality implies that the value of the linear-program relaxation is at least $\sum_{e \in U} y_e$. The theorem follows. ∎

40.4 GENERALIZATIONS AND VARIANTS

Greedy-set-cover and its analysis generalize naturally to *minimizing a linear function subject to a submodular constraint* [6] (Theorem 9.4).

Set-cover-2 generalizes to arbitrary covering problems with submodular cost [7], and has polylog-time parallel/distributed variants [8]. Broadly, it belongs to the class of *local-ratio* algorithms [9].

References

[1] D. S. Johnson. Approximation algorithms for combinatorial problems. *Journal of Computer and System Sciences*, **9**(3) (1974), 256–278.

[2] L. Lovász. On the ratio of optimal integral and fractional covers. *Discrete Math*, **13**(4) (1975), 383–390.

[3] V. Chvátal. A greedy heuristic for the set-covering problem. *Mathematics of Operations Research*, **4**(3) (1979), 233–235.

[4] D. S. Hochbaum. Approximation algorithms for the set covering and vertex cover problems. *SIAM Journal on Computing*, **11**(3) (1982), 555–556.

[5] R. Bar-Yehuda and S. Even. A linear-time approximation algorithm for the weighted vertex cover problem. *Journal of Algorithms*, **2**(2) (1981), 198–203.

[6] G. L. Nemhauser and L. A. Wolsey. *Integer and Combinatorial Optimization*, volume 18. Wiley, New York, 1988.

[7] C. Koufogiannakis and N. E. Young. Greedy Δ-approximation algorithm for covering with arbitrary constraints and submodular cost. *Algorithmica*, **66**(1) (2013), 113–152.

[8] C. Koufogiannakis and N. E. Young. Distributed algorithms for covering, packing and maximum weighted matching. *Distributed Computing*, **24**(1) (2011), 45–63.

[9] R. Bar-Yehuda, K. Bendel, A. Freund, and D. Rawitz. Local ratio: A unified framework for approximation algorithms. *ACM Computing Surveys*, **36**(4) (2004), 422–463.

CHAPTER 41

Approximation Schemes for Fractional Multicommodity Flow Problems

George Karakostas

CONTENTS

41.1 Multicommodity Flow Problems 1079
41.2 Maximum Concurrent Flow 1082
    41.2.1 Analysis of the Approximation Guarantee 1085
    41.2.2 Running Time 1088
    41.2.3 Explicit Case 1090
41.3 Minimum Cost Concurrent Flow 1092
41.4 Application: Maximum Concurrent Flow for Lossy Networks 1093

41.1 MULTICOMMODITY FLOW PROBLEMS

Flow problems are the basis of many optimization problems. Their efficient solution has been studied intensively ever since the breakthrough treatise by Ford and Fulkerson [1] established the field of network flows and introduced novel algorithmic ideas for calculating them (such as the famous Ford–Fulkerson augmenting paths algorithm, and the modeling of dynamic flows as static ones). Usually the input to such a problem consists of a directed network modeled as a graph $G = (V, E)$, an edge capacity function $u : E \to \mathbb{R}^+$ and a source–sink pair (s, t) (single commodity flow) or, more generally, a set of k source–sink pairs (s_i, t_i), $1 \leq i \leq k$ (multicommodity flow). We want to calculate *flows* f^i from s_i to t_i that would optimize an objective function, subject to flow conservation, and the constraint that the sum of flows through an edge cannot exceed the capacity of the edge. Each origin–destination pair (and its corresponding flow f^i) is usually called a *commodity*. More formally, given the graph G, the edge capacities u, and the k source–sink pairs above, the flows $f^i : E \to \mathbb{R}^+$ for $i = 1, \ldots, k$ must satisfy the following sets of constraints:

Capacity constraints: For all edges $e \in E$, we require $\sum_{i=1}^{k} f^i(e) \leq u(e)$.

Flow conservation: For all commodities $i = 1, \ldots, k$, and for all vertices $u \in V \setminus \{s_i, t_i\}$, we require
$$\sum_{v:(v,u)\in E} f^i(v, u) = \sum_{v:(u,v)\in E} f^i(u, v).$$

There may be additional constraints to be satisfied by the flows, such as lower bounds for the edge flows (instead of only upper bounds in the form of capacities), and integrality requirements for the routing of the flow through network paths (e.g., each commodity can use only a single path). In what follows, we will assume that there are no edge flow lower bounds, that is, they are all assumed to be 0, and each commodity can be split at every node without any restrictions, that is, the flows we need to compute are *fractional*.

The simplest multicommodity flow problem is the *maximum multicommodity flow* problem: the objective is the maximization of the total flow.

MAXIMUM MULTICOMMODITY FLOW (MAX-MF)

INPUT: A directed graph $G = (V, E)$, an edge capacity function $u : E \to \mathbb{R}^+$, and a set of k source–sink pairs (s_i, t_i), $1 \leq i \leq k$.

OUTPUT: Flows $f^i : E \to \mathbb{R}^+$ for $i = 1, \ldots, k$ that maximize $\sum_{i=1}^{k} \sum_{v:(s_i,v) \in E} f^i(s_i, v)$.

An additional constraint that flows must satisfy in many cases is the satisfaction of *demand* requirements. These are lower bounds on the total flow amount of a commodity routed over the network.

Demand constraints: Given demands $d_i \geq 0$ for $i = 1, \ldots, k$, we require that for all commodities i

$$\sum_{v:(s_i,v) \in E} f^i(s_i, v) \geq d_i.$$

Note that while the capacity and flow conservation constraints can always be satisfied, for example, by the zero flow, the addition of demands may render any flow infeasible for the given network. In what follows, a flow that satisfies its demand constraints (and any additional constraints we may introduce) will be called *feasible*.

A natural question that arises is the minimum-cost multicommodity problem: given a set of demands over a network with per unit of flow costs on its edges, we look for a routing of the demands with minimum total cost. More specifically, in addition to a network G, capacities u, commodities $i = 1, \ldots, k$, and demands d, we are also given edge costs $c : E \to \mathbb{R}^+$. If the flow through an edge $e \in E$ is $f(e) := \sum_{i=1}^{k} f^i(e)$, then the cost of the flow through e is $c(e)f(e)$, and the total cost of the flow on G is $\sum_{e \in E} c(e)f(e)$. Therefore, we can pose the minimum-cost multicommodity flow problem as the optimization problem of finding a feasible flow $f^i, i = 1, \ldots, k$ that minimizes $\sum_{e \in E} c(e)f(e)$. In fact, we can reduce this minimization problem to the feasibility question on a series of *budget-constrained* problems. In such problems, we are given a *budget* $B \geq 0$ and we ask for a flow routing that satisfies the demands, and its total cost is at most B.

Budget constraint: Given a budget $B \geq 0$, we require that a feasible flow satisfies

$$\sum_{e \in E} \left[c(e) \sum_{i=1}^{k} f^i(e) \right] \leq B.$$

BUDGET MULTICOMMODITY FLOW (BUDGET-MF)

INPUT: A directed graph $G = (V, E)$, an edge capacity function $u : E \to \mathbb{R}^+$, an edge cost function $c : E \to \mathbb{R}^+$, a set of k source–sink pairs (s_i, t_i), $1 \leq i \leq k$, and a set of demands $d_i \geq 0$ for $i = 1, \ldots, k$.

OUTPUT: Feasible flow $f^i : E \to \mathbb{R}^+$ for $i = 1, \ldots, k$, that is, a flow that satisfies the demand and budget constraints.

In what follows, we will always assume that all numbers given as input (capacities, edge costs, demands, etc.) are *integers*. If they are not, we first transform them by multiplying all of them with a sufficiently large number.

If we can solve BUDGET-MF, then we can obviously come very close to the solution of the minimum-cost multicommodity flow problem, by performing a binary search for the optimal budget B_{OPT} in the range $[0, DC]$, where $D = \sum_{i=1}^{k} d_i$ is the total demand, and $C = \sum_{e \in E} c(e)$. We will need to run BUDGET-MF at most $O(\log D + \log C)$ times (each time with a new budget) to get to within a (fixed) accuracy of the minimum cost.

Another question one can ask when a network and demands for commodities are given, is the following: what is the maximum fraction (throughput) of demands that can be routed on the given network? This problem is the *maximum concurrent flow problem*, and as before, it also has a budgeted (cost) version. The term *concurrent* means that we are trying to satisfy the *same* fraction from each commodity, introducing a notion of fairness in demand satisfaction. Also note that this fraction (or throughput) may be bigger than 1, in which case we are able to route multiples of the given demands. More formally, the maximum concurrent flow problem is defined as follows:

MAXIMUM CONCURRENT FLOW (MCF)

INPUT: A directed graph $G = (V, E)$, an edge capacity function $u : E \to \mathbb{R}^+$, a set of k source–sink pairs (s_i, t_i), $1 \le i \le k$, and a set of demands $d_i \ge 0$ for $i = 1, \ldots, k$.

OUTPUT: Maximum λ such that there is feasible flow that satisfies at least demand λd_i for each $i = 1, \ldots, k$.

The budgeted version is defined as follows:

BUDGETED CONCURRENT FLOW (BCF)

INPUT: A directed graph $G = (V, E)$, an edge capacity function $u : E \to \mathbb{R}^+$, an edge cost function $c : E \to \mathbb{R}^+$, a set of k source–sink pairs (s_i, t_i), $1 \le i \le k$, a set of demands $d_i \ge 0$ for $i = 1, \ldots, k$, and budget $B \ge 0$.

OUTPUT: Maximum λ such that there is feasible flow that satisfies at least demand λd_i for each $i = 1, \ldots, k$ and the budget constraint.

The output of MCF and BCF can be one of two possibilities: either we need to output just the value of λ we calculate, or we need to output λ *and* a feasible flow that achieves this λ. In the former case, the problem we solve will be called *implicit*, while in the latter it will be called *explicit*. In the explicit case, it is enough to enumerate only the non-zero flow on every edge and for every commodity. Simple examples of networks with n nodes and k commodities can be used to show that this explicit representation can be of size $\Omega(nk)$. One such example is shown in Figure 41.1, where each commodity is routed through a path of length $\Theta(n)$. This gives a trivial lower bound of $\Omega(nk)$ on the dependency of the running time of any algorithm that produces an explicit representation on k.

Note that we can assume that at most one commodity corresponds to a single source–sink pair of nodes, since we can combine commodities with the same origin and destination into a single commodity with demand equal to the summation of the individual demands. If a

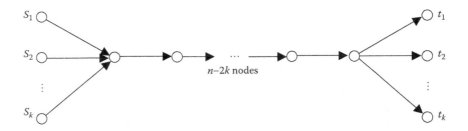

Figure 41.1 Simple example of k commodities, each with a flow path of length $\Theta(n)$.

fraction λ of the demand of this aggregated commodity is satisfied, then a fraction λ of the demand of each one of its constituent commodities is satisfied as well.

In what follows, G will always represent the underlying network (directed) graph for the flow routing, n will be its number of nodes and m its number of edges. We will assume that G is always connected, so $m = \Omega(n)$. Otherwise, we work on the connected components of G. We use the notation $\tilde{O}(f)$ to denote a quantity that is $O(f \log^{O(1)} n)$, that is, to hide polylogarithmic factors. We will first focus on polynomial time *approximation* algorithms for MCF and BCF, and then extend them to more general versions of these problems. By *approximation* we mean approximation by a positive *multiplicative* factor $\rho \leq 1$, that is, our algorithms will guarantee that if SOL and OPT are the values of the produced solution and the optimum value, respectively, then we have SOL $\geq \rho \cdot$ OPT. If for every positive constant $\varepsilon < 1$, we can present a polynomial time approximation algorithm with approximation factor $1 - \varepsilon$, then this family of algorithms is called a *polynomial time approximation scheme* (*PTAS*).* If, moreover, the PTAS running time depends *polynomially* on $1/\varepsilon$ then it is called a *fully polynomial time approximation scheme* (*FPTAS*).

Example 41.1 An approximation scheme with running time $O(\varepsilon^{-3} \cdot n^{2/\varepsilon})$ is a PTAS, while a scheme with running time $O(\varepsilon^{-3} \cdot n^2)$ is an FPTAS (and also a PTAS).

Note that, from a practical point of view, for small values of ε, that is, better approximations, an FPTAS is much more preferable than a PTAS. The approximation algorithms we will develop will be FPTASs. This is important for their practicality, since MCF and BCF (like all fractional multicommodity problems we defined) can be formulated as linear problems (LPs), and thus can be solved *exactly* in polynomial time by general LP algorithms such as the ellipsoid or interior point methods (see, e.g., [2]). We will also use the LP formulations, but in order to find good *approximate* solutions much faster than these general LP solvers.

41.2 MAXIMUM CONCURRENT FLOW

It is well known that any flow can be decomposed into a set of *path flows*, that is, flows that use a single path from the source to the sink. Let $\mathcal{P}_i$ be the set of paths between s_i and t_i in G, and let $\mathcal{P} := \cup_i \mathcal{P}_i$ be the set of all possible source–sink paths. Then the feasible flow f can be decomposed into path flows $\cup_{i=1}^{k} \{x(P) : P \in \mathcal{P}_i\}$, where $x(P)$ is the non-negative amount of flow that is routed through path P. We will solve MCF by computing path flows $x(P)$ and the maximum throughput λ. Then, we will output λ (in the implicit case), or λ and $f^i(e), e \in E$ (in the explicit case).

By using a variable λ for the throughput, and variables $x(P), P \in \mathcal{P}$ for the amount of flow sent along path P, for every $P \in \mathcal{P}$, the path flow LP formulation of

---
*Note that we define the approximation factor for a *maximization* problem. The definitions for a minimization problem are analogous, and the approximation factor is of the form $1 + \varepsilon$, with $\varepsilon > 0$.

MCF is the following:

$$\text{maximize } \lambda \text{ s.t.}$$
$$\sum_{P: e \in P} x(P) \leq u(e) \quad \forall e \in E$$
$$\sum_{P \in \mathcal{P}_i} x(P) \geq \lambda d(i) \quad \forall i = 1, \ldots, k \quad \text{(MCF)}$$
$$x(P) \geq 0 \quad \forall P \in \mathcal{P}$$
$$\lambda \geq 0$$

In MCF, the first set of constraints enforce the capacity constraints, and the second fact that at least a multiple (or fraction) of λ must be satisfied for each commodity. Since $x(P)$ are path flows, the flow conservation constraints are automatically satisfied. Note that MCF has an exponential number of variables $x(P)$ since the number of paths in $\mathcal{P}$ is exponential on n and m, but this will not bother us because we will never need to explicitly write MCF down.

The dual* LP of MCF has a variable $l(e)$ for each capacity constraint of the primal and a variable $z(i)$ for every commodity demand constraint:

$$\text{minimize } \sum_{e \in E} u(e) l(e) \text{ s.t.}$$
$$\sum_{e \in P} l(e) \geq z(i) \quad \forall i, \forall P \in \mathcal{P}_i$$
$$\sum_{i=1}^{k} d(i) z(i) \geq 1 \quad \text{(DMCF)}$$
$$l(e) \geq 0 \quad \forall e$$
$$z(i) \geq 0 \quad \forall i$$

Note that DMCF has a polynomial number of variables, but the number of constraints is exponential.

Let $D(l) := \sum_e u(e) l(e)$ be the quantity minimized by DMCF. We define $dist_i(l)$ as the distance of the shortest path from s_i to t_i in G under the length function l. Let $\alpha(l) := \sum_i d(i) dist_i(l)$.

Lemma 41.1 *Finding edge lengths $l(e)$ and $z(i)$ that minimize $D(l)$ under the constraints of DMCF is equivalent to computing lengths $l(e)$ for the edges and $z(i)$ that minimize $D(l)/\alpha(l)$.*

Proof. Let $l(e), z(i)$ be an optimal solution for DMCF, and $l'(e)$ an edge length function that minimizes $D(l)/\alpha(l)$. Note that by constraints $\sum_{e \in P} l(e) \geq z(i)$ for each commodity i, we have $dist_i(l) \geq z(i)$. Therefore, we have

$$\alpha(l) = \sum_i d(i) dist_i(l) \geq \sum_{i=1}^{k} d(i) z(i) \geq 1$$

and therefore

$$\frac{D(l')}{\alpha(l')} \leq \frac{D(l)}{\alpha(l)} \leq D(l).$$

Also, note that $(l'/\alpha(l'), dist_i(l)/\alpha(l'))$ is a feasible solution of DMCF. Hence,

$$D(l) \leq D\left(\frac{l'}{\alpha(l')}\right) = \frac{D(l')}{\alpha(l')}$$

and the lemma follows. ∎

Let $\beta := \min_l D(l)/\alpha(l)$. We will make the following assumption, which we will lift later.

---

*We assume basic knowledge of linear programming and its terminology, such as duality, the weak- and strong-duality theorems, and so on. A good reference is [3].

Assumption 41.1 *We assume that $\beta \geq 1$.*

The algorithm is described in Figure 41.2.

Input: Graph $G = (V, E)$, capacities $u(e)$, pairs (s_i, t_i) with demands $d(i)$,
$1 \leq i \leq k$, accuracy $w < 1$.

Output: Throughput λ.

Initialize $l(e) := \delta/u(e), \forall e, \quad x(P) := 0, \forall P$.
while $D(l) < 1$ **do**
 for $i = 1$ to $|\mathcal{S}|$ **do**
  $d'(c_q) := d(c_q), q = 1, \ldots, r$
  while $D(l) < 1$ and $d'(c_q) > 0$ for some q **do**
   P_{c_q} := shortest path in $\mathcal{P}_{c_q}$ using $l, q = 1, \ldots, r$ with $d'(c_q) > 0$
   f_{c_q} := $d'(c_q), q = 1, \ldots, r$ with $d'(c_q) > 0$
   σ := $\max\left\{1, \max_{e \in P_{c_1} \cup \ldots \cup P_{c_r}}\left\{\frac{\sum_{c_q : e \in P_{c_q}} f_{c_q}}{u(e)}\right\}\right\}$
   $\left.\begin{array}{l} f_{c_q} := f_{c_q}/\sigma \\ d'(c_q) := d'(c_q) - f_{c_q} \\ x(P_{c_q}) := x(P_{c_q}) + f_{c_q} \end{array}\right\} q = 1, \ldots, r$ with $d'(c_q) > 0$
   $l(e) := l(e)\left(1 + \varepsilon \cdot \frac{\sum_{c_q : e \in P_{c_q}} f_{c_q}}{u(e)}\right), \forall e \in P_{c_1} \cup \ldots \cup P_{c_r}$
  end while   /* end of step */
 end for   /* end of iteration */
end while   /* end of phase */
$x(P) := x(P)/\log_{1+\varepsilon}\frac{1+\varepsilon}{\delta}, \forall P$
$\lambda := \min_i \frac{\sum_{P \in P_i} x(P)}{d(i)}$
Output λ

Figure 41.2 Maximum concurrent flow FPTAS for $\beta \geq 1$.

In this algorithm, δ is a parameter that we will define exactly later. It computes variables x, λ for MCF and lengths $l(e), e \in E$ for the problem of minimizing $D(l)/\alpha(l)$ (or, equivalently, solving DMCF). Note that the algorithm has three nested loops. The iterations of each loop, from the outer to the inner, will be called *phases*, *iterations*, and *steps*. A new phase is entered if $D(l) < 1$ (for the current values of l). Each phase will run through $|\mathcal{S}|$ iterations, with $\mathcal{S}$ being the set of *sources* (notice that this is *not* necessarily the number of *commodities* k). In the ith iteration of the current phase, we consider all the commodities with the same source $s_i \in \mathcal{S}$. Let $c_1, c_2, \ldots, c_r$ be these commodities. Within the current iteration we try to route $d(c_q), q = 1, \ldots, r$ units of commodity c_q in a series of steps, until either all demand $d(c_q)$ is routed for all commodities c_q, or at some point $D(l) \geq 1$. Inside each step, commodities c_q that still have leftover demand ($d'(c_q) > 0$) route some of this demand on the single shortest path in the shortest path tree rooted at the common source s_i, which is computed by running Dijkstra's algorithm *once*, and using the current edge lengths l. The amount of flow routed on the shortest path tree is calculated so that no capacities are violated in the current step. Therefore, each phase goes through all distinct sources in a round-robin fashion in its iterations, and each iteration except possibly the last, routes the *whole* demand of each commodity; this may result in over-routing of flow (oversaturation of some capacities), but this is taken care of right after exiting the outermost loop, when variables $x(P)$ are appropriately scaled down by a factor of $\log_{1+\varepsilon}((1 + \varepsilon)/\delta)$. Finally, we calculate and output λ.

41.2.1 Analysis of the Approximation Guarantee

Let $l_{i,j,s}$ be the length function at the end of the sth step of the jth iteration of the ith phase. Initially $l_{1,1,0}(e) = \delta/u(e)$ for all edges e. At every step we compute the shortest path tree from s_j to the sinks of commodities $c_q, q = 1, \ldots, r$. Let $P_{i,j,s}^{c_q}$ be the path in this tree from s_j to the sink of commodity c_q, that is $P_{i,j,s}^{c_q}$ has length $\mathrm{dist}_{c_q} l_{i,j,s-1}$. It is crucial to our running time that the shortest path tree for all commodities with a common source s_j can be computed with only *one* call to Dijkstra's algorithm. Let $d_{i,j,s}^{c_q} > 0$ be the amount of commodity c_q that has not been routed yet at step s (obviously $d_{i,j,0}^{c_q} = d(c_q)$). Notice that we consider only commodities with *strictly* positive remaining demand, and we ignore commodities whose demand has already been completely routed. Then at step s we route $f_{i,j,s}^{c_q} = (d_{i,j,s-1}^{c_q})/\sigma$ units of each commodity c_q along path $P_{i,j,s}^{c_q}$, where σ is a scaling factor that ensures we do not push through an edge flow greater than its capacity. After this we have that $d_{i,j,s}^{c_q} = d_{i,j,s-1}^{c_q} - f_{i,j,s}^{c_q}$, and for every edge e on these paths we set

$$l_{i,j,s}(e) := l_{i,j,s-1}(e) \left(1 + \varepsilon \cdot \frac{\text{total new flow through } e}{u(e)}\right)$$

$$= l_{i,j,s-1}(e) \left(1 + \varepsilon \cdot \frac{\sum_{c_q : e \in P_{i,j,s}^{c_q}} f_{i,j,s}^{c_q}}{u(e)}\right).$$

Note that for every saturated edge e its length $l(e)$ increases by a factor of $1 + \varepsilon$, and that in each iteration, during each step except possibly the last one, at least one edge is saturated, that is, gets $u(e)$ units of flow. After the last step s $d_{i,j,s}^{c_q} = 0$ for all commodities c_q.

In the following calculations we abuse our notation a bit by writing P_{c_q} instead of $P_{i,j,s}^{c_q}$. After step s of the jth iteration of the ith phase, we have

$$D(l_{i,j,s}) = D(l_{i,j,s-1}) + \varepsilon \cdot \sum_{e \in P_{c_1} \cup \ldots \cup P_{c_r}} l_{i,j,s-1}(e) \sum_{q : e \in P_{c_q}} f_{i,j,s}^{c_q}$$

$$= D(l_{i,j,s-1}) + \varepsilon \cdot \sum_{e \in P_{c_1} \cup \ldots \cup P_{c_r}} \sum_{q : e \in P_{c_q}} l_{i,j,s-1}(e) \cdot f_{i,j,s}^{c_q}$$

$$= D(l_{i,j,s-1}) + \varepsilon \cdot \sum_{q=1}^{r} f_{i,j,s}^{c_q} \cdot \sum_{e \in P_{c_q}} l_{i,j,s-1}(e)$$

$$= D(l_{i,j,s-1}) + \varepsilon \cdot \sum_{q=1}^{r} f_{i,j,s}^{c_q} \cdot \mathrm{dist}_{c_q} l_{i,j,s-1}$$

where the third equality is a simple rearrangement of terms, and the fourth equality holds because P_{c_q} is by definition the shortest path for commodity c_q under the length function of the previous step.

Let s_{last} denote the last step of an iteration. By taking into account all the steps of the iteration we get

$$D(l_{i,j,s_{last}}) = D(l_{i,j,0}) + \varepsilon \cdot \sum_{s=1}^{s_{last}} \sum_{q=1}^{r} f_{i,j,s}^{c_q} \cdot \mathrm{dist}_{c_q} l_{i,j,s-1}$$

$$\leq D(l_{i,j,0}) + \varepsilon \cdot \sum_{q=1}^{r} \mathrm{dist}_{c_q} l_{i,j,s_{last}} \sum_{s=1}^{s_{last}} f_{i,j,s}^{c_q}$$

$$= D(l_{i,j,0}) + \varepsilon \cdot \sum_{q=1}^{r} \mathrm{dist}_{c_q} l_{i,j,s_{last}} d(c_q)$$

where the second inequality holds because l is a monotonically increasing function. Using the facts $D(l_{i,j,s_{last}}) = D(l_{i,j+1,0})$ and $l_{i,j,s_{last}} = l_{i,j+1,0}$, we have proven that

$$D(l_{i,j,0}) \leq D(l_{i,j-1,0}) + \varepsilon \cdot \sum_{q=1}^{r} \text{dist}_{c_q} l_{i,j,0} d(c_q).$$

Using the same arguments for the iterations, we can show that

$$D(l_{i,1,0}) \leq D(l_{i-1,1,0}) + \varepsilon \cdot \alpha(l_{i,1,0}). \tag{41.1}$$

By definition $\beta = \min_l (D(l)/\alpha(l)) \leq (D(l_{i,1,0})/\alpha(l_{i,1,0}))$, so from (41.1) we have

$$D(l_{i,1,0}) \leq \frac{D(l_{i-1,1,0})}{1 - \varepsilon/\beta}$$

and from the fact that $D(l_{1,1,0}) = m\delta$ because of the initialization of $l_{1,1,0}$, we have

$$D(l_{i,1,0}) \leq \frac{m\delta}{(1-\varepsilon/\beta)^{(i-1)}}$$
$$\leq \frac{m\delta}{1-\varepsilon/\beta} e^{\frac{\varepsilon(i-2)}{\beta-\varepsilon}}$$
$$\leq \frac{m\delta}{1-\varepsilon} e^{\frac{\varepsilon(i-2)}{\beta(1-\varepsilon)}}$$

where in the last inequality we use the hypothesis $\beta \geq 1$, and assume $i \geq 2$. The procedure stops after $D(l_{t,j,s}) \geq 1$ in the step s of iteration j in some phase t. Hence

$$1 \leq D(l_{t,j,s}) \leq D(l_{t+1,1,0}) \leq \frac{m\delta}{1-\varepsilon} e^{\frac{\varepsilon(t-1)}{\beta(1-\varepsilon)}}$$

where $D(l_{t+1,1,0})$ is computed using the edge lengths l right after we exit the last phase t. Therefore

$$\beta \leq \frac{\varepsilon(t-1)}{(1-\varepsilon)\ln\left(\frac{(1-\varepsilon)}{m\delta}\right)}. \tag{41.2}$$

Lemma 41.2 *If $\beta \geq 1$ the algorithm terminates after at most $t := \lceil \beta \log_{1+\varepsilon} \frac{1+\varepsilon}{\delta} \rceil$ phases.*

Proof. In order to bound the throughput λ produced by the algorithm, we first prove the following claim, that also provides the reason for scaling by σ in line 9 of the algorithm in Figure 41.2.

Claim 1 $\lambda > \frac{t-1}{\log_{1+\varepsilon}\left(\frac{(1+\varepsilon)}{\delta}\right)}.$

Proof. Since after each phase we route $d(i)$ units of commodity i, after $t-1$ phases we have routed $(t-1)d(i)$ units of flow. This flow may be infeasible (it may violate capacity constraints), hence in the end we scale it down to make it feasible. Since we scale down the flow through a path by σ in every step so that no edge is overflowed, for every $u(e)$ units of flow routed through edge e we increase the length $l(e)$ by a factor of at least $1+\varepsilon$. At the beginning $l_{1,1,0}(e) = \delta/u(e)$ and by the end of phase $t-1$ we have $l_{t-1,|S|,s_{last}}(e) = l_{t,1,0} < 1/u(e)$ (since $D(l_{t,1,0}) < 1$). Hence at the end of the algorithm $l_{t,j_{last},s_{last}}(e) < (1+\varepsilon)/u(e)$ (each step of the last phase increases $l(e)$ by at most a factor of $1+\varepsilon$, and the algorithm stops as soon as $D(l_{t,j_{last},s_{last}}) \geq 1$; at this point $D(l_{t,j_{last},s_{last}}) \leq 1+\varepsilon$). Therefore the total amount of flow through e in all t phases is strictly less than

$$\log_{1+\varepsilon} \frac{(1+\varepsilon)/u(e)}{\delta/u(e)} = \log_{1+\varepsilon} \frac{1+\varepsilon}{\delta}$$

times its capacity. So by scaling the flow by $\log_{1+\varepsilon}((1+\varepsilon)/\delta)$ we get the claimed *feasible* primal solution x, and we have routed a fraction of at least $(t-1)/(\log_{1+\varepsilon}(1+\varepsilon/\delta))$ of each demand.

This implies that
$$1 \le \frac{\beta}{\lambda} < \frac{\beta}{t-1} \log_{1+\varepsilon} \frac{1+\varepsilon}{\delta} \tag{41.3}$$
which, in turn, implies the lemma. ∎

From (41.3) and (41.2) we have
$$\frac{\beta}{\lambda} < \frac{\varepsilon \log_{1+\varepsilon}((1+\varepsilon)/\delta)}{(1-\varepsilon)\ln((1-\varepsilon)/m\delta)} = \frac{\varepsilon}{(1-\varepsilon)\ln(1+\varepsilon)} \cdot \frac{\ln((1+\varepsilon)/\delta)}{\ln((1-\varepsilon)/m\delta)}$$

By setting
$$\delta := \frac{1}{(1+\varepsilon)^{((1-\varepsilon)/\varepsilon)}} \cdot \left(\frac{1-\varepsilon}{m}\right)^{\frac{1}{\varepsilon}} \tag{41.4}$$
the dual-primal solution ratio becomes less than $(1-\varepsilon)^{-3}$ and we can pick ε so that $\beta/\lambda \le 1+z$ for any $z > 0$, or $\lambda \ge (1-w)\text{OPT}$ for any $1 > w > 0$. Hence we have proven the following.

Lemma 41.3 *The proposed algorithm for the concurrent multicommodity flow problem is an approximation scheme, provided $\beta \ge 1$.*

The algorithm in Figure 41.2 calculates an approximation of the *value* λ of a maximum concurrent flow. With the same algorithm and in the same running time we can also output an *implicit* representation of a flow that achieves this value: all we have to do is to store the shortest path tree computed in every step, together with the amount of flow routed through each path in this tree (the $x(P_{c_q})$'s in Figure 41.2). The problem with this representation is that it may produce many more paths than needed, because the same path may belong to many shortest path trees. In order to aggregate the flow $x(P)$ through path P we will need to be keeping track of the flow through each path. If we have a data structure that supports such update operations quickly, then maybe this representation of the flow may suffice for our needs. In the explicit case, we will show how to modify the algorithm in order to output the edge flow values as well without burdening the running time of our algorithm by much more than the time needed to just output all non-zero edge flows.

Next, we describe how to modify the algorithm, so that we get the same approximation guarantee even when Assumption 41.1 does not hold initially. In this case $\beta < 1$.

If $\zeta(i)$ is the maximum flow of commodity i that can be routed in G when no other commodity is routed, the value $\zeta = \min_i\{\zeta(i)/d(i)\}$ is an upper bound of the optimal solution for MCF (at best, all maximum *single* commodity flows for all commodities can be routed simultaneously). The solution that routes a fraction of $1/k$ of each flow $\zeta(i)$ is then a feasible solution, so ζ/k is a lower bound of the optimal solution. Hence for the optimal λ we have $\zeta/k \le \lambda \le \zeta$, or, because of Lemma 41.1,
$$\zeta/k \le \beta \le \zeta. \tag{41.5}$$

If we multiply the initial demands by k/ζ, the optimal throughput λ' of the new problem will be $\lambda' = k\lambda/\zeta$, or, equivalently, we will have from (41.5) that for the new scaled problem $1 \le \beta' \le k$. Therefore, we can now apply the algorithm of Figure 41.2 on this scaled problem, and output the throughput λ we compute multiplied by ζ/k.

The problem with this approach is that we do not want to calculate the $\zeta(i)$'s in order to calculate ζ, since flow computations are too expensive. Instead, we are going to use the following simple fact about single-commodity flows.

Fact 41.1 *Any single commodity flow on a network with m edges can be decomposed into at most m path flows.*

This is due to the fact that we can decompose the flow using a maximum capacity path on the edges that still carry flow, after we have removed the flow on the previous paths of the decomposition. Such a path will saturate at least one edge, and there are at most m of those. Each path of this decomposition will be the maximum remaining capacity path, where we start with initial edge capacities equal to the edge flow, and reduce the capacities accordingly after the next path of the decomposition has been found. Note that a saturating flow $\hat{\zeta}$ routed through a *maximum capacity* path in the original network is at least $(1/m)f$, where f is the flow we decomposed, that is, $\hat{\zeta} \geq (1/m)f$. Flow $\hat{\zeta}$ can be calculated by doing a binary search on the m (sorted) capacities: for each capacity C we test whether there is a source–sink path in the network we get from the original one after we remove all edges with capacity smaller than C in time $O(m)$, for an overall time of $O(m \log m)$. Therefore, by applying this insight to each commodity of MCF and its flow $\zeta(i)$, we get the following bounds

$$\hat{\zeta}(i) \geq \frac{1}{m}\zeta(i) \text{ for each commodity } i = 1,\ldots,k \tag{41.6}$$

in total time $O(min\{n,k\}m \log m)$ time, if we always group together commodities with a common source node. Let $\hat{\zeta} = \min_i \{\hat{\zeta}(i)/d(i)\}$. Note that $\hat{\zeta} \leq \zeta$. Therefore, we get from (41.5)

$$1 \leq \frac{\zeta}{\hat{\zeta}} \leq \frac{k\beta}{\hat{\zeta}} \leq \frac{k\zeta}{\hat{\zeta}} \leq km.$$

Hence, if we multiply the initial demands by $k/\hat{\zeta}$, we will have that for the new scaled problem $1 \leq \beta' \leq km$. We could now apply the algorithm of Figure 41.2. Unfortunately, for δ as in (41.4), the number of phases can be as large as $\lceil \beta/\varepsilon \log_{1+\varepsilon}(1+\varepsilon)m/(1-\varepsilon) \rceil$, and the new β' can be as large as km, affecting the number of phases (and, eventually, the running time of the algorithm). As a result, we do not run directly the algorithm, but initially we run it in the following manner, in order to first compute a 2-approximation of β': We set the constant δ (and the corresponding ε) so that the algorithm computes a 2-approximation of β. If the algorithm does not stop after $T = \lceil (2/\varepsilon) \log_{1+\varepsilon}\{((1+\varepsilon)m)/(1-\varepsilon)\} \rceil = O(\log m)$ phases, we know that $\beta \geq 2$. We double the demands of all commodities, so that β is halved and still $\beta \geq 1$, and continue running the algorithm for another T phases. If the algorithm does not stop we double again the demands and continue doing this until the algorithm ends. Since every time β is halved, this doubling can happen at most $\log(km)$ times, the total number of phases is at most $T \log(km) = O(\log^2 m + \log m \log k)$, and we have produced a 2-approximation $\hat{\beta}$ of the optimal β. Since $\beta \leq \hat{\beta} \leq 2\beta$, or, equivalently, $(\hat{\beta}/2) \leq \beta \leq 2(\hat{\beta}/2)$, if we multiply all the demands by $2/\hat{\beta}$ (as we did in (41.5)), only at most $O(\varepsilon^{-2} \log m)$ additional phases are needed to get an ε-approximation for any $0 < \varepsilon < 1$, because now $1 \leq \beta \leq 2$. After the end of the algorithm, we multiply back our computed flow* and λ by a factor $\hat{\beta}/2$.

41.2.2 Running Time

The number of iterations in each phase is bounded by the number of sources, which is at most $\min\{n,k\}$. Since we need $O(\log^2 m + \log m \log k)$ phases to reduce β to less than 2, and at most $T = O(\varepsilon^{-2} \log m)$ additional phases to get an ε-approximation, the total number of iterations is $O(\min\{n,k\} \log m (\log m + \log k + \varepsilon^{-2}))$. In order to compute the total running time, we need to calculate the number of steps in the algorithm. At every step

---

*That is, in case we want to explicitly output the edge flows. We will need this in the explicit case.

other than the last step in an iteration, the length of an edge increases by a factor of at least $1+\varepsilon$. At the beginning of the algorithm each edge has length $\delta/u(e)$ and at the end of the algorithm its length is at most $(1+\varepsilon)/u(e)$. So the number of steps exceeds the number of iterations by at most $m \log_{1+\varepsilon}((1+\varepsilon)/\delta) = O(\varepsilon^{-2} m \log m)$. Note that this estimate holds for both the first part of the algorithm that reduces β to β ≤ 2 and the second part that computes the final solution. Hence the total number of steps is at most $\tilde{O}(\varepsilon^{-2}(m+\min\{n,k\}))$ = $\tilde{O}(\varepsilon^{-2} m)$.*

Lemma 41.4 *Each step involves one run of Dijkstra's algorithm that takes time $O(n \log n + m)$ using Fibonacci heaps, and other computations that take at most $O(m)$ time, for a total of $O(m + n \log n)$ time per step.*

Proof. The only part of the inner while-loop of Figure 41.2 whose running time requires some explaining is the calculation of quantities

$$F(e) := \sum_{c_q : e \in P_{c_q}} f_{c_q}, \ \forall e \in P_{c_1} \cup \ldots \cup P_{c_r}$$

where P_{c_q} is the current shortest path for commodity c_q, $q = 1, \ldots, r$. Note that all these paths start from the same source and they form a shortest path tree. It will be easier for the exposition to assume that the sinks of the commodities c_q are the leaves of this tree (if a sink is at an internal node of the shortest path tree, we can connect this sink to this node via a new artificial edge of infinite capacity). For each one of these commodities we are routing f_{c_q} units of flow through the tree path P_{c_q}. The calculation of $F(e)$ is done as follows:

Step 0: Set $F(e) := f_{c_q}$ for all edges e that connect the sink of c_q to the tree.

Step 1: Repeat the following steps until all the tree nodes have been processed:

- Pick a tree node v which is connected to its children $u_1, \ldots, u_d$ via edges $e_1 = (v, u_1), \ldots, e_d = (v, u_d)$ and such that all $F(e_i)$, $i = 1, \ldots, d$ have already been calculated. Set $G(v) := \sum_{i=1}^{d} F(e_i)$.
- Let $e = (w, v)$ be the edge that connects v to its tree ancestor w. Set $F(e) := G(v)$.

It is clear that the running time of this procedure is $O(n)$, and once we have the $F(e)$'s, we can calculate easily σ and $l(e)$ for all the tree edges in time $O(n)$.

In addition, note that variables $x(P)$ do not need to be stored, and therefore we do not need to find a sophisticated data structure that would allow us efficient storage and updates of x. They are used (scaled in the end) to calculate the total demand of each commodity routed by the algorithm, in order to calculate λ. Instead, we can just keep a record of the total demand routed so far for each commodity, and use that (scaled by a factor of $\log_{1+\varepsilon}(1+\varepsilon)/\delta$) to calculate λ. ∎

Putting all these together with the scaling procedure we get an algorithm that runs in time $\tilde{O}(\varepsilon^{-2} m^2)$.

Theorem 41.1 *The algorithm described above is a FPTAS for the maximum concurrent multicommodity flow problem, that runs in $\tilde{O}(\varepsilon^{-2} m^2)$ time.*

---

*Recall that the $\tilde{O}$ notation hides the logarithmic factors in the running time bound.

41.2.3 Explicit Case

In most applications it is important to output an explicit description of a flow that achieves the near-optimal λ. Let $x_e(q)$ be the amount of flow of commodity $q \in \{1, \ldots, k\}$ that passes through edge $e \in E$. We would like our algorithm to output the explicit enumeration of $x_e(q)$, $\forall q, e$. As we have already seen, even for very simple examples of networks with n nodes and k commodities like the one in Figure 41.1 this explicit representation can be of size $\Omega(nk)$. This gives a trivial lower bound of $\Omega(nk)$ on the dependency of the running time of any algorithm that produces an explicit representation on the number of commodities k.

We will alter the algorithm of Figure 41.2 so that it computes *explicitly* a flow that achieves an almost optimal value for λ in time whose dependence on the number of commodities k is at most $O(nk)$ (times polylogarithmic factors). The crucial observation is that as long as in every step we route enough flow to either saturate an edge or so that no more flow remains to be routed by the current iteration (this will happen in the last step of this iteration), our analysis of correctness remains the same. We will route the *whole* remaining flow for a commodity until we get to a commodity whose remaining flow can be routed only partially and saturates an edge. The modified algorithm is shown in Figure 41.3.

Note that the only change in the algorithm is the way flow is distributed to paths during a step. The correctness analysis carries over to this modified algorithm exactly as is in the implicit case.

The modifications do not change the estimates for the number of phases and iterations within each phase. Therefore there are at most $O(\log m(\log m + \log k + \varepsilon^{-2}))$ phases. This results in at most $O(\min\{n,k\}\log m(\log m + \log k + \varepsilon^{-2}))$ iterations ($\min\{n,k\}$ iterations per phase). As was the case before the modification of the algorithm, each step will saturate an edge, unless it is the last step of an iteration. Therefore, the number of steps will exceed the number of iterations by the maximum possible edge saturations. Each such saturation increases the length function of the saturated edge by a factor at least $1+\varepsilon$. Just as before, the total number of steps is $\tilde{O}(\varepsilon^{-2}m)$. In every step we route as many *whole* remaining demands of commodities considered in the current iteration as possible, until an edge is saturated (if all commodities are wholly routed, the current iteration finishes). Hence the commodities routed during the current step are finished with (for the current iteration) except possibly the last one that may have some demand left to be routed in a subsequent step. We charge the path update cost (the cost of an iteration of the inner for-loop) to the commodity itself if the whole remaining demand for this commodity was routed through this path, and we charge the path cost to the saturated edge if only a part of the remaining demand for this commodity was routed through this path (resulting to the saturation of the edge). The cost for a single path update is at most $O(n)$. How many such updates are there? Note that inside a phase a commodity demand can be partially routed many times but can be wholly routed only once (this would be the last time we deal with this commodity in the current phase). Hence there are at most as many *whole* routings in the current phase as there are commodities (k), for a total of $O(k(\# \text{ of phases})) = O(k \log m(\log m + \log k + \varepsilon^{-2}))$ path updates. The total number of *partial* routings is at most the number of edge saturations, which is $O(\varepsilon^{-2} m \log m)$ as calculated in Section 41.2.2. In total we have at most $O(\varepsilon^{-2}(m+k)\log m + k \log m(\log m + \log k)) = \tilde{O}(\varepsilon^{-2}(m+k))$ path updates, for a total cost of $\tilde{O}(\varepsilon^{-2}(m+k)n)$.

The cost of the rest of the operations (Dijkstra's algorithm, etc.) can be estimated as in the implicit case. But in order not to inflate the running time of the algorithm by the update calculation for l and by the need to scale $x(P)$ in the end of the algorithm in Figure 41.2, we apply the following two further modifications:

- The update of the length $l(e)$ for every edge e in the modified algorithm presented in Figure 41.3 can be implemented by keeping track of the flow that was routed through every edge during one step. We can use one variable $f(e)$ for every edge e, that is

Input: Graph $G = (V, E)$, capacities $u(e)$, pairs (s_i, t_i) with demands $d(i)$,
$1 \leq i \leq k$, accuracy ε.

Output: Explicit solution x, λ.

Initialize $l(e) := \delta/u(e)$, $x_e(q) := 0$, $\forall e \in E, q = 1, \ldots, k$.
while $D(l) < 1$ **do**
 for $i = 1$ to $|\mathcal{S}|$ **do**
  $d'(c_q) := d(c_q), q = 1, \ldots, r$
  while $D(l) < 1$ and $d'(c_q) > 0$ for some q **do**
   $P_{c_q} := $ shortest path in $\mathcal{P}_{c_q}$ using $l, q = 1, \ldots, r$ with $d'(c_q) > 0$
   $u'(e) := u(e), \forall e \in P_{c_1} \cup \ldots \cup P_{c_r}$
   for $q = 1$ to r **do**
    $c := \min_{e \in P_{c_q}} u'(e)$
    if $d'(c_q) \leq c$ **then**
     $\left.\begin{array}{rcl} x_e(c_q) & := & x_e(c_q) + d'(c_q) \\ u'(e) & := & u'(e) - d'(c_q) \end{array}\right\} \forall e \in P_{c_q}$
     $f_{c_q} := d'(c_q)$
     $d'(c_q) := 0$
    else
     $x_e(c_q) := x_e(c_q) + c, \forall e \in P_{c_q}$
     $f_{c_q} := c$
     $d'(c_q) := d'(c_q) - c$
     break   /* out of the inner for-loop */
   end for
   $l(e) := l(e)(1 + \varepsilon \cdot \frac{\sum_{c_q:e \in P_{c_q}} f_{c_q}}{u(e)}), \forall e \in P_{c_1} \cup \ldots \cup P_{c_r}$
  end while /* end of step */
 end for /* end of iteration */
end while /* end of phase */
$x_e(q) := x_e(q) / \log_{1+\varepsilon} \frac{1+\varepsilon}{\delta}$, $\forall e \in E, q = 1, \ldots, k$
$\lambda := \min_i \frac{\text{flow}(s_i, t_i)}{d(i)}$
Output λ, x

Figure 41.3 Modified maximum concurrent flow FPTAS for $\beta \geq 1$ that produces an explicit flow.

updated every time the edge is in a flow path in the inner for-loop (at the beginning of the current step we set $f(e) := 0$). Then, the update of the length will be as follows:

$$l(e) := l(e)\left(1 + \varepsilon \frac{f(e)}{u(e)}\right).$$

- The last two lines of the algorithm scale down the solution so that it is feasible, and compute λ. Since we know the scaling factor even before the algorithm starts executing, we can scale down the amount by which the variable $x_e(q)$ is increased in the inner for-loop. If we do this then the second to last step of scaling x is not needed. Also it is very easy to keep track of the total flow routed so far for every commodity, and this can be used to calculate λ.

With these modifications the running time of all operations outside the inner for-loop is at most $\tilde{O}(\varepsilon^{-2}m^2)$, and when added to the cost of the inner for-loop which is $\tilde{O}(\varepsilon^{-2}(m + kn))$, the total running time is $\tilde{O}(\varepsilon^{-2}(m^2 + kn))$. Note that the commodity dependent part of the running time bound comes exclusively from the explicit calculation of the flow that achieves value λ.

Theorem 41.2 *The modified algorithm is a FPTAS for the maximum concurrent multi-commodity flow problem that runs in $\tilde{O}(\varepsilon^{-2}(m^2 + kn))$ time and outputs a throughput value together with the flow that achieves it.* ∎

41.3 MINIMUM COST CONCURRENT FLOW

In this section we extend the ideas for MCF to solving its budgeted version BCF. It is clear that if we have a FPTAS for this problem, we have a FPTAS for the minimum cost concurrent flow, by just using binary search in the same way we used it to solve the minimum cost multicommodity flow problem if we can solve BUDGET-MF at the beginning of this chapter. This search increases the running time by a factor of at most $\log M$, where M is the biggest number used to specify capacities, demands, or costs. Note that we can have different costs for different commodities, since there is a different primal variable for every commodity-path pair as the following LP formulation of BCF shows:

$$\begin{aligned}
\text{maximize } \lambda \text{ s.t.} & \\
\sum_{P:e\in P} x(P) \leq u(e) & \quad \forall e \in E \\
\sum_{P\in \mathcal{P}_i} x(P) \geq \lambda d(i) & \quad \forall i = 1,\ldots,k \\
\sum_{e\in E}\sum_{P:e\in P} c(e)x(P) \leq B & \\
x(P) \geq 0 & \quad \forall P \in \mathcal{P} \\
\lambda \geq 0 &
\end{aligned} \quad \text{(BCF)}$$

If ϕ is the dual variable associated with the budget onstraint in BCF, then its dual is

$$\begin{aligned}
\text{minimize } \sum_{e\in E} u(e)l(e) + B\phi \text{ s.t.} & \\
\sum_{e\in P} (l(e) + c(e)\phi) \geq z(i) & \quad \forall i, \forall P \in \mathcal{P}_i \\
\sum_{i=1}^{k} d(i)z(i) \geq 1 & \\
l(e) \geq 0 & \quad \forall e \\
z(i) \geq 0 & \quad \forall i \\
\phi \geq 0 &
\end{aligned} \quad \text{(DBCF)}$$

The algorithm is the same as in the maximum concurrent flow case, but instead of l the length function we use is $l' := l + c \cdot \phi$. Hence, in this case we have $D(l,\phi) := \sum_e l(e)u(e) + \phi B$, and $\alpha(l,\phi), \beta$ are defined as before but using l' as the edge distance.

Exercise 41.1 *Verify that Lemma 41.1 holds.*

The algorithm initializes $\phi = \delta/B$. The flow sent through at every step is the flow calculated above but scaled down (if necessary) so that its total cost is not more than B. In the explicit case, we route commodities *wholly* except possibly the last one considered, until (1) an edge becomes saturated, (2) the cost reaches B, or (3) all commodities are routed. Lengths l are updated as before, and the new ϕ becomes $\phi := \phi(1 + \varepsilon \cdot (\text{cost of flow sent})/B)$. Again, the algorithm ends when $D(l,\phi) \geq 1$. Since the analysis of the approximation guarantee and running time is almost identical to MCF, we leave its details as a series of exercises.

Exercise 41.2 *Prove that the algorithm achieves an $(1-\varepsilon)$-approximation for any $0 < \varepsilon < 1$ in both the implicit and explicit cases.*

Observe that at every step either the length l of an edge or ϕ increases by a factor of at least $1+\varepsilon$ (and, in the explicit case, we charge the path cost of partially routed commodities to the saturated edge or the *saturated cost* respectively). Also ϕ can be at most $(1+\varepsilon)/B$ at the end of the algorithm (recall that the algorithm ends when $D(l,\phi) \geq 1$ and $D(l,\phi) := \sum_e l(e)u(e)+\phi B$). For the case $\beta < 1$, we perform a binary search on the m capacities exactly as before, in order to find the maximum possible capacity of a path *that meets the budget*; the latter check can be done in every binary search step in time $O(m \log m)$ by running a shortest path computation.

Exercise 41.3 *Prove that the running time of the algorithm for the* MINIMUM COST CONCURRENT FLOW *problem is $\tilde{O}(\varepsilon^{-2} m^2 \log M)$ in the implicit case.*

Exercise 41.4 *Prove that the running time of the algorithm for the* MINIMUM COST CONCURRENT FLOW *problem is $\tilde{O}(\varepsilon^{-2}(m^2 \log M + kn))$ in the explicit case.* (Hint: Notice that the $\log M$ factor is due to the binary search for the right budget B. How many of these $O(\log M)$ iterations need to output the edge flow explicitly?)

41.4 APPLICATION: MAXIMUM CONCURRENT FLOW FOR LOSSY NETWORKS

A generalization of the flow problems discussed so far involves the introduction of one more network parameter, the *gain factor* $\gamma : E \to \mathbb{R}^+$. An edge e with gain factor $\gamma(e) > 0$ allows a fraction $\gamma(e)$ of the flow that enters e to exit. Hence, in case $\gamma(e) > 1$, $\gamma(e) = 1$ or $\gamma(e) < 1$, the edge increases, conserves or *leaks* the flow that passes through it. A network with a gain factor γ is a *generalized network* and the obvious extensions of flow problems to such networks are *generalized flow* problems. In the special case of $\gamma(e) \leq 1$ for all edges, the network is called *lossy*.

The analysis above applies directly to lossy networks, with two modifications: an appropriate (more general) definition of a shortest path, and the computation of a maximum capacity path, both of which we use when we compute estimates $\hat{\zeta}(i), i=1,\ldots,k$.

A simple modification of Dijkstra's algorithm works for a lossy network. Recall that our distances are still l, but now if we send one unit of flow from a source s to a sink t along an $s-t$ path, the amount of flow arriving at t is the product of the gain factors along the path, and this product is at most 1. Similarly, if a unit of flow is send from u to v on edge (u,v), then only $\gamma(u,v) \leq 1$ units arrive. This means that in order to get a unit of flow to v from u, we must send $1/\gamma(u,v)$ units from u. Hence, while, in the problems discussed so far, the unit of flow arriving to v from u experiences cost $l(u,v)$ (see the definition of α), in a lossy network it *experiences* cost $l(u,v)/\gamma(u,v)$. Therefore, in the case of lossy networks, the distance labels $\pi(v), v \in V$ calculated by Dijkstra's algorithm are now updated as follows: if we are currently relaxing edge (u,v) then

$$\pi(v) := \min\left\{\pi(v), \frac{\pi(u)+l(u,v)}{\gamma(u,v)}\right\}.$$

Note that the running time of the algorithm remains $O(m + n \log n)$.

Fact 41.1 continues to hold in the case of lossy networks, see [4]. We need to find the maximum capacity* path, and use the flow $\hat{\zeta}(i)$ we can send on this path as an estimate just like we did in (41.6). Again, a modified Dijkstra's algorithm can be used, where the labels $\pi(v), v \in V$ computed are the algorithm's current estimates for the maximum capacity of a

---

*Maximum capacity in this case means the path that can deliver the most flow to sink t, when we can send as much flow as we want from source s and always respecting the edge capacities.

path from source s to v for all $v \in V$. Then the capacity of a path reaching node v from node u through edge (u,v) is $\min\{u(u,v), \pi(u)\} \cdot \gamma(u,v)$. Therefore, if we are currently relaxing edge (u,v) then

$$\pi(v) := \max_{u:(u,v) \in E} \{\min\{u(u,v), \pi(u)\} \cdot \gamma(u,v)\}.$$

After the algorithm finishes, the estimates we output are $\hat{\zeta}(j) := \pi(t_j), j = 1, \ldots, k$. Note that the running time of the algorithm remains $O(m + n \log n)$, for a total of $O(\min\{n, k\}(m + n \log n))$ when we compute all estimates for all commodities.

By introducing these modifications to our algorithm for the maximum concurrent flow problem we get the following theorem.

Theorem 41.3 *There is a FPTAS that computes the maximum concurrent flow for lossy networks implicitly in time $\tilde{O}(\varepsilon^{-2} m^2)$ and explicitly in time $\tilde{O}(\varepsilon^{-2}(m^2 + nk))$.* ∎

Further Reading

A general book on flows is the one by Ahuja et. al. [5]. The presentation of the chapter was based on the work by Garg and Könemann [6], which achieved the same running time for minimum-cost multicommodity flow as the earlier work of Grigoriadis and Khachiyan [7], and subsequent improvements by Fleischer [8] and Karakostas [9]. The currently best running time of $\tilde{O}(\varepsilon^{-2} mn)$ for maximum multicommodity flow and $\tilde{O}(\varepsilon^{-2}(m+k)n \log M)$ for maximum- and minimum-cost concurrent flow problems was presented by Madry [10]; his algorithm uses a Monte Carlo (i.e., randomized) data structure for the calculation of approximate shortest paths. This is the culmination of a whole line of research starting with the seminal paper by Shahrokhi and Matula [11] (cf., e.g., [12–18]) based on Lagrangian relaxation and linear programming decomposition techniques that led to ever decreasing running times for multicommodity flow problems. The extension of these techniques to the case of generalized flows was done by Fleischer and Wayne [19].

References

[1] L. R. Ford, Jr., and D. R. Fulkerson. *Flows in Networks*. Princeton University Press, Princeton, NJ, 1962.

[2] A. Ben-Tal and A. Nemirovski. *Lectures on Modern Convex Optimization*. MPS-SIAM Series on Optimization, SIAM, 2001.

[3] C. H. Papadimitriou and K. Steiglitz. *Combinatorial Optimization: Algorithms and Complexity*. Dover Publications, New York, 1998.

[4] A. V. Goldberg, S. A. Plotking, and R. E. Tarjan. Combinatorial algorithms for the generalized circulation problem. *Mathematics of Operations Research*, **16** (1991), 351–379.

[5] R. K. Ahuja, T. L. Magnanti, and J. B. Orlin. *Network Flows: Theory, Algorithms, and Applications*. Prentice Hall, Englewood Cliffs, NJ, 1993.

[6] N. Garg and J. Könemann. Faster and simpler algorithms for multicommodity flow and other fractional packing problems. *SIAM Journal on Computing*, **37**(2) (2007), 630–652.

[7] M. D. Grigoriadis and L. G. Khachiyan. Approximate minimum-cost multicommodity flows in $\tilde{O}(knm/\epsilon^2)$ time. *Mathematical Programming*, **75** (1996), 477–482.

[8] L. Fleischer. Approximating fractional multicommodity flow independent of the number of commodities. *SIAM Journal on Discrete Mathematics*, **13** (2000), 505–520.

[9] G. Karakostas. Faster approximation schemes for fractional multicommodity flow problems. *ACM Transactions on Algorithms*, **4**(1) (2008).

[10] A. Madry. Faster approximation schemes for fractional multicommodity flow problems via dynamic graph algorithms. In *Proceedings of the 42nd ACM STOC*. Cambridge, MA, pp. 121–130, 2010.

[11] F. Shahrokhi and D. W. Matula. The maximum concurrent flow problem. *JACM*, **37** (1990), 318–334.

[12] P. Klein, S. Plotkin, C. Stein, and É. Tardos. Faster approximation algorithms for the unit capacity concurrent flow problem with applications to routing and finding sparse cuts. *SIAM Journal on Computing*, **23** (1994), 466–487.

[13] T. Leighton, F. Makedon, S. Plotkin, C. Stein, É. Tardos, and S. Tragoudas. Fast approximation schemes for multicommodity flow problems. *JCSS* **50** (1995), 228–243.

[14] M. D. Grigoriadis and L. G. Khachiyan. Fast approximation schemes for convex programs with many blocks and coupling constraints. *SIAM Journal on Optimization*, **4** (1994), 86–107.

[15] S. Plotkin, D. Shmoys, and É. Tardos. Fast approximation algorithms for fractional packing and covering problems. *Mathematics of Operations Research*, **20** (1995), 257–301.

[16] T. Radzik. Fast deterministic approximation for the multicommodity flow problem. In *Proceedings of the 6th ACM/SIAM SODA*. San Francisco, CA, pp. 486–492, 1995.

[17] D. Karger and S. Plotkin. Adding multiple cost constraints to combinatorial optimization problems, with applications to multicommodity flows. In *Proceedings of 27th ACM STOC*. Las Vegas, NV, pp. 18–25, 1995.

[18] P. Klein and N. Young. On the number of iterations for Dantzig-Wolfe optimization and packing-covering approximation algorithms. In *Proceedings of 7th Integer Programming and Combinatorial Optimization Conference*, 1610 (1999), 320.

[19] L. Fleischer and K.D. Wayne. Fast and simple approximation schemes for generalized flow. *Mathematical Programming Ser. A*, **91**(2) (2002), 215–238.

CHAPTER 42

Approximation Algorithms for Connectivity Problems

Ramakrishna Thurimella

CONTENTS

42.1 Introduction 1097
42.2 Definitions and Notation 1098
42.3 Simple 2-Approximation 1099
    42.3.1 Trivial Lower Bound 1099
    42.3.2 Non-Algorithmic Upper Bounds 1100
    42.3.3 Edge Connectivity 1100
    42.3.4 Vertex Connectivity 1100
    42.3.5 Time Complexity 1103
42.4 Linear-Time 2-Approximation 1104
    42.4.1 Time Complexity 1105
42.5 Beating 2 for the Edge Case 1105
    42.5.1 Khuller–Raghavachari Algorithm 1106
    42.5.2 Approximation Guarantee 1108
    42.5.3 Time Complexity 1110
42.6 Approximating Minimum-Size Spanning Subgraphs via Matching .. 1110
    42.6.1 Time Complexity 1112
42.7 Conclusion 1112

42.1 INTRODUCTION

The edge connectivity of an undirected graph G is the minimum number of edges that must be removed to disconnect G. For example, the edge connectivity of a tree is 1, and the edge connectivity of a simple cycle is 2. Similarly, the vertex connectivity of a graph is the minimum number of vertices that must be removed to disconnect it.

It is not hard to see why these concepts are fundamental and have applications to network reliability and distributed computing. For instance, a graph could be used to model a network such as the Internet where vertices correspond to routers and edges represent the connections between them.

In distributed computing, one of the measures of complexity is the *message* complexity—the number of messages that need to be exchanged in order to compute something in a distributed manner. A key building block for these algorithms is the broadcast operation. To broadcast, each vertex after receiving, or generating in the case of the vertex that wants to broadcast, the message can send out copies of that message on every outgoing link, that is, flood the network. This is expensive, costing m messages. To make this operation efficient, one can use a spanning tree and send the message out on every link of the tree. Since a spanning tree has only $n - 1$ edges, the number of messages used is also $n - 1$. For a dense graph, defined as $m = \Omega(n^2)$, the savings are substantial.

The problem with using a spanning tree is that it is too fragile: single vertex or edge failure causes the broadcast to fail. Hence, we seek a subgraph with high connectivity that has the fewest number of edges. Unfortunately, finding such a subgraph is NP-complete for all connectivity values greater than 1 [1]. Therefore, we turn to approximation algorithms for this problem—the focus of this chapter.

Given the importance of the problem, it was extensively studied [2–4]. It is not surprising that there exist many flavors of this basic problem and multitude of solutions. For example, each of the following assumptions and different combinations of them give rise to a number of variations: the presence or absence of directions/weights on edges, same connectivity between every pair of vertices—known as the *uniform* connectivity requirement—or possibly different connectivity values between vertex pairs, whether the graphs under consideration are simple or multigraphs, and finally if the special structure present for small constant values of connectivity, for example $k = 1, 2,$ or 3, can be exploited to design efficient algorithms.

The algorithms given in this chapter are limited for the most part to simple, undirected, unweighted graphs with uniform connectivity requirements. Rest of the chapter is organized as follows. Section 42.2 defines some common graph-theoretic terms that are used in this chapter; definitions that are specific to algorithms of this chapter are deferred to the respective sections. As a warm up, we present a simple 2-approximation in Section 42.3. Section 42.4 presents a different algorithm that also achieves the same factor, but runs in linear time. Next we show in Section 42.5 how to beat the factor of 2 for some specific values of connectivity. Section 42.6 presents the most technical result of this chapter, culminating in an algorithm that subsumes all the results from the previous sections. We end the chapter with some concluding remarks in Section 42.7.

42.2 DEFINITIONS AND NOTATION

For the most part, we use standard graph theory notation. Refer to Diestel [5] for definitions not covered here. The vertex set and edge set of a graph G are denoted by $V(G)$ and $E(G)$, respectively. The number of vertices and the number of edges are denoted by n and m, respectively. The degree of a vertex v in G is designated by $\delta_G(v)$.

We will only be dealing with undirected graphs. A *path* in a graph $G = (V, E)$ is a sequence of vertices $v_1, v_2, ..., v_p$ from V such that for all i, $1 \leq i < p$, $(v_i, v_{i+1}) \in E$. v_1 and v_p are called the *end vertices*. A *cycle* is a path that starts and ends at the same vertex. A *tree* is a connected graph that does not have any cycles. The *level* of a vertex v in a rooted tree with root r is the number of edges between the v and r. Thus, the root is at level 0, its children are at level 1, and so on. A *forest* is a collection of trees. For a graph $G = (V, E)$, a *maximal spanning forest* $F = (V, E_F)$ is a subgraph such that E_F has no cycles. Furthermore, E_F is *maximal* in the sense that $E_F \cup e$ contains a cycle for all edges $e \in E - E_F$. The cycle created by the addition of a nontree edge e to a maximal spanning forest is called the *fundamental cycle created by* e. Notice that when G is connected, F is a spanning tree and denoted by T depending on the context.

A *simple* path is a path in which if a vertex appears once, it cannot appear again. The *parent* of a vertex v in a rooted tree T with root r is the vertex that immediately follows v on the unique path from v to r in T. Two paths are *internally disjoint* if they share only the end vertices.

A *subgraph* $G_s = (V_s, E_s)$ of a graph $G = (V, E)$ is a graph whose vertex and edge sets are subsets of G, that is $V_s \subseteq V$ and $E_s \subseteq E$, and if $(x, y) \in E_s$, $x \in V_s$ and $y \in V_s$. A subgraph $G_s = (V_s, E_s)$ of $G = (V, E)$ is said to be a *spanning* subgraph if $V_s = V$ and $E_s \subseteq E$, that is G_s has the same vertex set as G, but not necessarily the same edge set. If $G_i = (V, E_i)$ and $G_j = (V, E_j)$ are spanning subgraphs of $G = (V, E)$, then $G_i + G_j$ is shorthand for the spanning subgraph $(V, E_i \cup E_j)$. Similarly, if $G_i = (V, E_i)$ is a spanning subgraphs of $G = (V, E)$, then $G - G_i$ refers to the spanning subgraph $(V, E - E_i)$.

A *connected component* of an undirected graph is a subgraph $C = (V_c, E_c)$ in which any two vertices are connected to each other by paths, and which is connected to no additional vertices in $V - V_c$.

Given $k > 0$, a connected graph $G = (V, E)$ with at least $k + 1$ vertices is called *k-edge* (respectively, *k-vertex*) *connected* if the deletion of any $k - 1$ edges (respectively, vertices) leaves the graph connected. If a connected graph contains a single vertex whose removal disconnects the graph, then that vertex is called an *articulation point*. A graph is called *biconnected* if it has no articulation points.

By connectivity, we mean both vertex and edge connectivity. An edge $e \in E$ in a k-connected graph (V, E) is *critical*, if $(V, E - \{e\})$ is not k-connected. In a *minimally* k-connected graph, every edge is critical.

For a subset of vertices $V' \subseteq V$, the subgraph G' induced by V' is (V', E') where $E' = \{(x, y) \in E \mid x \in V' \text{ and } y \in V'\}$. Also, the *neighborhood* of V', denoted $N(V')$, is the subset of vertices $\{y \mid \exists (x, y) \in E \text{ such that } x \in V' \text{ and } y \in V - V'\}$

For a connected graph $G = (V, E)$, a subset $Z \subset V$ is called a *vertex cut* if $G - Z$ has at least two connected components. The size of a vertex cut Z is defined by $|Z|$. We are generally interested in minimum size vertex cuts.

A *certificate* for the k-connectivity of G is a subgraph (V, E'), $E' \subseteq E$, that is k-connected if G is k-connected. A certificate for k-connectivity is *sparse* if it has $O(kn)$ edges. For example, a spanning tree is a sparse certificate for 1-connectivity of a connected graph. In our analysis of sparse certificate heuristics, we represent a k-connected subgraph of G that has the *optimum* number of edges by G^*.

42.3 SIMPLE 2-APPROXIMATION

In this section, we show how to find sparse certificates that have no more than twice the number of edges from the optimum. How do we know the optimum, given that the problem we are tackling is NP-complete? We do not. Instead, we bound the optimum from below and compute the ratio of upper and lower bounds. Obviously, the closer the lower and upper bounds, the better the approximation ratio.

42.3.1 Trivial Lower Bound

In order for a graph to be k-connected, notice that every vertex v must have degree at least k. Otherwise, if there exists a v such that $\delta(v) < k$, deleting the all edges incident on v, we can disconnect v from the rest of the graph. Similarly, $N(\{v\})$ constitutes a vertex cut whose size is less than k. From this observation, we can conclude that in order for a graph to be k-connected, each vertex must have degree at least k. In any graph, the sum of the degrees of all the vertices is twice the number of edges, as every edge is counted twice, once from each end. Therefore, in order for a graph to be k-connected, the sum of the degrees should be at least kn. In other words, every k-connected graph must have at least $(kn)/2$ edges. It is often referred to as the *degree lower bound*. This lower bound is sufficient to get a factor 2 approximation for the algorithms presented in the rest of this section.

42.3.2 Non-Algorithmic Upper Bounds

Note that any minimally k-edge connected graph has at most $k(n - k)$ edges [6–8]. This upper bound combined with the degree lower bound from Section 42.3.1 immediately yields a 2-approximation algorithm, as minimally k-edge connected graphs can be found in polynomial time.

Recent algorithmic work, the subject of the rest of this chapter, gives alternate, easy, and efficient methods for finding a k-connected spanning subgraph whose size (i.e., number of edges) is at most kn.

42.3.3 Edge Connectivity

Consider Algorithm 42.1 for finding edge certificates that operates iteratively. In the ith iteration, it computes a sparse certificate for i connectivity G_i. The desired k-connected subgraph G_k is the output at the end of the kth iteration.

Algorithm 42.1 Edge certificate

Require: $G = (V, E)$ and an integer $k > 0$
Ensure: A sparse certificate $G_k = (V, E_k)$ for k edge connectivity
1: **procedure** EC(G, k)
2: $\quad G_0 \leftarrow (V, \emptyset)$
3: $\quad$ **for** $i \leftarrow 1, k$ **do**
4: $\quad\quad$ Find a maximal spanning forest F_i in $G - G_{i-1}$
5: $\quad\quad G_i \leftarrow G_{i-1} + F_i$
6: $\quad$ **end for**
7: $\quad$ **return** G_k
8: **end procedure**

The following theorem proves the correctness of Algorithm 42.1 [9].

Theorem 42.1 *If G is k-edge connected, then G_k is k-edge connected.*

Proof. Denote the spanning forests found in the **for** loop of the algorithm by $F_1, F_2, \ldots, F_k$. Note that as $k > 0$ and G is connected, F_1 is a spanning tree. Hence G_k is connected. Assume for contradiction that G_k is not k-edge connected, but G is. Then, there must exist a set of edges K fewer than k in number whose removal disconnects G_k, but does not disconnect G. Let C_1 and C_2 be the two connected components of $G_k - K$. Since $|K| < k$, by the Pigeonhole Principle, there must exist an F_i, $1 \leq i \leq k$, such that F_i does not have any edges in common with K. Let e be an edge in G that has one end in C_1 and the other in C_2. Such an edge must exist as $E - K$ is connected. Now, adding e to F_i would not create a cycle, contradicting that each of the k spanning forests found in line 4 of Algorithm 42.1 is maximal. ∎

42.3.4 Vertex Connectivity

While Algorithm 42.1 is sufficient to find edge certificates, as shown in Figure 42.1, it can fail for vertex certificates. This is because a pair of vertices x and y are connected in maximal spanning forests F_i and F_j by paths P_i and P_j, respectively, where F_i and F_j are some maximal spanning forests found in Algorithm 42.1, then all we can say is that P_i and P_j are edge disjoint. But for vertex connectivity, we need these paths to be vertex disjoint.

However, by finding a specific kind of spanning forests in line 4 of Algorithm 42.1, we can make the same algorithm work for vertex certificates. For example, using breadth-first spanning forests, one can show that the resulting sparse subgraph preserves vertex

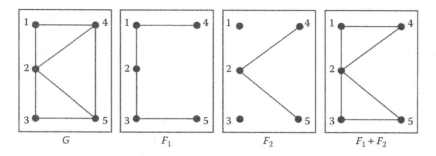

Figure 42.1 Algorithm edge certificate may not work for vertex certificates.

connectivity. We show in the rest of this section that the full power of breadth-first search (BFS) is not even needed. A less restrictive form of search called *scan-first* search (SFS) is sufficient. The algorithm presented in this section is due to Cheriyan et al. [10]. The earliest work that employed a similar notion is due to Doshi and Varman [11] though their method is limited to biconnected graphs.

Algorithm 42.2 Scan-first search

Require: A connected graph $G = (V, E)$ and a root $r \in V$
Ensure: A spanning tree $T = (V, E_T)$ where $E_T \subseteq E$
 1: **procedure**: SFS(G, r)
 2:   Initialize all vertices of G as unmarked and unscanned
 3:   $E_T \leftarrow \emptyset$
 4:   mark r
 5:   **while** there exist a marked, but unscanned vertex u **do**
 6:     **for** every unmarked neighbor v of u **do**
 7:       mark(v)
 8:       add (u, v) to E_T
 9:     **end for**   ▷ u is now considered scanned
10:   **end while**
11:   **return** $T = (V, E_T)$
12: **end procedure**

Starting from a given root r or an arbitrary root, BFS explores the graph in a systematic way satisfying the following two conditions:

1. When a vertex is being explored, all its unvisited neighbors are marked visited and added to a collection Q, and

2. The next vertex to be explored is taken from Q using the first-in-first-out (FIFO) order.

SFS is less restrictive than BFS. In SFS, we require only the first condition. In other words, SFS chooses the next vertex to be scanned from Q, but not necessarily in the FIFO order.

The spanning tree implied by a search, whether breadth-first or scan-first, is the one that results from each vertex v, $v \neq r$, choosing the edge (v, u) where u is the vertex that was responsible for marking v as visited. Such a tree is called BFS or SFS spanning tree as the case may be (see Algorithm 42.2).

Therefore, every BFS spanning tree is an SFS spanning tree, but the converse need not be true. The following property of BFS and SFS is noteworthy. The nontree edges with respect to a BFS tree are *all* cross edges (i.e., no back edges) and if (x, y) is one such nontree edge then $|level(x) - level(y)| \leq 1$. The nontree edges with respect to an SFS tree are also *all* cross edges, but the level difference can be arbitrary. The reason for the absence of back edges is the first condition. If a nontree edge (u, v), $level(u) < level(v)$, is present in a tree T, then while scanning u, its unvisited neighbor v was not added as a child in T. Hence T does not satisfy the first condition and it cannot be an SFS tree. It is also worth pointing out that not all spanning trees in which back edges are absent are not *necessarily* SFS trees, as the first condition still needs to be satisfied, i.e when a vertex is scanned *all* its unscanned neighbors must be added to the collection of vertices to be scanned. Equivalently, if a vertex u of a scan-first tree T has a nontree edge (u, v) incident on it, then the tree edge (t, v) incident on v must be such that t appears before u in the scan order that was used to construct T.

Here are some more examples that illustrate the difference between BFS and SFS spanning trees. Consider a graph that is a simple cycle C on 7 vertices $v_0, v_2, \ldots, v_6$. Denote the edge between $v_i, v_{(i+1) \bmod 7}$ as e_i. Then, $C - \{e_3\}$ is the only BFS tree starting from vertex v_0 whereas $C - \{e_1\}$, $C - \{e_2\}$, $C - \{e_3\}$, $C - \{e_4\}$, and $C - \{e_5\}$ are all valid SFS spanning trees.

Since the only vertex marked initially is r and at the time scanning r all neighbors of r get marked, the edges incident on r get added to the SFS spanning tree.

Proposition 42.1 *If r is the root of a SFS, the all edges incident on r are part of the spanning tree resulting from that search.*

We would like to prove the following theorem.

Theorem 42.2 *Let G be a k-vertex connected graph and G_k be $F_1 + F_2 + \ldots + F_k$ where F_i is a scan-first spanning forest in $G - G_{i-1}$, $1 \leq i \leq k$, where $G_0 = (V, \emptyset)$. Then, G_k is k-vertex connected.*

Before we present the proof, we propose the following fact which follows from the previous proposition.

Proposition 42.2 *If a vertex v is used as a root in F_i, then all edges incident on it are selected by G_i.*

We prove Theorem 42.2 along the same lines as Theorem 42.1, that is by contradiction, while keeping in mind a key difference between edge and vertex connectivity: assuming the size of the separator *minimal*, removal of an edge separator always results in two connected components. On the other hand, removal of a minimal vertex separator could result in more than two components. Such separators are called shredders [12]. This characteristic of vertex separators makes the proof considerably more technical. To keep the discussion simple, we will only prove the theorem for $k = 2$ and refer the reader to Cheriyan et al. [10] for the general case.

Proof of Theorem 42.2 Assume G is biconnected but G_2 is not. Then there exists an articulation point v in G_2. However, as G is biconnected, $G - \{v\}$ is connected. Therefore, there exists an edge $(x, y) \in E(G) - E(G_2)$, such that x and y are in different components in $G_2 - \{v\}$.

If v is the root of F_1, then by Proposition 42.1, all edges incident on v belong to F_1. But, since all paths between two vertices that belong to different components of $G_2 - \{v\}$ must go through v, as v is an articulation point, there is no path between x and y in $G - F_1$. But, F_2 is a maximal spanning forest in $G - F_1$ to which we can add (x, y) and not create a cycle, thus contradicting the maximality of F_2.

If on the other hand v is not the root of F_1, then let r be the root of F_1 and let C_1 be the component that contains r in $G_2 - \{v\}$. Pick any other component of $G_2 - \{v\}$ and call it C_2.

Notice that $G - \{v\}$ must contain an edge (x, y), from a vertex of C_1 to a vertex C_2, as we assumed G to be biconnected. Without loss of generality, let $x \in V(C_1)$ and $y \in V(C_2)$. We now show that adding (x, y) to F_2 does not create a cycle contradicting that F_2 is maximal. Notice that the SFS corresponding to F_1 starts in the component C_1 at root r and scans v before it scans any vertex of C_2. If the search does not proceed this way, there would be a path from a vertex of C_1 to a vertex in C_2 that does not go through v, contradicting our assumption that v is an articulation point in G_2. When scanning v, the spanning tree F_1 selects all edges (v, w) such that $w \in V(C_2)$. In other words, there is no path between x and y in $G - F_1$, that is adding (x, y) to F_2 does not create a cycle, a contradiction. ∎

42.3.5 Time Complexity

Algorithm 42.1 is very easy to implement, as formalized in the theorem below:

Theorem 42.3 *Given G and G_{i-1}, each maximal spanning forest F_i in $G - G_{i-1}$, $1 \leq i \leq k$, where $G_0 = (V, \emptyset)$ can be found in linear time sequentially. Therefore, a sparse certificate for k-edge connectivity G_k can be found in $O(k(m + n))$ time.*

Proof. Discard the edges of G_{i-1} from G and run any standard spanning forest algorithm [13]. Since finding a maximal spanning forest takes time linear in the size of the graph [13], the theorem follows. ∎

Algorithm 42.2 can be implemented to run in linear time, implying that a sparse certificate of k-connectivity can be found in $O(kn)$ time.

Theorem 42.4 *Given G and G_{i-1}, each maximal scan-first spanning forest F_i in $G - G_{i-1}$, $1 \leq i \leq k$, where $G_0 = (V, \emptyset)$ can be found in linear time sequentially. Therefore, a sparse certificate for k-vertex connectivity G_k can be found in $O(k(m + n))$ time.*

Proof. Discard the edges of G_{i-1} from G and run any linear-time breadth-first algorithm [13]. Since every BFS spanning forest is an SFS spanning forest, the theorem follows. ∎

Note that there are no efficient methods to execute BFS in parallel. Fortunately, SFS lends itself to an efficient parallel implementation [10]:

Theorem 42.5 *For graph G with m edges and n vertices, an SFS forest can be found in $O(\log n)$ time using using $C(n, m)$ processors on a CRCW PRAM, where $C(n, m)$ is the number of processors required to compute a spanning tree in each component in $O(\log n)$ time.*

Proof. One easy way to find an SFS tree T in parallel is to first find an arbitrary spanning tree T' rooted at some vertex r and rearrange the edges so that it becomes an SFS tree. First, label each vertex of T' with respect to preorder labeling. To obtain an SFS tree T from T', make each v, $v \neq r$, choose one edge (v, u) where u is a neighbor of v with the lowest preorder label. To see that T is indeed an SFS tree, treat the preorder labels as the order in which the vertices are scanned in some SFS. For an order to be a valid SFS order, when a vertex needs to be chosen to be scanned, it should be selected from the neighborhood of the vertices already scanned. Hence preorder is a valid scan-first order. With this interpretation, if a vertex v has a nontree edge (v, w) incident on v in T, then the preorder label of the parent of w in T is less than that of v, consistent with the meaning of scanning a vertex in that w should be attached to the first neighbor of w marks it.

The complexity bound follows because the preorder numbers and the minimum labeled neighbor can be found in $O(\log n)$ time using $(n + m)/\log n$ processors [14]. The spanning tree computation dominates the resource bounds as the processor bound is $C(n,m) = \Omega((n+m)/\log n)$ for $O(\log n)$ parallel time. ∎

42.4 LINEAR-TIME 2-APPROXIMATION

In this section, we present an algorithm that computes a sparse k-connected spanning subgraph that is within a factor 2 from the optimal in a *single scan* of the graph. This elegant algorithm is due to Nagamochi and Ibaraki [15].

We need a few definitions first. Let V_i denote $\{v_1, v_2, \ldots, v_i\}$. An ordering $\sigma = (v_1, v_2, \ldots, v_n)$ of vertices in G is called a maximum adjacency ordering (MA ordering, for short) if for all i, $2 \leq i < n$, v_{i+1} is one of the vertices from $N(V_i)$ that has the highest degree in the subgraph induced by $V_i \cup N(V_i)$.

An MA ordering can be found sequentially by starting with an arbitrary edge and designating its ends points as v_1 and v_2, thus defining V_2. Next, v_3 is selected from $N(V_2)$ based on the number of connections to the vertices in V_2. In general, V_i is extended to V_{i+1} by adding v_{i+1} where v_{i+1} is any vertex from $N(V_i)$ that has the maximum number edges to the vertices in V_i. Figure 42.2 illustrates MA labeling.

Next, we extend the MA labeling to edges. For each $2 \leq i \leq n$, assume there are j edges between $\{v_1, v_2, \ldots, v_{i-1}\}$ and v_i. Consider the neighbors of v_i in V_{i-1} in the increasing order of σ labeling and assign labels $e_{i,1}, e_{i,2}, \ldots, e_{i,j}$. Figure 42.1 on the right shows an example for edge labeling. For instance in Figure 42.2, the vertex labeled 4 has two edges to $V_3 = \{v_1, v_2, v_3\}$, to v_2 and v_3. Edge labeling would label (v_4, v_2) and (v_4, v_3) as $e_{4,1}$ and $e_{4,2}$, respectively.

Edge labeling defines a partition on E. We could use the second component of the subscript of an edge label as the partition number. Specifically, define a partition of E to be $(F_1, F_2, \ldots, F_n)$ where $F_i = \{e_{2,i}, e_{3,i}, \ldots, e_{n,i}\}$ for $i = 1, 2, \ldots, n$.

Note that some partitions F_i may be empty. See Figure 42.3 for an example. As before, define G_i to be $F_1 + F_2 + \cdots + F_i$ for $1 \leq i \leq n$.

Theorem 42.6 *Each (V, F_i), $1 \leq i \leq n$, is a maximal scan-first spanning forest in $G - G_{i-1}$.*

Proof: For every vertex v, there can be at most one edge $e_{v,i}$ present in F_i. Furthermore, there must be a unique vertex in every connected component C of F_i that does not have such an edge—the vertex with the smallest σ number in that component. Assume that there are n_c vertices in C. As every vertex v, with the exception of one, has a unique $e_{v,i}$ incident on it, C has $n_c - 1$ edges. Since any connected graph with n_c vertices and $n_c - 1$ edges is necessarily a tree, F_i is a spanning forest.

It remains to show that F_i is a scan-first spanning forest. The proof is similar to that of Theorem 42.5. Notice that the σ order is a valid scan-first order. This is because in SFS, starting from an arbitrary vertex, the next vertex to be scanned is chosen from the

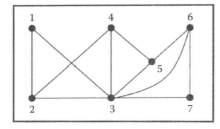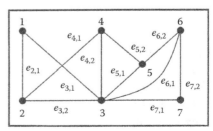

Figure 42.2 MA labeling example.

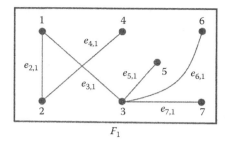
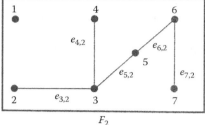

Figure 42.3 Edge partitions.

neighborhood of the vertices that have already been scanned and all its unmarked neighbors are marked. Interpreting the σ labeling as the scan-first order in $G - G_{i-1}$, the tree implied in each component of $G - G_{i-1}$ is an SFS tree. Therefore, F_i is a maximal scan-first spanning forest in $G - G_{i-1}$. ∎

Combining Theorem 42.2 with Theorem 42.6, we have the following corollary.

Corollary 42.1 *For any k, $F_1 + F_2 + \cdots + F_k$ is a sparse certificate for k-connectivity.*

42.4.1 Time Complexity

It turns out that MA labeling and the corresponding edge labeling can be computed in almost linear time, specifically in $O(m + n \log n)$ time [4]. Note that this bound is independent of k.

Theorem 42.7 *For graph G with m edges and n vertices, MA labeling can be computed in $O(m + n \log n)$ time.*

Proof. We will sketch an algorithm and allude to its implementation using Fibonacci heaps. (See Chapter 19 from [13] for a detailed description of Fibonacci heaps.) The algorithm, starting from an arbitrary vertex, scans one vertex at a time until all n vertices are processed. Assume inductively i vertices have been processed and an integer $d(v), 0 \leq d(v) \leq n - 1$, is associated with each of the $n - i$ unprocessed vertices v. The integer $d(v)$ represents the number of v's processed neighbors. The next vertex to be processed, $(i+1)$th vertex, is picked from one of the unprocessed vertices that has a maximum d value. Say, x is one such vertex. Processing vertex x entails incrementing the d value for all unprocessed neighbors of x and designating x as processed. Increment operation is performed once for each edge. Moving a vertex from unprocessed to processed involves using the `delete max` operation. This is done at most $n - 1$ times. Using Fibonacci heaps `increment key` can be performed in $O(1)$ amortized time and `delete max` takes $O(\log n)$ time. From these observations, the theorem follows. ∎

The above algorithm, while efficient, is inherently sequential as it processes one vertex at a time. However, Algorithm 42.2 lends itself to an efficient parallel implementation as proved in Theorem 42.5.

42.5 BEATING 2 FOR THE EDGE CASE

Khuller and Vishkin [16] gave the first approximation algorithm for the special case of $k = 2$ with a performance ratio less than 2. Their algorithm is based on depth-first search (DFS) and has a performance guarantee of $3/2$. A simple generalization of their algorithm has a performance guarantee of $2 - 1/k$ for all k. Unfortunately, for higher values of k, this expression approaches 2.

In this section we provide a different algorithm due to Khuller and Raghavachari [17]. This is the first algorithm that achieves an approximation factor 1.85 for *all* values of k, for unweighted k-edge connectivity. For smaller values of k, their bounds are actually better: 1.66, 1.75, and 1.733 for $k = 3$, 4, and 5, respectively. The structure of this algorithm is similar to the one shown in the previous sections where the connectivity of the solution is increased incrementally in stages.

42.5.1 Khuller–Raghavachari Algorithm

For the sake of clarity, we assume that k is even; if k is odd, first find a sparse $(k-1)$-connected spanning subgraph G_{k-1} by using the algorithm given in this section and add to it a maximal DFS spanning forest F_i from $G - G_{k-1}$, thus obtaining a k-edge connected spanning subgraph G_k of G. Proofs from this section can be used for the odd case with minor modifications.

Algorithm 42.3 KR edge certificate

Require: $G = (V, E)$ and an even integer $k > 0$
Ensure: A sparse certificate $G_k = (V, E_k)$ for k edge connectivity
1: **procedure:** KR(G, k)
2:   $G_0 \leftarrow (V, \emptyset)$
3:   **for** $i \leftarrow 0, k/2 - 1$ **do**
4:     Find a maximal DFS forest F_{2i+1} in $G - G_{2i}$ with arbitrary root(s)
5:     $G_{2i+1} \leftarrow G_{2i} + F_{2i+1}$
6:     Post-order label each edge of F_{2i+1}
7:     Build a forest F_{2i+2} as follows. Process each edge $e = (x, y)$, $level(x) < level(y)$,
8:       of F_{2i+1} in post-order and add b_e (see Definition 42.3) to F_{2i+2} if the number
9:       of edge-disjoint paths between x and y in $G_{2i+1} + F_{2i+2}$ is less than $2i + 2$.
10:      (Note that the edge (x, y) constitutes a path by itself.)
11:     $G_{2i+2} \leftarrow G_{2i+1} + F_{2i+2}$
12: **end for**
13: **return** G_k
14: **end procedure**

We use the following notation throughout this section.

Definition 42.1 *Let e be an edge in a DFS spanning forest F_{2i+1} with x and y as its ends points. We will assume that x is the parent of y in F_{2i+1}.*

Definition 42.2 *Let e be a critical edge in G_{2i+1}, that is G_{2i+1} is $2i + 1$ edge connected but $G_{2i+1} - \{e\}$ is not, and let K be a cut of size $2i$ in $G_{2i+1} - \{e\}$. Since G is connected, G_{2i+1} is connected for all i as it contains F_1, which is a spanning tree of G. Denote the two connected components that result when $K \cup \{e\}$ is deleted from G_{2i+1} by C_1 and C_2 where $x \in C_1$ and $y \in C_2$.*

See Figure 42.4 for an illustration of this definition.

Definition 42.3 *With every critical edge $e \in F_{2i+1}$, associate a back edge $b_e \in G - G_{2i+1}$ that satisfies two properties:*

- *The fundamental cycle C_e created by b_e in F_{2i+1} contains e, and*

- *Of all fundamental cycles that contain e, C_e contains a vertex v with a smallest $level(v)$ value.*

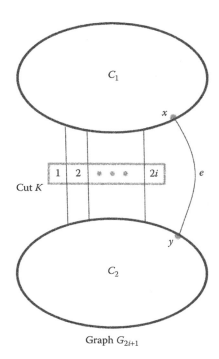

Figure 42.4 Illustration of Definition 42.2.

Definition 42.4 *Denote the set of edges that connect a vertex in C_1 to a vertex in C_2, excluding $K \cup \{e\}$, by B_e.*

In the following lemma, assume that G_{2i} is a $2i$ connected spanning subgraph of a k-connected graph G, for $k > (2i+1)$ and $i \geq 0$. Let $G_{2i+1} = G_{2i} + F_{2i+1}$ where F_{2i+1} is a maximal DFS spanning search forest in $G - G_{2i}$.

See Figure 42.5 for an illustration of these definitions.

Lemma 42.1 *If G_{2i+1} has an edge cut K' of size $2i + 1$, then there is exactly one edge e from F_{2i+1} in the cut, that is $K' \cap E(F_{2i+1}) = \{e\}$.*

Proof. We first show that K' must contain at least one edge from F_{2i+1}. Consider the two connected components C_1 and C_2 that result when K' is deleted from G_{2k+1}. Since G is k connected for some $k > (2i+1)$, there exists an edge (a,b) in $G - G_{2i+1}$ that connects a vertex from C_1 to a vertex in C_2. Since F_{2i+1} is maximal, every edge in $G - G_{2i+1}$, including (a,b), creates a cycle when added to F_{2i+1}. Let the path between a and b in F_{2i+1} be P. The path P cannot survive when the cut K' is deleted from G_{2i+1}, for otherwise C_1 would be connected to C_2 by P in $G_{2i+1} - K'$. Therefore, K' should contain at least one edge from F_{2i+1}. Denote K' as $K \cup \{e\}$ where $e \in F_{2i+1}$. To finish the proof, notice that K cannot contain any edges from F_{2i+1}, as it would mean that G_{2i} contains a cut smaller than $2i$. ∎

Corollary 42.2 *Let e, $e \in F_{2i+1}$, be a critical edge G_{2i+1} that belongs to a cut K of size $2i + 1$. Then, the edge b_e associated with each critical edge $e = (x,y)$ goes between a vertex from the subtree rooted at y, that is a descendant of y, and a vertex that is on the path from x to the root of F_{2i+1}, that is an ancestor of x.*

Proof. From the maximality of F_{2i+1}, we know that every edge from $G - G_{2i+1}$, including b_e, creates a cycle in F_{2i+1}. As F_{2i+1} is a DFS spanning forest, all such edges are back edges. From Lemma 42.1, we know that there is only one forest edge in K. Therefore, all edges from

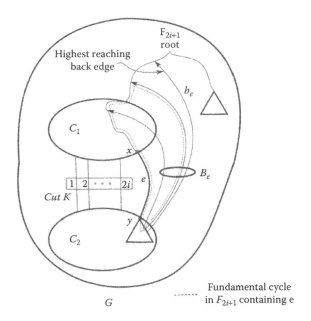

Figure 42.5 Illustration of Definitions 42.3 and 42.4.

$G - G_{2i+1}$ that go between C_2 and C_1 are back edges that connect a descendant of y to an ancestor of x. ∎

Theorem 42.8 *Let G be a graph which is at least k-edge connected for some $k > 0$. Then, Algorithm 42.3 with input G outputs a subgraph G_k of G that is k-edge connected.*

Proof. We prove this by induction on k. The theorem clearly holds for $k = 1$. Assume it holds for all values j, $j < k$. We will show that it holds for $j + 1$ as well. If j is even, then F_{j+1} is a maximal spanning forest in $G - G_j$. In this case, the fact that $G_j + F_{j+1}$ is $j + 1$ connected can be proved in a manner similar to the proof of Theorem 42.1.

Next consider the case when j is odd. Let $j = 2x + 1$ for some x. Then, F_{2x+2} is the set of back edges b_e for each critical edge e added in lines 7–9. We claim that every edge cut K in $G_j + F_{2x+2}$ is of size at least $2x + 2$. Note that the presence of a cut of size less than $2x + 1$ would violate the inductive hypothesis. If a cut K of size $2x + 1$ were to exist, one of the cut edges, say e, must come from F_{2x+1} by Lemma 42.1. This edge e would be a critical edge in G_j as it is part of a j cut. Therefore, the algorithm would add b_e to F_{2x+2} and make e non-critical. Hence no such cut K can exist in $G_j + F_{2x+2}$. ∎

42.5.2 Approximation Guarantee

Let b_e be a back edge added in lines 7–9 to F_{2i+2} corresponding to the critical edge e in G_{2i+1} and recall that B_e (see Definition 42.4) is the set of back edges whose fundamental cycles contain e.

Remark: It is important to note that F_{2i+2} changes as back edges associated with critical edges are added and it is this updated F_{2i+2} that is used when the criticality of a tree edge is checked in $G_{2i+1} + F_{2i+1}$ in lines 8–9.

Let d be another tree edge detected to be critical in the post-order corresponding to which b_d is added from the set of edges B_d whose fundamental cycles contains d (see Lemma 42.2).

Lemma 42.2 $B_d \cap B_e = \emptyset$.

Proof. Assume without loss of generality that e comes before d in the post-order traversal. Let $d = (s, t)$ and $e = (x, y)$. Then, by Corollary 42.2 the edges of B_d (resp. B_e) are all back edges and have one end point in the subtree rooted at t (resp. y). Therefore, if d is not an ancestor of e, the lemma follows immediately. Let us now consider the case when d is an ancestor of e. Assume, for contradiction, that there is an edge f that is in both B_d and B_e. Note that adding f to G_{2i+1} makes both d and e non-critical. But, when e is processed in lines 7–9, either f is added or another edge whose span includes the span of f is added, rendering d non-critical, contradicting the assumption that d was critical at the time it was examined in the updated G_{2i+1}. ∎

Recall that G^* denotes a subgraph of G with the optimum number of edges. The following lower bound is crucial for our analysis.

Lemma 42.3 *For any phase i, $0 \leq i < k/2$, $|G^*| \geq (k - 2i)|F_{2i+2}|$.*

Proof. At the beginning of phase i, G_{2i} is $2i$ connected. By the end of the iteration, the algorithm increases the connectivity by 2 and outputs G_{2i+2}. Corresponding to every edge $b_e \in F_{2i+2}$, there exists a critical edge e in G_{2i+1} that is part of a $2i+1$ cut $K \cup \{e\}$. Since G^* is k-connected, every cut is of size at least k. As $B_e \cup K \cup \{e\}$ is a cut that disconnects C_1 from C_2, $|B_e \cup K \cup \{e\}| \geq k$ (see Definition 42.2 for the definition of C_1 and C_2). That is, $|B_e \cup \{e\}| \geq k - 2i$. Furthermore, corresponding to every edge b_e of F_{2i+2} there is an edge e, $e \in F_{2i+1}$, and a set of back edges B_e, $B_e \subseteq G - G_{2i+1}$. By Lemma 42.2, the sets corresponding to two distinct edges from F_{2i+2} do not intersect with each other. Therefore G^* must contain at least $(k - 2i)|F_{2i+2}|$ edges. ∎

Using the bounds from Lemma 42.2, we can finally establish the following theorem.

Theorem 42.9 *The ratio of the number of edges of G_k output by Algorithm 42.3 to that of G^* is at most $(3 + \ln 2)/2$.*

Proof. As in the algorithm, we restrict our attention to even k; minor adaptation of the proof is required for the odd case. Note that since F_{2i+1} is a forest, it has at most n edges. This holds F_{2i+2} as well. Also, from the degree lower bound given Section 42.3.1, we know that $|G^*| \geq kn/2$. Thus,

$$\sum_{i=0}^{k/2-1} \frac{(|F_{2i+1}| + |F_{2i+2}|)}{|G^*|} \leq \frac{\sum_{i=0}^{k/2-1}|F_{2i+1}| + \sum_{i=0}^{(k/4-1)}|F_{2i+2}| + \sum_{i=k/4}^{(k/2-1)}|F_{2i+2}|}{\max\{kn/2, \max_i\{(k-2i)|F_{2i+2}|\}\}}$$

$$\leq \frac{3kn/4 + \sum_{i=0}^{(k/4-1)}|F_{2i+2}|}{\max\{kn/2, \max_i\{(k-2i)|F_{2i+2}|\}\}} \quad (42.1)$$

$$\leq \frac{3kn/4}{kn/2} + \frac{\sum_{i=0}^{(k/4-1)}|F_{2i+2}|}{\max_i\{(k-2i)|F_{2i+2}|\}}$$

$$\leq \frac{3}{2} + \frac{1}{2}\sum_{i=0}^{(k/4-1)} \frac{1}{(k/2-i)}$$

Using variable substitution we can rewrite the second term as

$$= \frac{3}{2} + \frac{1}{2}\sum_{x=k/4+1}^{k/2} \frac{1}{x}$$

Since $1/x$ is a monotonically decreasing function, we can set the upper bound of the sum using an integral (see Appendix A.2 from [13]) by reducing the bottom limit by 1:

$$\leq \frac{3}{2} + \frac{1}{2} \int_{x=k/4}^{k/2} \frac{1}{x} = \frac{3}{2} + \frac{1}{2}\left(\ln\left(\frac{x}{2}\right) - \ln\left(\frac{x}{4}\right)\right)$$

$$= \frac{3}{2} + \frac{\ln 2}{2} < 1.85 \qquad \blacksquare$$

As mentioned before this analysis assumes that k is even. By substituting $k = 4$ in (42.1), one can see that the bound is actually 1.75, which is somewhat better than 1.85. By a different analysis, it can be shown when k is 3 and 5, the approximation factors are 1.66 and 1.733, respectively, again better than 1.85.

42.5.3 Time Complexity

It turns out that analyzing the running time of Algorithm 42.3 is rather easy. There are $k/2$ iterations of the *for* loop. In each loop, we find a DFS spanning forest which takes linear time [13]. Similarly, post-order labeling can be performed in linear time [13]. Rest of the loop involves checking whether a given edge e is critical and finding b_e if it is determined that e is indeed critical. Karzanov and Timofeev show that in $O(n^2 i)$ time we can find all min-cuts of an i-connected graph [18]. Furthermore, they also show how to store all min-cuts using a compact tree-like representation. Given this representation, it is easy to check whether there exists a cut of size less than $2i + 2$ between x and y in $G_{2i+1} + F_{2i+2}$ in constant time. This will reveal whether e is critical. If it is, we can perform a DFS of the tree T containing the edge e in linear time and see which back edge b_e from the subtree rooted at y comes closest to the root of T. Therefore, the dominant step is the construction of the cut representation of Karzanov–Timofeev which takes $O(n^2 i)$ time. Summing this over $k/2$ iterations, we get a time bound of $O(k^2 n^2)$. Note that as back edges b_e are added, the cut in which b_e participates definitely gets destroyed. Also some other $(2i + 1)$-cuts might get destroyed by the addition of b_e. One assumption that is made in Khuller and Raghavachari [17], though not explicitly stated, is that the Karzanov–Timofeev cut representation can be updated in linear time under edge additions.

42.6 APPROXIMATING MINIMUM-SIZE SPANNING SUBGRAPHS VIA MATCHING

We conclude this chapter with the best-known algorithm for finding minimum-size k-connected spanning subgraphs. The algorithm presented in this section is extremely simple and for the vertex connectivity case, it meets or beats all other algorithms for all values of k. For undirected edge connectivity, it matches or improves the known bounds for values of $k \geq 3$. In particular, the algorithm presented in this section, due to Cheriyan and Thurimella [12], finds a k-vertex connected spanning subgraph of k-vertex connected graph, directed or otherwise, whose size is at most $1 + (1/k)$ times the optimal. For edge connectivity, the approximation factors are $1 + (2/(k+1))$ and $1 + (4/\sqrt{k})$ for undirected and directed graphs, respectively.

For clarity of exposition, we limit our discussion to the undirected, vertex case. Before we present the algorithm, the following background is required on generalized *matching*. Given a graph $G = (V, E)$, a matching M, $M \subseteq E$, is set of pairwise *non-adjacent* edges, that is, no two edges share a common vertex. A *maximum matching* is a matching M that contains the largest number of edges. Notice that in any matching M, the degree of any vertex v is at most 1. The maximum matching can be generalized wherein the degree of each vertex is at most d. Additionally, the degree requirement can be made non-uniform, that is each vertex

v can have its own degree constraint ranging from 1 to $\delta(v)$. This generalization is known as the b-matching problem.

A b-matching M of $G = (V, E)$ is really a subgraph with *most* number edges satisfying certain degree requirements. In particular, we are interested in a b-matching M where $\delta_M(v) \leq \delta_G(v) - (k-1)$. Denote $E - M$ by $\widetilde{M}$. Then $(V, \widetilde{M})$ is another subgraph of G with the *least* number of edges wherein $\delta_{\widetilde{M}}(v) \geq (k-1)$ for each v. Denote this subgraph by $\widetilde{M}$.

Algorithm 42.4 CT vertex certificate

Require: An integer $k > 0$ and k-vertex connected graph $G = (V, E)$
Ensure: A sparse certificate $G_k = (V, E_k)$ for k vertex connectivity
1: **procedure** CT(G, k)
2:    Find $\widetilde{M}$, a smallest subgraph of G where $\delta(v) \geq (k-1)$ for each $v \in G$
3:    $F \leftarrow \emptyset$
4:    **for** each edge $e = (u, v)$ in $G - \widetilde{M}$ **do**
5:      **if** (no. of vertex disjoint paths between u and v, including uv, in $\widetilde{M} + F$ is $\leq k$)
6:        **then** $F \leftarrow F \cup \{e\}$
7:    **end for**
8:    **return** $G_k = \widetilde{M} + F$
9: **end procedure**

Let us now analyze the algorithm above to see how many edges G_k contains in relation to the best possible G^*. We first show that $|F| < (n-1)$, a fact that follows easily from a theorem of Mader ([7]; [8], Theorem 1). (For an English translation of the proof of Mader's theorem see Lemma 1.4.4 and Theorem 1.4.5 in [6].)

Theorem 42.10 ([8], Theorem 1) *In a k-vertex connected graph, a cycle consisting of critical edges must be incident to at least one vertex of degree k.* ∎

Lemma 42.4 $|F| \leq (n-1)$

Proof. We claim that F cannot have any cycles. Assume otherwise. Since all edges of F are critical in $\widetilde{M} + F$, from Mader's theorem we know that this cycle must have a vertex v such that $\delta_{\widetilde{M}+F}(v) = k$. See Figure 42.6. Equivalently, $\delta_{\widetilde{M}}(v) < (k-1)$, contradicting that $\widetilde{M}$ is a subgraph of G in which every vertex has degree at least $(k-1)$. ∎

We now establish the performance guarantee.

Theorem 42.11 $\left((|\widetilde{M} + F|)/|G^*|\right) < (1 + (2/k))$

Proof. From the degree lower bound given in Section 42.3.1, we know that $|G^*| \geq kn/2$. Therefore

$$\frac{|\widetilde{M} + F|}{|G^*|} \leq \frac{|\widetilde{M}|}{|G^*|} + \frac{n-1}{(kn/2)} < \frac{|\widetilde{M}|}{|G^*|} + \frac{2}{k}$$

Since G^* is a subgraph in which every vertex v has $\delta(v) \geq k$, any subgraph of G that has the smallest number edges in which $\delta(v) \geq (k-1)$ will have no more than $|G^*|$ edges. Therefore, the first term is at most 1, thus establishing the theorem. ∎

Algorithm 42.4 performs significantly better and can be shown to have to an approximation ratio of $1 + (1/k)$ as mentioned at the beginning of this section. To show this improved bound, notice that there is some slack in our analysis in the proof of our theorem where we argue

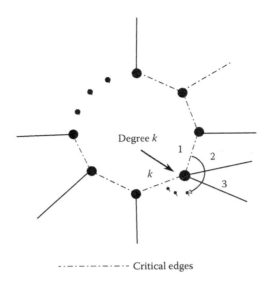

Figure 42.6 Illustration of Mader's theorem.

that the first term is at most 1, that is when we compare the degrees of vertices in G^* to those in $\widetilde{M}$.

It turns out that $|\widetilde{M}| \leq |G^*| - n/2$. Using this fact in conjunction with Lemma 42.4, we see that $|\widetilde{M} + F| < |G^*| + n/2$ which implies an approximation ratio of $1 + (1/k)$.

The fact $|\widetilde{M}| \leq |G^*| - n/2$ follows quite easily when G^* has a perfect matching M, because in that case $G^* - M$ would be a subgraph in which every vertex v has $\delta(v) \geq k - 1$. Proof of this fact is considerably technical when G^* does not have a perfect matching and is omitted here.

42.6.1 Time Complexity

Finding $\widetilde{M}$ is the same as finding a b-factor and removing the edges found, leaving a subgraph in which every vertex v satisfies $\delta(v) \geq k - 1$. Gabow and Tarjan [19] give a $O(m^{1.5} \log^2 n)$ algorithm for finding a b-factor.

To find a set of critical edges F whose addition to $\widetilde{M}$ would yield a k-connected subgraph, examine each edge $e = (u, v)$ of $G - \widetilde{M}$ in an arbitrary order (line 4). To see whether the if condition is satisfied in line 5, attempt to find $k + 1$ internally disjoint (vertex or edge as the case may be) paths between u and v. Each path can be found in linear time by an augmenting path algorithm [13]. Therefore each iteration of the for loop takes km time for a total of km^2 time. At termination, the subgraph $\widetilde{M} + F$ is k-vertex connected, and every edge $e \in F$ is critical.

The running time of finding F can be improved to $O(k^3 n^2)$ by first executing a linear-time preprocessing step to compute a sparse certificate of G for k-vertex connectivity and using that sparse certificate find k disjoint paths.

42.7 CONCLUSION

Approximation algorithms for graph connectivity are surprisingly elegant and have many useful applications. In this chapter, we have presented heuristics for finding k-connected spanning subgraphs of k-connected graphs. These polynomial-time heuristics are reasonably efficient and have *provable* performance guarantees. In some cases the approximation ratios come very close to the optimal. For example, the algorithm presented in the last section is provably within 5% from the optimal, that is $< 1.05 * \mathcal{OPT}$, if one is interested in finding a k-vertex connected subgraph of a k-vertex connected graph for $k = 20$.

References

[1] M. R. Garey and D. S. Johnson. *Computers and Intractability: A Guide to the Theory of NP-Completeness*. W. H. Freeman & Co., New York, 1979.

[2] D. S. Hochbaum, editor. *Approximation algorithms for NP-hard problems*. PWS Publishing Co., Boston, MA, 1997.

[3] H. Nagamochi and T. Ibaraki. Graph connectivity and its augmentation: applications of MA orderings. *Discrete App. Math.*, **123**(1–3) (2002), 447–472.

[4] H. Nagamochi. Graph algorithms for network connectivity problems. *J. Op. Res.*, **47**(4) (2004), 199–223.

[5] R. Diestel. *Graph Theory*. Graduate Texts in Mathematics. Springer-Verlag, Heidelberg/Berlin, Germany, 2006.

[6] B. Bollobás. *Extremal Graph Theory*. Dover Books on Mathematics. Dover Publications, Mineola, NY, 2004.

[7] W. Mader. Minimale n-fach kantenzusammenh ngende graphen. *Math. Ann.*, **191** (1971), 21–28.

[8] W. Mader. Ecken vom grad n in minimalen n-fach zusammenh ngenden graphen. *Arch. Math. (Basel)*, **23** (1972), 219–224.

[9] R. Thurimella. *Techniques for the Design of Parallel Graph Algorithms*. PhD thesis, University of Texas Austin, TX, 1989.

[10] J. Cheriyan, M. Kao, and R. Thurimella. Scan-first search and sparse certificates: An improved parallel algorithms for k-vertex connectivity. *SIAM J. Comput.*, **22**(1) (1993), 157–174.

[11] K. A. Doshi and P. J. Varman. Optimal graph algorithms on a fixed-size linear array. *IEEE Transac. Comput.*, **C-36**(4) (April 1987), 460–470.

[12] J. Cheriyan and R. Thurimella. Fast algorithms for k-shredders and k-node connectivity augmentation. *J. Algorithms*, **33**(1) (1999), 15–50.

[13] T. H. Cormen, C. E. Leiserson, R. L. Rivest, and C. Stein. *Introduction to Algorithms*. MIT Press, 3rd edition, Cambridge, MA, 2009.

[14] J. JáJá. *An Introduction to Parallel Algorithms*. Addison-Wesley Longman, Redwood City, CA, 1992.

[15] H. Nagamochi and T. Ibaraki. A linear-time algorithm for finding a sparse k-connected spanning subgraph of a k-connected graph. *Algorithmica*, **7**(5/6) (1992), 583–596.

[16] S. Khuller and U. Vishkin. Biconnectivity approximations and graph carvings. *J. ACM*, **41**(2) (March 1994), 214–235.

[17] S. Khuller and B. Raghavachari. Improved approximation algorithms for uniform connectivity problems. *J. Algorithms*, **21**(2) (1996), 434–450.

[18] A. V. Karzanov and E. A. Timofeev. Efficient algorithm for finding all minimal edge cuts of a nonoriented graph. *Cybernetics*, **22** (1986), 156–162.

[19] H. N. Gabow and R. E. Tarjan. Faster scaling algorithms for general graph matching problems. *J. ACM*, **38** (1991), 815–853.

CHAPTER 43

Rectilinear Steiner Minimum Trees

Tao Huang

Evangeline F. Y. Young

CONTENTS

43.1	Introduction	1115
	43.1.1 SMT Problem	1116
43.2	Steiner Ratio (RSMT vs. ESMT)	1116
	43.2.1 MST Problem	1116
	43.2.2 Euclidean Steiner Ratio	1117
	43.2.3 Rectilinear Steiner Ratio	1117
43.3	Heuristics	1120
	43.3.1 RMST-Based Heuristics	1120
	43.3.2 Iterated 1-Steiner	1121
	43.3.3 Batched Iterated 1-Steiner	1122
	43.3.4 FLUTE	1123
43.4	Exact Algorithms	1124
	43.4.1 FST Generation	1125
	43.4.2 FST Concatenation	1127
	43.4.2.1 Backtrack Search	1128
	43.4.2.2 Dynamic Programming	1128
	43.4.2.3 Integer Linear Programming	1129
43.5	Obstacle-Avoiding RSMT	1129
	43.5.1 Heuristics	1130
	43.5.1.1 Sequential Approach	1130
	43.5.1.2 Maze Routing–Based Approach	1130
	43.5.1.3 Connection Graph-Based Approach	1131
	43.5.2 Exact Algorithms	1131
43.6	Applications	1135
43.7	Summary	1136

43.1 INTRODUCTION

The Steiner minimum tree (SMT) problem asks for a shortest network that spans a set of given points in a metric space. The set of given points are usually referred to as *terminals* and new auxiliary Steiner points can be introduced so that the total length of the network can be reduced. The history of the SMT problem started with Fermat (1601–1665) who proposed the problem: given three points in a plane, find a fourth point such that the sum of its distances to the three given points is a minimum. Courant and Robbins [1] in their famous book *What Is Mathematics?* first named the problem after Steiner (1796–1863) who solved the problem of joining three villages by a system of roads having minimum total length. The popularity of this book has raised the research interests in the SMT problem. The formulation of the SMT problem is as follows.

43.1.1 SMT Problem

Given a set V of n terminals in the space L_p.* Find a shortest tree embedded in the space that spans V.

The original SMT problem considers the Euclidean space (i.e., L_2 space). The rectilinear Steiner tree problem (i.e., in L_1 space) is first attacked by Hanan [2]. The problem is equivalent to find a tree connecting all the terminals by using only horizontal and vertical lines. An optimal solution to this problem is called a rectilinear SMT (RSMT). Hanan proved that there is at least one RSMT that is contained in the Hanan grid. The Hanan grid can be obtained by constructing horizontal and vertical lines through each terminal and the intersections of these lines are thus candidate Steiner points. Although there is a finite number of candidate Steiner points in the Hanan grid, it is still a very difficult problem to select a subset of them to construct a RSMT. In fact, the RSMT problem is shown to be NP-complete by Garey and Johnson [3]. Moreover, they also showed that the Euclidean SMT (ESMT) problem is NP-hard.

This chapter discusses the RSMT problem, its properties, solutions, and applications. Section 43.2 of this chapter presents the Steiner ratio of the RSMT problem comparing with the ESMT problem. Section 43.3 introduces several heuristics for the RSMT problem. Section 43.4 describes the exact algorithms for the RSMT problem. In Section 43.5, a variant of the RSMT problem considering obstacles is discussed. Section 43.6 gives the applications of RSMT in very large-scale integration (VLSI) design. A conclusion is drawn in Section 43.7.

43.2 STEINER RATIO (RSMT VS. ESMT)

The RSMT problem is NP-complete which means that an efficient polynomial time algorithm for the problem may not exist. Therefore, finding a RSMT is usually of high computational cost. An alternative to the SMT is the minimum spanning tree (MST). The MST problem can be formulated as follows:

43.2.1 MST Problem

Given a set V of n terminals in the space L_p. Find a shortest tree embedded in the space that spans V using only edges connecting v_i and v_j where $v_i, v_j \in V$.

As we can see from the problem formulation, the only difference between the SMT problem and the MST problem is usage of the auxiliary points. Since the MST problem disallows any auxiliary point, it is much simpler than the SMT problem. In fact, the MST problem is polynomially solvable. Both Prim's algorithm and Kruskal's algorithm can find a MST in $O(n \log n)$ time. However, it is obvious that the length of a MST is always longer than its SMT counterpart, since the MST is also a candidate solution in the SMT problem. Let $|SMT(V)|$ and $|MST(V)|$ be the length of the SMT and MST over V, respectively,

$$|SMT(V)| \leq |MST(V)| \qquad (43.1)$$

---

*The distance between two points in the L_p space can be calculated by $d(u;v) = (|u_x - v_x|^p + |u_y - v_y|^p)^{1/p}$.

for any V. The question is, if we construct a MST instead of a SMT, how close can this approximation be. We define the Steiner ratio to be

$$\rho(L_p) = \inf_V \left\{ \frac{|SMT(V)|}{|MST(V)|} \right\} \quad (43.2)$$

where V is a set of points in L_p. That is, the Steiner ratio is the largest possible ratio between the length of a SMT and the length of a MST in the L_p space.

Since many of the SMT heuristics are based on improving a MST, their performances are closely related to the Steiner ratio. Therefore, it is of partical interest to determine the Steiner ratio. In this section, we will first give a brief introduction of the Euclidean Steiner ratio problem for a comparison with its rectilinear counterpart.

43.2.2 Euclidean Steiner Ratio

Early researches on the Steiner ratio in L_2 space mainly focus on special cases of V, or obtaining a lower bound of ρ.

Halton et al. [4] gave the first lower bound 0.5 on the Steiner ratio. Graham and Hwang [5] improved the low bound to 0.57. Later, the bound was pushed up to 0.74 by Chung and Hwang [6], and to 0.8 by Du and Hwang [7]. Finally, the lower bound was improved by Chung and Graham [8] to 0.824. Further improvement on the lower bound is possible, but it turns out that the decrease is marginal and it is not likely that the exact Steiner ratio can be found along this line.

Along another line, Gilbert and Pollak [9] verified that the Steiner ratio is $\sqrt{3}/2$ for special cases when the number of terminals $n = 3$. Based on a large number of simulations, they further conjectured that the Steiner ratio is $\sqrt{3}/2$ for general cases. This is known as the Steiner ratio Gilbert–Pollak conjecture. Pollak [10] proved that the conjecture is true for $n = 4$. By using a different approach, Yao et al. [11] also verified that the Steiner ratio is $\sqrt{3}/2$ for $n = 4$. They further extended this approach to prove the conjecture for $n = 5$. Rubinstein and Thomas [12] derived the variational approach and verified the conjecture for $n = 6$.

Finally, in 1990, Du and Hwang [13, 14] proposed some novel ideas and claimed that they proved the Steiner ratio Gilbert–Pollak conjecture. However, the proof is shown to be incorrect later by Ivanov and Tuzhilin [15]. Therefore, the Steiner ratio Gilbert–Pollak conjecture is still an open problem.

43.2.3 Rectilinear Steiner Ratio

The Steiner ratio in L_1 space was originally settled by Hwang [16] by first characterizing Steiner trees and then obtaining the Steiner ratio. Later, another simpler proof is provided by Salowe [17].

Both of these two approaches make use of the unique decomposition of a SMT into full Steiner trees (FSTs). Let V' be a set of points in the plane, and T be a SMT spanning V'. T is said to have a *full topology* if every point in V' is a leaf node in T. A terminal set V' is a *full set* if every SMT for V' has a full topology. An FST is a SMT that spans a full set of terminals. It can be easily verified that any SMT can be uniquely decomposed into a set of edge-disjoint FSTs by splitting at the terminals with degree* more than one. This fact brings out the importance of studying the characteristics of FSTs.

In the rectilinear plane, Hwang [16] first characterized the structures of FSTs. Two operations—*flipping* and *shifting* as shown in Figure 43.1—are defined. Shifting a

---

*The degree of a terminal is the number of edges connecting it.

Figure 43.1 (a) Shifting and (b) flipping.

line means moving a line between two parallel lines to a new position. Flipping an edge with two perpendicular lines meeting at a corner means moving these two lines to flip the corner to the opposite side diagonally. These two operations will not change the length of a RSMT. By using shiftings and flippings, Hwang developed a series of lemmas to reach Theorem 43.1.

Theorem 43.1 *For a full set of $n > 4$ terminals in the rectilinear plane, there exists a corresponding FST that either consists of a single line with $n - 1$ alternating incident segments, or a corner with $n - 3$ alternating segments incident to one leg and a single segment incident to the other leg.* ∎

The two FST structures described in Theorem 43.1 are shown in Figure 43.2. Hwang also showed that Theorem 43.1 holds for $n = 2, 3$, or 4. The only exception is when $n = 4$ and the four terminals are the endpoints of a cross as shown in Figure 43.3. By using Theorem 43.1, Hwang obtained the following theorem that determines a lower bound of the rectilinear Steiner ratio.

Theorem 43.2 *The rectilinear Steiner ratio $\rho(L_1) \geq 2/3$.*

Proof. Any RSMT can be partitioned into a set of edge-disjoint FSTs each of which corresponds to a full set. Therefore, we only need to establish the lower bound for each full set, because the sum of the length of the rectilinear minimum spanning trees (RMSTs) for each full sets is lower bounded by the length of RMST for the whole set of terminals.

Let V' be a full set of n terminals. The proof is by induction on n.

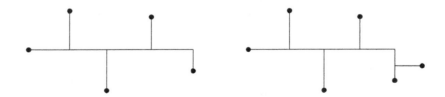

Figure 43.2 (a,b) Two generic forms for a FST when $n > 4$.

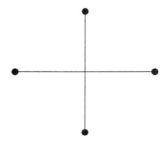

Figure 43.3 Only exception to Theorem 43.1.

If $n = 2$, then $RSMT(V') = RMST(V')$.

For $2 \leq n \leq 4$, let L and W be the length and width of the enclosing rectangle of V'. Obviously, $RSMT(V') \geq L + W$. Let R be the boundary of the enclosing rectangle and $|R| = L + W$. Since deleting any edge between two adjacent terminals on R yields a spanning tree of V', we have

$$|RMST(V')| \leq (1 - \frac{1}{4})|R| = \frac{3}{2}(L + W) \leq \frac{3}{2}|RSMT(V')|. \qquad (43.3)$$

Consider the case when $n > 4$. In any FST with $n > 4$, we can always find a subtree as shown in Figure 43.4 where $|ae| \leq |cg|$ and $|dh| \leq |bf|$. Let A be the set of terminals above a, B be the set of terminals below d, and $C = \{a, b, c, d\}$. Let s_A be the total length of the edges above ef, s_B be the total length of the edges below gh, and $s_C = |ef| + |bf| + |fg| + |gc| + |gh|$. Obviously, $|RSMT(V')| = s_A + s_B + s_C$. By considering the enclosing rectangle of C, we can show that $RMST(C) \leq (3/2)s_C$. Moreover, by inductive hypothesis, $RMST(A) \leq (3/2)s_A$ and $RMST(B) \leq (3/2)s_B$. Therefore,

$$|RMST(V')| \leq |RMST(A)| + |RMST(B)| + |RMST(C)| \leq \frac{3}{2}|RSMT(V')|. \qquad (43.4)$$

∎

It is also easy to verify that $2/3$ is an upper bound of the rectilinear Steiner ratio, by considering a set of four terminals with x and y coordinates $\{(1, 0), (-1, 0), (0, 1), (0, -1)\}$. The corresponding RSMT and RMST are shown in Figure 43.5. Clearly, the length of the RSMT is 4 and the length of the RMST is 6, which establish an upper bound of $2/3$. Note that by clustering more terminals arbitrarily close to any of the four terminals in Figure 43.5, the $2/3$ bound can be attained for any value of n. As a result, we have the following theorem for the rectilinear Steiner ratio.

Theorem 43.3 *The rectilinear Steiner ratio is $2/3$.* ∎

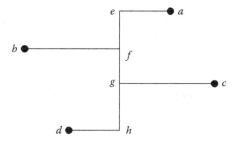

Figure 43.4 Subtree structure in a FST with $n > 4$.

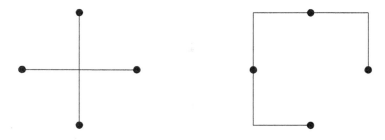

Figure 43.5 Example of a RSMT and a RMST of four terminals.

43.3 HEURISTICS

The RSMT problem is NP-complete. This means that efficient polynomial time exact algorithm may not exist. Therefore, many researches of the RSMT problem have been focused on the development of heuristics. Early heuristics are mainly based on improving over a RMST. Starting in 1990s, new class of RSMT heuristics that do not rely on the RMST construction has been proposed. Two typical examples are iterated one Steiner and batched iterated one Steiner. Recently, a look up table based algorithm called FLUTE is proposed. Comparing with the other heuristics, FLUTE can provide the best tradeoff between runtime and accuracy, and therefore is the state-of-the-art algorithm. In this section, a brief introduction to these approaches is presented.

43.3.1 RMST-Based Heuristics

In Section 43.2, we have shown that the rectilinear Steiner ratio is 2/3. It means that any heuristic based on improving over a RMST can guarantee a worst-case performance ratio of 3/2. Therefore, many RSMT heuristics in the literature use RMST-based strategies.

A RMST can be computed in $O(n\log n)$ time. The first RMST algorithm with this complexity is proposed by Hwang [18] and the algorithm is based on the construction of the rectilinear Voronoi diagram. Hwang showed that the rectilinear Voronoi diagram can be built in $O(n\log n)$ time. It can also be verified that a RMST can be computed in $O(n)$ time by using the Voronoi diagram, and therefore the complexity of finding a RMST is $O(n\log n)$. However, the computation of Voronoi diagram can be tedious. A simpler way is to use the nearest neighbors of each terminal. For each terminal we divide its surrounding area into eight regions separated by lines that intersect at a 45-degree angle, as shown in Figure 43.6. The following theorem is first proposed by Yao [19].

Theorem 43.4 *In a RMST, if two terminals v and u are connected, then v is the nearest to u in one of the eight regions of u.*

Proof. Assume the contrary that, in a RSMT, terminal v is connected to u, but v is not the closest point to u in one region. Let w be the terminal that is closer to u than v in the region. Therefore, $d(w,u) < d(v,u)$. It is also easy to verify that $d(w,v) < d(v,u)$. By deleting the edge connecting v and u, the RMST is divided into two subtrees with u and v belonging to different components. Despite which subtree w belongs to, a shorter spanning tree can be obtained by connecting either w and v or w and u, a contradiction. ∎

Theorem 43.4 shows that for the construction of RMST, only the edges connecting nearest neighbors in the eight regions need to be considered. Finding the nearest neighbor of all

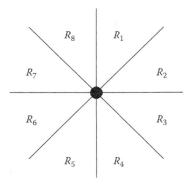

Figure 43.6 Eight regions of a terminal.

terminals in all eight regions can be done in $O(n\log n)$ time [20, 21]. Since there are at most $8n$ edges, a RMST can be therefore found in $O(n\log n)$ time by using either the Prim's or Kruskal's algorithm.

With a RMST as a starting point, a direct way to improve and obtain a RSMT is to remove overlapping segments by introducing Steiner points. These approaches are called Steinerization. Early overlap removal schemes all make use of simple heuristics. A pair of edges sharing a common terminal are chosen arbitrarily. If there is overlap, they are embedded by adding a Steiner point. This process terminates until all pairs of neighboring edges are explored. A comparison between different ways on selecting pairs of edges to process is done by Richards [22]. Later, Vijayan et al. [23] gave a polynomial time algorithm to find optimal embedding when starting with a special RMST called separable RMST. A RMST is separable if and only if for any pair of nonadjacent edges in the tree, any staircase layouts of the two edges will not intersect or overlap. They first gave a $O(n^2)$ time algorithm for the construction of a separable RMST. Based on the separable RMST, an $O(n)$ time optimal algorithm is proposed with the assumption that each edge has at most one corner (i.e., L-shaped). The algorithm starts by making a terminal as the root of the tree and solve the problem in a bottom-up fashion. The key observation is that the optimal solution of a subtree depends only on how the edge connecting the root node of the subtree and its parent is embedded. Since only L-shaped edges are considered, there are two options for embedding. Therefore, an $O(n)$ dynamic programming algorithm can find an optimal solution. Ho et al. further extended the algorithm to handle the case when each edge has at most two corners (i.e., Z-shaped). The difference is that there can be more embedding options for each subtree. Ho et al. showed that the corresponding dynamic programming algorithm has a time complexity of $O(n^7)$. Finally, they proved that the resulting RSMT after optimal Z-shaped embedding is also optimal when there is no restriction on edge shapes.

Another way to improve over a RMST is to add some new edges to replace longer ones repeatedly. These approaches are called edge-substitution. Borah et al. [24] proposed an edge-based heuristic that starts with a RMST and incrementally improves the cost by connecting a node* to a neighboring edge and removing the longest edge in the loop thus formed. The reduction in the cost of the tree due to this operation is the gain. The algorithm works in an iterative manner. In each iteration, a set of (node, edge) pairs are found and updates are applied to the tree starting from the (node, edge) pairs with the largest gain. Borah et al. showed that finding all possible (node, edge) pairs with positive gain can be done in $O(n\log n)$ time and applying the updates to the tree requires only $O(n)$ time. They further showed that a number of three iterations are sufficient in most cases. Therefore, the complexity of the algorithm is $O(n\log n)$. Zhou et al. [25] extended the edge-based heuristic by using a spanning graph [21]. A spanning graph is an undirected graph over the points that contain at least one MST. They showed that finding potential (node, edge) pairs in the spanning graph can be more efficient. They also proposed a simpler way to find the longest edge on the loop formed by connecting a node to an edge by using a merging binary tree. Although, the run time is dominated by the spanning graph and RMST generation, which take $O(n\log n)$ time, a good practical performance can be achieved.

43.3.2 Iterated 1-Steiner

While the RMST-based heuristics can guarantee a worst case performance ratio of 3/2, it is still a problem to find such a heuristic method with performance ratio strictly less than 3/2. Kahng and Robins [26] showed that the 3/2 bound is tight for a large number of RMST-based methods. Motivated by this fact, Kahng and Robins [27] proposed a heuristic called iterative

---

*A node can be a terminal or a Steiner point.

1-Steiner that does not, implicitly or explicitly, make use of a RMST. The algorithm is based on the answer to the following question. If at most one more Steiner point is allowed, what is the optimal Steiner tree and where should the Steiner point be placed? This is called the 1-Steiner problem.

In the Euclidean plane, Georgakopoulos and Papadimitriou [28] are the first to give an $O(n^2)$ algorithm to solve the 1-Steiner problem and Kahng and Robins adapted this method for the rectilinear plane. The algorithm makes use of the concept of nearest neighbor for the construction of RMST to partition the plane into $O(n^2)$ isodendral regions. An important property of isodendral regions is that introducing any point in a given region will result in a constant RMST topology. Therefore, after an $O(n^2)$ preprocessing step, updating the RMST to include a new point requires only constant time. Moreover, the optimal Steiner point in each region can also be determined in constant time. As a result, the 1-Steiner problem can be solved in $O(n^2)$ time by iterating through the isodendral regions and selecting the point with the lowest cost.

The iterative 1-Steiner heuristic works by iteratively calculating optimal 1-Steiner points and include them into the point set. Accepted Steiner points are deleted if they become useless, that is, if their degree becomes 1 or 2 in the tree. The algorithm terminates when no improvement can be achieved by adding new Steiner points or the maximum number of iterations has been reached. An example of the iterative 1-Steiner heuristic is shown in Figure 43.7. In [27], the maximum number of iterations is set to be the number of terminals n. Therefore, the overall time complexity of iterative 1-Steiner is $O(n^3)$.

43.3.3 Batched Iterated 1-Steiner

Kahng and Robins [27] proposed several variants to the iterative 1-Steiner. Among those variants, the most promising one makes use of batched processing to include Steiner points. Instead of adding one Steiner point per iteration, a maximal independent set of Steiner points are included.

The heuristic starts by evaluating every candidate Steiner points in the Hanan grid. By preprocessing the $O(n^2)$ isodendral regions as a planar subdivision, the planar region

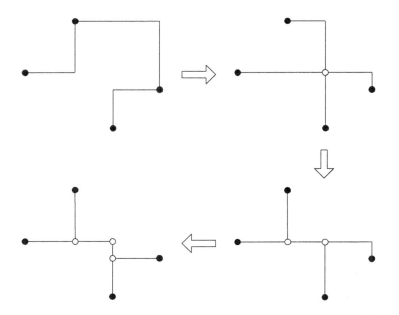

Figure 43.7 Example of the iterative 1-Steiner algorithm.

in which a given point lies can be determined in $O(\log n)$ time. This preprocessing requires $O(n^2 \log n)$ time. Since the MST of a planar-weighted graph can be maintained using $O(\log n)$ time per addition of a point, the RMST cost savings for all the candidate Steiner point can be calculated in $O(n^2 \log n)$ time. Then, the Steiner point candidates are sorted according to their gains on cost savings in decreasing order. Next, all the candidates are processed in order. Each candidate with a positive gain are added, as long as it is independent of all the Steiner points previously added in the same round. The criterion for independence is that no candidate is allowed to reduce the potential MST cost savings of any other candidate in the added set. This process iterates until no Steiner point can be included. The total time required for one iteration is $O(n^2 \log n)$. Since Steiner point candidates are added in batch, the number of iterations required grows much more slowly than the number of Steiner points considered. Empirical study showed that batched iterated 1-Steiner performs close to iterated 1-Steiner, but the computational cost is much lower.

Although batched iterated 1-Steiner can be implemented to run in $O(n^2 \log n)$ per iteration, the computational geometric methods have a large hidden constant and are also difficult to code. Therefore, an $O(n^4 \log n)$ implementation is used in [27]. A more efficient $O(n^3)$ implementation is later presented by Griffith et al. [29]. Experimental results showed that a speedup factor of three orders of magnitude over previous implementation can be achieved.

43.3.4 FLUTE

As will be discussed in Section 43.6, the RSMT problem has many applications in VLSI design. In VLSI circuits, many of the nets have a small number of terminals. Therefore, it is more important for RSMT algorithms to be simple and efficient for small problems. Based on this observation, Chu and Wong [30] proposed a RSMT algorithm called fast lookup table estimation (FLUTE).

Given a set of n terminals, the Hanan grid can be built by drawing horizontal and vertical lines through each terminal. Let x_i be the x-coordinates of the vertical grid lines such that $x_1 \leq x_2 \leq \ldots \leq x_n$, and y_i be the y-coordinates of the horizontal grid lines such that $y_1 \leq y_2 \leq \ldots \leq y_n$. Label the terminal in ascending order of the y-coordinates and let s_i be the rank of terminal i in ascending order of the x-coordinates. The sequence $s_1 s_2 \ldots s_n$ is called the position sequence. An example is shown in Figure 43.8 where the position sequence of the net is 3142. Let $v_i = y_{i+1} - y_i$ and $h_i = x_{i+1} - x_i$ be the distance between adjacent Hanan grid lines. Since a Steiner tree in the Hanan grid is a union of Hanan grid edges, the length of any Steiner tree can always be written as a linear combination of edge lengths in which every coefficient is a positive integer. For example, the length of the three Steiner trees can be expressed by $h_1 + 2h_2 + h_3 + v_1 + v_2 + 3v_3$, $h_1 + h_2 + h_3 + v_1 + 2v_2 + 3v_3$, and $h_1 + 2h_2 + h_3 + v_1 + v_2 + v_3$ (Figure 43.9). Therefore, a lookup table can be used to store the lengths of all possible Steiner trees as linear combinations of h_i and v_i. For simplicity, only the vectors of the coefficients are stored, for example, (1, 2, 1, 1, 1, 3), (1, 1, 1, 1, 2, 3), and (1, 2, 1, 1, 1, 1). It is also easy to find that some vectors are suboptimal, for example, the length induced by (1, 2, 1, 1, 1, 3) cannot be shorter than (1, 2, 1, 1, 1, 1). A vector that can potentially produce the optimal length is called a potentially optimal wirelength vector (POWV). For each POWV, a set of corresponding RSMTs called potentially optimal Steiner tree (POST) are also stored. A key observation is that, if two nets have the same position sequence, then every Steiner tree of one net is topologically equivalent to a Steiner tree of the other net. This means that nets with the same position sequence can be grouped together to share the set of POWVs and the following theorem can be stated.

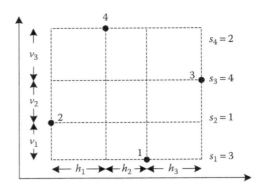

Figure 43.8 Example of the position sequence of a net.

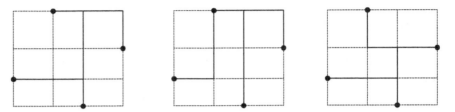

Figure 43.9 Example of different Steiner trees for a net.

Theorem 43.5 *The set of all nets with n terminals can be divided into n! groups according to the position sequence such that all nets in each group share the same set of POWVs.* ∎

FLUTE makes use of precomputed lookup table of POWVs and POSTs. Given a net, its position sequence is first determined and the corresponding POWVs are extracted from the table. The tree length of each POWV is computed according to the values of h_i and v_i and the POWV with the minimum length is selected. The corresponding POSTs are the RSMTs for the net.

The precomputation of the lookup table for small nets can be done by enumerating all possible Steiner trees in the Hanan grid. For larger nets, a boundary-compaction technique is proposed to efficiently generate all possible POWVs and POSTs. Some reductions are also applied to reduce the size of the lookup table. It is reported that the total table size is only 9.00 MB for all nets with up to 9 terminals.

FLUTE is able to generate optimal RSMTs for small nets (e.g., with up to 9 terminals) by using the lookup table. However, for large nets, the lookup table approach is impractical because of the high cost in both space and time. Therefore, a large net is divided into small nets with only the breaking terminals in common by using a net breaking heuristic. Each small net is then solved by using the lookup table and the resulting RSMTs are combined to form a RSMT for the original net. Finally, some refinement schemes are applied to eliminate overlapping segments or further reduce the length of the tree.

The total run time complexity of FLUTE is $O(n \log n)$. Empirical results on VLSI design showed that FLUTE is more accurate than the batched 1-Steiner heuristic and is almost as fast as a very efficient implementation of the Prim's RMST algorithm.

43.4 EXACT ALGORITHMS

In previous sections, we mentioned that at least one RSMT can be found in the Hanan grid graph. Therefore, exact algorithms for the Steiner problem in networks [31] can also be used to solve the RSMT problem. However, these approaches are considered to be less effective for the RSMT problem because they do not exploit the geometric of the problem. Therefore, in this section, we will focus on the geometric approaches.

In Section 43.2, we showed that any RSMT can be uniquely decomposed into a set of FSTs that have only two possible structures as shown in Figure 43.2. We refer these FST topologies as Hwang's topology in this section. Since FSTs are much simpler to construct than RSMTs, a straightforward strategy to construct RSMTs is to use a two-phase approach. The first phase is to generate a set of FSTs such that there is at least one RSMT composed of the FSTs in the set only. This phase is called the FST generation phase. In the second phase, a subset of FSTs with minimum total length are selected and combined such that all terminals are connected. This phase is called the FST concatenation phase.

43.4.1 FST Generation

Salowe and Warme [32] gave the first rectilinear FST generation algorithm. The algorithm generates FST by considering all pairs (a, b) of terminals as *backbone* in the Hwang's topology. The backbone is the complete corner in the Hwang's topology as described in Theorem 43.1. In the corner, the leg with alternating incident segments is called the long leg, and the other is called the short leg. For each pair (a, b), all candidate terminals that can be attached to the backbone are found. Then, the algorithm will recursively try to attach candidate terminals to the backbone and test if a FST can be formed. Some screening tests are developed to eliminate those FSTs that cannot be in any RSMT. The algorithm is able to generate FSTs for 100 terminals in a short time. However, it is impractical for larger instances because of the high computational cost. Later, Warme [33] improved this algorithm to handle 1000-terminal instance in hours.

The state-of-the-art rectilinear FST generation algorithm is presented by Zachariasen [34]. Let the root of a FST be the terminal incident to the long leg. For a given root z, the algorithm works by growing the long legs in four possible directions. For a given direction, the algorithm recursively tries to attach terminals to the long leg. A series of necessary conditions are used to prune away useless FSTs.

The *empty diamond property* states that no other points of the RSMT can lie in $L(u, v)$, where uv is a horizontal or vertical segment and $L(u, v)$ is an area on the plane such that all the points in this area are closer to both u and v than u and v are to each other. The empty diamond region of a segment is shown in Figure 43.10. This is because if there is a terminal w inside the empty region of segment uv, we can simply delete uv and connect either uw or vw to reduce the length of the tree. The empty diamond regions with respect to a FST are shown in Figure 43.11.

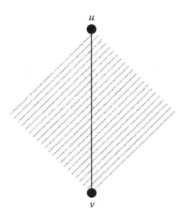

Figure 43.10 Empty diamond.

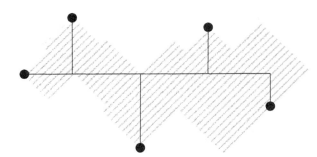

Figure 43.11 Empty diamond regions with respect to an FST.

Let uw and vw denote two perpendicular segments sharing a common endpoint w. The *empty corner rectangle property* states that no other points of the RSMT can lie in the interior of the smallest axis-aligned rectangle containing u and v. The empty corner rectangle region is shown in Figure 43.12. Assume that there is a terminal x inside the empty rectangle region. The unique path P from x to w in the RSMT visits either u or v first, or none of them, before reaching w. If P visits $u(v)$ first, we can delete $uw(vw)$ and add a vertical (horizontal) segment from x to a point on $vw(uw)$, forming a tree with shorter length. If P reaches neither u nor v before reaching w, we can delete uw or vw and add ux or vx depending on the location of x to obtain a shorter tree. The empty corner rectangle regions with respect to a FST are shown in Figure 43.13.

The *empty inner rectangle property* can be used to prune away useless FSTs. A FST can be transformed to its corner-flipped version by shifting segments and flipping corners as shown in Figure 43.14. The empty inner rectangle property states that no terminal can

Figure 43.12 Empty corner rectangle.

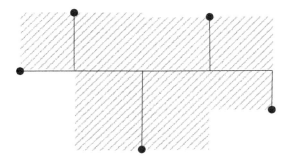

Figure 43.13 Empty corner rectangle regions with respect to an FST.

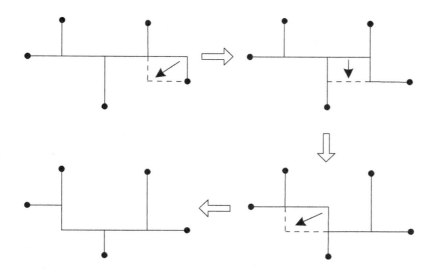

Figure 43.14 Transformation of an FST to its corner-flipped version.

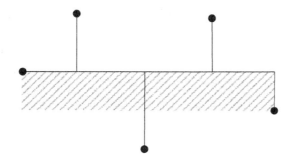

Figure 43.15 Empty inner rectangle in an FST.

be in between the backbone of the origin topology and that of the corner-flip topology (Figure 43.15). Assume that there is a terminal inside the empty inner rectangle region. We can shift some segments and flip some corners to align with the terminal such that splitting at this terminal will result in two smaller FSTs.

The *bottleneck Steiner distance*, which is analogous to that of the Steiner tree problem in networks, can also be used to eliminate useless rectilinear FSTs. Let $Tr(V)$ be a tree spanning the terminal set V. We use $\delta_{Tr(v_i v_j)}$ to denote the length of the longest edge on the unique path between v_i and v_j in $Tr(V)$. Let RMST(V) be a RMST of the terminal set V, then the bottleneck Steiner distance is equal to $\delta_{RMST(v_i v_j)}$. It can be proved that if RMST(V) and RSMT(V) are respectively a MST and a Steiner minimal tree on a set of vertices V, then $\delta_{RMST(v_i v_j)} \geq \delta_{RSMT(v_i v_j)}$ for any $v_i, v_j \in V$. Therefore, for a FST to be part of a RSMT, we require that $\delta_{RMST(v_i v_j)} \geq \delta_{FST(v_i v_j)}$ for any $v_i, v_j \in V$.

The above conditions are used to prune away those FSTs that cannot be part of any RSMT. Empirical study showed that most of the FSTs can be pruned away by one of these tests and the number of resulting FSTs grows almost linear with respect to the number of terminals. The algorithm is able to generate FSTs for 1000 terminals in less than a minute.

43.4.2 FST Concatenation

Let $F = \{f_1, f_2, \ldots, f_m\}$ be the set of FSTs generated in the first phase. The second phase is to select a subset such that all terminals are spanned. Different from the FST generation phase, the FST concatenation phase is purely combinatorial and metric-independent.

Therefore, early FST concatenation algorithms proposed for the ESMT problem can also be applied for the rectilinear case. These approaches include backtrack search, dynamic programming, and integer linear programming (ILP).

43.4.2.1 Backtrack Search

A straightforward way to combine FSTs is to use backtrack search. Starting from a single FST, recursively add new FSTs into the solution until the solution spans all terminals or it can be verified that the solution cannot be optimal. In these cases, the search backtracks to try to add some other FSTs.

Winter [35] proposed the first FST concatenation algorithm by backtrack search for the ESMT problem. Simple tests such as length test, degree tests, and cycle tests are employed during the search. The algorithm is able to solve, in a reasonable amount of time, problems with less than or equal to 15 terminals. Experimental results showed that, for the instances with more than 15 terminals, the computation time of the concatenation phase dominates that of the generation phase. Cockayne and Hewgill [36,37] presented an improved version of Winter's algorithm. Problem decomposition is applied to divide the initial concatenation problem into several subproblems. If the set of all FSTs can be divided into biconnected components, then each biconnected component corresponds to a subproblem on which concatenation can be done separately. They also proposed to use an incompatibility matrix to speedup the search. Two FSTs are incompatible if they cannot appear simultaneously in any of the SMTs (e.g., if they have more than one terminal in common, a cycle will be formed). This information is precomputed and stored in a matrix. The incompatibility matrix can be used to guide backtrack search. For example, only the FSTs that are compatible with every FST in the current solution can be added. This can significantly reduce the solution space with almost no computational overhead. In comparison with the savings in searching, the time required for computation of the incompatibility matrix is negligible. They reported a solvable range of 32 terminals. Salowe and Warme [32] proposed to select and add *the most promising* FST during the search. They also gave a more powerful graph decomposition theorem to decompose the problem. More recently, Winter and Zachariasen [38] improved FST compatibility and FST pruning substantially and report solutions for 140-terminal instances in Euclidean space.

43.4.2.2 Dynamic Programming

Ganley and Cohoon [39] presented a dynamic programming approach to combine FSTs. From Theorem 43.1, it is clear that any RSMT for any set of terminals is either a FST itself or it can be divided into two smaller RSMTs joining at a terminal. Therefore, dynamic programming is applicable. Subsets of terminals are processed in increasing order of their cardinality. For subsets of more than two terminals, the algorithm first tries to construct a FST according to Theorem 43.1. Then, several trees are produced by joining the RSMTs of every pair of disjoint subsets having exactly one terminal in common. Since the subsets are enumerated in increasing order of cardinality, the RSMTs of the smaller subsets are already computed and stored. Among all the generated trees, the one with minimum length is remembered in a lookup table. The time complexity of this algorithm is $O(n3^n)$. By proving that the number of candidate FSTs for a set of n terminals is at most $O(n1.62^n)$, Ganley and Cohoon improved the time complexity of the algorithm to $O(n^2 2.62^n)$. Based on this dynamic programming algorithm, Fößmeier and Kaufmann [40] make use of the empty region properties to reduce the number of candidate FSTs. An $O(n1.38^n)$ bound is derived which lead to an algorithm with $O(n^2 2.38^n)$ time complexity.

Although dynamic programming algorithms can provide the best theoretical worst-case time bound, their practical performance are inferior to the backtrack search.

43.4.2.3 Integer Linear Programming

Despite the substantial efforts made to improve the performance, backtrack search and dynamic programming algorithms can only handle problems with around 100 terminals. A breakthrough in the concatenation algorithm is achieved by Warme [33, 41] who observed that the FST concatenation problem is equivalent to find a MST in hypergraph and formulated the problem as an ILP.

Let V be the set of terminals to be connected and n be the number of terminals in the set. Let m be the number of FSTs in F. Each FST $f_i \in F$ is associated with a binary variable x_i indicating whether f_i is taken as a part of the RSMT. We use $|f_i|$ to denote the size of f_i, that is, the number of terminals connected by f_i, and use l_i to denote the length of f_i. In the following, $(A : B)$ means $\{f_i \in F : f_i \cap A \neq \emptyset \wedge f_i \cap B \neq \emptyset\}$. The ILP formulation is as follows.

Minimize:
$$\sum_{i=1}^{m} l_i \times x_i. \tag{43.5}$$

Subject to:
$$\sum_{i=1}^{m} x_i(|f_i| - 1) = n - 1, \tag{43.6}$$

$$\sum_{i: f_i \in (X:V-X)} x_i \geq 1 \quad \forall X \subset V \tag{43.7}$$

$$\sum_{i: f_i \cap X \neq \emptyset} x_i(|f_i \cap X| - 1) \leq |X| - 1 \quad \forall X \subset V \wedge |X| \geq 2. \tag{43.8}$$

In the ILP, the objective function (43.5) is to minimize the total length of selected FSTs. Constraint (43.6) is the *total degree constraint* that requires the right number of FSTs in order to span V. Constraints (43.7) are the *cutset constraints*. The constraints ensure that for any cut $(X : V - X)$ of the terminal set, there should be at least one selected FST to connect them. Constraints (43.8) are the *subtour elimination constraints* that eliminate any cycle in the solution. Since there is an exponential number of cutset constraints and subtour elimination constraints, they are considered in an incremental way and the ILP is solved by a branch-and-cut algorithm with the lower bound provided by linear programming (LP) relaxation, that is by relaxing integrality of variable x_i to $0 \leq x_i \leq 1$. At the beginning of the algorithm, only some simple constraints are considered. Other constraints are added by separation methods. The separation problems can be solved in polynomial time by finding minimum cuts in some graphs. It is shown in [41] that Warme's FST concatenation algorithm combined with Zachariasen's FST generation algorithm can solve instances with as many as 2000 terminals in a reasonable amount of time.

More recently, Polzin and Daneshmand [42] presented an efficient alternative for the concatenation phase. The set of FSTs are further decomposed into a set of edges. An algorithm which is originally designed for general graphs can then be applied to construct a RSMT. Polzin and Daneshmand showed that their algorithm, in most cases, is faster than Warme's algorithm. They claimed that the superiority is due to the sophisticated reduction techniques they developed to reduce the size of the problem instance.

43.5 OBSTACLE-AVOIDING RSMT

A more general version of the RSMT problem is to consider obstacles. We consider rectilinear obstacles that have all its boundary edges either horizontal or vertical. The obstacle-avoiding RSMT (OARSMT) problem asks for a rectilinear Steiner tree with minimum total length that connects all the given terminals in the presence of obstacles. No edge in the tree can intersect with any obstacle, but it can be point-touched at a corner or line-touched on an edge of an obstacle. The OARSMT problem is of practical interest because such obstacles exist in VLSI designs (e.g., macro cells, IP blocks, and pre-routed nets). The OARSMT problem is NP-complete as it is a generalization of the RSMT problem.

Analogous to the Hanan grid for the RSMT problem, Ganley and Cohoon [43] proposed the escape graph for the OARSMT problem. The escape graph consists of two types of segments. The first type is the segments that extend from the terminals in the vertical and horizontal directions, until an obstacle boundary is met. The second type of segments can be obtained by extending boundary segments of each obstacle until an obstacle boundary is met. An example of the escape graph is shown in Figure 43.16. It is proven in [43] that for any OARSMT problem, there is at least one optimal solution composed only of the escape segments in the escape graph. Therefore, by using the escape graph, the geometric OARSMT problem can be transferred into a graph problem.

43.5.1 Heuristics

Since the OARSMT problem is NP-complete, most of the previous works have been focused on the development of heuristics. These heuristics can be generally classified into three categories, namely sequential approach, maze routing–based approach, and connection graph based approach.

43.5.1.1 Sequential Approach

The sequential approach, also called the construction-by-correction approach, consists of two steps. In the first step, a RSMT is constructed without considering any of the obstacles. This step can be done by using any of the aforementioned RSMT algorithms. In the second step, edges that overlap with obstacles are found and replaced by edges going around the obstacles. Yang et al. [44] proposed a complicated 4-step heuristics to remove the overlaps in the second step. The sequential approach is popular in industry due to its simplicity and efficiency. However, this approach usually cannot provide solution with good quality because it lacks a global view of the obstacles.

43.5.1.2 Maze Routing–Based Approach

The maze-routing approach is originally proposed by Lee [45] for making connection between two points. Since then, several multiterminal variants have been proposed. Despite early works that incur unsatisfiable solution quality, recent developments on maze-routing demonstrate its effectiveness on the OARSMT problem. Hentschke et al. [46] presented AMAZE, a fast maze routing–based algorithm to build Steiner trees. The algorithm starts from a

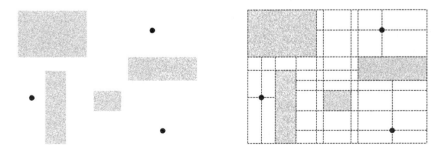

Figure 43.16 Escape graph.

particular terminal and grow the tree by connecting one terminal at a time by using A* search. Li and Young [47] proposed another maze routing-based approach for the OARSMT problem. Similar to Hentschke's algorithm, during the construction of the tree, terminals are added one by one to the existing tree. The key difference is that, in the work by Li and Young, instead of adding only one path between terminals, multiple paths will be kept and the path selection is delayed until all the terminals are reached. During this process, a number of candidate Steiner points can be generated. A MST is then constructed to connect all the Steiner points and the terminals. By deleting dangling Steiner points, an OARSMT can be obtained. Although this approach can provide solutions with high quality, the space and time complexities are relatively high which limit its applications to large-scale problems. Recently, Liu et al. [48] extended Li's work by using a simpler graph and showed a very competitive performance in both solution quality and run time.

43.5.1.3 Connection Graph-Based Approach

Most of the recent approaches on the OARSMT problem are graph-based algorithms where an OARSMT is built based on a connection graph (not necessary rectilinear) that captures the global blockage information. Shen et al. [49] proposed to use the obstacle-avoiding spanning graph. The obstacle-avoiding spanning graph can be formed by making connections between terminals and obstacle corners. Shen et al. showed that the graph contains only $O(n)$ edges and is much simpler than the escape graph. A MST in the graph can be easily found. The OARSMT can then be generated by rectilinearize and steinerize the MST. Lin et al. [50] extended Shen's approach by identifying many essential edges, which can lead to more desirable solutions in the construction of the obstacle-avoiding spanning graph. They proved the existence of a rectilinear shortest path between any two terminals in the new graph. With this property, their algorithm is able to find solutions with higher quality. However, the number of edges, in the worst case, is increased to $O(n^2)$. Therefore, the time complexity of their algorithm is $O(n^3)$. Long et al. [51] presented an efficient $O(n\log n)$ four-step algorithm to construct an OARSMT. They proposed a sparser graph model and efficient local and global refinements to improve the solution quality. Liu et al. [52] proposed another $O(n\log n)$ algorithm based on the generation of critical paths. Recently, Ajwani et al. [53] presented the FOARS, an FLUTE-based top-down approach for the OARSMT problem. They apply the obstacle avoiding spanning graph to partition the problem and construct the OARSMT by using the obstacle-aware version of FLUTE. The time complexity of their algorithm is also $O(n\log n)$.

43.5.2 Exact Algorithms

In comparison with heuristics, there has been relatively less research on exact algorithms for the OARSMT problem. Maze-routing [45] can give optimal solutions to two-terminal instances. Along with the escape graph, Ganley and Cohoon [43] presented a topology enumeration scheme to construct optimal three-terminal and four-terminal OARSMTs.

For multiterminal instances, a natural idea is to make use of the two-phase exact algorithm (i.e., generate FSTs in the first phase and then concatenate them in the second phase) which is originally proposed for the RSMT problem. However, this algorithm cannot be directly applied when obstacles exist in the plane. An example is shown in Figure 43.17. In the absence of obstacles, FST has some specific topology, as characterized by Hwang, which consists of a backbone and alternating incident segments connecting the terminals. In contrast, the structures of FSTs in the presence of obstacles can be very different. Therefore, the construction of FSTs in the presence of obstacles can itself be a difficult problem, which limits the application of the two-phase algorithm for the OARSMT problem.

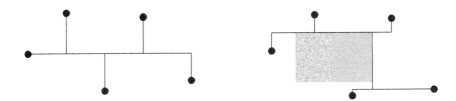

Figure 43.17 Example of FST in the presence of an obstacle.

Li et al. [54, 55] presented a pioneer work to extend the two-phase approach to solve the OARSMT problem. The key observation is that, by adding the so-called *virtual terminals*, the structures of FSTs can be greatly simplified. For each obstacle, four virtual terminals are added to its four corners as shown in Figure 43.18. We use T to denote the set of virtual terminals added. The direct impact of adding virtual terminals is that FSTs can be further decomposed into smaller FSTs by splitting at these virtual terminals. In Figure 43.19, the FST can be decomposed into a set of five smaller FSTs each of which is of simple structure. These smaller FSTs are called FSTs with blockages.

Let t be a rectilinear Steiner tree. A tree t' is equivalent to t if and only if t' can be obtained from t by shifting or flipping some edges which have no nodes on them. With the concept of equivalent trees, a FST f with blockage over a set of terminals $T_f \subseteq (V + T)$ can be defined as follows:

1. f is an OARSMT over T_f.

2. Every terminal in T_f has degree one in f and all its equivalent trees.

3. All the equivalent trees of f cannot contain forbidden edges as shown in Figure 43.20. (Otherwise, splitting can be done to further decompose the FST.)

With the definition, it can be easily verified that an OARSMT is a union of FSTs with blockages. An important theoretical result is that the structures of FSTs with blockages are the same as those of FSTs in the absence of obstacles. This indicates that, by adding virtual terminals, we can use the two-phase approach to construct an OARSMT efficiently.

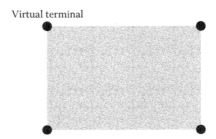

Figure 43.18 Locations of virtual terminals of an obstacle.

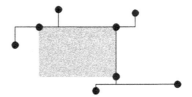

Figure 43.19 Decomposition of an FST.

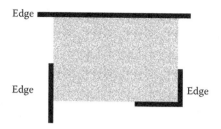

Figure 43.20 Forbidden edges in a FST with blockages.

In the first phase, we generate a sufficient set of FSTs with blockages. In the second phase, we identify and combine a subset of FSTs with minimum total length such that all real terminals are interconnected. For simplicity, we will use FSTs to denote FSTs with blockages in the following.

To generate FSTs with more than two terminals, a modified version of the Zachariasen's algorithm [34] is used. For the empty diamond property, when there are obstacles and virtual terminals, the points, which cannot lie in the empty region are the real terminals only. For the empty corner rectangle, we only need to consider those real terminals that can be projected on the two segments of the corner without intersecting with any obstacle.

To generate FSTs with exactly two terminals, a more efficient method is proposed. These FSTs can be divided into two types. The first type is FSTs connecting two real terminals. The second type is the FSTs connecting at least one virtual terminal. The first type of FSTs can be generated based on the OARMST. The second type of FSTs can be generated as follows. For each virtual terminal u, we divide its surrounding area into eight regions as shown in Figure 43.6. In every region, we find the real terminal v that has the shortest Manhattan distance (d_{uv}) from u and the rectangular area covered by u and v has no obstacles. Then, the edge connecting v and u is a two-terminal FST candidate. In this region, we also find those virtual terminals w with distance $d_{uw} \leq d_{uv}$ and the rectangular area covered by u and w is obstacle free. Then, the edge connecting u and w will also be included as a FST candidate.

For the FST concatenation phase, it can be formulated as an ILP. In the following, let F be the set of all FSTs found. Let V be the set of all real terminals and T be the set of all virtual terminals that survive after pruning. Let n be the number of real terminals, m be the number of FSTs in F and p be the number of virtual terminals. Each FST $f_i \in F$ is associated with a binary variable x_i indicating whether f_i is taken as a part of the OARSMT. Besides, there are binary variables y_i for $i = 1 \ldots p$ indicating whether virtual terminal $v_i \in T$ is connected in the OARSMT. We use $|f_i|$ to denote the size of f_i, that is, the number of terminals (including virtual ones) connected by f_i, and use l_i to denote the length of f_i. The ILP formulation is as follows.

Minimize:
$$\sum_{i=1}^{m} l_i \times x_i. \tag{43.9}$$

Subject to:
$$\sum_{i=1}^{m} x_i(|f_i| - 1) = n - 1 + \sum_{i=1}^{p} y_i, \tag{43.10}$$

$$2y_j \leq \sum_{i: t_j \in f_i} x_i \quad \forall t_j \in T, \tag{43.11}$$

$$4y_j \geq \sum_{i: t_j \in f_i} x_i \quad \forall t_j \in T, \tag{43.12}$$

$$\sum_{i:f_i\in(X:V+T-X)} x_i \geq 1 \qquad (43.13)$$

$$\forall X \text{ s.t. } (X \subseteq V+T) \wedge (X \cap V \neq V) \wedge (X \cap V \neq \emptyset)$$

$$\sum_{i:f_i\cap X\neq\emptyset} x_i(|f_i \cap X| - 1) \leq |X \cap V| + \sum_{i:t_i\in X} y_i - 1 \qquad (43.14)$$

$$\forall X \text{ s.t. } (X \subset V+T) \wedge (X \cap V \neq \emptyset) \wedge (|X| \geq 2)$$

$$\sum_{i:f_i\cap X\neq\emptyset} x_i(|f_i \cap X| - 1) \leq \sum_{i:t_i\in X} y_i - \max_{i:t_i\in X} y_i \qquad (43.15)$$

$$\forall X \text{ s.t. } (X \subseteq T) \wedge (|X| \geq 2)$$

Constraint (43.9) is the *total degree constraint*. $\sum_{i=1}^{p} y_i$ is added to indicate the number of selected virtual terminals. Constraints (43.11) and (43.12) bound the degree of any selected virtual terminal to be two, three, or four. Constraints (43.13) are the cutset constraints. We require $X \cap V \neq \emptyset$ and $X \cap V \neq V$, because we do not need to ensure the connectivity of the virtual terminals. Constraints (43.14) and (43.15) are the subtour elimination constraints. In (43.14), we consider those sets $X \cap V \neq \emptyset$. Since y_i tells whether t_i is selected, $|X \cap V| + \sum_{i:t_i\in X} y_i$ gives the exact number of selected terminals including virtual ones in X. In (43.15), we use $\sum_{i:t_i\in X} y_i$ to indicate the number of selected terminals in X. Since it is possible that the number of selected terminals in X is equal to zero, we do not simply subtract one from the right hand side of the inequality. Instead, the term $\max_{i:t_i\in X}(y_i)$ is used to ensure that the inequality is not binding when the number of selected terminals in X is zero.

Warme's branch-and-cut algorithm [33] is extended to solve the ILP. Efficient polynomial time separation algorithms are also developed to identify violated constraints. Experimental results showed that the proposed method is able to handle problems with hundreds of terminals in the presence of multiple obstacles, generating optimal solution in a reasonable amount of time. However, the performance is severely affected by the number of obstacles and all the solvable test cases contain less than one hundred obstacles. Moreover, the algorithm can only handle rectangular obstacles.

Recently, Huang and Young [56,57] extended this approach to handle complex rectilinear obstacles. Virtual terminals are added to the so-called *essential edges* of the obstacles. They proved that, after adding virtual terminals, the FSTs will follow four simple structures as shown in Figure 43.21. The first two structures are exactly the same as those in [16] and [55]. However, in the presence of complex obstacles, the FSTs have two additional structures. A main characteristic of these two additional structures is that the last corner connecting two Steiner points or one Steiner point and one terminal is blocked by some obstacles. The similarities in FSTs indicate that the two-phase approach can be used to solve the OARSMT

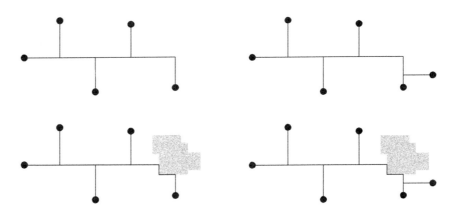

Figure 43.21 (a,b) FST structures in the presence of complex rectilinear obstacles.

problem in the presence of complex obstacles. Huang and Young further proposed to use an incremental way to handle obstacles. At the beginning of the algorithm, a RSMT without considering any obstacle is constructed. Then, check for obstacles that overlap with the solution. All such obstacles will be added into consideration and a new iteration begins to construct an OARSMT that avoid all these obstacles. This process iterates until no overlapping obstacle can be found. Empirical study showed that the algorithm is able to generate optimal solutions for test cases with up to two thousand obstacles.

43.6 APPLICATIONS

The RSMT problem has many applications in VLSI physical design. Although there are other applications such as heating system in building design [58]. They are not as popular as those in VLSI design.

In the VLSI physical design flow, one important step is routing. The specification of a routing problem usually consists of a set of modules, a netlist, and the area available for routing. Each module is with a set of terminals and has a fixed position. A netlist is a set of nets. Each net consists of a set of terminals that need to be made electronically equivalent (i.e., connected by wires). In modern VLSI design, there exist multiple routing layers, and each routing layer has a predefined direction (either horizontal or vertical) and routing capacity. Connectivity between layers can be achieved by vias. The objective of routing is to create an interconnection among the terminals of same nets such that the total wire length (i.e., routing resource) is minimized. For high performance design, it is also necessary to consider other requirements such as timing budget, signal integrity, and manufacturability issues. An example of the routing problem is shown in Figure 43.22.

In VLSI deign, routing is usually performed in two stages: global routing followed by detailed routing. The task of global routing is to first partition the routing region into tiles and then determine a loose tile-to-tile route for each net. In this stage, terminals within the same tile are assumed to be at the center of the tile. It is also common to represent a 3D routing problem as a 2D problem and perform layer assignment as a post-processing step. Therefore, the routing of a net can be realized by constructing a RSMT. A common approach for global routing algorithms is to first generate RSMTs for all the nets [59]. Since, RSMT only minimizes the wire length, it is possible that in some tiles, the number of wires may exceed the routing capacity creating some congested regions. In such cases, nets that are routed through the congested region will be ripped up and rerouted by using congestion-aware RSMT [60] or maze-routing algorithm. In the routing region, there can be routing obstacles such as macro cells, IP blocks, and pre-routed nets. Any route across an obstacle will lead to a violation to the capacity constraint. In such cases, OARSMT algorithms can be

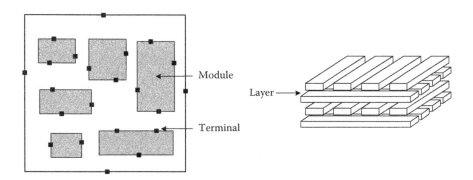

Figure 43.22 Example of routing problem.

used to deal with the obstacles. Given a global routing solution, detailed routing determines the actual geometric layout of each net (i.e., exact tracks, via position, and layer) within the assigned routing regions. In this stage, the RSMTs can also be used to guide the routing [61] to minimize the wire length and via usage.

Despite extensive applications in the routing stage, RSMTs can also find its application in an even earlier stage in VLSI design flow, such as floorplanning and placement. In floorplanning and placement, modules are not fixed and their positions are to be determined. A solution to the problem is a layout that specifies the location of each module such that there is no overlap. A good floorplanning or placement solution should be routable (i.e., be successfully routed in the later routing stage) by using the smallest amount of routing resources. This necessitates congestion and wire length estimations during floorplanning and placement. The estimation can be done by performing routing, but it is computationally too expensive. Therefore, using RSMTs as an approximation becomes an efficient alternative and is adopted by many estimation approaches [62]. Another target of floorplanning and placement is to achieve good timing. As deep submicron technology advances, interconnect delay is becoming increasingly dominant over transistor and logic delay. Timing estimation has to consider both interconnect and gate delays in order to be accurate. This requires actual topology of each net which is usually approximated by using RSMTs [63].

43.7 SUMMARY

In this chapter, we presented a survey of the RSMT problem. Ever since its first proposal, the problem has been of both theoretical and practical interests for nearly half a century. Substantial efforts have been made to develop efficient algorithms, prove performance bound of approximations, and solve the problem exactly. Being a premier application of the RSMT problem, the increasing demand on the design automation of VLSI has greatly promoted the research development of the problem. Recent algorithmic studies on the RSMT problem have resulted in a series of algorithms with excellent performance. For a single net with hundreds of terminals, these algorithms have demonstrated excellent tradeoffs between runtime and solution quality. As the process technology advances, the number of nets in a design can easily be millions and is still growing. Highly efficient RSMT algorithms are still in great demand. Besides minimizing the wire length, future research on RSMT should also be adapted to the new requirements of VLSI design, such as obstacle avoidance, timing constraints, signal integrity, and manufacturability issues.

References

[1] R. Courant and H. Robbins. *What is Mathematics?* Oxford University Press, New York, 1941.

[2] M. Hanan. On steiner minimal trees with rectilinear distance. *SIAM J. Appl. Math.*, **14** (1966), 225–265.

[3] M. Garey and D. Johnson. The rectilinear Steiner tree problem is NP-complete. *SIAM J. Appl. Math.*, **32** (1977), 826–834.

[4] J. H. Halton J. Beardwood, and J. M. Hammersley. The shortest path through many points. *Proc. Cambridge Phil. Soc.*, **55** (1959), 299–327.

[5] R. L. Graham and F. K. Hwang. Remarks on Steiner minimal trees I. *Bull. Inst. Math. Acad. Sinica*, **4** (1976), 177–182.

[6] F. R. K. Chung and F. K. Hwang. A lower bound for the Steiner tree problem. *SIAM J. Appl. Math.*, **34** (1978), 27–36.

[7] D. Z. Du and F. K. Hwang. A new bound for the Steiner ratio. *Trans. Am. Math. Soc.*, **278** (1983), 137–148.

[8] F. R. K. Chung and R. L. Graham. A new lower bound for the Steiner minimal trees. *Ann. N. Y. Acad. Sci.*, **440** (1985), 325–346.

[9] E. N. Gilbert and H. O. Pollak. Steiner minimal trees. *SIAM J. Appl. Math.*, **16** (1968), 323–345.

[10] H. O. Pollak. Some remarks on the Steiner problem. *J. Comb. Theory, Ser. A*, **24** (1978), 278–295.

[11] E. N. Yao, D. Z. Du, and F. K. Hwang. A short proof of a result of Pollak on Steiner minimal trees. *J. Comb. Theory, Ser. A*, **32** (1982), 356–400.

[12] J. H. Rubinstein and D. A. Thomas. The Steiner ratio conjecture for six points. *J. Comb. Theory, Ser. A*, **58** (1991), 54–77.

[13] D. Z. Du and F. K. Hwang. The Steiner ratio conjecture of Gilbert-Pollak is true. *Proc. Natl. Acad. Sci. USA*, **87** (1990), 9464–9466.

[14] D. Z. Du and F. K. Hwang. A proof of Gilbert-Pollak conjecture on the Steiner ratio. *Algorithmica*, **7** (1992), 121–135.

[15] A. O. Ivanov and A. A. Tuzhilin. The Steiner ratio Gilbert-Pollak conjecture is still open: Clarification statement. *Algorithmica*, **62** (2012), 630–632.

[16] F. K. Hwang. On Steiner minimal trees with rectilinear distance. *SIAM J. Appl. Math.*, **30** (1976), 104–114.

[17] J. S. Salowe. A simple proof of the planar rectilinear Steiner ratio. *Oper. Res. Lett.*, **12** (1992), 271–274.

[18] F. K. Hwang. An $O(nlogn)$ algorithm for rectilinear minimal spanning trees. *J. Assoc. Comput. Mach.*, **26** (1979), 177–182.

[19] A. C. C. Yao. On constructing minimal spanning trees in k-dimensional spaces and related problems. *SIAM J. Comput.*, **11** (1982), 721–736.

[20] L. J. Guibas and J. Stolfi. On computing all north-east nearest neighbor in the L_1 metric. *Inf. Process. Let.*, **17** (1983), 219–223.

[21] H. Zhou, N. Shenoy, and W. Nicholls. Efficient spanning tree construction without delaunay triangulation. *Inf. Process. Let.*, **81** (2002), 271–276.

[22] D. S. Richards. On the effectiveness of greed heuristics for the rectilinear steiner tree problem. Technical report, University of Virginia, Charlottesville, VA, 1991.

[23] G. Vijayan, J. M. Ho, and C. K. Wong. New algorithms for the rectilinear steiner tree problem. *IEEE Trans. Comput.-Aided Des.*, **9** (1990), 185–193.

[24] M. Borah, R. M. Owens, and M. J. Irwin. An edge-based heuristic for Steiner routing. *IEEE Trans. Comput.-Aided Des. Integr. Circuits Syst.*, **13** (1994), 1563–1568.

[25] H. Zhou. Efficient Steiner tree construction based on spanning graphs. In *Proc. Int. Symp. Phys. Des.*, pages 152–157. ACM, Monterey, CA, (2003).

[26] A. Kahng and G. Robins. On performance bounds for two rectilinear Steiner tree heuristics in arbitrary dimension. *IEEE Trans. Comput.-Aided Des. Integr. Circuits Syst.*, **11** (1992), 1462–1465.

[27] A. Kahng and G. Robins. A new class of iterative Steiner tree heuristics with good performance. *IEEE Trans. Comput.-Aided Des. Integr. Circuits Syst.*, **11** (1994), 893–902.

[28] G. Georgakopoulos and C. H. Papadimitriou. The 1-Steiner tree problem. **8** (1987), 122–130.

[29] J. Griffith, G. Robins, J. S. Salowe, and T. Zhang. Closing the gap: Nearoptimal Steiner trees in polynomial time. *IEEE Trans. Comput.-Aided Des. Integr. Circuits Syst.*, **13** (1994), 1351–1365.

[30] C. Chu and Y. C. Wong. FLUTE: Fast lookup table based rectilinear Steiner minimal tree algorithm for VLSI design. *IEEE Trans. Comput.-Aided Des. Integr. Circuits Syst.*, **27** (2008), 70–83.

[31] F. K. Hwang, D. S. Richards, and P. Winter. *The Steiner Tree Problem*. Number 53. Elsevier, Amsterdam, the Netherlands, 1992.

[32] J. S. Salowe and D. M. Warme. Thirty-five-point rectilinear Steiner minimal trees in a day. *Networks*, **25** (1995), 69–87.

[33] D. M. Warme. A new exact algorithm for rectilinear steiner minimal trees. Technical report, System Simulation Solutions, Alexandria, VA, 1997.

[34] M. Zachariasen. Rectilinear full Steiner tree generation. *Networks*, **33** (1999), 125–143.

[35] P. Winter. An algorithm for the Steiner problem in the euclidean plane. *Networks*, **15** (1985), 323–345.

[36] E. J. Cockayne and D. E. Hewgill. Exact computation of Steiner minimal trees in the plane. *Inf. Process. Lett.*, **22** (1986), 151–156.

[37] E. J. Cockayne and D. E. Hewgill. Improved computation of plane Steiner minimal trees. *Algorithmica*, **7** (1992), 219–229.

[38] P. Winter and M. Zachariasen. Euclidean Steiner minimum trees: an improved exact algorithm. *Networks*, **30** (1997), 149–166.

[39] J. L. Ganley and J. P. Cohoon. Optimal rectilinear Steiner minimal trees in $O(n^2 2.62^n)$ time. In *Proc. Canad. Conf. Comput. Geo.*, pages 308–313. Saskatoon, Saskatchewan, Canada, 1994.

[40] U. Fößmeier and M. Kaufmann. On exact solutions for the rectilinear Steiner tree problem part I: Theoretical results. *Algorithmica*, **26** (2000), 68–99.

[41] D. M. Warme, P. Winter, and M. Zachariasen. Exact algorithms for plane steiner tree problems: A computational study. In D. Z. Du, J. M. Smith, and J. H. Rubinstein, editors, *Advances in Steiner Trees*, pages 81–116. Kluwer Academic Publishers, Boston, MA, 2000.

[42] T. Polzin and S. V. Daneshmand. On Steiner trees and minimum spanning trees in hypergraphs. *Oper. Res. Lett.*, **31** (2003), 12–20.

[43] J. L. Ganley and J. P. Cohoon. Routing a multi-terminal critical net: Steiner tree construction in the presence of obstacles. In *Proc. of IEEE ISCAS*, pages 113–116. IEEE, London, 1994.

[44] Y. Yang, Q. Zhu, T. Jing, X. Hong, and Y. Wang. Rectilinear Steiner minimal tree among obstacles. In *Proc. Intl. Conf. ASIC*, pages 348–351, 2003.

[45] C. Y. Lee. An algorithm for connections and its application. *IRE Trans. Electron. Comput.*, **EC-10**(3) (1961), 346–365.

[46] R. Hentschke, J. Narasimham, M. Johann, and R. Reis. Maze routing Steiner trees with effective critical sink optimization. In *Proc. Int. Symp. Phys. Des.*, pages 135–142. ACM, Austin, TX, 2007.

[47] L. Li and Evangeline F. Y. Young. Obstacle-avoiding rectilinear Steiner tree construction. In *Proc. Int. Conf. Comput.-Aided Des.*, pages 523–528. IEEE, San Jose, CA, 2008.

[48] C. H. Liu, S. Y. Kuo, D. T. Lee, C. S. Lin, J. H. Weng, and S. Y. Yuan. Obstacle-avoiding rectilinear Steiner tree construction: A steiner-point-based algorithm. *IEEE Trans. on Comput.-Aided Des. Integr. Circuits Syst.*, **31** (2012), 1050–1060.

[49] Z. Shen, C. Chu, and Y. Li. Efficient rectilinear Steiner tree construction with rectilinear blockages. In *Proceedings ICCD*, pages 38–44. IEEE, San Jose, CA, 2005.

[50] C. W. Lin, S. Y. Chen, C. F. Li, Y. W. Chang, and C. L. Yang. Efficient obstacle-avoiding rectilinear Steiner tree construction. In *Proc. Int. Symp. Phys. Des.*, pages 380–385. ACM, Austin, TX, 2007.

[51] J. Y. Long, H. Zhou, and S. O. Memik. EBOARST: An efficient edge-based obstacle-avoiding rectilinear Steiner tree construction algorithm. *IEEE Trans. on Comput.-Aided Des. Integr. Circuits Syst.*, **27** (2008), 2169–2182.

[52] C. H. Liu, S. Y. Yuan, S. Y. Kuo, and Y. H. Chou. An $O(n \log n)$ path-based obstacle-avoiding algorithm for rectilinear Steiner tree construction. In *Proc. Des. Autom. Conf.*, pages 314–319. ACM, San Francisco, CA, 2009.

[53] G. Ajwani, C. Chu, and W. K. Mak. FOARS: FLUTE based obstacle-avoiding rectilinear Steiner tree construction. In *Proc. Int. Symp. Phys. Des.*, pages 27–34. ACM, San Francisco, CA, 2010.

[54] L. Li, Z. Qian, and Evangeline F. Y. Young. Generation of optimal obstacle-avoiding rectilinear Steiner minimum tree. In *Proc. Int. Conf. Comput.-Aided Des.*, pages 21–25, 2009.

[55] T. Huang, L. Li, and Evangeline F. Y. Young. On the construction of optimal obstacle-avoiding rectilinear Steiner minimum trees. *IEEE Trans. on Comput.-Aided Des. Integr. Circuits Syst.*, **30** (2011), 718–731.

[56] T. Huang and Evangeline F. Y. Young. Obstacle-avoiding rectilinear Steiner minimum tree construction: An optimal approach. In *Proc. Int. Conf. Comput.-Aided Des.*, pages 610–613. IEEE, San Jose, CA, 2010.

[57] T. Huang and Evangeline F. Y. Young. An exact algorithm for the construction of rectilinear Steiner minimum trees among complex obstacles. In *Proc. Des. Autom. Conf.*, pages 164–169. ACM, San Diego, CA, 2011.

[58] J. M. Smith and J. S. Liebman. Steiner trees, Steiner circuits and the interference problem in building design. *Eng. Opt.*, **4** (1979), 15–36.

[59] M. D. Moffitt, J. A. Roy, and I. L. Markov. The coming of age of (academic) global routing. In *Proc. Int. Symp. Phys. Des.*, pages 148–155. ACM, Portland, OR, 2008.

[60] M. Pan and C. Chu. FastRoute: A step to integrate global routing into placement. In *Proc. Int. Conf. Comput.-Aided Des.*, pages 464–471. IEEE, San Jose, CA, 2006.

[61] Y. Zhang and C. Chu. RegularRoute: An efficient detailed router with regular routing patterns. In *Proc. Int. Symp. Phys. Des.*, pages 45–52. ACM, Santa Barbara, CA, 2011.

[62] J. A. Roy and I. L. Markov. Seeing the forest and the trees: Steiner wirelength optimization in placement. *IEEE Trans. on Comput.-Aided Des. Integr. Circuits Syst.*, **26**(4) (2007), 632–644. ACM, San Diego, CA.

[63] H. Chen, C. Qiao, F. Zhou, and C. K. Cheng. Refined single trunk tree: A rectilinear Steiner tree generator for interconnect prediction. In *Proc. Int. Work. Sys. Interc. Pred.* pages 85–89, 2002.

CHAPTER 44

Fixed-Parameter Algorithms and Complexity

Venkatesh Raman

Saket Saurabh

CONTENTS

44.1	Introduction		1142
44.2	Algorithmic Techniques to Prove FPT		1144
	44.2.1	Kernelization	1145
		44.2.1.1 Vertex Cover	1146
		44.2.1.2 Kernel via Crown Decomposition—Max-SAT	1147
		44.2.1.3 Other Recent Upper Bounds	1149
		44.2.1.4 Kernelization Lower Bounds	1149
	44.2.2	Bounded Search Trees	1152
		44.2.2.1 Vertex Cover	1153
		44.2.2.2 Feedback Vertex Set	1153
		44.2.2.3 Vertex Cover above LP	1154
	44.2.3	Iterative Compression	1156
	44.2.4	Randomized Fixed-Parameter Algorithms	1157
		44.2.4.1 α-Covering Based Randomized Algorithms	1158
		44.2.4.2 Color Coding	1159
		44.2.4.3 Chromatic Coding	1160
	44.2.5	Important Separators	1162
		44.2.5.1 Important Vertex Separators in Undirected Graphs	1162
		44.2.5.2 Algorithm for Multiway Cut	1165
	44.2.6	Well-Quasi-Ordering	1166
	44.2.7	Bounded Treewidth Machinery	1168
	44.2.8	Subexponential Algorithms and Bidimensionality	1170
44.3	Ecology of Parameters		1171
	44.3.1	Parameterizing beyond the Guarantee Bounds	1171
	44.3.2	Structural Parameters	1172
	44.3.3	Backdoors to Satisfiability	1173
44.4	Parameterized Intractability		1173
	44.4.1	Example Reductions	1174
	44.4.2	Exponential Time Hypothesis and Stronger Lower Bounds	1178
		44.4.2.1 Exponential Time Hypothesis	1178
		44.4.2.2 Strong Exponential Time Hypothesis: $s_\infty=1$	1181
		44.4.2.3 Lower Bound on FPT Algorithms	1181
		44.4.2.4 $W[1]$-Hard Problems	1183
		44.4.2.5 Problems Parameterized by Treewidth	1183

44.5 FPT and Approximation 1185
    44.5.1 Approximation in FPT Time for *W*-Hard Problems 1186
    44.5.2 Approximation Parameterized by Cost 1186
44.6 Conclusions 1186

THE THEORY of parameterized computational complexity, pioneered by Downey and Fellows, is motivated by the observation that many NP-complete problems have as input several parameters, some of which are likely to be small in practice. While the classical notion of feasible computation is associated with an algorithm whose running time is a polynomial in the input size, parameterized complexity strengthens it by allowing exponential running time in the (small) parameters associated with the input.

Since the advent of this paradigm, a number of developments have happened in the last 30 years. For example, the theory has helped explain to some extent, why some problems are solvable reasonably well in practice despite being NP-complete. The theory has also led to the development of interesting new algorithmic techniques, new combinatorial, computational and complexity theoretic questions and answers. While showing NP-completeness has been a classical direction to show that a problem is unlikely to have a polynomial time algorithm, parameterized complexity and connections to exponential time hypothesis provide a method to show tighter lower bounds for such problems.

In this chapter, we survey the recent developments in the area starting from the basic notions. We highlight algorithmic techniques including iterated compression, kernelization, and separator–based algorithms. We also outline how the same problem can have different complexities when parameterized differently and also discuss connections to approximation. The notions of fixed-parameter intractability, including connections to the exponential time hypothesis on satisfiability, and lower bounds on kernelizations are also discussed.

44.1 INTRODUCTION

It is a widely held notion that polynomial-time computability captures feasible computation and NP-completeness identifies hard problems in this framework. However, the NP-hard problems can not be wished away and have to be handled algorithmically. One prominent approach to dealing with NP-hard optimization problems is to settle for polynomial-time computable (good) approximate solutions. Another, more classical, approach is to identify subclasses of instances of NP-hard problems, which are feasibly solvable. Both approaches have attained a reasonable degree of success [1–3]. Parameterized complexity can be considered as a refinement of the latter approach. Here exact solutions are sought for, and that the structure in the input is *parameterized*. Algorithms are designed and analyzed as a function of both the input size and the parameter.

It is not hard to see that useful parameters (besides the input size) are abound in an input in practice. Typical graph parameters include a measure of how close the graph is to a tree or a bipartite graph or a planar graph, the maximum degree of the graph, or the size of the solution sought for. An instance for the satisfiability problem has a number of parameters including the number of variables, the number of clauses, the number of 1s in a satisfying assignment sought for, the maximum number of variables in any clause, a measure of how close the instance is to a polynomially solvable instance and so on. In short, every input instance comes with a number of parameters besides the input size, and parameterized complexity provides a framework to perform a *multivariate* algorithmic complexity analysis.

The central notion of feasibility in parameterized complexity is *fixed-parameter tractability*. A computational problem with input x of size n, and a parameter k (k can, in practice represent multiple parameters) is said to be *fixed-parameter tractable* (FPT) if it can be solved by an algorithm running in time $f(k) + n^c$ where c is a constant independent of k, and f is a (typically exponential or worse) function of k alone (Algorithms with running time $f(k)n^c$ are also said to be fixed-parameter tractable, and both notions are equivalent). If the instances occurring in practice have some small parameters and if the problem is FPT under those parameterizations, then they can be solved well in practice.

Since the advent of the paradigm of parameterized complexity, there have been significant progress in obtaining practical algorithms for problems with multiple parameterization. The following is a sample of such algorithms (some of these have been discovered earlier than parameterized complexity came into picture as a paradigm).

- Simplex algorithm gives an $O(n^d)$ (assuming $d < n$) algorithm for the linear programming problem on d variables and n constraints. However, Megiddo [4] gave an algorithm for the problem that takes $O(2^{2^{O(d)}} n)$ time, which makes the problem FPT when parameterized by the number of variables.

- Given an undirected graph G, the achromatic number is the largest number of colors that can be assigned to the vertices of G so that adjacent vertices are assigned different colors and any two different colors are assigned to some pair of adjacent vertices. Given a graph G and an integer k, it is NP-complete to determine whether G has achromatic number at least k. However, it can be determined in $O(f(k) + |E(G)|)$ time whether G has achromatic number at least k for some function $f(k)$ [5].

  Note that such a result is not conceivable for the chromatic number problem as even for a fixed $k > 2$, it is NP-complete to check whether the graph has chromatic number k.

- Given an undirected graph G on n vertices, and an integer k, it is NP-complete [6] to determine whether G has a vertex cover of size at most k; however there is now an $O(kn + (1.2738)^k)$ [7] algorithm to answer this question.

  Furthermore, in polynomial (in n and k) time, one can reduce the graph G to a graph G' with at most $2k$ vertices such that G has a vertex cover of size at most k if and only if G' has one such. G' is said to be a *kernel* for vertex cover.

- Given an undirected graph G on n vertices, and an integer k, it is NP-complete [6] to determine whether G has a feedback vertex set (FVS; a subset of vertices whose deletion makes G acyclic) of size at most k; however there is now an $O(3.83^k n)$ [8] algorithm to answer this question. As in the case of vertex cover, undirected FVS problem has an $O(k^2)$-sized kernel.

  The corresponding version for the directed graph has an $O(4^k k! n)$ algorithm. It is still open whether the directed FVS has a polynomial-sized kernel.

The (first) linear programming example shows that the paradigm is applicable even for polynomially solvable problems. In the other examples above, the parameter is the solution size. The following are some examples where the parameter is different from the solution size.

- Consider the decision version of the maximum cut problem which asks whether a given undirected graph has a partition of the vertex set into two such that at least k edges go across the partition (such edges are said to form a cut).

  It can be shown that in any graph on m edges, it is easy to find a partition such that at least $m/2$ edges go across the partition. Hence the decision version of the problem becomes trivial if $k \leq m/2$ and for larger values of k, $m \leq 2k$ and hence any brute force algorithm becomes a FPT algorithm.

However a more natural parameterization is to parameterize above the guaranteed bound of $m/2$, and it is known [9] that it is FPT to determine whether the graph has a cut of size at least $m/2 + k$.

- For the satisfiability problem where the input is a propositional formula, we discussed a number of parameters earlier in this section. One parameter is how far (in terms of number of variables that need to be fixed) the instance is from an instance which is satisfiable by an all 0 assignment (also called 0-valid formula). It can be determined in $O(2.85^k + n^{O(1)})$ time [10] whether a given propositional formula in conjunctive normal form (CNF) form has at most k variables, setting of whom to 1 can result in a 0-valid formula.

- A natural parameter for graphs is how far the input is from a tree. This measure is captured by a notion called treewidth (see Section 44.2.7 for definitions). It is known that if the treewidth of a graph is w, then the independent set problem can be solved in $2^w + n^{O(1)}$ time [11] (see also Section 44.4.2). Hence independent set is FPT when parameterized by treewidth of the given graph.

Some of the FPT results above are obtained through techniques that have been developed during the last few years to deal with parameterized problems. Some of these techniques are elementary but powerful while others are based on the deep Robertson Seymour graph minor theorems and bounded treewidth machinery (see Section 44.2).

Parameterized versions of the dominating set problem and the clique problem, on the other hand, have only an $n^{\Omega(k)}$ algorithm where k is the size of the solution being sought and n is the size of the input. A completeness theory [12] in this framework explains this qualitative difference (of dominating set and clique compared to other problems mentioned before) that have been developed over the years to prove.

One of the main goals of this chapter is to survey some of the algorithmic techniques, and the lower bound and completeness theory. While we try to be as comprehensive as possible, due to the explosion of work in the area, and our own biases, we are bound to omit some of the research in the area. We refer to the recent monographs [11,13,14], surveys and the community wiki (http://fpt.wikidot.com/fpt-races) for other avenues to read more about the area.

Section 44.2 illustrates the algorithmic techniques through several examples. This section covers major algorithmic techniques including kernelization, bounded search trees, randomization, important separators, automata-based algorithms that use graph minor theory, and algorithms using bounded treewidth machinery. Section 44.3 discusses recent results on multiple parameterizations of the given input. In Section 44.4 we look at the completeness theory of which includes various complexity classes and parameterized reductions. We also discuss lower bound connections to exponential time hypothesis. In Section 44.5, we look at the relationship between fixed-parameter tractability and approximability of optimization problems. In Section 44.6, we conclude with recent developments.

Throughout this chapter, sometimes we refer to $f(k)n^{O(1)}$ as $O^*(f(k))$.

44.2 ALGORITHMIC TECHNIQUES TO PROVE FPT

Demonstrations of FPT sometimes uses novel approaches that shift the complexity burden onto the parameter. Some of these approaches run counter to our established practices of thought in designing polynomial time algorithms. In the parameterized setting, as Downey and Fellows [13] say, "the parameter can be 'sacrificed' in interesting ways."

44.2.1 Kernelization

Kernelization or preprocessing is a commonly used paradigm by practitioner while implementing algorithms. Parameterized complexity gives a formal framework to analyze such preprocessing algorithms. This is a paradigm where a lot of recent exciting progress has happened, in proving both upper and lower bounds.

The idea of this method is to reduce (but not necessarily solve) the given problem instance to an equivalent *smaller-sized* instance in time polynomial in the input size. Note, however that the input instance of a NP-hard problem can not always be reduced (in size of the input) in polynomial time, for we could use such a reduction algorithm repeatedly to actually solve the problem in polynomial time. However, the parameter plays a valuable secondary measure and kernelization talks about reducing the input size to a function of the *parameter* in polynomial time.

We now turn to the formal notion that captures the notion of kernelization, which is what most heuristics do when applied to a problem. A *data reduction rule* for a parameterized language L is a function $\phi : \Sigma^* \times \mathbb{N} \to \Sigma^* \times \mathbb{N}$ that maps an instance (x, k) of L to an equivalent instance (x', k') of L such that

1. ϕ is computable in time polynomial in $|x|$ and k;

2. $|x'| \leq |x|$.

Two instances of L are equivalent if $(x, k) \in L$ if and only if $(x', k') \in L$.

In general, a *kernelization algorithm* consists of a finite set of data reduction rules such that by applying the rules to an instance (x, k) (in some specified order) one obtains an instance (x', k') with the property that $|x'| \leq g(k)$ and $k' \leq g(k)$, for some function g only depending on k. Such a *reduced* instance is called a *problem kernel* and $g(k)$ is called the *kernel size*. Formally, this is defined as follows.

Definition 44.1 *Kernelization, Kernel* [15] *A kernelization algorithm for a parameterized problem $\Pi \subseteq \Sigma^* \times \mathbb{N}$ is an algorithm that, given $(x, k) \in \Sigma^* \times \mathbb{N}$, outputs, in time polynomial in $(|x| + k)$, a pair $(x', k') \in \Sigma^* \times \mathbb{N}$ such that (a) $(x, k) \in \Pi$ if and only if $(x', k') \in \Pi$ and (b) $|x'|, k' \leq g(k)$, where g is some computable function. The output instance x' is called the kernel, and the function g is referred to as the size of the kernel. If $g(k) = k^{O(1)}$, then we say that Π admits a polynomial kernel.*

It is important to mention here that the early definitions of kernelization required that $k' \leq k$. On an intuitive level this makes sense, as the parameter k measures the complexity of the problem—thus the larger the k, the harder the problem. This requirement was subsequently relaxed, notably in the context of lower bounds. An advantage of the more liberal notion of kernelization is that it is robust with respect to polynomial transformations of the kernel. However, it limits the connection with practical preprocessing. All the kernels mentioned in this survey respect the fact that the output parameter is at most the input parameter, that is, $k' \leq k$.

If we have a kernelization algorithm for a problem for which there is some (with any running time) algorithm to decide whether (x, k) is a YES instance, then clearly the problem is FPT, as the reduced instance x is simply a function of k (and independent of the input size n). However, a surprising result is that the converse is also true.

Lemma 44.1 [16] *If a parameterized problem Q is FPT via a computable function then it admits kernelization.*

Proof. Suppose that there is an algorithm deciding if $x \in Q$ in time $f(k) + |x|^c$ for some computable function f and constant c. If $|x|^c \geq f(k)$, then we run the decision algorithm on the instance in time $f(k)+|x|^c \leq 2|x|^c$. If the decision algorithm outputs *yes*, the kernelization algorithm outputs a constant size *yes* instance, and if the decision algorithm outputs *no*, the kernelization algorithm outputs a constant size *no* instance. On the other hand, if $|x|^c < f(k)$, then the kernelization algorithm outputs x. This yields a kernel of size $(f(k))^{1/c}$ for the problem. ■

Thus, in a sense, kernelizability can be another way of defining FPT. However, kernels obtained by this theoretical result are usually of exponential (or even worse) size, while problem-specific data reduction rules often achieve quadratic ($g(k) = O(k^2)$) or even linear-size ($g(k) = O(k)$) kernels. So a natural question for any concrete FPT problem is whether it admits polynomial-time kernelization to a problem kernel that is bounded by a polynomial function of the parameter ($g(k) = O(k^{O(1)})$).

We first illustrate the method of kernelization very briefly on two problems: vertex cover and Max-SAT. Then we outline recent research on upper bounds as well as lower bounds.

44.2.1.1 Vertex Cover

In this section we give two kernels for vertex cover, one of the quadratic size and the other having a linear number of vertices. Formally, the problem is defined as follows.

VERTEX COVER
Instance: A graph $G = (V, E)$, and a non-negative integer k.
Parameter: k.
Problem: Decide whether G has a vertex cover of size at most k.

Quadratic ($O(k^2)$) vertex kernel: Clearly if there is a vertex cover of size at most k, it should have all vertices of degree more than k in the graph (otherwise we need too many vertices to cover edges incident on any of them).

So the preprocessing algorithm [17] is to delete all vertices of degree more than k after including them in the solution, and to delete any isolated vertices created in the process. We repeat this step until no longer possible, at which time we have a graph with maximum degree at most k. If the resulting graph has more than k^2 edges, then the graph can not have a vertex cover of size at most k and so we stop giving a NO answer (or a trivial NO instance). Otherwise, the resulting graph is a *kernel* having at most k^2 edges and $2k^2$ vertices, and we are to find a vertex cover of size at most k minus the number of vertices we have already picked in the solution.

Remark 44.1 *Note that if we also want the subset of vertices in the solution (vertex cover) to induce a connected subgraph, then that is the connected vertex cover problem, and the above kernelization algorithm breaks down. In fact, it turns out that under complexity theoretic assumptions, connected vertex cover does not have polynomial-sized kernel. See Section 44.2.1 for lower bounds on kernel size.*

Linear $(O(k))$ vertex kernel: A kernel for the problem having linear number of vertices can be obtained using linear programming techniques.

The well-known integer linear programming (ILP) formulation for vertex cover is as follows.

ILP FORMULATION OF MINIMUM VERTEX COVER – ILPVC
Instance: A graph $G = (V, E)$.
Feasible Solution: A function $x : V \to \{0, 1\}$ satisfying edge constraints $x(u) + x(v) \geq 1$ for each edge $(u, v) \in E$.
Goal: To minimize $w(x) = \Sigma_{u \in V} x(u)$ over all feasible solutions x.

In the linear programming relaxation of the above ILP, the constraint $x(v) \in \{0, 1\}$ is replaced with $x(v) \geq 0$, for all $v \in V$. For a graph G, we call this relaxation $\mathrm{LPVC}(G)$. Clearly, every integer feasible solution is also a feasible solution to $\mathrm{LPVC}(G)$.

It is known [18] that

- There exists an optimum solution to the $\mathrm{LPVC}(G)$ where each variable takes value 0, 1/2 or 1, and that
- There exists an optimum integral solution that contains all the vertices having value 1 in the LP optimum and none of the vertices having value 0 in the LP optimum.

So, the preprocessing algorithm first solves the LP relaxation and returns a NO answer (or a trivial no instance) if the LP optimum value is more than k (as then the integral optimum can not be at most k). Then, it deletes all the vertices whose values are 1 or 0 in the LP optimum, after including in the solution, all vertices whose values are 1. The *kernel* is the induced subgraph on the vertices with LP value 1/2, and it is clear that there can not be more than $2k$ of them.

$2k - c \log k$ vertex kernel: Recently, a few papers [19–21] independently observed a kernel for vertex cover of size at most $2k - c \log k$ for some constant c. In Narayanaswamy et al. [21], an algorithm with running time $O^*(d^{k-LP})$ has been reported for the vertex cover, where LP is the optimum value of the LP relaxation, and d is a constant [22]. Hence if $k - LP \leq (c \log k)/2$, then one can solve the problem (and hence output a trivial kernel) in polynomial time. Otherwise $k - LP > (c \log k)/2$ which implies that $LP < k - (c \log k)/2$ which implies that the number of vertices (that remain in the kernel) with LP value 1/2 is at most $2k - c \log k$.

Note that the number of edges in the reduced graph remains $O(k^2)$, and it is known that this bound can not be improved to $O(k^{2-\epsilon})$ (see Section 44.2.1 on lower bounds for kernels). Kernels with smaller number of vertices remains open even for special classes of graphs (like planar graphs).

44.2.1.2 Kernel via Crown Decomposition—Max-SAT

Crown decomposition is a general kernelization technique that can be used to obtain kernels for many problems. The technique is based on the classical matching theorems of König [23] and Hall [24].

Definition 44.2 *A crown decomposition of a graph $G = (V, E)$ is a partitioning of V as C, H, and R, where C and H are nonempty and the partition satisfies the following properties.*

1. *C is an independent set.*
2. *There are no edges between vertices of C and R, that is, H separates C and R.*

3. Let E' be the set of edges between vertices of C and H. Then E' contains a matching of size $|H|$.

Set C can be seen as a crown put on head H of the remaining part R of the royal body. Let us remark that the fact that E' contains a matching of size $|H|$ implies that there is a matching of H into C, that is, a matching in the bipartite subgraph $G' = (C \cup H, E')$ saturating all the vertices of H.

The following lemma, which establishes that crown decompositions can be found in polynomial time, is the basis for kernelization algorithms using crown decompositions.

Lemma 44.2 (Crown lemma) [25] *Let G be a graph without isolated vertices and with at least $3k + 1$ vertices. There is a polynomial time algorithm that either finds a matching of size $k + 1$ in G or finds a crown decomposition of G.*

We demonstrate the application of crown decompositions on kernelization for Max-SAT.

Maximum satisfiability. Our next example concerns Max-SAT. We are interested in the following parameterized version of Max-SAT.

> MAX-SAT
> *Instance:* A CNF formula F, and a non-negative integer k.
> *Parameter:* k.
> *Problem:* Decide whether F has a truth assignment satisfying at least k clauses.

Theorem 44.1 [26] *Max-SAT admits a kernel with at most k variables and $2k$ clauses.*

Proof. Let F be a CNF formula with n variables and m clauses. If we assign values to the variables uniformly at random, linearity of expectation yields that the expected number of satisfied clauses is at least $m/2$. Since there has to be at least one assignment satisfying at least the expected number of clauses this means that if $m \geq 2k$ then (F, k) is a *yes* instance. In what follows we show how to give a kernel with $n < k$ variables. Whenever possible we apply a cleaning rule; if some variable does not occur in any clauses, remove the variable.

Let G_F be the *variable-clause* incidence graph of F. That is, G_F is a bipartite graph with bipartition (X, Y). The set X corresponds to the variables of F and Y corresponds to the clauses. For a vertex $x \in X$ we will refer to x as both the vertex in G_F and the corresponding variable in F. Similarly, for a vertex $c \in Y$ we will refer to c as both the vertex in G_F and the corresponding clause in F. In G_F there is an edge between a variable $x \in X$ and a clause $c \in Y$ if and only if either x, or its negation is in c. If there is a matching of X into Y in G_F, then there is a truth assignment satisfying at least $|X|$ clauses. This is true because we can set each variable in X in such a way that the clause matched to it becomes satisfied. Thus at least $|X|$ clauses are satisfied. Hence, in this case if $k \leq |X|$ then (F, k) is a *yes* instance. Else, $k > |X| = n$, which is the desired kernel. We now show that if F has at least $n \geq k$ variables, then we can in polynomial time, either reduce F to an equivalent smaller instance or find an assignment to the variables satisfying at least k clauses.

Suppose F has at least k variables. Using Hall's theorem and a polynomial time algorithm computing maximum-size matching, we can in polynomial time find either a matching of X into Y or an inclusion minimal set $C \subseteq X$ such that $|N(C)| < |C|$. If we found a matching we are done, as we can satisfy at least $|X| \geq k$ clauses. So suppose we found a set C as described. Let H be $N(C)$ and $R = V(G_F) \setminus (C \cup H)$. Clearly, $N(C) \subseteq H$, $N(R) \subseteq H$ and $G[C]$ is an independent set. Furthermore, for a vertex $x \in C$ we have that there is a matching of $C \setminus x$ into H since $|N(C')| \geq |C'|$ for every $C' \subseteq C \setminus x$. Since $|C| > |H|$, we have that the matching from $C \setminus x$ to H is in fact a matching of H into C. Hence (C, H, R) is a crown decomposition of G_F.

We prove that all clauses in H are satisfied in every truth assignment to the variables satisfying the maximum number of clauses. Indeed, consider any truth assignment t that does not satisfy all clauses in H. For every variable y in $C \setminus \{x\}$ change the value of y such that the clause in H matched to y is satisfied. Let t' be the new assignment obtained from t in this manner. Since $N(C) \subseteq H$ and t' satisfies all clauses in H, more clauses are satisfied by t' than by t. Hence t can not be an assignment satisfying the maximum number of clauses.

The argument above shows that (F, k) is a *yes* instance to Max-SAT if and only if $(F \setminus H, k - |H|)$ is. This gives rise to a simple reduction rule: remove H from F and decrease k by $|H|$. This completes the proof of the theorem. ∎

44.2.1.3 Other Recent Upper Bounds

For vertex cover, the rule that takes high degree vertices into the solution immediately reduces the maximum degree of the resulting graph, yielding a kernel. It turns out that such a (albeit more involved) degree reduction rule can be designed for the FVS problem in undirected graphs (which asks for a set of k vertices whose removal results in a forest), to obtain an $O(k^2)$ kernel for the problem [27]. Improving this bound (at least on the number of vertices) and the existence of a polynomial-sized kernel for the directed version of the problem are important open problems.

One of the earliest breakthrough results in kernelization was a linear kernel for the planar dominating set problem [28]. While the constants in the bound have seen some improvements, Bodlaender et al. [29] obtains a meta algorithmic results in kernelization that give linear kernels for a number of other problems in larger classes of graphs. For example, problems like vertex cover, FVS, and dominating set have linear kernels in a class of graphs called H-minor free graphs [29,30].

A recent breakthrough result in the area of kernelization is a randomized polynomial-sized kernel for the odd cycle traversal problem using representations of matroids. See [31, 32] for details. This approach was also used to develop randomized polynomial-sized kernel for vertex cover above LP (see Section 44.2.2.3 for the definition of the problem).

44.2.1.4 Kernelization Lower Bounds

Lemma 44.1 implies that a problem has a kernel if and only if it is FPT. However, we are interested in kernels that are as small as possible, and a kernel obtained using Lemma 44.1 has size that equals the dependence on k in the running time of the best-known FPT algorithm for the problem. The question is—can we do better? In particular, can we get polynomial-sized kernels for problems that admit FPT algorithms? The answer is that quite often we can, as we saw in the previous section, but it turns out that there are a number of problems, which are unlikely to have polynomial kernels. It is only very recently that a methodology to rule out polynomial kernels has been developed [15,33]. The existence of polynomial kernels are ruled out, in this framework, by linking the existence of a polynomial kernel to an unlikely collapse in classical complexity. These developments have deepened the connection between classical and parameterized complexity.

In this section we survey the techniques that have been developed to show kernelization lower bounds. To begin with, we consider the following problem.

> LONGEST PATH
> *Instance:* An undirected graph $G = (V, E)$ and a non-negative integer k.
> *Parameter:* k.
> *Problem:* Does G have a path of length k?

It is well known that the longest path problem can be solved in time $O(c^k n^{O(1)})$ using the method of color-coding [34] (see Section 44.2.4). Is it feasible that it also admits a polynomial kernel? We argue that intuitively this should not be possible. Consider a large set $(G_1, k), (G_2, k), \ldots, (G_t, k)$ of instances to the longest path problem. If we make a new graph G by just taking the disjoint union of the graphs $G_1, \ldots, G_t$ we see that G contains a path of length k if and only if G_i contains a path of length k for some $i \leq t$. Suppose the longest path problem had a polynomial kernel, and we ran the kernelization algorithm on G. Then this algorithm would in polynomial time return a new instance $(G' = (V', E'), k')$ such that $|V'| = k^{O(1)}$, a number potentially much smaller than t. This means that in some sense, the kernelization algorithm considers the instances $(G_1, k), (G_2, k), \ldots, (G_t, k)$ and in *polynomial time* figures out which of the instances are the most likely to contain a path of length k. However, at least intuitively, this seems almost as difficult as solving the instances themselves and since the longest path problem is NP-complete, this seems unlikely. We now formalize this intuition.

Definition 44.3 (Distillation [15])

- *An OR-distillation algorithm for a language $L \subseteq \Sigma^*$ is an algorithm that receives as input a sequence $x_1, \ldots, x_t$, with $x_i \in \Sigma^*$ for each $1 \leq i \leq t$, uses time polynomial in $\sum_{i=1}^{t} |x_i|$, and outputs $y \in \Sigma^*$ with (a) $y \in L \iff x_i \in L$ for some $1 \leq i \leq t$ and (b) $|y|$ is polynomial in $\max_{i \leq t} |x_i|$. A language L is OR-distillable if there is a OR-distillation algorithm for it.*

- *An AND-distillation algorithm for a language $L \subseteq \Sigma^*$ is an algorithm that receives as input a sequence $x_1, \ldots, x_t$, with $x_i \in \Sigma^*$ for each $1 \leq i \leq t$, uses time polynomial in $\sum_{i=1}^{t} |x_i|$, and outputs $y \in \Sigma^*$ with (a) $y \in L \iff x_i \in L$ for all $1 \leq i \leq t$ and (b) $|y|$ is polynomial in $\max_{i \leq t} |x_i|$. A language L is AND-distillable if there is an AND-distillation algorithm for it.*

Observe that the notion of distillation is defined for unparameterized problems. Bodlaender et al. [15] conjectured that no NP-complete language can have an OR-distillation or an AND-distillation algorithm.

Conjecture 1. (OR-Distillation conjecture [15]) *No NP-complete language L is OR-distillable.*

Conjecture 2. (AND-Distillation conjecture [15]) *No NP-complete language L is AND-distillable.*

One should notice that if any NP-complete language is distillable, then so are all of them. Fortnow and Santhanam [33] were able to connect the OR-distillation conjecture to a well-known conjecture in classical complexity. In particular they proved that if the OR-distillation conjecture fails, then $coNP \subseteq NP/poly$, implying that the *polynomial time hierarchy* [35] collapses to the third level, a collapse that is deemed unlikely. Until very recently, establishing a similar connection for the AND-distillation conjecture was one of the central open problems of the area. It is now established that both conjectures hold up to reasonable complexity-theoretic assumptions.

Theorem 44.2 [33,36]

- *If the OR-distillation conjecture fails, then $coNP \subseteq NP/poly$.*
- *If the AND-distillation conjecture fails, then $coNP \subseteq NP/poly$.* ∎

We are now ready to define the parameterized analogue of distillation algorithms and connect this notion to the Conjectures 1 and 2.

Definition 44.4 (Composition [15])

- *A composition algorithm (also called OR-composition algorithm) for a parameterized problem $\Pi \subseteq \Sigma^* \times \mathbb{N}$ is an algorithm that receives as input a sequence $((x_1, k), \ldots, (x_t, k))$, with $(x_i, k) \in \Sigma^* \times \mathbb{N}^+$ for each $1 \leq i \leq t$, uses time polynomial in $\sum_{i=1}^{t} |x_i| + k$, and outputs $(y, k') \in \Sigma^* \times \mathbb{N}^+$ with (a) $(y, k') \in \Pi \iff (x_i, k) \in \Pi$ for some $1 \leq i \leq t$ and (b) k' is polynomial in k. A parameterized problem is compositional (or OR-compositional) if there is a composition algorithm for it.*

- *An AND-composition algorithm for a parameterized problem $\Pi \subseteq \Sigma^* \times \mathbb{N}$ is an algorithm that receives as input a sequence $((x_1, k), \ldots, (x_t, k))$, with $(x_i, k) \in \Sigma^* \times \mathbb{N}^+$ for each $1 \leq i \leq t$, uses time polynomial in $\sum_{i=1}^{t} |x_i| + k$, and outputs $(y, k') \in \Sigma^* \times \mathbb{N}^+$ with (a) $(y, k') \in \Pi \iff (x_i, k) \in \Pi$ for all $1 \leq i \leq t$ and (b) k' is polynomial in k. A parameterized problem is AND-compositional if there is an AND-composition algorithm for it.*

Composition and distillation algorithms are very similar. The main difference between the two notions is that the restriction on output size for distillation algorithms is replaced by a restriction on the parameter size for the instance the composition algorithm outputs. We define the notion of the *unparameterized version* of a parameterized problem L. The mapping of parameterized problems to unparameterized problems is done by mapping (x, k) to the string $x\#1^k$, where $\# \notin \Sigma$ denotes the blank letter and 1 is an arbitrary letter in Σ. In this way, the unparameterized version of a parameterized problem Π is the language $\widetilde{\Pi} = \{x\#1^k \mid (x, k) \in \Pi\}$. The following theorem yields the desired connection between the two notions.

Theorem 44.3 [15,36] *Let Π be a compositional parameterized problem whose unparameterized version $\widetilde{\Pi}$ is NP-complete. Then, if Π has a polynomial kernel then $coNP \subseteq NP/poly$. Similarly, let Π be an AND-compositional parameterized problem whose unparameterized version $\widetilde{\Pi}$ is NP-complete. Then, if Π has a polynomial kernel, $coNP \subseteq NP/poly$.* ∎

We can now formalize the discussion from the beginning of this section.

Theorem 44.4 [15] *Longest path does not admit a polynomial kernel unless $coNP \subseteq NP/poly$.* ∎

Proof. The unparameterized version of longest path is known to be NP-complete [6]. We now give a composition algorithm for the problem. Given a sequence $(G_1, k), \ldots, (G_t, k)$ of instances we output (G, k) where G is the disjoint union of $G_1, \ldots, G_t$. Clearly G contains a path of length k if and only if G_i contains a path of length k for some $i \leq t$. By Theorem 44.3 longest path does not have a polynomial kernel unless $coNP \subseteq NP/poly$. ∎

An identical proof can be used to show that the longest cycle problem does not admit a polynomial kernel unless $coNP \subseteq NP/poly$. For many problems, it is easy to give AND-composition algorithms. For instance, the *disjoint union* trick yields AND-composition algorithms for the treewidth, pathwidth, and cutwidth problems (see Section 44.2.7 for definitions), among many others. Coupled with Theorem 44.3 this implies that these problems do not admit polynomial kernels unless $coNP \subseteq NP/poly$.

For some problems, obtaining a composition algorithm directly is a difficult task. Instead, we can give a reduction from a problem that provably has no polynomial kernel unless $coNP \subseteq NP/poly$ to the problem in question such that a polynomial kernel for the problem considered would give a kernel for the problem we reduced from. We now define the notion of *polynomial parameter transformations*.

Definition 44.5 [37] *Let P and Q be parameterized problems. We say that P is polynomial parameter reducible to Q, written $P \leq_{ppt} Q$, if there exists a polynomial time computable function $f : \Sigma^* \times \mathbb{N} \to \Sigma^* \times \mathbb{N}$ and a polynomial p, such that for all $(x, k) \in \Sigma^* \times \mathbb{N}$ (a) $(x, k) \in P$ if and only if $(x', k') = f(x, k) \in Q$ and (b) $k' \leq p(k)$. The function f is called polynomial parameter transformation.*

Proposition 44.1 [37] *Let P and Q be the parameterized problems and $\tilde{P}$ and $\tilde{Q}$ be the unparameterized versions of P and Q, respectively. Suppose that $\tilde{P}$ is NP-complete and $\tilde{Q}$ is in NP. Furthermore if there is a polynomial parameter transformation from P to Q, then if Q has a polynomial kernel then P also has a polynomial kernel.*

Proposition 44.1 shows how to use polynomial parameter transformations to show kernelization lower bounds. A notion similar to polynomial parameter transformation was independently used by Fernau et al. [38] albeit without being explicitly defined. We now give an example of how Proposition 44.1 can be useful for showing that a problem does not admit a polynomial kernel. In particular, we show that the path packing problem does not admit a polynomial kernel unless $coNP \subseteq NP/poly$. In this problem you are given a graph G together with an integer k and asked whether there exists a collection of k mutually vertex-disjoint paths of length k in G. This problem is known to be FPT [34] and is easy to see that for this problem the *disjoint union* trick discussed earlier does not directly apply. Thus we resort to polynomial parameter transformations.

Theorem 44.5 *Path packing does not admit a polynomial kernel unless $coNP \subseteq NP/poly$.*

Proof. We give a polynomial parameter transformation from the longest path problem. Given an instance (G, k) to longest path we construct a graph G' from G by adding $k - 1$ vertex disjoint paths of length k. Now G contains a path of length k if and only if G' contains k paths of length k. This concludes the proof. ∎

Dom et al. [39] provides a set of tools combined with colors, ids, compositions, and parameter preserving reductions to prove a large range of FPT problems do not have a polynomial-sized kernel. A further refinement to the lower bound technique was proved in Dell and van Melkebeek [40] where they show that problems like vertex cover and FVS who have polynomial-sized kernels can not have subquadratic ($O(k^{2-\epsilon})$)-sized kernels under the same complexity theoretic assumptions. See the works of Bodlaender et al. [41], Dell and Mark [42], Hermelin and Wu [43], and Kratsch [44] for variations of these techniques. ∎

44.2.2 Bounded Search Trees

Branching is a classical systematic way of exploring the search space of solutions to obtain exponential algorithms. The idea here is to first identify, in polynomial time, a small (typically a constant or some $f(k)$, but even logarithmically many is also fine) subset of elements of which at least one (or a subset of smaller size) must be in *any* feasible solution of the problem. Then we include one of them at a time and recursively solve the remaining problem with the reduced parameter value. Such search trees are analyzed by measuring the drop of the parameter in each branch. If we ensure that the parameter (or some measure bounded by a function of the parameter) drops in each branch, then we will be able to bound the depth of the search tree by a function of the parameter, resulting in a fixed-parameter algorithm.

We illustrate this technique with two parameterizations of vertex cover and FVS.

44.2.2.1 Vertex Cover

Let $G = (V, E)$ be the input to the vertex cover and k be the parameter. Our algorithm is based on the following two simple observations.

- For a vertex v, any vertex cover must contain either v or *all* of its neighbors $N(v)$.
- Vertex cover can be solved optimally in polynomial time when the maximum degree of a graph is at most 2.

So our algorithm recursively solves the problem by finding a vertex v of maximum degree in the graph and if $d(v) \geq 3$ then recursively branching on two cases by considering either v in the vertex cover or $N(v)$ in the vertex cover. When we consider two cases like this, we say we *branch according to v and $N(v)$*. And when the maximum degree of the graph is 2, we solve the problem in polynomial time.

The time complexity of the algorithm can be described by the following recurrence in k.

$$T(k) = \begin{cases} T(k-1) + T(k-3) + n^{O(1)} & \text{if } k \geq 2 \\ n^{O(1)} & \text{if } k \leq 1 \end{cases}.$$

The above recursive function bounds the size of the search tree and the time spent at each node in the tree. The above recursive function can be solved by finding the largest root of the characteristic polynomial $\lambda^k = \lambda^{k-1} + \lambda^{k-3}$. Using standard mathematical techniques (and/or symbolic algebra packages) the root is estimated to 1.466.

We could apply this branching step *after* applying the preprocessing steps outlined in Section 44.2.1 so that they are applied on a graph on at most $2k$ vertices.

This gives us the following theorem.

Theorem 44.6 *Vertex cover can be solved in $O(1.466^k k^{O(1)} + n^{O(1)})$.* ∎

This simple algorithm also suggests that one could branch on more complex structures (instead of one containing simply a vertex and its neighbors) and obtain better algorithms. A sequence of improvements have been made resulting in the current best bound of $O(1.2718^k + kn)$ [7].

44.2.2.2 Feedback Vertex Set

In this subsection we study the FVS problem which is formally defined below.

FEEDBACK VERTEX SET (FVS)
   Instance: An undirected graph $G = (V, E)$, and a non-negative integer k.
   Parameter: k.
   Problem: Decide whether G has a set of at most k vertices whose removal makes the graph acyclic.

We start with some simple reduction rules that simplifies the input instance. The following is well known in the literature on FVS problems.

Lemma 44.3 *Let G be an undirected multigraph. Perform the following steps as long as possible.*

1. *If G has a vertex of degree ≤ 1, remove it (along with the incident edge if any).*

2. *If G has a vertex x of degree 2 adjacent to vertices y and z, $y \neq x$ and $z \neq x$, short circuit by removing x and joining y and z by a new edge (even if y and z were adjacent earlier)*

Let G' be the resulting multigraph. Then G has a FVS of size at most k if and only if G' has a FVS of size at most k. ∎

Clearly the graph G' is such that each component of G' has minimum degree at least three unless that component is *either* an empty graph *or* a graph on one vertex with a self-loop (in which case that component has a FVS of size 1). In the last case we include that vertex in the FVS and remove it from the graph and decrease the parameter by 1. Thus, the reduced graph has minimum degree 3. Erdös and Pósa [45] observed that the girth of any undirected graph G with minimum degree at least 3 is bounded by $2 \log n + 1$. Given such a graph, one can find in $O(n)$ time a cycle of length at most $2 \log n$ by growing a breadth-first search (BFS) tree till the first non-tree edge is encountered.

Now clearly any FVS in the graph must have at least one of the vertices in this cycle. Once a vertex is included in the FVS, all edges incident on it can be removed. Thus iteratively pick one of the vertices of the cycle at a time, and recursively check whether the resulting graph has a FVS of size $k - 1$. Since the reduction in each step to remove vertices of degree 1 and 2, and to find the short cycle can be performed in $O(m + n)$ time, it can be verified that the entire algorithm takes $O((2 \lg n)^k n + m)$ time. Now, though this running time does not make the algorithm look like a fixed-parameter algorithm as the exponent k is on top of $\log n$, this should be allowed as a tractable algorithm under practical considerations. Fortunately,

$$(2 \log n)^k \leq (4k \log k)^k + n$$

for all n and $k \leq n$, and thus the algorithm is a fixed-parameter algorithm. This gives us the following theorem.

Theorem 44.7 *FVS can be solved in time $O((2 \lg n)^k n + m)$, or in $O((4k \log k)^k n + nm)$ time.* ∎

The bound in Theorem 44.7 has seen systematically improved [8,46–49], resulting in the current best bound of $O^*(3.83^k)$. The best-known algorithms use a clever combination of the iterated compression (see Subsection 44.2.3) and branching techniques. However, if we are willing to allow randomization then the problem admits an algorithm with running time $O^*(3^k)$ [19] (see also Subsection 44.2.4).

44.2.2.3 Vertex Cover above LP

Recall the ILP formulation of vertex cover discussed in Section 44.2.1.1.

If the minimum value of $\text{LPVC}(G)$ is $vc^*(G)$ then clearly the size of a minimum vertex cover is at least $vc^*(G)$. This leads to the following parameterization of vertex cover.

> **VERTEX COVER ABOVE LP**
> *Instance:* An undirected graph G, positive integers k and $\lceil vc^*(G) \rceil$, where $vc^*(G)$ is the minimum value of LPVC(G).
> *Parameter:* $k - \lceil vc^*(G) \rceil$.
> *Problem:* Does G have a vertex cover of size at most k?

Observe that since $vc^*(G) \geq m$, where m is the size of a maximum matching of G, we have that $k - vc^*(G) \leq k - m$.

Before we describe the algorithm we fix some notations. By the phrase an optimum solution to LPVC(G), we mean a feasible solution with $x(v) \geq 0$ for all $v \in V$ minimizing the objective function $w(x) = \sum_{u \in V} x(u)$. It is well known that for any graph G, there exists an optimum solution to LPVC(G), such that $x(u) \in \{0, 1/2, 1\}$ for all $u \in V$ [50]. Such a feasible optimum solution to LPVC(G) is called a half integral solution and can be found in polynomial time [50]. In this section we always deal with half integral optimum solutions to LPVC(G). Thus, by default whenever we refer to an *optimum solution* to LPVC(G) we will be referring to a *half integral optimum solution* to LPVC(G). Let $VC(G)$ be the set of all minimum vertex covers of G and $vc(G)$ denote the size of a minimum vertex cover of G. Let $VC^*(G)$ be the set of all optimal solutions (including non-half integral optimal solution) to LPVC(G). By $vc^*(G)$ we denote the value of an optimum solution to LPVC(G). We define $V_i^x = \{u \in V : x(u) = i\}$ for each $i \in \{0, 1/2, 1\}$ and define $x \equiv i$, $i \in \{0, 1/2, 1\}$, if $x(u) = i$ for every $u \in V$. Clearly, $vc(G) \geq vc^*(G)$ and $vc^*(G) \leq |V|/2$ since $x \equiv 1/2$ is always a feasible solution to LPVC(G). We also refer to the $x \equiv 1/2$ solution simply as the all $1/2$ solution.

To obtain an algorithm for this problem we will use the following well-known reduction rules that were also useful for linear vertex kernel for vertex cover.

Lemma 44.4 [18,51] *For a graph G, in polynomial time, we can compute an optimal solution x to LPVC(G) such that all $1/2$ is the unique optimal solution to LPVC($G[V_{1/2}^x]$). Furthermore, there is a minimum vertex cover for G which contains all the vertices in V_1^x and none of the vertices in V_0^x.* ∎

The Lemma 44.4 brings us to the following reduction rule.

Preprocessing Rule 44.1 Apply Lemma 44.4 to compute an optimal solution x to LPVC(G) such that all $1/2$ is the unique optimum solution to LPVC($G[V_{1/2}^x]$). Delete the vertices in $V_0^x \cup V_1^x$ from the graph after including V_1^x in the vertex cover we develop, and reduce k by $|V_1^x|$.

In the discussions in the rest of the section, we say that Preprocessing Rule 44.1 applies if all $1/2$ is not the unique solution to LPVC(G) and that it does not apply if all $1/2$ is the unique solution to LPVC(G).

Our algorithm for vertex cover above LP is same as the one described for vertex cover in Section 44.2.2.1. After the preprocessing rule is applied exhaustively, we pick an arbitrary vertex u in the graph and branch on it. In other words, in one branch, we add u into the vertex cover, decrease k by 1, and delete u from the graph, and in the other branch, we add $N(u)$ into the vertex cover, decrease k by $|N(u)|$, and delete $\{u\} \cup N(u)$ from the graph. The correctness of this algorithm follows from the soundness of the preprocessing rules and the fact that the branching is exhaustive.

In order to analyze the running time of our algorithm, we define a measure $\mu = \mu(G, k) = k - vc^*(G)$. We will first show that our preprocessing rules do not increase this measure.

Following this, we will prove a lower bound on the decrease in the measure occurring as a result of the branching, thus allowing us to bound the running time of the algorithm in terms of the measure μ. For each case, we let (G', k') be the instance resulting by the application of the rule or branch, and let x' be an optimum solution to LPVC(G').

Consider the application of Preprocessing Rule 1. We know that $k' = k - |V_1^x|$. Since $x' \equiv 1/2$ is the unique optimum solution to LPVC(G'), and G' comprises precisely the vertices of $V_{1/2}^x$, the value of the optimum solution to LPVC(G') is exactly $|V_1^x|$ less than that of G. Hence, $\mu(G, k) = \mu(G', k')$.

We now consider the branching step. Consider the case when we pick u in the vertex cover. In this case, $k' = k - 1$. We claim that $w(x') \geq w(x) - 1/2$. Suppose that this is not the case. Then, it must be the case that $w(x') \leq w(x) - 1$. Consider the following assignment $x'' : V \to \{0, 1/2, 1\}$ to LPVC(G). For every vertex $v \in V \setminus \{u\}$, set $x''(v) = x'(v)$ and set $x''(u) = 1$. Now, x'' is clearly a feasible solution and has a value at most that of x. But this contradicts our assumption that $x \equiv 1/2$ is the unique optimum solution to LPVC(G). Hence, $w(x') \geq w(x) - 1/2$, which implies that $\mu(G', k') \leq \mu(G, k) - 1/2$. Similarly, we can also show that in the other case we have that $\mu(G', k') \leq \mu(G, k) - 1/2$. We have thus shown that the preprocessing rules do not increase the measure $\mu(G, k)$ and the branching step results in a $(1/2, 1/2)$ decrease in $\mu(G, k) = \mu$, resulting in the recurrence $T(\mu) \leq 2T(\mu - 1/2)$ which solves to $4^\mu = 4^{k-vc^*(G)}$. Thus we get a $4^{(k-vc^*(G))}$ algorithm for vertex cover above LP.

Theorem 44.8 *Vertex cover above LP can be solved in time $O^*(4^{k-vc^*(G)})$.* ∎

A more involved application of the linear programming paradigm given above results in an algorithm running in time $O^*(2.32^k)$ for vertex cover above LP [22]. The above algorithm is not only an example of use of linear programming in parameterized algorithms but also an example of technique called *measure and conquer*. In a more sophisticated use of this technique, one comes up with a measure, which is a function of the input and the parameter to measure the progress in each branching step. We refer to Fomin et al. [52] for an introduction to measure and conquer as well as its application in obtaining non-trivial exact algorithms for vertex cover and dominating set.

44.2.3 Iterative Compression

This is a powerful technique that has helped solve several (minimization) problems FPT in the last decade. In the compression step, one gets hold of a solution of size larger than k (but still of size a function of k), and the goal is to look for a k-sized solution. Somehow having the larger sized solution provides a structure to the problem, which one can exploit. Now to get the larger sized solution, one approach is to appeal to an approximation algorithm for the problem. The other idea is to iterate on subgraphs of the graph. That is, we find a larger size solution to a subgraph of the problem (e.g., $k+1$ vertices form a vertex cover on any induced subgraph on $k+2$ vertices) and if we can not compress the solution to a k-sized solution for the subgraph, then we can return a NO answer. If we *can* compress for the subgraph, then we can include one more vertex to get a large solution for a larger subgraph now. This process stops when we have considered the entire subgraph.

We illustrate it again on the FVS problem.

Let G be an undirected graph and let $v_1, \ldots, v_n$ be an arbitrary ordering of its vertices. We use G_i to denote the graph induced on first i vertices, that is, $G_i = G[v_1, \ldots, v_i]$. Let S be the set containing the first $k+1$ vertices. This set is a FVS for the induced subgraph

G_{k+2} on the first $k + 2$ vertices. In the following compression step, we determine whether G_{k+2} has a FVS of size at most k. Note that if there is such a vertex subset, it will intersect S in some set $Y \subseteq S$ which could even be empty, but is of size at most k. We *guess* (i.e., try all possible such subsets) such a subset Y, delete it from the graph. Now the goal is to find a FVS of size at most $k - |Y|$ in $H = G_{k+2} \setminus Y$ and disjoint from $S \setminus Y$. We delete degree 0 and 1 vertices of H (as they can not be part of any solution). If there is a degree 2 vertex in $H \setminus S$ with at least one end point in $H \setminus S$, then delete this vertex by making its neighbors adjacent (even if they were adjacent before; the graph could become a multigraph now). It is easy to see that these reduction rules produce an equivalent instance.

Note that $H \setminus S$ is a forest as S is a FVS for H, and hence it has a vertex x whose degree is at most 1. Also x has at least two neighbors in $S \setminus Y$ (otherwise any of the reduction rules would have applied). If x has two neighbors in the same component of $G[S \setminus Y]$, then x must be in any FVS we are seeking, and so we include it in the solution, decrease the parameter by 1 and delete it from the graph and continue. Otherwise x *connects* at least two components of $G[S \setminus Y]$. Now branch by including x in the solution in one branch and excluding it in the other branch. In the branch we include k decreases by 1. In the branch we exclude it, x is added to $S \setminus Y$ (these are vertices which are excluded from the proposed FVS), and as x connects at least two components of $G[S \setminus Y]$, the number of components in $G[S \setminus Y]$ decreases by at least 1. So if we use a measure $\mu = k - |Y| + c$ where c is the initial number of components in $G[S \setminus Y]$, then μ decreases by at least 1 in each branch resulting in a $2^\mu = 2^{k-|Y|+c}$ algorithm for this step. Clearly, c is at most $k - |Y|$. Thus, the overall running time for the compression step is bounded by

$$\sum_{i=0}^{k} \binom{k+1}{i} 2^{2k-2i} n^d$$

is at most $5^{k+1} n^d$ where d is a constant. The n^d bound is the time spent in each branching step, which is primarily the time for deleting the selected vertices.

Now if there is no FVS of size at most k in H, then there is no such FVS in G and so we can return a NO answer. If we do get a FVS of size at most k in H, then we can add to that set the vertex $k + 3$ to get a new S, a FVS of size at most $k + 1$ in G_{k+3}, and we repeat the compression step on G_{k+3}. This process stops when we applied the compression step on G_n or we return a NO answer in between. Clearly this algorithm takes $5^k n^{d+1}$. This bound has been improved to $3.83^k n^c$ for some constant c in Cao et al. [8].

This idea of compression and iterating was first introduced by Reed et al. [53] to prove the odd cycle traversal problem (are there k vertices whose deletion results in a bipartite graph) FPT. This idea has been used to prove a number of minimization problems FPT. These include directed FVS [54] and almost-2-SAT [55]. See [56] for an application of iterated compression for the *above guarantee vertex cover* problem which asks whether there is a vertex cover of size at most k more than the size of the maximum matching in a given graph. See [57] for an application of iterated compression for determining whether a given perfect graph has cochromatic number (the minimum number of cliques and independent sets, the graph can be partitioned into) at most k.

44.2.4 Randomized Fixed-Parameter Algorithms

Randomization is a powerful paradigm in algorithms and complexity. It has also been useful in designing FPT algorithms. In this section we touch upon three commonly used approaches to design-randomized FPT algorithms.

44.2.4.1 α-Covering Based Randomized Algorithms

Consider the following simple randomized algorithm for the vertex cover problem on a graph on n vertices and m edges.

> Repeat for k steps (or until the resulting graph has no edges, whichever is earlier).
>
> - Choose an edge of the graph uniformly at random (i.e., each edge is chosen with probability $1/m$).
> - Choose one of the two vertices of the edge uniformly at random into the solution S, and delete it from the graph.

Vertex cover: Let X be a vertex cover G of size at most k, if exists. Then, it is clear that the probability that the above randomized algorithm will pick a vertex of X in one step, is at least $1/2$ (note that this applies even if we pick an arbitrary, but not a random, edge, in the first step), and hence will pick X in k steps is at least $1/2^k$. Thus the probability that the algorithm will pick a vertex cover of size at most k is at least $1/2^k$. Hence, if we just repeat the algorithm until it finds a vertex cover of size at most k, then this algorithm will find one such vertex cover in expected $O^*(2^k)$ time.

Feedback vertex set: It turns out that the above algorithm, after the standard preprocessing rules described in Lemma 44.3, also picks a feedback vertex cover of size at most k if there exists one, though in slightly worse expected time.

Let X be a FVS of size k. Note that $G[V \setminus X]$ is a forest, and hence has at most $n - k - 1$ edges. Thus, there are at least $m - n + k + 1 \geq n/2 + k + 1$ edges that have at least one end point in X (as the graph has minimum degree 3, $m \geq 3n/2$). That is, at least one-third of the edges has at least one end point in X, and hence our algorithm will pick these edges in the first step with probability at least $1/3$ and will pick a vertex in X with probability at least $1/6$. Hence, repeating these two steps k times, our algorithm will find X with probability at least $1/6^k$. Thus if we repeat these k steps until we find a set whose deletion results in a forest, the algorithm will take an expected $O^*(6^k)$ time. With a careful analysis, this can be made to run in $O^*(4^k)$ time [58]. Note that we needed the preprocessing rules to ensure this bound.

The algorithm described above not only works for vertex cover and FVS but for a large collection of problems. These problems are best described as follows. Let $\mathscr{G}$ be the set of all finite connected undirected graphs and let $\mathscr{L}$ be the family of all finite subsets of $\mathscr{G}$. Thus every element $\mathcal{F} \in \mathscr{L}$ is a finite set of connected graphs. Also assume that $\mathcal{F}$ is explicitly given. Then one can define the following $\mathcal{F}$-deletion problem.

> $\mathcal{F}$-DELETION
> *Instance:* A graph G and a non-negative integer k.
> *Parameter:* k.
> *Question:* Does there exist $S \subseteq V(G)$, $|S| \leq k$,
> such that $G \setminus S$ contains no graph from $\mathcal{F}$ as a minor?

See Section 44.2.6 for definitions of a minor.

Here, we would be only interested in the case when $\mathcal{F} \in \mathscr{L}$ contains at least one planar graph. This problem encompasses a number of the well-studied instances of $\mathcal{F}$-

deletion. For example, when $\mathcal{F} = \{K_2\}$, a complete graph on two vertices, this is the vertex cover problem. When $\mathcal{F} = \{C_3\}$, a cycle on three vertices, this is the FVS problem. Another fundamental problem, which is a special case of $\mathcal{F}$-deletion, is treewidth η-deletion or η-transversal which is to delete at most k vertices to obtain a graph of treewidth at most η. Since any graph of treewidth η excludes a $(\eta + 1) \times (\eta + 1)$ grid as a minor, we have that the set $\mathcal{F}$ of forbidden minors of treewidth η graphs contains a planar graph. Among other examples of $\mathcal{F}$-deletion that can be found in the literature on approximation and parameterized algorithms, are the cases of $\mathcal{F}$ being $\{K_{2,3}, K_4\}$, $\{K_4\}$, $\{\theta_c\}$, and $\{K_3, T_2\}$, which correspond to removing vertices to obtain an outerplanar graph, a series-parallel graph, a diamond graph, and a graph of pathwidth one, respectively. See [59] for more details.

It turns out that the algorithm described for vertex cover and FVS above works for $\mathcal{F}$-deletion when $\mathcal{F} \in \mathcal{L}$ contains a planar graph. But observe that to get the desired probability computation for FVS, it was important that the input graphs was preprocessed to have a minimum degree 3. We need similar kind of preprocessing for $\mathcal{F}$-deletion. Observe that we already have a notion of preprocessing in terms of kernelization. The goal of kernelization is to apply reduction rules such that the size of the reduced instance can be upper bounded by a function of the parameter. However, if we want to use preprocessing for approximation or FPT algorithms, it is not necessary that the size of the reduced instance has to be upper bounded. What we need is a preprocessing procedure that allows us to navigate the solution search space efficiently. Toward this a notion of α-cover is introduced in Fomin et al. [60]. For $0 < \alpha \leq 1$, we say that a vertex subset $S \subseteq V(G)$ is an α-cover, if the sum of vertex degrees $\sum_{v \in S} d(v)$ is at least $2\alpha|E(G)|$. For example, every vertex cover of a graph is also a 1-cover. The defining property of this preprocessing is that the equivalent simplified instance of the problem admits some optimal solution, which is also an α-cover. If we succeed with this goal, then for an edge selected uniformly at random, with a constant probability at least one of its endpoints belongs to some optimal solution. Using this as a basic step, we can construct FPT algorithms for $\mathcal{F}$-deletion. The preprocessing for these problems is captured by *protrusion reduction* described in Bodlaender et al. [29]. A detailed algorithm is described in Fomin et al. [60].

44.2.4.2 Color Coding

Alon et al. [34] developed a generic randomized technique called *color coding*, that works for a number of parameterized problems. We demonstrate this on the longest path problem. Note that the problem is NP-complete as it is simply the decision version of the well-known NP-complete Hamiltonial path problem. Finding a simple path directly involves keeping track of the vertices already visited, and this search may require n^k searches which is precisely what we are trying to avoid. The work around Alon et al. came up with, was to color the vertices randomly from 1 to k, and find a path on k vertices, if exists, whose colors are distinct. Now, finding such a path involves only keeping track of the colors and they argued (see [34] for details) that this can be done in $O(2^k m)$ time using simple dynamic programming techniques. Note, however that the graph may have a simple path of length k, but all its vertices may not be colored distinctly by this random coloring and hence we may miss finding it in the first step. But the probability of this happening is at most $1 - k!/k^k$ which is at most $1 - (e^{-k})$. That is, the algorithm finds a simple path of length k, if exists, with probability at least e^{-k}. Hence, if we repeat this algorithm until we find a path will all colors distinct, then in $O^*((2e)^k)$ expected time, we can find a simple path of length k if exists.

More generally, this algorithm can be modified to finding a subset of k vertices that form a cycle or a tree or a subgraph of treewidth at most constant. Furthermore, this algorithm can be derandomized by using functions from a (n, k) perfect hash family, which are of size $2^{O(k)} \log n$. See [34] for more details. Also see [61] for polynomial space implementation of algorithms based on color-coding.

This idea of color coding has been generalized to divide and color [62], chromatic-coding [63] and randomized monomial testing [64] to obtain faster parameterized algorithms for finding a k-sized subgraph of bounded treewidth in an input graph. Also see [65] for the best-known deterministic algorithm for finding a k-sized subgraph of bounded treewidth in an input graph.

44.2.4.3 Chromatic Coding

The idea of chromatic coding or color and conquer as it is sometimes referred to in literature, is intuitively based on the idea that in the yes instances of edge deletion/modification problems, the number of edges in the solution is bounded and therefore, using a sufficiently large number of colors and a random coloring of vertices, the edges in the solution set can be guaranteed to be properly colored with sufficiently large probability. Once this random coloring step is successfully completed, the problem instance seems to have a nice structure, which can be exploited for algorithmic applications. In particular, after a successful coloring, we can observe the following:

1. The subgraph induced on each color class is an element of the graph class for which the edge modification problem has to be solved.

2. All the solution edges go across color classes.

The next step is typically an algorithm to solve the problem on this colored instance. The next step of course would be to repeat the coloring procedure sufficiently large number of times to obtain a constant error probability. In Alon et al. [63], the authors also present a scheme to derandomize the algorithm using *Universal Coloring Families* at the cost of some increase in running time. The description and presentation of this section is taken from Ghosh et al. [66]. We demonstrate the paradigm by obtaining a subexponential time algorithm for the following problem.

SPLIT EDGE DELETION (SED)
  Instance: An undirected graph $G = (V, E)$, and a non-negative integer k.
  Parameter: k.
  Problem: Does there exist a set of edges of size at most k whose deletion from G results in a split graph?

This algorithm consists of three steps. In the first step, we reduce the instance (G, k) to an equivalent instance (G', k') with at most $O(k^2)$ vertices. In the second step, we color the vertices of the graph uniformly at random and we prove that with a sufficiently high probability, all the edges of some k-sized solution (if one exists) are non-monochromatic. Finally, we give an algorithm to check if a colored instance of split edge deletion (SED) has a non-monochromatic SED set of size at most k.

Kernelization. We first apply the kernelization algorithm described in Ghosh et al. [66] which, given an instance (G, k) of SED, in polynomial time, returns an equivalent instance (G', k') of SED such that the number of vertices in G' is $O(k^2)$ and $k' \leq k$. The instance (G', k') is called a *reduced* instance. In the rest of this section, we will assume that the given instance of SED is a reduced instance.

Probability of a good coloring. We now color the vertices of G independently and uniformly at random with $\sqrt{8k}$ colors and let A_c be the set of non-monochromatic edges. Suppose that $(G = (V, E), k)$ is a Yes instance and let $A \subseteq E$ be a solution to this instance. We now show that the probability of A being contained in A_c is at least $2^{-O(\sqrt{k})}$. We begin by estimating the probability of obtaining a proper coloring (making all the edges non-monochromatic) when applying the above random experiment on a graph with k edges.

Lemma 44.5 [63] *If the vertices of a graph on q edges are colored independently and uniformly at random with $\sqrt{8q}$ colors then the probability that G is properly colored is at least $(2e)^{-\sqrt{q/8}}$.* ∎

Now, since we colored each vertex of the graph G independently, the graph induced on the set A, of size at most k, will be properly colored with probability at least $2^{-O(\sqrt{k})}$, which gives us the following lemma.

Lemma 44.6 *Let $(G = (V, E), k)$ be a YES instance of SED which is colored by the random process described above, and let $A \subseteq E$ be a solution for this instance. The probability that no edge in A is monochromatic is at least $2^{-O(\sqrt{k})}$.* ∎

Solving a colored instance. We now present an algorithm to test if there is a *colorful* (all edges non-monochromatic) SED set in a given colored instance of SED. In the colored instance, every vertex is colored with one of $\sqrt{8k}$ colors. We start with the following simple observation.

Observation 44.1 *Let $G = (V_1 \bigcup V_2 \bigcup \ldots \bigcup V_t, E)$ be a t-colored graph. If there exists a colorful SED set A in G, then $G[V_i]$ is a split graph for every V_i.*

Proof. Since A is a colorful set, each of the subgraphs $G[V_i]$ is an induced subgraph of $G \setminus A$. But $G \setminus A$ is a split graph, and every induced subgraph of a split graph is also split. Hence, $G[V_i]$ is a split graph. ∎

We now proceed to the description of the algorithm. Suppose the given instance had a colorful SED set A. Observation 44.1 implies that $G[V_i]$ is a split graph and it remains a split graph in $G \setminus A$. Hence, we enumerate the split partitions of $G[V_i]$ for each i. Fixing a split partition for each $G[V_i]$ results in a *combined* split partition for the vertices in V. There are $O(k^2)$ split partitions for each V_i and $O(\sqrt{k})$ such sets. Hence, there are $k^{O(\sqrt{k})}$ many combined split partitions. Now, it simply remains to check if there is a combined split partition $(C \uplus I)$ such that the number of edges in the graph $G[I]$ is at most k and return YES if and only if there is one such combined split partition. Hence, we have the following lemma.

Lemma 44.7 *Given a colored instance (G, k) of SED of size $O(k^2)$, we can test if there is a colorful SED set of size at most k in time $2^{O(\sqrt{k} \log k)}$.* ∎

Combining Lemmas 44.6 and 44.7, we get the following theorem.

Theorem 44.9 *There is a randomized FPT algorithm for SED running in time $2^{O(\sqrt{k} \log k)} + n^{O(1)}$ with a success probability of at least $2^{-O(\sqrt{k})}$.* ∎

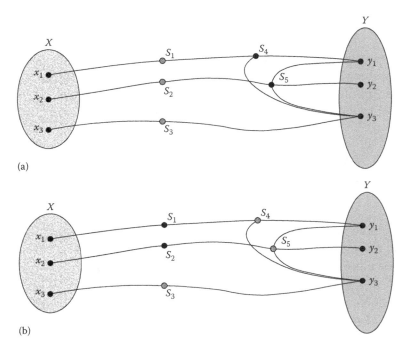

Figure 44.1 $S_1 = \{s_1, s_2, s_3\}$ and $S_2 = \{s_3, s_4, s_5\}$ are two $X - Y$ vertex separators. But S_2 also separates vertices of Y while S_1 does not.

44.2.5 Important Separators

Given a graph $G = (V, E)$ disjoint vertex sets X and Y, suppose we would like to find a minimum set of vertices in G whose removal disconnects X from Y and also disconnects every pair of vertices in Y. In such cases, if chosen carefully, the set of vertices which we choose to separate X from Y will also help us in separating some pairs of vertices in Y. Intuitively, *the closer the vertices are to Y, the better the chance of them separating vertices in Y* and hence some separators seem to be more *important* than others. The sets $S_1 = \{s_1, s_2, s_3\}$ and $S_2 = \{s_3, s_4, s_5\}$ (see Figure 44.1) are sets of the same size which separate X and Y. But by our intuition, S_2 is more *important* for us since it also separates the vertices in Y. This intuition was formalized in Marx [67] and was used to give an FPT algorithm for the multiway cut problem defined below. The same concept was used implicitly in Chen et al. [68] to give an improved algorithm for the same problem. The same concept was used by Chen et al. [54] to resolve the fixed-parameter tractability of the directed FVS problem and by Razgon and O'Sullivan [55] to prove the fixed-parameter tractability of the almost 2-Sat problem.

MULTIWAY CUT
    Instance: A graph $G = (V, E)$, $T \subseteq V$ of terminals and a non-negative integer k.
    Parameter: k.
    Question: Does there exist $S \subseteq V \setminus T$, $|S| \leq k$,
                    such that in $G \setminus S$ there is no path between any pair of vertices in T?

44.2.5.1 Important Vertex Separators in Undirected Graphs

Definition 44.6 *Let $G = (V, E)$ be an undirected graph and let $X \subseteq V$. We denote by $\delta(X)$ the vertices of $G \setminus X$ which have a neighbor in X. We define the function $\tilde{f} : 2^V \to \mathbb{N}$ as $\tilde{f}(X) = |\delta(X)|$.*

Definition 44.7 Let $G = (V, E)$ be an undirected graph, let $X \subseteq V$ and $S \subseteq V \setminus X$. We denote by $R_G(X, S)$ the set of vertices of G reachable from X in $G \setminus S$. We drop the subscript G if it is clear from the context.

Definition 44.8 Let Z be a finite set. A function $f : 2^Z \to \mathbb{R}$ is submodular if for all subsets A and B of Z, $f(A \cup B) + f(A \cap B) \leq f(A) + f(B)$.

Lemma 44.8 Let $G = (V, E)$ be an undirected graph and let $\tilde{f} : 2^V \to \mathbb{N}$ be a function defined as above. Then the function $\tilde{f}$ is submodular.

Definition 44.9 [67] Let $G = (V, E)$ be an undirected graph and let $X, Y \subset V$ be two disjoint vertex sets. A subset $S \subseteq V \setminus (X \cup Y)$ is called an $X - Y$ vertex separator in G if $R_G(X, S) \cap Y = \phi$ or in other words there is no path from X to Y in the graph $G \setminus S$. We denote by $\lambda_G(X, Y)$ the size of the smallest $X - Y$ vertex separator in G. An $X - Y$ separator S_1 is said to dominate an $X - Y$ separator S with respect to X if $|S_1| \leq |S|$ and $R(X, S_1) \supset R(X, S)$. If the set X is clear from the context, we just say that S_1 dominates S. An $X - Y$ vertex separator is said to be inclusionwise minimal if none of its proper subsets is an $X - Y$ vertex separator.

Proposition 44.2 If $R \supseteq X$ is any vertex set disjoint from Y such that $\delta(R) \cap Y = \phi$ then $\delta(R)$ is an $X - Y$ vertex separator.

Proof. This is because any path from X to Y in G must contain a vertex of $\delta(R)$. Consider a path P from $u \in X$ to $v \in Y$ in G. Since $u \in R$ and $v \notin R$, P must contain a vertex w which is outside R and is neighbor to a vertex in R implying that $w \in \delta(R)$. ∎

Definition 44.10 [67] Let $G = (V, E)$ be an undirected graph, $X, Y \subset V$ be disjoint vertex sets and $S \subseteq V \setminus (X \cup Y)$ be an $X - Y$ vertex separator in G. We say that S is an important $X - Y$ vertex separator if it is inclusionwise minimal and there does not exist another $X - Y$ vertex separator S_1 such that S_1 dominates S with respect to X. If $S \subset V$ is an important $X - Y$ vertex separator then the set $R(X, S)$ is called an important set and the subgraph $G[R(X, S)]$ is called an important component if it is connected.

Lemma 44.9 [67] Let $G = (V, E)$ be an undirected graph, $X, Y \subset V$ be disjoint vertex sets. There exists a unique important $X - Y$ vertex separator S^* of size $\lambda_G(X, Y)$.

Proof. Consider a minimum size $X - Y$ vertex separator of size $\lambda_G(X, Y)$. Since it is minimal, this separator is either important or there is another that dominates it. Hence, there is at least one important $X - Y$ vertex separator of size $\lambda_G(X, Y)$. Now we show that there can not be two such important $X - Y$ vertex separators.

Suppose S_1 and S_2 are two important $X - Y$ vertex separators of size $\lambda_G(X, Y)$ where $S_1 \neq S_2$ and let $R_1 = R(X, S_1)$ and $R_2 = R(X, S_2)$. We know that $R_1, R_2 \supset X$, and by the minimality of S_1 and S_2, $\delta(R_1) = S_1$ and $\delta(R_2) = S_2$. But $S_1, S_2 \cap Y = \phi$. Hence by Proposition 44.2 the sets $\delta(R_1 \cup R_2)$ and $\delta(R_1 \cap R_2)$ are also $X - Y$ vertex separators and hence $\tilde{f}(R_1 \cup R_2), \tilde{f}(R_1 \cap R_2) \geq \lambda_G(X, Y)$. By the submodularity of $\tilde{f}$ (Lemma 44.8), we have that

$$\underbrace{\tilde{f}(R_1)}_{=\lambda_G(X,Y)} + \underbrace{\tilde{f}(R_2)}_{=\lambda_G(X,Y)} \geq \underbrace{\tilde{f}(R_1 \cup R_2)}_{\geq \lambda_G(X,Y)} + \underbrace{\tilde{f}(R_1 \cap R_2)}_{\geq \lambda_G(X,Y)}$$

which implies that $\tilde{f}(R_1 \cup R_2) = \lambda_G(X, Y)$. But this contradicts our assumption that S_1 and S_2 were important $X - Y$ vertex separators since $\delta(R_1 \cup R_2)$ is an $X - Y$ vertex separator which dominates both S_1 and S_2.

Note: In future we will continue to refer to the unique smallest important $X - Y$ vertex separator as S^* without explicit reference to Lemma 44.9. ∎

Lemma 44.10 [69] *Let $G = (V, E)$ be an undirected graph, $X, Y \subset V$ be disjoint vertex sets and let S be an important $X - Y$ vertex separator. Then $R(X, S) \supseteq R(X, S^*)$.*

Proof. Suppose that this is not the case and let $R_1 = R(X, S)$ and $R_2 = R(X, S^*)$ where $S \neq S^*$. We know that $R_1, R_2 \supset X$ and the minimality of S and S^* implies that $\delta(R_1) = S$ and $\delta(R_2) = S^*$. But $S, S^* \cap Y = \phi$. Hence, by Proposition 44.2, the sets $\delta(R_1 \cup R_2)$ and $\delta(R_1 \cap R_2)$ are also $X - Y$ vertex separators and hence $\tilde{f}(R_1 \cup R_2), \tilde{f}(R_1 \cap R_2) \geq \lambda_G(X, Y)$. By the submodularity of $\tilde{f}$ (Lemma 44.8) we have that

$$\tilde{f}(R_1) + \underbrace{\tilde{f}(R_2)}_{=\lambda_G(X,Y)} \geq \tilde{f}(R_1 \cup R_2) + \underbrace{\tilde{f}(R_1 \cap R_2)}_{\geq \lambda_G(X,Y)}$$

which implies that $\tilde{f}(R_1 \cup R_2) \leq \tilde{f}(R_1)$. But this contradicts our assumption that S was an important $X - Y$ vertex separator since $\delta(R_1 \cup R_2)$ is an $X - Y$ vertex separator which dominates S. ∎

Lemma 44.11 *Let $G = (V, E)$ be an undirected graph, $X, Y \subset V$ be disjoint vertex sets and S be an important $X - Y$ vertex separator.*

   a. *S is a $\{v\} - Y$ vertex separator for every $v \in R(X, S)$.*

   b. *For every $v \in S$, $S \setminus \{v\}$ is an important $X - Y$ vertex separator in $G \setminus \{v\}$.*

   c. *If S is an $X' - Y$ vertex separator for some $X' \supset X$ such that X' is reachable from X in $G[X']$, then S is also an important $X' - Y$ vertex separator.*

Proof.

1. Suppose this were not the case. Since $v \in R(X, S)$ there is a path from X to v in $G \setminus S$. Now since there is also a path from v to Y, this implies the existence of a path from X to Y in $G \setminus S$. But, this is not possible since S is an $X - Y$ vertex separator.

2. Suppose $S' = S \setminus \{v\}$ is not an important $X - Y$ vertex separator in $G' = G \setminus \{v\}$. Then there is an $X - Y$ vertex separator S_1 in G' which dominates S' in G'. Consider the set $S_2 = S_1 \cup \{v\}$. Observe that S_2 is also an $X - Y$ vertex separator in G. This is because any path from X to Y which does not contain v exists in G' and hence must contain a vertex of S_2. Now, since S_1 dominates S' in G', S_2 dominates S in G which contradicts our assumption that S is an important $X - Y$ vertex separator.

3. Assume that this is not the case. Since S is a minimal $X - Y$ vertex separator, S is also a minimal $X' - Y$ vertex separator. Therefore, if S is not an important $X' - Y$ separator it must be the case that there is an $X' - Y$ vertex separator S_1 which dominates S with respect to X'. We will show that S_1 also dominates S with respect to X, which contradicts our assumption that S is an important $X - Y$ vertex separator. Clearly, $|S_1| \leq |S|$. Hence it is enough for us to show that $R(X, S) \subset R(X, S_1)$.

First we prove that $R(X, S) \subseteq R(X, S_1)$. Consider a vertex v in $R(X, S)$. Clearly $R(X', S) \supseteq R(X, S)$, which means that $v \in R(X', S)$ and since S_1 dominates S with respect to X', v is in $R(X', S_1)$. By our assumption we have that the vertices reachable from X in $G \setminus S_1$ and those reachable from X' in $G \setminus S_1$ are the same implying that $v \in R(X, S_1)$.

Now consider some vertex $u \in S \setminus S_1$. By the minimality of S, u has a neighbor w in $R(X, S)$. But w is also in $R(X, S_1)$ which implies that $u \in R(X, S_1)$ and hence $R(X, S) \subset R(X, S_1)$. ∎

The following lemma is implicit in Chen et al. [68].

Lemma 44.12 [68] *Let $G = (V, E)$ be an undirected graph, $X, Y \subset V$ be disjoint vertex sets of G. For every $k \geq 0$ there are at most 4^k important $X - Y$ vertex separators of size at most k.*

Proof. Given $G, X, Y, k \geq 0$ we define a measure $\mu(G, X, Y, k) = 2k - \lambda_G(X, Y)$. We prove by induction on $\mu(G, X, Y, K)$ that there are at most $2^{\mu(G, X, Y, k)}$ important $X - Y$ vertex separators of size at most k.

For the base case, if $2k - \lambda_G(X, Y) < k$ then $\lambda_G(X, Y) > k$ and hence the number of important separators of size at most k is 0. If $\lambda_G(X, Y) = 0$, it means that there is no path from X to Y and hence the empty set alone is the important $X - Y$ vertex separator. For the induction step, consider $G, X, Y, k \geq 0$ such that $\mu = \mu(G, X, Y, k) \geq k, \lambda_G(X, Y) > 0$ and assume that the statement of the Lemma holds for all G', X', Y', k' where $\mu(G', X', Y', k') < \mu$.

By Lemma 44.9, there is a unique important $X - Y$ vertex separator S^* of size $\lambda_G(X, Y)$. Since we have assumed $\lambda_G(X, Y)$ to be positive, S^* is non empty. Consider a vertex $v \in S^*$. Any important $X - Y$ vertex separator S either contains v or does not contain v. For any important $X - Y$ vertex separator S which contains v, $S \setminus \{v\}$ is an important $X - Y$ vertex separator in $G \setminus \{v\}$ (Lemma 44.11(b)). Hence the number of important $X - Y$ vertex separators containing v, of size at most k in G is at most the number of important $X - Y$ vertex separators of size at most $k - 1$ in $G \setminus \{v\}$. Observe that $\lambda_{G \setminus \{v\}}(X, Y) = \lambda_G(X, Y) - 1$ which implies that $\mu(G \setminus \{v\}, X, Y, k - 1) < \mu$ and by induction hypothesis, the number of important $X - Y$ vertex separators of size at most $k - 1$ in $G \setminus \{v\}$ is bounded by $2^{\mu - 1}$ which is also a bound on the number of important $X - Y$ vertex separators in G which have size at most k and contain v.

Now let S be an important $X - Y$ vertex separator of size at most k which does not contain v. By Lemma 44.10 we know that $R(X, S) \supseteq R(X, S^*)$ and by the minimality of S^*, v has a neighbor in $R(X, S)$ which implies that $R(X, S) \supseteq R(X, S^*) \cup \{v\}$. We now set $X' = R(X, S^*) \cup \{v\}$. By Lemma 44.11(c) we know that S is also an important $X' - Y$ vertex separator. Thus a bound on the number of important $X' - Y$ vertex separators of size at most k is also a bound on the number of important $X - Y$ vertex separators of size at most k which do not contain v. First note that $\lambda_G(X', Y) > \lambda_G(X, Y)$ since otherwise we would have an $X - Y$ vertex separator which dominates S^* with respect to X. Now, $\mu(G, X', Y, k) < \mu$ and by induction hypothesis, the number of important $X' - Y$ vertex separators of size at most k is bounded by $2^{\mu - 1}$.

Summing up the bounds we get that the number of important $X - Y$ separators of size at most k is bounded by $2 \cdot 2^{\mu - 1} = 2^\mu \leq 2^{2k}$. ∎

44.2.5.2 Algorithm for Multiway Cut

In this section we give a FPT algorithm for multiway cut. Next we show a lemma that proves that important separators are sufficient to separate a terminal from the set of terminals.

Lemma 44.13 [67] *Let $(G = (V, E), T, k)$ be an instance of multiway cut. If (G, T, k) is a YES instance then it has a solution $\hat{S}$ such that a minimal subset of $\hat{S}$ separating t_1 from $T \setminus \{t_1\}$ is an important $t_1 - T \setminus \{t_1\}$ vertex separator.*

Proof. Let $S \subseteq V$ be a minimal solution and let S_1 be a minimal subset of S such that $G \setminus S$ has no $t_1 - T \setminus \{t_1\}$ path. If S_1 is the empty set, it must be the case that there is no path from t_1 to $T \setminus \{t_1\}$ in G. By definition, S_1 is an important $t_1 - T \setminus \{t_1\}$ vertex separator and we are done by setting $\hat{S} = S$. Hence, we will assume that S_1 is non-empty. If S_1 is an important $t_1 - T \setminus \{t_1\}$ vertex separator, we are done by setting $\hat{S} = S$. Suppose that this is not the case.

Since S_1 is a minimal $t_1 - T \setminus \{t_1\}$ vertex separator which is not important, there is a $t_1 - T \setminus \{t_1\}$ vertex separator S_2 which dominates S_1. Set $\hat{S} = (S \setminus S_1) \cup S_2$. We claim that $\hat{S}$ is a solution of this instance, which satisfies the statement of the lemma. Clearly $|\hat{S}| \leq |S|$ and the minimal part of $\hat{S}$ separating t_1 from $T \setminus \{t_1\}$ is S_2 which by our assumption is an important $t_1 - T \setminus \{t_1\}$ vertex separator.

It remains for us to prove that $\hat{S}$ is a multiway cut of T. Suppose this is not so and let there be a path P from t_i to t_j in $G \setminus \hat{S}$. Since S was a multiway cut of T, P contains a vertex $v \in S_1 \setminus S_2$ and hence there is a path P' from v to T. Since S_1 was minimal, v has a neighbor u in $R(t_1, S_1)$. Now, $R(t_1, S_1) \subset R(t_1, S_2)$ and $v \notin S_2$. Hence, it must be the case that $v \in R(t_1, S_2)$. But then, S_2 is a $t_1 - T \setminus \{t_1\}$ vertex separator and there is a path from $v \in R(t_1, S_2)$ to T which is not possible by Lemma 44.11(a). This concludes the proof of the lemma. ∎

Theorem 44.10 *Multiway cut can be solved in time $4^{k^2} n^{O(1)}$.* ∎

Proof. The algorithm for multiway cut works as follows. Given a set T of terminals, let t_i be a terminal such that there is a path from t_i to $T \setminus \{t_i\}$. Now we try to separate t_i from the rest of terminals. By Lemma 44.13 we know that if there is a solution S that separates all pairs in T then there is one that contains an important $t_i - T \setminus \{t_i\}$ separator. So for the algorithm we enumerate all the $t_i - T \setminus \{t_i\}$ important separators using Lemma 44.12 (observe that the proof can be made algorithmic by making it a recursive algorithm). Since there is a path from t_i to $T \setminus \{t_i\}$ we have that each important separators has size at least one. For each important separator X, we delete X from G and try to separate the remaining terminals from each other as described above with at most $k - |X|$ vertices. Since in each branch k drops by at least one we have that the search tree has depth at most k. Since the branching factor is at most 4^k, we have that the branching algorithm runs in time $4^{k^2} n^{O(1)}$. ∎

A refined analysis of the algorithm described in Theorem 44.10 results in an algorithm with running time $4^k n^{O(1)}$. The best-known algorithm for multiway cut runs in time $2^k n^{O(1)}$ [70] and is based on ideas similar to the one described for the algorithm for veretx cover above LP. Marx and Razgon [69] showed that multicut is FPT using the ideas described in this section and a technique called *randomized selection of important separators*.

44.2.6 Well-Quasi-Ordering

The set of natural numbers is *well-ordered* since every two natural numbers are comparable, that is, for any two numbers a and b such that $a \neq b$, either $a < b$ or $b < a$. An unusual way to describe a well-ordered set is to say that it contains no *anti-chain* of length at least 2, where an anti-chain is a sequence $a_1, a_2, \ldots, a_k$ such that for every $i \neq j$ we have $a_i \neq a_j$ and neither $a_i < a_j$ nor $a_i > a_j$. We can relax the notion of well-ordering, and say that a set S is *well-quasi-ordered* under a relation $<$ if every anti-chain in S is finite.

Robertson and Seymour (2004) proved in their *graph minors* project that the set of graphs is well-quasi-ordered under the minor relation, thereby proving the *graph minor theorem* and resolving *Wagner's conjecture* [71]. This gives a very powerful and interesting way to prove a variety of parameterized problems FPT.

We start with some definitions. A graph H is a *minor* of a graph G, written $H \leq_m G$ if a graph isomorphic to H can be obtained from G by a sequence of the operations: (1) taking a subgraph, and (2) contracting an edge. While contracting an edge, its both end points become identified to a single vertex which is made adjacent to all the vertices that were adjacent to the end points of the edge.

A family of graphs $\mathcal{F}$ is closed under the minor order if $G \in \mathcal{F}$ and $H \leq_m G$ imply $H \in \mathcal{F}$.

Proposition 44.3 (Graph minor theorem [72]) *The set of graphs is well-quasi-ordered under the minor relation.*

Consider a graph class $\mathcal{G}$ that is closed under taking minors. That is, if $G \in \mathcal{G}$ and $H \leq_m G$ then $H \in \mathcal{G}$ as well. Consider the set of graphs $\mathcal{F}$ consisting of all graphs *not* in $\mathcal{G}$ such that all their minors are in $\mathcal{G}$. We call this set the *set of forbidden minors* of $\mathcal{G}$. Notice that every graph G that is not in $\mathcal{G}$ must have some minor in $\mathcal{F}$. Notice also that by definition, $\mathcal{F}$ is an antichain, and that by the graph minor theorem $\mathcal{F}$ is finite. Hence, to check whether G belongs to $\mathcal{G}$ it is sufficient to check for every $H \in \mathcal{F}$ whether G contains H as a minor. To this end, the following theorem, also from the graph minors project, is useful.

Proposition 44.4 [73] *For every fixed graph H there is an $O(n^3)$ time algorithm to check for an input graph G whether $H \leq_M G$.*

Combining the graph minors theorem with Proposition 44.4 yields that every minor closed graph class $\mathcal{G}$ can be recognized in $O(n^3)$ time, where the constant hidden in the big-Oh notation depends on $\mathcal{G}$. To put this in terms of parameterized algorithms—the problem of deciding whether G belongs to a minor closed class $\mathcal{G}$ is FPT parameterized by $\mathcal{G}$.

Corollary 44.1 [74] *Vertex cover is FPT.*

Proof. We prove that if a graph G has a vertex cover C of size at most k then so do all the minors of G. For any edge $e \in E(G)$, C is a vertex cover of $G \setminus e$. Similarly for any $v \in V(G)$, $C \setminus \{v\}$ is a vertex cover of $G \setminus v$. Finally, for an edge $uv \in E(G)$ let G_{uv} be the graph obtained by contracting the edge uv and let u' be the vertex obtained from u and v during the contraction. Then, every edge in $E(G)$ with u or v being one of its endpoints will have u' as its endpoint in G_{uv}. Hence, if neither u nor v is in C then C is a vertex cover of G_{uv}. If $u \in C$ or $v \in C$ then $(C \setminus \{u,v\}) \cup \{u'\}$ is a vertex cover of size at most k of G_{uv}. Hence, for a fixed integer $k \geq 0$ the class $\mathcal{C}_k$ of graphs that have a vertex cover of size at most k is closed under taking minors.

By Theorem 44.3 $\mathcal{C}_k$ has a finite set $\mathcal{F}_k$ of forbidden minors. By Proposition 44.4, for each $H \in \mathcal{F}_k$ we can decide whether G contains H as a minor in time $O(f(|H|)n^3)$ for some function $f: \mathbb{N} \to \mathbb{N}$. Let h be the size of the largest graph in $\mathcal{F}_k$. Then deciding whether a particular graph G is in $\mathcal{C}_k$ can be done in time $O(|\mathcal{F}_k|f(h)n^3)$. ∎

For a fixed minor closed graph class $\mathcal{G}$, consider the following problem. Input is graph G and integer k. The parameter is k and the objective is to determine whether there exists a set vertex set S of size at most k such that $G \setminus S \in \mathcal{G}$. The proof of Corollary 44.1 can easily be modified to show that for any fixed minor closed graph class $\mathcal{G}$ this problem is FPT.

Though this technique puts a lot of problems (some of which were not even known to be decidable earlier, see [74, 75] in FPT, there are two drawbacks with this approach. First, the minor testing algorithm of Robertson and Seymour has really huge constant of proportionality. The second, the more major, drawback is that the graph minor theorem is non-constructive. That is, the theorem simply guarantees the existence of a finite obstruction set without giving any means of identifying the elements of the set, the cardinality of the set, or even the order of the largest graph in the set. To handle these problems for which only an *existence* of a fixed-parameter algorithm is known, Downey and Fellows calls these problems nonuniformly FPT (these are problems for which there is a constant α and a sequence of algorithms Φ_x such that, for each $x \in N$, Φ_x computes L_x in time $O(n^\alpha)$).

Fortunately, for some of these problems, uniform fixed-parameter tractability is known using our earlier techniques. Furthermore, Fellows and Lungston [74–77] have developed some general techniques using self-reducibility and a graph theoretic generalization of Myhill Nerode theorem of formal language theory [76], to algorithmically construct the obstruction set along the way, for some of these problems thereby proving them uniformly FPT.

We refer to the survey of Downey and Thilikos [78] for a detailed overview of this method.

44.2.7 Bounded Treewidth Machinery

The notion of treewidth which is a measure of the graph to indicate how tree-like the graph is, was introduced by Robertson and Seymour [79].

Definition 44.11 *Let $G = (V, E)$ be a graph. A tree decomposition of G is a pair $(\{X_i | i \in I\}, T = (I, F))$ with $\{X_i | i \in I\}$ a family of subsets of V, and T a tree, with the following properties:*

- $\bigcup_{i \in I} X_i = V$.

- *Every edge $e = (v, w) \in E$, there is an $i \in I$ with $v \in X_i$ and $w \in X_i$.*

- *For every $v \in V$, the set $\{i | v \in X_i\}$ forms a connected subtree of T.*

The treewidth of a tree decomposition $(\{X_i | i \in I\}, T)$ is $max_{\{i \in I\}}(|X_i| - 1)$. The treewidth of G, denoted by treewidth(G), is the minimum treewidth of a tree decomposition of G, taken over all possible tree decompositions of G.

There are several alternative ways to characterize the class of graphs with treewidth $\geq k$, for example, as partial k-trees [80]. A large number of NP-complete (and other) graph problems can be solved in polynomial and even linear time when restricted to graphs with constant treewidth [81,82] especially if the tree decomposition is given.

Parameterized problems where the parameter is the treewidth of the input graph G are often tackled by doing dynamic programming over the tree decomposition of G. In particular it has been shown that independent set, vertex cover, and dominating set in graphs of treewidth at most k can be solved in time $O(2^k k^{O(1)} n)$, $O(2^k k^{O(1)} n)$, and $O(4^k n)$, respectively [11]. Many other problems admit fast dynamic programming algorithms in graphs of bounded treewidth.

A very useful tool for showing that a problem is FPT parameterized by treewidth is the celebrated Courcelle's theorem, which states that every problem expressible in Monadic second-order logic is FPT parameterized by the treewidth of the input graph. For a graph predicate ϕ expressed in MSO_2 let $|\phi|$ be the length of the MSO_2 expression for ϕ.

Theorem 44.11 (Courcelle's theorem [83]) *There is a function $f : \mathbb{N} \times \mathbb{N} \to \mathbb{N}$ and an algorithm that given a graph G together with a tree-decomposition of G of width t and a MSO_2 predicate ϕ decides whether $\phi(G)$ holds in time $f(|\phi|, t)n$.* ■

To apply Courcelle's theorem on a specific problem we need to show that the problem is expressible in monadic second-order logic. For example, consider the independent set problem. Here we are given as input a graph G, a tree-decomposition of G of width t and an integer k. The objective is to decide whether G has an independent set of size at least k. We formulate the independent set problem in MSO_2. That is, a graph G has an independent set of size k if and only if the following predicate holds:

$$\phi(G) = \exists v_1, v_2, \ldots, v_k \in V(G) : v_1 \neq v_2 \wedge \neg\mathbf{adj}(v_1, v_2), v_1 \neq v_3 \wedge \neg\mathbf{adj}(v_1, v_3),$$
$$\ldots v_2 \neq v_3 \wedge \neg\mathbf{adj}(v_2, v_3), \ldots v_{k-1} \neq v_k \wedge \neg\mathbf{adj}(v_{k-1}, v_k).$$

Observe that the length of the predicate ϕ depends only on k, and not on the size of the graph G. Hence, by Theorem 44.11, there is an algorithm that given a graph G together with a tree-decomposition G of width at most t and an integer k decides whether G has an independent set of size at least k in time $f(k, t)n$ for some function f. In fact, it is not really necessary that the tree-decomposition of G is given. Due to a result of Bodlaender [84], a tree-decomposition of width t of a graph G of treewidth t can be computed in time $f(t)n$ for some function f.

Theorem 44.12 (Bodlaender's theorem [84]) *There is a function $f : \mathbb{N} \to \mathbb{N}$ and an $f(t)n$ time algorithm that given a graph G and integer t decides whether G has treewidth at most t, and if so, constructs a tree-decomposition of width at most t.* ■

Hence, combining Theorems 44.11 and 44.12 yields that there is a function $f : \mathbb{N} \times \mathbb{N} \to \mathbb{N}$ and an algorithm that given a graph G of treewidth t and a MSO_2 predicate ϕ decides whether $\phi(G)$ holds in time $f(|\phi|, t)n$. For example, by the discussion in the previous paragraph there is an algorithm that given a graph G of treewidth t and an integer k decides whether G has an independent set of size at least k in time $f(t, k)n$ for some function f.

Theorem 44.11 has been generalized even further. For instance it has been shown that MSO_2-*optimization* problems are FPT parameterized by the treewidth of the input graph. In a MSO_2-*minimization* problem you are given a graph G and a predicate ϕ in MSO_2 that describes a property of a vertex (edge) set in a graph. The objective is to find a vertex (edge) set S of minimum size such that $\phi(G, S)$ holds. In a MSO_2-*maximization* problem the objective is to find a set S of maximum size such that $\phi(G, S)$ holds.

Theorem 44.13 [85,86] *There is a function $f : \mathbb{N} \times \mathbb{N} \to \mathbb{N}$ and an algorithm that given a graph G of treewidth t and a MSO_2 predicate ϕ finds a largest (smallest) set S such that $\phi(G, S)$ holds in time $f(|\phi|, t)n$.* ■

Turning our attention back to the independent set problem, we can observe that by applying Theorem 44.13 we can obtain a stronger result than from Theorem 44.11. In particular, the independent set can be expressed as a MSO_2-maximization problem as follows:

$$\max |S| \text{ s.t:} \phi(G, S) = \forall v_1, v_2 v_1 = v_2 \vee \neg\mathbf{adj}(v_1, v_2) \text{ holds}$$

Hence, by Theorem 44.13 there is an algorithm that given a graph G of treewidth t as input can find a maximum size independent set in time $f(t)n$ for some function f.

Another way to generalize Theorem 44.11 is to consider larger classes of graphs than graphs of bounded treewidth. In particular, if we restrict the predicate ϕ to MSO_1 logic, Theorem 44.13 can be extended to graphs of bounded cliquewidth.

Theorem 44.14 [87] *There is a function $f : \mathbb{N} \times \mathbb{N} \to \mathbb{N}$ and an algorithm that given a graph G and a clique-expression G of width t and a MSO_1 predicate ϕ finds a largest (smallest) set S such that $\phi(G, S)$ holds in time $f(|\phi|, t)n$.* ∎

There are natural problems expressible in MSO_2 but not in MSO_1. Examples include Hamiltonian cycle and max cut.

Treewidth concept has also been used to prove some parameterized problems FPT in the following way. For these parameterized problems, fixing the parameter k implies that the yes-instances (or sometimes the no-instances) have treewidth bounded by a function of k. We first use Theorem 44.12 for such problems to identify whether the treewidth of the given graph is that bounded function of k (otherwise it will be a no or yes instance appropriately) and if so we get the tree decomposition as well from Theorem 44.12. Given the tree decomposition, we apply the polynomial time algorithm for the problem for bounded treewidth graphs, thus obtaining a FPT algorithm.

We illustrate this technique through the longest cycle problem. That is, given an undirected graph G, and an integer parameter k, does G have a cycle of length at least k? We first need the following result.

Theorem 44.15 [88] *There exists an $O(c^k n^{O(1)})$ algorithm to find the longest cycle (or longest path) in a given graph G that is given together with a tree decomposition of G with a treewidth $\leq k$.* ∎

Now to find whether a given graph G has a cycle of length at least k, first grow a depth-first tree rooted at any vertex noting the depth-first number for each vertex. When a back edge is encountered, if the difference between the two DFS numbers is at least $k - 1$, we have already encountered a cycle of length at least k. Otherwise, once the DFS tree T is constructed, for all $v \in V$, let X_v be the set containing v and its at most $k - 2$ direct predecessors in T. Then $(\{X_v | v \in V\}, T)$ is a T-based tree decomposition of G of treewidth at most $k - 2$. Thus, we have that the *no* instances of the problem have treewidth at most $k - 2$. Thus (by using Theorem 44.12 or by the DFS tree method above) we can find a cycle of length at least k or find a tree decomposition of width at most $k - 2$. In the latter case, we apply the above theorem to test for the existence of a cycle of length k or more.

We refer to the book by Courcelle and Engelfriet [89] for an interplay between algebraic graph transformations and logic.

44.2.8 Subexponential Algorithms and Bidimensionality

Many parameterized problems are optimization problems parameterized by the objective function value. The objective function is said to be *bidimensional*, if the optimal value of an $n \times m$-grid is $O(nm)$, and if the optimal value of any minor H of G is at most the optimal value o¡/comment¿f G. A parameterized problem is *bidimensional* if the input is a graph G and an integer k which is the parameter, and the objective is to determine whether the optimal value of a bidimensional objective function on G is at most k. Most bidimensional problems exhibit nice algorithmic properties on planar graphs. In particular, it was proved by Demaine et al. [90] that any bidimensional problem which can be solved in time $2^{O(t+k)} n^{O(1)}$ on graphs of treewidth t admits a $2^{O(\sqrt{k})} n^{O(1)}$ time algorithm in planar graphs. The results proved in Demaine et al. [90] are for more general classes of graphs and more general classes of problems than considered here. We demonstrate how these ideas apply to vertex cover in planar graphs. We will need two propositions about treewidth and graph minors that we will use as black boxes. Notice that if a minor of G has treewidth at least t then the treewidth of G is also at least t.

Proposition 44.5 (Excluded grid theorem [91]) *Every planar graph G of treewidth at least t contains a $t/4 \times t/4$-grid as a minor.*

Proposition 44.6 [92] *The treewidth of planar graphs can be 3/2-approximated in polynomial time.*

We explain how to exploit the above propositions to show that vertex cover in planar graphs can be solved in time $2^{O(\sqrt{k})}$. As observed in Section 44.2.6, if a graph has a vertex cover of size at least k, then so do all its minors. Observe also that the size of a minimum vertex cover of a $k \times k$ grid is at least $k^2/2$.

Theorem 44.16 [90,93,94]. *There is an $2^{O(\sqrt{k})} n^{O(1)}$ time algorithm for vertex cover on planar graphs.*

Proof. On an input instance (G, k) we perform a test in polynomial time—we run the 3/2-approximation for treewidth on G, and let t be the width of the decomposition returned by the approximation algorithm. We have two possible outcomes, either $(2t/3 * 4)^2/2 = t^2/72 > k$ or not. If not, then we can find an optimal vertex cover in time $O(2^t n^{O(1)}) \leq 2^{O(\sqrt{k})} n^{O(1)}$ by applying the $2^t n^{O(1)}$ time dynamic programming algorithm for vertex cover in graphs of bounded treewidth [11]. If $t^2/72 > k$ then G contains a $(\sqrt{2k+1} \times \sqrt{2k+1})$ grid as a minor. The size of the minimum vertex cover of this grid is more than k and hence (G, k) is a no-instance. ∎

Algorithms of the above type can be given for any bidimensional problem which can be solved efficiently in graphs of bounded treewidth, and several survey papers have been written on bidimensionality [90,93–97].

44.3 ECOLOGY OF PARAMETERS

Parameterized complexity grew out of the realization that most problem inputs come with some natural parameters that are likely to be small. However, much of the early work on parameterized complexity focused on the solution size as the parameter (barring the exceptions of treewidth and some problems on strings; see [11]). Here we outline some recent paradigms that have departed from this parameterization.

44.3.1 Parameterizing beyond the Guarantee Bounds

For a number of optimization problems, there is some (upper or lower) bound for the optimum solution, and hence the problem becomes trivial for small values of k, the solution size (see examples below).

- In any graph on n vertices and maximum degree Δ, any vertex cover must have size at least n/Δ. Any dominating set must have size at least $n/(\Delta + 1)$.
- In any Boolean CNF formula on m clauses, there is an assignment that satisfies at least $m/2$ clauses.
- In any graph on m edges, there is a cut with at least $m/2$ edges; that is, there is a partition of the vertex set into two such that at least $m/2$ edges go between the two parts.

- In any planar graph on n vertices, there is an independent set of size at least $n/4$ (as the graph can be properly colored with 4 colors).

- In any graph, the size of the vertex cover is at least the size of the maximum matching, In particular, if the graph on n vertices has a perfect matching, then any vertex cover must have size at least $n/2$.

Hence the parameterized versions of these problems become trivial when k is small (e.g., for the parameterized version of the cut problem which asks where there is a cut of size at least k, the answer is trivially YES if $k \leq m/2$), and when k is large, the input itself becomes a *trivial* kernel, and the standard brute force algorithm becomes a FPT algorithm. But these algorithms are not practical, as k, for the situation when these algorithms are applied, is large.

To deal with this anomaly, Mahajan and Raman [9] introduced the notion of parameterizing beyond the guarantee. So the parameter is not the solution size, but the value of the solution minus the lower bound, which can range from small to moderate values for most instances.

At the other extreme, one can also obtain trivial upper bounds for the solution size for most of these problems, and one can parameterize away from those upper bounds (e.g., a cut of size at least $m-k$, a vertex cover of size at most $n-k$). Such parameterizations are similar to what Niedermierr [11] calls *the distance from triviality*. See [98] for more examples and some hardness results even for such parameterizations. See [99] for an extensive study on such parameterizations for a number of problems, in particular, for constraint satisfaction problems.

Most optimization problems can be formulated as ILP problems. And for such minimization problems, the optimum value of the linear programming relaxation gives a lower bound on the solution size. Cygan et al. [70] introduced the idea of parameterizing above this LP lower bound and Narayanaswamy et al. [21] and Lokshtanov et al. [22] have carried this forward to obtain impressive bounds for vertex cover above the maximum matching bound. In addition, by showing reductions from a number of problems (with standard parameterization) to this problem, Narayanaswamy et al. [21] and Lokshtanov et al. [22] obtain improved bounds for a number of vertex deletion problems including odd cycle traversal and split vertex deletion. See [31, 32] for randomized polynomial-sized kernel for these problems.

44.3.2 Structural Parameters

Another way to parameterize the *distance from triviality* is to look at the input instance itself and see how far it is from a *trivial* (or a simpler class of) input(s).

For example, graphs having small vertex covers are, in some sense, *close to* edgeless graphs. Graphs having small FVS are *close to* forests. Graphs having small odd cycle traversals are *close to* bipartite graphs. This suggests using some structural parameter (like the size of a vertex cover or a FVS or an odd cycle traversal) as the given parameter. For example, while treewidth as a measure of how tree-like the graph was, worked for generalizing the dynamic programming algorithms on trees, the size of the FVS can be another useful measure for how close to a forest the graph is.

An early work on this due to Cai [100] on graph coloring has been revived recently. See [101] for parameterized algorithms for problems parameterized by vertex cover or the size of the maximum leaf-spanning tree. Fellows et al. [102] has explored this in multiple directions including obtaining polynomial kernels for problems parameterized by other structural parameters.

44.3.3 Backdoors to Satisfiability

As was mentioned in the introduction, one area where parameterized complexity has played a major role is in explaining how practical solvers for satisfiability could solve, in a few seconds, instances having millions of variables while the problem is NP-complete and hence a feasible algorithm for general instances is unlikely. The answer lies in formally identifying hidden structures in practical input instances. A relevant notion in that direction is the notion of *a backdoor set*, a set of variables of a propositional formula such that fixing the truth value of the variables places the formula into some polynomially solvable class. See [103] for various parameterized complexity results for identifying small backdoor sets in different types of propositional formula.

44.4 PARAMETERIZED INTRACTABILITY

In this section we outline the parameterized interactability theory. How can we show that a problem does not have an algorithm with running time $f(k) \cdot n^{O(1)}$? One approach would be to show NP-hardness. For an example, consider coloring—given an input graph G and a positive integer k, test whether the graph G can be properly colored with at most k colors—parameterized by the number of colors. However, it is well known that the problem of testing whether a graph can be properly colored with at most 3 colors is NP-complete. Thus, this implies that coloring can not have even $n^{f(k)}$ algorithm leave alone an algorithm with running time $f(k) \cdot n^{O(1)}$. But with this method, we can not distinguish between problems that are solvable in time $n^{f(k)}$ from problems solvable in time $f(k) \cdot n^{O(1)}$. To be able to do this, Downey and Fellows [13] introduced the W-hierarchy. The hierarchy consists of a complexity class $W[t]$ for every integer $t \geq 1$ such that $W[t] \subseteq W[t+1]$ for all t. Downey and Fellows [13] proved that

$$\text{FPT} \subseteq W[1] \subseteq W[2] \ldots \subseteq W[t]$$

and conjectured that strict containment holds.

In particular, the assumption FPT $\neq W[1]$ is the fundamental complexity theoretic assumption in parameterized complexity. The reason for this is that the assumption is a natural parameterized analogue of the conjecture that P $\neq$ NP. The assumption P $\neq$ NP can be reformulated as "The non-deterministic turing machine problem can not be solvable in polynomial time." In this problem we are given a non-deterministic Turing machine M, a string s and an integer k coded in unary. The question is whether M can make its non-deterministic choices in such a way that it accepts s in at most k steps. The intuition behind the P $\neq$ NP conjecture is that this problem is so general and random that it is not likely to be in P. Similarly, one would not expect the problem parameterized by k to be FPT.

We start with a few simple definitions from parameterized complexity to formalize some of the notions. We mainly follow the notation of Flum and Grohe [14]. We describe decision problems as languages over a finite alphabet Σ.

Definition 44.12 *Let Σ be a finite alphabet.*

1. *A parameterization of Σ^* is a polynomial time computable mapping $\kappa : \Sigma^* \to \mathbb{N}$.*
2. *A parameterized problem (over Σ) is a pair (Q, κ) consisting of a set $Q \subseteq \Sigma^*$ of strings over Σ and a parameterization κ of Σ^*.*

For a parameterized problem (Q, κ) over alphabet Σ, we call the strings $x \in \Sigma^*$ the *instances* of Q or (Q, κ) and the number of $\kappa(x)$ the corresponding *parameters*.

A common way to obtain lower bounds is by reductions. A reduction from one problem to another is just a proof that a *too fast* solution for the latter problem would transfer to a too fast solution for the former. The specifics of the reduction varies based on what we mean by *too fast*. The next definition is of a kind of reduction that preserves fixed-parameter tractability.

Definition 44.13 *Let (Q, κ) and (Q', κ') be two parameterized problems over the alphabet Σ and Σ', respectively. An FPT reduction (more precisely FPT many-one reduction) from (Q, κ) to (Q', κ') is a mapping $R : \Sigma^* \to (\Sigma')^*$ such that*

1. *For all $x \in \Sigma^*$ we have $x \in Q$ if and only if $R(x) \in Q'$.*

2. *R is computable by an FPT-algorithm (with respect to κ).*

3. *There is a computable function $g : \mathbb{N} \to \mathbb{N}$ such that $\kappa'(R(x)) \leq g(\kappa(x))$ for all $x \in \Sigma^*$.*

It can be verified that FPT reductions work as expected: if there is an FPT reduction from (Q, κ) to (Q', κ') and $(Q', \kappa') \in$ FPT, then $(Q, \kappa) \in$ FPT as well.

The class $W[1]$ is the set of all parameterized problem that are FPT-reducible to the parameterized non-deterministic turning machine acceptance problem. A parameterized problem is said to be $W[1]$-*hard* if all problems in $W[1]$ FPT-reduce to it. The following theorem follows directly from Definition 44.13.

Theorem 44.17 [13] *Let P and Q be parameterized problems. If $P \leq_{FPT} Q$ and Q is in FPT then P is in FPT. Furthermore, if $P \leq_{FPT} Q$ and P is $W[1]$-hard then Q is $W[1]$-hard.* ∎

Theorem 44.17 implies that a $W[1]$-hard problem can not be in FPT unless FPT $= W[1]$. As mentioned above, it was proved in Downey and Fellows [13] that independent set is $W[1]$-*hard*. We now give two examples of FPT-reductions. First, we show that the multicolor clique problem is $W[1]$-hard.

44.4.1 Example Reductions

From an *engineering* viewpoint, how can we use the assumption that FPT $\subset W[1]$ to rule out $f(k) \cdot n^{O(1)}$ time algorithms for a particular problem? Downey and Fellows showed that the independent set problem is not in FPT unless FPT $= W[1]$. If we can show for a particular parameterized problem Π, that if Π is FPT then so is independent set, then this implies that $\Pi \notin$ FPT. The approach for using the multicolor clique problem in reductions is described in Fellows et al. [104], and has been proven to be very useful in showing hardness results in parameterized complexity.

MULTICOLOR CLIQUE
Instance: An undirected graph $G = (V[1] \cup V[2] \cdots \cup C[k], E)$ such that for every i the vertices of $V[i]$ induce an independent set, and a positive integer k.
Parameter: k.
Problem: Does there exist a k-sized clique C in G?

Theorem 44.18 *Multicolor clique is $W[1]$-hard.* ∎

Proof. We reduce from the independent set problem. Given an instance (G,k) to independent set we construct a new graph $G' = (V', E')$ as follows. For each vertex $v \in V(G)$ we make k copies of v in V' with the ith copy being colored with the ith color. For every pair $u,v \in V(G)$ such that $uv \notin E(G)$ we add edges between all copies of u and all copies of v with different colors. It is easy to see that G has an independent set of size k if and only if G' contains a clique of size k. This concludes the proof. ∎

One should notice that the reduction produces instances to multicolor clique with a quite specific structure. In particular, all color classes have the same size and the number of edges between every pair of color classes is the same. It is often helpful to exploit this fact when reducing from multicolor clique to a specific problem. We now give an example of a slightly more involved FPT-reduction.

Theorem 44.19 *Dominating set is $W[1]$-hard.* ∎

Proof. We reduce from the multicolor clique problem. Given an instance (G, k) to multicolor clique we construct a new graph G'. For every $i \leq k$ let V_i be the set of vertices in G colored i and for every pair of distinct integers $i, j \leq k$ let $E_{i,j}$ be the set of edges in $G[V_i \cup V_j]$. We start making G' by taking a copy of V_i for every $i \leq k$ and making this copy into a clique. Now, for every $i \leq k$ we add a set S_i of $k+1$ vertices and make them adjacent to all vertices of V_i. Finally, for every pair of distinct integers $i, j \leq k$ we consider the edges in $E_{i,j}$. For every pair of vertices $u \in V_i$ and $v \in V_j$ such that $uv \notin E_{i,j}$ we add a vertex x_{uv} and make it adjacent to all vertices in $V_i \setminus \{u\}$ and all vertices in $V_j \setminus \{v\}$. This concludes the construction. We argue that G contains a k-clique if and only if G' has a dominating set of size at most k.

If G contains a k-clique C then C is a dominating set of G'. In the other direction, suppose G' has a dominating set S of size at most k. If for some i, $S \cap V_i = \emptyset$ then $S_i \subseteq S$, contradicting that S has size at most k. Hence for every $i \leq k$, $S \cap V_i \neq \emptyset$ and thus S contains exactly one vertex v_i from V_i for each i, and S contains no other vertices. Finally, we argue that S is a clique in G. Suppose that $v_i v_j \notin E_{i,j}$. Then there is a vertex x in $V(G')$ with neighborhood $V_i \setminus \{u\}$ and $V_j \setminus \{v\}$. This x is not in S and has no neighbors in S contradicting that S is a dominating set of G'. ∎

In fact, it was an FPT-reduction from independent set to dominating set that was the starting point of the complexity part of parameterized algorithms and complexity. In 1989, Fellows realized that one could give a FPT-reduction from independent set to dominating set, but that it did not seem plausible to give a reduction in the other direction. Later, Downey and Fellows proved that the dominating set problem in fact is complete for the class $W[2]$ while independent set is complete for $W[1]$ [13]. It is therefore unlikely that an FPT-reduction from dominating set to independent set can exist. The theorem 44.19 is due to Downey and Fellows. The proof presented here is somewhat simpler than the original proof and due to Daniel Lokshtanov.

Even though it was shown by Downey and Fellows that dominating set is complete for $W[2]$ while independent set is complete for $W[1]$ [13], there are problems for which it is easier to give FPT-reductions from dominating set than from independent set. A typical example is the Steiner tree problem, parameterized by the number of *non-terminals* in the solution. In the Steiner tree problem we are given a connected graph G together with a subset X of $V(G)$ and an integer k. The vertices in the X are called *terminals*. The objective is to find a subtree T of G containing all terminals and at most k non-terminals. Such a tree is called a *Steiner tree* of G. Here we give a proof that the Steiner tree problem parameterized by the number of non-terminals in the Steiner tree is $W[2]$ hard. The proof of the following theorem first appeared in Bodlaender and Kratsch [105].

Theorem 44.20 [105] *The Steiner tree problem parameterized by $|V(T) \setminus X|$ is $W[2]$-hard.*

Proof. We reduce from the dominating set problem. For an instance (G, k) we build a graph G' as follows. We make two copies of $V(G)$, call them X' and N. The copy $x \in X'$ of each vertex $v \in V(G)$ is made adjacent to the copies in N of the vertices in $N(v)$. Finally the vertex set X is obtained from X' by adding a single vertex u_X and making it adjacent to all vertices in N. This concludes the construction of G'. We prove that G has a dominating set of size at most k if and only if G' has a Steiner tree with at most $|X| + k$ vertices.

In one direction, suppose G has a dominating set S on k vertices. Let S' be the copy of S in N. Then the graph $G'[X \cup S']$ is connected. Let T be a spanning tree of $G'[X \cup S']$, then T is a Steiner tree of G' with at most $|X| + k$ vertices. In the other direction, suppose G' has a Steiner tree T on at most $|X| + k$ vertices and let $S' = V(T) \cap N$. Then $|S'| \leq k$ and since X is an independent set in G' every vertex in X has a neighbor in S'. Thus, if we let S be the copy of S' in $V(G)$ then S is a dominating set of G of size at most k. ∎

Just to give a flavor of a *real parameterized completeness result*, we will now show that the dominating set problem is $W[2]$-complete for tournaments (complete directed graphs) [106,107], given that it is $W[2]$-complete for general directed graphs [13].

Theorem 44.21 [106,107] *Dominating set is $W[2]$-complete on tournaments.*

Proof. Given a directed graph $G = (V, E)$, and an integer k, we construct a tournament T using a FPT algorithm such that G has a dominating set of size k if and only if T has a dominating set of size $k + 1$.

Recall that a dominating set D in a directed graph G is a set of vertices of G such that for every other vertex of G, there is an outgoing edge from some vertex of D. A vertex x dominates a vertex y if $x = y$ or there is a directed edge from x to y. The set of vertices dominated by a set S of vertices is simply S union the set of outneighbors of vertices of S. First we will show that the (decision version of) the minimum dominating set problem is NP-complete for directed acyclic graphs.

Lemma 44.14 *Given a directed acyclic graph G, and an integer k, the problem of testing whether G has a dominating set of size k is NP-complete.*

Proof. Our reduction is from the same problem on general directed graphs. Let $G(V, E)$ be a directed graph. We construct a directed acyclic graph $\overline{G}$ from G as follows. $\overline{G}$ consists of two vertices v_o, and v_i for every vertex v of G with a directed edge from v_o to v_i. Furthermore, there is a directed edge from v_o to u_i for all vertices u dominated by v in G. There is a special vertex d which dominates all vertices of the form v_o. More specifically, the vertex set $V' = O \cup I \cup S$ where $O = \{v_o, v \in V\}$, $I = \{v_i, v \in V\}$ and $S = \{d\}$. The edge set E' of $\overline{G}$ is given by $E' = \{(u_o, v_i) | (u, v) \in E \text{ or } u = v\} \cup \{(d, u_o) | u_o \in O\}$.

$\overline{G}$ is acyclic because the edges are directed from the vertex d to the vertices in O and from some of the vertices in O to some vertices in I. Also if G has a dominating set D of size k, then the set $\{v_o | v \in D\} \cup \{d\}$ is a dominating set in $\overline{G}$ of size $k + 1$. Conversely if $\overline{G}$ has a dominating set D of size $k + 1$, then clearly $d \in D$ as no other vertex dominates it. Without loss of generality we can assume that no vertex of I is in D. Otherwise, choose instead the corresponding vertex of O in D if it is not already in D.

Thus we have a set of at most k vertices of O that dominates all of I. By our construction of E', the corresponding set $\{u | u_o \in D\}$ is a dominating set G of size at most k. ∎

Now, we complete the above construction to a weighted tournament and prove that the problem remains NP-complete on weighted tournaments.

Lemma 44.15 *Given a tournament with each vertex having a non-negative integer weight, and an integer k, the problem of testing whether G has a dominating set of weight (the sum of the weights of its vertices) k is NP-complete.*

Proof. Our reduction is from the problem on general directed graphs. We start from the directed graph $G(V, E)$ where we are interested in finding whether there is a dominating set of size k. We first construct the acyclic graph $\overline{G}$ obtained in the proof of Lemma 44.14. We assign a weight of 1 to the vertices in O and to the vertex d. The remaining vertices (those in I) get a weight of $k + 2$. To complete it into a tournament T, we have to specify the directions of the missing edges. They are arbitrary for those pairs of vertices in O or pairs of vertices in I. The remaining edges of the tournament are from I to O; that is, they are $\{(u_i, v_o) | (u_i, v_o) \notin E' \text{ and } (v_o, u_i) \notin E'\} \cup \{(u_i, d) | u_i \in I\}$.

Now, as before, if there is a dominating set of size k in G, then the corresponding vertices in O along with the vertex d give a dominating set of *weight $k + 1$* in T. Conversely if there is a dominating set D of weight $k + 1$ in T, then no vertex of I can be in D as each one of them has weight $k + 2$. Consequently the vertex d is in D, as no vertex of O dominates it. Thus a set of k vertices of O dominates all vertices of I. Hence by our construction of E', there is a dominating set of size k in G.

Now we convert the weighted tournament into an unweighted one by replacing each vertex of weight $k+2$ by a tournament M in which the size of the minimum dominating set is at least $k+2$. Such tournaments exist [108,109] and were used by Megiddo and Vishkin [110] to reduce the LOG²-CNF satisfiability problem to minimum dominating set problem in tournaments. The tournament M has $O(2^{2k}k^2)$ vertices and can be constructed in $O(2^{4k})$ time. Let $V(M)$ be the vertex set of M and $E(M)$ give the edge orientations of M. Then the new unweighted tournament T' has its vertex set $S \cup O \cup \overline{I}$ where S and O are as in Lemma 44.14 and $\overline{I} = \{[v, x] | v \in V(G) \text{ and } x \in V(M)\}$ where G is the original directed graph. Note that there is a copy of the tournament M for each element of O, that is, for every vertex of G. Let the vertices of G be labeled from 1 to n arbitrarily. As before, the edge directions for pairs of vertices in the set O are arbitrary, the vertex d dominates all vertices in O, and all vertices in $\overline{I}$ dominate d. The edge directions in each copy of M is dictated by the orientations of M. Between two copies of M (corresponding to vertices v and u of G, $v \neq u$), there is an edge $([v, x], [u, y])$ if and only if $(x, y) \in M$ or $x = y$ and $v < u$. And as before, there is an edge from $u_o \in O$ to all vertices $[v, x] \in \overline{I}$ (for a fixed v, and for all $x \in V(M)$) if and only if $(u, v) \in E(G)$ or $u = v$.

It can be easily seen that to dominate vertices of any one copy of M by vertices in $\overline{I}$, $k + 2$ vertices are required. Now it is easy to check that the original directed graph has a dominating set of size k if and only if the resulting tournament has a dominating set of size $k + 1$. Since we have reduced a general directed graph to a tournament consisting of $O(n2^{2k})$ vertices in FPT time and the resulting tournament has the desired property, the theorem follows. ∎

Corollary 44.2 *The minimum dominating set problem in tournaments is $W[2]$-complete.*

Papadimitriou and Yannanakis [111] have shown that the log dominating set problem for directed graphs is complete for the class LOGSNP introduced by them. This basically means if we can test whether a directed graph has a dominating set of size at most $k \leq \log n$, in polynomial time, then every problem in the class LOGSNP has a polynomial time algorithm. They have also shown by a reduction from a generic problem in LOGSNP that the tournament dominating set problem is also LOGSNP-complete. Since the function $f(k)$ used in our reduction is simply $2^{O(k)}$, this result also follows as a corollary.

Corollary 44.3 *The minimum dominating set problem in tournaments is LOGSNP-complete.*

Note that the number of vertices in the resulting tournament has grown exponential in k. If it had remained polynomial in k, the reduction would have proved that the tournament dominating set problem is NP-complete. But this is unlikely as the minimum dominating set problem can be solved in $n^{O(\log n)}$ time for tournaments, as every tournament has a dominating set of size at most $\log n$.

For a thorough introduction to the W-hierarchy we refer the reader to the books of Downey and Fellows [13] and Flum and Grohe [14].

44.4.2 Exponential Time Hypothesis and Stronger Lower Bounds

In this section we outline an alternate approach to proving lower bound on parameterized problems. This approach gives us more refined and tighter lower bounds on the form of running time of the algorithm. We first define the notion of subexponential time algorithms.

Definition 44.14 SUBEPT *is the class of parameterized problems (P, κ) where P can be solved in time $2^{\kappa(x)/s(\kappa(x))}|x|^{O(1)} = 2^{o(\kappa(x))}|x|^{O(1)}$. Here, $s(k)$ is a monotonically increasing unbounded function. A problem P in* SUBEPT *is said to have subexponential algorithms.*

A useful observation is that an *arbitrarily good* exponential time algorithm implies a subexponential time algorithm and vice versa.

Proposition 44.7 [14] *A parameterized problem (P, κ) is in SUBEPT if and only if there is an algorithm that for every fixed $\epsilon > 0$ solves instances x of P in time $2^{\epsilon \kappa(x)}|x|^c$ where c is independent of x and ϵ.*

The r-CNF-SAT problem is a central problem in computational complexity, as it is the canonical NP-complete problem. We will use this problem as a basis for our complexity assumptions.

> r-CNF-SAT
> *Instance:* A r-CNF formula F on n variables and m clauses.
> *Parameter 1:* n.
> *Parameter 2:* m.
> *Problem:* Decide whether there exists a $\{0, 1\}$ assignment to the variables of F such that it is satisfiable?

It is trivial to solve 3-CNF-SAT it time $2^n \cdot (n+m)^{O(1)}$. There are better algorithms for 3-CNF-SAT, but all of them have running time of the form $c^n \cdot (n+m)^{O(1)}$ for some constant $c > 1$ (the current best algorithm runs in time $O(1.30704^n)$ [112]. Our first complexity hypothesis, formulated by Impagliazzo et al. [113], states that every algorithm for 3-CNF-SAT has this running time, that is, the problem has no subexponential time algorithms.

44.4.2.1 Exponential Time Hypothesis

There is a positive real s such that 3-CNF-SAT with parameter n can not be solved in time $2^{sn}(n+m)^{O(1)}$ [113].

In particular, Exponential time hypothesis (ETH) states that 3-CNF-SAT with parameter n can not be solved in $2^{o(n)}(n+m)^{O(1)}$ time. We will use this assumption to show that several other problems do not have subexponential-time algorithms either. To transfer this hardness assumption to other problems, we need a notion of reduction that preserves solvability in subexponential time. It is easy to see that a polynomial-time FPT-reduction that increases the parameter only linearly (i.e., $\kappa'(R(x)) = O(\kappa(x))$ holds for every instance x) preserves subexponential-time solvability: if the target problem (Q', κ') is in SUBEPT, then so is the source problem (Q, κ). Most of the reductions in this survey are on this form. However, it turns out that sometimes a more general form of subexponential time reductions, introduced by Impagliazzo et al. [113], are required. Essentially, we allow the running time of the reduction to be subexponential and the reduction to be a Turing reduction rather than a many-one reduction:

Definition 44.15 *A SERF-T reduction from parameterized problem (A_1, κ_1) to a parameterized problem (A_2, κ_2) is a Turing reduction M from A_1 to A_2 that has the following properties.*

1. *Given an $\epsilon > 0$ and an instance x of A_1, M runs in time $O(2^{\epsilon \kappa_1(x)} |x|^{O(1)})$.*

2. *For any query $M(x)$ makes to A_2 with the input x',*

   a. $|x'| = |x|^{O(1)}$,

   b. $\kappa_2(x') = \alpha \kappa_1(x)$.

The constant α may depend on ϵ while the constant hidden in the $O()$-notation in the bound for $|x'|$ may not.

It can easily be shown that SERF-T reductions are transitive. We now prove that SERF-T reductions work as expected and indeed preserve solvability in subexponential time.

Proposition 44.8 *If there is a SERF-T reduction from (A_1, κ_1) to (A_2, κ_2) and A_2 has a subexponential time algorithm then so does A_1.*

Proof. By Proposition 44.7 there is an algorithm for (A_2, κ_2) that for every $\epsilon > 0$ can solve an instance x in time $O(2^{\epsilon \kappa_2(x)} |x|^c)$ for some c independent of x and ϵ. We show that such an algorithm also exists for (A_1, κ_1).

Given an $\epsilon > 0$ we need to make an algorithm running in time $O(2^{\epsilon \kappa_1(x)} |x|^{c'})$ for some c' independent of x and ϵ. We choose $\epsilon' = \epsilon/2$ and run the SERF-T reduction from (A_1, κ_1) to (A_2, κ_2) with parameter ϵ'. This reduction makes at most $O(2^{\epsilon' \kappa_1(x)} |x|^{O(1)})$ calls to instances x' of A_2, each with $|x'| \leq |x|^{O(1)}$ and $\kappa_2(x') \leq \alpha \kappa_1(x)$. Each such instance can be solved in time $2^{\epsilon \kappa_1(x)/2} |x|^{O(1)}$. Hence the total running time for solving x is $2^{\epsilon \kappa_1(x)} |x|^{c'}$ for some c' independent of x and ϵ. By Proposition 44.7 this means that (A_1, κ_1) is in SUBEPT.

Since every variable appears in some clause it follows that $n \leq rm$, and hence r-CNF-SAT with parameter m (the number of clauses) is SERF-T reducible to r-CNF-SAT with parameter n. However, there is no equally obvious SERF-T reduction from r-CNF-SAT with parameter n to r-CNF-SAT with parameter m. Nevertheless, Impagliazzo et al. [113] established such a reduction, whose core argument is called the *sparsification lemma* stated below.

Lemma 44.16 (Sparsification) [114] *For every $\epsilon > 0$ and positive integer r, there is a constant $C = O((r/\epsilon)^{3r})$ so that any r-CNF formula F with n variables, can be expressed as $F = \vee_{i=1}^{t} Y_i$, where $t \leq 2^{\epsilon n}$ and each Y_i is an r-CNF formula with every variable appearing in at most C clauses. Moreover, this disjunction can be computed by an algorithm running in time $2^{\epsilon n} n^{O(1)}$.* ∎

Lemma 44.16 directly gives a SERF-T reduction from r-CNF-SAT with parameter n to r-CNF-SAT with parameter m. Thus the following proposition is a direct consequence of the sparsification lemma.

Proposition 44.9 [113] *Assuming ETH, there is a positive real s' such that 3-CNF-SAT with parameter m can not be solved in time $O(2^{s'm})$. That is, there is no $2^{o(m)}$ algorithm for 3-CNF-SAT with parameter m.*

Proposition 44.9 has far-reaching consequences: as we shall see, by reductions from 3-CNF-SAT with parameter m, we can show lower bounds for a wide range of problems. Moreover, we can even show that several NP-complete problems are equivalent with respect to solvability in subexponential time. For an example, every problem in SNP and size-constrained SNP (see [113] for definitions of these classes) can be shown to have SERF-T reductions to r-CNF-SAT with parameter n for some $r \geq 3$. The SNP and size-constrained SNP problem classes contain several important problems such as r-CNF-SAT with parameter n and independent set, vertex cover, and clique parameterized by the number of vertices in the input graph. This gives some evidence that a subexponential time algorithm for r-CNF-SAT with parameter n is unlikely to exist, giving some credibility to ETH.

It is natural to ask how the complexity of r-CNF-SAT evolves as r grows. For all $r \geq 3$, define,

$$s_r = \inf\left\{\delta : \text{there exists an } O^*(2^{\delta n}) \text{ algorithm solving } r\text{-CNF-SAT} \text{ with parameter } n\right\}.$$

$$s_\infty = \lim_{r \to \infty} s_r.$$

Since r-CNF-SAT easily reduces to $(r+1)$-SAT it follows that $s_r \leq s_{r+1}$. However, saying anything else non-trivial about this sequence is difficult. ETH is equivalent to conjecturing that $s_3 > 0$. Impagliazzo et al. [113] present the following relationships between the s_r's and the solvability of problems in SNP in subexponential time. The theorem below is essentially a direct consequence of Lemma 44.16

Theorem 44.22 [113] *The following statements are equivalent*

1. *For all $r \geq 3$, $s_r > 0$.*
2. *For some r, $s_r > 0$.*
3. *$s_3 > 0$.*
4. *SNP $\not\subseteq$ SUBEPT.* ∎

The equivalence above offers some intuition that r-CNF-SAT with parameter n may not have a subexponential time algorithm and thus strengthens the credibility of ETH. The known NP-hardness proof combined with Proposition 44.9 implies the following theorem.

Theorem 44.23 *Assuming ETH, there is no $2^{o(n)}$ time algorithm for vertex cover, dominating set, Hamiltonian path, 3-COLORING, and independent set on a n vertex graph.* ∎

Similarly, we can show the following.

Theorem 44.24 *Assuming ETH, there is no $2^{o(\sqrt{n})}$ time algorithm for vertex cover, dominating set, Hamiltonian path, and independent set on a n vertex planar graph.* ∎

Impagliazzo et al. [113] and Calabro et al. [115] studied the sequence of s_r's and obtained the following results.

Theorem 44.25 [113,115] *Assuming ETH, the sequence $\{s_r\}_{r \geq 3}$ is increasing infinitely often. Furthermore, $s_r \leq s_\infty(1 - d/r)$ for some constant $d > 0$.* ∎

A natural question to ask is what is s_∞? As of today the best algorithms for r-CNF-SAT all use time $O(2^{n(1-c/r)})$ for some constant c independent of r and n. This, together with Theorem 44.25 hints at $s_\infty = 1$. The conjecture that this is indeed the case is known as the strong exponential time hypothesis.

44.4.2.2 Strong Exponential Time Hypothesis: $s_\infty = 1$

An immediate consequence of strong exponential time hypothesis (SETH) [113,115] is that strong exponential time hypothesis (SAT) with parameter n (here the input formula F could have arbitrary size clauses) can not be solved in time $(2 - \epsilon)^n(n + m)^{O(1)}$.

Now we give a few lower bounds based on ETH and SETH,

44.4.2.3 Lower Bound on FPT Algorithms

Cai and Juedes [116] were first to examine the existence of $2^{o(k)}$ or $2^{o(\sqrt{k})}$ algorithms for various parameterized problems solvable in time $2^{O(k)}$ or $2^{O(\sqrt{k})}$, respectively. They showed that for variety of problems assuming ETH, there is no $2^{o(k)}$ or $2^{o(\sqrt{k})}$ algorithms possible. In this section, we survey how ETH can be used to obtain lower bounds on the function f for various FPT problems.

Since $k \leq n$, a $2^{o(k)}n^c$ time algorithm directly implies a $2^{o(n)}$ time algorithm for vertex cover. However, by Theorem 44.24 we know that vertex cover does not have an algorithm with running time $2^{o(n)}$ unless ETH fails. This immediately implies the following theorem.

Theorem 44.26 [116] *Assuming ETH, there is no $2^{o(k)}n^{O(1)}$ time algorithm for vertex cover.* ∎

Similarly, assuming ETH, we can show that several other problems parameterized by the solution size, such as FVS or longest path do not have $2^{o(k)}n^{O(1)}$ time algorithms.

Similar arguments yield tight lower bounds for parameterized problems on special graph classes, such as planar graphs. As we have seen in the previous section, for many problems we can rule out algorithms with running time $2^{o(\sqrt{n})}$ even when the input graph is restricted to be planar. If the solution to such a problem is a subset of the vertices (or edges), then the problem parameterized by solution size can not be solved in time $2^{o(\sqrt{k})}n^{O(1)}$ on planar graphs, unless ETH fails.

Theorem 44.27 [116] *Assuming ETH, there is no $2^{o(\sqrt{k})}n^{O(1)}$ time algorithm for planar vertex cover.* ∎

Results similar to Theorem 44.27 are possible for several other graph problems on planar graphs. It is worth to mention that many of these lower bounds on these problems are tight. That is, many of the mentioned problems admit both $2^{O(k)}n^{O(1)}$ time algorithms on general graphs and $2^{O(\sqrt{k})}n^{O(1)}$ time algorithms on planar graphs.

Obtaining lower bounds of the form $2^{o(k)}n^{O(1)}$ or $2^{o(\sqrt{k})}n^{O(1)}$ on parameterized problems generally follows from the known NP-hardness reduction. However, there are several parameterized problems where $f(k)$ is *slightly superexponential* in the best-known running time: $f(k)$ is of the form $k^{O(k)} = 2^{O(k \log k)}$. Algorithms with this running time naturally occur when a search tree of height at most k and branching factor at most k is explored, or when all possible permutations, partitions, or matchings of a k element set are enumerated. Recently, for a number of such problems lower bounds of the form $2^{o(k \log k)}$ were obtained under ETH [117]. We show how such a lower bound can be obtained for an artificial variant of the clique problem. In this problem the vertices are the elements of a $k \times k$ table, and the clique we are looking for has to contain exactly one element from each row.

> $k \times k$ CLIQUE
> Input: A graph G over the vertex set $[k] \times [k]$.
> Parameter: k.
> Question: Is there a k-clique in G with exactly one element from each row?

Note that the graph G in the $k \times k$ clique instance has $O(k^2)$ vertices at most $O(k^4)$ edges, thus the size of the instance is $O(k^4)$.

Theorem 44.28 [117] *Assuming ETH, there is no $2^{o(k \log k)}$ time algorithm for $k \times k$ clique.* ∎

Lokshtanov et al. [117] first define other problems similar in flavor to $k \times k$ clique: basic problems artificially modified in such a way that they can be solved by brute force in time $2^{O(k \log k)} |I|^{O(1)}$. It is then shown that assuming ETH, these problems do not admit a $2^{o(k \log k)}$ time algorithm. Finally, combining the lower bounds on the variants of basic problems with suitable reductions one can obtain lower bounds for natural problems. One example is the bound for the closest string problem.

> CLOSEST STRING
> Input: Strings $s_1, \ldots, s_t$ over an alphabet Σ of length L each, an integer d.
> Parameter: d.
> Question: Is there a string s of length L such $d(s, s_i) \leq d$ for every $1 \leq i \leq t$?

Here $d(s, s_i)$ is the *Hamming distance* between the strings s and s_i, that is, the number of positions where s and s_i differ. Gramm et al. [118] showed that closest string is FPT parameterized by d: they gave an algorithm with running time $O(d^d \cdot |I|)$. The algorithm works over an arbitrary alphabet Σ (i.e., the size of the alphabet is part of the input). For fixed alphabet size, single-exponential dependence on d can be achieved: algorithms with running time of the form $|\Sigma|^{O(d)} \cdot |I|^{O(1)}$ were presented in Ma and Sun [119], Wang and Zhu [120], and Chen et al. [121]. It is an obvious question if the running time can be improved to $2^{O(d)} \cdot |I|^{O(1)}$, that is, single-exponential in d, even for arbitrary alphabet size. However, the following result shows that the running times of the cited algorithms have the best possible form.

Theorem 44.29 [117] *Assuming ETH, there is no $2^{o(d \log d)} \cdot |I|^{O(1)}$ or $2^{o(d \log |\Sigma|)} \cdot |I|^{O(1)}$ time algorithm for closest string.* ∎

Using similar methods one can also give tight running time lower bounds for the distortion problem. Here we are given a graph G and parameter d. The objective is to determine whether there exists a map f from the vertices of G to $\mathbb{N}$ such that for every pair of vertices u and v in G, if the distance between u and v in G is δ then $\delta \leq |f(u) - f(v)| \leq d\delta$. This problem belongs to a broader range of *metric embedding* problems where one is looking for a map from a complicated distance metric into a simple metric while preserving as many properties of the original metric as possible. Fellows et al. [122] give a $O(d^d n^{O(1)})$ time algorithm for distortion. The following theorem shows that under ETH the dependence on d of this algorithm can not be significantly improved.

Theorem 44.30 [117] *Assuming ETH, there is no $2^{o(d \log d)} \cdot n^{O(1)}$ time algorithm for distortion.* ∎

44.4.2.4 W[1]-Hard Problems

The complexity assumption ETH can be used not only to obtain running time lower bounds on problems that are FPT, but also on problems that are known to be $W[1]$-hard in parameterized complexity. For an example independent set and dominating set are known to be $W[1]$-complete and $W[2]$-complete, respectively. Under the standard parameterized complexity assumption that $FPT \neq W[1]$, this immediately rules out the possibility of having an FPT algorithm for clique, independent set, and dominating set. However, knowing that no algorithm of the form $f(k)n^{O(1)}$ exists, that these results do not rule out the possibility of an algorithm with running time, say, $n^{O(\log \log k)}$. As the best-known algorithms for these problems take $n^{O(k)}$ time, there is huge gap between the upper and lower bounds obtained this way.

Chen et al. [123] were the first to consider the possibility of showing sharper running time lower bounds for $W[1]$-hard problems. They show that lower bounds of the form $n^{o(k)}$ can be achieved for several $W[2]$-hard problems such as dominating set, under the assumption that $FPT \neq W[1]$. However, for problems that are $W[1]$-hard rather than $W[2]$-hard, such as independent set, we need ETH in order to show lower bounds. Later, Chen et al. [124,125] strengthened their lower bounds to also rule out $f(k)n^{o(k)}$ time algorithms (rather than just $n^{o(k)}$ time algorithms).

We outline one such lower bound result here and then transfer it to other problems using earlier reductions.

Theorem 44.31 [123,125] *Assuming ETH, there is no $f(k)n^{o(k)}$ time algorithm for clique or independent set.* ∎

Proof. We give a proof sketch. We will show that if there is an $f(k)n^{o(k)}$ time algorithm for clique, then ETH fails. Suppose that clique can be solved in time $f(k)n^{k/s(k)}$, where $s(k)$ is a monotone increasing unbounded function. We use this algorithm to solve 3-coloring on an n-vertex graph G in time $2^{o(n)}$. Let $f^{-1}(n)$ be the largest integer i such that $f(i) \leq n$. Function $f^{-1}(n)$ is monotone increasing and unbounded. Let $k := f^{-1}(n)$. Split the vertices of G into k groups. Let us build a graph H where each vertex corresponds to a proper 3-coloring of one of the groups. Connect two vertices if they are not conflicting. That is, if the union of the colorings corresponding to these vertices corresponds to a valid coloring of the graph induced on the vertices of these two groups, then connect the two vertices. A k-clique of H corresponds to a proper 3-coloring of G. A 3-coloring of G can be found in time $f(k)n^{k/s(k)} \leq n(3^{n/k})^{k/s(k)} = n3^{n/s(f^{-1}(n))} = 2^{o(n)}$. This completes the proof.

Since a graph G has a clique of size k if and only the complement of G has an independent set of size k. Thus, as a simple corollary to the result of clique, we get that independent set does not have any $f(k)n^{o(k)}$ time algorithm unless ETH fails. ∎

Using the parameter preserving reduction established in Theorems 44.18 and 44.19, we get the following theorem.

Theorem 44.32 *Assuming ETH, there is no $f(k)n^{o(k)}$ time algorithm for dominating set.* ∎

44.4.2.5 Problems Parameterized by Treewidth

It is well known that several graph problems parameterized by the treewidth of the input graph are FPT. See Table 44.1 for the time complexity of some known algorithms for problems parameterized by the treewidth of the input graph. Most of the algorithms on graphs of bounded treewidth are based on simple dynamic programming on the tree decomposition,

TABLE 44.1 $f(t)$ Bound in the Running Time of Various Problems Parameterized by the Treewidth of the Input Graph

Problem Name	$f(t)$ in the Best-Known Algorithms
Vertex cover	2^t
Dominating set	3^t
Odd cycle transversal	3^t
Partition into triangles	2^t
Max cut	2^t
Chromatic number	$2^{O(t \log t)}$
Disjoint paths	$2^{O(t \log t)}$
Cycle packing	$2^{O(t \log t)}$

although for some problems a recently discovered technique called fast subset convolution [126,127] needs to be used to obtain the running time shown in Table 44.1.

An obvious question is whether these algorithms can be improved. We can easily rule out the existence of $2^{o(t)}$ algorithm for many of these problems assuming ETH. Recall that, Theorem 44.23 shows that assuming ETH, the independent set problem parameterized by the number of vertices in the input graph does not admit a $2^{o(n)}$ algorithm. Since the treewidth of a graph is clearly at most the number of vertices, it is in fact a *stronger* parameter, and thus the lower bound carries over. Thus, we trivially have that independent set does not admit a subexponential algorithm when parameterized by treewidth. Along the similar lines we can show the following theorem.

Theorem 44.33 *Assuming ETH, independent set, dominating set, and odd cycle transversal parameterized by the treewidth of the input graph do not admit an algorithm with running time $2^{o(t)} n^{O(1)}$. Here, n is the number of vertices in the input graph to these problems.* ∎

For the problems chromatic number, cycle packing, and disjoint paths, the natural dynamic programming approach gives $2^{O(t \log t)} n^{O(1)}$ time algorithms. As these problems can be solved in time $2^{O(n)}$ on n-vertex graphs, the easy arguments of Theorem 44.33 can not be used to show the optimality of the $2^{O(t \log t)} n^{O(1)}$ time algorithms. However, as reviewed in Section 44.4.2, Lokshtanov et al. [117] developed a machinery for obtaining lower bounds of the form $2^{o(k \log k)} n^{O(1)}$ for parameterized problems and we can apply this machinery in the case of parameterization by treewidth as well.

Theorem 44.34 [19,117] *Assuming ETH, chromatic number, cycle packing, and disjoint paths parameterized by the treewidth of the input graph do not admit an algorithm with running time $2^{o(t \log t)} n^{O(1)}$. Here, n is the number of vertices in the input graph to these problems.* ∎

The lower bounds obtained by Theorem 44.33 are quite weak: they tell us that $f(t)$ can not be improved to $2^{o(t)}$, but they do not tell us whether the numbers 2 and 3 appearing as the base of exponentials in Table 44.1 can be improved. Just as we saw for exact algorithms, ETH seems to be too weak an assumption to show a lower bound that concerns the base of the exponent. Assuming the SETH, however, much tighter bounds can be shown. In Lokshtanov et al. [128] it is established that any non-trivial improvement over the best-known algorithms for a variety of basic problems on graphs of bounded treewidth would yield a faster algorithm for SAT.

Theorem 44.35 [128] *If there exists an $\epsilon > 0$ such that*

- *Independent set can be solved in $(2 - \epsilon)^{tw(G)} n^{O(1)}$ time, or*
- *Dominating set can be solved in $(3 - \epsilon)^{tw(G)} n^{O(1)}$ time, or*
- *Max cut can be solved in $(2 - \epsilon)^{tw(G)} n^{O(1)}$ time, or*
- *Odd cycle transversal can be solved in $(3 - \epsilon)^{tw(G)} n^{O(1)}$ time, or*
- *There is a $q \geq 3$ such that q-COLORING can be solved in $(q - \epsilon)^{tw(G)} n^{O(1)})$ time, or*
- *If partition into triangles can be solved in $(2 - \epsilon)^{tw(G)} n^{O(1)}$ time, then SETH fails.* ∎

Thus, assuming SETH, the known algorithms for the mentioned problems on graphs of bounded treewidth are essentially the best possible. We refer to Lokshtanov et al. [129] for more detailed survey on lower bounds based on ETH and SETH.

44.5 FPT AND APPROXIMATION

Though the study of parameterized complexity is not confined to parameterized versions of NP-complete problems alone, it could be considered as one way of coping with intractability of NP-complete problems. In this section, we will look at some relationship known between FPT and the other way of coping with NP-completeness - approximation.

A typical NP optimization problem Q is either a minimization or a maximization problem given by the 4-tuple (I, S, f, opt) where I is the set of input instances for the problem, $S(x)$ is the set of feasible solutions for the input $x \in I$, $f(x,y) \in N$ is the objective function value for each $x \in I$ and $y \in S(x)$, and $opt \in \{max, min\}$ so that $opt(x)$ is the optimum value of the instance x. See [130] for a precise definition of an NP optimization problem.

The natural parameterized version of the NP optimization problem is: given an instance $x \in I$, and an integer parameter k is there a $y \in S(x)$ of size at least (at most, for a minimization problem) k?

An NP optimization problem has a ratio r approximation algorithm if there is a polynomial time algorithm that for every instance x of the problem, produces a feasible solution y such that $\max\{f(x,y)/opt(x), opt(x)/f(x,y)\} \leq 1 + r$. The approximation algorithm is said to be an approximation scheme (PTAS) if given any $\epsilon > 0$, the algorithm runs in polynomial time and is an ϵ approximation algorithm. A PTAS is called a fully polynomial time approximation scheme (FPTAS) if its running time is polynomial in the input size as well as $1/\epsilon$. It is called an efficient polynomial time approximation scheme (EPTAS) if its running time is of the form $f(1/\epsilon) + n^c$ where f is some function of $1/\epsilon$ and n is the input size.

The following easy observation is due to Cai and Chen [131].

Theorem 44.36 *If an integer valued NP optimization problem has a fully polynomial time approximation scheme, then the corresponding parameterized problem is in FPT.*

Proof. Without loss of generality, let Q be an NP maximization problem having a FPTAS, and let Q_k be its corresponding parameterized problem. Given the parameterized instance x of Q, let y be the approximate solution obtained by the approximate scheme with the approximation ratio $1/2k$. That is, $opt(x)/y \leq 1 + 1/2k$. Now it is easy to check that $opt(x) > k$ if and only if $y > k$. Furthermore as the algorithm takes time polynomial in $2k$ and the size of x, Q_k is in FPT. ∎

This result shows that the decision versions of several knapsack-like and scheduling problems FPT [2].

It is easy to see that the above result can be strengthened to deduce the same result assuming the NP optimization problem has an EPTAS.

Cai and Chen [131] also observed that the parameterized versions of all maximization problems in MaxSNP are in FPT, from the result that the parameterized MaxcSat is FPT for any constant c, and that MaxcSat is MaxSNP-complete. Furthermore, using a branching algorithm similar to that for vertex cover, they could show that any minimization problem in the class $MINF^+\Pi_1$ [132] is in FPT.

The contrapositive consequences of these results are more interesting. For example, the parameterized dominating set problem (even for tournaments) is $W[2]$-hard implies that the minimum dominating set problem can not have FPTAS algorithm unless FPT=$W[2]$. Thus proving W-hardness could be one way of showing non-approximability of an optimization problem. One should also note that there are problems like the longest path problem that is hard to approximate [133] though its parameterized version is FPT as we saw in Section 2.4.

The notion of EPTAS suggests using the approximation ratio as a parameter. Marx [134] suggests the following additional ways in which approximation and parameterized complexity can be combined for better understanding of a problem.

44.5.1 Approximation in FPT Time for W-Hard Problems

This is a rather unexplored territory in fixed-parameter algorithms. The idea is, as in the case of classical approximation algorithms, to design an approximation algorithm in FPT time for a parameterized intractable problem. There are two views based on what we approximate.

- The parameter can be an instance parameter (e.g., genus of the graph, treewidth of the graph) and the value to be optimized is the solution size (which is known to be fixed-parameter intractable when parameterized by the instance parameter). For example, finding the minimum number of colors in a proper coloring of vertices in a bounded genus graph. This, for example, has a 2-approximation algorithm that runs in time $f(g)|x|$ where f is some function of the genus g, and $|x|$ is the input size.

- The parameter can be the solution size (like dominating set) which is known to be W-hard. And we would like an FPT 2-approximation algorithm to output that the given graph has no dominating set of size at most k or produce a dominating set of size at most $2k$. It is an open problem whether such an approximation algorithm exists for dominating set. See [135] for a related result.

44.5.2 Approximation Parameterized by Cost

Here the aim is to use the optimized value as the parameter and try to design an algorithm that finds the optimum value OPT or a small approximation to it using $f(OPT)|x|^c$ time. See [134] for some results and problems along these lines.

44.6 CONCLUSIONS

Parameterized complexity has emerged as an important practical direction to pursue for problems where a small range of parameter values is of particular interest. The study of parameterized complexity has given rise to novel and interesting algorithmic techniques to solve some difficult problems exactly. Kernelization, which started off as a direction to design fixed-parameter algorithms, has given rise to surprising connections to classical complexity theory. Iterated compression has turned out to be a novel technique that has been useful to show several long-standing problems FPT.

While NP-completeness is a useful direction to show problems that are unlikely have polynomial time algorithms, by using parameterized reductions, one can show stronger lower bounds. The large compendium of W-hard problems [13] gives support to the conjecture that they are unlikely to be FPT.

In this chapter, we have surveyed algorithmic techniques, hardness theory and various connections in parameterized complexity. For parameterized complexity applications in artifical intelligence, see [136] and in database theory, see [137].

For parameterized complexity applications in strings and biology, see [138]. For parameterized complexity applications in coding theory, see [139].

Given the pace of recent development in the area, we almost certainly have omitted some recent research areas, but we hope that this survey is comprehensive enough to motivate new entrants to the field. We also hope that development of new techniques in the area will bridge the theory and practice of computationally hard problems, and explain why some *heuristic* solutions to NP-complete problems work so well in practice.

References

[1] Dorit S. Hochbaum, editor. *Approximation Algorithms for NP-Hard Problems*. PWS Publishing, Boston, MA, 1997.

[2] Vijay V Vazirani. *Approximation Algorithms*. Springer, Berlin, Germany, 2004.

[3] Martin Charles Golumbic. *Algorithmic Graph Theory and Perfect Graphs*, volume 57. North Holland Publishing, New York, 2004.

[4] Nimrod Megiddo. Linear programming in linear time when the dimension is fixed. *J. ACM*, **31**(1) (1984), 114–127.

[5] Martin Farber, Gena Hahn, Pavol Hell, and Donald J. Miller. Concerning the achromatic number of graphs. *J. Comb. Theory, Ser. B*, **40**(1) (1986), 21–39.

[6] Michael R. Garey and David S. Johnson. *Computers and Intractability: A Guide to the Theory of NP-Completeness*. W. H. Freeman & Co., New York, 1990.

[7] Jianer Chen, Iyad A. Kanj, and Weijia Jia. Vertex cover: Further observations and further improvements. *J. Algorithms*, **41**(2) (2001), 280–301.

[8] Yixin Cao, Jianer Chen, and Yang Liu. On feedback vertex set: New measure and new structures. In *SWAT*, volume 6139 of *Lecture Notes in Computer Science*, pp. 93–104. Springer, Bergen, Norway, 2010.

[9] Meena Mahajan and Venkatesh Raman. Parameterizing above guaranteed values: Maxsat and maxcut. *J. Algorithms*, **31**(2) (1999), 335–354.

[10] Venkatesh Raman and Balsri Shankar. Improved fixed-parameter algorithm for the minimum weight 3-SAT. In *Proceedings of the Workshop on Algorithms and Computation*, volume 7748 of *Lecture Notes in Comput. Sci.*, pp. 265–273, 2013.

[11] Rolf Niedermeier. *Invitation to fixed-parameter algorithms*, volume 31 of *Oxford Lecture Series in Mathematics and Its Applications*. Oxford University Press, Oxford, 2006.

[12] Rodney G. Downey and Michael R. Fellows. Fixed-parameter tractability and completeness I: Basic results. *SIAM J. Comput.*, **24**(4) (1995), 873–921.

[13] Rodney G. Downey and Michael R. Fellows. *Fundamentals of Parameterized Complexity*. Monographs in Computer Science. Springer, New York, 1999.

[14] Jörg Flum and Martin Grohe. *Parameterized Complexity Theory*. Texts in Theoretical Computer Science. An EATCS Series. Springer-Verlag, Berlin, Germany, 2006.

[15] Hans L. Bodlaender, Rodney G. Downey, Michael R. Fellows, and Danny Hermelin. On problems without polynomial kernels. *J. Comput. Syst. Sci.*, **75**(8) (2009), 423–434.

[16] Rodney G. Downey, Michael R. Fellows, and Ulrike Stege. Computational tractability: The view from Mars. *Bull. Eur. Assoc. Theor. Comput. Sci. EATCS*, **69** (1999), 73–97.

[17] Jonathan F. Buss and Judy Goldsmith. Nondeterminism within P. *SIAM J. Comput.*, **22**(3) (1993), 560–572.

[18] George L. Nemhauser and Leslie E. Trotter. Vertex packings: Structural properties and algorithms. *Math. Program.*, **8** (1975), 232–248. doi:10.1007/BF01580444.

[19] Marek Cygan, Jesper Nederlof, Marcin Pilipczuk, Michal Pilipczuk, Johan M. M. van Rooij, and Jakub Onufry Wojtaszczyk. Solving connectivity problems parameterized by treewidth in single exponential time. In *FOCS*, pp. 150–159, IEEE FOCS, Palm Springs, CA, 2011.

[20] Michael Lampis. A kernel of order 2 k-c log k for vertex cover. *Inf. Process. Lett.*, **111**(23–24) (2011), 1089–1091.

[21] N. S. Narayanaswamy, Venkatesh Raman, M. S. Ramanujan, and Saket Saurabh. Lp can be a cure for parameterized problems. In *STACS*, pp. 338–349, 2012.

[22] Daniel Lokshtanov, N. S. Narayanaswamy, Venkatesh Raman, M. S. Ramanujan, and Saket Saurabh. Faster parameterized algorithms using linear programming. *CoRR*, abs/1203.0833, 2012.

[23] Dénes König. Über Graphen und ihre Anwendung auf Determinantentheorie und Mengenlehre. *Math. Ann.*, **77**(4) (1916), 453–465.

[24] Philip Hall. On representatives of subsets. *J. London Math. Soc.*, **10** (1935), 26–30.

[25] Benny Chor, Mike Fellows, and David W. Juedes. Linear kernels in linear time, or how to save k colors in $O(n^2)$ steps. In *Proceedings of the 30th Workshop on Graph-Theoretic Concepts in Computer Science*, Bad Honnef, Germany, volume 3353 of *Lecture Notes in Comput. Sci.*, pp. 257–269. Springer, 2004.

[26] Daniel Lokshtanov. New Methods in Parameterized Algorithms and Complexity. PhD thesis, University of Bergen, Norway, 2009.

[27] Stéphan Thomassé. A quadratic kernel for feedback vertex set. *ACM Transac. Algorithms*, **6**(2) (2010).

[28] Jochen Alber, Henning Fernau, and Rolf Niedermeier. Parameterized complexity: Exponential speed-up for planar graph problems. *J. Algorithms*, **52**(1) (2004), 26–56.

[29] Hans L. Bodlaender, Fedor V. Fomin, Daniel Lokshtanov, Eelko Penninkx, Saket Saurabh, and Dimitrios M. Thilikos. (Meta) kernelization. *IEEE Conference on Foundations of Computer Science*, pp. 629–638, 2009.

[30] Fedor V. Fomin, Daniel Lokshtanov, Saket Saurabh, and Dimitrios M. Thilikos. Linear kernels for (connected) dominating set on h-minor-free graphs. In *SODA*, pp. 82–93, 2012.

[31] Stefan Kratsch and Magnus Wahlström. Representative sets and irrelevant vertices: New tools for kernelization. *CoRR*, abs/1111.2195, 2011.

[32] Stefan Kratsch and Magnus Wahlström. Compression via matroids: a randomized polynomial kernel for odd cycle transversal. In *ACM-SIAM Symposium on Discrete Algorithms*, pp. 94–103, Kyoto, Japan, 2012.

[33] Lance Fortnow and Rahul Santhanam. Infeasibility of instance compression and succinct PCPs for NP. In *Proceedings of the 40th Annual ACM Symposium on Theory of Computing*, pp. 133–142. ACM, 2008.

[34] Noga Alon, Raphael Yuster, and Uri Zwick. Color-coding. *J. Assoc. Comput. Mach.*, **42**(4) (1995), 844–856.

[35] Larry J. Stockmeyer. The polynomial-time hierarchy. *Theor. Comp. Sci.*, **3** (1976), 1–22.

[36] Andrew Drucker. On the hardness of compressing an AND of SAT instances. Theory Lunch, Center for Computational Intractability, February 17, 2012. http://intractability.princeton.edu/blog/2012/03/theory-lunch-february-17/.

[37] Hans L. Bodlaender, Stéphan Thomassé, and Anders Yeo. Analysis of data reduction: Transformations give evidence for non-existence of polynomial kernels. Technical Report CS-UU-2008-030, Department of Information and Computer Sciences, Utrecht University, Utrecht, the Netherlands, 2008.

[38] Henning Fernau, Fedor V. Fomin, Daniel Lokshtanov, Daniel Raible, Saket Saurabh, and Yngve Villanger. Kernel(s) for problems with no kernel: On out-trees with many leaves. In *STACS*, pp. 421–432. Schloss Dagstuhl—Leibniz-Zentrum fuer Informatik, 2009.

[39] Michael Dom, Daniel Lokshtanov, and Saket Saurabh. Incompressibility through colors and ids. In *ICALP*, volume 5555 of *Lecture Notes in Comput. Sci.*, pp. 378–389, Springer, Rhodes, Greece, 2009.

[40] Holge Dell and Dieter van Melkebeek. Satisfiability allows no nontrivial sparsification unless the polynomial-time hierarchy collapses. In *ACM Symposium on Theory of Computing*, pp. 251–260, Cambridge, MA, 2010.

[41] Hans L. Bodlaender, Bart M. P. Jansen, and Stefan Kratsch. Cross-composition: A new technique for kernelization lower bounds. In *Proceedings of the 28th International Symposium on Theoretical Aspects of Computer Science*, volume 9 of *LIPIcs*, pp. 165–176. Schloss Dagstuhl—Leibniz-Zentrum fuer Informatik, 2011.

[42] Holger Dell and Dániel Marx. Kernelization of packing problems. In *SODA*, pp. 68–81, 2012.

[43] Danny Hermelin and Xi Wu. Weak compositions and their applications to polynomial lower bounds for kernelization. In *SODA*, pp. 104–113, 2012.

[44] Stefan Kratsch. Co-nondeterminism in compositions: a kernelization lower bound for a Ramsey-type problem. In *ACM-SIAM Symposium on Discrete Algorithms*, pp. 114–122, Kyoto, Japan, 2012.

[45] Pál Erdös and L. Pósa. On the maximal number of disjoint circuits of a graph. *Publ. Math. Debrecen*, **9** (1962), 3–12.

[46] Frank K. H. A. Dehne, Michael R. Fellows, Michael A. Langston, Frances A. Rosamond, and Kim Stevens. An $O(2^{O(k)} n^3)$ FPT algorithm for the undirected feedback vertex

set problem. In *Proceedings of the 11th Annual International Conference on Computing and Combinatorics*, volume 3595 of *Lecture Notes in Comput. Sci.*, pp. 859–869, Berlin, Germany, 2005. Springer.

[47] Jiong Guo, Jens Gramm, Falk Hüffner, Rolf Niedermeier, and Sebastian Wernicke. Compression-based fixed-parameter algorithms for feedback vertex set and edge bipartization. *J. Comput. Syst. Sci.*, **72**(8) (2006), 1386–1396.

[48] Venkatesh Raman, Saket Saurabh, and C. R. Subramanian. Faster fixed parameter tractable algorithms for finding feedback vertex sets. *ACM Transac. Algorithms*, **2**(3) (2006), 403–415.

[49] Jianer Chen, Fedor V. Fomin, Yang Liu, Songjian Lu, and Yngve Villanger. Improved algorithms for feedback vertex set problems. *J. Comput. Syst. Sci.*, **74**(7) (2008), 1188–1198.

[50] George L. Nemhauser and Leslie E. Trotter. Properties of vertex packing and independence system polyhedra. *Math. Program.*, **6** (1974), 48–61.

[51] Jean-Claude Picard and Maurice Queyranne. On the integer-valued variables in the linear vertex packing problem. *Math. Program.*, **12**(1) (1977), 97–101.

[52] Fedor V. Fomin, Fabrizio Grandoni, and Dieter Kratsch. A measure & conquer approach for the analysis of exact algorithms. *J. ACM*, **56**(5), (2009).

[53] Bruce Reed, Kaleigh Smith, and Adrian Vetta. Finding odd cycle transversals. *Oper. Research Lett.*, **32**(4) (2004), 299–301.

[54] Jianer Chen, Yang Liu, Songjian Lu, Barry O'Sullivan, and Igor Razgon. A fixed-parameter algorithm for the directed feedback vertex set problem. *J. ACM*, **55**(5), 2008.

[55] Igor Razgon and Barry O'Sullivan. Almost 2-SAT is fixed-parameter tractable. *J. Comput. Syst. Sci.*, **75**(8) (2009), 435–450.

[56] Venkatesh Raman, M. S. Ramanujan, and Saket Saurabh. Paths, flowers and vertex cover. In *Algorithms—ESA*, volume 6942 of *Lecture Notes in Comput. Sci.*, pp. 382–393. 2011.

[57] Pinar Heggernes, Dieter Kratsch, Daniel Lokshtanov, Venkatesh Raman, and Saket Saurabh. Fixed-parameter algorithms for cochromatic number and disjoint rectangle stabbing. In *SWAT, volume 6139 of Lecture Notes in Computer Science*, pp. 334–345. Springer, 2010.

[58] Ann Becker, Reuven Bar-Yehuda, and Dan Geiger. Random algorithms for the loop cutset problem. *J. Artif. Intell. Res.*, **12** (2000), 219–234.

[59] Fedor V. Fomin, Daniel Lokshtanov, Neeldhara Misra, Geevarghese Philip, and Saket Saurabh. Hitting forbidden minors: Approximation and kernelization. In *Proceedings of the 28th International Symposium on Theoretical Aspects of Computer Science*, volume 9 of *LIPIcs*, pp. 189–200. Schloss Dagstuhl—Leibniz-Zentrum fuer Informatik, Dortmund, Germany, 2011.

[60] Fedor V. Fomin, Daniel Lokshtanov, Neeldhara Misra, and Saket Saurabh. Planar f-deletion: Approximation, kernelization and optimal FPT algorithms. In *IEEE Foundations of Computer Science conference*, pp. 470–479, New Jersey, 2012.

[61] Omid Amini, Fedor V. Fomin, and Saket Saurabh. Counting subgraphs via homomorphisms. *SIAM J. Discrete Math.*, **26**(2) (2012), 695–717.

[62] Jianer Chen, Joachim Kneis, Songjian Lu, Daniel Mölle, Stefan Richter, Peter Rossmanith, Sing-Hoi Sze, and Fenghui Zhang. Randomized divide-and-conquer: Improved path, matching, and packing algorithms. *SIAM J. Comput.*, **38**(6) (2009), 2526–2547.

[63] Noga Alon, Daniel Lokshtanov, and Saket Saurabh. Fast fast. *ICALP*, **1** (2009), 49–58.

[64] Ioannis Koutis. Faster algebraic algorithms for path and packing problems. In *ICALP (1)*, volume 5125 of *Lecture Notes in Comput. Sci.*, pp. 575–586. Springer, 2008.

[65] Fedor V. Fomin, Daniel Lokshtanov, and Saket Saurabh. Efficient computation of representative sets with applications in parameterized and exact algorithms. *CoRR*, abs/1304.4626, 2013.

[66] Esha Ghosh, Sudeshna Kolay, Mrinal Kumar, Pranabendu Misra, Fahad Panolan, Ashutosh Rai, and M. S. Ramanujan. Faster parameterized algorithms for deletion to split graphs. In *SWAT*, volume 7357 of *Lecture Notes in Comput. Sci.*, pp. 107–118. Springer, 2012.

[67] Dániel Marx. Parameterized graph separation problems. *Theoret. Comput. Sci.*, **351**(3) (2006), 394–406.

[68] Jianer Chen, Yang Liu, and Songjian Lu. An improved parameterized algorithm for the minimum node multiway cut problem. *Algorithmica*, **55**(1) (2009), 1–13.

[69] Dániel Marx and Igor Razgon. Fixed-parameter tractability of multicut parameterized by the size of the cutset. *SIAM Journal on Computing*, **43**(2) (2014), 355–388.

[70] Marek Cygan, Marcin Pilipczuk, Michal Pilipczuk, and Jakub Onufry Wojtaszczyk. On multiway cut parameterized above lower bounds. In *IPEC*, number 7112 in *Lecture Notes in Comput. Sci.*, pp. 1–12. Springer, 2011.

[71] Klaus Wagner. Uber einer eigenschaft der ebener complexe. *Mathematische Annalen*, **14** (1937), 570–590.

[72] Neil Robertson and Paul D. Seymour. Graph minors. XX. Wagner's conjecture. *J. Comb. Theory, Ser. B*, **92**(2) (2004), 325–357.

[73] Neil Robertson and Paul D. Seymour. Graph minors. XIII. The disjoint paths problem. *J. Combin. Theory Ser. B*, **63**(1) (1995), 65–110.

[74] Michael R. Fellows and Michael A. Langston. Nonconstructive advances in polynomial-time complexity. *Inf. Process. Lett.*, **26**(3) (1987), 155–162.

[75] Michael R. Fellows and Michael A. Langston. Nonconstructive tools for proving polynomial-time decidability. *J. ACM*, **35**(3) (1988), 727–739.

[76] Michael R. Fellows and Michael A. Langston. An analogue of the Myhill-Nerode theorem and its use in computing finite-basis characterizations (extended abstract). In *FOCS*, pp. 520–525, IEEE, North Carolina, 1989.

[77] Michael R. Fellows and Michael A. Langston. On search, decision and the efficiency of polynomial-time algorithms (extended abstract). In *ACM Symposium on Theory of Computing*, pp. 501–512, Washington, DC, 1989.

[78] Rodney G. Downey and Dimitrios M. Thilikos. Confronting intractability via parameters. *Comput. Science Rev.*, **5**(4) (2011), 279–317.

[79] Neil Robertson and Paul D. Seymour. Graph minors. II. Algorithmic aspects of tree-width. *J. Algorithms*, **7**(3) (1986), 309–322.

[80] Jan van Leeuwen, editor. *Handbook of Theoretical Computer Science, Volume A: Algorithms and Complexity*. Elsevier and MIT Press, Cambridge, MA, 1990.

[81] Stefan Arnborg, Jens Lagergren, and Detlef Seese. Easy problems for tree-decomposable graphs. *J. Algorithms*, **12**(2) (1991), 308–340.

[82] Stefan Arnborg and Andrzej Proskurowski. Linear time algorithms for NP-hard problems restricted to partial k-trees. *Discrete Applied Math*, **23**(1) (1989), 11–24.

[83] Bruno Courcelle. The monadic second-order logic of graphs. III. Tree-decompositions, minors and complexity issues. *RAIRO Inform. Théor. Appl.*, **26**(3) (1992), 257–286.

[84] Hans L. Bodlaender. A linear-time algorithm for finding tree-decompositions of small treewidth. *SIAM J. Comput.*, **25**(6) (1996), 1305–1317.

[85] Richard B. Borie, R. Gary Parker, and Craig A. Tovey. Automatic generation of linear-time algorithms from predicate calculus descriptions of problems on recursively constructed graph families. *Algorithmica*, **7**(5/6) (1992), 555–581.

[86] Stefan Arnborg, Bruno Courcelle, Andrzej Proskurowski, and Detlef Seese. An algebraic theory of graph reduction. *J. ACM*, **40**(5) (1993), 1134–1164.

[87] Bruno Courcelle, Johann A. Makowsky, and Udi Rotics. Linear time solvable optimization problems on graphs of bounded clique-width. *Theory Comput. Syst.*, **33**(2) (2000), 125–150.

[88] Hans L. Bodlaender, Marek Cygan, Stefan Kratsch, and Jesper Nederlof. Solving weighted and counting variants of connectivity problems parameterized by treewidth deterministically in single exponential time. *CoRR*, abs/1211.1505, 2012.

[89] Bruno Courcelle and Joost Engelfriet. *Graph Structure and Monadic Second-Order Logic: A Language-Theoretic Approach*. Number 138 in *Encyclopedia of Mathematics and Its Application*. Cambridge University Press, Cambridge, 2012.

[90] Erik D. Demaine, Fedor V. Fomin, Mohammadtaghi Hajiaghayi, and Dimitrios M. Thilikos. Subexponential parameterized algorithms on graphs of bounded genus and H-minor-free graphs. *J. ACM*, **52**(6) (2005), 866–893.

[91] Neil Robertson, Paul D. Seymour, and Robin Thomas. Quickly excluding a planar graph. *J. Combin. Theory Ser. B*, **62**(2) (1994), 323–348.

[92] Paul D. Seymour and Robin Thomas. Call routing and the ratcatcher. *Combinatorica*, **14**(2) (1994), 217–241.

[93] Erik D. Demaine and MohammadTaghi Hajiaghayi. The bidimensionality theory and its algorithmic applications. *Comput. J.*, **51**(3) (2008), 292–302.

[94] Frederic Dorn, Fedor V. Fomin, and Dimitrios M. Thilikos. Subexponential parameterized algorithms. *Comput. Science Rev.*, **2**(1) (2008), 29–39.

[95] Fedor V. Fomin, Daniel Lokshtanov, Venkatesh Raman, and Saket Saurabh. Subexponential algorithms for partial cover problems. *Inf. Process. Lett.*, **111**(16) (2011), 814–818.

[96] Frederic Dorn, Fedor V. Fomin, Daniel Lokshtanov, Venkatesh Raman, and Saket Saurabh. Beyond bidimensionality: Parameterized subexponential algorithms on directed graphs. *Symposium on Theoretical Aspects of Computer Science*, **5** (2010), 251–262.

[97] Daniel Lokshtanov, Saket Saurabh, and Magnus Wahlström. Subexponential parameterized odd cycle transversal on planar graphs. *FSTTCS*, **18** (2012), 424–434.

[98] Meena Mahajan, Venkatesh Raman, and Somnath Sikdar. Parameterizing above or below guaranteed values. *J. Comput. Syst. Sci.*, **75**(2) (2009), 137–153.

[99] Gregory Gutin and Anders Yeo. Constraint satisfaction problems parameterized above or below tight bounds: A survey. In *The Multivariate Algorithmic Revolution and Beyond*, **7370** (2012), 257–286.

[100] Leizhen Cai. Linear time solvable optimization problems on graphs of bounded cliquewidth. *Discrete Applied Mathematics*, **127**(3) (2003), 415–429.

[101] Michael R. Fellows, Daniel Lokshtanov, Neeldhara Misra, Matthias Mnich, Frances A. Rosamond, and Saket Saurabh. The complexity ecology of parameters: An illustration using bounded max leaf number. *Theory Comput. Syst.*, **45**(4) (2009), 822–848.

[102] Michael R. Fellows, Bart M. P. Jansen, and Frances A. Rosamond. Towards fully multivariate algorithmics: Parameter ecology and the deconstruction of computational complexity. *Eur. J. Comb.*, **34**(3) (2013), 541–566.

[103] Serge Gaspers and Stefan Szeider. Backdoors to satisfaction. In *The Multivariate Algorithmic Revolution and Beyond*, **7370** (2012), 287–317.

[104] Michael R. Fellows, Danny Hermelin, Frances A. Rosamond, and Stéphane Vialette. On the parameterized complexity of multiple-interval graph problems. *Theor. Comput. Sci.*, **410**(1) (2009), 53–61.

[105] Hans L. Bodlaender and Dieter Kratsch. A note on fixed parameter intractability of some domination-related problems. 1994.

[106] Rodney G. Downey and Michael R. Fellows. Parameterized computational feasibility. In P. Clote and J. Remmel, editors, *Proceedings of the 2nd Cornell Workshop on Feasible Mathematics, Feasible Mathematics II*, pp. 219–244, Birkhauser, Boston, MA, 1995.

[107] Venkatesh Raman. Some hard problems in (weighted) tournaments. In Proceedings of the *5th National Seminar on Theoretical Computer Science*, Bombay, India, 1995.

[108] Ronald L. Graham and Joel H. Spencer. A constructive solution to a tournament problem. *Canad. Math. Bull.*, **14**(1) (1971), 45–48.

[109] Noga Alon and Joel H. Spencer. *The Probabilistic Method*, 3rd edition, July 2008, John Wiley.

[110] Nimrod Megiddo and Uzi Vishkin. On finding a minimum dominating set in a tournament. *Theoret. Comput. Sci.*, **61** (1988), 307–316.

[111] Christos H. Papadimitriou and Mihalis Yannakakis. On limited nondeterminism and the complexity of the v-c dimension. *J. Comput. Syst. Sci.*, **53**(2) (1996), 415–429.

[112] Timon Hertli. 3-SAT faster and simpler—Unique-SAT bounds for PPSZ hold in general. To appear in *FOCS, abs/1103.2165*, 2011.

[113] Russell Impagliazzo, Ramamohan Paturi, and Francis Zane. Which problems have strongly exponential complexity? *J. Comput. System Sci.*, **63**(4) (2001), 512–530.

[114] Chris Calabro, Russell Impagliazzo, and Ramamohan Paturi. A duality between clause width and clause density for sat. In *IEEE Conference on Computational Complexity*, pp. 252–260, Prague, Czech Republic, 2006.

[115] Chris Calabro, Russell Impagliazzo, and Ramamohan Paturi. The complexity of satisfiability of small depth circuits. In *IWPEC*, pp. 75–85, 2009.

[116] Liming Cai and David W. Juedes. On the existence of subexponential parameterized algorithms. *J. Comput. Syst. Sci.*, **67**(4) (2003), 789–807.

[117] Daniel Lokshtanov, Dániel Marx, and Saket Saurabh. Slightly superexponential parameterized problems. In *Proceedings of the 22nd Annual ACM-SIAM Symposium on Discrete Algorithms*, pp. 760–776, 2011.

[118] Jens Gramm, Rolf Niedermeier, and Peter Rossmanith. Fixed-parameter algorithms for closest string and related problems. *Algorithmica*, **37**(1) (2003), 25–42.

[119] Bin Ma and Xiaoming Sun. More efficient algorithms for closest string and substring problems. *SIAM J. Comput.*, **39**(4) (2009), 1432–1443.

[120] Lusheng Wang and Binhai Zhu. Efficient algorithms for the closest string and distinguishing string selection problems. In *Proceedings of the 3rd International Workshop, Frontiers in Algorithms*, Hefei, China, pp. 261–270, Springer, 2009.

[121] Zhi-Zhong Chen, Bin Ma, and Lusheng Wang. A fixed-parameter algorithm for the directed feedback vertex set problem. A three-string approach to the closest string problem, *Proceedings of 16th Annual International Conference on Computing and Combinatorics* 2010, Nha Treng, Vietnam, July 2010, Springer LNCS 6196, 449–458.

[122] Michael R. Fellows, Fedor V. Fomin, Daniel Lokshtanov, Elena Losievskaja, Frances A. Rosamond, and Saket Saurabh. Distortion is fixed parameter tractable. *ICALP*, **1** (2009), 463–474.

[123] Jianer Chen, Benny Chor, Mike Fellows, Xiuzhen Huang, David W. Juedes, Iyad A. Kanj, and Ge Xia. Tight lower bounds for certain parameterized NP-hard problems. *Inf. Comput.*, **201**(2) (2005), 216–231.

[124] Jianer Chen, Xiuzhen Huang, Iyad A. Kanj, and Ge Xia. On the computational hardness based on linear FPT-reductions. *J. Comb. Optim.*, **11**(2) (2006), 231–247.

[125] Jianer Chen, Xiuzhen Huang, Iyad A. Kanj, and Ge Xia. Strong computational lower bounds via parameterized complexity. *J. Comput. Syst. Sci.*, **72**(8) (2006), 1346–1367.

[126] Andreas Björklund, Thore Husfeldt, Petteri Kaski, and M. Koivisto. Fourier meets Möbious: Fast subset convolution. In *Proceedings of the 39th Annual ACM Symposium on Theory of Computing*, New York, 2007. ACM Press.

[127] Johan M. M. van Rooij, Hans L. Bodlaender, and Peter Rossmanith. Dynamic programming on tree decompositions using generalised fast subset convolution. In *ESA*, pp. 566–577, 2009.

[128] Daniel Lokshtanov, Dániel Marx, and Saket Saurabh. Known algorithms on graphs of bounded treewidth are probably optimal. In *Proceedings of the 22nd Annual ACM-SIAM Symposium on Discrete Algorithms*, pp. 777–789, ACM-SIAM, San Fansisco, CA, 2011.

[129] Daniel Lokshtanov, Dániel Marx, and Saket Saurabh. Lower bounds based on the exponential time hypothesis. *Bulletin of the EATCS*, **105** (2011), 41–72.

[130] Giorgio Ausiello, Pierluigi Crescenzi, Giorgio Gambosi, Viggo Kann, Alberto Marchetti Spaccamela, and Marco Protasi. *Complexity and Approximation: Combinatorial Optimization Problems and Their Approximability Properties*. Springer, 1999.

[131] Liming Cai and Jianer Chen. On fixed-parameter tractability and approximability of NP optimization problems. *J. Comput. Syst. Sci.*, **54**(3) (1997), 465–474.

[132] Phokion G. Kolaitis and Madhukar N. Thakur. Approximation properties of NP minimization classes. *J. Comput. Syst. Sci.*, **50**(3) (1995), 391–411.

[133] David R. Karger, Rajeev Motwani, and G. D. S. Ramkumar. On approximating the longest path in a graph. *Algorithmica*, **18**(1) (1997), 82–98.

[134] Dániel Marx. Parameterized complexity and approximation algorithms. *Comput. J.*, **51**(1) (2008), 60–78.

[135] Rodney G. Downey, Michael R. Fellows, Catherine McCartin, and Frances A. Rosamond. Parameterized approximation of dominating set problems. *Inf. Process. Lett.*, **109**(1) (2008), 68–70.

[136] Georg Gottlob and Stefan Szeider. Fixed-parameter algorithms for artificial intelligence, constraint satisfaction and database problems. *Comput. J.*, **51**(3) (2008), 303–325.

[137] Martin Grohe. Parameterized complexity for the database theorist. *SIGMOD Record*, **31**(4) (2002), 86–96.

[138] Hans L. Bodlaender, Rodney G. Downey, Michael R. Fellows, Michael Hallett, and Harold T. Wareham. Parameterized complexity analysis in computational biology. *Computer Applications in the Biosciences*, **11**(1) (1995), 49–57.

[139] Rodney G. Downey, Michael R. Fellows, Alexander Vardy, and Goeff Whittle. The parameterized complexity of some fundamental problems in coding theory. *SIAM J. Comput.*, **22**(2) (1999), 545–570.

Index

Note: Locator followed by '*f*' and '*t*' denotes figure and table in the text

0-1 network, 294
0-valid formula, 1144
1-extendable graph, 360–361
    ear decompositions of, 362
1-factorial connection, 220
1-minimal *P*-set, 423
1-port network, 249*f*
2-extendable bipartite graphs, 368
2-extreme set, 317
2-legged bad cycle, 566
2-legged cycle, 565
2-port network, 251*f*
2-ratio-cut, 857
2-visibility drawing of *G*, 541, 541*f*
3-approximate distance oracle, 1021–1022
3-asteroid, 717–718, 717*f*
3-connected bicritical graph, 362
3-legged cycle, 566
3-SAT problem, 441
4-chromatic graph, 460–461

A

Abelian group, 628
Accumulate-offset procedure, 551
Achromatic number, 463
Ackermann's functions, 696
Active pertinent vertex, 495
Active vertex, 493–494, 834
Acyclic chromatic number, 966
Acyclic-clustering of graph, 809
Acyclic directed graphs, 15
Acyclic graph, 8
Acyclic set, 768
Adjacency graph, 760, 1015
Adjacency list representation, 21–22
Adjacency matrix
    determinant of, 235
    of directed graph, 215–218
    representation, 21
    skew, 366

Adjacent edges, 4
Adjacent spanning tree structures, 133
Algebraic connectivity of graph, 243–245
Algebric notations, basic, 592
Algorithm
    for 3-colorability problem, 954
    AND-distillation, 1150
    arc-balancing, 110
    augmenting path. *See* Augmenting path algorithms
    BFM. *See* Bellman–Ford–Moore (BFM) algorithm
    biconnectivity, 66–68
    blocking flow, 104–107
    blossom, 358
    canonical labeling. *See* Canonical labeling algorithms
    capacity scaling, 115, 143–145, 144*f*
    chordal-recognition, 714
    computational complexity, 16–18, 21
    for convex grid drawing, 556–560, 556*f*, 557*f*, 558*f*, 559*f*
    cycle-canceling. *See* Cycle-canceling algorithm
    deferred acceptance, 405
    designs, 590
    Dijkstra's, 39–42, 1043
    Dinic's, 105
    Dinic's maximum flow, 294–295
    Edmonds', 25, 27, 377–379
    embedding, 529, 529*f*
    Even's vertex connectivity, 304*f*
    exact. *See* Exact algorithms
    fastest, 110
    Fastmincut, 1015
    Gabow's, 379–385
    graph search, 86
    GS, 405–406
    Hassin–Lorenz–Raz, 1035
    Hochbaum's pseudoflow, 110
    implementation in network, 590

Algorithm (*Continued*)
   interior point, 170
   Kruskal's, 22
   LARAC, 1042
   LARAC-BIN, 1049, 1050f
   LR, 1038
   MA ordering, 110
   matroid union, 907–910
   Matula's, 295–297, 297f
   maximally distant forests, 934–937
   network simplex, 115, 127, 130–143, 134f
   one-step greedy, 649
   optimal Chinese postman tour, 396–397
   outline for principal partition, 906
   out-of-kilter, 114, 129–130
   path finding, 73
   polynomial-time, 83, 147–150, 408
   preflow-push, 98–104
   primal-dual, 114, 129–130
   Prim's, 22, 24
   PSCSP, 1054
   pseudoflow push, 148
   pseudopolynomial-time, 83
   Rectangular-Draw, 566–567
   relaxation, 147
   shift, 546–551, 547f, 549f, 552f
   st-numbering, 71–74, 72f
   Stoer and Wagner's, 298–300, 301f, 302f–303f
   strong connectivity, 68–71
   successive shortest path, 114, 127–129, 129f
   transitive closure. *See* Transitive closure
   unit capacity maximum flow, 108
   with worst-case complexity, 83
Algorithmic graph, 590
All pairs shortest paths (APSP) problem, 1021
α-acyclic hypergraphs and their duals, 757–776
   bipartite graphs, 773–774
   dually chordal graphs, 770–773
   Graham's algorithm, 767–770
   hypergraph 2-coloring, 763–764
   hypergraph notions, 759–763
   hypertrees, 759–763, 773–774
   König property, 764–767, 766f
   matrix notions, 775–776
   maximum neighborhood orderings, 770–774
   other desirable properties, 775–776
   relational database theory, 757–759
   running intersection property, 759–763
   and tree structure, 764–767, 766f
α-covering based randomized algorithms, 1158–1159
Alternating group, 631
Alternating tree, 389, 390f
Amenable, 742
Analogous minimax theorem, 332
AND-distillation algorithm, 1150
Angle of vertex, 562
Anti-connected set, 710
Antipodal graph, 604
Approximation scheme, 1028
Arbitrary database, 758–759
Arboreal hypergraphs, 752
   hypertrees, 764–765
Arborescences, 13–14
Arboricity, 9
Arc, 502
   backward, 138
   blocking, 138, 141
   flow of, 572
   forward, 138
   infinite capacity, 867
Arc-at-a-time approach, 169
Arc-balancing algorithm, 110
Arc-coloring theorem, 265–267
Artemis graph, 741
Articulation set, 768
Asteroidal triple, 727
Atoms, chemical graph, 794
Attributes, 757
Augmented cubes (AQ_n)
   of dimensions, 617–618, 618f
   embedding into, 674–680
   Hamiltonicity property, 675
Augmenting cycle theorem, 1069
Augmenting path, 376
   to find, 377
Augmenting path algorithms, 85–98
   capacity-scaling, 89–91
      improvement in, 97
   generic. *See* Generic augmenting path algorithms
   maximum capacity, 89–91

residual capacity of, 85–86
residual network, 86f, 88f
shortest, 91–98
    admissible arcs, 91–94, 95f
    bad example for, 97f
    illustration, 92f
    worst-case improvements, 96–98
summary of, 98t
Automorphism, 5, 634, 644
Automorphism group of G, 5
AverageGain, 842
Axiom systems for matroids, 880–887
    base axioms, 882–884
    circuit axioms, 886–887
    independence axioms, 880–881
    rank axioms, 884–886
Azuma's, applications of, 950

B

Back edge, 60, 63, 493
Back edge traversal (BET), 527
Back neighbor vertex, 527
Backtrack search, 1128
Backward arcs, 138
Bad corner, 566
Balanced pancyclic graph, 678
Balance-factor of vertex, 668–669
Barnes' approach, 854–857
Barycentric mapping method, 552
Base axioms of matroid, 882–884
Batched iterated 1-Steiner, 1122–1123
Bellman–Ford–Moore (BFM) algorithm, 36–39, 1054
    negative cycle detection, 38
    shortest path tree, 38–39
Bend, 539
Bend-angle, 570
Bend-optimal orthogonal drawing of G, 569–570
Berge-acyclic graph, 802
Berge's alternating chain theorem, 373–377
Berge's deficiency theorem, 356–357
Bernoulli trial, 948
β-acyclic hypergraphs, 776–777
Bewildering diversity, 764
b-frugal chromatic number, 965
b-frugal k-coloring, 965
Biadjacency matrix, 364–365
Bicoloring problem, 763
Biconnected component of G, 490

Biconnectivity algorithm, 66–68
Bicritical graph, 363
    decomposition into bricks, 364f
Bicycle-based tripartition, 187
Big-Oh notation, 16
Big-Omega notation, 16
Bijective function composition, 630
Binary caterpillar, 666, 666f
Binary cube, 591
Binary de Bruijn multiprocessor network, 618, 619f
Binary groups, 629, 629t
Binary hypercubes, 640–641, 641f
Binary relation (R), 13
    transitive closure of, 30
Binary tree, 663
    double rooted complete, 664–665, 665f
Binet–Cauchy theorem, 203–204, 206, 241–242
Bipartite graph, 4, 663, 773–774, 786
    2-extendable, 368
    k-regular, 365
    matching in, 351–354, 403
    maximum matching in, 385–388
        flow-based approach, 388
        Hopcroft and Karp's approach, 386–388
    for optimal assignment problem, 392
    spectrum of complete, 234–235
Bipartite incidence graph, 762, 774, 787
Bipartite maximum matching algorithm, 389
Bipartite Q-critical graphs, 660, 661f
Bipartite vertex-edge, 762
Bipartition of G, 4, 831
    algorithm, 833
Birkhoff–von Neumann theorem, 354
Bit-wise addition, 629
Block diagonal matrix, 929
Block graphs, 802
Blocking arc, 138, 141
Blocking flow algorithms, 104–107, 107t
Blocking pair, 404, 406–407
    definition of, 412
Blocks of graph, 273–276, 275f
Blossom algorithm, 358
Blossom shrinking process, 378
b-matching problem, 1111
Boole's inequality (union theorem), 1018
Bottleneck Steiner distance, 1127

Boundary cycle, 527
Boundary path, external, 528
Box-orthogonal drawing of G, 540, 540f
Box-rectangular drawing of G, 540f, 541, 543, 543f
Branches of T, 8
Breadth-first search (BFS), 1019, 1101, 1154
Brick decomposition procedure, 362–364, 363f
Bridge, 570
Brittle graph, 735
Broadcasting problem, 649
Brute-force implementation, 529
Bubble sort generator, 636
Bubble sort graph, 642
Buckets, 695
Budget-constrained problems, 1080
Budgeted concurrent flow (BCF), 1081
Budget multicommodity flow (BUDGET-MF), 1080–1081
Buneman's four-point condition, 802–803
Butterfly graph, 619, 620f

C
CAD (computer-aided design), 830
Candidate list approach, 153
Canonical (s-t)-cut, 838
Canonical decomposition, 594
Canonical label, 998
Canonical labeling algorithms
    bound on failure rate of, 1005–1008
    Morgan's algorithm, 1001
Canonical name for graph, 999
Capacity-scaling algorithm, 89–91, 115, 143–145, 144f
    improvement in, 97
Cartesian product of graphs, 7, 592, 592f
Caterpillars, 666
Cayley coset graphs, 649
Cayley graph, 593–594, 594f, 627, 636–642
    abelian group, 638f
        binary hypercubes, 640–641, 641f
    d-dimensional toroid, 639
    degree four, 620–621, 621f
    n-node ring, 638, 638f
    of permutation groups, 642, 642f
    recursive structure of binary hypercube, 641, 641f
    supertoroid, 639–640, 640f
    symmetry in, 642–648

    graph symmetry, 643–647
    symmetry and Cayley graphs, 647–648
    toroid, 638–639, 639f
Cayley–Hamilton theorem, 231
Certified algorithms, 789
Characteristic graph, 792
Chebyschev's inequality, 948
Chemical graphs
    canonical naming algorithms for, 1000–1002
    chemical multigraph, 999–1000
    definition of, 998–999
    Faulon's algorithm, 1000
    random model for, 997–1008
Cheminformatics system, 998, 1000
Chernoff bound, 1018
Chernoff–Hoeffding bounds, 963
Chinese postman problem, 395–398
Chord, 544
Chordal bipartite graphs, 786–788
Chordal graphs, 710–715, 753–755
    characterization, 710–712
    optimization, 715
    recognition, 712–715
    subclasses of, 756–757
    and variants, 753–757
Chordal-recognition algorithm, 714
Chord cycle, 752
Chordless cycle, 710
Chromatically equivalent graphs, 459
Chromatically unique graph, 459
Chromatic coding, 1160–1162
    kernelization, 1160
    probability of good coloring, 1161
    SED problem, 1160
    solving colored instance, 1161
Chromatic number, 790
Chromatic polynomial, 458–459
Circuit axioms of matroid, 886–887
Circuit matrix, 196–197
    fundamental, 197
    submatrices of, 199–201
Circuits and cutsets
    fundamental, 177–179
    spaces of graph, 182–184
    dimensions of, 184–185
    orthogonality of, 186–187
    relationship between, 185–186
    spanning trees, 179–182

Circuit vector, 196
Clause graph, 735, 736f
Claw-free graph, 355, 426
Claw graph, 426, 710
Clique
    cutset, 711
    graph operator, 763
    hypergraph, 762
    number, 456, 459
    separator, 794
    trees, 752, 754
Clique cutset decomposition, 708
Clique net model, 859
Clique-width of graphs, 805–807
Closed-acyclic graph, 769
Closed walk, G, 6, 12
Closest string problem, 1182
Cluster of largest density, 838–839
C-nodes, 526
    creating first, 527–528
    creation in general, 528, 529f
    head, 528
Coarsening procedure, 848–849
Coates flow graph/Coates graph, 219, 219f
    associated with $n \times n$ matrix, 220
Coates' gain formula, 218–222
Cobase submatrix, maximum rank of, 933–934
Coda theorem optimization technique, 510–511
Code-optimization problem, 691
$C_o(G)$-component of G, 566
Cographs, 788–790
    optimization on, 790–791
Co-join operation, 788, 793
Collapsing vertex, 691
Coloop, 927, 939
Colorado School of Mines (CSM) mechanism, 1002
Color class, 449, 965
Color coding, 1159–1160
Column-oriented algorithm, 32
Combinatorial methods, 114
Combinatorial planar embedding, 490
Combinatorics
    in general and graph theory, 946
    mathematical statements in, 946
    probabilistic arguments in, 945–989
Commercial parallel computers, 589, 590t

Commodity, 1079
Commutative group, 628
Comparability graphs, 715–726, 716f, 717f
    characterization, 715–718
    optimization, 725–726
    recognition, 718–724
        modular decomposition, 720, 721f
        modular decomposition tree to transitive orientation, 721–722
        time, 722–724
        transitive orientation, 720
Comparability-recognition algorithm, 718–719
Complementary slackness optimality conditions, 122–123, 162, 166
Complement $\bar{G}$, 5
Complete bipartite graph, 4
Complete graph, 4
Complete k-admissible splitting, 318, 320, 325
Complete k-coloring, 463, 464f
Complete k-partite graph, 4
Completely scanned vertex, 59
Completeness theory, 1160
Computational complexity study, 353
Computation graph, 590
Computer-aided design (CAD), 830
Concentration inequality, 948
Concentration phenomena, 979
Concurrent defined, 1081
Conditional connectivity of graph, 283–286
Conditional domination number, 426
Conjunctive normal form (CNF), 763, 1144
Connected component of G, 490
Connected dominating sets, 429–430
Connected graph, 489–490
Connectedness of graph, 6–7
Connectivity augmentation problem, 315–316
Conservation law, 572
Constrained colorings, 965–968
Constrained shortest path (CSP) problem, 1035, 1042
    definition of, 1042
    and DUAL-RELAX CSP problems, 1042–1043
    ε-approximation algorithm for, 1035–1039
    feasible path, 1042

Constrained shortest path (CSP) problem
(*Continued*)
    GEN-LARAC for, 1056–1058
    LARAC algorithm for, 1042–1045
    PSCSP algorithm for, 1055f
Construction-by-correction approach, 1130
Contraction algorithm, global mincut, 1014–1016
Convex drawing of G, 539, 540f, 552–560
    canonical decomposition, 553–555, 553f, 554f, 555f
    convex grid drawing algorithm, 556–560, 556f, 557f, 558f, 559f
Convolution operation, 896–901
    connectedness for $f*g$, 899–901
    formal properties of, 897–899
    overview, 896
    polymatroid rank functions, 897
Cooling schedule, 865
Corner of rectangular drawing, 560, 560f
Cospanning tree, 8
Cost of link, 1052
Cost of path, 1052
Cost scaling approach, 1071
Cotree, 789
Courcelle's theorem, 1168–1169
Cover time of G, 258
χ-perfect graph, 790–791
Cramer's rule, 255
Critical value of $(f(\cdot), g(\cdot))$, 902
Critical values in principal partition, 843
Crossed cubes, embedding into, 680–681
Cross edge, 63
Crossing bi-supermodular function, 342
Crown decomposition, 1147–1148
Cube-connected cycles graph, 619, 620f
Cubes, enhanced, 681
Cubical dimension of G, 659
Cubical graphs, 5, 657–661, 662f, 662t
Cut-edges of graph, 273–276, 274f
Cut matrix, 193–196, 194f
    submatrices of, 199–201
Cutset, 9
    fundamental, 925
Cuts of G, 9
Cut vertex/vertices, 273–276, 274f, 490, 924
Cycle-canceling algorithm, 114, 125–127, 126f
Cycle free solutions, 131–132

Cycle, G, 489
Cyclic subgroup, 633
Cyclomatic number, 766

D
Dag (directed acyclic graph), 722
Dantzig–Wolfe decomposition method, 164–167
    optimality conditions, 165
    path flow
        complementary slackness conditions, 166–167
        reformulation with, 165
Dark edges, 374
Database scheme, 757
Data reduction rule, 1145–1146
Datum vertex, 192, 249
de Bruijn graph ($DG(d,k)$), 618–619, 619f
Decision graph, 562, 564
Decomposition of graphs, 788–794
    clique separator, 794
    cographs, 788–790
    modular, 793–794
    module properties, 791–792
    optimization on cographs, 790–791
Decomposition theorem, 793–794
Deferral costs, scheduling with, 118–119, 120f
Deferred acceptance algorithm, 405
Deficiency of G, 356
Degenerate iteration, 138
Degree four Cayley graph, 620–621, 621f
Degree lower bound, 1099
Degree matrix, 204
    of undirected graph, 253
Degree neighborhood canonical labeling, 1002f
Degree neighborhoods (DN) algorithm, 1001
Degree of vertex, 489
Degree-specified augmentation
    edge-connectivity augmentation
        of digraphs, 325
        of graph, 318–320
    local edge-connectivity augmentation, 321–323
Demand arcs, network, 117
Density ($Y(n)$), 970
Depth-first index (DFI), 493
Depth-first search (DFS), 492, 691, 974, 1011, 1105

of directed graph, 63–66, 65f
   forest, 64–65, 65f
   tree, 61
      of connected graph, 67
   of undirected graph, 59–63, 62f
Depth index, 135
Derivation chain, 48, 51
Detailed balance conditions, 865
DFS. *See* Depth-first search (DFS)
DFS tree, 493
δ-hyperbolic metric spaces, 810
   geodesic space, 811
Diakoptics, 931
Diamond vertex, 802
Digraphs
   edge-connectivity augmentation of, 323–325
   one-way pair in, 334
   vertex-connectivity augmentation of, 333–337
Dihedral group, 632
Dijkstra's algorithm, 39–42, 129, 1093–1094
   graph for illustrating, 40f
   illustration, 41f
Dilation of embedding, 645
Dilworth's theorem, 354, 708–709
Dimensional toroid, 639
Dinic's algorithm, 105
Dinic's maximum flow algorithm, 294–295
Dirac's theorem, 368
Directed acyclic graph (dag), 722
Directed edge, 490
Directed Eulerian graphs, 14
Directed Euler trails
   directed spanning trees and, 211–212
   procedure of constructing, 212
Directed graph/digraph, 11–12, 11f, 26f
   1-factor of, 216, 216f
   acyclic, 15
   adjacency matrix of, 215–218
   condensation, 12
   cut matrix of, 193
   DFS of, 63–66, 65f
   directed spanning trees in, 207–210
   disconnected, 12
   edge connectivity of, 298
   incidence matrix of, 925
   of network, 248f
   painting of, 266f
   paths and connections in, 12
   and relations, 13, 13f
   representing transitive closure, 30
   shortest paths, 35–44
      all pairs, 42–44
      with negative length edges, 37
      with no negative length edges, 39–42
      single source, 36–39
Directed Hamiltonian graphs, 14–15
Directed hypergraphs, 341–342
Directed path, 12
Directed spanning trees
   and directed Euler trails, 211–212
   in directed graph, 207–210
Directed trees, 13–14
Directed walk, 12
Direct product, G, 8, 634–635
Disconnected digraph, 12
Discrete probability
   space, 947
   tools from, 947–950
Disjoint paths, QoS constraints with
   algorithm analysis, 1069–1071
   computing upper and lower bounds, 1072–1073
   with delay constraint, 1067–1068
   MCDP problem. *See* Minimum cost disjoint paths (MCDP) problem
   minimizing computational complexity, 1071–1072
   minimizing path delays, 1068–1069
   model and problem definition, 1064–1065
   network flows, 1065–1066
   RDP problem, 1065
Disk hypergraph, 762
Distance-hereditary graphs
   minimum cardinality Steiner tree problem in, 799–801
   subclasses of, 794–803
      block graphs, 802
      γ-acyclic hypergraphs, 802–803
      ptolemaic graphs and bipartite, 801
Distance labels, 98–99, 105
Distance oracles, approximate
   3-approximate distance oracle, 1021–1023
   related work on, 1023
Distance-symmetric graph, 597

Distance transitive graph, 647
Distributed computing, 1097
Divisible graph, 744
Domatic number, 437–438
Dominating function (DF), 436
Dominating-induced matching problem, 808
Domination in graph
    algorithmic aspects, 441–442
    bounds, 428, 428t
    chain, 425
    conditional domination number, 426
    dominating sets
        connected, 429–430
        PDS, 430–432
        total, 427–428, 428f
    equivalence, 432–434
    factor, 434–435
    file server, 427
    global, 434–435
    hypercube, 431f
    other types of, 435–436
    parameters of Q_n, 430t
    in product graphs, 439–441
    types of, 426–427
    Vizing's conjecture, 439–440
Domination theory, 419
Dominator, 698
    algorithm, 701–702, 703f
    colorings, 466–467
    immediate, 698
    partition, 466
    partition number, 466
    in program graphs, 698–705, 698f
    tree, 698–699, 698f
Dominator chromatic number ($\chi_d(G)$), 466, 467f
Donath–Hoffman bound, 855
Double rooted complete binary tree, 664–665, 665f
Double starlike trees, 669
Doubly chordal graphs, 773
Doubly lexically order, 776, 782
Downhill groove, 865
Driving-point conductance, 250
Driving-point resistance, 250, 252
Dual combinatorial planar, 514
Dual graph, 481–484, 482f, 484f, 760
    construction of, 486f
Dual integrality property, 136

Duality, 485–487
    theory, 123
        strong, 124
        weak, 123–124
    Whitney's definition of, 485f
Dualization, 759
Dual-like graph, 542, 542f
Dually chordal graphs, 770–773
Dual matroids (M^*), 887–889
    principal partition of, 905
    union of matroids and union of, 912–914
Dual minimum cost flow problem, 123
DUAL-RELAX CSP, 1043
Dynamic programming approach, 1128
Dynamic trees, 97

E

ε-approximation algorithm for CSP problem, 1035–1039
Ear decompositions
    of 1-extendable graph, 362
    of matching-covered graph, 360–362
Ear vertices, 767
Ecology of parameters, 1171–1173
    backdoors to satisfiability, 1173
    beyond guarantee bounds, 1171–1172
    structural parameters, 1172
Edge addition method, 492
    nonplanarity minors, 500–502, 500f
Edge addition planarity testing algorithm
    efficient implementation, 502–511
        coda, 510–511
        external face management, 509–510
        forward edge lists, 503
        future pertinence management, 505
        graph storage and manipulation, 502–503
        merging, flipping, and embedding recovery, 505–509, 506f, 507f, 508f
        nonplanarity detection, 503–504
        pertinence management with walkup, 504
        sorted child lists, 503
        systematic walkdown invocation, 503–504
    isolating obstruction to planarity, 511–514, 513f
    overview, 493–495, 494f, 495f

proof of correctness, 499–502, 499f, 500f, 502f
top-level processing model, 495, 495f
visibility representation of planar graph, 514–522
   correctness and performance, 519–522
   horizontal positions of edges, 518–519, 518f, 519f
   vertical positions of vertices, 514–518
walkdown, 492, 494–498
   processing example, 497–498, 497f
   rules, 496–497
Edge case, beating 2 for, 1105–1110
   approximation guarantee, 1108–1110
   Khuller–Raghavachari algorithm, 1106–1108
   time complexity, 1110
Edge certificates algorithm, 1100, 1101f
Edge-chromatic number, 461
Edge-chromatic number $\chi_e(G)$, 353
Edge-coarsening procedure, 849
Edge coloring of G, 353, 461–463
Edge congestion, 654–655
Edge-connectivity of graphs, 276–279, 295–298, 1097, 1100
   augmentation of digraphs, 323–325
   augmentation of graph, 316–320
      constrained, 325–329
      degree-specified augmentation, 318–320
      local, 320–323
      simultaneous, 329
      variations and extensions, 320
   determining, 297f
   directed, 298
   undirected, 295–297
Edge contraction operation, 491
Edge covering number, 351
Edge-disjoint spanning trees, 672–673
Edge/edges, 4
   adjacent, 4
   back, 60, 63, 493
   computing horizontal positions of, 518–519
   cross, 63
   dark, 374
   of dimension, 592
   directed, 490
   forward, 63

   generator, 518
   light, 374
   marked, 50
   multiple, 4
   order, 518
   parallel, 592, 594
   short-circuit, 509
   subdivision, 491
Edge-induced subgraph, 5
Edge-intersection separator, 768
Edge-subhypergraphs, 759
Edge-substitution, 1121
Edge-symmetric graph, 597
Edge symmetry/transitivity, 646–647, 646f, 647f
Edge version of Menger's theorems, 292–293
Edge-weighted graph, 838
Edmonds' algorithm, 25, 27
   for optimum branching, 28–29
Edmonds' approach, 377–379
   blossom, 377–378, 379f
   correctness of, 378–379
   Gabow's implementation of, 383f
   illustration, 378f
Edmonds' optimum branching algorithm, 28–29
Edmonds' theorem, 308
Efficient domination problem, 807
Efficient edge domination problem, 808
Efficient polynomial time approximation scheme (EPTAS), 1185
Egalitarian cost, 410
Eigenvector-based clustering
   algorithm, 859–860
   theoretical motivation, 860
Electrical circuit theory, 177
Electrical network analysis, 923
   through topological transformation, 937–941
Elementary separators, 924
Empty corner rectangle property, 1126, 1126f
Empty diamond property, 1125, 1125f, 1126f
Empty inner rectangle property, 1126–1127, 1127f
END array, 379
Endpoints, 489
End vertices, 4
Epimorphism, 634

Eq-dominating set, structure of, 432, 433f
Eq-domination number, 432
Eq-irredundance number, 433
Equibipartite graph, 663
Equi-cofactor matrix, 204
Equipartite graph, 663
Equivalence chain, 432–433
Equivalence classes, 13
Equivalence domination, 432–434
Equivalence graph, 432
Equivalence number, 432
Equivalence relation, 13
Erdös method, 947
Escape graph, 1130, 1130f
Euclidean SMT (ESMT) problem, 1116
Euclidean Steiner ratio, 1117
Eulerian graph, 9, 396
    directed, 14
Eulerian trail, 9, 397
Euler's formula, 477–480
Even permutation, 631
Even's algorithm, 44–53
Even's st-numbering algorithm, 73–74
Even's vertex connectivity algorithm, 304f
Exact algorithms, 1124–1129, 1131–1135
    FST concatenation, 1127–1129
    FST generation, 1125–1127
Exact cover problem, 807
Exclusive-or addition, 629
Exclusive-or operation, 629
Exist vertex-disjoint paths, 671–672
Exponential time hypothesis (ETH), 954, 1178–1181
    lower bound on FPT algorithms, 1181–1183
    overview, 1178–1181
    problems parameterized by treewidth, 1183–1185
    SETH, 1181
    $W[1]$-hard problems, 1183
External edge cost, 845
External face, 490
    links, 509
    management, 509–510
Externally active vertex, 494
External private neighbor, 424

F
Faces, 490
Factor-critical graphs, structure of, 359–360
Factor domination number, 434
Fastest strongly polynomial time algorithm, 110
Fast lookup table estimation (FLUTE), 1120, 1123–1124
Fast subset convolution, 1184
Fault-free Hamilton cycle, 675
Fault tolerance, 273, 590
FBB (flow-balanced-bipartition), 868–869
$\mathcal{F}$-deletion problem, 1158–1159
Feasible drawing, 577
Feasible path, 1042
Feasible vertex labeling, 392
Feedback loops, 224
Feedback vertex set (FVS), 1153–1154, 1174
f-factor of graph, 368
Fibonacci code, 616
Fibonacci cube (Γ_n), 615–616, 617f
Fibonacci heaps, 30
Fibonacci tree, 669
Fiduccia–Mattheyses (F–M) algorithm, 831, 847
Fiedler value of G, 243
Finite group, 628
FIRST array, 380, 383
    at intermediate step, $384t$
First-in, first-out (FIFO) preflow-push algorithms, 103–104
    illustration, 104f
    worst-case complexity of, 103
Fission number of $\prod$, 940
Five color theorem, 455
Fixed-parameter tractable (FPT)
    algorithmic techniques, 1144–1171
        bounded search trees, 1152–1156
        bounded treewidth machinery, 1168–1170
        iterative compression, 1156–1157
        kernelization, 1161–1168
        randomized fixed-parameter algorithms, 1157–1162
        separators, 1162–1166
        subexponential algorithms and bidimensionality, 1170–1171
        well-quasi-ordering, 1166–1168
    and approximation, 1185–1186
    defined, 1143
    lower bound on, 1181–1183
    time for W-hard problems, 1186

Flip operation on planar graph, 492
    data structures for, 506f, 507f
Flow-balanced-bipartition (FBB), 868–869
Flow-based approach, 388
Flow decomposition algorithm, 1066
Flow network, 572, 838
Flow of arc, 572
Floyd's algorithm, 42–43, 397
F–M algorithm, 831, 847
Folded hypercube, 614
Forest, 8, 117
    building maximally distant, 934–937
    maximum distance between, 933
    of minimum size hybrid representation, 933
Forward arc list, 503
Forward arcs, 138
Forward edge, 63
Forward edge lists, 503
Forward path, 224
Foster's theorems, 263–265
    first theorem, 263–264
    second theorem, 264–265
Four color conjecture theorem, 455
Fractional domination number, 436
Fragment of G, 284
Freed-colors, 729
Frobenius theorem, 352
Frugal colorings of graph, 965–966
Full Steiner trees (FSTs), 1117
    concatenation, 1127–1129
        backtrack search, 1128
        dynamic programming, 1128
        integer linear programming, 1129
    decomposition of, 1132f
    forbidden edges in blockages, 1133f
    generation, 1125–1127
        empty corner rectangle property, 1126
        empty diamond property, 1125, 1125f, 1126f
        empty inner rectangle property, 1126–1127
    in presence of obstacle, 1132f
    structures, 1134f
Fully polynomial approximation scheme (FPAS), 1065
Fully polynomial time approximation scheme (FPTAS), 1028, 1038, 1082, 1185
    maximum concurrent flow, 1084f, 1091f

Fundamental circuit (f-circuit), 883, 925
Fundamental cutset, 925
Fusion–fission method, 937–939
    network N decomposed by, 939f
    network N to illustrate, 938f
    number of $\prod$, 940
Fusion number of $\prod$, 940
`FuturePertinentChild` member, 505

G
Gabow's algorithm, 379–385
    bipartite nature of, 389
    description of, 381
    implementation of Edmonds' algorithm, 383f
    outer vertex, 380
    procedures of, 380–381
Γ-acyclic hypergraphs, 794–803
Gaifman graph, 760
Gale–Shapley (GS) algorithm, 405–406
Gallai–Edmonds structure theorem, 357–359
    decomposition, 359
    partition, 358f
Gallai identities, 351
Generalized flow problems, 1093
Generalized-gain ($GGain$), 842
Generator edge, 518
Generator group, 635
Generic augmenting path algorithms, 85–89
    drawback of, 88–89
    implementations of, 89
    pathological instance for, 89f
Generic preflow-push algorithms, 99–102, 100f
GEN-LARAC method, 1055, 1057f
    convergence analysis of, 1058
Geodesic pancyclic graph, 679
Geometric improvement theorem, 1070
Γ-free matrices, 782–783, 782f
Global domination number, 434–435
Gomory–Hu cut-tree, 833
Gomory–Hu tree, 301–308
    construction, 306f
    defined, 308
    example, 307f
Good cycle, 566
Go-to-less programs, 692
Graham reduction, 767
Graham's algorithm, 767–770

Graph (G), 4, 489
  acyclic, 8
  acyclic colorings, 966–967
  after vertex splitting, 294f
  algebraic connectivity of, 243–245
  algorithms, 974
  automorphism, 5, 644
  biconnected components of, 67f
  canonical name of, 999
  circuits and cutsets spaces of, 182–184
    dimensions of, 184–185
    orthogonality of, 186–187
    relationship between, 185–186
  circuit vector of, 196
  components of, 6
  conditional connectivity of, 283–286
  criticality and minimality, 287–288
  critical subgraph of, 26f
  with degree sequence, 310f
  directed. *See* Directed graph/digraph
  domination number of, 420–423
  domination sequence of standard, 426t
  dual, 481–484, 482f, 484f
  edge connectivity of, 295–298, 316–320
  example, 4f
  expansion at vertex, 280f
  factor-critical, 359–360
  frugal colorings, 965
  fundamental circuit of, 178, 178f
  fundamental cutsets of, 179, 180f
  girth and chromatic number, 960–961
  global and local coloring, 961–962
  hierarchy of, 661, 662f, 662t
  interval. *See* Interval graphs
  isomorphism. *See* Graph isomorphism
  Kirchhoff Index of, 258–263
  λ_3-connected, 285f
  Laplacian spectrum of, 240–243
  list coloring of, 957–960
  local edge-connectivity augmentation of, 320–323
  matching in, 349
  with maximal irredundant set, 425f
  maximum matching in general, 377–385
  minors, 491
  multicoloring, 959
  with nodes, 261f
  nonplanar, 477, 477f
  operations on, 7–8
  order of, 4
  oriented, 241f
  pairs of, 440
  parameters, 10–11
  perfect matching in general, 354–355
  Petersen, 233, 234f, 350f, 481f
  planar. *See* Planar graph
  with prescribed degrees, 308–311
  principal partition of, 901–907
  principal sequence for, 918f
  representation on computer, 21–22
  restriction and contraction of, 889–890
  SCC of, 1017
  separator, 924
  simple, 490
  size of, 4
  small dominating sets in, 954
  spectrum/spectra of, 228–229
    complete bipartite graph, 234–235
    complete graph, 229
    product graphs, 236–240, 238f
    regular graphs, 230–234
  st-numbering of, 71–74, 72f
  storage and manipulation, 502–503
  structural results, 279–280
  subgraph of, 5
  Sylvester, 350f
  transitive closure of, 30f
  transitive orientation of, 44–53, 44f
  typical, 971
  underlying, 12
  undirected. *See* Undirected graph
  vertex-connectivity of, 329–333
  walk of, 6
Graph coloring theory, 449
  edge colorings, 461–463
  four color conjecture, 455
  greedy coloring, 451
  other variants of, 463–469
    achromatic number, 463–465
    complete colorings, 463–465
    dominator colorings, 466–467
    grundy, 465–466, 466f
    list colorings and choosability, 467–468
    total colorings, 469
  problems study, 455
  vertex colorings, 449–461
Graph drawing, styles

applications of, 542–543, 542f
convex drawing, 539, 540f, 552–560
    canonical decomposition, 553–555, 553f, 554f, 555f
    convex grid drawing algorithm, 556–560, 556f, 557f, 558f, 559f
grid drawing, 541, 541f
orthogonal drawing, 539–540, 540f, 567–577, 569f
    linear algorithm for bend-optimal drawing, 575–577
    and network flow, 570–575, 570f, 573f, 574f, 575f
planar drawing, 538, 539f
polyline drawing, 539, 539f
rectangular drawing, 540–541, 540f, 560–567, 560f, 561f
    linear algorithm, 564–567, 565f, 566f, 567f, 568f, 569f
    and matching, 562–564, 562f, 563f
straight line drawing, 539, 540f, 543–551
    canonical ordering, 544–546, 544f, 545f, 546f
    shift algorithm, 546–551, 547f, 549f, 552f
visibility drawing, 541, 541f
Graph/hypergraph partitioning variations, 866–871
    implementation, 869
    minimum cost balanced bipartition, 868–869
    net in flow-network, 867–868, 867f
    optimal minimum netcut algorithm, 868
    optimal replication for minimum k-cut partitioning, 869–871
        replication problem, 870–871
        vertex replication, 870
    r-balanced minimum netcut bipartition, 868
Graph isomorphism, 5
    algebraic methods for, 999
    chemical, 1000–1002
Graph Laplacian, 204
    Kirchhoff Index of, 258–263
Graph partition problems, 830–831
    class and combinatorial approaches, 832–837
        minimum cost multicut problem, 835–837
        multiway cut, 834–835
        Stoer, Wagner, and Frank algorithm, 833–835
    using multiple eigenvectors, 860–861
Graph search algorithm, 86
Graph symmetry, 643–647
    distance, 647
    edge, 645f, 646–647, 646f, 647f
    vertex, 645, 645f, 646f
Graph theory, 1063
    applications to, 951–971
    notation
        articulation point, 1099
        cycle created by edge, 1098
        end vertices, 1098
Greedy coloring, 451
Greedy set-cover algorithm, 1075–1077
    algorithm, 1076f
    analysis, 1075–1076
    generalizations and variants, 1077
    linear-program relaxation, 1076f
    vertex cover, 1076–1077
Grid drawing, 541, 541f
Gromov's hyperbolicity of graph, 753
Groups, algebraic system, 627–636
    automorphism, 634
    definitions and examples, 628–632
        binary groups, 629, 629t
        integer additive groups, 628, 629t
        integer multiplicative groups, 628, 629t
        matrix groups, 628
        symmetric permutation groups, 630–632, 630f, 631t, 632f, 633t
    generators of, 635–636
    homomorphism, 634
    isomorphism, 634
    operations on, 634–635
    subgroup, 633
Grundy colorings in graphs, 465–466, 466f
GYO algorithm. *See* Graham's algorithm

H

H_2O molecule, chemical graph, 998, 999f
Hadwiger's conjecture, 461
Hajós conjecture, 461
Hall, Philip, 352
Hall's approach and its variations, 849–854
Hall's condition, 352
Hall's theorem, 283, 352, 354, 1148–1149

Hamiltonian cycle, 9, 670
Hamiltonian graphs, 9–10
    directed, 14–15
Hamiltonian path, 9, 670
    ranking by, 16
Hamiltonicity property of AQ_n, 675
Hamming graph/generalized base-b cube, 618, 618f, 681–682
Hanan grid, 1116
Harary graphs, 278
Hassin–Lorenz–Raz (HLR) algorithm, 1035, 1038, 1038f
Havel's conjecture, 666
Helly graph, 785
    property, 760–761
Hereditary property, 423, 425
Heuristics, 1120–1124
    batched iterated 1-Steiner, 1122–1123
    connection graph-based approach, 1131
    flow-balanced-bipartition, 869
    FLUTE, 1123–1124
    iterated 1-Steiner, 1121–1122
    maze-routing approach, 1130–1131
    RMST-based, 1120–1121
    sequential approach, 1130
HIGHPT1 calculation, 693–698, 694f
Hilbert–Waring theorem, 969
Hinge vertex, 924
H-minor free graphs, 1149
Hochbaum's pseudoflow algorithm, 110
Hoffman–Wielandt inequality, 855–857
Holes, 710, 729, 796
Homeomorphic graph, 480
Homogeneous goods assumption, 158
Homomorphism, 464, 465f, 634, 964
Hopcroft and Karp's approach, philosophy of, 386–388
Horvert diagram, 514
Hospitals/Residents (HR) problem, 411–412, 412f
    with lower quotas, 412–413
    student-project allocation problem, 413–414
Hubs, 1000
Hungarian tree, 389
Hybrid analysis, 924
    proofs for topological, 931–932
    topological, 928–931
Hybrid rank, 933

Hydrocarbons C, class of, 1005
Hyperarcs, 341
Hyperbolicity of graph, 810
Hyperbolicity of quadruplet, 810
Hypercubes, 7, 591–612
    alternative definitions, 592–594
    basic properties of, 594–599, 595f
    characterizations, 599–612
        through convex sets, 608–609
        through edge colorings, 612
        through (0,2)-graphs, 599–600
        through intervals, 602–605
        through medians, 605–607
        through monotone properties, 609–612
        through projections, 607–608
        through splitting, 600–602
    definition, 591
    of dimension, 591, 591f
    embeddable generalized, 681–682
    embedding into, 656–673
        cubical graphs, 657–661, 662f, 662t
        cycles, paths, and trees, 670–673
        trees in, 662–669, 664f, 667f
    embedding into variants of, 674–682
        augmented cubes, 674–680
        crossed cubes, 680–681
        enhanced cubes, 681
        Hamming graphs, 681–682
        twisted cubes, 681
    optimal, 657
    paths in, 670–673
Hyperedges, 337, 831
Hypergraph, 831
    2-coloring, 763–764
    connectivity augmentation, 337–342
        detachments and augmentation, 340
        directed, 341–342
    disk, 762
    notions, 759–763
    reduced, 759
    totally balanced matrices and, 776–779
        hypergraph vs. b-acyclic hypergraph, 776–778
        matrices, 778–779
Hypergraph/graph partitioning variations, 866–871
    implementation, 869
    minimum cost balanced bipartition, 868–869

net in flow-network, 867–868, 867f
optimal minimum netcut algorithm, 868
optimal replication for minimum k-cut partitioning, 869–871
replication problem, 870–871
vertex replication, 870
r-balanced minimum netcut bipartition, 868
Hypergraph-theoretical background, 752
Hypertrees, 770–773, 773–774

I
i-component, 527
i-descendant, 527
Immediate dominator, 698
Implication class, 718
Inactive vertex, 494
Incidence matrix, 191–193, 192f, 249, 253
of directed graph, 925
reduced, 925
submatrices of, 199–201
In-degree matrix, 207–208, 210
Independence axioms of matroid, 880–881
Independence domination number, 424
Independence number of G, 351
lower bounds on, 954–955
Independent set, 424, 880
Indicator random variable, 948
Indivisible goods assumption, 158–159
Induced subgraph, 5, 460
Infinite capacity arcs, 867
Infinite group, 628
In-fragment, G, 334
Initial vertex, walk, 6
Inner angle, 562
Inner rectangular drawing of G, 561
Integer additive groups, 628, 629t
Integer linear programming (ILP), 1129, 1147
Integer multiplicative groups, 628, 629t
Integrality theorem, 87
Interconnection networks
computing
product of two matrices, 589–591, 589f
sum of n numbers, 588–589, 588f
definition, 588
graph embedding into hypercubes, 656–673
cubical graphs, 657–661, 662f, 662t
cycles, paths, and trees, 670–673
trees in hypercubes, 662–669, 664f, 667f
graph embedding into variants of hypercubes, 674–682
AQ_n, 674–680
crossed cubes, 680–681
enhanced cubes, 681
hamming graphs, 681–682
twisted cubes, 681
graph embeddings and quality measures, 653–656, 655f
hypercube-like, 612–623
AQ_n, 617–618, 618f
butterfly graph, 619, 620f
cube-connected cycles graph, 619, 620f
degree four Cayley graph, 620–621, 621f
$DG(d,k)$, 618–619, 619f
generalized base-b cube, 618, 618f
Γ_n, 615–616, 617f
hamming graph, 618, 618f
k-ary n-cube, 616–617
k-skip enhanced cube, 614–615, 615f
k-valent Cayley graph, 621–622, 622f
Möbius cube, 615, 616f
shuffle cube, 615, 617f
star graph, 622–623, 623f
twisted cube, 612–614, 614f
hypercubes, 591–612, 591f
alternative definitions, 592–594
characterizations of, 599–612
properties of, 594–599, 595f
problems, 590
topological properties, 612, 613t
Interior point algorithms, 170
Internal edge cost, 845
Internal embedding component, 527
Internal vertices, 12, 663
Interval analysis, 692
Interval graphs, 726–729
characterization, 727
optimization, 728–729
recognition, 728
Interval partitioning technique, 1028, 1031–1032, 1035
Inverse element, 628
Inverse exponential probability, 1012

i-path, 527
Irredundant set, 424
　graph with maximal, 425f
Isolated vertex, 4
Isomorphic graph, 644, 644f
Isomorphic permutations, 630
Isomorphism, 5, 634
Iterated 1-Steiner, 1121–1122
Iterative and multilevel partitioning algorithms, 845–849
　F–M algorithm, 847
　Kernighan–Lin algorithm, 847f, 847t
　　definition, 845
　　features of, 845–846
　　multilevel partitioning, 848–849, 848f
　　　coarsening, 849
　　　connectivity-based clustering algorithm, 849, 850f
　　　multilevel steps, 848–849
　　　V-cycle refinement, 849
Iterative averaging process, 864
Iterative move-based algorithms, drawback of, 847

J
Join decomposition, 794
Join operation, 788, 793
Jordan curves, 538

K
k-ary n-cube, 616–617
k-ary tree, 663
k-choosable/k-list colorable graph, 468
k-chromatic graph, 449, 461
k^*-connected graph, 287
k-cycle, 630
k-dimensional toroidal mesh, 284
k-distance dominating set, 435
k-distance domination number, 435
k-dominating set, 435
k-domination number, 435
k-edge connectivity augmentation problem, 317
k-edge connectivity of G, 297
　testing, 298f
Kekulé structures, 366–367
Kernelization, 1161–1168
　algorithm, 1145
　defined, 1145
　$\mathcal{F}$-deletion problem, 1159
　kernel via crown decomposition, 1147–1149
　lower bounds, 1165–1168
　other upper bounds, 1165
　vertex cover, 1162–1163
Kernighan–Lin (K–L) algorithm, 831, 847f, 847t
　definition, 845
　features of, 845–846
k-factor of graph, 368
k-fragment of G, 284
Khuller–Raghavachari algorithm, 1106–1108
Kirchhoff Index ($Kf(G)$) of G, 258–263
　formula for, 259–262
　topological formulas for network function, 262–263
Kirchhoff's current law (KCL), 177, 248–249, 923
Kirchhoff's voltage law (KVL), 248–249, 923
Kirchoff matrix of G, 240
Kirchoff's matrix-tree theorem, 241–242
Knapsack problem, 1031
Knob vertices, 767
Knotting graph, 716, 716f
König, Dénes, 349–350
König property, 764–767, 766f
König's edge coloring theorem, 350
König's theorem, 352, 462–463, 708–709, 915–916
κ-perfect graph, 790–791
k-regular bipartite graph, 365
k-restricted edge-connectivity of G, 284
k-restricted edge-cut graph, 284
Kruskal's algorithm, 22
　correctness of, 23
k-skip enhanced cube, 614–615, 615f
Kuhn–Munkres algorithm, 392–393
Kuratowski's graph, 477
Kuratowski subgraph, 511
　isolation, 491, 513
k-valent Cayley graph, 621–622, 622f

L
LABEL array, 380, 383
　at intermediate step, 384t
Lagrangian dual problem, 1043
Lagrangian function, 1053
Lagrangian multiplier problem, 163
Lagrangian relaxation, 163–164

Lagrangian Relaxation-based Aggregated
    Cost (LARAC) algorithm,
    1042–1044, 1044f
  convergence of, 1047
  equivalence of MCRT and, 1045
Laplacian eigenvalue of G, 244
  k-smallest, 857–858
  second smallest, 852–853
Laplacian matrix, 859–860
  of graph, 261f
  Moore–Penrose pseudoinverse of, 258
  properties, 850–852
Laplacian spectrum of graph, 240–243
LARAC-BIN algorithm, 1049, 1050f
Las Vegas algorithms, 1012
Lawrence Livermore National Laboratory
    (LLNL) mechanism, 1002
Layered networks, 105–106, 106f
$\mathcal{L}$-choosable/$\mathcal{L}$-list colorable graph, 467
Leaf node, 135
Leaf vertex, 466, 663
Least recently considered (LRC) rule, 146
Left multiplication of elements, 647
Leg-vertex, 564
Lempel algorithm, 44–53
Lempel–Even–Cederbaum algorithm, 492
Lempel–Even–Cederbaum planarity test,
    492
Length of edge, 35
Lexicographic breadth-first search
    (LexBFS), 712
Lexicographic product, G, 8
Light edges, 374
Linear algorithm for bend-optimal drawing,
    575–577
Linear multigrid, 674
Linear programming (LP) formulation, 1053,
    1129
Linear-time 2-approximation, 1104–1105
Linear-time algorithms, 538–539
Linear-time implementation, 529–530
Linear ($O(k)$) vertex kernel, 1147
Links, edge connections, 8, 528
LinMSOL, 806
List chromatic number, 468, 957
List colorings, 467–468
List set coloring, 959
Load of embedding, 645

Local edge-connectivity augmentation of G,
    320–323
  degree-specified augmentation, 321–323
  node-to-area augmentation problem, 323
Longest path problem, 1150
Loop vertex, 759
Lorenz and Raz (LR) algorithm, 1038
Lorenz–Raz approach, 1035–1039
Lossy networks, 1093–1094
Lovász local lemma (LLL), 950
  application of, 965
  non-symmetric form of, 950
Lovász's theta function, 708
LRC (least recently considered) rule, 146

M
Mader's theorem, 1112f
Malhotra, Kumar, and Maheshwari (MPM)
    algorithm, 107
Man-optimal stable matching, 406
Man-oriented GS algorithm, 405
MA (maximum adjacency) ordering of G,
    299
Marked edges, 50
Marked vertices, 50
Markov's inequality, 960, 972
Marriage theorem, 352
Mason's gain formula, 222–224
Mason's signal flow graph, 222, 223f
Master/coordinating problem, 166
Matching-covered G, ear decompositions of,
    360–362
Matching, G, 349
  applications of, 366–367
  in bipartite graphs, 351–354
  conditions for perfect, 355–356
  determinants, 364–366
  extension, 367–368
  in general graph, 354–355
  maximum, 356–357
  maximum matching in bipartite graphs,
    385–388
  maximum matching in general graphs,
    377–385
    Edmonds' approach, 377–379
    Gabow's algorithm, 379–385
  number, 351
  optimal stable, 410–411
  perfect, 364–366, 389–391, 391f
  permanents, 364–366

Matching pairwise edges, 710
MATE array, 380
Matrix groups, 628
Matrix reordering technique, 863
Matroid rank function, 916–917
Matroid(s)
    algorithm for construction of sequence of, 916–917
    axiom systems for, 880–887
    dual, 887–889
    examples of, 881–882
    matching, 882
    minors of, 889–896
    polygon, 881
    representability of, 915
    transversal, 915
    union. *See* Union of matroids
Matroid union algorithm, 907–910
    complexity of, 910–911
    example, 909f
Matroid union theorem, 911–912
Matula's algorithm, 295–297, 297f
Max-flow-min-cut algorithm, 833
Max-flow min-cut theorem, 87, 292, 354
Maximal clique, 710
Maximal independent sets, 861
Maximally independent subset, 882
Maximal matching vertex, 849
Maximizing multiplier, 1053
Maximum adjacency (MA) ordering, 110, 1104
Maximum capacity algorithm, 89–91
Maximum cardinality search (MCS), 712
    algorithm, 713
Maximum clique problem, 790
Maximum concurrent flow (MCF) problem, 1081, 1084f
Maximum flow problem
    assumptions, 82–83
    mathematical formulation, 81–82
    as minimum cost flow problem, 81, 81f
    with no feasible solution, 82f
    overview, 79–81
    preliminaries, 84–85
        flow across s–t cut, 84–85
        residual network, 84, 84f
    on unit capacity networks, 107–109

Maximum induced matching problem, 807
Maximum multicommodity flow (MAX-MF) problem, 1080
Maximum neighborhood ordering, 770–774
Maximum neighbor notion, 770
Maximum stable set problem, 790
Maximum weighted clique, 739
Maximum weight independent set (MWIS), 755
Maze-routing approach, 1130–1131
Mean cost of cycle, 127
Measure and conquer technique, 1156
Median, 605
Median graph, 605
Menger's theorem, 368
    and its applications, 280–283
    and maximum flows in network, 291–295
        connectivities in undirected graph, 294–295
        edge version of, 292–293
        vertex analog of, 293–294
Merge stack, 496
Mesh, 612, 614f
    of planar graph, 478
Metric embedding problems, 1182
Metric tree-like structures in G, 808–812
    hyperbolicity of graphs and embedding into trees, 810–812
    tree-breadth, 809
    tree-length, 808
    tree-stretch, 809–810
Metropolis algorithm, 865
    stimulated anneling, 866
Meyniel graph, 741
Mincut, global, 1014–1016
Minimal chord-path, 554
Minimal counter example, 440
Minimal dependent/non-independent subset, 882
Minimal DF (MDF), 436
Minimally noncomparability graph, 716
Minimal path, 1046
Minimal P-set, 423
Minimax theorem, 350, 352
Minimum cardinality Steiner tree problem, 799–801
Minimum cost concurrent flow, 1092–1093

Minimum cost disjoint paths (MCDP) problem
    algorithm for, 1066–1067, 1066f
    defined, 1064
Minimum cost flow problem, 572
    applications, 117–120
        distribution problems, 117
        optimal loading of hopping airplane, 118, 119f
        scheduling with deferral costs, 118–119, 120f
    defined, 115
    duality, 123–125
    optimality conditions, 120–123
        complementary slackness, 122–123
        negative cycle, 120–121
        reduced cost, 121–122
    overview
        notation and assumptions, 115–116
        residual networks, 116
        similarity assumption, 116
        tree, spanning tree, and forest, 117
    polynomial time algorithms for, 148t
Minimum-cost multicommodity problem, 1080–1081
Minimum cost multicut problem, 835–837
Minimum cost-restricted time combinatorial optimization (MCRT)
    algorithm, 1045f
    and LARAC algorithm, 1045
    NP-hard, 1044
    problem, 1042, 1044
Minimum cost-to-time ratio cycle algorithm, 1069
Minimum cut, 295
    problem, 85
Minimum mean cycle, 127
Minimum multicut, 832
Minimum multiway cut, 832
Minimum-size k-connected spanning subgraphs, 1110–1112
Minimum spanning tree (MST) problem, 1116–1117
Minimum spanning trees (MSTs), 1011
    linear time, 1012–1014
Minimum weighted coloring, 739
Minimum weight spanning tree, 22–25
    Kruskal's algorithm, 22–23
    Prim's algorithm, 22, 24
    unified version, 25
Minors of graphs, 926
    of dual vector spaces, 891–892
    of matroid, 893–896
    vector spaces, and matroids, 889–896
Möbius cube, 615, 616f
Modular decomposition of graphs, 720, 721f, 793–794
Modular function, 837–838, 897
Modules, 542
Molecule build algorithm, 1004
Molecules, chemical graph
    CH_3O, 1002f
    graph representation of, 998–999
    H_2O molecule, 998, 999f
Monadic second-order logic, 752, 804–806
Monomorphism, 634
Monte Carlo algorithms, 1012, 1017–1018
Monte Carlo-based metaheuristic optimization, 864–865
Moore-Penrose pseudoinverse, 258
    Kirchhoff Index using, 261–262
    trace of, 262
Morgan's algorithm, 1001
Multichip module (MCM), 542
Multicolor clique problem, 1174–1175
Multi commodity flow problem, 1079–1082
    applications, 159–161
        multi vehicle tanker scheduling, 160–161, 161f
        routing of multiple commodities, 160
    approaches to solve, 159
    approximation algorithms, 1082
    assumptions, 158–159
    dual of, 162
    explicit case, 1081, 1090–1092
    maximum concurrent flow, 1082–1092
        analysis of approximation guarantee, 1085–1088
        explicit case, 1090–1092
        running time, 1088–1089
    minimum cost concurrent flow, 1092–1093
    optimality conditions, 161–163
    overview, 157–158
Multigraph, 4
Multilevel partitioning, 848–849, 848f
    coarsening, 849

Multilevel partitioning (*Continued*)
    multilevel steps, 848–849
        coarsening phase, 848
        partitioning, 848
        refinement, 849
    V-cycle refinement, 849
Multilevel spectral partitioning, 861–862
    algorithm, 861–862
    eigenvalue-based hypergraph, 862–864
Multiple edges, 4
Multiprocessor computing system, 588
Multi-vehicle tanker scheduling, 160–161, 161*f*
Multiway cut, 834–835
    algorithm for, 1165–1166
    defined, 834
Mycielski's construction, graph, 456, 456*f*

N

Natural clusters, 849
Nauty algorithms, 1001
n-cube (Q_n), 7
Negative cycle optimality conditions, 120–121
Neighborhood hypergraph, 762
NEPS (noncomplete extended *P*-sum), 237–239
Nested point, 778
Nested sequence, 843
Netcut bipartition, 867
Netlist, 831
Network
    analysis through topological transformation, 937–941
        fusion–fission method, 937–939, 938*f*, 939*f*
        solution of node fusion–fission problem, 939–941
    design and analysis of, 590
    element, 247–248, 248*f*
    flow
        algorithms, 105
        disjoint paths for, 1065–1066
        problems, 114–117
    hypercube-like interconnection, 612–623
        AQ_n, 617–618, 618*f*
        butterfly graph, 619, 620*f*
        cube-connected cycles graph, 619, 620*f*
        degree four Cayley graph, 620–621, 621*f*
        $DG(d,k)$, 618–619, 619*f*
        generalized base-*b* cube, 618, 618*f*
        Γ_n, 615–616, 617*f*
        Hamming graph, 618, 618*f*
        k-ary *n*-cube, 616–617
        k-skip enhanced cube, 614–615, 615*f*
        k-valent Cayley graph, 621–622, 622*f*
        Möbius cube, 615, 616*f*
        shuffle cube, 615, 617*f*
        star graph, 622–623, 623*f*
        twisted cube, 612–614, 614*f*
    resistance, 247–249
        element, 248*f*
        no-gain property of, 265–267
        topological formulas for functions of, 249–253
    scale-free, 998
    and system theory, 224
    topological degree of freedom of, 932–933
Network flow theory, 831–832
Network simplex algorithm, 115, 127, 130–143, 134*f*
    arc
        entering, 137–138
        leaving, 138, 141*f*
    computing node potentials and flows, 135–137
    cycle free solutions, 131–133
    spanning tree solutions, 131–133
    spanning tree structure
        initial, 133–134
        maintaining, 134–135
    steps, 135
    strongly feasible spanning trees, 140–143
    termination, 140
    updating tree, 138–140
Next-to-optimal hypercube, 657
$n \times n$ matrix, 204, 207, 215
No congestion assumption, arc, 158
Node collections, 831
Node conductance matrix, 250, 257
Node congestion, 655
Node-disjoint paths algorithms, 1067
Node-disjoint tree paths, 530
Node fusion–fission problem, 938
    solution of, 939–941

Node potentials and flows, computing, 135–136
Node ring, 638, 638f
Node splitting process, 692
Node-to-area augmentation problem, 323
No-gain property of resistance networks, 265–267
Nonabelian group, 628
Non-algorithmic upper bounds, 1100
Non-bipartite knotting graphs, 717–718, 717f
Noncommutative group, 628
Nondegenerate iteration, 138
Nonempty bipartite graph, 462–463
Non-Hamilton cycles, 670
Nonisomorphic graph, 644, 644f
Nonplanar drawing of G, 538, 539f
Nonplanar graphs, basic, 477, 477f
Nonplanarity detection, 503–504
Non-reducible graph, 692–693, 693f
Nonsingular matrix, 628
Non-transitively orientable graph, 48, 48f
Nontrivial modules, 792
Nordhaus–Gaddum type results, 436–437
    for domatic number, 438
Normal subgroup, 633
Notation, 710
Notion of graph isomorphism, 643–644
NP-complete problem, 17–18, 441–442, 759, 868, 1027
    solve in polynomial time, 1027
NP optimization problem, 1185
Null graph, 4
Number theory, applications to, 951–971

O

Obstacle-avoiding RSMT (OARSMT) problem, 1129–1135
    exact algorithms, 1131–1135
    heuristics, 1130–1131
Obstacle-avoiding spanning graph, 1131
Odd chord, 757
Odd permutation, 631
Ohm's law, 248
One-dimensional multigrid, 674
One-step greedy algorithm, 649
The On-Line Encyclopedia of Integer Sequences, 670
Open circuit resistance matrix, 251
Open directed Euler trail, 14

Operations on graphs, 7–8
Optimal assignment problem, 392–394, 394f
Optimal Chinese postman tour, 396
Optimal Lagrange multipliers, 163
Optimal loading of hopping airplane, 118, 119f
Optimal node potentials, 122
Optimal stable matchings, 410–411
Optimal vertex coloring, 729
Optimum branching, graph, 25–30, 29f
Oracle test algorithm, 1055f
Order-n Fibonacci code, 616
Orientation arrow, 247
Orientation of graph, 12, 962–964
Oriented chromatic numbers, bounds on, 964–965
Oriented coloring of G, 964
Orthogonal drawing of G, 539–540, 540f, 567–577, 569f
    linear algorithm for bend-optimal drawing, 575–577
    and network flow, 570–575, 570f, 573f, 574f, 575f
Orthogonality relation, graph, 197–199
Outer angle, 562
OUTER array, 380, 383–385
Outer cycle, 544
Outer edge, 544
Outer rectangle, 566
Outer vertex, 544
Out-fragment, G, 334
Out-of-kilter algorithm, 114, 129–130
Overfull graph, 463

P

Packing number of hypergraph, 764
Painting theorem, 265
Paired-dominating sets (PDS), 430–432
Pairwise comparable elements, 718
Pairwise incomparable elements, 718
Pairwise intersecting hypergraph, 760
Parallel edges, 592, 594
Parallel face walking optimization technique, 504
Parallel nodes, 793
Parameterized intractability, 1173–1185
    example reductions, 1174–1178
    exponential time hypothesis, 1178–1181
    stronger lower bounds, 1181–1183
Parameters of graph, 10–11

Parametric search-based CSP (PSCSP)
    algorithm, 1054
Parity
    of binary string, 629
    defined, 631
Partial hypercube, 661, 662f
Partial k-trees, 752
Partial subhypergraph, 759
Partite sets, 4
Partition associate, 839
Partitioning problem
    definitions, 831–832
    overview of, 830–831
Partitioning procedure, 831
Partitions principal sequence, 840–841
Path addition approach, 492, 525
Path finding algorithm, 73
Path, walk, 6
PC tree data structure, 525
Pendant vertex, 4, 466
Perfect 2-matching of G, 354
Perfect code, 807
Perfect elimination graphs, 752
Perfect graphs, 707–744
    χ-bounded graphs, 744
    chordal graphs, 710–715
        characterization, 710–712
        optimization, 715
        recognition, 712–715
    comparability graphs, 715–726, 716f, 717f
        characterization, 715–718
        optimization, 725–726
        recognition, 718–724
    interval graphs, 726–729
        characterization, 727
        optimization, 728–729
        recognition, 728
    notation, 710
    overview, 707–709
    perfectly contractile graphs, 741
    perfectly orderable graphs, 733–741
        characterization, 735
        optimization, 738–741
        recognition, 735–738, 736f, 737f
    recognition of perfect graphs, 742–743
    theorem, 708
    weakly chordal graphs, 729–733
        characterization, 729–730
        optimization, 732–733
        recognition, 731–732
        remarks, 733
Perfect graph theorem, 459
Perfectly orderable graphs, 733–741
    characterization, 735
    optimization, 738–741
    recognition, 735–738, 736f, 737f
Perfect matching decision graph, 564
Perfect subgraph, 708, 734
Permutation graph, 328
Permutation group, 630–632
Permutation matrix, 228, 354
PertinentEdge member, 504
PertinentRoots member, 504
Pertinent vertex
    future, 494–495
    management with walkup, 504
Petersen, Dane Julius, 349
Petersen graph, 233, 234f, 350f, 481f, 637, 637f
Pfaffians, method of, 365–366
Pieces (nonempty subsets), 332
Pigeonhole principle, 283
Pivot cycle, 138
    procedure for identifying, 139f
Pivot operation, 133
    updating node potentials in, 139f
Planar embedding graph, 475, 476f
    regions of, 477, 478f
Planar graph, 455, 456f, 468, 475–477
    cycle view, 526, 526f
    drawing, 490, 538, 539f
    embedding of, 477
    meshes of, 478
    nonplanar embedding, 490, 490f
    obstructions, 491, 491f
    planar embedding, 490, 490f
    representation, 570
    tree view, 526, 526f
    visibility representation of, 514–522
        correctness and performance, 519–522
        horizontal positions of edges, 518–519, 518f, 519f
        vertical positions of vertices, 514–518
Planarity, 485–487
    isolating obstruction to, 511–514, 513f
    Kuratowski's characterization, 480–481
    testing based on PC-trees

creating C-nodes in general, 528, 529f
creating first C-node, 527–528
embedding algorithm, 529, 529f
linear-time implementation, 529–530
S&H planarity test, 526
Plane graph, 539, 541f
Plant/model nodes, 117
Plant nodes, 117
P-nodes, 526
Pnueli algorithm, 44–53
Poisson trial, 948
Polygonal chain, 539
Polygon matroid, 881
Polyline drawing of G, 539, 539f
Polymatroid rank functions ($f(\cdot)$), 897, 901
Polynomial parameter transformations, 1152
Polynomial time algorithm, 83, 147–150, 148t, 408
Polynomial time approximation algorithms, 1027, 1028
Polynomial time approximation scheme (PTAS), 1082
Polynomial time dual simplex algorithm, 146
Positive semidefinite matrix, 245
Potentially optimal Steiner tree (POST), 1123
Potentially optimal wirelength vector (POWV), 1123
Potential vector, 925
PQ-tree, 492, 525, 728
Predecessor index, 134–135
Preflow-push algorithms, 83, 98–104
  FIFO, 103–104
  generic, 99–102, 100f
  summary of, 105t
Preliminaries, G, 924–928
Pre-multiplier algorithm, 147
Preprocessing rule, 1155–1156
Price-directive decomposition method, 164, 167
Primal-dual algorithm, 114, 129–130
Primal graph, 760
Primal integrality property, 137
Prime graph, 792
  simple structure, 793
Prime nodes, 793
Prim's algorithm, 22, 24
Principal partition of graph, 901–907, 933
  characterization, 903

of contraction and restriction, 904
and density of sets, 905–906
of dual, 905
notes, 906–907
outline of algorithm for, 906
overview, 901
proof of properties of, 902–903
properties of, 901–904
relevant properties of, 843
Principal partition problem, 932–934
  forests
    maximum distance between, 933
    of minimum size hybrid representation, 933
  lattice of, 940–941
  maximum rank of cobase submatrix, 933–934
  Shannon switching game, 933
  topological degree of freedom of network, 932–933
Principal sequence of $(f(\cdot), g(\cdot))$, 902
Principal structure of $h(\cdot)$, 902
Principle of deferred decision, 1013
Printed circuit board (PCB), 537
Probabilistic arguments
  application of, 946, 951
  in combinatorics, 945–989
Probabilistic method
  application of, 968–670
  basic philosophy, 950–951
Probability, 947
  inverse exponential, 1012
  inverse polynomial, 1012
  threshold, 973
  union law of, 947
Probability theory, 946
  tools of, 950
Problem kernel, 1145
PROC-EDMONDS procedure, 380–381
Procedure compute-flows, 136, 137f
Procedure compute-potentials, 135, 136f
PROC-LABEL procedure, 381–382
PROC-REMATCH procedure, 381–382
Product graphs, domination in, 439–441
Product group, 635
Production arcs, network, 117
Production–distribution model, 117, 118f
Program graphs, 691–705
  dominators in, 698–705, 698f

Program graphs (*Continued*)
    overview, 691
    reducibility of, 691–698, 692f, 693f
Proper face, 490
Proper subgraph, 5
Pruning sequences, 797
Pruning trees, 797
P-set, 423
Pseudo-code of GS algorithm, 405
Pseudoflow push algorithms, 148
Pseudo graph, 4
Pseudopolynomial time algorithm, 83
Pseudo-vertex, 27–28
Ptolemaic graphs, 801

Q
Quadratic ($O(k^2)$) vertex kernel, 1146
Quality of service (QoS), 1063
Quasi-strongly connected digraph, 12
Quotient graph, 721, 721f

R
r-admissible S-detachment, 340
Rado's theorem, 916
Random digraphs
    being strongly connected, 987–988
    emergence of strongly connected component, 988–989
    exposition of, 984–989
    induced acyclic subgraphs, 985–986
    induced acyclic tournaments, 984–985
    induced tournaments, 986–987
Random graph models, 999–1000
    for chemistry, 1002–1005
    definitions, 999–1000, 1002–1003
    molecule build model, 1004–1005
    random pair model, 1003
Random graphs
    algorithm for, 1000
    being connected, 973–974
    concentration of invariants, 979–984
    diameter of, 977–978
    emergence of giant component, 974–977
    existence of triangles, 972–973
    exposition of, 971–984
    scale-free networks, 998
    theory of, 947
Randomized algorithm, 998
Randomized fixed-parameter algorithms, 1157–1162
    chromatic coding, 1160–1162
    color coding, 1159–1160
    α-covering based, 1158–1159
Randomized polynomial (RP), 1012
Randomness phenomenon, 946
Random variable, 963
Random walks on undirected graph, 254–258
Rank axioms of matroid, 884–886
Ratio-cut
    bipartitioning using eigenspectrum, 857
    defined, 857
    problem, 842
Ratio-optimization problems, 842
Rayleigh quotient iterations, 861–862
r-CNF-SAT problem, 1178–1179
Reachability matrix, 30
Realizer method, 543
Reasonable canonical labeling algorithm, 1005–1008
Rectangular-Draw algorithm, 566–567
Rectangular drawing of G, 540–541, 540f, 560–567, 560f, 561f
    linear algorithm, 564–567, 565f, 566f, 567f, 568f, 569f
    and matching, 562–564, 562f, 563f
Rectangular dual plane graph, 561
Rectangular grid drawing, 541, 541f
Rectilinear minimum spanning trees (RMSTs), 1118
    based heuristics, 1120–1121
    separable, 1121
Rectilinear SMT (RSMT), 1116
Rectilinear Steiner ratio, 1117–1119
    flipping, 1118
    shifting, 1117–1118
Rectilinear Voronoi diagram, 1120
Reduced cost optimality conditions, 121–122
Reduced hypergraphs, 759
Reduced incidence matrix, 925
Reducibility, 691
    algorithm, 695
    concept of, 18, 18f
    of program graphs, 691–698, 692f, 693f
Reduction order, 697
Reference vertex, 192, 206, 249
Reflexive directed graph, 13
Reflexive relation, 13

Regular graph, 5
    spectra of line graphs of, 232–233
    spectrum of complement of, 231–232
Regular labeling, 562–563
Relational database, 757
Relational database scheme, 757
Relation schemes, 757
Relaxation algorithm, 147
RELAX-CSP problem, algebraic study of, 1046–1049
Relaxed multicut problem, 836
Representative boundary cycle (RBC), 527
Representative graph, 760
Representative vertex, 1019
Residual network, concept of, 84, 84f, 91f
Residual sequence, 308
Resistance networks, 247–249
    element, 248f
    no-gain property of, 265–267
    topological formulas for functions of, 249–253
Resource-allocation problem, 168–169
Resource-directive decomposition method, 159, 167–169
Restricted coarsening procedure, 849
Restricted disjoint paths (RDP) problem, 1065
    algorithm RDP1, 1068f
    algorithm RDP2, 1069f
        analysis, 1069, 1071
        minimizing computational complexity, 1071
        pseudopolynomial algorithms, 1071
    algorithm RDP3, 1072f
Restricted edge-connectivity $\lambda'(G)$, 283
Restricted edge-cut of G, 283
Restricted master problem, 166
Restricted shortest path (RSP) problem, 1064–1065
Retailer/model nodes, 117
Retailer nodes, 117
Right coset, 633
Rigid-circuit graphs, 752
Rooted tree, 663
Rounding technique, 1028–1031
Round-robin arc-balancing algorithm, 110
Round-robin tournament, 15
Routing process, 590
    global, 1135–1136

Row-oriented transitive closure algorithm, 33, 33f
Running intersection property, 768
Rural Hospitals Theorem, 412

S
Sachs' theorem, 232–233
Saran–Vazirani multiway cut, 835, 835f
Satisfiability (SAT) problem, 17–18
Satisfiable truth assignment, 441
Scaled cost concept, 842
Scaled pseudopolynomial plus (SPPP) algorithm, 1036f
    complexity of, 1037
Scale-free networks, 998
Scaling techniques, 143
Scan-first search (SFS), 1101–1102
s^*-critical of G, 287
Search trees, 1152–1156
    feedback vertex set, 1153–1154
    vertex cover, 1153
    vertex cover above LP, 1154–1156
Self-complementary graph, 5
Self-loop//loop, 4, 927, 939
Semidirect product, 634–635
Semi-dominator, 699
Separable RMST, 1121
Separation technique, 1028, 1032–1033
Separators, 710, 1162–1166
    algorithm for multiway cut, 1165–1166
    in undirected graphs, 1162–1165
Sequential edge order, 490
Series edges, 480
Series insertion, 480, 480f
Series merger, 480, 480f
Series nodes, 793
Series reduction, 491
Set functions, coverings of, 337–342
Set splitting problem, 763
Sex-equality cost, 410
S&H algorithm, 542
Shannon switching game, 933
Shift algorithm, 546–551, 547f, 549f, 552f
Shift method, 543
Short-circuit edges, 509
Shortest augmenting path algorithms, 91–98
    admissible arcs, 91–94, 95f
    bad example for, 97f
    illustration, 92f
    worst-case improvements, 96–98

Shortest path problem, 648–649
Short separator, 768
S&H planarity test, 526
Shredders, 1102
Shuffle cube, 615, 617f
Signal flow graph theory, 215
Signal nets, 831
Simple 2-approximation, 1099–1104
    edge connectivity, 1100
        non-algorithmic upper bounds, 1100
        time complexity, 1103–1104
        trivial lower bound, 1099
        vertex connectivity, 1100–1103
Simple efficient approximation (SEA) algorithm, 1039f
Simplex algorithm, 1143
Simplex method, 114
Simulated annealing and graph bisection, 864–866
Single biconnected component, 528
Singleton biconnected component of G, 493
Sink nodes, 572
Skew adjacency matrix, 366
SMI (stable marriage with incomplete lists), 406, 407f
SMT (stable marriage with ties), 407, 408f, 409f
SMTI (stable marriage with ties and incomplete lists), 409–410, 409f
SNUMBER, 697, 697f
Soft vertex, 735
Sorted DFS child list, 503
Source nodes, 572
Spanning 2-trees, number of, 205–207
Spanning binomial trees, 676
Spanning subgraph, 5
Spanning tree, 8
    minimum weight, 22–25
    number of, 203–205
    solutions, 131–132
    strongly feasible, 140–143
    structure, 132
        maintaining, 134–135
        obtaining initial, 133–134
Spectral approaches for partitioning, 849–864
    eigenvector-based clustering algorithm, 859–860
        theoretical motivation, 860

graph bisection, 854
Hall's approach and its variations, 849–854
    multilevel spectral partitioning algorithm, 861–862
        eigenvalue-based hypergraph, 862–864
    multiple eigenvectors
        spectral bound involving, 857–859
        using, 860–861
    partitioning into k blocks, 854–857
    ratio-cut bipartitioning using eigenspectrum, 857
Spectral bipartitioning algorithm, 854, 860
Spectrum of graph, 228–229
    of cycle, 229–230
    Laplacian, 240–243
Sperner hypergraph, 762, 767
Spider head, 793–794
Split decomposition, 794
Split edge deletion (SED) problem, 1160
Split graph, 756
Split nodes, 867
Split partition, 756
SpMxV (sparse matrix with vector), 830
Square vertices, 498
Stable marriage problem, 403–404, 404f
    SMI, 406–407, 407f
    SMT, 407–408, 408f, 409f
    SMTI, 409–410, 409f
Stable matching, 404, 407f, 410–411
    man-optimal, 406
    other variants, 412–414
        HR problem with lower quotas, 412–413
        SPA problem, 413–414
        woman-pessimal, 106
Stable roommates (SR) problem, 411
Stable set, 740
Stalling, 146
Standard representative matrix, 881, 887
Star chromatic number, 967
Star coloring, 967
Star-delta transformation, 264f
Star generator, 636
Star graph, 622–623, 623f, 642
Starlike trees, 669
Start vertex, 691
s–t cut, 85, 292
ST-DHG algorithm, 799

ST-edge-connectivity, augmenting, 335–337
Steinerization, 1121
Steiner minimum tree (SMT) problem, 1115–1116
Steiner ratio (RSMT *vs.* ESMT), 1116–1119
    Euclidean, 1117
    MST problem, 1116–1117
    Rectilinear, 1117–1119
Steiner ratio Gilbert–Pollak conjecture, 1117
Steiner tree problem, 1175–1176
Stereographic projection, 476f, 477
st-numbering of graph, 71–74, 72f
Stoer and Wagner's algorithm, 298–300, 301f
    example, 302f–303f
Stoer–Wagner and Frank algorithm, 833–835
Stopping vertex, 494
Straight line drawing of G, 539, 540f, 543–551
    canonical ordering, 544–546, 544f, 545f, 546f
    shift algorithm, 546–551, 547f, 549f, 552f
Strong connectivity algorithm, 68–71
Strong duality theorem, 124
Strong elimination orderings, 779–781
Strong exponential time hypothesis (SETH), 1181
Strongly chordal graph, 779–786
    elimination orderings of, 779–781, 780f
    Γ-free matrices and, 782–783, 782f
    as sun-free chordal graph, 783–786, 784f
Strongly connected components (SCC), 1017
    decremental algorithm for, 1019–1020
Strongly connected digraph, 12
Strongly feasible spanning trees, 140–143
Strongly stable matching, 408
Strong perfect graph conjecture, 460
Strong perfect graph theorem, 708–709
Strong product, G, 7
Structural solvability, matroids, 907
Student-Project Allocation (SPA) problem, 413–414
Student-Project Allocation problem with preferences over Projects (SPA-P), 414

Subgraph
    edge-induced, 5
    homeomorphic, 491
    induced, 5, 460
    minimum-size k-connected spanning, 1110–1112
    problem, 656
    weight of, 22
Subgroup, 633
Subhypergraph, 759
Submatrix, rank of cobase, 933–934
Submodular function, 832, 837–838, 841, 897
    sum of $(h(\cdot))$, 902
Submodular inequality, 304–305, 331
Submodular optimization-based partitioning algorithms, 837–844
    finding densest cluster, 838–839
    improving ratio-cut, 842–844
        algorithm to, 843–844
        justification, 844
    principal lattice of partitions, 839–841
Subtree matrix, 775
Successive shortest path algorithm, 114, 127–129, 129f
Sum-free subset constant, 970
Sun-free chordal graph, 785
Super connectivity, 680
Super edge-connectivity, 680
Supermodular function, 837–838, 897
Super spanning, 679–680
Super-stable matching, 408
Supertoroid, 639–640, 640f
Super vertex-cut, 680
Support vertex, 466
Suurballe–Tarjan algorithm, 1064
Sylvester graph, 350f
Symmetric directed graph, 13
Symmetric group, 630
Symmetric permutation groups, 630–632, 630f, 631t, 632f, 633t
Symmetry in Cayley graphs, 642–648
    analyzing, 647–648
    graph symmetry, 643–647
        distance symmetry, 647
        edge symmetry, 645f, 646–647, 646f, 647f
        vertex symmetry, 645, 645f, 646f

Systematic walkdown invocation, 503–504
System of distinct representatives (SDR), 283, 354

T

Tarjan's algorithm, 692–693, 695–698
Tarjan's st-numbering algorithm, 73–74
Tellegen's theorem, 925
Terminal nodes, 527
Terminal path, 528
Terminal vertex, walk, 6
Thick spider graph, 794
Thin spider graph, 793–794
Thomassen's theorem, 564–565
Thread index, 135
Threshold probability, 973
Tight cut decomposition procedure, 363
Time complexity, 1103–1105
Time reductions, 710
Timetable scheduling problem, 394–395
Topological hybrid analysis procedure, 928–931
    proofs for, 931–932
Topological sorting, 15
Toroid, 638–639, 639f
Torus, 612, 614f
Total chromatic number, 469
Total colorings, 469
Total graph $T(G)$, 469
Totally disconnected graph, 4
Totally unimodular matrices, 202–203
Tournament, graph, 15–16
Traceable graph, 356
Trail, walk, 6
Transfer resistances, 252
Transitive closure, 30–35
    of binary relation, 30
        directed graph representing, 30
        estimating the size of, 1016–1018
        of G, 30f
        row-oriented, 33, 33f
        of Warshall's algorithm, 32
Transitive directed graphs, 13
Transitively orientable graph, 720, 791
Transitive orientation algorithm, 44–53, 44f
    illustration of, 47f
Transportation arcs, network, 117
Transportation problem, 117
Transposition graph, complete, 642

Transversal matroids, 915
Transversal number, 764
Tree (T), 8–9, 117
    decomposition, 804
        breadth of, 809
        length of, 808
        width of, 804
    edge, 59, 493
    entering arc, 137–138
    height, 663
    in hypercubes, 662–669, 664f, 666f, 667f, 670–673
    indices, 134f
    leaving arc, 138, 141f
    schema, 765
    strongly feasible spanning, 140–143
    updating, 138–140
Tree-distortion, 811–812
Tree-stretch, 809–810
Tree-structured graph
    α-acyclic hypergraphs and their duals, 757–776
        bipartite graphs, 773–774
        dually chordal graphs, 770–773
        Graham's algorithm, 767–770
        hypergraph 2-coloring, 763–764
        hypergraph notions, 759–763
        hypertrees, 770–774
        König property, 764–767, 766f
        matrix notions, 775–776
        maximum neighborhood orderings, 770–774
        other desirable properties, 767–770
        relational database theory, 757–759
        running intersection property, 767–770
        and tree structure, 764–767, 766f
    algorithmic implications, 752–753
    balanced hypergraph and matrices, 776–779
        hypergraph $vs.$ b-acyclic hypergraph, 776–778
        matrices, 778–779
    chordal bipartite graphs, 786–788
    chordal graph
        strongly, 779–786
        and variants, 753–757
    clique-width of graphs, 805–807
    complexity of problems, 807–808
    decomposition of graph, 788–794

clique separator, 794
cographs, 788–790
modular, 793–794
module properties, 791–792
optimization on cographs, 790–791
distance-hereditary graph, 794–798
minimum cardinality Steiner problem in, 799–801
subclasses of, 801–803
metric tree-like structures, 808–812
related graph classes, 752–753
treewidth of graphs, 803–805
Tree t-spanner, 810
Tree t-spanner problem, 810
Treewidth, 752, 803, 1144
of graph, 803–805
machinery, 1168–1170
Triangle-free graph, 456–457
Triangulated graph, 542, 542f, 752
Trivial lower bound, 1099
t-uniform hypergraph, 338
Tuples, 757–758
Turtles graph, 793–794
Tutte's method, 208
Tutte's theorem, 354–355, 357
Twin arcs, 502–503
Twisted cube, 612–614, 614f, 681
Twofold transposal matrix, 760
Two-phase exact algorithm, 1131

U

Unbounded clique-width, 806
Underlying graph, 12
Undirected edge, 490
Undirected graph, 44, 527
degree matrix of, 253
determining connectivities in, 294–295
DFS of, 59–63, 62f
edge connectivity of, 295–297
global minimum cut in, 298–300
random walks on, 254–258
vertex connectivity in, 300–301
Unilaterally connected digraph, 12
Unimodular matrices, totally, 202–203
Union of matroids, 907–919
algorithm. *See* Matroid union algorithm
circuits and coloops, 912
example, 918–919
and matroid intersection, 914–915
applications of, 915–916

matroid union theorem, 911–912
and union of dual matroids, 912–914
Unique games conjecture (UGC), 410
Uniquely k-colorable graph, 454
Unit capacity bridging arc, 867
Unit capacity maximum flow algorithm, 108
Unit capacity networks, maximum flow, 107–109
Unit capacity simple networks, 109
Unit element, 628
Universal Coloring Families, 1160
Universal relation, 758–759
Unstable blocking pair, 404
Upper eq-domination number, 432
Upper eq-irredunance number, 433
Upper fractional domination number, 436

V

van der Waerden's Conjecture, 365
V-cycle refinement, 849
Vector spaces
minors of dual, 891–892
representative matrices of minors of, 892
restriction and contraction of, 890–891
Verification algorithm, 17
Vertex addition approach, 525
Vertex addition planarity test, 492
Vertex addition processing model, 514
Vertex analog of Menger's theorems, 293–294
Vertex-angle, 570
Vertex-connectivity of G, 276–279, 329–333, 1100–1103
algorithm, 304f
augmentation of digraphs, 333–337
in undirected graphs, 300–301
Vertex-disjoint, 649
directed circuits, 223–224
Vertex-induced subgraph, 5
Vertex-subhypergraphs, 759
Vertex-symmetric graph, 597
Vertex-transitive graph, 284
Vertex transitivity
consequences of, 648–649
broadcasting, 649
shortest path problem, 648–649
illustration of, 645f
Vertex/vertices, 4
angle of, 562
back neighbor, 527

Vertex/vertices (*Continued*)
- balance-factor of, 668–669
- collections, 831
- colorings, 449–461
- computing vertical positions of, 514–518
  - localized sense of up and down, 515
  - making localized settings, 515–516, 516*f*, 517*f*
  - postprocessing to generate vertex order, 516–518
- cover, 351
- covering number, 351
- diamond, 802
- expansion of graph at, 280*f*
- gain in cost, 845
- inactive, 494
- internal, 12, 663
- isolated, 4
- kernel
  - linear, 1147
  - quadratic, 1146
- least ancestor of, 505
- levels, 598
- lowpoint of, 505
- marked, 50
- order, 516
- orientation, 490
- partitioning, 722
- replication, 870
- square, 498
- subsets, 607
- symmetry, 645, 645*f*, 646*f*
- virtual, 493

Vertical vertex order, 516
Very large-scale integrated circuits (VLSI), 830, 1123, 1135
- cells, 831
- routing
  - detailed, 1136
  - global, 1135–1136
  - stage, 1135–1136, 1135*f*
- RSMTs floorplanning and placement, 1136

Virtual vertex, 493
Visibility drawing of G, 541, 541*f*
Visitation detection optimization technique, 504

Vizing's conjecture, 439–440
VLSI floorplanning problem, 542
Voltage vector, 925

W

$W[1]$-hard problems, 1183
`Walkdown`, 492, 495–498, 504
- processing example, 497–498, 497*f*
- rules, 496–497

Walk, graph, 6, 489
- length of, 12
- random, 254–258

Warren's algorithm, 34–35, 34*f*
Warshall's algorithm, 30
- illustration of, 31*f*
- transitive closure of, 32
- Warren's modification of, 33–34

w-container, 679
Weak β-cycles, 777
Weak duality theorem, 123–124
Weakly chordal (wc) graph, 729–733
- characterization, 729–730
- optimization, 732–733
- optimization algorithm, 733
- recognition algorithm, 731–732
- remarks, 733

Weakly connected, digraph, 12
Weakly geodesic pancyclic, 679
Weakly stable matching, 408
Weakly triangulated graph, 729
Weighted chromatic number, 739
Weighted clique number, 739
Weighted-polar Dilworth theorem, 336
Weighted vertex-cover problem, 1076
Well quasi ordering, 1166–1168
Whitney's definition of duality, 485*f*
Winter's algorithm, 1128
Woman-pessimal stable matching, 406
Worst-case analysis, 21
Wrapped butterfly graph, 619, 620*f*

X

X-tree, 674

Z

Zero error probabilistic polynomial (ZPP), 1012
Zero-one optimization models, 107